Harry Smith

CRC HANDBOOK OF

ENERGY EFFICIENCY

CRC HANDBOOK OF

ENERGY EFFICIENCY

Edited by

FRANK KREITH
RONALD E. WEST

CRC Press
Boca Raton New York London Tokyo

Aquiring Editor:	Norm Stanton
Project Editor:	Jennifer L. Richardson
Marketing Manager:	Susie Carlisle
Direct Marketing Manager:	Arline Massey
Typesetter:	Kathy Johnson
Prepress:	Kevin Luong and Greg Cuciak
Cover Designer:	Denise Craig
Manufacturing:	Sheri Schwartz

Library of Congress Cataloging-in-Publication Data

CRC handbook of energy efficiency / edited by Frank Kreith, Ronald E. West.
p. cm.
Includes bibliographical references and index.
ISBN 0-8493-2514-5 (alk. paper)
1. Energy conservation--Handbooks, manuals, etc. I. Kreith, Frank. II. West, Ronald E.
TJ163.3.C731994
621′.042--dc20

95-39603
CIP

International Standard Book Number 0-8493-2514-5
Library of Congress Card Number 95-39603
Printed in the United States of America 1 2 3 4 5 6 7 8 9 0
Printed on acid-free paper

Foreword

The real story behind energy efficiency is productivity, industrial competitiveness, jobs, and a clean environment. The importance of these elements to the future health of our economy cannot be overstated.

Today's energy bill for seven major U.S. materials and process industries is $100 billion annually. Pollution control costs are $50 billion a year. These dollars can be reprogrammed into more productive uses if we incorporate pollution prevention and energy efficiency into industry at the wellhead, rather than downstream.

Getting the word out is key to making more people aware of the huge benefits to be had from using our energy resources more wisely in every way. This handbook may serve to inspire readers to move enthusiastically toward a lean, green energy future.

The Honorable Hazel R. O'Leary

Editors

Dr. Frank Kreith currently serves as the ASME Legislative Fellow for Energy and Environment at the National Conference of State Legislatures. In this capacity, he provides technical assistance on energy management, utility restructuring, waste disposal, and environmental protection to legislators and their staffs in all fifty state governments.

Prior to joining NCSL in 1988, Dr. Kreith was Chief of Thermal Research and Senior Research Fellow at the Solar Energy Research Institute, currently the National Renewable Energy Laboratory. During his tenure at SERI, he participated in the Presidential Domestic Energy Review, served as one of the energy advisors to the Governor of Colorado, was the Editor of the ASME Journal of Solar Energy Engineering, and Associate Editor of Energy, the International Journal.

From 1951 to 1977, Dr. Kreith taught engineering at the University of California, Lehigh University, and the University of Colorado. He is the author of textbooks on heat transfer, nuclear power, solar energy, and energy management. He is the recipient of the Charles Greeley Abbot Award from ASES, the Max Jakob Award from ASME-AIChE, and the Ralph Coats Roe Medal from ASME for "significant contributions to a better appreciation of the engineer's worth to society...through provision of information to legislators about energy and conservation".

Dr. Kreith has served as consultant and advisor on energy planning all over the world. His assignments included consultancies to NATO, U.S. Agency for International Development, and the United Nations.

Dr. Kreith resides with his wife in Boulder, Colorado.

Dr. Ronald E. West is Professor Emeritus of Chemical Engineering at the University of Colorado-Boulder. He retired in 1995 after 38 years on the faculty there. He earned B.S.E., M.S.E., and Ph.D. degrees in chemical engineering at the University of Michigan. His professional interests are eclectic, but center on the technically and economically efficient use of energy and material resources.

Professor West first worked on solar energy in 1960 and continued to do so on and off throughout his career. For many years, he was a consultant with the Solar Energy Research Institute (now the National Renewable Energy Laboratory). He served as Fulbright Professor at the Bosphorus University, Istanbul, Turkey; the University of the Aegean, Izmir, Turkey; and the Budapest Technical University, Hungary. Among the several books he co-edited are: *Economics of Solar Energy and Conservation Systems* (with F. Kreith, CRC Press, 1980); *Economic Analysis of Solar Thermal Energy Systems*, (with F. Kreith, MIT Press, 1988); and *Implementation of Solar Thermal Technology* (with R. W. Larson, MIT Press, 1996).

Contributors

Anthony F. Armor, M.S.
Director, Fossil Power Plants
Electric Power Research Institute
Palo Alto, California

Barbara Atkinson, M.S.
Staff Research Associate
Energy Analysis Program
Lawrence Berkeley National Laboratory
Berkeley, California

Bruce Bailey, Ph.D.
AWS Scientific, Inc.
Albany, New York

Stanley H. Boghosian, M.S.
Principal Research Associate
Energy Analysis Program
Lawrence Berkeley National Laboratory
Berkeley, California

Louise K. Bonadies, M.B.A.
Energy Industry Specialist
Energy Information Administration
U.S. Department of Energy
Washington, D.C.

Daryl R. Brown, B.S.
Staff Engineer
Technology Systems Analysis
Pacific Northwest National Laboratory
Richland, Washington

Jerald A. Caton, Ph.D.
Texas A&M University
Department of Mechanical Engineering
College Station, Texas

Robert Clear, Ph.D.
Building Technologies Program
Lawrence Berkeley National Laboratory
Berkeley, California

Anibal T. de Almeida, Ph.D.
Department of Electrical Engineering
University of Coimbra
Coimbra, Portugal

Andrea Denver, M.S.
Energy Analysis Program
Lawrence Berkeley National Laboratory
Berkeley, California

David M. Eggleston, Ph.D
Principal
DME Engineering
Midland, Texas

Dennis Elliot, M.S.
Senior Scientist—Wind Energy
Wind Technology Division
National Renewable Energy Laboratory
Golden, Colorado

Marjorie A. Franklin, B.S.
President
Franklin Associates
Prairie Village, Kansas

Randy C. Gee, P.E.
Industrial Solar Technology, Principal Inc.
Golden, Colorado

Steve Greenberg, M.S.
Facilities Department
Lawrence Berkeley National Laboratory
Berkeley, California

Vahab Hassani, Ph.D.
Senior Engineer
Buildings and Energy Systems
National Renewable Energy Laboratory
Golden, Colorado

Steve Hauser
Center Director
Market Partnership
National Renewable Energy Laboratory
Golden, Colorado

Roger S. Hecklinger, P.E.
Technical Director
Roy F. Weston, Inc.
Valhalla, New York

Lawrence J. Hill, Ph.D.
Energy Division
Oak Ridge National Laboratory
Oak Ridge, Tennessee

Eric Hirst, Ph.D
Energy Division
Oak Ridge National Laboratory
Oak Ridge, Tennessee

David A. Jump, Ph.D.
Visiting Staff Scientist
Energy and Environment Division
Lawrence Berkeley National Laboratory
Berkeley, California

Jonathan Koomey, Ph.D.
Energy Analysis Program
Energy and Environment Division
Lawrence Berkeley National Laboratory
Berkeley, California

Florentin Krause, Ph.D.
Energy Analysis Program
Energy and Environment Division
Lawrence Berkeley National Laboratory
Berkeley, California

Jan F. Kreider, Ph.D., P.E.
Professor and Director
Joint Center for Energy Management
University of Colorado
Boulder, Colorado

Frank Kreith, P.E
Consulting Engineer
Boulder, Colorado

Jerrold H. Krenz, Ph.D.
Associate Professor Emeritus
Electrical and Computer Engineering
University of Colorado
Boulder, Colorado

Andy S. Kydes, Ph.D.
Senior Technical Analyst
Energy Information Administration
U.S. Department of Energy
Washington, D.C.

J.D. Lutz, B.S.
Staff Research Associate
Energy Analysis Program
Lawrence Berkeley National Laboratory
Berkeley, California

William Marion, M.S.
Senior Scientist
Resource Assessment Program
National Renewable Energy Laboratory
Golden, Colorado

Eugene L. Maxwell, Ph.D.
Senior Scientist
Resource Assessment Program
National Renewable Energy Laboratory
Golden, Colorado

E. Kenneth May, P.E.
Principal
Industrial Solar Technology
Golden, Colorado

James E. McMahon, Ph.D.
Energy Analysis Program
Lawrence Berkeley National Laboratory
Berkeley, California

Jeffrey H. Morehouse, Ph.D.
Associate Professor
Department of Mechanical Engineering
University of South Carolina
Columbia, South Carolina

Julie Phillips, M.S.
J.A. Phillips and Associates
Boulder, Colorado

T. Agami Reddy, Ph.D.
Assistant Director
Energy Systems Laboratory
Texas Engineering Experiment Station
Texas A&M University
College Station, Texas

David S. Renné, Ph.D.
Senior Scientist
Resource Assessment Program
National Renewable Energy Laboratory
Golden, Colorado

Wesley M. Rohrer, Jr.
Associate Professor Emeritus
Mechanical Engineering Department
University of Pittsburgh
Pittsburgh, Pennsylvania

Gregory J. Rosenquist, M.S.
Staff Research Associate
Energy Analysis Program
Lawrence Berkeley National Laboratory
Berkeley, California

Rosalie T. Ruegg
Chief, Economic Office
Advanced Technology Program
National Institute of Standards and Technology
Gaithersburg, Maryland

Martin D. Rymes, M.S.
Staff Scientist
Resource Assessment Program
National Renewable Energy Laboratory
Golden, Colorado

Marc Schwartz
Wind Technology Division
National Renewable Energy Laboratory
Golden, Colorado

Ramesh K. Shah, Ph.D.
Department of Mechanical Engineering
University of Kentucky
Lexington, Kentucky

Max Sherman, Ph.D.
Senior Scientist
Energy and Environment Division
Lawrence Berkeley National Laboratory
Berkeley, California

Leslie Shown, M.A.
Energy Analysis Program
Lawrence Berkeley National Laboratory
Berkeley, California

Scott Sitzer, M.A.
Director, Energy Supply and Conversion Division
Energy Information Administration
U.S. Department of Energy
Washington, D.C.

Craig B. Smith, Ph.D.
Sr. Vice President
Daniel, Mann, Johnson, & Mendenhall
Los Angeles, California

S. Somasundaram, Ph.D.
Staff Engineer
Energy Sciences
Pacific Northwest National Laboratory
Richland, Washington

Kirtan K. Trivedi, Ph.D.
Principal Process Engineer
Process Technology
Brown and Root Petroleum and Chemicals
Alhambra, California

Isaac Turiel, Ph.D.
Staff Scientist
Energy and Environment Division
Lawrence Berkeley National Laboratory
Berkeley, California

W. Dan Turner, Ph.D.
Texas A&M University
Department of Mechanical Engineering
College Station, Texas

Lorin L. Vant-Hull, Ph.D.
Physics Department
University of Houston
Houston, Texas

Charles O. Velzy, P.E.
Consultant
Lyndonville, Vermont

Ronald E. West, Ph.D.
Professor Emeritus
Department of Chemical Engineering
University of Colorado
Boulder, Colorado

Contents

Section I General Principles

Section II Energy Conservation

Section III Renewable Energy

Introduction

"The future belongs to the efficient"
—Admiral James D. Watkins
Former U.S. Secretary of Energy

This handbook is essentially an updated and expanded version of a three volume series titled *Economics of Solar Energy and Conservation Systems* which we edited in the late 1970s and was published by CRC Press in 1980. The update is obviously required because a great deal of research and development in the energy field has taken place in the past 15 years. The expansion is necessary because the energy climate in the U.S. has changed enormously since the original handbook was prepared. At that time, the country's energy policy included the optimistic projection of President Carter that by the year 2000 20% of our energy use would be provided by renewable sources. President Carter's projections have not been fulfilled, however, and a large fraction of the solar energy industry, developed with tax incentives and other financial stimulants provided under the Ford and Carter Administrations, has gone out of business. The Arab oil embargo, which stimulated the interest in renewable energy in the 1970s, has collapsed and fuel prices have plummeted from the heights they reached in the latter part of that decade. The energy market has become more competitive and new ground work for future energy policy was laid in the 1992 National Energy Policy Act (NEPA), which was signed into law by President Bush.

Although energy efficiency, integrated resource planning, and consideration of social costs of energy were important facets of NEPA, its most revolutionary aspect may well be the breakup of the monopolies that electric utilities enjoyed for many years. This is expected to lead to a restructuring of utilities. Under the NEPA regulations, utilities will have to transport electric power from independent producers at competitive prices or divest themselves of the transmission lines. At the same time, new generation technologies have been developed which can provide electric power considerably cheaper than large nuclear generators. Once placed into the grid, these new technologies may also offer severe competition to conservation and renewable energy. Hence, the coverage of this handbook has been expanded to include these new developments in the energy field, as well as the results of research and development in the conservation and renewable technology areas.

The handbook is divided into three sections. The first covers general principles. Chapter 1 gives an overview of the U.S. energy system. Chapter 2 covers the basic elements of thermodynamics, heat transfer, and fluid flow necessary for an understanding of specific technologies. Economic methods to compare various options are covered in Chapter 3. The next five chapters deal essentially with the environment for energy planning created by NEPA. Chapter 4 discusses integrated resource planning, Chapter 5 discusses the social costs of energy, a cost that the public has to bear in the form of adverse health effects and environmental degradation resulting from energy generation. Chapter 6 covers the high efficiency generation technologies that are expected to be used in the next decade and Chapter 7 gives an outlook for the U.S. demand, supply, and prices of energy sources. Finally, Chapter 8 deals with the changes that are expected to occur in the utility industry as a result of restructuring.

Section II covers the various options available for reducing energy consumption using conservation methods. Chapters 9, 10, and 11 cover thermal energy conservation, electrical energy conservation, and HVAC control systems in buildings, while Chapter 12 deals with newly developed energy efficient technologies for motors, lighting, appliances, and air conditioning. Chapters 13–17 deal with energy conservation potential in industry. Chapter 13 describes recuperator, regenerator, and storage technologies, while Chapter 14 deals with thermal energy management in industry. Chapter 15 is devoted to pinch technology of heat exchanger networks because this has great potential for saving energy in future industrial design. Electrical power management in industry is covered in Chapter 16 and cogeneration is covered in Chapter 17.

Section III deals with renewable energy technologies. Chapter 18 describes the availability of renewable resources, including solar radiation, wind energy, and biomass. Chapter 19 deals with active solar heating systems and Chapter 20 with passive solar heating, cooling, and daylighting possibilities. Chapter 21 covers solar thermal power and industrial heat, while Chapter 22 presents the basics of photovoltaics. Chapters 23 and 24 cover wind power and waste-to-energy combustion, respectively.

This handbook does not cover nuclear energy technologies, transportation, geothermal, or hydropower. We debated whether or not nuclear energy should be included in this handbook. Our decision not to include nuclear energy is based upon the objectives set for this handbook, in particular, the handbook has a timeframe of approximately ten to fifteen years and deals with the energy technologies that are likely to be of significance during that period. On December 13, 1994, Gary Fields wrote in USA Today "an era in nuclear energy ended Monday when the Tennessee Valley Authority announced that it would not complete the last three nuclear reactors being built in the U.S.A". Tennessee Valley Authority (TVA) Chairman, Craven Crowell, said in Knoxville that, "given the TVA board's commitment to limit TVA's debt, we believe it no longer makes sense to finish construction by ourselves on plants that are significantly short of completion". He continued, "today we are still building nuclear plants and piling up debts guided by policies that are decades out of date. The TVA Board believes that it is time for a change — time to recognize that policies conceived decades ago are no longer adequate as we approach the 21st Century".

At present, 109 nuclear power plants are operating in the U.S., producing 20% of the country's electricity, which constitutes about 4% of the country's total energy consumption. As shown in detail in this handbook, advances in natural gas powered turbines, wind energy, and co-generation, as well as niche applications for other solar energy technologies, combined with conservation measures promise an adequate supply of energy for the next decade and beyond at considerably less cost than current nuclear power technology. Phillip Baney, President and Chief Executive Officer of the Nuclear Energy Institute in Washington, D.C., argues that American manufacturers have designed advanced nuclear plants that promise greater safety and better performance. However, there are no indications that current plans of utilities in this country include the construction of any new nuclear power plants in the U.S. Since, under current regulations, the time between conception, financing, permitting, and construction of a nuclear power plant is in excess of seven or eight years, there appears to be no likelihood that nuclear power will play a role in the next fifteen years. This does not mean that nuclear power is not likely to find increased application and use in the third millennium, but for the goals and perspective of this publication, it is not a topic of relevance.

The issues involved in energy efficiency for the transportation sector of our economy are quite different from those related to electric power, building comfort, and industrial heat and power. Those issues could not be treated adequately under the auspices of a handbook such as this, but deserve a handbook of their own. Rather than give a superficial treatment to the transportation issues, we decided to delegate this task to specialists in that field for a future book.

The editors have received help and support from many different people in the preparation of this handbook. The editors have benefitted from the suggestions of many reviewers, including Dr. Charles Gay, Director of the National Renewable Energy Laboratory (NREL) (Photovoltaics); Dr. Ralph Overend, Acting Director of the Industrial Technology Division at NREL (Biomass); Dr. T.A. Reddy, Texas A&M University (Solar Radiation); Mr. Mathew Brown, Principal Scientist at the National Conference of State Legislators (Utility Restructuring); Dr. Arthur Rosenfeld, Senior Advisor, Energy Efficiency at the U.S. Department of Energy (Thermal Energy Conservation in Buildings); and Professor David DiLaura, University of Colorado, and Professor Robert W. Jones, Arizona State University (Passive Heating, Cooling, and Daylighting). In addition to the above-mentioned reviewers, the editors are pleased to acknowledge the invaluable assistance rendered by Ms. Bev Weiler, the Editorial Assistant throughout the development of this project and by Ms. Jennifer Richardson, Project Editor.

The very existence of this handbook is a result of the diligent work of the contributors from academia, national laboratories, and industry. Their names are listed with each chapter and the editors dedicate this volume to them in the hopes that their contributions will lead to a more energy efficient nation in the future.

Frank Kreith
Ronald E. West
Boulder, Colorado 1996

Nomenclature

The symbols in this list are used in several chapters. When additional symbols are used in a chapter, they are defined in the text.

- A — Area m^2 (ft^2)
- $\overline{B}$ — Monthly-averaged, daily beam radiation, $kWh/m^2 \cdot day$ ($Btu/day \cdot ft^2$)
- B — Daily total beam radiation, $kWh/m^2 \cdot day$ ($Btu/day \cdot ft^2$)
- C — Cost, $
- C_{se} — Average annual cost of delivered solar enrgy or energy savings by conservation, $/GJ ($/MBtu)
- c_p — Specific heat at constant pressure, $kJ/kg \cdot K$, ($Btu/lb°F$)
- c_v — Specific heat at constant volume, $kJ/kg \cdot K$, ($Btu/lb°F$)
- COP — Coefficient of performance of a heat pump or refrigeration system
- CR — Concentration ratio
- CRF — Capital recovery factor
- d — Diameter, m (ft)
- d_H — Hydraulic diameter, m (ft)
- D — Daily total diffuse (scattered) radiation, $kWh/m^2 \cdot day$ ($Btu/day \cdot ft^2$)
- $E_{b\lambda}$ — Spectral emissive power of a black body at λ, $W/m^2 \cdot \mu$ ($Btu/h \cdot ft^2 \cdot \mu$)
- e — Specific internal energy; eccentricity of earth orbit
- E — Energy
- F — Fin efficiency; force
- F' — Plate efficiency of a flat plate collector
- F'' — Flow factor of a flat plate collector
- F_{ij} — Radiation shape factor between surfaces i and j
- F_R — Heat removal factor of a flat plate collector
- f — Fanning friction factor; frequency
- f_s — Fraction of energy demand delivered by solar system
- g — Gravitational acceleration, m/s^2 (ft/sec^2)
- g_c — Inertia proportional factor occurring in relation Force = (mass × acceleration)/g_c
- Gr_L — Grashof number based on length dimension L
- G — Mass flow per unit area (= ρv), $kg/m^2 \cdot h$ ($lb/ft^2 \cdot h$)
- $\overline{H}$ — Monthly-averaged, total radiation on a horizontal surface on earth $kWh/m^2 \cdot day$ ($Btu/ft^2 \cdot day$)
- $\overline{H}_o$ — Monthly-average, total radiation on a horizontal surface outside the atmosphere $kWh/m^2 \cdot day$ ($Btu/ft^2 \cdot day$)
- h — Specific enthalpy, kJ/kg (Btu/lb); Planck's constant; altitude

h_c — Convection heat transfer coefficient between a surface and a fluid, $W/m^2 \cdot K$ ($Btu/h \cdot ft^2{}^\circ F$)
I — Insolation, defined as the instantaneous, hourly or annual solar radiation on a surface, W/m^2 ($Btu/h \cdot ft^2$)
I_o — Solarrar constant, W/m^2 ($Btu/h \cdot ft^2$)
i — Interest rate, percent; incidence angle, rad (deg)
k — Thermal conductivity $W/m \cdot k$ ($Btu/h \cdot ft^\circ F$); Boltzmann's constant
L — Length, m (ft); latitude; rad (deg); thermal load or demand
l — Length, m (ft)
m — Mass, kg (lb_m)
$\dot{m}$ — Mass flow rate, kg/s (lb_m/h)
n — Index of refraction; number index
NTU — Number of transfer units (dimensionless)
Nu_L — Nusselt number based on length dimension L
P — Power, W (hp)
p — Pressure, Pa = N/m^2 (lb/in^2)
Pr — Prandtl number
PV — Present value
PWF — Present worth factor
q — heat flux, Wm^{-2}
$\dot{Q}$ — Rate of heat flow, W (Btu/hr)
Q — Quantity of energy or heat, kJ or kWh (Btu)
r — Radius, m (ft)
R — Thermal resistance, ($K \cdot m^2/W$) ($^\circ F \cdot ft^2 \cdot h/Btu$); tilt factor; gas constant in $pv = RT$
Ra — Rayleigh number
Re_L — Reynolds number based on length dimension L
S — Surface area, m^2 (ft^2)
s — Specific entropy, $kJ/kg \cdot K$ ($Btu/lb^\circ R$)
T — Temperature, K or °C (°F or °R); tax or tax rate
t — Time, s (h); thickness, m (ft)
U — Overall heat transfer coefficient, $W/m^2 \cdot K$ ($Btu/h \cdot ft^2{}^\circ F$)
u — Specific internal energy, kJ/kg (Btu/lb_m)
V — Volume, m^3 (ft^3); Voltage, V
v — Velocity, m/s (ft/sec); specific volume, m^3/kg (ft^3/lb_m)
w — Width, m (ft)
W — Humidity ratio, kg water/kg dry air, (lb water/lb dry air); Work, kWh (ft lb_f)
z — Zenith angle, deg; altitude above mean sea level, m (ft)

Subscripts

a — air
ab — absorbent, ambient
ann — annual
atm — atmospheric
aux — auxiliary
b — beam; black-body
c — collector; convection
d — diffuse
eff — effective
f — fluid; fin; fuel
h — horizontal surface
i — incident
in — inlet or inside
k — conduction
max — maximum
min — minimum
n — normal
o — reference or standard; extraterrestrial
opt — optimum
out — outlet or outside
r — reflected
rad — radiation
s — solar; surface
terr — terrestrial
t*, T — total
u — useful
w — wall; water
y — yearly

z — zenith

∞ — environmental conditions

Greek Symbols (May also be used as Subscripts)

α — absorptance; solar altitude angle, deg

β — volumetric thermal expansion coefficient collector tilt angle from horizontal plane

γ — specific heat ratio; azimuth, deg; wavelength, m

δ — boundary layer thickness, m (ft); declination, deg

Δ — difference

ε — emittance

E — heat-exchanger effectiveness

η — efficiency; effectiveness

θ,ϕ — angles

λ — wavelength; mean free path, m (ft)

μ — dynamic viscosity, kg/m · s (lb_m/ft · h)

ν — frequency; kinematic viscosity, m^2/s (ft^2/s)

ϱ — reflectance; density, kg/m^3 (lb_m/ft^3)

σ — Stefan-Boltzman constant

τ — transmittance; shear stress, N/m^2 (lb_f/m^2)

ω — solid angle

$\dot{\omega}$ — angular velocity, s^{-1}

Prefixes (factor, prefix, symbol)

10^{12} tera T

10^{9} giga G

10^{6} mega M

10^{3} kilo k

10^{-2} centi c

Section I

General Principles

1

U.S. Energy System

Frank Kreith
Consulting Engineer

Ronald E. West
University of Colorado

Introduction

Energy is a mainstay of modern industrial society. The energy sources on which the United States energy infrastructure is based are mostly fossil fuels, namely petroleum, natural gas and coal, hydropower, and nuclear energy. American energy use can be divided into three end-use segments: transportation, residential and commercial buildings, and industrial uses. Each of these sectors consumes about one-third of the total energy used in this country. The amount of money spent on energy in 1993 was estimated to be about $505 billion or about 8% of the gross domestic product (GDP). In addition to the actual cost of energy, there are environmental and health impacts, which are often referred to as social costs. Including these social costs, the 1994 energy bill was conservatively estimated to be about $577 billion. Obviously, greater energy use efficiency and reduction in the cost of energy can have beneficial impacts on the competitiveness of industry, the environment, and the health and well-being of the American people.

Our current energy structure is based upon a legal and commercial environment in which utilities have had a virtual monopoly over the production, transmission, and sale of electric power and, to a certain extent, natural gas. This traditional energy infrastructure developed more than 20 years ago in an era when information, communication, and computations were expensive, while the cost for labor, fuel, waste disposal, and materials was low. Over the past two decades the environment has been altered as a result of new developments in computer and information technology, increasing population, and social changes. The conditions under which the old infrastructure developed have changed; consequently, this is an opportune time to reconsider the structure of the energy system and plan for a future that provides energy more efficiently with lower environmental impacts and minimum overall cost. All of these trends indicate that energy efficiency will become increasingly important and may be a key to the future economic health of this country.

The U.S. energy system is quite complex. As shown in Figures 1.1 and 1.2, the United States consumed 88.5 quadrillion Btus (or quads) of energy in 1994. Our energy demand depends on the energy service requirements of individuals and industry, as well as on the technologies used to convert fuel and power into heat, light, mobility, and other services. Energy supply is determined

0-8493-2514-5/96/$0.00+$.50

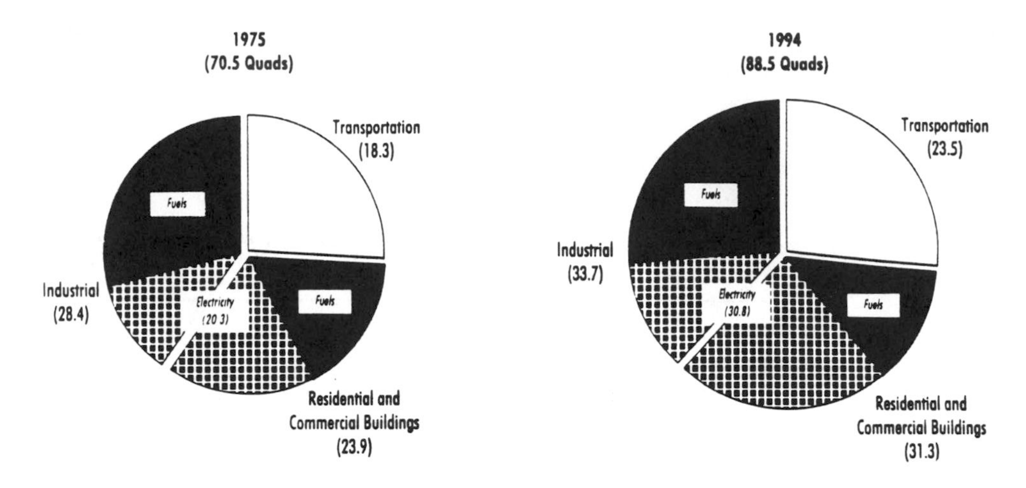

FIGURE 1.1 U.S. consumption of primary energy by end-use sector, 1975 and 1994. *Source:* Department of Energy, 1995.

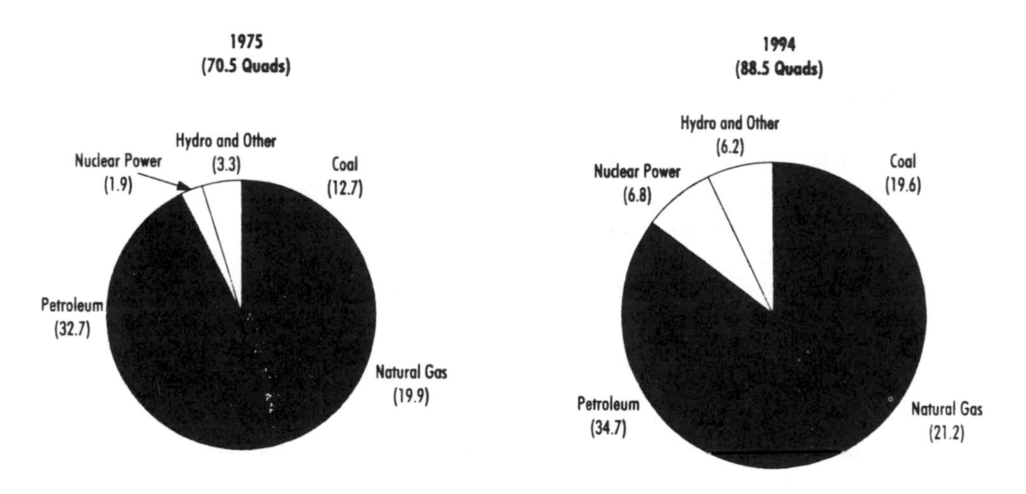

FIGURE 1.2 U.S. consumption of primary energy by energy source, 1975 and 1994. *Source:* Department of Energy, 1995.

by the natural resources (renewable and nonrenewable) that provide the energy and the technologies used to extract, convert, and distribute the energy product to end users. As shown in Figure 1.1, total U.S. energy consumption increased by more than 25% in the past two decades, with the expanded use of electricity in residential and commercial buildings contributing the most to this growth. The energy sources that fueled the growing demand for electricity in the last 20 years were mostly nuclear power and coal, as shown in Figure 1.2. Oil imports also increased by more than one-third to meet the increasing need for fuel in the transportation sector at a time of declining domestic production.

Figure 1.3 depicts how the United States spent money for energy products in 1993, the last year for which complete data are available. The figure shows the energy products in various stages of the market as they are extracted or imported as raw materials (primary energy), then refined and converted into useful products such as gasoline and electricity, and finally distributed and sold to consumers such as businesses and households. The figure also shows that the lion's share of the total energy expenditure went to residential and commercial building uses ($203 billion), followed by transportation ($181 billion), and industrial uses ($121 billion).

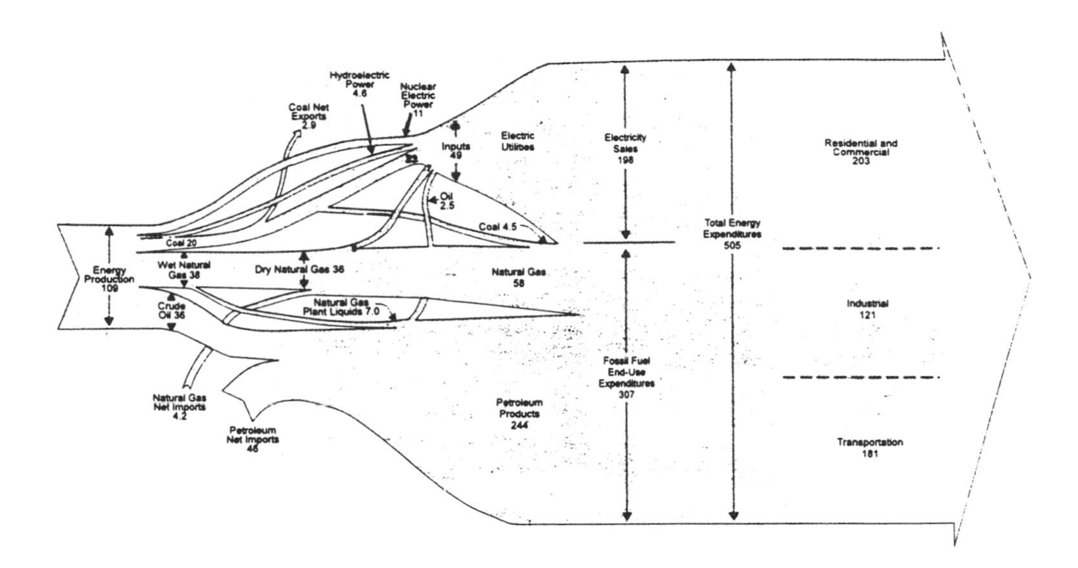

FIGURE 1.3 U.S. energy expenditures, 1993 (billion dollars). *Source:* Department of Energy, 1995.

1.1 1992 National Energy Policy Act

Energy supply and demand markets have changed considerably over the past 20 years. On the supply side, greater competition among suppliers has led to expanded production of coal, natural gas, and renewable energy. On the demand side, many industries have become more efficient and the U.S. economy has shifted away from energy-intensive industry. Thus, the amount of energy used to produce a dollar's worth of GDP, called the energy intensity of the economy, has declined. This trend is expected to continue well into the next century.

Despite an improvement in energy efficiency over the past two decades, North America remains the most energy-intensive region of the world. With increasing global competition, it is becoming imperative that U.S. energy efficiency continue to improve. Past improvements in U.S. energy intensity have not been uniform, as illustrated in Figure 1.4. This figure shows the change in energy intensity between 1975 and 1991 for the most important end use sectors. The energy intensities of industry and residential buildings have shown the greatest improvement in energy efficiency. The need for further improvement has been incorporated into the provisions of the 1992 National Energy Policy Act (EPACT), which aims to ensure continued improvement in energy efficiencies in all sectors for the future. Under this Act, future utility energy planning will have to follow an integrated approach that requires comparisons of various technologies to provide specified services. This procedure, referred to as integrated resource planning (IRP), is discussed in more detail in Chapter 4. Essentially it requires a comparison of conservation of energy with production of energy on a level playing field. It also recommends that the social and environmental costs of energy production be included in the planning process. Chapter 5 provides an overview of the social costs of energy production, which should be considered in a fair comparison between production and conservation. Chapter 6 describes the new high-efficiency electrical generation technologies that are entering the market and their economic prospects. Chapter 7, contributed by experts from the Energy Information Administration, gives estimates of the supply and cost of energy sources for the next decade.

Another facet of the 1992 EPACT, combined with Executive Order 636 passed by the Federal Energy Regulatory Commission in 1992, was designed to break up the monopolistic structure of the electric and gas utilities. Under these new provisions, utilities will have to transport electric power and natural gas from independent producers and their own plants at comparable prices or

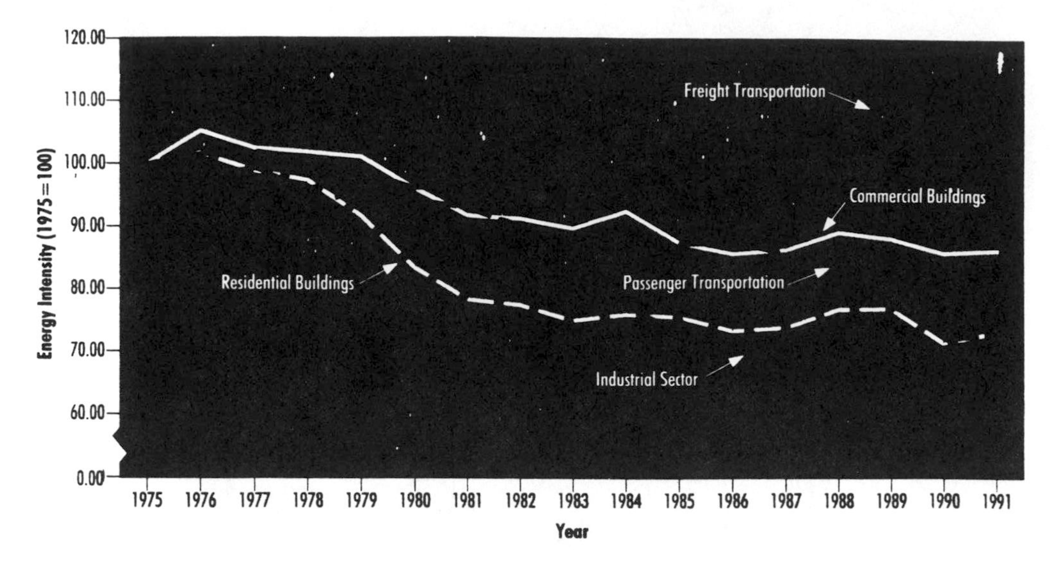

FIGURE 1.4 Changes in U.S. energy intensity by end-use sector, 1975–1991. *Source:* Department of Energy, 1995.

divest themselves of the transmission lines required to transport gas and electricity. These provisions, coupled with recent developments of more efficient electric conversion and cogeneration technologies using natural gas as fuel, and the current low prices of natural gas, will increase competition in the utility business enormously. Chapter 8 describes the processes of utility restructuring and the role of state governments in ensuring a smooth transition of a previously regulated monopoly to a future free market competitive energy industry and its potential impact on inefficient and expensive power producers such as nuclear power plants.

1.2 Industrial Energy Uses

Our standard of living and industrial productivity depends on a reliable supply of competitively priced energy. In 1993, industrial facilities purchased more than $100 billion worth of energy, while the average family spent more than $2,000 per year on energy for its various activities. The country also spent $56 billion for energy imports, which accounted for almost 10% of the total cost of imported products.

American industry needs energy to generate heat and steam for industrial processes, electrical power for motors, machines, lighting and cooling factories and buildings, and moving materials from place to place. Oil and natural gas are also needed as feedstocks for manufacturing plastics and other materials.

In 1991, U.S. industry consumed about 23 quads (quadrillion Btu) of energy, which is equivalent to about 37% of the country's total energy use if electric generation losses are included. The breakdown of energy consumption by major industry is shown in Figure 1.5. It should be noted that six major industries (petroleum refining, food processing, metals processing, pulp and paper manufacturing, glass and ceramics manufacturing, and chemical industry) accounted for about 78% of all industrial energy use and 87% of all energy used in U.S. manufacturing. The cost of energy for the industrial sector was about $100 billion/year. Obviously, any measures that can reduce energy use in industry will have beneficial results for the American economy. In particular, materials and process industries have great opportunities to reduce the cost of their products by improving their energy efficiency.

Chapter 13 deals with the design of recuperators and regenerators that can be used to reclaim waste energy in industrial processes and with the opportunities for thermal storage. Chapter 14

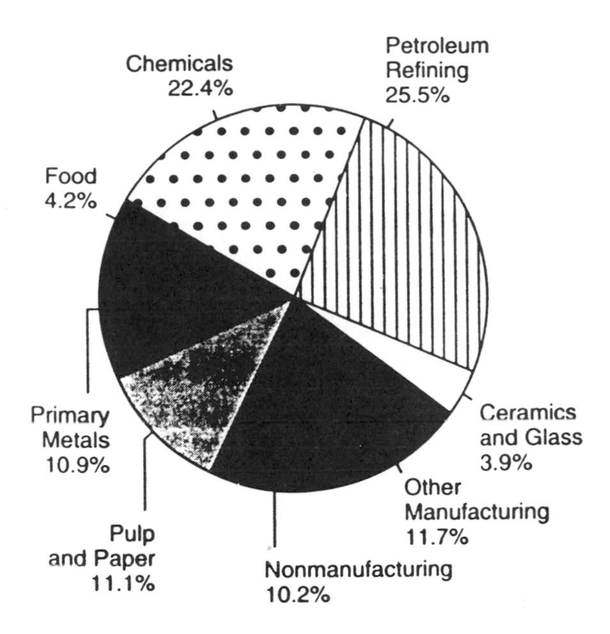

FIGURE 1.5 Energy consumption by major industry, 1991.

describes conservation by improved thermal energy management in industry. Chapter 15 deals with a specific aspect of thermal energy management, namely the design of heat exchanger networks. Improving the network heat exchanger design can save as much as 30% of the energy used in the chemical industry according to the author. Chapter 16 describes how electrical power management in industry can be used to improve energy efficiency. Also, some of the material in Chapters 9 and 10, which deal with energy conservation in buildings, Chapter 11, which describes improved methods for HVAC control systems, and Chapter 12, which summarizes energy efficient technologies, such as motors and lighting, can contribute to a more energy-efficient industrial sector in the U.S. energy system.

1.3 Energy Use in Buildings

Energy use in buildings accounts for 35% of the total primary energy consumption in the United States and 42% of the total energy costs, and produces 35% of all U.S. carbon emissions. Utility bills are a substantial part of a family budget, so residential energy efficiency affects the kinds of housing people can afford as well as their comfort level. Energy use in commercial buildings represents a cost to business, and the comfort and lighting in them have a substantial bearing on employee productivity. Recent studies have shown that the overall potential for reducing energy use in buildings through cost-effective investments is substantial. These energy savings can be achieved simultaneously with overall enhancement of indoor air quality, occupant comfort, and worker productivity. Cost savings for energy efficiency measures in buildings are estimated at $13 billion per year by the year 2000 (in 1992 dollars) and 38 billion by 2010. In addition, several hundred thousand new jobs can be created through energy efficiency measures in buildings, according to recent DOE studies.

Energy use in U.S. buildings has increased from about 22 quads in 1970 to about 30 quads in 1989 and has continued to rise since. Residential and commercial buildings account for about

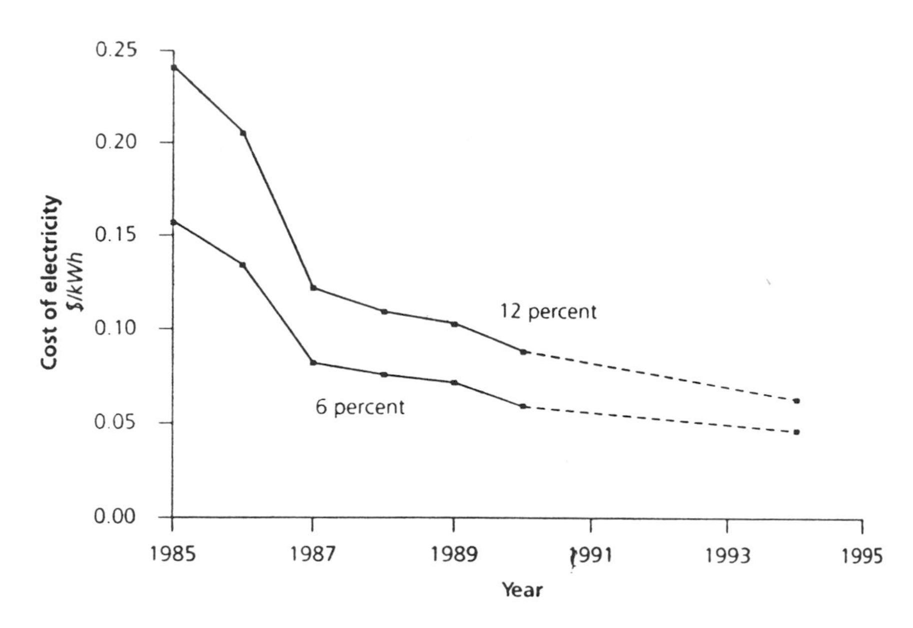

FIGURE 1.6 Building energy use: two future scenarios. *Source:* OTA, 1992.

one-third of U.S. energy consumption at an annual cost of $170 billion. Using commercially available cost-effective technologies, building energy consumption could be reduced up to one-third by 2015, compared to a business as usual projection as shown in Figure 1.6, according to an Office of Technology Assessment study released in 1992 (OTA 1992). The growth in building energy consumption has been attributed to increases in population and the number of households and offices. In addition, increased service demands for more air conditioning, larger houses, and more electrical equipment, such as computers, has been a factor. These increases have been ameliorated by more energy-efficient building shells, energy-efficient appliances, improved building designs, and stabilization in the annual energy use per square foot in the commercial sector.

Building energy efficiency measures include many technologies, such as improved lighting, energy-efficient appliances, increased insulation, and reducing unnecessary infiltration, as well as high-efficiency windows. Among the most significant energy efficiency measures are improvements in HVAC systems. The reason for the significance of HVAC is that air conditioning is the fastest-growing energy use in the country and most air conditioning systems rely on electric power for their operation and control.

According to OTA, building energy use will continue to grow in a business-as-usual scenario (that is, assuming no change in current policy) at a moderate pace, reaching roughly 42 quads by 2015. In an alternative perspective, assuming that all energy-efficient technologies with a positive net present value to the consumer are implemented, energy use in buildings could actually decrease to about 28 quads by 2015. Some of the most important cost-effective energy conservation technologies are listed in Table 1.1 with their estimated paybacks in years. Although many conservation technologies have a high initial cost, this summary shows that energy savings more than repay the initial outlay in a reasonable time.

Figure 1.7 shows a breakdown of the 17 quads of energy used by residential buildings in 1988. Increased demand for particularly energy-intensive services, such as central air conditioning and clothes dryers, have contributed to the large percentage of electric power use in the residential sector.

Figure 1.8 shows a breakdown of the 13 quads of energy used by commercial buildings. About two-thirds of the energy use was in the form of electricity, mainly for space heating, cooling, and lighting. The most energy-intensive commercial buildings are those used for offices, health care,

TABLE 1.1 Cost-Effectiveness of Selected Energy Efficient Technologies

Technology	Typical payback (years)
Additional insulation	6 to 7
Compact fluorescent lamps	Less than 2
Condensing gas furnace—95% + efficient	4 to 7
Electronic ballasts for commercial lighting	3 to 4
Improved burner head for oil furnaces	2 to 5
Residential duct repair	Less than 2
Highly efficient room air conditioner	6 to 7
Water heater tank insulation	Less than 1

Source: OTA 1992.

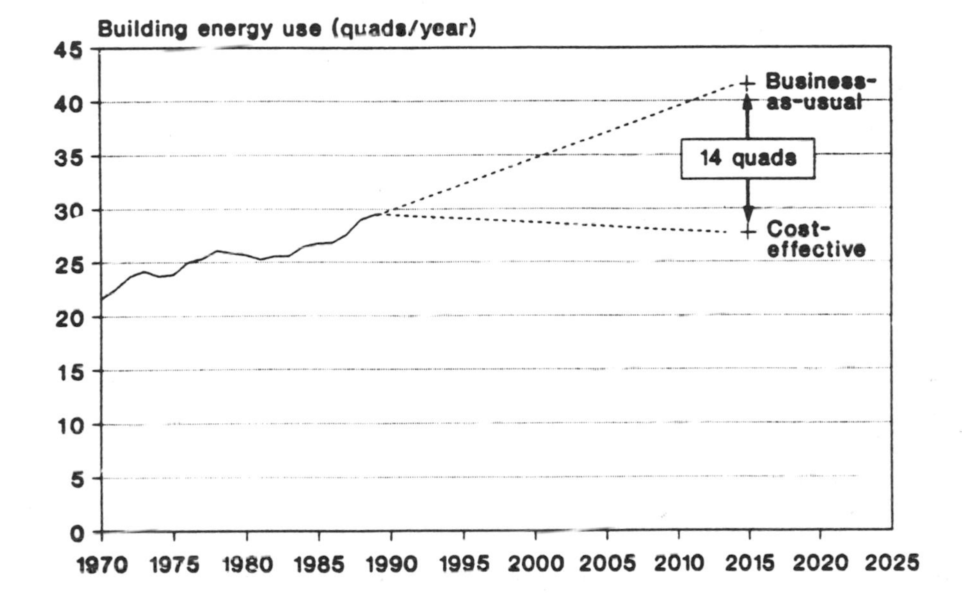

FIGURE 1.7 Residential sector energy use by end use and fuel type, 1988. *Source:* OTA, 1992.

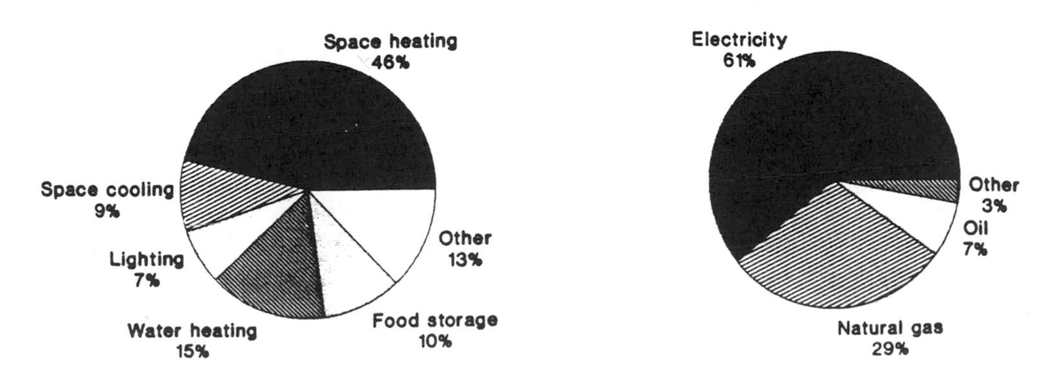

FIGURE 1.8 Commercial sector energy use by end use and fuel type, 1988. *Source:* OTA, 1992.

and food. The most important factor in the rapid growth (an average annual rate of 2.3%) is the increase in the number of commercial buildings. However, better building shells (windows and insulation), better appliances (furnaces, air conditioners, and refrigerators), and better building design have stabilized the energy intensity (energy use per square foot per year) in the commercial sector.

Several modeling efforts by various organizations and government agencies have attempted to estimate the potential energy savings possible. Although there is considerable uncertainty in these modeling efforts, there is general agreement that significant opportunities for increased energy efficiency in U.S. buildings exist through the use of cost-effective technologies described in considerable detail in subsequent chapters of this book.

1.4 Energy Use and Environment

It is well recognized that energy use and environmental quality are intimately related. Greater energy efficiency reduces the amount of fuel needed to perform a given task and thereby also reduces pollution and improves environmental quality. The U.S. relies largely on regulations that focus on specific energy technologies such as power plants, buildings, and automobiles to reduce the adverse environmental impact of energy use. To some extent, these regulations have led to improvements in energy efficiency. Implementation of the Clean Air Act Amendments of 1990 will further reduce the emission of many pollutants from the use of fossil fuels (e.g., nitrogen oxides and sulfur oxides), which contribute to acid rain, but low energy prices over the past few years have also blunted the incentive to invest in energy conservation measures and renewable conversion systems that could improve energy use efficiency in the future.

A sustainable energy future requires that we maximize energy productivity. Getting the most out of the energy we use will keep costs of energy services such as light, heat, power, and mobilities at affordable levels. At the same time, maximizing energy productivity also reduces pollution and global warming. Despite the many benefits that can be derived from improved energy efficiency, there are some barriers to its widespread implementation. The foremost of these is the high initial investment required to implement conservation and energy efficiency technologies. This is especially a problem in public institutions, which do not have the financial freedom to allocate funds to long-term energy efficiency investments. Moreover, state and local governments interested in energy efficiency programs often do not have access to technical support systems and financial borrowing opportunities. Finally, there is often a lack of reliable information, and with this comes a distrust of the methods used to estimate future cost savings that would accrue from investments in energy efficiency.

1.5 Renewable Energy Technologies

According to a survey conducted in 1994 (Farhar, 1994), the American public prefers and is also willing to pay for the implementation of energy efficiency, conservation, and renewable energy systems. Public concern about the environment has increased during the past decade and, according to Farhar, "the public prefers policies supportive of energy efficiency and the use of renewables over other energy options and policies that foster environmental protection and improvement...." Finally, the poll suggests that the public wants a broad national agenda of sustainable development, with "energy efficiency and the use of renewable energy increasingly institutionalized." Similar conclusions have been drawn by other surveys in the U.S. and other parts of the world.

Despite public preference, market penetration of renewable energy technologies has been relatively modest. Although it is generally recognized that our fossil fuel supply is limited and that increasing dependence on petroleum imports undermines the energy security of the country, the high initial cost of renewable technologies has limited their deployment. Moreover, despite scientific advances in research, with the exception of wind and photovoltaics, there has been little improvement in the performance and cost of renewable energy conversion systems because technology transfer has

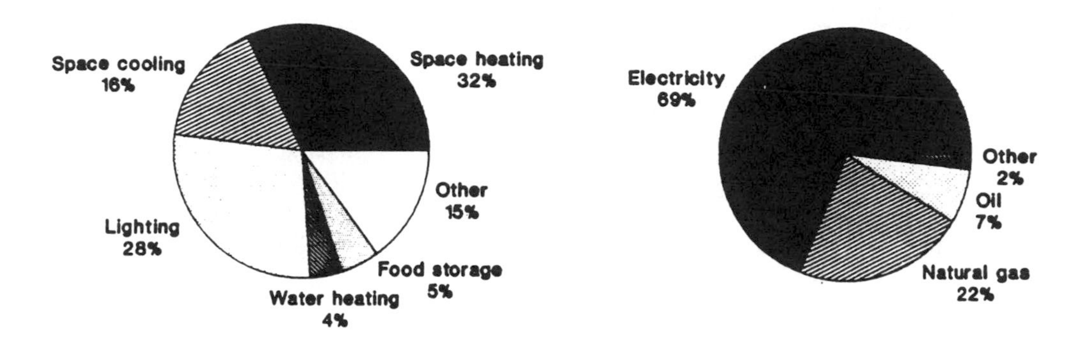

FIGURE 1.9 Declining cost of wind energy from wind turbines in California (all taxes have been neglected). *Source:* Cavallo et al., 1993.

been lacking. Nevertheless, opportunities for renewable technologies, especially in wind, bioconversion, hydropower, and solar thermal energy, are promising and should lead to greater deployment in the next 10 to 20 years.

Wind Power

Results from field operation of close to 20,000 wind turbines in Europe and California during the past decade have demonstrated the viability of wind turbine technology. The reliability of the turbines and operating and maintenance procedures in wind farms are satisfactory. In areas with a good wind source, defined as 450 W/m^2 of wind power density at hub level, the current generation of wind turbines can generate electricity at a cost below 6¢/kWh. Availability of the wind resource is covered in Chapter 18B. As shown in Figure 1.9, the cost of wind-generated electricity in California has decreased substantially during the last decade and indications are that this trend will continue (Cavallo et al., 1993). Also, installed capital costs have decreased and the DOE projects that, within the next decade, wind turbine technology will deliver electric power at less than 3¢/kWh. If this goal can be achieved, wind-generated electricity will be fully competitive in good wind climates with electricity generated by fossil fuels and, as shown in Chapter 23 wind-generated electric power will become a larger source of electricity in the next decade.

Photovoltaics

The situation is quite different for photovoltaic (PV) systems as is discussed in Chapter 22. In the 1960s, PV systems were largely used to power spacecraft and calculators. In the 1970s, PV systems were used to power remote industrial loads, such as telecommunications repeaters, offshore navigation aids, and cathodic protection systems. In the early 1980s, some fairly large centralized utility PV power plants, built with federal financial support, raised expectations that PV systems could compete with fossil fuels in utility applications, but the high cost of PV power, combined with a reduction in oil prices, made utility applications of PV systems economically unattractive, and there is little expectation that PV systems will be used within the next decade for centralized utility power plants.

At the present time, photovoltaics cannot compete with conventional electric power on an energy cost basis. While conventional electric power can today be produced for between 4 and 6 ¢/kWh, photovoltaic DC energy costs are between 20 and 40 ¢/kWh. Although the cost of photovoltaic cells and modules has decreased considerably over the last 30 years because of improved manufacturing, the elements beyond the modules, called the balance of system (BOS) components, which include array structures to support the modules, inverters, power conditioning circuitry for control and modification of the output, and means of storing energy, have not similarly decreased. At the present time, the cost of BOS items is about equal to the cost of the PV module, although the BOS fractional cost can vary from one-third to two-thirds of the total installed cost of the system. The cost of batteries for energy storage in photovoltaic systems, including maintenance and replacement cost, is roughly 10 ¢/kWh for present day lead-acid battery technology. Consequently, even if the entire photovoltaic system cost nothing, PV electric power could still not compete with conventional electric power in situations where battery storage is required.

There are, however, niche applications for photovoltaic systems in situations where either no power is available or the cost of building an extension line to the location where the power is needed is prohibitively expensive. Such applications occur for lighting, refrigeration, or water pumping in remote locations and offer near-term opportunities for small PV systems.

Solar Thermal

Solar thermal technology can be divided into three segments:

- Passive solar technologies for heating, cooling and lighting, as covered in Chapter 20.
- Flat-plate solar collectors suitable for domestic hot water and swimming pool heating at temperatures below 180°F as covered in Chapter 19.
- Intermediate- and high-temperature technologies that require the concentration of solar radiation as covered in Chapters 21A and B. Intermediate temperatures can be delivered by nontracking compound parabolic concentrating collectors (CPCs), but for temperatures above 250°F, tracking is required.

During the late 1970s, flat-plate collector technology, assisted by government financial support in the form of tax credits, developed to the point where it was poised for serious market penetration. In 1980, flat-plate domestic hot water heaters were selling at a price that achieved, without tax credits, a payback of 5 to 6 years in favorable solar climates with a system life in excess of 20 years. With federal tax credits, the payback period was less than 4 years by 1980. By the end of that decade, more than 1.5 million buildings and more than 100,000 swimming pools had solar systems installed. Unfortunately, the cessation of federal support caused a major disruption for the budding solar industry and a majority of suppliers ceased operation, either due to bankruptcy or falling profits. However, overseas, use of solar hot water systems is growing. In Israel, for example, 3% of the country's total energy needs are met by flat-plate solar collectors for domestic hot water installed on the roofs of buildings. Similar opportunities exist in the U.S. The availability of the solar resource is covered in Chapter 18A.

In the U.S., the leading solar thermal technology today is the one-axis tracking parabolic trough, which focuses radiation on a tube through which a heat-absorbing fluid passes. The hot fluid is circulated through a boiler or preheater, where its thermal energy is used to generate steam that generates electric power with a conventional turbine. More than 350 MW of power from parabolic trough solar thermal systems are providing energy for the Southern California Edison Company's utility grid. The U.S. Department of Energy has set an ambitious goal for parabolic trough technology to provide an additional 440 MW in the United States and an equal amount of power overseas by the year 2000 (Sheinkopf, 1994).

Central receiver technology is considered by many solar experts the most promising approach to cost effective solar electric power. It uses a field of mirrors to focus the sun's energy on a central receiver that is mounted on a tower. An experimental 10 MW central receiver power plant, Solar One, was built and operated in California during the 1980s. The plant has now been refurbished with upgraded heliostats and the water–steam system has been replaced by a new and more efficient nitrate salt receiver, storage, and steam generation system. The current project, called Solar Two, is a cooperative venture between the U.S. Department of Energy, industry, and a consortium of eight utilities led by Southern California Edison Company. The primary purpose of Solar Two is to validate the projections for the cost, performance, and reliability of future commercial plants.

A parabolic tracking dish, using a Sterling engine to generate power in the range of 5 to 25 kW, is in the R&D stage for small-scale application in isolated regions (Mancini et al., 1994). According to Sandia National Laboratory, 25 kW electric systems should be commercially available in 1998 or 1999 and electric power cost is projected to be between $0.06 and $0.11 per kWh.

The prospects for solar thermal applications are intimately tied to the cost of natural gas and coal. Although for the immediate future economic viability may limit development, in the long term, solar thermal systems are expected to find a variety of applications.

Hydropower

Hydropower is an old and well-established technology. Its current domestic installed capacity of 72,000 MW accounts for almost half of the total energy contribution from renewable energy sources in the United States. Hydropower uses the kinetic energy of flowing water to turn a turbine, which is coupled to a generator that produces electricity. Although most hydropower facilities use large impoundment dams, hydropower can also be generated on a small scale by diverting a portion of a stream or river. Such diversion projects may require a dam, but the dams are usually much smaller and less obtrusive than impoundments dams.

Hydropower technology is also used to store energy. During low-load periods, excess electrical supplies can be routed to a pumped storage facility, which stores potential energy by pumping water from a lower reservoir to another reservoir at a higher elevation. During peak-load periods, the water is allowed to return from the upper reservoir to the lower reservoir, turning a turbine and generating electricity in the process.

Hydropower plants have a rich history and played a major role in spurring industrial development in the 19th century. By the 1930s, hydropower provided 30% of the nation's generating capacity. However, the growth of other generation sources slowly eroded the hydropower capacity share to its current 12%. The most favorable hydropower sources in the U.S. have been tapped, but there is still potential for obtaining additional capacity through a combination of new site development, development of generating capability at preexisting impoundments, and equipment upgrade at existing plants. There is also significant potential for development of small hydropower facilities throughout the country with some existing dams.

However, hydropower development has slowed in recent years because of environmental concerns and more stringent regulatory and operating requirements. As a result of the Electric Consumers Protection Act (ECPA), enacted in 1986, the time and cost of licensing hydroelectric projects has escalated. Many older hydropower projects will require relicensing during the 1990s, exposing these projects to greater scrutiny and a potential loss of capacity.

Biomass

Biomass-based energy systems utilize wood, agricultural and wood waste, municipal waste, and landfill methane gas as fuels. In all its energy uses, biomass currently supplies more than 3% of total U.S. energy needs and provides almost 10,000 MW of electric generating capacity. Wood fuels provide the bulk of this generation (66%), followed by municipal waste (24%), agricultural waste (5%), and landfill gas (5%). The availability biomass resources is covered in Chapter 18C.

Wood is the leading biomass energy resource used for power generation, primarily because of its use as a boiler fuel in the lumber and in the pulp and paper industries. Wood has environmental advantages in terms of emissions of carbon dioxide, a greenhouse gas. Although the burning of a tree releases carbon dioxide, an equal amount of carbon dioxide is removed from the atmosphere when the tree grows. Thus, as long as the trees that are burned are replaced by growing new trees, the net emission of carbon dioxide is small. The lumber industry satisfies close to 75% of its energy needs through direct wood combustion, while the pulp and paper industry has achieved a 55% aggregate fuel contribution from wood. Many of these companies use cogeneration systems for power generation. The Electric Power Research Institute estimates that more than 6,000 MW of nonutility, wood-fired generating capacity was in place at the end of 1991. While biomass resources are present in all 50 states, handling and transportation costs present serious economic barriers.

Moreover, combustion efficiencies in existing power plants is less than for fossil fuels. Many experts believe that the development and utilization of short-rotation woody crops are necessary for any significant expansion of wood resources to occur.

Municipal waste is the second largest source of biomass power, generating more than 2,000 MW of electricity and providing steam for industrial uses. More than 580,000 tons of municipal waste are generated in the United States each day, with three-quarters or more of this total going to landfills. With landfills charging higher costs and adopting stricter regulations, many localities have turned to waste-to-energy (WTE) systems as a disposal alternative. WTE technologies are covered in Chapter 24. An estimated 15 to 20% of municipal waste is burned for energy. Several industry sources have predicted that from one-third to one-half of the nation's municipal waste could be combusted for energy generation.

Agricultural waste plants are the third largest biomass generators, producing another 575 MW nationwide. These plants use such diverse feedstocks as bagasse (sugarcane residue), rice hulls, rice straw, nut shells, crop residues, and prunings from orchards and vineyards.

Finally, more than 100 power plants in 31 states burn landfill-generated methane. The existing landfill gas energy projects have a combined electricity generating capacity of about 350 MW. According to the General Accounting Office, the EPA estimates that potential energy production from landfill gases ranges from 800 to, at most, 5,000 MW of additional electricity generating capacity. At the same time, the EPA cautions that energy production from methane faces barriers, such as the low price of competing fuels, limited and unstable markets, and regulatory constraints (GAO, 1994).

Geothermal

Geothermal resources can be used for power generation or for heating. They exist as either dry steam or as hot water. Dry steam, which is a rare resource, can be routed directly to a turbine to generate power. For power generation from hot water, there are two primary conversion technologies: flash plants (for resource temperatures above 175°C), which rely on flashing the hot water to steam, and binary plants (for resource temperatures of 100°C to 175°C), which use the heat of the hot water to vaporize a "working fluid," usually an organic compound. These technologies are currently used to generate electricity from geothermal resources in California, Hawaii, Nevada, and Utah. In 1990, 62 geothermal electric power plants were in place with a total generating capacity of slightly more than 2,350 MW.

Geothermal energy is also found in the form of geopressured brines at depths of 3 km to 6 km or more. The technology has been developed to use this resource, but it is not cost-effective.

Relatively high-temperature resources for geothermal electric power generation exist only in limited amounts in the Western, U.S. Low-temperature resources (<130°C) are more widespread across the country. These resources can be used for application such as heat pumps, aquaculture, industrial processes, and domestic hot water. It is estimated that there are 130,000 direct-use installations with a total thermal installed capacity of 2,100 MW, providing annually about 19 trillion Btu. The fastest growing direct-use application is geothermal (ground source) heat pumps (NREL, 1993).

1.6 Energy Efficiency and Conservation

The First Law of Thermodynamics states that energy can neither be created nor destroyed. Hence, this law alone is of little help in determining how effectively a given energy resource is used to perform a given task. The process of mining primary energy resources (such as fossil or nuclear) or building hydropower plants and converting these sources into a usable form of energy is highly complex. To perform a given task for which energy is needed requires first conversion of a primary resource into a form suitable to perform the task. For example, to provide domestic hot water for washing and other household tasks requires raising the temperature of water from a lower to a higher level. The heat required to raise the temperature could be provided by a natural gas-powered furnace or an electric resistance heater. The former converts between 75 and 80% of the primary energy into heat, while the latter has an efficiency of less than 30%. Despite the large efficiency advantage of the natural gas system, many new homes have electric heating systems because the initial investment for the electric heater is considerably less than that for the gas furnace. Although the savings in heating bills from the gas furnace would repay the difference in initial investment in 2 or 3 years, the energy choice of many American consumers ignores the levelized cost of the system in favor of up-front savings. This short-term outlook also exists in many industries and is responsible for most of the inefficient energy designs in buildings and industrial installations.

Thermodynamicists have realized for some time the shortcomings of First Law approaches and have recently developed the theoretical underpinnings for using the Second Law of Thermodynamics to demonstrate specific ways and means to improve energy efficiency (Bejan et al., 1996). The approach, called Exergy Analysis, has been combined with economic principles into a branch of engineering called **Exergoeconomics** to provide principles for energy efficient system design. Exergoeconomics can be used as a tool for exergy-based cost minimization. The approach recognizes that it is not energy but exergy, or availability, that is the resource of value, the commodity that "fuels" possess and which the consumer is willing to pay for. Recent books such as *Thermal Design and Optimization* (Bejan et al., 1996) have introduced Second Law methods to pinpoint and quantify losses and inefficiencies in energy systems as well as exergy-based optimization methods, but until exergy pricing is used by industry and government and is widely accepted as the basis for pricing in the real world of economics, it will be necessary to analyze systems with energy concepts and First Law optimization processes, the approach taken in this book.

Generation of electricity from fossil, nuclear, or solar thermal resources requires a power cycle with a high-temperature reservoir and a low-temperature sink, usually the environment. The process and technology that converts a primary resource to a useful form, such as shaft power, heat, or electricity, performs this conversion with an efficiency that is usually defined as the energy output divided by the energy input. The higher this efficiency, the less primary resource is required to convert its latent internal energy content into a useful form.

Except for nuclear power, hydroelectric power, and a small amount of biomass, essentially all primary sources used today are fossil fuels. Fossil fuels are limited in supply, and their anticipated future cost and availability are described in Chapter 7. There is general agreement that a transition from fossil fuels to renewable energy sources (including solar, hydro, wind, biomass, breeder nuclear, and fusion) is inevitable and that energy conservation measures will make the transition easier. However, the time scale of such a shift is being intensely debated. In this handbook, a time frame of the next 10 to 15 years has been chosen and the energy-efficiency perspectives of those technologies that are expected to have a major impact in this period are treated.

In the introduction to this handbook, U.S. Secretary of Energy Hazel O'Leary points out that in addition to the technical advantages of using conservation and renewable energy sources to reduce the need of fossil and nuclear fuels, there are also some social gains. Renewable energy and conservation systems are attractive not only because of their economic and environmental benefits, but also because of the potential for job generation. For example, as shown in Figure 1.10, according to a study by the World Watch Institute, direct employment from solar and wind technologies per kWh/year is on the order of 250 and 540 people respectively, compared to 116 from coal-fired power plants (Kreith, 1994). In an era when generation of new jobs is important for the economic well-being of the country, the enhanced job generation potential of renewable and conservation technologies make these technologies socially and politically attractive for the future.

Broadly speaking, higher energy efficiency can be achieved by two approaches:

1. Improving the conversion efficiency of a primary resource into a form of energy suitable to perform a given task
2. Reducing the energy required to perform a specific task.

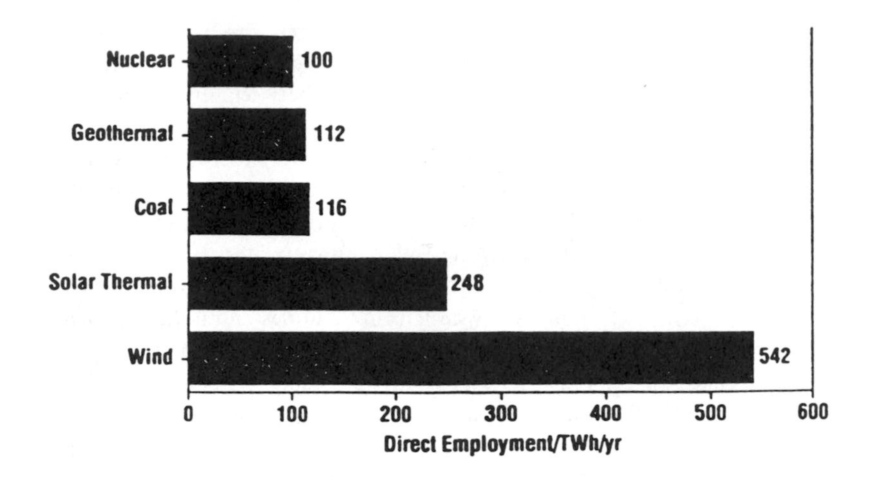

FIGURE 1.10 Jobs from electricity generation. *Source:* Worldwide Institute, 1991. Paul Gipe & Associates, taken from Kreith, 1994.

The first approach to achieving high energy efficiency is to utilize technologies that can achieve a high conversion efficiency with a given energy resource. For example, if a steam or gas turbine is used as the electric generation technology, the higher the turbine inlet temperature, the higher the conversion efficiency. Turbine efficiencies approach 40%, and the maximum temperature is usually limited by metallurgical and structural considerations. But there may be other limitations. The upper temperature of a working fluid in a nuclear electric power plant is limited by safety concerns, and thermal conversion efficiency cannot exceed the order of 30%. On the other hand, a gas turbine cogeneration plant that not only utilizes a high-temperature working fluid but also provides heat from the thermal energy that is ordinarily discarded into the atmosphere or a body of water can achieve efficiencies between 60 and 70% with cogeneration. Obviously, for a given cost of primary energy, a cogeneration plant can deliver a great deal more useful power and/or heat than an energy conversion system that generates electricity and thus only utilizes a smaller portion of the available energy. Modern energy technologies that can achieve high energy efficiencies by use of high-temperature gas turbines and cogeneration conversion are described in Chapters 6 and 17.

The other side of the coin of achieving energy efficiency is to reduce the amount of heat or power required for a given task. This is commonly called energy conservation. In January 1990, a study by the staffs of five national laboratories (Oak Ridge National Laboratory, the Argonne National Laboratory, the Lawrence Berkeley National Laboratory, the Pacific Northwest Laboratories, and the Idaho National Engineering Laboratory) asked the question: "Energy efficiency: How far can we go? (Carlsmith et al., 1990)" The review began by noting that in 15 years following the Arab oil embargo in 1973, energy use in the U.S. increased by only 8% while gross national product increased 46%. At the end of this period, the U.S. was producing a dollar's worth of goods and services with 26% less energy than in 1973. The reduction came mostly from increases in energy efficiency and conservation. However, overall energy utilization efficiency has hardly changed since oil prices dropped precipitously in 1986. Nonetheless, detailed studies of available energy technologies point to largely unrealized opportunities for more efficient energy use. A study by the Electric Power Research Institute (EPRI) has shown that by replacing existing electric devices with more efficient ones, the U.S. could cut electricity consumption by over 31%. Figure 1.11 illustrates EPRI's estimates of the opportunities for energy savings by replacing existing devices, such as motors, lighting, and water heating, with currently available, more efficient technologies.

The 1990 review by the national laboratories considered only end-use consumption of energy (residential, commercial, transportation, and industrial sectors) but omitted the significant

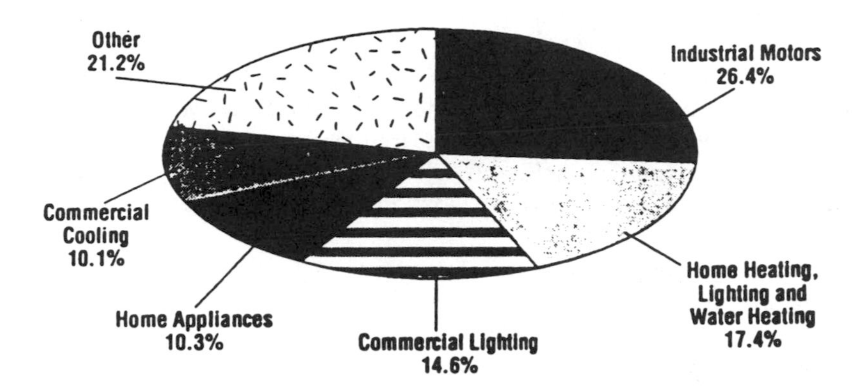

FIGURE 1.11 Energy Saving Technologies. Replacing existing devices with more efficient ones could cut U.S. electricity consumption over time by 31.3%. *Source:* Electric Power Research Institute, taken from Kreith, 1994.

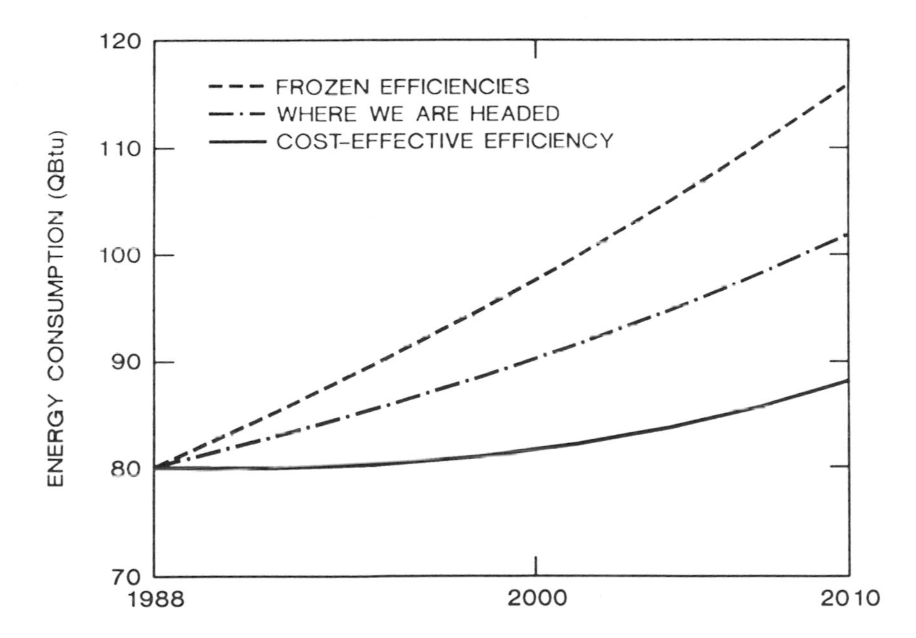

FIGURE 1.12 Energy consumption projections reflecting energy efficiency scenarios. *Source:* Carlsmith et al., 1990).

efficiency improvements possible in the generation, transmission, and distribution of power. The study first estimated likely future trends in end-use energy efficiency, using an assumed course of events in population growth, economic growth, and fuel prices. The study then compared this business-as-usual scenario with a scenario that utilizes all cost-effective efficiency measures. This baseline scenario, titled "Where We Are Headed," is shown by the dash-dot line in Figure 1.12. The solid line in this figure shows the large gains that would be obtained if advantage were taken of efficient cost-effective technology in all new installation and replacement. The energy consumption with each of these two scenarios was then compared with the energy consumption that would result if there were no improvements in efficiencies in the various end uses. The comparisons displayed in Figure 1.12 clearly show the enormous opportunities for improving energy efficiency.

References

Bejan, A., G. Tsatsarouis, and M. Moran. 1996. *Thermal Design and Optimization,* Wiley, New York.

Carlsmith, R.S., W.U. Chandler, J.E. McMahon, and D.J. Santini, 1990. *Energy Efficiency: How Far Can We Go,* Oak Ridge National Laboratory/Martin Marietta, ORNL/TM-11441, January.

Cavallo, A.J., S.M. Hock, and D.R. Smith. 1993. "Wind Energy: Technology and Economics," in *Renewable Energy —Sources for Fuel and Electricity,* Johansson, T.B., et al., Eds., Island Press, Washington, DC.

DOE. 1995. *Sustainable Energy Strategy.* U.S. Department of Energy. Washington, D.C. (Energy Information Administration).

EIA. 1992. *Manufacturing Energy Consumption 1991.* U.S. Department of Energy. Washington, D.C.

Farhar, B. 1994. Trends in U.S. Public Perception and Preference on Energy and Environmental Policy, *Annual Review of Energy and Environment,* vol. 19, p. 211.

GAO. 1994. *Energy Potential of Municipal Solid Waste is Limited,* GAO Report to the Chairman, Subcommittee on Investigations and Oversight, Committee on Science, Space, and Technology, House of Representatives, GAO/RCED-94-200, September.

Kreith, F. 1994. "The Role of IRP and Conservation in Electric Utility Transition," in *The Electric Industry in Transition,* pp. 219–236, Pub. Util. Rep., Inc. and New York State Energy Research and Development Authority, Albany, NY.

Mancini, T.R., J.M. Chavez, and G.J. Kolb. 1994. "Solar Thermal Power, Today and Tomorrow", *Mech. Eng.,* pp. 74–79, August.

NREL. 1993. *Profiles in Renewable Energy,* DOE/CH10093-206, DE 93000081, Washington, DC, produced by the National Renewable Energy Laboratory, Golden, CO 80401, October.

OTA. 1992. U.S. Congress, Office of Technology Assessment, Building Energy Efficiency, OTA-E-518, Washington, DC, U.S. Government Printing Office, May.

Sheinkopf, K. 1994. "High Temperature Future"; *Independent Energy,* pp. 68–71, September.

2

Fundamentals of Thermodynamics, Heat Transfer, and Fluid Mechanics

Vahab Hassani

Steve Hauser
National Renewable Energy Laboratory

Introduction

Design and analysis of energy conversion systems require an in-depth understanding of basic principles of thermodynamics, heat transfer, and fluid mechanics. **Thermodynamics** is that branch of engineering science that describes the relationship and interaction between a system and its surroundings. This interaction usually occurs as a transfer of energy, mass, or momentum between a system and its surroundings. Thermodynamic laws are usually used to predict the changes that occur in a system when moving from one equilibrium state to another. The science of **heat transfer** complements the thermodynamic science by providing additional information about the energy that crosses a system's boundaries. Heat-transfer laws provide information about the mechanism of transfer of energy as heat and provide necessary correlations for calculating the rate of transfer of energy as heat. The science of **fluid mechanics**, one of the most basic engineering sciences, provides governing laws for fluid motion and conditions influencing that motion. The governing laws of fluid mechanics have been developed through a knowledge of fluid properties, thermodynamic laws, basic laws of mechanics, and experimentation.

In this chapter, we will focus on the basic principles of thermodynamics, heat transfer, and fluid mechanics that an engineer needs to know to analyze or design an energy conversion system.

Because of space limitations, our discussion of important physical concepts will not involve detailed mathematical derivations and proofs of concepts. However, we will provide appropriate references for those readers interested in obtaining more detail about the subjects covered in this chapter. Most of the material presented here is accompanied by examples that we hope will lead to better understanding of the concepts.

2.1 Thermodynamics

During a typical day, everyone deals with various engineering systems such as automobiles, refrigerators, microwaves, and dishwashers. Each engineering system consists of several components, and a system's optimal performance depends on each individual component's performance and interaction with other components. In most cases, the interaction between various components of a system occurs in the form of energy transfer or mass transfer. Thermodynamics is an engineering science that provides governing laws that describe energy transfer from one form to another in an engineering system. In this chapter, the basic laws of thermodynamics and their application for energy conversion systems are covered in the following four sections. The efficiency of the thermodynamic cycles and explanations of some advanced thermodynamic systems are presented in the succeeding two sections. Several examples have been presented to illustrate the application of concepts covered here.

Energy and the First Law of Thermodynamics

In performing engineering thermodynamic analysis, we must define the *system* under consideration. After properly identifying a thermodynamic system, everything else around the system becomes that system's *environment*. Of interest to engineers and scientists is the *interaction* between the system and its environment.

In thermodynamic analysis, systems can either consist of specified matter (**control mass**, CM) or specified space (**control volume**, CV). In a control-mass system, energy—but not mass—can cross the system boundaries while the system is going through a thermodynamic process. Control-mass systems may be called **closed systems** because no mass can cross their boundary. On the other hand, in a control-volume system—also referred to as an **open system**—both energy and matter can cross the system boundaries. The shape and size of CVs need not necessarily be constant and fixed; however, in this chapter, we will assume that the CVs are of fixed shape and size. Another system that should be defined here is an **isolated system**, which is a system where no mass or energy crosses its boundaries.

The energy of a system consists of three components: kinetic energy, potential energy, and internal energy. The **kinetic** and **potential energy** of a system are macroscopically observable. **Internal energy** is associated with random and disorganized aspects of molecules of a system and is not directly observable. In thermodynamic analysis of systems, the energy of the whole system can be obtained by adding the individual energy components.

Conservation of Energy—The First Law of Thermodynamics

The First Law of Thermodynamics states that energy is conserved: it cannot be created or destroyed, but can change from one form to another. The energy of a closed system can be expressed as

$$E = me + \frac{mu^2}{2g_c} + \frac{mgz}{g_c}, \tag{2.1}$$

where E is the total energy of the system, e is its internal energy per unit mass, and the last two terms are the kinetic energy and potential energy of the system, respectively. The proportionality

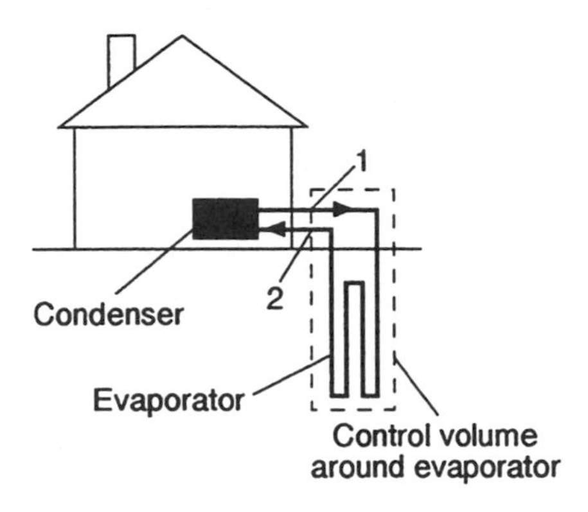

FIGURE 2.1 Geothermal-based (ground-source) heat pump.

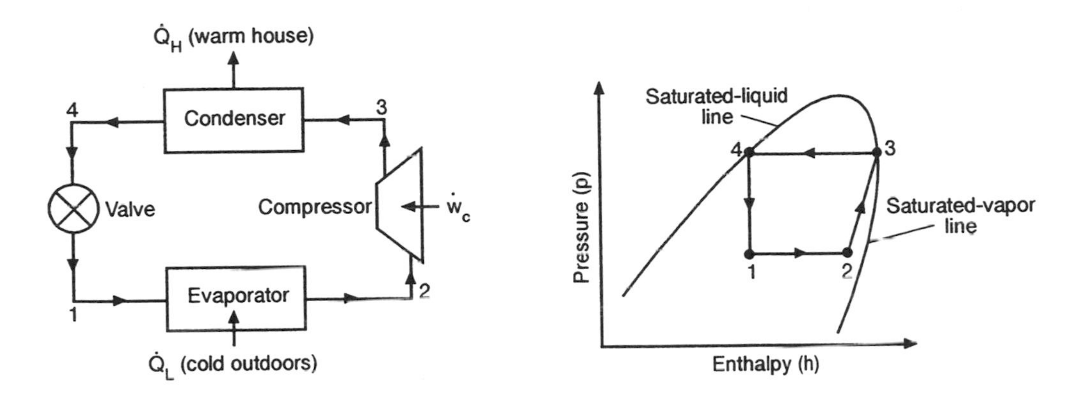

FIGURE 2.2 Thermodynamic cycle and p–h diagram for heat pump (heating mode).

constant g_c is defined in the nomenclature (listed at the end of this chapter) and is discussed in the text following Eq. (2.73). When a system undergoes changes, the energy change within the system can be expressed by a general form of the energy-balance equation:

Energy stored in the system	=	Energy entering the system	–	Energy leaving the system	+	Energy generated in the system (e.g., chemical reactions)

For example, consider the geothermal-based heat pump shown in Figure 2.1. In this heat pump, a working fluid (R-22, a common refrigerant used with geothermal heat pumps, which is gaseous at room temperature and pressure) is sealed in a closed loop and is used as the transport medium for energy. Figure 2.2 presents a simple thermodynamic cycle for a heat pump (heating mode) and an associated pressure-enthalpy (p–h) diagram. The saturated vapor and liquid lines are shown in Figure 2.2, and the region between these two lines is referred to as the wet region, where vapor and liquid coexist. The relative quantities of liquid and vapor in the mixture, known as the quality of the mixture (x), is usually used to denote the state of the mixture. The **quality of a mixture** is defined as the ratio of the mass of vapor to the mass of the mixture. For example, in 1 kg of mixture with quality x, there are x kg of vapor and $(1 - x)$ kg of liquid. Figure 2.2 shows that the working fluid leaving the evaporator (point 2) has a higher quality than working fluid entering the evaporator (point 1). The working fluid in Figure 2.2 is circulated through the closed loop and undergoes several phase changes. Within the evaporator, the working fluid absorbs heat from the surroundings (geothermal resource) and is vaporized. The low-pressure gas (point 2) is then directed into the compressor, where its pressure and temperature are increased by compression. The hot compressed gas (point 3) is then passed through the condenser, where it loses heat to the surroundings (heating up the house). The cool working fluid exiting the condenser is a high-pressure liquid (point 4), which then passes through an expansion device or valve to reduce its pressure to that of the evaporator (underground loop).

Specifically, consider the flow of the working fluid in Figure 2.1 from point 1 to point 2 through the system shown within the dashed rectangle. Mass can enter and exit this control-volume system. In flowing from point 1 to 2, the working fluid goes through the evaporator (see Figure 2.2). Assuming no accumulation of mass or energy, the First Law of Thermodynamics can be written as

$$e_2 + \frac{u_2^2}{2g_c} + \frac{gz_2}{g_c} + p_2 v_2 = e_1 + \frac{u_1^2}{2g_c} + \frac{gz_1}{g_c} + p_1 v_1 + \frac{\dot{Q} - \dot{w}}{\dot{m}}, \quad (2.2)$$

where $\dot{m}$ is the mass-flow rate of the working fluid, $\dot{Q}$ is the rate of heat absorbed by the working fluid, $\dot{w}$ is the rate of work done on the surroundings, v is the specific volume of the fluid, p is the pressure, and the subscripts 1 and 2 refer to points 1 and 2. A mass-flow energy-transport term, pv, appears in Eq. (2.2) as a result of our choice of control-volume system. The terms e and pv can be combined into a single term called **specific enthalpy,** $h = e + pv$, and Eq. (2.2) then reduces to

$$\frac{1}{2g_c}\left(u_2^2 - u_1^2\right) + \frac{g}{g_c}\Delta z_{2\text{-}1} + \Delta h_{2\text{-}1} = \frac{\dot{Q} - \dot{w}}{\dot{m}}. \quad (2.3)$$

For a constant-pressure process, the enthalpy change from temperature, T_1 to temperature T_2 can be expressed as

$$\Delta h_{2\text{-}1} = \int_{T_1}^{T_2} c_p \; dT = \bar{c}_p\left(T_2 - T_1\right), \quad (2.4)$$

where $\bar{c}_p$ is the mean specific heat at constant pressure.

Entropy and the Second Law of Thermodynamics

In many events, the state of an isolated system can change in a given direction, whereas the reverse process is impossible. For example, the reaction of oxygen and hydrogen will readily produce water, whereas the reverse reaction (electrolysis) cannot occur without some external help. Another example is that of adding milk to hot coffee. As soon as the milk is added to the coffee, the reverse action is impossible to achieve. These events are explained by the Second Law of Thermodynamics, which provides the necessary tools to rule out impossible processes by analyzing the events occurring around us with respect to time. Contrary to the First Law of Thermodynamics, the Second Law is sensitive to the *direction* of the process.

To better understand the second law of thermodynamics, we must introduce a thermodynamic property called **entropy** (symbolized by S, representing total entropy, and s, representing entropy per unit mass). The entropy of a system is simply a measure of the degree of molecular chaos or disorder at the microscopic level within a system.

The more disorganized a system is, the less energy is available to do useful work; in other words, energy is required to create order in a system. When a system goes through a thermodynamic process, the natural state of affairs dictates that entropy be produced by that process. In essence, the Second Law of Thermodynamics states that, in an isolated system, entropy can be produced, but it can never be destroyed.

$$\Delta S = S_{\text{final}} - S_{\text{initial}} \geq 0 \quad \text{for isolated system.} \quad (2.5)$$

Thermodynamic processes can be classified as reversible and irreversible processes. A **reversible process** is a process during which the net entropy of the system remains unchanged. A reversible process has equal chances of occurring in either a forward or backward direction because the net entropy remains unchanged. The absolute incremental entropy change for a closed system of fixed mass in a reversible process can be calculated from

$$dS = \frac{dQ}{T}, \tag{2.6}$$

where dS is the increase in entropy, dQ is the heat transferred, and T is the absolute temperature. However, the net change in entropy for all the participating systems in the reversible process must equal zero; thus,

$$\Delta S = \sum_{\text{all systems}} dS = \sum_{\text{all systems}} \frac{dQ}{T} = 0. \tag{2.7}$$

We emphasize that most real processes are not reversible and the entropy of a real process is not usually conserved. Therefore, Eq. (2.6) can be written in a general form as

$$dS \geq \frac{dQ}{T}, \tag{2.8}$$

where the equality represents the reversible process. A reversible process in which $dQ = 0$ is called an **isentropic process**. It is obvious from Eq. (2.6) that for such processes, $dS = 0$, which means that no net change occurs in the entropy of the system or its surroundings.

Application of the Thermodynamic Laws to Energy Conversion Systems

We can now employ these thermodynamic laws to analyze thermodynamic processes that occur in energy conversion systems. Among the most common energy conversion systems are heat engines and heat pumps. In Figure 2.3, the *solid* lines indicate the operating principle of a **heat engine,** where energy, Q_H, is absorbed from a high-temperature thermal reservoir and is converted to work, w, by using a turbine, and the remainder, Q_L, is rejected to a low-temperature thermal reservoir. The **energy-conversion efficiency** of a heat engine is defined as

$$\eta_{\text{heat engine}} = \frac{\text{desired output energy}}{\text{required input energy}} = \frac{w}{Q_H}. \tag{2.9}$$

In the early 1800s, Nicholas Carnot showed that to achieve the maximum possible efficiency, the heat engine must be completely reversible (i.e., no entropy production, no thermal losses due to friction). Using Eq. (2.7), Carnot's heat engine gives

$$\Delta S = \frac{Q_H}{T_H} - \frac{Q_L}{T_L} = 0 \tag{2.10}$$

or

$$\frac{Q_H}{Q_L} = \frac{T_H}{T_L}. \tag{2.11}$$

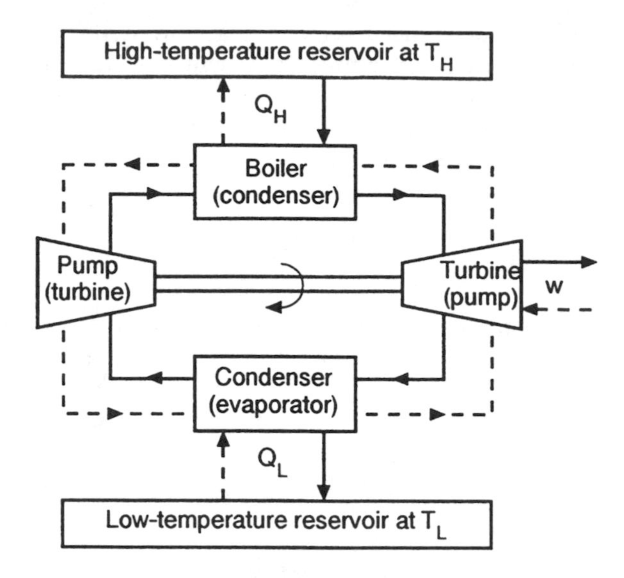

FIGURE 2.3 Principle of operation of a heat engine (solid lines and upper terms) and heat pump (dashed lines and lower terms in parentheses).

An energy balance gives

$$w = Q_H - Q_L. \tag{2.12}$$

Therefore, the maximum possible efficiency is

$$\eta_{\text{rev}} = \frac{Q_H - Q_L}{Q_H} = 1 - \frac{Q_L}{Q_H} = 1 - \frac{T_L}{T_H}. \tag{2.13}$$

In real processes, however, due to entropy production, the efficiency is

$$\eta \leq 1 - \frac{T_L}{T_H}. \tag{2.14}$$

A **heat pump** is a heat engine with the reverse thermodynamic process. In heat pumps, work input allows for thermal energy transfer from a low-temperature reservoir to a high-temperature reservoir as shown in Figure 2.3 by *dashed* lines. Energy (heat), Q_L, is absorbed by a working fluid from a low-temperature reservoir (geothermal resource or solar collectors), then the energy content (temperature and pressure) of the working fluid is increased as a result of input work, *w*. The energy, Q_H, of the working fluid is then released to a high-temperature reservoir (e.g., a warm house). The efficiency of a heat pump is defined as

$$\eta_{\text{heat pump}} = \frac{\text{desired output energy}}{\text{required input energy}} = \frac{Q_H}{w} = \frac{Q_H}{Q_H - Q_L}. \tag{2.15}$$

The efficiency of a heat pump is often expressed as **coefficient of performance** (COP). The COP of a Carnot (or reversible) heat pump can be expressed as

$$COP = \frac{T_H}{T_H - T_L}. \tag{2.16}$$

Heat engines and heat pumps are broadly discussed by Sandord [1962], Reynolds and Perkins [1977], Wood [1982], Karlekar [1983], and Van Wylen and Sonntag [1986].

Efficiencies of Thermodynamic Cycles

To evaluate and compare various thermodynamic cycles (or systems), we further define and employ the term *efficiency.* The operating efficiency of a system reflects irreversibilities that exist in the system. To portray various deficiencies or irreversibilities of existing thermodynamic cycles, the following thermodynamic efficiency terms are most commonly considered.

$$\text{Mechanical efficiency} \quad \eta_m = \frac{w_{\text{act}}}{w_{\text{rev}}}, \tag{2.17}$$

which is the ratio of the actual work of a system to that of the same system under reversible process. Note that the reversible process is not necessarily an adiabatic process (which would involve heat transfer across the boundaries of the system).

$$\text{Isentropic Efficiency} \quad \eta_s = \frac{w_{\text{act}}}{w_{\text{isent}}}, \tag{2.18}$$

which is the ratio of actual work to the work done under an isentropic process.

$$\text{Relative efficiency} \quad \eta_r = \frac{w_{\text{rev}}}{w_{\text{isent}}}, \tag{2.19}$$

which is the ratio of reversible work to isentropic work.

$$\text{Thermal efficiency} \quad \eta_T = \frac{\dot{w}_{\text{out}}}{\dot{Q}_{\text{in}}}, \tag{2.20}$$

which is the ratio of the net power output to the input heat rate. Balmer [1990] gives a comprehensive discussion on the efficiency of thermodynamic cycles.

Some Thermodynamic Systems

The most common thermodynamic systems are those used by engineers in generating electricity for utilities and for heating or refrigeration/cooling purposes.

Modern power systems employ after Rankine cycles, and a typical **Rankine cycle** is shown in Figure 2.4(a). In this cycle, the working fluid is compressed by the pump and is sent to the boiler where heat Q_H is added to the working fluid, bringing it to a saturated (or superheated) vapor state. The vapor is then expanded through the turbine, generating shaft work. The mixture of vapor and liquid exiting the turbine is condensed by passing through the condenser. The fluid coming out of the condenser is then pumped to the boiler, closing the cycle. The enthalpy-entropy (h–s) diagram for the Rankine cycle is shown in Figure 2.4(b). The dashed line 3→4 in Figure 2.4(b) represents actual expansion of the steam through the turbine, whereas the solid line 3→4s represents an isentropic expansion through the turbine.

In utility power plants, the heat source for the boiler can vary depending on the type of generating plant. In geothermal power plants, for example, geothermal brine at temperatures as high as 380°C is pumped from geothermal resources located several hundred meters below the earth's surface, and the brine's energy is transferred to the working fluid in a boiler.

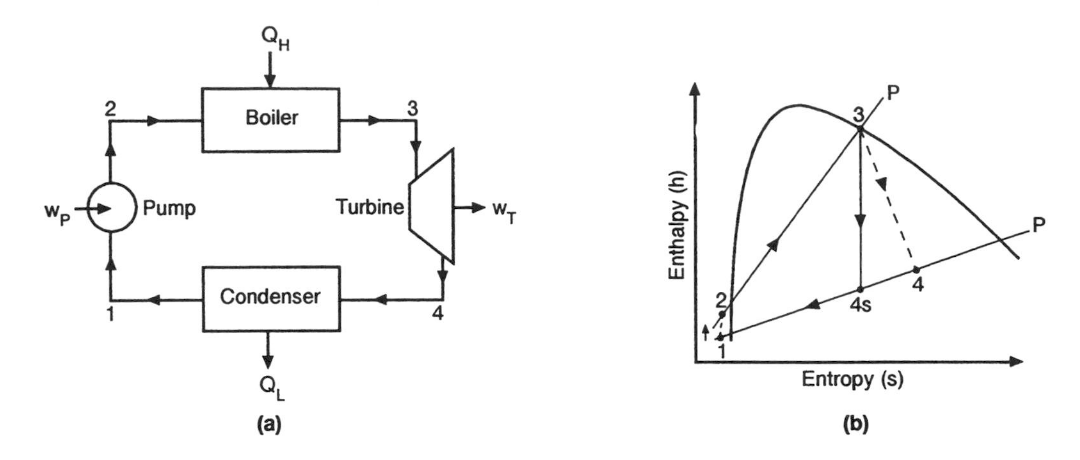

FIGURE 2.4 Typical Rankine cycle (a) and its *h–s* diagram (b).

The other commonly used thermodynamic cycle is the **refrigeration cycle** (heat-pump cycle). As stated earlier, a heat engine and a heat pump both operate under the same principles except that their thermodynamic processes are reversed. Figures 2.2 and 2.3 provide detailed information about the heat-pump cycle.

Modified Rankine Cycles

Modifying the Rankine cycle can improve the output work considerably. One modification usually employed in large central power stations is introducing a **reheat process** into the Rankine cycle. In this modified Rankine cycle, as shown in Figure 2.5(a), steam is first expanded through the first stage of the turbine. The steam discharging from the first stage of the turbine is then reheated before entering the second stage of the turbine. The reheat process allows the second stage of the turbine to have a greater enthalpy change. The enthalpy-versus-entropy plot for this cycle is shown in Figure 2.5(b), and this figure should be compared to Figure 2.4(b) to further appreciate the effect of the reheat process. Note that in the reheat process, the work output per pound of steam increases; however, the efficiency of the system may be increased or reduced depending on the reheat temperature range.

Another modification also employed at large power stations is called a **regeneration process**. The schematic representation of a Rankine cycle with a regeneration process is shown in Figure 2.6(a), and the enthalpy-versus-entropy plot is shown in Figure 2.6(b). In this process, a portion of the steam (at point 6) that has already expanded through the first stage of the turbine is extracted and is mixed in an open regenerator with the low-temperature liquid (from point 2) that is pumped from the condenser back to the boiler. The liquid coming out of the regenerator at point 3 is saturated liquid that is then pumped to the boiler.

Example 2.1

A geothermal heat pump, shown in Figure 2.7, keeps a house at 24°C during the winter. The geothermal resource temperature is –5°C. The amount of work required to operate the heat pump for a particular month is 10^6 kilojoules (kJ). What is the maximum heat input to the house during that 1–month period?

Solution:

The energy balance for the system gives

$$w_i + Q_L = Q_H. \quad (2.21)$$

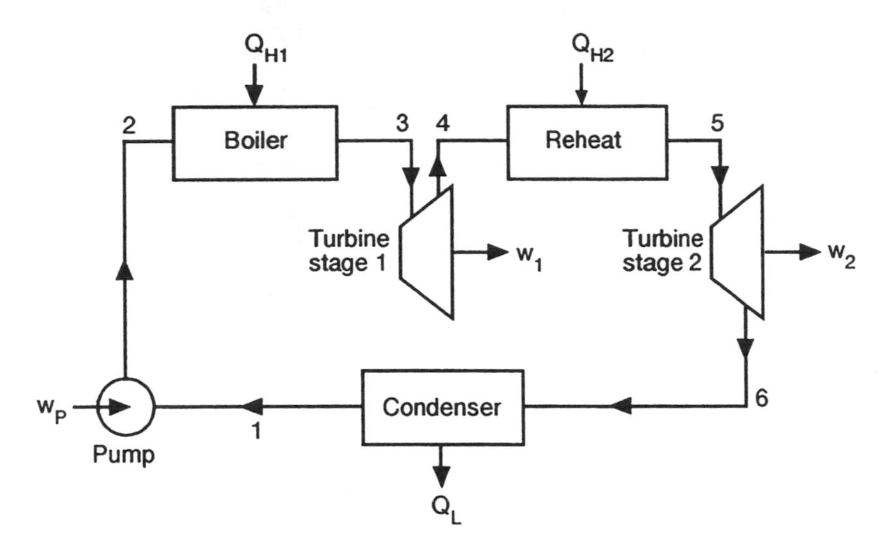

FIGURE 2.5(a) Rankine cycle with a reheat process.

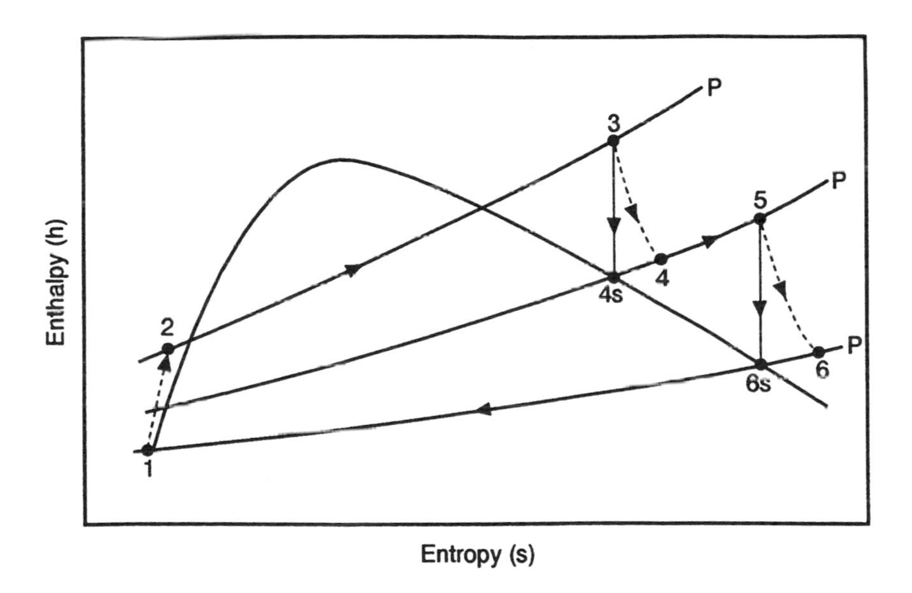

FIGURE 2.5(b) The *h–s* plot for the modified Rankine cycle of Figure 2.5(a).

Eq. (2.7) gives the entropy change for the system:

$$\Delta S = \frac{Q_H}{T_H} - \frac{Q_L}{T_L}, \tag{2.22}$$

where, from Eq. (2.22), we can get an expression for Q_L:

$$Q_L = \frac{Q_H T_L}{T_H} - T_L \Delta S. \tag{2.23}$$

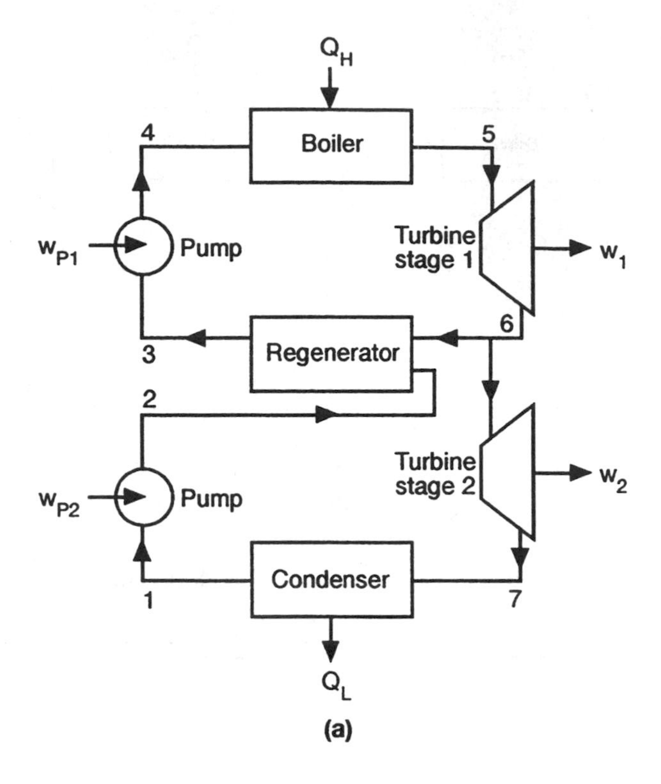

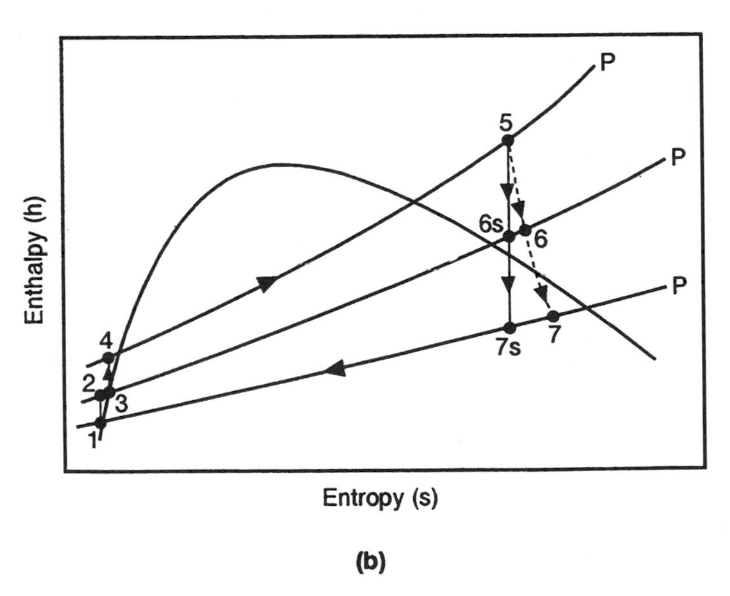

FIGURE 2.6 (a) Rankine cycle with regeneration process, and (b) its *h–s* diagram.

Substituting for Q_L from Eq. (2.21), we get an expression for Q_H:

$$Q_H = \frac{1}{1-\dfrac{T_L}{T_H}}\left(w_i - T_L \Delta S\right).$$

The maximum Q_H is obtained when $\Delta S = 0$; therefore,

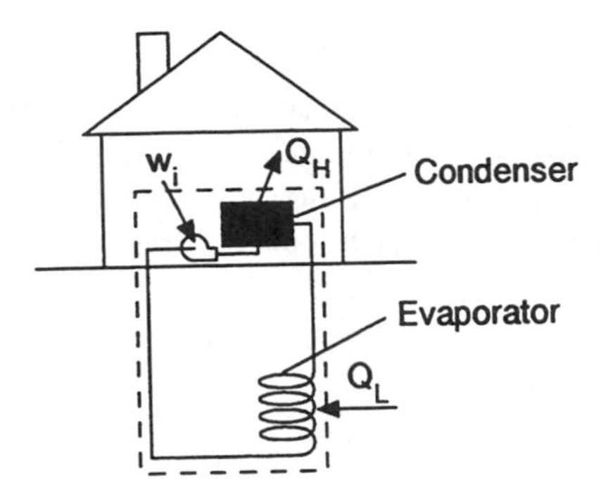

FIGURE 2.7 Ground-source heat pump.

$$Q_H \le \frac{1}{1-\frac{T_L}{T_H}} w_i, \tag{2.24}$$

and substituting the actual values yields

$$Q_H \le \frac{1}{1-\frac{268.15\text{ K}}{297.15\text{ K}}} \times 10^6 \text{ kJ}, \ or \ Q_H \le 10^7 \text{ kJ}.$$

Example 2.2

Calculate the maximum COP for the heat pump of Example 2.1.

Solution:

$$COP_{\text{heat pump}} = \frac{Q_H}{Q_H - Q_L} = \frac{10^7 \text{ kJ}}{10^6 \text{ kJ}} = 10.$$

Example 2.3

A simple heat-pump system is shown in Figure 2.2. The working fluid in the closed loop is R-22. The *p–h* diagram of Figure 2.2 shows the thermodynamic process for the working fluid. The following data represent a typical operating case.

$$T_1 = T_2 = -5°C \ (23°F)$$

$$T_3 = T_4 = 24°C \ (75°F)$$

$$x_1 = 0.17$$

(a) Determine the COP for this heat pump assuming isentropic compression, $s_2 = s_3$.

(b) Determine the COP by assuming a compressor isentropic efficiency of 70%.

Solution:

First, the thermodynamic properties at each station can be found using the *ASHRAE Handbook of Fundamentals* [1993].

State Point #1:

The evaporation of working fluid R-22 occurs at a constant pressure (between points 1 and 2). This pressure can be obtained from the saturated liquid/vapor Table of Properties for R-22 at $T_1 = -5°C$ (23°F), which is $p_1 = 422$ kPa (61.2 psia). At point 1, the quality is $x_1 = 0.17$. Therefore, the enthalpy and entropy at this point can be obtained from:

$$h_1 = h_{f_1} + x_1 \; h_{fg_1} \quad \text{and} \quad s_1 = s_{f_1} + x_1 \; s_{fg_1},$$

where $h_{f_1} = 39.36$ kJ/kg, $h_{fg_1} = h_{g_1} - h_{f_1} = 208.85$ kJ/kg, $s_{f_1} = 0.1563$ kJ/kg K, and $s_{fg_1} = s_{g_1} - s_{f_1} = 0.7791$ kJ/kg K.

The quantities listed are read from the Table of Properties for R-22. Using these properties, we obtain:

$$h_1 = 39.36 \text{ kJ/kg} + 0.17 \; (208.85 \text{ kJ/kg}) = 74.86 \text{ kJ/kg},$$

$$s_1 = 0.1563 \text{ kJ/kg K} + 0.17 \; (0.7791 \text{ kJ/kg K}) = 0.2887 \text{ kJ/kg K}.$$

We then find the state properties at point 3, because they will be used to find the quality of the mixture at point 2.

State Point #3:

At point 3, the working fluid is saturated vapor at $T_3 = 24°C$ (75°F). From the Table of Properties, the pressure, enthalpy, and entropy at this point are $p_3 = 1{,}014$ kPa (147 psia), $h_3 = 257.73$ kJ/kg, and $s_3 = 0.8957$ kJ/kg K.

State Point #2:

The temperature at this point is $T_2 = -5°C$ (23°F), and because we are assuming an isentropic compression, the entropy is $s_2 = s_3 = 0.8957$ kJ/kg K. The quality of the mixture at point 2 can be calculated from

$$x_2 = \frac{s_2 - s_{f_2}}{s_{fg_2}},$$

where $s_{fg_2} = s_{g_2} - s_{f_2}$, and the quantities s_{g_2} and s_{f_2} can be obtained from the Table of Properties at $T_2 = -5°C$ (23°F).

Note that the saturation properties for points 1 and 2 are the same because both points have the same pressure and temperature. Therefore,

$$s_{f_2} = s_{f_1} = 0.1563 \text{ kJ/kg K},$$

$$h_{f_2} = h_{f_1} = 39.36 \text{ kJ/kg},$$

$$s_{fg_2} = s_{fg_1} = 0.7791 \text{ kJ/kg K},$$

$$h_{fg_2} = h_{fg_1} = 208.85 \text{ kJ/kg},$$

$$x_2 = \frac{s_2 - s_{f_2}}{s_{fg_2}} = \frac{0.8957 \text{ kJ/kg K} - 0.1563 \text{ kJ/kg K}}{0.7791 \text{ kJ/kg K}} = 0.95.$$

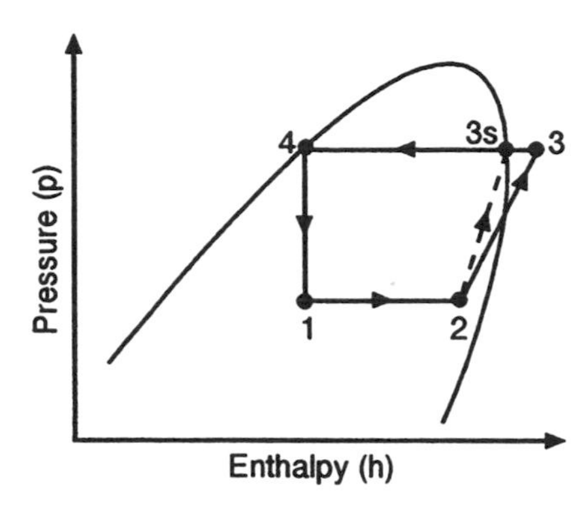

FIGURE 2.8 The *p–h* diagram for a heat-pump cycle with a 70% isentropic efficiency for the compressor.

Knowing the quality at point 2, the enthalpy at point 2 can be calculated:

$$h_2 = h_{f_2} + x_2\ h_{fg_2} = 39.36\ \text{kJ/kg} + 0.95\ (208.85\ \text{kJ/kg}) = 237.77\ \text{kJ/kg}.$$

State Point #4

At point 4, we have saturated liquid at $T_4 = 24°\text{C}$ (75°F). Therefore, from the Table of Properties, $s_4 = 0.2778$ kJ/kg K and $h_4 = 74.16$ kJ/kg.

(a) The coefficient of performance for a heat pump is

$$\text{COP} = \frac{\text{rate of energy transfer to house}}{\text{compressor shaft power}}$$

$$= \frac{h_3 - h_4}{h_3 - h_2} = \frac{183.57\ \text{kJ/kg}}{19.96\ \text{kJ/kg}} = 9.2.$$

(b) If the isentropic efficiency is 70%, the *p–h* diagram is as shown in Figure 2.8. The isentropic efficiency for the compressor is defined as

$$\eta_s = \frac{h_{3s} - h_2}{h_3 - h_2}.$$

Using this relationship, h_3 can be calculated as follows:

$$h_3 = \frac{h_{3s} - h_2}{\eta_s} + h_2 = \frac{257.73\ \text{kJ/kg} - 237.77\ \text{kJ/kg}}{0.70} + 237.77\ \text{kJ/kg} = 266.28\ \text{kJ/kg}.$$

Therefore, the COP is

$$\text{COP} = \frac{h_3 - h_4}{h_3 - h_2} = \frac{266.28\ \text{kJ/kg} - 74.16\ \text{kJ/kg}}{266.28\ \text{kJ/kg} - 237.77\ \text{kJ/kg}} = 6.7.$$

Advanced Thermodynamic Power Cycles

Over the past 50 years, many technological advances have improved the performance of power plant components. Recent developments in exotic materials have allowed the design of turbines

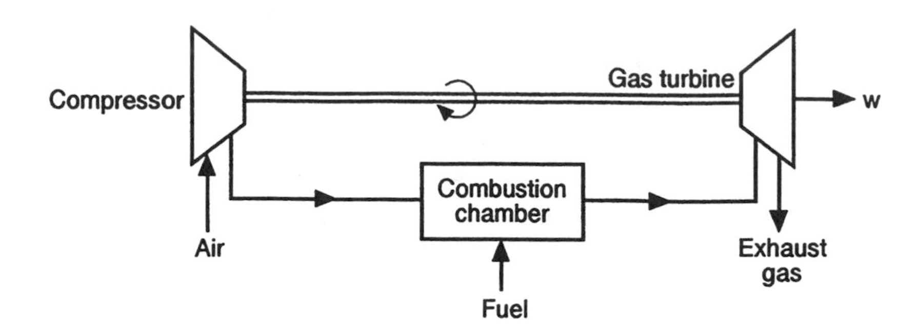

FIGURE 2.9 A basic gas-turbine or Brayton-cycle representation.

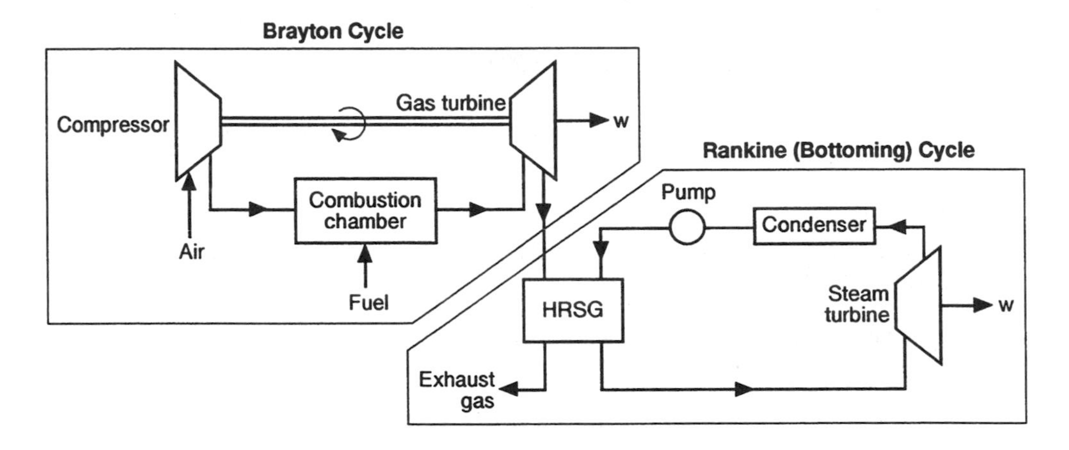

FIGURE 2.10 A combined cycle known as the steam-and-gas-turbine cycle.

that can operate more efficiently at higher inlet temperatures and pressures. Simultaneously, innovative thermodynamic technologies (processes) have been proposed and implemented that take advantage of improved turbine isentropic and mechanical efficiencies and allow actual operating thermal efficiencies of a power station to approach 50%. These improved technologies include (1) modification of existing cycles (reheat and regeneration), and (2) use of combined cycles. In the previous section, we discussed reheat and regeneration techniques. In the following paragraphs, we give a short overview of the combined-cycle technologies and discuss their operation.

The basic gas-turbine or **Brayton cycle** is shown in Figure 2.9. In this cycle, ambient air is pressurized in a compressor and the compressed air is then forwarded to a combustion chamber, where fuel is continuously supplied and burned to heat the air. The combustion gases are then expanded through a turbine to generate mechanical work. The turbine output runs the air compressor and a generator that produces electricity.

The exhaust gas from such a turbine is very hot and can be used in a bottoming cycle added to the basic gas-turbine cycle to form a **combined cycle.** Figure 2.10 depicts such a combined cycle where a **heat-recovery steam generator** (HRSG) is used to generate steam required for the bottoming (Rankine) cycle. The high-temperature exhaust gases from the gas-turbine (Brayton) cycle generate steam in the HRSG. The steam is then expanded through the steam turbine and condensed in the condenser. Finally, the condensed liquid is pumped to the HRSG for heating. This combined cycle is referred to as a **steam and gas turbine cycle.**

Another type of bottoming cycle proposed by Kalina [1984] uses a mixture of ammonia and water as a working fluid. The multicomponent mixture provides a boiling process that does not occur at a constant temperature; as a result, the available heat is used more efficiently. In addition,

Kalina employs a distillation process or working-fluid preparation subsystem that uses the low-temperature heat available from the mixed-fluid turbine outlet. The working-fluid mixture is enriched by the high-boiling-point component; consequently, condensation occurs at a relatively constant temperature and provides a greater pressure drop across the turbine. The use of multi-component working fluids in Rankine cycles provides variable-temperature boiling; however, the condensation process will have a variable temperature as well, resulting in system inefficiencies. According to Kalina, this type of bottoming cycle increases the overall system efficiency by up to 20% above the efficiency of the combined-cycle system using a Rankine bottoming cycle. The combination of the cycle proposed by Kalina and a conventional gas turbine is estimated to yield thermal efficiencies in the 50 to 52% range.

2.2 Fundamentals of Heat Transfer

In Section 2.1, we discussed thermodynamic laws and through some examples showed that these laws are concerned with interaction between a system and its environment. Thermodynamic laws are always concerned with the **equilibrium** state of a system and are used to determine the amount of energy required for a system to change from one equilibrium state to another. These laws do not quantify the mode of the energy transfer or its rate. Heat transfer relations, however, complement thermodynamic laws by providing **rate equations** that relate the heat transfer rate between a system and its environment.

Heat transfer is an important process that is an integral part of our environment and daily life. The heat-transfer or heat-exchange process between two media occurs as a result of a temperature difference between them. Heat can be transferred by three distinct modes: conduction, convection, and radiation. Each one of these heat transfer modes can be defined by an appropriate rate equation presented below:

Fourier's Law of Heat Conduction—represented here by Eq. (2.25) for the one-dimensional steady-state case:

$$\dot{Q}_{cond} = -kA\frac{dT}{dx}. \tag{2.25}$$

Newton's Law of Cooling—which gives the rate of heat transfer between a surface and a fluid:

$$\dot{Q}_{conv} = hA\Delta T, \tag{2.26}$$

where h is the average heat-transfer coefficient over the surface with area A.

Stefan-Boltzmann's Law of Radiation—which is expressed by the equation:

$$\dot{Q}_r = A_1 F_{1-2}\sigma\left(T_1^4 - T_2^4\right). \tag{2.27}$$

Conduction Heat Transfer

Conduction is the heat-transfer process that occurs in solids, liquids, and gases through molecular interaction as a result of a temperature gradient. The energy transfer between adjacent molecules occurs without significant physical displacement of the molecules. The rate of heat transfer by conduction can be predicted by using Fourier's law, where the effect of molecular interaction in the heat-transfer medium is expressed as a property of that medium and is called the **thermal conductivity**. The study of conduction heat transfer is a well-developed field where sophisticated analytical and numerical techniques are used to solve problems.

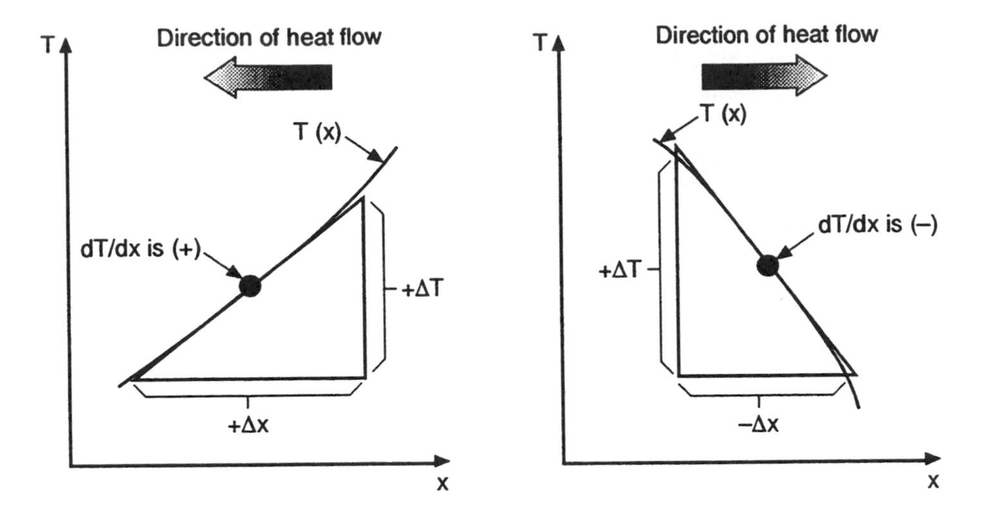

FIGURE 2.11 The sign convention for conduction heat flow.

In this section, we discuss basics of steady-state one-dimensional conduction heat transfer through homogeneous media in cartesian and cylindrical coordinates. Some examples are provided to show the application of the fundamentals presented, and we also discuss fins or extended surfaces.

One-Dimensional Steady-State Heat Conduction

Fourier's law, as represented by Eq. (2.25), states that the rate of heat transferred by conduction is directly proportional to the temperature gradient and the surface area through which the heat is flowing.

The proportionality constant k is the thermal conductivity of the heat-transfer medium. Thermal conductivity is a thermophysical property and has units of W/m K in the SI system, or Btu/h ft °F in the English system of units. Thermal conductivity can vary with temperature, but for most materials it can be approximated as a constant over a limited temperature range. A graphical representation of Fourier's law is shown in Figure 2.11.

Eq. (2.25) is only used to calculate the rate of heat conduction through a one-dimensional homogenous medium (uniform k throughout the medium). Figure 2.12 shows a section of a plane wall with thickness L, where we assume the other two dimensions of the wall are very large compared to L. One side of the wall is at temperature T_1, and the other side is kept at temperature T_2, where $T_1 > T_2$. Integrating Fourier's law with constant k and A, the rate of heat transfer through this wall is

$$\dot{Q} = kA\frac{T_1 - T_2}{L}, \tag{2.28}$$

where k is the thermal conductivity of the wall.

The Concept of Thermal Resistance

Figure 2.12 also shows the analogy between electrical and thermal circuits. Consider an electric current I flowing through a resistance R_e, as shown in Figure 2.12. The voltage difference $\Delta V = V_1 - V_2$ is the driving force for the flow of electricity. The electric current can then be calculated from

$$I = \frac{\Delta V}{R_e}. \tag{2.29}$$

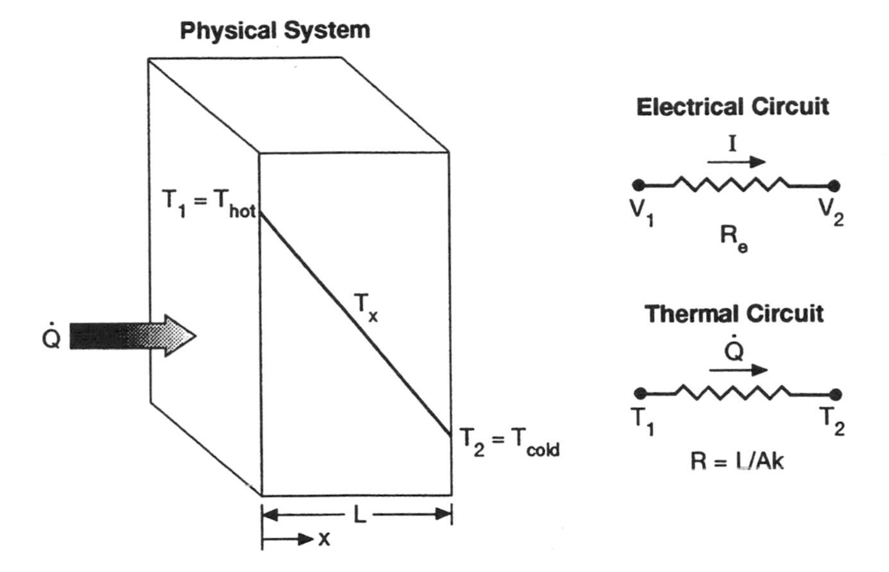

FIGURE 2.12 Analogy between thermal and electrical circuits for steady-state conduction through a plane wall.

Like electric current flow, heat flow is governed by the temperature difference, and it can be calculated from

$$\dot{Q} = \frac{\Delta T}{R}, \tag{2.30}$$

where, from Eq. (2.28), $R = L/Ak$ and is called **thermal resistance**. Following this definition, the thermal resistance for convection heat transfer given by Newton's Law of Cooling becomes $R = 1/(hA)$. Thermal resistance of composite walls (plane and cylindrical) has been discussed by Kakac and Yener [1988], Kreith and Bohn [1993], and Bejan [1993]. The following example shows how we can use the concept of thermal resistance in solving heat-transfer problems.

Example 2.4

One wall of a house, shown in Figure 2.13, has a thickness of 0.30 m and a surface area of 11 m^2. The wall is constructed from a material (brick) that has a thermal conductivity of 0.55 W/m K. The outside temperature is –10°C, while the house temperature is kept at 22°C. The convection heat-transfer coefficient is estimated to be h_o = 21 W/m^2 K in the outside and h_i = 7 W/m^2 K in the inside. Calculate the rate of heat transfer through the wall, as well as the surface temperature at either side of the wall.

Solution:

The conduction thermal resistance is

$$R_{t,\text{cond}} = \frac{L}{Ak} = \frac{0.3 \text{ m}}{11 \text{ m}^2 \times 0.55 \text{ W/m K}} = 0.0496 \frac{\text{K}}{\text{W}}.$$

Note that the heat-transfer rate per unit area is called **heat flux** and is given by

$$q'' = \frac{\dot{Q}}{A} = \frac{T_1 - T_2}{L/k}.$$

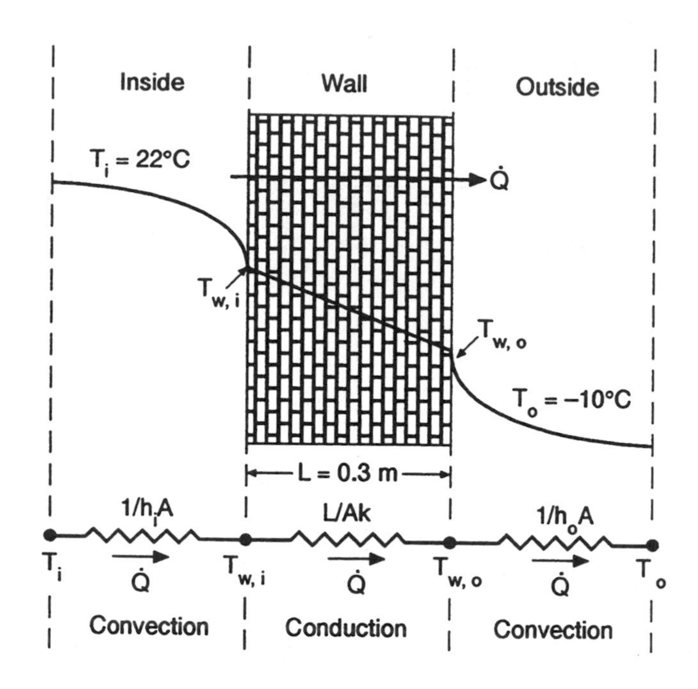

FIGURE 2.13 Heat loss through a plane wall.

In this case, the resistance to heat transfer over a 1–m by 1–m area of the wall is

$$R_{t,\text{cond}} = L/k = \frac{0.3 \text{ m}}{0.55 \text{ W/m K}} = 0.5455 \frac{\text{m}^2 \text{ K}}{\text{W}}.$$

The convection resistances for inside and outside, shown in Figure 2.13, are

$$R_{i,\text{conv}} = \frac{1}{h_i A} = \frac{1}{7 \text{ W/m}^2 \text{ K} \times 11 \text{ m}^2} = 0.0130 \frac{\text{K}}{\text{W}}, \text{ and}$$

$$R_{o,\text{conv}} = \frac{1}{h_o A} = \frac{1}{21 \text{ W/m}^2 \text{ K} \times 11 \text{ m}^2} = 0.0043 \frac{\text{K}}{\text{W}}.$$

Note that the highest resistance is provided by conduction through the wall. The total heat flow can be calculated from

$$\dot{Q} = \frac{\Delta T}{\Sigma R} = \frac{T_i - T_o}{\frac{1}{h_i A} + \frac{L}{kA} + \frac{1}{h_o A}} = \frac{295.15 \text{ K} - 263.15 \text{ K}}{0.013 \frac{\text{K}}{\text{W}} + 0.0496 \frac{\text{K}}{\text{W}} + 0.0043 \frac{\text{K}}{\text{W}}} = 478.3 \text{ W}.$$

The surface temperatures can then be calculated by using the electric analogy depicted in Figure 2.13. For the inside surface temperature,

$$\dot{Q} = \frac{T_i - T_{w,i}}{\frac{1}{h_i A}},$$

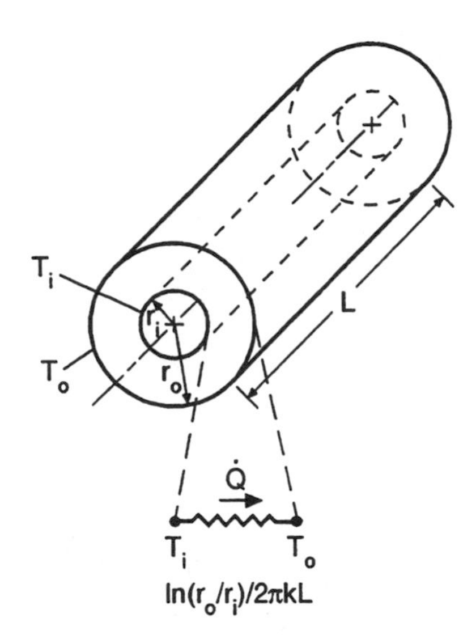

FIGURE 2.14 Conduction through hollow cylinders.

or

$$T_{w,i} = T_i - \frac{\dot{Q}}{h_i A} = 295.15\ \text{K} - \frac{478.3\ \text{W}}{77\ \text{W/K}} = 288.94\ \text{K} = 15.79°C.$$

Similarly, for the outside surface temperature:

$$T_{w,o} = \frac{\dot{Q}}{h_o A} + T_o = \frac{478.3\ \text{W}}{231\ \text{W/K}} + 263.15\ \text{K} = 265.22\ \text{K} = -7.93°C.$$

Conduction Through Hollow Cylinders

A cross section of a long hollow cylinder with internal radius r_i and external radius r_o is shown in Figure 2.14. The internal surface of the cylinder is at temperature T_i and the external surface is at T_o, where $T_i > T_o$. The rate of heat conduction in a radial direction is calculated by

$$\dot{Q} = \frac{2\pi Lk(T_i - T_o)}{\ln(r_o/r_i)}, \tag{2.31}$$

where L is the length of the cylinder that is assumed to be long enough so that the end effects may be ignored. From Eq. (2.31) the resistance to heat flow in this case is

$$R = \frac{\ln(r_o/r_i)}{2\pi kL}. \tag{2.32}$$

Equation (2.31) can be used to calculate the heat loss through insulated pipes, as presented in the following example.

Example 2.5

The refrigerant of the heat pump discussed in Example 2.3 is circulating through a thin-walled copper tube of radius r_i = 6 mm, as shown in Figure 2.15. The refrigerant temperature

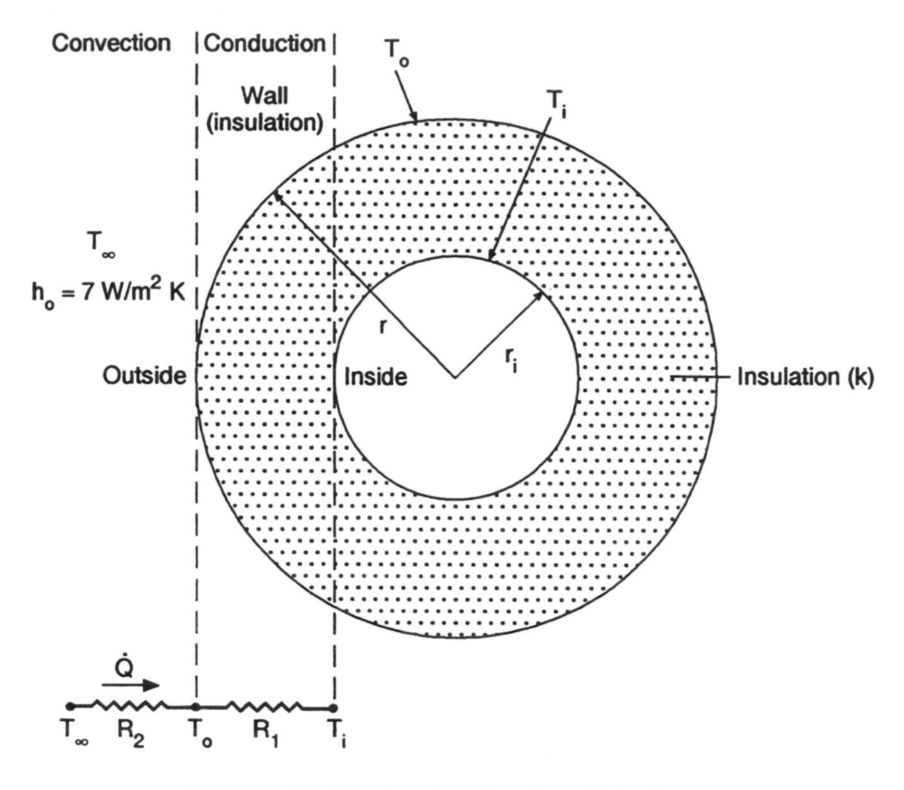

FIGURE 2.15 Heat loss through a pipe with insulation.

is T_i, ambient temperature is T_∞, and $T_i < T_\infty$. The outside convection heat-transfer coefficient is $h_o = 7$ W/m² K.

(a) If we decide to insulate this tube, what would be the optimum thickness of the insulation?

(b) Show the behavior of heat flow through the tube at different insulation thicknesses such as 0, 3, 6, 10, 15, and 20 mm, and plot the results for $\dot{Q}/L$ versus radius r. Assume an insulation material with thermal conductivity $k = 0.06$ W/m K.

Solution:

(a) In thermal analysis of radial systems, we must keep in mind that there are competing effects associated with changing the thickness of insulation. Increasing the insulation thickness increases the conduction resistance; however, the area available for convection heat transfer increases as well, resulting in reduced convection resistance. To find the optimum radius for insulation, we first identify the major resistances in the path of heat flow. Our assumptions are that (1) the tube wall thickness is small enough that conduction resistance can be ignored, (2) heat transfer occurs at steady state, (3) insulation has uniform properties, and (4) radial heat transfer is one-dimensional.

The resistances per unit length are

$$R_1 = \frac{\ln(r/r_i)}{2\pi k} \quad \text{and} \quad R_2 = \frac{1}{2\pi r h_o},$$

where r, the outer radius of insulation, is unknown. The total resistance is

$$R_t = R_1 + R_2 = \frac{\ln(r/r_i)}{2\pi k} + \frac{1}{2\pi r h_o}$$

and the rate of heat flow per unit length is

$$\dot{Q} = \frac{T_\infty - T_i}{R_t}.$$

The optimum thickness of the insulation is obtained when the total heat flow is minimized or when the total resistance is maximized. By differentiating R_t with respect to r, we obtain the condition under which R_t is maximum (or minimum). Therefore,

$$\frac{dR_t}{dr} = \frac{1}{2\pi r k} - \frac{1}{2\pi r^2 h_o} = 0,$$

from which we obtain $r = k/h_o$. To determine if R_t is maximum or minimum at $r = k/h_o$, we take the second derivative and find its sign at $r = k/h_o$.

$$\frac{d^2R_t}{dr^2} = -\frac{1}{2\pi k(k/h_o)^2} + \frac{1}{\pi h_o(k/h_o)^3} = \frac{1}{2\pi k^3/h_o^2} > 0.$$

Therefore, R_t is a minimum at $r = k/h_o$, which means that the heat flow is maximum at this insulation radius. An optimum radius of insulation does not exist; however, the radius obtained in this analysis is referred to as the **critical radius**, r_c, and this radius should be avoided when selecting insulation for pipes.

(b) For this example the critical radius is $r_c = k/h_o = (0.06 \text{ W/m K})/(7 \text{ W/m}^2 \text{ K}) = 0.0086 \text{ m} = 8.6$ mm, and $r_i = 6$ mm, so $r_c > r_i$. This means that by adding insulation, we will increase the heat loss from the tube. Using the expression for R_t, we can plot the total resistance versus the insulation thickness as shown in Figure 2.16. Note that the minimum total resistance occurs at an insulation thickness of about 0.025 m (corresponding to the r_c calculated earlier). Also note that as the insulation thickness is increased, the conduction resistance increases; however, the convection resistance decreases as listed in Table 2.1. This table shows that the total resistance for zero insulation is almost equal to the total resistance with an insulation thickness of 6 mm.

Convection Heat Transfer

Energy transport (heat transfer) in fluids usually occurs by the motion of fluid particles. In many engineering problems, fluids come into contact with solid surfaces that are at different temperatures than the fluid. The temperature difference and random/bulk motion of the fluid particles result in an energy transport process known as convection heat transfer. Convection heat transfer is more complicated than conduction because the motion of the fluid, as well as the process of energy transport, must be studied simultaneously. Convection heat transfer can be created by external forces such as pumps and fans in a process referred to as **forced convection**. In the absence of external forces, the convection process may result from temperature or density gradients inside the fluid; in this case, the convection heat-transfer process is referred to as **natural convection**. We will discuss this type of convection in more detail in the next section. There are other instances where a heat-transfer process consists of both forced and natural convection modes and they are simply called **mixed-convection** processes.

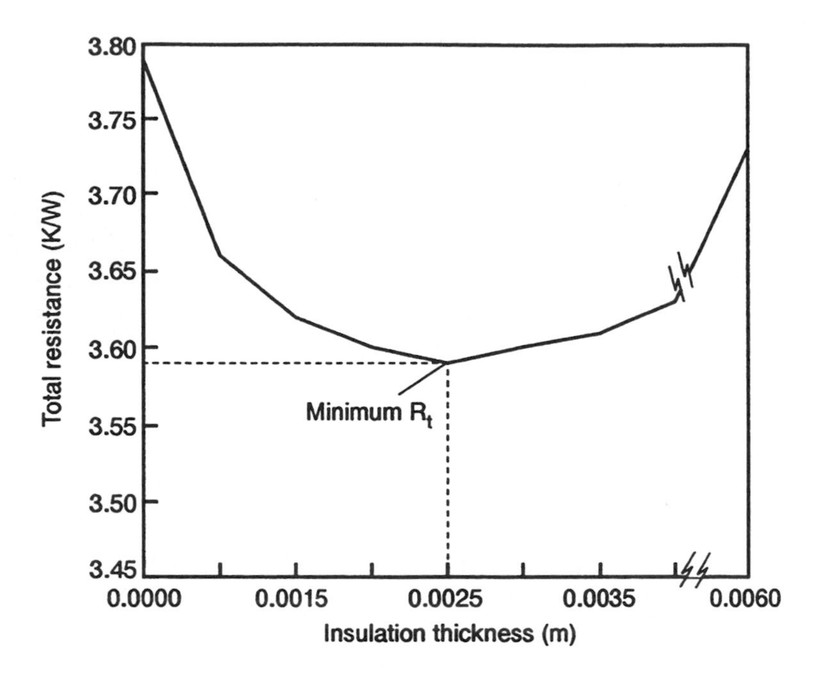

FIGURE 2.16 Total resistance versus insulation thickness for an insulated tube.

TABLE 2.1 Effect of Insulation Thickness on Various Thermal Resistances

Insulation Thickness (m)	Outer Radius, r (m)	Convection Resistance, R_2, (K/W)	Conduction Resistance, R_1, (K/W)	Total Resistance, R_t, (K/W)
0	0.0060	3.79	0	3.79
0.0010	0.0070	3.25	0.41	3.66
0.0015	0.0075	3.03	0.59	3.62
0.0020	0.0080	2.84	0.76	3.60
0.0025	0.0085	2.67	0.92	3.59
0.0030	0.0090	2.53	1.07	3.60
0.0035	0.0095	2.39	1.22	3.61
0.0040	0.0100	2.27	1.36	3.63
0.0060	0.0120	1.89	1.84	3.73

The main unknown in the convection heat-transfer process is the heat-transfer coefficient (see Eq. 2.26). Figure 2.17 serves to explain the convection heat-transfer process by showing the temperature and velocity profiles for a fluid at temperature T_∞ and bulk velocity u_∞ flowing over a heated surface. As a result of viscous forces interacting between the fluid and the solid surface, a region known as **velocity boundary layer** is developed in the fluid next to the solid surface. In this region the fluid velocity is zero at the surface and increases to the bulk fluid velocity u_∞. Because of the temperature difference between the fluid and the surface, a region known as **temperature boundary layer** also develops next to the surface, where the temperature at the fluid varies from T_w (surface temperature) to T_∞ (bulk fluid temperature). The velocity-boundary-layer thickness δ and temperature-boundary-layer thickness δ_t and their variation along the surface are shown in Figure 2.17.

Depending on the thermal diffusivity and kinematic viscosity of the fluid, the velocity and temperature boundary layers may be equal or may differ in size. Because of the no-slip condition at the solid surface, the fluid next to the surface is stationary; therefore, the heat transfer at the interface occurs only by conduction.

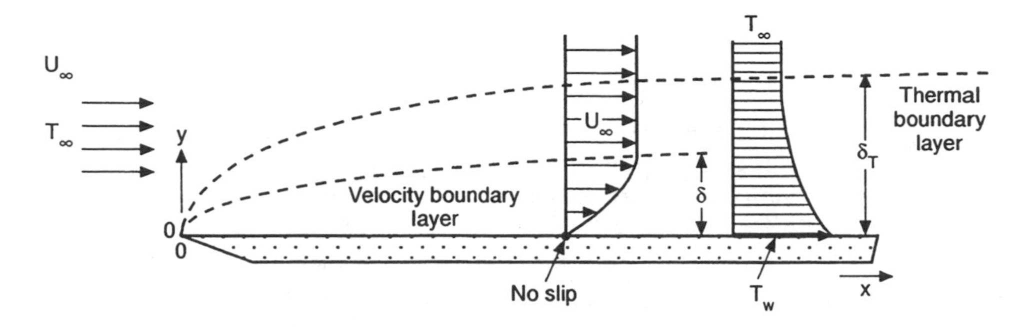

FIGURE 2.17 Temperature and velocity profiles for convection heat-transfer process over a heated surface.

If the temperature gradient were known at the interface, the heat exchange between the fluid and the solid surface could be calculated from Eq. (2.25), where k in this case is the thermal conductivity of the fluid and dT/dx (or dT/dy in reference to Figure 2.17) is the temperature gradient at the interface. However, the temperature gradient at the interface depends on the macroscopic and microscopic motion of fluid particles. In other words, the heat transferred from or to the surface depends on the nature of the flow field.

Therefore, in solving convection problems, engineers need to determine the relationship between the heat transfer through the solid-body boundaries and the temperature difference between the solid-body wall and the bulk fluid. This relationship is given by Eq. (2.26), where h is the convection coefficient averaged over the solid surface area. Note that h depends on the surface geometry and the fluid velocity, as well as on the fluid's physical properties. Therefore, depending on the variation of the above quantities, the heat-transfer coefficient may change from one point to another on the surface of the solid body. As a result, the *local* heat-transfer coefficient may be different than the *average* heat-transfer coefficient. However, for most practical applications, engineers are mainly concerned with the *average* heat-transfer coefficient, and in this section we will use only average heat-transfer coefficients unless otherwise stated.

Natural-Convection Heat Transfer

Natural-convection heat transfer results from density differences within a fluid. These differences may result from temperature gradients that exist within a fluid. When a heated (or cooled) body is placed in a cooled (or heated) fluid, the temperature difference between the fluid and the body causes heat flow between them, resulting in a density gradient inside the fluid. As a result of this density gradient, the low-density fluid moves up and the high-density fluid moves down. The heat-transfer coefficients (and consequently, the rate of heat transfer in natural convection) are generally less than that in forced convection because the driving force for mixing of the fluid is less in natural convection.

Natural-convection problems can be divided into two categories: **external natural convection** and **internal natural convection**. Natural-convection heat transfer from the external surfaces of bodies of various shapes has been studied by many researchers. Experimental results for natural-convection heat transfer are usually correlated by an equation of the type

$$\mathrm{Nu} = \frac{hL}{k} = f\left(\mathrm{Ra}\right), \quad \textbf{(2.33)}$$

where the **Nusselt number**, Nu, provides a measure of the convection heat transfer occurring between the solid surface and the fluid. Knowing Nu, the convection heat-transfer coefficient, h, can be calculated. Note that in Eq. (2.33), k is the fluid conductivity and Ra is the **Rayleigh number**, which represents the ratio of buoyancy force to the rate of change of momentum. The Rayleigh number is given by

$$\mathrm{Ra} = \frac{g\beta(T - T_\infty)L^3}{\nu\alpha}, \tag{2.34}$$

where β is the coefficient of thermal expansion equal to $1/T$ (T is the absolute temperature expressed in Kelvin) for an ideal gas, L is a characteristic length, ν is the kinematic viscosity of the fluid, and α is its thermal diffusivity. A comprehensive review of the fundamentals of natural-convection heat transfer is provided by Raithby and Hollands [1985]. Table 2.2 gives correlations for calculating heat transfer from the external surfaces of some common geometries.

Experiments conducted by Hassani and Hollands [1989], Sparrow and Stretton [1985], Yovanovich and Jafarpur [1993], and others have shown that the external natural-convection heat transfer from bodies of arbitrary shape exhibit Nu–Ra relationships similar to regular geometries such as spheres and short cylinders. An extensive correlation for predicting natural-convection heat transfer from bodies of arbitrary shape was developed by Hassani and Hollands [1989]; this correlation covers the conduction limit, as well as laminar and turbulent convection flow regimes.

Internal natural-convection heat transfer occurs in many engineering problems such as heat loss through building walls, electronic equipment, double-glazed windows, and flat-plate solar collectors. Some of the geometries and their corresponding Nusselt numbers are listed in Table 2.3. Anderson and Kreith [1987] provide a comprehensive summary of natural-convection processes that occur in various solar thermal systems.

The natural-convection heat transfer for long concentric horizontal cylinders and concentric spheres has been studied by Raithby and Hollands [1985]. Their proposed correlations are listed in Table 2.3, where D_o and D_i represent the diameters of outer cylinder (or sphere) and inner cylinder (or sphere), respectively. The Rayleigh number is based on the temperature difference across the gap and a characteristic length defined as $b = (D_o - D_i)/2$. The **effective thermal conductivity** k_{eff} in their correlation is the thermal conductivity that a stationary fluid in the gap must have to transfer the same amount of heat as the moving fluid. Raithby and Hollands also provide correlations for natural-convection heat transfer between long eccentric horizontal cylinders and eccentric spheres.

Example 2.6

One component of the total heat loss from a room is the heat loss through a single-pane window in the room, as shown in Figure 2.18. The inside temperature is kept at $T_i = 22°C$, and the outside temperature is $T_o = -5°C$. The window height H is 0.5 m, and its width is 2 m. The weather is calm, and there is no wind blowing. Assuming uniform glass temperature T_w, calculate the heat loss through the window.

Solution:

The air flow pattern next to the window is shown in Figure 2.18. When warm room air approaches or contacts the window, it loses heat and its temperature drops. Because this cooled air next to the window is denser and heavier than the room air at that height, it starts moving down and is replaced by warmer room air at the top of the window. A similar but opposite air movement occurs at the outside of the window. The total heat loss can be calculated from

$$\dot{Q} = h_i A(T_i - T_w) = h_o A(T_w - T_o), \tag{2.35}$$

where h_i and h_o are the average natural-convection heat-transfer coefficients for inside and outside, respectively. Using the correlation recommended by Fujii and Imura [1972] for a vertical plate with constant temperature T_w (from Table 2.2) and substituting for angle of inclination $\theta = 0$, we get

$$\overline{Nu_H} = 0.56\ \mathrm{Ra}_H^{1/4},$$

TABLE 2.2 Natural-Convection Correlations for External Flows

Configuration	Correlation	Restrictions	Source
Vertical plate with constant T_w, T_∞	$\bar{Nu}_L = \left\{0.825 + \frac{0.387\ Ra_L^{1/6}}{[1 + (0.492/Pr)^{9/16}]^{8/27}}\right\}^2$	$10^{-1} < Ra_L < 10^{12}$ All Pr	Churchill and Chu [1975]
	$\bar{Nu}_L = 0.56\ (Ra_L \cos\theta)^{1/4}$	$10^5 < Ra_L \cos\theta < 10^{11}$ $0 \le \theta \le 89$	Fujii and Imura [1972]
Horizontal plate with hot surface facing upward or cold surface facing downward	$\bar{Nu}_L = 0.54\ Ra_L^{1/4}$ $\bar{Nu}_L = 0.15\ Ra_L^{1/3}$ $\left(L = \frac{\text{surface area}}{\text{perimeter}}\right)$	$10^5 < Ra_L < 10^7$ $10^7 < Ra_L < 10^{10}$ $Pr > 0.5$	McAdams [1954] and Incropera and DeWitt [1990]
Horizontal plate with hot surface facing downward or cold surface facing upward	$\bar{Nu}_L = 0.27\ Ra_L^{1/4}$ $\left(L = \frac{\text{surface area}}{\text{perimeter}}\right)$	$10^5 < Ra_L < 10^{10}$ $Pr > 0.5$	McAdams [1954] and Incropera and DeWitt [1990]

TABLE 2.2 (continued) Natural-Convection Correlations for External Flows

Configuration	Correlation	Restrictions	Source
Horizontal cylinders	$\bar{Nu}_D = \left\{0.6 + \frac{0.387\ Ra_D^{1/6}}{\left[1 + (0.559/Pr)^{9/16}\right]^{8/27}}\right\}^2$	$10^{-5} < Ra_D < 10^{12}$ All Pr	Churchill and Chu [1975]
Vertical cylinders of height L	$\bar{Nu}_L = 0.68 + \frac{0.67\ Ra_L^{1/4}}{\left[1 + (0.492/Pr)^{9/16}\right]^{4/9}}$	$1 < Ra_L < 10^9$	Churchill and Chu [1975]
		$\frac{D}{L} \geq 35 \left(\frac{Pr}{Ra_L}\right)^{1/4}$	McAdams [1954]
	$\bar{Nu}_L = 0.13\ (Ra_L)^{1/3}$	$Ra_L > Pr \times 10^9$	
Sphere	$\bar{Nu}_D = 2 + \frac{0.589\ Ra_D^{1/4}}{[1 + (0.469/Pr)^{9/16}]^{4/9}}$	$Pr \geq 0.7$ $Ra_D < 10^{11}$	Churchill [1983]
Other immersed bodies such as cubes, bispheres, spheroids			Hassani and Hollands [1989]

Note: Ra is Rayleigh number, Pr is Prandtl number, and Nu is Nusselt number.

TABLE 2.3 Natural-Convection Correlations for Internal Flows

Configuration	Correlation	Restrictions	Source
Space enclosed between two horizontal plates heated from below	$\overline{Nu}_L = 1 + 1.44\left[1 - \frac{1708}{Ra_L}\right]^* + \left[\left(\frac{Ra_L}{5830}\right)^{1/3} - 1\right]^*$	Air, $1700 < Ra_L < 10^8$	Hollands et al. [1975]
	$\overline{Nu}_L = 1 + 1.44\left[1 - \frac{1708}{Ra_L}\right]^* + \left[\left(\frac{Ra_L}{5830}\right)^{1/3} - 1\right]^* + 2.0\left[\frac{Ra_L^{1/3}}{140}\right]^{[1 - \ln(Ra_L^{1/3}/140)]}$	Water, $1700 < Ra_L < 3.5 \times 10^9$	Hollands et al. [1975]
	The quantities contained between the parenthesis with asterisk, $(\ )^*$, must be set equal to zero if they become negative.		
Space enclosed between two vertical plates heated from one side	$\overline{Nu}_L\ (90°) = 0.22\left(\frac{H}{L}\right)^{-1/4}\left(\frac{Pr}{0.2 + Pr}\ Ra_L\right)^{0.28}$	$2 < \frac{H}{L} < 10$, $Pr < 10$, $Ra_L < 10^{10}$	Catton [1978]
	$\overline{Nu}_L\ (90°) = 0.18\left(\frac{Pr}{0.2 + Pr}\ Ra_L\right)^{0.29}$	$1 < \frac{H}{L} < 2$, $10^{-3} < Pr < 10^5$; $10^3 < \frac{Ra_L Pr}{0.2 + Pr}$	
	$\overline{Nu}_L = 0.42\ Ra_L^{0.25}\ Pr^{0.012}\left(\frac{H}{L}\right)^{-0.3}$	$10 < \frac{H}{L} < 40$, $1 < Pr < 2 \times 10^4$; $10^4 < Ra_L < 10^7$	MacGregor and Emery [1969]

TABLE 2.3 (continued) Natural-Convection Correlations for Internal Flows

Configuration	Correlation	Restrictions	Source
Natural convection in inclined enclosures. (Stably statified fluid, $\theta = 180°C$; Cellular convection, $\theta = 0°C$)	$\overline{Nu}_L(\theta) = 1 + \left[Nu_L(90°) - 1\right] \sin\theta$	$90° < \theta < 180°$, air	Arnold et al. [1976], Catton (1978), Arnold et al. [1974], and Ayyaswamy and Catton [1973]. Hollands et al. [1976]
	$\overline{Nu}_L(\theta) = \overline{Nu}_L(90°)\,(\sin\theta)^{1/4}$	$\theta^* < \theta < 90$, air	
	$\overline{Nu}_L(\theta) = \left[\dfrac{\overline{Nu}_L(90°)}{\overline{Nu}_L(0°)}(\sin\theta^*)^{1/4}\right]^{\theta/\theta^*}$	$0° < \theta < \theta^*$, $\dfrac{H}{L} < 10$, air	
	$\overline{Nu}_L(\theta) = 1 + 1.44\left(1 - \dfrac{1708}{Ra_L\cos\theta}\right)^* \times \left[1 - \dfrac{(\sin 1.8\theta)^{1.6} \times 1708}{Ra_L\cos\theta}\right] + \left[\left(\dfrac{Ra_L\cos\theta}{5830}\right)^{1/3} - 1\right]^*$	$0° < \theta < \theta^*$, $\dfrac{H}{L} > 10$, air	
	where: H/L = 1, 3, 6, 12, > 12; θ^* = 25°, 53°, 60°, 67°, 70°		
	The quantities contained between the parentheses with asterisk, ( )*, must be set equal to zero if they become negative.		
Spherical cavity interior (Diameter D)	$\overline{Nu}_D = C(Gr_D Pr)^n$	$Gr_D Pr$ = 10^4-10^9: C = 0.59, n = 1/4; 10^9-10^{12}: C = 0.13, n = 1/3	Kreith [1970]

H/L	1	3	6	12	> 12
θ^*	25°	53°	60°	67°	70°

$Gr_D Pr$	C	n
10^4-10^9	0.59	1/4
10^9-10^{12}	0.13	1/3

Configuration	Correlation	Restrictions	Source
Long concentric cylinders	$\frac{k_{eff}}{k} = 0.386 \left[\frac{\ln(D_o/D_i)}{b^{3/4}(1/D_i^{3/5} + 1/D_o^{3/5})^{5/4}} \right] \times \left(\frac{Pr}{0.861 + Pr} \right)^{1/4} Ra_b^{1/4}$ where $b = (D_o - D_i) / 2$	$0.70 \le Pr \le 6000$ $10 \le \left[\frac{\ln(D_o/D_i)}{b^{3/4}(1/D_i^{3/5} + 1/D_o^{3/5})^{5/4}} \right]^4 Ra_b \le 10^7$	Raithby and Hollands [1974]
Concentric spheres	$\frac{k_{eff}}{k} = 0.74 \left[\frac{b^{1/4}}{D_o D_i (D_i^{-7/5} + D_o^{-7/5})^{5/4}} \right] \times \left(\frac{Pr}{0.861 + Pr} \right)^{1/4} Ra_b^{1/4}$ where $b = (D_o - D_i) / 2$	$0.70 \le Pr \le 4200$ $10 \le \left[\frac{b}{(D_o D_i)^4 (D_i^{-7/5} + D_o^{-7/5})^5} \right] Ra_b \le 10^7$	Raithby and Hollands [1974]

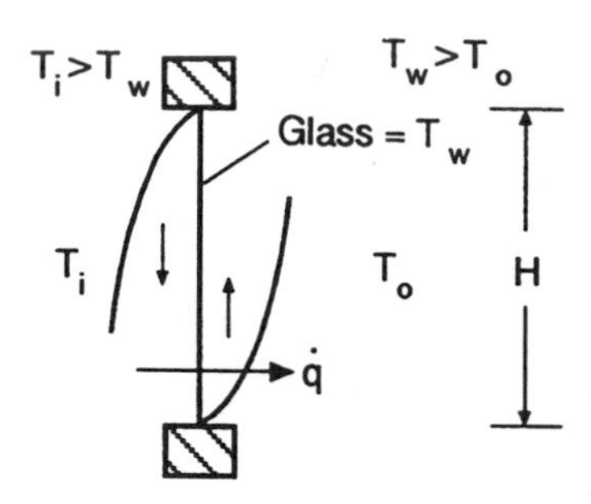

FIGURE 2.18 Heat loss through a single-pane window.

where

$$\text{Ra}_H = \frac{g\beta\ \Delta T\ H^3}{\nu\alpha},$$

and H is the height of the window pane. Note that the temperature difference in the expression for Rayleigh number depends on the medium for which the heat-transfer coefficient is sought. For example, for calculating h_i, we write

$$\overline{\text{Nu}}_{H,i} = \frac{h_i H}{k} = 0.56\left[\frac{g\beta(T_i - T_w)H^3}{\nu\alpha}\right]^{1/4}, \tag{2.36}$$

and for calculating h_o, we write

$$\overline{\text{Nu}}_{H,o} = \frac{h_o H}{k} = 0.56\left[\frac{g\beta(T_w - T_o)H^3}{\nu\alpha}\right]^{1/4}. \tag{2.37}$$

Note that all the properties in Eqs. (2.36) or (2.37) should be calculated at film temperature $T_f = (T_i + T_w)/2$ or $T_f = (T_o + T_w)/2$. To calculate $\dot{Q}$ and air properties, we need to know T_w. To estimate T_w, we assume that air properties over the temperature range of interest to this problem do not change significantly (refer to air property tables to verify this assumption). Using this assumption, we find the ratio between Eqs. (2.36) and (2.37) as

$$\frac{h_o}{h_i} = \left(\frac{T_w - T_o}{T_i - T_w}\right)^{1/4}, \tag{2.38}$$

which provides a relationship between h_o, h_i, and T_w. Another equation of this kind can be obtained from Eq. (2.35):

$$\frac{h_o}{h_i} = \frac{T_i - T_w}{T_w - T_o}. \tag{2.39}$$

Solving Eqs. (2.38) and (2.39), we can show that

$$T_w = \frac{T_o + T_i}{2} = 8.5°\text{C}.$$

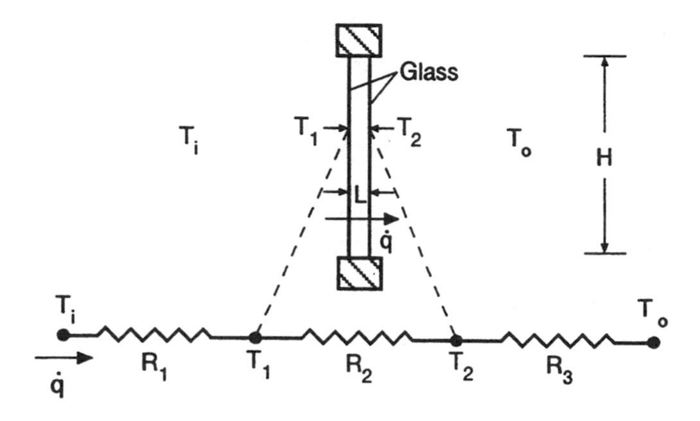

FIGURE 2.19 Heat loss through a double-pane window.

Now, by calculating h_i or h_o and substituting into Eq. (2.35), the total heat transfer can be calculated. In this case, we choose to solve for h_o. Therefore, the air properties should be calculated at

$$T_f = \frac{T_o + T_w}{2} = 1.75°\text{C}.$$

Air properties at $T_f = 1.75$°C are $k = 0.0238$ W/m K, $\nu = 14.08 \times 10^{-6}$ m²/s, $\alpha = 19.48 \times 10^{-6}$ m²/s, and $\beta = 1/T_f = 0.00364$ K⁻¹. Using these properties, the Rayleigh number is

$$\text{Ra}_{H,o} = \frac{g\beta\,(T_w - T_o)\,H^3}{\nu\alpha} = 220 \times 10^6.$$

From Eq. (2.37), we obtain

$$h_o = \frac{k}{H}\,(0.56)\,\text{Ra}_{H,o}^{1/4} = 3.25\ \text{W/m}^2\ \text{K},$$

and

$$\dot{Q} = h_o A(T_w - T_o)$$
$$= 3.25\ \text{W/m}^2\ \text{K} \times 1\ \text{m}^2 \times (281.65\ \text{K} - 268.15\ \text{K}) = 43.87\ \text{W}.$$

Example 2.7

The single-pane window of the previous example is replaced by a double-pane window as shown in Figure 2.19. The outside and inside temperatures are the same as in Example 2.6 ($T_i = 22$°C, $T_o = -5$°C). The glass-to-glass spacing is $L = 20$ mm, the window height $H = 0.5$ m, and the width is 2 m. Find the heat loss through this window and compare it to the heat loss through the single-pane window. Ignore conduction resistance through the glass.

Solution:

The thermal circuit for the system is shown in Figure 2.19. Temperatures T_1 and T_2 are unknown and represent the average glass temperature (i.e., we assume that the glass temperature is uniform over the entire surface). As in Example 2.6, we first need to estimate temperatures T_1 and T_2. The rate of heat transfer can be expressed as

$$\dot{Q} = h_1 A(T_i - T_1) = h_2 A(T_1 - T_2) = h_3 A(T_2 - T_o). \quad \textbf{(2.40)}$$

The heat-transfer coefficients h_1 and h_3 for natural-convection heat transfer between the glass surface and interior/exterior can be calculated using Eqs. (2.36) and (2.37), and the ratio between h_1 and h_3 is

$$\frac{h_1}{h_3} = \left(\frac{T_i - T_1}{T_2 - T_o}\right)^{1/4}. \quad \textbf{(2.41)}$$

Another relationship between h_1, h_3, T_1, and T_2, is obtained from Eq. (2.40):

$$\frac{h_1}{h_3} = \left(\frac{T_2 - T_o}{T_i - T_1}\right). \quad \textbf{(2.42)}$$

Solving Eqs. (2.41) and (2.42), we get:

$$T_1 + T_2 = T_i + T_o. \quad \textbf{(2.43)}$$

We need an additional equation that provides a relationship between T_1 and T_2, and we obtain this equation from the correlation that expresses the natural-convection heat transfer in the enclosed area of the double-pane window. We choose the correlation recommended by MacGregor and Emery [1969] from Table 2.3:

$$h_2 = \frac{k}{L}\ 0.42\ \mathrm{Ra}_L^{0.25}\left(\frac{H}{L}\right)^{-0.3} \quad \text{for } \mathrm{Pr} = 0.72, \quad \textbf{(2.44)}$$

where

$$\mathrm{Ra}_L = \frac{g\beta(T_1 - T_2)L^3}{\nu\alpha}.$$

Using some mathematical manipulations, Eq. (2.44) can be written as

$$h_2 = \frac{k}{L} 0.16 \left(\frac{T_1 - T_2}{T_i - T_o}\right)^{1/4} \left(\frac{L}{H}\right)^{3/4} \left(\frac{g\beta(T_i - T_o)H^3}{\nu\alpha}\right)^{1/4}, \quad \textbf{(2.45)}$$

where $L/H = 0.04$. The heat-transfer coefficient h_1 can be calculated from Eq. (2.36) (used to calculate h_i) and can be written as

$$h_1 = \frac{k}{H} 0.56 \left(\frac{T_i - T_2}{T_i - T_o}\right)^{1/4} \left(\frac{g\beta(T_i - T_o)H^3}{\nu\alpha}\right)^{1/4}, \quad \textbf{(2.46)}$$

where the Rayleigh number has been written in terms of $(T_i - T_o)$ instead of $(T_i - T_1)$. Substituting for L/H, the ratio between h_2 and h_1 is

$$\frac{h_2}{h_1} = \frac{0.63\left(T_1 - T_2\right)^{1/4}}{\left(T_i - T_1\right)^{1/4}}. \tag{2.47}$$

Note that in finding h_2/h_1, we have assumed that the properties do not change much in the temperature range of interest. From Eq. (2.40), we have

$$\frac{h_2}{h_1} = \frac{T_i - T_1}{T_1 - T_2}. \tag{2.48}$$

Therefore, solving Eqs. (2.47) and (2.48), we obtain T_1 in terms of T_i and T_o:

$$T_1 = 0.71\ T_i + 0.29\ T_o. \tag{2.49}$$

Substituting for T_i and T_o, we obtain $T_1 = 14.2$°C, and substituting for T_1, T_i, and T_o in Eq. (2.43), we get $T_2 = 2.8$°C. Knowing T_1 and T_2, we can calculate Ra_L. To calculate Ra_L, we should obtain air properties at $T_f = (T_1 + T_2)/2 = 8.5$°C, which are $k = 0.0244$ W/m K, $\nu = 14.8 \times 10^{-6}$ m²/s, $\alpha = 20.6 \times 10^{-6}$ m²/s, and $\beta = 1/T_f = 0.00355$ K^{-1}. Therefore,

$$Ra_L = \frac{9.81\ \text{m/s}^2 \times 0.00355\ \text{K}^{-1} \times \left(287.35\ \text{K} - 275.95\ \text{K}\right) \times \left(0.02\right)^3\ \text{m}^3}{14.8 \times 10^{-6}\ \text{m}^2/\text{s} \times 20.6 \times 10^{-6}\ \text{m}^2/\text{s}} = 10{,}420,$$

and h_2 from Eq. (2.44) is

$$h_2 = 1.97\ \text{W/m}^2\ \text{K},$$

and

$$\dot{Q} = h_2\ A\left(T_1 - T_2\right) = 22.46\ \text{W}.$$

Comparing the heat loss to that of Example 2.6 for a single-pane window, we note that the heat loss through a single-pane window is almost twice as much as through a double-pane window for the same inside and outside conditions.

Forced-Convection Heat Transfer

Forced-convection heat transfer is created by auxiliary means such as pumps and fans or natural phenomena such as wind. This type of process occurs in many engineering applications such as flow of hot or cold fluids inside ducts and various thermodynamic cycles used for refrigeration, power generation, and heating or cooling of buildings. As with natural convection, the main challenge in solving forced-convection problems is to determine the heat-transfer coefficient.

The forced-convection heat-transfer processes can be divided into two categories: **external-flow forced convection** and **internal-flow forced convection**. External forced-convection problems are important because they occur in various engineering applications such as heat loss from external walls of buildings on a windy day, from steam radiators, from aircraft wings, or from a hot wire anemometer. To solve these problems, researchers have conducted many experiments to develop correlations for predicting the heat transfer. The experimental results obtained for external forced-convection problems are usually expressed or correlated by an equation of the form

$$\mathrm{Nu} = f(\mathrm{Re})\ g(\mathrm{Pr}),$$

where f and g represent the functional dependance of the Nusselt number on the Reynolds and Prandtl numbers. The Reynolds number is a nondimensional number representing the ratio of inertia to viscous forces, and the **Prandtl number** is equal to ν/α, which is the ratio of momentum diffusivity to thermal diffusivity.

Table 2.4a lists some of the important correlations for calculating forced-convection heat transfer from external surfaces of common geometries. Listed in Table 2.4a is the correlation for the forced-convection heat-transfer to or from a fluid flowing over a bundle of tubes, which is relevant to many industrial applications such as the design of commercial heat exchangers. Figures 2.20 and 2.21 show different configurations of tube bundles in cross-flow whose forced-convection correlations are presented in Table 2.4a.

Forced-convection heat transfer in confined spaces is also of interest and has many engineering applications. Flow of cold or hot fluids through conduits and heat transfer associated with that process is important to both chemical and mechanical engineering processes. The heat transfer associated with internal forced convection can be expressed by an equation of the form

$$\mathrm{Nu} = f(\mathrm{Re})\ g(\mathrm{Pr})\ e(x/D_H),$$

where $f(\mathrm{Re})$, $g(\mathrm{Pr})$, and $e(x/D_\mathrm{H})$ represent the functional dependance on Reynolds number, Prandtl number, and x/D_H, respectively. The functional dependance on x/D_H becomes important for short ducts in laminar flow. The quantity D_H is called the **hydraulic diameter** of the conduit and is defined as

$$D_H = 4 \times \frac{\text{flow cross-sectional area}}{\text{wetted perimeter}} \tag{2.50}$$

and is used as the characteristic length for Nusselt and Reynolds numbers.

Fully developed laminar flow through ducts of various cross section has been studied by Shah and London [1978], and they present analytical solutions for calculating heat transfer and friction coefficients. Solving internal tube-flow problems requires knowledge of the nature of the tube-surface thermal conditions. Two special cases of tube-surface conditions cover most engineering applications: constant tube-surface heat flux and constant tube-surface temperature. The axial temperature variations for the fluid flowing inside a tube are shown in Figure 2.22. Figure 2.22(a) shows the mean fluid-temperature variations inside a tube with constant surface heat flux. Note that the mean fluid temperature, $T_\mathrm{m}(x)$, varies linearly along the tube. Figure 2.22(b) shows the mean fluid-temperature variations inside a tube with constant surface temperature. Some of the recommended correlations for forced convection of incompressible flow inside tubes and ducts are listed in Table 2.5.

Now that we have reviewed both natural- and forced-convection heat-transfer processes, it is useful to compare the order of magnitude of the heat-transfer coefficient for both cases. Table 2.6 provides some approximate values of convection heat-transfer coefficients.

Example 2.8

A solar-thermal central-receiver system is depicted in Figure 2.23. In this system, solar radiation is reflected from tracking mirrors onto a stationary receiver. The receiver consists of a collection of tubes that are radiatively heated, and a working fluid (coolant) flows through them; the heat

TABLE 2.4a Forced-Convection Heat-Transfer Correlations for External Flows*

Configuration	Correlation	Restrictions	Source
Flat plate in parallel flow	$\bar{Nu}_x = 0.664\ Re_x^{1/2}\ Pr^{1/3}$	Laminar $Pr \geq 0.6$	Incropera and DeWitt [1990]
Flat plate in parallel flow	$Nu_x = 0.0296\ Re_x^{4/5}\ Pr^{1/3}$	Turbulent, local, $0.6 < Pr < 60$, $Re_x < 10^8$	Incropera and DeWitt [1990]
Circular cylinder in cross flow	$\bar{Nu}_D = C\ Re_D^m\ Pr^n \left(\frac{Pr_\infty}{Pr_s}\right)^{1/4}$ n=0.36 for $Pr > 10$ n=0.37 for $Pr \leq 10$ Re_D / C / m 1–40 / 0.75 / 0.4 40–1000 / 0.51 / 0.5 10^3–2×10^5 / 0.26 / 0.6 2×10^5–10^6 / 0.076 / 0.7	$0.7 < Pr < 500$ $1 < Re_D < 10^6$ Properties at T_∞	Zukauskas [1972]
Non-circular cylinder in cross flow in a gas	$\bar{Nu}_D = C\ Re_D^m\ Pr^{1/3}$ For C and m, see Table 4b.	$0.4 < Re_D < 4 \times 10^5$ $Pr \geq 0.7$	Jakob [1949]
Short cylinder in a gas	$\bar{Nu}_D = 0.123\ Re_D^{0.651} + 0.00416 \left(\frac{D}{L}\right)^{0.85} Re_D^{0.792}$	$7 \times 10^4 < Re_D < 2.2\times10^5$ $L/D < 4$	Quarmby and Al-Fakhri [1980]

TABLE 2.4a (continued) Forced-Convection Heat-Transfer Correlations for External Flows*

Configuration	Correlation	Restrictions	Source
Sphere in a gas or liquid	$\bar{Nu}_D = 2 + \left(0.4\, Re_D^{1/2} + 0.06\, Re_D^{2/3}\right) Pr^{0.4} \left(\frac{\mu_\infty}{\mu_s}\right)^{1/4}$	$3.5 < Re_D < 7.6 \times 10^4$ $0.7 < Pr < 380$ $1.0 < \frac{\mu_\infty}{\mu_s} < 3.2$ Properties at T_∞	Whitaker [1972]
Tube bundle in cross-flow	$\bar{Nu}_D = Pr^{+0.36} \left(\frac{Pr}{Pr_s}\right)^{1/4} C \left(\frac{S_T}{S_L}\right)^n Re_D^m$ C, m, n: 0.8, 0.4, 0: $10 < Re_D < 100$, in-line 0.9, 0.4, 0: $10 < Re_D$ 100, staggered 0.27, 0.63, 0: $1000 < Re_D < 2 \times 10^5$, in-line $S_T/S_L \geq 0.7$ 0.35, 0.60, 0.2: $1000 < Re_D < 2 \times 10^5$, staggered $S_T/S_L < 2$ 0.40, 0.60, 0: $1000 < Re_D < 2 \times 10^5$, staggered $S_T/S_L \geq 2$ 0.021, 0.84, 0: $Re_D > 2 \times 10^5$, in-line 0.022, 0.84, 0: $Re_D > 2 \times 10^5$, staggered $Pr > 1$ $\bar{Nu}_D = 0.019\, Re_D^{0.84}$ — $Re_D > 2 \times 10^5$, staggered $Pr = 0.7$	See Figures 20 and 21 Properties at T_∞	Zukauskas [1972]
Flow over staggered tube bundle, gas and liquid	$\bar{Nu}_D = 0.0131\, Re_D^{0.883}\, Pr^{0.36}$	$4.5 \times 10^5 < Re_D < 7 \times 10^6$ $Pr > 0.5$ $S_T/D = 2$, $S_L/D = 1.4$	Achenbach [1989]

*All properties calculated at $(T_\infty + T_s)/2$ unless otherwise stated under the column "condition." Properties with the subscript "s" are calculated at T_s (surface temperature).

TABLE 2.4b Constants for Noncircular Cylinders in Cross Flow of a Gas

Configuration	Re_D	C	m
Square Flow direction → (D)	$5 \times 10^3 - 10^5$	0.246	0.588
Flow direction → (D)	$5 \times 10^3 - 10^5$	0.102	0.675
Hexagon Flow direction → (D)	$5 \times 10^3 - 1.95 \times 10^4$ $1.95 \times 10^4 - 10^5$	0.160 0.0385	0.638 0.782
Flow direction → (D)	$5 \times 10^3 - 10^5$	0.153	0.638
Vertical plate Flow direction → (D)	$4 \times 10^3 - 1.5 \times 10^4$	0.228	0.731

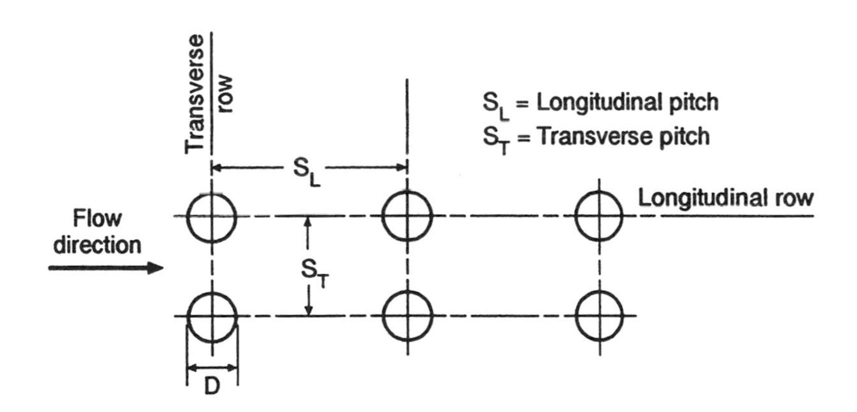

FIGURE 2.20 In-line tube arrangement for tube bundle in cross-flow forced convection.

absorbed by the working fluid is then used to generate electricity. Consider a central-receiver system that consists of several horizontal circular tubes each with an inside diameter of 0.015 m. The working fluid is molten salt that enters the tube at 400°C at a rate of 0.015 kg/s. Assume that the average solar flux approaching the tube is about 10,000 W/m^2.

(a) Find the necessary length of the tube to raise the working-fluid temperature to 500°C at the exit.

(b) Determine the tube-surface temperature at the exit.

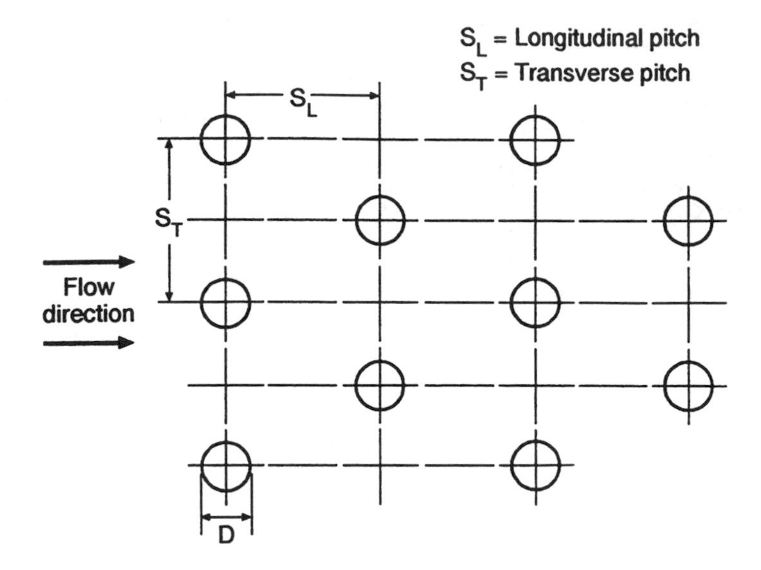

FIGURE 2.21 Staggered tube arrangement for tube bundle in cross-flow forced convection.

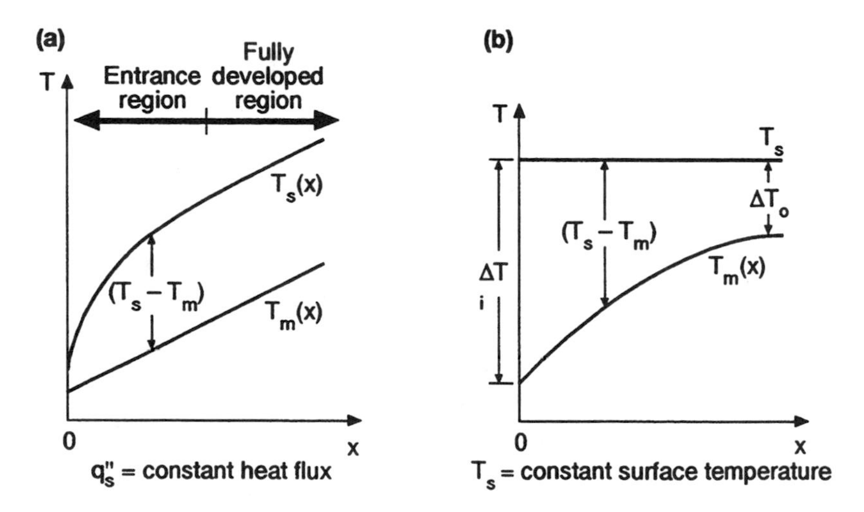

FIGURE 2.22 Axial fluid temperature variations for heat transfer in a tube for (a) constant surface heat flux, and (b) constant surface temperature.

Solution:

We will assume steady-state conditions, fully developed flow, and incompressible flow with constant properties. The axial temperature variations for heat transfer in a tube for constant surface temperature are shown in Figures 2.22(a) and 2.22(b), respectively.

(a) The specific heat of molten salt at $T_m = (T_i + T_o)/2 = 450°C$ is $c_p = 1{,}520$ J/kg K. The total heat transferred to the working fluid is $q''A_t = \dot{m}c_p\ (T_o - T_i)$, where q'' is the solar flux and $A_t = \pi DL$ is the surface area of the tube (assuming that the solar flux is incident over the entire perimeter of the tube). Therefore,

$$L = \frac{\dot{m}c_p\left(T_o - T_i\right)}{q''\pi D} = \frac{0.015\ \text{kg/s} \times 1520\ \text{J/kg K} \times \left(773\ \text{K} - 673\ \text{K}\right)}{10^4\ \text{W/m}^2 \times \pi \times 0.015\ \text{m}} = 4.8\ m$$

TABLE 2.5 Forced-Convection Correlations for Incompressible Flow Inside Tubes and Ducts*,†

Configuration	Correlation	Restrictions	Source
Fully developed laminar flow in long tubes:			
a. With uniform wall temperature	$\overline{\mathrm{Nu}}_D = 3.66$	$\mathrm{Pr} > 0.6$	Kays and Perkins [1985]
b. With uniform heat flux	$\overline{\mathrm{Nu}}_D = 4.36$	$\mathrm{Pr} > 0.6$	Incropera and DeWitt [1990]
c. Friction factor (liquids)	$f = \left(\frac{64}{\mathrm{Re}_D}\right)\left(\frac{\mu_s}{\mu_b}\right)^{0.14}$		
d. Friction factor (gas)	$f = \left(\frac{64}{\mathrm{Re}_D}\right)\left(\frac{T_s}{T_b}\right)^{0.14}$		
Laminar flow in short tubes and ducts with uniform wall temperature	$\overline{\mathrm{Nu}}_{D_H} = 3.66 + \frac{0.0668\ \mathrm{Re}_{D_H}\ \mathrm{Pr}\frac{D_H}{L}}{1+0.045\left(\mathrm{Re}_{D_H}\ \mathrm{Pr}\frac{D_H}{L}\right)^{0.56}}\left(\frac{\mu_b}{\mu_s}\right)^{0.14}$	$100 < \mathrm{Re}_{D_H}\ \mathrm{Pr}\frac{D_H}{L} < 1500$ $\mathrm{Pr} > 0.7$	Hausen [1983]
Fully developed turbulent flow through smooth, long tubes and ducts:			
a. Nusselt number	$\overline{\mathrm{Nu}}_{D_H} = 0.027\ \mathrm{Re}_{D_H}^{0.8}\ \mathrm{Pr}^{0.33}\left(\frac{\mu_b}{\mu_s}\right)^{0.14}$	$6\times 10^3 < \mathrm{Re}_{D_H} < 10^7$ $0.7 < \mathrm{Pr} < 10^4$ $60 < L/D_H$	Sieder and Tate [1936]
b. Friction factor	$f = \frac{0.184}{\mathrm{Re}_{D_H}^{0.2}}$	$10^4 < \mathrm{Re}_{D_H} < 10^6$	Kays and London [1984]

* All physical properties are evaluated at the bulk temperature T_b except μ_s, which is evaluated at the surface temperature T_s.

† Incompressible flow correlations apply to gases and vapors when average velocity is less than half the speed of sound (Mach number < 0.5).

TABLE 2.6 Order of Magnitude of Convective Heat-Transfer Coefficients h_c

	W/m² K	Btu/h ft²°F
Air, free convection	6–30	1–5
Superheated steam or air, forced convection	3–300	5–50
Oil, forced convection	60–1,800	10–300
Water, forced convection	300–18,000	50–3,000
Water, boiling	3,000–60,000	500–10,000
Steam, condensing	6,000–120,000	1,000–20,000

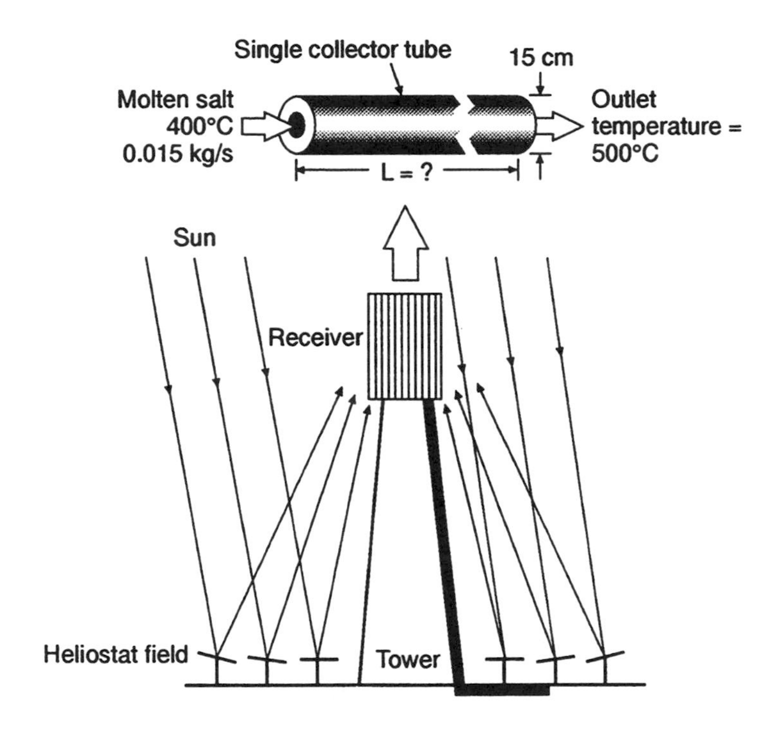

FIGURE 2.23 A solar-thermal central-receiver system.

(b) Molten salt properties at $T_o = 500°C$ are $\mu = 1.31 \times 10^{-3}$ Ns/m², $k = 0.538$ W/m K, and Pr = 3.723. The tube-surface temperature can be obtained from $q'' = h(T_s - T_o)$, where h is the local convection coefficient at the exit. To find h, the nature of the flow must first be established by calculating the Reynolds number:

$$\text{Re} = \frac{uD}{\nu} = \frac{4\dot{m}}{\pi\mu D} = \frac{4 \times 0.015 \ \text{kg/s}}{\pi \times 1.31 \times 10^{-3} \ \text{Ns/m}^2 \times 0.015 \ \text{m}} = 972.$$

Because Re < 2,300, the flow inside the tube is laminar. Therefore, from Table 2.5, $\text{Nu}_D = hD/k = 4.36$, and

$$h = \frac{\text{Nu}_D k}{D} = \frac{4.36 \times 0.538 \ \text{W/m K}}{0.015 \ \text{m}} = 156.4 \ \text{W/m}^2 \ \text{K}.$$

The surface temperature at the exit is

$$T_s = \frac{q''}{h} + T_o = \frac{10^4 \ \text{W/m}^2}{156.4 \ \text{W/m}^2 \ \text{K}} + 773 \ \text{K} = 836.9 \ \text{K}.$$

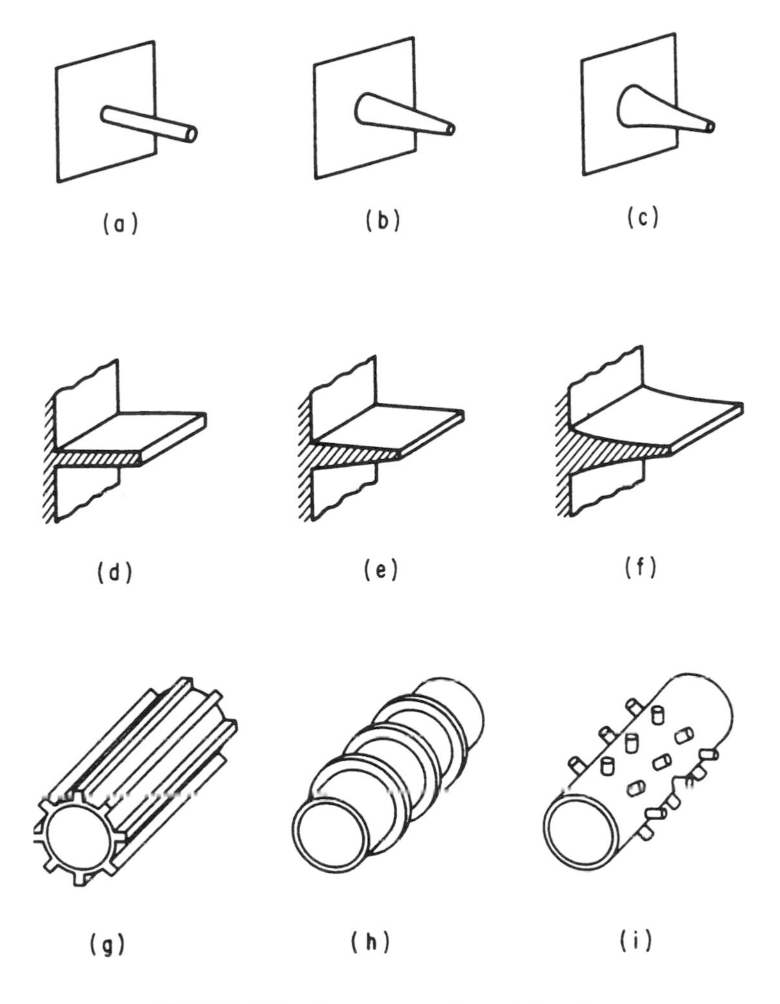

FIGURE 2.24 Various types of extended surfaces.

Extended Surfaces or Fins

According to Eq. (2.25), the rate of heat transfer by conduction is directly proportional to the heat flow area. To enhance the rate of heat transfer, we can increase the effective heat-transfer surface area. Based on this concept, extended surfaces or fins are widely used in industry to increase the rate of heat transfer for heating or cooling purposes. Various types of extended surfaces are shown in Figure 2.24. The simplest type of extended surface is the fin with a uniform cross section, as shown in Figure 2.24(d). The temperature distribution and fin heat-transfer rate can be found by solving a differential equation that expresses energy balance on an infinitesimal element in the fin as given by

$$\frac{d^2T(x)}{dx^2}-\frac{hP}{kA}\left[T(x)-T_\infty\right]=0, \tag{2.51}$$

where P is the cross-sectional perimeter of the fin, k is the thermal conductivity of the fin, A is the cross-sectional area of the fin, and h is the mean convection heat-transfer coefficient between the fin and its surroundings. To solve Eq. (2.51), we need two boundary conditions: one at $x = 0$ (base of the fin) and the other at $x = L$ (tip of the fin). The boundary condition used at the base of the fin is usually $T(x = 0) = T_b$, the temperature of the main body to which the fin is attached. The second boundary condition at the tip of the fin ($x = L$) may take several forms:

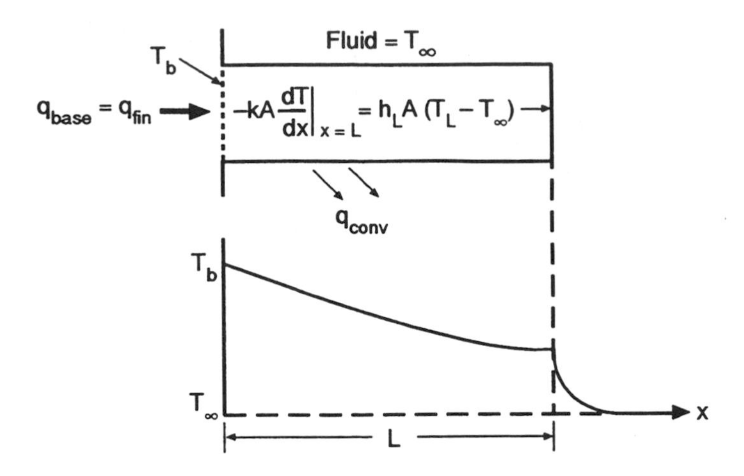

FIGURE 2.25 Schematic representation of temperature distribution in a fin with boundary condition 4 at its tip.

1. The fin temperature approaches the environment temperature:

$$T \approx T_{\infty} \quad at \quad x = L.$$

2. There is no heat loss from the end surface of the fin (insulated end):

$$\frac{dT}{dx} = 0 \quad at \quad x = L.$$

3. The fin-end surface temperature is fixed:

$$T = T_L \quad at \quad x = L.$$

4. There is convection heat loss from the end surface of the fin:

$$-k \left.\frac{dT}{dx}\right|_{x=L} = h_L \left(T_L - T_{\infty}\right).$$

Using the boundary condition at $x = 0$ along with one of the four boundary conditions for $x = L$, we can solve Eq. (2.51) and obtain the temperature distribution for a fin with a uniform cross section. Knowing the temperature distribution of the fin, the fin heat-transfer rate q_{fin} can be obtained by applying Fourier's law at the base of the fin:

$$g_{fin} = g_{base} = -kA \left.\frac{dT}{dx}\right|_{x=0} = -kA \left.\frac{d\theta}{dx}\right|_{x=0}, \tag{2.52}$$

where A is the cross-sectional surface area of the fin and $\theta(x) = T(x) - T_{\infty}$. Figure 2.25 is a schematic representation of the temperature distribution in a fin with boundary condition 4. Table 2.7 lists equations of temperature distribution and rate of heat transfer for fins of uniform cross section with all four different tip boundary conditions.

Fins or extended surfaces are used to increase the heat-transfer rate from a surface. However, the presence of fins introduces an additional conduction resistance in the path of heat dissipating from the base surface. If a fin is made of highly conductive material, its resistance to heat conduction

TABLE 2.7 Equations for Temperature Distribution and Rate of Heat Transfer for Fins of Uniform Cross Section*

Case	Tip Condition ($x = L$)	Temperature Distribution (θ/θ_b)	Fin Heat-Transfer Rate (q_{fin})
1	Infinite fin ($L \rightarrow \infty$): $\theta(L) = 0$	e^{-mx}	M
2	Adiabatic: $\left.\frac{d\theta}{dx}\right\|_{x=L} = 0$	$\frac{\cosh m(L-x)}{\cosh mL}$	$M \tanh mL$
3	Fixed temperature: $\theta(L) = \theta_L$	$\frac{(\theta_L/\theta_b)\sinh mx + \sinh m(L-x)}{\sinh mL}$	$M\frac{\cosh mL - (\theta_L/\theta_b)}{\sinh mL}$
4	Convection heat transfer: $h\theta(L) = k\left.\frac{d\theta}{dx}\right\|_{x=L}$	$\frac{\cosh m(L-x) + (h/mk)\sinh m(L-x)}{\cosh mL + (h/mk)\sinh mL}$	$M\frac{\sinh mL + (h/mk)\cosh mL}{\cosh mL + (h/mk)\sinh mL}$

* $\theta \equiv T - T_\infty$; $\theta_b \equiv \theta(0) = T_b - T_\infty$; $m^2 \equiv \frac{hP}{kA}$; $M \equiv \sqrt{hPkA}\ \theta_b$

is small, creating a small temperature gradient from the base to the tip of the fin. As a result, the fin ideally dissipates heat at a temperature almost equal to the base temperature. However, fins show a temperature distribution similar to that shown in Figure 2.25. Therefore, the thermal performance of fins is usually assessed by calculating **fin efficiency**.

The efficiency of a fin is a measure of its effectiveness in transferring heat and is defined as the ratio of the actual heat loss to the maximum heat loss that would have occurred if the total surface of the fin were at the base temperature, that is,

$$\eta_{fin} = \frac{q_{fin}}{q_{max}} = \frac{q_{fin}}{hA_f(T_b - T_\infty)}, \tag{2.53}$$

where A_f is the total surface area of the fin, and q_{fin} for fins with uniform cross section is obtained from Table 2.7.

Radiation Heat Transfer

Thermal radiation is a heat-transfer process that occurs between any two objects that are at different temperatures. All objects emit thermal radiation by virtue of their temperature. Scientists believe that the thermal radiation energy emitted by a surface is propagated through the surrounding medium either by electromagnetic waves or is transported by photons. In a vacuum, radiation travels at the speed of light C_0 (3×10^8 m/s in a vacuum); however the **speed of propagation** c in a medium is less than C_0 and is given in terms of index of refraction of the medium, as in Eq. (2.54). The radiation wavelength depends on the source frequency and refractive index of the medium through which the radiation travels, according to the equation

$$c = \lambda\nu = \frac{C_0}{n}, \tag{2.54}$$

where n = index of refraction of the medium
$C_0 = 3 \times 10^8$ m/s (9.84×10^8 ft/s)
λ = wavelength, m (ft)
ν = frequency, s^{-1}

Thermal radiation can occur over a wide spectrum of wavelengths, namely between 0.1 and 100 μm. The spectral distribution and the magnitude of the emitted radiation from an object depends strongly on its absolute temperature and the nature of its surface. For example, at the surface temperature of the sun, 5,800 K, most energy is emitted at wavelengths below 3 μm. However, the earth (at a surface temperature of about 290 K) emits its energy at wavelengths longer than 3 μm. This particular radiation-process property has caused environmental concerns such as global warming (or the greenhouse effect) in recent years. Global warming is a result of the increased amount of carbon dioxide in the atmosphere. This gas absorbs radiation from the sun at shorter wavelengths but is opaque to emitted radiation from the earth at longer wavelengths, thereby trapping the thermal energy and causing a gradual warming of the atmosphere, as in a greenhouse.

A perfect radiator—called a **blackbody**—emits and absorbs the maximum amount of radiation at any wavelength. The amount of heat radiated by a blackbody is

$$\dot{Q}_r = \sigma A T_b^4, \tag{2.55}$$

where σ = the Stefan-Boltzmann constant = 5.676×10^{-8} W/m² K⁴ (or 0.1714×10^{-8} Btu/h ft² °R⁴)
T_b = absolute temperature of the blackbody, K (°R)
A = surface area, m² (ft²)

The spectral (or monochromatic) **blackbody emissive power** according to Planck's Law is

$$E_{b\lambda}(T) = \frac{C_1 \lambda^{-5}}{e^{C_2/\lambda T} - 1}, \tag{2.56}$$

where $E_{b\lambda}(T)$ = spectral emissive power of a blackbody at absolute temperature T, $\frac{W}{m^3}\left(\frac{Btu}{h\,ft^2\mu}\right)$
λ = wavelength, m (μ)
T = absolute temperature of blackbody, K (°R)
C_1 = constant, 3.7415×10^{-16} W m² $\left(1.187 \times 10^8 \frac{Btu\mu^4}{hft^2}\right)$
C_2 = constant, 1.4388×10^{-2} m K (2.5896×10^4 μ °R)

The spectral blackbody emissive power for different temperatures is plotted in Figure 2.26, which shows that as the temperature increases, the emissive power and the wavelength range increase as well. However, as temperature increases, the wavelength at which maximum emissive power occurs decreases. Wien's Displacement Law provides a relationship between the maximum wavelength λ_{max} and the absolute temperature at which $E_{b\lambda}$ is maximum:

$$\lambda_{max}\, T = 2.898 \times 10^{-3} \text{ mK} = 5216.4\ \mu\ °\text{R}.$$

To obtain the total emissive power of a blackbody, we integrate the spectral emissive power over all wavelengths:

$$E_b = \int_0^\infty E_{b\lambda}\, d\lambda = \sigma T_b^4. \tag{2.57}$$

Equation (2.57) is the same as Eq. (2.55) except that it is expressed per unit area. At a given temperature T_b, the quantity E_b of Eq. (2.57) is the area under the curve corresponding to T_b in Figure 2.26.

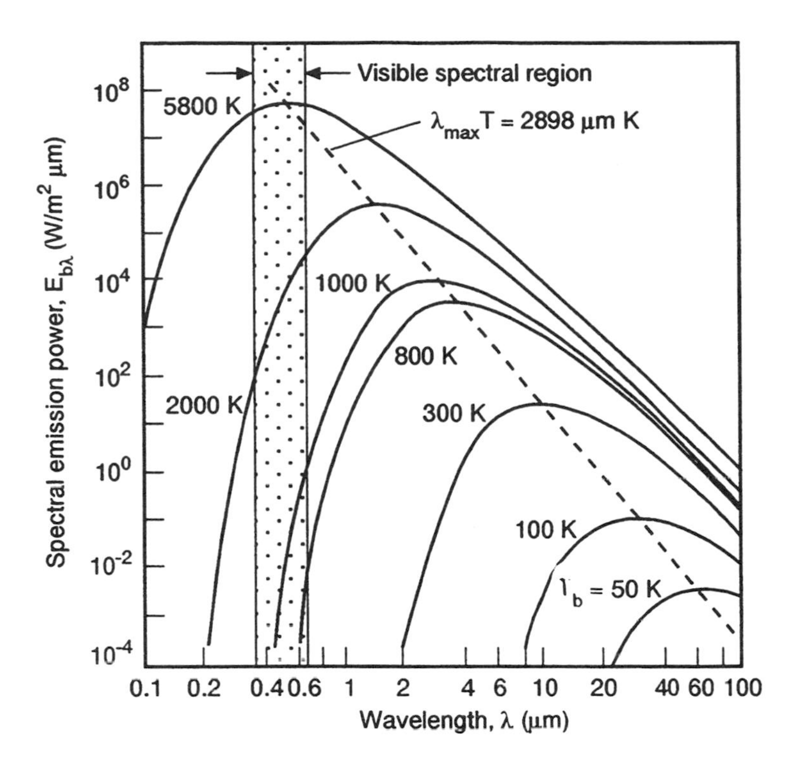

FIGURE 2.26 Spectral blackbody emissive power for different temperatures.

Engineers sometimes encounter problems where it is necessary to find the fraction of the total energy radiated from a blackbody in a finite interval between two specific wavelengths λ_1 and λ_2. This fraction for an interval from 0 to λ_1 can be determined from:

$$B(0 \rightarrow \lambda_1) = \frac{\int_0^{\lambda_1} E_{b\lambda}\ d\lambda}{\int_0^{\infty} E_{b\lambda}\ d\lambda} = \frac{\int_0^{\lambda_1} E_{b\lambda}\ d\lambda}{\sigma T_b^4}.$$

This integral has been calculated for various λT quantities, and the results are presented in Table 2.8. The fraction of total radiation from a blackbody in a finite wavelength interval from λ_1 to λ_2 can then be obtained from

$$B(\lambda_1 \rightarrow \lambda_2) = \frac{\int_0^{\lambda_2} E_{b\lambda}\ d\lambda - \int_0^{\lambda_1} E_{b\lambda}\ d\lambda}{\sigma T_b^4} = B(0 \rightarrow \lambda_2) - B(0 \rightarrow \lambda_1)$$

where quantities $B(0\rightarrow\lambda_2)$ and $B(0\rightarrow\lambda_1)$ can be read from Table 2.8.

Radiation Properties of Objects

When radiation strikes the surface of an object, a portion of the total incident radiation is reflected, a portion is absorbed, and if the object is transparent, a portion is transmitted through the object, as depicted in Figure 2.27.

TABLE 2.8 Blackbody Radiation Functions

λT ($mK \times 10^3$)	$B(0 \rightarrow \lambda)$	λT ($mK \times 10^3$)	$B(0 \rightarrow \lambda)$
0.2	0.341796×10^{-26}	6.2	0.754187
0.4	0.186468×10^{-11}	6.4	0.769234
0.6	0.929299×10^{-7}	6.6	0.783248
0.8	0.164351×10^{-4}	6.8	0.796180
1.0	0.320780×10^{-3}	7.0	0.808160
1.2	0.213431×10^{-2}	7.2	0.819270
1.4	0.779084×10^{-2}	7.4	0.829580
1.6	0.197204×10^{-1}	7.6	0.839157
1.8	0.393499×10^{-1}	7.8	0.848060
2.0	0.667347×10^{-1}	8.0	0.856344
2.2	0.100897	8.5	0.874666
2.4	0.140268	9.0	0.890090
2.6	0.183135	9.5	0.903147
2.8	0.227908	10.0	0.914263
3.0	0.273252	10.5	0.923775
3.2	0.318124	11.0	0.931956
3.4	0.361760	11.5	0.939027
3.6	0.403633	12	0.945167
3.8	0.443411	13	0.955210
4.0	0.480907	14	0.962970
4.2	0.516046	15	0.969056
4.4	0.548830	16	0.973890
4.6	0.579316	18	0.980939
4.8	0.607597	20	0.985683
5.0	0.633786	25	0.992299
5.2	0.658011	30	0.995427
5.4	0.680402	40	0.998057
5.6	0.701090	50	0.999045
5.8	0.720203	75	0.999807
6.0	0.737864	100	1.00000

The fraction of incident radiation which is reflected is called the **reflectance (or reflectivity)** ρ, the fraction transmitted is called the **transmittance (or transmissivity)** τ, and the fraction absorbed is called the **absorptance (or absorptivity)** α. There are two types of radiation reflections: specular and diffuse. A **specular reflection** is one in which the angle of incidence is equal to the angle of reflection, whereas a **diffuse reflection** is one in which the incident radiation is reflected uniformly in all directions. Highly polished surfaces such as mirrors approach the specular reflection characteristics, but most industrial surfaces (rough surfaces) have diffuse reflection characteristics. By applying an energy balance to the surface of the object as shown in Figure 2.27, the relationship between these properties can be expressed as

$$\alpha + \rho + \tau = 1. \tag{2.58}$$

The relative magnitude of each one of these components depends on the characteristics of the surface, its temperature, and the spectral distribution of the incident radiation. If an object is opaque ($\tau = 0$), it will not transmit any radiation. Therefore

$$\alpha + \rho = 1 \text{ and } \tau = 0 \quad \text{for an opaque object.} \tag{2.59}$$

If an object has a perfectly reflecting surface (a good mirror), then it will reflect all the incident radiation, and

$$\rho = 1, \ \alpha = 0, \text{ and } \tau = 0 \quad \text{for a perfectly reflective surface.} \tag{2.60}$$

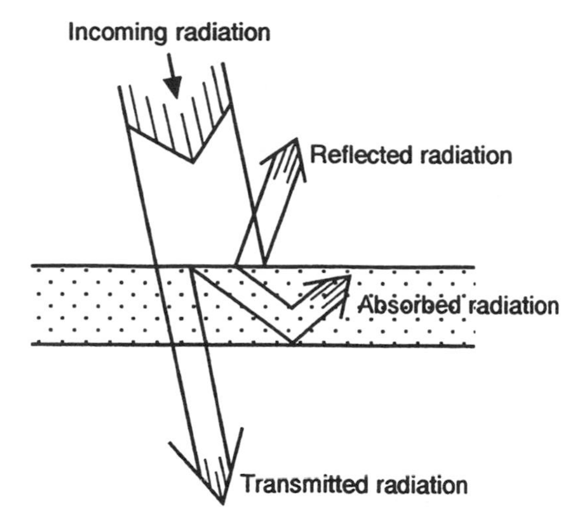

FIGURE 2.27 Schematic of reflected, transmitted, and absorbed radiation.

The **emissivity**, ε, of a surface at temperature T is defined as the ratio of total energy emitted to the energy that would be emitted by a blackbody at the same temperature T:

$$\varepsilon = \frac{E(T)}{\sigma T^4}, \tag{2.61}$$

where $E(T)$ represents the radiation energy emitted from the surface. For a blackbody, Eq. (2.61) gives $\varepsilon_b = 1$. The absorptivity for a blackbody is also equal to unity; therefore, $\varepsilon_b = \alpha_b = 1$.

A special type of surface called a gray surface or **graybody** is a surface with spectral emissivity and absorptivity that are both independent of the wavelength. Therefore, for a graybody, $\bar{\alpha} = \alpha_\lambda = \bar{\varepsilon} = \varepsilon_\lambda$ where $\bar{\varepsilon}$ and $\bar{\alpha}$ are the average values of emissivity and absorptivity, respectively. In many engineering problems, surfaces are not gray surfaces. However, one can employ graybody assumptions by using suitable $\bar{\alpha}$ and $\bar{\varepsilon}$ values.

Table 2.9 provides emissivities of various surfaces at several wavelengths and temperatures. A more extensive list of experimentally measured radiation properties of various surfaces has been provided by Gubareff et al. [1960] and Kreith and Bohn [1993]; note that the listed quantities in Table 2.9 are hemispherical emissivities. Detailed directional and spectral measurements of radiation properties of surfaces are limited in the literature. Because of the difficulties in performing these detailed measurements, most of the tabulated properties are averaged quantities, such as those presented in Table 2.9. Properties averaged with respect to wavelength are termed *total* quantities, and properties averaged with respect to direction are termed *hemispherical* quantities. Hemispherical spectral emissivity of a surface is the ratio of (1) the spectral radiation emitted by a unit surface area of an object into all directions of a hemisphere surrounding that area to (2) the spectral radiation emitted by a unit surface area of a blackbody (at the same temperature) into all directions of that hemisphere.

The Radiation Shape Factor (View Factor)

In this section, we will only deal with surfaces that have diffuse reflection characteristics, because most real surfaces used in different industries can be assumed to have diffuse reflection characteristics. In solving radiation problems, we must find out how much of the radiation leaving one surface is being intercepted by another surface.

The **radiation shape** factor F_{1-2} is defined as

$$F_{1-2} = \frac{\text{diffuse radiation leaving surface } A_1 \text{ and being intercepted by surface } A_2}{\text{total diffuse radiation leaving surface } A_1}.$$

TABLE 2.9 Hemispherical Emissivities of Various Surfaces[a]

	Wavelength and Average Temperature				
Material	9.3 μm 310 K	5.4 μm 530 K	3.6 μm 800 K	1.8 μm 1700 K	0.6 μm Solar ~6,000 K
			Metals		
Aluminum					
polished	~0.04	0.05	0.08	~0.19	~0.30
oxidized	0.11	~0.12	0.18		
24-ST weathered	0.40	0.32	0.27		
surface roofing	0.22				
anodized (at 1,000°F)	0.94	0.42	0.60	0.34	
Brass					
polished	0.10	0.10			
oxidized	0.61				
Chromium					
polished	~0.08	~0.17	0.26	~0.40	0.49
Copper					
polished	0.04	0.05	~0.18	~0.17	
oxidized	0.87	0.83	0.77		
Iron					
polished	0.06	0.08	0.13	0.25	0.45
cast, oxidized	0.63	0.66	0.76		
galvanized, new	0.23			0.42	0.66
galvanized, dirty	0.28			0.90	0.89
steel plate, rough	0.94	0.97	0.98		
oxide	0.96		0.85		0.74
molten				0.3–0.4	
Magnesium	0.07	0.13	0.18	0.24	0.30
Molybdenum filament			~0.09	~0.15	~0.20[b]
Silver					
polished	0.01	0.02	0.03		0.11
Stainless steel					
18-8, polished	0.15	0.18	0.22		
18-8, weathered	0.85	0.85	0.85		
Steel tube					
oxidized		0.94			
Tungsten filament	0.03			~0.18	0.35[c]
Zinc					
polished	0.02	0.03	0.04	0.06	0.46
galvanized sheet	~0.25				
		Building and Insulating Materials			
Asbestos paper	0.93	0.93			
Asphalt	0.93		0.90		0.93
Brick					
red	0.93				0.70
fire clay	0.90		~0.70	~0.75	
silica	0.90		0.75	0.84	
magnesite refractory	0.90			~0.40	
Enamel, white	0.90				
Marble, white	0.95		0.93		0.47
Paper, white	0.95		0.82	0.25	0.28
Plaster	0.91				
Roofing board	0.93				
Enameled steel, white				0.65	0.47
Asbestos cement, red				0.67	0.66

TABLE 2.9 (continued) Hemispherical Emissivities of Various Surfaces[a]

	Wavelength and Average Temperature				
Material	9.3 μm 310 K	5.4 μm 530 K	3.6 μm 800 K	1.8 μm 1700 K	0.6 μm Solar ~6,000 K
		Paints			
Aluminized lacquer	0.65	0.65			
Cream paints	0.95	0.88	0.70	0.42	0.35
Lacquer, black	0.96	0.98			
Lampblack paint	0.96	0.97		0.97	0.97
Red paint	0.96				0.74
Yellow paint	0.95		0.50		0.30
Oil paints (all colors)	~0.94	~0.90			
White (ZnO)	0.95		0.91		0.18
		Miscellaneous			
Ice	~0.97[d]				
Water	~0.96				
Carbon					
T-carbon, 0.9% ash	0.82	0.80	0.79		
filament	~0.72			0.53	
Wood	~0.93				
Glass	0.90				(Low)

[a] Since the emissivity at a given wavelength equals the absorptivity at that wavelength, the values in this table can be used to approximate the absorptivity to radiation from a source at the temperature listed. For example, polished aluminum will absorb 30% of incident solar radiation.
[b] At 3,000 K.
[c] At 3,600 K.
[d] At 273 K.
Sources: Fischenden and Saunders [1932]; Hamilton and Morgan [1962]; Kreith and Black [1980]; Schmidt and Furthman [1928]; McAdams [1954]; Gubareff et al. [1960].

For example, consider two black surfaces A_1 and A_2 at temperatures T_1 and T_2, as shown in Figure 2.28. The radiation leaving surface A_1 and reaching A_2 is

$$\dot{Q}_{1\to 2} = A_1 F_{1-2} E_{b1}, \quad \textbf{(2.62)}$$

and the radiation leaving surface A_2 and reaching surface A_1 is

$$\dot{Q}_{2\to 1} = A_2 F_{2-1} E_{b2}, \quad \textbf{(2.63)}$$

From Eqs. (2.62) and (2.63), we can calculate the net radiation heat exchange between these two black surfaces:

$$\Delta\dot{Q}_{1-2} = A_1 F_{1-2} E_{b1} - A_2 F_{2-1} E_{b2}.$$

Shape factors for some geometries that have engineering applications are presented in Table 2.10. For more information and an extensive list of shape factors, refer to Siegel and Howell [1972].

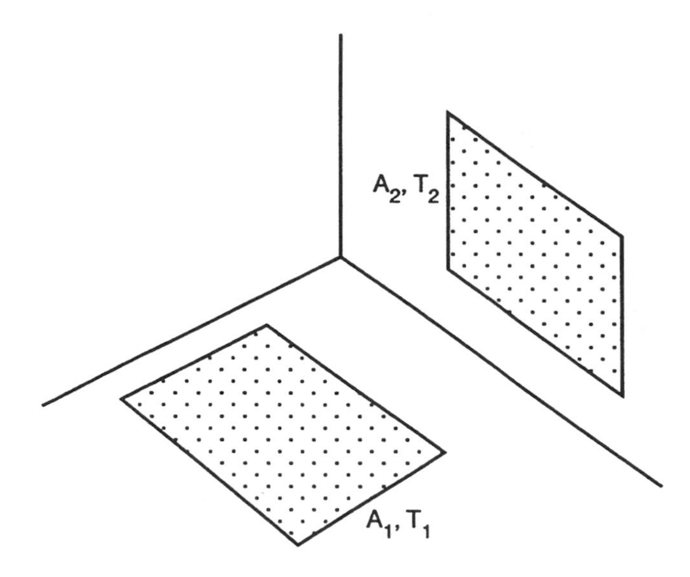

FIGURE 2.28 Sketch illustrating the nomenclature for shape factor between the two surfaces A_1 and A_2.

Example 2.9

A flat-plate solar collector with a single glass cover is shown in Figure 2.29. The following quantities are known:

The solar irradiation, $G_s = 750$ W/m^2
The absorptivity of the cover plate to solar radiation, $\alpha_{cp,s} = 0.16$
The transmissivity of the cover plate to solar radiation, $\tau_{cp} = 0.84$
The emissivity of the cover plate to longwave radiation, $\varepsilon_{cp} = 0.9$
The absorptivity of the absorber plate to solar radiation, $\alpha_{ap,s} = 1.0$
The emissivity of the absorber plate to longwave radiation, $\varepsilon_{ap} = 0.1$
The convection coefficient between the absorber plate and the cover plate, $h_i = 2$ W/m^2 K
The convection coefficient between the cover plate and ambient, $h_o = 5$ W/m^2 K
The absorber-plate temperature, $T_{ap} = 120°C$
The ambient air temperature $T_\infty = 30°C$
The effective sky temperature, $T_{sky} = -10°C$

Using this information, calculate the useful heat absorbed by the absorber plate.

Solution:

We will assume the following:

- Steady-state conditions
- Uniform surface heat-flux and temperature for the cover plate and the absorber plate
- Opaque, diffuse-gray surface behavior for longwave radiation
- Well-insulated absorber plate

To find the useful heat absorbed by the absorber plate, perform an energy balance on a unit area of the absorber plate, as in Figure 2.30:

$$\alpha_{ap,s}\ \tau_{cp,s}\ G_s = \dot{Q}_{\text{conv},i} + \dot{Q}_{\text{rad},ap-cp} + \dot{Q}_u, \tag{2.64}$$

where $\dot{Q}_{\text{conv,i}} = h_i\,(T_{ap} - T_{cp})$ is the convection heat exchange between the absorber plate and the cover plate and $\dot{Q}_{\text{rad,ap-cp}} = \sigma\left(T_{ap}^4 - T_{cp}^4\right)/\left(1/\varepsilon_{ap} + 1/\varepsilon_{cp} - 1\right)$ is the heat exchange by radiation between

TABLE 2.10 Minicatalog of Geometric View Factors

Configuration	Geometric View Factor
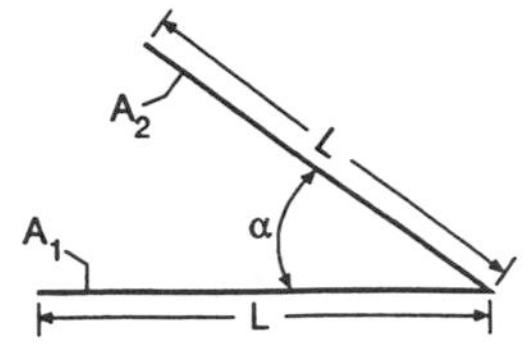 A_2, L, α, A_1, L	Two infinitely long plates of width L, joined along one of the long edges: $$F_{1\text{-}2} = F_{2\text{-}1} = 1 - \sin\frac{\alpha}{2}$$
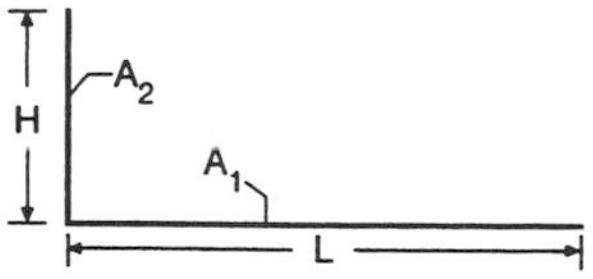 H, A_2, A_1, L	Two infinitely long plates of different widths (H, L), joined along one of the long edges and with a 90° angle between them: $$F_{1\text{-}2} = \frac{1}{2}\left[1 + x - (1 + x^2)^{1/2}\right]$$ where $x = H/L$
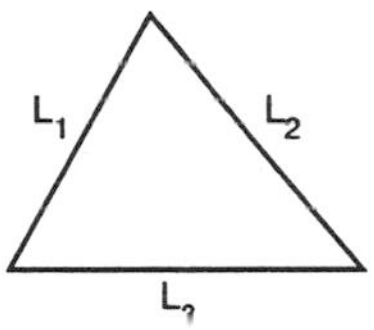 L_1, L_2, L_3	Triangular cross-section enclosure formed by three infinitely long plates of different widths (L_1, L_2, L_3): $$F_{1\text{-}2} = \frac{L_1 + L_2 - L_3}{2L_1}$$
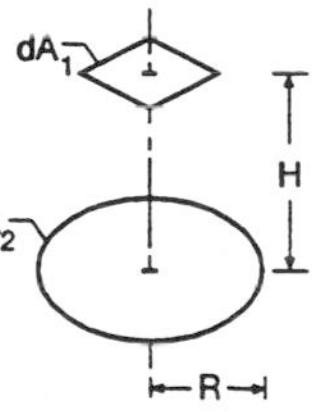 dA_1, H, A_2, R	Circular disk and plane element positioned on the disc centerline: $$F_{1\text{-}2} = \frac{R^2}{H^2 + R^2}$$
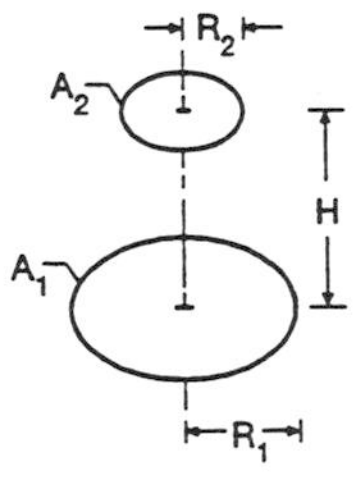 R_2, A_2, H, A_1, R_1	Parallel discs positioned on the same centerline: $$F_{1\text{-}2} = \frac{1}{2}\left\{X - \left[X^2 - 4\left(\frac{x_2}{x_1}\right)^2\right]^{1/2}\right\}$$ where $$x_1 = \frac{R_1}{H},\ x_2 = \frac{R_2}{H},\ \text{and}\ X = 1 + \frac{1 + x_2^2}{x_1^2}$$

TABLE 2.10 (continued) Minicatalog of Geometric View Factors

Configuration	Geometric View Factor
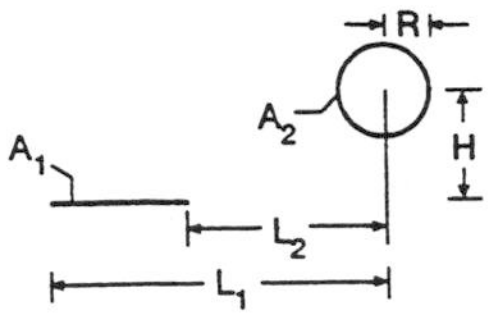	Infinite cylinder parallel to an infinite plate of finite width $(L_1 - L_2)$: $$F_{1-2} = \frac{R}{L_1 - L_2}\left(\tan^{-1}\frac{L_1}{H} - \tan^{-1}\frac{L_2}{H}\right)$$
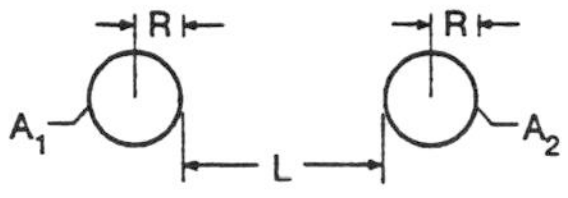	Two parallel and infinite cylinders: $$F_{1-2} = F_{2-1} = \frac{1}{\pi}\left[\left(X^2 - 1\right)^{1/2} + \sin^{-1}\left(\frac{1}{X}\right) - X\right]$$ where $X = 1 + \frac{L}{2R}$
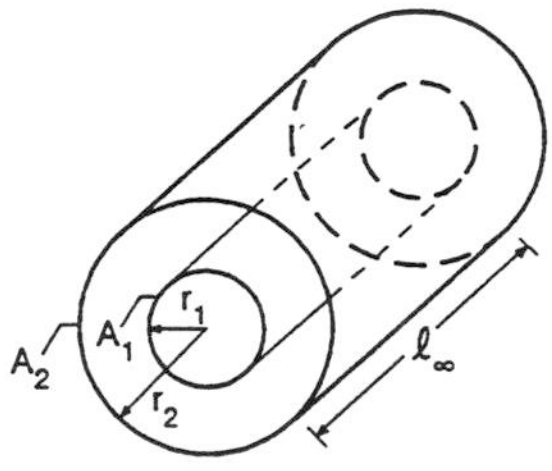	Concentric cylinders of infinite length: $$F_{1-2} = 1$$ $$F_{2-1} = \frac{R_1}{R_2}$$ $$F_{2-2} = 1 - \frac{R_1}{R_2}$$
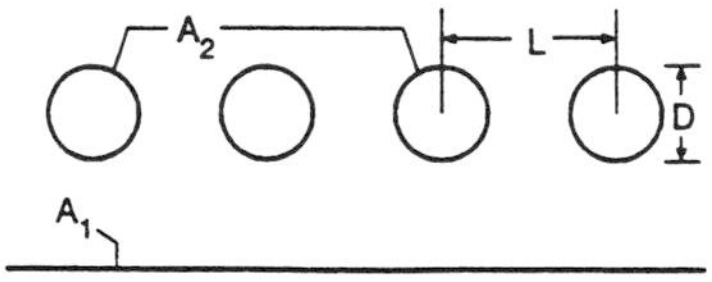	Row of equidistant infinite cylinders parallel to an infinite plate: $$F_{1-2} = 1 - (1 - x^2)^{1/2} + x\tan^{-1}\left(\frac{1 - x^2}{x^2}\right)^{1/2}$$ where $x = D/L$

TABLE 2.10 (continued) Minicatalog of Geometric View Factors

Configuration	Geometric View Factor
	Sphere and disc positioned on the same centerline: $F_{1\text{-}2} = \frac{1}{2}\left[1 - \frac{1}{\sqrt{1+x^2}}\right]$ where $x = \frac{R_2}{H}$
	Sphere and a sector of disk positioned on the same centerline: $F_{1\text{-}2} = \frac{\alpha}{4\pi}\left[1 - \frac{1}{\sqrt{1+x^2}}\right]$ where $x = \frac{R_2}{H}$
	Concentric spheres: $F_{1\text{-}2} = 1$ $F_{2\text{-}1} = \left(\frac{R_1}{R_2}\right)^2$ $F_{2\text{-}2} = 1 - \left(\frac{R_1}{R_2}\right)^2$

Sources: Howell (1982) and Siegel and Howell (1972)

them. Note that the shape factor between two parallel plates is equal to one. The left-hand side of Eq. (2.64) represents the solar irradiation transmitted through the cover plate and absorbed by the absorber plate. Substituting for $\dot{Q}_{conv,i}$ and $\dot{Q}_{rad,ap\text{-}cp}$ in Eq. (2.64), we obtain (for $\alpha_{ap,s} = 1$)

$$\tau_{cp,s}\, G_s = h_i\left(T_{ap} - T_{cp}\right) + \frac{\sigma\left(T_{ap}^4 - T_{cp}^4\right)}{1/\varepsilon_{ap} + 1/\varepsilon_{cp} - 1} + \dot{Q}_u. \qquad \textbf{(2.65)}$$

To find q_u from Eq. (2.65), T_{cp} should be known, which is obtained from an energy balance on the cover plate, as in Figure 2.31:

$$\alpha_{cp,s}\, G_s + \dot{Q}_{conv,i} + \dot{Q}_{rad,ap-cp} = \dot{Q}_{conv,o} + \dot{Q}_{rad,cp-sky}, \qquad \textbf{(2.66)}$$

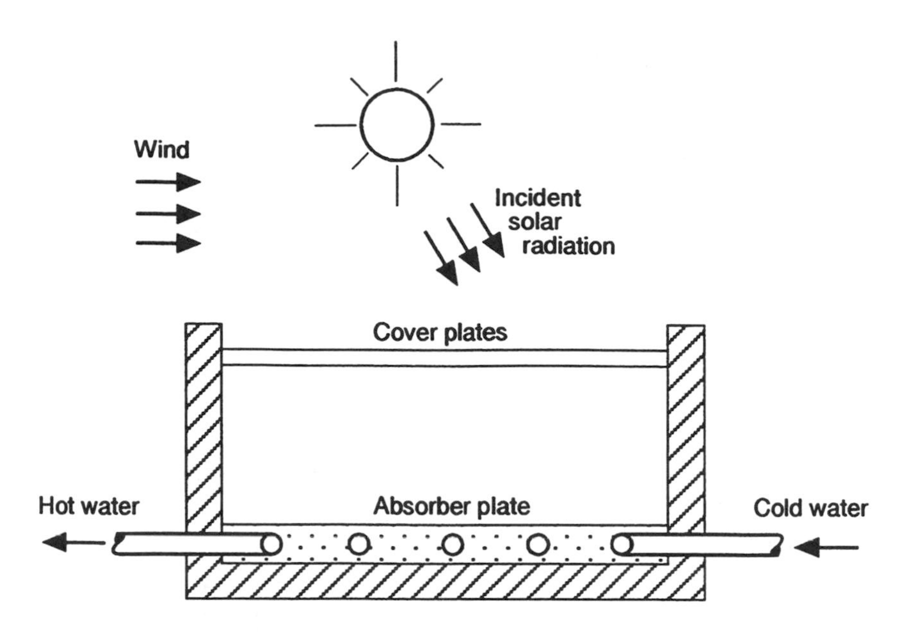

FIGURE 2.29 Flat-plate solar collector with a single glass cover.

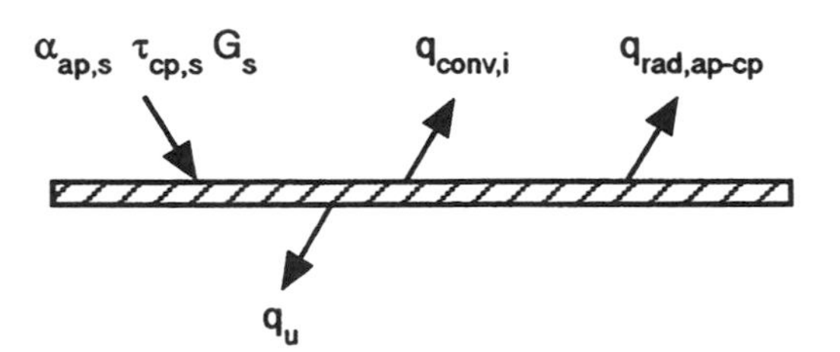

FIGURE 2.30 Energy balance on a unit area of the absorber plate.

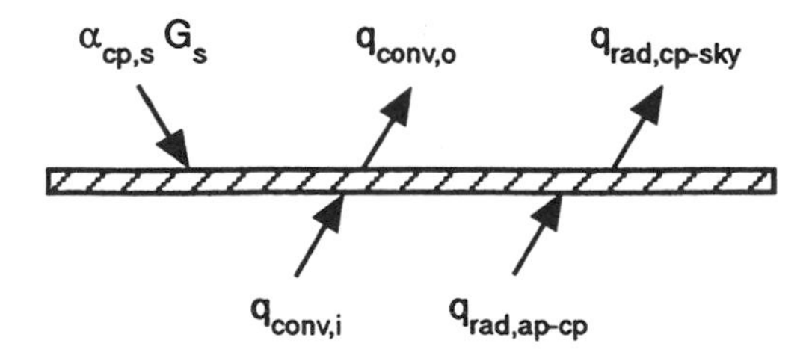

FIGURE 2.31 Energy balance on a unit area of the cover plate.

where $\dot{Q}_{conv,o} = h_o\,(T_{cp} - T_\infty)$ is the heat loss by convection and $\dot{Q}_{rad,cp\text{-}sky} = \varepsilon_{cp}\sigma\left(T_{cp}^4 - T_{sky}^4\right)$ is the heat exchange by radiation between the cover plate and sky. Equation (2.66) can be written as

$$\alpha_{cp,s}\; G_s + h_i\left(T_{ap} - T_{cp}\right) + \frac{\sigma\left(T_{ap}^4 - T_{cp}^4\right)}{1/\varepsilon_{ap} + 1/\varepsilon_{cp} - 1} = h_o\left(T_{cp} - T_\infty\right) + \varepsilon_{cp}\,\sigma\left(T_{cp}^4 - T_{sky}^4\right). \quad \textbf{(2.67)}$$

Substituting for known quantities in Eq. (2.67), T_{cp} is calculated to be $T_{cp} = 44.6°C$. Substituting for T_{cp} and other known quantities in Eq. (2.65), $\dot{Q}_u$ is 402.5 W/m².

2.3 Fundamentals of Fluid Mechanics

The distribution of fluids by pipes, ducts, and conduits is essential to most engineering processes and systems. The fluids encountered in engineering processes are gases, vapors, liquids, or mixtures of liquid and vapor (2–phase flow). This section briefly reviews certain basic concepts of fluid mechanics that are often encountered in analyzing engineering systems.

Fluid flowing through a conduit will encounter shear forces that result from viscosity of the fluid. The fluid undergoes continuous deformation when subjected to these shear forces. Furthermore, as a result of shear forces, the fluid will experience pressure losses as it travels through the conduit.

Viscosity, μ, is a property of fluid best defined by Newton's Law of Viscosity:

$$\tau = \mu \frac{du}{dy}, \tag{2.68}$$

where τ is the frictional shear stress, and du/dy represents the measure of the motion of one layer of fluid relative to an adjacent layer. The following observation will help to explain the relationship between viscosity and shear forces. Consider two very long parallel plates with a fluid between them, as shown in Figure 2.32. Assume a uniform pressure throughout the fluid. The upper plate is moving with a constant velocity u_0, and the lower plate is stationary. Experiments show that the fluid adjacent to the moving plate will adhere to that plate and move along with the plate at a velocity equal to u_0, whereas the fluid adjacent to the stationary plate will have zero velocity. The experimentally verified velocity distribution in the fluid is linear and can be expressed as

$$u = \frac{y}{\ell} u_0, \tag{2.69}$$

where ℓ is the distance between the two parallel plates. The force necessary to keep the upper plate moving at a constant velocity of u_0 should be large enough to overcome (or balance) the frictional forces in the fluid. Again, experimental observations indicate that this force is proportional to the ratio u_0/ℓ. One can conclude from Eq. (2.69) that u_0/ℓ is equal to the rate of change of velocity, du/dy. Therefore, the frictional force per unit area (shear stress), τ, is proportional to du/dy, and the proportionality constant is μ, which is a property of the fluid known as viscosity. The quantity μ is a measure of the viscosity of the fluid and depends on the temperature and pressure of the fluid. Equation (2.68) is analogous to Fourier's Law of Heat Conduction given by Eq. (2.25). Fluids that do not obey Newton's Law of Viscosity are called non-Newtonian fluids. Fluids with zero viscosity are known as inviscid or ideal fluids. Molasses and tar are examples of highly viscous liquids; water and air on the other hand, have very low viscosities. The viscosity of a gas increases with temperature, but the viscosity of a liquid decreases with temperature. Reid, Sherwood, and Prausnitz [1977] provide a thorough discussion on viscosity.

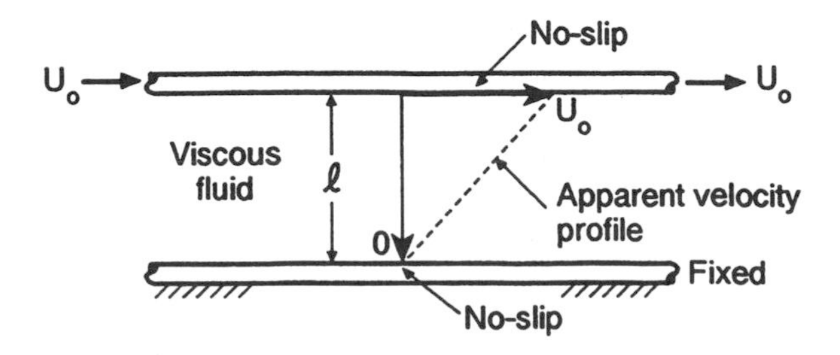

FIGURE 2.32 A fluid sheared between two parallel plates.

Flow Characteristics

The flow of a fluid may be characterized by one or a combination of the following descriptor pairs: laminar/turbulent, steady/unsteady, uniform/nonuniform, reversible/irreversible, rotational/irrotational. In this section, however, we will focus our attention only on laminar and turbulent flows.

In **laminar flow**, fluid particles move along smooth paths in layers, with one layer sliding smoothly over an adjacent layer without significant macroscopic mixing. Laminar flow is governed by Newton's Law of Viscosity. Turbulent flow is more prevalent than laminar flow in engineering processes. In **turbulent flow**, the fluid particles move in irregular paths, causing an exchange of momentum between various portions of the fluid; adjacent fluid layers mix and this mixing mechanism is called eddy motion. In this type of flow, the velocity at any given point under steady-state conditions fluctuates in all directions about some time-mean value. Turbulent flow causes greater shear stresses throughout the fluid, producing more irreversibilities and losses. An equation similar to Newton's Law of Viscosity may be written for turbulent flows:

$$\tau = (\mu + n)\,\frac{du}{dy}, \tag{2.70}$$

where the factor n is the eddy viscosity, which depends on the fluid motion and density. Unlike the fluid viscosity, μ, the eddy viscosity is not a fluid property and is determined through experiments.

The type of flow is primarily determined by the value of a nondimensional number known as a Reynolds number, which is the ratio of inertia forces to viscous forces given by

$$\mathrm{Re} = \frac{\rho\, u_{\mathrm{avg}}\, D_H}{\mu}, \tag{2.71}$$

where u_{avg} is the average velocity and D_H is the hydraulic diameter defined by Eq. (2.50). The value of the Reynolds number can be used as the criterion to determine whether the flow is laminar or turbulent. In general, laminar flow occurs in closed conduits when $\mathrm{Re} < 2{,}100$; the flow goes through transition when $2{,}100 < \mathrm{Re} < 6{,}000$ and becomes turbulent when $\mathrm{Re} > 6{,}000$.

For fluid flow over flat plates, laminar flow is generally accepted to occur at $\mathrm{Re}_x = \rho u x/\mu < 3 \times 10^5$, where x is the distance from the leading edge of the plate and u is the free-stream velocity. Note that if the flow approaching the flat plate is turbulent, it will remain turbulent from the leading edge of the plate forward.

When a fluid is flowing over a solid surface, the velocity of the fluid layer in the immediate neighborhood of the surface is influenced by viscous shear; this region of the fluid is called the **boundary layer.** Boundary layers can be laminar or turbulent depending on their length, the fluid viscosity, the velocity of the bulk fluid, and the surface roughness of the solid body.

Analysis of Flow Systems

Most engineering problems require some degree of system analysis. Regardless of the nature of the flow, all fluid-flow situations are subject to the following relations:

1. Newton's Law of Motion, $\Sigma F = \dfrac{1}{g_c}\dfrac{d(mu)}{dt}$
2. Conservation of mass
3. The First and Second Laws of Thermodynamics
4. Boundary conditions such as zero velocity at a solid surface.

In an earlier section, the First Law of Thermodynamics was applied to a system shown in Figure 2.1. With some modifications, the same energy balance can be applied to any fluid-flow system. For example, a term representing the frictional pressure losses should be added to the left-hand side of Eq. (2.2), as expressed by the following equation:

$$e_2 + \frac{u_2^2}{2g_c} + \frac{gz_2}{g_c} + \frac{p_2}{\rho_2} + \frac{\dot{w}_f}{\dot{m}} = e_1 + \frac{u_1^2}{2g_c} + \frac{gz_1}{g_c} + \frac{p_1}{\rho_1} + \frac{\dot{Q} + \dot{w}}{\dot{m}}, \tag{2.72}$$

where $\dot{w}_f$ represents the frictional pressure losses and is the rate of work done on the fluid (note the sign change from $-\dot{w}$ to $+\dot{w}$ in Eq. (2.72), because the work is done on the fluid). In the remainder of this section, we will focus on obtaining an expression for $\dot{w}_f$ and analyzing different sources of frictional pressure losses.

Using Newton's Law of Motion, the weight of a body, w, can be defined as the force exerted on the body as a result of the acceleration of gravity, g,

$$w = \frac{g}{g_c} m. \tag{2.73}$$

In the English system of units, 1 lbm weighs 1 lbf at sea level because the proportionality constant g_c is numerically equal to the gravitational acceleration (32.2 ft/s²). However, in the SI system, 1 kg of mass weighs 9.81 N at sea level because $g_c = 1$ kg m/N s² (or $g_c = 10^3$ kg m³/kJ s²) and g = 9.81 m/s².

Equation (2.73) can be used to determine the static pressure of a column of fluid. For example, a column of fluid at height z that experiences an environment or atmospheric pressure of p_0 over its upper surface will exert a pressure of p at the base of the fluid column given by

$$p = p_0 + \rho z \frac{g}{g_c}, \tag{2.74}$$

where ρ is the density of the fluid. The base pressure as expressed by Eq. (2.74) is a function of fluid height or fluid head and does not depend on the shape of the container. Knowing the fluid head is very important, especially in specifying a pump, as it is common practice to specify the performance of the pump in terms of fluid head. Therefore, we can calculate the required mechanical power from

$$\dot{w}_{\text{pump}} = zg\ \dot{m}. \tag{2.75}$$

Equation (2.75) expresses the pump power at 100% efficiency; in reality, however, mechanical pumps have efficiencies of less than 100%. Therefore, the required mechanical power $\dot{w}$ is

$$\dot{w} = \frac{\dot{w}_{\text{pump}}}{\eta_{\text{pump}}}. \tag{2.76}$$

A pump used in a system is expected to overcome various types of pressure losses such as frictional pressure losses in the piping; pressure losses due to fittings, bends, and values; and pressure losses due to sudden enlargements and contractions. All these pressure losses should be calculated for a system and summed up to obtain the total pressure drop through a system.

The frictional pressure losses in the piping are caused by the shearing force at the fluid-solid interface. Through a force balance, we can obtain the frictional pressure loss of an incompressible fluid in a pipe between two points as

$$p_1 - p_2 = 4f\ \frac{L}{D}\ \frac{\rho u^2}{2g_c}, \tag{2.77}$$

where L is the length of the pipe between points 1 and 2, D is the pipe diameter, u is the average fluid velocity in the pipe, and f is the dimensionless friction factor. For laminar flow inside a pipe, the friction factor is

$$f = \frac{16}{\mathrm{Re}_{D_H}}, \tag{2.78}$$

where the Reynolds number is based on the hydraulic diameter D_H. The friction factor for turbulent flow depends on the surface roughness of the pipe and on the Reynolds number. The friction factor for various surface roughnesses and Reynolds numbers is presented in Figure 2.33, which is called the **Moody diagram.** The relative roughnesses of the various commercial pipes are given in Figure 2.34.

Pressure losses due to fittings, bends, and valves are generally determined through experiments. This type of pressure loss can be correlated to the average fluid velocity in the pipe by

$$\Delta p_b = k_b \frac{\rho u^2}{2g_c}, \tag{2.79}$$

where k_b is a pressure-loss coefficient obtained from a handbook or from the manufacturer, and u is the average fluid velocity in the pipe upstream of the fitting, bend, or valve. For typical values of k_b, refer to Perry, Perry, Chilton, and Kirkpatrick [1963], Freeman [1941], and the *Standards of Hydraulic Institute* [1948].

Pressure losses due to sudden enlargement of the cross section of the pipe can be calculated using

$$\Delta p = \alpha\left(1 - \frac{A_s}{A_L}\right)\frac{\rho u^2}{2g_c} = k_e \frac{\rho u^2}{2g_c}, \tag{2.80}$$

where A_s/A_L is the ratio of the cross-sectional area of the smaller pipe to that of the larger pipe, α is the nondimensional pressure-loss coefficient ($\alpha = 1$ for turbulent flow and 2 for laminar flow), and u is the average fluid velocity in the smaller pipe. Note that a gradual increase in pipe cross section will have little effect on pressure losses. In case of sudden contraction of pipe size, the pressure drop can be calculated from

$$\Delta p = 0.55\,\alpha\left(1 - \frac{A_s}{A_L}\right)\frac{\rho u_c^2}{2g_c} = k_c \frac{\rho u_c^2}{2g_c}, \tag{2.81}$$

where A_s/A_L and α are as defined for Eq. (2.80), and u_c is the average fluid velocity in the smaller pipe (contraction). Adding the various pressure losses, the total pressure loss in a system can be calculated from

$$\frac{\dot{w}_f}{\dot{m}} = \Sigma\frac{\Delta p}{\rho} = 4f\frac{L}{D}\frac{u^2}{2g_c} + \Sigma k_b \frac{u^2}{2g_c} + k_e \frac{u^2}{2g_c} + k_c \frac{u_c^2}{2g_c}. \tag{2.82}$$

For a system under consideration, a pump must be chosen that can produce sufficient pressure head to overcome all the losses presented in Eq. (2.82). For system engineering applications, Eq. (2.82) can be simplified to

$$\frac{\dot{w}_f}{\dot{m}} = \Sigma\frac{2fLu^2}{g_c D}, \tag{2.83}$$

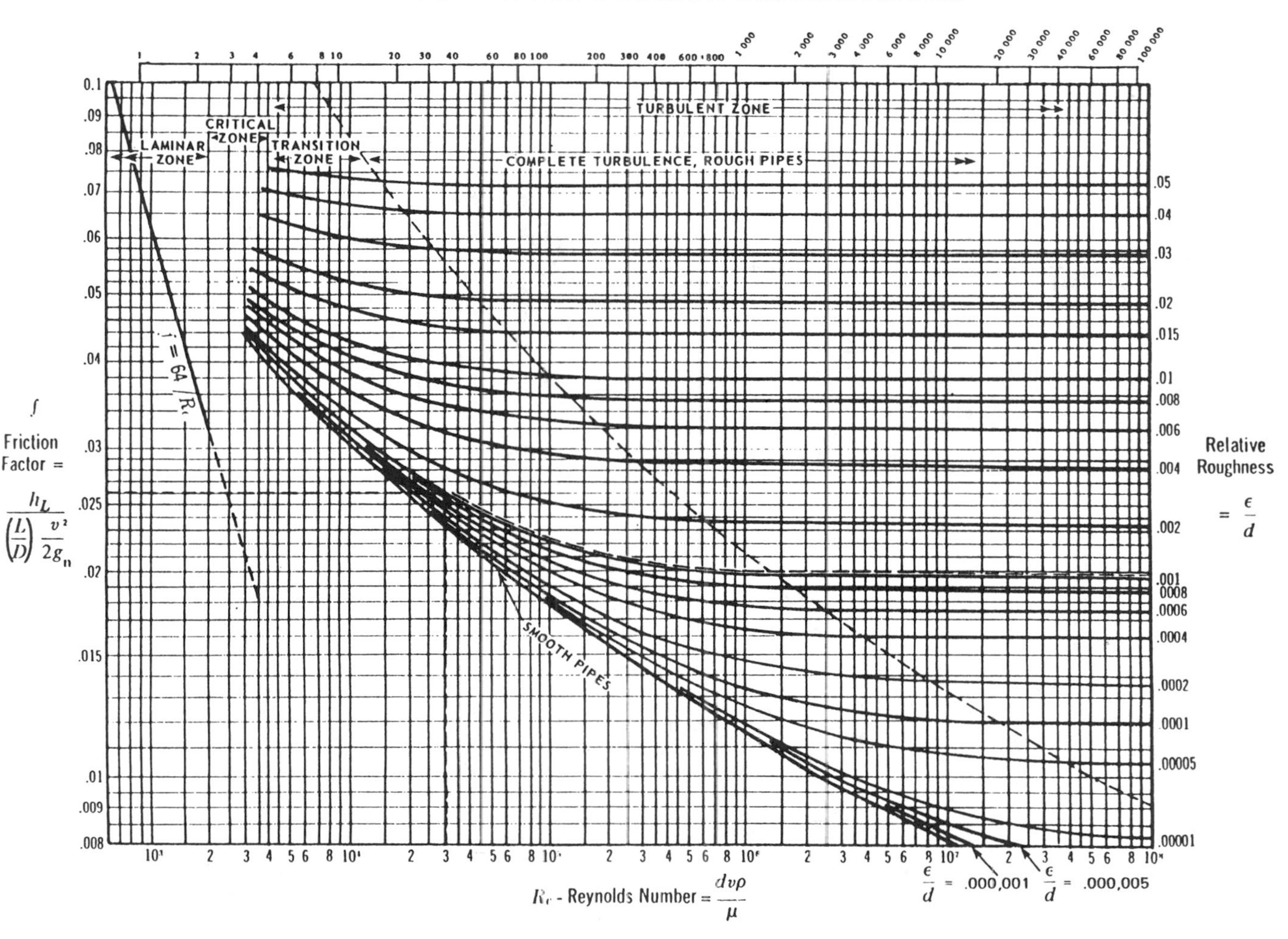

FIGURE 2.33 Friction factors for various surface roughness and Reynolds numbers. Data extracted from *Friction Factor for Pipe Flow* by L.F. Moody (1944), with permission of the publisher, The American Society of Mechanical Engineers.

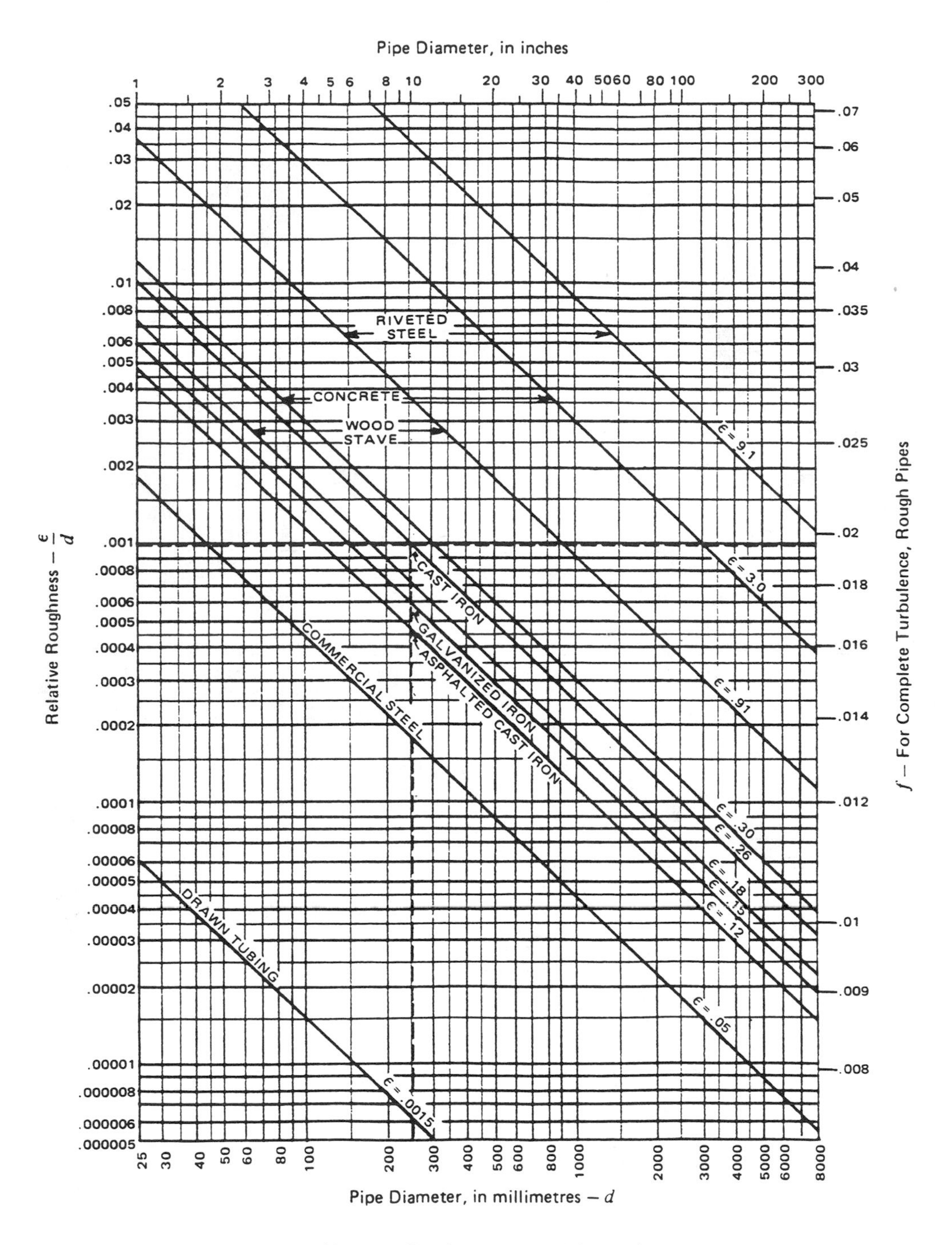

FIGURE 2.34 Relative roughness of commercial pipe. Data extracted from *Friction Factor for Pipe Flow* by L.F. Moody (1944), with permission of the publisher, The American Society of Mechanical Engineers.

where f is as defined for Eq. (2.77), u is the average velocity inside the conduit, and D is the appropriate diameter for the section of the system under consideration. The summation accounts for the effect of changes in pipe length, diameter, and relative roughness. The length L represents not only the length of the straight pipe of the system, but also, equivalent lengths of straight pipe

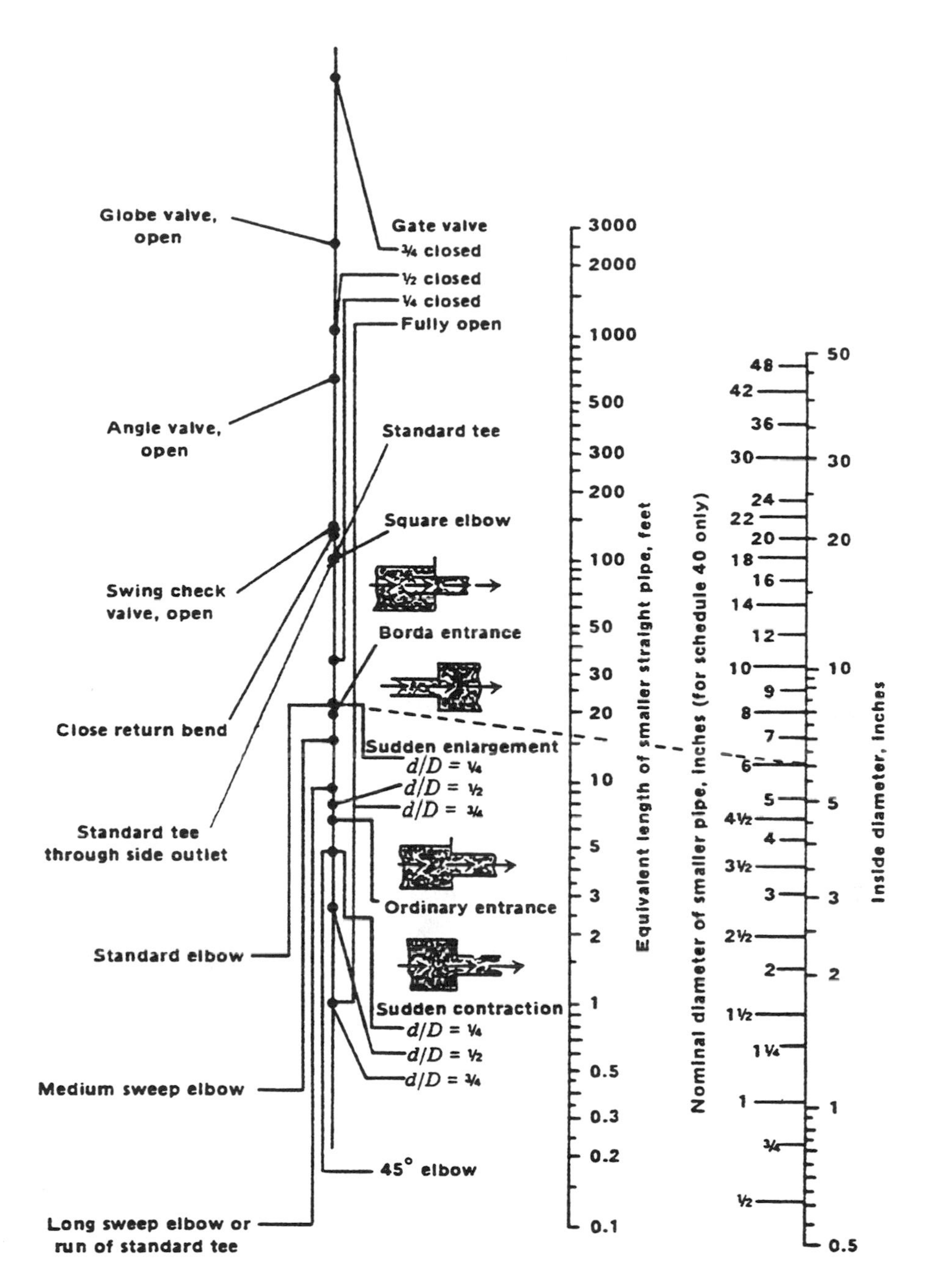

FIGURE 2.35 Equivalent lengths for friction losses. Data extracted from *Flow of Fluids through Valves, Fittings and Pipe,* Publication 410M (1988), with permission of the publisher, Crane Company.

that would have the same effects as the fittings, bends, valves, and sudden enlargements or contractions. Figure 2.35 provides a nomogram to determine such equivalent lengths.

Example 2.10

Figure 2.36 shows a system layout for a small solar collector where water at 35°C (95°F) is pumped from a tank (surface-area heat exchanger) through three parallel solar collectors and back to the tank. The water flow rate is 0.9 m^3/min (23.8 gal/min). All the piping is 1-in. Sch 40 steel pipe (cross-sectional area = 0.006 ft^2 = 5.57 × 10^{-4} m^2, with inside diameter = 1.049 in. = 0.0266 m).

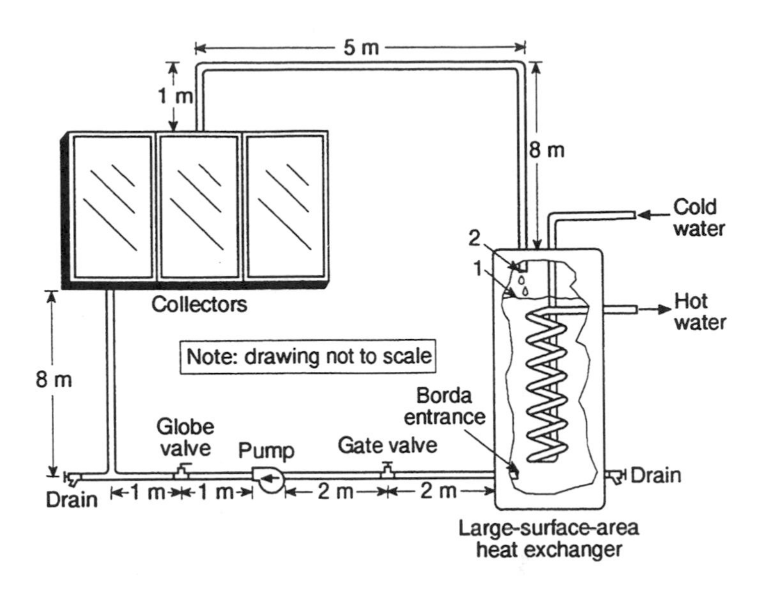

FIGURE 2.36 Layout of a small solar-collector system.

The pressure drop through each solar collector is estimated to be 1.04 kPa (0.15 psi) for a flow rate of 0.03 m^3/min (7.9 gal/min). Find the appropriate pump size for this system using the lengths and fittings specified in Figure 2.36. Assume a pump efficiency of about 75% and that the heat gain through the collectors is equal to the change in the internal energy of the water from point 1 to point 2.

Solution:

To find the pump 1 size, we apply an energy-balance similar to Eq. (2.72) between points 1 and 2 shown in Figure 2.36. Point represents the water free-surface in the tank, whereas point 2 represents the water inlet to the tank after the water has circulated through the collectors. Because both points (1 and 2) have the same pressure, we have

$$\frac{\Delta p_{2-1}}{\rho} = \frac{p_2}{\rho_2} - \frac{p_1}{\rho_1} = 0.$$

Similarly, because there is no significant height difference between these two points, we have

$$\frac{g}{g_c} \Delta z_{2-1} = \frac{gz_2}{g_c} - \frac{gz_1}{g_c} = 0.$$

The velocity at the water free-surface (point 1) is $u_1 = 0$. For point 2, the velocity is

$$u_2 = \frac{q_2}{A_2} = \frac{0.09 \text{ m}^3/\text{min}}{5.57 \times 10^{-4} \text{ m}^2} = 161.6 \text{ m/min} = 2.7 \text{ m/s}.$$

Therefore,

$$\frac{\Delta u_{2-1}^2}{2g_c} = \frac{u_2^2}{2g_c} - \frac{u_1^2}{2g_c} = \frac{(2.7 \text{ m/sec})^2}{2\left(10^3 \dfrac{\text{kg}}{\text{kJ}} \dfrac{\text{m}^2}{\text{sec}^2}\right)} = 3.7 \times 10^{-3} \text{ kJ/kg}.$$

Also note that

$$\Delta e_{2-1} = e_2 - e_1 = \frac{\dot{Q}}{\dot{m}}.$$

Therefore, Eq. (2.72) reduces to

$$\frac{\Delta u_{2-1}^2}{2g_c} + \frac{\dot{w}_f}{\dot{m}} = \frac{\dot{w}}{\dot{m}}. \tag{2.84}$$

The frictional pressure losses $\dot{w}_f$ should be determined for the whole system between points 1 and 2. Equation (2.83) can be used to determine $\dot{w}_f$; however, the total equivalent length should be determined first. The total straight piping in the system is

$$L_s = 2\text{ m} + 2\text{ m} + 1\text{ m} + 1\text{ m} + 8\text{ m} + 1\text{ m} + 5\text{ m} + 8\text{ m} = 28\text{ m}.$$

Using Figure 2.35, the equivalent lengths for bends and valves are obtained as follows:

Borda entrance: 0.79 m (2.6 ft)
Open gate valve: 0.18 m (0.6 ft)
Open globe valve: 7.90 m (26.0 ft)
Standard tee: 1.80 m (5.9 ft)
Standard elbow: 0.82 m (2.7 ft)

Therefore, with two standard elbows in this system, the equivalent length for bends, elbows, and valves becomes L_b = 0.79 m + 0.18 m + 7.9 m + 1.8 m + 2 (0.82 m) = 12.31 m, and the total equivalent of 1-in. Sch 40 pipe is $L = L_b + L_s$ = 40.31 m. To calculate the friction factor, we must calculate the Reynolds number. Assuming an average fluid density of ρ = 988 kg/m³ and an absolute viscosity of $\mu = 555 \times 10^{-6}$ Ns/m², the Reynolds number is

$$\text{Re} = \frac{\rho u D_H}{\mu} = \frac{988\ \text{kg/m}^3 \times 2.7\ \text{m/s} \times 0.0266\ \text{m}}{555 \times 10^{-6}\ N\,\text{s/m}^2} = 128 \times 10^3.$$

From Figure 2.34, the relative roughness of the pipe obtained is e/D = 0.0018, and by using Figure 2.33 (the Moody diagram), the friction factor is $f \approx 0.006$. Substituting in Eq. (2.83), the work required to overcome the frictional losses is obtained from

$$\frac{\dot{w}_f}{\dot{m}} = \frac{2fLu^2}{g_c D} + \text{Fr}_c,$$

where Fr_c is the required work to overcome pressure loss through the collectors. Since the collectors are in parallel, the total pressure loss is equal to the pressure drop through each collector. Therefore,

$$\text{Fr}_c = 1.04\ kN/\text{m}^2 \times \frac{1}{988\ \text{kg/m}^3} = 0.0011\ \text{kJ/kg},$$

and

$$\frac{\dot{w}_f}{\dot{m}} = \frac{2 \times 0.006 \times 40.31\ \text{m} \times (2.7\ \text{m/s})^2}{10^3\ \frac{\text{kg m}^2}{\text{kJ s}^2} \times 0.0266\ \text{m}} + 0.0011\ \text{kJ/kg} = 0.134\ \text{kJ/kg}.$$

Substituting for $\frac{\Delta u_{2-1}^2}{2g_c}$ and $\dot{w}_f$ in Eq. (2.84), we can calculate the input power to the pump (the mass flow rate of the fluid is 0.494 kg/s):

$$\frac{\dot{w}}{\dot{m}} = \frac{\Delta u_{2-1}^2}{2g_c} + \frac{\dot{w}_f}{\dot{m}} = 3.7 \times 10^{-3} \text{ kJ/kg} + 0.134 \text{ kJ/kg} = 0.138 \text{ kJ/kg}.$$

With a 75% efficiency, the actual mechanical energy required will be

$$w_{\text{act}} = \frac{0.138 \text{ kJ/kg}}{0.75} = 0.184 \text{ kJ/kg}.$$

The appropriate pump size is

$$\dot{w}_{\text{act}} = 0.184 \text{ kJ/kg} \times 0.494 \text{ kg/s} = 0.091 \text{ kJ/s} = 0.12 \text{ hp}.$$

2.4 Heat Exchangers

A **heat exchanger** is a device designed to transfer heat between two fluids. Heat exchangers are often used to transfer heat from its source (e.g., a boiler or solar collector) through a working fluid to the point of use (e.g., an engine or turbine). They are particularly important for improving overall process efficiency of energy-efficient systems. Heat exchangers can be expensive and must be designed carefully to maximize effectiveness and minimize cost. Depending on their application, heat exchangers can have different shapes, designs, and sizes. The major types of heat exchangers include boilers, condensers, radiators, evaporators, cooling towers, regenerators, and recuperators. All heat exchangers are identified by their geometric shape and the direction of flow of the heat-transfer fluids inside them. Figure 2.37 depicts some heat exchangers used in industry. In the following paragraphs, we describe the operating principles of some of the heat exchangers that have extensive use in industrial processes—direct-contact heat exchangers, regenerators, and recuperators.

A **direct-contact heat exchanger** is designed so that two fluids are physically brought into contact, with no solid surface separating them. In this type of heat exchanger, fluid streams form a mutual interface through which the heat transfer takes place between the two fluids. Direct-contact heat exchangers are best used when the temperature difference between the hot and cold fluids is small. An example of a direct-contact heat exchanger is a cooling tower, where water and air are brought together by letting water fall from the top of the tower and having it contact a stream of air flowing upward. Kreith and Boehm [1987] provide more information about direct-contact heat exchange.

Regenerators are heat exchangers in which the hot and cold fluids flow alternately through the same space. As a result of alternating flow, the hot fluid heats the core of the heat exchanger, where the stored heat is then transferred to the cold fluid. Regenerators are used most often with gas streams, where some mixing of the two streams is not a problem and where the cost of another type of heat exchanger would be prohibitive. For example, heat recovery in very energy-efficient homes is often done with "air-to-air" regenerators to maintain an acceptable quality of air inside the homes.

The **recuperator** is the heat exchanger encountered most often. It is designed so that the hot and cold fluids do not come into contact with each other. Heat is exchanged from one fluid to a solid surface by convection, through the solid by conduction, and from the other side of the solid surface to the second fluid by convection.

In Section 2.2, we described these heat-transfer processes and developed some simple equations that are applied here to determine basic equipment performance. Designing a heat exchanger also requires estimating the pressure flow losses that can be carried out, based on the information

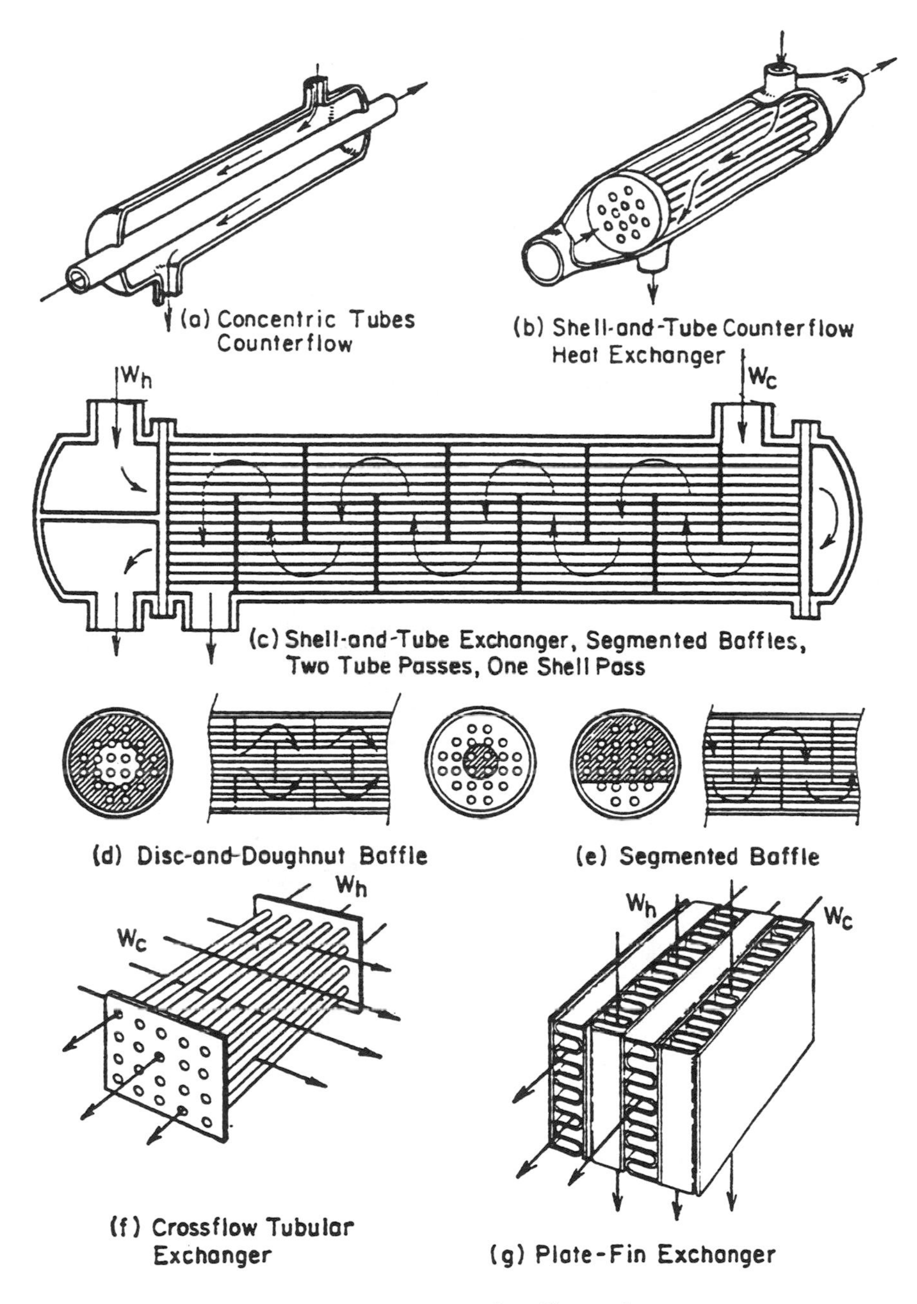

FIGURE 2.37 Some examples of heat exchangers.

provided in Section 2.3. Finally, appropriate materials must be selected and a structural analysis done; Fraas and Ozisik [1965] provide a good discussion of these topics. More detailed discussion on heat exchangers and their application to renewable technologies is found in Chapter 13a of this handbook.

Heat-Exchanger Performance

The performance of a heat exchanger is based on the exchanger's ability to transfer heat from one fluid to another. Calculating the heat transfer in heat exchangers is rather involved because the temperature of one or both of the fluids is changing continuously as they flow through the exchanger. There are three main flow configurations in heat exchangers: parallel flow, counter flow, and cross flow. In **parallel-flow** heat exchangers, both fluids enter from one end of the heat exchanger flowing in the same direction and they both exit from the other end. In **counter-flow** heat exchangers, hot fluid enters from one end and flows in an opposite direction to cold fluid entering from the other end. In **cross-flow** heat exchangers, baffles are used to force the fluids to move perpendicular to each other, to take advantage of higher heat-transfer coefficients encountered in a cross-flow configuration. Figure 2.38 shows the temperature variation of the fluids inside the heat exchanger for a parallel-flow and a counter-flow heat exchanger. In parallel-flow heat exchangers, the temperature difference ΔT_i between the two fluids at the inlet of the heat exchanger is much greater than ΔT_o, the temperature difference at the outlet of the heat exchanger. In counter-flow heat exchangers, however, the temperature difference between the fluids shows only a slight variation along the length of the heat exchanger. Assuming that the heat loss from the heat exchanger is negligible, usually the case in a practical design, the heat loss of the hot fluid should be equal to the heat gain by the cold fluid. Therefore, we can write

$$\dot{Q} = \dot{m}_c c_{p,c} \left(T_{co} - T_{ci}\right) = \dot{m}_h \; c_{p,h} \left(T_{hi} - T_{ho}\right), \tag{2.85}$$

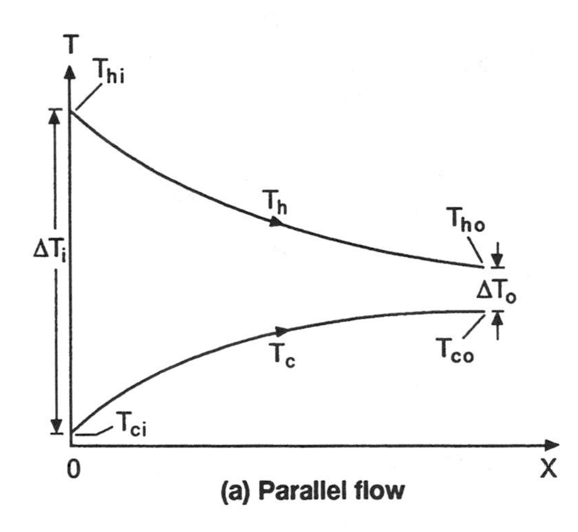

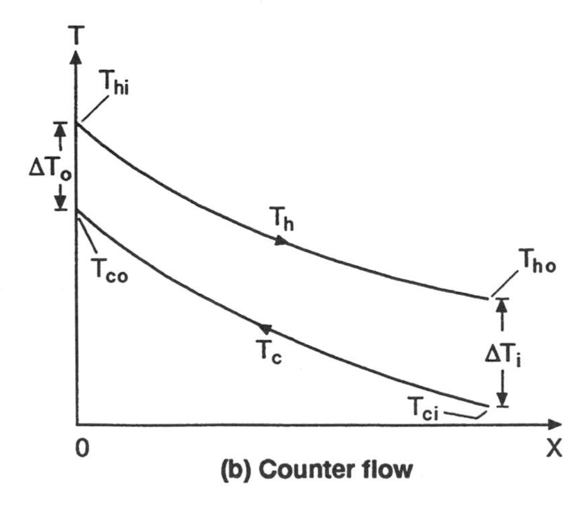

FIGURE 2.38 Fluid temperature variation for (a) parallel-flow configuration, and (b) counter-flow configuration.

where the subscripts c and h refer to cold and hot fluids, respectively. Note that in heat-exchanger analysis, the terms $\dot{m}_c c_{p,c}$ and $\dot{m}_h c_{p,h}$ are called the **capacity rates** of the cold and hot fluids, respectively, and are usually represented by C_c and C_h.

The heat-transfer rate between a hot and a cold fluid can be written as

$$\dot{Q} = UA\ \Delta T_m, \tag{2.86}$$

where U is the overall heat-transfer coefficient (and is assumed to be constant over the whole surface area of the heat exchanger) and ΔT_m is an appropriate mean temperature difference to be defined later. In fact, the overall heat-transfer coefficient is not the same for all locations in the heat exchanger, and its local value depends on the local fluid temperatures as was shown in Example 2.8. For most engineering applications, designers of heat exchangers are usually interested in the *overall* average heat-transfer coefficient. Common practice is to calculate the overall heat-transfer coefficient based on some kind of mean fluid temperature.

Expanding on the definition of thermal resistance described earlier, the heat transfer in a heat exchanger can be expressed as

$$\dot{Q} = UA\ \Delta T_m = \frac{\Delta T_m}{\sum_{i=1}^{n} R_i}, \tag{2.87}$$

where ΣR_i represents the total thermal resistance to heat transfer between fluid streams in the heat exchanger. For example, consider the simple case of heat transfer inside a shell-and-tube heat exchanger, where a hot fluid at T_h is flowing inside a steel tube with inside radius r_i and outside radius r_o as shown in Figure 2.14. The cold fluid at T_c is flowing in the shell side over the steel tube, where the convection heat-transfer coefficient between the cold fluid and the exterior of steel tube is h_o. For this case, the total resistance to heat transfer can be written as

$$\frac{1}{UA} = \sum_{i=1}^{5} R_i = \frac{1}{2\pi r_i L h_i} + R_{f,i} + \frac{\ln(r_o/r_i)}{2\pi k_p L} + R_{f,o} + \frac{1}{2\pi r_o L h_o}, \tag{2.88}$$

where L is the length of the heat exchanger, k_p is the thermal conductivity of steel, and h_i and h_o are the convection heat-transfer coefficients of hot and cold fluid sides, respectively. Terms $R_{f,i}$ and $R_{f,o}$ represent the fouling resistances on the cold and hot heat-transfer surfaces. The overall heat-transfer coefficient can be based on either the hot surface area (in this case, $A_i = 2\pi r_i L$) or on the cold surface area ($A_o = 2\pi r_o L$). Therefore, the numerical value of U will depend on the area selected; however, it is always true that $UA \equiv U_i A_i \equiv U_o A_o$.

Table 2.11 gives some typical values of overall heat-transfer coefficients that are useful in preliminary system analysis and design. For all but the simplest heat exchangers, designing the best heat exchanger for a given application involves using a model that accurately sums the temperature difference and the resistance over the entire surface of the heat exchanger. In our computer age, most engineers use very sophisticated computer models for designing heat exchangers. These computer models incorporate the most accurate algorithms for a myriad of applications. A reasonably good estimate of heat-exchanger performance can be calculated by hand by using one of various readily available handbooks (e.g., *Handbook of Heat Exchanger Design* [1983]).

The other important term in Eq. (2.86) for calculating the heat-transfer rate is the mean temperature difference ΔT_m. The mean temperature difference for a heat exchanger depends on its flow configuration and the degree of fluid mixing in each flow stream.

TABLE 2.11 Approximate Overall Heat-Transfer Coefficients for Preliminary Estimates

Heat-Transfer Duty	Overall Heat-Transfer Coefficient (W/m² K)
Steam to water	
Instantaneous heater	2,200–3,300
Storage-tank heater	960–1,650
Steam to oil	
Heavy fuel	55–165
Light fuel	165–330
Light petroleum distillate	275–1,100
Steam to aqueous solutions	550–3,300
Steam to gases	25–275
Water to compressed air	55–165
Water to water, jacket water coolers	825–1,510
Water to lubricating oil	110–330
Water to condensing oil vapors	220–550
Water to condensing alcohol	250–660
Water to condensing Freon-12	440–830
Water to condensing ammonia	830–1,380
Water to organic solvents, alcohol	275–830
Water to boiling Freon-12	275–830
Water to gasoline	330–500
Water to gas, oil, or distillate	200–330
Water to brine	550–1,100
Light organics to light organics	220–420
Medium organics to medium organics	110–330
Heavy organics to heavy organics	55–220
Heavy organics to light organics	55–330
Crude oil to gas oil	170–300

For a simple single-pass heat exchanger with various temperature profiles (e.g., parallel flow, counter flow, and constant surface temperature), the mean temperature of Eq. (2.87) can be calculated from

$$\Delta T_m = \frac{\Delta T_i - \Delta T_o}{\ln\left(\Delta T_i / \Delta T_o\right)}, \tag{2.89}$$

where ΔT_i represents the greatest temperature difference between the fluids and ΔT_o represents the least temperature difference, and only if the following assumptions hold:

1. U is constant over the entire heat exchanger.
2. The flow of fluids inside the heat exchanger is in steady-state mode.
3. The specific heat of each fluid is constant over the entire length of the heat exchanger.
4. Heat losses from the heat exchanger are minimal.

The mean temperature difference ΔT_m given by Eq. (2.89) is known as the **logarithmic mean temperature difference** (LMTD).

Example 2.11

A stream of lubricating oil at initial temperature of 115°C and flow rate of 2 kg/s is to be cooled to 70°C in a shell-and-tube heat exchanger. A stream of cold water at a flow rate of 2 kg/s and initial temperature of 20°C is used as the cooling fluid in the heat exchanger. Calculate the heat-exchanger area required for this task by employing first a counter-flow and then a parallel-flow heat-exchanger arrangement. The overall heat-transfer coefficient is $U = 900$ W/m² K, and the specific heat of the oil is $c_{p,h} = 2.5$ kJ/kg K.

Solution:

First, we use Eq. (2.85) to calculate the water outlet temperature. The specific heat of water can be assumed to be constant over the temperature range of interest, and it is $c_{p,c} = 4.182$ kJ/kg K.

$$\dot{m}_c c_{p,c}(T_{co} - T_{ci}) = \dot{m}_h c_{p,h}(T_{hi} - T_{ho})$$

or

$$T_{co} = \frac{\dot{m}_h c_{p,h}}{\dot{m}_c c_{p,c}}(T_{hi} - T_{ho}) + T_{ci}.$$

$$T_{co} = \frac{2 \text{ kg/s } (2.5 \text{ kJ/kg K})}{2 \text{ kg/s } (4.182 \text{ kJ/kg K})}(115°\text{C} - 70°\text{C}) + 20°\text{C}.$$

$$T_{co} = 46.9°\text{C}.$$

The total heat transferred from hot fluid to the cold fluid is

$$\dot{Q} = \dot{m}_c \; c_{p,c} \; (T_{co} - T_{ci}) = 2 \text{ kg/s } (4.182 \text{ kJ/kg K}) \; (46.9°\text{C} - 20\text{C}) = 225 \text{ kJ/s}.$$

For a *counter-flow* arrangement, the temperature differences are shown in Figure 2.39(a). The greatest temperature difference is $\Delta T_i = 68.1$°C, and the least temperature difference is $\Delta T_o =$ 50°C. Using Eq. (2.89), the mean temperature can be calculated

$$\Delta T_m = \frac{\Delta T_i - \Delta T_o}{\text{ln}(\Delta T_i/\Delta T_o)} = \frac{68.1°\text{C} - 50°\text{C}}{\text{ln}(68.1/50)} = 58.58°\text{C}.$$

The heat-exchanger surface area for counter-flow arrangement can be obtained from Eq. (2.86):

$$\dot{Q} = UA \; \Delta T_m,$$

or

$$A = \frac{\dot{Q}}{U \; \Delta T_m} = \frac{225 \times 10^3 \text{ J/s}}{900 \text{ W/m}^2 \text{ K} \times 58.58°\text{C}} = 4.27 \text{ m}^2.$$

A similar procedure can be followed for the *parallel-flow* arrangement. The water outlet temperature T_{co} and the total heat transfer calculated earlier still hold for this arrangement. However, the temperature differences are as shown in Figure 2.39(b). The mean temperature for this arrangement is

$$\Delta T_m = \frac{(\Delta T_i - \Delta T_o)}{\text{ln}(\Delta T_i/\Delta T_o)} = \frac{95°\text{C} - 23.1°\text{C}}{\text{ln}(95/23.1)} = 50.84°\text{C},$$

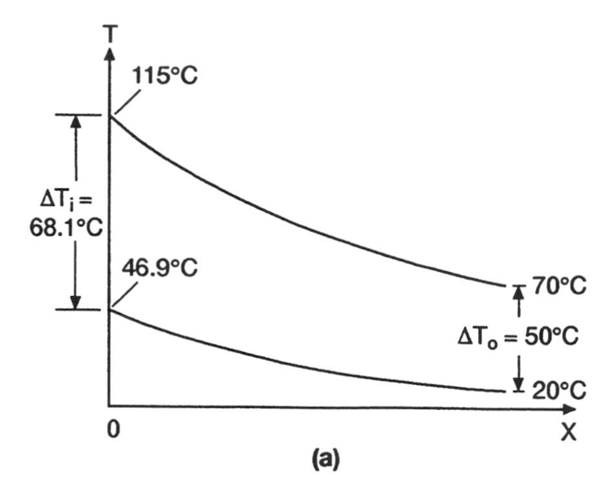

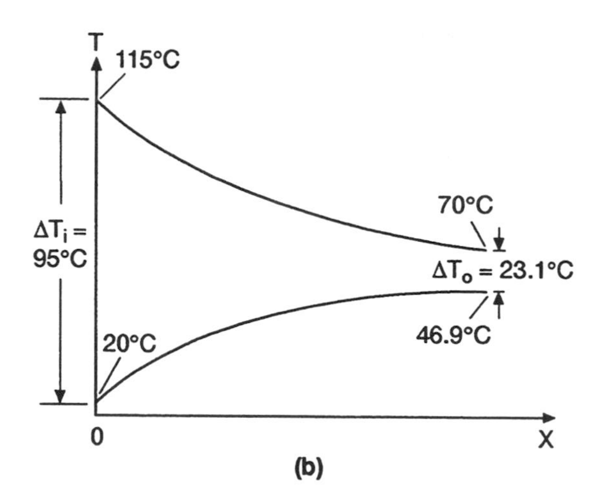

FIGURE 2.39 Temperature differences of Example 2.11 for (a) counter-flow arrangement, and (b) parallel-flow arrangement.

and the area required is

$$A = \frac{\dot{Q}}{U\ \Delta T_m} = \frac{225 \times 10^3\ \text{J/s}}{900\ \text{W/m}^2\ \text{K} \times 50.84°\text{C}} = 4.92\ \text{m}^2.$$

Therefore, the heat-exchanger surface area required for parallel flow is more than that required for the counter-flow arrangement if all the other conditions are assumed to be the same. Consequently, whenever possible, it is advantageous to use the counter-flow arrangement because it will require less heat-exchanger surface area to accomplish the same job. In addition, as seen from Figure 2.39, with the counter-flow arrangement, the outlet temperature of the cooling fluid may be raised much closer to the inlet temperature of the hot fluid.

The LMTD expression presented by Eq. (2.89) does not hold for more complex flow configurations such as cross flow or multipass flows. To extend the LMTD definition to such configurations, a correction factor is defined as

$$F = \frac{\Delta T_m}{\Delta T_{m,cF}}, \tag{2.90}$$

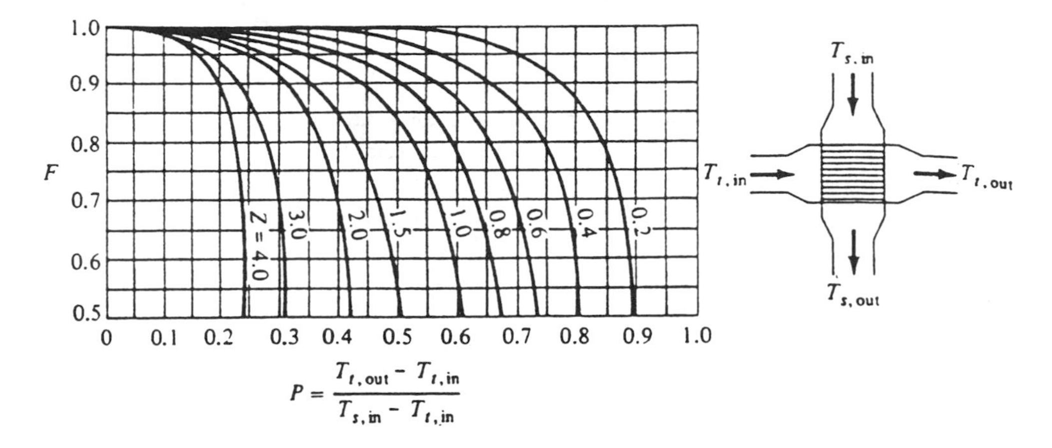

FIGURE 2.40 Correction factor for counter-flow LMTD for cross-flow heat exchangers with the fluid on the shell side mixed, the other fluid unmixed, and one pass through tube. Data extracted from "Mean Temperature Difference in Design" by Bowman et al. (1940) in *ASME Proceedings,* with permission of the publisher, The American Society of American Engineers.

where $\Delta T_{m,cF}$ is calculated from Eq. (2.89) for a counter-flow configuration. Bowman et al. [1940] provide charts for calculating the correction factor F for various flow configurations of heat exchangers. A sample of their charts is shown in Figure 2.40 for one fluid mixed and the other fluid unmixed. The term *unmixed* means that a fluid stream passes through the heat exchanger in separated flow channels or passages with no fluid mixing between adjacent flow passages. Note that the correction factor F in Figure 2.40 is a function of two dimensionless parameters Z and P defined as

$$Z = \frac{T_{hi} - T_{ho}}{T_{co} - T_{ci}} \equiv \frac{C_c}{C_h} \tag{2.91}$$

and

$$P = \frac{T_{co} - T_{ci}}{T_{hi} - T_{ci}}, \tag{2.92}$$

where the term Z is the ratio of the capacity rates of the cold and hot streams, and the term P is referred to as the temperature effectiveness of the cold stream. Kays and London [1984] provide a comprehensive representation of F charts.

Heat-Exchanger Design Methods

Heat-exchanger designers usually use two well-known methods for calculating the heat-transfer rate between fluid streams—the UA-LMTD and the effectiveness-NTU (number of heat-transfer units) methods.

UA-LMTD Method

In this method, the relationship between the total heat-transfer rate, the heat-transfer area, and the inlet and outlet temperatures of the two streams is obtained from Eqs. (2.87) and (2.90). Substituting Eq. (2.90) in Eq. (2.87) gives

$$\dot{Q} = F \times \Delta T_{m,cF} \times UA. \tag{2.93}$$

For a given heat-exchanger configuration, one can calculate UA by identifying heat-transfer resistances and summing them as in Eq. (2.88), calculating $\Delta T_{m,cF}$ from Eq. (2.89), reading the F value from an appropriate chart, and substituting them in Eq. (2.93) to find the heat-transfer rate. The UA-LMTD method is most suitable when the fluid inlet and outlet temperatures are known or can be determined readily from an energy-balance expression similar to Eq. (2.85). There may be situations where only inlet temperatures are known. In these cases, using the LMTD method will require an iterative procedure. However, an alternative is to use the effectiveness-NTU method described in the following section.

Effectiveness-NTU Method

In this method, the capacity rates of both hot and cold fluids are used to analyze the heat-exchanger performance. We will first define two dimensionless groups that are used in this method—the number of heat-transfer units (NTU) and the heat exchanger effectiveness ε.

The NTU is defined as

$$\mathrm{NTU} = \frac{UA}{C_{\min}}, \tag{2.94}$$

where $C_{\min}$ represents the smaller of the two capacity rates C_c and C_h. NTU is the ratio of the heat-transfer rate per degree of mean temperature-difference between the fluids, Eq. (2.86), to the heat-transfer rate per degree of temperature change for the fluid of minimum heat-capacity rate. NTU is a measure of the physical size of the heat exchanger: the larger the value of the NTU, the closer the heat exchanger approaches its thermodynamic limit.

The heat-exchanger effectiveness is defined as the ratio between the actual heat-transfer rate $\dot{Q}$ and the maximum possible rate of heat that thermodynamically can be exchanged between the two fluid streams. The actual heat-transfer rate can be obtained from Eq. (2.85). To obtain the maximum heat-transfer rate, one can assume a counter-flow heat exchanger with infinite surface area, where one fluid undergoes a temperature change equal to the maximum temperature-difference available, $\Delta T_{\max} = T_{hi} - T_{ci}$. The calculation of $\dot{Q}_{\max}$ is based on the fluid having the smaller capacity rate $C_{\min}$, because of the limitations imposed by the Second Law of Thermodynamics (see Bejan [1993] for more detail). Therefore,

$$\dot{Q}_{\max} = C_{\min}\left(T_{hi} - T_{ci}\right), \tag{2.95}$$

where Eq. (2.95) is not limited to counter-flow heat exchangers and can be applied equally to other configurations. Therefore, the effectiveness can be expressed as

$$\varepsilon = \frac{\dot{Q}}{\dot{Q}_{\max}} = \frac{C_c\left(T_{co} - T_{ci}\right)}{C_{\min}\left(T_{hi} - T_{ci}\right)}, \tag{2.96}$$

or

$$\varepsilon = \frac{C_h\left(T_{hi} - T_{ho}\right)}{C_{\min}\left(T_{hi} - T_{ci}\right)}. \tag{2.97}$$

Knowing the effectiveness of a heat exchanger, one can calculate the actual rate of heat transfer by using Eq. (2.96) or from

$$\dot{Q} = \varepsilon C_{\min}\left(T_{hi} - T_{ci}\right). \tag{2.98}$$

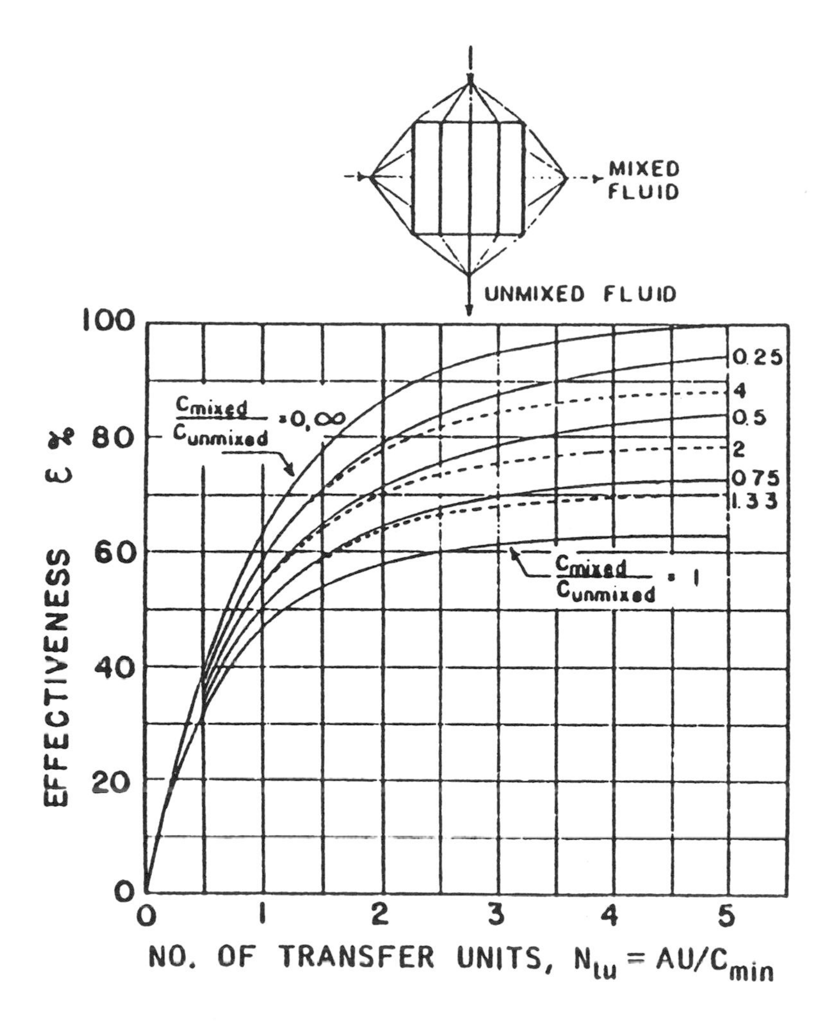

FIGURE 2.41 Heat-transfer effectiveness as a function of number of heat-transfer units and capacity-rate ratio; cross-flow exchanger with one fluid mixed. Data extracted from *Compact Heat Exchangers* by W.M. Kays and A.L. London (1984), with permission of the publisher, McGraw-Hill.

Expressions for the effectiveness of heat exchangers with various flow configurations have been developed and are given in heat-transfer texts (e.g., Bejan [1993], Kreith and Bohn [1993]). For example, the effectiveness of a counter-flow heat exchanger is given by

$$\varepsilon = \frac{1-\exp\left[-\text{NTU}\,(1-R)\right]}{1\text{-R}\,\exp\left[\text{-NTU}\,(1-R)\right]}, \tag{2.99}$$

where $R = C_{min}/C_{max}$. The effectiveness of a parallel-flow heat exchanger is given by

$$\varepsilon = \frac{1-\exp\left[-\text{NTU}\,(1+R)\right]}{1+\text{R}}. \tag{2.100}$$

The effectiveness for heat exchangers of various flow configurations has been evaluated by Kays and London [1984] and is presented in a graph format similar to the one shown in Figure 2.41. In this figure, the heat-exchanger effectiveness has been plotted in terms of NTU and R. Note that for an evaporator and a condenser, $R = 0$ because the fluid remains at a constant temperature during the phase change.

The two design-and-analysis methods just described are equivalent, and both can be equally employed for designing heat exchangers. However, the NTU method is preferred for rating problems where at least one exit temperature is unknown. If all inlet and outlet temperatures are known, the UA-LMTD method does not require an iterative procedure and is the preferred method.

Example 2.12

A schematic representation of a hybrid central receiver is shown in Figure 2.42. In this system, molten nitrate salt is heated in a central receiver to temperatures as high as 1,050°F (565°C). The molten salt is then passed through a heat exchanger, where it is used to preheat combustion air for a combined-cycle power plant. For more information about this cycle, refer to Bharathan et al. (1995) and Bohn et al. [1995]. The heat exchanger used for this purpose is shown in Figure 2.43. The plates of the heat exchanger are made of steel and are 2 mm thick. The overall flow is a counter-flow arrangement where the air and molten salt both flow through duct-shape passages (unmixed). The shell side, where the air flow takes place, is baffled to provide cross flow between the lateral baffles. The baseline design conditions are

Air flow rate:	0.503 kg/s per passage (250 lbm/s)
Air inlet temperature:	340°C (≈650°F)
Air outlet temperature:	470°C (≈880°F)
Salt flow rate:	0.483 kg/s per passage (240 lbm/s)
Salt inlet temperature:	565°C (≈1050°F)
Salt outlet temperature:	475°C (≈890°F)

Find the overall heat-transfer coefficient for this heat exchanger. Ignore the fouling resistances.

Solution:

The overall heat-transfer coefficient for this problem can be calculated from

$$\frac{1}{UA} = \frac{1}{h_s A} + \frac{t}{k_p A} + \frac{1}{h_a A}, \tag{2.101}$$

where h_s and h_a are the convection heat-transfer coefficients on the salt side and on the air side, respectively, k_p is the conductivity of the steel plate, and t is its thickness. To obtain the overall heat-transfer coefficient, we first identify all thermal resistances to heat transfer between the salt stream and the air stream. These resistances are shown in Figure 2.43, and they consist of convection heat-transfer resistance between the salt stream and the steel plate $(1/h_s A)$, conduction heat-transfer resistance through the steel plate $(t/k_p A)$, and the convection heat-transfer resistance between the steel plate and the air stream $(1/h_a A)$. Next, we calculate the convection heat-transfer coefficients h_s and h_a. We first consider the *air side*. The air properties are calculated at

$$T_a = \frac{340°\text{C} + 470°\text{C}}{2} = 405°\text{C},$$

where we obtain the following values from the Table of Properties for air: $\mu = 32.75 \times 10^{-6}$ N s/m^2, $k = 0.0485$ W/m K, $\rho = 0.508$ kg/m^3, and Pr = 0.72. To obtain h_a, we first calculate the Reynolds number and then choose an appropriate expression for Nu. The Reynolds number is

$$\text{Re} = \frac{\rho u D_H}{\mu},$$

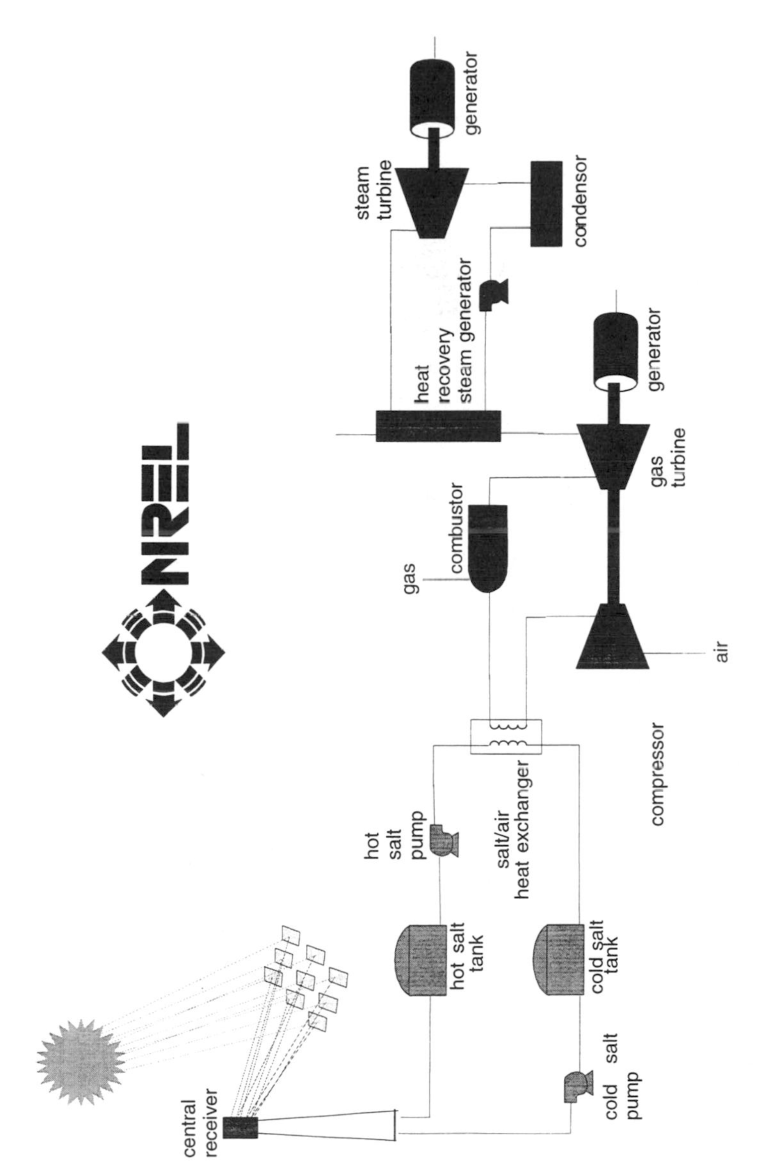

FIGURE 2.42 Schematic representation of a hybrid central-receiver concept developed at the National Renewable Energy Laboratory (see Bharathan et al. (1995) and Bohn et al. [1995]).

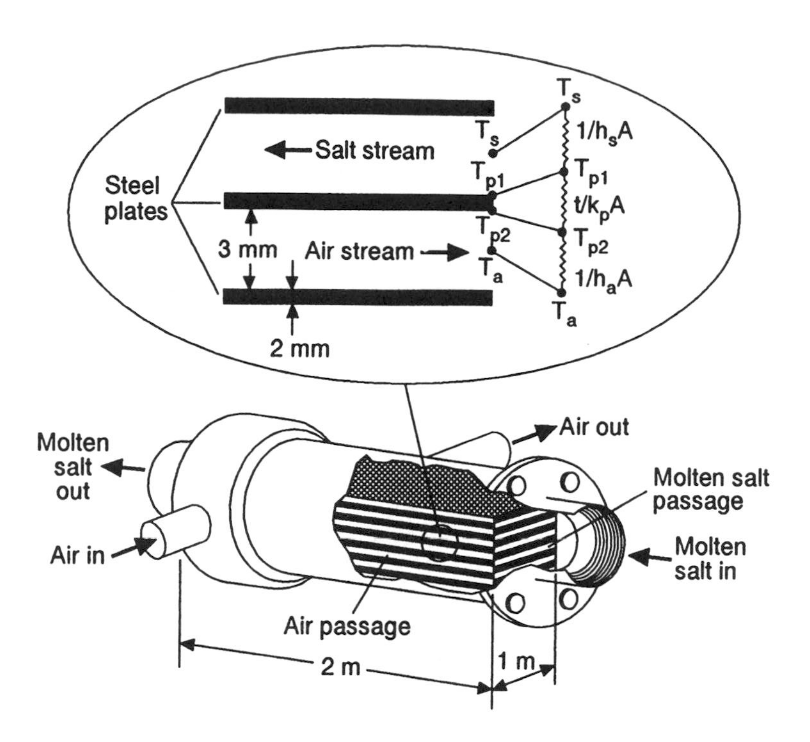

FIGURE 2.43 Molten-salt-to-air heat exchanger used to preheat combustion heat.

where D_H is the hydraulic diameter given by Eq. (2.50). The flow area for air is $A_a = 2 \text{ m} \times 0.003 \text{ m} = 0.006 \text{ m}^2$; therefore,

$$D_H = \frac{4\ A_a}{P_{\text{wet}}} = \frac{4 \times 0.006 \text{ m}^2}{2\left(2 \text{ m} + 0.003 \text{ m}\right)} = 0.006 \text{ m}\,.$$

The mass flow rate for the air is

$$\dot{m} = \rho u\ A_a = 0.503 \text{ kg/s per passage}\,.$$

Therefore, with $A_a = 0.006 \text{ m}^2$, we obtain

$$\rho u = \frac{0.503 \text{ kg/s}}{0.006 \text{ m}^2} = 83.83 \text{ kg/m}^2 \text{ s}\,.$$

Substituting for D_H and ρu in the expression for the Reynolds number, we obtain

$$\text{Re}_{\text{air}} = \frac{83.83 \text{ kg/m}^2\text{s} \times 0.006 \text{ m}}{32.75 \times 10^{-6} \text{ N s/m}^2} \approx 15{,}400\,.$$

From Table 2.5, we choose

$$\text{Nu}_{D_H} = 0.027\ \text{Re}_{\text{air}}^{0.8}\ \text{Pr}^{0.33}\left(\frac{\mu_b}{\mu_s}\right)^{0.14}$$

$$= 0.027\ \left(15{,}400\right)^{0.8}\ \left(0.72\right)^{0.33} = 54.18\,.$$

The term $(\mu_b/\mu_s)^{0.14}$ represents the ratio of air viscosities at bulk and surface temperatures, respectively; this term is very close to unity and therefore is ignored here. The Nusselt number in terms of the heat-transfer coefficient is

$$\mathrm{Nu}_{D_H} = \frac{h_a D_H}{k} \quad or \quad h_a = \frac{\mathrm{Nu}_{D_H} k}{D_H}.$$

Therefore,

$$h_a = \frac{54.18 \times 0.0485 \text{ W/m K}}{0.006 \text{ m}} = 438 \text{ W/m}^2 \text{ K}.$$

We follow a similar procedure for the *salt side*:

$$T_s = \frac{565°\text{C} + 475°\text{C}}{2} = 520°\text{C}.$$

The salt properties at this average temperature are $\mu = 1.25 \times 10^{-3}$ N s/m², $k = 0.543$ W/m K, and $\rho = 1756$ kg/m³. The flow area for the salt side is $A_s = 1 \text{ m} \times 0.003 \text{ m} = 0.003 \text{ m}^2$, and the hydraulic diameter is

$$D_H = \frac{4\ A_s}{P_{\text{wet}}} = \frac{4 \times 0.003 \text{ m}^2}{2\ (1 \text{ m} + 0.003 \text{ m})} = 0.006 \text{ m}.$$

The mass flow rate for the salt is

$$\dot{m} = \rho u A_s = 0.483 \text{ kg/s per passage}.$$

Therefore,

$$\rho u = \frac{0.483 \text{ kg/s}}{0.003 \text{ m}^2} = 161 \text{ kg/m}^2 \text{ s}.$$

Now, we can calculate the Reynolds number for the salt side:

$$\mathrm{Re}_{\text{salt}} = \frac{\rho u D_H}{\mu} = \frac{161 \text{ kg/m}^2\text{s} \times 0.006 \text{ m}}{1.25 \times 10^{-3} \text{ N s/m}^2} \approx 773.$$

Therefore, the flow on the salt side is laminar, and the Nusselt number for a laminar flow inside a duct with very large aspect ratio is $\mathrm{Nu}_{D_H} \approx 8$, as given by Shah and London [1978]. Following a similar procedure as for the air side, we calculate

$$h_s = \frac{\mathrm{Nu}_{D_H} k}{D_H} = \frac{8 \times 0.543 \text{ W/m K}}{0.006 \text{ m}} = 724 \text{ W/m}^2 \text{ K}.$$

The thickness of the steel plate is $t = 0.002$ mm, and its thermal conductivity at about 450°C is $k_p = 22$ W/m K. Now, by substituting for all these in Eq. (2.101), we can obtain the overall heat-transfer coefficient for this heat exchanger. Note that the heat-flow area A is the same for convection and conduction terms in Eq. (2.101); therefore, it can be canceled from both sides of the equation. Therefore,

$$\frac{1}{U} = \frac{1}{724\ \text{W/m}^2\ \text{K}} + \frac{0.002\ \text{m}}{22\ \text{W/m K}} + \frac{1}{365\ \text{W/m}^2\ \text{K}}$$

or

$$U = 266\ \text{W/m}^2\ \text{K}.$$

2.5 Nomenclature

A	heat-transfer surface area, m^2 (ft^2)
c	speed of propagation of radiation energy, m/s (ft/s)
C_c	capacity rate of cold fluid equal to $\dot{m}_c(c_p)_c$
C_h	capacity rate of hot fluid equal to $\dot{m}_h(c_p)_h$
C_0	speed of light in a vacuum, 3×10^8 m/s (9.84×10^8 ft/s)
c_p	specific heat, kJ/kg K (Btu/lbm R)
D_H	hydraulic diameter, defined by Eq. (2.50)
E	total energy of a thermodynamic system, kJ (Btu)
e	internal energy of a thermodynamic system, kJ/kg (Btu/lbm)
E_b	total emissive power of a blackbody, W/m^2 (Btu/h ft^2)
$E_{b\lambda}$	spectral blackbody emissive power, given by Eq. (2.56), W/m^3 (Btu/h ft^2 μ)
f	dimensionless friction factor, see Eq. (2.77)
$F_{1\text{-}2}$	radiation shape factor between surfaces 1 and 2
g	gravitational acceleration, 9.806 m/s^2 (32.17 ft/s^2)
g_c	Newton constant, equal to 32.17 ft $\times$ lbm/(lbf $\times$ s^2), or 10^3 kg $\times$ m^2/(kJ $\times$ s^2)
h	enthalpy, kJ/kg (Btu/lbm), or convection heat-transfer coefficient, W/m^2 K (Btu/h ft^2 °F)
h_f	enthalpy of saturated liquid, kJ/kg (Btu/lbm)
h_g	enthalpy of saturated vapor, kJ/kg (Btu/lbm)
k	thermal conductivity, W/m K (Btu/h ft °F)
k_b	pressure-loss coefficient due to bends and fittings, see Eq. (2.79)
k_c	pressure-loss coefficient due to sudden contraction of pipe, see Eq. (2.81)
k_e	pressure-loss coefficient due to sudden enlargement of pipe, see Eq. (2.80)
k_p	conductivity of steel plate, W/m K (Btu/h ft °F)
L	length of the circular cylinder or heat exchanger, m (ft)
m	mass of a system, kg (lbm)
$\dot{m}$	mass-flow rate of working fluid, kg/s (lbm/s)
n	index of refraction of a medium
Nu	Nusselt number, see Eq. (2.33)
p	pressure, N/m^2 (lbf/in^2)
Pr	Prandtl number, equal to ν/α
Q_H	heat absorbed from a high-temperature reservoir, kJ (Btu)
Q_l	heat rejected to a low-temperature reservoir, kJ (Btu)
$\dot{Q}$	rate of energy (heat) transfer, W (Btu/h)
$\dot{q}''$	heat flux
$\dot{Q}_r$	heat-transfer rate by radiation, W (Btu/h)
R	thermal resistance, see Eq. (2.30), m^2 K/W (h ft^2°F/Btu)
Ra	Rayleigh number, defined by Eq. (2.34)
Re	Reynolds number, equal to uD_H/ν
$R_{f,i}$	fouling resistance of cold (inner) heat transfer surface, m^2 K/W (h ft^2°F/Btu)
$R_{f,o}$	fouling resistance of hot (outer) heat transfer surface, m^2 K/W (h ft^2°F/Btu)
r_i	inner radius of cylinder, m (ft)
r_o	outer radius of cylinder, m (ft)
K_s	specific

S	entropy, kJ/K (Btu/R)
s_f	entropy of saturated liquid, kJ/kg K (Btu/lbm R)
s_g	entropy of saturated vapor, kJ/kg K (Btu/lbm R)
T	temperature, °C (°F)
T_H	temperature of higher-temperature reservoir, K (R)
T_L	temperature of low-temperature reservoir, K (R)
U	overall heat transfer coefficient, see Eq. (2.88), W/m²K (Btu/h ft²°F)
u	velocity, m/s (ft/s)
v	specific volume (volume per unit mass), m³/kg (ft³/lbm)
w	work or energy, kJ (Btu)
$\dot{w}$	rate of work (done on the surroundings), kJ/s (Btu/h)
w_{act}	actual work of a system, kJ (Btu)
$\dot{w}_f$	frictional pressure losses, kJ/s (Btu/h)
w_{isent}	work done under an isentropic process, kJ (Btu)
w_{rev}	work of a system under reversible process, kJ (Btu)
x	distance along x axis, m (ft)
y	distance along y axis, m (ft)
z	distance along z axis, or elevation of a thermodynamic system, m (ft)

Greek Symbols

a	thermal diffusivity, m²/s (ft²/h), or absorptivity of a surfacc
$\bar{\alpha}$	average absorptivity of a surface
b	thermal expansion coefficient, K^{-1} (R^{-1})
d	velocity boundary-layer thickness, mm (in.)
δ_T	temperature boundary-layer thickness, mm (in.)
Δ	difference
ΔT_i	equal $T_{ho} - T_{ci}$ for counter-flow, and $T_{hi} - T_{ci}$ for parallel-flow heat exchanger, see Figure 2.38
ΔT_o	equal to $T_{hi} - T_{co}$ for counter-flow, and $T_{ho} - T_{co}$ for parallel-flow heat exchanger, see Figure 2.38
e	cmissivity of a surface
$\bar{\varepsilon}$	average absorptivity of a surface
η_m	mechanical efficiency, defined by Eq. (2.17)
η_s	isentropic efficiency, defined by Eq. (2.18)
η_r	relative efficiency, defined by Eq. (2.19)
η_{rev}	energy conversion efficiency for a reversible system
η_T	thermal efficiency, defined by Eq. (2.20)
l	wavelength of radiation energy, m (ft)
μ	viscosity of fluid, Ns/m²
n	kinematic viscosity, m²/s (ft²/h), or frequency, s^{-1}
r	reflectivity of a surface, or density (mass per unit volume), kg/m³ (lbm/ft³)
s	Stefan-Boltzmann constant, $Wm^{-2} K^{-4}$ (Btu/h ft² R⁴)
t	transmissivity of a surface, or frictional shear stress N/m² (lbf/ft²)
u	thermal radiation source frequency, s^{-1}

References

Achenbach, E. 1989. Heat transfer from a staggered tube bundle in cross-flow at high Reynolds numbers, *Int. J. Heat Mass Transfer,* 32:271–280.

Anderson, R.S. and Kreith, F. 1987. Natural convection in active and passive solar thermal systems, *Advances in Heat Transfer,* Vol. 18, pp. 1–86. Academic Press, New York.

Arnold, J.N., Bonaparte, P.N., Catton, I., and Edwards, D.K. 1974. *Proceedings 1974 Heat Transfer Fluid Mech. Inst.,* Stanford University Press, Stanford, CA.

Arnold, J.N., Catton, I., and Edwards, D.K. 1976. Experimental investigation of natural convection in inclined rectangular regions of differing aspect ratios, *J. Heat Transfer,* 98:67–71.

ASHRAE Handbook of Fundamentals, 1993. Robert A. Parsons, Ed., I-P edition, American Society of Heating, Ventilating, and Air Conditioning Engineers, Atlanta, GA.

Ayyaswamy, P.S. and Catton, I. 1973. The boundary layer regime for natural convection in a differentially heated tilted rectangular cavity, *J. Heat Transfer,* 95:543–545.

Balmer, R.T. 1990. *Thermodynamics,* West Publishing Company, St. Paul, MN.

Bejan, A. 1993. *Heat Transfer,* John Wiley & Sons, Inc., New York.

Bharathan, D., Bohn, M.S., and Williams, T.A. 1995. Hybrid Solar Central Receiver for Combined Cycle Power Plant, U.S. Patent 5,417,052.

Bohn, M.S., Williams, T.A., and Price, H.W. 1995. "Combined Cycle Power Tower," Presented at *Int. Solar Energy Conference,* ASME/JSME/JSES joint meeting, Maui, HI.

Bowman, R.A., Mueller, A.C., and Nagle, W.M. 1940. "Mean Temperature Difference in Design," *Trans. ASME,* Vol. 62, pp. 283–294.

Carnot, S. 1960. *Reflections on the Motive Power of Fire (and Other Papers on the Second law of Thermodynamics by E. Clapeyron and R. Clausius),* ed. E. Mendoza. Dover, New York.

Catton, I. 1978. Natural convection in enclosures. In *Proceedings Sixth International Heat Transfer Conference, Toronto,* Vol. 6, pp. 13–31. Hemisphere, Washington, D.C.

Churchill, S.W. 1983. Free convection around immersed bodies. In *Heat Exchanger Design Handbook,* ed. E.U. Schlünder, Section 2.5.7. Hemisphere, New York.

Churchill, S.W. and Chu, H.H.S. 1975. Correlating equations for laminar and turbulent free convection from a vertical plate, *Int. J. Heat Mass Transfer,* 18:1323–1329.

Fischenden, M. and Saunders, O.A. 1932. *The Calculation of Heat Transmission.* His Majesty's Stationary Office, London.

Fraas, A.P. and Ozisik, M.N. 1965. *Heat Exchanger Design,* John Wiley & Sons, Inc., New York.

Freeman, A. 1941. *Experiments upon the Flow of Water in Pipes and Pipe Fittings,* American Society of Mechanical Engineers, New York.

Fujii, T. and Imura, H. 1972. Natural convection heat transfer from a plate with arbitrary inclination, *Int. J. Heat Mass Transfer,* 15:755–767.

Gubareff, G.G., Janssen, E.J., and Torborg, R.H. 1960. *Thermal Radiation Properties Survey,* Honeywell Research Center, Minneapolis, MN.

Hamilton, D.C. and Morgan, W.R. 1962. Radiant Interchange Configuration Factors, NACA TN2836, Washington, D.C.

Hassani, A.V. and Hollands, K.G.T. 1989. On natural convection heat transfer from three-dimensional bodies of arbitrary shape, *J. Heat Transfer,* 111:363–371.

Hausen, H. 1983. *Heat Transfer in Counter Flow, Parallel Flow and Cross Flow,* McGraw-Hill, New York.

Hollands, K.G.T., Raithby, G.D., and Konicek, L.J. 1975. Correlation equations for free convection heat transfer in horizontal layers of air and water, *Int. J. Heat Mass Transfer,* 18:879–884.

Hollands, K.G.T., Unny, T.E., Raithby, G.D., and Konicek, L.J. 1976. Free convection heat transfer across inclined air layers, *J. Heat Transfer,* 98:189–193.

Howell, J.R. 1982. *A Catalog of Radiation Configuration Factors,* McGraw-Hill, New York.

Incropera, F.P. and DeWitt, D.P. 1990. *Introduction to Heat Transfer,* 2nd ed., Wiley, New York.

Jakob, M. 1949. *Heat Transfer,* Vol. 1, Wiley, New York.

Kakac, S. and Yener, Y. 1988. *Heat Conduction,* 2nd ed. Hemisphere, Washington, D.C.

Kalina, A.I. 1984. Combined-cycle system with novel bottoming cycle, *J. of Eng. for Gas Turbines and Power,* 106:737–742.

Karlekar, B.V. 1983. *Thermodynamics for Engineers,* Prentice-Hall, Englewood Cliffs, NJ.

Kays, W.M. and London, A.L. 1984. *Compact Heat Exchangers,* 3rd ed., McGraw-Hill, New York.

Kays, W.M. and Perkins, K.R. 1985. Forced convection, internal flow in ducts. In *Handbook of Heat Transfer Applications,* eds. W.R. Rosenow, J.P. Hartnett, and E.N. Ganic, Vol. 1, Chap. 7. McGraw-Hill, New York.

Kreith, F. 1970. Thermal design of high altitude balloons and instrument packages, *J. Heat Transfer,* 92:307–332.

Kreith, F. and Black, W.Z. 1980. *Basic Heat Transfer,* Harper & Row, New York.

Kreith, F. and Boehm, eds. 1987. *Direct Condenser Heat Transfer,* Hemisphere, New York.

Kreith, F. and Bohn, M.S. 1993. *Principles of Heat Transfer,* 5th ed., West Publishing Company, St. Paul, MN.

McAdams, W.H. 1954. *Heat Transmission,* 3rd ed., McGraw-Hill, New York.

MacGregor, R.K. and Emery, A.P. 1969. Free convection through vertical plane layers: moderate and high Prandtl number fluid, *J. Heat Transfer,* 91:391.

Moody, L.F. 1944. Friction factors for pipe flow, *Trans. ASME,* 66:671.

Perry, J.H., Perry, R.H., Chilton, C.H., and Kirkpatrick, S.D. 1963. *Chemical Engineer's Handbook,* 4th ed., McGraw-Hill, New York.

Planck, M. 1959. *The Theory of Heat Radiation,* Dover, New York.

Quarmby, A. and Al-Fakhri, A.A.M. 1980. Effect of finite length on forced convection heat transfer from cylinders, *Int. J. Heat Mass Transfer,* 23:463–469.

Raithby, G.D. and Hollands, K.G.T. 1974. A general method of obtaining approximate solutions to laminar and turbulent free convection problems. In *Advances in Heat Transfer,* Academic Press, New York.

Raithby, G.D. and Hollands, K.G.T. 1985. Natural convection. In *Handbook of Heat Transfer Fundamentals,* 2nd ed., eds. W.M. Rosenow, J.P. Hartnett, and E.N. Ganic, McGraw-Hill, New York.

Reid, R.C., Sherwood, T.K., and Prausnitz, J.M. 1977. *The Properties of Gases and Liquids,* 3rd ed., McGraw-Hill, New York.

Reynolds, W.C. and Perkins, H.C. 1977. *Engineering Thermodynamics,* 2nd ed., McGraw Hill, New York.

Sanford, J.F. 1962. *Heat Engines,* Doubleday, Garden City, NY.

Schlünder, E.U., ed. 1983. *Handbook of Heat Exchanger Design,* Vol. 1. Hemisphere, Washington, D.C.

Schmidt, H. and Furthman, E. 1928. Ueber die Gesamtstrahlung fäster Körper. Mitt. Kaiser-Wilhelm-Inst. Eisenforsch., Abh. 109, Dusseldorf.

Shah, R.K. and London, A.L. 1978. *Laminar Flow Forced Convection in Ducts,* Academic Press, New York.

Sieder, E.N. and Tate, C.E. 1936. Heat transfer and pressure drop of liquids in tubes, *Ind. Eng. Chem.,* 28:1429.

Siegel, R. and Howell, J.R. 1972. *Thermal Radiation Heat Transfer,* 3rd ed., Hemisphere, New York.

Sparrow, E.M. and Stretton, A.J. 1985. Natural convection from variously oriented cubes and from other bodies of unity aspect ratio, *Int. J. Heat Mass Transfer,* 28(4):741–752.

Standards of Hydraulic Institute Tentative Standards, Pipe Friction, 1948. Hydraulic Institute, New York.

Van Wylen, G.J. and Sonntag, R.E. 1986. *Fundamentals of Classical Thermodynamics,* 3rd ed., Wiley, New York.

Whitaker, S. 1972. Forced convection heat transfer correlations for flow in pipes, past flat plates, single cylinders, single spheres, and for flow in packed beds and tube bundles, *AIChE J.,* 18:361–371.

Wood, B.D. 1982. *Applications of Thermodynamics,* 2nd ed., Addison-Wesley, Reading, MA.

Yovanovich, M.M. and Jafarpur, K. 1993. Models of laminar natural convection from vertical and horizontal isothermal cuboids for all Prandtl numbers and all Rayleigh numbers below 10^{11}, *Fundamentals of Natural Convection,* ASME, HTD, Vol. 264, pp. 111–126.

Zukauskas, A.A. 1972. Heat transfer from tubes in cross flow, *Advances in Heat Transfer,* Vol. 8, pp. 93–106. Academic Press, New York.

3

Economic Methods*

Rosalie T. Ruegg
National Institute of Standards and Technology

Introduction

Comparing investments in energy conservation and renewable energy systems that both reduce energy consumption and lower total lifetime building costs, with buying more electric power or heat is facilitated by the use of economic evaluation methods. Guidelines for making economically efficient energy-related decisions are needed by all sectors of the energy community.

The purpose of this chapter is to provide an introduction to some basic methods that are helpful in designing and sizing cost-effective systems, and in determining if it is economically efficient to invest in specific energy conservation or renewable energy projects. The targeted audience includes analysts, architects, engineers, designers, builders, codes and standards writers, and government policy makers—collectively referred to as the "design community."

The focus is on microeconomic methods for measuring cost-effectiveness and computing rates of return for individual projects or groups of projects under **uncertainty**. The chapter does not treat macroeconomic methods and national market penetration models for measuring economic impact of energy conservation and renewable energy investments on the national economy. It provides sufficient guidance for computing measures of economic performance for relatively simple investment choices, and it provides the fundamentals for dealing with complex investment decisions.

*This chapter is a contribution of the National Institute of Standards and Technology and therefore is not subject to copyright.

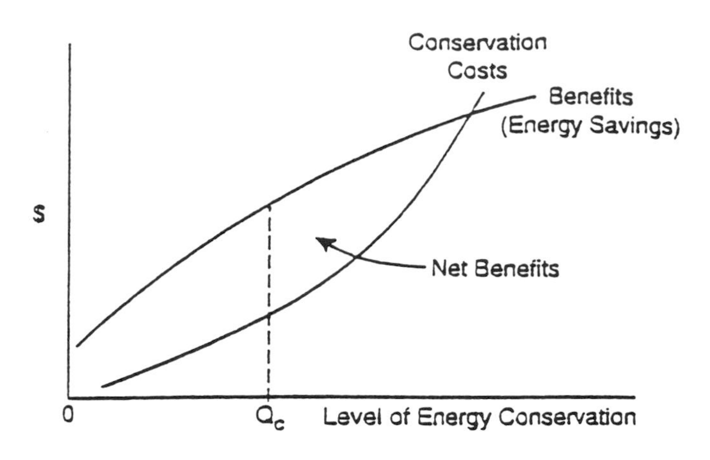

FIGURE 3.1 Maximizing net benefits.

3.1 Making Economically Efficient Choices*

Economic evaluation methods can be used in a number of ways to increase the **economic efficiency** of energy-related decisions. There are methods that can be used to obtain the largest possible savings in energy costs for a given energy budget; there are methods that can be used to achieve a targeted reduction in energy costs for the lowest possible conservation/renewable energy investment; and there are methods that can be used to determine how much it pays to spend on energy conservation and renewable energy in order to lower total lifetime costs, including both **investment costs** and energy cost savings.

The first two ways of using economic evaluation methods (i.e., to obtain the largest savings for a fixed budget and to obtain a targeted savings for the lowest budget) are more limited applications than the third, which aims at minimizing total building costs or maximizing **net benefits** (net savings) from expenditure on energy conservation and renewables. As an example of the first, building owners may budget a specific sum of money for the purpose of retrofitting their buildings for energy conservation. As an example of the second, designers may be required by state or federal building standards or codes to reduce the design energy loads of new buildings below some specified level. As an example of the third, designers or builders may be required by their clients to include in their buildings those energy conservation and renewable energy features that will pay off in terms of lower overall building costs over the long run.

Note that economic efficiency is not necessarily the same as engineering thermal efficiency. For example, one furnace may be more "efficient" than another in the engineering technical sense if it delivers more units of heat for a given quantity of fuel than another. Yet it may not be economically efficient if the first cost of the higher output furnace outweighs its fuel savings. The focus in this chapter is on economic efficiency, not technical efficiency.

Economic efficiency is illustrated conceptually in Figures 3.1 through 3.3. In these three figures, the physical quantity of energy conservation (e.g., the number of units of insulation) is measured on the horizontal axis, and dollar costs are measured on the vertical axis.

Figure 3.1 shows the level of energy conservation, Q_c, that maximizes net benefits from energy conservation, that is, the level that is most profitable over the long run. Note that it is the point at which the curves are most distant.

*This section is based on a treatment of these concepts provided by Marshall and Ruegg in *Economics of Solar Energy and Conservation Systems*, ed. Kreith and West, CRC Press, 1980.

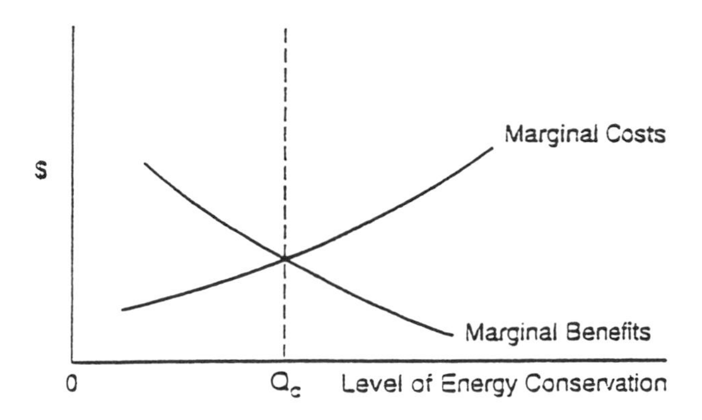

FIGURE 3.2 Equating marginal benefits and marginal costs.

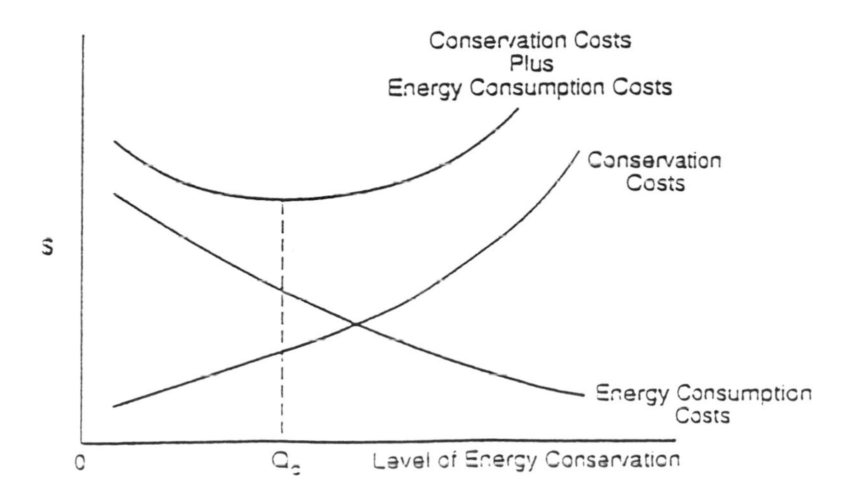

FIGURE 3.3 Minimizing life-cycle costs.

Figure 3.2 shows how "marginal analysis" can be used to find the level of conservation, Q_c, that will yield the largest net benefits. It depicts changes in the total benefits and cost curves (i.e., the derivatives of the curves in Figure 3.1) as the level of energy conservation is increased. The point of intersection of the marginal curves coincides with the most profitable level of energy conservation indicated in Figure 3.1. This is the point at which the cost of adding one more unit of conservation is just equal to the corresponding benefits in terms of energy savings (i.e., the point at which "marginal costs" and "marginal benefits" are equal). To the left of the point of intersection, the additional benefits from increasing the level of conservation by another unit are greater than the additional costs, and it pays to invest more. To the right of the point of intersection, the costs of an addition to the level of conservation exceed its benefits and the level of total net benefits begins to fall, as shown by Figure 3.1. Figure 3.3 shows that the most economically efficient level of energy conservation, Q_c, is that for which the total cost curve is at a minimum.

The most economically efficient level of conservation is indicated in Figures 3.1 to 3.3 as Q_c. Three different approaches to finding Q_c are illustrated: finding the maximum difference between benefits and costs; finding the point where marginal benefits equal marginal costs; and finding the lowest life-cycle costs.

3.2 Economic Evaluation Methods*

There are a number of closely related, commonly used methods for evaluating economic performance. These include the **life-cycle cost** method, net benefits (net present worth) method, benefit/cost (or savings-to-investment) ratio method, internal rate-of-return method, overall rate-of-return method and payback method. All of these methods are used when the important effects can be measured in dollars. If incommensurable effects are critical to the decision, it is important that they also be taken into account. But, since only quantified effects are included in the calculations for these economic methods, unquantified effects must be treated outside the models. Brief treatments of the methods are provided; some additional methods are identified but not treated. For more comprehensive treatments, see Ruegg and Marshall (1990) and other sources listed at the end of the chapter.

Life-Cycle Cost (LCC) Method

The life-cycle costing method sums, for each investment alternative, the costs of acquisition, maintenance, repair, replacement, energy, and any other monetary costs (less any positive amounts, such as salvage value) that are affected by the investment decision. The **time value of money** must be taken into account for all amounts, and the amounts must be considered over the relevant period. All amounts are usually measured either in **present value** or annual value dollars. This is discussed later under "Discounting" and "Discount Rate". At a minimum, for comparison, the investment alternatives should include a "base-case" alternative of not making the conservation or renewable investment and at least one case of an investment in a specific conservation or renewable system. Numerous alternatives may be compared. The alternative with the lowest life-cycle cost that meets the investor's objectives and constraints is the preferred investment.

Following is a formula for finding the life-cycle costs of each alternative:

$$\mathrm{LCC}_{\mathrm{A1}} = I_{\mathrm{A1}} + E_{\mathrm{A1}} + M_{\mathrm{A1}} + R_{\mathrm{A1}} - S_{\mathrm{A1}} \tag{3.1}$$

where

$\mathrm{LCC}_{\mathrm{A1}}$ = life-cycle cost of alternative A1
I_{A1} = present value investment costs of alternative A1
E_{A1} = present value energy costs associated with alternative A1
M_{A1} = present value nonfuel operating and maintenance cost of A1
R_{A1} = present value repair and replacement costs of A1
S_{A1} = present value resale (or salvage) value less disposal cost associated with alternative A1

The LCC method is particularly useful for decisions that are made primarily on the basis of cost effectiveness, such as whether a given conservation or renewable energy investment will lower total cost, (i.e., the sum of owning and operating costs). It can be used to compare alternative designs or sizes of systems. The method, if used correctly, can be used to find the overall cost-minimizing combination of energy conservation investments and energy supply investments within a given facility. However, it cannot be used to find the best investment in general.

Net Benefits (NB) or Net Savings (NS) Method

The net benefits method finds the excess of benefits over costs, where all amounts are discounted for their time value. (If costs exceed benefits, net losses result.) The NB method is also often called the "net present worth" or "net present value" method.

*These methods are treated in detail in Ruegg and Marshall *Building Economics: Theory and Practice*, Chapman and Hall, New York, NY, 1990.

When this method is used for evaluating a cost-reducing investment, the cost savings are the benefits, and it is often called the "net savings" (NS) method.

Following is a formula for finding the net benefits (net savings) from an investment, such as an investment in energy conservation or renewable energy systems:

$$NB_{A1:A2} = \sum_{t=0}^{N} \left(B_t - C_t\right) / \left(1+d\right)^t \tag{3.2}$$

where

$NB_{A1:A2}$ = net benefits, that is present value benefits (savings) net of present value costs for alternative A1 as compared with alternative A2

B_t = benefits in year t, which may be defined to include energy savings

C_t = costs in year t associated with alternative A1 as compared with a mutually exclusive alternative A2

d = **discount rate**

The NB method is particularly suitable for decisions made on the basis of long-run profitability. The NB (NS) method is also useful for deciding whether to make a given investment and for designing and sizing systems. The method is useful for combining conservation and renewable energy projects in an efficiently integrated system.

Benefit-to-Cost Ratio (BCR) or Savings-to-Investment Ratio (SIR) Method

This method divides benefits by costs, or savings by investment. When used to evaluate energy conservation and renewable energy systems, benefits are in terms of energy cost savings. Usually the numerator of the SIR ratio is constructed as energy savings, net of maintenance and repair costs, and the denominator as the sum of investment costs and replacement costs less salvage value (capital cost items). However, depending on the objective, sometimes only initial investment costs are placed in the denominator and the other costs are subtracted in the numerator, or sometimes only the investor's equity capital is placed in the denominator. Like the two preceding methods, this method is based on discounted cash flows.

Unlike the two preceding methods, which provided a performance measure in dollars, this method gives the measure as a dimensionless number. The higher the ratio, the more dollar savings realized per dollar of investment.

Following is a commonly used formula for computing the ratio of savings-to-investment costs:

$$SIR_{A1:A2} = \sum_{t=0}^{N} \left(CS_t \left(1+d\right)^{-t}\right) \Big/ \sum_{t=0}^{N} \left(I_t \left(1+d\right)^{-t}\right) \tag{3.3}$$

where

$SIR_{A1:A2}$ = savings-to-investment ratio for alternative A1 relative to mutually exclusive alternative A2

CS_t = Cost savings (excluding those investment costs in the denominator) plus any positive benefits of alternative A1 as compared with mutually exclusive alternative A2

I_t = additional investment costs for alternative A1 relative to A2

Note that the particular formulation of the ratio with respect to the placement of items in the numerator or denominator can affect the outcome. One should use a formulation appropriate to the decision maker's objectives.

The ratio method can be used to determine whether or not to accept or reject a given investment on economic grounds. If applied incrementally, it can also be used for design and size decisions and other choices among mutually exclusive alternatives. A primary application of the ratio method is to set funding priorities among projects competing for a limited budget. When it is used in this way, it should be supplemented by the NB or NS method when project costs are "lumpy," making it impossible to fully allocate the budget by taking projects in order according to the size of their ratios.

Internal Rate-of-Return (IRR) Method

The **internal rate-of-return** method solves for the interest rate for which dollar savings are just equal to dollar costs over the relevant period; that is, the rate for which net savings are zero. This interest rate is the rate of return on the investment. It is compared to the investor's minimum acceptable rate of return to determine if the investment is desirable. Unlike the preceding three techniques, the internal rate of return does not call for the inclusion of a prespecified cost of capital in the computation, but, rather, solves for a rate.

The rate of return is typically calculated by a process of trial and error, by which various compound rates of interest are used to discount cash flows until a rate is found for which the net value of the investment is zero. The approach is the following: compute NB or NS using Eq. (3.2), except substitute a trial interest rate in place of the discount rate, d, in the equation. A positive NB or NS means that the actual solution rate (the IRR) is greater than the trial rate; a negative NB or NS means that the IRR is less than the trial rate. Based on the information, try another rate. By a series of iterations, find the rate at which NB or NS approaches zero.

Computer algorithms, graphical techniques, and, for simple cases, discount-factor tabular approaches, are often used to facilitate IRR solutions (Ruegg and Marshall, 1990, pp. 71–72).

Expressing economic performance as a rate of return can be desirable for ease in comparing the returns on a variety of investment opportunities, since returns are often expressed in terms of annual rates of return. The IRR method is useful for accepting or rejecting given investments. For designing or sizing projects, the IRR method, like the SIR, must be applied incrementally.

It is a widely used method, but it is often misused. Shortcomings include the possibility of no solution or of multiple solution values and failure to give a measure of overall yield associated with the project over the study period, with the attendant possibility of overstating profitability due to an implicit reinvestment rate assumption (Ruegg, 1991). A measure of the overall rate of return that makes the reinvestment rate assumption explicit is recommended over the IRR (Ruegg and Marshall, 1990, pp. 79–91).

Overall Rate-of-Return (ORR) Method

The ORR method, like the IRR, expresses economic performance in terms of an annual rate of return over the period of study. But unlike the IRR, the ORR provides for an explicit reinvestment rate on interim receipts and a unique solution value. The explicit reinvestment rate makes it possible to express net cash flows (excluding investment costs) as a terminal amount. The ORR is then easily computed with a closed-form solution as shown in Eq. (3.4).

$$\mathrm{ORR}_{\mathrm{A1:A2}} = \left[\left[\sum_{t=0}^{N}\left(B_t - \overline{C}_t\right)(1+r)^{N-t}\right] \Big/ \sum_{t=0}^{N}\left[\overline{I}_t\,(1+d)^{-t}\right]\right]^{1/N} - 1 \qquad \textbf{(3.4)}$$

where

$\mathrm{ORR}_{\mathrm{A1:A2}}$ = overall rate of return on a given investment alternative A1 relative to a mutually exclusive alternative over a designated study period

B_t = benefits from a given alternative relative to a mutually exclusive alternative A2 over time period t

$\overline{C}_t$ = costs (excluding that part of investment costs on which the return is to be maximized) associated with a given alternative relative to a mutually exclusive alternative A2 over time t

r = the reinvestment rate at which net returns can be reinvested, usually set equal to the discount rate

d = the discount rate

N = the length of the study period

$\overline{I}_t$ = investment costs in time t on which the return is to be maximized

The ORR is recommended as a substitute for the IRR because it avoids the limitations and problems of the IRR. It can be used for deciding whether or not to fund a given project, for designing or sizing projects (if it is used incrementally), and for budget allocation decisions.

Discounted Payback (DPB) Method

This evaluation method measures the elapsed time between the point of an initial investment and the point at which accumulated savings, net of other accumulated costs, are sufficient to offset the initial investment, taking into account the cost of capital. (If the cost of capital is not included, the technique is called "simple payback.") For the investor who requires a rapid turnover of investment funds, the shorter the length of time until the investment pays off, the more desirable the investment. However, a shorter payback time does not always mean a more economically efficient investment, making this a method that must be used with care.

To determine the **discounted payback** period, find the minimum solution value of Y (year in which payback occurs) in the equation, such that the following equality condition is satisfied.

$$\sum_{t=1}^{y_0}\left(B_t - C_t'\right)/(1+d)^t = I_0 \tag{3.5}$$

where

B_t = benefits associated in period t with one alternative as compared with a mutually exclusive alternative

C_t' = costs in period t (not including initial investment costs) associated with an alternative as compared with a mutually exclusive alternative in period t

I_0 = initial investment costs of an alternative as compared with a mutually exclusive alternative, where the initial investment cost comprises total investment costs

DPB is often—correctly—used as a supplementary measure when project life is uncertain. It is used to identify feasible projects when the investor's time horizon is constrained. It is used as a supplementary measure in the face of uncertainty to indicate how long capital is at risk. It is a rough guide for accept/reject decisions. It is also overused and misused. Since it indicates the time at which the investment just breaks even, it is not a reliable guide for most complex investment decisions where the objective is to choose the most profitable investment alternatives.

Other Economic Evaluation Methods

A variety of other methods have been used to evaluate the economic performance of energy systems, but these tend to be hybrids of those presented here. One of these is the *required revenue method*, which computes a measure of the before-tax revenue in present or annual-value dollars required to cover the costs on an after-tax basis of an energy system (Ruegg and Short, 1988, pp. 22–23). Another is the levelized cost of energy (LCOE) method, which finds the annualized cost per unit of load for an energy system (Ruegg and Short, 1988, p. 26). Mathematical programming methods have been used to evaluate the optimal size or design of projects, as well as other mathematical and statistical techniques.

3.3 Risk Assessment*

Decisions about energy systems occur under conditions of uncertainty. Using the evaluation methods alone, without techniques for accounting for risk, results in deterministic answers. The decision maker then lacks information about the degree of risk associated with the decision.

Risk assessment provides decision makers with information about the "**risk exposure**" inherent in a given decision, that is, the probability that the outcome will be different from the "best-guess" estimate. Risk assessment is also concerned with the "**risk attitude**" of the decision maker, that is, the willingness of the decision maker to take a chance on an investment of uncertain outcome. Risk assessment techniques are typically used in conjunction with the evaluation methods outlined earlier; the risk assessment techniques are generally not used as stand alone evaluation techniques.

The risk assessment techniques range from simple and partial to complex and comprehensive. Though none takes the risk out of making decisions, the techniques, if used correctly, can help the decision maker make more informed choices in the face of uncertainty.

This chapter provides an overview of the following probability-based risk assessment techniques:

- Expected Value Analysis
- Mean-Variance Criterion and Coefficient of Variation
- Risk-Adjusted Discount Rate Technique
- Certainty Equivalent Technique
- Simulation Analysis
- Decision Analysis

It also treats one widely used non-probability-based technique for treating uncertainty: **Sensitivity Analysis.** There are other techniques that are used to assess risks and uncertainty (e.g., the Mathematical/Analytical Technique, and various combinations of probability-based techniques, and break-even analysis, a non-probability-based technique), but those are not treated here.**

Expected Value (EV) Analysis

Expected Value Analysis provides a simple way of taking into account uncertainty about input values, but it does not provide an explicit measure of risk in the outcome. It is helpful in explaining and illustrating risk attitudes.

How to Calculate EV. An "expected value" is the sum of the products of the dollar value of alternative outcomes and their probabilities of occurrence. That is, where a_i $(i = 1, \ldots, n)$ indicates the value associated with alternative outcomes of a decision, and p_i indicates the probability of occurrence of alternative a_i, the expected value (EV) of the decision is calculated as follows:

$$EV = a_1p_1 + a_2p_2 + \cdots + a_np_n \quad (3.6)$$

Example of EV Analysis. The following simplified example illustrates the combining of expected value analysis and net present value analysis to support a purchase decision.

Assume that a not-for-profit organization must decide whether to buy a given piece of energy-saving equipment. Assume that the unit purchase price of the equipment is $100,000, the yearly operating cost is $5,000 (obtained by a fixed-price contract), and both costs are known with certainty. The annual energy cost savings, on the other hand, are uncertain, but can be estimated

*Material adapted from the author's contribution to the *Engineering Handbook,* ed. Richard Dorf, CRC Press, 1995.

**For a treatment of these other techniques see Ruegg and Marshall, 1990.

TABLE 3.1 Expected Value (EV) Example

Year	Equipment Purchase ($1000)	Operating Cost ($1000)	Energy Savings a_1 ($1000)	p_1	a_2 ($1000)	p_2	PV* Factor	PV ($1000)
0	−100	—	—	—	—	—	1.00	−100
1		−5	25	0.8	50	0.2	0.93	23
2		−5	30	0.8	60	0.2	0.86	27
3		−5	30	0.7	60	0.3	0.79	27
4		−5	30	0.6	60	0.4	0.74	27
5		−5	30	0.8	60	0.2	0.68	21
					Expected Net Present Value			25

* Present value calculations are based on a discount rate of 8%. Probabilities sum to 1.0 in a given year.

in probabilistic terms as shown in Table 3.1 in the columns headed a_1, p_1, a_2, and p_2. The present value calculations are also given in Table 3.1.

If the equipment decision were based only on net present value calculated with the "best-guess" energy savings (column a_1), the equipment purchase would be found to be uneconomic. But if the possibility of greater energy savings is taken into account by using the expected value of savings rather than the best guess, the conclusion is that over repeated applications the equipment is expected to be **cost-effective**. The expected net present value of the energy-saving equipment is $25,000 per unit.

Advantages and Disadvantages of the EV Technique. An advantage of the technique is that it predicts a value that tends to be closer to the actual value than a simple "best-guess" estimate over repeated instances of the same event, provided, of course, that the input probabilities can be estimated with some accuracy.

A disadvantage of the EV technique is that it expresses the outcome as a single-value measure, such that there is no explicit measure of risk. Another is that the estimated outcome is predicated on many replications of the event, with the expected value, in effect, a weighted average of the outcome over many like events. But the expected value is unlikely to occur for a single instance of an event. This is analogous to a single coin toss: the outcome will be either heads or tails, not the probabilistic-based weighted average of both.

Expected Value and Risk Attitude. Expected values are useful in explaining risk attitude. Risk attitude may be thought of as a decision maker's preference between taking a chance on an uncertain money payout of known probability versus accepting a sure money amount. Suppose, for example, a person were given a choice between accepting the outcome of a fair coin toss where heads means winning $10,000 and tails means losing $5,000 and accepting a certain cash amount of $2,000. Expected value analysis can be used to evaluate and compare the choices. In this case the expected value of the coin toss is $2,500, which is $500 more than the certain money amount. The "risk-neutral" decision maker will prefer the coin toss because of its higher expected value. The decision maker who prefers the $2,000 certain amount is demonstrating a "risk-averse" attitude. On the other hand, if the certain amount were raised to $3,000 and the first decision maker still preferred the coin toss, he or she would be demonstrating a "risk-taking" attitude. Such tradeoffs can be used to derive a "utility function" that represents a decision maker's risk attitude.

The risk attitude of a given decision maker is typically a function of the amount at risk. Many people who are risk averse when faced with the possibility of significant loss, become risk neutral, or even risk taking, when potential losses are small. Since decision makers vary substantially in their risk attitudes, there is a need to assess not only risk exposure (i.e., the degree of risk inherent in the decision) but also the risk attitude of the decision maker.

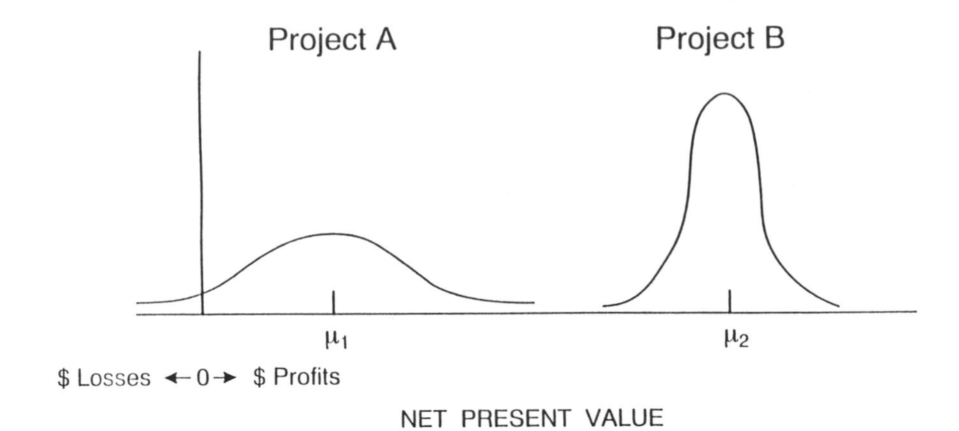

FIGURE 3.4 Stochastic dominance as demonstrated by mean-variance criterion.

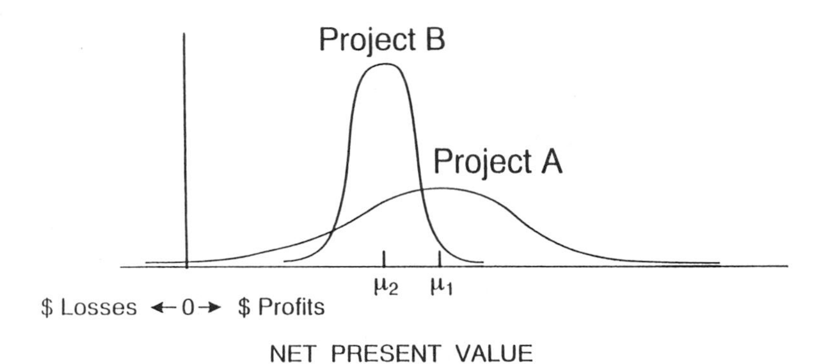

FIGURE 3.5 Inconclusive results from mean-variance criterion.

Mean-Variance Criterion (MVC) and Coefficient of Variation (CV)

These techniques can be useful in choosing among risky alternatives if the mean outcomes and standard deviations (variation from the mean) can be calculated.

Consider a choice between two projects—one with higher mean net benefits and a lower standard deviation than the other. This situation is illustrated in Figure 3.4. In this case, the project whose probability distribution is labeled B can be said to have stochastic dominance over the project labeled A. Project B is preferable to Project A both on grounds that its output is likely to be higher and that it entails less risk of loss. But what if Project A, the alternative with higher risk, has the higher mean net benefits, as illustrated in Figure 3.5? If this were the case, the MVC would provide inconclusive results.

When there is not stochastic dominance of one project over the other(s), it is helpful to compute the coefficient of variation (CV) to determine the relative risk of the alternative projects. The CV indicates which alternative has the lower risk per unit of project output. Risk-averse decision makers will prefer the alternative with the lower CV, other things being equal. The CV is calculated as follows:

$$CV = \sigma/\mu \tag{3.7}$$

where

CV = coefficient of variation
σ = standard deviation
μ = mean

TABLE 3.2 Risk-Adjusted Discount Rate (RADR) Example

Year	Costs ($M)	Revenue ($M)	PV Costs* ($M)	PV Revenue* ($M)	NPV ($M)
0	80		80	—	–80
1	5	20	4	17	13
2	5	20	4	14	10
3	5	20	4	12	8
4	5	20	3	10	7
5	5	20	3	9	6
6	5	20	3	7	4
7	5	20	2	6	4
			Total NPV		–28

* Costs are discounted with a discount rate of 12%; revenue with a discount rate of 18%.

The principal advantage of these techniques is that they provide quick, easy-to-calculate indications of the returns and risk exposure of one project relative to another. The principal disadvantage is that the MVC does not provide a clear indication of preference when the alternative with the higher mean output has the higher risk, or vice versa.

Risk-Adjusted Discount Rate (RADR) Technique

The RADR technique takes account of risk through the discount rate. If a project's benefit stream is riskier than that of the average project in the decision maker's portfolio, a higher than normal discount rate is used; if the benefit stream is less risky, a lower than normal discount rate is used. If costs are the source of the higher-than-average uncertainty, a lower than normal discount rate is used, and vice versa. The greater the variability in benefits or costs, the greater the adjustment in the discount rate.

The RADR is calculated as follows:

$$RADR = RFR + NRA + XRA \tag{3.8}$$

where

RADR = risk-adjusted discount rate
RFR = risk-free discount rate, generally set equal to the treasury bill rate
NRA = "normal" risk adjustment to account for the average level of risk encountered in the decision maker's operations
XRA = extra risk adjustment to account for risk greater or less than normal risk

An example of using the RADR technique is the following: a company is considering an investment in a new type of alternative energy system with high payoff potential and high risk on the benefits side. The discount rate normally used by the company to discount its costs, computed as the weighted marginal cost of capital, is 12%. The Treasury bill rate, taken as the risk-free rate, is 8%. The company uses a risk adjustment factor of 5% to account for the average level of risk encountered in its operations. This investment is twice as risky as the company's average investment, such that it will add an extra risk adjustment of 5%. Hence, the RADR is 18%. The projected cost and revenue streams and the discounted present values are shown in Table 3.2. Because net present value is estimated to be a loss of $28 million, the company would not accept the project.

Advantages of the RADR technique are that it provides a way to account for both *risk exposure* and *risk attitude*. Moreover, RADR does not require any additional steps for calculating net present value once a value of the RADR is established. The disadvantage is that it provides only an approximate adjustment. The value of the RADR is typically a rough estimate based on sorting investments into risk categories and adding a "fudge factor" to account for the decision maker's risk attitude. It generally is not a fine-tuned measure of the inherent risk associated with variations in cash flows. Further, it typically is biased toward investments with short payoffs because it applies a constant RADR over the entire time period even though risk may vary over time.

Certainty Equivalent (CE) Technique

The CE Technique adjusts investment cash flows by a factor that will convert the measure of economic worth to a "certainty equivalent" amount—the amount a decision maker will find equally acceptable to a given investment with an uncertain outcome. Central to the technique is the derivation of the Certainty Equivalent Factor (CEF), which is used to adjust net cash flows for uncertainty.

Risk exposure can be built into the CEF by establishing categories of risky investments for the decision maker's organization and linking the CEF to the coefficient of variation of the returns—greater variation translating into smaller CEF values. The procedure is as follows:

1. Divide the organization's portfolio of projects into risk categories. Examples of investment risk categories for a private utility company might be the following: low-risk investments—expansion of existing energy systems and equipment replacement; moderate-risk investments—adoption of new, conventional energy systems; and high-risk investments—investment in new alternative energy systems.
2. Estimate the coefficients of variation (see the section on the CV technique) for each investment-risk category (e.g., on the basis of historical risk-return data).
3. Assign CEFs by year according to the coefficients of variation, with the highest-risk projects being given the lowest CEFs. If the objective is to reflect only risk exposure, set the CEFs such that a risk-neutral decision maker will be indifferent between receiving the estimated certain amount and the uncertain investment. If the objective is to reflect risk attitude as well as risk exposure, set the CEFs such that the decision maker with his or her own risk preference will be indifferent.

To apply the technique, proceed with the following steps:

4. Select the measure of economic performance to be used—such as the measure of net present value (i.e., net benefits).
5. Estimate the net cash flows and decide in which investment-risk category the project in question fits.
6. Multiply the yearly net cash flow amounts by the appropriate CEFs.
7. Discount the adjusted yearly net cash flow amounts with a risk-free discount rate (a risk-free discount rate is used because the risk adjustment is accomplished by the CEFs).
8. Proceed with the remainder of the analysis in the conventional way.

In summary, the certainty equivalent net present value is calculated as follows:

$$\mathrm{NPV}_{\mathrm{CE}} = \sum_{t=0}^{N} \left[\left(\mathrm{CEF}_t \left(B_t - C_t \right) \right) / \left(1 + \mathrm{RFD} \right)^t \right], \quad (3.9)$$

where

$\mathrm{NPV}_{\mathrm{CE}}$ = net present value adjusted for uncertainty by the CE technique
B_t = estimated benefits in time period t
C_t = estimated costs in time period t
RFD = risk-free discount rate

TABLE 3.3 Certainty Equivalent (CE) Example (Investment-Risk Category: High-Risk — New-Alternative Energy System)

Yearly Net Cash Flow	($M)	CV	CEF	RFD Discount Factors*	NPV ($M)
1	−100	0.22	0.76	0.94	−71
2	−100	0.22	0.76	0.89	−68
3	20	0.22	0.76	0.84	13
4	30	0.22	0.76	0.79	18
5	45	0.22	0.76	0.75	26
6	65	0.22	0.76	0.70	35
7	65	0.22	0.76	0.67	33
8	65	0.22	0.76	0.63	31
9	50	0.22	0.76	0.59	22
10	50	0.22	0.76	0.56	21
				Total NPV	60

* The RFD is assumed equal to 6%.

Table 3.3 illustrates the use of this technique for adjusting net present value calculations for an investment in a new, high-risk alternative energy system. The CEF is set at 0.76 and is assumed to be constant with respect to time.

A principal advantage of the CE Technique is that it can be used to account for both risk exposure and risk attitude. Another is that it separates the adjustment of risk from discounting and makes it possible to make more precise risk adjustments over time. A major disadvantage is that the estimation of CEF is only approximate.

Simulation Technique

Simulation entails the iterative calculation of the measure of economic worth from probability functions of the input variables. The results are expressed as a probability density function and as a cumulative distribution function. The technique thereby enables explicit measures of risk exposure to be calculated. One of the economic evaluation methods treated earlier is used to calculate economic worth; a computer is employed to sample repeatedly—hundreds of times—and make the calculations.

Simulation can be performed by the following steps:

1. Express variable inputs as probability functions. Where there are interdependencies among input values, multiple probability density functions, tied to one another, may be needed.
2. For each input for which there is a probability function, draw randomly an input value; for each input for which there is only a single value, take that value for calculations.
3. Use the input values to calculate the economic measure of worth and record the results.
4. If inputs are interdependent, such that input X is a function of input Y, first draw the value of Y, then draw randomly from the X values that correspond to the value of Y.
5. Repeat the process many times until the number of results is sufficient to construct a probability density function and a cumulative distribution function.
6. Construct the probability density function and cumulative distribution function for the economic measure of worth, and perform statistical analysis of the variability.

The strong advantage of the technique is that it expresses the results in probabilistic terms, thereby providing explicit assessment of risk exposure. A disadvantage is that it does not explicitly treat risk attitude; however, by providing a clear measure of risk exposure, it facilitates the implicit incorporation of risk attitude in the decision. The necessity of expressing inputs in probabilistic terms and the extensive calculations are also often considered disadvantages.

Decision Analysis

Decision Analysis is a versatile technique that enables both risk exposure and risk attitude to be taken into account in the economic assessment. It diagrams possible choices, costs, benefits, and probabilities for a given decision problem in "decision trees," which are useful in understanding the possible choices and outcomes.

Although it is not possible to capture the richness of this technique in a brief overview, a simple decision tree, shown in Figure 3.6, is discussed to give a sense of how the technique is used. The decision problem is whether to lease or build a facility. The decision must be made now, based on uncertain data. The decision tree helps to structure and analyze the problem. The tree is constructed left to right and analyzed right to left. The tree starts with a box representing a decision juncture or node—in this case, whether to lease or build a facility. The line segments branching from the box represent the two alternative paths: the upper one the lease decision and the lower one the build decision.

An advantage of this technique is that it helps to understand the problem and to compare alternative solutions. Another advantage is that, in addition to treating risk exposure, it can also accommodate risk attitude by converting benefits and costs to utility values (not addressed here). A disadvantage is that the technique as typically applied does not provide an explicit measure of the variability of the outcome.

Sensitivity Analysis

A technique for taking into account uncertainty that does not require estimates of probabilities is sensitivity analysis. It tests the sensitivity of economic performance to alternative values of key factors about which there is uncertainty. Although sensitivity analysis does not provide a single answer in economic terms, it does show decision makers how the economic viability of a conservation project changes as fuel prices, discount rates, time horizons, and other critical factors vary.

Figure 3.7 illustrates the sensitivity of fuel savings realized by a solar energy heating system to three critical factors: time horizons (zero to 25 years), discount rates (D equals 0, 5, 10 and 15%), and energy escalation rates (E equals 0, 5, 10, and 15%). The present value of savings is based on yearly fuel savings valued initially at $1,000.

Note that, other things being equal, the present value of savings increase with time, but less with higher discount rates and more with higher escalation rates. The huge impact of fuel price escalation is most apparent when comparing the top line of the graph ($D = 0.10$, $E = 0.15$) with the line next to the bottom ($D = 0.10$, $E = 0$). The present value of savings at the end of 25 years is approximately $50,000 with a fuel escalation rate of 15%, and only about $8,000 with no escalation, other things equal. Whereas the quantity of energy saved is the same, the dollar value varies widely, depending on the escalation rate.

This example graphically illustrates a situation frequently encountered in the economic justification of energy conservation and solar energy projects: The major savings in energy costs, and thus the bulk of the benefits, accrue in the later years of the project and are highly sensitive to both the assumed rate of fuel-cost escalation and the discount rate. The two rates set equal will be offsetting.

3.4 Building Blocks of Evaluation

Beyond the formula for the basic evaluation methods and risk assessment techniques, the practitioner needs to know some of the "nuts-and-bolts" of carrying out an economic analysis. He or she needs to know how to structure the evaluation process; how to choose a method of evaluation; how to estimate dollar costs and benefits; how to perform discounting operations; how to select a study period; how to choose a discount rate; how to adjust for inflation; how to take into account taxes; how to treat residual values; and how to reflect assumptions and constraints, among other things. This section provides brief guidelines for these topics.

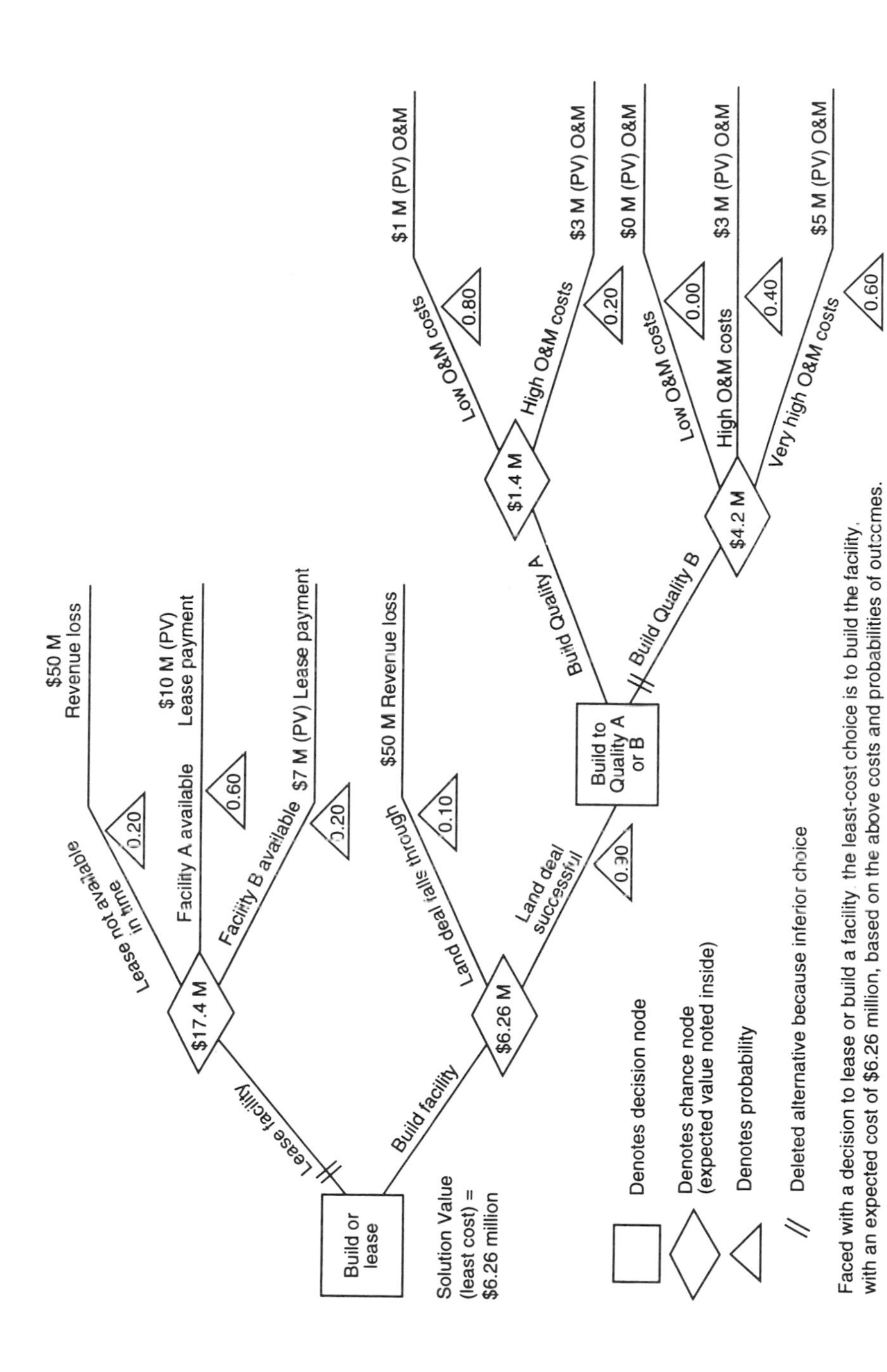

FIGURE 3.6 Decision tree: build versus lease.

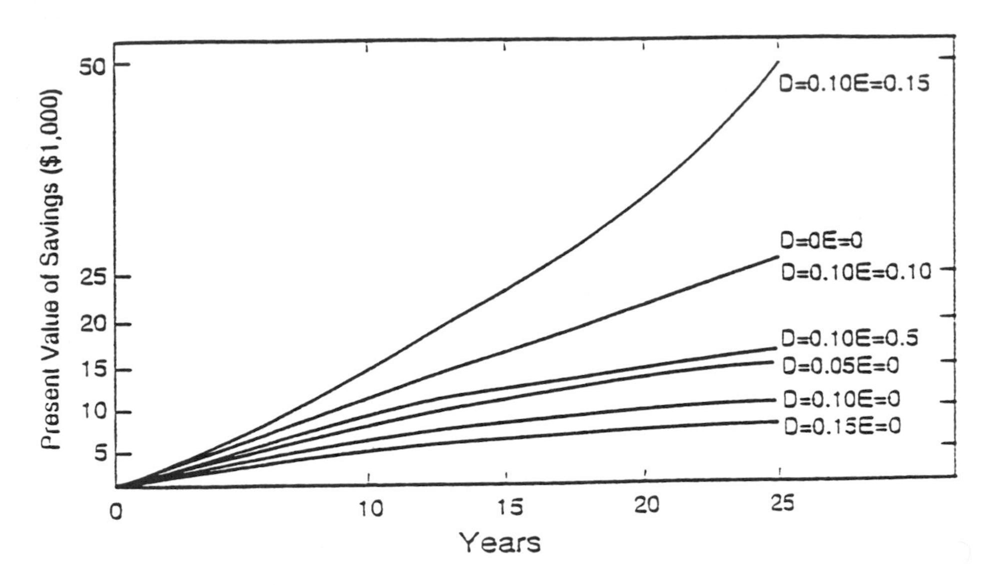

FIGURE 3.7 Sensitivity of present value energy savings to time horizons, discount rates, and energy price escalation rates.

Structuring the Evaluation Process and Selecting a Method of Evaluation

A good starting point for the evaluation process is to define the problem and the objective. Identify any constraints to the solution and possible alternatives. Consider if the best solution is obvious, or if economic analysis and risk assessment are needed to help make the decision. Select an appropriate method of evaluation and a risk assessment technique. Compile the necessary data and determine what assumptions are to be made. Apply the appropriate formula(s) to compute a measure of economic performance under risk. Compare alternatives and make the decision, taking into account any incommensurable effects that are not included in the dollar benefits and costs. Take into account the risk attitude of the decision maker if it is relevant.

Although the five evaluation methods given earlier are similar, they are also sufficiently different that they are not always equally suitable for evaluating all types of energy investment decisions. For some types of decisions, the choice of method is more critical than for others. The choice is usually not critical, for example, in simple "accept-reject" investment decisions where the problem is to determine if a given conservation or renewable energy investment will save more than it costs. Any of the five techniques will usually work in this case provided the correct criterion is used.

Accept/Reject Criteria:

- LCC technique—LCC must be lower as a result of the energy conservation or renewable energy investment than without it.
- NB (NS) technique—NB (NS) must be positive as a result of the investment.
- B/C (SIR) technique—B/C (SIR) must be greater than one.
- IRR technique—the IRR must be greater than the investor's minimum attractive rate of return.
- DPB technique—the number of years to achieve DPB must be less than the project life or the investor's time horizon, and there are no cash flows after payback is achieved that would reverse payback.

The choice of evaluation methods, on the other hand, is usually important for determining the priority to assign to conservation or renewable energy investments competing for a limited budget.

For this purpose, either the B/C (SIR) ratio technique or the IRR technique is recommended. A ratio or rate-of-return measure is preferred because they reflect the return per dollar spent and can be used to rank investment projects in order of their return. This facilitates selecting a combination of investment projects that will result in the largest total return for a given conservation budget.

In the case where a fast turnaround on investment funds is required, DPB is recommended. The other methods, although more comprehensive and accurate for measuring an investment's lifetime profitability, do not indicate the time required for recouping the investment funds.

For determining the economically efficient size of a conservation investment, any of the methods (except DPB) will usually work, provided they are used correctly. However, the LCC or NB (NS) method is usually recommended for this purpose, because they are less likely to be misapplied. As long as LCC falls or NB (NS) rises as more conservation or renewable energy is added, it pays to increase the investment. The B/C (SIR) and IRR methods also can be used for efficient sizing of an investment, but only if they are applied to incremental investments, rather than to the total investment.

Discounting

Some or all of investment costs in energy conservation or renewable energy systems are incurred near the beginning of the project and are treated as "first costs." The benefits, on the other hand, typically accrue over the life span of the project in the form of yearly energy savings. To compare benefits and costs that accrue at different points in time, it is necessary to put all cash flows on a time-equivalent basis. The method for converting cash flows to a time-equivalent basis is often called "discounting."

The value of money is time dependent for two reasons: first, inflation or deflation can change the buying power of the dollar, and second, money can be invested over time to yield a return over and above inflation. For these two reasons, a given dollar amount today will be worth more than that same dollar amount a year later. For example, suppose a person were able to earn a maximum of 10% interest per annum risk-free. He or she would require $1.10 a year from now in order to be willing to forego having $1.00 today. If the person were indifferent between $1.00 today and $1.10 a year from now, then the 10% rate of interest would indicate that person's time preference for money. The higher the time preference, the higher the rate of interest required to make future cash flows equal to a given value today. The rate of interest for which an investor feels adequately compensated for trading money now for money in the future is the appropriate rate to use for converting present sums to future equivalent sums and future sums to present equivalent sums (i.e., the rate for discounting cash flows for that particular investor). This rate is often called the "discount rate."

To evaluate correctly the economic efficiency of an energy conservation or renewable energy investment, it is necessary to convert the various expenditures and savings that accrue over time to a lump-sum, time-equivalent value in some base year (usually the present), or to annual values. The remainder of this section illustrates how to discount various types of cash flows.

Discounting is illustrated by Figure 3.8 in a problem of installing, maintaining, and operating a heat pump. Discounting has also been built into Eqs. (3.2) to (3.5). The calculations would be required, for example, for comparing the life-cycle costs of a heat pump to those of an alternative heating/cooling system.

The life-cycle cost calculations are shown for two reference times. The first is the present, and it is therefore called a present value. The second is based on a yearly time scale and is called an annual value. These two reference points are the most common in economic evaluations of investments. When the evaluation methods are derived properly, each time basis will give the same relative ranking of conservation investment priorities.

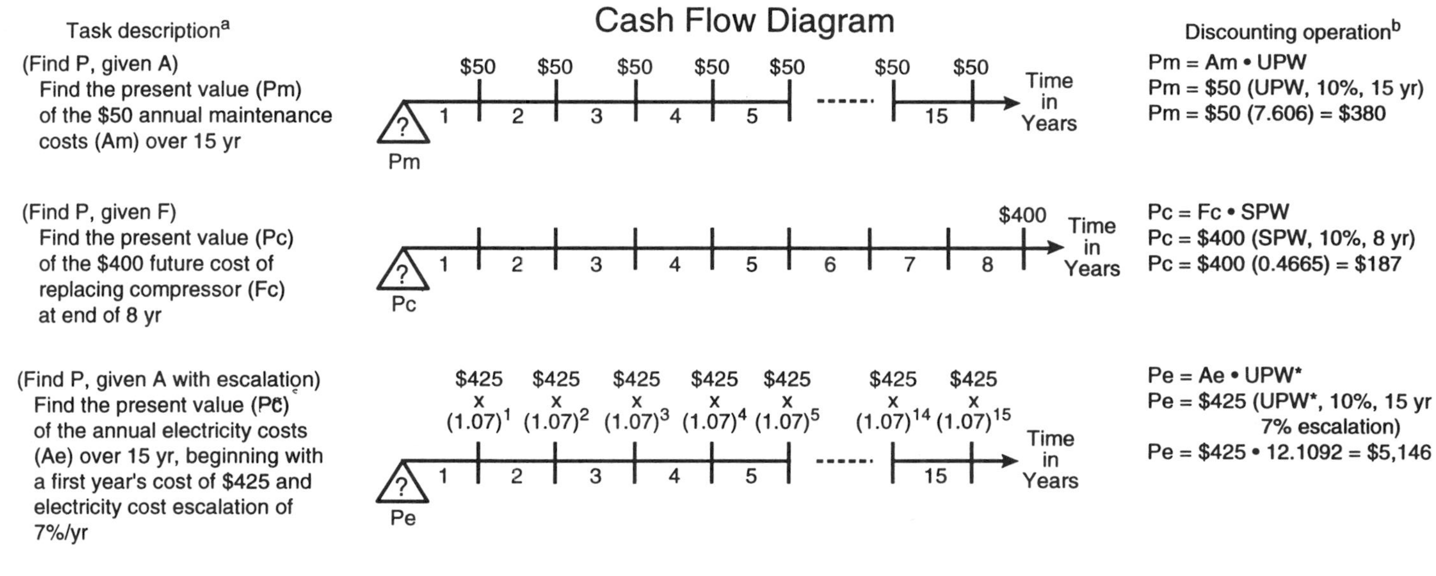

FIGURE 3.8 Determining present value life-cycle costs: heat pump example.

The assumptions for the heat pump problem—which are given only for the sake of illustration and not to suggest actual prices—are as follows:

1. The residential heat pump (not including the ducting) costs $1,500 to purchase and install.
2. The heat pump has a useful life of 15 years.
3. The system has annual maintenance costs of $50 every year over its useful life, fixed by contractual agreement.
4. A compressor replacement is required in the eighth year at a cost of $400.
5. The yearly electricity cost for heating and cooling is $425, evaluated at the outset, and increased at a rate of 7% per annum due to rising electricity prices.
6. The discount rate (a nominal rate that includes an inflation adjustment) is 10%.
7. No salvage value is expected at the end of 15 years.

The life-cycle costs in the sample problem are derived only for the heat pump and not for alternative heating/cooling systems. Hence, no attempt is made to compare alternative systems in this discounting example. To do so would require similar calculations of life-cycle costs for other types of heating/cooling systems. Total costs of a heat pump system include costs of purchase and installation, maintenance, replacements, and electricity for operation. Using first the present as the base-time reference point, we need to convert each of these costs to the present before summing them. If we assume that the purchase and installation costs occur at the base reference point, the present, the $1,500 is already in present value terms.

Figure 3.8 illustrates how to convert the other cash flows to present values. The first task is to convert the stream of annual maintenance costs to present value. The maintenance costs, as shown in the cash flow diagram of Figure 3.8, are $50 per year, measured in current dollars (i.e., dollars of the years in which they occur). The triangle indicates the value to be found. We follow the practice here of compounding interest at the end of each year. The present refers to the beginning of year one.

The discounting operation for calculating the present value of maintenance costs (last column of Figure 3.8) is to multiply the annual maintenance costs times the uniform present worth (UPW) factor. The UPW is a multiplicative factor computed from the formula given in Table 3.4, or taken from a look-up table of factors that have been published in many economics textbooks. UPW factors make it easy to calculate the present values of a uniform series of annual values. For a discount rate of 10% and a time period of 15 years, the UPW factor is 7.606. Multiplying this factor by $50 gives a present value maintenance cost equal to $380. Note that the $380 present value of $50 per year incurred in each of 15 years is much less than simply adding $50 for 15 years (i.e., $750). Discounting is required to achieve correct statements of costs and benefits over time.

The second step is to convert the one-time future cost of compressor replacement, $400, to its present value. The operation for calculating the present value of compressor replacement is to multiply the **future value** of the compressor replacement times the single-payment present worth factor (SPW), which can be calculated from the formula in Table 3.4, or taken from a discount factor look-up table. For a discount rate of 10% and a time period of 15 years, the SPW factor is 0.4665. Multiplying this factor by $400 gives a present value cost of the compressor replacement of $187, as shown in the last column of Figure 3.8. Again note that discounting makes a significant difference in the measure of costs. Failing to discount the $400 would result in an overestimate of cost in this case of $213.

The third step is to convert the annual electricity costs for heating and cooling to present value. A year's electricity costs, evaluated at the time of installation of the heat pump, are assumed to be $425. Electricity prices, for purposes of illustration, are assumed to rise at a rate of 7% per annum. This is reflected in Table 3.4 by multiplying $425 times $(1.07)^t$ where $t = 1, 2, \ldots, 15$. The electricity cost at the end of the fourth year, for example, is $\$425(1.07)^4 = \$55$.

The discounting operation for finding the present value of all electricity costs (shown in Figure 3.8) is to multiply the initial, yearly electricity costs times the appropriate UPW* factor. (An

TABLE 3.4 Discount Formulas

Standard Nomenclature	Use When	Standard Notation	Algebraic Form
Single Compound Amount Formula	Given *P*; to find *F*	(SCA),*i*%,*N*)	$F = P(1+i)^N$
Single Present Worth Formula	Given *F*; to find *P*	(SPW,*i*%,*N*)	$P = F\frac{1}{(1+i)^N}$
Uniform Compound Amount Formula	Given *A*; to find *F*	(UCA,*i*%,*N*)	$F = A\frac{(1+i)^N - 1}{i}$
Uniform Sinking Fund Formula	Given *F*; to find *A*	(USF,*i*%,*N*)	$A = F\frac{i}{(1+i)^N - 1}$
Uniform Capital Recovery Formula	Given *P*; to find *A*	(UCR,*i*%,*N*	$A = P\frac{i(1+i)^N}{(1+i)^N - 1}$
Uniform Present Worth Formula	Given *A*; to find *P*	(UPW,*i*%,*N*)	$P = A\frac{(1+i)^N - 1}{i(1+i)^N}$
Uniform Present Worth Modified	Given *A* escalating at a rate *e*; to find *P*	(UPW*,*i*%,*e*,*N*)	$P = A\frac{(1+e)\left[1-\left(\frac{1+e}{1+i}\right)^N\right]}{(i-e)}$

Note: P = a present sum of money; F = a future sum of money, equivalent to P at the end of N periods of time at an interest of i; i - an interest rate; N = number of interest periods; A = an end-of-period payment (or receipt) in a uniform series of payments (or receipts) over N periods at i interest rate, usually annually; e = a rate of escalation in A in each of N periods.

asterisk following UPW denotes that a term for price escalation is included.) The UPW or UPW* discount formulas in Table 3.4 can also be used to obtain present values from annual costs or multiplicative discount factors from look-up tables can be used. For a period of 15 years, a discount rate of 10%, and an escalation rate of 7%, the UPW* factor is 12.1092. Multiplying the factor by $425 gives a present value of electricity costs of $5,146. Note once again that failing to discount (i.e., simply adding annual electricity expenses in current prices) would overestimate costs by $1,229 ($6,376 – $5,146). Discounting with a UPW factor that does not incorporate energy price escalation would underestimate costs by $1,913 ($5,146 – $3,233).

The final operation described in Figure 3.8 is to sum purchase and installation cost and the present values of maintenance, compressor replacement, and electricity costs. Total life-cycle costs of the heat pump in present value terms are $7,213. This is one of the amounts that a designer would need for comparing the cost effectiveness of heat pumps to alternative heating/cooling systems.

Only one discounting operation is required for converting the present value costs of the heat pump to annual value terms. The total present value amount is converted to the total annual value simply by multiplying it by the uniform capital recovery factor (UCR)—in this case the UCR for 10% and 15 years. The UCR factor, calculated with the UCR formula found in Table 3.4, is 0.13147. Multiplying this factor by the total present value of $7,213 gives the cost of the heat pump as $948 in annual value terms. The two figures—$7,213 and $948 per year—are time-equivalent values, made consistent through the discounting.

Figure 3.8 provides a model for the designer who must calculate present values from all kinds of benefit or cost streams. Most distributions of values occurring in future years can be handled with the SPW, the UPW, or the UPW* factors.

Discount Rate

Of the various factors affecting the net benefits of conservation and renewable energy investments, the discount rate is one of the most dramatic. A project that appears economic at one discount rate will often appear uneconomic at another rate. For example, a project that yields net savings at a 6% discount rate might yield net losses if evaluated with a 7% rate.

As the discount rate is increased, the present value of any future stream of costs or benefits is going to become smaller. High discount rates tend to favor projects with quick payoffs over projects with benefits deferred further in the future.

The discount rate should be set equal to the rate of return available on the next best investment opportunity of similar risk to the project in question, that is, it should indicate the opportunity cost of the investor. The discount rate may be formulated as a "real rate" exclusive of general price inflation or as a "nominal rate" inclusive of inflation. The former should be used to discount cash flows that are stated in **constant dollars**. The latter should be used to discount cash flows stated in **current dollars**.

Inflation

Inflation is a rise in the general price level. Although all prices cannot be expected to rise or fall at the same time and at the same rate, it is possible to measure the average price increases in categories of goods and services.

In evaluating the economic performance of energy conservation and energy investments, a recommended approach is to remove price inflation (as indicated by average price increases in the economy) from the estimates of benefits and costs, and to perform the evaluation in constant dollar values. This does not mean that future amounts will necessarily remain unchanged; rather, only changes above or below the rate of general price inflation will be registered.

Since dollar benefits from conservation and renewable energy vary directly with fuel prices, assumptions regarding the change in fuel prices over time tend to have a significant impact on the predicted benefits of a project. Among the energy price projections that are available are those provided by the U.S. Department of Energy (DOE); see Chapter 7. The National Institute of Standards and Technology (NIST), acting as the technical arm of DOE, has published annual tables of UPW* factors incorporating DOE projections of energy prices by sector of the economy, by type of energy, and by region of the country (Petersen, 1994). These projections are differentials relative to the rate of general price inflation.

Alternatively, inflation may be handled in analyses by including it in all estimates of costs and benefits, but in this case, discounting is performed with a "nominal" discount rate that reflects both the real changes in the value of money and the expected inflation rate.

Time Horizon

The time horizon is the period of analysis. For conservation and renewable investments it is the length of time over which costs and benefits are calculated. The time horizon is often called the "**life cycle**" in life-cycle cost analysis.

The selection of a time horizon is usually based on the personal time perspective of the investor. There is no fixed time horizon that will be appropriate for all investment projects. When the investor plans to hold a facility indefinitely, the life of the building, energy system, or conservation investment is often used as the time horizon.

Two common concepts of life are the "useful life" and "**economic life.**" The useful life is the period over which the investment has some value; that is, the investment continues to conserve or provide energy during this period. Economic life is the period during which the conservation investment in question is the least-cost way of meeting the requirement. Often economic life is shorter than useful life. A common time horizon should be used for comparing alternatives.

The actual selection of a time horizon will depend on the objectives and perspective of the decision maker. A speculative builder who plans to build for immediate sale, for example, may view the relevant time horizon as that short period of ownership from planning and acquisition of property to the first sale of the building. Although the useful life of a solar domestic hot water heating system, for example, might be 20 years, the speculative builder might operate on the basis of a 2–year time horizon if the property is expected to change hands within that period. Only if the speculator expects to gain the benefit of those energy savings through a higher selling price for the building, will the higher first cost of the solar energy investment likely be economic.

If an analyst is performing an economic analysis for a particular client, that client's time horizon is the relevant period. If an analyst is performing an analysis in support of public investment or policy decisions, the life of the building or system is typically the appropriate time horizon.

Taxes and Subsidies

Taxes and subsidies should be taken into account in economic evaluations because they may affect the economic viability of an investment, the return to the investor, and the optimal size of the investment. Taxes, which may have positive and negative effects, include, but are not limited to, income taxes, sales taxes, property taxes, excise takes, capital gain taxes, depreciation recapture taxes, tax deductions, and tax credits.

Subsidies are inducements for a particular type of behavior or action. They include grants—cash subsidies of specified amounts; government cost sharing; loan interest reductions, and tax-related subsidies. Income tax credits for conservation or renewable energy expenditures provide a subsidy by allowing specific deductions from the investor's tax liability. Property tax exemptions eliminate the property taxes that would otherwise add to annual costs. Income tax deductions for energy conservation or renewable energy expenses reduce annual tax costs. The imposition of higher taxes on nonrenewable energy sources raises their prices and encourages conservation and renewable energy investments.

Residual Values

Residual values may arise from salvage (net of disposal costs) at the end of the life of systems and components, from reuse values when the purpose is changed, and from remaining value when assets are sold prior to the end of their lives. The present value of residuals can generally be expected to decrease, other things equal, as (1) the discount rate rises, (2) the building or equipment deteriorates, and (3) the time horizon lengthens.

To estimate the residual value of energy conservation or renewable energy systems and components, it is helpful to consider the amount that can be added to the selling price of a building because of those systems. It might be assumed that a building buyer will be willing to pay an additional amount equal to the capitalized value of energy savings over the remaining life of the conservation or renewable investment. If the time horizon is the same as the useful life, there will be no residual value.

Summary

There are multiple methods of economic performance and multiple techniques of risk analysis that can be selected and combined to improve decisions in energy conservation and renewable energy investments. Economic performance can be stated in a variety of ways depending on the problem and preferences of the decision maker: as dollar net benefits or net savings, as a schedule of life-cycle costs, as a rate of return, as years to payback, or as a ratio. To reflect the reality that most decisions are made under conditions of uncertainty, risk assessment techniques can be used to reflect the risk exposure of the project and the risk attitude of the decision maker. Rather than expressing results in single, deterministic terms, they can be expressed in probabilistic terms, thereby revealing the likelihood that the outcome will differ from the best-guess answer. These methods and techniques can be used to decide whether or not to invest in a given energy conservation or alternative energy system; to determine which system design or size is economically efficient; to find the combination of components and systems that are expected to be cost-effective; to estimate how long before a project will break even; and to decide which energy-related investments are likely to provide the highest rate of return to the investor. They support the goal of achieving economic efficiency—which may differ from technical efficiency.

3.5 Defining Terms

Benefit/Cost (B/C) or Savings-to-Investment (SIR) Ratio—A method of measuring the economic performance of alternatives by dividing present value benefits (savings) by present value costs.

Constant Dollars—Values expressed in terms of the general purchasing power of the dollar in a base year. Constant dollars do not reflect price inflation or deflation.

Cost Effective—The least-cost alternative for achieving a given level of performance.

Current Dollars—Values expressed in terms of actual prices of each year (i.e., current dollars reflect price inflation or deflation).

Discount Rate—Based on the opportunity cost of capital, this minimum acceptable rate of return is used to convert benefits and costs occurring at different times to their equivalent values at a common time.

Discounted Payback Period—The time required for the discounted annual net benefits derived from an investment to pay back the initial investment.

Discounting—A technique for adjusting the opportunity cost of capital by converting cash flows that occur over time to equivalent amounts at a common point in time.

Economic Efficiency—Maximizing net benefits or minimizing costs for a given level of benefits (i.e., "getting the most for your money").

Economic Life—That period of time over which an investment is considered to be the least-cost alternative for meeting a particular objective.

Future Value (Worth)—The value of a dollar amount at some point in the future, taking into account the opportunity cost of capital.

Internal Rate of Return—A method of measuring the economic performance of an investment by solving for the interest rate that will equate total discounted benefits with total discounted costs and then comparing that solution rate with the minimum acceptable rate of return on an investment of comparable risk.

Investment Costs—The sum of the planning, design, and construction costs necessary to obtain an asset.

Life Cycle—The period of time between the starting point and cutoff date for analysis, over which the costs and benefits of an asset or project are incurred.

Life-Cycle Cost—The total of all relevant costs associated with an asset or project over the time horizon.

Net Benefits—Present value benefits minus present value costs.

Present Value (Worth)—Past, present, or future case flows all expressed as a lump sum amount as of the present time, taking into account the time value of money.

Risk Assessment—As applied to economic decisions, the body of theory and practice that helps decision makers assess their risk exposures and risk attitudes in order to increase the probability that they will make economic choices that are best for them.

Risk Attitude—The willingness of decision makers to take chances on investments of uncertain, but predictable, outcome. Risk attitudes may be classified as risk averse, risk neutral, and risk taking.

Risk Exposure—The probability of investing in a project whose economic outcome is less favorable than what is considered economically desirable.

Sensitivity Analysis—A non-probability-based technique for reflecting uncertainty that entails testing the outcome of an investment by altering one or more system parameters from the initially assumed values.

Time Value of Money—The willingness of people to trade present dollar amounts for future dollar amounts.

Uncertainty—As used in the context of this chapter, a state of not being certain about the values of variable inputs to an economic analysis.

References

Marshall, H.E. and Ruegg, R.T. 1980. "Principles of Economics Applied to Investments in Energy Conservation and Solar Energy Systems." In *Economics of Solar Energy and Conservation Systems,* ed. F. Kreith and R. West, pp. 123–173. CRC Press, Boca Raton, FL.

Marshall, H.E. and Ruegg, R.T. 1980. *Simplified Energy Design Economics,* ed. F. Wilson. National Bureau of Standards Special Publication 544.

Petersen, S.R. 1994. *Energy Price Indexes and Discount Factors for Life-Cycle Cost Analysis, 1995.* National Institute of Standards and Technology Report. NISTIR 85-3273-9.

Ruegg, R.T. 1991. The Internal Rate of Return and the Reinvestment Rate Controversy: An Attempt at Clarification. In *Persistent Problems in Investment Appraisal,* ed. J.B. Weaver and H.C. Thorne, AICE Symposium Series, 285(87):52–61.

Ruegg, R.T. and Marshall, H.E. 1990. *Building Economics: Theory and Practice,* Chapman and Hall, New York, NY.

Ruegg, R.T. and Short, W. 1988. Economic Methods. In *Economic Analysis of Solar Thermal Energy Systems,* ed. R. West and F. Kreith, pp. 18–83. The MIT Press, Cambridge, MA.

For Further Information

There are a number of computer programs on the market for applying risk analysis techniques. An overview of a selection of software is provided by Ruegg and Marshall, 1990, Appendix D; however, it should be recognized that any assessment of software is quickly dated.

For video tutorials on life-cycle cost analysis, risk analysis, and choosing among economic evaluation methods, see the video training series, *Least-Cost Energy Decisions for Buildings,* Parts I, II, and III, respectively. The series was produced by the National Institute of Standards and Technology, and is available in VHS, with workbooks, through Video Transfer, Inc., 5709-B Arundel Avenue, Rockville, MD 20852. (301–881–0270).

4

Electric-Utility Integrated Resource Planning: Practice and Possibilities

Eric Hirst
Oak Ridge National Laboratory

Author's Note—The electricity industry has changed dramatically since I completed this chapter in mid-1995. As the structure, operation, and regulation of the U.S. electricity industry evolves, so too has my thinking about resource planning. Therefore, the current chapter should be read primarily as an archival piece that explains how electric utilities and their regulators met customer electric energy-service needs from the mid-1980s through the mid-1990s. See Hirst (1996) for a discussion of my perspective on IRP as of mid-1996. Recognize, however, that resource planning within the electricity sector will continue to evolve.

Abstract—Integrated resource planning helps utilities and state regulatory commissions consistently assess a broad range of demand and supply resources to meet customer energy-service needs cost-effectively. Key characteristics of this planning approach include explicit and fair treatment of a wide variety of demand and supply options, consideration of the environmental and other social costs of providing energy services, public participation in the development of the resource plan, and analysis of the uncertainties associated with different external factors and resource options.

This chapter presents suggestions to utilities on how to conduct such planning and what to include in their resource-planning reports. The chapter also discusses the roles that state regulatory commissions play in requiring utilities to develop resource plans, in requiring utilities to involve outsiders in plan preparation, and in regulatory review and approval of these plans. Because the electricity industry is undergoing substantial changes, this chapter also speculates on the possible future roles and activities of utilities and regulatory commissions with respect to resource planning.

Introduction

During the past several years, more state public utility commissions (PUCs) have adopted rules that require electric utilities to prepare integrated resource plans. Both because such planning represents good business practice and because of these rules, more and more utilities throughout the U.S. now prepare such plans on a regular basis. This chapter offers guidelines to utilities on how to conduct the analyses that support integrated resource planning (IRP) and what to present in the IRP report. See Goldman et al. (1993) for similar information on resource planning for gas utilities.

The next two sections provide background on the electric-utility industry and explain the basics of IRP. Sections 4.3 through 4.7 discuss technical aspects of IRP, including load forecasting, assessment of supply and demand resources, resource integration, environmental costs, and uncertainty analysis. Sections 4.8 through 4.11 discuss short-term action plans, public participation in IRP development, and the final report on a utility's resource plan (Hirst 1992). Section 4.12 describes the roles that PUCs can play to encourage good planning. Finally, Section 4.13 comments on the changes that might take place, both within utilities and PUCs, as the electricity industry changes during the next few years.

4.1 Historical Background

Until the early 1970s, resource planning at electric utilities was straightforward. Electricity demand was growing steadily, and technological developments were lowering electricity prices in both real and nominal terms. Since then, the utility industry has been shaken by wrenching changes. Perhaps most important, the era of declining costs ended and was replaced by one of increasing costs during the 1970s and early 1980s. As a consequence, proceedings before PUCs, which had been largely routine during prior decades, became stormy and contentious. Utilities argued for substantial rate increases to pay for escalating capital and fuel costs, while consumer groups opposed these frequent and large price increases.

Inflation and the sudden increase in oil prices (and, as a consequence, other fossil-fuel prices) after the 1973–1974 oil embargo had major effects on electricity prices (Energy Information Administration 1993). Many utilities found it necessary to petition their PUCs repeatedly for price increases. Opposition, especially from newly formed citizens' groups, became intense as more groups learned to use PUC proceedings to pursue their objectives (e.g., protection of low-income customers, protection of the physical environment, or opposition to nuclear plants). The second oil-price increase in 1979–1980 reinforced pressures on PUCs.

One consequence of these changes was that PUCs found it necessary to delve more deeply into the details of utility management. This, in turn, made it easier for intervenor groups to challenge utility decisions. The increasingly complex and adversarial PUC proceedings led to calls for reform of utility regulation.

During the past few years, real electricity prices have stabilized or declined but utilities continue to face fundamental changes. These changes include (1) deregulation of electricity generation; (2) greater access by utilities and others to the transmission systems of other utilities; (3) competition for retail customers from other fuels, self-generation, and from other electricity suppliers; (4) reductions in the costs of renewable resources; (5) growing use of demand-side management (DSM) programs as capacity and energy resources; (6) increased concern with the environmental consequences of electricity production; (7) growing public opposition to construction of power plants and transmission lines; and (8) uncertainty about future load growth, fossil-fuel prices and availability, the costs and construction times of facilities needed to meet future energy needs, and possible regulatory delays associated with permission to construct these facilities.

Utilities are also concerned about the low cost at which independent power producers (IPPs) can construct and operate new power plants (especially cogeneration and gas-fired units). Utility prices reflect each utility's historical costs, including the costs of any excess capacity; high equity capitalization rates; social programs like low-income weatherization; and research and development. Because IPPs typically do not face these costs, they can often underprice traditional utilities. Indeed, IPP prices reflect primarily current fuel costs and the latest generation technologies. Recently, IPP costs have been lower than utility historical costs, especially for utilities with expensive nuclear plants in their rate bases. This gap between the incremental cost of power from IPPs and the historical utility cost presents a strong incentive to large electricity consumers and resellers (e.g., municipal, cooperative, and investor-owned distribution utilities) to buy from IPPs. Similar cost gaps exist between vertically-integrated utilities that have no excess capacity or cost-overrun plants and utilities with one or both of these problems.

As alternative supply and demand options became more viable, many people considered the traditional approach to utility planning, with its narrow focus on utility-built power plants and its inadequate treatment of uncertainty, no longer adequate. By the mid-1980s many people were calling for a new paradigm for utility resource planning, one that accounted for the availability of DSM and renewable-energy technologies, the changing nature of the utility industry, the public's concern with environmental quality, and the increase in the number of parties interested in utility resource acquisitions. In addition, a fundamental truth was recognized, that customers buy electricity not as an end in itself but as a means to provide energy services, such as comfortable temperatures, light, and motive power. This search for a new planning paradigm produced integrated resource planning (Cavanagh 1986; Hirst 1992; Bauer and Eto 1992).

However, the call for IRP was not the only reform being considered for the utility industry. The argument that economies of scale in electricity generation could be captured by *competitive* bulk-power markets as well as by monopoly utilities was gaining acceptance. The excess generating capacity that existed during the 1980s led to vigorous trading of power among utilities. The efficiency gains of such trades showed that the transmission system could be used to support competitive bulk-power markets.

During the late 1980s and early 1990s, regulators began implementing both models. State regulators led in the implementation of IRP and the Federal Energy Regulatory Commission (FERC) led in encouraging competitive wholesale power markets. These two futures conflict in some ways, but they are clearly not contradictory. Indeed, the Energy Policy Act of 1992 encouraged both futures. As a result, utilities and their regulators will need to identify IRP activities that are consistent with evolving competition.

4.2 IRP Basics

IRP is a process by which utilities and PUCs can consistently assess various demand and supply resources to meet customer energy-service needs at the lowest economic or societal cost. IRP involves deliberations among utility planners and executives, PUCs, customers, and other interested citizen groups. These deliberations are intended to lead to the development of a plan that will provide reliable and low-cost electric-energy services to customers, financial stability for the utility, a reasonable return on investment for investors, and protection of the environment. (IRP practice may differ from its principles. In particular, some utilities favor IRP principles but oppose what they consider to be "regulatory IRP," a process they believe is cumbersome, expensive, and time consuming.)

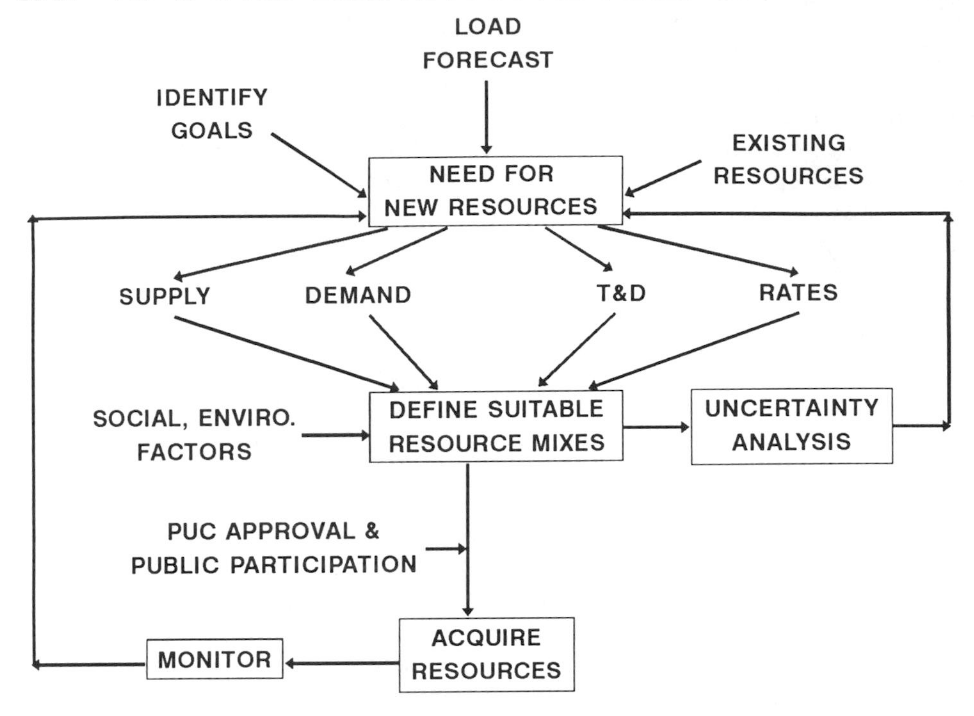

FIGURE 4.1 The activities involved in integrated resource planning.

Typically, a utility begins its IRP process by identifying its goals and the key issues that the resource plan must address (Figure 4.1). Corporate goals often concern customer service, returns to shareholders, maintenance of low electricity prices, and protection of the physical environment. Specific issues might involve forthcoming decisions on an aging power plant that could be retired, repowered, or restored to full service; DSM programs that might be expanded or modified; a response to a recent PUC order requiring the utility to conduct a competitive bidding process to acquire new resources; efforts to restore the utility's bond rating to a higher level; and so on.

Next the utility develops alternative load forecasts. Then the utility assesses the costs and remaining lifetimes of its existing resources and identifies the need for additional resources. Here, "resources" refers to any method used to meet customer energy-service needs, including conventional and renewable power plants, contracts to buy electricity from other organizations, and programs that improve the efficiency or timing of customer electricity use.

The utility then assesses a broad array of alternatives that could satisfy the need for more electric-energy services. Such alternatives might include power-supply options, DSM, transmission and distribution (T&D) additions, and pricing options. Supply resources include modifications to existing power plants that extend their lifetimes or increase their output, purchase of power from other utilities and from nonutility companies, as well as the construction of new power plants.

Utility DSM programs might include (1) promotion of new lighting systems, motors, and other equipment to improve energy efficiency and/or (2) direct control of customer loads at critical times. These DSM programs constitute resources that can substitute for power plants and perhaps also for transmission lines and distribution systems. T&D resources include thicker conductors (which lower losses and permit more of the generation output to reach customers) and additional lines that provide access to more generating units. Pricing options include time-of-use rates that encourage customers to shift load from onpeak to offpeak periods as well as overall price changes that affect overall electricity use.

Different combinations of these supply and demand resources are then analyzed to see how well they meet future electricity needs and how expensive they are. These analyses are repeated time and again to test various resource portfolios for their resilience against different uncertainties. These analyses test the effects of uncertainty about the external environment (e.g., local economic growth and fossil-fuel prices) and about the costs and performances of different resources. Such uncertainty analysis helps to identify a mix of resource options that meets the growing demand for electricity, is consistent with the utility's corporate goals, avoids exposure to undue risks, and satisfies other environmental and social criteria.

The utility prepares a formal report based on the preceding analyses and on suggestions arising from public involvement. That report presents the preferred resource plan and the justification for that plan. After acceptance or approval by the PUC, the plan is implemented and resources are acquired. Although the PUC formally reviews the plan and various nonutility parties participate in its preparation, the utility has the ultimate responsibility for its development and implementation.

While the plan is in force, the utility monitors changes in fuel prices, electricity demand, DSM participation, and a host of other factors and modifies the plan as events and opportunities warrant. Although resource planning is an ongoing process, only once every few years does the utility issue a formal plan along the lines discussed here.

In summary, IRP differs from traditional utility planning in two ways. First, it calls upon each utility to consider a broad array of ways to meet customer energy-service needs. In particular, IRP requires utilities to look through customer demands for kWh and kW to the demands for the underlying energy services. Second, IRP generally calls upon each utility to consult extensively with all parties that want to participate in the planning process. The expectation is that the IRP data, analysis, and process will lead to decisions that are more widely accepted and that reduce total risk. IRP, because of its extensive public involvement, may also shift risks from one party to another; indeed, utilities expect that their IRP activities will lead to the shifting of some risks from utility shareholders to customers. These IRP attributes are also key elements for planning in a competitive environment.

4.3 Energy and Demand Forecasts

The forecasts of annual electricity use and peak demand are typically the starting point for resource planning. These forecasts, when compared with the utility's existing and committed resources, determine the amounts, timing, and types of future resources that the utility will need during its planning period.

Utilities use two types of methods to develop their forecasts, econometric and end-use. Econometric models are characterized by modest data requirements and by their statistical foundation. While econometric models are aggregate, end-use models are disaggregate. End-use models take an engineering approach and estimate customer-class electricity use from the individual end-use details.

Several organizations have developed forecasting models that combine the best features of the econometric and end-use approaches. These engineering/economic models include the behavioral features of the econometric models (e.g., sensitivity of electricity use to the price of electricity) and the detail of end-use models (e.g., disaggregation of electricity use by end use and building type).

Because forecasts of electricity use and demand are so important in IRP, the forecasts should be detailed (e.g., by building type, end use, and technology). Such detail is needed so that the utility can link its forecasts with its assessment, planning, and evaluation of DSM programs. Specifically, the forecasts should show the effects on electricity use of:

- Past, present, and possible new utility DSM programs
- Government efficiency standards and DSM programs
- Market forces (e.g., changes in electricity and fuel prices, incomes and economic activity, and energy-using technologies)

These details will ensure that the forecasts appropriately account for changes in energy use caused by different factors, with no under- or double-counting of effects.

The load-forecasting process used by Wisconsin Electric (1991) demonstrates these points well. The utility first developed a forecast without any DSM programs. It then developed forecasts that included nonutility DSM, existing utility DSM programs, and all DSM programs. Thus, the difference between the first two forecasts is the effect of nonutility DSM programs, the difference between the second and third forecasts is the effect of existing DSM programs, and the difference between the last two forecasts is the effect of new (planned) utility DSM programs.

PacifiCorp's (1992) resource plan shows the link between the load forecasts and the amount of energy-efficiency resources that can be obtained. Higher load growth is caused by increases in the number of new customers (i.e., construction of new buildings and factories) and by greater use of electricity in existing facilities. Both factors more than double the potential for cost-effective energy savings in going from PacifiCorp's low to high load forecasts.

The energy and load forecasts should be linked together. Some utilities develop detailed (i.e., end-use) forecasts of annual energy use but produce a peak-load forecast with a single-equation model. Worse yet, this simple model is not linked to annual energy use. In such situations, certain changes (for example in the federal efficiency standards for air conditioners) would have no effect on the forecast of peak demand. However, such changes could dramatically affect future demands at the time of system peak. One utility developed separate forecasts of annual energy use for 14 residential end uses, 12 commercial building types, and 14 industrial sectors. The peak-demand forecasts, however, were based only on three years of data on "historic load factors." Thus, estimates of future peaks were based on a single number, the company's system load factor. On the other hand, New England Electric System (NEES) (1992) produced detailed forecasts of peak demands (e.g., for nine commercial building types and sixteen manufacturing groups).

4.4 Assessment of Supply Resources

The utility should next examine a broad range of supply alternatives, including those that use new technologies (e.g., clean-coal and renewables). The generic categories include existing utility-owned power plants and contracts; life extension, repowering, or fuel-switching of utility-owned plants; utility construction of new power plants; purchases from other organizations (including other utilities and IPPs); and T&D improvements. These alternatives include different technologies, fuels, and ownership of the resources. A key element in review of both supply and demand alternatives is to consider a sufficiently wide array to be sure that potentially attractive options are not overlooked.

Central Maine Power (1991) emphasized stewardship of existing resources (especially the company's hydroelectric facilities and older fossil-fired power plants). Such a focus on deployment and preservation of existing resources (both power plants and contracts) is especially important for a utility that faces long-term surpluses.

Some utilities with substantial generating capacity beyond that needed to meet native loads are active in wholesale markets. To illustrate, Public Service Company of New Mexico (1992) sold 38% of its output to other utilities in 1991 and is selling portions of its "excess" generating units to other utilities. Utilities should analyze the implications on both retail customers and shareholders of these wholesale transactions and asset sales.

Renewable resources, including solar-thermal, wind, geothermal, and photovoltaics, are rarely considered seriously by utilities. Some utilities, including the Sacramento Municipal Utility District, PacifiCorp, and NEES, are making commitments to renewable resources (Lowell 1994). These utilities are pursuing renewables because of their economics, environmental benefits, contribution to resource diversity, contribution to risk reduction (e.g., from a tax on CO_2), or help in lowering the long-term costs of renewable resources.

Several utilities considered seriously the purchase of energy and capacity from other sources, including IPPs (Kliman 1994). Such resources can be obtained through direct negotiation or through bidding programs. Puget Power (1992) issued its first request for proposals in 1989. Contracts were subsequently signed for 153 MWa of supply projects plus 10 MWa of energy-efficiency projects. The company's 1992 plan called for 100 to 200 MWa in the late 1990s, based on responses to its second request for proposals (RFP), which was issued in September 1991.

Increased competition among power providers and improved planning may appear inconsistent to some people. To the extent that utilities rely on markets to determine the prices, types, and amounts of new resources, some argue that careful planning is not needed. An alternative perspective suggests that completion of an integrated resource plan is a necessary prerequisite to the preparation of an RFP to acquire resources competitively. Careful planning is needed to determine the types of resources that are needed, when these resources are needed, and the maximum prices that the utility can pay for such resources.

In this approach, the utility first develops a long-term resource plan. Information from the resource plan is used to structure an RFP that specifies what types of demand and supply proposals will be acceptable to the utility. In other words, the values of avoided energy and capacity costs, year by year, are crucial outputs of the resource plan that are needed as inputs to bidding. Resources offered at costs below the utility's avoided costs are attractive; others are not. The utility then receives bids and uses the information gained from the bids and subsequent contract negotiations as input (e.g., on the amounts and costs of nonutility power likely to be available) to the next IRP. This approach is essentially the one chosen by Puget Power (1992).

In an alternative approach, the utility would conduct a solicitation early in the development of its IRP. Early submissions of bidder responses would allow the utility to integrate the analysis of these IRP resources into analysis of its power plants and DSM programs. The final IRP would then include those resources the utility selected in its solicitation (including contracts to be signed pursuant to the solicitation). Public Service Company of Colorado (1993) used this approach to coordinate bidding and IRP.

For several reasons, utilities are increasingly coordinating their T&D planning with their resource planning. The public is concerned about electromagnetic fields and the public is often opposed to construction of new transmission and distribution lines. The greater number of generating companies (discussed above) increases the complexity of power transactions and flows, and new generation sources that are distributed (e.g., renewables) complicate the local balancing of supplies and demands.

Utilities can improve T&D-system efficiencies by replacing components, such as transformers and conductors, with components having lower losses; modifying operating conditions; and reconfiguring systems to, for example, reduce the distance between substations and their loads. Unfortunately, few utilities include T&D improvements in their resource plans. Some utilities provide lengthy lists of existing and planned transmission facilities (e.g., showing voltage, type of structure, location, and miles of line), these numbers, absent strategic analysis, have little to do with resource planning.

Green Mountain Power's (1991) IRP analyzed the loss-reduction opportunities on the company's T&D system. In addition to identifying the sources of these losses, the plan discusses several strategies for cost-effective loss reduction. PacifiCorp's (1992) plan shows T&D efficiency improvements that amount to 1.1% of annual sales and 2.4% of the winter peak by the year 2011.

Pacific Gas & Electric analyzed the use of DSM programs to defer costly expansion of its distribution system in rapidly growing areas (Orans et al. 1992). It found that aggressive DSM programs could defer construction of a new substation and reconductoring of a local transmission line. These deferrals could cut total investment costs in its Delta district by one-third, from $112 million to $76 million.

PacifiCorp (1994) divided its multistate service area into six load and resource regions; its analysis accounted explicitly for the transmission links and constraints among these regions (Figure 4.2). Two of these regions, California and the desert southwest, are entirely wholesale markets for PacifiCorp.

PACIFICORP EXAMINED LOADS AND RESOURCES IN SIX REGIONS

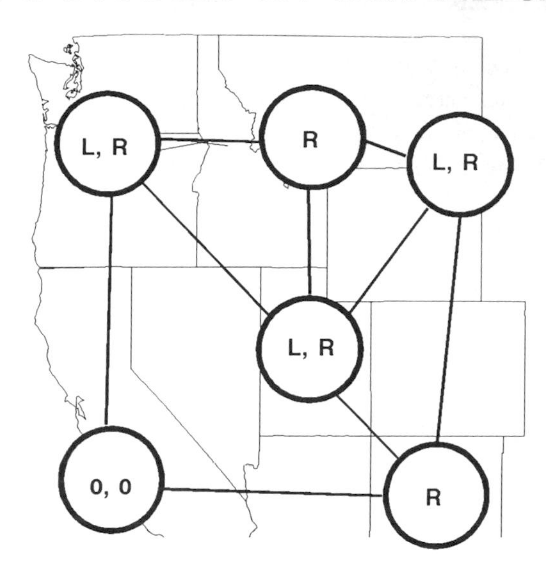

FIGURE 4.2 Map of the Pacific Northwest showing the regions that PacifiCorp used in its IRP to analyze power flows and transmission constraints among its load (L) and resource (R) centers.

The analysis of many options in the resource integration phase is complicated and time consuming. Therefore, utilities use a screening process to eliminate from integration those resources that are clearly inferior to others. The supply options discussed above should be screened for their construction characteristics (costs and time to license and build), operating characteristics (fixed and variable operating and maintenance costs, availability and likely capacity factors, environmental effects, and plant lifetime), and transmission additions required to connect the source to the utility's system. These supply alternatives should be compared to each other with screening curves or other methods to identify those that provide capacity or energy at the lowest cost (in $/kW or ¢/kWh). The more expensive options can then be dropped from consideration. However, utilities should not prematurely drop resources because they are not *the* least-cost options; what appears to be cost-ineffective during resource screening may, in the subsequent integration and uncertainty-analysis stages, turn out to be very attractive.

Potomac Electric (1992) considered 25 supply technologies in its screening process. Each technology was scored on its commercial status, technical status, licensing and leadtime risks, cost risk, capital costs, and operating costs. The ratings for each resource on these six characteristics were used to develop an overall score for each resource, which then allowed the company to rank the resources. This approach goes beyond the simple screening discussed above.

4.5 Assessment of Demand-Side Resources

The utility should, in parallel with its assessment of supply options, conduct a thorough assessment of energy-efficiency, load-management, and fuel-switching resources. These DSM programs are, in later steps, compared with supply options to develop a portfolio that best meets customer needs. A key requirement here is that the utility assessment of DSM resources be comparable to and consistent with its assessment of supply resources.

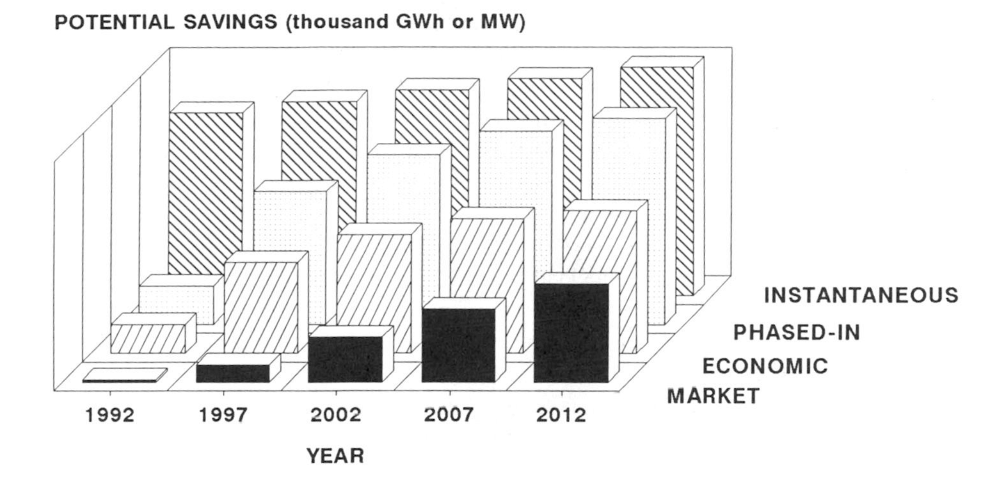

FIGURE 4.3 Schematic showing the differences among technical (both instantaneous and phased-in), economic, and market potentials for a utility DSM program.

Utility DSM programs represent a modest, but growing, source of energy and capacity resources. To illustrate, in 1993 these programs cut annual electricity sales by 1.5% and potential peak demand by 6.8% (Hadley and Hirst 1995).

Utilities should examine DSM resources in all customer classes, all major end uses, and a variety of current and emerging technologies. DSM resources that are slightly more expensive than supply resources under baseline conditions should be retained because they may later turn out to be attractive as the integration and uncertainty analyses proceed. For example, Wisconsin Electric (1991) screened DSM options with a benefit/cost ratio of 0.75 "because it is possible that an option may be cost-effective after more information is gathered or because the option's positive externality impacts indicate that it should be implemented anyway."

Assessment of DSM options should begin with a review of the company's DSM programs. This review should include program description, utility budgets, program participation rates, estimated energy and load effects (and the basis for these estimates), and program cost-effectiveness. The utility should show what evaluations and market research support its knowledge about the process and performance of existing programs.

The utility should next discuss new program possibilities. This assessment should consider the technical, economic, and market potentials for energy efficiency, load management, and fuel switching for each customer class (Figure 4.3). Such assessments begin with a careful analysis of existing and projected electricity-use trends by customer class, end use, and technology. The need for such information is one important reason to use end-use forecasting methods rather than econometric approaches.

New programs can include modifications of existing programs (e.g., to gain more participation from target markets or to change financial incentives) and initiation of new programs (new end uses, technologies, or market segments). DSM options (e.g., electric heat pumps, high-efficiency lighting systems, and industrial cogeneration) should be combined into program designs because that is what the utility delivers to its customers. It is not enough to analyze the costs and electricity savings of high-efficiency lights and motors for commercial buildings. The combination of these measures and the utility's delivery system (e.g., marketing approach and audit delivery) is what is relevant.

Because the resource-integration process is complicated, the utility first needs to screen the candidate DSM technologies and programs to reduce these lists to manageable size. The California PUC and California Energy Commission (1987) developed standard tests that assess the economics of DSM programs from different perspectives. These perspectives include participating customers, nonparticipating customers (the rate-impact measure), utility revenue requirements, all customers (the total resource cost test), and society as a whole. The plan should state which tests the utility used and the sensitivity of the results to the input assumptions. Assumptions concerning program costs, participation rates, and marginal energy and capacity costs are especially important. These marginal costs should reflect, in addition to avoided generation costs, those costs associated with investments in T&D. Joskow and Marron (1992) and Eto et al. (1994) note that utilities typically ignore some of the costs of DSM programs and often overestimate program energy and demand reductions, leading to an optimistic view of DSM-program cost effectiveness.

The utility's IRP should examine its load-building programs as well as those programs aimed at load reductions. In particular, the utility should show whether these programs reduce electricity prices to customers in the short term with no adverse long-term increases in electricity costs or prices (e.g., because construction of a new power plant must begin sooner or more pollution will be emitted than otherwise would occur). Some utilities are not explicit about their load-building efforts. One utility's resource plan noted the "substantial" contribution of its DSM programs to energy savings. Only in the Appendix to the report, however, was it shown that more than half of these savings were from cogeneration (a source that most would not classify as DSM). The Appendix also showed that these "savings" were more than offset by increases in electricity use caused by load-building programs. Thus, the net effect of the company's DSM activities was to *increase* electricity use, rather than to decrease it as the company claimed in the body of its IRP report.

The level of electricity prices and the structure of the rate tariff—i.e., monthly customer charge, demand charge(s), and energy charge(s)—affect the amount and timing of customer electricity use. Therefore, the level of prices can be considered a baseload resource and the structure of prices (especially time-of-use prices and interruptible rates) can be considered a peaking resource. However, most utilities do not treat pricing in their resource plan. Also, rate tariffs (especially the energy and demand charges) should be consistent with the utility's estimates of long-run avoided costs.

4.6 Resource Integration

Integration of the supply and demand resources involves specification of the criteria to be used in assessing resource portfolios, development of alternative resource portfolios designed to address different objectives, analytical integration of resources, explicit treatment of uncertainty, assurance that the results of the analysis are internally consistent, presentation of avoided energy and capacity costs, explicit consideration of reliability and reserve margins, and candid treatment of the environmental costs of electricity production.

The selection of resource portfolios (analogous to the screening and selection of individual options) can be based on different criteria (e.g., minimize revenue requirements, capital costs, or average electricity prices; ensure adequate reserve margins and the ability to meet high load growth; maintain certain financial ratios; or reduce environmental effects of electricity production). The utility should clearly specify what criteria it used in selecting individual resources and choosing among alternative resource mixes.

For example, the Lower Colorado River Authority (LCRA) (1994) used eight factors in assessing alternative resource portfolios: environment (CO_2 and NO_x emissions), present worth of utility revenue requirements over 45 years, reserve margin, technology experience for each resource mix, percentage of resources that are distributed (small and close to load centers), debt service coverage, 10-year increase in rates, and fuel risk. These eight factors were assigned weights according to the perspectives of the environment, investors, LCRA, and LCRA's wholesale customers. LCRA developed a dozen resource portfolios, combining different amounts of energy efficiency, load management,

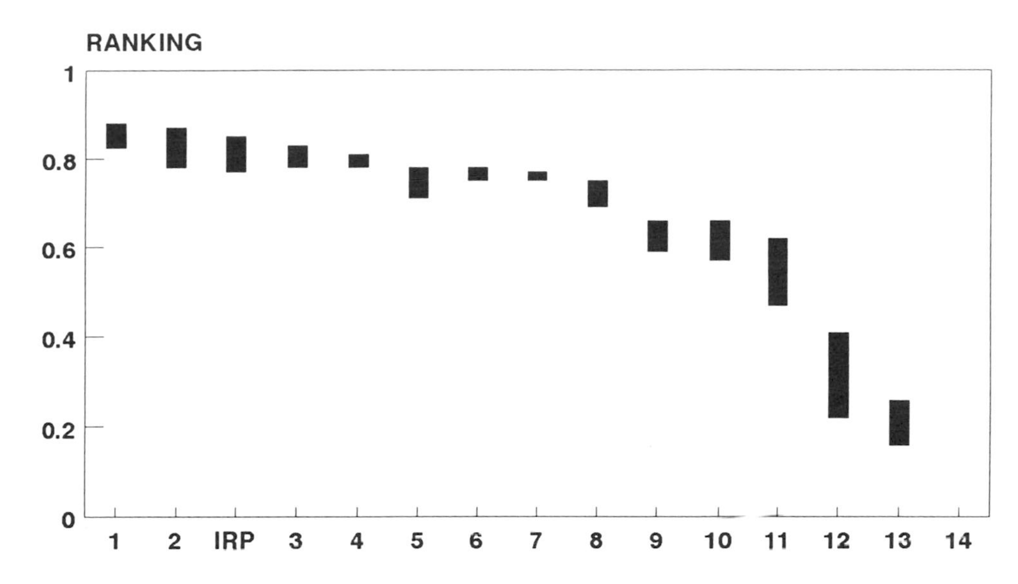

FIGURE 4.4 The Lower Colorado River Authority's ranking of various resource portfolios against eight factors.

distributed resources (diesel generators and fuel cells), wind, gas-fired combustion turbines (CTs), and coal plants. Figure 4.4 shows the aggregate ranking for each portfolio against the eight factors.

The utility should develop, analyze, and report the results on alternative resource portfolios designed to meet different goals. To illustrate, the utility could analyze portfolios intended to minimize revenue requirements, total resource costs, societal costs, electricity prices, utility capital costs, or utility use of oil. Comparisons among these portfolios will show clearly the types of tradeoffs that must be made among competing objectives.

PacifiCorp (1994) examined 155 different futures. These cases included five load forecasts, three gas-price forecasts, various resource portfolios, and environmental sensitivities (NO_x and CO_2).

Niagara Mohawk (1991) developed ten resource portfolios for three scenarios reflecting the need for different amounts of new resources. These scenarios represent different assumptions on load growth and the amount of nonutility generation that the utility must use. DSM options were grouped into "moderate, aggressive, and very aggressive DSM blocks based on the economic merits of individual programs." These DSM blocks were combined with various supply options to meet future resource needs.

A utility's preferred resource plan is unlikely to be the "least cost" plan under any particular conditions. Rather, it will reflect a balance among competing interests. As Potomic Electric (1992) noted, "the 'least-cost' resource mix may not necessarily be the one with the lowest expected cost when compared to all other feasible resource plans." Its preferred plan differed slightly from the base-case plan in that the preferred plan will acquire more DSM resources and defer by one year the additions of several gas-fired CTs and combined cycle units.

Reported results on a utility's preferred plan should include the annual MW and GWh contributions from different resources, revenue requirements (including capital costs, fuel costs, and other cost components), electricity prices, emissions, reserve margin, and measures of utility financial performance (e.g., interest coverage and percent of construction internally funded).

Utilities typically use one of two approaches to assess alternative resource portfolios. One method involves mathematical optimization, in which a dynamic programming model automatically selects

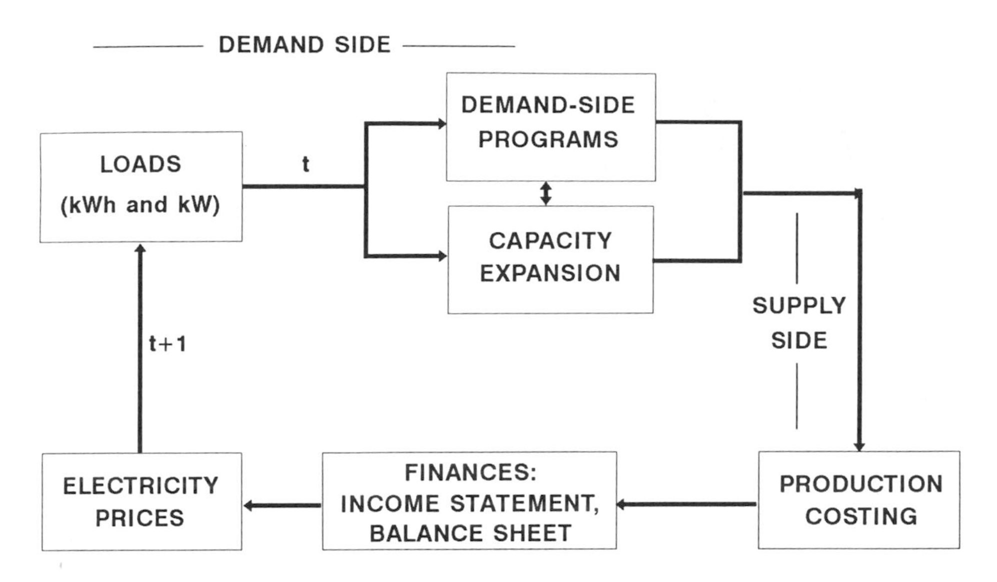

FIGURE 4.5 Preferred approach used to analytically integrate demand and supply resources; t refers to the year of analysis.

the mix of resource options that meets the stated objective function (typically the lowest net present value of revenue requirements over a stated planning horizon). The primary strength of these models is their ability to identify the *least-cost* mix of resources. However, these models are complicated, difficult to run, and their solutions depend strongly on the input constraints and assumptions.

The other approach uses a simulation model, for which the user must provide the resource mix to be tested. This approach requires substantial interaction between the analyst and the model. On the other hand, one can never be sure that the *least-cost* mix of resources has been identified; therefore, substantial trial and error is required with this approach to identify suitable resource portfolios.

No matter what type of modeling approach a utility uses, DSM resources should be treated in a fashion that is both substantively and analytically consistent with the treatment of supply resources. Demand and supply resources should compete head to head (Figure 4.5). The plan must show how the process integrates load forecasting, DSM resources, supply resources, finances, rates, and the important feedbacks among these components (especially between rates and future loads).

Hill (1991) compared three methods often used to integrate demand and supply resources: (1) a sequential approach in which DSM resources are selected first (e.g., Wisconsin Electric); (2) a sequential approach in which supply resources are selected first (e.g., Duke Power); and (3) a simultaneous approach in which supply and demand resources are chosen together (e.g., NEES). His results suggest that the simultaneous approach yields the resource mix with the lowest total cost. Thus, this approach is both theoretically appealing and practically important.

The integration process must recognize the degree of dispatchability (control that the system operator has over the generating unit) and intermittency of each resource. For example, nuclear units are generally treated as "must run," because it is expensive and difficult to vary their output quickly. Combustion turbines, on the other hand, are very flexible machines. Some renewable resources, such as wind and photovoltaics, are intermittent and require special consideration in modeling to account properly for their capacity value (Logan et al. 1995).

AVOIDED COST OF DSM DEPENDS ON LOAD SHAPE

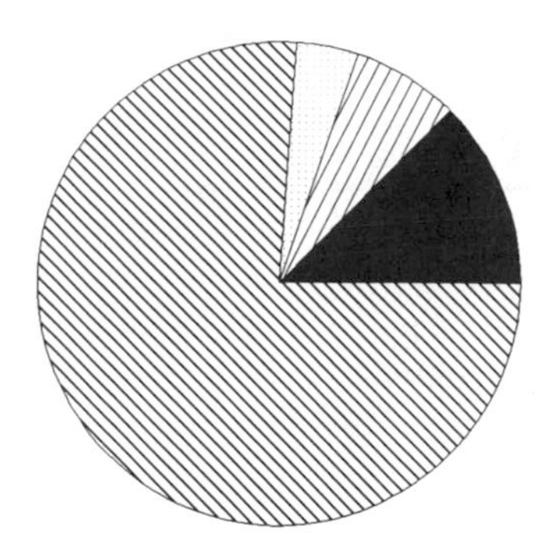

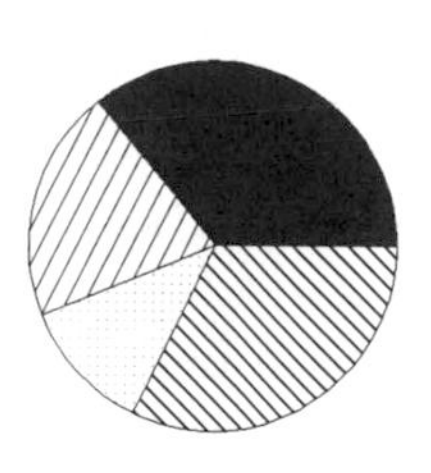

FIGURE 4.6 Georgia Power's avoided costs for two hypothetical DSM programs.

At the end of the integration process, utilities should compare the initial assumptions on future electricity prices used as inputs to the load forecast with the prices produced by the integration process. The utility may need to iterate through the process (i.e., develop new load forecasts and resource portfolios) if the two sets of prices differ materially. Ensuring consistency between the avoided costs used to screen resources and those that result from the IRP process may also require iteration of the analysis (Potomac Electric Power Company 1992) .

Because avoided energy and capacity costs (for generation, transmission, and distribution) play a vital role in assessing resource cost-effectiveness, the utility should report its estimates of these costs on an annual basis for the entire analysis period. Some utilities include only the costs of a CT in their avoided costs. Unless the utility has no plans to add baseload capacity any time during the planning period, this assumption understates avoided costs. In essence, this approach ignores the energy-related capital costs of a baseload plant (e.g., the difference in cost per kW of a baseload coal plant and a CT). Ignoring these costs undervalues DSM programs aimed at improving energy efficiency.

Avoided capacity costs have several components. Because different resources (especially location-specific DSM programs and renewables) affect these costs differently, utilities should be explicit about the details of avoided cost. Figure 4.6 shows, as an example, the avoided generation, T&D, and energy costs associated with two types of DSM programs. One program cuts demand by 1 kW for every hour of the year (equivalent to a baseload power plant with 100% capacity factor). The second program also cuts demand by 1 kW, but for only 900 hours during the summer peaking season (equivalent to a peaking unit with a capacity factor of 10%). The avoided costs associated with these two programs are quite different.

4.7 Environmental Costs

Because electricity production, transmission, and distribution have substantial environmental effects, utility resource plans must address these impacts. Not surprisingly, utilities use different methods to assess these impacts (Eco Northwest 1993). The simplest approach is to ignore externalities and to assume that compliance with all government environmental regulations yields zero

TABLE 4.1 Results of PacifiCorp's Analysis of the Effects of Different Environmental-Externality Costs

Environmental level (cost)	Net present value of revenue (billion $)	Electricity price in 2011 (1991-¢/kWh)	Percent increase in emissions to 2011		
			CO_2	SO_2	NO_x
1	31.33	4.51	32.2	12.4	–8.3
2	31.31	4.51	30.0	12.1	–12.0
3	31.31	4.52	29.0	12.7	–11.7
4	31.73	4.58	26.1	11.3	–12.6

externalities. Another approach is to characterize and describe qualitatively the environmental effects of different resource options. A more complicated approach is to rank and weigh the individual air, water, and terrestrial impacts of individual options. Finally, some utilities quantify and monetize the emissions associated with resource options. This approach requires quantification of the emissions (e.g., X tons of sulfur dioxide emitted per million Btu of coal) and monetization of these emissions (e.g., Y $ of environmental damage per ton of sulfur dioxide). These monetary values (typically expressed in ¢/kWh) reflect the damages imposed on society by the emissions from a particular resource.

A few utilities have specific environmental goals. For example, NEES (1992) committed "to continuously reduce the environmental impact of its electric service by reducing net air emissions including greenhouse gases from its operations by an estimated 45%" by the year 2000. The NEES resource plan shows the resources to be acquired and how they will help the utility meet the 45%-reduction goal.

PacifiCorp (1992) conducted a sensitivity analysis with respect to the external costs of different resources. Four sets of costs were tested for SO_2, NO_x, particulates, and CO_2. Using these externality values in the selection of resources showed how different estimates of environmental costs affected the resource portfolio chosen, utility revenue requirements, and electricity price. As the environmental level (cost) increased, so did the amount of renewables and gas-fired generation. Except for environmental level 4, the revenues, prices, and even emissions differed only slightly (Table 4.1).

Consumers Power (1992) as part of its scenario analysis, considered two global-warming cases. These scenarios included a tax on CO_2 ($30/ton beginning in 2000) or a cap on CO_2 emissions (a requirement to reduce emissions to its 1990 level beginning in the year 2000). By 2007, the emissions-limit case included four times as much DSM as the base case, a 28% reduction in carbon emissions, and a 4% increase in revenue requirements from 1993 to 2007.

Because the estimates of monetized environmental costs vary so much, many utilities are reluctant to use this approach in resource planning. An alternative to monetization is to conduct a tradeoff analysis. Andrews (1991) examined the New England Power Pool system from 1990 to 2010. His analysis shows the tradeoffs between revenue requirements and emissions for different resource portfolios (Figure 4.7). Use of low-sulfur fuels plus DSM programs reduces both costs and emissions. Fuel switching lowers emissions but not costs, while DSM lowers costs but not emissions. PacifiCorp (1994) used this approach in its most recent IRP.

4.8 Uncertainty Analysis

A thorough analysis of a variety of plausible future conditions and the options available to deal with them is essential to a good resource plan. Such an analysis would use one or more of the techniques shown in Table 4.2. These techniques should be used to assess uncertainties about both the utility's external environment and factors at least partly under the utility's control. Utilities, in their uncertainty analysis, must focus on the key decisions they are considering for the next few years.

Uncertainties about the external environment include economic growth, inflation rates, fossil-fuel prices, and regulation. The analysis should also consider uncertainties about the costs and

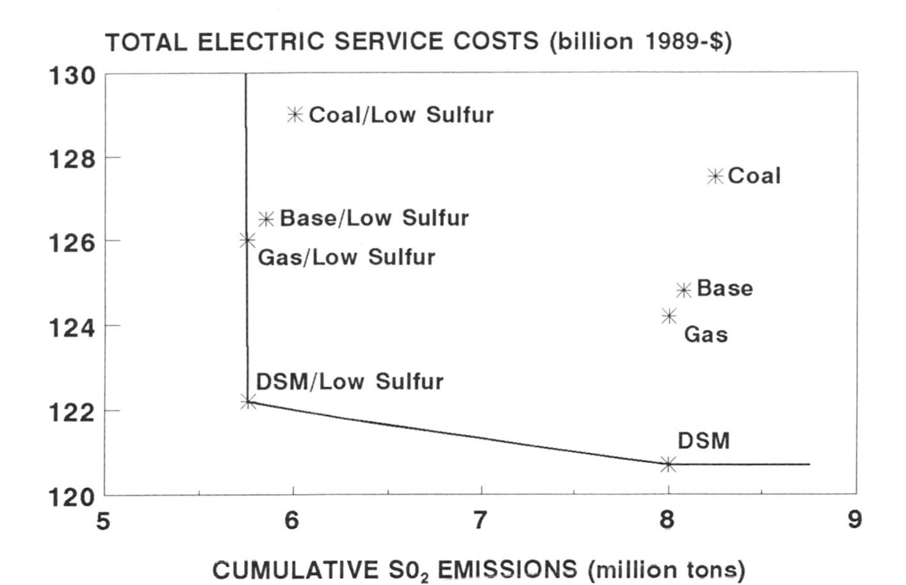

FIGURE 4.7 Tradeoffs between sulfur emissions and revenue requirements for various resource portfolios for the New England Power Pool. The solid curve shows the envelope of preferred solutions.

TABLE 4.2 Analytical Techniques Used to Treat Uncertainty

Scenario	Alternative, internally consistent, futures are constructed, and then resource options are identified to meet each future.
Sensitivity	Preferred plan (combination of options) is identified. Key factors are then varied to see how the plan responds to these variations.
Portfolio	Multiple plans are developed, each of which meets different corporate goals.
Probabilistic	Probabilities are assigned to different values of key uncertain variables, and outcomes are identified that are associated with the different values of the key factors in combination.
Worst-case	Utility creates a plan to meet an extreme set of conditions (e.g., high load growth and high fuel prices) and later learns that it faces an entirely different set of conditions (e.g., low load growth and low fuel prices). The utility then adjusts its resource acquisitions to meet the newly perceived conditions.

performance of different demand and supply resources. The analysis should show how utility resource-acquisition *decisions* are affected by these different assumptions and show the *effects* of these uncertainties and decisions on the customer and utility costs. Differences among resources in unit size, construction time, capital cost, and operating performance should be considered for how they affect the uncertainties faced by utilities.

PacifiCorp (1992) developed different, scenario-specific mixes of CTs, renewable resources, DSM programs, cogeneration, and coal plants. Its scenarios included, in addition to the medium-high load forecast, electrification (leading to higher load growth), loss of major generating resources, high natural-gas prices, and a tax on CO_2. For each scenario, results were presented on the amounts of each resource acquired; utility operating revenues; average electricity prices; and emissions of sulfur dioxide, nitrogen oxides, and carbon dioxide.

Southern California Edison's (SCE) (1991) San Onofree Nuclear Generation Station 1 (SONGS1) needed \$125 million of capital improvements to continue operating. To test the value of this

CAPACITY FACTOR KEY UNCERTAINTY IN SCE ANALYSIS OF SONGS1

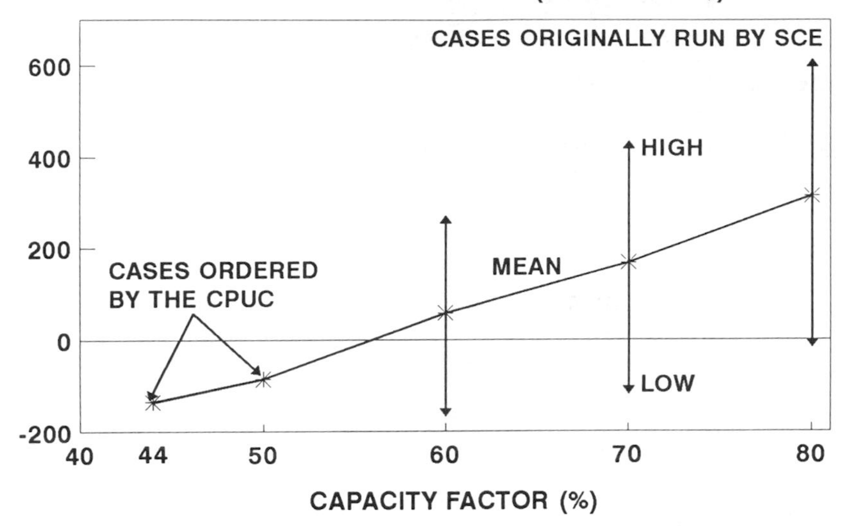

FIGURE 4.8 Results of the uncertainty analysis conducted by Southern California Edison on the benefits of extending the lifetime of its San Onofree Nuclear Generating Station 1.

investment, SCE analyzed 32 cases with different assumptions on the values of environmental externalities, future natural gas prices, DSM, the capacity factor of SONGS1, and the capital costs of extending the life of SONGS1 (Figure 4.8). The utility's analysis suggested that the continued operation of SONGS1 would be cost-effective.

The California PUC directed SCE to expand its uncertainty analysis to show the effects of assuming lower capacity factors based on the plant's historical performance. SCE analyzed cases with capacity factors of 44 and 50% (compared with the 60, 70, and 80% values used earlier). For these additional cases, the net benefits were generally negative. Ultimately, SCE decided to shut down the plant.

This uncertainty analysis illustrates three critical points about this part of IRP. First, the analysis should focus on key decisions that the utility will face during the next few years. A reasoned treatment of the key uncertainties that the utility can influence is far more valuable than an exhaustive treatment of all uncertainties with little regard for their importance. Second, the input assumptions must be reasonable and encompass a broad enough range to show the effects of uncertainties. Third, including nonutility parties in the uncertainty analysis can help with the first two issues. Portland General Electric (1992) analyzed the economics of retrofitting its Trojan nuclear plant with scenario and probabilistic analyses; it reached the same conclusion that SCE did and decided to prematurely retire the plant.

New England Electric (1993) used options theory in its latest IRP. This method, based on one used in financial markets, allowed the utility to determine the benefits of waiting to take action. In many cases, delaying a decision can yield benefits because more information becomes available during that time.

The links between the results of uncertainty analysis and the utility's resource-acquisition decisions must be demonstrated. The uncertainty analysis should also demonstrate the robustness of the selected resource plan. For example, Georgia Power (1992) examined the possibility that a non-combustion turbine (something other than a peaking unit) would be needed before the year 2000. It examined uncertainties related to alternative financial, environmental, load-growth, fuel-cost, and generating-unit characteristics to assess these possibilities. The results "point overwhelmingly to CTs as the generation additions for the mid-1990s," lending confidence to its decision to build several CTs this decade. The mix of resources selected should be able to withstand the shocks of different futures and should minimize the risks associated with various adverse outcomes (e.g., rapid increases in oil prices or a moratorium on nuclear power).

4.9 Action Plans and Progress Reports

The utility's action plan is, in many ways, the "bottom line" of the resource plan, because it reflects the utility's commitment to specific actions. The action plan must be consistent with the long-term resource plan to assure that what is presented as appropriate for the long haul is actually implemented, and implemented in an efficient manner.

The action plan should be specific and detailed so that readers can judge the utility's commitment to different actions. Specific tasks should be identified along with milestones and budgets. The action plan should present the utility's expected accomplishments during the next few years, including the number of participants and the reductions in annual energy use, summer peak, and winter peak for each DSM program.

The Portland General Electric (PGE) (1990) action plan included nine tables showing actions planned for 1991 and 1992 to capture DSM lost opportunities by market segment, discretionary DSM resources by market segment, DSM technology assessment and pilot programs, T&D system actions, enhancements to existing utility-owned resources, acquisition of new supply resources, development of renewable resources, development of customer resources, and monitoring of key indicators (e.g., oil and gas prices). Each table included specific actions, timing, program impact, and estimated budget. For example, the table on new supply resources committed PGE to "implement a competitive bid solicitation to coincide with a projected need for power in 1996" with an estimated program impact of 200 MW.

The action plan should also discuss the data and analysis activities, such as model development, data collection, and resource assessments, needed to prepare for the next resource plan. The LCRA (1991) action plan, for example, committed to the preparation of a "combustion turbine contingency plan" to help decide when to begin construction of a CT and a comparable "demand-side 'fall back' plan."

Both the utility and the readers of the IRP report should recognize that not all the projects presented in the action plan will be completed as specified. As circumstances and opportunities change, the utility should respond accordingly. Thus, the action plan is the utility's plan as of a certain date. Because changes will occur (e.g., local economic growth, environmental regulations, or experience with DSM programs), the plan should indicate how, and under what circumstances, the utility will revise its action plan.

The utility should prepare annual progress reports that explain its activities in implementing the action plan during the past year. These interim reports should be prepared during the off years. For example, if a utility files long-term resource plans in 1993 and again in 1996, then progress reports should be filed in 1994 and 1995.

These progress reports should show accomplishments in resource acquisition and in collection and analysis of additional data and development of improved analytical methods. These reports should also indicate when (and why) plans are changed. For example, Duke Power's (1991) progress report explained why it had delayed the online dates for the first of several planned CTs.

4.10 Public Participation

Because the interests of all stakeholders in a utility's resource plan are not identical, the ways in which these interests will be affected by utility actions will differ. Therefore, utilities should seek advice from different groups as they develop resource plans (Schweitzer et al. 1994). In addition,

TABLE 4.3 Spectrum of Issues Addressed by DSM Collaboratives

Least difficult
Identify potential DSM technologies
Package DSM measures into programs
Screen measures and programs for cost-effectiveness
Design evaluation plans
Select cost-effectiveness tests for screening
Select annual budgets for DSM programs
Design incentives to encourage utilities to run DSM programs
Decide how to treat environmental externalities
Decide whether and how to examine fuel switching
Most difficult

utilities should report results for various resource plans along enough dimensions so that different groups can assess the plan's effects on them.

Unfortunately, some utilities treat public involvement as one-way communication in which they "sell" their plan to others. But without two-way communication between the utility and its customers and interest groups, a plan is in danger of ignoring community needs. Equally important, a plan developed solely by utility personnel is likely to lack the breadth that outside perspectives can give. Accordingly, the plan should present evidence that the utility sought ideas from its customers and other interested parties. Energy experts from local universities, the state energy office, the PUC, environmental groups, and organizations representing industrial customers should be consulted as the plan is being developed. So too should trade allies, such as building contractors, appliance dealers and wholesalers, nonutility generators, and fuel suppliers. In order of increasing participation, utilities can hold information meetings, conduct surveys, establish advisory groups, or do joint planning (collaboratives) to solicit public input and involvement in its resource planning.

Utilities in the Pacific Northwest invite nonutility experts, customers, and other interested stakeholders to participate in plan development and review. The PGE (1992) planning process included a 40-member Technical Advisory Group (TAG), which met eight times, plus a 25-member Public Policy Group, which met seven times. Puget Power, PacifiCorp, Public Service Company of Colorado, and the Tennessee Valley Authority (TVA) used technical advisory groups similar to PGE's. TVA provided funds to hire consultants to examine issues that its IRP Review Group felt warranted independent analysis. The Review Group selected consultants to examine TVA's load forecasting methods and results, and TVA's assumptions concerning future operation of its nuclear units.

Some utilities use collaborative working groups, especially for issues related to the planning of DSM programs. These collaboratives typically involve, in addition to the utility, environmental groups, the state's consumer advocate, the state energy office, PUC staff, and representatives of different customer groups (typically large industrial and low-income residential). These groups study and agree on appropriate approaches to different program-design and policy issues, including those shown in Table 4.3 (Raab and Schweitzer 1992).

Finally, the utility should document its responses to the comments offered by nonutility parties during plan development. One utility, which ran a comprehensive and extensive public-involvement process, nevertheless angered some of the participants. Because the utility provided no feedback on their comments, these participants felt that their inputs were largely ignored.

4.11 The IRP Report

Although resource planning is an ongoing process, utilities should periodically publish formal reports on their plans. The primary purpose of an IRP report is to help utility executives decide which resources to acquire, what amounts to acquire, and when to acquire those resources. The report provides a forum for the utility to present its vision of the future and how it plans to meet that future. The report documents the utility's decisions and helps the PUC and public to review and understand the basis for these decisions. Thus, the report must be useful both within and outside the utility.

Preparing a document that serves the needs of different readers is difficult. Some utility plans are so detailed and complicated that only the most technically sophisticated readers can understand what the plan contains. And some utilities publish short, glossy documents that present little information; the lack of detail frustrates readers interested in how the utility developed its plan.

The utility's plan might best be presented in more than one volume. Several utilities prepare three sets of documents: an executive summary, a plan, and one or more volumes of technical appendixes. The Northwest Power Planning Council (1991) published a 50-page summary, backed up by almost 1000 pages of details bound in two additional volumes.

The full report should discuss the goals of the utility's planning process, explain the process used to produce the plan, present load forecasts, compare existing resources with future loads to identify the need for additional resources, document the demand and supply resources considered, describe alternative resource portfolios, show the preferred long-term resource plan, and present the short-term actions to be taken in line with the long-term plan. The report should also explain the technical aspects of the planning process as described above.

4.12 PUC Role in IRP

Public utility commissions play important roles in IRP. They require utilities to prepare such plans, they establish the rules that determine the frequency and nature of such planning activities, they identify the key public-policy goals that the utility plans must meet, they may specify minimum requirements for public involvement, and they review or approve utility plans (Hirst et al. 1992).

Commissions may specify what economic test(s) the utilities should use in analyzing alternative resources and resource portfolios. As examples, these tests could focus on minimizing electricity prices, energy-service costs, or the total societal costs of energy services. The latter measure would include, in addition to the direct economic costs that customers face, the indirect costs of electricity production and consumption. These indirect costs could include air pollution from power plants, electromagnetic fields from transmission lines, and the benefits of increased employment caused by the utility's energy-efficiency programs. As of early 1994, 29 PUCs required utilities to explicitly account for environmental externalities in their resource planning (Fang and Galen 1994).

PUCs might also specify a preference for certain types of resources and resource-acquisition approaches. For example, the commission might require a minimum "setaside" for renewable resources to ensure that utilities acquire some of these not-quite-cost-effective resources that are environmentally benign. Commissions might also require utilities to conduct competitive auctions for supply and/or demand resources as a precondition to approval of any utility-built power plants.

PUCs can also influence utility inclusion of nonutility parties in development and review of the plan. In particular, PUCs might require a utility to conduct a broad public-participation process as the utility prepares its resource plan. In addition, the PUC might conduct formal hearings upon receipt of the utility plan, allowing additional opportunity for interested parties to comment on the plan. Such proceedings, both the informal and formal ones, can serve as useful mutual education sessions, and they can narrow the scope of disagreement for later determination by the PUC. But the ultimate decisions are made by the regulators, who decide on the allocation of risks between the utility and its customers.

The PUC role in IRP will, in general, play out in formal proceedings before the PUC. These proceedings could be long and complicated, because many parties may choose to participate and because the issues to be decided are complicated. Utilities will want some assurance from the regulators that the large sums of money they may commit to implement their resource plan will

TABLE 4.4 Comparison of Alternative Regulatory Approaches

	Pro (benefits)	Con (concerns)
IRP	Public participation	Cumbersome process
	Includes externalities	Insufficient cost-cutting incentives
	Comprehensive planning, includes DSM and renewables	DSM and renewables may raise prices
Markets	Cost minimization	Markets may not be competitive
	Lower regulatory costs	Possible unintended effects
	Greater customer choice	Loss of societal benefits

be recovered from customers. Intervenors will want a detailed defense of these planned expenditures. Fundamentally, the debates will center on which party (customers or the utility) will bear the risks associated with resource-acquisition decisions. A key issue for all parties is the extent to which PUCs will pre-approve utility actions in an approved IRP.

4.13 IRP in the Future

The preceding sections described today's IRP principles and practice. However, competitive forces sweeping the electric-utility industry will affect IRP practices both within utilities and PUCs (Tonn et al. 1994; Stalon and Hirst 1994). The industry is changing from one of cooperating vertically-integrated utilities to one of competitive generating companies free of pricing regulation and selling across federally regulated transmission networks to state, provincial, and locally regulated distribution companies (and perhaps to large industrial customers). Thus, generation, transmission, distribution, and customer services are being deintegrated. In this sense, the electricity industry may come to look somewhat like the natural-gas industry. Although subsequent discussions focus on the investor-owned sector, these competitive forces will affect consumer-owned utilities as well.

Because utility assets have long lives, investments made today will be part of the firm's portfolio in the new industry structure, whether it takes ten years to develop or only three years. Utility executives, consequently, will likely resist any investments or expenditures that might put any of the separate sectors at a competitive disadvantage in the forthcoming industry structure. These concerns will constrain current IRP practices and will likely lead to new forms of IRP.

The speed of the transition from regulation to competition will depend in part on the extent to which wholesale markets (especially transmission access, pricing, and investment) are resolved equitably. Concerns about market abuses (e.g., utilities that unfairly control transmission networks) may slow the transition to a competitive electricity industry and, therefore, enhance the importance of IRP for a longer time than would otherwise occur.

In addition to the forces of competition, utilities face increasingly stringent environmental constraints. Future regulations of carbon dioxide, nitrogen oxides, small particulates, and air toxics; limits on electromagnetic fields surrounding T&D facilities; or local opposition to construction of any power-system facilities may be quite different from those that occur today. However, the direction of change is clear—stricter controls on emissions of all kinds. These environmental constraints will also affect the evolution and practice of IRP.

Table 4.4, which compares IRP and markets, shows that each approach has advantages and disadvantages. While markets may yield lower direct and administrative costs, they may also provide fewer societal benefits and afford fewer opportunities for public involvement in resource selection.

IRP Within the Utility

Because IRP represents sound business planning, the underlying IRP principles will be unchanged from today's situation. However, because the utility will be a much different entity than today's vertically integrated company, IRP will differ in its details (Table 4.5).

TABLE 4.5 Likely Characteristics of IRP in a Competitive Environment

Use shorter time frames (5–10 years instead of 20+ years)
Focus on contracts rather than on power-plant construction
Emphasize T&D more than in the past
Emphasize low load forecasts rather than high forecasts
Treat electricity pricing as a resource

Generation companies (Gencos) will likely be competitive and largely unregulated. Transmission companies will be regulated by FERC (probably with heavy reliance on industry self-regulation), and local distribution companies (Discos) will be regulated by state PUCs. The utility for which IRP will be most applicable is the Disco. This company, although it may own transmission and generating assets, will rely primarily on other organizations for these outputs. Thus, its access to generation and transmission will be more through contracts than through ownership and operation. And, as with today's natural gas industry, these contracts will likely be of much shorter duration than either the life of the underlying physical assets or today's contracts. Discos will likely own a portfolio of contracts, including some as long as 10 years and many that price electricity at short-term spot prices. In essence, Discos will accept some price volatility to reduce their exposure to the risks and costs of excess generating capacity.

IRP will continue to involve utility assessment of a broad array of supply, demand, T&D, and pricing resources to meet future energy needs. A major difference from traditional practice will be the general exclusion of new power plants from this mix of potential resources. Therefore, utility planning will involve more contract law and less detailed engineering expertise. Utilities will pay less attention to the technical details of power-supply options and pay more attention to the markets for bulk power. Discos will consider distributed generation as an alternative to expanding distribution systems.

Electricity pricing will be increasingly important, both for its resource value and as a marketing tool. Both wholesale and retail services will be increasingly unbundled and separately priced. Discos and their customers will purchase separately energy, capacity, reliability, and the other attributes that are now all bundled into simple tariffs. Increasingly, prices rather than large physical assets (i.e., capacity intended to provide substantial reserve margins) will be used to allocate generating and transmission capacity. Furthermore, markets will determine the prices for power that Discos buy, and with unbundled prices and services, prices will approximate competitive market prices (i.e., prices will tend to equal short-run marginal costs of generation and transmission). For example, when capacity is tight, prices will increase substantially. Thus, society is likely to experience volatile prices that are, on average, lower than regulated (but stable) prices would be.

The planning horizon for IRP will likely change within utilities. Long time horizons will continue to dominate the planning of T&D systems (the remaining natural monopolies). Planning horizons will shorten dramatically when Discos plan power-acquisition strategies. Instead of the traditional 20+ year outlook, Disco will probably look no more than 5 to 10 years ahead. Four factors drive this change:

- Because utilities will contract for power rather than build their own plants, they need not be concerned with the costs and benefits of these facilities over their full 40-year lifetimes.
- Because of retail competition, utilities will have a much tougher time forecasting the likely demands of those who buy power from them. However, to plan for expansion of the distribution system, Discos will still need to forecast total system loads and the locations of those loads.
- An increasingly competitive environment will force utilities to use shorter time periods (equivalent to higher discount rates) in assessing the financial benefits and costs of different investments.
- Discos will be increasingly reluctant to accept the risks associated with excess capacity, in part because state regulators will not be able to shift these costs to retail customers.

Utility load forecasting will be different than in the past if retail competition becomes important. In particular, utilities will need to forecast the likely demands for T&D (wheeling) services separate from the demands for energy and capacity. The former forecast will determine utility investments in its "wires" business, while the latter forecast will determine its acquisition of generating resources.

For example, if large industrial customers and resellers are free to choose other suppliers, the utility's load forecast may not include their demands. Exit and entry from the local utility system may require advance notice, which simplifies somewhat the forecasting problem. Beyond that, utilities are likely to focus more on their low load forecasts, whereas historically they focused more on their high forecasts. Attention to the low forecast will be a consequence of the Disco's desire to avoid the risk of paying for excess capacity and to provide only enough resources to meet the needs of its current customers in an effort to keep retail prices as low as possible. Historically, utilities emphasized reliability and assurance of adequate power supplies to meet the possible loads of both existing and potential new customers. In the future, the total generating capacity will be determined by market incentives (i.e., by investor forecasts of electricity prices). No utility or group of utilities will determine the amount of generating capacity to make available or the appropriate reserve margin for a region.

In other respects, forecasting will be more complicated, because electricity services and prices are unbundled. Therefore, utilities will try to forecast demands for energy, power, reliability, energy-efficiency services, power quality, and other services separately.

For at least three reasons, utilities in the future may devote more attention to T&D planning within their IRP processes. First, transmission and distribution will remain natural monopoly functions, and these functions require a single control center for the dispatch of power plants and transportation of power from generators to consumers. As noted above, utilities will own less generation and will increasingly buy power from geographically dispersed sources. To some extent, the local Disco may be responsible for providing transmission paths from these remote generating stations to the step-down transformers that define the input points to the Disco. Second, transmission-owning utilities will be required by FERC regulations to provide transmission access to other parties; this may involve construction of new facilities. Finally, Discos will continue to build distribution facilities to meet the loads of all end (retail) users in its geographic service area, regardless of who owns the power plants that provide the electricity.

The extent to which Discos continue to encourage the use of DSM and renewables will depend on the extent of retail wheeling and the ability of state regulators to maintain a utility's monopoly franchise (Hirst 1994). If retail wheeling is widespread, Discos may find it difficult to spread the costs of DSM and renewable resources across all customers. If state regulators impose a fee on distribution services, however, that money could be used to provide DSM and renewable resources to benefit all customers (California Public Utilities Commission 1995). Experience may show that the transaction costs of buying power from nonregulated merchants is so high that residential and small commercial customers remain full-service customers of the local Disco. This will permit the Disco to run DSM programs that benefit those "core" customers.

If the regulated monopoly franchise shrinks, Discos may increasingly use the rate impact measure, rather than the total resource cost test, to design DSM programs. The total resource cost test leads to a selection of programs that reduce energy-service costs, but may raise electricity prices. The rate impact measure limits DSM to those programs that both reduce costs and prices. It is unclear how much less DSM would be achieved if the rate impact measure is the primary economic criterion used to screen DSM programs. Some recent IRP reports show greater use of the rate impact measure and, as a consequence, less reliance on DSM as an energy resource; compare, for example, the 1992 and 1995 Georgia Power Company resource plans.

However, state regulators might limit retail wheeling to ensure that the Disco can continue to acquire these environmentally benign resources. In this case, utility auctions could include special provisions to ensure acquisition of these resources. These special provisions might include setasides, green pricing, or explicit consideration of environmental externalities. However, regulators may not be able to make such decisions on their own. Given competition among states for industrial plants and the economic and political power of large customers, state legislators and regulators will be pressured to grant these large customers access to competitive markets. In any case, Disco concerns about competition will place strong pressure on them to cut the costs of DSM programs and focus increasingly on programs that meet customer needs and are paid for directly by participating customers.

In summary, IRP will differ from today's model in five key respects (Table 4.5). First, competition is likely to shorten the time horizon over which Discos plan and acquire new resources. Although utilities historically built plants expected to last 40 years (and Discos will likely continue that practice for their distribution systems), Discos will likely sign power contracts with a range of lifetimes, and the longer contracts are likely to be for 10 years or less.

Second, the distribution utility assessment of supply resources will generally focus on the purchase of energy and capacity from other parties. Because the Disco increasingly will be a buyer, rather than a producer, of electricity, it needs to know less about the details of production processes and more about electricity markets. As short-term power markets mature, pressures for long-term contracts at defined prices will come only from those who want protection from market-induced price fluctuations. Long-term contracts will be neither necessary nor useful to ensure adequate supplies of power. Such contracts will, furthermore, have no effect on the short-term prices for power in competitive markets.

Third, utility planning will focus more on their transmission and distribution systems, because this is where the bulk of their capital investments will go. Also, T&D planning will be more complicated because it must encompass a larger number and variety of wholesale and retail transactions than was true in the past.

Fourth, because Discos may no longer have an "obligation to serve" all customers, they will focus their planning to meet the low end of their load forecasts. Historically, utilities focused more on the high end, to ensure that sufficient capacity was available for reliability purposes. Forecasting will also involve separate analyses of the needs for additional T&D investments.

Fifth, pricing of electricity services will be increasingly unbundled and complicated. Distribution utilities, as a consequence, will pay much more attention to the temporal and spatial determinants of their costs to serve different types of customers. These costs, combined with information on customer value, will be used to set unbundled electricity prices.

The extent to which resource planning remains *integrated* will depend on industry structure. If competition focuses on wholesale markets and Discos retain much of their retail-monopoly franchise, IRP will change in its details but not in its overall concepts. However, if retail wheeling is widespread, the integrated part of IRP will disappear as the planning responsibilities are spread among competitive generation companies, transmission monopolies, distribution monopolies, and competitive customer-service companies.

The PUC Role in IRP

The role of the PUC will be substantially different (and diminished) from the typical PUC involvement in IRP today. Because the only entity regulated at the state level will be the Disco (which may not own generation), PUCs will have less influence than they did in the past on construction and operation of generation and transmission facilities.

This loss of PUC influence over generation is important because about two-thirds of the total costs of major U.S. investor-owned electric utilities are in generation. Thus, the fraction of today's retail electricity dollar attributable to Disco investments and activities is small; the bulk of the capital and operating expenses is concentrated in the increasingly competitive generating sector and, therefore, beyond the reach of state regulators.

PUCs will determine the "rules of the road" for the Disco's resource procurements but will have little to say about specific acquisition decisions. For example, PUCs may approve the structure of utility requests for proposals, including the price and nonprice factors included in the scoring system as well as their weights. However, PUCs may or may not review and approve the utility's selection of resources acquired in response to its competitive solicitation. The PUC role in setting the structure of Disco acquisitions will focus on DSM and renewables. However, commissions will have a more difficult time requiring utilities to acquire power from sources that are not lowest in price but, for example, are more environmentally benign. (If state laws or regulations require consideration of environmental externalities, PUCs may be able to impose these requirements on all electricity suppliers, both utilities and independent power producers regardless of location). This difficulty will occur because of competitive pressures at the retail level, many of which will focus on price. In addition, PUCs will increasingly use incentive regulation in lieu of the traditional cost-of-service regulation to cap prices or revenues. Such changes, although they may increase utility use of IRP principles, will further reduce the PUC role in IRP.

Discos will continue to prepare and submit to the PUC resource plans on a regular basis (e.g., once every two or three years). These reports will be less detailed than their early 1990s counterparts because PUCs will impose fewer regulatory requirements on IRP filings. PUCs will continue to review these plans, may conduct public hearings (but with less litigation than now occurs), and may even "approve" such plans. However, the plans and their approvals will focus more on resource-acquisition criteria and strategies than on specific resources.

Because power-plant construction will not be an issue, PUCs might be able to render decisions on utility plans more quickly than they do now. PUC consideration of T&D investments, however, may lengthen the process, especially if a proposed transmission line is to be built in one state for the benefit of producers and consumers in another state.

In summary, PUCs will likely play a smaller role in Disco resource planning. Largely, the PUC role will be (1) to preserve the key social obligations of utilities; (2) to ensure that the more important environmental and broad economic benefits of DSM and renewables are maintained; and (3) to ensure that the Disco develops a balanced resource portfolio for its "core customers" for whom it retains an obligation to serve.

Conclusions

Integrated resource planning is a powerful and flexible way for utilities to plan for and manage the resources needed to provide their customers with desired energy services at a reasonable cost. IRP includes a broad array of supply and demand resources, explicit treatment of uncertainty, consideration of environmental costs as well as direct economic costs, and public involvement. Because of these features, IRP is likely to yield a better mix of resources and fewer controversies among the utility, its regulators, and the public than would traditional planning approaches. By considering a large number and variety of resource options, IRP ensures that the plan adopted will be lower in cost, more environmentally benign, subject to fewer risks, and more acceptable to the public than would a plan that was developed entirely within the utility and that considered only a few supply options.

During the past few years, IRP has expanded along several dimensions. First, more utilities in more parts of the country are preparing and presenting integrated resource plans. Second, the technical quality of these plans is improving. Finally, the plans are addressing additional important topics, such as fuel switching, bidding for resources, and explicit treatment of environmental externalities.

IRP has probably had a slight downward pressure on electricity *costs* and a slight upward pressure on electricity *prices*. This effect is a consequence of the emphasis within IRP on DSM programs as low-cost alternatives to construction of new power plants. In the future, IRP is likely to have virtually no effect on the price of electricity. This conclusion is driven by two beliefs. First, prices will be determined primarily by competitive markets, not by regulation. Second, utility DSM programs will focus more on providing services for which each participating customer will pay.

IRP historically led to slightly greater fuel diversity. This increase in fuel diversity is a consequence of the environmental concern and the explicit attention to uncertainty that characterize IRP. In the future, however, fuel-choice decisions will likely be based primarily on costs in a competitive market with little influence from IRP. However, tighter controls on air-pollution emissions and other environmental constraints will likely lead to greater use of IRP principles and will affect resource selections.

In summary, IRP has encouraged utilities and their regulators to think much more expansively about ways to meet the energy-service needs of customers than they otherwise would have. IRP will continue to be a positive force in the future, although many of its benefits will be captured by emerging competitive markets.

References

C. J. Andrews 1991, *The Marginality of Regulating Marginal Investments: Why We Need a Systemic Perspective on Environmental Externality Adders*, Energy Laboratory, MIT, Cambridge, MA, June.

D. C. Bauer and J. H. Eto 1992, "Future Directions: Integrated Resource Planning," *Proceedings ACEEE 1992 Summer Study on Energy Efficiency in Buildings, Volume 8—Integrated Resource Planning*, pp. 8.1-8.16, American Council for an Energy-Efficient Economy, Washington, D.C., August.

California Public Utilities Commission and California Energy Commission 1987, *Standard Practice Manual, Economic Analysis of Demand-Side Management Programs*, P-400-87-006, San Francisco and Sacramento, CA, December.

California Public Utilities Commission 1995, *Working Group Report, Options for Commission Consideration*, OIR 94-04-031, San Francisco, CA, February 22.

R. Cavanagh 1986, "Least-Cost Planning Imperatives for Electric Utilities and their Regulators," *The Harvard Environmental Law Review 10*(2).

Central Maine Power Company 1991, *Energy Resource Plan*, Augusta, ME, September.

Consumers Power 1992, *1992 Integrated Resource Planning Report*, Jackson, MI.

Duke Power Company 1991, *Least Cost Integrated Resource Planning 1991, Short-Term Action Plan*, Charlotte, NC, April.

Eco Northwest 1993, *Environmental Externalities and Electric Utility Regulation*, prepared for the National Association of Regulatory Utility Commissioners, Washington, D.C., September.

Energy Information Administration 1993, *Changing Structure of the Electric Power Industry 1970-1991*, DOE/EIA-0562, U.S. Department of Energy, Washington, D.C., March.

J. Eto, E. Vine, L. Shown, R. Sonnenblick, and C. Payne 1994, *The Cost and Performance of Utility Commercial Lighting Programs*, LBL-34967, Lawrence Berkeley National Laboratory, Berkeley, CA, May.

J. M. Fang and P. S. Galen 1994, *Issues and Methods in Incorporating Environmental Externalities into the Integrated Resource Planning Process*, NREL/TP-461-6684, National Renewable Energy Laboratory, Washington, D.C., November.

Georgia Power Company 1992, *1992 Integrated Resource Plan*, Atlanta, GA, January.

Georgia Power Company 1995, *1995 Integrated Resource Plan*, Atlanta, GA, January.

C. Goldman, C. A. Comnes, J. Busch, and S. Wiel 1993, *Primer on Gas Integrated Resource Planning*, LBL-34144, Lawrence Berkeley National Laboratory, Berkeley, CA, December.

Green Mountain Power Corp. 1991, *Green Mountain Power Corporation Integrated Resource Plan*, So. Burlington, VT, October.

S. Hadley and E. Hirst 1995, *Utility DSM Programs from 1989 Through 1998: Continuation or Crossroads?*, ORNL/CON-405, Oak Ridge National Laboratory, Oak Ridge, TN, February.

L. J. Hill 1991, *Comparison of Methods to Integrate DSM and Supply Resources in Electric-Utility Planning*, ORNL/CON-341, Oak Ridge National Laboratory, Oak Ridge, TN, December.

E. Hirst 1992, *A Good Integrated Resource Plan: Guidelines for Electric Utilities and Regulators*, ORNL/CON-354, Oak Ridge National Laboratory, Oak Ridge, TN, December.

E. Hirst 1994, *Electric-Utility DSM Programs in a Competitive Market*, ORNL/CON-384, Oak Ridge National Laboratory, Oak Ridge, TN, April.

E. Hirst 1996, "Is There a Future for Electric-Industry IRP?", *Proceedings of the 1996 ACEEE Summer Study on Energy Efficiency in Building*, Volume 7, American Council for an Energy-Efficient Economy, Washington, D.C., August 1996.

E. Hirst, B. Driver, and E. Blank 1992, *Trial By Fire: A Sensible Integrated Resource Planning Rule for Electric Utilities*, Land and Water Fund of the Rockies, Boulder, CO, December.

P. Joskow and D. Marron 1992, "What Does a Negawatt Really Cost? Evidence from Utility Conservation Programs," *The Energy Journal 13*(4), 41-75.

M. Kliman 1994, "Competitive Bidding for Independent Power, Developments in the USA," *Energy Policy 22*(1), 41-56, January.

D. M. Logan, C. A. Neil, A. S. Taylor, and P. Lilienthal 1995, "Integrated Resource Planning with Renewable Resources," *The Electricity Journal* **8**(2), 56-66, March.

J. B. Lowell 1994, "Acquiring Renewables Through IRP: The NEES Experience," *Proceedings Fifth National Conference on Integrated Resource Planning*, pp. 111-116, National Association of Regulatory Utility Commissioners, Washington, D.C., May.

Lower Colorado River Authority 1991, *1991 Integrated Electric Resource Plan*, Austin, TX, September.

Lower Colorado River Authority 1994, *1993 Integrated Resource Plan*, Austin, TX, February.

New England Electric 1992, *Integrated Least Cost Resource Plan for the Fifteen Year Period 1992-2006*, Westborough, MA, April 30.

New England Electric 1993, *NEESPLAN 4, Preparing for the Next Century*, Westborough, MA, September.

Niagara Mohawk Power Corp. 1991, *1991 Integrated Electric Resource Plan*, Syracuse, NY, September.

Northwest Power Planning Council 1991, *1991 Northwest Conservation and Electric Power Plan*, Portland, OR, April.

R. Orans, C. K. Woo, J. N. Swisher, B. Wiersma, and B. Hori 1992, *Targeting DSM for Transmission and Distribution Benefits: A Case Study of PG&E's Delta District*, EPRI TR-100487, Energy and Environmental Economics and Pacific Gas and Electric Company, prepared for the Electric Power Research Institute, Palo Alto, CA, May.

PacifiCorp 1992, *Balanced Planning for Growth, Resource and Market Planning Program RAMPP-2*, Portland, OR, May.

PacifiCorp 1994, *Positioning for Competition and Uncertainty, Resource and Market Planning Program RAMPP-3*, Portland, OR, April.

Portland General Electric 1990, *The 1990 Integrated Resource Plan, A Least Cost Approach*, Portland, OR, October.

Portland General Electric 1992, *1992 Integrated Resource Plan*, Portland, OR, November.

Potomac Electric Power Company 1992, *Integrated Least-Cost Resource Plan: 1992 Energy Plan*, Washington, D.C., March.

Public Service Company of Colorado 1993, *Integrated Resource Plan, 1993, A Balanced Approach to Meeting Customers' Future Electricity Needs*, Denver, CO, October.

Public Service Company of New Mexico 1992, *Annual Report 1991*, Albuquerque, NM.

Puget Power 1992, *Integrated Resource Plan 1992-1993*, Bellevue, WA, April.

J. Raab and M. Schweitzer 1992, *Public Involvement in Integrated Resource Planning: A Study of Demand-Side Management Collaboratives*, ORNL/CON-344, Oak Ridge National Laboratory, Oak Ridge, TN, February.

M. Schweitzer, M. English, S. Schexnayder, and J. Altman 1994, *Energy Efficiency Advocacy Groups: A Study of Selected Interactive Efforts and Independent Initiatives*, ORNL/CON-377, Oak Ridge National Laboratory, Oak Ridge, TN, March.

Southern California Edison Company 1991, *Southern California Edison Company's (U 338-E) Energy Policy and Resource Plan for the Biennial Resource Plan Update*, OII-89-07-004, Rosemead, CA, August and October.

C. Stalon and E. Hirst 1994, *Effects of Electric-Utility Integrated Resource Planning*, Oak Ridge National Laboratory, Oak Ridge, TN, June.

B. Tonn, E. Hirst, and D. Bauer 1994, *IRP and the Electricity Industry of the Future: Workshop Results*, ORNL/CON-398, Oak Ridge National Laboratory, Oak Ridge, TN, September.

Wisconsin Electric Power Company 1991, *Advance Plan 6, Technical Support Document D1, Supplemental Information*, Milwaukee, WI, March.

5

Introduction to Social Externality Costs*

Jonathan Koomey

Florentin Krause
Lawrence Berkeley National Laboratory

Introduction

According to Griffin and Steele (1986), external costs exist when "the private calculation of benefits or costs differs from society's valuation of benefits or costs." Pollution represents an external cost because damages associated with it are borne by society as a whole and are not reflected in market transactions.

The goals of this chapter are modest. It does not attempt a systematic review of the current state of externalities analysis, because such reviews have been recently completed by other authors (CEC 1994, CECA 1993, ECO Northwest 1993, OTA 1994, Weil 1991). Instead, it serves as an introduction for the interested but uninformed reader to some of the key issues in assessing environmental externality costs and gives references for those readers wishing to investigate further.

Many analysts have attempted to quantify societal costs of pollution and other externalities associated with fossil fuel combustion, and some regulatory bodies have even attempted to crudely incorporate externality costs into investment decisions (Cohen et al. 1990, Hashem and Haites 1993). Efforts to incorporate externalities have generally been confined to the regulated sectors of the energy system (electricity and, to a lesser extent, natural gas). Unfortunately, estimates of externality costs are often based on quite different assumptions, making comparisons difficult. Uncertainties in such estimates are large, and can even span orders of magnitude.

5.1 Framework for Understanding Externalities

Exploitation of any energy source generates externalities, defined as societal costs that are not reflected in market transactions. Figure 5.1 (Holdren 1981) shows a detailed listing of stages of energy sources, from exploration to end use. It also shows phases of each stage, from research to dismantling. A comprehensive analysis of external costs must treat each and every stage in the process, which makes any such calculation inherently difficult.

*The work described in this paper was funded by the Assistant Secretary for Energy Efficiency and Renewable Energy of the U.S. Department of Energy, under Contract No. DE-AC03-76SF0098.

0-8493-2514-5/96/$0.00+$.50

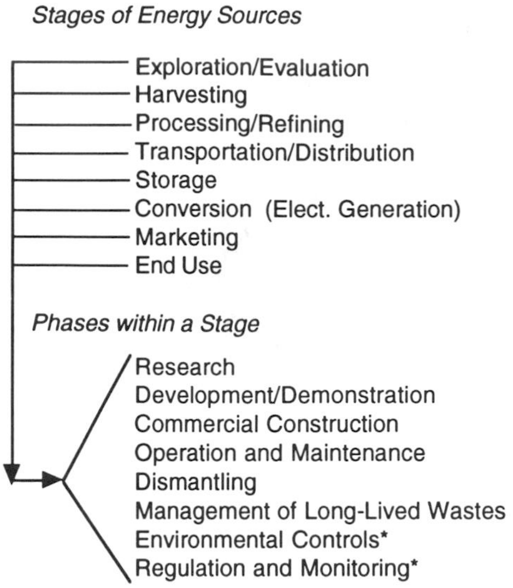

Stages of Energy Sources

Exploration/Evaluation
Harvesting
Processing/Refining
Transportation/Distribution
Storage
Conversion (Elect. Generation)
Marketing
End Use

Phases within a Stage

Research
Development/Demonstration
Commercial Construction
Operation and Maintenance
Dismantling
Management of Long-Lived Wastes
Environmental Controls*
Regulation and Monitoring*

FIGURE 5.1 Steps in energy production, processing, and use. *Source:* Holdren, John P. 1981.

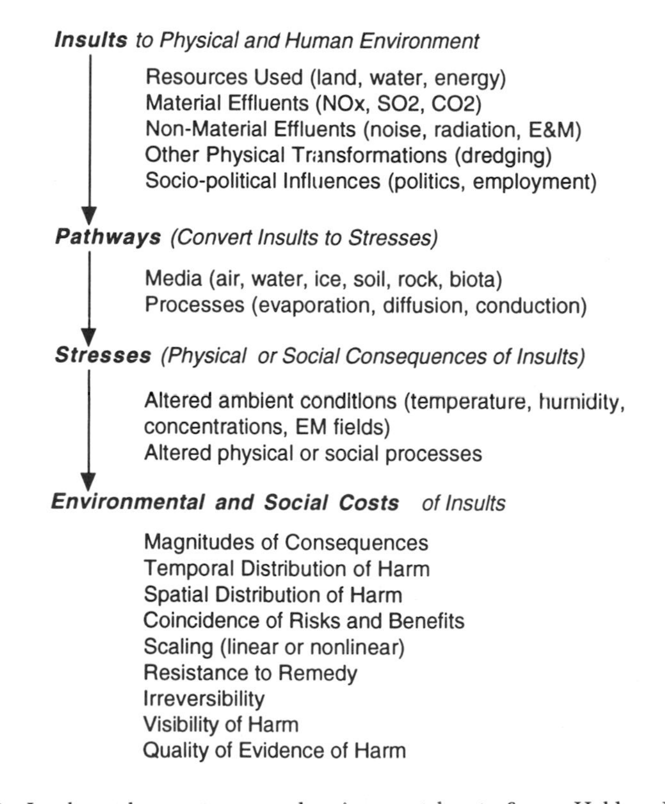

FIGURE 5.2 Insults, pathways, stresses, and environmental costs. *Source:* Holdren, John P. 1981.

Figure 5.2 (Holdren 1981) presents insults, pathways, stresses, and costs. Insults are humankind's physical and chemical intrusions into the natural world. Pathways are those mechanisms by which insults are converted to stresses. Stresses, defined as changes in ambient conditions (social, political, or environmental), then lead directly to societal costs.

TABLE 5.1 Environmental Insults from Fossil Fuels

	All Fuels	Natural Gas	Oil	Coal
Exploration/ harvesting	CO_2, CH_4, N_2O, NO_x, CO, ROG, HCs, particulates, trace metals, thermal pollution	drilling accidents, drilling sludge disposal	drilling accidents, SO_2, drilling sludge disposal	mining injuries, land degradation, SO_2
Processing/ refining	CO_2, CH_4, N_2O, NO_x, CO, ROG, HCs, particulates, trace metals, thermal pollution	refinery accidents, refinery waste disposal	SO_2, refinery accidents, refinery waste disposal	SO_2
Transport/ distribution	CO_2, CH_4, N_2O, NO_x, CO, ROG, HCs, particulates, trace metals, thermal pollution	pipeline accidents, LNG explosions	pipeline and tanker accidents, oil spills, SO_2	train accidents, SO_2
Conversion/ marketing/ end use	CO_2, CH_4, N_2O, NO_x, CO, ROG, HCs, particulates, trace metals, thermal pollution		ash disposal, SO_2	ash disposal, SO_2

Note: ROG = Reactive Organic Gases, HC = hydrocarbons.

TABLE 5.2 Environmental Insults from Existing Nuclear Power, Hydroelectric, and Wind Generation

	Nuclear Power	Hydroelectric	Wind
Exploration/ harvesting	mining accidents, radioactive tailing disposal, land degradation, indirect fossil fuel emissions (from fuel used in harvesting)	N/A	
Processing/ refining	processing accidents, indirect fossil fuel emissions	N/A	
Transport/ distribution	truck accidents, risk of proliferation, indirect fossil fuel emissions	N/A	
Conversion/ marketing/end use	risk of catastrophic accidents, creation of low- and high-level radioactive wastes	may inhibit fish migration	may kill birds, noise pollution
Decommissioning	disposal of low- and high-level radioactive wastes,* indirect fossil-fuel emissions	concrete disposal	

* All U.S. nuclear reactors are charged an annual fee to cover decommissioning and disposal of radioactive wastes. However, neither a disposal site or disposal method has yet been chosen, and no large reactor has ever been decommissioned. It is therefore unknown if the actual costs will correspond to the value of this fee.

Table 5.1 lists environmental and social insults attributable to fossil fuel combustion, and Table 5.2 shows those insults attributable to nuclear, hydroelectric, and wind generation. To understand how Figure 5.2 relates to such insults, consider the case of sulfur dioxide. SO_2 (the insult) is emitted from oil and coal combustion as a gas (this is the first pathway). Some of the SO_2 is converted, through chemical reactions in the atmosphere, to sulfuric acid, some of which then falls in rain into lakes and watersheds (another pathway). Some of this sulfuric acid is neutralized by buffering cations in the water and soil (a third pathway). The altered acidity of the lakes is the stress. The costs (social, economic, and environmental consequences) are the destruction of fish and other wildlife, mobilization of aluminum, damage to trees, and reduction in recreational value of the forest (Harte 1985).

While it is often possible to quantify the size of the insult, the pathways may be so numerous or complicated that only the crudest approximations are possible. Even if it is possible to confidently predict stresses from a given insult, translating those stresses into societal costs is problematic.

Calculating Externality Costs

In general external costs can be crudely characterized by Eq. (5.1)

$$\textit{Externality cost} = \textit{Size of insult} \times \textit{value of environmental damage per unit of insult} \quad \textbf{(5.1)}$$

where

Externality cost = total external cost to society, in dollars:
Size of insult is expressed in physical units (lb emitted or hectares degraded).
Value of environmental damage (VED) is expressed in dollars per physical unit of insult.

Externality costs must be normalized to some common unit of service for consistent comparison. This unit is *delivered* kWh, which includes transmission and distribution losses. For direct fuel consumption, the unit of service is a million BTU (MMBtu).

Air pollution and climate effects tend to dominate most analyses of fossil fuel externalities. Such external costs, which vary with power plant fuel consumption, can be characterized by Eq. (5.2), which is a variation of Eq. (5.1):

$$\textit{Externality cost}\left(\frac{¢}{\mathrm{kWh}}\right) = \mathrm{EF}\left(\frac{\mathrm{lb}}{\mathrm{Btu}}\right) \times \mathrm{HR}\left(\frac{\mathrm{Btu}}{\mathrm{kWh}}\right) \times \mathrm{VED}\left(\frac{¢}{\mathrm{lb}}\right) \quad \textbf{(5.2)}$$

where

EF = Emission factor, in lb/Btu of fuel consumed,
HR = Heat rate of power plant, in Btu/kWh,*
VED = Value of environmental damage, in ¢/lb.

The emission factor relates the particular insult to the amount of fuel burned. The heat rate characterizes the first pathway by which the insult is converted from its fuel-related state to a form that impinges upon the natural environment. The marginal damage cost relates the insult to the social costs. It embodies a relationship between the insult and the stresses that depends on assumptions about geography, dose response, weather, biotic interactions, population density, postcombustion pathways, and myriad other factors.

EF and HR are physical parameters that can be measured, while VED can be calculated using direct cost estimation, abatement costs, or some combination of both. VED is an important parameter for regulatory policy analysis, but it is usually difficult to calculate. It should always be stated explicitly, along with the large number of assumptions needed to calculate such a value.

Environmental Insults from Energy Efficiency and Supply Technologies

Consistent comparisons require that environmental insults from both energy efficiency and supply technologies must be included in externality assessments. Emissions from supply technologies are both direct (from the combustion of fossil fuels) and indirect (from the construction of the equipment and the extraction, processing, and the transport of the fuel). Emissions from efficiency technologies are generally only of the indirect type. On balance, increasing the efficiency of end use reduces emissions and other externalities.

*For direct fuel use, this term becomes the inverse of combustion efficiency.

Energy Supply Technologies

Direct and indirect emissions for fossil fuels are calculated by DeLuchi et al. (1987b), Unnasch et al. (1989), Fritsche et al. (1989), Meridian Corp. (1989), and San Martin (1989). For a complete treatment of both direct and indirect emissions of carbon dioxide, NO_x and SO_2 associated with the latest fossil-fired cogeneration and district heating technologies, see Krause et al. (1994a). The net emissions from cogeneration vary by cogeneration fuel, cogeneration technology, and boiler fuel, and are rarely analyzed.

Indirect emissions for nuclear power are calculated by Meridian Corp. (1989), San Martin (1989), and Fritsche et al. (1989), while emissions and other environmental insults for nuclear are calculated by Ottinger et al. (1990) and Krause et al. (1994b). Meridian Corp. (1989), Krause et al. (1995), and San Martin (1989) also show indirect emissions for renewable power sources.

Efficiency Technologies

Feist (1988) investigates 11 different insulating materials and wall compositions that are widely used in Germany. He finds, based on process analyses for the manufacture and installation of alternative insulating systems by Marmé and Seeberger (1982), that for wall insulation thicknesses now typically applied in retrofits or new buildings in Europe (5–15 cm), indirect primary energy consumption can be neglected, since it amounts to less than 5% of the direct primary energy savings associated with installing the insulation. An analysis of efficiency technologies in the U.S. context (Anderson 1987) came to a similar conclusion.

Methods of Calculating the Value of Emissions Reductions

There are two basic approaches to calculating the value of incremental emissions reductions: "direct damage estimation" and "cost of abatement." Direct damage estimation involves calculating damages that can be definitively linked to emissions of a particular pollutant, in dollar terms (Hohmeyer 1988, Ottinger et al. 1990). For instance, Cavanagh et al. (1982) monetize and tally the human health and environmental effects due to coal consumption in new power plants. These effects include premature human deaths, increased health costs, potential famine induced by global warming, and other effects. Direct estimation is extremely difficult, even when there are relatively few pathways. Some of the most important effects are impossible to quantify, while others depend on pathways that we do not fully understand.

Cost of abatement approaches typically use the cost of pollution controls imposed by regulatory decisions as a proxy for the true externality costs imposed by a pollutant (Chernick and Caverhill 1989, Marcus 1989). This approach (sometimes called "revealed preferences") assumes that regulators' choices embody society's preferences for pollution control, that the marginal costs of mitigation are known, and that these marginal mitigation costs are incurred solely to reduce emissions of a single pollutant (i.e., that there are no other benefits to a pollution reduction investment).

If society's preferences are changing rapidly, abatement cost calculations may be misleading, because society's preferences for pollution control may not accurately represent its present preferences. If mitigation measures have multiple or incommensurate benefits, revealed preference calculations become difficult. For instance, the cost of an energy conservation measure cannot be used to estimate the true value of mitigating SO_2 emissions, since the conservation measure avoids power plants, reduces fuel use, and eliminates other pollutants (Krause and Koomey 1989). In contrast, the cost of flue-gas desulfurization equipment or the price premium of low-sulfur oil over high-sulfur oil can be used without modification in abatement cost analysis, because the cost of these mitigation measures is incurred solely to reduce sulfur emissions.

Estimates from Other Studies

Table 5.3 shows the value of incremental emissions reductions from a variety of studies (in $/lb of pollutant). The difference in assessments of pollutant value reflect different geographic and

TABLE 5.3 Value of Environmental Damage (1989$/lb)

	SO_2 $/lb	NO_x $/lb	CO_2 $/lb C**	ROG $/lb	CH_4 $/lb	N_2O $/lb	Particulates $/lb
EPRI (1987) rural PA, WV							
Low	0.21	0.02	—	—	—	—	—
High	0.85	0.23	—	—	—	—	—
Best estimate	0.48	0.07	—	—	—	—	—
EPRI (1987) (Sub)urban*							
Low	0.48	0.02	—	—	—	—	—
High	2.31	0.23	—	—	—	—	—
Best estimate	1.27	0.07	—	—	—	—	—
Hohmeyer (1988)							
Low	0.233	0.292	—	0.233	—	—	0.233
High	1.244	1.555	—	1.244	—	—	1.244
Chernick and Caverhill (1989)	0.92	1.58	0.042	—	0.37	—	>2.63
Schilberg et al (1989)							
Outside CA	0.50	1.35	0.027	0.33	0.19	1.85	—
CA outside SCAQMD	0.90	9.40	0.027	0.57	0.19	1.85	—
CA inside SCAQMD	9.15	12.25	0.027	8.75	0.19	1.85	—
CEC Staff (1989) in CA	5.75	5.80	0.013	1.65	—	—	3.9
Implied by NY PSC (1989)	0.48	0.94	0.0015	—	—	—	1.01
MA DPU (1990)	0.75	3.25	0.040	2.65	0.11	1.98	2.00
NV PSC	0.78	3.40	0.040	0.59	0.11	2.07	2.09
Pace University	2.03	0.82	0.026	—	—	—	1.19
Minnesota interim values							
Low	0.00	0.03	0.009	0.50	—	—	0.07
High	0.13	0.69	0.021	0.50	—	—	1.00

* NV PSC and Pace University values taken from Weil (1991). Massachusetts Department of Public Utilities values taken from MA DPU (1990). Minnesota interim values are taken from MN PUC (1994), and are adjusted from 1994$ to 1989$ using the consumer price index, which increased 19% over this 5-year period. Other values as reported in Koomey (1990a).

** Values for CO_2 are expressed in $ per lb of carbon.

environmental circumstances, as well as other factors. The assessments of the value of NO_x reductions in California are substantially higher than those estimated by Chernick for New England and implied by the NY Public Service Commission's bidding system. California has some of the strictest air pollution controls in the nation, which reflect its severe NO_x-related ozone problems. Chernick's, Schilberg's, and the California Energy Commissions (CEC) estimates of the value of CO_2 reductions are based on proxy approaches, while the value implied in the Con Ed/NY PSC system (which is less than 12% of the other estimates) is based on cautious initial regulatory response to the global warming problem, and not on explicit analysis.

Damage Costs vs. Cost of Abatement

California is arguably at the leading edge of externality policy. Table 5.4 shows a summary of the externality values calculated in the context of California's electricity planning process, as embodied in the *1992 Electricity Report* (CEC 1993). The table compares damage cost and control cost methods for estimating externalities in ten distinct regions of California. The estimates vary as a function of population density, geographic and meteorological conditions, stringency of emissions regulations, and other factors. Except for particulates with a diameter of less than 10 microns (PM10) in six of the ten regions, control cost methods yield higher externality values than do damage cost methods.

5.2 Pitfalls in Externality Analysis

Holdren (1980) identifies pitfalls in calculating total societal costs associated with energy technologies that affect both direct estimation and cost of abatement approaches. These include

TABLE 5.4 Value of Incremental Emissions Reductions in California (1989$/lb)

Air Basin or District	Valuation Method	SO_x	NO_x	ROG	PM10*	CO
South Coast	Damage functions	3.71	7.24	3.46	23.81	0.00
	Control costs	9.90	13.20	9.45	2.85	4.65
Ventura County	Damage functions	0.75	0.82	0.14	2.05	0.00
	Control costs	3.10	8.25	10.55	0.90	Attainment
Bay Area	Damage functions	1.74	3.67	0.05	12.20	0.00
	Control costs	4.45	5.20	5.10	1.30	1.10
San Diego	Damage functions	1.34	2.78	0.05	7.11	0.00
	Control costs	1.80	9.15	8.75	0.50	0.55
San Joaquin Valley	Damage functions	0.75	3.24	1.86	1.88	0.00
	Control costs	8.90	4.55	4.55	2.60	1.60
Sacramento Valley	Damage functions	0.75	3.04	2.06	1.09	0.00
	Control costs	4.80	4.55	4.55	1.40	2.50
North Coast	Damage functions	0.75	0.40	0.23	0.28	0.00
	Control costs	1.50	3.00	1.75	0.45	Attainment
North Central Coast	Damage functions	0.75	0.98	0.40	1.43	0.00
	Control costs	1.50	4.55	4.55	0.45	Attainment
South Central Coast	Damage functions	0.75	0.82	0.14	2.05	0.00
	Control costs	1.50	4.55	4.55	0.45	Attainment
Southeast Desert	Damage functions	0.75	0.22	0.08	0.34	0.00
	Control costs	9.85	3.00	1.75	2.85	1.45

* PM10 = particulates less than 10 microns in diameter.
Source: Taken from CEC (1993).

(1) inconsistent boundaries; (2) confusing average and marginal effects; (3) illusory precision; (4) environmental stochasticity; and (5) "confusing things that are merely countable with things that really count."

1. **Inconsistent Boundaries.** Boundaries must be drawn consistently to ensure that comparisons between estimates of external costs are fair. This principle implies that the service delivered by competing resources be identical, that all relevant stages of each resource be included in the comparison, and that geographic boundaries be drawn to include all external effects.
2. **Average vs. Marginal Comparisons.**Hohmeyer (1988) calculates costs of externalities from the existing power supply mix in West Germany. While this calculation is useful to show total societal costs from power production, it will almost certainly be misleading to use these embedded externality costs per kWh to calculate the cost of externalities from either *new* power plants or from *marginal existing capacity*, both of which may have characteristics quite different from average existing plants.
3. **Illusory Precision.** There are often large uncertainties in specifying the size of insults, in translating insults through pathways to stresses, in converting stresses to consequences, and in valuing consequences. To ignore such uncertainty by specifying single-point estimates to many significant figures can be quite misleading, since it creates the illusion that the estimates are certain. To avoid misunderstandings, externality cost estimates should be assigned appropriate error bounds. Such uncertainty creates a quandary for regulators, since most regulatory determinations *must* be in terms of point estimates. Analysts can best serve regulators by making the uncertainties explicit and understandable.
4. **Stochasticity.** Environmental and social systems are often characterized by *stochasticity*, or probabilistic variability about some mean value. The most interesting and important interactions between human societies and the natural world occur when one or both of these systems are far from their respective mean values. Overzealous averaging of important parameters may disguise damages that occur only under extreme conditions.

   For instance, calculations of damages from ambient air pollutant concentrations may yield vastly different results depending on how the concentrations are averaged over time. Damages may not be linearly related to pollutant concentration, and may only occur if concentrations exceed some threshold value. Calculating damages based on the annual average pollutant concentration might be misleading for these reasons. Daily or hourly averages sorted by concentration would give a more accurate picture.

5. **What Is Countable versus What Counts.** Analysts often focus on those things that are amenable to quantitative treatment. Yet the probabilities, consequences, and risk-adjusted expected costs of many important external effects (such as nuclear sabotage, nuclear proliferation, or global warming) may be difficult or impossible to quantify, and may also be irreversible once the damages are incurred. Since the "facts" are uncertain or nonexistent, and may not become certain before decisions need to be made, such costs can only be valued through the political process. In that circumstance, it is especially crucial that analysts' value judgements be made explicit.

5.3 Using Externality Costs in the Context of Policy

Some states have not explicitly monetized externality values, but have expressed their preferences for low-polluting resources (such as efficiency and renewables) by increasing the cost of conventional resources by some fixed percentage (15% in Wisconsin, 10% in Iowa) for the purposes of resource planning. While such approximations push resource choices toward the less polluting resources, basing externalities on a percentage of the busbar cost of the resource can lead to perverse results (Koomey 1990a). Damages from pollutants are not, in general, correlated with resource costs, but are strongly related to pollutant emissions, local topography, population density, and other physical characteristics of the surrounding area.

There are huge uncertainties in assessing externality costs related to greenhouse gas emissions, and the exact values of these costs are probably unknowable. In spite of such uncertainties, it is clear that all emissions that contribute to global warming should be treated similarly. Carbon, which is the most important contributor to the global warming problem, is by no means the only one. Radiatively active trace gases like methane (CH_4), nitrous oxide (N_2O), and chloroflourocarbons (CFCs) should all be assigned the same externality cost per unit of global warming contribution. The appendices in Krause et al. (1989) explain how to convert concentrations of the other gases into *equivalent CO_2 concentrations*, which can then be used to assign these gases externality costs (once the appropriate cost for CO_2 has been determined). Others have also derived "warming factors" that can be used to achieve the same result [e.g., Unnasch et al. (1989) and DeLuchi et al. (1987a)].

An important consideration for policy makers in this area is that "getting prices right" is not the end of the story. Many market failures affecting energy use will still remain after external costs are incorporated (Fisher and Rothkopf 1989, Koomey 1990b, Levine et al. 1994, Sanstad et al. 1993). They are amenable to a variety of nonprice policies, including efficiency standards, and incentive, information, and research and development programs. For more discussion of the policy issues surrounding energy tax and nonprice policies, see Krause et al. (1993).

Conclusions

Estimates of the consequences of technological choices (including, but not limited to, estimates of externality costs) will always be inaccurate because many effects are spread geographically and chronologically, and the causal links are extremely complicated. Our understanding of these links will always lag behind our ability to alter them, as David Bella (1979) points out:

> Changes can be accomplished one at a time as if they were essentially in isolation from each other. Moreover, only a small part of the environment and only a few environmental properties

must be understood in order to produce a change. In contrast, to foresee the consequences of change requires that one examine the combined effect of many changes.

As in all areas of life, externality policy must be made in the face of imperfect information. We can make action easier by looking for common ground and by undertaking policies that have multiple benefits. We must be prepared to experiment, to change course in response to new information, and to learn from our mistakes. To not incorporate externalities in prices is to implicitly assign a value of zero, a number that is demonstrably wrong.

Acknowledgments

We wish to thank the following colleagues for their substantive assistance with this chapter and related work: Frank Kreith, Olar Hohmeyer, Dick Ottinger, and Beth Schwehr. The authors alone are responsible for the analysis presented here and any errors contained within.

References

Anderson, Kent B. 1987. *Conservation versus Energy Supply: An Economic and Environmental Comparison of Alternatives for Space Conditioning of New Residences.* Ph.D. Thesis, Energy and Resources Group, University of California, Berkeley.

Bella, David A. 1979. "Technological Constraints on Technological Optimism." *Technological Forecasting and Social Change.* vol. 14, p. 15.

Cavanagh, Ralph, Margie Gardner, and David Goldstein. 1982 "Part IIIA2E: Environmental Costs." In *A Model Electric Power and Conservation Plan for the Pacific Northwest.* Northwest Conservation Act Coalition.

CEC. 1993. *1992 Electricity Report.* California Energy Commission. P104-92-001. January.

CEC. 1994. *Staff Report on Internalizing Externalities.* California Energy Commission. Docket #93-ER-94. September 15.

CEC. 1989. Energy Facility Siting and Environmental Protection Division. 1989. *Valuing Emission Reductions for Electricity Report 90.* California Energy Commission. Staff Issue Paper #3R, Docket #88-ER-8. November 21.

CECA. 1993. *Incorporating Environmental Externalities Into Utility Planning: Seeking a Cost-Effective Means of Assuring Environmental Quality.* Consumer Energy Council of America. July.

Chernick, Paul, and Emily Caverhill. 1989. *The Valuation of Externalities From Energy Production, Delivery, and Use: Fall 1989 Update.* A Report by PLC, Inc. to the Boston Gas Co. December 22.

Cohen, S.D., J.H. Eto, C.A. Goldman, J. Beldock, and G. Crandall. 1990. *A Survey of State PUC Activities to Incorporate Environmental Externalities into Electric Utility Planning and Regulation.* Lawrence Berkeley National Laboratory. LBL-28616. May.

DeLuchi, Mark A., Robert A. Johnston, and Daniel Sperling. 1987a. *Transportation Fuels and the Greenhouse Effect.* University-Wide Energy Research Group, University of California. UER-182. December.

DeLuchi, Mark A., Daniel Sperling, and Robert A. Johnston. 1987b. *A Comparative Analysis of Future Transportation Fuels.* Institute of Transportation Studies, University of California, Berkeley. UCB-ITS-RR-87-13. October.

ECO Northwest. 1993. *Environmental Externalities and Electric Utility Regulation.* Prepared for the National Association of Regulatory Utility Commissioners, Washington, DC. September.

EPRI. 1987. *TAG-Technical Assessment Guide: Vol. 4: Fundamentals and Methods, End Use.* EPRI. Electric Power Research Institute. P-4463-SR, vol. 4, August.

Feist, W. 1988. *Improving the Thermal Integrity of Buildings: State of the Art versus Current Standards.* Institut fuer Wohnen und Umwelt, Darmstadt.

Fisher, Anthony C., and Michael H. Rothkopf. 1989. "Market Failure and Energy Policy: A Rationale for Selective Conservation." *Energy Policy*. vol. 17, no. 4. p. 397.

Fritsche, U., L. Rausch, and K.H. Simon. 1989. *Umweltwirkungsanalyse von Energiesystemen: Gesamtemissions-Modell Integrieter Systeme (GEMIS)*. Oeko-Institut Darmstadt und Wissenschaftliches Zentrum GHS Kassel. August.

Griffin, James M., and Harry B. Steele. 1986. *Energy Economics and Policy*. 2nd ed. Orlando, FL: Academic Press College Division.

Harte, John. 1985. *Consider a Spherical Cow: A Course in Environmental Problem Solving*. Los Altos, CA: William Kaufmann, Inc.

Hashem, Julie, and Erik Haites. 1993. *Status Report: State Requirements for Considering Environmental Externalities in Electric Utility Decision Making*. Prepared for the Association of DSM Professionals (ADSMP) by Barakat and Chamberlin, Inc. January.

Hohmeyer, Olav. 1988. *Social Costs of Energy Consumption: External Effects of Electricity Generation in the Federal Republic of Germany*. Berlin: Springer-Verlag.

Holdren, John P. 1980. *Integrated Assessment for Energy-Related Environmental Standards: A Summary of Issues and Findings*. Lawrence Berkeley National Laboratory. LBL-12779. October.

Holdren, John P. 1981. "Energy and Human Environment: The Generation and Definition of Environmental Problems." In *The European Transition from Oil: Societal Impacts and Constraints on Energy Policy*. Chapter V. Edited by G.T. Goodman, L.A. Kristoferson and J.M. Hollander. London: Academic Press.

Koomey Jonathan. 1990a. *Comparative Analysis of Monetary Estimates of External Environmental Costs Associated with Combustion of Fossil Fuels*. Lawrence Berkeley National Laboratory. LBL-28313. July.

Koomey, Jonathan. 1990b. *Energy Efficiency Choices in New Office Buildings: An Investigation of Market Failures and Corrective Policies*. Ph.D. Thesis, Energy and Resources Group, University of California, Berkeley.

Krause, Florentin, Wilfred Bach, and Jon Koomey. 1989. *Energy Policy in the Greenhouse*. Volume 1. *From Warming Fate to Warming Limit: Benchmarks to a Global Climate Convention*. El Cerrito, CA: International Project for Sustainable Energy Paths.

Krause, Florentin, Eric Haites, Richard Howarth, and Jonathan Koomey. 1993. *Energy Policy in the Greenhouse*. Volume II, part 1. *Cutting Carbon Emissions—Burden or Benefit? The Economics of Energy-Tax and Non-Price Policies*. El Cerrito, CA: International Project for Sustainable Energy Paths.

Krause, Florentin, Jonathan Koomey, Hans Becht, David Olivier, Giuseppe Onufrio, and Pierre Radanne. 1994a. *Energy Policy in the Greenhouse*. Volume II, Part 3C. *Fossil Generation: The Cost and Potential of Low-Carbon Resource Options in Western Europe*. El Cerrito, CA: International Project for Sustainable Energy Paths.

Krause, Florentin, Jonathan Koomey, David Olivier, Pierre Radanne, and Mycle Schneider. 1994b. *Energy Policy in the Greenhouse*. Volume II, Part 3E. *Nuclear Power: The Cost and Potential of Low-Carbon Resource Options in Western Europe*. El Cerrito, CA: International Project for Sustainable Energy Paths.

Krause, Florentin, Jonathan Koomey, and David Oliver. 1995. *Energy Policy in the Greenhouse*. Volume II, Part 3D. *Renewable Power: The Cost and Potential of Low Carbon Resource Options in Western Europe*. El Cerrito, CA: International Project for Sustainable Energy Paths.

Krause, Florentin, and Jonathan G. Koomey. 1989. *Unit Costs of Carbon Savings from Urban Trees, Rural Trees, and Electricity Conservation: A Utility Cost Perspective*. Lawrence Berkeley National Laboratory. LBL-27311. Presented at the Conference on Urban Heat Islands, February 23–24, 1989. Berkeley, California. July.

Levine, Mark D., Eric Hirst, Jonathan G. Koomey, James E. McMahon, and Alan H. Sanstad. 1994. *Energy Efficiency, Market Failures, and Government Policy*. Lawrence Berkeley National Laboratory. LBL-35376. March.

MA DPU. 1990. *Implementing an Environmental Externality Evaluation Method for Electric Utility Planning and Resource Acquisition in Massachusetts (excerpts from MA DPU order 89-239).* Presenter: Henry Yoshimura, Massachusetts Department of Public Utilities. August 31.

Marcus, William B. 1989. *Prepared Testimony of William B. Marcus on Marginal Cost and Revenue Allocation.* San Francisco, CA: California Public Utilities Commission.

Marmé, W. and J. Seeburger. 1982. "Der Primaerenergicinhalt von Baustoffen." *Bauphysik.* vol. 4, no. 5, p. 155 and no. 6, p. 208.

Meridian Corp. 1989. *Energy System Emissions and Material Requirements.* The Deputy Assistant Secretary for Renewable Energy, U.S. Department of Energy. February.

MN PUC. 1994. In the *Matter of the Quantification of Environmental Costs Pursuant to laws of Minnesota 1993, Chapter 356, Section 3.* Minnesota Public Utilities Commission. Docket No. E-999/CI-93-583, 1994 Minn. PUC LEXIS; 150 P.U.R. 4th 130. March 1.

NY PSC, New York Public Service Commission. 1989. *Order Issuing a Final Environmental Impact Statement—Case 88-E-246—Proceeding on Motion of the Commission (established in Opinion No. 88-15) as to the Guidelines for Bidding to Meet Future Electric Needs of Consolidated Edison Company of New York, Inc.* July 19.

OTA, Office of Technology Assessment. 1994. *Background Paper: Studies of the Environmental Costs of Electricity.* U.S. Government Printing Office. OTA-ETI-134. September.

Ottinger, Richard L., David R. Wooley, Nicholas A. Robinson, David R. Hodas, Susan E. Babb, Shepard C. Buchanan, Paul L. Chernick, Emily Caverhill, Alan Krupnick, Winston Harrington, Seri Radin, and Uwe Fritsche. 1990. *Environmental Costs of Electricity.* New York, NY: Oceana Publications, Inc., for the Pace University Center for Environmental and Legal Studies.

San Martin, Robert L. 1989. *Environmental Emissions from Energy Technology Systems: The Total Fuel Cycle.* U.S. Department of Energy. Spring 1989.

Sanstad, Alan H., Jonathan G. Koomey, and Mark D. Levine. 1993. *On the Economic Analysis of Problems in Energy Efficiency: Market Barriers, Market Failures, and Policy Implications.* Lawrence Berkeley National Laboratory. LBL-32652. January.

Schilberg, G.M., J.A. Nahigian, and W.B. Marcus. 1989. *Valuing Reductions in Air Emissions and Incorporation into Electric Resource Planning: Theoretical and Quantitative Aspects (re: CEC Docket 88-ER-8).* JBS Energy, Inc., for the Independent Energy Producers. August 25.

Unnasch, Stefan, Carl B. Moyer, Douglas D. Lowell, and Michael D. Jackson. 1989. *Comparing the Impact of Different Transportation Fuels on the Greenhouse Effect.* Report to the California Energy Commission. P500-89-001. April 1989.

Weil, Stephen. 1991. "The New Environmental Accounting: A Status Report." *The Electricity Journal.* vol. 4, no. 9, p. 46. November.

6

Generation Technologies Through the Year 2005

Anthony F. Armor
Electric Power Research Institute

Introduction

This chapter reviews the status and likely application of competing generation technologies, particularly those with near-term potential. Capacity additions in the United States in the next 10 years will be based on gas, coal, and to some extent on renewables. Repowering of older plants will likely be increasingly attractive.

Gas turbine-based plants will dominate in the immediate future. The most advanced combustion turbines achieve more than 40% efficiency in simple cycle mode and greater than 50% lower heating value (LHV) efficiency in combined cycle mode. In addition, combustion turbine/combined cycle (CT/CC) plants offer siting flexibility, phased construction, and capital costs between $400/kW and $800/kW. These advantages, coupled with adequate natural gas supplies and the assurance of coal gasification backup, make this technology a prime choice for green field and repowered plants.

There is also good reason why the pulverized coal plant may still be a primary choice for many generation companies. Scrubbers have proved to be more reliable and effective than early plants indicated. Up to 99% SO_2 removal efficiency is possible. By the year 2000, 30 GW of U.S. coal-fired generation will likely be equipped with flue gas desulfurization (FGD) systems. Also, the pulverized-coal (PC) plant has the capability for much improved heat rate (about 8,500 Btu/kWh) even with full flue gas desulfurization.

0-8493-2514-5/96/$0.00+$.50

Atmospheric and pressurized fluidized bed combustion (FBC), offer reductions in both SO_2 and NO_x and also permit the efficient combustion of low-rank fuels. In the United States, there are now over 150 operating units for power generation and ten vendors of FBC boilers, four of which offer units up to 250 MW in size.

Gasification power plants exist at the 100 MW and 160 MW levels and are planned up to 450 MW. Much of the impetus is now coming from the DOE clean coal program, where three gasification projects are in progress and four more are planned.

In small unit sizes, often suitable for distributed generation, technical progress will be made (although large-scale applications still remain modest) in renewables (solar, wind, biomass) and in fuel cells. Fuel cells promise high efficiencies, low emissions, and compact plants.

Capital cost will remain a determining issue in the application of all these possible generation options.

Overall Industry Needs

The U.S. electric utility industry consists of a network of small and large utilities, both private and public—more than 3,000 in all. They generate more than 700 GW of electric power—by far the greatest concentration of electric power production in the world, about equal to the next five countries combined. However, although advanced generation technologies are beginning to find their way into the power industry, most installed capacity is 20 or more years old and equipment efficiency reflects this vintage.

The National Energy Strategy of 1991 set general U.S. policy for the future, and one specific directive was to enhance the efficiency of generation, transmission, and use of electricity. There are two key drivers for this directive: one is to reduce emissions of undesired air, water, and ground pollutants, and the other is to conserve our fossil fuels. The recent energy tax, proposed by the current administration, is another aspect of the desire to move to a more energy-conscious mode of operation, which will clearly give impetus to the renewables as significant (although probably not major) future power sources and will encourage efficiency advances in both fossil and nuclear plants.

6.1 Management of Existing Fossil Plant Assets

Utilities are now looking at power plants in a more profit-focused manner, treating them as company assets, to be invested in a way that maximizes the company bottom line or the profit for the utility. As the average age of fossil unit inches upward, this question is often being asked by utility executives, "If I invest a dollar into this plant to improve heat rate, availability, or some other plant performance measure, will this produce more in base profit to my company than investing say in some other plant, or building new capacity, or buying power from outside?" One important aspect of this new business strategy concerns the "use" that is being made of any particular plant, since increased plant usage implies more company value for that plant.

Here are some measures of plant utilization:

Heat rate is one. It is of course the quantifiable measure of how efficiently we can convert fuel into MW. It is inherently limited by cycle and equipment design and by how we operate the plant. In a simple condensing cycle the heat rate of a fossil fuel plant cannot fall much below 8,500 Btu/kWh, even with supercritical cycles, and double reheat of the feedwater.

Capacity factor (CF) is a measure of use that tells us how the plant is loaded over the year. Few fossil units achieve 90% capacity factor these days, and this has an impact on the measured heat rate of the unit. Under ideal conditions for effective asset management, and apart from downtime for maintenance, CF should be close to 100% for the purposes of getting the most out of the plant asset. Market conditions and the reserve margins of the utility often dictate otherwise.

Cost of electricity is a determining factor in how units are dispatched. Electricity cost is largely dictated by fuel cost, which typically makes up 70% of the cost of operating a power plant. It is interesting that none of the top ten U.S. units in heat rate makes the top ten in electricity cost.

Finally, I have suggested a term called **energy efficiency**, which describes how well a plant utilizes the basic feedstock (coal, oil or gas). If we can produce other products from a fossil plant besides electricity, the value of that asset goes up, and of course the "effective" heat rate drops significantly.

Using these measures we are seeing many, perhaps most, of the major U.S. generation companies take a close look at their plant assets to judge their bottom-line value to the company. In the growing competitive generation business, an upgrade or maintenance investment in a power plant will be determined largely on the return on investment the company can expect at the corporate level. In order to achieve these corporate goals, it is necessary to have a good handle on equipment life and the probability of failures. Also needed are options for improving heat rate, for increasing output (by repowering), and for improving plant productivity to make the assets competitive.

6.2 Clean Coal Technology Development

At an increasing rate in the last few years, innovations have been developed and tested aimed at reducing emissions through improved combustion and environmental control in the near term, and in the longer term by fundamental changes in how coal is preprocessed before converting its chemical energy to electricity. Such technologies are referred to as clean coal technologies—described by a family of precombustion, combustion/conversion, and postcombustion technologies. They are designed to provide the coal user with added technical capabilities and flexibility and at lower net cost than current environmental control options. They can be categorized as follows:

- **Precombustion**, in which sulfur and other impurities are removed from the fuel before it is burned.
- **Combustion**, in which techniques to prevent pollutant emissions are applied in the boiler while the coal burns.
- **Postcombustion**, in which the flue gas released from the boiler is treated to reduce its content of pollutants.
- **Conversion**, in which coal, rather than being burned, is changed into a gas or liquid that can be cleaned and used as a fuel.

Coal Cleaning

Cleaning of coal to remove sulfur and ash is well established in the United States with more than 400 operating plants, mostly at the mine. Coal cleaning removes primarily pyritic sulfur (up to 70% SO_2 reduction is possible) and in the process increases the heating value of the coal, typically by about 10% but occasionally by 30% or higher. Additionally, if slagging is a limiting item, increased MW may be possible, as at one station which increased generation from 730 MW to 779 MW. The removal of organic sulfur, chemically part of the coal matrix, is more difficult, but may be possible using microorganisms or through chemical methods, and research is underway. Finally, heavy metal trace elements can be removed also, conventional cleaning removing (typically) 30 to 80% of arsenic, mercury, lead, nickel, and antimony.

Pulverized-Coal-Fired Plants

Built in 1959, Eddystone 1 at PECO Energy was, and still is, the supercritical power plant with the highest steam conditions in the world. Main steam pressure was 5,000 psi when built, and steam

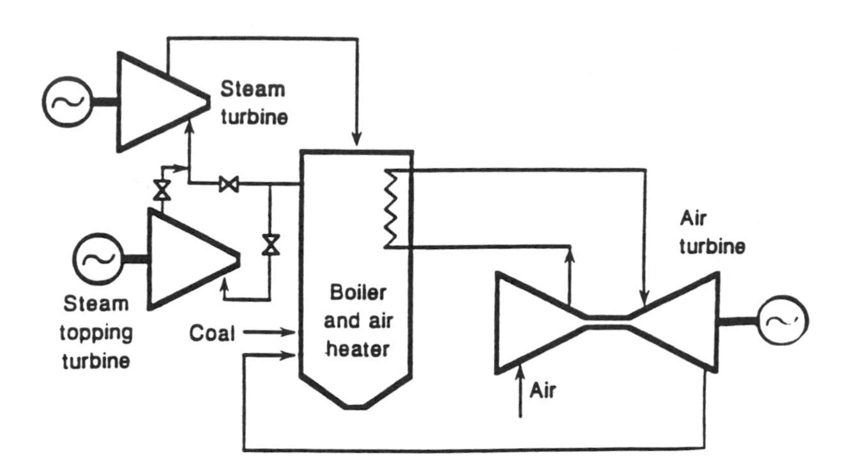

FIGURE 6.1 A pulverized-coal combined cycle with topping steam turbine has a projected heat rate of 7,200 Btu/kWh. The air turbine uses 1,800°F air, or 2,300°F air with supplemental firing. The topping turbine uses steam at 1,300°F.

temperature 1,200°F for this double reheat machine. PECO Energy will continue to operate Eddystone I to the year 2010, an impressive achievement for a prototype unit.

But the most efficient pulverized-coal-fired plant of the future is likely to be a combined cycle plant, perhaps with a topping steam turbine, as shown in Figure 6.1. With a 1,800°F air turbine and 1,300°F topping steam turbine, the heat rate of this cycle is about 7,200 Btu/kWh—very competitive with any other proposed advanced cycles in the near term.

So although there has been a perception that the pulverized-coal power plant has come to the end of the road and that advanced coal technologies will quite soon make the PC plant with a scrubber, whose efficiency hovers around 35%, obsolete, this perception is premature. There is good reason why the PC plant is still a primary choice for many electricity generators, since scrubbers have proved to be more reliable and effective than early plants indicated. Up to 99% SO_2 removal efficiency is possible. By the year 2000 60 GW of U.S. coal-fired generation will likely be equipped with FGD systems.

6.3 Emissions Control

Worldwide about 40% of electricity is generated from coal. Installed coal-fired generating capacity, more than 1,000 GW, is largely made up of 340 GW in North America, 220 GW in Western Europe, Japan, and Australia, 250 GW in Eastern Europe and the former USSR, and 200 GW in China and India. In the decade to the year 2000, 190 GW of new coal-fired capacity will likely be added. The control of particulates, sulfur dioxides, and nitrogen oxides from those plants is one of the most pressing needs of today and of the future. This is accentuated when the impact of carbon dioxide emissions, with its contribution to global warming, is considered. To combat these concerns, a worldwide move toward environmental retrofitting of older fossil-fired power plants is underway, focused largely on sulfur dioxide scrubbers and combustion or postcombustion optimization for nitrogen oxides.

Sulfur Dioxide Removal

When it is a matter of retrofitting an existing power plant, no two situations are identical: fuels, boiler configurations, and even space available for new pollution control equipment all play a role in the decision on how a utility will meet new emission reduction requirements. For example, a decision to install a sorbent injection technology rather than flue gas desulfurization (FGD) for SO_2 reduction may depend not only on the percentage reduction required but also on the space

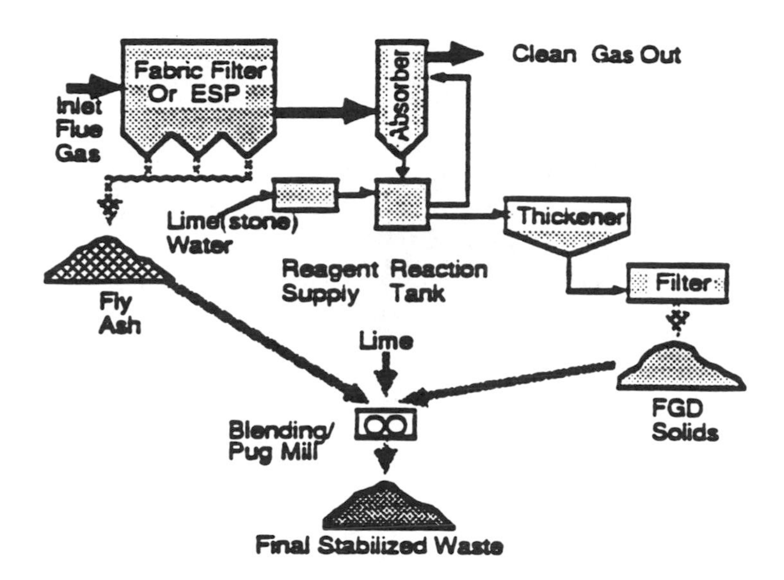

FIGURE 6.2 The conventional lime/limestone wet scrubber is the dominant system.

constraints of the site and on the capacity factor of the plant (with a lower capacity factor, the lower capital cost of sorbent injection is advantageous compared to FGD). Most utilities, though, have been selecting either wet or dry scrubbing systems for desulfurization. Generically these can be described as in the following sections.

Conventional Lime/Limestone Wet Scrubber. By 1994 the United States already had more than 280 flue gas desulfurization (FGD) systems in operation on 95,000 MW at utility stations; now the U.S. experience is approaching 1,000 unit years.

The dominant system is the wet limestone design, limestone being one-quarter the cost of lime as a reagent. In this system (Figure 6.2) the limestone is ground and mixed with water in a reagent preparation area. It is then conveyed to a spray tower called an absorber, as a slurry of 90% water and 10% solids, and sprayed into the flue gas stream. The SO_2 in the flue gas is absorbed in the slurry and collected in a reaction tank, where it combines with the limestone to produce water and calcium sulfate or calcium sulfite crystals. A portion of the slurry is then pumped to a thickener where these solids/crystals settle out before going to a filter for final dewatering. Mist eliminators installed in the system ductwork at the spray tower outlet collect slurry/moisture entrained in the flue gas stream. Calcium sulfate is typically mixed with flyash (1:1) and lime (5%), and disposed of in a landfill.

Various improvements can be made to this basic process, including the use of additives for performance enhancement and the use of a hydrocyclone for dewatering, replacing the thickener, and leading to a salable gypsum byproduct. The Chiyoda-121 process (Figure 6.3) reverses the classical spray scrubber and bubbles the gas through the liquid. This eliminates the need for spray pumps, nozzle headers, separate oxidation towers, and thickeners. Bechtel has licensed this process in the United States, the first commercial installation is at the University of Illinois on a heating boiler, and a DOE clean coal demonstration is underway. The waste can be sold as gypsum or disposed of in a gypsum stack.

Spray Drying. Spray drying (Figure 6.4) is the most advanced form of dry SO_2 control technology. Such systems tend to be less expensive than wet FGD but typically remove a smaller percentage of the sulfur (90% compared with 98%). They are used when burning low-sulfur coals and utilize fabric filters for particle collection, although recent tests have shown applicability to high-sulfur coals also.

Spray driers use a calcium oxide reagent (quicklime), which, when mixed with water, produces a calcium hydroxide slurry. This slurry is injected into the spray drier, where it is dried by the hot

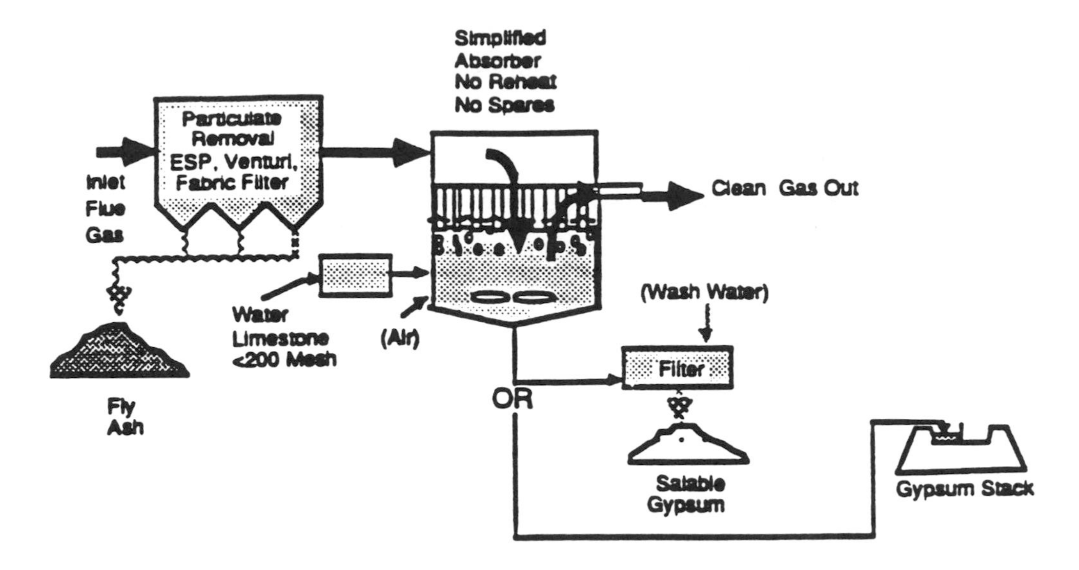

FIGURE 6.3 The Chiyoda-121 scrubber simplifies the process by bubbling the flue gas through the liquid, eliminating some equipment needs.

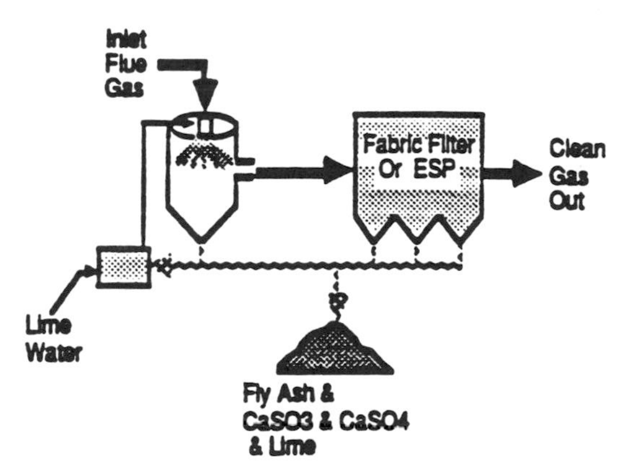

FIGURE 6.4 Spray driers use a calcium oxide reagent mixed with water, which is dried by the flue gas. A dry product is collected in a fabric filter.

flue gas. As the drying occurs, the slurry reacts to collect SO_2. The dry product is collected at the bottom of the spray tower and in the downstream particulate removal device, where further SO_2 removal may take place. It may then be recycled to the spray drier to improve SO_2 removal and alkali utilization.

For small, older power plants with existing electrostatic precipitators (ESPs), the most cost-effective retrofit spray dry configuration locates the spray dryer and fabric filter downstream of the ESP, separating in this manner the spray dryer and fly ash waste streams. The fly ash can then be sold commercially.

Control of Nitrogen Oxides

Nitrogen oxides can be removed either during or after coal combustion. The least expensive option and the one generating the most attention in the United States is combustion control, first through adjustment of the fuel/air mixture, and second through combustion hardware modifications.

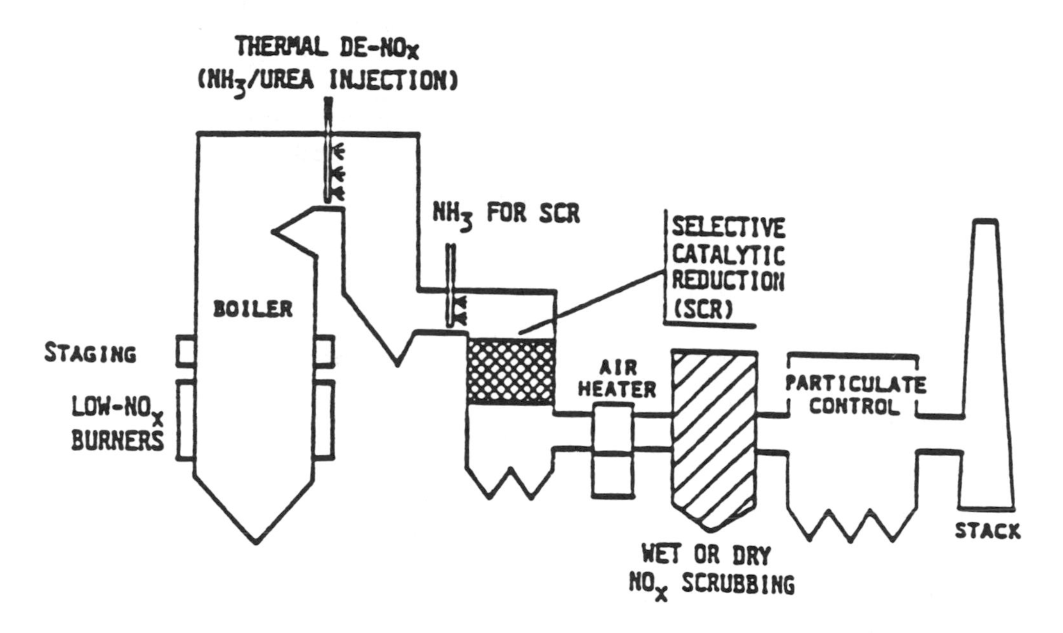

FIGURE 6.5 Control options for NO_x include operational, hardware, and postcombustion modifications.

Postcombustion processes seek to convert NO_x to nitrogen and water vapor through reactions with amines such as ammonia and urea. Selective catalytic reduction (SCR) injects ammonia in the presence of a catalyst for greater effectiveness. So the options (Figure 6.5) can be summarized as

Operational changes. Reduced excess air and biased firing, including taking burners out of service.

Hardware combustion modifications. Low NO_x burners, air staging, and fuel staging (reburning).

Postcombustion modifications. Injection of ammonia or urea into the convection pass, selective catalytic reduction (SCR), and wet or dry NO_x scrubbing (usually together with SO_2 scrubbing).

Low NO_x burners can reduce NO_x by 50%, and SCR by 80%, but the low NO_x burner option is much more cost-effective in terms of cost per ton of NO_x removed. Reburning is intermediate in cost per removed ton and can reduce NO_x by 50%, or 75% in conjunction with low NO_x burners.

Fluidized-Bed Plants. Atmospheric fluidized-bed boilers (see Figure 6.6) offer reductions in both SO_2 and NO_x, and also permit the efficient combustion of low-rank fuels. In the United States there are now over 150 operating units generating over 5,000 MW, and ten vendors of FBC boilers, four of which offer units up to 250 MW in size. The rapid growth in number of FBC units and capacity is shown in Figure 6.7.

The demonstration projects at TVA (Shawnee, 160 MW) and Northern States Power (Black Dog, 133 MW) are examples of successful bubbling-bed units. The Shawnee unit now routinely operates at 85 to 90% capacity factor, having overcome initial fuel feed problems. The Black Dog unit has been dispatched in a daily cycling mode and has successfully fired a blend of coal and petroleum coke. The focus of AFBC in the United States is now on circulating fluid beds. In fact, more than 70% of operating fluid-bed boilers in the United States are of the circulating type. The CFBC project at Nucla (Tri-state) has been successful in demonstrating the technology at the 110 MW level, and commercial plants have now reached 250 MW in size. Most fluidized-bed units are being installed by independent power producers in the 50 to 100 MW size range, where the inherent SO_2 and NO_x advantages over the unscrubbed PC plant have encouraged installations even in such traditional noncoal arenas as California.

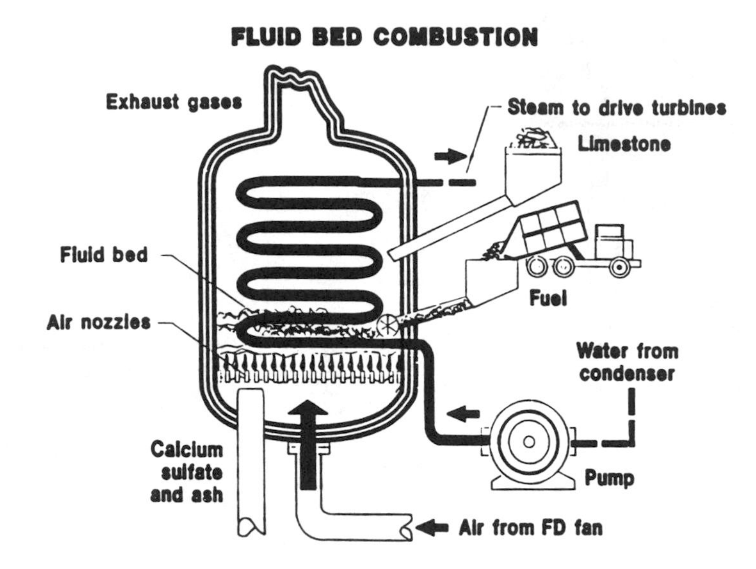

FIGURE 6.6 The addition of limestone or Dolomite to the combustion chamber allows the coal limestone mixture to be burned in a suspended bed, fluidized by an underbed air supply. The sulfur in the coal reacts with the calcium to produce a solid waste of calcium sulfate. The combustion temperature is low (1,500°F), reducing NO_x emissions.

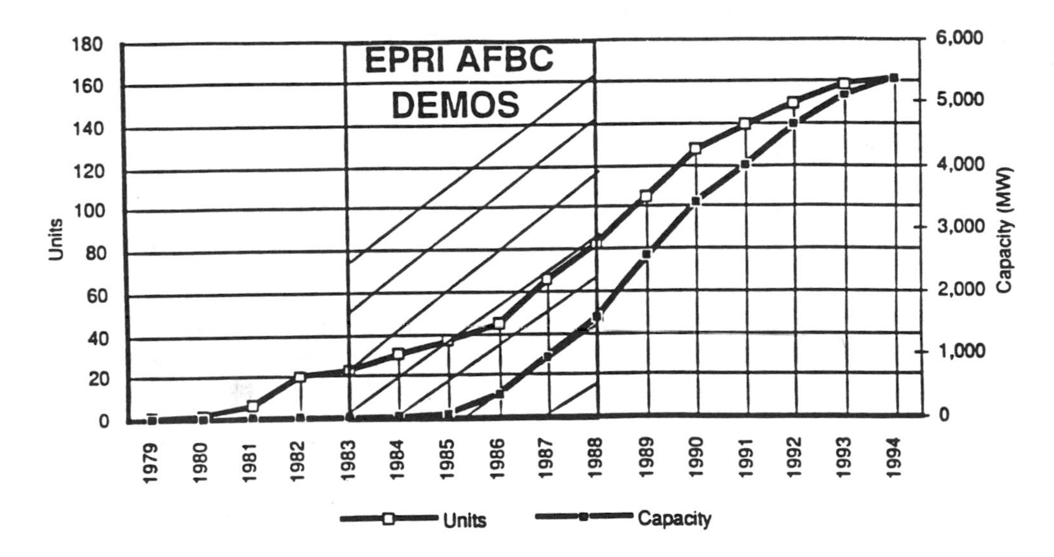

FIGURE 6.7 The growth in number and size of AFBC plants has been significant since the demonstration of 100+ MW sizes in U.S. utility plants.

Pressurized Fluidized Bed Boilers. The future for fluidized-bed utility boilers is evident in the move by several countries (Sweden, Japan, Spain, Germany, United States) toward the pressurized fluidized-bed combined cycle design. The 80 MW units at Vaertan (Sweden) and Escatron (Spain) and the 70 MW unit at Tidd (American Electric Power) are of a commercial size, and larger units up to 350 MW are under development. The modular aspect of the PFBC unit is a particularly attractive feature leading to short construction cycles and low-cost power. This was particularly evident in the construction of the Tidd plant, which first generated power from this combined cycle (Figure 6.8) on November 29, 1990. This $185 million project is partially funded

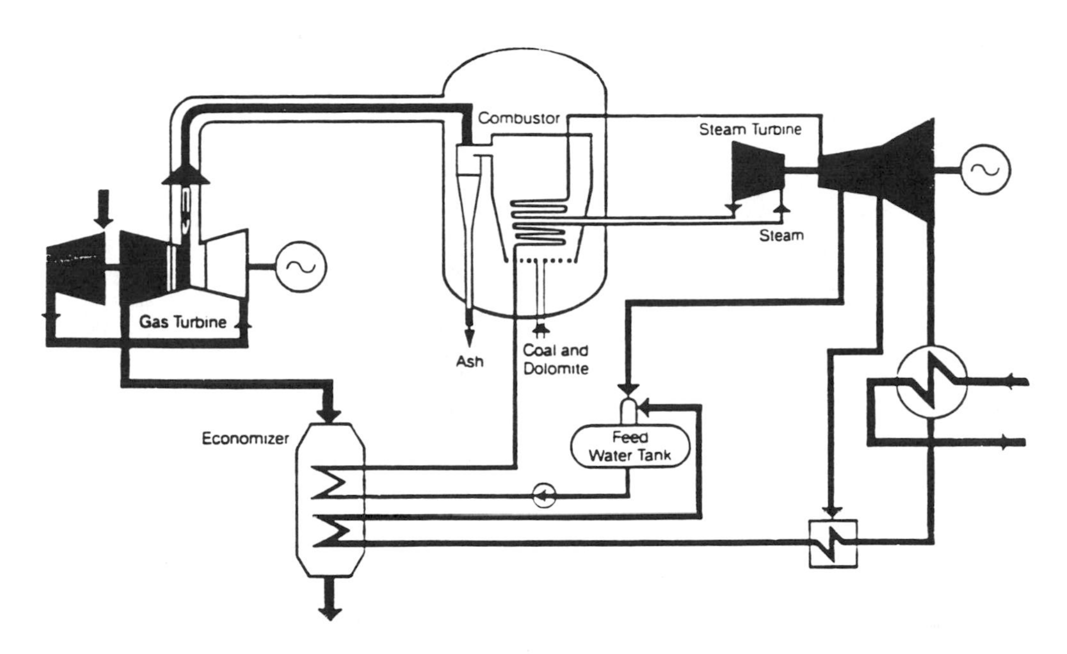

FIGURE 6.8 Pressurized, fluidized-bed combustor with combined cycle. This 70 MW system is now in operation at the Tidd plant of American Electric Power.

by DOE. The heat rate and capital cost of the PFBC plant are forecast to reach very competitive levels, which, when combined with shortened construction schedules, will position the technology for a role in future generation plans, particularly where modular additions have advantages.

6.4 Combustion Turbine Plants

Combustion turbine-based plants comprise the fastest growing technology in power generation. Between now and 2002, natural gas-fired combustion turbines and combined cycles burning gas will account for 50 to 70% of the 900 to 1,000 GW of new generation to be ordered worldwide. GE forecasts that combustion turbines and combined cycles will account for 45% of new orders globally, and for 66% of new U.S. orders. Almost all of these CT and CC plants will be gas-fired, leading a major expansion of gas for electricity generation.

DOE estimates that utilities will install 23 GW of new gas-fired CT capacity in the United States between 1990 and 2000 (Figure 6.9). While not broken out, nonutility generation will also be almost exclusively CT and CC. Worldwide there are even more striking examples. In the United Kingdom, 100% of all new generation ordered or under construction is gas-fired combined cycle.

International Vendor and Technology Base

Both the CT/CC market and vendor base are broadly dispersed today. Siemens and ABB have committed to manufacturing as well as marketing in the United States. Siemens opened a manufacturing plant in Milwaukee, which shipped (to an Asian customer) the first U.S.-assembled machine in May, 1992. ABB, through its acquisitions in the United States, has extensive manufacturing facilities which could be converted to combustion turbines. In a merging of CT expertise, European Gas Turbines is a close associate of General Electric. Also, Rolls-Royce and Westinghouse have recently concluded a technology sharing agreement. This competitive market fosters rapid introduction of new updated technology, and therefore displacement of less attractive, older designs.

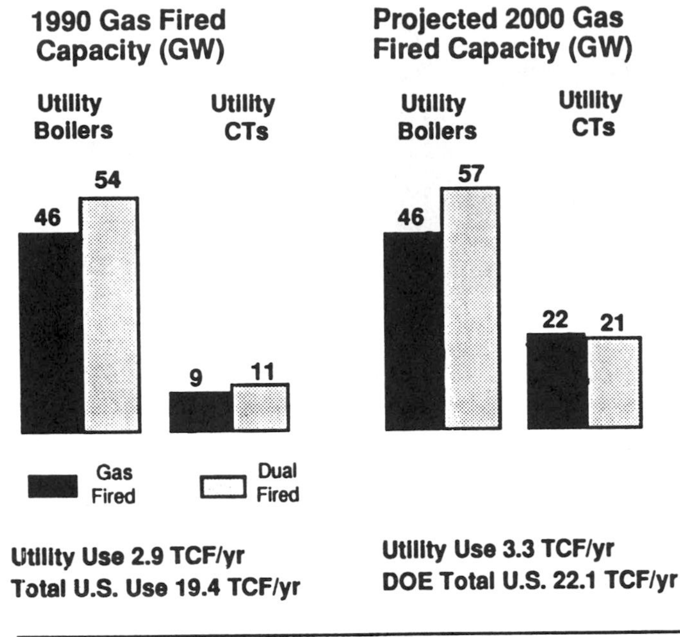

FIGURE 6.9 Combustion turbine additions in the United States will likely total more than 20 GW in the decade before the year 2000.

Aircraft Technology

In the 1960s, gas turbines derived from military jet engines formed a major source of utility peaking generation capacity. Fan-jets, though, which replaced straight turbojets, were much more difficult and expensive to convert to utility use, and the resulting aeroderivative turbines have been little used. The main reason for the high cost was the need to replace the fan and add a separate power turbine.

As bypass ratios, and hence fan power, have increased, the most recent airline fan-jets can be converted to utility service without adding a separate power turbine. Further, modifications of these engines, with intercooling and possibly reheat, appear to be useful for advanced power cycles such as chemical recuperation and the humid air turbine. EPRI is currently managing a major project with a utility consortium to convert advanced aircraft fan-jet engines to high-pressure-ratio, natural gas-fired utility gas turbines. Commercial readiness by about the year 1999, very high efficiency (ultimately 60%, HHV, is expected), very low NO_x emissions, low water usage, and low capital cost are specific project goals. Three leading developers of large fan jets—the aircraft divisions of General Electric, Rolls Royce, and United Technologies—have been contracted to analyze the electric power derivatives of these fan jets in several high efficiency cycles. One utility has estimated a saving of up to $1 billion/yr in the post-2000 time frame due to high-efficiency aeroderivatives.

Humidified Air Power Plants

A new class of combustion turbine-based approaches are termed humidified air power plants. In these combustion turbine cycles the compressor exit air is highly humidified prior to combustion. This reduces the mass of dry air needed and the energy required to compress it.

The continuous plant cycle for this concept is termed the Humid Air Turbine (HAT). This cycle using, for example, extensive modification of the TPM FT4000, has been calculated to have a heat rate on natural gas about 5% better than the latest high-technology combined cycle. The HAT cycle is adaptable to coal gasification leading to the low emissions and high efficiency characteristics of gasification combined cycle plants (discussed later) but at a low capital cost, since the steam turbine bottoming cycle is eliminated.

Gasification Plants

One option of particular importance is that of coal gasification (Figure 6.10). After the EPRI Coolwater demonstration in 1984 at the 100 MW level, the technology has moved ahead in the United States largely through demonstrations under the CCT program. Overseas, the 250 MW Buggenham plant in Holland was operating in 1994, and the PSI/Destec 265 MW and TECO 260 MW clean coal demos should both operate in 1996. Beyond this, there is a 300 MW plant scheduled for Endesa, Spain and a 330 MW unit for RWE in Germany (Figure 6.11).

Gasification-based plants have among the lowest emissions of pollutants of any central station fossil technology. With the efficiency advantages of combined cycles, CO_2 emissions are also lower. Fuel flexibility is an additional benefit, since the gasifier can accommodate a wide range of coals, plus petroleum coke. Integrated gasification combined cycle (IGCC) plants permit a hedge against long-term increases in natural gas prices. Natural gas-fired combustion turbines can be installed initially, and gasifiers at a later time when a switch to coal becomes prudent.

Concurrent with the advances in gasification are efficiency improvements in combustion turbines. The new F-type CTs operate at 2,300°F, and 2,500°F machines are likely by the turn of the century. This makes the IGCC a very competitive future option.

A Look at the Future of Combustion Turbines

Combustion turbines and combined cycles will grow steadily more important in all generation regimes: peaking, mid-range, and base load. Between now and 2000 they will account for the majority of new generation ordered and installed. If the present 2,300°F firing temperature machines operate reliably and durably, CT and CC plants will begin to retire older steam plants and uneconomic nuclear plants. With no clear rival other than fuel cells, which are now emerging, CT technology may dominate fossil generation, and new advanced CT cycles—with intercooling, reheat, possibly chemical recuperation, and most likely humidification—should result in higher efficiencies and lower capital costs. Integrated gasification which guarantees a secure bridge to coal in the near term, will come into its own as gas prices rise under demand pressure. By 2015, coal through gasification will be the economic fuel for a significant fraction of new base-load CT/CC generation. The rate at which these trends develop depends in large measure on the degree of deregulation and competition in the industry.

6.5 Distributed Generation

A new approach to meeting cost, environmental, and customer service objectives is distributed generation, which creates an integrated delivery network employing small, modular generators as well as central stations.

Such modular generators—typically hundreds of kW to tens of MW—can normally be sited, permitted, and assembled quickly, and even relocated should needs change. Strategic placement in the subtransmission and distribution system provides frequency control, voltage regulation, and local spinning reserve capacity, and may allow deferral or avoidance of T&D capacity upgrades (Figure 6.12).

Some distributed generating technologies—fuel cells in particular—operate quietly with extremely low emissions and are thus well suited for environmentally sensitive areas. Moreover, placement of distributed generators near customer sites enables utilities to offer more reliable service to facilities with critical power needs. Such tailored energy services may prove paramount in meeting competitive challenges from other power producers.

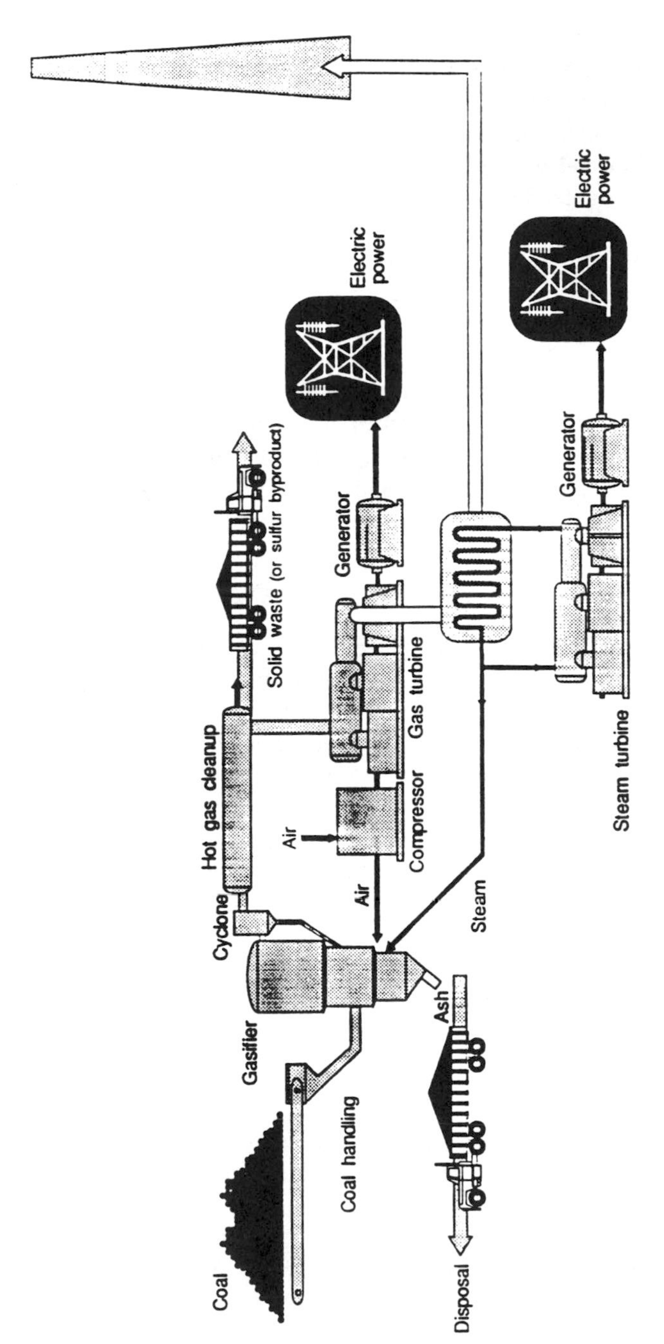

FIGURE 6.10 Gasification combined cycle.

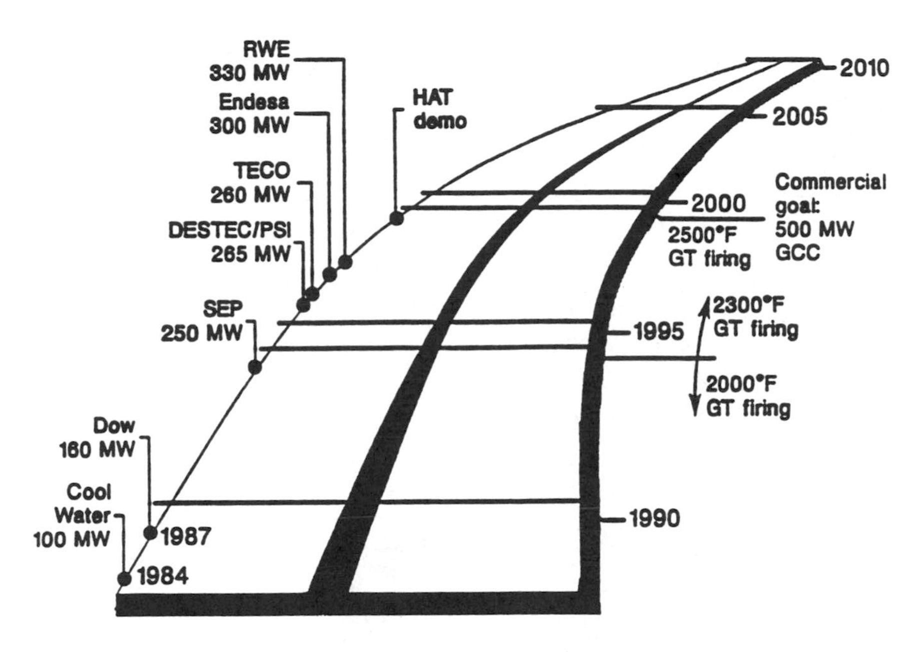

FIGURE 6.11 Gasification power plant time line.

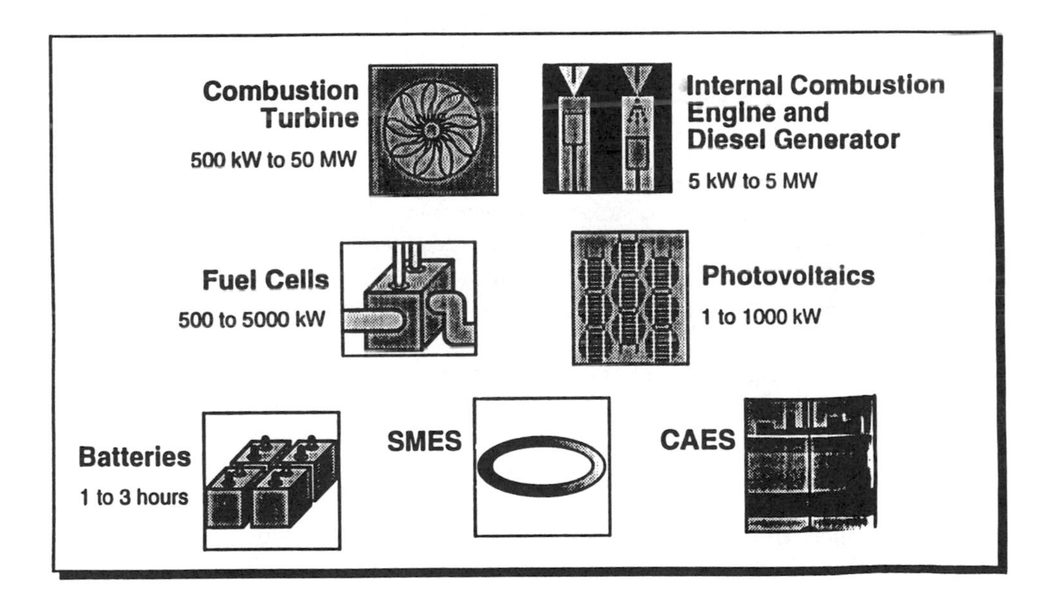

FIGURE 6.12 Distributed generation and energy storage technologies.

Commercial technologies for distributed generation include internal combustion engines and small CTs. Emerging technologies include fuel cells, markedly improved CTs, and improved reciprocating engines.

Fuel Cells

Used in space craft since the 1950s, fuel cells convert hydrocarbon fuels to electricity without combustion. Like a continuous battery, the fuel cell has electrodes in an electrolyte medium, which serves to carry electrons released at one electrode to an absorbing electrode. Hydrogen and air are

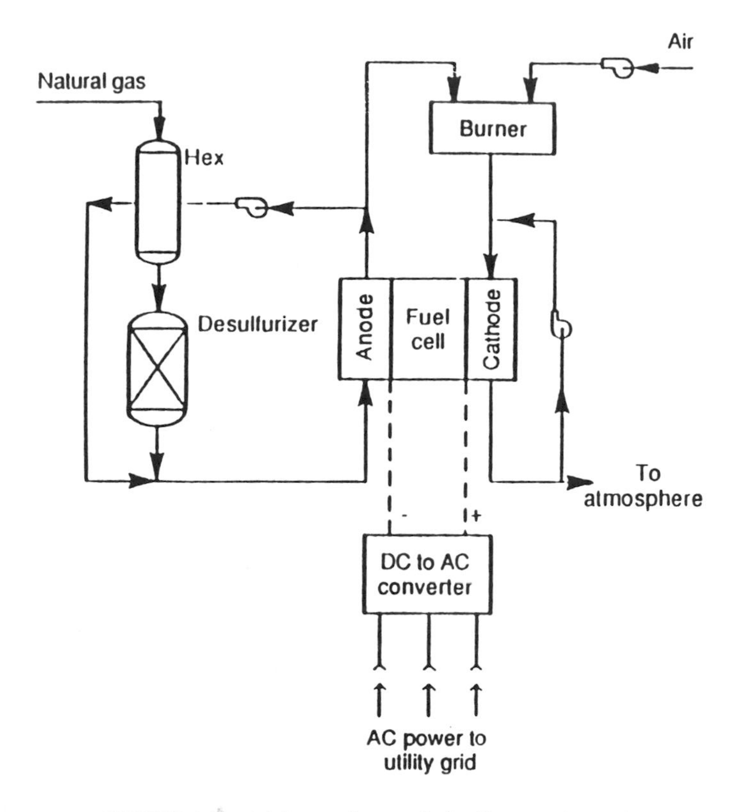

FIGURE 6.13 Molten carbonate fuel cell generating unit.

supplied to the fuel cell, and DC-electricity-produced by-products are water and carbon dioxide. Electrolytes used are phosphoric acid, molten carbonate, and solid oxide (Figure 6.13).

While precommercial fuel cells based on phosphoric acid technology are now operating successfully in Japan, the units most likely to be used in the United States apply more advanced carbonate and solid oxide designs, which offer higher efficiencies, greater compactness, and potentially lower cost.

Tokyo Electric Power Company (TEPCO) first operated phosphoric acid fuel cells at utility scale in 1983. TEPCO is currently running the world's largest fuel cell, an 11 MW phosphoric acid model, in addition to several smaller units.

Although currently fired by natural gas, carbonate fuel cells are expected to ultimately rely on coal gasification as a primary energy source. The carbonate fuel cell combined with a gasifier is the cleanest, most efficient coal power plant now envisioned. Figure 6.14 shows that an Integrated Gasification Molten Carbonate Fuel Cell (IGMCFC) has minimum atmospheric emissions.

EPRI studies indicate that the near-term market for 2 MW carbonate fuel cells, which are expected to be commercially available by 1997, is conservatively 12,000 to 14,000 MW by the year 2005. Earliest uses will probably include

- Supplying power to dense urban areas facing severe siting and environmental constraints.
- Supporting substations to minimize upgrade costs.
- Making use of more cogeneration opportunities in industrial parks.

The prospect of two U.S. manufacturers offering commercial molten carbonate fuel cell (MCFC) units in the next 3 to 5 years represents the culmination of more than a decade of fuel cell R&D

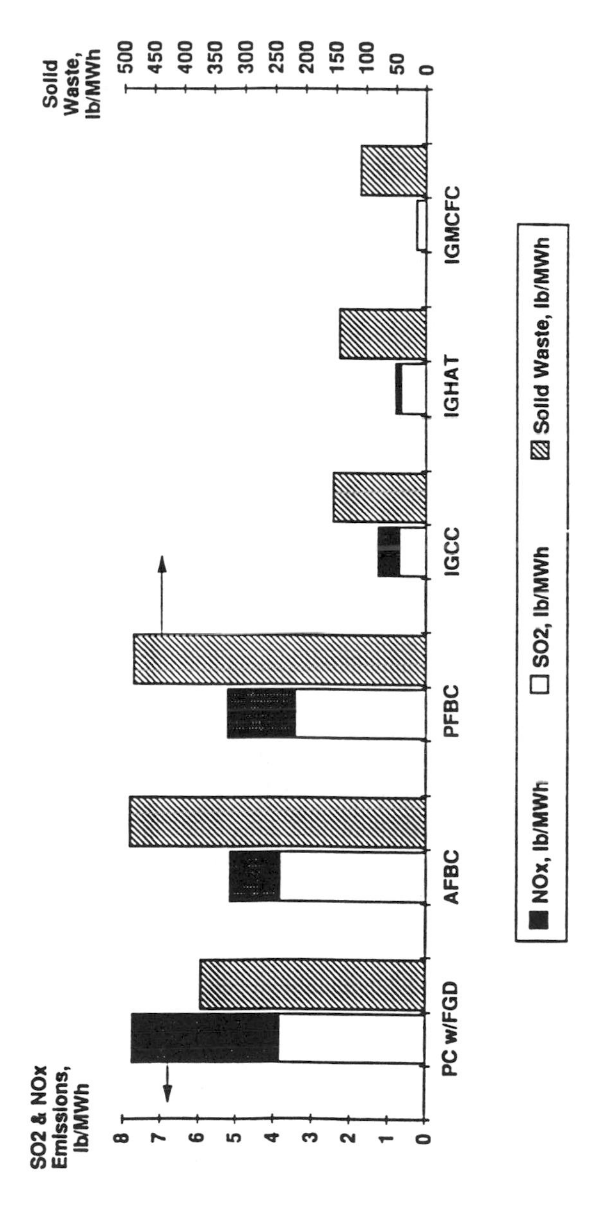

FIGURE 6.14 Emissions and solid waste from coal-based technologies are seen to be lowest with the gasification plants. IGCC (Integrated Gasification Combined Cycle), IGHAT (Integrated Gasification Humid Air Turbine), and IGMCFC (Integrated Gasification Molten Carbonate Fuel Cell).

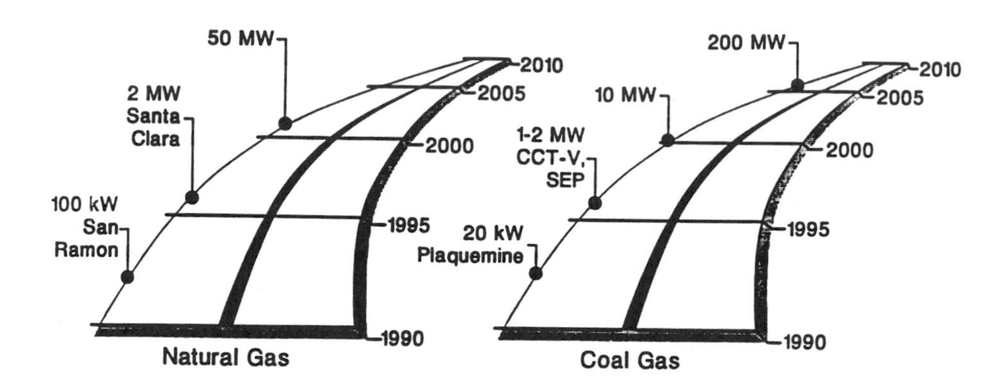

FIGURE 6.15 Time line for molten carbonate fuel cells shows that 10 to 15 MW units are possible by the turn of the century.

at EPRI. A utility "buyers group" has placed advance orders for nearly 100 MW of initial commercial units in exchange for future royalties. As demand grows, the economies of standardization and mass production will reduce unit costs substantially, expanding cost-effective application (Figure 6.15).

If production cost targets are met, MCFC technology could already be cost-effective in some instances. Working with EPRI, the Los Angeles Department of Water & Power, Central and South West Services, and Oglethorpe Power evaluated the benefits of prototype 2 MW fuel cells in various distributed applications. Projected savings ranged from 2 to 71 mills/kWh.

EPRI is providing technical management for a consortium of utilities and manufacturers that will demonstrate a commercial prototype of a modular 2 MW molten carbonate fuel cell. This quiet, efficient unit will serve customers of the City of Santa Clara, California. Ground-breaking ceremonies were held in April 1994; commercial generation will begin in 1996.

Capital and Operating Costs of Power Plants

Table 6.1 lists typical costs for constructing and operating fossil-fuel plants as of 1993. Costs will vary according to type of coal burned, the size of the unit, the plant location in the United States, and the extent of environmental control employed. In this table, Pittsburgh bituminous coal is assumed, which has 7% ash, 2% sulfur, 5% moisture, and a higher heating value of 13,395 Btu/lb. Wet scrubbers for pulverized coal plants use limestone forced oxidation. Fixed O&M costs include labor, maintenance, and overhead. Variable O&M costs are largely consumables: water, chemicals, and other materials. Fuel costs, not listed explicitly, will be affected by the plant efficiency, listed in the table. For comparison with the coal-burning plants, a gas-fired combined cycle unit is included.

6.6 Nuclear Power

Nuclear power plants represent about 100 GW of capacity in the United States, and some 300 GW worldwide. Environmental benefits can be seen in terms of elimination of acid rain precursors and in the absence of any greenhouse gas emissions. In recent years, the progress in nuclear plant implementation has slowed, and in the United States has been halted, by public opposition, largely resulting from concerns over plant safety and nuclear waste disposal. Underway are evolutionary nuclear plants with advanced safety features; increased standardization, modularity, and simplicity; and simplified operation and maintenance procedures to reduce the risk of human error. In the time frame of the next 10 years, it is unlikely that such advanced nuclear plants will have significant impact on the U.S. generation picture.

TABLE 6.1 Cost Projections for Representative Generation Technologies for a Plant in the Northeast U.S.

Plants	Pulverized Coal Plants: Subcritical Wet Scrubber 300 MW	Pulverized Coal Plants: Supercritical Wet Scrubber 400 MW	Atmospheric FBC Circulating No Scrubber 200 MW	Pressurized FBC Combined Cycle 340 MW
Capital cost (1993)* ($/kW)	1,607	1,600	1,805	1,318
Nonfuel O&M costs				
Variable (mills/kWh)	3.0	2.8	5.4	1.8
Fixed ($/kW yr)	46.6	43.1	37.0	37.6
Efficiency (%)	36	39	35	41

Plants	Coal Gasification Combined Cycle 500 MW	Coal Gasification Humid Air Turbine 500 MW	Coal Gasification Molten Carbonate Fuel Cell 400 MW	Gas-Fired Combined Cycle 225 MW
Capital cost (1993) ($/kW)	1,648	1,447	2,082	595
Nonfuel O&M costs				
Variable (mills/kWh)	0.5	1.3	1.1	0.4
Fixed ($/kW yr)	49.9	40.4	57.2	26.5
Efficiency (%)	42	42	50	46

* Costs of new plants are likely to reduce, in real terms, over the next 10 years due to technology developments and increased worldwide competition for markets in the developing countries. New technologies (PFBC, IGCC, fuel cells) will lower capital costs as production capacity grows.

Source: "Technical Assessment Guide," EPRI TR-102275-V1R7 June, 1993. Electric Power Research Institute, Palo Alto, CA.

Conclusions

The generation picture for the United States can be divided into two phases: the next 10 years, and after 2005. In the near term gas-fired new additions will dominate, led by more efficient and reliable combustion turbines. Coal-fired plants will be added, but in smaller unit sizes than those which characterized the last 20 years, with 400 MW being a likely upper limit. Coal remains the fuel of choice for most existing generation though, providing the majority of generated power.

Beyond 2005, distributed generation, perhaps in the form of fuel cells will increasingly appear at strategic load points in the electric system. In addition, the gas turbine plants of the 1990s will be adapted for coal gasification as natural gas becomes scarcer and more expensive, and the 30- to 40-year-old fossil plants, which have been the backbone of the utility industry, will be retired as more efficient new generation begins to dominate.

References

1. Armor, A.F., "Generation Technologies Through the Year 2005," The Electric Industry in Transition, *Public Utilities Reports, Inc.*, New York, December 1994.
2. National Energy Strategy, 1st ed. 1991/1992, Washington D.C., February 1991.
3. Couch, G., "Advanced Coal Cleaning Technology," IEACR/44, London, IEA Coal Research, December 1991.
4. Torrens, I.M., "Developing Clean Coal Technologies," *Environment*, 32:6 (11-33), July/August, 1990.

5. Blunden, W.E., *Colorado-UTE's Nucla Circulating AFBC Demonstration Project,* EPRI Report CS-5831, Palo Alto, February, 1989.
6. Armor, A.F., Touchton, G.L., Cohn, A., "Powering The Future: Advanced Combustion Turbines and EPRI's Program," EPRI Coal Gasification Conference, San Francisco, California, October 1992.
7. Carpenter, L.K., Dellefield, R.J., "The U.S. Department of Energy PFBC Perspective," EPRI Fluidized Bed Combustion for Power Generation Conference, Atlanta, Georgia, May 17-19, 1994.
8. Manaker, A.M., *TVA 160-MWe Atmospheric Fluidized-Bed Combustion Demonstration Project,* EPRI Report TR-100544, Palo Alto, December 1992.
9. Olesen, C., "Pressurized Fluidized-Bed Combustion for Power Generation," in EPRI CS-4028 *Proceedings: Pressurized Fluidized-Bed Combustion Power Plants,* May, 1985.
10. U.S. Department of Energy, Clean Coal Technology Demonstration Program, DOE/FE-0299P, Washington, D.C., March 1994.
11. Douglas, J., "IGCC: Phased Construction for Flexible Growth," *EPRI Journal,* Palo Alto, California, September, 1986.
12. Lamarre, L., "The Vision of Distributed Generation," *EPRI Journal,* Palo Alto, California, April/May 1993.

7

The Outlook for U.S. Energy Supply, Demand, and Prices Through 2010*

Andy S. Kydes

Scott Sitzer

Louise K. Bonadies

U.S. Department of Energy

*This product was prepared as an account of work sponsored by an agency of the United States Government. Neither the United States Government nor any agency thereof, nor any of their employees, makes any warranty, express or implied, or assumes any legal liability or responsibility for the accuracy, completeness, or usefulness of any information, apparatus, product, or process disclosed, or represents that its use would not infringe on privately owned rights. Reference herein to any specific commercial product, process, or service by trade name, trademark, manufacturer, or otherwise does not necessarily constitute or imply its endorsement, recommendation, or favoring by the United States Government or any agency thereof. The views and opinions of the authors expressed herein do not necessarily state or reflect those of the United States Government or any agency thereof.

Authors' Note — Since this chapter was prepared, the Energy Information Administration has issued a new energy outlook based on new information concerning the primary assumptions affecting demand, supply, and prices. Among other changes, these assumptions tended to reduce fossil fuel price projections while making only minor differences to forecasts of energy consumption and production. In addition, the new outlook extended the forecast horizon through 2015. A full description of the new forecast is available in the *Annual Energy Outlook 1996*[1b]. Please see Appendix B for a table that summarizes the projections appearing in that publication.

Introduction

This chapter is adapted from the Energy Information Administration's *Annual Energy Outlook 1995*,[1] which provides projections of energy demand, supply, and prices through the year 2010. These projections are based on the National Energy Modeling System, which is developed and maintained by the Energy Information Administration's Office of Integrated Analysis and Forecasting to provide projections of domestic energy-economy markets in the midterm time period and to perform policy analyses. The forecasts and projections appearing in this chapter are based on the reference case published in the *Annual Energy Outlook 1995*.

The National Energy Modeling System

The National Energy Modeling System has a time horizon of 20 years, the midterm time period in which the structure of the economy and the nature of energy markets are sufficiently understood that it is possible to represent considerable structural and regional detail. The national model results are presented each year in the *Annual Energy Outlook*. Regional and other detailed results are presented annually in the Energy Information Administration's *Supplement to the Annual Energy Outlook*.[2] Detailed documentation of the modeling system is available in a series of documentation reports.* A synopsis of the National Energy Modeling System, the model components, and the interrelationships of the modules is presented in *The National Energy Modeling System: An Overview*.[3]

For each fuel and consuming sector, the National Energy Modeling System balances the energy supply and demand, accounting for the economic competition between the various energy fuels and sources. It is organized and implemented as a modular system. The modules represent each of the fuel supply markets, conversion sectors, and end-use consumption sectors of the energy system. The modeling system also includes macroeconomic and international modules. Each model component also represents the impact and cost of legislation and environmental regulations that affect that sector and reports key emissions. The model reflects all current legislation and environmental regulations, such as the Clean Air Act Amendments of 1990, and the costs of compliance with other regulations. For the *Annual Energy Outlook 1995*, the model includes an analysis of the impacts of the provisions of the Climate Change Action Plan.

*NEMS documentation reports are available from the National Energy Information Center (202/586–8800).

7.1 Historical Overview

Energy Consumption Trends

U.S. energy consumption increased from 66 quadrillion British thermal units (Btu) in 1970 to 87 quadrillion Btu in 1993, the reference year in the *Annual Energy Outlook 1995* and the most recent year for which annual data were available when the forecast was developed.* During this period energy consumption fluctuated in response to dramatic changes in oil prices, economic growth rates, and environmental concerns. Following the oil price shock in 1973, higher energy prices led to increases in energy efficiency. In the 1970s, the composition of energy end-use demand also reflected a shift away from petroleum and natural gas toward electricity generated by other fuels.

Consumption of coal increased more than consumption of any other energy source between 1970 and 1993, growing from 12 quadrillion Btu in 1970 to 20 quadrillion Btu in 1993. Between 1970 and 1993, petroleum consumption increased from 30 quadrillion Btu to 34 quadrillion Btu. Natural gas consumption began to decline in the 1970s due to uncertainties about supply and regulatory restrictions, falling below 17 quadrillion Btu in 1986, before turning upward again. At 21 quadrillion Btu in 1993, natural gas consumption nearly regained its 1970 level of 22 quadrillion Btu. Consumption of nuclear electric power increased from less than 1 quadrillion Btu in 1970 to nearly 7 quadrillion Btu in the early 1990s.

Energy Production Trends

Energy production increased from 62 quadrillion Btu in 1970 to 66 quadrillion Btu in 1993. For a decade coal has accounted for the largest share of domestic energy production, contributing 20 quadrillion Btu in 1993. Dry natural gas production totaled 19 quadrillion Btu, crude oil production totaled 15 quadrillion Btu, and natural gas plant liquids contributed another 2.4 quadrillion Btu in 1993. As petroleum consumption outpaced U.S. domestic production, expanded imports filled the gap.

Nuclear electric power production grew from less than 1 quadrillion Btu in 1970 to more than 6 quadrillion Btu in 1990, and approached 7 quadrillion Btu in each of the next 3 years. Hydroelectric power production remained relatively steady, hovering around 3 quadrillion Btu for most of the period between 1970 and 1993.

7.2 Forecast Highlights

Energy Consumption

Total U.S. energy consumption is projected to increase at an average annual rate of 1% through 2010, with oil and natural gas showing the largest consumption increases. Total U.S. energy consumption increases from 87 quadrillion Btu in 1993 to nearly 104 quadrillion Btu in 2010 (Table 7.1).

Petroleum consumption grows at an average annual rate of 1.1% through 2010, reaching 41 quadrillion Btu. The largest share of petroleum consumption, approximately two-thirds, is in the transportation sector, with a 1.4% annual growth rate (Table 7.2). Increased vehicle efficiency over the projection period is offset by an increase in vehicle-miles traveled.

Natural gas consumption increases at an average rate of 1.2% a year, led by a rapid increase in gas-fired electricity generation.** Consumption of gas for generation in 2010 is 1.8 quadrillion Btu

*Historical data are taken from the Energy Information Administration, *Annual Energy Review 1993*, DOE/EIA-0384(93) (Washington, DC, July 1994).

**Electricity generation includes all electric power generators except cogenerators, which produce electricity as a byproduct of other processes.

TABLE 7.1 Total Energy Supply and Disposition Summary (quadrillion Btu per year, unless otherwise noted)

Supply, Disposition, and Price	Reference Case 1992	1993	2000	2005	2010	Annual Growth 1993–2010 (percent)
			Production			
Crude Oil and Lease Condensate	15.19	14.50	11.33	10.92	11.42	–1.4
Natural Gas Plant Liquids	2.36	2.41	2.57	2.69	2.81	0.9
Dry Natural Gas	18.37	18.90	19.65	20.53	21.51	0.8
Coal	21.59	20.23	22.08	23.21	24.51	1.1
Nuclear Power	6.61	6.52	6.96	6.97	6.36	–0.1
Renewable Energy/Other[1]	7.14	7.06	7.92	8.43	9.25	1.6
Total	71.25	69.62	70.51	72.75	75.86	0.5
			Imports			
Crude Oil[2]	13.25	14.79	19.14	19.72	19.53	1.6
Petroleum Products[3]	3.73	3.79	5.18	7.10	8.12	4.6
Natural Gas	2.18	2.32	3.17	3.33	3.97	3.2
Other Imports[4]	0.41	0.50	0.71	0.73	1.00	4.2
Total	19.57	21.40	28.20	30.87	32.62	2.5
			Exports			
Petroleum[5]	2.00	2.11	1.97	1.68	1.64	–1.5
Natural Gas	0.21	0.15	0.21	0.27	0.31	4.5
Coal	2.68	1.96	2.22	2.54	2.89	2.3
Total	4.89	4.21	4.40	4.49	4.83	0.8
Discrepancy[6]	–0.32	0.46	0.31	0.24	0.22	N/A
			Consumption			
Petroleum Products[7]	33.56	33.71	36.89	39.30	40.82	1.1
Natural Gas	20.15	20.81	22.78	23.76	25.30	1.2
Coal	18.87	19.43	20.14	21.01	21.97	0.7
Nuclear Power	6.61	6.52	6.96	6.97	6.36	–0.1
Renewable Energy/Other[8]	6.43	6.80	7.83	8.32	9.41	1.9
Total	85.61	87.27	94.61	99.37	103.88	1.0
Net Imports — Petroleum	14.99	16.47	22.36	25.13	26.02	2.7
			Prices (1993 dollars per unit)			
World Oil Price (dollars per barrel)[9]	18.70	16.12	19.13	21.50	24.12	2.4
Gas Wellhead Price (dollars per thousand cubic feet)[10]	1.80	2.02	2.14	3.02	3.39	3.1
Coal Minemouth Price (dollars per ton)	21.57	19.85	20.34	21.55	22.77	0.8

[1] Includes utility and nonutility electricity from hydroelectric, geothermal, wood and wood waste, municipal solid waste, other biomass, wind, photovoltaic and solar thermal sources; nonelectric energy from renewable sources, such as active and passive solar systems, groundwater heat pumps, and wood; alcohol fuels from renewable sources; and, in addition to renewables, liquid hydrogen, methanol, supplemental natural gas, and some domestic inputs to refineries.

[2] Includes imports of crude oil for the Strategic Petroleum Reserve.

[3] Includes imports of finished petroleum products, imports of unfinished oils, alcohols, ethers, and blending components.

[4] Includes coal, coal coke (net), and electricity (net).

[5] Includes crude oil and petroleum products.

[6] Balancing item. Includes unaccounted for supply, losses, and gains.

[7] Includes natural gas plant liquids, crude oil consumed as a fuel, and nonpetroleum based liquids for blending, such as ethanol.

TABLE 7.1 (continued) Total Energy Supply and Disposition Summary (quadrillion Btu per year, unless otherwise noted)

[8] Includes utility and nonutility electricity from hydroelectric, geothermal, wood and wood waste, municipal solid waste, other biomass, wind photovoltaic and solar thermal sources; nonelectric energy from renewable sources, such as active and passive solar systems, groundwater heat pumps, and wood; alcohol fuels from renewable sources; and, in addition to renewables, net coal coke imports, net electricity imports, methanol, and liquid hydrogen.

[9] Average refiner acquisition cost for imported crude oil.

[10] Represents lower 48 onshore and offshore supplies.

Btu = British thermal unit.

N/A = Not applicable.

Note: Totals may not equal sum of components due to independent rounding. Figures for 1992 and 1993 may differ from published data due to internal conversion factors.

Sources: 1992 natural gas values: Energy Information Administration (EIA), *Natural Gas Annual 1992, Volume 1,* DOE/EIA-0131(92)/1 (Washington, DC, November 1993). 1992 coal minemouth prices: EIA, *Coal Production 1992,* DOE/EIA-0118(92) (Washington, DC, October 1993). Other 1992 values: EIA, *Annual Energy Review 1993,* DOE/EIA-0384(93) (Washington, DC, July 1994). 1993 natural gas price: EIA, *Natural Gas Monthly,* DOE/EIA-0130(94/6) (Washington, DC, June 1994). 1993 coal minemouth price: EIA, *Coal Industry Annual 1993,* DOE/EIA-0584(93) (Washington, DC, December 1994). Other 1993 values: EIA, *Annual Energy Review 1993,* DOE/EIA-0384(93) (Washington, DC, July 1994). *Projections:* EIA, AEO95 National Energy Modeling System run AEO95B.D1103942.

higher than in 1993. Because of its competitive price, gas consumption also grows significantly in the industrial sector, with much of the increment used for cogeneration.

Coal consumption for electricity generation—nearly 90% of total coal use in 2010—grows at an average annual rate of 0.8%. Total coal use grows by 0.7% a year. Consumption of renewables, including hydropower, increases from 6.6 to 8.9 quadrillion Btu between 1993 and 2010. In 2010, 5.9 quadrillion Btu of renewables is used for electricity generation, including cogeneration. The balance is used for dispersed heating and cooling and for blending into vehicle fuels.

The *Outlook* projects increased use of nuclear power for generation through 2005 due to increasing capacity utilization and the addition of two new units.* However, by 2010, retirements of existing plants cause a slight reduction in nuclear generation relative to 1993 levels.

Demand for electricity grows at an average annual rate of 1.1%, lower than the economic growth rate of 2.2%, primarily because of increased performance standards for new energy-using equipment. Electricity use is the largest source of energy growth in the buildings sectors. Industrial demand for electricity also grows faster than total energy use in that sector, as new electricity-using technologies continue their penetration.

Coal remains the primary fuel for generation, but natural gas use has a higher annual growth rate of 2.8%. Gas-fired units are cleaner than coal-fired units, are less costly to construct, and have shorter construction lead times. The share of coal in generation declines slightly between 1993 and 2010, while the natural gas share increases.

Consumption of renewable energy sources for generation also increases. Hydropower and geothermal remain the primary and secondary renewable sources, about 62 and 16%, respectively, in 2010. Wind is the fastest-growing renewable generation source, at an average 13% annual rate, as a result of technology improvements.

Energy Production

Domestic crude oil production continues to decline through most of the forecast period, at an average annual rate of 1.4%. With production declining and demand increasing over most of the

*One of the two new units anticipated was Watts Bar 2. After the *Annual Energy Outlook 1995* analysis was completed, the Tennessee Valley Authority announced that it was halting construction work on the unit and was considering the options of converting the plant to alternative fuel or seeking a financial partner to complete construction.

TABLE 7.2 Energy Consumption by End-Use Sector and Source (quadrillion Btu per year)

Sector and Source	Reference Case 1992	1993	2000	2005	2010	Annual Growth 1993–2010 (percent)
Residential						
Distillate Fuel	0.86	0.88	0.79	0.75	0.73	–1.1
Kerosene	0.06	0.07	0.06	0.06	0.05	–1.1
Liquefied Petroleum Gas	0.38	0.39	0.34	0.31	0.29	–1.8
Natural Gas	4.82	5.11	5.17	5.04	5.04	–0.1
Coal	0.06	0.06	0.06	0.06	0.05	–0.2
Renewable Energy[1]	0.69	0.72	0.72	0.73	0.74	0.2
Electricity	3.19	3.39	3.61	3.73	3.94	0.9
Total	10.07	10.61	10.75	10.69	10.85	0.1
Commercial						
Distillate Fuel	0.46	0.42	0.39	0.37	0.35	–1.0
Kerosene	0.01	0.01	0.01	0.01	0.01	–0.1
Motor Gasoline[2]	0.08	0.07	0.07	0.07	0.08	0.5
Residual Fuel	0.19	0.17	0.17	0.17	0.17	–0.1
Natural Gas	2.87	2.98	3.04	3.04	3.06	0.2
Other[3]	0.15	0.15	0.15	0.15	0.15	–0.1
Renewable Energy[4]	0.01	0.01	0.03	0.03	0.03	5.1
Electricity	2.86	2.96	3.22	3.34	3.50	1.0
Total	6.64	6.77	7.08	7.17	7.34	0.5
Industrial[5]						
Distillate Fuel	1.14	1.12	1.24	1.33	1.39	1.3
Liquefied Petroleum Gas	1.86	1.82	2.04	2.24	2.40	1.6
Motor Gasoline[2]	0.19	0.19	0.21	0.23	0.24	1.4
Petrochemical Feedstocks	1.19	1.17	1.31	1.42	1.52	1.5
Residual Fuel	0.39	0.38	0.49	0.54	0.56	2.2
Other Petroleum[6]	3.86	3.78	3.86	3.98	4.07	0.4
Natural Gas[7]	8.84	9.09	10.32	10.71	11.28	1.3
Metallurgical Coal	0.87	0.84	0.77	0.67	0.59	–2.1
Steam Coal	1.65	1.66	1.87	2.01	2.11	1.4
Net Coal Coke Imports	0.03	0.02	0.02	0.03	0.04	5.5
Renewable Energy	2.09	2.12	2.45	2.67	2.84	1.7
Electricity	3.32	3.35	3.68	4.01	4.24	1.4
Total	25.44	25.55	28.25	29.83	31.28	1.2
Transportation						
Distillate Fuel	3.81	3.86	4.59	5.02	5.43	2.0
Jet Fuel[8]	3.00	3.04	3.74	4.13	4.46	2.3
Motor Gasoline[2]	13.70	13.88	14.97	15.44	15.51	0.7
Residual Fuel	1.08	1.10	1.32	1.51	1.67	2.5
Liquefied Petroleum Gas	0.02	0.02	0.08	0.16	0.24	16.3
Other Petroleum[9]	0.20	0.20	0.23	0.24	0.25	1.2
Pipeline Fuel Natural Gas	0.61	0.63	0.67	0.67	0.67	0.4
Compressed Natural Gas	0.00	0.01	0.16	0.29	0.43	26.3
Renewables (ethanol)[10]	0.00	0.00	0.01	0.03	0.06	35.2
Liquid Hydrogen	0.00	0.00	0.00	0.00	0.00	N/A
Methanol[11]	0.00	0.00	0.01	0.03	0.05	30.8
Electricity	0.06	0.06	0.09	0.13	0.18	6.5
Total	22.48	22.79	25.86	27.66	28.94	1.4
Electric Generators[12]						
Distillate Fuel	0.01	0.01	0.01	0.01	0.05	10.2
Residual Fuel	0.97	1.07	0.93	1.26	1.31	1.2

TABLE 7.2 (continued) Energy Consumption by End-Use Sector and Source (quadrillion Btu per year)

Sector and Source	Reference Case 1992	1993	2000	2005	2010	Annual Growth 1993–2010 (percent)
Natural Gas	3.01	3.00	3.42	4.01	4.82	2.8
Steam Coal	16.21	16.78	17.36	18.19	19.13	0.8
Nuclear Power	6.61	6.52	6.96	6.97	6.36	–0.1
Renewable Energy/Other[13]	3.61	3.92	4.60	4.80	5.65	2.2
Total	30.42	31.30	33.27	35.23	37.32	1.0
		Primary Energy Consumption				
Distillate Fuel	6.29	6.29	7.02	7.49	7.95	1.4
Kerosene	0.09	0.09	0.08	0.08	0.07	–0.9
Jetfuel[8]	3.00	3.04	3.74	4.13	4.46	2.3
Liquefied Petroleum Gas	2.33	2.29	2.53	2.77	2.99	1.6
Motor Gasoline[2]	13.97	14.14	15.25	15.75	15.83	0.7
Petrochemical Feedstocks	1.19	1.17	1.31	1.42	1.52	1.5
Residual Fuel	2.63	2.72	2.90	3.47	3.70	1.8
Other Petroleum[14]	4.05	3.98	4.07	4.21	4.31	0.5
Natural Gas	20.15	20.81	22.78	23.76	25.30	1.2
Metallurgical Coal	0.87	0.84	0.77	0.67	0.59	–2.1
Steam Coal	18.00	18.59	19.37	20.34	21.38	0.8
Net Coal Coke Imports	0.03	0.02	0.02	0.03	0.04	5.5
Nuclear Power	6.61	6.52	6.96	6.97	6.36	–0.1
Renewable Energy/Other[15]	6.41	6.78	7.81	8.29	9.37	1.9
Total	85.61	87.26	94.61	99.37	103.88	1.0
Electricity Consumption (all sectors)	9.43	9.77	10.60	11.21	11.86	1.1

[1] Includes electricity generated by the sector for self-use from hydroelectric, geothermal, wood and woodwaste, municipal solid waste, other biomass, wind, photovoltaic and solar thermal sources, and nonelectric energy from renewable sources, such as solar thermal water heaters, groundwater heat pumps, and wood.

[2] Includes ethanol (blends of 10 percent or less) and ethers blended into gasoline.

[3] Includes liquefied petroleum gas and coal.

[4] Includes commercial sector electricity cogenerated using wood and wood waste, municipal solid waste, and other biomass; nonelectric energy from renewable sources, such as active solar and passive solar systems, geothermal heat pumps, and solar water heating systems.

[5] Fuel consumption includes consumption for cogeneration.

[6] Includes petroleum coke, asphalt, road oil, lubricants, still gas, and miscellaneous petroleum products.

[7] Includes lease and plant fuel.

[8] Includes naphtha and kerosene type.

[9] Includes aviation gas and lubricants.

[10] Only E85 (85 percent ethanol).

[11] Only M85 (85 percent methanol).

[12] Includes consumption of energy by all electric power generators except cogenerators and generators with standard industrial classification other than 49, both of which produce electricity as a byproduct of other processes.

[13] Includes hydroelectric, geothermal, wood and wood waste, municipal solid waste, other biomass, wind, photovoltaic and solar thermal sources, plus net electricity imports.

[14] Includes unfinished oils, natural gasoline, motor gasoline blending compounds, aviation gasoline, lubricants, still gas, asphalt, road oil, and miscellaneous petroleum products.

[15] Includes electricity generated for sale to electric utilities and for self use from renewable sources, nonelectric energy from renewable sources, net electricity imports, liquid hydrogen, and methanol.

Btu = British thermal unit.

N/A = Not applicable.

Note: Totals may not equal sum of components due to independent rounding. Figures for 1992 and 1993 may differ from published data due to internal conversion factors in the AEO95 National Energy Modeling System.

TABLE 7.2 (continued) Energy Consumption by End-Use Sector and Source (quadrillion Btu per year)

Sources: 1992 natural gas lease, plant, and pipeline fuel values: Energy Information Administration (EIA), *Natural Gas Annual 1992 Volume 1,* DOE/EIA-0131(92)/1 (Washington, DC, November 1993). Other 1992 values: EIA, *State Energy Data Report 1992,* DOE/EIA-0214(92) (Washington, DC, May 1994) and Office of Coal, Nuclear, Electric, and Alternative Fuels estimates. 1993 natural gas lease, plant, and pipeline fuel values: EIA, *Natural Gas Monthly,* DOE/EIA-0130(94/6) (Washington, DC, June 1994). 1993 transportation sector compressed natural gas consumption: EIA, AEO95 National Energy Modeling System run AEO95B.D1103942. 1992 and 1993 coal consumption is estimated from EIA, *State Energy Data Report 1992,* DOE/EIA-0214(92) (Washington, DC, May 1994). 1992 and 1993 metallurgical consumption is estimated from this source using unpublished data. 1992 and 1993 residential and commercial coal consumption tonnages are from EIA, *Quarterly Coal Report, October–December 1993,* DOE/EIA-0121(93/4Q) (Washington, DC, May 1994) and have been converted to quadrillion Btu using *State Energy Data Report 1992* thermal conversion factors. Other 1993 values: EIA, *Short-Term Energy Outlook,* DOE/EIA-0202(94/3Q) (Washington, DC, August 1994) with adjustments to end-use sector consumption levels for consumption of natural by electric wholesale generators based on EIA, AEO95 National Energy Modeling System run AEO95B.D1103942. *Projections:* EIA, AEO95 National Energy Modeling System run AEO95B.D1103942.

forecast, the share of petroleum consumption met by net imports reaches 59% in 2010 (in terms of barrels per day), compared with 44% in 1993. Increases in production of coal, natural gas, and renewables also offset the decline in crude oil production.

A progression of recent environmental regulations and initiatives—including the Clean Air Act Amendments of 1990, the Energy Policy Act of 1992, and the Climate Change Action Plan—are projected to increase demand for low-sulfur coal and natural gas. Coal production increases by an average of 1.1% a year through 2010, to meet the demand for coal in both domestic and export markets. Growth in natural gas consumption is met by increases in both production and imports, primarily from Canada. U.S. production of natural gas increases at an average annual rate of 0.8% between 1993 and 2010, satisfying more than half the growth in consumption by 2010.

Renewable energy production grows significantly throughout the forecast period—at 1.6% a year—primarily for electricity generation. Conventional hydropower increases only slightly, but other renewable sources, including ethanol for transportation uses, increase more rapidly.

Nuclear power production increases through the middle part of the next decade, then declines as existing plants are retired. By 2010, because of improved performance of existing units, nuclear production is only slightly below the 1993 level.

Fuel Prices

Nominal prices of primary fuels (Table 7.3) are projected to rise faster than the general rate of inflation through 2010, despite the dampening effects on prices of increases in Organization of Petroleum Exporting Countries (OPEC) crude oil production and capacity, expanded estimates of economically recoverable reserves for oil and gas, and improved labor productivity for coal. Larger reserves and improved productivity result, in part, from advances in technology.

The average world oil price is projected to increase from the 1993 level of $16.12 to $24.12 per barrel (in 1993 dollars) in 2010. Wellhead prices of natural gas increase more slowly over most of the forecast, reflecting the effects of technology improvements on oil and gas production. The average price in 2010 is projected to reach $3.39 per thousand cubic feet, compared with $2.02 in 1993. Delivered prices rise even more slowly than the wellhead price, because the average costs of transmission and distribution decline over most of the projection period. Projected minemouth coal prices rise at an average annual rate of 0.8%, reflecting trends of increasing productivity and lower demand for coal in the electric utility sector and for export.

Electricity prices are projected to rise at an average annual growth rate of only 0.3% a year through 2010, due to lower input fuel prices and reduced requirements for new capacity. Recent

TABLE 7.3 Energy Prices by End-Use Sector and Source (1993 dollars per million Btu)

Sector and Source	Reference Case					Annual Growth 1993–2010 (percent)
	1992	1993	2000	2005	2010	
Residential	12.52	12.48	12.72	13.73	14.53	0.9
Primary Energy	6.28	6.32	6.41	7.18	7.42	0.9
Petroleum Products	7.99	7.72	8.97	9.54	10.06	1.6
Distillate Fuel	6.86	6.55	7.70	8.24	8.71	1.7
Liquefied Petroleum Gas	10.55	10.40	11.96	12.72	13.58	1.6
Natural Gas	5.87	6.00	5.88	6.71	6.92	0.8
Electricity	24.62	24.30	23.93	24.66	25.67	0.3
Commercial	12.54	12.25	12.53	13.49	14.25	0.9
Primary Energy	4.86	4.93	5.00	5.79	6.07	1.2
Petroleum Products	5.16	4.96	5.76	6.32	6.84	1.9
Distillate Fuel	4.89	4.61	5.52	6.05	6.49	2.0
Residual Fuel	2.66	2.67	3.06	3.48	3.89	2.2
Natural Gas[1]	4.86	5.02	4.93	5.79	6.03	1.1
Electricity	22.66	21.63	21.47	22.25	23.15	0.4
Industrial[2]	5.22	5.02	5.28	5.95	6.42	1.5
Primary Energy	3.42	3.32	3.67	4.32	4.77	2.2
Petroleum Products	4.66	4.34	4.99	5.57	6.19	2.1
Distillate Fuel	4.99	4.78	5.52	6.06	6.53	1.8
Liquefied Petroleum Gas	5.04	4.85	7.18	7.91	8.69	3.5
Residual Fuel	2.40	2.35	2.93	3.34	3.75	2.8
Natural Gas[3]	2.62	2.79	2.96	3.80	4.17	2.4
Metallurgical Coal	1.84	1.77	1.71	1.74	1.76	0.0
Steam Coal	1.51	1.45	1.45	1.48	1.53	0.3
Electricity	15.40	14.53	14.34	14.78	15.22	0.3
Transportation	8.11	7.76	9.06	9.52	9.70	1.3
Primary Energy	8.09	7.74	9.04	9.49	9.66	1.3
Petroleum Products	8.09	7.74	9.06	9.51	9.68	1.3
Distillate Fuel[4]	8.22	8.05	8.89	9.22	9.49	1.0
Jet Fuel[5]	4.63	4.28	5.80	6.46	6.85	2.8
Motor Gasoline[6]	9.33	8.93	10.54	11.07	11.21	1.3
Residual Fuel	2.24	2.07	2.76	3.17	3.59	3.3
Natural Gas[7]	4.03	4.65	6.36	8.34	8.67	3.7
Electricity	15.63	14.85	15.34	15.55	15.73	0.3
Total End-Use Energy	8.22	8.05	8.61	9.24	9.64	1.1
Primary Energy	7.90	7.68	8.32	8.90	9.25	1.1
Electricity	20.72	20.08	19.78	20.30	21.04	0.3
	Electric Generators[8]					
Fossil Fuel Average	1.64	1.61	1.65	1.88	2.05	1.4
Petroleum Products	2.53	2.48	2.91	3.33	3.78	2.5
Distillate Fuel	4.80	4.72	5.27	5.53	5.91	1.3
Residual Fuel	2.52	2.46	2.90	3.30	3.70	2.4
Natural Gas	2.40	2.57	2.59	3.38	3.73	2.2
Steam Coal	1.45	1.39	1.39	1.45	1.50	0.5
	Average Price to all Users[9]					
Petroleum Products	7.04	6.73	7.91	8.35	8.65	1.5
Distillate Fuel[4]	7.19	7.02	7.97	8.39	8.74	1.3
Jet Fuel	4.63	4.28	5.80	6.46	6.85	2.8
Liquefied Petroleum Gas	6.09	5.94	8.13	8.90	9.74	3.0
Motor Gasoline[6]	9.33	8.93	10.52	11.05	11.20	1.3
Residual Fuel	2.40	2.30	2.85	3.26	3.67	2.8
Natural Gas	3.79	3.97	3.94	4.74	5.01	1.4
Coal	1.47	1.41	1.40	1.45	1.50	0.4
Electricity	20.72	20.08	19.78	20.30	21.04	0.3

TABLE 7.3 (continued) Energy Prices by End-Use Sector and Source (1993 dollars per million Btu)

[1] Excludes independent power producers.
[2] Includes cogenerators.
[3] Excludes uses for lease and plant fuel.
[4] Includes federal and state taxes on diesel fuel and excludes county and local taxes.
[5] Kerosene-type jet fuel.
[6] Average price for all grades. Includes federal and state taxes and excludes county and local taxes.
[7] Compressed natural gas used as a vehicle fuel.
[8] Includes all electric power generators except cogenerators, which produce electricity as a byproduct of other processes.
[9] Weighted averages of end-use fuel prices are derived from the prices shown in each sector and the corresponding sectoral consumption.
Btu = British thermal unit.

Note: 1992 and 1993 figures may differ from published data due to internal rounding in the AEO95 National Energy Modeling System.

Sources: 1992 prices for gasoline, distillate, and jet fuel are based on prices in the Energy Information Administration (EIA), *Petroleum Marketing Annual 1992,* DOE/EIA-0487(92) (Washington, DC, July 1993). 1993 prices for gasoline, distillate, and jet fuel are based on prices in various 1993 issues of EIA, *Petroleum Marketing Monthly,* DOE/EIA-0380(93/1-12) (Washington, DC, 1993). 1992 and 1993 prices for all other petroleum products are derived from the EIA, *State Energy Price and Expenditure Report 1991,* DOE/EIA-0376(91) (Washington, DC, September 1993). 1992 residential and transportation natural gas delivered prices: EIA, *Natural Gas Annual 1992,* Volume 1, DOE/EIA-0131(92)/1 (Washington, DC, November 1993). 1993 residential natural gas delivered prices: EIA *Natural Gas Monthly,* DOE/EIA-0130(94/6) (Washington, DC, June 1994). Other 1992 and 1993 natural gas delivered prices: EIA, AEO95 National Energy Modeling System run AEO95B.D1103942. 1992 coal sectoral prices are from EIA, *State Price and Expenditure Report 1992,* DOE/EIA-0376(92) (Washington, DC, December 1994) except for the Utility sector which is from EIA, *Cost and Quality of Fuels for Electric Utility Plants 1993,* DOE/EIA-0191(93) (Washington, DC, July 1994). Values for 1993 coal prices have been estimated from EIA, *State Price and Expenditure Report 1992* using consumption quantities aggregated from EIA, *State Energy Data Report 1992,* DOE/EIA-0214(92) (Washington, DC, May 1994). 1992 residential electricity prices derived from EIA, *Short Term Energy Outlook,* DOE/EIA-0202(94/3Q) (Washington, DC, August 1994). 1992 and 1993 electricity prices for commercial, industrial, and transportation: EIA, AEO95 National Energy Modeling System run AEO95B.D1103942. *Projections:* EIA, AEO95 National Energy Modeling System run AEO95B.D1103942.

trends to deregulate and unbundle the electricity market have the potential to contribute to even lower prices in the future.

Energy Intensity

From 1970 to the mid-1980s, U.S. energy intensity declined—both energy use per capita and use per dollar of gross domestic product (GDP). Energy use per capita began to increase in the mid-1980s, as energy prices dropped. With projections of low prices and increasing demand for energy services, that trend is expected to continue. Energy use per person is likely to increase very slightly, at an average rate of 0.2% per year through 2010 (Table 7.4).

Energy use per dollar of GDP is projected to continue declining from 1993 through 2010, at an average annual rate of 1.2%. The expected rate of decline is slower than that seen in the 1970s and early 1980s (nearly 2%), when the economy shifted to less energy-intensive industries and increasingly efficient technologies. Low energy prices and the growth of more energy-intensive industries contribute to the slower rate of decline in the projection period.

The Energy Policy Act of 1992 (EPACT) and the National Appliance Energy Conservation Act of 1987 (NAECA) mandate additional efficiency standards for new energy-using equipment in the buildings sectors (residential and commercial) and for motors in the industrial sector. As demands for energy services and prices increase, additional investment in new technologies is anticipated, leading to further declines in energy intensity.

TABLE 7.4 Macroeconomic Indicators (billion 1987 dollars, unless otherwise noted)

	Reference Case					Annual Growth 1993–2010
Indicators	1992	1993	2000	2005	2010	(percent)
GDP Implicit Price Deflator (index 1987 = 1.000)	1.211	1.242	1.491	1.755	2.074	3.1
Real Gross Domestic Product	4986	5136	6126	6852	7485	2.2
Real Consumption	3342	3453	4035	4399	4748	1.9
Real Investment	733	820	1175	1385	1547	3.8
Real Government Spending	945	939	1003	1077	1143	1.2
Real Exports	578	598	938	1297	1625	6.1
Real Imports	611	674	1025	1306	1578	5.1
Real Disposable Personal Income	3633	3701	4371	4764	5140	2.0
Index of Manufacturing Gross Output (index 1987 = 1.000)	1.077	1.097	1.306	1.469	1.598	2.2
AA Utility Bond Rate (percent)	8.55	7.43	8.41	8.24	7.99	N/A
90-Day U.S. Government Treasury Bill Rate (percent)	3.43	3.00	4.66	4.53	4.37	N/A
Real Yield on Government 10-Year Bonds (percent)	3.62	3.23	4.13	3.50	3.51	N/A
Real 90-Day U.S. Government Treasury Bill Rate (percent)	0.54	0.44	1.61	1.17	1.01	N/A
Real Utility Bond Rate (percent)	5.66	4.87	5.36	4.87	4.64	N/A
Energy Intensity (thousand Btu per 1987 dollar of GDP)	17.17	16.99	15.44	14.50	13.88	−1.2
Consumer Price Index (1982 = 1.00)	1.40	1.45	1.81	2.19	2.64	3.6
Employment Cost Index (1987 = 1.00)	1.12	1.15	1.44	1.75	2.14	3.7
Unemployment Rate (percent)	7.55	7.42	6.28	6.00	6.47	N/A
U.S. Trade-Weighted Exchange Rate	0.97	1.00	0.95	0.90	0.87	N/A
		Million Units				
Truck Deliveries, Light-Duty	4.50	5.20	6.35	7.17	7.05	1.8
Unit Sales of Automobiles	8.38	8.71	9.59	9.92	10.17	0.9
		Millions of People				
Population with Armed Forces Overseas	255.8	258.4	275.6	287.1	298.9	0.9
Population (aged 16 and over)	196.3	198.2	212.8	223.8	235.4	1.0
Employment, Non-Agriculture	107.9	107.9	120.4	128.9	133.5	1.3

GDP = Gross domestic product.
Btu = British thermal unit.
N/A = Not available.
Note: Totals may not equal sum of components due to independent rounding.
Sources: 1992 and 1993: Data Resources Incorporated (DRI), DRI Trend0294. *Projections:* Energy Information Administration, AEO95 National Energy Modeling System run AEO95B.D1103942.

Macroeconomic Activity

The rate of growth of the economy, measured by the growth in GDP, is a key determinant of the growth in demand for energy. Associated economic factors, such as interest rates and disposable income, strongly influence various elements of the supply and demand for energy. These forecasts also reflect reactions to energy markets by the aggregate economy, such as a slowdown in economic growth resulting from increasing energy prices. The key macroeconomic indicators used are shown in the Table 7.4 and are discussed in Section 7.3, "Key Assumptions of the Forecast."

International Petroleum Markets

Changes in the world oil price and the supply prices of petroleum products for import to the United States are reflected in these forecasts. These values are summarized in Table 7.5. Some of the major assumptions underlying the forecast, including the level of oil production by OPEC countries, non-OPEC production, and worldwide regional economic growth rates and the associated regional demand for oil are discussed next.

7.3 Key Assumptions of the Forecast

The National Energy Modeling System

The National Energy Modeling System (NEMS) was used to generate the projections in this chapter. NEMS uses a unified modeling system to forecast alternative energy futures in the middle term—a time horizon of 20 years, in which the economy and the nature of energy markets are understood sufficiently well to allow for considerable structural and regional detail. Like its predecessors at the Energy Information Administration (EIA), NEMS incorporates a market-based approach to energy analysis, balancing energy supply and demand for each fuel and consuming sector, and taking into account the economic competition among energy sources.

Economic Growth

Economic growth rates, historically and in the reference case, affect all market activities, including energy supply and demand. In the reference case, the economy's output, as measured by real GDP, is expected to grow by 2.2% a year between 1993 and 2010. GDP grows at a slightly declining rate over the forecast period, continuing a long-term trend. The long-run potential for economic growth depends on growth in the labor force, capital stock, and labor productivity. The slowing of the economic growth rate in large part reflects demographic changes.

The labor force grows by 1.2% a year on average over the forecast horizon, with population growth of 0.9% a year. The labor force participation rate increases to a peak in 2005, then declines as "baby boom" cohorts reach retirement age. In the last 15 years of the forecast, the labor force grows more slowly. At the same time, there is a slight decline in the rate of productivity growth. Consequently, the GDP growth rate slowly declines, averaging 2.5% a year from 1995 to 2000 and 1.8% from 2005 to 2010.

Productivity growth is key to achieving the long-run GDP growth rate of 2.2%. Productivity is expected to increase on average by 1.1% a year, with fixed business investment and research and development expenditures in the private sector contributing to real output gains. Savings as a share of GDP increases over time as the government deficit is reduced. Increasing the pool of funds available for investment is an important ingredient for boosting long-term economic growth.

Overall manufacturing growth in the forecast averages 2.2% annually from 1993 to 2010. Energy-intensive industries grow at a slower average rate of 1.7% a year. Thus, their share of manufacturing output declines. Other, non-energy-intensive manufacturing industries grow at a faster rate, 2.6% a year, because two major non-energy-intensive manufacturing industries—electronic equipment and industrial machinery—grow by more than 3.0% annually.

The six major energy-intensive industries are chemicals, food processing, paper, petroleum refining, primary metals, and stone, clay, and glass. Among those six industries, chemicals show the fastest growth—2.5% annually. The refining industry has the slowest growth rate, averaging only 0.9% a year from 1993 through 2010.

TABLE 7.5 International Petroleum Supply and Disposition Summary (million barrels per day, unless otherwise noted)

	Reference Case					Annual Growth 1993–2010
Supply and Disposition	1992	1993	2000	2005	2010	(percent)
World Oil Price (1993 dollars per barrel)[1]	18.70	16.12	19.13	21.50	24.12	2.4
		Production[2]				
U.S. (50 states)	9.71	9.53	8.24	8.20	8.58	–0.6
Canada	2.08	2.10	2.17	2.42	2.34	0.6
OECD Europe[3]	5.08	5.17	6.38	5.22	4.67	–0.6
OPEC	26.21	27.02	36.01	42.68	46.67	3.3
Rest of the World[4]	11.56	11.67	12.84	12.05	11.77	0.1
Total Production	54.65	55.48	65.65	70.58	74.03	1.7
Net Eurasian Exports	1.26	1.33	1.24	1.42	1.60	1.1
Total Supply	55.70	56.78	66.89	72.00	75.63	1.7
		Consumption				
U.S. (50 states)	17.03	17.24	18.88	20.10	20.89	1.1
U.S. Territories	0.21	0.22	0.26	0.29	0.30	1.9
Canada	1.64	1.69	1.94	2.02	2.08	1.2
Japan	5.35	5.45	7.02	7.67	8.05	2.3
Australia & New Zealand	0.82	0.84	0.96	1.03	1.10	1.6
OECD Europe	13.59	13.48	15.87	16.64	16.97	1.4
Rest of the World[4]	17.36	18.16	22.27	24.56	26.55	2.3
Total Consumption	56.00	57.08	67.19	72.30	75.93	1.7
Discrepancy	0.30	0.30	0.30	0.30	0.30	N/A
		Consumption Aggregations				
OECD	38.64	38.92	44.92	47.75	49.38	1.4
OPEC	4.91	5.05	5.89	6.51	7.18	2.1
Rest of the World[4]	12.45	13.11	16.38	18.05	19.37	2.3
Non-OPEC Production[4]	28.44	28.47	29.63	27.90	27.36	–0.2
		OPEC Summary				
Market Share	0.47	0.47	0.54	0.59	0.61	1.5
Production Capacity[5]	28.90	31.30	38.40	44.10	47.75	2.5
Capacity Utilization	0.91	0.86	0.94	0.97	0.98	0.7

[1] The average cost to domestic refiners of imported crude oil.
[2] Includes production of crude oil (including lease condensates), natural gas plant liquids, other hydrogen and hydrocarbons for refinery feedstocks, alcohol, liquids produced from coal and other sources, and refinery gains.
[3] OECD Europe includes the unified Germany.
[4] Does not include Eurasia.
[5] Maximum sustainable production capacity.
OECD = Organization for Economic Cooperation and Development —Australia, Austria, Belgium, Canada, Denmark, Finland, France, Germany, Greece, Iceland, Ireland, Italy, Japan, Luxembourg, the Netherlands, New Zealand, Norway, Portugal, Spain, Sweden, Switzerland, Turkey, the United Kingdom, and the United States (including territories).
OPEC = Organization of Petroleum Exporting Countries — Algeria, Gabon, Indonesia, Iran, Iraq, Kuwait, Libya, Nigeria, Qatar, Saudi Arabia, the United Arab Emirates, and Venezuela.
Eurasia = Albania, Bulgaria, China, Czech Republic, Hungary, Poland, Romania, Slovak Republic, the Former Soviet Union, and the Former Yugoslavia.
N/A = Not applicable.
Note: Totals may not equal sum of components due to independent rounding.
Sources: 1992 and 1993: Energy Information Administration (EIA), *Short-Term Energy Outlook*, DOE/EIA-0202(94/3Q) (Washington, DC, August 1994). *Projections:* EIA, AEO95 National Energy Modeling System run AEO95B.D1103942.

World Oil Prices

In the reference case, the world oil price is about $19.10 per barrel in 2000 and about $24.10 in 2010. The prices reflect the assumption that OPEC member nations, particularly in the Persian Gulf, will produce sufficient oil to keep real price increases moderate.

Variation in OPEC production levels is one of the key assumptions used to generate the world oil price path for the reference case. Most oil market analyses agree that OPEC will continue capacity expansion after the year 2000 to meet growing demand.[4] The reference case projects OPEC production of 47 million barrels a day by 2010. Many estimates have OPEC capacity peaking and stabilizing at a level of 45 to 50 million barrels a day.[4] Two reasons are given. First, capacity additions after 2000 must offset production declines from old fields in the Middle East (some of them "super-giant" complexes). Second, OPEC producers are not inclined to develop fields that contain heavy crude oils, and those fields make up a substantial part of total reserves. Whereas there is considerable potential to expand capacity in such fields, particularly in Saudi Arabia and Venezuela, OPEC argues that worldwide downstream refining capacity is not sophisticated enough to upgrade heavier crude oils into lighter products.

Non-OPEC oil production in the reference case rises slightly through 2000, then declines in the 2000–2010 decade. Production in mature producing areas, such as the United States, declines over time, as is normal for older oil fields. Offsetting the declines are increases in production from areas with expanded capacity or new discoveries, including Latin America and the North Sea.

Economic growth is a major determinant of the *demand* for oil throughout the world, and the fastest growth is projected for the developing countries. Particularly high growth rates are expected for the Pacific Rim. Demand growth of 2.5% a year is expected for developing countries and 1.5% for countries in the Organization for Economic Cooperation and Development (OECD).

Legislation

The reference case assumes that all federal, state, and local laws and regulations in effect as of August 15, 1994, will remain unchanged through 2010. The impacts of pending or proposed legislation and sections of existing legislation for which funds have not been appropriated are not reflected.

The projections include the provisions of the Climate Change Action Plan (CCAP), a set of 44 actions designed to achieve carbon stabilization in the United States by 2000, relative to 1990. Of the 44 actions, 13 are not related to energy fuels and are not incorporated in the analysis. Emissions in the early 1990s have grown more rapidly than projected at the time the plan was formulated, and the forecasts of continued moderate prices make it more difficult to achieve stabilization. Funding for many of the CCAP programs had been curtailed in budget negotiations at the time of the analysis, and their full impact is not reflected Restoration of government investment in carbon mitigation programs or more rapid adoption of voluntary programs could lead to lower emissions levels.

Other major pieces of recent federal legislation included in the forecasts are the Omnibus Budget Reconciliation Act of 1993 (which adds 4.3 cents per gallon to the federal tax on highway fuels), the Clean Air Act Amendments of 1990 (CAAA90), and EPACT. The provisions of EPACT focus primarily on reducing energy demand. Minimum building efficiency standards are required for federal buildings and other new buildings that receive federally backed mortgages. Efficiency standards for electric motors, lights, and other equipment are required, and owners of automobile and truck fleets must phase in vehicles that do not rely on petroleum products. CAAA90 requires a phased reduction in vehicle emissions of regulated pollutants, primarily through the use of reformulated gasoline. Electric utilities are required by CAAA90 to reduce annual emissions of sulfur dioxide to less than 9 million short tons a year in 2000 and after.

Caveats of the Forecast

Despite the complexity of NEMS, a state-of-the-art energy-economy forecasting system, and the care and effort that have gone into developing and documenting the model, there is clearly uncertainty inherent in the forecast due to three fundamental sources: (a) uncertainty in the principle assumptions and parameters of the model, (b) uncertainty that the functional form selected may not be the best in representing the behavior of the agents being modeled, and (c) regulatory uncertainty. Consequently, potential users of this forecast who intend to use the reference case for capital investment purposes should understand the great uncertainties inherent in any forecast. User beware.

7.4 Energy Demand

Energy Consumption

In the reference case, the United States is projected to consume 104 quadrillion Btu of primary energy resources in 2010, 19% more than in 1993. (Primary energy consumption includes fuels used directly by consumers, in addition to energy used for electricity generation and distribution.) GDP is 46% higher in 2010 than in 1993, and the population is 16% larger. Thus, on a per-capita basis, energy use is projected to increase slightly, while the overall energy intensity of the economy continues to decline.

Between 1970 and 1980, the transportation sector accounted for most of the increase in end-use energy consumption. Other sectors showed relatively little growth in end-use energy consumption, but because they relied increasingly on electricity, there was a 15% increase in primary energy use. Between 1980 and 1993, energy prices (adjusted for inflation) were either stable or declined, and end-use energy consumption increased in all sectors.

Between 1993 and 2010, the residential and commercial sectors show little growth in end-use energy consumption. Energy efficiency improvements, in part mandated by legislation, largely offset energy consumption increases due to increases in the number of residential and commercial buildings. The industrial and transportation sectors, where growth in energy use is more sensitive to growth in the economy, account for about 90% of the projected increase in end-use energy consumption between 1993 and 2010 in the reference case.

During the 1970s, electricity consumption grew rapidly as electricity-based technologies such as air conditioning became commonplace in homes, workplaces, and commercial establishments. Between 1980 and 1993, electricity consumption grew more slowly. Lower growth rates are expected for electricity consumption throughout the forecast, as a result of improved efficiency and, for some applications, market saturation.

In the transportation sector, petroleum demand grew modestly during the 1970s. Rising fuel prices and new federal vehicle efficiency standards dampened the growth in petroleum consumption between 1980 and 1993. Demand for transportation fuels rises slowly in the forecast, as a larger vehicle population and increased travel in the context of stable oil prices more than offset improvements in average fleet efficiency.

Regulation of natural gas markets during most of the 1970s had the effect of limiting the availability of natural gas. As a result, natural gas use declined. Between 1980 and 1993, as gas supplies became more certain and regulations on end uses were removed, natural gas markets expanded—a trend that is expected to continue over the forecast. The largest increase is in the industrial sector, due to increasing use of natural gas for heat and power, chemical feedstocks, and cogeneration. Natural gas meets 26% of end-use energy requirements by 2010, at an annual consumption level of 20.5 quadrillion Btu.

Household Energy Expenditures

On the forecast, average household expenditures for energy increase by 12% through 2010. Most of the increase is for motor gasoline through 2000. Higher equipment efficiencies associated with recent standards are expected to dampen growth in energy expenditures for home use.

Climate Change Action Plan

The CCAP was developed by the Clinton Administration in response to a worldwide commitment to stabilize greenhouse gas concentrations in the atmosphere at levels that would prevent "dangerous anthropogenic interference with the climate system." The goal of the plan was to stabilize greenhouse gas emissions in 2000 at their 1990 levels. Emissions include carbon dioxide, methane, nitrous oxide, and other gases. Energy use is the primary source of carbon emissions and constitutes 87% of the total.

These projections account for carbon releases from fuel combustion and related activities. However, emissions of other gases from other sources, such as methane from landfills and agriculture, are not considered. Moreover, NEMS does not provide a net emissions balance that takes into account carbon "sinks," such as forests, which remove carbon from the atmosphere.

Estimates of energy consumption in the reference case include a significant decline in the growth rate of carbon emissions as a result of anticipated improvements in end-use equipment and building shell efficiencies.

Residential Demand

Over the past two decades, electricity's share of residential energy demand has doubled, as electrically powered services such as air conditioning have become commonplace in American households. In 1970, fewer than 40% of new single-family homes were equipped with central air conditioning, compared with 78% in 1993. In 1985, as many as 30% of new single-family homes had electric air-source heat pumps as the primary space heating devices. With competition from natural gas increasing, only 24% of new homes used heat pumps in 1993. Tighter federal appliance efficiency standards are also expected to slow future growth in residential electricity demand to 0.9% a year from 1993 to 2010.

Natural gas is expected to continue to account for about 50% of residential energy use throughout the forecast, virtually the same share as in 1993. While the number of households using gas rises, overall gas demand is stable as a result of substantial gains in end-use efficiency.

Distillate fuel use has fallen considerably over the past 20 years because of a shrinking customer base and improved efficiency of oil heating equipment. Further declines are projected, with declines in new distillate-fueled installations and conversions of existing units to natural gas.

Space and water heating are the two most energy-intensive services in the residential sector, accounting for over two-thirds of residential energy use in the forecast. Space heating continues to dominate energy use in the sector, but a 19% decline in average heating energy intensity between 1993 and 2010 is expected to reduce overall energy requirements for space heating. In addition, refrigerators and freezers are projected to use nearly 31% less total electricity in 2010 than in 1993.

Growth in households and appliances is a major determinant of residential energy consumption. Air conditioning and natural gas heaters grow substantially during the forecast period, as most new homes are built with them and additional units are added through retrofits of existing stock. Freezers and distillate-fueled heaters decline in use.

The reference case incorporates the effects of current policies aimed at increasing residential end-use efficiency, such as minimum efficiency standards for appliances, new building codes, and "Golden Carrot" programs to promote manufacturing innovations.

Commercial Demand

Projected energy use trends in the commercial sector feature an increasing market share for electricity, a stable share for natural gas, and declining shares for petroleum and other fuels (Table 7.2). Growth in electricity consumption is expected to slow considerably from the pace over the past two decades, with significant efficiency gains attributed to recent equipment standards, government programs aimed at increasing efficiency, and lower growth in commercial floorspace. Much of the growth in electricity is due to continued penetration of office equipment.

During the past 10 years, the share of energy supplied by natural gas in the commercial sector has ranged between 40 and 44%, with increases in gas-heated floorspace offset by the improved efficiency of gas-using equipment. Natural gas continues to be the fuel of choice for space heating and water heating, but no new end uses are expected to increase its share.

Distillate's share has dropped by half since 1970, but no further major declines in its current market share (about 7%) are anticipated. The stability of the distillate share is attributable to the historically low levels of its projected price.

Renewable energy currently accounts for only a small part of commercial energy use (0.2%). Its share is expected to more than double by 2010, with anticipated gains in solar technologies.

Between 1993 and 2010, new commercial floorspace is added at an average rate of 1% a year in the reference case. Commercial energy use is expected to grow at half that rate. Only two of the major commercial end uses of energy—office personal computers and miscellaneous electric office equipment—have projected growth rates in excess of floorspace growth. For the other end uses, which are assumed to be at full market penetration, increases in energy consumption are the net of consumption by new equipment resulting from floorspace additions, offset by savings resulting from retrofits with more energy-efficient equipment in existing floorspace.

Consumption of electricity for lighting remains practically unchanged over the forecast period, implying an average efficiency increase of over 1% annually. Most of the efficiency gain comes from the replacement of existing equipment with more efficient lighting. The forecast for lighting includes the estimated effects of the Environmental Protection Agency's Green Lights program. Participants in Green Lights will replace existing equipment as long as the payback of the project is greater than the prime rate plus 6 percentage points. If energy consumption for lighting grew at the same rate as floorspace, total electricity consumption in 2010 would be about 240 trillion Btu, or roughly 7% higher than in the reference case.

In the reference case, efficiency gains in the commercial sector can be attributed largely to equipment available for purchase today. The most significant opportunities for energy savings in the commercial sector are provided by technologies that use electricity and natural gas.

Industrial Demand

Overall energy use in the industrial sector—which includes the agriculture, mining, and construction (nonmanufacturing) industries in addition to traditional manufacturing industries—has not grown over the past two decades. However, there has been a shift in consumption toward electricity (early in the period) and natural gas (more recently). In the forecast, industrial use of purchased electricity grows at a significantly slower rate than was seen in the early 1980s, reflecting significant efficiency gains in electricity-based processes as well as increased reliance on self-generated electricity.

Natural gas is used both as a raw material (feedstock) in the chemical industry and as a fuel in other industrial applications. Between 1993 and 2010, gas consumption grows at an average annual rate of 1.3%, reaching 11.3 quadrillion Btu (including lease and plant fuel) by 2010. High prices and concerns over possible supply shortages have curtailed the use of oil as an industrial boiler fuel, but there have been recent increases in the use of petroleum as a feedstock for petrochemicals, as a byproduct fuel in refineries, and as a motor fuel in the nonmanufacturing industries. This trend is expected to continue. The use of coal for steelmaking, which dropped radically during the

1970s, is not expected to recover. As a boiler fuel, coal faces competition from natural gas, but its use in industrial boilers increases by 1.4% a year in the forecast because projected coal prices are more stable than either natural gas or oil prices.

The six energy-intensive industries—chemicals, food processing, paper, petroleum refining, primary metals, and stone, clay, and glass—accounted for 67% of all manufacturing energy consumption in 1993. By 2010, however, their share of manufacturing energy use is expected to fall to 64%, as their output (the value of manufactured products) grows more slowly than that of the other, non-energy-intensive manufacturing industries.

Energy conservation is expected to continue its important role in industries that rely heavily on energy, either as a fuel or as a feedstock. In the reference case, manufacturing output from the energy-intensive industries increases by 33% between 1993 and 2010, while their energy consumption increases by only 18%. The output from other manufacturing industries, such as transportation equipment, increases by 54% between 1993 and 2010. Energy expenditures in these industries, however, typically represent a small portion of total production costs. Nonetheless, over the same period, energy use in the non-energy-intensive manufacturing industries is projected to increase by only 40%.

The major use of energy in the nonmanufacturing industries is to fuel off-road equipment, such as farm tractors, bulldozers, and coal-handling equipment. Between 1993 and 2010, output from these industries increases by 35%, and energy consumption increases by 24%.

Aggregate industrial energy intensity (thousand Btu per dollar of output) decreased by 2.5% a year between 1970 and 1987, reflecting the use of more energy-efficient technologies and a shift toward less energy-intensive products. Intensity rose briefly in the late 1980s, when energy prices fell and output from energy-intensive industries rose slightly. Between 1993 and 2010, industrial energy consumption and output are projected to grow at average annual rates of 1.2 and 2.1%, respectively, in the reference case, with a 0.9% average annual decline in energy intensity. Substantially higher output growth rates are expected for key non-energy-intensive manufacturing industries, compared with the growth rates expected for energy-intensive industries such as iron and steel production (Table 7.6).

TABLE 7.6 Energy Intensity and Industry Growth

Industry	1993 Energy Intensity (thousand Btu per dollar)	1993–2010 Industry Growth (percent)
Cement	129.0	1.4
Iron and Steel	30.5	0.2
Transportation Equipment	0.9	2.4
Industrial Machinery	1.2	3.5

Transportation Demand

Transportation Energy Use Increases, Petroleum Fuels Dominate

Motor gasoline consumption increases by 0.7% annually between 1993 and 2010 in the reference case, a total increase of 1.6 quadrillion Btu. Gasoline's share of transportation energy use declines, as the overall fuel efficiency of conventional light-duty vehicles continues to improve and sales of alternative-fuel vehicles increase. In 2010, alternative fuels displace some 465,000 barrels of oil per day.

Distillate fuel oil is used mostly to transport freight. As such, its use follows trends in industrial output. In the reference case, distillate use is expected to increase by an average of 2% a year between 1993 and 2010.

Total jet fuel consumption increases at an average annual rate of 2.3% between 1993 and 2010, due to a 3.9% annual increase in commercial air travel. Fuel efficiency gains in new aircraft, increasing reliance on more fuel-efficient (per passenger mile) jumbo jets, and declining military use of jet fuel account for the difference in growth rates.

Light-Duty Vehicles Lead Gains in Transportation Fuel Efficiency

Average fuel efficiency continues to improve in the forecast, but at a slower pace than during the 1980s. Automobiles show the greatest average efficiency improvement between 1993 and 2010 at 1% a year, or about one-third the rate of improvement between 1980 and 1990. Contributing to the slower efficiency gains for light-duty vehicles are relatively low and stable projected gasoline prices and increases in disposable per-capita income, both of which tend to reduce consumer demand for fuel economy and increase demand for larger, better-performing cars. New care fuel economy in the reference case is only 18% higher in 2010 than it was in 1993.

Even smaller efficiency gains are projected for large trucks and locomotives, because their stock turnover is slower and there is limited potential for new technologies to improve fuel efficiencies. Aircraft stock efficiency improves at an average annual rate of 0.7%, as more wide-body aircraft are purchased.

Energy Use for Transportation Rises

Light-duty vehicles (cars, vans, and light trucks) currently account for more than one-half of all transportation energy consumption, and their dominant role continues in the forecast. Fuel use by light-duty vehicles increases on average by 1.0% annually between 1993 and 2010, slightly higher than the rate of population growth (0.9%). Fuel use by light trucks grows rapidly because of a 2.4% annual growth rate in vehicle-miles traveled by small commercial trucks.

Fuel consumption by freight trucks is projected to grow by 1.9% a year. Air transport fuel consumption is expected to grow by 2.2% a year, mainly because of a 3.9% annual increase in travel demand.

Personal Travel Grows More Slowly Than in the Past

Personal travel, consisting primarily of air travel and highway travel in light-duty vehicles, continues to grow with increasing population, personal income, and GDP. Other modes of personal travel, such as commuter rail and intercity bus, have historically accounted for an estimated 4% of passenger miles, and their share is assumed to remain stable over the forecast. Implementation of CCAP provisions that promote telecommuting, adoption of a transportation system efficiency strategy, and reform of tax subsidies for employer-provided parking lead to a 1.3% reduction in vehicle miles traveled by 2000.

Highway travel by light-duty vehicles, which grew at an annual rate of 3.4% from 1980 to 1990, increases at a more moderate 1.8% annual rate from 1993 to 2010. Contributing factors include slower growth of the driving-age population and demographic aging trends. Air travel grows at an annual rate of 3.9% from 1993 to 2010, substantially below the 5.8% annual growth between 1980 and 1990, reflecting the gradual maturation of the industry.

Freight transport by truck grows steadily from 1993 to 2010, driven by the growth of industrial output. Truck transport has the highest annual growth rate, 2.0%, but is well below the 4.1% annual growth rate recorded between 1980 and 1990. Rail and waterborne freight services maintain modest annual growth rates in the forecast, 1.3% and 1.1%, respectively.

Legislative Mandates Drive Sales of Alternative-Fuel Vehicles

Practically all sales of alternative-fuel vehicles (AFVs) before 1997 result from legislation, but by 2010, about one-fifth of AFV sales are based on market competition. Federal legislation mandates increasing use of AFVs by government and by energy utilities, and more sales are expected as a result of California's Low Emissions Vehicle Program (LEVP). CAAA90 includes a provision for states to join the pilot LEVP, which requires 2% of vehicles sold within the state in 1997 to be capable of using alternative fuels. By 2003, 10% of sales must be zero emission (electric) vehicles. The projections assume that California, New York, and Massachusetts—which currently account for 21% of all vehicle sales in the country—will participate. The expected regional distribution of AFV sales reflects regional fuel prices and availability, as well as regional legislation.

Approximately 3% of all new light-duty vehicles are expected to be capable of running on alternative fuels by 2000. The share of AFV sales rises to 9% by 2010, when the total stock of AFVs is expected to equal about 6% of the U.S. light-duty vehicle fleet. AFVs are expected to displace 465,000 barrels of oil per day in 2010, reducing transportation sector carbon emissions by approximately 7.3 million metric tons.

Among the AFV technologies, compressed natural gas, liquefied petroleum gas (propane), and electric vehicles claim the largest market shares in 2010. Almost 60% of all AFVs are purchased for use in commercial fleets, where natural gas and propane technologies are preferred. Electric, methanol, and ethanol vehicles are used primarily in nonfleet applications for personal use.

7.5 Electricity

Electricity markets in the United States are in the midst of a period of rapid change. The age of the domination of the vertically integrated system, where one utility controlled the generation, transmission, and distribution of electricity to each customer, is waning. The final outcome of the changes occurring is still unclear.

Currently, most state regulators, utilities, nonutility power producers, and their customers continue to struggle to define the future of the industry. The degree to which the market will, or should, move toward open competition is unclear.

One indication of the changes occurring is the rise of nonutility generators in wholesale power markets. From the beginning of this century through the 1970s, the economies of scale associated with large generating facilities led the industry to be dominated by large, vertically integrated utilities. During the 1970s, the capacity of nonutility generators actually declined, while utility generating capacity grew by more than 240 gigawatts. Electric utilities continued to dominate through the 1980s, but nonutilities, spurred by the passage of the Public Utilities Regulatory Policies Act of 1978 (PURPA), increased their capacity substantially. Their growth continued into the early 1990s, when nonutilities accounted for approximately half of all new capacity additions.

Electricity Market Restructuring: Moving Toward Competition

There are clear trends toward competition in electricity markets. The wholesale power market is becoming increasingly competitive. In many states, when a need for new resources is identified, the competition among utility and nonutility plants, utility demand-side management (DSM) programs, and DSM programs provided by energy service companies is fierce. Many utilities have instituted integrated resource planning (IRP) programs, which attempt to weigh the costs and benefits of all available resource options, and many utilities are using competitive bidding processes to acquire new resources.

Large electricity consumers, mainly industrial customers, are pressuring utilities to reduce their prices. Industrial customers have many options for meeting their energy needs, including self-generation, cogeneration, fuel switching, and relocation. Disparities among the industrial rates of neighboring utilities, and among utilities and nearby nonutility generators, have led industrial customers to demand lower rates from franchised utilities or access to cheaper power supplies. In some cases, utilities have offered lower rates in the form of economic development or interruptible rates. Utilities are also taking steps to reduce costs and increase competitiveness, such as consolidating with other utilities, reducing staff, and buying out uneconomical contracts with nonutilities.

The use of traditional cost-based ratemaking is also under review in some states. Concerned that cost-of-service-based pricing does not provide sufficient incentive to utilities to reduce their costs and provide electricity at the lowest possible rates, state commissions continue to investigate alternative methodologies, such as performance-based rate mechanisms and price caps.

States have been taking a variety of approaches to test the competitive waters. Many, often through the use of IRP programs, are continuing efforts to open wholesale markets to all resource options. The Federal Energy Regulatory Commission (FERC) is contributing to the effort through its implementation of Section 211 of EPACT, which gives FERC authority to order owners of transmission capacity to provide services to all requesters.

Michigan and California have taken even more aggressive steps to open both wholesale and retail electricity markets. In Michigan, regulators have decided to test open retail competition in a small experiment, allowing the customers of the state's two largest utilities to purchase power from other utilities or nonutilities for up to 60 megawatts of capacity. Regulators in California have stepped away from other states by proposing to give all customers access to any electricity supplier. Their proposal contains a timetable for moving from traditional franchised cost-of-service regulation to performance-based regulation, with all consumers being able to choose their suppliers by 2002.

The debate on the merits of increased competition is continuing, and numerous areas of contention remain. Among the most important are how best to ensure transmission access for wholesale suppliers, the potential impacts of consumer choice on utility systems, and how non-production expenditures, such as those associated with DSM, IRP, fuel diversity, and environmental compliance programs, can be recovered in a more competitive marketplace. Through individual cases, FERC is establishing criteria for acceptable transmission tariffs, but some argue that it may be necessary to force utilities to divest their transmission assets in order to ensure truly open access. If customers are permitted to leave their utilities and shop for other power providers, utilities could be left with underutilized, uneconomical assets. There is disagreement about whether the costs of such "stranded assets" should be recovered from the departing customers, the utility's remaining ratepayers, or stockholders.

With the 1992 passage of EPACT, nonutilities are expected to continue to play an important role in meeting growth in the demand for electricity through 2010. However, utilities are also taking steps to increase their competitiveness. Through consolidations, staffing reductions, and other cost-cutting measures, utilities are preparing for a more competitive future.

While utilities and nonutilities try to keep up with the evolving structure of the electricity market, they are also faced with slower growth in demand than in the past. Historically, the demand for electricity has been linked with economic growth. This positive relationship is expected to continue, but the magnitude is uncertain.

During the 1960s, electricity demand grew by over 7% a year, nearly twice the rate of economic growth. In the 1970s and 1980s, however, the ratio of electricity demand growth to economic growth declined to 1.5 and 1.0, respectively. Throughout the forecast, this slowing trend is expected to continue. Several factors have contributed to this trend, including increasing market penetration of electric appliances, improvements in equipment efficiency brought about by consumer expectations of higher energy prices, utility investments in DSM programs, and legislation establishing more stringent equipment efficiency standards. For example, by 1991, nearly 70% of homes in the United States, the majority of those in warmer climates, had air conditioning systems.

There are factors that could mitigate the slowing of electricity demand growth seen in these projections. New electric appliances seem to appear almost daily. Only a few years ago, no one could have foreseen the growth in home computers, facsimile machines, copiers, and security systems, all powered by electricity. If new uses of electricity are more substantial than currently expected, they could partially offset the efficiency gains shown in these projections.

Electricity Demand

From 1993 to 2010, the annual growth rate for electricity sales is projected to be about 1.1% (Table 7.7), well below the projected GDP rate of 2.2% per year and well below the 3.2% per year experienced between 1970 and 1993.

Several factors contribute to the projected decrease in electricity demand growth in the forecast. By complying with EPACT, businesses and municipalities are expected to improve energy efficiency through the installation of energy-efficient lighting and appliances, increased building efficiency, and support of energy efficiency in process-related industries. In addition, National Appliance

TABLE 7.7 Electricity Supply, Disposition, and Prices (billion killowatthours, unless otherwise noted)

Supply, Disposition, and Prices	Reference Case 1992	1993	2000	2005	2010	Annual Growth 1993–2020 (percent)
	Generation by Fuel Type					
Electric Utilities						
Coal	1576	1639	1689	1761	1825	0.6
Petroleum	89	100	80	111	118	1.0
Natural Gas	264	259	301	342	387	2.4
Nuclear Power	619	610	652	653	596	–0.1
Pumped Storage	–3	–2	–2	–2	–2	–0.8
Renewable Sources[1]	253	277	303	306	317	0.8
Total	2797	2883	3022	3171	3241	0.7
Nonutilities (excluding cogenerators)[2]						
Coal	2	5	9	12	44	13.4
Petroleum	1	1	0	0	0	–10.4
Natural Gas	16	21	36	53	90	8.8
Renewable Sources[1]	40	45	64	73	101	4.9
Total	60	73	109	139	236	7.1
Cogenerators[3]						
Coal	46	49	55	58	60	1.3
Petroleum	8	10	38	41	44	9.2
Natural Gas	144	151	177	196	216	2.2
Renewable	35	35	38	38	39	0.5
Other[4]	3	3	4	5	5	3.0
Total	237	248	313	338	365	2.3
Sales to Utilities	102	111	138	148	159	2.1
Generation for Own Use	135	137	175	190	206	2.4
Net Imports	28	28	35	33	56	4.1
	Electricity Sales by Sector					
Residential	936	994	1058	1094	1156	0.9
Commercial	837	868	945	979	1026	1.0
Industrial	973	983	1079	1174	1242	1.4
Transportation	18	18	25	38	52	6.5
Total	2763	2862	3108	3285	3475	1.1
	End-Use Prices (1993 cents per kilowatthour)[5]					
Residential	8.4	8.3	8.2	8.4	8.8	0.3
Commercial	7.7	7.4	7.3	7.6	7.9	0.4
Industrial	5.3	5.0	4.9	5.0	5.2	0.3
Transportation	5.3	5.1	5.2	5.3	5.4	0.3
All Sectors Average	7.1	6.8	6.7	6.9	7.2	0.3
	Price Components (1993 cents per kilowatthour)					
Capital Component	2.9	2.8	2.6	2.5	2.5	–0.5
Fuel Component	1.2	1.2	1.2	1.4	1.4	1.1
Operation and Maintenance Component	2.8	2.7	2.7	2.8	2.7	0.0
Wholesale Power Cost	0.1	0.1	0.2	0.3	0.5	7.0
Total	7.1	6.8	6.7	6.9	7.2	0.3

[1] Includes conventional hydroelectric, geothermal, wood, wood waste, municipal solid waste, other biomass, solar and wind power.

[2] Electricity was produced solely for sale to an electric utility or another end user, and there is no business activity at the site (standard industrial classification 49).

TABLE 7.7 (continued) Electricity Supply, Disposition, and Prices (billion kilowatthours, unless otherwise noted)

[3] Includes generation and cogeneration at facilities whose primary function is not electricity production (standard industrial classification other than 49). Includes sales to utilities and generation for own use.

[4] Other includes methane, propane, and blast furnace gas.

[5] Prices represent average revenue per kilowatthour.

Note: Totals may not equal sum of components due to independent rounding.

Sources: 1992 commercial and transportation sales: Total transportation plus commercial sales comes from Energy Information Administration (EIA), *State Energy Data Report 1992,* DOE/EIA-0214(92) (Washington, DC, May 1994), but individual sectors do not match because sales taken from commercial and placed in transportation, because Oak Ridge National Laboratories, *Transportation Energy Data Book 14* (May 1994) indicates the transportation value should be higher. 1992 and 1993 generation by electric utilities, nonutilities, and cogenerators, net electricity imports, residential sales, and industrial sales: EIA, *Annual Energy Review 1993,* DOE/EIA-0384(93) (Washington, DC, July 1994). 1992 residential electric prices derived from EIA, *Short Term Energy Outlook,* DOE/EIA-0202(94/3Q) (Washington, DC, August 1994). *1992 and 1993 electricity prices for commercial, industrial, and transportation, price components, and projections:* EIA, AEO95 National Energy Modeling System run AEO95B.D1103942.

Energy Conservation Act of 1987 (NAECA) requires appliance manufacturers to meet efficiency standards for certain appliances before they can be marketed.

DSM programs are also expected to increase throughout the 1990s. By granting customers rebates for installing energy-efficient appliances, utilities provide incentives to lower energy consumption. In charging reduced rates for off-peak service, utilities can delay the need for new capacity by promoting the efficient use of currently available capacity. DSM programs are expected to reduce the demand for electricity by 73 billion kilowatthours in 1997, relative to the level that would have been reached in their absence.

New Capacity Additions

In the reference case, utilities are expected to add 78 gigawatts of new capacity and retire 53 gigawatts between 1993 and 2010, for a net gain of 25 gigawatts (Table 7.8). Nonutilities add 41 gigawatts of new net capacity, accounting for 62% of the total net capacity added. Nearly 16 gigawatts of new net capacity is supplied by cogenerators during this period.

Gas-fired combined-cycle and combustion turbine technologies provide a significant share of the expected new capacity. These technologies have the advantages of relatively low initial capital cost, high efficiency, and low emissions. New coal and renewable capacity will also be used for baseload and peak requirements, respectively. Gas turbines, while adding the most capacity, essentially meet peak requirements. Most electricity requirements will be met by coal-fired and gas combined-cycle capacity.

Electricity Fuel Shares

Flexibility Makes Gas-Fired Capacity More Attractive

Before building new capacity, utilities are expected to use other options to meet demand growth—life extension and repowering of existing plants, imported power from Canada and Mexico, demand-side management, and purchases from cogenerators. Even so, assuming an average plant capacity of 300 megawatts, a projected 450 new plants with a total of 135 gigawatts of capacity will be needed by 2010 to meet growing demand and to offset retirements.

Of the new capacity needed, 61% is projected to be gas-fired or oil- and gas-fired combined-cycle or combustion turbine technology. Both technologies are designed primarily to supply peak and intermediate capacity, whereas combined-cycle technology can also be used to meet baseload requirements.

TABLE 7.8 Electricity Generating Capability (thousand megawatts)

Net Summer Capability[1]	Reference Case					Annual Growth 1993–2010 (percent)
	1992	1993	2000	2005	2010	
	Electric Utilities					
Capability						
Coal Steam	301.0	300.7	297.1	301.1	306.4	0.1
Other Fossil Steam[2]	141.8	141.3	130.4	124.3	120.4	–0.9
Combined Cycle	9.2	9.7	17.7	23.4	27.2	6.2
Combustion Turbine/Diesel	49.2	49.2	65.9	71.0	77.8	2.7
Nuclear Power	99.0	99.0	101.3	101.3	88.7	–0.6
Pumped Storage	19.1	19.1	20.0	20.0	20.0	0.3
Renewable Sources[3]	76.6	76.8	78.1	78.9	81.6	0.4
Total	695.9	695.7	710.5	720.0	722.1	0.2
Cumulative Planned Additions[4]						
Coal Steam	0.00	0.00	5.63	11.98	13.74	N/A
Other Fossil Steam[2]	0.00	0.00	0.00	0.00	0.00	N/A
Combined Cycle	0.00	0.61	7.22	11.79	12.38	19.3
Combustion Turbine/Diesel	0.00	0.14	16.20	22.41	22.77	35.2
Nuclear Power	0.00	1.15	3.49	3.49	3.49	6.7
Pumped Storage	0.30	0.30	1.29	1.29	1.29	9.0
Renewable Sources[3]	0.19	0.20	1.14	1.15	1.15	10.9
Total	0.49	2.40	34.97	52.11	54.82	20.2
Cumulative Unplanned Additions[4]						
Coal Steam	0.00	0.00	0.02	0.34	6.68	N/A
Other Fossil Steam[2]	0.00	0.00	0.00	0.00	0.00	N/A
Combined Cycle	0.00	0.00	1.42	2.56	5.79	N/A
Combustion Turbine/Diesel	0.00	0.00	0.74	1.23	8.05	N/A
Nuclear Power	0.00	0.00	0.00	0.00	0.00	N/A
Pumped Storage	0.00	0.00	0.00	0.00	0.00	N/A
Renewable Sources[3]	0.00	0.00	0.13	0.52	3.07	N/A
Total	0.00	0.00	2.32	4.66	23.59	N/A
Cumulative Total Additions	0.49	2.40	37.29	56.77	78.41	22.8
Cumulative Retirements[5]	6.50	8.71	29.26	39.86	59.57	12.0
	Nonutilities (excludes cogenerators)[6,7]					
Capability						
Coal Steam	0.47	0.62	1.82	2.10	7.11	15.5
Other Fossil Steam[2]	0.21	0.21	0.25	0.25	0.25	0.9
Combined Cycle	2.02	2.03	5.29	7.00	10.48	10.1
Combustion Turbine/Diesel	1.28	1.33	2.80	3.97	13.72	14.7
Nuclear Power	0.00	0.00	0.00	0.00	0.00	N/A
Pumped Storaged	0.00	0.00	0.00	0.00	0.00	N/A
Renewable Sources[3]	8.61	9.11	13.45	15.47	22.07	5.3
Total	12.60	13.30	23.61	28.79	53.62	8.5
	Cogenerators[7,8]					
Capacity						
Coal	6.64	6.90	8.03	8.37	8.72	1.4
Petroleum	2.89	3.36	5.96	6.37	6.81	4.2
Natural Gas	20.83	21.06	24.12	26.74	29.52	2.0
Renewables	6.66	6.92	7.41	7.46	7.51	0.5
Other	0.00	0.00	0.03	0.03	0.03	N/A
Total	37.01	38.25	45.54	48.96	52.59	1.9
Cumulative Additions[4,7]	5.86	7.80	25.40	34.00	62.46	13.0

TABLE 7.8 (continued) Electricity Generating Capability (thousand megawatts)

[1] Net summer capability is the steady hourly output that generating equipment is expected to supply to system load (exclusive of auxiliary power), as demonstrated by tests during summer peak demand.

[2] Includes oil-, gas-, and dual-fired capability.

[3] Includes conventional hydroelectric, geothermal, wood, wood waste, municipal solid waste, other biomass, solar and wind power.

[4] Cumulative additions after December 31, 1992. Nonzero utility planned additions in 1992 indicate units operational in 1992, but not supplying power to the grid.

[5] Cumulative total retirements from 1990.

[6] Electricity was produced solely for sale to an electric utility or another end user, and there is no business activity at the site (standard industrial classification 49).

[7] Nameplate capacity is reported for nonutilities on Form EIA-867, "Annual Power Producer Report." Nameplate capacity is designated by the manufacturer. The nameplate capacity has been converted to the net summer capacity based on historic relationships.

[8] Incudes generators and cogenerators at facilities whose primary function is not electricity production (standard industrial classification other than 49).

N/A = Not applicable.

Notes: Totals may not equal sum of components due to independent rounding. Net summer capacity has been estimated for nonutility generators for AEO95. Net summer capacity is used to be consistent with electric utility capacity estimates. Electric utility capacity is the most recent data available as of August 15, 1994. Therefore, capacity estimates may differ from other Energy Information Administration sources.

Sources: Net summer capacity at electric utilities in 1992 and 1993, and planned additions: Energy Information Administration (EIA), Form EIA-860, "Annual Electric Generator Report." Net summer capacity for nonutilities and cogeneration in 1992 and 1993 and planned additions estimated based on EIA, Form EIA-867, "Annual Nonutility Power Producer Report." *Projections:* EIA, AEO95 National Energy Modeling System run AEO95B.D1103942.

Through 2010, the equivalent of 97 plants with a total of 29 gigawatts of new planned and additional unplanned coal-steam capacity are projected to come online, and the equivalent of 52 plants totaling 16 gigawatts are expected to retire. After 2005, coal-steam plants will compete effectively with gas-fired plants for baseload capacity. More than 46% of the new coal-fired capacity will be added after 2005.

Except for units in the construction pipeline, no additional nuclear or hydroelectric capacity is expected. Nuclear plants are assumed to retire, with no life extension, as their current operating licenses expire. Increases in hydroelectric capacity that result from repowering are expected to be offset by retirements.

Coal Is Projected to Remain the Dominant Fuel for Electricity Generation

As they have since early in this century, coal-fired power plants are expected to remain the dominant source of electricity through 2010. In 1993, coal plants accounted for 41% of the generating capacity in the United States and for 53% of the electricity generated. Rising environmental concerns about coal plants, combined with their relatively long construction lead times and the availability of economical natural gas, make it unlikely that many new coal plants will be built before 2000. However, slow demand growth and the huge investment in existing plants will keep coal in its dominant position.

The large investment in existing plants will also make nuclear power an important source of electricity through 2010. In recent years, the performance of nuclear power plants has improved substantially, and two units now under construction are expected to become operable in the near term.* As a result, nuclear generation is projected to increase through 2006. After 2006, however, nuclear generation is expected to decline as older units are retired.

*One of the two new units anticipated was Watts Bar 2. After the *Annual Energy Outlook 1995* analysis was completed, the Tennessee Valley Authority announced that it was halting construction work on the unit and was considering the options of converting the plant to alternative fuel or seeking a financial partner to complete construction.

In percentage terms, generation from gas-fired power plants shows the largest increase in the forecast. As a result, by 2010, gas-fired generation by utilities, nonutilities, and cogenerators overtakes nuclear power as the nation's second-largest source of electricity. Generation from oil-fired plants remains fairly small throughout the forecast.

Gas-Fired Generation Increases for All Types of Generators

The future for natural gas in electricity is bright. Between 1993 and 2010, gas-fired generation is expected to increase by over 60%. Increased use of gas by utilities, nonutilities, and cogenerators is projected to raise its share of total generation to 18% by 2010. New combustion turbine and combined-cycle plants, with efficiencies approaching 50%, make gas-fired plants competitive with other resource options. Their high efficiencies relative to other resource options, such as conventional coal-fired power plants (around 35%), partially offset their higher fuel costs. Only after 2005 do rising gas prices begin to make gas-fired plants less economical.

Financially, combustion turbines and combined-cycle plants are attractive because of their modularity and relatively low capital costs. Unlike with traditional steam plants, the per-kilowatt costs and thermal efficiencies of small turbine and combined-cycle plants are similar to those for larger plants. Thus, it is economical to add them in small increments, reducing the financial burden on the utility or nonutility and allowing capacity to be added slowly as demand grows.

Gas-fired plants are also attractive for environmental reasons, since gas produces much lower carbon and sulfur emissions during combustion than do coal and oil. Because sulfur dioxide emissions from gas-fired plants are near zero, there is no need for operators to purchase the emission allowances required for coal- and oil-fired plants under CAAA90.

Sulfur Dioxide Emissions Cap Goes Into Effect

In response to CAAA90, utilities and nonutilities have begun taking steps to reduce sulfur dioxide emissions below the established ceilings. Relatively "dirty" plants must take action by 1995, while other affected plants have until 2000. The goal is to reduce annual emissions below 9 million short tons, compared with 14.8 million in 1993. Because utilities can hold allowances for use in future years, and because temporary allowances are issued between 2000 and 2010, the 9-million-ton cap is not reached until 2010 or later.

In each phase of the CAAA90, affected facilities are issued annual permits, or allowances, to emit a fixed amount of sulfur dioxide in the permit year or any year thereafter. Each operator must ensure that there are sufficient allowances on hand to cover the year's emissions. Compliance options include fuel switching (to lower sulfur fuels) or blending, purchasing allowances, and installing flue gas desulfurization equipment (scrubbers). Utilities have reported plans to switch or blend fuels at most of their units affected by Phase I.* They have already begun actions to modify older plants that burn bituminous coal to enable them to burn lower sulfur subbituminous coal. For other affected units, utilities have reported plans to purchase allowances and add scrubbers. After 2000, as the restrictions tighten, more operators are expected to use scrubber retrofits to stay within compliance limits (Table 7.9).

TABLE 7.9 Scrubber Retrofits and Allowance Costs, 2000–2010

Industry	2000	2005	2010
Cumulative Retrofits From 1993 (gigawatts of capacity)	18.4	18.4	22.2
Allowance Costs (1993 dollars per ton of sulfur dioxide)	23.8	22.1	23.5

*For more information on the electricity industry's CAAA90 compliance plans, see *Electric Utility Phase I Acid Rain Compliance Strategies for the Clean Air Act Amendments of 1990*, DOE/EIA-0582 (Washington, DC, March 1994).

Nuclear Power

The reference case forecast assumes that all nuclear units will operate to the end of their current license terms, with 19 units (13.7 gigawatts) retiring by 2010. Two units under construction, Watts Bar 1 and 2, were assumed to be completed by 1997, and no newly order plants become operational during the forecast period. Given these assumptions, 93 nuclear units are projected to produce 16% of total electricity generation in 2010.

Because the average age of nuclear units is currently less than 20 years, the performance of older reactors is not well established. By 2010 most operable units will have been in service for more than 20 years; however, their expected lifetimes are uncertain. The early retirements of Yankee Rowe and San Onofre 1 in 1992 and Trojan in 1993 occurred because the utilities faced costly repairs when future competitiveness was uncertain. Also, the lack of a permanent waste repository may require increased on-site spent fuel storage capability for continued operation at some units.

The Nuclear Regulatory Commission has recognized that nuclear reactors can operate safely beyond the initial license period. There is a process in place for utilities to extend their operating licenses for up to 20 years past the current 40-year license term. Also, two advanced designs have recently received final design approval, which, under the new streamlined design certification process, will allow future orders of the same type to be placed without a complete design safety review each time. Given the challenges facing the nuclear industry, there is considerable uncertainty in the assumptions for the nuclear forecast. Other possible assumptions could allow for early retirements, different completion schedules for units under construction, and possible new orders of advanced reactors.

Electricity Prices

Electricity prices are expected to remain steady despite rising fuel prices. Between 1993 and 2010, average electricity prices are expected to remain nearly unchanged, rising by a scant 0.4 cents per kilowatthour (Table 7.7). Although residential prices rise slightly more, the average household electric increases by only $3 to $4 (1993 dollars) per month by 2010. Rising natural gas prices place upward pressure on electricity prices, but stable coal prices, slowing expenditures on new and existing plants, and steady operations and maintenance costs nearly offset their impact.

From 1993 to 2010, natural gas prices to electric generators rise by 45%, from $2.63 to $3.82 per thousand cubic feet. Over the same period, utilities and nonutilities (including cogenerators) increase their generation from natural gas by 49 and 78%, respectively. By 2010, purchases of natural gas are expected to account for 35% of total utility fossil fuel expenditures, up from 23% in 1993 (excluding purchases by cogenerators). Dependence on natural gas plants is expected to increase because of their relatively low construction costs, high efficiencies, and low sulfur and carbon emissions. These factors partially offset their higher fuel costs relative to those of coal plants.

Coal prices to electric utilities are projected to increase very little, rising by only 8% between 1993 and 2010. Large domestic coal reserves and improvements in mining productivity combine to keep coal prices stable. Because coal plants provide over half the electricity generated, stable coal prices contribute to stable electricity prices.

Utility capital costs, associated with recovery of investments in power plants and transmission and distribution facilities (including annual depreciation expenses, return on investment, and taxes), are expected to grow more slowly than electricity sales, reducing their impact on electricity prices. In contrast, purchases of power from wholesale suppliers are expected to grow in importance. Factors contributing to these trends include the abundant generating capacity that exists today, increasing reliance on economical wholesale suppliers for new resources as needed, and construction of relatively inexpensive natural gas combustion turbine and combined-cycle plants.

Most regions of the country have sufficient capacity in place to meet expected demand growth for many years. As a result, growth in generating capacity will lag sales growth for the next decade or so, before picking up when existing capacity is fully utilized. From 1993 through 2010, generating capacity grows at an annual rate of only 0.6%, while electricity sales grow by 1.1% a year.

The need for utilities to make capital investments should also be reduced in two other ways. First, when new generating resources are needed, utilities are expected, in large part, to look to wholesale suppliers. Thus, the suppliers will make the needed capital investments. Second, building natural gas combustion turbine and combined-cycle plants, which are less expensive to build than other generating options, reduces capital requirements.

Electricity: Challenges for the Future

As discussed, the U.S. electric utility industry is in a period of transition. Among the issues facing the industry, which could significantly affect the outlook for the future, are the continued development of IRP and DSM programs by utilities, recent legislation affecting nonutility generators, new environmental regulations, the emergence of new generating technologies, changes in nuclear plant refurbishment and retirement options, and state policies affecting the electric power market.

Many utilities have developed IRP programs to evaluate the resource options available to meet the demand for electricity. These programs attempt to make all options equally accessible and allow for the participation of all interested stakeholders in the evaluation of costs and benefits. In some cases, utilities and their public utility commissions (PUCs) have attempted to adjust for factors not normally captured in traditional cost-benefit comparisons, such as including emission adders to account for environmental externalities associated with particular technologies or evaluating the impacts on the local economy of choosing one technology over another.

Utility investments in DSM programs have also increased dramatically during the 1990s. Utilities reported to EIA that DSM programs reduced their peak demand by 17.7 gigawatts in 1992, and they reported plans to continue investing heavily in DSM programs through 2000. With the passage of EPACT, however, utility DSM programs may be co-opted by the Act's stringent appliance efficiency standards as consumers purchase new appliances. The future of utility programs targeted at the same appliances is therefore ambiguous. The new standards may lead utilities to refocus their programs on other areas, and continued technological development may create new opportunities for DSM programs

EPACT also contains provisions with potentially significant impacts on the development of nonutility generators and the flow of electricity trade. EPACT creates a class of generators, referred to as exempt wholesale generators (EWGs), which can develop non-rate-based generating systems and market the power from them to utilities. EPACT also guarantees EWGs greater access to utility transmission systems. These provisions will lead to an increase in nonutility generation and, to some degree, a restructuring of the electricity industry.

The possibility of new environmental legislation and the continued development of advanced generating technologies also present challenges for the future. While the CCAP primarily involves voluntary compliance, more stringent carbon reduction regulations are possible if the current approach does not achieve the desired reductions. Efforts to develop generating technologies with reduced environmental impacts are underway. New technologies, particularly for generation from renewable fuels, might play an important role in reducing the emissions associated with electricity generation.

In the area of nuclear power, the Advanced Light-Water Reactor Program, a joint initiative by the U.S. Department of Energy (DOE) and the nuclear supply industry, is a current priority. The goal of the program is to develop a standardized nuclear plant design for commercial orders. Four plant types are involved, two "evolutionary" and two "midsized." The current schedule calls for design certification in 1996 and 1997, respectively. It is still unlikely, however, that any new orders for nuclear plants will be placed until a number of difficult issues are satisfactorily resolved: concerns about disposal of radioactive waste, public concerns with safety, concerns about economic and financial risks, uncertainty about future power plant performance, and uncertainty in the licensing and regulatory processes.

At the federal and state levels, initiatives that will change the way electric power is marketed are being proposed and implemented. Their goal is to lower the cost of electricity to all classes of consumers through increased competition in the electric power industry. The initiatives will grant utilities the flexibility to compete for market share, ensure nonutilities a fair opportunity to compete, and give consumers direct access to electricity markets. There is still much uncertainty about how effective such measures will be in achieving lower costs for consumers. In many cases, the initiatives introduce potential problems, such as increased risk, higher cost of capital, conflict between state and federal authority, multiple ownership of transmission systems, and cost allocation.

7.6 Oil and Natural Gas

Although the share of total U.S. energy *production* captured by oil and gas declines in the reference case, the combined oil and gas share of total energy *consumption* increases. Domestic production of dry natural gas increases from 18.9 quadrillion Btu in 1993 to 21.5 in 2010, but that increase is more than offset by a drop in crude oil production (including natural gas plant liquids) from 16.9 to 14.2 quadrillion Btu. Oil production falls to 13.6 quadrillion Btu in 2005, as the depletion of lower-cost existing resources continues, then rebounds to the 2010 level as prices rise and technological advances, especially for enhanced oil recovery (EOR), reduce production costs.

Net petroleum imports are needed to fill the widening gap between domestic production and consumption. In 2010, 59% of U.S. domestic petroleum consumption is met by imports, compared with 44% in 1993.

A significant increase in natural gas consumption is driven by its expanding role as a boiler fuel for electric generators. (Electric generators here include all electric power generators except cogenerators, which produce electricity as a byproduct of other processes.) Natural gas also continues to be the leading source of end-use energy for space heating. Natural gas is further expected to capture the largest share of the alternative-fuel vehicle market in 2010. In total, natural gas consumption grows in the forecast from 20.8 quadrillion Btu in 1993 to 25.3 quadrillion Btu in 2010. Net imports of natural gas total only 3.7 quadrillion Btu in 2010, and thus they satisfy a much smaller percentage of consumption (15%) than is the case for oil imports.

There has been a steady downward trend in most forecasts of oil and natural gas wellhead prices over the past 5 years. The revisions have been based primarily on reassessments of the resource base, reevaluations of improvements in exploration and production technology, changing expectations with regard to the global oil supply/demand balance, and revised estimations of the effects of increased competition following ongoing restructuring of the natural gas industry.

Oil and Gas Production

Wellhead Prices for Oil and Gas Rise in the Forecast

Domestic oil and natural gas prices are projected to reverse recent declines (Table 7.10). Domestic oil prices are determined largely by the international market, whereas domestic natural gas prices are determined largely by competition in North American energy markets. Unlike oil, natural gas is not easily transported between the United States and countries outside North America.

The prices of both fuels increase over most of the forecast in response to rising domestic demand and resource depletion. Natural gas prices rise despite increased competition and technological advances that reduce the cost of finding and developing reserves. The world oil price interacts with domestic natural gas markets in complex ways. A positive, direct effect occurs on gas *supply* when oil prices, and hence levels of oil drilling, increase. Nearly one-sixth of domestic gas production (associated and dissolved natural gas) is currently a coproduct of oil production. On the other hand, a negative, indirect effect on gas *supply* also occurs, because increases in oil prices increase the profitability of oil drilling relative to gas drilling. The change in relative profitability induces

TABLE 7.10 Forecast of World Oil Price and Domestic Natural Gas Wellhead Price

Forecast	1993	2000	2005	2010
World Oil Price (1993 dollars per barrel)	16.12	19.13	21.50	24.12
Natural Gas Wellhead Price (1993 dollars per thousand cubic feet)	2.02	2.14	3.02	3.39

drillers to shift exploration and development investments in the direction of oil, the increasingly profitable opportunity. Crude oil prices can also affect *demand* for natural gas, as oil and gas compete as substitute fuels in some end-use sectors. The competition affects natural gas demand, and thus natural gas prices, in the forecast period.

Gas Prices Increase More Dramatically Than Oil Prices

Lower 48 natural gas wellhead prices in the reference case are projected to grow at an average annual rate of 3.1% between 1993 and 2010 (Table 7.11). Lower 48 wellhead prices for crude oil in the reference case grow more slowly than gas prices, at an average rate of 2.4% a year over the 1993–2010 period (Table 7.12). The average natural gas price comes much closer to attaining its historic peak than does the crude oil price.

The historic peak for average annual lower 48 wellhead prices was reached in 1983 for natural gas ($3.58 per million Btu in constant 1993 dollars) and in 1981 for oil ($9.15 per million Btu). Reasons for the closer approach of the gas price to its historic peak include faster growth in demand for gas and the fact that, over most of the historical period, the price of gas was regulated at artificially low levels. It should be noted, however, that relative trends in wellhead prices are not necessarily equivalent to relative trends in end-use prices. End-use prices are the primary determinant of fuel selection in end-use applications where technologies provide a choice among competing fuels. Natural gas end-use prices increase less than the wellhead price because transmission and distribution margins decline over most of the forecast. In contrast, petroleum product prices (in particular, prices for light products) increase more than the oil wellhead price, because markups from the wellhead to the end user increase during the forecast.

Drilling for Oil and Gas Increases

Rising prices (combined with lower finding and operating costs) generally lead to more drilling for both natural gas and crude oil. Projected drilling statistics include the expected values of both exploratory and developmental well completions. The number of successful natural gas wells grows at an average annual rate of 3.9% over the 1993–2010 period.

The number of successful oil wells grows at a faster average annual rate (6.7%) over the same period, reaching significantly higher drilling levels by 2010. In 1993, gas drilling exceeded oil drilling by 10%; by 2010, oil drilling exceeds gas drilling by 43%. For both natural gas and oil, the exploratory share of total wells rises between 1993 and 2010. The exploratory share of total natural gas wells rises from 4.5 to 10.5% over the period. The exploratory share of total oil wells rises less sharply—from 4.7 to 7.7%. The diminishing size of new discoveries requires less developmental drilling relative to exploratory drilling. The stronger relative growth in oil well completions occurs because the prospective profitability of new oil projects increases more than the prospective profitability of new natural gas projects over the forecast period.

Gas Reserve Additions Are Expected To Exceed Oil Reserve Additions

Higher levels of natural gas drilling produce significant increases in annual reserve additions, continuing the trend of the past two decades. Projected reserve additions for both gas and oil include expected volumes from new field discoveries, extensions, and revisions. Lower 48 reserve

TABLE 7.11 Natural Gas Prices, Margins, and Revenues (1993 dollars per thousand cubic feet, unless otherwise noted)

Prices, Margins, and Revenue	Reference Case 1992	1993	2000	2005	2010	Annual Growth 1993–2010 (percent)
			Source Price			
Average Lower 48 Wellhead Price[1]	1.80	2.02	2.14	3.02	3.39	3.1
Average Import Price	1.88	2.01	2.01	3.16	3.49	3.3
Average[2]	1.81	2.02	2.12	3.04	3.41	3.1
			Delivered Prices			
Residential	6.05	6.19	6.06	6.92	7.13	0.8
Commercial	5.01	5.18	5.08	5.97	6.22	1.1
Industrial[3]	2.70	2.87	3.05	3.92	4.30	2.4
Electric Generators[4]	2.45	2.63	2.65	3.45	3.82	2.2
Transportation[5]	4.15	4.80	6.56	8.60	8.94	3.7
Average[6]	3.90	4.09	4.06	4.88	5.16	1.4
			Transmission & Distribution Margins by Sector[7]			
Residential	4.25	4.17	3.94	3.89	3.72	–0.7
Commercial	3.21	3.15	2.96	2.94	2.82	–0.7
Industrial[3]	0.89	0.85	0.93	0.88	0.90	0.3
Electric Generators[4]	0.65	0.61	0.53	0.41	0.41	–2.3
Transportation[5]	2.35	2.78	4.44	5.56	5.53	4.1
Average[6]	2.10	2.07	1.94	1.85	1.75	–1.0
			Transmission and Distribution Revenue (billion 1993 dollars)			
Residential	19.86	20.66	19.75	19.00	18.21	–0.7
Commercial	8.92	9.12	8.74	8.64	8.36	–0.5
Industrial[3]	6.61	6.48	8.07	7.96	8.51	1.6
Electric Generators[4]	1.91	1.79	1.78	1.63	1.93	0.5
Transportation[5]	0.00	0.02	0.69	1.59	2.30	31.5
Total	37.30	38.07	39.03	38.81	39.32	0.2

[1] Represents lower 48 onshore and offshore supplies.

[2] Quantity-weighted average of the average lower 48 wellhead price and the average price of imports at the U.S. border.

[3] Includes consumption by cogenerators.

[4] Includes all electric power generators except cogenerators, which produce electricity as a byproduct of other processes.

[5] Compressed natural gas used as a vehicle fuel.

[6] Weighted average price and margin. Weights used are the sectoral consumption values excluding lease, plant, and pipeline fuel.

[7] Within the table, "transmission and distribution" margins equal the difference between the delivered price and the source price (average of the wellhead price and the price of imports at the United States border) of natural gas, and thus, reflect the total cost of bringing natural gas to market. When the term "transmission and distribution" margins is used in today's natural gas market, it generally does not include the cost of independent natural gas marketers or costs associated with aggregation of supplies, provisions of storage, and other services. As used here, the term includes the cost of all services and the cost of pipeline fuel used in compressor stations.

Note: Totals may not equal sum of components due to independent rounding.

Sources: 1992 residential delivered price, transportation delivered price, average lower 48 wellhead price, and average import price: Energy Information Administration (EIA), *Natural Gas Annual 1992, Volume 1*, DOE/EIA-0131(92)/1 (Washington, DC, November 1993). 1993 residential delivered price, average lower 48 wellhead price, and average import price: EIA, *Natural Gas Monthly*, DOE/EIA-0130(94/6) (Washington, DC, June 1994). *Other 1992 values, other 1993 values, and projections:* EIA, AEO95 National Energy Modeling System run AEO95B.D1103942.

TABLE 7.12 Petroleum Product Prices (1993 cents per gallon unless otherwise noted)

Sector and Fuel	Reference Case 1992	1993	2000	2005	2010	Annual Growth 1993–2010 (percent)
World Oil Price (dollars per barrel)	18.70	16.12	19.13	21.50	24.12	2.4
	Delivered Sector Product Prices					
Residential						
Distillate Fuel	95.1	90.8	106.7	114.3	120.7	1.7
Liquefied Petroleum Gas	91.0	89.8	103.3	109.8	117.2	1.6
Commercial						
Distillate Fuel	67.8	64.0	76.6	83.9	90.1	2.0
Residual Fuel	39.8	40.0	45.8	52.1	58.3	2.2
Residual Fuel (dollars per barrel)	16.72	16.80	19.25	21.87	24.47	2.2
Industrial[1]						
Distillate Fuel	69.2	66.3	76.6	84.1	90.5	1.8
Liquefied Petroleum Gas	43.5	41.9	62.0	68.3	75.0	3.5
Residual Fuel	36.0	35.1	43.9	50.0	56.1	2.8
Residual Fuel (dollars per barrel)	15.11	14.74	18.45	20.99	23.58	2.8
Transportation						
Distillate Fuel[2]	114.0	111.6	123.3	127.8	131.6	1.0
Jet Fuel[3]	62.5	57.8	78.2	87.3	92.5	2.8
Motor Gasoline[4]	116.7	111.7	129.5	135.9	137.7	1.2
Residual Fuel	33.6	31.0	41.4	47.4	53.8	3.3
Residual Fuel (dollars per barrel)	14.11	13.00	17.37	19.90	22.59	3.3
Electric Generators[5]						
Distillate Fuel	66.5	65.5	73.0	76.8	81.9	1.3
Residual Fuel	37.7	36.9	43.4	49.5	55.4	2.4
Residual Fuel (dollars per barrel)	15.82	15.49	18.24	20.77	23.26	2.4
Refined Petroleum Product Prices[6]						
Distillate Fuel	99.8	97.4	110.5	116.4	121.2	1.3
Jet Fuel	62.5	57.8	78.2	87.3	92.5	2.8
Liquefied Petroleum Gas	52.6	51.2	70.1	76.8	84.1	3.0
Motor Gasoline	116.7	111.7	129.2	135.7	137.5	1.2
Residual Fuel	35.9	34.4	42.7	48.8	54.9	2.8
Residual Fuel (dollars per barrel)	15.08	14.47	17.94	20.48	23.06	2.8
Average	92.9	89.4	104.2	109.6	113.0	1.4

[1] Includes cogenerators.

[2] Includes federal and state taxes on diesel fuel and excludes county and local taxes.

[3] Kerosene-type jet fuel.

[4] Average price for all grades. Includes federal and state taxes and excludes county and local taxes.

[5] Includes all electric power generators except cogenerators, which produce electricity as a byproduct of other processes.

[6] Weighted averages of end-use fuel prices are derived from the prices in each sector and the corresponding sectoral consumption.

Sources: 1992 prices for gasoline, distillate, and jet fuel are based on prices in the Energy Information Administration (EIA), *Petroleum Marketing Annual 1992,* DOE/EIA-0487(92) (Washington, DC, July 1993). 1993 prices for gasoline, distillate, and jet fuel are based on prices in various 1993 issues of EIA, *Petroleum Marketing Monthly,* DOE/EIA-0380(93/1-12) (Washington, DC, 1993). 1992 and 1993 prices for all other petroleum products are derived from EIA, *State Energy Price and Expenditures Report: 1991,* DOE/EIA-0376(91) (Washington, DC, September 1993). *Projections:* EIA, AEO95 National Energy Modeling System run AEO95B.D1103942.

additions generally increase to about the level of their recent peaks, but lower 48 reserves fall at a 0.7% annual rate as production exceeds reserve additions.

Higher levels of crude oil drilling also lead to significant increases in annual reserve additions, reversing the generally declining trend of the past two decades. However, reserve additions per new well are lower for oil than for gas. Lower 48 oil reserve additions increase on average by 5.2% a year. Still, lower 48 oil reserves fall at an average annual rate of 1.6%, because oil production generally exceeds reserve additions.

Oil Production Declines, Gas Production Rises over the Forecast Horizon

Despite increasing reserve additions, domestic natural gas production generally declined over the 1970–1986 period. This unusual relationship reflected supply and demand imbalances that arose from distorted price signals, generally associated with the past regulatory environment affecting natural gas. Gas production has increased since 1986, largely because of increasing industry deregulation and rising demand. In the forecast, relatively abundant natural gas is expected to be available from lower-cost sources, allowing production to increase steadily to meet rising demand. In contrast, oil production is projected to continue its historic decline through 2005, increasing slightly thereafter in response to rising prices and improvements in drilling technology (Table 7.13).

Oil production includes lease condensate (a mixture of hydrocarbons recovered as a liquid from natural gas through the normal process of condensation in lease or field separation facilities). It does not include natural gas plant liquids (those hydrocarbons in natural gas that are separated from the gas through the processes of absorption, condensation, adsorption, or other methods in gas processing or cycling plants). Natural gas production comprises nonassociated (NA) and associated-dissolved (AD) production. NA natural gas is not in contact with significant quantities of crude oil in a reservoir. AD natural gas consists of the combined volume of natural gas which occurs in crude oil reservoirs either as free gas (associated) or as gas in solution with crude oil (dissolved).

Future domestic oil and natural gas production depends on many uncertain factors—the size and geologic distribution of remaining resources; technological advances in exploration, development, and production; and prices, costs, and other factors affecting industry profitability and expectations. The resource estimates underlying the production projections assume that, given technological innovation, economically recoverable domestic oil and natural gas resources (measured as of 1990) will increase (Table 7.14).

Economically recoverable resources are those volumes considered to be of sufficient size and quality for their production to be commercially profitable by current conventional or nonconventional technologies, under specified economic conditions. Proved reserves are the estimated quantities that analysis of geological and engineering data demonstrate with reasonable certainty to be recoverable in future years from known reservoirs under existing economic and operating conditions. Unproved resources comprise inferred reserves and undiscovered resources. Inferred reserves are that part of expected ultimate recovery from known fields in excess of cumulative production plus current reserves. Undiscovered resources are located outside oil and gas fields in which the presence of resources has been confirmed by exploratory drilling; they include resources from undiscovered pools within confirmed fields, when they occur as unrelated accumulations controlled by distinctly separate structural features or stratigraphic conditions.

The current resource forecast uses data and assumptions that reflect the demonstrated successes of the industry during the recent period of low prices. Two major reasons account for the improved outlook: technology and productivity. An upward revision to the impact of technological advances affecting recoverable resource potential from higher-cost recovery (e.g., tight gas and coalbed methane) and the frontier areas of the lower 48 offshore and Alaska was incorporated into the analysis. This change is based on continued review of the relevant literature and discussions with DOE and other experts. Technological advances expected by 2010 are assumed to raise estimated economically recoverable oil and gas resources by 38.6 and 58.5% compared to the volumes based on 1990 technology.

Lower 48 natural gas production in the reference case is projected to grow at an average annual rate of 0.8% over the 1993–2010 period (Table 7.15), whereas lower 48 crude oil production is projected to *decline* at an average annual rate of 0.8% (Table 7.16). Historically, for both natural gas and oil, major price and drilling peaks have had only limited impact on contemporaneous levels of new reserve additions and production. This is in large part because higher levels of drilling

TABLE 7.13 Oil and Gas Supply

Production and Supply	Reference Case 1992	1993	2000	2005	2010	Annual Growth 1993–2010 (percent)
			Crude Oil			
Lower 48 Average Wellhead Price[1] (1993 dollars per barrel)	17.91	15.36	18.47	20.58	22.92	2.4
Production (millions barrels per day)[2]						
U.S. Total	7.17	6.85	5.35	5.16	5.39	–1.4
Lower 48 Onshore	4.39	4.18	3.53	3.70	3.96	–0.3
Conventional	3.71	3.55	2.86	2.91	3.01	–1.0
Enhanced Oil Recovery	0.68	0.62	0.67	0.79	0.95	2.5
Lower 48 Offshore	1.07	1.09	0.66	0.63	0.66	–2.9
Alaska	1.71	1.58	1.16	0.83	0.77	–4.2
U.S. End of Year Reserves (billion barrels)	24.97	23.30	17.59	16.12	16.97	–1.8
			Natural Gas			
Lower 48 Average Wellhead Price[1] (1993 dollars per thousand cubic feet)	1.80	2.02	2.14	3.02	3.39	3.1
Production (trillion cubic feet)[3]						
U.S. Total	17.84	18.35	19.08	19.94	20.88	0.8
Lower 48 Onshore	12.74	12.96	13.81	14.72	15.88	1.2
Associated-Dissolved[4]	2.18	2.21	1.55	1.67	1.80	–1.2
Nonassociated	10.55	10.75	12.26	13.04	14.08	1.6
Conventional	8.51	8.68	9.70	10.28	11.05	1.4
Unconventional	2.04	2.07	2.57	2.77	3.02	2.3
Tight Sands	1.35	1.36	1.29	1.43	1.84	1.8
Coal Bed Methane	0.54	0.55	1.13	1.18	1.01	3.6
Devonian Shale	0.15	0.15	0.15	0.16	0.17	0.7
Lower 48 Offshore	4.69	4.98	4.76	4.73	4.53	–0.6
Associated-Dissolved[4]	0.58	0.61	0.37	0.36	0.39	–2.6
Nonassociated	4.11	4.37	4.40	4.38	4.14	–0.3
Alaska	0.41	0.41	0.50	0.49	0.47	0.9
U.S. End of Year Reserves (trillion cubic feet)	165.02	167.78	145.67	147.48	152.48	–0.6
Supplemental Gas Supplies (trillion cubic feet)[5]	0.12	0.13	0.12	0.12	0.08	–3.0
Lower 48 Wells Completed (thousands)	23.08	23.06	28.07	44.53	57.75	5.5

[1] Represents lower 48 onshore and offshore supplies.
[2] Includes lease condensate.
[3] Market production (wet) minus extraction losses.
[4] Gas which occurs in crude oil reserves either as free gas (associated) or as gas in solution with crude oil (dissolved).
[5] Synthetic natural gas, propane air, coke oven gas, refinery gas, biomass gas, air injected for Btu stabilization, and manufactured gas commingled and distributed with natural gas.

Note: Totals may not equal sum of components due to independent rounding.

Sources: 1992 lower 48 onshore, lower 48 offshore, Alaska crude oil production: Energy Information Administration (EIA), *Petroleum Supply Annual 1992*, DOE/EIA-0340(92)/1 (Washington, DC, May 1993). 1992 U.S. crude oil reserves, and U.S. natural gas reserves: EIA, *U.S. Crude Oil, Natural Gas, and Natural Gas Liquids Reserves*, DOE/EIA-0216(92) (Washington, DC, October 1993). 1992 crude oil lower 48 average wellhead price: EIA, *Annual Energy Review 1992*, DOE/EIA-0384(92) (Washington, DC, June 1993). 1992 natural gas lower 48 average wellhead price, Alaska, total natural gas production, and supplemental gas supplies: EIA, *Natural Gas Annual 1992*, DOE/EIA-0131(92) (Washington, DC, November 1993). 1992 and 1993 total wells completed: EIA, Office of Integrated Analysis and Forecasting. 1993: Lower 48 onshore, lower 48 offshore, Alaska crude oil production: EIA, *Petroleum Supply Annual 1993*, DOE/EIA-0340(93) (Washington, DC, June 1994). 1993 natural gas lower 48 average wellhead price, total natural gas production: *Natural Gas Monthly*, DOE/EIA-0130(94/06) (Washington, DC, June 1994). Other 1992 and 1993 values: EIA, Office of Integrated Analysis and Forecasting. Figures for 1992 and 1993 may differ from published data due to internal conversion factors within the AEO95 National Energy Modeling System. *Projections:* EIA, AEO95 National Energy Modeling System run AEO95B.D1103942.

TABLE 7.14 Economically Recoverable Oil and Gas Resources in 1990, Measured Under Different Technology Assumptions

	Crude Oil (billion barrels)		Natural Gas (trillion cubic feet)	
Resources	1990 Technology	2010 Technology	1990 Technology	2010 Technology
Proved	26.3	26.3	169.4	169.4
Unproved	85.7	128.9	851.9	1,449.2
U.S. Total	112.0	155.2	1,021.2	1,618.6

tend to lead to the drilling of less promising prospects and because production levels tend to be proportional to the overall stock of reserves rather than annual reserve additions. The projections reflect a continuation of those basic relationships.

The continuing increase in domestic natural gas production for most of the forecast is partly attributable to increases in onshore conventional production and increasing use of unconventional gas recovery (UGR) technologies. UGR consists principally of production from reservoirs with low permeability (tight sands) but also includes methane from coal seams and gas from Devonian shales. The trends in natural gas production reflect the advantage of relatively lower costs associated with conventional production and the reduction in unit costs of UGR production due to technological advances. The gradual decline in offshore production reflects the impact of relatively higher costs and resource depletion.

The increasing levels of domestic oil production from 2005 to 2010 are attributable primarily to increasing production by EOR methods. Despite technological advances that improve recovery, conventional oil production in the lower 48 onshore regions is expected to decline from 1993 levels as a result of the maturity of the oil resource base.

Natural Gas Market Competitive Issues

Over the past decade (1985–1995), the focus of federal policy initiatives has been to promote competition in the procurement of natural gas supplies by deregulating wellhead prices and restructuring the natural gas interstate pipeline industry. The concept of relying on market forces for the pricing of natural gas services by deregulating services in workably competitive markets is expected to expand into other industry segments. Although refinement of recent federal initiatives continues, attention is shifting to promoting competition in the gathering and distribution segments of the industry. The goal of the policy changes is to modify the market to allow pricing signals to flow freely between the wellhead and the burnertip.

Industry restructuring is leading to an environment in which the prices of services are based on their commercial value. This is a major change for the industry and contrasts with the old environment, where service prices were often based on the cost of placing and maintaining physical assets in service. As the market continues to change direction, the industry may find that it has surplus plant, leading to a significant amount of stranded investment. The level of physical assets that were used to provide bundled sales service, including assets related to storage, production, product extraction, transmission, gathering, and upstream capacity needed for operational integrity, may no longer be required to meet the demand for various services in an unbundled environment. The costs associated with these assets are all categorized as stranded costs. This could bring about a number of writedowns, internal restructuring, sales of assets, or a new round of "transition costs."

In the implementation of Order 636, the term "transition costs" referred to the costs associated with the transition from a bundled to an unbundled environment and included costs incurred in reforming contracts with gas producers (gas supply realignment costs), unrecovered gas costs

TABLE 7.15 Natural Gas Supply and Disposition (trillion cubic feet per year, unless otherwise noted)

Supply and Disposition	Reference Case 1992	1993	2000	2005	2010	Annual Growth 1993–2010 (percent)
Production						
Dry Gas Production[1]	17.84	18.35	19.08	19.94	20.88	0.8
Supplemental Natural Gas[2]	0.12	0.13	0.12	0.12	0.08	–3.0
Net Imports	1.93	2.13	2.90	3.00	3.60	3.1
Canada	2.03	2.14	2.65	2.69	2.71	1.4
Mexico	–0.09	–0.04	–0.01	0.00	0.17	N/A
Liquefied Natural Gas	–0.01	0.03	0.25	0.30	0.71	21.5
Total Supply	19.88	20.61	22.09	23.06	24.55	1.0
Consumption by Sector						
Residential	4.68	4.96	5.01	4.89	4.89	–0.1
Commercial	2.78	2.89	2.95	2.94	2.97	0.2
Industrial[3]	7.41	7.61	8.68	9.01	9.50	1.3
Electric Generators[4]	2.95	2.94	3.34	3.92	4.72	2.8
Lease and Plant Fuel[5]	1.17	1.20	1.33	1.38	1.44	1.1
Pipeline Fuel	0.59	0.61	0.65	0.65	0.65	0.4
Transportation[6]	0.00	0.01	0.15	0.29	0.42	26.3
Total	19.57	20.21	22.12	23.08	24.59	1.2
Discrepancy[7]	0.31	0.40	–0.03	–0.02	–0.03	N/A

[1] Market production (wet) minus extraction losses.

[2] Synthetic natural gas, propane air, coke oven gas, refinery gas, biomass gas, air injected for Btu stabilization, and manufactured gas commingled and distributed with natural gas.

[3] Includes consumption by cogenerators.

[4] Includes all electric power generators except cogenerators, which produce electricity as a byproduct of other processes.

[5] Represents natural gas used in the field gathering and processing plant machinery.

[6] Compressed natural gas used as vehicle fuel.

[7] Balancing item. 1992 and 1993 values reflect net storage injections plus natural gas lost as a result of converting flow data measured at varying temperatures and pressures to a standard temperature and pressure, and the merger of different data reporting systems which vary in scope, format, definition, and respondent type.

N/A = Not applicable.

Note: Totals may not equal sum of components due to independent rounding. Figures for 1992 and 1993 may differ from published data due to internal conversion factors in the AEO95 National Energy Modeling System.

Sources: 1992 supply values and consumption as lease, plant, and pipeline fuel: Energy Information Administration (EIA), *Natural Gas Annual 1992, Volume 1,* DOE/EIA-0131(92)/1 (Washington, DC, November 1993) with adjustments to end-use sector consumption levels based on Form EIA-867, "Annual Nonutility Power Producer Report." Other 1992 consumption: EIA, *State Energy Data Report 1992,* DOE/EIA-0214(92) (Washington, DC, May 1994) with adjustments to end-use sector consumption levels based on Form EIA-867, "Annual Nonutility Power Producers Report." 1993 supply values and consumption as lease, plant, and pipeline fuel: EIA, *Natural Gas Monthly,* DOE/EIA-0130(94/6) (Washington, DC, June 1994) with adjustments to end-use sector consumption levels for consumption of natural gas by electric wholesale generators based on EIA, AEO95 National Energy Modeling System run AEO95B.D1103942. 1993 transportation sector consumption: EIA, AEO95 National Energy Modeling System run AEO95B.D1103942. Other 1993 consumption: EIA, *Short-Term Energy Outlook,* DOE/EIA-0202(94/3Q) (Washington, DC, August 1994) with adjustments to end-use sector consumption levels for consumption of natural gas by electric wholesale generators based on EIA, AEO95 National Energy Modeling System run AEO95B.D1193942. *Projections:* EIA, AEO95 National Energy Modeling System run AEO95B.D1103942.

(account 191) remaining when the purchased gas adjustment mechanism was terminated, stranded costs for assets no longer needed in an unbundled environment, and new facilities costs for new assets required because of unbundling, such as electronic bulletin boards. In the context of this discussion, "transition costs" refer to the costs associated with surplus pipeline capacity that becomes available as long-term transportation agreements expire.

The effects of the increasingly competitive market are evident in the observed changes in the components of the end-use prices over the past decade. All end-use sectors have seen some reduction in the cost of supplies and the cost of getting those supplies to the burnertip. Changes

TABLE 7.16 Petroleum Supply and Disposition Balance (million barrels per day, unless otherwise noted)

Supply and Disposition	Reference Case 1992	1993	2000	2005	2010	Annual Growth 1993–2010 (percent)
			Crude Oil			
Domestic Crude Production[1]	7.17	6.85	5.35	5.16	5.39	−1.4
Alaska	1.71	1.58	1.16	0.83	0.77	−4.2
Lower 48 States	5.46	5.26	4.19	4.33	4.62	−0.8
Net Imports	5.99	6.69	8.70	8.97	8.88	1.7
Other Crude Supply[2]	0.25	0.08	0.00	0.00	0.00	N/A
Total Crude Supply	13.41	13.61	14.05	14.13	14.27	0.3
Natural Gas Plant Liquids	1.70	1.74	1.85	1.93	2.03	0.9
Other Inputs[3]	0.21	0.19	0.23	0.23	0.23	1.1
Refinery Processing Gain[4]	0.77	0.77	0.80	0.81	0.83	0.5
Net Product Imports[5]	0.94	0.93	1.76	2.82	3.34	7.8
Total Primary Supply[6]	17.03	17.24	18.69	19.92	20.71	1.1
			Refined Petroleum Products Supplied			
Motor Gasoline[7]	7.27	7.48	8.10	8.36	8.41	0.7
Jet Fuel[8]	1.45	1.47	1.81	1.99	2.15	2.3
Distillate Fuel[9]	2.98	3.04	3.34	3.56	3.78	1.3
Residual Fuel	1.09	1.08	1.27	1.51	1.61	2.4
Other[10]	4.24	4.17	4.37	4.66	4.92	1.0
Total	17.03	17.24	18.87	20.09	20.88	1.1
			Refined Petroleum Products Supplied			
Residential and Commercial	1.12	1.14	1.00	0.95	0.92	−1.3
Industrial[11]	4.55	4.60	4.81	5.14	5.39	0.9
Transportation	10.94	11.08	12.66	13.45	13.98	1.4
Electric Generators[12]	0.42	0.41	0.41	0.55	0.59	2.1
Total	17.03	17.24	18.87	20.10	20.88	1.1
Discrepancy[13]	0.00	0.00	−0.18	−0.17	−0.17	N/A
World Oil Price (1993 dollars per barrel)[14]	18.70	16.12	19.13	21.50	24.12	N/A
Domestic Refinery Distillation Capacity	15.5	15.3	15.7	15.8	15.9	0.2
Capacity Utilization Rate (percent)	88.0	92.0	89.7	90.0	90.0	−0.1

[1] Includes lease condensate.

[2] Strategic petroleum supply stock additions plus unaccounted for crude oil plus crude stock withdrawals minus crude products supplied.

[3] Includes alcohols, ethers, petroleum product stock withdrawals, domestic sources of blending components, and other hydrocarbons.

[4] Represents volumetric gain in refinery distillation and cracking processes.

[5] Includes net imports of finished petroleum products, unfinished oils, other hydrocarbons, alcohols, ethers, and blending components.

[6] Total crude supply plus natural gas plant liquids plus other inputs plus refinery processing gain plus net petroleum imports.

[7] Includes ethanol and ethers blended into gasoline.

[8] Includes naphtha and kerosene type.

[9] Includes distillate and kerosene.

[10] Includes aviation gasoline, liquefied petroleum gas, petrochemical feedstocks, lubricants, waxes, asphalt, road oil, still gas, special naphthas, petroleum coke, crude oil product supplied, and miscellaneous petroleum products.

[11] Includes consumption by cogenerators.

[12] Includes all electric power generators except cogenerators, which produce electricity as a byproduct of other processes.

[13] Balancing item. Includes unaccounted for supply, losses and gains.

[14] Average refiner acquisition cost for imported crude oil.

N/A = Not applicable.

Note: Totals may not equal sum of components due to independent rounding.

Sources: 1992: Energy Information Administration (EIA), *Petroleum Supply Annual 1992,* DOE/EIA-0340(92) (Washington, DC, May 1993). 1993: EIA, *Petroleum Supply Annual 1993,* DOE/EIA-0340(93) (Washington, DC, June 1994). *Projections:* EIA, AEO95 National Energy Modeling System run AEO95B.D1103942.

TABLE 7.17 Components of Natural Gas End-Use Prices (1993 dollars per thousand cubic feet)

Price Component	1984	1993
Wellhead Price	2.73	2.02
Markup to Citygate	1.67	1.19
LDC Distribution Markup		
Residential	2.96	2.98
Commercial	2.18	1.97
End-Use Price		
Residential	8.35	6.19
Commercial	7.57	5.18

in the components of the residential and commercial end-use prices are shown in Table 7.17. Benefits of industry restructuring could spread further downstream, resulting in lower end-use prices, if similar changes occur in the industry's distribution segment.

The lowering of wellhead prices and wellhead-to-citygate markups since 1984, coupled with little change in distribution markups, has caused a significant increase in the distribution share of the core market end-use price. Many states and local distribution companies (LDCs) are currently investigating restructuring of the distribution segment of the industry to move toward competitive pricing—either through deregulation or through performance-based ratemaking—and a number of possible scenarios have been suggested. Such changes in the distribution segment of the industry could improve the competitive position of natural gas at the burnertip and increase its use in the residential and commercial sectors, which traditionally have been supplied by the LDCs. Examples are New Jersey's proposal to unbundle gas utility services to all customers and Boston Gas Company's proposal to consolidate all Massachusetts LDCs into a single utility.

In November 1993, the New Jersey Board of Regulatory Commissioners approved a set of guidelines involving the unbundling of gas utility services to all commercial, industrial, and electricity generation customers.[5] This would effectively remove the regulated LDCs from the gas supply business for all nonresidential customers. The LDCs were given an April 1, 1994, deadline to file tariffs for new competitive natural gas services. All have filed, and the approval process is ongoing. The Board intends to conduct a review of "the successes of this deregulation effort in the future and consider how the benefits can be extended to residential customers." No date for the review has been set, and it is expected that it will not occur for at least 2, and more likely 3, years.

Boston Gas Company, the largest gas distribution company in New England, has submitted an analysis paper (known as a "white paper") to the Massachusetts Department of Public Utilities (DPU) detailing substantial cost savings and service improvements that could be achieved through the consolidation of the 10 gas distribution companies in Massachusetts into 1 large utility. This has been evaluated by the DPU, and a mergers and acquisitions order has been released with comments invited. The Massachusetts DPU is also looking into incentive ratemaking for gas and electric utilities, and in September 1994 issued a Notice of Inquiry (NOI) inviting comment.

The New Jersey and Massachusetts proposals illustrate how industry restructuring could move the roles of LDCs in two different directions. Since FERC Order 636 is bringing system supply costs in line with off-system purchases, LDCs may play a larger role as supply aggregators in the future. Individual customers could go off-system (purchase supplies from an entity other than the pipeline) prior to Order 636, but LDCs were bound by contracts with the pipelines limiting their ability to go off-system for supply. Order 636 provided the opportunity to renegotiate the contracts and unbundle the supply from the transportation service, putting all pipeline customers on a level playing field with comparable service and access to receipt and delivery points. Thus, the costs of supply offered by the LDCs are coming in line with off-system purchases. Consolidation in the industry could occur either through mergers, as proposed by Boston Gas, or by LDCs pooling together in a fashion similar to the electric utility power pools. (A pool is a grouping of companies for the common benefit of all member companies.) Some municipal gas distributors have grouped together to form cooperatives for the purpose of procuring natural gas supplies. In contrast, the New Jersey proposal extends the unbundling concept to the distribution segment of the industry, increasing the opportunity for natural gas marketing firms to expand their base of customers and intensifying competition among suppliers of natural gas.

Changes are also anticipated in the gathering end of the industry. A May 1994 FERC ruling that gathering falls under FERC jurisdiction only if performed by a regulated interstate pipeline has encouraged pipelines to transfer their gathering facilities to existing affiliates, create new affiliates to handle their gathering facilities, or sell their facilities to nonaffiliates. The rate of those activities will most likely accelerate, bringing competition to yet another segment of the natural gas market. Although FERC will no longer have direct jurisdiction over the transfers it approves, it maintains a built-in safeguard in the right to step in if there appears to be an abuse of the pipeline/gathering relationship.

Both Domestic and Foreign Suppliers Are Expected to Gain from an Expanding Gas Market

Between 1993 and 2000, foreign and domestic producers capture equal shares in the growth of 1.5 trillion cubic feet in the U.S. natural gas market. Domestic producers fare better after 2000, capturing 73% of the market growth of 2.5 trillion cubic feet between 2000 and 2010.

Total gas consumption in the industrial, electric generator, and vehicle market sectors increases by more than 4 trillion cubic feet by 2010 in the reference case (Table 7.18). The market expansion is driven primarily by the demand for electricity (including industrial cogeneration) and the requirements for alternative-fuel vehicles. Residential and commercial consumption remains flat, as conservation and efficiency improvements offset the growth in the number of customers.

TABLE 7.18 Reference Case Natural Gas Consumption (trillion cubic feet per year)

Sector	1993	2000	2010
Residential	4.96	5.01	4.89
Commercial	2.89	2.95	2.97
Industrial	7.61	8.68	9.50
Electric Generators	2.94	3.34	4.72
Transportation	0.01	0.15	0.42
Total	18.40	20.14	22.49

Gas Transmission and Distribution Revenues Stabilize, Margins Decline

Transmission and distribution revenues stabilize over the forecast period, in contrast to their nearly continuous growth through the early 1980s and the 17.8% decline between 1983 and 1993. The leveling of revenues reflects a balance between cost decreases resulting from improved alignment of services with customer needs and technology improvements, and cost increases resulting from the investment needed to support market expansion.

Unlike revenues, the average transmission and distribution margin (revenue divided by consumption)* continues the decline that began after 1983, when margins were at their peak. In the forecast, the average margin declines from $2.07 per thousand cubic feet in 1993 to $1.75 in 2010.

*Transmission and distribution margins equal the difference between the delivered price and the source price (average of the wellhead price and the price of imports at the U.S. border) of natural gas, and thus reflect the total cost of bringing natural gas to market. When the term "transmission and distribution margins" is used in today's natural gas market, it generally does not include the cost of independent natural gas marketers or costs associated with aggregation of supplies, provisions of storage, and other services. As used here, the term includes the cost of all services and the cost of pipeline fuel used in compressor stations.

Greater throughput, a large base of depreciated plant, industry automation, and the growing number of end users that typically use nonfirm services all contribute—along with increased competition—to the decline in the average transmission and distribution margin.

Lower Margins Dampen Wellhead Price Increases at the Burnertip

The industry restructuring begun in 1984 has allowed competition to place downward pressure on transmission and distribution margins in most end-use sectors. The exception is the vehicle natural gas market, where margins are projected to rise as they reflect motor fuels taxes and changes in service as the market moves from demonstration programs to commercial multiuser refueling stations. Nevertheless, natural gas still retains a significant price edge over motor gasoline.

In contrast with petroleum products, the relative fuel prices favor coal over natural gas in the electric generator sector. Although they have higher fuel costs, gas plants currently operate more efficiently than coal plants, generally cost less to build and operate, and have additional advantages in siting, permitting, construction time, and load-following flexibility. Toward the end of the forecast period, coal's price advantage begins to outweigh other factors, and coal is generally projected to be chosen over natural gas for new electric generator builds.

Natural Gas Pipelines

The nation's gas pipeline network is aging. Most of the system was constructed before the 1972 peak in natural gas consumption, although newer segments have been added to meet shifting regional supply and demand patterns. With proper maintenance and new technology, it is possible to extend the useful life of the existing transmission network. Costs for routine maintenance are accounted for in the pipeline operation and maintenance expenses included in the reference case forecast; however, significant additional investment may be needed to extend the life of aging plants or to refurbish and replace pipe as it approaches the end of its useful life. On the basis of a survey of major pipeline companies, conducted as part of the study published in *The Potential for Natural Gas in the United States* (December 1992), the National Petroleum Council (NPC) has estimated that the industry could be faced with an average annual capital investment of $1.7 billion (1991 dollars) in replacement or refurbishment expenses through 2010.

Industry research and development expenditures, operation and maintenance activities, and passage of the Pipeline Safety Act of 1992 (PSA) provide ample evidence that pipeline safety has long been of concern to legislators and the pipeline industry. A key provision of the PSA requires that all new and replacement pipelines accommodate internal inspection devices known as "smart pigs"—electronic devices that are sent through the pipeline to inspect for structural weaknesses. In 1992, after a New Jersey pipeline explosion, the Department of Transportation's Research and Special Programs Administration (RSPA) expanded the definition of "replacement," as referenced in the PSA, to include any line section of pipe requiring replacement of any portion of the pipe or other component of the line. The new definition could require replacement of many miles of pipe whenever a small section is repaired. As a result of industry contention that the ruling would have an adverse financial impact on the industry, the RSPA has indefinitely suspended enforcement of the "line section" replacement ruling pending further investigation. The American Gas Association (AGA) and others have argued that the additional costs to the industry could be as high as $100 million annually.

New interregional capacity for interstate pipeline facilities will be required to support the expansion of natural gas markets and regional shifts in supply. Much of the expected capacity expansion occurs in the early years of the forecast. Some of the projected interregional capacity may be in the form of additional storage facilities. Between 40 and 50% of the projected pipeline capacity additions by 2010 occur between 1993 and 1996. These additions largely reflect industry plans to expand pipeline capacity as documented in current regulatory filings. It is assumed that over the forecast period there will be no change in load profiles in the end-use sectors.

Pipeline capacity and utilization continue to increase in many regions to support emerging supply sources, such as Canadian imports and the East South Central and Mountain regions. Continuing production declines in traditional producing regions (for example, the West South Central region) result in no new capacity beyond planned additions, as well as decreases in the capacity utilization of pipelines exiting those regions.

Domestic Oil Markets

Refined Products Make Up a Growing Share of Petroleum Imports

Imports of both crude oil and refined petroleum products are projected to increase in the forecast. Refined products represent a growing share of petroleum imports because expansions in domestic refining will not keep pace with growing domestic consumption. Large, new refineries are not expected to be built in the United States because of the time and costs associated with obtaining permits and meeting environmental regulations—the Clean Water Act, the Comprehensive Environmental Response, the Compensation and Liability Act, the Resource Conservation and Recovery Act, and the Oil Pollution Act, as well as anticipated regulations related to limiting pollution at refinery sites. Investment funds will also be limited, as refiners make large investments to comply with CAAA90.

In the reference case, crude oil inputs to U.S. refineries in 2010 are only 0.6 million barrels a day higher than 1993 levels, because refinery utilization rates remain stable at around 90%. A growing share of the crude oil processed in U.S. refineries is imported, as domestic crude oil production declines over time.

Petroleum Continues to be a Major Source of U.S. Energy Consumption

Despite programs to encourage the use of alternative fuels, about 40% of the growth in domestic energy consumption over the forecast is supplied by petroleum products. Light products (including diesel fuel, heating oil, jet fuel, gasoline, kerosene, and liquefied petroleum gases), distilled from crude oil at lower temperatures, represent about 75% of the growth in petroleum consumption. Transportation fuels continue to account for around two-thirds of U.S. petroleum use.

Environmental Regulations Change the U.S. Gasoline Market

The makeup of U.S. gasoline consumption will change significantly as CAAA90 requirements continue to be phased in. Starting in 1995, cleaner burning "reformulated gasoline" will be sold in the nine metropolitan areas with the most severe ozone pollution and in other areas, predominantly in the Northeast, that choose to impose the requirement.* In California, beginning in 1996, all gasoline sold must be "reformulated" according to standards set by the California Air Resources Board. After 1996, reformulated gasoline will make up some 40% of the gasoline consumed in the United States. About one-eighth of the reformulated gasoline must also meet preexisting higher oxygen standards in areas with high carbon monoxide levels.

Higher oxygen content is one characteristic that sets reformulated gasoline apart from conventional gasoline. Oxygen is added to gasoline by blending with "oxygenates," including methyl tertiary butyl ether (MTBE), ethyl tertiary butyl ether (ETBE), and ethanol, which offset a small portion of the petroleum content of gasoline. Moreover, a recent U.S. Environmental Protection Agency (EPA) ruling, the Renewable Oxygenate Standard (currently pending legal review), requires that 15% of reformulated gasoline use renewable oxygenates for blending. The requirement will be stepped up to 30% in 1996. In the reference case, approximately 0.57 million barrels of petroleum a day (about 6% of gasoline) is offset by blending with renewable and nonrenewable oxygenates in 2010.

*The affected areas are Baltimore, Chicago, Hartford, Houston, Los Angles, Milwaukee, New York City, Philadelphia, and San Diego. The 1995 opt-in areas are in the following states: Connecticut, Delaware, Kentucky, Maine, Massachusetts, Maryland, New Hampshire, New Jersey, New York, Pennsylvania, Rhode Island, Texas, Virginia, and the District of Columbia. Additional 1996 opt-ins include Atlanta and areas of Wisconsin.

Prices for Lighter Products May Rise More Sharply Than Heavy Product Prices

Prices of lighter petroleum products, including heating oil, diesel, gasoline, jet fuel, and liquefied petroleum gases (LPGs), increase in the forecast relative to the prices of heavier products, including residual fuel. Growth in the consumption of lighter products, which are distilled from crude oil at lower boiling ranges, will require investment in conversion processes that turn the heavier streams into lighter ones.

Compared with prices in the early 1990s, the prices of lighter products will bear an additional 3 to 5 cents per gallon as a result of refinery compliance with emissions, health, and safety regulations. The prices of heavier petroleum products will not be affected because they compete closely with other products, such as natural gas, and are therefore more price-sensitive.

Requirements for reformulated gasoline will increase the cost of producing and distributing gasoline. In the forecast, reformulated gasoline has a national price premium of 4 to 7 cents a gallon over conventional gasoline. In the Northeast, where reformulated gasoline use will be heavily concentrated, the premium ranges between 4 and 6 cents a gallon. Higher price premiums, between 10 and 14 cents a gallon, will be seen on the West Coast as a result of the mandate for reformulated gasoline in California beginning in 1996. Relative to federal requirements, the California law places tighter limits on the sulfur and olefin contents of reformulated gasoline, which make it more costly to produce.

International Oil Markets

Persian Gulf Oil, Lighter Products Have Larger Shares of U.S. Imports

The 1.1% annual growth rate of oil consumption in the reference case translates into increased U.S. imports of both crude oil and refined products. Crude oil imports are projected to grow by 1.6% a year, and refined products by a vigorous 4.8%.

OPEC is expected to account for approximately one-half of total U.S. petroleum imports for the remainder of this decade. After 2000, however, the OPEC share increases to just under 60%. The Persian Gulf share of U.S. imports from OPEC increases more dramatically, from today's 44% to more than 65% in 2010. Crude oil imports from the North Sea increase throughout the 1990s but then begin to decline as North Sea production ebbs. Significant quantities of crude oil continue to be imported from both Canada and Mexico.

Most additions to worldwide refining capacity over the next 15 years will be outside the United States, with significant capacity increases expected in the Caribbean Basin, Middle East, and Far East. U.S. imports of refined products from each of these regions are expected to increase, with the most dramatic gain for the Caribbean Basin. Traditionally, significant volumes of residual fuel oil have been imported from Caribbean Basin exporters; however, lighter products are expected to make up a larger share of imports from the region as demand for them increases.

Declining Quality of Crude Oil Supply Will Challenge Refiners Worldwide

The declining quality of world crude oil production over the forecast period presents challenges to the refining industry. The production of light, low-sulfur oils, so valued by industrialized nations for their robust yield of light products, is expected to peak around the turn of the century, then drops off with the decline of North Sea fields. OPEC members with significant light, low-sulfur crude oil production (Algeria, Libya, Nigeria, Gabon, and Indonesia) are not expected to maintain current output levels through 2010.

A large volume of the reserves in the Persian Gulf region consists of light, high-sulfur crude oils, which make up the largest share of world production throughout the forecast. A substantial drop in their share is expected, however, as giant Middle East oil fields mature and decline. The production losses are likely to be replaced by Middle East heavier crude oils, although some analysts are uncertain about the willingness of Persian Gulf producers to expand production capacity when world oil prices are low.

Crude oil reserves, especially in the Middle East, can satisfy world petroleum demand well into the next century. The real challenge faces the refining industry. With demand for lighter products increasing and more stringent product specifications resulting from environmental regulations, refining becomes more complex and more expensive. Additions to distillation capacity will be needed worldwide merely to keep pace with demand, and significant downstream capacity will have to be added so that heavier crude oils can be upgraded.

Oil and Gas: Challenges for the Future

Natural Gas Markets

Far-reaching changes have occurred in the U.S. natural gas industry over the past decade as a result of deregulation of many aspects of the industry, and significant future changes are anticipated. The reference case assumes modest changes from current market trends, but the gas market of the future may deviate substantially from that pattern.

Currently, interstate pipeline companies are reassessing their markets and positioning themselves for the new environment. The strategic position a pipeline company could take depends on the market position of its parent company and its physical relationships to suppliers and consumers in existing and potential markets. Companies only offering transportation service need to be profitable transporting gas, while integrated firms may use gas transportation as an entree to their other businesses. Slightly higher margins may be seen in the case of companies with captive LDCs as customers, provided that natural gas remains competitive at the burnertip. Companies only offering transportation service may face competition from alternative transportation routes. They will strive to provide the lowest rates possible by cutting operating costs. A key issue that must be taken into account in efforts to cut costs is that of pipeline safety and refurbishment. Ongoing market developments may change the capacity expansion and utilization picture reflected in the forecasts. Requirements for new capacity could be reduced and utilization increased if traditional users of transportation services adopt a portfolio strategy that includes more storage and some noncore services. DSM programs, as well as pricing changes (including straight fixed variable rate design and incremental pricing of new capacity), may levelize transmission loads and increase pipeline capacity utilization.

Other challenges lie in future LDC developments, the outcome of electric utility deregulation, and the market for alternative-fuel vehicles. Prices for compressed natural gas could be considerably lower than those presented in the forecasts if advances in technology lower dispensing costs, if incentives such as a reduction in the motor fuels tax are provided, or if the industry provides favorable transportation rates.

Petroleum Markets

Producers and distributors of petroleum products will also be facing a number of long-term challenges over the forecast period. First, under the Air Toxics title of CAAA90, many petroleum refineries will be required to install "maximum achievable control technology" (MACT) to prevent releases of hazardous air pollutants at the facilities. After the EPA promulgates the MACT standards in 1995, refineries will have 3 to 5 years to comply.

The gradual decline in the quality of crude oil inputs, coupled with relatively flat demand for residual fuel oil, will present another challenge. U.S. refineries will need to alter refinery configurations and invest in conversion units to handle crude oils with higher sulfur contents and lower gravities.

Another challenge will result from continued pressure to produce environmentally friendly products. CAAA90 requirements for cleaner burning fuels have been phased in over the past 5 years. Reformulated gasoline, which will begin to be used in 1995, will evolve over the forecast period. Certification of reformulated gasoline is currently based on a uniform set of EPA standards described as the "simple model." The simple model includes specifications for the content of aromatics, benzene, oxygen, and sulfur and on Reid vapor pressure (Rvp). Beginning in 1998, reformulated gasoline must be certified by a results-oriented "complex formula," based on achieving EPA emissions parameters. Initial requirements for a 15% reduction of volatile organic compounds (VOCs) and air toxics relative to baseline 1990 gasoline will be stepped up after 2000.

Combined with federal clean fuel requirements, state gasoline requirements will multiply the logistical problems already complicating the distribution of motor fuels. California will require its own version of reformulated gasoline statewide beginning in 1996. As an alternative to federal reformulated gasoline, a number of states have proposed further restrictions on Reid vapor pressure as a means of reducing pollution in ozone nonattainment areas. Refiners and distributors will have to produce and handle multiple octane grades of conventional, oxygenated, reformulated, and reformulated high-oxygen gasoline; during the summer, all gasoline blends must meet regional requirements for Reid vapor pressure. Reformulated blendstock for oxygenate blending (RBOB) will be delivered to terminals for possible blending with ethanol and other oxygenates before delivery at the pump. The greatly increased number of gasoline products and blendstocks can be expected to restrict the flexibility of the marketing system and increase the potential for distribution problems. Testing and recordkeeping will play increasingly vital roles.

7.7 Coal

Currently accounting for a greater share of U.S. primary energy production than any other fuel, coal maintains its lead position in energy production throughout the forecast horizon. The continued growth of coal consumption for electricity generation and the response of the coal industry and electric utilities to CAAA90 are the two major determinants of the coal forecasts. In the industrial sector, steam coal demand will increase in certain energy-intensive process industries, reflecting higher levels of output and greater use of cogeneration in those industries. However, domestic coking coal consumption will decline, primarily as a result of changes in domestic steelmaking technology. Coal exports rise in the forecast as a result of increases in steam coal imports in Europe.

The average minemouth price of U.S. coal is expected to be only slightly higher by 2010 than in 1993, reflecting the interplay of changes in coal mine capacity utilization rates,* continued gains in labor productivity, and the cost impacts of opening new mines.

Coal Production

Annual coal production rises to 1,137 million tons** by 2010, an increase of 192 million tons from 1993. The slower growth in coal production over the forecast relative to the preceding 20 years (1.1% a year compared with 2.3% a year) is primarily the result of a smaller projected increase in electricity coal demand.

Between 1993 and 2010, western coal production increases by 18 percent, from 429 million tons in 1993 to 508 million tons in 2010 (Table 7.19). Production from mines east of the Mississippi River rises by 22 percent, from 516 million tons in 1993 to 629 million tons in 2010.

In the forecast, the increase in western production is based primarily on the ability of western producers to satisfy increased demand for low-sulfur coal in the electricity sector. Growth in eastern production results from increases in electricity coal demand, a recovery in U.S. coal exports, and a return to more normal coal supply conditions after 1993.

Eastern production declined by 72 million tons in 1993, mainly due to a strike against the Bituminous Coal Operators Association. As a result of strike-related disruptions in coal supply,

*Capacity utilization is defined as the ratio of production to capacity.

**Throughout this chapter, tons refers to short tons (2,000 pounds).

TABLE 7.19 Coal Supply, Disposition, and Prices (million short tons per year, unless otherwise noted)

Supply, Disposition, and Prices	Reference Case 1992	1993	2000	2005	2010	Annual Growth 1993–2010 (percent)
			Production[1]			
East of the Mississippi	589	516	589	591	629	1.2
West of the Mississippi	409	429	439	485	508	1.0
Total	998	945	1027	1076	1137	1.1
			Net Imports			
Imports	4	7	13	14	15	4.3
Exports	103	75	87	100	115	2.5
Total	−99	−67	−74	−86	−100	2.4
Total Supply[2]	899	877	953	990	1037	1.0
			Consumption by Sector			
Residential and Commercial	6	6	7	6	6	0.2
Industrial[3]	74	75	86	93	97	1.6
Coke Plants	32	31	29	25	22	−2.1
Electric Generators[4]	780	814	832	868	913	0.7
Total	892	926	954	992	1039	0.7
Discrepancy and Stock Change[5]	6	−49	0	−2	−1	N/A
Average Minemouth Price (1993 dollars per short ton)	21.57	19.85	20.34	21.55	22.77	0.8
			Delivered Prices (1993 dollars per short ton)[6]			
Industrial	33.62	32.23	31.32	31.90	33.25	0.2
Coke Plants	49.15	47.44	45.96	46.64	47.26	0.0
Electric Generators	30.11	28.60	29.06	30.38	31.43	0.6
Average[7]	31.10	29.54	29.78	30.94	31.94	0.5
Average Price to All Users (1993 dollars per million Btu)	1.47	1.41	1.40	1.45	1.50	0.4
Exports[8]	42.39	41.41	41.03	42.62	43.52	0.3

[1] Includes anthracite, bituminous coal, and lignite.
[2] Production plus net imports plus net storage withdrawals.
[3] Includes consumption by cogenerators.
[4] Includes all electric power generators except cogenerators, which produce electricity as a byproduct of other processes.
[5] Balancing item: the sum of production, net imports, and net storage minus total consumption.
[6] Weighted average excludes residential and commercial prices; sectoral prices weighted by consumption tonnage.
[7] Weighted average excluded residential and commercial prices.
[8] Free-alongside-ship (f.a.s.) price at U.S. port-of-exit.

N/A = Not applicable.

Btu = British thermal unit.

Note: Totals may not equal sum of components due to independent rounding.

Sources: 1992 production and minemouth price: Energy Information Administration (EIA), *Coal Production 1992*, DOE/EIA-0118(92) (Washington, DC, October 1993). 1992 imports, exports, consumption, and other prices: EIA, *Quarterly Coal Report October-December 1993*, DOE/EIA-0121(93/4Q) (Washington, DC, May 1994). 1993 production and minemouth price: EIA, *Coal Industry Annual 1993*, DOE/EIA-0584(93) (Washington, DC, December 1994). 1993 imports, exports, consumption, and other prices: EIA, *Quarterly Coal Report October-December 1993*, DOE/EIA-0121(93/4Q) (Washington, DC, May 1994). *Projections:* EIA, AEO95 National Energy Modeling System run AEO95B.D1103942.

electric utilities drew heavily from their coal stockpiles to meet electricity demand* and U.S. exports declined sharply. In the forecast, eastern production recovers rapidly after 1993, primarily due to replenishment of utility coal stocks and increases in domestic coal demand. U.S. coal exports recover gradually after 1994.

Coal Demand

Electricity Generation Accounts for Nearly All the Increase in Coal Use

Domestic coal demand rises by 113 million tons in the forecast, from 926 million tons in 1993 to 1,039 million tons in 2010. Most of the increase is caused by growth in coal use for electricity generation. Coal demand in the other end-use sectors, taken as a whole, increases by 13 million tons.

Coal consumption for electricity generation (excluding cogenerators) rises from 814 million tons in 1993 to 913 million tons in 2010. By region, electricity coal consumption increases by 54 million tons west of the Mississippi River and by 45 million tons east of the Mississippi River, benefiting coal suppliers in both the East and the West. Increases in coal-fired generation come from a combination of increased utilization of existing generating capacity and additions of new capacity. The average utilization rate for existing coal-fired plants increases from 63% in 1993 to 68% in 2010. Together, utilities, nonutilities, and cogenerators increase net coal-fired generating capacity by 14 gigawatts over the forecast period. One gigawatt of coal-fired generating capacity corresponds to roughly 2.5 million tons of electric utility coal consumption, varying with changes in the average utilization rate of coal-fired plants and the heat content of coal consumed. Coal consumption (in tons) per kilowatthour of generation is higher for lower-rank coals, such as lignite and subbituminous, than for higher-rank bituminous coal.

Coal Maintains the Largest Share of Electricity Generation

Although coal maintains its fuel cost advantage over both oil and natural gas, gas-fired combined cycle is the most economical choice for new power generation through 2005 in terms of total generating costs (capital, operating, and fuel). Between 2000 and 2010, rising natural gas prices and a growing need for baseload generation result in an increase in coal-fired capacity.**

Through 2000, increases in coal consumption for electricity generation result mainly from increased utilization of existing plants. During this period, more coal-fired capacity is retired than built, resulting in a net capacity reduction of 1 gigawatt. Although coal-fired generation increases substantially, its share of total generation declines from 53% in 1993 to 51% in 2000, as gas, nuclear, and renewable fuels are used for additional generation. Increased nuclear generation between 1993 and 2000 is attributable mostly to improved operating performance of existing plants.

Between 2000 and 2010, 15 gigawatts (net of retirements) of new coal capacity is added, and coal fuels 44% of the new generation required to satisfy demand. As a result, coal's share of total generation is further eroded, falling to 50% by 2010. Generation from natural gas and renewable fuels continues to increase during the period, but expected retirements of nuclear plants result in a decline in nuclear generation.

Rising Industrial Coal Use Is Offset by Falling Demand for Coking Coal

In the sectors other than electricity, an increase in industrial steam coal consumption of 22 million tons between 1993 and 2010 is partly offset by a decline of 9 million tons in coking coal consumption. The higher consumption forecast for industrial steam coal results primarily from increased use of coal in the chemical and food-processing industries. In addition to higher levels of output projected

*At the end of 1993, stocks at electric utilities were 41 million tons lower than at the end of 1992. Energy Information Administration, *Quarterly Coal Report, October–December 1993,* DOE/EIA-0121(93/4Q) (Washington, DC, May 1994), Table 54.

**Baseload capacity represents the generating equipment normally operated to serve loads on an around-the-clock basis.

for those industries, the increased use of coal for cogeneration (the production of both electricity and usable heat for industrial processes) also contributes to the overall increase.

A projected decline in domestic consumption of coking coal results from the displacement of raw steel production from integrated steel mills (which use coal coke both for energy and as a raw material input) by increased steel production from minimills (which use electric arc furnaces, and thus bypass the use of coal coke) and increases imports of semifinished steels. Also contributing to the decrease is a reduction in the amount of coke required per ton of pig iron produced, based on energy efficiency improvements and increased supplemental fuel injections (mostly pulverized coal) to blast furnaces.

Coal consumption in the residential and commercial sectors remains constant, accounting for less than 1% of total U.S. coal demand over the forecast.

Coal Exports

U.S. coal exports rise in the forecast from 75 million tons in 1993 to 115 million tons. in 2010, because of higher demand for steam coal imports for electricity generation in Europe. U.S. exports of metallurgical coal change little, falling from 50 million tons in 1993 to 42 million tons in 2000, then rising to 53 million tons by 2010. World metallurgical coal trade declines slightly over the forecast.

U.S. coal exports to Europe increase from 38 million tons in 1993 to 70 million tons in 2010. Coal imports to Europe from all sources are projected to rise by 69 million tons, from 168 million tons in 1993 to 237 million tons in 2010. Coal demand for electricity generation is expected to rise in Europe, while reduced producer subsidies curtail European coal production. However, increased use of natural gas and environmental considerations restrain the growth in coal consumption for electricity generation and, consequently, the need for additional imports of steam coal.

U.S. coal exports to Asia increase by only 5 million tons, from 19 million tons in 1993 to 24 million tons in 2010. Coal imports to Asia from all sources rise in the forecast by 174 million tons, from 211 million tons in 1993 to 385 million tons in 2010. The large increase is based on current plans to add substantial amounts of coal-fired generating capacity, together with the expectation that much of the new capacity will be fueled by imported coal. Most of the increased exports to Asia should originate from Australia, South Africa, and Indonesia.

Coal Prices

On average, the minemouth price of domestic coal rises by 0.8% a year between 1993 and 2010. Gains in coal-mining productivity, an abundant coal reserve base, and lingering excess production capacity in the predominantly high-sulfur coal regions help stabilize U.S. average coal prices. Coal mining productivity has varied considerably in the past and is particularly difficult to forecast. Recent history shows that U.S. coal mining productivity declined by an average of 3.2% a year between 1970 and 1978 and then increased by 6.7% a year between 1978 and 1993. In the reference case, labor productivity is assumed to increase by an average rate of 3.9% a year through 2010.

Between 1993 and 2000, the minemouth price of coal declines in both Appalachia and the interior, with continuing improvements in labor productivity and lingering excess mine capacity for high-sulfur coal. During the same period, the average price in the West rises in response to increased demand for the region's low-sulfur coal. Prices rise in all three supply regions after 2000 as a result of reserve depletion and higher-capacity utilization levels, which are only partially offset by slower growth in labor productivity.

Uncertainties for the Future

Two key areas of uncertainty that affect the U.S. coal outlook are environmental concerns and the ability of the coal industry to keep prices competitive with those for other fuels.

Environmental Issues

Regulations to control utility emissions of air toxics, permitting requirements, and legislation or policy initiative to restrain greenhouse gas emissions could significantly affect coal's future in electricity generation. Title III of CAAA90 requires the EPA to submit its findings and recommendations on utility emissions to Congress in 1995. A decision to regulate emissions could require utilities to install equipment for removing air toxics from combustion gases. On the supply side, regulations could result in some interregional switching to coals with lower levels of toxic trace elements or affect the amount and degree of preparation required.

Air toxics (or hazardous air pollutants) are those pollutants that are hazardous to human health or the environment. Recent field studies conducted by the U.S. Department of Energy and the Electric Power Research Institute have been undertaken to determine and evaluate emissions of air toxics from coal-fired power plants. Air toxic emissions being evaluated include antimony, arsenic, beryllium, cadmium, chromium, cobalt, lead, manganese, nickel, mercury, chlorine/hydrochloric acid (HCl), selenium, benzene, toluene, formaldehyde, polycyclic aromatic hydrocarbons, and dioxins. With the exception of three of the pollutants, preliminary findings indicate that hazardous pollutants emitted by coal-fired power plants are (1) effectively removed from the combustion gases with electrostatic precipitators (ESP) or fabric filters or (2) low from the perspective of both presence in the flue gas and health risk impact. Mercury, selenium, and chlorine/HCl are not effectively controlled with conventional particulate control devices because they are relatively volatile at stack gas conditions.[6]

In addition to obtaining sulfur dioxide emission allowances as specified by CAAA90, electricity producers face permitting requirements under the Prevention of Significant Deterioration (PSD) program and for nonattainment areas. The Clean Air Act Amendments of 1977 established three types of regions with respect to National Ambient Air Quality Standards (NAAQS): attainment, nonattainment, and unclassifiable. Attainment and unclassifiable areas were made subject to a new PSD program for air quality and were further divided into three classes: Class I, Class II, and Class III. Of the three, Class I areas are the most pristine areas and are afforded the greatest degree of protection. They include international parks, national wilderness areas, national memorial parks larger than 5,000 acres, and national parks larger than 6,000 acres that were established as of August 7, 1977. Because of the general nature of Class I areas, the siting of a new coal-fired power plant within their boundaries is not likely to occur. In Class II areas, all new major power plants or major modifications of plants are required to use Best Available Control Technology (BACT). The BACT requirement is at least as stringent as a new source performance standard, but only as strict as the Lowest Achievable Emission Rate (LAER) requirement. In addition to technology-related requirements, proposed new plants in Class II areas face a maximum allowable increment test for particulate matter, sulfur dioxide, and nitrogen oxides, and are subject to an adverse impact test regarding key air-quality-related values for nearby Class I areas. At present, no PSD areas are classified as Class III.[7]

For nonattainment areas, new power plants are required to use control technologies that meet specifications for LAER. In addition, new sources in nonattainment areas must obtain offsets and meet various other requirements as well. How greatly the requirements limit additions of new coal-fired capacity depends on such factors as growth in electricity demand, costs and dependability of pollution control technologies, the ability to model environmental impacts accurately, remaining emissions increments in PSD areas, and the cost of environmental assessment activities.

In regard to greenhouse gas emissions, the CCAP provides environmental agencies, regulatory bodies, and electric utilities with guidelines and incentives for increasing the use of low-carbon fuels, such as natural gas and renewables, and for reducing growth in electricity demand. Success in meeting these objectives would reduce demand for both coal and oil in the electricity sector. For stabilizing greenhouse gas emissions in the long term, the CCAP states that measures must be taken to "ensure that a constant stream of improved technologies is available and that market conditions are favorable to their adoption." As a result, the prospects for increased reliance on coal in the nation's energy mix may hinge on the success of elements in DOE's Clean Coal Technology Demonstration Program, as well as on other federal and state initiatives.

Current Clean Coal Technology projects include the development of such advanced coal technologies as gasification combined cycle and pressurized fluidized-bed combustion, which have the potential to reduce carbon emissions through improvements in conversion efficiencies. These advanced coal plants offer conversion efficiencies in the range of 40 to 45%, which corresponds to a reduction in carbon emissions per unit of generation of between 17 and 27% relative to current units fired with pulverized coal.[8] State funding for the development of clean coal technologies is provided by governmental agencies such as the Pennsylvania Energy Development Authority and the Ohio Coal Development Office.

Coal Prices

In addition to labor productivity, other key areas of uncertainty in the coal price forecasts relate to estimates of reserve depletion and state actions aimed at bringing about full-cost pricing of energy. As new reserves are opened to mining, incremental costs related to differences in geologic conditions of new mines versus existing mines can raise mining costs, even when factor input costs, labor productivity, and technology remain constant. In the forecast, the effects of reserve depletion come into play as new mines are opened to meet increased demand and to replace capacity lost when existing mines are retired. The effects on future coal prices depend on the rate at which existing capacity is retired, growth in both domestic and foreign demand, and the availability and geological characteristics of coal reserves for new mines.

In recent years, several states have issued proposed rules and regulations specifying that estimated costs and benefits of factors such as pollution externalities, economic development, and social distributional effects be included in the prices consumers pay for energy. Currently, the following states have issued proposed rules and regulations aimed at bringing about full-cost pricing of energy: California, New York, Massachusetts, Nevada, New Jersey, and Wisconsin.[9] While consideration of environmental costs favors cleaner burning fuels such as natural gas and renewable fuels over coal, it is unclear how consideration of other externalities will affect relative fuel prices. Additional uncertainties relate to the ability of states to implement full-cost pricing policies and how widespread such policies may become. The forecasts account for the costs of environmental externalities for electric capacity planning decisions for regions with established externality costs.

7.8 Renewables

Interest in renewables, spurred both by government actions (including subsidies, incorporated into the Energy Policy Act of 1992, and set-asides for new capacity, as mandated by some state PUCs) and by the perceived environmental costs associated with fossil fuel consumption, has been strong in recent years. However, with projections of only moderate or low growth in oil, gas, and coal prices, the future of renewable energy in the absence of strong government support is far from certain. This section briefly examines the major renewable energy sources for both electricity and end-use demand, and provides some indication of future growth prospects and some of the hindrances to those prospects.

Renewables in Electricity Generation

Conventional Hydroelectric Power

Conventional hydropower, used primarily for baseload generation, now generates as much electricity as about 100 medium-sized (500-megawatt) coal plants. According to the reported plans of electric utilities, hydroelectric generation will grow slightly (about 0.6% annually) through 2010, mainly as a result of turbine repowering. Hydropower is expected to generate 306 billion kilowatt-hours of electricity by 2010 (Table 7.20). New hydropower developments will face increased environmental constraints, and FERC relicensing requirements will offset capacity increases in many cases.

FERC now considers relicensing of hydroelectric plants on a "cumulative impacts" basis, weighing the environmental and water use impacts of existing and proposed projects on an entire river basin and modifying existing licenses when it finds the cumulative negative impacts of all dams in a given river basin unacceptable. In May 1994, the U.S. Supreme Court held that states may impose conditions on hydroelectric operations, such as minimum stream flow requirements, as part of their authority under the Clean Water Act of 1977. That interpretation could limit hydroelectric generation at both existing and proposed projects such as states require FERC to impose license conditions that operators release water over spillways instead of through turbines, or otherwise moderate their operations. Possibilities include reducing generation or changing the timing of generation, both of which would reduce hydropower's economic value.

Because of hydropower's zero fuel cost, it is used by electricity producers whenever available. However, variations in precipitation can limit its availability, and generation from the same installed capacity can vary by 5% or more from one year to another as a result of variations in annual water flows.

Wind and Solar

Much of the wind energy market before 2000 will result from legislated set-asides, which add about a gigawatt of capacity, bringing the total to 3 gigawatts in 2000. Strong growth is expected, especially after 2005, as improved technology, higher fuel prices, increased capacity needs, and externality costs combine to make wind energy more attractive. Wind, which accounted for less than 3 billion kilowatthours of electricity in 1993, should provide about 25 billion kilowatthours by 2010, an annual growth rate of 13%, the highest expected for the renewable sources of electricity generation.

Wind, along with solar, is an "intermittent" source of electricity, which means that the resource is available only at certain times of the day and year, and only with a distribution that is less certain than other sources. Although the general pattern of wind resources is known, the uncertainty associated with those resources at any given point in time makes them unsuitable for baseload generation. Consequently, unless storage is available, wind's primary benefit is as a "fuel saver," mainly for natural gas. A consequence of that fact is that the average capacity factor for wind is low compared with most other sources of electricity. Average utilization in 1993 was just under 20%. As new, more efficient turbines move into the marketplace, the utilization rate is expected to increase to almost 29% by 2010.

Grid-connected solar-powered electricity generation, still a tiny share of electricity generation overall, is expected to grow from 0.9 billion kilowatthours in 1993 to almost 5 billion kilowatthours by 2010, an annual growth rate of 10.5%. The cost of solar thermal generating units remains high by comparison with other sources of electricity; consequently, it is expected to make inroads primarily in those areas with the most abundant solar resources, such as California and the desert Southwest. Nevertheless, there are current indications that solar photovoltaics may be on the verge of becoming a more attractive alternative for generating electricity. A recent announcement of a joint project by DOE and Enron for a 100-megawatt facility in Nevada, together with other indications of technological breakthroughs, could lead to somewhat more market acceptance for solar photovoltaics as an electricity source. In addition, photovoltaics have significant advantages as a remote site electricity source, where costs of connection to distribution lines are prohibitive or the costs of transporting and using other fuels are too high or environmentally problematic. While such applications are not specifically addressed in these projections, the advent of lower-cost photovoltaic units could conceivably create additional markets where access to the national transmission infrastructure is not currently available.

TABLE 7.20 Renewable Energy (quadrillion Btu per year, unless otherwise noted)

Electric and Nonelectric	Reference Case 1992	1993	2000	2005	2010	Annual Growth 1993–2010 (percent)
	Electric Utilities and Nonutilities[1] (excluding cogenerators)					
Capability (gigawatts)						
Conventional Hydropower	76.55	76.74	79.27	79.65	79.81	0.2
Geothermal[2]	2.90	2.96	3.42	3.44	4.57	2.6
Municipal Solid Waste	2.34	2.59	3.51	4.35	5.14	4.1
Biomass/Other Waste[3]	1.38	1.50	1.75	1.95	2.73	3.6
Solar	0.34	0.34	0.54	0.79	1.37	8.5
Wind[4]	1.75	1.76	3.05	4.18	10.04	10.8
Total	85.25	85.89	91.55	94.34	103.65	1.1
Generation (billion kilowatthours)[1]						
Conventional Hydropower	248.77	275.73	303.10	305.22	306.30	0.6
Geothermal[2]	16.68	17.32	20.55	20.45	28.21	2.9
Municipal Solid Waste	17.17	17.75	23.58	29.75	35.71	4.2
Biomass/Other Waste[3]	6.84	7.46	11.09	12.45	17.90	5.3
Solar	0.75	0.90	1.50	2.52	4.90	10.5
Wind[4]	2.92	3.05	6.33	9.37	25.22	13.2
Total	293.13	322.21	366.14	379.76	418.24	1.5
Consumption						
Conventional Hydropower	2.56	2.84	3.12	3.14	3.16	0.6
Geothermal[2]	0.35	0.36	0.52	0.57	0.83	5.0
Municipal Solid Waste	0.28	0.29	0.38	0.48	0.58	4.2
Biomass/Other Waste[3]	0.09	0.09	0.13	0.14	0.19	4.1
Solar	0.01	0.01	0.02	0.03	0.05	10.5
Wind[4]	0.03	0.03	0.07	0.10	0.26	13.2
Total	3.32	3.63	4.24	4.46	5.06	2.0
	Cogenerators[5]					
Capacity (gigawatts)						
Conventional Hydropower	0.81	0.97	1.29	1.29	1.29	1.7
Municipal Solid Waste	0.31	0.31	0.51	0.51	0.51	3.0
Biomass/Other Waste	5.49	5.60	5.62	5.66	5.71	0.1
Total	6.61	6.87	7.41	7.46	7.51	0.5
Generation (billion killowatthours)						
Conventional Hydropower 3.16	3.16	3.24	4.48	4.48	4.48	1.9
Municipal Solid Waste	1.46	1.54	1.89	1.89	1.89	1.2
Biomass/Other Waste	30.20	30.70	31.71	31.97	32.24	0.3
Total	34.81	35.48	38.08	38.34	38.61	0.5
Consumption						
Conventional Hydropower	0.03	0.03	0.03	0.03	0.03	N/A
Municipal Solid Waste	0.02	0.03	0.03	0.03	0.03	1.2
Biomass/Other Waste	0.64	0.64	0.72	0.78	0.82	1.5
Total	0.69	0.69	0.78	0.84	0.88	1.4
	Nonelectric					
Nonelectric Renewable Energy Consumption						
Geothermal[6]	0.01	0.01	0.02	0.03	0.04	7.3
Biofuels[7]	2.03	2.09	2.32	2.48	2.61	1.3
Solar Thermal[8]	0.06	0.06	0.08	0.08	0.09	2.3
Ethanol	0.08	0.07	0.13	0.19	0.23	7.8
Total	2.18	2.22	2.55	2.78	2.96	1.7

TABLE 7.20 (continued) Renewable Energy (quadrillion Btu per year, unless otherwise noted)

Electric and Nonelectric	Reference Case					Annual Growth 1993–2010 (percent)
	1992	1993	2000	2005	2010	
Total Renewable Energy Consumption[9]						
Conventional Hydropower	2.60	2.87	3.16	3.18	3.19	0.6
Geothermal	0.36	0.37	0.54	0.60	0.86	5.0
Municipal Solid Waste	0.30	0.31	0.41	0.52	0.61	4.0
Biofuels/Other Waste	2.76	2.82	3.17	3.40	3.61	1.5
Solar	0.07	0.07	0.09	0.11	0.14	4.2
Wind	0.03	0.03	0.07	0.10	0.26	13.2
Ethanol	0.08	0.07	0.13	0.19	0.23	7.8
Total	6.19	6.55	7.57	8.08	8.91	1.8

[1] Grid connected only.
[2] Includes hydrothermal resources only (hot water and steam).
[3] Does not include projections for energy crops.
[4] Includes horizontal-axis wind turbines only.
[5] Includes generators and cogenerators at facilities whose primary function is not electricity production (standard industrial classification 49). In general, biomass and other waste facilities are cogenerators while the remaining renewables produce only electricity.
[6] Residential and commercial ground-source heat pumps.
[7] Residential and industrial wood and wood waste.
[8] Residential and commercial water heating.
[9] Actual heat rates used to determine fuel consumption for all renewable fuels except, hydropower, solar, and wind. Consumption at hydroelectric, solar, and wind facilities determined using the fossil fuel equivalent of 10,302 Btu per kilowatthour.
N/A = Not available.
Btu = British thermal unit.
Notes: Totals may not equal sum of components due to independent rounding. Net summer capacity has been estimated for nonutility generators for AEO95. Net summer capacity is used to be consistent with electric utility capacity estimates. Electric utility capacity is the most recent data available as of August 15, 1994. Additional retirements are also determined based on the size and age of the units. Therefore, capacity estimates may differ from other Energy Information Administration sources.
Sources: 1992 and 1993 electric utility capacity: Energy Information Administration (EIA), Form EIA-860, "Annual Electric Utility Report." 1992 and 1993 nonutility and cogenerator capacity: Form EIA-867, "Annual Nonutility Power Producer Report." 1992 ethanol, 1992 generation, and 1993 generation: EIA, *Annual Energy Review,* DOE/EIA-0384(93) (Washington, DC, July 1994). 1992 nonutility consumption other than ethanol: EIA, *State Energy Data Report 1992,* DOE/EIA-0214(92) (Washington, DC, May 1994). 1993 ethanol: EIA, *Petroleum Supply Annual,* 1993, DOE/EIA-0340(93/1) (Washington, DC, June 1994). 1993 nonutility consumption other than ethanol: EIA, *Short-Term Energy Outlook,* DOE/EIA-0202(94/3Q) (Washington, DC, August 1994). *Projections:* EIA, AEO95 National Energy Modeling System run AEO95B.D1103942.

Geothermal

Currently exploitable geothermal resources (hot water and steam) are limited to the western United States, where capacity from geothermal plants is projected to grow by about 1.5 gigawatts by 2010. Most of the expected growth occurs after 2005 as demand for new capacity begins to grow. Total geothermal electricity generation in 1993 was 17 billion kilowatthours, generated from almost 3 gigawatts of capacity. By 2010, geothermal is expected to account for 28 billion kilowatthours of electricity, a growth rate of 2.9% annually. While this will marginally increase geothermal's share of total generation, its status as a largely western resource will prevent it from becoming a more important source of electricity nationwide unless further technological breakthroughs, such as hot-dry-rock technology, become commercialized.

Geothermal is generally a baseload application. Utilization in 1993 was 67%, reflecting the high output from California's Geysers project. By 2010, utilization is expected to rise to 70%, as new technology combined with increasing demand makes the generation of power from geothermal sources more economical.

Municipal Solid Waste

Municipal solid waste (MSW) generation capacity is projected to grow at a rate just over 4% a year. MSW plants serve a dual purpose: they are a source of baseload generation, and they provide a means for the disposal of MSW. In 1993, MSW generation was 18 billion kilowatthours, slightly more than that obtained from geothermal sources. By 2010, MSW is expected to generate 36 billion kilowatthours, a doubling in 17 years. Utilization rates should reach nearly 80%, with growth in the economy providing the material necessary to run these dual-purpose plants.

However, legal issues could affect the use of MSW as an energy source, with plants seeking to obtain guaranteed fuel supplies through local ordinances that direct the flow of waste. Environmental issues could also have adverse effects on MSW plants. There have been instances of local opposition to new incinerators, which are perceived to contribute adversely to air pollution. In addition, further mandated recycling or growth in alternative uses of combustible waste (such as in the pulp and paper industry) could reduce the amount of waste available. Thus, while the economics of MSW are reasonably promising under certain conditions, these concerns could inhibit the growth over the forecast horizon.

Biomass

Generation from biomass (wood) is projected to grow only slightly before 2005, because new competitive biomass technologies are as yet unproven and conventional fuel prices remain relatively low. After 2005, however, the market for biomass energy begins to grow, driven by expected technology enhancements, slightly higher prices for conventional fuels, and increased need for new capacity.

Production of electricity from wood was just over 7 billion kilowatthours in 1993. By 2010, generation should total 18 billion kilowatthours, an annual rate of growth of 5.3%. Much of the growth is expected from cogeneration in the pulp and paper industries, which have access to significant supplies of wood and waste wood. One of the inhibiting factors in the growth of wood use for electricity generation is the expense of long-distance transportation, as well as the perception that wood is not as environmentally benign as other renewable sources. In addition, the wood resource is a highly regional one. However, the promise of "energy crops" could greatly increase the market potential for biomass, depending on the ability of suppliers to provide a low-cost, marketable fuel source. The projections in this forecast do not assume that energy crops will be available in the 2010 time frame. However, if other sources of electricity, such as natural gas, are seen as increasingly expensive, or further environmental regulations inhibit the use of, for example, coal as a fuel source, then energy crops could greatly increase the attractiveness of biomass in the post-2010 period.

Nonelectric Renewable Energy Uses

Wood

Projections for wood use include steam production in the industrial sector and heating in the residential sector. The primary industries that use wood for energy are the pulp and paper and lumber industries. Growth in wood use in the industrial sector is a direct function of the growth in demand for wood-based products. Wood use in the residential sector is affected by changes in the housing stock and, to some degree, by fossil fuel and electricity prices.

Wood consumption is by the far the largest contributor among the nonelectric renewable energy categories in the forecast. In 1993, wood use accounted for almost 97% of total nonelectric renewable energy consumption (excluding ethanol), and it accounts for nearly 90% of the projected growth in the use of these renewables over the forecast period. Nevertheless, wood consumption for heat and steam production is expected to increase relatively slowly, from 2.09 quadrillion Btu in 1993 to 2.61 quadrillion Btu in 2010, at an average annual rate of 1.3% a year. Furthermore, in contrast to most renewable energy applications, the use of woodstoves in the residential sector is expected to decline slightly.

Geothermal

Projections of geothermal energy use other than for electricity generation are limited to ground-source heat pumps. Ground-source heat pumps include a buried heat exchanger to permit the extraction of ground heat. Because ground temperatures remain relatively stable throughout the year, from about 42 to 77 degrees Fahrenheit (depending on the region of the country), earth energy can be more effective than air in providing cooling in the summer and heating in winter. Growth in the use of ground-source heat pumps is likely to be the greatest for new construction in the residential and commercial markets.

Geothermal energy use for ground-source heat pumps increases rapidly over the forecast period, from about 10 trillion Btu in 1993 to about 40 trillion Btu in 2010, increasing at an average annual rate of over 7% a year. However, ground-source heat pumps are expected to remain a small share of the overall heating and cooling market.

Solar

Similarly, solar thermal energy use for water heating in the residential and commercial sectors is expected to expand through 2010. Energy consumption for water heating is expected to grow from around 60 trillion Btu in 1993 to about 90 trillion Btu in 2010.

Overall, these nonelectric renewable energy uses are expected to increase more slowly than electricity applications, growing at an average rate of 1.4% a year through 2010 and continuing to account for less than one-third of total renewable energy consumption throughout the forecast.

References

1a. Energy Information Administration, *Annual Energy Outlook 1995,* DOE/EIA-0383(95) (Washington, DC, January 1995).

1b. Energy Information Administration, *Annual Energy Outlook 1996,* DOE/EIA-0383(96) (Washington D.C. January 1996). The full report may be accessed at the EIA's World Wide Web Site, address http://www.eia.doe.gov.

2. Energy Information Administration, *Supplement to the Annual Energy Outlook 1995,* DOE/EIA-0554(95) (Washington, DC, February, 1995).
3. Energy Information Administration, *The National Energy Modeling System: An Overview,* DOE/EIA-0581 (Washington, DC, May 1994).
4. I. Ismail, "Future Growth in OPEC Oil Production Capacity and the Impact of Environmental Measures." Presentation to the Sixth Meeting of the International Energy Workshop (Vienna, Austria, June 1993).
5. *Foster Natural Gas Report,* December 16, 1993, p. 24.
6. P. Chu and C. Schmidt, "Hazardous Air Pollutant Emissions from Coal Fired Power Plants," in *Eleventh Annual International Pittsburgh Coal Conference,* Pittsburgh, PA, September 12–16, 1994, ed. Shiao-Hung Chiang (Pittsburgh, PA: University of Pittsburgh, 1994), pp. 551–556.
7. A.F. Loeb and T.J. Elliott, "PSD Constraints in Utility Planning: A Review of Recent Visibility Litigation," *Natural Resources Journal,* Vol. 34 (Spring 1994), pp. 1–40.
8. U.S. Department of Energy, Office of Fossil Energy, *Clean Coal Technology: The New Coal Era,* DOA/FE-0193P (Washington, DC, June 1990).
9. T.F. Torries and V.J. Norton, "Implications of Full Cost Pricing of Fossil Energy," in *Eleventh Annual International Pittsburgh Coal Conference,* Pittsburgh, PA, September 12–16, 1994, ed. Shiao-Hung Chiang (Pittsburgh, PA: University of Pittsburgh, 1994), pp. 1598–1603.
10. Lawrence Berkeley National Laboratory, *U.S. Residential Appliance Energy Efficiency: Present Status and Future Direction.*
11. Richard F. Mast et al., U.S. Department of the Interior, Geological Survey and Minerals Management Service, *Estimates of Undiscovered Conventional Oil and Gas Resources in the United States—A Part of the Nation's Energy Endowment* (Washington, DC, 1989).

12. Larry W. Cooke, U.S. Department of the Interior, Minerals Management Service, *Estimates of Undiscovered, Economically Recoverable Oil and Gas Resources for the Outer Continental Shelf, Revised as of January 1990,* OCS Report MMS 91-0051 (Washington, DC, July 1991).
13. National Petroleum Council, Committee on Natural Gas, *The Potential for Natural Gas in the United States, Volume II, Source and Supply* (Washington, DC, December 1992).
14. William L. Fisher et al., Oil Resources Panel convened by the U.S. Department of Energy, *An Assessment of the Oil Resource Base of the United States* (Washington, DC, October 1992).
15. Potential Gas Committee, *Potential Supply of Natural Gas in the United States (December 31, 1992)* (Potential Gas Agency, Colorado School of Mines, May 1993).
16. Pacific Northwest Laboratory, *An Assessment of the Available Windy Land Area and Wind Energy Potential in the Contiguous United States* (PNL-7789), prepared for the U.S. Department of Energy under Contract DE-AC06-76RLO 1830 (August 1991).
17. Energy Information Administration, Form EIA-861, "Annual Electric Utility Report," and Form EIA-867, "Annual Nonutility Power Producer Report."
18. U.S. Bureau of the Census, U.S. Department of Commerce, *Current Construction Reports, Series C25 Characteristics of New Housing: 1992* (Washington, DC, 1993).
19. National Renewable Energy Laboratory, "Baseline Projections of Renewables Use in the Buildings Sector," prepared for the U.S. Department of Energy under Contract DE-AC02-83CH10093 (December 1992).

Appendix A. Major Assumptions for the Reference Case

Buildings Sector Assumptions

The buildings sector includes both residential and commercial structures. The National Appliance Energy Conservation Act of 1987 (NAECA), the Energy Policy Act of 1992 (EPACT), and the Climate Change Action Plan (CCAP) contain provisions that impact future buildings sector energy use. The provisions with the most significant effect are minimum equipment efficiency standards. These standards require that new heating, cooling and other specified energy-using equipment meet minimum energy efficiency levels, which change over time. The manufacture of equipment that does not meet the standards is prohibited.

Residential Assumptions

The NAECA minimum standards[10] for the major types of equipment in the residential sector are

- Heat pumps—a 10.0 minimum seasonal energy efficiency ratio for 1992
- Room air conditioners—an 8.6 energy efficiency ratio in 1990
- Gas/oil furnaces—a 0.78 annual fuel utilization efficiency in 1992
- Refrigerators—a standard of 976 kilowatthours per year in 1990, decreasing to 691 kilowatthours per year in 1993
- Electric water heaters—a 0.88 energy factor in 1990
- Natural gas water heaters—a 0.54 energy factor in 1990

Building codes relevant to CCAP are represented by an increase in the shell integrity of new construction over time. By the year 2000, heating and cooling shell efficiency in new construction improves by 7% relative to 1994.

Other programs which could have a major impact on residential energy consumption are the EPA Green Programs. These programs, which are cooperative efforts between the EPA and energy appliance manufacturers, encourage the development and production of highly energy-efficient equipment. One of the best known examples of these programs is the "golden carrot refrigerator," a very efficient design that is projected to be available by 1998 and to consume less than two-thirds of the energy specified in the 1993 standard.

Commercial Assumptions

Minimum 1994 equipment efficiency standards for the commercial sector are mandated in the EPACT legislation.* Minimum standards for representative equipment types produced after January 1, 1994, are

- Central air-conditioning heat pumps—a 9.7 seasonal energy efficiency rating
- Gas-fired forced-air furnaces—a 0.8 annual fuel utilization efficiency standard
- Fluorescent reflector lamps—a 75.0 lumens per watt lighting efficacy standard

The CCAP programs recognized in the reference case include enhanced efficiency standards for central air-conditioning units, the expansion of the EPA Green Lights and Energy Star Buildings programs, and improvements to building shells from advanced insulation methods and technologies. The minimum efficiency standard for air-conditioning units is assumed to rise to a seasonal energy efficiency rating of 10.0 in 1998. The EPA Green Programs are designed to facilitate cost-justified retrofitting of equipment by providing participants with information and analysis as well as participation recognition. Retrofitting behavior is captured in the commercial module via discount parameters for controlling cost-based equipment retrofit decisions for various market segments. To model programs such as Green Lights, which target particular end uses, the reference case version of the commercial module includes end-use-specific segmentation of discount rates. Existing building shell efficiency is assumed to increase by 1% over the 1990 average by the year 2000, saving a proportional amount of space heating and cooling energy.

Industrial Sector Assumptions

Compared to the buildings sectors, relatively few regulations target industrial sector energy use. The electric motor standards in EPACT require a 10% increase in efficiency above 1992 efficiency levels for motors sold after 1997.** These standards have been incorporated into the Industrial Demand Module through the analysis of process efficiencies for new industrial processes. These standards are expected to lead to significant improvements in efficiency, since it has been estimated that electric motors account for about 60% of industrial process electricity use.

Climate Change Action Plan

Several programs included in the CCAP target the industrial sector. The intent of these programs is to reduce greenhouse gas emissions by lowering industrial energy consumption. It was estimated that full implementation of these programs would reduce industrial electricity consumption by 55 billion kilowatthours and nonelectric consumption by 370 trillion Btu by 2000. However, the programs were not fully funded. The energy savings were revised in proportion to the funding received. Consequently, electricity consumption is reduced by 6 billion kilowatthours, and nonelectric energy consumption is reduced by 43 trillion Btu. The nonelectric energy is assumed to be steam coal.

Transportation Sector Assumptions

The transportation sector accounts for two-thirds of the nation's oil use and has been subject to regulations for many years. The Corporate Average Fuel Economy (CAFE) standards, which mandate average miles per gallon standards for manufacturers, continue to be widely debated. The reference case projections assume that there will be no further increase in the CAFE standards from the current 27.5 miles per gallon standard. This assumption is consistent with the overall policy that only current legislation is assumed.

*National Energy Policy Act of 1992, P.L. 102–486, Title I, Subtitle C, Sections 122 and 124.
***Ibid.*, Title II, Subtitle C, Section 342.

EPACT requires that centrally fueled automobile fleet operators—federal, state, and local governments, and fuel providers (e.g., gas and electric utilities)—purchase a minimum fraction of alternative-fuel vehicles.* Federal fleet purchases of alternative-fuel vehicles must reach 50% of their total vehicle purchases by 1998 and 75% by 2002. Purchases of alternative-fuel vehicles by state and local governments must realize 20% of total purchases by 2002 and 70% by 2005. Private fuel companies are required to purchase 30% alternative-fuel vehicles in 1996, increasing to 90% by 1999.

In addition to these requirements, the State of California has adopted a Low Emission Vehicle Program, which requires that 10% of all new vehicles sold by 2000 meet the "zero emissions requirements." At present, only electric-dedicated vehicles meet these requirements. Both Massachusetts and New York have also adopted this program. Other states could "opt in" for adoption, but these projections assume that only the three states that have formally adopted the California program will participate.

The projections assume that these regulations represent minimum requirements for alternative-fuel vehicle sales; consumers are allowed to purchase more of the vehicles, should vehicle cost, fuel efficiency, range, and performance characteristics make them desirable. In fact, the projections indicate that more than the minimum will be purchased.

Projections for both vehicle-miles traveled** and ton-miles traveled† are calculated endogenously and are based on the assumption that modal shares—for example, personal automobile travel versus mass transit—remain stable over the forecast and track recent historical patterns. Other important factors affecting the forecast of vehicle-miles traveled are personal disposable income per capita; the ratio of miles driven by females to miles driven by males in the total driving population, which increases from 56% in 1990 to 70% by 2010; and the proportion of the driving-age population over the age of 60, which increases from 21.6% in 1990 to 24.1% in 2010.

Climate Change Action Plan

Four CCAP programs focus on transportation energy use: (1) reform the federal subsidy for employer-provided parking; (2) adopt a transportation system efficiency strategy; (3) promote telecommuting; and (4) develop fuel economy labels for tires. The combined assumed effect of the federal subsidy, system efficiency, and telecommuting policies is a 1.3% reduction in vehicle-miles traveled (190 trillion Btu). The fuel economy tire labeling program improves new fuel efficiency by 4% among vehicles that switch to low rolling resistance tires, resulting in a reduction in fuel consumption of 40 trillion Btu.

Electricity Assumptions

Characteristics of Generating Technologies

The costs and performance of new generating technologies are important factors in determining the future mix of capacity. There are 22 fossil, renewable, and nuclear generating technologies included in these projections. Technologies represented include those currently available as well as those that are assumed to be commercially available within the horizon of the forecast. Capital cost estimates and operational characteristics, such as efficiency of electricity production expressed by the percent heat rate, are used for decision making where it is assumed that the selection of new plants to be built is based on least cost. The levelized lifetime cost, including fuel costs, is evaluated and is used as the basis for selecting plants to be built.

**Ibid.*, Title III, Section 303, and Title V, Sections 501 and 507.

**Vehicle-miles traveled are the miles traveled yearly by light-duty vehicles.

†Ton-miles traveled are the miles traveled and their corresponding tonnage for freight modes, such as trucks, rail, air, and shipping.

Regulation of Electricity Markets

It is assumed that electricity producers will comply with CAAA90, which mandates a limit of 8.95 million short tons of sulfur dioxide emissions by 2000. Utilities are assumed to comply with the limits on sulfur emissions by retrofitting units with flue gas desulfurization (FGD) equipment, transferring or purchasing sulfur emission allowances, operating high-sulfur coal units at a lower capacity utilization rate, or switching to low-sulfur fuels. The costs for FGD equipment average approximately $179 per kilowatt, in 1993 dollars, although the costs vary widely across the regions. It is also assumed that the market for trading emission allowances will be allowed to operate without regulation and that the states will not further regulate the selection of coal to be used.

The provisions of EPACT include revised licensing procedures for nuclear plants and the creation of exempt wholesale generators.* These entities are included among nonutility producers and are assumed to have a capital structure that is highly leveraged compared with that of investor-owned regulated utilities.

Prices for electricity are assumed to be regulated at the state level. Prices for the residential, commercial, industrial, and transportation sectors are developed by classifying costs into four categories: fuel, fixed operation and maintenance, variable operation and maintenance, and capital. These costs are allocated to each of the four customer classes using the proportion of sales to the class and the contribution of each class to system peak load requirements. These allocated costs are divided by the sales to each sector to obtain electricity prices to the sector.

Energy Efficiency and Demand-Side Management

Improvements in energy efficiency induced by growing energy prices, new appliance standards, and utility DSM programs are represented in the end-use demand models. Appliance choice decisions are a function of the relative costs and performance characteristics of a menu of technology options. Utilities have reported plans to increase their expenditures on DSM programs to more than $4 billion per year by 1997.

Nuclear Power

It is assumed that two nuclear generating units currently under construction will be operational by 2010: Watts Bar 1 in 1995 and Watts Bar 2 in 1997. Bellefonte 1 and 2 are assumed not to be completed. These four units are owned by the Tennessee Valley Authority (TVA). The TVA is in the process of developing an Integrated Resource Plan for completion by late 1995 to determine long-term energy needs in the region and the most economical way to meet them. In recent months the chairman of the TVA has made several statements that make the completion of these nuclear units highly uncertain. In particular, he stated that completion of the unfinished units may not be economically feasible and has had cost estimates developed for the conversion of the Bellefonte units to coal or natural gas facilities. In addition, the TVA has, for the first time, issued requests for proposals totaling 4 gigawatts of power, suggesting that purchased power may be a preferable option to the completion of the nuclear units.**

It is assumed that no newly ordered nuclear power plants will be operational through 2010 for the following reasons:

- Concerns about the disposal of radioactive waste
- Public concerns about safety
- Concerns about economic and financial risk
- Uncertainty in the licensing and regulatory processes.

*National Energy Policy Act of 1992, P.L. 102–486, Title VII, Subtitle A, Section 711, and Title XXVIII, Sections 2801 and 2802.

**One of the two new units anticipated was Watts Bar 2. After the *Annual Energy Outlook 1995* analysis was completed, the Tennessee Valley Authority announced that it was halting construction work on the unit and was considering the options of converting the plant to alternative fuel or seeking a financial partner to complete construction.

In the reference case, nuclear units are assumed to operate until their license expiration, typically 40 years from the date of first operation. The average nuclear capacity factor is expected to increase from 71% currently to 74% by 2000 and remain at that level through 2010. Capacity factor assumptions are developed at a regional level, based on historical performance of individual units.

Oil and Gas Supply Assumptions

Domestic Oil and Gas Economically Recoverable Resources

The projections are based on analyses of estimates of the economically recoverable resource base from the U.S. Geological Survey and the Minerals Management Service of the U.S. Department of the Interior, the National Petroleum Council, the Office of Fossil Energy of the U.S. Department of Energy, and the Potential Gas Committee.[11–15] Economically recoverable resources are those volumes considered to be of sufficient size and quality for their production to be commercially profitable by current conventional or nonconventional techniques, under specified economic conditions. Estimates were developed on a regional basis. Total unproved oil resources are assumed to be 86 billion barrels with 1990 technology and 129 billion barrels with 2010 technology. Total unproved gas resources are assumed to be 852 trillion cubic feet with 1990 technology and 1,449 trillion cubic feet with 2010 technology.

The CCAP includes a program promoting the capture of methane from coal mining activities to reduce carbon emissions. That methane would be marketed as part of the domestic natural gas supply. The assumption for this program is that it begins in 1995, reaching a maximum annual production level of 19.1 billion cubic feet by 2000. The volumes of recoverable methane from the program are not included in the economically recoverable gas estimates discussed previously in this appendix.

Technological Improvements Affecting Recovery and Costs

Productivity improvements are simulated by assuming that the recoverable resource target will expand and the effective cost of supply activities will be reduced. The projections assume that the total volumes of unproved domestic oil and natural gas resources that are economically recoverable will increase over the 1990–2010 period at average annual rates of roughly 2.1 and 2.6%, respectively, in response to technological innovation, as indicated by the volumes cited earlier. The increase is due to both the development and deployment of new technologies, such as three-dimensional seismology and horizontal drilling and completion techniques. Drilling, operating, and lease equipment costs are expected to decline at assumed rates that vary somewhat by cost and fuel categories, ranging from roughly 1 to 3%, with most of them generally at 2%.

Leasing and Drilling Restrictions

The projections of crude oil and natural gas supply assume that current restrictions on leasing and drilling will continue to be enforced throughout the forecast period. At present, drilling is prohibited along the entire East Coast, the west coast of Florida, and the West Coast except for the area off Southern California. In Alaska, drilling is prohibited in a number of areas including the Arctic National Wildlife Refuge. The projections also assume that coastal leasing and drilling activities will be reduced in response to the restrictions of CAAA90, which require that offshore drilling sites within 25 miles of the coast, with the exception of areas off Texas, Louisiana, Mississippi, and Alabama, meet the same clean air requirements as onshore drilling sites.

Gas Supply from Alaska and Imports

The Alaska Natural Gas Transportation System is assumed to come online no earlier than 2005 and only after the border price reaches $3.64 (1993 dollars) per thousand cubic feet. Pipeline import volumes from Canada are constrained by the pipeline design capacity, which is assumed to increase from 2.3 trillion cubic feet in 1990 to 4.2 trillion cubic feet in 2010. The liquefied natural gas facilities at Everett, Massachusetts, and Lake Charles, Louisiana (the only ones currently in operation) have an operating capacity of 311 billion cubic feet. The facilities at Cove Point, Maryland, and Elba Island, Georgia, are assumed to reopen when economically justified, but not before 1996 and 1998, respectively, expanding total liquefied natural gas operating capacity to 794 billion cubic feet.

Natural Gas Transmission and Distribution Assumptions

The projections reflect the assumptions that the provisions of Order 636 have been fully implemented, and that all interstate pipeline companies have completed the switch from modified fixed variable (MFV) to straight fixed variable (SFV) rate design. Approved transition costs are assumed to be consistent with the revised cost estimate published by FERC in the November 1993 General Accounting Office (GAO) report. Gas supply realignment (GSR) costs are recovered from 1994 through 1998, with 90% assigned to firm markets and 10% to interruptible markets, as stipulated in FERC Order 636. Account 191 costs are collected in 1994 and 1995 from firm customers only.

Consistent with the industry restructuring, the methodology employed in solving for the market equilibrium assumes that marginal costs are the basis for determining market-clearing prices for noncore markets and that average cost-of-service rates are the basis for core market prices.

Firm transportation rates for pipeline services are calculated assuming that the costs of new pipeline capacity will be rolled into the existing rate base (although the test for determining whether or not to build new capacity is based on incremental rates). Distribution markups to firm service customers are based on historical data and are assumed to decline by 1% per year. Although the market is perceived to be changing from recent history, this assumption is consistent with current regulatory policy and with EIA's definition of a reference case.

In determining interstate pipeline tariffs, capital expenditures for refurbishment over and above that included in operations and maintenance costs are not considered, nor are potential future expenditures for pipeline safety. (Refurbishment costs include any expenditures for repair and/or replacement of existing pipe.)

Prices for use in compressed natural gas (CNG) vehicles are phased from EIA's *Natural Gas Annual* historical transportation prices to what is assumed to be a retail market price in 2005. The linear phase-in begins in 1994 and continues through 2005, after which the price is assumed to be the firm industrial price plus a markup to cover the cost of dispensing the fuel (plus taxes). The phase-in period represents the transition from a market where users must obtain and dispense their own supplies (as in the case of fleet vehicles) to a market in which retail outlets are readily available to all customers. Federal taxes of $0.49 (1993 dollars) per thousand cubic feet plus corresponding state taxes are levied starting in 1994.

Provisions of the CCAP are assumed to have no impact on the transmission and distribution segment of the industry. Although regulatory changes that are recommended in the CCAP may be considered by the FERC in the near future, they go beyond the current FERC regulatory policy and thus are not considered in the reference case.

Petroleum Market Assumptions

The petroleum refining and marketing industry is assumed to incur large environmental costs to comply with CAAA90 and other regulations. Investments related to reducing emissions at refineries are represented as an average annualized expenditure. Costs identified by the National Petroleum Council* are allocated among the prices of liquefied petroleum gases, gasoline, distillate, and jet fuel, assuming they are recovered in the prices of light products. The lighter products, such as gasoline and distillate, are assumed to bear a greater amount of these costs because demand for these products is less price-responsive than the demand for the heavier products.

*Estimated from National Petroleum Council, *U.S. Petroleum Refining—Meeting Requirements for Cleaner Fuels and Refineries*, Volume I (Washington, DC, August 1993).

Petroleum product prices also include additional costs resulting from requirements for new fuels, including oxygenated and reformulated gasolines and low-sulfur diesel. These additional costs are determined in the representation of refinery operations by incorporating specifications and demands for these fuels. Demands for traditional, reformulated, oxygenated, and high-oxygen reformulated gasolines are disaggregated from composite gasoline consumption on the basis of market share assumptions for each Census Division. The expected oxygenated gasoline market shares assume wintertime participation of 39 carbon monoxide nonattainment areas and year-round participation of Minnesota beginning in 1995.

Starting in 1995, reformulated gasoline is assumed to be consumed in the nine serious ozone nonattainment areas required by CAAA90 and in areas in 12 states and the District of Columbia that had opted into the program as of January 1994.* Nonattainment areas in Wisconsin will join the program in June 1995 and are assumed to opt in beginning in 1996, along with Atlanta, Georgia. The state of Georgia, which had been considering joining the reformulated gasoline program, has recently adopted tighter restrictions on Reid vapor pressure (Rvp) in the Atlanta area in lieu of reformulated gasoline.

Reformulated gasoline reflects "Simple Model" standards between 1995 and 1997 and meets the "Complex Model" definition beginning in 1998 as required by the EPA. The reference case projections also reflect California's statewide requirement for severely reformulated gasoline beginning in 1996. In accordance with the Renewable Oxygenate Standard, renewable oxygenates, such as ethanol and ethyl tertiary butyl ether (ETBE), are blended into 15% of reformulated gasoline in 1995 and 30% starting in 1996. Throughout the forecast, traditional gasoline is blended according to 1990 baseline specifications, to reflect CAAA90 "antidumping" requirements aimed at preventing traditional gasoline from becoming more polluting.

On the basis of recent tax trend analyses, the reference case assumes that state taxes on gasoline, diesel, jet fuel, M85, and E85 will increase with inflation, while federal taxes remain at 1994 levels. This differs from previous forecasts, which assumed that both state and federal taxes would increase with inflation.

Coal Market Assumptions

Resource Base

Estimates of recoverable coal reserves are based on the EIA Demonstrated Reserve Base (DRB) of in-ground coal resources of the United States. Resource estimates from the DRB are correlated with coal quality data from other sources to create a Coal Reserves Data Base. Estimates are developed on a regionally disaggregated basis.

In certain coal-producing regions, the DRB estimates have been augmented by a portion of inferred resources. The extent of augmentation varies by state and coal type, based on the recency of DRB estimates and the amount of inferred coal that meets criteria related to seam thickness, depth, and overburden. The purpose of this change is to represent expected additions to demonstrated reserves that would occur in later years of the forecast. The effect of the change is to reduce somewhat minemouth and delivered prices in that period.

Productivity

Technological advances in the coal industry, such as continuous mining, contribute to increases in productivity, as measured in average tons of coal per miner per hour. Productivity improvements are assumed to continue, but to decline in magnitude over the forecast horizon. Different rates of improvement are assumed by region and by mine type, surface or deep. On a national basis, labor

*Required areas are Baltimore, Chicago, Hartford, Houston, Los Angeles, Milwaukee, New York City, Philadelphia, and San Diego. The 1995 opt-in areas are in the following states: Connecticut, Delaware, Kentucky, Maine, Massachusetts, Maryland, New Hampshire, New Jersey, New York, Rhode Island, Texas, Virginia, and the District of Columbia. Additional 1996 opt-ins are Atlanta (Georgia) and areas of Wisconsin.

productivity is assumed to improve at a rate of 3.9% per year, declining from an annual rate of 7.7% in 1993 to 2.2% in 2010.

Renewable Fuels Assumptions

Energy Policy Act of 1992

The Renewable Fuels Module incorporates the provisions of EPACT that support the development of renewable energy forms. EPACT provides a renewable electricity production credit of 1.5 cents per kilowatthour for electricity produced by wind, applied to plants that become operational between January 1, 1994, and June 30, 1999.* The credit extends for 10 years after the date of initial operation. EPACT also includes provisions that allow an investment tax credit of 10% for solar and geothermal technologies that generate electric power.† This credit is included as a 10% reduction to the capital costs in the Renewable Fuels Module.

State Mandates

The reference case includes estimates of mandated capacities (net summer capability) for renewables, as follows:

- Wind—California (U.S. supply only), 927 megawatts; Minnesota, 380 megawatts; New York, 6 megawatts
- Geothermal—California, 159 megawatts
- Biomass wood—Minnesota, 119 megawatts

Several energy supply actions under the CCAP encourage states to promote the demonstration and use of renewable energy systems.

Renewable Resources

The major source of renewable energy for electricity generation is hydroelectric power. Environmental and other restrictions are assumed to limit the growth of hydroelectric power, which grows slightly. The total resources for most other renewables are theoretically large—for example, the amount of sunlight. However, total resources are not always the relevant measure. Regional characterization is required in order to properly represent the resource. For example, although the capability to produce solar thermal energy is present in all regions of the United States, it is assumed that solar energy technologies will penetrate first in those regions where its economics are most favorable. Wind energy resource potential, although large, is constrained by land-use and environmental factors that result in the exclusion of some land areas within suitable wind classes. The geographic distribution of available wind resources is based on a resource assessment study by the Pacific Northwest Laboratory.[16] Geothermal energy is limited geographically to regions in the western United States with hydrothermal resources of hot water and steam. Capacity in 1992 totaled 2.9 gigawatts.[17] Although the potential for biomass is large, transportation costs limit the amount of the resource that is economically producible, since biomass fuels have a low Btu value per weight of fuel. Municipal solid waste resources are limited by the amount of the waste that is disposed of by other methods, such as recycling or landfills, and the impact of waste minimization as a strategy for managing the waste problem.

Nonelectric Renewable Energy

The forecast for wood consumption in the residential sector is based on EIA's Residential Energy Consumption Survey (RECS) and data from the *Characteristics of New Housing: 1992,* published by the Bureau of the Census.[18] The RECS data provide a benchmark for Btu of wood use in 1990.

*National Energy Policy Act of 1992, P.L. 102–486, Title XIX, Section 1914.
†*Ibid.,* Section 1916.

The census data are used to develop the forecasts of new housing units utilizing wood. Wood consumption is then computed by multiplying the number of homes that use wood for main and secondary space heating by the amount of wood used. Ground-source (geothermal) heat pump consumption is also based on the latest RECS and census data; however, the measure of geothermal energy consumption is represented by the amount of primary energy displaced by using a geothermal heat pump in place of an electric resistance furnace. Solar thermal consumption for water heating is also represented by displaced primary energy relative to electric water heaters.

Exogenous projections of active and passive solar technologies and geothermal heat pumps in the commercial sector are based on projections from the National Renewable Energy Laboratory.[19] Industrial use of renewable energy is primarily the use of wood and wood byproducts in the paper and lumber industries as well as a small amount of hydropower for electricity generation.

Appendix B. Total Energy Supply and Disposition Summary, Annual Energy Outlook 1996 Reference Case

(Quadrillion Btu per Year, Unless Otherwise Noted)

	Reference Case						
Supply, Disposition, and Prices	1993	1994	1995	2000	2005	2010	2015
Production							
Crude Oil and Lease Condensate	14.50	14.10	13.84	11.96	11.10	11.51	12.30
Natural Gas Plant Liquids	2.49	2.47	2.43	2.54	2.74	2.97	3.20
Dry Natural Gas	18.97	19.41	19.47	20.24	21.89	23.52	25.72
Coal	20.23	22.01	21.89	22.59	23.95	24.94	26.14
Nuclear Power	6.52	6.84	7.02	7.09	6.93	6.52	4.63
Renewable Energy[1]	6.40	6.26	6.68	7.01	7.38	7.78	8.51
Other[2]	0.54	0.99	0.45	0.44	0.50	0.50	0.58
Total	69.64	72.08	71.79	71.87	74.50	77.79	81.08
Imports							
Crude Oil[3]	14.76	15.33	16.33	19.67	21.13	21.05	20.87
Petroleum Products[4]	3.73	3.92	3.53	4.03	5.23	6.16	6.40
Natural Gas	2.39	2.60	2.78	3.30	3.48	3.84	4.42
Other Imports[5]	0.50	0.67	0.62	0.69	0.59	0.61	0.56
Total	21.38	22.53	23.25	27.70	30.43	31.65	32.25
Exports							
Petroleum[6]	2.12	2.00	2.28	2.12	1.88	1.77	1.91
Natural Gas	0.15	0.16	0.17	0.27	0.31	0.32	0.33
Coal	1.96	1.88	1.88	2.16	2.37	2.66	3.14
Total	4.23	4.04	4.33	4.56	4.56	4.74	5.38
Discrepancy[7]	0.59	−1.43	−0.12	0.06	0.01	0.07	N/A
Consumption							
Petroleum Products[8]	33.83	34.56	34.88	36.88	39.12	40.68	41.69
Natural Gas	20.80	21.36	21.95	23.00	24.79	26.76	29.52
Coal	19.55	19.65	19.66	20.67	21.83	22.55	23.27
Nuclear Power	6.52	6.84	7.02	7.09	6.93	6.52	4.63
Renewable Energy[1]	6.40	6.27	6.68	7.01	7.38	7.78	8.52
Other[9]	0.29	0.46	0.40	0.42	0.33	0.40	0.40
Total	87.38	89.14	90.60	95.07	100.3	104.69	108.02
Net Imports — Petroleum	16.37	17.5	17.58	21.58	24.48	25.44	25.36
Prices (1994 dollars per unit)							
World Oil Price (dollars per barrel)[10]	16.48	15.52	16.81	19.27	21.86	23.70	25.43
Gas Wellhead Price (dollars per Mcf)[11]	2.09	1.88	1.60	1.89	1.99	2.15	2.57
Coal Minemouth Price (dollars per ton)	19.85	19.41	18.54	17.44	17.68	17.43	17.39

(Quadrillion Btu per Year, Unless Otherwise Noted) (continued)

[1] Includes utility and nonutility electricity from hydroelectric, wood and wood waste, municipal solid waste, other biomass, wind, photovoltaic and solar thermal sources; non-electric energy from renewable sources, such as active and passive solar systems, and wood; and both the ethanol and gasoline components of E85, but not the ethanol components of blends less than 85 percent. Excludes nonmarketed renewable energy. See Table 18 for selected nonmarketed residential and commercial renewable energy.

[2] Includes liquid hydrogen, methanol, supplemental natural gas, and some domestic inputs to refineries.

[3] Includes imports of crude oil for the Strategic Petroleum Reserve.

[4] Includes imports of finished petroleum products, imports of unfinished oils, alcohols, ethers, and blending components.

[5] Includes coal, coal coke (net), and electricity (net).

[6] Includes crude oil and petroleum products.

[7] Balancing item. Includes unaccounted for supply, losses, gains, and net storage withdraws.

[8] Includes natural gas plant liquids, crude oil consumed as a fuel, and nonpetroleum based liquids for blending, such as ethanol.

[9] Includes net electricity imports, methanol, and liquid hydrogen.

[10] Average refiner acquisition cost for imported crude oil.

[11] Represents lower 48 onshore and offshore supplies.

Btu = British thermal unit.

Mcf = Thousand cubic feet.

N/A = Not applicable.

Note: Totals may not equal sum of components due to independent rounding. Figures for 1993 and 1994 may differ from published data due to internal conversion factors.

Sources: 1993 natural gas values: Energy Information Administration (EIA), *Natural Gas Annual 1993,* DOE/EIA-0131(93) (Washington, D.C., October 1994). 1993 coal minemouth prices: EIA, *Coal Industry Annual 1993,* DOE/EIA-0584(93) (Washington, D.C., December 1994). Other 1993 values: EIA, *Annual Energy Review 1994,* DOE/EIA-0384(94) (Washington, D.C., July 1995). 1994 natural gas supply and price: EIA, *Natural Gas Monthly,* DOE/EIA-0130(95/6) (Washington, D.C., June 1995). 1994 coal minemouth prices: EIA, *Coal Industry Annual 1994,* DOE/EIA-0584(94) (Washington, D.C., December 1995). 1994 coal production and exports: EIA, *Monthly Energy Review,* DOE/EIA-0035(95/08) (Washington, D.C., August 1995). Other 1994 values: EIA, *Annual Energy Review 1994,* DOE/EIA-0384(94) (Washington, D.C., July 1995). **Projections:** EIA, AEO96 National Energy Modeling Systems, run AEO96B.D101995C.

8

Electric Utility Restructuring

Lawrence J. Hill
Oak Ridge National Laboratory

Introduction

Current public policy promotes "restructuring" of industries historically subject to federal and state regulation, relying more on market forces and less on regulation. This restructuring has drastically changed the amount of economic activity subject to economic regulation. Currently, less than 10% of total U.S. output is subject to economic regulation, in comparison with nearly 20% two decades ago. The deregulated industries include transportation (airlines, railroads, and trucking), communications, finance, and energy.

The energy industries are among the most recent to undergo change. In the natural gas industry, the Federal Energy Regulatory Commission's (FERC's) Order 636 in April 1992 was the culmination of a series of orders designed to restructure that industry. As a result, natural gas pipelines are now required to separate gas sales from transportation, allowing open access to pipelines for both gas producers and users.

In the electric industry, public policy at the federal level promotes entry of firms into wholesale electric markets, allowing these firms to compete with vertically integrated, monopoly electric utilities in the business of generating electricity. An early piece of legislation was the Public Utility Regulatory Policies Act (PURPA) of 1978, which required electric utilities to purchase power from certain "qualifying," nonutility power generators. The Energy Policy Act of 1992 (EPACT) amended the Public Utility Holding Company Act of 1935 (PUHCA), allowing the creation of "exempt wholesale generators" to compete in the electric-generating business and also allowing electric utilities to construct power plants anywhere where they choose. EPACT also requires that electric transmission-line owners allow access to their lines without discrimination. FERC implemented this portion of EPACT, requiring that electric transmission-line owners post tariffs that allow open access to their lines.[1]

The objective of these federal initiatives is to increase the economic efficiency of electric markets, ultimately lowering the cost of producing electricity and the price that customers pay for it. Electric-utility mergers are a by-product of these federal initiatives. In expectation of increased competition, many electric utilities are merging to improve their combined efficiency and lower their production

0-8493-2514-5/96/$0.00+$.50

costs to compete with lower-cost electricity producers. Recently announced mergers include those of Northern States Power Company and Wisconsin Energy Corporation, forming Primergy Corporation; Southwestern Public Service Company and Public Service Company of Colorado; and Baltimore Gas and Electric Company and Potomac Electric Power Company. More announcements are expected in the future.

The restructuring of the electric industry is different from that of other industries restructured in the past because state-level policy-making will be important in determining the type of restructuring to take place. At the federal level, FERC has the authority to shape wholesale electric markets, but it is policy-making by state regulators and legislators that will ultimately shape the structure of electric industries in states and regions. It is very likely that the market structures of electric industries in the 48 contiguous states will not be the same. If competitive markets are the policy objectives for state policymakers, they will have to manage the transition from the current structure to the one they are trying to attain. There are many considerations in managing this transition, including the strandable benefits and commitments that may arise as a result of restructuring and the need to consider changes in economic regulation to allow electric utilities to compete more effectively with nonutility entrants into electric markets.

In this chapter, we examine the current structure of the electric industry, its regulation, some future possibilities for the structure of the industry, and some important issues that state policymakers must address in the restructuring process. In Section 8.1 we examine the current structure of the industry and its regulation. In Section 8.2 we look at federal policy-making toward electric markets in more detail. Then, we discuss some future wholesale and retail market possibilities in Sections 8.3 and 8.4. We conclude by discussing some important state public-policy issues, including strandable benefits, strandable commitments, and changes in economic regulation.

8.1 Organization and Regulation of the Electric Industry

The United States has one of the most decentralized and diverse electric industries in the world. It consists of electric utilities, nonutility electricity generators (NUGs), and net imports of electricity from Canada and Mexico. Currently, electric utilities dominate the industry. Of the 803,143 megawatts (MW) of total capacity in the industry at the end of 1993, electric utilities owned 745,009 MW, or 93%. NUGs owned the remaining 58,134 MW.

However, new increments of generating capacity are being constructed more and more by NUGs. In 1994, for example, NUGs accounted for three-fifths of the new capacity in the industry. This share is likely to increase in the future as a result of federal policy initiatives (discussed later).

There were 3,212 electric utilities in the electric industry at the end of 1993 (Table 8.1). These utilities were (1) investor-owned with shares of stock traded on stock exchanges, (2) publicly owned by subnational governments (i.e., state-, county-, or city-owned), (3) owned by cooperatives in rural areas, or (4) publicly owned by the federal government. In Table 8.1, we show the total amount of generating capacity in the industry at the end of 1993 by ownership type. By only including the capacity of electric utilities, we exclude the 58,134 MW of capacity owned by NUGs.

The electric industry in most states is dominated by the 254 investor-owned electric utilities (IOUs). The data in Table 8.1 show that IOUs account for more than 75% of the total capacity owned by electric utilities. Federally owned utilities, including the Tennessee Valley Authority, own less than 10%. The IOUs generally are vertically integrated, franchise-monopoly electric utilities regulated by state regulatory commissions in return for an exclusive franchise to sell electricity in prespecified geographical areas. State regulators approve the rates and quality and terms of electric service in regulatory proceedings for the IOUs.

Most of the IOUs are vertically integrated. In Figure 8.1, we characterize a vertically integrated electric utility. (The arrows in Figure 8.1 indicate the flow of power in an electric system.) The utility generates its own electric power, dispatches and transmits it, and distributes it to retail customers. For the most part, technology shaped this vertical structure. Historically, there have

TABLE 8.1 U.S. Electric Utilities — Number and Capacity by Ownership Type, 1993

		Capacity	
Ownership	Number	Amount	%
Investor-Owned	254	576,005	77.3
State/Municipal	2,007	76,824	10.3
Federal	10	66,129	8.9
Cooperatives	941	26,052	3.5
Total	3,212	745,010	100.0

Source: Edison Electric Institute (1994) and Energy Information Administration (1995).

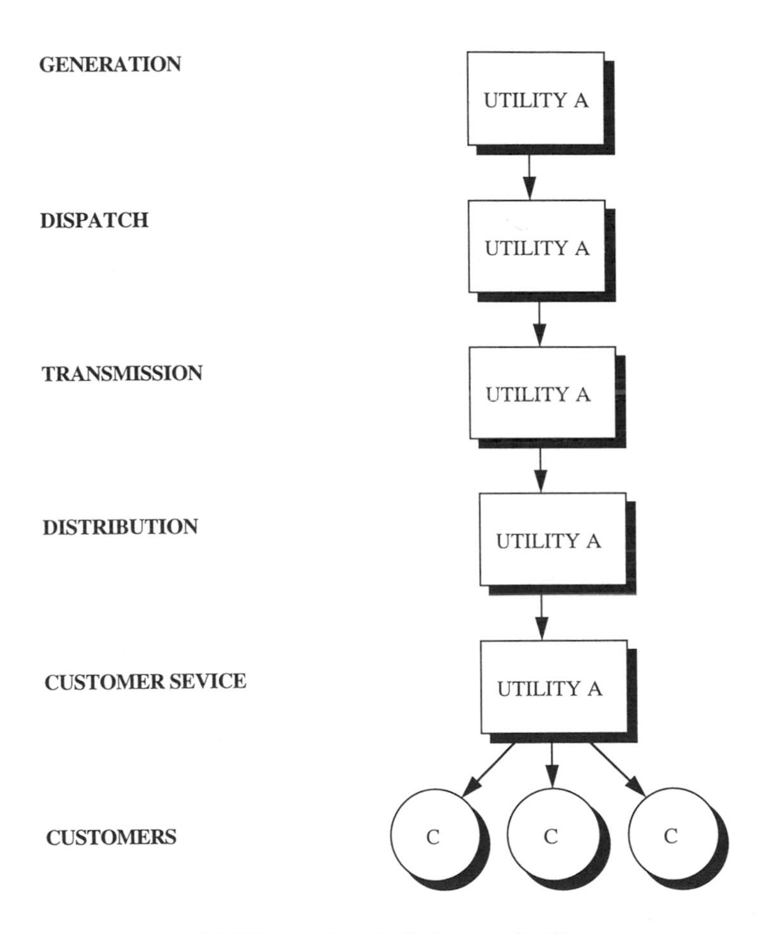

FIGURE 8.1 A vertically integrated utility.

been economies of scale and scope associated with electric-power delivery, so it was uneconomic for policymakers to utilities to compete with each other.

The vertically integrated utility in Figure 8.1 is a theoretical construct, showing the major vertical stages of the industry. In real-world settings, the overwhelming majority of electric utilities are not totally integrated from generation through transmission and distribution. Most utilities purchase some of their power requirements on spot markets or through long-term contracts and, therefore, are only partially integrated.

In Figure 8.2, we show a typical investor-owned electric utility (Utility A) engaging in typical market transactions. (The broken lines indicate contracts between different entities.) Utility A is partially integrated through five primary utility functions—generation, power dispatch, transmission, distribution, and customer service. In its customer-service function, it runs demand-side management programs.

Utility A engages in three types of market transactions on a regular basis. First, *coordination transactions* are short term in nature. Utility A buys energy in bulk power markets under bilateral contracts. The typical bulk power market involves the sale of surplus energy: utilities with surpluses selling to capacity-deficit utilities at different periods of the day, week, or season.

Second, *requirements transactions* are related to the capacity needs of some utilities, such as municipally owned ones ("muny" in Figure 8.2), that do not own all of their capacity requirements. An investor-owned utility, for example, may serve all or a portion of the capacity requirements of the municipal utility. Although we show Utility A selling capacity requirements to a municipally owned utility in Figure 8.2, the purchasing utility could be any ownership form requiring generating capacity such as other investor-owned utilities or rural electric cooperatives.

Third, *wholesale wheeling transactions* are becoming more commonplace after passage of EPACT. In this type of transaction, a power generator other than a franchise-monopoly utility (a NUG in Figure 8.2) contracts to sell power to a wholesale power supplier (Utility B) using Utility A's transmission lines.

The economic regulation of the electric industry characterized in Figure 8.2 reflects the industry's diversity and decentralization. In Figure 8.3, we show the regulation of the industry. FERC regulates the wholesale power market: (1) coordination transactions, (2) requirements transactions, and (3) wholesale wheeling transactions. Regulatory jurisdiction over proposed retail wheeling transactions is contentious. Although FERC claims jurisdiction over these types of transactions because they involve electricity transmission to a large extent and FERC regulates transmission lines, some states also claim jurisdiction because retail-wheeling transactions are sales made to retail customers, transactions over which state regulatory commissions have jurisdiction. In Figure 8.3, retail sales are made by Utilities A and B and the "MUNY".

In contrast to the regulation of IOUs, few state regulatory commissions have jurisdiction over the ratemaking of state and municipal electric systems. Approximately 50% of municipal utilities are under the direct control of their governing legislative body, and the other half are under the jurisdiction of an independent power board. For municipals under the jurisdiction of a board, 25% are controlled by elected boards and the remainder are controlled by boards that are appointed by the mayor, by the city's governing board, or by the mayor with approval of the city governing board.

Although not shown in Figure 8.3 for reasons of clarity, rural electric cooperatives also make retail sales. Regulatory authority over rural electric cooperatives is at three levels. The Rural Utilities Service (RUS), formerly the Rural Electrification Administration, has overall responsibility to ensure the financial soundness of the cooperatives. At the state level, less than one-half of state regulatory commissions have economic jurisdiction over cooperatives. For those cooperatives not under state regulatory jurisdiction, regulation is based on an RUS policy regarding financial soundness.

Also for reasons of clarity, we do not show federal power projects in Figure 8.3. Federal power operations are divided into five power marketing agencies and the Tennessee Valley Authority (TVA). The TVA is a government corporation and is more autonomous than the power marketing agencies. The TVA's rate level and rate structure are set internally, outside the jurisdiction of federal and state regulatory bodies. The rate structures and rate levels of the five power marketing agencies, however, are reviewed and approved by FERC. State commissions do not have jurisdiction over the operation of these federal projects.

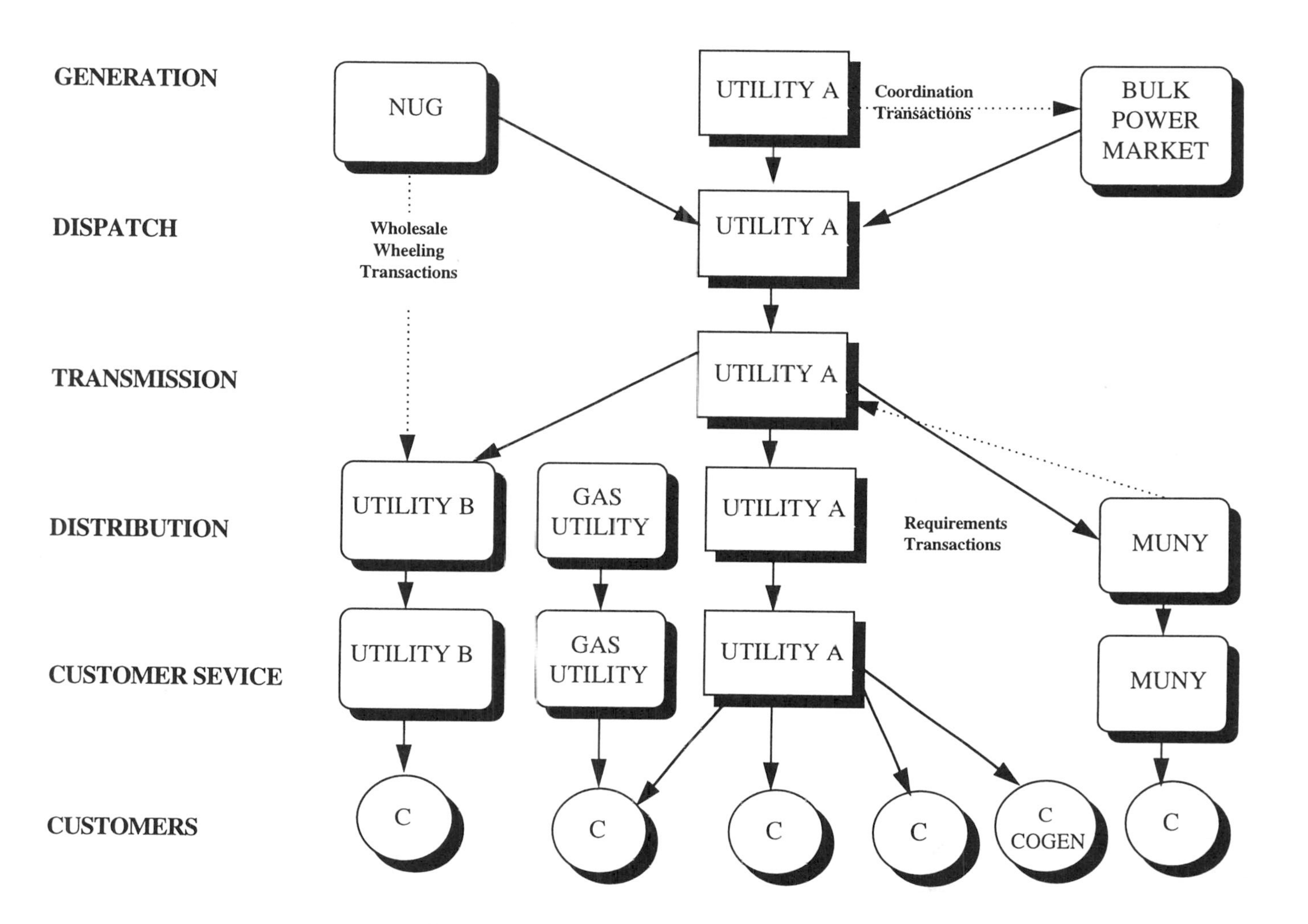

FIGURE 8.2 A typical investor-owned electric utility.

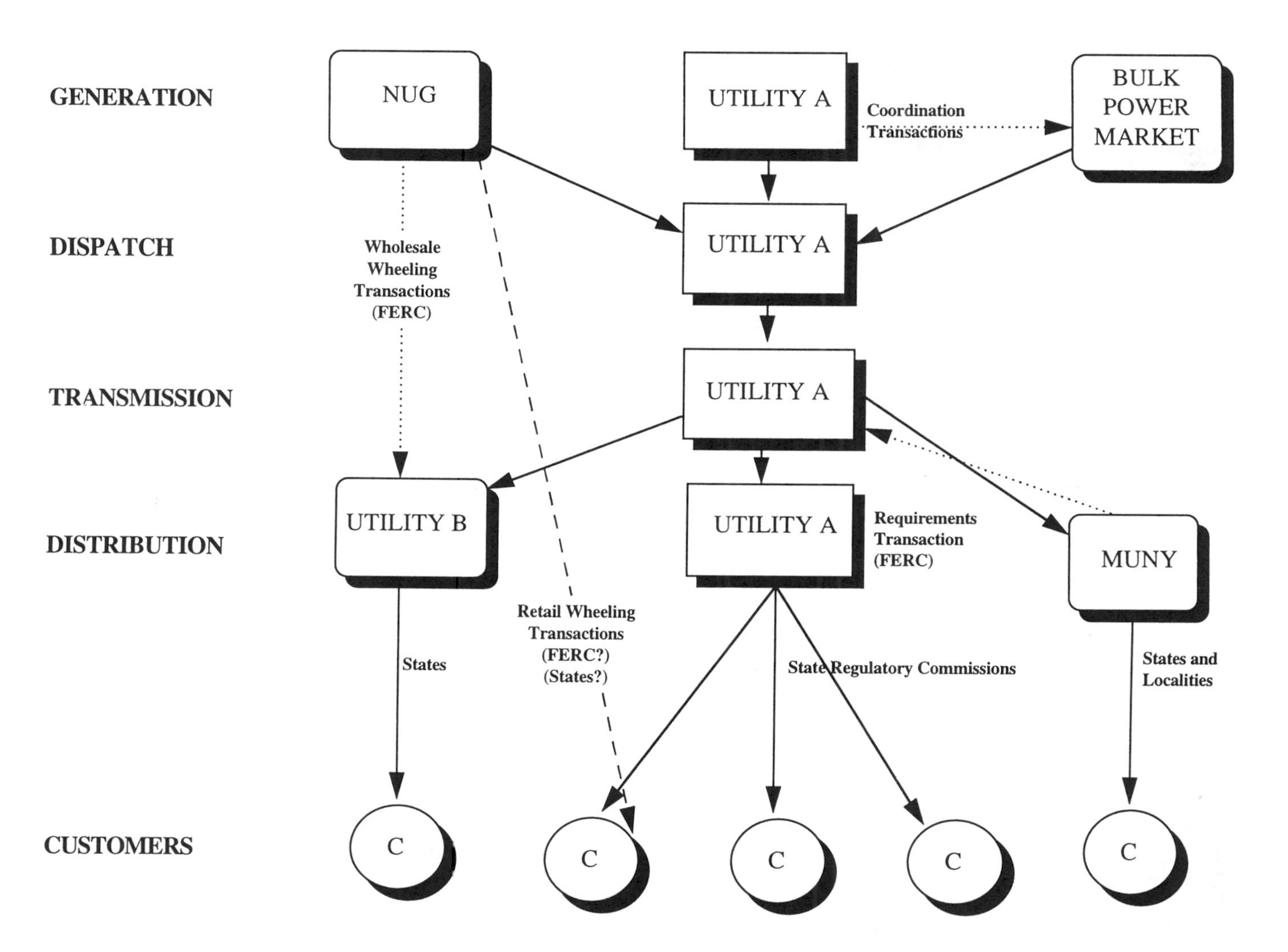

FIGURE 8.3 Regulatory jurisdiction in the electric industry.

8.2 Federal Initiatives to Promote Competition

To varying degrees, investor-owned utilities face both wholesale and retail competition today, and the degree of wholesale competition is likely to increase in the future as a result of federal initiatives to promote competition. These initiatives include federal legislation and implementing rulings by FERC. We review these initiatives.

The Public Utility Holding Company Act (PUHCA) was enacted in 1935 in response to the consolidation movement in the electric power industry. In this movement, electric utilities were controlled more and more by holding companies. At the time of PUHCA's enactment, for example, three holding companies accounted for nearly one-half of electricity production in the United States. Because these companies operated across state lines, state regulatory commissions were powerless to control them. PUCHA was enacted to correct this jurisdictional problem and curb the alleged market abuses of holding companies.

Under PUHCA, the Securities and Exchange Commission confined holding companies to a single geographical area. Under this "one-area" rule, holding companies could control just one electric system. However, a holding company could control more than one system if three conditions were met: (1) control of multiple systems was required to preserve economies of operation; (2) the electric systems were located in one state, in adjoining states, or in contiguous foreign countries; and (3) the resulting holding company was not so large as to impair efficient management or regulation.

The result of these restrictions was to confine utility ownership of generating plants to specific areas. A utility on the East Coast, for example, could not construct and operate a power plant on the West Coast. Therefore, competition in the wholesale power market was limited. Under PUHCA's "death clause," holding companies that did not operate as an integrated power system were abolished.

The Energy Policy Act of 1992 (EPACT) amended PUHCA with respect to utility ownership of generating plants. EPACT allows formation of another type of nonutility generator, the exempt wholesale generator (EWG). EWGs can be created by any type of business firm to compete with existing electric utilities. In fact, as a result of EPACT electric utilities themselves can now construct and operate generating plants anywhere in the United States and foreign countries.

As a result of EPACT, the structure of the electric industry is likely to change significantly over the next two decades. EPACT facilitates competition in the electric generating business, allowing business firms—including subsidiaries of electric utilities—to build power plants outside of the traditional regulated environment and sell power to local distribution companies. Additionally, EPACT requires electric transmission-line owners to allow open access to their lines. This provision requires owners to provide nondiscriminatory use of the lines by nonowners.

FERC is in the process of implementing the transmission-access provision of EPACT. First, to further promote competition in wholesale electricity markets in the aftermath of EPACT, FERC adopted a new policy in July 1993 encouraging the creation of regional transmission groups (RTGs) to improve regional transmission-line coordination and planning. This policy reflects FERC's belief that transmission markets will develop regionally over multiple states and that transmission arrangements freely entered into by affected parties are superior to arrangements imposed on participants from Washington.

Second, FERC's 1996 rulemaking on open access of transmission lines moves these lines toward common-carrier status. In this rulemaking, FERC requires transmission-line owners to file open-access tariffs and prevents them from self-dealing or using the lines to their own advantage.

In Figure 8.4, we show how these initiatives may relieve a bottleneck for potential entrants into the electric-generating business. Prior to these federal initiatives, Utility A was the dominant producer of electricity and was vertically integrated downstream into distribution. EPACT was enacted to increase the number of entrants to power generation. In Figure 8.4, these are the NUGs

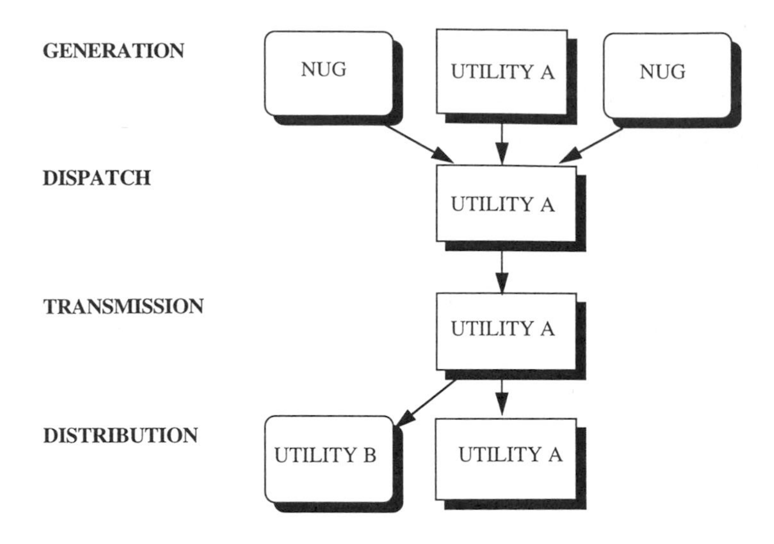

FIGURE 8.4 Vertical control in wholesale electric markets.

participating in the generating market. A potential barrier to these NUGs entering the market is access to the transmission system. If NUGs cannot get access to Utility A's transmission lines, they are not able to compete with Utility A for Utility B's business. FERC's objective in its open-access rulemaking is to eliminate the bottleneck by requiring Utility A to open up its transmission lines to all existing and potential generators (the NUGs in Figure 8.4) without discrimination. Open-access tariffs are filed in Washington, and every existing or potential generator—including the transmission line's owners—would have comparable access to the lines under the tariffs.

8.3 Some Wholesale Market Possibilities

These federal initiatives will directly affect the organization of wholesale electricity markets throughout the country. We discuss some possibilities in this section. These initiatives may also indirectly affect the organization of retail markets in many parts of the country, prompting state policymakers to change retail electricity markets in response to the changes in wholesale markets. We discuss retail markets in the following section.

FERC's requirement of arm's-length transactions between an electric utility's generating plants and its transmission lines may lead to radical restructuring of wholesale power markets. FERC requires "functional unbundling" of generation and transmission by electric utilities—that is, an arm's-length separation of generation and transmission. To comply with this requirement, utilities may reorganize into functions. For example, many utilities may form generation, transmission, and distribution subsidiaries, each under separate management, to ensure arm's-length transactions. Each of the three subsidiaries operates independently, but all are under the same ownership and central management.

Depending on business conditions, many utilities may go a step further and divest themselves of a portion of their assets. Also known as "corporate unbundling" or "structural unbundling," divestiture means a change in ownership of a portion of a utility's assets. For example, a utility may sell all or a portion of its generating assets to another company.

We illustrate the difference between functional and structural unbundling in Figure 8.5, where we show some wholesale market possibilities with power pools. A power pool refers to two or more utilities or NUGs that allow their electric-generating units to be dispatched by a single entity so that power can be produced more cheaply. The utilities in a power pool retain control over much of their own planning, including decisions on constructing new power plants.

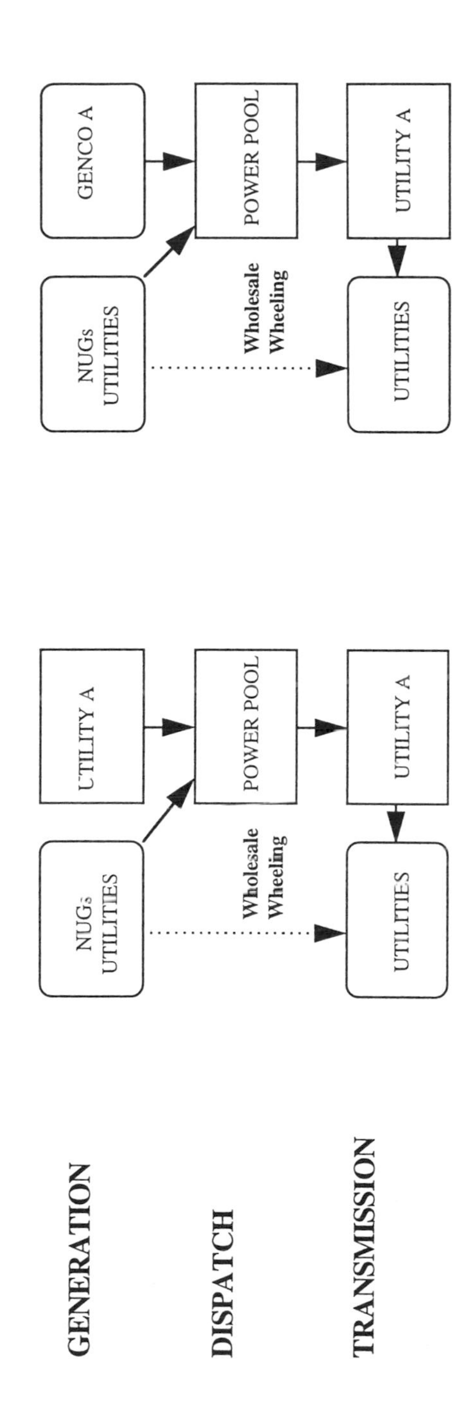

FIGURE 8.5 Wholesale power pooling arrangements.

In the market structure on the left in Figure 8.5, Utility A functionally unbundles and joins a power pool such as that in New England [the New England Power Pool (NEPOOL)].* In this case, the only major change for Utility A is more economic dispatch of its power. Utility A still owns both its generating and transmission assets.

As shown on the right of Figure 8.5, Utility A could spin off or divest itself of its generating assets into a new company, Genco A, that would join a regional pool. In this case, Utility A retains ownership of its transmission lines but is no longer involved in the business of generating electricity.

A more radical solution to exploit the economies of dispatching the power plants of multiple utilities is to create a "poolco". A poolco is an independent, regulated monopoly in business solely to promote a competitive electric market. It matches the power offered by electricity generators to the electricity demand requirements of the territory it serves. A poolco is a more radical solution than a power pool because power plants are not automatically dispatched and used in a poolco as in a power pool. Like other utilities and NUGs in the region, a typical utility would "offer" power to the poolco at fixed intervals during each day (e.g., in half-hour increments). If the prices at which it offers to supply its power are competitive with other offers to the pool, the poolco purchases Utility A's power for resale to wholesale purchasers.

We show three possible poolco arrangements in Figure 8.6, emphasizing Utility A and excluding its power-supply competitors. (The broken-lined boxes indicate control by the poolco.) In the diagram on the left, Utility A unbundles functionally, retaining its generating and transmission assets in separate subsidiaries, but participates in the poolco for dispatching purposes. It also runs its own transmission lines. In the middle arrangement, Utility A divests its generating assets to form a new company (Genco A), retaining ownership of its transmission and distribution assets, but consigns operation of its transmission system to the poolco. The poolco has exclusive control over the use of the lines. In this situation, Utility A is compensated for the use of its transmission lines, but does not control them. Further, Genco A offers power to the poolco. In the third arrangement on the right, Utility A divests itself of its generating (Genco A), transmission (Transco A), and distribution functions (Disco A, not shown), separating into three companies. Again, the poolco controls dispatch and transmission in this arrangement, even though it does not own the transmission lines.

What do electric markets look like after these changes? The number of possibilities is large. We show two in Figures 8.7 and 8.8. In Figure 8.7, Utility A divests itself of its generating assets, retaining the "wires" portion of the former vertically integrated company. The divested generating assets, organized as Genco A, join a power pool with other utilities and NUGs in the surrounding area. The electric-generating market is competitive—and not regulated. The wires portion of the industry—transmission and distribution—is still a natural monopoly and regulated as it was in the past. However, instead of cost-of-service regulation, a performance-based approach is used for transmission and distribution. (Performance-based regulation will be discussed later.)

In Figure 8.8, Utility A divests itself of its transmission assets rather than its generating assets. Utility A now consists of generating and distribution assets. A poolco is formed in the region, and Utility A offers capacity and energy to this pool. The poolco controls—but does not own—Utility A's divested transmission assets. The transmission system is a separate company (Transco A). The poolco compensates the owners of Transco A for use of their lines in the wholesale market.

Although not explicitly shown in Figure 8.8, the NUGs and utilities participating in the poolco have two opportunities to hedge against the risk of price fluctuations. First, they can contract for differences between the price of power charged by the poolco and the price in a separate contract made between a generator and purchaser of power. For example, Utility B and a NUG can enter into a fixed-price contract for power (the dotted line in Figure 8.8). Any time the poolco price (i.e., the market price) deviates from the contract price, Utility A and the NUG settle the difference

*NEPOOL consists of the generating plants owned by multiple electric utilities in the Northeast, voluntarily pooling the dispatch of these plants to promote more efficient electricity generation and lower electricity prices than would be obtained if each utility dispatched their plants individually.

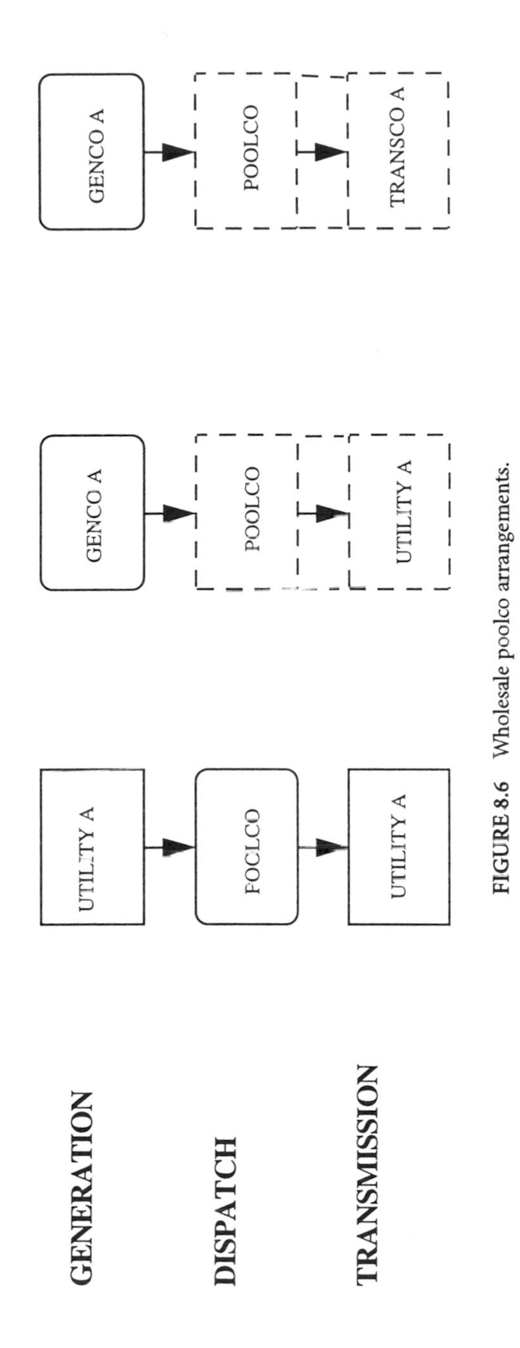

FIGURE 8.6 Wholesale poolco arrangements.

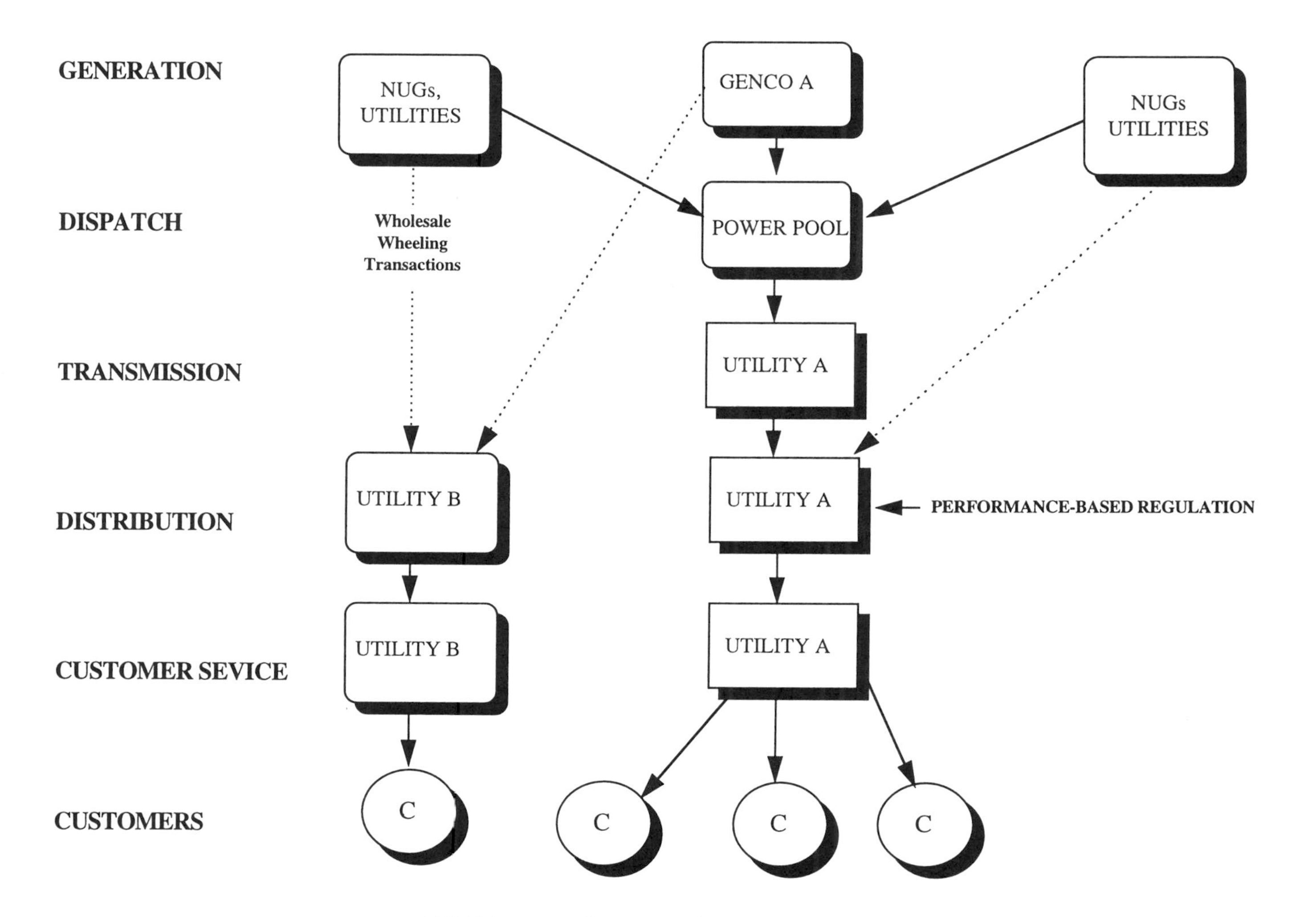

FIGURE 8.7 An electricity market with a power pool.

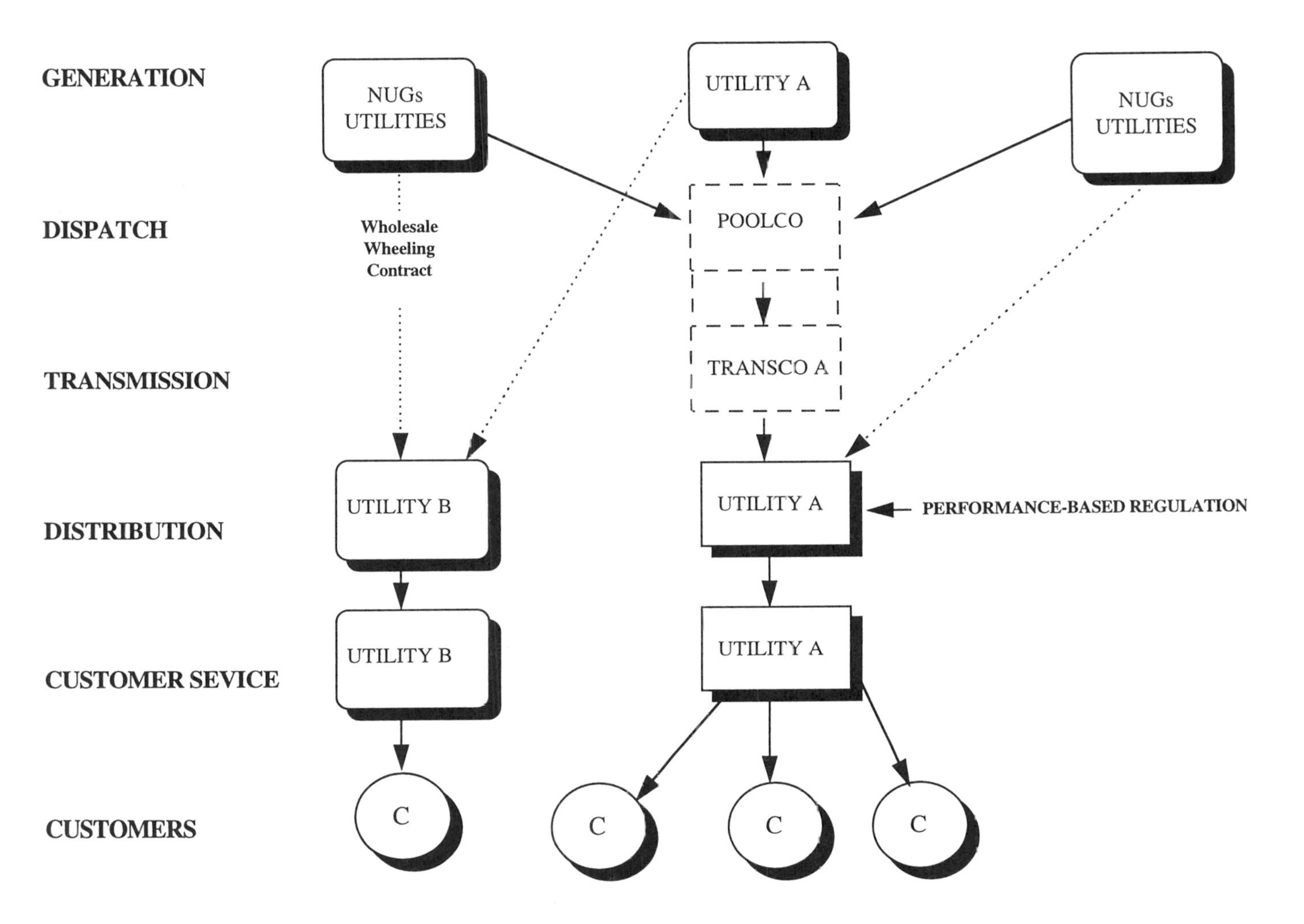

FIGURE 8.8 An electricity market with a poolco.

between themselves. Second, a futures market for electricity will exist. As in other futures markets, a buyer can specify a given quantity of electricity to be delivered to a given location at a specified time, typically between 1 and 18 months in the future.

Figures 8.7 and 8.8 show two possible market structures in a restructured electric industry. The examples are not exhaustive of the range of possibilities. All of these structures have one feature in common, however. The former franchise-monopoly utility is not the sole supplier of electricity. It competes with others.

8.4 Retail Markets

Wholesale competition in regional electric markets is inevitable. Although state policymakers can play a significant role in shaping the type of wholesale electric market (e.g., power pool, poolco, or neither), the institutions to promote wholesale competition were created at the federal level. The situation is much different in retail markets, where state policymakers have considerable discretion in shaping their structure.

There are many possibilities for increasing the number of participants in retail markets for state policymakers to consider. The possibilities include municipalizing the distribution systems of local utilities, encouraging the formation of electricity brokers and energy service companies, and engaging in bilateral contracting. We illustrate these possibilities in Figure 8.9. To facilitate the presentation, we do not include a detailed characterization of a wholesale market in Figure 8.9.

In Figure 8.9, Utility A still distributes electricity, performs customer-service functions, and owns the wires, but faces much more competition for its heretofore captive customers. Some customers of Utility A may contract with a separate electricity generator for their power requirements. This retail wheeling transaction is illustrated on the far right of the figure. The contract is made with a generator other than Utility A, and the power is transmitted and distributed over Utility A's lines.

A Consumer Service District option, originally proposed in Massachusetts, is illustrated on the left. In this form of municipalization, local governments contract for electric power supplies directly from power generators, leasing the facilities owned by the monopoly utility.

The other transactions illustrated in Figure 8.9 are various combinations of brokers, energy service companies, and Utility A. In one type of transaction, a broker—or "aggregator"—serves as a conduit for a group of customers to get direct access to power generation and the economies associated with large-scale purchases. Energy service companies (ESCOs) may perform a similar function by aggregating the demands of smaller-volume users and also performing some electricity-service functions such as running demand-side management programs.

Finally, in addition to these new mechanisms to provide electricity to end-use customers, Utility A still faces more traditional types of competition for its formerly captive customers. In Figure 8.9, gas utilities may be able to compete effectively for the energy service needs of some customer groups, depending on the relative prices of competing fuels and electricity. Depending on the price of electricity, some of Utility A's customers may opt to cogenerate their own electricity.

8.5 Some Public-Policy Issues

An emerging competitive marketplace will result in other public-policy issues that must be addressed by state policymakers. For example, as the price of electricity is determined more by markets than by regulatory proceedings, less attention will be given to energy efficiency programs, renewable energy, and environmental quality. State policymakers must address these benefits stranded as a result of more competition. Also, if competition emerges and electricity prices decline as a result, some of the past investments made by electric utilities may no longer be economic. These "strandable commitments" must be addressed by policymakers because utilities made these commitments with regulatory approval and may face bankruptcy if not compensated. Finally, the

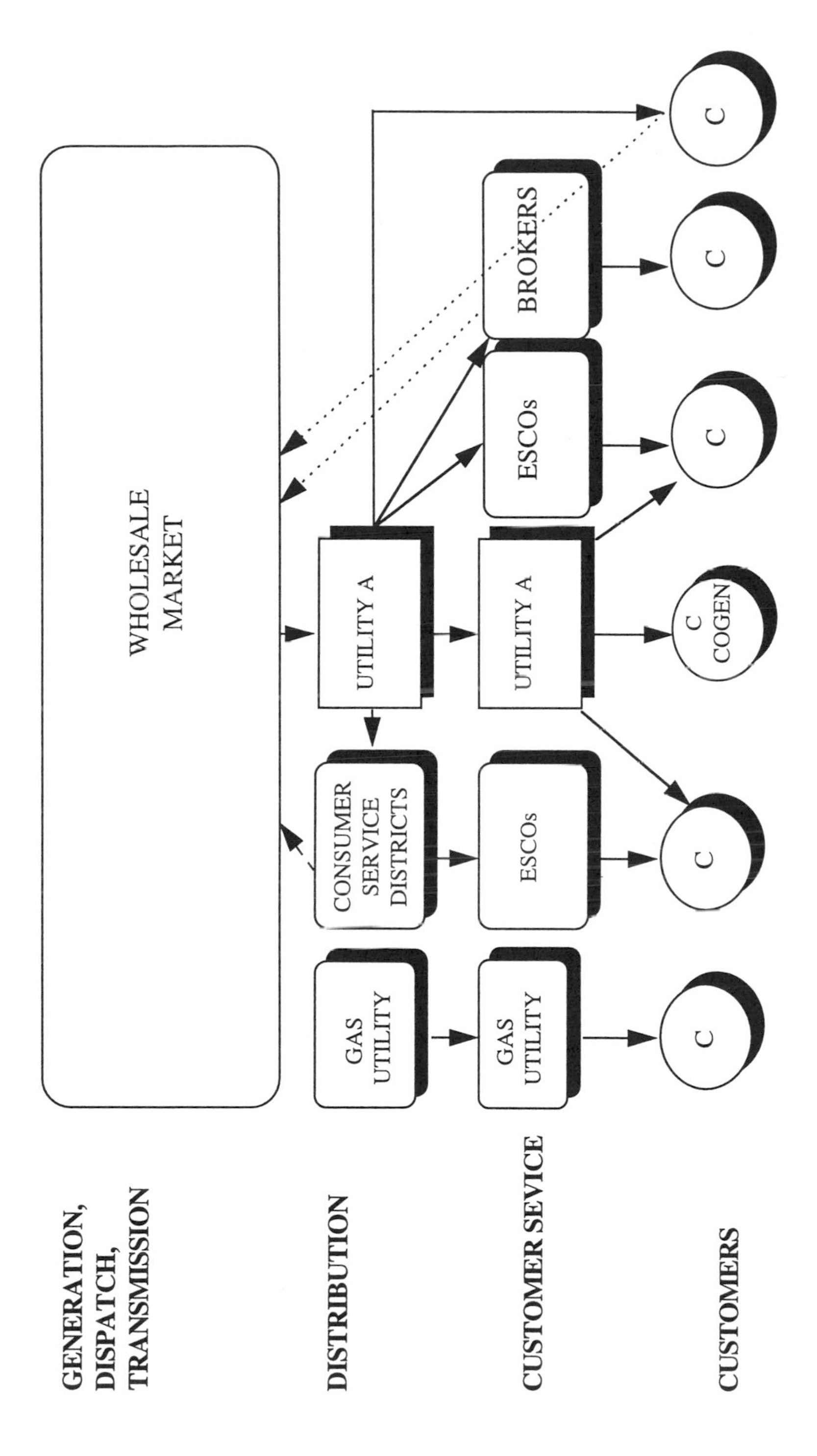

FIGURE 8.9 Examples of retail market transactions.

way in which most utilities are currently regulated may not be the most effective way if competition increases in the industry. We discuss each of these policy issues in turn.

Strandable Benefits

Strandable benefits are social "goods" supported by existing state electric-industry regulation that may be jeopardized in restructured electric markets. Looking across states, these benefits include such activities as energy efficiency programs, environmental protection, resource diversity through use of renewable energy forms, and low-income energy-assistance programs.

The reason that these "benefits" could become stranded is that planning and decision-making by electric utilities could change drastically in restructured electric markets. In many current electric planning processes, the decision to purchase power from neighboring utilities or NUGs, construct conventional or renewable generating plants, or provide other social benefits such as running energy conservation programs is made through an integrated resource planning process. This decision-making process is likely to change in the future in many jurisdictions because the price of electricity charged to customers will no longer be determined by this process. Rather, it will be determined more and more in markets. Renewables, DSM programs, and environmental quality will no longer be a part of the process of determining resource choice. If these public policy goals are still to be pursued, other mechanisms may have to be developed to implement them.

There are mechanisms other than the IRP process to provide these benefits in competitive markets. Although some of them could be implemented through the private sector, the most encouraging are those implemented through the public sector. For example, state policymakers could offer incentives for the adoption of renewable energy forms to encourage fuel diversity and promote environmental quality. The incentives could be in the form of reduced taxes, favorable rates, or subsidies for financing costs. States can go further than providing incentives. New York, for example, requires a 300 MW set-aside of various renewables for a demonstration program. In Minnesota, electric utilities that operate nuclear power plants must construct 225 MW of wind energy by the end of 1998 and an additional 200 MW by the end of 2002. Biomass plants with 50 and 75 MW of capacity must also be constructed by these respective dates. At the federal level, emissions taxes are efficient mechanisms to promote environmental quality. However, the political feasibility of enacting such taxes is questionable.

Strandable Commitments

As competition emerges in the electric industry, some of the past financial commitments made by electric utilities—and approved by their regulatory commissions—may not be competitive in a marketplace with lower electricity prices. Although these "strandable commitments" take several forms, the majority of them are likely to be investments in physical assets, especially high-cost electric-generating plants. The important state policymaking issue is to decide who should pay for strandable commitments.

The amount of strandable commitments varies by utility, state, and region, depending on (1) the degree of competition in regional utility markets, (2) the types of power plants used to generate electricity, (3) the cost of these plants, (4) the amount and cost of purchased fuel and power, and (5) policies established by state commissions. Because future electricity prices are not known with certainty and there are different ways to estimate the amount of strandable commitments, estimates of the total amount in the industry vary from as low as $20 billion to as high as $500 billion. The range between $50 and $150 billion is the most plausible.

Strandable commitments could arise at either the wholesale or retail level. Figure 8.10 illustrates the two possibilities. In the wholesale wheeling transaction on the left, assume that a NUG sells power to Utility B at 3¢/kWh using Utility A's transmission lines. Assuming that Utility B originally purchased that electricity from Utility A for 6¢/kWh prior to the NUG entering the market, the amount of Utility A's capacity used to service Utility B is now "stranded," or uneconomic. Further, if the 3¢/kWh

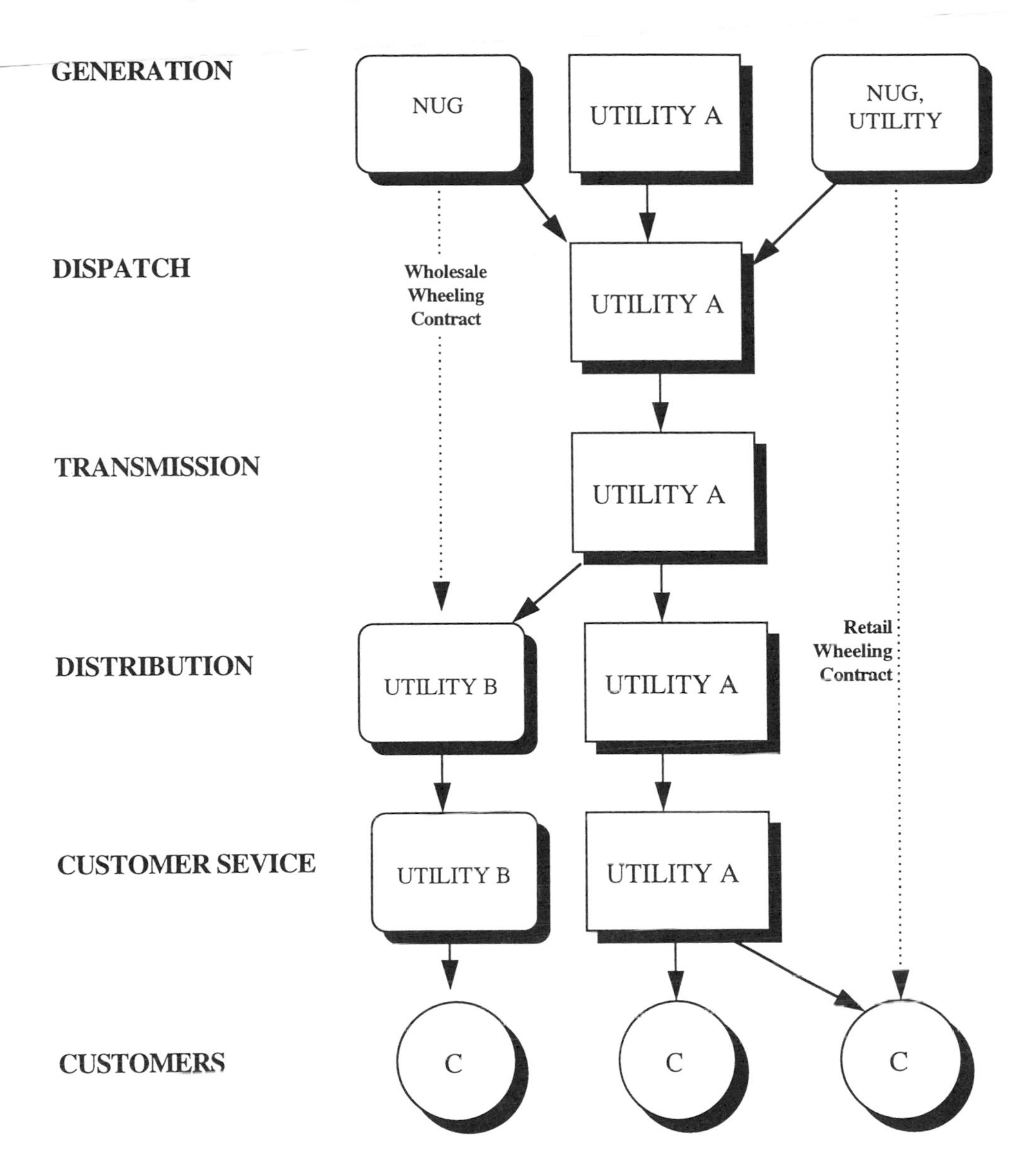

FIGURE 8.10 Wholesale and retail wheeling transactions.

that Utility B pays for electricity becomes the "market price" of electricity, the amount of stranded commitments is based on the difference between Utility A's cost (6¢/kWh) and the market price.

Depending on state policymaking, electric utilities may also face more competition at the retail level. An example of retail competition is provided on the right side of Figure 8.10. A similar type of stranded commitment could arise from retail transactions. In this case, the NUG contracts directly with an end-use customer using Utility A's transmission and distribution system. Utility A's generating capacity, built to satisfy the needs of customers that are now purchasing their electricity from the NUG, is stranded. If the NUG sells the power to Utility A's former retail customer at 4¢/kWh and this is the market price of electricity, Utility A's stranded commitments are calculated as the difference between its total cost of serving the customer and the market price of 4¢/kWh.

Both FERC and state regulatory commissions have jurisdiction over strandable commitments. FERC has jurisdiction over transmission rates and has already ruled on several cases involving wholesale stranded commitments. Further, in its Rulemaking on Stranded Assets, FERC restated its authority. It argued that it has jurisdiction over any assets stranded as a result of wholesale

TABLE 8.2 Estimated Strandable Commitments — 114 Investor-Owned Electric Utilities, By NERC Region

NERC Region (States)	Stranded Amount ($\$10^6$)	% of Equity
Northeast (CT, ME, MA, NY, NH, RI, VT)	29,544	163
West (AZ, CA, CO, ID, MT, NV, NM, ND, SD, TX, UT, WA, WY)	28,863	109
East Central (IN, KY, MI, OH, PA)	20,164	90
Southwest (AR, KS, LA, MS, MO, OK)	14,384	118
Mid-Atlantic (DE, MD, NJ, PA, VA)	13,303	67
Southeast (AL, FL, GA, MS, NC, SC)	11,261	43
Texas	10,307	89
Mid-America (IL, MO, WI)	5,984	48
Mid-Continent (IA, MN, ND, SD)	632	14

Source: Moody's Investor Service, "Stranded Costs will Threaten Credit Quality of U.S. Electrics," New York, August 1995.

transactions (the transaction on the left in Figure 8.10), leaving states to deal with strandable commitments resulting from retail transactions. However, in dollar terms, the vast majority of strandable commitments are likely to arise from retail transactions.

In Table 8.2, we provide an estimate by Moody's Investor Service of the amount of assets that are potentially strandable for 114 investor-owned utilities. The estimates take into consideration both investment in generating plant and deferred assets over a 10-year period. The results are organized by the nine North American Electric Reliability Council (NERC) regions. The regions do not exactly correspond to state boundaries.

In terms of dollar values and percentage of equity, the Northeast has the largest amount of strandable commitments. Three utilities in the Northeast have strandable commitments in excess of three times their equity. Of the 114 utilities in the study, 34 have strandable commitments in excess of their equity. Fifty-seven have strandable commitments in excess of 50% of their equity.

There is no accepted formula for allocating stranded commitments to affected parties. Therefore, political compromise will determine the share paid by affected parties. This sharing is crucial because the stock and bond ratings of electric utilities are likely to be affected if their stockholders are required to absorb a significant share of any stranded costs.

Besides the shareholders of a utility, other groups that may absorb portions of strandable commitments include

- The customers departing from a utility's system who cause the commitments to be stranded
- Remaining or "core" customers on a utility's system
- New nonutility sources of electric power for customers exiting the system (e.g., nonutility electricity generators)
- All taxpayers through new taxes

Irrespective of any proposed sharing of strandable commitments, it is likely that they will still affect the credit ratings of electric utilities. Moody's concludes in part:

> we are skeptical that regulators will allow utilities to recover all of their stranded costs. Furthermore, even if regulators are accommodating, economic and competitive realities will probably preclude full recovery. The utilities are likely to absorb a significant portion of these costs, which will exert pressure on their credit ratings.

Thus far, strandable commitments have not generally been addressed at the state level. At the federal level, FERC recommended that business firms causing stranded assets in wholesale transactions compensate the affected utility. Using this rule for the wholesale transaction in Figure 8.10, Utility B must compensate Utility A for the amount of stranded commitments.

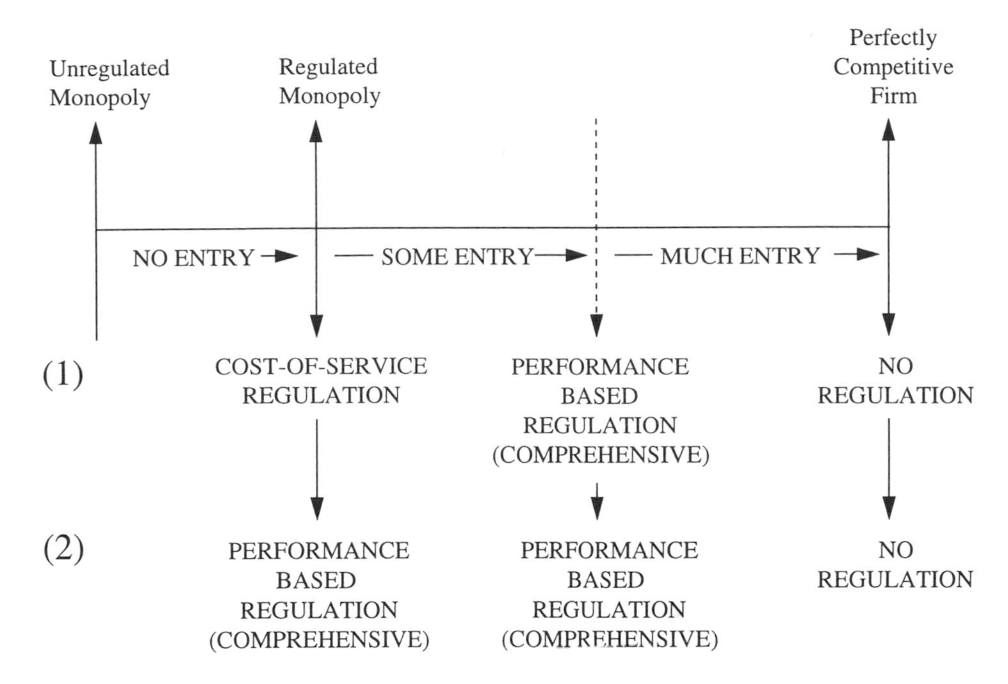

FIGURE 8.11 Market stages for electric utilities.

Changes in Regulation

At the top of Figure 8.11, we show a range of public-policy approaches toward electric utilities, using the degree of entry into electricity markets as the organizing metric. At one extreme, electric utilities could be left unregulated. At the other, they could participate in competitive markets.

As shown in Scenario 1 of Figure 8.11, state regulators have generally used cost-of-service regulation (COSR) for utilities that do not face competition. Concisely, COSR allows electric utilities to recover all prudently incurred operating costs plus a fair return on their investment in electric plant when setting rates. Three features of cost-of-service regulation make it attractive for regulators. First, rates are based on accounting costs, information that utilities gather as a matter of course. Second, because utilities are allowed to recover their costs, they are likely to maintain and even improve service quality. Third, the regulatory process allows a fair return on the franchise monopoly's investment in plant and equipment, neither confiscating it over time nor allowing an excess return.

Then why replace COSR with another approach based on performance when more entry is allowed in electric markets as shown in Scenario 1 of Figure 8.11? The problems with the cost-of-service approach are the incentives and opportunities given to electric utilities. They are not the type of incentives that characterize a competitive market, and that balance the additional risk of operating in a competitive market. Under COSR, incentives *to operate* "efficiently" in the sense of an unregulated firm are minimal.

Incentives and opportunities *to invest* wisely in utility plant and equipment are also distorted under COSR. In contrast to a competitive market, COSR does not generally reward utilities with a higher return for making an especially good investment in plant and equipment or penalize them for making an especially bad investment. The return that a utility earns on a highly successful investment is generally the same as its return on a less successful one. The opportunity to prosper—or, alternatively, to go bankrupt—is generally not part of COSR. Performance-based regulation (PBR) is viewed as a mechanism to substitute for COSR and eliminate these distorted incentives.

The degree of competition in electric markets will not be uniform across the country. Despite federal initiatives, state policymakers have a lot of discretion to shape their markets. In some areas of the country, electric utilities have been facing competition for years, and state policymakers have taken actions to allow utilities to better compete in the emerging competitive marketplace. In other parts of the country, utilities face very little competition, and are not likely to face much more in the future.

The speed at which markets become competitive will also vary. Again, policymaking at the state level will determine the speed of the transition from a regulated-monopoly market to a competitive one.

For policymakers in states where utilities are facing—or will soon face—competition, two time frames must be considered. First, there is the transitional phase from a regulated-monopoly electric market to a competitive one. By severing the relationship between costs and prices and providing more pricing flexibility, PBR is one way that regulators can give utilities the ability to compete effectively. We show this case in Scenario 1 of Figure 8.11. PBR is a step across the spectrum from COSR. At some point when entry into the industry is sufficiently free that electric utilities will suffer financially if they cannot compete with new entrants on an equal basis, PBR should be considered as the regulatory model for electric utilities. Many would argue that PBR should be introduced even in regulated-monopoly electric markets (Scenario 2 in Figure 8.4) because COSR does not generally provide incentives for utilities to reduce costs or innovate.

The second important time frame for policymakers is at a point in time when regulation is no longer needed because markets are competitive. In a competitive market, there are no barriers to exit and entry. Firms have the incentive to lower costs and become more efficient because lower costs translate into higher profits—at least in the short run. In the long run, firms have the incentive to innovate, lower costs, and thereby gain a competitive advantage over their rivals. Competitive firms can immediately raise or lower their prices consistent with market conditions. Regulation is replaced by the market.

However, only the generating portion of the industry will be competitive in most electricity markets. Regulation will still be required for residual natural monopolies: the transmission and distribution functions. The challenge for regulators at this point in time is to develop a regulatory model that accommodates the specific features of the "wires" portion of the industry. Is PBR the proper approach for transmission and distribution only? If so, how must PBR be adapted to meet the special circumstances of transmission and distribution? If not, will COSR provide the proper incentives to reduce costs?

Reference

1. Federal Energy Regulatory Commission, *Final Rule: Promoting Wholesale Competition Through Open Access Non-Discriminatory Transmission Services and Recovery of Stranded Costs by Public Utilities and Transmitting Utilities,* Docket Nos. RM95-8-000 and RM94-7-001, Washington, DC, April 24, 1996.

Section II

Energy Conservation

9

Thermal Energy Conservation in Buildings

Max Sherman

David Jump
Lawrence Berkeley National Laboratory

9.1 The Indoor Environment

The building sector is an important part of the energy picture. In the United States about 40% of all U.S. energy expenses are attributable to buildings. While buildings may consume about one-third of the fuel resources in the country, they consume over 65% of the electricity.

As mentioned in Chapter 1, residential buildings accounted for 17 quads of energy at a cost of 104 billion dollars, while commercial buildings (offices, stores, schools, and hospitals) accounted for about 13 quads and 68 billion dollars of energy consumption in 1989. A breakdown by sector reveals where most of the energy is used in buildings and provides background for improving building energy efficiency. Figure 1.7 shows the breakdown of energy consumption by end use fuel type for residential buildings, while Figure 1.8 provides the same kind of information for the 13 quads used in commercial buildings. There are great opportunities for reducing energy consumption in buildings and Figure 1.6 shows the Office of Technology Assessment's estimate of energy growth in "a business-as-usual scenario," and in a scenario where all energy-efficient technologies with a positive net present value are implemented. The OTA's estimate suggests that by 2010, more than 10 quads of energy could be saved with improved energy efficiency.

The energy consumed in residential and commercial buildings provides many services, including weather protection, thermal comfort, communications, facilities for daily living, esthetics, a healthy work environment, and so on. Since in a modern society, people spend the vast majority of their time inside buildings, the quality of the indoor (or built) environment is important to their comfort, and good thermal performance of buildings is important for energy efficiency as well as worker productivity in commercial buildings.

0-8493-2514-5/96/$0.00+$.50

This chapter first discusses issues that influence the quality of the indoor environment in terms of the occupants' comfort, health, and productivity. Most all of the facets of providing an acceptable indoor environment involve energy-intensive services. Factors influencing space conditioning energy needs, such as the building envelope's thermal properties and ventilation requirements, are then discussed. Next, the means of providing space conditioning to the building interior is examined. This section focuses on the thermal distribution system; the actual heating and cooling equipment is covered in Chapter 12C. Where applicable, opportunities for energy conservation through improved efficiency are demonstrated. Finally, some building simulation tools are reviewed to provide the reader with the means necessary to complete the complicated analysis of building energy consumption.

Thermal Comfort

The most fundamental building service is to protect the occupants of the building from the outdoor environment. The building structure keeps the wind and rain out, but energy is required to provide an acceptable thermal environment for the occupants. The amount of energy required depends in part on the optimum comfort conditions required by the occupants activity levels and clothing.

ASHRAE Standard 55 describes thermal comfort as "that condition of mind which expresses satisfaction with the thermal environment." The thermal environment variables that influence thermal comfort are the air temperature, mean radiant temperature, relative air velocity, and water vapor pressure in ambient air. Two other important variables are the person's activity level and clothing.

A heat balance of the body provides a basis for the development of an equation for thermal comfort. The equation provides a means of determining all possible acceptable ranges of the environment variables for given values of activity level and clothing. Tabulated values of activity level (met) and clothing (clo) as well as diagrams charting the solution to the comfort equation are found in Fanger (1970) and in Chapter 8 of the ASHRAE Handbook of Fundamentals (1993). It is of interest to note that Fanger found no significant differences in thermal comfort conditions between sexes, age groups or geographic location.

While the comfort equation shows how the environmental variables are combined to obtain optimal thermal comfort, it does not indicate the preferred comfort conditions of persons. This is the function of the predicted mean vote (PMV), which provides information on the degree of discomfort experienced in a thermal environment. The PMV is a seven-point scale from –3 (cold) to +3 (hot), with 0 as the neutral level. Models for predicting the PMV are also given in the referenced sources just noted. A last indicator of the degree of satisfaction with the thermal environment is the predicted percentage of dissatisfied people (PPD). As shown in Figure 9.1 (Fanger, 1970), the PPD has a minimum of 5% for the PMV = 0. This shows that it is impossible to satisfy a large group of people in the same conditions. The PPD rises sharply as the PMV moves away from zero in both directions.

As an example, suppose it is desired to find the degree of satisfaction of persons in an overly vented, windowless meeting room. The people are clothed in normal business suits (clo = 1.0), the relative air velocity is 0.5 m/s, and the relative humidity is 40%.

For a windowless meeting room the air temperature is set equal to the mean radiant temperature. From Figure 9.2(c) (Fanger, 1970), for sedentary persons, $T_{air} = T_{MR} = 24.8$°C. From Table 9.1 (Fanger 1970), the PMV is approximately 0.08, which is only slightly warmer than the optimum satisfaction level. To find out what fraction of people would express dissatisfaction with the environment, the PPD is determined to be 22%. Thus approximately 1 in 5 people in the room may report discomfort with the thermal conditions.

Thermal comfort is one of a number of design constraints when determining space conditioning energy needs. Other constraints are summarized in the next sections.

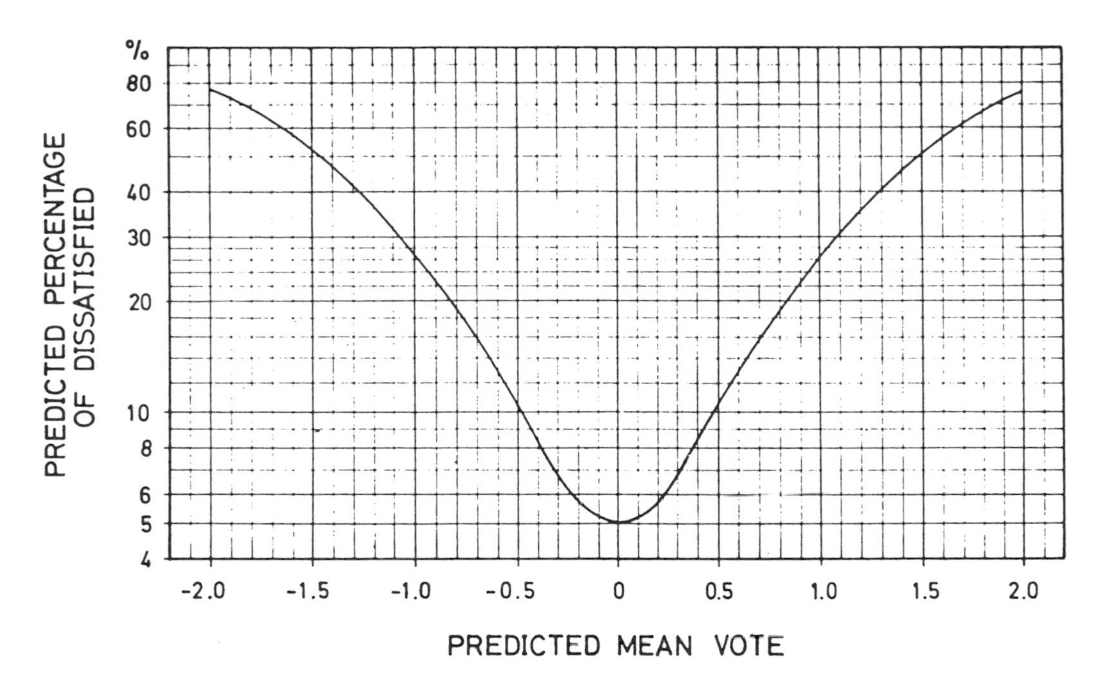

FIGURE 9.1 Predicted Percentage of Dissatisfied people (PPD) as a function of Predicted Mean Vote (PMV). *Source:* Fanger, P.O., *Thermal Comfort Analysis and Applications in Environmental Engineering,* McGraw-Hill, New York, 1970. With permission.

Lighting and Visual Comfort

After the HVAC system, lighting accounts for the largest energy consumption of the building services. Building occupants require sufficient light levels to go about their normal activities. Just as the building envelope protects the occupants from weather, it also reduces the amount of usable sunlight. The desire to continue activities at any time or place necessitates the use of electric lighting.

In the perimeter of buildings and throughout small buildings natural lighting or daylighting can provide a high-quality visual environment. Care must be taken in the design phase to control the admission of daylight as it varies over the day and throughout seasons. This is done primarily with the use of exterior overhangs, fins, awnings, blinds, and interior shades, drapes, or blinds. In the core areas of buildings, electric lighting must be continuously provided to meet the occupants' lighting requirements.

Achieving visual comfort requires more than providing average light levels. Glare from high-intensity sources, poor color rendition, or flickering can all cause discomfort or reduce visual performance.

Indoor Air Quality

Good indoor air quality may be defined as air that is free of pollutants that cause irritation, discomfort or ill health to occupants, or premature degradation of the building materials, paintings, and furnishings or equipment. Thermal conditions and relative humidity also impact the perception of air quality in addition to their effects on thermal comfort. Focus on indoor air quality issues increased as reduced-ventilation energy-saving strategies, and consequently increased pollution levels, were introduced. A poor indoor environment can manifest itself as a sick building in which some occupants experience mild illness symptoms during periods of occupancy. More serious pollutant problems may result in long-term or permanent ill-health effects.

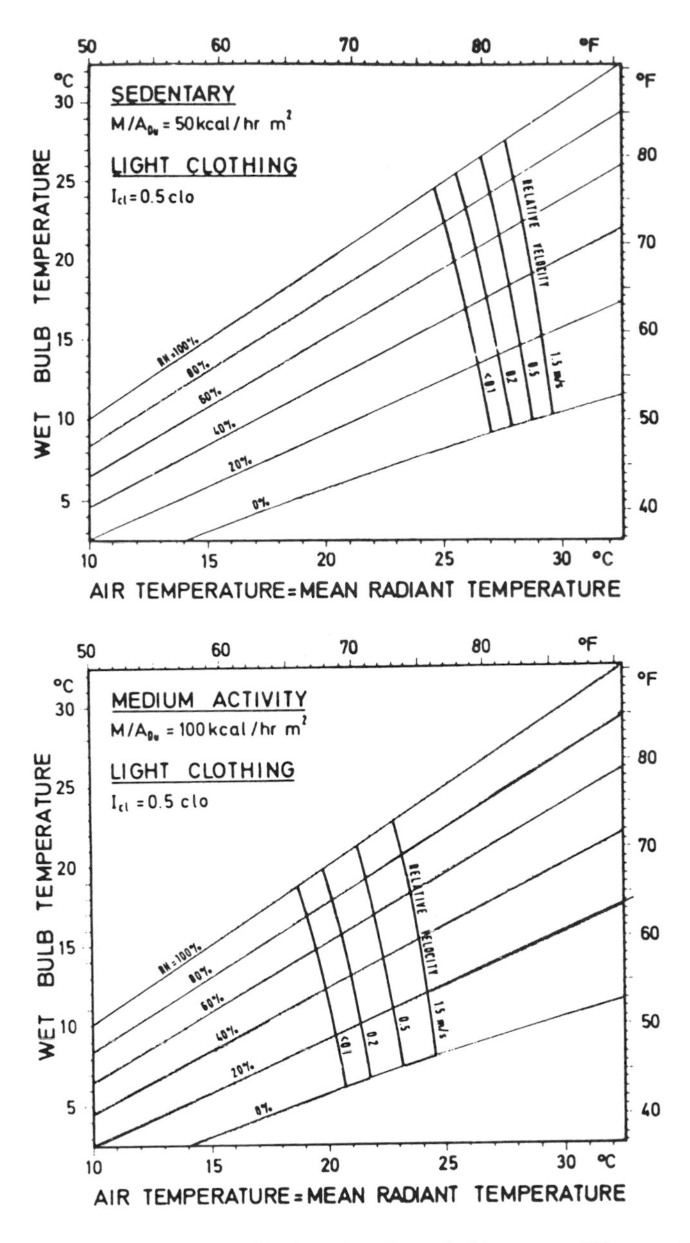

FIGURE 9.2 Comfort lines for persons with light and medium clothing at two different activity levels. *Source:* Fanger, P.O., *Thermal Comfort Analysis and Applications in Environmental Engineering,* McGraw-Hill, New York, 1970. With permission.

An almost limitless number of pollutants may be present in a space, of which many are at immeasurably low concentrations and have largely unknown toxicological effects. Sources of indoor air pollutants in the home and in offices and their typical concentrations are given in Table 9.2. The task of identifying and assessing the risk of individual pollutants has become a major research activity in the past 20 years. Some pollutants can be tolerated at low concentrations, while irritation and odor often provide an early warning of deteriorating conditions. Health-related air quality standards are typically based on risk assessment and are specified in terms of a maximum permitted exposure, which is determined by exposure time and pollutant concentration. Higher concentrations of pollutants are normally permitted for shorter term exposures.

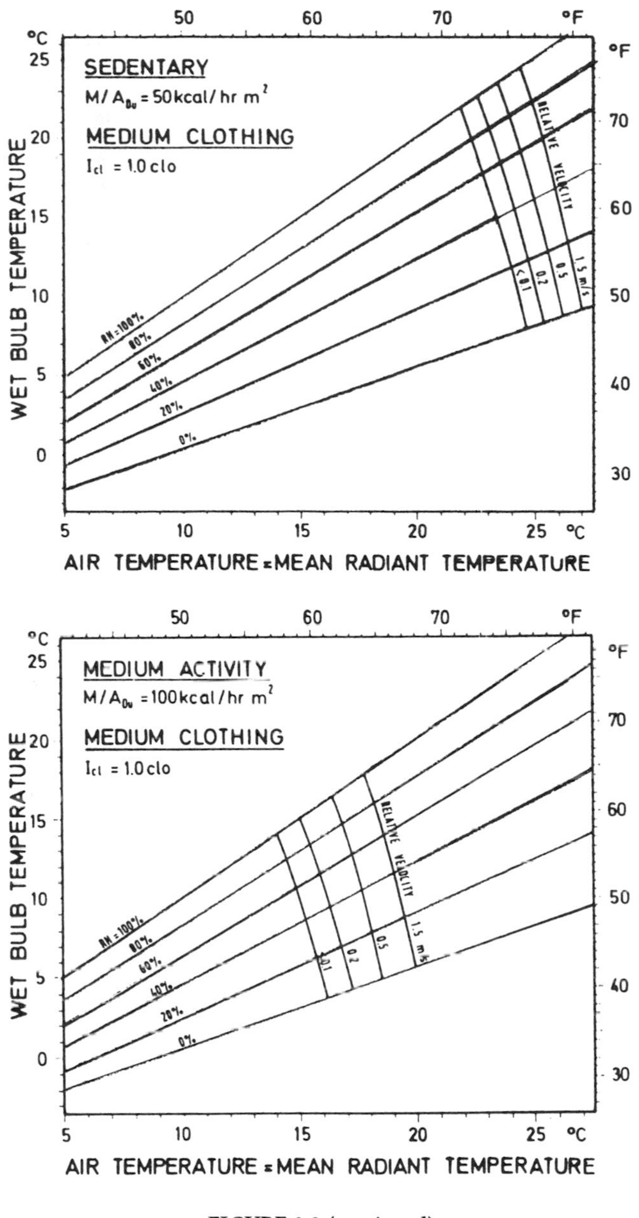

FIGURE 9.2 (continued)

Air quality needs for comfort are highly subjective and dependent on circumstances. Some occupations allow higher exposures than would be allowed for the home or office. Health-related air quality standards are normally set at minimum safety requirements and may not necessarily provide for adequate comfort or energy efficiency at work or in the home.

Pollution-free environments are a practical impossibility. Optimum indoor air quality relies on an integrated approach to managing exposures by the removal and control of pollutants and ventilating the occupied space. It is often useful to differentiate between unavoidable pollutants (such as human bioeffluents) over which little source control is possible, and avoidable pollutants (such as emissions of volatile organic compounds) for which control is possible. Whole-building ventilation usually provides an effective measure to deal with the unavoidable emissions, but source control is the preferred and sometimes only practical method to address avoidable pollutant sources. Examples of source control are given in Table 9.3.

TABLE 9.1 Predicted Mean Vote

Activity Level kcal/m² hr	Clothing clo	Ambient Temp. °C	Relative Velocity (m/s) <0.10	0.10	0.15	0.20	0.30	0.40	0.50	1.00	1.50
50	0.50	23.	−1.10	−1.10	−1.33	−1.51	−1.78	−1.99	−2.16		
		24.	−0.72	−0.74	−0.95	−1.11	−1.36	−1.55	−1.70	−2.22	
		25.	−0.34	−0.38	−0.56	−0.71	−0.94	−1.11	−1.25	−1.71	−1.99
		26.	0.04	−0.01	−0.18	−0.31	−0.51	−0.66	−0.79	−1.19	−1.44
		27.	0.42	0.35	0.20	0.09	−0.08	−0.22	−0.33	−0.68	−0.90
		28.	0.80	0.72	0.59	0.49	0.34	0.23	0.14	−0.17	−0.36
		29.	1.17	1.08	0.98	0.90	0.77	0.68	0.60	0.34	0.19
		30.	1.54	1.45	1.37	1.30	1.20	1.13	1.06	0.86	0.73
50	1.00	20.	−0.85	−0.87	−1.02	−1.13	−1.29	−1.41	−1.51	−1.81	−1.98
		21.	−0.57	−0.60	−0.74	−0.84	−0.99	−1.11	−1.19	−1.47	−1.63
		22.	−0.30	−0.33	−0.46	−0.55	−0.69	−0.80	−0.88	−1.13	−1.28
		23.	−0.02	−0.07	−0.18	−0.27	−0.39	−0.49	−0.56	−0.79	−0.93
		24.	0.26	0.20	0.10	0.02	−0.09	−0.18	−0.25	−0.46	−0.58
		25.	0.53	0.48	0.38	0.31	0.21	0.13	0.07	−0.12	−0.23
		26.	0.81	0.75	0.66	0.60	0.51	0.44	0.39	0.22	0.13
		27.	1.08	1.02	0.95	0.89	0.81	0.75	0.71	0.56	0.48
100	0.50	14		−1.08	−1.16	−1.31	−1.53	−1.71	−1.85	−2.32	
		16.		−0.69	−0.79	−0.92	−1.12	−1.27	−1.40	−1.82	−2.07
		18.		−0.31	−0.41	−0.53	−0.70	−0.84	−0.95	−1.31	−1.54
		20.		0.07	−0.04	−0.14	−0.29	−0.40	−0.50	−0.81	−1.00
		22.		0.46	0.35	0.27	0.15	0.05	−0.03	−0.29	−0.45
		24.		0.83	0.75	0.68	0.58	0.50	0.44	0.23	0.10
		26.		1.21	1.15	1.10	1.02	0.96	0.91	0.75	0.65
		28.		1.59	1.55	1.51	1.46	1.42	1.38	1.27	1.21
100	1.00	8.		−0.95	−1.02	−1.11	−1.26	−1.36	−1.45	−1.71	−1.86
		10.		−0.68	−0.75	−0.84	−0.97	−1.07	−1.15	−1.38	−1.52
		12.		−0.41	−0.48	−0.56	−0.68	−0.77	−0.84	−1.05	−1.18
		14.		−0.13	−0.21	−0.28	−0.39	−0.47	−0.53	−0.72	−0.83
		16.		0.14	0.06	0.00	−0.10	−0.16	−0.22	−0.39	−0.49
		18.		0.41	0.34	0.28	0.20	0.14	0.09	−0.06	−0.14
		20.		0.68	0.61	0.57	0.50	0.44	0.40	0.28	0.20
		22		0.96	0.91	0.87	0.81	0.76	0.73	0.62	0.56

Source: Fanger, P.O., *Thermal Comfort Analysis and Applications in Environmental Engineering*, McGraw-Hill, New York, 1970.

Productivity

The indoor environment contains and is affected by a variety of other issues that have indirect effects on thermal conservation in buildings. One of these that is of primary importance is worker productivity. Productivity is the workers' efficiency in performing their duties and responsibilities, which ultimately result in the economic well-being of the organization. In commercial buildings productivity has traditionally been viewed as the monetary return on employee compensation. Efforts to increase worker productivity have evolved from improving job satisfaction by various means (Stokes, 1978) and improving worker incentives (Lawler and Porter, 1967), to focusing on factors that negatively influence productivity, such as poor indoor environments.

Poor indoor environments can be generally described in three categories: inadequate thermal comfort, unhealthy environments, and poor lighting. Manifestations of poor productivity can be characterized by worker illness, absenteeism, distractions to concentration, and drowsiness or lethargy at work as well as by defects and mistakes in manufacturing and routine office work, and so forth. Primarily because of inadequate productivity measures, direct relationships between productivity and environmental factors are difficult to quantify (Daisey, 1989).

TABLE 9.2 Principal Indoor Pollutants, Sources, and Typical Concentrations (Samet et al., 1988). A cfu is a colony forming unit.

Pollutant	Source	Concentrations
Respirable Particles	Tobacco smoke, unvented kerosene heaters, wood and coal stoves, fireplaces, outside air, occupant activities, attached facilities	>500 μg/m³ bars, meetings, waiting rooms with smoking 100–500 μg/m³ smoking sections of planes 10 to 100 μg/m³ homes 1,000 μg/m³ burning food or fireplaces
NO,NO_2	Gas ranges and pilot lights, unvented kerosene and gas space heaters, some floor heaters, outside air	25 to 75 ppb homes with gas stoves 100 to 500 ppb peak values for kitchens with gas stoves or kerosene gas heaters
CO	Gas ranges, pilot lights, unvented kerosene and gas space heaters, tobacco smoke, back drafting water heater, furnace, or wood stove, attached garages, street level intake vents, gasoline engines	> 50 ppm when oven used for heating > 50 ppm attached garages, air intakes 2 to 15 ppm cooking with gas stove 2 to 10 ppm heavy smoking in homes and other locations
CO_2	People, unvented kerosene and gas space heaters, tobacco smoke, outside air	320 to 400 ppm outdoor air 2,000 to 5,000 ppm crowded indoor environment, inadequate ventilation
Infectious, allergenic, irritating biological materials	Dust mites and cockroaches, animal dander, bacteria, fungi, viruses, pollens	> 1,000 cfu/m³ homes with mold problems, offices with water damage 500 + 200 cfu/m³ homes and offices without obvious problems
Formaldehyde	Urea Formaldehyde Foam Insulation (UFFI), glues, fiberboard, pressed board, plywood, particle board, carpet backing fabrics	0.1 to 0.8 ppm homes with UFFI 0.5 ppm average in mobile homes
Radon and radon daughters	Ground beneath a home, domestic water, some utility natural gas	1.5 pCi/l estimated average in homes > 8 pCi/l in 3 to 5% homes
Volatile organic compounds: benzene, styrene, tetrachloroethylene, dichlorobenzene, methylene chloride, chloroform	Outgassing from water, plasticizers, solvents, paints, cleaning compounds, mothballs, resins, glues, gasoline, oils, combustion, art materials, photocopiers, personal care products	Typical concentrations of selected compounds: benzene — 15 μg/m³; 1,1,1 trichloroethylene — 20 μg/m³; chloroform — 2 μg/m³; tetrachloroethylene — 5 μg/m³; styrene — 2 μg/m³; m,p-dichlorobenzene — 4 μg/m³; m,p-xylene — 15 μg/m³
Semivolatile organics: chlorinated hydrocarbons, DDT heptachlor, chlordane, polycyclic compounds	Pesticides, transformer fluids, germicides, combustion of wood, tobacco, kerosene and charcoal, wood preservatives, fungicides, herbicides, insecticides	limited data
Asbestos	Insulation on building structural components, asbestos plaster around pipes and furnaces	> 1,000 ng/m³ when friable asbestos, otherwise no systematic measurements

TABLE 9.3 Methods of Controlling Sources of Indoor Pollution (Nero, 1992)

Use of building materials, furnishings, and consumer products with low emissions rates
Physical removal of emitted pollutant
Isolating, encapsulating, or controlling emission sources
Local venting of pollutants at the point of emission (e.g., range hood, substructure radon control system)

Examination of the cost of improving energy efficiency in buildings reveals that while significant energy cost savings are being achieved through retrofits, the relative savings may be dwarfed by savings due to increased worker productivity. Romm and Browning (1994) presented data based upon a national survey of office building stock in the United States showing that while energy costs are roughly 1.8$/ft^2 yr, the office workers' salaries amount to approximately 130$/ft^2 yr. As the authors state, "a 1% gain in productivity is equivalent to the entire annual energy cost." The point, often overlooked when considering an energy efficiency measure's cost-effectiveness, is that increased worker productivity can dramatically reduce the payback time of the retrofit.

Envelope Thermal Properties

Some of the most important properties of building materials are their strength, weight, durability, and cost. In terms of energy conservation, their most important properties are their ability to absorb and transmit heat. The materials' thermal properties govern the rate of heat transfer between the inside and outside of the building, the amount of heat that can be stored in the material, and the amount of heat that is absorbed into the surface by heat conduction and radiation. The rate of heat transfer through the building materials in turn determines the magnitude of heat losses and gains in the building. This information is important in order to determine the proper and most efficient design of space heating equipment required to maintain the desired indoor environmental conditions.

Heat loss and gain through the building envelope is a complex process involving four main mechanisms: heat conductance through solid and porous parts of the building envelope; heat convection from air to walls, ceilings, floors, and exteriors; solar radiation absorbed on exterior surfaces and transmitted through windows; and heat transport through ventilation or infiltration of air. This section discusses the thermal properties associated with the first three mechanisms; the next section discusses infiltration and ventilation issues. The importance of moisture control will also be addressed.

In order to take full advantage of different materials' thermal properties for energy conservation purposes, it is necessary first to determine the nature of building loads in each building sector. Residential, lightly loaded small commercial buildings and warehouses typically have low internal loads (e.g., heat from appliances, office equipment, lights, people, etc.), high infiltration loads, and high envelope transmission loads. The heat losses in these buildings are roughly proportional to the indoor-outdoor temperature difference. Depending on orientation and shading, solar heat gains can also be large. In large commercial, industrial, and institutional buildings, envelope transmission loads are relatively lower than in houses and affect only the peripheral zones, not the building core. In these buildings, internal loads are dominant.

A complete demonstration of a building load analysis would be too ambitious for the purposes of this chapter. There exist several outstanding examples of such analyses, of which the ASHRAE "Handbook of Fundamentals" (1993) is recommended to the reader. Because of the complexity of the heat transfer processes in a building; its dynamic nature; the impacts of building design, climate, and orientation, and so on; annual energy analyses are normally carried out by computer simulation. However, peak load and equipment sizing is still often carried out manually. The end of this chapter reviews various computer simulation tools of current use in the analysis of buildings.

The objective of this section is to offer some basic understanding of the heat transfer process in order to demonstrate how the appropriate use of building materials such as insulation, windows and paint can result in increased building energy efficiency.

Above-Grade Opaque Surfaces

Steady-state heat transfer through the walls, floors, and ceilings of a building depends on the indoor-outdoor temperature difference and the heat transmittance through each envelope component. Nothing can be done about the weather, and indoor conditions are constrained by occupant thermal comfort, but the conductance of the building envelope can be advantageously manipulated. Equation 9.1 shows the calculation required to determine envelope transmission heat losses, where the summation is taken over each component of the building envelope that separates the interior from the exterior.

$$Q_{tr} = \sum_i (UA)_i \Delta T \tag{9.1}$$

where

ΔT is the indoor-outdoor air temperature difference, in K (°F)
A is the component's surface area, in m^2 (ft^2)
U is the thermal transmittance of the component, in W/m^2K (Btu/hr ft^2°F)

The inverse of the transmittance U, is the component's resistance to heat flow, $R = 1/U$. Thermal resistance is analogous to electrical resistance when the heat transfer is one-dimensional, which is often the case in buildings. Thus, a very good approximate method of determining wall resistance to heat flow, for example, is to use electric circuit analogs. This is particularly useful when analyzing composite walls, ceilings, or floors made up of supporting framework, insulation, interior wall-board, exterior facing and so on, where the total resistance to heat flow can be determined from the individual resistances of each component. As an example, consider the composite wall of Figure 9.3(a), which represents the structure of a wall in a house, with insulation between wood studs on 46 cm (18 inches) centers. The interior wallboard is gypsum and the exterior facing is Douglas fir. The wall is represented by the electrical circuit shown in Figure 9.3(b).

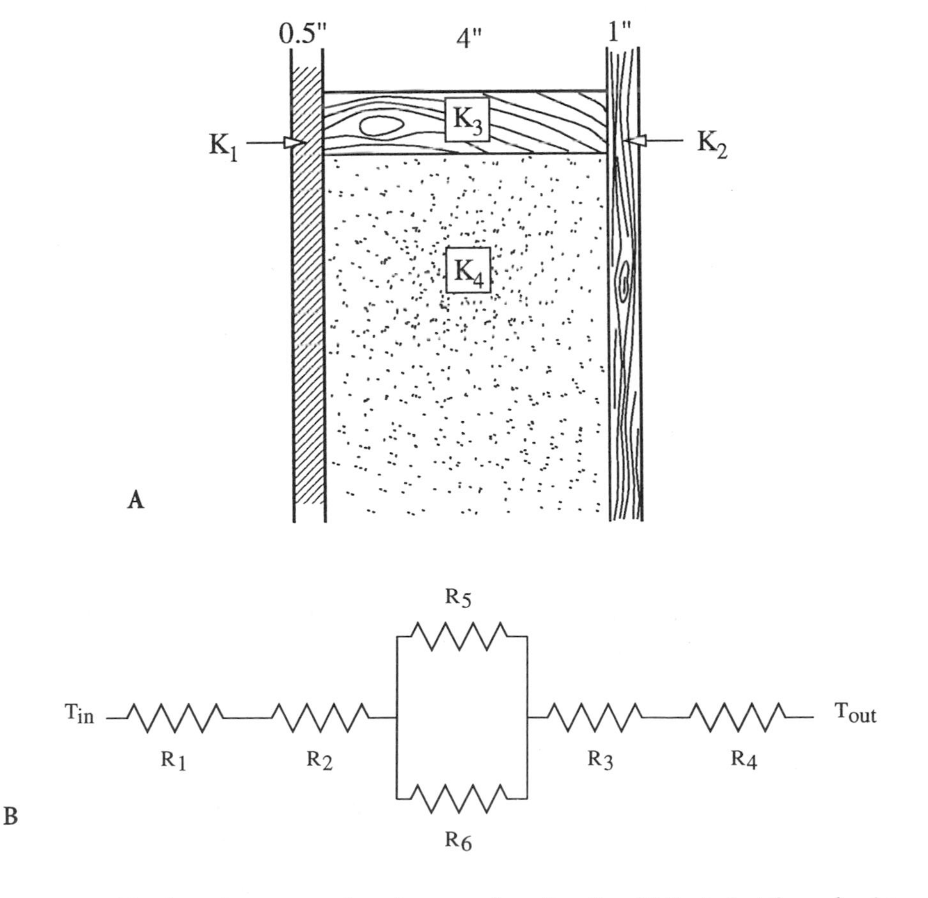

FIGURE 9.3 (a) Schematic representation of a composite wall section. (b) Equivalent thermal resistance for wall section of (a).

The total thermal resistance of the wall is given by

$$R_{tot} = R_1 + R_2 + (R_5 R_6)/(R_5 + R_6) + R_3 + R_4$$

where $R_1 = 1/h_i,\ R_2 = a/k_1,\ R_3 = c/k_2,\ R_4 = 1/h_o,\ R_5 = b/(0.11 \cdot k_3),\ R_6 = 1/(0.89 \cdot k_4)$

R_1 and R_4 are convective resistances, while all others are conductive resistances. R_5 and R_6 have been corrected for the fractional amount of area perpendicular to heat flow that they occupy. To show the effect of insulation, the calculation of the total wall resistance will be done first by assuming the insulation space is occupied by air. The calculation will be repeated with fiberglass installed in the insulation space.

Using thermal conductivity data from Table 9.4, and assuming typical values for the convective coefficients as shown:

$$h_i = 7.5\ \text{W/m}^2\text{K},\ h_o = 15\ \text{W/m}^2\text{K},$$
$$k_1 = 0.48\ \text{W/m K},\ a = 0.0127\ \text{m}$$
$$k_2 = 0.11\ \text{W/m K},\ c = 0.0254\ \text{m}$$
$$k_3 = 0.17\ \text{W/m K},\ b = 0.1016\ \text{m}$$

for the airspace, the ASHRAE *Handbook of Fundamentals,* (1993) Chapter 22, Table 2 gives R_6 = 0.44K m²/W. Using these values, the total thermal resistance is

$$R_{tot} = 0.01 + 0.06 + (5.43 \cdot 0.44)/(5.43 + 0.44) + 0.23 + 0.07 = 0.78\ \text{Km}^2/\text{W}\ (4.43°\text{F ft}^2\ \text{hr/Btu})$$

If the air space is filled with cellulose insulation, k_4 = 0.043 W/m K, R_4 = 0.1016/(.89·0.042) = 2.72 K m²/W, and

$$R_{tot} = 0.01 + 0.06 + (5.43 \cdot 2.72)/(5.43 + 2.72) + 0.23 + 0.07 = 2.18\ \text{K m}^2/\text{W}\ (12.38°\text{F ft}^2\ \text{hr/Btu})$$

which is more than double the thermal resistance of the wall with air between the studs.

The wood frame in the previous example acts as a thermal bridge between the wallboard and external facing. To further reduce heat losses a 2.54 cm (1 inch) layer of open-cell rigid foam can be added between the wood frame and exterior facing. This would result in a new total resistance of

$$R_{tot} = 2.18 + 0.0254/0.033 = 2.95\ \text{K m}^2/\text{W}\ (16.75°\text{F ft}^2\ \text{hr/Btu})$$

which is a 35% improvement. It is common practice in cold climates and in superinsulated houses to use an external insulation layer. However adding more insulation to the house exterior is not always beneficial. Adding insulation decreases envelope heat transmission losses; however, the percentage reduction diminishes quickly with increased insulation. After a certain point, the energy cost savings do not justify the cost of the added insulation. This point represents the economic optimum insulation thickness, and is determined by minimizing the life cycle and installation costs. This optimum varies by climate zone.

The application of insulation in internally loaded buildings must be analyzed on a case-by-case basis. For example, in a commercial building in a cool climate with outdoor temperatures below room temperatures more than half the time, adding insulation to the envelope would unnecessarily increase cooling costs. The optimum insulation thickness depends on the amount of internal and solar gains, hours of use, and so on. These factors vary between zones of the building.

TABLE 9.4 Thermal Properties of Some Common Building Materials (Sources: ASHRAE Handbook of Fundamentals, 1993; Holman, 1976)

Material	Density kg/m³ (lb/ft³)	Conductivity W/m K (Btu/hr ft°F)	Specific Heat J/kg K (Btu/lb°F)	Emissivity Ratio
		Wallboard		
Douglas fir plywood	140 (8.7)	0.11 (0.06)	2,720 (0.65)	
Gypsum board	1,440 (90)	0.48 (0.27)	840 (0.20)	
Particle board	800 (50)	0.14 (0.08)	1,300 (0.31)	
		Masonry		
Red brick	1,200 (75)	0.47 (0.27)	900 (0.21)	0.93
White brick	2,000 (125)	1.10 (0.64)	900 (0.21)	
Concrete	2,400 (150)	2.10 (121)	1,050 (0.25)	
		Hardwoods		
Oak	704 (44)	0.17 (0.10)	1,630 (0.39)	0.09 (planed)
Birch	704 (44)	0.17 (0.10)	1,630 (0.39)	
Maple	671 (42)	0.16 (0.09)	1,630 (0.39)	
Ash	642 (40)	0.15 (0.09)	1,630 (0.39)	
		Softwoods		
Douglas fir	559 (35)	0.14 (0.08)	1,630 (0.39)	
Redwood	420 (26)	0.11 (0.06)	1,630 (0.39)	
Southern pine	615 (38)	0.15 (0.09)	1,630 (0.39)	
Cedar	375 (23)	0.11 (0.06)	1,630 (0.39)	
		Other		
Steel (mild)	7,830 (489)	45.3 (26.1)	500 (0.12)	0.12 (cleaned)
Aluminum Alloy 1100	2,740 (171)	221 (127.7)	896 (0.21)	0.09 (commercial sheet)
Bronze	8,280 (517)	100 (57.8)	400 (0.10)	
Rigid foam insulation	32.0 (2.0)	0.033 (0.02)		
Glass (soda-lime)	2,470 (154)	1.0 (0.58)	750 (0.18)	0.94 (smooth)

The overall conductance of the building can be found by analyzing each component separately, then summing over all components. The process is cumbersome because the bookkeeping of all conductivities, thicknesses, resistances, and so on must be accurate. However, calculated values have been shown to agree well with measured data if the quality of installation is high.

Heat transmission through the building envelope is one of the major loss mechanisms in residences. Increasing the thermal resistance of the envelope by adding insulation reduces space heat loss on a long-term basis. This increases the building's thermal efficiency and also improves the occupants' thermal comfort by providing a more constant indoor temperature. Insulating walls and ceilings also keeps the inside surface of the exterior wall above the dewpoint temperature, thereby preventing condensation.

There are many types of insulation materials. Table 9.5 lists some of the various types of insulation, the various forms available, and the approximate thermal resistance per unit thickness.

While each type of insulation shows a high R-value, there are some disadvantages to some insulation types that must be mentioned. Both cellulose and perlite will pack down and lose their insulation value when they get wet. Rock wool and fiberglass irritate the skin. Polystyrene is moisture resistant, but combustible. Urea formaldehyde and urethane give off noxious gases during fire, even though they're fire resistant. They require specialized equipment to inject the foam into wall cavities, which makes them the most expensive type. If not installed correctly, they'll leave a lingering odor.

TABLE 9.5 Available Building Envelope Insulation

Insulation Type	Blanket	Batt	Loose Fill	Rigid Panels	Formed-in-Place	R-Value/thickness m² K/W/m (°F ft² hr/Btu/in.)
Cellulose	√	√	√			25.7 (3.7)
Rock wool and fiberglass	√	√	(pellets)	(semirigid board)		22.9 (3.3)
Perlite	√					18.7 (2.7)
Polystyrene				√		24.3 (3.5)
Urea formaldehyde, urethane					√	31,2. 36.7 (4.5, 5.3)

TABLE 9.6 Appropriate Application of Insulation Forms (Jones, 1979)

Batts	Between joists, on unfinished attic floors or basement/crawl space ceilings
Blankets	Same as batts, but with longer continuous coverage
Loose fill	Poured in unfinished attic floor, useful around obstructions and hard-to-reach corners
Rigid boards	Interior or exterior basement walls
Foam	Between framing studs in wall, virtually anywhere in building

Reflective-type insulations are not mentioned in Table 9.5. These insulations generally have smooth and shiny surfaces and are installed with this surface facing the source of heat. Reflective insulations are generally installed over exposed studs in walls, attics, and floors, often enclosing an air space underneath. The thermal resistance of reflective insulations so installed depends on the orientation of the insulation and direction of heat flow. Some blanket and batt insulations have reflective backing, which also serves as a vapor barrier.

Currently the highest thermal resistance per unit thickness insulation is the blown polymer foam types as shown in Table 9.5. These foam insulations contain chlorofluorocarbons (CFCs), which have been identified with the depletion of the earth's ozone layer. Use of CFCs will be discontinued in 1996 in the United States. The performance of blown foam insulations with CFC substitutes is expected to decrease by up to 25%.

Insulation technology is still evolving. A new type of insulation showing great promise is gas-filled panels (GFPs). These insulations are made up of a thin walled baffle structure with low-emittance coatings. The minimal solid construction prevents heat conduction; the baffle structure and coating reduce convection and radiation heat transfer. High-performance GFPs use low-conductivity gases such as argon and krypton in place of air in the baffle spaces. Originally designed for refrigerators and freezers, GFPs can be applied as a building insulation. The R-value per inch for argon filled units, R7.5/in, is twice as high as fiberglass insulation, and more expensive krypton filled units have achieved R12/in. Projected costs in 1992 for 3-inch R22 GFPs is $0.60/ft². GFPs are in development (Griffith and Arasteh, 1992).

The various forms of insulation are convenient to use in different parts of the building. Table 9.6 shows where each form is typically installed.

Figure 9.4 shows a map of the United States roughly divided into climate zones. The recommended R-value has of residential ceiling, wall and floor insulation corresponding to the zones on the map are given in Table 9.7.

The thermal mass of a building material is the product of the material's specific heat and density. Judicious use of a building's thermal mass can also be used to increase its energy efficiency. For example, in winter a building's thermal mass can store heat from solar radiation during the day and be made to release its heat to the interior at night, thereby reducing the need for space heating. These strategies are reviewed in the chapter on passive solar energy. Peak power shifting can also be accomplished with the use of a building's thermal mass by cooling the mass overnight and circulating indoor air over it to provide cooling during the day. Such strategies make use of lower

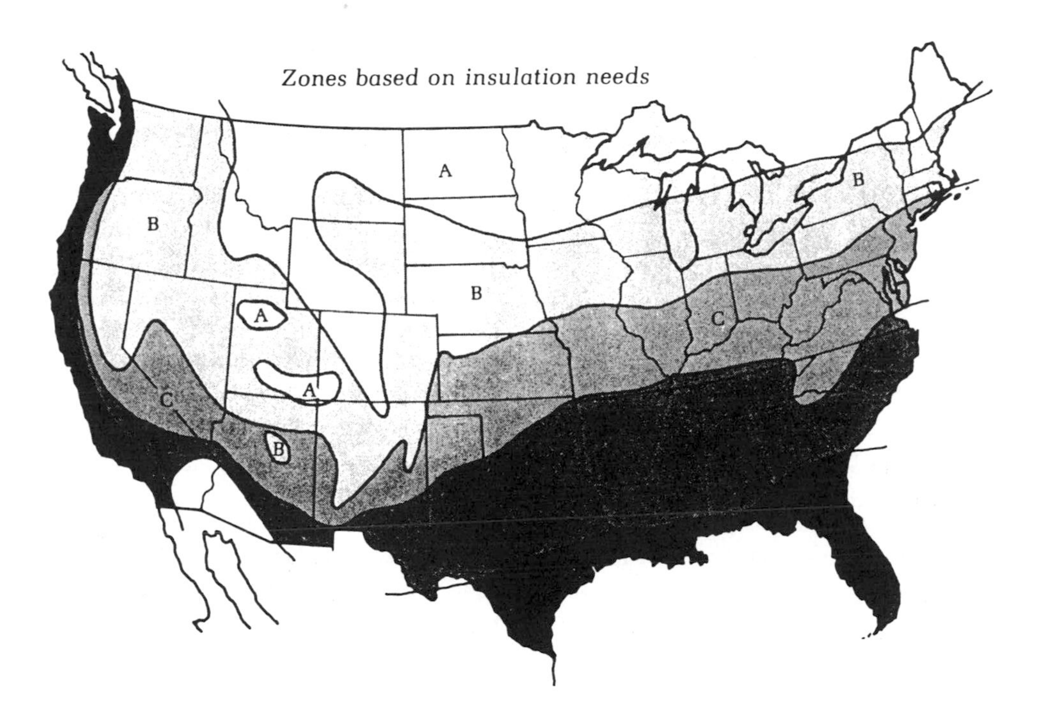

FIGURE 9.4 U.S. zones based on insulation needs. *Source:* Jones, P., *How to Cut Heating and Cooling Costs*, Butterick Publishing, New York, 1979.

TABLE 9.7 Recommended R-Value for U.S. Zones (Jones, 1979)

Zone	Ceiling	Wall	Floor
A	R-38	R-19	R-22
B	R-33	R-19	R-22
C	R-30	R-19	R-19
D	R-26	R 19	R-13
E	R-26	R-13	R-11

utility rates during the night. Examples of materials with a high thermal mass are concrete, masonry, and cinder block. Gypsum board also serves as thermal mass in houses, though to a lesser degree. To use a buildings thermal mass effectively, it should have a large exposed surface area to the interior.

Foundations and Basements

Foundations and basements of buildings raise other issues besides those of the rest of the building envelope. Unlike the ambient air, the ground temperature does not undergo large daily temperature swings. Problems with foundations and basements arise primarily in winter. For houses with uninsulated concrete slab foundations, the floor of the house can become very cold and water vapor diffusing through carpets can condense on the slab surface. This causes damage to the floors and carpets. The same problems can occur on uninsulated basement floors and walls. Also, pipes can freeze in uninsulated basements and the space is generally unusable. Moisture diffusion through concrete and cinder block and exposed soil in basements can be a problem. These problems can be mitigated by the proper placement of insulation and vapor barriers.

Insulating foundations and basements also has energy benefits. In a handbook on foundations and basements from Oak Ridge National Laboratory (ORNL, 1988), it was shown that both interior and exterior insulated masonry basement walls saved approximately the same amount of energy, while basement walls insulated only from the ground level to the frost line saved less energy. In regard to moisture problems, a vapor barrier between the soil and foundation can be used to keep condensing water vapors outside the building shell.

Albedo and Shading

In summer, solar radiation striking roofs and walls of houses and other buildings can be absorbed and transmitted to the building interior. For houses, attic temperatures can soar as a result of sunshine being absorbed on the roof and transmitted to the attic as heat. This increases the heat transmittance through the ceiling. One strategy to reduce ceiling heat gains is to increase attic insulation levels appropriately. In warmer climates such as the sun belt of the United States, attic insulation levels are being increased to the same levels as those found in the coldest climates of Table 9.7. This strategy can be expensive. Another strategy for preventing solar heat gains from penetrating the building envelope is to reflect the incoming solar radiation from building surfaces or by shading the surfaces from the sun with deciduous trees. These strategies are much more cost-effective.

Reflecting solar gains from building surfaces is a method with some precedence in history. In ancient Greece, whitewashed building walls reflected solar radiation to keep the interiors cool. This practice is being revisited in modern times. Recent studies have shown that increasing the albedo of the building roofs and walls can reduce air conditioner energy consumption. It was shown that a 22% reduction in the air conditioning bill was achieved by increasing the albedo from 0.2 to 0.6 of a residential roof in Sacramento (Akbari, et al. 1992).

Albedo is a measure of a surface's reflectance. Generally the higher the surface's reflectance, the lower its emissivity, but this does not always hold for paints, which can reflect visible light but absorb infrared radiation. The scale for both reflectivity and emissivity is from 0 to 1. Many paint manufacturers have begun listing the reflective properties on their products. Surface roughness and color also have an impact: rougher, darker surfaces generally have a lower albedo. Costs for retrofitting buildings with higher-albedo paints are minimal if the paint is used at the time of regular building maintenance.

Planting deciduous trees (i.e., broad-leaved trees that lose their leaves in the fall) near a house can provide the needed shading of buildings in summer, while allowing needed solar gains into the building in winter. There are many beneficial side effects as well: trees cool their surroundings by evapotranspiration, absorbing heat from the ambient air and evaporating up to 100 gallons of water per day, trees can filter air pollution, provide a windbreak for houses, reduce street noise, prevent soil erosion and provide habitat for wildlife. A study of the cooling effect of trees has shown that annual cooling costs of air conditioning were reduced by 10% by the addition of shade trees (Akbari, et al. 1994).

Air conditioning energy and cost savings can be further increased by combining the benefits of the use of shade trees, shrubs, and other greenery with raised albedo building surfaces and other surfaces in the urban environment as well. Figure 9.5 shows the typical albedo of various surfaces found in communities. When combining the shading properties of trees with high albedo surfaces, researchers estimate cooling costs to be reduced by 15 to 35% for the whole community. An excellent information source on the cooling effect of trees and high-albedo surfaces is listed in the bibliography (USEPA, 1992).

Windows

Windows, or building fenestration systems, influence building energy use in four ways. Heat is transmitted through windows in the same manner as through walls. Windows absorb and transmit direct and scattered solar radiation into the building, where it is absorbed on surfaces and convected to the inside air. Air leakage around windows increases infiltration loads. Windows also let in visible light, reducing the need for electric lighting.

Heat transmittance through building window systems is generally larger than through the opaque part of the building shell. A typical single pane of window glass has an insulating value of

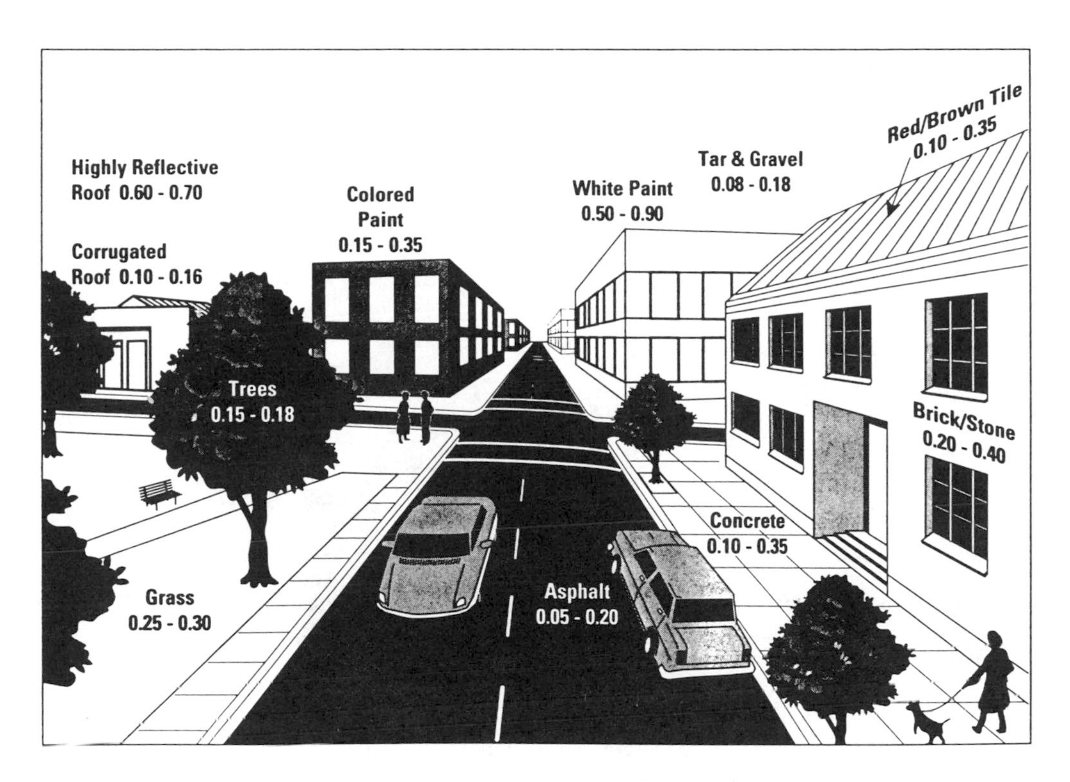

FIGURE 9.5 Surface albedo values in a typical urban environment. *Source:* United States Environmental Protection Agency, "Cooling Our Communities: A Guidebook on Tree Planting and Light-Colored Surfaces," USEPA Report No. 22p-2001, Office of Policy Analysis, Climate Change Division, Washington, DC, January 1992.

R-1. This value is low in comparison to typical R-15 wall thermal resistances. Adding a second or third pane of glass doubles or triples the R-value respectively. Alternatively, filling the gap space between the window panes with argon gas and adding a transparent low emissivity coating can increase the resistance to R4. A triple layer window with two low-emissivity coatings and argon gas fill can achieve R8. Naturally, windows become more expensive with each step taken to increase the R-value.

The overall heat conductance of windows is made up of three parts: through the center of the glass, through the edge of the glass and through the window frame, as in Eq. (9.2)

$$U_0 = \left(U_{cg}A_{cg} + U_{eg}A_{eg} + U_f A_f\right) / A_{pf} \tag{9.2}$$

where U is the heat conductance, A is the area of the surface, cg refers to the center of the glass, eg refers to the edge of the glass (defined by the area 2½ inches from the frame), f refers to the frame and A_{pf} refers to the opening area of the wall. This methodology of determining the window U value is preferred because of the different combinations of window and frame technologies possible in fenestration systems.

Table 5 in the ASHRAE Handbook of Fundamentals (1993), Chapter 27, lists computed U-values for various combinations of single-, double-, or triple-pane windows with glass spacer type, gap width, frame material, low-*e* coatings and gas-filled spaces. A similar methodology is used by the National Fenestration Rating Council (NFRC, 1995) which publishes a directory listing for over 20,000 commercially available windows with their U-values. Both sources of information are recommended as starting points for design purposes.

Windows also allow radiative heat transmission through the glass. The amount of solar radiation passing through the glass during the day depends on the window's optical properties. Generally,

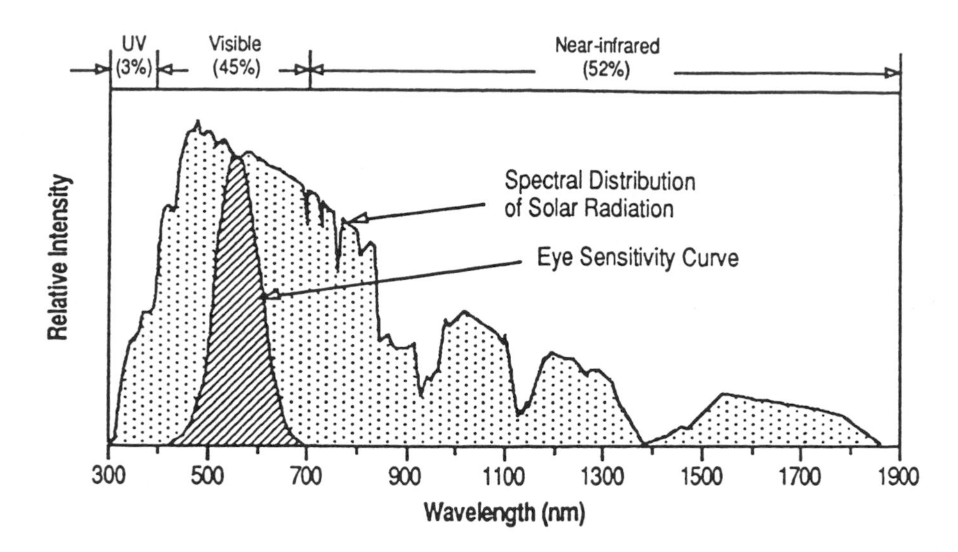

FIGURE 9.6 Solar spectrum. *Source:* Davids, B.J., "Taking the Heat out of Sunlight — New Advances in Glazing Technology for Commercial Buildings," Proc. ACEEE Summer Study on Energy Efficiency in Buildings, August 26–September 1, Asilomar Conference Center, Pacific Grove, CA, 1990.

clear glass is not spectrally selective and allows most incident solar radiation to pass through, with a small amount of absorptance (depending on thickness) and with about 8% reflected from each layer of glass. However, most architectural-quality glass is opaque to long wave radiation from surfaces at temperatures below 120°C. This tends to produce the greenhouse effect, in which radiation passing through the glass is absorbed on interior room surfaces and then re-radiated with long wave radiation to other interior surfaces, which warm the room air by convection. This re-radiated energy is unable to exit through the glass.

Figure 9.6 shows the relative intensity of solar radiation as a function of wavelength. Also shown in the figure is the region the human eye perceives as visible light. To make optimum use of daylighting, an ideal spectrally selective coating on the surface of a window would block out all solar radiation with wavelengths below about 450 nm and above about 690 nm. Such a coating would pass virtually all visible light through the window while blocking out approximately 50% of all solar radiation. Multilayer metal and dielectric coatings on glass approach this behavior, allowing more than 80% of visible light through and reflecting most of the rest of the sun's spectrum. Some types of green, blue-green, and blue absorbing glasses also provide spectral selectivity, although they are generally not as effective as reflective coatings.

Some of the nonreflected radiation is absorbed in the glass; the rest is transmitted through it. Part of the absorbed energy is conducted to the inside surface. Thus, the solar heat gain from a window is the sum of the transmitted radiation and the fraction of absorbed radiation that is conducted to the inside surface:

$$q_i = E_t(\tau_s + N_i\alpha_s) = E_t \times \text{SHGC} \tag{9.3}$$

where E_t is the total solar radiation incident on the window, τ_s is the glass transmittance, N_i is the fraction conducted inward and α_s is the glass absorbance. The fraction of the total radiation that finds its way to the interior is called the Solar Heat Gain Coefficient (SHGC). The SHGC and the window's overall U-value should be used in any description of a window's energy performance.

The total rate of heat transfer through a window is the sum of the normal heat transmittance due to the indoor-outdoor temperature difference, the solar gain, and an infiltration term as shown in Eq. (9.4):

$$q_{tot_i} = \text{SHGC}_i E_t + U_0 \Delta T + KQ_{\text{inf}} \Delta T \quad \mathbf{(9.4)}$$

Cracks around window frames or loose sliding window sashes are sites for air infiltration in buildings. This topic is covered in the next section. In building design or retrofit, appropriate window selection depends on the climate and nature of the building. For heating-dominated buildings, it is desirable to have a high transmittance over the entire solar spectrum, but a high reflectivity (low emissivity) over the long wavelength infrared portion of the spectrum. This arrangement allows most solar radiation into the building, but traps low-temperature radiant heat inside the building. Many commercially available windows with low-emittance (low-e) coatings approach this performance. The opposite performance is desirable for cooling dominated buildings. The ideal in this case is to have a high reflectance (low emittance) over all solar wavelengths outside the visible portion of the spectrum. Depending on the spectral selectivity of the window, this would eliminate from 50 to 70% of the solar radiation from entering the building, without loss in light transmission. Windows approximating this ideal behavior are available. Many absorbing glass and reflective coatings also block some portion of the visible spectrum which may be desirable to control glare from direct sun. These glazings may also change the color-rendering properties of the transmitted light. This must also be considered in the selection of window systems.

Windows provide daylight, which aids visual comfort and reduces the need for artificial lighting, thus saving energy. An accurate evaluation of daylighting is beyond the purposes of this chapter. It has been addressed in Chapter 20, "Passive Solar Energy." As a rule of thumb, properly designed windows can provide daylight adequate for typical indoor tasks for a depth 2½ times the height of the window, based on normal sill height. In typical office buildings, if the effective aperture (visible transmittance of glass times fractional area of wall that is glazed) of a building is in the range of 0.2 to 0.35, daylighting can provide approximately 50% of annual electric lighting needs in perimeter zones adjacent to windows. In a skylighted building, an effective aperture of 0.04 can provide 50 to 70% of lighting needs. For more on the subject, the reader is referred to "Recommended Practices of Daylighting" (IES, 1979).

Infiltration and Ventilation

Ventilation is the building service most associated with controlling the indoor air quality to provide a healthy and comfortable indoor environment. In large buildings ventilation is normally supplied through mechanical systems, but in smaller buildings such as single-family homes it is principally supplied by leakage through the building envelope (i.e., *infiltration*). Most U.S. buildings with mechanical ventilation systems also use the system for thermal energy distribution, but this function is described in the following section, and we restrict our discussion here to ventilation of *outdoor* air.

Ventilation is the process by which clean air is provided to a space. It is needed to meet the metabolic requirements of occupants and to dilute and remove pollutants emitted within a space. Usually ventilation air must be conditioned by heating or cooling it to maintain thermal comfort and then it becomes an energy liability. Ventilation energy requirements can exceed 50% of the space conditioning load; thus excessive or uncontrolled ventilation can be a major contributor to energy costs and global pollution. Thus, in terms of cost, energy, and pollution, efficient ventilation is vital, but inadequate ventilation can cause comfort or health problems for the occupants.

Mechanically Dominated Ventilation

Most medium and large buildings are ventilated by mechanical systems designed to bring in outside air, filter it, supply it to the occupants, and then exhaust an approximately equal amount of stale air.

Ideally these systems should be based on criteria that can be established at the design stage. After the systems are designed and installed, attempts to mitigate problems may lead to considerable expense and energy waste, and may not be entirely successful. The key factors that must be included

TABLE 9.8 Design Considerations for Mechanical Ventilation

Code requirement, regulations or standards	ASHRAE Standard 55
Ventilation strategy and system sizing	ASHRAE Standard 62-1989
Climate and weather variations	ASHRAE *Handbook of Fundamentals*, ch. 24 (1993)
Air distribution, diffuser location, and local ventilation	ASHRAE *Handbook of Fundamentals*, chs. 23, 25 and 26 (1993), ASHRAE *Handbook of Systems and Equipment*, ch. 17 (1992)
Location of outdoor air inlets and outlets	ASHRAE *Handbook of Fundamentals*, ch. 14 (1993)
Ease of operations and maintenance	Equipment Manufacturer, ASHRAE *Handbook of Aplications*, chs. 32–38 (1991)
Impact of system on occupants (e.g., acoustics and vibration)	ASHRAE *Handbook of Fundamentals*, chs. 7, 37 (1993)

in the design of ventilation systems are given in Table 9.8, along with suggested sources for more information.

These factors differ for various building types and occupancy patterns. For example, in office buildings pollutants tend to come from occupancy, office equipment, and automobile fumes. Occupant pollutants typically include metabolic carbon dioxide emission, odors, and sometimes smoking. When occupants (and not smoking) are the prime source, carbon dioxide acts as a surrogate and can be used to cost-effectively modulate the ventilation, forming what is known as a *demand controlled ventilation* system.

Schools are dominated by high occupant loads, transient occupancy, and high levels of metabolic activity. Design ventilation in hospitals must aim at providing fresh air to patient areas, combined with clean room design for operating theaters. Ventilation in industrial buildings poses many special problems, which frequently have to be assessed on an individual basis. Contaminant sources are varied, but often well-defined, and limiting values are often determined by occupational standards. Poorly designed, operated or maintained ventilation systems, rather than the ventilation rate itself, may cause sick building syndrome (SBS). The causes of SBS were summarized earlier.

Infiltration

Infiltration is the process of air flowing in (or out) of leaks in the building envelope, thereby providing ventilation in an uncontrolled manner. All buildings are subject to infiltration, but it is more important in smaller buildings. In larger buildings there is less surface area to leak for a given amount of building volume, so the same leakage matters less. More important, the pressures in larger buildings are usually dominated by the mechanical ventilation system and the leaks in the building envelope have only a secondary impact on the ventilation rate. However, infiltration in larger buildings may affect thermal comfort, control, and system balance.

In low-rise residential buildings (most typically, single-family houses) infiltration is the dominant force. In these buildings mechanical systems contribute little (intentionally) to the ventilation rate.

Infiltration is made up of two parts: weather-induced pressures and envelope leakage. Since little of practical import can be done about the weather, it is the envelope leakage, or *air tightness*, that is the variable factor in understanding infiltration. Virtually all knowledge about the air tightness of buildings comes through making *fan pressurization* measurements, done most typically with a *blower door.*

Blower door is the popular name for a device that is capable of pressurizing or depressurizing a building and measuring the resultant air flow and pressure. The name comes from the common utilization of the technology, where there is a fan (i.e., blower) mounted in a door; the generic term is "fan pressurization." Blower door technology was first used around 1977 to test the tightness of building envelopes (Blomsterberg, 1977), but the diagnostic potential of the technology soon became apparent. Blower doors helped uncover hidden bypasses that accounted for a much greater percentage of building leakage than did the presumed culprits of window, door, and electrical outlet leakage. The use of blower doors as part of retrofitting and weatherization became known as *House Doctoring*. This led to the creation of instrumented audits and computerized optimizations (Blasnik and Fitzgerald, 1992). A brief description of a typical blower door test follows in the measurements section.

While it is understood that blower doors can be used to measure air tightness, the use of blower door data alone cannot be used to estimate real-time air flows under natural conditions or to estimate the behavior of complex ventilation systems. However, a rule of thumb relating blower door data to seasonal air change data exists (Sherman, 1987): namely, the seasonal air exchange can be estimated from the flow required to pressurize the building to 50 Pascal divided by 20. Ventilation and infiltration air flows are generally measured with a tracer gas, as described in the measurements section.

A more accurate description of infiltration rates can be found by separating the leakage characteristics of the building from the driving forces, which are wind- and temperature-induced pressures on the building shell. Modeling the leakage data from blower door tests as orifice flow, Sherman and Grimsrud (1980) developed the LBL infiltration model, Eq. (9.5):

$$Q = L\left(f_s \Delta T + f_w V_w^2\right)^{0.5} \tag{9.5}$$

Here Q is the volumetric air flow rate, L is the effective leakage area of the house at 4 Pa, ΔT is the indoor-outdoor temperature difference, V_w is the time-averaged wind speed, and f_s and f_w are the stack and wind coefficients, respectively, as determined from a fit of the data. The model was validated by Sherman and Modera (1984) and incorporated into the ASHRAE "Handbook of Fundamentals" (1989). Much of the subsequent work on quantifying infiltration is based on that model, including ASHRAE Standard 119 (1988) and ASHRAE Standard 136 (1993).

Blower doors are still used to find and fix the leaks. A common method of locating leak sites is to hold a smoke source near the leak and to watch where the smoke exits the house. Depending on the leak site, different methods are used to stop the leakage.

Often, the values generated by blower door measurements are used to estimate infiltration for both indoor air quality and energy consumption analyses. These estimates in turn are used for comparison to standards or to provide program or policy decisions. Each specific purpose has a different set of associated blower door issues.

Compliance with standards, for example, requires that the measurement protocols be clear and easily reproducible, even if this reduces accuracy. Public policy analyses are more concerned with getting accurate aggregate answers than reproducible individual results. Measurements that might result in costly actions are usually analyzed conservatively, but "conservatively" for IAQ is diametrically opposed to "conservatively" for energy conservation.

Because infiltration depends on the weather, buildings that have much of it can have quite variable ventilation rates. Determining when there is insufficient infiltration to provide adequate indoor air quality or energy-wasteful excess infiltration is not a simple matter. The trade-off in determining optimal levels depends on various economic and climactic factors.

Individual variations notwithstanding, Sherman and Matson (1993) have shown that the stock of housing in the United States is significantly overventilated from infiltration and that there are 2 exajoules (1.9 quads) of potential annual savings that could be captured. Much of this savings could be captured by simple tightening, but a significant portion requires installation of a ventilation system or strategy to assure adequate ventilation levels.

Natural Ventilation

Natural ventilation is a strategy suitable for use in mild climates or during mild parts of the year. As commonly interpreted, natural ventilation is the use of operable parts of the building envelope (e.g., windows, etc.) to allow natural airflow at the discretion of the occupants.

Natural ventilation shares many of the same properties as infiltration: it depends on weather for driving forces; it is a function of the leakage area of the buildings; and so on. The distinguishing feature of natural ventilation, however, is that it is under the control of the building occupants.

From the point of view of the HVAC designer, natural ventilation is quite bothersome, because a conservative ventilation designer cannot count on it, but one must consider its potential effects on the building load. From the perspective of the occupants, however, natural ventilation gives them more control of their environment and usually makes it more acceptable. Studies have shown that those in naturally ventilated buildings tend to suffer less sick building syndrome and less respiratory disease (e.g., colds) than buildings that are fully mechanically ventilated.

The designs of new commercial buildings have been curtailing the availability of natural ventilation as an option by removing operable windows. Natural ventilation still dominates in the residential sector.

Ventilative Cooling

In dwellings, natural ventilation serves more than just a means to provide clean air; it serves to cool the building and its occupants and reduce the requirement for mechanical cooling. Fans are used to assist ventilative cooling by increasing the air change rates. These *whole house fans* are of much larger capacity than is needed for ventilation. A standard such fan may provide as much as 10 air changes per hour, compared with the ventilation requirement of no more than half an air change per hour.

Ventilative cooling removes internally generated heat as long as the outdoor temperature is less than the indoor temperature. When thermally massive elements are included in the structure, night ventilation can be used to store "coolth" and reduce the cooling requirements the following day.

The air motion caused by ventilative cooling provides additional cooling to the occupants of the building by removing more heat and lowering the apparent temperatures. Increased air motion raises the upper limit of acceptable air temperature from a thermal comfort perspective and therefore also reduces cooling demand.

In commercial buildings ventilative cooling is accomplished principally by the use of an *economizer.* Economizers are nothing more than dampers that allow outdoor air to be used instead of recycled (i.e., return) air in the building's thermal distribution system. This is usually done when there is a cooling load and the outdoor air is cooler than the indoor air.

A simple economizer works in lieu of the cooling system of the building. Because of the large internal loads of some commercial buildings it can be necessary to supply some mechanical cooling and also use outdoor air. Such a system is called an *integrated economizer* because it can do both. Not all systems are capable of running in this way. In buildings using cooling towers, *water-side economizers* can effectively use evaporative cooling to substitute for chiller operation.

Heat Recovery

Other than periods when ventilative cooling is useful, considerable energy is lost from a building through the departing air stream. When air change is dominated by infiltration, little can be done to recapture this energy, but if the departing air stream is centrally collected, a variety of methods for recovering or recycling the waste heat become possible (AIVC TN 45, 1994).

Ventilation heat recovery is the process by which sensible and latent heat is recovered from the air stream. Methods have been developed for air-to-air systems, in which the energy in the departing air stream is transferred directly to the entering air stream. Heat recovery can also be accomplished with heat-pump systems, in which heat from the exhaust air is pumped to another system such as the domestic hot water.

While the heat recovery process can be efficient at collecting the heat, benefits must always be weighed against the energy needed to drive the recovery, the capital costs of the equipment, and the maintenance. Induced losses such as infiltration or duct leakage must be understood. Without careful design and construction of the building envelope the system total performance can be considerably impaired and in some cases could increase the total energy costs.

The efficiency of all systems can be defined in terms of the proportion of outgoing ventilation energy that is recovered. Quoted efficiencies can be quite high (e.g., 65–75%), and the attractiveness therefore quite strong. For various reasons, however, field studies do not always come up to expectations. Basically, if poor attention is given to planning and installation, then the level of heat recovery can be quite disappointing.

Measurement Methods

A brief description of a blower door test follows. The reader is referred to ASTM Standard E779 for a complete description. Shown in Figure 9.7 is a sketch of a fan mounted in a doorway of a single-family house. A means of measuring the pressure difference between the house and outside is provided in this case by a digital manometer. Other pressure measuring instruments are acceptable as long as they are accurate over the measured pressure range: 0 to 100 Pascal. Volumetric air flow through the fan must also be determined. The air flow through most blower door fans is calibrated against fan speed in revolutions per minute or the pressure drop across the fan (range of approximately 0 to 500 Pa). The latter method is shown in Figure 9.7. The test is performed after ensuring that all windows and fireplace dampers in the house are closed. Some protocols also require exhaust vents in the kitchen and bathrooms to be sealed. The general procedure is to depressurize the house in steps of about 12 Pascal to about 50 Pascal, recording the house pressure and fan air flow at each step. The air flow direction through the fan is then reversed and the procedure is repeated for house pressurization measurements.

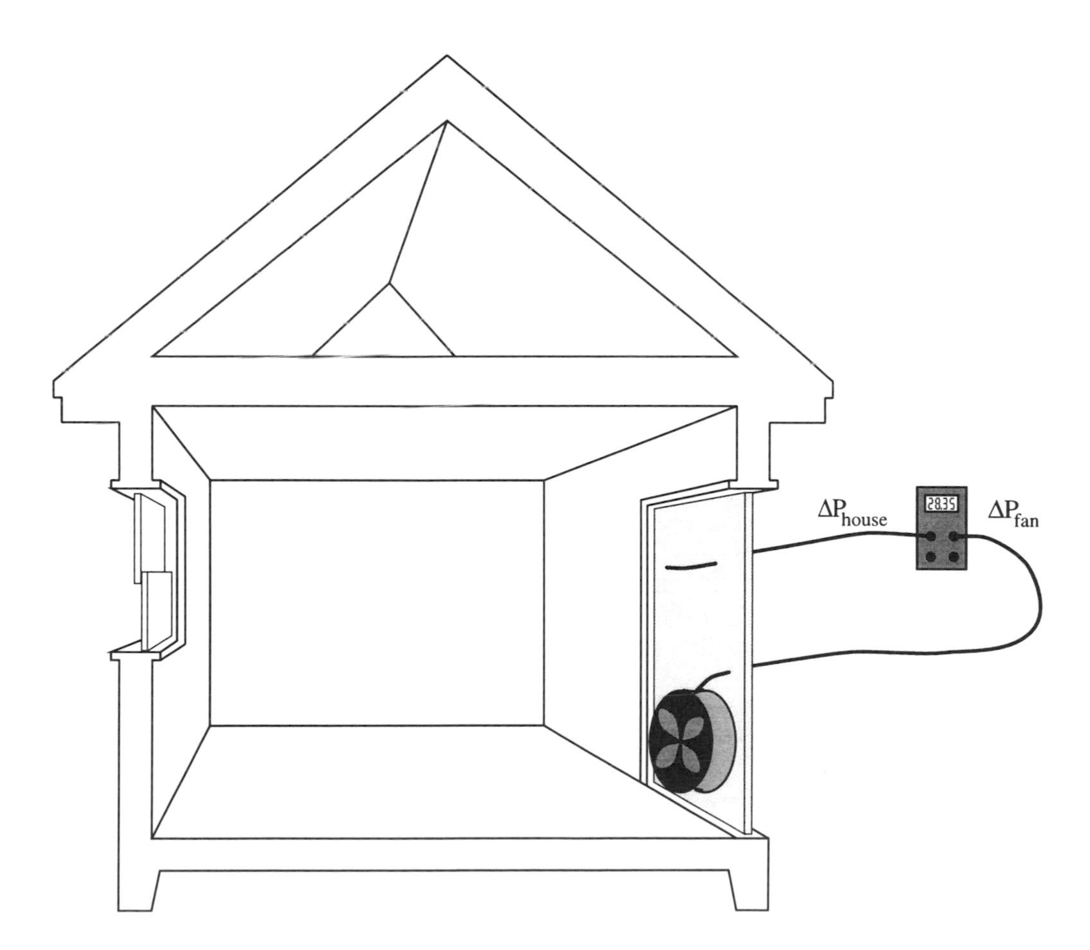

FIGURE 9.7 Sketch of a typical blower door test setup.

The air flow is plotted against the house pressure and a power law relationship of the type $Q = c(\Delta P)^n$ is fitted to the data (c and n are determined from the chosen curve fitting procedure). Using this relationship and modeling the house leakage as orifice flow, the effective leakage area of the house is determined by Eq. (9.6):

$$\mathrm{ELA} = CQ_r\left(\rho/\left(2\Delta P\right)\right)^{0.5} \tag{9.6}$$

where ELA is the effective leakage area at the reference pressure ΔP_r, usually 4 Pa, Q_r is the reference pressure air flow, ρ is the air density, and C is a conversion constant. At 4 Pascal, the ELA is a good estimate of the equivalent area of holes in the envelope that provide the same leakage. Note that the ELA at 4 Pa is determined by extrapolation of measured data.

Other indications of a houses air tightness are often used. Two of these are CFM50 and ACH50. CFM50 is the flow in cubic feet per minute at a house pressure of 50 Pa. ACH 50 is the air change rate at 50 Pa, which is obtained by dividing CFM50 by the house volume in cubic feet. Unlike ELA measurements, which require an entire range of air flow and house pressure data, both CFM50 and ACH50 can be determined from one measurement. This simplifies testing. On the other hand, normalizing ELAs with house floor area enables comparisons between houses—to determine if one house is unusually leaky, for example. These air tightness indications are used for comparisons to standards, to provide background for program or policy decisions, or to estimate the energy load caused by the infiltration.

Tracer gas techniques have become widely used to measure the ventilation rates in buildings. The basic principle involved is that of conservation of mass (of tracer gas) as expressed in the continuity equation. By monitoring the injection and concentration of the tracer, one can infer the exchange of air. Although there is only one continuity equation, there are many different experimental injection strategies and analytical approaches. These different techniques may result in different estimates of infiltration due to uncertainties and biases of the procedures.

An ideal tracer gas is one which is inert, safe, mixes well, does not adhere to surfaces, and its concentration in air is easily measurable. A mass balance of tracer gas within the building, assuming the outdoor concentration is zero, takes the form of Eq. (9.7):

$$\frac{dc(t)}{dt} = F(t) - Q(t)c(t) \tag{9.7}$$

where

V is the volume of the interior space
$c(t)$ is the tracer gas concentration at time t
$dc(t)/dt$ is the time rate of change of tracer concentration
$F(t)$ is the tracer injection rate at time t
$Q(t)$ is the airflow rate of the building at time t

The development of Eq. (9.7) made use of a number of assumptions: the airflow out of the building accounts for the removal of tracer gas, not chemical reactions or absorption onto surfaces, and the tracer gas concentration is uniform throughout the interior space.

There are a number of methods to use in determining infiltration and ventilation rates with a tracer gas. The choice of method in a given situation will depend on the practical details of the experiment as well as the reason for measuring the air change in the first place. A standardized method (ASTM 1990) is the decay method, which requires the least time and usually the least preparation.

In the decay method, a small amount of tracer gas is mixed with the interior air. The injection is then stopped and the concentration of tracer gas is monitored as it decays over time. The air change rate is then determined from the solution of Eq. (9.7):

$$c(t) = c_0 e^{-\text{ACH}t} \tag{9.8}$$

where c_0 is the concentration at time = 0 and ACH = Q/V is the air change rate.

The advantages of the decay method are numerous: the injection rate F need not be measured, the concentration of tracer gas can be measured on site or collected in sample containers and analyzed elsewhere, and the test can be performed with a minimum of equipment and time. Disadvantages include errors introduced by the nonuniform mixing of tracer gas with air, large uncertainties in the air change rate unless the precision of the measuring equipment is high, and biased estimates of the average air change rate.

Other experimental techniques, such as the constant concentration and constant injection techniques are summarized in the ASHRAE *Handbook of Fundamentals*, Chapter 23 (1993). The constant concentration technique can be both accurate and precise, but it requires the most equipment as well as sophisticated control systems and real-time data acquisition. The constant injecting technique (without charge-up) can be considered a somewhat simpler version of the constant concentration technique, in that no active control of the injection rate is needed.

As more detailed information is required for both energy and indoor air quality purposes, researchers are turning to complex, multizone tracer strategies. Both single-gas and multigas techniques are being utilized, but only multigas techniques are capable of uniquely determining the entire matrix of air flows.

Tracer gases are used for a wide range of diagnostic techniques including leak detection and atmospheric tracing. In cases (e.g., mechanically ventilated rooms) in which ventilation rates are known, tracer gases can be used to measure the ventilation efficiency within the zone. Age of air concepts are often used to describe the spatial variation of ventilation. Sandberg (1983) summarizes the definitions and some of the tracer techniques for determining the efficiency (e.g., by seeding inlet streams or monitoring exhaust streams). Further discussion of intrazonal air flows and ventilation efficiency is beyond the scope of this chapter. The reader is referred to Chapter 23 of the 1993 ASHRAE *Handbook of Fundamentals*.

9.2 Review of Thermal Distribution Systems

Thermal distribution systems are the ductwork, piping or other means used to transport heat or cooling effect from the heating or cooling equipment (furnaces, boilers, compressors, etc.) to the building space where it is needed. this section focuses on the distribution system connected to the heating/cooling equipment, rather than on the equipment itself. For a review of the heating/cooling equipment, refer to Chapter 12C. Energy efficiency research in buildings has been primarily focused on the building shell, lighting, appliances, or the space-conditioning equipment. Although the need for improved efficiency in thermal distribution systems has been often cited (Modera, 1989, Cummings and Tooley, 1989), this need has received more attention only in recent years.

Thermal distribution systems are primarily characterized by two transport mediums, air and water. Andrews and Modera (1991) classify the type of distribution system used in residential and small commercial buildings and estimate the potential energy savings possible. In residential applications they found that 85% of the primary energy used for space conditioning was in forced-air systems, with the remainder in hydronic systems. In small commercial buildings the authors reported that 69% of all small commercial buildings in 1986 were heated or cooled with forced air systems, and that the fraction was continually increasing. The focus of this section will be on forced air systems, because this type of system is the largest fraction in buildings and because they have the most potential for efficiency improvements.

There are three primary modes of energy loss associated with air distribution. One is direct air leakage from the ducts to unconditioned spaces or to outside via holes or cracks in the ductwork. This mechanism is mainly a function of the quality and durability of the duct installation. Another is heat conduction through the duct walls resulting from inadequate insulation. The third mode is increased infiltration resulting from increased pressures across building shell leakage sites, pressures which are generated by the operating forced air system. Infiltration is also increased when the system is not operating because the leakage sites in the duct system add to the buildings overall leakage area.

The magnitude of the energy loss depends on many factors: level and location of duct leakage sites, insulation level, location of ducts, space conditioning system sizing, and climate region are a few examples. Typically, the inefficiencies of ducts are at their worst at the time of day when they are needed most. This is due to the extreme temperature differences between the ducts and their surroundings, which increase the conduction losses through the duct walls and worsen return-side leakage losses. This is true in both residential and commercial buildings where the ducts can be located in attics or on rooftops, for example. Because the demand for space conditioning is highest during these peak hours, inefficient ducts exacerbate the problem electric utilities have in meeting the power demand.

Measurement Methods

Measurement techniques available to use in the analysis of distribution efficiencies and to characterize the existing stock of forced air systems are outlined here. Duct leakage area (DLA) is an important parameter used to characterize both direct duct leakage losses and additional envelope leakage area. The most common measurement methods are listed in Table 9.9.

TABLE 9.9 Duct Leakage Area Measurement Methods

No.	Primary Equipment	Result	Accuracy	Reference
1	Blower door	DLA (to outside)	low	(Proctor et al., 1993)
2	Blower door + flow hood	DLA (to outside)	medium	(Proctor et al., 1993)
3	Duct pressurizing fan	DLA (total)	high	(Energy Conservatory, 1995)
4	Duct fan + blower door	DLA (to outside)	high	(Energy Conservatory, 1995)

The blower door subtraction method utilizes two blower door tests, a normal blower door test as described in a previous section, and a second test with the duct system sealed from the house by taping over the duct registers. The DLA is then determined from the difference between the ELAs measured in each test. This method yields duct leakage to outside values only, as the total leakage area of ducts may also include leakage sites to inside the conditioned space. The accuracy of this method is low because the DLA is determined by the subtraction of two large numbers.

In the blower door plus flow hood measurement, the method is similar to a normal blower door test except that a flow capture hood is placed over one unsealed register, usually a return register, and the airflow into the ducts is measured during the test. Simultaneous measurements of house and duct pressure are made, and with the measured flow through the flow capture hood an orifice model is used to determine the DLA. This method has been shown to be more accurate than the subtraction technique.

The third measurement of DLA is the most versatile and is gaining acceptance as the preferred method in field measurements: the fan pressurization technique. The procedure is similar to the determination of the house ELA. In this method, all registers are taped except one, where a fan is connected. The fan has been previously calibrated for volumetric flow rate. With this technique, airflow and duct pressurization measurements are made and the total duct leakage of the supply side, return side, or entire system is determined. Measurements of duct leakage to the outside are made with the help of a blower door. Besides being the most accurate of the three methods, this method has advantages when sealing ducts in retrofit applications.

Actual duct air leakage rates during system operation can be measured by individually measuring the fan air flow rate and air flow through the supply and return registers. The volumetric leakage rate of supply air is then the difference between the fan air flow rate and the sum of the air flow out the supply registers. The method is similar on the return side. Fan air flow rates can be measured with a tracer gas technique (ASTM Standard E741). Air flow rates out of registers can be measured with a flow capture hood or a modified version of a flow capture hood as done in Jump and Modera (1994).

An indication of duct conduction losses can be obtained by measuring temperatures in the supply and return plenums and in each supply register. The percentage of energy lost due to duct conduction is the temperature difference between the supply plenum and register temperature divided by the difference between the supply and return plenum temperatures. Measurement of the distribution system's impact on the building infiltration rate can be measured by the tracer gas or blower door techniques outlined previously.

Alternatively, a duct systems efficiency can be determined by an electric coheating method. This method compares the energy used by the heating equipment (furnace, heat-pump, etc.) and duct system to maintain a set indoor temperature with the energy consumed by electric heaters placed inside the conditioned space to maintain the same indoor temperature. This method requires extensive data acquisition and measurement equipment, but has revealed much insight on the factors that influence a duct system's energy impacts. An example of this method is found in Palmiter and Francisco (1994).

As research into the energy efficiency of air distribution systems has only recently begun, development of standard methods of testing to determine distribution system efficiencies is not yet complete. ASHRAE is currently sponsoring the development of such tests. There are three test pathways currently in development. The first is the design pathway, and it will rely on computer modeling of the building, equipment, and duct system. The other two pathways involve field measurements. The research pathway is intended to obtain the most complete description of the in-situ performance of a duct system. The diagnostic pathway is intended to rely on field measurements which may be obtained quickly and at minimal cost. Each field measurement pathway will contain descriptions of the recommended procedures to analyze an air distribution system's performance. First versions of the ASHRAE test methods should be available by 1997.

Residential Ducts

Forced air heating and cooling systems are used in approximately 50% of existing single family housing in the United States (Andrews and Modera, 1991). In the Northeast and Midwest, 52% of all homes have forced air systems, with 44% of those homes having ducts in unconditioned spaces such as attics or crawl spaces, and 50% having them in partly conditioned spaces, such as basements. In the southern and western United States, 46% of all homes have forced air systems, with 82% of their ducts located in unconditioned spaces.

Field work on existing housing has revealed the potential for efficiency improvements. Modera et al. (1992) measured DLAs (at 4 Pa) of 0.90 cm^2/m^2 (normalized for house floor area) in 19 California houses built before 1980, and 0.92 cm^2/m^2 in post-1980 construction (12 houses). Jump and Modera (1994) measured 1.57 cm^2/m^2 leakage area (at 25 Pa) in 13 houses in Sacramento, California. In terms of conduction losses, Modera et al. reported that 23% of the energy delivered to the air at the coil was lost due to conduction before it arrived at the registers. In the Sacramento houses, Jump and Modera reported a 13% loss. In terms of distribution system induced air infiltration, Gammage (1986) reported an average increase of 80% in the infiltration rate of 31 Tennessee houses when the air handler fan was operating. In five Florida houses, Cummings and Tooley (1989) determined that the infiltration rate tripled when the air handler fan was on and internal doors in the house were open, and that rate further tripled when internal doors were closed. The former effect was attributed to duct system leakage, while the latter effect was attributed to pressure imbalances in the house due to inadequate return-air pathways.

Palmiter and Francisco (1994) measured the system and delivery efficiencies in 24 all-electric homes in the Pacific Northwest. They found that delivery efficiencies averaged 56% for 22 of those homes, which had ducts in unconditioned spaces. The corresponding system efficiency was 71%. In the 2 homes with 50% or more of the ducts in conditioned spaces, they found that delivery efficiencies were 67%, with system efficiencies of 98% on average.

The majority of new single-family housing construction is housing with solid concrete foundations. This practice usually results in duct placement in the attic. This is the most unfavorable location for the distribution system because of the extreme temperature differences that exist between ducts and attics, particularly in summer. In designing homes for energy efficiency, care must be taken to either seal and insulate the ducts well, or locate them in a less harsh environment, preferably inside the conditioned space. Efficiency-conscious home builders use techniques such as fan pressurization to verify minimum leakage levels in the duct system. Stum (1993) has other advice for duct installation in new construction.

Efficiency improvements in existing housing is mainly accomplished by duct sealing and insulation retrofits. Monitoring of duct system retrofits has shown a reduction of space conditioning energy use of up to 20% for the houses studied (Jump and Modera, 1994; Palmiter and Francisco, 1994). The average cost per house in the Jump and Modera study was $600. This cost is approximately 1½ times as much as adding R19 attic insulation.

In new construction, the savings potential is much larger. Actual energy savings are much larger if care is taken at the time of installation, ensuring airtight duct systems and adequate insulation levels. Other efficiency improvements which can be incorporated into new construction are the inclusion of zoned air distribution systems and installing the ducts inside the conditioned space. In the latter case, energy losses from the duct system will not be lost to outside the home.

Commercial Buildings

Commercial building air distribution systems have received far less attention as compared to residential systems. Small commercial buildings have received some attention because of their similarities with residential systems. Andrews and Modera (1991) determined that 69% of small commercial buildings used forced air systems for heating and cooling.

Large commercial building air distribution systems are characterized by extensive networks of ducts, mixing boxes, dampers, in-line fans, and controls. They operate at higher pressures and serve many different building zones and often the building's ventilation system is included. Unlike residential systems, the duct location in large commercial buildings is not typically in a severe environment. Two sources of energy losses in these systems come from warming the conditioned air with the in-line fans and leaky air dampers. Efficiency problems with large commercial duct systems has not received as much attention as residential and small commercial systems.

Advanced Systems

A new application of an existing technology is being developed for energy-efficient cooling, thermal comfort, and high indoor air quality in commercial buildings (Feustel, 1993). These systems provide cooling effect through radiative exchange between humans and water-cooled ceilings or ceiling panels. There are several advantages to radiant hydronic systems. The first is that water is a far better thermal medium than air. Water systems in general do not leak, and when they do, the problem is quickly noted and repaired. Second, the preferred thermal comfort arrangement of a cool head and warm feet is maintained in these systems. Overhead plenum space need no longer be provided. High indoor air quality is maintained by the continual supply of fresh air, eliminating the need to mix fresh and recirculated air as is the case in all-air systems. Such systems are more common in Europe, but some studies have shown their high application potential in the U.S. (Feustel, 1993; Feustel and Stetiu, 1995).

9.3 Tools

A variety of tools and reference materials are available to help estimate the energy flows discussed in this chapter. The four volumes of the ASHRAE Handbook, for example, contain fundamentals of HVAC design as well as applications and examples. This section will focus on computer-based tools for design and analysis.

Whole-Building Simulation Tools

Computer-based building energy simulation tools allow architects, engineers, and researchers much needed flexibility in analyzing a buildings energy performance while preserving accuracy. Development of the various tools has been underway for the past 30 years. The oil shortages of the 1970s greatly accelerated their development. These tools serve a variety of functions, from relatively uncomplicated algorithms that are used to predict a building's peak loads and system requirements to comprehensive room-by-room analysis packages that yield information about the load impacts of specific building components, the performance of heating or cooling equipment, and life cycle costs.

The most comprehensive building energy simulation programs in use today are based on the same modeling strategy. They first determine the loads in each room, considering all heat entering the room as positive, then the required system heat extraction (or addition) rates necessary to maintain the indoor temperature and finally the physical plant size and performance requirements. The flow diagram is shown in Figure 9.8. In the past, each step of the calculations is performed separately—meaning, for example, that the output of the load module cannot be changed by the procedures of the systems module, even though in reality systems operation may affect the building loads. The modules were separated because the simultaneous solution of coupled loads and system and plant modules would require an enormous amount of computer resources (e.g., memory and calculation time). With recent advances in computers, newer simulation programs are able to account for the interaction between components.

The two major public domain software programs, BLAST and DOE-2, follow the algorithm outlined in Figure 9.8. However, each differs in the way the loads are determined. BLAST determines the building loads using a heat balance method, while DOE-2 uses a weighting factor method. The heat balance method is the more fundamental of the two methods, using the conservation of energy principle to calculate room heat transfers. This method requires fewer assumptions than the weighting factor method but more calculations. The basic procedure is to write a heat transfer equation for each surface in the room and also for the room air mass. A matrix solution routine is employed to solve the set of simultaneous equations for the surface temperatures and air temperature. The convective heat flows from each surface are then determined. An overall heat balance on the interior surfaces, interior heat sources, and infiltration determines the required heat that the air conditioning system must remove. This is used as an input to the next module.

The weighting factor method is simpler but more flexible than the heat balance method. It is performed in two steps. In the first step, instantaneous heat gains and losses through the building shell are calculated using the response factor method (RFM). The RFM is the most computationally advantageous way to calculate heat conduction through multilayer, massive walls under time varying conditions. The heat gains and losses through the building shell and internal heat sources are combined under a set of weighting factors to determine the cooling or heating load. The weighting factors are a set of Z-transforms that were obtained for each heat gain by solving the differential equations of heat transfer. The weighting factor method assumes that the heat transfer processes can be linearized and that the system properties are constant. As long as these assumptions are valid, the method yields reasonable results. One advantage of the weighting factor method is that it is able to model somewhat floating room temperatures. A more obvious advantage is that it takes much less computational resources than the heat balance method.

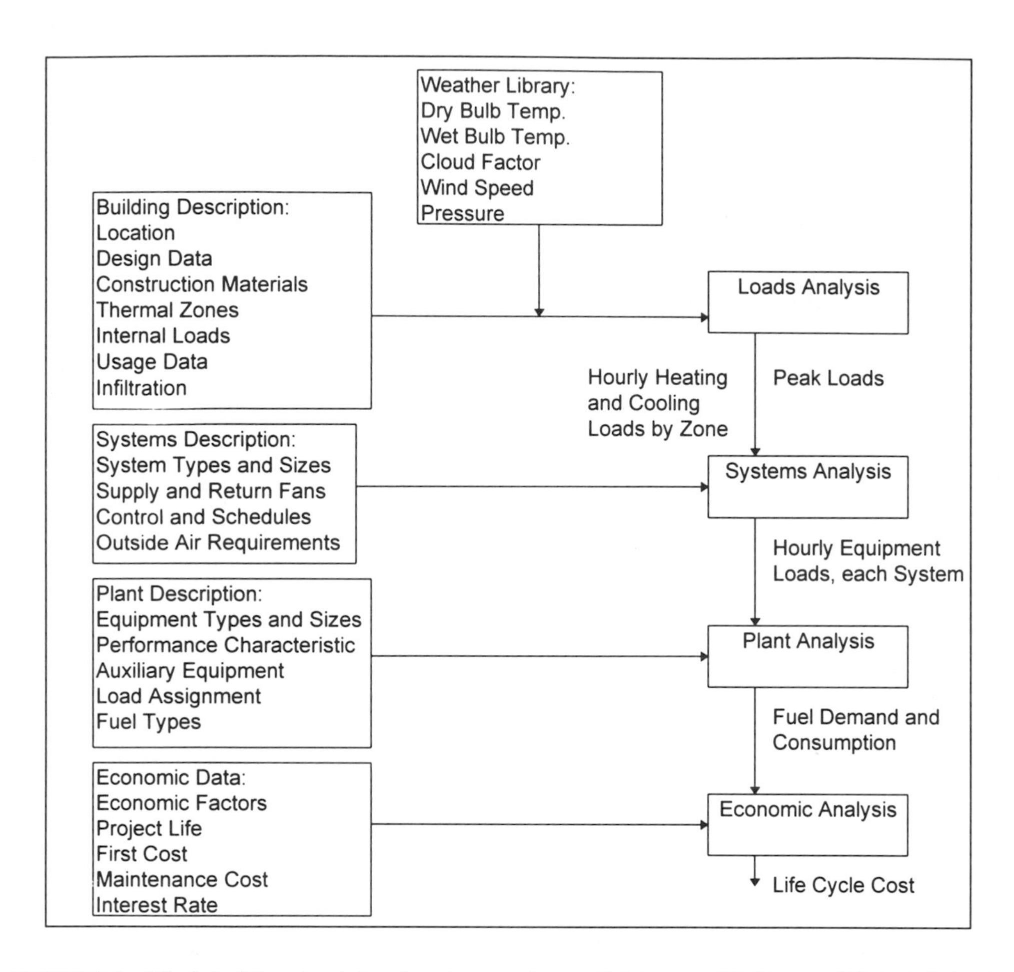

FIGURE 9.8 Whole building simulation flow diagram. *Source:* J.M. Ayres and E. Stamper, "Historical Development of Building Energy Calculations," *ASHRAE Journal*, Vol. 37, No. 2, February 1995. With permission.

The system simulation module next predicts the performance of the chosen HVAC system. Both DOE-2 and BLAST have extensive libraries of different HVAC systems. The user must choose the HVAC system and the various components and control features for each type of system used. The plant simulation module determines the part-load operating conditions under which the plant will operate. The plant libraries contain polynomial equations to approximate the part-load behavior of the selected plant. Most plant information is general, however specific plant performance data can be used, but this information must be supplied by the manufacturer.

Both BLAST and DOE-2, as well as many of the other energy simulation tools of Table 9.10 are now being used as the energy analysis engines behind user-friendly computer interfaces. This is an attempt to make these valuable tools more accepted among architects and design engineers, as well as researchers. A more recent innovation involves linking these engines with CAD software. Table 9.10 contains a short selection and description of building energy simulation software packages that are currently available.

Air Flow and Air Quality Simulation Tools

Air flow models are used to simulate the rates of incoming and outgoing air flows for a building with known leakage under given weather and shielding conditions. Air flow models can be divided into two main categories, single-zone models and multizone models. Single-zone models assume that the structure can be described by a single, well-mixed zone. The major application for this

TABLE 9.10 Whole Building Simulation Tools

Program	Description*	Users†	CAD	Developer
ASG Energy	L	ACEPR	yes	University of Oregon School of Architecture, Pacific Northwest Laboratories
AXCESS	LSP	ERC	no	AXCESS Consulting and Training, Roswell GA
BLAST	LSPEU	ACEPR	some versions	U.S. Army Construction Engineering Research Laboratory, University of Illinois
Builder Guide	LSEU	ACEPR	no	National Renewable Energy Laboratory, Golden, CO
DOE-2	LSPEU	ACEPR	some versions	Lawrence Berkeley National Laboratory, Berkeley, CA
FSEC 2.1	LSP	ER	no	Florida Solar Energy Center, Cape Canaveral, FL
HAP	LSP	ACEP	some versions	Carrier Corporation
SERIRES	LSU	ER	no	National Renewable Energy Laboratory, Golden, CO
TARP	LU	R	no	National Institute of Standards and Technology, Gaithersburg, MD
TRACE	LSP	ACEP	some versions	The Trane Company
TRYNSYS	LSP	ACER	no	Solar Energy Laboratory, University of Wisconsin

* L = loads analysis, S = systems analysis, P = plant analysis, E = economics analysis, U = residential solar analysis

† A = architects, C = energy consultants, E = design engineers, P = HVAC professionals, R = building energy researchers

model type is the single-story, single-family house with no internal partitions. A large number of buildings, however, have structures that would characterize them more accurately as multizone structures. Therefore, more detailed models have been developed, which also take internal partitions into account.

Multizone airflow network models deal with the complexity of flows in a building by recognizing the effects of internal flow restrictions. They require extensive information about flow characteristics and pressure distribution.

As for their single-zone counterparts, these models are based on the mass-balance equation. Unlike the single-zone approach, where there is only one internal pressure to be determined, the multizone model must determine one pressure for each of the zones. This adds considerably to the complexity of the numerical solving algorithm, but by the same token, the multizone approach offers great potential in analyzing infiltration and ventilation airflow distribution.

The advantage of multizone models, besides being able to simulate air flows for larger buildings, is their ability to calculate mass flow interactions between the different zones. Understanding the air mass flow in buildings is important for several reasons:

- Exchange of outside air with inside air necessary for building ventilation
- Energy consumed to heat or cool the incoming air to inside comfort temperature
- Air needed for combustion
- Airborne particles and germs transported by air flow in buildings
- Smoke distribution in case of fire

Literature reviews undertaken in 1985 and 1992 (Feustel and Kendon, 1985; Feustel and Dieris, 1992) showed a large number of different multizone airflow models. One of the first models found was already published in 1970 (Jackman, 1970). Newer airflow models (e.g., CONTAM 93 [Walton, 1993] and COMIS [Feustel and Raynor-Hoosen, 1990]) include pollutant transport models, are more user-friendly and run on personal computers. Furthermore, faster and more robust solvers guarantee shorter calculation times and allow integration of all kinds of flow resistance (e.g., large vertical openings with two-way flows) besides the basic crack flow resistance. The limits of zones and flow paths per zone a model can handle depends on the computer storage.

One of the most bothersome exercises for the modeler is to provide the characteristic parameters for the flow resistance and the outside pressure field for different wind directions. In COMIS, the wind pressure distribution for rectangular shaped buildings can be calculated, using an algorithm derived from a parametric study based on wind tunnel results (Feustel and Raynor-Hoosen, 1990).

Thermal Distribution Simulation Tools

Air Distribution Systems

Air distribution system leakage, conduction losses, and the associated impact on whole-building air infiltration has received little attention in building energy simulation tools. The widely known building simulation program DOE 2.1 used a simple efficiency multiplier of approximately 0.9 (Lawrence Berkeley National Laboratory, 1984) to compensate for duct energy losses. Other simulation tools such as the Thermal Analysis Research Program (TARP) (Walton, 1983) and TRNSYS (University of Wisconsin, 1978) ignored the space conditioning equipment and associated distribution system when calculating building thermal loads.

In more recent times, researchers have begun to consider air distribution systems and their impact on building thermal loads in simulations. An ASHRAE special projects committee (Jakob et al., 1986) showed a reduction of up to 40% in the overall system efficiency in the heating mode. Parker et al. (1993) used a simulation tool, FSEC 2.1 (Kerestecioglu and Gu, 1990), coupled with a detailed duct model to predict duct system impacts on building loads and associated energy consumption. Details considered in the duct model were duct leakage flows based upon duct leakage areas and operating pressures, infiltration impacts across the return ducts and building envelope, and heat storage and heat transfer losses across the duct walls. They found that the impacts of duct leakage were of the largest magnitude and that electrical demand during summer peak hours were significantly increased.

Modera and Jansky (1992) developed a simulation tool to analyze air distribution system energy impacts in residences. The tool is based upon the DOE-2 thermal simulation code (Birdsall et al. 1990), the COMIS airflow network code (Feustel and Raynor-Hoosen, 1990), and a duct performance model developed specifically for the simulation tool. The duct performance model calculates the combined impacts of duct leakage and conduction on duct performance and also acts as the interface between COMIS and DOE-2. One of the major findings of their study was the identification of a thermal siphon loop with a heat exchange rate of more than four times larger than that due to system-off duct leakage. Modera and Treidler (1995) improved the simulation tool in order to look at the thermal siphon effect in more detail and improve the modeling of the duct thermal mass and its effects on duct losses, and model duct impacts on multispeed space conditioning equipment. They estimated thermal siphon loads to be between 5 and 16% of the heating load, and that duct thermal mass effects decrease the energy delivery efficiency of the duct system by 1 to 6%. Most significantly, multispeed air conditioners were shown to be more sensitive to duct efficiency than single-speed equipment, because their efficiency decreases with increasing building load. Subsequent field measurements have shown the simulation tool to be accurate.

Hydronic Systems

Energy savings and peak load impacts of radiant hydronic systems have not been studied as systematically. Stetiu and Feustel (1995), developed a simulation tool to perform sensitivity analyses of nonresidential radiantly cooled buildings. The model is based on a methodology for describing and solving the dynamic, nonlinear equations that correspond to complex physical systems as found in buildings. Accurate simulation of the dynamic performance of hydronic radiant cooling systems is described. The model calculates loads, heat extraction rates, room air temperature and room surface temperature distributions, and can be used to evaluate issues such as thermal comfort, controls, system sizing, system configuration, and dynamic response. The authors present favorable comparisons with available field data.

Window Thermal Analysis and Daylighting/Fenestration Tools

Fenestration software programs are used to determine the windows thermal performance and daylighting capabilities. The software can be either stand-alone or used as a front end of a whole building energy simulation program. To facilitate their use in the building design stage, window analysis software often has CAD-compatible inputs for geometric details or graphics displays.

TABLE 9.11 Sample of Fenestration Software

Software	Analysis	CAD	Building Simulation	Platform	Developer
ADELINE	thermal/ lighting	yes	yes	DOS	Lawrence Berkeley National Laboratory, International Energy Agency
AGI	lighting	yes	no	DOS	Lighting Analysis Corporation
CALA	lighting	yes	no	DOS	Holophane Corporation
Building Design Advisor	thermal/ lighting	yes	yes	WINDOWS	Lawrence Berkeley National Laboratory
Lumen-Micro	lighting	yes	no	DOS	Lighting Technologies Inc.
LUXICON	lighting	yes	no	WINDOWS	Cooper Lighting
Radiance	lighting	yes	no	UNIX	Lawrence Berkeley National Laboratory
RESFEN	thermal	no	yes	DOS	Lawrence Berkeley National Laboratory
SUPERLITE	lighting	no	no	DOS	Lawrence Berkeley National Laboratory, International Energy Agency
WINDOW 4.1	thermal	no	no	DOS	Lawrence Berkeley National Laboratory

Software developers are continually improving the user interface to make the programs more accessible to building designers, who may not have time or design fees to use less friendly tools.

The thermal performance of a window is characterized by its overall *U*-value, solar heat gain coefficient, and shading coefficient. These factors depend on the number of panes, gas-filled spaces, low-emittance coatings, frame material and construction, and so on. Window thermal performance software does the required bookkeeping and calculations to determine the parameters important to a window's impact on the thermal loads in a building. Such programs can be used to design and develop new window products and compare the performance characteristics of different types of windows.

The daylighting capabilities of windows are important to consider in the design phase. Some fenestration programs are designed to demonstrate a window's ability to illuminate a space. Some of these programs can calculate the interreflection of light from surfaces in the space and present the results in high-resolution photorealistic graphical displays using a variety of ray-tracing and radiosity techniques. This offers a significant advantage over traditional software that provides simple numerical results or isolux contours. The results tell the designer where the room is under- or overilluminated and where glare problems may exist. These programs can also be integrated into whole building energy simulation programs and thus present a complete picture of the impacts of the windows on the building's energy efficiency and comfort. Table 9.11 presents a sample of available lighting or fenestration software. A more complete review of the lighting or daylighting design software can be found in (IESNA, 1995).

Conclusion

This chapter presented the issues that determine the quality of the indoor environment and the energy issues that affect them. Many facets of building properties and energy services that affect comfort levels of occupants, their health, and their productivity were reviewed. These included conductive, convective, and radiant heat transmission through the building envelope, ventilation and infiltration, and thermal distribution. It was demonstrated how improvements in the building envelope or thermal distribution system can provide the same services, but much more efficiently. Technical advances in construction materials, insulation, windows, and paint provide the means to control building loads or use them to advantage. This reduces the requirements of heating or cooling in the building, thus reducing energy consumption, operating costs, and peak power demand. Whole building simulation programs allow the building designer or retrofitter to evaluate an energy service or shell technology's impact on a building's energy efficiency. These tools give architects the information and means necessary to evaluate a proposed building's thermal and visual comfort, heating and cooling equipment, energy budgets, design cost-effectiveness, and so on. The benefit to society is reduced pressure on limited natural resources, independence from foreign fuel supplies, less demand for new power plants, and reduced air pollution and groundwater contamination.

References

Akbari, H., S. Bretz, H. Taha, D. Kurn, and J. Hanford, "Peak Power and Cooling Energy Savings of High-Albedo Roofs," Lawrence Berkeley National Laboratory Report LBL-34411, Berkeley, CA, 1992.

Akbari, H., S. Bretz, H. Taha, D. Kurn, and J. Hanford, "Peak Power and Cooling Energy Savings of Shade Trees," Lawrence Berkeley National Laboratory Report LBL-34411, Berkeley, CA, 1994.

AIVC, "A Review of Building Airtightness and Ventilation Standards," TN 30, Air Infiltration and Ventilation Centre, UK, 1990.

AIVC, "Air-to-Air Heat Recovery in Ventilation," TN 45, Air Infiltration and Ventilation Centre, UK, 1994.

ASHRAE Standard 119, "Air Leakage Performance for Detached Single-Family Residential Buildings," American Society of Heating, Refrigerating and Air conditioning Engineers, 1988.

ASHRAE Standard 62, "Air Leakage Performance for Detached Single-Family Residential Buildings," American Society of Heating, Refrigerating and Air conditioning Engineers, 1989.

ASHRAE Handbooks, American Society of Heating, Refrigerating and Air conditioning Engineers 1989–1993.

ASHRAE Standard 55, "Thermal Environmental Conditions for Human Occupancy," American Society of Heating, Refrigerating and Air conditioning Engineers, 1992.

ASHRAE Standard 136, "A Method of Determining Air Change Rates in Detached Dwellings," American Society of Heating, Refrigerating and Air conditioning Engineers, 1993.

ASTM STP 1067, "Air Change Rate and Airtightness in Buildings, American Society of Testing and Materials," M.H. Sherman, Ed., 1990.

ASTM Standard E741-83, "Standard Test Method for Determining Air Leakage Rate by Tracer Dilution," *ASTM Book of Standards,* American Society of Testing and Materials, Vol. 04.07, 1994.

ASTM Standard E779-87, "Test Method for Determining Air Leakage by Fan Pressurization," *ASTM Book of Standards,* American Society of Testing and Materials, Vol. 04.07, 1991.

ASTM Standard E1186-87, "Practices for Air Leakage Site Detection in Building Envelopes," *ASTM Book of Standards,* American Society of Testing and Materials, Vol. 04.07, 1991.

Andrews, J.W. and M.P. Modera, "Energy Savings Potential for Advanced Thermal Distribution Technology in Residential and Small Commercial Buildings," Prepared for the Building Equipment Division, Office of Building Technologies, U.S. Dept. of Energy, Lawrence Berkeley National Laboratory Report, LBL-31042, 1991.

Andrews, J.W. and M.P. Modera, "Thermal Distribution in Small Buildings: A Review and Analysis of Recent Literature," Brookhaven National Laboratory Report, BNL-52349, September 1992.

Ayres, J.M. and E. Stamper, "Historical Development of Building Energy Calculations," *ASHRAE Journal,* Vol. 37, No. 2, February 1995.

Birdsall, B., W.F. Buhl, K.L. Ellington, A.E. Erdem, and F.C. Winkelmann, "Overview of the DOE-2 Building Energy Analysis Program, Version 2.1D," Lawrence Berkeley National Laboratory Report LBL-19735 Rev.1, Lawrence Berkeley National Laboratory, Berkeley, CA, July 1992.

Blasnik, M. and J. Fitzgerald, "In Search of the Missing Leak," *Home Energy,* Vol. 9, No. 6, November/December 1992.

Blomsterberg, A., "Air Leakage in Dwellings," Dept. Bldg. Constr. Report No. 15, Swedish Royal Institute of Technology, 1977.

Cummings, J.R. and J.J. Tooley Jr., "Infiltration and Pressure Differences Induced by Forced Air Systems in Florida Residences," *ASHRAE Trans.*, Vol. 95, Pt. 2, 1989.

Daisey, J.M., "Buildings of the 21st Century: A Perspective on Health and Comfort and Work Productivity," presented at the International Energy Agency's Workshop on Buildings of the 21st Century: "Developing Innovative Research Agendas," Gersau, Switzerland, May 16–18, 1989.

Davids, B.J., "Taking the Heat Out of Sunlight—New Advances in Glazing Technology for Commercial Buildings," Proc. ACEEE Summer Study on Energy Efficiency in Buildings, August 26–September 1, Asilomar Conference Center, Pacific Grove, CA, 1990.

Energy Conservatory, "Minneapolis Duct Blaster Operation Manual," The Energy Conservatory, 5158 Bloomington Ave, S, Minneapolis, MN 55417, 1995.

Fanger, P.O., *Thermal Comfort Analysis and Applications in Environmental Engineering,* McGraw-Hill, New York, 1970.

Feustel, H.E., "Hydronic Radiant Cooling—Preliminary Performance Assessment, Lawrence Berkeley National Laboratory Report, LBL-33194, 1993.

Feustel, H.E. and C. Stetiu, "Hydronic Radiant Cooling—Preliminary Assessment," *Energy and Buildings,* Vol. 22, 1995.

Feustel, H.E. and V.M. Kendon, "Infiltration Models for Multicellular Structures—A Literature Review," *Energy and Bulidings,* Vol. 8, 1985.

Feustel, H.E. and A. Raynor-Hoosen, "Fundamentals of the Multizone Airflow Model COMIS," Technical Note 29, Air Infiltration and Ventilation Centre, Warwick, UK, 1990, also Lawrence Berkeley National Laboratory Report LBNL-28560, 1990.

Feustel, H.E. and J. Dieris, "A Survey of Airflow Models for Multizone Structures," *Energy and Buildings,* Vol. 18, 1992.

Gammage, R.B., A.R. Hawthorne, and D.A. White, "Parameters Affecting Air Infiltration and Air Tightness in Thirty-One East Tennessee Homes," In *Measured Air Leakage of Buildings, ASTM STP 904,* H.R. Trechsel and P.L. Lagus, Eds., pp. 61–69, American Society for Testing and Materials, Philadelphia, 1986.

Griffith, B. and D. Arasteh, "Gas Filled Panels: A Thermally Improved Building Insulation," Proc. ASHRAE/DOE/BTECC Conference: Thermal Performance of the Exterior Envelopes of Buildings V, Clearwater Beach, FL, December 1992.

Holman, J.P., *Heat Transfer,* 4th Edition, McGraw-Hill, New York, 1976.

IES, "Recommended Practices of Daylighting," IES RP-5, Illuminating Engineering Society of North America, 120 Wall St., 17th Floor, New York, NY 10005, 1979.

IESNA, "1995 Software Summary, Lighting Design and Applications," Vol. 25, No. 7, Illuminating Engineering Society of North America, 120 Wall St., 17th Floor, New York, NY 10005, July 1995.

Jackman P.J. "A Study of Natural Ventilation of Tall Office Buildings," *J. Inst. Heat. Vent. Eng.,* 38, 1970.

Jakob, F.E., D.W. Locklin, P.E. Fisher, L.J. Flanigan, and R.A. Cudnick, "SP43 Evaluation of Systems Options for Residential Forced Air Heating," *ASHRAE Trans.,* Vol. 92, Pt. 2, Atlanta, GA, 1986.

Jones, P., *How to Cut Heating and Cooling Costs,* Butterick Publishing, New York, 1979.

Jump, D.A. and M.P. Modera, "Energy Impacts of Attic Duct Retrofits in Sacramento Houses," Proc. ACEEE 1994 Summer Study, Pacific Grove, CA, August 28–September 3, 1994, American Council for an Energy Efficient Economy, 1001 Connecticut Ave., NW, Suite 801, Washington, DC 20036, 1994.

Kerestecioglu, A. and L. Gu, "Theoretical and Computational Investigation of Heat and Moisture Transfer in Buildings: Evaporation and Condensation Theory," *ASHRAE Trans.,* Vol. 96, Pt. 1, 1990.

Lawler, E.E. III and L.W. Porter, "The Effect of Performance on Job Satisfaction," *Industrial Relations,* Vol. 7, pp. 20–28, 1967.

Lawrence Berkeley National Laboratory, "DOE 2.1 Supplement, Version 2.1C," Building Energy Simulation Group, Lawrence Berkeley National Laboratory, Berkeley, CA, 1984.

Modera, M.P., "Residential Duct System Leakage: Magnitude, Impacts, and Potential for Reduction," *ASHRAE Trans.*, Vol. 95, Pt. 2, 1989, also LBL-26575.

Modera, M.P., D.J. Dickerhoff, R.E. Jansky, and B.V. Smith, "Improving the Energy Efficiency of Residential Air Distribution Systems in California—Final Report: Phase 1," Lawrence Berkeley National Laboratory Report, LBL-30886, 1991.

Modera, M.P. and E.B. Treidler, "Improved Modeling of HVAC System/Envelope Interactions in Residential Buildings," Proc. 1995 ASME International Solar Energy Conference (March 19–24), 1995.

Modera, M.P. and R. Jansky, "Residential Air-Distribution Systems: Interactions with the Building Envelope," Lawrence Berkeley National Laboratory Report LBL-31311, UC-350, Lawrence Berkeley National Laboratory, Berkeley, CA, July 1992.

Modera, M.P., J.C. Andrews, and E. Kweller, "A Comprehensive Yardstick for Residential Thermal Distribution Efficiency," Proc. ACEEE 1992 Summer Study, Pacific Grove, CA, August 30–September 5, 1992, American Council for an Energy Efficient Economy, 1001 Connecticut Ave., NW, Suite 801, Washington, DC 20036, 1992.

Modera, M.P. and D.A. Jump, "Field Measurement of the Interactions Between Heat Pumps and Duct Systems in Residential Buildings," Proc. 1995 ASME International Solar Energy Conference (March 19–24), 1995; also LBL-36047, Hawaii.

National Fenestration Rating Council (NFRC), "Certified Products Directory," 4th Edition, Silver Springs, MD, January 1995.

Nero, A.V. Jr., "Personal Methods of Controlling Exposure to Indoor Air Pollution," *Principles and Practice of Environmental Medicine,* New York, Plenum Medical Book Company, 1992.

Oak Ridge National Laboratory (ORNL), *Building Foundation Design Handbook,* ORNL/SUB/86-72143/1, Oak Ridge National Laboratory, Oak Ridge, TN, 37831, May 1988.

Palmiter, L.E. and P.W. Francisco, "Measured Efficiency of Forced-Air Distribution Systems in 24 Homes," Proc. ACEEE 1994 Summer Study, Pacific Grove, CA, August 28–September 3, 1994, American Council for an Energy Efficient Economy, 1001 Connecticut Ave. NW, Suite 801, Washington, DC 20036, 1992.

Parker, D., P. Fairey, and L. Gu, "Simulation of the Effects of Duct Leakage and Heat Transfer on Residential Space-Cooling Energy Use," *Energy and Buildings,* Vol. 20, No. 2, 1993.

Proctor, J., M. Blasnik, B. Davis, T. Downey, M.P. Modera, G. Nelson, and J.J. Tooley, Jr., "Leak Detectors: Experts Explain the Techniques," *Home Energy,* Vol. 10, No. 5, pp. 26–31, September/October 1993.

Romm, J.J. and W.D. Browning, "Greening the Building and the Bottom Line: Increasing Productivity Through Energy Efficient Design," Proc. ACEEE 1994 Summer Study on Energy Efficiency in Buildings, Panel 9, Demonstrations and Retrofits, Pacific Grove, CA, September 1994.

Samet, J.M., M.C. Marbury, and J.D. Spengler, "Health Effects and Sources of Indoor Air Pollution. Part II," *Am. Rev. Respir. Dis.,* Vol. 137, pp. 221–242, 1988.

Sandberg, M. and M. Sjoberg, "The Use of Moments for Assessing Air Quality in Ventilated Rooms," *Building & Environment,* Vol. 18, p. 181, 1983.

Sherman, M.H., "Estimation of Infiltration from Leakage and Climate Indicators," *Energy and Buildings,* Vol. 10, No. 1, p. 81, 1987.

Sherman, M.H., and D.T. Grimsrud, "The Measurement of Infiltration using Fan Pressurization and Weather Data," Proc. First International Air Infiltration Centre Conference, London, England. Lawrence Berkeley National Laboratory Report, LBL-10852, October 1980.

Sherman, M.H., and N.E. Matson, "Ventilation-Energy Liabilities in U.S. Dwellings," Proc. 14th AIVC Conference, pp. 23–41, 1993, LBL Report No. LBL-33890, 1994, Copenhagen, Denmark.

Sherman, M.H., and M.P. Modera, "Infiltration Using the LBL Infiltration Model," Special Technical Publication No. 904, "Measured Air Leakage Performance of Buildings," pp. 325–347. ASTM, Philadelphia, PA, 1984.

Stetiu, C. and H.E. Feustel, "Development of a Model to Simulate the Performance of Hydronic Radiant Cooling Ceilings," presented at the ASHRAE Summer Meeting, San Diego, CA, June 1995.

Stokes, B., "Worker Participation—Productivity and the Quality of Work Life," Worldwatch Paper 25, Worldwatch Institute, December 1978.

Stum, K., "Guidelines for Designing and Installing Tight Duct Systems," *Home Energy*, Vol. 10, No. 5, pp. 55–59, September/October 1993.

Treidler, E.B. and M.P. Modera, "Peak Demand Impacts of Residential Air-Conditioning Conservation Measures," Proc. ACEEE 1994 Summer Study, Pacific Grove, CA, August 28–September 3, 1994, American Council for an Energy Efficient Economy, 1001 Connecticut Ave. NW, Suite 801, Washington, DC 20036, 1992.

United States Congress, Office of Technology Assessment, "Building Energy Efficiency," OTA-E-518, Washington DC: U.S. Government Printing Office, May, 1992.

United States Environmental Protection Agency, "Cooling Our Communities: A Guidebook on Tree Planting and Light-Colored Surfaces," USEPA Report No. 22P-2001, Office of Policy Analysis, Climate Change Division, Washington, DC, January 1992.

University of Wisconsin, "TRNSYS, A Transient Simulation Program," University of Wisconsin Experiment Station, Report 38-9, Madison WI, 1978.

Walton, G.N., "Thermal Analysis Research Program," National Bureau of Standards, NSBIR 83-2655, Washington DC, 1983.

Walton, G. "CONTAM 93, User Manual," NISTIR 5385, National Institute of Standards and Technology, 1993, Chapter 53. "Thermal Insulation and Airtightness Building Regulations," Royal Ministry of Local Government and Labor, Norway, 27 May, 1987.

10

Electrical Energy Management in Buildings

Craig B. Smith
Daniel, Mann, Johnson, & Mendenhall

10.1 Principal Electricity Uses in Buildings

Introduction: The Importance of Energy Efficiency in Buildings

A typical building is designed for a 40-year economic life. This implies that the existing inventory of buildings—with all their good and bad features—is turned over very slowly. Today we know it is cost-effective to design a high degree of energy efficiency into new buildings, because the savings on operating and maintenance costs will repay the initial investment many times over. Many technological advances have occurred in the last two decades, resulting in striking reductions in the energy usage required to operate buildings safely and comfortably. An added benefit of these developments is the reduction in air pollution, which has occurred as a result of generating less electricity.

There are hundreds of building types, and buildings can be categorized in many ways—by use, type of construction, size, or thermal characteristics, to name a few. For simplicity, two designations will be used here: residential and nonresidential.

The residential category includes features common to single-family dwellings, apartments, and hotels. The nonresidential category includes a major emphasis on office buildings, as well as a less detailed discussion of features common to retail stores, hospitals, restaurants, and laundries. There is an estimated 5 million commercial buildings totalling 65 billion square feet in the United States today. Most of this space is contained in buildings larger than 10,000 square feet. Industrial facilities are not included here, but are discussed in Chapters 14 and 16. The extension to other types is either obvious or can be pursued by referring to the literature listed in the references.

The approach taken in this chapter is to list two categories of specific strategies that are cost-effective methods for conserving electricity. The first category includes those measures that can be implemented at low capital cost using existing facilities and equipment in an essentially unmodified state. The second category includes technologies that require retrofitting, modification of existing

0-8493-2514-5/96/$0.00+$.50

TABLE 10.1 Electricity End Use in the Residential/Commercial Sectors—1992 (10^{15} Btu)

End Use	Residential Sector	Commercial	Total	Percent
Spacing heating	0.33	0.06	0.39	6.4
Space cooling	0.51	0.26	0.77	12.6
Water heating	0.34	0.03	0.37	6.1
Refrigeration	0.52	*	0.52	8.5
Cooking	0.15	*	0.15	2.5
Clothes dryers	0.18	—	0.18	3.0
Freezers	0.14	—	0.14	2.3
Lighting	0.30	1.20	1.50	24.5
Office equipment	—	0.41	0.41	6.7
Other uses	0.72	0.95	1.67	27.4
Total	3.19	2.91	6.10	100

* Included in "other uses".
Source: Ref 1, Table A4, A5 pp. 60, 62.

equipment, or new equipment or processes. Generally, moderate to substantial capital investments are also required.

Residential Uses

Total energy use by the residential sector has remained essentially constant since the oil embargo, despite the fact that the number of households has increased by about 20%. This is the best measure of the success of energy efficiency in the residential sector. Electricity use in the residential and commercial sectors has actually increased since 1975. Current values are shown in Table 10.1. The most significant end uses are heating, ventilating, and air conditioning (HVAC) and lighting, followed by refrigeration, office equipment, and water heating. The increase in office electricity use is one significant change in the 15 years since this first edition of this book.

HVAC energy use accounts for approximately 20% of total electricity use. Electricity is used in space heating and cooling to drive fans and compressors, as a direct source of heat (resistance heating), or as an indirect heat source (heat pumps). Heat pumps are discussed elsewhere (Chapter 12).

Lighting, which accounts for about 20% of U.S. electricity use (7% of total U.S. energy use), amounts to 9% of the electricity used in the residential sector. This represents a decrease from 10% in the 1980s, due to the introduction of more efficient lamps, principally compact fluorescents. The bulk of residential lighting is incandescent, and offers substantial opportunities for improved efficiency.

Water heating amounts to about 11% of residential electricity use. Electricity use for this purpose currently occurs in regions where there is cheap hydroelectricity or where alternative fuels are not available. Solar water heating (discussed in Chapter 19) is another alternative that is used on a limited basis.

Refrigerators are another important energy use in the residential sector, accounting for about 16% of electricity. The vast majority of refrigerators in use today are electric, and efficiency has increased during the past two decades as new standards have been implemented.

Cooking and other appliances account for the balance of electricity use in the residential sector (15% of electricity).

Nonresidential Uses

For the commercial sector as a whole, lighting accounts for over 40% of electricity use, followed by office equipment (14%) and HVAC (11%).

In nonresidential buildings where space conditioning is used, HVAC is one of the major electricity end uses. There are exceptions of course—in energy-intensive facilities such as laundries, the process energy will be most important. Electricity is used to run fans, pumps, chillers, and cooling towers. Other uses include electric resistance heating (for example, in terminal reheat systems) or electric boilers.

Nonresidential lighting is generally next in importance to HVAC for total use of electricity, except in those nonresidential facilities with energy-intensive processes. In a typical office building, lighting consumes approximately 25% of the total energy used in the building and approximately 50% of the electricity. Lighting is generally fluorescent, with a growing use of high-intensity-discharge (e.g., high-pressure sodium or metal halide) lamps, and with a small fraction of incandescent lamps. Incandescent lamps are largely used in older buildings or for decorative or esthetic applications.

Water heating is another energy use in nonresidential buildings, but here circulating systems (using a heater, storage tank, and pump) are more common. Many possibilities exist for using heat recovery as a source of hot water.

Refrigeration is an important use of energy in supermarkets and several other types of nonresidential facilities. It is now common practice to include heat recovery units on commercial refrigeration systems.

Computers are an increasingly common feature of office buildings and many nonresidential facilities. For the larger units, specially designed rooms with temperature and humidity control are required. As a rule of thumb, the electricity used by the computer must be at least doubled, since cooling must be provided to remove the heat from the equipment, lights, and personnel. The trend toward office automation has increased the installation of other types of office equipment, including facsimile machines, copiers, computer printers, typewriters, electronic mail, and telephone message systems.

In nonresidential facilities, the balance of the electricity use is for elevators, escalators, and miscellaneous items.

10.2 Strategies for Electricity End-Use Management

Setting Up an Energy Management Program

The general procedure for establishing an energy management program in buildings involves five steps:

- Review historical energy use.
- Perform energy audits.
- Identify energy management opportunities.
- Implement changes to save energy.
- Monitor the energy management program, set goals, review progress.

Each step will be described briefly.

Review of Historical Energy Use

Utility records can be compiled to establish electricity use for a recent 12-month period. These should be graphed on a form (see Figure 10.1) so that annual variations and trends can be evaluated. By placing several years (e.g., last year, this year, and next year projected) on the form, past trends can be reviewed and future electricity use can be compared with goals. Alternatively, several energy forms can be compared, or energy use vs. production determined (meals served, for a restaurant; kilograms of laundry washed, for a laundry; etc.).

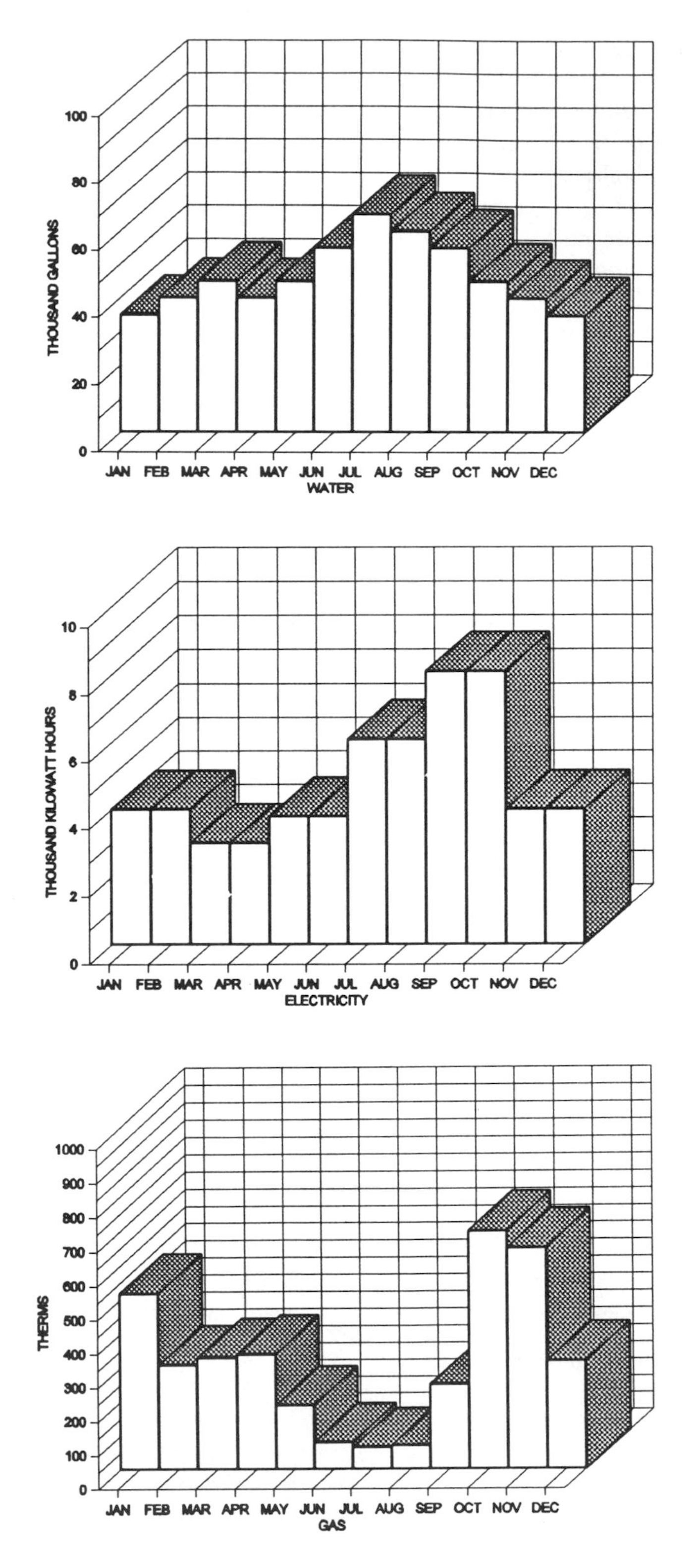

FIGURE 10.1 Sample graph: historical energy use in an office building.

Building Description

- Name:_______________ Age: ___years heating degree days ___
- Location: ____
- No. of floors ____ Gross floor area ___ m² (ft²) Net floor area ___m² (ft²)
- Percentage of surface area which is glazed ___% cooling degree days ___
- Type of air conditioning system; heating only___ evaporative ____ dual duct ____ other (describe) ______________________________
- Percentage breakdown of lighting equipment: Incandescent _____%

Building Mission

- What is facility used for:
- Full time occupancy (employees) ____ persons
- Transient occupancy (visitors or public) _____ persons
- Hours of operations per year ____
- Unit of production per year ____ Unit is___________

Installed Capacity

- Total installed capacity for lighting ___ kW
- Total installed capacity of electric drives greater than 7.5 kW (10 hp) (motors, pumps, fans, elevators, chillers, etc.)___hp x 0.746 = ___ kW
- Total steam requirements ___lbs/day or ___kg/day
- Total gas requirements ___ ft³/day or BTU/hr or ___m³/day
- Total other fuel requirements ______________________________

Annual Energy End Use

Energy Form	x Conversion		kBTU/yr	Metric Units		Conversion		MJ/yr
• Electricity	___kWh/yr x 3.41	=	_____	_____ kWh/yr	x	3.6	=	_____
• Steam	___lb/yr x 1.00	=	_____	_____ kg/yr	x	2.32	=	_____
• Natural gas	___cf/yr x 1.03	=	_____	_____ m3/yr	x	38.4	=	_____
• Oil	___gls/yr x {#2 139, #6 150}	=	_____	_____ ℓ/yr	x	{#2 38.9, #6 41.8}	=	_____
• Coal	___tons/yr x 24,000	=	_____	_____ kg/yr	x	28.0	=	_____
• Other	___ x		_____	_____	x		=	_____
	Totals		_____					_____

Energy Use Performance Factors (EUPF's) for Building

- EUPF 1 = MJ/yr(kBTU/yr) ÷ Net floor area = ______ MJ/m²yr(kBTU/ft² yr)
- EUPF 2 = MJ/yr(kBTU/yr) ÷ Average annual occupancy = ______ MJ/person · yr (kBTU/person · yr)
- EUPF 3 = MJ/yr(kBTU/yr) ÷ Annual units of production = ______ MJ/unit · yr (kBTU/unit · yr)

FIGURE 10.2 Building energy survey form.

Perform Energy Audit

Figures 10.2 and 10.3 are data sheets used in performing an energy audit of a building. The Building Energy Survey Form, Figure 10.2, provides a gross indication of how energy is used in the building in meeting the particular purpose for which it was designed. This form would not be applicable to single-family residences, but it could be used with apartments. It is primarily intended for commercial buildings.

Figure 10.3 is a form used to gather information concerning energy used by each piece of equipment in the building. When totaled, the audit results can be compared with the historical energy use records plotted on Figure 10.1. The energy audit results show a detailed breakdown and permit identification of major energy using equipment items.

Another way to perform energy audits is to use a microcomputer and a commercially available database program to record the data and make the calculations. If the work load is extensive, the database can include look-up tables of frequently used electrical loads, utility rates, and other essential information to automate the process. Teams of engineers with portable computers have been used to rapidly survey and collect the energy data from large commercial facilities. See Figure 10.4 for an example of an audit result.

Facility Name____________ Date____________ By________ Sheet ___ of ___
Location____________ Period of Survey: 1 day 1 wk 1 mo 1 yr
Symbols: K = 10^3 M = 10^6 Notes____________

Conversion Factors

Multiply	by	to get
kWh	3.6	MJ
Btu/hr	0.000293	kW
hp	0.746	kW

Item No.	Equipment Description	Power			Est. Hrs Use Per Period	kWh	Conv. Factor	Total Energy Use Per Period (MJ)
		Name Plate Rating (Btu/hr, kW, hp, etc.)	Conv. Factor to kW	Est.% Load (100%, 50%, etc.)				

FIGURE 10.3 Energy audit data sheet.

Still another way of making an energy audit is to use commercially available computer software that estimates energy use for typical building occupancies based on size, type, climate zone, and other identifying parameters. These programs are not as accurate as an actual survey but can be used as a preliminary screening criteria to select the buildings worthy of more detailed investigations.

Identify Energy Management Opportunities

An overall estimate should be made of how effectively the facility uses its energy resources. This is difficult to do in many cases, because so many operations are unique. An idea can be obtained, however, by comparing similar buildings having similar climates. Table 10.2 shows representative values and indicates the range in performance factors that is possible.

Next, areas or equipment which use the greatest amounts of electricity should be examined. Each item should be reviewed and these questions asked:

- Is this actually needed?
- How can the same equipment be used more efficiently?
- How can the same purpose be accomplished with less energy?
- Can the equipment be modified to use less energy?
- Would new, more efficient equipment be cost effective?

Implement Changes

Once certain actions to save energy have been identified, an economic analysis will be necessary to establish the economic benefits and to determine if the cost of the action is justified. (Refer to Chapter 3 for guidance.) Those changes which satisfy the economic criteria of the building owner (or occupant) will then be implemented. Economic criteria might include a minimum return on investment (e.g., 25%), a minimum payback period (e.g., 2 years) or a minimum benefit-cost ratio (e.g., 2.0).

XYZ Corporation

Lighting Energy Savings Summary				Prepare by:	
Existing Annual kWh:	181,828 kWh	Existing kW Draw:	36.37 kW	Annual Energy $$ Saved:	$12,094
Proposed Annual kWh:	95,234 kWh	Proposed kW Draw:	18.52 kW	Estimated PRE-REBATE Cost:	$13,592
Annual kWh Savings:	86,594 kWh	KW Savings:	17.85 kW		
% kWh Savings:	47.6%	% kW Savings:	49.1%		

Lighting Inventory, Recommendations, and Savings

		Existing				Recommended			Savings			Estimated	
Item #	Location	Weekly Hours	Qty	Fixtures	Watts/ Fix	Qty	Fixture	Watts/ Fix	kW	Annual kWh	Annual Energy $	Unit Cost	Total Cost
1	Presidents Office	60	8	75 Watt INC Spotlight	75	8	18 Watt CFL/SI/Ref.	18	0	1368	$219	$22	$176
2	Presidents Office	60	6	2-F40T12(40W)/STD	96	6	2-F32T8(32W)/ELEC	61	0	630	$101	$48	$288
3	V.P. Office	60	8	75 WATT INC Spotlight	75	8	18 Watt CFL/SI/Ref.	18	0	1,368	$219	$22	$176
4	V.P. Office	60	4	2-F40T12(40W)/STD	96	4	2-F32T8(32W)/ELEC	61	0	420	$67	$48	!92
5	Night Lighting	168	4	2-F40T12(40W)/STD	96	4	2-F32T8(32W)/ELEC	61	0	1176	$145	$48	$192
6	Women's Restroom Mirror	60	12	25 Watt INC	25		None						
7	Women's Restroom	60	6	2-F40T12(34W)/U/STD	94	6	2-F40T12(34W)/U/ELEC	60	0	612	$98	$48	$288
8	Men's Restroom	60	6	2-F40T12(34W)/U/STD	94	6	2-F40T12(34W)/U/ELEC	60	0	612	$98	$48	$288
9	Main Office Area	60	56	3-F40T12(40W)/2-Class 1 1 1	136	56	3-F32T8(32W)/1-ELEC	90	2	7728	$1,240	$48	$2,688
10	Storage Room	25	1	100 Watt INC	100	1	28 Watt PL CFL/SI	30	0	88	$19	$32	$32
11	Parking Garage	168	26	500 Watt Quartz	350	26	175 Watt MH	205	3	3,166	$3,906	$200	$5,200
12	Parking Garage	168	8	2-F96T12(75W)/STD	173	8	20F96T8(50W)/ELEC	104	0	4637	$572	$60	$480
13	Physical Plant	80	28	2-F96T12(215W)/WHO/STD	450	28	2-F96T12(95W)/HO/ELEC	166	8	3,180	$4,741	$110	$3,080
14	Physical Plant	80	162	100 Watt INC	100	16	28 Watt PL CFL/SI	30	1	4480	$668	$32	$512
Total									1	8,659	$12,094	N/A	$13,592

FIGURE 10.4 Sample energy audit results.

TABLE 10.2 Typical Energy Use Performance Factors (EUPFs)

Facility Type	EUPF #1		EUPF #2	
	KBtu/ft²·yr	MJ/m²·yr	KBtu/Person·yr	GJ/person·yr
Small office building (300 m²)	40	455	12	12.7
Engineering office (1,000 m²)	30	341	14	14.8
Elementary school (4,000 m²)	70	796	5	5.3
Office building (50,000 m²)	100	1,138	50	53

Monitor the Program, Establish Goals

This is the final—and perhaps most important—step in the program. A continuing monitoring program is necessary to ensure that energy savings do not gradually disappear as personnel return to their old ways of operation, equipment gets out of calibration, needed maintenance is neglected, and so on. Also, setting goals (they should be realistic) provides energy management personnel with targets against which they can gauge their performance and the success of their programs.

Summary of Energy Management Programs

The foregoing has been outlined in two tables to provide a step-by-step procedure for electrical energy management in buildings. Table 10.3 is directed at the homeowner or apartment manager; Table 10.4 has been prepared for the commercial building owner or operator. Industrial facilities are treated separately; refer to Chapter 16.

One problem is performing the energy audit is determining the energy used by each item of equipment. In many cases published data are available—as in Table 10.5 for residential appliances. In other cases, engineering judgments must be made, the manufacturer consulted, or instrumentation provided to actually measure energy use.

Electricity-Saving Techniques by Category of End Use

In the discussion that follows, strategies for saving energy that can be implemented in a short time at zero or low capital cost are discussed. Retrofit and new design strategies are then described. The ordering of topics corresponds approximately to their importance in terms of building energy use. Energy use specifically for a process (e.g., heating) is excluded except as it relates to buildings and their occupants.

Residential HVAC

Residential HVAC units using electricity are generally heat pumps, refrigeration systems, and electrical resistance heaters. Heaters range from electric furnace types, small radiant heaters, duct heaters, and strip or baseboard heaters, to embedded floor or ceiling heating systems. Efficiency for heating is usually high, since there are no stack or flue losses and the heater transfers heat directly into the living space.

Cooling systems range from window air conditioning to central refrigeration or heat pump systems. Evaporative coolers are also used in some climates.

Principle operational and maintenance strategies for existing equipment include

- System maintenance and cleanup
- Thermostat calibration and setback
- Time clocks, night cool-down
- Improved controls and operating procedures
- Heated or cooled volume reduction
- Reduction of infiltration and exfiltration losses

TABLE 10.3 An Energy Management Plan for the Homeowner or Apartment Manager

First Step: Review Historical Data

1. Collect utility bills for a recent 12-month period.
2. Add up the bills and calculate total kWh, total $, average kWh (divide total by 12), and average $, and note the months with the lowest and highest kWh.
3. Calculate a seasonal variation factor (svf) by dividing the kWh for the greatest month by the kWh for the lowest month.

Second Step: Perform Energy Audits

4. Identify all electrical loads greater than 1 kW (1000 W). Refer to Table 10.5 for assistance. Most electrical appliances have labels indicating the wattage. If not, use the relation $W = V \times A$.
5. Estimate the number of hours per month each appliance is used.
6. Estimate the percentage of full load (pfl) by each device under normal use. For a lamp, it is 100%; for water heaters and refrigerators, which cycle on and off, about 30%, for a range, about 25% (only rarely are *all* burners *and* the oven used), etc.
7. For each device, calculate kWh by multiplying: kW × hours/month × pfl = kWh/month.
8. Add up all kWh calculated by this method. The total should be smaller than the average monthly kWh calculated in 2.
9. *Note:* if the svf is greater than 1.5, the load shows strong seasonal variation (e.g., summer air conditioning, winter heating, etc.). If this is the case, make two sets of calculations, one for the lowest month (when the fewest loads are operating) and one for the highest month.
10. Make a table listing the wattage of each lamp and the estimated numbers of hours of use per month for each lamp. Multiply watts times hours for each, sum, and divide by 1,000. This gives kWh for the lighting loads. Add this to the total shown.
11. Add the refrigerator, television, and all other appliances or tools that use 5 kWh per month or more.
12. By this process you should now have identified 80 to 90% of electricity using loads. Other small appliances which are used infrequently can be ignored. The test is to now compare with the average month (high or low month if svf is greater than 1.5). If your total is too high, you have overestimated the pfl or the hours of use.
13. Now rank each appliance in descending order of kWh used per month. Your list should read approximately like this:

First:	Heating (in cold climates); air conditioning would be first in hot climates.
Second:	Waterheating
Third:	Lighting
Fourth:	Refrigeration
Fifth:	Cooking
Sixth:	Television
Seventh to last:	All others

Third Step: Apply Energy Management Principles

14. Attack the highest-priority loads first. There are three general things which can be done: (1) reduce kW (smaller lamps, more efficient appliances); (2) reduce pfl ("oven cooked" meals, change thermostats, etc.); (3) reduce hours of use (turn lights off, etc.). Refer to the text for detailed suggestions.

Fourth Step: Monitor Program, Calculate Savings

15. After the energy management program has been initiated, examine subsequent utility bills to determine if you are succeeding.
16. Calculate savings by comparing utility bills. (*Note:* since utility rates are rising, your utility bills may not be any lower. In this case it is informative to calculate what your bill would have been without the energy management program.)

System maintenance is an obvious but often neglected energy-saving tool. Dirty heat transfer surfaces decrease in efficiency. Clogged filters increase pressure drops and pumping power. Inoperable or malfunctioning dampers can waste energy and prevent proper operation of the system.

In residential systems heating and cooling is generally controlled by the room thermostat. As a first step the calibration of the thermostat should be checked, since these low-cost devices can be inaccurate by as much as ±5°C. Several manufacturers now offer replacement thermostats with night setback controllers. Thus two set points are provided; one for daytime use when heating to

TABLE 10.4 An Energy Management Plan for Commercial Building Operator

First Step: Review Historical Data

1. Collect utility bills for a recent 12-month period.
2. Add up the bills and calculate total kWh, total $, average kWh (divide total by 12), and average $, and note the months with the lowest and highest kWh.
3. Calculate a seasonal variation factor (svf) by dividing the kWh for the greatest month by the kWh for the lowest month.
4. Prepare a graph of historical energy use (see Figure 10.1).

Second Step: Perform Energy Audits

5. Evaluate major loads. In commercial buildings loads can be divided into four categories:
   a. HVAC (fans, pumps, chillers, heaters, cooling towers)
   b. Lighting
   c. Office equipment and appliances (elevators, typewriters, cash registers, copy machines, hot water heaters, etc.)
   d. Process equipment (as in laundries, restaurants, bakeries, shops, etc.)

   Items a, b, and c are common to all commercial operations and will be discussed here. Item d overlaps with industry and the reader should also refer to Chapters 14, 15, and 16. Generally items a, b, and d account for the greatest use of electricity and should be examined in that order.
6. In carrying out the energy audit, focus on major loads. Items that together comprise less than 1% of the total connected load in kW can often be ignored with little sacrifice in accuracy.
7. Use the methodology described above and in Chapter 16 for making the audit.
8. Compare audit results with historical energy use. If 80 to 90% of the total (according to the historical records) has been identified, this is generally adequate.

Third Step: Formulate the Energy Management Plan

9. Secure management commitment. The need for this varies with the size and complexity of the operation. However, any formal program will cost something, in terms of salary for the energy coordinator as well as (possibly) an investment in building modifications and new equipment. At this stage it is very important to project current energy usage and costs ahead for the next 3 to 5 years, make a preliminary estimate of potential savings (typically 10 to 50% per year), and establish the potential payback or return on investment in the program.
10. Develop a list of energy management opportunities (EMOs); e.g., install heat recovery equipment in building exhaust air), estimate the cost of each EMO, and also the payback. Methods for economic analysis are given in Chapter 3. For ideas and approaches useful for identifying EMOs, refer to the text.
11. Communicate the plan to employees, department heads, equipment operators, etc. Spell out who will do what, why there is a need, what are the potential benefits and savings. Make the point (if appropriate) that "the energy you save may save your job." If employees are informed, understand the purpose, and realize that the plan applies to everyone, including the president, cooperation is increased.
12. Set goals for department managers, building engineers, equipment operators, etc., and provide monthly reports so they can measure their performance.
13. Enlist the assistance of all personnel in: (1) better "housekeeping and operations" (e.g., turning off lights, keeping doors closed) and (2) locating obvious wastes of electricity (e.g., equipment operating needlessly, better methods of doing jobs).

Fourth Step: Implement Plan

14. Implementation should be done in two parts. First, carry out operational and housekeeping improvements with a goal of, say, 10% reduction in electricity use at essentially no cost and no reduction in quality of service or quantity of production. Second, carry out those modifications (retrofitting of buildings, new equipment, process changes) that have been shown to be economically attractive.
15. As changes are made it is important to continue to monitor electricity usage to determine if goals are being realized. Additional energy audits may be justified.

Fifth Step: Evaluate Progress, Management Report

16. Compare actual performance to the goals established in Item 12. Make corrections for weather variations, increases or decreases in production or number of employees, addition of new buildings, etc.
17. Provide a summary report of energy quantities and dollars saved, and prepare new plans for the future.

TABLE 10.5 Residential Energy Usage— Typical Appliances

Electric Appliances	Power (watts)	Typical use (kWh/year)	Typical use (GJe/year)
	Home Entertainment		
Radio	10	10	0.036
Radio/record player	90	90	0.324
Color television	100	229	0.79
Compact disk player	12	6	0.022
Video cassette recorder	30	15	0.054
Personal computer	125	63	0.23
Computer printer	250	25	0.09
Fax machine	65	75	0.27
	Food Preparation		
Blender	300	1	0.0036
Broiler	1,140	85	0.31
Carving knife	92	8	0.03
Coffee maker	1,200	140	0.50
Deep fryer	1,448	83	0.30
Dishwasher	1,201	363	1.31
Egg cooker	516	14	0.05
Frying pan	1,196	100	0.36
Hot plate	1,200	90	0.32
Mixer	127	2	0.0072
Oven, microwave (only)	1,450	190	0.68
Range with oven	12,200	700	2.52
With self-cleaning oven	12,200	730	2.63
Roaster	1,333	60	0.22
Sandwich grill	1,161	33	0.12
Toaster	1,146	39	0.14
Trash compactor	400	50	0.18
Waffle iron	1,200	20	0.07
Waste dispenser	445	7	0.03
	Food Preservation		
Freezer			
Manual defrost, upright 26 ft³	—	1,000	3.60
Refrigerator-freezer			
Automatic defrost 18 ft³	—	700	2.52
	Laundry		
Clothes dryer	4,856	993	3.57
Iron (hand)	1,100	60	0.22
Washing machine (automatic)	512	103	0.37
Water heater	2,475	4,219	15.19
(quick recovery)	4,474	4,811	17.32
	Housewares		
Clock	2	17	0.06
Floor polisher	305	15	0.05
Sewing machine	75	11	0.04
Vacuum cleaner	630	46	0.17
	Comfort Conditioning		
Air cleaner	50	216	0.78
Air conditioner (room)	600	600	2.16
Bed covering	177	147	0.53

TABLE 10.5 (continued) Residential Energy Usage—Typical Appliances

Electric Appliances	Power (watts)	Typical use (kWh/year)	Typical use (GJe/year)
Dehumidifier	257	377	1.36
Fan (attic)	370	291	1.05
Fan (circulating)	88	43	0.15
Fan (rollaway)	171	138	0.50
Fan (window)	200	170	0.61
Heater (portable)	1,322	176	0.63
Heating pad	65	10	0.04
Humidifier	177	163	0.59
	Health and Beauty		
Germicidal lamp	20	141	0.51
Hair dryer	1,000	40	0.14
Heat lamp (infrared)	250	13	0.05
Shaver	15	0.5	0.0013
Sun lamp	279	16	0.06
Tooth brush	1.1	1.0	0.0036
Vibrator	10	1.0	0.0036

approximately 18°C (65°F) is recommended, the other for night time when temperatures are permitted to drop to 10 to 13°C (50 to 55°F). Depending on temperature conditions, hours of use, and so on, correct temperature settings and nighttime temperature setback can save 5 to 10% of the annual heating bill.

A similar approach can be used with air conditioners. Thermostats should be set to 24°C (75°F) or higher. Be careful that this setting does not cause heating to occur. Time clocks are sometimes useful for heating or cooling control—for example, for shutting systems off at night or for starting heating or cooling systems early in the morning.

Sometimes simple changes in controls or operating procedures will save energy. In cooling, use night air for summer cool-down. When the outside air temperature is cool, turn off the refrigeration unit and circulate straight outside air. If fan units have more than one speed, use the lowest speed that provides satisfactory operation. Check the balance of the system and the operation of dampers and vents to ensure that heating and cooling is provided in the correct quantities where needed.

Energy savings can be achieved by reducing the volume of the heated or cooled space. This can be accomplished by closing vents, doors, or other appropriate means. Usually it is not necessary to heat or cool an entire residence; the spare bedroom is rarely used, halls can be closed off, and so on.

A major cause of energy wastage is infiltration and exfiltration. In a poorly "sealed" residence, infiltration of cold or hot air will increase heating or cooling energy use. In many residences this is the major loss of heated or cooled air. If ventilation rates are excessive due to high-fan speeds, the dwelling can be overpressurized, increasing the leakage of heated or cooled air from the structure.

Check for open doors and windows, open fireplace dampers, inadequate weather stripping around windows and doors, and any other openings which can be sealed. Caution must be exercised to provide adequate ventilation, however. Standards vary, depending on the type of occupancy and whether smoking is allowed or not. In nonsmoking areas the standard is typically 5 to 10 ft^3/min; higher rates would be required for certain industrial processes, kitchens, and so on.

In retrofit or new designs, two general strategies should be observed:

- Reduce heating and cooling needs to the minimum level practical by proper selection of site, dwelling orientation on the site and dwelling design and construction.
- Provide the most efficient heating/cooling system possible. Generally, this will require provisions for heat storage and heat recovery.

These two considerations are particularly important for installations where solar heating is being considered.

In retrofit or new design projects the following techniques will save energy:

- Site selection and building orientation
- Building envelope design
- Selection of efficient heating/cooling equipment

Site selection and building orientation is not always under the control of the owner/occupant. Where possible, though, select a site sheltered from temperature extremes and wind. Orient the building (in cold climates) with a maximum southerly exposure to take advantage of direct solar heating in winter. Use earth berms to reduce heat losses on northerly exposed parts of the building. Deciduous trees provide summer shading but permit winter solar heating.

Building envelope design can improve heat absorption and retention in winter and summer coolness. The first requirement is to design a well-insulated, thermally tight structure. Guidelines for accomplishing this depend on climate. Consult the literature, local contractors, or your utility representative to determine recommendations for your area. Windows are an important source of heat gain and loss. The heat loss for single-pane glazing is around 5 to 7 W/m^2 °C. For double glazing, the comparable value is in the range of 3 to 4 W/m^2 °C, while for triple glazing it is 2 to 3 W/m^2 °C.

In general, the most efficient electric heating and cooling system is the heat pump. Common types are air-to-air heat pumps, either a single package unit (similar to a window air conditioner) or a split system in which the air handling equipment is inside the building and the compressor and related equipment is outdoors. Commercially available equipment demonstrates a wide range of efficiency. Heating performance is measured in terms of a Heating Seasonal Performance Factor (HSPF), in BTUs of heat added per watt-hour of electricity input. Typical values are 6.0 to 10.0. Cooling performance of residential air conditioners is measured in terms of a Seasonal Energy Efficiency Ratio (SEER), which describes the ratio of cooling capacity to electrical power input. Typical values are 9.0 to 12.0. In purchasing new equipment, selection of equipment with the highest HSPF and SEER should be considered.

Sizing of equipment is important, since the most efficient operation generally occurs at or near full load. Selection of oversized equipment is thus initially more expensive and will also lead to greater operating costs.

The efficiency of heat pumps declines as the temperature difference between the heat source and heat sink decreases. Since outside air is generally the heat source, heat is most difficult to get when it is most needed. For this reason heat pumps often have electrical backup heaters for extremely cold weather.

An alternative approach is to design the system using a heat source other than outside air. Examples include heated air, such as is exhausted from a building; a deep well, providing water at a constant year-round temperature; or a solar heat source. There are a great many variations on solar heating and heat pump combinations.

Nonresidential HVAC

HVAC systems in nonresidential installations may involve package rooftop or ground-mounted units, or a central plant. Although the basic principles are similar to those discussed earlier in connection with residential systems, the equipment is larger and control more complex.

Efficiency of many existing HVAC systems can be improved. Modifications can reduce energy use by 10 to 15%, often with building occupants unaware that changes have been made.

The basic function of HVAC systems is to heat, cool, dehumidify, humidify, and provide air mixing and ventilating. The energy required to carry out these functions depends on the building design, its duty cycle (e.g., 24 hours per day use, as in a hospital, vs. 10 hours per day in an office), the type of occupancy, the occupants' use patterns and training in using the HVAC system, the type of HVAC equipment installed, and finally, daily and seasonal temperature and weather conditions to which the building is exposed.

A complete discussion of psychometrics, HVAC system design, and commercially available equipment types is beyond the scope of this chapter (see also Chapter 11).

Energy management strategies will be described in three parts:

- Equipment modifications (control, retrofit, and new designs)
  - Fans
  - Chillers
  - Ducts and dampers
  - Pumps
- Economizer systems and enthalpy controllers
- Heat recovery techniques

Equipment Modifications (Control, Retrofit, and New Designs)

Fans. All HVAC systems involve some motion of air. The energy needed for this motion can make up a large portion of the total system energy used. This is especially true in moderate weather, when the heating or cooling load drops off but the distribution systems often operate at the same level.

Control—Simple control changes can save electrical energy in the operation of fans. Examples include turning off large fan systems when relatively few people are in the building or stopping ventilation half an hour before the building closes. The types of changes that can be made will depend upon the specific facility. Some changes involve more sophisticated controls, which may already be available in the HVAC system.

Retrofit—The capacity of the building ventilation system is usually determined by the maximum cooling or heating load in the building. This load has been changing due to reduced outside air requirements, lower lighting levels, and wider acceptable comfort ranges. As a result it is now feasible to decrease air flow in many existing commercial buildings.

The volume rate of air flow through the fan, Q, varies directly with the speed of the impeller's rotation. This is expressed as follows for a fan whose speed is changed from N_1 to N_2.

$$Q_2 = (N_2/N_1) \times Q_1 \tag{10.1}$$

The pressure developed by the fan, P, (either static or total) varies as the square of the impeller speed.

$$P_2 = (N_2/N_1)^2 \times P_1 \tag{10.2}$$

The power needed to drive the fan, H, varies as the cube of the impeller speed.

$$H_2 = (N_2/N_1)^3 \times H_1 \tag{10.3}$$

The result of these laws is that for a given air distribution system (specified ducts, dampers, etc.), if the air flow is to be doubled, eight (2^3) times the power is needed. Conversely, if the air flow is to be cut in half, one-eighth [$(1/2)^3$] of the power is required. This is useful in HVAC systems because even a small reduction in air flow (say 10%) can result in significant energy savings (27%).

The manner in which the air flow is reduced is critical in realizing these savings. Maximum savings are achieved by sizing the motor exactly to the requirements. Simply changing pulleys to provide the desired speed will also result in energy reductions according to the cubic law. The efficiency of existing fan motors tends to drop off below the half-load range.

If variable volume air delivery is required, it may be achieved through inlet vane control, outlet dampers, variable speed drives (VSDs), controlled pitch fans, or cycling. Energy efficiency in a retrofit design is best obtained with variable speed motors or controlled pitch fans. This can be seen by calculating the power reduction which would accompany reduced flow using different methods of control, as noted below. Numbers in the table are the percentages of full-flow input power:

Fans				Pumps	
% Flow	Inlet Vanes	Dampers	VSDs	Throttle Valve	VSDs
100	102	103	102	101	103
90	86	98	76	96	77
80	73	94	58	89	58
70	64	88	43	83	41
60	56	81	31	77	30
50	50	74	22	71	19
40	46	67	15	65	13
30	41	59	9	59	8

New design—The parameters for new design are similar to those for fan retrofit. It is desirable, when possible, to use a varying ventilation rate which will decrease as the load decreases. A system such as variable air volume incorporates this in the interior zones of a building. In some cases there will be a trade-off between power saved by running the fan slower and the additional power needed to generate colder air. The choices should be determined on a case-by-case basis.

Pumps. Pumps are found in a variety of HVAC applications such as chilled water, heating hot water, and condenser water loops. They are another piece of peripheral equipment that can use a large portion of HVAC energy, especially at low system loads.

Control—The control of pumps is often neglected in medium and large HVAC systems, where it could significantly reduce the demand. A typical system would be a three-chiller installation in which only one chiller is needed much of the year. Two chilled water pumps in parallel are designed to handle the maximum load through all three chillers. Even when only one chiller is on, both pumps are used. By manual adjustments two chillers could be by-passed and one pump turned off. All systems should be reviewed in this manner to ensure that only the necessary pumps operate under normal load conditions.

Retrofit—Pumps follow laws similar to fan laws, the key being the cubic relationship of power to the volume pumped through a given system. Small decreases in flow rate can save significant portions of energy. In systems in which cooling or heating requirements have been permanently decreased, flow rates may be reduced also. A simple way to do this is by trimming the pump impeller. The pump curve must be checked first, however, because pump efficiency is a function of the impeller diameter, flow rate, and pressure rise. It would be ensured that after trimming, the pump will still be operating in an efficient region. This is roughly the equivalent of changing fan pulleys in that the savings follow the cubic law of power reduction.

Another common method for decreasing flow rates is to use a "throttle" (pressure reducing) valve. The result is equivalent to that of the discharge damper in the air-side systems, as shown earlier. The valve creates an artificial use of energy, which can be responsible for much of the work performed by the pumps. Variable speed drives are more efficient.

New design—In a variable load situation, common to most HVAC systems, more efficient systems with new designs are available, rather than the standard constant volume pump. (These may also apply to some retrofit situations.)

One option is the use of several pumps of different capacity so that a smaller pump can be used when it can handle the load and a larger pump used the rest of the time. This can be a retrofit modification as well when a back-up pump provides redundancy. Its impeller would be trimmed to provide the lower flow rate.

Another option is to use variable speed drive pumps. While their initial cost is greater, they offer an improvement in efficiency over the standard pumps. The economic desirability of this or any similar change can be determined by estimating the number of hours the system will operate under various loads.

Package Units. Great improvements in package air-conditioner efficiency have been made in the last two decades. Today room air conditioning units have Energy Efficiency Ratios (EERs) in the range of 8.0 to 9.0. The Federal National Appliance Energy Conservation Act set standards in 1992 that required room air conditioners to have an EER of 9.0 (for sizes 8,000 to 13,999 BTU.) High-efficiency rooftop units and split-system heat pumps are also available with seasonal energy efficiency ratios (SEERs) in the range of 8.0 to 10.0 and cooling capacities of 5 tons, 10 tons, and higher. The Federal standards are SEER = 10.0 (1992) and a heating seasonal performance factor (HSPF) of 6.8 (1992). The higher initial cost of these units is almost always justified by operating savings. In addition, many utilities offer rebates of $20 to $50 per ton for installing the more efficient units.

Chillers. Chillers are often the largest single energy user in the HVAC system. The chiller cools the water used to extract heat from the building and outside air. By optimizing chiller operation, the performance of the whole system is improved.

Two basic types of chillers are found in commercial and industrial applications: absorption and mechanical chillers. Absorption units boil water, the refrigerant, at a low pressure through absorption into a high-concentration lithium bromide solution. Mechanical chillers cool through evaporation of a refrigerant, such as freon, at a low pressure after it has been compressed, cooled, and passed through an expansion valve.

There are three common types of mechanical chillers. They have similar thermodynamic properties, but use different types of compressors. Reciprocating and screw-type compressors are both positive displacement units. The centrifugal chiller uses a rapidly rotating impeller to pressurize the refrigerant.

All of these chillers must reject heat to a heat sink outside the building. Some use air-cooled condensers, but most large units operate with evaporative cooling towers. Cooling towers have the advantage of rejecting heat to a lower temperature heat sink because the water approaches the ambient wet-bulb temperature whereas air-cooled units are limited to the dry-bulb temperature. This results in a higher condensing temperature, which lowers the efficiency of the chiller. Air-cooled condensers are used because they require much less maintenance than cooling towers.

Mechanical cooling can also be performed by Direct Expansion (DX) units. These are very similar to chillers except that they cool the air directly instead of using the refrigerant as a heat transfer medium. They eliminate the need for chilled water pumps and also reduce efficiency losses associated with the transfer of the heat to and from the water. DX units must be located close (~30 m) to the ducts they are cooling, so they are typically limited in size to the cooling required for a single air handler. A single large chiller can serve a number of distributed air handlers. Where the air handlers are located close together, it can be more efficient to use a DX unit.

Air-cooled chillers today have efficiencies of 1.2 kW/ton or greater, while water-cooled systems have declined to around 0.6 to 0.7 kW/ton for units in the 300-ton capacity range.

Controls—Mechanical chillers operate on a principle similar to the heat pump. The objective is to remove heat from a low-temperature building and deposit it in a higher-temperature atmosphere. The lower the temperature rise the chiller has to face, the more efficiently it will operate. It is useful, therefore, to maintain as warm a chilled water loop and as cold a condenser water loop as possible.

Energy can be saved by using lower-temperature water from the cooling tower to reject the heat. However, as the condenser temperature drops, the pressure differential across the expansion valve drops, starving the evaporator of refrigerant. Many units with expansion valves therefore operate at a constant condensing temperature, usually 41°C (105°F), even when more cooling is available from the cooling tower. Field experience has shown that in many systems, if the chiller is not fully loaded, it can be operated with a lower cooling tower temperature.

Retrofit—Where a heat load exists and the wet-bulb temperature is low, cooling can be done directly with the cooling tower. If proper filtering is available, the cooling tower water can be piped directly into the chilled water loop. Often a direct heat exchanger between the two loops is preferred to protect the coils from fouling. Another technique is to turn off the chiller but use its refrigerant to transfer heat between the two loops. This "thermocycle" uses the same principles as a heat pipe and only works on chillers with the proper configuration.

A low wet-bulb temperature during the night can also be utilized. It requires a chiller that handles low condensing temperatures and a cold storage tank. This technique may become even more desirable as time-of-day or demand pricing for electricity increases.

New design—In the purchase of a new chiller, an important consideration should be the load control feature. Since the chiller will be operating at partial load most of the time, it is important that it be able to do so efficiently.

In addition to control of single units, it is sometimes desirable to use multiple compressor reciprocating chillers. This allows some units to be shut down at partial load. The remaining compressors operate near full load, usually more efficiently.

If a new chiller is being installed to replace an old unit or to retire equipment that uses environmentally unacceptable chlorinated fluorocarbon (CFC) refrigerants, this is a good opportunity to install a high-efficiency chiller. Currently R-11 and R-12 refrigerants are being phased out; non-CFC refrigerants are defined as R-22, R-123, R-134a, and ammonia.

Commonly, in commercial and industrial buildings, a convenient source of heat for a heat pump is the building exhaust air. This is a constant source of warm air available throughout the heating season. A typical heat pump design could generate hot water for space heating from this source at around 32 to 35°C (90 to 95°F). Heat pumps designed specifically to use building exhaust air can reach 66°C (150°F).

Another application of the heat pump is a continuous loop of water traveling throughout the building with small heat pumps located in each zone. Each small pump can both heat and cool, depending upon the needs of the zone. This system can be used to transfer heat from the warm side of a building to the cool side. A supplemental cooling tower and boiler are included in the loop to compensate for net heating or cooling loads.

A double-bundle condenser can be used as a retrofit design for a centralized system. This creates the option of pumping the heat higher to the cooling tower or into the heating system hot duct. Some chillers can be retrofitted to act as heat pumps. Centrifugal chillers will work much more effectively with a heat source warmer than outside air (exhaust air, for example). The compression efficiency of the centrifugal chiller falls off as the evaporator temperature drops.

Because they are positive displacement machines, reciprocating and screw-type compressors operate more effectively at lower evaporator temperatures. They can be used to transfer heat across a larger temperature differential. Multistage compressors increase this capacity even further.

Ducting-Dampers. **Controls**—In HVAC systems using dual ducts, static pressure dampers are often placed near the start of the hot or cold plenum run. They control the pressure throughout the entire distribution system and can be indicators of system operation. Often in an overdesigned system, the static pressure dampers may never open more than 25%. Fan pulleys can be changed to slow the fan and open the dampers fully, eliminating the previous pressure drop. The same volume of air is delivered with a significant drop in fan power.

Retrofit—Other HVAC systems use constant-volume mixing boxes for balancing that create their own pressure drops as the static pressure increases. An entire system of these boxes could be overpressurized by several inches of water without affecting the air flow, but the required fan power would increase. (One inch of water pressure is about 250 N/m^2 or 250 Pa.) These systems should be monitored to ensure that static pressure is controlled at the lowest required value. It may also be desirable to replace the constant-volume mixing boxes with boxes without volume control to eliminate their minimum pressure drop of approximately 1 in. of water. In this case, static pressure dampers will be necessary in the ducting.

Leakage in any dampers can cause a loss of hot or cold air. Neoprene seals can be added to blades to slow leakage considerably. If a damper leaked more than 10%, it may be less costly to replace the entire damper assembly with effective positive-closing damper blades rather than to tolerate the loss of energy.

New design—In the past small ducts were installed because of their low initial cost despite the fact that the additional fan power required offset the initial cost on a life cycle basis. New ASHRAE guidelines set a maximum limit on the fan power that can be used for a given cooling capacity. As a result the air system pressure drop must be low enough to permit the desired air flow. In small buildings this pressure drop is often largest across filters, coils, and registers. In large buildings the duct runs may be responsible for a significant fraction of the total static pressure drop, particularly in high-velocity systems.

Systems. The use of efficient equipment is only the first step in the optimum operation of a building. Equal emphasis should be placed upon the combination of elements in a system and the control of those elements.

Control—Many systems use a combination of hot and cold to achieve moderate temperatures. Included are dual duct, multizone, and terminal reheat systems, and some induction, variable air volume, and fan coil units. Whenever combined heating and cooling occurs, the temperatures of the hot and cold ducts or water loops should be brought as close together as possible while still maintaining building comfort.

This can be accomplished in a number of ways. Hot and cold duct temperatures are often reset on the basis of the temperature of the outside air or the return air. A more complex approach is to monitor the demand for heating and cooling in each zone. For example, in a multizone building, the demand of each zone is transferred back to the supply unit by electric or pneumatic signals. At the supply end hot and cold air are mixed in proportion to this demand. The cold air temperature should be just low enough to cool the zone calling for the most cooling. If the cold air were any colder, it would be mixed with hot air to achieve the right temperature. This creates an overlap in heating and cooling not only for the zone but for all the zones, because they would all be mixing in the colder air.

If no zone calls for total cooling, then the cold air temperature can be increased gradually until the first zone requires cooling. At this point, the minimum cooling necessary for that multizone configuration is performed. The same operation can be performed with the hot air temperature until the first zone is calling for heating only.

Note that simultaneous heating and cooling is still occurring in the rest of the zones. This is not an ideal system, but it is a first step in improving operating efficiency.

The technique for resetting hot and cold duct temperatures can be extended to the systems that have been mentioned. It may be performed automatically, with pneumatic or electric controls, or manually. In some buildings it will require the installation of more monitoring equipment (usually only in the zones of greatest demand), but the expense should be relatively small and the payback period short.

Nighttime temperature setback is another control option that can save energy without significantly affecting the comfort level. Energy is saved by shutting off or cycling fans. Building heat loss may also be reduced because the building is cooler and no longer pressurized.

In moderate climates complete night shutdown can be used with a morning warm-up period. In colder areas where the overall night temperature is below 4°C (40°F), it is usually necessary to provide some heat during the night. Building setback temperature is partially dictated by the capacity of the heating system to warm the building in the morning. In some cases it may be the mean radiant temperature of the building rather than air temperature that determines occupant comfort.

Some warm-up designs use “free” heating from people and lights to help attain the last few degrees of heat. This also provides a transition period for the occupants to adjust from the colder outdoor temperatures.

In some locations during the summer, it is desirable to use night air for a cool-down period. This “free cooling” can decrease the temperature of the building mass, which accumulates heat during the day. In some buildings, with high heat content (such as libraries or buildings that have thick walls), a long period of night cooling may decrease the building mass temperature by a degree or two. This represents a large amount of cooling that the chiller will not have to perform the following day.

Retrofit—Retrofitting HVAC systems may be an easy or difficult task depending upon the possibility of using existing equipment in a more efficient manner. Often retrofitting involves control or ducting changes that appear relatively minor but will greatly increase the efficiency of the system. Some of these common changes, such as decreasing air flow, are discussed elsewhere in this chapter. This section describes a few changes appropriate to particular systems.

Both dual duct and multizone systems mix hot and cold air to achieve the proper degree of heating or cooling. In most large buildings the need for heating interior areas is essentially non-existent, due to internal heat generation. A modification that adjusts for this is simply shutting off air to the hot duct. The mixing box then acts as a variable air volume box, modulating cold air according to room demand as relayed by the existing thermostat. (It should be confirmed that the low volume from a particular box meets minimum air requirements.)

Savings from this modification come mostly from the elimination of simultaneous heating and cooling. Since fans in these systems are likely to be controlled by static pressure dampers in the duct after the fan, they do not unload very efficiently and represent only a small portion of the savings.

Economizer Systems and Enthalpy Controllers. The economizer cycle is a technique for introducing varying amounts of outside air to the mixed air duct. Basically it permits mixing warm return air at 24°C (75°F) with cold outside air to maintain a preset temperature in the mixed air plenum (typically 10 to 15°C, 50 to 60°F). When the outside temperature is slightly above this set point, 100% outside air is used to provide as much of the cooling as possible. During very hot outside weather, minimum outside air will be added to the system.

A major downfall of economizer systems is poor maintenance. The failure of the motor or dampers may not cause a noticeable comfort change in the building, because the system is often capable of handling the additional load. Since the problem is not readily apparent, corrective maintenance may be put off indefinitely. In the meantime the HVAC system will be working harder than necessary for any economizer installation.

Typically, economizers are controlled by the dry-bulb temperature of the outside air rather than its enthalpy (actual heat content). This is adequate most of the time but can lead to unnecessary cooling of air. When enthalpy controls are used to measure wet-bulb temperatures, this cooling can be reduced. However, enthalpy controllers are more expensive and less reliable.

The rules that govern the more complex enthalpy controls for cooling-only applications are as follows:

- When outside air enthalpy is greater than that of the return air or when outside air dry-bulb temperature is greater than that of the return air, use minimum outside air.
- When the outside air enthalpy is below the return air enthalpy and the outside dry-bulb temperature is below the return air dry-bulb temperature but above the cooling coil control point, use 100% outside air.

- When the outside air enthalpy is below the return air enthalpy and the outside air dry-bulb temperature is below the return air dry-bulb temperature and below the cooling coil controller setting, the return and outside air are mixed by modulating dampers according to the cooling set point.

These points are valid for the majority of cases. When mixed air is to be used for heating and cooling, a more intricate optimization plan will be necessary, based on the value of the fuels used for heating and cooling.

Heat Recovery. Heat recovery is often practiced in industrial processes which involve high temperatures. It can also be employed in HVAC systems. Systems are available that operate with direct heat transfer from the exhaust air to the inlet air. These are most reasonable when there is a large volume of exhaust air—for example, in once-through systems and when weather conditions are not moderate.

Common heat recovery systems are broken down into two types, regenerative and recuperative. Regenerative units use alternating air flow from the hot and cold stream over the same heat storage/transfer medium. This flow may be reversed by dampers or the whole heat exchanger may rotate between streams. Recuperative units involve continuous flow; the emphasis is upon heat transfer through a medium with little storage. See Chapter 13A for more information.

The rotary regenerative unit, or heat wheel, is one common heat recovery device. It contains a corrugated or woven heat storage material that gains heat in the hot stream. This material is then rotated into the cold stream, where the heat is given off again. The wheels can be impregnated with a desiccant to transfer latent as well as sensible heat. Purge sections for HVAC applications can reduce carryover from the exhaust stream to acceptable limits for most installations. The heat transfer efficiency of heat wheels generally ranges from 60 to 85% depending upon the installation, type of media, and air velocity. For easiest installation the intake and exhaust ducts should be located near each other.

Another system that can be employed with convenient duct location is a plate-type air-to-air heat exchanger. This system is usually lighter though more voluminous than heat wheels. Heat transfer efficiency is typically in the 60 to 75% range. Individual units range from 1,000 to 11,000 SCFM and can be grouped together for greater capacity. Almost all designs employ counterflow heat transfer for maximum efficiency.

Another option to consider for nearly contiguous ducts is the heat pipe. This is a unit that uses a boiling refrigerant within a closed pipe to transfer heat. Since the heat of vaporization is utilized, a great deal of heat transfer can take place in a small space. Heat pipes are often used in double-wide coils, which look very much like two steam coils fastened together. The amount of heat transferred can be varied by tilting the tubes to increase or decrease the flow of liquid through capillary action. Heat pipes cannot be "turned off," so bypass ducting is often desirable. The efficiency of heat transfer ranges from 55 to 75% depending upon the number of pipes, fins per inch, air face velocity, and so on.

Runaround systems are also popular for HVAC applications, particularly when the supply and exhaust plenums are not physically close. Runaround systems involve two coils (air-to-water heat exchangers) connected by a piping loop of water or glycol solution and a small pump. The glycol solution is necessary if the air temperatures in the inlet coils are below freezing.

Standard air-conditioning coils can be used for the runaround system, and some equipment manufacturers supply computer programs for size optimization. Precaution should be used when the exhaust air temperature drops below 0°C (32°F), which would cause freezing of the condensed water on its fins. A three-way bypass valve will maintain the temperature of the solution entering the coil at just above 0°C (32°F). The heat transfer efficiency of this system ranges from 60 to 75% depending upon the installation.

Another system similar to the runaround in layout is the desiccant spray system. Instead of using coils in the air plenums, it uses spray towers. The heat transfer fluid is a desiccant (lithium chloride) which transfers both latent and sensible heat—desirable in many applications. Tower capacities range from 7,700 to 92,000 SCFM; multiple units can be used in large installations. The enthalpy recovery efficiency is in the range of 60 to 65%.

Thermal Energy Storage. Thermal energy storage (TES) systems are used to reduce the on-peak electricity demand caused by large cooling loads. TES systems utilize several different storage media, with chilled water, ice, or eutectic salts being most common. Chilled water requires the most space, with the water typically being stored in underground tanks. Ice storage systems can be above-ground insulated tanks with heat exchanger coils that cause the water to freeze or can be one of several types of ice-making machines. See Chapter 13B for more information on thermal storage.

In a typical system a chiller operates at off-peak hours to make ice (usually at night). Since the chiller can operate for a longer period of time than during the daily peak, it can have a smaller capacity. Efficiency is greater at night, when the condensing temperature is lower than it is during the day. During daytime operation, chilled water pumps circulate water through the ice storage system and extract heat. Systems can be designed to meet the entire load or to meet a partial load with an auxiliary chiller as a backup.

This system reduces peak demand and can also reduce energy use. With ice storage, it is possible to deliver water at a lower temperature than is normally done. This means that the chilled water piping can be smaller and the pumping power reduced. A low-temperature air distribution system will allow smaller ducts and lower-capacity fans to deliver a given amount of cooling. Careful attention must be paid to the system design to ensure occupant comfort in conditioned spaces. Many electric utilities offer substantial rebates to customers installing TES systems, with amounts in the range of $250 to 500 per ton (or per kW of deferred peak demand).

Water Heating

Residential water heaters typically range in size from 76 liters (20 gal) to 303 liters (80 gal). Electric units generally have one or two immersion heaters, each rated at 2 to 6 kW depending on tank size. Energy input for water heating depends on the temperature at which water is delivered, the supply water temperature, and standby losses from the water heater, storage tanks, and piping. The National Appliance Energy Conservation Act of 1992 established these energy efficiency (EF) standards for water heaters: electric: 88.4% EF; natural gas: 52.5% EF.

In single-tank residential systems, major savings can be obtained by

- Thermostat temperature setback to 60°C (140°F)
- Time clock control
- Supplementary tank insulation
- Hot water piping insulation

The major source of heat loss from electric water heaters is standby losses through the tank walls and from piping, since there are no flame or stack losses in electric units. The heat loss is proportional to the temperature difference between the tank and its surroundings. Thus, lowering the temperature to 60°C will result in two savings: (1) a reduction of the energy needed to heat water and (2) a reduction in the amount of heat lost. Residential hot water uses do not require temperatures in excess of 60°C; for any special use that does, it would be advantageous to provide a booster heater to meet this requirement when needed, rather than maintain 100 to 200 liters of water continuously at this temperature with associated losses.

When the tank is charged with cold water, both heating elements operate until the temperature reaches a set point. After this initial rise, one heating element thermostatically cycles on and off to maintain the temperature, replacing heat that is removed by withdrawing hot water or that is lost by conduction and convection during standby operation.

Experiments indicate that the heating elements may be energized only 10 to 20% of the time, depending on the ambient temperature, demand for hot water, water supply temperature, and so on. By carefully scheduling hot water usage, this time can be greatly reduced. In one case a residential water heater was operated for 1 hour in the morning and 1 hour in the evening. The morning cycle provided sufficient hot water for clothes washing, dishes, and other needs. Throughout the day the water in the tank, although gradually cooling, still was sufficiently hot for incidental needs. The evening heating cycle provided sufficient water for cooking, washing dishes, and bathing. Standby losses were eliminated during the night and much of the day. Electricity use was cut to a fraction of the normal amount. This method requires the installation of a time clock to control the water heater. A manual override can be provided to meet special needs.

Supplementary tank insulation can be installed at a low cost to reduce standby losses. The economic benefit depends on the price of electricity and the type of insulation installed. However, paybacks of a few months up to a year are typical on older water heaters. Newer units have better insulation and reduced losses. Hot water piping should also be insulated, particularly when hot water tanks are located outside or when there are long piping runs. If copper pipe is used, it is particularly important to insulate the pipe for the first 3 to 5 meters where it joins the tank, since it can provide an efficient heat conduction path.

Since the energy input depends on the water flow rate and the temperature difference between the supply water temperature and the hot water discharge temperature, energy use is reduced by reducing either of these two quantities. Hot water demand can be reduced by cold water clothes washing and by providing hot water at or near the use temperature to avoid the need for dilution with cold water. Supply water should be provided at the warmest temperature possible. Since reservoirs and underground piping systems are generally warmer than the air temperature on a winter day in a cold climate, supply piping should be buried, insulated, or otherwise kept above the ambient temperature.

Solar systems are available today for heating hot water. Simple inexpensive systems can preheat the water, reducing the amount of electricity needed to reach the final temperature. Alternatively, solar heaters (some with electric backup heaters) are also available, although current costs correspond to 5- to 10-year paybacks.

Heat pump water heaters may save as much as 25 to 30% of the electricity used by a conventional electric water heater. Some utilities have offered rebates of several thousand dollars to encourage customers to install heat pump water heaters.

Heat recovery is another technique for preheating or heating water, although opportunities in residences are limited. This is discussed in more detail under commercial water heating. Apartments and larger buildings use a combined water heater/storage tank, a circulation loop, and a circulating pump. Cold water is supplied to the tank, which thermostatically maintains a preset temperature, typically 71°C (160°F). The circulating pump maintains a flow of water through the circulating loop, so hot water is always available instantaneously upon demand to any user. This method is also used in hotels, office buildings, and so on.

Adequate piping and tank insulation is even more important here, since the systems are larger and operate at higher temperature. The circulating hot water line should be insulated, since it will dissipate heat continuously otherwise.

Commercial/industrial hot water systems offer many opportunities for employing heat recovery. Examples of possible sources of heat include air compressors, chillers, heat pumps, refrigeration systems, and water-cooled equipment. Heat recovery permits a double energy savings in many cases. First, recovery of heat for hot water or space heating reduces the direct energy input needed for heating. The secondary benefit comes from reducing the energy used to dissipate waste heat to a heat sink (usually the atmosphere). This includes pumping energy and energy expended to operate cooling towers and heat exchangers. Solar hot water systems are also finding increasing use. It is interesting that the prerequisites for solar hot water systems also permit heat recovery. Once the hot water storage capacity and backup heating capability have been provided for the solar hot water system, it is economical to tie in other sources of waste heat (e.g., water jackets on air compressors).

Lighting

There are seven techniques for improving the efficiency of lighting systems:

- Delamping
- Relamping
- Improved controls
- More efficient lamps and devices
- Task-oriented lighting
- Increased use of daylight
- Room color changes, lamp maintenance

The first two techniques and possibly the third are low in cost and may be considered operational changes. The last four items generally involve retrofit or new designs. See Chapters 12B and 20 for more information.

The first step in reviewing lighting electricity use is to perform a lighting survey. An inexpensive hand-held lightmeter can be used as a first approximation; however, distinction must be made between raw intensities (lux or footcandles) recorded in this way and *equivalent sphere illumination (ESI)* values.

Many variables can affect the "correct" lighting values for a particular task: task complexity, age of employee, glare, and so on. For reliable results consult a lighting specialist or refer to the literature and publications of the Illuminating Engineering Society.

The lighting survey indicates those areas of the building where lighting is potentially inadequate or excessive. Deviations from adequate illumination levels can occur for several reasons: overdesign, building changes, change of occupancy, modified layout of equipment or personnel, more efficient lamps, improper use of equipment, dirt buildup, and so on. Once the building manager has identified areas with potentially excessive illumination levels, he or she can apply one or more of the seven techniques listed earlier. Each of these will be described briefly.

Delamping refers to the removal of lamps to reduce illumination to acceptable levels. With incandescent lamps, bulbs are removed. With fluorescent or high-intensity-discharge lamps, ballasts account for 10 to 20% of total energy use and should be disconnected after lamps are removed.

Fluorescent lamps often are installed in luminaires in groups of two or more lamps where it is impossible to remove only one lamp. In such cases an artificial load (called a "phantom tube") can be installed in place of the lamp that has been removed.

Relamping refers to the replacement of existing lamps by lamps of lower wattage or increased efficiency. Low-wattage fluorescent tubes are available that require 15 to 20% less wattage (but produce 10 to 15% less light). In some types of high-intensity-discharge lamps, a more efficient lamp can be substituted directly. However, in most cases, ballasts must also be changed. Table 10.6 shows typical savings by relamping.

Improved controls permit lamps to be used only when and where needed. For example, certain office buildings have all lights for one floor on a single contactor. These lamps will be switched on at 6 A.M., before work begins, and are not turned off until 10 P.M., when maintenance personnel finish their cleanup duties.

Energy usage can be cut by as much as 50% by installing individual switches for each office or work area, installing time clocks, installing occupancy sensors, using photocell controls, or instructing custodial crews to turn lights on as needed and turn them off when work is complete.

There is a great variation in the efficacy (a measure of light output per electricity input) of various lamps. Incandescent lamps have the lowest efficacy, typically 5 to 30 lm/W. Wherever possible, fluorescent lamps should be substituted for incandescent lamps. This not only saves energy but offers substantial economic savings as well, since fluorescent lamps last 10 to 50 times longer. Fluorescent lamps have efficacies in the range of 30 to 70 lm/W.

TABLE 10.6 Typical Relamping Opportunities

Change Office Lamps (2,700 h per year) From	To	Energy Savings/Cost Savings kWh	GJe	5¢ kWh
		To save annually		
1–300 W incandescent	1–100 W mercury vapor	486	5.25	$24.30
2–100 W incandescent	1–40 W fluorescent	400	4.32	20.00
7–150 W incandescent	1–150 W sodium vapor	2,360	25.5	$118.00
Change industrial lamps (3,000 h per year)				
1–300 W incandescent	2–40 W fluorescent	623	6.73	$31.15
1–100 W incandescent	2–215 W fluorescent	1,617	17.5	80.85
3–300 W incandescent	1–250 W sodium vapor	1,806	19.5	90.30
Change store lamps (3,300 h per year)				
1–300 W incandescent	2–40 W fluorescent	685	7.40	$34.25
1–200 W incandescent	1–100 W mercury vapor	264	2.85	13.20
2–200 W incandescent	1–175 W mercury vapor	670	7.24	33.50

Compact fluorescent lamps are available as substitutes for a wide range of incandescent lamps. They range in wattage from 5 to 25 watts with efficacies of 26 to 58 lm/W and will replace 25- to 100-watt incandescent lamps. In addition to the energy savings, they have a 10,000-hour rated life and do not need to be replaced as often as incandescent lamps. Exit lights are good candidates for compact fluorescent lamps. Conversion kits are available to replace the incandescent lamps. Payback is rapid because of the lower energy use and lower maintenance cost, since these lamps are normally on 24 hours a day. There are also light-emitting-diode exit lights that are very energy efficient.

Still greater improvements are possible with high-intensity-discharge lamps such as mercury vapor, metal halide, and high-pressure sodium lamps. Although they are generally not suited to residential use (high light output and high capital cost) they are increasingly being used in commercial buildings. They have efficacies in the range of 40 to 100 lm/W.

Improved ballasts are another way of saving lighting energy. A comparison of the conventional magnetic ballast with improved ballasts shows the difference:

Type lamp	2 lamp 40 W	F40 T-12 CW	2 lamp 32 W	F32 T-8
Ballast type	Standard magnetic	Energy-efficient magnetic	Electronic	Electronic
Input watts	96	88	73	64
Efficacy	60 lm/W	65 lm/W	78 lm/W	90 lm/W

The best performance comes from electronic ballasts, which operate at higher frequency. In addition to the lighting energy savings, there are additional savings from the reduced air-conditioning load due to less heat output from the ballasts. The environmental benefit for the electronic ballast described earlier, as estimated by the U.S. Environmental Protection Agency, is a reduction in CO_2 production of 150 lb/year, a reduction of 0.65 lb/year of SO_2, and a reduction of 0.4 lb/year of NO_x.

Task-oriented lighting is another important lighting concept. In this, lighting is provided for work areas in proportion to the needs of the task. Hallways, storage areas, and other nonwork areas receive less illumination. This approach can be contrasted to the so-called "uniform illumination" method sometimes used in office buildings. The rationale for uniform illumination was based on the fact that the designer could never know that exact layout of desks and equipment in advance, so uniform illumination was provided. This also accommodates revisions in the floor plan. Originally, electricity was cheap and any added cost was inconsequential.

Daylighting was an important element of building design for centuries before the discovery of electricity. In certain types of buildings and operations today, daylighting can be utilized to at least reduce (if not replace) electric lighting. Techniques include windows, an atrium, skylights, and so on. There are obvious limitations such as those imposed by the need for privacy, 24-hour operation, and building core locations with no access to natural light.

The final step is to review building and room color schemes and decor. The use of light colors can substantially enhance illumination without modifying existing lamps.

An effective lamp maintenance program also has important benefits. Light output gradually decreases over lamp lifetime. This should be considered in the initial design and when deciding on lamp replacement. Dirt can substantially reduce light output; simply cleaning lamps and luminaries more frequently can gain up to 5 to 10% greater illumination, permitting some lamps to be removed.

Reflectors are available to insert in fluorescent lamp fixtures. These direct and focus the light onto the work area, yielding a greater degree of illumination. Alternatively, in multiple lamp fixtures, one lamp can be removed and the reflector keeps the level of illumination at about 75% of the previous value.

Refrigeration

The refrigerator, at 100 to 200 kWh/month, is among the top six residential users of electricity. In the last 50 years the design of refrigerator/freezers has changed considerably, with sizes increasing from 0.14 to 0.28 m (5 to 10 ft^3) to 0.34 to 0.68 m^3 today. At the same time, the energy input increased, from roughly 210 to 490 W/m^3 (6 to 14 W/ft^3). Following the oil embargo efforts were initiated to increase refrigerator efficiency. Today, a typical residential refrigerator/freezer combination uses about half the energy required prior to the embargo. For example, an 18.1 ft^3 refrigerator-freezer with automatic defrost now uses only 50 to 90 kWh/month, with more improved versions being developed.

Heat losses in refrigerators arise from a variety of sources. The largest losses are due to heat losses through the walls. Additional energy is used in modern refrigerators to provide for automatic defrosting (so-called frost-free operation) and to prevent condensation on the exterior of the refrigerator. These functions are conveniences and are not essential to the operation of the refrigerator. Some types have a "power saver" switch by which the antisweat heaters can be turned off during dry weather.

Since much of the energy used by a refrigerator depends on its design, care should be used in selection. Efficiency should be an important consideration. Units are available with added insulation and efficient compressors that decrease annual energy use by 25 to 45%. Information on the energy efficiency of specific models is available from the Association of Home Appliances Manufacturers.

Recent research efforts have been directed at designing a high-efficiency refrigerator that would be 25% more efficient than the 1993 Federal energy efficiency standards and that would not use CFC refrigerants. The 1993 standards are already aggressively tighter than the 1990 standards, so this is a tough goal:

National Appliance Energy Conservation Act Standards

• Refrigerator freezers:	960 kWh/year (1990)
	688 kWh/year (1993)
• Freezers:	706 kWh/year (1990)
	533 kWh/year (1993)

The U.S. Environmental Protection Agency, in collaboration with a group of American utilities, formed a consortium to sponsor a design competition to develop a "Super Efficient Refrigerator" with a prize of $30 million. The competition was won by Whirlpool Corporation, which will deliver the new refrigerators in 1995.

Purchase of a new, more efficient unit is not a viable option for many individuals who have a serviceable unit and do not wish to replace it. In this case, the energy management challenge is to obtain the most effective operation of the existing equipment. More efficient operation of refrigeration equipment can be achieved by

- Better insulation
- Disconnecting or reducing operation of automatic defrost and antisweat heaters
- Providing a cool location for the refrigerator coils (or reducing room temperature) and cleaning coils frequently
- Reducing the number of door openings
- Increasing temperature settings
- Precooling foods before refrigerating

Studies have been made of refrigerator operation based on these approaches by the author and his colleagues. Typical results will be described briefly.[3]

A 0.4 m^3 refrigerator (14 ft^3) refrigerator/freezer was studied in a home. The refrigerator was located in the kitchen, where the ambient temperature ranged from 19 to 22°C (66 to 72°F). The refrigerator temperature averaged 4.4C (40°F) and the freezer –13.9°C (7°F).

Measurements indicated that in normal operation the compressor would cycle on for 7 to 8 minutes and then turn off for roughly an equal amount of time. This meant the compressor was on, on the average, 50% of the time. Under these conditions normal electricity use was about 0.17 kWh per hour or 4 kWh per day. Better insulation and disconnecting antisweat heaters and defrost heaters could potentially reduce this by 50%. However, it is generally not practical to do this in the home.

Providing a cool location (such as a basement) can potentially save 5 to 10% by reducing the heat sink temperature to which the refrigerator must transfer heat. Calculations indicate a potential savings of 3.6%/°C(2%/°F).

Holding the number of door openings to a minimum also saves energy—estimated to be 0.3% per opening. As an extreme case, the door of the experimental refrigerator described above was opened every 5 minutes for a 10-second interval, resulting in a 46% increase in energy use (to 0.25 kWh/h).

An effective means for saving energy is to increase temperature settings. This will save about 7 to 9%/°C (4 to 5%/°F). The correct temperature depends on the foods to be stored and the length of storage; for most household purposes 4°C for the refrigerator and –14°C for the freezer is adequate.

Precooling foods will also save refrigeration energy. Hot foods should be cooled to room temperature before placing in the refrigerator, to take advantage of as much "free" cooling as possible.

The same general concepts apply to commercial refrigeration systems. Commercial refrigeration systems are found in supermarkets, liquor stores, restaurants, hospitals, hotels, schools, and other institutions—about one-fifth of all commercial facilities. Systems include walk-in dairy cases, open refrigerated cases, and freezer cases. In a typical supermarket, lighting, HVAC, and miscellaneous uses account for half the electricity use, and refrigerated display cases, compressors, and condenser fans account for the other half. Thus commercial refrigeration can be an important element of electric energy efficiency.

Power requirements vary with case temperature. Low-temperature freezers (–25 to –35°F) require 2.0 to 5.0 kW/ton. Medium-temperature systems (+10 to +20°F) require 1.0 to 2.5 kW/ton, while high-temperature cases (+25 to +40°F) require 0.75 to 1.5 kW/ton.

It is common practice in some types of units to have the compressor and heat exchange equipment located far from the refrigerator compartment. In such systems, a cool location should be selected, rather than locating the compressor next to other equipment that gives off heat. Many of the newer commercial refrigerators now come equipped with heat recovery systems, which recover compressor heat for space conditioning or water heating.

Walk-in freezers and refrigerators lose energy through openings; refrigerated display cases have direct transfer of heat. Covers, strip curtains, air curtains, glass doors, or other thermal barriers can help mitigate these problems. The most efficient light sources should be used in large refrigerators and freezers; every watt eliminated saves 3 watts. Elimination of 1 watt of electricity to produce light also eliminates 2 additional watts required to extract the heat. Other improvements can reduce energy use, including high-efficiency motors, variable speed drives, more efficient compressors, and improved refrigeration cycles and controls.

Energy efficient ice machines are available, with either water- or air-cooled condensers. They have capacities ranging from 75 lb (34 kg) to 1,200 lb (545 kg) per 24-hour period. The water-cooled units are about 20% more efficient and produce ice for from 4.7 kWh/100 lb of ice in the larger sizes to 11.3 kWh/100 lb for the smaller machine.

Cooking

Cooking accounts for only about 1% of total U.S. energy use but constitutes about 6% of residential electricity use. Improvements in energy use efficiency for cooking can be divided into three categories:

- More efficient use of existing appliances
- Use of most efficient existing appliances
- More efficient new appliances

The most efficient use of existing appliances can lead to substantial reductions in energy use. Although slanted toward electric ranges and appliances, the following observations also apply to cooking devices using other sources of heat.

First, select the right size equipment for the job. Do not heat excessive masses or large surface areas that will needlessly radiate heat. Second, optimize heat transfer by ensuring that pots and pans provide good thermal coupling to the heat sources. Flat-bottomed pans should be used on electric ranges. Third, be sure that pans are covered to prevent heat loss and to shorten cooking times. Fourth, when using the oven, plan meals so that several dishes are cooked at once. Use small appliances (electric fry pans, "slow" cookers, toaster ovens, etc.) whenever they can be substituted efficiently for the larger appliances such as the oven.

Different appliances perform similar cooking tasks with widely varying efficiencies. For example, the electricity used and cooking time required for common foods items can vary by as much as ten to one in energy use and five to one in cooking times, depending on the method. As an example, four baked potatoes require 2.3 kWh and 60 minutes in an oven (5.2 kW) of an electric range, 0.5 kWh and 75 minutes in a toaster oven (1.0 kW), and 0.3 kWh and 16 minutes in a microwave oven (1.3 kW). Small appliances are generally more efficient when used as intended. Measurements in a home indicated that a pop-up toaster cooks two slices of bread using only 0.025 kWh. The toaster would be more efficient than using the broiler in the electric range oven, unless a large number of slices of bread (more than 17 in this case) were to be toasted at once.

Cooking several dishes at once can save energy. A complete meal, consisting of a canned ham (2.3 kg), frozen peas (0.23 kg), four yams, and a pineapple upside-down cake (23 cm × 23 cm) were cooked separately using a toaster oven and an electric range, together in an electric oven, and separately in a microwave oven. Cooking separately using the toaster oven and range required 5.2 kWh; cooking together in the oven took 2.5 kWh, and the microwave required 1.2 kWh.

If new appliances are being purchased, select the most efficient ones available. Heat losses from a typical oven approach 1 kW, with insulation accounting for about 50%; losses around the oven door edge and through the window are next in importance. These losses are reduced in certain models. Self-cleaning ovens are normally manufactured with more insulation. Careful design of heating elements can also contribute to better heat transfer.

Microwave cooking is highly efficient for many types of foods, since the microwave heat is deposited directly in the food. Energy input is minimized because there is no need to heat the cooking utensil. Although many common foods can be prepared effectively using a microwave oven, different methods must be used as certain foods are not suitable for microwave cooking. Convection ovens and induction cooktops are two new developments that may also reduce cooking energy use.

Commercial cooking operations range from small restaurants and cafes, where methods similar to those described earlier for residences are practiced, to large institutional kitchens in hotels and hospitals, and finally, to food processing plants.

Many of the same techniques apply. Microwave heating is finding increasing use in hotel and restaurant cooking. Careful scheduling of equipment use and provision of several small units rather than a single large one will save energy. For example, in restaurants, grills, soup kettles, bread warmers, and so on often operate continuously. Generally it is unnecessary to have full capacity during off-peak hours; one small grill might handle mid-morning and mid-afternoon needs, permitting the second and third units to be shut down. The same strategy can be applied to coffee warming stations, hot plates, and so on.

In food processing plants where food is cooked and canned, heat recovery is an important technique. Normally, heat is rejected via cooling water at some step in the process. This heat can be recovered and used to preheat products entering the process, decreasing the amount of heating which eventually must be done.

Residential Appliances

A complete discussion of energy management opportunities associated with all the appliances found in homes is beyond the scope of this chapter. However, several of the major ones will be discussed and general suggestions applicable to the others will be given.

Clothes Drying. Electric clothes dryers operate most efficiently when fully loaded. To remove 97% of the moisture in a typical load of clothes (2 to 7 kg) requires 0.85 to 1.2 kWh/kg. Operating with one-third to one-half a load costs roughly 10 to 15% in energy efficiency.

Measurements were made on a 6 kW dryer used in a residence. The dryer used 0.3 kWh to heat up before cycling off the first time. Electricity used to dry typical loads was found to range from 0.67 kWh/kg of clothes ("permanent press" items dried at a low setting) to 1.13 kWh/kg of heavy towels (dried at a "hot" setting). The average energy input for three different types of loads was 0.88 kWh/kg.

Theoretically, it should take an energy input of 0.73 kWh/kg of water removed. Averaging results from three typical loads for the dryer discussed earlier showed that it took 1.52 kWh/kg of water removed. Therefore about 50% of the energy input is dissipated as waste heat and does not go into water vaporization.

Locating clothes dryers in heated spaces could save 10 to 20% of the energy used by reducing energy needed for heating up. Another approach is to save up loads and do several loads sequentially, so the dryer does not cool down between loads.

The heavier the clothes the greater the amount of water they hold. Mechanical water removal (pressing, spinning, wringing) generally requires less energy than electric heat. Therefore, be certain the washing machine goes through a complete spin cycle (0.1 kWh) before putting clothes in the dryer.

Solar drying, which requires a clothesline (rope) and two poles or trees, has been practiced for millennia and is very sparing of electricity. The chief limitation of course, is inclement weather.

Clothes Washing. Electric clothes washers are designed for typical loads of 3 to 7 kg. Most of the energy used in clothes washing is for hot water; the washer itself only requires 1 to 2% of the total energy input. Typical machines use 1 to 2 kWh/kg of clothes. This is the total energy input, which is the sum of the machine energy plus the water heating energy. The machine energy is typically 0.0025 to 0.05 kWh/kg.[3]

For example, in a study conducted in a residence, a 68-liter washer used 0.16 kWh per load. Hot water usage was 68 liters for the wash cycle plus 34 liters during the rinse cycle. Total water heating energy was 23 MJ, equivalent to 6.5 kWh per load for hot water. These data indicate that the major opportunity for energy management in clothes washing is the use of cold water for washing. Under normal household conditions it is not necessary to use hot water. Clothes are just as clean (in terms of bacteria count) after a 20°C wash as after a 50°C wash. If there is concern for sanitation (e.g., a sick person in the house), authorities recommend use of chlorine bleach. If special cleaning is required, such as removing oil or grease stains, hot water (50°C) and detergent will emulsify oil and fat. There is no benefit in a hot rinse. A secondary savings can come from using full loads. Surveys indicate that machines are frequently operated with partial loads, even though a full load of hot water is used. Conversion to cold water washing (5 loads per week, electric water heater) would save approximately 1,700 kWh per year, or $50 to 100 per year.

Dishwashers. The two major energy uses in electric dishwashers are the hot water and the dry cycle. The volume of hot water used ranges from 45 to 61 liters and can be varied on some machines depending on the load. The temperature requirement for dishwashing sometimes dictates the hot water temperature for the entire residence. Water at 55 to 60°C is needed for dishwashing, although most other residential functions can be satisfied at 38 to 43°C. Some dishwashers accept lower-temperature water and use a booster heater to provide the proper temperature. This is a more efficient approach than maintaining the entire hot water system at the higher temperature with its attendant losses.

Since commercially available machines vary widely, the characteristics of the particular machine are important. In a study of one unit in a residence, the cycle used

	Electricity (kWh)	Hot Water (liters)
Prewash	0.035 (6%)	21 (29%)
Wash	0.125 (20%)	51 (71%)
Dry	0.450 (74%)	—
Totals	0.610 (100%)	72 (100%)

By eliminating the prewash (e.g., by rinsing dishes in cold water prior to placing in the dishwasher) 6% of the electricity and 20% of the hot water can be saved. By eliminating the dry cycle (e.g., opening the door following the wash cycle and letting the dishes air dry), 74% of the electricity can be saved.

Commercial low-temperature dishwasher operate at 140°F rather than 180°F as used in high-temperature models. The low-temperature unit injects a chlorine sanitizer in the final rinse to sanitize dishware. This eliminates the need for an electric heating element of several kilowatts to boost the rinse water temperature.

Television. The transition from black and white to color television roughly doubled or tripled electrical power requirements (from typically 100 to 240 W). More recently, however, the substitution of solid-state electronics for vacuum tube technology reduced energy use by 40 to 50%. This decline has been offset in recent years by the proliferation of home entertainment devices including VCRs, home stereos, compact disk players, and electronic games. Depending on the type and size, a television and other home entertainment devices can account for 100 to 500 kWh per year in a typical residence.

Cable television is growing in popularity due to special programming, improved reception, etc. Some home receiving units have electronic amplifying and demodulating units, which require electricity. These units require 5 to 10 W and unless unplugged or switched off will use 40 to 80 kWh per year.

General Suggestions for Residential Appliances and Electrical Equipment. Many electrical appliances (blenders, floor polishers, hand tools, mixers, etc.) perform unique functions that are difficult to duplicate. This is their chief value. In addition, their relatively infrequent use (a few hours per year) leads to annual energy use in the range of 1 to 10 kWh per year. These appliances are generally insignificant when measured against annual electricity use of 5,000 to 15,000 kWh per year.

Attention should be focused on those appliances that use more than a few percent of annual electricity use. General techniques for energy management include the following:

- Reduce use of equipment (where feasible).
- Reduce losses (e.g., better insulation).
- Substitute for the same function (e.g., heat recovery).
- Perform maintenance to improve efficiency (e.g., clean filters).
- Reduce connected load (e.g., use smaller motors).
- Schedule use for off-peak hours (evenings).

The last point requires further comment. Some utilities now offer "time-of-day" rates which include a premium charge for usage that occurs "on-peak" (when the greatest demand for electricity occurs). Even if there is no direct economic benefit to the user, there is an indirect benefit. By scheduling energy-intensive activities for off-peak hours (clothes washing and drying in the evening for example) the user helps the utility reduce its peaking power requirement, thereby reducing generating costs and helping ensure lower utility rates for all customers.

Computers and Office Equipment

A wide variety of equipment can be found in commercial buildings, depending on the size and function. Process equipment within buildings (e.g., clothes washers in laundries, printing presses, refrigerated display cases, etc.) will not be discussed due to the great diversity of these items.

Excluding process equipment, major energy-using equipment in commercial buildings generally includes HVAC systems, lighting, and "other" equipment. Since energy management options for HVAC and lighting have already been described, the discussion here will be directed at "other" equipment, which includes

- Office equipment (e.g., typewriters and calculators)
- Computers and peripherals
- Facsimile machines, electronic mail
- Vending machines and water coolers
- Copy machines
- Elevators and escalators

To assess the relative importance of these loads, consider a 15-story office building using approximately 65,000 kWh/week excluding HVAC. Half (32,000 kWh/week) of this total was attributable to lighting. Of the balance, roughly 16,000 kWh/week (25%) was used by a computer center, 12,000 kWh/week (18%) was for miscellaneous power (typewriters, duplicating machines, water coolers, etc.), 4,000 kWh/week (6%) was for elevators, and 1,000 kWh/week (1%) was for miscellaneous exhaust fans, pumps, emergency lights, and so on.

Computing equipment varies from small minicomputers using a few hundred watts to large central computer systems using hundreds of kilowatts. The computer center just described, which handled accounting and data processing for a large city, had 90 kW of computing equipment installed. This included about 35 separate equipment items, ranging from display stations rated at 200 W up to the central processing unit (512K memory) rated at 16 kW. Included were card readers (0.5 kW), printers (1.5 kW), disk storage units (3.4 kW), and card punches (1.6 kW), etc.

In this example, special air-conditioning equipment was installed in the computing facility to maintain proper temperature and humidity for the computer equipment. Total heat removal capacity of approximately 60 kW (200,000 Btu/h) was required. According to the computer equipment supplier, acceptable environmental conditions ranged from 16 to 32°C with relative humidity in the range of 20 to 80%. Due to operational constraints, an effort was made to avoid either the high or lower limit on humidity. Equipment to both humidify and dehumidify was thus made part of the separate HVAC system which served the computer facility.

Several energy management opportunities were discovered in this facility. First, an energy audit revealed that overlapping control ranges on the humidify/dehumidify system permitted both units to be on simultaneously. In fact this mode of operation was observed during the audit. A minor modification to the control system would eliminate this duplication of energy waste and pay back the investment in 6 months.

The facility operates three shifts, 7 days per week. There were 10 key punches, which were left on throughout the day shift; many were left on throughout the second and third shift so that personnel could avoid the delay caused by equipment "warmup." By allocating one or two machines to "quick turnaround" jobs, many of the other machines could be turned off except when needed for production work.

Undoubtedly other energy management opportunities could be found in similar computer facilities. These examples suggest some of the considerations involved in evaluating energy use in computing equipment.

General office equipment includes electric typewriters (150 W), coffeemakers (1.2 to 6 kW), teletypes (0.5 kW), vending machines and calculators (10 to 100 W), tape recorders (100 W), and so forth. The energy management opportunities with these types of equipment are more restricted, since they are generally designed to use energy only when operating. One obvious strategy is to ensure that all equipment is turned off when not in use. Another is to size equipment with the right capacity to do the job, avoiding overcapacity, which increases both energy and demand charges.

The four elevators in the building had a total capacity of 14.2 kW. During off hours one or more could be shut down. Efforts were made to encourage employees to walk on short trips—"one floor up, two floors down"—to avoid excessive use of the elevators. Overall energy use was small and no major savings opportunities were possible. In general the most likely opportunity with elevators, escalators, and similar equipment is to shut them down during off hours or other times when they are not needed.

There is a host of miscellaneous equipment in buildings that contributes only a small percentage of total energy use, is used infrequently, and is an unlikely candidate for improved efficiency. Its interaction with the building should be examined, however. For example, in one building stairwells and toilets were exhausted separately to the atmosphere, resulting in large and continuous heat losses. By adding heat recovery to these ventilation systems, a substantial savings resulted.

Closing Remarks

This chapter has discussed the management of electrical energy in buildings. Beginning with a discussion of energy use in buildings, I next outlined the major energy using systems and equipment, along with a brief description of their features which influence energy use and waste. A systematic methodology for implementing an energy management program was then described. The procedure has been implemented in a wide variety of situations including individual homes, commercial buildings, institutions, multinational conglomerates, and cities, and has worked well in each case. Following the discussion of how to set up an energy management program, a series of techniques for saving electricity in each major end use was discussed. The emphasis has been on currently available, cost-effective technology. Undoubtedly other techniques are available, but I have concentrated on those that I know will work in today's economy, for the typical energy consumer.

The first edition of this book was published in 1980. Much of the data in the first edition dated from the 1975–1980 time frame, when the initial response to the oil embargo of 1973 was gathering momentum and maturing. It is remarkable to return to those data and look at the progress that has been made. Most projections at that time predicted U.S. energy use in excess of 100 quads (106 GJ) by 1992; instead, we find that it is 85 quads (90 GJ). As noted earlier in this chapter, a significant growth of the residential sector has occurred in the intervening two decades, but efficiency improvements have kept energy use at roughly the mid-1970s value. The improvement in energy efficiency in lighting, refrigerators, air conditioning, and other devices has been truly remarkably. Today the local hardware or home builder's store has supplies of energy efficient devices that were beyond imagination in 1973.

This is a remarkable accomplishment, technically and politically, given the diversity of the residential/commercial market. Besides the huge economic savings this has meant to millions of homeowners, apartment dwellers, and businesses, think of the environmental benefits associated with avoiding the massive additional amounts of fuel, mining, and combustion which otherwise would have been necessary.

I fear that this important message will be lost in the rush and clamor to deregulate utilities and to curtail government programs. Let us hope that the economic and environmental benefits we have obtained and can continue to reap are clear to the public and that further actions will be encouraged by the reappearance of this book.

Acknowledgments

I am grateful to my colleague T. Cosentino for her editorial support and assistance with the manuscript.

References

1. Energy Information Administration, *Annual Energy Outlook 1994—with Projections to 2010,* U.S. Department of Energy, (Washington D.C.: U.S. Government Printing Office, January 1994).
2. U.S. Congress, Office of Technology Assessment, *Industrial Energy Efficiency,* OTA-E-560, (Washington D.C.: U.S. Government Printing Office, August 1993).
3. Smith, Nancy J., Energy management and appliance efficiency in residences, in *Energy Use Management—Proceedings of the International Conference,* Vol. 1, Pergamon Press, New York, 1977.

11

Heating, Ventilating, and Air Conditioning Control Systems

Jan F. Kreider
University of Colorado—Boulder

Introduction—The Need for Control

This chapter describes the essentials of control systems for heating, ventilating, and air conditioning (HVAC) of buildings designed for energy conserving operation. Of course, there are other renewable and energy conserving systems that require control. The principles described herein for buildings also apply with appropriate and obvious modification to these other systems. For further reference, the reader is referred to several standard references in the list at the end of this chapter.

HVAC system controls are the information link between varying energy demands on a building's primary and secondary systems and the (usually) approximately uniform demands for indoor environmental conditions. Without a properly functioning control system, the most expensive, most thoroughly designed HVAC system will be a failure. It simply will not control indoor conditions to provide comfort.

The HVAC designer must design a control system that

- Sustains a comfortable building interior environment
- Maintains acceptable indoor air quality
- Is as simple and inexpensive as possible and yet meets HVAC system operation criteria reliably for the system lifetime
- Results in efficient HVAC system operation under all conditions

A considerable challenge is presented to the HVAC system designer to design a control system that is both energy conserving and reliable. One of the reasons for inadequate control operation

0-8493-2514-5/96/$0.00+$.50

historically is inadequate design or unclear assignment of responsibility for control system design. In this chapter devoted entirely to controls we will describe the rudiments of control design from the point of view of the HVAC system designer. The reader is encouraged to do additional study in the following references on the subject: ASHRAE (1987), Haines (1987), Honeywell (1988), and Letherman (1981).

In order to achieve proper control based on the control system design, the HVAC system itself must be constructed and calibrated according to the mechanical system drawings. These must include properly sized primary and secondary systems. In addition, air stratification must be avoided, proper provision for control sensors is required, freeze protection is necessary in cold climates, and proper attention must be paid to minimizing energy consumption subject to reliable operation and occupant comfort.

The principal, ultimate controlled variable in buildings is zone temperature (and to a lesser extent air quality in some buildings). Therefore, we will focus our discussion in this chapter on temperature control. Of course, the control of zone temperature involves many other types of control within the primary and secondary HVAC systems, including boiler and chiller control, pump and fan control, liquid and air flow control, humidity control, and auxiliary system control (for example thermal storage control). This chapter discusses only *automatic* control of these subsystems. Honeywell (1988) defines an automatic control system as "a system that reacts to a change or imbalance in the variable it controls by adjusting other variables to restore the system to the desired balance."

Figure 11.1 shows a familiar control problem. It is necessary to maintain the water level in the tank under varying outflow conditions. The float operates a valve that admits water to the tank as the tank is drained. This simple system includes all the elements of a control system:

Sensor—float; reads the controlled variable, the water level
Controller—linkage connecting float to valve stem; senses difference between full tank level and operating level and determines needed position of valve stem
Actuator (controlled device)—internal valve mechanism; sets valve (the final control element) flow in response to level difference sensed by controller.
Controlled system characteristic—water level; this is often termed the **controlled variable**

This system is called a **closed loop** or **feedback** system because the sensor (float) is directly affected by the action of the controlled device (valve). In an open loop system the sensor operating the controller does not directly sense the action of the controller or actuator. An example would be a method of controlling the valve based on an external parameter such as the time of day, which may have an indirect relation to water consumption from the tank.

There are four common methods of control of which Figure 11.1 shows but one. In the next section we will describe each with relation to an HVAC system example.

11.1 Modes of Feedback Control

The four common modes of relating the error (difference between desired set point and sensed value of controlled variable—see Figure 11.1) to the corrective action to be taken by the controller are

Two-position
Proportional
Integral
Derivative

The latter three are usually used in a variety of combinations with one another. Figure 11.2(a) shows a steam coil used to heat air in a duct. The simple control system shown includes an air temperature sensor, a controller that compares the sensed temperature to the set point, a steam

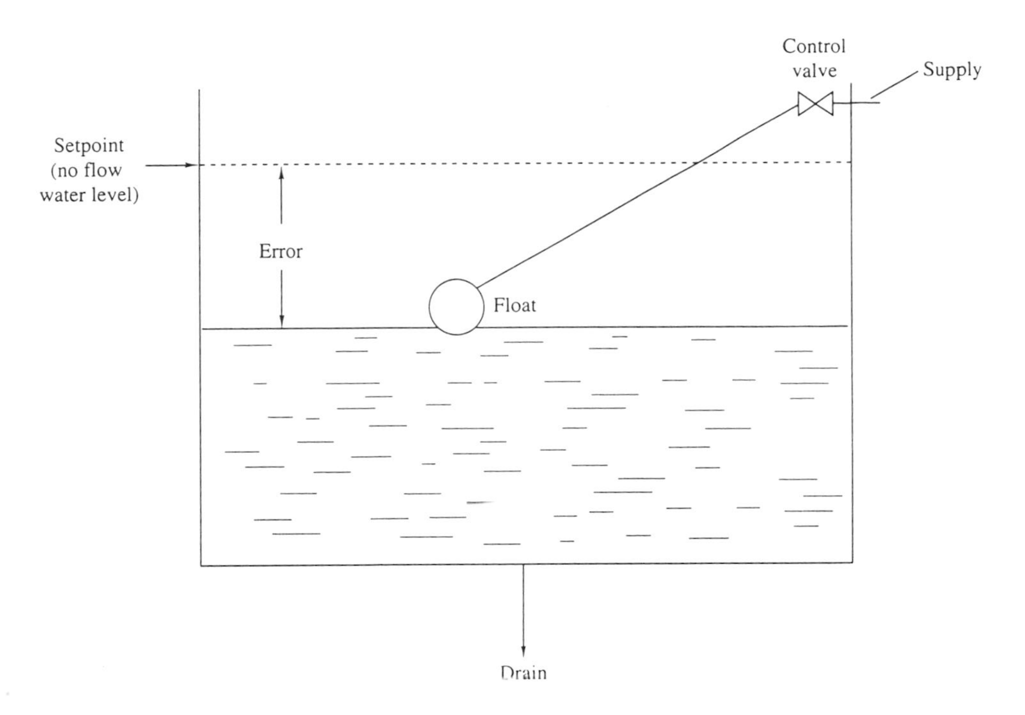

FIGURE 11.1 Simple water level controller. The set point is the full water level; the error is the difference between the full level and the actual level.

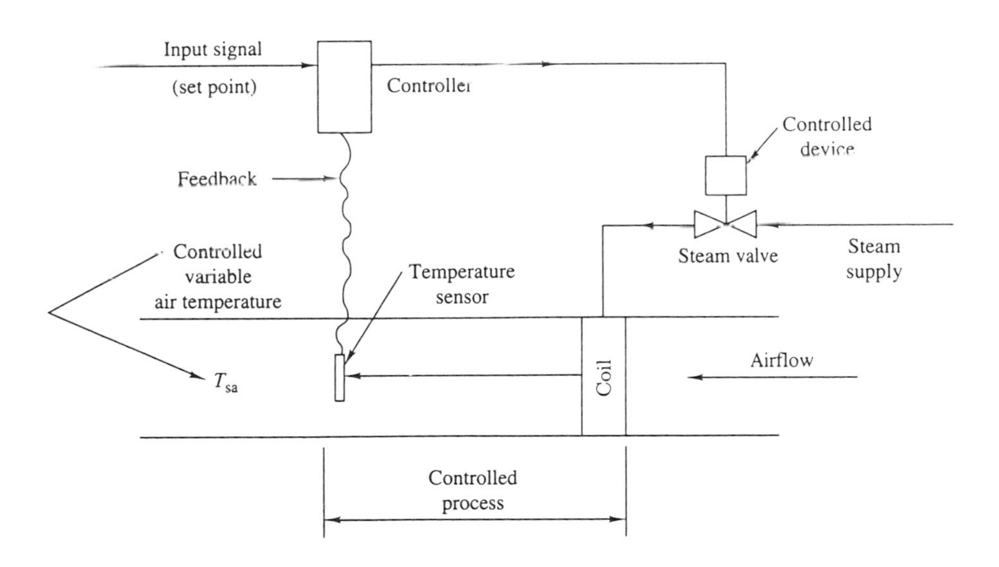

FIGURE 11.2(a) Simple heating coil control system showing the process (coil and short duct length), controller, controlled device (valve and its actuator) and sensor. The set point entered externally is the desired coil outlet temperature.

valve controlled by the controller, and the coil itself. We will use this example system as point of reference when discussing the various control system types. Figure 11.2(b) is the **control diagram** corresponding the physical system shown in Figure 11.2(a).

Two-position control applies to an actuator that is either fully open or fully closed. In Figure 11.2(a) the valve is a two-position valve if two-position control is used. The position of the

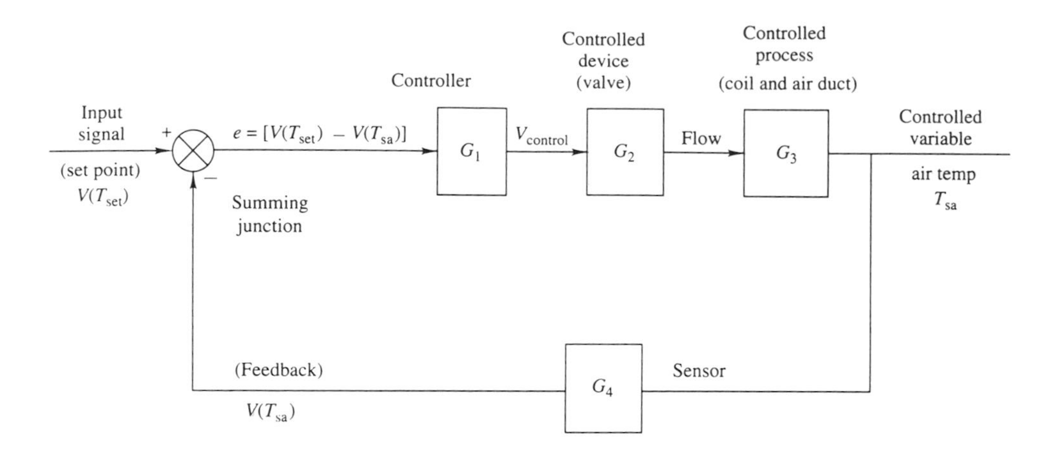

FIGURE 11.2(b) Equivalent control diagram for heating coil. The *G*'s represent functions relating the input to the output of each module. Voltages *V* represent both temperatures (set point and coil outlet) and the controller output to the valve in electronic control systems.

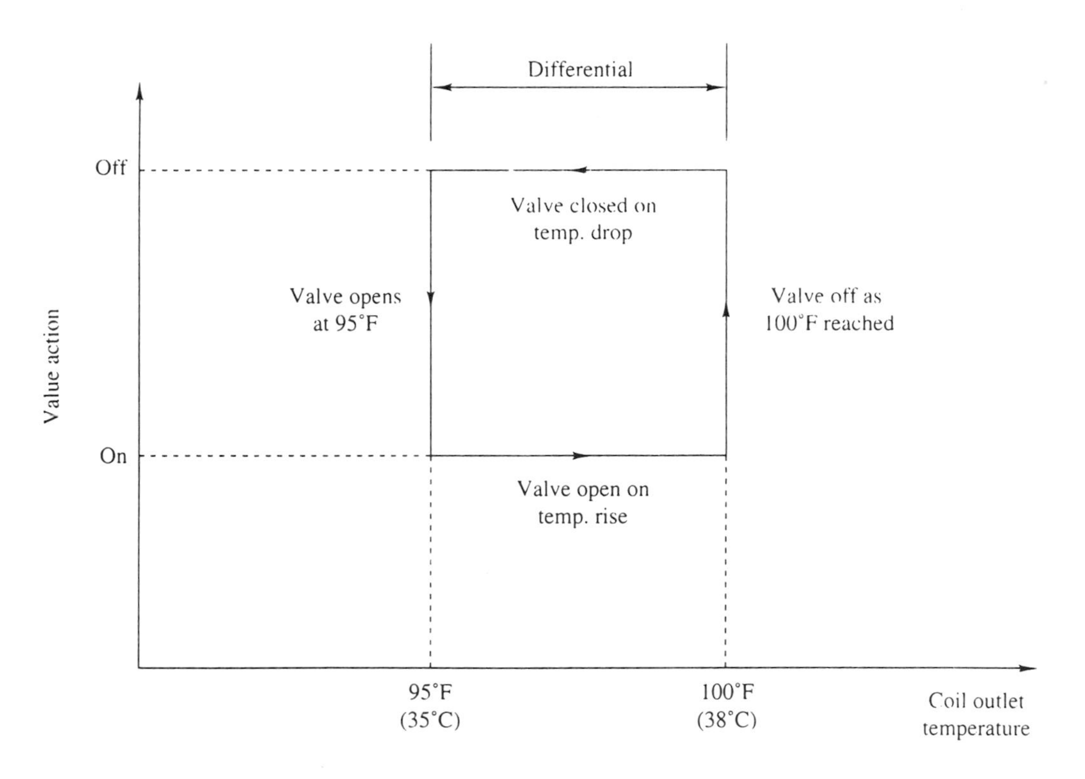

FIGURE 11.3 Two-position (on-off) control characteristic.

valve is determined by the value of the coil outlet temperature. Figure 11.3 depicts two-position control of the valve. If the air temperature drops below 95°F the valve opens and remains open until the air temperature reaches 100°F. The differential is usually adjustable, as is the temperature setting itself. Two-position control is the least expensive method of automatic control and is suitable for control of HVAC systems with large time constants. Examples include residential space and water heating systems. Systems that are fast reacting should not be controlled using this approach, since overshoot and undershoot may be excessive.

Proportional control corrects the controlled variable in proportion to the difference between the controlled variable and the set point. For example, a proportional controller would make a 10% increase in coil heat rate in Figure 11.2 if a 10% decrease in coil outlet air temperature were sensed. The proportionality constant between the error and the controller output is called the **gain** K_p. Equation (11.1) shows the characteristic of a proportional controller:

$$V = V_0 + K_p e \qquad \textbf{(11.1)}$$

where

- V is the controller output; V is used since in electronic controls the controller output is often a voltage
- V_0 is the constant value of controller output when no error exists
- e is the **error**; in the case of the steam coil it is the difference between the sensed air temperature and the needed air temperature, the set point

$$e = T_{set} - T_{sensed} \qquad \textbf{(11.2)}$$

A progressively lower coil air outlet temperature results in a greater error e and hence a greater control action—a greater steam flow rate.

The **throttling range** (ΔV_{max}) is the total change in the controlled variable that is required to cause the actuator or controlled device to move between its limits. For example, if the nominal temperature of a zone is 72°F and the heating controller throttling range is 6°F, then the heating control undergoes its full travel between a zone temperature of 69°F and 75°F. This control, whose characteristic is shown in Figure 11.4, is **reversing acting**; that is, as temperature (controlled variable) increases, the heating valve position decreases.

The throttling range is inversely proportional to the gain as shown in Figure 11.4. Beyond the throttling range the system is out of control. In actual hardware one can set the set point and either the gain or the throttling range (most common) but not both of the latter. Proportional control by itself is not capable of reducing the error to zero, since an error is needed to produce the capacity required for meeting a load as we will see in the following example. This unavoidable value of the error in proportional systems is called the **offset**. From Figure 11.4 it is easy to see that the offset is larger for systems with smaller gains. There is a limit to which one can increase the gain to reduce offset, because high gains can produce control instability.

Example 11.1 Proportional Gain Calculation

If the steam heating coil in Figure 11.2(a) has a heat output that varies from 0 to 20 kW as the outlet air temperature varies from 35 to 45°C in an industrial process, what is the coil gain and what is the throttling range? Find an equation relating the heat rate at any sensed air temperature to the maximum rate in terms of the gain and set point.

Given:

$$\dot{Q}_{max} = 20 \text{ kW}$$
$$\dot{Q}_{max} = 0 \text{ kW}$$
$$T_{max} = 45°\text{C}$$
$$T_{min} = 35°\text{C}$$

Figure: See Figure 11.2(a).

Assumptions: Steady-state operation.

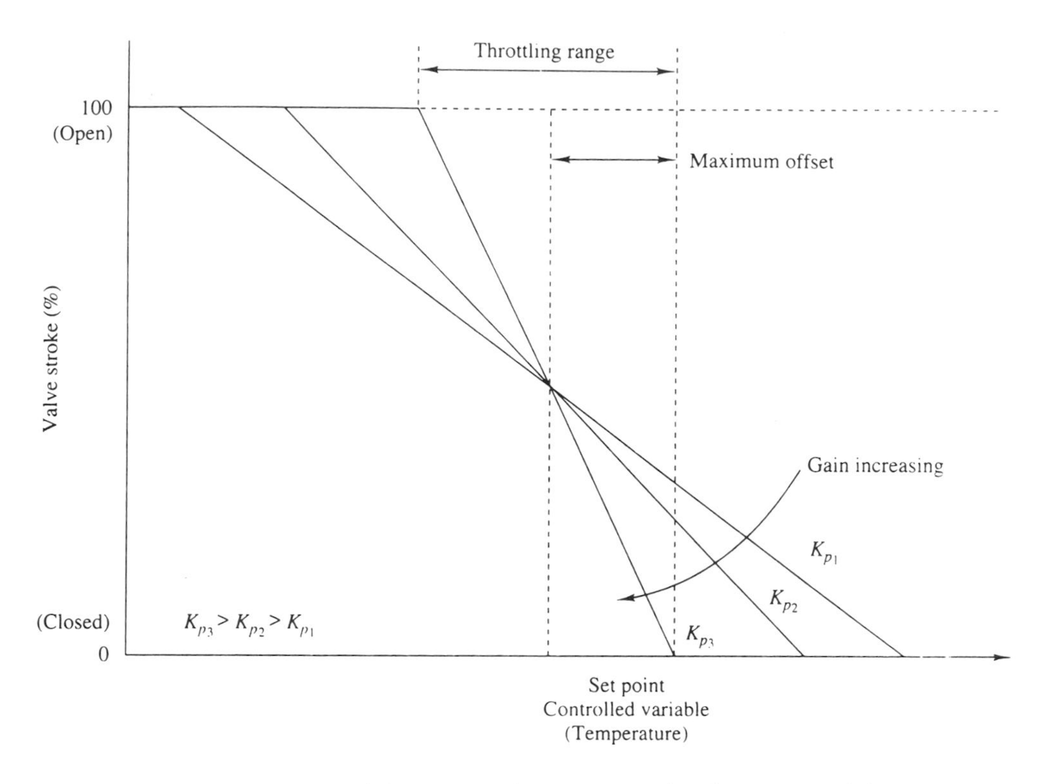

FIGURE 11.4 Proportional control characteristic showing various throttling ranges and the corresponding proportional gains K_p. This characteristic is typical of a heating coil temperature controller.

Find: K_p, ΔV_{max}

Solution: The throttling range is the range of the controlled variable (air temperature) over which the controlled system (heating coil) exhibits its full capacity range. The temperature varies from 90°C to 100°C; therefore the throttling range is

$$\Delta V_{max} = 45°C - 35°C = 10°C \quad \textbf{(11.3)}$$

The proportional gain is the ratio of the controlled system (coil) output to the throttling range. For this example the gain is

$$K_p = \frac{\dot{Q}_{max} - \dot{Q}_{min}}{\Delta V_{max}} = \frac{(20-0)\ \text{kW}}{10\ \text{K}} = 2.0\frac{\text{kW}}{\text{K}} \quad \textbf{(11.4)}$$

The controller characteristic can be found by inspecting Figure 11.4. It is assumed that the average air temperature (40°C) occurs at the average heat rate (10 kW). The equation of the straight line shown is

$$\dot{Q} = K_p\left(T_{set} - T_{sensed}\right) + \frac{\dot{Q}_{max}}{2} = K_p e + \frac{\dot{Q}_{max}}{2} \quad \textbf{(11.5)}$$

The quantity $(T_{set} - T_{sensed})$ is the **error** e and is the signal to the control system that a valve opening change is needed to meet the desired set point.

Inserting the numerical values we have

$$\dot{Q} = 2.0\frac{\text{kW}}{\text{K}}\left(40 - T_{\text{sensed}}\right) + 10 \text{ kW} \tag{11.6}$$

Comments: In an actual steam coil control system it is the steam valve that is controlled directly to indirectly control the heat rate of the coil. This is typical of many HVAC system controls in that the desired control action is achieved indirectly by controlling another variable that in turn accomplishes the desired result. That is why the controller and controlled device are often shown separately as in Figure 11.2(b).

This example illustrates in a simple system why proportional control always requires an error signal, with the result that there is always an **offset** (i.e., the desired set point and actual temperature are always different) with this control mode. Equation (11.5) shows that for any coil heat output other than $\dot{Q}_{max}/2$ an error must exist. This error is the offset and is smaller for higher coil gains.

Proportional control is used with stable, slow systems that permit the use of a narrow throttling range and resulting small offset. Fast-acting systems need wide throttling ranges in order to avoid instability and large offsets result.

Integral control is often added to proportional control to eliminate the offset inherent in proportional only control. The result, proportional plus integral control is identified by the acronym **PI**. Initially the corrective action produced by a PI controller is the same as for a proportional-only controller. After the initial period a further adjustment due to the integral term reduces the offset to zero. The rate at which this occurs depends on the time scale of the integration. In equation form the PI controller is modeled by

$$V = V_0 + K_p e + K_i \int e dt \tag{11.7}$$

in which K_i is the integral gain constant. It has units of reciprocal time and is the number of times that the integral term is calculated per unit time. This is also known as the reset rate; **reset** control is an older term used by some to identify integral control.

The integral term in Eq.(11.7) has the effect of adding a correction to the output signal V as long as the error term exists. The continuous offset produced by the proportional-only controller can thereby reduced to zero because of the integral term. The time scale (K_p/K_i) of the integral term is of the order of 1 to 60 minutes. PI control is used for fast-acting systems for which accurate control is needed. Examples include mixed air controls, duct static pressure controls, and coil controls. Because the offset is eventually eliminated with PI control, the throttling range can be set rather wide to insure stability under a wider range of conditions than good control would permit with only proportional control.

Derivative control is used to speed up the action of PI control. When derivative control is added to PI control, the result is called PID control. The derivative term added to Eq.(11.7) generates a correction signal proportional to the time rate of change of error. This term has little effect on a steady proportional system with uniform offset (time derivative is zero) but initially, after a system disturbance, produces a larger correction more rapidly. Equation (11.8) includes the derivative term in the mathematical model of the PID controller

$$V - V_0 + K_p e + K_i \int e\, dt + K_d \frac{de}{dt} \tag{11.8}$$

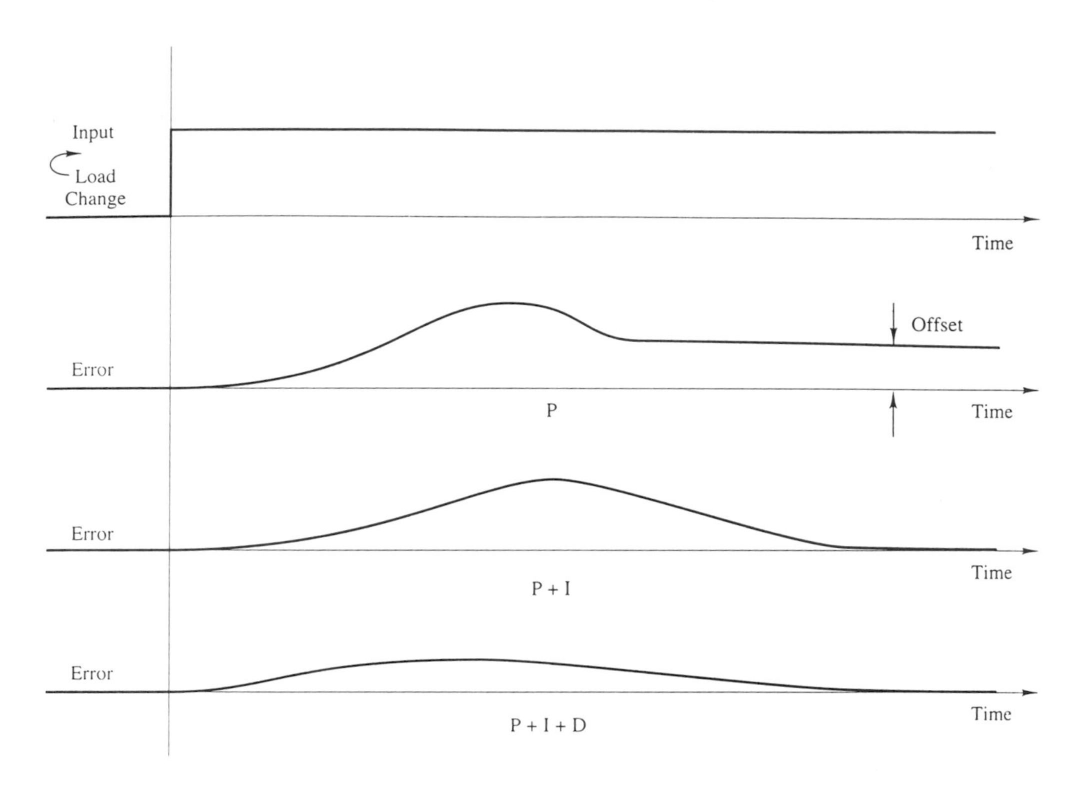

FIGURE 11.5 Performance comparison of P, PI, and PID controllers when subjected to a uniform, input step change.

in which K_d is the derivative gain constant. The time scale (K_d/K_p) of the derivative term is typically in the range of 0.2 to 15 minutes. Since HVAC systems do not often require rapid control response, the use of PID or PD control is less common than use of PI control. Since a derivative is involved, any noise in the error (i.e., sensor) signal must be avoided to maintain good control. One application in buildings where PID control has been effective is in duct static pressure control, a fast-acting subsystem that has a tendency to be unstable otherwise.

Figure 11.5 compares the reaction of the three systems discussed earlier to a step change in load on a coil. The proportional system never achieves the set point. Since the integral term generates a correction proportional to the area under the error curve, it slowly reduces the error to zero but the maximum error is greater than for the PID approach, which takes action faster to reduce the error than does the PI system alone. The PD system (not shown) never achieves zero error, since the zero derivative of the constant offset produces no control action.

11.2 Basic Control Hardware

In this section we describe the various physical components needed to achieve the actions required by the control strategies of the previous section. Since there are two fundamentally different control approaches—pneumatic and electronic—the following material is so divided. Sensors, controllers, and actuators for principal HVAC applications are described.

Pneumatic Systems

The first, widely adopted automatic control systems used compressed air as the operating medium. Although a transition to electronic controls is occurring, about 50% of the sales of one large controls company in the early 1990s were pneumatic equipment. Pneumatic controls use compressed air

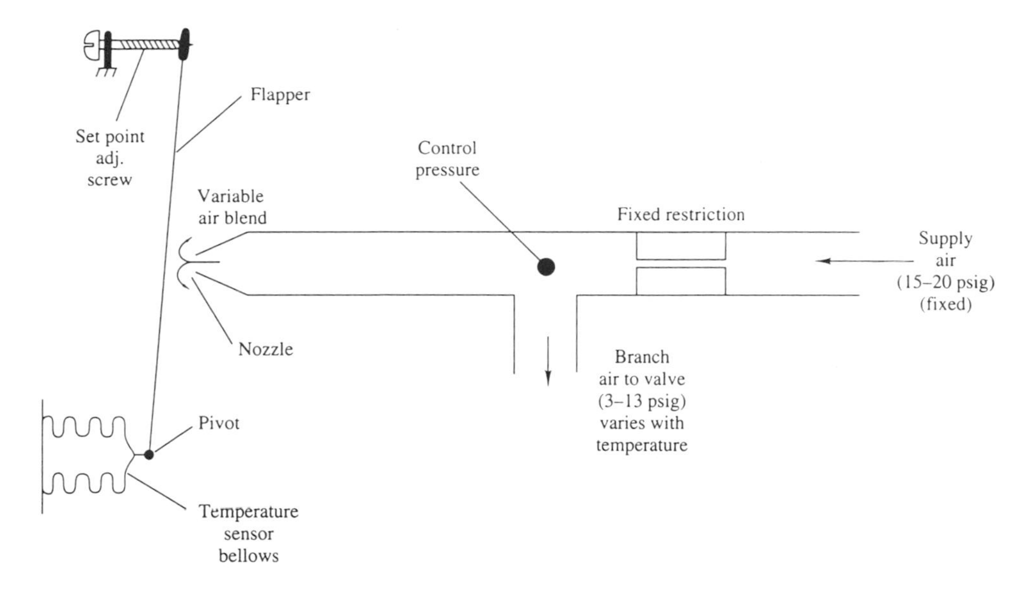

FIGURE 11.6 Drawing of pneumatic thermostat showing adjustment screw used to change temperature setting.

(approximately 20 psig in the U.S.) for operation of sensors and actuators. This section provides an overview of how these devices operate.

Since temperature is the most common parameter controlled, the most common pneumatic **sensor** is the temperature sensor. Figure 11.6 shows one method of sensing temperature and producing a control signal from it. Main supply air is supplied by a compressor to the zone thermostat. An amount of this air is bled from the nozzle depending on the position of the flapper and the size of the restrictor (diameters of the order of thousandths of an inch, [hundredths of a millimeter] are typical). The pressure in the branch line that controls the valve ranges between 3 and 13 psig (20–90 kPa) typically. In simple systems this pressure from a thermostat could operate an actuator such as a control valve for a room heating unit. In this case the thermostat is both the sensor and the controller—a rather common configuration. The consumption of air in this sensor is quite small.

Many other temperature sensor approaches can be used. For example, the bellows shown in Figure 11.6 can be eliminated and the flapper can be made of a bimetallic strip. As temperature changes, the bimetal strip changes curvature, opening or closing the flapper/nozzle gap. Another approach uses a remote bulb filled with either liquid or vapor that pushes a rod (or a bellows) against the flapper to control the pressure signal. This device is useful if the sensing element must be located where direct measurement of temperature by a metal strip or bellows is not possible, such as in a water stream or high-velocity ductwork. The bulb and connecting capillary size may vary considerably by application.

Pressure sensors may use either bellows or diaphragms to control branch line pressure. For example, the motion of a diaphragm may replace that of the flapper in Figure 11.6 to control the bleed rate. A bellows similar to that shown in the same figure may be internally pressurized to produce a displacement that can control air bleed rate. A bellows produces significantly greater displacements than a single diaphragm.

Humidity sensors in pneumatic systems are made from materials that change size with moisture content. Nylon or other synthetic hygroscopic fibers that change size significantly (i.e., 1–2%) with humidity are commonly used. Since the dimensional change is relatively small on an absolute basis, mechanical amplification of the displacement is used. The materials that exhibit the desired property include nylon, hair, and cotton fibers. Since the properties of hair vary with age, the more

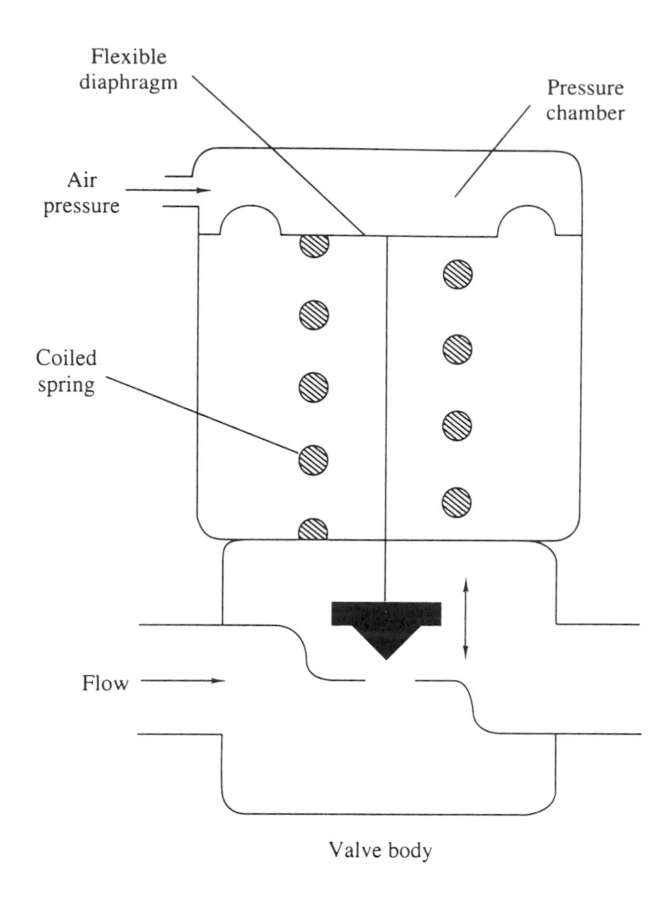

FIGURE 11.7 Pneumatic control valve showing counterforce spring and valve body. Increasing pressure closes the valve.

stable material, nylon, is preferred and most widely used (Letherman, 1981). Humidity sensors for electronic systems are quite different and are discussed in the next section.

An **actuator** converts pneumatic energy to motion—either linear or rotary. It creates a change in the controlled variable by operating control devices such as dampers or valves. Figure 11.7 shows a pneumatically operated control valve. The valve opening is controlled by the pressure in the diaphragm acting against the spring. The spring is essentially a linear device. Therefore, the motion of the valve stem is essentially linear with air pressure. However, this does not necessarily produce a linear effect on flow as discussed later. Figure 11.8 shows a pneumatic damper actuator. Linear actuator motion is converted into rotary damper motion by the simple mechanism shown.

Pneumatic **controllers** produce a branch line (see Figure 11.6) pressure that is appropriate to produce the needed control action for reaching the set point. Such controls are manufactured by a number of control firms for specific purposes. Classifications of controllers include the sign of the output (direct or reverse acting) produced by an error, by the control action (proportional, PI, or two-position), or by number of inputs or outputs. Figure 11.9 shows the essential elements of a dual-input, single-output controller. The two inputs could be heating system supply temperature and outdoor temperature sensors used to control the output water temperature setting of a boiler in a building heating system. This is essentially a boiler **temperature reset** system that reduces heating water temperature with increasing ambient temperature for better system control and reduced energy use.

The air supply for pneumatic systems must produce very clean, oil-free, dry air. A compressor producing 80 to 100 psig is typical. Compressed air is stored in a tank for use as needed, avoiding

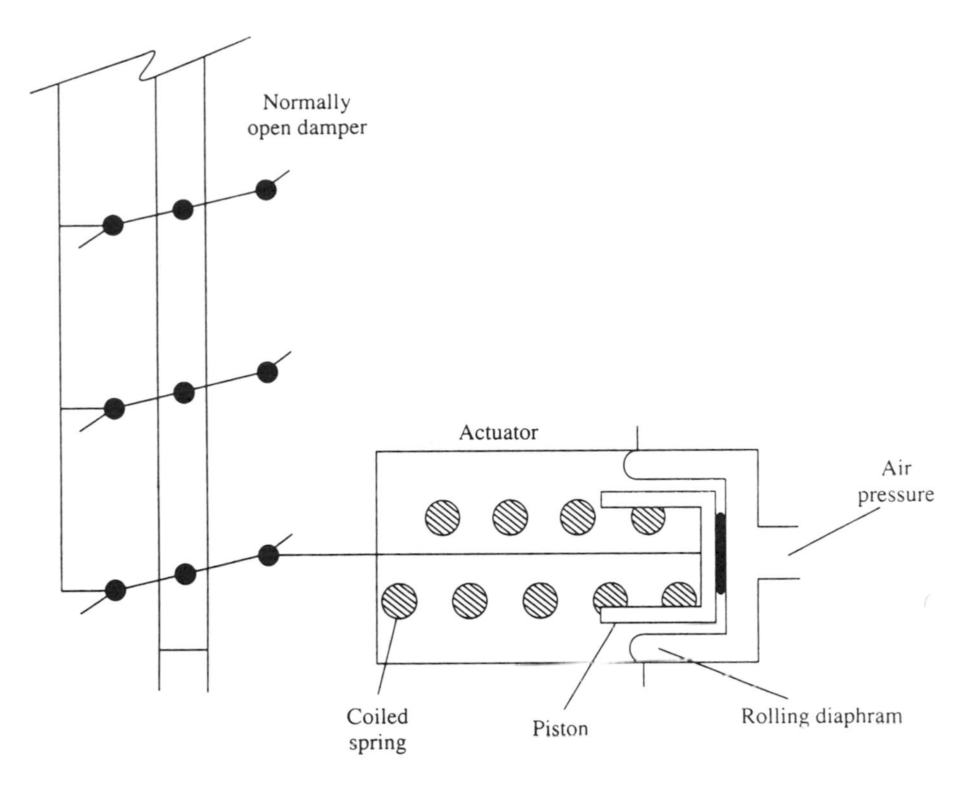

FIGURE 11.8 Pneumatic damper actuator. Increasing pressure closes the parallel blade damper.

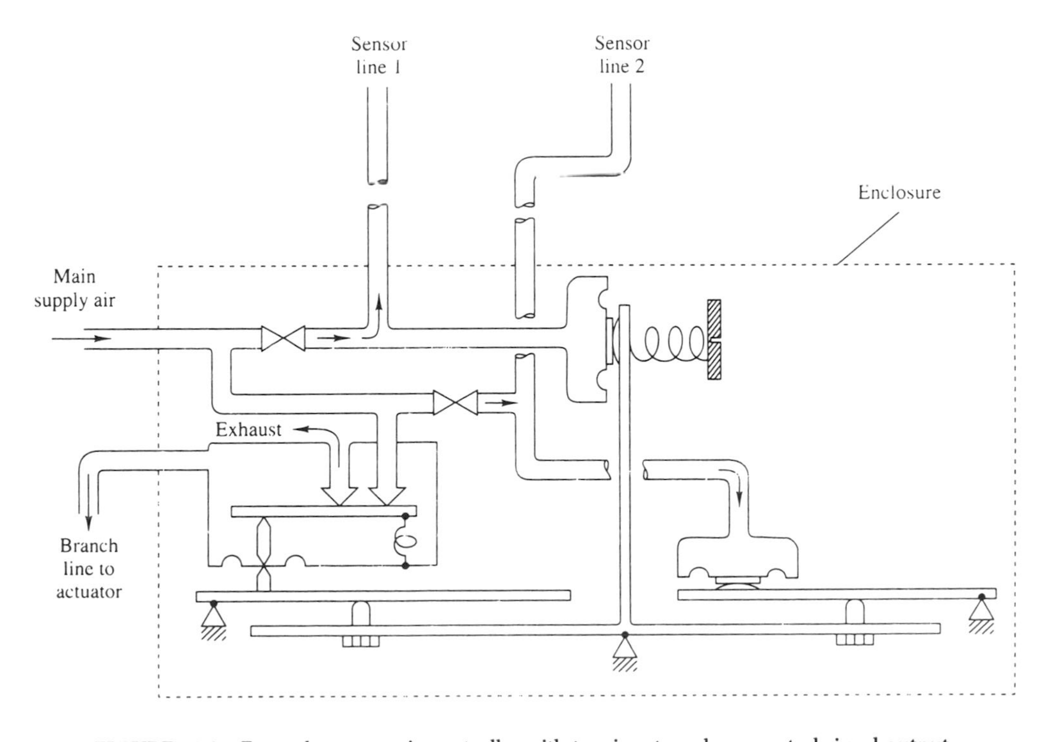

FIGURE 11.9 Example pneumatic controller with two inputs and one control signal output.

continuous operation of the compressor. The air system should be oversized by 50 to 100% of estimated, nominal consumption. The air is then dried to avoid moisture freezing in cold control lines in air handling units and elsewhere. Dried air should have a dew point of –30°F or less in severe heating climates. In deep cooling climates the lowest temperature to which the compressed air lines are exposed may be the building cold air supply. Next, the air is filtered to remove water droplets, oil (from the compressor), and any dirt. Finally the air pressure is reduced in a pressure regulator to the control system operating pressure of approximately 20 psig. Control air piping uses either copper or nylon (in accessible locations).

Electronic Control Systems

Electronic controls are used increasingly more widely in HVAC systems in commercial buildings. Their precise control, flexibility, compatibility with microcomputers, and reliability are advantages but they may have increased cost per control loop. With the continuous decrease in microprocessor cost and associated increase in capability, the cost penalty, if any, is expected to virtually disappear, especially when calculated on a per function basis. In this section we survey the sensors, actuators, and controllers used in modern electronic control systems for buildings.

Direct digital control (DDC) enhances the previous analog-only electronic system with digital features. Modern DDC systems use analog sensors (converted to digital signals within a computer) along with digital computer programs to control HVAC systems. The output of this microprocessor-based system can be used to control electronic, electrical, or pneumatic actuators or a combination. DDC systems have the advantage of reliability and flexibility. For example, it is easier to accurately set control constants in computer software than by adjustments at a controller panel with a screwdriver. DDC systems offer the option of operating energy management systems (EMSs) and HVAC diagnostic, knowledge-based systems, since the sensor data used for control is very similar to that used in EMSs. Pneumatic systems do not offer this ability. Figure 11.10 shows a schematic diagram of a DDC controller. The entire control system must include sensors and actuators not shown in this controller-only drawing.

Temperature measurements for DDC applications are made by three principal methods:

Thermocouples
Resistance temperature detectors (RTDs)
Thermistors

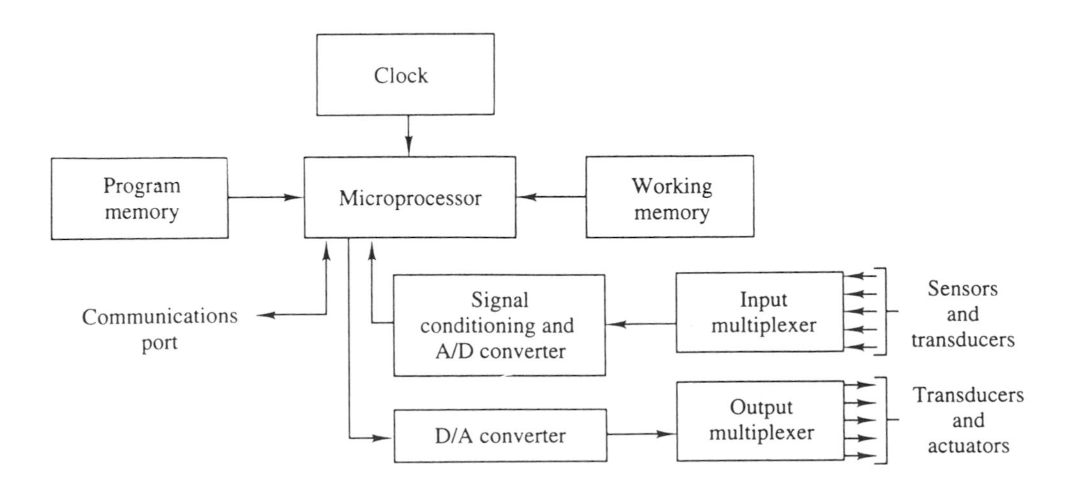

FIGURE 11.10 Block diagram of DDC controller.

Each has its advantages for particular applications. Thermocouples consist of two dissimilar metals chosen to produce a measurable voltage at the temperature of interest. The voltage output is low (millivolts) but a well-established function of the junction temperature. Except for flame temperature measurements thermocouples produce voltages too small to be useful in most HVAC applications (for example, a type J thermocouple produces only 5.3 mV at 100°C).

RTDs use small, responsive sensing sections constructed from metals whose resistance-temperature characteristic is well established and reproducible. To first order

$$R = R_0(1 + kT) \tag{11.9}$$

where

R is the resistance, ohms
R_0 is the resistance at the reference temperature (0°C), ohms
k is the temperature coefficient of resistance, °C^{-1}
T is the RTD temperature, °C

This equation is easy to invert to find the temperature as a function of resistance. Although complex higher order expressions exist, their use is not needed for HVAC applications.

Two common materials for RTDs are platinum and Balco (a 40% nickel, 60% iron alloy). The nominal values of k, respectively, are 3.85×10^{-3} and 4.1×10^{-3}°C^{-1}.

Resistance is measured indirectly by measurements of voltage and current. Therefore, the controller must supply current to the RTD. The current can cause self-heating and consequent errors. These are avoidable by using higher resistance RTDs. In addition, lead wire resistance can cause lack of accuracy for the class of platinum RTDs whose nominal resistance is only 100 ohms because the lead resistance of 1 to 2 ohms is not negligible in comparison to that of the sensor itself.

Thermistors are semiconductors with the property that resistance is a strong, but nonlinear function of temperature given approximately by Eq. (11.10)

$$R = Ae^{(B/T)} \tag{11.10}$$

A is related to the nominal value of resistance at the reference temperature and is of the order of 0.06 to 0.07 for a nominal 10 Kohm thermistor. The exponential coefficient B (a weak function of temperature) is of the order of 5,400 to 7,200 R (3,000–4,000 K). The nonlinearity inherent in thermistors can be reduced by connecting a properly selected fixed resistor in parallel with it. The resulting linearity is desirable from a control system design viewpoint. Thermistors can have a problem with long-term drift and aging; the designer and control manufacturer should consult on the most stable thermistor design for HVAC applications.

Humidity measurements are needed for control of economizers or may also be needed to control special environments as such as clean rooms, hospitals, and areas housing computers. Relative humidity, dew point, and humidity ratio are all indicators of the moisture content of air. An electrical, capacitance-based approach is preferred. The high dielectric constant of water absorbed into a polymer causes a significant change in capacitance. This is used along with a local electronic circuit to produce a linear voltage signal with humidity. If not saturated by excessive exposure to very high humidities, these devices produce reproducible signals without excessive hysteresis (Huang, 1991). Response times are of the order of seconds if the local air velocity over the sensor is sufficiently high. Another solid-state device for humidity measurement uses the variation in resistance of a thin film of lithium chloride. The resistance change is significant but also depends on temperature. The most common sensor type used presently employs the capacitance approach.

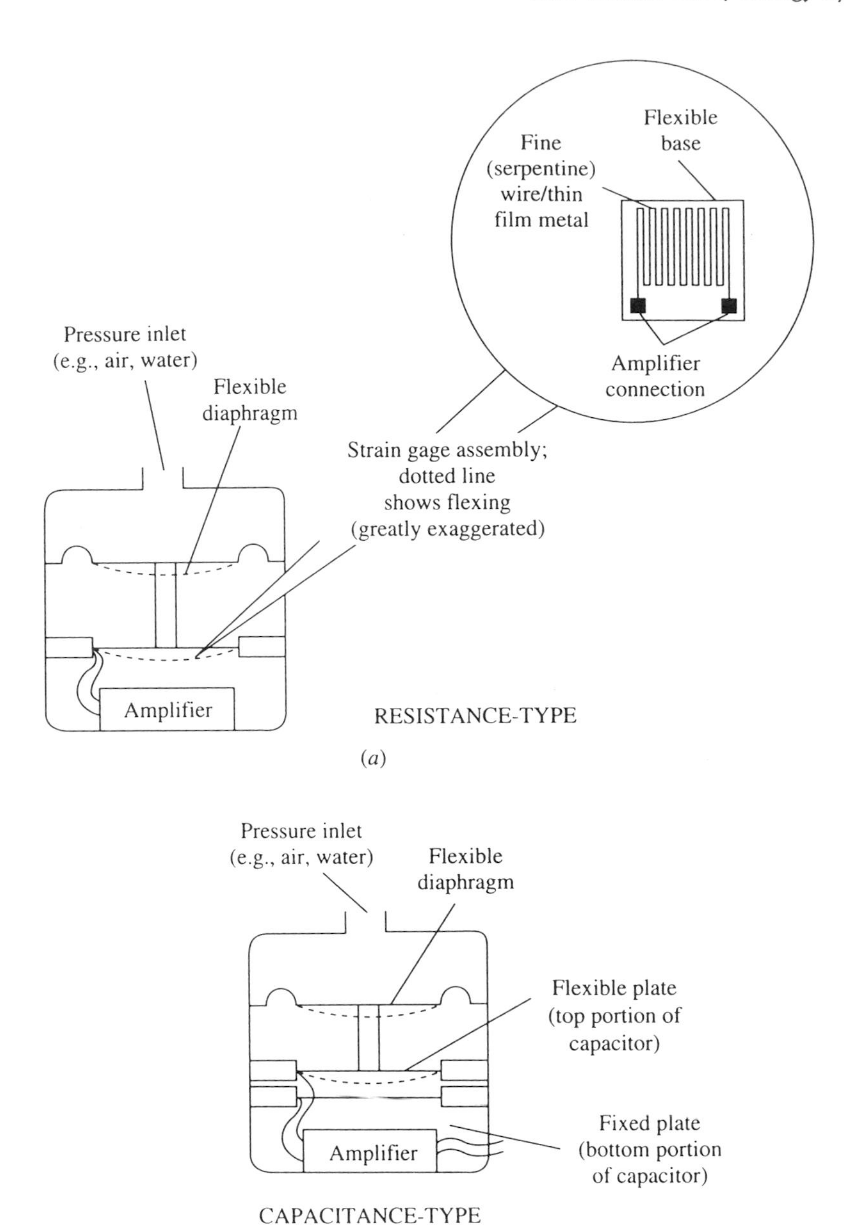

FIGURE 11.11 Resistance- and capacitance-type pressure sensors.

Pressure measurements are made by electronic devices that depend on a changing of resistance or capacitance with imposed pressure. Figure 11.11 shows a cross-sectional drawing of each. In the resistance type, stretching of the membrane lengthens the resistive element, thereby increasing resistance. This resistor is an element in a Wheatstone bridge; the resulting bridge voltage imbalance is linearly related to the imposed pressure. The capacitative type of unit has a capacitance between a fixed and a flexible metal that decreases with pressure. The capacitance change is amplified by a local amplifier that produces an output signal proportional to pressure.

Flow measurement or indication is needed in DDC systems. Pitot tubes (or arrays of tubes) and other flow measurement devices can be used to measure either air or liquid flow in secondary HVAC systems. Air flow information is important for proper control of building pressure, variable air volume (VAV) system control, and outside air supply. Water flow rates are needed for chiller and boiler control and for secondary system liquid loop control. In some cases the quantitative value of flow is not needed, only a knowledge that flow exists. Sensors for this, called flow switches, are electromechanical switches that change from open to closed (or vice versa) upon the existence of flow. Similar switches are used in some control system designs to sense damper position (open or closed).

Temperature, humidity, and pressure **transmitters** are often used in HVAC systems. They amplify signals produced by the basic devices described in the preceding paragraphs and produce an electrical signal over a standard range thereby permitting standardization of this aspect of DDC systems. The standard ranges are

Current—4 to 20 mA (DC)
Voltage—0 to 10 volts (DC)

Although the majority of transmitters produce such signals, the noted values are not universally used.

Figure 4.10 shows the elements of a DDC controller. The heart of the controller is the microprocessor, which can be programmed in either a standard or system-specific language. Control algorithms (linear or not), sensor calibrations, output signal shaping, and historical data archiving can all be programmed as the user requires. A number of firms have constructed controllers on standard personal computer platforms. It is beyond the scope of this book to describe the details of programming HVAC controllers, since each manufacturer uses a different approach. However, the essence of any DDC system is the same as shown in the figure. Honeywell (1988) discusses DDC systems and their programming in more detail.

Actuators for electronic control systems include

Motors—operate valves, dampers
Variable speed controls—pump, fan, chiller drives
Relays and motor starters—operate other mechanical or electrical equipment (pumps, fans, chillers, compressors), electrical heating equipment
Transducers—for example, convert electrical signal to pneumatic (EP transducer)
Visual displays—not actuators in the usual sense but used to inform system operator of control and HVAC system function

Pneumatic and DDC systems both have advantages and disadvantages. Pneumatics have the advantage of lower cost, inherently modulating actuators and sensors, explosion-proof components, and diagnostic simplicity. Disadvantages include the need for a compressor producing clean and dry air, the cost of air piping, and the need for regular component calibration. DDC systems can be very precise (limited by sensor and actuator accuracy), can accommodate complex control algorithms and scheduling, can accept easy changes to control constants, can produce data usable by EMSs, and can allow central control of a group of buildings. Present disadvantages include cost (this penalty is decreasing as discussed earlier) and training needs of maintenance personnel, who may be more familiar with pneumatic controls.

11.3 Basic Control System Design Considerations

This section discusses selected topics in control system design including control system zoning, valve and damper selection, and control logic diagrams. The following section shows several HVAC system control design concepts.

The ultimate purpose of an HVAC system control system is to control zone temperature (and secondarily air motion and humidity) to conditions that assure maximum comfort and productivity of the occupants. From a controls viewpoint the HVAC system is assumed to be able to provide comfort conditions if controlled properly. Basically a zone is a portion of a building that has loads that differ in magnitude and timing sufficiently from other areas so that separate portions of the secondary HVAC system and control system are needed to maintain comfort.

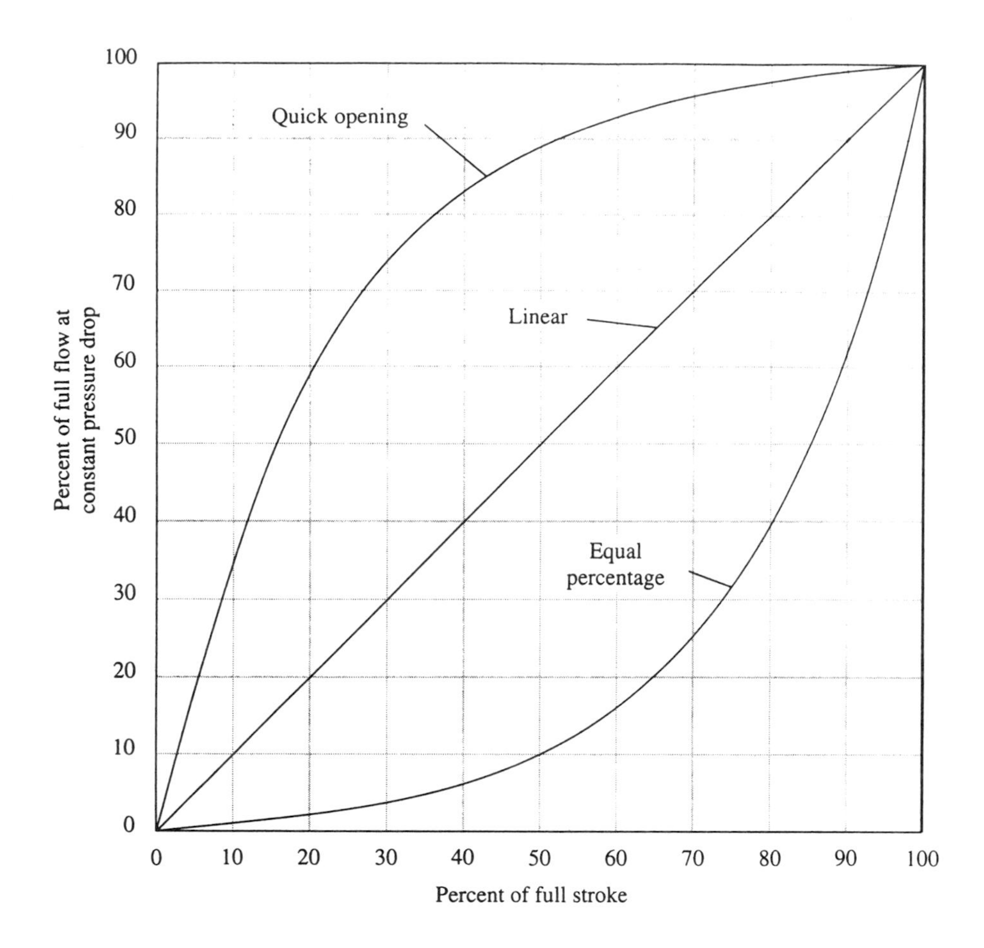

FIGURE 11.12 Quick-opening, linear, and equal percentage valve characteristics.

Having specified the zones, the designer must select the thermostat (and other sensors, if used) location. Thermostat signals are either passed to the central controller or used locally to control the amount and temperature of conditioned air or coil water introduced into a zone. The air is conditioned either locally (e.g., by a unit ventilator or baseboard heater) or centrally (e.g., by the heating and cooling coils in the central air handler). In either case a flow control actuator is controlled by the thermostat signal. In addition, airflow itself may be controlled in response to zone information in VAV systems. Except for variable speed drives used in variable volume air or liquid systems, flow is controlled by valves or dampers. The design selection of valves and dampers is discussed next.

Steam and Liquid Flow Control

The flow through valves such as that shown in Figure 11.7 is controlled by valve stem position, which determines the flow area. The variable flow resistance offered by valves depends on their design. The flow characteristic may be linear with position or not. Figure 11.12 shows flow characteristics of the three most common types. Note that the plotted characteristics apply only for **constant valve pressure drop**. The characteristics shown are idealizations of actual valves. Commercially available valves will resemble but not necessarily exactly match the curves shown.

The **linear** valve has a proportional relation between volumetric flow $\dot{V}$ and valve stem position z.

$$\dot{V} = kz \tag{11.11}$$

The flow in **equal percentage** valves increases by the same fractional amount for each increment of opening. In other words, if the valve is opened from 20% to 30% of full travel the flow will increase by the same percentage as if the travel had increased from 80% to 90% of its full travel. However, the absolute volumetric flow increase for the latter case is much greater than for the former. The equal percentage valve flow characteristic is given by

$$\dot{V} = Ke^{(kz)} \tag{11.12}$$

in which k and K are proportionality constants for a specific valve. Quick-opening valves do not provide good flow control but are used when rapid action is required with little stem movement for on/off control.

Example 11.2 Equal Percentage Valve

A valve at 30% of travel has a flow of 4 gal/min. If the valve opens another 10% and the flow increases by 50% to 6 gal/min what are the constants in Eq. (11.12)? What will be the flow at 50% of full travel?

Figure: See Figure 11.15.

Assumptions: Pressure drop across the valve remains constant.

Find: k, K, $\dot{V}_{50}$

Solution: Equation (11.12) can be evaluated at the two flow conditions. If the results are divided by each other we have

$$\frac{\dot{V}_2}{\dot{V}_1} = \frac{6}{4} = e^{k(z_2 - z_1)} = e^{k(0.4-0.3)} \tag{11.13}$$

In this expression the travel z is expressed as a fraction of the total travel and is dimensionless. Solving this equation for k gives the result

$$k = 4.05 \text{ (no units)}$$

From the known flow at 30% travel we can find the second constant K

$$K = \frac{4\frac{\text{gal}}{\text{min}}}{e^{4.05\times 0.3}} = 1.19\frac{\text{gal}}{\text{min}} \tag{11.14}$$

Finally, the flow is given by

$$\dot{V} = 1.19e^{4.05z} \tag{11.15}$$

At 50% travel, the flow can be found from this expression

$$\dot{V}_{50} = 1.19e^{4.05\times 0.5} = 9.0\frac{\text{gal}}{\text{min}} \tag{11.16}$$

Comments: This result can be checked, since the valve is an equal percentage valve. At 50% travel the valve has moved 10% beyond its 40% setting at which the flow was 6 gal/min. Another 10% stem movement will result in another 50% flow increase from 6 gal/min to 9 gal/min, confirming the solution.

The plotted characteristics of all three valve types assume constant pressure drop across the valve. In an actual system the pressure drop across a valve will not remain constant, but if the valve is to maintain its control characteristics, the pressure drop across it must be the majority of the entire loop pressure drop. If the valve is designed to have a full open pressure drop equal to that of the balance of the loop, good flow control will exist. This introduces the concept of valve **authority**, defined as the valve pressure drop as a fraction of total system pressure drop:

$$A \equiv \frac{\Delta p_{\mathrm{v,\ open}}}{\left(\Delta p_{\mathrm{v,\ open}} + \Delta p_{\mathrm{system}}\right)} \tag{11.17}$$

For proper control the full open valve authority should be at least 0.50. If the authority is 0.5 or more, control valves will have installed characteristics not much different from those shown in Figure 11.12. If not, the valve characteristic will be distorted at low flow, since the majority of the system pressure drop will be dissipated across the valve.

Valves are further classified by the number of connections or ports. Figure 11.13 shows sections of typical **two-way** and **three-way** valves. Two-port valves control flow through coils or other HVAC equipment by varying valve flow resistance as a result of flow area changes. As shown, the flow must oppose the closing of the valve. If not, near closure the valve would slam shut or oscillate, both of which cause excessive wear and noise. The three-way valve shown in the figure is configured in the **diverting** mode. That is, one stream is split into two depending on the valve opening. The three-way valve shown is double seated (single-seated three-way valves are also available); it is therefore easier to close than a single-seated valve, but tight shutoff is not possible.

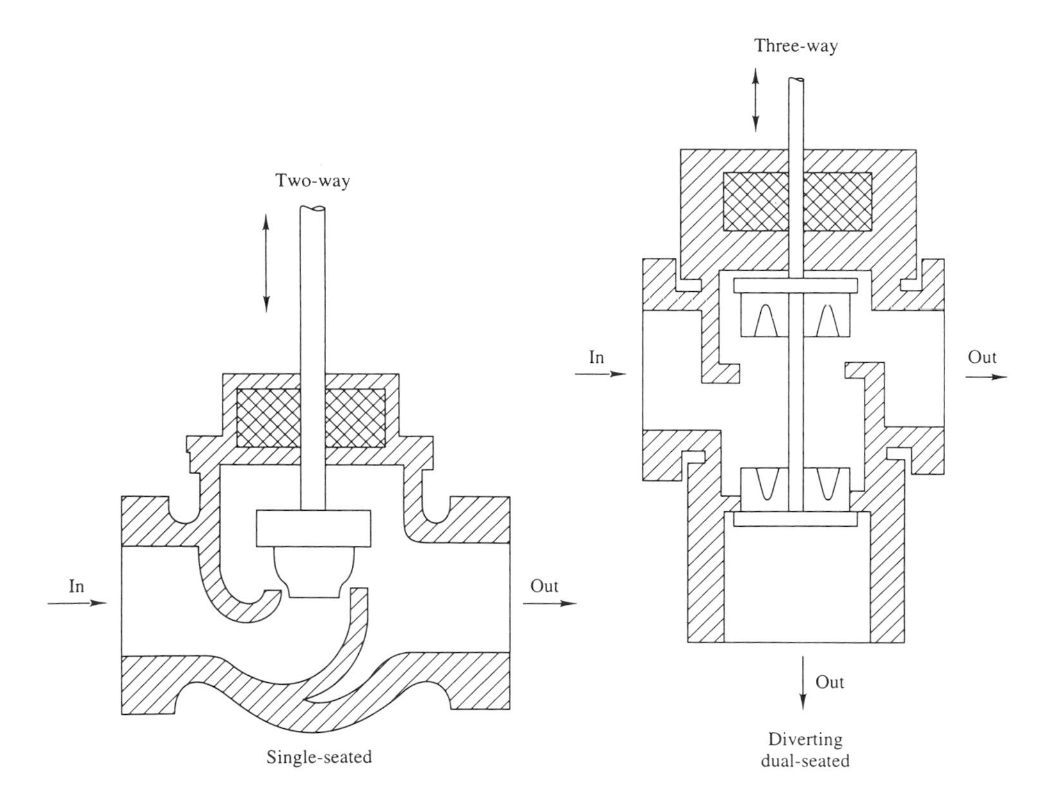

FIGURE 11.13 Cross-sectional drawings of direct-acting, single-seated two-way valve and dual-seated, three-way diverting valve.

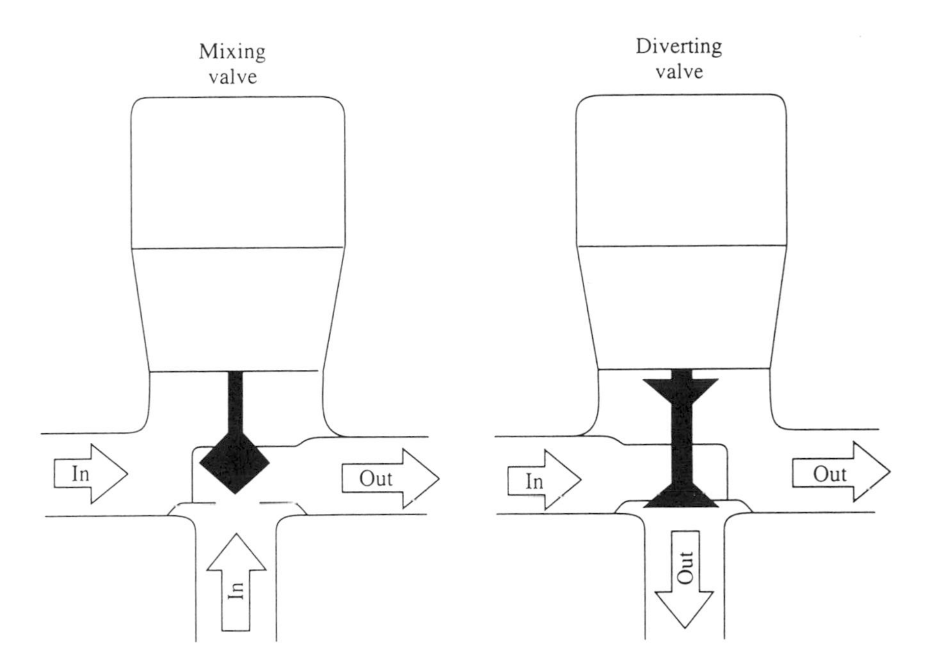

FIGURE 11.14 Three-way mixing and diverting valves. Note the significant difference in internal construction. Mixing valves are more commonly used.

Three-way valves can also be used as **mixing** valves. In this application two streams enter the valve and one leaves. Mixing and diverting valves *cannot be used interchangeably*, since their internal design is different to ensure that they can each seat properly. Particular attention is needed by the installer to be sure that connections are made properly; arrows cast in the valve body show the proper flow direction. Figure 11.14 shows an example of three-way valves for both mixing and diverting applications.

Valve flow capacity is denoted in the industry by the dimensional flow coefficient C_v defined by

$$\dot{V}\left(\frac{\text{gal}}{\text{min}}\right) = C_v\sqrt{\Delta p(psi)} \tag{11.18}$$

C_v is specified as the flow rate of 60°F water that will pass through the fully open valve if a pressure difference of 1.0 psi is imposed across the valve. If SI units (m^3/s and Pa) are used, the numerical value of C_v is 17% larger than in USCS units. Once the designer has determined a value of C_v, manufacturer's tables can be consulted to select a valve for the known pipe size. If a fluid other than water is to be controlled, the C_v found from Eq. (11.18) should be multiplied by the square root of the fluid's specific gravity.

Steam valves are sized using a similar dimensional expression

$$\dot{m}\left(\frac{\text{lb}}{\text{hr}}\right) = 63.5C_v\sqrt{\frac{\Delta p(psi)}{v\left(\frac{\text{ft}^3}{\text{lb}}\right)}} \tag{11.19}$$

in which v is the steam specific volume. If the steam is highly superheated, multiply C_v found from Eq. (11.19) by 1.07 of every 100°F of superheat. For wet steam multiply C_v by the square root of the steam quality. Honeywell (1988) recommends that the pressure drop across the valve to be used in the equation be 80% of the difference between steam supply and return pressures (subject to the sonic flow limitation discussed below). Table 11.1 can be used for preliminary selection of control valves for either steam or water.

The type of valve (linear or not) for a specific application must be selected so that the controlled system is as nearly linear as possible. Control valves are very commonly used to control the heat transfer rate in coils. For a linear system, the combined characteristic of the actuator, valve, and coil should be linear. This will require quite different valves for hot water and steam control, for example, as we shall see.

Figure 11.15 shows the part load performance of a hot water coil used for air heating; at 10% of full flow the heat rate is 50% of its peak value. The heat rate in a cross-flow heat exchanger increases roughly in exponential fashion with flow rate, a highly nonlinear characteristic. This heating coil nonlinearity follows from the longer water residence time in a coil at reduced flow and the relatively large temperature difference between air being heated and the water heating it.

However, if we were to control the flow through this heating coil by an equal percentage valve (positive exponential increase of flow with valve position), the combined valve plus coil characteristic would be roughly linear. Referring to Figure 11.15 we see that 50% of stem travel corresponds to 10% flow. The third graph in the figure is the combined characteristic. This near-linear subsystem is much easier to control than if a linear valve were used with the highly nonlinear coil. Hence the rule: use equal percentage valves for heating coil control.

Linear, two-port valves are to be used for steam flow control to coils, since the transfer of heat by steam condensation is a linear, constant temperature process—the more steam supplied the greater the heat rate in exact proportion. Note that this is a completely different coil flow characteristic than for hot water coils. However, steam is a compressible fluid and the sonic velocity sets the flow limit for a given valve opening when the pressure drop across the valve is more than 60% of the steam supply line absolute pressure. As a result, the pressure drop to be used in Eq. (11.19) is the *smaller* of (1) 50% of the absolute stream pressure upstream of the valve and (2) 80% of difference between the steam supply and return line pressures. The 80% rule gives good valve modulation in the subsonic flow regime (Honeywell, 1988).

Chilled water control valves should also be linear, since the performance of chilled water coils (smaller air-water temperature difference than in hot water coils) is more similar to steam coils than to hot water coils.

Either two- or three-way valves can be used to control flow at part load through heating and cooling coils as shown in Figure 11.16. The control valve can either be controlled from coil outlet water or air temperature. Two- or three-way valves achieve the same local result at the coil when used for part load control. However, the designer must consider effects on the balance of the secondary system when selecting the valve type.

In essence, the two-way valve flow control method results in variable flow (tracking variable loads) with constant coil water temperature change whereas the three-way valve approach results in roughly constant secondary loop flow rate but smaller coil water temperature change (beyond the local coil loop itself). Since chillers and boilers require that flow remain within a narrow range, the energy and cost savings that could accrue due to the two-way-valve, variable volume system are difficult to achieve in small systems unless a two-pump, primary/secondary loop approach is employed. If this dual-loop approach is not used, the three-way valve method is required to maintain constant boiler or chiller flow. In large systems a primary/secondary design with two-way valves is preferred.

The location of the three-way valve at a coil must also be considered by the designer. Figure 11.16(b) shows the valve used downstream of the coil in a mixing, bypass mode. If a balancing valve is installed in the bypass line and set to have the same pressure drop as the coil, the local coil loop will have the same pressure drop for both full and zero coil flow. However, at

TABLE 11.1 Quick Sizing Chart for Control Valves

	Steam Capacity, lb/h						Water Capacity, gal/min						
	Vacuum Return Systems*			Atmospheric Return Systems			Differential Pressure, psig						
C_v	2-psi Supply Press. 3.2-psi Press. Drop†	5-psi Supply Press. 5.6-psi Press. Drop†	10-psi Supply Press. 9.6-psi Press. Drop†	2-psi Supply Press. 1.6-psi Press. Drop†	5-psi Supply Press. 4.0-psi Press. Drop†	10-psi Supply Press. 8.0-psi Press Drop†	2	4	6	8	10	15	20
0.33	7.7	11.0	16.0	5.4	9.3	14.6	0.41	0.66	0.81	0.93	1.04	1.27	1.47
0.63	14.6	20.9	30.5	10.4	17.7	27.8	0.89	1.26	1.54	1.78	1.99	2.4	2.81
0.73	17.0	24.3	35.4	12	20.5	32.2	1.0	1.46	1.78	2.06	2.3	2.8	3.25
1.0	23.0	33.2	48.5	16.4	28	44	1.4	2.0	2.44	2.82	3.16	3.9	4.46
1.6	37.09	53.1	77.6	26.8	45	70.6	2.25	3.2	3.9	4.51	5.06	6.2	7.13
2.5	58.25	82.9	121.2	41.9	70.25	110.25	3.53	5.0	6.1	7.05	7.9	9.68	11.15
3.0	69.9	99.5	145.5	50.2	84.3	132.3	4.23	6.0	7.32	8.46	9.48	11.61	13.38
4.0	93.2	132.2	194.0	67	112.4	177.4	5.6	8.0	9.76	11.28	12.6	15.5	17.87
5.0	116.2	165.2	242.5	82.7	140.5	220.5	7.1	10.0	12.2	14.1	15.8	19.4	22.3
6.0	139	200	291.0	99	168	265	8.5	12.0	14.6	16.92	18.9	23.2	27.0
6.3	146	209	311.5	104	177	278	8.9	12.6	15.4	17.78	19.9	24.4	28.1
7.0	162	233	339.5	115	196	309	9.9	14.0	17.1	19.74	22.1	27.1	31
8.0	186.5	264.4	388.0	131.2	224.8	352.8	11.3	16.0	19.5	22.56	25.3	31.6	35.7
10.0	232	332	485.0	164	281	441	14.1	20	24.4	28.2	31.6	38.7	44.6
11.0	256	366	533.5	181	309	486	15.5	22	27	31.02	34.4	42.5	49
13.0	303	434	630.5	213.7	365.3	373.3	18.3	27	31.7	36.7	41.1	50.3	58
14.0	326	465	679.0	232	393	617	19.7	28	34	39	44	54	62
15.0	349.3	497.6	727.5	246	421.5	661.5	21.1	30	36.6	42.3	47.4	58	66.9
16.0	370.9	531	776.0	268	450	706	22.5	32	39	45.1	50.6	62	71.3
18.0	419	597	873.0	301	505	794	25	36	44	51	57	70	80
20.0	466	664	970.0	335	562	882	28	40	49	56	63	77	89
23.0	541	763	1,115	385	646	1,014	32	46	56	65	73	89	103
25.0	582.5	829	1,212	419	702.5	1,102.5	35.3	50	61	70.5	79	96.8	111.5
27.0	628.2	896	1,309	452.5	758.7	1,190.7	38.1	54	65.9	76.1	85.3	104.5	120.4
30.0	699	995	1,455	502	843	1,323	42.3	60	73.2	84.6	94.8	116.1	133.8
38.0	885	1,257	1,833	636	1,069	1,676	53	76	93	107	120	147	169
40.0	932	1,322	1,940	670	1,124	1,764	56	80	97.6	112.8	126	155	178.7
50.0	1,162	1,652	2,425	827	1,405	2,205	71	100	122	141	158	194	223
56.0	1,305	1,851	2,716	938	1,574	2,469	79	112	137	158	177	217	250

TABLE 11.1 (continued) Quick Sizing Chart for Control Valves

	Steam Capacity, lb/h						Water Capacity, gal/min						
	Vacuum Return Systems*			Atmospheric Return Systems			Differential Pressure, psig						
	2-psi Supply Press.	5-psi Supply Press.	10-psi Supply Press.	2-psi Supply Press.	5-psi Supply Press.	10-psi Supply Press.							
C_v	3.2-psi Press. Drop†	5.6-psi Press. Drop†	9.6-psi Press. Drop†	1.6-psi Press. Drop†	4.0-psi Press. Drop†	8.0-psi Press Drop†	2	4	6	8	10	15	20
63.0	1,460	2,090	3,056	1,043	1,770	2,778	89	126	154	178	199	244	281
75.0	1,748	2,481	3,637	1,230	2,107	3,307	106	150	183	212	237	290	335
80.0	1,865	2,644	3,880	1,312	2,248	3,528	113	160	195	225.6	253	316	357
90.0	2,096	2,980	4,365	1,476	2,529	3,969	127	180	220	254	284	348	401
97.0	2,229	3,204	4,703	1,590	2,725	4,277	137	196	231	274	307	375	432
100.0	2,330	3,319	4,850	1,640	2,816	4,410	141	200	244	282	316	387	446
105.0	2,442	3,481	5,092	1,722	2,950	4,630	148	210	256	296	332	406	468
130.0	3,030	4,340	6,305	2,137	3,653	5,733	183	270	317	367	411	503	580
150.0	3,493	4,976	7,275	2,460	4,215	6,615	211	300	366	423	474	580	699
160.0	3,709	5,310	7,760	2,680	4,500	7,060	225	320	390	451	560	620	713
170.0	3,960	5,642	8,245	2,788	4,777	7,497	240	340	415	479	537	658	758
190.0	4,450	6,310	9,215	3,116	5,339	8,379	268	360	464	536	600	735	847
244.0	5,670	7,930	11,834	4,001	6,856	10,760	344	488	595	688	771	944	1,088
250.0	5,825	8,290	12,125	4,190	7,025	11,025	353	500	610	705	790	968	1,115
270.0	6,282	8,960	13,095	4,525	7,587	11,907	381	540	659	761	853	1,045	1,204
300.0	6,990	9,950	14,550	5,025	8,430	13,230	423	600	732	846	948	1,161	1,338
350.0	8,160	11,590	16,975	5,860	9,835	15,435	494	700	854	987	1,106	1,355	1,561
360.0	8,380	11,910	17,460	6,030	10,116	15,876	508	720	878	1,015	1,137	1,393	1,606
430.0	10,010	14,225	20,855	7,200	12,083	18,963	606	860	1,049	1,213	1,359	1,664	1,918
480.0	11,180	15,860	23,280	8,045	13,408	21,168	677	960	1,171	1,353	1,517	1,858	2,141
640.0	14,910	21,180	31,040	10,496	17,984	28,224	902	1,280	1,561	1,805	2,022	2,477	2,854
760.0	17,700	25,120	36,860	12,464	21,356	33,516	1,071	1,520	1,854	2,143	2,401	2,941	3,390
1,000.0	23,300	33,190	48,500	16,400	28,160	44,100	1,410	2,000	2,440	2,820	3,160	3,870	4,460
1,200.0	27,150	39,790	58,200	19,680	33,720	52,920	1,692	2,400	2,928	3,384	2,792	4,644	5,352
1,440.0	33,290	47,160	69,840	23,616	40,464	63,504	2,030	2,880	3,514	4,061	4,550	5,573	6,422

* Assuming a 4-in through 8-in vacuum.

† Pressure drop across fully open valve taking 80% of the pressure difference between supply and return main pressures.

Source: "Engineering Manual of Automatic Control," Honeywell, Inc., Minneapolis, MN, 1988.

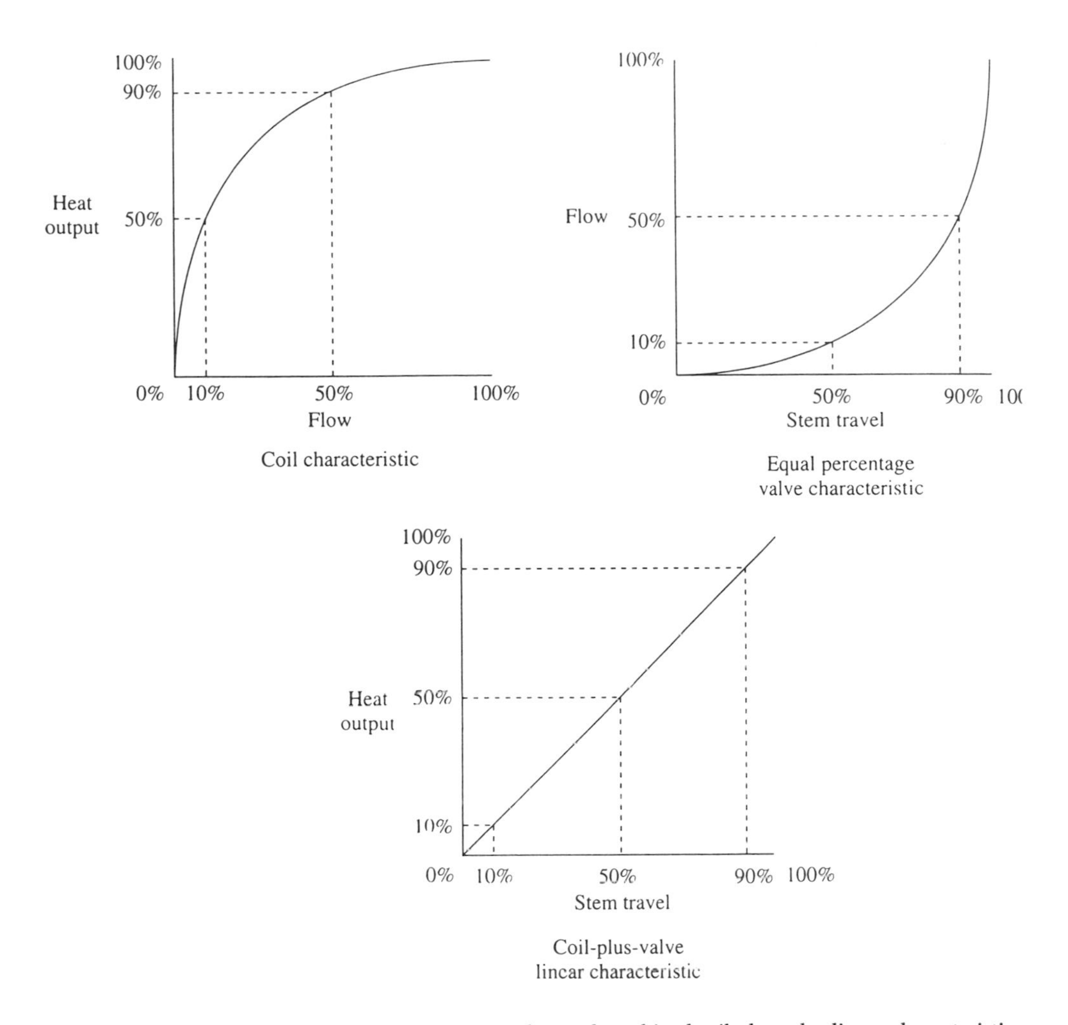

FIGURE 11.15 Heating coil, equal percentage valve, and combined coil plus valve linear characteristic.

the valve mid-flow position, overall flow resistance is less, since two parallel paths are involved, and the total loop flow increases to 25% more than that at either extreme.

Alternatively, the three-way valve can also be used in a diverting mode as shown in part (c) of the figure. In this arrangement essentially the same considerations apply as for the mixing arrangement discussed earlier.* However, if a circulator (small pump) is inserted as shown in Figure 11.16(d), the direction of flow in the branch line changes and a mixing valve is used. The reason that pumped coils are used is that control is improved. With constant coil flow, the highly nonlinear coil characteristic shown in Figure 11.15 is reduced, since the residence time of hot water

*A little-known disadvantage of three-way valve control has to do with the *conduction* of heat from a closed valve to a coil. For example, the constant flow of hot water through two ports of a *closed* three-way heating coil control valve keeps the valve body hot. Conduction from the closed, hot valve mounted close to a coil can cause sufficient air heating to actually decrease the expected cooling rate of a downstream cooling coil during the cooling season. Three-way valves have a second practical problem; installers often connect three-way valves incorrectly given the choice of three pipe connections and three pipes to be connected. Both of these problems are avoided by using two-way valves.

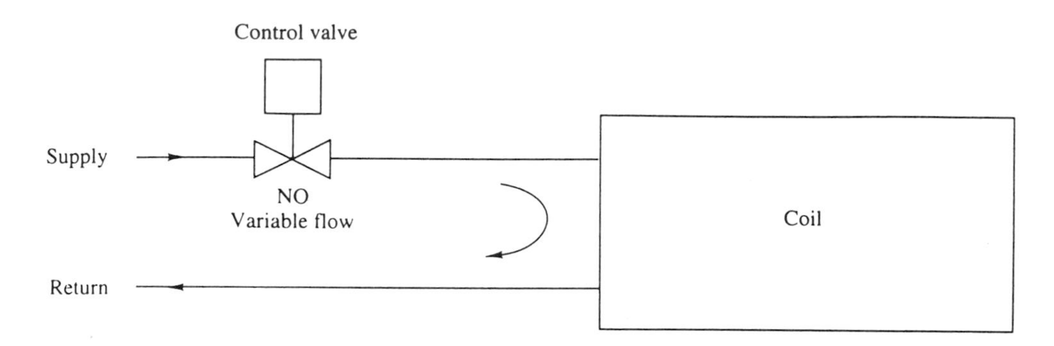

FIGURE 11.16 Various control valve piping arrangements: (a) two-way valve; (b) three-way mixing valve; (c) three-way diverting valve; (d) pumped coil with three-way mixing valve.

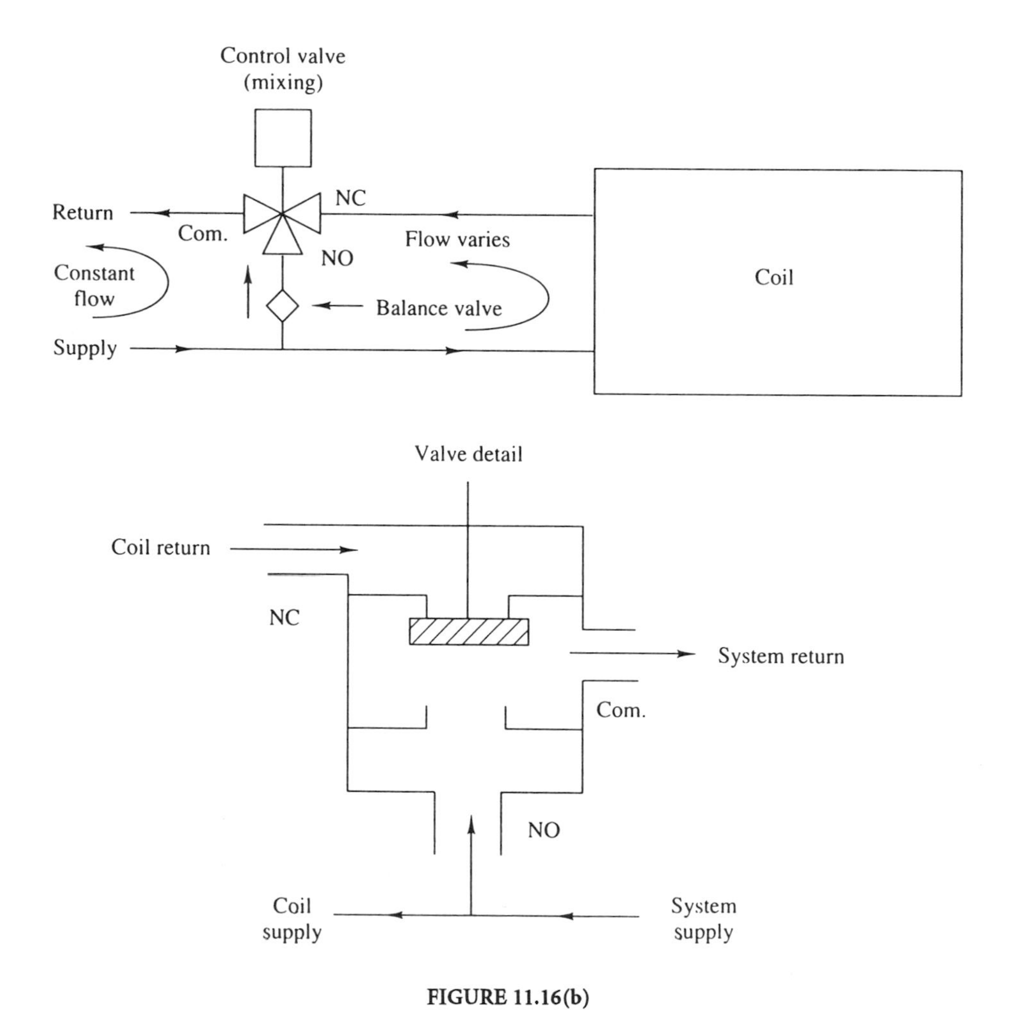

FIGURE 11.16(b)

in the coil is constant independent of load. However, this arrangement appears to the external secondary loop the same as a two-way valve. As load is decreased, flow into the local coil loop also decreases. Therefore, the uniform secondary loop flow normally associated with three-way valves is not present unless the optional bypass is used.

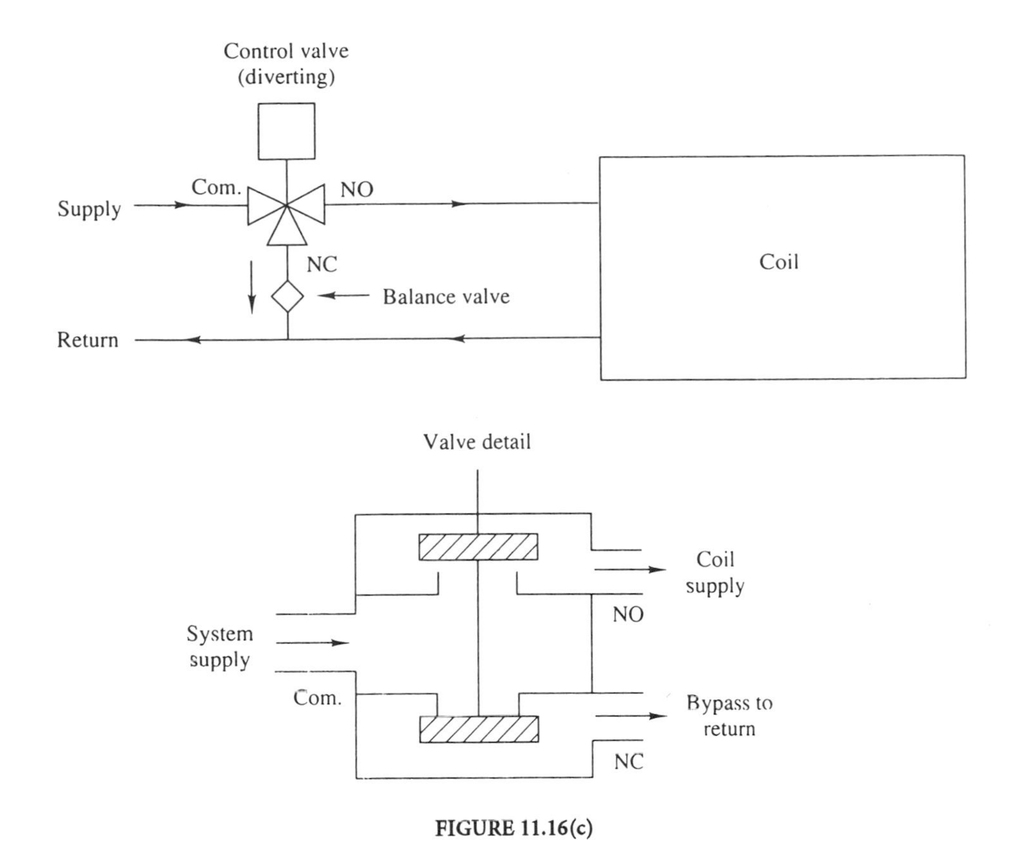

FIGURE 11.16(c)

For HVAC systems requiring precise control, high-quality control valves are required. The best controllers and valves are of "industrial quality." The additional cost for these valves compared to conventional building hardware results in more accurate control and longer lifetime.

Air Flow Control

Dampers are used to control airflow in secondary HVAC air systems in buildings. In this section we discuss the characteristics of dampers used for flow control in systems where constant speed fans are involved. Figure 11.17 shows cross sections of the two common types of dampers used in commercial buildings. Parallel blade dampers use blades that all rotate in the same direction. They are most often applied to two position locations—open or closed. Use for flow control is not recommended. The blade rotation changes airflow direction, a characteristic that can be useful when airstreams at different temperatures are to be effectively blended.

Opposed blade dampers have adjacent counterrotating blades. Air flow direction is not changed with this design, but pressure drops are higher than for parallel blading. Opposed blade dampers are preferred for flow control. Figure 11.18 shows the flow characteristics of these dampers to be closer to the desired linear behavior. The parameter α on the curves is the ratio of system pressure drop to fully open damper pressure drop.

A common application of dampers controlling the flow of outside air uses two sets in a **face and bypass** configuration as shown in Figure 11.19. For full heating all air is passed through the coil and the bypass dampers are closed. If no heating is needed in mild weather, the coil is bypassed (for minimum flow resistance and fan power cost, flow through fully open face and bypass dampers can be used if the preheat coil water flow is shut off). Between these extremes, flow is split between the two paths. The face and bypass dampers are sized so that the pressure drop in full bypass mode (damper pressure drop only) and full heating mode (coil plus damper pressure drop) is the same.

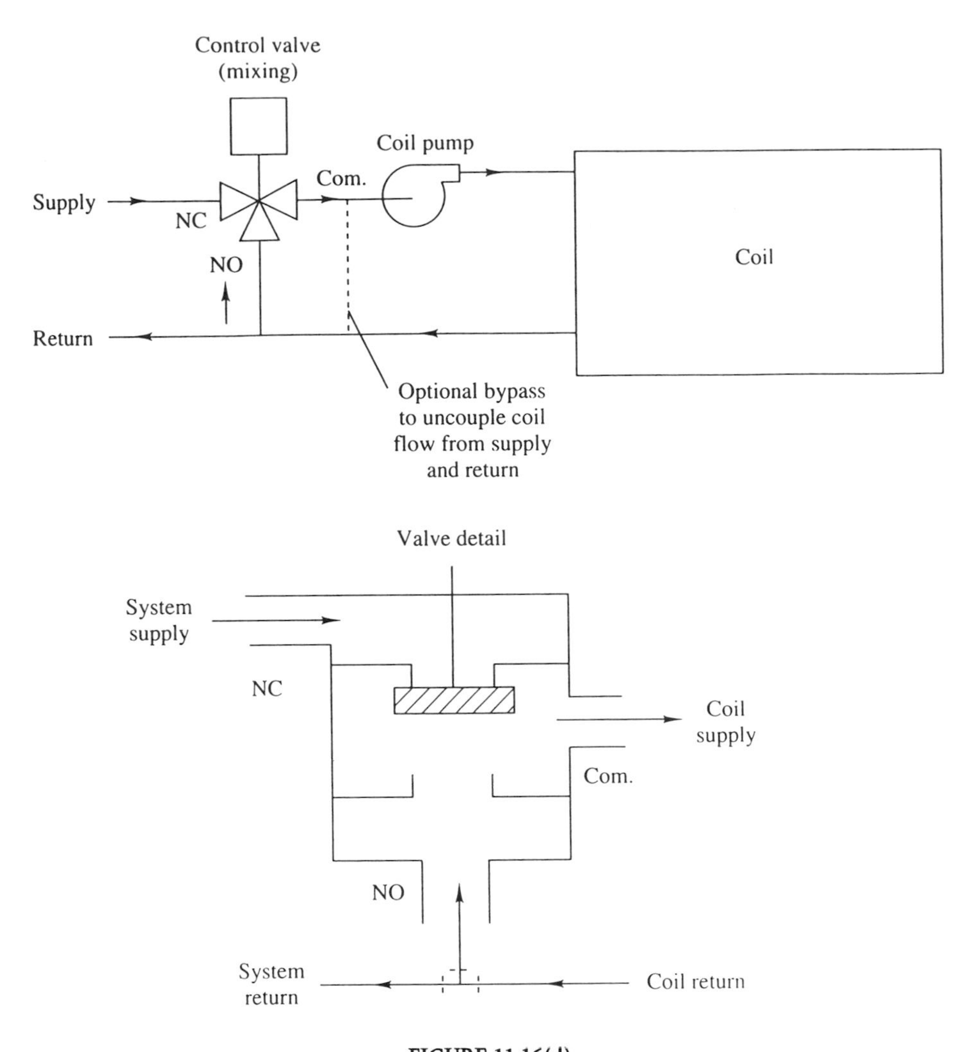

FIGURE 11.16(d)

11.4 Example HVAC System Control Systems

Several widely used control configurations for specific tasks are described in this section. These have been selected from the hundreds of control system configurations that have been used for buildings. The goal of this section is to illustrate how control components described previously are assembled into systems and what design considerations are involved. For a complete overview of HVAC control system configurations see ASHRAE (1987), Grimm and Rosaler (1990), and Honeywell (1988). The illustrative systems in this section are drawn in part from the latter reference.

In this section we will discuss seven control systems in common use. Each system will be described using a schematic diagram, and its operation and key features will be discussed in the accompanying text.

Outside Air Control

Figure 11.20 shows a system for controlling outside and exhaust air from a central air handling unit equipped for economizer cooling when available. In this and the following diagrams the following symbols are used

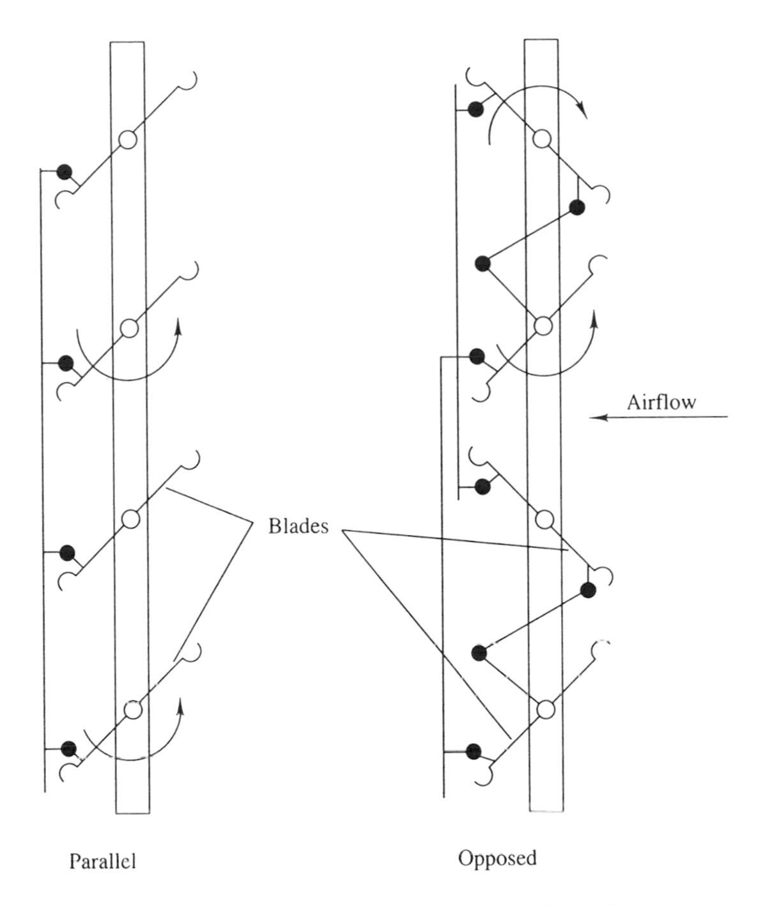

FIGURE 11.17 Diagram of parallel and opposed blade dampers.

C—cooling coil
DA—discharge air (supply air from fan)
DX—direct-expansion coil
EA—exhaust air
H—heating coil
LT—low-temperature limit sensor or switch, must sense the lowest temperature in the air volume being controlled
M—motor or actuator (for damper or valve), variable speed drive
MA—mixed air
NC—normally closed
NO—normally open
OA—outside air
PI—proportional plus integral controller
R—relay
RA—return air
S—switch
SP—static pressure sensor used in VAV systems
T—temperature sensor; must be located to read the average temperature representative of the air volume being controlled

This system is able to provide the minimum outside air during occupied periods, to use outdoor air for cooling when appropriate by means of a temperature-based economizer cycle, and to operate

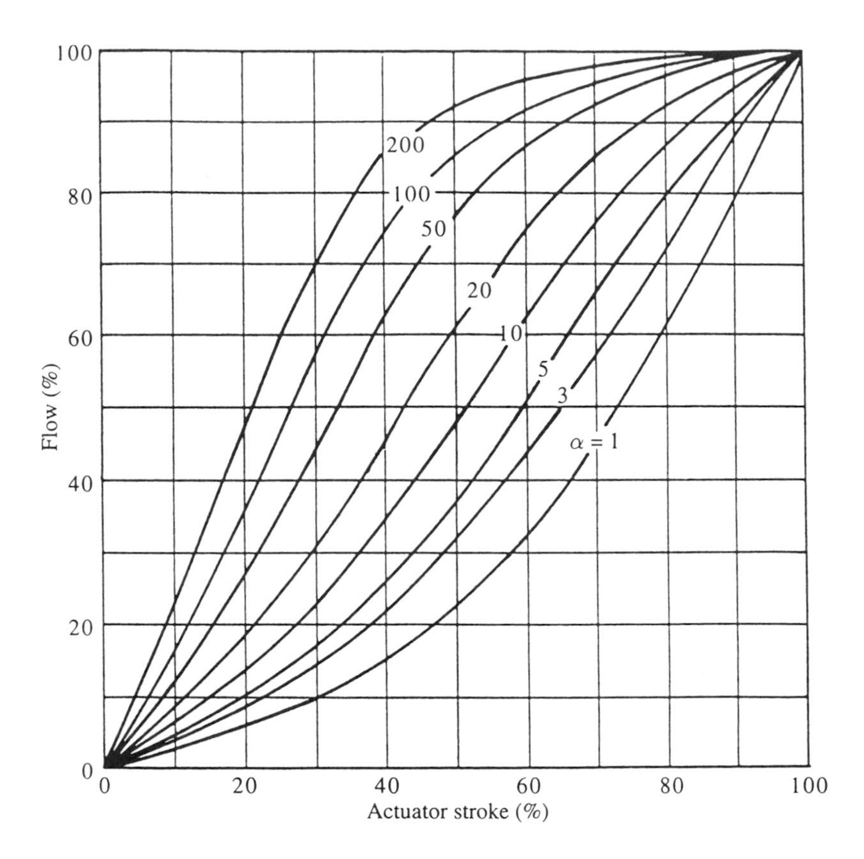

FIGURE 11.18 Flow characteristics of opposed blade dampers. The parameter α is the ratio of system resistance (not including the damper) to damper resistance. The ideal linear damper characteristic is achieved if this ratio is about 10 for opposed blade dampers.

fans and dampers under all conditions. The numbering system used in the figure indicates the sequence of events as the air handling system begins operation after an off period:

1. The fan control system turns on when the fan is turned on. This may be by a clock signal or a low-temperature space condition.
2. The space temperature signal determines if the space is above or below the set point. If above, the economizer feature will be activated and control the outdoor and mixed air dampers. If below, the outside air damper is set to its minimum position.
3. The mixed air PI controller controls both sets of dampers (OA/RA and EA) to provide the desired mixed air temperature.
4. When the outdoor temperature rises above the cutoff point for economizer operation, the outdoor air damper is returned to its minimum setting.
5. Switch S is used to set the minimum setting on outside and exhaust air dampers manually. This is ordinarily done only once during building commissioning and flow testing.
6. When the supply fan is off, the outdoor air damper returns to its NC position and the return air damper returns to its NO position.
7. When the supply fan is off, the exhaust damper also returns to its NC position.
8. Low temperature sensed in the duct will initiate a freeze-protect cycle. This may be as simple as turning on the supply fan to circulate warmer room air. Of course, the OA and EA dampers remain tightly closed during this operation.

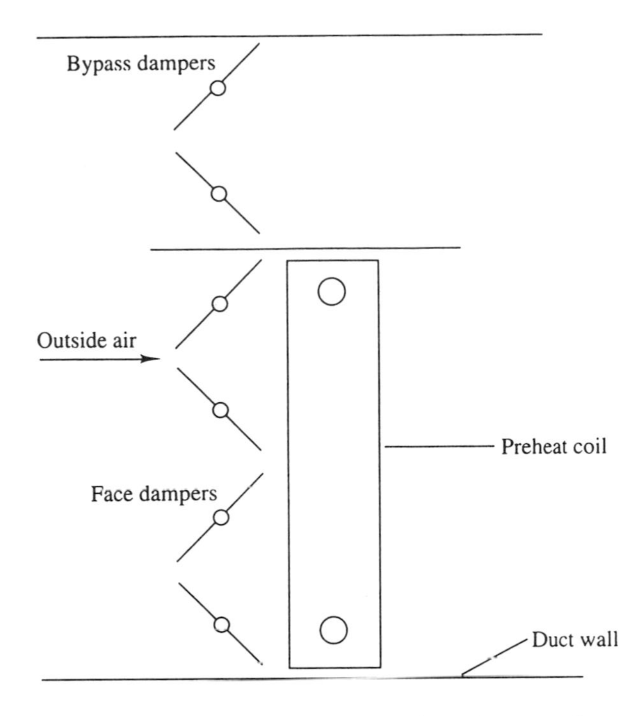

FIGURE 11.19 Face and bypass dampers used for preheating coil control.

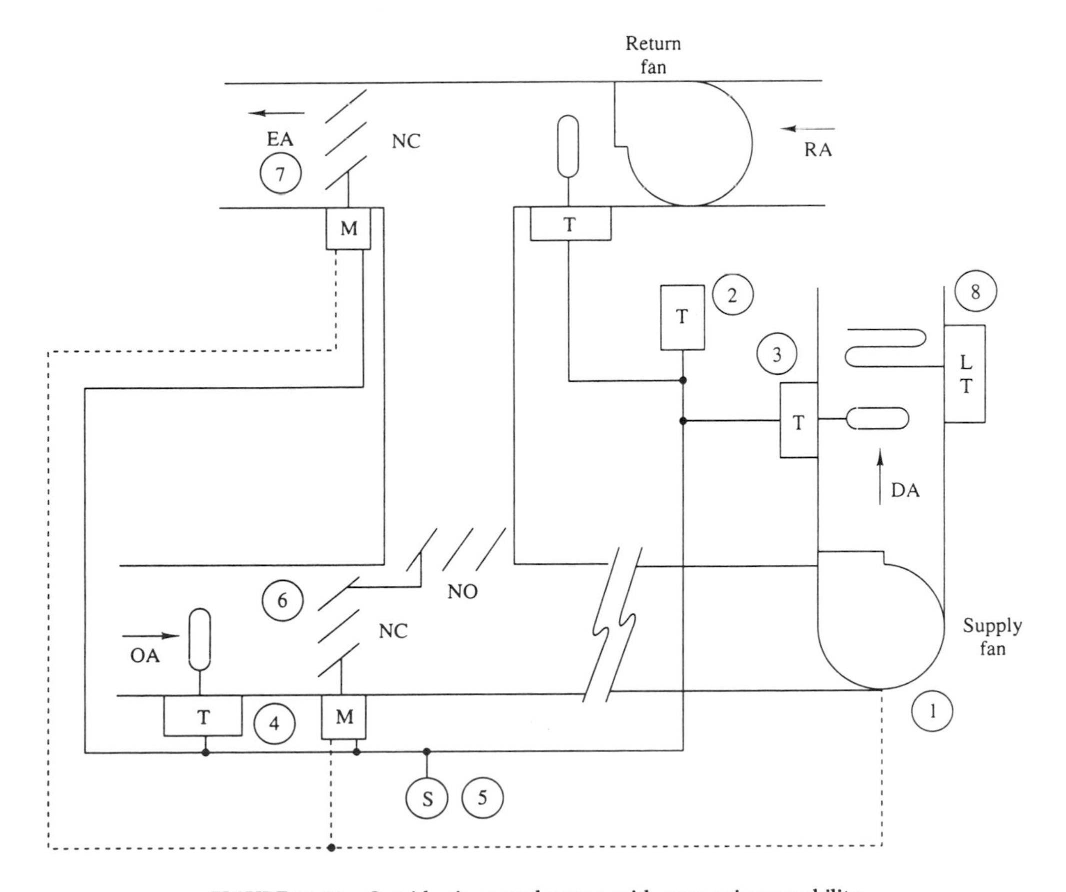

FIGURE 11.20 Outside air control system with economizer capability.

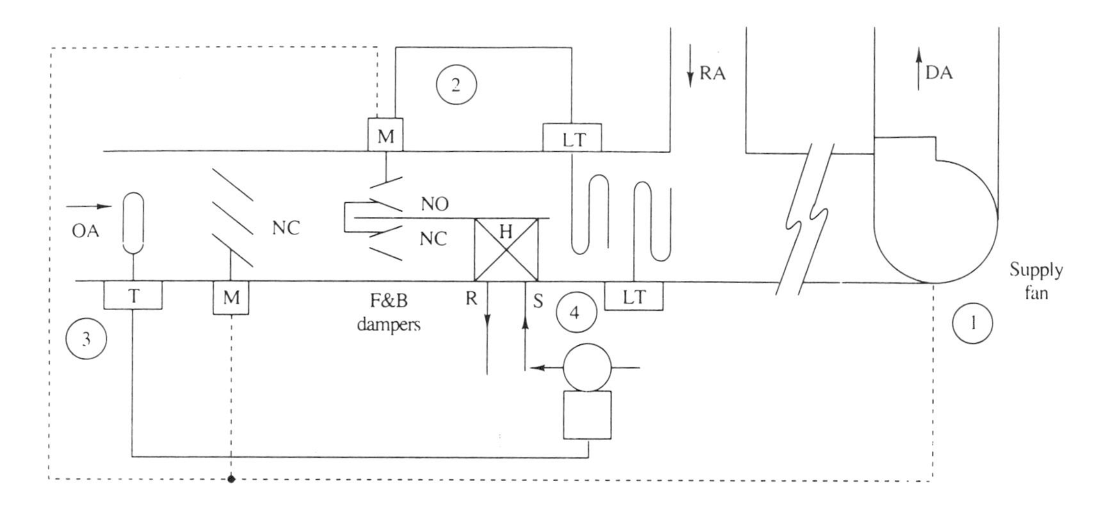

FIGURE 11.21 Preheat control system. Counter flow of air and hot water in the preheat coil results in the highest heat transfer rate.

Heating Control

If the minimum air setting is large in the preceding system, the amount of outdoor air admitted in cold climates may require preheating. Figure 11.21 shows a preheating system using face and bypass dampers. (A similar arrangement is used for direct-expansion [DX] cooling coils.) The equipment shown is installed upstream of the fan in Figure 11.20. This system operates as follows:

1. The preheat subsystem control is activated when the supply fan is turned on.
2. The preheat PI controller senses temperature leaving the preheat section. It operates the face and bypass dampers to control the exit air temperature between 45 and 50°F.
3. The outdoor air sensor and associated controller controls the water valve at the preheat coil. The valve may be either a modulating valve (better control) or an on-off valve (less costly).
4. The low-temperature sensors (LTs) activate coil freeze protection measures including closing dampers and turning off the supply fan.

Note that the preheat coil (as well as all coils in this section) is connected so that the hot water (or steam) flows counter to the direction of airflow. Counter flow provides a higher heating rate for a given coil than does parallel flow. Mixing of heated and cold bypass air must occur upstream of the control sensors. Stratification can be reduced by using sheet metal **air blenders** or by propeller fans in the ducting. The preheat coil should be located in the bottom of the duct. Steam preheat coils must have adequately sized traps and vacuum breakers to avoid condensate buildup that could lead to coil freezing at light loads.

The face and bypass damper approach enables air to be heated to the required system supply temperature without endangering the heating coil. (If a coil were to be as large as the duct—no bypass area—it could freeze when the hot water control valve cycles open and closed to maintain discharge temperature.) The designer should consider pumping the preheat coil as shown in Figure 11.19(d) to maintain water velocity above the 3 ft/s needed to avoid freezing. If glycol is used in the system, the pump is not necessary but heat transfer will be reduced.

During winter, in heating climates, heat must be added to the mixed air stream to heat the outside air portion of mixed air to an acceptable discharge temperature. Figure 11.22 shows a common heating subsystem controller used with central air handlers. (It is assumed that the mixed air temperature is kept above freezing by action of the preheat coil, if needed.) This system has

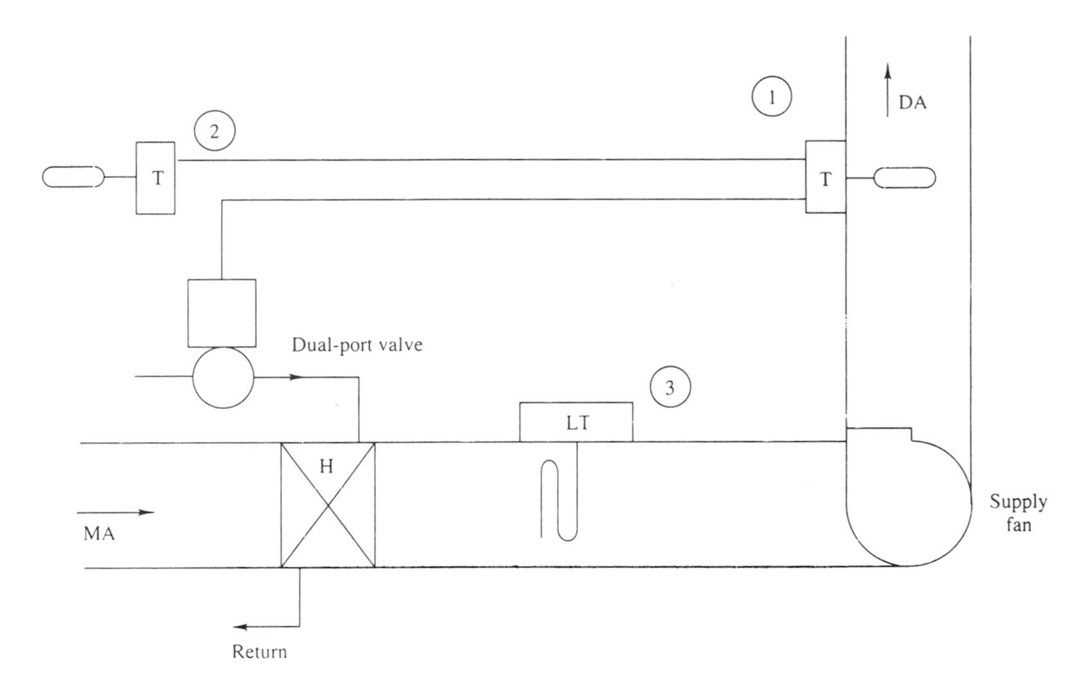

FIGURE 11.22 Heating coil control subsystem using two-way valve and optional reset sensor.

the added feature that coil discharge temperature is adjusted for ambient temperature, since the amount of heat needed decreases with increasing outside temperature. This feature, called coil discharge **reset**, provides better control and can reduce energy consumption. The system operates as follows:

1. During operation the discharge air sensor and PI controller controls the hot water valve.
2. The outside air sensor and controller resets the **set point** of the discharge air PI controller up as ambient temperature drops.
3. Under sensed low-temperature conditions freeze-protection measures are initiated as discussed earlier.

Reheating at zones in VAV or other systems uses a system similar to that just discussed. However, boiler water temperature is reset and no freeze protection is normally included. The air temperature sensor is the zone thermostat for VAV reheat, not a duct temperature sensor.

Cooling Control

Figure 11.23 shows the components in a cooling coil control system for a single-zone system. Control is similar to that for the heating coil discussed above except that the zone thermostat (not a duct temperature sensor) controls the coil. If the system were a central system serving several zones, a duct sensor would be used. Chilled water supplied to the coil partially bypasses and partially flows through the coil, depending on the coil load. The use of three- and two-way valves for coil control has been discussed in detail earlier. The valve NC connection is used as shown so that valve failure will not block secondary loop flow.

Figure 11.24 shows another common cooling coil control system. In this case the coil is a direct-expansion (DX) refrigerant coil and the controlled medium is refrigerant flow. DX coils are used when precise temperature control is not required, since the coil outlet temperature drop is large whenever refrigerant is released into the coil because refrigerant flow is not modulated, it is most commonly either on or off. The control system sequences as follows:

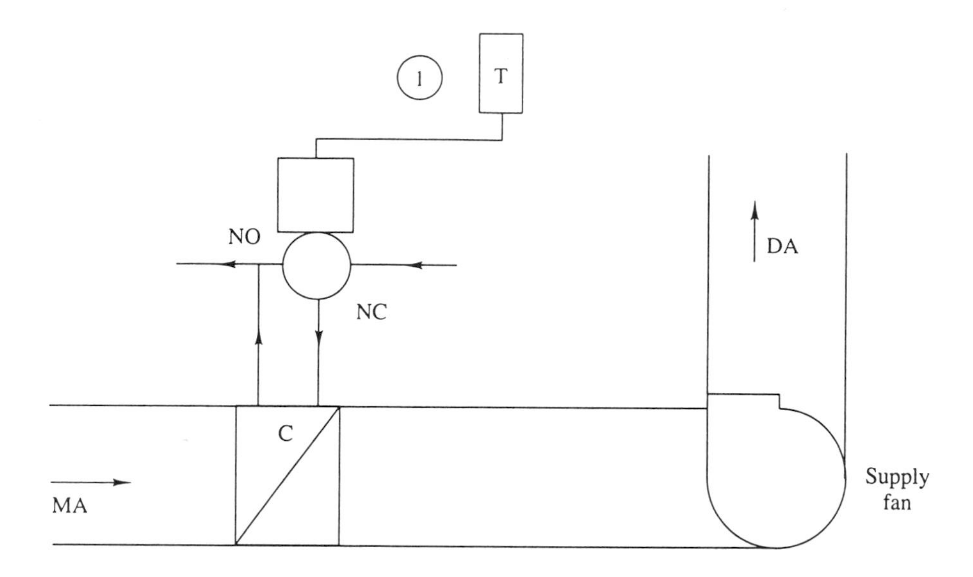

FIGURE 11.23 Cooling coil control subsystem using three-way diverting valve.

1. The coil control system is energized when the supply fan is turned on.
2. The zone thermostat opens the two-position refrigerant valve for temperatures above the set point and closes it in the opposite condition.
3. At the same time, the compressor is energized or de-energized. The compressor has its own internal controls for oil control and pumpdown.
4. When the supply fan is off, the refrigerant solenoid valve returns to its NC position and the compressor relay to its NO position.

At light loads bypass rates are high and ice may build up on coils. Therefore, control is poor at light loads with this system.

Complete Systems

The preceding five example systems are actually control subsystems that must be integrated into a single control system for the HVAC system's primary and secondary systems. In the remainder of this section we will describe briefly two complete HVAC control systems widely used in commercial buildings. The first is a fixed volume system, whereas the second is a VAV system.

Figure 11.25 shows a constant volume, central system air handling system equipped with supply and return fans, heating and cooling coils, and economizer for a single-zone application. If the system were to be used for multiple zones the zone thermostat shown would be replaced by a discharge air temperature sensor. This fixed volume system operates as follows:

1. When the fan is energized the control system is activated.
2. The minimum air setting is set (usually only once during commissioning as described earlier).
3. The OA temperature sensor supplies a signal to the damper controller.
4. The RA temperature sensor supplies a signal to the damper controller.
5. The damper controller positions the dampers to use outdoor or return air depending on which is cooler.
6. The mixed-air low-temperature controller controls the outside air dampers to avoid excessively low-temperature air from entering the coils. If a preheating system were included, this sensor would control it.

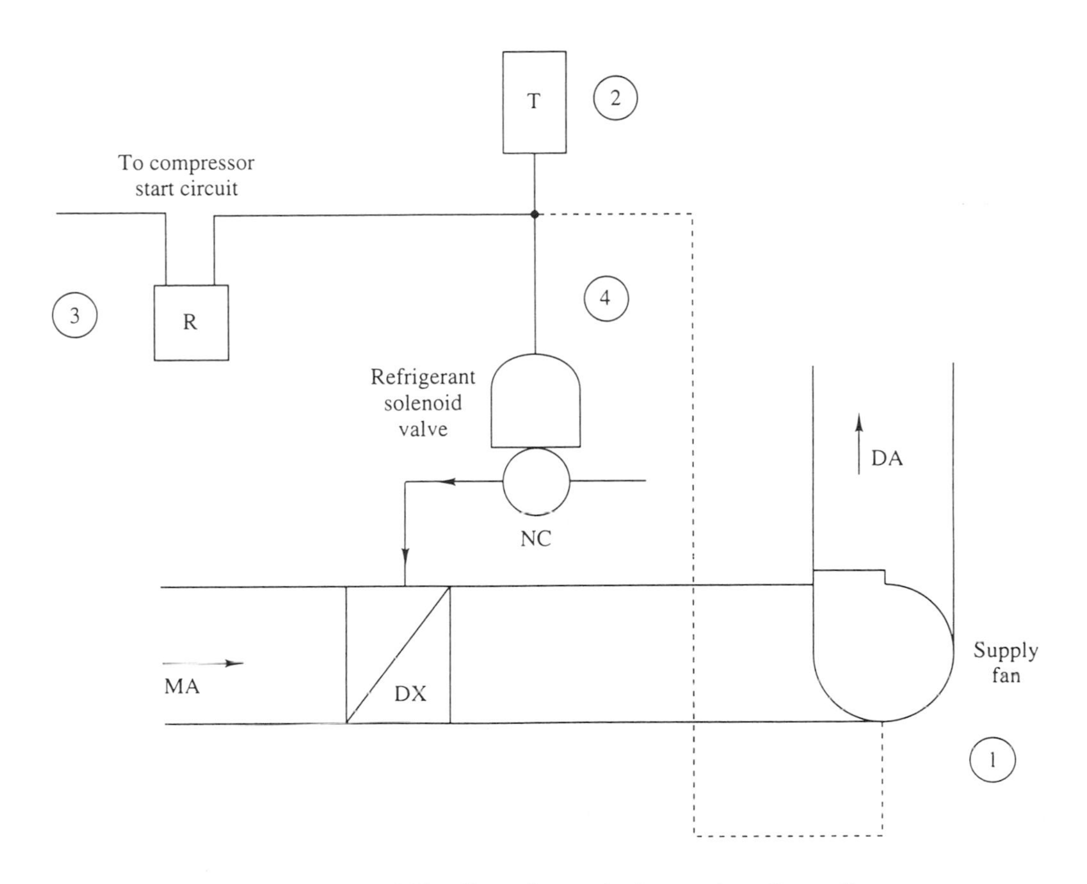

FIGURE 11.24 DX cooling coil control subsystem (on-off control).

7. Optionally the space sensor could reset the coil discharge air PI controller.
8. The discharge air controller controls the
   a. Heating coil valve
   b. Outdoor air damper
   c. Exhaust air damper
   d. Return air damper
   e. Cooling coil valve after the economizer cycle upper limit is reached
9. The low-temperature sensor initiates freeze protection measures as described previously.

A method for reclaiming either heating or cooling energy is shown by dashed lines in Figure 11.25. This so-called "runaround" system extracts energy from exhaust air and uses it to precondition outside air. For example, the heating season exhaust air may be at 75°F while outdoor air is at 10°F. The upper coil in the figure extracts heat from the 75°F exhaust and transfers it through the lower coil to the 10°F intake air. To avoid icing of the air intake coil, the three-way valve controls this coil's liquid inlet temperature to a temperature above freezing. In heating climates the liquid loop should also be freeze protected with a glycol solution. Heat reclaiming systems of this type can also be effective in the cooling season, when outdoor temperatures are well above the indoor temperature.

A VAV system has additional control features including a motor speed (or inlet vane) control and a duct static pressure control. Figure 11.26 shows a VAV system serving both perimeter and interior zones. It is assumed that the core zones always require cooling during the occupied period. The system shown has a number of options and does not include every feature present in all VAV

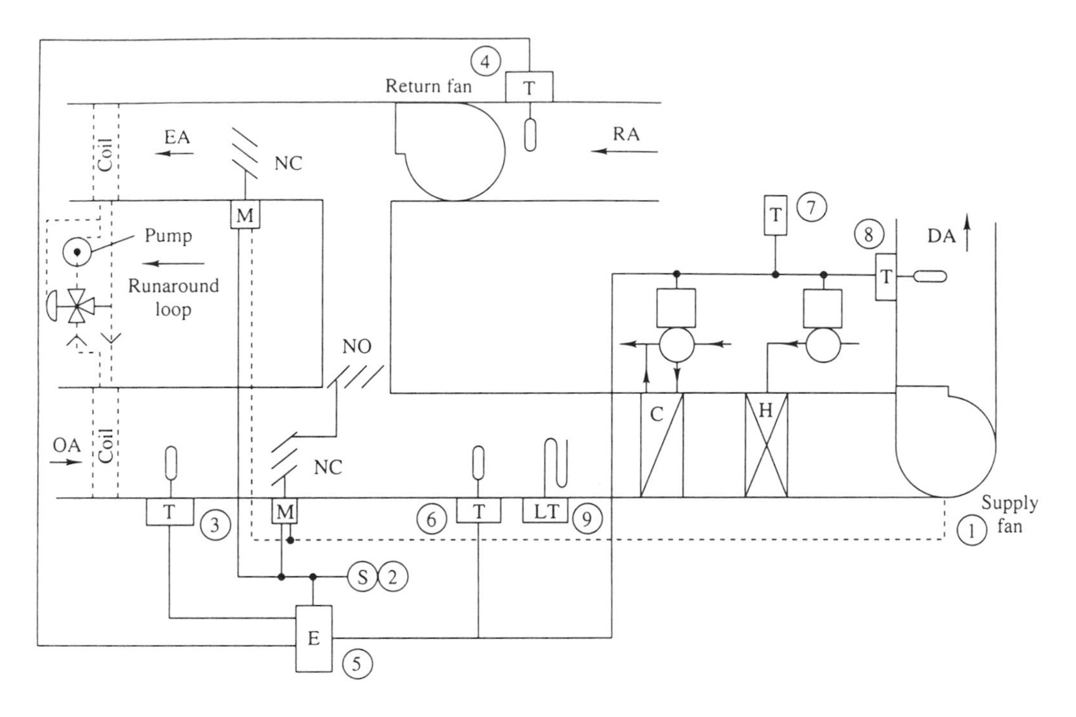

FIGURE 11.25 Control for a complete, fixed volume HVAC system. Optional runaround heat recovery system is shown to left in dashed lines.

systems. However, it is representative of VAV design practice. The sequence of operation is as follows:

1. When the fan is energized the control system is activated. Prior to activation during unoccupied periods the perimeter zone baseboard heating is under control of room thermostats.
2. Return and supply fan interlocks are used to prevent pressure imbalances in the supply air ductwork.
3. The mixed air sensor controls the outdoor air dampers (and/or preheat coil not shown) to provide proper coil air inlet temperature. The dampers will typically be at their minimum position at about 40°F.
4. The damper minimum position controls the minimum outdoor air flow.
5. As the upper limit for economizer operation is reached, the OA dampers are returned to their minimum position.
6. The return air temperature is used to control the morning warmup cycle after night setback (option present only if night setback is used).
7. The outdoor air damper is not permitted to open during morning warmup by action of the relay shown.
8. Likewise, the cooling coil valve is de-energized (NC) during morning warmup.
9. All VAV box dampers are moved full open during morning warmup by action of the relay override. This minimizes warmup time. Perimeter zone coils and baseboard units are under control of the local thermostat.
10. During operating periods the PI static pressure controller controls both supply and return fan speeds (or inlet vane positions) to maintain approximately 1.0 in. WG of static pressure at the pressure sensor location (or optionally to maintain building pressure). An additional pressure sensor (not shown) at the supply fan outlet will shut down the fan if fire dampers or other dampers should close completely and block air flow. This sensor overrides the duct static pressure sensor shown.

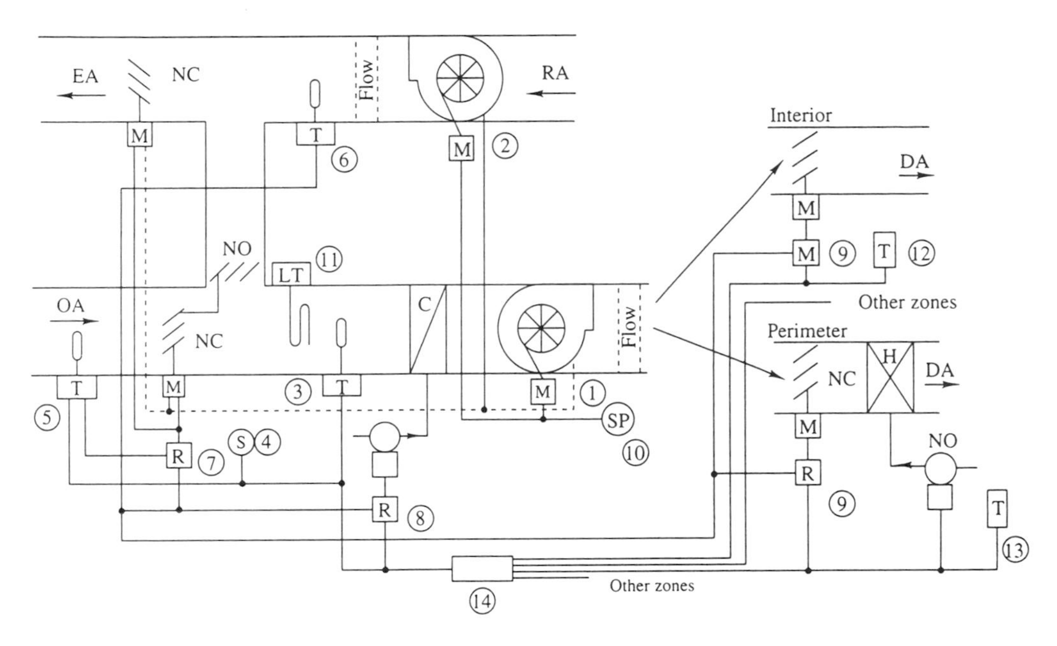

FIGURE 11.26 Control for complete, VAV system. Optional supply and return flow stations shown with dashed lines.

11. The low-temperature sensor initiates freeze-protection measures.
12. At each zone room thermostats control VAV boxes (and fans if present); as zone temperature rises the boxes open more.
13. At each perimeter zone room, thermostats close VAV dampers to their minimum settings and activate zone heat (coil and/or perimeter baseboard) as zone temperature falls.
14. The controller, using temperature information for all zones (or at least for enough zones to represent the characteristics of all zones), modulates outdoor air dampers (during economizer operation) and the cooling control valve (the above economizer cycle cutoff) to provide air sufficiently cooled to meet the load of the warmest zone.

The duct static pressure controller is critical to the proper operation of VAV systems. The static pressure controller must be of PI design, since a proportional-only controller would permit duct pressure to drift upward as cooling loads drop due to the unavoidable offset in P-type controllers. In addition, the control system should position inlet vanes closed during fan shutdown to avoid overloading on restart.

Return fan control is best achieved in VAV systems by an actual flow measurement in supply and return ducts as shown by dashed lines in the figure. The return air flow rate is the supply rate less local exhausts (fume hoods, toilets, etc.) and exfiltration needed to pressurize the building.

VAV boxes are controlled locally, assuming that adequate duct static pressure exists in the supply duct and that supply air is at an adequate temperature to meet the load (this is the function of the controller described in item 14). Figure 11.27 shows a *local* control system used with a series-type, fan-powered VAV box. This particular system delivers a constant flow rate to the zone, to assure proper zone air distribution, by action of the airflow controller. Primary air varies with cooling load as shown in lower part of the figure. Optional reheating is provided by the coil shown.

Other Systems

This section has not covered the control of central plant equipment such as chillers and boilers. Most primary system equipment controls are furnished with the equipment and as such do not

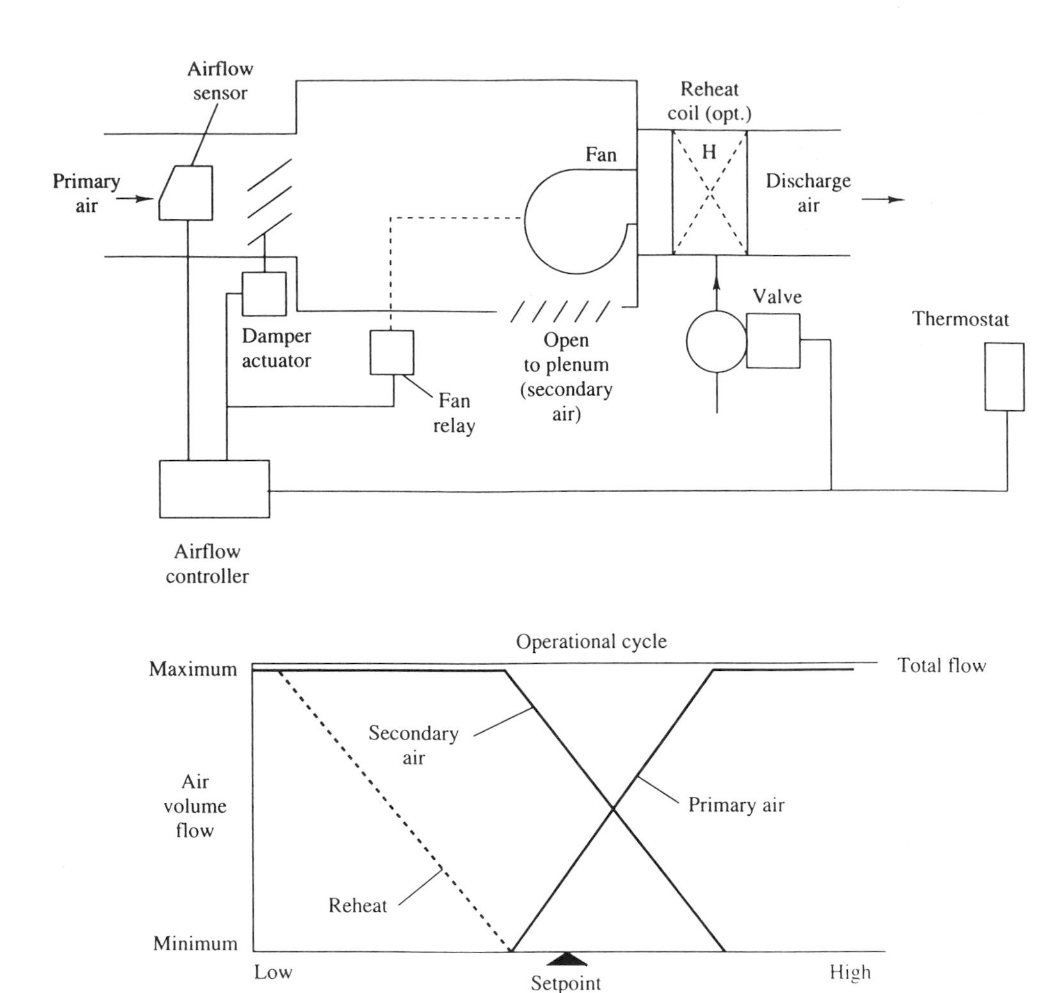

FIGURE 11.27 Series-type, fan-powered VAV box control subsystem and primary flow characteristic. The total box flow is constant at level identified as "maximum" in the figure. The difference between primary and total air flow is secondary air recirculated through the return air grille. Optional reheat coil requires air flow shown by dashed line.

offer much flexibility to the designer. However, Braun et al. (1989) have shown that considerable energy savings can be made by properly sequencing cooling tower stages on chiller plants and by properly sequencing chillers themselves in multiple chiller plants.

Fire and smoke control are important for life safety in large buildings. The design of smoke control systems is controlled by national codes. The principal concept is to eliminate smoke from the zones where it is present while keeping adjacent zones pressurized to avoid smoke infiltration. Some components of space conditioning systems (e.g., fans) can be used for smoke control, but HVAC systems are generally not smoke control systems by design.

Electrical systems are primarily the responsibility of the electrical engineer on a design team. However, HVAC engineers must make sure that the electrical design accommodates the HVAC control system. Interfaces between the two occur where the HVAC controls activate motors on fans or chiller compressors, pumps, electrical boilers, or other electrical equipment.

In addition to electrical specifications, the HVAC engineer often conveys electrical control logic using a **ladder diagram**. An example is shown in Figure 11.28 for the control of the supply and return fans in a central system. The electrical control system is shown at the bottom and operates on low voltage (24 or 48 VAC) from the control transformer shown. The supply fan is started manually by closing the "start" switch. This activates the motor starter coil labeled 1M, thereby closing the three contacts labeled 1M in the supply fan circuit. The fourth 1M contact (in parallel with the start switch) holds the starter closed after the start button is released.

The hand-off-auto switch is typical and allows both automatic and manual operation of the return fan. When switched to the "hand" position the fan starts. In the "auto" position the fans will operate only when the adjacent contacts 3M are closed. Either of these actions activates the relay coil 2M, which in turn closes the three 2M contacts in the return fan motor starter. When either fan produces actual airflow a flow switch is closed in the ducting, thereby completing the circuit to the pilot lamps L. The fan motors are protected by fuses and thermal overload heaters. If motor current draw is excessive, the heaters shown in the figure produce sufficient heat to open the normally closed thermal overload contacts.

This example ladder diagram is primarily illustrated and is not typical of an actual design. In a fully automatic system both fans would be controlled by 3M contacts actuated by the HVAC control system. In a fully manual system the return fan would be activated by a fifth 1M contact, not by the 3M automatic control system.

1.5 Advanced Control System Design Topics—Neural Networks

Neural networks offer considerable opportunity to improve control achievable in standard PID systems. This section provides a short introduction to this novel approach to control.

Neural Network Introduction

An artificial neural network is a massively parallel, dynamic system of interconnected, interacting parts based on some aspects of the brain. Neural networks are considered to be intuitive because they learn by example rather than by following programmed rules. The ability to "learn" is one of the key aspects of neural networks. A neural network consists of several layers of neurons that are connected to each other. A "connection" is a unique information transport link from one sending to one receiving neuron. The structure of part of an NN is schematically shown in Figure 11.29. Any number of input, output, and "hidden layer" (only one hidden layer is shown) neurons can be used. One of the challenges of this technology is to construct a net with sufficient complexity to learn accurately without imposing a burden of excessive computational time.

The neuron is the fundamental building block of a network. A set of inputs is applied to each. Each element of the input set is multiplied by a *weight*, indicated by the W in the figure, and the products are summed at the neuron. The symbol for the summation of weighted inputs is termed INPUT and must be calculated for each neuron in the network. In equation form this process for one neuron is

$$\text{INPUT} = \sum_i O_i W_i + B \qquad \textbf{(11.20)}$$

where O_i are inputs to a neuron—i.e., outputs of the previous layer
W_i are weights
B are the biases

After INPUT is calculated, an activation function F is applied to modify it, thereby producing the neuron's output as described shortly.

Artificial networks have been trained by a wide variety of methods (McClelland and Rumelhart, 1988). **Back-propagation** is one systematic method for training multilayer neural networks. The weights of a net are initiated with small random numbers. The objective of training the network

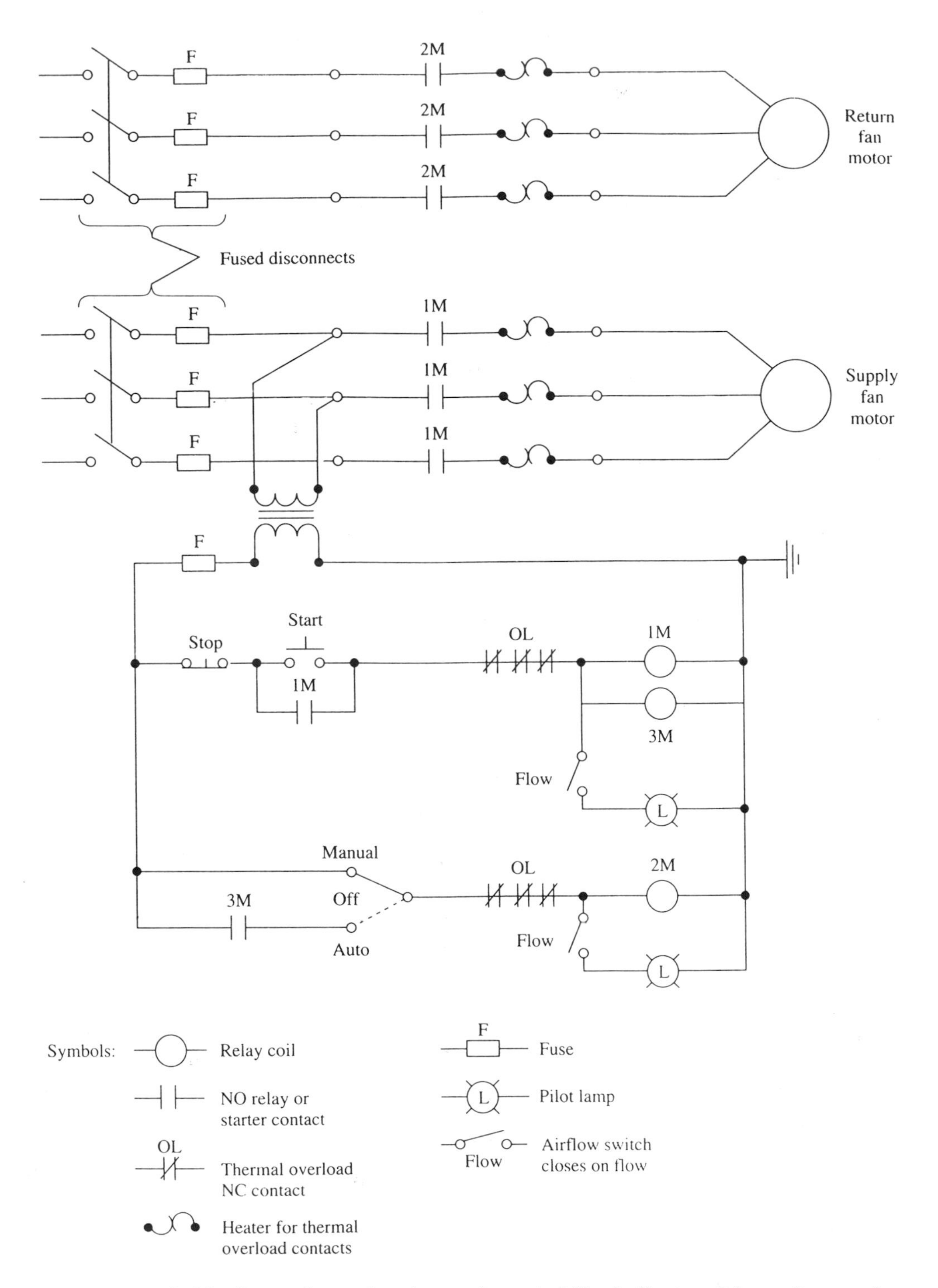

FIGURE 11.28 Ladder diagram for supply and return fan control. Hand-off-auto switch permits manual or automatic control of the return fan.

is to adjust the weights iteratively so that application of a set of inputs produces the desired set of outputs matching a training data set. Usually a network is trained with a training data set that consists of many input-output pairs; these data are called a training set. Training the net using back-propagation requires the following steps:

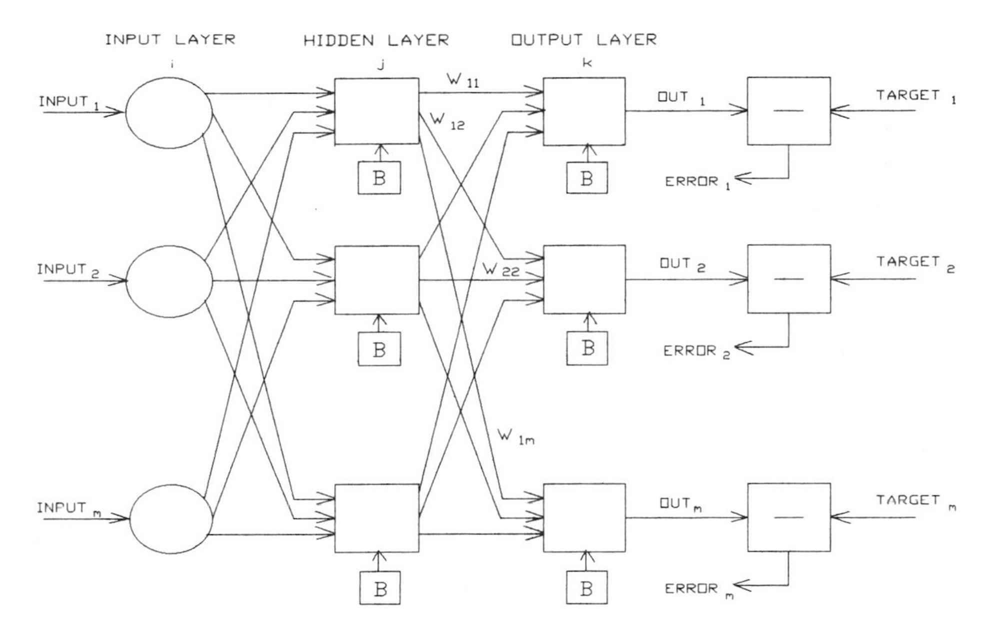

FIGURE 11.29 Schematic diagram of a neural network showing input layer, hidden layers, and output along with target training valves. Hidden and output layers consist of connected neurons; the input layer does not contain neurons.

1. Select a training pair from the **training set** and apply the input vector to the network input layer.
2. Calculate the output of the network, OUT_i.
3. Calculate the error $ERROR_i$ the network output and the desired output (the **target vector** from the training pair).
4. Adjust the weights of the network in a way that minimizes the error.
5. Repeat steps 1 through 4 for each vector in the training set until the error for the entire set is lower than the user specified, preset training tolerance.

Steps 1 and 2 are the "forward pass." The following expression describes the calculation process in which an **activation function** *F* is applied to the weighted sum of inputs INPUT as follows.

$$\text{OUT} = F(\text{INPUT}) = F(\Sigma O_i W_i + B) \tag{11.21}$$

where *F* is the activation function
B is the bias of each neuron

The activation function used for this work was selected to be

$$F(\text{INPUT}) = \frac{1}{(1 + e^{-\text{INPUT}})} \tag{11.22}$$

This is referred to as a sigmoid function and is shown in Figure 11.30. It has a value of 0.0 when INPUT is a large negative number and a value of 1.0 for large and positive INPUT, making a smooth transition between these limiting values. The bias *B* is the activation threshold for each neuron. The bias avoids the tendency of a sigmoid function to get "stuck" in the saturated, limiting value area.

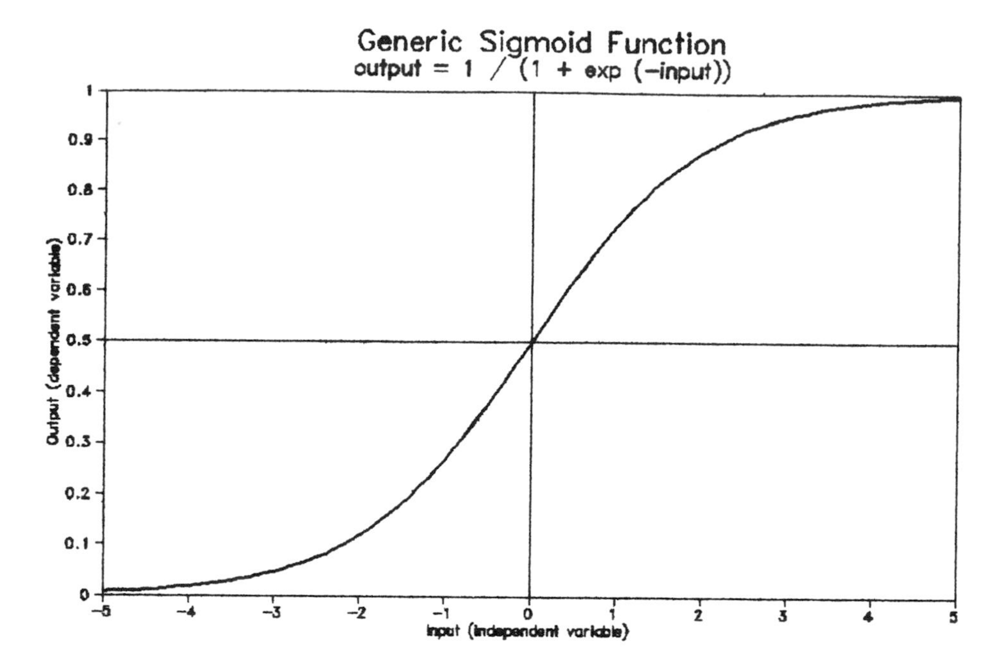

FIGURE 11.30 Sigmoid function used to process the weighted sum of network inputs.

Steps 3 and 4 comprise the "reverse pass" in which the **delta rule** is used as follows: For each neuron in the output layer, the previous weight $W(n)$ is adjusted to a new value $W(n+1)$ to reduce the error by the following rule:

$$W(n+1) = W(n) + (\eta\delta)\text{OUT} \tag{11.23}$$

where $W(n)$ is the previous value of a weight
$W(n+1)$ is the weight after adjusting
η is the training rate coefficient
δ is calculated from

$$\delta = \left(\frac{\partial \text{INPUT}}{\partial \text{OUT}}\right)(\text{TARGET} - \text{OUT}) = \text{OUT}(1 - \text{OUT})(\text{TARGET} - \text{OUT}) \tag{11.24}$$

in which the derivative has been calculated from Eqs. (11.21) and (11.22), and TARGET (see Fig. 11.29) is the training set target value. This method of correcting weights bases the magnitude of the correction on the error itself.

Of course, hidden layers have no target vector: therefore, back-propagation trains these layers by propagating the output error back through the network layer by layer, adjusting weights at each layer. The **delta rule** adjustment δ is calculated from

$$\delta_j = \text{OUT}(1 - \text{OUT})\Sigma(\delta_{j+1} W_{j+1}) \tag{11.25}$$

where δ_j and δ_{j+1} belong to the jth and $(j+1)$th hidden layers, respectively (being numbered with increasing values from left to right in Figure 11.29). This overall method of adjusting weights belongs to the general class of steepest descent algorithms. The weights and biases after training contain meaningful system information; before training the initial, random biases and random weights have no physical meaning.

Commercial Building Adaptive Control Example

A proof of concept experiment in which neural networks (NNs) were used for both local and global control of a commercial building HVAC system was conducted in the JCEM laboratory in which full-scale testing of multizone HVAC systems can be done repeatably. Data collected in the laboratory were used to train NNs for both the components and the full systems involved (Curtiss, 1993a, 1993b). Any neural network based controller will be useful only if it can perform better than a conventional PID controller. Figures 11.31 and 11.32 show typical results for the PID and NN control of a heating coil. The difficulty that the PID controller experienced is due to the highly nonlinear nature of the heating coil. A PID controller tuned at one level of load is unable to control acceptably at another whereas the NN controller does not have this difficulty. With the NN controller we see excellent control—minimal overshoot and quick response to the set-point changes.

In an affiliated study Curtiss et al. (1993b) showed that NNs offered a method for global control of HVAC systems as well. The goal of such controls could be to reduce energy consumption as much as possible while meeting comfort conditions as a constraint. Energy savings of over 15% were achieved by the NN method vs. standard PID control.

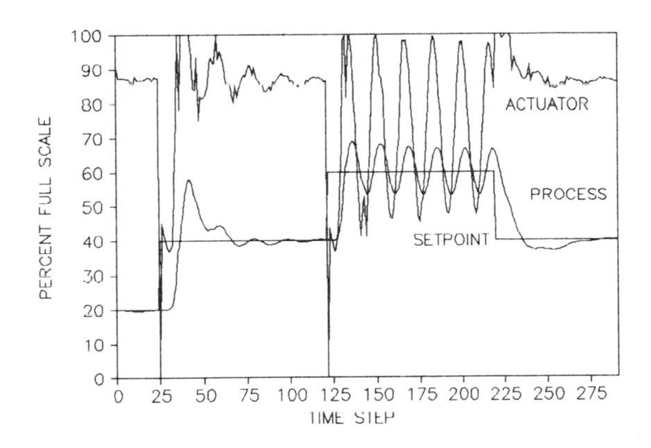

FIGURE 11.31 PID controller response to step changes in coil load. Proportional gain of 2.0 *Source:* Curtiss et al., 1993a.

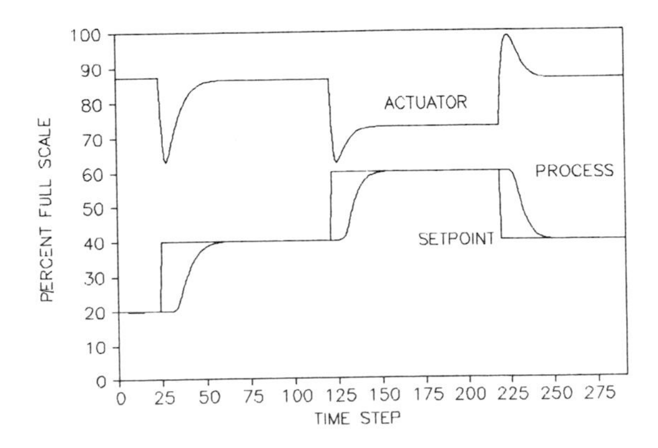

FIGURE 11.32 NN controller with learning rate of 1.0 and window of 15 time steps. *Source:* Curtiss et al., 1993a.

Summary

This chapter has introduced the important features of properly designed control systems for HVAC applications. Sensors, actuators, and control methods have been described. The method for determining control system characteristics either analytically or empirically has been discussed.

The following rules (ASHRAE, 1987) should be followed to ensure that the control system is as energy efficient as possible. Neural networks offer one method for achieving energy efficient control.

1. Operate HVAC equipment only when the building is occupied or when heat is needed to prevent freezing.
2. Consider the efficacy of night setback visà vis building mass. Massive buildings may not benefit from night setback due to overcapacity needed for the morning pickup load.
3. Do not supply heating and cooling simultaneously. Do not supply humidification and dehumidification at the same time.
4. Reset heating and cooling air or water temperature to provide only the heating or cooling needed.
5. Use the most economical source of energy first, the most costly last.
6. Minimize the use of outdoor air during the deep heating and cooling seasons subject to ventilation requirements.
7. Consider the use of "dead-band" or "zero-energy" thermostats.
8. Establish control settings for stable operation to avoid system wear and to achieve proper comfort.

References

ASHRAE, 1987: *HVAC Systems and Applications*, ASHRAE, Atlanta, GA.

J.E. Braun, et al., 1989: "Applications of Optimal Control to Chilled Water Systems without Storage," *ASHRAE Transactions*, Vol. 95, Pt. 1.

P.S. Curtiss, J.F. Kreider, and M.J. Brandemuehl, 1993a: "Adaptive Control of HVAC Processes Using Predictive Neural Networks," *ASHRAE Transactions*, Vol. 99, Pt. 1.

P.S. Curtiss, M.J. Brandemuehl, and J.F. Kreider, 1993b: "Energy Management in Central HVAC Plants using Neural Networks," *ASHRAE Transactions*, Vol. 100, Pt. 1.

A.L. Dexter and P. Haves, 1989: "A Robust Self-Tuning Predictive Controller for HVAC Applications," *AHSRAE Transactions*, Vol. 95, Pt. 2.

N.R. Grimm and R.C. Rosaler, 1990: *Handbook of HVAC Design*, McGraw-Hill, New York.

R.W. Haines, 1987: *Control Systems for Heating, Ventilating and Air Conditioning*, 4th ed., Van Nostrand Reinhold, New York.

Honeywell, Inc., 1988: *Engineering Manual of Automatic Control*, Minneapolis, MN.

P.H. Huang, 1991: "Humidity Measurements and Calibration Standards," *ASHRAE Transactions*, Vol. 97, Pt. 2.

K.M. Letherman, 1981: *Automatic Controls for Heating and Air Conditioning*, Pergamon Press, New York.

J.L. McClelland and D.E. Rumelhart, 1988: *Exploration in Parallel Distributed Processing*, MIT Press, Cambridge, MA.

W.F. Stoecker and P.A. Stoecker, 1989: *Microcomputer Control of Thermal and Mechanical Systems*, Van Nostrand Reinhold, New York.

D.M. Underwood, 1989: "Response of Self-Tuning Single Loop Digital Controllers to a Computer Simulated Heating Coil," *ASHRAE Transactions*, Vol. 95, Pt. 2.

12A

Energy Efficient Technologies: Electric Motor Systems Efficiency

Anibal T. de Almeida
University of Coimbra

Steve Greenberg
Lawrence Berkeley National Laboratory

Introduction

Motor systems are by far the most important type of electrical load, ranging from small fractional-hp motors incorporated in home appliances to multimegawatt motors driving pumps and fans in power plants. Motors consume over half of the total electricity, and in industry they are responsible for about two-thirds of the electricity consumption. In the commercial and residential sectors motors consume slightly less than half of the electricity. The cost of powering motors is immense; roughly $90 billion a year in the United States alone. There is a vast potential for saving energy and money by increasing the efficiency of motors and motor systems.

12A.1 Motor Types

Motors produce useful work by causing the shaft to rotate. Motors have a rotating part, the rotor, and a stationary part, the stator. Both parts produce magnetic fields, either through windings excited by electric currents or through the use of permanent magnets. It is the interaction between these two magnetic fields that is responsible for the torque generation, as shown in Figure 12A.1.

There are a wide variety of electric motors, based on the type of power supply (AC or DC) that feeds the windings as well as on different methods and technologies to generate the magnetic fields in the rotor and in the stator. Figure 12A.2 presents the most important types of motors.

0-8493-2514-5/96/$0.00+$.50

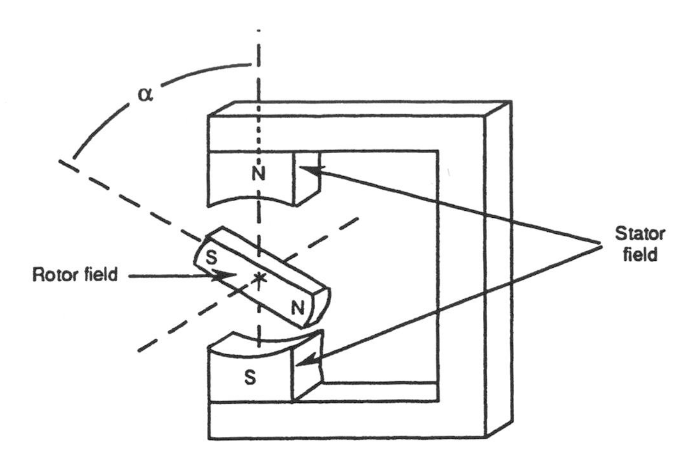

FIGURE 12A.1 Torque generation in a motor. The generated torque is proportional to the strength of each magnetic field and depends on the angle α between the two fields. Mathematically, torque equals $\mathbf{B}_{rotor}| \times |\mathbf{B}_{sta\text{-}tor}| \times |\sin \alpha$, where **B** refers to a magnetic field.(Reprinted from *Energy-Efficient Motor Systems: A Handbook on Technology, Program and Policy Opportunities,* 1992, American Council for an Energy Efficient Economy, Washington, D.C. With permission.)

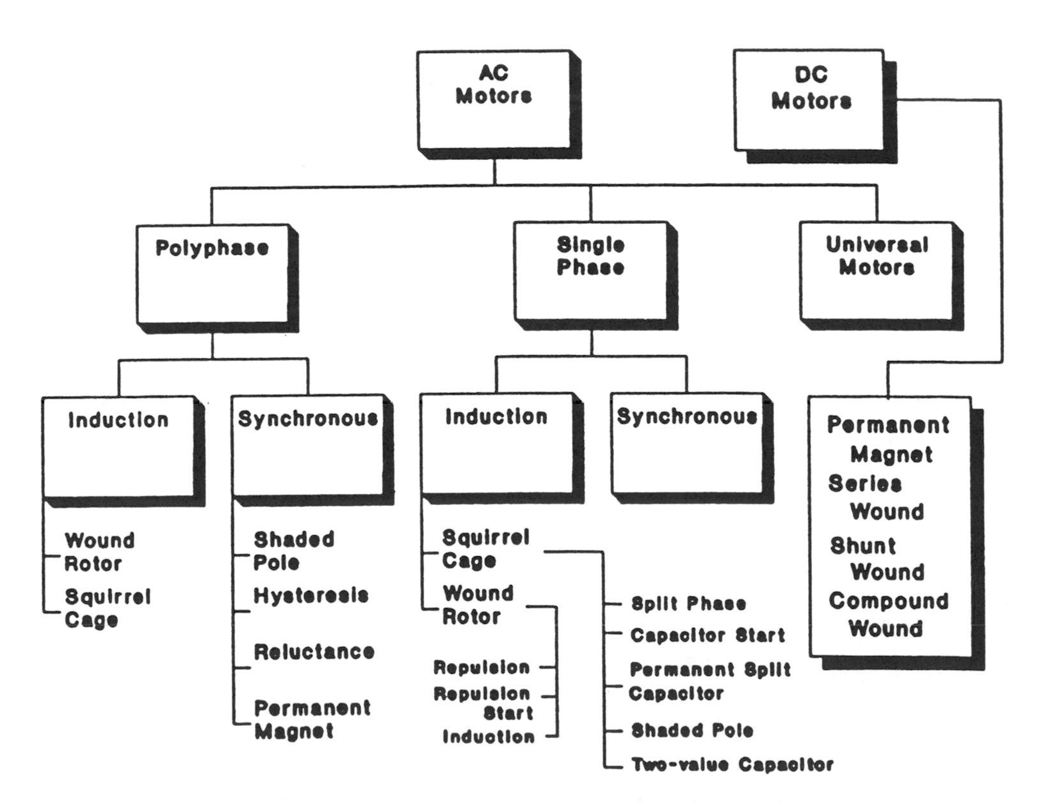

FIGURE 12A.2 Motor types. *Source:* EPRI, 1992a.

Because of their low cost, high reliability, and fairly high efficiency, most of the motors used in large home appliances, industry, and commercial buildings are induction motors. Figure 12A.3 shows the operating principle of a three-phase induction motor.

Synchronous motors are used in applications requiring constant speed, high operating efficiency and controllable power factor. Efficiency and power factor are particularly important above 1000 hp. Although DC motors are easy to control, both in terms of speed and torque, they are

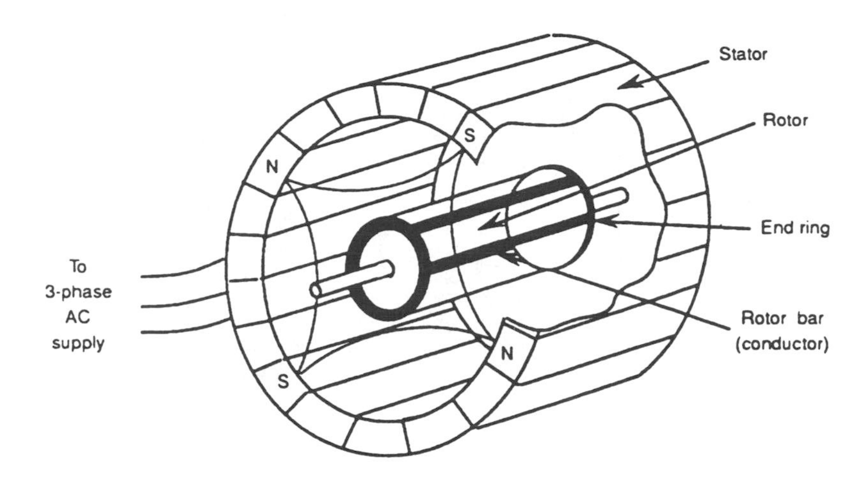

FIGURE 12A.3 Operation of a four-pole induction motor. Rotating magnetic field is created in the stator by AC currents carried in stator winding. Three-phase voltage source results in the creation of north and south magnetic poles that revolve or "move around" the stator. The changing magnetic field from the stator induces current in the rotor conductors, in turn creating the rotor magnetic field. Magnetic forces in the rotor tend to follow the stator magnetic fields, producing rotary motor action. (Reprinted from *Energy-Efficient Motor Systems: A Handbook on Technology, Program and Policy Opportunities*, 1992, American Council for an Energy Efficient Economy, Washington, D.C. With permission.)

expensive to produce and have modest reliability. DC motors are used for some industrial and electric traction applications, but their importance is dwindling.

12A.2 Motor Systems Efficiency

The efficiency of a motor-driven process depends upon several factors, which may include

- Motor efficiency
- Motor speed controls
- Proper sizing
- Power supply quality
- Distribution losses
- Transmission
- Maintenance
- Driven equipment (pump, fan, etc.) mechanical efficiency

It must be emphasized that the design of the process itself influences the overall efficiency (units produced/kWh or service produced/kWh) to a large extent. In fact, in many systems the largest opportunity for increased efficiency is in improved use of the mechanical energy (usually in the form of fluids or solid materials in motion) in the process. Comprehensive programs to address motor-system energy use start with the process and work back toward the power line, optimizing each element in turn. Outlining such a program is beyond the scope of this discussion; see (e.g.,) Baldwin (1989) for an example of the benefits that propagate all the way back to the power plant.

Motor Efficiency

Figure 12A.4 shows the distribution of the losses of an induction motor as a function of the load. At low loads the core magnetic losses (hysteresis and eddy currents) are dominant, whereas at higher loads the copper resistive ("Joule" or I^2R) losses are the most important. Mechanical losses are also present in the form of friction in the bearings and windage.

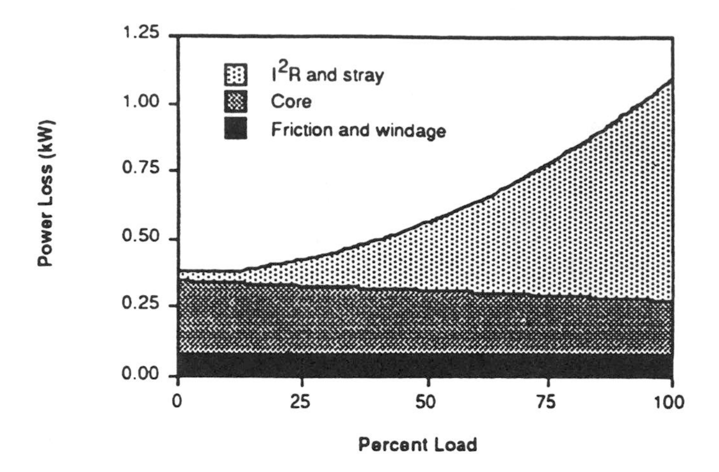

FIGURE 12A.4 Variation of losses with load for a 10 hp motor. (Reprinted from *Energy-Efficient Motor Systems: A Handbook on Technology, Program and Policy Opportunities,* 1992, American Council for an Energy Efficient Economy, Washington, D.C. With permission.)

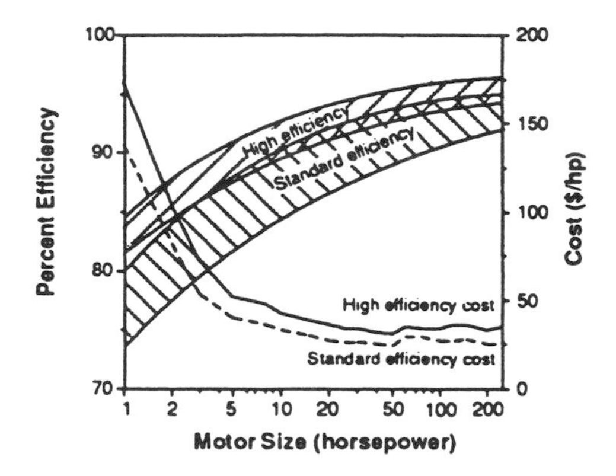

FIGURE 12A.5 Ranges for full load efficiency versus size, and costs (average per hp trade prices) versus size for NEMA Design B standard and high-efficiency 1,800-rpm three-phase induction motors. Width of efficiency bands reflects variation among manufacturers. Note that in some cases there is considerable overlap (one manufacturer's "energy-efficient" motor may be less efficient than another's "standard" motor. (Reprinted from *Energy-Efficient Motor Systems: A Handbook on Technology, Program and Policy Opportunities,* 1992, American Council for an Energy Efficient Economy, Washington, D.C. With permission.)

Energy-Efficient Motors (EEMs)

After World War II and until the early 1970s, there was a trend to design inefficient motors that minimized the use of raw materials (copper, aluminum, and silicon steel). These induction motors had lower initial costs and were more compact than previous generations of motors, but their running costs were higher. When electricity prices started escalating rapidly in the mid-1970s, most of the large motor manufacturers added energy-efficient motors (EEMs) to their product lines. EEMs feature optimized design, more generous electrical and magnetic circuits, and higher-quality materials (Baldwin, 1989).

Incremental efficiency improvements are still possible with the use of superior materials (e.g., amorphous silicon steel) and optimized computer-aided design techniques. EEMs offer an efficiency improvement that typically ranges from 6% for a 5 hp motor to 3% for a 150 hp motor, as shown in Figure 12A.5. EEMs normally carry a price premium of around 15 to 25% in relation to standard motors, which translates into a price premium of $8 to 12/hp. In new applications, and for motors with a large number of operating hours, the paybacks are normally under 2 years.

Efficiency of Rewound Motors

When a motor fails, the user has the option of either having the motor rebuilt, buying a new standard motor, or buying an EEM. The price premium of buying an EEM instead of rebuilding the motor is now in the range of $25 to 35/hp, leading to long paybacks, normally more than 4 years. Although motor rebuilding is a low-cost alternative, the efficiency of a rebuilt motor can be substantially decreased by the use of improper methods for stripping the old winding. On average, the efficiency of a motor decreases by about 1% each time the motor is rewound.

The use of high temperatures (above 350°C) can damage the interlaminar insulation and distort the magnetic circuit with particular impact on the air gap shape, leading to substantially higher core and stray losses. Before any motor is rewound, it should be checked for mechanical damage and the condition of the magnetic circuit should be tested with an electronic iron core loss meter. Techniques are available to remove the old windings, even the ones coated with epoxy varnish, that do not exceed 350°C (Dreisilker, 1987).

Recent Motor Developments

In the low-horsepower range, the induction motor is being challenged by new developments in motor technology such as permanent magnet and switched reluctance motors, which are as durable as induction motors and have higher efficiency. These advanced motors do not have losses in the rotor and feature higher torque and power/weight ratio. In fractional-horsepower motors, such as the ones used in home appliances, the efficiency improvements can reach 10 to 15%, compared with single-phase induction motors. Compared to the shaded-pole motors commonly used in small fans, improved motor types can more than double motor efficiency.

Permanent-Magnet Motors

Over the last few decades, there has been substantial progress in the area of permanent-magnet materials. Figure 12A.6 shows the relative performance of several families of magnetic materials. High-performance permanent-magnet materials, such as neodymium-iron-boron alloys, with a large energy density and moderate cost, offer the possibility of achieving high efficiency and compact lightweight motors.

In modern designs, the permanent magnets are used in the rotor. The currents in the stator windings are switched by semiconductor power devices based on the position of the rotor, normally detected by Hall sensors, as shown in Figure 12A.7. The rotor rotates in synchronism with the rotating magnetic field created by the stator coils, leading to the possibility of accurate speed control. Because these motors have no brushes and with suitable control circuits can be fed from a DC supply, they are sometimes called brushless DC motors.

Switched Reluctance Motors

Reluctance motors are synchronous motors whose stator windings are commutated by semiconductor power switches to create a rotating field. The rotor has no windings, being made of iron with salient poles. The rotor poles are magnetized by the influence of the stator rotating field. The attraction between the magnetized poles and the rotating field creates a torque that keeps the rotor moving at synchronous speed. Figure 12A.8 shows the structure of a switched reluctance motor.

Switched reluctance motors have higher efficiency than induction motors, are simple to build, and are robust. If they are mass-produced, their price can compete with induction motors. Switched reluctance motors can also be used in high-speed applications (above the 3,600 rpm possible with induction or synchronous motors operating on a 60 Hz AC supply) without the need for gears.

Motor Speed Controls

AC induction and synchronous motors are essentially constant-speed motors. Most motor applications would benefit if the speed could be adjusted to the process requirements. This is especially

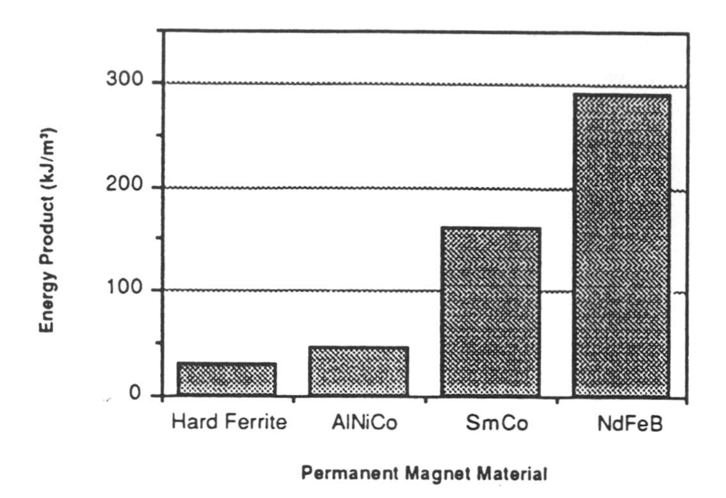

FIGURE 12A.6 The evolution of permanent-magnet materials, showing the increasing magnetic energy density ("energy product"). Ferrites were developed in the 1940s; AlNiCos (aluminum, nickel, and cobalt) in the 1930s. The rare-earth magnets were developed beginning in the 1960s (samarium-cobalt) and in the 1980s (neodymium-iron-boron). The higher the energy density, the more compact the motor design can be for a given power rating. (Reprinted from *Energy-Efficient Motor Systems: A Handbook on Technology, Program and Policy Opportunities,* 1992, American Council for an Energy Efficient Economy, Washington, D.C. With permission.)

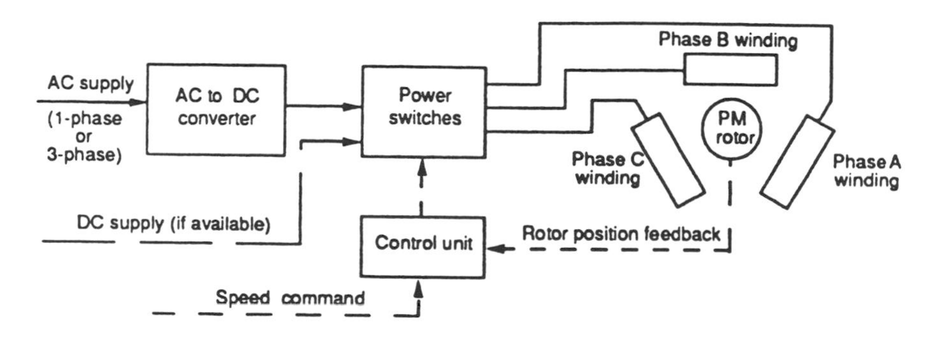

FIGURE 12A.7 Schematic of electronically commutated permanent-magnet (PM) motor. The motor is at right, composed of three sets of stator windings arranged around the permanent-magnet rotor. AC power is first converted to DC, then switched to the windings according to the signals provided by the control unit, which responds to both the desired speed ("speed command") and rotor position feedback. If a DC supply is available, it can be used in place of the AC supply and converter. The function of the commutator and brushes in the conventional DC motor is replaced by the control unit and power switches. The PM rotor follows the rotating magnetic field created by the stator windings. The speed of the motor is easily changed by varying the frequency of switching. (Reprinted from *Energy-Efficient Motor Systems: A Handbook on Technology, Program and Policy Opportunities,* 1992, American Council for an Energy Efficient Economy, Washington, D.C. With permission.)

true for new applications, where the processes can be designed to take advantage of the variable speed. The potential benefits of speed variation include increased productivity and product quality, less wear in the mechanical components, and substantial energy savings.

In many pump, fan, and compressor applications, the mechanical power grows roughly with the cube of the fluid flow; to move 80% of the nominal flow only half of the power is required. Fluid-flow applications are therefore excellent candidates for motor speed control.

Conventional methods of flow control have used inefficient throttling devices such valves, dampers, and vanes. These devices have a low initial cost but introduce high running costs due to

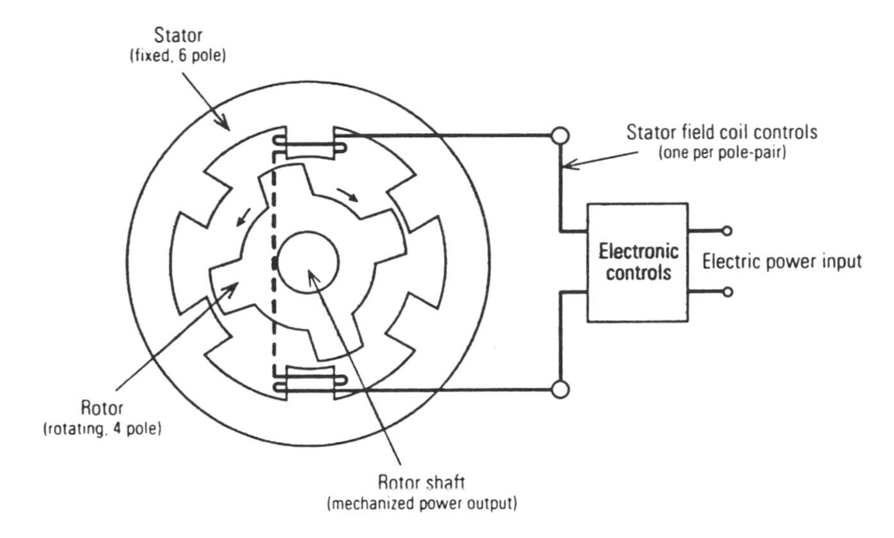

FIGURE 12A.8 Schematic view of a switched reluctance motor. Configuration shown is 6/4 pole, with only one of the three sets of stator windings shown for clarity. Switching times are controlled by microprocessors with custom programming. *Source:* Lovins and Howe, 1992.

their inefficiency. Figure 12A.9 shows the relative performance of different techniques to control flow produced by a fan.

Motor system operation can be improved through the use of several speed-control technologies, such as those covered in the following three sections.

Mechanical and Eddy-Current Drives

Mechanical speed-control technologies include hydraulic transmissions, adjustable sheaves, and gearboxes. Eddy-current drives work as induction clutches with controlled slip (Magnusson, 1984).

Both mechanical drives and eddy drives have relatively low importance. They suffer from low efficiency, bulkiness, limited flexibility, and limited reliability when compared with other alternatives; mechanical drives may require regular maintenance.

Mechanical and eddy drives are not normally used as a retrofit due to their space requirements. Their use is more and more restricted to the low-horsepower range, where their use may be acceptable due to the possible higher cost of ASDs.

Multispeed Motors

In applications where only a few operating speeds are required, multispeed motors may provide the most cost-effective solution. These motors are available with a variety of torque-speed characteristics (variable torque, constant torque, and constant horsepower) (Andreas, 1992) to match different types of loads. Two-winding motors can provide up to four speeds, but they are normally bulkier (one frame size larger) than single-speed motors for the same horsepower rating. Pole-amplitude modulated (PAM) motors are single-winding, two-speed, squirrel cage induction motors that provide a wide range of speed ratios (Pastor, 1986). Because they use a single winding they have the same frame size as single-speed motors for the same horsepower and are thus easy to install as a retrofit. PAM motors are available with a broad choice of speed combinations (even ratios close to unity), being especially suited and cost-effective for those fan and pump applications that can be met by a two-speed duty cycle.

Electronic Adjustable-Speed Drives (ASDs)

Electronic ASDs (Bose, 1986) convert the fixed-frequency power supply (50 or 60 Hz), normally first to a DC supply and then to a continuously variable frequency/variable voltage (Figure 12A.10).

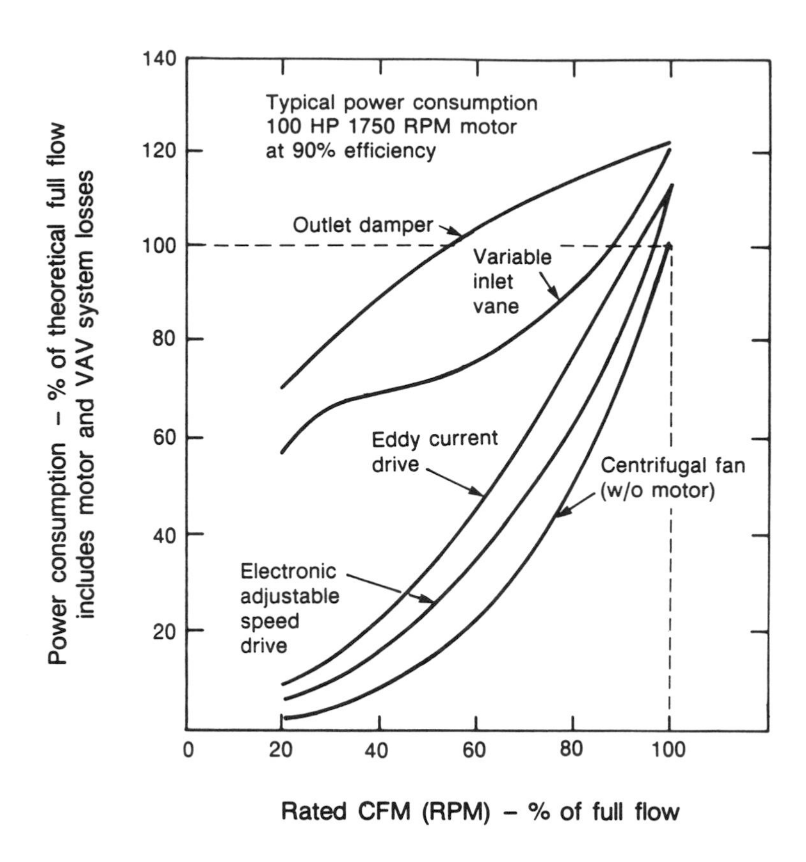

FIGURE 12A.9 Comparison of several techniques for varying air flow in a variable-air-volume (VAV) ventilation system. The curve on the lower right represents the power required by the fan itself, not including motor losses. Electronic ASDs are the most efficient VAV control option, offering large savings compared to outlet dampers or inlet vanes, except at high fractions of the rated fan speed. *Source:* Greenberg et al., 1988.

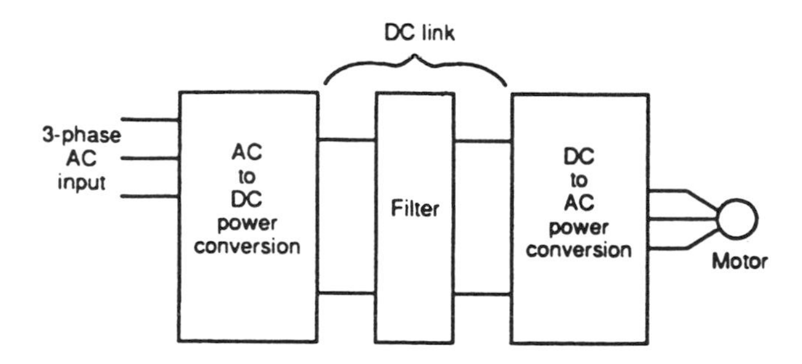

FIGURE 12A.10 General inverter power circuit with motor load. (Reprinted from *Energy-Efficient Motor Systems: A Handbook on Technology, Program and Policy Opportunities,* 1992, American Council for an Energy Efficient Economy, Washington, D.C. With permission.)

ASDs are thus able to continuously change the speed of AC motors. Electronic ASDs have no moving parts (sometimes with the exception of a cooling fan), presenting high reliability and efficiency and low maintenance requirements. Because ASDs are not bulky and have flexible positioning requirements, they are easy to retrofit.

Electronic ASDs are the dominant motor speed-control technology at the present and for the foreseeable future. Developments in the past two decades in the areas of microelectronics and power electronics make possible the design of efficient, compact, and increasingly cost-competitive electronic ASDs. As ASDs control the currents/voltages fed to the motor through power semiconductor

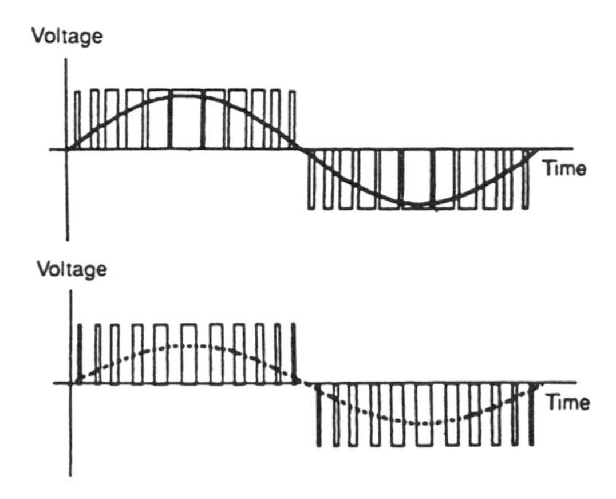

FIGURE 12A.11 Pulse-width modulation. Output voltage is varied by changing the width of the voltages pulses. Output frequency is varied by changing the length of the cycle. (Reprinted from *Energy-Efficient Motor Systems: A Handbook on Technology, Program and Policy Opportunities,* 1992, American Council for an Energy Efficient Economy, Washington, D.C. With permission.)

switches, it is possible to incorporate motor protection features, soft-start, and remote control at a modest cost.

Across the range of motor applications, no single ASD technology emerges as a clear winner when compared with other ASD types. Pulse-width modulation (PWM) voltage-source inverter ASDs dominate in the low to medium horsepower range (up to a few hundred horsepower) due to their lower cost and good overall performance. Figure 12A.11 shows how the variable-frequency/variable-voltage waveform is synthesized by a PWM ASD.

In the range above a few hundred horsepower the choice of ASD technology depends on several factors including the type of motor, horsepower, speed range, and control requirements (Greenberg et al., 1988). Table 12A.1 presents a general classification of the most widely used adjustable-speed motor drive technologies.

Motor System Oversizing

Motor systems are often oversized as a result of the compounding of successive safety factors in the design of a system (Smeaton, 1988). The magnetic, friction, and windage losses are practically constant as a function of the load. Therefore, motors that are oversized (working all the time below 50% of capacity) present not only lower efficiency but also a poor power factor (NEMA, 1977), as shown in Figure 12A.12. The efficiency drops significantly when a motor operates lightly loaded (below 40% for a standard motor). The power factor drops continuously from full load. The decrease in performance is especially noticeable in small motors and standard-efficiency motors.

It is therefore essential to size new motors correctly and to identify motors that run grossly underloaded all the time. In the last case, the economics of replacement by a correctly sized motor should be considered. In medium or large industrial plants, where a stock of motors is normally available, oversized motors may be exchanged for the correct size versions.

Power Quality

Electric motors, and in particular induction motors, are designed to operate with optimal performance when fed by symmetrical three-phase sinusoidal waveforms with the nominal voltage value. Deviations from these ideal conditions may cause significant deterioration of the motor efficiency and lifetime. Possible power quality problems include voltage unbalance, undervoltage or overvoltage, and harmonics and interference. Harmonics and interference can be caused by, as well as affect, motor systems.

TABLE 12A.1 Adjustable-Speed Motor Drive Technologies

Technology	Applicability (R = Retrofit; N = New)	Cost[2]	Comments
		Motors	
Multispeed (incl PAM[1]) Motors	Fractional–500 hp PAM: fractional–2,000 + hp R,N	1.5 to 2 times the price of single-speed motors	Larger and less efficient than 1-speed motors. PAM more promising than multiwinding. Limited number of available speeds.
Direct-Current Motors	Fractional–10,000 hp N	Higher than AC induction motors	Easy speed control. More maintenance required.
		Shaft-Applied Drives (on Motor Output)	
Mechanical			
Variable-Ratio Belts	5–125 hp N	\$350–\$50/hp (for 5–125 hp)	High efficiency at part load. 3:1 speed range limitation. Requires good maintenance for long life.
Friction Dry Disks	Up to 5 hp N	\$500–\$300/hp	10:1 speed range. Maintenance required.
Hydraulic Drive	5–10,000 hp N	Large variation	5:1 speed range. Low efficiency below 50% speed.
Eddy-Current Drive	Fractional–2,000 + hp N	\$900–\$60/hp (for 1 to 150 hp)	Reliable in clean areas. Relatively long life. Low efficiency below 50% speed.
		Wiring-Applied Drives (on Motor Input)	
Electronic Adjustable Speed Drives			
Voltage-Source Inverter	Fractional–1,000 hp R,N	\$500–\$80/hp (for 1 to 300 hp)	Multimotor capability. Can generally use existing motor. PWM[3] appears most promising.
Current-Source Inverter	100–100,000 hp R,N	\$200–\$30/hp (for 100 to 20,000 hp)	Larger and heavier than VSI. Industrial applications, including large synchronous motors.
Others	Fractional–100,000 hp R,N	Large variation	Includes cycloconverters, wound rotor, and variable voltage. Generally for special industrial applications.

[1] PAM means Pole Amplitude Modulated.

[2] The prices are listed from high to low to correspond with the power rating, which is listed from low to high. Thus, the lower the power rating, the higher the cost per horsepower.

[3] PWM means Pulse Width Modulation.

Voltage Unbalance

Induction motors are designed to operate at their best with three-phase balanced sinusoidal voltages. When the three-phase voltages are not equal, the losses increase substantially. Phase unbalance is normally caused by an unequal distribution of the single-phase loads (such as lighting) on the three phases or by faulty conditions. An unbalanced supply can be mathematically represented by two balanced systems rotating in opposite directions. The system rotating in the opposite direction to the motor induces currents in the rotor that heat the motor and decrease the torque. Even a modest phase unbalance of 2% can increase the losses by 25% (Cummings et al., 1985). When a phase unbalance is present, the motor must be derated according to Figure 12A.13.

Voltage Level

When an induction motor is operated at above or below the rated voltage, its efficiency and power factor change. If the motor is underloaded, a voltage reduction may be beneficial, but for a properly sized motor the best overall performance is achieved at the rated voltage. The voltage fluctuations

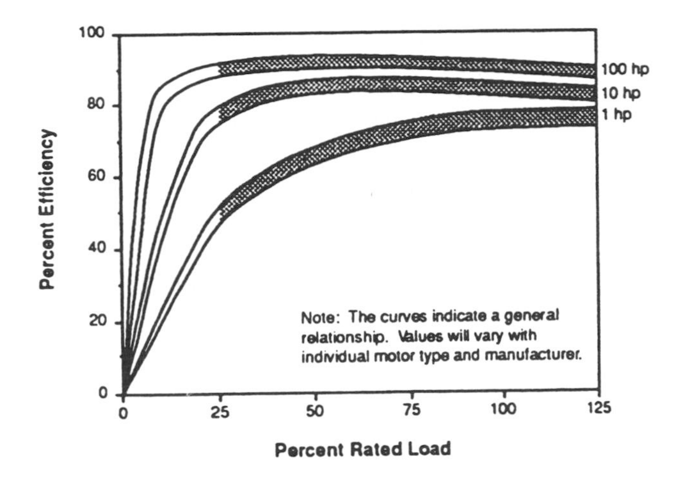

FIGURE 12A.12 Typical efficiency vs. load curves for 1800 rpm, three-phase 60 Hz Design B squirrel cage induction motors. (Reprinted from *Energy-Efficient Motor Systems: A Handbook on Technology, Program and Policy Opportunities*, 1992, American Council for an Energy Efficient Economy, Washington, D.C. With permission.)

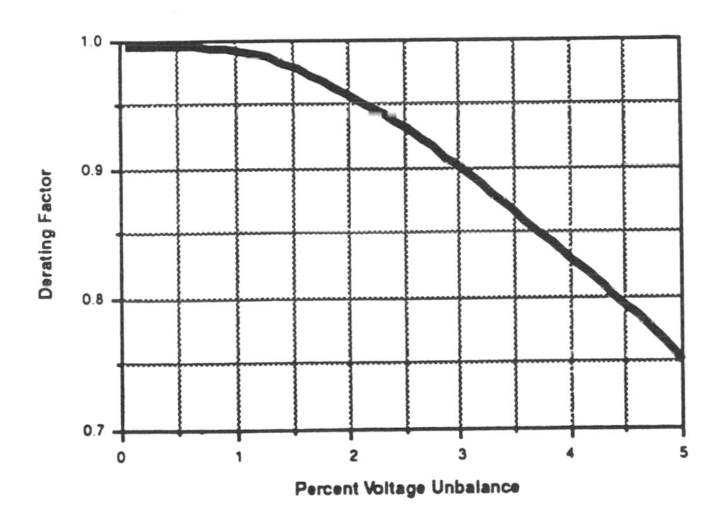

FIGURE 12A.13 Derating factor due to unbalanced voltage for integral-horsepower motors. (Reprinted from *Energy-Efficient Motor Systems: A Handbook on Technology, Program and Policy Opportunities*, 1992, American Council for an Energy Efficient Economy, Washington, D.C. With permission.)

are normally associated with ohmic (*IR*) voltage drops or with reactive power (poor power factor) flow in the distribution network (see "Distribution Losses").

Harmonics and Electromagnetic Interference

When harmonics are present in the motor supply, they heat the motor and do not produce useful torque. This in turn affects the motor lifetime and causes a derating of the motor capacity. This is also true when motors are supplied by ASDs, which generate the harmonics themselves. The use of EEMs can alleviate these problems due to their higher efficiency and thermal capacity; there are also motors specially designed for use with ASDs known as inverter-duty motors.

Reduction of harmonics is also important for the benefit of other consumer and utility equipment. Harmonics, caused by nonlinear loads such as the semiconductor switches in ASDs, should be reduced to an acceptable level as close as possible to the source. The most common technique uses inductive/capacitive filters at the ASD input circuit to provide a shunt path for the harmonics and to perform power factor compensation.

IEEE Standard No. 519 (IEEE, 1992) contains guidelines for harmonic control and reactive power compensation of power converter. The cost of the harmonic filter to meet this standard is typically around 5% of the cost of the ASD.

ASD power semiconductor switches operate with fast switching speeds to decrease energy losses. The fast transitions in the waveforms contain high-frequency harmonics, including those in the radio-frequency range. These high-frequency components can produce interference through both conduction and radiation. The best way to deal with EMI is to suppress it at the source. Radiated EMI is suppressed through shielding and grounding of the ASD enclosure. Proper ASD design, the use of a dedicated feeder, and the use of a low-pass input filter (an inductor; often called a "line reactor") will normally suppress conducted EMI.

Distribution Losses

Cable Sizing

The currents supplied to the motors in any given installation will produce Joule (I^2R) losses in the distribution cables and transformers of the consumer. Correct sizing of the cables not only allows a cost-effective minimization of those losses, it also helps to decrease the voltage drop between the transformer and the motor. The use of the National Electrical Code for sizing conductors leads to cable sizes that prevent overheating and allow adequate starting current to the motors, but can be far from being an energy-efficient design. For example, when feeding a 100 hp motor located at 150 meters from the transformer with a cable sized using NEC, 6% of the power will be lost in heating the cable (Lovins et al., 1989). Considering a 2-year payback, it is normally economical to use a cable with a cross section 2 to 4 times the one required by the NEC.

Reactive Power Compensation

In most industrial consumers, the main reason for a poor power factor is the widespread application of oversized motors. Correcting oversizing can thus contribute in many cases to a significant improvement of the power factor.

Reactive power compensation, through the application of correction capacitors, not only reduces the losses in the network but also allows full use of the power capacity of the power system components (cables, transformers, circuit breakers, etc.). In addition, voltage fluctuations are reduced, thus helping the motor to operate closer to its design voltage.

Mechanical Transmissions

The transmission subsystem transfers the mechanical power from the motor to the motor-driven equipment. To achieve overall high efficiency it is necessary to use simple, properly maintained transmissions with low losses. The choice of transmission depends on many factors including speed ratio desired, horsepower, layout of the shafts, type of mechanical load, and so forth.

Transmission types available include direct shaft couplings, gearboxes, chains, and belts. Belt transmissions offer significant potential for savings. About a third of motor transmissions use belts (Lovins et al., 1989). Several types of belts can be used, such as V-belts, cogged V-belts, and synchronous belts.

V-belts have an efficiency in the 90 to 96% range. V-belt losses are associated with flexing, slippage, and a small percentage due to windage. With wear, the V-belt stretches and needs retensioning; otherwise the slippage increases and the efficiency drops. Cogged V-belts have lower flexing losses and have better gripping on the pulleys, leading to a 2 to 3% efficiency improvement when compared with standard V-belts.

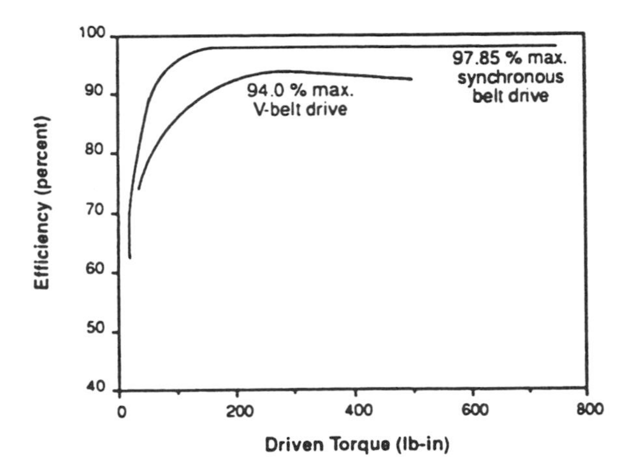

FIGURE 12A.14 Efficiency versus torque for V-belts and synchronous belts in a typical application. (Reprinted from *Energy-Efficient Motor Systems: A Handbook on Technology, Program and Policy Opportunities,* 1992, American Council for an Energy Efficient Economy, Washington, D.C. With permission.)

Synchronous belts can be 98 to 99% efficient, as they have no slippage and have low flexing losses; they typically last over twice as long as V-belts, leading to savings in avoided replacements that more than offset their extra cost. Figure 12A.14 shows the relative performance of V-belts and synchronous belts. The efficiency gains increase with light loads.

Maintenance

Regular maintenance (such as inspection, adjustment, cleaning, filter replacement, lubrication, and tool sharpening) is essential to maintain peak performance of the mechanical parts and to extend their operating lifetime. Both under- and overlubrication can cause higher friction losses in the bearings and shorten the bearing lifetime. Additionally, overgreasing can cause the accumulation of grease and dirt on the motor windings, leading to overheating and premature failure.

The mechanical efficiency of the driven equipment (pump, fan, cutter, etc.) directly affects the overall system efficiency. Monitoring wear and erosion in this equipment is especially important, as its efficiency can be dramatically affected. For example, in chemical process industries the erosion of the pump impeller will cause the pump efficiency to drop sharply; a dull cutter will do the same to a machine tool.

Cleaning the motor casing is also relevant because its operating temperature increases as dust and dirt accumulates on the case. The same can be said about providing a cool environment for the motor. The temperature increase leads to an increase of the windings' resistivity and therefore to larger losses. An increase of 25°C in the motor temperature increases the Joule losses by 10%.

12A.3 Energy-Saving Applications of ASDs

Typical loads that may benefit from the use of ASDs include those covered in the following four sections:

Pumps and Fans

In many pumps and fans where there are variable-flow requirements, substantial savings can be achieved, as the power is roughly proportional to the cube of the flow (and thus speed of the motor). The use of ASDs instead of throttling valves with pumps shows behavior similar to that for fans in Figure 12A.9.

Centrifugal Compressors and Chillers

Centrifugal compressors and chillers can take advantage of motor controls in the same way as other centrifugal loads (pumps and fans). The use of wasteful throttling devices or the on-off cycling of the equipment can be largely avoided, resulting in both energy savings and extended equipment lifetime.

Conveyors

The use of speed controls in both horizontal and inclined conveyors allows the matching of speed to material flow. As the conveyor friction torque is constant, energy savings are obtained when the conveyor is operated at reduced speed. In long conveyors, such as are found in power plants and in the mining industry, the benefits of soft-start without the need for complex auxiliary equipment are also significant.

High-Performance Applications

AC motors have received much attention in recent years as a proposed replacement for DC motors in high-performance speed-control applications, where torque and speed must be independently controlled. Induction motors are much more reliable, more compact, more efficient, and less expensive than DC motors. As induction motors have no carbon brush commutation, they are especially suitable for corrosive and explosive environments. In the past, induction motors have been difficult to control, as they behave as complex nonlinear systems. However, the appearance on the market of powerful and inexpensive microprocessors has made it possible to implement in real time the complex algorithms required for induction motor control.

Field-oriented control, also called vector control, allows accurate control of the speed and torque of induction motors, in a way similar to DC motor control (Leonhard, 1984). The motor current and voltage waveforms, together with motor position feedback, are processed in real time, allowing the motor current to be decomposed into a field-producing component and a torque-producing component. The vector control operation principle is represented in Figure 12A.15 and is being applied to a wide variety of high-performance applications described next.

Rolling mills were one of the strongholds of DC motors, due to the accurate speed and torque requirements. With present ASD technology, AC drives can outperform DC drives in all technical aspects (reliability, torque/speed performance, maximum power, efficiency), and are capable of accurate control down to zero speed.

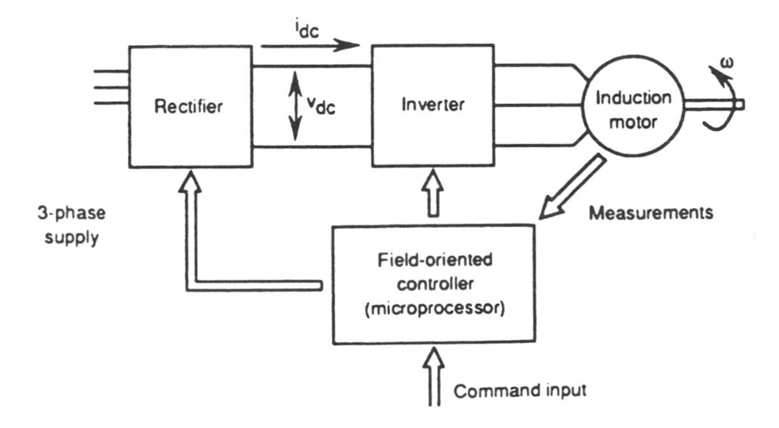

FIGURE 12A.15 Schematic of a vector-control drive (also known as a field-oriented control). (Reprinted from *Energy-Efficient Motor Systems: A Handbook on Technology, Program and Policy Opportunities*, 1992, American Council for an Energy Efficient Economy, Washington, D.C. With permission.)

The availability of large-diameter, high-torque, and low-speed AC drives makes them suitable for use in applications such as ball mills and rotary kilns without the need for gearboxes. This area was also a stronghold of DC drives. Again, AC drives have the capability to offer superior performance in terms of reliability, power density, overload capability, efficiency, and dynamic characteristics.

AC traction drives can also feature regenerative braking. AC traction drives are already being used in trains, rapid transit systems, and ship propulsion, and are the proper choice for the electric automobile.

DC drives have traditionally been used with winders in the paper and steel industry in order to achieve the constant-tension requirements as the winding is performed. Sometimes the constant tension is obtained by imposing friction, which wastes energy. AC drives can also be used to replace DC drives, saving energy and achieving better overall performance.

The use of field-oriented AC drives in machine tools and robotics allows the stringent requirements regarding dynamic performance to be met. Positioning drives can produce a peak torque up to 10 times the rated torque and make possible the adjustment of the speed to very low values.

In machine tools, AC drives can provide accurate higher spindle speeds than DC drives without the need for gearboxes and their associated losses. The ASDs can also adjust the voltage level when the spindle drive is lightly loaded, providing further savings. In robotics, the higher power density and superior dynamics of AC drives are important advantages.

12A.4 Energy and Power Savings Potential; Cost-Effectiveness

The energy and peak power savings potential for any motor-related technology depends on a number of factors, including the characteristics of the motor, the motor drive, the driven equipment, and the load (Nadel et al., 1992). Since all of this information is seldom available, it is difficult to determine the effect of even a single application; it is far more difficult to determine the savings potential for diverse applications, across all sectors, for an entire nation.

This section estimates the energy and power savings that could be realized nationwide in the United States through the application of these technologies in the residential, commercial, and industrial sectors. Table 12A.2 lists the major motor-driven end uses, estimates of their energy use as a percentage of total national electricity use (based on Lovins et al., 1989, and EEI, 1987) and the potential energy savings from ASDs expressed as a fraction of existing use.

Potential Savings in the Residential Sector

The primary motor technology for realizing energy and power savings in the residential sector is the electronic adjustable-speed drive (ASD). Heat pumps, air conditioners, forced-air furnaces, and washing machines with ASDs have already been introduced into the market. Other appliances, such as refrigerators, freezers, heat pump water heaters, and evaporative coolers, are also potential candidates for adjustable-speed controls. Most of the energy-saving potential of ASDs in the home is associated with the use of refrigerant compressors for cooling or heating (as in heat pumps, air conditioners, refrigerators, and freezers). In all of these applications, ASDs can reduce energy consumption by matching the speed of the compressor to the instantaneous thermal load.

Several improvements in home appliance technology that are likely to become common over the next few years will complement the use of ASDs:

- *High-efficiency compressor and fan motors.* The use of permanent-magnet and reluctance motors can increase the efficiency of the motor by 5 to 15%, when compared with conventional squirrel cage single-phase induction motors; as noted earlier, even larger savings are possible with many small fan motors. Permanent-magnet AC motors are used in the latest ASD-equipped furnace and heat pump.

TABLE 12A.2 Estimated Motor Electricity Usage and Potential ASD Savings

Sector	End Use	Usage (% of Total U.S. Usage)	Potential ASD Savings (% of Usage for Each End Use)
Residential	Refrigeration	6.9	20
	Space heating	1.4	25
	Air conditioning	3.8	20
	Other	0.9	0
	Residential total	13.1	19
Commercial	Air conditioning	7.5	15
	Ventilation	2.2	25
	Refrigeration	1.3	25
	Other	0.1	20
	Commercial total	11.3	18
Industrial	Pumps	7.7	25
	Blowers and fans	4.5	25
	Compressors	3.9	25
	Machine tools	2.1	10
	Other integral hp	2.8	0
	DC drives	2.5	15
	Fractional hp	1.1	0
	HVAC	0.6	15
	Industrial total	25.4	18
Public and Miscellaneous	Muni. water works	1.7	15
	Electric utilities	5.4	25
	Other	0.2	0
	Pub. & misc. total	7.3	22
Totals		57.1	19

- *Rotary and scroll compressors.* The use of rotary (in small applications) or scroll compressors in place of reciprocating compressors can take full advantage of the speed variation potential of ASDs.
- *Larger heat exchangers with improved design.* Improved heat exchangers increase the efficiency by decreasing the temperature difference of the compressor thermal cycle.

Potential Savings in the Commercial Sector

An estimated 47% of commercial electricity use is for motor-driven end uses (Lovins et al., 1989; EEI, 1987); the percentage for peak power is higher. The savings potential of ASDs in air-conditioning and ventilation applications was estimated by running DOE-2 computer simulations on two representative building types in five U.S. cities (Eto and de Almeida, 1988). Comparisons were made between the cooling and ventilation systems with and without ASDs. The results indicate ventilation savings of approximately 25% for energy and 6% for peak power, and cooling savings of about 15% and 0%, respectively. In Table 12A.1 we assumed these energy results can be applied nationwide.

The estimated energy savings for refrigeration are shown in Table 12A.1, assuming 25% savings in energy; an estimated 10% savings in peak power should be attainable. Other motor efficiency measures (discussed in Section 12A.2) combined can capture approximately 10% more energy and demand savings.

Potential Savings in the Industrial and Public/Miscellaneous Sectors

In Table 12A.2, the fluid moving end-use savings are estimated at 25%, except for municipal water works, where system requirements limit the savings. The machine tool savings are due to a combination of the abilities of ASDs to eliminate gearboxes and to act as a variable-voltage control to increase motor efficiency at light loads. The DC drive savings assume the DC drive is replaced with an AC motor and ASD. HVAC load is mostly chillers, so savings are taken to be the same as for commercial cooling. The other integral and fractional-hp categories probably have savings potential as well, but not enough is known about them to make an estimate.

As most industries are nonseasonal with flat load profiles during operating hours, the peak savings are similar to energy savings. When other motor efficiency measures (see Section 12A.2) are combined, approximately 10% more energy and demand savings can be obtained, resulting in a total of 28%.

Cost-Effectiveness of ASDs

The price of ASD equipment, in terms of dollars/horsepower, is a function of the horsepower range, the type of AC motor used, and the additional control and protection facilities offered by the electronic ASD. ASD installation costs vary tremendously depending on whether the application is new or retrofit, available space, weather protection considerations, labor rates, and so on. Thus there is a huge range of installed costs possible for any given ASD size, and the costs listed here necessarily have large uncertainties.

A market survey (PEAC, 1987), modified by a compilation of case studies (Nadel et al., 1992; EPRI, 1992b) and general market trends (Hoffman, 1992) showed the following typical dollar values for equipment and installation of drives for induction motors (the higher $/hp in each range corresponds to the small end of the hp range):

Installed Cost ($/hp)	Size Range (hp)
1,000–380	7.5–50
380–280	50–200
280–200	200–1,000
200–130	1,000–2,500
130–80	2,500–20,000

Mass production of smaller motors with built-in ASDs has been very successful in bringing down the cost of Japanese variable-speed heat pumps, with reported incremental costs of $25/hp for the ASD (Abbate, 1988).

To determine whether an adjustable-speed drive is cost-effective for any given application, the following need to be taken into account:

- First cost (acquisition and installation)
- System operating load profile (number of hours per year at each level of load)
- Cost of electricity
- Maintenance requirements
- Reliability
- Secondary benefits (less wear on equipment, less operating noise, regeneration capability, improved control, soft-start, and automatic protection features)
- Secondary problems (power factor, harmonics, and interference)

A careful analysis should weigh the value of the benefits offered by each option against the secondary problems, such as power quality, that may impose extra costs for filters and power factor correction capacitors. Figure 12A.16 applies this general approach in developing a schematic showing ASD cost-effectiveness, once the net costs are known.

Comparing the cost of conserved energy to the cost of electricity is a crude way to assess the cost effectiveness of energy efficiency measures. More accurate calculations would account for the

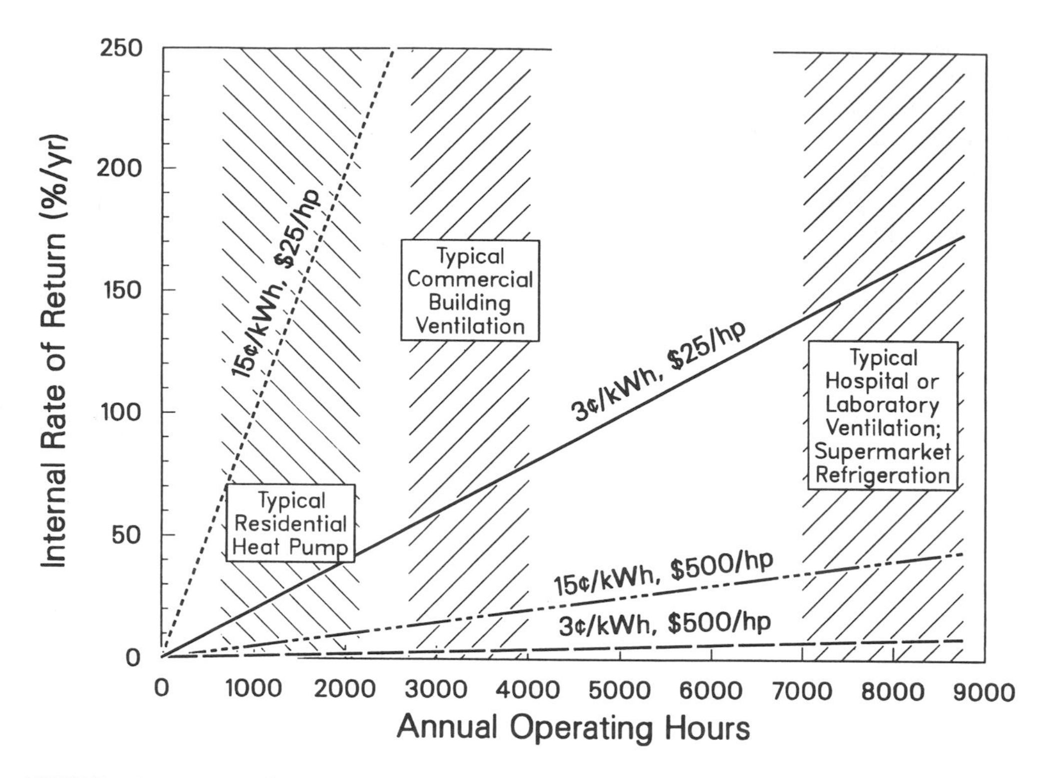

FIGURE 12A.16 Cost-effectiveness of adjustable-speed drive applications for various combinations of initial ASD cost, electricity prices, and annual operating hours. Typical energy savings of 20% are assumed here, although actual savings depend on the application. *Source:* Greenberg et al., 1988.

time at which conservation measures save energy relative to the utility system peak demand, and relate these "load shape characteristics" to baseload, intermediate, and peaking supply resources. See Koomey et al. (1990a and 1990b) for more details.

References

Abbate, G. 1988. "Technology Developments in Home Appliances." In *Demand-Side Management and Electricity End-Use Efficiency,* A. de Almeida and A. Rosenfeld, eds. Boston: Kluwer Academic Publishers.

Andreas, J. 1992. *Energy-Efficient Electric Motors: Selection and Application.* New York: Marcel Dekker.

Baldwin, S.F. 1989. "Energy-Efficient Electric Motor Drive Systems." In *Electricity: Efficient End-Use and New Generation Technologies, and Their Planning Implications,* T.B. Johansson et al., eds. Lund, Sweden: Lund University Press.

Bose, B. 1986. *Power Electronics and AC Drives.* Englewood Cliffs, NJ: Prentice Hall.

BPA/EPRI. 1993. *Electric Motor Systems Sourcebook.* Olympia, WA: Bonneville Power Administration, Information Clearinghouse.

Cummings, P., Dunki-Jacobs, J., and Kerr, R. 1985. "Protection of Induction Motors Against Unbalanced Voltage Operation," *IEEE Transactions on Industry Applications,* Vol. IA-21, No. 4.

Dreisilker, H. 1987. "Modern Rewind Methods Assure Better Rebuilt Motors." *Electrical Construction and Maintenance,* 84. August.

EEI, 1987. *Statistical Yearbook 1987.* Washington, DC: Edison Electric Institute.

EPRI, 1992a. "Electric Motors: Markets, Trends and Application." EPRI report TR-100423. Palo Alto, CA: Electric Power Research Institute.

EPRI, 1992b. "Adjustable Speed Drive Directory," 3rd ed. Palo Alto, CA: Electric Power Research Institute.

EPRI, 1993. "Applications of AC Adjustable Speed Drives." Palo Alto, CA: Electric Power Research Institute.

Eto, J. and de Almeida, A. 1988. "Saving Electricity in Commercial Buildings with Adjustable Speed Drives." *IEEE Transactions on Industry Applications,* Vol. IA-24, No. 3, pp. 439–443.

Greenberg, S., Harris, J.H., Akbari, H., and de Almeida, A. 1988. "Technology Assessment: Adjustable Speed Motors and Motor Drives." Lawrence Berkeley National Laboratory Report LBL-25080. Berkeley, CA: University of California Lawrence Berkeley National Laboratory.

Hoffman, J. 1992. "Building on Success: EPA Initiatives for Energy Efficiency." E-Source Members' Forum. Snowmass, CO: E-Source. October 2.

IEEE. 1992. "IEEE Guide for Harmonic Control and Reactive Compensation of Static Power Converters." IEEE Standard 519. New York: Institute of Electrical and Electronics Engineers.

Jarc, D. and Schieman, R. 1985. "Power Line Considerations for Variable Frequency Drives." *IEEE Transactions on Industry Applications,* Vol. IA-21, No. 5.

Koomey, J., Rosenfeld, A.H., and Gadgil, A.K. 1990a. "Conservation Screening Curves to Compare Efficiency Investments to Power Plants: Applications to Commercial Sector Conservation Programs." In *Proceedings of the 1990 ACEEE Summer Study on Energy Efficiency in Buildings.* Asilomar, CA: American Council for an Energy-Efficient Economy.

Koomey, J., Rosenfeld, A.H., and Gadgil, A.K. 1990b. "Conservation Screening Curves to Compare Efficiency Investments to Power Plants." *Energy Policy,* Vol. 18, No. 8. p. 774.

Leonhard, W. 1984. *Control of Electrical Drives.* New York: Springer-Verlag.

Lovins, A., Neymark, J., Flanigan, T., Kiernan, P., Bancroft, B., and Shepard, M. 1989. *The State of the Art: Drivepower.* Snowmass, CO: Rocky Mountain Institute.

Lovins, A. and Howe, B. 1992. "Switched Reluctance Motor Systems Poised for Rapid Growth." Boulder, CO: E-Source.

Magnusson, D. 1984. "Energy Economics for Equipment Replacement." *IEEE Transactions on Industry Applications,* Vol. IA-20, No. 2.

Moore, T. 1988. "The Advanced Heat Pump: All the Comforts of Home ... and Then Some." *EPRI Journal.* Palo Alto, CA: Electric Power Research Institute. March.

Nadel, S., Shepard, M., Greenberg, S., Katz, G., and de Almeida, A. 1992. *Energy-Efficient Motor Systems—A Handbook on Technologies, Programs, and Policy Opportunities.* Washington, DC: American Council for an Energy-Efficient Economy.

NEMA. 1977. NEMA Standards Publication MG10-1977, "Energy Management Guide for Selection and Use of Polyphase Motors." Washington, DC: National Electrical Manufacturers' Association.

Pastor, C.E. 1986. "Motor Application Considerations—Single Speed, Multi-Speed, Variable Speed." Oakland, CA: Pacific Gas and Electric Company Conference. April 30–May 2.

PEAC (Power Electronics Applications Center). 1987. *ASD Directory,* 2nd ed. Knoxville, TN: Power Electronics Applications Center.

Smeaton, R. 1988. *Motor Application and Maintenance Handbook.* New York: McGraw-Hill.

Taylor, R.W. 1987. "Terminal Characteristics of Power Semiconductor Devices." *TechCommentary,* Vol. 1, No. 2. Knoxville, TN: Power Electronics Applications Center.

12B

Energy Efficient Technologies: Energy Efficient Lighting Technologies and Their Applications in the Commercial and Residential Sectors

Barbara Atkinson

Andrea Denver

James E. McMahon

Leslie Shown

Robert Clear
Lawrence Berkeley National Laboratory

Introduction

Lighting is an important electrical end use in every sector and building type across the United States and accounts for approximately one-fifth of national electricity use (see Table 12B.1). Researchers estimate that an additional 3 to 4% of national electricity use can be attributed indirectly to lighting because of the heat that lighting systems produce and the extra air conditioning energy that is therefore required to cool buildings.

In 1990, approximately 38% of commercial, 11% of residential, and 9% of industrial electricity consumption was attributable to lighting measures. The commercial sector consumes the majority of the electricity used for lighting in the United States. In 1990, the commercial sector consumed approximately 60% of total lighting electricity while the residential and industrial sectors consumed approximately 20% and 16%, respectively. In this chapter, we focus primarily on lighting in the commercial and residential sectors.

TABLE 12B.1 Lighting Electricity Use in the United States in 1990

Sector	Total Electricity Use (terawatt-hours)	Lighting Electricity Use (terawatt-hours)	Lighting as Percentage of Sectoral Total
Residential	921	103	11%
Commercial	840	316	38%
Industrial	856	81	9%
Other	92	15	16%
Total	2,709	515	19%

Sources: Atkinson et al., 1992; U.S. Department of Energy, 1993c; and EIA, 1993.

There is great potential for saving electricity, reducing the emission of greenhouse gases associated with electricity production, and reducing consumer energy costs through the use of more efficient lighting technologies as well as advanced lighting design practices and control strategies. According to Atkinson et al. (1992), using the cost-effective lighting technologies that are already available, electricity consumption for commercial interior lighting could be reduced by 50 to 60%. Similarly, electricity consumption for residential interior and exterior lighting could be reduced by 20 to 35%. Over the next 35 years (1995 to 2030), these savings from energy-efficient technologies in the commercial sector would have a net present value of $50 to 100 billion, while the savings from energy-efficient technologies in the residential sector would have a net present value of $35 to 40 billion.* New, efficient technologies that enter the market during this time period can further reduce energy use and increase financial savings.

In this chapter, we provide an overview of both traditional and newer, more efficient lighting technologies. Our discussion includes

- Variations in lighting energy consumption among different building types
- The design of energy-efficient lighting systems
- The designs, applications, and efficacies of various lighting technologies (including lamps, ballasts, fixtures, and controls)
- The operation of energy-efficient lighting systems
- The current production of lighting technologies
- The cost-effectiveness of selected lighting technologies

See Chapter 20 for information on daylighting.

12B.1 Variations in Lighting Energy Consumption Among Different Building Types

The amount of energy consumed for lighting in a given building, and thus the potential for energy conservation, depends on a variety of factors: the lighting level, the number of hours per year that the lighting equipment is in use, the type of equipment used, and the design of the lighting system. In general, lighting levels depend on the type of building and the type of activities within the building. Because of changes in building codes over time, the age of a building can also be a factor in its lighting level.** The annual hours of use for lighting equipment in a given building depend largely on the type of activity and the use of lighting controls.

*These values of savings account for changes in equipment costs as well as changes in energy costs.

**For example, between the early 1970s and late 1980s, the lighting levels suggested by the Illuminating Engineering Society (IES) decreased by 34% in retail buildings, 21% in office buildings, 17% in schools, and 15% in hospitals (Mills and Piette, 1992).

Commercial and industrial lighting consumption are described in terms of energy use intensity (EUI), which has units of kilowatt-hours per square meter per year (kWh/m²/yr).* Table 12B.2 provides EUIs and annual operating hours for major commercial building types in 1992.** For most building types, the annual operating hours are approximately equal to the number of hours that lighting equipment is in use; however, in some building types (e.g., lodging), many of the lamps are not in use during all of a facility's operating hours. As indicated in Table 12B.2, EUIs and operating hours can vary dramatically by building type. In a warehouse, for example, the average energy use intensity is very low and operating hours average less than 13 hours per day. In contrast, the average energy use intensity for grocery stores is more than three times as high and operating hours average almost 17 hours per day.

TABLE 12B.2 Lighting EUIs and Annual Hours of Use in the Commercial Sector, 1992

Commercial Building Types	Energy Use Intensity (existing stock, interior lighting) [kWh/m²/yr (kWh/ft²/yr)]	Annual Operating Hours
Small office	47.3 (4.4)	2,867
Large office	56.0 (5.2)	3,465
Restaurant	53.8 (5.0)	5,035
Retail	42.0 (3.9)	3,711
Grocery	76.4 (7.1)	6,001
Warehouse	24.7 (2.3)	4,651
School	24.7 (2.3)	—
College	57.0 (5.3)	—
Education	—	3,083
Health	64.6 (6.0)	7,218
Lodging	68.9 (6.4)	8,529
Assembly	22.6 (2.1)	3,639
Miscellaneous	26.9 (2.5)	6,012

Source: Regional Economic Research (RER), Del Mar, CA, 1995. RER data are based on (1) Energy Information Administration (EIA) "Commercial Buildings Energy Consumption Survey (CBECS) 1992," Washington, DC, 1994 and (2) "CBECS, Commercial Buildings Characteristics 1992," DOE/EIA-0246(92), Washington, DC, 1994.

Variations in energy use and operating hours within a building type or sector are probably as important as the variations between types or sectors. This is particularly true for the residential sector, where there are significant variations in the number of people per household, the size of the house or apartment, the type of lighting equipment used, and individual use characteristics. On average, lighting energy use per household in the United States is estimated to be approximately 1,100 kWh/year (Atkinson et al., 1992), but surveys show that it can range from less than 1,000 to as much as 3,000 kWh/year or more.

12B.2 The Design of Energy-Efficient Lighting Systems

A lighting system is an integral part of a building's architectural design and interacts with the shape of each room, its furnishings, and the level of natural light. Energy efficiency is an important component of lighting system design; however, lighting designers must also consider economics, aesthetics, and consumer preference. To improve the lighting efficiency in a building, a lighting designer must understand the user's lighting tastes and needs, the most efficient technologies available to meet these needs, and how individual lighting components function together as a system.

*One kWh/m²/yr equals 0.093 kWh/ft²/yr.

**The average EUI for industrial uses in 1992 was about 41 kWh/m²/yr (3.8 kWh/ft²/yr).

Efficient, high-quality lighting design includes

- Attention to task and ambient lighting
- Effective use of daylighting
- Effective use of lighting controls
- Use of the most cost-effective and efficacious technologies

Because people require less light in surrounding areas than they do where they perform visual tasks, it is usually both unnecessary and inefficient for an entire space to be lit at a level that is appropriate for visual tasks. For this reason, lighting designers practice *task-ambient lighting* design. For visual comfort and ease of visual transition between task and ambient spaces, the "ambient" lighting in a room should be at least one-third as bright as the lighting of the task areas (U.S. DOE, 1993a). A common task-ambient lighting strategy is to design the overall lighting system to provide an appropriate ambient level of light and then add task lights (e.g., desk lamps) in areas where people are working.

Effective use of *daylighting* is also an important component of lighting system design. After decades of overdependence on artificial light, many lighting designers are returning to the use of sunlight to illuminate interior spaces. For lighting designers to make good use of natural light, however, requires more than the simple addition of multiple windows. Light pouring in through windows can create glare and cause other spaces to appear very dark by comparison; in addition, windows that are too large can allow too much heat loss or gain. The challenges to successful daylighting are to admit only as much light as needed, to distribute it evenly, and to avoid glare. The effective use of daylighting can be greatly enhanced by the overall architectural design of a building. For example, more sunlight is available to a building design that maximizes surface area (e.g., a building that is U-shaped or has an interior courtyard). In addition, skylights, wide windowsills, reflector systems, louvers, blinds, and other innovations can be used to bounce natural light farther into a building. The use of window glazes can limit heat transmission while permitting visible light to pass through a window or skylight.

Efficient lighting design depends on the careful selection of cost-effective and efficient lighting technologies. *Lighting control systems* and permanent dedicated fixtures are important components of efficient lighting systems. In order to complement other efficiency improvements, lighting designers can use lighting controls to reduce lighting when it is not needed. For example, lighting energy is saved when occupancy sensors turn off the lights in an empty room or daylighting controls dim the fluorescent lamps as the level of natural light in a room increases. Dimming systems can also be used to maintain a constant light level as a system ages, which saves energy when lamps are new. In order to ensure the persistence of energy savings, lighting designers can install permanent lighting fixtures that are dedicated to efficient lamps. For example, if an office retrofit consists of substituting screw-in compact fluorescent lamps (CFLs) for incandescent lamps in incandescent fixtures, rather than replacing the incandescent system with CFLs and hard-wired CFL fixtures, the more efficient CFLs could be replaced with incandescents when they fail and the opportunity for ongoing energy savings would be lost.

Lighting design that promotes energy-efficient lighting technologies can also influence the design and energy use of a building's cooling system. Because efficient lighting systems produce less heat, the air conditioning systems installed in new buildings with efficient lighting can have lower capacities. Consequently, less money is spent on air conditioning systems as well as cooling energy.* The cost-effectiveness of different lighting systems is discussed in Section 12B.6.

*More energy may be required to heat a building when lighting electricity consumption is reduced, but in most climates cooling savings offset this heating penalty and net cost savings are accrued.

12B.3 The Design, Application, and Efficacy of Various Lighting Technologies

Lighting system components can be broken into four basic categories:

- Lamps
- Ballasts
- Fixtures
- Lighting controls

In this section, we describe the most common and the most efficacious lighting technologies within each category as well as where and how different technologies are used. In addition, we discuss some of the most promising design options for further efficiency improvements in lighting technologies.*

Lamps

Because the purpose of a lamp is to produce light, and not just radiated power, there is no direct measure of lamp efficiency. Instead, a lamp is rated in terms of its **efficacy**, which is the ratio of the amount of light emitted (lumens) to the power (watts) drawn by the lamp. The unit used to express lamp efficacy is lumens per watt (LPW). The theoretical limit of efficacy is 683 LPW and would be produced by an ideal light source emitting monochromatic radiation with a wavelength of 555 nanometers. The lamps that are currently on the market produce from a few percent to almost 25% of the maximum possible efficacy.** The efficacies of various light sources are depicted in Figure 12B.1. Lamps also differ in terms of their cost; size; color; lifetime; **optical controllability**; dimmability; **lumen maintenance*****; reliability; simplicity and convenience in use, maintenance, and disposal; and environmental effects (e.g., emission of noise, radio interference, and ultraviolet [UV] light).

The color properties of a lamp are described by its color temperature and its color rendering index. **Color temperature**, expressed in degrees Kelvin (K), is a measure of the color appearance of the light of a lamp. The concept of color temperature is based on the fact that the emitted radiation spectrum of a blackbody radiator depends on temperature alone. The color temperature of a lamp is the temperature at which an ideal blackbody radiator would emit light that is closest in color to the light of the lamp. Lamps with low color temperatures (3,000 K and below) emit "warm" white light that appears yellowish or reddish in color. Incandescent and warm-white fluorescent lamps have a low color temperature. Lamps with high color temperatures (3,500 K and above) emit "cool" white light that appears bluish in color. Cool-white fluorescent lamps have a high color temperature.

The **color rendering index** (**CRI**) of a lamp is a measure of how surface colors appear when illuminated by the lamp compared to how they appear when illuminated by a reference source of the same color temperature. For color temperatures above 5,000 K, the reference source is a standard daylight condition of the same color temperature; below 5,000 K, the reference source is

*The technical information in this section was obtained from many sources including Turiel et al., 1995; Atkinson et al., 1992; U.S. DOE, 1993a; Audin et al., 1994; and manufacturer catalogs.

**The efficacies of fluorescent lamp/ballast combinations reported in this section are based on data compiled by Oliver Morse from tests by the National Electrical Manufacturer's Association; all comparisons assume SP41 lamps. Efficacies of incandescent lamps are based on manufacturer catalogs. Efficacies of high-intensity discharge lamps are based on manufacturer catalogs and an assumed ballast factor of 0.95.

***Over time, most lamps continue to draw the same amount of power but produce fewer lumens. The lumen maintenance of a lamp refers to the extent to which the lamp sustains its lumen output, and therefore efficacy, over time.

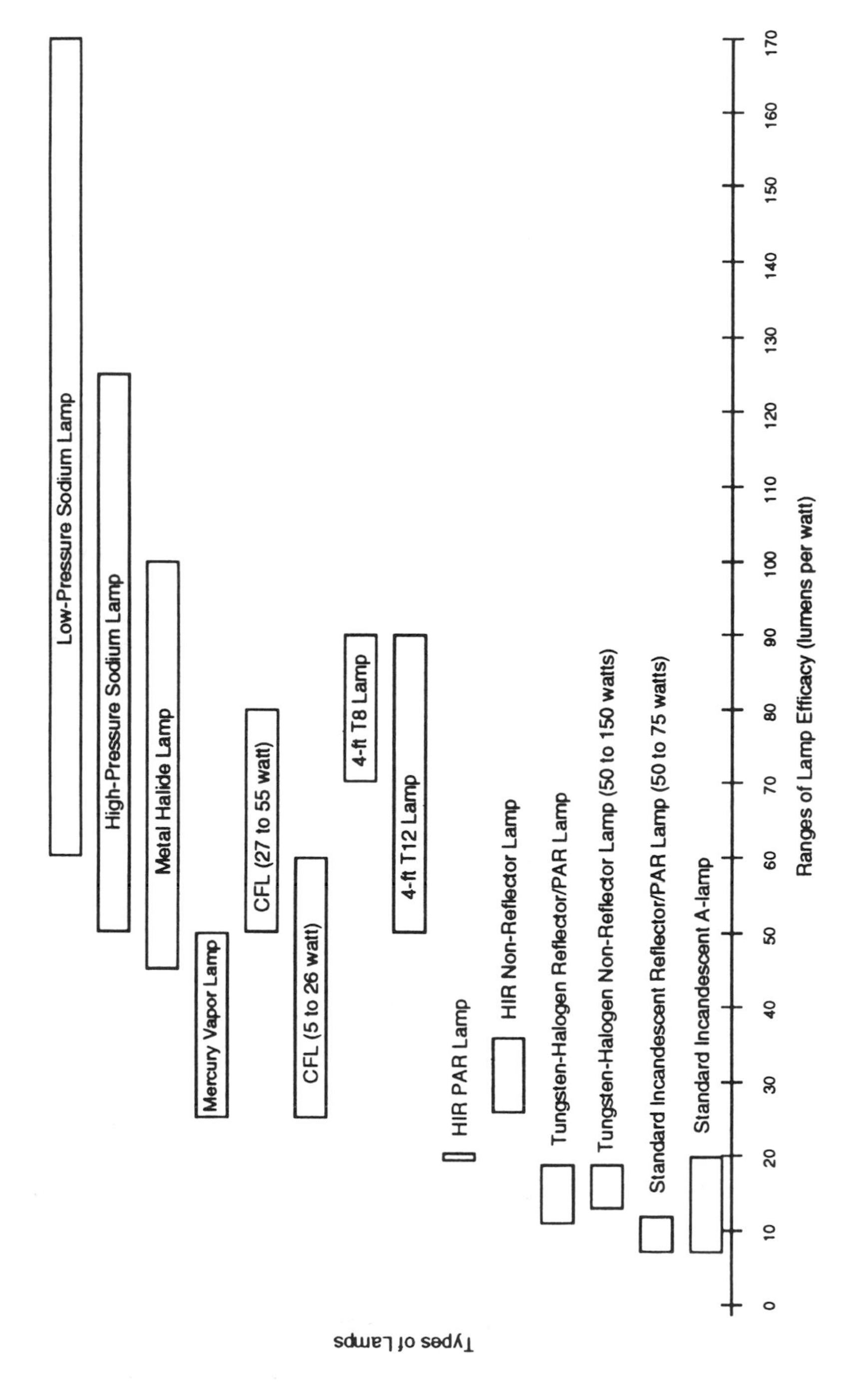

FIGURE 12B.1 Ranges of lamp efficacy. (Ballast losses are included for all discharge lamps.) *Sources:* U.S. DOE, 1993a; Morse, 1994; and manufacturer's catalogs.

a blackbody radiator. A lamp's CRI indicates the difference in the perceived color of objects viewed under the lamp and under the reference source. There are 14 differently colored test samples, 8 of which are used in the calculation of the general CRI index.

The CRI is measured on a scale that has a maximum value of 100 and is an average of the results for the 8 colors observed. A CRI of 100 indicates that there is no difference in perceived color for any of the test objects; a lower value indicates that there are differences. CRIs of 70 and above are generally considered to be good, while CRIs of 20 and below are considered to be quite poor. Most incandescent lamps have CRIs equal to or approaching 100. Low-pressure sodium lamps have the lowest CRI of any common lighting source; their light is essentially monochromatic.

The optical controllability of a lamp describes the extent to which a user can direct the light of the lamp to the area where it is desired. Optical controllability depends on the size of the light-emitting area, which determines the beam spread of the light emitted. In addition, controllability depends on the fixture in which the lamp is used. Incandescent lamps emit light from a small filament area: they are almost point sources of light, and their optical controllability is excellent. In contrast, fluorescent lamps emit light from their entire phosphored area: their light is extremely diffuse, and their controllability is poor.

Because of the many different characteristics and the variety of applications by which a lamp can be judged, no one type of source dominates the lighting market. The types of lamps that are commonly available include incandescent, fluorescent, and high-intensity discharge (HID).

Incandescent Lamps

The *incandescent lamp* was invented independently by Thomas Edison in the United States and Joseph Swan in England in the late 1800s. An incandescent lamp produces light when electricity heats the lamp filament to the point of incandescence. In modern lamps the filament is made of tungsten. Because 90% or more of an incandescent's emissions are in the infrared (thermal) rather than the visible range of the electromagnetic spectrum, incandescent lamps are less efficacious than other types of lamps.

The two primary types of standard incandescent lamps are general service and reflector/PAR (parabolic aluminized reflector) lamps. General-service lamps (also known as "A-lamps) are the pear-shaped, common household lamps (see Figure 12B.2). Reflector lamps, such as flood or spot lights, are generally used to illuminate outdoor areas or highlight indoor retail displays and artwork. They are also commonly used to improve the optical efficiency of downlights (discussed later). Downlights are used where controlling glare or hiding the light source is important.

In spite of the fact that they are the least efficacious lamps on the market today, standard incandescent lamps are used for almost all residential lighting in the United States and are also common in the commercial sector. Although the prevalence of incandescent lamps in the residential sector may be partially due to historical precedence and inertia, these lamps do have advantages that to some extent counterbalance their relatively poor efficacies: they have excellent CRIs and a warm color; they are easily dimmed, inexpensive, small, lightweight, and can be used with inexpensive fixtures, and, in a properly designed fixture, they permit excellent optical control. In addition, incandescent lamps make no annoying noises, emit little or no harmful radiation, and contain essentially no toxic chemicals. They are simple to install, maintain, and dispose of. The characteristics and applications of A-lamps and reflector lamps are summarized in Tables 12B.3 and 12B.4.

Although they account for less than 3% of the incandescent market, *tungsten-halogen* and now *tungsten-halogen infrared-reflecting lamps* are the incandescents that offer the greatest opportunity for energy savings. Practical halogen lamps were developed in the late 1950s, and, especially since the late 1980s, their popularity has increased dramatically. According to U.S. census data, production of general-use halogen lamps more than doubled between 1984 and 1993 (from 3.9 million to 8.6 million lamps). Halogen lamps produce bright white light and have color temperatures and CRIs that are similar to, or slightly higher than, those of standard incandescents. In addition, they have longer lives, can be much more compact, are slightly more efficacious, and have better lumen maintenance than standard incandescent lamps.

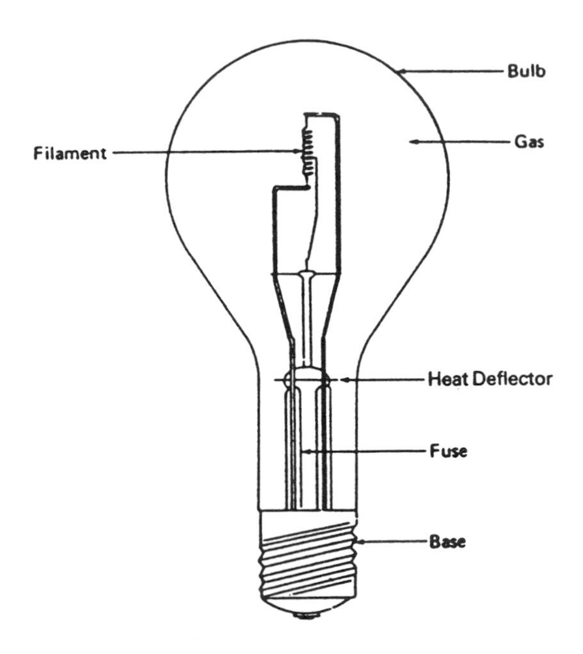

FIGURE 12B.2 Typical incandescent A-lamp. *Source:* Atkinson et al., 1992.

TABLE 12B.3 Characteristics and Applications of Standard Incandescent A-Lamps

Available wattages	15 to 250 watts
Efficacy	10 to 19 LPW (for lamps of 40 to 250 watts)
Rated lifetime	750 to 2,500 hours
Color rendition	Excellent, CRI ≈ 100
Color temperature	Warm, 2,500 to 3,000 K
Lumen maintenance	Very good (light output typically declines by 10 to 20% over rated lamp life)
Optical controllability	Excellent (point source)
Typical uses	In the U.S., standard incandescent lamps are used for most residential lighting; they are also common in commercial establishments such as restaurants and lodging where they are used, in part, to make the customer feel at home. In retail establishments, incandescents are often used for display lighting.
Technologies for which these lamps are energy-efficient alternatives	Not applicable
Notes	These lamps are dimmable. When dimmed even slightly, the life of an incandescent lamp increases dramatically.

Like standard incandescent lamps, tungsten-halogen lamps produce light when electricity heats the tungsten filament to the point of incandescence. In a standard incandescent lamp, tungsten evaporating from the filament deposits on the glass envelope. Generally, tungsten-halogen lamps use a quartz envelope rather than a glass envelope, which allows the lamp to operate at a much higher temperature. In place of the normal inert gas fill, the tungsten-halogen lamps use a small amount of halogen gas. The halogen gas in the lamp reacts with the tungsten that deposits on the quartz envelope to make a volatile tungsten-halide compound; because tungsten-halide vapor is not stable at the temperature of the filament, the vapor dissociates and deposits the tungsten back onto the filament. The cycle is then repeated. This cycle does not necessarily return the tungsten to the same portion of the filament from which it evaporated, but it does substantially reduce net evaporation of tungsten and thus prolong the life of the filament.

Both the efficacy and the lifetime of an incandescent lamp depend upon the temperature at which the filament is operated. Increasing the filament temperature increases the amount of energy radiated

TABLE 12B.4 Characteristics and Applications of Standard Incandescent Reflector/PAR Lamps

Available wattages	15 to 500 watts
Efficacy	8 to 12 LPW (for lamps of 50 to 75 watts)
Rated lifetime	Typically 2,000 hours
Color rendition	Excellent, CRI ≈ 100
Color temperature	Warm, 2,500 to 3,000 K
Lumen maintenance	Very good (light output typically declines by 10 to 20% over rated lamp life)
Optical controllability	Excellent (point source)
Typical uses	Floodlights and spotlights are generally used to illuminate outdoor areas or to highlight retail displays and artwork. They are also used in downlights in the commercial sector.
Technologies for which these lamps are energy-efficient alternatives	Not applicable
Notes	These lamps are dimmable. When dimmed even slightly, the life of an incandescent lamp increases dramatically.

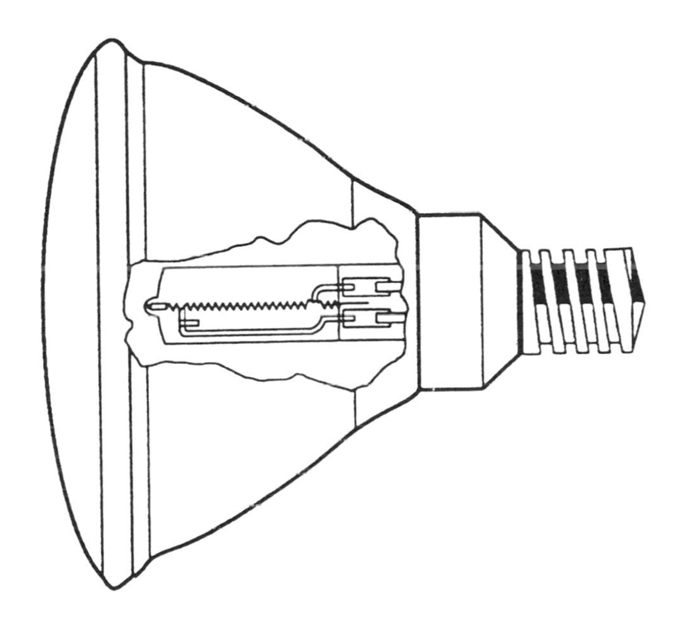

FIGURE 12B.3 Cross section showing a tungsten-halogen capsule within a reflector lamp. *Source:* U.S. DOE, 1993a.

as visible light and thus improves lamp efficacy. At the same time, increasing the temperature increases the evaporation rate of the filament and thus reduces expected lamp life. In the tungsten-halogen lamp, the net reduction of evaporation due to the halogen cycle is usually partially offset by running the filaments slightly hotter than in a standard incandescent lamp, so most tungsten-halogen lamps are designed to have slightly better efficacies and lifetimes than comparable standard lamps.

Tungsten-halogen lamps meant for general service are referred to as "capsule" lamps and have two envelopes: an inner quartz capsule that is the actual halogen lamp and an outer glass envelope that protects the user from the very high operating temperatures of the quartz envelope. Figure 12B.3 shows a cross section of a tungsten-halogen capsule lamp. Tungsten-halogen lamps are efficacious replacements for incandescents where lamp use is moderate and the other features of an incandescent (e.g., optical control, dimmability, color) are desired. The characteristics and applications of tungsten-halogen capsule lamps are summarized in Table 12B.5.

TABLE 12B.5 Characteristics and Applications of Tungsten-Halogen (T-H) Capsule Lamps

Available wattages	Nonreflector lamps: 50 to 100 watts; Reflector/PAR lamps: 35 to 1000 watts
Efficacy	Nonreflector lamps: 17 to 19 LPW; Reflector/PAR lamps: 11 to 19 LPW
Rated lifetime	Nonreflector lamps: 2,000 hours; Reflector/PAR lamps: 2,000 to 6,000 hours
Color rendition	Excellent, CRI ≈ 100
Color temperature	Warm, 2,850–3,050 K
Lumen maintenance	Excellent (light output typically declines by less than 10% over rated lamp life)
Optical controllability	Excellent (point source)
Typical uses	Both reflector and nonreflector T-H lamps are popular with lighting designers in the residential and commercial sectors. Generally, T-H reflector/PAR lamps are used in the same way that standard incandescent reflector/PAR lamps are used. Floodlights and spotlights are typically used to illuminate outdoor areas or to highlight retail displays and artwork.
Technologies for which these lamps are energy-efficient alternatives	Lower-wattage T-H reflector/PAR lamps can be used to replace standard incandescent reflector/PAR lamps (the light output of a 90-watt T-H reflector is comparable to that of a 120-watt standard incandescent reflector lamp). The nonreflector T-H lamps can be used to replace standard incandescents where lamp use is moderate and the other features of an incandescent (e.g., optical control, dimmability, color) are desired.
Notes	T-H lamps are dimmable. Like standard incandescents, when dimmed even slightly, the life of a T-H lamp increases dramatically. However, it is recommended to periodically run T-H lamps at full power to clean the tungsten from the bulb wall.

TABLE 12B.6 Characteristics and Applications of Tungsten-Halogen Infrared-Reflecting (HIR) PAR Lamps

Available wattages	50 to 100 watts for PAR lamps
Efficacy	≈20 LPW for 50- and 60-watt PAR lamps
Rated lifetime	3,000 hours for 50- and 60-watt PAR lamps
Color rendition	Excellent, CRI ≈ 100
Color temperature	Warm, 2,850 to 2,925 K
Lumen maintenance	Excellent (light output typically declines by less than 10% over rated lamp life)
Optical controllability	Excellent (point source)
Typical uses	HIR PAR lamps are used in the same way that standard incandescent reflector/PAR lamps are used. Floodlights and spotlights are typically used to illuminate outdoor areas or to highlight retail displays and artwork.
Technologies for which these lamps are energy-efficient alternatives	Like the T-H reflector/PAR lamps, HIR lamps can be used to replace standard incandescent reflector/PAR lamps. HIRs are the most efficacious PAR lamps on the market.
Notes	High-wattage (350- and 900-watt) nonreflector HIRs are also available for highly specialized uses (such as stage lighting). They have efficacies ranging from 26 to 36 LPW. HIR lamps are dimmable. Like standard incandescents, when dimmed even slightly, the life of a T-H lamp increases dramatically. However, it is recommended to periodically run T-H lamps at full power to clean the tungsten from the bulb wall.

Even more efficacious than the standard tungsten-halogen lamp is the tungsten-halogen infrared-reflecting (HIR) lamp. Because approximately 90% of the energy radiated by incandescent lamps is in the form of heat (infrared radiation), their efficacy can be improved by reflecting the infrared portion of the spectrum back onto the lamp filament. HIR lamps use a selective, reflective, thin-film coating on the halogen-filled capsule or on the reflector surface. The coating transmits visible light but reflects much of the infrared radiation back to the filament, so it takes less electricity to heat it. HIR lamps have been available for a number of years as high-wattage double-ended quartz lamps, and are only recently (1994) becoming widely available as PAR lamps. The characteristics and applications of HIR PAR lamps are summarized in Table 12B.6.

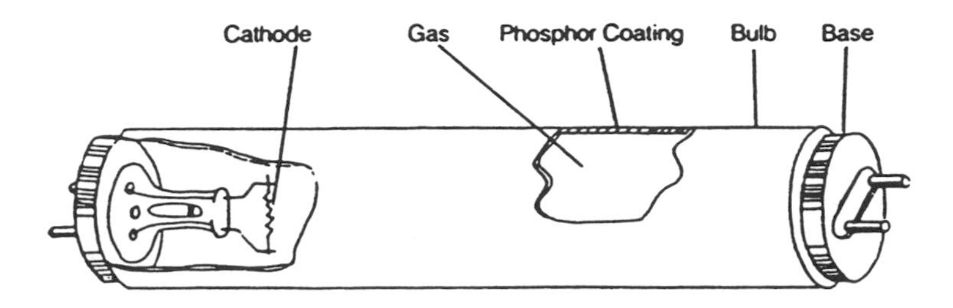

FIGURE 12B.4 Typical full-size fluorescent lamp. *Source:* Atkinson et al., 1992.

HIR lamps have been promoted to residential- and commercial-sector customers primarily as low-wattage reflector lamps. In general, HIR lamps have a small market share due to their high cost. Researchers are working to further improve the efficacy of HIR technology. General-service HIR lamps have been developed as prototypes but are not yet commercially available.

It is expected that tungsten-halogen and HIR reflector lamps will rapidly gain a large share of the market starting in November 1995, when the U.S. Energy Policy Act (EPACT) will increase the minimum lamp efficacies required for all incandescent reflector lamps (spotlights and floodlights). The minimum required efficacies will range from 10.5 LPW (for lamps of 40 to 50 watts) to 15 LPW (for lamps of 156 to 205 watts). Standard incandescent reflector lamps (50 to 75 watts) have efficacies of only 8 to 12 LPW, and altering their design to meet the EPACT requirements could result in lamps with only half of their current operating lifetimes.

Full-Size Fluorescent Lamps

The scientific principles of fluorescent lighting were understood by the mid-1800s, and the first practical fluorescent lamps were produced in the United States in the late 1930s. *Fluorescent lamps* came into general use in the 1950s. When a rapid-start version was introduced in the 1960s, their popularity increased dramatically.* In a fluorescent lamp, gaseous mercury atoms within a phosphor-coated lamp tube are excited by an electric discharge. As the mercury atoms return to their ground state, ultraviolet radiation is emitted. This UV radiation excites the phosphor coating on the lamp tube and causes it to fluoresce, thus producing visible light. Fluorescent lamps are far more efficacious than incandescent lamps. The efficacy of a fluorescent lamp system depends upon the lamp length and diameter, the type of phosphor used to coat the lamp, the type of **ballast** used to drive the lamp, the number of lamps per ballast, the temperature of the lamp (which depends on the fixture and its environment), and a number of lesser factors.

Fluorescent lamps have long lives and fairly good lumen maintenance. While the standard-phosphor [cool-white (CW) and warm-white (WW)] lamps have CRIs of 50 to 60, the new rare earth phosphor lamps have CRIs of 70 to 80. The majority of lighting used in the commercial sector is fluorescent. Fluorescent lighting is also common in the industrial sector. The small amount of full-size fluorescent lighting in the residential sector is primarily found in kitchens, bathrooms, and garages. The most common fluorescent lamps are tubular and 4 feet (1.2 meters) in length (see Figure 12B.4).

Lamp tubes with a diameter of 1.5 inches (38 mm) are called T12s, and tubes that are 1 inch (26 mm) in diameter are called T8s. The 8 and 12 refer to the number of eighths of an inch in the diameter of the lamp tube. Lamp tubes are available in other diameters as well. Four-foot T12s are available in 32, 34, and 40 watts. The specified wattage of a lamp refers to the power draw of the lamp alone, even if lamp operation requires a ballast, as do all fluorescent and high-intensity discharge light sources. The ballast typically adds another 10 to 20% to the power draw, thus reducing system efficacy. Ballasts are discussed in greater detail later. The characteristics and applications of 4-foot, standard phosphor T12 lamps are summarized in Table 12B.7.

*In a rapid-start fluorescent lamp, start time is nearly instantaneous and system reliability is improved.

TABLE 12B.7 Characteristics and Applications of 4-Foot Full-Size Fluorescent T12 Lamps

Available wattages	32, 34, 40 watts
Efficacy	Combined with a high-efficiency magnetic ballast, 60 to 70 LPW
Rated lifetime	≈20,000 hours
Color rendition	50 to 90 (50 to 60 for CW and WW lamps)
Color temperature	3,000–7,500 K (typically, 3,000–4,150 K for CW and WW lamps)
Lumen maintenance	Fairly good (light output typically declines by 20 to 25% over rated lamp life)
Optical controllability	Poor (very diffuse light)
Typical uses	The majority of lighting used in the commercial sector is fluorescent. Fluorescent lighting is also used in the industrial sector. In the residential sector, full-size fluorescent lighting is used in some kitchens, bathrooms, and garages.
Technologies for which these lamps are energy-efficient alternatives	These lamps are good replacements for incandescent lamps providing general ambient light where the color of the light is not a high priority and a point source of light is not needed (e.g., garages, storage areas)
Notes	Rare earth phosphor T12s are also available; their CRIs resemble those of the T8s described in Table 12B.8.

Even the smallest, least efficacious fluorescent lamps (≈30 LPW for a 4-watt lamp) are more efficacious than the most efficacious incandescent lamps (≈20 LPW). Of the full-size fluorescent lamps available today, *rare earth phosphor lamps* are the most efficacious. In these lamps, rare earth phosphor compounds are used to coat the inside of the fluorescent lamp tube. Rare earth phosphor lamps are also called triphosphor lamps because they are made with a mixture of three rare earth phosphors that produce visible light of the wavelengths to which the red, green, and blue retinal sensors of the human eye are most sensitive. These lamps have improved color rendition as well as efficacy. Fluorescent lamps with diameters of 1 inch (26 mm) and smaller use triphosphors almost exclusively. Rare earth coatings can also be used for lamps of larger diameter.

The most efficacious of the fluorescent lamps available today are T8 lamps operating with electronic ballasts. The efficacy of two 32-watt T8 lamps operating with a single electronic ballast is about 90 LPW, approximately 30% more efficacious than the more standard lighting system consisting of two 40-watt T12 lamps and a high-efficiency magnetic ballast. T12 lamps are also available with rare earth phosphors, and can attain over 80 LPW when used with a two-lamp electronic ballast. The characteristics and applications of 4-foot T8 lamps are summarized in Table 12B.8.

Certain 8-foot (2.4-meter), standard-wattage, standard-phosphor T12s were outlawed by the Energy Policy Act in 1994. When 40-watt standard-phosphor T12s are outlawed in November 1995, many of them are likely to be replaced with rare earth phosphor lamps (both T8s and T12s). T8 lamps are rapidly gaining market share and can be effectively used to replace less efficacious lamps in areas where maintaining lighting levels and improving lighting quality are a priority.* Newer building designs may benefit from the new higher-efficacy rare earth phosphor lamps. Lower-wattage T12 lamps can be used in place of the banned T12s and will provide the same color and about the same efficacy, at a somewhat lower wattage and light level. These lamps can be effectively used to replace the T12s in older buildings, which are often overlit by today's lighting standards.

In spite of their much greater efficacy, fluorescent lamps have several disadvantages when compared to incandescent lamps. Fluorescent lamps can be dimmed, but only with special equipment that costs much more than the dimming controls used for incandescent lamps. Standard fluorescent lamps are much bigger than incandescent lamps of equivalent output and are much harder to control optically. In addition, fluorescent lamps contain trace amounts of mercury, a toxic metal, and emit more ultraviolet light than incandescent lamps. The ballast equipment that drives the lamps is sometimes noisy and may emit radio interference.

*Between 1992 and 1993, production of T8s increased by almost 60% and the percentage of full-size fluorescents that is T8s rose from 5% to 8% (Audin et al., 1994). See the commercial retrofit example in Section 12B.6.

TABLE 12B.8 Characteristics and Applications of 4-Foot Full-Size Fluorescent T8 Lamps

Available wattages	32 watts
Efficacy	For two T8s and a single electronic ballast, ≈90 LPW
Rated lifetime	Typically 12,000 to 20,000 hours
Color rendition	Typically 75–84
Color temperature	2,800–7,500 K
Lumen maintenance	Very good (light output typically declines by 10 to 12% over rated lamp life)
Optical controllability	Poor (very diffuse light but slightly better than a T12)
Typical uses	The typical uses of T8s are the same as the uses of T12s. The majority of lighting used in the commercial sector is fluorescent. Fluorescent lighting is also used in the industrial sector. In the residential sector, full-size fluorescent lighting is used in some kitchens, bathrooms, and garages.
Technologies for which these lamps are energy-efficient alternatives	These lamps are most often replacements for less efficient fluorescents (T12s).
Notes	Some T8 lamps are now available with CRIs above 90, especially those with high color temperature; however, these lamps are less efficacious than lamps with lower CRIs.

Although standard-phosphor (halophosphor) fluorescent lamps are only slightly more expensive than incandescent lamps with equivalent lumens, the newer rare earth phosphor lamps are considerably more extensive at this time.

Circular and Compact Fluorescent Lamps:

Circular fluorescent lamps in 20- to 40-watt sizes have been available for many years, but have always had a fairly small market. Essentially, a circular lamp is a standard fluorescent lamp tube (as described earlier) that has been bent into a circle. Although they have a more compact geometry than a straight tube, circular lamps are still moderately large (16.5 to 41 cm in diameter). *Compact fluorescent lamps* (CFLs), which are substantially smaller than standard fluorescent lamps, were developed in the late 1970s and introduced to the U.S. market in the early 1980s. In a CFL, the lamp tube is smaller in diameter and is bent into two to six sections (see Figure 12B.5). CFLs have much higher power densities per phosphor area than standard fluorescents, and their design was therefore dependent on the development of rare earth phosphors, which could hold up much better than standard phosphors at high power loadings. CFLs are available as both screw-in replacements for incandescent lamps and as pin-base lamps for hard-wired fixtures. They may be operated with separate ballasts or purchased as integral lamp/ballast units.

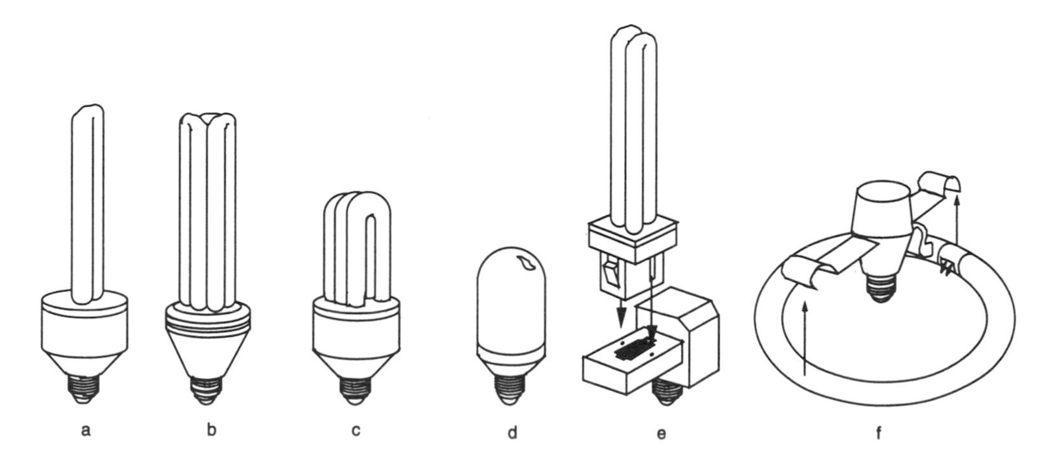

FIGURE 12B.5 A variety of compact fluorescent and circular lamps: (a) twin-tube integral, (b and c) triple-tube integrals, (d) integral model with glare-reducing casing, (e) modular quad tube and ballast, and (f) modular circline and ballast. *Source:* Katherine Falk, courtesy of *Home Energy* magazine, 1995.

TABLE 12B.9 Characteristics and Applications of Compact Fluorescent Lamps (CFL)

Available wattages	5 to 55 watts
Efficacy	Combined with an electronic ballast, 60 to 70 LPW; less when combined with a magnetic ballast
Rated lifetime	Typically 9,000–12,000 hours
Color rendition	Typically 82 to 85
Color temperature	≈2,700 to 5,000K
Lumen maintenance	Light output typically declines by 10 to 30% over the rated lamp life (the larger the lamp, the less lumen depreciation)
Optical controllability	Fair
Typical uses	In the residential sector, CFLs are sometimes used in place of incandescents. In the commercial sector, where more fixtures are hard-wired for CFLs, they are used in downlights, wall sconces, exit signs, and 2-foot by 2-foot troffers.
Technologies for which these lamps are energy-efficient alternatives	On average, a CFL consumes about one-third to one-fourth of the energy consumed by an incandescent lamp to produce an equal number of lumens, and has a lamp life up to 10 times as long. In the residential sector, screw-in CFLs may be substituted for incandescents in porch lights, table lamps, floor lamps, wall sconces, and ceiling fixtures. In the commercial sector, CFLs can be useful for replacing incandescents in downlights, wall sconces, storage rooms, and closets.

CFLs are much more efficacious than the incandescent lamps that they are often used to replace. On average, a CFL consumes only about one-third to one-fourth of the energy consumed by an incandescent lamp and has a lamp life up to 10 times longer. They are available in a variety of sizes, shapes, and wattages, and typically have CRIs in the 80s. A CFL with an electronic ballast may have an efficacy of 60 or even 70 LPW. Initially, the large size of CFLs relative to incandescent bulbs limited their use, especially in the retrofit of existing household fixtures. Although CFLs are now close in size to incandescent lamps and provide better optical control than full-size fluorescents, they are not point sources and cannot be used where very fine optical control is required. The characteristics and applications of CFLs are summarized in Table 12B.9.

The number of CFLs sold in the United States has increased significantly in recent years. U.S. sales of CFLs increased from approximately 200,000 units in 1982 (Audin et al., 1994) to more than 33 million units in 1993. CFLs have been used in the commercial sector for more than 10 years. Initially, they were used to replace incandescent bulbs in fixtures that were designed for incandescent lamps. Over time, however, more and more fixture manufacturers have begun to design commercial-sector fixtures specifically for CFLs. In the commercial sector, where more fixtures are hard-wired for CFLs, they are common in downlight fixtures, wall sconces, exit signs, and 2-foot by 2-foot troffers.

Although utility incentives and improved dedicated fixtures are increasing the CFL market share in the residential sector, CFLs have been slow to gain popularity in the average household. The lack of CFLs used in households is at least partially attributable to their limited availability in local supermarkets and hardware stores, their high price, and consumer uncertainty about how CFL watts and lumens correspond to the watts and lumens of a standard incandescent lamp.* In the residential sector, screw-in CFLs may be substituted for incandescents in table lamps, floor lamps, ceiling fixtures, and porch lights (CFL use in porch lights is primarily a southern states application because some CFLs are not able to start at low temperatures). CFL fixtures for the residential sector are becoming available in local lighting stores and in home improvement chains.

CFLs are available with integrated or modular electronic ballasts. Although electronic ballasts are more expensive than magnetic ballasts, a lamp used with an electronic ballast is generally 20% more efficacious than a lamp used with a magnetic ballast. In addition, electronic ballasts start more quickly, weigh less, make less noise, have no flicker, and can be made much smaller than magnetic ballasts. Because the life of the ballast is longer than the life of a lamp, CFLs with integrated

*Typically, multiplying the watts of a CFL lamp by three or four will give the approximate wattage of its incandescent counterpart.

ballasts (which must be disposed of at the end of the lamp life) are more expensive than modular units. Recently, modular units have become readily available. As CFLs with electronic ballasts become less expensive, shorter in length, and higher in lumen output, they should become more attractive to the residential consumer.

Some hard-wired, electronic-ballast CFLs are now manufactured with *integrated dimming ballasts.* CFLs with dimming ballasts can be more energy efficient because they can be controlled by daylight sensors, occupancy sensors, and manual dimmers. Introduction of dimmable CFL ballasts to the residential sector will allow the replacement of more incandescent lamps with CFLs.

High-Intensity Discharge Lamps

Like fluorescent lamps, *high-intensity discharge (HID) lamps* produce light by discharging an electrical arc through a mixture of gases. In contrast to fluorescent lamps, HID lamps use a compact "arc tube" in which both temperature and pressure are very high. Compared to a fluorescent lamp, the arc tube in an HID lamp is small enough to permit compact reflector designs with good light control. Consequently, HID lamps are both compact and powerful. There are currently three common types of HID lamps available: mercury vapor (MV), metal halide (MH), and high-pressure sodium (HPS).

Because of their very high light levels, except in the smallest lamps, and their substantial cost ($15 to $80 for lamps up to 400 watts), HID lamps are most often used for exterior applications such as street lighting and commercial, industrial, and residential floodlighting (sports and security lighting). They are also used in large, high-ceilinged, interior spaces such as industrial facilities and warehouses, where good color is not typically a priority. Occasionally, HID lamps are used for indirect lighting in commercial offices. Interior residential applications are rare because of high cost, high light level, and the fact that HID lamps take several minutes to warm up to full light output. If there is a momentary power outage, the lamps must cool down before they will restrike. Some HID lamps are now available with dual arc tubes or parallel filaments. Dual arc tubes eliminate the restrike problem and a parallel filament gives instantaneous light output both initially and on restrike, but at a cost of a high initial power draw.

The *mercury vapor lamp* was the first HID lamp developed. Including ballast losses, the efficacies of MV lamps range from about 25 to 50 LPW. Uncoated lamps have a bluish tint and very poor color rendering (CRI ≈ 15). Phosphor-coated lamps emit more red, but are still bluish, and have a CRI of about 50. Because of their poor color rendition, these lamps are only used where good color is not a priority. Although MV lamps generally have rated lifetimes in excess of 24,000 hours, light output can diminish significantly after only 12,000 hours (Audin et al., 1994). The characteristics and applications of MV lamps are summarized in Table 12B.10. Both metal halide and high-pressure sodium HID lamps have higher efficacies than mercury vapor lamps and have consequently replaced them in most markets (see Figure 12B.6).

Including ballast losses, *metal halide lamps* range in efficacy from 46 to 100 LPW. They produce a white to slightly bluish-white light and have CRIs ranging from 65 to 70. Lamp lifetimes range from only 3,500 hours to 20,000 hours, depending on the type of MH lamp. Lower-wattage metal halides (particularly the 70-watt and 100-watt) are now available with CRIs of 65 to 75 and color temperatures of 3,200 K to 4,200 K. Good lumen maintenance, longer life, reduced maintenance costs, and the fact that they blend more naturally with fluorescent sources have made MH lamps a very good replacement in the commercial sector for 300-watt and 500-watt PAR lamps. New fixtures utilizing these lamps, particularly 1-foot by 1-foot recessed lensed troffers (downlights), are becoming common in lobbies, shopping malls, and retail stores. The characteristics and applications of MH lamps are summarized in Table 12B.11.

Including ballast losses, *high-pressure sodium lamps* have efficacies ranging from 50 LPW for the smallest lamps to 124 LPW for the largest lamps. Standard HPS lamps emit a yellow light and have very poor color rendition. Like MV lamps, HPS lamps are only used where good color is not a priority. The rated lifetimes of HPS lamps rival those of MV lamps and typically exceed 24,000 hours. The characteristics and applications of HPS lamps are summarized in Table 12B.12.

TABLE 12B.10 Characteristics and Applications of Mercury Vapor (MV) Lamps

Available wattages	40 to 1,000 watts
Efficacy	Including ballast losses, MV efficacies range from about 25 LPW for the smallest sizes (50 to 100 watts) to 50 LPW for the largest sizes (400 to 1,000 watts)
Rated lifetime	Typically more than 24,000 hours (see lumen maintenance below)
Color rendition	Uncoated, CRI ≈ 15; phosphor-coated, CRI ≈ 50. These lamps produce a bluish light.
Color temperature	Uncoated, ≈5,700 K; phosphor-coated, 3,300 K to 3,900 K
Lumen maintenance	In spite of their very long lifetimes, light output typically declines by 25 to 40% after only 12,000 hours
Optical controllability	Fair
Typical uses	For many years, mercury vapor lamps were the most cost-effective light source for many industrial and outdoor uses where color was not a priority. Although many building owners have retrofitted to the more efficacious MH and HPS lamps, MV lamps are still in use because of their relatively low cost and the fact that the switchover to another type of HID lamp requires installation of new ballasts (and sometimes new fixtures). Because of their low cost, home improvement centers often promote MV lamps for residential dusk-to-dawn yard lighting.
Technologies for which these lamps are energy-efficient alternatives	Because they bring out the greens in foliage, unlike other HID sources, MV lamps are most useful for replacing incandescents in landscape lighting. For security lighting, other HID sources are recommended, since they are much more efficacious.
Notes	Both metal halide and high-pressure sodium HID lamps have higher efficacies than mercury vapor lamps and have consequently replaced them in most markets

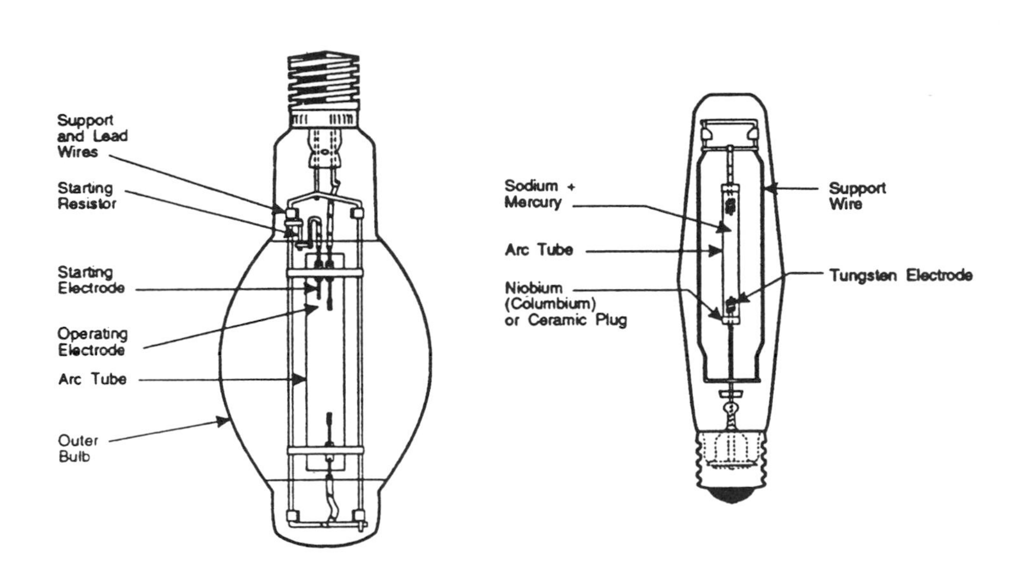

FIGURE 12B.6 High-intensity discharge lamps: metal halide (left) and high-pressure sodium (right). *Source:* U.S. DOE, 1993a.

Ongoing improvements in HID technologies include introduction of high-CRI, white HPS lamps and low-wattage (32 to 35 watts) HPS and MH lamps, as well as the gradual improvement in rated lifetimes for all HID products. These improvements will increase the number of interior applications for HID lamps.

Low-Pressure Sodium Lamps

A *low-pressure sodium (LPS) lamp* produces light by discharging an electrical arc through vaporized sodium. The arc tube is typically enclosed in an outer bulb at a high vacuum or in a vacuum flask. LPS lamps have rated lamp lives ranging from 14,000 to 18,000 hours. Because they require more power as they age, their light output is almost constant over the lamp life.

TABLE 12B.11 Characteristics and Applications of Metal Halide (MH) Lamps

Available wattages	32 to 1,500 watts
Efficacy	Including ballast losses, MH efficacies range from 46 to 100 LPW. The 1,000-watt MH lamp is the most efficacious.
Rate lifetime	3,500 hours for compact arc lamps; 5,000 to 10,000 hours for the smallest and largest lamps; and on up to 20,000 hours for the 400-watt lamps
Color rendition	CRIs range from 65 to 70
Color temperature	3,000 to 4,400 K; MH lamps produce a white to slightly bluish-white light and tend to change color slightly as they age
Lumen maintenance	Light output typically declines by as much as 20% after 12,000 hours
Optical controllability	Good (better in the lower-wattage lamps)
Typical uses	MH lamps were developed in the 1960s for industrial and outdoor lighting. They are now suitable for many applications. Higher-wattage lamps are used in gymnasiums and for exterior security lighting. Low-wattage MH lamps are becoming increasingly common in lobbies, shopping malls and retail stores.
Technologies for which these lamps are energy-efficient alternatives	MH lamps can be used to replace the less efficacious MV lamps. They are also useful for replacing incandescents, especially where ceiling heights are more than 10 feet.
Notes	MH lamps are sensitive to operating position. Many lamp models have substantially shorter lifetimes and poorer efficacies when operated in a horizontal, rather than vertical, orientation.

TABLE 12B.12 Characteristics and Applications of High-Pressure Sodium (HPS) Lamps

Available wattages	35 to 1,000 watts
Efficacy	Including ballast losses, high-pressure sodium lamps have efficacies ranging from 50 LPW for the smallest lamps to 124 LPW for the largest lamps. The 35- and 70-watt reflector lamps are approximately two-thirds as efficacious as the nonreflector lamps.
Rated lifetime	The rated lifetimes of HPS lamps rival those of MV lamps, and typically exceed 24,000 hours.
Color rendition	CRI ≈ 22
Color temperature	1,900 to 2,100 K; standard HPS lamps emit a yellow light
Lumen maintenance	Of the HID lamps, HPS lamps have the best lumen maintenance; light output declines by approximately 20% after 18,000 hours.
Optical controllability	Good
Typical uses	HPS lamps were developed in the late 1960s and promoted as an efficacious light source for outdoor, security, and industrial lighting. Most street lighting today is done with HPS lamps. HPS lamps are also used for outdoor lighting and in large, high-ceilinged, interior spaces such as industrial facilities and warehouses
Technologies for which these lamps are energy-efficient alternatives	HPS lamps can be used to replace the less efficacious MV lamps. HPS lamps are most useful in the industrial sector and out of doors, where high light levels are often needed and good color rendering is not a priority.
Notes	New higher-pressure HPS lamps are now available with color temperatures ranging from 2,200 to 2,800 K, and CRIs of 65 to more than 70. These lamps have lower efficacies (35 to 85 LPW) and lower rated lifetimes (10,000 to 15,000 hours) than the standard HPS lamps. Color rendering in these higher-CRI lamps may deteriorate over time.

This is the most efficacious light source on the market—with a magnetic ballast, the 180-watt lamp has an efficacy of approximately 140 LPW. However, the use of LPS lamps is extremely limited because their light is monochromatic (yellow), making them inappropriate for situations in which even moderate color rendition is needed. In addition, LPS lamps are very large and thus hard to control optically.

One situation in which LPS lamps are highly useful is as street lighting in areas where astronomical observatories are located. The monochromatic light is simpler to filter out and allows observatory equipment to be used more effectively. Because of its very high efficacy, LPS lighting would seem to be ideal for outdoor uses such as street lighting. However, for security reasons,

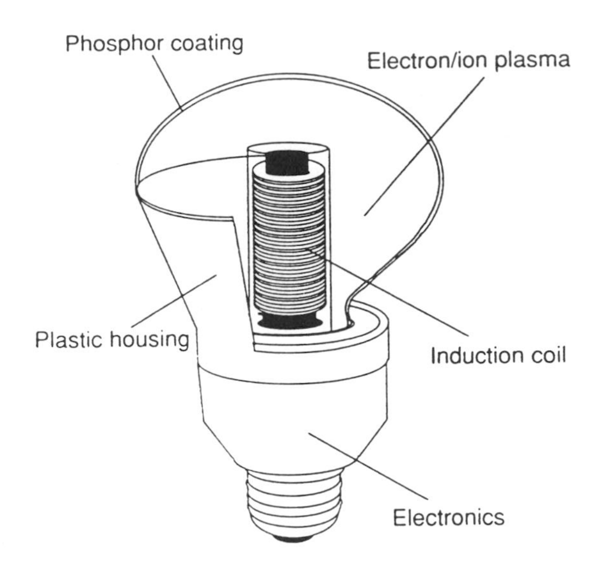

FIGURE 12B.7 Electrodeless induction lamp. *Source:* Katherine Falk, courtesy of *Home Energy* magazine, 1995.

many cities do not permit LPS street lighting (Audin et al., 1994). The monochromatic light makes it very difficult to discern the color of cars, people's clothing, and other objects.

New Developments in Efficient Lamps

Electrodeless Induction Lamp. The *electrodeless induction lamp* (see Figure 12B.7) was introduced to the U.S. market in 1994 and is a modification of a standard compact fluorescent lamp. In an induction lamp, the power supply converts ordinary 60 Hz current into radio-frequency power that is fed into an electrical coil. The coil excites a gas plasma inside the bulb, releasing UV radiation that strikes the rare earth phosphor coating of the bulb and is converted into visible light. Although lamp efficacy is only about 50 LPW, the induction lamp has a very long lamp life (potentially, up to 60,000 hours). Because the lamp has no filament, lamp life is limited only by the degradation of the lamp phosphors. The induction lamp has a CRI of approximately 82 and a color temperature of 3,000 K.

The very long life of induction lamps could make them extremely useful in many applications where maintenance is expensive and disruptive. In addition, because of their small size, reflector induction lamps permit more precise optical control than standard CFLs. In the United States, lower-wattage induction lamps are presently available only in reflector versions. Currently, they are produced by only one manufacturer and are intended to replace 75-watt incandescent reflector lamps in the commercial sector. They cost approximately $20 per lamp. Induction lamps will be used primarily in the commercial and industrial sectors in the near-term.

Sulfur Lamp. The *sulfur lamp* is a very new high-intensity discharge source. The lamp's system consists of a power source that feeds radio-frequency or microwave radiation to a small, rotating quartz sphere containing sulfur and a mixture of noble gases. The radiation creates molecular emissions in the sulfur gas and visible light is produced. In preliminary demonstrations, 3.5 kW lamps have shown efficacies of 80 LPW. A 1,000-watt, microwave-driven sulfur lamp is expected to be available in late 1995, and its preliminary specifications indicate an efficacy of about 100 LPW.

The advantages of the lamp are its long life (it has no electrodes or phosphors), lack of mercury in the lamp, white color, high color rendition, quick start and restrike, precise optical control from a small point source, and dimmability. These lamps are most likely to be used to light large spaces such as stadiums, and may replace standard HID lamps for tasks requiring a lot of light. Another

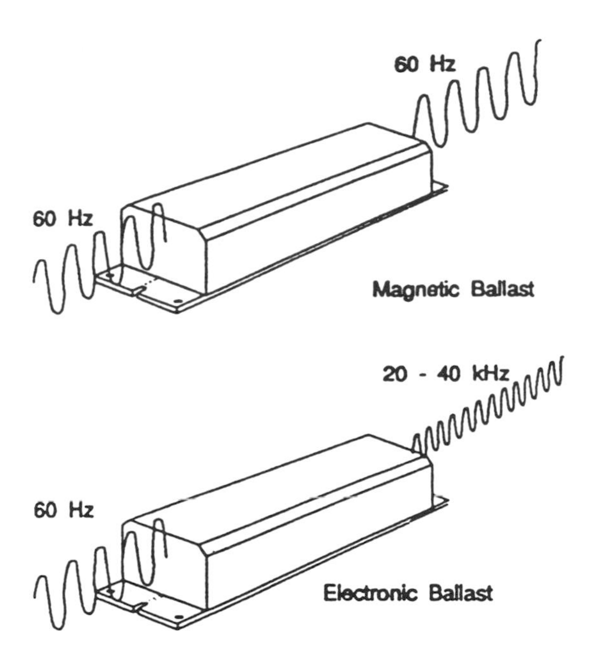

FIGURE 12B.8 Typical magnetic and electronic ballasts. *Source:* Atkinson et al., 1992.

potential application is in tunnel lighting. The small size of the sulfur lamp makes it possible to efficiently couple it to a light pipe. The combination of a long-life sulfur lamp and a light pipe could substantially reduce maintenance costs in tunnels while maintaining high efficacies. Prototypes of the sulfur lamp/light pipe combination are now operating at a Department of Energy office building and the Air and Space Museum in Washington, DC.

Ballasts

Because both fluorescent and HID lamps (discharge lamps) have a low resistance to the flow of electric current once the discharge arc is struck, they require some type of device to limit current flow. A lamp ballast is an electrical device used to control the current provided to the lamp. In most discharge lamps, a ballast also provides the high voltage necessary to start the lamp. Older "preheat" fluorescent lamps require a separate starter, but these lamps are becoming increasingly uncommon. In many HID ballasts, the ignitor used for starting the lamp is a replaceable module.

The most common types of ballasts are magnetic core-coil and electronic high-frequency ballasts (see Figure 12B.8). A *magnetic core-coil ballast* uses a transformer with a magnetic core coiled in copper or aluminum wire to control the current provided to a lamp. Magnetic ballasts operate at an input frequency of 60 Hz and operate lamps at the same 60 Hz. An *electronic high-frequency ballast* uses electronic circuitry rather than magnetic components to control current. Electronic ballasts use standard 60 Hz power but operate lamps at a much higher frequency (20,000 to 60,000 Hz). Both magnetic and electronic ballasts are available for most fluorescent lamp types.

Of the ballasts that are currently available for fluorescent lamps, the most efficient options are the electronic ballast and the cathode cut-out ballast. Because an electronic ballast is more efficient than a standard core-coil magnetic ballast in transforming the input power to lamp requirements, and because fluorescent lamps are more efficient when operated at frequencies of 20,000 Hz or more, a lamp/ballast system using an electronic rather than magnetic ballast is more efficacious. Where there are two lamps per ballast, electronic ballast systems are approximately 20% more efficacious than magnetic ballast systems; where there is only one lamp per ballast, electronic ballast systems are almost 40% more efficacious than magnetic ballast systems.

In addition, electronic ballasts eliminate flicker, weigh less than magnetic ballasts, and operate more quietly. Since electronic ballasts are packaged in "cans" that are the same size as magnetic ballasts, they can be placed in fixtures designed to be used with magnetic ballasts. Electronic ballasts are widely available, but their use is limited because of their high cost and some technological limitations for certain applications. Fluorescent electronic ballasts are now widely available for standard commercial-sector applications. They have also become available for small circular and CFL lamps. Electronic dimming ballasts have been on the market since the early 1990s. Dimmable fluorescent ballasts allow savings from daylighting in offices and other building types.

In 1988, federal standards mandated that ballast manufacturers could no longer produce standard magnetic ballasts, effective in 1990.* Although standard magnetic ballasts are no longer available, manufacturers now produce high-efficiency magnetic ballasts and cathode cut-out ballasts (also known as hybrid ballasts). The high-efficiency magnetic ballast is built with low-loss metal and denser windings, and is approximately 10 to 15% more efficacious than the standard magnetic ballast.

The *cathode cut-out (hybrid) ballast* is a modified magnetic ballast. It uses an electronic circuit to remove the filament power after the discharge has been initiated for rapid-start lamps. Cathode cut-out ballasts use approximately 5 to 10% less energy than energy-efficient magnetic ballasts. Used with 8-foot lamps, cathode cut-out ballasts can meet, and sometimes exceed, the efficacy of electronic ballasts. Cathode cut-out ballasts are available commercially, but account for only a small share of the market.

Almost all ballasts used for HID lamps are magnetic, and a number of different types are available. The various types differ primarily in how well they tolerate voltage swings and, in the case of HPS lamps, the increased voltage required to operate the lamp as it ages. Although some ballasts do tolerate wide voltage swings ("regulating" ballasts), they are relatively inefficient. Non-regulating ballasts are more efficient but are less tolerant of voltage swings. Electronic ballasts are becoming available for some MH lamps, but they do not seem to significantly improve system efficacy. A positive feature of the current MH electronic ballasts is that they do maintain an almost constant lamp color over the lamp life.

Lighting Fixtures

A lighting fixture is a housing for securing lamp(s) and ballast(s) and for controlling light distribution to a specific area. The function of the fixture is to distribute light to the desired area without causing glare or discomfort. The distribution of light is determined by the geometric design of the fixture as well as the material of which the reflector and/or lens is made. The more efficient a fixture is, the more light it emits from the lamp(s) within it. Although a lighting fixture is sometimes referred to as a luminaire, the term "luminaire" is most commonly used to refer to a complete lighting system including a lamp, ballast, and fixture.

Types of fluorescent lighting fixtures that are commonly used in the nonresidential sectors include recessed troffers, pendant-mounted indirect fixtures and indirect/direct fixtures, and surface-mounted commercial fixtures such as wraparound, strip, and industrial fixtures.

Most offices are equipped with *recessed troffers*, which are direct (downward) fixtures and emphasize horizontal surfaces. Many forms of optical control are possible with recessed luminaires. In the past, prismatic lenses were the preferred optical control because they offer high luminaire efficiency and uniform illuminance in the work space. Electronic offices have become increasingly common, however, and the traditional direct lighting fixtures designed for typing and other horizontal tasks have become less useful because they tend to cause reflections on video display terminal (VDT) screens.

No lighting system reduces glare entirely, but some fixtures and/or components can reduce the amount of glare significantly. Because the glossy, vertical VDT screen can potentially reflect bright

*Because ballasts have long lifetimes, many standard ballasts are still in use in today's building stock.

spots on the ceiling, and because VDT work is usually done with the head up, existing fixtures are sometimes replaced with indirect or direct/indirect fixtures, which produce light that is considered more visually comfortable. Most indirect lighting systems are suspended from the ceiling. They direct light towards the ceiling where the light is then reflected downward to provide a calm, diffuse light. Some people describe the indirect lighting as similar to the light on an overcast day, with no shadows or highlights. Generally, indirect lighting does not cause bright reflections on VDT screens. A *direct/indirect fixture* is suspended from the ceiling and provides direct light as well as indirect. These fixtures combine the high efficiency of direct lighting systems with the uniformity of light and lack of glare produced by indirect lighting systems.

A *wraparound fixture* has a prismatic lens that wraps around the bottom and sides of the lamp, and is always surface-mounted rather than recessed. Wraparound fixtures are less expensive than other commercial fixtures and are typically used in areas where lighting control and distribution are not a priority. *Strip and industrial fixtures* are even less expensive, and are typically used in places where light distribution is less important such as large open areas (grocery stores, for example) and hallways. These are open fixtures in which the lamp is not hidden from view.

The most common incandescent fixture in the nonresidential sector is the *downlight*, also known as a recessed can fixture. Fixtures designed for CFLs are available to replace downlight fixtures in areas where lighting control is less critical. Other open and enclosed incandescent fixtures are also used in these sectors. For outdoor lighting and street lighting, HID lamps are used in reflector and refractor fixtures, including square, rectangular, and cobra head fixtures. In the residential sector, a large variety of fixtures are used for incandescent lamps. Often, these fixtures are selected based on their aesthetic appeal rather than their energy efficiency. Attractive CFL fixtures for ceiling, wall, floor, and table lamps, as well as outdoor lamps, are now available.

Much research is focused on improving the efficiency of fluorescent lighting fixtures.* Among the most promising efficiency options are specular reflector fixtures and improvements in the thermal operating conditions of both standard and compact fluorescent fixtures.

Researchers have been striving to improve the optical design of the fixtures. Currently, one technique used to improve optical efficiency is through the use of specular reflectors. *Specular reflector fixtures* use reflective materials such as anodized aluminum, silver, and multiple dielectric coating to improve reflectivity and reduce reflection within the fixture. In new fixture designs the specular reflectors can be shaped and matched with lenses to give essentially the same light pattern as a standard fixture. However, in retrofit applications, where a reflector is simply inserted into an existing fixture, the specular reflectors tend to focus light more precisely and distribute it more narrowly. This can be a problem in that the installation can end up with insufficient overlap of light from one fixture to the next, so that the lighting uniformity is very poor. Ideally, the higher optical efficiency of the reflector system may allow one to use fewer lamps in a fixture, or fewer fixtures, and still achieve the same lighting level and quality associated with standard fixtures. Because of their high cost, the market is small for efficient fluorescent fixtures with reflectors.

Researchers are also striving to design more efficient lighting fixtures by optimizing the *thermal operating conditions* of both full-size and compact fluorescent systems. Lamps in most full-size or compact fluorescent fixtures operate below their optimum efficacy because their operating temperature is too high. The addition of vents to fixtures and thermal bridging are among the strategies being used to decrease the minimum temperature of the lamp wall.

Lighting Controls

Lighting controls include a wide range of technologies that electronically and/or mechanically control the use of lights in a building. Control systems range from simple mechanical clocks to sophisticated building energy management systems that control the lighting in a building as well

*For the most part, incandescent fixtures are designed for aesthetics rather than efficiency. Most of the HID fixtures available today are highly efficient.

as the heating, ventilation, and air conditioning systems. Lighting control systems include programmable timers, occupancy sensors, daylighting controls, lumen maintenance control, and dimmers for incandescent and compact fluorescent lamps. In the commercial sector, both simple and complex control systems are used. Although lighting control systems in the residential sector are typically simple, the systems installed in the newly emerging "smart houses" are quite complex.

Of the control systems available today, integrated workstation sensors and energy management systems are two of the most promising efficiency options. An *integrated workstation sensor* allows users to control lighting, electric heating and cooling equipment, and other electrical equipment (such as plug loads) for individual workstations or spaces. For example, user lighting controls might include dimmer switches for area and task lighting as well as daylight sensors. From their workspace, users can adjust lighting and HVAC controls according to their preference. Occupancy sensors automatically shut down electrical equipment when the space is unoccupied, and system memory allows the equipment to come back on at the same level when the occupant returns. Comprehensive, automated, *building energy management systems* are user-programmable and can control equipment for several energy end uses including lighting, HVAC, security, and safety systems. A well-designed energy management system may offer greater energy savings than individual controls for single end uses; the "systems approach" is becoming more common in both new construction and retrofits of existing buildings.

12B.4 Efficient Lighting Operation

In addition to high-quality design and the use of efficient lighting technologies, commissioning and maintenance of lighting systems play important roles in maximizing energy savings.* The commissioning process is necessary because lighting systems (particularly lighting controls) often do not perform exactly as they were designed to perform. Although often ignored, maintenance is essential for maximizing the efficient performance of lighting fixtures.

The presence of dirt on a luminaire can significantly reduce its light output. In a very clean environment (high-quality offices, clean rooms, laboratories), luminaire dirt is estimated to reduce light output over a period of 3 years by 10 to 20%, depending upon the type of luminaire used. In a moderately dirty environment (mill offices, paper processing, light machining), dirt on luminaries can reduce light output by as much as 45% in 3 years. In a very dirty environment, light output can be reduced by 50% in less than 2 years. Dirt on the surfaces in a room can also reduce the available light by limiting the ability of surfaces to interreflect from one to another. Consequently, cleaning not only makes the lighting system and room look better, it also increases the efficiency of the lighting system.

Group relamping (replacing all lamps in an area or building simultaneously rather than as each lamp burns out) is another strategy for maintaining high-efficiency fluorescent and HID lighting systems and often saves time and money as well.** As mentioned earlier, most lamps become less efficacious and produce fewer lumens as they age. Therefore, if a lighting system is designed to maintain a minimum light level over many years, it must be designed to produce more than the minimum light level when first installed. Group relamping is usually done when the lamps have operated for 60 to 80% of their rated lifetimes (Audin et al., 1994). Even though the lamps are retired before the end of their useful lives, early relamping can greatly reduce the amount of initial overdesign that is necessary and results in a more efficient lighting system. Compared to the high cost of having a person replace one lamp at a time, group relamping can often result in substantial

*"Commissioning involves reviewing design documentation, verifying installation and testing of equipment and system performance, training building operators, and analyzing the operation of an energy-saving strategy" (Mills and Piette, 1992).

**Group relamping is less important for incandescent lighting, since incandescents have shorter lifetimes and less lumen depreciation.

labor cost savings. In addition, while replacing lamps, a maintenance crew has a convenient opportunity to clean lamp lenses and reflectors.

12B.5 Current Production

As a whole, the U.S. lamp industry is highly concentrated—almost all of the market is shared among a few firms. Lighting manufacturers are typically multinational corporations serving markets around the world. The U.S. ballast industry is also highly concentrated, with almost all ballasts produced by only a few firms. It should be noted that these are not the same companies that dominate the lamp market; most lamp-producing companies do not produce ballasts and vice versa. U.S. fixture manufacturers have seen a series of acquisitions over the past few decades and many firms are now subsidiaries of larger firms. Five or so fixture manufacturers account for more than half of the market share. In contrast to the concentrated markets for lamps, ballasts, and fixtures, there are many firms involved in the lighting controls market. The technologies in this group are highly varied; for example, some firms manufacture very simple timers while others manufacture highly complex control systems for whole buildings. Table 12B.13 shows, for different types of lamps, the number that were shipped by manufacturers in 1974, 1984, and 1993, and the value of lamp shipments in 1993.

Although the growth in total incandescent lamp shipments closely matched the growth in the number of households between 1974 and 1993, there were a number of substantial changes within the incandescent lamp market during that time. Shipments of large incandescent lamps (>150 watts) in 1993 were less than a quarter of large incandescent shipments in 1974. Most of this decrease appears to have been compensated for by reflector lamp shipments, which more than doubled between 1974 and 1993. In addition, between 1974 and 1993, shipments of decorative lamps more than doubled and shipments of tungsten-halogen lamps increased by more than a factor of five.

Shipments of fluorescent and high-intensity discharge lamps have increased faster than the growth in population, number of households, and commercial floorspace. In the commercial sector, many high-wattage standard incandescents have been replaced with full-size fluorescent lamps and reflector/PAR lamps for interior applications and with HID lamps for outdoor applications. In the industrial sector, many high-wattage incandescents have been replaced with HID lamps. In 1974, virtually all HID lamps were mercury vapor. By 1993, MV lamps accounted for only about 21% of the HID market. Metal halide and high-pressure sodium lamps accounted for 32% and 47% of the market, respectively. Overall, shipments of HID lamps approximately tripled between 1974 and 1993.

Like the lamp market, the ballast market has evolved considerably in recent years. As can be seen in Table 12B.14, electronic ballasts accounted for less than 3% of total ballast sales in 1987 but more than one-third of totals by 1993. Ultimately, electronic ballasts are expected to replace magnetic ballasts for many fluorescent lighting applications. The dramatic increase in the sale of electronic ballasts in recent years has been accompanied by a significant decrease in price.

12B.6 The Cost-Effectiveness of Efficient Lighting Technologies

As mentioned in the introduction, the use of cost-effective design practices and lighting technologies could reduce the energy consumed in the U.S. for commercial interior lighting by 50 to 60% and reduce the energy consumed by residential interior and exterior lighting by 20 to 35%. However, consumer choices regarding what types of efficient lighting technologies to purchase depend not only on the reliability, lighting quality, and energy-saving potential of the alternative system, but also on the additional cost of the more efficient technologies.

TABLE 12B.13 The Size and Value of U.S. Lamp Shipments

	1974 Shipments 1,000s of units	1984 Shipments 1,000s of units	1993 Shipments 1,000s of units	Value of 1993 Shipments $1,000s
	Incandescent Lamps[1]			
Standard incandescents (15–150 W)	1,104,515	1,298,545	1,447,521	183,825
Large incandescents (>150 W)	53,722	26,365	12,517	16,866
Reflector incandescents (including PAR)	52,023	83,551	114,664	293,527
Three-way incandescents	44,627	58,877	60,413	55,987
Decorative incandescents	37,840	49,787	79,991	39,111
Tungsten-halogen, general lighting	2,600	3,863	8,596	42,705
Tungsten-halogen, other	1,001	1,720	10,015	49,893
	Fluorescent Lamps			
Standard fluorescents >30 W[2] (excluding linear T8 lamps)	170,724	269,561	346,083	375,589
Linear T8s	N/A	N/A	43,839	80,895
Standard fluorescents ≤30 W[3]	40,135	43,822	38,085	75,294
Compact fluorescent lamps	N/A	N/A	33,441	136,040
Circular lamps	6,127	8,605	7,818	19,201
All other fluorescents[4]	67,543	80,250	90,337	273,610
	High-Intensity Discharge Lamps[5]			
HID general lighting	8,742	16,279	25,213	302,510
High-pressure sodium	N/A	N/A	11,681	127,913
Mercury vapor	N/A	N/A	5,403	41,645
Metal halide	N/A	N/A	8,129	132,952

[1] These numbers refer to U.S. shipments and do not include imports. The lamp categories reported in this table do not include specialty lamps such as traffic, streetlighting, rough service, photo flash, miniature, and sun lamps.

[2] The numbers in this category in 1974 and 1984 represent lamps greater than or equal to 40 watts.

[3] The numbers in this category in 1974 and 1984 represent lamps less than 40 watts.

[4] Includes slimline and high-output lamps.

[5] Includes display lamps, but not special-purpose lamps or lamps of nonstandard voltages.

Source: Bureau of the Census, 1976, 1985, 1994.

TABLE 12B.14 Magnetic and Electronic Ballast Sales, 1987–1993

	Total Sales of Ballasts	
Year	Magnetic (millions of units)	Electronic (millions of units)
1987	49.5	1.0
1988	51.6	1.2
1989	52.4	1.5
1990	51.4	3.0
1991	49.6	8.0
1992	48.6	13.5
1993	44.3	23.0

Source: U.S. DOE, 1994.

TABLE 12B.15 Example of a Residential Retrofit

Residential Retrofit	Existing Lamp	Replacement Lamp
Lamp type	75-watt standard incandescent A-lamp	20-watt CFL with integral electronic ballast
CRI	100	82
Cost per lamp	$0.50	$15.00
Incremental cost per lamp	—	$14.50
Lumens	1,220	1,180
Lamp wattage	75	20
Watts saved	—	55
Rated lamplife	750 hours	10,000 hours
Operating hours per year	1,100 (3 h/day)	1,100 (3 h/day)
Price of electricity[1]	8.4¢/kWh	8.4¢/kWh
Electricity savings per year	—	61 kWh
Reduction in the consumer's annual cost of electricity	—	$5.12
Reduction in the consumer's annual equipment cost[2]	—	$0.74
Total annual consumer savings	—	$5.86
Simple payback time	—	2.5 years

[1] Average cost of residential electricity in the United States, estimated by the U.S. Department of Energy in December 1993.

[2] The incandescent lamp would have to be replaced about every 8 months. The annual replacement cost is therefore about $0.74.

To fully assess the financial costs and benefits of an alternative lighting system, a sophisticated analysis is required. In addition to the cost of equipment and installation, one should consider energy, relamping, and maintenance costs over the life of the lighting system, as well as disposal costs.

Two examples of simplified cost-effectiveness calculations for efficient retrofits in the residential and commercial sectors are shown below in Tables 12B.15 and 12B.16. The cost-effectiveness of the more efficient lighting systems is expressed in terms of a simple payback time. The simple payback time for an efficient lighting measure is the number of years required for the customer to recoup the initial incremental investment in that measure through the resulting reduction in energy bills and lamp replacement costs. The payback time is calculated by dividing the increased purchase price of the efficiency measure (incremental cost) by the dollar value of the annual energy and lamp replacement savings attributed to the measure.*

In the example of a residential retrofit outlined in Table 12B.15, a 20-watt CFL with an integral electronic ballast is substituted for a 75-watt standard incandescent lamp. Both the CFL and the incandescent lamp produce about 1,200 lumens. The electronic ballast ensures that the CFL will come on almost as quickly as the incandescent lamp would have. As can be seen in Table 12B.15, the cost of the CFL lamp is much higher than the incandescent ($15.00 compared to $0.50), but the financial savings associated with the CFL make the simple payback time only 2 1/2 years. Because these savings are calculated using an average residential electricity rate, savings could be much more significant in areas with high electricity costs (e.g., the Northeast and California). In addition, the cost of these lamps is expected to decline over time, in which case their payback time will decrease.

Payback times in the commercial and industrial sectors tend to be shorter than in the residential sector because labor is not considered to be a cost in the residential sector (household occupants change their own lamps rather than paying someone to do it for them) and because annual lighting hours are typically less in the residential sector.

*A more precise calculation of cost effectiveness would incorporate a discount rate and would consider time-of-day pricing instead of using an average energy price per kilowatt-hour. See Koomey et al. (1990a and 1990b) for more details.

TABLE 12B.16 Example of a Commercial Retrofit

Commercial Retrofit	Existing Lamp/Ballast	Option 1 Replacement Lamp/Ballast	Option 2 Replacement Lamp/Ballast
Lamp type	34-watt, 4-ft T12	32-watt, 4-ft T8	32-watt, 4-ft T8
Rated lamplife	20,000 hours	20,000 hours	20,000 hours
CRI	62	84	84
Ballast type (factor)	Two 2-lamp efficient magnetic (0.87)	One 4-lamp rapid-start electronic (0.88)	One 3-lamp rapid-start electronic (0.88)
Fixture type	Lensed recessed troffer	Lensed recessed troffer	Lensed recessed troffer
Number of lamps per fixture	4	4	3
Mean lumens per lamp	2,280	2,650	2,650
Wattage[1]	137	111	86
Full cost of retrofit[2]	$57.00	$55.00	$65.00
Incremental cost of retrofit at the end of ballast life	—	$200	$8.00
Incremental cost of retrofit-early replacement[3]	—	$42.24	$52.24
Watts saved	—	26	51
Operating hours per year	3,465	3,465	3,465
Price of electricity[4]	7.6¢/kWh	7.6¢/kWh	7.6¢/kWh
Electricity savings per year	—	90 kWh	177 kWh
Reduction in the consumer's annual cost of electricity	—	$6.84	$13.45
Simple payback time— early replacement	—	6.1 years	3.9 years
Simple payback time at the end of ballast life	—	None— the cost of the retrofit is less expensive	0.6 years
Application correction factor	0.83	0.91	0.91
Approximate light level relative to pre-retrofit system[5]	—	127%	96%

[1] Wattages and the application correction factors used to calculate approximate light levels were taken from U.S. DOE, 1993a; calculations assume a closed static recessed fixture.

[2] Retrofit costs includes lamps, ballasts, and labor; Option 2 cost also includes a new raceway for properly positioning the three lamps in the fixture.

[3] Full cost of retrofit less cost of replacing existing lamps.

[4] Based on EIA, 1995.

[5] Number of lamps per fixture times mean lumens per lamp time application correction factor (U.S. DOE, 1993a).

In the example of the commercial retrofit outlined in Table 12B.16, T8 lamps are substituted for T12s. Because they are common luminaires in commercial buildings today, we assume that the baseline luminaire in the retrofit example is a 2-foot by 4-foot recessed troffer with four 4-foot, 34-watt T12 lamps in two 2-lamp efficient magnetic ballasts. Two retrofit options are presented.

Option 1 replaces the baseline lamp/ballast system with four 4-foot, 32-watt T8 lamps and a four-lamp, rapid-start electronic ballast. This option will result in about 25% more light on the task area. Assuming the original fixtures were installed using the manufacturer's recommended spacing, this would be too much light for most computer-oriented offices. As can be seen in Table 12B.16, for every fixture retrofitted, the customer would save $6.84 in annual electricity costs. If the retrofit occurs before the end of the ballast life (early replacement), the simple payback time is a little more than 6 years. However, if the retrofit occurs at the end of the ballast life, the more efficient option is actually less expensive than replacing the baseline equipment.

In Option 2, three T8s are installed in a three-lamp, rapid-start, electronic ballast. In addition, a new raceway is required for properly positioning the three lamps in the fixture. Option 2 decreases the light level by a very small amount and would be appropriate where lighting levels are a little higher than necessary. Slightly less light will reach the task area. As can be seen in Table 12B.16, for every fixture retrofitted, the customer would save $13.45 in annual electricity costs. If the retrofit occurs before the end of the ballast life (early replacement), the simple payback time is approximately 4 years. However, if the retrofit occurs at the end of the ballast life, the payback time decreases to less than a year.

Conclusion

In the United States, the development and promotion of efficient lighting technologies is encouraged in a number of ways. Research and development is an ongoing process for manufacturers of lighting equipment and they regularly refine existing technologies and introduce new products to the market. Developments in the electronics industry are likely to foster innovations in lighting technologies as well. In addition, during the last decade, utility DSM incentive and information programs have been particularly active in promoting efficient lighting technologies. Although utility deregulation will probably reduce the involvement of utilities in the promotion of efficient technologies, utilities are likely to continue to be advocates of energy efficiency. Regulations are also important in encouraging energy efficiency. Within states, lighting levels are regulated through building codes; at the federal level, component regulations for lamps and ballasts are required through manufacturing codes.

In addition to establishing labeling and efficiency requirements for some types of lamps and ballasts, the federal government promotes lighting efficiency through procurement policies. By executive order, energy-using equipment purchased by the government must be energy efficient. Federal laboratories are involved in efficiency-related research and market transformation programs and have been especially active in developing and promoting electronic ballasts. In addition, the U.S. Environmental Protection Agency sponsors the highly successful Green Lights program. Green Lights is a voluntary, nonregulatory program aimed at reducing air pollution. The program provides design assistance to large businesses that are interested in dramatically reducing their lighting energy consumption and helps them to set savings goals.

In this chapter, we have provided an overview of energy-efficient lighting design practices as well as both traditional and newer, more efficient lighting technologies. Lighting is an important electrical end use in all sectors and building types in the United States, and accounts for approximately one-fifth of national electricity use. Through the use of more efficient lighting technologies as well as advanced lighting design practices and control strategies, there is significant potential for saving electricity, reducing consumer energy costs, and reducing the emission of greenhouse gases associated with electricity production. In addition, efficient lighting technologies and design can improve the quality of light in both the home and workplace.

Acknowledgments

This work was supported by the Assistant Secretary for Energy Efficiency and Renewable Energy, Office of Buildings Technology, Office of Codes and Standards, of the U.S. Department of Energy, under Contract No. DE-AC03-76SF00098. The opinions expressed in this paper are solely those of the authors and do not necessarily represent those of Ernest Orlando Lawrence Berkeley National Laboratory, or the U.S. Department of Energy. We would like to thank Katherine Falk of *Home Energy* magazine for permission to use her drawings of lamps in this report.

12B.7 Glossary

Ballast: Because both fluorescent and HID lamps (discharge lamps) have a low resistance to the flow of electric current once the discharge arc is struck, they require some type of device to limit current flow. A lamp ballast is an electrical device used to control the current provided to the lamp. In most discharge lamps, a ballast also provides the high voltage necessary to start the lamp.

Color Rendering Index (CRI): The color rendering index (CRI) of a lamp is a measure of how surface colors appear when illuminated by the lamp compared to how they appear when illuminated by a reference source of the same color temperature. For color temperatures above 5,000 K, the reference source is a standard daylight condition of the same color temperature; below 5,000 K, the reference source is a blackbody radiator. A lamp's CRI indicates the difference in the perceived color of objects viewed under the lamp and under the reference source; there are 14 differently colored test samples, 8 of which are used in the calculation of the general CRI index. The CRI is measured on a scale that has a maximum value of 100 and is an average of the results for the 8 colors observed. A CRI of 100 indicates that there is no difference in perceived color for any of the test objects; a lower value indicates that there are differences. CRIs of 70 and above are generally considered to be good; CRIs of 20 and below are considered to be quite poor.

Color temperature: The color of a lamp's light is described by its color temperature, expressed in degrees Kelvin (K). The concept of color temperature is based on the fact that the emitted radiation spectrum of a blackbody radiator depends on temperature alone. The color temperature of a lamp is the temperature at which an ideal blackbody radiator would emit light that is the same color as the light of the lamp. Lamps with low color temperatures (3,000 K and below) emit "warm" white light that appears yellowish or reddish in color. Incandescent and warm-white fluorescent lamps have a low color temperature. Lamps with high color temperatures (3,500 K and above) emit "cool" white light that appears bluish in color. Cool-white fluorescent lamps have a high color temperature.

Efficacy: Because the purpose of a lamp is to produce light, and not just radiated power, there is no direct measure of lamp efficiency. Instead, a lamp is rated in terms of its efficacy, which is the ratio of the amount of light emitted (lumens) to the power (watts) drawn by the lamp. The unit used to express lamp efficacy is lumens per watt (LPW).

Lumen maintenance: Over time, lamps continue to draw the same amount of power but produce fewer lumens. The lumen maintenance of a lamp refers to the extent to which the lamp sustains its lumen output, and therefore efficacy, over time.

References

* Atkinson, B., McMahon, J., Mills, E., Chan, P., Chan, T., Eto, J., Jennings, J., Koomey, J., Lo, K., Lecar, M., Price, L., Rubinstein, F., Sezgen, O., and Wenzel, T. 1992. *Analysis of Federal Policy Options for Improving U.S. Lighting Energy Efficiency: Commercial and Residential Buildings.* Lawrence Berkeley Laboratory, Berkeley, CA. LBL-31469.

* Audin, L., Houghton, D., Shepard, M., and Hawthorne, W. 1994. *Lighting Technology Atlas.* E-Source, Snowmass, CO.

Bureau of the Census, U.S. Department of Commerce. 1976. "Electric Lamps: Summary for 1975." Series: MQ-36B(75)-5.

Bureau of the Census, U.S. Department of Commerce. 1985. "Electric Lamps: Summary for 1985." Series: MQ-36B(85)-5.

Bureau of the Census, U.S. Department of Commerce. 1994. "Electric Lamps: Summary for 1993." Series: MQ-36B(94)-1.

Buildings Technologies Program, Energy and Environment Division, Lawrence Berkeley National Laboratory. 1991. *Technology Reviews: Envelope and Lighting Technology to Reduce Electric Demand, A Multi-Year Research Project for the California Institute for Energy Efficiency.* Lawrence Berkeley National Laboratory, Berkeley, CA.

Byrne, J. 1994. "Energy-Efficient Lighting for the Home." *Home Energy.* 11(6):53–60.

Code of Federal Regulations: Energy, 10, Parts 400 to 499. 1995. Office of the Federal Register National Archives and Records Administration, Washington, DC.

Economopoulos, O. 1992. *Lighting Reference Guide,* Fifth Edition. Ontario Hydro, Toronto, Ontario.

Energy Information Administration (EIA). 1992. *Energy Consumption Series: Lighting in Commercial Buildings.* EIA, Washington, DC. DOE/EIA-0555(92)/1.

Energy Information Administration (EIA). 1993. *Monthly Energy Review, September 1993.* EIA, Washington, DC. DOE/EIA-0035(93/09).

Energy Information Administration (EIA). 1995. *Annual Energy Outlook 1995 with Projections to 2010.* EIA, Washington, DC. DOE/EIA-0383(95).

Falk, Katherine, 1995. Personal communication. Drawings courtesy of *Home Energy* magazine.

* Illuminating Engineering Society of North America. 1993. *Lighting Handbook,* Eighth Edition. Ed. M. Rhea. Illuminating Engineering Society of North America, New York.

Koomey, J., Rosenfeld, A., and Gadgil, A. 1990a. *Conservation Screening Curves to Compare Efficiency Investments to Power Plants: Applications to Commercial Sector Conservation Programs.* In Proceedings of the 1990 ACEEE Summer Study on Energy Efficiency in Buildings, Asilomar, CA. American Council for an Energy Efficient Economy, Washington, DC.

Koomey, J., Rosenfeld, A., and Gadgil, A. 1990b. Conservation Screening Curves to Compare Efficiency Investments to Power Plants. *Energy Policy.* 18(8):774.

*Leslie, R. and Conway, K. 1993. *The Lighting Pattern Book for Homes.* Lighting Research Center, Rensselaer Polytechnic Institute, Troy, NY.

Mills, E. and Piette, M. 1992. Advanced Energy-Efficient Lighting Systems: Progress and Potential. *Energy.* 18(2):75–97.

Morse, O. Personal communication, November, 1994.

Plexus Research, Inc. and Scientific Communications, Inc. 1993. *1992 Survey of Utility Demand-Side Management Programs,* Volume 1. Electric Power Research Institute (EPRI), Palo Alto, CA. EPRI TR-102193s.

*Turiel, I., Atkinson, B., Boghosian, S., Chan, P., Jennings, J., Lutz, J., McMahon, J., and Rosenquist, G. 1995. *Evaluation of Advanced Technologies for Residential Appliances and Residential and Commercial Lighting.* Lawrence Berkeley National Laboratory, Berkeley, CA. LBL-35982.

*U.S. Department of Energy (DOE). 1993a. *Advanced Lighting Guidelines: 1993.* Final Report. Prepared for the U.S. Department of Energy, California Energy Commission, and the Electric Power Research Institute. Eley Associates, San Francisco, CA.

U.S. Department of Energy (DOE). 1993b. *Technical Support Document: Energy Efficiency Standards for Consumer Products: Room Air Conditioners, Water Heaters, Direct Heating Equipment, Mobile Home Furnaces, Kitchen Ranges and Ovens, Pool Heaters, Fluorescent Lamp Ballasts and Television Sets.* Volume 1: *Methodology* DOE, Washington, DC, DOE/EE-0009.

U.S. Department of Energy (DOE). 1993c. *Technical Support Document: Energy Efficiency Standards for Consumer Products: Room Air Conditioners, Water Heaters, Direct Heating Equipment, Mobile Home Furnaces, Kitchen Ranges and Ovens, Pool Heaters, Fluorescent Lamp Ballasts and Television Sets.* Volume 2: *Fluorescent Lamp Ballasts, Television Sets, Room Air Conditioners, & Kitchen Ranges and Ovens.* DOE, Washington, DC. DOE/EE-0009.

U.S. Department of Energy (DOE), Office of Building Technologies. 1993d. *CORE Databook.* DOE, Washington, DC.

U.S. Department of Energy (DOE). 1994. Comments for the Department of Energy Workshop, April 5–7, on behalf of the National Electrical Manufacturers Association, Ballast Section, p. 11.

*If you are interested in more detailed discussions of energy-efficient lighting design strategies and technologies, we suggest consulting the references that have an asterisk before them as well as requesting information from the Green Lights program at the following address: Green Lights, U.S. EPA, Air and Radiation (6202-J), Washington, DC, 20460.

12C

Energy Efficient Technologies: Appliances, Heat Pumps, and Air Conditioning

James McMahon
Isaac Toriel
Gregory J. Rosenquist
J.D. Lutz
Stanley H. Boghosian
Leslie Shown
Lawrence Berkeley National Laboratory

Introduction

Although individual household appliances do not consume large amounts of energy, appliance energy consumption in the aggregate is significant. Residential energy consumption accounts for approximately 35% of electricity use and 25% of natural gas use in the United States. In total, U.S. households in 1990 spent more than $110 billion on energy and the average U.S. household spent almost $1,200 on its energy bill (EIA, 1993). We focus our appliance discussion on electricity and natural gas because, together, they account for more than 90% of the primary energy consumed by the residential sector* (Turiel et al., 1995). In Table 12C.1, 1990 consumption of gas and electricity is shown in exajoules (10^{18} joules) for the household appliances discussed in this chapter.

Reducing the energy consumption of residential appliances depends both on replacing older appliances with the much more efficient models that are now available and on continuing to design even more energy-efficient appliances. National energy efficiency standards for appliances have

*Throughout the chapter, when we refer to electricity consumption, we refer not to the amount of electricity consumed by the end user, but to the amount of energy required to generate that electricity in total (the primary energy). Incorporating the generational and distributional inefficiencies of electricity typically increases the actual number of Btu consumed by about a factor of three.

TABLE 12C.1 The Energy Consumption of U.S. Household Appliances in 1990

Appliance	Fuel	Primary Exajoules (EJ) Consumed	Percentage of Residential Electricity Consumption	Percentage of Residential Natural Gas Consumption	Percentage of Residential Energy Consumption	Household Saturation Rate (%)[1]
Refrigerator and refrigerator-freezer	Electricity	1.45	13.7	—	8.0	114
Freezer	Electricity	0.41	3.9	—	2.3	36
Water heater[2]	Electricity	1.67	15.7	—	9.3	36
	Natural gas	1.48	—	26.6	8.2	58
Furnace for central heating	Natural gas	2.30	—	41.4	12.7	38
Boiler for central heating	Natural gas	0.61	—	10.9	3.4	10
Heat pump for central space heating	Electricity	0.14	1.3	—	0.8	6
Heat pump for central air conditioning	Electricity	0.18	1.7	—	1.0	—
Central air conditioning	Electricity	0.80	7.5	—	4.4	28
Room air conditioning	Electricity	0.36	3.4	—	2.0	32
Oven	Electricity	0.37	3.5	—	2.0	57
	Natural gas	0.21	—	3.8	1.2	41
Cooktop	Electricity	0.34	3.2	—	1.9	59
	Natural gas	0.24	—	4.3	1.3	41
Microwave oven	Electricity	0.19	1.8	—	1.1	70
Clothes washer[3]	Electricity and natural gas	0.46	2.8	2.9	2.5	81
Clothes dryer	Electricity	0.51	4.8	—	2.8	54
	Natural gas	0.05	—	0.9	0.3	14
Total for above appliances		11.77	63.3	90.8	65.2	—
Other appliances[4]	Electricity	3.89	36.7	—	21.6	—
	Natural gas	0.51	—	9.2	2.8	—
	Other[5]	1.88	—	—	10.4	—
Total for all appliances		18.05	10.61 Ej elec	5.56 EJ gas	100.0	—

[1] The percentage of houses that own a particular appliance is called a "saturate rate."
[2] Water heating for clothes washers and dishwashers is not included.
[3] Includes energy used to heat water.
[4] Other appliances include lighiting, electric and oil central space heating, room space heating, dishwashers, televisions, and pool heaters.
[5] Includes oil, kerosene, and wood fuel use.

driven efficiency improvements over the last 5 years, and appliances have become significantly more efficient as a result. Further improvement of appliance efficiency represents a large, untapped technological opportunity. Research indicates that appliance efficiency could be improved by 30 to 60%, based on technologies that could be mass produced by 1998 (Turiel et al., 1995).

12C.1 The Most Common Types of Appliances, How They Work, and How Much Energy They Use

Refrigerator-Freezers and Freezers

Refrigerators, refrigerator-freezers, and freezers keep food cold by transferring heat from the air in the appliance cabinet to the outside. A refrigerator is a well-insulated cabinet used to store food at 0°C or above; a refrigerator-freezer is a refrigerator with an attached freezer compartment that stores food below −13°C; and a standalone freezer is a refrigerated cabinet to store and freeze foods at −18°C or below (DOE, 1988). Almost all refrigerators are fueled by electricity. The refrigeration system includes an evaporator, a condenser, and a compressor. The system uses a vapor compression cycle, in which the refrigerant changes phase (from liquid to vapor and back to liquid again) while circulating in a closed system. The refrigerant (usually CFC-12) absorbs or discharges heat as it changes phase. Although most refrigerants and insulating materials now contain chlorofluorocarbons (CFCs), all models sold after January 1, 1996, will be CFC-free.

In 1990, almost all households had a refrigerator-freezer and approximately 14% of households had more than one. Thirty-five percent of households had standalone freezers. Together, refrigerators, refrigerator-freezers, and freezers account for more than 10% of residential energy consumption and almost 18% of residential electricity consumption. Ambient temperature is a significant determinant of the energy consumed by these appliances (Meier, 1995). Extensive research has shown that user behavior has relatively little effect on the energy consumption of these appliances (Hanford et al., 1994).

Water Heaters

A water heater is an appliance that is used to heat potable water for use outside the heater upon demand. Water heaters supply water to sinks, bathtubs and showers, dishwashers, and clothes washing machines. Most water heaters in the United States are storage water heaters, which continuously maintain a tank of water at a thermostatically controlled temperature. The most common storage water heaters consist of a cylindrical steel tank that is lined with glass in order to prevent corrosion. Most hot water tanks manufactured today are insulated with polyurethane foam and wrapped in a steel jacket. Although some use oil, almost all storage water heaters are fueled by natural gas or electricity.

Rather than storing water at a controlled temperature, instantaneous water heaters heat water as it is being drawn through the water heater. Both gas-fired and electric instantaneous water heaters are available. Although instantaneous water heaters are quite popular in Europe, they are not commonly used in the United States. No heaters of this type are manufactured in the United States.

Like refrigerators, water heaters are present in almost all households. Approximately 58% of households have gas-fired water heaters, and approximately 36% have electric water heaters. Excluding water-heating for dishwashers and clothes washers, gas and electric water heaters together account for more than 17% of residential energy consumption. Hot water use varies significantly from household to household. This variation is mostly due to differences in household size and occupant behavior. Climate also affects energy consumption; more energy is required to heat water in colder parts of the country (Hanford et al., 1994).

Space-Conditioning Systems

Furnaces and Boilers

Furnaces and boilers are major household appliances used to provide central space heating. Both fuel-burning and electric furnaces and boilers are available. A typical *gas furnace* installation is composed of the following basic components: (1) a cabinet or casing; (2) heat exchangers; (3) a system for obtaining air for combustion; (4) a combustion system including burners and controls; (5) a venting system for exhausting combustion products; (6) a circulating air blower and motor; and (7) an air filter and other accessories (ASHRAE, 1992).* In an *electric furnace*, the casing, air filter, and blower are very similar to those used in a gas furnace. Rather than receiving heat from fuel-fired heat exchangers, however, the air in an electric furnace receives heat from electric heating elements. Controls include electric overload protection, contactor, limit switches, and a fan switch (ASHRAE, 1992). Furnaces provide heated air through a system of ducts leading to spaces where

*Furnaces that burn oil and liquid petroleum gas (LPG) are also available, though not as common.

heat is desired. In a *boiler system*, hot water or steam is piped to terminal heating units placed throughout the household. Boilers can be fueled by natural gas, oil, coal, wood, or LPG, or can be powered by electricity. The boiler itself is typically a pressurized heat exchanger of cast iron, steel, or copper in which water is heated.

The majority of U.S. households use natural gas for space heating. Natural gas was the primary space-heating fuel in approximately 55% of U.S. households in 1990, and electricity was the primary space-heating fuel in 23% of households (EIA, 1993). More than half of residential natural gas consumption, and more than 9% of residential electricity consumption is attributable to central space heating.* The amount of energy consumed by a household for space heating varies significantly with climate and various household characteristics. For example, energy use for space heating is higher in the Northeast and Midwest, in single-family rather than multifamily homes, and in larger homes (EIA, 1993).

Central and Room Air Conditioners

A central air conditioning (AC) system is an appliance designed to provide cool air to an enclosed space. Typically, central AC systems consist of an indoor unit and an outdoor unit. The outdoor unit contains a compressor, condenser (outdoor heat exchanger coil), condenser fan, and condenser fan motor; the indoor unit consists of an evaporator (indoor conditioning coil) and a flow control device (a capillary tube, thermostatic expansion valve, or orifice) residing either in a forced-air furnace or an air handler. Refrigerant tubing connects the two units. A central AC system provides conditioned air by drawing warm air from the space and blowing it through the evaporator. In passing through the evaporator, the air gives up its heat content to the refrigerant. The conditioned air is then delivered back to the space (via a ducted system) by the blower residing in the furnace or air handler. The compressor takes the vaporized refrigerant coming out of the evaporator and raises it to a temperature exceeding that of the outside air. The refrigerant then passes on to the condenser (outside coil), where the condenser fan blows outside air over it, gives up its heat to the cooler outside air, and condenses. The liquid refrigerant is then taken by the flow control device and its pressure and temperature are reduced. The refrigerant reenters the evaporator, where the refrigeration cycle is repeated.

Unlike the two-unit, central AC system, a room air conditioner is contained within one cabinet and is mounted in a window or a wall so that part of the unit is outside and part is in. The two sides of the cabinet are typically separated by an insulated divider wall in order to reduce heat transfer. The components in the outdoor portion of the cabinet are the compressor, condenser, condenser fan, fan motor, and capillary tube. The components in the indoor portion of the cabinet are the evaporator and evaporator fan. The fan motor drives both the condenser and evaporator fans. A room AC provides conditioned air in the same manner described for a central AC system. Almost all AC systems are powered by electricity.

Approximately 66% of U.S. households have some sort of AC system. Central AC systems (including heat pumps, which are discussed later) are used in 34% of households; room ACs are used in 32% of households. More than 8% of residential energy consumption and approximately 13% of residential electricity consumption is attributable to space cooling. Energy use for space cooling is higher in warmer parts of the United States; in single-family rather than multifamily homes, and in larger homes (EIA, 1993).

Heat Pumps

Unlike air conditioners, which provide only space cooling, heat pumps use the same equipment to provide both space heating and cooling. A heat pump draws heat from the outside air into a building during the heating season and removes heat from a building to the outside during the cooling season. An *air-source heat pump* contains the same components and operates in the same way as a central AC system but is able to operate in reverse as well, in order to provide space

*Fuel oil was the primary space-heating fuel in 11% of households, LPG in 5%, wood in 4%, and kerosene in 1%.

heating. In providing space heat, the indoor coil acts as the condenser while the outdoor coil acts as the evaporator. When the outside air temperature drops below 2°C during the heating season, the available heat content of the outside air significantly decreases; in this case, a heat pump will utilize supplementary electric-resistance backup heat. A *ground-source heat pump* operates on the same principle as air-source equipment except that heat is rejected or extracted from the ground instead of the air. Since ground temperatures do not vary over the course of a day or a year as much as the ambient air temperature, more stable operating temperatures are achieved. The ground loop for a ground-source heat pump is a closed system that uses a pressurized, sealed piping system filled with a water/antifreeze mixture. The indoor mechanical equipment of a ground-source system includes a fan coil unit with an indoor coil, a compressor, and a circulation pump for the ground loop. Almost all heat pumps are powered by electricity. Heat pumps were used in approximately 6% of U.S. households in 1990. The energy consumption of heat pumps varies according to the same user characteristics discussed above for furnaces and AC systems.

Cooking

Cooktops and Conventional Ovens

Cooktops and ovens are used to cook or heat different types of food. A cooktop is a horizontal surface on which food is cooked or heated from below; a conventional oven is an insulated, cabinetlike appliance in which food is surrounded by heated air. When a cooktop and an oven are combined in a single unit, the appliance is referred to as a *range*. Both gas and electric ranges are available. Cooktops and ovens are present in almost all households. Almost 60% of households use electric cooktops and ovens, and the remaining 40% of households use gas cooktops and ovens. Together, these cooking appliances account for approximately 7% of residential energy consumption. Energy consumption over time depends primarily on the cooking habits of the user.

Microwave Ovens

In a microwave oven, microwaves directed into the oven cabinet cause water molecules inside the food to vibrate. Movement of the water molecules heats the food from the inside out. The number of households with microwave ovens has increased dramatically in recent years. In 1980, only 14% of households had microwave ovens; by 1990, 79% of households had microwave ovens (EIA, 1992). Almost 2% of residential electricity consumption was attributable to microwave ovens in 1990. Almost a quarter of U.S. households use microwave ovens to cook half or more of their food; these households tend to be younger, have fewer persons, and have more education than households that are less dependent on their microwave ovens (EIA, 1992).

Clothes Washers

A clothes washer is an appliance that is designed to clean fabrics by using water, detergent, and mechanical agitation. The clothes are washed, rinsed, and spun within the insulated cabinet of the washer. Top-loading washers move clothes up and down, and back and forth, typically about a vertical axis. Front-loading machines move clothes around a horizontal axis. Electricity is used to power an electric motor that agitates and spins the clothes, as well as a pump that controls both hot and cold water flowing into and out of the machine. A separate water heater is used to heat the water used in the washer.

Approximately 81% of households have clothes washers. Washers consume approximately 2.6% of total residential energy. Most of the clothes washers sold in the United States are top-loading, vertical-axis machines. The majority of energy used for clothes washing (85 to 90%) is used to heat the water. User behavior significantly affects the energy consumption of clothes washers. The user can adjust the amount of water used by the machine to the size of the load, and thereby save water and energy. Choosing to wash with warm water rather than hot water reduces energy consumption per cycle by approximately 50%. Similarly, rinsing with cold water rather than warm can reduce energy consumption per cycle by approximately 20% (Levine et al., 1992). Energy consumption over time will depend on how frequently the washer is used. A 1989 estimate of the number of washer cycles per household was 380 per year (Proctor and Gamble, 1989).

Clothes Dryers

A clothes dryer is an appliance that is designed to dry fabrics by tumbling them in a cabinetlike drum with forced-air circulation. The source of heated air may be powered either by electricity or natural gas. The motors that rotate the drum and drive the fan are powered by electricity. Approximately 53% of households have electric clothes dryers, and 16% have gas dryers. Clothes dryers account for more than 3% of residential energy consumption. Energy consumption over time will depend on how frequently the dryer is used. A 1989 estimate of the number of dryer cycles per household was 359 per year (Proctor and Gamble, 1989).

12C.2 Current Production

Table 12C.2 shows the number of each type of appliance that was shipped by manufacturers in 1992. Shipment trends are relatively stable for appliances found in most households such as refrigerators, water heaters, clothes washers, and clothes dryers; some variation is attributable to changing levels of new construction in the residential sector. Shipments of room air conditioners are less stable and tend to increase substantially during unusually hot weather. As mentioned earlier, shipments of microwaves have steadily increased over the last 15 years. Also shown in Table 12C.2 is the extent to which the four largest manufacturers of each appliance controlled the market for that appliance in 1991. For example, 70% of all freezers shipped to merchants were manufactured by one company and 95% of all freezers shipped were manufactured by one of the top four companies. In comparison to freezers, the production of heat pumps and central air conditioners is much more disaggregated among several manufacturers.

12C.3 Efficient Designs

Federal standards requiring increased efficiency for residential appliances, earlier state standards, and utility programs have improved appliance efficiency dramatically since the 1970s. For example, compared with new models in 1972, a new freezer in 1990 was more efficient by 95% and a new central air conditioner or heat pump was more efficient by 40 to 45%. Because of the slow turnover rate of appliances, however, the older, less efficient equipment remains in use for a long time. Compared to 1972 models, the freezers in use in 1990 were only 15% more efficient, and the air conditioners and heat pumps in use were only 15 to 21% more efficient. "If everyone could be persuaded to replace appliances 10 years and older with new 1990 appliances … 7.3 percent of total residential end-use consumption per year" could be saved (EIA, 1993).

Promising design options for further improving the efficiency of residential appliances are discussed next. Potential increases in appliance efficiency are expressed relative to the existing efficiency standard for each appliance.*

Refrigerator-Freezers and Freezers

Vacuum-Panel Insulation

The use of vacuum-panel insulation (VPI) can significantly reduce heat gain in a refrigerated cabinet and thereby decrease the amount of energy necessary to maintain a refrigerator or freezer

*Information regarding efficient appliances is taken from Turiel et al. (1995) unless otherwise noted. Energy consumption comparisons are based on appliance consumption at the site rather than the source.

TABLE 12C.2 The Size and Concentration of the Appliance Market

Appliance	1992 Shipments	Market Share Controlled by Top Manufacturers in 1991 (%)			
Number of manufacturers		1	2	3	4
Refrigerator-freezers	7,760,800[1]	35	60	77	89
Freezers (chest and upright)	1,691,000	70	86	92	95
Gas and oil water heaters	4,241,000[2]	28	56	74	88
Electric water heaters	3,399,000[2]				
Gas furnaces	2,106,900	20	35	50	64
Electric furnaces	290,400	NA	NA	NA	NA
Boilers	321,900[3]	NA	NA	NA	NA
Central air conditioners	3,095,600	19	33	47	60
Room air conditioners	2,910,000	28	50	69	77
Heat pumps	913,500	19	33	47	60
Electric ranges	3,574,000	43	61	78	90
Gas ranges	2,614,000	25	49	71	92
Microwave ovens	8,207,000[4]	23	42	58	69
Clothes washers	6,515,000	51	69	85	97
Gas clothes dryers	1,154,000	52	66	80	92
Electric clothes dryers	3,563,000	52	69	84	98

[1] Includes imports and exports of units 6.5 cubic feet or more.
[2] Includes both residential and small commercial.
[3] Includes both residential and commercial.
[4] 1991 shipments; information from *Appliance*, 1992.
Sources: Shipment information in this table was taken from U.S. DOE, 1994; market share information was taken from *Appliance*, 1992.

at a low temperature. When using VPI, a partial vacuum is created within the walls of the insulation panels. Because air is conductive, the amount of heat transfer from the outside air to the refrigerated cabinet is reduced as the amount of air within the panels is reduced. Evacuated panels are filled with low-conductivity powder, fiber, or aerogel in order to prevent collapse. Energy savings associated with the use of vacuum panel insulation range from 10 to 20% depending on the size of the panels, the percentage of the panel area that is insulated with VPI, the resistivity of the panels, and edge losses. A limited number of VPI refrigerators and freezers can be found on the market. Additional testing and development of VPI is necessary before manufacturers can begin to mass produce VPI appliances.

Improved Fan Motors

The evaporator and condenser fans of large refrigerators are powered by motors. The most common and inexpensive motor used for this purpose is a shaded-pole motor. Large efficiency gains are possible in refrigerators and freezers by switching to electronically commutated motors (ECMs), also known as brushless permanent-magnet motors, which typically demand less than half as much power as shaded-pole motors. Although ECM motors are more expensive, energy savings of approximately 9% can be achieved when they are substituted for standard refrigerator motors. At present, insufficient supplies of ECMs are produced to supply the U.S. market. At least two companies plan to manufacture large quantities of ECMs in the near future.

Water Heaters

Gas-Fired Condensing Storage Water Heater

The amount of heat extracted from the fuel used to fire a gas appliance can be increased by condensing the combustion products in the flue gases. In a condensing storage water heater, the flue is lengthened by coiling it around the inside of the water tank. The flue exit is located near the bottom of the tank where the water is coolest. Because the flue gases are relatively cool, a plastic venting system may be used. A drain must be installed in condensing systems. Energy savings associated with the use of a gas-fired condensing water heater are approximately 34% compared to the 1990 efficiency standard. At this time, the high incremental cost (≈$1,700) of the water heater results in a payback time that exceeds the typical lifetime of a water heater. It is reasonable to assume that the incremental cost could be reduced to approximately $750 in the next few years; in that case, the water heater would be a cost-effective investment. When such a heater is used in an integrated water heater/furnace system (described later), the payback time is reduced to approximately 12 years. Currently, only one company produces condensing storage water heaters for residential use; these are sold as combined water heater/space-heating systems. No sales data are available for these models, but shipments are assumed to be very low.

EPRI/E-Tech Heat Pump

Heat pumps used with water heaters capture heat from the surrounding air or recycle waste heat from air conditioning systems and then transfer the heat to the water in the storage tank. In this way, less energy is used to bring the water to the desired temperature. The EPRI/E-Tech heat pump water heater was developed collaboratively by the Electric Power Research Institute (EPRI) and a private corporation. The heat pump is a separate unit that can be attached to a water heater. Water is circulated out of the water heater storage tank, through the heat pump, and back to the storage tank. The pump is small enough to sit on the top of a water heater but could be anywhere nearby. Research indicates that this technology uses 60 to 70% less energy than conventional electric resistance water heaters. This model is smaller, more efficient, and less costly than conventional residential heat pump water heaters. Field and lab tests have been completed for several prototypes. Extensive field monitoring is planned by EPRI for the next few years. This model is currently available for sale, and developers are actively promoting its distribution through utility company rebate programs.

Solar Water Heaters

Technological improvements in the last decade have improved the quality and performance of both passive and active solar water heaters. Research indicates that, in general, solar water heaters use 60% less energy than conventional electric resistance water heaters. There are several types of solar water heaters commercially available today.

Space-Conditioning Systems

Furnaces

Gas-Fired Condensing Furnaces. The efficiency of a conventional gas furnace can be increased by using an additional heat exchanger to capture the heat of the flue gases before they are expelled to the outside. The secondary heat exchanger is typically located at the outlet of the circulating air blower, upstream of the primary heat exchanger. A floor drain is required for the condensate. A gas-fired condensing furnace uses approximately 17% less energy than a standard gas furnace. Condensing furnaces have been on the market since the 1980s. They are now very popular, and constitute 20% of all central furnace sales. They are particularly popular in colder areas of the United States, where heating bills are high. An early technical problem, corrosion of the secondary heat exchanger, has been resolved by the industry.

Integrated Water Heaters and Furnaces. Traditionally, water heating and space heating have required two separate appliances—a hot water heater and a furnace. The processes of heating water and air, however, are very similar; consequently, the average home has redundant controls, burners, and flues. Combining a water heater and a furnace into a single system can provide both space heating and hot water with less fuel and less complexity, and at a lower cost. Integrated water and space heating is most cost-effective when installed in new buildings because vent and gas connections are necessary for only one appliance rather than two.

There are two basic systems that integrate space heating and water heating. The more common system, called a *combo system* by the residential heating and cooling industry, uses a water heater to provide heat for both space heating and water heating. The water heater is connected by piping to one or more fan coil air handlers in the house. In the space-heating mode, a pump circulates hot water from the heater to the fan coil unit(s) and the heat is then distributed by ducting from the fan coil unit(s). In the water-heating mode, hot water is simply drawn from the water heater in the usual manner. There are only two manufacturers of components for these systems, although the components (most important, the fan coil air handlers) are marketed under a number of trade names and companies. The second type of system is referred to here as an *integrated system*. It consists of a boiler, circulating pump, hot water storage tank, and two heat exchangers. One heat exchanger is a fan coil air handler for space heating, and the other is located in the hot water tank for water heating. When space heat is called for, the boiler heats the heat transfer fluid (water or a glycol solution) and the pump circulates the fluid through the fan coil. For hot water, the boiler heats the fluid and the pump circulates the fluid through the interconnecting pipe to a heat exchanger in the hot water tank. There are only one or two manufacturers of this type of system.

The efficiencies of these integrated systems are determined largely by the hot water heating component of the system. Compared to a system using a standard water heater or boiler, an integrated system using a condensing water heater or condensing boiler can reduce energy consumption by as much as 25%. Two primary manufacturers produce residential fan coil units designed for operation with domestic hot water heaters. Currently, there is only one manufacturer of condensing gas water heaters. The condensing gas heater is marketed as both a standalone water heater and as a part of an integrated appliance. At present only one manufacturer produces integrated systems.

Central and Room Air Conditioners

Electric Variable-Speed Air Conditioning. Variable-speed central air conditioners use electronically commutated motors (ECMs). These motors are more efficient than the induction motors used in a single-speed system. In addition, the speed of the ECMs can be varied to match system capacity more precisely to a building load. Cycling losses, which are associated with a system that is continually turned off and on in order to meet building load conditions, are thus reduced. Unlike induction motors, ECMs retain their efficiency at low speeds; consequently, energy use is also reduced at low-load conditions. A variable-speed AC system uses approximately 40% less energy than a standard single-speed AC system. Although these AC systems are now available from two major manufacturers, they accounted for less than 1% of sales in 1994.

Electric Two-Speed Air Conditioning. Although two-speed induction motors are not as efficient as variable-speed ECMs, they are less expensive. Like variable-speed air conditioners, two-speed air conditioners reduce cycling losses. When two-speed induction motors are used to drive the compressor and fans, the system can operate at two distinct capacities. Cycling losses are reduced because the air conditioner can operate at a low speed to meet low building loads. In some models, two-speed compressors are coupled with variable-speed indoor blowers to improve system efficiency further. A two-speed AC system reduces energy consumption by approximately one-third. Although these AC systems are available from three major manufacturers, they accounted for less than 1% of sales in 1994.

Room Air Conditioners. The most efficient room air conditioners have relatively large evaporator and condenser heat-exchanger coils, high-efficiency rotary compressors, and permanent split-capacitor fan motors. Compared to standard room ACs, efficient room ACs reduce energy consumption by approximately 28%. Efficient room ACs are available from only two manufacturers and accounted for less than 1% of sales in 1994.

Heat Pumps

Variable-Speed and Two-Speed Heat Pumps. Like central air conditioners, heat pumps can be made more efficient by the use of two-speed and variable-speed motors (see the earlier discussion of efficient central air conditioners). Both two-speed and variable-speed air-source heat pumps are available; variable-speed ground-source heat pumps are not commercially available at this time. Compared to standard models, two-speed air-source heat pumps reduce energy consumption by approximately 27%, variable-speed air-source heat pumps reduce energy consumption by 35%, and two-speed ground-source heat pumps reduce energy consumption by 46%. Variable-speed and two-speed air-source heat pumps are made by the same companies that make variable-speed and two-speed AC systems. Approximately three manufacturers produce efficient air-source heat-pumps, and they accounted for less than 2% of all heat pump sales in 1994. Approximately four or five manufacturers produce two-speed ground-source heat pumps, and they accounted for less than 1% of heat pump sales in 1994.

Gas-Fired Heat Pumps. Currently, all residential heat pumps are electric, but researchers have been developing gas heat pumps. The Gas Research Institute (GRI) and a private corporation have jointly developed a natural gas, engine-driven, variable-speed heat pump in which the compressor is driven by an internal combustion spark-ignition engine and heat is recovered in the space-heating mode. This engine-driven heat pump was put on the market in 1994. In addition, the DOE has been funding the development of a gas-fired ammonia-water absorption-cycle heat pump. The unit is in full-scale testing and a major manufacturer has been selected to produce a number of units for field testing. Gas-driven heat pumps have the potential to reduce heat pump energy consumption by approximately 35 to 45%.

Distribution Systems. When assessing the efficiency of a space conditioning system, it is important to consider the efficiency of the distribution system as well as the appliance. Pipes carrying hot water for heating can lose up to 10% of their heat through conduction; heat losses can be reduced to less than 5% when the pipes are well insulated. More important, it is not uncommon for air ducts to have distribution losses of 20 to 40% due to conduction as well as leakage. Better insulation as well as more careful duct sealing can reduce these losses. In general, the most effective strategy for reducing distribution losses is to include the distribution system in the conditioned space so that any losses due to conduction or air leakage go directly into the space to be conditioned. This requires careful attention by the architect in the design of new buildings and is typically very expensive as a retrofit measure.

Cooking

Pilotless Ignitions for Gas Ranges

Gas cooktops and ovens that are not equipped with electrical power cords use standing pilot ignition systems. The natural gas consumption of cooktops and ovens can be reduced by replacing the pilot ignition with an electronic ignition system that consumes a negligible amount of electricity. The electronic ignition used for a *gas cooktop* is an intermittent ignition system in which a spark igniter is activated by an electronic control module when the burner valve is turned to the start position. A spark-ignition system is also available for a *gas oven.* When the thermostat knob of the oven is set to a specific temperature, a control module activates a spark igniter, which lights a pilot, and the pilot in turn ignites the oven burner. The pilot will burn until the thermostat is turned off. In a gas range, the control module controls both the cooktop and oven ignition systems. Because the spark igniter is activated for an extremely short time period, electricity consumption is negligible. Almost all pilotless gas ovens in households today use a carbide "glo" igniter. The glo igniter must continually draw power (approximately 380 watts), however, to keep the burner ignited. Although gas energy is saved, the electrical energy consumption of a glo ignition system is significant; consequently, the electronic ignition system is far more efficient. The use of electronic ignition rather than a standard pilot in a gas cooktop and oven can reduce a range's energy consumption by almost half.

All major range manufacturers offer gas ranges with electronic spark-ignition systems. Most manufacturers use glo ignition systems in ovens because they are less expensive than spark-ignition systems. In almost all gas ranges equipped with pilotless ignitions, the cooktop will use a spark-ignition system while the oven will use a glo ignition system. Spark-ignition systems are available for ovens, but only a small number are manufactured.

Induction Cooktop

The energy efficiency of electric cooktops can be improved with the use of induction heating elements. In an induction heating element, a flat spiral inductor is located just beneath a glass-ceramic panel. A high-frequency current supplied to the inductor causes it to generate a magnetic field that passes through the ceramic-glass panel unaffected. The cooking vessel must be made of a ferromagnetic material, and when it is placed on the panel above the inductor, the magnetic field from the inductor causes it to heat up. Thus, the vessel essentially becomes the heating element. Temperature is controlled by a sensor placed between the inductor and the panel. The sensor also enables the inductor to heat only objects of 4 inches or more in diameter. This prevents any small metal objects, such as forks or spoons, from accidentally being heated. In addition, because the glass-ceramic panel is unaffected by the magnetic field, it remains relatively cool, preventing any accidental burns. The primary advantages of an induction element are that the heat source is precisely controlled, response time is fast, it is easy to clean, and it can heat vessels that are not flat. Compared to an electric coil-type cooktop, an induction cooktop can reduce energy consumption by approximately 8%. A few small manufacturers that cater to the high-end market offer induction cooktops. In the past, two major manufacturers also offered induction cooktops but discontinued the product because of low sales.

Microwave Ovens: Improved Power Supply to the Magnetron

In order to operate the magnetron in microwave ovens, a transformer is used to increase the input voltage from 120 volts to 4,000 volts. By reducing power losses through the controller and transformer, the efficiency of a microwave oven can be improved. Transformers of higher efficiency than those currently used in microwaves have been tested in the laboratory. Microwave manufacturers may be able to make use of this technology in the future. More efficient transformers might reduce the energy consumption of a microwave oven by as much as 12%.

Clothes Washers

Horizontal-Axis Washers

Although horizontal-axis clothes washers dominate the European market, the vast majority of clothes washers sold in the United States are top-loading, vertical-axis machines. Horizontal-axis washers, in which the tub spins around a horizontal axis, use much less water than their vertical-axis counterparts, and less hot water is therefore required from water heaters. As mentioned earlier, the majority of energy used for clothes washing is used for heating water, so a significant amount of energy can be saved by using horizontal-axis washers. Research has indicated that horizontal-axis washers are more than twice as efficient as vertical-axis washers of comparable size. Currently, only two U.S. manufacturers produce horizontal-axis washers, though not in great numbers ($\approx$100,000 per year). Two more manufacturers plan to produce horizontal-axis washers by 1996. Many are produced in Europe, and a small number are imported to the U.S.

High-Spin-Speed Washers

Clothes washers can be designed so that less energy is required to dry clothes after they have been washed. Extracting water from clothes mechanically in a clothes washer uses approximately 70 times less energy than extracting the water with thermal energy in an electric clothes dryer. Thus, by increasing the speed of a washer's spin cycle, one can reduce the energy required to dry clothes. Generally, U.S. clothes washers have spin speeds of about 550 revolutions per minute (rpm); a number of European washers have spin speeds of 1,000 rpm and do not create excessive noise or wrinkling (Levine et al., 1992).

Because gas clothes dryers require so much less energy than electric dryers, energy savings are much more significant when a high-spin-speed washer is used with an electric dryer. In a vertical-axis clothes washer, an increase in spin speed from 550 rpm to 850 rpm reduces moisture retention from 65% to 41%. In an electric dryer, this reduces the energy consumption by more than 40%. In a horizontal-axis washer, an increase in spin speed from 550 rpm to 750 rpm reduces moisture retention from 65% to 47%; in an electric dryer, energy consumption is reduced by more than 30%. No high-spin-speed washers were manufactured in the United States in 1994. Washers with greater than 1,000 rpm spin speeds are common in Europe, but these washers have smaller tubs than U.S. models. Researchers are working to design a high-spin-speed washer with a larger tub. EPRI is currently working with a manufacturer to develop a high-spin-speed, horizontal-axis washer.

Clothes Dryers

Microwave Dryers

In conventional clothes dryers, hot air passes over wet clothes and vaporizes the surface water. During the later stages of drying, the surface dries out and heat from the hot air must be transferred to the interior, where the remaining moisture resides. In contrast, in microwave drying, water molecules in the interior of a fabric absorb electromagnetic energy at microwave wavelengths, thereby heating the water and allowing it to vaporize. In addition to using less energy for drying, microwave dryers can extend the life of clothes because fabric temperature will be lower and the action of the tumbler will be reduced or eliminated. Research indicates that microwave dryers are more efficient than conventional electric dryers by 20 to 26%.

Several U.S. appliance manufacturers have experimented with microwave clothes dryers, and a few small companies have built demonstration machines. EPRI has been working with two California companies to provide industry with generic data that could be used in the development of new microwave products; one of their goals is to quantify the hazards (e.g., arcing and heating of metals, such as zippers, possibly leading to dryer fires) and to develop designs that would minimize or eliminate those hazards. EPRI, through the Industrial Advisor Committee, will be placing 12 test units in homes around the country in 1995.

Heat Pump Dryers

A heat pump dryer is essentially a clothes dryer and an air conditioner packaged as one appliance. In a heat pump dryer, exhaust heat energy is recovered by recirculating all the exhaust air back to the dryer; the moisture in the recycled air is removed by a refrigeration-dehumidification system. A drain is required to remove the condensate; because washers and dryers are usually located side by side, a drain is generally easily accessible. Research has indicated that heat pump dryers can be 50 to 60% more efficient than conventional electric dryers. No one is selling these at this time. One U.S. company has successfully tested several prototypes; several European companies also have test models.

12C.4 Cost-Effectiveness of Efficient Designs

Consumer choices regarding what appliances to purchase depend not only on how much energy can be saved but also the additional cost of the more efficient appliance. The simple payback time for an energy-efficient appliance is the number of years required for the consumer to recoup the initial incremental investment in an energy-efficiency measure through the resulting reduction in energy bills. The payback time is calculated by dividing the increased purchase price of the appliance by the dollar savings per year attributed to the more efficient appliance. Table 12C.3 shows the reduction in energy use, increase in purchase cost, and simple payback time for each of the efficiency options mentioned earlier.

TABLE 12C.3 Price Efficiency for Energy-Efficient Appliances

Appliance Efficiency Option	Reduction in Energy Use[1] (%)	Increase in Purchase Price ($)	Simple Payback Time[2] (years)
Refrigerators and Freezers			
Vacuum panel insulation	17	100	10
Improved fan motors	9	25	5
Water Heaters			
Gas: condensing	34	1,672	27
EPRI/E-tech heat pump	65	720	3
Solar	≈60	≈2,500	10
Furnaces[3]			
Gas-fired condensing furnaces	17	≈740	12
Integrated water heater and condensing gas furnaces	≈20	≈1,700	12
Central and Room AC			
Variable-speed central AC	41	4,860	38
Two-speed central AC	33	2,860	27
Efficient room AC	28	449	18
Heat Pumps			
Air-source variable-speed heat pump	35	4,800	13
Air-source two-speed heat pump	27	2,800	10
Ground-source two-speed heat pump	46	4,490	9
Cooking Equipment			
Electronic ignition for gas ranges	47	170	7
Induction cooktop	8	700	352
Microwave: improved power supply to magnetron	12	5	2
Clothes Washers			
Horizontal-axis (compared to electric water heater//gas water heater)	61//63	200	2.3//3.0
High spin-speed (vertical axis)	35	95	4
Clothes Dryers			
Microwave dryers (compared to electric)	23	160	10
Heat pump dryers (compared to electric)	65	300	6

[1] Compared to existing efficiency standard for each appliance.
[2] Energy prices used in the calculation of these payback times were 8.7¢/kWh and 69¢/therm.
[3] Energy savings for efficient furnaces are higher in colder climates and payback times can be less than 10 years.
Source: This table is based on information in Turiel et al., 1995.

Conclusion

Residential appliances consume significant amounts of electricity and natural gas. Consumers typically spend more money annually to operate their appliances than to buy new appliances. The energy consumption of each household is divided among a number of end uses. In this section, the basic engineering principles of the major appliances and space conditioning equipment have been described, and alternative, energy-efficient designs for each have been identified.

Significant potential energy savings are possible beyond the models typically sold in the marketplace today. Among the various end uses, energy savings range from 8 to 65%. Many of these efficient appliances appear to be cost-effective at currently projected manufacturing costs, with simple payback times that are shorter than the typical appliance lifetimes of 10 to 20 years. If costs of efficiency improvements decrease, or if future energy prices increase, more of the potential energy savings that are already technically possible will become increasingly economically attractive. In addition, future research is likely to identify additional technological opportunities to save energy.

Acknowledgments

This work was supported by the Assistant Secretary for Energy Efficiency and Renewable Energy, Office of Buildings Technology, Office of Codes and Standards, of the U.S. Department of Energy, under Contract No. DE-AC03-76SF00098. The opinions expressed in this paper are solely those of the authors and do not necessarily represent those of Ernest Orlando Lawrence Berkeley National Laboratory, or the U.S. Department of Energy.

References

A Portrait of the U.S. Appliance Industry 1992. *Appliance*. September, pp. 41–47.

American Society of Heating, Refrigerating and Air-Conditioning Engineers, Inc. (ASHRAE). 1992. *1992 ASHRAE Handbook: Heating, Ventilating, and Air-Conditioning Systems and Equipment, Inch-Pound Edition*. ASHRAE, Atlanta, GA.

Bancroft, B., Shepard, M., Lovins, A., and Bishop, R. 1991. *The State of the Art: Water Heating*. E-Source, Snowmass, CO.

Energy Information Administration (EIA). 1993. *Household Energy Consumption and Expenditures 1990*. DOE/EIA-0321(90).

Energy Information Administration (EIA). 1992. *Housing Characteristics 1990*. DOE/EIA-0314(90).

Gregerson, J., George, K., Shepard, M., Webster, L., and Davia, D. 1994. *Residential Appliances: Technology Atlas*. E-Source, Snowmass, CO.

Hanford, J.W., Koomey, J.G., Stewart, L.E., Lecar, L.E., Brown, R.E., Johnson, F.X., Hwang, R.J., and Price, L.K. 1994. *Baseline Data for the Residential Sector and Development of a Residential Forecasting Database*. Lawrence Berkeley National Laboratory, Berkeley, CA. LBL-33717.

Levine, M.D., Geller, H., Koomey, J., Nadel, S., and Price, L. 1992. *Electricity End-Use Efficiency: Experience with Technologies, Markets, and Policies Throughout the World*. Lawrence Berkeley National Laboratory, Berkeley, CA. LBL-31885.

Meier, Alan. 1995. "Refrigerator Energy Use in the Laboratory and the Field." *Energy and Buildings*, 22(3):p 233-243.

Proctor and Gamble. 1989. *Comments on Notice of Proposed Rulemaking by Proctor and Gamble*. Cincinnati, Ohio.

Turiel, I., Atkinson, B., Boghosian, S., Chan, P., Jennings, J., Lutz, J., McMahon, J., and Rosenquist, G. 1995. *Evaluation of Advanced Technologies for Residential Appliances and Residential and Commercial Lighting*. Lawrence Berkeley National Laboratory, Berkeley, CA. LBL-35982. Draft Report.

U.S. Department of Energy (DOE). 1988. *Technical Support Document: Energy Conservation Standards for Consumer Products: Refrigerators, Furnaces, and Television Sets*. DOE/CE-0239.

U.S. Department of Energy (DOE). 1993. *Technical Support Document: Energy Efficiency Standards for Consumer Products: Room Air Conditioners, Water Heaters, Direct Heating Equipment, Mobile Home Furnaces, Kitchen Ranges and Ovens, Pool Heaters, Fluorescent Lamp Ballasts and Television Sets*. Volume 2: *Fluorescent Lamp Ballasts, Television Sets, Room Air Conditioners, and Kitchen Ranges and Ovens*. DOE/EE-0009.

U.S. Department of Energy (DOE). 1994. *Core Data Book, Office of Building Technologies*.

13A

Recuperators, Regenerators and Storage: Recuperators, Regenerators and Compact Heat Exchangers

Ramesh K. Shah
University of Kentucky

Introduction

A heat exchanger is a device to provide for change of mutual thermal energy (enthalpy) levels between two or more fluids, between a solid surface and a fluid, or between solid particulates and a fluid in thermal contact without external heat and work interactions. The fluids may be single

0-8493-2514-5/96/$0.00+$.50

compounds or mixtures. Typical applications involve heating or cooling of a fluid stream of concern, evaporation or condensation of single or multicomponent fluid stream, and heat recovery or heat rejection from a system. In other applications, the objective may be to heat, cool, condense, vaporize, sterilize, pasteurize, fractionate, distill, concentrate, crystallize, or control process fluid. In some heat exchangers, the fluids transferring heat are in direct contact. In other heat exchangers, heat transfer between fluids takes place through a separating wall or into and out of a wall in a transient manner. In most heat exchangers, the fluids are separated by a heat transfer surface, and ideally they do not mix. Such exchangers are referred to as *direct transfer type*, or simply *recuperators*. In contrast, exchangers in which there is an intermittent flow of heat from the hot to cold fluid—via heat storage and heat rejection through the exchanger surface or matrix—are referred to as *indirect transfer type* or *storage type exchangers*, or simply *regenerators*.

A heat exchanger consists of heat exchanging elements such as a core or a matrix containing the heat transfer surface, and fluid distribution elements such as headers, manifolds, tanks, inlet and outlet nozzles or pipes, or seals. Usually there are no moving parts in a heat exchanger; however, there are exceptions such as a rotary regenerator, in which the matrix is mechanically driven to rotate at some design speed.

The heat transfer surface is a surface of the exchanger core that is in direct contact with fluids and through which heat is transferred by conduction in a recuperator. The portion of the surface which also separates the fluids is referred to as *primary* or *direct surface*. To increase heat transfer area, appendages known as fins may be intimately connected to the primary surface to provide *extended*, *secondary*, or *indirect surface*. Thus, the addition of fins reduces the thermal resistance on that side and thereby increases the net heat transfer from the surface for the same temperature difference.

Heat exchangers may be classified according to transfer process, construction, flow arrangement, surface compactness, number of fluids, and heat transfer mechanisms as shown in Figure 13A.1 (Shah, 1981; Shah and Mueller, 1988) or according to process function as shown in Figure 13A.2 (Shah and Mueller, 1988). Further general descriptions of heat exchangers are provided in Walker (1990), Saunders (1988), and Hewitt (1989). In the following section, waste heat recovery exchangers are further described, since this information is not readily available.

A gas-to-fluid heat exchanger is referred to as compact heat exchanger if it incorporates heat transfer surface having a surface area density above about 700 m^2/m^3 (213 ft^2/ft^3) on at least one of the fluid sides that usually has gas flow. It is referred to as a laminar flow heat exchanger if the surface area density is above about 3,000 m^2/m^3 (914 ft^2/ft^3), and as a micro heat exchanger if the surface area density is above about 10,000 m^2/m^3 (3,050 ft^2/ft^3). A liquid/two-phase heat exchanger is referred to as compact heat exchanger if the surface area density on any one fluid side is above about 400 m^2/m^3 (122 ft^2/ft^3). A typical process industry shell-and-tube exchanger has a surface area density of less than 100 m^2/m^3 on one fluid side with plain tubes, and 2 or 3 times that with the high-fin-density, low-finned tubing. Plate-fin, tube-fin, and rotary regenerators are examples of compact heat exchangers for gas flows on one or both fluid sides, and gasketed and welded plate heat exchangers are examples of compact heat exchangers for liquid flows.

In this chapter, waste heat recovery exchangers are described first, followed by exchanger heat transfer and pressure drop analysis. Next, theoretical results/insights and empirical correlations for nondimensional heat transfer and flow friction characteristics of exchanger surfaces are presented. Overall design methodology and step-by-step design procedures for the exchanger rating and sizing problems are then outlined. Finally, the flow maldistribution and fouling problems and design considerations are summarized.

There is ever-increasing use of compact heat exchangers for advanced power cycles, and hence the focus in this chapter is on compact heat exchangers. Readers are referred to excellent works of Singh and Soler (1984), Palen (1987), Saunders (1988), Hewitt (1989), Yokell (1990), and Hewitt et al. (1994) for design information on shell-and-tube and other heat exchangers.

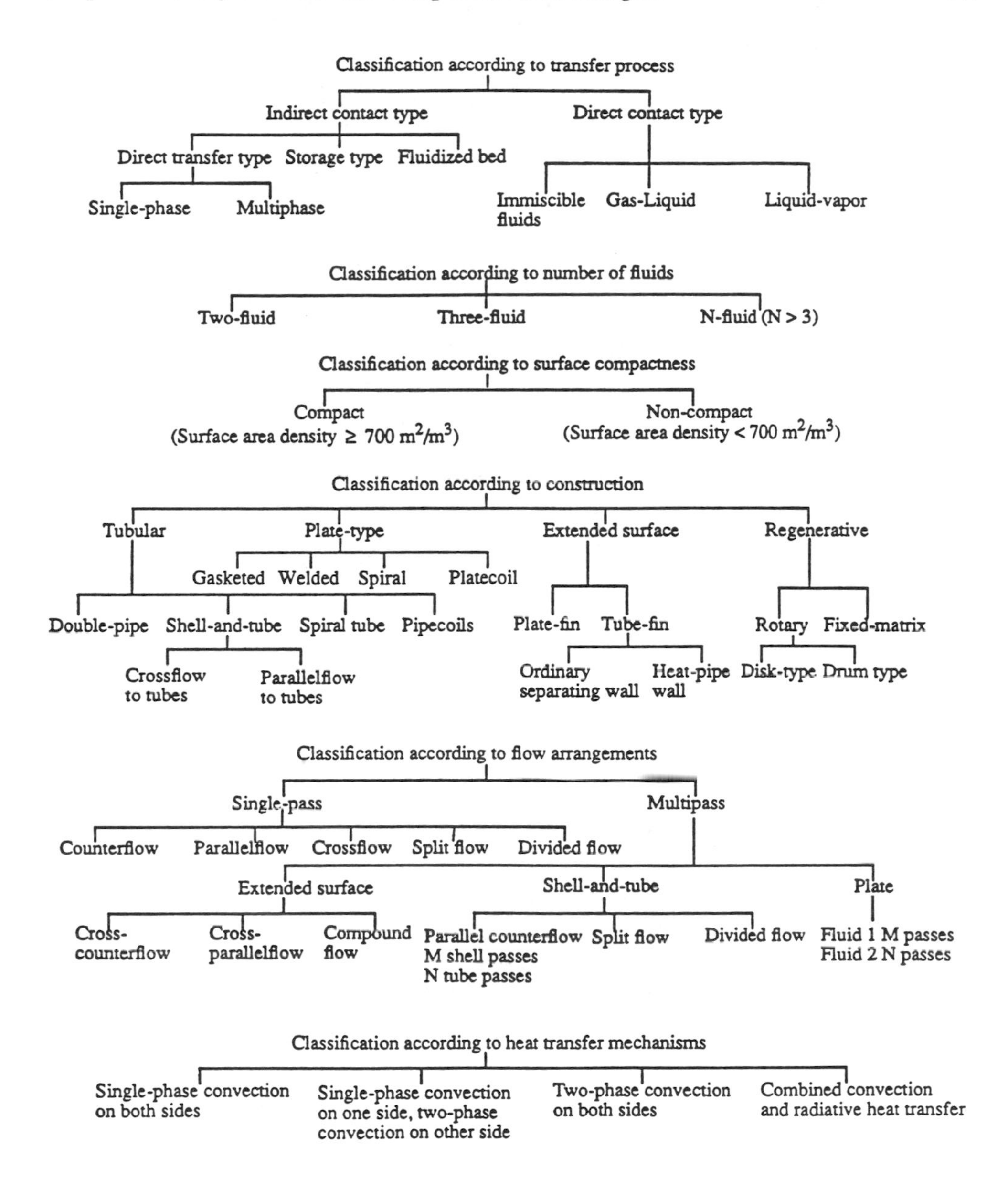

FIGURE 13A.1 General classification of heat exchangers. *Source:* Modified from Shah and Mueller, 1988.

13A.1 Waste Heat Recovery Exchangers

From the energy conservation and operating cost reduction point of view, the thermal energy recovery from exhaust gases and combustion products is getting common in many industries. Depending upon the temperature range of the hot gases, the waste heat recovery may be classified as low-temperature (lower than 230°C or 450°F), medium-temperature (between 230 and 650°C or 450 and 1,200°F), and high-temperature (greater than 650°C or 1,200°F) recovery. If the recovered thermal energy is transferred back in the same process (as by preheating incoming air with the exhaust gas in the same process), this is referred to as internal heat recovery. If the recovered heat is used in another process (such as heating of water for space heating from the exhaust gas), this is referred to as external heat recovery (Meunier, 1991). In addition to hot gases as the source of waste heat, the other sources of waste heat could be process steam, process liquids/solids, and

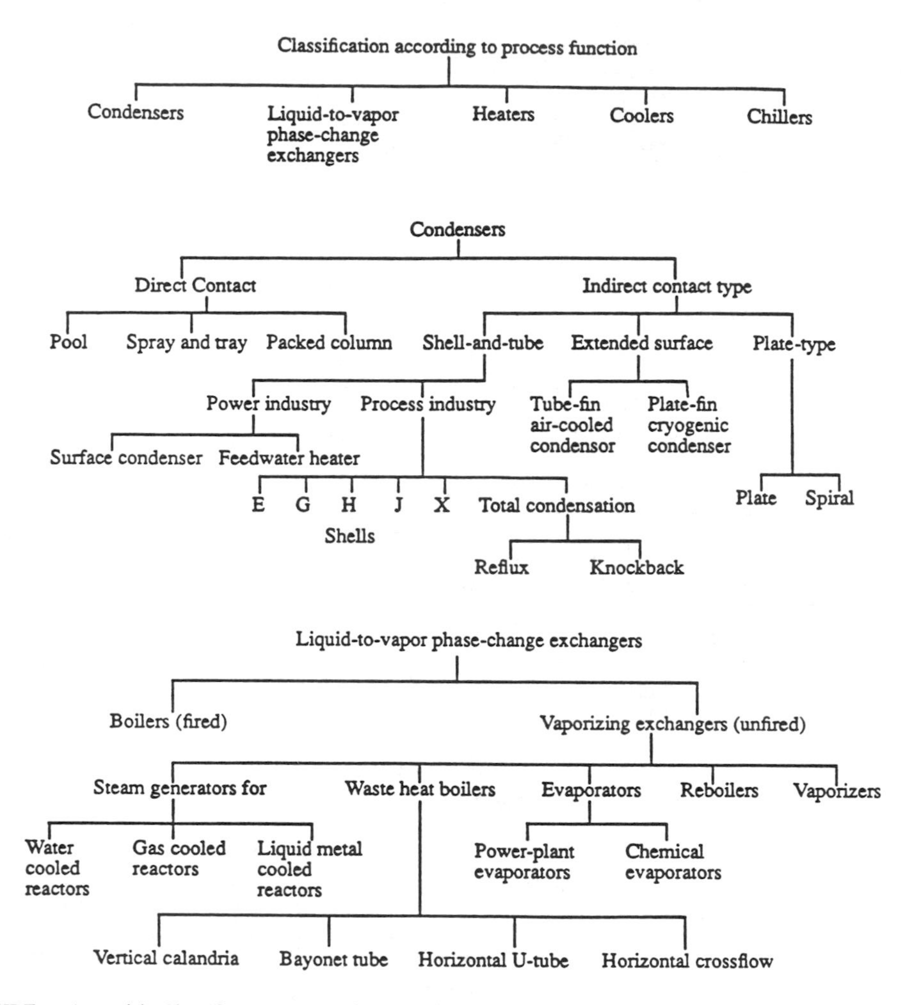

FIGURE 13A.2 (a) Classifications according to the process functions, (b) Classification of condensers, (c) Classification of liquid-to-vapor phase-change exchangers. *Source:* Modified from Shah and Mueller, 1988.

exhaust air. Waste heat can be utilized in many ways, such as preheating combustion air or boiler water, generating electrical and mechanical power or process steam, heating general process liquids and solids or viscous and corrosive liquids, and heating, ventilation, and refrigeration applications. Early waste heat recovery exchangers were extensions of then existing exchangers, and were generally not optimized. Now new designs and materials are more often introduced, and designs are optimized based on the service requirements and understanding of heat transfer phenomena. In general, to preserve strength and resist oxidation or corrosion is an engineering challenge in high-temperature waste heat recovery, and to mitigate the problems of fouling and corrosion is a challenge in low-temperature waste heat recovery. Waste heat recovery exchangers may be classified as gas-to-gas, gas-to-liquid, and liquid-to-liquid recovery exchangers as shown in Figure 13A.3; these three types of exchangers are primarily used in high-, medium-, and low-temperature waste heat recovery applications respectively. It should be emphasized that if the desired maximum temperature of recovered heat is only low to medium, the most cost-effective method to recover thermal energy from a high-temperature waste heat stream is to dilute it and operate the equipment at lower temperatures.

Process efficiency must be improved before considering waste heat recovery from any stream/system; Richlen (1990) also describes five factors for the selection of waste heat recovery exchangers: usability (of available waste heat), temperature, fouling, corrosion characteristics (of the waste heat stream), and quantity (flow rate of waste heat stream and desired exchanger effectiveness).

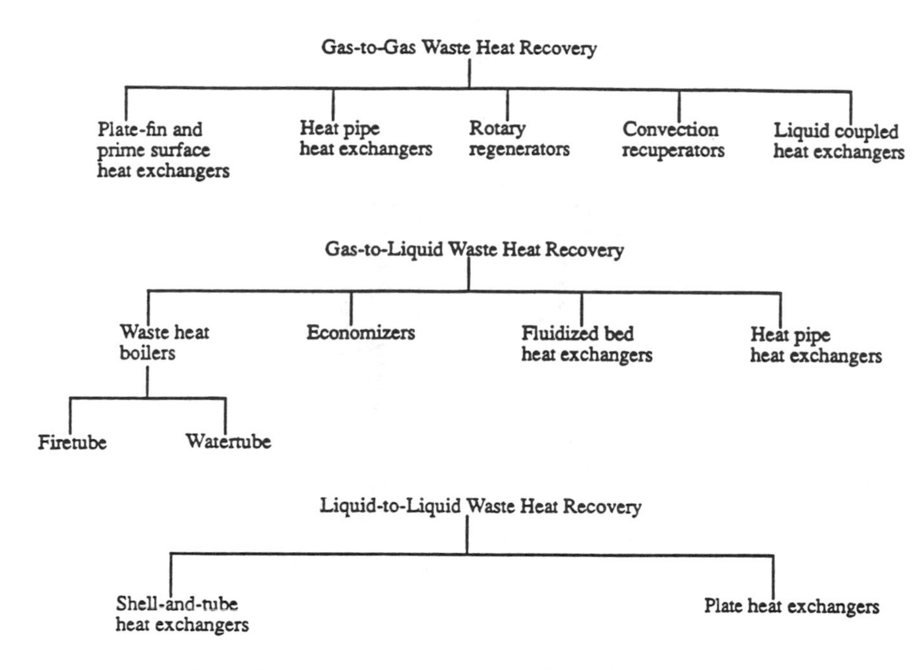

FIGURE 13A.3 Classification of waste heat recovery exchangers.

Comprehensive information is provided by Richlen and Parks (1992) and Henriette (1991) on ceramic heat exchangers used for high- and medium-temperature waste heat recovery. These include tube-in-shell types, bayonet tubes, plain tubular, crossflow plate-fin, and fluidized-bed heat exchangers. Scaccia and Theoclitus (1980) also discuss high-temperature and other heat exchangers.

Gas-to-Gas Waste Heat Recovery Exchangers

Gas-to-gas waste heat recovery exchangers may be categorized as plate-fin and primary surface exchangers, heat pipe heat exchangers, rotary regenerators, radiation and convection recuperators, and runaround coils. Each type of exchanger has its own niche applications. The plate-fin and heat-pipe exchangers and rotary regenerators are described by Shah (1991a). The primary surface exchangers (flat plates with some sort of bumps/ribs to provide desired plate spacing) are usually of brazed constructions for higher operating temperature (up to about 800°C or 1,500°F) applications. Plate-fin exchangers are used for low- and medium-temperature applications, except for the ceramic plate-fin units (Richlen and Parks, 1992). Both plate-fin and primary surface exchangers are used for relatively low heat duty and low pressure drop applications.

A *metallic radiation recuperator* consists of two concentric metal tubes as shown in Figure 13A.4 with the hot exhaust (flue) gas flowing through the central duct and the air to be preheated flowing in the outer annulus (Goldstick and Thumann, 1986; Meunier, 1991; Reay, 1979). The pressure drop on the hot gas side is generally extremely small, and this recuperator can act as part of a chimney or flue with the length of the recuperator being up to 50 m (165 ft) long, thus producing natural draft and eliminating a need for a fan. The diameter of radiation recuperators varies from 0.25 m (10 inches) to 3 m (10 ft) depending upon the application. In the recuperator, the majority of the heat is transferred from the hot gas to the inner wall by radiation, thus providing air preheating at temperatures up to the maximum usable by the best burners ordinarily available (Boyen, 1976). However, heat transfer to air in the outer annulus takes place by convection, since the air is transparent to infrared radiation. Air and gas flowing in counterflow directions is the most desirable flow arrangement from a thermal performance point of view. However, in high-temperature applications, a parallelflow arrangement may be used in order to maintain low wall

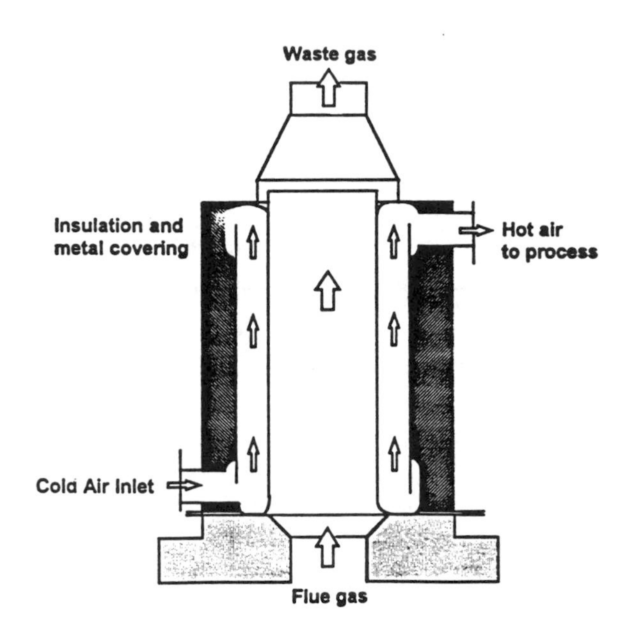

FIGURE 13A.4 A metallic radiation recuperator.

temperatures at the hot gas inlet. Sometimes, the recuperator is made up of two sections, a parallelflow arrangement in the first short section at the hot gas inlet, and then counterflow in the second section, where the inner wall temperatures are within specified limits. Typical exchanger effectiveness is 40% or lower. Smooth inner walls prevent the deposition of particles and hence can be used for dirty gases. An expansion bellows or a flexible part is provided at the cold end to allow for thermal expansion. This type of recuperator is best suited for continuous operation such as in steel plants and glass-melting furnaces. The inner wall of metallic recuperators is made up of high-nickel stainless steels with the operating temperature limits of 1,053°C or 1,925°F. For operating temperatures of up to 1,500°C (2,800°F), ceramic tube recuperators have been developed. In this design, ceramic tubes are connected to two headers and hot gas flows normal to the tubes (Goldstick and Thumann, 1986; Reay, 1979). Typical exchanger effectiveness for ceramic recuperators is 70 to 85%.

Convection recuperators consist of one or more tube bundles of plain or finned tubes with the flue gas flowing normal to the tubes and air to be preheated flowing in the tubes. The size, type, and form of the recuperator depends upon the heat duty, operating temperature range, and fouling characteristics of the gas. For low-temperature applications (250°C or 500°F), a glass tube crossflow recuperator is ideal for highly corrosive or fouling gases such as in the textile industry (Reay, 1979) or preheating combustion air for nonutility boilers firing heavy fuel oil. For high-temperature applications (1,000°C or 1,900°F or less), either drawn tubes (for high-pressure applications) or cast tubes (for low-pressure applications) with or without fins are used. Ceramic tube recuperators are used for higher operating temperatures of up to about 1,450°C (2,600°F) (Richlen and Parks, 1992).

A *runaround coil* (liquid-coupled indirect heat exchanger) consists of two exchangers (coils) connected to each other by a circulating liquid that transfers heat from the hot fluid to the cold fluid. The hot and cold fluids in general are gas and air in many waste heat recovery applications, as shown in Figure 13A.5, although they can be gas and liquid or liquid and liquid. Essentially, each unit is a gas-to-liquid exchanger with extended surface on the gas side and both units are connected by the liquid loop with a pump. This system is generally not used in furnaces but is used in a heat recovery system without rerouting the ductwork for the waste gas and cold air, since the cost and space requirement for a liquid pipe is much less than the cost of air ducting. Thus the coils in the inlet air and exhaust gas ducts (separated by a dashed line in Figure 13A.5) can be many meters away or even on a different floor. Since hot gas and cold air are significantly far apart,

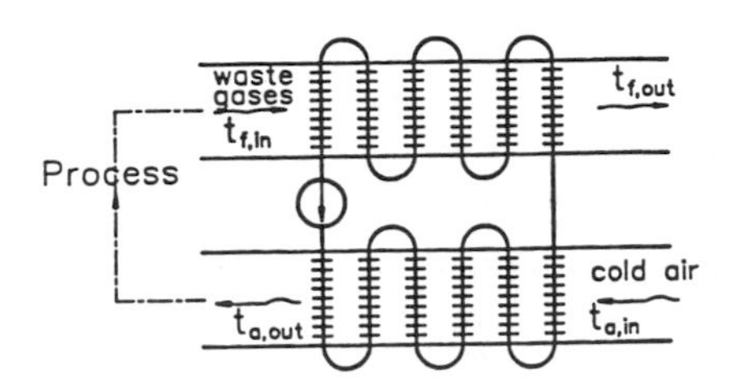

FIGURE 13A.5 A runaround coil.

no cross contamination is possible. The liquid circulating pump, a very reliable component, is the only moving part in the system. Water is used as a circulating liquid for temperatures lower than 180°C (360°F) and heat transfer fluids (oils, Dowtherm, etc.) are used for temperatures up to 300°C (570°F). The runaround coils are used in HVAC applications and drying applications; they have also been used in a number of industrial applications where high- and low-temperature fluids cannot be brought close together and the risk of mixing them must be minimized.

Gas-to-Liquid Waste Heat Recovery Exchangers

These exchangers may be categorized as waste heat boilers, economizers, thermal fluid heaters, fluidized-bed heat exchangers, and heat-pipe heat exchangers (Reay, 1979). Economizers and thermal fluid heaters are used for low- to medium-temperature waste heat recovery, while the rest can be used for either medium- or high-temperature waste heat recovery. These exchangers are briefly described next. Waste heat boilers are used in recovering heat from the flue gases from gas turbines, incinerators, furnaces, and so on, or for cooling and controlling chemical process gases by generating steam. In contrast, economizers produce hot water by using the waste heat. The steam generated by a waste heat boiler can be used for industrial processes, power generation, or space heating. A combination boiler that operates both by firing fuel and by removing heat from a gas stream is also referred to as a waste heat boiler. Waste heat boilers may be categorized as firetube and watertube units. In a *firetube boiler*, the hot gas flows through the tubes, and a water-steam mixture flows outside the tubes. In a *watertube boiler*, the water-steam mixture flows through the tubes and the hot gas flows outside the tubes. Firetube boilers are designed with natural circulation, and watertube boilers with natural or forced circulation.

Firetube boilers are used to recover or remove heat from relatively high-pressure, high-temperature process gases. For example, in the steam-methane reforming process, the high-pressure gas leaves the reformer at several hundred kPa (several atmospheric pressures) and at 870 to 980°C (1,600–1,800°F). The boiler resembles a shell-and-tube type heat exchanger. The firetube boilers are generally limited to steam pressure of under 6.9 MPa gauge (1,000 psig). They are usually more economical than the watertube boilers if the flow area required is less than 0.4 m^2 (4 ft^2), since construction is relatively uncomplicated. The heat exchanger effectiveness of firetube boilers usually ranges from 65 to 75% (Eriksen, 1980). Refer to Csathy (1980) and Hinchley (1979) for further design details.

Watertube boilers are used to recover heat from flue gases at near atmospheric pressure. Nearly all watertube boilers have finned tubes; plain tubes are used for those watertube boilers having gases above 760°C (1,400°F). The heat exchanger effectiveness for watertube boilers with fins usually ranges from 90 to 95%, and that for watertube boilers without fins from 65 to 75% (Eriksen, 1980). In natural circulation boilers, thermosiphon action or natural circulation results from the density difference between the steam–water mixture in the steaming area and the water in the downcomer. In this case, watertubes are either vertical or inclined. When horizontal tubes are used in the boiler with vertical gas flow, forced circulation is mandatory, since buoyant force is not sufficient to sustain natural circulation. Forced circulation is achieved with pumps. Natural circulation boilers cost less, are easier to construct, and require minimum maintenance. The circulation ratio (water mass flow:steam mass flow) in natural circulation boilers varies from 15:1 to 25:1; the circulation ratio for forced convection boilers is restricted to 5:1 to 7:1 to minimize pump size and power consumption. Refer to Csathy (1980) and Hinchley (1979) for further design details. A number of special waste heat boiler designs are available, including (1) integral burners to burn liquid or solid waste products; (2) integral Ljungstrom rotary regenerators to save fuel; or (3) trailer-mounted boilers and a special packaged boiler in modest sizes for niche applications (Watts et al. 1984).

Economizers and Water Heaters

An economizer in most applications is used with boilers to preheat the boiler feedwater with the hot flue gases that leave the boiler. In applications other than boilers, it is referred to as a process fluid or water heater. The use of economizers has spread to other applications (such as for process heat, waste incineration, process water heating, etc.) to heat water or raise steam using the heat from the hot flue gas. In a direct contact economizer, liquid is sprayed to capture the latent heat and the remaining sensible heat. An economizer is an individually finned tube bundle, with gas flowing outside normal to the finned tubes and water inside the tubes. Finned tubes are rugged, heavy, usually made up of steel, and able to withstand flue gas temperatures of up to 900°C (1,650°F). At the hot gas inlet side, water hammer can occur due to formation of steam pockets. The other operating problem is that if the hot gas temperature drops below the dew point of condensation of acids in the gas, it can corrode the steel tube. As a rule of thumb, 1% fuel consumption is saved for every 6°C (11°F) rise in the feedwater temperature (Reay, 1979).

Thermal Fluid Heaters

These heaters operate on the same principle as a central domestic warm-water system, except that the water is replaced by a high-temperature organic heat transfer fluid (operating between –50 and +400°C or –60 and +750°F). They circulate throughout the plant while receiving or rejecting heat through many processes. Thermal fluid heaters are used to conserve energy in three ways: they use the combustion of waste products for firing the burners, they are used to heat and cool the plant in the same building, and they are used for heat recovery (Reay, 1979). For further details refer to Boyen (1976). Fluid heaters are firetube or liquid-tube types, like the waste heat boilers, and both systems can be fired or could use waste heat sources. Fired-fluid heaters are sometimes referred to as furnaces.

Fluidized-Bed Heat Exchangers

In this exchanger, one side of a two-fluid exchanger is immersed in a bed of finely divided solid material such as a tube bundle immersed in a bed of sand or coal particles. If the upward fluid velocity on the bed side is low, the solid particles will remain fixed in position and the fluid will flow through the interstices of the bed. If the upward fluid velocity is high, the solid particles will be carried away with the fluid. At a proper value of the fluid velocity, the upward drag force is slightly higher than the weight of the bed particles. As a result, the solid particles will float with an increase in the bed volume, and the bed behaves as a liquid. This characteristic of the bed is referred to as fluidized condition. Then the fluid pressure drop through the bed remains almost constant, independent of the flow rate. A strong mixing of the solid particles occurs. This results in an isothermal temperature for the total bed (gas and particles) with an apparent thermal conductivity of the solid particles as infinity. Very high heat transfer coefficients compared to particle-free or dilute-phase particle gas flows are achieved on the fluidized side. For example, the typical heat transfer coefficient of 60 W/m^2 K for flow normal to a tube bank is increased to between 150 and 500 W/m^2 K depending upon the particle size. A plain or finned tube bundle or heat pipes are embedded in the fluidized bed. Water, steam, or heat transfer fluid in the tubes is heated in the fluidized bed with waste heat recovery from hot gases flowing over the fluidized bed. This exchanger is used for heating boiler feedwater, process tanks, washing machines, dyeing and printing plant fluids, space heating, and domestic hot water services (Reay, 1979).

Gas–Liquid Heat-Pipe Heat Exchangers

This exchanger is used when there is an absolute requirement that the gas and liquid do not mix even with small leaks. Because of the higher heat transfer coefficients with liquids, the heat-pipe portion (condenser) in the liquid is unfinned and that (evaporator) in the gas stream is finned. Otherwise, the principle of operation is the same as that for the gas-to-gas heat-pipe heat exchanger (Shah, 1991a).

Liquid-To-Liquid Waste Heat Recovery

Liquid-to-liquid or condensing fluid-to-liquid waste heat recovery is common in many industrial processes. The exchangers used most commonly are tubular and plate types, particularly shell-and-tube and plate heat exchangers. Whenever liquids and condensing fluids are at high pressures and temperatures and with or without significant fouling, shell-and-tube construction may be preferred. These exchangers are discussed by Shah and Mueller (1989) and Shah (1991a); see also Figures 13A.1 and 13A.2 for different types of tubular, plate-type, and other exchangers available for liquid-to-liquid and condensing fluid-to-liquid waste heat recovery applications.

13A.2 Exchanger Heat Transfer and Pressure Drop Analysis

In this section, starting with the thermal circuit associated with a two-fluid exchanger, ε-NTU, *P*-NTU, and MTD methods used for an exchanger analysis are presented, followed by the fin efficiency concept and various expressions. Finally, pressure drop expressions are outlined for various single-phase exchangers.

Thermal Circuit

In order to develop relationships among variables for various exchangers, consider the counterflow exchanger of Figure 13A.6 as an example. Two energy conservation differential equations for a two-fluid exchanger with any flow arrangement are

$$dq = q''\,dA = -C_h dT_h = \pm C_c dT_c \tag{13A.1}$$

where the ± sign depends upon whether dT_c is increasing or decreasing with increasing dA. The overall rate equation on a local basis is

$$dq = q''\,dA = U(T_h - T_c)_{\text{local}}\,dA = U\,\Delta T\,dA \tag{13A.2}$$

Integration of Eqs. (13A.1) and (13A.2) across the exchanger surface area results in

$$q = C_h(T_{h,i} - T_{h,o}) = C_c(T_{c,o} - T_{c,i}) \tag{13A.3}$$

and

$$q = UA\,\Delta T_m = \Delta T_m / R_o \tag{13A.4}$$

Here ΔT_m is the true mean temperature difference dependent upon the exchanger flow arrangement and degree of fluid mixing within each fluid stream. The inverse of the overall thermal conductance *UA* is referred to as the overall thermal resistance R_o as follows (see Figure 13A.7).

$$R_o = R_h + R_{s,h} + R_w + R_{s,c} + R_c \tag{13A.5}$$

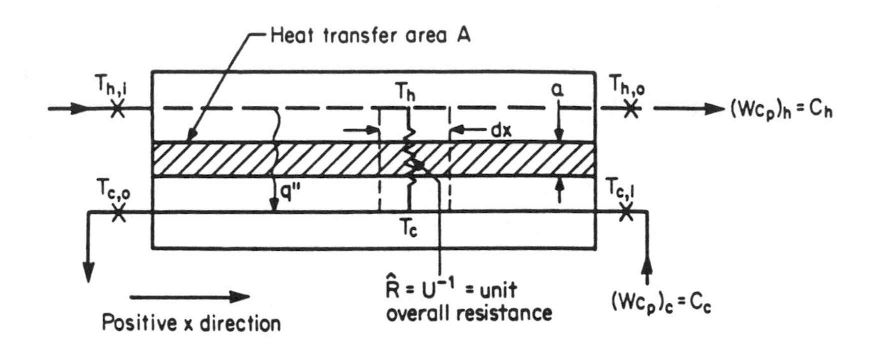

FIGURE 13A.6 Nomenclature for heat exchanger variables.

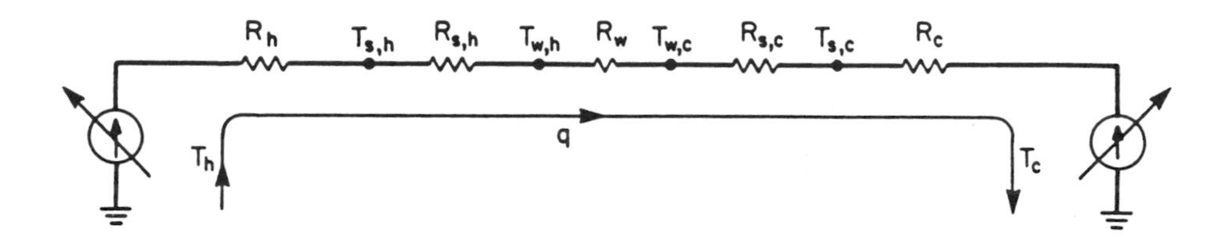

FIGURE 13A.7 Thermal circuit for heat transfer in an exchanger.

where the subscripts *h*, *c*, *s*, and *w* denote hot, cold, scale (or fouling), and wall respectively. In terms of the overall and individual heat transfer coefficients, Eq. (13A.5) is represented as

$$\frac{1}{UA} = \frac{1}{(\eta_o hA)_h} + \frac{1}{(\eta_o h_s A)_h} + R_w + \frac{1}{(\eta_o h_s A)_c} + \frac{1}{(\eta_o hA)_c} \tag{13A.6}$$

where η_o = the total surface effectiveness of an extended (fin) surface and is related to the fin efficiency η_f, fin surface area A_f and the total surface area A as follows:

$$\eta_o = 1 - \frac{A_f}{A}(1 - \eta_f) \tag{13A.7}$$

The wall thermal resistance R_w of Eq. (13A.5) is given by

$$R_w = \begin{cases} \delta / A_w K_w & \text{for a flat wall} \\ \dfrac{\ln(d_o/d_i)}{2\pi k_w L N_t} & \text{for a circular tube with a single-layer wall} \\ \dfrac{1}{2\pi L N_t}\left[\sum_j \dfrac{\ln(d_{j+1}/d_j)}{k_{w,j}}\right] & \text{for a circular tube with a multiple-layer wall} \end{cases} \tag{13A.8}$$

If there is any contact or bond resistance present between the fin and tube or plate on the hot or cold fluid side, it is included as an added thermal resistance on the right-hand side of Eq. (13A.5) or (13A.6). For a heat-pipe heat exchanger, additional thermal resistances associated with the heat pipe should be included on the right-hand side of Eq. (13A.5) or (13A.6); these resistances are evaporator resistance at the evaporator section of the heat pipe, viscous vapor flow resistance inside the heat pipe (very small), internal wick resistance at the condenser section of the heat pipe, and condensation resistance at the condenser section.

If one of the resistances on the right-hand side of Eq. (13A.5) or (13A.6) is significantly higher than the otheı resistances, it is referred to as the controlling resistance. A reduction in the controlling thermal resistance will have much more impact in reducing the exchanger surface area (A) requirement compared to the reduction in A due to the reduction in other thermal resistances.

UA of Eq. (13A.6) may be defined in terms of hot or cold fluid side surface area or wall conduction area as

$$UA = U_h A_h = U_c A_c = U_w A_w \quad \textbf{(13A.9)}$$

When R_w is negligible, $T_{w,h} = T_{w,c} = T_w$ of Figure 13A.7 is computed from

$$T_w = \frac{T_h + \left[(R_h + R_{s,h}) / (R_c + R_{s,c}) \right] T_c}{1 + \left[(R_h + R_{s,h}) / (R_c + R_{s,c}) \right]} \quad \textbf{(13A.10)}$$

When $R_{s,h} = R_{s,c} = 0$, Eq. (13A.10) reduces to

$$T_w = \frac{T_h / R_h + T_c / R_c}{1/R_h + 1/R_c} = \frac{(\eta_o hA)_h T_h + (\eta_o hA)_c T_c}{(\eta_o hA)_h + (\eta_o hA)_c} \quad \textbf{(13A.11)}$$

ε-NTU, *P*-NTU, and MTD Methods

If we consider the fluid outlet temperatures or heat transfer rate as dependent variables, they are related to independent variable/parameters of Figure 13A.6 as follows.

$$T_{h,o},\ T_{c,o} \text{ or } q = \phi\{T_{h,i},\ T_{c,i},\ C_c,\ C_h,\ U,\ A,\ \text{flow arrangement}\} \quad \textbf{(13A.12)}$$

Six independent and three dependent variables of Eq. (13A.12) for a given flow arrangement can be transferred into two independent and one dependent dimensionless groups; three different methods are presented in Table 13A.1 based on the choice of three dimensionless groups. The relationship among three dimensionless groups is derived by integrating Eqs. (13A.1) and (13A.2) across the surface area for a specified exchanger flow arrangement. Such expressions are presented later, in Table 13A.3, for industrially most important flow arrangements. Now we will briefly describe the three methods.

The ε-NTU Method

In this method, the heat transfer rate from the hot fluid to the cold fluid in the exchanger is expressed as

$$q = \varepsilon C_{\min} (T_{h,i} - T_{c,i}) \quad \textbf{(13A.13)}$$

Here the exchanger effectiveness ε is an efficiency factor. It is a ratio of the actual heat transfer rate from the hot fluid to the cold fluid in a given heat exchanger of any flow arrangement to the maximum possible heat transfer rate $q_{\max}$ thermodynamically permitted by the Second Law of Thermodynamics. The $q_{\max}$ is obtained in a *counterflow* heat exchanger (recuperator) of *infinite surface area* operating with the fluid flow rates (heat capacity rates) and fluid inlet temperatures

TABLE 13A.1 General Functional Relationships and Dimensionless Groups for ε-NTU, ε-NTU, and LMTD Methods

ε-NTU method	*P*-NTU method*
$q = \varepsilon C_{\min}(T_{h,i} - T_{c,i})$	$q = P_1 C_1 \lvert T_{1,i} - T_{2,i} \rvert$
$\varepsilon = \phi(\text{NTU}, C^*, \text{flow arrangment})$	$P_1 = \phi(\text{NTU}_1, R_1, \text{flow arrangement})$
$\varepsilon = \dfrac{C_h(T_{h,i} - T_{h,o})}{C_{\min}(T_{h,i} - T_{c,i})} = \dfrac{C_c(T_{c,o} - T_{c,i})}{C_{\min}(T_{h,i} - T_{c,i})}$	$P = \dfrac{T_{1,o} - T_{1,i}}{T_{2,i} - T_{1,i}}$
$\text{NTU} = \dfrac{UA}{C_{\min}} = \dfrac{1}{C_{\min}} \int_A U\,dA$	$\text{NTU}_1 = \dfrac{UA}{C_1} = \dfrac{\lvert T_{1,o} - T_{1,i} \rvert}{\Delta T_m}$
$C^* = \dfrac{C_{\min}}{C_{\max}} = \dfrac{(\dot{m}c_p)_{\min}}{(\dot{m}c_p)_{\max}}$	$R = \dfrac{C_1}{C_2} = \dfrac{T_{2,i} - T_{2,o}}{T_{1,o} - T_{1,i}}$

MTD method*
$q = UAF\,\Delta T_{lm}$
$\text{LMTD} = \Delta T_{lm} = \dfrac{\Delta T_1 - \Delta T_2}{\ln(\Delta T_1/\Delta T_2)}$
$\Delta T_1 = T_{h,i} - T_{c,o} \quad \Delta T_2 = T_{h,o} - T_{c,i}$
$F = \phi(P, R, \text{flow arrangement})$
$F = \dfrac{\Delta T_m}{\Delta T_{lm}}$
P and R are defined in the P-NTU method.

* Although *P*, *R*, and NTU are defined on Fluid 1 side, it must be emphasized that all the results of the *P*-NTU and MTD methods are valid if the definitions of *P*, NTU, and *R* are consistently based on C_t, C_s, C_h, or C_c.

equal to those of an actual exchanger (constant fluid properties are idealized). As noted in Table 13A.1, the exchanger effectiveness ε is a function of NTU and C^* in this method. The number of transfer units NTU is a ratio of the overall conductance *UA* to the smaller heat capacity rate $C_{\min}$. NTU designates the dimensionless "heat transfer size" or "thermal size" of the exchanger. Other interpretations of NTU are given by Shah (1983). The heat capacity rate ratio C^* is simply a ratio of the smaller to the larger heat capacity rate for the two fluid streams. Note that $0 < \varepsilon < 1$, $0 < \text{NTU} < \infty$ and $0 \le C^* \le 1$.

The *P*-NTU Method

This method represents a variant of the ε-NTU method. The ε-NTU relationship is different depending upon whether the shell fluid is the $C_{\min}$ or $C_{\max}$ fluid in the (stream unsymmetric) flow arrangements commonly used for shell-and-tube exchangers. In order to avoid possible errors and confusion, an alternative is to present the temperature effectiveness *P* as a function of NTU and *R*, where *P*, NTU, and *R* are defined consistently either for the Fluid 1 side or the Fluid 2 side; in Table 13A.1, they are defined for the Fluid 1 side (regardless of whether that side is hot or cold fluid side), and the Fluid 1 side is clearly identified for each flow arrangement in Table 13A.3; it is the shell side in a shell-and-tube exchanger. Note that

$$q = P_1 C_1 |T_{1,i} - T_{2,i}| = P_2 C_2 |T_{2,i} - T_{1,i}| \tag{13A.14}$$

$$P_1 = P_2 R_2 \text{ or } P_2 = P_1 R_1 \tag{13A.15}$$

$$NTU_1 = NTU_2 R_2 \text{ or } NTU_2 = NTU_1 R_1 \tag{13A.16}$$

and

$$R_1 = 1/R_2 \tag{13A.17}$$

The MTD Method

In this method, the heat transfer rate from the hot fluid to the cold fluid in the exchanger is given by

$$q = UAF\Delta T_{lm} \tag{13A.18}$$

Here the log-mean temperature difference correction factor F is a ratio of true (actual) mean temperature difference (MTD) to the log-mean temperature difference (LMTD) where

$$\text{LMTD} = \Delta T_{lm} = \frac{\Delta T_1 - \Delta T_2}{\ln(\Delta T_1/\Delta T_2)} \tag{13A.19}$$

Here ΔT_1 and ΔT_2 are defined as

$$\Delta T_1 = T_{h,i} - T_{c,o} \quad \Delta T_2 = T_{h,o} - T_{c,i} \quad \text{for all flow arrangements except for parallelflow} \tag{13A.20}$$

$$\Delta T_1 = T_{h,i} - T_{c,i} \quad \Delta T_2 = T_{h,o} - T_{c,o} \quad \text{for parallelflow} \tag{13A.21}$$

The LMTD represents a true mean temperature difference for a counterflow arrangement under the idealizations listed next. Thus, the LMTD correction factor F represents a degree of departure for the MTD from the counterflow LMTD; it does not represent the effectiveness of a heat exchanger. It depends on two dimensionless groups P_1 and R_1 or P_2 or R_2 for a given flow arrangement.

The relationships among the dimensionless groups of the ε-NTU, P-NTU and MTD methods are presented in Table 13A.2. The closed-form formulas for industrially important exchangers are presented in terms of P_1, NTU$_1$ and R_1 in Table 13A.3. These formulas are valid under the following idealizations.

1. The heat exchanger operates under steady-state conditions, that is, constant flow rate, and fluid temperatures (at the inlet and within the exchanger) independent of time.
2. Heat losses to the surroundings are negligible.
3. There are no thermal energy sources and sinks in the exchanger walls or fluids.
4. In counterflow and parallelflow exchangers, the temperature of each fluid is uniform over every flow cross section. From the temperature distribution point of view, in crossflow exchangers each fluid is considered mixed or unmixed at every cross section depending upon the specifications. For a multipass exchanger, the foregoing statements apply to each

TABLE 13A.2 Relationships between Dimensionless Groups of the *P*-NTU and MTD Methods and Those of the ε-NTU Method

$$P_1 = \frac{C_{min}}{C_1}\varepsilon = \begin{cases} \varepsilon & \text{for } C_1 = C_{min} \\ \varepsilon C^* & \text{for } C_1 = C_{max} \end{cases}$$
$$R_1 = \frac{C_1}{C_2} = \begin{cases} C^* & \text{for } C_1 = C_{min} \\ 1/C^* & \text{for } C_1 = C_{max} \end{cases}$$
$$\text{NTU}_1 = \text{NTU}\frac{C_{min}}{C_1} = \begin{cases} \text{NTU} & \text{for } C_1 = C_{min} \\ \text{NTU } C^* & \text{for } C_1 = C_{max} \end{cases}$$
$$F = \frac{\text{NTU}_{cf}}{\text{NTU}} = \frac{1}{\text{NTU}(1-C^*)}\ln\frac{1-C^*\varepsilon}{1-\varepsilon} \xrightarrow[C^*=1]{} \frac{\varepsilon}{\text{NTU}(1-\varepsilon)}$$
$$F = \frac{1}{\text{NTU}_1(1-R_1)}\ln\left[\frac{1-R_1P_1}{1-P_1}\right] \xrightarrow[R_1=1]{} \frac{P_1}{\text{NTU}_1(1-P_1)}$$

pass depending upon the basic flow arrangement of the passes; the fluid is considered mixed or unmixed between passes.

5. Either there are no phase changes in the fluid streams flowing through the exchanger or the phase changes (condensation or boiling) occur under one of the following conditions: (a) Phase change occurs at a constant temperature as for a single component fluid at constant pressure; the effective specific heat for the phase-changing fluid is infinity in this case, and hence $C_{max} \to \infty$. (b) The temperature of the phase changing fluid varies linearly with heat transfer during the condensation or boiling. In this case, the effective specific heat is constant and finite for the phase-changing fluid.
6. The specific heat of each fluid is constant throughout the exchanger so that the heat capacity rate on each side is treated as constant.
7. The velocity and temperature at the entrance of the heat exchanger on each fluid side are uniform.
8. For an extended surface exchanger, the overall extended surface temperature effectiveness η_o is considered uniform and constant.
9. The individual and overall heat transfer coefficients are constant throughout the exchanger, including the case of phase-changing fluid in idealization 5.
10. The heat transfer area is distributed uniformly on each fluid side. In a multipass unit, heat transfer surface area is equal in each pass.
11. For a plate-baffled shell-and-tube exchanger, the temperature rise per baffle pass is small compared to the overall temperature rise along the exchanger; that is, the number of baffles is large.
12. The fluid flow rate is uniformly distributed through the exchanger on each fluid side in each pass. No stratification, flow bypassing, or flow leakages occur in any stream. The flow condition is characterized by the bulk (or mean) velocity at any cross section.
13. Longitudinal heat conduction in the fluid and in the wall is negligible.

Idealizations 1 to 4 are necessary in a theoretical analysis of steady-state heat exchangers. Idealization 5 essentially restricts the analysis to single-phase flow on both sides or on one side with a dominating thermal resistance. For two-phase flows on both sides, many of the foregoing idealizations are not valid, since mass transfer in phase change results in variable properties and variable flow rates of each phase, and the heat transfer coefficients vary significantly. As a result, the heat exchanger cannot be analyzed using the theory presented here.

TABLE 13A.3 P_1-NTU_1 Formulas and Limiting Values of P_1 for $R_1 = 1$ and $\mathrm{NTU}_1 \to \infty$ for Various Exchanger Flow Arrangements

Flow arrangement	Eq. no.	General formula	Value for $R_1 = 1$	Value for $\mathrm{NTU}_1 \to \infty$
Counterflow exchanger, stream symmetric	1.1.1	$P_1 = \dfrac{1 - \exp[-\mathrm{NTU}_1(1 - R_1)]}{1 - R_1 \exp[-\mathrm{NTU}_1(1 - R_1)]}$	$P_1 = \dfrac{\mathrm{NTU}_1}{1 + \mathrm{NTU}_1}$	$P_1 \to 1$ for $R_1 \le 1$ $P_1 \to 1/R_1$ for $R_1 \ge 1$
	1.1.2	$\mathrm{NTU} = \dfrac{1}{(1 - R_1)} n \left[\dfrac{1 - R_1 P_1}{1 - P_1}\right]$	$\mathrm{NTU}_1 = \dfrac{P_1}{1 - P_1}$	$\mathrm{NTU}_1 \to \infty$
	1.1.3	$F = 1$	$F = 1$	$F = 1$
Parallel flow exchanger, stream symmetric	1.2.1	$P_1 = \dfrac{1 - \exp[-\mathrm{NTU}_1(1 + R_1)]}{1 + R_1}$	$P_1 = \dfrac{1}{2}[1 - \exp(-2\mathrm{NTU}_1)]$	$P_1 \to \dfrac{1}{1 + R_1}$
	1.2.2	$\mathrm{NTU}_1 = \dfrac{1}{1 + R_1} \ln\left[\dfrac{1}{1 - P_1(1 + R_1)}\right]$	$\mathrm{NTU}_1 = \dfrac{1}{2} \ln\left[\dfrac{1}{1 - 2P_1}\right]$	$\mathrm{NTU}_1 \to \infty$
	1.2.3	$F = \dfrac{(R_1 + 1)\ln\left[\dfrac{1 - R_1 P_1}{1 - P_1}\right]}{(R_1 - 1)\ln[1 - P_1(1 + R_1)]}$	$F = \dfrac{2P_1}{(P_1 - 1)\ln(1 - 2P_1)}$	$F \to 0$
Single-pass crossflow exchanger, both fluids unmixed, stream symmetric	2.1	$P_1 = 1 - \exp(\mathrm{NTU}_1) - \exp[-(1 + R_1)\mathrm{NTU}_1] \cdot \sum_{n=1}^{\infty} R_1^n P_n(\mathrm{NTU}_1)$ $P_n(y) = \dfrac{1}{(n + 1)!} \sum_{j=1}^{n} \dfrac{(n + 1 - j)}{j!} y^{n+j}$	same as Eq. (2.1) with $R_1 = 1$	$P_1 \to 1$ for $R_1 \le 1$ $P_1 \to \dfrac{1}{R_1}$ for $R_1 \ge 1$

* Table condensed from R. K. Shah and A. Pignotti: *Basic Thermal Design of Heat Exchangers*, National Science Foundation Report, Int-8601771, 1988. In this table, all variables, except P_1, R_1, NTU_1, and F, are local or dummy variables not necessarily related to similar ones defined on pp. **2-4** to **2-4**.

TABLE 13A.3 (continued) P_1-NTU_1 Formulas and Limiting Values of P_1 for $R_1 = 1$ and $NTU_1 \rightarrow \infty$ for Various Exchanger Flow Arrangements

Flow arrangement	Eq. no.	General formula	Value for $R_1 = 1$	Value for $NTU_1 \rightarrow \infty$
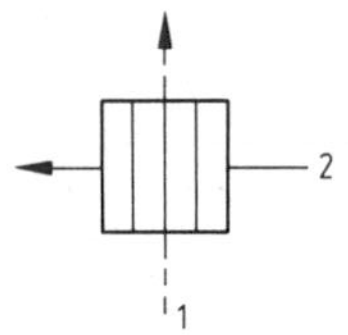 Single-pass crossflow exchanger, fluid 1 unmixed, fluid 2 mixed	2.2.1	$P_1 = [1 - \exp(-KR_1)]/R_1$ $K = 1 - \exp(-NTU_1)$	$P_1 = 1 - \exp(-K)$	$P_1 \rightarrow \frac{1 - \exp(-R_1)}{R_1}$
	2.2.2	$NTU_1 = \ln\left[\frac{1}{1 + \frac{1}{R_1}\ln(1 - R_1 P_1)}\right]$	$NTU_1 = \ln\left[\frac{1}{(1 + \ln(1 - P_1))}\right]$	$NTU_1 \rightarrow \infty$
	2.2.3	$F = \frac{\ln[(1 - R_1 P_1)/(1 - P_1)]}{(R-1)\ln\left[1 + \frac{1}{R_1}\ln(1 - R_1 P_1)\right]}$	$F = \frac{P_1}{(P_1 - 1)\ln[1 + \ln(1 - P_1)]}$	$F \rightarrow 0$
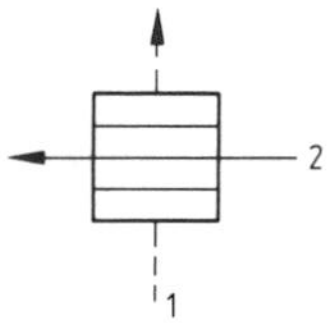 Single-pass crossflow exchanger, fluid 1 mixed, fluid 2 unmixed	2.3.1	$P_1 = 1 - \exp(-K/R_1)$ $K = 1 - \exp(-R_1 NTU_1)$	$P_1 = 1 - \exp(-K)$ $K = 1 - \exp(-NTU_1)$	$P_1 \rightarrow 1 - \exp(-1/R_1)$
	2.3.2	$NTU_1 = \frac{1}{R_1}\ln\left[\frac{1}{1 + R_1\ln(1 - P_1)}\right]$	$NTU_1 = \ln\left[\frac{1}{(1 + \ln(1 - P_1))}\right]$	$NTU_1 \rightarrow \infty$
	2.3.3	$F = \frac{\ln[(1 - R_1 P_1)/(1 - P_1)]}{(1 - 1/R)\ln[1 + R_1\ln(1 - P_1)]}$	$F = \frac{P_1}{(P_1 - 1)\ln[1 + \ln(1 - P_1)]}$	$F \rightarrow 0$
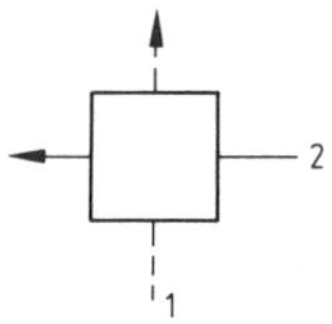 Single-pass crossflow exchanger, both fluid mixed, stream symmetric	2.4	$P_1 = \left[\frac{1}{K_1} + \frac{R_1}{K_2} - \frac{1}{NTU_1}\right]^{-1}$ $K_1 = 1 - \exp(-NTU_1)$ $K_2 = 1 - \exp(-R_1 NTU_1)$	$P_1 = \left[\frac{2}{K_1} - \frac{1}{NTU_1}\right]^{-1}$	$P_1 \rightarrow \frac{1}{1 + R_1}$

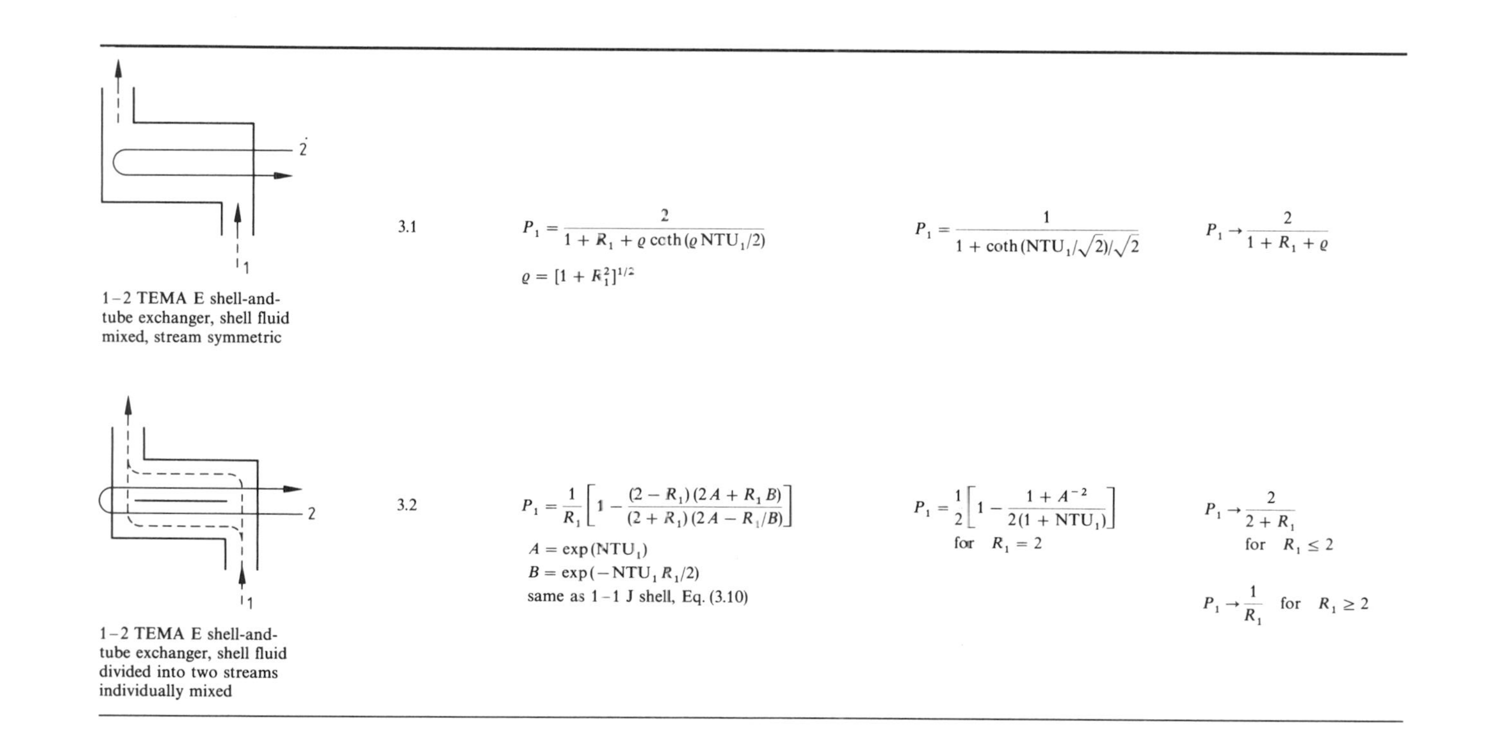

1–2 TEMA E shell-and-tube exchanger, shell fluid mixed, stream symmetric	3.1	$P_1 = \dfrac{2}{1 + R_1 + \varrho \coth(\varrho \, \mathrm{NTU}_1/2)}$ $\varrho = [1 + R_1^2]^{1/2}$	$P_1 = \dfrac{1}{1 + \coth(\mathrm{NTU}_1/\sqrt{2})/\sqrt{2}}$	$P_1 \to \dfrac{2}{1 + R_1 + \varrho}$
1–2 TEMA E shell-and-tube exchanger, shell fluid divided into two streams individually mixed	3.2	$P_1 = \dfrac{1}{R_1}\left[1 - \dfrac{(2 - R_1)(2A + R_1 B)}{(2 + R_1)(2A - R_1/B)}\right]$ $A = \exp(\mathrm{NTU}_1)$ $B = \exp(-\mathrm{NTU}_1 R_1/2)$ same as 1–1 J shell, Eq. (3.10)	$P_1 = \dfrac{1}{2}\left[1 - \dfrac{1 + A^{-2}}{2(1 + \mathrm{NTU}_1)}\right]$ for $R_1 = 2$	$P_1 \to \dfrac{2}{2 + R_1}$ for $R_1 \leq 2$ $P_1 \to \dfrac{1}{R_1}$ for $R_1 \geq 2$

TABLE 13A.3 (continued) P_1-NTU_1 Formulas and Limiting Values of P_1 for $R_1 = 1$ and $NTU_1 \to \infty$ for Various Exchanger Flow Arrangements

Flow arrangement	Eq. no.	General formula	Value for $R_1 = 1$	Value for $NTU_1 \to \infty$
1–3 TEMA E shell-and-tube exchanger, shell and tube fluids mixed, one parallel-flow and two counterflow passes	3.3	$P_1 = \frac{1}{R_1}\left[1 - \frac{C}{AC + B^2}\right]$ $A = X_1(R_1 + \lambda_1)(R_1 - \lambda_2)/2\lambda_1 - X_3\delta - X_2(R_1 + \lambda_2)(R_1 - \lambda_1)/2\lambda_2 + 1/(1 - R_1)$ $B = X_1(R_1 - \lambda_2) - X_2(R_1 - \lambda_1) + X_3\delta$ $C = X_2(3R_1 + \lambda_1) - X_1(3R_1 + \lambda_2) + X_3\delta$ $X_i = \exp(\lambda_i NTU_1/3)/2\delta$ $\delta = \lambda_1 - \lambda_2$ $\lambda_1 = -\frac{3}{2} + \left[\frac{9}{4} + R_1(R_1 - 1)\right]^{1/2}$ $\lambda_2 = -\frac{3}{2} - \left[\frac{9}{4} + R_1(R_1 - 1)\right]^{1/2}$ $\lambda_3 = R_1$	same as Eq. (3.3) with $R_1 = 1$ $A = -\exp(-NTU_1)/18 - \exp(NTU_1/3)/2 + (NTU_1 + 5)/9$	$P_1 \to 1$ for $R_1 \le 1$ $P_1 \to \frac{1}{R_1}$ for $R_1 \ge 1$
1–4 TEMA E shell-and-tube exchanger, shell and tube fluids mixed	3.4	$P_1 = 4[2(1 + R_1) + \varrho A + R_1 B]^{-1}$ $A = \coth(\varrho NTU_1/4)$ $B = \tanh(R_1 NTU_1/4)$ $\varrho = [4 + R_1^2]^{1/2}$	$P_1 = 4[4 + \sqrt{5}A + B]^{-1}$ $A = \coth(\sqrt{5}NTU_1/4)$ $B = \tanh(NTU_1/4)$	$P_1 \to \frac{4}{2(1 + R_1) + \varrho + R_1}$
Limit of 1-*n* TEMA E shell-and-tube exchanger, for large N. Shell and tube fluids mixed, stream symmetric	3.5	Eq. (2.4) applies in this limit	same as for Eq. (2.4)	same as for Eq. (2.4)

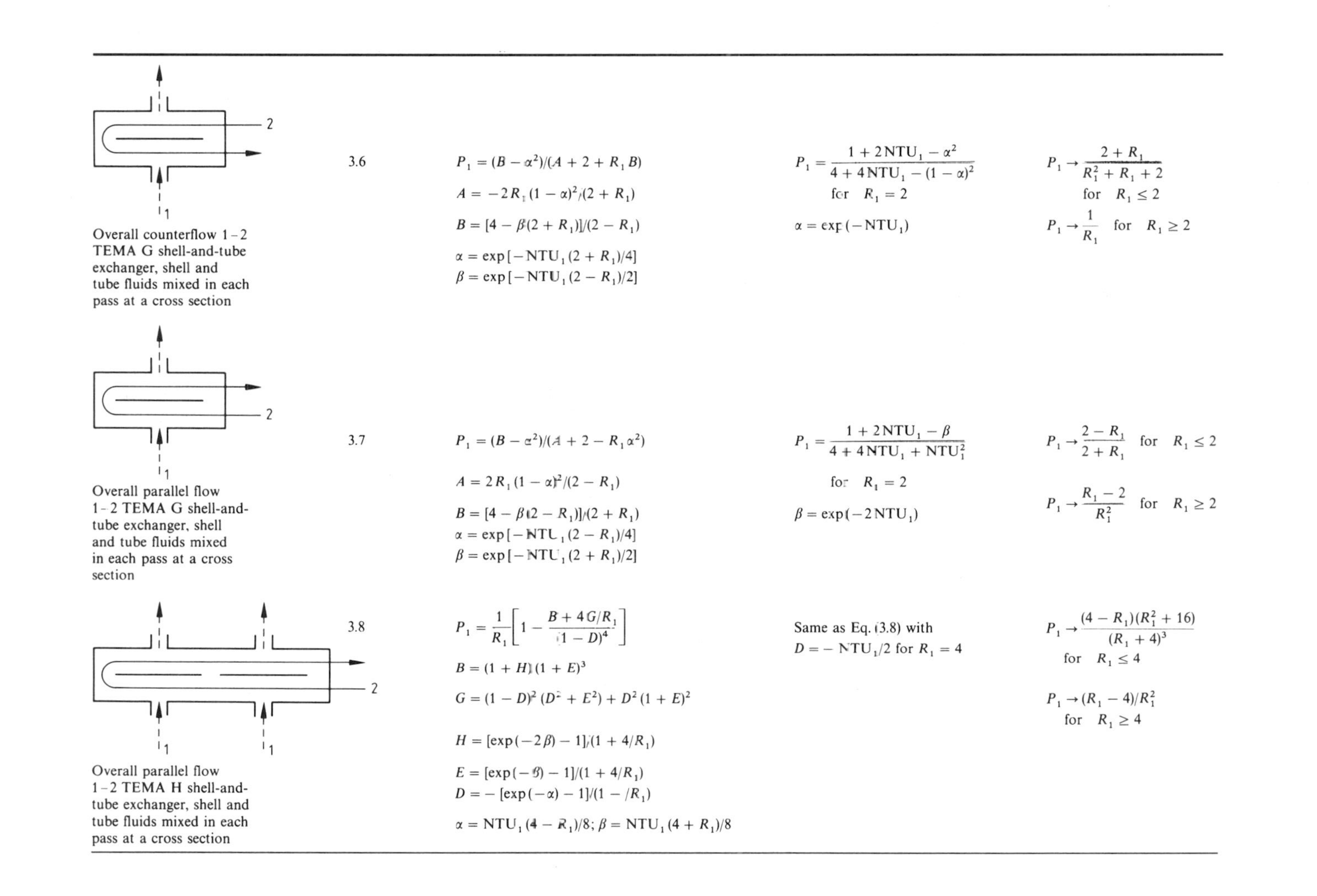

Flow arrangement	Eq.	General formula	Value for special case	Value for $\text{NTU}_1 \to \infty$
Overall counterflow 1-2 TEMA G shell-and-tube exchanger, shell and tube fluids mixed in each pass at a cross section	3.6	$P_1 = (B - \alpha^2)/(A + 2 + R_1 B)$ $A = -2R_1(1-\alpha)^2/(2+R_1)$ $B = [4 - \beta(2 + R_1)]/(2 - R_1)$ $\alpha = \exp[-\text{NTU}_1(2 + R_1)/4]$ $\beta = \exp[-\text{NTU}_1(2 - R_1)/2]$	$P_1 = \dfrac{1 + 2\text{NTU}_1 - \alpha^2}{4 + 4\text{NTU}_1 - (1-\alpha)^2}$ for $R_1 = 2$ $\alpha = \exp(-\text{NTU}_1)$	$P_1 \to \dfrac{2 + R_1}{R_1^2 + R_1 + 2}$ for $R_1 \le 2$ $P_1 \to \dfrac{1}{R_1}$ for $R_1 \ge 2$
Overall parallel flow 1-2 TEMA G shell-and-tube exchanger, shell and tube fluids mixed in each pass at a cross section	3.7	$P_1 = (B - \alpha^2)/(A + 2 - R_1\alpha^2)$ $A = 2R_1(1-\alpha)^2/(2 - R_1)$ $B = [4 - \beta(2 - R_1)]/(2 + R_1)$ $\alpha = \exp[-\text{NTU}_1(2 - R_1)/4]$ $\beta = \exp[-\text{NTU}_1(2 + R_1)/2]$	$P_1 = \dfrac{1 + 2\text{NTU}_1 - \beta}{4 + 4\text{NTU}_1 + \text{NTU}_1^2}$ for $R_1 = 2$ $\beta = \exp(-2\text{NTU}_1)$	$P_1 \to \dfrac{2 - R_1}{2 + R_1}$ for $R_1 \le 2$ $P_1 \to \dfrac{R_1 - 2}{R_1^2}$ for $R_1 \ge 2$
Overall parallel flow 1-2 TEMA H shell-and-tube exchanger, shell and tube fluids mixed in each pass at a cross section	3.8	$P_1 = \dfrac{1}{R_1}\left[1 - \dfrac{B + 4G/R_1}{(1-D)^4}\right]$ $B = (1 + H)(1 + E)^3$ $G = (1 - D)^2(D^2 + E^2) + D^2(1 + E)^2$ $H = [\exp(-2\beta) - 1]/(1 + 4/R_1)$ $E = [\exp(-\beta) - 1]/(1 + 4/R_1)$ $D = -[\exp(-\alpha) - 1]/(1 - /R_1)$ $\alpha = \text{NTU}_1(4 - R_1)/8;\ \beta = \text{NTU}_1(4 + R_1)/8$	Same as Eq. (3.8) with $D = -\text{NTU}_1/2$ for $R_1 = 4$	$P_1 \to \dfrac{(4 - R_1)(R_1^2 + 16)}{(R_1 + 4)^3}$ for $R_1 \le 4$ $P_1 \to (R_1 - 4)/R_1^2$ for $R_1 \ge 4$

TABLE 13A.3 (continued) P_1-NTU_1 Formulas and Limiting Values of P_1 for $R_1 = 1$ and $NTU_1 \to \infty$ for Various Exchanger Flow Arrangements

Flow arrangement	Eq. no.	General formula	Value for $R_1 = 1$	Value for $NTU_1 \to \infty$
Overall counterflow 1–2 TEMA H shell-and-tube exchanger, shell and tube fluids mixed in each pass at a cross section	3.9	$P_1 = \frac{1}{R_1}\left[1 - \frac{(1-D)^4}{B - 4G/R_1}\right]$ $B = (1 + H)(1 + E)^2$ $G = (1 - D)^2(D^2 + E^2) + D^2(1 + E)^2$ $H = [1 - \exp(-2\beta)]/(4/R_1 - 1)$ $E = [1 - \exp(-\beta)]/(4/R_1 = 1)$ $D = [1 - \exp(-\alpha)]/(4/R_1 + 1)$ $\alpha = NTU_1(4 + R_1)/8$ $\beta = NTU_1(4 - R_1)/8$	same as Eq. (3.9) with $H = NTU_1$ $E = NTU_1/2$ for $R_1 = 4$	$P_1 \to \left[R_1 + \frac{(4 - R_1)^3}{(4 + R_1)(R_1^2 + 16)}\right]$ for $R_1 \le 4$ $P_1 \to \frac{1}{R_1}$ for $R_1 \ge 4$
1–1 TEMA J shell-and-tube exchanger, shell and tube fluids mixed	3.10	$P_1 = \frac{1}{R_1}\left[1 - \frac{(2 - R_1)(2A + R_1B)}{(2 + R_1)(2A - R_1/B)}\right]$ $A = \exp(NTU_1)$ $B = \exp(-NTU_1 R_1/2)$ same as Eq. (3.2)	$P_1 = \frac{1}{2}\left[1 - \frac{1 + A^{-2}}{2(1 + NTU_1)}\right]$ for $R_1 = 2$	$P_1 \to \frac{2}{2 + R_1}$ for $R_1 \le 2$ $\to \frac{1}{R_1}$ for $R_1 \ge 2$
1–2 TEMA J shell-and-tube exchanger, shell and tube fluids mixed	3.11	$P_1 = [1 + \frac{R_1}{2} + \lambda B - 2\lambda CD]^{-1}$ $B = (A^\lambda + 1)/(A^\lambda - 1)$ $C = \frac{A^{(1+\lambda)/2}}{\lambda - 1 + (1 + \lambda)A^\lambda}$ $D = 1 + \frac{\lambda A^{(\lambda-1)/2}}{A^\lambda - 1}$ $A = \exp(NTU_1)$ $\lambda = (1 + R_1^2/4)^{1/2}$	same as Eq. (3.11) with $R_1 = 1$	$P_1 \to [1 + \frac{R_1}{2} + \lambda]^{-1}$

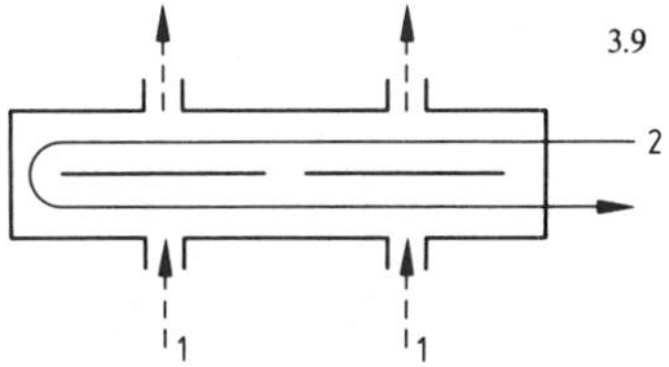

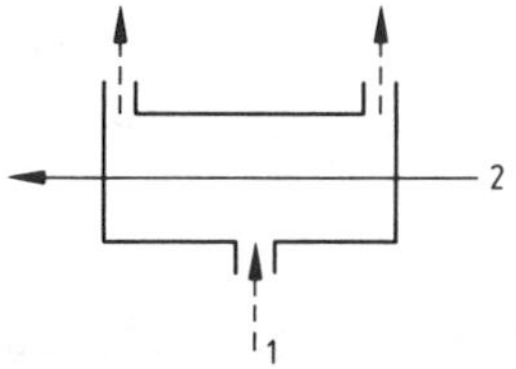

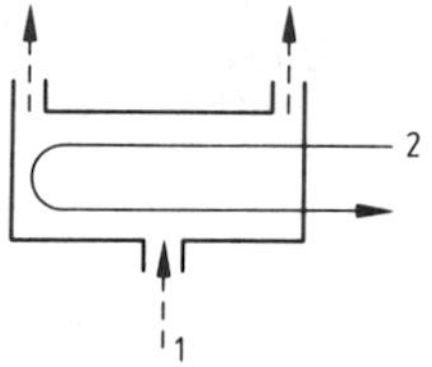

1–4 TEMA J shell-and-tube exchanger, shell and tube fluids mixed	3.12	$P_1 = \left[1 + \frac{R_1}{4}\left(\frac{1+3E}{1+E}\right) + \lambda B - 2\lambda C D\right]^{-1}$ $B = \frac{A^\lambda + 1}{A^\lambda - 1}$ $C = \frac{A^{(1+\lambda)/2}}{\lambda - 1 + (1+\lambda)A^\lambda}$ $D = 1 + \frac{\lambda A^{(\lambda-1)/2}}{A^\lambda - 1}$ $A = \exp(\mathrm{NTU}_1)$ $E = \exp(R_1\,\mathrm{NTU}_1/2)$ $\lambda = (1 + R_1^2/16)^{1/2}$	same as Eq. (3.12) with $R_1 = 1$	$P_1 \to [1 + \frac{3R_1}{4} + \lambda]^{-1}$
Limit of $1-n$ TEMA J shell-and-tube exchangers for large N. Shell and tube fluids mixed	3.13	Eq. (2.4) applies in this limit	same as for Eq. (2.4)	same as for Eq. (2.4)
	4.1.1	$P_1 = 1 - \prod_{i=1}^{n}(1 - P_{1A_i})$	same as Eq. (4.1.1)	
	4.1.2	$\frac{1}{R_1} = \sum_{i=1}^{n} \frac{1}{R_1 A_i}$	$1 = \sum_{i=1}^{n} \frac{1}{R_{1A_i}}$	
Parallel coupling of n exchangers. Fluid 2 split arbitrarily into n streams	4.1.3	$\mathrm{NTU}_1 = \sum_{i=1}^{n} \mathrm{NTU}_{A_i}$	same as Eq. (4.1.3)	

TABLE 13A.3 (continued) P_1-NTU_1 Formulas and Limiting Values of P_1 for $R_1 = 1$ and $NTU_1 \rightarrow \infty$ for Various Exchanger Flow Arrangements

Flow arrangement	Eq. no.	General formula	Value for $R_1 = 1$	Value for $NTU_1 \rightarrow \infty$
	4.2.1	$P_1 = \dfrac{\prod_{i=1}^{n}(1 - R_1 P_{1A_i}) - \prod_{i=1}^{n}(1 - P_{1A_i})}{\prod_{i=1}^{n}(1 - R_1 P_{1A_i}) - R_1 \prod_{i=1}^{n}(1 - P_{1A_i})}$	$P_1 = \dfrac{\sum_{i=1}^{n} \dfrac{P_{1A_i}}{1 - P_{1A_i}}}{1 + \sum_{i=1}^{n} \dfrac{P_{1A_i}}{1 - P_{1A_i}}}$	same as Eq. (1.1.1) (counterflow)
	4.2.2.	$R_1 = R_{1A_i}, \quad i = 1, \ldots, n$	$1 = R_{1A_i}, \quad i = 1, \ldots, n$	same as Eq. (4.2.2)
	4.2.3	$NTU_1 = \sum_{i=1}^{n} NTU_{1A_i}$	same as Eq. (4.2.3)	$NTU_{1A_i} \rightarrow 0$ $i = 1, \ldots, n$
Series coupling of n exchangers, overall counterflow arrangement.	4.2.4	$F = \dfrac{1}{NTU_1} \sum_{i=1}^{n} NTU_{1A_i} F_{A_i}$	same as Eq. (4.2.4)	same as Eq. (4.2.4)
	4.3.1	$P = \dfrac{1}{1 + R_1}\left\{1 - \prod_{i=1}^{n}[1 - (1 + R_1) P_{1A_i}]\right\}$	$P_1 = \dfrac{1 - \prod_{i=1}^{n}[1 - 2P_{1A_i}]}{2}$	
Series coupling of n exchangers, overall parallel flow arrangement; stream symmetric if all A_i are stream symmetric	4.3.2	$R_1 = R_{1A_i}, \quad i = 1, \ldots, n$	$1 = R_{1A_i}, \quad i = 1, \ldots, n$	
	4.3.3	$NTU_1 = \sum_{i=1}^{n} NTU_{1A_i}$	same as Eq. (4.3.3)	

* In this table all variables except P_1, R_1, NTU_1, and F are local or dummy variables not necessarily related to similar ones defined in nomenclature and the text. Reprinted from Shah and Mueller (1988). With permission.

If idealization 6 is not valid, divide the exchanger into small segments until the specific heats can be treated as constant. Idealizations 7 and 8 are primarily important for compact heat exchangers and will be discussed on pages 493 and 467-471 respectively.

Idealization 9 has been discussed in detail by Shah (1993). The overall heat transfer coefficient can vary because of variations in local heat transfer coefficients due to two effects: (1) change in heat transfer coefficients in the exchanger as a result of changes in the fluid properties or radiation due to rise or drop of fluid temperatures, and (2) change in heat transfer coefficients in the exchanger due to developing thermal boundary layers; this is referred to as the *length effect.* The first effect due to fluid property variations (or radiation) consists of two components: (i) distortion of velocity and temperature profiles at a given flow cross section due to fluid property variations; this effect is usually taken into account by the so-called property ratio method, with the correction scheme of Eqs. (13A.54) and (13A.55), and (ii) variations in the fluid temperature along the axial and transverse directions in the exchanger depending upon the exchanger flow arrangement; this effect is referred to as the *temperature effect.* The resultant axial changes in the overall mean heat transfer coefficient can be significant; the variations in U_{local} could be nonlinear depending upon the type of the fluid. The effect of varying U_{local} can be taken into account by evaluating U_{local} at a few points in the exchanger and subsequently integrating U_{local} values by Simpson or Gauss method (Shah, 1993). The temperature effect can increase or decrease mean U slightly or significantly depending upon the fluids and applications. The length effect is important for developing laminar flows for which high heat transfer coefficients are obtained in the thermal entrance region. However, in general it will have a less impact on the overall heat transfer coefficient because the other thermal resistances in series in an exchanger may be controlling. The length effect reduces then overall heat transfer coefficient compared to the mean value calculated conventionally (assuming uniform mean heat transfer coefficient on each fluid side). It is shown that this reduction is up to about 14% for the worst case (Shah, 1993).

Idealization 10 is generally true for individual passes due to the manufacturing considerations. However, in a multipass unit, it is possible to have different surface areas in different passes due to the design considerations. This effect has been investigated by Roetzel and Spang (1987, 1989) for 1-2, 1-3, and 1-2N TEMA E exchangers, by Spang et al. (1991) for 1-N split flow TEMA G exchangers, by Xuan et al. (1991) for 1-N divided flow TEMA J exchangers, and by Băclić et al. (1988) for two-pass cross-counterflow exchangers.

For Idealization 11, Shah and Pignotti (1995) have shown that the following are the specific number of baffles beyond which the influence of the finite number of baffles on the exchanger effectiveness is not significantly larger than 2%: $N_b \geq 10$ for the 1-1 TEMA E counterflow exchanger; $N_b \geq 6$ for the 1-2 TEMA E exchanger for $\text{NTU}_s \leq 2$, $R_s \leq 6$; $N_b \geq 9$ for the 1-2 TEMA J exchanger for $\text{NTU}_s \leq 2$, $R_s \leq 5$; $N_b \geq 5$ for the 1-2 TEMA G exchanger for $\text{NTU}_s \leq 3$, all R_s; $N_b \geq 11$ for the 1-2 TEMA H exchanger for $\text{NTU}_s \leq 3$, all R_s.

If any of these idealizations are not valid for a particular exchanger application, the best solution is to work directly with either Eqs. (13A.1) and (13A.2) or their modified form by including a particular effect, and to integrate them over a small exchanger segment in which all of the idealizations are valid.

Fin Efficiency and Extended Surface Efficiency

Extended surfaces have fins attached to the primary surface on one or both sides of a two-fluid or a multifluid heat exchanger. Fins can be of a variety of geometries—plane, wavy, or interrupted, and can be attached to the inside, the outside, or both sides of circular, flat, or oval tubes, or parting sheets. Fins are primarily used to increase the surface area (when the heat transfer coefficient on that fluid side is relatively low) and consequently to increase the total rate of heat transfer. In addition, enhanced fin geometries also increase the heat transfer coefficient compared to that for a plain fin. Fins may also be used on the high heat transfer coefficient fluid side in a heat exchanger primarily for structural strength purpose (for example, for high-pressure water flow through a flat tube) or to provide a thorough mixing of a highly viscous liquid (such as for laminar oil flow in a flat or a round tube). Fins are attached to the primary surface by brazing, soldering, welding, adhesive bonding, or mechanical expansion, or they are extruded or integrally connected to the tubes. Major categories of extended surface heat exchangers are plate-fin (Figure 13A.8) and tube-fin (Figure 13A.9). Note that shell-and-tube exchangers sometimes employ individually finned tubes—low finned tubes (similar to Figure 13A.9a but with low height fins) (Shah, 1985).

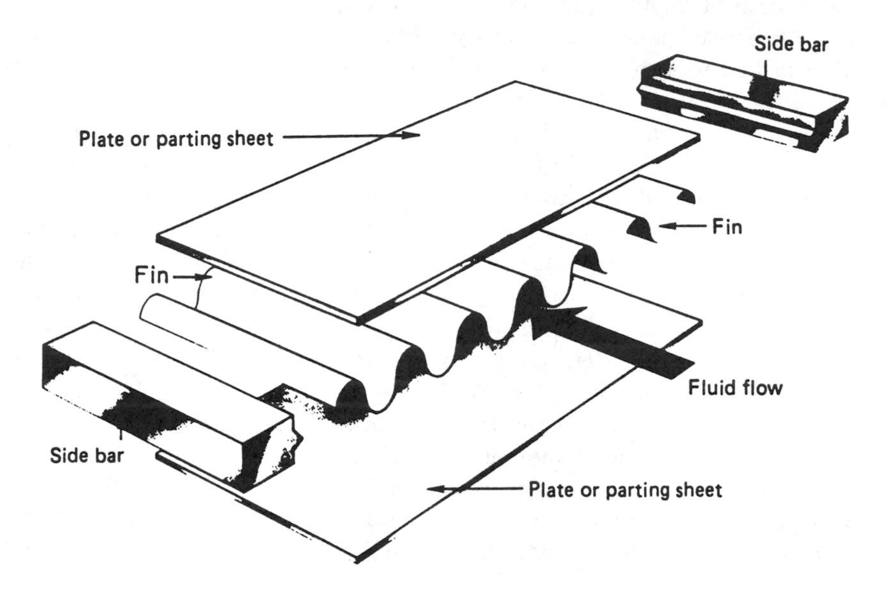

FIGURE 13A.8 Typical components on one fluid side of a plate-fin exchanger.

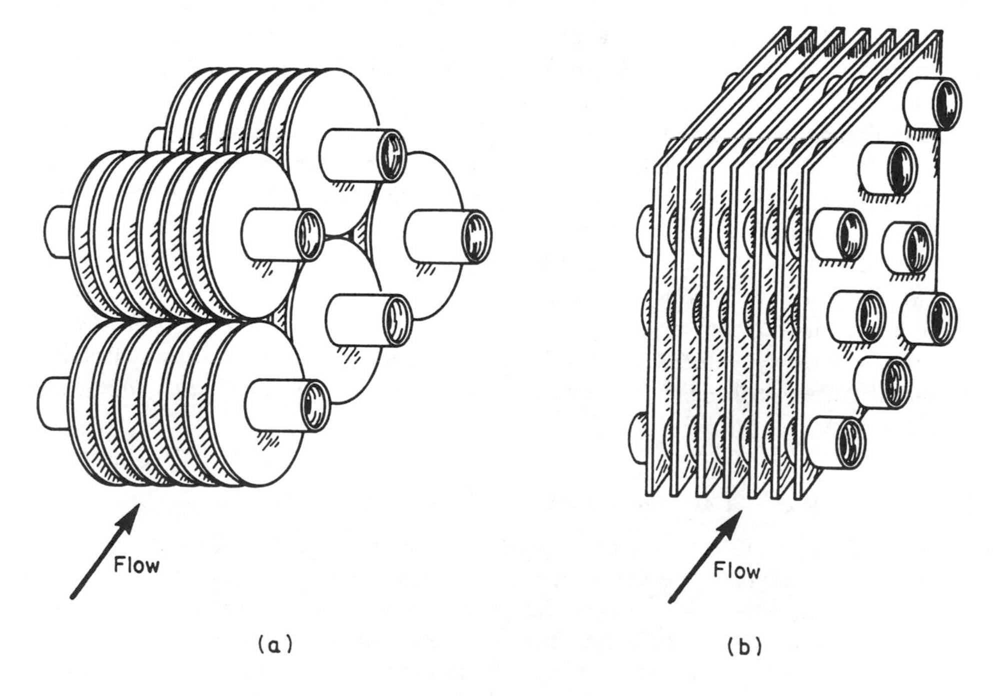

FIGURE 13A.9 Types of tube-fin exchangers: (a) individually finned tubes, (b) flat (continuous) fins on an array of tubes; flat fins are shown as plain fins, but they can be wavy, louvered or interrupted.

TABLE 13A.4 Fin Efficiency Expressions for Plate-Fin and Tube-Fin Geometries of Uniform Fin Thickness

Geometry	Fin efficiency formula
	$m_i = \left[\frac{2h}{k_f \delta_i}\left(1 + \frac{\delta_i}{l_{ef}}\right)\right]^{1/2}$ $E_i = \frac{\tanh(m_i l_i)}{m_i l_i}$ $i = 1, 2, 3, 4$
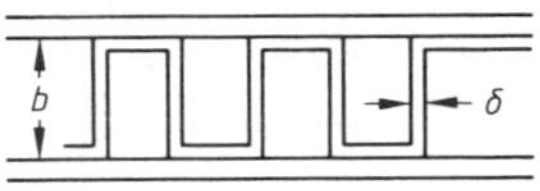 Plain, wavy, or offset strip fin of rectangular cross section	$\eta_f = E_1$ $l_1 = \frac{b}{2} - \delta_1 \quad \delta_1 = \delta$
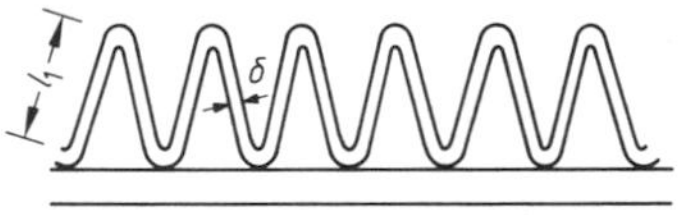 Triangular fin heated from one side	$\eta_f = \dfrac{hA_1(T_0 - T_a)\frac{\sinh(m_1 l_1)}{m_1 l_1} + q_e}{\cosh(m_1 l_1)\left[hA_1(T_0 - T_a) + q_e \frac{T_0 - T_a}{T_1 - T_a}\right]}$
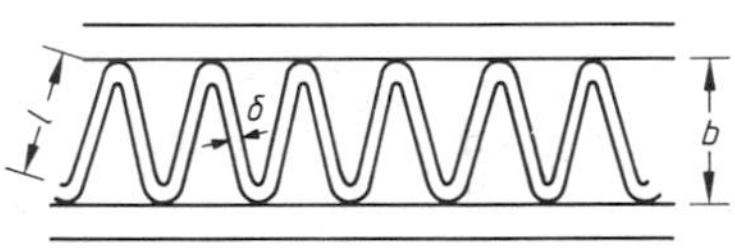 Plain, wavy, or louver fin of triangular cross section	$\eta = E_1$ $l_1 = \frac{l}{2} \quad \delta_1 = \delta$
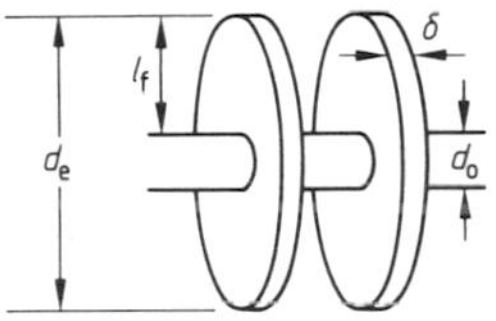 Circular fin	$\eta_f = \begin{cases} a(ml_e)^{-b} & \text{for } \Phi > 0.6 + 2.257(r^*)^{-0.445} \\ \frac{\tanh \Phi}{\Phi} & \text{for } \Phi \leq 0.6 + 2.257(r^*)^{-0.445} \end{cases}$ $a = (r^*)^{-0.246} \quad \Phi = ml_e(r^*)^{\exp(0.13ml_e - 1.3863)}$ $b = \begin{cases} 0.9107 + 0.0893 r^* & \text{for } r^* \leq 2 \\ 0.9706 + 0.17125 \ln r^* & \text{for } r^* > 2 \end{cases}$ $m = \left(\frac{2h}{k_f \delta}\right)^{1/2} \quad l_e = l_f + \frac{\delta}{2} \quad r^* = \frac{r_e}{r_o}$
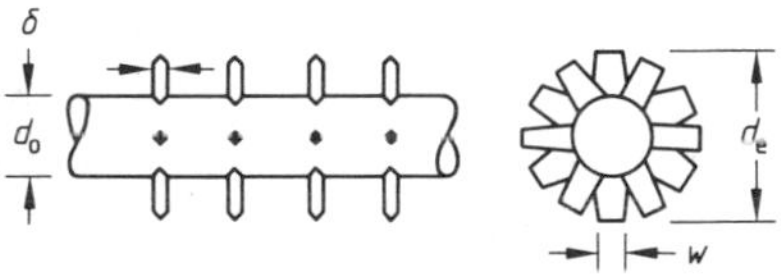 Studded fin	$\eta_f = \frac{\tanh(ml_e)}{ml_e}$ $m = \left[\frac{2h}{k_f \delta}\left(1 + \frac{\delta}{w}\right)\right]^{1/2} \quad l_e = l_f + \frac{\delta}{2} \quad l_f = \frac{(d_e - d_o)}{2}$
Rectangular fin over circular tubes	See the text.

The concept of fin efficiency accounts for the reduction in temperature potential between the fin and the ambient fluid due to conduction along the fin and convection from or to the fin surface depending upon the fin cooling or heating situation. The fin temperature effectiveness or fin efficiency is defined as the ratio of the actual heat transfer rate through the fin base divided by the maximum possible heat transfer rate through the fin base that would be obtained if the entire fin were at the base temperature (i.e., its material thermal conductivity were infinite). Since most of the real fins are "thin," they are treated as one-dimensional (1D) with standard idealizations used for the analysis (Huang and Shah, 1992). This 1D fin efficiency is a function of the fin geometry, fin material thermal conductivity, heat transfer coefficient at the fin surface, and the fin tip boundary condition; it is not a function of the fin base or fin tip temperature, ambient temperature, and heat flux at the fin base or fin tip in general. Fin efficiency formulas for some common fins are presented in Table 13A.4 (Shah, 1985).

The fin efficiency for flat fins (Figure 13A.9b) is obtained by a sector method (Shah, 1985). In this method the rectangular or hexagonal fin around the tube (Figure 13A.10a and b) or its smallest symmetrical section is divided into n sectors. Each sector is then considered as a circular fin with the radius $r_{e,i}$ equal to the length of the centerline of the sector. The fin efficiency of each sector is

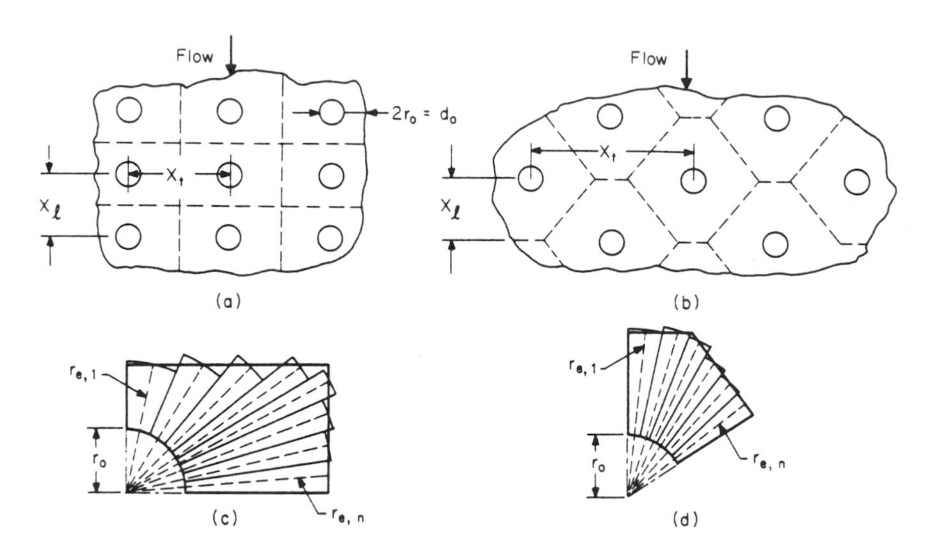

FIGURE 13A.10 Flat Fin over (a) an inline and (b) a staggered tube arrangement; the smallest representative segment of the fin for (c) an inline and (d) a staggered tube arrangement.

subsequently computed using the circular fin formula of Table 13A.4. The fin efficiency η_f for the whole fin is then the surface area weighted average of $\eta_{f,i}$ of each sector.

$$\eta_f = \frac{\sum_{i=1}^{n} \eta_{f,i} A_{f,i}}{\sum_{i=1}^{n} A_{f,i}} \tag{13A.22}$$

Since the heat flow seeks the path of least thermal resistance, actual η_o will equal to or higher than that calculated by Eq. (13A.22); hence Eq. (13A.22) yields a somewhat conservative value of η_f.

The η_f values of Table 13A.4 or Eq. (13A.22) are not valid in general when the fin is thick, is subject to variable heat transfer coefficients or variable ambient fluid temperature, or has a temperature depression at the base. For a thin rectangular fin of constant cross section, the fin efficiency as presented in Table 13A.4 is given by

$$\eta_f = \frac{\tanh(ml)}{(ml)} \tag{13A.23}$$

where $m = [2h(1 + \delta_f/l_f)/k_f\delta_f)^{1/2}$. For a thick rectangular fin of constant cross section, the fin efficiency (a counterpart of Eq. (13A.23) is given by (Huang and Shah, 1992)

$$\eta_f = \frac{(Bi^+)^{1/2}}{KBi} \tanh\left[K(Bi^+)^{1/2}\right] \tag{13A.24}$$

where $Bi^+ = Bi/(1 + Bi/4)$, $Bi = (h\delta_f/2k_f)^{1/2}$, $K = 2H/\delta_f$. Equation (13A.23) is accurate (within 0.3%) for a "thick" rectangular fin of having $\eta_f > 80\%$; otherwise use Eq. (13A.24) for a thick fin.

The nonuniform heat transfer coefficient over the fin surface can lead to a significant error in η_f (Huang and Shah, 1992) compared to that for a uniform h over the fin surface. However, generally h is obtained experimentally by considering a constant (uniform) value of h over the fin surface. Hence, such experimental h will not introduce significant errors in η_f while designing a heat exchanger, particularly for $\eta_f > 80\%$. However, one needs to be aware of the impact of nonuniform h on η_f if the heat exchanger test conditions and design conditions are significantly different. Nonuniform ambient temperature has less than a 1% effect on the fin efficiency for $\eta_f > 60\%$ and hence can be neglected. The longitudinal heat conduction effect on the fin efficiency is less than 1% for $\eta_f > 10\%$ and hence can be neglected. The fin base temperature depression increases the total heat flow rate through the extended surface compared to that having no fin base temperature depression, and hence neglecting this effect provides a conservative approach for the extended surface heat transfer. Refer to Huang and Shah (1992) for further details on the foregoing effects and modifications to η_f for rectangular fins of constant cross sections. Refer to Srinivasan and Shah (1995) for fin efficiency of extended surfaces for two-phase flows.

In an extended surface heat exchanger, heat transfer takes place from both the fins ($\eta_f < 100\%$) and the primary surface ($\eta_f = 100\%$). In that case, the total heat transfer rate is evaluated through a concept of total surface effectiveness or extended surface efficiency η_o defined as

$$\eta_o = \frac{A_p}{A} + \eta_f \frac{A_f}{A} = 1 - \frac{A_f}{A}\left(1 - \eta_f\right) \quad \textbf{(13A.25)}$$

where A_f is the fin surface area, A_p is the primary surface area, and $A = A_f + A_p$. In Eq. (13A.25), heat transfer coefficients over the finned and unfinned surfaces are idealized to be equal. Note that $\eta_o \geq \eta_f$ and η_o is always required for the determination of thermal resistances of Eq. (13A.5) in heat exchanger analysis.

Pressure Drop Analysis

Usually a fan, blower, or pump is used to flow fluid through individual sides of a heat exchanger. Due to potential initial and operating high cost, low fluid pumping power requirement is highly desired for gases and viscous liquids. The fluid pumping power $\wp$ is approximately related to the core pressure drop in the exchanger as (Shah, 1985).

$$\wp = \frac{\dot{m}\Delta p}{\rho} \approx \begin{cases} \dfrac{1}{2g_c}\dfrac{\mu}{\rho^2}\dfrac{4L}{D_h}\dfrac{\dot{m}^2}{D_h A_o} f\,\mathrm{Re} & \text{for laminar flow} \quad \textbf{(13A.26a)} \\[2ex] \dfrac{0.046}{2g_c}\dfrac{\mu^{0.2}}{\rho^2}\dfrac{4L}{D_h}\dfrac{\dot{m}^{2.8}}{A_o^{1.8}D_h^{0.2}} & \text{for turbulent flow} \quad \textbf{(13A.26b)} \end{cases}$$

It is clear from Eqs. (13A.26a) and (13A.26b) that the fluid pumping power is strongly dependent upon the fluid density ($\wp \propto 1/\rho^2$), particularly for low-density fluids in laminar and turbulent flows, and upon the viscosity in laminar flow. In addition, the pressure drop itself can be an important consideration when blowers and pumps are used for the fluid flow, since they are head limited. Also, for condensing and evaporating fluids, the pressure drop affects the heat transfer rate. Hence, the pressure drop determination in the exchanger is important.

The pressure drop associated with a heat exchanger consists of (1) core pressure drop and (2) the pressure drop associated with the fluid distribution devices such as inlet and outlet manifolds, headers, tanks, nozzles, ducting, and so on, which may include bends, valves, and fittings. This second Δp component is determined from Idelchik (1994) and Miller (1978). The core pressure drop may consist of one or more of the following components depending upon the exchanger construction: (i) friction losses associated with fluid flow over heat transfer surface, which usually consists of skin friction, form (profile) drag, and internal contractions and expansions, if any; (ii) the momentum effect (pressure drop or rise due to fluid density changes) in the core, (iii) pressure drop associated with sudden contraction and expansion at the core inlet and outlet, and (iv) the gravity effect due to the change in elevation between the inlet and outlet of the exchanger. The gravity effect is generally negligible for gases. For vertical flow through the exchanger, the pressure drop or rise ("static head") due to the elevation change is given by

$$\Delta p = \pm \frac{\rho_m g L}{g_c} \tag{13A.27}$$

Here the "+" sign denotes vertical upflow (i.e., pressure drop) and the "–" sign denotes vertical downflow (i.e., pressure rise). The first three components of the core pressure drop are now presented for plate-fin, tube-fin, plate, and regenerative heat exchangers.

Plate-Fin Heat Exchangers

For the plate-fin exchanger (Figure 13A.8), all three components are considered in the core pressure drop evaluation as follows.

$$\frac{\Delta p}{p_i} = \frac{G^2}{2g_c} \frac{1}{p_i \rho_i} \left[\left(1 - \sigma^2 + K_c\right) + f \frac{L}{r_h} \rho_i \left(\frac{1}{\rho}\right)_m + 2\left(\frac{\rho_i}{\rho_o} - 1\right) - \left(1 - \sigma^2 - K_e\right) \frac{\rho_i}{\rho_o} \right] \tag{13A.28}$$

where f is the Fanning fraction factor, K_c and K_e are flow contraction (entrance) and expansion (exit) pressure loss coefficients (see Figure 13A.11), and σ is a ratio of minimum free flow area to frontal area. K_c and K_e for four different entrance flow passage geometries are presented by Kays and London (1984). The entrance and exit losses are important at low values of σ and L (short cores), high values of Re, and for gases; they are negligible for liquids. The values of K_c and K_e apply to long tubes for which flow is fully developed at the exit. For partially developed flows, K_c is lower and K_e is higher than that for fully developed flows. For interrupted surfaces, flow is never fully developed boundary layer type. For highly interrupted fin geometries, the entrance and exit losses are generally small compared to the core pressure drop and the flow is well mixed; hence, K_c and K_e for Re $\rightarrow \infty$ should represent a good approximation. The mean specific volume v_m or $(1/\rho)_m$ in Eq. (13A.28) is given as follows: For liquids with any flow arrangement, or for a perfect gas with $C^* = 1$ and any flow arrangement (except for parallelflow),

$$\left(\frac{1}{\rho}\right)_m = v_m = \frac{v_i + v_o}{2} = \frac{1}{2}\left(\frac{1}{\rho_i} + \frac{1}{\rho_o}\right) \tag{13A.29}$$

where v is the specific volume in m^3/kg.

For a perfect gas with $C^* = 0$ and any flow arrangement,

$$\left(\frac{1}{\rho}\right)_m = \frac{\tilde{R}}{p_{\text{avg}}} T_{lm} \tag{13A.30}$$

Here $\tilde{R}$ is the gas constant in J/(kg K), $p_{\text{avg}} = (p_i + p_o)/2$, and $T_{\text{lm}} = T_{\text{const}} + \Delta T_{\text{lm}}$ where T_{const} is the mean average temperature of the fluid on the other side of the exchanger; the log-mean temperature difference ΔT_{lm} is defined in Table 13A.1. The core frictional pressure drop in Eq. (13A.28) may be approximated as

$$\Delta p \approx \frac{4fLG^2}{2g_c D_h} \left(\frac{1}{\rho}\right)_m \tag{13A.31}$$

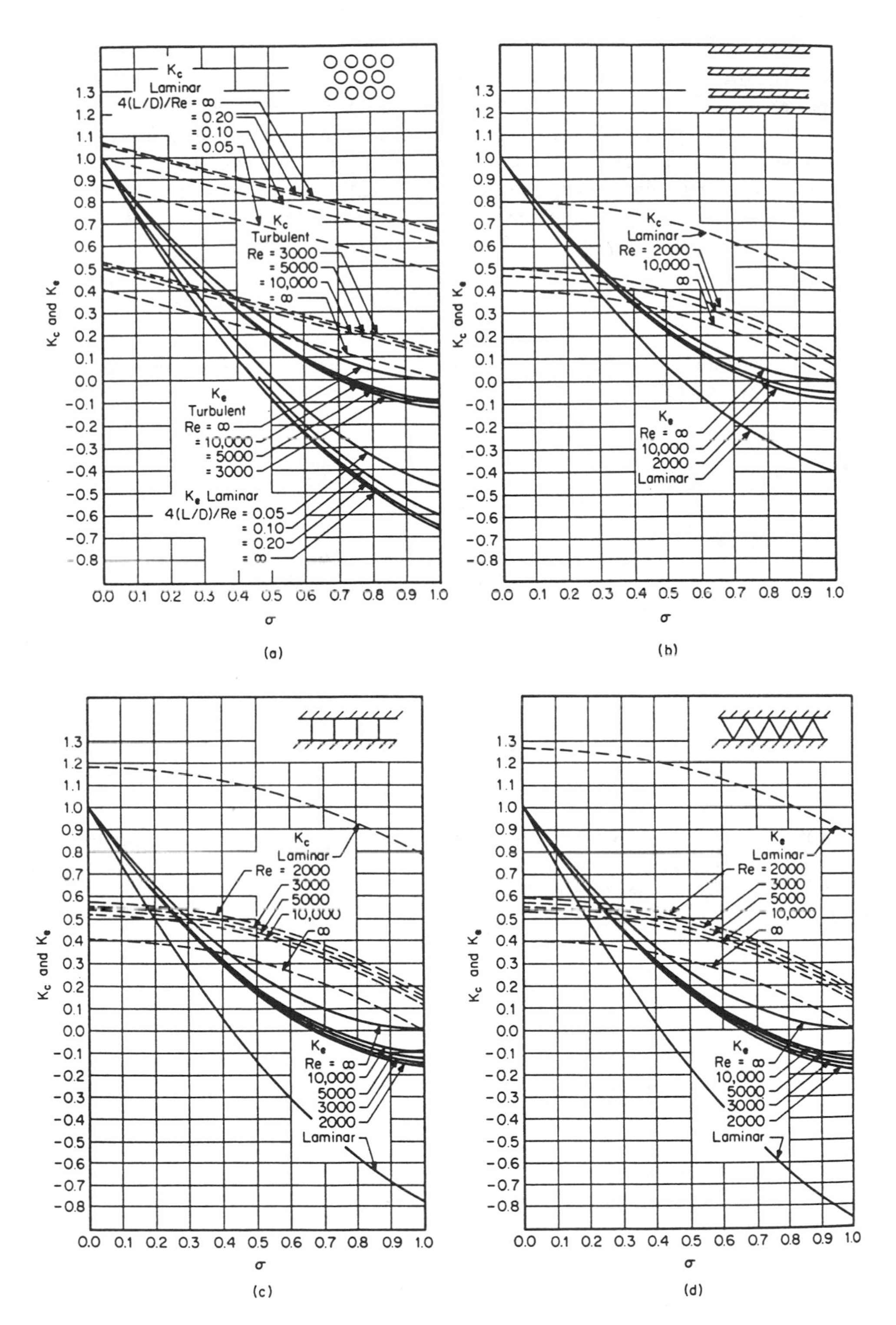

FIGURE 13A.11 Entrance and exit pressure loss coefficients: (a) circular tubes, (b) parallel plates, (c) square passages, and (d) triangular passages. For each of these flow passages shown in the inset, the fluid flows perpendicular to the plane of the paper into the flow passages. *Source:* Kays and London, 1984.

Tube-Fin Heat Exchangers

The pressure drop inside a circular tube is computed using Eq. (13A.28) with proper values of f factors (see equations in Table 13A.7) and K_c and K_e from Figure 13A.11 for circular tubes.

For flat fins on an array of tubes (see Figure 13A.9b), the components of the core pressure drop, such as those in Eq. (13A.28), are the same with the following exception: the core friction and momentum effect take place within the core with $G = \dot{m}/A_o$, where A_o is the minimum free flow area within the core, and the entrance and exit losses occur at the leading and trailing edges of the core with the associated flow area A_o' so that

$$\dot{m} = GA_o = G'A_o' \quad \text{or} \quad G'\sigma' = G\sigma \tag{13A.32}$$

where σ' is the ratio of free flow area to frontal area at the fin leading edges. The pressure drop for flow normal to a tube bank with flat fins is then given by

$$\frac{\Delta p}{p_i} = \frac{G^2}{2g_c}\frac{1}{p_i\rho_i}\left[f\frac{L}{r_h}\rho_i\left(\frac{1}{\rho}\right)_m + 2\left(\frac{\rho_i}{\rho_o} - 1\right)\right] + \frac{G'^2}{2g_c}\frac{1}{p_i\rho_i}\left[\left(1 - \sigma'^2 + K_c\right) - \left(1 - \sigma'^2 - K_e\right)\frac{\rho_i}{\rho_o}\right] \tag{13A.33}$$

For individually finned tubes as shown in Figure 13A.9a, flow expansion and contraction take place along each tube row, and the magnitude is of the same order as that at the entrance and exit. Hence, the entrance and exit losses are generally lumped into the core friction factor. Equation (13A.28) for individually finned tubes then reduces to

$$\frac{\Delta p}{p_i} = \frac{G^2}{2g_c}\frac{1}{p_i\rho_i}\left[f\frac{L}{r_h}\rho_i\left(\frac{1}{\rho}\right)_m + 2\left(\frac{\rho_i}{\rho_o} - 1\right)\right] \tag{13A.34}$$

Regenerators

For regenerator matrices having cylindrical passages, the pressure drop is computed using Eq. (13A.28) with appropriate values of f, K_c, and K_e. For regenerator matrices made up of any porous material (such as checkerwork, wire, mesh, spheres, copper wools, etc.), the pressure drop is calculated using Eq. (13A.34) in which the entrance and exit losses are included in the friction factor f.

Plate Heat Exchangers

Pressure drop in a plate heat exchanger consists of three components: (1) pressure drop associated with the inlet and outlet manifolds and ports, (2) pressure drop within the core (plate passages), and (3) pressure drop due to the elevation change. The pressure drop in the manifolds and ports should be kept as low as possible (generally < 10%, but it is found as high as 25 to 30% or higher in some designs). Empirically, it is calculated as approximately 1.5 times the inlet velocity head per pass. Since the entrance and exit losses in the core (plate passages) cannot be determined experimentally, they are included in the friction factor for the given plate geometry. The pressure drop (rise) caused by the elevation change for liquids is given by Eq. (13A.27). Hence, the pressure drop on one fluid side in a plate heat exchanger is given by

$$\Delta p = \frac{1.5G^2 N_p}{2g_c\rho_i} + \frac{4fLG^2}{2g_c D_e}\left(\frac{1}{\rho}\right)_m + \left(\frac{1}{\rho_o} - \frac{1}{\rho_i}\right)\frac{G^2}{g_c} \pm \frac{\rho_m g L}{g_c} \tag{13A.35}$$

where N_p is the number of passes on the given fluid side, and D_e is the equivalent diameter of flow passages (usually twice the plate spacing). Note that the third term on the right-hand side of the equality sign of Eq. (13A.35) is for the momentum effect which generally is negligible for liquids.

13A.3 Heat Transfer and Flow Friction Correlations

Accurate and reliable surface heat transfer and flow friction characteristics are a key input to the exchanger heat transfer and pressure drop analyses or to the rating and sizing problems (Shah, 1985). After presenting the associated nondimensional groups, we will present important analytical solutions and empirical correlations for some important exchanger geometries.

Dimensionless Groups

Heat transfer characteristics of an exchanger surface are presented in terms of the Nusselt number, Stanton number, or Colburn factor vs. the Reynolds number, x^*, or the Graetz number. Flow friction characteristics are presented in terms of the Fanning friction factor vs. Re or x^+. These and other important dimensionless groups used in presenting and correlating internal flow forced convection heat transfer are summarized in Table 13A.5 with their definitions and physical meanings. Where applicable, the hydraulic diameter D_h is used as a characteristic dimension in all dimensionless groups. A number of different definitions are used in the literature for some of the dimensionless groups; the user should pay particular attention to the specific definitions used in any research paper before using specific results. This is particularly true for the Nusselt number, where many different temperature differences are used in the definition of h, and for f, Re, and other dimensionless groups having characteristic dimensions different from D_h.

Analytical Solutions

Flow passages in most compact heat exchangers are complex with frequent boundary layer interruptions; some heat exchangers (particularly the tube side of shell-and-tube exchangers and highly compact regenerators) have continuous flow passages. The velocity and temperature profiles across the flow cross section are generally fully developed in the continuous flow passages, whereas they develop at each boundary layer interruption in an interrupted surface and may reach a periodic fully developed flow. The heat transfer and flow friction characteristics are generally different for fully developed flows and developing flows. Analytical results for developed and developing flows for simple flow passage geometries follow. For complex surface geometries, the basic surface characteristics are primarily obtained experimentally (Shah, 1985); the pertinent correlations are presented in the next subsection.

Analytical solutions for developed and developing velocity/temperature profiles in constant cross section circular and noncircular flow passages are important when no empirical correlations are available, extrapolations are needed for empirical correlations, or in the development of empirical correlations. Fully developed laminar flow solutions are applicable to highly compact regenerator surfaces or highly compact plate-fin exchangers, with plain uninterrupted fins. Developing laminar flow solutions are applicable to interrupted fin geometries and plain uninterrupted fins of "short" lengths, and turbulent flow solutions to not-so-compact heat exchanger surfaces.

The heat transfer rate in laminar duct flow is very sensitive to the thermal boundary condition. Hence, it is essential to carefully identify the thermal boundary condition in laminar flow. The heat transfer rate in turbulent duct flow is insensitive to the thermal boundary condition for most common fluids ($Pr > 0.7$); the exception is liquid metals ($Pr < 0.03$). Hence, there is generally no need to identify the thermal boundary condition in turbulent flow for all fluids except liquid metals.

Fully developed laminar flow analytical solutions for some duct shapes of interest in compact heat exchangers are presented in Table 13A.6 for three important thermal boundary conditions denoted by the subscripts H1, H2, and T (Shah and London, 1978; Shah and Bhatti, 1987). Here, H1 denotes constant axial wall heat flux with constant peripheral wall temperature, H2 denotes constant axial and peripheral wall heat flux, and T denotes constant wall temperature. The following observations may be made from this table: (1) There is a strong influence of flow passage geometry on Nu and fRe. Rectangular passages approaching a small aspect ratio exhibit the highest Nu and

TABLE 13A.5 Important Dimensionless Groups for Internal Flow Forced Convection Heat Transfer and Flow Friction, Useful in Heat Exchanger Design

Dimensionless Groups	Definitions and Working Relationships	Physical Meaning and Comments
Reynolds number	$\mathrm{Re} = \frac{\rho u_m D_h}{\mu} = \frac{G D_h}{\mu}$	a flow modulus, proportional to the ratio of flow momentum rate ("inertia force") to viscous force
Fanning friction factor	$f = \frac{\tau_w}{\left(\rho u_m^2 / 2g_c\right)}$	the ratio of wall shear (skin frictional) stress to the flow kinetic energy per unit volume; commonly used in heat transfer literature
	$f = \Delta p^* \frac{r_h}{L} = \frac{\Delta p}{\left(\rho u_m^2 / 2g_c\right)} \frac{r_h}{L}$	
Apparent Fanning friction factor	$f_{\mathrm{app}} = \Delta p^* \frac{r_h}{L}$	includes the effects of skin friction and the change in the momentum rates in the entrance region (developing flows)
Incremental pressure drop number	$K(x) = \left(f_{\mathrm{app}} - f_{fd}\right) \frac{L}{r_h}$	represents the excess dimensionless pressure drop in the entrance region over that for fully developed flow
	$K(\infty) = \text{constant for } x \rightarrow \infty$	
Darcy friction factor	$f_D = 4f = \Delta p^* \frac{D_h}{L}$	four times the Fanning friction factor; commonly used in fluid mechanics literature
Euler number	$\mathrm{Eu} = \Delta p^* = \frac{\Delta p}{\left(\rho u_m^2 / 2g_c\right)}$	the pressure drop normalized with respect to the dynamic velocity head; commonly used in the Russian literature
Dimensionless axial distance for the fluid flow problem	$x^+ = \frac{x}{D_h \mathrm{Re}}$	the ratio of the dimensionless axial distance (x/D_h) to the Reynolds number; useful in the hydrodynamic entrance region
Nusselt number	$\mathrm{Nu} = \frac{h}{k/D_h} = \frac{q'' D_h}{k(T_w - T_m)}$	the ratio of the convective conductance h to the pure molecular thermal conductance k/D_h
Stanton number	$St = \frac{h}{G c_p}$	the ratio of convection heat transfer (per unit duct surface area) to amount virtually transferable (per unit of flow cross-sectional area); no dependence upon any geometric characteristic dimension
	$St = \frac{\mathrm{Nu}}{\mathrm{Pe}} = \frac{\mathrm{Nu}}{\mathrm{Re\,Pr}}$	
Colburn factor	$j = \mathrm{St\,Pr}^{2/3} = \left(\mathrm{Nu\,Pr}^{-1/3}\right)/\mathrm{Re}$	a modified Stanton number to take into account the moderate variations in the Prandtl number for $0.5 \lesssim \mathrm{Pr} \lesssim 10.0$ in turbulent flow
Prandtl number	$\mathrm{Pr} = \frac{\nu}{\alpha} = \frac{\mu c_p}{k}$	a fluid property modulus representing the ratio of momentum diffusivity to thermal diffusivity of the fluid
Péclet number	$\mathrm{Pe} = \frac{\rho c_p u_m D_h}{k} = \frac{u_m D_h}{\alpha}$ $= \mathrm{Re\,Pr}$	proportional to the ratio of thermal energy convected to the fluid to thermal energy conducted axially within the fluid; the inverse of Pe indicates relative importance of fluid axial heat conduction
Dimensionless axial distance for the heat transfer problem	$x^* = \frac{x}{D_h \mathrm{Pe}} = \frac{x}{D_h \mathrm{Re\,Pr}}$	useful in describing the thermal entrance region heat transfer results
Graetz number	$Gz = \frac{\dot{m} c_p}{kL} = \frac{\mathrm{Pe}\,P}{4L} = \frac{P}{4C_h} \frac{1}{x^*}$	conventionally used in the chemical engineering literature related to x^* as shown when the flow length in Gz is treated as a length variable
	$Gz = \pi / \left(4x^*\right)$ for a circular tube	

Source: Shah and Mueller, 1988. With permission.

TABLE 13A.6 Solutions for Heat Transfer and Friction for Fully Developed Laminar Flow Through Specified Ducts

Geometry ($L/D_h > 100$)		Nu_{H1}	Nu_{H2}	Nu_T	fRe	$\frac{j_{H1}}{f}$ [a]	$K(\infty)$ [b]	L^+_{hy} [c]
Sine (2b, 2a)	$\frac{2b}{2a} = \frac{\sqrt{3}}{2}$	3.014	1.474	2.39	12.630	0.269	1.739	0.04
Triangle 60°C (2b, 2a)	$\frac{2b}{2a} = \frac{\sqrt{3}}{2}$	3.111	1.892	2.47	13.333	0.263	1.818	0.04
Square (2b, 2a)	$\frac{2b}{2a} = 1$	3.608	3.091	2.976	14.227	0.286	1.433	0.090
Hexagon		4.002	3.862	3.34	15.054	0.299	1.335	0.086
Rectangle (2b, 2a)	$\frac{2b}{2a} = \frac{1}{2}$	4.123	3.017	3.391	15.548	0.299	1.281	0.085
Circle		4.364	4.364	3.657	16.000	0.307	1.25	0.056
Rectangle (2b, 2a)	$\frac{2b}{2a} = \frac{1}{4}$	5.331	2.94	4.439	18.233	0.329	1.001	0.078
Rectangle (2b, 2a)	$\frac{2b}{2a} = \frac{1}{6}$	6.049	2.93	5.137	19.702	0.346	0.885	0.070
Rectangle (2b, 2a)	$\frac{2b}{2a} = \frac{1}{8}$	6.490	2.94	5.597	20.585	0.355	0.825	0.063
Parallel plates	$\frac{2b}{2a} = 0$	8.235	8.235	7.541	24.000	0.386	0.674	0.011

[a] $j_{H1}/f = Nu_{H1}\, Pr^{-1/3}/(f\,Re)$ with $Pr = 0.7$. Similarly, values of j_{H2}/f and j_T/f may be computed.
[b] $K(\infty)$ for sine and equilateral triangular channels may be too high [40]; $K(\infty)$ for some rectangular and hexagonal channels is interpolated based on the recommended values in Ref. [40].
[c] L^+_{hy} for sine and equilateral triangular channels is too low [40], so use with caution. L^+_{hy} for rectangular channels is based on the faired curve drawn through the recommended value in [40]. L^+_{hy} for a hexagonal channel is an interpolated value.

Source: Shah and London, 1978. With permission.

fRe. (2) The thermal boundary conditions have a strong influence on the Nusselt numbers. (3) As $Nu = hD_h/k$, a constant Nu implies the convective heat transfer coefficient h independent of the flow velocity and fluid Prandtl number. (4) An increase in h can be best achieved either by reducing D_h or by selecting a geometry with a low aspect ratio rectangular flow passage. Reducing the hydraulic diameter is an obvious way to increase exchanger compactness and heat transfer, or D_h can be optimized using well-known heat transfer correlations based on design problem specifications. (5) Since $f\mathrm{Re}$ = constant, $f \propto 1/\mathrm{Re} \propto 1/u_m$. In this case, it can be shown that $\Delta p \propto u_m$. Many additional analytical results for fully developed laminar flow ($\mathrm{Re} \leq 2{,}000$) are presented by Shah and London (1978) and Shah and Bhatti (1987). The entrance effects, flow maldistribution, free convection, property variation, fouling, and surface roughness all affect fully developed analytical solutions. In order to account for these effects in real plate-fin plain fin geometries having fully-developed flows, it is best to reduce the magnitude of the analytical Nu by at least 10% and to increase the value of the analytical fRe by 10% for design purposes.

The initiation of transition flow (the lower limit of the critical Reynolds number, Re_{crit}) depends upon the type of entrance (e.g., smooth vs. abrupt configuration at the exchanger flow passage entrance). For a sharp square inlet configuration, Re_{crit} is about 10 to 15% lower than that for a rounded inlet configuration. For most exchangers, the entrance configuration would be sharp. Some information on Re_{crit} is provided by Ghajar and Tam (1994).

Transition flow and fully developed turbulent flow Fanning friction factors are given by Bhatti and Shah (1987) as

$$f = A + B\mathrm{Re}^{-1/m} \tag{13A.36}$$

where

$$\begin{array}{llll} A = 0.0054, & B = 2.3 \times 10^{-8}, & m = -2/3 & \text{for } 2{,}100 \le \mathrm{Re} \le 4{,}000 \\ A = 0.00128, & B = 0.1143, & m = 3.2154 & \text{for } 4{,}000 \le \mathrm{Re} \le 10^7 \end{array}$$

Equation (13A.36) is accurate within ± 2% (Bhatti and Shah, 1987). The transition flow and fully developed turbulent flow Nusselt number correlation for a circular tube is given by Gnielinski as reported in Bhatti and Shah (1987) as

$$\mathrm{Nu} = \frac{(f/2)(\mathrm{Re} - 1{,}000)\mathrm{Pr}}{1 + 12.7(f/2)^{1/2}(\mathrm{Pr}^{2/3} - 1)} \tag{13A.37}$$

which is accurate within about ±10% with experimental data for $2{,}300 \le \mathrm{Re} \le 5 \times 10^6$ and $0.5 \le \mathrm{Pr} \le 2{,}000$. For higher accuracies in turbulent flow, refer to the correlations by Petukhov et al. reported by Bhatti and Shah (1987). Churchill (1977) (see also Bhatti and Shah, 1987) provides a correlation for laminar, transition, and turbulent flow regimes in a circular tube for $2{,}100 < \mathrm{Re} < 10^6$ and $0 < \mathrm{Pr} < \infty$.

A careful observation of accurate experimental friction factors for all noncircular smooth ducts reveals that ducts with laminar $f\mathrm{Re} < 16$ have turbulent f factors lower than those for the circular tube, whereas ducts with laminar $f\mathrm{Re} > 16$ have turbulent f factors higher than those for the circular tube (Shah and Bhatti, 1988). Similar trends are observed for the Nusselt numbers. If one is satisfied within ± 15% accuracy, Eqs. (13A.36) and (13A.37) for f and Nu can be used for noncircular passages with the hydraulic diameter as the characteristic length in f, Nu, and Re; otherwise refer to Table 13A.7 for more accurate results for turbulent flow.

For hydrodynamically and thermally developing flows, the analytical solutions are boundary condition dependent (for laminar flow heat transfer only) and geometry dependent. The reader may refer to Shah and London (1978), Shah and Bhatti (1987), and Bhatti and Shah (1987) for specific solutions. The hydrodynamic entrance lengths for developing laminar and turbulent flow are given by Shah and Bhatti (1987) and Bhatti and Shah (1987) as

$$\frac{L_{hy}}{D_h} = \begin{cases} 0.0565\ \mathrm{Re} & \text{for laminar flow } (\mathrm{Re} \le 2{,}100) \\ 1.359\ \mathrm{Re}^{1/4} & \text{for turbulent flow } (\mathrm{Re} \ge 10^4) \end{cases} \tag{13A.38}$$

Experimental Correlations

Analytical results presented in the preceding section are useful for well-defined constant cross-sectional surfaces with essentially unidirectional flows. The flows encountered in heat exchangers

TABLE 13A.7 Fully Developed Turbulent Flow Friction Factors and Nusselt Numbers (Pr > 0.5) for Technically Important Smooth-Walled Ducts

Duct geometry and characteristic dimension	Recommended correlations[a]
 Circular $D_h = 2a$	friction factor correlation for $2300 \leq Re \leq 10^7$ [41]: $$f = A + \frac{B}{Re^{1/m}}$$ where $A = 0.0054$, $B = 2.3 \times 10^{-8}$, $m = -2/3$ for $2100 \leq Re \leq 4000$ and $A = 1.28 \times 10^{-3}$. $B = 0.1143$, $m = 3.2154$ for $4000 \leq Re \leq 10^7$ [41]. Nusselt number correlation by Gnielinski for $2300 \leq Re \leq 5 \times 10^6$: $$Nu = \frac{(f/2)(Re - 1000)\,Pr}{1 + 12.7(f/2)^{1/2}(Pr^{2/3} - 1)}$$
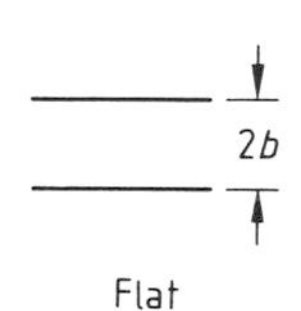 Flat $D_h = 4b$	use circular duct f and Nu correlations. Predicted f are up to 12.5% lower and predicted Nu are within $\pm 9\%$ of the most reliable experimental results.
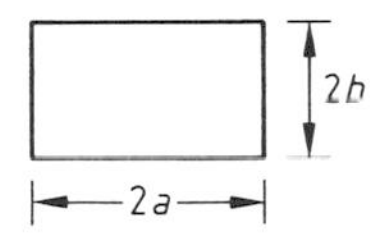 Rectangular $D_h = \frac{4ab}{a+b}$, $\alpha^* = \frac{2b}{2a}$ $\frac{D_l}{D_h} = \frac{2}{3} + \frac{11}{24}\alpha^*(2 - \alpha^*)$	f factors: (1) substitute D_l for D_h in the circular duct correlation, and calculate f from the resulting equation. (2) Alternatively, calculate f from $f = (1.0875 - 0.1125\,\alpha^*)\, f_c$ where f_c is the friction factor for the circular duct using D_h. In both cases, predicted f factors are within $\pm 5\%$ of the experimental results. Nusselt numbers: (1) With uniform heating at four walls, use circular duct Nu correlation for an accuracy of $\pm 9\%$ for $0.5 \leq Pr \leq 100$ and $10^4 \leq Re \leq 10^6$. (2) With equal heating at two long walls, use circular duct correlation for an accuracy of $\pm 10\%$ for $0.5 < Pr \leq 10$ and $10^4 \leq Re \leq 10^5$. (3) With heating at one long wall only, use circular duct correlation to get approximate Nu values for $0.5 < Pr < 10$ and $10^4 \leq Re \leq 10^6$. These calculated values may be up to 20% higher than the actual experimental values.
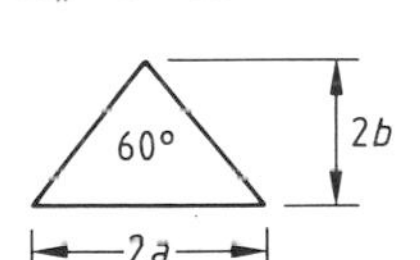 Equilateral triangular $D_h = 2\sqrt{3}\,a = 4b/3$ $D_l = \sqrt{3}\,a = 2b/3\sqrt{3}$	use circular duct f and Nu correlations with D_h replaced by D_l. Predicted f are within $+3\%$ and -11% and predicted Nu within $+9\%$ of the experimental values.
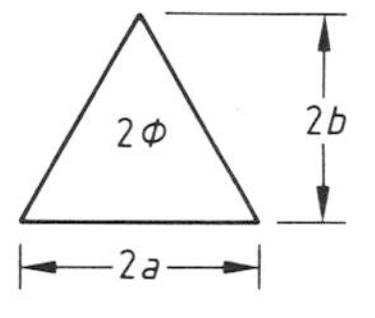 Isosceles triangular $D_h = \frac{4ab}{a + \sqrt{a^2 + b^2}}$ $\frac{D_g}{D_h} = \frac{1}{2\pi}\left[3 \ln \cot \frac{\theta}{2} + 2 \ln \tan \frac{\varphi}{2} - \ln \tan \frac{\theta}{2}\right]$ where $\theta = (90° - \varphi)/2$	for $0 < 2\varphi < 60°$, use circular duct f and Nu correlations with D_h replaced by D_g; for $2\varphi = 60°$, replace D_h by D_l (see above); and for $60° < 2\varphi \leq 90°$ use circular duct correlations directly with D_h. Predicted f and Nu are within $+9\%$ and -11% of the experimental values. No recommentations can be made for $2\varphi > 90°$ due to lack of the experimental data.

TABLE 13A.7 (continued) Fully Developed Turbulent Flow Friction Factors and Nusselt Numbers (Pr > 0.5) for Technically Important Smooth-Walled Ducts

Duct geometry and characteristic dimension	Recommended correlations[a]
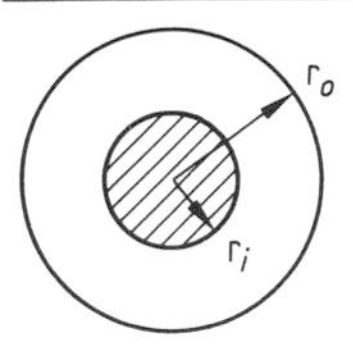 Concentric annular $D_h = 2(r_o - r_i)$, $r^* = \frac{r_i}{r_o}$ $\frac{D_l}{D_h} = \frac{1 + r^{*2} + (1 - r^{*2})/\ln r^*}{(1 - r^*)^2}$	f Factors: (1) Substitute D_l for D_h in the circular duct correlation, and calculate f from the resulting equation. (2) Alternatively, calculate f from $f = (1 + 0.0925\,r^*)\,f_c$ where f_c is the friction factor for the circular duct using D_h. In both cases, predicted f factors are within $\pm 5\%$ of the experimental results. Nusselt Numbers: In all the following recommendations, use D_h with a wetted perimeter in Nu and Re: (1) Nu at the outer wall can be determined from the circular duct correlation within the accuracy of about $\pm 10\%$ regardless of the condition at the inner wall. (2) Nu at the inner wall cannot be determined accurately regardless of the heating/cooling condition at the outer wall.

[a] The friction factor and Nusselt number corrlations for the circular duct are the most reliable and agree with a large amount of the experimental data within $\pm 2\%$ and $\pm 10\%$ respectively. The correlations for all other duct geometries are not as good as those for the circular duct on an absolute basis.

Source: Bhatti and Shah, 1987. With permission.

are generally very complex, having flow separation, reattachment, recirculation, and vortices. Such flows significantly affect Nu and f for the specific exchanger surfaces. Since no analytical or accurate numerical solutions are available, the information is derived experimentally, Kays and London (1984) and Webb (1994) present most of the experimental results reported in the open literature. In the following, empirical correlations for only some important surfaces are summarized due to space limitations.

Bare Tubebanks

One of the most comprehensive correlations for crossflow over a plain tube bank is presented by Zukauskas (1987) as shown in Figures 13A.12 and 13A.13 for inline (90° tube layout) and staggered arrangements (30° tube layout) respectively. These results are valid for the number of tube rows above about 16. For other inline and staggered tube arrangements, a correction factor χ is obtained from the inset of these figures. Zukauskas (1987) also presented the mean Nusselt number $Nu_m = h_m d_o / k$ as

$$Nu_m = F_c (Nu_m)_{16\ \text{rows}} \tag{13A.39}$$

Values of Nu_m for 16 or more tube rows are presented in Table 13A.8 for inline (90° tube layout) and staggered (30° tube layout) arrangements. For all expressions in Table 13A.8, fluid properties in Nu, Re_d and Pr are evaluated at the bulk mean temperature and for Pr_w at the wall temperature. The tube row correction factor F_c is presented in Figure 13A.14 as a function of the number of tube rows N for inline and staggered tube arrangements.

Plate-Fin Extended Surfaces

Offset Strip Fins. This is one of the most widely used enhanced fin geometries (Figure 13A.15) in aircraft, cryogenics, and many other industries that do not require mass production. This surface has one of the highest heat transfer performance relative to the friction factor. Extensive analytical, numerical, and experimental investigations have been conducted over the last 50 years. The most

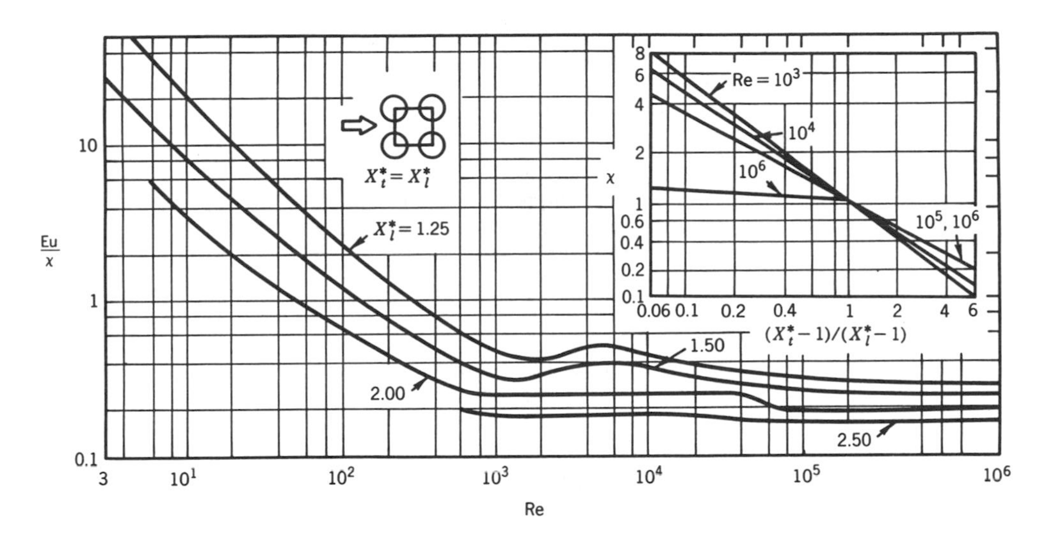

FIGURE 13A.12 Friction factors for inline tube arrangement; (a) $X_l^* = 1.25$, (b) $X_l^* = 1.5$, (c) $X_l^* = 2.0$ and (d) $X_l^* = 2.5$; $X_l^* = x_l/d_0$. *Source:* Zukauskas, 1987. With permission.

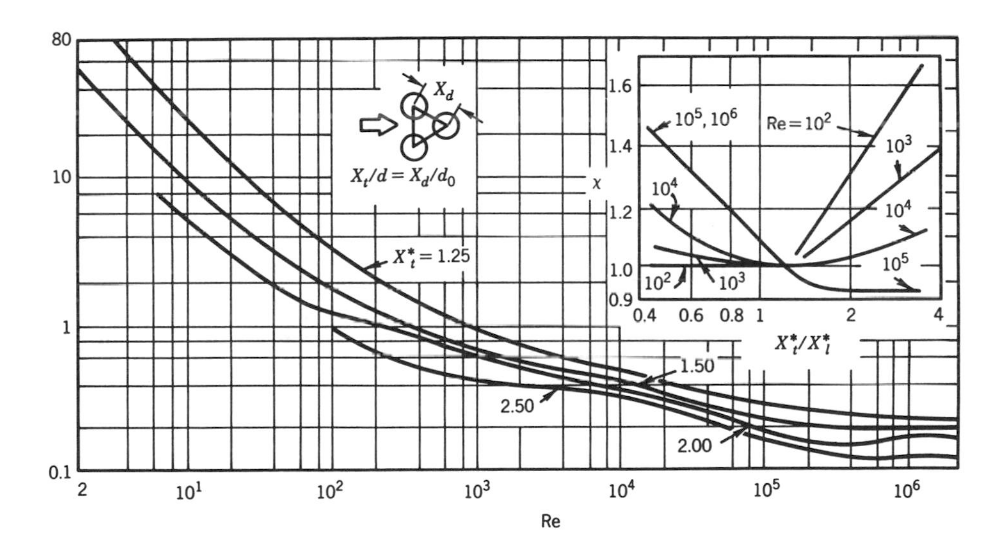

FIGURE 13A.13 Friction factors for staggered tube arrangement; (a) $X_l^* = 1.25$, (b) $X_l^* = 1.5$, (c) $X_l^* = 2.0$ and (d) $X_l^* = 2.5$; $X_l^* = X_l/d_o$. *Source:* Zukauskas, 1987. With permission.

comprehensive correlations for j and f factors for the offset strip fin geometry is provided by Manglik and Bergles (1995) as follows:

$$j = 0.6522\,\mathrm{Re}^{-0.5403}\left(\frac{s}{h'}\right)^{-0.1541}\left(\frac{\delta_f}{l_f}\right)^{0.1499}\left(\frac{\delta_f}{s}\right)^{-0.0678}\left[1+5.269\times10^{-5}\,\mathrm{Re}^{1.340}\left(\frac{s}{h'}\right)^{0.504}\left(\frac{\delta_f}{l_f}\right)^{0.456}\left(\frac{\delta_f}{s}\right)^{-1.055}\right]^{0.1} \quad \textbf{(13A.40)}$$

$$f = 9.6243\,\mathrm{Re}^{-0.7422}\left(\frac{s}{h'}\right)^{-0.1856}\left(\frac{\delta_f}{l_f}\right)^{0.3053}\left(\frac{\delta_f}{s}\right)^{-0.2659}\left[1+7.669\times10^{-8}\,\mathrm{Re}^{4.429}\left(\frac{s}{h'}\right)^{0.920}\left(\frac{\delta_f}{l_f}\right)^{3.767}\left(\frac{\delta_f}{s}\right)^{0.236}\right]^{0.1} \quad \textbf{(13A.41)}$$

TABLE 13A.8A Mean Nusselt Number Correlations for Inline Tube Bundles for $N > 16$

Recommended Correlations	Range of Re
$Nu = 0.9\ Re^{0.4}Pr^{0.36}(Pr/Pr_w)^{0.25}$	10^0–10^2
$Nu = 0.52\ Re^{0.5}Pr^{0.36}(Pr/Pr_w)^{0.25}$	10^2–10^3
$Nu = 0.27\ Re^{063}Pr^{0.36}(Pr/Pr_w)^{0.25}$	10^3–2×10^5
$Nu = 0.033\ Re^{0.8}Pr^{0.4}(Pr/Pr_w)^{0.25}$	2×10^5–2×10^6

Source: Zukauskas, 1987.

TABLE 13A.8B Mean Nusselt Number Correlations for Staggered Tube Bundles for N > 16

Recommended Correlations	Range of Re
$Nu = 1.04\ Re^{0.4}Pr^{0.36}(Pr/Pr_w)^{0.25}$	10^0–5×10^2
$Nu = 0.71\ Re^{0.5}Pr^{0.36}(Pr/Pr_w)^{0.25}$	5×10^2–10^3
$Nu = 0.35\ (X_t^*/X_1^*)^{0.2}Re^{0.6}Pr^{0.36}(Pr/P_w)^{0.25}$	10^3–2×10^5
$Nu = 0.31\ (X_t^*/X_1^*)^{0.2}Re^{0.8}Pr^{0.36}(Pr/Pr_w)^{0.25}$	2×10^5–2×10^6

Source: Zukauskas, 1987.

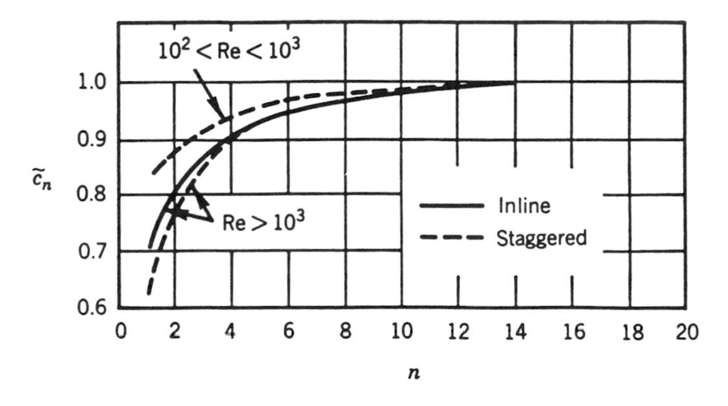

FIGURE 13A.14 A correction factor F_c to account for the tube-row effective for heat transfer for flow normal to bare tube banks. *Source*: Zukauskas, 1987. With permission.

where

$$D_h = 4A_o \big/ \left(A/l_f\right) = 4sh'l_f \big/ \left[2\left(sl_f + h'l_f + \delta_f h'\right) + \delta_f s\right] \tag{13A.42}$$

Geometrical symbols in Equation (13A.42) are shown in Figure 13A.15.

These correlations predict the experimental data of 18 test cores within ±20% for $120 \le Re \le 10^4$. Although all the experimental data for these correlations are obtained for air, the j factor takes into consideration minor variations in the Prandtl number, and the correlations should be valid for $0.5 < Pr < 15$.

Louver Fins. Louver or multilouver fins are extensively used in the auto industry due to their mass production manufacturability and hence lower cost. It has generally higher j and f factors than those for the offset strip fin geometry, and the increase in the friction factors is in general

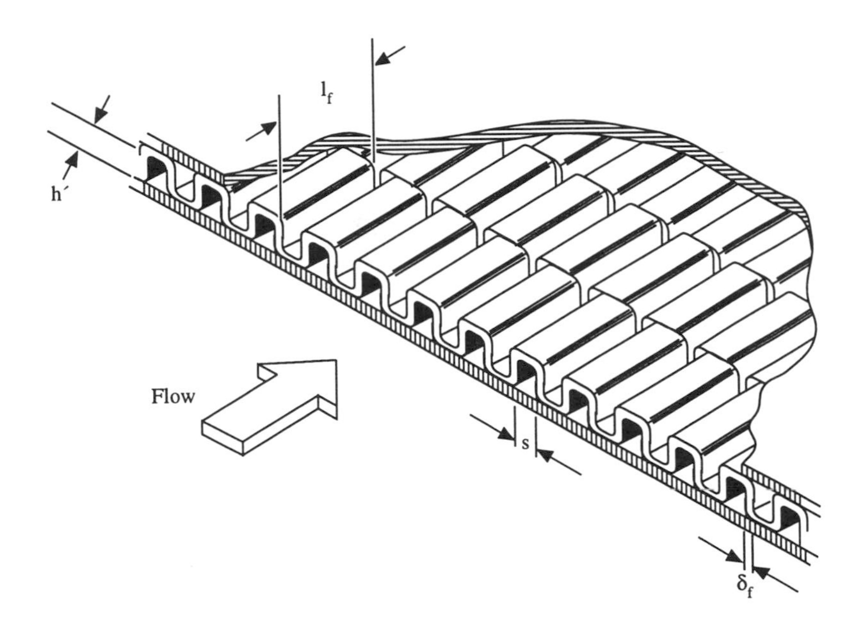

FIGURE 13A.15 An offset strip fin geometry.

higher than the increase in the j factors. However, the exchanger can be designed for higher heat transfer and the same pressure drop compared to that with the offset strip fins by a proper selection of exchanger frontal area, core depth, and fin density. Published literature and correlations on the louver fins are summarized by Webb (1994) and Cowell et al. (1995), and the understanding of flow and heat transfer phenomena is summarized by Cowell et al. (1995). Because of a lack of systematic studies reported in the open literature on modern louver fin geometries, no correlation can be recommended for the design purpose.

Tube-Fin Extended Surfaces

Two major types of tube-fin extended surfaces are (1) individually finned tubes, and (2) flat fins (also sometimes referred to as plate fins) with or without enhancements/interruptions on an array of tubes as shown in Figure 13A.9. Extensive coverage of the published literature and correlations for these extended surfaces is provided by Webb (1994), Kays and London (1984), and Rozenman (1976). Empirical correlations for some important geometries are summarized next.

Individually Finned Tubes. This fin geometry, helically wrapped (or extruded) circular fins on a circular tube as shown in Figure 13A.9a, is commonly used in process and waste heat recovery industries. The following correlation for j factors is recommended by Briggs and Young (see Webb, 1994), for individually finned tubes on staggered tubebanks.

$$j = 0.134\,\mathrm{Re}_d^{-0.319}\left(s/l_f\right)^{0.2}\left(s/\delta_f\right)^{0.11} \tag{13A.43}$$

where l_f is the radial height of the fin, δ_f the fin thickness, $s = p_f - \delta_f$ is the distance between adjacent fins, and p_f is the fin pitch. Equation (13A.43) is valid for the following ranges: $1{,}100 \le \mathrm{Re}_d \le 18{,}000$, $0.13 \le s/l_f \le 0.63$, $1.01 \le s/\delta_f \le 6.62$, $0.09 \le l_f/d_o \le 0.69$, $0.011 \le \delta_f/d_o \le 0.15$, $1.54 \le X_t/d_o \le 8.23$, fin root diameter d_0 between 11.1 and 40.9 mm, and fin density N_f ($= 1/p_f$) between 246 and 768 fins per meter. The standard deviation of Eq. (13A.43) with experimental results is 5.1%.

For friction factors, Robinson and Briggs (see Webb, 1994) recommended the following correlation.

$$f_{tb} = 9.465\ \mathrm{Re}_d^{-0.316} \left(X_t/d_0\right)^{-0.927} \left(X_t/X_d\right)^{0.515} \tag{13A.44}$$

Here $X_d = (X_t^2 + X_\ell^2)^{1/2}$ is the diagonal pitch, and X_t and X_ℓ are the transverse and longitudinal tube pitches, respectively. The correlation is valid for the following ranges: $2{,}000 \le \mathrm{Re}_d \le 50{,}000$, $0.15 \le s/l_f \le 0.19$, $3.75 \le s/\delta_f \le 6.03$, $0.35 \le l_f/d_0 \le 0.56$, $0.011 \le \delta_f/d_0 \le 0.025$, $1.86 \le X_t/d_o \le 4.60$, $18.6 \le d_o \le 40.9$ mm, and $311 \le N_f \le 431$ fins per meter. The standard deviation of Eq. (13A.44) with correlated data is 7.8%.

For crossflow over low-height finned tubes, Rabas and Taborek (1987), Ganguli and Yilmaz (1987), and Chai (1988) have assessed the pertinent literature. A simple but accurate correlation for heat transfer is given by Ganguli and Yilmaz (1987) as

$$j = 0.255\ \mathrm{Re}_d^{-0.3}\left(d_e/s\right)^{-0.3} \tag{13A.45}$$

A more accurate correlation for heat transfer is given by Rabas and Taborek (1987). Chai (1988) provides the best correlation for friction factors:

$$f_{tb} = 1.748\ \mathrm{Re}_d^{-0.233}\left(\frac{l_f}{s}\right)^{0.552}\left(\frac{d_o}{X_t}\right)^{0.599}\left(\frac{d_o}{X_l}\right)^{0.1738} \tag{13A.46}$$

This correlation is valid for $895 < \mathrm{Re}_d < 713{,}000$, $20 < \theta < 40°$, $X_t/d_0 < 4$, $N \ge 4$, and θ is the tube layout angle. It predicts 89 literature data points within a mean absolute error of 6%; the range of actual error is from –16.7 to 19.9%.

Flat Plain Fins on a Staggered Tubebank. This geometry, as shown in Figure 13A.9b, is used in the air-conditioning/refrigeration industry as well as where the pressure drop on the fin side prohibits the use of enhanced flat fins. An inline tubebank is generally not used unless very low fin side pressure drop is the essential requirement. Heat transfer correlation for Figure 13A.9b flat plain fins on staggered tubebanks is provided by Gray and Webb (see Webb, 1994) as follows for four or more tube rows.

$$j_4 = 0.14\ \mathrm{Re}_d^{-0.328}\left(X_t/X_l\right)^{-0.502}\left(s/d_o\right)^{0.031} \tag{13A.47}$$

For the number of tube rows N from 1 to 3, the j factor is lower and is given by

$$\frac{j_N}{j_4} = 0.991\left[2.24\ \mathrm{Re}_d^{-0.092}\left(N/4\right)^{-0.031}\right]^{0.607(4-N)} \tag{13A.48}$$

Gray and Webb hypothesized the friction factor consisting of two components, one associated with the fins and the other associated with the tubes as follows:

$$f = f_f\frac{A_f}{A} + f_t\left(1-\frac{A_f}{A}\right)\left(1-\frac{\delta_f}{p_f}\right) \tag{13A.49}$$

where

$$f_f = 0.508\ \mathrm{Re}_d^{-0.521}\left(X_t/d_0\right)^{1.318} \tag{13A.50}$$

and f_t (defined the same way as f_f) is the Fanning friction factor associated with the tube and can be determined from Eu of Figure 13A.13 as $f_t = \mathrm{Eu}N(X_t - d_0)/\pi d_0$. Equation (13A.49) correlated 90% of the data for 19 heat exchangers within ±20%. The ranges of dimensionless variables of Eqs. (13A.49) and (13A.50) are $500 \le \mathrm{Re} \le 24{,}700$, $1.97 \le X_t/d_o \le 2.55$, $1.7 \le X_t/d_0 \le 2.58$, and $0.08 \le s/d_0 \le 0.64$.

13A.4 Exchanger Design Methodology

The problem of heat exchanger design is complex and multidisciplinary (Shah, 1991b). The major design considerations for a new heat exchanger include process/design specifications, thermal and hydraulic design, mechanical design, manufacturing and cost considerations, and trade-offs and system-based optimization, as shown in Figure 13A.16 (with possible strong interactions among these considerations as indicated by double-sided arrows). The thermal and hydraulic design has mainly analytical solutions; the structural design has also to some extent analytical solutions. Most of the other major design considerations involve qualitative and experience-based judgments, trade-offs, and compromises. Therefore, there is no unique solution to designing a heat exchanger for given process specifications. Further details on this design methodology are given by Shah (1991b).

Two of the most important heat exchanger design problems are rating and sizing. Determination of heat transfer and pressure drop performance of either an existing exchanger or an already sized exchanger is referred to as the rating problem. The objective here is to verify the vendor's specifications or to determine the performance at off-design conditions. The rating problem is also sometimes referred to as the performance problem. In contrast, the design of a new or existing type exchanger is referred to as the sizing problem. In a broad sense, this means the determination of the exchanger construction type, flow arrangement, heat transfer surface geometries and materials, and physical size of an exchanger to meet the specified heat transfer and pressure drops. However, from the viewpoint of quantitative thermal-hydraulic analysis, we will consider that the selection of the exchanger construction type, flow arrangement, and materials has already been made. Thus in the sizing problem, we will determine the physical size (length, width, height) and surface areas on each side of the exchanger. The sizing problem is also sometimes referred to as the design problem.

The step-by-step solution procedures for the rating and sizing problems for counterflow and crossflow single-pass plate-fin heat exchangers have been presented with a detailed illustrative example by Shah (1981). Shah (1988a) presented further refinements in these procedures as well as step-by-step solution procedures for two-pass cross-counterflow plate-fin exchangers, and single-pass crossflow and two-pass cross-counterflow tube-fin exchangers. Also, step-by-step solution procedures for the rating and sizing problems for rotary regenerators (Shah, 1988b), heat-pipe heat exchangers (Shah and Giovannelli, 1988), and plate heat exchangers (Shah and Wanniarachchi, 1991) are available. As an illustration, the step-by-step solution procedures will be covered here for a single-pass crossflow exchanger.

Rating Problem for a Crossflow Plate-Fin Exchanger

We will present here a step-by-step solution procedure for the rating problem for a crossflow plate-fin exchanger. Inputs to the rating problem for a two-fluid exchanger are the exchanger construction, flow arrangement and overall dimensions, complete details on the materials and surface

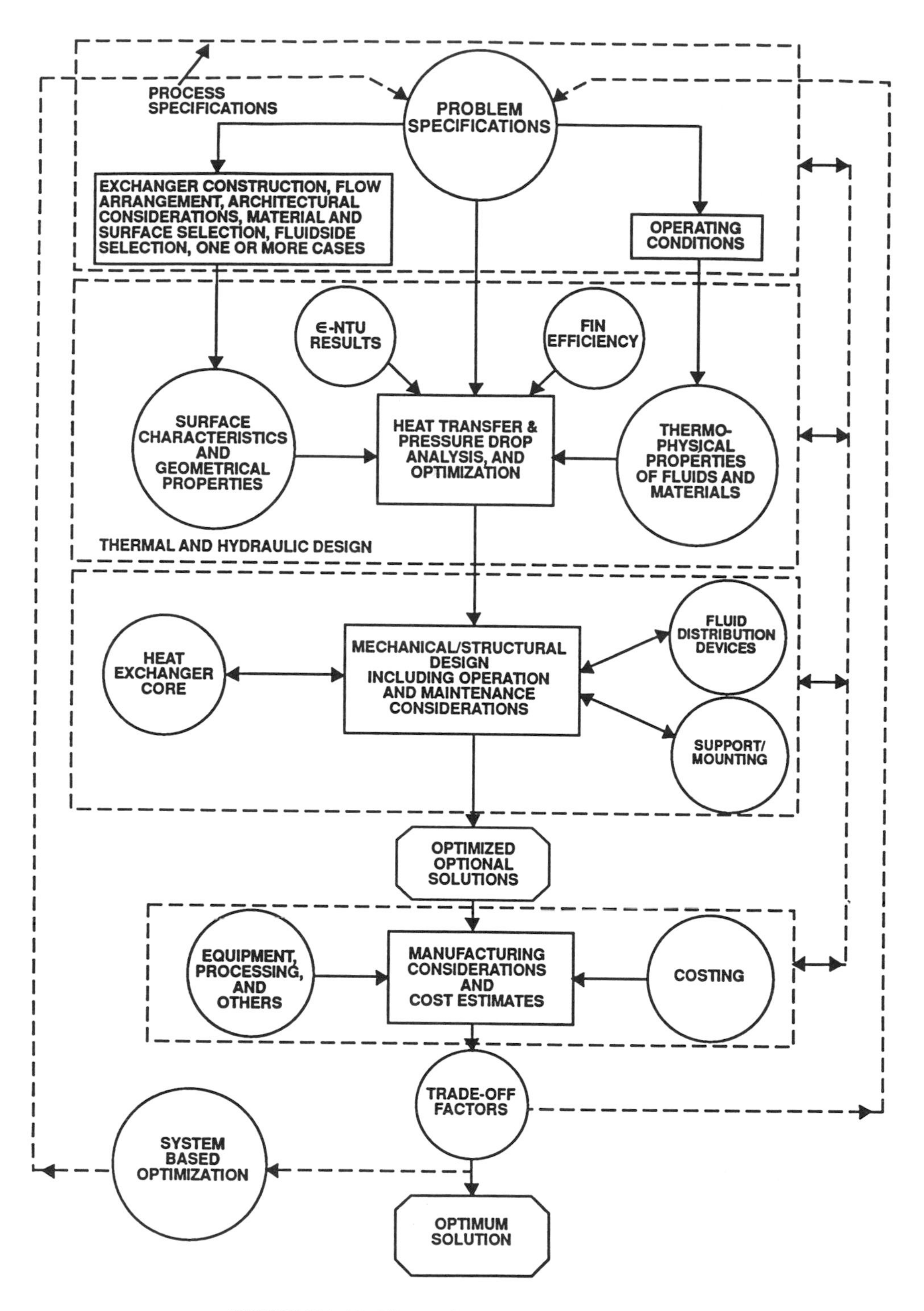

FIGURE 13A.16 Heat exchanger design methodology.

geometries on both sides including their nondimensional heat transfer and pressure drop characteristics (j and f vs. Re), fluid flow rates, inlet temperatures, and fouling factors. The fluid outlet temperatures, total heat transfer rate, and pressure drops on each side of the exchanger are then determined as the rating problem solution.

1. Determine the surface geometrical properties on each fluid side. This includes the minimum free flow area A_o, heat transfer surface area A (both primary and secondary), flow lengths L, hydraulic diameter D_h, heat transfer surface area density β, the ratio of minimum free flow area to frontal area σ, fin length and fin thickness δ for fin efficiency determination, and any specialized dimensions used for heat transfer and pressure drop correlations.
2. Compute the fluid bulk mean temperature and fluid thermophysical properties on each fluid side. Since the outlet temperatures are not known for the rating problem, they are estimated initially. Unless it is known from the past experience, assume an exchanger effectiveness of 60 to 75% for most single-pass crossflow exchangers or 80 to 85% for single-pass counterflow exchangers. For the assumed effectiveness, calculate the fluid outlet temperatures.

$$T_{h,o} = T_{h,i} - \varepsilon\left(C_{\min}/C_h\right)\left(T_{h,i} - T_{c,j}\right) \quad \mathbf{(13A.51)}$$

$$T_{c,o} = T_{c,i} + \varepsilon\left(C_{\min}/C_c\right)\left(T_{h,i} - T_{c,i}\right) \quad \mathbf{(13A.52)}$$

Initially, assume $C_c/C_h = \dot{m}_c/\dot{m}_h$ for a gas-to gas exchanger, or $C_c/C_h = \dot{m}_c c_{p,c}/\dot{m}_h c_{p,h}$ for a gas-to-liquid exchanger with very approximate values of c_p's for the fluids in question.

For exchangers with $C^* > 0.5$ (usually gas-to-gas exchangers), the bulk mean temperatures on each fluid side will be the arithmetic mean of the inlet and outlet temperatures on each fluid side (Shah, 1981). For exchangers with $C^* > 0.5$ (usually gas-to-gas exchangers), the bulk mean temperature on the $C_{\max}$ side will be the arithmetic mean of inlet and outlet temperatures; the bulk mean temperature on the $C_{\min}$ side will be the log-mean average temperature obtained as follows.

$$T_{m,C_{\min}} = T_{m,C_{\max}} \pm \Delta T_{lm} \quad \mathbf{(13A.53)}$$

where ΔT_{lm} is the log-mean temperature difference based on the terminal temperatures (see Eq. 13A.19); use the plus sign if the $C_{\min}$ side is hot, otherwise use the negative sign.

Once the bulk mean temperature is obtained on each fluid side, obtain the fluid properties from thermophysical property books or from handbooks. The properties needed for the rating problem are μ, c_p, k, Pr and ρ. With this c_p, one more iteration may be carried out to determine $T_{h,o}$ or $T_{c,o}$ from Eq. (13A.51) or (13A.52) on the $C_{\max}$ side, and subsequently T_m on the $C_{\max}$ side, and refine fluid properties accordingly.
3. Calculate the Reynolds number Re = GD_h/μ and/or any other pertinent dimensionless groups (from the basic definitions) needed to determine the nondimensional heat transfer and flow friction characteristics (e.g., j or Nu and f) of heat transfer surfaces on each side of the exchanger. Subsequently, compute j or Nu and f factors. Correct Nu (or j) for variable fluid property effects (Shah, 1981) in the second and subsequent iterations from the following equations.

$$\text{For gases,} \quad \frac{Nu}{Nu_{cp}} = \left[\frac{T_w}{T_m}\right]^{n'} \quad \frac{f}{f_{cp}} = \left[\frac{T_w}{T_m}\right]^{m'} \quad \mathbf{(13A.54)}$$

$$\text{For liquids,} \quad \frac{Nu}{Nu_{cp}} = \left[\frac{\mu_w}{\mu_m}\right]^{n'} \quad \frac{f}{f_{cp}} = \left[\frac{\mu_w}{\mu_m}\right]^{m'} \quad \mathbf{(13A.55)}$$

TABLE 13A.9A Property Ratio Method Exponents of Eqs. (13A.54) and (13A.55) for Laminar Flow

Fluid	Heating	Cooling
Gases	$n' = 0.0$, $m' = 1.00$ for $1 < T_w/T_m < 3$	$n' = 0.0$, $m' = 0.81$ for $0.5 < T_w/T_m < 1$
Liquids	$n' = -0.14$, $m' = 0.58$ for $\mu_w/\mu_m < 1$	$n' = -0.14$, $m' = 0.54$ for $\mu_w/\mu_m > 1$

TABLE 13A.9B Property Ratio Method Correlations or Exponents of Eqs. (13A.54) and (13A.55) for Turbulent Flow

Fluid	Heating	Cooling
Gases	$\text{Nu} = 5 + 0.012\,\text{Re}^{0.83}(\text{Pr} + 0.29)\,(T_w/T_m)^n$ $n = -[\log_{10}(T_w/T_m)]^{1/4} + 0.3$ for $1 < T_w/T_m < 5$, $0.6 < \text{Pr} < 0.9$, $10^4 < \text{Re} < 10^6$ and $L/D_h > 40$	$n' = 0$
	$m' = -0.1$ for $1 < T_w/T_m < 2.4$	$m' = -0.1$ (tentative)
Liquids	$n' = -0.11^*$ for $0.08 < \mu_w/\mu_m < 1$	$n' = -0.25^*$ for $1 < \mu_w/\mu_m < 40$
	$f/f_{cp} = (7 - \mu_m/\mu_w)/6^\dagger$ or $m' \simeq 0.25$ for $0.35 < \mu_w/\mu_m < 1$	$m' = 0.24^\dagger$ for $1 < \mu_w/\mu_m < 2$

* Valid for $2 \leq \text{Pr} \leq 140$, $10^4 \leq \text{Re} \leq 1.25 \times 10^5$.
† Valid for $1.3 \leq \text{Pr} \leq 10$, $10^4 \leq \text{Re} \leq 2.3 \times 10^5$.
Source: Shah, 1981.

where the subscript *cp* denotes constant properties and m' and n' are empirical constants provided in Table 13A.9. Note that T_w and T_m in Eqs. (13A.54) and (13A.55) and in Tables 13A.9A and B are absolute temperatures.

4. From Nu or j, compute the heat transfer coefficients for both fluid streams.

$$h = \text{Nu}\; k/D_h = jGc_p\; \text{Pr}^{-2/3} \tag{13A.56}$$

Subsequently, determine the fin efficiency η_f and the extended surface efficiency η_o

$$\eta_f = \frac{\tanh ml}{ml} \quad \text{where} \quad m^2 = \frac{h\tilde{P}}{k_f A_k} \tag{13A.57}$$

where $\tilde{P}$ is the wetted perimeter of the fin surface;

$$\eta_o = 1 - (1 - \eta_f) A_f / A \tag{13A.58}$$

Also calculate the wall thermal resistance $R_w = a/A_w k_w$. Finally compute the overall thermal conductance *UA* from Eq. (13A.6), knowing the individual convective film resistances, wall thermal resistances, and fouling resistances, if any.

5. From the known heat capacity rates on each fluid side, compute $C^* = C_{min}/C_{max}$. From the known *UA*, determine $\text{NTU} = UA/C_{min}$. Also calculate the longitudinal conduction parameter λ. With the known NTU, C^*, λ, and the flow arrangement, determine the exchanger effectiveness ε from either closed-form equations of Table 13A.3 or tabular/graphical results from Kays and London (1984).

6. With this ε, finally compute the outlet temperatures from Eqs. (13A.51) and (13A.52). If these outlet temperatures are significantly different from those assumed in Step 2, use these outlet temperatures in Step 2 and continue iterating Steps 2 to 6, until the assumed and computed outlet temperatures converge within the desired degree of accuracy. For a gas-to-gas exchanger, most probably one iteration may be sufficient.
7. Finally, compute the heat duty from

$$q = \varepsilon C_{\min}\left(T_{h,i} - T_{c,i}\right) \tag{13A.59}$$

8. For the pressure drop calculations, we first need to determine the fluid densities at the exchanger inlet and outlet, ρ_i and ρ_o, for each fluid. The mean specific volume on each fluid side is then computed from Eq. (13A.29). Next, the entrance and exit loss coefficients, K_c and K_e are obtained from Figure 13A.11 for known σ, Re, and the flow passage entrance geometry. The friction factor on each side is corrected for the variable fluid properties using Eq. (13A.54) or (13A.55). Here, the wall temperature T_w is computed from

$$T_{w,h} = T_{m,h} - \left(R_h + R_{s,h}\right)q \tag{13A.60}$$

$$T_{w,c} = T_{m,c} + \left(R_c + R_{s,c}\right)q \tag{13A.61}$$

where the various resistance terms are defined by Eq. (13A.6). The core pressure drops on each fluid side are then calculated from Eq. (13A.28). This then completes the procedure for solving the rating problem.

Sizing Problem for Plate-Fin Exchangers

As defined earlier, we will concentrate here to determine the physical size (length, width, and height) of a single-pass crossflow exchanger for specified heat duty and pressure drops. More specifically inputs to the sizing problem are surface geometries (including their nondimensional heat transfer and pressure drop characteristics), fluid flow rates, inlet and outlet fluid temperatures, fouling factors, and pressure drops on each side.

For the solution to this problem, there are four unknowns: two flow rates or Reynolds numbers (to determine correct heat transfer coefficients and friction factors) and two surface areas for the two-fluid crossflow exchanger. The following four equations (Eqs. 13A.62, 13A.64 and 13A.66) are used to solve iteratively the surface areas on each fluid side: *UA* in Eq. (13A.62) is determined from NTU computed from the known heat duty or ε and C^*; *G* in Eq. (13A.64) represents two equations, for Fluid 1 and 2 (Shah, 1988a); and the volume of the exchanger in Eq. (13A.66) is the same based on the surface area density of Fluid 1 and 2 sides.

$$\frac{1}{UA} \approx \frac{1}{\left(\eta_o hA\right)_h} + \frac{1}{\left(\eta_o hA\right)_c} \tag{13A.62}$$

Here we have neglected the wall and fouling thermal resistances. This equation in nondimensional form is given by

$$\frac{1}{\mathrm{NTU}} = \frac{1}{\mathrm{ntu}_h\left(C_h/C_{\min}\right)} + \frac{1}{\mathrm{ntu}_c\left(C_c/C_{\min}\right)} \tag{13A.63}$$

$$G = \left[\frac{2g_c\Delta p}{\mathrm{Deno}}\right]^{1/2} \tag{13A.64}$$

where

$$\text{Deno} = \frac{f}{j}\frac{\text{ntu}}{\eta_o}\text{Pr}^{2/3}\left(\frac{1}{\rho}\right)_m + 2\left(\frac{1}{\rho_o} - \frac{1}{\rho_i}\right) + \left(1 - \sigma^2 - K_e\right)\frac{1}{\rho_o} \quad \textbf{(13A.65)}$$

$$V = \frac{A_1}{\alpha_1} = \frac{A_2}{\alpha_2} \quad \textbf{(13A.66)}$$

In the iterative solutions, the first time one needs ntu_h and ntu_c to start the iterations. These can be either determined from the past experience or by estimations. If both fluids are gases or both fluids are liquid, one could consider that the design is "balanced," that is, the thermal resistances are distributed approximately equally on the hot and cold sides. In that case, $C_h = C_c$, and

$$\text{ntu}_h \approx \text{ntu}_c \approx 2\,\text{NTU} \quad \textbf{(13A.67)}$$

Alternatively, if we have liquid on one side and gas on the other side, consider 10% thermal resistance on the liquid side; that is,

$$0.10\left(\frac{1}{UA}\right) = \frac{1}{\left(\eta_o hA\right)_{\text{liq}}} \quad \textbf{(13A.68)}$$

Then from Eqs. (13A.62) and (13A.63) with $C_{\text{gas}} = C_{\min}$, we can determine the ntu on each side as follows.

$$\text{ntu}_{\text{gas}} = 1.11\,\text{NTU}, \quad \text{ntu}_{\text{liq}} = 10C^*\,\text{NTU} \quad \textbf{(13A.69)}$$

Also note that initial guesses of η_o and j/f are needed for the first iteration to solve Eq. (13A.65). For a good design, consider $\eta_o = 0.80$ and determine approximate value of j/f from the plot of j/f vs. Re curve for the known j and f vs. Re characteristics of each fluid side surface. The specific step-by-step design procedure is as follows.

1. In order to compute the fluid bulk mean temperature and the fluid thermophysical properties on each fluid side, determine the fluid outlet temperatures from the specified heat duty.

$$q = \left(\dot{m}c_p\right)_h\left(T_{h,i} - T_{h,o}\right) = \left(\dot{m}c_p\right)_c\left(T_{c,o} - T_{c,i}\right) \quad \textbf{(13A.70)}$$

   or from the specified exchanger effectiveness using Eqs. (13A.51) and (13A.52). For the first time, estimate the values of c_p's.

   For exchangers with $C^* \geq 0.5$, the bulk mean temperature on each side will be the arithmetic mean of inlet and outlet temperatures on each side. For exchangers with $C^* < 0.5$, the bulk mean temperature on the $C_{\max}$ side will be the arithmetic mean of the inlet and outlet temperatures on that side, the bulk mean temperature on the $C_{\min}$ side will be the log-mean average as given by Eq. (13A.53). With these bulk mean temperatures, determine c_p and iterate one more time for the outlet temperatures if warranted. Subsequently, determine μ, c_p, k, Pr, and ρ on each fluid side.
2. Calculate C^* and ε (if q is given), and determine NTU from the ε-NTU expression, tables or graphical results for the selected flow arrangement (in this case, it is unmixed–unmixed crossflow, Table 13A.3). The influence of longitudinal heat conduction, if any, is ignored in the first iteration since we don't know the exchanger size yet.

3. Determine ntu on each side by the approximations discussed with Eqs. (13A.67) and (13A.69) unless it can be estimated from the past experience.
4. For the selected surfaces on each fluid side, plot j/f vs. Re curve from the given surface characteristics, and obtain an approximate value of j/f. If fins are employed, assume $\eta_o =$ 0.80 unless a better value can be estimated.
5. Evaluate G from Eq. (13A.64) on each fluid side using the information from Steps 1 to 4 and the input value of Δp.
6. Calculate Reynolds number Re, and determine j and f on each fluid side from the given design data for each surface.
7. Compute h, η_f, and η_o using Eqs. (13A.56) to (13A.58). For the first iteration, determine U_1 on Fluid 1 side from the following equation derived from Eqs. (13A.6) and (13A.66).

$$\frac{1}{U_1} = \frac{1}{(\eta_o h)_1} + \frac{1}{(\eta_o h_s)_1} + \frac{\alpha_1/\alpha_2}{(\eta_o h_s)_2} + \frac{\alpha_1/\alpha_2}{(\eta_o h)_2} \tag{13A.71}$$

where $\alpha_1/\alpha_2 = A_1/A_2$, $\alpha = A/V$, V is the exchanger total volume, and subscripts 1 and 2 denote Fluid 1 and 2 sides. For a plate-fin exchanger, the α's are given by (Shah, 1981; Kays and London, 1984)

$$\alpha_1 = \frac{b_1\beta_1}{b_1 + b_2 + 2a} \qquad \alpha_2 = \frac{b_2\beta_2}{b_1 + b_2 + 2a} \tag{13A.72}$$

Note that the wall thermal resistance in Eq. (13A.71) is ignored in the first iteration. In second and subsequent iterations, compute U_1 from

$$\frac{1}{U_1} = \frac{1}{(\eta_o h)_1} + \frac{1}{(\eta_o h_s)_1} + \frac{\delta A_1}{k_w A_w} + \frac{A_1/A_2}{(\eta_o h_s)_2} + \frac{A_1/A_2}{(\eta_o h)_2} \tag{13A.73}$$

where the necessary geometry information A_1/A_2 and A_1/A_w is determined from the geometry calculated in the previous iteration.

8. Now calculate the core dimensions. In the first iteration, use NTU computed in Step 2. For subsequent iterations, calculate longitudinal conduction parameter λ (and other dimensionless groups for a crossflow exchanger). With known ε, C^*, and λ, determine the correct value of NTU using either a closed-form equation or tabulated/graphical results (Shah and Mueller, 1985), Determine A_1 from NTU using U_1 from the previous step and known C_{min}.

$$A_1 = \text{NTU } C_{min}/U_1 \tag{13A.74}$$

and hence

$$A_2 = (A_2/A_1)A_1 = (\alpha_2/\alpha_1)A_1 \tag{13A.75}$$

A_o from known $\dot{m}$ and G is given by

$$A_{o,1} = (\dot{m}/G)_1 \quad A_{o,2} = (\dot{m}/G)_2 \tag{13A.76}$$

so that

$$A_{fr,1} = A_{o,1}/\sigma_1 \quad A_{fr,2} = A_{o,2}/\sigma_2 \tag{13A.77}$$

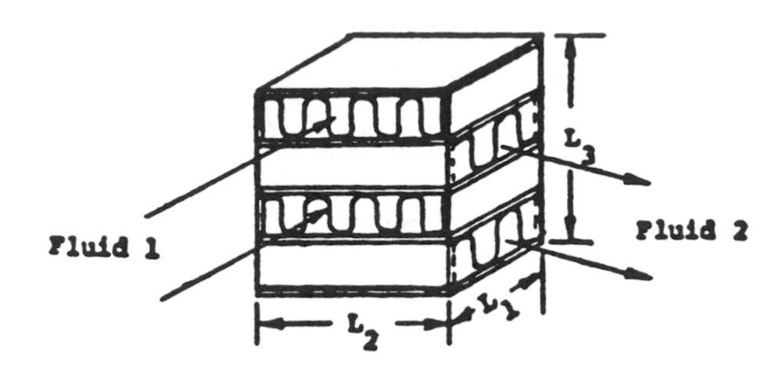

FIGURE 13A.17 A single-pass crossflow exchanger.

where σ_1 and σ_2 are generally specified for the surface or can be computed for plate-fin surfaces from (Shah, 1981; Kays and London, 1984):

$$\sigma_1 = \frac{b_1\beta_1 D_{h,1}/4}{b_1 + b_2 + 2a} \quad \sigma_2 = \frac{b_2\beta_2 D_{h,2}/4}{b_1 + b_2 + 2a} \tag{13A.78}$$

Now compute the fluid flow lengths on each side (see Figure 13A.17) from the definition of the hydraulic diameter of the surface employed on each side.

$$L_1 = \left(\frac{D_h A}{4A_o}\right)_1 \quad L_2 = \left(\frac{D_h A}{4A_o}\right)_2 \tag{13A.79}$$

Since $A_{fr,1} = L_2 L_3$ and $A_{fr,2} = L_1 L_3$, we can obtain

$$L_3 = \frac{A_{fr,1}}{L_2} \text{ or } L_3 = \frac{A_{fr,2}}{L_1} \tag{13A.80}$$

Theoretically, L_3's calculated from both expressions of Eq. (13A.80), should be identical. In reality, they may differ slightly due to the round-off error. In that case, consider an average value for L_3.

9. Now compute the pressure drop on each fluid side, after correcting f factors for variable property effects, in a manner similar to Step 8 of the Rating Problem for a Crossflow Plate-Fin Exchanger.
10. If the calculated values of the Δp's are within and close to input specifications, the solution to the sizing problem is completed. Finer refinements in the core dimensions such as integer numbers of flow passages, and so on may be carried out at this time. Otherwise, compute the new value of G on each fluid side using Eq. (13A.28) in which Δp is the input specified value, and f, K_c, K_e, and the geometrical dimensions are from the previous iteration.
11. Repeat (iterate) Steps 6 to 10 until both heat transfer and pressure drops are met as specified. It should be emphasized that since we have imposed no constraints on the exchanger dimensions, the procedure will yield L_1, L_2, and L_3 for the selected surfaces such that the design will meet the heat duty and pressure drops on both fluid sides exactly.

13A.5 Flow Maldistribution

In previously presented heat transfer (ε-NTU, MTD, etc. methods) and pressure drop analyses, it is presumed that the fluid is uniformly distributed through the core. In practice, flow maldistribution does occur to some extent and often severely, and may result in a significant reduction in

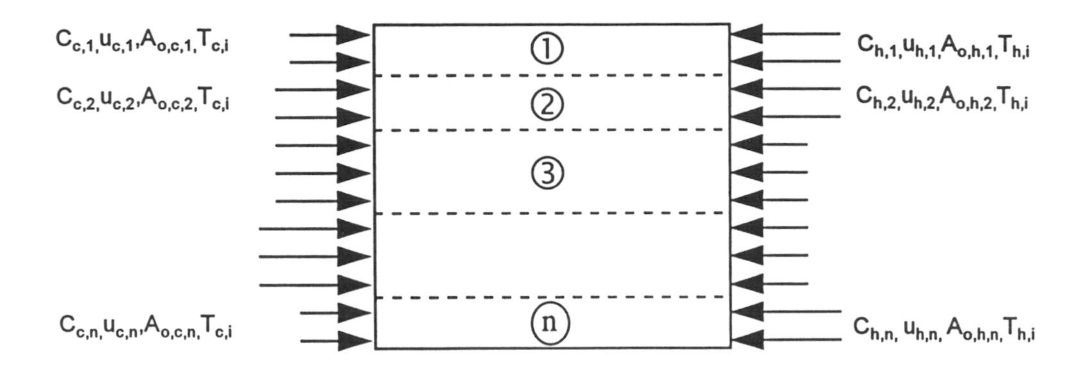

FIGURE 13A.18 Idealized flow nonuniformity in a counterflow exchanger.

exchanger heat transfer performance and an increase in the pressure drop. Hence, it may be necessary for the designer to take into account the effect of flow maldistribution, causing undesirable performance deterioration upfront while designing a heat exchanger.

Some maldistributions are geometry induced (i.e., the result of exchanger fabrication conditions, such as header design or manufacturing tolerances, or the duct geometry/structure upstream of the exchanger); others are the result of exchanger operating conditions. Gross, passage-to-passage, and manifold-induced flow maldistributions are the examples of the former category; flow maldistributions induced by viscosity, natural convection, and density difference are of the latter category. Flow maldistributions associated with two-phase and multiphase flow are too complex and beyond the scope of this chapter. The analysis methods and results for some of these flow maldistributions for single-phase flows are summarized next.

Gross Flow Maldistribution

In gross flow maldistribution, the nonuniform distribution of the fluid flow is on the macroscopic level. It results due to the shape of the inlet header, upstream flow conditions, or gross blockage (caused by brazing, soldering, or other manufacturing considerations) of any part of the exchanger. Therefore, gross flow maldistribution is independent of the exchanger heat transfer surface geometry configuration and its microscopic nonuniformity. Gross flow maldistribution generally results in a *significant increase* in exchanger pressure drop and *some reduction* in heat transfer.

A design procedure for calculating the reduction in heat exchanger effectiveness due to one-dimensional nonuniform flow distribution on only one side of the exchanger has been introduced by Shah (1981). The flow maldistribution in this procedure is characterized by an n-step velocity function relating the flow nonuniformity as n adjacent subexchangers, each having a uniform flow distribution unique from the rest. This method has been applied to a counterflow, a parallelflow, and a single-pass unmixed–mixed crossflow exchanger (for the case where the nonuniform side is the unmixed side) with the maldistributed side having either the hot or cold fluid. For all other exchanger flow arrangements, a numerical analysis is essential.

In the case of a counterflow or parallelflow exchanger, the flow nonuniformity can be on hot or cold sides as shown in Figure 13A.18. The analysis for the exchanger of Figure 13A.18 is straightforward, computing temperature effectiveness of individual subexchangers based on one fluid side only, and computing the total heat transfer rate by adding the q's of the subexchangers.

A typical influence of gross flow maldistribution on the counterflow exchanger effectiveness is shown in Figure 13A.19 (Shah, 1985). The subject exchanger contains a two-step velocity maldistribution on the hot side with $C^* = 1$, $u_{max}/u_m = 0.5$ and 2.0, and $A_{o,1}/A_o = 0.50$, 0.35, and 0.25. It is idealized that the flows are fully developed laminar, hence h and U are constant. Otherwise, the variations in h should be evaluated at respective flow velocities (i.e., Reynolds numbers), and the

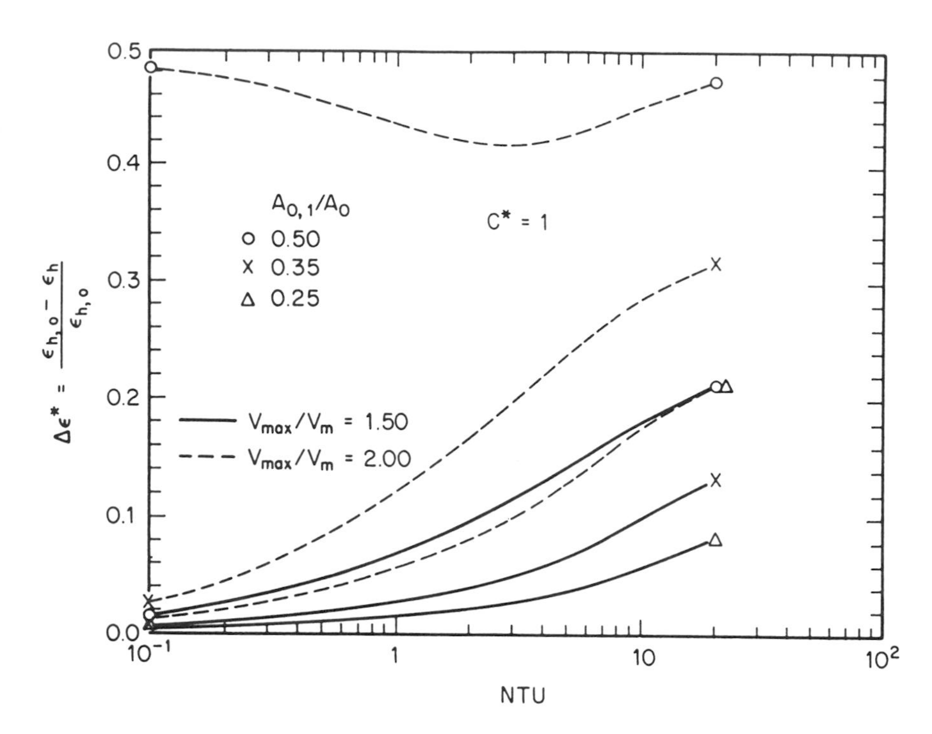

FIGURE 13A.19 Influence of gross flow maldistribution on the counterflow exchanger effectiveness. *Source:* Shah, 1985.

subsequent new values of U (or UA) should be calculated. From Figure 13A.19, it can be seen that for particular values of u_{max}/u_m and exchanger NTU, the greatest reduction in exchanger effectiveness occurs when the velocity function is equally distributed over flow area (e.g., $A_{o,1}/A_o = A_{o,2}/A_o = 0.5$). Also, as expected, the greater the maldistribution the greater is the loss in the exchanger effectiveness.

Although not demonstrated here, it can be shown that an n-step ($n > 2$) velocity maldistribution results in less deterioration in ε compared to that for the two-step velocity maldistribution for the same u_{max}/u_m ratio for the same exchanger (Shah, 1981).

In a series of papers, Chiou investigated numerically the effect of various flow maldistributions on a single-pass unmixed–unmixed crossflow exchanger, and some of these results are summarized by Mueller and Chiou (1987). Chiou (1982) reported a lower reduction in the exchanger effectiveness for nonuniform inlet temperature than for nonuniform inlet velocity for a single-pass unmixed–unmixed crossflow exchanger. Cichelli and Boucher (1956), Fleming (1966), and Chowdhury and Sarangi (1985) analyzed the case of tube-side maldistribution in a counterflow shell-and-tube exchanger, while Mueller (1977) considered flow maldistributions on both tube and shell sides. For a rotary regenerator, Kutchey and Julien (1974) investigated gross flow maldistribution, while Köhler (1974) investigated the influence of flow path geometry and manufacturing tolerances.

No analysis is available for an increase in the pressure drop due to gross flow maldistribution. Since such nonuniform flow is associated with poor header design or gross core blockage, the static pressure distributions at the core inlet and outlet faces may not be uniform. Hence, no simple modeling for the pressure drop evaluation is possible. As a conservative approach, it is recommended that the core pressure drop be evaluated based on the highest-velocity component of the flow maldistribution.

Passage-to-Passage Flow Maldistribution

Neighboring passages in a compact heat exchanger are geometrically never identical because of manufacturing tolerances. It is especially difficult to control precisely the passage size when small dimensions are involved (e.g., a rotary regenerator with $D_h = 0.5$ mm or 0.020 in.). Since differently sized and shaped passages exhibit different flow resistances in laminar flow and the flow seeks a path of least resistance, a nonuniform flow results through the matrix. This passage-to-passage flow nonuniformity can result in a *significant penalty* in heat transfer performance with only a small compensating effect of *reduced* pressure drop. This effect is especially important for laminar flows in continuous cylindrical passages, but is of lesser importance for interrupted surfaces (where transverse flow mixing can occur) or for turbulent flow.

The theoretical analysis for this flow maldistribution for low-Re laminar flow surfaces has been carried out by London (1970) for a two-passage model and extended by Shah and London (1980) for an n-passage model. In the latter model, there are n different-size passages of the same basic shape, either rectangular or triangular. In Figure 13A.20, a reduction in NTU for rectangular passages is shown when 50% of the flow passages are large ($c_2 > c_r$) and 50% of the passages are small ($c_1 < c_r$) compared to the reference or nominal passages. The results are presented for the passages having a nominal aspect ratio α^* of 1, 0.5, 0.25, and 0.125 for the (H1) and (T) boundary conditions and for a reference NTU of 5.0. Here, a percentage loss in NTU and the channel deviation parameter δ_c are defined as

$$\mathrm{NTU}^*_{\mathrm{cost}} = \left(1 - \frac{\mathrm{NTU}_e}{\mathrm{NTU}_r}\right) \times 100 \qquad \textbf{(13A.81)}$$

$$\delta_c = 1 - \frac{c_1}{c_r} \qquad \textbf{(13A.82)}$$

where NTU_e is the effective NTU when two-passage model passage-to-passage nonuniformity is present, and NTU_r is the reference or nominal NTU. It can be seen from Figure 13A.20 that a 10% channel deviation (which is common for a highly compact surface) results in 10 and 21% reduction in $\mathrm{NTU}_{\mathrm{H1}}$ and $\mathrm{NTU}_{\mathrm{T}}$, respectively, for $\alpha_r^* = 0.125$ and $\mathrm{NTU}_r = 5.0$. In contrast, a gain in the pressure drop due to passage to passage nonuniformity is only 2.5% for $\delta_c = 0.10$ and $\alpha_r^* = 0.125$, as found from Figure 13A.21. Here $\Delta p^*_{\mathrm{gain}}$ is defined as

$$\Delta p^*_{\mathrm{gain}} = \left(1 - \frac{\Delta p_{\mathrm{actual}}}{\Delta p_{\mathrm{nominal}}}\right) \times 100 \qquad \textbf{(13A.83)}$$

The results of Figures 13A.20 and 13A.21 are also applicable to an n-passage model in which there are n different-size passages in a normal distribution about the nominal passage size. The channel deviation parameter needs to be modified for this case to

$$\delta_c = \left[\sum_{i=1}^{n} \chi_i \left(1 - \frac{c_i}{c_r}\right)^2\right]^{1/2} \qquad \textbf{(13A.84)}$$

Here χ_i is the fractional distribution of the ith shaped passage. For $n = 2$, Eq. (13A.84) reduces to Eq. (13A.82). For triangular passages, c in Eq. (18A.84) is replaced by r_h. The following observations may be made from Figure 13A.20 and additional results presented by Shah and London (1980): (1) The loss in NTU is more significant for the (T) boundary condition than for the (H1) boundary condition. (2) The loss in NTU increases with higher nominal NTU. (3) The loss in NTU is much more significant compared to the gain in Δp at a given δ_c. (4) The deterioration in performance is the highest for the two-passage model compared to the n-passage model with $n > 2$ for the same value of $c_{\max}/c_r$.

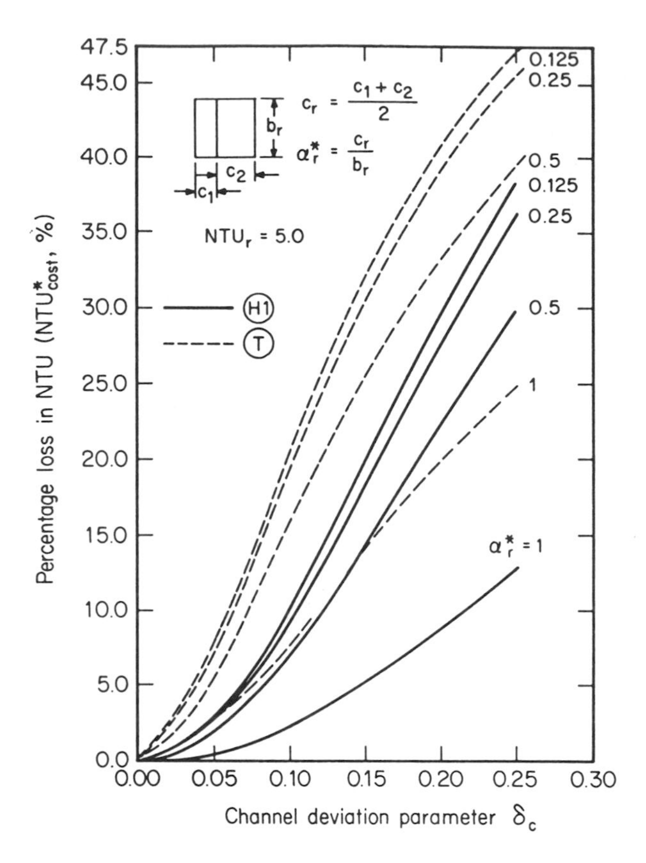

FIGURE 13A.20 Percentage loss in NTU for two-passage nonuniformities in rectangular passages.

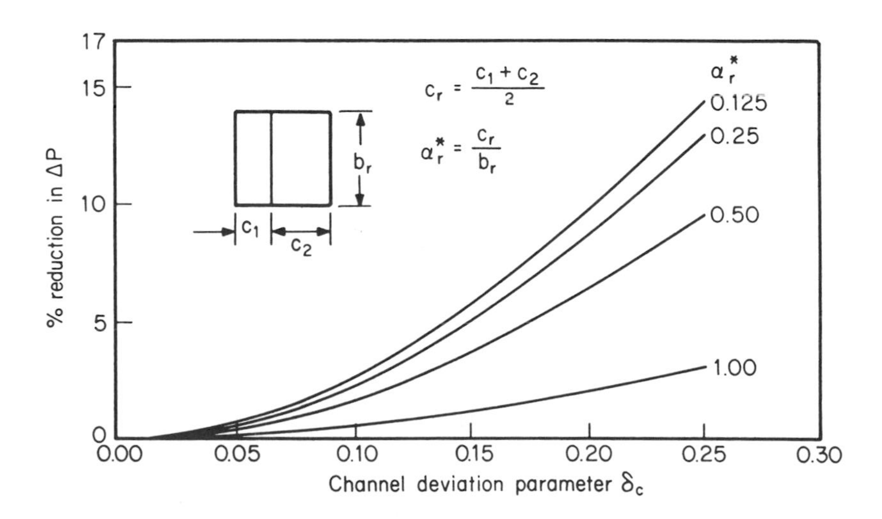

FIGURE 13A.21 Percentage reduction in Δp for two-passage nonuniformities in rectangular passages.

Manifold-Induced Flow Maldistribution

Two of the most commonly used manifold systems connecting manifolds to the branches (particularly for plate heat exchangers) are the parallelflow (S or Z) and reverse-flow (U) systems as shown in Figure 13A.22 for a single-pass exchanger. Since the flow length (and subsequently the flow resistance) is different for fluid particles going through different branches and inlet/outlet manifolds,

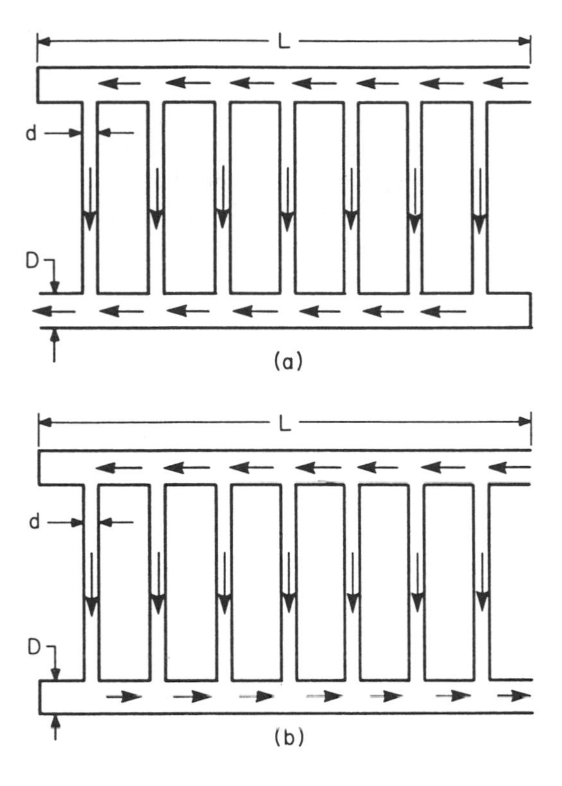

FIGURE 13A.22 (a) Parallelflow and (b) reverse-flow manifold systems.

and the imposed pressure drop (between inlet and outlet of the exchanger) is the same, it results in flow maldistribution regardless of perfect flow passages (branch geometry) and the type of the flow through the manifolds and branches (such as laminar, transition, or turbulent). Bajura and Jones (1976), Majumdar (1980) and Shen (1992) have investigated this flow maldistribution, and qualitative conclusions of the first two investigators are summarized below due to space limitation; for some quantitative results, refer to Shah (1985).

- A design rule of thumb is to limit the ratio of the flow area of lateral branches (exchanger core) to the flow area of the inlet header (area of pipe before lateral branches) A_o^* to less than unity to minimize flow maldistributions.
- More uniform flow distribution through the core is achieved by a reverse-flow (U) manifold system in comparison to a parallelflow (S) manifold system.
- In a parallelflow (S) manifold system, the maximum flow occurs through the last port, and in the reverse-flow (U) manifold system the first port.
- The influence of the manifold pipe friction parameter F is less significant than that of the flow area ratio A_o^*.
- Flow distribution becomes more uniform with a higher branch pressure loss coefficient K.
- The flow area of a combining-flow header should be larger than that for the dividing-flow header for a more uniform flow distribution through the core in the absence of heat transfer within the core. If there is heat transfer in the lateral branches (core), the fluid densities will be different in the inlet and outlet headers, the flow areas should be adjusted for this density change, and then the flow area of the combing header should be larger than that calculated previously.
- Flow reversal is more likely to occur in parallelflow (S flow) systems that are subject to poor flow distribution.

Viscosity-Induced Flow Maldistribution

Whenever one or both fluids flowing in a heat exchanger are liquids and operate in the laminar flow region, there exists a possibility of viscosity-induced flow instability and maldistribution, and thus exchanger performance deterioration. The viscosity-induced flow instability is a result of large changes in fluid viscosity within the exchanger and is found to be present when the viscous liquid is being cooled; it is not present for the case of liquid being heated, although flow maldistribution will be present. For the case of liquid cooling, the liquid viscosity increases significantly with the decreasing temperature. Thus moderate temperature differences between parallel passages (or tubes) within an exchanger can result in large viscosity differences between these adjacently flowing streams. Because of the direct proportionality between the pressure drop and the product of flow rate and viscosity ($\Delta p \propto m\mu$) in laminar flow, flow will be maldistributed in tubes having different heat transfer rates (and hence resultant different viscosities for the tube fluid), but will have the same pressure drop across each tube. The end result is that slower fluid streams are forced to flow slower and faster fluid streams to flow faster. This trend normally continues until a new equilibrium is reached, but unfortunately the resulting flow nonuniformity of this new exchanger condition will generally cause undesirable performance deterioration.

Mueller (1974) and Putman and Rohsenow (1985) addressed the problem of viscosity-induced flow maldistribution from the flow instability point of view and not from the heat transfer performance deterioration point of view. Mueller (1974) provided a simple analysis of the pressure drop–mass flow rate relationship in a single tube laminar flow cooler. From an extension of this analysis, he proposed a method for determining a point (maximum pressure drop or flow rate) beyond which it would be safe to assume flow instability of this kind will not occur in a multitubular exchanger.

When the exchanger is operating outside the region of viscous instability, the deterioration in heat transfer performance and the pressure drop "gain" can be calculated using the same procedure as that for the passage-to-passage flow maldistribution. In this case, the flow rate ratio depends on the viscosity ratio, whereas it depends on the ratios of the friction factor, hydraulic diameter, and flow area of two different flow passages in the case of passage-to-passage flow maldistribution (London, 1970).

13A.6 Fouling in Heat Exchangers

Fouling, Its Effect and Mechanisms

Fouling refers to undesired accumulation of solid material (byproducts of the heat transfer processes) on heat exchanger surfaces, which results in additional thermal resistance to heat transfer, thus reducing exchanger performance. The fouling layer also blocks the flow passage/area and increases surface roughness, which either reduces the flow rate in the exchanger or increases the pressure drop or both. The foulant deposits may be loose, such as magnetite particles, or hard and tenacious, such as calcium carbonate scale; other deposits may be sediment, polymers, cooking or corrosion products, inorganic salts, biological growth, and so on. Depending upon the fluids, operating conditions, and heat exchanger construction, the maximum fouling layer thickness on the heat transfer surface may result in a few hours to a number of years.

Fouling could be very costly depending upon the nature of fouling and the applications. It increases capital costs: (a) oversurfacing the heat exchanger, (b) provisions for cleaning, and (c) use of special materials and constructions/surface features. It increases maintenance costs: (a) cleaning techniques, (b) chemical additives, and (c) troubleshooting. It may cause a loss of production: (a) reduced capacity, and (b) shutdown. It increases energy losses: (a) reduced heat transfer, (b) increased pressure drop, and (c) dumping dirty streams. Fouling promotes corrosion, severe plugging, and eventual failure of uncleaned heat exchangers. In a fossil-fired exhaust environment, gas-side fouling produces a potential fire hazard in heat exchangers.

The following are the major fouling mechanisms:

- *Crystallization or Precipitation Fouling.* This results from the deposition/formation of crystals of dissolved substances from the liquid onto the heat transfer surface due to solubility changes with temperatures beyond the saturation point. If the deposited layer is hard and tenacious, it is often referred to as scaling. If it is porous and mushy, it is called sludge.
- *Particulate Fouling.* This results from the accumulation of finely divided substances suspended in the fluid stream onto the heat transfer surface. If the settling occurs due to gravity, it is referred to as *sedimentation fouling.*
- *Chemical Reaction Fouling.* This is defined as the deposition of material produced by chemical reaction (between reactants contained in the fluid stream) in which the heat transfer surface material does not participate.
- *Corrosion Fouling.* This results from corrosion of the heat transfer surface that produces products fouling the surface and/or roughens the surface, promoting attachment of other foulants.
- *Biological Fouling.* This results from the deposition, attachment, and growth of biological organisms from liquid onto a heat transfer surface. Fouling due to microorganisms refers to *microbial fouling,* and fouling due to macroorganisms refers to *macrobial fouling.*
- *Freezing Fouling.* This results from the freezing of a single component liquid or higher-melting-point constituents of a multicomponent liquid onto a subcooled heat transfer surface.

Biological fouling occurs only with liquids, since there are no nutrients in the gases. Also, crystallization fouling is not too common with gases, since most gases contain a few dissolved salts (mainly in mists) and even fewer inverse-solubility salts. All other types of fouling occur in both liquid and gas. More than one mechanism is usually present in many fouling situations, often with synergetic results. Liquid-side fouling generally occurs on the exchanger side where the liquid is being heated, and gas-side fouling occurs where the gas is being cooled; however, reverse examples can be found.

Importance of Fouling

Fouling in liquids and two phase flows has a significant detrimental effect on heat transfer with some increase in pressure drop. In contrast, fouling in gases reduces heat transfer somewhat (5–10% in general) in compact heat exchangers but increases pressure drop significantly (up to several hundred percent). For example, consider $U = 1{,}400$ W/m^2 K as in a process plant liquid-to-liquid heat exchanger. Hence, $R = 1/U = 0.00072$ m^2 K/W. If the fouling factors $(r_{s,h} + r_{s,c})$ together amount to 0.00036 (considering a typical TEMA value of the fouling factor as 0.00018), 50% of the transfer area requirement A for given q is chargeable to fouling. However, for gas flows on both sides of an exchanger, $U \approx 280$ W/m^2 K, and the same fouling factor of 0.00036 would represent only about 10% of the total surface area. Thus one can see a significant impact on heat transfer surface area requirement due to fouling in heat exchangers having high U values (such as having liquids or phase-change flows).

Considering the core frictional pressure drop, Eq. (13A.31), as the main pressure drop component, the ratio of pressure drops of fouled and cleaned exchanger is given by

$$\frac{\Delta p_F}{\Delta p_c} = \frac{f_F}{f_c}\left(\frac{D_{h,C}}{D_{h,F}}\right)\left(\frac{u_{m,F}}{u_{m,C}}\right)^2 = \frac{f_F}{f_c}\left(\frac{D_{h,C}}{D_{h,F}}\right)^5 \qquad \textbf{(13A.85)}$$

where the term after the second equality sign is for a circular tube and the mass flow rates under fouled and clean conditions remain the same. Generally $f_F > f_C$ due to the fouled surface being rough. Thus although the effect of fouling on the pressure drop is usually neglected, it can be

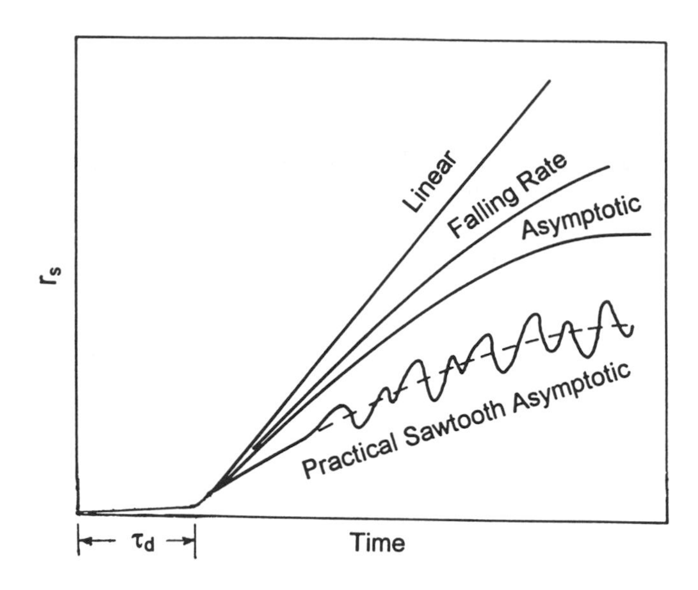

FIGURE 13A.23 Characteristic fouling curves.

significant, particularly for compact heat exchangers with gas flows. If we consider $f_F = f_C$, and the reduction in the tube inside diameter due to fouling by only 10 and 20%, the resultant pressure drop increase will be 69 and 205% respectively according to Eq. (13A.85), regardless of whether the fluid is liquid or gas!

Accounting of Fouling in Heat Exchangers

Fouling is an extremely complex phenomena characterized by a combined heat, mass, and momentum transfer under transient condition. Fouling is affected by a large number of variables related to heat exchanger surfaces, operating conditions, and fluids. Fouling is time dependent and zero at $\tau = 0$. After the induction or delay period τ_d, the fouling resistance is pseudo linear, a falling rate, or asymptotic, as shown by the smooth lines in Figure 13A.23, and the actual transient behavior is as represented by the dashed line for one asymptotic behavior.

Fouling is characterized by all or some of the following sequential events: initiation, transport, attachment, removal, and aging (Epstein, 1983). Research efforts are concentrated on quantifying these events by semi-theoretical models (Epstein, 1978) with a very limited success on specific fouling situations. Hence, the current heat exchanger design approach is to use a constant (supposedly an asymptotic) value of the fouling factor $r_s = 1/h_s$. Equation (13A.6) presented earlier includes the fouling resistances on the hot and cold sides for a nontubular extended surface exchanger. Here $1/h_s = r_s$ is generally referred to as the *fouling factor*. Fouling factors for some common fluids are presented in Tables 13A.10 and 13A.11.

The specification of fouling effects in a process heat exchanger is usually represented in the following form, wherein the combined fouling factor $r_{s,t}$ is the sum of the fouling factors on the hot and cold sides.

$$\text{Combined fouling factor} \quad r_{s,t} = \frac{1}{U_c} - \frac{1}{U_F} \qquad \textbf{(13A.86)}$$

$$\text{Cleanliness factor} \quad \text{CF} = U_F / U_C \qquad \textbf{(13A.87)}$$

TABLE 13A.10 Fouling Factors for Various Fluid Streams Used in Heat Exchangers

Fluid	Fouling Factors (m^2 K/W)
Water Type	
Seawater (43°C maximum outlet)	0.000275 to 0.00035
Brackish water (43°C maximum outlet)	0.00035 to 0.00053
Treated cooling tower water (49°C maximum outlet)	0.000175 to 0.00035
Artificial spray pond (49°C maximum outlet)	0.000175 to 0.00035
Closed loop treated water	0.000175
River water	0.00035 to 0.00053
Engine jacket water	0.000175
Distilled water or closed cycle condensate	0.00009 to 0.000175
Treated boiler feedwater	0.00009
Boiler blowdown water	0.00035 to 0.00053
Liquids	
No. 2 fuel oil	0.00035
No. 6 fuel oil	0.0009
Transformer oil	0.000175
Engine lube oil	0.000175
Refrigerants	0.000175
Hydraulic fluid	0.000175
Industrial organic HT fluids	0.000175 to 0.00035
Ammonia	0.000175
Ammonia (oil bearing)	0.00053
Methanol solutions	0.00035
Ethanol solutions	0.00035
Ethylene glycol solutions	0.00035
MEA and DEA solutions	0.00035
DEG and TEG solutions	0.00035
Stable side draw and bottom products	0.000175 to 0.00035
Caustic solutions	0.00035
Gas or Vapor	
Steam (non-oil-bearing)	0.0009
Exhaust steam (oil-bearing)	0.00026 to 0.00035
Refrigerant (oil-bearing)	0.00035
Compressed air	0.000175
Ammonia	0.000175
Carbon dioxide	0.00035
Coal flue gas	0.00175
Natural gas flue gas	0.00090
Acid gas	0.00035 to 0.00053
Solvent vapor	0.000175
Stable overhead products	0.000175
Natural Gas & Petroleum Streams	
Natural gas	0.000175 to 0.00035
Overhead products	0.000175 to 0.00035
Lean oil	0.00035
Rich oil	0.000175 to 0.00035
Natural gasoline and liquefied petroleum gases	0.000175 to 0.00035
Oil Refinery Streams	
Crude and vacuum unit gases and vapors	
Atmospheric tower overhead vapors	0.00017
Light naphthas	0.00017
Vacuum overhead vapors	0.00035

TABLE 13A.10 (continued) Fouling Factors for Various Fluid Streams Used in Heat Exchangers

Fluid	Fouling Factors (m^2 K/W)
Crude and vacuum liquids	
Crude oil	
Gasoline	0.00035
Naphtha and light distillates	0.00035 to 0.00053
Kerosene	0.00035 to 0.00053
Light gas oil	0.00035 to 0.00053
Heavy gas oil	0.00053 to 0.0009
Heavy fuel oil	0.00053 to 0.00123
Vacuum tower bottoms	0.00176
Atmospheric tower bottoms	0.00123
Cracking and coking unit streams	
Overhead vapors	0.00035
Light cycle oil	0.00035 to 0.00053
Heavy cycle oil	0.00053 to 0.0007
Light coker gas oil	0.00053 to 0.0007
Heavy coker gas oil	0.00070 to 0.0009
Bottoms slurry oil (1.5 m/s minimum)	0.00053
Light liquid products	0.00035
Catalytic reforming, hydrocracking and hydrodesulfurization streams	
Reformer charge	0.00026
Reformer effluent	0.00026
Hydrocharger charge and effluent	0.00035
Recycle gas	0.000175
Liquid product over 50°C (API)	0.000175
Liquid product 30 to 50°C (API)	0.00035
Light ends processing streams	
Overhead vapors and gases	0.000175
Liquid products	0.000175
Absorption oils	0.00035 to 0.00053
Alkylation trace acid streams	0.00035
Reboiler streams	0.00035 to 0.00053

Source: Chenoweth, 1988.

$$\text{Percentage oversurface} \quad \%OS = \left(\frac{A_F}{A_C} - 1\right)100 \tag{13A.88}$$

Here the subscripts F and C denote fouled and clean exchanger values. From Eq. (13A.6) with $A_h = A_c = A$, $\eta_o = 1$, $\Delta t_{m,F} = \Delta t_{m,C}$, it can be shown that

$$\frac{A_F}{A_C} = \frac{U_C}{U_F} = 1 + U_C r_{s,t} \tag{13A.89}$$

where $r_{s,t} = r_{s,h} + r_{s,c}$. In heat exchanger design, constant (supposedly and asymptotic) values of $r_{s,h}$ and $r_{s,c}$ are used. Accordingly, extra heat transfer surface area is provided to take into account the deleterious effect of fouling. Thus, the heat exchanger will be "oversized" for the initial clean condition, "correctly sized" for asymptotic fouling (if it occurs in practice), and "undersized" just before the cleaning operation for nonasymptotic fouling.

TABLE 13A.11 Fouling Factors and Design Parameters for Finned Tubes in Fossil-Fuel Exhaust Gases

Type of Flue Gas	Fouling Factor, m^2 K/W	Minimum Spacing Between Fins, m	Maximum Gas Velocity to Avoid Erosion, m/s
Clean Gas (cleaning devices not required)			
Natural Gas	0.0000881–0.000528	0.00127–0.003	30.5–36.6
Propane	0.000176–0.000528	0.00178	
Butane	0.000176–0.000528	0.00178	
Gas Turbine	0.000176		
Average Gas (provisions for future installation of cleaning devices)			
No. 2 Oil	0.000352–0.000704	0.00305–0.00384	25.9–30.5
Gas Turbine	0.000264		
Diesel Engine	0.000528		
Dirty Gas (cleaning devices required)			
No. 6 Oil	0.000528–0.00123	0.00457–0.00579	18.3–24.4
Crude Oil	0.000704–0.00264	0.00508	
Residual Oil	0.000881–0.00352	0.00508	
Coal	0.000881–0.00881	0.00587–0.00864	15.2–21.3

Source: Weierman, 1982.

Influence of Operating and Design Variables

Based on operational experience and research over the last several decades, many variables have been identified that have a significant influence on fouling. Most important variables are summarized next.

Flow Velocity

Flow velocity is one of the most important variables affecting fouling. Higher velocities increase fluid shear stress at the fouling deposit-fluid interface and increase the heat transfer coefficient, but at the same time increase pressure drop and fluid pumping power, may erode the surface, and may accelerate the corrosion of the surface by removing the protective oxide layer. The fouling build-up in general is inversely proportional to $u_m^{1.5}$. For water, the velocity should be kept above 2 m/s to suppress fouling, and the absolute minimum should be above 1 m/s to minimize fouling.

Surface Temperature

Higher surface temperatures promote chemical reaction, corrosion, crystal formation (with inverse solubility salts), and polymerization, but reduce biofouling for temperatures above the optimum growth, avoid potential freezing fouling, and avoid precipitation of normal solubility salts. It is highly recommended that the surface temperature be maintained below the reaction temperature; it should be kept below 60°C for cooling tower water.

Tube Material

The selection of the tube material is important from the corrosion point of view, which in turn could increase crystallization and biological fouling. Copper alloys can reduce certain biofouling, but its use is limited by environmental concerns for river, ocean, and lake waters. Many other variables affect fouling. Discussing them is beyond our scope here, but the reader may refer to TEMA (1988).

Fouling Control and Cleaning Techniques

Control of fouling should be attempted first before any cleaning method is attempted. For gas-side fouling, one should verify that fouling exists, identify the sequential event that dominates the foulant accumulation, and characterize the deposit. For liquid-side fouling, fouling inhibitors/additives should be employed while the exchanger is in operation, such as using antidispersant polymers to prevent sedimentation fouling, "stabilizing" compounds to prevent polymerization and chemical reaction fouling, corrosion inhibitors to prevent corrosion fouling, biocide/germicides to prevent biofouling, and softeners, acids, and polyphosphates to prevent crystallization fouling.

If the foulant control is not effective, the exchanger must be cleaned either on-line or off-line. On-line cleaning includes: flow-driven brushes/sponge balls inside tubes, power-driven rotating brushes inside tubes, acoustic horns/mechanical vibrations for tube banks with gases, sootblowers, and shutting the cold gas supply, flowing hot gas, or reversing the fluids. Off-line cleaning methods without dismantling the exchanger include chemical cleaning (circulate acid/detergent solutions), circulating particulate slurry (such as sand and water), and thermal cleaning to melt frost layers. Off-line cleaning with a heat exchanger opened includes high-pressure stream or water cleaning, thermal baking of an exchanger, and rinsing of small heat exchanger modules removed from the container of modular exchangers.

Concluding Remarks

Heat exchangers play a critical and dominant role in energy conservation, conversion, recovery, and utilization, and in the economic development of new energy sources as well as in many solutions to environmental problems. Compact heat exchangers have the major advantages of having high heat fluxes, small volumes and packaging, and total lower cost. With the advancement of manufacturing technologies, many innovative and new designs of compact heat exchangers have emerged in recent years that are capable of withstanding either extreme high pressures or high temperatures, but generally cannot handle heavy fouling due to small flow passages that cannot be cleaned mechanically. Hence, compact heat exchangers are being applied to many niche applications, replacing shell-and-tube and other heat exchangers. In this article, a comprehensive review is made of thermal and hydraulic design aspects of single-phase compact heat exchangers. A good number of references are also provided for further study of many types of heat exchangers.

13A.7 Nomenclature

- A total heat transfer area (primary + fin) on one fluid side of a heat exchanger—A_p, primary surface area; A_f, fin surface area, m^2
- A_{fr} frontal area on one side of an exchanger, m^2
- A_k total wall cross-sectional area for heat conduction in fin or for longitudinal conduction in the exchanger, m^2
- A_o minimum free flow area on one fluid side of a heat exchanger, m^2
- b plate spacing, $h' + \delta_f$, m
- C flow stream heat capacity rate with a subscript c or h, $\dot{m}c_p$, W/°C
- C^* heat capacity rate ratio, C_{min}/C_{max}, dimensionless
- c_p specific heat of fluid at constant pressure, J/kg K
- D_h hydraulic diameter of flow passages, $4A_oL/A$, m
- d_e fin tip diameter of an individually finned tube, m
- d_i, d_o tube inside and outside diameters, respectively, m
- Eu N-row average Euler number, $\Delta p/(\rho u_m^2\ N/2g_c)$, $\rho\ \Delta p\ g_c/NG^2/2)$, dimensionless
- F log-mean temperature difference correction factor, dimensionless
- f Fanning friction factor, $\rho\ \Delta p\ g_c\ D_h/(2LG^2)$, dimensionless

f_{tb}	average Fanning friction factor per tube row for crossflow over a tubebank outside, $\rho\ \Delta p\ g_c/(2NG^2)$, dimensionless
G	mass velocity based on the minimum free flow area, $\dot{m}/A_o$, kg/m² s
g	gravitational acceleration, m²/s
g_c	proportionality constant in Newton's Second Law of Motion, $g_c = 1$ and dimensionless in SI units
H	fin length for heat conduction from primary surface to either fin tip or midpoint between plates for symmetric heating, m
h	heat transfer coefficient, W/m² K
h'	height of the offset strip fin (see Figure 13A.15), m
j	Colburn factor, $\mathrm{NuPr}^{-1/3}/\mathrm{Re}$, $\mathrm{StPr}^{2/3}$, dimensionless
k	fluid thermal conductivity, W/m K
k_f	thermal conductivity of the fin material, W/m K
k_w	thermal conductivity of the matrix (wall), material W/m K
L	fluid flow (core or tube) length on one side of an exchanger, m
l	fin length for heat conduction from primary surface to the midpoint between plates for symmetric heating, see Table 13A.4 for other definitions of l, m
l_f	offset strip fin length or fin height for individually finned tubes, l_f represents the fin length in the fluid flow direction for an uninterrupted fin with $l_f = L$ in most cases, m
m	fin parameter, 1/m
N	number of tube rows
N_f	number of fins per meter, 1/m
N_t	total number of tubes in an exchanger
NTU	number of heat transfer units, UA/C_{min}, it represents the total number of transfer units in a multipass unit, $\mathrm{NTU}_s = UA/C_{shell}$, dimensionless
Nu	Nusselt number, hD_h/k, dimensionless
ntu_c	number of heat transfer units based on the cold side, $(\eta_o hA)_c/C_c$, dimensionless
ntu_h	number of heat transfer units based on the hot side, $(\eta_o hA)_h/C_h$, dimensionless
$\dot{m}$	mass flow rate, kg/s
P	temperature effectiveness of one fluid, dimensionless
$\wp$	fluid pumping power, W
Pr	fluid Prandtl number, $\mu c_p/k$, dimensionless
p	fluid static pressure, Pa
Δp	fluid static pressure drop on one side of heat exchanger core, Pa
p_f	fin pitch, m
q	heat duty, W
q_e	heat transfer rate (leakage) at the fin tip, W
q''	heat flux, W/m²
R	heat capacity rate ratio used in the P-NTU method, $R_1 = C_1/C_2$, $R_2 = C_2/C_1$, dimensionless
R	thermal resistance based on the surface area A, compare Eqs. (13A.5) and (13A.6) for definitions of specific thermal resistances, K/W
Re	Reynolds number, GD_h/μ, dimensionless
Re_d	Reynolds number, $\rho u_m d_o/\mu$, dimensionless
r_h	hydraulic radius, $D_h/4$, A_oL/A, m
r_s	fouling factor, $1/h_s$, m² K/W
St	Stanton number, h/Gc_p, dimensionless
s	distance between adjacent fins, $p_f - \delta_f$, m
T	fluid static temperature to a specified arbitrary datum, °C
T_s	ambient temperature, °C
T_o	fin base temperature, °C
T_ℓ	fin tip temperature, °C

U overall heat transfer coefficient, W/m^2 K
u_m mean axial velocity in the minimum free flow area, m/s
V heat exchanger total volume, m^3
X_d diagonal tube pitch, m
X_ℓ longitudinal tube pitch, m
X_t transverse tube pitch, m
α ratio of total heat transfer area on one side of an exchanger to the total volume of an exchanger, A/V, m^2/m^3
β heat transfer surface area density, a ratio of total transfer area on one side of a plate-fin heat exchanger to the volume between the plates on that side, m^2/m^3
ε heat exchanger effectiveness, represents an overall exchanger effectiveness for a multipass unit, dimensionless
δ wall thickness, m
δ_f fin thickness, m
η_f fin efficiency, dimensionless
η_o extended surface efficiency, dimensionless
λ longitudinal wall heat conduction parameter based on the total conduction area, $\lambda = k_w A_{k,t}/C_{min}L$, $\lambda_c = k_w A_{k,c}/C_c L$, $\lambda_h = k_w A_{k,h}/C_h L$, dimensionless
μ fluid dynamic viscosity, Pa s
ρ fluid density, kg/m^3
σ ratio of free flow area to frontal area, A_o/A_{fr}, dimensionless

Subscripts

C clean surface value
c cold fluid side
F fouled surface value
f fin
h hot fluid side
i inlet to the exchanger
o outlet to the exchanger
s scale or fouling
w wall or properties at the wall temperature
1 one section (inlet or outlet) of the exchanger
2 other section (outlet or inlet) of the exchanger

References

Băclić, B.S., Romie, F.E., and Herman, C.V. 1988. The Galerkin Method for Two-Pass Crossflow Heat Exchanger Problem, *Chem. Eng. Comm.*, Vol. 70, pp. 177–198.

Bajura, R.A. and Jones Jr., E.H. 1976. Flow Distribution Manifolds, *J. Fluid Eng., Trans. ASME,* Vol. 98, pp. 654–666.

Bhatti, M.S. and Shah, R.K. 1987. Turbulent and Transition Flow Convective Heat Transfer in Ducts. In *Handbook of Single-Phase Convective Heat Transfer,* ed. S. Kakaç, R.K. Shah, and W. Aung, Chap. 4, Wiley, New York.

Boyen, J.L. 1976. *Practical Heat Recovery,* Wiley, New York.

Chai, H.C. 1988. A Simple Pressure Drop Correlation Equation for Low Finned Tube Crossflow Heat Exchangers, *Int. Commun. Heat Mass Transfer,* Vol. 15, pp. 95–101.

Chenoweth, J. 1988. Final Report, HTRI/TEMA Joint Committee to Review the Fouling Section of TEMA Standards HTRI, Alhambra, CA.

Chiou, J.P. 1982. The Effect of Nonuniformities of Inlet Temperatures of Both Fluids on the Thermal Performance of Crossflow Heat Exchanger, ASME Paper 82-WA/HT-42.

Chowdhury, K. and Sarangi, S. 1985. The Effect of Flow Maldistribution on Multipassage Heat Exchanger Performance, *Heat Transfer Eng.*, Vol. 6, No. 4, pp. 45–54.

Churchill, S.W. Nov. 7, 1977. Friction-Factor Equation Spans All Fluid Flow Regimes, *Chem. Eng.*, Vol. 84, No. 24, pp. 91-92.

Cichelli, M.T. and Boucher, D.F. 1956. Design of Heat Exchanger Heads for Low Holdup, AIChE, *Chem. Eng. Prog.*, Vol. 52, No. 5., pp. 213–218.

Cowell, T.A., Heikal, M.R., and Achaichia, A. 1995. Flow and Heat Transfer in Compact Louvered Fin Surfaces, *Exp. Thermal and Fluid Sci.*, Vol. 10, pp. 192-199.

Csathy, D. Feb. 4, 1980. Latest Practice in Industrial Heat Recovery, presented at The Energy-Source Technology Conference and Exhibition, New Orleans.

Epstein, N. 1978. Fouling in Heat Exchangers, In *Heat Transfer 1978*, Vol. 6, pp. 235–254, Hemisphere, New York.

Epstein, N. 1983. Thinking about Heat Transfer Fouling: A 5 × 5 Matrix, *Heat Transfer Eng.*, Vol. 4, No. 1, 43–56.

Eriksen, V.L. 1980. Waste Heat Recovery Exchangers. In *Compact Heat Exchangers—History, Technological Development and Mechanical Design Problems*, Book No. G00183, HTD-Vol. 10, pp. 181–185, ASME, New York.

Fleming, R.B. 1966. The Effect of Flow Distribution in Parallel Channels of Counterflow Heat Exchangers, *Advances in Cryogenic Engineering*, pp. 352–362.

Ganguli, A. and Yilmaz, S.B. 1987. New Heat Transfer and Pressure Drop Correlations for Crossflow over Low-Finned Tube Banks, *AIChE Symp. Ser. 257*, Vol. 83, pp. 9–14.

Ghajar, A.J. and Tam, L.M. 1994. Heat Transfer Measurements and Correlations in the Transition Region for a Circular Tube with Three Different Inlet Configurations, *Exp. Thermal Fluid Sci.*, Vol. 8, pp. 79–90.

Goldstick, R. and Thumann, A. 1986. *Principles of Waste Heat Recovery*, The Fairmont Press, Atlanta, GA.

Henriette, J. 1991. Ceramic Heat Exchangers. In *Industrial Heat Exchangers*, ed. J-M. Buchlin, Lecture Series No. 1991-04, von Kármán Institute for Fluid Dynamics, Belgium.

Hewitt, G.F. 1989. Coordinating Editor. *Hemisphere Handbook of Heat Exchanger Design*, Hemisphere, New York.

Hewitt, G.F., Shires, G.L., and Bott, T.R. 1994. *Process Heat Transfer*, CRC Press, Boca Raton, Florida.

Hinchley, P. 1979. Specifying and Operating Reliable Waste-Heat Boilers, *Chem. Eng.*, Vol. 86, Aug. 13, No. 17, pp. 120–134.

Huang, L.J. and Shah, R.K. 1992. Assessment of Calculation Methods for Efficiency of Straight Fins of Rectangular Profiles, *Int. J. Heat and Fluid Flow*, Vol. 13, pp. 282–293.

Idelchik, I.E. 1994. *Handbook of Hydraulic Resistance*, 3rd Edition, CRC Press, Boca Raton, FL.

Kays, W.M. and London, A.L. 1984. *Compact Heat Exchangers*, 3rd Edition, McGraw-Hill, New York.

Köhler, M. 1974. The Influence of Flow Path Geometry and Manufacturing Tolerances on Gas Turbine Regenerator Efficiency, SAE Paper No. 740183.

Kutchey, J.A. and Julien, H.L. 1974. The Measured Influence of Flow Distribution on Regenerator Performance, *SAE Trans.*, Vol. 83, SAE Paper No. 740164.

London, A.L. 1970. Laminar Flow Gas Turbine Regenerators—The Influence of Manufacturing Tolerances, *J. Eng. Power*, Vol. 92A, pp. 45–56.

Majumdar, A.K. 1980. Mathematical Modeling of Flows in Dividing and Combining Flow Manifold, *Applied Mathematical Modeling*, Vol. 4, pp. 424–432.

Manglik, R.M. and Bergles, A.E. 1995. Heat Transfer and Pressure Drop Correlations for the Rectangular Offset-Strip-Fin Compact Heat Exchanger, *Exp. Thermal and Fluid Sci.*, Vol. 10, pp. 171-180.

Meunier, H. 1991. Heat Exchangers and Regenerators for High Temperature Waste Gases. In *Industrial Heat Exchangers*, ed. J-M. Buchlin, Lecture Series No. 1991-04, von Kármán Institute for Fluid Dynamics, Belgium.

Miller, D.S. 1978. Internal Flow Systems, *BHRA Fluids Engineering Series,* Vol. 5, British Hydromechanics Research Association, Cranfield, United Kingdom.

Mueller, A.C. 1974. Criteria for Maldistribution in Viscous Flow Coolers, *Heat Transfer 1974,* Vol. 5, pp. 170–174.

Mueller, A.C. 1977. An Inquiry of Selected Topics on Heat Exchanger Design, *AIChE Symp. Ser.* 164, Vol. 73, pp. 273–287.

Mueller, A.C. and Chiou, J.P. 1987. Review of Various Types of Flow Maldistribution in Heat Exchangers, Book No. H00394, HTD-Vol. 75, pp. 3–16, *ASME,* New York.

Palen, J.W., Editor, 1987. *Heat Exchanger Sourcebook,* Hemisphere Publishing Corp., Washington, DC.

Putnam, G.R. and Rohsenow, W.M. 1985. Viscosity Induced Nonuniform Flow in Laminar Flow Heat Exchangers, *Int. J. Heat Mass Transfer,* Vol. 28, pp. 1031–1038.

Rabas, T.J. and Taborek, J. 1987. Survey of Turbulent Forced-Convection Heat Transfer and Pressure Drop Characteristics of Low-Finned Tube Banks in Cross Flow, *Heat Transfer Eng.*, Vol. 8, No. 2, pp. 49–62.

Reay, D.A. 1979. *Heat Recovery Systems,* E. & F.N. Spon, London.

Richlen, S.L. 1990. Factors in Choosing Heat Recovery Systems, *Eng. Dig.,* 18, No. 12, pp. 34–36.

Richlen, S.L. and Parks, W.P. Jr. 1992. Heat Exchangers. In *Engineered Materials Handbook,* Vol. 4: *Ceramics and Glasses,* pp. 978–986, ASM International, Materials Park, Ohio 44073.

Roetzel, W. and Spang, B. 1987. Analytisches Verfahren zur thermisen Berechnung mehrgängiger Rohrbünädelwreübertrager, Fortschr.-Ber. VDI, Reihe 19, Nr. 18.

Roetzel, W. and Spang, B. 1989. Thermal Calculation of Multipass Shell and Tube Heat Exchangers, *Chem. Eng. Res. Des.,* Vol. 67, pp. 115–120.

Rozenman, T. 1976. Heat Transfer and Pressure Drop Characteristics of Dry Cooling Tower Extended Surfaces, Part I: Heat Transfer and Pressure Drop Data, Report BNWL-PFR 7-100; Part II: Data Analysis and Correlation, Report BNWL-PFR 7-102, Battelle Pacific Northwest Laboratories, Richland, WA.

Saunders, E.A.D. 1988. *Heat Exchangers: Selection, Design & Construction,* Longman Scientific & Technical, Essex, UK.

Scaccia, C. and Theoclitus, G. Oct. 6, 1980. Selecting Heat Exchangers: Types, Performance and Applications, *Chem. Eng.* 87, pp. 121–132.

Shah, R.K. 1981. Compact Heat Exchangers. In *Heat Exchangers: Thermal-Hydraulic Fundamentals and Design,* ed. S. Kakaç, A.E. Bergles, and F. Mayinger, pp. 111–151, Hemisphere Publishing Corp., Washington, DC.

Shah, R.K. 1983. Heat Exchanger Basic Design Methods. In *Low Reynolds Number Flow Heat Exchangers,* ed. S. Kakaç, R.K. Shah, and A.E. Bergles, pp. 21–72, Hemisphere Publishing Corp. Washington, DC.

Shah, R.K. 1985. Compact Heat Exchangers. In *Handbook of Heat Transfer Applications,* 2nd Edition, ed. W.M. Rohsenow, J.P. Hartnett, and E.N. Ganić, Chap. 4, Part 3, McGraw-Hill, New York.

Shah, R.K. 1988a. Plant-Fin and Tube-Fin Heat Exchanger Design Procedures. In *Heat Transfer Equipment Design,* ed. R.K. Shah, E.C. Subbarao, and R.A. Mashelkar, pp. 255–266, Hemisphere Publishing Corp., Washington, DC.

Shah, R.K. 1988b. Counterflow Rotary Regenerator Thermal Design Procedures. In *Heat Transfer Equipment Design,* edited by R.K. Shah, E.C. Subbarao, and R.A. Mashelkar, pp. 267–296, Hemisphere Publishing Corp., Washington, DC.

Shah, R.K. 1991a. Industrial Heat Exchangers—Functions and Types. In *Industrial Heat Exchangers,* ed. J-M. Buchlin, Lecture Series No. 1991-04, von Kármán Institute for Fluid Dynamics, Belgium.

Shah, R.K. 1991b. Multidisciplinary Approach to Heat Exchanger Design. In *Industrial Heat Exchangers,* ed. J-M. Buchlin, Lecture Series No. 1991-04, von Kármán Institute for Fluid Dynamics, Belgium.

Shah, R.K. 1993. Nonuniform Heat Transfer Coefficients for Heat Exchanger Thermal Design. In *Aerospace Heat Exchanger Technology 1993,* ed. R.K. Shah and A. Hashemi, pp. 417–445, Elsevier Science, Amsterdam, Netherlands.

Shah, R.K. and Bhatti, M.S. 1987. Laminar Convective Heat Transfer in Ducts. In *Handbook of Single-Phase Convective Heat Transfer,* edited by S. Kakaç, R.K. Shah and W. Aung, Chapter 3, Wiley, New York.

Shah, R.K. and Bhatti, M.S. 1988. Assessment of Correlations for Single-Phase Heat Exchangers. In *Two-Phase Flow Heat Exchangers: Thermal-Hydraulic Fundamentals and Design,* ed. S. Kakaç, A.E. Bergles, and E.O. Fernandes, pp. 81–122, Kluwer Academic Publishers, Dordrecht, The Netherlands.

Shah, R.K. and Giovannelli, A.D. 1988. Heat Pipe Heat Exchanger Design Theory. In *Heat Transfer Equipment Design,* ed. R.K. Shah, E.C. Subbarao, and R.A. Mashelkar, pp. 609–653, Hemisphere Publishing Corp., Washington, DC.

Shah, R.K. and London, A.L. 1978. *Laminar Flow Forced Convection in Ducts,* Supplement 1 to *Advances in Heat Transfer,* Academic Press, New York.

Shah, R.K. and London, A.L. 1980. Effects of Nonuniform Passages on Compact Heat Exchanger Performance, *ASME J. Eng. Power,* Vol. 102A, pp. 653–659.

Shah, R.K. and Mueller, A.C. 1985. Heat Exchanger Basic Thermal Design Methods. In *Handbook of Heat Transfer and Applications,* 2nd Edition, edited by W.M. Rohsenow, J.P. Hartnett, and E.N. Ganić, Chapter 4, Part I, pp. 4-1 to 4-77, McGraw-Hill, New York.

Shah, R.K. and Muller, A.C. 1988. Heat Exchange. In *Ullmann's Encyclopedia of Industrial Chemistry, Unit Operations II,* Vol. B3, Chapter 2, VCH Publishers, Weinheim, Germany.

Shah, R.K. and Pignotti, A. 1995. The Influence of a Finite Number of Baffles on the Shell-and-Tube Heat Exchanger Performance, Proc. 2nd ISHMT-ASME Heat and Mass Transfer Conference, Surathkal, India, Tata-McGraw-Hill Publishing, New Delhi, pp. 931-938; to be published in *Heat Transfer Eng.*, Vol. 18, 1997.

Shah, R.K. and Wanniarachchi, A.S. 1991. Plate Heat Exchanger Design Theory. In *Industrial Heat Exchangers,* ed. J.M. Buchlin, Lecture Series No. 1991-04, von Kármán Institute for Fluid Dynamics, Belgium.

Shen, P.I. 1992. The Effect of Friction on Flow Distribution in Dividing and Combining Flow Manifolds, *ASME J. Fluids Eng.,* Vol. 114, pp. 121–123.

Singh, K.P. and Soler, A.I. 1984. *Mechanical Design of Heat Exchangers and Pressure Vessel Components,* Arcturus Publishers, Cherry Hill, NJ.

Spang, B., Xuan, Y., and Roetzel, W. 1991. Thermal Performance of Split-Flow Heat Exchangers, *Int. J. Heat Mass Trans.,* Vol. 34, pp. 863–874.

Srinivasan, V. and Shah, R.K. 1995. Fin Efficiency of Extended Surfaces in Two-Phase Flow, Proc. Int. Symp. Two-Phase Flow Modelling and Experimentation. Rome, Italy.

TEMA, 1988. *Standards of the Tubular Exchanger Manufacturers Association,* 7th Edition, Tubular Exchanger Manufacturers Association, New York.

Walker, G. 1990. *Industrial Heat Exchangers—A Basic Guide,* 2nd Edition, Hemisphere, Washington, D.C.

Watts, R.L., Dodge, R.E., Smith, S.A., and Ames, K.R. March 1984. Identification of Existing Waste Heat Recovery and Process Improvement Technologies, Report No. PNL-4910, UC95f, Pacific Northwest Laboratory, Richland, Washington.

Webb, R.L. 1994. *Principles of Enhanced Heat Transfer,* Wiley, New York.

Weierman, R.C. 1982. Design of Heat Transfer Equipment for Gas-Side Fouling Service, Workshop on an Assessment of Gas-side Fouling in Fossil Fuel Exhaust Environments, ed. W.J. Marner and R.L. Webb, JPL Publ. 82-67, Jet Propulsion Lab., Calif. Inst. of Technology, Pasadena, CA.

Xuan, Y., Spang, B., and Roetzel, W. 1991. Thermal Analysis of Shell and Tube Exchangers with Divided-Flow Pattern, *Int. J. Heat Mass Trans.,* Vol. 34, pp. 853–861.

Yokell, S. 1990. *A Working Guide to Shell-and-Tube Heat Exchangers,* McGraw-Hill, New York.

Zukauskas, A. 1987. Convective Heat Transfer in Cross Flow. In *Handbook of Single-Phase Convective Heat Transfer,* edited by S. Kakaç, R.K. Shah, and W. Aung, Chapter 6, Wiley, New York.

13B

Recuperators, Regenerators, and Storage: Thermal Energy Storage Applications in Gas-Fired Power Plants

D.R. Brown

S. Somasundaram
Pacific Northwest National Laboratory

Natural gas–fired systems are becoming increasingly popular generating technologies in utility and industrial settings. The popularity of gas–fired systems stems from several factors, including low construction cost, short construction period, improving efficiency, moderate fuel cost, and relatively benign environmental impacts. Applications include both electric power and cogeneration systems based on simple-cycle and combined-cycle plant configurations.

Despite these attractive features, other characteristics of gas turbines and cogeneration applications are at odds with the needs of electric utilities. In general, electricity demand varies with the time of day, the day of the week, and the season. Most U.S. electric utilities experience peak demands in the summer, corresponding to weather-driven demands for space cooling. Cogeneration systems, particularly for industrial applications, are often designed around a fixed thermal load, which translates into a fixed electric power output as well. While the combined production of thermal and electric energy via cogeneration results in high efficiency, the electric power is not always produced when it is needed the most. In addition, all combustion turbines suffer from reduced generating capacity and higher heat rates at high ambient temperatures, or precisely when the demand for power is usually the greatest.

Thermal energy storage (TES) can be applied to address both of these problems. Turbine inlet air can be cooled prior to compression to offset the effects of high ambient temperatures. TES can be used to reduce the size and cost of the chiller unit and reduce on-peak parasitic power consumption for electrically driven chillers. For cogeneration systems, TES can be used to decouple the generation of power from the gas turbine from the generation of steam by capturing and storing thermal energy from the turbine exhaust gases.

13B.1 Cogeneration with Thermal Energy Storage

Concept Overview

Cogeneration is playing an increasingly important role in providing efficiently produced power and thermal energy for commercial and industrial applications. However, the number of useful applications of conventional cogeneration systems—cogeneration systems without thermal energy storage (TES)—is limited by the temporal mismatch between the demands for electricity and thermal energy. Increasingly, utilities are requiring cogenerators to provide dispatchable power, whereas most industrial thermal loads are relatively constant. Diurnal TES can decouple the generation of electricity from the production of thermal energy to provide the following benefits:

- Power can be produced on demand, at various levels, while meeting a constant thermal load.
- The generating capacity of the gas turbine will be several times larger than for a conventional cogeneration facility. For example, a cogeneration plant with a TES system sized for an 8-hour peak demand period would provide 30 MW of peaking capacity whereas to a similar conventional cogeneration facility would provide 10 MW of baseload capacity.
- Maximum cogeneration efficiency is achieved by firing all natural gas in the combustion turbine. If a conventional cogeneration system is required to shut down due to lack of electricity demand, the steam demand would have to be met via direct firing of natural gas in a separate boiler.

The concept for integrating TES in a natural gas–fired cogeneration plant is shown in Figure 13B.1. The facility consists of (1) a gas turbine, (2) a heat recovery storage medium heater, (3) TES medium and storage tank, (4) a medium-heated steam generator, and (5) interconnecting piping. In contrast, a conventional cogeneration facility would consist of a gas turbine and a heat recovery steam generator (HRSG). The gas turbine is operated during peak demand periods to produce electricity and heat the storage media. Cold storage media is pumped from the cold storage tank, through the heat recovery medium heater, and back to the warm storage tank. Depending on the storage medium used, hot and cold storage may be in separate tanks or combined in a single tank with hot and cold sections effectively separated by the difference in fluid densities. Hot medium is pumped to the steam generator as required to meet the steam demand schedule.

Equipment Descriptions and Storage Options

An important objective in designing a cogeneration plant with TES is to develop arrangements that minimize the impact of including TES on the design and layout of the cogeneration plant. Therefore, there should be substantial similarity between a conventional cogeneration plant and a cogeneration plant with TES. As described in the concept overview, the HRSG in a conventional cogeneration plant is replaced by a medium heater, medium storage, and medium-heated steam generator. The gas turbine technology remains the same, but there would be multiple units or a single larger unit. Similarities and differences between the TES and conventional system components are described next, along with the principal storage media options.

Medium Heater

The medium heater replaces the HRSG in a conventional cogeneration plant. Compared to an HRSG of the same thermal duty, the medium mass flow rate will be higher because the thermal storage medium does not go through a phase change as water does in the HRSG. On the other hand, the medium operating pressure is much lower than for water/steam because the medium need only overcome frictional losses during circulation. Heat transfer area requirements are likely to be similar because the dominant resistance to heat transfer is on the gas side. As a rule of thumb,

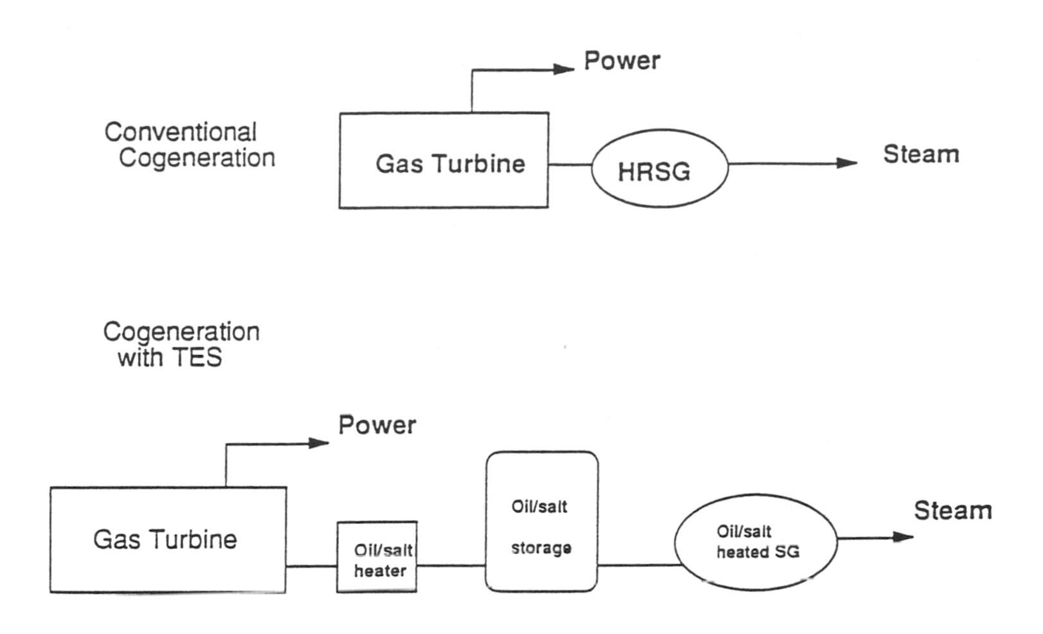

FIGURE 13B.1 Cogeneration with thermal energy storage.

sizing calculations for medium heaters can be based on an overall heat transfer coefficient of 150 W/m^2 °C. The preferred medium heater design could be similar to an HRSG (i.e., it could employ a series of finned-tube banks within a shell). A more likely alternative, with low-pressure media, is to use a plate-type heat transfer surface, which should result in a more compact and less costly design than the finned-tube arrangement. For molten salt storage media, it may be possible to use direct heat exchange between the exhaust gas and the salt. Direct-contact heat exchange would dramatically decrease the cost of the heat recovery salt heater, but further development is required to prove this concept.

Oil/Rock Storage

The oil/rock storage system consists of a heat transfer oil and river rock storage medium. The oil and rock are contained in one or more carbon steel tanks, depending on the size of the TES system. The tank or tanks are insulated to reduce heat loss, and appropriate foundations and miscellaneous support equipment are included. A substantial fraction of the tank volume is filled with the inexpensive rock; the remaining volume is filled with the more costly oil. The oil, which is about a quarter of the storage volume, stores about 20% of the thermal energy as sensible heat, while the rest of it is stored in the rock. Hot oil is added to or removed from the top of the tank, while cool oil is added to or removed from the bottom of the tank. This arrangement maintains a density-driven segregation (thermocline) between the hot oil in the top of the tank and the cooler, denser oil in the bottom of the tank. The thermal storage capacity of the tank is determined by the temperature range of the heat transfer oil. Many different heat transfer oils are available, but Caloria HT-43 (an Exxon product) is a proven, low-cost option for large-scale TES systems. To avoid significant degradation, Caloria HT-43 should be limited to an upper temperature of about 290°C. Thus, this medium is applicable for serving most process steam needs, but its upper temperature limit is too low to service combined-cycle cogeneration applications. Oil/rock storage has been successfully demonstrated for solar thermal applications at the Solar I test facility near Barstow, California.

Molten Nitrate Salt Storage

Molten nitrate salt is an excellent storage medium for high-temperature TES applications. The preferred molten salt for TES is a mixture of sodium nitrate (60 wt%) and potassium nitrate (40 wt%) that can operate without significant degradation up to about 570°C. This mixture freezes at about 240°C, so freeze protection measures must be employed. Typically, systems are designed for a minimum operating temperature of about 290°C to provide a working safety margin against freezing. The minimum operating temperature limits the amount of waste heat that can be recovered from a combustion turbine's exhaust, because the exhaust can only be cooled to about 315°C.

Molten nitrate salt TES has been extensively investigated for solar thermal power generation applications. Investigations have included bench-scale testing, detailed design studies, and relatively small-scale system demonstrations. Based on the results of these investigations, the Solar I test facility is being converted to use a molten salt working fluid. The retrofit system will include a 105 MWht (megawatt-hour, thermal) molten salt storage system and a 35 MWt (megawatt, thermal) molten-salt-heated steam generator. Presuming success for the molten salt retrofit suggests that molten nitrate salt TES is technically ready for large-scale cogeneration applications as well.

Typically, molten nitrate salt systems use separate hot and cold salt tanks. A single storage tank with a thermocline is also possible. However, it is difficult to maintain the thermocline because of radiative heat transfer between the hot and cold regions of the tank. In addition, the hot tank is much more expensive than the cold tank, so eliminating the cold tank does not significantly reduce salt storage costs.

Combined Oil/Rock and Molten Salt Storage

The advantages of both storage concepts just described can be retained by using a combination of molten nitrate salt TES for high-temperature storage and oil/rock TES for lower-temperature storage. Selection of the specific storage concept will depend on the characteristics of the thermal load. If the thermal load requires high-temperature thermal energy, molten nitrate salt TES or a combined molten nitrate salt and oil/rock TES can be chosen. Alternatively, if the thermal load requires temperatures below 290°C, oil/rock TES would be the preferred option.

Alternative Storage Media

Several ternary mixtures of salts offer wider operating temperature ranges than the "60/40" mixture of sodium and potassium nitrates described earlier, but at a higher cost. Hitec (a product of Coastal Chemical) is a mixture of sodium nitrate (7 wt%), potassium nitrate (53 wt%), and sodium nitrite (40 wt%) and can safely operate between 180°C and 450°C. Thus, Hitec salt will allow greater recovery of exhaust gas energy than 60/40 salt but does not have the higher temperature capability important for combined-cycle power production applications.

Two other ternary mixtures have also been moderately investigated for TES, but are even more expensive than Hitec. Alternative nitrate salts that can operate between 120°C and 510°C have been identified. One option is a mixture of $LiNO_3$, $NaNO_3$, and KNO_3 with proportions of 30, 18, and 52%, respectively, that melts at 120°C and can operate at up to 510°C. Another is a mixture of $Ca(NO_3)_2$, $NaNO_3$, and KNO_3 with proportions of 30, 24, and 46%, respectively, that melts at 160°C and can also operate at up to 510°C. These latter two ternary nitrate salts would allow maximum heat recovery from the turbine exhaust and can be operated at temperatures nearly as high as 60/40 salt, but are much more expensive. At current (1994) media prices, a combined oil/rock and 60/40 molten salt storage system would be more cost-effective.

Steam Generator

Steam is generated from hot media using conventional shell and tube technology. Generally, the higher pressure steam would be on the tube side and the medium would be on the shell side. The steam generator system will usually consist of two or three separate heat exchangers: (1) a preheater for raising the temperature of the feedwater to the saturation temperature, (2) an evaporator for generating saturated steam, and (3) a superheater, if steam is being generated for power production. Heat transfer occurs more readily in the steam generator than in the medium heater; overall heat transfer coefficients near 850 W/m^2 °C are typical.

Preliminary Feasibility Assessment

The feasibility of a generic cogeneration system with TES has been evaluated by Somasundaram, Brown, and Drost (1992). The procedures, assumptions, and results of this reference were updated and are summarized here to further illustrate the concept and the general approach for conducting a site-specific investigation. Although the designs developed are presumed to be reasonable and workable, they were not optimized. Hence, more advantageous versions of each TES/cogeneration system likely exist. For example, alternative approach temperatures between the turbine exhaust and storage medium or between the storage medium and steam may reduce the total system cost.

System Alternatives

The reference systems are a conventional standalone boiler plant and a conventional cogeneration system with a gas turbine and an HRSG. Both systems were designed to supply the same constant, 24 hr/day, steam load. The conventional cogeneration system also produces electricity on a baseload schedule. Six different cogeneration systems with TES were evaluated, based on the combination of three different storage media (oil/rock, 60/40 molten salt and Hitec molten salt) and two different gas turbine operating schedules (8 and 12 hr/day). Limiting gas turbine operation to less than 24 hr/day increases the size of gas turbine required to meet the daily thermal duty. For example, if the gas turbine only operates 8 hr/day it will need to be three times as large as a turbine operating 24 hr/day. The additional cost is offset by the increased power output and higher value of peaking and intermediate duty power compared to baseload power.

Design Conditions

Rudimentary design specifications were developed for each major system component to define the cost and performance basis. In general, equipment was sized to meet the steam requirements shown in Table 13B.1. Other key design and performance assumptions are presented in Table 13B.2.

TABLE 13B.1 Process Steam Requirements

Flow Rate:	181,400 kg/hr; 24 hr/day; 320 days/year
Supply Conditions:	saturated vapor @ 690 kPA (164°C)
Return Conditions:	49°C saturated liquid

TABLE 13B.2 Design and Performance Assumptions

- Natural Gas–Fired Systems
- Gas Turbine Heat Rate = 12,130 kJ/kWh
- Gas Turbine Exhaust Temperature = 538°C
- Overall Heat Transfer Coefficients
  - 150 W/m^2 °C for HRSG and storage media heaters
  - 850 W/m^2 °C for media-heated steam generators
- Storage Media Cycle Temperatures
  - 60/40 salt: 288–510°C
  - oil/rock: 149–288°c
  - Hitec salt: 177–454°C
- TES System Losses
  - 1% thermal loss
  - 2% parasitic pumping power

Steam Generator Sizing

Steam generators include conventional gas turbine heat recovery steam generators and steam generation from thermal storage media. Process steam supply and return conditions specified in Table 13B.1 define the economizer and boiler duties and water/steam inlet and exit temperatures. For an HRSG, sizing of these units depends on the gas turbine exhaust temperature and the stack reject temperature after heat recovery. In general, lowering the reject temperature increases the waste heat recovery fraction and reduces the size of the gas turbine required, but results in larger, more costly heat exchangers. The minimum reject temperature is limited by the boiler pinch point.

The first step for sizing storage-media-heated steam generators was to select the operating temperature range from within the upper and lower temperature limits of each medium. In general, the temperature range should be as large as possible to minimize storage costs. A higher upper temperature will reduce steam generator costs but increase medium heater costs. Boiler pinch point limitations must also be considered in setting the lower media temperature. Thus, the media operating temperature range affects all TES charging and discharging equipment as well as the TES unit itself. The specific temperature range for each medium type was shown in Table 13B.2. Once the temperature range was established, the design procedure was the same as that described for the HRSG.

Storage Sizing

Thermal storage capacity (kWht or kWh-thermal) is independent of the medium type because the same total energy must be transferred in the steam generator and the storage efficiency is essentially the same. Thermal losses for large (500,000 to 2,500,000 kWht) storage systems such as those required for the systems evaluated in this example are less than 1%. In addition, pumping the storage medium to and from the medium heater, storage, and steam generator is estimated to consume about 2% of gross plant power production. The required storage capacity is directly proportional to the steam generation energy and the number of hours the steam generator is operated from storage.

Medium Heater Sizing

Medium heater thermal duties and medium temperatures were established as part of the medium heated steam generator sizing process. Design considerations and procedures are similar to that described earlier for the HRSG sizing. In general, the minimum gas turbine exhaust stack reject temperature is limited by the minimum medium temperature.

Gas Turbine Sizing

The required gas turbine generating capacity depends on its heat rate and the stack reject temperature (after thermal recovery in the HRSG or media heater). Gas turbines were assumed to have a heat rate of 12,130 kJ/kWh based on the higher heating value (HHV) of natural gas. If inputs and outputs are both expressed in kWh, the equivalent heat rate is 3.369 (i.e., 3.369 kWh of fuel energy is converted into 1.000 kWh of electricity and 2.369 kWh of waste heat). Thus, the ratio of electric energy to exhaust energy is 1/2.369, or more generally 1/(HR–1), where "HR" is the heat rate in kWh of fuel energy power kWh of electricity based on the HHV of the fuel.

Not all of the 2.369 kWh of waste heat is recoverable, however. The HHV heat rate is calculated based on cooling the combustion products to 25°C and condensing all water vapor. Practically none of the water vapor is condensed, however, so the heat rate based on the fuel's lower heating value (LHV), which excludes the heat of condensation, is more useful for establishing the recoverable energy in the exhaust. All waste heat was assumed to leave the turbine in its exhaust gas at a temperature of 538°C. Recoverable energy in the exhaust is measured relative to 25°C, the reference temperature for measuring the energy input from the fuel. Therefore, the heat recovery fraction becomes (538–TR)/(538–25), where TR equals the stack reject temperature after heat recovery. The waste heat recovery fraction can be combined with the electric/exhaust energy ratio to produce the equation presented below defining the relationship between gas turbine generating capacity and the waste heat recovery rate. The maximum capacity of an individual gas turbine unit was limited to about 150 MW, resulting in either two or three parallel gas turbine and heat recovery trains for TES systems.

$$kW = kWt \cdot 513/\left[(HR-1)\cdot(538-TR)\right]$$

Note: HR in the equation is the heat rate in kWh of fuel energy per kWh of electricity based on the LHV of the fuel. For natural gas, LHV = 0.9 · HHV.

Equipment Sizes

Equipment sizes for the six cogeneration systems with TES are shown in Table 13B.3.

TABLE 13B.3 System Component Ratings and Sizes

Characteristic	Turbine Operating Period = 12 hours/day		
Hitec salt	oil/rock	60/40 salt	Hitec salt
Medium temperature range, °C	139	222	278
Turbine capacity, MW	2 × 100	2 × 150	2 × 100
Medium heater size, m²	2 × 6790	2 × 31,600	2 × 17,360
Charging piping capacity, MWt	265	265	265
Storage capacity, MWht	1,562	1,562	1,562
Discharging piping capacity, MWt	129	129	129
Steam generator size, m²	3,365	643	1,122
	Turbine Operating Period = 8 hours/day		
Hitec salt	oil/rock	60/40 salt	Hitec salt
Medium temperature range, °C	139	222	278
Turbine capacity, MW	2 × 150	3 × 150	2 × 150
Medium heater size, m²	2 × 10,180	3 × 31,600	2 × 26,030
Charging piping capacity, MWt	398	398	398
Storage capacity, MWht	2,083	2,083	2,083
Discharging piping capacity, MWt	129	129	129
Steam generator size, m²	3,365	643	1,122

Economic Analysis

The economic evaluation was conducted by calculating and comparing the levelized cost of electricity produced by the alternative systems being considered. Levelized cost analysis combines initial cost, annually recurring cost, and system performance characteristics with financial parameters to produce a single figure of merit (the levelized cost) that is economically correct and can be used to compare the projected costs of alternative steam and electric power production options. The specific methodology used was that defined in Brown, et al. (1987). The economic assumptions used to calculate the levelized costs are listed in Table 13B.4.

In general, a levelized cost analysis solves for the revenue required to exactly cover all costs associated with owning and operating a facility, including return on investment. Typically, the required revenue is expressed per unit of production, e.g., $/kWh or $/kg steam. For cogeneration systems, there are two revenue producing products, electricity and steam. Increasing the revenue associated with electricity decreases the revenue required from steam and vice versa. Levelized electricity costs were calculated by setting the value of cogenerated steam at the cost of steam produced from a stand-alone boiler.

TABLE 13B.4 Economic Assumptions

Parameter	Assumption
System economic life	30 years
System depreciable life	20 years
Nominal discount rate	9.3%/year
General inflation rate	3.1%/year
Capital inflation rate	3.1%/year
O&M inflation rate	3.1%/year
Natural gas cost in 1994	$2.45/GJ
Natural gas inflation rate	6.5%/year
Combined state and federal income tax rate	39.1%
Property tax and insurance rate	2.0%
System construction period	2 years
Price year	1994
First year of system operation	1998

Capital Cost Estimates

Capital cost estimating equations were developed for the following energy production systems or energy production system components:

- Boiler plant
- Gas turbine power plant
- Combined-cycle power plant
- Heat recovery steam generator
- Oil and salt heaters
- Oil and salt piping
- Oil/rock, 60/40 salt, and Hitec salt media
- Oil-heated and salt-heated steam generators

The fundamental basis for each equation is documented in Somasundaram, Brown, and Drost (1992), but it should be noted that the figures have been updated from this reference. All cost equations represent the completed construction cost, including indirect costs and contingency, but do not include allowances for startup and working capital, which were calculated separately. All cost data are in 1994 dollars.

Boiler plant capital cost

$$= \$17{,}900{,}000 \cdot (S/150)^{0.76}$$

S = net steam generating capacity, Mg/hr.
Cost increases by 1% for each 690 kPa above 690 kPa.

Gas turbine power plant capital cost

$$= \$43{,}200{,}000 \cdot (P/100)^{0.70}$$

P = turbine generating capacity, MW.

Combined-cycle power plant capital cost

$$= \$5{,}040{,}000 \cdot (P)^{0.60}$$

P = power plant generating capacity, MW.

Heat recovery steam generator capital cost

Heat transfer surface cost:

$HX = 5490 \cdot [k_1 \cdot (C_1/1.9)^{0.8} + k_2 \cdot (C_2/1.9)^{0.8} + \ldots + k_n * (C_n/1.9)^{0.8}]$
C = conductance, kJ/s °C, for each section of the heat exchanger
k = 1 for intermediate temperature surfaces
k = 2 for high-temperature (e.g., superheater) surfaces
k = 2 for condensing surfaces

Heat transfer surface pressure drop correction factor:

$\text{HX factor} = 0.8764 \cdot (1.194 \cdot C_t/G)^{0.28} - 0.67$
C_t = the sum of individual conductances for each section of the heat exchanger.
G = exhaust gas flow rate, kg/s.
Steam flow component cost:
$S = 10{,}646 \cdot W$
W = heated fluid flow rate, kg/s

Exhaust gas flow component cost:

$E = 459 \cdot (2.2 \cdot G)^{1.2}$

Total heat recovery steam generator installed initial capital cost:
$\text{Cost} = 2 \cdot (HX \cdot \text{HX factor} + S + E)$
Cost increases by 1% for each 690 kPa above 690 kPa.

Oil heater capital cost

$= \$199 \cdot (10.76 \cdot A)^{0.889}$

A = bare plate surface area, m^2

Molten salt heater capital cost

$= \$112 \cdot (10.76 \cdot A)^{0.951}$

A = bare plate surface area, m^2

Molten salt piping capital cost

$= \$6{,}000 \cdot MWt_{278}$

MWt_{278} = the thermal flow rate for molten salt systems cycling through a 278°C temperature range. If an alternate temperature range is used, the actual MWt must be adjusted to use this estimating formula. For example, if the temperature range were only 222°C, MWt_{278} would equal the actual MWt times 278/222.

Oil piping capital cost

$= \$12{,}000 \cdot MWt_{167}$

MWt_{167} = the thermal flow rate for oil systems cycling through a 167°C temperature range. If an alternate temperature range is used, the actual MWt must be adjusted to use this estimating formula. For example, if the temperature range was only 139°C, MWt_{167} would equal the actual MWt times 167/139.

Note: One of the preceding two equations (depending on the medium type) needs to be applied twice: once for storage charging piping and once for storage discharging piping.

Molten salt storage capital cost

60/40 Salt Storage:

Hardware capital cost = \$8,500,000 · $(MWht_{278}/1,500)^{0.52}$ for $MWht_{278} \leq 1,500$.
Hardware capital cost = \$2,540,000 + \$3,973 · $(MWht_{278})$ for $1,500 < MWht_{278} < 3,000$
Hardware capital cost = \$4,820 · $(MWht_{278})$ for $MWht_{278} \geq 3,000$
Medium capital cost = \$5,300 $(MWht_{278})$

Hitec Salt Storage:

Hardware capital cost = \$8,500,000 · $(MWht_{278}/1,563)^{0.52}$ for $MWht_{278} \leq 1,563$.
Hardware capital cost = \$4,630 · $(MWht_{278})$ for $MWht_{278} \geq 3,125$
Hardware capital cost = \$2,540,000 + \$3,816 · $(MWht_{278})$ for $1,563 < MWht_{278} < 3,125$
Medium capital cost = \$13,040 $(MWht_{278})$

The three alternative hardware cost equations for molten salt storage reflect the transition from single to multiple hot tanks above 1,500 $MWht_{278}$ and single to multiple cold tanks above 3,000 $MWht_{278}$ for 60/40 salt. The transition points for Hitec are slightly greater, as indicated in the Hitec cost equations. The 278 subscript identifies the cost equations as being valid for thermal capacities calculated for a 278°C temperature range (e.g., from 288°C to 566°C for 60/40 salt or 176°C to 454°C for Hitec). MWht requirements associated with other temperature ranges must be adjusted to a 278°C basis before using the cost estimating equations as shown next.

$$MWht_{278} = MWht_x \cdot (278/x)$$

For example, if $x = 139$, then $MWht_{278} = MWht_{139} \cdot (278/139)$. This adjustment accounts for the doubling of the physical size of a 139°C range storage system that would be required to achieve the same thermal capacity as a 278°C range storage system.

Oil/rock storage capital cost

Hardware capital cost = \$199,000 · $(MWht_{167})^{0.3851}$ for $MWht_{167} < 1,500$
Hardware capital cost = \$2,218 · $(MWht_{167})$ for $MWht_{167} \geq 1,500$
Media capital cost = \$2,152 $(MWht_{167})$

The two alternative hardware cost equations reflect the transition from single to multiple tanks above 1,500 $MWht_{167}$. The "167" subscript identifies the cost equations as being valid for thermal capacities calculated for a 167°C temperature range (e.g., from 121°C to 288°C). MWht requirements associated with other temperature ranges must be adjusted to a 167°C basis before using the cost estimating equations as shown next.

$$MWht_{167} = MWht_x \cdot (167/x)$$

For example, if $x = 83$, then $MWht_{167} = MWht_{83} \cdot (167/83)$. This adjustment accounts for the doubling of the physical size of a 83°C range storage system that would be required to achieve the same thermal capacity as a 167°C range storage system.

Steam generator capital cost

Baseline heat exchanger initial capital cost

= \$780,000 · $(HTA/1,394)^{0.75}$

HTA = heat transfer surface area, m^2.

The baseline cost presumes carbon steel tubes and shell. Cost multiplier for carbon steel tubes and stainless steel shell or carbon steel shell and stainless steel tubes = 1.7.

Cost multiplier for stainless steel shell and tubes = 3.
Cost multiplier for evaporator = 1.5.
Add 1% for each 690 kPa increase in steam pressure above 690 kPa.

Note: the baseline heat exchanger cost equation and multipliers must be evaluated based on the heat transfer area of each steam generator component, that is, for the preheater, evaporator, superheater (if included), and reheater (if included).

Startup and working capital cost estimates must be added to the costs estimated with the previous equations to obtain a total cost. Startup costs include operator training, equipment checkout, minor changes in equipment, extra maintenance to get the system on-line, and fuel consumption incurred after the plant is constructed, but prior to regular operation. Working capital represents a "revolving account" used to pay for the procurement of current expenses and an investment in spare parts. The cost relations used for estimating startup and working capital follow.

$$\begin{aligned}\text{Startup capital cost} &= 0.02 \cdot \text{total system construction cost}\\ &\quad + 1/12 \cdot \text{total annual O\&M}\\ &\quad + 1/52 \cdot \text{total annual fuel}\end{aligned}$$

$$\begin{aligned}\text{Working capital cost} &= 0.005 \cdot \text{total system construction cost}\\ &\quad + 1/6 \cdot \text{total annual O\&M}\\ &\quad + 1/6 \cdot \text{total annual fuel}\end{aligned}$$

Operation and Maintenance Cost Estimates

O&M costs include fuel, operating labor, maintenance labor and materials, consumable supplies, and overhead. Nonfuel O&M cost estimating relations were developed for each of the system cost elements described earlier.

Steam generation O&M

Fixed labor:

$= 0.07 \cdot (S/150)^{-0.5} \cdot$ construction capital

S = net steam generating capacity, Mg/hr.

Fixed maintenance materials:

$= 0.0085 \cdot$ construction capital

Variable maintenance materials:

$= 3 \cdot 10^{-6} \cdot$ annual operating hours at full capacity $\cdot$ construction capital

Consumable supplies:

$= 0.33 \cdot$ annual steam production, Mg

Gas Turbine Power Plant O&M

Fixed annual O&M cost:

$= \$69{,}500 \cdot P^{0.61}$

P = power plant generating capacity, MW

Variable O&M cost:

$= \$0.00021 \cdot$ power output, kWh

TABLE 13B.5 Relative System Levelized Electricity Costs

Systems Evaluated	Daily Power Production Period	
	8 hr/day	12 hr/day
Simple-cycle power plant with stand alone boiler	0.96	0.85
Combined-cycle power plant with stand alone boiler	1.00	0.81
Simple-cycle cogeneration plant with stand alone boiler	0.68	0.56
Simple-cycle cogeneration plant with oil/rock TES	0.65	0.55
Simple-cycle cogeneration plant with Hitec TES	0.77	0.64
Simple-cycle cogeneration plant with 60/40 salt TES	0.87	0.72

Combined-Cycle Power Plant O&M

Fixed annual O&M cost:

$= \$1{,}480{,}000 \cdot P^{0.263}$

P = power plant generating capacity, MW

Variable O&M cost:

$= \$0.00082 \cdot$ power output, kWh

Annual O&M costs for the remaining components were estimated as 10% of the hardware costs. Note that the 10% factor was not applied to TES media capital; media maintenance was assumed to be negligible at the operating temperatures being considered.

Results

The results of the economic analyses for the different system configurations considered are presented in Table 13B.5. Each system is designed to produce 181,400 kg/hr of steam 24 hr/day, 320 days/year. Power output varies according to the generating capacities listed in Table 13B.3 for the cogeneration plants with TES. The generating capacity for the simple cycle power plant and cogeneration plant with a standalone boiler is 97 MW. The combined-cycle power plant was presumed to have a generating capacity of 160 MW. Each plant generates electricity during the daily power production period. The simple cycle and combined cycle power plants operate independent from their stand alone boilers, which operate 24 hr/day. For the simple cycle cogeneration plant with a stand alone boiler, steam is supplied by the cogeneration plant during the daily power production period and from the boiler during the balance of the day. Steam is supplied continuously by the cogeneration plants with TES.

The results show the significant economic advantage of cogeneration over separate power and steam production, regardless of whether TES is used or not. Another obvious trend is the lower cost for the longer daily power production period, which results from greater utilization of fixed capital investment.

Oil/rock is shown to be preferred over the two salts for this application. Higher-temperature storage is inherently more expensive than lower-temperature storage due to increased costs of containment. For example, stainless steel is required at salt storage temperatures, while carbon steel will suffice at oil/rock storage temperatures. In addition, thicker and/or more expensive insulation is required for the higher temperatures. High-temperature storage is not required for the relatively low-temperature process steam application investigated. The higher temperature capability of salts would be required for combined-cycle cogeneration applications, where maintenance of the thermodynamic availability in the gas turbine exhaust gases is important.

The lower operating temperature of the oil/rock medium also results in a lower-cost medium heater due to the higher approach temperatures that are possible. While approach temperatures for the steam generator are higher for the two salt media, the cost impact is much less. The medium

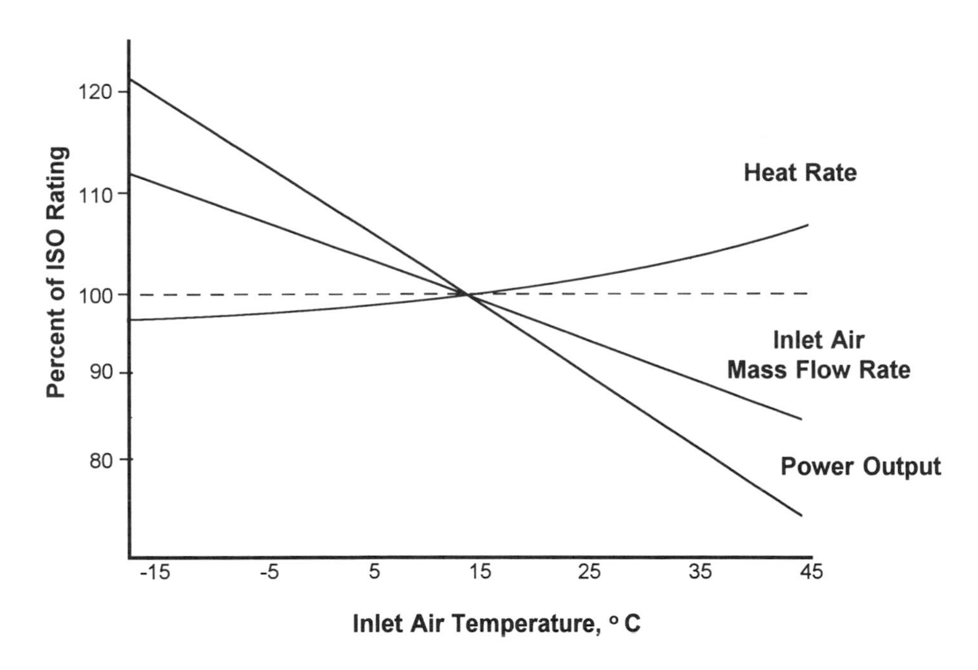

FIGURE 13B.2 Effect of inlet air temperature on combustion turbine performance.

heater is much larger because it has a greater thermal duty and a much lower overall heat transfer coefficient. Hence, the net advantage again exists for the oil/rock TES system.

Cogeneration with oil/rock TES produces a moderately lower electricity cost than cogeneration with a standalone boiler, but a significantly lower cost compared to separately producing electricity and steam. In addition, the cogeneration plant with oil/rock TES generates 2 to 3 times as much electricity during the daily power production period. This difference has added importance when utilities are seeking additional intermediate duty capacity and attractive cogeneration applications are in short supply. Finally, another benefit not captured in the cost comparison is the reduced emissions of the cogeneration plant with oil/rock TES that result from its higher efficiency.

13B.2 Gas Turbine Inlet Air Cooling

Concept Overview

Combustion turbines are constant volume machines (i.e., air intake is limited to a fixed volume of air regardless of ambient air conditions). As air temperature rises its density falls. Thus, although the volumetric flow rate remains constant, the mass flow rate is reduced as air temperature rises. Power output is also reduced as air temperature rises because power output is proportional to mass flow rate. The conversion efficiency of the gas turbine also falls as air temperature rises because more power is required to compress the warmer air.

The impact of compressor inlet air temperature on mass flow rate, power output, and conversion efficiency is shown in Figure 13B.2 for a typical "industrial" type of turbine. Per custom, conversion efficiency is reported as a heat rate, which is the amount of fuel energy consumed per kWh of electricity produced. Thus, a rise in the heat rate is consistent with a drop in conversion efficiency. In general, the relationships are linear with temperature, or nearly so.

The performance curves show the dramatic effect of temperature on turbine performance and the opportunity to improve performance via inlet air cooling. Cooling the inlet air improves both the power output and the heat rate, but the impact on power output is greater. Furthermore, the positive impacts on power output and heat rate increase with higher ambient temperatures, when

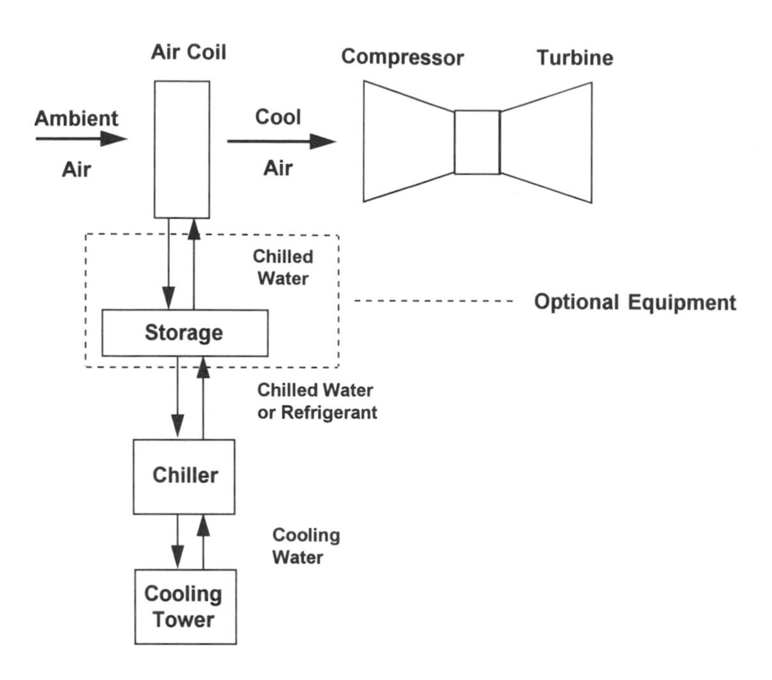

FIGURE 13B.3 Generic inlet air cooling system.

utilities typically experience the highest demand for power. Therefore, incremental power production is greatest at the time it is most valued. *Usually, the primary objective of gas turbine inlet air cooling is to increase peak power output. The heat rate improvement is a significant, but secondary benefit.*

The relative importance of the two impacts depends mostly on the value of incremental kW and kWh production, the number of operating hours per year, and ambient air temperatures during the operating periods. A greater number of annual operating hours will increase the importance of the heat rate impact, while fewer hours will emphasize the increase in power output. Applications with a higher ratio of average to peak ambient operating temperature will also increase the importance of the heat rate impact, all else equal.

In theory, power output could be further increased by cooling below the temperature range indicated in Figure 13B.2. In practice, all turbines are designed around a maximum thrust level, which sets a useful limit on the inlet air temperature. In particular, the maximum power output of "aeroderivative" types of turbines often occurs around the freezing point and may actually decline below this temperature. Cooling near or below the freezing point also presents concerns associated with icing, which will be discussed later.

A block diagram of a generic gas turbine inlet air cooling system, with and without storage, is shown in Figure 13B.3. The basic building blocks are the chiller, its cooling tower, the air coil, and interconnecting piping. Cold fluid from the chiller is pumped through the air coil, where the coolant is heated and returned to the chiller, while the inlet air is cooled prior to entering the compressor. The cooling tower provides cooling water to the chiller condenser. Alternatively, an evaporative condenser might be used with some types of chillers. Including storage and its associated piping loop increases the number of system components but allows the chiller and cooling tower components to be downsized, assuming that cooling is not conducted 24 hr/day. Storage also significantly reduces on-peak parasitic power consumption for electrically driven chiller systems.

The fundamental benefits of inlet air cooling were described earlier. The most obvious cost is the up-front investment and periodic maintenance of the inlet air cooling system hardware. The energy used to drive the chiller may also result in a significant expense, although the cost will vary significantly depending on whether the chiller is thermally or electrically driven and the source of thermal or electric energy. Inclusion of an air cooling coil within the inlet duct to the compressor will incur an additional pressure loss, with negative consequences to power output and heat rate, but the impacts will generally be less than 0.5%. These and other potential effects of inlet air cooling are discussed in more detail in subsequent sections.

The previous two paragraphs described the use of refrigerative systems for cooling turbine inlet air. A commonly employed alternative, besides not cooling the inlet air at all, is to use evaporative cooling. Direct contact evaporative cooling, accomplished by passing the inlet air through a wetted medium, can be particularly effective in drier climates. Any consideration of refrigerative inlet air cooling should also consider evaporative cooling as an option. Note that evaporative and refrigerative approaches should be considered independently. Evaporative cooling followed by refrigerative cooling would not reduce the refrigerative cooling load. It would only substitute latent load for sensible load.

System and Component Options

For a concept that may seem relatively straightforward, a plethora of options complicate the inlet air cooling component selection and system design process. The unique conditions existing for any potential application make it difficult to establish useful rules of thumb. In particular, it is difficult, if not foolhardy, to suggest a single configuration of components that will always be the optimal choice. Nevertheless, some rules of thumb can be developed to at least narrow the field.

Gas Turbine Applications

Gas turbines are employed in simple-cycle and combined-cycle electric power and cogeneration plants. Key differences in these applications for inlet air cooling systems are the presumed value of low-temperature thermal energy and the number of operating hours per day. For simple-cycle electric power plants, the energy in the exhaust gases is free (except for the cost of collection and assuming other uses of this energy are not perceived), while steam extracted from a combined-cycle or cogeneration power plant results in foregone electricity or steam sales, which has value. Simple-cycle electric plants tend to operate only a few hours per day, while combined-cycle and cogeneration plants tend to operate at least half the day. The inlet air cooling system may operate for fewer hours than the plant operates, of course, which adds to the number of system possibilities.

The availability of free thermal energy in the form of gas turbine exhaust suggests that a thermally driven chiller may be preferred over an electrically driven chiller in that situation. The use of storage, however, can significantly reduce the size and cost of the electrically driven chiller and allow it to be driven by low-valued off-peak electricity. While storage could be employed to reduce the size and cost of the thermally driven chiller, there would be no free energy to drive the chiller during off-peak hours. For combined-cycle and cogeneration plants operating most of the day, the availability of low-valued thermal energy during off-peak hours makes storage a more reasonable option for thermally driven systems. Assignment of value to the thermal energy may also justify the expense of a more efficient thermally driven chiller, whereas efficiency is of little concern when the energy is free. Again, broad rules of thumb regarding preferred systems are difficult to establish.

Inlet Air Temperature

The inlet air cooling system must be designed to avoid icing at the compressor inlet or anywhere in the air intake structure. Ice fragments sucked into the compressor can cause serious structural damage. Icing is a potential problem anytime the ambient air temperature drops near the freezing mark. Compressor bleed air is commonly used to either internally heat compressor inlet surfaces or is directly injected into the inlet air stream. Exhaust from the turbine has also been routed back through the inlet air structure within closed heat exchanger surfaces, and electric heat tracing has been employed as well.

The potential icing problem is exacerbated for inlet air cooling systems because warm ambient air will almost always be saturated after passing through the inlet air cooling coils. When the air is drawn into the mouth of the compressor, its velocity increases and its temperature drops further as air enthalpy is transformed into kinetic energy in an adiabatic process. Condensate icing can occur if the temperature drops below freezing. Equation 13B.1 describes the relationship between air velocity and temperature. A temperature drop of about 5°C is common. Therefore, the design inlet air temperature should be at least 5°C to avoid potential icing problems altogether. The temperature depression will vary by a few degrees for specific machinery, however, and should be calculated, rather than relying on this general rule of thumb.

$$V^2/2\,g = C_p \cdot dT \tag{13B.1}$$

where V = velocity
g = gravitational constant
C_p = heat capacity
dT = temperature drop

The design inlet temperature may also be affected by the capabilities of the chiller. The minimum chilled water temperature available from lithium bromide absorption chillers is about 6°C. Therefore, practical approach temperatures at the air coil will limit the design inlet air temperature to about 10°C for this type of chiller. Mechanical vapor compression chillers can cool to much lower temperatures, but the cost per unit of cooling capacity increases and the efficiency decreases as evaporator temperatures are lowered. In general, it will be cost-effective to cool the inlet air as low as the chilling technology allows without incurring potential icing problems.

Chillers

The chiller is the central piece of hardware in an inlet air cooling system. Its selection will determine or limit other equipment and design options. The two basic options are thermally driven absorption chillers and mechanically driven vapor compression chillers. Other thermal and mechanical refrigeration options exist, but are more expensive to own and/or operate.

Lithium bromide absorption chillers are commonly used to provide chilled water where steam or hot water is available. In these systems, lithium bromide is the absorbent and water is the refrigerant. A lithium bromide chiller operates at subatmospheric pressure in a hermetic vessel. Therefore, its evaporator cannot be directly placed in the path of the turbine inlet air. Thus, chilled water (limited to a minimum temperature of about 6°C) is produced and would be piped through a coil to cool the inlet air. Single-stage and double-stage units are available, with coefficients of performance (COPs) typically about 0.6 for the former and 1.2 for the latter. The double-stage unit does cost about 1/3 more than a single-stage unit and also requires a higher steam pressure. Therefore, a single-stage unit would be preferred if turbine exhaust gas heat is free.

Aqua–ammonia is another absorption chilling option. In these systems, water is the absorbent and ammonia is the refrigerant. Aqua–ammonia systems have the capability of serving refrigeration applications at temperatures well below 0°C, but this is of little value if inlet air cooling is limited to 5°C. Furthermore, the COP of an aqua-ammonia system is no better than a single-stage lithium bromide system, while its initial cost is several times higher. Therefore, lithium bromide is generally preferred to aqua–ammonia.

The preferred cooling technology for industrial refrigeration applications is mechanical vapor compression with ammonia refrigerant and a screw compressor. The preference for ammonia stems from its superior heat transfer properties and high COP. Centrifugal compressors using R-22 have been widely used as water chillers for space cooling applications. Due to ozone depletion concerns, R-134a may become more common for this application in the future. Safety concerns have precluded the use of ammonia for space cooling applications. The superior performance of ammonia makes it the preferred refrigerant for gas turbine inlet air cooling. Centrifugal compressors using carbon-based refrigerants may offer lower first cost, but at lower efficiency and generally with less durability than the screw compressors designed for industrial refrigeration applications.

The mechanical vapor compression chiller can be powered by either electric motors, steam turbines, or gas-fired engines. Electric motors are the most common choice, their advantages being lower first cost and simpler installation. For inlet air cooling, electric motors offer the flexibility of using off-peak electricity to run the chiller and charge a storage system, even when the gas turbine is not running. On the other hand, the availability of low-cost or essentially free steam while the gas turbine is running may make the steam turbine option attractive. Gas-fired engines would also offer the system charging flexibility that electric motors provide.

Air Coils

In general, the inlet air is cooled by passing it across a set of coils containing either refrigerant or a secondary coolant such as water. The type of inlet air cooling coil used will depend on the type of chiller selected and whether or not storage is used. As described earlier, lithium bromide absorption chillers will always be coupled with an air/water coil. Vapor compression chillers offer the option of evaporating the refrigerant directly in the coils, which reduces the air temperature for a given air coil approach temperature and also reduces equipment costs and pumping power compared to chilled water circulation systems. If storage is included with a vapor compression system, either chilled water or a mixture of chilled water and ice will be generated by evaporating refrigerant in the storage vessel. Chilled water would subsequently be pumped to the air coil.

As noted in the overview, the inlet air coil must be designed to minimize air pressure drop and its negative effects on turbine performance. As a general rule of thumb, pressure drop should probably be limited to 250 Pa; 125 Pa is a common design point. Pressure drops of this magnitude should result in less than a 0.5% decline in power output and 0.5% increase in heat rate for the turbine. An additional consideration is the potential carryover of condensed water droplets into the air stream, with possible negative effects as the droplets impact the compressor blades. User and/or vendor experience should be consulted to determine if demister pads will be required.

Storage

Incorporating storage into the inlet air cooling system may be desirable to downsize the chiller and heat rejection (cooling tower) components and significantly reduce on-peak parasitic power consumption for electrically driven systems. The chiller is usually the most expensive system component, so reducing its size and cost at the expense of adding storage and related piping can be cost-effective. Reducing on-peak parasitic power consumption is also important because increasing peak power output is usually the primary objective of gas turbine inlet air cooling. Storage is not desirable in all situations, however. An obvious example is when inlet air cooling is conducted 24 hr/day. In general, storage will be more attractive with fewer daily operating hours. Storage also makes little sense for systems with thermally driven chillers that operate for the same number of hours as the power plant. In this situation, there would be no low-cost or free thermal energy available to run the chiller. If natural gas were used instead, much or all of the chiller downsizing benefit would be offset by increased energy costs.

Chilled water or ice (really a mixture of water and ice) are the most obvious and effective storage media for inlet air cooling systems. Both are applicable to diurnal storage, and ice storage is applicable to weekly storage cycles as well. Seasonal storage of ice via engineered ice or snow ponds, or chilled water in naturally occurring aquifers would also be possible, but these concepts suffer from site-specific limitations and have had limited successful application to date. Eutectic salts are another possibility, but the salts are more expensive than water, suffer availability losses on charge and discharge, and also suffer from limited application experience. Direct storage of refrigerant has thermodynamic appeal, but vapor storage is a difficult problem.

Steel or concrete cylindrical tanks can be used for water or ice storage. External insulation should be adequately thick to avoid condensation. Chilled water is added and removed from the bottom of water storage tanks, while warm water is added or removed from the top. This procedure forms a thermally stratified tank with a thermocline that rises on charge and falls on discharge. The preferred ice-making method uses a harvesting approach that periodically passes hot refrigerant from the compressor through the evaporator to release ice from the evaporator surface. The ice falls from the evaporator and makes a pile within the tank. Several evaporators are used to aid in distributing the ice. The alternative approach is to build up logs of ice around evaporator coils that run back and forth throughout the tank. Although the defrost cycle increases the effective cooling load by about 15%, the ice harvester is less costly to build because it requires much less evaporator surface and refrigerant inventory.

Selection of the storage medium depends partly on the chiller type. Lithium bromide absorption chillers can only use water storage. Either water or ice storage is possible for vapor compression chillers. The principal advantage of ice is its greater chill storage density. The principal advantage of water is the mechanical simplicity of its storage system. Ice storage will generally allow the inlet air to be cooled to a lower temperature than will water storage, but ice generation requires a lower chiller evaporator temperature, which results in poorer chiller efficiency and higher chiller cost. Limited operating experience to date suggests that ice storage systems with vapor compression systems are generally preferred.

Auxiliary Components

In addition to the chiller, air coil, and storage tank, several auxiliary components are required to make a complete inlet air cooling system. While the majority of the cost and energy consumption for an inlet air cooling system is associated with the three components described earlier, the following auxiliary components require significant additional investment and add nontrivial energy consumption to the system. For vapor compression chillers using storage, the refrigerant would be evaporated in a water chiller or ice generator, depending on the type of storage selected. Without storage, the refrigerant is evaporated in the air coil. Evaporators are included in the hermetically sealed lithium bromide units, which produce chilled water directly. Chilled water circulation equipment (pumps, pipe, valves, etc.) is required for any system with water or ice storage or a lithium bromide chiller. Refrigeration circulation equipment is required for any system with a vapor compression chiller. Steam and condensate circulation equipment will be required for lithium bromide chillers. In addition, a heat recovery steam generator would be required to generate steam from simple-cycle turbine exhaust. In general, a cooling tower and cooling water circulation equipment will be required to provide cooling water to chiller condensers. Alternatively, heat rejection for vapor compression chillers can be served by evaporative condensers.

Preliminary Feasibility Assessment Approach

As noted in the concept overview, the primary objective of gas turbine inlet air cooling is to increase the power generating capacity when the ambient air temperature is high. Heat rate improvement is an additional, but secondary benefit. The components described earlier represent conventional hardware, so technical feasibility is not much of an issue. The real issue is economic feasibility for a set of site-specific conditions, with the key conditions being meteorology, gas-turbine application (simple-cycle or combined-cycle, electric power production only or cogeneration), plant operating schedule, inlet air cooling schedule, and the value of incremental electric power and energy production.

General Approach

The general assessment approach compares the benefit of incremental power and energy production with the cost of owning and operating the inlet air cooling system. With costs and benefits denominated in dollars, economic feasibility can be determined by calculating the economic decision factor (e.g., net present value, internal rate of return, payback period) used by the investing organization. As always, care must be taken in selecting the alternatives to be evaluated. Refrigerative inlet air cooling options should be compared to evaporative inlet air cooling, or simply building a larger turbine if the proposed application is for a new turbine rather than an existing one.

The value of incremental power and energy production may vary depending on ownership perspective. From a utility perspective, value may be defined by the alternative to inlet air cooling for providing additional peaking capacity. The most likely alternative would be to build a new turbine without inlet air cooling. The value of incremental power would be equal to the expected fixed costs (capital and fixed O&M), while the value of incremental energy would be equal to the expected variable costs (fuel and variable O&M). From an industrial perspective, the value of incremental power and energy will usually be set more specifically by the prevailing contract with the utility for power purchases or sales. This contract will also likely contain requirements that will dictate the period the inlet air cooling system will need to operate to achieve the maximum credit for increased generating capacity.

System Selection and Design Conditions

The next step in the evaluation process is to select the inlet air cooling systems to be investigated. Consideration of application characteristics and the rules of thumb described earlier should narrow the field, but several potentially valid options will likely still exist. Obviously, the details of the evaluation process will vary depending on the type of system and component options selected.

Power plant performance curves will be needed to size and assess the performance of the inlet air cooling system. At a minimum, the relationship between the inlet air temperature and plant power output, plant heat rate, and gas turbine inlet air mass flow rate must be established. Note that cooling the inlet air to the gas turbine will affect its exhaust temperature and flow rate as well, thus having an impact on cogeneration and combined-cycle plants. Therefore, performance curves for a simple-cycle turbine are inadequate for these applications. The pressure drop incurred by the air coil (and an HRSG, if one is added to a simple-cycle turbine to produce steam for the chiller) also affects power output, heat rate, and inlet air mass flow rate, but the impact is generally small enough to ignore for a preliminary analysis.

Historical weather data (dry bulb temperature and humidity) must be acquired for the prospective site to establish the cooling load, size equipment, and calculate component performance. Hourly weather data is preferable and is available for 200 U.S. cities in data files known as a typical meteorological year. The peak hourly air cooling load is set by the maximum ambient temperature occurring during the on-peak period, the design inlet air temperature, and the air flow rate at the design inlet temperature. If daily or weekly storage is used, the weather data would have to be surveyed to determine the maximum daily or weekly loads too.

Equipment Sizing

Equipment "sizing" can proceed once the design air cooling loads have been calculated. Chiller and storage sizing depends on whether storage is employed or not, whether storage occurs on a weekly or daily cycle, and whether storage is designed to provide chiller "load shifting" or "load leveling." In a load-shifting system, the chiller and storage system would be sized such that the chiller never operates during the peak demand period, but would run continuously at full load during off-peak hours on the peak demand day or week. Load-shifting minimizes on-peak parasitic energy consumption for electrically driven vapor compression systems. In a load-leveling system, the chiller and storage system would be sized such that the chiller would run continuously, 24 hr/day, on the peak demand day or week. Load-shifting minimizes the size of the chiller.

In addition to the air cooling load, other cooling loads and losses need to be considered when sizing the chiller and storage components. As inlet air cooling increases the turbine shaft output on a hot day, it is possible the generator capacity may be exceeded. As a rule of thumb, supplementary cooling equal to about 5% of the air cooling load must be applied to the generator to maintain generating capability commensurate with shaft output. Thermal losses in the storage tank must also be included for proper sizing and performance assessment of the chiller and storage components. Thermal losses through the storage tank walls will usually be about 1%. The overall thermal efficiency of water storage is only about 85%, however, due to losses across the thermocline.

Defrost heat added to ice harvesting storage systems is also a significant problem, and typically adds about 15% to the gross input required by the chiller. Note that the physical size of the ice storage tank is not increased by adding the defrost heat, but chiller size and energy consumption are.

Auxiliary equipment sizes can be calculated once the air coil, chiller, and storage components have been sized. Chilled water and refrigerant circulation equipment size is directly proportional to the peak hourly air load. Cooling tower and cooling water circulation equipment size is directly proportional to the peak chiller capacity and the heat rejection rate per unit of capacity, which depends on the type of chiller. If a heat recovery steam generator (HRSG) is added to supply steam to an absorption chiller, the sizing of the HRSG and steam/condensate circulation equipment will be proportional to the chiller capacity and its COP at peak design conditions.

Equipment capital costs can be estimated from the sizing specifications. Annual operating and maintenance costs can be estimated from equipment sizes, equipment capital costs, and/or annual operating hours.

System Performance

The performance evaluation must calculate the power output and fuel consumption of the gas turbine, with and without inlet air cooling, plus the parasitic power consumption of inlet air cooling system components. The latter includes compressor power for vapor compression chillers, pump power for absorption chillers, pump power for the various fluid circulation systems, and fan power for the cooling tower.

The most accurate approach would be to aggregate performance calculated for each operating hour of the year, using a computer model capable of reading TMY weather tapes. Such a model could also be programmed to calculate cooling loads and size system equipment. The desirability of developing or obtaining a model with these capabilities increases as the number of systems to be considered increases, system complexity increases, and/or the planned operating hours per year increases. For simple-cycle applications operating only a few hundred hours per year, the whole focus of the study will likely be on the incremental cost per kW of generating capacity. In this situation, the analysis can be simplified considerably by focusing on performance for the design conditions only. Another alternative simpler than an hourly simulation model would couple a design condition analysis of incremental cost per kW with an estimate of incremental kWh and fuel consumption based on the average weather conditions during the inlet air cooling system operating period. This approach may be more accurate than it first seems because the relationship between temperature and gas turbine power output, heat rate, and mass flow rate is nearly linear. Annual parasitic power consumption can be approximated by multiplying the peak hourly parasitic load by the ratio of the annual cooling load to peak hourly cooling load. Whichever analytical approach is taken, it is important to segregate the parasitic loads into on-peak and off-peak components. On-peak parasitics result in both capacity and energy cost penalties, while off-peak parasitics should only be charged for their energy cost.

Example Application

One of the first applications of refrigerative inlet air cooling using thermal energy storage was installed in 1991 by Lincoln Electric System (LES) of Lincoln, Nebraska. This project has been described in detail in Ebeling et al. (1992) and is summarized here to further illustrate the concept. The Lincoln Electric application was a simple-cycle industrial frame combustion turbine with a generating capacity of about 63 MW at ISO conditions. The turbine was used to meet peak electric demands during hot, humid summer conditions.

LES became interested in inlet air cooling because it was faced with rising peak demand and the need for additional generating capacity. Inlet air cooling offered low-cost incremental capacity that could be designed and installed relatively quickly. The primary benefit of inlet air cooling for LES was a substantial increase in generating capacity. At their 38.6°C design day temperature (with

34% relative humidity), generating capacity is 53.1 MW with no cooling, 57.8 MW with evaporative cooling, and 67.1 MW with refrigerative cooling. The reduction in heat rate occurring at the lower inlet temperature is an important, but secondary, benefit to LES. The heat rate was lowered by about 4%, with the resulting fuel cost savings offsetting the cost of energy to run the chiller.

The LES system uses an electrically driven vapor compression chiller with ammonia refrigerant and a screw-type compressor. Heat rejection is accomplished with an evaporative condenser. The chiller operates during off-peak hours to charge an ice storage system using an ice harvesting evaporator design. During turbine operations water circulates from the bottom of the storage tank, through the inlet air cooling coils, and back to the storage tank, where it is sprayed onto the top of the remaining ice.

Storage was incorporated into the system to minimize on-peak parasitic power consumption and to reduce the size of the chiller. Weekly storage, which further reduces the size of the chiller, but increases the size of the storage tank, was determined to be the economic optimum. Ice harvesting was chosen over ice-on-tube to reduce the investment in evaporator tubing and refrigerant inventory. Ice-on-tube evaporators also suffer from reduced ice generating efficiency as ice builds up and insulates the tube.

The inlet air cooling system operates 4 hr/day, 5 days/week. This schedule was selected based on the requirements for accredited peaking capacity within LES's power generating pool. Peak cooling loads were calculated based on the design day conditions noted earlier, and a design inlet air temperature of 4.4°C. The latter represents the minimum temperature that will avoid condensate icing at the compressor bell mouth. Design day conditions were used to size the air coil and chilled water circulation equipment, while slightly lower temperature conditions were used to calculate the weekly load and size the chiller and storage equipment. The resulting air cooling loads are about 50 GJ/hr or 1,000 GJ/week. The refrigeration system consists of one compressor, three evaporators, one evaporative condenser, plus a liquid receiver and suction accumulator. The rated capacity of the compressor is 7.6 GJ/hr, with each evaporator 1/3 this size.

The air cooling coils are spiral-wound finned tubes constructed from copper (tubes) and aluminum (fins). Tube diameter is 1.6 cm with a wall thickness of 0.064 cm. Fin thickness is 0.05 cm with 3.15 fins/cm. The air coil is designed to cool inlet air from 38.6°C to 4.4°C with water entering at 1.1°C and leaving at 7.8°C. Air-side pressure drop is 125 kPa. Water condensing from the inlet air is collected and routed to site drainage.

The ice storage tank was constructed from cast in-place concrete and installed partially below grade. Insulation with an R value of 10 was applied to the exterior. Its height and diameter are approximately 9 m and 24 m, respectively. The required tank volume (4,360 m^3) was calculated to allow for the maximum active weekly ice charge, a reserve inventory equal to about 10% of the maximum weekly charge, and 10% of freeboard volume.

Circulating chilled water pumps draw water from the bottom of the storage tank via a loop header. The loop header is strategically perforated to draw water from various locations and equalize ice melt within the tank. Water is pumped through buried piping to the air coil and back to the tank. Return water is distributed through a spray system designed to wet the entire ice surface equally. Part of the water returning from the air coil is routed to the generator to supplement its cooling, prior to returning to the storage tank. A smaller water circulation loop pumps water from the bottom of the tank up to the evaporators on top of the tank when ice is being generated.

Project capital costs for the LES inlet air cooling system are presented in detail in Table 13B.6. The refrigeration system, storage tank, and inlet air coil (including its structure) make up the majority of the total cost. Refrigerative inlet air cooling increased the generating capacity by 14 MW compared to no inlet air cooling, and by 9.3 MW compared to evaporative cooling. On-peak parasitic power consumption was approximately 0.6 MW for the refrigerative inlet air cooling system. Therefore, the cost of incremental generating capacity would be $165/kW if the evaporative cooler did not exist. With the existing evaporative cooler, the cost of incremental generating capacity rises to $254/kW.

TABLE 13B.6 Lincoln Electric System Project Capital Costs

Component	Cost, $1,000s	Component	Cost, $1,000s
	Refrigeration System		
Compressor	101	Motor control center	65
Evaporator	290	Ammonia	5
Evaporative condenser	60	Transportation	8
Receiver	12	Installation	40
Accumulator	14	Startup and training	4
Piping	50	Design	20
Wiring	35	Refrigeration building	65
	Tank Construction		
Ice storage	227	Excavation and backfill	35
Insulation	30	Other	39
Pump pit	75		
	Controls		
PLC and cabinets	109	Installation and other	64
Conduit and wiring	35	Calibration and startup	10
Instrumentation	43	Site electrical	76
Valves and operators	23		
	Site Mechanical		
Pipe and fittings	150	Hangers and supports	11
Valves	16	Other	71
Piping specials	39	Startup and testing	5
Air cooling coils	135	Circulating water pumps	63
	Inlet Structure		
Concrete structure	65	Roofing, doors, and other	37
Steel	87		
	Total project capital cost	2,213	

References

Brown, D.R., J.A. Dirks, M.K. Drost, G.E. Spanner, and T.A. Williams. 1987. *An Assessment Methodology for Thermal Energy Storage Evaluation.* PNL-6372, Pacific Northwest Laboratory, Richland, Washington.

Ebeling, J., R. Halil, D. Bantam, B. Bakenhus, H. Schreiber, and R. Wendland. 1992. "Peaking Gas Turbine Capacity Enhancement Using Ice Storage for Compressor Inlet Air Cooling." 92-GT-265. *Proceedings, 1991 International Gas Turbine and Aeroengine Congress and Exposition,* American Society of Mechanical Engineers, New York.

Somasundaram, S., D.R. Brown, and M.K. Drost. 1992. *Evaluation of Diurnal Thermal Energy Storage Combined with Cogeneration Systems* PNL-8298. Pacific Northwest Laboratory, Richland, Washington.

14

Thermal Energy Management in Industry

Wesley M. Rohrer, Jr.
University of Pittsburgh

Introduction

The Potential for Fuel Conservation

The potential for fuel conservation in the industrial sector of the economy is still considered to be large. Many companies, particularly the large corporations with abundant technical resources, have energy management programs that date back to the time of the first OPEC oil embargo. The early programs of the U.S. Department of Commerce enlisted the cooperation of the big corporations in conserving scarce fuels and helped the industrial and commercial trade organizations develop programs to monitor fuel consumption within their membership and to organize information on conservation to be used within the group. However, Ross and Williams in 1977 estimated from their Second-Law analyses of the consumption for energy in 1973 that the whole economy could do better by 42% and that the potential in industry amounted to 10.43 quads (1 quad = 10^{15} Btu = 2.93×10^{11} kWh = 1.055×10^{18} J) of a total of 29.65 or 35% of the total use. As shown in Table 14.1, the big potential was in improved housekeeping (37%) and in cogeneration of steam and electricity (25%). Waste-heat utilization contributed only 12% of the total potential savings. The most significant aspect of their study was the large savings realizable with minimum capital requirement. Since some industries use up to one-third of their total energy purchases for space conditioning, it can be seen that a large potential existed that is not considered in Table 14.1.

Figure 14.1 shows that industrial energy consumption was at all-time highs in 1973 through 1979. In the following period, from 1980 through 1986, industrial consumption fell as energy prices escalated rapidly, but energy use increased again through 1990 as energy prices fell toward previous levels. Figure 14.2 shows rather conclusively that the drop in energy consumption from

0-8493-2514-5/96/$0.00+$.50

TABLE 14.1 Hypothetical Potential for Energy Conservation in the Industrial Sector

Industrial Sector	Potential Savings (10^{18} J)
Improve housekeeping measures (better management practices with no changes in capital equipment)	4.06
Use fossil fuel instead of electric heat in direct-heat applications	0.18
Adopt steam/electric cogeneration for half of process steam	2.73
Use heat recuperators or regenerators in half of direct-heat applications	0.78
Generate electricity from bottoming cycles in half of direct-heat application	0.52
Recycle aluminum in urban refuse	0.11
Recycle iron and steel in urban refuse	0.12
Use organic wastes in urban refuse for fuel	0.74
Savings from reduced throughput at petroleum refineries	0.92
Reduced field and transport losses associated with reduced use of natural gas	0.84
Total savings	11.00
Total demand in 1973	31.28×10^{18} J
Hypothetical energy demand with savings—20.28×10^{18} J	

Source: From Ross, M.H. and Williams, R.H., 1977. With permission.

1980 to 1985 was due to vigorous efforts to conserve energy. The average increase in energy use efficiency was a respectable 25.1%, more than half of the potential for conservation estimated by Ross and Williams in 1977. However, more recent data (EIA, 1995) show that from 1985 to 1991 overall industrial energy use efficiency decreased by 1.0%. In several industrial groups the potential is seen to remain high, and in all other groups surveyed, there remains some potential for energy savings. The lesson seems to be that energy conservation is driven largely by market forces. This does not mean that government and industry trade group programs have been ineffectual, but rather that the technical content of those programs is not fully utilized until the economics of energy use begins to pinch profits. The secret to a successful project is in carefully planning and organizing for it. The technical solutions are available to everyone. The energy problem must be viewed as an overall systems problem, and a systems analysis must be employed with everything considered that the systems approach entails. A random or scatter-gun approach, no matter how enthusiastically and energetically entered into, cannot produce satisfactory results. Only a carefully derived synthesis of good management practice and good engineering solutions will produce optimum results. As essentially presented in the EPIC Guide (Gatts et al., 1974), the following elements are necessary for success:

- Gaining management commitment
- Setting up the organization
- Carrying out the energy audit
- Educating employees
- First-level economic studies
- Engineering studies
- Second-level economic analyses
- Investment analysis
- Installation of energy conservation measures
- Developing and maintaining the energy information system
- Installation of submeters
- Exploiting program successes

Each of these essentials will be discussed in greater detail in the remainder of this chapter.

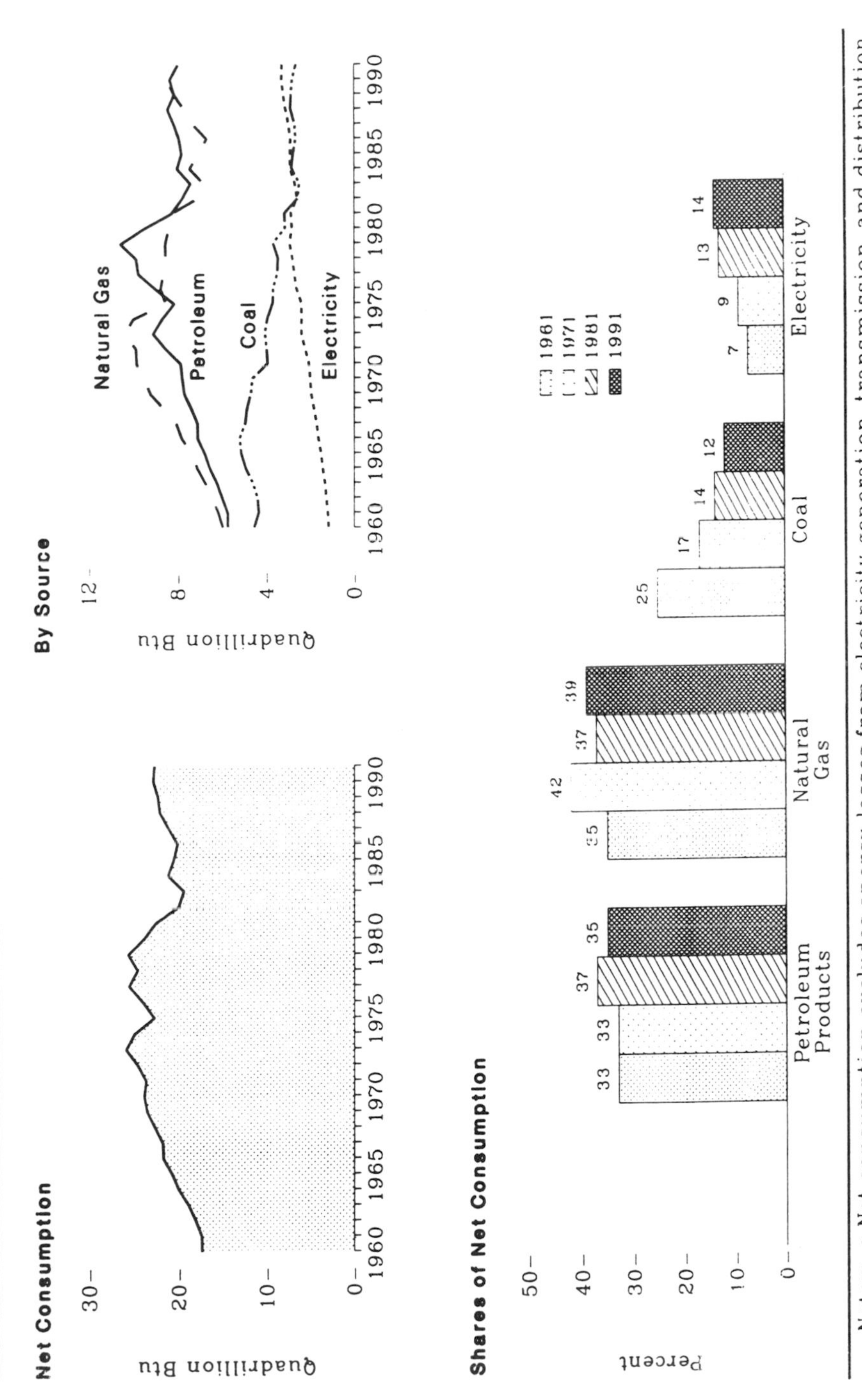

Notes: ● Net consumption excludes energy losses from electricity generation, transmission, and distribution. Electricity includes hydroelectric power generated by the industrial sector. ● Because vertical scales differ, graphs should not be compared.
Source: Table 12.

FIGURE 14.1 Industrial energy consumption 1949–1993. *Source:* EIA, 1994.

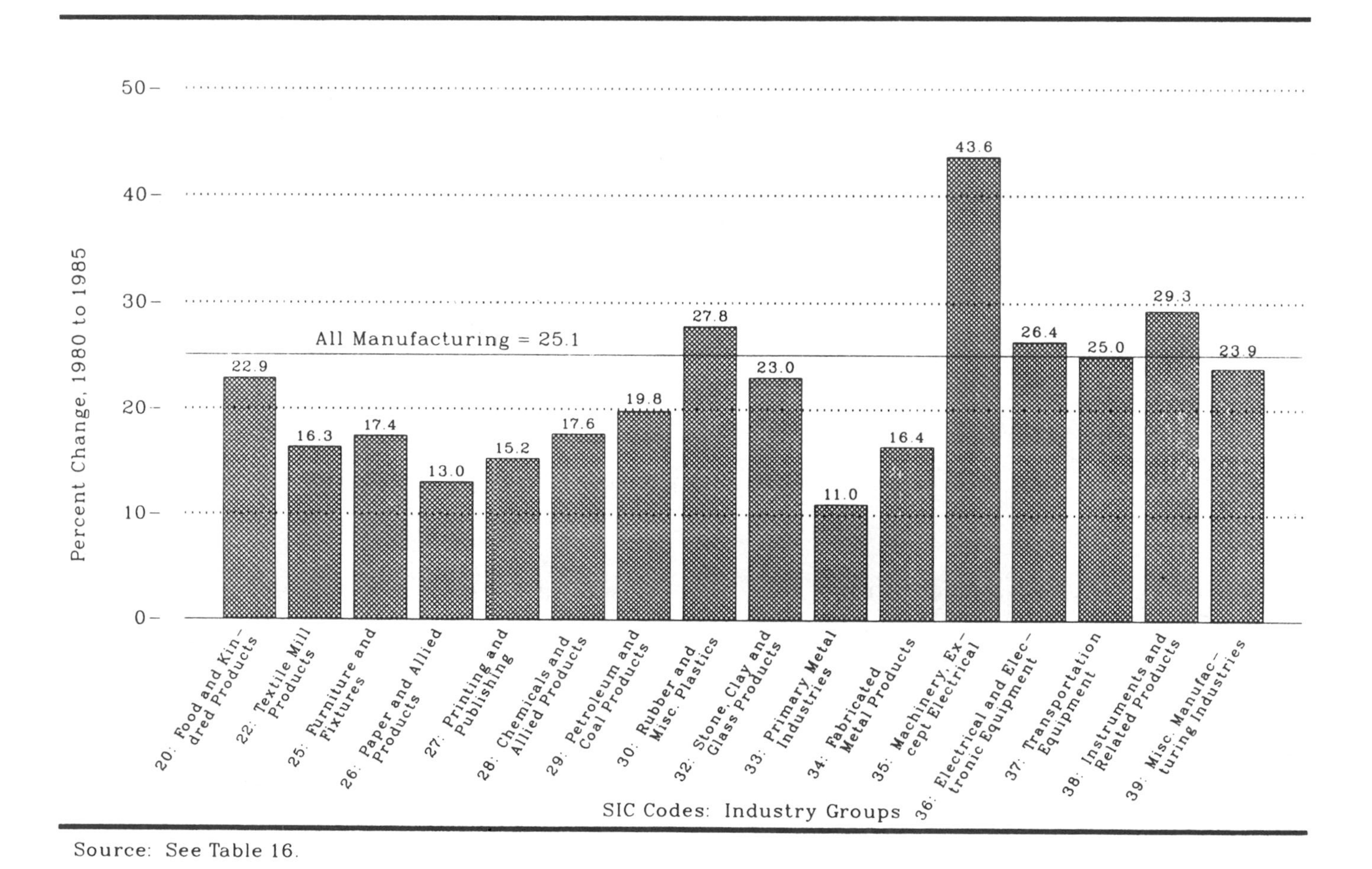

FIGURE 14.2 Manufacturing sector energy efficiency by industry group. *Source:* EIA, 1994.

Gaining Commitments to the Energy Conservation Program

The greatest impediment to a successful program in energy conservation is lack of commitment at all levels of the organizational structure. Generally the difficulties in gaining the necessary commitments are related to disbelief in the seriousness of the national energy problem, or to a belief that the only solution to that problem is in a more vigorous program of exploration for conventional fuels or in future technologic breakthroughs, which will result in boundless energy supplies. It is unlikely that large additions to the world's oil and gas reserves will be forthcoming, and the fulfillment of the promise of abundant energy from nuclear fission and fusion plants seems today to be much farther away than it seemed in 1973. World population growth and the continuing industrialization of third world countries point inevitably to future energy crises. Furthermore, environmental concerns will result in large increases in the costs of using fossil fuels. Here we are speaking not only of the direct costs of reducing industrial emissions, but also of the increased cost of electric energy due to more stringent regulations of emissions from electric utility plants. Fortunately, in many cases, all or part of the costs of regulating air and water pollution can be offset by appropriate energy conservation programs. The energy situation in the long term can only worsen, but it can be alleviated by utilizing available fuel supplies in the most efficient manner consistent with environment requirements. Energy conservation programs will require the full cooperation of everyone, but in the industrial setting the highest priority, of course, rests with top management, for without top-level commitment the necessary resources cannot be made available and the program will fail by starvation. Furthermore, a great part of the necessary cooperation from all lower echelons can be secured in some measure by management dictum, although the potential for success is enhanced by the addition of every single willing and enthusiastic soul. Horror stories are legion concerning mighty programs being made impotent by the stubborn resistance of one supposedly minor character who believes that the energy problem has been contrived for someone else's personal gain. To counter these attitudes and arguments one must use as inducement the beneficial effects of energy conservation on the economic self-interests of everyone concerned. Using the methods developed here, management and labor must be convinced that energy conservation measures are indeed economically feasible, avoid costs, and thereby put money in the bank—money that is just as good as that derived from profits on product sold. Furthermore, this additional income is earned regularly and perpetually, and as fuel costs rise increases in value. Additionally, the savings in fuel may help to forestall future fuel allocation cuts or to avoid exceeding present fuel allocations. During the hard winter of 1976–1977 one plant in Northwest Pennsylvania, by taking Draconian steps in energy conservation, was actually able to maintain full production on maintenance fuel levels and thus emerge from the emergency with their business unscathed, while neighboring plants lost several weeks' production. Thus, good energy management can be sold on the basis that (1) overall profits go up, (2) the possibility of maintaining full production is increased, (3) jobs are saved, and (4) a step is taken toward the fulfillment of national energy goals. Investment in cost containment programs has not hitherto been a popular management tool. However, a growing list of case studies involving successful energy conservation programs is educating us to the acceptance of that possibility.

Organizing for Energy Conservation Programs

The most important organizational step which will effect the success of an energy conservation (e.c.) program is the appointment of one person who has full responsibility for its operation. Preferably that person should report directly to the top management position and be given substantial discretion in directing technical and financial resources within the bounds set by the level of management commitment. It is difficult to stress enough the importance of making the position of plant energy coordinator a full-time job. Any diversion of interest and attention to other aspects of the business is bound to badly affect the e.c. program. The reason is that the greatest opportunity for conservation is in improved housekeeping. The improvement and maintenance of good housekeeping procedures is an exceedingly demanding job and requires a constant attention and a dedication to detail that is rarely found in corporate business life. The coordinator should be energetic, enthusiastic, dedicated, and political.

The second step is the appointment of the plant e.c. committee. This should consist of one group of persons who are able to and have some motivation for cutting fuel costs and a second group who have the technical knowledge or access to data needed for the program department managers or their assistants. Union stewards and/or other labor representatives often make up the first group, while the second should include the maintenance department head, a manager of finance or data storage, some engineers, and a public relations person. The coordinator should keep up to date on the energy situation daily, convene the committee weekly, and present a definitive report to top management at least monthly and at other times when required by circumstance. It is suggested also that several subcommittees be broken out of the main committee to consider such important aspects as capital investments, employee education, operator-training programs, external public relations, and so on. The committee will define strategy, provide criticism, publish newsletters and press releases, carry out employee programs, argue for the acceptance of feasible measures before management, represent the program in the larger community, and be as supportive as possible to the energy coordinator. This group has the most to risk and the most to gain. They must defend their own individual interests against the group but at the same time must cooperate in making the program successful and thus be eligible for rewards from top management for their good work and corporate success.

As the e.c. program progresses to the energy audit and beyond, it will be necessary to keep all employees informed as to its purposes, its goals, and how its operation will impact on plant operations and employee routine, comfort, and job performance. The education should proceed through written and oral channels as best benefits the organizational structure. Newsletters, posters, and employee meetings have been used successfully.

In addition to general education about energy conservation, it may prove worthwhile to offer specialized courses for boiler and mechanical equipment operators and other workmen whose jobs can affect energy utilization in the plant. The syllabuses should be based on thermodynamic principles applied to the systems involved and given on an academic level consistent with the workmen's backgrounds. Long-range attempts to upgrade job qualifications through such training can have very beneficial effects on performance. The courses can be given by community colleges, by private enterprises, or by in-house technical staff, if available.

The material presented here on organization is based on the presumption that a considerable management organization already exists and that sufficient technical and financial resources exist for support of the energy conservation program as outlined. Obviously, very small businesses cannot operate on this scale, and some are so small that energy conservation cannot be made a realistic goal. However, we have found companies with energy costs below $50,000 annually with capabilities for carrying out effective energy conservation efforts.

Setting Energy Conservation Goals

It is entirely appropriate and perhaps even necessary to select an energy conservation goal for the first year of the program very early in the program. The purpose is to gain the advantage of the competitive spirit of those employees that can be aroused by a target goal. Unfortunately, the true potential for conservation and the investment costs required to achieve it are not known until the plant energy audit is completed and a detailed study made of the data. Furthermore, a wide variety of energy-use patterns exists even with a single industry. A number of industry studies have shown that energy consumption per unit of production does not fit a normal population curve. Individual plants are often clustered around low, average, and high values for consumption with two or three peaks in the distribution curve. However, it may be of interest to know where your plant stands vis-à-vis national averages in your sector of industry. Table 14.2 lists the primary energy consumption

TABLE 14.2 Primary Energy Consumption Data for Nine U.S. Industries

	Notes	Primary Energy Consumption (GJ/t product)	Breakdown of Primary Energy Use by Type of Resource (%)					Total Energy Used in 1970 in Making Product (10^{15} J)
			Coal	Oil	Gas	Purchased Electricity	Derivative Fuels	
Low-density polyethylene resin	a,b	108.73	0	23.6	67.3	18.2	(9.1)	212
High-density polyethylene resin	a,b	103.09	0	28.1	73.1	8.4	(9.6)	79
Polystyrene resin	a,b	136.56	1.1	100.4	27.1	6.9	(35.5)	208
Polyvinyl-chloride resin	a,b	96.44	9.1	19.4	55.6	23.4	(7.5)	138
Petroleum refinery products	a,e,d	(0.51)	—	—	—	—	—	1,841
Portland cement—wet process	a,b	9.35	30.4	13.7	39.9	16.0	—	399
Portland cement—dry process	a,b	8.43	42.6	8.0	32.4	17.0	—	236
Primary copper	a,b	130.07	10.1	13.5	38.4	38.0	—	179
Primary aluminum	a,c	201.50	0.5	15.1	9.0	72.2	3.2	728
Raw steel	a,b	22.35	81.1	6.6	13.5	8.4	(9.6)	2,667
Glass containers	a,b	21.12	35.8	7.3	48.8	14.5	(6.4)	216
Newsprint	a,b	25.53	6.6	12.8	13.5	67.1	—	77
Writing paper	a,b	28.46	19.3	27.8	23.5	29.4	—	60
Corrugated containers	a,b	25.28	26.2	26.2	42.1	6.9	(1.4)	228
Folding boxboard	a,b	25.47	17.1	25.8	40.6	16.5	—	22
Virgin styrene butadiene rubber	a,b	155.41	0.1	47.8	53.9	9.7	(11.5)	210

Note: The total energy figures quoted in this table correspond to the primary energy consumption shown in the first column of numbers. These consumption figures were multiplied by the total tonnage of product manufactured in 1970. Total energy use represented by these products was $7{,}604 \times 10^{15}$ J in 1970.

a. The figures shown are based on average industrial practice in the U.S. during 1970–1971.

b. For all process steps other than alumina smelting and petroleum refining, electric energy was derived from the following mix of primary energy sources: 48.5% coal, 16.1% oil, 26.8% natural gas, 2.6% nuclear fuels, 6.0% hydroelectric. Taking generation and transmission losses into account, this is equivalent to 10,286 Btu/kWh.

c. For alumina smelting, electric energy was derived from the following mix of primary energy sources: 39.0% coal, 12.9% oil, 21.5% natural gas, 2.1% nuclear fuels, 24.5% hydroelectric.

d. For petroleum refining, the primary energy consumption in generating electricity was taken as equivalent to 12,000 Btu/kWh, which is derived from the breakdown given in Note b with the exclusion of the hydroelectric contribution.

e. Petroleum refining industry data expressed as million Btu per barrel of crude oil processed. Data taken from typical refinery calculations. Breakdown by resource type not determined for overall industry. Typically, all energy is derived from oil, gas, and purchased electricity. Electricity use in the refinery is typically 11%. Total energy use is based on 0.44 million Btu/bbl and the 1970 total of crude runs to stills, 3.967×10^9 bbls *(API Annual Statistical Review)*.

Source: FEA. 1975. With permission.

data for the nine most energy-intensive industries for the census year 1970. The notes following the table explain some of the rationales used and list several data sources.

Current data for the nine industry groups listed in Table 14.2 is not available in the same form. However, Table 14.3 presents the "energy intensity ratios" for 20 groups of manufacturers identified by their SIC code numbers. The percentage energy intensity change is also presented for three periods: 1980 to 1985, 1985 to 1988, and 1980 to 1988. In Table 14.2 the energy intensity factor has units of gigajoules per ton of product, GJ/ton, while in Table 14.3 the "energy intensity ratio" has units of kiloBtu per constant (1982) dollar of product shipped. Although different in scale, the two energy intensity indices measure essentially the same entity. From 1980 to 1985 the average industrial energy intensity dropped an average of 23.4%, from 1985 to 1988 it dropped by only 4.3%, with an overall improvement in energy use efficiency of 26.7% from 1980 to 1988.

It is suggested that an intuitive selection be made, perhaps in the range of 5 to 25% for the first year. More realistic goals will be forthcoming as the work progresses. For every additional time period the potential savings and the required investment will be known with increasing certainty. Thus, in subsequent years, goals will be set and will be expected to be achieved as a matter of course, just as are other management goals such as those for production and sales.

The Question of Alternate Fuels

In an economy with abundant supplies of all types of fuels, the choice of fuel would depend first on the requirements of process and product quality and secondly on cost. In a fuel-scare economy the choice may be reduced to what is available regardless of cost, perhaps with sharply reduced technical specifications. Experience has shown that it is usually more economic to pay higher energy costs than to shut down or even reduce production. Thus many managers are considering the use of all electric production facilities, even though the present price ratio of electric to fossil fuels ranges from 3:1 to 15:1 on an energy basis, depending on local market conditions and annual consumption levels.

The reasons for the large price differentials are related mainly to the attainable thermal efficiency of fixed-station power plants, which average about 35% in this country. However the price difference may be partly offset by a potentially more efficient use of electrically derived heat over that from combustion processes. For example, the substitution of radiant electric heat for a gas-fired oven could involve the following statistics: First-Law efficiency of gas ovens is 10%; First-Law efficiency of electric ovens is 85%. The ratio of cost of electric energy to gas-derived energy is 4:1; the assumed electric generating station efficiency is 35%.

$$\begin{array}{l}\text{Percentage fuel savings}\\ \text{using electricity}\end{array} = 100\left(\frac{\dfrac{1}{0.1} - \dfrac{1}{0.85 \times 0.35}}{\dfrac{1}{0.1}}\right) = 66\%$$

On the other hand the cost of the electrical energy would be:

$$\left(\frac{1}{0.85}\Big/\frac{1}{0.10}\right) \times 4 = 0.47$$

times that of the gas-derived energy. The use of electrical energy for heating purposes is sometimes in poor taste from a thermodynamic outlook and usually costs more per unit of heat delivered, but it does have some demonstrable advantages. It alleviates the most pressing national energy problem because it substitutes (in most areas of the nation) coal as a fuel for scarcer liquid and gaseous fuels. In some cases coal can be used as a local industrial fuel, but in the majority of plants in this country coal is unacceptable because of process requirements, environmental restrictions, or the availability of small-scale coal-fired equipment. This is not to say that the present outlook

TABLE 14.3 Manufacturing Energy Intensity by Industry Group, 1980, 1985, and 1988

SIC[b] Code	Industry Group	Energy Intensity Ratios[c]			Energy Intensity Change[a] (percent)		
		1980	1985	1988	1980 to 1985	1985 to 1988	1980 to 1988
20	Food and kindred products	3.52	2.72	2.98	22.8	–9.4	15.5
21	Tobacco products	NA	NA	NA	NA	NA	NA
22	Textile mill products	5.69	4.80	4.85	15.6	–0.9	14.8
23	Apparel and other textile mill products	NA	NA	NA	NA	NA	NA
24	Lumber and wood products	NA	NA	NA	NA	NA	NA
25	Furniture and fixtures	1.87	1.55	1.72	16.5	–10.1	8.0
26	Paper and allied products	15.92	13.96	12.86	12.3	7.9	19.2
27	Printing and publishing	NA	NA	NA	NA	NA	NA
28	Chemicals and allied products	14.91	12.40	11.34	16.8	8.6	24.0
29	Petroleum and coal products	5.32	4.87	5.67	8.3	–16.3	–6.6
30	Rubber and misc. plastics products	4.29	3.10	3.22	27.7	–3.8	25.0
31	Leather and leather products	NA	NA	NA	NA	NA	NA
32	Stone, clay, and glass products	21.53	16.74	16.74	22.3	0	22.2
33	Primary metal industries	16.30	14.64	14.37	10.2	1.8	11.8
34	Fabricated metal products	2.74	2.33	2.42	15.2	–4.0	11.8
35	Industrial machinery and equipment	1.66	0.95	0.77	43.2	18.7	53.8
36	Electronic and other electric equipment	1.67	1.25	1.18	24.9	5.4	29.0
37	Transportation equipment	1.51	1.15	1.06	23.6	8.0	29.7
38	Instruments and related products	1.60	1.19	1.16	26.0	2.5	27.9
39	Misc. manufacturing industries	1.71	1.36	1.35	20.3	1.1	21.1
	Total manufacturing	5.78	4.43	4.23	23.4	4.3	26.7

[a] A decrease in the energy intensity ratio results in an increase in energy efficiency represented by a positive value.
[b] Standard Industrial Classification.
[c] Thousand Btu per constant (1982) dollar of value of shipments and receipts.
NA = Not available.
Note: Data for 1985 are different from previously published data due to deflator adjustments.
Sources: EIA, 1992.

for coal as an industrial fuel will continue indefinitely. Should suitable coal-fired equipment with the capability of removing sulfur and/or its oxides from the liquid and gaseous emissions become available to industrial users, or should the environmental constraints on sulfur oxide emissions be relaxed, then coal can be expected to take over much more of the industrial fuel market. In the meantime each enterprise might well review its position vis-à-vis alternate fuel sources and make provisions for more flexibility in fuel use. Such flexibility should pay off in enhanced ability to stay in production and to use the cheapest available fuel, as market factors cause relative changes in per-unit fuel costs. Recently, small-scale industrial boilers have been introduced that can burn gaseous, liquid, or solid fuels. This is clearly a step in the right direction. Furthermore, more attention needs to be given to the possibility of using as fuels locally generated waste which is presently being discarded. We are referring here to such materials as waste paper and paperboard, used lubricants, and contaminated liquid solvents, which can often be used as boiler fuels when trifuel burners have been installed.

14.1 The Energy Audit and Energy Information

Carrying Out the Energy Audit

Rationale for the Energy Audit

The energy audit consists of a complete study of the consumption and costs of all purchased energy, how it is distributed throughout the plant, how it is used in individual systems and processes, the potential for energy saving and where it exists, and the capital costs involved in recovering that potential. Additionally, it provides the basis for creating a very inexpensive energy information system (EIS), which can and should be incorporated into the present management information system (MIS). It represents in effect the total body of knowledge needed by management to institute an effective energy conservation program. The particular uses to which the audit can be applied are

- To inform management of the growing cost of energy and to provide motivation for carrying out an energy conservation program as a cost-containment measure
- To inform the engineering staff of the plant's energy utilization characteristics so that intelligent conservation measures may be planned
- To provide management with the information needed to make wise investment decisions concerning energy conservation measures
- To provide part of the basis for planning and installing alternate fuels
- To provide base-line energy consumption data to which future energy consumption can be compared
- To provide the basis for a perpetual energy information system (EIS) that can be integrated with existing management information systems
- To uncover poor housekeeping practices that can be quickly remedied to give almost immediate energy and cost savings

In other words, the energy audit provides the rationale for an energy management program, as well as all the information for design, implementation, and evaluation of energy conservation measures, and in addition provides the base upon which to build the EIS which is needed to complete the MIS.

Mobilizing for the Energy Audit

The first step to be taken is always to construct a retrospective record of energy purchased monthly from all sources for a consecutive 2-year period. Monthly consumption and monthly costs are recorded. This includes separate records for electricity, natural gas, each grade of oil, propane, steam at each pressure purchased, and each waste fuel. It does not include steam generated on site or waste heat recovered, but does include by-products used as waste fuel. Purchased water is usually included because it is closely associated with energy utilization and is a controllable expense. Along with the energy consumption data, one must gather month-by-month measures of weather severity and level of plant activity. The most useful weather data are the monthly degree-days for the heating season. The level of plant activity can be measured in a number of ways, and one cannot predetermine the most useful measure. In the brewing industry barrels of beer produced suffices. In complicated fabrication industries in which highly diversified products are manufactured and then assembled, direct labor hours are often a better measure. The chronology of records is often a big problem in constructing the consumption audit. Billing dates for electric and natural gas bills lag behind the consumption and the records must be compensated. For oil, propane, and coal, one usually has records of delivery dates and quantity delivered. Depending upon the ratio of the rate of use to the storage volume, the delivery dates may or may not be a significant measure of monthly use. Happy is the auditor who finds records of monthly inventory of these fuels or better still meter records.

The next step is to construct process flow charts for the total operation. These should be arranged so that indications of energy added to the product can be entered at each process stage. These allow the quick determination of the most energy-intensive operations, on which one should presumably concentrate first, and permit studies of possible waste-heat interchange and process modifications for purposes of energy conservation.

The third step is to derive an in-plant distribution tree for each source of energy purchased. This should be done by plant, building, department, production line, and unit process. This step is probably the most difficult one in the audit but is most important for the successful operation of your program. If submetering is done at any of the levels, the time and effort required is correspondingly reduced. Furthermore, as one suffers through the attempt to derive a realistic in-plant distribution, pain should motivate him or her toward mounting a program to add submetering for all large energy flows. The reason that such a program is so important has to do with establishing responsibility for energy utilization within the plant. If top management has no way of knowing how much energy departments A, B, and C consume, then they can hardly be in a position to reward the managers of those departments for good energy management or punish them for poor energy management. If manager A is aware that no way exists to become a hero by way of improved energy management, that manager will expend as little effort as possible on that activity, for any improvements that he or she brings about may be credited more or less equally to all of the departments.

The next activity to be taken is to produce energy balances on each integrated system and unit process that consumes appreciable amounts of energy. In order to prevent undue emphasis on arbitrating the division of large and small energy users, it is best to select the 5 or 10 or 20 largest energy users as your first focus for analysis. The information that must be provided for each system includes mass flow rates, identification of materials, and the temperature for each entering and exiting stream. This should include the flows of raw materials as well as finished product. In addition, any surface heat losses must be estimated or measured. Time-of-use profiles must also be derived when possible. An alternative to typical daily, weekly, or monthly time profiles is an estimate of the annual load factor for each system or process unit studied.

A data bank that stores the information from the initial audit must be established and fed data on at least a monthly basis. This energy information system should be maintained in perpetuity. The data bank may be computerized or kept by hand. If it is complex enough to be computerized, the data acquisition, handling, and retrieval system can be used for carrying out the numerical analysis required by the audit and the associated energy conservation program. It is relatively easy and much less expensive to develop the energy information system while constructing the energy audit than it is to establish it later. A well-designed EIS can be maintained at a fraction of the cost of making annual energy audits. A detailed log of energy conservation activities must be maintained. The log must include changes in scheduling, production techniques, operating set points,

ENERGY SAVING SURVEY

Department: ____________

Date: ____________

Surveyed by: ____________

Fuel Gas or Oil Leaks	Steam Leaks	Compressed Air Leaks	Condensate Leaks	Water Leaks	Damaged or Lacking Insulation	Excess Lighting	Excess Utility Usage	Equipment Running & Not Needed	Burners Out of Adjustment	Leaks of or Excess of HVAC	Location	Date Corrected

FIGURE 14.3 Energy saving survey. *Source:* Gatts et al., 1974.

additions or replacement of equipment, and retrofit of energy conservation features to existing equipment.

During the energy audit one must make an energy-saving survey. This consists of a walking plant tour designed to uncover poor housekeeping practices that result in energy waste. The defects which one looks for are fuel, steam, compressed air, or water leaks; uninsulated, poorly insulated, or damaged insulation on building, pipes, or equipment; excess use of any utility, particularly due to lighting or power in use when not needed; heat loss due to broken or open windows and open doors; control temperatures set too high or too low, and fuel burners in need of maintenance or poorly adjusted. A sample form is shown in Figure 14.3. A recapitulation of the content of the energy survey appears here.

- Retrospective survey of energy consumption and cost for each source of purchased energy and monthly measures of weather severity and level of plant activity
- Process flow charts
- In-plant energy distribution
- Heat balances and time-of-use profiles for all integrated systems and unit processes
- An energy information system
- Energy conservation activity log
- Energy-saving survey

The contents missing from this list, which are often considered to be part of the audit, are the engineering and economic analyses leading to specification of energy conservation measures requiring capital expenditures. This author has chosen to regard these analyses as integral parts of the energy conservation program of which the audit is also an integral part but not to specifically include it in the content of the audit itself. The reason is that energy conservation is much more a matter of good management practices than it is of capital spending. Thus it deemphasizes the importance of capital spending by making the feasibility and engineering studies separate and discretionary activities. The topics of engineering and economic studies are treated later in this chapter.

Data Collection

Data collection is always a task of considerable magnitude. A summary outline of the data requirements is given next.

First a retrospective, month-by-month record of the consumption and cost of every form of energy purchased or sold out of plant for at least a 2-year period is needed. This includes energy in the form of electricity, natural gas, fuel oils, gasoline, propane, coal, purchases steam, and waste or by-product fuels. Separate records should be used for each form of energy, including each grade of fuel oil and each steam pressure. The data for purchased energy are usually derived from utility bills and other financial records. Gas and electric utilities often keep computerized records, which are available for audit purposes. However, bulk purchases of oil, propane, and coal cause problems unless monthly inventories are taken of fuel storage. If this has not been the practice, it should be instituted at the earliest possible time. Audit records for electricity should include monthly peak demands, power factor multipliers, and any other available data that affect the cost of energy and at the same time are susceptible to some control.

Second, monthly measures of production level and of weather severity are required for the same period of time as the purchased energy records. Measures of production level are needed in order to assess the effects of production on energy use. It has been found, however, that the best measure to use for this purpose differs widely from industry to industry and perhaps from plant to plant. Measures of production (or of business activity) range from tonnage produced to direct labor hours used per month. In general, it is found that basic industries such as steel, cement, oil, and coal can use monthly units of production and more complex assembly-type producers can best use direct labor hours. Once the proper measure is found, then records should be kept of the total energy use per unit of activity. These will then constitute a measure of efficiency of fuel utilization for the plant.

The best measures of weather severity are degree-days for the heating season and for the cooling season. These are available monthly from the National Oceanographic and Atmospheric Administration.

Third, a complete inventory must be kept of energy-consuming equipment with as much information as possible about energy and material flows, entrance and exit pressures, and temperatures. This information should preferably be derived from actual inspections and from experienced operators and engineers.

Fourth, time-of-use profiles for all systems and unit processes and the corresponding operating conditions should also be kept.

A set of audit forms that were reproduced from Rohrer (1980) is appended to this section. These are shown as Figures 14.4 through 14.7.

Analyzing the Gross Energy Consumption Data

This author's approach to wringing out the energy consumption data is to plot it chronologically in as many ways as seems useful for each case. This is illustrated with some plots of both raw and derived data. In Figure 14.8 there is a block diagram of the energy distribution within a commercial laundry and in Table 14.4 the annual energy and cost data.

This rather atypical facility uses no direct-fired laundry equipment. Instead, natural gas is used as a boiler fuel to supply steam-heated washers, dryers, and ironers, for firing a hot-water furnace for space heating, and to supply four direct gas-fired air make-up heaters. Figure 14.9 shows the monthly energy consumption for a 2-year period with the gas and electric energy used converted to thermal units. The seasonal variation is quite apparent, but the base load is some 55% of the peak consumption. The base load is shown by the dashed lines on the figure as 2.796 TJ per month. The average annual gas consumption is known from utility bills to be 40.778 TJ. The difference between the total and the base load is the heating load, where annual heating energy = $40.778 - 2.796 \times 12 = 7.226$ TJ.

19__/19__

1	2	3	4	5	6	7	8	9	10	11	12	13	14	15	16
Month	Demand MCF	Demand Charge Rate	Demand Charge	MCF	MMBTU	Net Energy Charge	Pch'd Gas Surcharge	Tax Surcharge	Gross Energy Charge	Total Cost	Year-to-Date MCF	Year-to-Date Cost	Last Year Cost	% Diff	Cost/MMBTU
Total															
19__/19__															
Total															

FIGURE 14.4 Historical quantity and cost for natural gas. *Source:* Rohrer, 1980.

19__/19__

1	2	3	4	5	6	7	8	9	10	11	12	13	14	15	16	17
Month	Actual Demand KW	Power Factor Multiplier	Billing Demand KW	Demand Charge Rate	Demand Charge	KWH	MMBTU	Net Energy Charge	Fuel Surcharge Rate	Tax Surcharge Rate	Gross Energy Charge	Total Cost of Electricity	Year-to-Date Cost	Last Year Cost	% Diff	Cost MMBTU
Total																
19__/19__																
Total																

FIGURE 14.5 Historical quantity and cost data for electricity. *Source:* Rohrer, 1980.

The plant operates on one 10-hr split shift a day for 5 days a week. Thus the total production time is 2,600 hr per year and

$$\text{Base load consumption} = \frac{2796 \times 12}{2600} = 13 \ GJ/hr$$

Since the rated input of each of the 250 hp fire-tube boilers is 11.1 GJ/hr, a single boiler will hardly be able to maintain this base load, which averages 16% overload year round. Thus a study is

19__/19__

1	2	3	4	5	6	7	8	9	10
	DEGREE-DAYS		SIZE OF FACILITY			LEVEL OF ACTIVITY			
Month	Heat	Cool	sq. ft.	cu. ft.	Production Units	Operating Cost	Operating Level		
Total									
19__/19__									
Total									

FIGURE 14.6 Major factors affecting energy use. *Source:* Rohrer, 1980.

19__/19__

1	2	3	4	5	6	7	8	9	10	11	12	13	14	15
Month	Elec.	Natural Gas	#__Oil	#__Oil	Coal	Steam __ PSIG	Steam __ PSIG	Water	Waste	Other	Total	Year-to-Date	Last Year	% Diff.
Total														
Last Year's Total														
% Diff														

FIGURE 14.7 Summary of all energy, quantity. *Source:* Rohrer, 1980.

required to determine the proper load distribution for the boilers in order to minimize fuel consumption for each system energy requirement. Figure 14.10 is a plot of the total energy consumption and cost on a monthly basis. Costs seem to follow consumption closely except for the last 6 months of 1976, where costs seem to rise faster than consumption compared to the previous 18 months.

Figure 14.11 shows monthly peak demand and the corresponding power factors. The peak monthly demands at the end of 1976 were eventually attributed to an electrical hookup permitted to be used by the contractor installing a new continuous washer. Large variation in the power

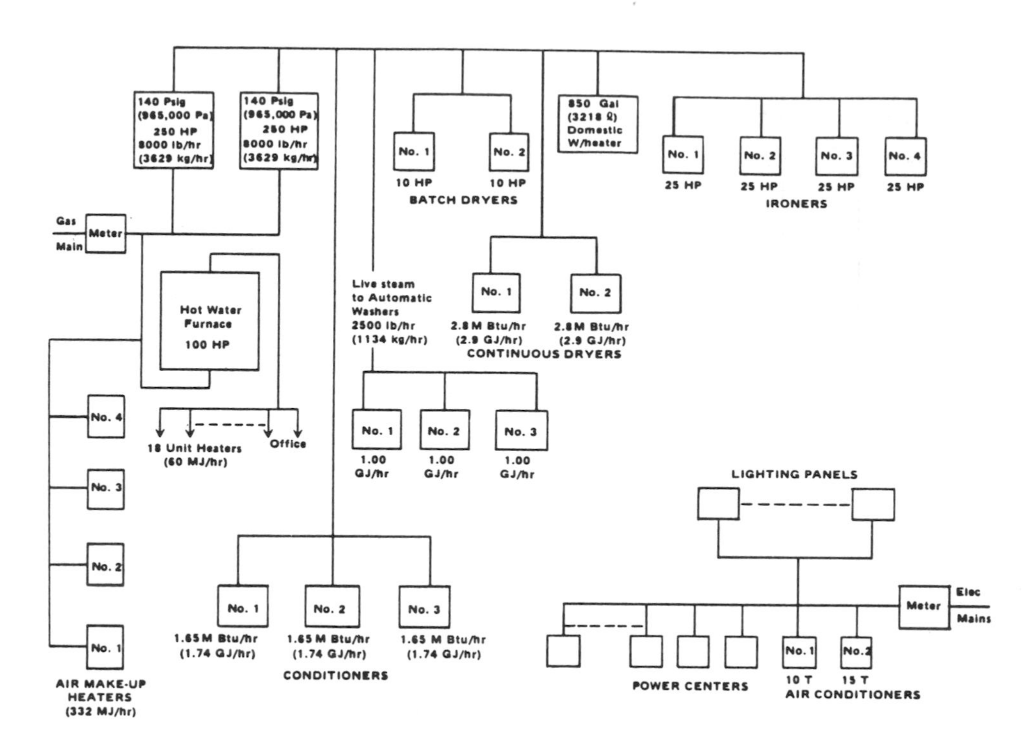

FIGURE 14.8 Laundry process diagram.

TABLE 14.4 Annual Energy and Cost Data for Laundry

Total annual gas consumption	40.8 TJ
Total annual electric consumption	5.0 TJ
Gross annual energy consumption	45.8 TJ
Production	4,445,200 kg
Degree-days (°C), heating	3,786
Annual energy cost (1993$)	$301,570
Average unit energy cost—all sources	$6.59
Per unit energy consumption	10,311 kJ/kg

factor multiplier usually signals a defect in a system, which should be corrected immediately to end an unnecessary expense. In fact, any plant load with power factor less than 0.95 should be compensated for. In most parts of the country the cost of capacitors is less than the expense of paying increased demand charges. On the other hand, increases in demand are usually caused by increased connected capacity and may not be susceptible to control.

Demand charges can be as much as half of the total electrical utility bill. Ways to reduce them may be suggested by a study of daily demand variations. The utility will usually provide continuous records of demand over periods as long as a week at no cost to the consumer.

Per unit costs of gas and electricity are plotted in the graph of Figure 14.12.

It is instructive to view the energy use per unit of production activity seen in Figure 14.13. The laundry data are plotted as energy consumed per kilogram of laundry processed. This index goes up in the winter due to the heating load but otherwise is quite constant, which indicates a constant production rate. One more often sees a variation of the sort shown for a commercial heat treatment plant in Figure 14.14. Energy per unit of product treated is shown to be inverse to the quantity processed. This points out the advantages in energy efficiency of keeping your plant at full capacity. In the case of this job shop and in most other businesses, production is scheduled to fit product

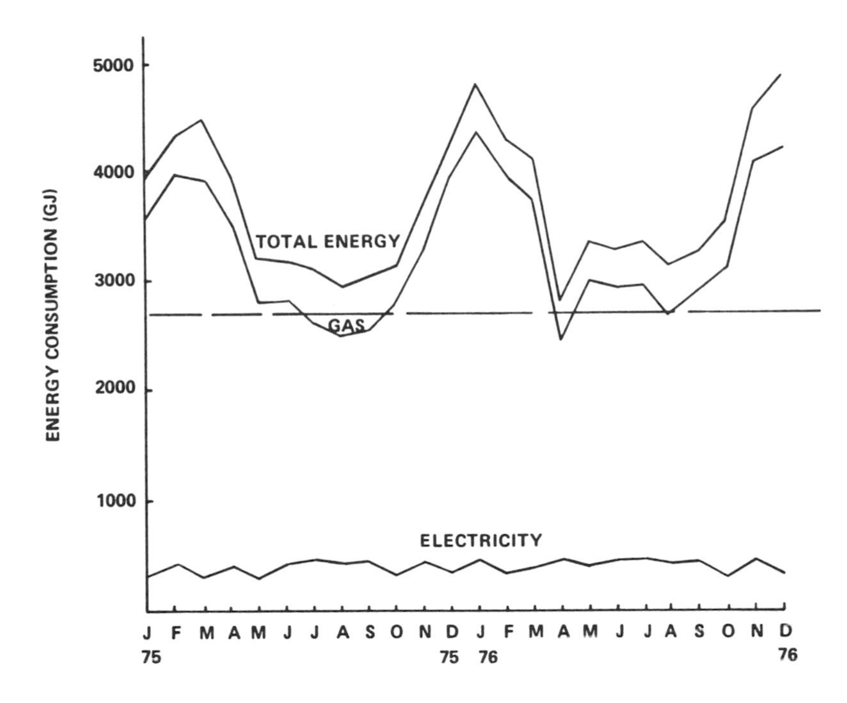

FIGURE 14.9 Monthly energy consumption for laundry.

demand. Businesses that have seasonal demand, such as breweries, can study the implications of leveling production and stockpiling to meet peak demands. Some considerations that control the economics are startup and shutdown losses, electrical demand costs, incremental costs for refrigerated storage, overtime costs for labor, and a host of others.

Creating Unit Energy Balances

Up to this point, the discussion has been concentrated mainly on the gross consumption audit and the information about the plant that can be derived from it. Although the benefits derived from it can be large, the gross consumption audit is really just the beginning of the whole program. The biggest and the most difficult portion is still ahead. At this point an understanding of the grand design must be developed, both of the plant and its processes that lead from the receipt of raw materials to the shipping of the final products. The more thoroughly the plant operation is studied, the greater the amount of knowledge that can be derived about smaller and still smaller parts of the overall scheme. The reason for carrying out the energy audit is to produce an understanding that is as complete as possible of the ways that energy is involved in the activity of the plant, particularly as it is essential to the plant's primary purposes.

The laundry referred to in the previous section will be used as an example of how one proceeds with distributing the gross energy consumption over unit processes within the plant. Table 14.5 is an equipment list with nameplate ratings, excluding the process boilers and hot-water furnace.

From Figure 14.9 the annual heating load had been previously estimated to be 7.227 TJ. Subtracting this from the total natural gas consumption we get (40.776 − 7.227) TJ or 33.549 TJ as the estimated process load. Note that this is also the boiler input, since no other direct-fired gas equipment is used.

Because of the energy losses from liquid and solid surfaces, very little if any space heating is required during production hours. Therefore almost all heating during that time is due to the heating of make-up air. However, the ventilation is reduced to a low value during the remaining 14 hours of the day, so that during off-production hours the largest energy loads are the unit space heaters and the office heating system.

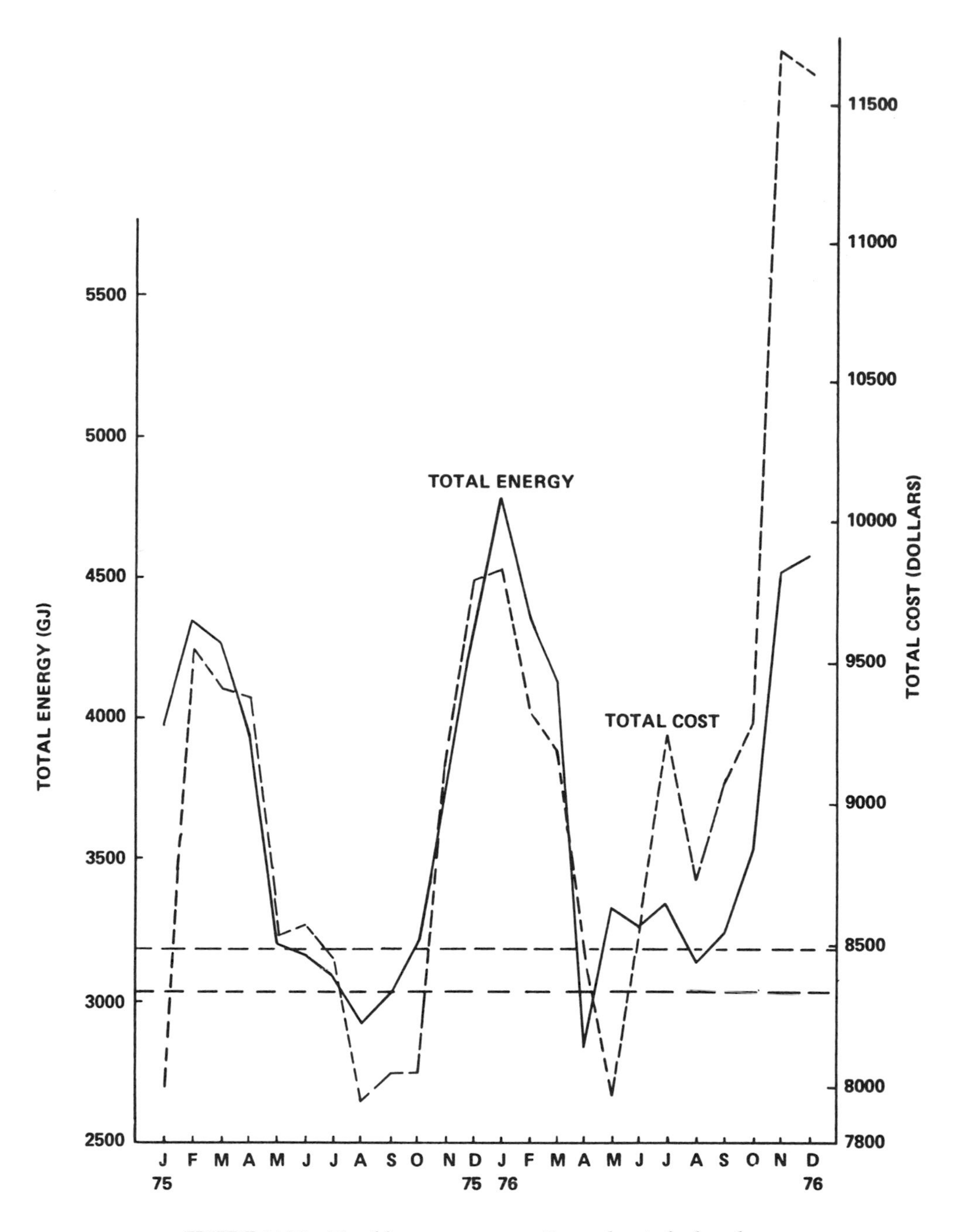

FIGURE 14.10 Monthly energy consumption and costs for laundry.

An air-flow survey for the plant gave the following information:

Ventilation	905.6 m³/min
Continuous dryer exhaust	277.3
Batch dryer exhaust (average)	244.8
Boiler	148.6
Furnace	29.7
Total outdoor air	1,606 m³/min

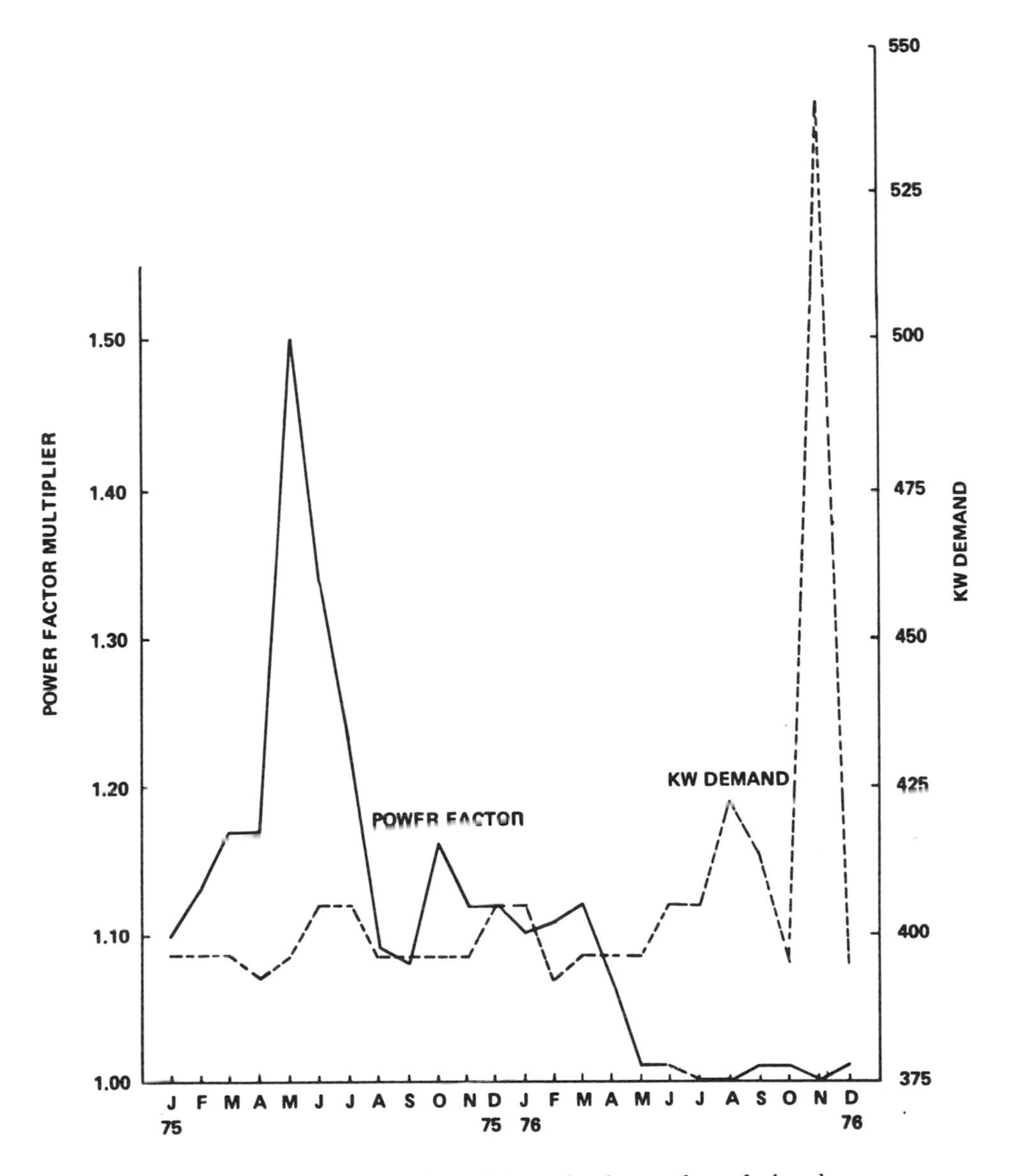

FIGURE 14.11 Monthly electrical demand and power factor for laundry.

The average outdoor temperature during the heating season is –3°C. Therefore, the approximate output of the make-up heaters in a heating season, Q_v is

$$Q_v = 1{,}606 \ \text{m}^3/\text{min} \times 60 \times 10 \ \frac{\text{min}}{\text{day}} \times 1.2 \ \frac{\text{kg air}}{\text{m}^3}$$

$$\times 0.519 \ \frac{\text{kJ}}{\text{kg °C}} \times 3{,}786 \ \text{DD} \Big/ \left(10^6 \ \frac{\text{kJ}}{\text{GJ}}\right) = 2{,}273 \ \text{GJ}$$

The annual input to the heaters, Q_h, assuming a 60% efficiency, is

$$Q_h = 2.273/0.60 = 3.788 \ \text{TJ}$$

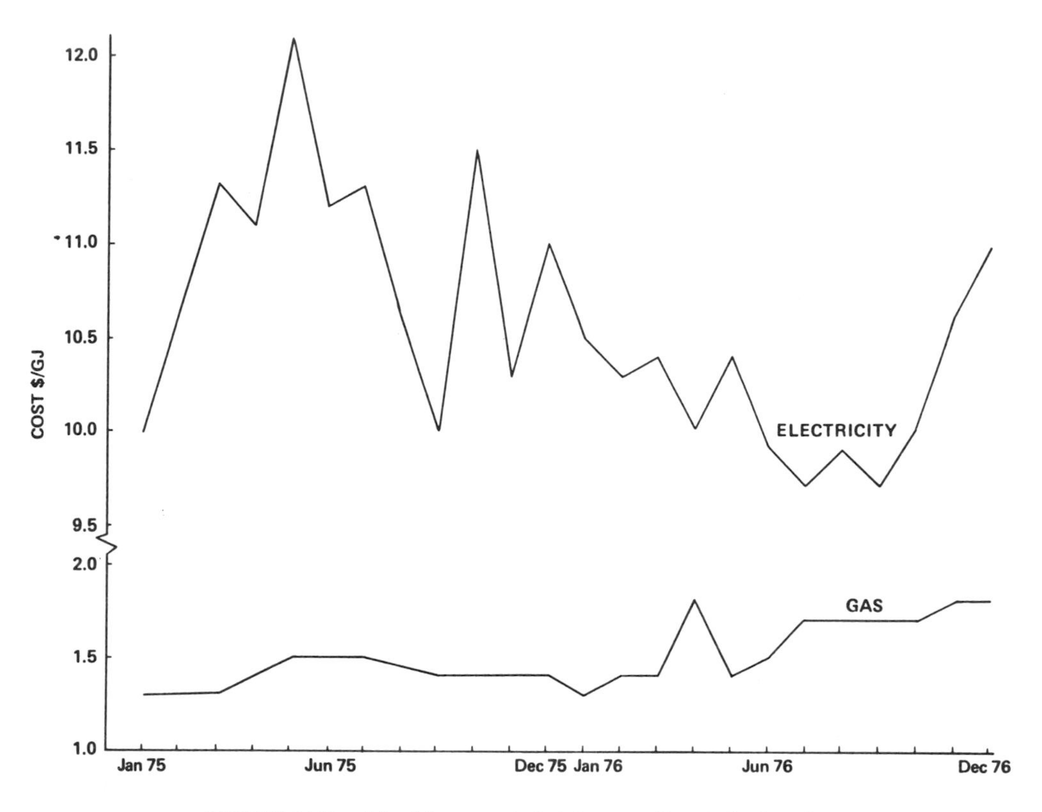

FIGURE 14.12 Monthly gas and electric per unit costs for laundry.

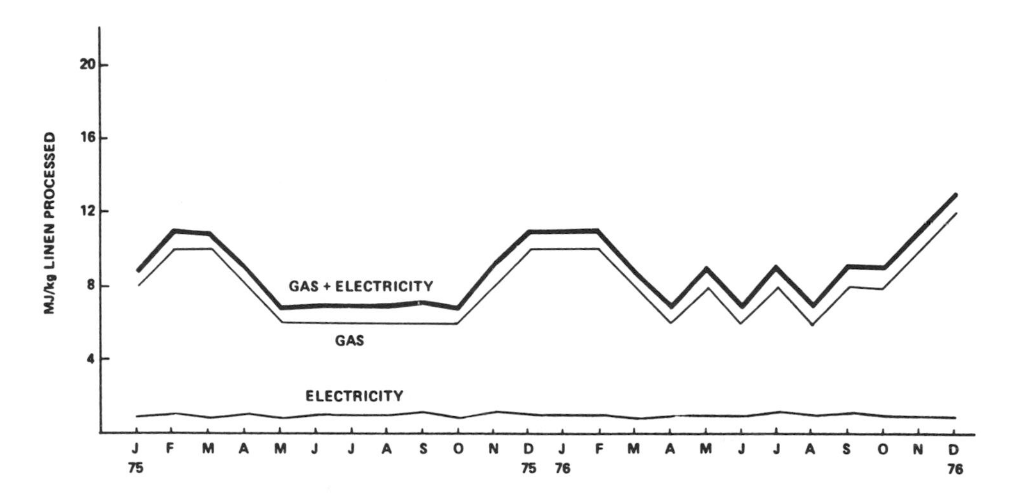

FIGURE 14.13 Per unit energy consumption for a laundry.

Since the total heating load was 7.227 TJ, the annual furnace input is 7.227 – 3.788 = 3.439 TJ. That is, half the heating load is make-up air heating and the other half is due to heat leaks. The total process heating load

$$q_p = (40.776 - 7.227) = 33.549 \ \text{TJ/year}$$

For 80% boiler efficiency, the heat supplied to the steam, q_{st}, is $q_{st} = 0.80 \times 33.349 = 26.839$ TJ.

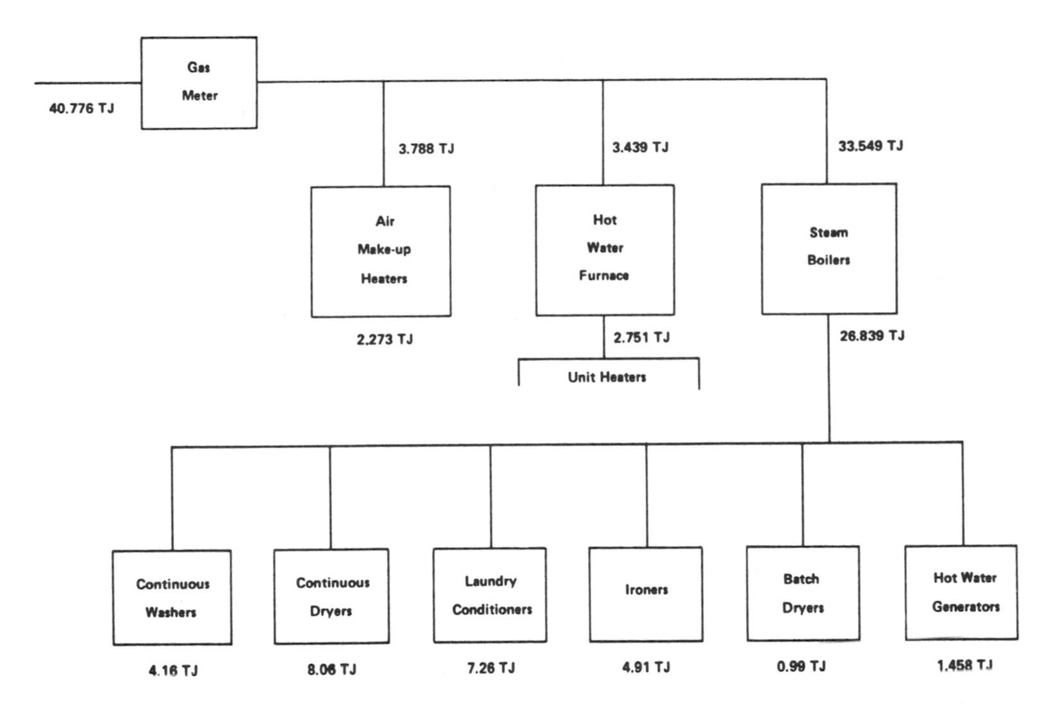

FIGURE 14.14 Energy distribution over thermal process units in laundry.

TABLE 14.5 Laundry Equipment List

Equipment	Number	Energy Source	Input Rating	Output Rating
Air make-up heaters	4	Natural gas	332 MJ/hr	—
Batch dryers	2	Steam	10 BHP	—
Automatic washers	3	Steam	1.15×10^6 kJ/hr	—
Conditioners	3	Steam	1.75×10^6 kJ/hr	—
Continuous dryers	2	Steam	3.0×10^6 kJ/hr	—
Flat ironers	4	Steam	25 BHP	—
Unit heaters	18	Steam	60 MJ/hr	—
Domestic hot water heater	1	Steam	527,500 kJ/hr	—
50-lb batch-washer motors	1	Electric	—	1 hp
100-lb batch-washer motors	2	Electric	—	2.5 hp
Batch-dryer motors	2	Electric	—	2.5 hp
Automatic-washer motors	3	Electric	—	2.5 hp
Conditioner fans	3	Electric	—	1 hp
Continuous dryer fans	2	Electric	—	1.5 hp
Ironer motors	4	Electric	—	¼ hp
Return pump	1	Electric	—	1.5 hp
Make-up air fans	4	Electric	—	3 hp
Ventilating fans	8	Electric	—	¾ hp
Air conditioner	1	Electric	—	10 ton
Air conditioner	1	Electric	—	15 ton

The only batch processes are the flat ironers and the batch washers and dryers. The ironers are never valved off, and when not in use they heat the plant. The batch washers average five batches of 488 kg of linen each day using approximately 8.345 L of hot water for each kilogram or 20,363 L per day. Each of the 70 employees uses an average of 37.85 L per day for another 2650 L or a total of 23,013 L per day. The heat used to heat this water, q_{hw}, is

$$q_{hw} = 23{,}013\ \text{L} \times 1\ \text{kg/L} \times 4.186\ \text{kJ/kg}$$
$$-{}^\circ\text{C} \times (71.1 - 12.8)^\circ\text{C} / (10^6\ \text{kJ/GJ}) = 5.62\ \text{GJ}$$

or 0.0056 × 52 × 5 = 1.458 TJ per year. That leaves (26.839 – 1.458) = 25.381 TJ per year. This is distributed among all other steam-heated units in direct proportion to the nameplate ratings. This allows the preparation of a new schematic diagram of the heating processes shown in Figure 14.14. The same procedure is used to distribute the electrical load but will not be carried out in this example. The crude estimating done here to distribute the total energy load over the unit processes is necessary so that approximate heat balances for the large-unit energy users can be derived. The heat balances will then aid in the design of economic waste-heat recovery systems or other measures that will increase the efficiency of energy utilization. The crude estimating is unnecessary when all large process units are submetered.

In-Plant Metering

Submetering reduces the work and time required for an energy audit; indeed, it does much more than that. Because meters are tools for assessing production control and for measuring equipment efficiency, they can contribute directly to energy conservation and cost containment. Furthermore, submetering offers the most effective way of evaluating the worth of an energy conservation measure. Too many managers accept a vendor's estimate of fuel savings after buying a recuperator. They may scan the fuel bills for a month or two after the purchase to get an indication of savings—usually in vain—and then relax and accept the promised benefit without ever having any real indication that it exists. It may well be that, in fact, it does not yet exist. The equipment may not be adjusted correctly or it may be operated incorrectly, and there is no way of knowing without directly metering the fuel input. It is estimated that at least 2.5% waste is recoverable by in-plant metering.

Oil meters are just as effective as gas meters used in the same way and are even less expensive on an energy-flow basis. Electric meters are particularly helpful in monitoring the continued use of machines or lighting during shutdown periods and for evaluating the efficacy of lubricants and the machineability of feed stock. The use of in-plant metering can have its dark side too. The depressing part is the requirement for making periodic readings. It does not stop even there. Someone must analyze the readings so that something can be done about them. If full use is to be made of the information contained in meter readings, it must be incorporated into the energy information portion of the management information system. At the very least each subreading must be examined chronologically to detect malfunctions or losses of efficiency. Better still, a derived quantity such as average energy per unit of production should be examined.

In-plant metering need not be as accurate as that required for billing. It is often possible to buy out-of-tolerance billing meters from your utility for a fraction of the cost of a new meter. If these are installed by plant personnel during slack periods, the overall cost can be very modest indeed.

14.2 The Technology of Energy Conservation

Overview

Energy conservation can be accomplished in two general ways categorized as "belt-tightening" and "leak-plugging." Belt-tightening consists of those measures that represent some sacrifice or restraint on normal operations such as lowered living space temperatures, less ventilating air flow, lower lighting levels, the elimination of aesthetic architectural features that are energy intensive, the purchase of smaller and less comfortable company autos, and the substitution of telephone conferences for meetings requiring air travel. This category can be considered almost an emergency category, since it requires complete employee cooperation and can have negative effects on morale and organization harmony if resisted. On the other hand, the range of conditions for human comfort are quite wide (ASHRAE, 1989) and physiological and psychological adjustments can be made given enough time and proper communications. Furthermore, a great deal of waste is presently encountered in situations where excess lighting, ventilation, and temperature levels produce human discomfort that can easily be alleviated. However, the real impact of energy conservation efforts over the long range will come about by leak-plugging measures. These are measures that reduce the amount of heat loss to the surroundings, thus increasing the efficiency of utilization of industrial heat. Obvious examples include better building and piping insulation, repair of gas, oil, and steam leaks, use of waste heat and waste fuels, and improved control of combustion processes. These and other measures will be discussed in greater detail later in this section.

Improving Housekeeping

Excellent housekeeping is a measure of good management, yet it is relatively rare in industry. That is so because of the dedication to purpose and the attention to detail that is required to accomplish it. However, it is well worth striving for because, in terms of energy management, it pays off so well. Since the major part of industrial energy waste results from poor housekeeping, the biggest savings can be had by improving it. An equally attractive feature of housekeeping improvement is that it involves little or no capital expenditure. The combination of big savings at small cost should prove irresistible, but one must face up to the demands on management attention that are entailed.

The procedure requires that one person or a small party make a detailed inspection of the plant looking of energy leaks. When found, these are duly noted in an inspection form and orders for corrective action are issued as soon as possible following the inspection. Figure 14.3 is an example of a suitable inspection form.

The initial inspection is just the beginning of a perpetual routine that is used to evaluate the corrective actions generated by previous inspections and to find new energy leaks as soon as possible after they occur. The best person to carry out this part of the program is the plant energy coordinator. It would not be unreasonable for him or her to dedicate the beginning of every working day to a thorough plant inspection. The danger is in making it so routine that it becomes a purely perfunctory checkoff.

The EPIC Guide[5] provides a detailed list of energy leaks to look for. A sampling of these is

- Leaks in air, gas, oil, steam, or water piping
- Deteriorated insulation
- Broken or poor-fitting windows and doors
- Electric motors running when not in use
- Lights on in unused areas
- Overheated/overcooled, overventilated, and overlighted spaces
- Gas torches lighted when not in use
- Unused transformers, motor generators, or rectifiers excited
- Furnaces and boilers idling unnecessarily long
- Steam coils heated when not in use
- Doors and windows of conditioned spaces left open

Many of the corrective actions that are suggested face potential opposition from the work force. The opposition can be overcome with patient explanation, estimates of potential annual savings, and offers to change back if the corrective action disrupts effective plant operation.

A 1.59-mm hole in a gas line at 413,700 Pa will waste 2.3 TJ of energy worth approximately $6,400 at $2.80/GJ.

One 1.59-mm hole in a compressed air line operating at 689,000 Pa will waste 62,826 m^3 per year, which requires 6060 kWh costing $315 at an electric rate of $.052/kWh.

As much as 0.01 TJ per year costing over $28 can be wasted by 1 m^2 of missing insulation from a 20-cm steam main operating at 138,000 Pa.

Five 100 W fixtures more than required can waste more than 0.014 TJ per year and cost $208 in excess electrical billing when the rate is $0.052/kWh.

Combustion Control

The stoichiometric equation for the combustion of methane, the principal constituent of natural gas, with air is

$$CH_4 + 2\,O_2 + 7.52\,N_2 \rightarrow CO_2 + 2\,H_2\,O + 7.52\,N_2 \tag{14.1}$$

The stoichiometric equation is the one representing the exact amount of air necessary to oxidize the carbon and hydrogen in the fuel to carbon dioxide and water vapor. However, it is necessary to provide more than the stoichiometric amount of air since the mixing of fuel and air is imperfect in the real combustion chamber. Thus the combustion equation for hydrocarbon fuels becomes

$$\begin{aligned} &C_x\,H_y + \phi\left(x + y/4\right) O_2 + 3.76\,\phi\left(x + y/4\right) N_2 \\ &\rightarrow xCO_2 + y/2\ H_2O + (\phi - 1)(x + y/4)\ O_2 \\ &+3.76\ \phi(x + y/4)\ N_2 \end{aligned} \tag{14.2}$$

Note that for a given fuel nothing in the equation changes except the parameter ϕ, the equivalence ratio, as the fuel–air ratio changes.

As ϕ is increased beyond the optimal value for good combustion, the stack losses increase and the heat available for the process decreases. As the equivalence ratio increases for a given flue temperature and a given fuel, more fuel must be consumed to supply a given amount of heat to the process.

The control problem for the furnace or boiler is to provide the minimum amount of air for good combustion over a wide range of firing conditions and a wide range of ambient temperatures. The most common combustion controller uses the ratio of the pressure drops across orifices, nozzles, or Venturis in the air and fuel lines. Since these meters measure volume flow, a change in temperature of combustion air with respect to fuel, or vice-versa, will affect the equivalence ratio of the burner. Furthermore, since the pressure drops across the flow meters are exponentially related to the volume flow rates control dampers must have very complicated actuator motions. All the problems of ratio controllers are eliminated if the air is controlled from an oxygen meter. These are now coming into more general use as reasonably priced, high-temperature oxygen sensors become available. It is possible to control to any set value of percentage oxygen in the products; that is,

$$\%\ O_2 = \frac{\phi - 1}{\dfrac{x + y/2}{x + y/4} + 4.76\ \phi - 1} \tag{14.3}$$

Figure 14.15 is a nomograph from the Bailey Meter Company (Dukelow, 1974) that gives estimates of the annual dollar savings resulting from the reduction of excess air to 15% for gas-, oil-, or coal-fired boilers with stack temperatures from 300 to 700°F. The fuel savings are predicted on the basis that as excess air is reduced, the resulting reduction in mass flow of combustion gases results in reduced gas velocity and thus a longer gas residence time in the boiler. The increased residence time increases the heat transfer from the gases to the water. The combined effect of lower

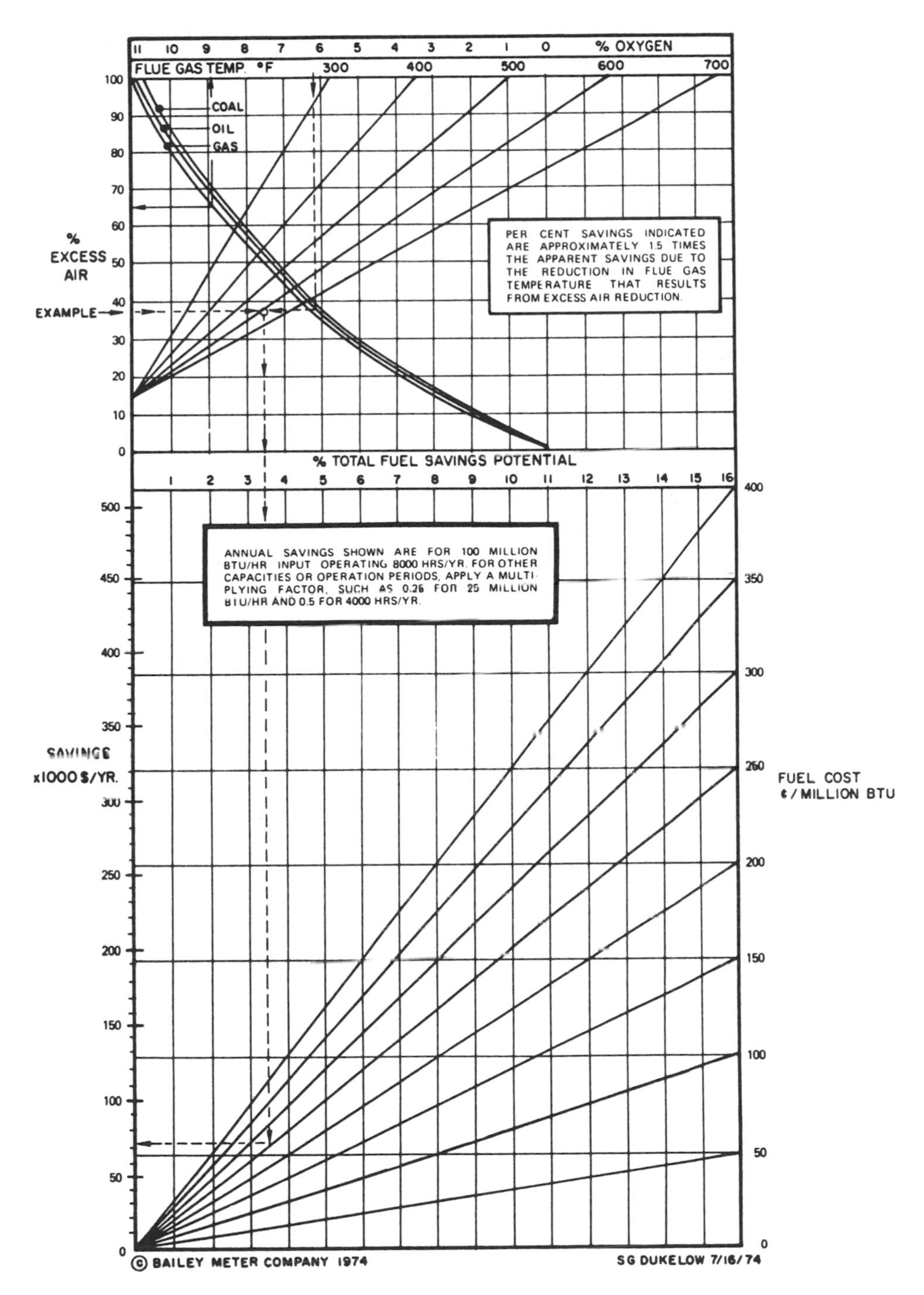

FIGURE 14.15 Nonograph for estimating savings from adjustment of burners. (From Dukelow, S.G., Bailey Meter Company, 29801 Euclid Avenue, Wickliffe, Ohio. With permission.)

exhaust gas flows and increased heat exchange effectiveness is estimated to be 1.5 times greater than that due to the reduced mass flow alone.

As an example assume the following data pertaining to an oil-fired boiler: Entering the graph at the top abscissa with 6.2% O_2, we drop to the oil fuel line and then horizontally to the 327°C (620°F) flue gas temperature line. Continuing to the left ordinate we can see that 6.2% O_2

corresponds to 37.5% excess air. Dropping vertically from the intersection of the flue gas temperature line and the excess air line we note a 3.4% total fuel savings. Fuel costs are

Burner capacity	63 GJ/hr (60 × 10^6 Btu/hr)
Annual operating hours	6,200
Fuel cost	$0.19/L ($0.72/gal)
Heating value fuel	42.36 MJ/L (152,000 Btu/gal)
Percent O_2 in exhaust gases	6.2%
Stack temperature	327°C (620°F)

$$10^9\ \mathrm{J/GJ}\left(10^6\ \mathrm{Btu/million\ Btu}\right) \times \frac{\$0.19/\mathrm{L}}{42{,}365{,}000\ \mathrm{J/L}} \left(\frac{\$0.72/\mathrm{gal}}{152{,}000\ \mathrm{Btu/gal}}\right)$$

$$= \$4.48/\mathrm{GJ}\ \left(\$4.74/\mathrm{million\ Btu}\right)$$

Continuing the vertical line to intersect the $2.60/million Btu and then moving to left ordinate shows a savings of $140,000 per year for 8,000 hr of operation, 100 × 10^6 Btu/hr input and $5.00/million Btu fuel cost. Adjusting that result for the assumed operating data:

$$\text{Annual savings} = \frac{6{,}200}{8{,}000} \times \frac{60\times 10^6}{100\times 10^6} \times \frac{4.74}{5.00} \times \$140{,}000$$

$$= \$61{,}715 \text{ per year}$$

This savings could be obtained by installing a modern oxygen controller, an investment with approximately a 1-year payoff, or from heightened operator attention with frequent flue gas testing and manual adjustments. Valuable sources of information concerning fuel conservation in boilers and furnaces are given by FEA (1977).

Waste-Heat Management

Waste heat as generally understood in industry is the energy rejected from any process to the atmospheric air or to a body of water. It may be transmitted by radiation, conduction, or convection, but often it is contained in gaseous or liquid streams emitted to the environment. Almost 50% of all fuel energy used in the United States is transferred as waste heat to the environment, causing thermal pollution as well as chemical pollution of one sort or another. It has been estimated that half of that total may be economically recoverable for useful heating functions.

What must be known about waste heat streams in order to decide whether they can become useful? Here is a list along with a parallel list of characteristics of the heat load that should be matched by the waste-heat supply.

Waste-Heat Supply
Quantity
Quality
Temporal availability of supply
Characteristics of fluid
Heat Load
Quantity required
Quality required
Temporal availability of load
Special fluid requirements

Let us examine the particular case of a plant producing ice-cream cones. All energy quantities are given in terms of 15.5°C reference temperature. Sources of waste heat include

- Products of combustion from 120 natural gas jets used to heat the molds in the carousel-type baking machines. The stack gases are collected under an insulated hood and released to the atmosphere through a short stack. Each of six machines emits 236.2 m^3/hr of stack gas at 160°C. Total source rate is 161,400 kJ/hr or 3,874 MJ/day for a three-shift day.
- Cooling water from the jackets, intercoolers, and aftercoolers of two air compressors used to supply air to the pneumatic actuators of the cone machines; 11.36 L/min of water at 48.9°C is available. This represents a source rate of 96 MJ/hr. The compressors run an average of 21 hr per production day. Thus this source rate is 2,015 MJ/day.
- The water chillers used to refrigerate the cone batter make available—at 130°F—264 MJ/hr of water heat. This source is available to heat water to 48.9°C using desuperheaters following the water chiller compressors. The source rate is 6,330 MJ/day.
- 226 m^3/min of ventilating air is discharged to the atmosphere at 21.2°C. This is a source of rate of less than 22.2 MJ/hr or 525 MJ/day.

Uses for waste heat include

- 681 L/hr of hot water at 82.2°C is needed for cleanup operations during 3 hours of every shift or during 9 of every 24 hours. Total daily heat load is 4,518 MJ.
- Heating degree-days total in excess of 3,333 annually. Thus any heat available at temperatures above 21.1°C can be utilized with the aid of runaround systems during the 5½-month heating season. Estimated heating load per year is 4,010 GJ.
- Total daily waste heat available—12.74 GJ/day
- Total annual waste heat available—3.19 TJ/year
- Total annual worth of waste heat (at $2.80/GJ for gas)—$8,900.00
- Total daily heat load—this varies from a maximum of 59.45 GJ/day at the height of the heating season to the hot-water load of 4.52 GJ/day in the summer months.

Although the amount of waste heat from the water chillers is 40% greater than the load needed for hot-water heating, the quality is insufficient to allow its full use, since the hot water must be heated to 82°C and the compressor discharge is at a temperature of 54°C.

However the chiller waste heat can be used to preheat the hot water. Assuming 13°C supply water and a 10° heat exchange temperature approach, the load that can be supplied by the chiller is

$$\frac{49-13}{82-13} \times 4.52 = 2.36 \text{ GJ/day}$$

Since the cone machines have an exhaust gas discharge of 3.87 GJ/day at 160°C, the remainder of the hot-water heating load of 2.17 GJ/day is available. Thus a total saving of 1,129 GJ/year in fuel is possible with a cost saving of $3,161 annually based on $2.80/GJ gas. The investment costs will involve the construction of a common exhaust heater for the cone machines, a desuperheater for each of the three water chiller compressors, a gas-to-liquid heat exchanger following the cone-machine exhaust heater, and possibly an induced draft fan following the heat exchanger, since the drop in exhaust gas temperature will decrease the natural draft due to hot-gas buoyancy.

It is necessary to almost match four of four characteristics. Not exactly, of course, but the closer the match the more economic the waste-heat recovery. The term quality used in the list really means thermodynamic availability of the waste heat. Unless the energy of the waste stream is sufficiently hot, it will be impossible to even transfer it to the heat load, since spontaneous heat transfer occurs only from higher to lower temperature.

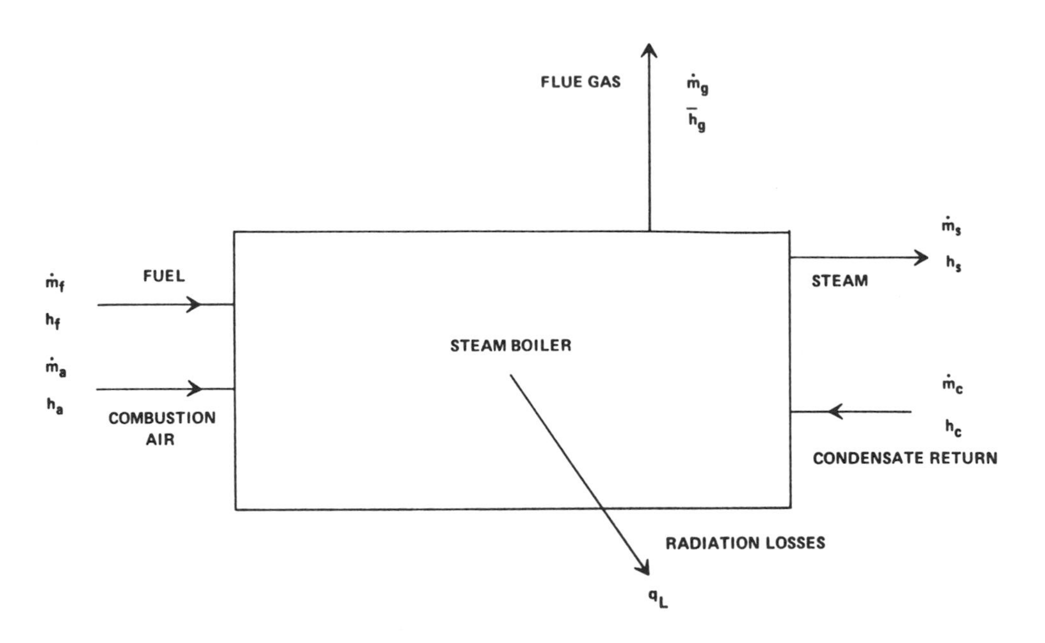

FIGURE 14.16 Heat balance on steam boiler.

The quantity and quality of energy available from a waste-heat source or for a heat load are studied with the aid of a heat balance. Figure 14.16 shows the heat balance for a steam boiler. The rates of enthalpy entering or leaving the system fluid streams must balance with the radiation loss rate from the boiler's external surfaces. Writing the First-Law equation for a steady-flow–steady-state process

$$\dot{q}_L = \dot{m}_f\ h_f + \dot{m}_a\ h_a + \dot{m}_c\ h_c - \dot{m}_s\ h_s - \dot{m}_g\ h_g \tag{14.4}$$

and referring to the heat-balance diagram, one sees that the enthalpy flux $\dot{m}_g$, h_g, leaving the boiler in the exhaust gas stream, is a possible source of waste heat. A fraction of that energy can be transferred in a heat exchanger to the combustion air, thus increasing the enthalpy flux $\dot{m}_g\ h_a$ and reducing the amount of fuel required. The fraction of fuel that can be saved is given in the equation

$$\frac{\dot{m}_f - \dot{m}_{f'}}{\dot{m}_f} = 1 - \left[\frac{K_1 - (1+\phi)\overline{C}_p\ T_g}{K_1 - (1+\phi')\overline{C}_p'\ T_g'}\right] \tag{14.5}$$

where the primed values are those obtained with waste-heat recovery. K_1 represents the specific enthalpy of the fuel-air mixture, $h_f + \phi\ h_a$, which is presumed to be the same with or without waste-heat recovery, ϕ is the molar ratio of air to fuel, and $\overline{C}_p$ is the specific heat averaged over the exhaust gas components. Figure 14.17, which is derived from Equation 14.5, gives possible fuel savings from using high-temperature flue gas to heat the combustion air in industrial furnaces.

It should be pointed out that the use of recovered waste heat to preheat combustion air, boiler feedwater, and product to be heat treated, confers special benefits not necessarily accruing when the heat is recovered to be used in another system. The preheating operation results in less fuel being consumed in the furnace, and the corresponding smaller air consumption means even smaller waste heat being ejected from the stacks.

Table 14.6 shows heat balances for a boiler with no flue gas-heat recovery, with a feedwater preheater (economizer) installed and with an air preheater, respectively.

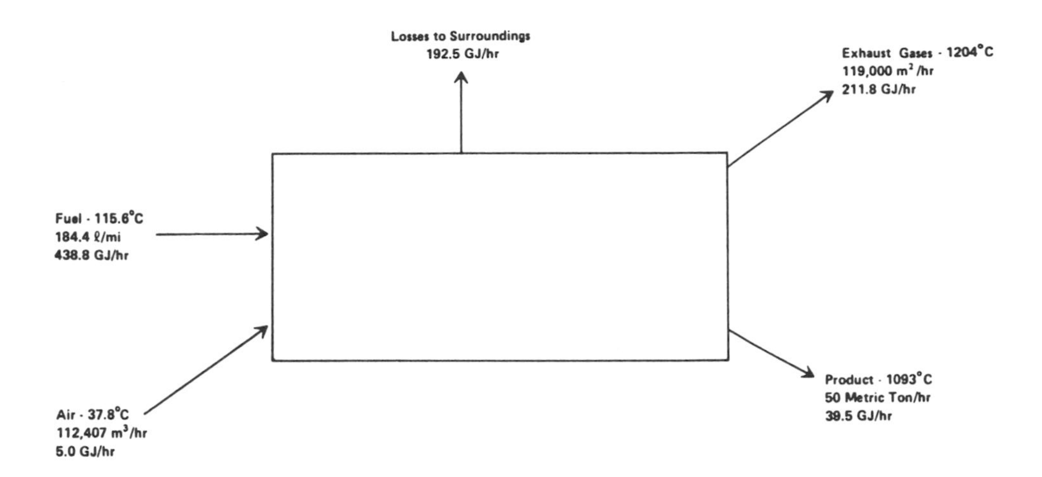

FIGURE 14.17 Heat balance for a simple continuous steel tube furnace.

It is seen that air preheater alone saves 6% of the fuel and the economizer saves 9.2%. Since the economizer is cheaper to install than the air preheater, the choice is easy to make for an industrial boiler. For a utility boiler, both units are invariably used in series in the exit gas stream.

Figure 14.18 is an example of a heat balance for a high-temperature industrial steel tube furnace. A comparison with the boiler diagrams reveals an extremely large difference in efficiencies. The tube furnace extracts only $39.47 \times 10^9/43.88 \times 10^{10}$ or 9% efficiency, while the simple boiler shows an efficiency of 84%.

Table 14.7 is an economic study using 1993 prices for fuel, labor, and equipment. At that time it was estimated that a radiation recuperator fitted to a fiberglass furnace would cost $410,000 and effect a savings of $267,170/year, making for a payoff period of approximately 1¼ years. This assumes of course that the original burners, combustion control system, and so on, could be used without modification. It is unlikely that this would be the case for such a high-temperature system. However, one could estimate with a good deal of confidence that the cost of additional equipment would not exceed the cost of the recuperator itself. This would mean a maximum payoff period of 28 months, which still offers a very interesting investment opportunity.

At this point, we can take some time to relate waste-heat recovery to the combustion process itself. We can first state categorically that the use of preheated combustion air improves combustion conditions and efficiency at all loads, and in a newly designed installation permits a reduction in size of the boiler or furnace. It is true that the increased mixture temperature that accompanies air preheat results in some narrowing of the mixture stability limits, but in all practical furnaces this is of small importance.

In many cases low-temperature burners may be used for preheated air, particularly if the air preheat temperature is not excessive and if the fuel is gaseous. However, the higher volume flow of preheated air in the air piping may cause large enough increases in pressure drop to require a larger combustion air fan. Of course larger diameter air piping can prevent the increased pressure drop, since air preheating results in reduced quantities of fuel and combustion air. For high preheat temperatures alloy piping, high-temperature insulation, and water-cooled burners may be required. Since many automatic combustion-control systems sense volume flows of air and/or fuel, the correct control settings will change when preheated air is used. Furthermore, if air preheat temperature varies with furnace load, then the control system must compensate for this variation with an auxiliary temperature-sensing control. On the other hand, if control is based on the oxygen content of the flue gases, the control complications arising from gas volume variation with temperature is obviated. This is the preferred control system for all furnaces, and only cost prevents

TABLE 14.6 Heat Balances for a Steam Generator

	Input Streams				Output Streams			
Case	Name	Temperature (°C)	Flow Rate	Energy	Name	Temperature (°C)	Flow Rate	Energy (GJ/hr)
Without economizer or air preheater	Natural gas	26.7	3,611 m³/hr	134.78 GJ/hr	Steam	185.6	45,349 kg/hr	126.23
	Air	26.7	35,574 m³/hr	560.32 MJ/hr	Flue gas	372.2	41,185 m³/hr	19.49
	Make-up water	10.0	7,076 kg/hr	297.25 MJ/hr	Surface losses	—	—	1.21
	Condensate return	82.2	41,277 kg/hr	14.21 GJ/hr	Blow down	185.6	2,994 kg/hr	2.36
With air preheater	Natural gas	26.7	3,395 m³/hr	126.72 GJ/hr	Steam	185.6	45,359 kg/hr	126.23
	Air	232.2	32,114 m³/hr	478.90 GJ/hr	Flue gas	260.0	35,509 m³/hr	11.42
	Make-up water	10	70,767 kg/hr	297.25 MJ/hr	Surface losses	—	—	1.21
	Condensate return	82.2	41,277 kg/hr	14.21 GJ/hr	Blow down	185.6	2,994 kg/hr	2.36
With economizer	Natural gas	26.7	3,278 m³/hr	122.37 GJ/hr	Steam	185.6	45,359 kg/hr	126.23
	Air	26.7	31,013 m³/hr	462.48 MJ/hr	Flue gas	176.7	34,281 m³/hr	7.08
	Make-up water	101	7,076 kg/hr	297.25 GJ/hr	Surface loss	—	—	1.21
	Condensate water	82.2	41,277 kg/hr	12.21 GJ/hr	Blow down	185.6	2,994 kg/hr	2.36

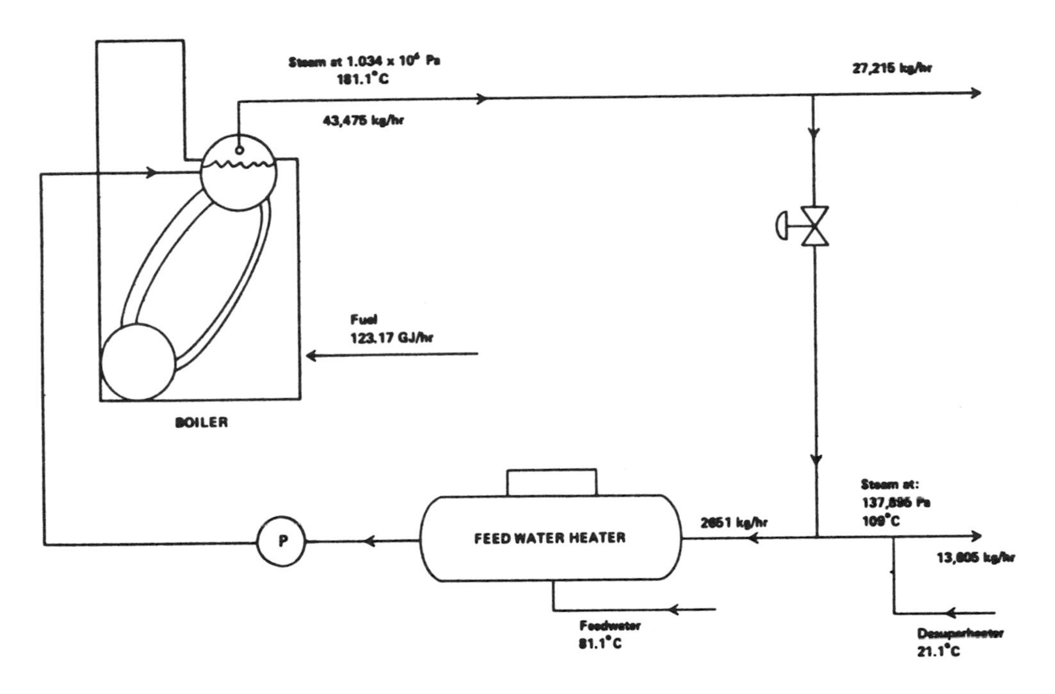

FIGURE 14.18 Steam plant schematic before adding electrical generation.

TABLE 14.7 Cost–Fuel Savings Analysis of a Fiberglass Furnace Recuperator (costs as per 1993)

Operation	Continuous
Fuel input	19.42 GJ/hr
Fuel	No. 3 fuel oil
Furnace temperature	1,482°C
Flue gas temperature entering recuperator	1,204°C
Air preheat (at burner)	552°C
Fuel Savings = 37.4%	
Q = 7.26 GJ/hr or 173.6 L/hr of oil	
Fuel cost savings estimation	
per GJ = 4.60	
per hour = $33.40	
per year (8,000 hr) = $267,170	
Cost of recuperator	$210,700
Cost of installation, related to recuperator	$199,400
Total cost of recuperator installation	$410,100
Approximate payback time 1½ years	

its wide use in small installations. Burner operation and maintenance for gas burners is not affected by preheating, but oil burners may experience accelerated fuel coking and resulting plugging from the additional heat being introduced into the liquid fuel from the preheated air. Careful burner design, which may call for water cooling or for shielding the fuel tip from furnace radiation, will always solve the problems. Coal-fired furnaces may use preheated air up to temperatures that endanger the grates or burners. Again, any problems can be solved by special designs involving water cooling if higher air temperatures can be obtained and/or desired.

The economics of waste-heat recovery today range from poor to excellent depending upon the technical aspects of the application as detailed earlier, but the general statement can be made that, at least for most small industrial boilers and furnaces, standard designs and/or off-the-shelf heat exchangers prove to be the most economic. For large systems one can often afford to pay for special designs, construction, and installations. Furthermore, the applications are often technically constrained by material properties and space limitations, and as shall be seen later, always by economic considerations. For further information on heat exchangers see Chapter 13.

Heating, Ventilating, and Air Conditioning

Heating, ventilating, and air conditioning (HVAC), while not usually important in the energy intensive industries, may be responsible for the major share of energy consumption in the light manufacturing field, particularly in high-technology companies and those engaged primarily in assembly.

Because of air pollution from industrial processes, many HVAC systems require 100% outside ventilating air. Furthermore, ventilating air requirements are often much in excess of those in residential and commercial practice (Hayashi et al., 1985). An approximate method for calculating the total heat required for ventilating air in Btu per heating season is given by

$$E_v\ (\text{kJ}) = 60 \times 24 \left(\frac{\text{min}}{\text{day}}\right) \times (1.2 \times 0.519) \left(\frac{\text{kJ}}{\text{m}^3 - \text{K}^\circ}\right) \times \text{SCMM} \times \text{DD} \tag{14.6}$$

$$= 896.8 \times \text{SCMM} \times \text{DD}$$

where SCMM = standard cubic meter per minute of total air entering plant including unwanted infiltration; DD = heating degree-days (C).

This underestimates the energy requirement, because degree-days are based on 18.33°C reference temperature and indoor temperatures are ordinarily held 1.6 to 3.9° higher. For a location with 3,333 degree-days each year the heating energy given by Eq. 14.13 is about 17% low.

Savings can be effected by reducing the ventilating air rate to the actual rate necessary for health and safety and by ducting outside air into direct-fired heating equipment such as furnaces, boilers, ovens, and dryers. Air infiltration should be prevented through a program of building maintenance to replace broken windows, doors, roofs, and siding, and by campaigns to prevent unnecessary opening of windows and doors.

Additional roof insulation is often economic, particularly because thermal stratification makes roof temperatures much higher than average wall temperatures. Properly installed vertical air circulators can prevent the vertical stratification and save heat loss through the roof. Windows can be double glazed, storm windows can be installed, or windows can be covered with insulation. Although the benefits of natural lighting are eliminated by this measure, it can be very effective in reducing infiltration and heat transfer losses.

Waste heat from ventilating air itself, from boiler and furnace exhaust stacks, and from air-conditioning refrigeration compressors can be recovered and used to preheat make-up air. Consideration should also be given to providing spot ventilation in hazardous locations instead of increasing general ventilation air requirements.

As an example of the savings possible in ventilation air control, a plant requiring 424.5 CMM outside air flow is selected. A gas-fired boiler with an energy input of 0.0165 GJ is used for heating and is supplied with room air.

$$\text{Combustion air} = \frac{16.5 \times 10^6}{37{,}281} \times \frac{12}{60}$$

$$= 88.52\ \text{CMM}$$

for a fuel with 37,281 kJ/m^3 heating value and an air fuel ratio of 12 m^3 air per m^3 fuel. The number of annual degree-days was 3,175.

A study showed that the actual air supplied through infiltration and air handlers was 809 m^3/min. An outside air duct was installed to supply combustion air for the boiler and the actual ventilating air supply was reduced to the required 424 m^3/min. The fuel saving that resulted using Eq. was

$$896.8\ (809-424)\ 3{,}175=1{,}096.2\ GJ$$

worth $3,070 in fuel at $2.80/GJ for natural gas.

Modifications of Unit Processes

The particular process used for the production of any item affects not only the cost of production but also the quality of the product. Since the quality of the product is critical in customer acceptance and therefore in sales, the unit process itself cannot be considered a prime target for the energy conservation program. That does not say that one should ignore the serendipitous discovery of a better and cheaper way of producing something. Indeed, one should take instant advantage of such a situation, but that clearly is the kind of decision that management could make without considering energy conservation at all.

Optimizing Process Scheduling

Industrial thermal processing equipment tends to be quite massive compared to the product treated. Therefore, the heat required to bring the equipment to steady-state production conditions may be large enough to make start-ups fuel intensive. This calls for scheduling this equipment so that it is in use for as long periods as can be practically scheduled. It also may call for idling the equipment (feeding fuel to keep the temperature close to production temperature) when it is temporarily out of use. The fuel rate for idling may be between 10 and 40% of the full production rate for direct-fired equipment. Furthermore, the stack losses tend to increase as a percentage of fuel energy released. It is clear that overfrequent start-ups and long idling times are wasteful of energy and add to production costs. The hazards of eliminating some of that waste through rescheduling must not be taken lightly. For instance, a holdup in an intermediate heating process can slow up all subsequent operations and make for inefficiency down the line. The breakdown of a unit that has a very large production backlog is much more serious than that of one having a smaller backlog. Scheduling processes in a complex product line is a very difficult exercise and perhaps better suited to a computerized PERT program than to an energy conservation engineer. That does not mean that the possibilities for saving through better process scheduling should be ignored. It is only a warning to move slowly and take pains to find the difficulties that can arise thereby.

A manufacturer of precision instruments changed the specifications for the finishes of over half of his products, thereby eliminating the baking required for the enamel that had been used. He also rescheduled the baking process for the remaining products so that the oven was lighted only twice a week instead of every production day. A study is now proceeding to determine if electric infrared baking will not be more economic than using the gas-fired oven.

Cogeneration of Process Steam and Electricity

In-plant (or on-site) electrical energy cogeneration is nothing new. It has been used in industries with large process steam loads for many years, both in the U.S. and Europe. It consists of producing steam at a higher pressure than required for process use, expanding the high-pressure steam through a back-pressure turbine to generate electrical energy, and then using the exhaust steam as process steam. Alternatively, the power system may be a diesel engine that drives an electrical generator. The diesel engine exhaust is then stripped of its heat content as it flows through a waste-heat boiler where steam is generated for plant processes. A third possibility is operation of a gas

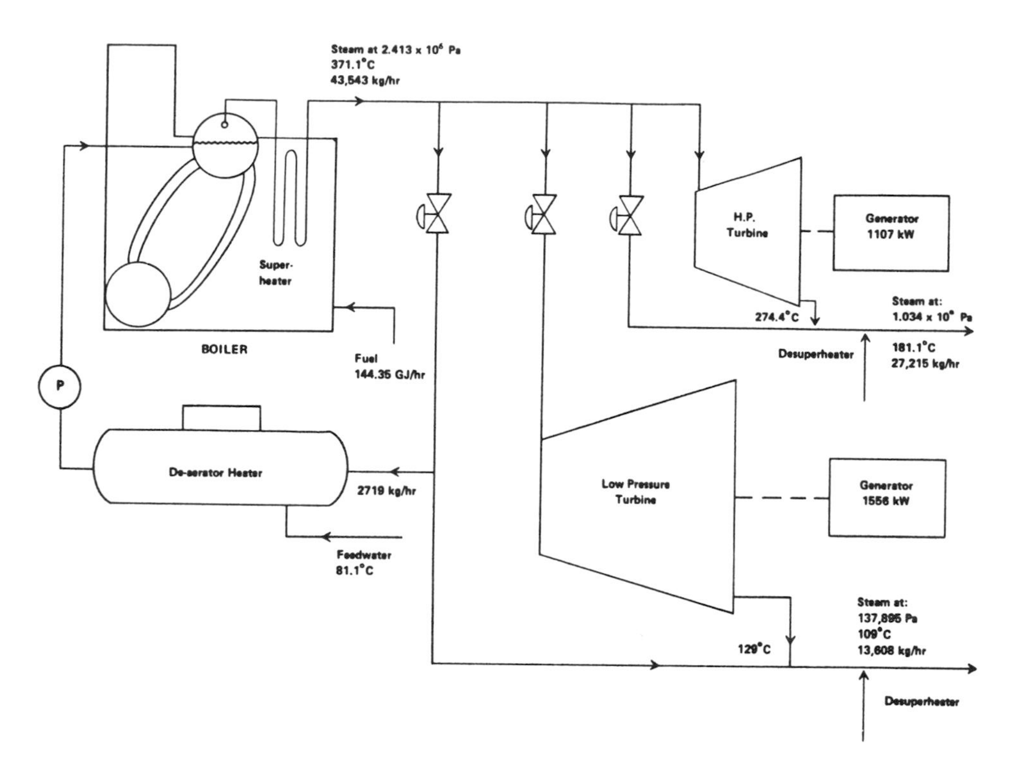

FIGURE 14.19 Steam plant schematic after installing electrical generation.

turbine generator to supply electric power and hot exhaust gases, which produce process steam in a waste-heat boiler. As will be seen later, the ratio of electric power to steam-heat rate varies markedly from one of these types of systems to the next. In medium to large industrial plants the cogeneration of electric power and process steam is economically feasible provided certain plant energy characteristics are present. In small plants or in larger plants with small process steam loads, cogeneration is not economic because of the large capital expenditure involved. Under few circumstances is the in-plant generation of electric power economic without a large process steam requirement. A small industrial electric plant cannot compete with an electric utility unless the generation required in-plant exceeds the capacity of the utility. In remote areas where no electric utility exists, or where its reliability is inferior to that of the on-site plant, the exception can be made.

Cogeneration if applied correctly is not only cost effective, it is fuel conserving. That is, the fuel for the on-site plant is less than that used jointly by the utility to supply the plant's electric energy and that used on-site to supply process steam. Figures 14.18 and 14.19 illustrate the reasons for and the magnitude of the savings possible. However, several conditions must be met in order that an effective application be possible. First, the ratio of process steam heat rate to electric power must fall close to these given in the table below:

Heat engine type	E_{steam}/E_{elect}
Steam turbine	2.3
Gas turbine	4.0
Diesel engine	1.5

The table is based upon overall electric plant efficiencies to 30, 20, and 40% respectively, for steam turbine, gas turbine, and diesel engine. Second, it is required that the availability of the steam load coincide closely with the availability of the electric load. If these temporal availabilities are out of phase, heat-storage systems will be necessary and the economy of the system altered. Third, it is necessary to have local electric utility support. Unless backup service is available from your utility, the cost of building in redundancy is too great. This may be the crucial factor in the majority of cases. The subject of cogeneration and the available technology is covered in Chapter 17.

Data Acquisition and Control Systems for Energy Management

Data acquisition is essential in energy conservation for at least three reasons: (1) base-line operational data is an absolute requirement for understanding the size and timing of energy demands for each plant, division, system, and component, and this information is basic for the design of the energy conservation strategy; (2) continuing data acquisition during the course of the energy conservation effort is necessary to calculate the gains made in energy use efficiency and to measure the success of the program; and (3) effective automatic control depends upon the accurate measurement of the controlled variables and the system operational data. More information on control systems can be found in Chapter 11.

With small, manually controlled systems, data acquisition is possible using indicating instruments and manual recording of data. With large systems, or almost any size automatically controlled system, manual data collection is impractical. The easy availability and the modest price of personal digital computers (PCs) and their present growth in speed and power as their price declines makes them the preferred choice for data acquisition equipment. The only additions needed to the basic PC are the video monitor; a mass data storage device (MSD), analog to digital (A/D) interface cards; software for controlling the data sampling, data storage, and data presentation; and a suitable enclosure for protecting the equipment from the industrial environment.

That same computer is also suitable for use as the master controller of an automatic control system. Both functions can be carried on simultaneously, with the same equipment, with a few additions necessary for the control part of the system. These additional electronic components include multiplexers to increase the capacity of the A/D cards; direct memory access (an addition to the A/D boards), which speeds operation by allowing the data transfer to bypass the CPU and programmable logic controllers (PLCs); and digital input/output cards (I/Os), both used for controlling the equipment controllers. The most critical parts of either the data acquisition or the control system are the software. The hardware can be ordered off the shelf, but the software must be either written from scratch or purchased and modified for each particular system in order to achieve the reliable, high performance operation that is desired. Although many proprietary program languages exist for the PCs and PLCs used for control purposes, BASIC is the most widely used. A schematic diagram of a typical data acquisition and control system is shown in Figure 14.20.

A major application has been in the control of mechanical and electrical systems in commercial and industrial buildings. These have been used to control lighting, electric demand, ventilating fans, thermostat setbacks, air-conditioning systems, and the like. These same computer systems can also be used in industrial buildings with or without modification for process control. Several manufacturers of the computer systems will not only engineer and install the system but will maintain and operate it from a remote location. Such operations are ordinarily regional. For large systems in large plants, one may be able to have the same service provided in-plant. However, there are also many small standalone analog control systems that can be used advantageously for the control of simple processes. Examples of these are a temperature controller using a thermocouple output to control the temperature of a liquid storage tank; a level controller, which keeps the liquid level in a tank constant by controlling a solenoid valve; and an oxygen trim system for a small boiler, which translates the measurement of the oxygen concentration in the exhaust stream into the jack shaft position, which regulates the combustion air flow to the burner.

After installing a demand limiter on an electric-arc foundry cupola, the manager was able to reduce the power level from 7,100 kW to 4,900 kW with negligible effect on the production time and no effect on product quality. The savings in demand charges alone were $4,400 per month with an additional savings in energy costs.

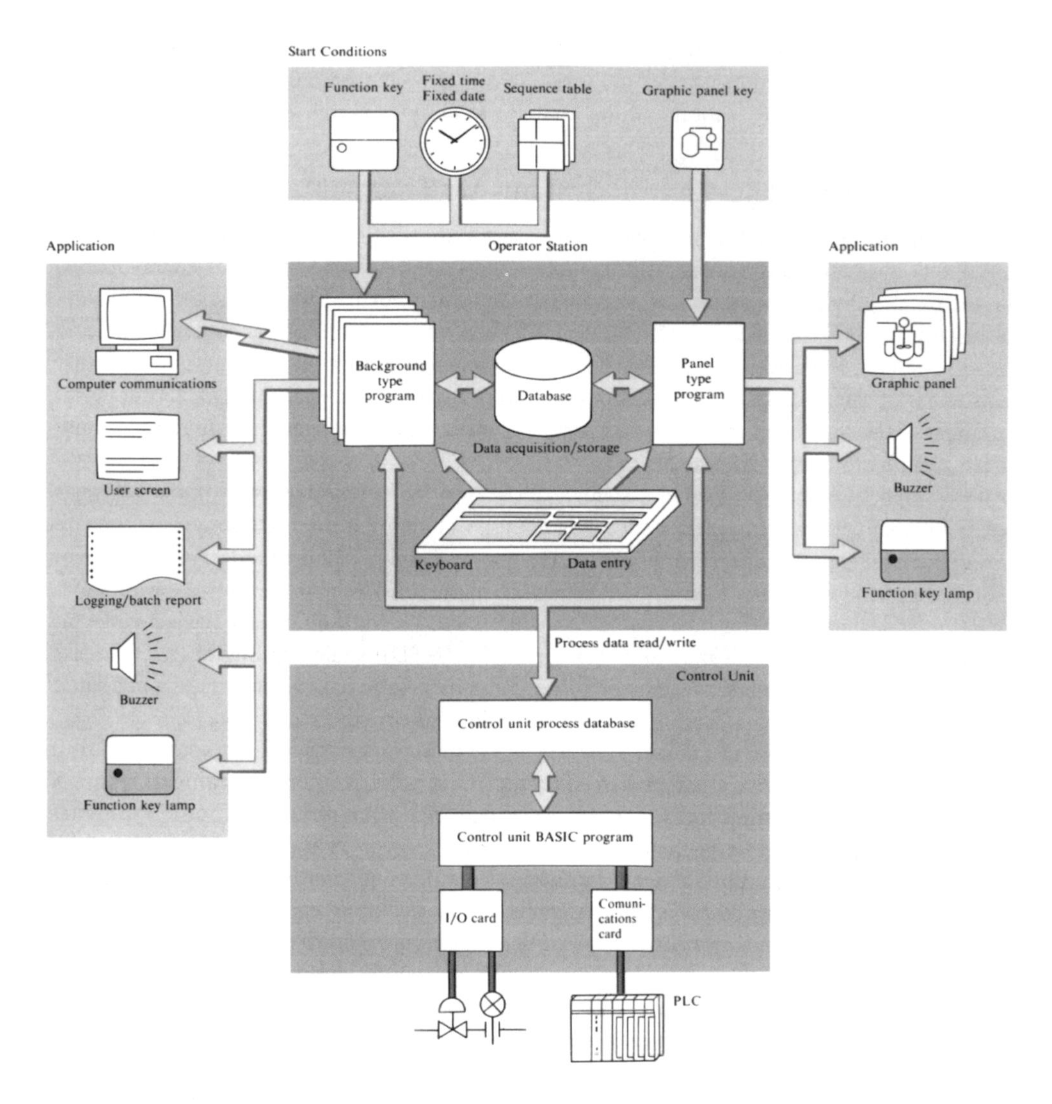

FIGURE 14.20 Typical data acquisition and control system.

Commercial Options in Waste-Heat Recovery Equipment

The equipment that is useful in recovering waste heat can be categorized as heat exchangers, heat storage systems, combination heat storage–heat exchanger systems, and heat pumps.

Heat exchangers certainly constitute the largest sales volume in this group. They consist of two enclosed flow paths and a separating surface that prevents mixing, supports any pressure difference between the fluids of the two fluids, and provides the means through which heat is transferred from the hotter to the cooler fluid. These are ordinarily operated at steady-state–steady-flow condition. The fluids may be gases, liquids, condensing vapors, or evaporating liquids, and occasionally fluidized solids. For more information, see Chapter 13.

Radiation recuperators are high-temperature combustion-air preheaters used for transferring heat from furnace exhaust gases to combustion air. As seen in Figure 14.21 they consist of two concentric cylinders, the inner one as a stack for the furnace and the concentric space between the inner and outer cylinders as the path for the combustion air, which ordinarily moves upward and therefore parallel to the flow of the exhaust gases. With special construction materials these can handle 1,355°C furnace gases and save as much as 30% of the fuel otherwise required. The main

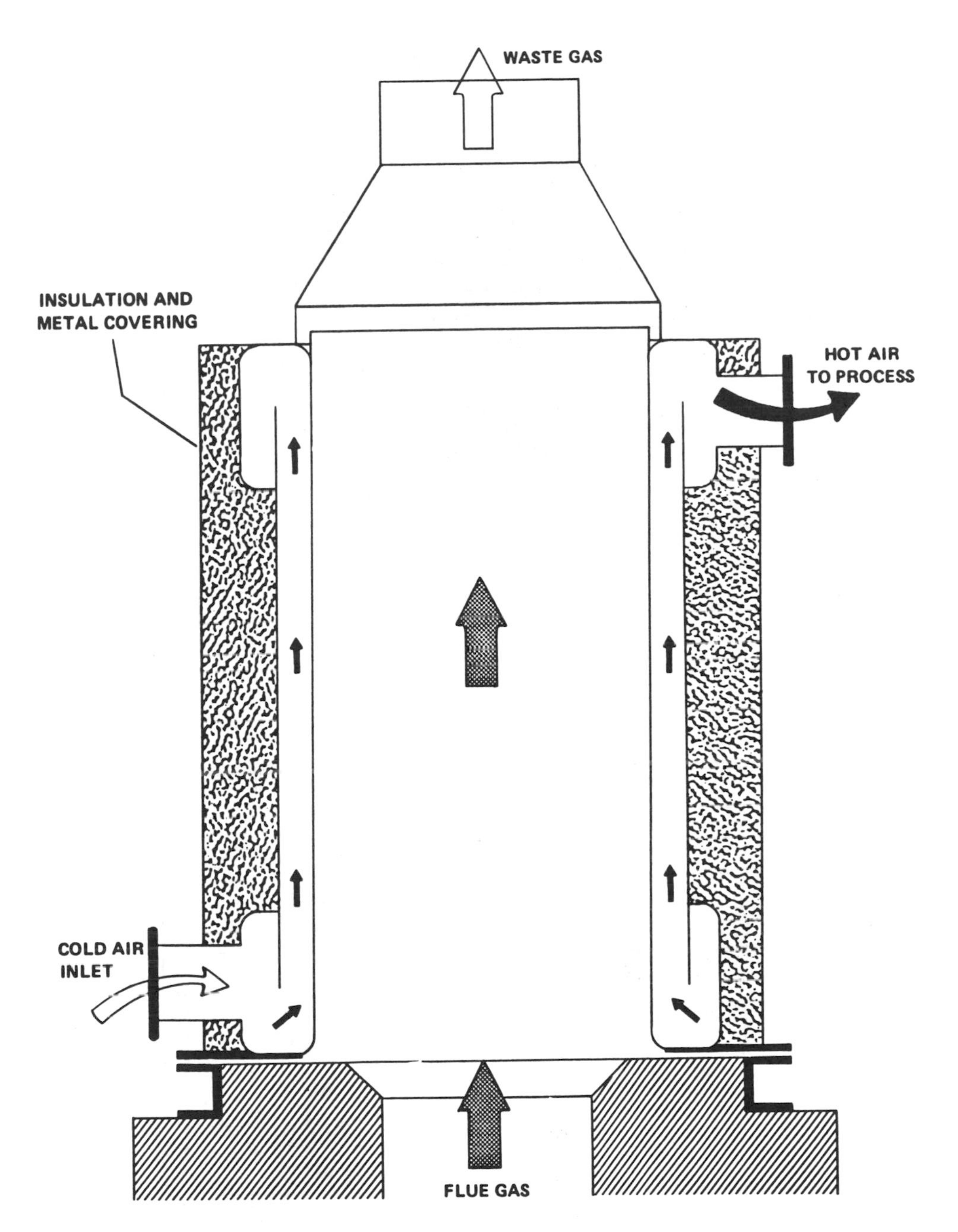

FIGURE 14.21 Metallic radiation recuperator.

problem in their use is damage due to overheating for reduced air flow or temperature excursions in the exhaust gas flow.

Convective air preheaters are corrugated metal or tubular devices that are used to preheat combustion air in the moderate temperature range (121°C to 649°C) for ovens, furnaces, boilers, and gas turbines, or to preheat ventilating air from sources as low in temperature as 21°C. Figures 14.22 and 14.23 illustrate typical construction. These are often available in modular design so that almost any capacity and any degree of effectiveness can be obtained by multiple arrangements. The biggest problem is keeping them clean.

Economizer is the name traditionally used to describe the gas-to-liquid heat exchanger used to preheat the feedwater in boilers from waste heat in the exhaust gas stream. These often take the

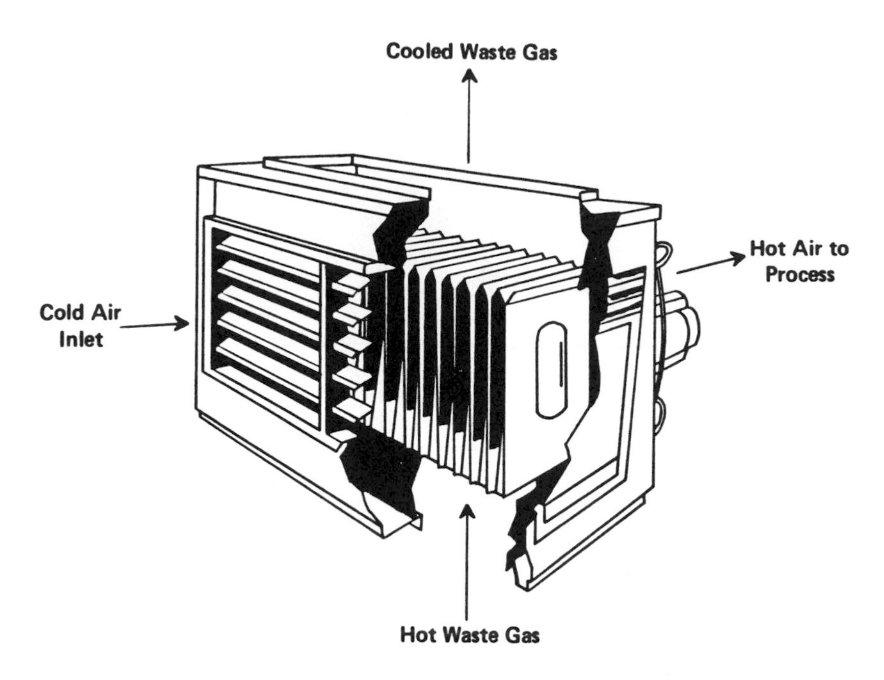

FIGURE 14.22 Air preheater.

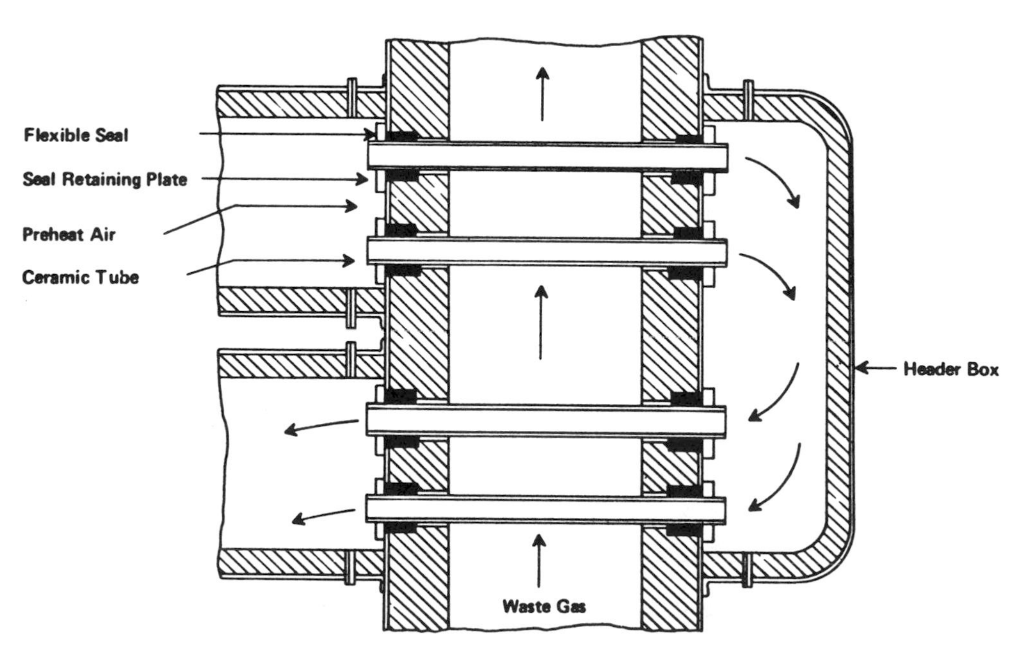

FIGURE 14.23 Ceramic tube recuperator.

form of loops, spirals, or parallel arrays of finned tubing through which the feedwater flows and over which the exhaust gases pass. They are available in modular form to be introduced into the exhaust stack or into the breeching. They can also be used in reverse to heat air or other gases with waste heat from liquid streams.

A more recent development is the use of condensing economizers, which are placed in the exhaust stream following high-temperature economizers. They are capable of extracting an additional 6 to 8% of the fuel input energy from the boiler exhaust gases. However, they are only used under certain restricted conditions. Obviously, the cooling fluid must be at a temperature below

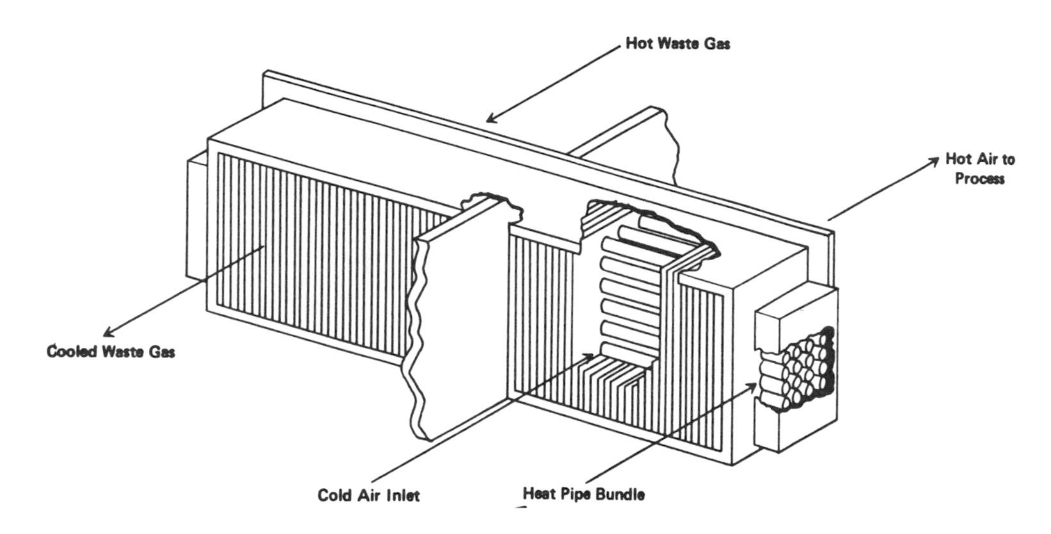

FIGURE 14.24 Heat pipe recuperator.

the dew point of the exhaust stream. This condition is often satisfied when boilers are operated with 100% make-up water. A second, less restrictive condition is that the flue gases be free of sulfur oxides. This is normally the case for natural gas fired boilers. Otherwise the economizer tubes will be attacked by sulfurous and/or sulfuric acid. Acid corrosion can be slowed down markedly by the use of all–stainless steel construction, but the cost of the equipment is increased significantly.

Heat-pipe arrays are often used for air-to-air heat exchangers because of their compact size. Heat-transfer rates per unit area are quite high. A disadvantage is that a given heat pipe (that is, a given internal working substance) has a limited temperature range for efficient operation. The heat pipe transfers heat from the hot end by evaporative heating and at the cold end by condensing the vapor. Figure 14.24 is a sketch of an air preheater using an array of heat pipes.

Waste-heat boilers are water-tube boilers, usually prefabricated in all but the largest sizes, used to produce saturated steam from high-temperature waste heat in gas streams. The boiler tubes are often finned to keep the dimensions of the boiler smaller. They are often used to strip waste heat from diesel-engine exhausts, gas-turbine exhausts, and pollution-control incinerators or afterburn ers. Figure 14.25 is a diagram of the internals of a typical waste-heat boiler. Figure 14.26 is a schematic diagram showing a waste heat boiler for which the evaporator is in the form of a finned-tube economizer. Forced water circulation is used giving some flexibility in placing the steam drum and allowing the use of smaller tubes. It also allows the orientation of the evaporator to be either vertical or horizontal. Other advantages of this design are the attainment of high boiler efficiencies, a more compact boiler, less cost to repair or retube, the ability to make superheated steam using the first one or more rows of downstream tubes as the superheater, and the elimination of thermal shock, since the evaporator is not directly connected to the steam drum.

Heat storage systems, or regenerators, once very popular for high-temperature applications, have been largely replaced by radiation recuperators because of the relative simplicity of the latter. Regenerators consist of twin flues filled with open ceramic checkerwork. The high-temperature exhaust of a furnace flowing through one leg of a swing valve to one of the flues heated the checkerwork while the combustion air for the furnace flowed through the second flue in order to preheat it. When the temperatures of the two masses of checkerwork were at proper levels, the swing valve was thrown and the procedure was continued, but with reversed flow in both flues. Regenerators are still used in some glass- and metal-melt furnaces, where they are able to operate in the temperature range 1,093 to 1,649°C. It should be noted that the original application of the regenerators was to achieve the high-melt temperatures required with low-heating-value fuel.

A number of ceramic materials in a range of sizes and geometric forms are available for incorporation into heat-storage units. These can be used to store waste heat in order to remedy

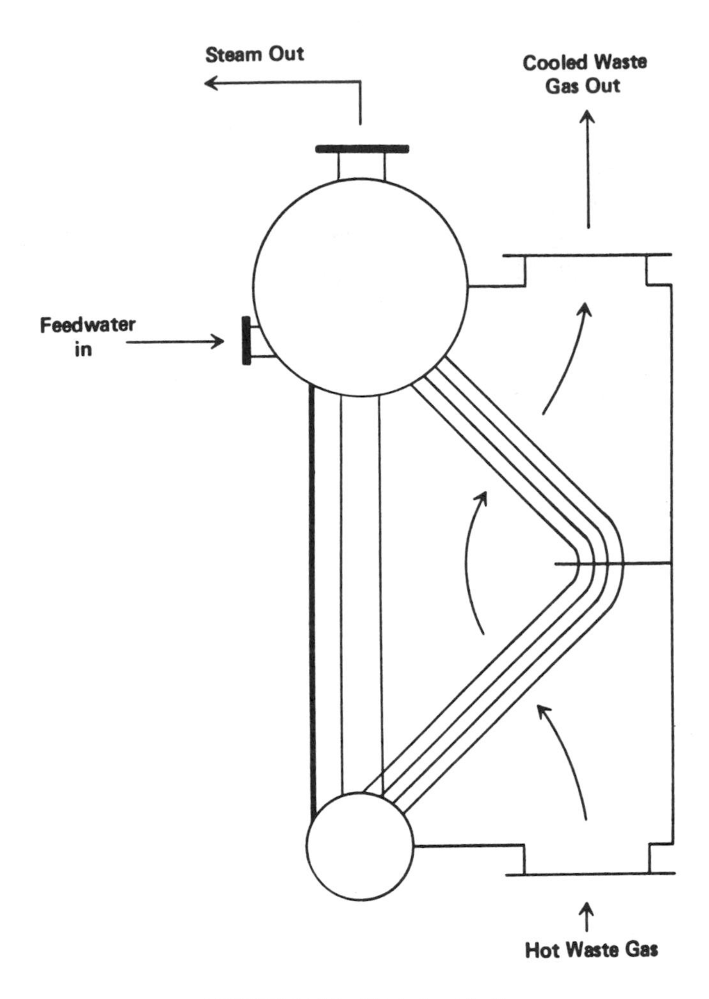

FIGURE 14.25 Waste-heat boiler.

time discrepancies between source and load. A good example is the storage of solar energy in a rock pile so that it becomes available for use at night and on cloudy days. Heat storage, other than for regenerators in high-temperature applications, has not yet been used a great deal for waste-heat recovery but will probably become more popular as more experience with it accumulates.

Combination heat-storage unit–heat exchangers called heat wheels are available for waste-heat recovery in the temperature range 0 to 982°C. The heat wheel is a porous flat cylinder that rotates within a pair of parallel ducts, as can be observed in Figure 14.28. As the hot gases flow through the matrix of the wheel they heat one side of it, which then gives up that heat to the cold gases as it passes through the second duct. Heat-recovery efficiencies range to 80%. In low- and moderate-temperature ranges the wheel is composed of an aluminum or stainless steel frame filled with a wire matrix of the same material. In the high-temperature range the material is ceramic. In order to prevent cross-contamination of the fluid streams, a purge section, cleared by fresh air, can be provided. If the matrix of the wheel is covered with a hygroscopic material, latent heat as well as sensible heat can be recovered. Problems encountered with heat wheels include freeze damage in winter, seal wear, and bearing maintenance in high-temperature applications.

The heat pump is a device operating on a refrigeration cycle that is used to transfer energy from a low-temperature source to a higher-temperature load. It has been highly developed as a domestic heating plant using energy from the air or from a well but has not been used a great deal for industrial applications. The COP (or the ratio of heat delivered to work input) for an ideal Carnot refrigeration cycle equals $T_H/(T_H - T_L)$, where T_H is the load temperature and T_L is the source

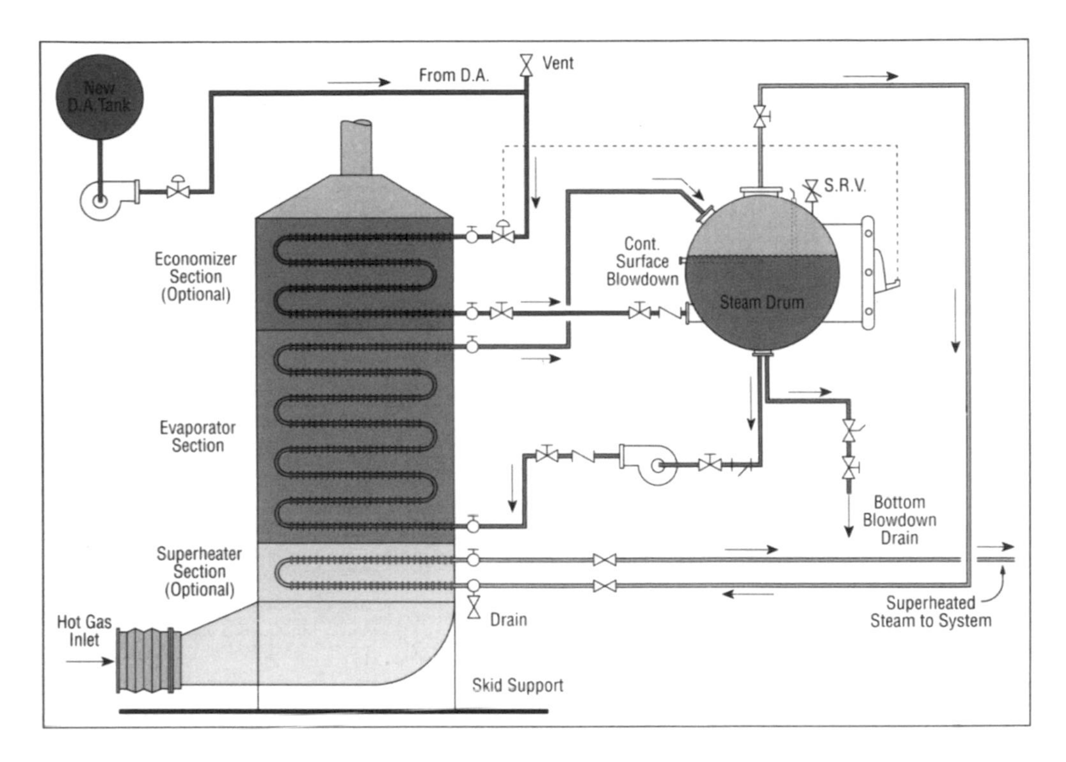

FIGURE 14.26 Schematic diagram of a finned tube waste-heat boiler. *Source:* Courtesy Cannon Technology.

temperature. It is obvious that when the temperature difference $T_H - T_L$ becomes of the order of T_H, the heat could be derived almost as cheaply from electric resistance heating. However for efficient refrigeration machines and a small temperature potential to overcome, the actual COP is favorable to moderate-cost waste energy. The heat pump can be used to transfer waste heat from and to any combination of liquid, gas, and vapor.

14.3 Sample Report on an Energy Study in a Foundry

The following report is synthesized from several actual energy studies made in gray iron foundries. The characteristics of any one of those locations are not recognizable from the description of the fictitious Iron Brothers Foundry. The energy consumption, fuel costs, and investment costs proposed for the waste-heat recovery systems are well within the wide range possible for those systems. However, no actual quotations from vendors have been used in the following example.

Iron Brothers Foundry located at 145th Street and West Newland in Cleveland, Ohio is a foundry wholly owned by Smith and York Turbine Company. It employs 320 production workers on three shifts. In addition there are 60 management and office workers. Production in 1993 was 13,600 metric tons of castings, down 27% from 1975. Production in 1994 is estimated to be up 30% over 1993. In 1993 total energy consumption was 312,800 GJ at a cost of $1,313,500 or an average unit cost of $4.20 per GJ. The plant is housed in three adjacent buildings totaling 156,000 ft^2. Annual sales are in the vicinity of $22 million. Annual degree-days (C) in 1993 were slightly over 3,889, up 11% over 1992.

Figure 14.28 gives us the production flow chart in block diagram form. Iron and steel scrap and carbon are the raw materials. The gas-fired cupola produces iron during one shift only. This is held in an electric furnace until the molds are ready for pouring in the second and third shifts. Sand is used to form the molds, many of which are dried and thermally conditioned in gas-fired ovens. After cooling, the castings are ground and machined, heat treated, assembled if necessary, painted if necessary, and packed for shipping. An electric motor–driven air compressor supplies

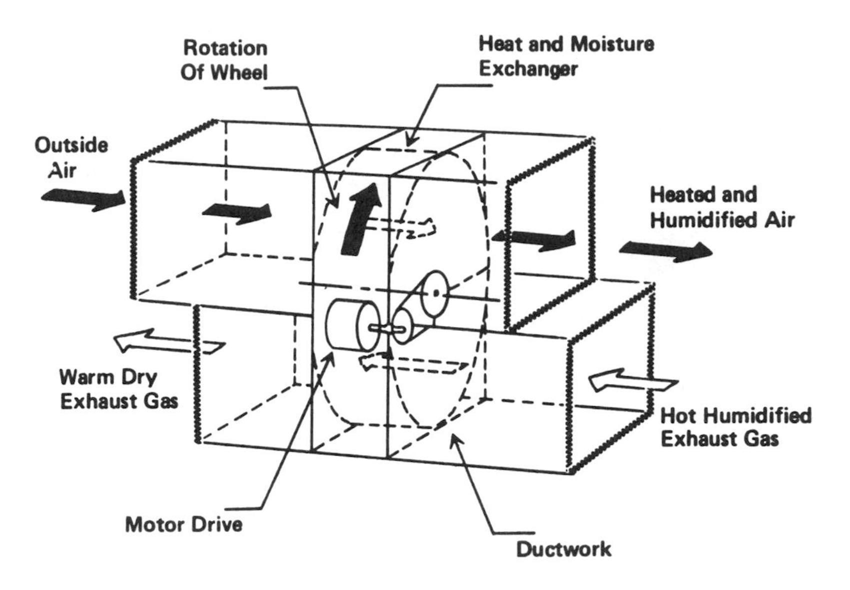

FIGURE 14.27 Heat wheel.

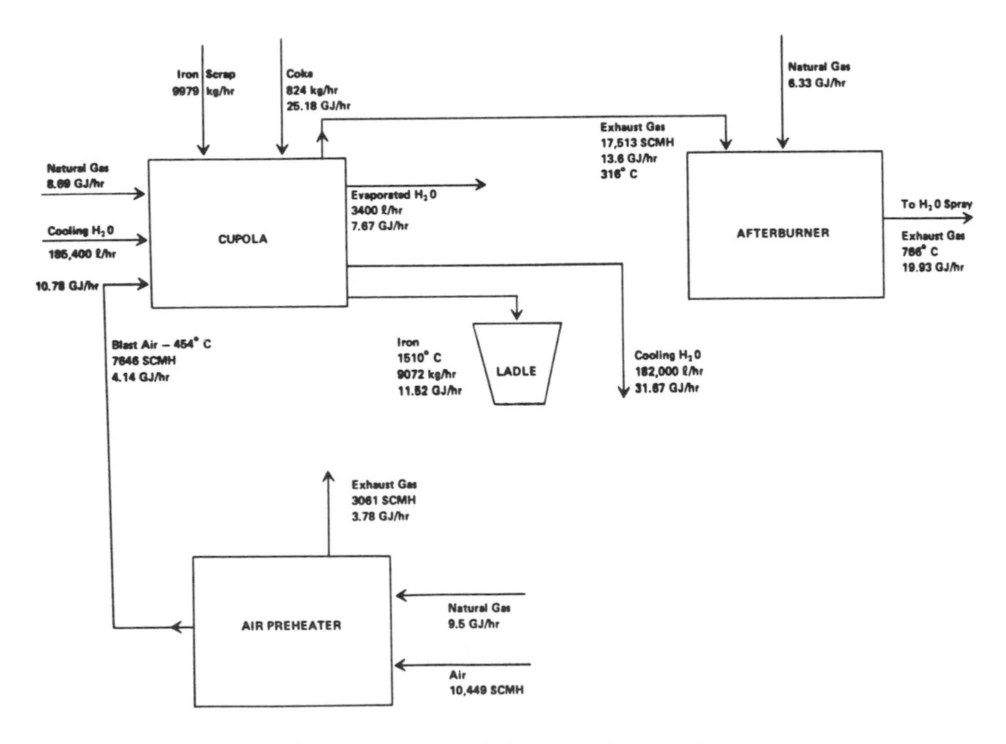

FIGURE 14.28 Block diagram of iron cupola.

air for packing and compressing molds, shaking out castings, and for driving pneumatic hand tools and mold-preparation machines. Electricity is used to drive fans and blowers for combustion air, to supply heat for the holding furnace, for miscellaneous drive motors, and for lighting. Table 14.8 shows the estimated electrical consumption division among the major user systems.

Space and hot-water heating is accomplished with natural gas. An average of 23.06 GJ is used per ton of product, with 12% derived from coke, 75% from natural gas, and 13% from electricity.

Figures 14.29 and 14.30 show, respectively, the monthly consumption and monthly cost of natural gas and electric energy. The consumption of gas, although obviously seasonal, shows a

TABLE 14.8 Distribution of Electrical Energy over Major Systems

Type System	Total Capacity (hp or kW; 1 hp = 0.746 kW)	% Total Load
Collector and combustion air fans	1,025 hp	35.5
Air compressors	450 hp	33.2
Electric furnace	545 kW	20.4
Mold shakers	225 hp	5.0
Lighting	65 kW	4.0
Miscellaneous		1.9
Total		100.0

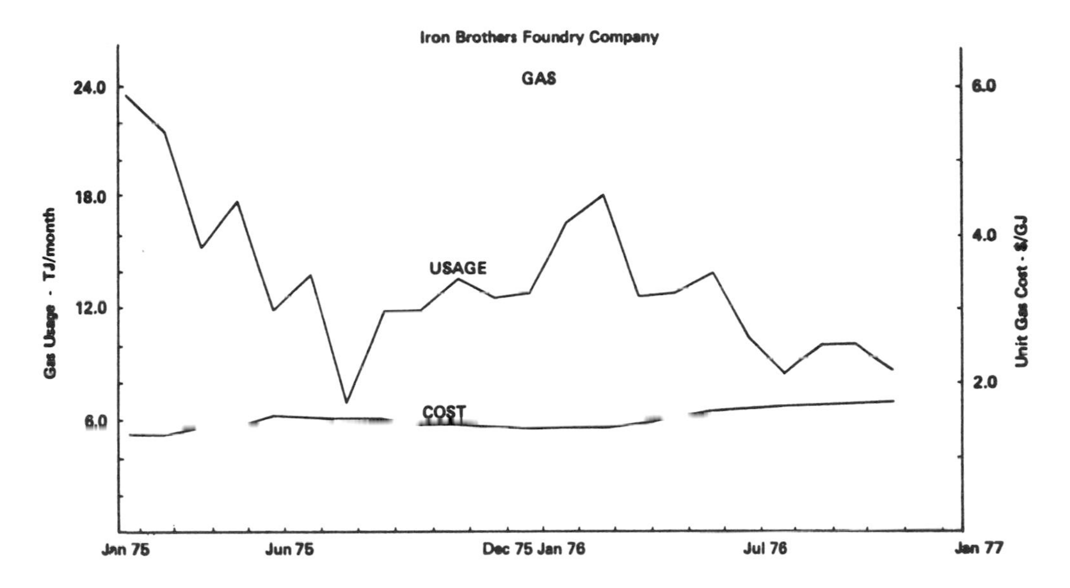

FIGURE 14.29 Foundry natural gas usage and unit costs.

drop in 1993 over that used in 1992, which is consistent with the lower production in 1993. Figure 14.31 shows clearly the relationship between gas consumption and heating degree days. The use of electric energy is not seasonally dependent and is not clearly production dependent. In fact, electric consumption increased 6% in the first 10 months of 1993 over the same period of 1992, while production dropped some 27%. Figure 14.32 is a graph of monthly production and monthly electric billing demand. It shows that in December 1992 electric demand rose over 500 kW above the average of 1992 and continued at that level over the greater part of the year. It is assumed that this increase in both electric billing demand and in the electric energy consumption at the beginning of 1993 was due to the installation of air-pollution control systems. Figure 14.33 graphs total monthly energy consumption and total monthly energy costs as well as unit energy cost by month. Total consumption on the average is dropping while the cost is rising, giving a definite upward trend to unit energy costs. These are seen to increase from a low of $3.20 per GJ to $5.00 per GJ. Figure 14.34 is an attempt to plot energy efficiency and energy cost efficiency on a monthly basis. The graphs indicate a strong seasonal influence on gas consumption. There is also a strong inverse relationship between total energy efficiency and production, which can be seen by comparing the curves of Figure 14.34 with the monthly production curves of Figure 14.32. The introduction of the Kleen-Air System for cupola emissions control raised the energy requirements and increased the energy costs per unit of product. During the latter part of 1993 the average unit cost for gas and electric energy was $4.70 per GJ. This results in a cost for utilities of $95.36 per metric ton. At $192.00 per metric ton for metallurgical coke, the total energy cost per metric ton of iron is $112.45.

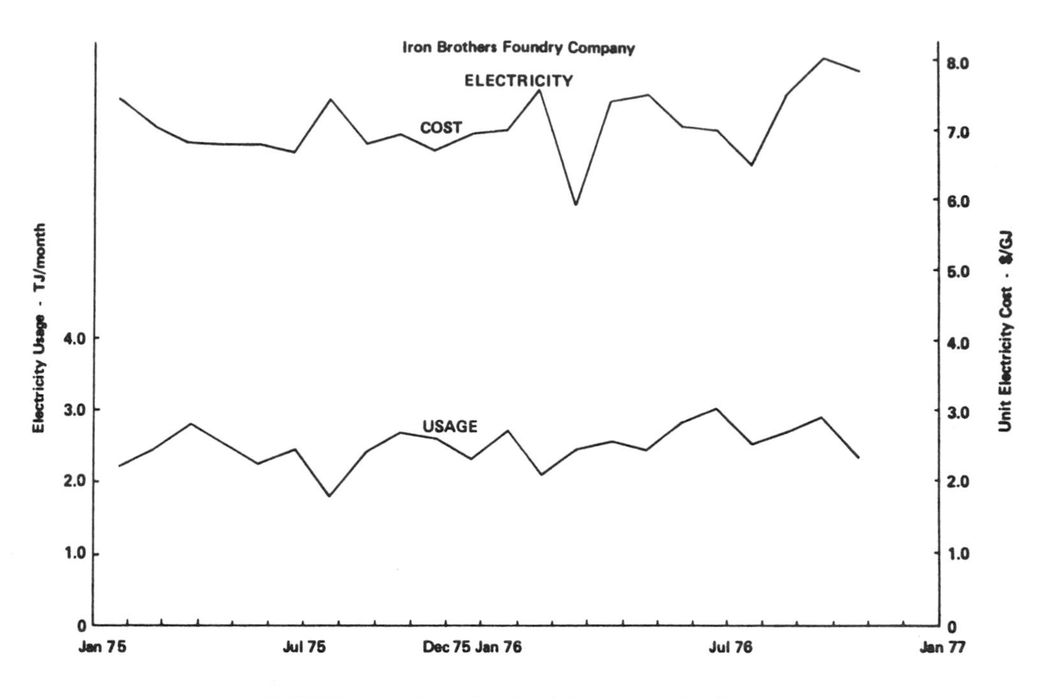

FIGURE 14.30 Foundry electricity usage and unit costs.

14.4 Energy Conservation Opportunity

The following paragraphs comprise a short list of measures that may provide significant energy savings and at the same time promise a suitable return on the required investment. It must be emphasized, however, that more detailed engineering and economic studies should be undertaken to provide a firmer basis for making investment decisions.

On the block diagram, Figure 14.35, we look for the energy intensive parts of the operation. Fifty-seven percent of the natural gas is consumed in the eight heat-treat furnaces, 21% in the boiler used exclusively for space heating, 11.7% in the cupola, and 8.6% in the mold-baking ovens. Less than 2.0% is used for other purposes throughout the plant. On the basis that the biggest energy users have the biggest potential for energy conservation, we look at the operation of the heat-treating furnaces first.

Recuperation of a Heat-Treating Furnace

A battery of eight front-loading bottom-fired furnaces operate on staggered 5½-hr treatment cycles, four cycles per day. New ratio controllers have recently been installed to replace the original controls. The air–fuel ratio is maintained in an inverse ratio to the firing rate so that the volume of combustion products are more or less constant with varying temperature. This is necessary in order to retain the proper flow path and velocities of the hot gases in the furnace so that the castings are heated uniformly at each control temperature.

After the castings are loaded into the furnace on steel pallets by a forklift truck, the furnace goes on high fire for approximately 45 min to reach 893°C, which is then maintained constant for 1¾ hr. At the end of the austinitizing process, the castings are quenched in a water pool and returned to the furnaces for drawing at 693°C for 2½ hr under a low-fire (25 to 30%) condition. During periods of production holdup the furnace is maintained at a standby temperature of 621°C. Each of the eight identical furnaces has five burners with a total rating of 5.3 GJ/hr. The average firing rate is 47%. The availability factor for the furnaces is 87.5%. Production hours run 8,000 hr annually.

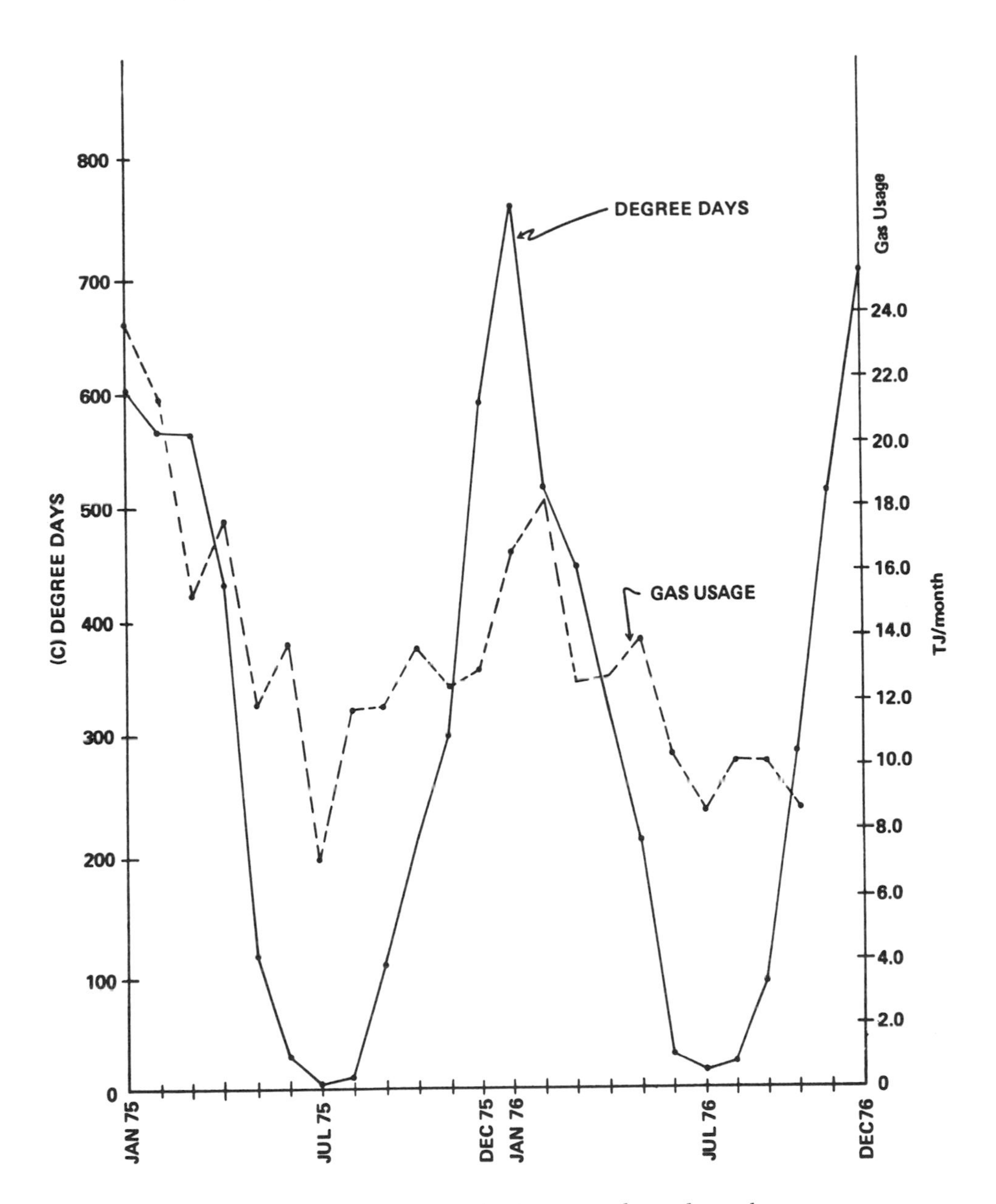

FIGURE 14.31 Iron Brothers Foundry Company—degree-days and gas usage.

Figure 14.36 is an annual heat-balance diagram of the eight furnaces. It shows an enthalpy flux of 60.53 TJ per year for the exhaust gases, or 52.6% of the 133.7 TJ of fuel-input energy. To heat the iron 5.75 TJ or 4.3% are required, –401 GJ per year or 0.3% for heating outside air to reference temperature (15.6°C) while the remainder, equaling 57.49 TJ per year or 43% of the fuel-input energy, is assumed to be surface heat loss and radiation losses from the interior through cracks and door openings.

Studies showed that recuperation of the exhaust gases using ceramic heat wheels could represent the most economic heat-recovery solution. It was decided that the furnace battery be divided into two groups of four and that two heat wheels be employed each to provide the combustion air for four adjacent furnaces. With a heat recovery efficiency of 70% representing an annual saving of 42.30 TJ, a fuel reduction of 32% was predicted. Because of the problem of gas flow and temperature

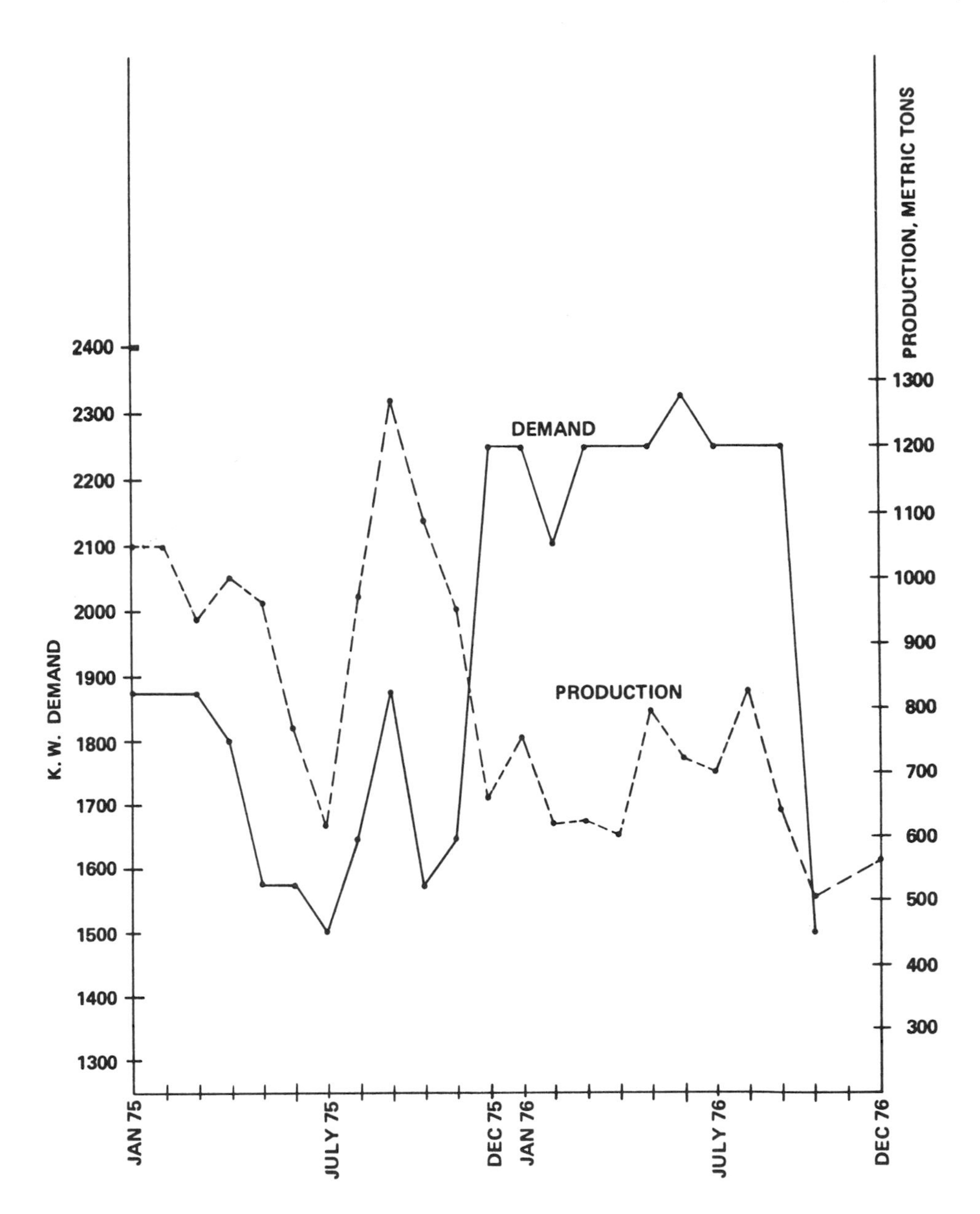

FIGURE 14.32 Iron Brothers Foundry Company—kW demand and production..

uniformity, no additional furnace efficiency was assumed because of the longer gas residence time in the furnace. Instead, it was assumed that the excess air quantities would have to be increased to compensate for the smaller fuel flows. However it was recognized that recuperation could result in shortened full-fire periods in the cycle. Table 14.9 is a list of specifications for the heat wheels. Since the combustion air fan had been overdesigned as evidenced by considerable throttling at the air damper occurring under all observed firing conditions, it was concluded that the fan was capable of overcoming the additional pressure drop caused by the heat wheels. However, the burners and the air piping would have to be replaced by high-temperature burners and stainless steel air piping, respectively.

At 70% recovery efficiency from the enthalpy flux of 60.4 TJ per year the energy savings are

$$\text{Savings} = 0.7 \times 60.4\ \text{TJ/year} = 42.3\ \text{TJ/year}$$

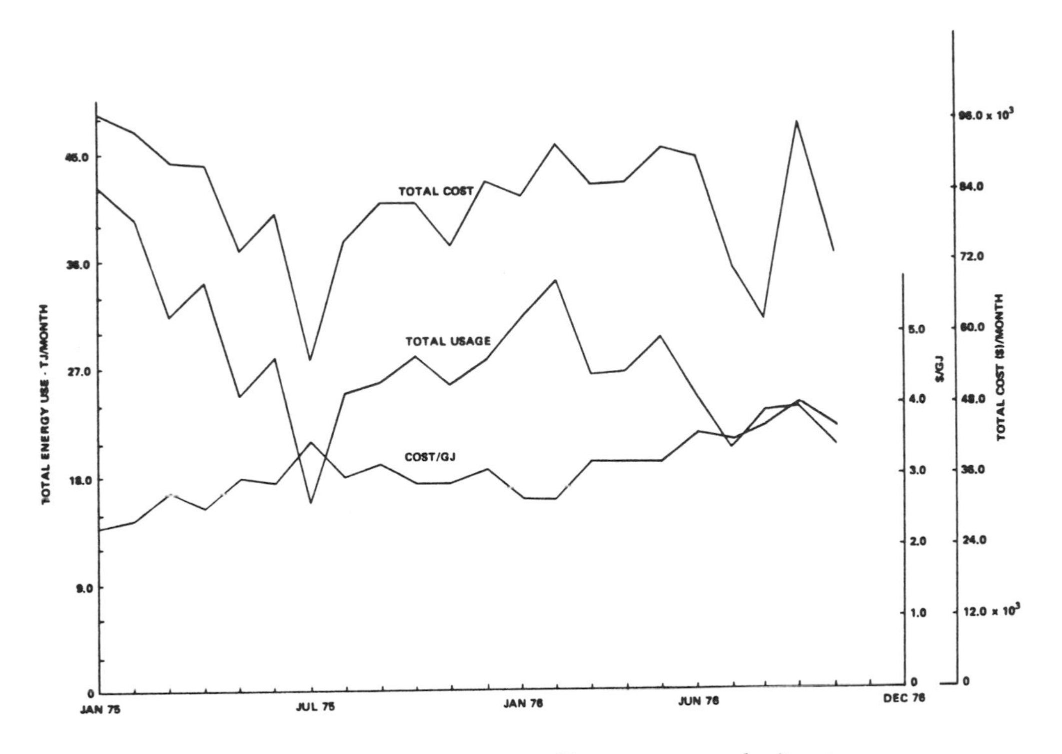

FIGURE 14.33 Foundry total monthly energy usage and unit costs.

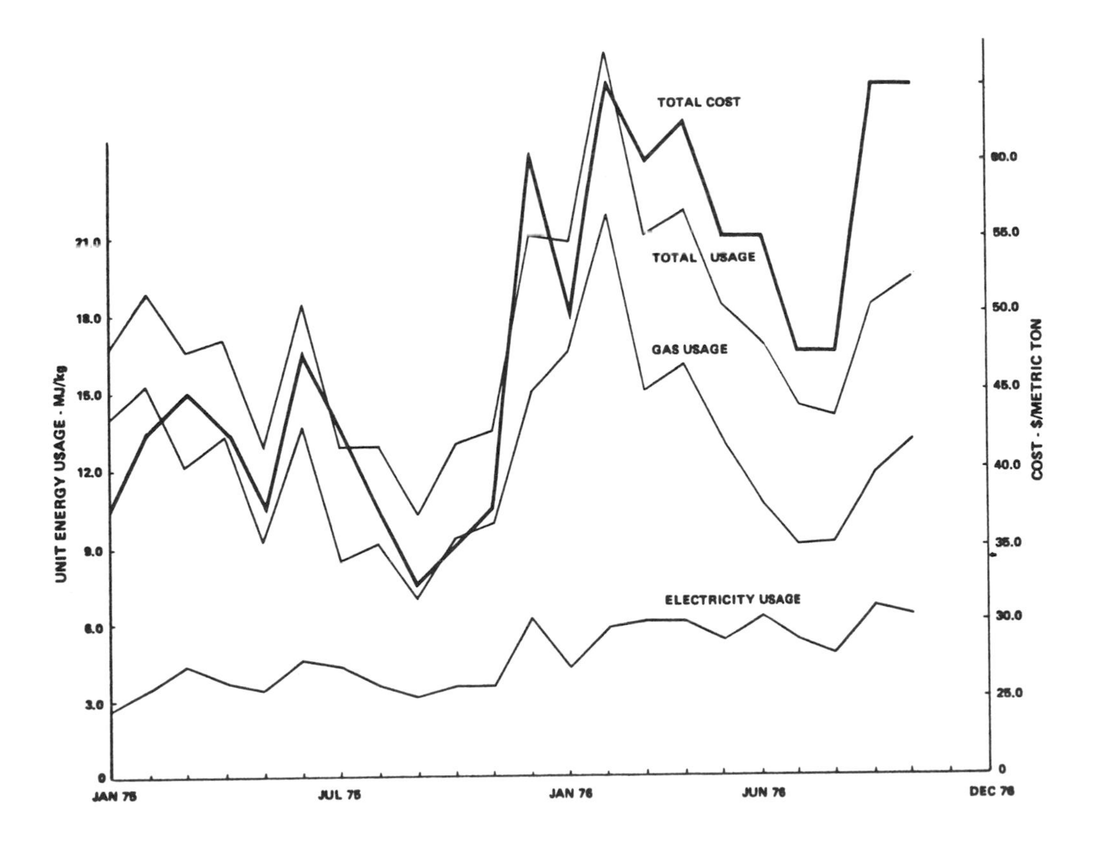

FIGURE 14.34 Energy use and cost per production unit.

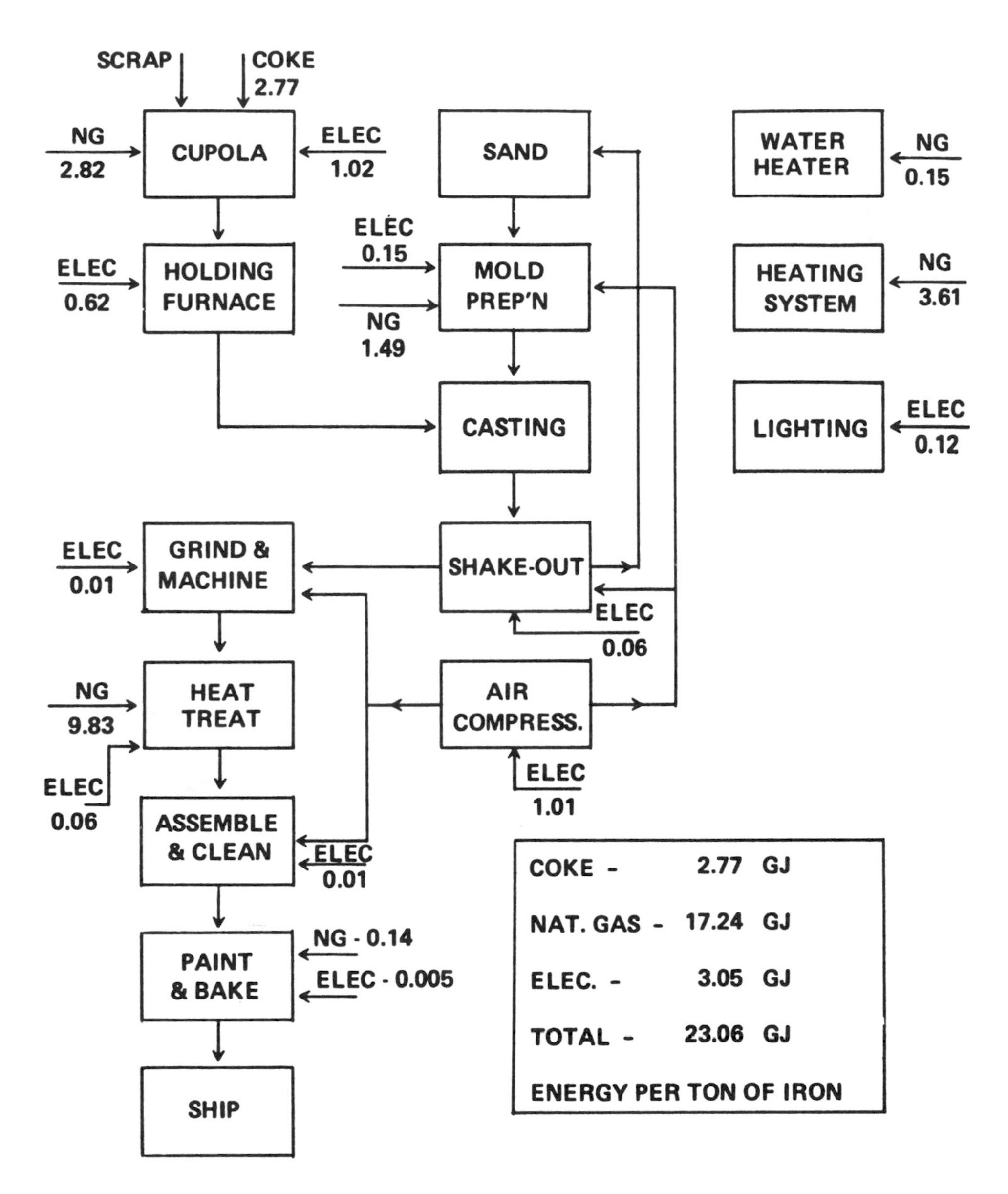

FIGURE 14.35 Unit energy consumption in GJ/per metric ton of product. Numbers preceeded by ELEC (electricity) and NG (natural gas) showing inputs to systems have units of GJ/metric ton.

The life of the units is expected to be 8 years. The following natural gas prices are interpolated from projections in Chapter 7.

Year	Energy Price (1993$/GJ)	Savings (1993$)
1995	2.86	120,980
1996	2.89	122,250
1997	2.91	123,100
1998	2.93	123,940
1999	2.96	125,200
2000	3.13	132,400
2001	3.30	139,600
2002	3.47	146,800

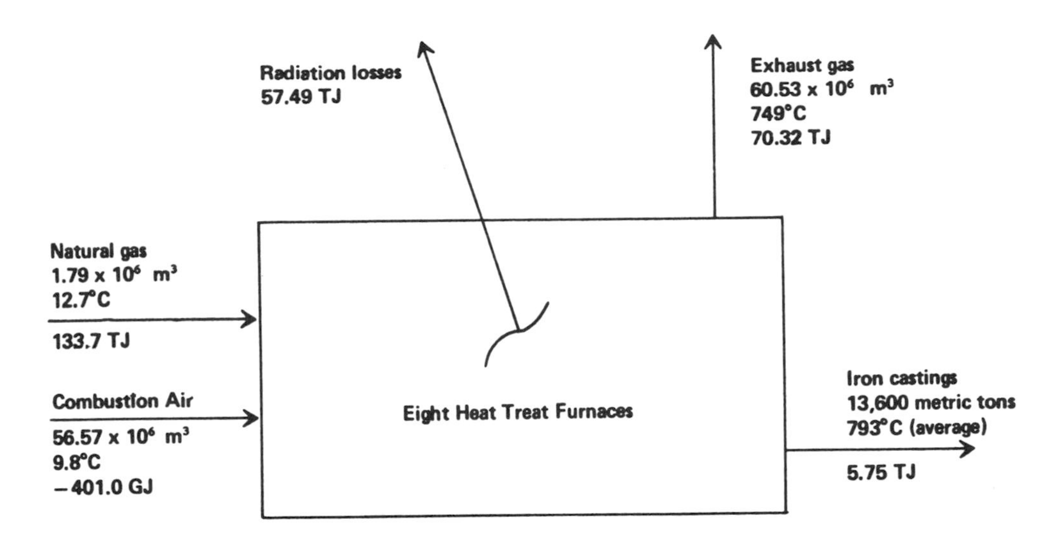

FIGURE 14.36 Annual heat balance on foundry treatment furnaces—Iron Brothers Foundry Company.

TABLE 14.9 Specifications for the Heat Wheel Recuperator

Exhaust gas flow	60 SCM/min
Average hot gas temperature	749°C
Maximum hot gas temperature	916°C
Pressure drop—hot side	3 cm H_2O
Cold air flow	51 SCM/min
Combustion air temperature	624°C
Pressure drop—cold side	1.25 cm H_2O
Energy recovery efficiency	70%

The costs of the project are estimated to be heat wheel, \$105,900 each × 2 = \$211,800; high-temperature burners, \$2,460 × 40 = \$98,400; installation (including ducting and piping), \$524,800; engineering, \$49,000—giving a total investment of \$884,000. By trial and error, it is found that a discount rate of 3.47%/year makes the present worth of the savings equal to the present worth of the investment (net present worth equals zero). This is equivalent to earning at 3.47% per year, not an attractive investment. Although the effects of the investment and the savings on income taxes have not been included, because these two effects tend to cancel one another, they do not change the result significantly. See Chapter 3 for more detail on methods of economic analysis.

Recuperation of Cupola

The second largest energy consumer is the cupola. The melting furnace is a hot-blast, continuous production, water-cooled cupola with external air heaters, an afterburner to accelerate combustion of CO following the air sweep, a spray cooler and a dust-collecting system. The operating data are given in Table 14.10.

$$\frac{\text{heat in steel}}{\text{preheat + energy coke}} = \left(9{,}091\ \text{kg/hr} \times 0.7\ \text{kJ/kg °C}\right)$$

$$\times \left(\left(1{,}510 - 27\right)\ \text{°C}\right) / \left(9 \times 10^6 + 822\ \text{kg coke/hr} \times 30{,}500\ \ \text{kJ/kg}\right)$$

The system appears to be 27.6% efficient. That is not a low efficiency for a furnace operating at 1,510°C. On the other hand, one cannot help but notice that the heat content of the coke is

TABLE 14.10 Cupola Operating Data

Capacity	13,636 kg/hr
Production	9.091×10^3 kg/hr
Temperature of iron	1,510°C
Coke consumption	826 kg/hr
Preheater fan capacity	127.5 SCMM
Hot-blast temperature	482°C
Preheater capacity	9 GJ/hr
Exhaust gas temperature	316°C
Afterburner capacity	2 each—3.17 GJ/hr
Temperature—following afterburning	454°C one burner
	607°C two burners
Composition of exhaust gas	
CO	15.65%
CO_2	11.50%
Air-to-carbon mass ratio	1.90
Exhaust gas to carbon mass ratio	9.22

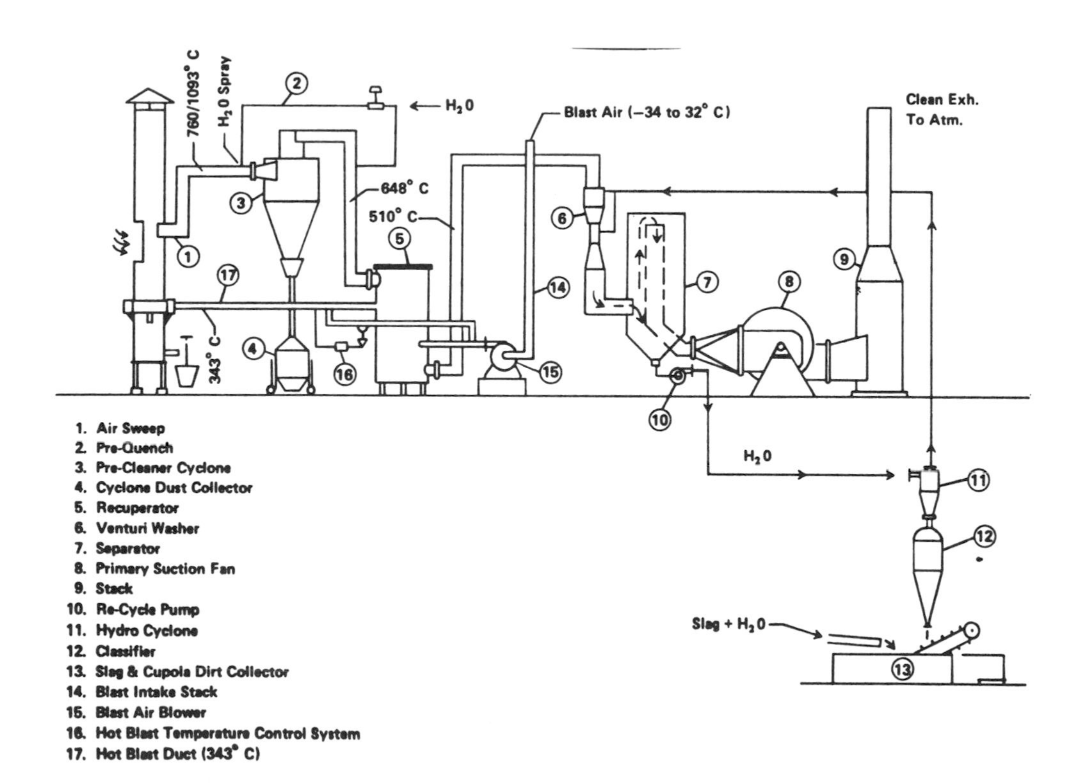

FIGURE 14.37 Cupola with recuperator installed. *Source:* Courtesy of Clean Air Engineering, Inc. Palatine, Ill. With permission.

many times that required to supply the energy needed to carry out the reaction. It is presumed, however, that the air blast and the raw gas feed increase the production rate and perhaps provide a more uniform product. If so, the cost does not seem excessive, for the total expense for the additional 18.39 GJ/hr is about $52 or just $5.20 per ton of product. If this expense is indeed justified, then we need to give thought in how we may recover some of the waste heat that leaves in the various fluid streams. Some 31.7 GJ/hr is available in the cooling water at a low temperature

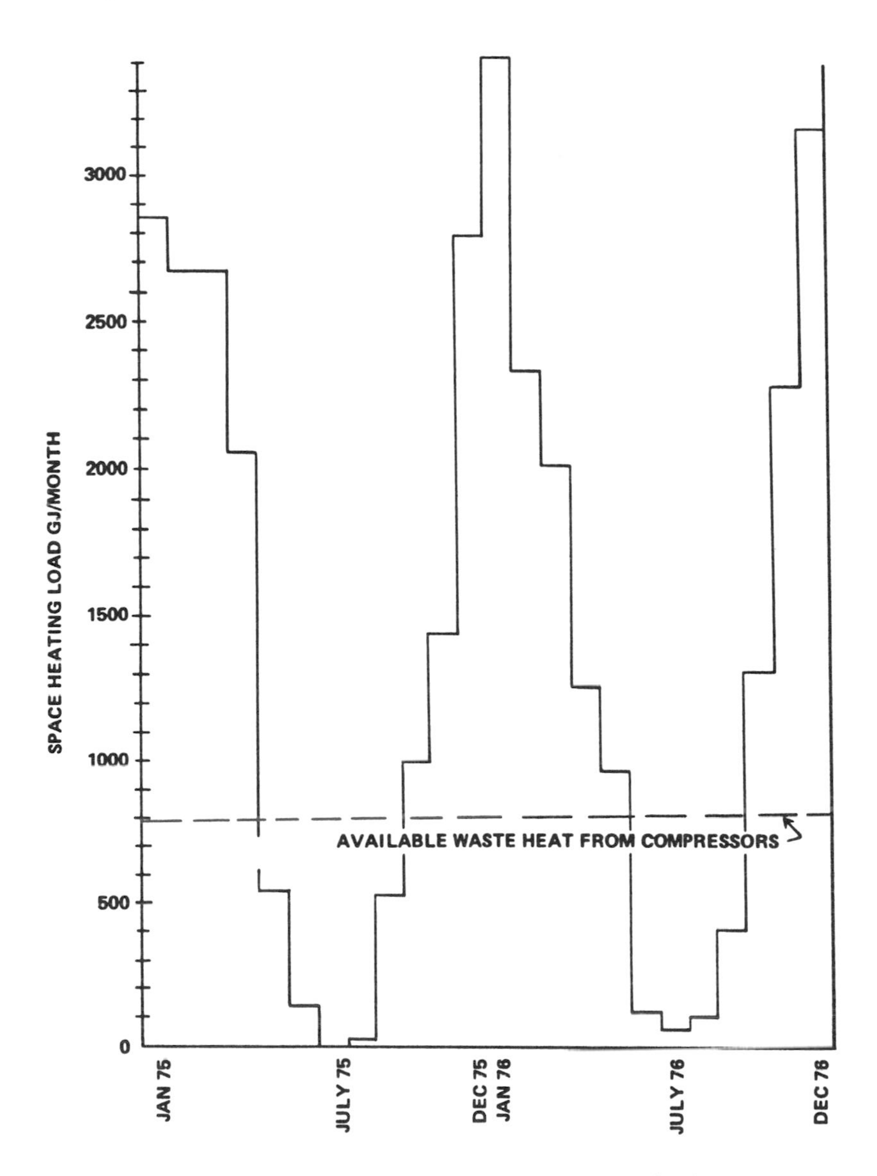

FIGURE 14.38 Space heating load—Iron Brothers Foundry Company.

and 19.93 GJ/hr in the exhaust gas streams. Some additional waste heat might become available if high-temperature insulation is added to the preheaters and afterburners. The following two figures illustrate ways to recover some of that waste heat. Figure 14.38 illustrates the retrofit installation of a recuperator, which will recover 50% of the exhaust gas energy for preheating combustion air. The recuperator is a finned-tube-and-shell heat exchanger with automatic internal soot blowing because of the dirty gas stream. Note that a cyclone dust separator is required upstream of the heat exchanger to remove the large particulate load. The rest of the equipment is essentially unchanged from the present system. The specifications for the recuperator are given in Table 14.11. The annual energy savings based on a 22% recovery of the waste heat, as derived from the specifications, will be

$$19.93 \text{ GJ/hr} \times 0.22 \times 1{,}950 \text{ hr/year} = 8{,}573 \text{ GJ/year}$$

TABLE 14.11 Recuperator Specifications

Clean air volume	127.5 SCMM
Clean air temperature	21°C
Clean air outlet temperature	482°C
Clean air inlet pressure	114-cm water column*
Dirty gas volume	292 SCMM
Dirty gas inlet temperature	766°C
Dirty gas inlet pressure	−10-cm water column*

* Pressure in height of water column.

The additional costs and investment required are, in 1993$

Recuperator	$132,100
Piping and fittings	9,200
Installation	93,800
Engineering	21,200
Total investment	256,300
Additional fan energy, $/yr	735
Project period, years	5
Operating hours/year	1,950
Electric rate, $/kWh	0,052
Salvage value, $	20,000

Natural gas prices are interpolated from projections of Chapter 7.

Year	Natural gas price, 1993$/GJ	Savings 1993$
1	2.86	24,519
2	2.89	24,776
3	2.91	24,947
4	2.93	25,119
5	2.96	25,376
	Total savings	$124,737

The savings resulting from this design are far less than the investment. Thus, even without looking at the economics in more detail, it is clear that this is a highly unfavorable investment. The main reason is that the cupola operates only 22% of the time during a year. A lesser reason is that a high efficiency of energy recovery is not achievable because of the very dirty exhaust gas.

Heat Recovery from Cupola Air Compressor Cooling Water

The steel cupola is cooled with 3,028 L of water, which is recirculated after cooling in a cooling tower. It enters the cooling tower at 57°C and is cooled to 29°C. The heat recovery could be

$$3{,}028 \text{ L/min} \times 60 \text{ min/hr} \times 1 \text{ kg/L} \times 4.185 \text{ kJ/kg °C}$$
$$\times (57.2 - 29.4) \text{ °C} = 21.1 \text{ GJ/hr}$$

Since the cupola operates 1,950 hr per year, the monthly waste heat available is

$$\frac{1{,}950 \times 21.1}{12} = 3{,}429 \text{ GJ}$$

This is sufficient heat to take care of the total maximum space heating load (excluding heating make-up air), but because it is only available during the melting shifts each day, it can supply only 27% of the heat required for space heating. The cooling towers would be required to operate during those times when less heat was required. Since the air compressor cooling water potentially can supply about 30% of the base heating load over the year, a combination system utilizing the compressor and cupola cooling water is worth investigating. This is particularly appropriate since the compressor cooling water flow exceeds the normal evaporation loss from the cupola cooling system.

The monthly space heating load is shown in the graph of Figure 14.38 where the constant supply of waste heat from the compressors is shown as a dashed line.

Waste heat from the compressor intercoolers averages 44.3 kJ/min/hp as 49°C cooling water.

$$450 \text{ hp} \times 44.3 \text{ kJ/min} \times 60 \text{ min/hr} \times 8{,}000 \text{ hr/year}$$

$$= 9{,}570 \text{ GJ annually}$$

This waste heat is presently carried to the sewer with the 133.5 L/min or 8,010 L/hr of cooling water. Since 3,400 L/hr of cupola cooling water is evaporated into the air, we propose the system of Figure 14.39, where the cooling water in the sump at 57°C is supplemented with the 49°C cooling water from the compressor.

Because the cupola cooling water is very dirty and because the make-up air preheater that constitutes the principal waste-heat load for the system would freeze during the coldest weather, the system consists of a runaround secondary system where a water–glycol mixture picks up heat in coils immersed in the sump and the heated liquid is circulated to the preheater coils and to unit space heaters. Any waste heat that is not used by the space heating system must be removed by the existing cooling tower. Because of the time discrepancies in availability of the cupola waste heat and the heating load, the present heating system must be retained and additional hot-water unit-space heaters purchased for use in the runaround loop. If the immersed coils of that loop have a heat exchange effectiveness of 75%, then the compressor waste heat can supply a maximum of 0.75 × 9,570/12 or 598 GJ per month. From the space heating graph of Figure 14.39, it was determined that 76.7% of the compressors' contribution to the heating system could be utilized, or 5,505 GJ per year. The remaining heating load of

$$(18{,}462 - 5{,}505) \text{ GJ} = 12{,}957 \text{ GJ/year}$$

can be partially supplied by the cupola. Since the cupola operates only about 22% of the time, then it can supply approximately 22% of that load or 12,957 × 0.22 = 2,850 GJ per year. The total waste-heat utilization is then (5,505 + 2,950) GJ = 8,455 GJ per year.

The investment cost is estimated as

	Cost in 1993$ at beginning of 1995	Present worth in 1993$ at beginning of 1995
Heat exchangers	25,276	25,276
Pipe and fittings	54,065	54,065
Controls	3,040	3,040
Pumps	8,426	8,426
Unit heaters	33,702	33,702
Installation labor	40,807	40,807
Engineering	10,017	11,119
Total	175,333	176,435

The annual savings and their present worth are estimated next. The project life is taken as 10 years. Natural gas prices are interpolated from Chapter 7 projections. A discount rate of 0.0788 was used to give a net present worth of zero.

Year	Natural Gas Price 1993$/GJ	Savings 1993$	Present Worth Factor @ 7.88%/yr	Present Worth at Beginning of 1995 in 1993$
1995	2.86	23,895	0.9270	22,150
1996	2.89	24,146	0.8592	20,747
1997	2.91	24,313	0.7965	19,365
1998	2.93	24,480	0.7383	18,073
1999	2.96	24,731	0.6844	16,925
2000	3.13	26,151	0.6344	16,589
2001	3.30	27,572	0.5880	16,213
2002	3.47	28,992	0.5451	15,803
2003	3.64	30,412	0.5053	15,366
2004	3.81	31,833	0.4683	14,909
Total		266,525		176,139

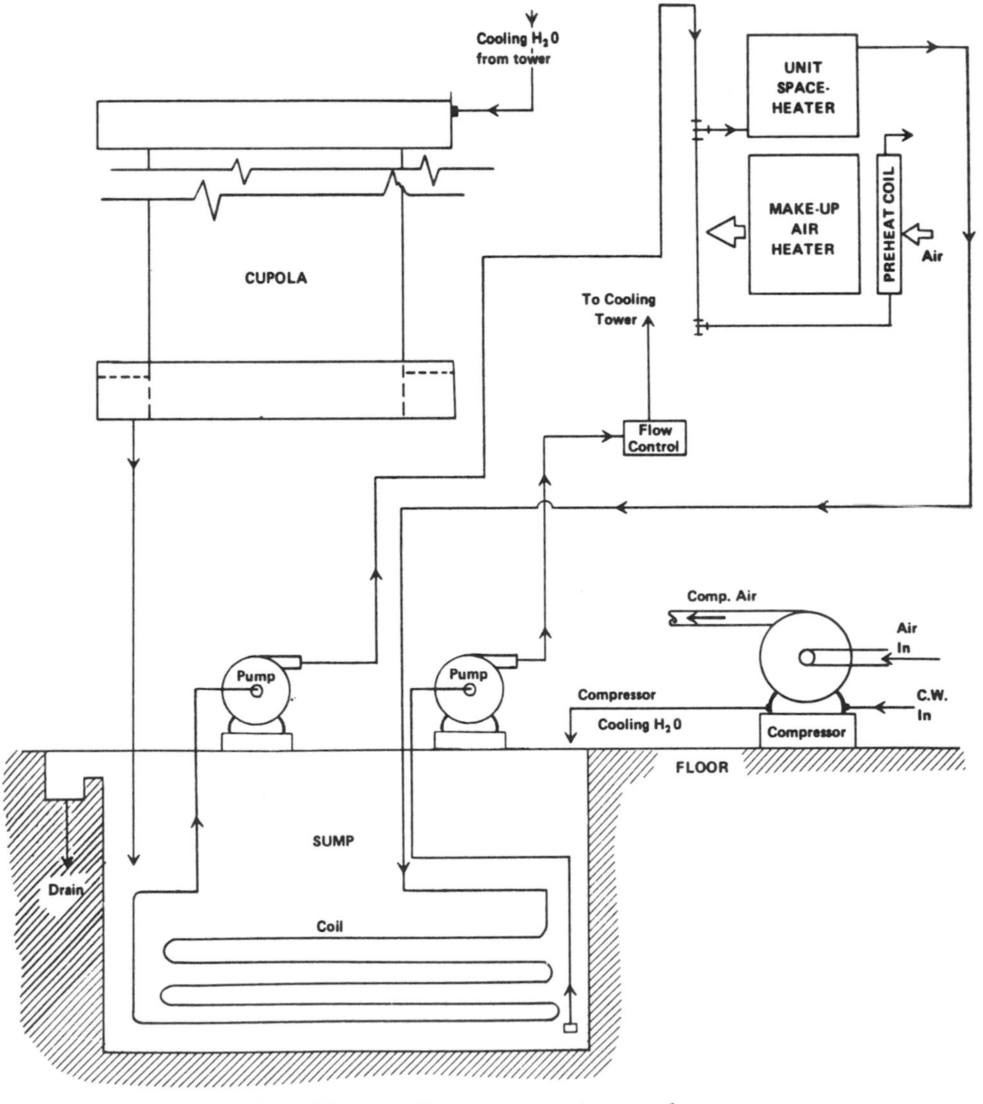

FIGURE 14.39 Heating system using waste heat.

This analysis indicates an earning rate of 7.88% per year. When the effects of depreciation and savings (reduced expenses) on income taxes are incorporated into the analysis (at 34% per year), the discounted cash flow earning rate becomes 6.01% per year. This return is moderately attractive and would have to be compared to alternative investments before a decision could be made.

Summary of Example

The economic analyses provided in the preceeding sections indicate the following results:

Project	Project Life, years	Simple Payout Time, years	Discounted Cash-Flow Rate of Return, %/yr Before Income Tax	After Income Tax
Recuperation of heat-treat furnaces	8	7	3.47	—
Recuperation of cupola	5	10 (loss in 5 years)	(loss)	(loss)
Waste heat for space heating	10	7	7.88	6.01

Of these, only the third is of even moderate economic interest.

Rohrer (1980) reported analyses for these same projects at the beginning of 1979. At that time, each project showed a better economic outcome than in the current analysis (though recuperation of the cupola even then showed a loss). The explanation for this unfavorable change in the economics is that the prices of energy have not escalated as much over the intervening years as have the costs of process equipment and labor. Thus, the investment has increased more than the savings and the economics look worse.

References

ASHRAE (American Society of Heating, Refrigeration and Air-Conditioning Engineers). 1989. *ASHRAE Handbook of Fundamentals.* Atlanta, GA: ASHRAE.

Dukelow, S.G. 1974. Nomograph. Wickliffe, OH: Bailey Meter Company.

EIA (Energy Information Administration). 1990. *Annual Energy Review 1989.* Washington, DC: U.S. Department of Energy.

EIA. 1992. *Annual Energy Review 1991.*.Washington, DC: U.S. Department of Energy.

EIA. 1994. *Annual Energy Review 1993.* Washington, DC: U.S. Department of Energy.

EIA. 1995. *Changes in Energy Intensity in the Manufacturing Sector 1985–1991.* Washington, DC: U.S. Department of Energy.

FEA (Federal Energy Administration). 1975. *The Potential for Energy Conservation in Nine Selected Industries.* NTIS PB-243-611/AS. Washington, DC: FEA.

FEA. 1977. *Assessment of the Potential for Energy Conservation Through Improved Boiler Efficiency,* Vol. I. Washington, DC: FEA.

Gatts, R., R. Massey, and J. Robertson. 1974. *Energy Conservation Program Guide for Industry and Commerce.* NBS Handbook 115. Washington, DC: National Bureau of Standards.

Hayashi, T., R.H. Howell, M. Shibata, and K. Tsuji. 1985. *Industrial Ventilation and Air Conditioning.* Boca Raton, FL: CRC Press.

Rohrer, W.M., Jr. 1980. "Industrial Energy Management." In *Economics of Solar Energy and Conservation Systems,* Vol. III, F. Kreith and R.E. West, eds. Boca Raton, FL: CRC Press.

Ross, M.H. and R.H. Williams. 1977. "The Potential for Fuel Conservation." *Technol. Rev.* 79(4), 49.

15

Process Energy Efficiency: Pinch Technology

Kirtan K. Trivedi
Brown and Root Petroleum and Chemicals

0-8493-2514-5/96/$0.00+$.50

Introduction

Many industrial processes require that heat be added to and removed from streams within the process. This is notably the case in the chemical process industry. Raw materials are heated or cooled to the appropriate reaction temperature. Heat may be added or removed to carry out the reaction at the specified condition of temperature and pressure. The product separation is achieved by heat addition (to a reboiler in a distillation column) and removal (from a condenser of the distillation column). The product may be heated or cooled to the proper temperature for storage and transportation. Thus, at all times heat is either added to and removed from a variety of process streams by utilities or by heat exchange between process streams. In an integrated plant this is normally achieved with a *heat exchanger network* (HEN). A large fraction of the capital cost of many chemical process plants is attributed to heat recovery networks.

The aim of a designer is to synthesize an optimal configuration for a process, in which a proper trade-off between the capital invested and the operating cost of the plant are achieved. The capital cost depends on the type, number, and size of units utilized to satisfy the design objectives. A substantial part of the operating cost usually depends upon the utilities consumed. To reduce these costs, the process designer aims for an economic combination having near the theoretical minimum number of heat exchanger units and recovers the maximum possible heat with them. An obvious way to minimize utility consumption is to recover heat by exchanging it between hot process streams that need to be cooled and cold process streams that need to be heated. Furthermore, the designer should also investigate the operability of the final design.

These objectives can be achieved by synthesizing a good heat exchanger network. However, a very large number of alternatives exist. Because of this, it is highly rewarding to synthesize quickly and systematically the best possible alternatives. This is now possible with the aid of "pinch technology."

Pinch technology refers to a large and growing set of methods for analyzing process energy requirements in order to find economically optimal and controllable designs. Considerable development has taken place in pinch technology during the last two decades, mainly due to the efforts of Linnhoff and co-workers. Pinch technology is proved to be effective and is successfully applied to process integration[1] that encompasses overall plant integration and includes heat exchanger networks, heat and power integration or cogeneration, and thermal integration of distillation columns. To date, there are 65 concepts used by this technology. However, due to the limited scope of this chapter, only the most important concepts are discussed here.

Industrial applications of this technology include capital cost reduction, energy cost reduction, emissions reduction, operability improvement, and yield improvement for both new process design and revamp process design. Imperial Chemical Industries (ICI), where this technology was first developed, reported an averaged energy savings of about $11 million/year (about 30% of annual consumption) in processes previously thought optimized.[2] The payback time was typically in the order of 12 months. Union Carbide showed even better results.[3] Studies conducted by Union Carbide on nine projects showed an average savings of 50% with an average payback period of 6 months. Savings in energy cost of about MM$8/yr was achieved by Union Carbide on these nine projects. BASF reports energy savings of 25% obtained by application of pinch technology to their Ludwigshafen site in Germany.[4] Over a period of 3 years about 150 projects were undertaken by BASF. Fraser and Gillespie[5] reported energy savings of 10% with a payback period of 2 years by applying pinch technology to the Caltex Refinery situated in Milnerton, Cape, South Africa. The energy consumption before the study was 100 MW for the whole refinery. A newsfront article in *Chemical Engineering*[6] gives more details about the experience of various companies in using pinch technology and the benefits obtained.

15.1 Fundamental Principles and Basic Concepts

Pinch technology is based on thermodynamic principles. This section reviews important thermodynamic principles and some basic concepts.

Temperature–Enthalpy Diagram

Whenever there is a temperature change occurring in a system, the enthalpy of the system changes. If a stream is heated or cooled, the amount of heat absorbed or liberated can be measured by the amount of the enthalpy change. Thus,

$$Q = \Delta H$$

For sensible heating or cooling at constant pressure where $CP = mc_p$,

$$\Delta H = CP \ \Delta T \tag{15.1}$$

For latent heating or cooling,

$$\Delta H = m\lambda$$

If we assume that the temperature change for latent heating or cooling is 1°C, then

$$CP = m\lambda$$

Equation (15.1) enables us to represent a heating or cooling process on a temperature–enthalpy (T-H) diagram. The abscissa is the enthalpy (H) and the ordinate is the temperature (T). The slope of the line is $(1/CP)$. Figure 15.1(a) shows a cold stream being heated from 20°C to 80°C with $CP = 2.0$ kW/°C.

As enthalpy is a relative function, the stream can be drawn anywhere on the enthalpy scale as long as it is between its starting and target temperatures and has the same enthalpy change. Figure 15.1(b) shows such a case. Thus, one of the important advantages of representing a stream on the temperature enthalpy plot is that the stream can be moved horizontally between the same temperature intervals. Figure 15.1(c) shows two different streams in the same temperature interval.

Two streams can be easily added on the temperature–enthalpy plot to represent a composite of the two streams. Figure 15.2 shows how to obtain the composite of two streams on this plot. This feature will be used in later sections to predict the minimum utility requirement for multistream problems.

A heat exchanger is represented by two heat curves on the temperature–enthalpy diagram as shown in Figure 15.3. This figure also shows how we will represent heat exchangers in a grid diagram form and the conventional form. In the grid diagram form, the hot stream goes from left to right and the cold stream goes from right to left i.e., the temperature of the hot stream decreases as it goes from left to right and the temperature of the cold stream increases as it goes from right to left. So, if the flow is counter current in the exchanger, then the temperature will decrease from left to right. The exchanger is represented by two circles connected by a vertical line. Heaters and coolers are represented by a single circle with a H or C inside the circle. Beneath the circle, the heat load is noted. The advantages of the grid diagram will become apparent when we discuss the design of heat exchanger networks. The exchanger has a temperature difference at the hot end and another at the cold end of the heat exchanger. The smaller of these two temperature differences is called Δt_{min}.

Second Law of Thermodynamics and Exergy Analysis

According to the Second Law of Thermodynamics, heat can be transferred from a higher temperature to a lower temperature only. Design of low-temperature processes uses exergy analysis that is based on this law. For a further discussion on the Second Law of Thermodynamics see Chapter 2.

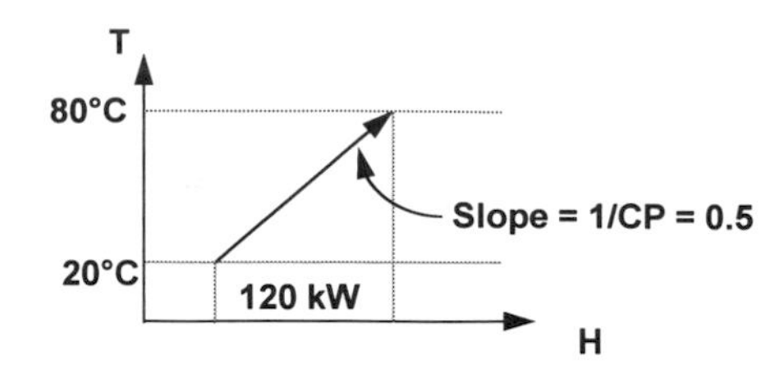

(a) Representation of a cold stream on the T-H diagram.

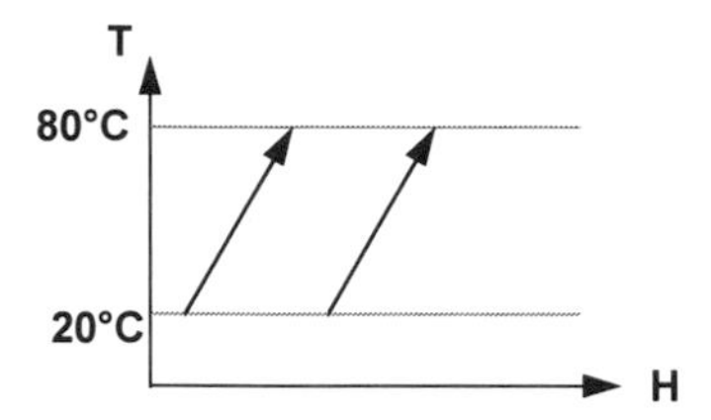

(b) A stream can be moved horizontally on the T-H diagram in a given temperature interval.

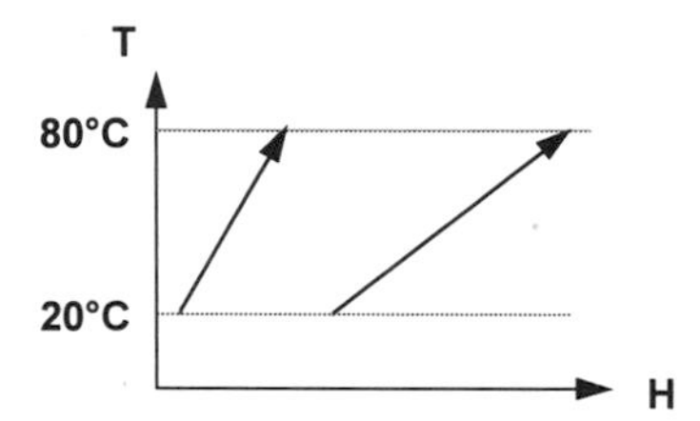

(c) Two different streams in the same temperature interval.

FIGURE 15.1 The temperature–enthalpy diagram.

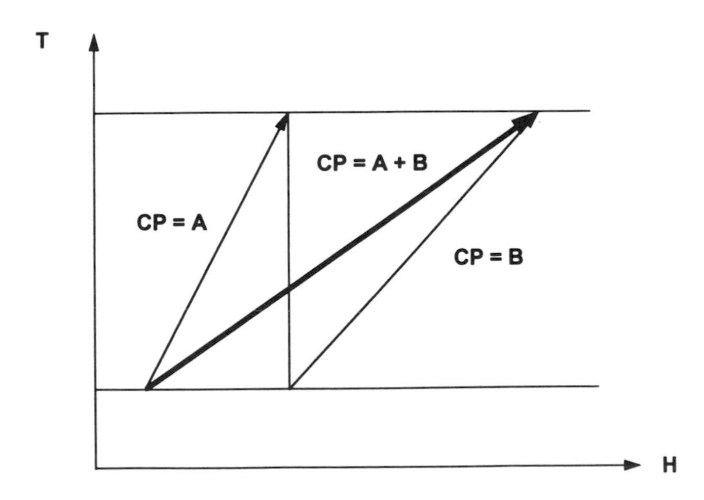

FIGURE 15.2 Composite stream obtained from two different streams in the same temperature interval.

Some Definitions

A *match* between a hot and a cold stream indicates that heat transfer is taking place between the two streams. A match between two streams is physically achieved via a heat exchanger *unit*. The number of heat exchanger units impacts the plot plan and determines the installed piping and cost including foundations.

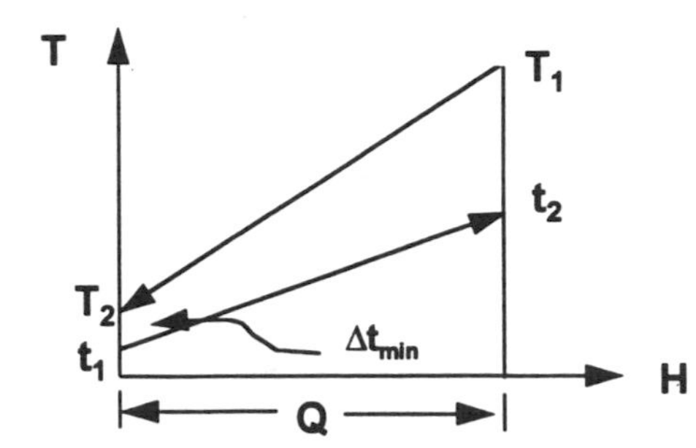

(a) Representation of an exchanger on the T-H diagram.

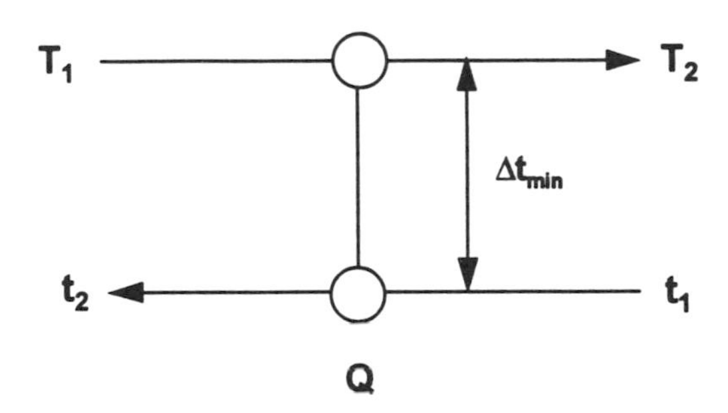

(b) Grid diagram representation of an exchanger.

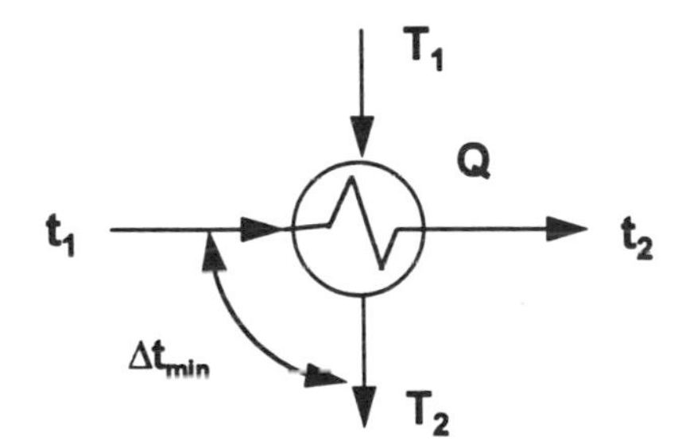

(c) Conventional representation of an exchanger.

FIGURE 15.3 Heat exchanger representation.

For reasons of fouling, thermal expansion, size limitation, cleaning, improved heat transfer coefficients, and so on, most process heat exchangers are shell-and-tube types with 1 shell pass and 2 tube passes or with multiple passes. Often what appears as a single match between two streams in a heat exchanger network representation is actually installed as several 1-2 exchangers in series or parallel. The term *shell* will be used to represent a single 1-2 shell-and-tube heat exchanger. See Chapter 13 for more discussion on 1-2 shell-and-tube heat exchangers.

15.2 Software

Pinch technology is a mature state-of-the-art technology for process integration and design. While various programs have been written by different researchers at various universities, only a handful of commercial programs are available for pinch technology applications. Table 15.1 shows the commercially available programs for process integration.

TABLE 15.1 Commercially Available Computer Programs for Pinch Technology

Software Name	Marketed By
Supertarget	Linnhoff-March Inc., 9800 Richmond Av. Suite 560, Houston, TX 77042 Phone: (713) 787 6861 Fax: (713) 787 6865
Advent	Aspen Technology Inc., Ten Canal Park, Cambridge, MA 02141 Phone: (617) 577 0100 Fax: (617) 577 0303
Heatnet	Britain's National Engineering Laboratory East Kilbride Scotland
HEXTRAN	Simulation Sciences Inc. 601 S. Valencia Ave., Brea, CA 92621 Phone: (714) 579 0354 Fax: (714) 579 0412

15.3 Heat Exchanger Network Design Philosophy

The objective of the heat exchanger network synthesis problem is to design a network that meets an economic criterion such as minimum total annualized cost. The total annualized cost is the sum of the annual operating cost (which consists mainly of energy costs) and annualized capital cost. The capital cost of a network primarily depends on the total surface area, the number of shells, and the number of units that will be installed. The capital cost will also depend on the individual type of heat exchangers and their design temperature, pressure, and material of construction.

If we include the pressure drop incurred in a heat exchanger, then the capital and operating cost of the pumps and compressors will also have to be taken into account. Being limited in the scope of this chapter we will not include stream pressure drop constraints.

Figure 15.4 shows the various variables that affect the optimization of a heat exchanger network. To make the network operable or flexible, extra cost may be incurred. This cost may be in the form of added equipment, extra utilities, or additional area on some of the exchangers in the network.

From this discussion, it can be seen that the synthesis of heat exchanger networks is a multivariable optimization problem. The design can never violate the laws of thermodynamics. Hence, the philosophy adopted in pinch technology is to establish targets for the various optimization variables based on thermodynamic principles. These targets set the boundaries and constraints for the design problem. Further, targets help us identify the various trade-offs between the optimization parameters. They help us obtain a bird's eye view of the solution space and identify the optimal values of the optimization parameters. Once the optimal values are identified, the network can be designed at these values. This approach will always lead to an optimal design.

To understand the interaction between the different optimization variables consider the heat curves with a small Δt_{min} for a simple-two stream system shown in Figure 15.5(a). Three different sections can be easily identified on this diagram. Section 1 represents the cold utility requirement, Section 2 represents the process–process heat exchange, and Section 3 represents the hot utility requirement. From this set of heat curves we can calculate the utility requirements for the system, and for each section we can calculate the exchanger area required, as we know the duty for each section and the terminal temperatures. Further, we can deduce that each section will require one

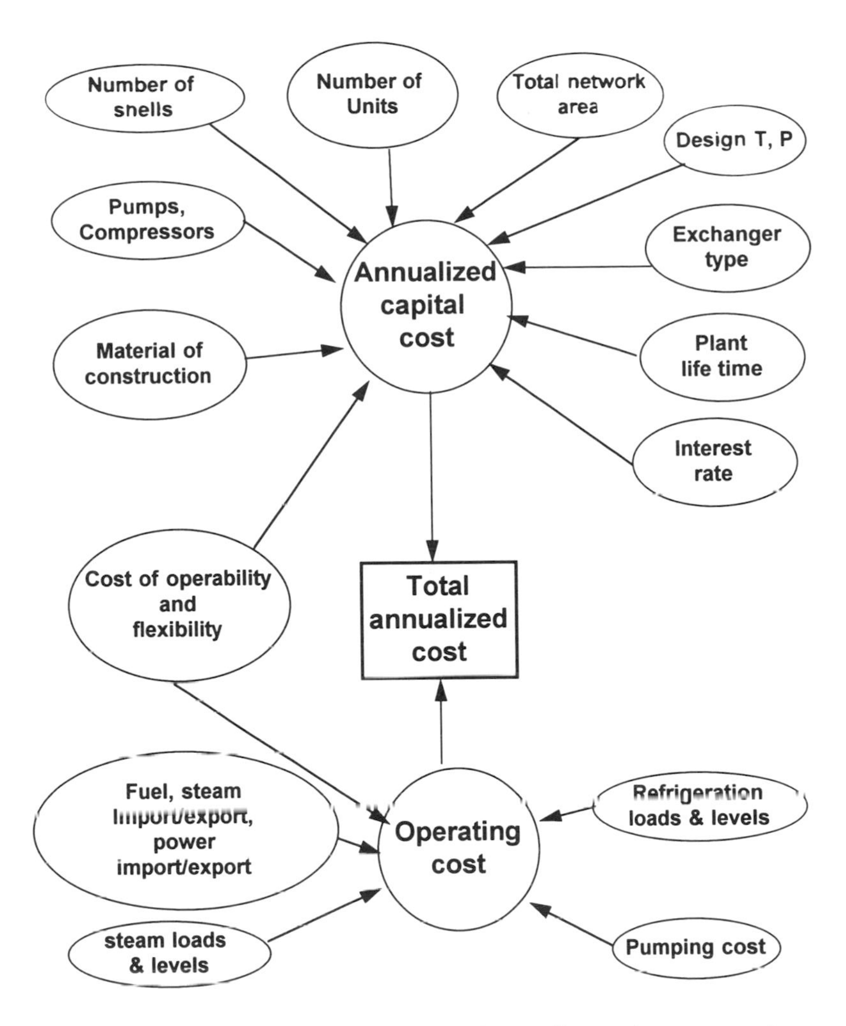

FIGURE 15.4 Optimization variables in the design of heat exchanger networks.

unit. We can use Bell's method[7] to estimate the number of shells required for each unit that represents the different sections. In Section 2, two shells are required. The heat curves for the hot and cold streams establish targets for energy, area, units, and shells. These are the major components that contribute towards the annualized cost of the network.

Now for the same system, let us increase the Δt_{min}. The new set of heat curves is shown in Figure 15.5(b). The preceding exercise can now be repeated to obtain targets for energy, area, units, and shells. When we increase the value of Δt_{min}, the utility requirement will increase and the total network area required will decrease. The number of units for this system remains the same, but the number of shells required decreases. Section 2 now only requires one shell. Thus, as the value of Δt_{min} increases, the utility cost increases and the capital cost decreases. This indicates that Δt_{min} is the single variable that fixes the major optimization variables shown in Figure 15.4. Hence, we can reduce the multivariable optimization problem to a single-variable optimization problem. This single variable is Δt_{min}. At the optimum value of Δt_{min}, the other optimization variables—that is, number of shells, number of units, total network area requirement, and the hot and cold utility requirements—will also be optimal. For a multistream problem the same conclusion can be easily derived. Hence, for the multistream problem optimization discussed in the next sections, we will develop methods to optimize the value of Δt_{min}.

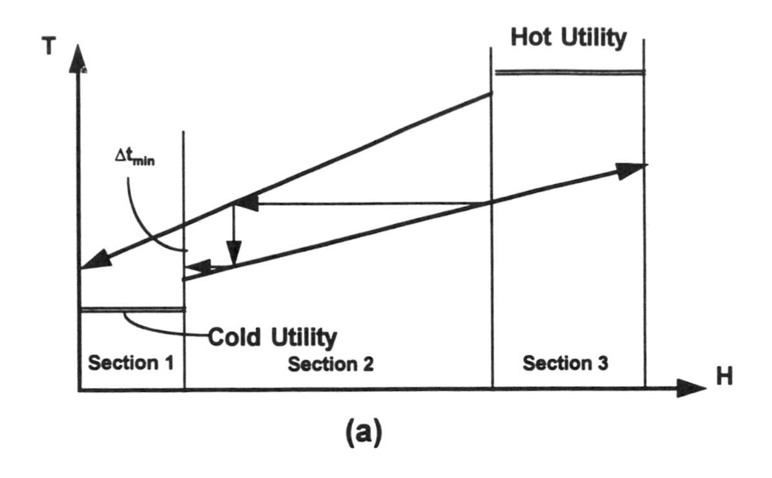

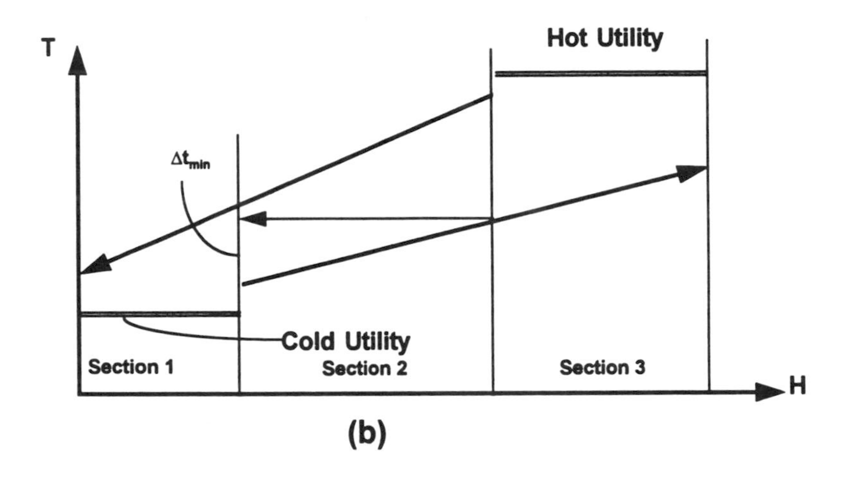

FIGURE 15.5 Effects of Δt_{min} on the energy, area, units, and shells required for a heat exchange system.

15.4 Design Problem

Consider the plant flow sheet shown in Figure 15.6. The feed is heated to the reaction temperature. The reactor effluent is further heated, and the products are separated in a distillation column. The reboiler and condenser use external utilities for control purposes. The overhead and bottoms products are cooled and sent for further processing. We shall use this problem to illustrate the use of pinch technology concepts.

Table 15.2 lists the starting and target temperatures of all streams involved in the flowsheet. It also shows the *CP* values. While pressure drop constraints on the individual streams determine the film heat transfer coefficients, we shall not take into account this constraint. Instead, we use (unrealistically) the same heat transfer coefficient value of 1 kW/m^2 °C for all streams. Our objective is to design an optimum heat exchanger network for this process using the economic basis outlined in Table 15.3. Further, let Δt_{min} denote the minimum temperature difference between any hot stream and any cold stream in any exchanger in the network. It is desired that the final network should be able to control the target temperature of all the streams at the design temperature.

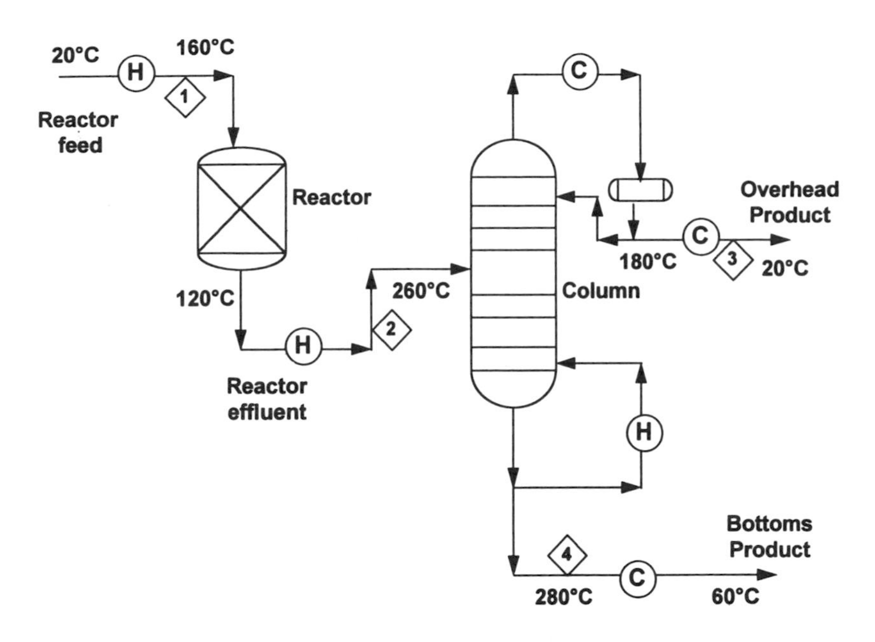

FIGURE 15.6 A typical plant flowsheet.

TABLE 15.2 Stream Data for the Flowsheet in Figure 15.6

Stream Number	Stream Name	T_s (°C)	T_t (°C)	CP (kW/°C)
1	Reactor Feed	20	160	40
2	Reactor Effluent	120	260	60
3	Overhead Product	180	20	45
4	Bottoms Product	280	60	30

TABLE 15.3 Economic Basis for the Flowsheet in Figure 15.6

Heat Exchangers	
Installed cost of shell ($)/(m²)	= 10,000 $A^{0.6}$
Utility Data	
Cost of using hot oil	= 68 ($/kW yr), ($2.36/MMBtu)
Cooling water	= 2.5 ($/kW yr), ($0.09/MMBtu)
Temperature range of hot oil	= 320–310°C
Temperature range of cooling water	= 10–20°C
Plant Data	
Interest rate	= 10%/year
Lifetime	= 5 years
Operation time	= 8000 hr/yr
Calculated capital recovery factor, CRF (see Chapter 3 for details)	= 0.2638/yr

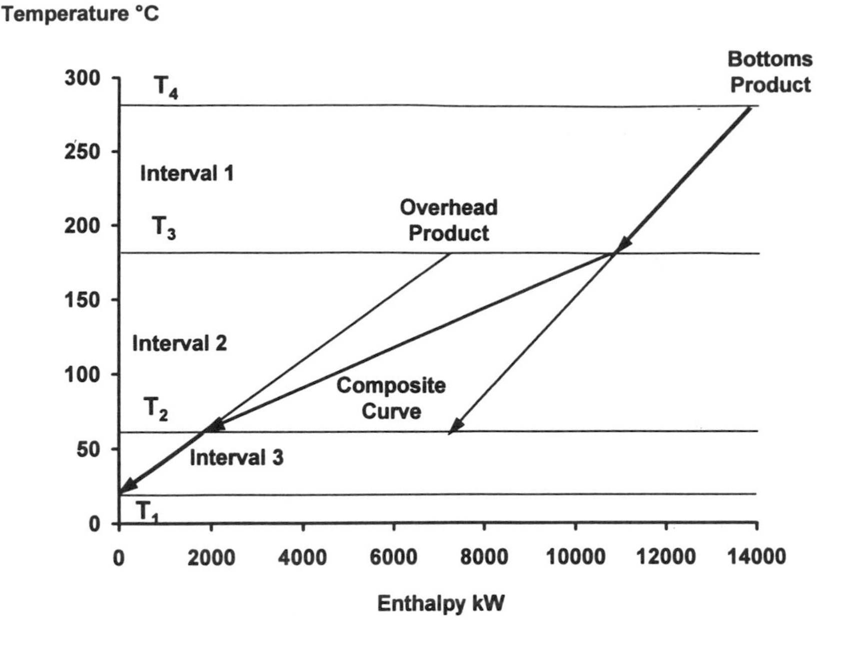

FIGURE 15.7 Construction of the hot composite curve.

15.5 Targets for Optimization Parameters

Energy Targets

To find the energy targets we need to plot the composite curves.

Composite Curves

Let us plot the heat curves for all the hot streams on the T–H diagram (see Fig. 15.7). We can divide the diagram into a number of temperature intervals, defined by the starting and target temperatures for all the streams. Between two adjacent temperatures we can calculate the total heat content of all streams that are present in this temperature interval. For example, between 180°C and 60°C, the sum total of the heat available is calculated as

$$\Delta H = (T_3 - T_2) \sum_j CP_j$$

$$\Delta H_2 = (180 - 60)(40 + 30) = 8,400 \text{ kW}$$

A composite curve that represent the heat content of all the hot streams is obtained by summing the heat available in each of these temperature intervals as shown in Figure 15.7. Similarly, a composite curve for all the cold streams can be obtained. These two composite curves can be used as the heat curves for the whole process.

To obtain the energy target, the hot composite curve is fixed and the cold composite curve is moved horizontally until the shortest vertical distance between the two curves is equal to the value of $\Delta t_{\min}$ (see Figure 15.8). The overshoot of the cold composite curve is the minimum hot utility requirement (Q_{hmin})and the overshoot of the hot composite curve is the minimum cold utility requirement (Q_{cmin}).[8] For $\Delta t_{\min}$ = 30°C, the minimum hot utility requirement is 4,750 kW and the cold utility requirement is 4,550 kW.

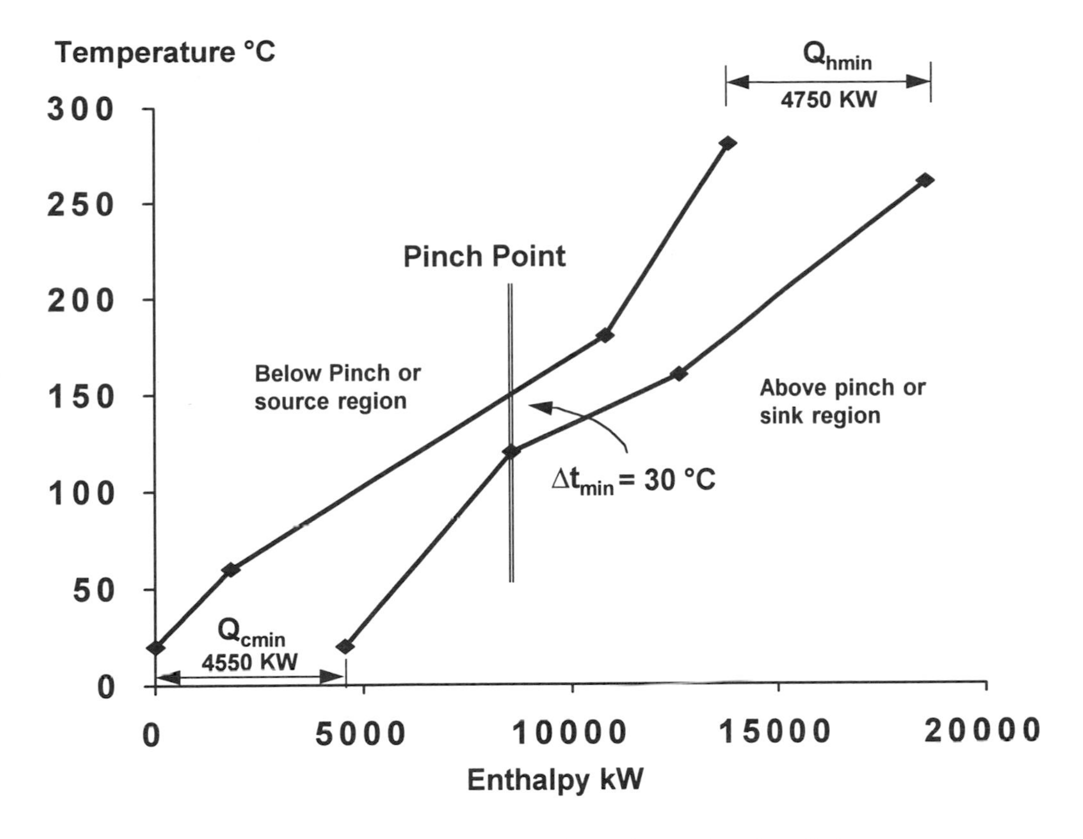

FIGURE 15.8 Composite curves for the flowsheet in Figure 15.6.

The Problem Table

The preceding procedure for obtaining the energy targets using composite curves can be time-consuming. An alternative method based on thermodynamic principles is developed. Hohmann[8] called it the feasibility table. Linnhoff and Flower[9] independently developed the problem table. The problem table algorithm is easy and involves no trial and error.

The algorithm consists of the following steps:

- Select a value of Δt_{min}. Since we have already established the energy targets using composite curves for Δt_{min} = 30°C, we shall use that value here.
- Convert the actual terminal temperatures into interval temperatures as follows:

  for the hot streams: $T_{int} = T_{act} - \Delta t_{min}/2$
  for the cold streams: $T_{int} = T_{act} + \Delta t_{min}/2$

  where T_{int} = interval temperature
  T_{act} = actual stream temperature

  The interval temperatures now have the allowance for Δt_{min}. This modification guarantees that for a given T_{int}, the actual temperature difference between the hot and cold stream will always be greater than or equal to Δt_{min}.

- All the interval temperatures for both the hot and cold streams are sorted in descending order and the duplicate intervals are removed (see Table 15.4).
- For each interval, an enthalpy balance is made. The enthalpy balance for interval i is calculated using the following equation:

TABLE 15.4 Generation of Temperature Intervals for Δt_{min} = 30°C

Stream No.	Actual Temperature (°C) T_s	Actual Temperature (°C) T_t	Interval Temperature (°C) T_s	Interval Temperature (°C) T_t	Interval Number	Ordered Interval Temperatures (°C)
1	20	160	35	175	1	275
2	120	260	135	275	2	265
3	180	20	165	5	3	175
4	280	60	265	45	4	165
					5	135
					6	45
					7	35
					8	5

TABLE 15.5 The Problem Table

Stream No.		1	2	3	4					
Type		Cold	Cold	Hot	Hot					
CP		40	60	45	30					
Temperature Interval	Interval Number					ΔT_{int}	$\Sigma CP_{cj} - \Sigma CP_{hj}$	ΔH_{int}	Heat Cascade	Corrected Heat Cascade
Hot utility–275	0								0	4,750
275–265	1					10	60	600	–600	4,150
265–175	2					90	30	2,700	–3,300	1,450
175–165	3					10	70	700	–4,000	750
165–135	4					30	25	750	–4,750	0
135–45	5					90	–35	–3,150	–1,600	3,150
45–35	6					10	–5	–50	–1,550	3,200
35–5	7					30	–45	–1,350	–200	4,550
5–Cold utility	8									4,550

$$\Delta H_i = \left(T_{int,i} - T_{int,i+1}\right)\left(\sum_{j}^{N_{streams}} CP_{cj} - \sum_{j}^{N_{streams}} CP_{hj}\right)$$

where ΔH_i = net heat surplus or deficit in interval i
CP_c = mass specific heat of a cold stream
CP_h = mass specific heat of a hot stream

For the example problem this calculation is shown in Table 15.5. With each interval the enthalpy balance will indicate that there is either heat deficit or surplus or the interval is in heat balance. A heat surplus is negative and a heat deficit is positive.

The Second Law of Thermodynamics only allows us to cascade heat from a higher temperature to a lower temperature. Thus, a heat deficit from any interval can be satisfied in two possible ways—by using an external utility or by cascading the surplus heat from a higher temperature interval. If no external utility is used, the heat cascade column of Table 15.5 can be constructed by subtracting the net heat surplus or deficit in an interval from the total cascade of the previous interval. All intervals in this heat cascade have negative heat flows. This is thermodynamically impossible in any interval (it would mean heat flowing from lower to higher temperature). To correct this situation, we take the largest negative flow in the cascade and supply that amount of external hot utility at the highest temperature interval. This modification will make the heat cascade feasible (i.e., none of the intervals will have negative heat flows). The amount of external heat

For the example under consideration, the minimum hot utility required is 4,750 kW and the minimum cold utility required is 4,550 kW. These are the same targets obtained from the composite curves. It is clear, however, that the problem table algorithm is an easier, quicker, and more exact method for setting the energy targets. One can set up a spreadsheet to implement the problem table algorithm. Further, once the energy targets are obtained, the composite curves can be drawn to visualize the heat flows in the system. It should be noted that the absolute minimum utility targets for a fixed flowsheet are determined by the case of $\Delta t_{min} = 0$.

Capital Cost Targets

To find the optimal value of Δt_{min} ahead of design, we need to set the targets for the capital and energy costs. As seen before, the capital cost target for a given Δt_{min} depends on the total area of the network, the number of shells required in the network, and the number of units required in the network. We will now establish these targets.

Target for Minimum Total Area for the Network

The composite curves can be divided into different enthalpy intervals at the discontinuities in the hot and cold composite curves. Between any two adjacent enthalpy intervals, if the hot streams in that interval transfer heat only to the cold streams in that interval and vice versa, then we can say that vertical heat transfer takes place along the composite curves.[10] This mode of heat transfer models the pure counter-current heat exchange.

The equation for establishing the minimum area target assumes a complex network called the "spaghetti" network. This network models the vertical heat transfer on the composite curves. In this network, for any enthalpy interval defined by the discontinuities in the composite curves, each hot stream is split into the number of cold streams in that enthalpy interval. Each cold stream in the enthalpy interval is also split into the number of hot streams in that interval. A match is made between each hot stream and each cold stream. Figure 15.9 shows this network for an enthalpy interval. The area target is obtained by the following equation:

$$A_{min} = \sum_{i}^{N_{intervals}} \frac{1}{\Delta T_{LMTDi}} \left(\sum_{j}^{N_{streams}} \frac{q_j}{h_j} \right)_i$$

where A_{min} = total minimum area for the network
ΔT_{LMTDi} = ΔT_{LMTD} for enthalpy interval i
q_j = heat content of stream j in enthalpy interval i
h_j = film plus fouling heat transfer coefficient of stream j in enthalpy interval i

The equation gives a minimum total surface area for any system where the process streams have uniform heat transfer coefficients. Townsend and Linnhoff[10] claim that for nonuniform heat transfer coefficients the equation gives a useful approximation of the minimum area, with errors typically within 10%.

Minimum Number of Units Target

The minimum number of units is given by:[11]

$$u = N + L - s$$

where u = number of units including heaters and coolers
N = number of streams including utilities
L = number of loops
s = number of separate networks

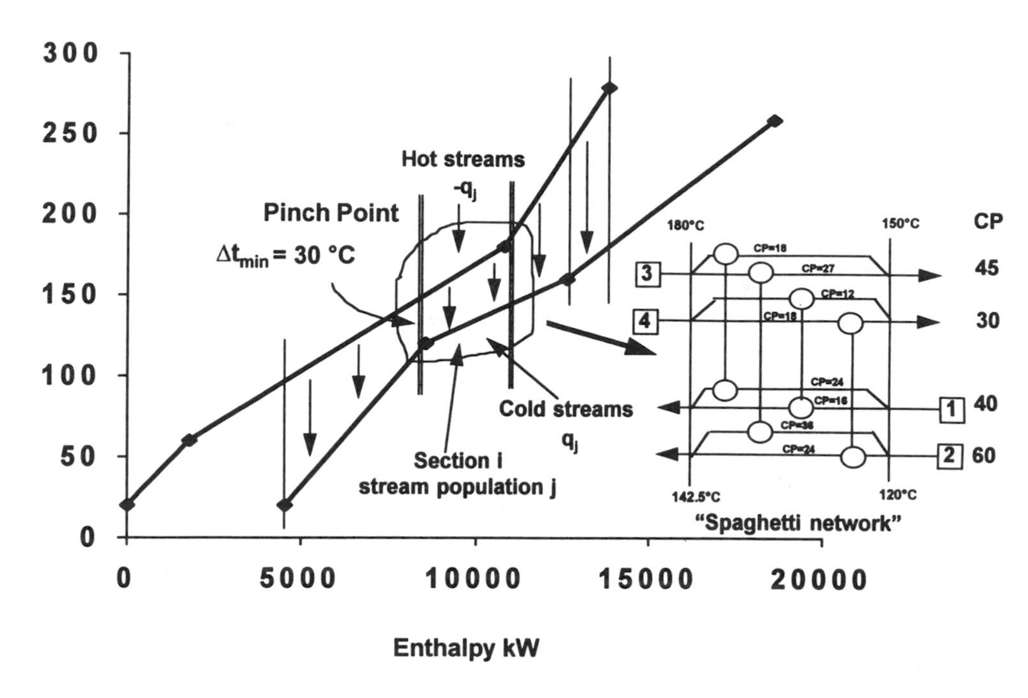

FIGURE 15.9 Enthalpy intervals for area targeting and the spaghetti network that mimics vertical heat transfer.

A loop is a closed path through the network, and its effect is to increase the number of units. For a network to have the minimum number of units, the number of loops should be zero. Figure 15.10 shows a loop on the grid diagram. Normally for a given stream system, only a single network exists and if the number of loops is assumed to be zero, then the equation can be reduced to[8]

$$u = N - 1$$

Figure 15.10 illustrates the occurrence of independent networks. The total network consists of two independent subnetworks. Thus, for the system shown in this figure, $N = 5$, $L = 1$, $s = 2$, and $u = 4$.

Target for Minimum Number of Shells

Trivedi et al.[12] have developed a method for estimating the total number of shells required for the heat exchanger network. The method starts by setting the energy targets for the selected value of Δt_{min}. The hot and cold utilities are added to the process stream system and the composite curves for the process, and the utility streams are constructed. No external utility requirement will be needed for this set of composite curves. Hence, these are called balanced composite curves.[13] The method estimates the number of shells based on the balanced composite curves. The method involves two steps:

Step 1: Estimate the total number of shells required by the cold process streams and the cold utility streams. See Figure 15.11(a).

- Commencing with a cold stream target temperature, a horizontal line is drawn until it intercepts the hot composite curve. From that point a vertical line is dropped to the cold composite curve. This section, defined by the horizontal line, represents a single exchanger shell in which the cold stream under consideration gets heated without the possibility of

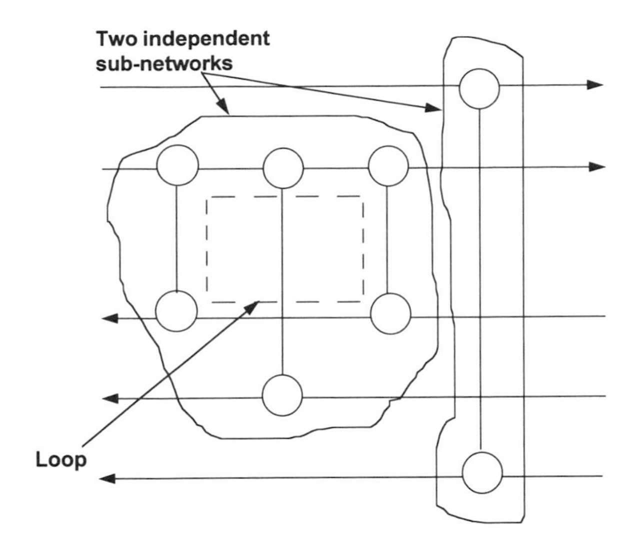

FIGUkE 15.10 Loops and independent subnetworks in a heat exchanger network.

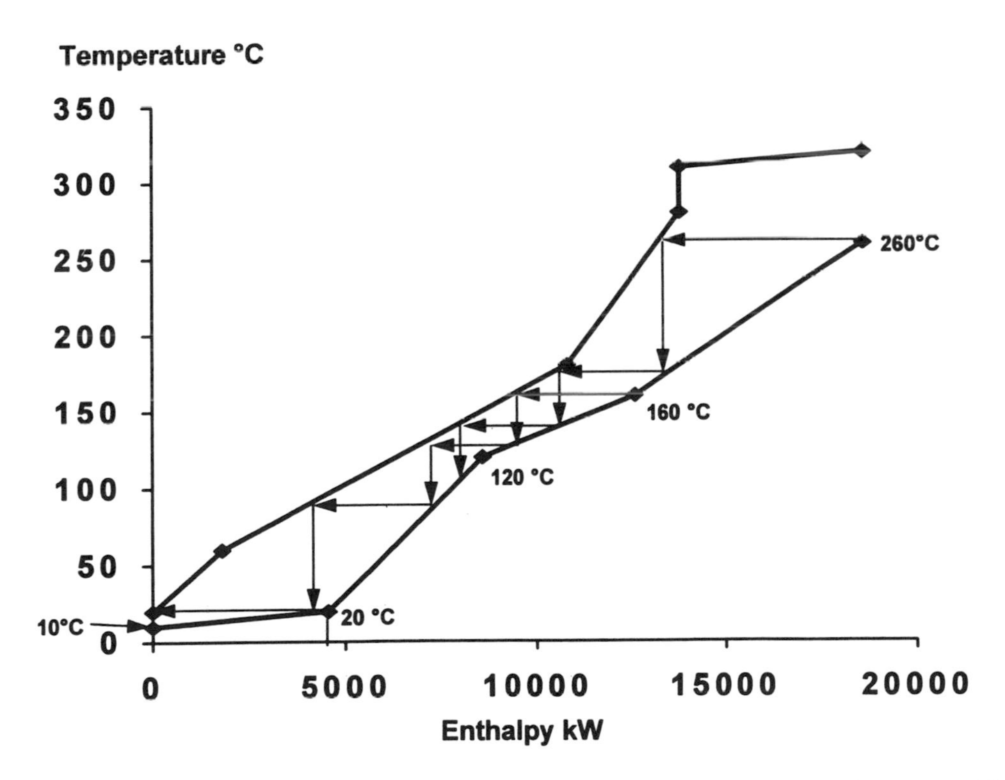

FIGURE 15.11(a) Shell targeting for cold streams on the balanced composite curves. Cold streams will require 8 shells.

a temperature cross. In this section, the cold stream will have at least one match with a hot stream. Thus, this section implies that the cold stream will require at least one shell. Further, it ensures that log mean temperature (LMTD) correction factor, $F_t \geq 0.8$.

- Repeat the procedure until a vertical line intercepts the cold composite curve at or below the starting temperature of that particular stream.

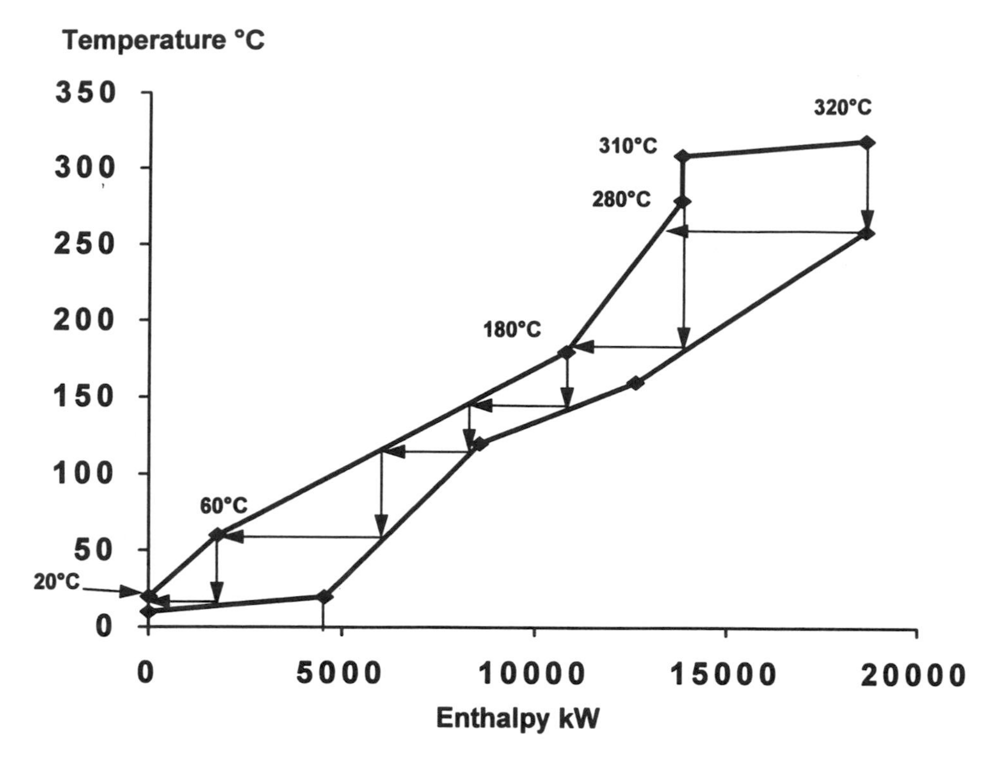

FIGURE 15.11(b) Shell targeting for hot streams on the balanced composite curves. Hot streams will require 9 shells.

- The number of horizontal lines will be the number of shells the cold stream is likely to require to reach its target temperature.
- Repeat the procedure for all the cold streams including the cold utility streams.
- The sum of the number of shells for all the cold streams is the total number of shells required by the cold streams to reach their respective target temperatures.

Step 2: Estimate the total number of shells required by the hot process streams and the hot utility streams. See Figure 15.11(b).

- Starting from the hot stream initial temperature, drop a vertical line on the balanced composite curve until it intercepts the cold composite curve. From this point construct the horizontal and vertical lines until a horizontal line intercepts the hot composite curve at or below the hot stream target temperature.
- The number of horizontal lines will be the number of shells required by the hot streams for heat exchange in the network.
- Repeat the procedure for all the hot streams including the hot utility streams.
- The sum of the number of shells required by the hot streams will be the total number of shells required by the hot streams to reach their respective target temperatures.

The quasi-minimum number of shells required in the network would be the larger of either the total number of shells required by the hot streams or the total number of shells required by the cold streams.

Ahmad and Smith[14] have pointed out that this procedure works for the problems whose composite curves are wide. They suggest that the procedure be applied independently to the above and below pinch subnetworks (above and below pinch subnetworks are discussed in the section "Pinch Point and Network Design"). They have also proposed another method for finding the number of shells in a network. Their method is more sophisticated and is not within the scope of this chapter.

Optimum Δt_{min} Value

The procedures presented so far establish, for a selected Δt_{min} value, targets for minimum energy, minimum area, minimum number of units, and minimum number of shells. These targets along with the cost data can be translated into capital and energy cost for the network. The targets can be evaluated at different values of Δt_{min} to obtain a bird's-eye view of the solution space and the optimal value of Δt_{min} ahead of design. This philosophy was first proposed by Hohmann[8] and later developed by Linnhoff and Ahmad[15] in what is now known as "supertargeting."

For the process flowsheet illustrated in Figure 15.6, the different targeting curves are shown in Figure 15.12(a)–(f). It is seen from the total annualized cost curve, Figure 15.12(f), that the optimal region is very flat. While selecting a value of Δt_{min} from the optimal region, a couple of points should be kept in mind. Different values of Δt_{min} lead to different topologies. Hence, we should take into account different factors that can affect the final cost. They are the wideness of the composite curves and the problem constraints that have significance in the network synthesis and refinement. The optimal value of Δt_{min} selected for this problem is 40°C.

15.6 The Pinch Point

On the composite curves there are one or more enthalpy values for which the two composite curves are Δt_{min} apart. For the example under consideration (see Figure 15.8), this occurs at the hot stream temperature of 150°C and cold stream temperature 120°C. This is also the fourth temperature interval in the problem table (see Table 15.5). In this interval, the heat cascade has zero heat flow (i.e., no heat is transferred across this interval when minimum hot utility is used). This interval, identified by a zero in the corrected heat cascade column of Table 15.5, is referred to as the pinch point.[11,16] The significance of the pinch point is now clear—for the minimum external utility requirement do not transfer heat across the pinch point. Any extra amount of external heat that is put into the system above the minimum will be transferred across the pinch point and will be removed by the cold utility.[17]

Cross Pinch Principle[17]

For a given value of Δt_{min}, if the network is using Q_h units of hot utility and if Q_{hmin} is the minimum energy target, then

$$Q_h = Q_{h\min} + \alpha$$

If the network uses Q_c units of cold utility and if Q_{cmin} is the minimum energy target, then

$$Q_c = Q_{c\min} + \alpha$$

where α = the amount of cross pinch heat transfer.

The Pinch Point and Its Significance

The pinch point divides the stream system into two independent subsystems. The subsystem above the pinch point is a net heat sink, and the subsystem below the pinch point is a net heat source. Thus, for a system of hot and cold streams, to design a network that meets the minimum utility targets the following rules set by the pinch principle should be followed:[17]

- Do not transfer heat across the pinch point.
- Do not use hot utility below the pinch point.
- Do not use cold utility above the pinch point.

We shall use these principles to design the network for the optimal value of Δt_{min}.

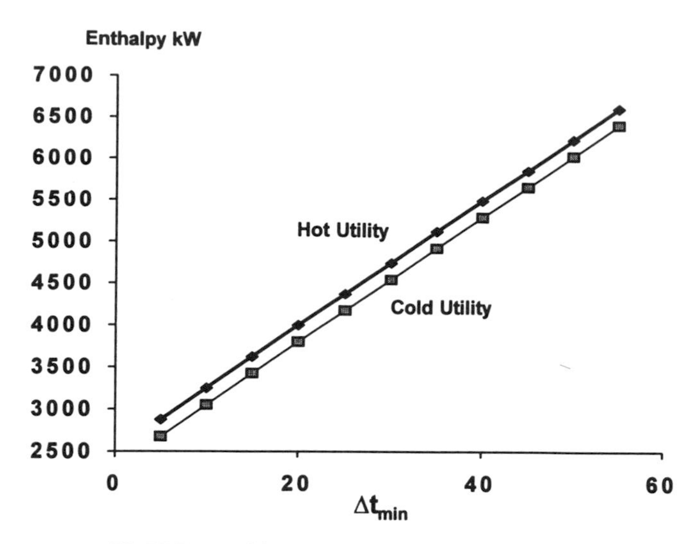

FIGURE 15.12(a) Minimum energy targeting plot.

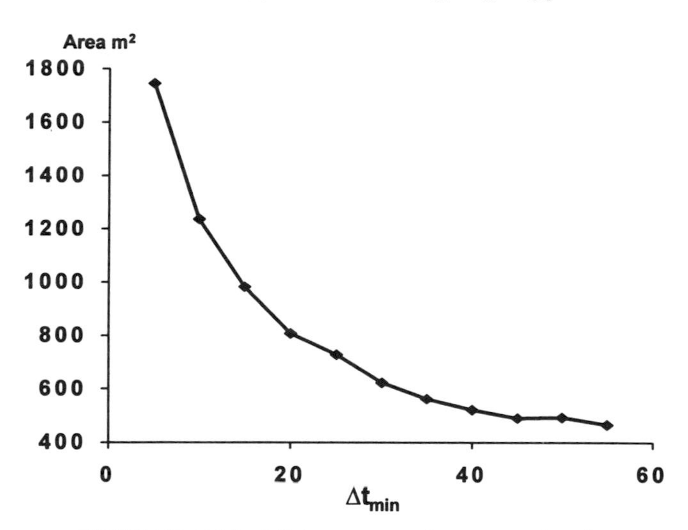

FIGURE 15.12(b) Minimum network area targeting plot.

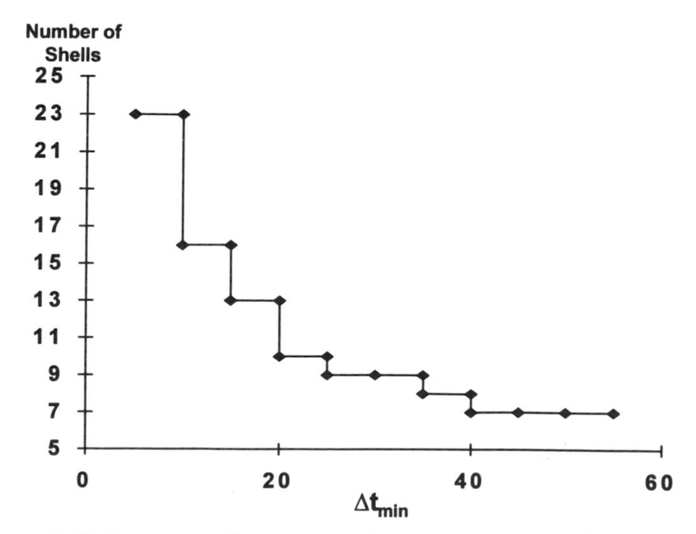

FIGURE 15.12(c) Targeting plot for minimum number of shells.

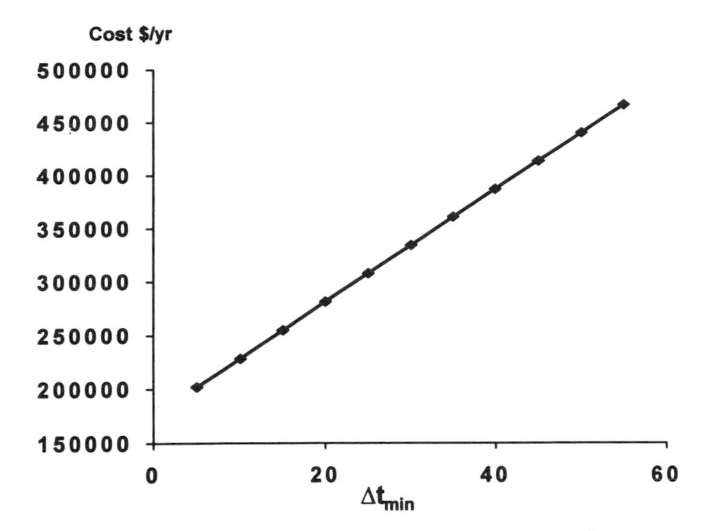

FIGURE 15.12(d) Minimum energy cost targeting plot.

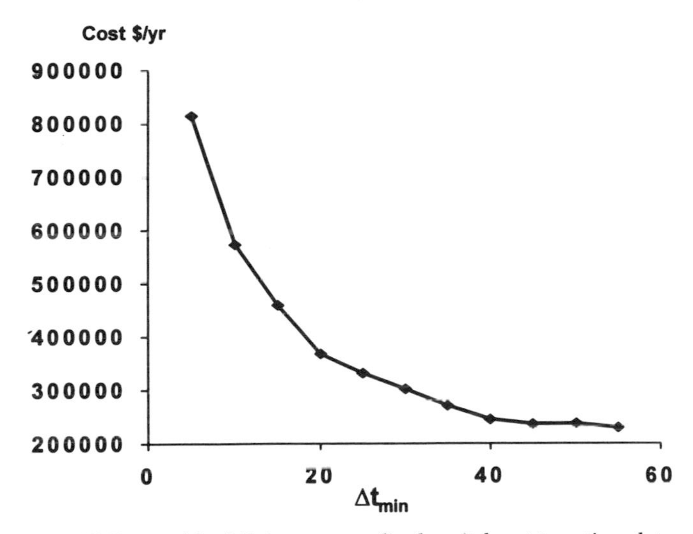

FIGURE 15.12(e) Minimum annualized capital cost targeting plot.

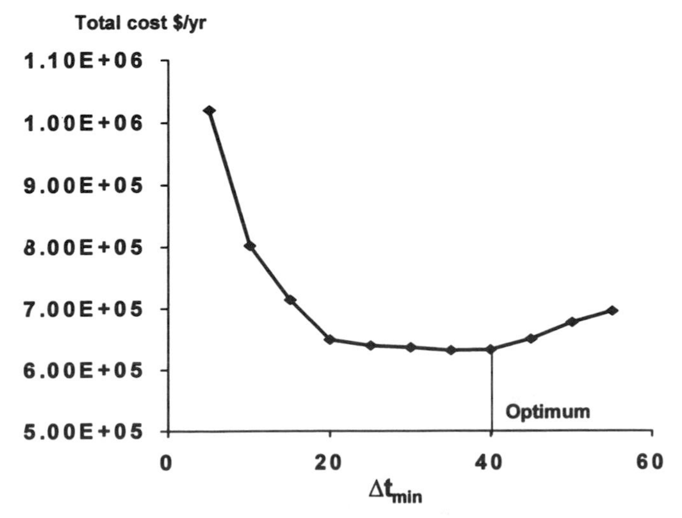

FIGURE 15.12(f) Targeting plot for total annualized cost.

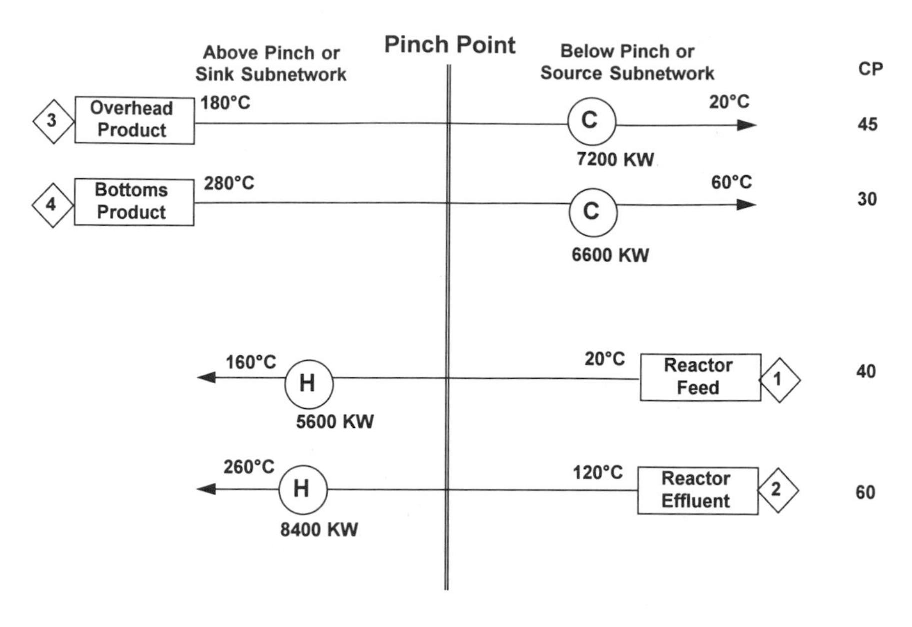

FIGURE 15.13 Grid diagram representation of the network in the flowsheet of Figure 15.6.

15.7 Network Design

Network Representation on the Grid Diagram

If we attempt to design the network using the conventional flowsheet format, any changes in the design may lead to redrawing the flowsheet. Hence, we shall use the grid diagram to represent and design the network. Figure 15.13 shows the grid diagram for the flowsheet shown in Figure 15.6. On this grid diagram we can place a heat exchanger match between two streams without redrawing the whole stream system. The grid representation reflects the counter-current nature of heat transfer that makes it easier to check temperature feasibility of the match that is placed. Furthermore, we can easily represent the pinch point on the grid diagram as shown in Figure 15.13.

Pinch Point and Network Design

We now continue with the design of the example problem. Figure 15.8 shows the composite curves with Δt_{min} = 30°C for the process flowsheet. It also shows the pinch point. The pinch point is the most constrained region of the composite curves. From optimization principles, we know that for a constrained problem, the optimal solution is located at the point formed by the intersection of multiple constraints. Hence, if we can satisfy the constraints at that point then we are guaranteed an optimal design. Thus, we should start the design where the problem is most constrained, that is, the pinch point.[17] The pinch point divides the problem into two independent subnetworks (see Figures 15.8 and 15.3)—an above the pinch or the sink subnetwork that requires only hot utility and a below the pinch or the source subnetwork that requires only cold utility.

Pinch Design Rules and Maximum Energy Recovery

Let us make some observations at the pinch point. Immediately above the pinch point,[17]

$$\Sigma CP_h \leq \Sigma CP_c$$

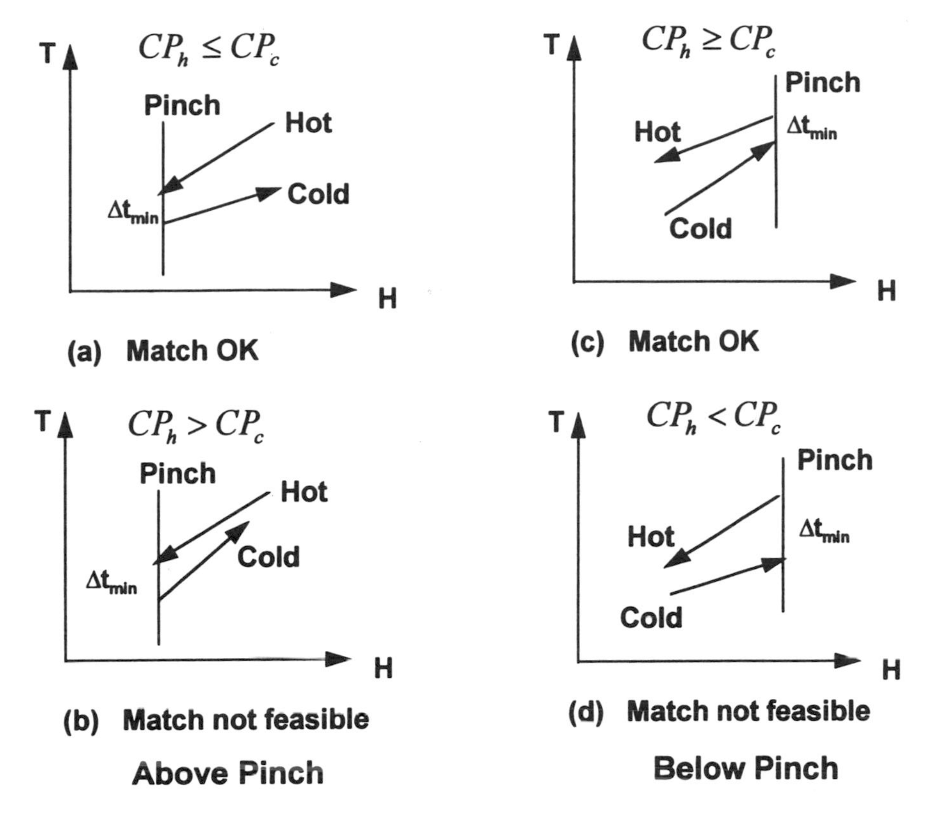

FIGURE 15.14 CP matching rules at the pinch point. (Reproduced by permission from the Institute of Chemical Engineers from *The Chemical Engineer,* Nov. 1987, page 30.)

and immediately below the pinch point,

$$\Sigma CP_h \ge \Sigma CP_c$$

This condition defines the pinch point, that is, the constraint which the designer has to satisfy. Thus, each and every match in the sink subnetwork at the pinch point should be placed such that $CP_h \le CP_c$. Similarly, each and every match in the source subnetwork at the pinch point should be placed such that $CP_c \le CP_h$. Figure 15.14 shows these conditions on the *T–H* diagram.

Two situations arise where the above condition may not be satisfied. The first condition is shown in Figure 15.15. The solution is to split the hot stream. Another situation is shown in Figure 15.16. Here we will have to split the cold stream.

Consider the situation shown in Figure 15.17, which shows the stream population at the pinch point in a sink subnetwork. Here, the number of hot streams is greater than the number of cold streams. Two matches can be easily placed, but for the third stream there is no cold stream to match unless we split a cold stream. Thus, immediately at the pinch for a sink subnetwork, the stream population constraint tells us that the number of hot streams should be less than or equal the number of cold streams. Applying the same logic to the source subnetwork immediately at the pinch point, the number of cold streams should be less than or equal to the number of hot streams. If this condition is not satisfied, then split a hot stream. Figure 15.18[18] shows the algorithms that are developed for placing matches immediately at the pinch point for both the above and below pinch subnetworks.

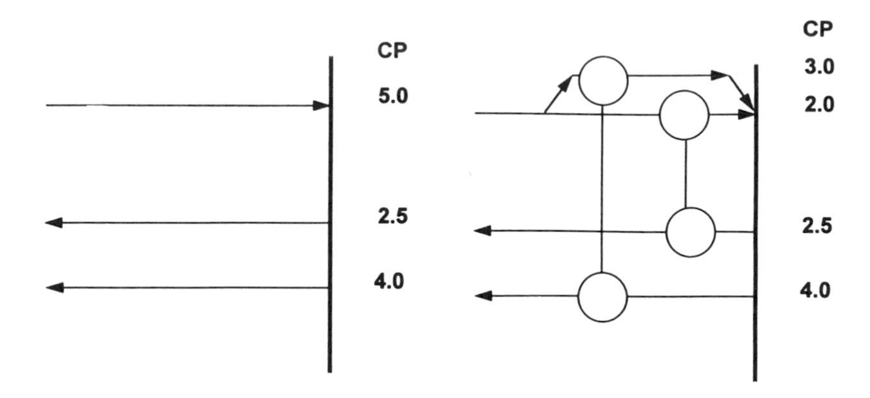

FIGURE 15.15 Stream splitting when CP rule is not satisfied.

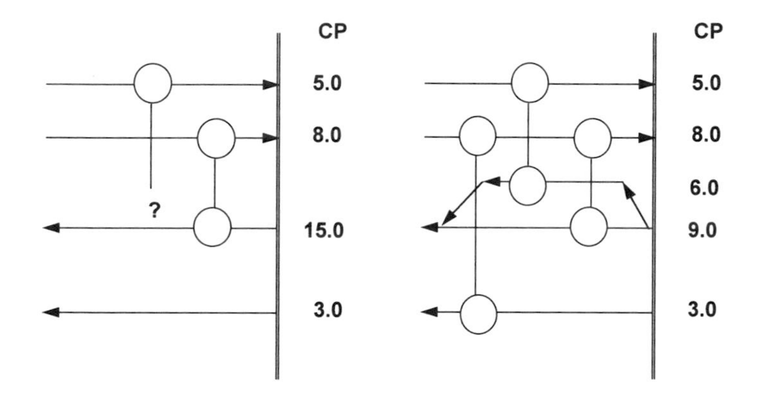

FIGURE 15.16 Stream splitting at the pinch point when CP rule is not satisfied for all hot streams above the pinch point.

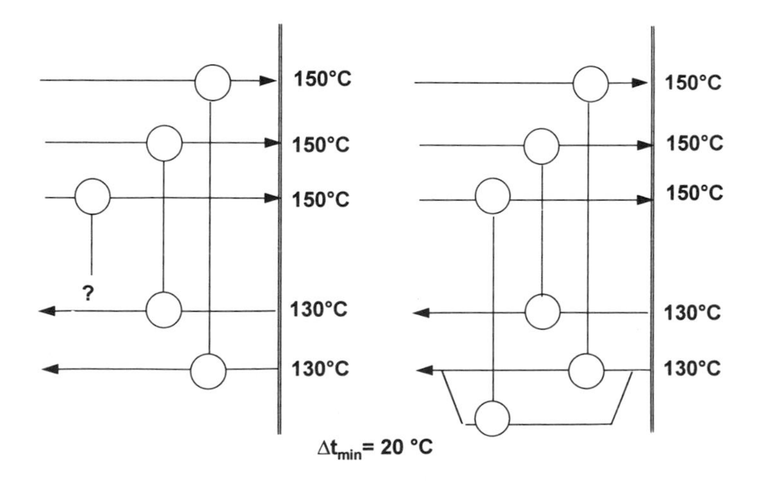

FIGURE 15.17 Stream population at the pinch point may force a stream to split.

The matches and stream splitting can be easily identified with the *CP* table.[17] Consider the stream system shown in Figure 15.19. The *CP* table is shown in the same figure. The first row in this table contains the conditions that need to be satisfied for the subnetwork under consideration. The *CP* values of the hot and cold streams are arranged in a descending order in the columns. The stream

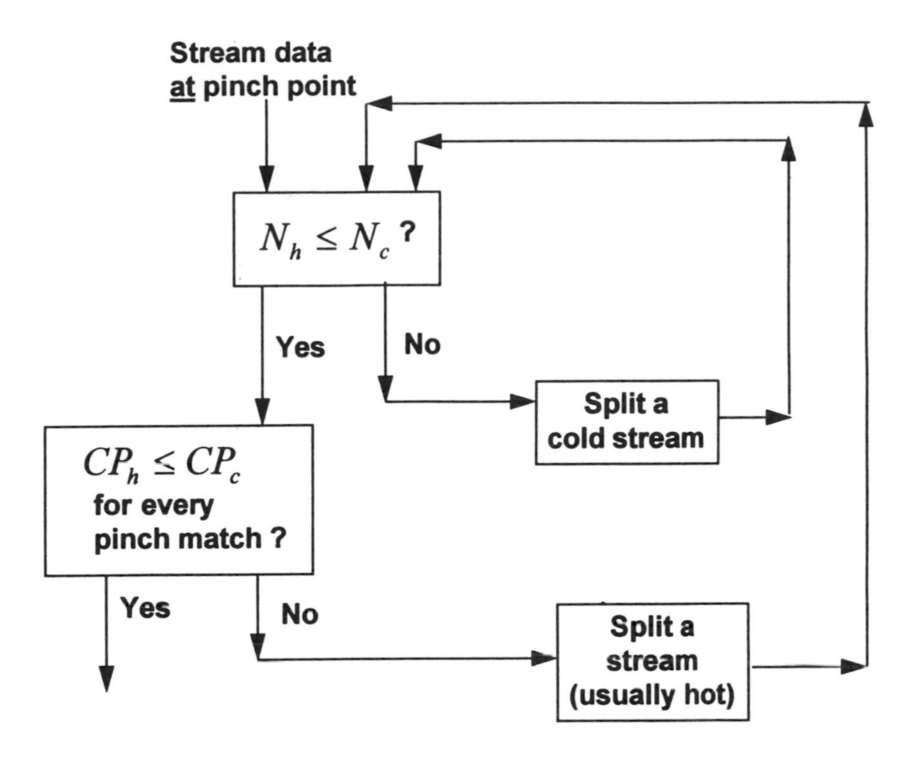

FIGURE 15.18(a) Algorithm for sink subnetwork design at the pinch point. (Reproduced with permission from B. Linhoff.)

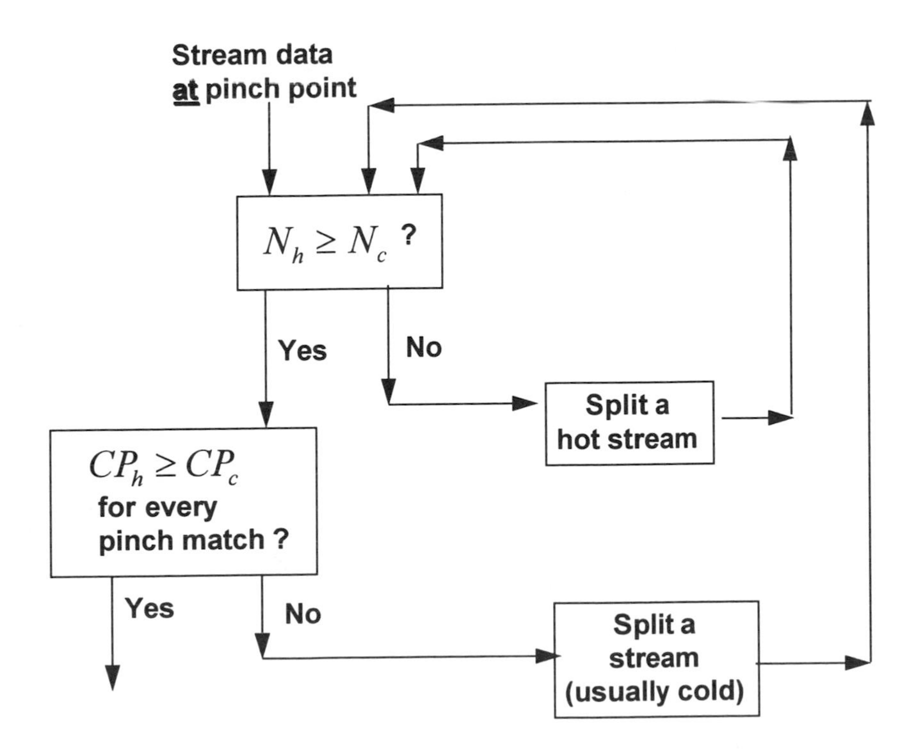

FIGURE 15.18(b) Algorithm for source subnetwork design at the pinch point. (Reproduced with permission from B. Linhoff.)

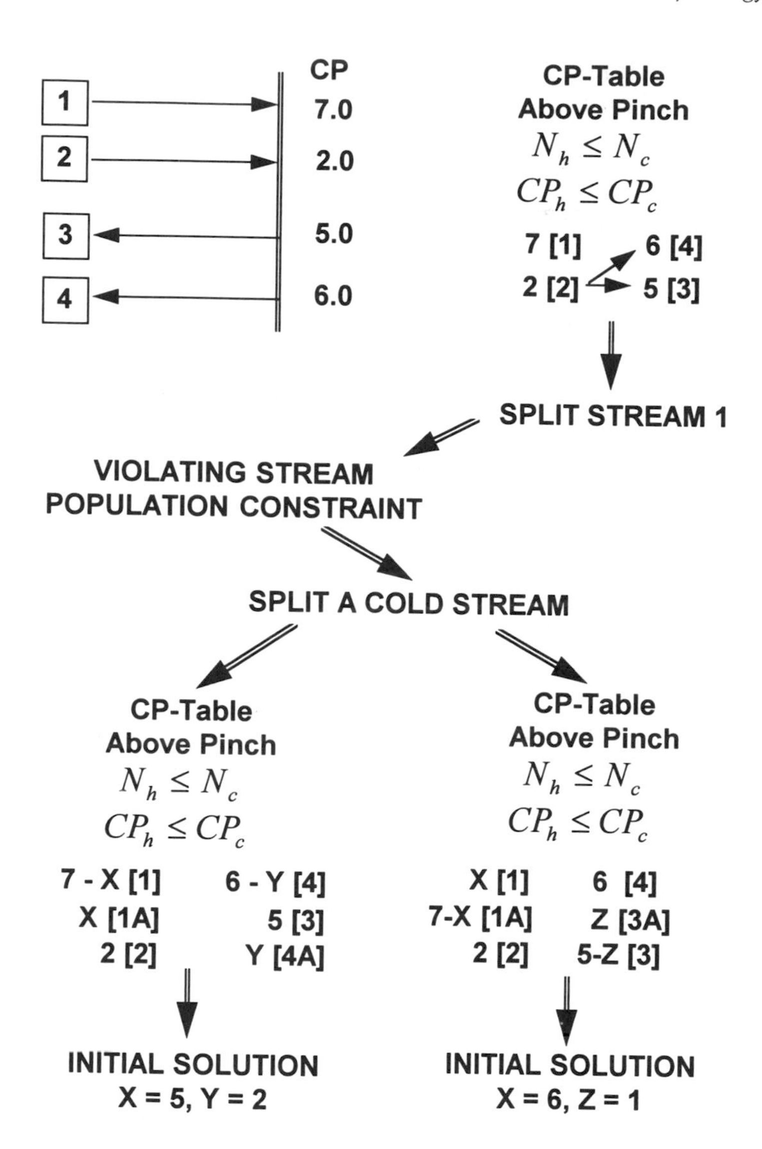

FIGURE 15.19 Identifying matches and stream splitting using the CP table.

numbers are shown in brackets adjacent to the *CP* values. For the sink subnetwork under consideration, the *CP* value for the hot streams is listed in the left column and the *CP* value for the cold stream is listed in the right column. There is no feasible match for hot stream 1. Hence, it will have to be split. Hot stream 2 can match with either stream 3 or 4. Once we split hot stream 1 we violate the stream population constraint. To satisfy this constraint we will have to split a cold stream. Either stream 3 or 4 can be split. The designer can use his or her judgment and decide which stream should be split depending on controllability and other physical constraints. For example, it is not advisable to split a stream that may be a two-phase flow.

Two different designs will be obtained depending on which stream is split. The hot stream is split into two streams having *CP* values of X and $7 - X$. In the first alternative, we split stream 4 into streams having *CP* values of Y and $6 - Y$. In the second option we split stream 3 into streams having *CP* values of Z and $5 - Z$. To find the initial values of X, Y, and Z it is recommended that all matches except for one are set for *CP* equality. Thus, for the first option set $X = 5$ and $Y = 2$. For the second option set $X = 6$ and $Z = 1$. This is an initial solution. The values of X, Y, and Z

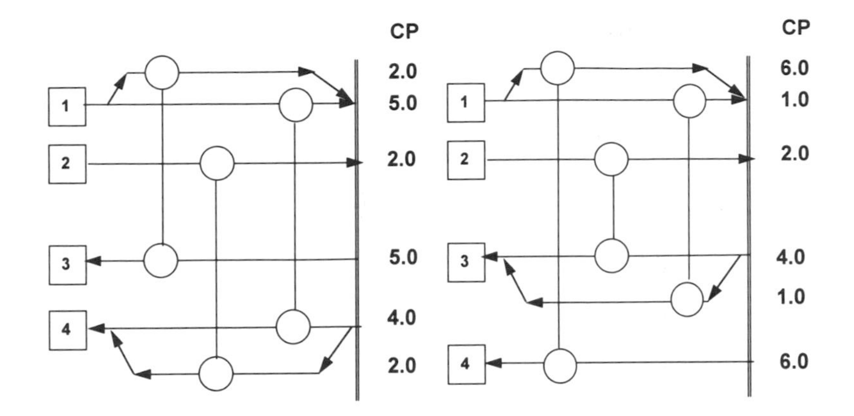

FIGURE 15.20 Two different topologies for the problem in Figure 15.19.

can be adjusted to obtain an optimal design. The two different topologies obtained for this example are shown in Figure 15.20.

Once the matches are identified at the pinch, the heat loads for these matches are fixed to maximize the heat exchange to the limit of heat load of either the hot stream or the cold stream. This will eliminate that stream from the analysis. Fixing the heat loads for match with this heuristic will also help us minimize the number of units and thus the installation cost.[17]

Matches away from the pinch point do not have to satisfy the condition just outlined, as the system is not constrained in this region. Matches are easily identified away from the pinch point. The network obtained will satisfy the energy targets set. Such a network is called maximum energy recover (MER) network.[17]

ΔT–T_c Plot

To appropriately place matches in the network that utilizes the correct driving forces, the driving force plot is a very useful design tool.[19] A plot of the temperature difference ΔT between the hot composite curve and the cold composite curve vs. the cold composite temperature is derived from the composite curves as shown in Figure 15.21. Different versions of this plot can also be used such as the ΔT–T_h plot or T_h–T_c plot. To meet the area targets established one must place the matches such that the ΔT–T_c plot of the exchanger must fall in line with the ΔT–T_c plot obtained from the composite curves. This is a good guiding principle when the heat transfer coefficients of all the streams are in the same order of magnitude.

Figure 15.22 shows two matches placed in the network. The match in Figure 15.22(a) is a good match, as it makes the correct utilization of the available driving forces. The match in Figure 15.22(b) is a bad match, as it uses more ΔT than available. If a match uses more driving force than is available, it will force some other match in the network to use less driving force than available. This results in an overall increase in the total surface area of the network. Badly placed matches do not waste energy but waste surface area.[3]

Remaining Problem Analysis

As we have noted, the rules that are established for design guidelines can identify multiple options. The option that satisfies the targets established ahead of design will be the optimal design. To help screen different options, the Remaining Problem Analysis (RPA) technique[19] should be used. RPA evaluates the impact of the design decision made on the remaining problem. Once a matching decision is made, the remaining stream data excluding the match is retargeted. The targets for the

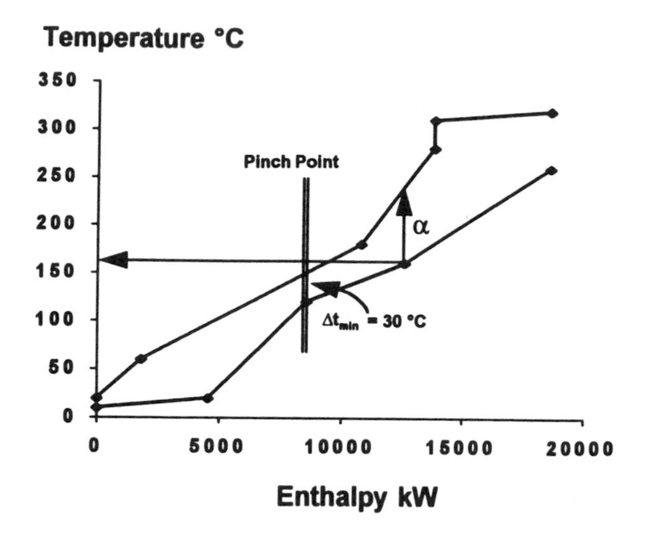

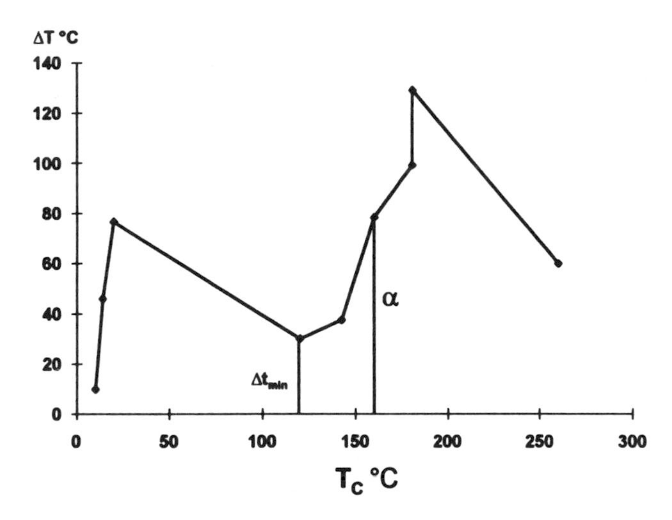

FIGURE 15.21 Construction of the ΔT–T_c plot.

remaining problem should be in line with the original targets. In case they are not, look for other options such as decreasing the heat duty on the match or finding a new match.

Figure 15.23 shows how RPA is undertaken. For the original problem the targets are

Q_{hmin}	= 30 MW
Q_{cmin}	= 20 MW
Minimum area	= 1,000 m²
Minimum number of shells	= 15

A match is placed that has a duty of 40 MW, has a 200 m² area, and needs 3 shells (see Figure 15.23b). The targets for the remaining stream data that excluded the match are

Q_{hmin}	= 40 MW
Q_{cmin}	= 30 MW
Minimum area	= 900 m²
Minimum number of shells	= 14

Thus, by placing the selected match, 10 MW of extra energy will be required, 100 m² more area will be required, and two extra shells will be needed. If we place a new match as shown in

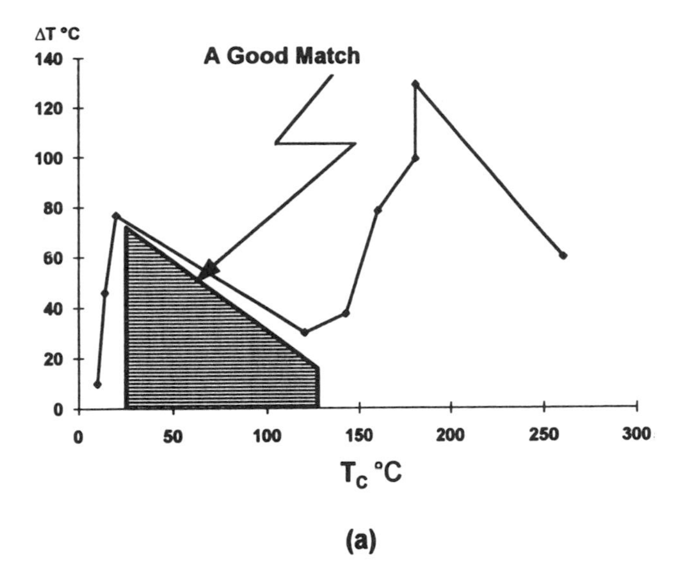

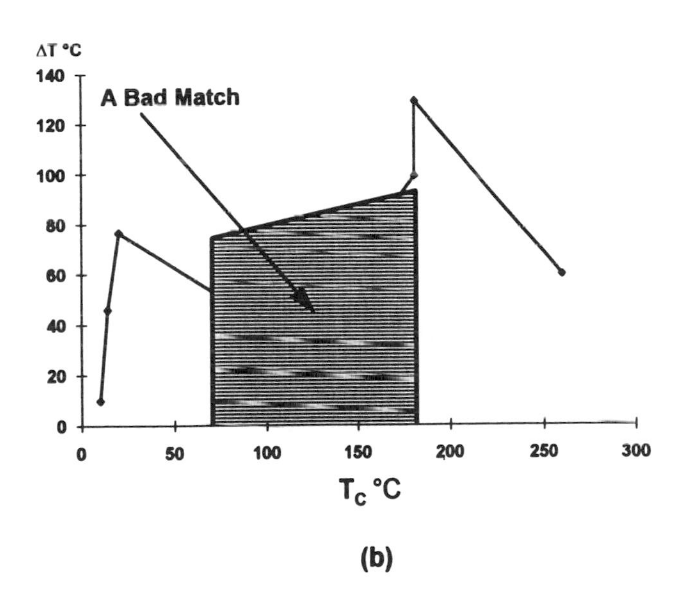

FIGURE 15.22 Identifying good and bad matches using the ΔT–T_c plot.

Figure 15.23(c) the remaining problem targets are in line with the original targets (i.e., only a small penalty is associated with this match). Hence this match should be selected.

The procedure is repeated after placing each and every match. This way the designer gets continuous feedback on the impact of the matching decisions and the heat load assigned to the match. The final design will meet all the targets set before we start the design.

Local Optimization—Energy Relaxation

The optimal region is generally very flat, and the optimal value of Δt_{min} is not a single point but a region. The targeting exercise helps identify this region. Once a value of Δt_{min} is selected, and the network is designed using the methodology outlined earlier, there is still some scope for further optimization. This is achieved by a process called energy relaxation.[17]

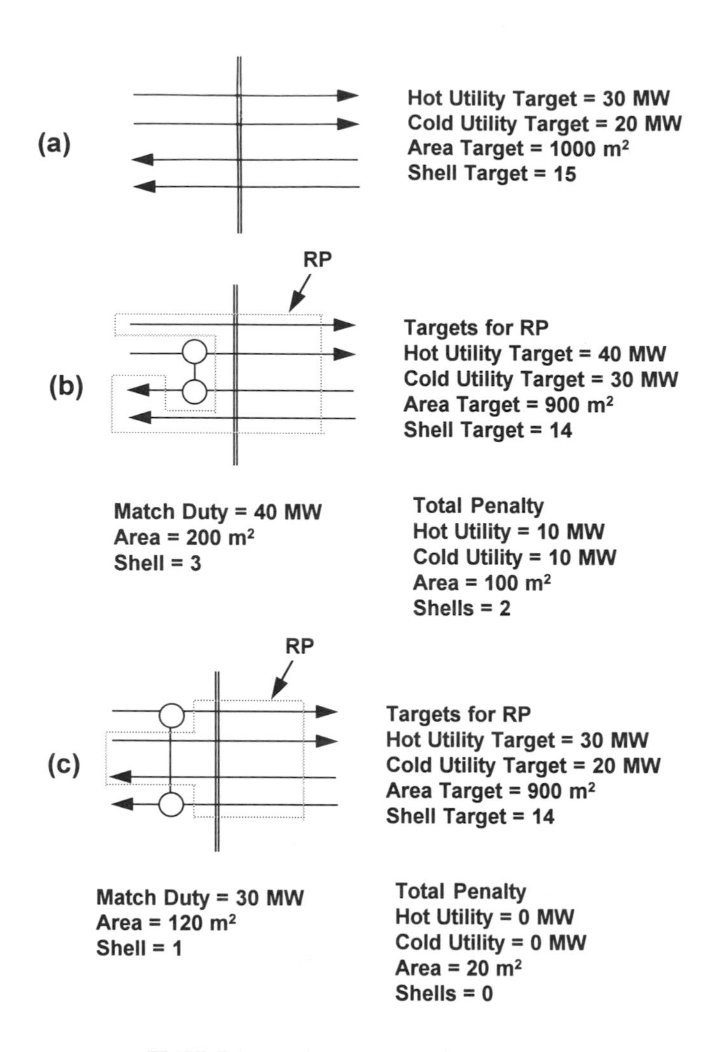

FIGURE 15.23 Remaining problem analysis.

The number of units in the design obtained using this procedure will generally be greater than the target value. This is due to the presence of loops. If we can eliminate the loops in a network, then the heat transfer area is concentrated on fewer matches and hence will decrease the piping and foundation requirements. This will tend to decrease the capital cost of the network. However, some energy penalty may be incurred.

Identifying loops in a large network cannot be done visually as we had done in the example problem. Various algorithms exist to identify loops in a given network.[20,21] Trivedi et al.[22] have proposed LAPIT (Loop and Path Identification Tree) for identifying loops and paths present in the network. A path is the connection between a heater and a cooler via process–process matches.

Loops can be broken by removal of unit in a loop and redistributing the load of the unit among the remaining units of the loop. Some exchangers may result having a very small Δt_{min} when a loop is broken. The Δt_{min} across such exchangers can be increased by increasing the utility consumption along a path that consists of a heater, the unit having a small value of Δt_{min}, and a cooler.

Loop breaking is a very complex optimization process. Trivedi et al.[22,23] have proposed a detail method that systematically breaks loops and identifies options available at each step of the process. The method is based on loop network interaction and load transfer analysis (LONITA)[23] and on a best-first-search procedure.[22]

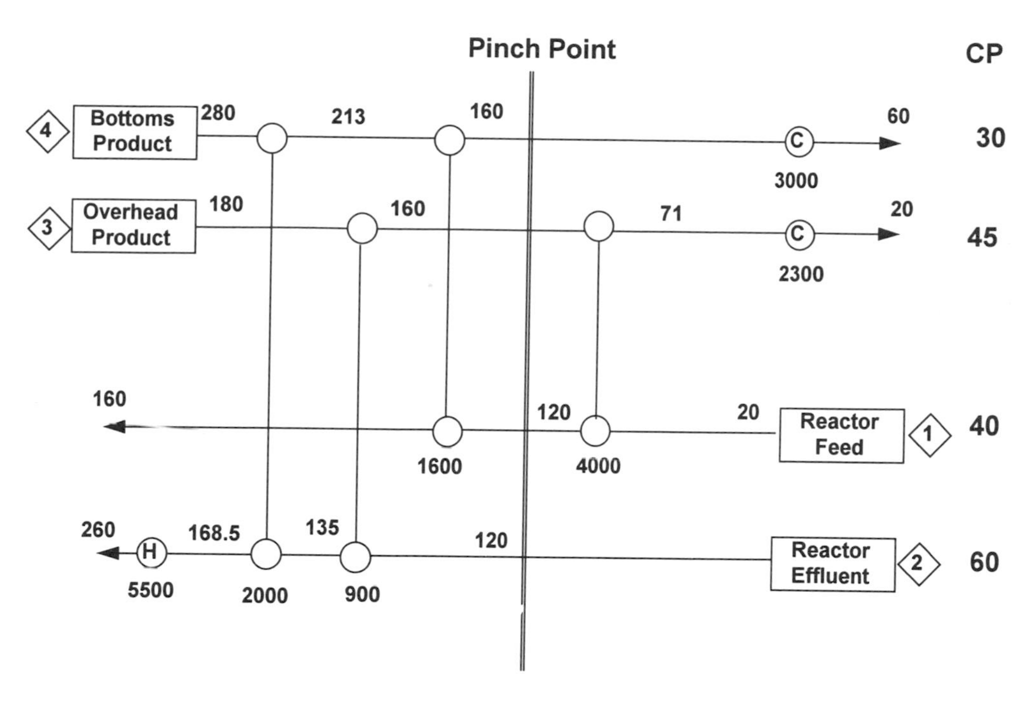

FIGURE 15.24 Pinch design for the flowsheet in Figure 15.6. Total annualized cost is $720,800.

Summary of the Design Procedure

The pinch design procedure can now be summarized. Establish the value of the optimum Δt_{min} using the targeting techniques. This identifies the region in which the design should be initialized. Using the pinch design procedure, design the network. Reduce the number of units using loop breaking and energy relaxation techniques outlined earlier.

Example

The principles and procedures just discussed are used for the design of the network for the flowsheet of Figure 15.6. The value of the optimum Δt_{min} is 40°C. The final network MER design is shown in Figure 15.24. Note that the total annualized cost of this network is only about 4% higher than the target value of $692,700. This design is in the neighborhood of the predicted optimum.

The design after energy relaxation is shown in Figure 15.25. This design has one less unit than the MER design in Figure 15.24. Hence, this will reduce piping, installation, and maintenance costs. The total annualized cost of the new design is about 2% higher than the MER design. Since we are using simplified models for estimating costs, the cost of the design in Figure 15.25 is within the errors of most cost estimates. However, this design will have a lower cost when detailed costs are evaluated, as it has a lower number of units.

Table 15.6 compares the target values of the various optimization parameters with actual design values for Figures 15.24 and 15.25.

15.8 The Dual Temperature Approach Design Method

Concept

The pinch design procedure fixes the Δt_{min}, designs a network, and then undertakes local optimization. During the process of local optimization either the constraint of Δt_{min} across exchangers

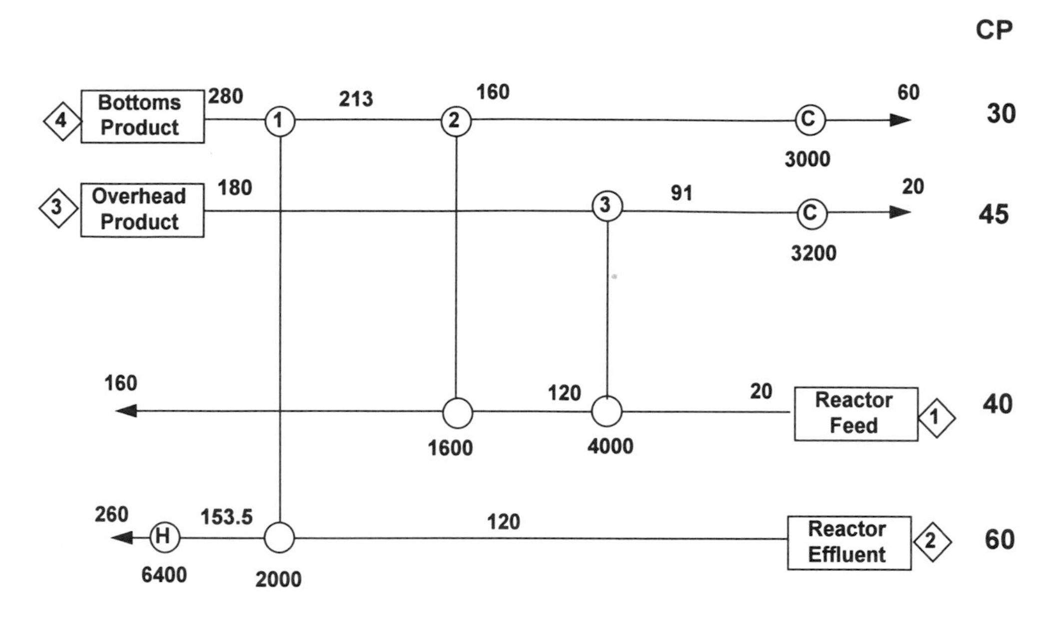

FIGURE 15.25 Energy relaxation of the pinch design in Figure 15.24. Total annualized cost is $773,700. Δt_{Nmin} = 52°C and Δt_{Emin} = 40°C.

TABLE 15.6 Comparison of Target and Design Values for Different Network Optimization Variables at Δt_{min} = 40°C

		Figure 15.24 MER Design		Figure 15.25 Design	
	Target		% Deviation from Target		% Deviation from Target
Total network area, m²	687	779	13	708	3
Total hot utility consumption, kW	5,500	5,500		6,400	16
Total cold utility consumption, kW	5,300	5,300		6,200	17
Number of units	7	7		6	
Number of shells	8	9		7	
Energy cost, $/yr	387,250	387,250		450,700	16
Annualized capital cost $/yr	305,407	333,576	9	282,996	–7
Total annualized cost, $/yr	692,657	720,826	4	733,696	6

may be relaxed or the energy consumption may be increased. Thus, the final network will have a Δt_{min} less than the Δt_{min} on the composite curves corresponding to the final energy consumption.

The final network for the flowsheet in Figure 15.6 using pinch design is shown in Figure 15.25. In this network the energy consumption corresponds to a Δt_{min} of 52°C on the composite curves. However, the minimum approach temperature is 40°C across exchanger 2. The dual-temperature approach design method for the heat exchanger network synthesis is based on this observation.

The network is characterized by two approach temperatures. One approach temperature is called the network minimum approach temperature Δt_{Nmin}. This is the minimum approach temperature on the composite curves and thus fixes the energy consumption for the network. The other approach temperature is called the exchanger minimum approach temperature Δt_{Emin}. This is the minimum approach temperature that any exchanger will have in the network. In the network of Figure 15.25, Δt_{Nmin} = 52°C and Δt_{Emin} = 40°C. Generally, $\Delta t_{Emin} \leq \Delta t_{Nmin}$. When $\Delta t_{Emin} = \Delta t_{Nmin}$ the network has a global minimum approach temperature just like the Δt_{min} parameter of the pinch design procedure. The dual-temperature design procedure starts with a fixed value of Δt_{Nmin} and Δt_{Emin}. This is in contrast to the pinch design method, which utilizes a global value of Δt_{min} (i.e.,

$\Delta t_{Nmin} = \Delta t_{Emin}$) and then relaxes the energy consumption of the network (with the obvious implication of increase in Δt_{min}).

The concept of using dual approach temperatures for the design of the network is very powerful as the designer retains complete control over the energy consumption of the network. This is in contrast with the pinch design method where energy consumption increases following loop breaking and is determined by the new topology and not by the designer. Another advantage in using the dual-temperature method is the reduction in the number of shells and units. As well, unnecessary stream splits may be avoided, making a more practical network with better operability characteristics. This may result in a substantial decrease (~20%) in the capital cost for a fixed operating cost, representing millions of dollars savings in industrial applications.[24]

The design procedure proposed by Trivedi et al.[25] incorporates the best features of both the pinch design method and the dual temperature approach design procedure. Their design procedure is based on the concept of a pseudo-pinch point.

Location of the Pseudo-Pinch Point

Initially two sets of composite curves are generated using Δt_{Nmin} and Δt_{Emin}, with Δt_{Nmin} selected such that $\Delta t_{Nmin} \geq \Delta t_{Emin}$. Both sets will have different energy consumption denoted as *EC*. As $\Delta t_{Emin} \leq \Delta t_{Nmin}$, $EC(\Delta t_{Nmin}) \geq EC(\Delta t_{Emin})$.

Let us define the energy difference α (see Figure 15.26a and b) as

$$\alpha = EC\left(\Delta t_{N\min}\right) - EC\left(\Delta t_{E\min}\right)$$

As the network is designed employing two approach temperatures, a new interpretation of the problem could be this: As Δt_{Emin} is less than Δt_{Nmin}, an amount of energy α traverses the Δt_{Emin} conventional pinch point. This can be compensated by energy carried upward across the pinch because heat exchange is allowed at temperature differences as low as Δt_{Emin} between streams. The maximum amount of this upward carriage that can be achieved is the value of α defined earlier. This represents additional flexibility in stream matching by providing for upward and downward carriage, which is permitted by the reduced Δt_{Emin} between streams, while the total energy requirement remains fixed at $EC(\Delta t_{Nmin})$, which in turn is defined by the network approach temperature Δt_{Nmin}.

A pseudo-pinch point (actually a set of stream temperatures) is defined so that the stream temperatures at this point allow the problem to be partitioned as shown in Figure 15.27. Thus, the two parts of the network are in enthalpy balance with the utility consumption determined by Δt_{Nmin}.

The pseudo-pinch point is defined based on the observation that the changes in slope of the composite curve that occur at the conventional pinch point require (at least) one hot or cold stream entering at a real pinch point temperature.[26] This stream's starting temperature is chosen to determine the pseudo-pinch point. Let the subscripts *pp* denote the pseudo-pinch temperatures. To determine the pseudo-pinch point temperatures with α units of heat transferred across the conventional pinch point at Δt_{Emin}, various strategies for allocating the α units are proposed by Trivedi et al.[25] Generally the following heuristics are used:

1. Pseudo-pinch point temperatures having $\Delta T = \Delta t_{Emin}$: If a hot stream with a starting temperature T_{hs} determines the conventional pinch point for Δt_{Nmin} or Δt_{Emin}, then this stream is also assumed to be at the pseudo-pinch point with all the cold stream temperatures given by $T_{hs} - \Delta T_{Emin}$. Likewise, if a cold stream entrance determines the conventional pinch point, then the pseudo-pinch temperatures of all the hot streams are given by $T_{cs} + \Delta t_{Emin}$. It is possible that different streams may determine the conventional pinch point for both Δt_{Emin} and Δt_{Nmin}. In such a situation different topologies will result and all possible configurations should be investigated.

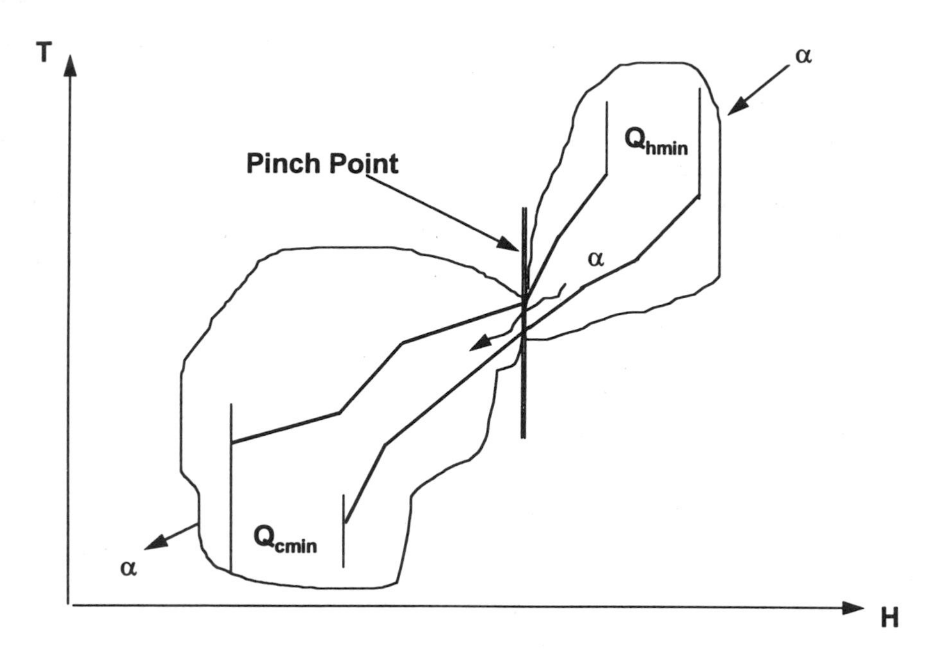

FIGURE 15.26(a) Energy relaxation with heat transfer across the pinch point.

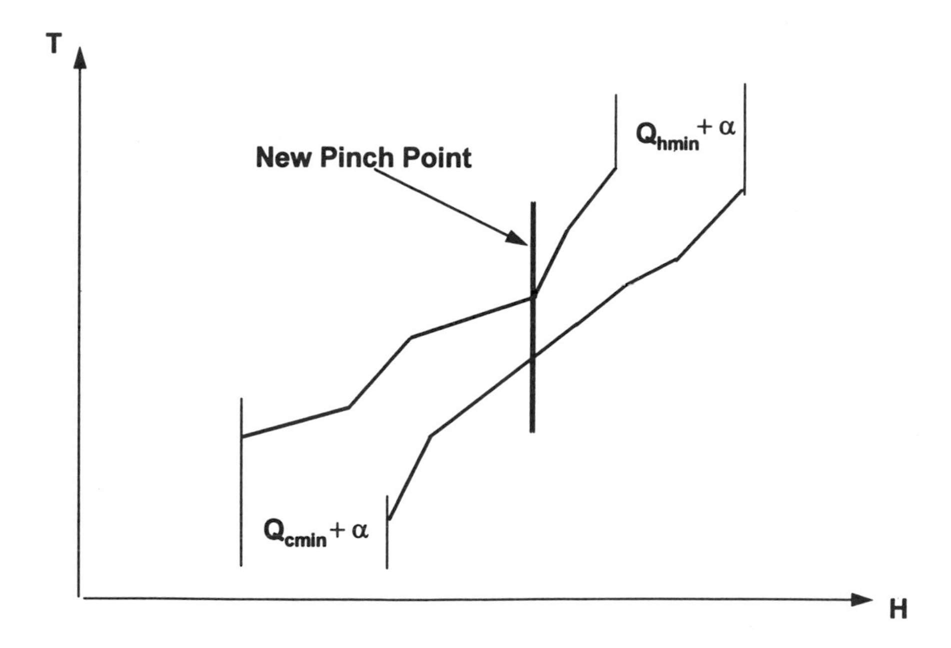

FIGURE 15.26(b) New pinch point with minimum energy consumption plus (α).

2. Pseudo-pinch point temperatures with $\Delta T > \Delta t_{Emin}$: The α units of heat carried across the conventional pinch point will increase the pseudo-pinch point ΔT's for some stream combinations. Assuming that a hot stream determines the conventional pinch point T_{hspp}, various strategies are used. Generally, if there are N other hot streams with starting temperature $T_{hsj} \geq T_{hspp}$, then $\alpha_j = \alpha/N$ is allocated to each of these streams so that their pseudo-pinch point temperatures are increased by α/NCP_j.

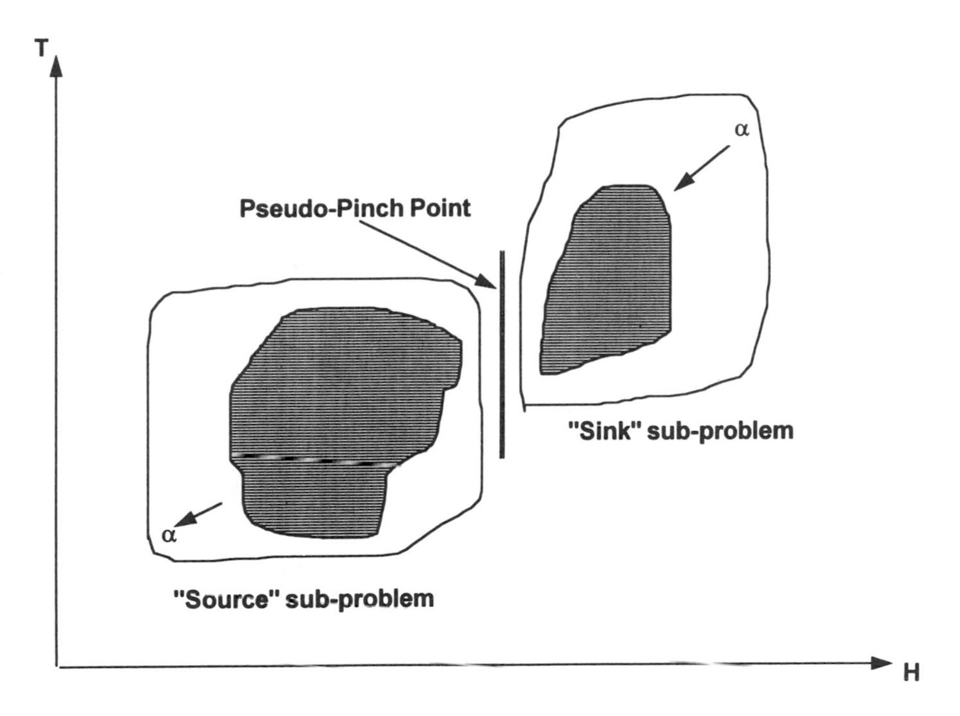

FIGURE 15.27 Pseudo-pinch division of the system.

Design of the Sink Subproblem

Feasibility Criteria at the Pseudo-Pinch

As the problem constraints are relaxed at the conventional pinch point by passing heat across it, a wide variety of network topologies can be generated. The method commences the design at the pseudo-pinch point. As there are only a few essential matches to be made at the pseudo-pinch point, all the options available to the designer are readily identified. The designer, utilizing process knowledge and experience, can now select the necessary matches. These could include imposed and constrained matches required for a safe, controllable, and practical network or any other preferences the designer may have. The feasibility criteria for stream matching and splitting follow.

Number of Process Streams and Branches. In a conventional pinch situation, the stream population is compatible with a minimum utility design only if a pinch match is found for each hot stream above the conventional pinch point. The situation is illustrated in Figure 15.17, where stream splitting is unavoidable. However, for the pseudo-pinch situation, splitting may not be necessary as demonstrated in Figure 15.28. This results from the relaxation of the conventional pinch temperatures resulting in an increase in the available driving forces for the pseudo-pinch matches.

It is exceedingly difficult to determine in advance if stream splitting is necessary. Hence, two approaches are suggested. The first approach is identical to the pinch design method. Stream splits may be removed once the initial design is generated. This will be discussed later. In the second approach, we place the pseudo-pinch matches and see if the unmatched streams at the pseudo-pinch can be satisfied. If the matching streams are violating Δt_{Emin}, then a stream is split and the pseudo-pinch matching is repeated.

Temperature Feasibility Criteria. Four temperature profiles are possible for a match at the pseudo-pinch point as shown in Figure 15.29. These profiles are for the sink subproblem. The

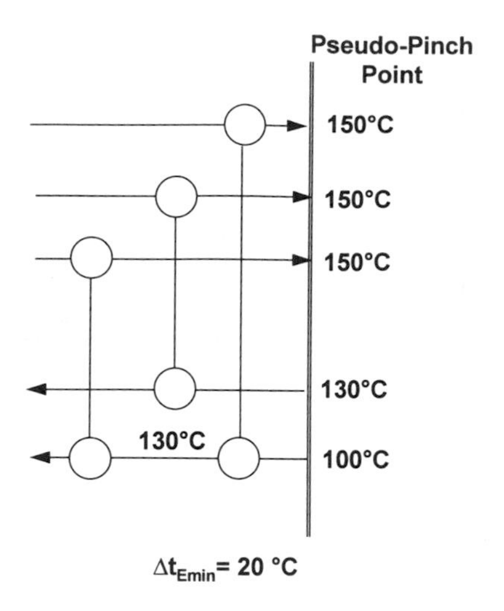

FIGURE 15.28 Stream splitting may not be necessary in pseudo-pinch problems.

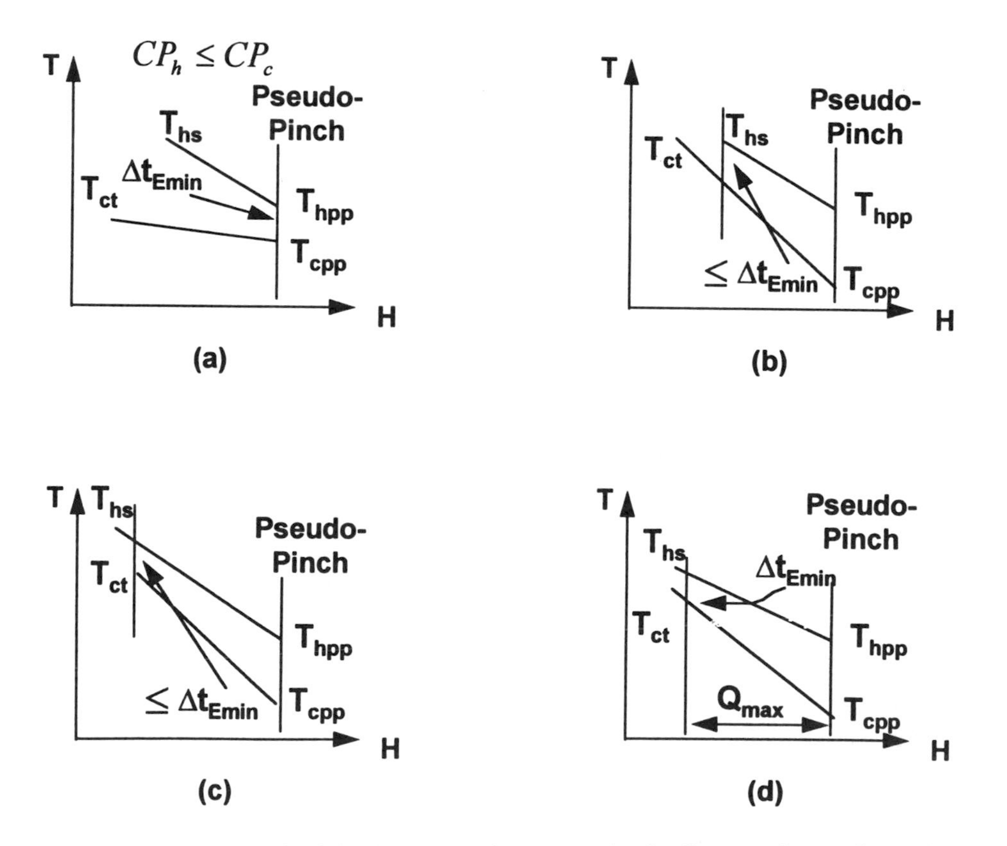

FIGURE 15.29 Temperature feasibility for stream splitting immediately adjacent to the pseudo-pinch point for the sink subproblem.

match illustrated in Figure 15.29(a) possesses the same characteristics as that of an above conventional pinch match. However, for a pseudo-pinch, other matches of the type illustrated in Figure 15.29(b), (c), and (d) are also possible. Trivedi et al.[25] have outlined the various criteria to be used at the pseudo-pinch that determine the maximum heat load Q_{max} for a match between a hot and a cold stream.

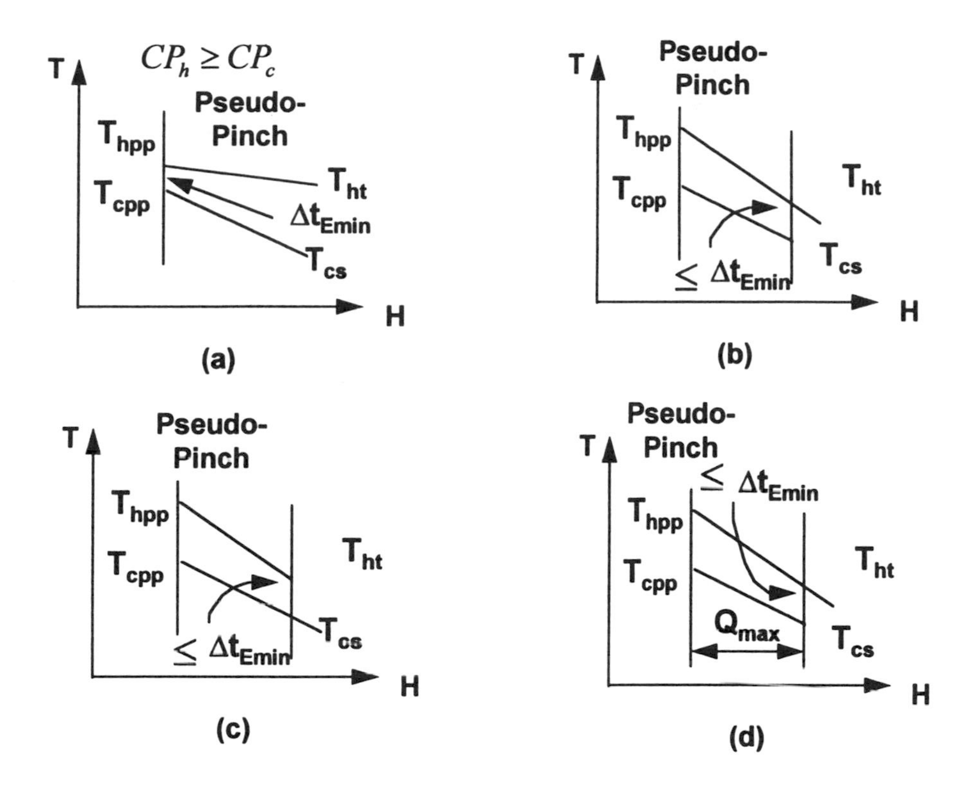

FIGURE 15.30 Temperature feasibility for stream splitting immediately adjacent to the pseudo-pinch point for the source subproblem.

Design of the Source Subproblem

For the source subproblem analogous feasibility criteria can readily be developed. These depend both on the stream populations and temperature feasibility shown in Figure 15.30. The complete design procedure is summarized in Figure 15.31(a) and (b).

Design of the Remaining Network

Once the pseudo-pinch matches are placed, the remaining problem can be designed employing the rules mentioned earlier depending on the subproblem. It should be noted that the remaining problem is still in enthalpy balance and only requires either a hot or cold utility, depending on the subproblem. The utilities can be placed on either end or at an appropriate temperature level. (The rules outlined by Trivedi et al.[25] can also be employed in selecting the nonpinch matches when the pinch design method is employed.)

Once an initial topology is synthesized, its capital cost can be further reduced by decreasing the number of units and removing stream splits keeping the energy consumption constant. A linear program can be used to remove stream splits.[25] To reduce the number of units, LONITA can be used.

Example

We shall illustrate the dual temperature design method for the flowsheet shown in Figure 15.6. A final topology obtained by employing the pinch design method is illustrated in Figure 15.25. In this design Δt_{Emin} = 40°C and Δt_{Nmin} = 52°C. For Δt_{Emin} = 40°C, the pinch point is determined by the entrance of cold stream 2. Hence, the pseudo-pinch temperature for all the hot streams is 160°C and for the cold stream 2 is 120°C. As the energy consumption of the network corresponds

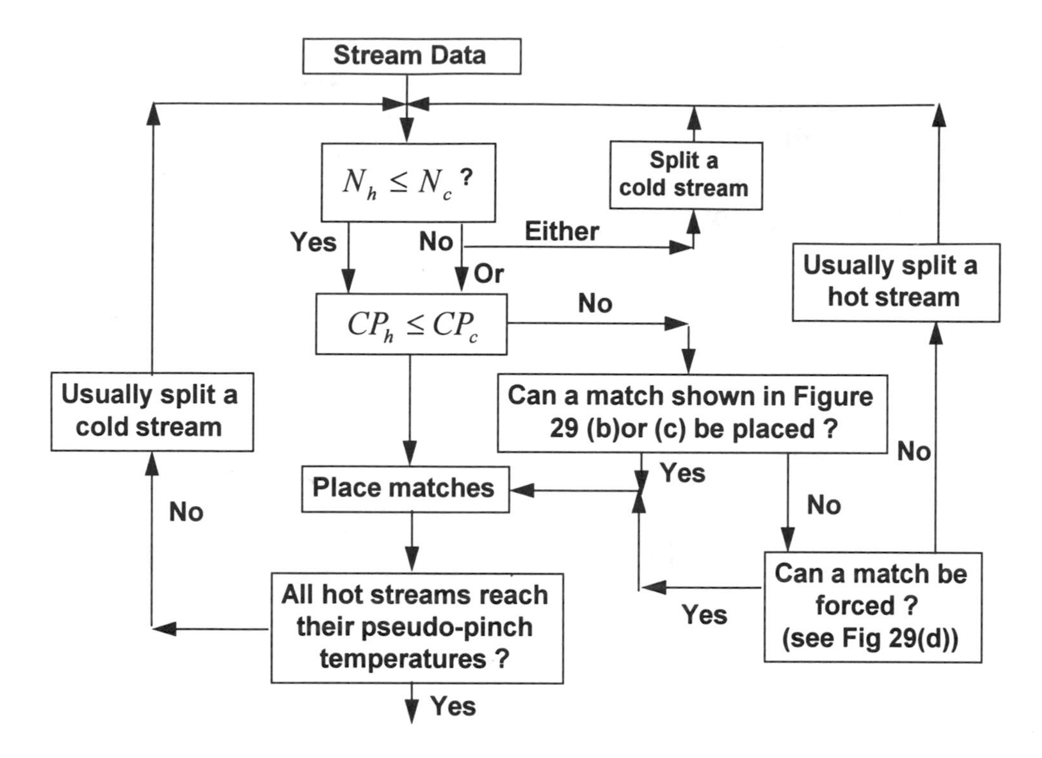

FIGURE 15.31(a) Above pseudo-pinch design algorithm.

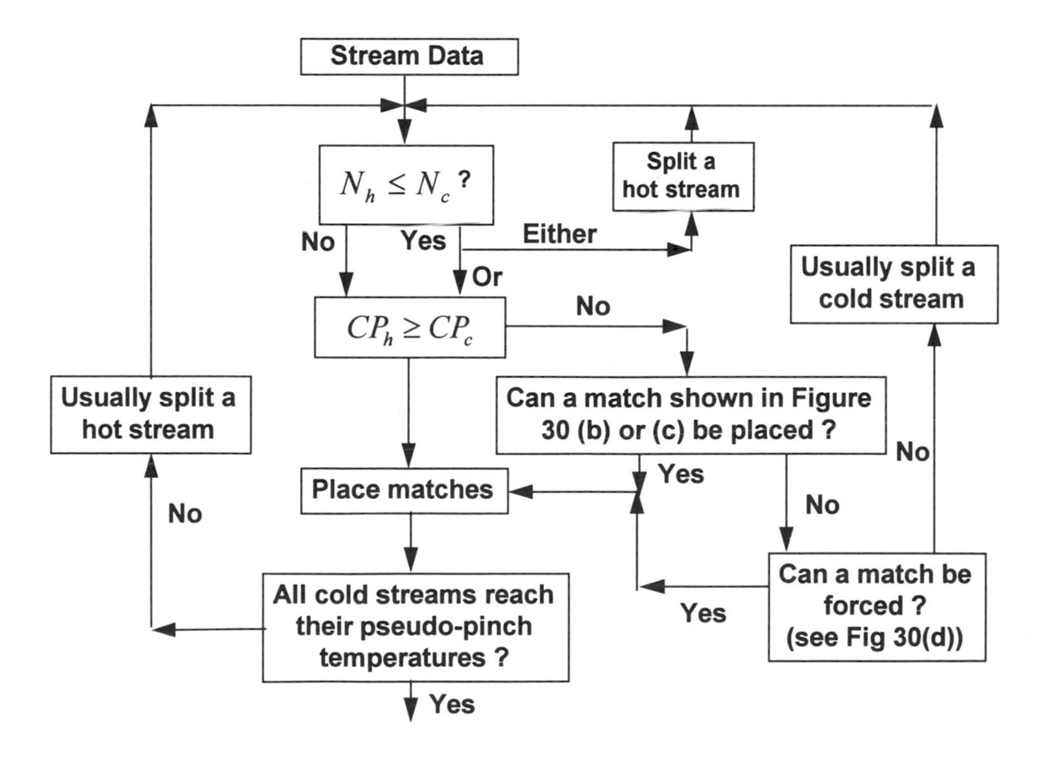

FIGURE 15.31(b) Below pseudo-pinch design algorithm.

to Δt_{Nmin} = 52°C, α = 900 kW. This additional heat load is carried by stream 1, and hence the pseudo-pinch temperature of cold stream 1 is 97.5°C (120-(900/40)).

Above Pseudo-Pinch Design

The two possible designs are illustrated in Figure 15.32(a) and (b). It is interesting to note that the topology of Figure 15.32(b) will not be identified if the pinch design method is employed.

Below Pseudo-Pinch Design

Again two designs are possible as illustrated in Figure 15.32(c) and (d). The matches suggested in Figure 15.32(c) will not be identified by the use of pinch design rules.

The above and below pseudo-pinch designs can be combined to create four different fixed-energy designs. Combining Figure 15.32(a) and (c) and breaking the loop gives the same design as shown in Figure 15.25. The other designs are illustrated in Figure 15.33(a–c). The capital cost of the designs obtained so far are within 3% of the one shown in Figure 15.24. As the cost of all these networks is very close, we will choose the final design based on other factors such as operability, controllability, and flexibility. A detailed analysis may have to be undertaken on all these designs.

The design shown in Figure 15.33(a) has utility matches for all the streams. This design will probably be the most controllable design. Hence it is selected. The final process flow diagram for the flowsheet in Figure 15.6 is shown in Figure 15.33(d). Observe that there is a heater on the reactor feed before it enters the reactor. This heater controls the reactor feed temperature. Similarly, the heater on the reactor effluent stream to the distillation column controls the distillation column feed enthalpy. The coolers on the products will control the final product temperatures.

15.9 Criss-Cross Mode of Heat Transfer

The pinch design procedure, the ΔT–T_c plot, and the remaining problem analysis try to achieve the targets set ahead of design. The area target is based on the assumption that vertical heat transfer takes place in the network. This target is good if the film heat transfer coefficients are the same for all the streams. In majority of the processes, phase changes will be occurring, and streams will be in different phases. Thus, a process may consist of streams that have, at times, film heat transfer coefficients that differ by an order of magnitude. Such a situation is shown in Figure 15.34, where there are two gas streams and two liquid streams. A network that criss-crosses on the composite curves may have a smaller area than predicted by vertical heat transfer mode.[27]

A network will generally have different types of exchangers. Further, each exchanger will have a different material of construction and design temperature and pressure. These factors contribute to the capital cost of the network. Thus each stream match will have different coefficients in the cost equation. To optimize the network one will have to undertake criss-crossing on the composite curves.

Stream matching constraints may also require criss-crossing on the composite curve. For example, a match between two streams may be prohibited to maintain process safety or to avoid incurring excessive piping costs associated with a particular match. When constraints are imposed, extra utilities may be required. As a result, one may have to use cold utilities above the pinch point, use hot utilities below the pinch point, or transfer heat across the pinch point. This is totally in contrast to the pinch design rules.

All these scenarios force the designer to criss-cross on the composite curves. Various design methods are proposed in the literature to obtain optimal designs using the criss-cross mode of heat transfer in the network and they are not within the scope of this work.

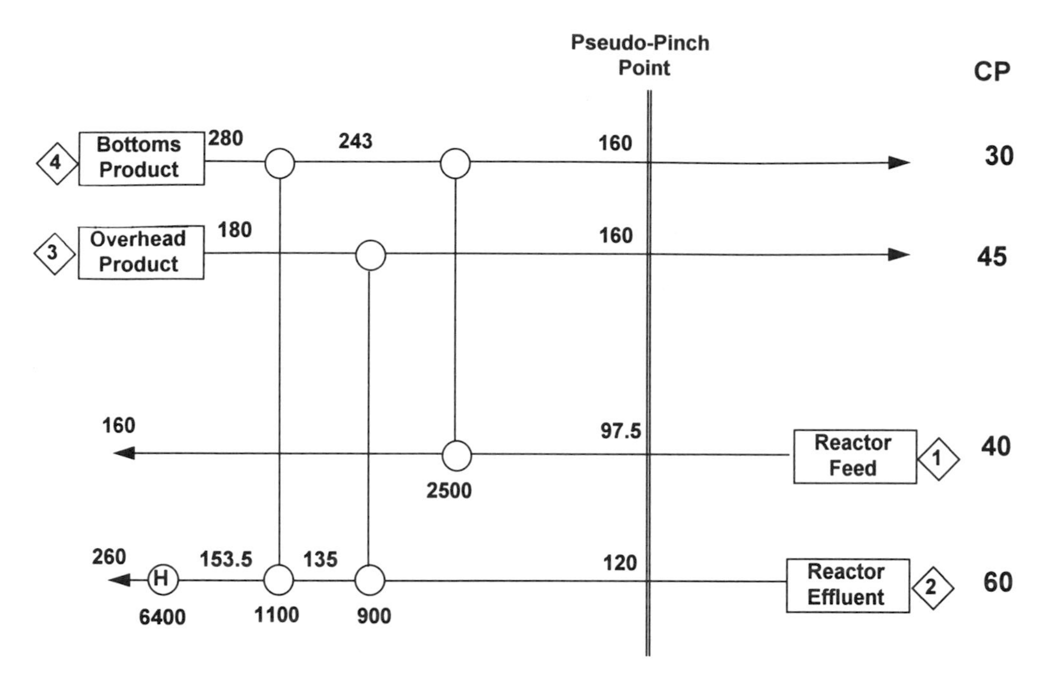

FIGURE 15.32(a) Above pseudo-pinch design—1.

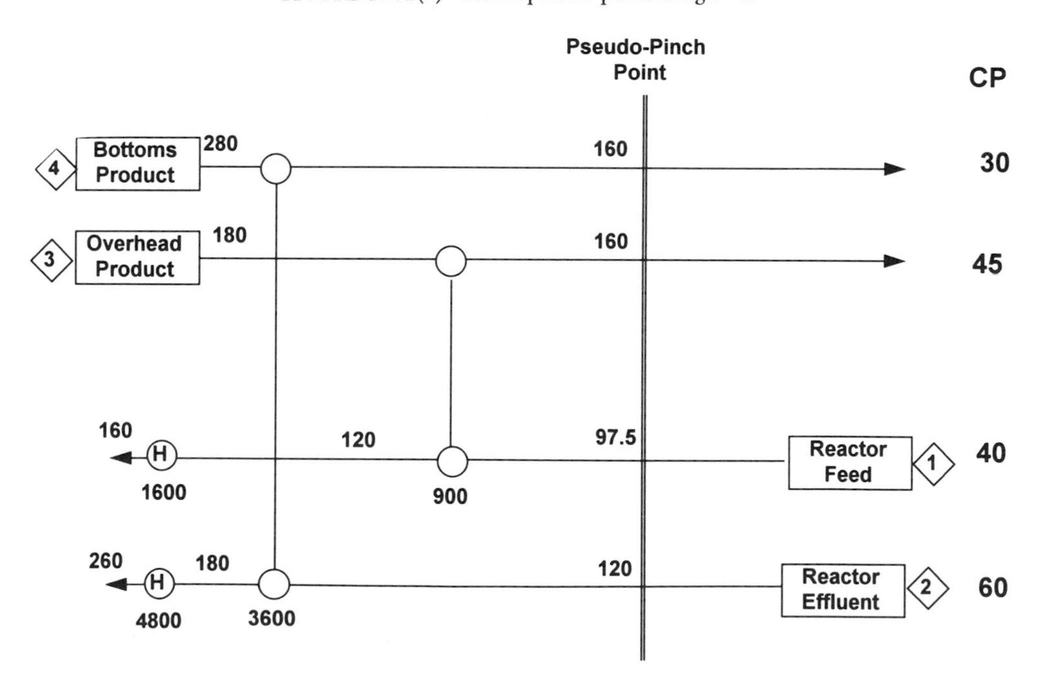

FIGURE 15.32(b) Above pseudo-pinch design—2.

15.10 Selection of Utility Loads and Levels

The operating cost depends on the amount and type of utilities used. In a complex process such as an ethylene plant, there would be about four or five steam levels and about seven or eight refrigeration levels. High-pressure steam generated within the process is let down to other steam levels via steam turbines that generate power. Steam from these levels is used for heating the process.

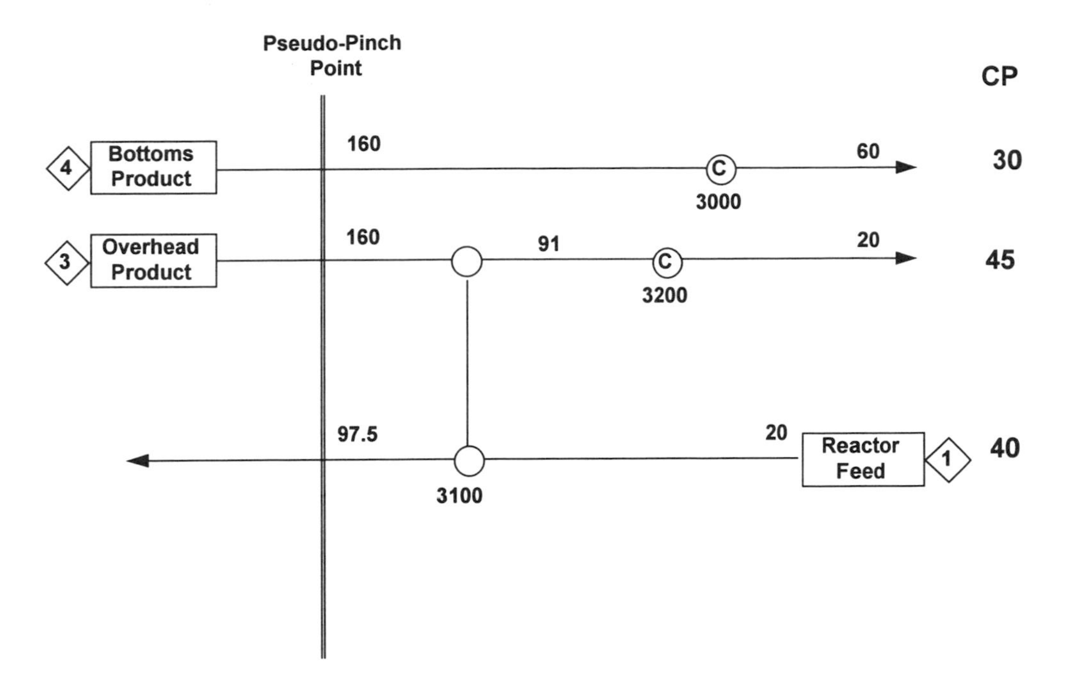

FIGURE 15.32(c) Below pseudo-pinch design—1.

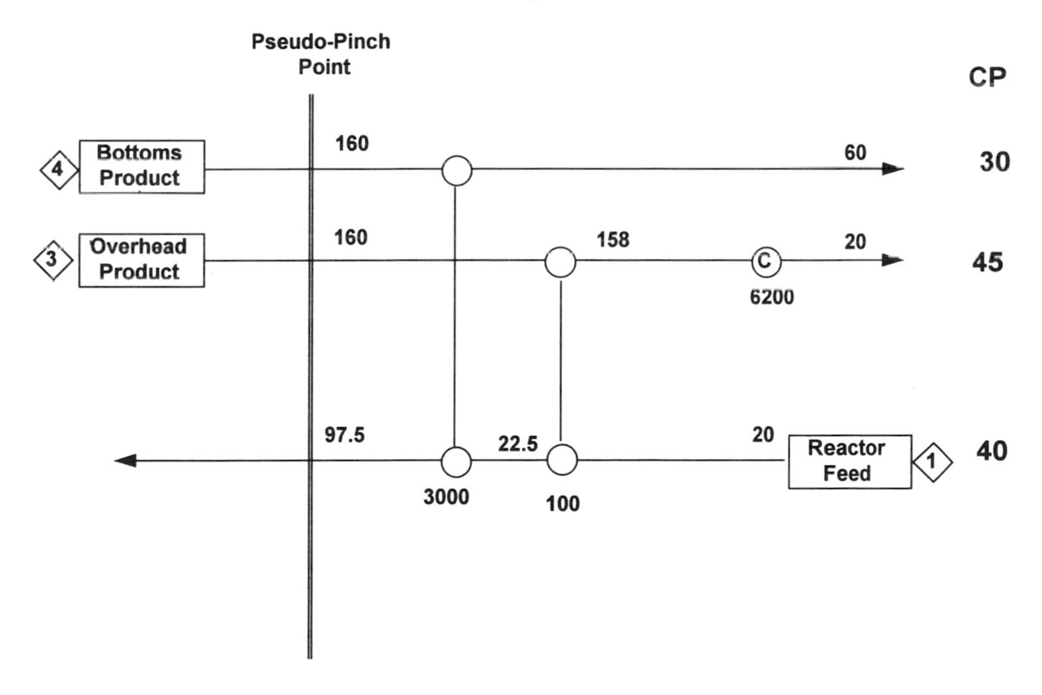

FIGURE 15.32(d) Below pseudo-pinch design—2.

The power generated from the turbines is used by the pumps and the compressors to generate various levels of refrigeration needed. The question we face is what pressure levels of steam to use and what is the load on each level? Also, what are the best temperature levels of refrigeration and what will be their respective loads?

Pinch technology helps us answer these questions in a very simple manner using the Grand Composite Curves (GRCC).[28] The GRCC is the curve that shows the heat demand and supply within each temperature interval. This curve is derived from the problem table (refer to Table 15.5).

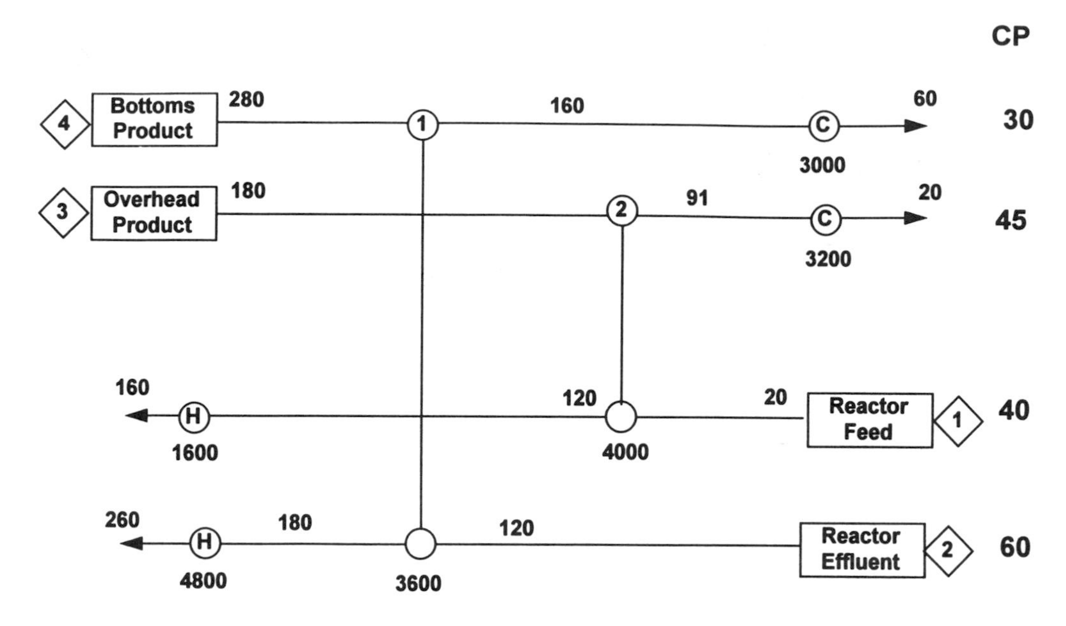

FIGURE 15.33(a) Pseudo-pinch design—2. Total annualized cost is \$743,400.

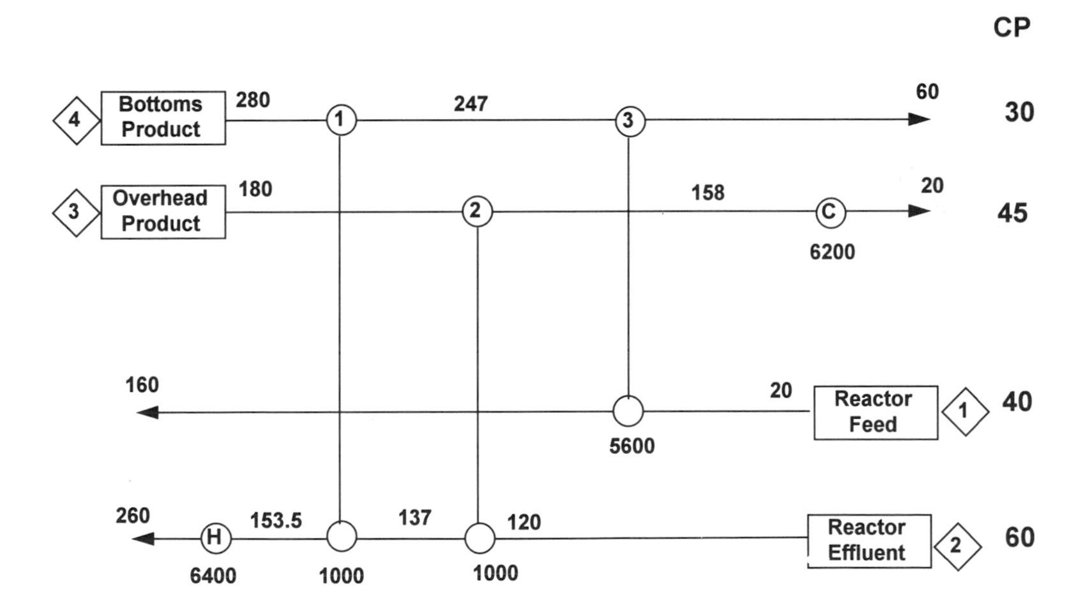

FIGURE 15.33(b) Pseudo-pinch design—3. Total annualized cost is \$727,600.

In the problem table, we had modified the stream starting and target temperatures depending on the value of Δt_{min}. From now on we shall refer to these modified temperatures as shifted temperatures. The heat flows in the intervals between two adjacent shifted-temperatures can be plotted on the shifted-temperature–enthalpy plot. Figure 15.35 shows the heat cascade and how the grand composite curve is developed for the flowsheet shown in Figure 15.6. The grand composite curve gives us a graphical representation of heat flows taking place in the system. At the pinch point the heat flow is zero.

The grand composite curve is piecewise linear. The slope of this curve also changes from interval to interval. A line with a positive slope indicates that the system in that region needs external heat. A line with a negative slope indicates that there is surplus heat available within that temperature

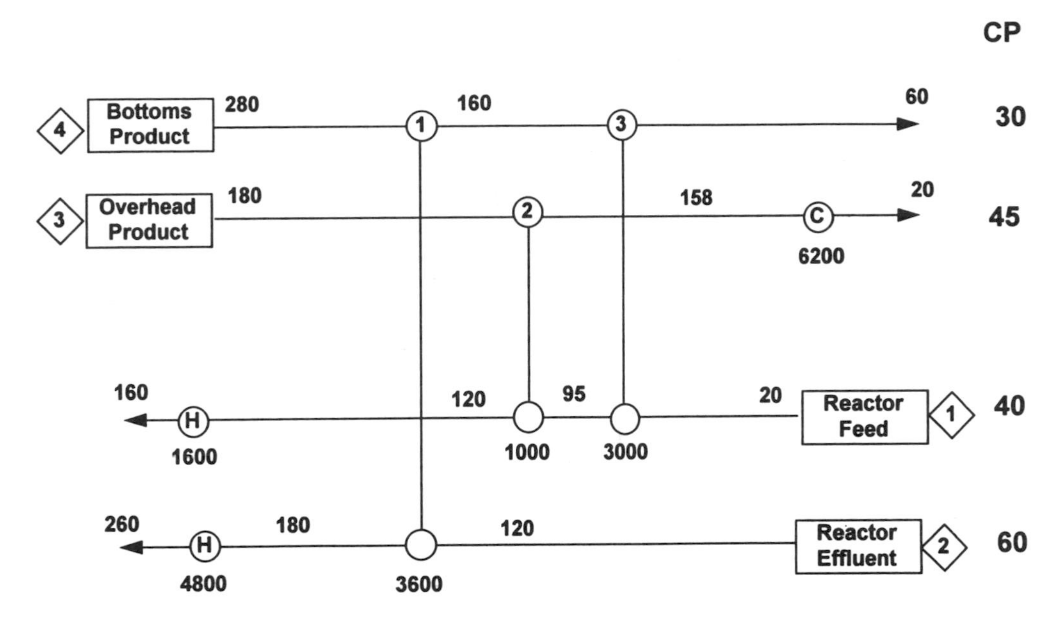

FIGURE 15.33(c) Pseudo-pinch design—4. Total annualized cost is $740,100.

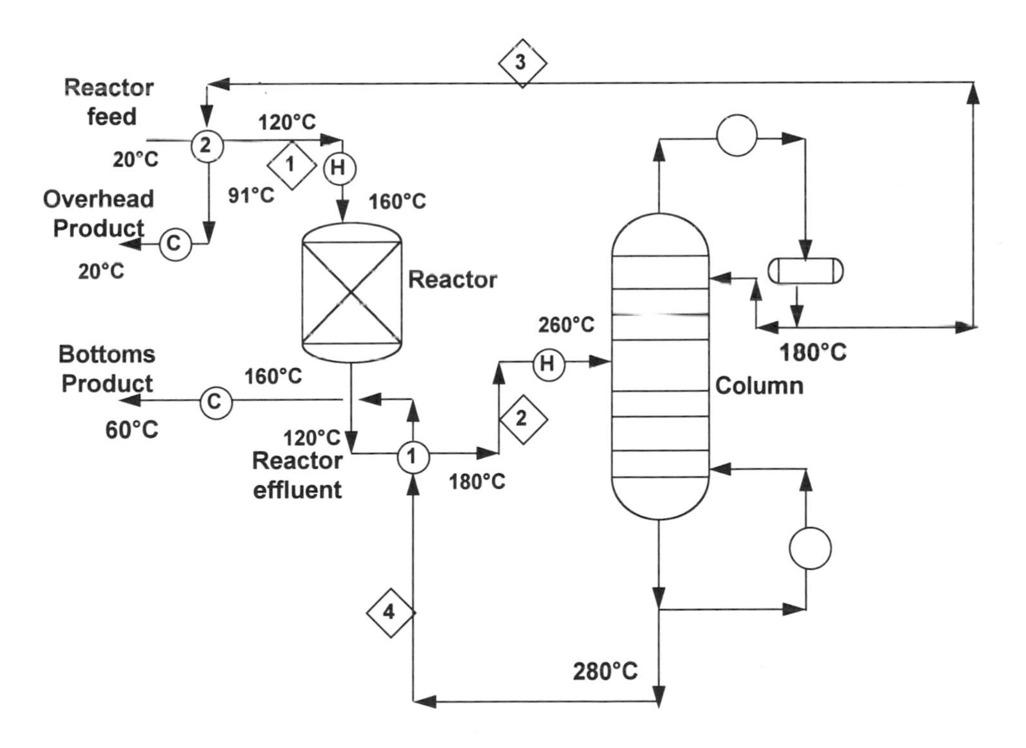

FIGURE 15.33(d) Final flow diagram for the plant in Figure 15.6.

interval that can be cascaded down within the system and used at a lower temperature interval. It is very clear from the grand composite curve that the above pinch region is a heat sink and the below pinch region is a heat source.

To further explain the importance of the grand composite curve and the kind of information that can be extracted from it, consider the curve shown in Figure 15.36. Consider the section *AB* that is between the shifted temperature interval of 295°C and 280°C. This section demands external

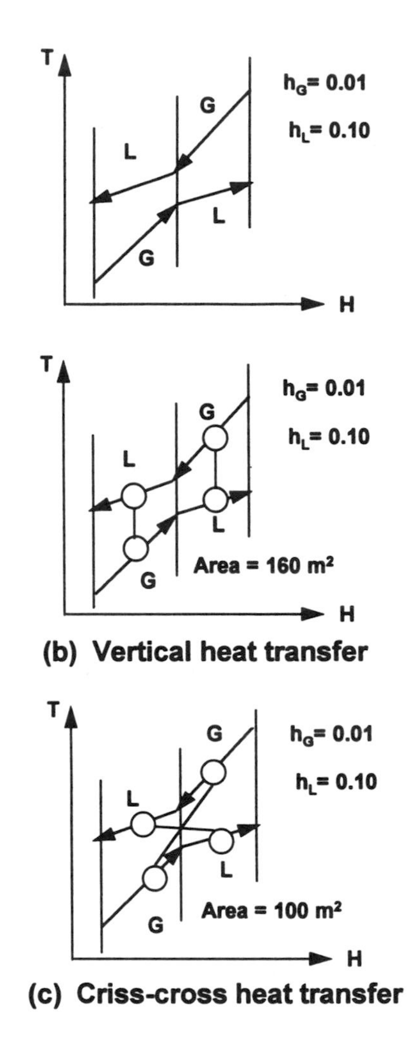

FIGURE 15.34 Criss-cross heat transfer will reduce the network area when heat transfer coefficients are very different.

heat of 2,000 kW. Hence we should place a hot utility such as high-pressure steam or hot oil to supply the heat.

Next, consider section *BC* of the GRCC between the shifted temperature intervals of 280°C and 250°C. This section has a heat surplus, so we can use it elsewhere in the system. We can drop a vertical line from *B* to meet the GRCC at point *D*. Section *CD* of the GRCC, which is a heat deficit region of the system, can now be satisfied by the heat surplus from the BC section. This section of the GRCC is called a pocket, and the process is self-sufficient with respect to energy in this region. The section *DE* now needs external heat. This heat can be supplied at any temperature ranging from the highest available temperature to a minimum temperature corresponding to point *D*.

Following the same logic, *EFG* is a pocket. In section *GHI* one hot utility level can be used at a shifted-temperature level of point *G* with a total duty of 3,000 kW, or two levels can be used—one at 160°C with a duty of 1,500 kW, and the other at the shifted temperature level of point *H* (i.e., 140°C with a duty of 1,500 kW). The choice is dictated by the trade-off between the power requirement, capital investment, and complexity of the design. Using only one level will make the design of the utility system simpler and the capital cost of the heat exchanger smaller due to higher-temperature approaches. On the other hand, if there is demand for power, then using two levels will produce more power.

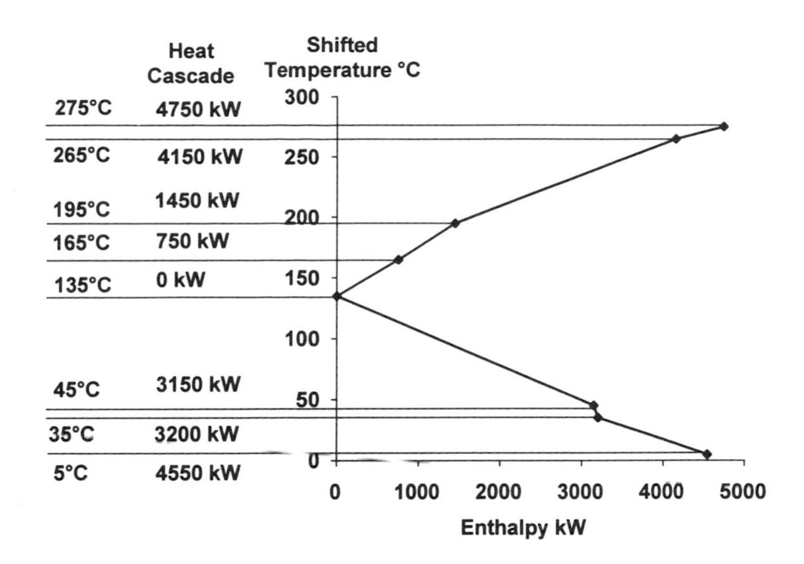

FIGURE 15.35 Grand composite curve for the flowsheet of Figure 15.6.

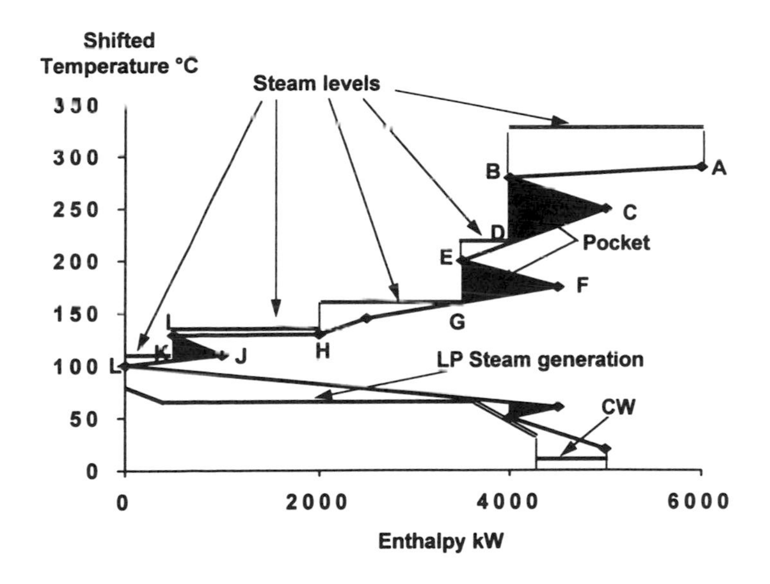

FIGURE 15.36 Selection of utility loads and levels.

A similar economic trade-off will be required for supplying the external heating requirement in section *KL* of the composite curve. Point L is the pinch point. Below point *L* heat needs to be rejected into a cooling utility such as an air cooler or cooling water. Also, in the below pinch section of the process we can address the question, Is it possible to raise steam at some temperature? If so, how much?

For example, in the process grand composite curve shown in Figure 15.36, we want to find out how much low-pressure superheated steam can be generated. The saturation temperature of the low pressure steam is 70°C and boiler feed water is available at 30°C. The superheat is 10°C. Using a simple trial-and-error procedure we can find out how much steam will be generated. Assume the amount of steam that is generated. Develop a heat curve for the low-pressure steam generation on the shifted-temperature scale. As generation of low-pressure steam will be a cold stream, the temperature of the stream will be increased by $\Delta t_{min}/2$. Keep on increasing the amount of steam

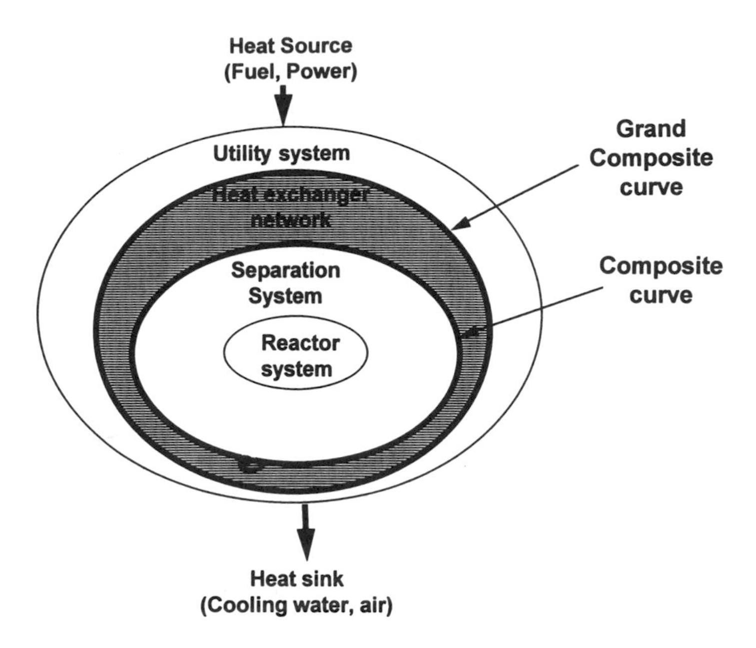

FIGURE 15.37 Onion diagram. (Reproduced with permission from B. Linnhoff.)

generated till the steam generation heat curve touches the process grand composite curve at any point. The selection of refrigeration levels uses the same technique.

Once the utility levels are decided, introduce them into the stream data and obtain the balanced composite curves. The number of pinch points will increase. In addition to the original process pinch point, each utility level will introduce at least one pinch point. A balanced grid diagram that includes all the utility streams and all the pinch points identified on the balanced composite curve can now be used along with the network design algorithms to develop a network that achieves the target set. Advances have made the task of selecting and optimizing multiple utilities easier.[29,30]

15.11 Process Integration

Until now we have discussed the design and optimization of heat exchanger networks. But heat exchanger networks are a part of a whole process. Processes consist of reactors, distillation columns, utilities, and so on. For the process to be optimally designed, all the unit operations should be properly integrated. Generally each unit operation is individually optimized. It is a misconception to believe that if each individual unit is individually optimized then the resulting process is also optimized. Each unit operation interacts with the other operations in the process. Hence, for the process to be optimized, each unit operation should be properly integrated.

Figure 15.37 shows the onion diagram proposed by Linnhoff et al.[18] that represents the hierarchy of process design. Generally, the reactors are designed first and that fixes the heat and material balance for the whole process. The products, by-products, and reactants are then separated using the most common unit operation—distillation. A number of columns may be used if a large number of components need to be separated. To achieve the separation and to carry out the reaction in the reactor, the heat exchanger network, heats and cools the process streams. External heating and cooling is supplied by the utility system. The utility system may consist of a simple boiler system or a cogeneration unit that produces the necessary steam for the process and the power requirement. All of these layers interact. An optimal process design must take into account these interactions and must properly integrate them.

Process integration uses principles that we have already established. Process integration starts by optimizing stand alone distillation columns with respect to reflux ratio, feed conditioning,

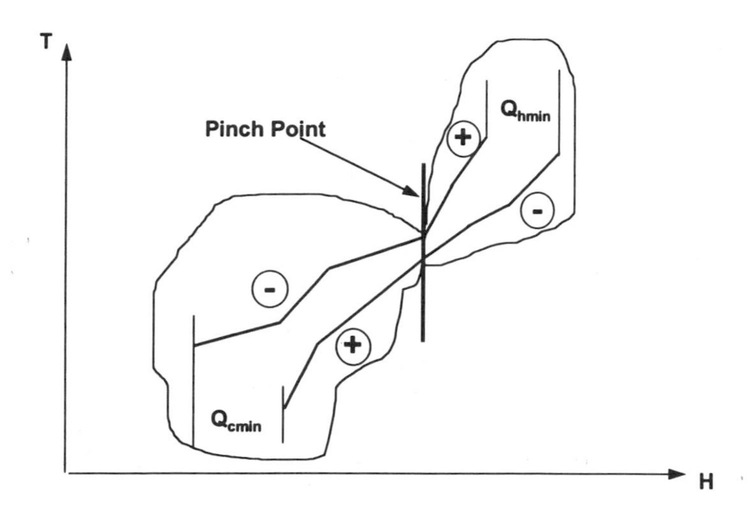

FIGURE 15.38 The plus/minus principle. (Reproduced with permission of the American Institute of Chemical Engineers, copyright ©, 1984. All rights reserved.)

side-reboilers, and side-condensers.[31] To properly integrate the distillation column with the background process, it should be placed above or below but not across the background process pinch point.[32] The same principle applies for the appropriate integration of heat engines. Heat pumps should be placed across the pinch point.[28] Linnhoff and Leur[33] have developed procedures for integrating furnaces with the rest of the process.

15.12 Process Modification

In the above pinch region, hot streams should be modified such that they can transfer more heat to the cold streams. The cold streams should be modified such that they require less heat. Both of these modifications will decrease the amount of external hot utility requirement. Similarly, in the below pinch region, the cold streams should be modified such that they require more heat and the hot streams should be modified such that they transfer less heat. These modifications will reduce the amount of cold utility requirement. This principle is called the plus/minus principle[3] and is illustrated in Figure 15.38.

15.13 Shaftwork Targets

Recently, Linnhoff and Dhole[34] proposed shaftwork targets based on exergy analysis of the stream data for a plant. The shaftwork targeting procedure calculates the change in the total shaftwork requirement of the system due to any changes in the base case. They use exergy composite and grand composite curves that are obtained from the composite and grand composite curves by changing the temperature axis to the Carnot factor. Shaftwork targets are very important in the design of low-temperature processes.

15.14 Sitewide Integration

Generally, individual processes are located in a site. Sites consist of multiple units. For example a refinery may consist of a crude unit, vacuum unit, naphtha hydrotreater, diesel hydrotreater, fluidized catalytic cracking (FCC) unit, visbreaker unit, coker unit, and so on. A petrochemical complex may consist of an ethylene unit, polyethylene unit, and so on. All of these units will have one utility system that provides hot and cold utilities as well as power to the whole site. Each unit will have steam demands at different temperatures. At the same time, each unit may be producing steam at different levels. It is possible to directly integrate different units, but that may cause political problems or piping problems. Hence, different units are generally integrated indirectly through the utility system. The problem that sitewide integration addresses is what is the correct level of steam for a site and what is the trade-off between heat and power. Should power be imported or cogenerated?[35]

The most important benefit of a sitewide analysis is that correct pricing for different levels of steam are no longer required. The impact of any modification in terms of fuel or power can be easily evaluated. There is no need for "cost of steam," as energy pricing is only done with respect to either fuel or power at the battery limit.

Sitewide analysis has increased the understanding of global emissions associated with any processing industry. To minimize the emissions associated with a process, sharper separations are required. Since distillation is the workhorse of the chemical industry, obtaining sharp separation requires increasing, the reboiler and condenser duties or adding additional processes that require additional heating and cooling. External energy is obtained by burning fuel, and when fuel is burned emissions are generated. If extra power is used, then the emissions in the utility company will increase. Thus, to decrease the emissions in the process we might end up increasing the emissions associated with burning the fuel. The net effect might be that we increase the global emissions.[1] Recently, Smith and Delaby[36] have established techniques to target for CO_2 emissions associated with energy. Combining these CO_2 targeting techniques with sitewide analysis will help in trading off different emissions.[37]

15.15 Data Extraction

Process integration studies start from a base-case flowsheet. This flowsheet may be existing or may be developed from the designer's experience. To conduct pinch analysis properly, it is important to extract the flowrate, temperature, and heat duty data correctly.

Stream target and starting temperatures should be chosen so that we do not generate the original flowsheet.[18] To illustrate this, consider the flowsheet shown in Figure 15.39. If we extract the data as two streams, then we might end up with the original flowsheet. If the drum temperature is not important and the water flow rate is small then we can consider the path from the reactor to the column to be one stream and we stand a chance for finding new matches. The drum and the pump can then be kept at a natural break point in the system.

While extracting data, extra care should be taken when streams are mixing nonisothermally.[18] Consider the system shown in Figure 15.40. Stream A is being cooled to the mixed temperature, and stream B is being heated to the mixed temperature. This happens due to mixing the streams. The mixed stream is then heated to a higher temperature. If we extract the data as shown in Figure 15.40(a) and if the pinch temperature is 70°C then we are inherently transferring heat across the pinch point by mixing the two streams. In the process of mixing we are cooling stream A and then subsequently heating it up again. The correct way to extract the data is shown in Figure 15.40(b).

When a stream is split and the split streams have two different target temperatures then each stream is considered as a separate stream. However during the design phase we can use the stream splitting and mixing technique to eliminate one unit.[18,38] See Figure 15.41.

15.16 Procedure for Optimization of an Existing Design*

Process design generally starts with an existing flowsheet. The flowsheet is modified to meet the new requirements defined by the project. Figure 15.42 shows the procedure proposed by Trivedi et al.[39] that utilizes all the concepts of pinch technology for optimizing an existing design. The following sections discuss various steps of this procedure.

*Reproduced with permission from Brown and Root, Inc.

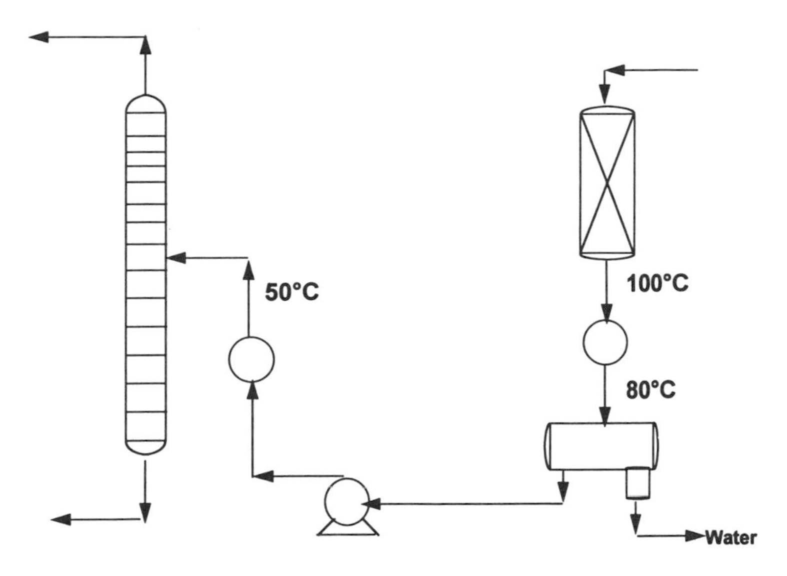

FIGURE 15.39 Flowsheet for data extraction example.

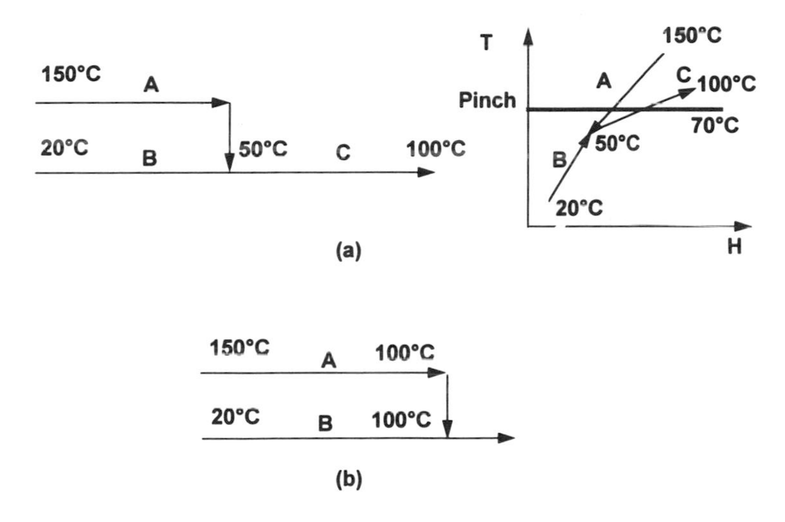

FIGURE 15.40 Data extraction for streams that are mixing.

Problem Definition

From the existing process flow diagram and the project requirements develop a problem definition and scope of the project. At this early stage in the design process, identify all the key project constraints. These constraints may be utility temperature levels (if the construction of the plant is going to be in an existing site), environmental constraints, budget constraints, schedule constraints, and so on.

Conceptual Flow Design

On the basis of the problem definition and project constraints, develop a new conceptual flow diagram. This flow diagram need not be detailed. However, it should at least contain the reactors, the distillation columns, and the associated conceptual utility system. Furthermore, it should also contain key operating parameters such as feed conditions to distillation columns, operating temperature, and pressure of distillation columns. This flow diagram becomes the basis for optimization.

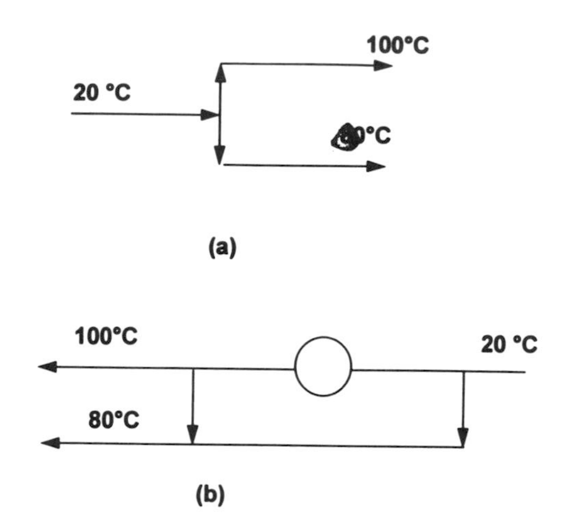

FIGURE 15.41 Stream bypassing and mixing can save a unit.

Marginal Cost of Utilities

Develop a simulation model of the conceptual utility system to evaluate the marginal cost of various utility levels. With this model it is possible to evaluate the interactions between the various utility levels. Use a reference utility level as the basis for calculating the savings associated with the other levels. The reference utility level may be the fuel fired to generate steam or export/import steam at a certain level.

Simulate and Optimize Distillation Columns

In this step, simulate the distillation columns using the conditions set by the problem definition and the conceptual flow diagram. Optimize these columns on a stand alone basis using the procedures outlined by Dhole and Linnhoff.[31]

Stream Data Extraction

The conceptual flow diagram and simulation of the distillation columns establish the heat and material balance for the entire process. Simulation will also generate the heat curves for all the hot and cold streams. Extract stream data from the process flow diagram using the guidelines established earlier.

Targeting

Assume a value of Δt_{min}. Set overall energy targets using the composite curves. Divide the problem into above ambient temperature and below ambient temperature subproblems. The temperature of cooling water will generally decide this division point. Construct the grand composite curve (GRCC) for the above ambient temperature subproblem. For the below ambient temperature subproblem, construct the process exergy grand composite curve (EGCC). Fine-tune the column optimization undertaken previously.

Optimize Total Annualized Cost

From the grand composite curve, optimize the steam levels and their corresponding loads. From these utility loads and levels, design the utility system and estimate its capital cost. Now construct the composite curves that include the utilities. These composite curves are the balanced composite

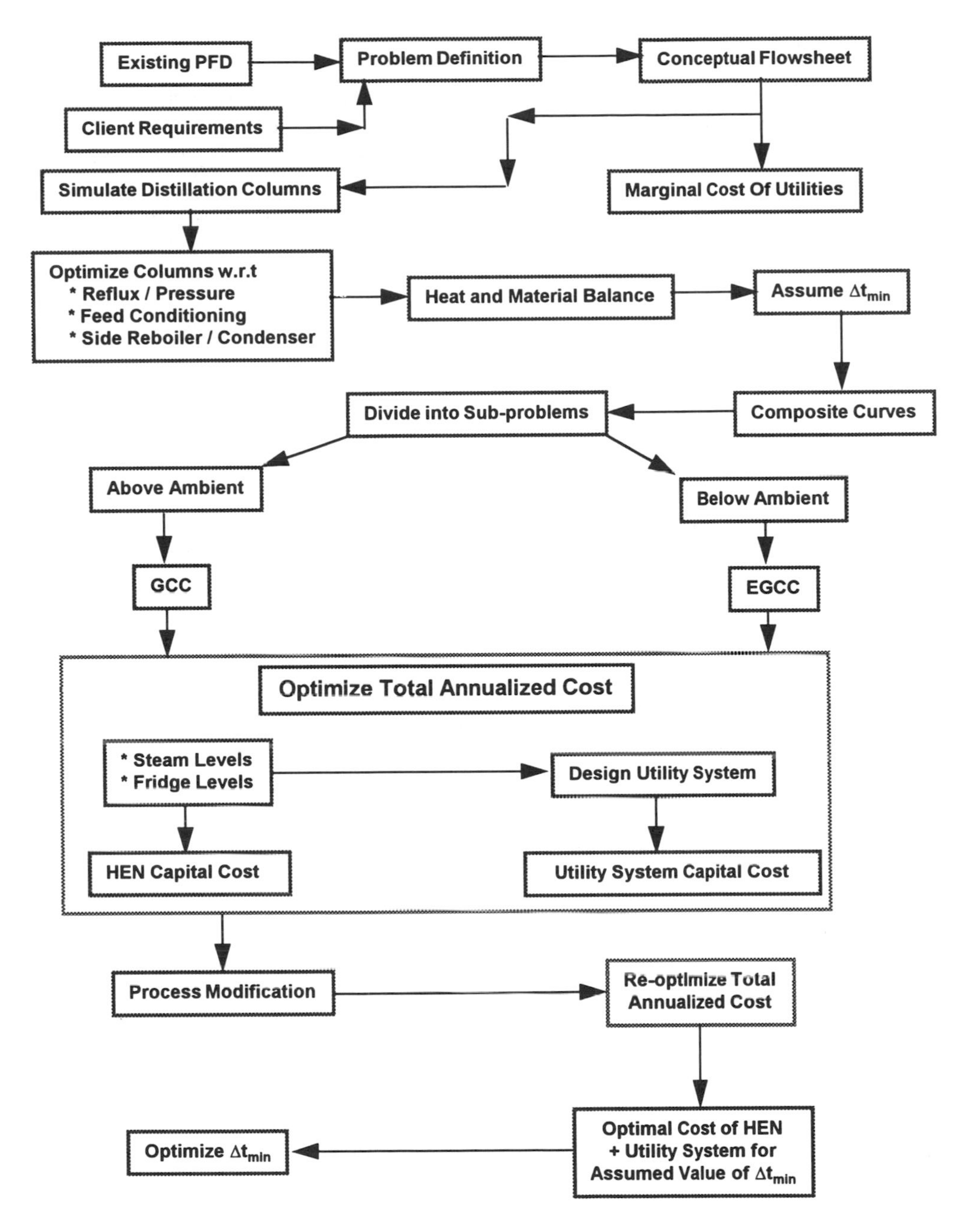

FIGURE 15.42 Proposed new procedure. (Reproduced with permission from Brown and Root, Inc.)

curves. Estimate the capital cost of the heat exchanger network from these curves. Calculate the utility cost from the utility loads and levels. Then estimate the total annualized cost for the heat exchanger network and the utility system. Optimize this cost by changing the utility loads and levels.

Process Modification

Use the balanced composite curves to identify process modifications that can potentially reduce the total annualized cost. Changing the operating pressure of a distillation column or modifying the flow rate of a stream are some of the potential process modifications that can be made. The total annualized cost needs to be reoptimized if the utility system design changes significantly.

Optimize the Value of Δt_{min}

At this stage we have the optimal total annualized cost targets for the heat exchanger network and the utility system for the assumed value of Δt_{min}. Assuming another value of Δt_{min} will give another optimized total annualized capital cost. Repeat the exercise over a range of Δt_{min} values to find the optimal value. Design the heat exchanger network at the optimal value of Δt_{min}.

Trivedi et al.[39] have reported a case study that discusses the application of the above outlined approach to optimize a state-of-the-art Ethylene process. They achieved 12% savings in energy consumption. The capital cost increased by 2% with a simple payback period of 2 years. (Typical specific energy consumption of an ethylene plant for cracking ethane through naphtha ranges from 3,000 to 5,000 Kcal/kg. The total world production capacity of ethylene in 1989 was 61 million MTY.[40])

15.17 Recent Developments

Recent developments have been made to solve different process design problems using pinch technology principles and concepts. We shall discuss these developments briefly.

Normally plants have many variables that change during the course of time. For example, depending on market conditions the throughput may change or the feedstock can change. Process conditions can change from summer to winter operations. Catalytic reactors lose their effectiveness and heat exchangers get fouled over time. The final process design should be able to handle such changes. Sensitivity analysis and multiple base case design procedures are developed that address the issue of flexibility.[8,41]

The pressure drop in a heat exchanger network is also important. Polley et al.[42] have developed methods that account for stream pressure drops during the targeting and detail design phase. Stream pressure drop constraints are very important during revamp projects. Pinch technology concepts are also applied to retrofit projects, heat integrate batch processes, minimize waste water, integrate evaporator systems, and design power cycles.[1]

References

1. Linnhoff, B., "Use Pinch Analysis to Knock Down Capital Costs and Emissions," *Chem. Eng. Prog.*, 90(9), 32, Aug. (1994).
2. Linnhoff, B. and J.A. Turner, "Heat Recovery Networks: New Insights Yield Big Savings," *Chem. Eng.*, 56, Nov. 2 (1981).
3. Linnhoff, B. and D.R. Vredeveld, "Pinch Technology Has Come of Age," *Chem. Eng. Prog.*, 80(7), 33 (1984).
4. Korner, H., "Optimal Use of Energy in the Chemical Industry," *Chem. Eng. Tech.*, 60(7), 511 (1988).
5. Fraser, D.M. and N.E. Gillespie, "The Application of Pinch Technology to Retrofit Energy Integration of an Entire Oil Refinery," *Trans. IChemE*, 70, Part A, 395, July (1992).
6. Samdani, G. and S. Moore, "Pinch Technology: Doing More With Less," *Chem. Eng.*, 43, July (1993).
7. Bell, K.J., "Estimate S & T Exchanger Design Fast," *The Oil & Gas Journal*, 59, Dec. 4 (1978).
8. Hohmann, E.C., "Optimum Networks for Heat Exchange," Ph.D. thesis, University of Southern California, Los Angeles, CA (1971).
9. Linnhoff, B. and J.R. Flower, "Synthesis of Heat Exchanger Networks," *AIChE J*, 633, July (1978).
10. Townsend, D.W. and B. Linnhoff, "Surface Area Targets for Heat Exchanger Networks," IChemE 11th Annual Research Meeting, Bath University, U.K., April (1984).
11. Linnhoff, B., D.R. Mason, and I. Wardle, "Understanding Heat Exchanger Networks," *Comp. & Chem. Eng.*, 3, 295 (1979).

12. Trivedi, K.K., J.R. Roach, and B.K. O'Neill, "Shell Targeting in Heat Exchanger Networks," *AIChE J*, 33(12), 2087 (1987).
13. Linnhoff, B., "The Process/Utility Interface," paper presented at the Second International Meeting, "National Use of Energy," Liege, Belgium, March (1986).
14. Ahmad, S. and R. Smith, "Targets and Design for Minimum Number of Shells in Heat Exchanger Networks," *Chem. Eng. Res. Des. Devel.*, 67, 481, Sept. (1989).
15. Linnhoff, B. and S. Ahmad, "SUPERTARGETING: Optimum Synthesis of Energy Management Systems," *Journal of Energy Resources Technology*, 111, 121 (1989).
16. Umeda, T., J. Itoh, and K. Shiroko, "Heat Exchange System Synthesis," *Chem. Eng. Prog.*, 70, July (1978).
17. Linnhoff, B. and E. Hindmarsh, "The Pinch Design Method for Heat Exchanger Networks," *Chem. Eng. Sci.*, 38(5), 745 (1983).
18. Linnhoff, B., D.W. Townsend, D. Boland, G.F. Hewitt, B.E.A. Thomas, A.R. Guy, and R.H. Marshland, "Users Guide on Process Integration for the Efficient Use of Energy," IChemE, Rugby, U.K. (1982).
19. Linnhoff, B. and S. Ahmad, "Cost Optimum Heat Exchanger Networks—Part I: Minimum Energy and Capital Using Simple Models for Capital Cost," *Comp. & Chem. Eng.*, 14(7), 729 (1990).
20. Su, L.J., "A Loop Breaking Evolutionary Method for the Synthesis of Heat Exchanger Networks," M.S. thesis, Sever Institute of Washington University (1979).
21. Forder, G.J. and H.P. Hutchison, "The Analysis of Chemical Flowsheets," *Chem. Eng. Sci.*, 24, 771 (1969).
22. Trivedi, K.K., B.K. O'Neill, J.R. Roach, and R.M. Wood, "A Best-First Search Method for Energy Relaxation," *Engineering Optimization*, 16, 291 (1990).
23. Trivedi, K.K., B.K. O'Neill, J.R. Roach, and R.M. Wood, "Systematic Energy Relaxation in MER Heat Exchanger Networks," *Comp. & Chem. Eng.*, 14(6), 601 (1990).
24. O'Reilly, M., Personal View, *The Chemical Engineer*, 410, 46 (1985).
25. Trivedi, K.K., B.K. O'Neill, J.R. Roach, and R.M. Wood, "A New Dual-Temperature Design Method for the Synthesis of Heat Exchanger Networks," *Comp. & Chem. Eng.*, 13(6), 667 (1989).
26. Grimes, L.E., M.D. Rychener, and A.W. Westerberg, "The Synthesis and Evolution of Networks of Heat Exchanges that Feature the Minimum Number of Units," *Chem. Eng. Commu.*, 14, 339 (1982).
27. Townsend, D.W., "Surface Area and Capital Cost Targets for Process Energy Systems," Ph.D. thesis, UMIST, Manchester, U.K. (1989).
28. Townsend, D.W. and B. Linnhoff, "Heat and Power Networks in Process Design: Part I: Criteria for Placement of Heat Engines and Heat Pumps in Process Networks; Part II: Design Procedure for Equipment Selection and Process Matching," *AIChE J*, 29(5), 742 (1983).
29. Hall, S.G., "Targeting for Multiple Utilities in Pinch Technology," Ph.D. thesis, UMIST, Manchester, U.K. (1989).
30. Parker, S.J., "Supertargeting for Multiple Utilities," Ph.D. thesis, UMIST, Manchester, U.K. (1989).
31. Dhole, V.R. and B. Linnhoff, "Distillation Column Targets," *Comp. & Chem. Eng.*, 17(5/6), 549 (1993).
32. Linnhoff, B., H. Dunford, and R. Smith, "Heat Integration of Distillation Columns into Overall Processes," *Chem. Eng. Sci.*, 38(8), 1175 (1983).
33. Linnhoff, B. and J. de Leur, "Appropriate Placement of Furnaces in the Integrated Process," paper presented at IChemE Symposium—Understanding Process Integration—II, UMIST, Manchester, U.K. (1988).
34. Linnhoff, B. and V.R. Dhole, "Shaftwork Targets for Low Temperature Process Design," *Chem. Eng. Sci.*, 47(8), 2081 (1992).
35. Raissi, K., "Total Site Integration," Ph.D. thesis, UMIST, Manchester, U.K. (1994).

36. Smith, R. and O. Delaby, "Targeting Flue Gas Emissions," *Trans, IChemE,* 69, Part A, 492 (1992).
37. Linnhoff, B. and V.R. Dhole, "Targeting for CO_2 Emissions for Total Sites," *Chem. Ing. Tech.,* 16, 256 (1993).
38. Wood, R.M., R.J. Wilcox and I.E. Grossmann, "A Note of the Minimum Number of Units for Heat Exchanger Network Synthesis," *Chem. Eng. Commun.,* 39, 371 (1985).
39. Trivedi, K.K., K.H. Pang, H.R. Klavers, D.L. O'Young, and B. Linnhoff, "Integrated Ethylene Process Design using Pinch Technology," presented at AIChE Meeting, Atlanta, April (1994).
40. Ma, J.L. James, Ethylene Supplement E, Report 29E, SRI International, Menlo Park, CA, Oct. (1991).
41. Kotjabasakis, E. and B. Linnhoff, "Sensitivity Tables for the Design of Flexible Processes, Part I: How Much Contigency in Heat Exchange Networks is Cost-Effective?" *Chem. Eng. Res. & Des.,* 64, 197 (1986).
42. Polley, G.T., M.H. Panjeh Shahi, and F.O. Jegede, "Pressure Drop Consideration in the Retrofit of Heat Exchanger Networks," *Trans of IChemE,* 68, Part A, 211 (1990).

16

Electrical Power Management in Industry

Craig B. Smith
Daniel, Mann, Johnson, & Mendenhall

16.1 The Industrial Sector

Introduction—The Significance of Industrial Energy Use

Industrial energy use continues to dominate the U.S. energy economy, but with remarkable changes since the 1973 oil embargo. At that time about 26 quads (1 quad = 10^{15} BTU) were used directly in industry. Adding electricity conversion losses of roughly 6 quads, the total was about 32 quads, or 42% of total U.S. energy use.

Industrial energy use declined in the mid-1970s, rose to approximately the 1973 values in 1979, declined again, and only now—20 years later—has returned to the 1973 level.[2] During that same period, industrial output increased by 35%, while the energy intensity (BTU input per constant dollar value of output) declined from around 32,500 BTU/$ to approximately 22,500 BTU/$, or a 30% reduction. Electricity now accounts for about 3,000 BTU/$ of this intensity.

Today, industrial energy use including conversion losses associated with electricity is slightly less than 39% of the U.S. total (see Table 16.1). Some of the savings in the industrial sector have resulted from a shift to less energy-intensive products, but efficiency improvements are estimated to be responsible for one-half to two-thirds of the savings.[2] Electricity now accounts for nearly 15% of industrial end-use energy, up from 12.5% in 1975. When primary fuels used to generate electricity are included, it accounts for 32.4% of industrial energy use. Because of this, the efficient use of electricity in industry is of great importance in both economic and environmental terms.

0-8493-2514-5/96/$0.00+$.50

TABLE 16.1 Estimated U.S. Energy Use, 1992

	Used as Fuel	%	Allocated Electricity Coversion Losses	Used as Electricity	Total	%
Residential	6.85	8.1	6.77	3.19	16.81	19.8
Commercial	3.87	4.6	6.17	2.91	12.95	15.2
Industrial	22.13	26.0	7.33	3.29	32.75	38.6
Transportation	22.37	26.3	0.02	0.01	22.40	26.4
Utilities	29.69	35.0	—	—	—	—
	84.91	100%	20.29	9.40	84.91	100%

Source: Data from EIA (1994), Table A2, p. 56.

TABLE 16.2 Industrial Electricity by End Use, 1987

	GWh	%
Motor drivers	570,935	67.6
Electrolysis	98,193	11.6
Process heat	83,008	9.8
Lighting	84,527	10.0
Other	8,452	1.0
Total	845,115	100%

Source: Compiled from OTA (1993b), p. 90.

TABLE 16.3 Major Industrial Users of Electricity, 1987

		GWh
1.	Primary metals	146,410
2.	Chemicals	141,191
3.	Paper	83,219
4.	Mining	55,676
5.	Food	47,213
6.	Agriculture	44,541
7.	Petroleum	41,444
8.	Transportation equipment	37,560
9.	Stone, clay, and glass	34,019
10.	Industrial machinery	33,194
11.	Other	180,648
	Total	845,115

Source: Compiled from OTA (1993b), p. 90.

The Importance of Electricity to Industry

Electricity use in industry is primarily for electric drives, electrolysis, electric heat, air conditioning, space heating, lighting, and refrigeration. Tables 16.2 and 16.3 list the relative importance of electricity in various industrial processes and show which industries account for 80% of industrial energy use. The important ones will be discussed in more detail.

16.2 Electricity End Uses in Industry

Electric Drives

Electric drives of one type or another use 68% of industrial electricity. Examples include electric motors, machine tools, compressors, refrigeration systems, fans, and pumps. Improvements in these applications would have a significant effect on reducing industrial electrical energy.

Motor efficiency can be improved in some cases by retrofit (modifications, better lubrication, improved cooling, heat recovery), but generally requires purchasing of more efficient units. Particularly for small motors (<10 kW), manufacturers today supply a wide range of efficiency. Greater efficiency in a motor requires improved design, more costly materials, and generally greater first cost. Losses in electric drive systems may be divided into four categories:

	Typical Efficiency (%)
Prime mover (motor)	10–95
Coupling (clutches)	80–99
Transmission	70–95
Mechanical load	1–90

Each category must be evaluated to determine energy management possibilities. In many applications the prime mover will be the most efficient element of the system. Table 16.4 shows typical induction motor data, illustrating the improvement in efficiency. Note that both efficiency and power factor decrease with partial load operation, which means that motors should be sized to operate at or near full load ratings.

Manufacturers have introduced new high-efficiency electric motors in recent years. Many utilities offer rebates of $5 to $20 per horsepower for customers installing these motors.

Electrolysis

Industrial uses of electrolysis include electrowinning, electroplating, electrochemicals, electrochemical machining, fuel cells, welding, and batteries. An indirect application of electrolysis is corrosion—which results in a substantial loss of materials which embody energy. It has been estimated that corrosion damage amounts to approximately 1% of the U.S. gross national product.

A major use of electrolytic energy is in electrowinning—the electrolytic smelting of primary metals such as aluminum and magnesium. Current methods require on the order of 13 to 15 kWh/kg; efforts are under way to improve electrode performance and reduce this to 10 kWh/kg. Recycling now accounts for one-third of aluminum production; this requires only about 5% of the energy required to produce aluminum from ore.

Electrowinning is also an important low-cost method of primary copper production. Another major use of electrolysis is in the production of chlorine and sodium hydroxide from salt brine. Electroplating and anodizing are two additional uses of electricity of great importance. Electroplating is basically the electrodeposition of an adherent coating upon a base metal. It is used with copper, brass, zinc, and other metals. Anodizing is roughly the reverse of electroplating, with the workpiece (aluminum) serving as the anode. The reaction progresses inward from the surface to form a protective film of aluminum oxide on the surface.

TABLE 16.4 Typical Electric Motor Data

	Full Load Efficiency (%)	
Size (hp)	1975	1993
1	76	82.5–85.5
2	80	84.0–86.5
5	84	87.5–89.5
10	85	89.5–91.7
20	85	91.0–93.0
40	85	93.0–94.5
100	91	94.1–95.4
200	90	95.0–96.2

Fuel cells are devices for converting chemical energy to electrical energy directly through electrolytic action. Currently they represent a small use of energy, but research is directed at developing large systems suitable for use by electric utilities for small dispersed generation plants. Batteries are another major use of electrolytic energy, ranging in size from small units with energy storage in the joule or fractional joule capacity up to units proposed for electric utility use that will store 18×10^9 J (5 MWh). Electroforming, etching, and welding are forms of electrolysis used in manufacturing and material shaping. The range of applications for these techniques stretches from microcircuits to aircraft carriers. In some applications, energy for machining is reduced and reduction of scrap also saves energy. Welding has benefits in the repair and salvage of materials and equipment, reducing the need for energy to manufacture replacements.

Electric Process Heat

Electricity is widely used as a source of process heat due to ease of control, cleanliness, wide range in unit capacities (watts to megawatts), safety, and low initial cost. Typical heating applications include resistance heaters (metal sheath heaters, ovens, furnaces), electric salt bath furnaces, infrared heaters, induction and high-frequency resistance heating, dielectric heating, and direct arc electric furnaces.

Electric arc furnaces in the primary metals industry are a major use of electricity. Typical energy use in a direct arc steel furnace is about 2.0 kWh/kg. Electric arc furnaces are used primarily to refine recycled scrap steel. This method uses about 40% of the energy required to produce steel from iron ore using basic oxygen furnaces. Energy savings can be achieved by using waste heat to preheat scrap iron being charged to the furnace.

Glass making is another process which uses electric heat. An electric current flows between electrodes placed in the charge, causing it to melt. Electric motors constitute a small part of total glass production. Major opportunities for improved efficiency with electric process heat applications in general include improved heat transfer surfaces, better insulation, heat recovery, and improved controls.

HVAC

Heating, ventilating, and air conditioning (HVAC) is an important use of energy in the industrial sector. The environmental needs in an industrial operation can be quite different from residential or commercial operations. In some cases, strict environmental standards must be met for a specific function or process. More often, the environmental requirements for the process itself are not limiting, but space conditioning is a prerequisite for the comfort of production personnel. Chapters 10 and 11 have a more complete discussion of energy management opportunities in HVAC systems.

Lighting

Industrial lighting needs range from low-level requirements for assembly and welding of large structures (such as shipyards) to the high levels needed for manufacture of precision mechanical and electronic components such as integrated circuits. Lighting uses about 20% of U.S. electrical energy and 7% of all energy. Of all lighting energy, about 20% is industrial, with the balance being residential, commercial, and miscellaneous. Although there are differences in equipment types and sizes of systems, energy management opportunities in industrial lighting systems are similar to those in residential/commercial systems (see Chapter 10).

16.3 Energy and Power Management in Industry

Setting Up an Industrial Power Management Program

The effectiveness of energy utilization varies with specific industrial operations because of the diversity of the products and the processes required to manufacture them. The organization of

TABLE 16.5 Elements of an Energy Management Program

Phase 1:	Management commitment
1.1	Commitment by management to an energy management program
1.2	Assignment of an energy management coordinator
1.3	Creation of an energy management committee of major plant and department representatives
Phase 2:	Audit and analysis
2.1	Review of historical patterns of fuel and energy use
2.2	Facility walk-through survey
2.3	Preliminary analyses, review of drawings, data sheets, equipment specifications
2.4	Development of energy audit plans
2.5	Conduct facility energy audit, covering
	a. Processes
	b. Facilities and equipment
2.6	Calculation of annual energy use based on audit results
2.7	Comparison with historical records
2.8	Analysis and simulation step (engineering calculations, heat and mass balances, theoretical efficiency calculations, computer analysis and simulation) to evaluate energy management options
2.9	Economic analysis of selected energy management options (life-cycle costs, rate of return, benefit-cost ratio)
Phase 3:	Implementation
3.1	Establish energy effectiveness goals for the organization and individual plants
3.2	Determine capital investment requirements and priorities
3.3	Establish measurement and reporting procedures, install monitoring and recording instruments as required
3.4	Institute routine reporting procedures ("energy tracking" charts) for managers and publicize results
3.5	Promote continuing awareness and involvement of personnel
3.6	Provide for periodic review and evaluation of overall energy management program

personnel and operations involved also varies. Consequently, an effective conservation program should be custom designed for each company and its plant operations. There are some generalized guidelines, however, for initiating and implementing an energy management program. Many large companies have already instituted energy management programs and have realized substantial savings in fuel utilization and costs. Smaller industries and plants, however, often lack the technical personnel and equipment to institute and carry out effective programs. In these situations, reliance on external consultants may be appropriate to initiate the program. Internal participation, however, is essential for success. A well-planned, organized, and executed energy management program requires a strong commitment by top management.

Assistance also can be obtained from local utilities. Utility participation would include help in getting the customer started on an energy management program, technical guidance, or making information available. Some utilities today have active programs that include training of customer personnel or provision of technical assistance. Table 16.5 summarizes the elements of an effective energy management program. These will now be discussed in more detail.

Phase I: Management Commitment

A commitment by the directors of a company to initiate and support a program is essential. An energy coordinator is designated and an energy management committee is formed. The committee should include personnel representing major company activities utilizing energy. A plan is formulated to set up the program with a commitment of funds and personnel. Realistic overall goals and guidelines in energy savings should be established based on overall information in the company records, projected activities, and future fuel costs and supply. A formal organization as described earlier is not an absolute requirement for the program; smaller companies will simply give the energy management coordination task to a staff member.

Phase 2: Audit and Analysis

Energy Audit of Equipment and Facilities. Historical data for the facility should be collected, reviewed, and analyzed. The review should identify gross energy uses by fuel types, cyclic trends, fiscal year effects, dependence on sales or work load, and minimum energy use ratios. Historical data are graphed in a form similar to the one shown in Chapter 10. Historical data assist in planning a detailed energy audit and alert the auditors as to the type of fuel and general equipment to expect. A brief facility walk-through is recommended to establish the plant layout, major energy uses, and primary processes or functions of the facility.

The energy audit is best performed by an experienced or at least trained team, since visual observation is the principal means of information gathering and operational assessment. A team would have from three to five members, each with a specific assignment for the audit. For example, one auditor would check the lighting, another the HVAC system, another the equipment and processes, another the building structure (floor space, volume, insulation, age, etc.), and another the occupancy use schedule, administration procedures, and employees' general awareness of energy management.

The objectives of the audit are to determine how, where, when, and how much energy is used in the facility. In addition, the audit helps to identify opportunities to improve the energy use efficiency and its operations. Some of the problems encountered during energy audits are determining the rated power of equipment, determining the effective hours of use per year, and determining the effect of seasonal, climatic, or other variable conditions on energy use. Equipment ratings are often obscured by dust or grease (unreadable nameplates). Complex machinery may not have a single nameplate listing the total capacity, but several giving ratings for component equipment. The effect of load is also important because energy use in a machine operating at less than full load may be reduced along with a possible loss in operating efficiency.

The quantitative assessment of fuel and energy use is best determined by actual measurements under typical operational conditions using portable or installed meters and sensing devices. Such devices include light meters, ammeters, thermometers, air flow meters, recorders, and so on. In some situations, sophisticated techniques such as infrared scanning or thermography are useful. The degree of measurement and recording sophistication naturally depends on available funds and the potential savings anticipated. For most situations, however, nameplate and catalog information are sufficient to estimate power demand. Useful information can be obtained from operating personnel and their supervisors—particularly as it relates to usage patterns throughout the day. A sample form that can be used for recording audit data is shown in Chapter 10 (Figure 10.3).

The first two columns of the form are self-explanatory. The third column is used for the rated capacity of the device, (e.g., 5 kW). The sixth column is used if the device is operated at partial load. Usage hours (column 7) are based on all work shifts, and are corrected to account for the actual operating time of the equipment. The last three columns are used to convert energy units to a common basis (e.g., MJ or Btu).

Data recorded in the field are reduced easily either by hand or computer. The advantage of using a computer is that uniform results and summaries can be obtained easily in a form suitable for review or for further analysis. Computer analysis also provides easy modification of the results to reflect specific management reporting requirements or to present desired comparisons for different energy use, types of equipment, and so on.

A Special Case: Energy Audit of a Process. In some manufacturing and process industries it is of interest to determine the energy content of a product. This can be done by a variation of the energy audit techniques described earlier. Since this approach resembles classical financial accounting, it is sometimes called *energy accounting*. In this procedure the energy content of the raw materials is determined in a consistent set of energy units. Then, the energy required for conversion to a product is accounted for in the same units. The same is done for energy in the waste streams and the by-products. Finally, the net energy content per unit produced is used as a basis for establishing efficiency goals.

In this approach, all materials used in the product or used to produce it are determined. Input raw materials used in any specific period are normally available from plant records. Approximations of specific energy content for some materials can be found in the literature or can be obtained from the U.S. Department of Commerce or other sources. The energy content of a material includes that due to extraction and refinement as well as any inherent heating value it would have as a fuel prior to processing. Consequently, nonfuel-type ores in the ground are assigned zero energy and petroleum products are assigned their alternate value as a fuel prior to processing in addition to the refinement energy. The energy of an input metal stock would include the energy due to extraction, ore refinement to metal, and any milling operations.

Conversion energy is an important aspect of the energy audit, since it is under direct control of plant management. All utilities and fuels coming into the plant are accounted for. They are converted to consistent energy units (joules or Btu) using the actual data available on the fuels or using approximate conversions.

Electrical energy is assigned the actual fuel energy required to produce the electricity. This accounts for power conversion efficiencies. A suggested approach is to assume (unless actual values are available from your utility) that 10.8 MJ (10,200 Btu) is used to produce 3.6 MJe (1 kWh), giving a fuel conversion efficiency of $3.6 \div 10.8 = 0.33$ or 33%.

The energy content of process steam includes the total fuel and electrical energy required to operate the boiler as well as line and other losses. Some complexities are introduced when a plant produces both power and steam, since it is necessary to allocate the fuel used to the steam and power produced. One suggested way to make this allocation is to assume that there is a large efficient boiler feeding steam to a totally condensing vacuum turbine. Then, one must determine the amount of extra boiler fuel that would be required to permit the extraction of steam at whatever pressure while maintaining the constant load on the generator. The extra fuel is considered the energy content of the steam being extracted.

Waste disposal energy is that energy required to dispose of or treat the waste products. This includes all the energy required to bring the waste to a satisfactory disposal state. In a case where waste is handled by a contractor or some other utility service, it would include the cost of transportation and treatment energy.

If the plant has by-products or coproducts, then energy credit is allocated to them. A number of criteria can be used. If the by-product must be treated to be utilized or recycled (such as scrap), then the credit would be based on the raw material less the energy expended to treat the by-product for recycle. If the by-product is to be sold, the relative value ratio of the by-product to the primary product can be used to allocate the energy.

Analysis of Audit Results, Identification of Energy Management Opportunities. Often the energy audit will identify immediate energy management opportunities, such as unoccupied areas that have been inadvertently illuminated 24 hours per day, equipment operating needlessly, and so on. Corrective housekeeping and maintenance action can be instituted to achieve short-term savings with little or no capital investment.

An analysis of the audit data is required for a more critical investigation of fuel waste and identification of the potential for conservation. This includes a detailed energy balance of each process, activity, or facility. Process modification and alternatives in equipment design should be formulated, based on technical feasibility and economic and environmental impact. Economic studies to determine payback, return on investment, and net savings are essential before making capital investments.

Phase 3: Implementation

At this point goals for saving energy can be established more firmly and priorities set on the modifications and alterations to equipment and the process. Effective measurement and monitoring procedures are essential in evaluating progress in the energy management program. Routine reporting procedures between management and operations should be established to accumulate information on plant performance and to inform plant supervisors of the effectiveness of their operation. Time-tracking charts of energy use and costs can be helpful. Involving employees and recognizing their contributions facilitate the achievement of objectives. Finally, the program must be continually reviewed and analyzed with regard to established goals and procedures.

Electric Load Analysis

The energy audit methodology is a general tool that can be used to analyze energy use in several forms and over a short or long period of time. Another useful technique, particularly for obtaining a short-term view of industrial electricity use, is an analysis based on evaluation of the daily load curve. Normally this analysis uses metering equipment installed by the utility and therefore available at the plant. However, special metering equipment can be installed if necessary to monitor a specific process or building.

For small installations both power and energy use can be determined from the kilowatt hour meter installed by the utility. Energy in kWh is determined by

$$E = (0.001)(K_h P_t C_t N) \text{ kWh} \quad \textbf{(16.1)}$$

where E = electric energy used, kWh; K_h = meter constant, watt hours/revolution; P_t = potential transformer ratio; C_t = current transformer ratio; and N = number of revolutions of the meter disk. (The value of K_h is usually marked on the meter. P_t and C_t are usually 1.0 for small installations.) To determine energy use, the meter would be observed during an operation and the number of revolutions of the disk counted. Then the equation can be used to determine E.

To determine the average load over some period p (hours), determine E as earlier for time p and then use the relation that

$$L = E/p \text{ kW} \quad \textbf{(16.2)}$$

where E is in kWh, p is in hours, and L is the load in kW. Larger installations will have meters with digital outputs or strip charts. Often these will provide a direct indication of kWh and kW as a function of time. Some also indicate the reactive load (kVARs) or the power factor.

The first step is to construct the daily load curve. This is done by obtaining kWh readings each hour using the meter. The readings are then plotted on a graph to show the variation of the load over a 24-hour period. Table 16.6 shows a set of readings obtained over a 24-hour period in the XYZ manufacturing plant located in Sacramento, California and operating one shift per day. These readings have been plotted in Figure 16.1.

Several interesting conclusions can be immediately drawn from this figure:

- The greatest demand for electricity occurs at 11:00.
- Through the lunch break the third highest demand occurs.
- The ratio of the greatest demand to the least demand is approximately 3:1.
- Only approximately 50% of the energy used actually goes into making a product (54% on-shift use, 46% off-shift use).

When presented to management, these facts were of sufficient interest that a further study of electricity use was requested. Additional insight into the operation of a plant (and into the cost of purchase of electricity) can be obtained from the load analysis. Following a brief discussion of electrical load parameters, a load analysis for the XYZ Company will be described.

TABLE 16.6 Kilowatt Hour Meter Readings for XYZ Manufacturing Company

Time Meter Read	Elapsed kWh	Notes Concerning Usage	Percentage of Total Usage
1:00 (A.M.)	640		
2:00	610		
3:00	570		
4:00	570	7 hr preshift use	
5:00	640		
6:00	770		
7:00	1,120		
Subtotal	4,920		17%
8:00	1,470		
9:00	1,700		
10:00	1,790		
11:00	1,850		
	(Peak)	9 hr on shift use	
12:00 (noon)	1,830		
13:00 (1 P.M.)	1,790		
14:00 (2 P.M.)	1,790		
15:00 (3 P.M.)	1,760		
16:00 (4 P.M.)	1,690		
Subtotal	15,670		54%
17:00 (5 P.M.)	1,470		
18:00 (6 P.M.)	1,310		
19:00 (7 P.M.)	1,210		
20:00 (8 P.M.)	1,090	8 hr postshift use	
21:00 (9 P.M.)	960		
22:00 (10 P.M.)	800		
23:00 (11 P.M.)	730		
24:00 (12 A.M.)	640		
Subtotal	8,210		29%
Grand totals	28,800		100%

Any industrial electrical load consists of lighting, motors, chillers, compressors, and other types of equipment. The sum of the capacities of this equipment, in kW, is the *connected* load. The actual load at any point in time is normally less that the connected load since every motor is not turned on at the same time; only part of the lights may be on at any one time, and so on. Thus the load is said to be diversified, and a measure of this can be gotten by calculating a *diversity factor:*

$$DV = \left(D_{ml} + D_{m2} + D_{m3} + \ldots\right) / \left(D_{\max}\right) \tag{16.3}$$

where D_{m1}, D_{m2}, and so on = sum of maximum demand of individual loads in kW and D_{max} = maximum demand of plant in kW. If the individual loads do not occur simultaneously (usually they do not), the diversity factor will be greater than unity. Typical values for industrial plants are 1.3 to 2.5.

If each individual load operated to its maximum extent simultaneously, the *maximum* demand for power would be equal to the connected load and the diversity factor would be 1.0. However, as pointed out earlier, this does not happen except for special cases. The demand for power varies over time as loads are added and removed from the system. It is usual practice for the supplying utility to specify a demand interval (usually 0.25, 0.5, or 1.0 hr) over which it will calculate the demand and compute the demand charge using the relationship

$$D = E/p \text{ kW} \tag{16.4}$$

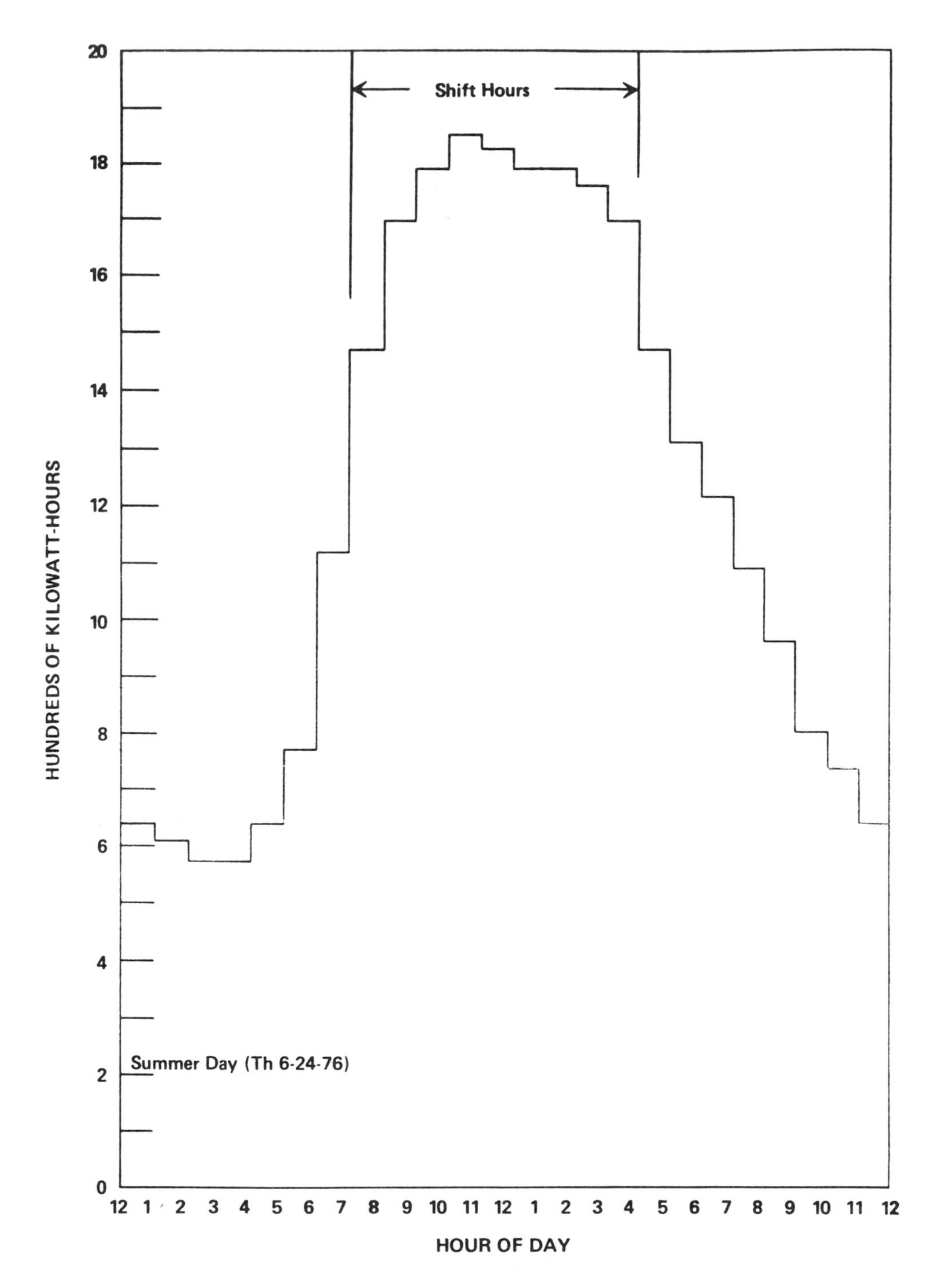

FIGURE 16.1 Daily load curve for XYZ company.

where D = demand in kW, E = kilowatt hours used during p, and p = demand interval in hours. The demand calculated in this manner is an average value, being greater than the lowest instantaneous demand during the demand interval but less than the maximum demand during the interval.

Utilities are interested in *peak demand*, since this determines the capacity of the equipment they must install to meet the customer's power requirements. This is measured by a demand factor, defined as

$$DF = D_{max}/CL \quad (16.5)$$

where D_{max} = maximum demand in kW and CL = connected load in kW. The demand factor is normally less than unity; typical values range from 0.25 to 0.90.

Since the customer normally pays a premium for the maximum load placed on the utility system, it is of interest to determine how effectively the maximum load is used. The most effective use of the equipment would be to have the peak load occur at the start of the use period and continue unchanged throughout it. Normally, this does not occur, and a measure of the extent to which the maximum demand is sustained throughout the period (a day, month, or year) is given by the *hours use of demand:*

$$\text{HUOD} = E/D_{max} \text{ hours} \quad (16.6)$$

where HUOD = hours use of demand in hours; E = energy used in period p, in kWh; D_{max} = maximum demand during period p, in kW; and p = period over which HUOD is determined—for example, 1 day, 1 month, or 1 year (p is always expressed in *hours*).

The *load factor* is another parameter that measures the plant's ability to use electricity efficiently. In effect it measures the ratio of the average load for a given period of time to the maximum load which occurs during the same period. The most effective use results when the load factor is as high as possible once E or HUOD has been minimized (it is always less than one). The load factor is defined as

$$LF = E/(D_{max})(p) \quad (16.7)$$

where LF = load factor (dimensionless); E = energy used in period p in kWh; D_{max} = maximum demand during period p in kW; and p = period over which load factor is determined (e.g., 1 day, 1 month, or 1 year) in hours. Another way to determine LF is from the relation

$$LF = \text{HUOD}/p \quad (16.8)$$

Still another method is to determine the average load, L = kWh/p during p divided by p and then use the relation

$$LF = L/D_{max} \quad (16.9)$$

These relations are summarized for convenience in Table 16.7.

Returning to the XYZ plant, the various load parameters can now be calculated. Table 16.8 summarizes the needed data and the results of the calculations. The most striking thing shown by the calculations is the hours use of demand, equal to 15.6. This is a surprise, since the plant is only operating one shift. The other significant point brought out by the calculations is the low load factor.

An energy audit of the facility was conducted and the major loads were evaluated. The audit results revealed a number of energy management opportunities whereby both loads (kW) and energy use (kWh) could be reduced. The audit indicated that inefficient lighting (on about 12 hr

TABLE 16.7 Summary of Load Analysis Parameters

Formulas	Definitions	
$E = \frac{K_h P_t C_t N}{1{,}000}$	E	= Electric energy used in period p, kWh
	E_{max}	= Maximum energy used during period p, kWh
$L = \frac{E}{p}$	K_h	= Meter constant, watt hours/revolution
	P_t	= Potential transformer ratio
$DV = \frac{D_{m1} + D_{m2} + D_{m3}}{D_{max}}$	C_t	= Current transformer ratio
	N	= Number of revolutions of the meter disk
$D = \frac{E}{p} D_{max} = \frac{E_{max}}{p}$	L	= Average load, kW
$DF = \frac{D_{max}}{CL}$	p	= Period of time used to determine load, demand, electricity use, etc., normally 1 hour, day, month, or year; measured in hours
$HUOD = \frac{E}{D_{max}}$	DV	= Diversity factor, dimensionless
	D_{max}	= Maximum demand in period p, kW
$LF = \frac{E}{(D_{max})(p)} = \frac{HUOD}{p} = \frac{L}{D_{max}}$	D_{m1}, D_{m2}, etc.	= Maximum demand of individual load, kW
	D	= Demand during period p, kW
	DF	= Demand factor for period p, dimensionless
	CL	= Connected load, kW
	HUOD	= Hours use of demand during period p, hours
	LF	= Load factor during period p, dimensionless

TABLE 16.8a Data for Load Analysis of XYZ Plant

p	= 24 hr
E	= 28,800 kWh/day
E_{max}	= 1,850 kWh
CL	= 2,792 kW
$D_{m1,2,3,etc.}$	= 53, 62, 144, 80, 700, 1,420 kW

per day) could be replaced in the parking lot. General office lighting was found to be uniformly at 100 fc; by selective reduction and task lighting the average level could be reduced to 75 fc. The air conditioning load would also be reduced. Improved controls could be installed to automatically shut down lighting during off-shift and weekend hours (the practice had been to leave the lights on). Some walls and ceilings were selected for repainting to improve reflectance and reduced lighting energy. It was found that the air conditioning chillers operated during weekends and off-hours; improved controls would prevent this. Also, the ventilation rates were found to be excessive and could be reduced. In the plant, compressed air system leaks, heat losses from plating tanks, and on-peak operation of the heat treat furnace represented energy and load management opportunities.

The major energy management opportunities were evaluated to have the following potential savings, with a total payback of 5.3 months, as shown in Table 16.8b: the average daily savings of electricity amounted to approximately 4,400 kWh/day. This led to savings of $80,000 per year, with the cost of the modification being $36,000.

TABLE 16.8b Sample Calculations for XYZ Plant

1. $D_{max} = \frac{E_{max}}{p} = \frac{1{,}850 \text{ kWh}}{1 \text{ hr}} = 1{,}850 \text{ kW}$

2. $DV = \frac{D_{m1} + D_{m2} + D_{m3} + \ldots}{D_{max}} = \frac{2{,}459}{1{,}850} = 1.33$

3. $DF = \frac{D_{max}}{CL} = \frac{1{,}850}{2{,}792} = 0.66$

4. $\underset{\text{(daily)}}{\text{HUOD}} = \frac{E}{D_{max}} = \frac{28{,}800 \text{ kWh/day}}{1{,}850 \text{ kW}} = 15.6 \text{ hr/day}$

5. $\underset{\text{(daily)}}{LF} = \frac{\text{HUOD}}{p} = \frac{15.6}{24.0} = 0.65$

The load parameters after the changes were made can be found:

$$D_{max} = \frac{1{,}850 - 235}{1 \text{ hr}} = 1{,}615 \text{ kW}$$

$$\text{HUOD} = \frac{24{,}400 \text{ kWh/day}}{1{,}615} = 15.1 \text{ hr/day}$$

$$LF = \frac{15.1}{24} = 0.63$$

Calculated savings:	Savings kW	Savings kWh/yr
More efficient parking lot lighting	16	67,000
Reduce office lighting	111	495,000
Office lighting controls to reduce off-shift use	—	425,000
Air conditioning controls and smaller fan motor	71	425,000
Compressed air system repairs and reduction of heat losses from plating tanks	—	200,000
Shift heat treat oven off-peak	37	—
Totals	235	1,612,000

The revised electricity use was found to be:

	kWh	%
Preshift	4,400	18
On shift	14,400	59
Postshift	5,600	23
Totals	24,400	100

This can be compared to the original situation (Table 16.6). See also Figure 16.2, which shows the daily load curve after the changes have been made. The percentage of use on shift is now higher. Note that D_{max} has been improved significantly (reduced by 13%); the HUOD has improved slightly (about

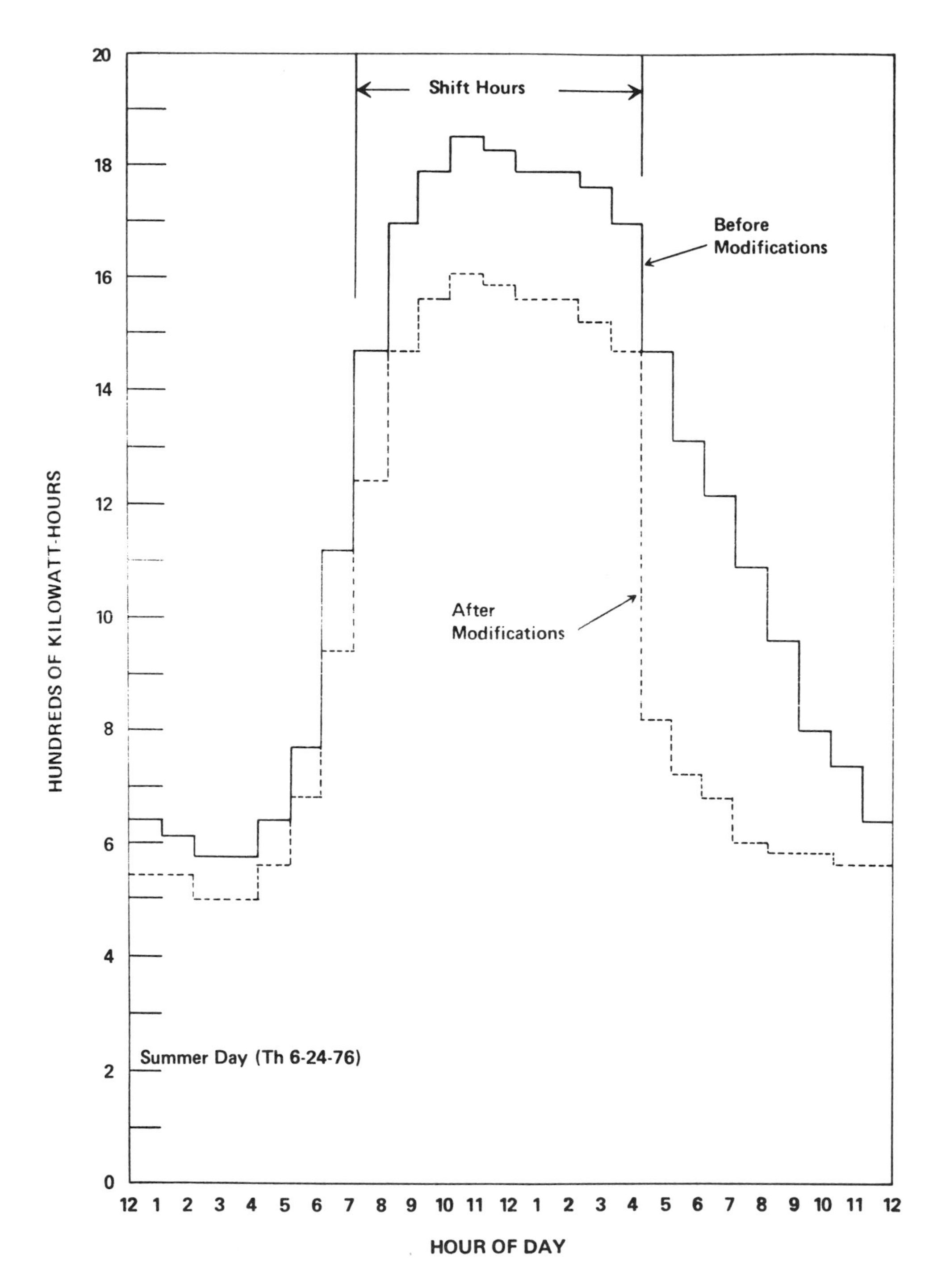

FIGURE 16.2 Daily load curve for XYZ company after modifications.

3% lower now); and the LF is slightly lower. Further improvements are undoubtedly still possible in this facility; they should be directed first at reducing nonessential uses, thereby reducing HUOD.

So far the discussion has dealt entirely with power and has neglected the reactive component of the load. In the most general case the apparent power in kVA that must be supplied to the load is the sum of the active power in kW and the *reactive power* in KVAR (the reader who is unfamiliar with these terms should refer to a basic electrical engineering text):

$$|S| = \sqrt{P^2 + Q^2} \tag{16.10}$$

where S = apparent power in kVA; P = active power in kW; and Q = reactive power in KVAR. In this notation the apparent power is a vector of magnitude S and angle θ where θ is commonly referred to as the phase angle and given as

$$\theta = \tan^{-1}(\text{KVAR/kW}) \tag{16.11}$$

Another useful parameter is the *power factor*, given by

$$pf = \cos\theta \tag{16.12}$$

The power factor is also given by

$$pf = |P| / |S| \tag{16.13}$$

The power factor is always less than or equal to unity. A high value is desirable because it implies a small reactive component to the load. A low value means the reactive component is large.

The importance of the power factor is related to the reactive component of the load. Even though the reactive component does not dissipate power (it is stored in magnetic or electric fields), the switch gear and distribution system must be sized to handle the current required by the apparent power, or the vector sum of the active and reactive components. This results in a greater capital and operating expense. The operating expense is increased due to the standby losses that occur in supplying the reactive component of the load.

The power factor can be improved by adding capacitors to the load to compensate for part of the inductive reactance. The benefit of this approach depends on the economics of each specific case and generally requires a careful review or analysis.

These points can be clarified with an example. Consider the distribution system shown in Figure 16.3. Four loads are supplied by a 600 A bus. Load A is a distant load that has a large reactive component and a low power factor (pf = 0.6). To supply the active power requirement of 75 kW, an apparent power of 125 kVA must be provided and a current of 150 A is required.

The size of the wire to supply the load is dictated by the current to be carried and voltage drop considerations. In this case, #3/0 wire that weighs 508 lb per 1,000 ft and has a resistance of 0.062 Ω per 1,000 ft is used. Since the current flowing in this conductor is 150 A, the power dissipated in the resistance of the conductor is

$$P = \sqrt{3}\ i^2 r = (1.732)(150^2)(0.124)\ \text{W}$$

$$\text{P} = 4.8\ \text{kW}$$

Similar calculations can be made for load B, which uses #1 wire, at 253 lb per 1,000 ft and 0.12 Ω/1,000 ft.

Now the effect of the power factor is visible. Although the active power is the same for both load A and load B, load A requires 150 A vs. 90 A for load B. The i^2r standby losses are also higher for load A as opposed to load B. The installation cost to service load B is roughly half that of load A, due to the smaller wire, switches, and conduit sizes possible. Note the greater power dissipation due to the long conduit run of load B compared to load D. For large loads that are served over long distances and operate continuously, consideration should be given to using larger wire sizes to reduce standby losses.

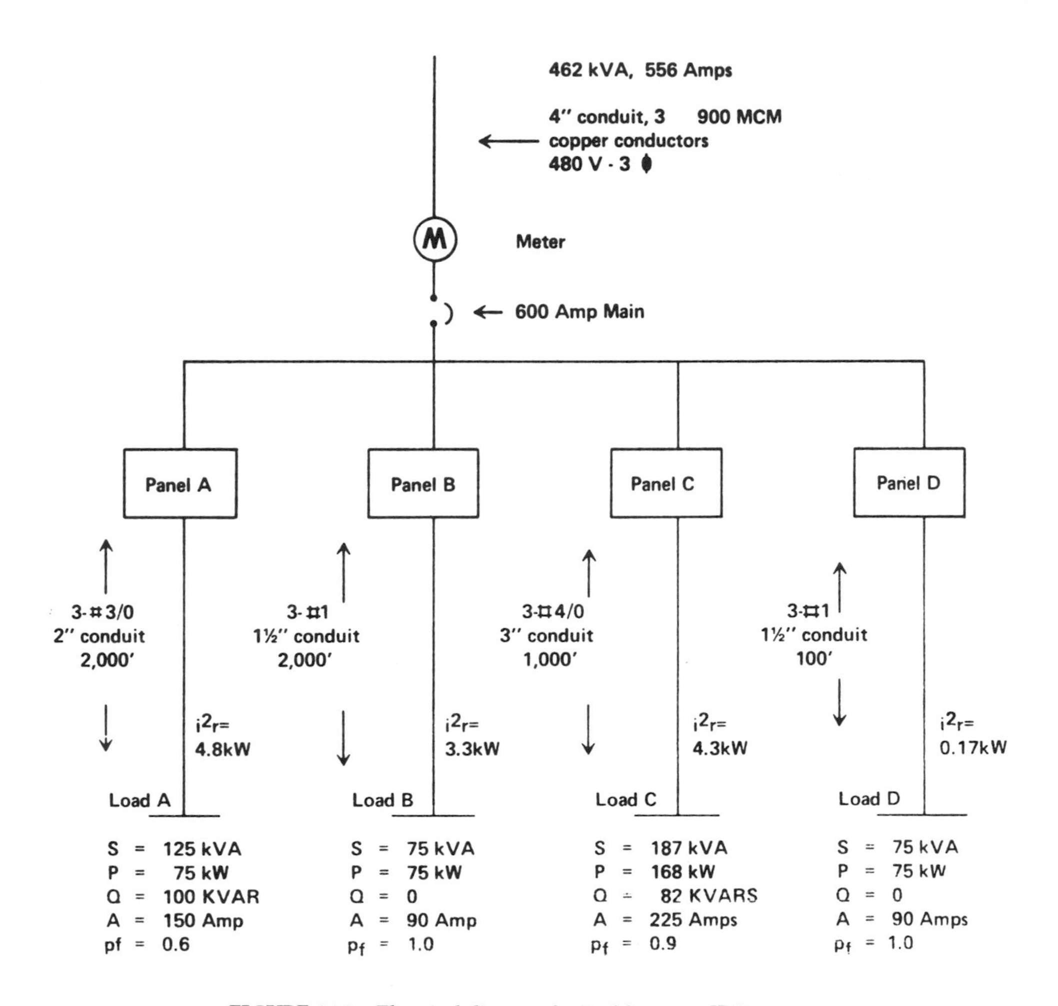

FIGURE 16.3 Electrical diagram for Building 201, XYZ company.

An estimate of the annual cost of power dissipated as heat in these conduit runs can be made if the typical operating hours of each load are known:

Load	Line Losses in kW	Operating Hours/yr	kWh/year
A	4.8	2,000	9,600
B	3.3	2,000	6,600
C	4.3	4,000	17,200
D	0.17	2,000	340
Total			33,740

At an average cost of 6¢/kWh (includes demand and energy costs), the losses in the distribution system alone are $2,022 per year. Over the life of the facility, this is a major expense for a totally unproductive use of energy.

Energy Management Strategies for Industry

Energy management strategies for industry can be grouped into three categories:

- Operational and maintenance strategies
- Retrofit or modification strategies
- New design strategies

The order in which these are listed corresponds approximately to increased capital investment and increased implementation times. Immediate savings at little or no capital cost can generally be achieved by improved operations and better maintenance. Once these "easy" savings have been realized, additional improvements in efficiency will require capital investments.

Electric Drives and Electrically Driven Machinery

About 68% of industrial electricity use is for electrical-motor-driven equipment. Integral horsepower (<0.75 kW) motors are more numerous. Major industrial motor loads are, in order of importance, pumps, compressors, blowers, fans, miscellaneous integral motor applications including conveyors, DC drives, machine tools, and fractional horsepower applications.

Numerous examples of pumping in industry can be observed. These include process pumping in chemical plants, fluid movement in oil refineries, and cooling water circulation. An example of compressors is in the production of nitrogen and oxygen, two common chemicals. Large amounts of electricity are used to drive the compressors, which supply air to the process.

The typical industrial motor is a polyphase motor rated at 11.2 kW (15 hp) and having a life of about 40,000 hr. The efficiencies of electric motors have increased recently as a result of higher energy prices, conservation efforts, and new government standards (the 1992 Energy Policy Act). High-efficiency motors cost roughly 20% to 30% more than standard motors, but this expense is quickly repaid for motors that see continuous use.

Most efficient use of motors requires that attention be given to the following:

Optimum power—Motors operate most efficiently at rated voltage. Three-phase power supplies should be balanced; an unbalance of 3% can increase losses 25%.[1]

Good motor maintenance—Provide adequate cooling, keep heat transfer surfaces and vents clean, and provide adequate lubrication. Improved lubrication alone can increase efficiency a few percentage points.

Equipment scheduling—Turn equipment off when not in use; schedule large motor operation to minimize demand peaks.

Size equipment properly—Match the motor to the load and to the duty cycle. Motors operate most efficiently at rated load.

Evaluate continuous vs. batch processes—Sometimes a smaller motor operating continuously will be more economical.

Power factor—Correct if economics dictate savings. Motors have the best power factor at rated load.

Retrofit or new designs permit use of more efficient motors. For motors up to about 10 to 15 kW (15 to 20 hp) there are variations in efficiency. Select the most efficient motor for the job. Check to verify that the additional cost (if any) will be repaid by the savings that will accrue over the life of the installation.

In addition to reviewing the electric drive system, consider the power train and the load. Friction results in energy dissipation in the form of heat. Bearings, gears, and belt drives all have certain losses, as do clutches. Proper operation and maintenance can reduce energy wastage in these systems and improve overall efficiency.

Material shaping and forming, such as is accomplished with machine tools, requires that electrical energy be transformed into various forms of mechanical energy. The energy expenditure is related to the material and to the depth and speed of the cut. By experimenting with a specific process, it is possible to establish cutting rates that are optimum for the levels of production required and are most efficient in terms of energy use. Motors are not the only part of the electric drive system that sustains losses. Other losses occur in the electric power systems that supply the motor. Electric power systems include substations, transformers, switching gear, distribution systems, feeders, power and lighting panels, and related equipment. Possibilities for energy management include the following:

Use highest voltages that are practical. For a given application, doubling the voltage cuts the required current in half and reduces the i^2r losses by a factor of four.

Eliminate unnecessary transformers. They waste energy. Proper selection of equipment and facility voltages can reduce the number of transformers required and cut transformer losses. Remember, the customer pays for losses when the transformers are on his side of the meter. For example, it is generally better to order equipment with motors of the correct voltage, even if this costs more, than to install special transformers.

Energy losses are an inherent part of electric power distribution systems. This is primarily due to i^2r losses and transformers. The end-use conversion systems for electrical energy used in the process also contribute to energy waste. Proper design and operation of an electrical system can minimize energy losses and contribute to the reduction of electricity bills. Where long feeder runs are operated at near-maximum capacities, check to see if larger wire sizes would permit savings and be economically justifiable.

The overall power factor of electrical systems should be checked for low power factor. This could increase energy losses and the cost of electrical service, in addition to excessive voltage drops and increased penalty charges by the utility. Electrical systems studies should be made and consideration should be given to power factor correction capacitors. In certain applications as much as 10 to 15% savings can be achieved in a poorly operating plant.

Check load factors. This is another parameter that measures the plant's ability to use electrical power efficiently. It is defined as the ratio of the actual kWh used to the maximum demand in kW times the total hours in the time period. A reduction in demand to bring this ratio closer to unity without decreasing plant output means more economical operation. For example, if the maximum demand for a given month (200 hr) is 30,000 kWe and the actual kWh is 3.6×10^6 kWh, the load factor is 60%. Proper management of operations during high demand periods, which may extend only 15 to 20 minutes, can reduce the demand during that time without curtailing production. For example, if the 30,000 kWe could be reduced to 20,000 kWe, this would increase the load factor to about 90%. Such a reduction could amount to a $20,000 to $50,000 reduction in the electricity bill.

Reduce peak loads wherever possible. Many nonessential loads can be shed during the demand peak without interrupting production. These loads would include such items as air compressors, heaters, coolers, and air conditioners. Manual monitoring and control is possible but is often impractical because of the short periods of time that are normally involved and the lack of centralized control systems. Automatic power demand control systems are available.

Provide improved monitoring or metering capability, submeters, or demand recorders. While it is true that meters alone will not save energy, plant managers need feedback to determine if their energy management programs are taking effect. Often the installation of meters on individual processes or buildings leads to immediate savings of 5 to 10% by virtue of the ability to see how much energy is being used and to test the effectiveness of corrective measures.

Fans, Blowers, Pumps

Simple control changes are the first thing to consider with these types of equipment. Switches, timeclocks, or other devices can insure that they do not operate except when needed by the process. Heat removal or process mass flow requirements will determine the size of fans and pumps. Often there is excess capacity, either as a result of design conservatism or because of process changes subsequent to the installation of equipment. The required capacity should be checked, since excess capacity leads to unnecessary demand charges and decreased efficiency.

For fans, the volume rate of air flow Q varies in proportion to the speed of the impeller:

$$Q = c_f N \ m^3/s \tag{16.14}$$

where Q = air flow in m^3/s; c_f = a constant with units m^3/r; and N = fan speed, r/s. The pressure developed by the fan varies as the square of the impeller speed. The important rule, however, is that the power needed to drive the fan varies as the cube of the speed:

$$P = P_c N^3 \text{ W} \tag{16.15}$$

where P = input power in watts and P_c = a constant with units W s^3/r^3.

The cubic law of pumping power indicates that if the air flow is to be doubled, eight (2^3) times as much power must be supplied. Conversely, if the air flow is to be cut in half, only one eighth ($1/2^3$) as much power must be supplied. Air flow (and hence power) can be reduced by changing pulleys or installing smaller motors.

Pumps follow laws similar to fans, the key being the cubic relationship of power to the volume pumped through a given system. Small decreases in flow rate, such as might be obtained with a smaller pump or gotten by trimming the impeller, can save significant amounts of energy.

Variable speed drives (VSDs) are another technique for reducing process energy use. VSDs permit fans, blowers, and pumps to vary speed depending on process requirements. This can lead to significant savings on noncontinuous processes. Recent improvements in solid-state electronics have caused the price of VSDs to drop substantially. This is another technology that is supported by utility rebates in many areas.

Air Compressors

Compressed air is a major energy use in many manufacturing operations. Electricity used to compress air is converted into heat and potential energy in the compressed air stream. Efficient operation of compressed air systems therefore requires the recovery of excess heat where possible, as well as the maximum recovery of the stored potential energy.

Efficient operation is achieved in these ways:

Select the appropriate type and size of equipment for the duty cycle required. Process requirements vary, depending on flow rates, pressure, and demand of the system. Energy savings can be achieved by selecting the most appropriate equipment for the job. The rotary compressor is more popular for industrial operations in the range of 20 to 200 kW, even though it is somewhat less efficient than the reciprocal compressor. This has been due to lower initial cost and reduced maintenance. When operated at partial load, reciprocating units can be as much as 25% more efficient than rotary units. However, newer rotary units incorporate a valve that alters displacement under partial load conditions and improves efficiency. Selection of an air-cooled vs. a water-cooled unit would be influenced by whether water or air was the preferred medium for heat recovery.

Proper operation of compressed air systems can also lead to improved energy utilization. Obviously, air leaks in lines and valves should be eliminated. The pressure of the compressed air should be reduced to a minimum. The percentage saving in power required to drive the compressor at a reduced pressure can be estimated from the fan laws described previously. For example, suppose the pressure were reduced to one half the initial value. Since pressure varies as the square of the speed, this implies the speed would be 70.7% of the initial value. Since power varies as the cube of the speed, the power would now be $0.707^3 = 35\%$ of the initial value. Of course, this is the theoretical limit; actual compressors would not do as well, and the reduction would depend on the type of compressor. Measurements indicate that actual savings would be about half the theoretical limit; reducing pressure 50% would reduce brake horsepower about 30%. To illustrate this point further, for a compressor operating at 6.89×10^5 N/m^2 (100 psi) and a reduction of the discharge pressure to 6.20×10^5 N/m^2 (90 psi), a 5% decrease in brake horsepower would result. For a 373 kW (500 hp) motor operating for 1 year, the 150,000 kWh savings per year would result in about $7,500 per year at current electric power costs.

The intake line for the air compressor should be at the lowest temperature available. This normally means outside air. The reduced temperature of air intake results in a smaller volume of air to be compressed. The percentage horsepower saving relative to a 21°C (70°F) intake air temperature is about 2% for each 10°F drop in temperature. Conversely, input power increases by about 2% for each 10°F increase in intake air temperature.

Leakage is the greatest efficiency offender in compressed air systems. The amount of leakage should be determined and measures taken to reduce it. If air leakage in a plant is more than 10% of the plant demand, a poor condition exists. The amount of leakage can be determined by a simple test during off-production hours (when air-using equipment is shut down) by noting the time that the compressor operates under load compared with the total cycle. This indicates the percentage of the compressor's capacity that is used to supply the plant air leakage. Thus if the load cycle compared with the total cycle were 60 seconds compared with 180 seconds, the efficiency would be 33%, or 33% of the compressor capacity is the amount of air leaking in m^3/min (ft^3/min).

Recover heat where feasible. There are sometimes situations where water-cooled or air-cooled compressors are a convenient source of heat for hot water, space heating, or process applications. As a rough rule of thumb, about 300 J/m^3 min of air compressed (~10 Btu/ft^3 min) can be recovered from an air-cooled rotary compressor.

Substitute electric motors for air motor (pneumatic) drives. Electric motors are far more efficient. Typical vaned air motors range in size from 0.15 to 6.0 kW (0.2 to 8 hp), cost $300 to $1,500, and produce 1.4 to 27 N m (1 to 20 ft lb) of torque at 620 kN/m^2 (90 psi) air pressure. These are used in manufacturing operations where electric motors would be hazardous, or where light weight and high power are essential. Inefficiency results from air system leaks and the need (compared to electric motors) to generate compressed air as an intermediate step in converting electric to mechanical energy.

Review air usage in paint spray booths. In paint spray booths and exhaust hoods, air is circulated through the hoods to control dangerous vapors. Makeup air is constantly required for dilution purposes. This represents a point of energy rejection through the exhaust air.

Examination should be made of the volumes of air required in an attempt to reduce flow and unnecessary operation. Possible mechanisms for heat recovery from the exhaust gases should be explored using recovery systems.

Electrolytic Operations

Electrolysis is an important industrial use of electricity, particularly in the primary metals industry, where it is used in the extraction process for several important metals. Energy management opportunities include

Improve design and materials for electrodes. Evaluate loss mechanisms for the purpose of improving efficiency.

Examine electrolysis and plating operations for savings. Review rectifier performance, heat loss from tanks, and the condition of conductors and connections.

Welding is another electrolytic process. Alternating current welders are generally preferable when they can be used, since they have a better power factor, better demand characteristics, and more economical operation.

Welding operations can also be made more efficient by the use of automated systems which require 50% less energy than manual welding. Manual welders deposit a bead only 15 to 30% of the time the machine is running. Automated processes, however, reduce the no-load time to 40% or less. Different welding processes should be compared in order to determine the most efficient process. Electroslag welding is suited only for metals over 1 cm (0.5 in.) thick but is more efficient than other processes.

Two other significant applications of electrolysis of concern to industry are batteries and corrosion. Batteries are used for standby power, transportation, and other applications. Proper battery maintenance, and improved battery design contribute to efficient energy use.

Corrosion is responsible for a large loss of energy-intensive metals every year and thus indirectly contributes to energy wastage. Corrosion can be prevented and important economies realized, by use of protective films, cathodic protection, and electroplating or anodizing.

Electric Process Heat and Steam Systems

Inasmuch as approximately 40% of the energy utilized in industry goes toward the production of process steam, it presents a large potential for energy misuse and fuel waste from improper maintenance and operation. Even though electrically generated steam and hot water is a small percentage of total industrial steam and hot water, the electrical fraction is likely to increase as other fuels increase in price. This makes increased efficiency even more important. For example:

Steam leaks from lines and faulty valves result in considerable losses. These losses depend on the size of the opening and the pressure of the steam, but can be very costly. A hole 0.1" in diameter with steam at 200 psig can bleed $1,000 to $2,000 worth of steam (500 GJ) in a year.

Steam traps are major contributors to energy losses when not functioning properly. A large process industry might have thousands of steam traps, which could result in large costs if they are not operating correctly. Steam traps are intended to remove condensate and non-condensable gases while trapping or preventing the loss of steam. If they stick open, orifices as large as 6 mm (0.25 in.) can allow steam to escape. Such a trap would allow 1,894 GJ/year (2,000 MBtu/year) of heat to be rejected to the atmosphere on a 6.89×10^5 N/m (100 psi) pressure steam line. Many steam traps are improperly sized, contributing to an inefficient operation. Routine inspection, testing, and a correction program for steam valves and traps are essential in any energy program and can contribute to cost savings.

Poor practice and design of steam distribution systems can be the source of heat waste up to 10% or more. It is not uncommon to find an efficient boiler or process plant joined to an inadequate steam distribution system. Modernization of plants results from modified steam requirements. The old distribution systems are still intact, however, and can be the source of major heat losses. Large steam lines intended to supply units no longer present in the plant are sometimes used for minor needs, such as space heating and cleaning operations, that would be better accomplished with other heat sources.

Steam distribution systems operating on an intermittent basis require a start-up warming time to bring the distribution system into proper operation. This can extend up to 2 or 3 hr, which puts a demand on fuel needs. Not allowing for proper ventilating of air can also extend the startup time. In addition, condensate return can be facilitated if it is allowed to drain by gravity into a tank or receiver and is then pumped into the boiler feed tank.

Proper management of condensate return. Proper management can lead to great savings. Lost feedwater must be made up and heated. For example, every 0.45 kg (1 lb) of steam that must be generated from 15°C feedwater instead of 70°C feedwater requires an additional 1.056×10^5 J (100 Btu) more than 1.12 MJ (1,063 Btu) required or a 10% increase in fuel. A rule of thumb is that a 1% fuel saving results for every 5°C increase in feedwater temperature. Maximizing condensate recovery is an important fuel saving procedure.

Poorly insulated lines and valves due either to poor initial design or a deteriorated condition. Heat losses from a poorly insulated pipe can be costly. A poorly insulated line carrying steam at 400 psig can lose ~1,000 GJ/year (10^9 Btu/year) or more per 30 m (100 ft) of pipe. At steam costs of $2.00/GJ, this translates to a $2,000 expense per year.

Improper operation and maintenance of tracing systems. Steam tracing is used to protect piping and equipment from cold weather freezing. The proper operation and maintenance of tracing systems will not only insure the protection of traced piping but also saves fuel.

Occasionally these systems are operating when not required. Steam is often used in tracing systems and many of the deficiencies mentioned earlier apply (e.g., poorly operating valves, insulation, leaks).

Reduce losses in process hot water systems. Electrically heated hot water systems are used in many industrial processes for cleaning, pickling, coating, or etching components. Hot or cold water systems can dissipate energy. Leaks and poor insulation should be repaired.

Electrical Process Heat

Industrial process heat applications can be divided into four categories: direct-fired, indirect-fired, fuel, or electric. Here we shall consider electric direct-fired installations (ovens, furnaces) and indirect-fired (electric water heaters and boilers) applications. Electrical installations use metal sheath resistance heaters, resistance ovens or furnaces, electric salt bath furnaces, infrared heaters, induction and high-frequency resistance heaters, dielectric heaters, and direct arc furnaces. From the housekeeping and maintenance point of view, typical opportunities would include:

Repair or improve insulation. Operational and standby losses can be considerable, especially in larger units. Remember that insulation may degrade with time or may have been optimized to different economic criteria.

Provide finer controls. Excessive temperatures in process equipment waste energy. Run tests to determine the minimum temperatures that are acceptable, then test instrumentation to verify that it can provide accurate process control and regulation.

Practice heat recovery. This is an important method, applicable to many industrial processes as well as HVAC systems and so forth. It is described in more detail in the next section.

Heat Recovery

Exhaust gases from electric ovens and furnaces provide excellent opportunities for heat recovery. Depending on the exhaust gas temperature, exhaust heat can be used to raise steam or to preheat air or feedstocks. Another potential source of waste heat recovery is the exhaust air that must be rejected from industrial operations in order to maintain health and ventilation safety standards. If the reject air has been subjected to heating and cooling processes, it represents an energy loss inasmuch as the makeup air must be modified to meet the interior conditions. One way to reduce this waste is through the use of heat wheels or similar heat exchange systems.

Energy in the form of heat is available at a variety of sources in industrial operations, many of which are not normally derived from primary heat sources. Such sources include electric motors, crushing and grinding operations, air compressors, and drying processes. These units require cooling in order to maintain proper operation. The heat from these systems can be collected and transferred to some appropriate use such as space heating or water heating.

The heat pipe is gaining wider acceptance for specialized and demanding heat transfer applications. The transfer of energy between incoming and outgoing air can be accomplished by banks of these devices. A refrigerant and a capillary wick are permanently sealed inside a metal tube, setting up a liquid-to-vapor circulation path. Thermal energy applied to either end of the pipe causes the refrigerant to vaporize. The refrigerant vapor then travels to the other end of the pipe, where thermal energy is removed. This causes the vapor to condense into liquid again, and the condensed liquid then flows back to the opposite end through the capillary wick.

Industrial operations involving fluid flow systems that transport heat such as in chemical and refinery operations offer many opportunities for heat recovery. With proper design and sequencing of heat exchangers, the incoming product can be heated with various process steams. For example, proper heat exchanger sequence in preheating the feedstock to a distillation column can reduce the energy utilized in the process.

Many process and air conditioning systems reject heat to the atmosphere by means of wet cooling towers. Poor operation can contribute to increased power requirements.

Water flow and air flow should be examined to see that they are not excessive. The cooling tower outlet temperature is fixed by atmospheric conditions if operating at design capacity. Increasing the water flow rate or the air flow will not lower the outlet temperature.

The possibility of utilizing heat that is rejected to the cooling tower for other purposes should be investigated. This includes preheating feedwater, heating hot water systems, space heating, and other low-temperature applications. If there is a source of building exhaust air with a lower wet bulb temperature, it may be efficient to supply this to a cooling tower.

Power Recovery

Power recovery concepts are an extension of the heat recovery concept described earlier. Many industrial processes today have pressurized liquid and gaseous streams at 150 to 375°C (300 to 700°F) that present excellent opportunities for power recovery. In many cases high-pressure process stream energy is lost by throttling across a control valve.

The extraction of work from high-pressure liquid streams can be accomplished by means of hydraulic turbines (essentially diffuser-type or volute-type pumps running backward). These pumps can be either single or multistage. Power recovery ranges from 170 to 1,340 kW (230 to 1,800 hp). The lower limit of power recovery approaches the minimum economically justified for capital expenditures at present power costs.

Heating, Ventilating, Air Conditioning Operation

The environmental needs in an industrial operation can be quite different from those in a residential or commercial structure. In some cases strict environmental standards must be met for a specific function or process. More often the environmental requirements for the process itself are not severe; however, conditioning of the space is necessary for the comfort of operating personnel, and thus large volumes of air must be processed. Quite often opportunities exist in the industrial operation where surplus energy can be utilized in environmental conditioning. A few suggestions follow:

Review HVAC controls. Building heating and cooling controls should be examined and preset.

Ventilation, air, and building exhaust requirements should be examined. A reduction of air flow will result in a savings of electrical energy delivered to motor drives and additionally reduce the energy requirements for space heating and cooling. Because pumping power varies as the cube of the air flow rate, substantial savings can be achieved by reducing air flows where possible.

Do not condition spaces needlessly. Review air conditioning and heating operations, seal off sections of plant operations that do not require environmental conditioning, and use air conditioning equipment only when needed. During nonworking hours the environmental control equipment should be shut down or reduced. Automatic timers can be effective.

Provide proper equipment maintenance. Insure that all equipment is operating efficiently. (Filters, fan belts, and bearings should be in good condition.)

Use only equipment capacity needed. When multiple units are available, examine the operating efficiency of each unit and put operations in sequence in order to maximize overall efficiency.

Recirculate conditioned (heated or cooled) air where feasible. If this cannot be done, perhaps exhaust air can be used as supply air to certain processes (e.g., a paint spray booth) to reduce the volume of air that must be conditioned.

For additional energy management opportunities in HVAC systems, see Chapter 10.

Lighting

Industrial lighting needs range from low-level requirements for assembly and welding of large structures (such as shipyards) to the high levels needed for manufacture of precision mechanical and electronic components (e.g., integrated circuits). There are four basic housekeeping checks that should be made:

Is a more efficient lighting application possible? Remove excessive or unnecessary lamps.

Is relamping possible? Install lower-wattage lamps during routine maintenance.

Will cleaning improve light output? Fixtures, lamps, and lenses should be cleansed periodically.

Can better controls be devised? Eliminate turning on more lamps than necessary. For modification, retrofit, or new design, consideration should be given to the spectrum of high-efficiency lamps and luminaires that are available. For example, high-pressure sodium lamps are finding increasing acceptance for industrial use, with savings of nearly a factor of five compared to incandescent lamps. See Chapter 10 for additional details.

The New Electrotechnologies

Electricity has certain characteristics that make it uniquely suitable for industrial processes. These characteristics include electricity's suitability for timely and precise control; its ability to interact with materials at the molecular level; the ability to apply it selectively and specifically, and the ability to vary its frequency and wavelength so as to enhance or inhibit its interaction with materials. These aspects may be said to relate to the *quality* of electricity as an energy form. It is important to recognize that different forms of energy have different qualities in the sense of their ability to perform useful work. Thus, although the Btu content of two energy forms may be the same, their ability to transform materials may be quite different.

New electrotechnologies based on the properties of electricity are now finding their way into modern manufacturing. In many cases the introduction of electricity reduces manufacturing costs, improves quality, reduces pollution, or has other beneficial results. Some examples include

- Microwave heating
- Induction heating
- Plasma processing
- Magnetic forming
- RF drying and heating
- Ion nitriding
- Infrared drying
- UV drying and curing
- Advanced finishes
- Electron beam heating

Microwave heating is a familiar technology that exhibits the unique characteristics of electricity described earlier. First it is useful to review how conventional heating is performed to dry paint, anneal a part, or remove water. A source of heat is required, along with a container (oven, furnace, pot, etc.) to which the heat is applied. Heat is transferred from the container to the work piece by conduction, radiation, convection, or a combination of these. There are certain irreversible losses associated with heat transfer in this process. Moreover, since the container must be heated, more energy is expended than is really required. Microwave heating avoids these losses due to the unique characteristics of electricity.

Timely control. There is no loss associated with the warmup or cool-down of ovens. The heat is applied directly when needed.

Molecular interaction. By interacting at the molecular level, heat is deposited directly in the material to be heated, without having to preheat an oven, saving the extra energy required for this purpose and avoiding the losses that result from heat leakage from the oven.

Selective application. By selectively applying heat only to the material to be heated, parasitic losses are avoided. In fact, the specificity of heat applied this way can improve quality by not heating other materials.

Selective wavelength and frequency. A microwave frequency is selected that permits the microwave energy to interact with the material to be heated, and not with other materials. Typically the frequency is greater than 2,000 MHz.

Microwave heating was selected for this discussion, but similar comments could be made about infrared, ultraviolet, dielectric, induction, or electron beam heating. In each case the frequency or other characteristics of the energy form are selected to provide the unique performance required.

Ultraviolet (UV) curing (now used for adhesive and finishes) is another example. The parts to be joined or coated can be prepared and the excessive adhesive removed without fear of prehardening. Then the UV energy is applied, causing the adhesive to harden.

Induction heating is another example. It is similar to microwave heating except that the energy is applied at a lower frequency. Induction heating operates on the principal of inducing electric currents to flow in materials, heating them by the power dissipated in the material. The method has several other advantages. In a conventional furnace, the work piece has to be in the furnace for a sufficient time to reach temperature. Because of this, some of the material is oxidized and lost as scale. In a typical high-temperature gas furnace this can be 2% of the throughput. Additional product is scrapped as a result of surface defects caused by uneven heating and cooling. This can amount to another 1% of throughput. Induction heating can reduce these losses by a factor of four.

The fact that electricity can be readily controlled and carries with it a high information content through digitization or frequency modulation also offers the potential for quantum improvements in efficiency. A slightly different example is the printing industry.

Today the old linotype technology has been replaced by electronic processes. The lead melting pots that used to operate continuously in every newspaper plant have been removed, eliminating a major energy use and an environmental hazard. Books, magazines, and newspapers can be written, composed, and printed entirely by electronic means. Text is processed by computer techniques. Camera-ready art is prepared by computers directly or prepared photographically and then optically scanned to create digital images. The resulting electronic files can be used in web offset printing by an electronic photochemical process. The same information can be transmitted electronically, via satellite, to a receiving station at a remote location where a high-resolution fax machine reconstitutes the image. This method is being used to simultaneously and instantaneously distribute advertising copy to multiple newspapers, using a single original. Previously, to insert an advertisement in 25 newspapers, 25 sets of photographic originals would have to be prepared and delivered, by messenger, air express, or mail, to each newspaper.

Some of the other applications of the new electrotechnologies include RF drying of plywood veneers, textiles, and other materials; electric infrared drying for automobile paint and other finishes; electric resistance melting for high purity metals and scrap recovery; and laser cutting of wood, cloth, and other materials.

General Industrial Processes

The variety of industrial processes is so great that detailed specific recommendations are outside the scope of this chapter. Useful sources of information are found in trade journals, vendor technical bulletins, and manufacturers' association journals. These suggestions are intended to be representative, but by no means do they cover all possibilities.

In machining operations, eliminate unnecessary operations and reduce scrap. This is so fundamental from a purely economic point of view that it will not be possible to find significant improvements in many situations. The point is that each additional operation and each increment of scrap also represents a needless use of energy. Machining itself is not particularly energy intensive. Even so, there are alternate technologies that can not only save energy but reduce material wastage as well. For example, powder metallurgy generates less scrap and is efficient if done in induction-type furnaces.

Use stretch forming. Forming operations are more efficient if stretch forming is used. In this process sheet metal or extrusions are stretched 2 to 3% prior to forming, which makes the material more ductile so that less energy is required to form the product. The finished part is stronger, so thinner sheets and lighter extrusions can be used.

Use alternate heat treating methods. Conventional heat treating methods such as carburizing are energy intensive. Alternate approaches are possible. For example, a hard surface can be produced by induction heating, which is a more efficient energy process. Plating, metallizing, flame spraying, or cladding can substitute for carburizing, although they do not duplicate the fatigue strengthening compressive skin of carburization or induction hardening.

Use **alternative painting methods.** Conventional techniques using solvent-based paints require drying and curing at elevated temperature. Powder coating is a substitute process in which no solvents are used. Powder particles are electrostatically charged and attracted to the part being painted so that only a small amount of paint leaving the spray gun misses the part and the overspray is recoverable. The parts can be cured rapidly in infrared ovens, which require less energy than standard hot air systems. Water-based paints and high-solids coatings are also being used and are less costly than solvent-based paints. They use essentially the same equipment as the conventional solvent paint spray systems so that the conversion can be made at minimum costs. New water-based emulsion paints contain only 5% organic solvent and require no afterburning. High-solids coatings are already in use commercially for shelves, household fixtures, furniture, and beverage cans, and require no afterburning. They can be as durable as conventional finishes and are cured by either conventional baking or ultraviolet exposure.

Substitute for energy intensive processes such as hot forging. Hot forging may require a part to go through several heat treatments. Cold forging with easily wrought alloys may offer a replacement. Lowering the preheat temperatures may also be an opportunity for savings. Squeeze forging is a relatively new process in which molten metal is poured into the forging dye. The process is nearly scrap free, requires less press power, and promises to contribute to more efficient energy utilization.

Movement of materials through the plant creates opportunities for saving energy. Material transport energy can be reduced by

Combining processes or relocate machinery to reduce transport energy. Sometimes merely relocating equipment can reduce the need to haul materials.

Turning off conveyors and other transport equipment when not needed. Look for opportunities where controls can be modified to permit shutting down of equipment not in use.

Using gravity feeds wherever possible. Avoid unnecessary lifting and lowering of products.

Demand Management

The cost of electrical energy for medium to large industrial and commercial customers generally consists of two components. One component is the *energy charge,* which is based on the cost of fuel to the utility, the average cost of amortizing the utility generating plant, and on the operating and maintenance costs experienced by the utility. Energy costs for industrial users in the United States are typically in the range of 0.04 to 0.06 $/kWh.

The second component is the *demand charge,* which reflects the investment cost the utility must make to serve the customer. Besides the installed generating capacity needed, the utility also provides distribution lines, transformers, substations, and switches whose cost depends on the size of the load being served. This cost is recovered in a demand charge which typically is 2 to 10 $/kW month.

Demand charges typically account for 10 to 50% of the bill, although wide variations are possible depending on the type of installation. Arc welders, for example, have relatively high demand charges, since the installed capacity is great (10 to 30 kW for a typical industrial machine) and the energy use is low.

From the utility's point of view it is advantageous to have its installed generating capacity operating at full load as much of the time as possible. To follow load variations conveniently, the utility operates its largest and most economical generating units continuously to meet its base load, and then brings smaller (and generally more expensive) generating units on line to meet peak load needs.

Today consideration is being given to time-of-day or *peak load pricing* as a means of assigning the cost of operating peak generating capacity to those loads that require it. See Chapter 5 for a discussion of utility pricing strategies. From the viewpoint of the utility, *load management* implies maintaining a high capacity factor and minimizing peak load demands. From the customer's viewpoint, *demand management* means minimizing electrical demands (both on and off peak) so as to minimize overall electricity costs.

Utilities are experimenting with several techniques for load management. Besides rate schedules that encourage the most effective use of power, some utilities have installed remotely operated switches that permit the utility to disconnect nonessential parts of the customer's load when demand is excessive. These switches are actuated by a radio signal, through the telephone lines, or over the power grid itself through a harmonic signal (ripple frequency) that is introduced into the grid.

Customers can control the demand of their loads by any of several methods:

- Manually switching off loads ("load shedding")
- Use of timers and interlocks to prevent several large loads from operating simultaneously
- Use of controllers and computers to control loads and minimize peak demand by scheduling equipment operation
- Energy storage (e.g., producing hot or chilled water during off-peak hours and storing it for use on peak)

Demand can be monitored manually (by reading a meter) or automatically using utility-installed equipment or customer-owned equipment installed in parallel with the utility meter. For automatic monitoring, the basic approach involves pulse counting.

The demand meter produces electronic pulses, the number of which is proportional to demand in kW. Demand is usually averaged over some interval (e.g., 15 minutes) for calculating cost. By monitoring the pulse rate electronically, a computer can project what the demand will be during the next demand measurement interval, and can then follow a preestablished plan for shedding loads if the demand set point is likely to be exceeded.

Computer control can assist in the dispatching of power supply to the fluctuating demands of plant facilities. Large, electrically based facilities are capable of forcing large power demands during peak times that exceed the limits contracted with the utility or cause penalties in increased costs. Computer control can even out the load by shaving peaks and filling in the valleys, thus minimizing power costs. In times of emergency or fuel curtailment, operation of the plant can be programmed to provide optimum production and operating performance under prevailing conditions. Furthermore, computer monitoring and control provide accurate and continuous records of plant performance.

It should be stressed here that many of these same functions can be carried out by manual controls, time clocks, microprocessors, or other inexpensive devices. Selection of a computer system must be justified economically on the basis of the number of parameters to be controlled and the level of sophistication required. Many of the benefits described here can be obtained in some types of operations without the expense of a computer.

Closing Remarks

Electrical energy is increasing in importance in the industrial sector as conventional processes have been electrified and new electrotechnologies introduced. The reason for this is that electricity is a high-quality energy form and offers opportunities for innovations in manufacturing and industry that have the potential for jumps in productivity and product quality. This stems from the unique nature of electricity and the way it interacts with materials at the molecular level, as well as its controllability and flexibility. Flexibility arises from the fact that electrical energy can be converted to many different energy forms and can be supplied at different wavelengths depending on the application.

As the introduction to this chapter showed, electricity now accounts for almost 15% of industrial energy use. The increased use of electricity is certainly in part responsible for the fact that industrial output has increased by 30% during the last 20 years while the energy input has declined and only now is approaching the 1973 level. The electrical share is likely to increase further as new technologies are introduced and there is greater use of automation and electronic controls. While electricity causes little pollution at the end use, the generation of electricity involves hydropower, the combustion of fossil fuels, or nuclear power, and their associated environmental concerns. The efficient use of electrical power is therefore vital to avoid wasting natural resources and to minimize environmental and economic costs.

The approach taken in this chapter is to outline the important industrial applications of electricity (electric drives, electrolysis, electric process heat, HVAC, and lighting) and then to describe the process of establishing an industrial power management program. Methods for evaluating electrical loads in an industrial facility are illustrated by analyzing a typical daily load curve. This approach helps the energy manager establish priorities.

The balance of the chapter describes specific energy management strategies for each of the major electricity end uses, beginning with electric drives and including fans and pumps, air compressors, electrolytic operations, electric process heat, and other topics. There is also a discussion of new electrotechnologies and general industrial processes where electricity plays a role. The strategies include better housekeeping and maintenance; more efficient equipment, such as high-efficiency electric motors, ballasts, and lamps; improved controls; new processes; heat recovery; reduction of losses, and other techniques.

We now live in a world where trade barriers are dropping and there is increased international competition in every field. Customers demand and receive high quality products, goods, and services at competitive prices. The efficient use of electricity will continue to be vital to maintaining a competitive edge in many industries.

Acknowledgments

I am grateful to my colleague T. Cosentino for her editorial support and assistance with the manuscript.

References

1. Energy Information Administration (EIA). 1994. *Annual Energy Outlook 1994—with Projections to 2010,* U.S. Department of Energy, (Washington D.C.: U.S. Government Printing Office, January).
2. U.S. Congress, Office of Technology Assessment (OTA). 1993a. *Industrial Energy Efficiency,* OTA-E-560, (Washington D.C.: U.S. Government Printing Office, August).
3. U.S. Congress, Office of Technology Assessment (OTA). 1993b. *Energy Efficiency—Challenges and Opportunities for Electric Utilities,* OTA-E-561 (Washington D.C.: U.S. Government Printing Office, September).

17

Cogeneration

Jerald A. Caton
W. Dan Turner
Texas A&M University

Introduction

Background

The term "cogeneration" as used in this chapter is defined as the combined production of electrical power* and useful thermal energy by the sequential use of a fuel or fuels. Each term in this definition is important. Combined means that the production processes of the electric power and thermal energy are linked, and often are accomplished in a series or parallel fashion. Electrical power is

*Although the power output of a cogeneration system could be mechanical power as well as electrical power, the majority of systems produce electrical power. This chapter will consider only cogeneration systems that produce electrical power.

0-8493-2514-5/96/$0.00+$.50

the electricity produced by an electrical generator, which is most often powered by a prime mover such as a steam turbine, gas turbine, or reciprocating engine. Thermal energy is that product of the process which provides heating or cooling. Forms of this thermal energy include hot exhaust gases, hot water, steam, and chilled water. Useful means that the energy is directed at fulfilling an existing need for heating or cooling. Simply exhausting hot exhaust gases does not meet the definition of useful thermal energy. In other words, cogeneration is the production of electrical power and the capture of coexisting thermal energy for useful purposes.

The importance of cogeneration is monetary and energy savings. Any facility that uses electrical power and needs thermal energy is a candidate for cogeneration. Although many considerations are involved in determining if cogeneration is feasible for a particular facility, the basic consideration is if the saving on thermal energy costs is sufficient to justify the capital expenditures for a cogeneration system. Facilities that may be considered for cogeneration include those in the industrial, commercial, and institutional sectors.

The technology for cogeneration is for the most part available and exists over a range of sizes: from less than 100 kW to over 100 MW. The major equipment requirements include a prime mover, electrical generator, electrical controls, heat recovery systems, and other typical power plant equipment. These components are well developed, and the procedures to integrate these components into cogeneration systems are well established.

In addition to the economic and technical considerations, the application of cogeneration systems involves an understanding of the governmental regulations and legislation on electrical power production and on environmental impacts. With respect to electrical power production, certain governmental regulations were passed during the late 1970s that remove barriers and provide incentives to encourage cogeneration development. Finally, no cogeneration assessment would be complete without an understanding of the financial arrangements that are possible.

Scope of the Overall Chapter

The objective of this chapter is to provide a complete overview of the important aspects necessary to understand cogeneration. Specifically, this chapter includes information on the technical equipment and components, technical design issues, regulatory considerations, economic evaluations, financial aspects, computer models and simulations, and future technologies. As briefly described earlier, a thorough discussion of cogeneration includes many aspects of engineering, economics, law, finance, and other topics. The emphasis of this chapter is on the technical and engineering considerations. The descriptions of the economic, legal, and governmental aspects are provided for completeness, but should not be considered a substitute for consulting with appropriate specialists such as attorneys, accountants, and bankers.

This chapter is divided into sections on basic cogeneration systems and terminology, technical components, design issues, regulatory considerations, economic evaluations, and financial aspects. Also, two case studies are presented that illustrate the application of cogeneration to an industrial and an institutional facility. Finally, the chapter concludes with some comments on the future technologies of cogeneration. This introductory section ends with a brief review of the history of cogeneration.

History of Cogeneration

At the beginning of the twentieth century, electrical power generation was in its infancy. Most industrial facilities generated all their own electrical power and often supplied power to nearby communities. They used the thermal energy that was available during the electrical power production to provide or supplement process or building heat. These industrial facilities, therefore, were the first "cogenerators." The dominant prime mover at this time was the reciprocating steam engine, and the low-pressure exhaust steam was used for heating applications.

Between the early 1920s and through the 1970s, the public electric utility industry grew rapidly because of increasing electrical power demands. Coincident with this rapid growth was a general reduction in the costs to produce the electrical power mainly due to the economies of scale, more efficient technologies, and decreasing fuel costs. During this period, industry often abandoned their own electrical power generation because of (1) the decreasing electrical rates charged by public utilities, (2) income tax regulations, which favored expenses instead of capital investments, (3) increasing costs of labor, and (4) the desire of industry to focus on their product rather than the side issue of electrical power generation. Estimates are available that suggest that industrial cogenerated electrical power decreased from about 25 to 9% of the total electrical power generated in the country between the years of 1954 and 1976. Since about the mid-1980s, this percentage has been fairly constant at about 5%. For example, at the end of 1992, 5.1% of the total U.S. electrical capacity was due to cogeneration systems.

During the 1960s and 1970s, the natural gas industry promoted a "total energy" concept of cogeneration. This effort was not very successful due to the relatively poor economics (e.g., relatively inexpensive electricity and expensive fuels), and the lack of governmental regulations to ease the interface with public utilities.

In late 1973 and again in 1979, America experienced major "energy crises," which were largely a result of reduced petroleum imports. Between 1973 and 1983, the prices of fuels and electrical power increased by a factor of about five (5). Any facility purchasing electrical power began to consider (or reconsider) the economic savings associated with cogeneration. These considerations were facilitated by federal regulations that were enacted to ease or remove barriers to cogeneration. In 1978, the government passed the National Energy Act (NEA), which included several important pieces of legislation. The NEA included the Fuel Use Act, the Natural Gas Policy Act, and the Public Utility Regulatory Policies Act (PURPA). Each of these acts had a direct impact on cogeneration, but PURPA was the most significant. In particular, PURPA defined cogeneration systems to include those power plants that supplied a specified fraction of their input energy as useful thermal output in addition to a mechanical or electrical output. This legislation is discussed in more detail in later sections of this chapter.

In addition, other regulatory legislation passed beginning in the late 1950s and continuing through the early 1990s impacts the installation of cogeneration systems. In particular, federal legislation directed at managing air and water quality significantly affects the installations of cogeneration systems. For managing air pollution, the original legislation is the Air Quality Act of 1967, which was amended with the Clean Air Act Amendments in 1970, 1977, and 1990. The main legislative basis for managing water pollution is the Federal Water Pollution Control Act of 1956, as amended by the Water Quality Act of 1965, the Federal Water Pollution Control Act amendments of 1972, and the Clean Water Act of 1977. These Acts and other legislation, and their impacts on the development of cogeneration projects, are discussed in a subsequent section of this chapter.

As the years of the twenty-first century near, cogeneration will continue to experience progressive growth because of the energy and monetary savings. New technology will become available, and new regulations and legislation will be enacted. The future of cogeneration will build on the accomplishments and successes of the past, and will continue to increase in importance.

17.1 Basic Cogeneration Systems

Advantages of Cogeneration

To illustrate the fuel and monetary savings of a cogeneration system, the following simple comparison is presented. Figure 17.1 shows the details of this comparison. The power and thermal needs to be satisfied are 30 energy units* of electrical power and 34 energy units of heat. These power and thermal needs may be satisfied by either a cogeneration system or by conventional systems. Conventional systems for this example would be an electric utility and a standard boiler.

*For this example, an energy unit is defined as any consistent unit of energy such as kW, MW, Btu, MMBtu, or Btu/hr. The use of arbitrary units avoids confusion and the need for unit conversions.

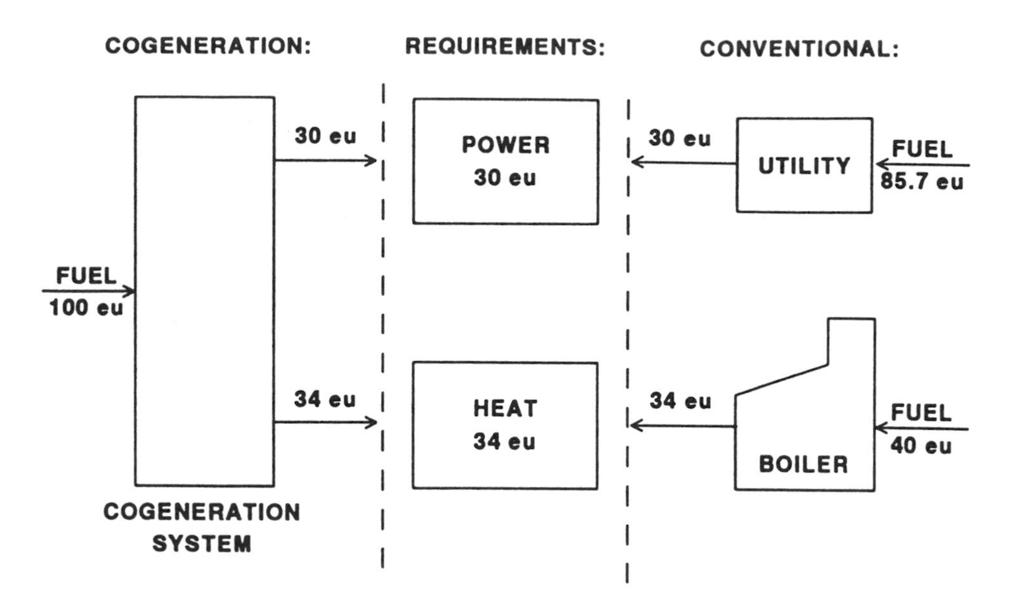

FIGURE 17.1 A schematic illustration of the advantages of cogeneration relative to conventional systems.

At the left of Figure 17.1 is the cogeneration system, and at the right of Figure 17.1 is the conventional system. All the components of the cogeneration system are represented by the box. Assuming an electrical generation efficiency of 30% for the cogeneration system, then 30 energy units are produced with 100 energy units of fuel energy. This cogeneration system then supplies 34 energy units of heat. These are conservative numbers for the electrical and thermal outputs of the cogeneration system, and many systems may have higher efficiencies. The overall thermodynamic efficiency* of the cogeneration system is defined as

$$\eta_0 = (P + T)/F \tag{17.1}$$

where P represents the power, T represents the thermal or heat energy rate, and F represents the fuel input rate (all in consistent units). This overall efficiency also may be found as the sum of the individual efficiencies of the electrical and thermal production. For this example, the cogeneration system has an overall efficiency of

$$\eta_0 = (30 + 34)/100 = 64\% \tag{17.2}$$

The conventional systems, diagrammed at the right of Figure 17.1, satisfy the power and thermal needs by the use of an electric utility and a boiler. In this case, the utility is assumed to be able to deliver the 30 energy units of electrical power with a plant efficiency of 35% (which is on the high side for most utilities). This results in a fuel input of 85.7 energy units. The boiler supplies the 34 energy units of heat with an 85% efficiency, which requires a fuel input of 40 energy units. The overall efficiency is, therefore,

$$\eta_0 = (P + T)/F = (30 + 34)/(85.7 + 40) = 51\% \tag{17.3}$$

*This overall thermodynamic efficiency should not be confused with the "PURPA" efficiency, which is a legislated definition and is described in a later subsection.

This simple example demonstrates the thermodynamic advantage of a cogeneration system relative to conventional systems for accomplishing the same objectives. In this example, the cogeneration system had an overall efficiency of 64% compared to 51% for the conventional systems. Compared to the conventional systems, this is an absolute increase of 13% and a relative improvement of 25% (based on the 51%).

To determine the fuel and monetary savings of the cogeneration system relative to the conventional systems, the output power will be assumed to be 50 MW. This is a power level that would be typical for a large university or hospital. The associated thermal output rate for this example would be

$$T = (34/30)\ 50\ \text{MW} = 56.7\ \text{MW} \tag{17.4}$$

This may be converted to units of MMBtu/hr (i.e., million Btu/hr)

$$T = 56.7\ \text{MW}\,(3.41\ \text{MMBtu/hr/MW}) = 193.4\ \text{MMBtu/hr} \tag{17.5}$$

To determine the fuel inputs, the power must be converted to consistent units:

$$P = 50\ \text{MW}\,(3.41\ \text{MMBtu/hr/MW}) = 170.7\ \text{MMBtu/hr} \tag{17.6}$$

Now the fuel inputs are:

$$(\text{Fuel})_{\text{cogen}} = (P+T)/(\eta_0)_{\text{cogen}} = (170.7+193.4)/0.64 = 569\ \text{MMBtu/hr} \tag{17.7}$$

$$(\text{Fuel})_{\text{conv}} = (P+T)/(\eta_0)_{\text{conv}} = (170.7+193.4)/0.51 = 714\ \text{MMBtu/hr} \tag{17.8}$$

The fuel savings is the difference between these two numbers:

$$(\text{Fuel Savings}) = (\text{Fuel})_{\text{conv}} - (\text{Fuel})_{\text{cogen}} = 714 - 569 = 145\ \text{MMBtu/hr} \tag{17.9}$$

If the plant operates 6,000 hours per year (this is about 68% of the time on average, also a conservative estimate), then the fuel energy savings per year is 870,000 MMBtu. If the fuel price is \$2.00/MMBtu, then the monetary saving is $\$1.74 \times 10^6$ per year. Although this is a simple economic analysis, the general trends are relevant. A later section of this chapter will describe more comprehensive economic analyses used in actual assessments of cogeneration applications.

In summary, the use of cogeneration is an effective way to more efficiently use fuels. The energy released from the fuels is used to both produce electrical power as well as provide useful thermal energy. The savings in energy and money can be substantial, and such systems have been shown to be technically and economically feasible in a wide range of applications.

Topping and Bottoming Cycles

A cogeneration system may be classified as either a topping cycle system or a bottoming cycle system. Figure 17.2 is a schematic illustration of a topping cycle system. As shown, a prime mover uses fuel to power an electrical generator to produce electricity. This electricity may be used completely on-site or may be tied into an electrical distribution network for sale to the local utility or other customers. The hot exhaust gases are directed to a heat recovery boiler (HRB)* to product steam or hot water. This steam or hot water is used on-site for process or building heat.

*Many other terms for this boiler are common: for example, waste heat boiler (WHB) and heat recovery steam generator (HRSG).

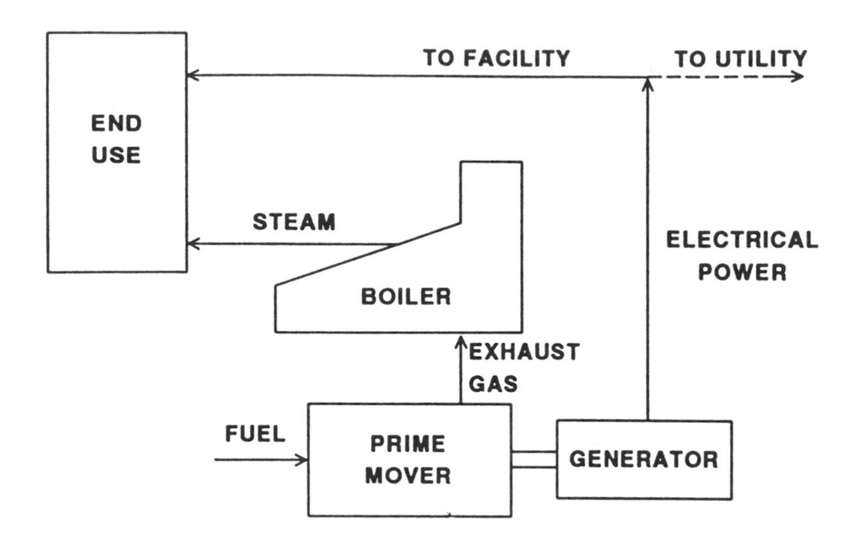

FIGURE 17.2 A schematic illustration of a cogeneration topping-cycle system.

This cogeneration system is classified as a topping cycle because the electrical power is first generated at the higher (top) temperatures associated with the fuel combustion process and then the rejected or exhausted energy is used to produce useful thermal energy (such as the steam or hot water in this example). Figure 17.3 shows the thermodynamic states of the exhaust gases for this process on a temperature–entropy diagram. For this process, as energy is removed from the combustion gases, the temperature and entropy decrease (since the process involves energy removal). As shown, the top or high-temperature gases are used first to produce the electrical power and then the lower-temperature gases (exhaust) are used to produce useful thermal energy. The majority of cogeneration systems are based on topping cycles.

The other classification of cogeneration systems is bottoming cycle systems. Figure 17.4 is a schematic illustration of a bottoming cycle system. As shown, the high-temperature combustion gases are used first in a high-temperature thermal process (such as high-temperature metal treatment) and then the lower-temperature gases are used in a special low-temperature cycle to produce electrical power. Figure 17.5 shows the thermodynamic states of the exhaust gases for this process on a temperature–entropy diagram. After the energy is removed at the high temperatures, the energy available at the bottom or lower temperatures is then used to produce electrical power.*

Bottoming-cycle cogeneration systems have fewer applications than topping-cycle systems and must compete with waste heat recovery systems such as feedwater heaters, recuperators, and process heat exchangers. One of the difficulties with bottoming cycle systems is the low-temperature electrical power producing cycle. One example, depicted in Figure 17.4, is a low-temperature Rankine cycle. The low-temperature Rankine cycle is a power cycle similar to the conventional steam Rankine cycle, but a special fluid such as an organic substance (like a refrigerant) is used in place of water. This fluid vaporizes at a lower temperature than water, so this cycle is able to utilize the low-temperature energy. These cycles are generally much less efficient than conventional power cycles, often involve special equipment, and use more expensive working fluids. The majority of this chapter on cogeneration is based on topping cycle systems.

*Other definitions of bottoming cycles are common. These other definitions often include any second use of the energy. For example, some authors refer to the steam turbine in a combined cycle as using a bottoming cycle. The more precise thermodynamic definition employed here is preferred, although fewer applications meet this definition.

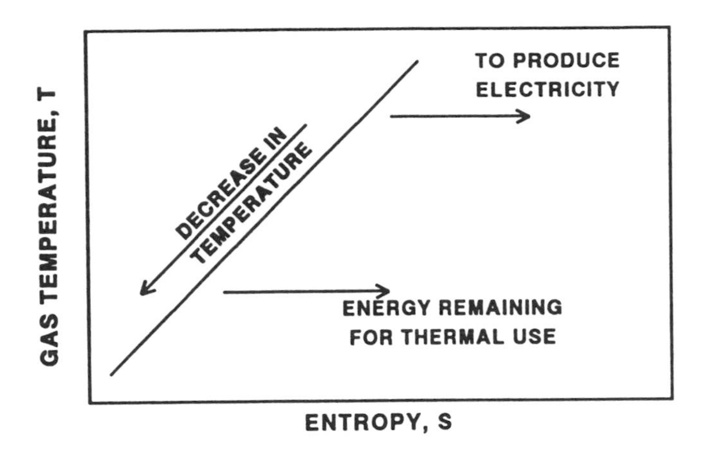

FIGURE 17.3 The gas temperature as a function of entropy for a topping-cycle system.

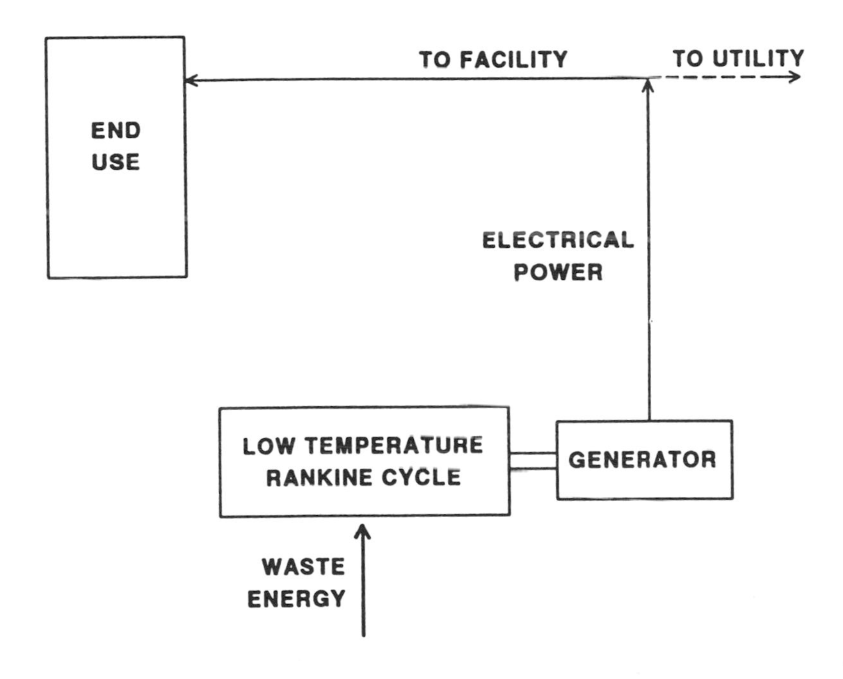

FIGURE 17.4 A schematic illustration of a cogeneration bottoming-cycle system.

Combined Cycles

One power plant configuration that is based on a form of a topping cycle and is widely used in industry and by electrical utilities is known as a combined cycle. Typically in this configuration, a gas turbine is used to generate electricity and the exhaust gas is ducted to a heat recovery steam generator. The steam then is ducted to a steam turbine, which produces additional electricity. Such a combined cycle gas turbine power plant is often denoted as CCGT. In a cogeneration application, some steam would then need to be used to satisfy a thermal requirement. As might be expected, combined cycles have high power-to-heat ratios and high electrical efficiencies. Current designs have electrical efficiencies of up to 55% depending on the equipment, location, and details of the specific application. These current designs for combined cycle plants result in gas turbine power of between 1.5 to 3.5 times the power obtained from the steam turbine. These plants are most

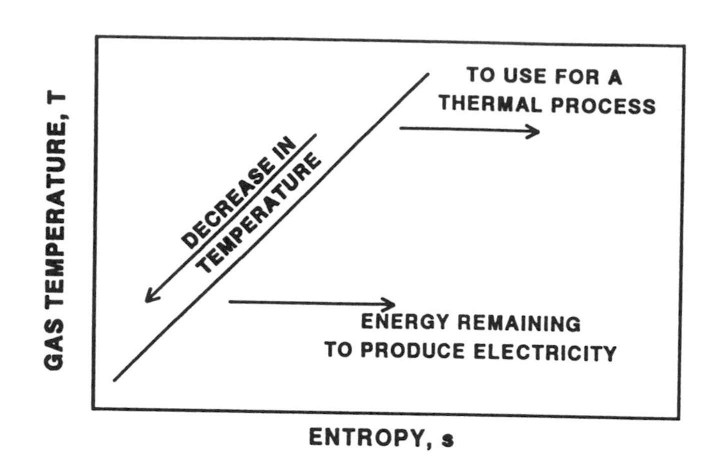

FIGURE 17.5 The gas temperature as a function of entropy for a bottoming-cycle system.

often base load systems operating more than 6,000 hours per year. More details on gas turbines and steam turbines are provided in the following sections on the prime movers.

Applications of Cogeneration Systems

General. Cogeneration systems may involve different types of equipment and may be designed to satisfy specific needs at individual sites. On the other hand, many sites have similar needs and packaged (pre-engineered) cogeneration systems may satisfy these needs and are more economical than custom-engineered systems. The following are examples of cogeneration systems in three different economic sectors.

Cogeneration systems are found in all economic sectors of the world. For convenience, cogeneration systems are often grouped into one of three sectors: (1) industrial, (2) institutional, and (3) commercial. The types and sizes of the cogeneration systems in these sectors overlap to varying degrees, but these sectors are nonetheless convenient for describing various applications of cogeneration. This section will provide examples of the variety of applications that exist. Obviously, these examples are not inclusive, but are intended to illustrate the breadth of possibilities.

Industrial Sector. Compared to the other economic sectors, the industrial sector includes the oldest, largest, and greatest number of cogeneration systems. As mentioned in the preceding history, cogeneration was first used by industry in the early 1900s to supply both electrical and thermal needs in an efficient manner. Many industries have a rich and continuous history of cogeneration applications. The industrial sector is dominant in cogeneration for several reasons. Industrial facilities often operate continuously, have simultaneous electrical and thermal requirements, and already have a power plant and operating staff. Those industries that are particularly energy intensive, such as the petrochemical and paper and pulp industries, are significant cogenerators. Many of these industries are cogenerating hundreds of megawatts of electrical power at one site. In addition, of course, mid-sized and smaller industries also use cogeneration.

Institutional Sector. The institutional sector includes a wide range of largely not-for-profit enterprises including universities, colleges, other schools, government building complexes, hospitals, military bases, and not-for-profit institutes. Many of these enterprises operate for a majority of the day, if not continuously. Some, such as hospitals, may already have provisions for emergency backup power and have the staff to operate a cogeneration system. Although not as large as the largest industrial cogenerators, a large hospital complex or university may require 50 or more megawatts of cogenerated electrical power.

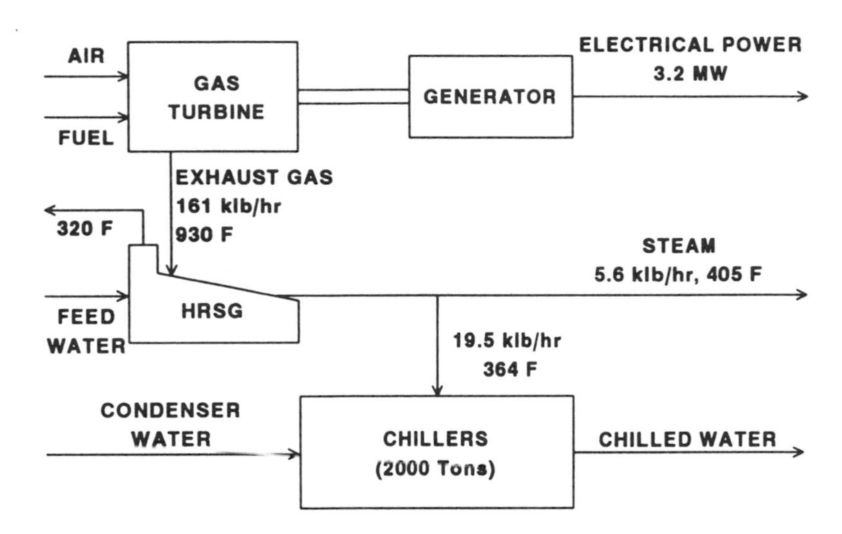

FIGURE 17.6 A schematic diagram of the topping-cycle cogeneration system installed in 1986 at Rice University in Houston, Texas.

Figure 17.6 is a schematic of a cogeneration system installed in 1986 at Rice University in Houston, Texas. As shown, this is a topping-cycle cogeneration system that uses a 3.2 MW gas turbine, a heat recovery steam generator, and two 1,000-ton absorption chillers. Because this system was so successful, a second cogeneration system using a combined cycle was installed in 1989. It included a 3.7 MW gas turbine, a 400 kW steam turbine, and a 1,500-ton absorption chiller. These two cogeneration systems supply the university with 90% of its electrical power, heating, and cooling requirements.

Commercial Sector. The commercial sector includes a wide range of for-profit enterprises including businesses, hotels, motels, apartment and housing complexes, restaurants, shopping centers, laundries, and laboratories. Generally, this sector includes the smallest cogenerators, and unless the electrical rates are unusually high, the economics are often less favorable than for the other sectors. Since many of these facilities share similar characteristics, pre-engineered or packaged cogeneration systems are often possible. These units are less expensive than customized systems. Packaged cogeneration systems are discussed in more detail in a subsequent section of this chapter.

17.2 Equipment and Components

Cogeneration systems consist of several major pieces of equipment and many smaller components. This section will describe and provide guidance on the selection of this equipment and components. The following discussion is grouped into four subsections: (1) prime movers, (2) electrical equipment, (3) heat recovery devices, and (4) absorption chillers.

Prime Movers

Prime movers include those devices that convert fuel energy into rotating shaft power to drive electrical generators. The prime movers that are used most often in cogeneration systems are steam turbines, gas turbines, and reciprocating engines. Each of these prime movers is described here. Important distinctions between the prime movers are the fuels that they may use, their combustion processes, their overall thermal efficiency, and the type, amount, and temperature of their rejected thermal energy. In cogeneration applications, a significant parameter for each type of prime mover is the ratio of the rate of supplied thermal energy and the output power. This ratio is called the heat-to-power ratio, and is unitless (i.e., kW/kW or Btu hr/Btu hr). Knowing the value of the heat-to-power ratio assists in matching a particular prime mover to a particular application. This matching is discussed in a later section.

Steam Turbines. Steam turbines are widely used in power plants throughout industry and electric utilities. Steam turbines use high-pressure, high-temperature steam from a boiler. The steam flows through the turbine, forcing the turbine to rotate. The steam exits the turbine at a lower pressure and temperature. A major difference of the steam turbine relative to the reciprocating engines and gas turbines is that the combustion occurs externally in a separate device (boiler). This allows a wide range of fuels to be used including solid fuels such as coal or solid waste materials. The exit steam, of course, can be used for thermal heating or to supply the energy to an absorption chiller.

Steam turbines are available in a multitude of configurations and sizes. This description will highlight only the major possibilities, but more complete information is available elsewhere (Wood, 1982). A major distinction is whether the machine is a condensing or noncondensing (back-pressure) steam turbine. Condensing steam turbines are steam turbines designed so that the steam exits at a low pressure (less than atmospheric) such that the steam may be condensed in a condenser at near ambient temperatures. Condensing steam turbines provide the maximum electrical output, and hence, are most often used by central plants and electric utilities. Since the exiting steam possesses little available energy, the application of condensing steam turbines for cogeneration is negligible.

Noncondensing steam turbines are steam turbines designed such that the exiting steam is at a pressure above atmospheric. These steam turbines are also referred to as back-pressure steam turbines. The exiting steam possesses sufficient energy to provide process or building heat. Either type of steam turbine may be equipped with one or more extraction ports so that a portion of the steam may be extracted from the steam turbine at pressures between the inlet and exit pressures. This extracted steam may be used for higher temperature heating or process requirements.

Noncondensing steam turbines are available in a wide range of outputs beginning at about 50 kW and increasing to over 100 MW. Inlet steam pressures range from 150 to 2,000 psig, and inlet temperatures range from 500 to 1,050°F. Depending on the specific design and application, the heat-to-power ratio for steam turbines could range from 4 to over 10. The thermal efficiency increases with size (or power level) from typically 8 to 20%. Although the major source of thermal energy is the exit or extracted steam, the boiler exhaust may be a possible secondary source of thermal energy in some cases.

Gas Turbines. As with steam turbines, stationary gas turbines are major components in many power plants. Stationary gas turbines share many of the same components with the familiar aircraft gas turbines. In fact, both stationary (or industrial) and aircraft gas turbines are used in cogeneration systems. This brief description will highlight the important characteristics of gas turbines as applied to cogeneration.

Many configurations, designs, and sizes of gas turbines are available. The simple-cycle gas turbine uses no external techniques such as regeneration to improve its efficiency. The thermal efficiency of simple cycle gas turbines may therefore be increased by the use of several external techniques, but the designs and configurations become more complex. Many of these modifications to the simple cycle gas turbine are directed at using the energy in the exhaust gases to increase the electrical output and efficiency. Of course, such modifications will decrease the available energy in the exhaust. For cogeneration applications, therefore, the most efficient gas turbine may not always be the appropriate choice.

Gas turbines are available in a wide range of outputs beginning at about 100 kW and increasing to over 100 MW. Depending on the specific design, the heat-to-power ratio for gas turbines could range from about 1 to 3. The design point thermal efficiency increases with size (or power level) and complexity from typically 20 to 45%. The higher thermal efficiencies, compared to steam turbines, is the reason the heat-to-power ratio is lower than for steam turbines. These high efficiencies are for full-load (design point) operation. At part load, a gas turbine's efficiency decreases rapidly. As mentioned earlier, the use of regeneration, intercooling, reheating, and other modifications are used to improve the overall performance of the simple cycle gas turbine.

Due to the large amount of excess air (the total air mass may be on the order of 100 times the fuel mass) used in the combustion process of gas turbines, the exiting exhaust gas contains a relatively high concentration of nitrogen and oxygen. Hence, the gas turbine exhaust may be characterized as mostly heated air and is nearly ideal for process or heating purposes. Gas turbines may use liquid fuels such as jet fuel or kerosene, or they may use gaseous fuels such as natural gas or propane. The highest performance is possible with liquid fuels, but the lowest emissions have been reported for natural gas operation.

Power ratings for gas turbines are provided for continuous and intermittent duty cycles. The continuous ratings are slightly lower than the intermittent ratings to provide long life and good durability. In both cases, the power ratings are provided for a set of standard operating conditions known as ISO conditions. ISO conditions include 1 atm, 59°F (15°C), sea level, and no inlet or exhaust pressure losses. For a specific application, the local conditions must be used to adjust the rated power to reflect the actual operating conditions. Most manufacturers will provide potential users with their recommended adjustments, which depend on the difference between the local and standard conditions. In particular, the local ambient inlet air temperature affects the gas turbine output in a significant manner. As the ambient air temperature increases, the performance of a gas turbine decreases due to the lower air density.

Gas turbines require more frequent and specialized maintenance than steam turbines. Major overhauls are required at 20,000 to 75,000 hours of operation, depending on the service duty and the manufacturer. Aircraft gas turbines are designed to have their "hot section" returned to the manufacturer and replaced with a conditioned unit to minimize downtime, although the overhaul is more expensive. Industrial gas turbines are designed to be overhauled in the field, and this is generally a less expensive procedure.

Reciprocating Engines. A third category of prime movers for cogeneration systems is internal combustion (IC), reciprocating engines. Although rotary engines could be used in cogeneration systems, no significant applications are known. The remaining discussion will focus on reciprocating engines. These engines are available in several forms. Probably the most common form of the reciprocating engineer is the typical spark-ignited, gasoline engine used in automobiles. For cogeneration applications, the spark-ignited gasoline engine has been converted to operate in a stationary, continuous mode with fuels such as natural gas. The majority of reciprocating engines for mid- to large-sized cogeneration systems are stationary diesel engines operating either with diesel fuel or in a dual-fuel mode with natural gas. These engines share some common characteristics for cogeneration applications and have some distinctive features as well.

Power ratings for reciprocating engines are similar to those for gas turbines in that both continuous and intermittent duty cycle ratings are provided. As with the gas turbines, these power ratings are provided for a set of standard conditions for ambient temperature, pressure, and elevation. The standard power ratings need to be adjusted for the local conditions at the site of the installation. For cogeneration applications, reciprocating engines are available in many power levels and designs. These power levels range from less than 50 kW to over 200 MW. Some manufacturers even offer "mini" cogeneration systems with outputs as low as 6 kW.

For those reciprocating, internal combustion engines that are liquid cooled (the majority of the engines considered here), the cooling liquid is a secondary source of thermal energy. Although not at the high temperatures of exhaust gas, this energy can be used to produce hot water or low

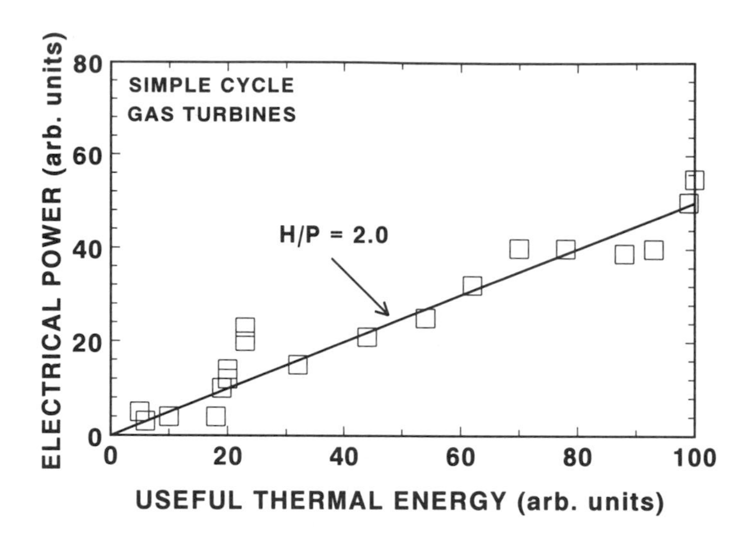

FIGURE 17.7 The electrical power as a function of the useful thermal energy (in arbitrary units) for a number of simple-cycle gas turbines. The average heat-to-power ratio for this family of gas turbines is 2.0, and is shown as the solid line.

pressure steam. Several designs are available for recovering the energy in the cooling liquid. These designs use one or more direct or indirect heat exchangers to generate the hot water or low pressure steam. Liquid-to-liquid heat exchangers can have high efficiencies, and most of this energy is recoverable (but at relatively low temperatures). Other energy sources from reciprocating engines include oil-coolers and turbocharger after-coolers. This energy is usually at temperatures below 160°F, and would only be practical to recover for low-temperature requirements. Another benefit of the reciprocating engine is that the maintenance and repair is less specialized than for gas turbines. On the other hand, the maintenance may be more frequent and more costly.

Heat-to-Power Characteristics. As mentioned earlier, an important feature in the selection of a prime mover for a cogeneration system is the heat-to-power ratio. This ratio may be a fairly constant characteristic of a family of a particular type of prime mover. Figures 17.7 and 17.8 show the electrical power output as a function of the potential available thermal energy rate output for a specific family of gas turbines and a specific family of reciprocating engines, respectively. The symbols represent the power and thermal output for individual gas turbines and engines in the two "families." The solid lines are the linear best fits to this data and represent the average heat-to-power (H/P) ratio for each family. As shown, the data does scatter some about the linear line, but the characteristic for each prime mover family is a good average. Also as shown, the average heat-to-power ratio is 2.0 for the gas turbine and 0.5 for the reciprocating engine. These are representative values for these two types of prime movers. These characteristic heat-to-power ratios will be used here in matching the prime mover to a specific application.

Electrical Equipment

The electrical equipment for cogeneration systems includes electrical generators, transformers, switching components, circuit breakers, relays, electric meters, controls, transmission lines, and related equipment. In addition to the equipment that supports electrical production, cogeneration systems may need equipment to interconnect with an electric utility to operate in parallel for the use of backup (emergency) power or for electrical sales to the utility. This section will briefly highlight some of the aspects regarding electrical generators and interconnections, but will not be able to review the other important electrical considerations.

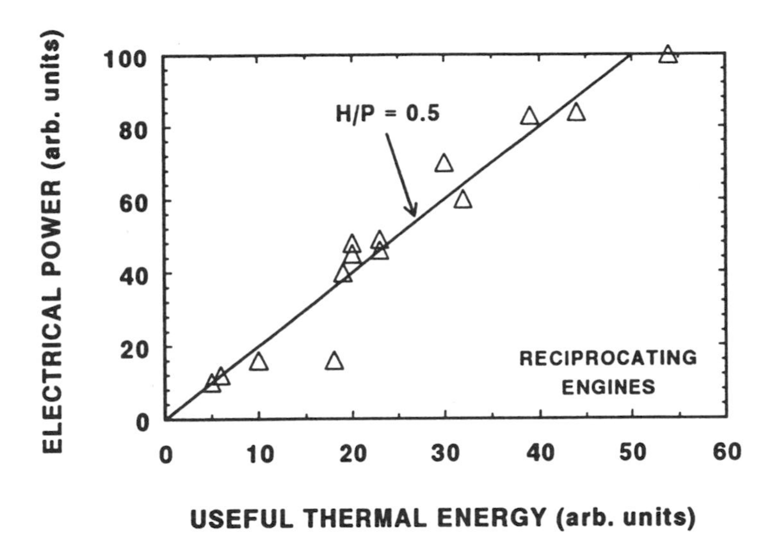

FIGURE 17.8 The electrical power as a function of the useful thermal energy (in arbitrary units) for similar diesel engines. The average heat-to-power ratio for this family of reciprocating engines is 0.5, and is shown as the solid line.

The electric generator is a device for converting the rotating mechanical energy of a prime mover to electrical energy (i.e., electricity). The basic principle for this process, known as the Faraday effect, is that when an electrically conductive material such as a wire moves across a magnetic field, an electric current is produced in the wire. This can be accomplished in a variety of ways, so there are several types of electric generators. The frequency of the generator's output depends on the rotational speed of the assembly.

An important feature of generators is that they require a magnetic field to operate. The source of the energy for this magnetic field serves to distinguish the two major types of generators. If the generator is connected to an electric source and uses that source for the magnetic field, it is called an induction generator. In this case, the generator operates above the synchronous speed, and cannot operate if the external current (usually from the electric utility) is not available. On the other hand, if the magnetic field is generated internally using a small alternator, the generator is called a synchronous generator and operates at the synchronous speed. Synchronous generators can operate independent of the external electric grid.

Most often the manufacturer of the prime mover will provide the prime mover and generator as an integrated, packaged assembly (called a gen-set). Performance characteristics of generators include power rating, efficiency, voltage, power factor, and current ratings. Each of these performance characteristics must be considered when selecting the proper generator for a given application. Electric generators may have conversion efficiencies of between about 50 to over 98%. The efficiency increases in this range as the generator increases in size and power rating. Only the largest electric generators (say, on the order of 100 MW) can attain efficiencies of 98%.

In many cases, the cogeneration system will need to be interconnected to the utility. Although a cogeneration system could be isolated from the electric grid, there are several reasons why the system may need to be interconnected. An interconnection is necessary for receiving supplementary or backup (emergency) electric power. Also, if the cogenerator elects to sell excess power, such an interconnection would be needed. The interconnection equipment includes relays, circuit breakers, fuses, switches, transformers, synchronizers, and meters. The specific equipment and design for the interconnection largely is dictated by the utility for safety and compatibility reasons. The utility has specific responsibilities for the integrity of the electric grid and for maintaining the electrical power quality.

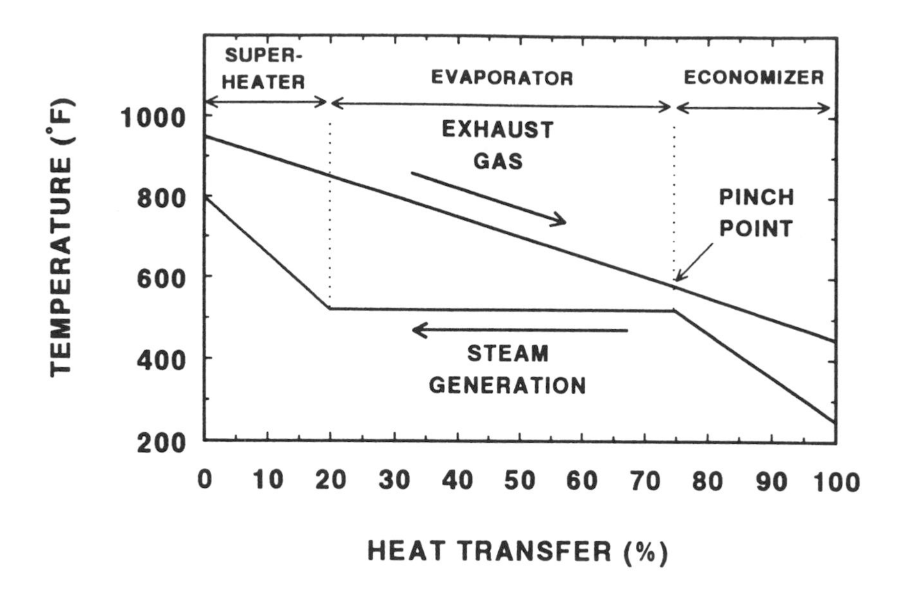

FIGURE 17.9 The temperatures of the exhaust gas and water/steam as a function of the heat transfer coordinate for the steam generation process of an HRSG.

Heat Recovery Equipment

The primary heat recovery equipment used in cogeneration systems includes several types of steam and hot-water production facilities. In addition, absorption chillers could be considered in this section, but for organizational reasons, chillers will be discussed in the following section. Several configurations of heat recovery devices are available. As mentioned earlier, these devices may be referred to as "heat recovery steam generators" or HRSGs. HRSGs are often divided into the following categories: (1) unfired, (2) partially fired, and (3) fully fired. An unfired HRSG is essentially a convective heat exchanger. A partially fired HRSG may include a "duct burner," which often uses a natural gas burner upstream of the HRSG to increase the exhaust gas temperature. A fully fired HRSG is basically a boiler that simply uses the exhaust gas as preheated air.

For most of these heat recovery devices, exhaust gas flows up through the device and exits at the top. Energy from the exhaust gas is used to heat and vaporize the water, and to superheat the steam. Figure 17.9 shows this process on a temperature diagram. The top line shows the exhaust gas temperature decreasing from left to right as energy is removed from the gas to heat the water. The lower line represents the water heating up from right to left in the diagram. The lower-temperature exhaust is used to preheat the water to saturation conditions in the economizer. The intermediate-temperature exhaust is used to vaporize (or boil) the water to form saturated steam. Finally, the highest-temperature exhaust is used to superheat the steam.

The temperature difference between the exhaust gas and the water where the water first starts to vaporize is referred to as the pinch point temperature difference. This is the smallest temperature difference in the HRSG and may limit the overall performance of the heat recovery device. Since the rate of heat transfer is proportional to the temperature difference, the greater this difference the greater the heat transfer rate. On the other hand, as this temperature difference increases the steam flow rate must decrease and less of the exhaust gas energy will be utilized. To use smaller temperature differences and maintain higher heat transfer rates, larger heat exchanger surfaces are required. Larger heat transfer surface areas result in higher capital costs. These, then, are the types of trade-offs that must be decided when incorporating a heat recovery device into a cogeneration system design.

The proper selection of HRSG equipment depends on the prime mover, the required steam conditions, and other interdependent factors. Some of these considerations are described next. Most of these considerations involve a trade-off between increased performance and higher initial capital cost.

The "back-pressure" of the HRSG unit affects the overall system performance. As the back-pressure decreases, the HRSG efficiency increases, but the cost of the HRSG unit increases. Successful units often have 10 to 15 inches water pressure of back-pressure. As mentioned earlier, the selection of the "pinch point" temperature difference affects the performance and cost. Typical pinch point temperature differences are between 30 and 80°F. High-efficiency, high-cost units may use a temperature difference as low as 25°F.

The final outlet stack gas temperature must be selected so as to avoid significant acid formation. This requires the outlet stack temperature to be above the water condensation temperature. This is typically above 300°F. The amount of sulfur in the fuel affects this decision, since the sulfur forms sulfuric acid. As the amount of sulfur in the fuel increases, the recommended stack gas outlet temperature increases.

The final consideration is the steam temperature and pressure. This is a complex decision that depends on many factors such as the application for the steam, the source of the exhaust gas, the exhaust gas temperature and flow rate, and inlet water condition and temperature.

Absorption Chillers

Absorption chillers may use the thermal energy from cogeneration systems to provide cooling for a facility. This section will briefly review the operation of absorption chillers and their application to cogeneration systems. Absorption chillers use special fluids and a unique thermodynamic cycle that produces low temperatures without the requirement of a vapor compressor, which is used in mechanical chillers. Instead of the vapor compressor, an absorption chiller uses liquid pumps and energy from low-temperature sources such as hot water, steam, or exhaust gas.

Absorption chillers utilize fluids that are solutions of two components. The basic principle of the operation of absorption chillers is that once the solution is pumped to a high pressure, low-temperature energy is used to vaporize one component from the solution. This component serves as the "refrigerant" for this cycle. Examples of these solutions are (1) water and ammonia, (2) lithium bromide and water, and (3) lithium chloride and water. In the first case the ammonia serves as the refrigerant, and in the latter two cases the water serves as the refrigerant.

For cogeneration applications, the important feature of absorption chillers is that they use relatively low-temperature energy available directly or indirectly from the prime mover and produce chilled water for cooling. The use of absorption chillers is particularly advantageous for locations where space and water heating loads are minimal during a good part of the year. For these situations, the thermal output of a cogeneration system can be used for heating during the colder part of the year and, using an absorption chiller, for cooling during the warmer part of the year. Furthermore, by not using electric chillers, the electric loads are more constant throughout the year. In warm climates, absorption chillers are often an important, if not an essential, aspect of technically and economically successful cogeneration systems.

Some machines are designed as indirect-fired units using hot water or steam. As examples of typical numbers, a single-stage unit could use steam at 250°F to produce a ton of cooling for every 18 pounds of steam flow per hour. A dual-stage unit would need 365°F steam to produce a ton of cooling for every 10 pounds of steam flow per hour. If hot-water is available, a ton of cooling could be produced for every 220 pounds of 190°F hot water per hour.

Other machines use the exhaust gas directly and are called direct-fired units. In these cases, the exhaust gas temperature needs to be 550 to 1,000°F. The higher the exhaust temperature, the less energy (or exhaust gas flow) is needed per ton of cooling. For example, for 1,000°F exhaust gas, a ton of cooling requires 77 pounds per hour of flow whereas for 550°F exhaust gas a ton of cooling requires 313 pounds per hour of flow.

17.3 Technical Design Issues

Selecting and Sizing the Prime Mover

The selection of a prime mover for a cogeneration system involves the consideration of a variety of technical and nontechnical issues. Technical issues, which often dominate the selection process, include the operating mode or modes of the facility, the required heat-to-power ratio of the facility, the overall power level, and any special site considerations (e.g., low noise). Other issues, which may play a role in the selection process, include matching existing equipment and utilizing the skills of existing plant personnel. Of course, the final decision is often dominated by the economics.

Steam turbines and boilers usually are selected for a cogeneration system if the fuel of choice is coal or another solid fuel. Occasionally, for very large (>50 MW) systems operating at base load, a steam turbine system may be selected even for a liquid or gaseous fuel. Also, steam turbines and boilers would be selected if a high heat-to-power ratio is needed. Steam turbines also may be selected for a cogeneration system in certain specialized cases. For example, a large pressure reduction valve in an existing steam system could be replaced with a steam turbine and thereby provide electrical power and thermal energy. In many applications, however, steam turbines are selected to be used in conjunction with a gas turbine in a combined cycle power plant to increase the power output. Combined cycle gas turbine power plants were described in an earlier section.

Gas turbines are selected for many cogeneration systems where the required heat-to-power ratio and the electrical power need are high. Also, gas turbines are the prime mover of choice where minimal vibration or a low weight-to-power ratio (such as for a roof installation) is required. Reciprocating engines are selected where the heat-to-power ratio is modest, the temperature level of the thermal energy is low, and the highest electrical efficiency is necessary for the economics. Additionally, reciprocating engines may be selected if the plant personnel are more suited to the operation and maintenance of these engines.

Selecting the appropriate size prime mover involves identifying the most economic cogeneration operating mode. This is accomplished by first obtaining the electrical and thermal energy requirements of the facility. Next, various operating modes are considered to satisfy these loads. By conducting a comprehensive economic analysis, the most economic operating mode and prime mover size can be identified. The process of matching the prime mover and the loads is described next.

Matching Electrical and Thermal Loads

To properly select the size and operating mode of the prime mover, the electric and thermal loads of the facility need to be obtained. For the most thorough "matching," these loads are needed on an hourly, daily, monthly, and yearly basis. Figure 17.10 is an example of the hourly electrical and thermal loads of a hypothetical facility for a typical work day in the winter. As shown, the heating and power demands begin to increase at 6:00 A.M. as the day's activities begin. The heating load is shown to peak shortly after 8:00 A.M. and then decrease for the remainder of the day. The electrical load remains nearly constant during the working part of the day, and is somewhat lower during the evening and nighttime hours.

Figure 17.11 shows the month totals for the electrical and thermal loads for the same hypothetical facility. For the summer months, the heating and electrical loads are minimum. In addition, this figure shows dotted and dashed lines, which represent the "base loads" for the electrical and heating loads, respectively. The base loads are the minimum loads during the year and form a floor or base for the total loads. Often a cogeneration system may be sized so as to provide only the base loads. In this case, auxiliary boilers would provide the additional heating needed during the days where the heating needs exceeded the base amount. Similarly, electrical power would need to be purchased to supplement the base power provided by the cogeneration system.

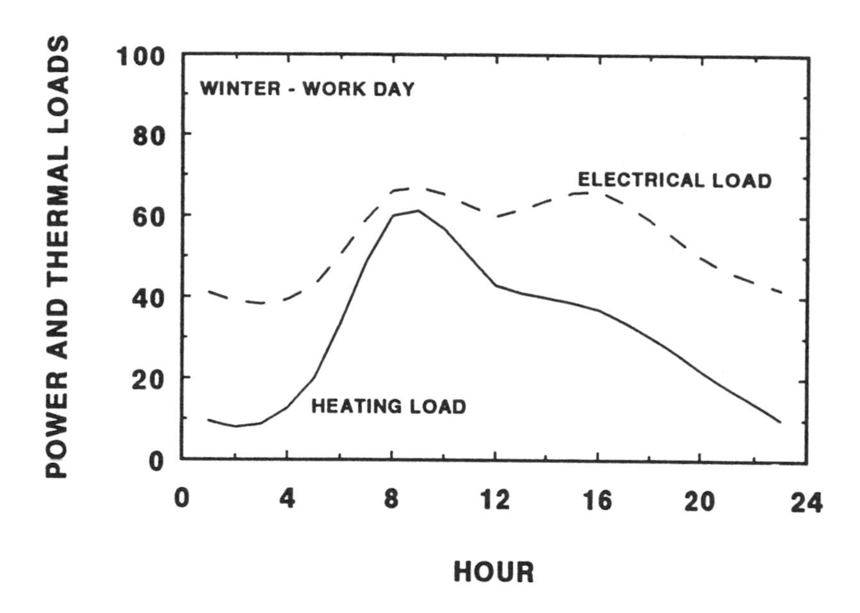

FIGURE 17.10 The power and thermal loads (in arbitrary units) as a function of the hour of the day for a hypothetical facility for one work day in the winter.

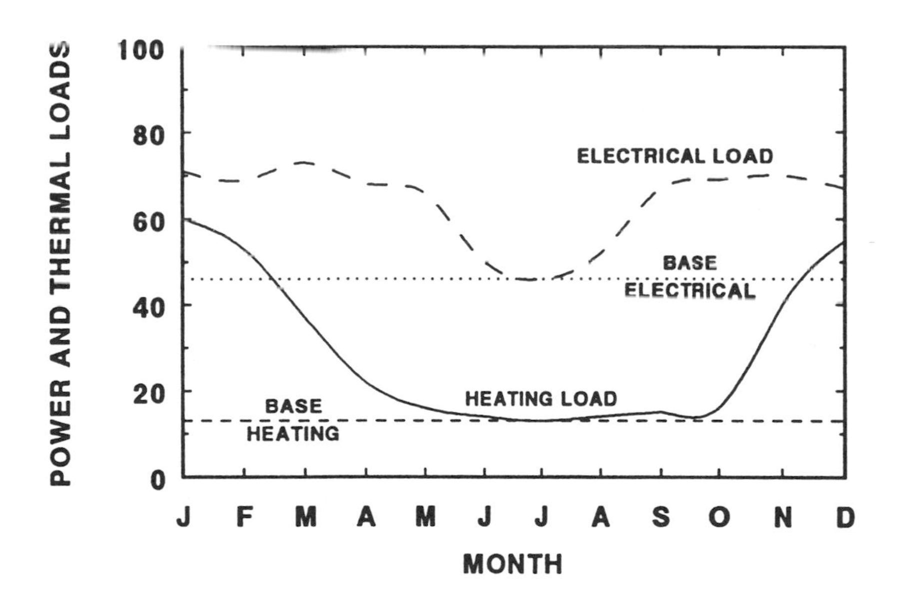

FIGURE 17.11 The total power and thermal loads (in arbitrary units) as a function of the month of the year for a hypothetical facility.

Several options exist in matching the facility's electrical power and thermal needs to a prime mover for a cogeneration system. To illustrate this matching procedure, consider Figure 17.12, which shows electrical power as a function of thermal energy. The three dashed lines are the heat-to-power characteristics of three families of prime movers. In this example, the heat-to-power ratio increases from A to B to C. The facility's needed electrical power and thermal energy are shown as the horizontal and vertical dotted lines, respectively. This needed power and thermal energy is often the base load of the facility as explained earlier.

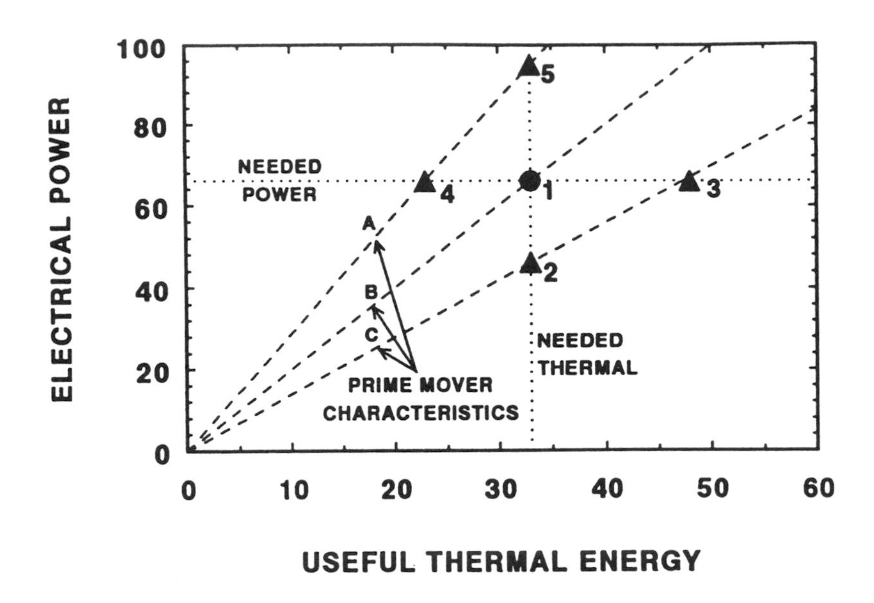

FIGURE 17.12 The electrical power as a function of the useful thermal energy (in arbitrary units).

Five operating points are identified:

1. Point 1 represents an exact match between the facility's needed power and thermal energy, and the power and thermal energy available from a prime mover with the heat-to-power ratio characteristic B. Such an exact match is not likely or desired, so the other operating points must be considered.
2. Point 2 represents a case where a prime mover is selected with a characteristic C such that the thermal needs are matched but the electrical power supplied is too low. In this case, the facility would need to purchase supplementary electrical power usually from the local electrical utility. This is a common case that is often the best economic and operational choice.
3. Point 3 represents a case where a prime mover is selected with a characteristic C such that the electrical power needs are matched, but excess thermal energy is available. Generally, this is not an economical choice because the overall system efficiency is low. For certain situations, however, this case might be selected—for example, if future thermal loads are expected to increase or if a customer can be identified to purchase the excess thermal energy.
4. Point 4 represents a case where a prime mover is selected with a characteristic A such that the electrical power needs are matched but the available thermal energy does not satisfy the needs of the facility. In this case, supplementary boilers are often used to supply the additional thermal energy. This may be the best economic choice, particularly if existing boilers are available and the desire is to size the cogeneration system such that all the thermal energy is used.
5. Point 5 represents a case where a prime mover is selected with a characteristic A such that the thermal needs are matched but the electrical power supplied is too high. This would only be economical if an external customer was available to purchase the excess electrical power. This is often the situation with third-party cogenerators, who own and operate a cogeneration system and sell the thermal energy and electrical power to one or more customers.

These cases represent distinct operating mode selections. In practice, other technical and nontechnical aspects must be considered in the final selection. These cases are presented to illustrate the global decision process.

The possible overall operating modes for a cogeneration power plant are often categorized into one of three classes. (1) The plant may operate as a base load system with little or no variation in power output. Base load plants operate in excess of 6,000 hours per year. Power needs above the base load are typically provided by interconnections to a local utility or by an auxiliary power plant. (2) The plant may operate as an intermediate system for 3,000 to 4,000 hours per year. These systems are less likely than base load systems, but if the economics are positive, may have application for facilities that are not continuously operated such as some commercial enterprises. (3) Finally, a third class of plant is a peaking system, which operates only for 1,000 hours or less per year. Utility plants often use peaking systems to provide peaking power during periods of high electrical use. For cogeneration applications, peaking units may be economical where the cost of the electricity above a certain level is unusually high. These units are sometimes referred to as peak shaving systems.

Packaged Systems

In general, facilities with low electrical power needs cannot utilize customized cogeneration systems because of the relatively high initial costs that are associated with any size system. These initial costs include at least a portion of the costs related to the initial design, engineering, and related development and installation matters. Also, smaller facilities often do not have the specialized staff available to develop and operate complex power plants.

To solve some of these above problems, pre-engineered, factory-assembled, "packaged" cogeneration systems have been developed. Packaged cogeneration systems range from less than 50 kW to over 1 MW. Some larger units (say, 20 MW or larger) using gas turbines or reciprocating engines are sometimes described as packaged cogeneration systems, and are, at least to some degree, "packaged." For example, a gas turbine and generator may be factory assembled, tested, and skid mounted for shipping. These units, although packaged, do not completely eliminate the problems of project development, installation, and operation. The remainder of this section will consider only the smaller packaged cogeneration systems.

The major advantage of packaged cogeneration systems is that the initial engineering, design, and development costs can be spread over many units, which reduces the capital cost (per kW) for these systems. Other advantages of packaged cogeneration systems include factory assembly and testing of the complete system. If there are any problems, they can be fixed while the system is still at the manufacturer's plant. The standard design and reduced installation time result in short overall implementation times. In some cases, a packaged cogeneration system could be operational within a few months after the order is received. This short implementation time reduces the project's uncertainty, which eases making decisions and securing financing.

Another advantage of packaged cogeneration systems is that the customer only interacts with one manufacturer. In some cases, the packaged cogeneration system manufacturer will serve as project engineer and take the project from initial design to installation to operation. The customer often may decide to purchase a turn-key system. This provides the customer with little uncertainty and places the burden of successful project completion on the manufacturer. Also, the manufacturer of a packaged cogeneration system will have experience interacting with regulating boards, financing concerns, and utilities, and may assist the customer in these interactions.

The major disadvantage of packaged cogeneration systems is that the system is not customized for a specific facility. This may mean some compromise and lack of complete optimization. Specialized configurations may not be available. Also, beyond a certain size, packaged cogeneration systems are simply not offered and a customized unit is the only alternative.

Although the initial capital costs of packaged cogeneration systems are low (on a per kW basis), the other site-specific costs could be relatively high and result in unsatisfactory economics. These other costs depend on the selected fuel, the available space at the site, and the compatibility of existing electrical and mechanical facilities. In particular, the interconnection costs associated with the electric utility and the mechanical systems can be prohibitively high. The electric interconnection costs depend on the electric utility's requirements. The fuel system costs depend on whether the selected fuel is already available on site at the appropriate conditions (such as gas pressure for gaseous fuels). Other miscellaneous costs include those for services and assistance with engineering, permitting, zoning, financing, legal review, and construction oversight.

In summary, packaged cogeneration systems offer a cost-effective solution for facilities with low power and thermal requirements such as small to medium hospitals, schools, hotels, and restaurants. The majority of current packaged cogeneration systems use reciprocating engines and standard components. They are factory assembled and tested, and are often sold as turn-key installations. Although the packaged cogeneration systems' costs can be attractive, each facility must be assessed for total costs to determine the economic feasibility. Future developments of packaged cogeneration systems will include even more efficient and cost-effective engines, more advanced microprocessor control systems, and more effective integration of the various components.

17.4 Regulatory Considerations

This section includes brief overviews of the relevant federal regulations on electric power generation and the related environmental constraints.

Federal Regulations Related to Cogeneration

As mentioned earlier, the passage of PURPA, the Public Utility Regulatory Policies Act, in 1978, helped make cogeneration a much more attractive option for electric power generation for a variety of facilities. PURPA is one of the most controversial bills relating to power generation that has been passed by Congress, and even today remains at the center of controversy. PURPA was one of five separate acts comprising the National Energy Act passed by Congress in 1978:

1. National Energy Conservation Policy Act (NECPA)
2. Natural Gas Policy Act (NGPA)
3. Powerplant and Industrial Fuel Use Act (FUA)
4. Energy Tax Act (ETA)
5. Public Utility Regulatory Policies Act (PURPA)

The federal government, and for that matter the entire nation, was concerned about the oil crisis in 1973, the rapidly increasing price of energy, the increased dependence on foreign oil, and our lack of concern for energy efficiency and the use of renewable energy. The National Energy Act addressed most of those concerns, and PURPA, in particular, dealt with independent power generation, energy efficiency, and the use of renewables for power generation. Although all five acts had an impact on our nation's energy consumption and energy use patterns, PURPA had the greatest overall impact and led to a resurgence of interest in cogeneration. Prior to the passage of PURPA, the three most common barriers to cogeneration were (Laderoute, 1980) (1) no general requirement by electric utilities to purchase electric power from cogenerators, (2) discriminatory backup power for cogenerators, and (3) fear on the part of the cogenerator that they might become subject to the same state and federal regulations as electric utilities. The passage of PURPA helped to remove these obstacles to development of cogeneration facilities (Spiewak, 1980, 1987).

Although PURPA was passed in 1978, it was not until 1980 that the Federal Energy Regulatory Commission (FERC) issued its final rulemakings and orders on PURPA. In the National Energy Act, FERC was designated as the regulatory agency for implementation of PURPA. The regulations dealing with PURPA are contained in Part 292 of the FERC regulations, and Sections 201 and 210 are the two primary sections relevant to small power production and cogeneration (FERC, 1978).

Section 201 contains definitions of cogeneration and sets annual efficiency standards for new topping cycle* cogeneration facilities that use oil or natural gas.† For a cogenerating facility to qualify for the privileges and exclusions specified in PURPA, the facility must meet these legislated standards. These standards define a legislated or artificial "efficiency," which facilities must equal or exceed to be considered a "qualified facility" (QF). A qualified facility is eligible to use the provisions outlined in PURPA regarding nonutility electric power generation. This legislated "efficiency" is defined as

$$\eta_{PURPA} = \left(P + \tfrac{1}{2}T\right)/F \tag{17.10}$$

where P is the electrical energy output, T is the useful thermal energy, and F is the fuel energy used. Table 17.1 lists the standards for the PURPA efficiencies. These standards state that a facility must produce at least 5% of the site energy in the form of useful thermal energy, and must meet the efficiency standards in Table 17.1. Values for the thermal fraction and the PURPA efficiency are based upon projected or estimated annual operations.

TABLE 17.1 Required Efficiency Standards for Qualified Facilities (QFs)

If the Useful Thermal Energy Fraction Is	The Required η_{PURPA} Must Be
≥5.0%	≥45.0%
≥15.0%	≥42.5%

The purpose of introducing the "artificial" standards was to insure that useful thermal energy was produced on-site in sufficient quantities to make the cogenerator more efficient than the electric utility. Any facility that meets or exceeds the required efficiencies will be more efficient than any combination of techniques producing electrical power and thermal energy separately. Section 201 also put limitations on cogenerator ownership; that is, electric utilities could not own a majority share of a cogeneration facility, nor could any utility holding company, nor a combination thereof. This section also defined the procedures for obtaining QF status.

Section 210 of the PURPA regulations specifically addressed major obstacles to developing cogeneration facilities. In fact, the regulations in Section 210 not only leveled the playing field, they tilted it in favor of the cogenerator. The principal issues in Section 210 include the following legal obligations of the electric utility toward the cogenerator:

- Obligation to purchase cogenerated energy and capacity from QFs.
- Obligation to sell energy and capacity to QFs.
- Obligation to interconnect.
- Obligation to provide access to transmission grid to "wheel" to another electric utility.
- Obligation to operate in parallel with QFs.
- Obligation to provide supplementary power, back-up power, maintenance power, and interruptible power.

*Since bottoming cycle cogeneration facilities do not use fuel for the primary production of electrical power, these facilities are only regulated when they use oil or natural gas for supplemental firing. The standards states that during any calendar year the useful power output of the bottoming cycle cogeneration facility must equal or exceed 45% of the energy input if natural gas or oil is used in the supplementary firing. The fuels which are used first in the thermal process prior to the bottoming cycle cogeneration facility are not taken into account for satisfying PURPA requirements.

†For topping cycle cogeneration facilities using energy sources other than oil or natural gas (or facilities installed before 13 March 1980), no minimum has been set for efficiency.

Section 210 also exempted QFs from utility status and established a cost basis for purchase of the power from QFs. FERC specified that the price paid to the QF must be determined both on the basis of the utility's avoided cost for producing that energy and, if applicable, on the capacity deferred as a result of the QF power (e.g., cost savings from not having to build a new power plant). Other factors, such as QF power dispatchability, reliability, and cooperation in scheduling planned outages, could also be figured into the price paid to the QFs by the electric utilities. The state public utility commissions were responsible for determining the value of these avoided cost rates.

PURPA also gave special consideration to power produced from small (less than 80 MW) power production facilities (SPPFs). These must be fueled by biomass, geothermal, wastes, renewable resources, or any combination thereof. H.R. 4808, passed by the 101st Congress, lifted the 80 MW size restriction on wind, solar, geothermal cogeneration, and some waste facilities.* These SPPFs could become QFs and have the same status as a large gas-fired industrial cogeneration facility. Facilities between 1 kW and 30 MW may receive exemptions from the Federal Power Act (FPA) and the Public Utility Holding Company Act (PUHCA), and SPPFs using biomass are exempt from PUHCA and certain state restrictions but not from the FPA.

PURPA impacted all states, but not all electric utilities were impacted equally. In Texas, for instance, PURPA had an enormous impact on Houston Lighting and Power (HL&P). In a hearing before the House subcommittee on Energy and Commerce held on June 11, 1986, HL&P discussed both their participation with cogeneration facilities and their problems.** HL&P has more cogenerated power in their service area than any electric utility in the U.S., and they are well qualified to speak on the utility/cogenerator's issues. From HL&P's perspective, two of the principal issues are

1. Obligation to serve; that is, their exclusive rights to the electric customers in their service area and hence their obligation to plan for long-term needs.
2. Profit regulation; that is, electric utilities are constrained to maximum profits, while cogenerators have no such constraints.

Another example from the same House Committee Hearing came from a very small utility in Idaho, Idaho Power, which was faced with a deluge of small hydropower producers after PURPA was enacted. They had 47 projects totaling 82.5 MW on line within 8 years, and another 85 projects in work totaling 230 MW. Idaho Power has surplus hydro capacity and is forced to buy cogenerated power while their surplus hydro is spilled over the dam. The cogenerated power is more expensive and is less reliable. The small hydro systems, in particular, are more susceptible to flooding and washouts.**

The other extreme is the view taken by the Electricity Consumers Resource Council (ELCON), which has membership from the largest industrial cogenerators in the United States. ELCON testified at the same 1986 hearing,** arguing that cogenerated power, particularly for the large industrials, was more efficient, reliable, and in the public's best interest. The testimony in the 1986 hearing was typical of many of the issues and includes text and documentation from dozens of individuals and firms speaking for and against PURPA.

Regulatory Developments of the 1990s

New and restored terms in the power generation area were introduced in the early 1990s: NUGs, IPPs, and EWGs. The NUGs are nonutility generators, and the IPPs are independent power producers. An EWG is an exempt wholesale generator, an entity that wishes to build a new non-rate-based power plant under PUHCA and is only in the business of selling wholesale power. A revised PUHCA and a 1992 FERC ruling gave more open access to transmission lines and limited what a utility could charge third-party transmission users for that access.

*Serial 101-160, Hearing of the Committee on Energy and Commerce, U.S. House of Representatives, June 14, 1990.

**Serial 99-146, Hearing of the Committee on Energy and Commerce, U.S. House of Representatives, June 11, 1986.

The Energy Policy Act of 1992 enacted into law many of FERC's rules on EWGs and defined their legal status. In addition to opening up transmission line access, the Energy Act also opened the possibility of retail wheeling, an area strongly favored by industry and heavily opposed by the electric utilities. An example of retail wheeling might be an industrial cogenerator in the Texas Gulf Coast area selling power to an industry in the Dallas area. Two or more electric utilities would be affected on their transmission lines, and the Dallas-area utility would lose the revenue from the customer in their service territory. From industry's viewpoint, they would like the ability to buy power wherever they could purchase it cheaper. The Energy Policy Act of 1992, some feel, will ultimately lead to a complete deregulation of the electric utility industry. Carried to the ultimate extreme, consumers may be able to buy their electricity similar to the current selection of telephone long distance service. The local electric utility will be an electric transmission distributor just like the local telephone company in this deregulated arena.

The California Public Utility Commission held hearings in August 1994 on a proposal to have individual consumers choose their own electric company by 2002. Large companies might have that option as early as 1996. California is the state leading the way "for the now inevitable downfall of regulated electric utility monopolies..."* Part of the problem lies with the huge differential in electricity prices throughout the lower 48. Prices to consumers range from under 3¢ per kWh in some areas of the Pacific Northwest to over 10¢/kWh in the Chicago area, Central Vermont, and Southern California Edison. Electric utilities with big investments in nuclear plants such as El Paso Electric are facing huge capital debts and are forced to charge high electric rates. Consumers are unhappy because they feel they are having to pay the price for possible mismanagement by the monopolistic utility. At this time, no one is certain if the National Energy Policy Act of 1992 will eventually lead to a complete deregulation of the electric utility industry, and if it does, what the outcomes will be. Some argue that prices will go down and services will improve. Others argue that the lights may not always come on.

Finally, major regulatory questions still exist and need to be resolved to allow cogeneration to maintain a contribution to the nation's electric power generation. For example, do transmission line owners charge the same transmission rates as they would charge themselves? Do they provide the same level of services? Who pays for "stranded costs," which arise when a major customer dumps its local utility in favor of another supplier? Someone will have to pay for unused generating capacity. Only time will tell how all these questions will be answered.** See Chapter 8 for a more complete discussion of these issues.

Environmental Considerations, Permitting, Water Quality

As described in the introduction, air and water regulations have been legislated since the 1950s. The Clean Air Act Amendments of 1977 established a new source of performance standards and placed environmental enforcement in the hands of the Environmental Protection Agency (EPA). The National Energy Policy Act of 1992 and the revised Clean Air Act Amendments (1990) placed even more stringent regulations on environmental pollutants. Concerns over acid rain, global warming, and the depletion of the ozone layer have prompted these tougher regulations.

The Clean Air Act Amendment of 1990 also produced new challenges and opportunities for power generation—trading and selling of emissions. Certain areas in the United States were designated as nonattainment zones where no new emissions sources could be located. Industries could, however, trade or even sell emissions (i.e., so much per ton of CO_2 or SO_2). This could become very important if a cogenerator wanted to locate a plant in a nonattainment zone and could buy emissions from a company that reduced production or somehow cut emissions from their plant.

**USA Today,* August 5, 1994, "Electricity Consumers May Get Power of Choice," p. 10A.

***Wall Street Journal,* August 8, 1994, "Electric Power Brokers Create New Breed of Business."

A number of certifications are required to get a cogeneration plant approved. These include not only a FERC certificate but also various state permits. Since each state will set its own permitting requirements, it is not possible to generalize what is required on each application, but the following information will likely be required:

1. Nature of pollutant source.
2. Type of pollutant.
3. Level of emissions.
4. Description of process technology.
5. Process flow diagram.
6. Certification that the "Best Available Control Technology" (BACT) is being used.

Different requirements must be met if the proposed site is in an attainment or nonattainment zone. An attainment area will require sufficient modeling of ambient air conditions to insure that no significant deterioration of existing air quality occurs. This is PSD modeling or "prevention of significant deterioration." For a nonattainment area, no new emissions can be added unless they are offset by the removal of existing emissions. This has led to the selling or trading of emissions by industrial facilities and utilities.

Water Quality and Solid Waste Disposal. Water quality is usually not a problem with natural gas–fired cogeneration plants but will probably be monitored and will require both state and federal permits if the wastewater is discharged into a public waterway. In extreme conditions it may be necessary to control both wastewater temperature and pH.

Solid waste disposal is not a problem with either natural gas or oil-fired cogeneration plants but could be a major problem for a coal-fired, coal gasification, or waste-to-energy cogeneration system. In some states the bottom ash from waste-to-energy plants was considered hazardous waste and the ash disposal cost per ton was more expensive than the refuse disposal cost. In 1994, bottom ash was declared by the EPA to be nonhazardous and therefore was allowed for land filling in standard landfills. Again, all states have different standards for both the quality of the water discharged and the requirements for solid waste disposal, and all project planners should check with the appropriate regulatory agencies in the state where the project is planned to be sure what air, water, and solid waste (if applicable) permits are required.

17.5 Economic Evaluations

This section will present a brief summary of the important aspects of economic evaluations of cogeneration systems. This section is not intended to be a complete presentation of engineering economics, which would be beyond the scope of this chapter. Rather, those aspects of economic evaluations that are specific or especially important to cogeneration systems will be described. These aspects include the baseline (or "business as usual") financial considerations, cogeneration economics including capital and operating costs, and the available economic and engineering computer programs. The two case studies presented at the end of this chapter will further illustrate some of the economic principles discussed in this section.

Baseline Considerations

Because a cogeneration project is capital intensive, it is extremely important to have accurate data to determine current and estimated (projected) future energy consumption and costs. In a large industrial plant with known process steam needs and 15-minute electric data available, it is usually fairly straightforward to establish the base-line case (i.e., the no-cogeneration case). For universities, hospitals, small manufacturing plants, and so on, the data are not always available. Typically whole campus/facility electrical data are available on a monthly basis only (from utility bills), and hourly electrical profiles may have to be "constructed" from monthly energy and demand data. Hourly thermal data also may not exist. Boiler operators may have daily logs, but these are typically not in electronic form. Constructing accurate hourly thermal energy profiles can involve hours of tedious work pouring over graphs and boiler operator logs. In the worst-case scenario, only monthly gas bills may be available, and it may be necessary to construct hourly thermal profiles from monthly bills, boiler efficiencies, and Btu content of the fuel. Since the monthly dollars have to be matched, there is often a great deal of trial and error involved before there is a match between assumed hourly energy profiles and monthly energy consumption and costs from the utility bills.

While a first-cut energy and cost analysis may be done on a monthly basis for an industrial facility that operates 7 days a week, 24 hours a day, that is not the case for a university, a commercial building complex, or a one- or two-shift manufacturing facility. Energy loads are highly variable, and hourly analyses are required. The possible exception may be hospitals, where 7-day-a-week operation is normally the case. However, there will often be a significant difference between weekday and weekend loads, which would require some sort of "day typing" in the analysis.

Assuming a good energy profile (hourly preferred) can be constructed, the "business as usual" scenario costs are determined. This is the annual cost of doing business without the cogeneration system, including annual purchased electrical energy sales, electrical demand charges, boiler fuel costs, and boiler maintenance. That total number is the baseline to which the cogeneration economics are compared.

Cogeneration Economics

If a simple payback approach is used, the cogeneration system costs divided by the savings over the baseline costs, would be the simple payback:

$$\text{Annual savings} = \text{Baseline costs} - \text{Cost with cogeneration system}$$

$$\text{Simple payback} = \frac{\text{Cogeneration system capital costs}}{\text{Annual savings}} \qquad \textbf{(17.11)}$$

For many analyses, a simple payback approach is often enough. Once an acceptable payback period is defined, the cogeneration decision can be made. If the simple payback period is fairly short (2–3 years), then small variations in energy prices will have little effect on the decision. As the payback period lengthens (e.g., 4–8 years), then other factors should be examined. The time value of money has to be considered, as well as projected energy rates, and projected changes in energy needs for the facility. See Chapter 3 for more information.

A typical scenario for a project financed over 15 to 20 years might require an economic analysis that would include the following steps:

1. Determine current energy costs and current thermal and electrical energy requirements.
2. Project energy costs into the 15- to 20-year future using forecasts from the local utility or from the state public utility commission.
3. Project thermal and electrical needs into the future, considering possible growth (or declines) in energy requirements.

Note: Any projections of future thermal and electrical loads should carefully consider the impact of energy conservation on the total energy needs. Energy conservation is extremely important in the cost analysis because it represents dollars that do not have to be spent. The term "negawatt" has been applied to demand-side management programs as a means of expressing a decrease in electrical demand at a facility.

4. Project increasing maintenance costs associated with an expected increase in thermal loads (if appropriate).

TABLE 17.2 Possible Scenarios to Consider for Cogeneration System Sizing

Scenario	Result	Application
Baseline cogeneration system that meets a portion of the facility's electrical and thermal requirements.	Requires production of supplemental thermal energy and purchase of additional electrical energy.	Universities/commercial buildings/thermal energy plants that have variable loads and do not want to get into the power sales business.
Cogeneration system where the thermal load exceeds the equivalent electrical output of the generator–thermal load sizing.	Electrical energy will have to be purchased from the local utility.	Industrial facility or manufacturing plant with relatively constant thermal load.
Cogeneration system designed to match electrical loads.	Thermal energy to be produced by a boiler.	Industrial facility with fairly high and constant electrical loads and lower, but variable, thermal loads.
Cogeneration system sized to meet high thermal loads, but with lower equivalent electrical requirements.	Excess electrical power sales to the local utility.	Industrial or university facility that has lower or variable electrical needs, which require electricity sales to make the project economical.

Summing these costs will provide a more realistic "business as usual" case over the expected financing period of the cogeneration system.

Once the base case is determined, various sizing and operating scenarios for the cogeneration system should then be evaluated. These scenarios include (1) sizing a base-load cogeneration system and, if necessary, purchasing additional required electrical energy and producing supplemental required thermal energy, (2) following the thermal load profile and purchasing electrical energy needs, (3) following the electrical load profile and producing supplemental thermal energy, and (4) following the thermal profile with excess electricity produced. Note that with any potential scenario considered, all the thermal energy produced by the cogeneration system is used. It is the use of all the thermal energy that generally makes a cogeneration system feasible. If electricity production were the primary output, the cogenerator could not compete with the local electric utility. It is the simultaneous production of and need for the thermal energy that makes cogeneration attractive.

The next step is to determine the costs associated with a cogeneration system. This will depend on the type and size of system selected—diesel engine/generator, dual fuel diesel engine/generator, natural gas–fired engine/generator, or gas turbine/generator. As described in a previous section, a variety of possibilities exist in sizing the cogeneration system for a particular application (see "Matching Electrical and Thermal Loads"). Ideally, the cogeneration system could be sized to match the electrical and thermal loads exactly; however, seldom is there an exact match. Table 17.2 summarizes four possible outcomes.

Operating and Capital Costs

In any scenario, the savings and income from the cogeneration system must be determined as well as the additional costs such as the additional fuel, and the maintenance. Steps for determining the savings and income of the cogeneration system include

a. Savings in electricity generated (avoided purchase) = kWh × ¢/kWh.
b. Savings in demand charges (decreased demand) = kW × $/kW.
c. Savings in HRSG-produced steam.
d. Income from excess electrical power sales (if applicable).
e. Income from excess thermal energy sales (if applicable.

There will be additional costs for the cogeneration system. These include

i. Turbine/generator annual maintenance—use manufacturer's recommended fees, typically expressed in terms of ¢/kWh of electricity produced. The range may be from 0.4¢/kWh for gas turbines to as much as 1.5¢/kWh for small engines.
ii. Fuel cost for the prime mover (fuel oil and/or natural gas) at cost per million Btu.
iii. Maintenance cost for HRSG—use manufacturer's recommended fee schedule, typically around $1/klb of steam.
iv. Any electrical utility charges for standby power, maintenance energy or power, etc., expressed in $/kW or ¢/kWh.

As an example of the savings and costs of a cogeneration system, see Table 17.4 in the following case study of the Austin State Hospital. This table provides detailed cost and savings data for that application.

Finally, the system or capital costs must be determined. A rough rule of thumb for estimating these system costs is as follows:

• Major equipment packages (combustion turbine and HRSG)	40%
• Balance of plant equipment	25%
• Engineering and construction	15%
• Other (interest during construction, permitting, contingency, etc.)	20%
	100%

Any cogeneration analysis is very complicated, and the results will vary greatly depending on interest rates, cost of fuel, permitting requirements, cost of electricity sales, and other factors.

Cogeneration Software Programs

Computer models and simulations of various aspects of cogeneration systems have been developed to assist in the design and economic evaluation of actual applications. Because cogeneration systems are combinations of major equipment, which are often operated in complex fashions, detailed calculations are necessary to complete accurate analyses of the complete systems. Computer programs are therefore often necessary to perform these calculations in a time-effective, repetitive manner. These computer programs are available in a variety of formats from the very simple to the highly sophisticated. Examples of the types of analyses that these programs perform include (1) simple ("first-cut") technical and economic feasibility assessments, (2) complete technical and economic evaluations, (3) preliminary engineering design, (4) detailed engineering design, (5) complete detailed engineering and economic evaluations, (6) assessments of various financial arrangements, and (7) large-scale economic forecasting (Hu, 1985). Many computer programs contain combinations of these aspects, and many may even include other features.

In general, these programs require data on the power and thermal loads of a facility being considered for a cogeneration system. These load profiles may be needed on an hourly, daily, monthly, or yearly basis depending on the level of the program. In addition, these programs may need information on one or more of the following: the electrical rates for purchases (including standby or emergency power) and sales, fuel rates, required steam and chilled water conditions, prime mover performance data, operating and maintenance expenses, plant availability, economic life of the major equipment, emission regulations, and economic parameters such as interest rates, tax rate structures, depreciation schedules, financing charges, and escalation rates. Many of these programs allow several operating modes to be investigated: following the thermal loads, following the electrical loads, maintaining a user-specific power or thermal level, or some combination of these operating modes. The outputs of these computer programs generally include the amount of electric power and thermal energy produced, the operating costs, and any revenues. Often these programs may provide additional economic outputs such as simple payback periods, undiscounted and discounted cash flows, internal rates of return, and net present values of cash flows.

These computer programs are designed to enable the user to assess the benefits and costs associated with a cogeneration system operated in one of several modes. These programs allow the user to determine which cogeneration system maximizes economic benefits within the given constraints by selecting different systems, sizes, operating modes, and other design parameters. The programs may yield simple payback, net present value, or other economic indicators to determine the success of a given cogeneration system design relative to meeting the facility's thermal and electrical power needs. In some cases, the computer programs will be able to select the optimum configuration for a given facility. But as these programs become more complex, generally they become less user friendly.

Although a complete list of cogeneration computer programs is beyond the scope of this chapter, a few representative programs will be mentioned. CELCAP (Civil Engineering Laboratory Cogeneration Analysis Program), developed by the U.S. Navy, is a simulation program that uses hourly electrical and thermal profiles to predict the cogeneration economics (Lee, 1988). CELCAP is primarily used as a screening tool, and was designed to permit rapid consideration of cogeneration alternatives for U.S. Naval facilities. The DEUS model was developed by the General Electric Company for the Electric Power Research Institute (EPRI) in 1982 (General Electric Company, 1982). DEUS contains a cogeneration data base that includes performance and cost information for nine generic systems. COGEN2 was developed by Mathtech, Inc., for EPRI, and is a complex, optimization program (Mathtech, 1980). COGEN2 is best used by experienced users who have a thorough understanding of cogeneration systems. OASIS (Optimization and Simulation of Integrated Systems) is similar to COGEN2 (Argonne National Laboratory, 1981). Finally, COGEN-Micro (Fisher and Schmidt, 1983) was configured for microcomputers to determine optimal size systems based on a plant' energy demand conditions. Many other commercial and noncommercial (Fennell, 1993) programs are available for different purposes.

Final Comments on Economic Evaluations

Economic evaluations for assessing the feasibility of cogeneration projects is an important aspect of any development. Although many of the considerations are similar to other engineering economic evaluations, several aspects are unique to cogeneration projects. Obtaining detailed electric and thermal energy data is a prerequisite to completing accurate economic assessments. Once the data are available, the savings and income of a cogeneration project must be compared to the additional costs of operating the system. If the net savings will "pay back" the capital expense of the cogeneration system in a reasonable time, then the project may make economic sense. A variety of sizes and operating modes of the cogeneration system should be examined to determine the most economical scenario. Computer programs are available to help in this process. Other detailed considerations that are often necessary in a thorough economic evaluation include tax and financing arrangements, but these considerations are beyond the scope of this chapter.

17.6 Financial Aspects

Overall Considerations

Cogeneration facilities can be financed by a variety of options. The traditional approach to financing is owner financing; however, cogeneration facilities are expensive to construct, and a company may choose not to use its own capital. When deciding on the most favorable financial arrangement, companies often consider the following factors:

- Shortage of capital.
- Effect on credit rating (if borrowing the money to finance the plant).
- Impact on balance sheet.
- Desire to fix savings (minimize risk).
- Inability to utilize available tax benefits.
- Lack of interest in operating or owning the plant (not main line of business).

Financing is critical to the success of a cogeneration facility, and it is best to determine, early on, the financing method to be used. Various financial structures are described below. There may be other, more creative ways, to finance projects, including combinations of these, but most financing will fall under one of the following:

1. Conventional ownership and operation.
2. Joint venture partnership.
3. Lease.
4. Third-party ownership.
5. Guaranteed savings.

Conventional Ownership and Operation (100% Ownership)

The heat load owner has two basic financing options in a conventional owner/operate structure: (1) to fund the project internally from profits in other areas of the business, or (2) to fund part of the project from internal sources and borrow the remainder from a conventional lending institution. Figure 17.13 shows one such financial structure.

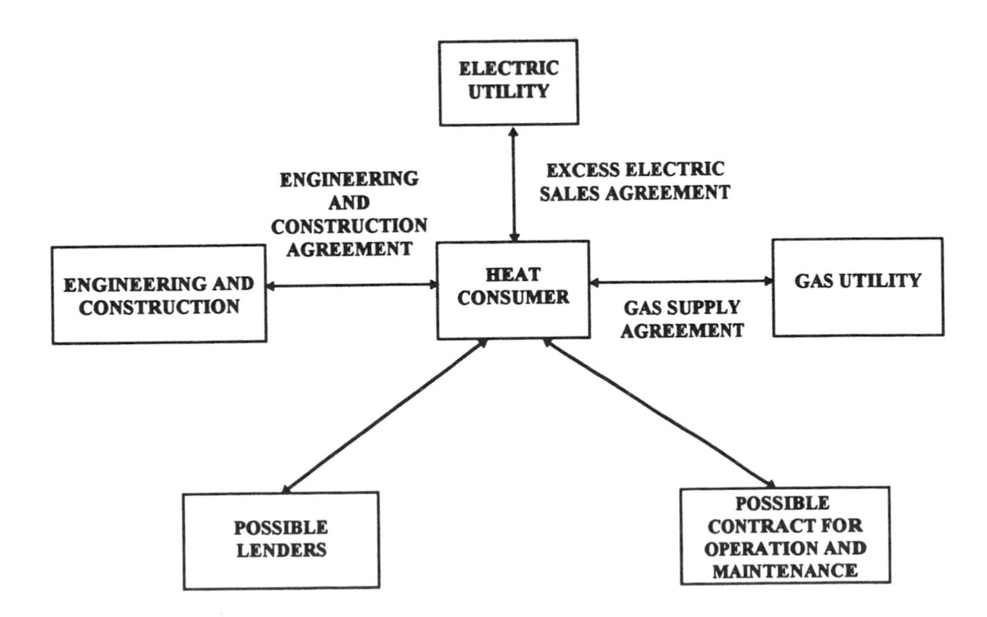

FIGURE 17.13 A schematic illustration of one financial structure for conventional ownership and operation of a cogeneration facility.

Most businesses have a minimum internal rate of return on equity that they require for any investment. They may not be willing to fund any project that does not meet the internal hurdle rate using 100% equity (internal) financing. With 100% internal financing the company avoids the problems of arranging external financing (perhaps having to add partners). If there is a marginal return on equity, however, the company will not finance the project, especially if the money could be used to expand a product line or create a new product that could provide a greater return on equity.

Because the cost of borrowed money is typically lower than a business's own return on equity requirements, the combination of partial funding internally and conventional borrowing is often used. By borrowing most of the funds, the internal funds can be leveraged for other projects, thus magnifying the overall return on equity.

External contract issues are simpler in conventional ownership. Contracts will be required for the gas supply, for excess power sales to the local utility, possible O&M agreements, plus any agreements with possible lenders, but no contracts may be needed for the steam and electricity if all is used internally.

Joint Venture

One alternative to 100% ownership is to share ownership. Figure 17.14 shows the outline of one possible arrangement of a joint venture project. PURPA regulations allow electric utilities to own up to a 50% share in a cogeneration facility. Profits shared with a utility in a partial ownership role are unregulated. Other partners might include a gas utility or a major equipment vendor, such as a major gas turbine producer.

The major advantages of a joint venture are the sharing of risks and credit. The disadvantages are that profits are also shared and that contract complexity increases. Since the joint venture company is the owner, steam and power sales agreements, in addition to the other contracts required under conventional ownership, have to be signed with the heat consumer.

Leasing

In this financing arrangement, a company will build the cogeneration plant with the agreement that the steam host or heat consumer will lease the plant from the builder. In this model, the heat consumer is still heavily involved with the project construction but keeps at arm's length from the financing, since the lessor will have put that package together. These deals are often more difficult to put together and could take longer to develop than either joint venture or conventional ownership financing. Figure 17.15 is a schematic of one such lease agreement approach.

Third-Party Ownership

In third-party ownership, the heat consumer is distanced from both the financing and construction of the cogeneration facility. A third party arranges the finances, develops the project, and arranges for gas supply, power sales for the excess power produced, steam sales to the heat consumer, and O&M agreements. Under the 1992 National Energy Policy Act, the third party may also be able to enter into a power sales contract with the heat consumer as well. Figure 17.16 is a schematic of one type of third-party arrangement. The lessee, shown in Figure 17.16, can be a minimally capitalized entity that is primarily in the business of plant operation and handling energy sales once the plant is built. The contracts for firm energy sales are key to this type of financing operation. This type of financing option can be complex, requiring more development time, so development costs could be much higher.

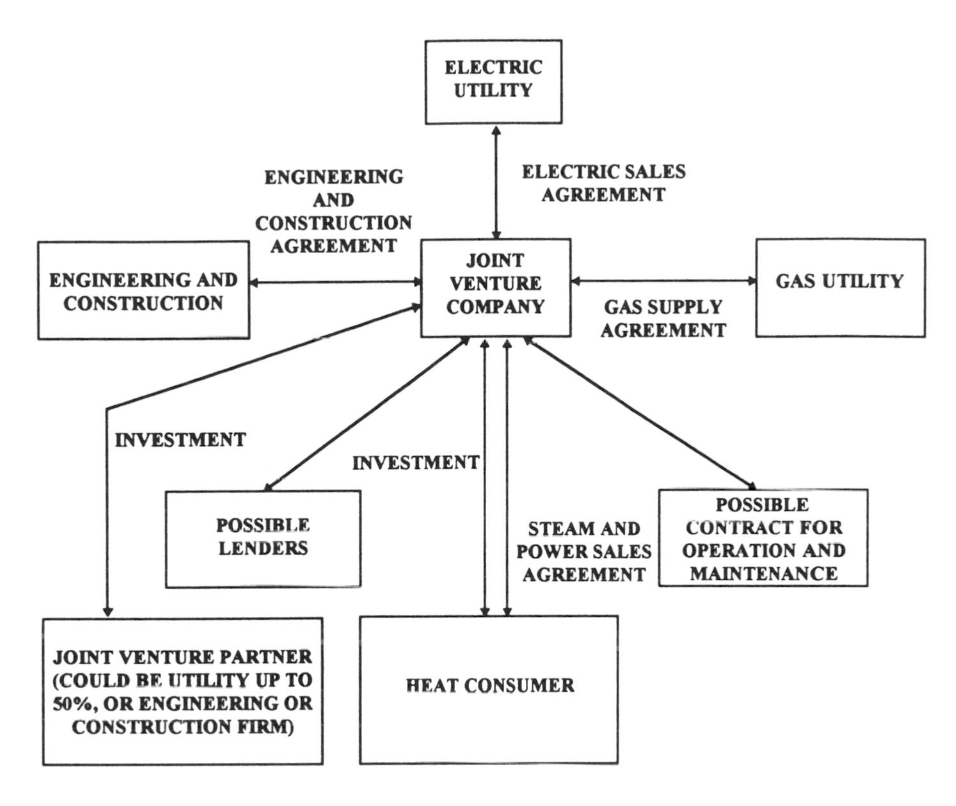

FIGURE 17.14 A schematic illustration of one financial structure for a joint venture ownership of a cogeneration facility.

Guaranteed Savings Contracts

These types of contracts are more common in smaller sized cogeneration systems, and typically may use packaged facilities. Figure 17.17 is a schematic of one possible guaranteed savings arrangement. A developer may pay all costs for construction of a cogeneration facility and then operate and maintain the system. Further, a guarantee of savings would be made to the owner over a specified length of time. Typically, a guaranteed savings contract will run from 5 to 10 years, with an average of 8 years, and will guarantee a fixed savings per year. The "split" between the guaranteed savings contractor usually ranges from a 75/25 division to a 95/5 division, with a "typical" split of 85/15; that is, 85% of the savings to the guaranteed savings contractor and 15% to the owner. Any additional savings above the guarantee are often split between the guaranteed savings contractor and the owner.

Guaranteed savings contracts have a distinct advantage to the owner in that all the capital and most of the risks are taken by the guaranteed savings contractor. The owner is guaranteed a percentage savings each year. These types of contracts should be considered when (1) the owner does not have capital to invest in a cogeneration system, or (2) when the owner does not have the necessary experience in-house to operate/manage such a project. If the capital is available, and if the in-house expertise exists, the company would be ahead financially to do the project itself, as the net return would be much higher.

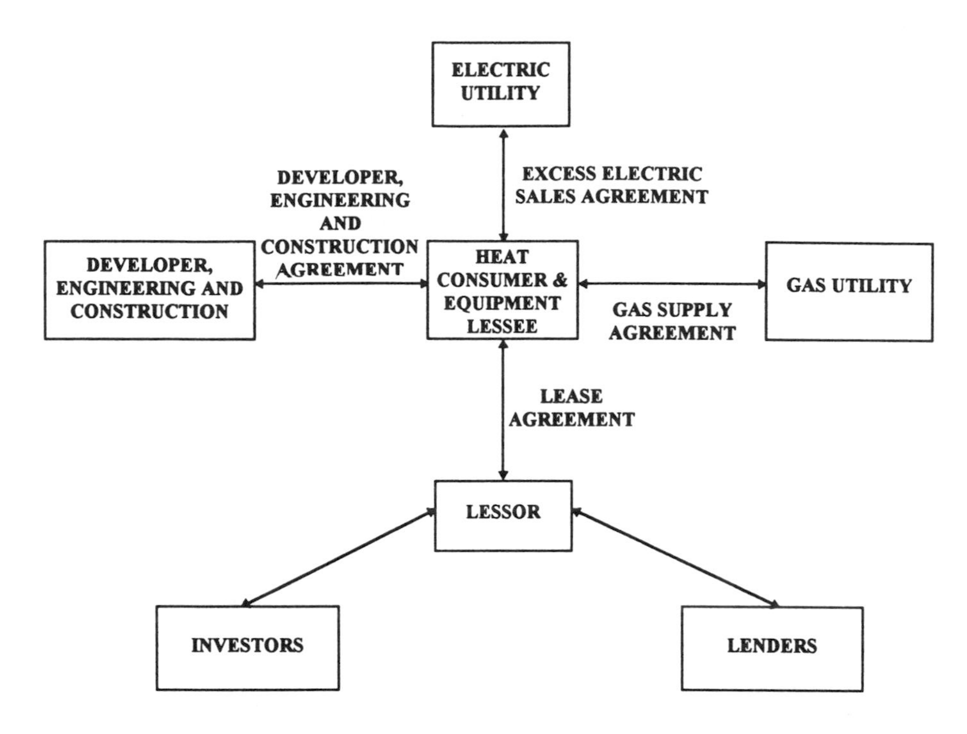

FIGURE 17.15 A schematic illustration of one financial structure for a lease arrangement for a cogeneration facility.

Final Comments on Financial Aspects

Financing arrangements are a crucial aspect of most cogeneration developments. These arrangements may range from simple to highly complex. They are affected by internal factors such as ownership arrangements, credit ratings, and risk tolerance. In addition, these financial aspects are affected by external factors such as the financial and credit markets, tax laws, and cogeneration regulations. A variety of initial financial arrangements have been outlined in this section to illustrate the nature of these arrangements. Much more detailed arrangements are possible and often necessary, but these are beyond the scope of this chapter.

17.7 Case Studies

Two case studies will be presented. One is a state hospital where a small gas turbine was installed, and the other is an industrial application which used a large gas turbine.

Introduction to the Austin State Hospital (ASH) Case Study

At the request of the Texas State Energy Office, and to counter rising costs of energy, 15 potential sites were examined by Texas A&M University's Energy Systems Laboratory (ESL) in 1985. ESL faculty and graduate students conducted preliminary feasibility studies of possible cogeneration sites at state universities, state hospitals, prisons, and the Capitol Complex. Not all the feasibility studies proved economically promising, and these were not considered for further study. Several

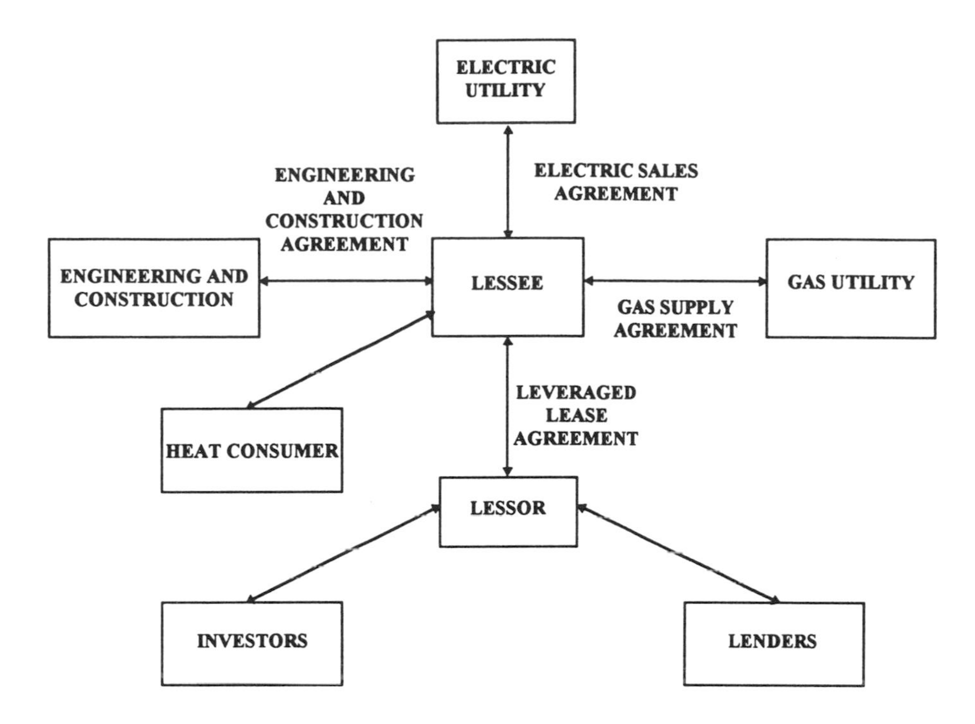

FIGURE 17.16 A schematic illustration of one financial structure for third-party ownership and operation of a cogeneration facility.

of the agencies pursued the most feasible cogeneration systems, and four sites have installed systems. These four sites and the corresponding cogeneration system are

Austin State Hospital; Austin, Texas:	1 MW gas turbine
Texas Tech University; Lubbock, Texas:	900 kW steam turbine
Southwest Texas State University; San Marcos, Texas:	6.8 MW dual-fuel diesel engine
University of Texas—Dallas; Dallas, Texas:	3.8 MW dual-fuel diesel engine

Several more of the Texas state agencies and universities are still considering cogeneration, but as of 1995, these were the only ones identified in the 1985 ESL studies to have installed systems. The University of Texas at Austin and Texas A&M University already had cogeneration systems operating, and two private universities, Baylor (3.5 MW) and Rice (7 MW), have installed gas turbine cogeneration systems. Texas Tech University has a third-party cogeneration system on its campus supplying thermal and electrical energy to the university, and the University of Texas at San Antonio has a cogeneration system to supply peaking power in the summer to the campus power plant. The small-size cogeneration system case study will focus on the Austin State Hospital.

The Austin State Hospital Case Study

The Austin State Hospital is a facility operated by the Texas Mental Health and Mental Retardation (MHMR) state agency. It was one of 15 sites selected for a potential cogeneration site and was analyzed in 1985. A large laundry and cafeteria provided most of the thermal loads needed for the project, but there were also significant heat loads during the winter. In 1985, the ESL study recommended gas turbines totaling 1.5 MW be installed, at an estimated cost of $1.5 million. An

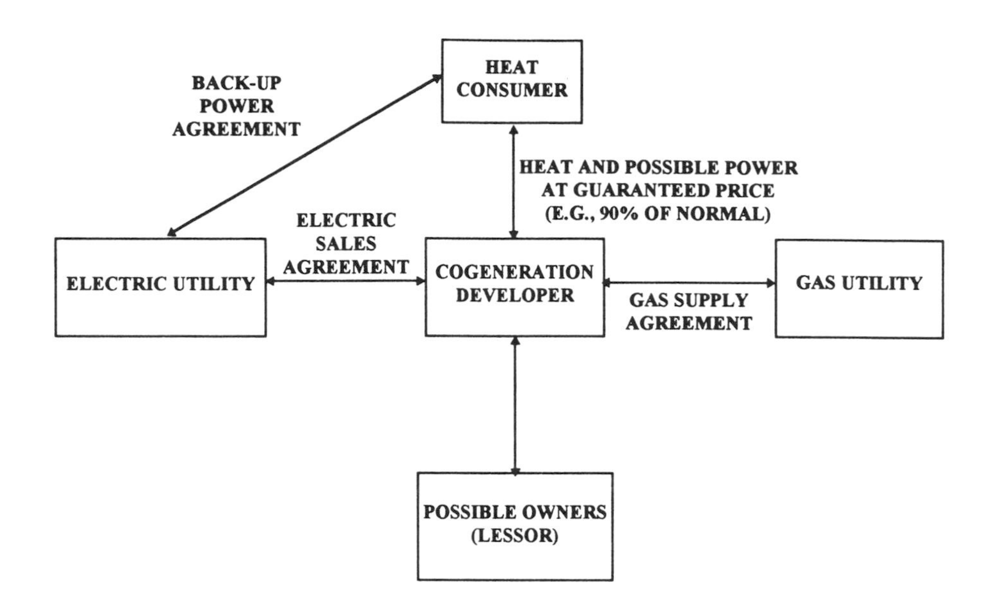

FIGURE 17.17 A schematic illustration of one financial structure for a guaranteed savings arrangement of a cogeneration facility.

engineering study was then initiated in 1986–1987, and steam data were taken to determine the steam loads more accurately. The consulting engineering study recommended a 1 MW system at an estimated cost of $1.3 million. A $1.5 million grant project application was submitted to and approved by DOE. Specifications were prepared, the project was bid, and all bids came in too high. The project was then stalled for nearly 2 years until additional funding was secured and a long-term gas contract could be signed with the General Land Office of the State of Texas. The project was rebid in 1990, and a contractor was selected (Caton et al., 1991).

The 1 MW gas turbine, heat recovery steam generator, and controls were completed in May 1992, at a cost of over $1.8 million, or slightly more than $1,800/kW. The system supplies steam to the Austin State Hospital complex and provides base load electricity for the facility. The cogeneration system was projected to save $225,000 per year, for a simple payback of 8 years. A long-term gas contract with the Texas General Land Office (GLO) assured the long-term economics of the system. The GLO established a maximum gas cost escalation rate of 5% per year for the duration of the payback period. Since DOE was providing a $1.5 million grant for the system, the benefit of the State of Texas was large. The state contribution of $300,000 would be paid back in less than 2 years. The ESL has been monitoring the system performance, and Table 17.3 summarizes 2 years of operation of the cogeneration system.

Shortly after the installation of the cogeneration system was completed (in May 1992), lightning struck the transformer and power line into the Austin State Hospital and burned out many of the controls of the cogeneration system. This caused several weeks of downtime during that summer. Additional start-up problems were attributed to the waste heat system. Overall, the cogeneration system has performed exceptionally well, with a turbine/generator availability of 96.8% for a 2-year period, July 1992 through June 1994. This 96.8% availability includes downtime for scheduled maintenance. Excluding scheduled maintenance downtime, the turbine/generator availability is nearly 100%.

TABLE 17.3 Energy Usage for the Austin State Hospital Cogeneration System

Date	Purchased Electricity (kWh/mo)	Cogenerated Electricity (kWh/mo)	Total Electricity (kWh/mo)	HRSG Heating (MMBtu/mo)
Aug. 1992	994,909	629,714	1,624,623	3,851
Sep. 1992	1,159,110	395,537	1,554,647	1,182
Oct. 1992	694,243	655,740	1,349,983	4,030
Nov. 1992	475,120	550,046	1,025,166	3,449
Dec. 1992	303,951	721,222	1,025,173	5,168
Jan. 1993	251,741	704,480	956,221	4,862
Feb. 1993	268,584	663,623	932,207	4,755
Mar. 1993	346,492	698,975	1,045,467	4,910
Apr. 1993	541,877	557,241	1,099,118	3,947
May 1993	601,993	649,361	1,251,354	5,066
Jun. 1993	795,057	601,731	1,396,788	4,923
Jul. 1993	880,342	629,613	1,509,955	4,913
Aug. 1993	887,767	641,948	1,529,715	5,221
Sep. 1993	756,969	610,222	1,367,191	5,016
Oct. 1993	605,129	588,025	1,193,154	4,812
Nov. 1993	284,797	684,472	969,269	5,077
Dec. 1993	305,742	703,319	1,009,061	5,535
Jan. 1994	320,072	676,443	996,515	4,708
Feb. 1994	246,905	656,497	903,402	4,643
Mar. 1994	326,711	661,360	988,071	4,841
Apr. 1994	409,269	641,899	1,051,168	5,009
May 1994	515,273	655,417	1,170,690	5,437
Jun. 1994	866,959	582,881	1,449,840	5,220
Jul. 1994	794,524	608,172	1,402,696	2,981
Total	13,633,536	15,167,938	28,801,474	109,457

Problems with the waste heat boiler reduced its availability to approximately 92% over the same period. This has reduced the expected steam recovery below the assumed availability. This is not the reason, however, for the lower than projected savings. The City of Austin changed the type of electrical service and increased the demand charges in 1992. An extra standby charge adds approximately $30,000 annually to the bill. Without this charge, the cogeneration system would be saving nearly $210,000 annually, almost exactly the annual amount predicted by the savings analysis. Table 17.4 is a summary based on actual metered data for the ASH cogeneration system for 23 months of operation, August 1992 through June 1994. The annual dollar savings are approximately $181,000 per year, which gives a simple payback of approximately 10 years. Since the bulk of the money was provided by a federal grant, the state portion has already been repaid. Therefore, the cogeneration project is saving Texas over $180,000 annually at today's electricity and gas prices.

Large Industrial Cogeneration Plant Case Study

This case study was selected to illustrate the scale of industrial cogeneration plants and to describe the variety of technical possibilities associated with these large plants. This material was originally presented by Wohischiegel et al. (1985). Figure 17.18 illustrates the process flow of the plant. In the conventional mode of operation (base case), process steam would be produced in a boiler and electricity would be purchased from the local electrical utility. In this case study, the plant electrical load is 45 MW. The steam loads are high in this application, which is a prerequisite for any potential cogeneration system. The process uses 725,000 lb/hr of steam at 600 psig, 300 psig, and 5 psig. The boiler efficiency is 84%.

Several decisions have to be made. Should a new system be sized for the plant electrical load with supplemental firing of a boiler for the additional thermal, or should the cogeneration plant provide all the thermal needs, which would require electricity to be sold to the grid? Supplemental firing in a duct burner may also be necessary to get the process steam required. One of the first

TABLE 17.4 Summary of Savings Calculations for the Austin State Hospital

Electrical Energy Cost	$0.02609/kWh
Average Demand Cost	$11.69/kW
Natural Gas Fuel Cost	$3.19/MCF

(August 1992–June 1994)

Energy	Usage	Savings (per year)
a. Electricity	14,599,766 kWh	$198,209
b. Demand (avoided	23,190 kW	141,183
c. Heating (HRSG)	106,477 MMBtu	229,565
d. Maintenance of conventional boiler (not required)	at $1.10/klb of steam	57,496
Total		$626,452

Energy Costs for the Cogeneration System	Expenses (per year)
Natural gas for the gas turbine	$332,394
Standby charge	30,600
Maintenance of the cogeneration system ($0.004/kWh)	30,365
Maintenance of HRSG ($1.00/klb)	52,261
Total	$445,620
Net Savings	$181,000/year

scenarios to consider is a combustion turbine sized for the electrical loads (45 MW). In this case, the steam produced in the HRSG would be approximately 300,000 lb/hr. Even with supplemental firing the match with the thermal load would not be good. Therefore, the decision was made to consider larger gas turbines, sell excess power, and produce more thermal energy from the cogeneration system.

Three (3) possible cogeneration configurations were considered for supplying the required thermal loads, the required electrical loads, and producing excess electrical power for sales. All three of the cogeneration cases were sized to produce 90 MW of electrical power, but each case results in increasing amounts of process steam. The base case and the three cogeneration cases were

Base case (business as usual)—fired boilers for process steam and purchase electricity.

Case A—90 MW combustion turbine, excess electrical power sales, remaining thermal energy produced in a fired-boiler, single-pressure HRSG.

Case B—90 MW combustion turbine, excess electrical power sales, supplemental firing in a duct burner, no process steam in a fired-boiler, single-pressure HRSG.

Case C—90 MW combustion turbine, excess electrical power sales, supplemental firing in a duct burner, no process steam in a fired-boiler, dual-pressure HRSG.

Each of these scenarios was then analyzed to determine the savings relative to the base case, and the simple payback and net present value (NPV) was determined.* Table 17.5 summarizes the comparison of the various options.

For each of the cogeneration cases, the hot exhaust gas from the gas turbine is used in the HRSG to generate steam, so less or no fuel is used in the original boiler. For case A, the total fuel consumption has increased by 634 million Btu/hr and the overall thermal efficiency has been reduced from the boiler efficiency of 84% to a combined efficiency of 72%. The lower efficiency

*The numbers and paybacks for these scenarios were based on 1985 prices (Wohischiegel et al., 1985). Current prices for capital equipment, natural gas, electricity sales, insurance, maintenance, and other items would have to be used for economic analyses of any system being considered today.

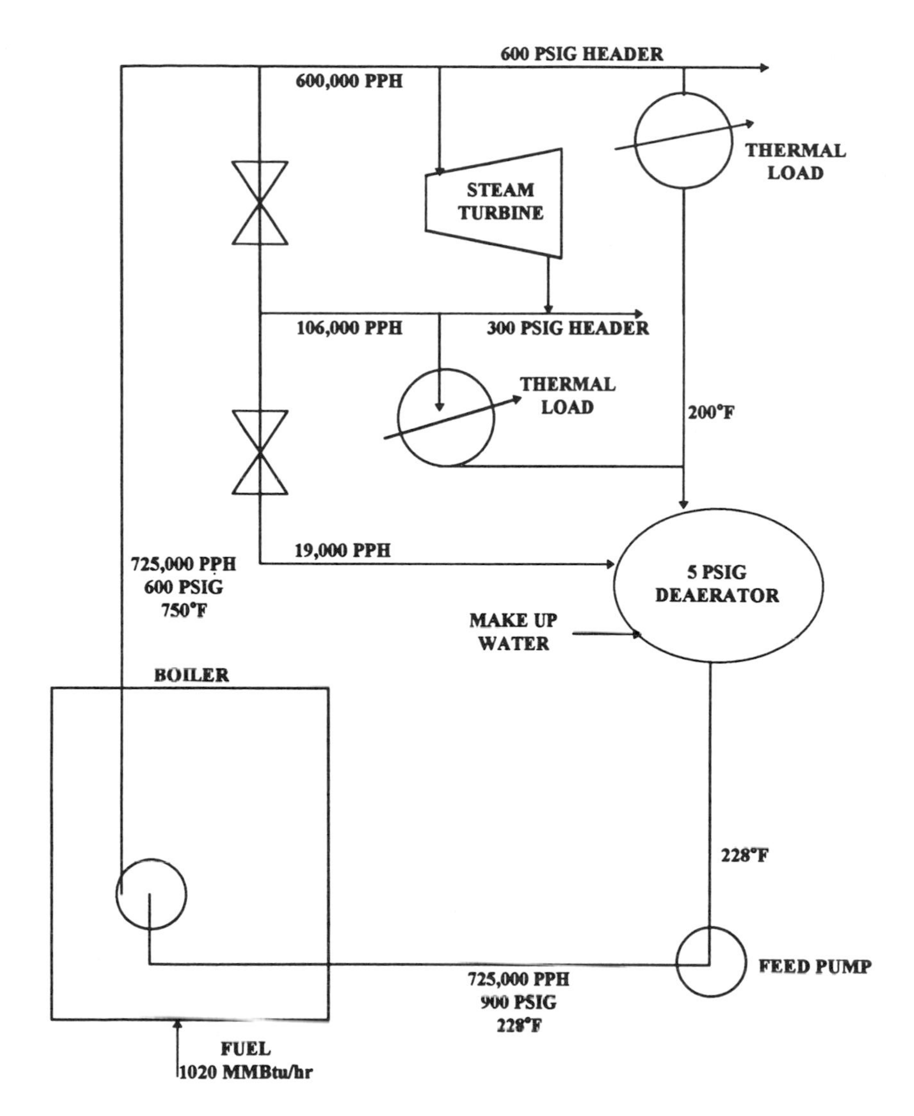

FIGURE 17.18 A schematic illustration of an industrial application process flow diagram.

for the cogeneration case reflects the use of the fuel to generate electrical power. Although the efficiency is lower, the cogeneration cases generate a more valuable commodity, electrical power, and this is reflected in the better economics (described later). Case B uses supplementary duct firing to generate more steam and eliminates the need for the original boiler. This improves the overall efficiency to 76%, because the duct burners are highly efficient. Finally, case C uses a dual-pressure HRSG for even greater efficiencies.

Table 17.5 also presents a summary of the economic results for this case study. Although the capital cost increased as the cogeneration systems became more complex from case A through case C, the NPV, simple payback, and internal rate of return all increased. In these applications, based on analyses completed in 1985, all cogeneration cases had excellent paybacks and high rates of return. The best scenarios favored supplemental firing (duct burners), and because of the price being paid for electricity sales (5.5¢ per kWh in 1985), the best scenario was to produce as much electricity as possible, which was the situation for case C.

This industrial case study was selected to depict a large gas turbine application, to illustrate the variety of technical choices that may be considered, and to provide an example of economic evaluation techniques. Complete details of this study are presented by Wohischiegel et al. (1985).

TABLE 17.5 Comparison of Three Cogeneration Systems for the Industrial Gas Turbine Case Study (1985 dollars)

	Summary of Results						
	Fuel Use (MMBtu/hr)				Efficiency		
Case	Boiler	Combustion Turbine	Duct Burner	Increased Fuel Used	HRSG (%)	η (%)	Cogeneration Heat Rate (Btu/kWh)
Base	1,020	N/A	N/A	N/A	N/A	84	N/A
Case A	491	1,163	0	634	77	72	6,356
Case B	0	1,163	417	560	90	76	5,614
Case C	0	1,163	406	549	96	79	4,937

	Summary of Economic Results			
Case	Capital Cost ($1,000)	NPV ($1,000)	Simple Payback (years)	Internal Rate of Return (%)
Base	N/A	N/A	N/A	N/A
Case A	49,900	107,624	2.11	47.1
Case B	51,400	118,511	2.09	48.6
Case C	55,970	142,533	2.02	52.0

Actual decisions on the exact system configuration are of course constrained by other site-specific considerations such as the importance of separating the electrical power production from the thermal production or emission limitations and regulations.

17.8 Future of Cogeneration

The future of cogeneration will include refined and better versions of the technology already described. In addition, several advanced cogeneration cycles are currently in various stages of development. These include (1) the Cheng cycle, (2) the Kalina cycle, (3) fuel cells, (4) coal gasification combined cycles, and (5) magnetohydrodynamic applications.

Cheng Cycle

The Cheng Cycle for optimizing the performance of gas turbine cogeneration systems was first identified by Dr. Dan Yu Cheng in 1974. It uses the steam produced in the HRSG for reinjection into the combustion chamber to provide additional power from the gas turbine. Since the gas turbine is a mass flow device, combining the steam with the combustion gases will increase the total electrical power produced. This cycle has an advantage over the traditional combined cycle (Rankine cycle) because of its simplicity. There is no need for steam turbine condensers and pumps. Part of the steam produced in the HRSG can also be used for process steam. Following electrical loads or following process steam loads is possible using steam injection in the Cheng Cycle. Substantial increases in both power output and generating efficiency are possible.

An added benefit from the steam injection is reduced NO_x emissions from the system. Strasser (1991) summarizes the results of several Cheng cycle installations in the United States and abroad, the oldest of which was an installation at San Jose State University in San Jose, California. That particular site had accumulated over 46,000 operating hours through November 1991.

Kalina Cycle

The Kalina cycle has been proposed by Dr. Alex Kalina as a means of improving the overall efficiency of combined cycle plants. Its application to cogeneration systems would be to replace the conventional Rankine cycle power generation with a more efficient Kalina cycle. Dr. Kalina proposed the use of an ammonia–water mixture as the working fluid instead of water. The advantages include a 10 to 20% increased thermal efficiency over a Rankine cycle with the same boundary conditions.

Ibrahim and Kovach (1993) describe a power generation application using the Kalina cycle, and Zervos et al. (1992) describe a 3 MW demonstration plant that uses the Kalina cycle. The complexity of the plant increases with the use of the NH_3–H_2O mixture. The conventional boiler is replaced by a vapor generator, the steam turbine by a vapor turbine, and the condenser by a distillation/condensation system.

In the DOE demonstration project (Zervos et al., 1992), the liquid ammonia concentration is 70% by weight as it leaves the distillation/condensation system and enters the simulated waste heat boiler. The superheated vapor then enters a specially designed vapor turbine, which drives a generator and produces electrical power. The Kalina cycle demonstration project operates between 2,000 and 20 psia with ammonia mixture concentrations varying from 30 to 70%. The vapor turbine inlet conditions include a pressure of 1,600 psia, a temperature of 960°F, an enthalpy of 1,221 Btu/lb, an ammonia concentration of 70%, and a flow rate of 31,000 lb/hr. The distillation/condensation system consists of a low-pressure condenser, vaporizer, preheater, and pump; an intermediate pressure shell-and-tube heat exchanger, falling film evaporator, flash drum, and pump; and a high-pressure condenser, preheater, pump, and solids removal system. Obviously, the management of the distillation/condensation system is crucial to the operation of a Kalina cycle system.

The vapor turbine has to withstand the high-temperature corrosion of the ammonia–water mixture and is made of Type 316 stainless steel. Pressurized steam is injected into the labyrinth seals to contain the ammonia–water mixture and prevent its leakage during the vapor expansion phase in the turbine. This demonstration plant using the Kalina cycle began operation in December 1991.

Fuel Cells

Fuel cells, in theory, should be ideal cogeneration devices, capable of very high overall efficiencies. According to Cameron (1990), fuel cell devices that have an overall efficiency of 80 to 90% are possible, with roughly a 50-50 split between thermal energy and electrical energy. An electrical power production efficiency of 65% may ultimately be demonstrated.

The fuel cell concept is very old, invented more than 150 years ago in England (Itoh, 1990). An electrochemical reaction produces direct-current electricity continuously if fed by air and a fuel, from which oxygen and hydrogen are produced. Since fuel cells are electrochemical devices, there is a minimum of rotating machinery, little noise, and no vibration in their operation. Waste heat is recovered from the chemical reaction.

Fuel cells received a big boost by the United States' manned spacecraft flights in the 1960s, and while the United States and Japan continue to be the leaders in fuel cell development, other countries, including Italy, The Netherlands, Norway, Spain, Canada, Korea, Switzerland, Germany, the United Kingdom, France, Belgium, Denmark, India, the People's Republic of China, and Russia have major R&D programs or active demonstration plants (Hirschenhofer, 1990). Japan is by far the most active in fuel cell research, and at the present rate of R&D spending will surpass the United States in the 1990s as the leader in fuel cell technology.

The three most common fuel cell types under development for terrestrial electric power applications are phosphoric acid (PAFC); a mixture of molten lithium, sodium, and potassium carbonate (molten-carbonate fuel cell, MCFC); and solid zirconium oxide stabilized with yttrium oxide (solid-oxides fuel cell, SOFC). The PAFC is available commercially as a small cogeneration unit.

Table 17.6 is a summary of current and projected fuel cell technology (Itoh, 1990). The first generation fuel cells are the alkali fuel cell and the phosphoric acid fuel cell. The second- and third-generation fuel cells, MCFC and SOFC, are high-temperature cells, operating at temperatures up to 650°C and 1,000°C, respectively. Large quantities of high-temperature thermal energy can be produced in the fuel cell operation or by burning hydrogen produced for the fuel cell. These higher operating temperatures are in contrast to the PAFC which typically operates in the 170 to 222°C range, providing hot water or low-pressure steam as the thermal energy by-product.

TABLE 17.6 Types of Fuel Cells and Their Characteristics

	First Generation		Second Generation	Third Generation
	Alkali Fuel Cell	Phosphoric Acid Fuel Cell	Molten-Carbonate Fuel Cell	Solid-Oxide Fuel Cell
Electrolyte	Potassium hydroxide	Phosphoric acid aqueous solution	Lithium carbonate and potassium carbonate	Zirconium oxide stabilized with ytrium oxide
Ion that mediates reaction	OH^-	H^+	CO_3^{--}	O^{--}
Operating temperature	Room temperature to 100°C	170–220°C	To 650°C	To 1,000°C
Fuel	Refined hydrogen	Natural gas, methanol, and light-distillate oil (such as naphta)	Natural gas, methanol, petroleum, and coal	Natural gas, methanol, petroleum, and coal
System generation efficiency	To 60%	40–50%	45–60%	50% or more
Major materials for cell	Synthetic resins	Carbon group	Nickel and stainless steel	Ceramics
Features	No noble-metal catalyst is needed.	Waste heat can be recovered.	CO can be tolerated.	CO can be tolerated.
	Many types of materials can be adapted.		Media of nickel group can be used.	No liquid medium is required.
	CO_2 is eliminated from fuel air.		Waste heat can be recovered.	Waste heat can be recovered.
Applicable areas	Special applications such as spacecraft.	Substitute for mid-size thermal power plant; distributed power generation (at substations, and so on); on-site cogeneration unit at factory, restaurant, hospital, hotel.	Substitute for large or mid-size thermal power plant; distributed power generation at substations; cogeneration at factories, and so on.	

Source: Adapted from Itoh, 1990.

Coal Gasification Combined Cycle Applications

The United States Clean Coal Technology (CCT) program was established in 1984 with the set-aside of $750 million for research and development of commercial processes to accelerate the use of clean coal technologies for power production. The Clean Air Act amendments of 1990 prescribed reductions in acid rain effects and emissions, aimed primarily at the electric utility industry. Coal gasification projects, as well as fluidized-bed combustion, have been the focus of the DOE CCT program (Hemenway et al., 1990).

The purpose of the coal gasification project is to produce a syngas that can be burned in a combustion turbine, hopefully at a price which is competitive with natural gas. The obstacles are coal cleanup, including desulfurization and particulate removal, development of components capable of operating at higher temperatures, disposal of solids and particulates from the process, high-temperature cleaning of the gas stream, and optimization of combustor design for the lower-temperature gas.

Sulfur removal is a must for these units to meet the standards of the clean air act for electric power production. There are various approaches used to remove sulfur, including the addition of limestone or dolomite to the gasifier. Particulate cleanup is usually accomplished by a series of cyclone separators and particle filters. The tricky problem with the gas cleanup is that it should be accomplished in the heated gas stream, which may be at a temperature of 1,000°F or more.

Waste heat from the combustion turbine exhaust is routed through the HRSG for steam production. The steam may then be piped to a steam turbine to produce additional electrical power or used partly for process heating. If some stage in the gasification process requires a fixed temperature to be maintained, it is possible that additional steam could be produced in the gasifier that could also be used for process heating or for additional power production.

Ruth and Bedick (1992) summarize five major integrated gasification combined cycle (IGCC) applications sponsored by the DOE. All of the projects are scheduled for start-up/demonstration in the 1995–1996 calendar years. The five activities are (1) Combustion Engineering IGCC Repowering, (2) Tampa Electric IGCC, (3) Pinion Pine IGCC, (4) Toms Creek IGCC Demonstration, and (5) Wabash River Coal Gasification Repowering Project.

Other authors have looked at the economics of obtaining and marketing the byproducts produced by the gasification process, including methanol, acetic acid, nitric acid, and formaldehyde, as a means of enhancing revenues (Baumann et al., 1992). Several scenarios were analyzed for cost effectiveness depending on the type of coproduct produced. Paffenbarger (1991) proposed a variation of the integrated gasification combined cycle plant by producing a coproduct, methanol, which could be stored on-site. His approach is to use the syngas from the gasification process as the base load IGCC system and then burn the methanol, along with the syngas, in the combustor for peaking purposes. His argument is that the gasification process could always operate at a steady load and therefore would not have to be cycled to meet a utility's peaking demand, which is always a problem in a coal-fired plant. The liquid methanol in the storage system could be used when needed to provide increased power production.

Magnetohydrodynamic Applications

Magnetohydrodynamic (MHD) generation utilizes the movement of a hot, electrically conducting gas through a magnetic field to produce a direct current. Hot air (preheated by the ionizing gases downstream of the MHD generator) is mixed with the fuel and combusted. The hot gases then enter the strong magnetic field in the MHD generator. Since very high temperatures (>3,000 K) are usually necessary for ionization, some "seeding" with cesium or potassium may be necessary. The current flow in the ionized gas is picked up by electrodes and flows through the external load. The hot ionized gases are cooled by the combustion air in an air preheater and then are channeled to a boiler where the steam production occurs. The cogeneration is accomplished by the direct current produced in the MHD generator and by the steam produce in the boilers. If this were a repowering application in an electric utility plant, conventional boilers would be used but would be unfired if the hot gas temperatures and flow rates were high enough to produce the required steam flow.

The Electric Power Research Institute (EPRI) reported in 1978 (Pomeroy et al., 1978) that the MHD technology was capable of capturing the entire electric utility base load, when developed to maturity. That development has been slow, however. The potential improvement in cycle efficiency to a typical electric utility plant is 15 to 30%, raising the maximum thermal efficiency from 38 to 45–54%. Large MHD systems appear to have the best potential, but this may not be the way electric utilities would use MHD systems (Chapman and Attig, 1990). Of all the advanced technologies, MHD is the farthest away from large-scale demonstration. Problems with low net electrical power output, high-temperature corrosion, and thermal stress have proven difficult to overcome. There are also practical problems in retrofitting an MHD system to existing power boilers. These include adding refractory to boiler tube surfaces because of the reducing atmosphere of the gases, additional particulate removal, the effect of more particulates on convective surfaces, and perhaps the need to redesign for secondary combustion air.

Summary and Conclusions

The importance of cogeneration is both monetary and energy savings. Any facility that uses electrical power and has thermal energy needs is a candidate for cogeneration. Facilities that may be considered for cogeneration include those in the industrial, commercial, and institutional sectors. The technology for conventional cogeneration systems is for the most part available and exists over a range of sizes: from less than 100 kW to over 100 MW. The major equipment requirements include a prime mover, electrical generator, electrical controls, heat recovery systems, and other typical power plant equipment. These components are well developed, and the procedures to integrate these components into cogeneration systems are well established.

In addition to the economic and technical considerations, the application of cogeneration systems involves an understanding of the governmental regulations and legislation on electrical power production and on environmental impacts. With respect to electrical power production, certain governmental regulations were passed during the late 1970s that remove barriers and provide incentives to encourage cogeneration development. Finally, no cogeneration assessment would be complete without an understanding of the financial arrangements that are possible.

This chapter has provided a brief survey of the important aspects necessary to understand cogeneration. Specifically, this chapter has included information on the technical equipment and components, technical design issues, regulatory considerations, economic evaluations, financial aspects, and future cogeneration technologies. The chapter included two case studies to illustrate the range of technologies available for cogeneration systems, the types of considerations involved in selecting a cogeneration system, and the overall economic evaluations that may be completed.

As the years of the twenty-first century near, cogeneration will continue to experience progressive growth because of both the energy and monetary savings. In the future, new technology will become available, and new regulations and legislation will be enacted. The future of cogeneration will build on the accomplishments and successes of the past, will continue to increase in importance, and will continue to contribute to the efficient use of our nation's energy resources.

References

Argonne National Laboratory. 1981. "OASIS CODE Application to Proposed Argonne National Laboratory Cogeneration Plant," ANL/CNSV-TM-67, Argonne, IL, June 1981.

Baumann, P.D., Epstein, M., and Kern, E.E. 1992. "Coal Gasification-Based Integrated Coproduction Energy Facilities," 1992 ASME Cogen-Turbo Conference, IGTI, Vol. 7, pp. 69–74.

Cameron, D.S. 1990. "World Developments of Fuel Cells," *International Journal of Hydrogen Energy*, Volume 15, No. 9, pp. 669–675.

Caton, J.A., Muraya, N., and Turner, W.D. 1991. "Engineering and Economic Evaluations of a Cogeneration System for the Austin State Hospital," Proceedings of the 26th Intersociety Energy Conversion Engineering Conference, Boston, MA, Paper No. 910444, Vol. 5, pp. 438–443, August.

Chapman, J.N. and Attig, R.C. 1990. "MHD Repowering of a Coal Power Plant—An Option for Reduced Environmental Emissions and Improved Heat Rate," Proceedings of the Intersociety Energy Conversion Engineering Conference, Reno, NV, Vol. 2, pp. 431–436.

Fennell, Steve R. 1993. "Technical and Economic Evaluations of Cogeneration Systems Using Computer Simulations," *MSME*, May.

FERC Regulations. 1978. Part 292—Regulations Under Sections 201 and 210 of the Public Utility Regulatory Policies Act of 1978 with Regard to Small Power Production and Cogeneration.

Fisher, D.B. and Schmidt, P.S. 1983. "Analysis of In-Plant Cogeneration with a Microcomputer," Proceedings of the 1983 Industrial Energy Technology Conference, Houston, TX, April 1983.

General Electric Company. 1982. "DEUS Computer Evaluation Model—Program Descriptive Manual," Final report to EPRI, EM-2776, Vol. 1, Palo Alto, CA, November 1982.

Hemenway, A., Williams, W.A., and Huber, D.A. 1991. "Effects of the Clean Air Act Amendments of 1990 on the Commercialization of Fluidized Bed Technology," ASME Fluidized Bed Combustion, pp. 219–224.

Hirschenhofer, J.H. 1990. "Latest International Activities in Fuel Cells," Proceedings of the Intersociety Energy Conversion Engineering Conference, Reno, NV, Vol. 2, pp. 176–184.

Hu, S. David. 1985. *Cogeneration,* Reston Publishing Company, Inc., Reston, Virginia.

Ibrahim, M. and Kovach, R.M. 1993. "A Kalina Cycle Application for Power Generation," *Energy,* Volume 18, No. 9, pp. 961–969.

Itoh, N. 1990. "New Tricks for an Old Power Source," *IEEE Spectrum,* July 1990, pp. 40–43.

Laderoute, C.D. 1980. "The FERC's PURPA Cogeneration Rules: Economics, Rate Design, and Policy Aspects," Annual UMM DNR 7th Conference Energy Proceedings, University of Missouri, Rolla, pp. 114–122.

Lee, T.Y. Richard. 1988. "Cogeneration System Selection Using the Navy's CELCAP Code," *Energy Engineering,* pp. 4–24, Vol. 85, No. 5, 1988.

Mathtech. 1980. "Advanced Cogeneration Technology Economic Optimization Study (ACTEOS)," Final report prepared for Jet Propulsion Laboratory, NASA, JPL Contract 955559, December 15, 1980.

Paffenbarger, J.A. 1991. "A GCC Power Plant with Methanol Storage for Intermediate-Load Duty," *Journal of Engineering for Gas Turbines and Power,* Volume 113, pp. 151–157.

Pomeroy, B.D. et al. 1978. "Comparative Study and Evaluation of Advanced Cycle Systems," EPRI Report No. AF0-664, Vol. 1, General Electric, Schenectady, New York.

Ruth, L.K. and Bedick, R.C. 1992. "Research and Development Efforts at the Department of Energy (DOE) Supporting Integrated Gasification Combined Cycle (IGCC) Demonstrations, 1992 ASME Cogen-Turbo Conference, IGTI, Vol. 7, pp. 87–94.

Spiewak, S. 1980, "Regulation of Cogeneration," McGraw-Hill Conference on Industrial Power—Electrical Rates, Reliability, and Energy Management, October 7–8.

Spiewak, Scott A. 1987. *Cogeneration and Small Power Production Manual,* The Fairmont Press, Atlanta, GA.

Strasser, A. 1991. "The Cheng Cycle Cogeneration System: Technology and Typical Applications." ASME Cogen-Turbo Conference, IGTI, Vol. 6, pp. 419–428.

Wohischiegel, M.V., Myers, G., and Marcellino, A. 1985. "Flexibility and Economics of Combustion Turbine-Based Cogeneration Systems," from *Planning Cogeneration Systems,* ed. Dilip R. Limaye, The Fairmont Press, Atlanta, GA, Chapter 9, pp. 119–143.

Wood, B.D. 1982. *Applications of Thermodynamics,* 2nd ed., Addison-Wesley, Reading, MA.

Zervos, N.G., Leibowitz, H.M., and Robinson, K.S., 1992. "Startup and Demonstration Experience of the Kalina Cycle Demonstration Plant," ASME Cogen-Turbo Conference, IGTI, Vol. 7, pp. 187–191.

Section III

Renewable Energy

18A

Availability of Renewable Resources: Solar Radiation

David S. Renné
Eugene L. Maxwell
Martin D. Rymes
William Marion
National Renewable Energy Laboratory

Julie Phillips
J.A. Phillips & Associates

Introduction

Solar radiation is the electromagnetic energy emitted by the sun. The amount of solar radiation available for terrestrial solar energy projects is that portion of the total solar radiation that reaches the earth's surface. In this chapter, the terms "insolation"; "incident solar radiation"; and "irradiance" are used interchangeably to define solar radiation incident on the earth per unit of area per unit of time as measured in watts per square meter (W/m^2) or watt-hours per square meter per 24-hour day ($Wh/m^2/day$).

The amount of solar radiation that reaches the earth's surface at any given location and time depends on many factors. They include time of day, season, latitude, surface albedo, the translucence of the atmosphere, and the weather.

This chapter describes practical methods for estimating the amount of solar radiation at specific locations. Section 18A.1, "Extraterrestrial Solar Radiation," discusses the measurement and modeling of solar energy outside the earth's atmosphere. Section 18A.2, "Terrestrial Solar Radiation," describes the factors affecting solar radiation at the earth's surface and provides a general description of considerations important in estimating the available resource. Section 18A.3, "Solar Radiation Data for the United States," describes the National Solar Radiation Data Base and presents insolation maps of the nation's solar resource. Section 18A.4, "Solar Design Tools," describes publications available from the National Renewable Energy Laboratory that provide estimates of the U.S. solar resource and also gives equations needed to estimate insolation on both horizontal and tilted surfaces. Section 18A.5, "International Solar Radiation Data," describes data bases that may be useful in estimating solar resources in foreign countries.

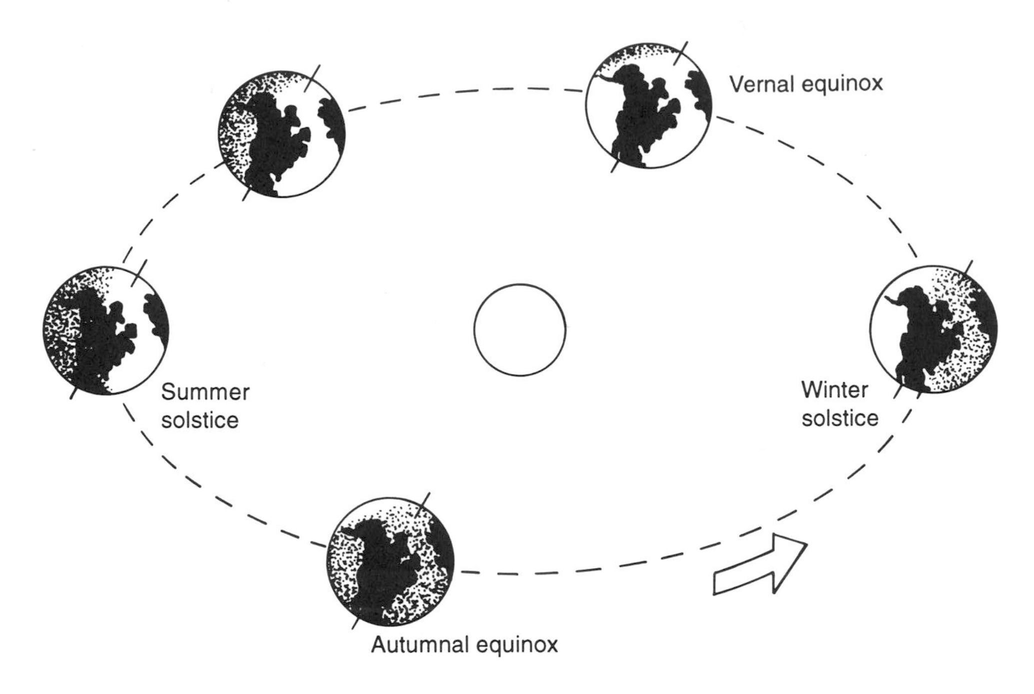

FIGURE 18A.1 The earth's elliptical orbit around the sun.

18A.1 Extraterrestrial Solar Radiation

Extraterrestrial radiation is the amount of solar energy above the earth's atmosphere. Also known as *top-of-the-atmosphere* radiation, this parameter indicates the amount of solar energy that would fall on the earth in the absence of an atmosphere.

The sun's energy is nearly constant, varying from year to year by less than 1%. The small interannual variations are associated primarily with the 22-year sunspot cycle. However, extraterrestrial solar radiation also varies due to the earth's elliptical orbit around the sun. Extraterrestrial radiation increases by about 7% from July 4 to January 3, at which time the earth reaches the point in its orbit closest to the sun. Figure 18A.1 shows a diagram of the earth's orbit around the sun.

Calculations of extraterrestrial solar radiation typically include an eccentricity correction factor, E_0, to account for this variation. For many engineering applications, the following hand calculations is adequate (Duffie and Beckman, 1980):

$$E_0 = 1 + 0.033 \cos\left(\frac{360n}{365}\right) \tag{18A.1}$$

where n = the day of the year counted from January 1, such that $1 \leq n \leq 365$.

The earth's axis is tilted 23.5° with respect to the plane of its orbit around the sun. Tilting results in longer days in the northern hemisphere from the spring equinox (approximately March 23) to the autumnal equinox (approximately September 22) and longer days in the southern hemisphere during the other 6 months. On the March and September equinoxes, the sun is directly over the equator, both poles are equidistant from the sun, and the entire earth experiences 12 hours of daylight and 12 hours of darkness.

In the temperate latitudes ranging from 23.5° to 66.5° north and south, variations in insolation are pronounced. At a latitude of 40° N, for example, the average daily total extraterrestrial solar radiation (on a horizontal surface) varies from 3.94 kilowatt-hours (kWh) per square meter (m^2) in December to 11.68 kWh/m^2 in June. Therefore, it is necessary to know the location of the sun in the sky to determine the extraterrestrial radiation at a specific location and time.

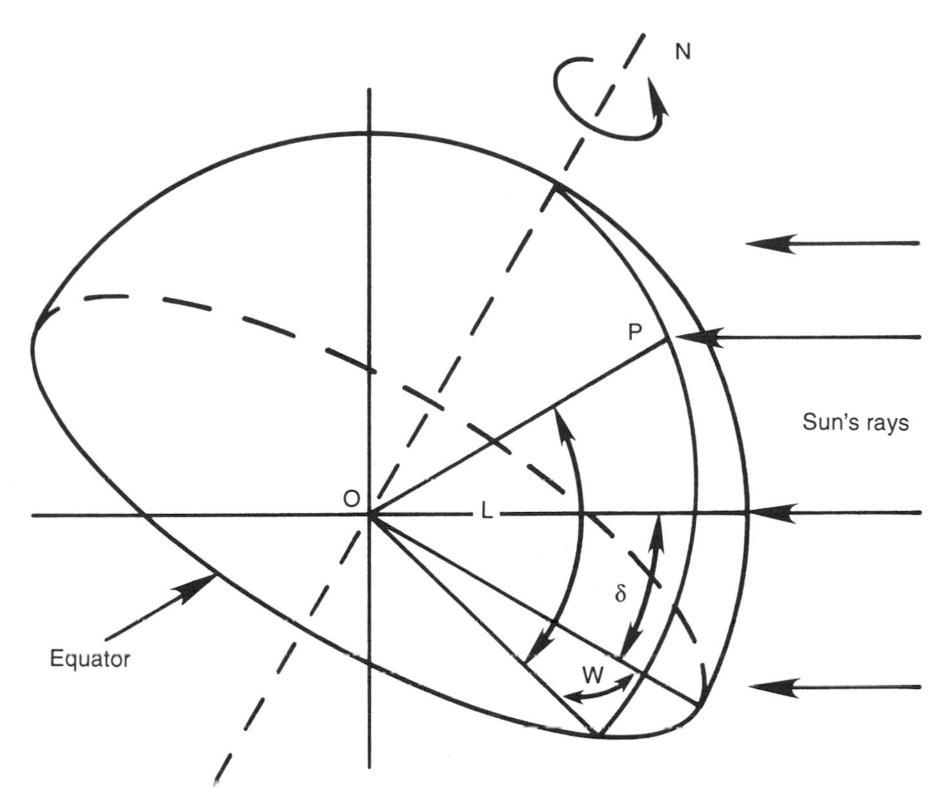

FIGURE 18A.2 Solar geometry: latitude, hour angle, and the sun's declination.

The sun's position relative to a location can be determined if a location's latitude *(L)*, a location's hour angle *(W)*, and the sun's declination are known. These three parameters are shown graphically in Figure 18A.2. Latitude is the angular distance north or south of the earth's equator, measured in degrees along a meridian.

The hour angle is measured in the earth's equatorial plane. It is the angle between the projection of a line drawn from the location to the earth's center and the projection of a line drawn from the center of the earth to the sun's center. Thus, at solar noon, the hour angle is zero. At a specific location, the hour angle expresses the time of day with respect to solar noon, with one hour of time equal to 15° of hour angle. By convention, the westward direction from solar noon is positive.

The sun's declination is the angle between a projection of the line connecting the center of the earth with the center of the sun and the earth's equatorial plane. It varies from –23.45° on the winter solstice (December 21), to +23.45° on the summer solstice (June 22). The approximate declination for any given day is given by (Cooper, 1969):

$$\delta = 23.45 \sin\left[360\left(\frac{284+n}{365}\right)\right] \quad \textbf{(18A.2)}$$

Several other angles are used to identify the location of the sun in the sky. These angles include the solar azimuth angle (γ), the solar zenith angle (θ_Z), the solar altitude angle (α), and the sunset hour angle W_s. The solar azimuth angle is the compass direction. The solar zenith angle, as shown in Figure 18A.3, is the angle of the sun from the vertical, or zenith. The solar altitude angle is the angle of the sun with the horizon. Therefore, $\alpha + \theta_Z = 90°$.

If the solar zenith angle is known, the extraterrestrial radiation (H_0) can be calculated as follows:

$$H_0 = I_{SC} \cdot E_0 \cdot \cos(\theta_z) \quad \textbf{(18A.3)}$$

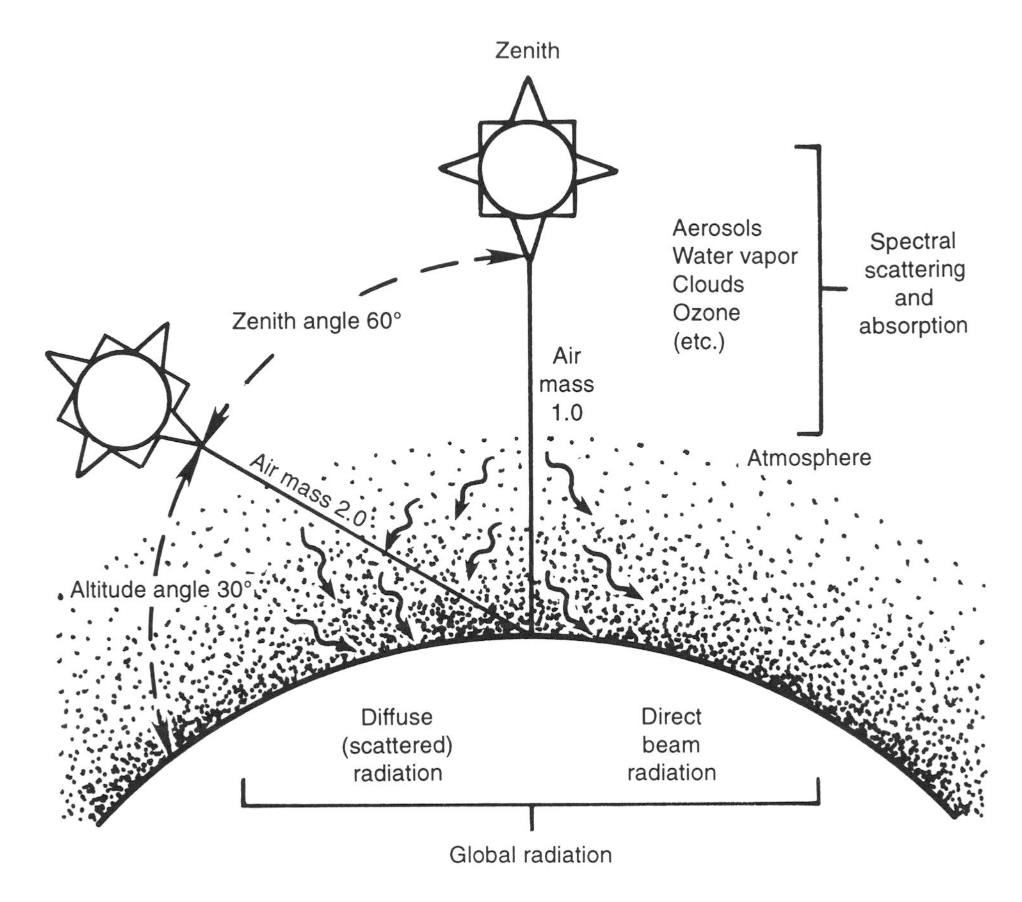

FIGURE 18A.3 Solar geometry: solar altitude and solar zenith angles.

where

I_{SC} = the solar constant
E_0 = the correction factor for the eccentricity of earth's orbit

The value of the solar constant (I_{SC}) endorsed by the World Radiation Center in DavosDorf, Switzerland, is 1,367 W/m^2. The eccentricity correction factor (E_0), calculated using Eq. (18A.1), allows for the daily adjustment of the solar constant.

The relationship among the solar zenith angle, the solar altitude angle, the sun's declination, the hour angle, and latitude is given as follows:

$$\cos\,\theta_z = \sin\,\alpha = \cos\,\delta\,\cos\,W\,\cos\,L + \sin\,\delta\,\sin\,L \tag{18A.4}$$

This equation can be used to compute the sunset hour angle. At sunset, $\sin\,\alpha = 0$, therefore:

$$\cos\,W_S = -\tan\,\delta\,\tan\,L \tag{18A.5}$$

Because day length can be easily derived from Eq. (18A.5), W_S is also a measure of the position of the earth in its orbit around the sun. This relationship is given by:

$$\text{day length} = \frac{2}{15}\cos^{-1}\left(-\tan\,\delta\,\tan\,L\right) \tag{18A.6}$$

18A.2 Terrestrial Solar Radiation

Measuring or calculating the total global solar radiation striking the earth's surface is far more complicated than determining extraterrestrial radiation. Changing atmospheric conditions, topography, and the changing position of the sun interact to moderate the solar resource at a given time and location. The problem is further complicated for the solar engineer because the vast majority of solar radiation measurements (and extrapolations based on them) have been made on horizontal surfaces rather than on the tilted surfaces typically employed by efficient solar energy collection. Solar designers must use reference tables that translate horizontal data into directional data for tilted surfaces or make those calculations themselves. Section 18A.4, "Solar Design Tools," describes such calculations and provides information on obtaining data manuals that describe the available solar resource for tilted surfaces in specific locations.

The total solar radiation on earth consists of (1) *direct beam radiation,* which comes to the surface of a direct line from the sun, and (2) *diffuse radiation* from the sky, created when part of the direct beam radiation is scattered by atmospheric constituents (e.g., clouds and aerosols). Concentrating solar collectors rely almost entirely on direct beam radiation, whereas other collectors, including passive solar buildings, capture both direct beam and diffuse radiation. Radiation reflected from the surface in front of a collector may also contribute to total solar radiation.

The atmosphere through which solar radiation passes is quite variable and acts as a dynamic filter, absorbing and scattering solar radiation. Low and mid-level cumulus and stratus clouds are generally opaque, blocking the direct beam of the sun. In contrast, high, thin cirrus clouds are usually translucent, scattering the direct beam, but not totally blocking it.

On sunny days with clear skies, most of the solar radiation is direct beam radiation and diffuse radiation only accounts for about 5 to 20% of the total. Under overcast skies, diffuse radiation, which is scattered out of the direct beam by gases, aerosols, and clouds, accounts for 100% of the solar radiation reaching the earth's surface. Under clear skies, the total instantaneous solar radiation at the planet's surface at midday can exceed 1,000 W/m^2. In contrast, instantaneous midday radiation on a dark, overcast day can be less than 100 W/m^2.

The impact of clouds on the solar resource can be seen in Figure 18A.4, which shows the inverse relationship between average monthly opaque cloud cover and average daily solar energy for Fresno, California. Fresno experiences relatively little seasonal variation in atmospheric turbidity and only small variations in atmospheric water vapor. Therefore, the city's seasonal variations in average daily solar radiation are due almost entirely to variations in cloud cover (Maxwell, Marion, and Myers, 1995).

Aerosols, including dust, smoke, pollen, and suspended water droplets reduce the transmittance or amount of solar radiation reaching the planet's surface. These factors are influenced by climate and seasonal changes, as demonstrated in Figure 18A.5, which shows seasonal variations in the aerosol optical depth for six cities with different climates. If the annual mean aerosol optical depth and the amplitude of seasonal variations are known for a particular location, the aerosol optical depth for any day of the year can be estimated as follows (Maxwell and Myers, 1992):

$$\tau_A = A \sin\left[\left(360n/365\right) - \phi\right] + C \tag{18A.7}$$

where

τ_A = the aerosol optical depth on day n
A = the amplitude of seasonal variations in τ_A
n = the day of the year (1 to 365)
ϕ = the phase factor that determines the days when the maximum and minimum values occur. For most locations in the northern hemisphere, $\phi = 90°$
c = the annual mean aerosol optical depth.

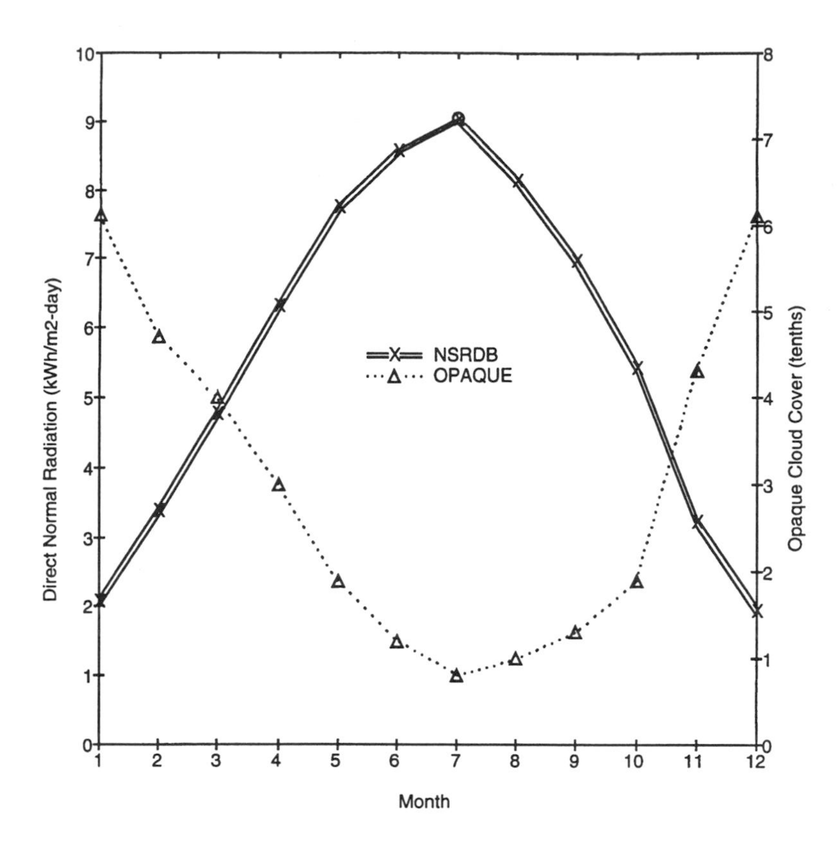

FIGURE 18A.4 Comparison of direct normal solar radiation and opaque cloud cover for Fresno, CA. (Maxwell, Marion, and Myers, 1995).

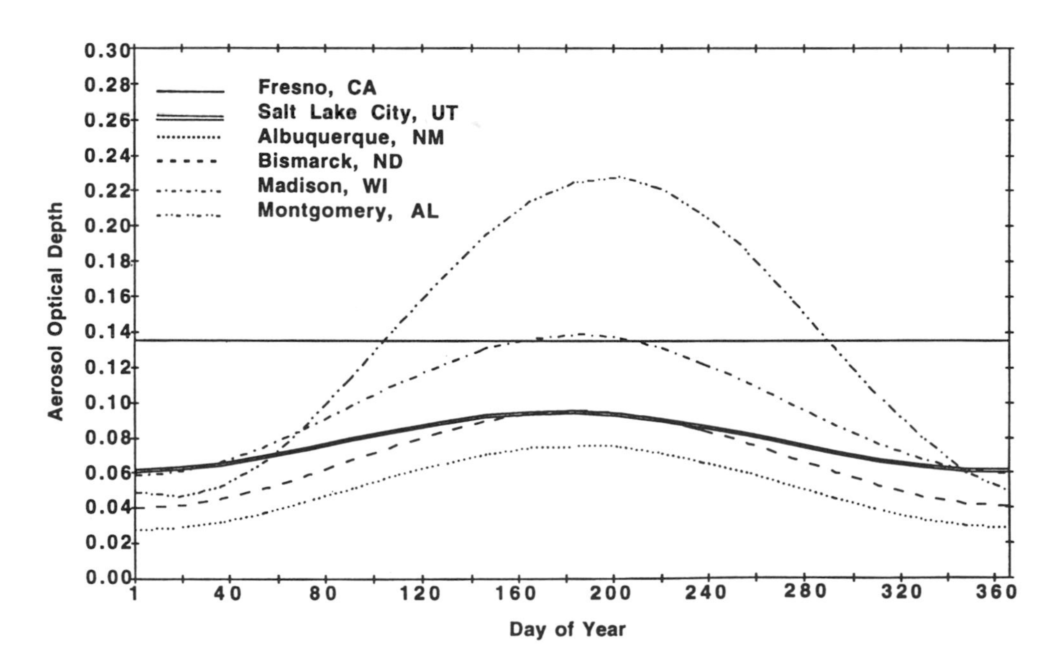

FIGURE 18A.5 Seasonal variation of aerosol optical depth for six locations in the United States. (Maxwell and Myers, 1992).

The monthly average clearness index ($\overline{K}$) is often used to quantify the relative transmittance of the atmosphere. The clearness index is (Iqbal, 1983:249)

$$\overline{K} = \frac{\overline{H}}{\overline{H}_0} \quad \textbf{(18A.8)}$$

where

$\overline{H}$ = the monthly average daily totals of the global horizontal radiation (see Section 18A.3)
$\overline{H}_0$ = the average daily total amount of extraterrestrial radiation incident upon a horizontal surface during that month

The albedo of a surface affects the total solar radiation available to a tilted collector. For example, with snow on the ground and scattered clouds overhead, surface to cloud reflections can greatly increase diffuse solar radiation. Under such circumstances, the total global radiation on the collector can increase. Any highly reflective surface material such as Utah's salt flats or light-colored sand will have similar effects. Engineers planning a solar project in highly reflective areas should measure the average albedo in the region and work with a model that incorporates these surface reflection effects.

Seasonal and diurnal changes in insolation impact resource availability. Insolation is normally greater at midday, when the path of the sun's rays through the atmosphere is shortest, than in the early morning or late afternoon, when sunlight must pass through more of the atmosphere to reach the planet's surface, as shown in Figure 18A.3. The longer the path through the atmosphere, the greater will be the absorption and rescattering of solar radiation. In the temperate and polar latitudes, daily total solar radiation is much higher during summer months because there are more daylight hours.

Some solar energy applications may require spectral information in addition to the parameters discussed earlier. The spectral distribution of extraterrestrial solar radiation, with wavelengths ranging from 0.3 μm to 3.0 μm, is shown in Figure 18A.6. Figure 18A.6 also shows an example of a similar spectral distribution after solar radiation has passed through the atmosphere and is absorbed by particulates, ozone, water vapor, and other gases. Since many photovoltaic devices are optimized for specific wavelengths in the solar spectrum, spectral information is necessary for evaluating their performance. Such information is available from "The American Society for Testing and Materials: Standard for Terrestrial Solar Spectral Irradiance Tables at Air Mass 1.5 for a 37° Tilted Surface, ASTM Standard E892-82." The Standard is found in *Annual Book of ASTM Standards,* Volume 12.02, Section 12, 1984, pp. 712–719 and is available from The American Society of Testing and Materials in Philadelphia, Pennsylvania.

Even with the best instruments and calibration methods available today, solar radiation data is no more certain than ±5% and the uncertainty is usually much higher. Solar designers should take this uncertainty into account in predicting the performance of solar energy systems.

18A.3 Solar Radiation Data for the United States

Solar designers and engineers can obtain basic information about the nation's solar resource from the National Solar Radiation Data Base (NSRDB-Vol. 1, 1992), which was developed as part of a national resource assessment project conducted by the National Renewable Energy Laboratory for the U.S. Department of Energy. A series of solar resource maps of the United States has been produced from this data base. The data base and maps are described in this section.

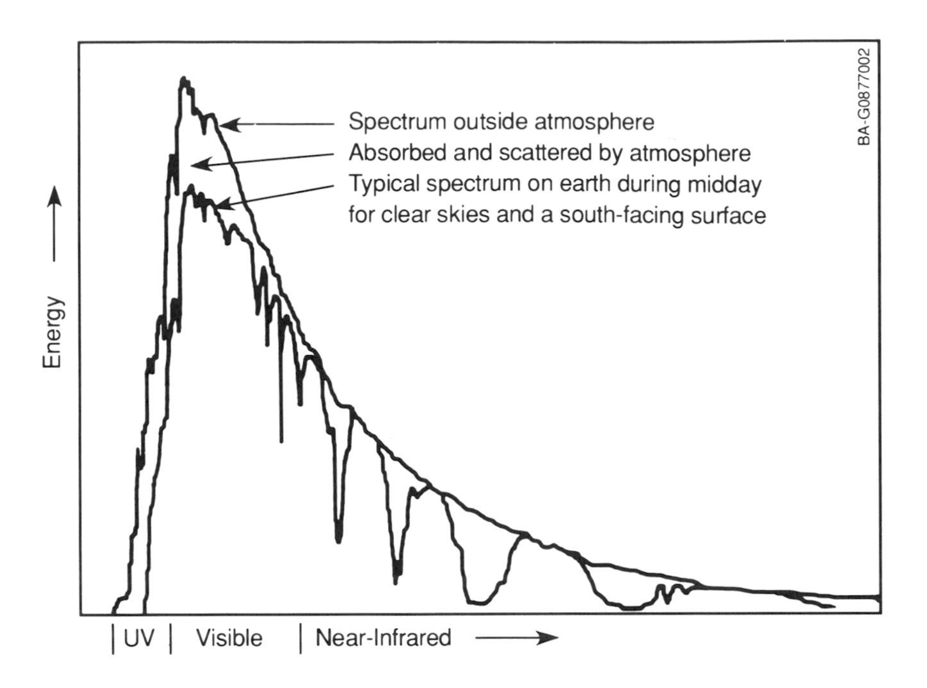

FIGURE 18A.6 Comparison of solar spectral distributions in space and on earth.

National Solar Radiation Data Base

The National Renewable Energy Laboratory completed the National Solar Radiation Data Base in 1992 to assist designers, planners, and engineers with solar resource assessments. The data base contains 30 years of hourly values of measured or modeled solar radiation together with meteorological information from 239 stations across the United States, covering the period from 1961 through 1990. Figure 18A.7 shows the locations of these stations and indicates whether each location was a primary station, which measured solar radiation data for at least 1 year, or a secondary station. Solar radiation data for secondary stations were derived from computer models. In most cases, primary and secondary stations were located at National Weather Service stations that collected hourly or three-hourly meteorological data throughout the period from 1961 to 1990. Table 18A.1 shows the solar radiation and meteorological elements included in the data base.

All data are referenced to local standard time. The solar radiation elements represent the total energy received during the hour preceding the designated time. The position of the sun in the sky and the earth's position in orbit were calculated at the midpoint of the hour preceding the designated time. Meteorological elements are the values observed at the designated time.

The most comprehensive products available from the data base are serial hourly data in either a synoptic or TD-3282 format. The synoptic format presents all solar radiation and meteorological data for 1 hour in a line of data. The next line contains data for the next hour.

In the TD-3282 format, each line of data contains a day's worth of data (24 hours) for one element. The next line contains a day of data for the next element. This format is more flexible in that it allows the user to tailor data files to a specific project's requirements. Both synoptic and TD-3282 formats present the degree of uncertainty for global horizontal, direct normal, and diffuse horizontal radiation data. They also indicate whether these elements were measured or modeled.

Other data base products include (1) hourly, daily, and quality statistics for the solar radiation elements, (2) daily statistics for the meteorological elements, and (3) persistence statistics for daily total solar radiation. The statistical summaries include hourly, monthly, annual, and 30-year

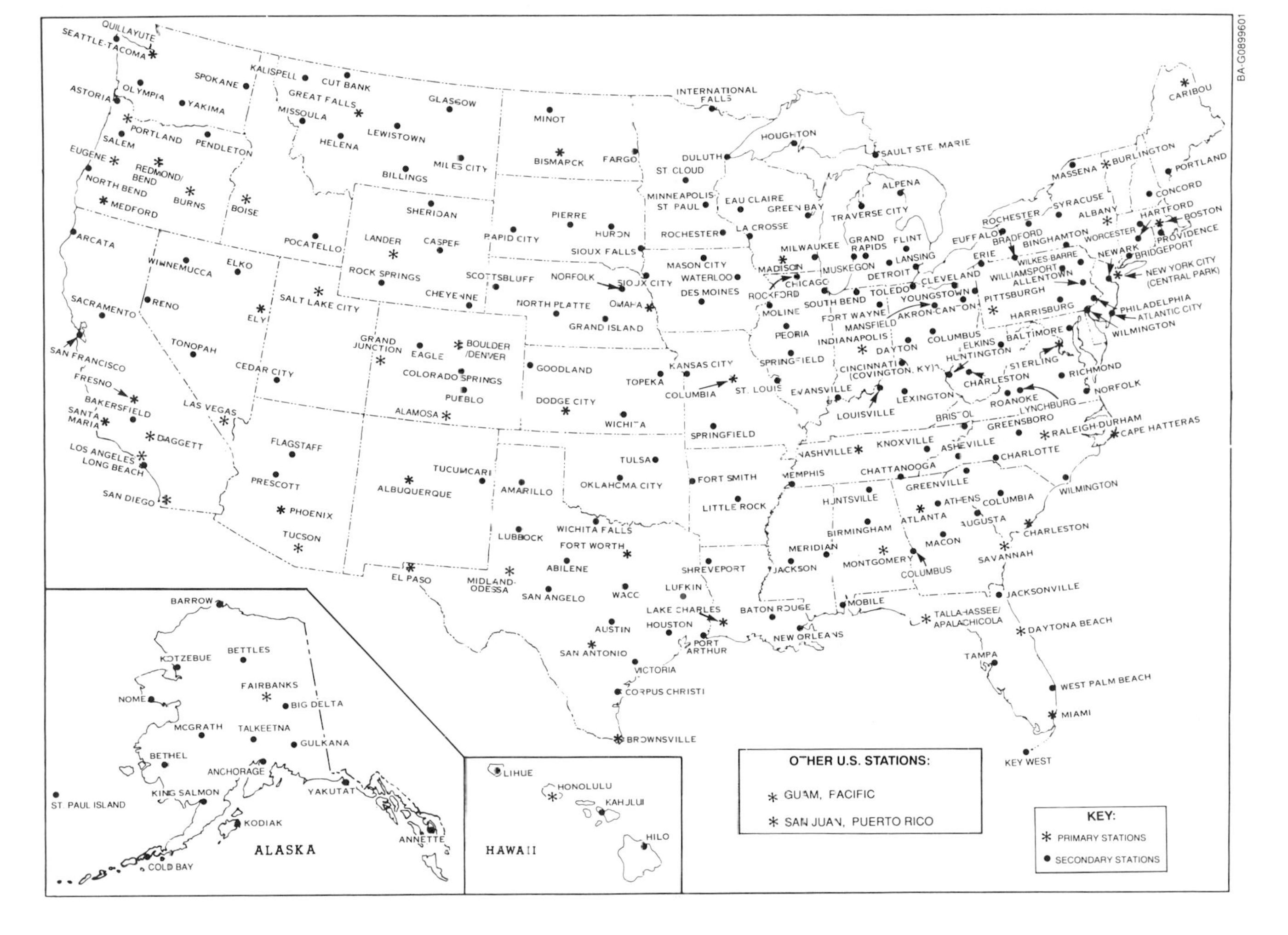

FIGURE 18A.7 Map showing the locations of 239 stations in the National Solar Radiation Data Base.

TABLE 18A.1 National Solar Radiation Data Base

Hourly Solar Radiation Elements	Meteorological Elements
Global horizontal radiation (in Wh/m^2)	Total sky cover (in tenths)
Direct normal radiation (in Wh/m^2)	Opaque sky cover (in tenths)
Diffuse horizontal radiation (in Wh/m^2)	Dry-bulb temperature (in °C)
Extraterrestrial radiation (in Wh/m^2)	Dew-point temperature (in °C)
Direct normal ETR (in Wh/m^2)	Wind direction (in degrees)
	Wind speed (in m/s)
	Horizontal visibility (in km)
	Ceiling height (in km)
	Present weather
	Total precipitable water (in cm)
	Aerosol optical depth
	Snow depth (in cm)
	Days since last snowfall

averages and their standard deviations. The persistence statistics show the number of times the daily solar energy persisted above or below set thresholds for periods from 1 to 15 days during the 30-year period. Summary characteristics of specific measurement sites are also available. The statistical products and the synoptic and TD-3282 data formats are completely described in a user's manual (NSRDB-Vol. 1, 1992).

The National Solar Radiation Data Base is available in the synoptic format on a 3-disk set of CD-ROMs from:

User Services
National Climatic Data Center
Federal Building
Asheville, NC 28801-2696
Telephone: 704-259-0682
Facsimile: 704-259-0876

The user's manual is included on the CD-ROMs, but a hard copy is also available. Other products developed from the data base are also available from the National Climatic Data Center. These products include:

- Hourly Statistics Files on Floppy Disks: Statistics include means, standard deviations, and cumulative frequency distributions of hourly solar radiation (Wh/m^2) for global, direct, and diffuse elements for each station-year-month, each station-year, and for the 30-year period.
- Daily Statistics Files on Floppy Disks: Statistics include means and standard deviations of daily-total radiation (Wh/m^2) for global, direct, and diffuse elements for each station-year-month, each station-year, and for the 30-year period. Statistics for meteorological elements include
    - Mean daily temperature in °C
    - Mean daily temperature during daylight hours in °C
    - Maximum daily temperature in °C
    - Minimum daily temperature in °C
    - Mean wind speed
    - Mean total and opaque sky cover
    - Mean total precipitable water in cm
    - Mean aerosol optical depth
    - Mean heating and cooling degree days
- Persistence Files of Daily-Total Solar Radiation on Floppy Disks.

Solar Resource Maps of the United States

The National Renewable Energy Laboratory is using the data base to develop solar radiation resource maps of the United States. The maps are included in a solar atlas scheduled for publication by NREL in 1996.

Solar resource maps of the United States present solar radiation throughout the year on (1) horizontal surfaces and (2) tilted surfaces, as shown in Figures 18A.8 to 18A.14. Figure 18A.8(a) shows the total solar radiation on a horizontal surface for the continental United States, whereas Figure 18A.8(b) shows the total solar radiation on a south-facing surface tilted at an angle equal to the latitude. A comparison of these two maps clearly shows that the collectible portion of the annual solar resource can be increased over much of the nation by using collectors mounted on a tilted surface.

Figures 18A.9 to 18A.14 show a bimonthly comparison between radiation on a horizontal surface and radiation on a tilted surface throughout the year. Examination of these maps shows that south-facing, tilted solar collectors harness more of the solar resource during those times of the year when the sun is close to the equator or south of it. During the late spring and early summer months in the northern hemisphere, when the sun is relatively far north of the equator, collectors on a horizontal surface receive more solar radiation than those tilted at latitude.

Figure 18A.15 shows the annual direct normal, or beam, solar radiation available to a concentrating dish collector that tracks the sun. The large solar resource in the southwestern United States reflects the relatively cloud-free, desert climate typical of the region. By comparing Figures 18A.15 and 18A.8(b), the relative contributions of direct beam and diffuse radiation to the total global resource available to a south-facing, tilted surface can be estimated.

The maps included in this section are spatial interpolations based on values obtained from the National Solar Radiation Data Base. Though useful for identifying general trends, they should be used with caution for site-specific resource evaluations because variations in insolation not reflected in the maps can exist, introducing uncertainty into resource estimates.

The Renewable Resource Data Center

Many of the data base product described in this chapter can be accessed via the Internet from the Renewable Resource Data Center (RReDC) located at the National Renewable Energy Laboratory. The data center contains information developed by the Laboratory's Resource Assessment Program and allows users to comment on the program's products and services. It is accessible through the Laboratory's home page at http://www.nrel.gov or directly at http://rredc.nrel.gov.

18A.4 Solar Design Tools

Since producing the solar radiation data base, the National Renewable Energy Laboratory has developed new design tools to assist people interested in using the data base. This section describes new typical meteorological year data sets (called TMY2s) and a new solar radiation data manual. The TMY2s present a single, representative year of resource data for each of the 239 stations in the data base. The data manual customizes solar radiation values for common types of solar collectors, including fixed flat-plate and concentrating.

The section also presents a basic set of equations required to determine solar radiation on a horizontal surface and then translate this information into data for tilted surfaces.

Typical Meteorological Years 2

A typical meteorological year (TMY) is a data set of hourly values of solar radiation and meteorological elements for a 1-year period. A TMY consists of months selected from different individual

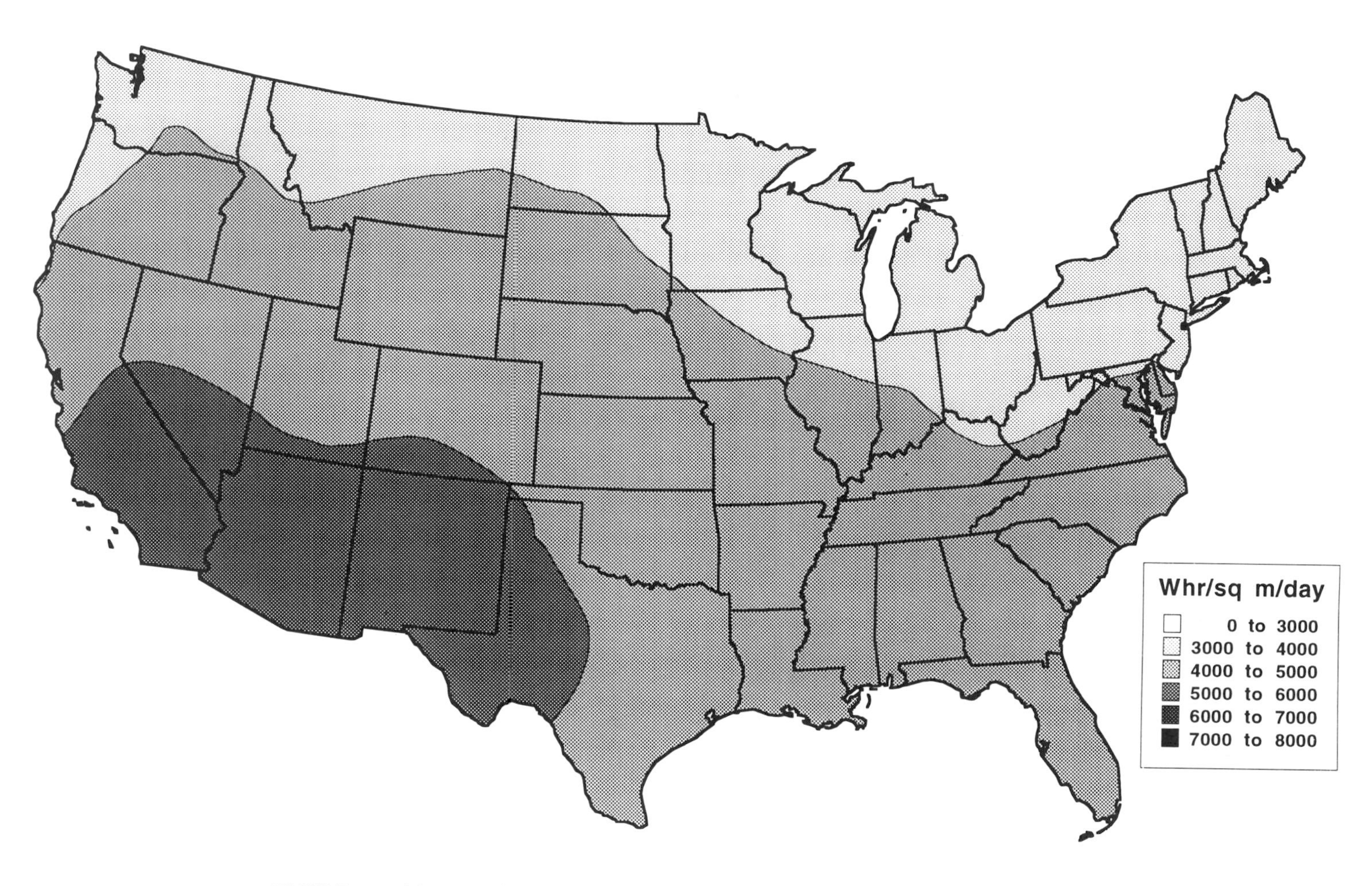

FIGURE 18A.8(a) Annual average daily total global horizontal solar radiation in the United States.

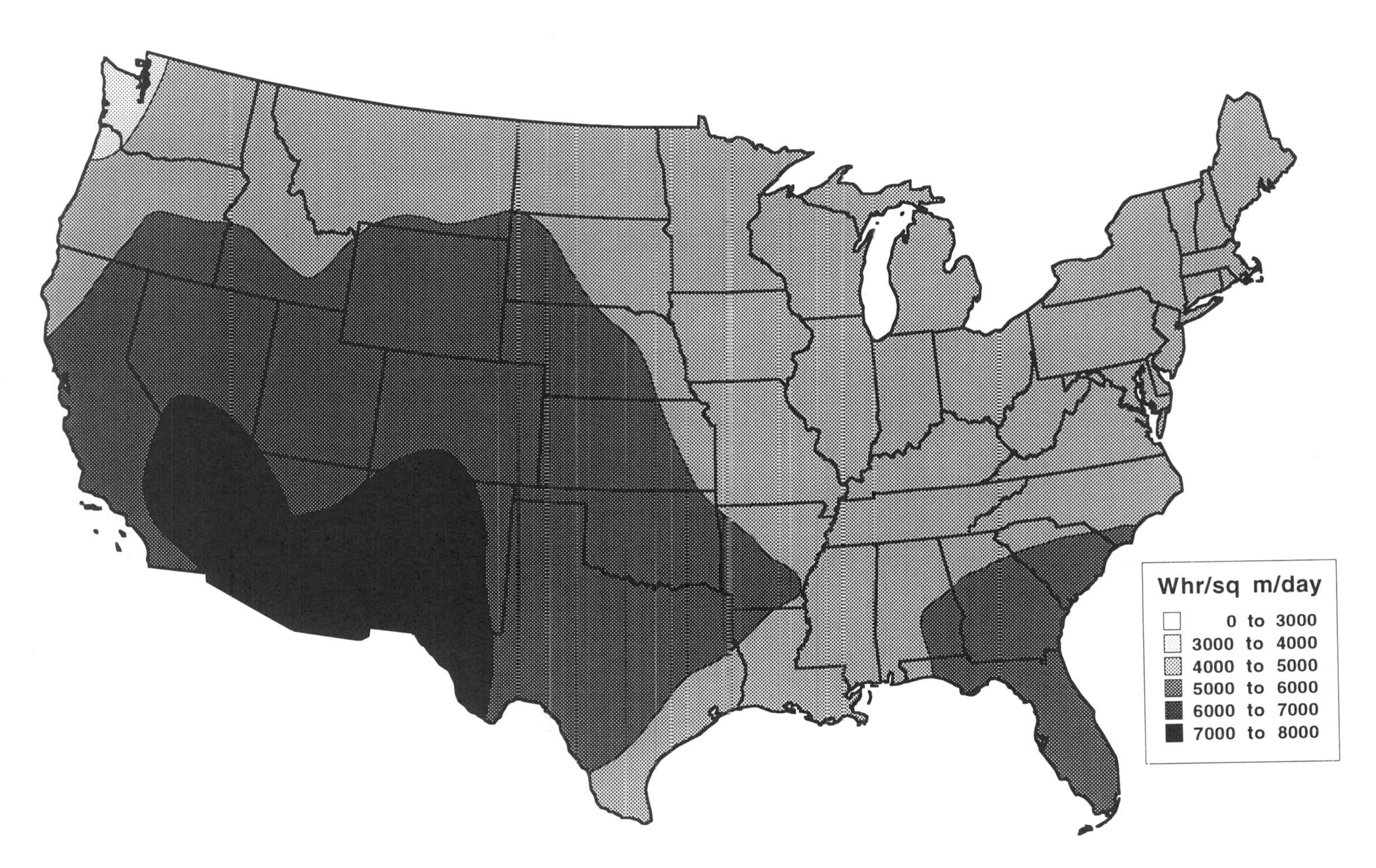

FIGURE 18A.8(b) Annual average daily total global solar radiation on a south-facing surface with tilt equal to latitude.

FIGURE 18A.9(a) Average daily total global horizontal solar radiation during January in the United States.

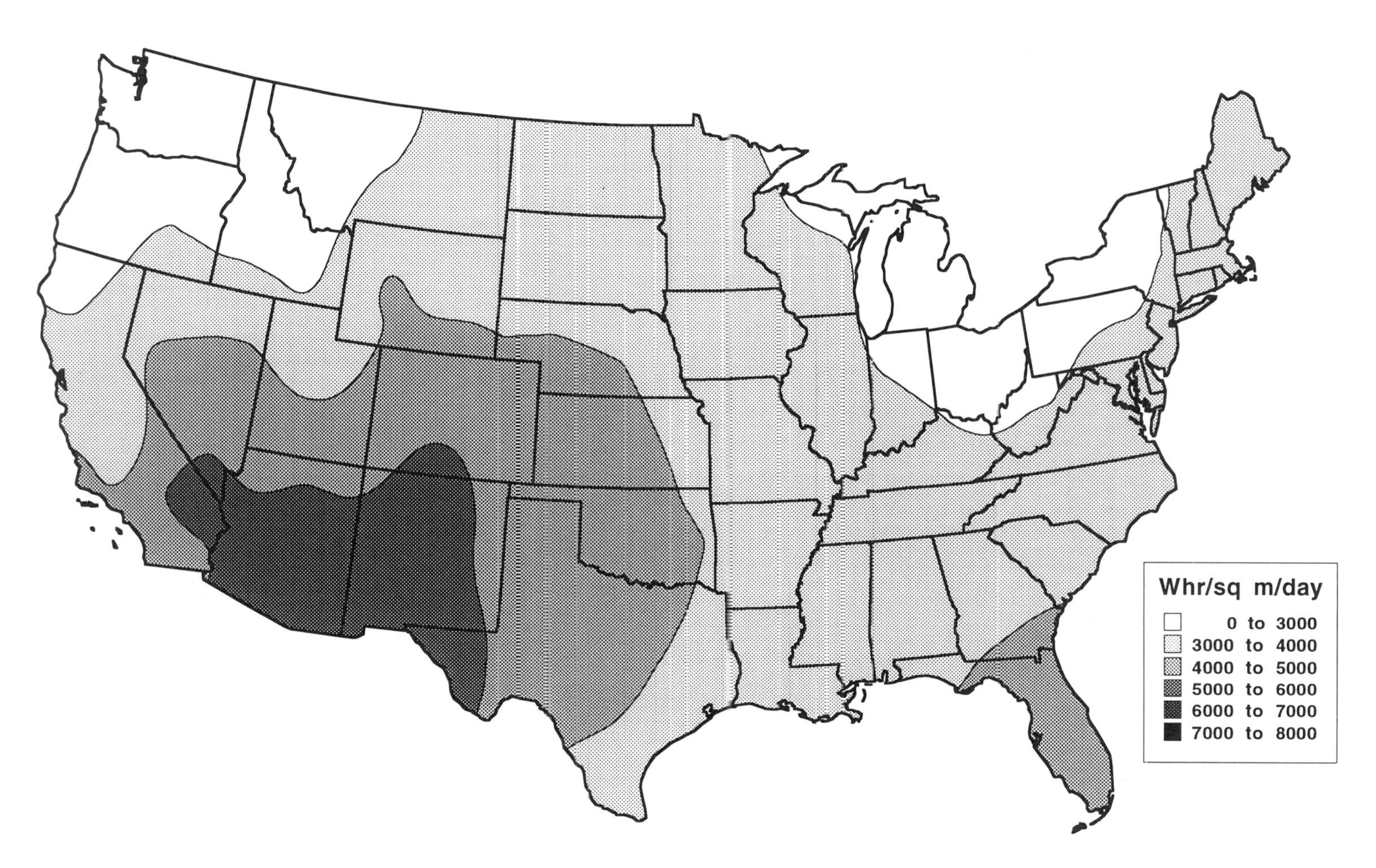

FIGURE 18A.9(b) Average daily total global solar radiation during January on a south-facing surface, with tilt equal to latitude.

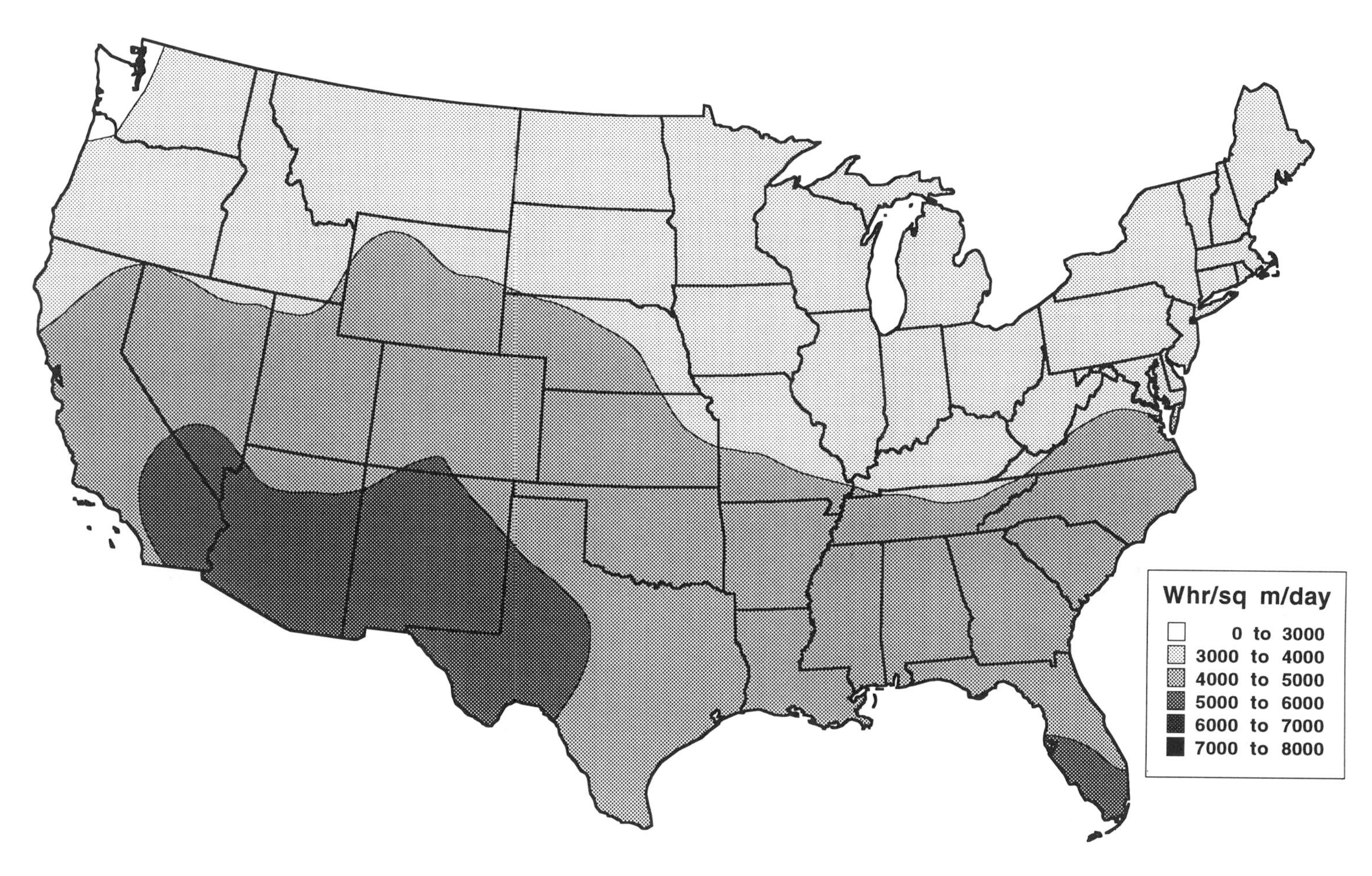

FIGURE 18A.10(a) Average daily total global horizontal solar radiation during March in the United States.

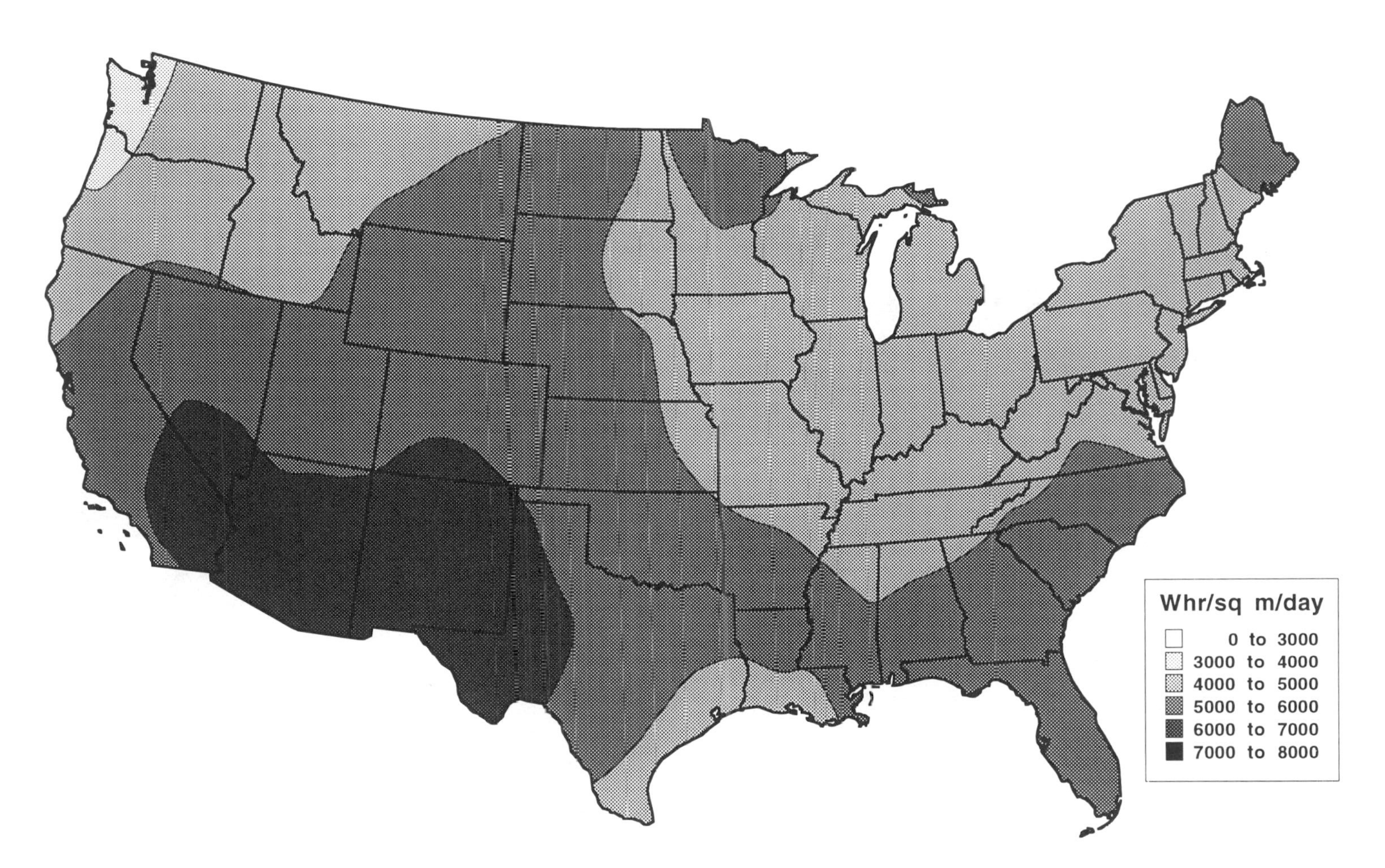

FIGURE 18A.10(b) Average daily total global solar radiation during March on a south-facing surface, with tilt equal to latitude.

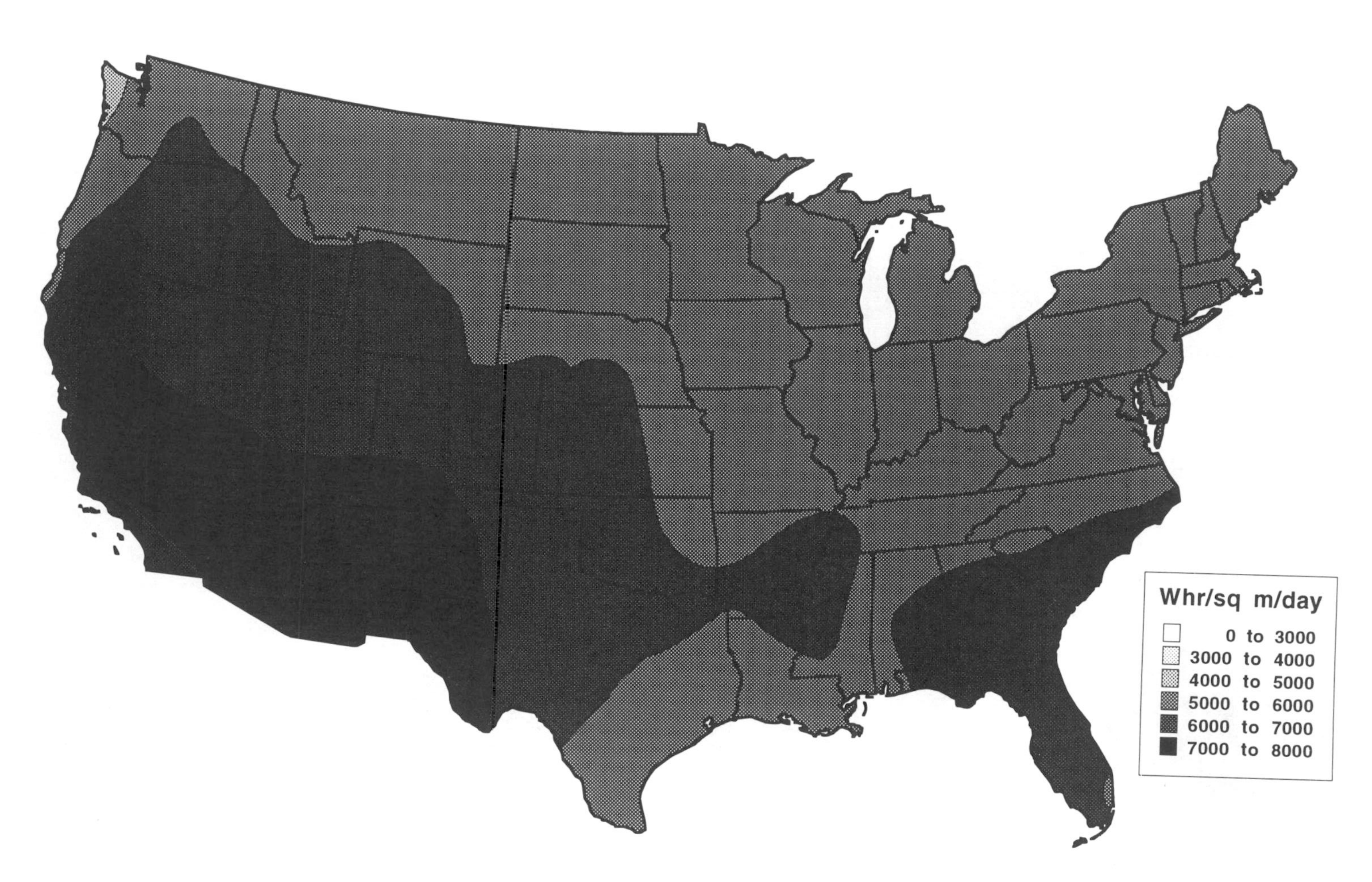

FIGURE 18A.11(a) Average daily total global horizontal solar radiation during May in the United States.

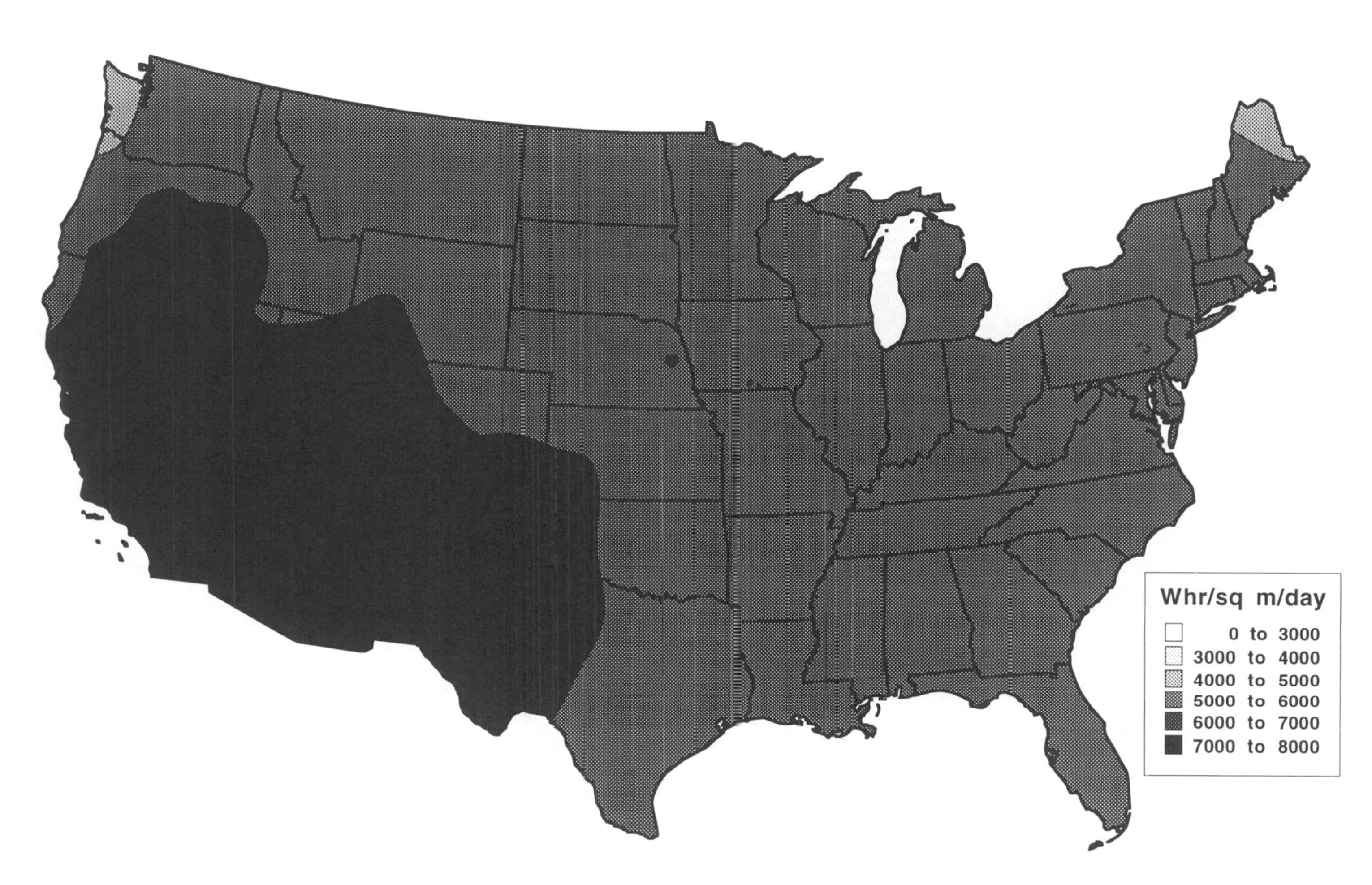

FIGURE 18A.11(b) Average daily total global solar radiation during May on a south-facing surface, with tilt equal to latitude.

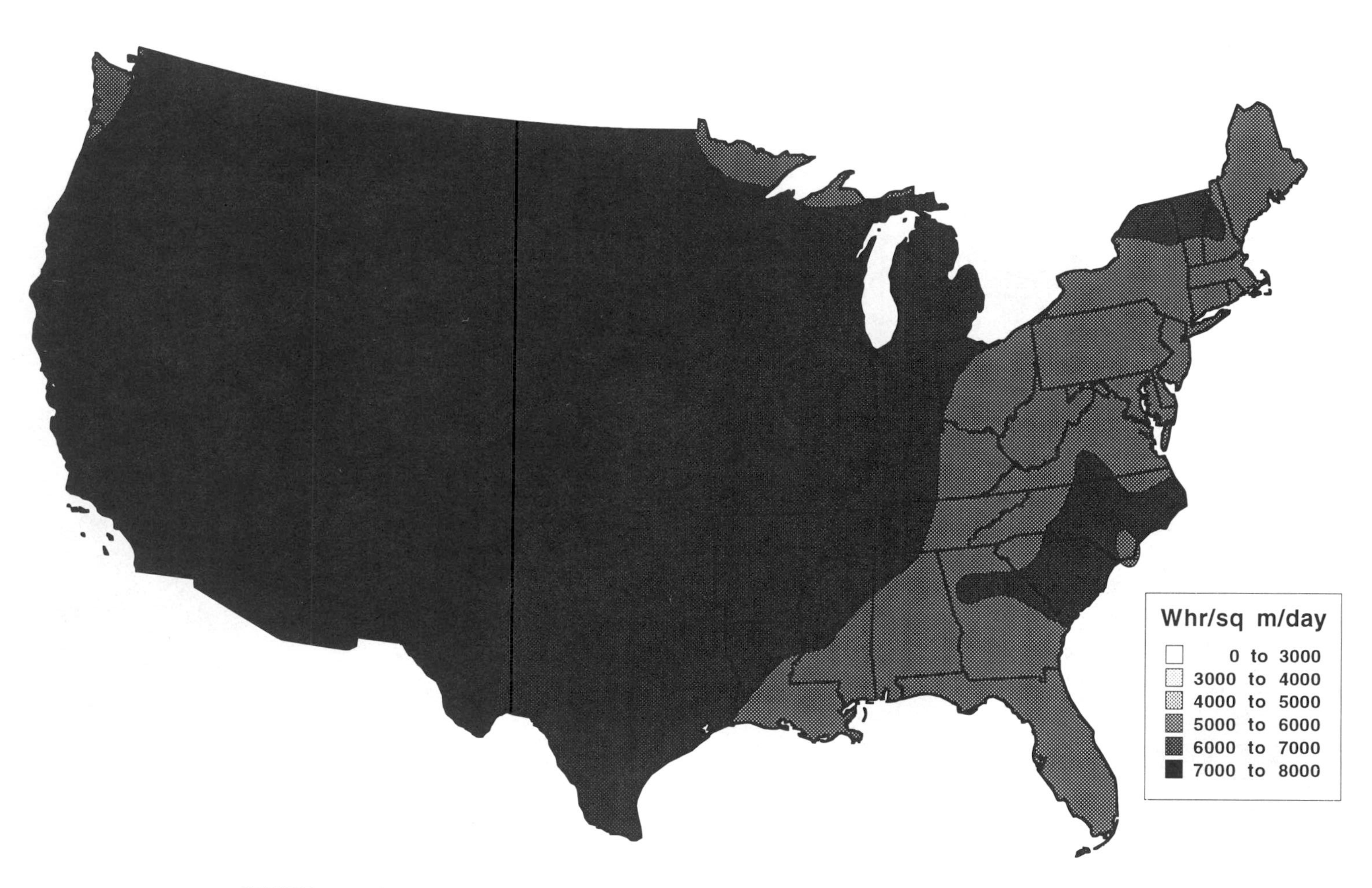

FIGURE 18A.12(a) Average daily total global horizontal solar radiation during July in the United States.

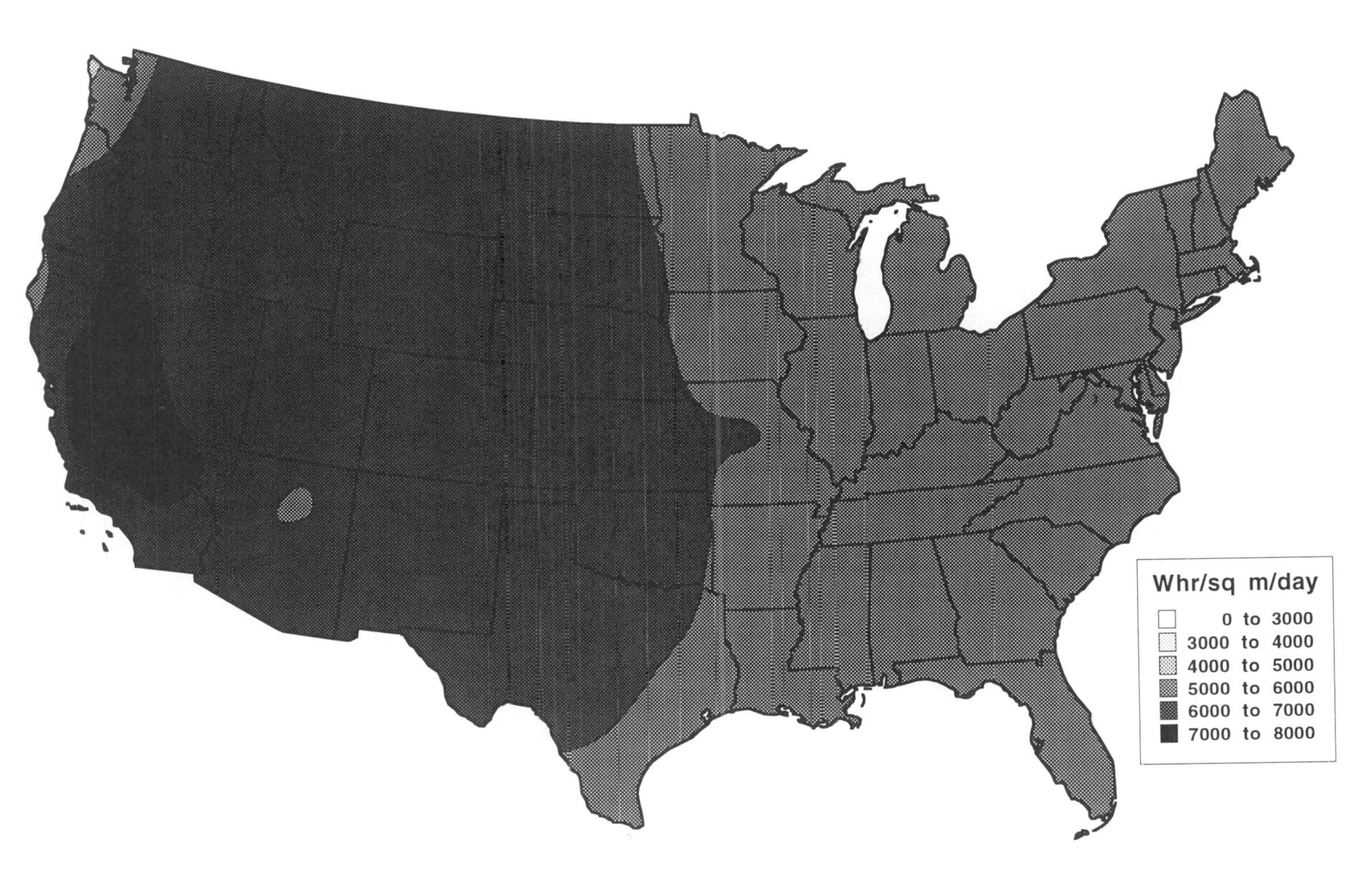

FIGURE 18A.12(b) Average daily total global solar radiation during July on a south-facing surface, with tilt equal to latitude.

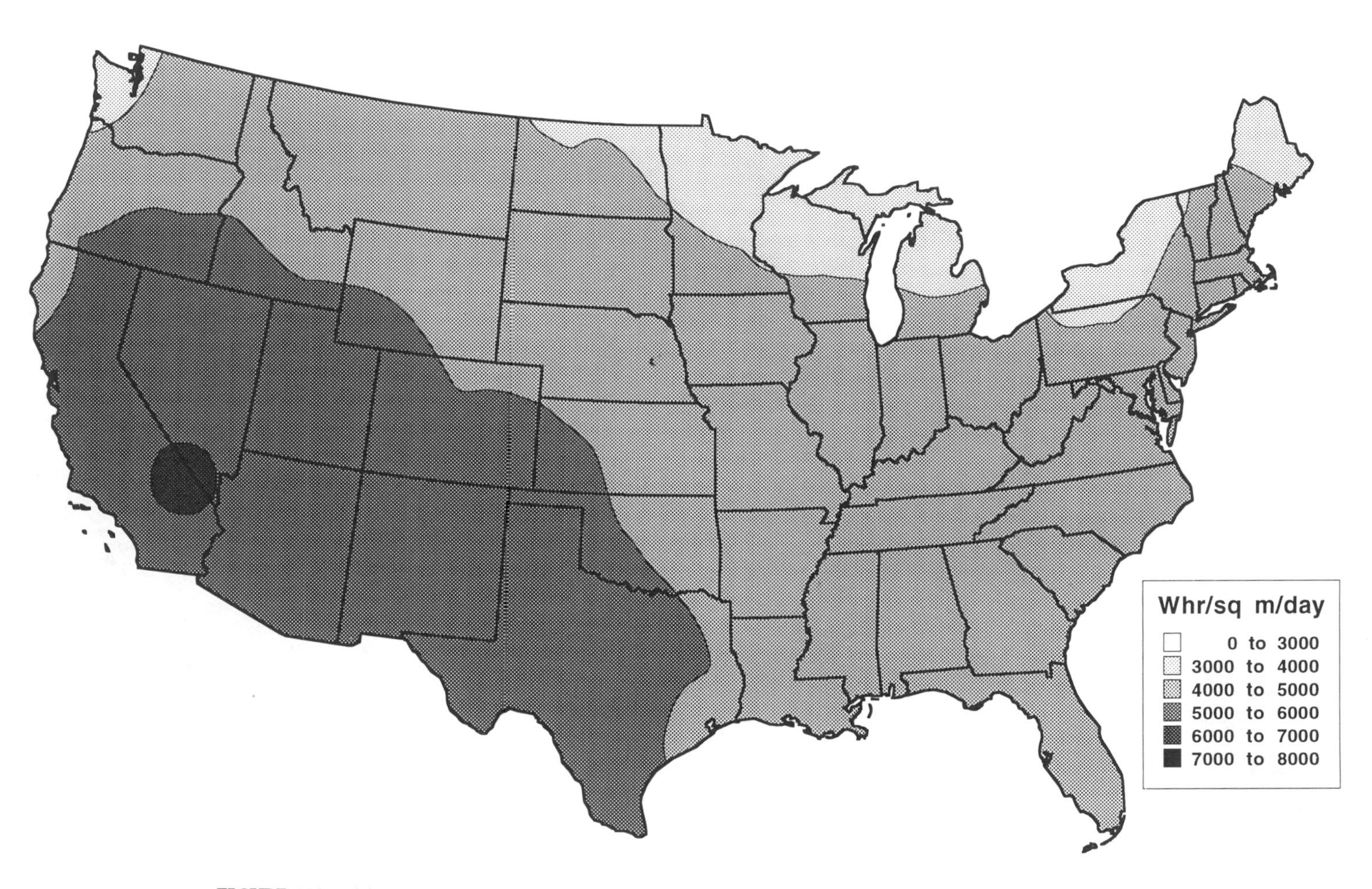

FIGURE 18A.13(a) Average daily total global horizontal solar radiation during September in the United States.

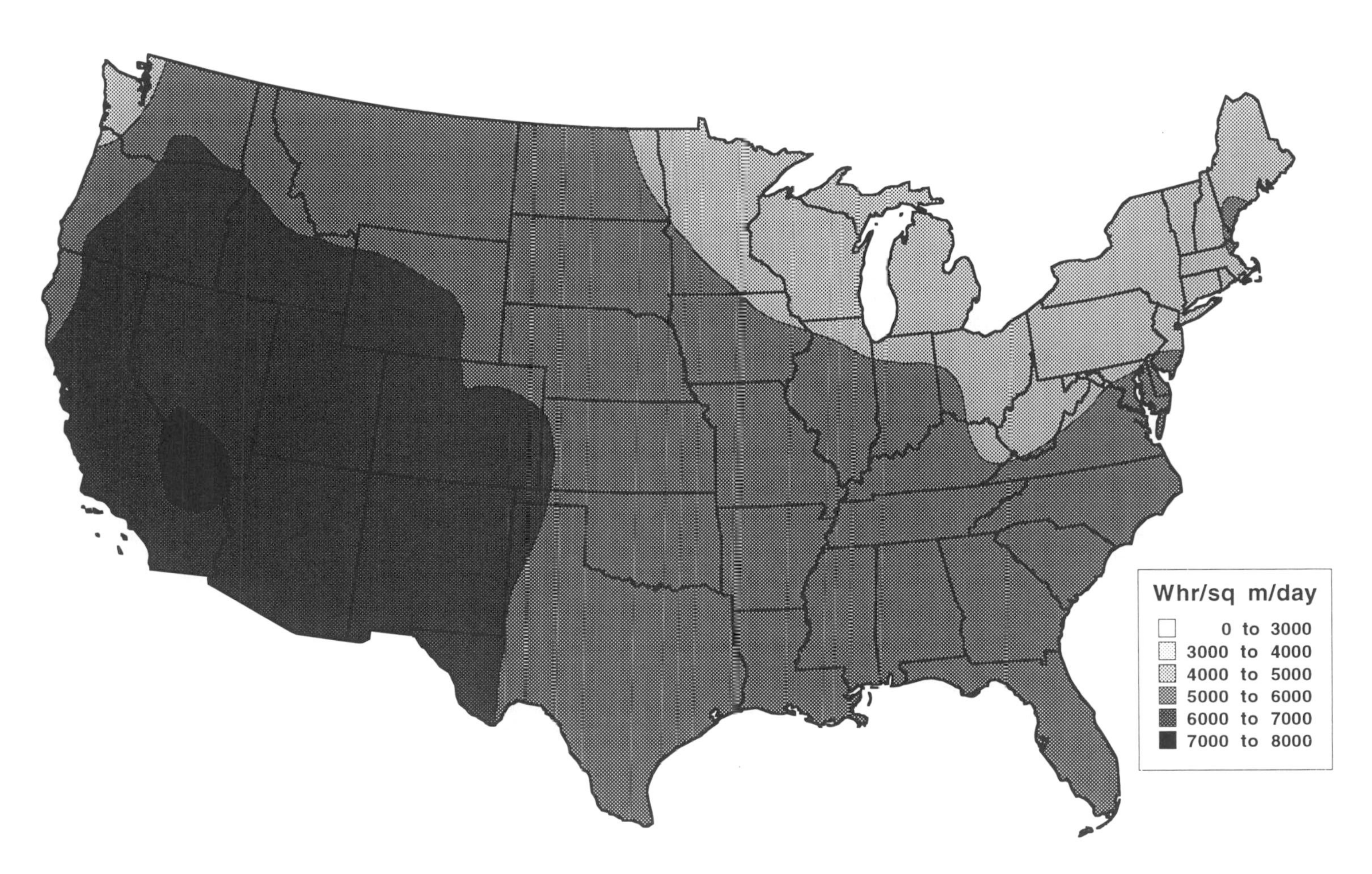

FIGURE 18A.13(b) Average daily total global sclar radiation during September on a south-facing surface, with tilt equal to latitude.

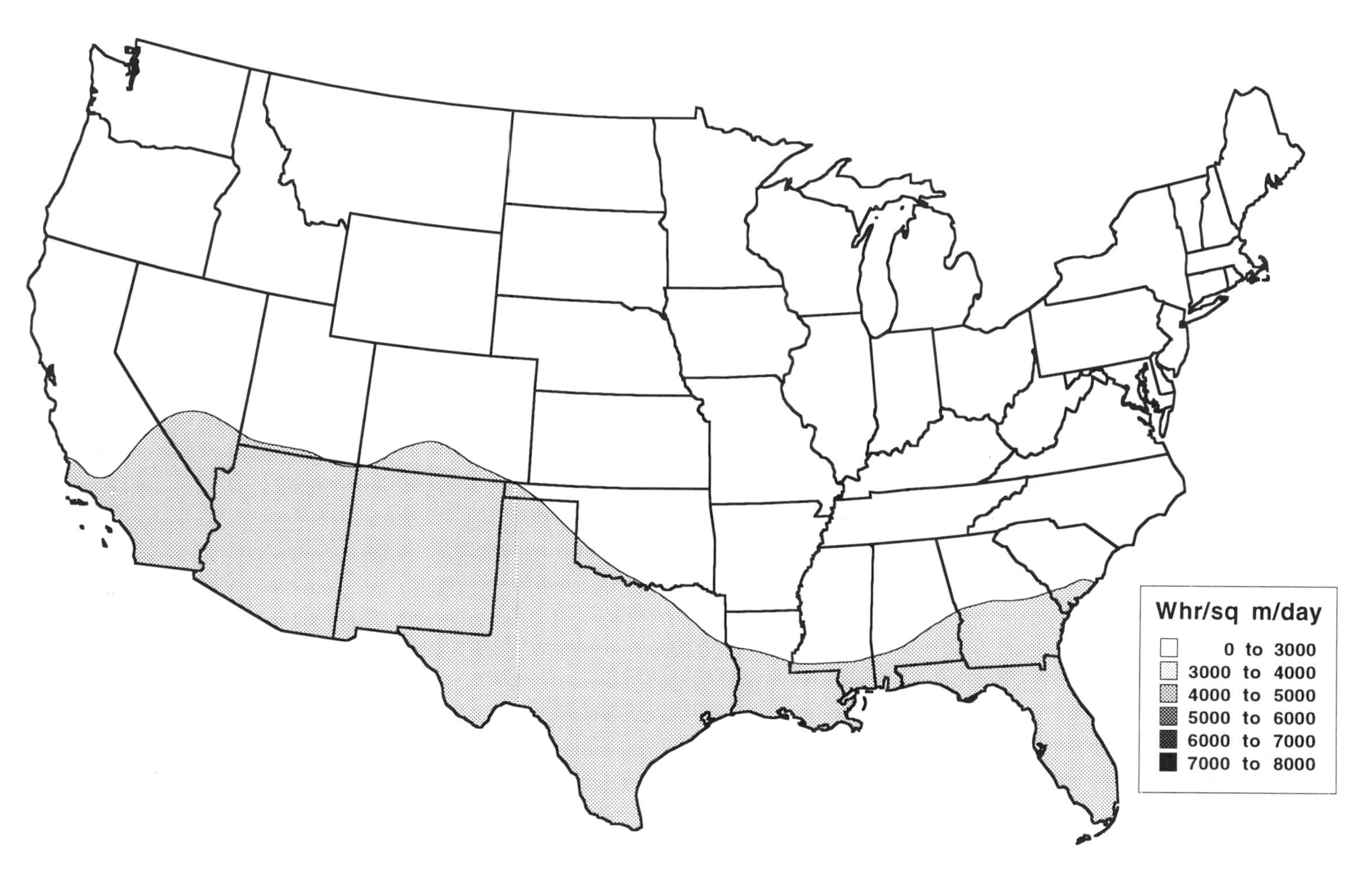

FIGURE 18A.14(a) Average daily total global horizontal solar radiation during November in the United States.

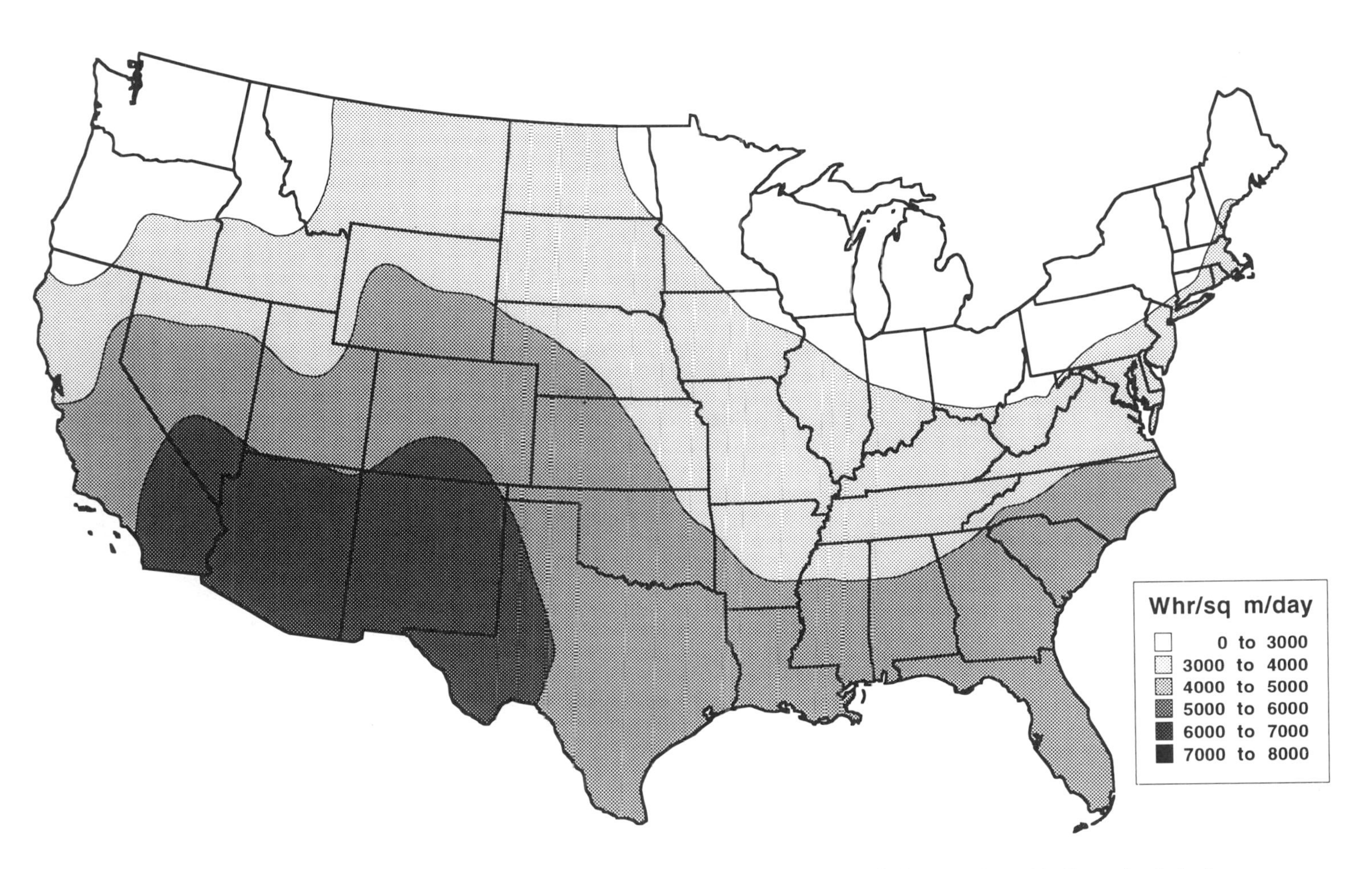

FIGURE 18A.14(b) Average daily total global solar radiation during November on a south-facing surface with tilt equal to latitude.

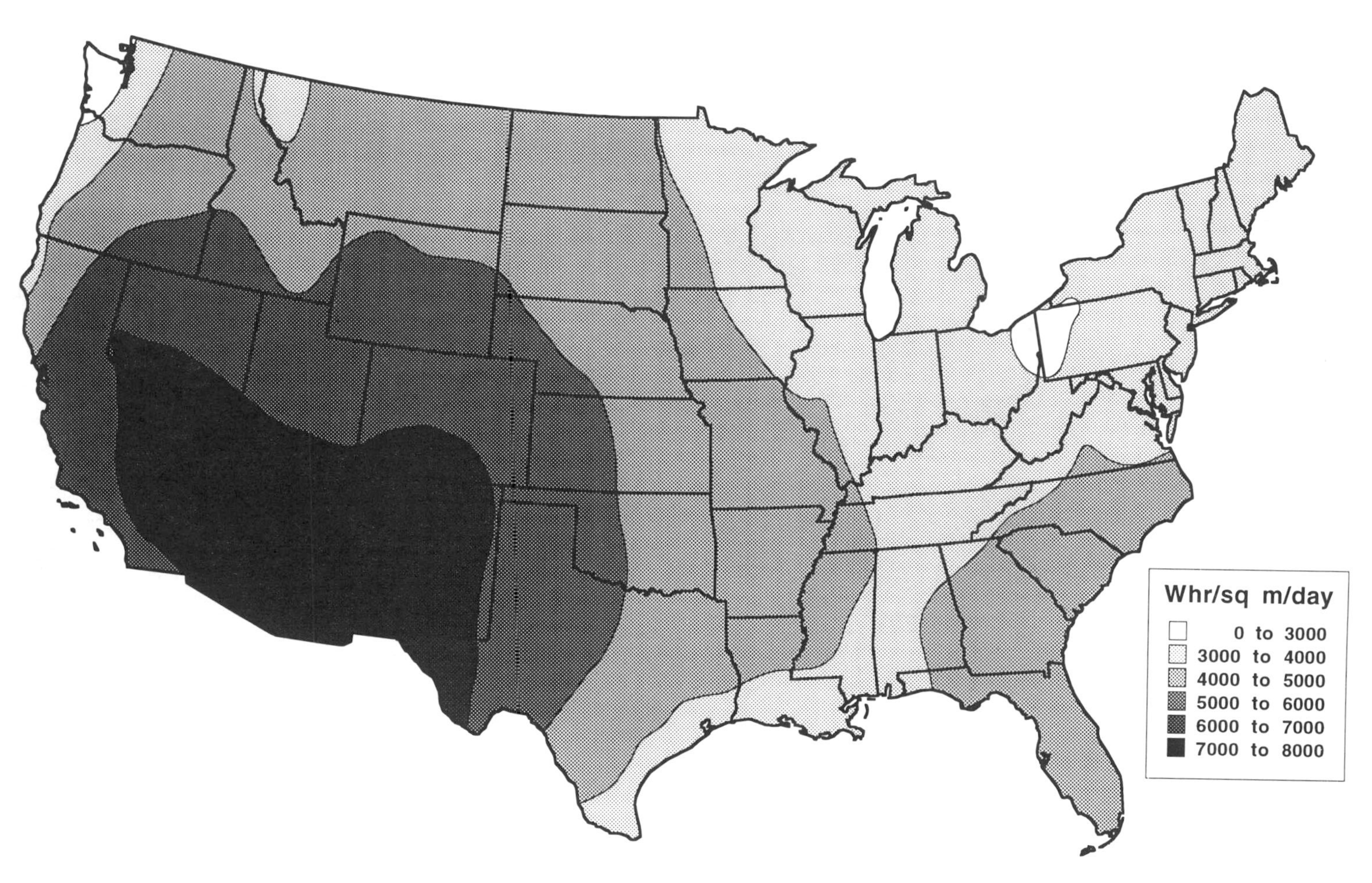

FIGURE 18A.15 Annual average daily total direct beam solar radiation in the United States.

years and linked together to form a complete year. For example, if 30 years of data are available, all 30 Januarys would be examined and the one judged most typical would be selected for inclusion in the TMY. The remaining months in the TMY would be chosen in a similar fashion, with particular attention paid to five statistical elements: global horizontal radiation, direct normal radiation, dry bulb temperature, dew point temperature, and wind speed.

The TMY2 data sets were derived from Version 1.1 of the National Solar Radiation Data Base, and are designated as TMY2s to distinguish them from an earlier TMY developed in the late 1970s for the earlier SOLMET/ERSATZ solar radiation data base.

Designers and planners often find such a representative data set useful in predicting the performance and economic viability of a solar energy system. For instance, the TMY2s were designed for computer simulations of solar energy conversion and building systems. They provide a standard for hourly data for solar radiation and other meteorological elements that supports performance comparisons of system types and configurations for one or more locations.

A TMY is not specific to any particular year and should not be used as an indicator of actual conditions expected over the next year, or even 5 years. Rather, a TMY represents typical conditions over a long time period such as 30 years. Because TMY data represent typical rather than extreme conditions at a specific location, they should be used with caution for component and system design. System hardware should be designed to withstand a location's worst case conditions, rather than typical conditions.

The TMY2 data sets are available via the Internet from the National Renewable Energy Laboratory's Renewable Resource Data Center (RreDC) at http://rredc.nrel.gov. TMY2 data sets for all 239 stations may also be obtained on a CD-ROM from

NREL Document Distribution Service
1617 Cole Boulevard
Golden, CO 80401-3393
Telephone: 303-275-4363
Facsimile: 303-275-4053
Internet: sally_evans@nrel.gov

A printed user's manual (Marion and Urban, 1995) is also available from the document distribution center.

Solar Radiation Data Manual

The *Solar Radiation Data Manual for Flat-Plate and Concentrating Collectors* (Marion and Wilcox, 1994), published by the National Renewable Energy Laboratory, provides information on the solar resource for 239 stations in the United States and its territories. Researchers developed the manual's data from the hourly values of direct beam and diffuse horizontal solar radiation values in the National Solar Radiation Data Base.

The manual converts raw data regarding insolation into a form directly useful to solar system designers and engineers. It contains estimates of the resource available to flat-plate collectors mounted at fixed tilt angles of 0°, latitude minus 15°, latitude, latitude plus 15°, and 90°. The manual also shows solar irradiation for flat-plate and concentrating collectors, including parabolic trough designs, with 1-axis tracking and 2-axis tracking. Figure 18A.16 shows drawings of 5 collector types for which resource estimates were made.

Data include monthly and hourly averages for 14 solar collectors for the period of 1961 to 1990. Data tables for each station are presented on a single page, which also contains a graph highlighting the variability of the solar resource at that particular station. Figure 18A.17 shows a sample data page from the manual for Fresno, California. Each data page also contains a table listing climatic conditions such as average temperatures, average daily minimum and maximum temperatures,

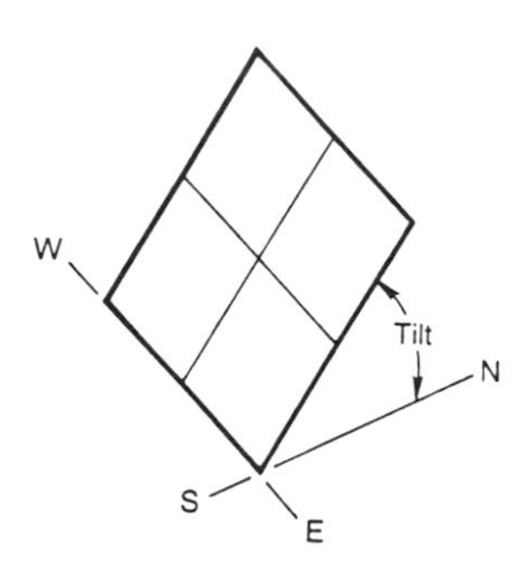

Flat-plate collector facing south at fixed tilt

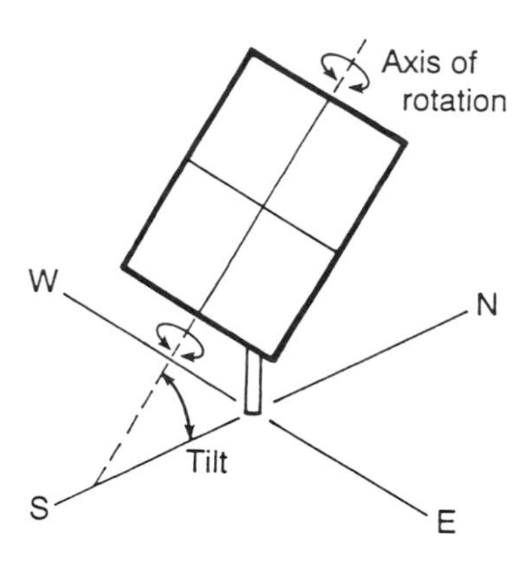

One-axis tracking flat-plate collector with axis oriented north-south

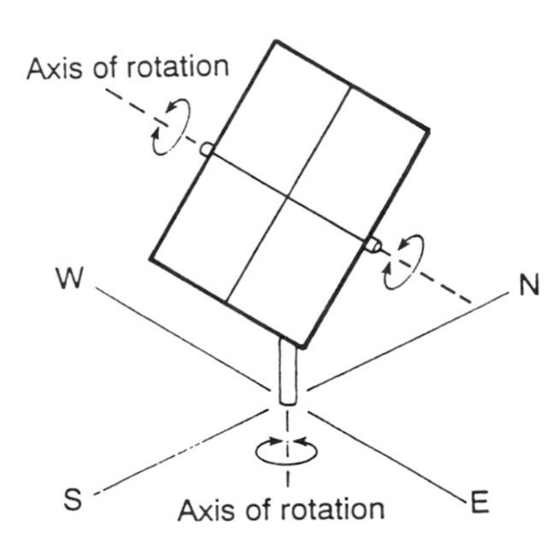

Two-axis tracking flat-plate collector

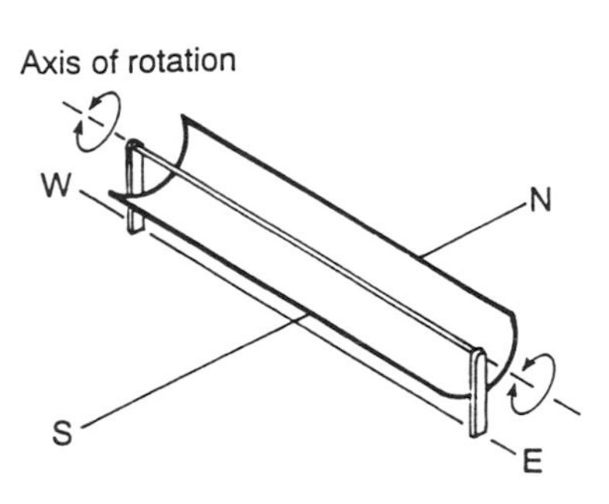

One-axis tracking parabolic trough with axis oriented east-west

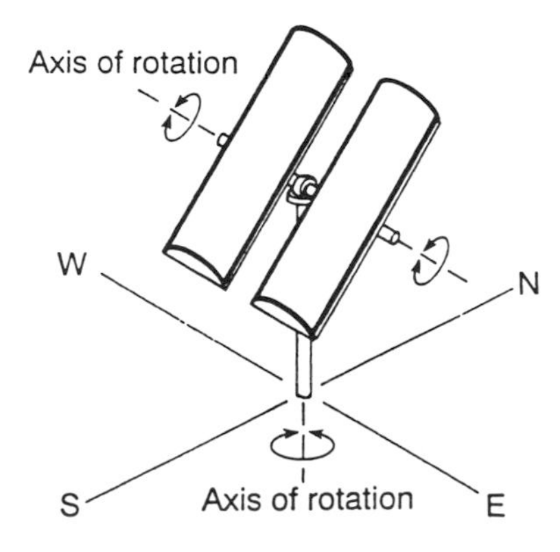

Two-axis tracking concentrator

FIGURE 18A.16 Flat plate and concentrating solar collector configurations.

record minimum and maximum temperatures, average heating and cooling degree days, average relative humidity, and average wind speed.

The manual is organized alphabetically by state and, within each state, alphabetically by station. It explains how to interpret data in the tables and discusses the methods used to determine the

Fresno, CA

WBAN NO. 93193

LATITUDE: 36.77° N
LONGITUDE: 119.72° W
ELEVATION: 100 meters
MEAN PRESSURE: 1004 millibars

STATION TYPE: Primary

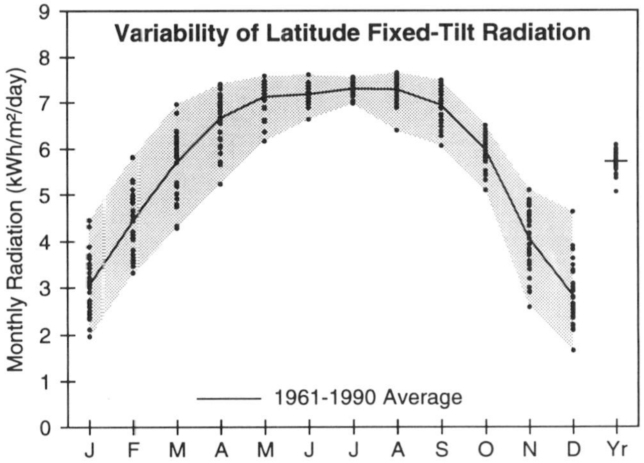

Variability of Latitude Fixed-Tilt Radiation

Monthly Radiation (kWh/m²/day)

——— 1961-1990 Average

J F M A M J J A S O N D Yr

Solar Radiation for Flat-Plate Collectors Facing South at a Fixed Tilt (kWh/m²/day), Uncertainty ±9%

Tilt (°)		Jan	Feb	Mar	Apr	May	June	July	Aug	Sept	Oct	Nov	Dec	Year
0	Average	2.1	3.2	4.7	6.3	7.5	8.1	8.0	7.2	5.9	4.3	2.7	1.9	5.2
	Min/Max	1.7/2.7	2.5/3.9	3.7/5.5	5.1/7.0	6.6/8.0	7.4/8.7	7.6/8.4	6.3/7.6	5.2/6.3	3.9/4.6	2.0/3.2	1.4/2.6	4.7/5.4
Latitude -15	Average	2.8	4.1	5.5	6.8	7.6	7.8	7.9	7.5	6.8	5.5	3.6	2.5	5.7
	Min/Max	1.9/3.9	3.1/5.3	4.2/6.6	5.4/7.5	6.6/8.1	7.2/8.3	7.5/8.2	6.6/7.9	5.9/7.3	4.8/6.0	2.4/4.5	1.6/3.9	5.1/6.0
Latitude	Average	3.1	4.4	5.7	6.7	7.1	7.2	7.3	7.3	6.9	6.0	4.1	2.8	5.7
	Min/Max	2.0/4.5	3.3/5.8	4.3/6.9	5.2/7.4	6.2/7.6	6.6/7.6	7.0/7.6	6.4/7.7	6.1/7.5	5.1/6.5	2.6/5.1	1.7/4.6	5.1/6.1
Latitude +15	Average	3.2	4.5	5.6	6.2	6.3	6.1	6.3	6.6	6.7	6.1	4.2	3.0	5.4
	Min/Max	2.0/4.8	3.3/6.1	4.1/6.9	4.8/6.9	5.4/6.7	5.7/6.4	6.0/6.5	5.8/6.9	5.8/7.2	5.1/6.6	2.6/5.4	1.7/5.0	4.7/5.8
90	Average	2.8	3.7	4.0	3.6	3.0	2.6	2.7	3.5	4.4	4.8	3.7	2.6	3.4
	Min/Max	1.6/4.3	2.6/5.1	2.9/4.9	2.9/4.0	2.7/3.2	2.5/2.8	2.6/2.9	3.2/3.6	3.8/4.8	3.9/5.3	2.2/4.8	1.4/4.7	2.9/3.7

FIGURE 18A.17(a) Monthly average daily total solar radiation data for flat plate and concentrating collectors located in Fresno, California (Marion and Wilcox, 1994).

Solar Radiation for 1-Axis Tracking Flat-Plate Collectors with a North-South Axis (kWh/m²/day), Uncertainty ±9%

Axis Tilt (°)		Jan	Feb	Mar	Apr	May	June	July	Aug	Sept	Oct	Nov	Dec	Year
0	Average Min/Max	2.7 1.8/3.9	4.3 3.3/5.7	6.4 4.5/8.0	8.5 6.5/10.0	10.2 8.7/11.3	11.0 9.8/12.2	11.2 10.1/11.8	10.1 8.0/10.9	8.3 7.1/9.3	6.2 5.0/6.8	3.7 2.3/4.7	2.4 1.5/3.9	7.1 6.0/7.6
Latitude -15	Average Min/Max	3.2 2.0/4.8	5.0 3.7/6.7	7.0 4.9/8.9	9.0 6.8/10.5	10.4 8.8/11.5	10.9 9.8/12.1	11.2 10.2/11.8	10.4 8.3/11.3	9.1 7.7/10.0	7.1 5.7/7.8	4.4 2.6/5.6	2.9 1.6/4.9	7.5 6.3/8.1
Latitude	Average Min/Max	3.4 2.1/5.2	5.2 3.9/7.1	7.2 4.9/9.1	8.9 6.7/10.5	10.1 8.5/11.1	10.5 9.3/11.6	10.8 9.8/11.4	10.3 8.1/11.1	9.2 7.8/10.2	7.4 6.0/8.2	4.7 2.8/6.1	3.1 1.7/5.4	7.6 6.3/8.2
Latitude +15	Average Min/Max	3.5 2.0/5.5	5.3 4.0/7.3	7.1 4.8/9.0	8.6 6.4/10.1	9.5 8.0/10.5	9.8 8.7/10.9	10.1 9.1/10.7	9.8 7.7/10.6	9.0 7.6/10.0	7.5 6.0/8.3	4.9 2.8/6.4	3.2 1.7/5.8	7.4 6.0/8.0

Solar Radiation for 2-Axis Tracking Flat-Plate Collectors (kWh/m²/day), Uncertainty ±9%

Tracker		Jan	Feb	Mar	Apr	May	June	July	Aug	Sept	Oct	Nov	Dec	Year
2-Axis	Average Min/Max	3.6 2.1/5.5	5.3 4.0/7.3	7.2 4.9/9.1	9.0 6.8/10.6	10.5 8.9/11.6	11.2 9.9/12.4	11.4 10.3/12.1	10.5 8.3/11.4	9.2 7.8/10.2	7.5 6.0/8.3	4.9 2.8/6.4	3.3 1.7/5.8	7.8 6.5/8.4

Direct Beam Solar Radiation for Concentrating Collectors (kWh/m²/day), Uncertainty ±8%

Tracker		Jan	Feb	Mar	Apr	May	June	July	Aug	Sept	Oct	Nov	Dec	Year
1-Axis, E-W Horiz Axis	Average Min/Max	1.8 0.5/3.4	2.7 1.7/4.4	3.6 2.0/5.2	4.6 2.7/5.9	5.7 4.4/6.8	6.4 4.6/7.4	6.6 5.7/7.3	5.9 4.2/6.7	5.1 4.0/5.9	4.3 2.9/5.0	2.7 1.2/4.0	1.7 0.5/3.8	4.3 3.2/4.8
1-Axis, N-S Horiz Axis	Average Min/Max	1.3 0.4/2.6	2.5 1.5/4.1	4.1 2.1/6.0	5.9 3.6/7.9	7.6 5.8/9.0	8.4 5.8/10.0	8.8 7.2/9.9	7.8 5.2/9.0	6.3 4.8/7.4	4.3 2.8/5.2	2.2 0.9/3.3	1.2 0.4/2.7	5.1 3.7/5.8
1-Axis, N-S Tilt=Latitude	Average Min/Max	1.9 0.6/3.7	3.3 2.0/5.3	4.7 2.5/6.9	6.2 3.8/8.2	7.3 5.7/8.8	7.9 5.4/9.4	8.4 6.8/9.4	7.9 5.3/9.1	6.9 5.3/8.1	5.4 3.5/6.3	3.1 1.3/4.6	1.8 0.5/4.0	5.4 3.9/6.2
2-Axis	Average Min/Max	2.1 0.6/4.0	3.4 2.0/5.4	4.8 2.5/6.9	6.3 3.8/8.4	7.8 6.0/9.3	8.6 5.9/10.2	9.0 7.3/10.1	8.1 5.4/9.4	7.0 5.3/8.2	5.4 3.5/6.4	3.2 1.3/4.8	1.9 0.6/4.3	5.6 4.1/6.4

Average Climatic Conditions

Element	Jan	Feb	Mar	Apr	May	June	July	Aug	Sept	Oct	Nov	Dec	Year
Temperature (°C)	7.6	10.7	12.8	16.2	20.6	24.8	27.7	26.8	23.6	18.4	12.0	7.4	17.4
Daily Minimum Temp Daily Maximum Temp	3.0 12.3	4.7 16.5	6.3 19.2	8.5 23.9	12.1 29.0	15.8 33.7	18.4 37.0	17.7 35.9	14.9 32.3	10.4 26.5	5.8 18.2	2.8 12.1	10.1 24.7
Record Minimum Temp Record Maximum Temp	-7.2 25.6	-4.4 26.7	-3.3 32.2	0.0 37.8	2.2 41.7	6.7 43.3	10.0 44.4	9.4 43.9	2.8 43.9	-2.8 38.9	-3.3 31.7	-7.8 24.4	-7.8 44.4
HDD, Base 18.3°C CDD, Base 18.3°C	332 0	214 0	174 4	101 38	19 88	0 196	0 291	0 263	4 163	47 51	190 0	338 0	1420 1093
Relative Humidity (%) Wind Speed (m/s)	83 2.2	77 2.5	69 3.0	57 3.4	47 3.8	42 3.8	39 3.4	45 3.1	50 2.8	58 2.3	74 2.1	84 2.1	60 2.9

FIGURE 18A.17(b)

data values. Manual data are also available on floppy disks in ASCII format, which can be imported into a spreadsheet program. Manual and disks are available from

NREL Document Distribution Service
1617 Cole Boulevard
Golden, CO 80401-3393
Telephone: 303-275-4363
Facsimile: 303-275-4053
Internet: sally_evans@nrel.gov

Estimating Daily and Hourly Radiation from Monthly Averages

The only extant solar radiation data for developing countries is often monthly averages of daily total global horizontal radiation. Given sufficient time and money, an engineer could conduct a multi-year measurement program. Lacking such resources and time, a statistical correlation approach such as that developed by Reddy (1987) can be used. This approach assumes that correlations between monthly averages and daily and hourly distributions at locations having measured data can be used to derive daily and hourly data for locations exhibiting similar clearness indices. This section summarizes Reddy's approach.

Horizontal Surfaces

The first step in estimating daily and hourly values of solar radiation on horizontal surfaces is the calculation of the clearness index (K) using available monthly average daily total global horizontal data and Eq. (18A.8). $\overline{K}$ can be used to predict the monthly average daily total horizontal diffuse radiation ($\overline{H}_d$). The monthly average daily total horizontal beam radiation can then be obtained from the global solar radiation by subtraction.

The correlation between the clearness index and the average daily total horizontal diffuse radiation takes account of the effect of seasonal variation by including the sunset hour angle, W_s. The correlation for a clearness index in the range of $0.3 \leq \overline{K} \leq 0.8$ is

$$\frac{\overline{H}_d}{\overline{H}} = \begin{cases} 1.391 - 3.560\ \overline{K} + 4.189\ \overline{K}^2 - 2.137\ \overline{K}^3 & \text{for } W_S \leq 81.4° \\ 1.311 - 3.022\ \overline{K} + 3.427\ \overline{K}^2 - 1.821\ \overline{K}^3 & \text{for } W_S > 81.4° \end{cases} \quad \textbf{(18A.9)}$$

where W_s = the monthly mean value of the sunset hour angle in degrees on a horizontal surface.

When subjected to statistical analysis, different locations with the same mean monthly clearness index, $\overline{K}$, often display cumulative frequency curves that are more or less identical. This gives rise to the long-term implication that the monthly average distribution of the daily total global horizontal radiation may be independent of location and month. This relationship is most consistent for temperate regions. Thus, if one assumes that the mean monthly clearness index, $\overline{K}$, is known and that the random variable K will also have known minimum and maximum values, then a generalized probability function can be represented as follows:

$$P(K) = C\left[\frac{K_{max} - K}{K_{max}}\right]^n e^{\gamma K} \quad \text{for } K_{min} \leq K \leq K_{max} \quad \textbf{(18A.10a)}$$

where C and γ can be calculated from

$$\int_{K_{min}}^{K_{max}} P(K)dK = 1 \ \ and \ \ \int_{K_{min}}^{K_{max}} KP(K)dK = \overline{K} \quad \textbf{(18A.10b)}$$

and K_{max} can be calculated as follows:

$$K_{max} = 0.6313 + 0.267\,\overline{K} - 11.9\left(\overline{K} - 0.75\right)^8 \tag{18A.10c}$$

The monthly average daily total diffuse radiation, H_d, can be predicted from the daily total horizontal global radiation and the clearness index using the following relationship, which includes the sunset angle W_s to account for seasonal variation. For $W_s < 81.4°$,

$$\frac{H_d}{H} = 1.0 - 0.2727\,K + 2.449\,K^2 - 11.9541\,K^3 + 9.3879\,K^4 \quad \text{for } K < 0.715 \tag{18A.11a}$$

and

$$\frac{H_d}{H} = 0.143 \quad \text{for } K \geq 0.715 \tag{18A.11b}$$

For $W_s \geq 81.4°$,

$$\frac{H_d}{H} = 1.0 + 0.2832\,K - 2.5557\,K^2 + 0.8448\,K^3 \quad \text{for } K < 0.722 \tag{18A.11c}$$

and

$$\frac{H_d}{H} - 0.175 \quad \text{for } K \geq 0.722 \tag{18A.11d}$$

The hourly horizontal global radiation can also be predicted from daily total horizontal global radiation using a statistical correlation. The correlation is possible if average diurnal solar radiation is assumed to be symmetrical about solar noon. In the relationship, r is defined as the ratio of the monthly mean hourly global radiation to the monthly mean daily global radiation on a horizontal surface (i.e., $r = \overline{I}/\overline{H}$). The statistical correlation is expressed as

$$r(W) = \frac{\pi}{24}(a + b\cos W)\left[\frac{(\cos W - \cos W_s)}{\left(\sin W_s - \frac{\pi}{180} W_s \cos W_s\right)}\right] \tag{18A.12}$$

where

a = 0.409 + 0.5016 sin(W_s – 60)
b = 0.6609 – 0.4767 sin(W_s – 60)
W_s = the sunset hour angle
W = the hour angle (in degrees) corresponding to the midpoint of the hour

The correlation is valid for long-term averages and may be used for determining monthly mean values. It can also be applied to individual days so long as the days are clear. However, curve fitting is typically better when monthly time scales are employed.

Similarly, it is possible to predict hourly horizontal diffuse radiation. In this case, r_d is defined as the ratio of the monthly mean hourly diffuse radiation to the monthly mean daily total diffuse radiation on a horizontal surface (i.e., $r_d = \bar{I}_d/\bar{H}_d$). The corresponding relationship is given by

$$r_d(W) = \frac{\pi}{24}\left[\frac{(\cos W - \cos W_s)}{\left(\sin W_s - \frac{\pi}{180} W_s \cos W_s\right)}\right] \tag{18A.13}$$

This correlation can also be applied to clear days, but works better for monthly time scales.

If the daily total global radiation on a horizontal surface *(H)* is known, then the hourly diffuse radiation (I_d) can be estimated by first calculating the average global diffuse radiation (H_d) from Eq. (18A.11) and then using Eq. (18A.13). Alternatively, one can determine the hourly diffuse radiation (I_d) by calculating the hourly global radiation *(I)* from Eq. (18A.12) and then using the following relationships:

$$\frac{I_d}{I} = 1.0 - 0.09\,K \quad \text{for } K \le 0.22 \tag{18A.14a}$$

$$\frac{I_d}{I} = 0.9511 - 0.1604\,K + 4.388\,K^2 - 16.638\,K^3 + 12.336\,K^4 \quad \text{for } 0.22 < K \le 0.8 \tag{18A.14b}$$

$$\frac{I_d}{I} = 0.165 \quad \text{for } K > 0.8 \tag{18A.14c}$$

where K is the hourly clearness index = I/I_0.

Eq. (18A.14) should give the most consistent results because the interdependence between the hourly global radiation and the hourly diffuse radiation is generally greater than that between the hourly diffuse radiation and the average global diffuse radiation.

Tilted Surfaces

Radiation incident on a tilted surface is composed of beam radiation, diffuse radiation from the sky, and radiation reflected from the ground. An estimate of the beam radiation on a tilted surface can be obtained by first subtracting the diffuse radiation on a horizontal surface from the global horizontal radiation and then multiplying the resulting number by a geometrical conversion factor. Diffuse and reflected radiation values with respect to the tilted surfaces are obtained by multiplying their horizontal values by the fraction of the total hemisphere intersected by the inclined plane.

The conversion factor for hourly global radiation (I_T) on a surface with a tilt angle β is

$$I_T = (I - I_d) r_{b,T} + I_d\left(\frac{1 + \cos\beta}{2}\right) + I\rho\left(\frac{1 - \cos\beta}{2}\right) \tag{18A.15}$$

where

ρ = the surface albedo
β = the tilt angle of the inclined surface with respect to the horizontal
$r_{b,T}$ = the ratio of hourly beam radiation on the tilted surface to that on a horizontal surface

Similarly, the daily total global radiation (H_T) on a tilted surface can be determined as follows:

$$H_T = (H - H_d)R_{b,T} + H_d\left(\frac{1+\cos\beta}{2}\right) + H\rho\left(\frac{1-\cos\beta}{2}\right) \quad \textbf{(18A.16)}$$

where $R_{b,T}$ = the ratio of the daily beam radiation on the tilted surface to that on the horizontal surface and ρ and β are as previously.

18A.5 International Solar Radiation Data

The sparsity of solar radiation measurements in some countries has led to the use of satellite data for estimating solar radiation. An international solar radiation data base from a satellite-derived surface radiation budget developed by the National Aeronautics and Space Administration's Langley Research Center has recently become available. The data set uses a model to convert satellite images to daily total or monthly averages of the daily total surface solar radiation estimates with a grid resolution of 280 km. The data set covering the period from 1985 to 1988 is currently available on CD-ROM from

The Langley DAAC
User and Data Services Office
NASA Langley Research Center, Mail Stop 157B
Hampton, VA 23681-0001
Telephone: 804-864-8656
Facsimile: 804-864-8807
e-mail: userserv@eosdis.larc.nasa.gov

Historically, one of the most commonly used ground-based international data sets has been one developed by the University of Massachusetts at Lowell. In this set, many published reports were consolidated into a data base of monthly average daily total global horizontal solar radiation. The major source for this information has been the conversion of sunshine duration records to global horizontal estimates. The data base is available from

The University of Lowell Research Foundation
450 Aiken Street
Lowell, MA 01854

The World Radiation Data Center (WRDC) in St. Petersburg, Russia, is another well-known international source of solar radiation measurements. The center was established in 1964 under the auspices of the World Meteorological Organization. Since then, it has received summarized solar radiation data from as many as 1,250 stations throughout the world. Most of the measurement data consist of daily total global horizontal radiation, but some measurements of direct beam and diffuse radiation are also included.

The center's role has been to assess the quality of data sent in by different countries and to archive it. For many years, the center has produced quarterly reports providing data summaries from each station submitting data. With the assistance of the U.S. Department of Energy and the National Renewable Energy Laboratory, the center has recently upgraded its computer system. As a result, its archive is becoming available on the Internet at the following address: http://wrdc-mgo.nrel.gov.

References

"The American Society for Testing and Materials: Standard for Terrestrial Solar Spectral Irradiance Tables at Air Mass 1.5 for a 37° Tilted Surface, ASTM Standard E892-82" in *Annual Book of ASTM Standards,* Volume 12.02, Section 12, 1984, pp. 72–719, The American Society of Testing and Materials, Philadelphia, PA.

Cooper, P.I., The Absorption of Solar Radiation on Solar Stills, *Solar Energy,* 12(3), 333–346, 1969.

Duffie, J.A. and Beckman, W.A., *Solar Engineering of Thermal Processes,* Wiley, New York, 1980.

Iqbal, Muhammad, *An Introduction to Solar Radiation,* Academic Press, Toronto, New York, 1983.

Kreider, Jan F. and Kreith, Frank, *Solar Energy Handbook,* McGraw-Hill, New York, 1981.

Maxwell, E.L. and Myers, D.R., Daily Estimates of Aerosol Optical Depth for Solar Radiation Models, *Proc. 1992 Annu. Conf. American Solar Energy Soc.,* June 1992, Cocoa Beach, FL, 323–327.

Maxwell, E.L., Marion, W., and Myers, D.R., Comparisons of NSRDB and SOLMER Data, in NSRDB-Vol. 2, *National Solar Radiation Data Base (1961–1990): Final Technical Report,* Ch. 10, pp. 216–260, National Renewable Energy Laboratory, Golden, CO, 1995.

Marion, William and Wilcox, Stephen, *Solar Radiation Data Manual for Flat-Plate and Concentrating Collectors,* National Renewable Energy Laboratory, Golden, CO. NREL/TP-463-5607. (Available from: NREL Document Distribution Service, 1617 Cole Boulevard, Golden, CO 80401-3393) 1994.

Marion, William and Urban, Ken, *User's Manual for TMY2s,* Derived from the 1961–1990 National Solar Radiation Data Base, National Renewable Energy Laboratory, Golden, CO. NREL/SP-463-7668 1995.

NSRDB-Vol. 1, National Solar Radiation Data Base (1961–1990): *User's Manual,* National Renewable Energy Laboratory, Golden, CO and National Climatic Data Center, Asheville, NC, 1992

Reddy, T. Agami, *The Design and Sizing of Active Solar Thermal Systems,* Clarendon Press, Oxford, 1987.

18B

Availability of Renewable Resources: Wind Energy

Dennis Elliott

Marc Schwartz
National Renewable Energy Laboratory

Bruce Bailey
AWS Scientific, Inc

Julie Phillips
J.A. Phillips & Associates

Wind is a renewable energy resource with the potential to generate significant amounts of electricity for both grid-connected utilities and standalone applications around the world. Since the oil embargos of the 1970s, there have been major strides in wind technology development, and wind energy costs have become competitive with other generation technologies in regions with good wind resources. In mid-1996, there were more than 5,000 MW of nameplate installed capacity worldwide. Currently, there are no technical barriers to using wind as a resource for as much as 20% of the world's total electrical generation.

Wind resource assessment is crucial in determining the feasibility of wind energy development. Whether the development objective is a single wind turbine or a utility wind power plant consisting of hundreds of turbines, a thorough understanding of a proposed site's wind resource is necessary. Resource assessments are typically conducted in conjunction with an evaluation of a proposed site. Site evaluation considers vegetation patterns, land forms, accessibility, existing land use, land ownership, zoning ordinances, permitting requirements, environmental impact, and transmission access. This section provides an overview of wind resource assessment and highlights site evaluation strategies.

The siting of a wind power facility is complex and requires an understanding of how to gather and interpret wind resource data. The power in the wind is a function of the cube of the wind speed. Therefore, inaccuracies in projecting long-term annual average wind speeds can lead to significant errors in energy production estimates. Readers are encouraged to use this chapter in conjunction with the resources listed in Section 18B.6, and, if undertaking the development of a wind project for the first time, to enlist the assistance of experienced professionals. Names of wind energy meteorologists and other wind energy consultants are available from

0-8493-2514-5/96/$0.00+$.50

The American Wind Energy Association (AWEA)
122 C Street, NW
Washington, DC 20001
Telephone: (202) 383-2500
Facsimile: (202) 383-2505

Wind Resource Assessment

Three stages comprise a wind resource assessment program. The first stage, prospecting for candidate sites, is also known as preliminary area identification. For prospecting, wind developers typically rely on existing wind resource information to narrow the search region to one or more promising locations. Existing wind data, wind resource and wind electric potential maps, topographic maps, and site evaluations characterize this effort. Section 18B.1 discusses preliminary area identification.

The second stage is area wind resource evaluation, in which wind measurement programs are undertaken in one or more areas under serious consideration for wind power development. Wind developers use such measurements to determine the quality of the resource, compare different candidate sites, estimate the performance of specific turbines, analyze the economic potential of wind development, and make preliminary determinations of turbine placements. Section 18B.2 provides an account for area wind resource evaluation.

Micrositing is the final stage of wind resource assessment. By carefully studying the small-scale variability of the wind resource at a particular location, the developer can position the wind turbines to maximize the overall energy output. The goal of micrositing is for each turbine to get maximum wind exposure with minimal interference from other turbines or obstructions. Section 18B.3 provides a general discussion of micrositing.

18B.1 Prospecting

Prospecting for wind energy development sites has much in common with exploring for new mineral or oil deposits. The first step is to identify areas within a fairly large region such as a utility service area, a county, or even a multistate region that are likely to have good wind resources. One way to accomplish this is to use existing wind data in conjunction with topographic maps. The largest source of wind information is the National Climatic Data Center, which archives data from weather stations operated by the National Weather Service, the U.S. Military, the Federal Aviation Administration, and the Coast Guard:

National Climatic Data Center
Climate Services Division
151 Patton Ave. Rm. 120
Asheville, NC 28801-5001
Telephone: 704-271-4800
Facsimile: 704-271-4876

Other sources include regional climate centers, which are listed in Section 18B.4, the U.S. Forest Service, universities, air quality monitoring networks, utilities, state energy and natural resource agencies, and various other organizations. The American Wind Energy Association can assist readers in contacting specific information sources, many of which can also be accessed via the World Wide Web. In addition, the U.S. Department of Energy has synthesized information from many of these sources into a national wind resource atlas, which will be discussed in the next section.

In evaluating historical wind data, it is important to recognize their limitations. Little information was collected for the purpose of wind energy assessment, and many stations were located near or in cities in relatively flat terrain. The records can provide a good general description of the wind

TABLE 18B.1 Wind Power Density Classes

Wind Power Class	10 m (33 ft)[a]		30 m (98 ft)[a]		50 m (164 ft)[a]	
	Wind Power Density, W/m²	Speed,[b] m/s (mph)	Wind Power Density, W/m²	Speed,[b] m/s (mph)	Wind Power Density, W/m²	Speed,[b] m/s (mph)
1						
	100	4.4 (9.8)	160	5.1 (11.4)	200	5.6 (12.5)
2						
	150	5.1 (11.5)	240	5.9 (13.2)	300	6.4 (14.3)
3						
	200	5.6 (12.5)	320	6.5 (14.6)	400	7.0 (15.7)
4						
	250	6.0 (13.4)	400	7.0 (15.7)	500	7.5 (16.8)
5						
	300	6.4 (14.3)	480	7.4 (16.6)	600	8.0 (17.9)
6						
	400	7.0 (15.7)	640	8.2 (18.3)	800	8.8 (19.7)
7						
	1000	9.4 (21.1)	1600	11.0 (24.7)	2000	11.9 (26.6)

[a] Vertical extrapolation of wind power density and wind speed are based on the 1/7 power law.

[b] Mean wind speed is estimated assuming a Rayleigh distribution of wind speeds and standard sea-level air density. The actual mean wind speed may differ from these estimated values by as much as 20%, depending on the actual wind speed distribution and elevation above sea level.

resource within a large area but typically do not contain sufficient information to pinpoint candidate sites for wind development.

The Wind Energy Resource Atlas

The use of wind power classes to describe the magnitude of a location's wind resource was first defined in conjunction with the preparation of the U.S. Department of Energy's 1987 *Wind Energy Resource Atlas of the United States*. The atlas is currently available through the American Wind Energy Association and is an excellent source of regional wind resource estimates for the United States and its territories. In the atlas, the magnitude of a wind resource is expressed in terms of seven wind power classes, rather than as a function of wind speed.

The wind power classes range from Class 1 (for winds containing the least energy) to Class 7 (for winds containing the greatest energy). The classes are based on average "wind power density," which is expressed in watts per square meter (W/m²) of the area swept by a turbine rotor. This term incorporates the combined effects of the time variation of wind speed and the dependence of wind power on both air density and the cube of the wind speed. Each wind class is based on a range of average wind speeds at specified heights above the ground.

Table 18B.1 shows the wind power classes in terms of their mean wind power density and the mean wind speed at 10 m (33 ft), 30 m (98 ft), and 50 m (164 ft) above the ground. The 30-m and 50-m heights correspond to the range of hub heights of many turbines currently operating or under development.

Areas designated as Class 5 or above are well suited to wind power generation. Class 4 regions, which are abundant throughout the nation's midsection, are just now becoming economical for power generation. Class 3 areas could potentially be developed for wind power at some time in the near future, particularly with expected improvements in advanced wind turbine technology. Class 2 areas are marginal, and Class 1 areas are unsuitable for wind power development. The U.S. map of the annual average wind resource is shown in Figure 18B.1.

The atlas displayed wind resource values on analyzed and gridded national and state maps with a resolution of 1/4° latitude by 1/3° longitude (an area of approximately 300 square miles) in the contiguous United States. The size of the grid cells in Alaska was 1/2° latitude by 1° longitude, and in Hawaii the grid cells were 1/8° latitude by 1/8° longitude. The gridded maps also indicate the

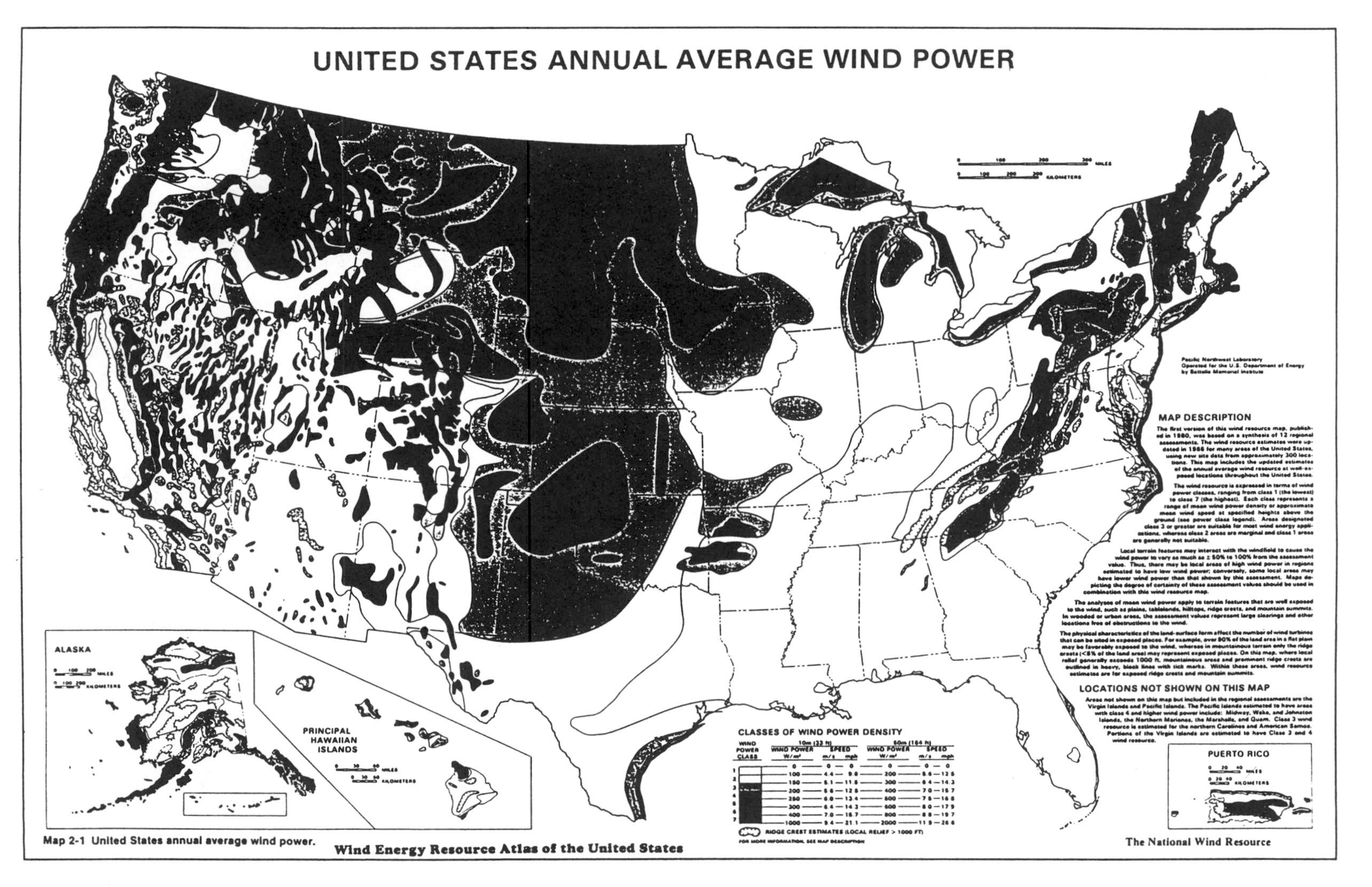

FIGURE 18B.1 Annual average wind power density in the United States.

certainty of the wind resource estimates and their areal distribution. However, they do not account for the variability in mean wind speed on a local scale; rather, they indicate regions where a high wind resource is likely to exist. In regions with complex terrain, for example, a Class 2 area could contain specific sites with a more energetic wind resource. Correspondingly, it should not be assumed that every part of a Class 6 region consistently experiences high winds.

Certainty ratings were assigned to each grid cell to indicate the degree of confidence in the resource estimate. The abundance and quality of both the historical and new wind data, the complexity of the terrain, and the geographical variability of the wind resource were all considerations. The highest degree of confidence (rating 4) was given to grid cells with relatively simple terrain for which there was abundant historical data. Regions with complex terrain or for which little data was available received a low certainty rating of 1.

Gridded seasonal maps of the wind resource are presented in Figures 18B.2 through 18B.5. These maps show the intra-annual variability of the wind resource in different regions of the United States. In general, winter and spring have the highest wind resource and summer the lowest—except for wind corridors in the western United States, where a summertime maximum prevails.

The U.S. Wind Electric Potential

Following the publication of the atlas, researchers produced estimates of the U.S. wind electric potential, that is, the amount of wind generated electricity a region could produce. In developing wind electric potential estimates for all 48 contiguous states, researchers excluded urban areas, half the nation's forested ridge crests, 30% of agricultural lands, and 10% of range lands. Parks, wilderness areas, wildlife refuges, and other environmentally sensitive areas were also excluded from wind energy development estimates.

Gridded maps of the United States have been prepared showing the wind electric potential from (1) available lands with Class 4 and above resources and (2) available lands with Class 3 and above resources. The U.S. wind electric potentials for lands with Class 4 and above resources are shown in Figure 18B.6. The wind electric potential maps enhance the information available from the atlas and are currently available from

The National Renewable Energy Laboratory
Wind Technology Division
1617 Cole Boulevard
Golden, Colorado 80401-3393
Telephone: (303) 384-6900
Facsimile: (303) 384-6901

Wind electric potential information sheets for individual states are available from the American Wind Energy Association.

Wind Information for Specific Sites

Occasionally, a wind prospector will discover existing data sets for locations near a site under consideration for wind development. To evaluate the significance of this data, it will be necessary to determine (1) the exact location of the measurement station, (2) the local topography, (3) the anemometer height and exposure, (4) the type of observation (i.e., instantaneous or hourly average), and (5) the time period during which data were gathered.

If the terrain surrounding the station (and including the proposed site) is relatively flat, the station data may well be representative of the site. However, if the surrounding terrain is complex, it may not be possible to reliably extrapolate the existing information beyond the immediate vicinity of the station. For example, most airport measurements are now taken adjacent to the runways, where winds blow unobstructed. Such measurements would provide no information about wind flows in and around nearby canyons, ridges, or other landforms. Because buildings alter natural

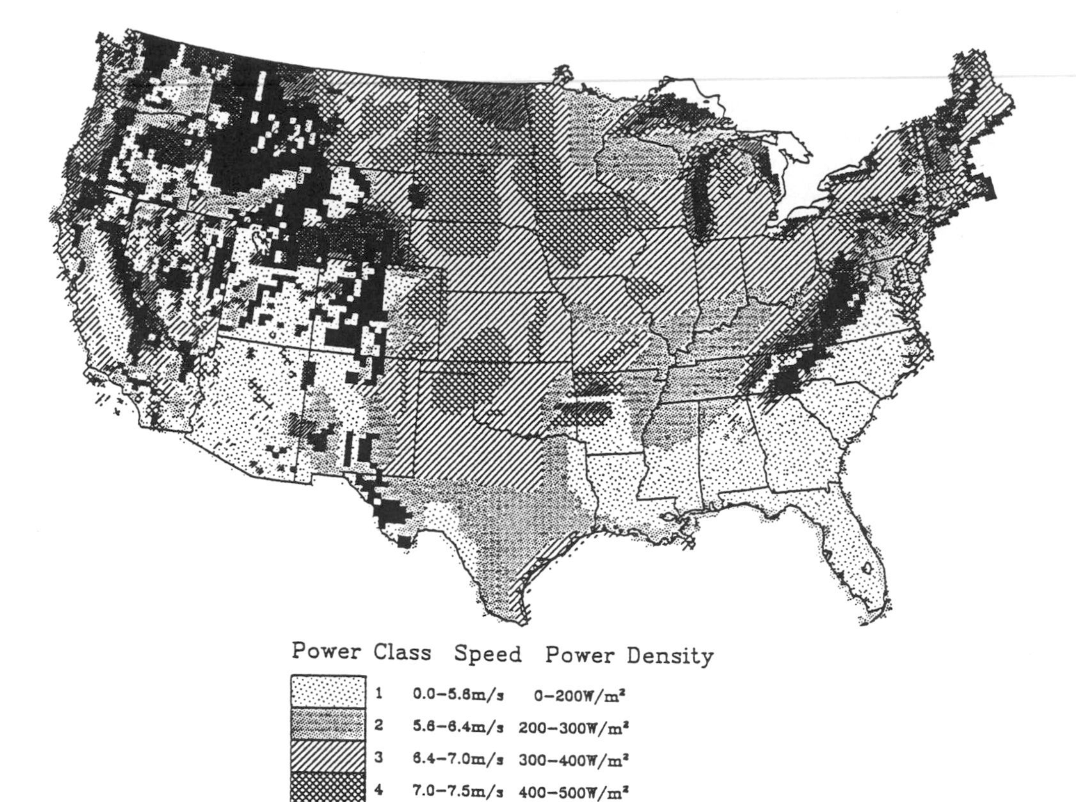

FIGURE 18B.2(a) Winter wind power density in the United States at 50 m.

wind flows, rooftop measurements are often considered unreliable for wind resource assessment purposes.

The next factor to consider is anemometer placement. The height of airport anemometers typically ranges from 6 m to 15 m (20 ft to 50 ft). If data from one station are to be compared with similar data from another station, it is a good idea to extrapolate all wind speed data to a standard reference height such as 30 m or 40 m as shown in the next section.

To be most useful to a wind prospector, existing data sets should contain information for at least a full year, although the reliability of the wind resource estimates would be significantly increased with 3 to 5 years of data. A *time series* of hourly wind speed and wind direction is the most useful format for the data. This format can be analyzed for several different wind characteristics, including the frequency distribution of wind speed and direction, as well as the diurnal, monthly, seasonal, and annual variability of these parameters. In many instances, wind data summaries may already be available.

Variations in Wind Speed

Wind speeds and wind patterns vary according to atmospheric circulation patterns, time of day, season, and in response to large-scale weather systems such as fronts and storms. Wind speed also varies with the wind's height above ground, a phenomenon known as wind shear.

Wind speed increases when air flows around and over hills and ridges, making such locations (particularly if they have strong prevailing winds) excellent candidates for wind power development.

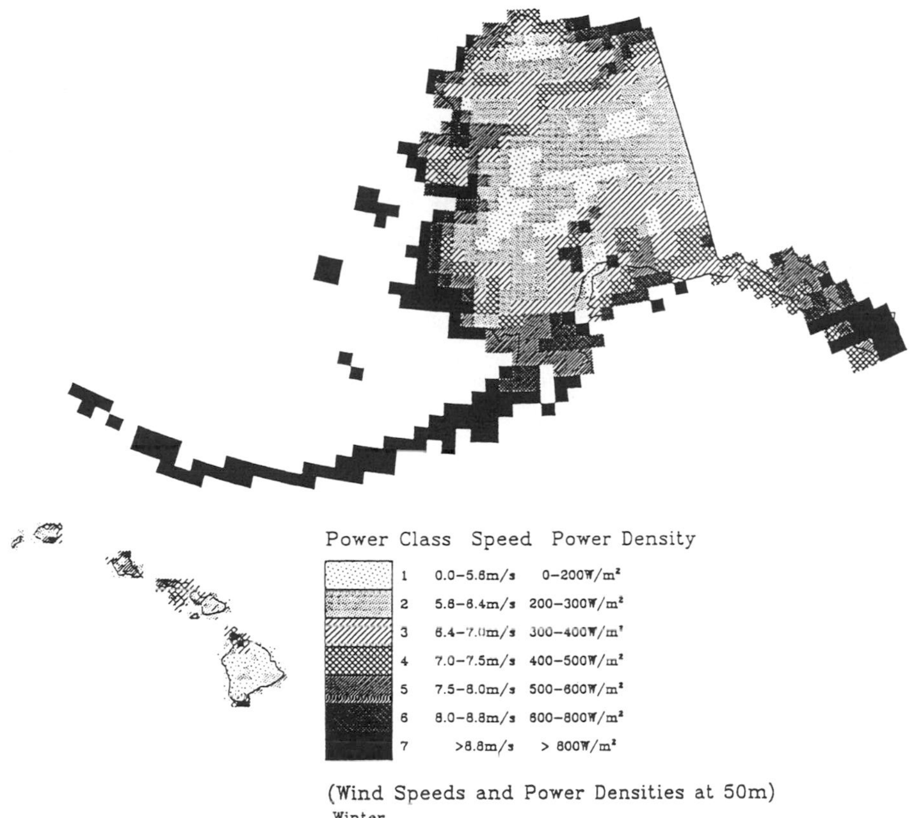

FIGURE 18B.2(b)

Depending upon the slope and orientation of such terrain to the wind flow, the wind speed at the top of a hill can be 50 to 100% greater than at a location off the crest of the hill.

Most historical wind data were collected at heights near 10 m (33 ft), but the hub height for many of today's wind turbines ranges from 30 m to 50 m(98 ft to 164 ft). At greater heights, not only does wind speed increase, winds also become more constant. Thus, resource assessments should include wind speed measurements at the hub height of turbines planned for installation at the site. But, if wind speed data are not available at the planned hub height of the turbines, z_2, wind speeds can be estimated from another height, z_1, with the *power law* equation:

$$V_2 = V_1 \left[\frac{z_2}{z_1} \right]^{\alpha} \tag{18B.1}$$

where

V_2	= the unknown speed at height z_2
V_1	= the known wind speed at the known height, z_1
α	= the wind shear exponent

The wind shear exponent is often assigned a value of 0.143, known as the *1/7 power law,* to predict wind profiles in a well-mixed atmosphere over flat, open terrain. This was the wind shear value used in the *Wind Energy Resource Atlas* to adjust the mean wind speed and wind power density to the 10-m (33-ft) and 50-m (164-ft) reference levels. Wind shear exponent values of 0.16 or higher may more accurately reflect many inland sites with higher surface roughness, while values of 0.25 or greater may apply to obstructed rural and urban areas. Higher exponent values

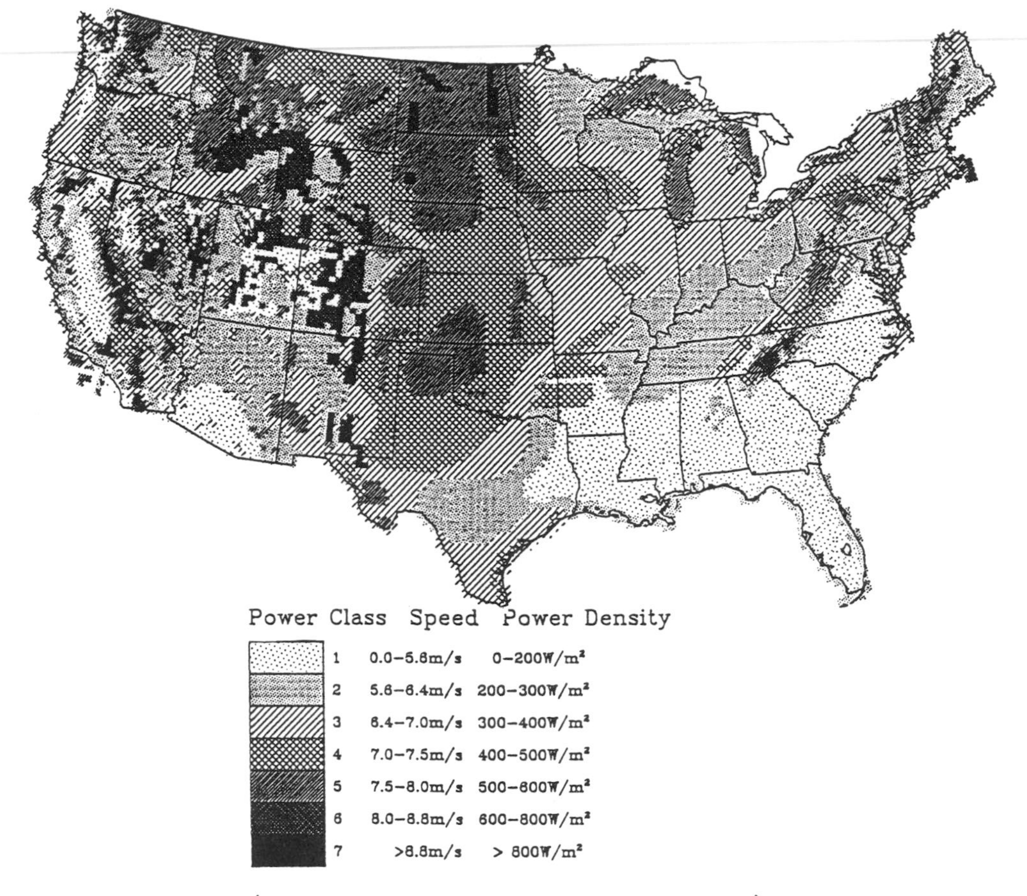

FIGURE 18B.3(a) Spring wind power density in the United States at 50 m.

are normally observed over vegetated surfaces and when wind speeds are light. In complex terrain, using the power law equation to estimate wind speeds at higher altitudes from data gathered relatively close to the ground is not reliable.

Figure 18B.7 shows how wind speed and power varies with height for different wind shear exponent values. Because the wind power varies as the cube of the wind speed, the differences in wind power as a function of height and wind shear exponent are considerably greater than the differences in wind speed.

Topography

Topographic maps are an excellent resource for wind prospecting. Because they allow a prospector to identify attractive terrain features, they help streamline the process of identifying candidate wind development sites. Topographic maps at various scales are available from the U.S. Geological Survey. The most detailed are 1:24,000 scale (1 inch = 2,000 feet).

Topographic screening can identify features likely to experience higher winds than the general surroundings. The process is invaluable for areas for which there is little or no relevant historical wind speed data. Landforms often associated with better wind resources include:

- Exposed ridges and mountain summits
- High-elevation plains and plateaus
- Gaps, passes, and gorges (locations where funneling of local winds is common)

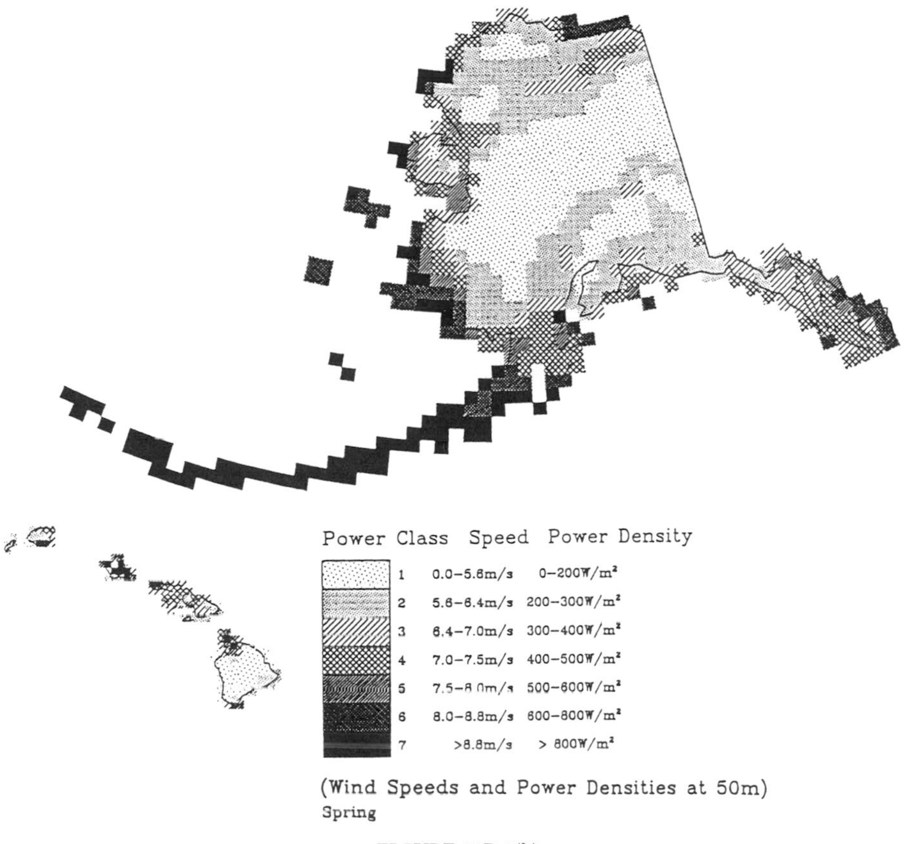

FIGURE 18B.3(b)

- Long valleys extending down from mountains
- Coastlines and immediate inland strips with a minimum of relief and vegetation
- Upwind and crosswind corners of islands.

Topographic maps also contain useful information about locations to avoid in siting a wind power system. Low-wind areas include valleys perpendicular to the prevailing winds, sheltered basins, short or narrow valleys and canyons, areas immediately upwind of mountains, and areas with many surface obstructions such as hilly or forested terrain. In addition, areas immediately downwind of higher terrain, the lee sides of ridges and steeply sloped terrain should be avoided because they frequently have excessive turbulence. Turbulence has historically been associated with severe damage to wind turbines.

Field Surveys

As soon as one or more candidate sites are identified using the methods just described, the next step is to visit each site to confirm the wind resource and terrain features. Wind prospectors should arrive at each site with a topographic map of the area, a Global Positioning System (GPS) to record the latitude and longitude, and a video or still camera. Use the map (and cameras, if desired) to record information about the following site characteristics:

- Exact location and elevation above sea level
- Available land area
- Existing land use
- Land ownership

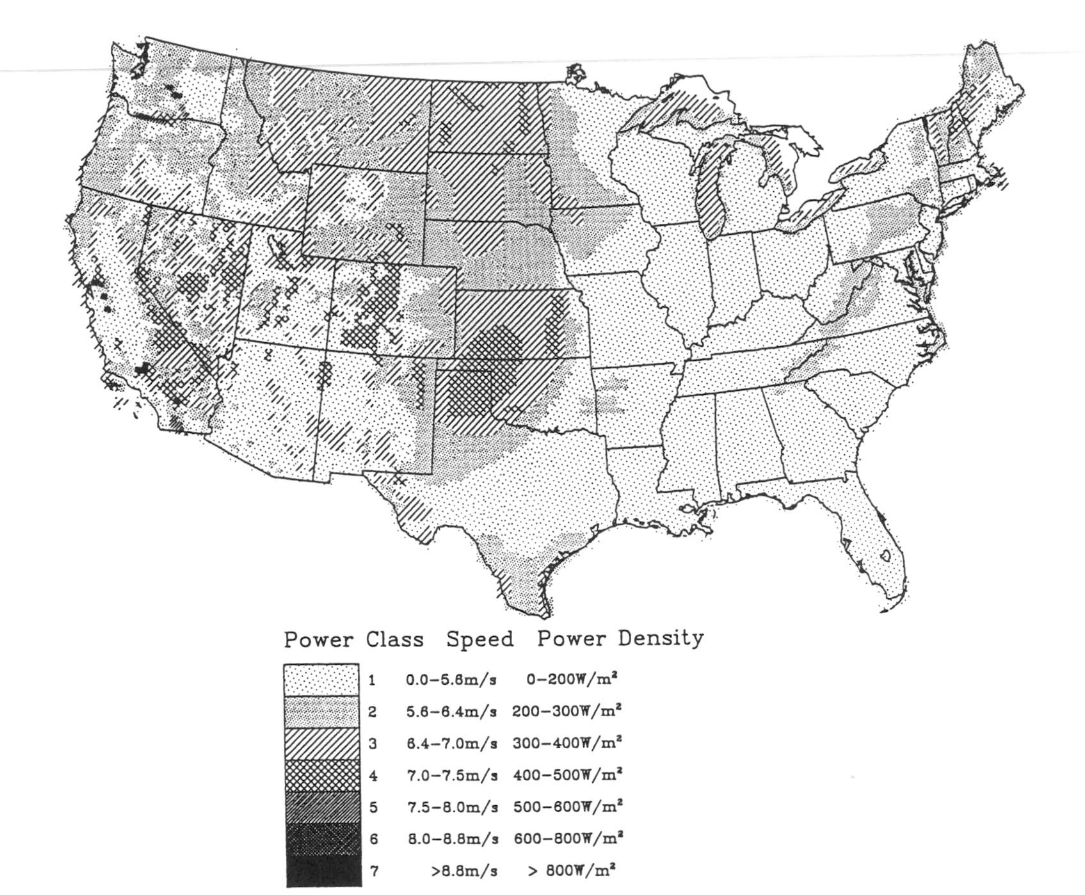

(Wind Speeds and Power Densities at 50m)

Summer

FIGURE 18B.4(a) Summer wind power density in the United States at 50 m.

- Location of any obstructions
- Trees deformed by persistent strong winds
- Accessibility into the site
- Visual impact if wind turbines are installed
- Cellular phone service (for data transfers)
- Possible locations for wind monitoring instruments

In areas where wind data are sparse, the terrain, vegetation, or local residents may hold clues to the nature of the resource. For instance, certain trees and bushes record evidence of persistently strong winds in their form. Branches may be missing or shortened on the windward side of trees, and entire trees may be permanently bent over by the wind. The use of trees as wind speed indicators is limited to those areas where winds blow consistently from the same direction because strong winds that blow from different directions may not deform foliage. Terrain features, such as playas (flattened bottoms of undrained desert basins) and sand dunes, may also indicate strong winds from a specific direction. In addition, local residents are usually familiar with seasonal wind patterns and the relative windiness of one site compared to another, and may even be able to recommend an area for consideration.

Regional Wind Prospecting: An Example

Identifying prospective wind development sites involves knowledge of the wind resource and topographic features. Suppose an entrepreneur desires to build a wind generating facility somewhere

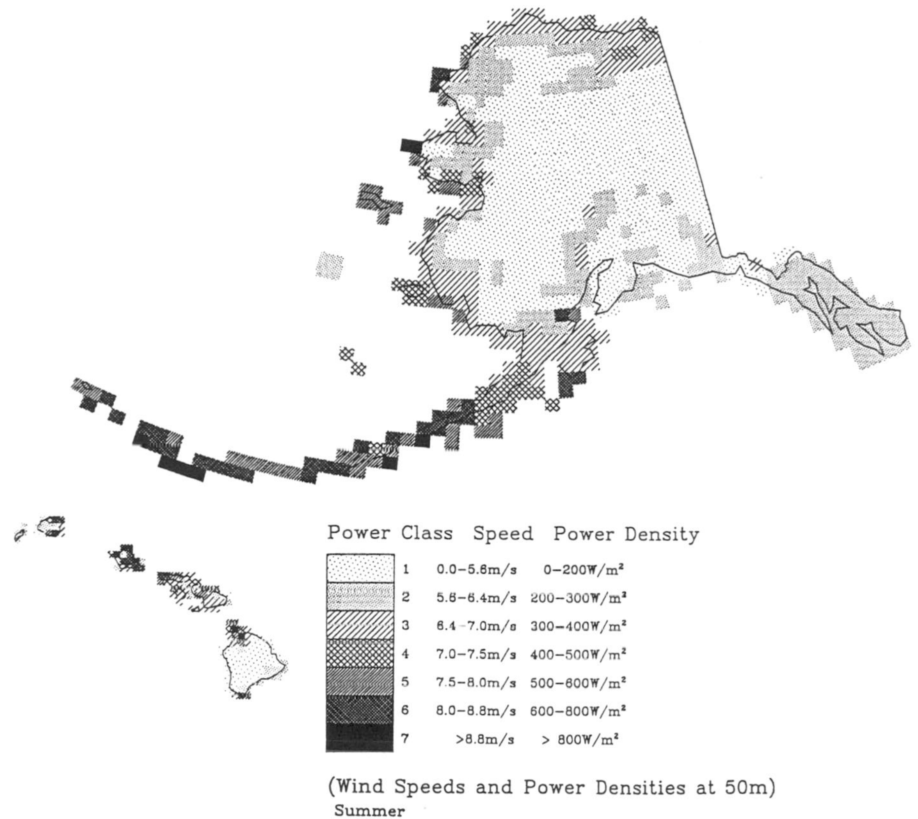

FIGURE 18B.4(b)

in the United States. A glance at the U.S. wind resource map (Figure 18B.1) in the *Wind Energy Resource Atlas of the United States* reveals that Wyoming contains abundant winds. A look at the state's gridded wind resource map (shown in Figure 18B.8) reveals numerous areas with good-to-excellent (Class 4 to Class 6) wind resources.

How can one narrow the search?

The atlas, which summarizes data from the National Climatic Data Center and other sources, for all states, in Appendices C and E, reveals that existing wind data for Wyoming (Table 18B.2) are somewhat limited given the size of the state. The U.S. Wind Electric Potential Map (Figure 18B.6) shows that most of the state's wind electric potential lies in the southern half of the state. Topographic and land-use maps reveal why: much of northern Wyoming's wind resources are located in relatively inaccessible mountain ranges, national parks, and designated wilderness areas.

Topographic maps reveal the complex terrain in southern Wyoming. In several areas, numerous small mountain ranges abut potentially windy bluffs, ridges, and exposed plains. While the state appears to be sparsely populated, it still has several population centers and urban areas in Utah and Colorado are not far from the border. Several areas in the state appear sufficiently promising to schedule field inspections along with preliminary site evaluations.

18B.2 Area Wind Resource Evaluation

The second stage of wind resource assessment, area wind resource evaluation, consists of wind measurement programs that characterize the wind resource within areas identified as potential development sites during prospecting. The objectives of this stage most commonly include

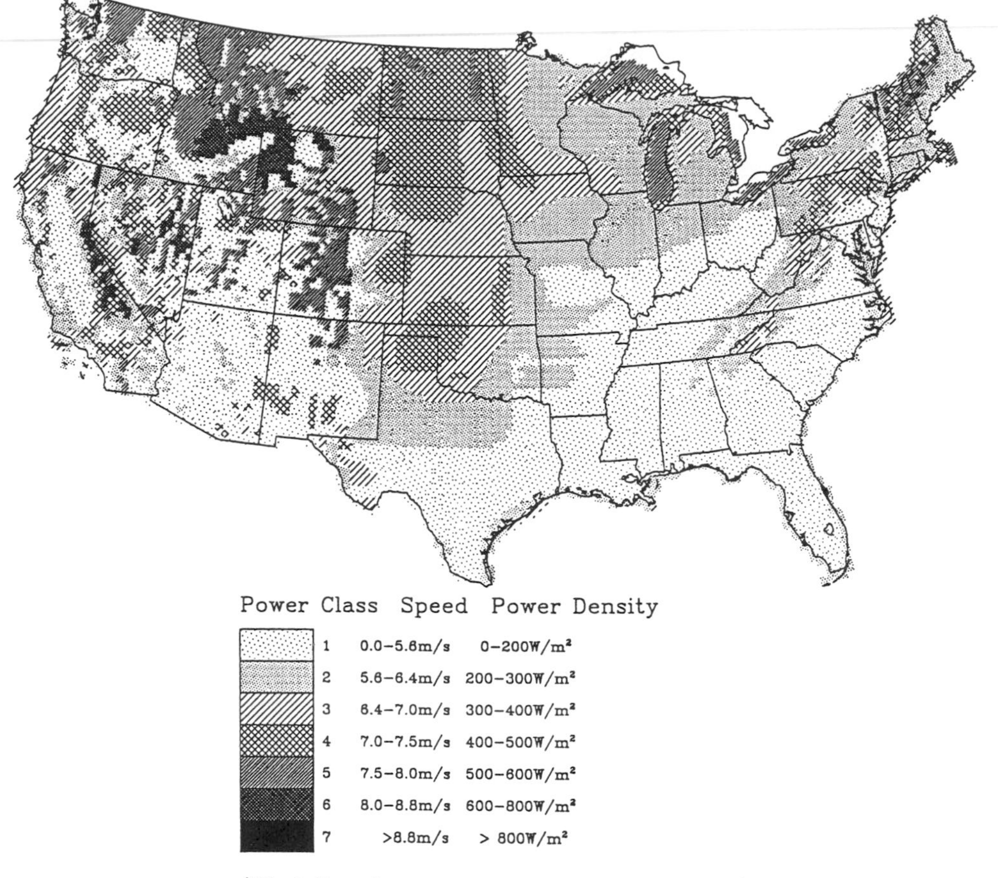

FIGURE 18B.5(a) Autumn wind power density in the United States at 50 m.

- Determining whether an area's wind resource is sufficient to justify more comprehensive and site-specific studies
- Comparing the wind resource in different areas
- Obtaining data for estimating the performance and economic viability of selected wind turbines
- Screening for potential wind turbine installation sites

Planning

A *measurement plan* is necessary before undertaking a wind monitoring program. The plan specifies the tower type, sensors, data acquisition equipment, duration of the program, and data processing procedures. Its purpose is to ensure that the program will meet the objectives of the specific wind energy development project. For example, a short-duration program employing a few towers with anemometers set at a standard height would meet the objective of screening an area to locate the windiest sites. In contrast, the complete wind characterization necessary for wind power plant development would require a much longer-term effort using a variety of sensors mounted on multilevel towers strategically placed within an area.

Measurement plans should be documented in writing, then reviewed and approved by project participants prior to implementation. The plan should specify the following:

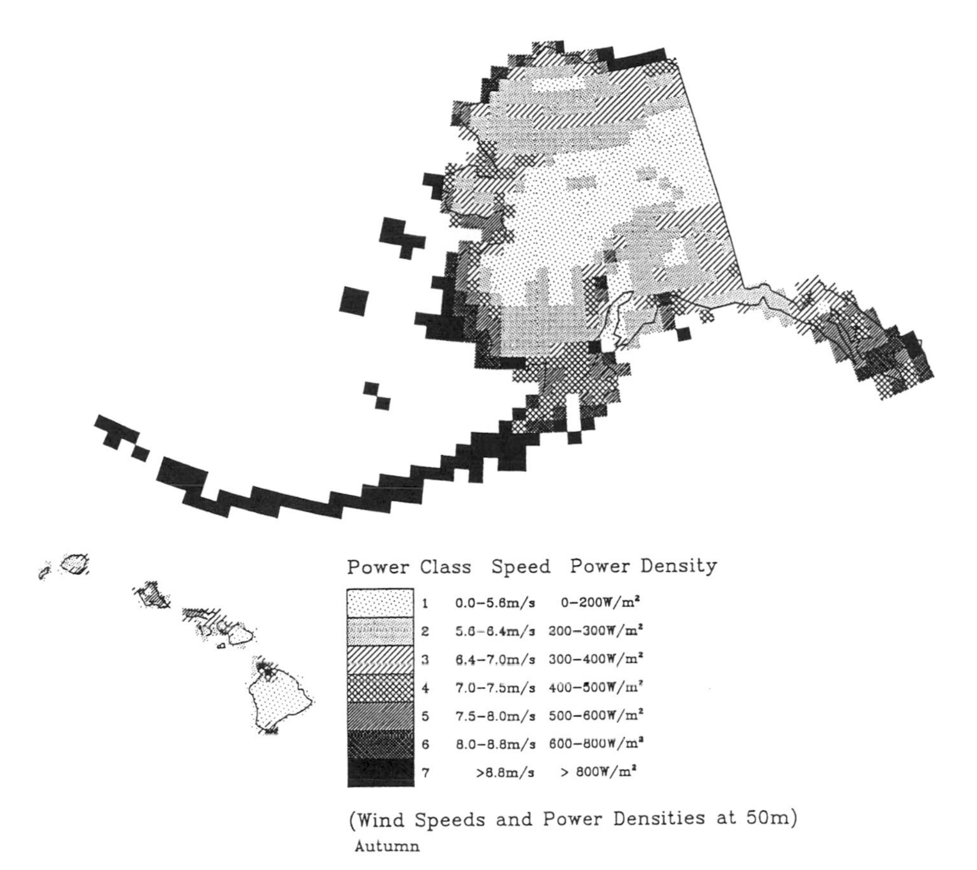

FIGURE 18B.5(b)

- Measurement parameters
- Equipment type, quality, and cost
- Number and location of monitoring stations
- Sensor measurement heights
- Minimum measurement accuracy, duration, and data recovery
- Data sampling and recording intervals
- Data storage format
- Data handling and processing procedures
- Quality control measures
- Final data products

The monitoring plan must include a specified location for the monitoring tower, which should be placed in a location that is representative of the site. Placing the tower in either the windiest or least windy portion of a site will not reflect the site's overall wind conditions. It is also important to place the tower as far away as possible from obstructions to the wind.

Siting a meteorological tower near trees or buildings can introduce errors into the analysis of a site's wind characteristics. As shown in Figure 18B.9, an obstruction can reduce wind speed, increase turbulence, and decrease the amount of energy in the wind as measured by instruments downwind from the obstruction. Obstructions to the airflow will alter the perceived magnitude of the wind resource, wind shear, and turbulence levels. If sensors must be near an obstruction, place them at a horizontal distance no closer than 10 times the height of the obstruction in the prevailing wind direction.

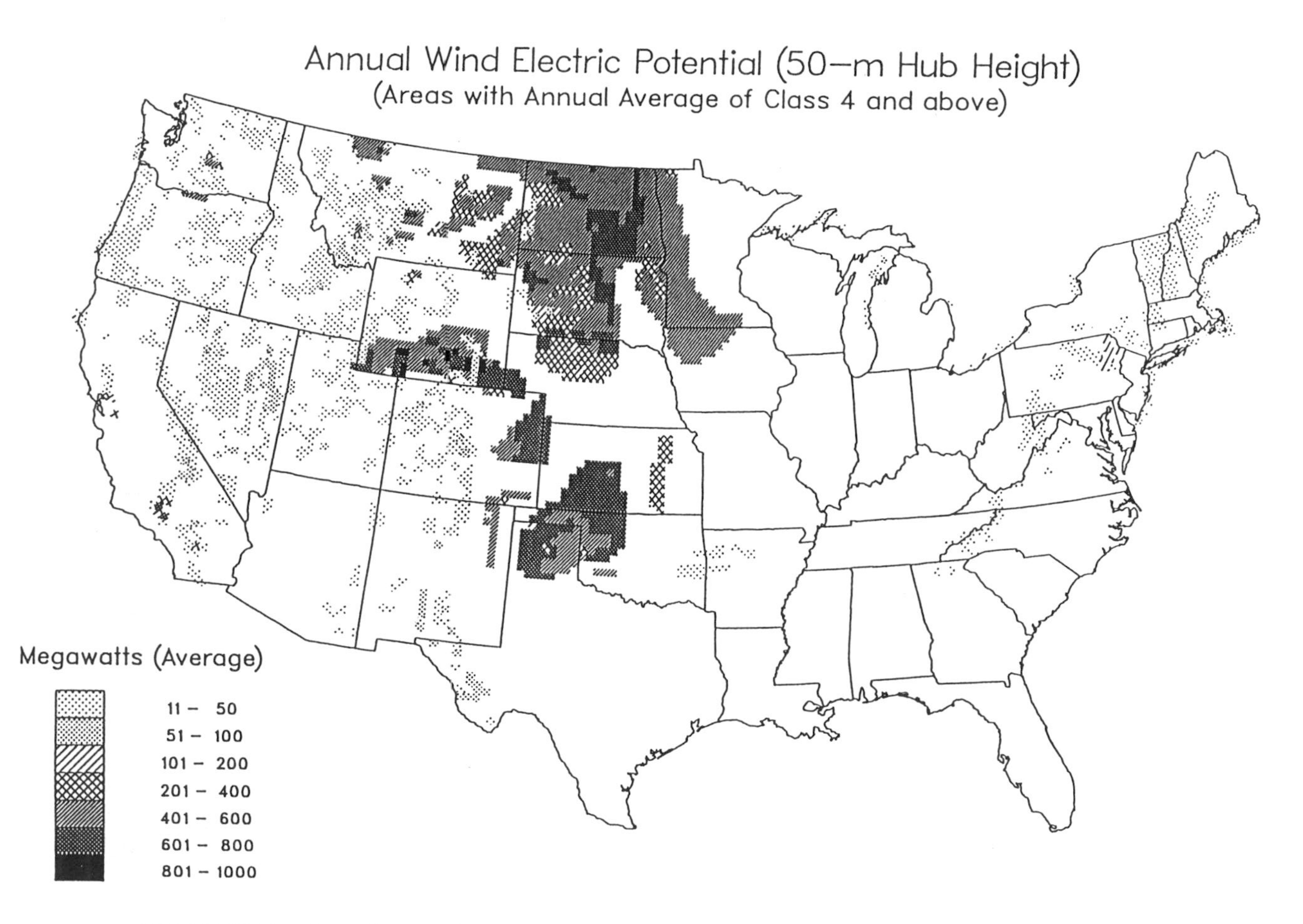

FIGURE 18B.6 Map of U.S. wind electric potential from areas with wind power Class 4 and above at 50 m.

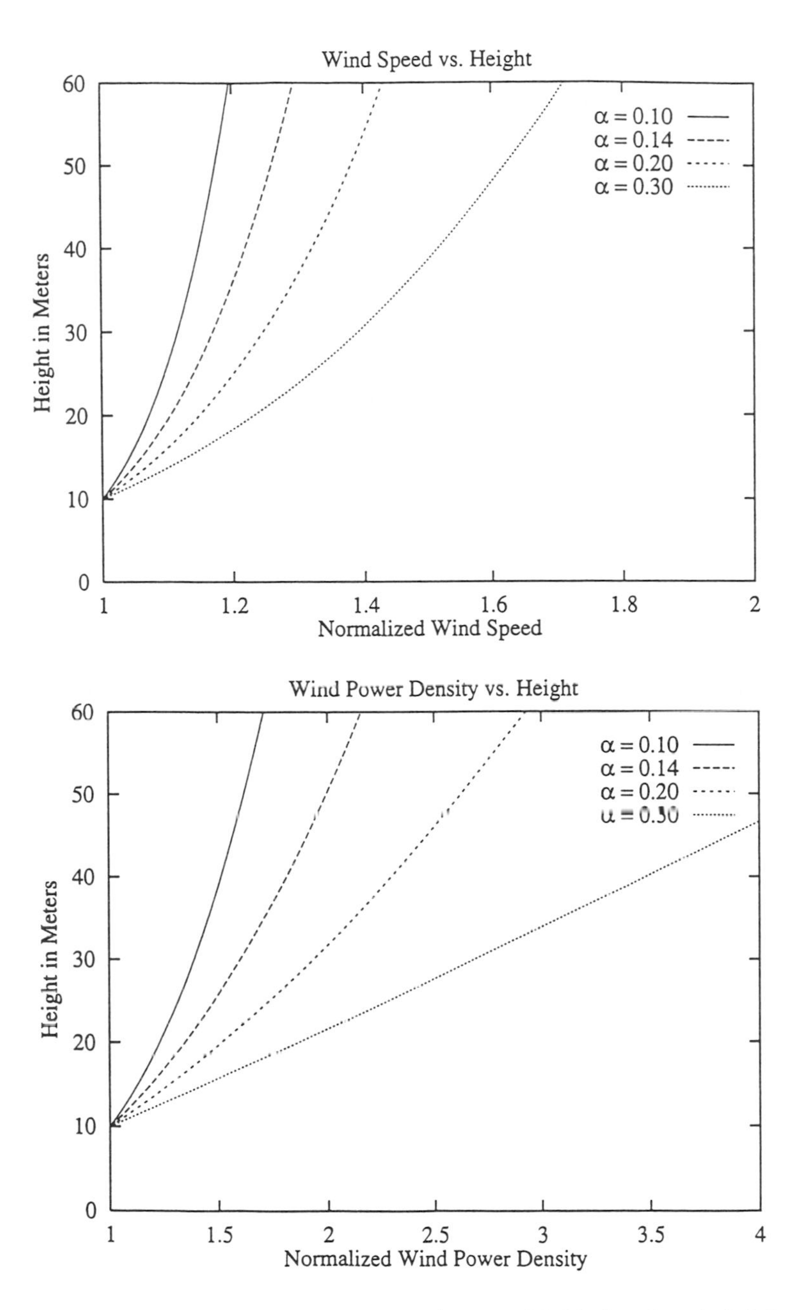

FIGURE 18B.7 Variation of wind speed and wind power density with height for selected wind shear exponent values.

Ideally, wind measurement programs should take place for a minimum of 2 years. This time period is normally sufficient to reliably determine the diurnal and seasonal variability of the wind resource within candidate sites. With the aid of a well-correlated, long-term reference station, such as an airport, the annual variability of the wind can also be determined. The data recovery for all measured parameters should be at least 90% during the duration of the program, with data gaps kept to less than 1 week.

If a large organization such as a utility is responsible for the monitoring program, then a *monitoring strategy* should also be developed for carrying out the measurement plan. The core of a sound monitoring strategy is good management, qualified staff, and adequate resources. The

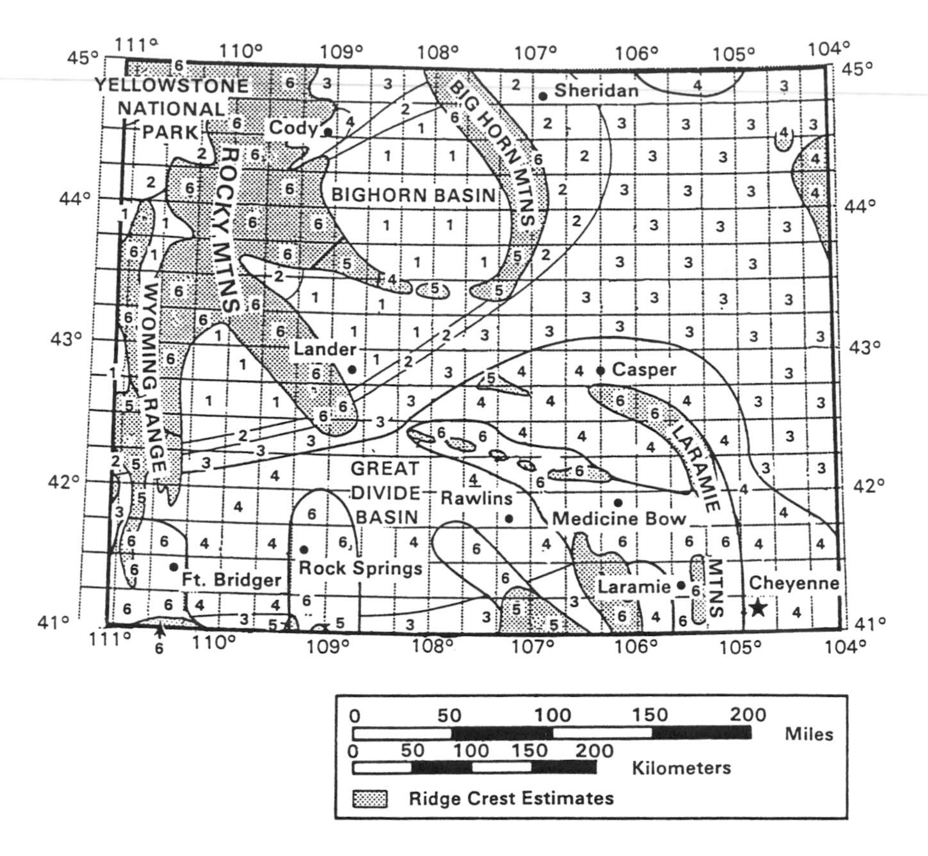

FIGURE 18B.8 Wyoming annual average wind power classes.

project team should include at least one person with prior field measurement experience and team members with skills in computer-based data processing, analysis, and interpretation.

Data Validation and Processing

There are two parts to data validation: data screening and data verification. Data screening uses a series of validation routines (algorithms) to screen all data for questionable and erroneous values, known as suspect values. A suspect value deserves scrutiny, but is not necessarily erroneous. For instance, a severe thunderstorm can cause an unusually highly hourly wind speed within an otherwise average day. Data screening typically produces a report (printout) listing any suspect values and the particular algorithm each value failed. There are many possible causes of erroneous data, including faulty, or damaged sensors; loose wire connections; broken wires, damaged mounting hardware, datalogger malfunctions; static discharges; sensor calibration drift; and icing conditions.

Data verification requires a case-by-case decision on what to do with suspect values: (1) retain them as valid; (2) reject them as invalid; or (3) replace them with redundant, valid values, if available. This decision is often a judgment call best made by someone familiar with the monitoring equipment and local weather conditions.

After validation, a data set is ready for processing to evaluate a site's wind resource. Processing involves different calculations and combining selected values into useful subsets that are readily translated into summary tables or performance graphs. It is useful to produce these data summaries monthly. Wind data is most useful when presented in tabular or graphic form. Tables 18B.3 and 18B.4 and Figures 18B.10 through 18B.12 present a list of sample data reports.

TABLE 18B.2 Summarized Wind Data for Wyoming

St	Station Name	WBAN Code	TYP	Lat DD.MM	Long DDD.MM	Elev (M)	Period		OBS	ANEM Ht (M)	LOC	Mean Wind Speed (m/s) and Wind Power Density (Watt/M**2)									
							Start YYMMDD	End YYMMDD				Annual Spd	Annual Pow	Winter Spd	Winter Pow	Spring Spd	Spring Pow	Summer Spd	Summer Pow	Autumn Spd	Autumn Pow
WY	Casper	24089	W	42.55	−106.28	1622	500311	580129	A	24.4	R	6.1	255	7.8	437	5.9	246	5.0	130	5.9	215
WY	Casper	24089	W	42.55	−106.28	1622	580130	640811	A	8.8	R	5.6	191	6.6	255	5.7	205	4.6	107	5.7	203
WY	Casper	24089	W	42.55	−106.28	1622	640812	781231	A	6.1	G	5.7	194	7.1	329	5.7	191	4.6	95	5.4	163
WY	Casper/Ward.	24016	W	42.55	−106.20	1617	480101	500311	A	15.2	R	6.4	365	8.5	719	6.3	278	5.0	147	5.3	213
WY	Cheyenne	24018	W	41.09	−104.49	1871	480101	540811	A	12.2	R	5.9	237	7.1	395	6.2	264	4.6	98	5.5	191
WY	Cheyenne	24018	W	41.09	−104.49	1871	540812	571001	A	22.3	R	6.3	287	7.0	394	7.0	380	4.9	114	6.2	275
WY	Cheyenne	24018	W	41.09	−104.49	1871	571002	641231	A	10.1	G	6.2	234	7.1	339	6.6	274	5.3	120	5.9	203
WY	Cheyenne	24018	W	41.09	−104.49	1871	650101	781231	A	10.1	G	5.8	199	6.7	301	6.3	246	4.7	92	5.4	158
WY	Douglas	24019	W	42.45	−105.22	1486	480101	541231	A	17.7	B	5.6	285	6.4	475	6.2	323	5.1	165	4.9	205
WY	Fort Bridger	24118	W	41.24	−110.25	2136	480101	541231	A	18.3	U	6.6	313	7.6	472	6.7	317	5.9	206	6.3	273
WY	Lander	24021	W	42.49	−108.44	1697	480101	620331	C	9.8	R	2.9	55	2.5	52	3.3	70	3.2	50	2.7	45
WY	Lander	24021	W	42.49	−108.44	1697	620401	730919	A	9.8	R	3.2	59	2.8	67	3.5	72	3.5	53	2.9	42
WY	Lander	24021	W	42.49	−108.44	1697	730920	781231	A	9.8	G	3.1	45	2.7	40	3.6	61	3.5	50	2.7	32
WY	Laramie	24022	W	41.19	−105.41	2217	480101	541231	A	19.5	U	6.2	255	7.2	395	6.6	277	5.5	155	5.6	195
WY	Moorcraft	24088	W	44.16	−104.57	1300	500101	520731	A	9.8	R	4.1	128	3.6	103	4.4	143	4.1	115	4.5	157
WY	Rawlins	24057	W	41.48	−107.12	2067	550101	641231	A	8.5	R	5.2	189	6.2	282	5.6	214	4.0	94	4.9	167
WY	Rock Springs	24027	F	41.36	−109.04	2056	480101	600726	A	15.2	R	5.7	242	6.2	319	6.3	281	5.0	147	5.5	220
WY	Rock Springs	24027	F	41.36	−109.04	2056	600727	781231	A	6.1	G	4.9	137	5.4	191	5.3	161	4.3	86	4.5	111
WY	Sheridan	24029	W	44.46	−106.58	1203	480101	580609	A	11.6	R	3.8	111	3.7	124	4.3	139	3.6	75	3.7	103
WY	Sheridan	24029	W	44.46	−106.58	1203	580610	640902	A	12.8	R	3.6	104	3.5	115	4.3	137	3.3	70	3.4	100
WY	Sheridan	24029	W	44.46	−106.58	1203	640903	781231	A	6.1	G	3.5	74	3.5	83	4.1	103	3.2	47	3.3	64
WY	Sinclair	24031	W	41.48	−107.03	1999	480101	510228	A	8.8	R	6.5	402	7.7	647	6.7	347	5.2	191	6.3	377
DOE Sites																					
WY	Bridger Butte			41.17	−110.29	2290	800901	820930		45.7		8.4	589	9.4	821*	8.5	570	7.4	396	8.4	629
										30.0		8.2	542	9.1	737*	8.3	525	7.4	392	8.2	575
										9.1		7.0	371	7.7	467*	7.3	376	6.4	272	7.0	398

* Mean wind speed and power are based on less than 75% data recovery for the period.

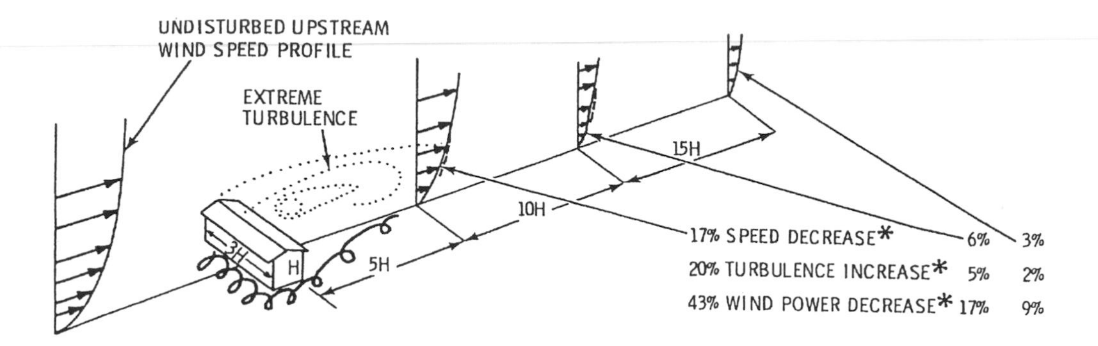

FIGURE 18B.9 Impact of obstructions on airflow.

TABLE 18B.3 Recommended Monthly Data Reports

Report Products	Presentation	Reporting Heights
Mean hourly diurnal wind speed	Graph/Table	10 m, 25 m, and 40 m
Joint wind speed and direction frequency distribution (16 sectors)	Table	25 m and 40 m
Wind speed frequency distribution: 0.5 m/s (1 mph) bins	Graph/Table	10 m, 25 m, and 40 m, and 10 m and 40 m
Mean hourly diurnal temperature	Graph/Table	3 m
Mean hourly wind shear	Graph/Table	Between 25 m and 40 m
Mean turbulence intensity	Graph/Table	10 m, 25 m, and 40 m
Wind rose	Graph	25 m and 40 m
Mean hourly wind power density	Graph/Table	10 m, 25 m, and 40 m

The wind shear exponent, turbulence intensity, and wind power density are not internal processing functions of most dataloggers. Unless a fully programmable datalogger has been used in data gathering, these parameters will need to be calculated to obtain hourly and monthly averages. Calculations for these three parameters are described next.

Wind shear is the change in wind speed with change in height. The wind shear exponent must be determined for each site from Eq. (18B.1) because its magnitude is site specific. Using the 1/7 power law is not recommended.

Wind turbulence is characterized by rapid disturbances in both wind speed and wind direction. It is an important part of wind resource evaluation because high turbulence can not only lower wind turbine power output, it can also cause mechanical stress on turbine components. The most commonly used indicator of turbulence is the standard deviation (σ) of wind speed. Normalizing this value with the mean wind speed yields the turbulence intensity *(TI)*. Thus,

$$TI = \frac{\sigma}{\overline{V}} \quad \textbf{(18B.2)}$$

where

σ = the standard deviation of wind speed
$\overline{V}$ = the mean wind speed

The *TI* value allows for an overall assessment of a site's turbulence. Values of $TI \leq 0.10$ indicate low levels of turbulence; values >0.10 but ≤ 0.25 indicate moderate levels of turbulence; and values >0.25 indicate high levels of turbulence.

TABLE 18B.4 Sample Hourly Wind Data Summary Report

Hour	40-m Wind Speed (m/s)	25-m Wind Speed (m/s)	10-m Wind Speed (m/s)	Wind Shear (40 m/25 m)	40-m Wind Power Density (W/m²)	40-m Turbulence Intensity	3-m Temp (°C)
0:00	5.8	5.1	4.8	0.26	175	0.13	13.3
1:00	5.7	5.1	4.8	0.27	175	0.13	15.6
2:00	5.7	5.0	4.7	0.29	166	0.13	16.7
3:00	5.5	4.9	4.6	0.27	159	0.13	16.9
4:00	5.5	4.8	4.6	0.28	158	0.13	17.2
5:00	5.5	4.8	4.6	0.28	155	0.13	17.2
6:00	5.4	4.7	4.5	0.28	150	0.13	17.4
7:00	5.5	4.9	4.6	0.26	160	0.14	18.9
8:00	5.7	5.1	4.9	0.21	186	0.14	19.3
9:00	5.9	5.4	5.2	0.19	201	0.14	22.2
10:00	6.1	5.7	5.4	0.16	217	0.14	23.5
11:00	6.3	5.9	5.6	0.16	232	0.14	26.0
12:00	6.4	6.0	5.7	0.15	249	0.15	28.6
13:00	6.4	6.0	5.7	0.15	252	0.14	29.6
14:00	6.5	6.0	5.7	0.17	261	0.14	31.3
15:00	6.4	5.9	5.7	0.17	245	0.14	30.9
16:00	6.1	5.6	5.3	0.20	215	0.14	30.2
17:00	5.9	5.3	5.0	0.22	189	0.13	26.0
18:00	5.7	5.1	4.9	0.23	177	0.13	24.2
19:00	5.8	5.2	4.9	0.24	184	0.13	22.9
20:00	5.8	5.1	4.8	0.26	180	0.13	20.2
21:00	5.7	5.1	4.8	0.27	173	0.13	18.7
22:00	5.6	5.0	4.7	0.27	163	0.13	17.3
23:00	5.7	5.0	4.8	0.27	164	0.13	15.4
Average	5.9	5.3	5.0	0.23	191	0.14	21.7

Wind power density corresponds to the magnitude of a site's wind resource. Hence, the parameter is an excellent indicator of wind energy potential. Wind power density (WPD) is defined as the wind power available per unit area swept by a turbine rotor. If hourly wind data are available, the average WPD in watts/m² is given by the following equation:*

$$\text{WPD} = \frac{1}{2n}\sum_{i=1}^{n} \rho_i V_i^3 \tag{18B.3}$$

where

n = the number of hourly records in the averaging interval
ρ_i = the air density (kilograms/m³; ($\rho = P/RT$)
V_i = the wind speed (m/s)at the ith hourly average record

If barometric pressure and temperature have not been measured, air density can be estimated as a function of site elevation *(z)* as follows:

$$\rho = 1.225 - \left(1.194 \times 10^{-4}\right)z \tag{18B.4}$$

*Equations (18B.2 to 18B.6) were taken from the *Wind Energy Resource Atlas of the United States,* pp. 145–146.

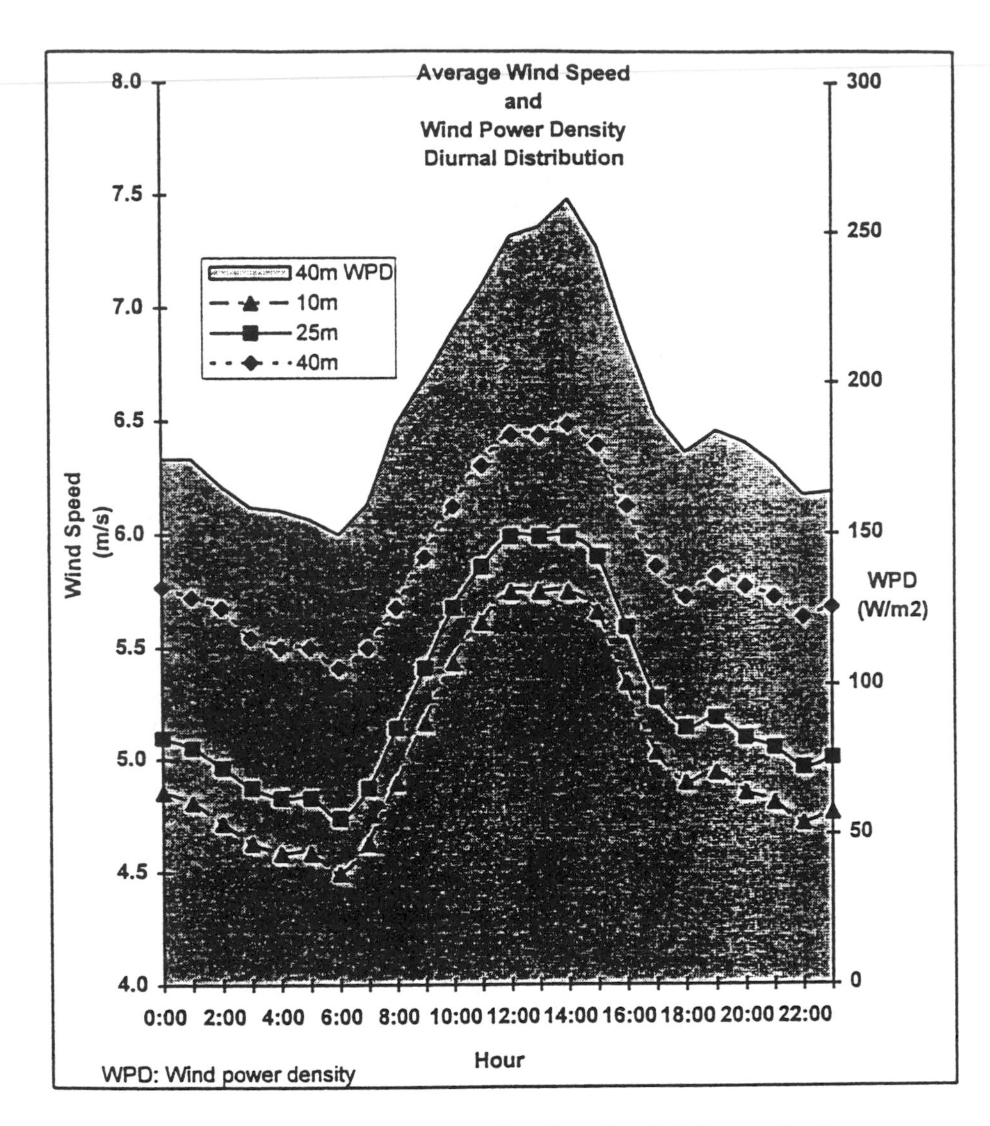

FIGURE 18B.10 Diurnal pattern of wind speed and wind power density.

The wind power density equation should only be used for instantaneous ($n = 1$) or multiple hourly average wind speed values where $n > 1$. It should not be used for a single long-term average such as yearly. The following example shows why this is true:

Suppose that the average measured wind speed at a site is 6.7 m/s (15 mph) over a 2-hour period. First, calculate the WPD using this average ($n = 1$). Assume an air pressure of 101,325 Pa and an air temperature of 288 K. The WPD is 184 W/m^2. Now suppose that the average wind speed for the first hour was 4.5 m/s (10 mph) and for the second hour 8.9 m/s (20 mph). A calculation of the WPD using both average values ($n = 2$) yields an average wind power density of 246 W/m^2. The correct calculation reveals that there is nearly 34% more wind power available at the site than indicated by using the single, 2-hour average wind speed.

However, if only summarized wind data is available, an average WPD can be estimated using:

$$WPD = \frac{1}{2}\bar{\rho}\sum_{j=1}^{c} f_j V_j^3 \tag{18B.5}$$

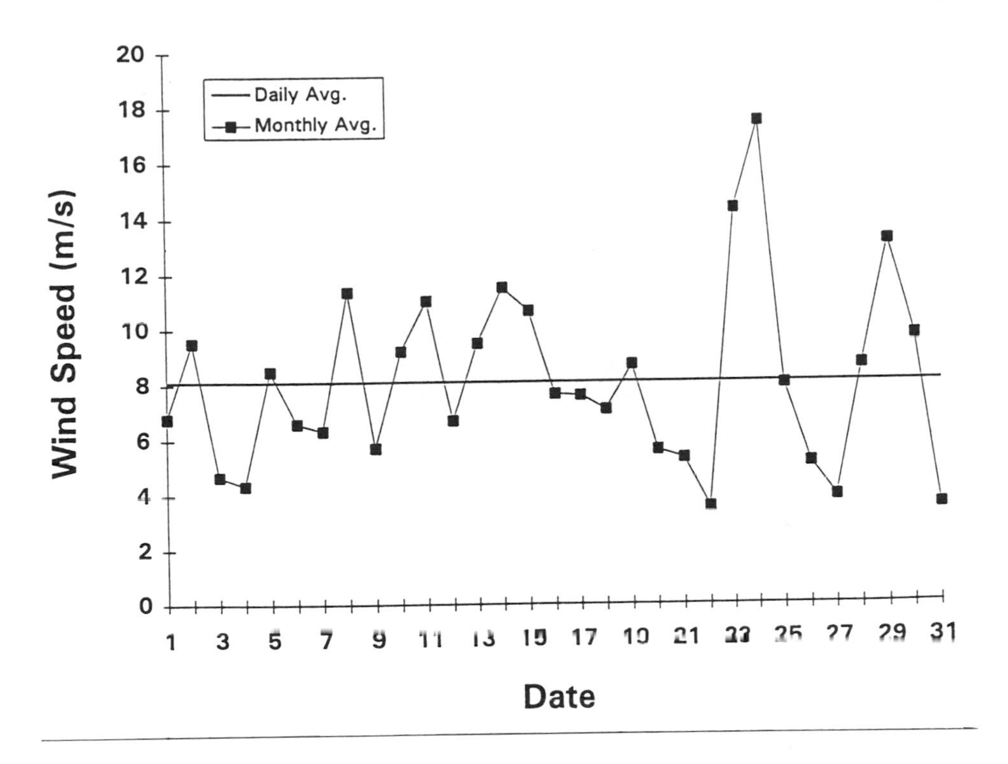

FIGURE 18B.11 Sample monthly report of daily average wind speed.

where

$\overline{\rho}$ = the mean air density
c = the number of wind speed classes
f_j = the frequency of occurrence of winds in the jth class
V_j = the median wind speed of the jth class

In those cases in which only wind data that have not been summarized are available, the mean WPD can be roughly estimated from the annual average wind speed ($\overline{V}$), as follows:

$$WPD = 0.955\overline{\rho}\overline{V}^3 \quad \textbf{(18B.6)}$$

Equation 18B.6 assumes that the wind speed frequency distribution follows a Rayleigh distribution, which derives from the *Weibull distribution*. The Weibull distribution gives the proportion of time, *t(v)*, for which the wind speed exceeds the value *v*. It describes the distribution of wind speeds at any given site:

$$t(V) = \exp\left[\left(\frac{-V}{V_c}\right)^k\right] \quad \textbf{(18B.7)}$$

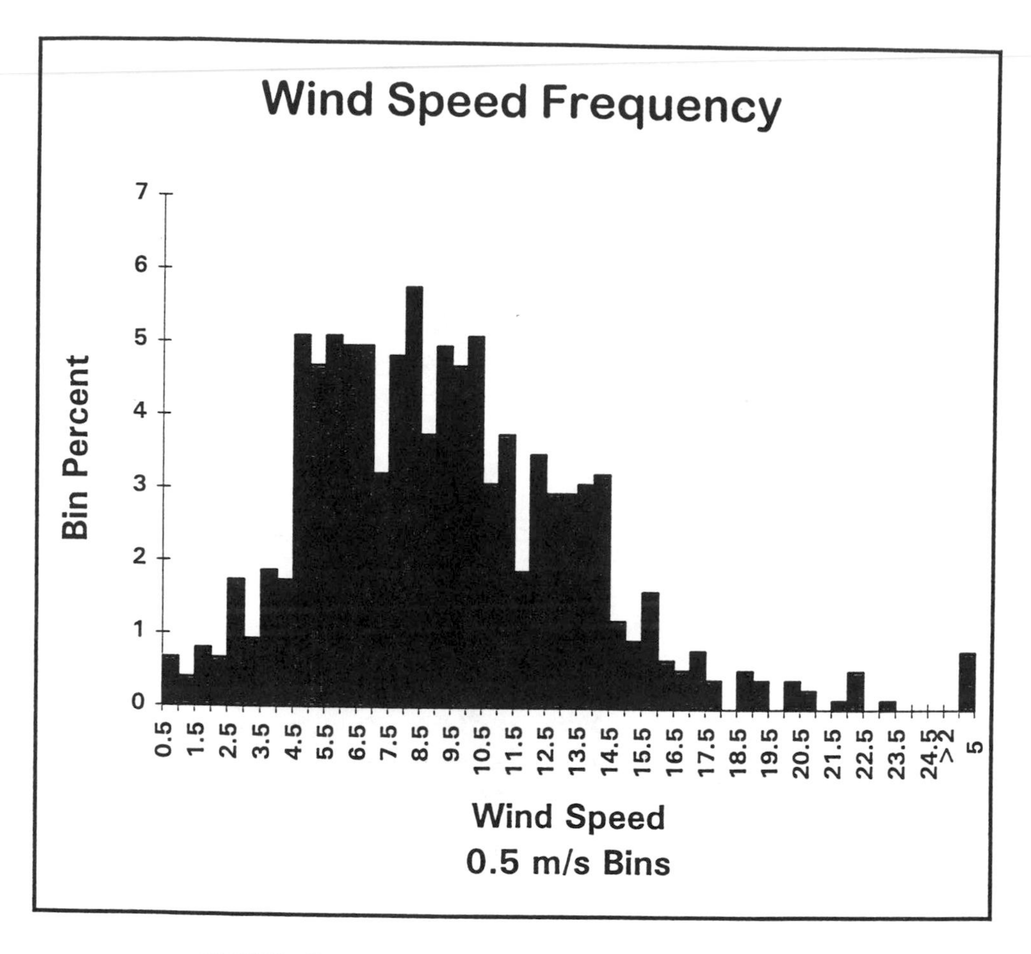

FIGURE 18B.12 Sample monthly report of wind speed frequency.

where

V_c = the site's characteristic wind speed, or Weibull scaling parameter
k = the Weibull shape parameter

Weibull statistics provide a convenient way to summarize information about wind regimes for a given period of time. For instance, the Weibull shape parameter usually ranges from 1 to 4, with the higher values often associated with trade winds. The lower the Weibull shape parameter, the broader is the distribution of wind speeds at a station. When the shape parameter is 2, the function is known as the *Rayleigh distribution*. Because wind regimes in many areas of the continental United States often approximate a Rayleigh distribution, wind prospectors can use this distribution when better data are not available, by using Eq. (18B.6). If the Weibull shape parameter k is known, or can be estimated, then the mean wind power density can be estimated more precisely than in Eq. (18B.6) by using:

$$WPD = \frac{1}{2}\bar{\rho}\bar{V}^3\left[\frac{\Gamma\left(1+\frac{3}{k}\right)}{\Gamma^3\left(1+\frac{1}{k}\right)}\right] \quad \mathbf{(18B.8)}$$

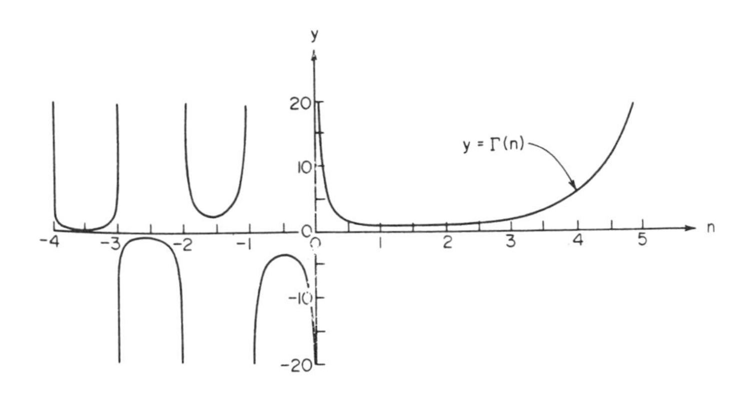

Scheme 18B.1 Gamma function.

where

k = the Weibull shape parameter

Γ = the Gamma function,* $\Gamma(n) = \int_0^\infty x^{n-1} e^{-x}\, dx$

18B.3 Micrositing

Micrositing is the third stage of wind resource assessment. Its objective is to quantify the small-scale variability of the wind resource over the terrain targeted for development as a result of the area wind resource evaluation. Micrositing is used to position wind turbines on a parcel of land to maximize the energy output of the wind power plant. Two considerations are crucial in micro-siting: (1) siting the turbines to take maximum advantage of natural variations in the local wind flow, and (2) siting the turbines to minimize their exposure to wind disturbances induced by other operating turbines.

Micrositing requires a denser network of wind monitoring towers than resource evaluation. The methods and procedures given for area wind resource evaluation can also provide information necessary for micrositing. In most instances, however, micrositing will require more extensive wind measurements than does a general site evaluation. Micrositing requires time series averages and frequency distributions of wind speed and direction, including diurnal and seasonal variability. In addition, it is also wise to characterize wind shear and turbulence intensity within the site. Methods for calculating these parameters are found in "Variations in Wind Speed and Data Validation and Processing."

When siting a single wind turbine, wind measurements should be taken as close as possible to the proposed installation site. Unfortunately, there is no simple method for determining the number of monitoring stations needed for a larger facility. One anemometer per 20 turbines may be sufficient on relatively flat terrain. However, one anemometer for every 3 turbines might be required in more complex terrain. If cost permits, the ideal solution is to install as many measurement stations as practical. Alternatively, a small number of fixed measurement stations could be augmented with portable anemometry.

*The gamma function is shown in Scheme 18B.1. Numerical values can be obtained from tables such as Table 1005, Page 260 in *Tables of Integrals and Other Mathematical Data*, by H.W. Dwight, 4th ed., MacMillan, New York, 1964.

Within a wind power plant, wind turbines are often grouped in clusters, or arrays. Grouping the turbines takes advantage of the site's windiest locations and helps maximize energy production. One aspect of micrositing is determining the optimum spacing for individual turbines in an array.

Spacing turbines too closely together can reduce overall energy production and, in some cases, lead to early turbine failure. Wakes produced by operating turbines may interact to produce excessive turbulence and disrupt the wind flow. Differences in wind flow within an array are influenced not only by turbine wakes, but also by the terrain, vegetation, turbine spacing, and meteorological conditions. Because so many of these factors are site specific, there are few general guidelines for turbine layouts.

In some cases, wind flow modeling can help reduce the time and expense of the wind measurements and decrease the risks inherent in wind turbine siting. Two types of models can be used in micrositing: (1) numerical, or computer, modeling and (2) physical modeling. In physical modeling, a scale model of an area is studied in a wind tunnel. Both types of models approximate the terrain, vegetation, and atmospheric conditions of a wind development site. However, modeling only works well when validated with field data and should be used primarily to augment conventional siting techniques.

For example, field measurement studies used in conjunction with wind tunnel simulations and computer models do offer some insights into expected wake and array effects of proposed turbine layouts. Experimental data suggest that turbine wakes have the greatest impact on wind speed when the ambient wind speed and turbulence are low and the atmosphere is stable. In addition, spacing turbines at least 3 rotor diameters apart within a row and spacing rows a minimum of 8 to 10 diameters apart will reduce the impact of wake effects of most locations.*

Computer and physical models can also assist wind engineers in designing turbine arrays in complex areas. Computer models provide relatively simple assessments of the performance of array designs, but may not provide reliable estimates of turbulence within the array or the interaction of wakes from multiple turbines. These models have been verified only for small, two- to four-row arrays, so their general reliability for use with larger arrays is still unknown. Wind tunnel simulations of array designs have also been used to a limited extent. However, their reliability has not yet been verified with field data.

Given the complexity and importance of micrositing and that there is as much art as science in it, the novice wind developer should hire experts to assist with this process. In addition, many wind developers continue to monitor wind conditions after the turbines are installed. Ongoing measurements allow turbine operators to evaluate machine performance as a function of wind conditions and air density, monitor local variations in the wind, analyze impacts of wake and array losses on overall performance, improve the long-term record of wind conditions at the site, and determine if the siting process was successful and that wind turbine output estimates were accurate. Because turbine siting is so dependent on specific site characteristics, follow-up siting studies are critical for improving future efforts.

18B.4 International Wind Resource Assessment

Wind developers are interested in more than just the U.S. wind resource. Industrialized nations (particularly in Europe), developing countries, and remote military installations represent sizeable markets for the U.S. wind industry. International wind resource assessments are key to capitalizing on wind energy's large export potential. This section highlights considerations in undertaking a wind resource assessment for a foreign country.

*Dennis Elliott, *Status of Wake and Array Loss Research*, presented at AWEA's Windpower '91 Conference, Palm Springs, CA, September 1991.

There is no single formula for conducting a wind resource assessment that applies to every country. Weather, topography, and in-country technical expertise all vary from one country to another. More importantly, the existence of reliable wind data cannot be assumed. With these considerations in mind, one place to begin is with the world wind resource map created by DOE's Pacific Northwest Laboratory in 1981.

The map, shown in Figure 18B.13, indicates wind power classes with a color scale, in which the darker shade of blue, the greater is the wind energy potential. The map reflects the known broad-scale global wind patterns and shows clearly the greater complexity of wind patterns over land as compared with those over the sea. However, much greater regional and local variability exists than can be shown on the map. For this reason, analysis of existing regional and local wind data and new wind measurements is necessary.

Since 1993, organizations such as the National Renewable Energy Laboratory, the Pacific Northwest Laboratory, Sandia National Laboratories, the U.S. Department of Energy, the U.S. Agency for International Development, and the American Wind Energy Association have provided technical assistance for wind resource assessments to developing countries, including Mexico, Indonesia, the Caribbean Islands, the former Soviet Union, Brazil, Chile, and Argentina.

Resource assessments in these countries have focused on the development of rural wind power applications. As part of the effort in Mexico, a new, two-pronged strategy for wind resource assessment in developing countries was developed.

Maps of the areas suitable for both utility-scale wind power plant development and rural applications were prepared for Mexico. Both maps are shown in Figure 18B.14. Note that a new wind power classification was used for the development of the rural applications map. The reasoning behind this change was that the level of wind resource needed to make a rural wind system economically viable is significantly less than that required for a utility-scale wind development. The rural power classifications are based on wind power densities at a height of 30 m (98 ft), the hub height of turbines used in many rural wind systems. The rural power classes range from 1 to 4, with power classes 3 and 4 best suited for rural wind power development. In contrast, the utility-scale classification is derived from the 7-power classification system used in the *Wind Energy Resource Atlas of the United States,* described in Section 18B.1.

18B.5 Federal Wind Resource Assessment Programs

During 1994 and 1995, the U.S. Department of Energy's National Wind Energy Program began supporting selected wind resource assessment programs to encourage the development of wind power in the U.S. Program participants receive guidance, technical assistance, and co-funding. The programs include:

The *Utility Wind Resource Assessment Program (U*WRAP),* sponsored by the National Wind Coordinating Committee, a consortium of leaders from the wind industry, the utility industry, and federal and state government agencies. The cost-shared program assists utilities in conducting detailed wind resource measurements.

The *Cooperative Networks for Renewable Resource Measurements (CONFRRM)* program, which supports measurement programs for solar radiation and wind resource evaluations. The multi-year program is helping to establish many wind measurement stations. The stations provide high quality wind data to the National Renewable Energy Laboratory for public distribution through the laboratory's Renewable Resource Data Center.

The *Sustainable Technology Energy Partnerships (STEP)* program, which provides funding for pilot projects undertaken by state energy offices in partnership with industry, academia, and utilities. Eligible pilot projects include bird population studies, wind technology deployment, and wind resource assessments.

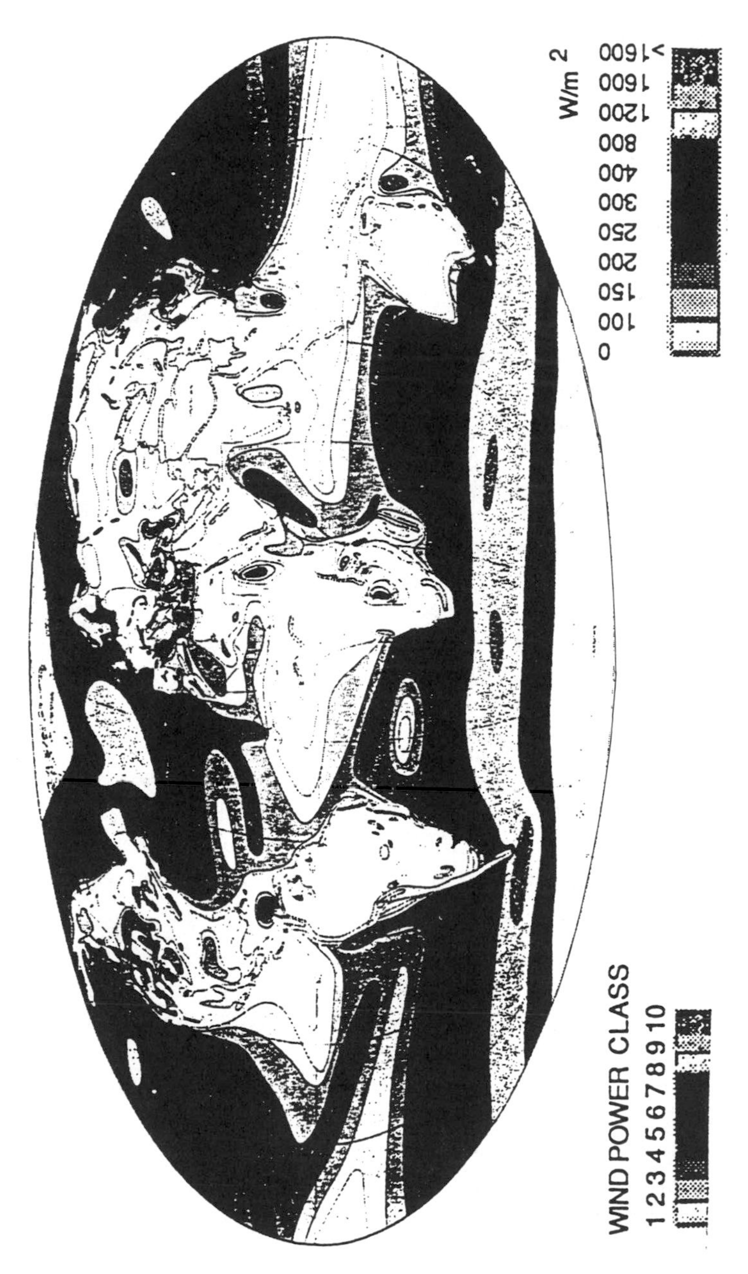

FIGURE 18B.13 Global wind resource map.

MEXICO—PRELIMINARY WIND RESOURCE MAP FOR UTILITY-SCALE APPLICATIONS

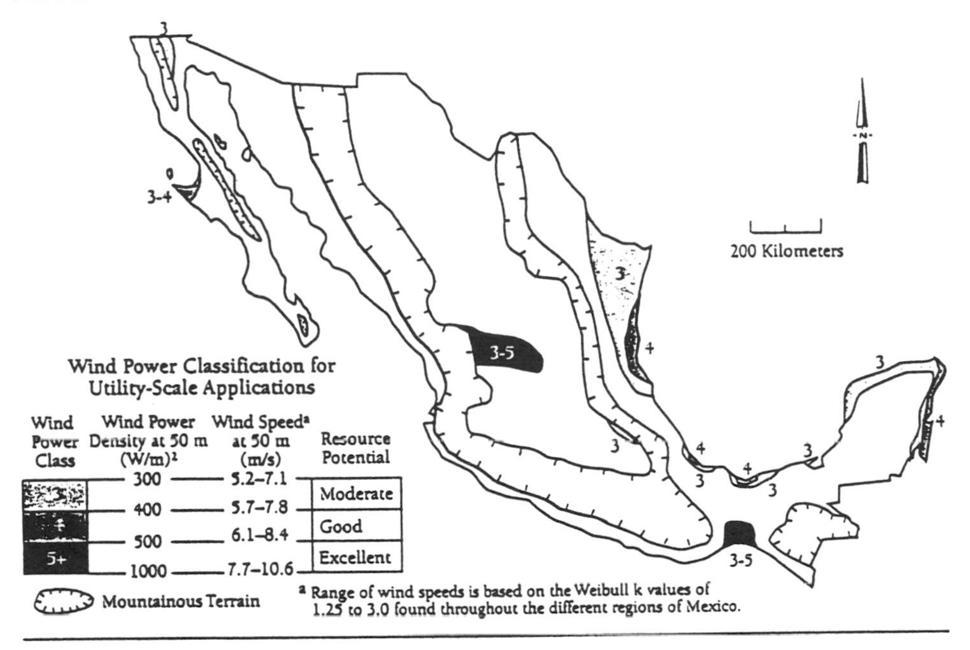

MEXICO—PRELIMINARY WIND RESOURCE MAP FOR RURAL POWER APPLICATIONS

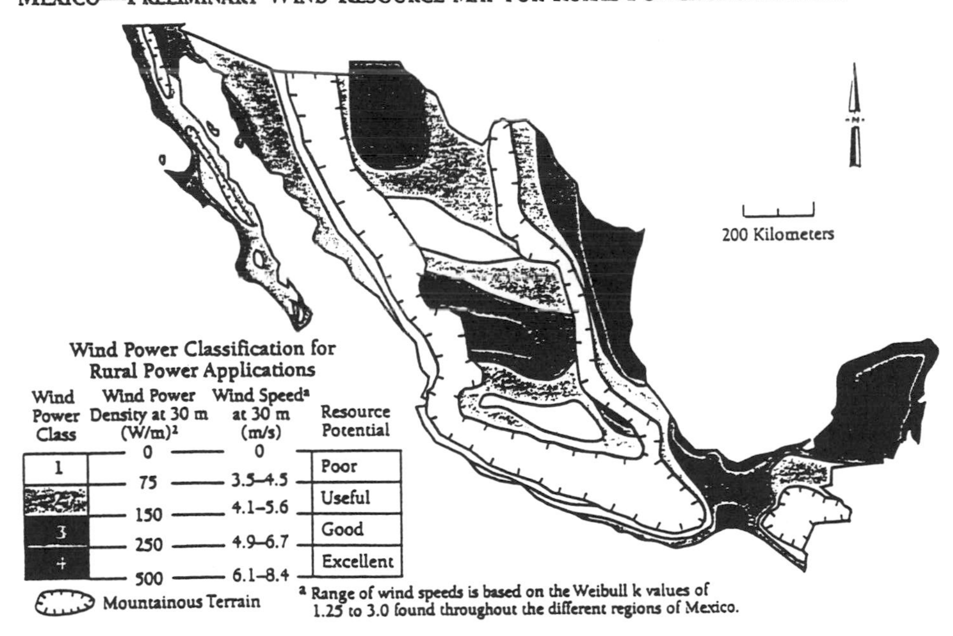

FIGURE 18B.14 Comparison of Mexico wind resource maps for utility-scale and rural power applications.

18B.6 Contacts for Wind Resource Information

Organizations

The American Wind Energy Association (AWEA)
122 C Street, NW
Washington, DC 20001
Telephone: (202) 383-2500
Facsimile: (202) 383-2505

National Climatic Data Center
Climate Services Division
151 Patton Avenue, Room 120
Asheville, NC 28801-5001
Telephone: (704) 271-4800
Facsimile: (704) 271-4876

The National Renewable Energy Laboratory
Wind Technology Division
1617 Cole Boulevard
Golden, CO 80401-3393
Telephone: (303) 384-6963
Facsimile: (303) 384-6901

Regional Climate Centers

High Plains Climate Center
Room 236, L.W. Chase Hall
University of Nebraska—Lincoln
Lincoln, NB 68583-0728
Telephone: (402) 472-6709
Facsimile: (402) 472-6614

Midwest Regional Climate Center
2204 Griffiths Drive
Champagne, IL 61820-7495
Telephone: (217) 244-8226
Facsimile: (217) 244-0220

Northeastern Regional Climate Center
1123 Bradfield Hall
Cornell University
Ithaca, NY 14853
Telephone: (607) 255-1751
Facsimile: (607) 255-2106

Southeast Regional Climate Center
1201 Main Street, Suite 1100
Columbia, SC 29201
Telephone: (803) 737-0849
Facsimile: (803) 765-9080

Southern Regional Climate Center
260 Howe-Russell Geoscience Complex
Department of Geography
Louisiana State University
Baton Rouge, LA 70803
Telephone: (504) 388-5021
Facsimile: (504) 388-2912

Western Regional Climate Center
P.O. Box 60220
Reno, NV 89506-0220
Telephone: (702) 677-3105
Facsimile: (702) 677-3131

General References

Characteristics of the Wind (Walter Frost and Carl Aspliden, Chapter 8 in *Wind Turbine Technology*, edited by David A. Spera, 1994: available from The American Society of Mechanical Engineers, 345 East 47th Street, New York, NY 10017).

Recommended Practice for the Siting of Wind Energy Conversion Systems (American Wind Energy Association (AWEA), AWEA Standard 8.2—1993: available from AWEA, 122 C Street NW, Washington DC 20002).

Standard Procedures for Meteorological Measurements at a Potential Wind Turbine Site (document number 8,1, 1986: available from the American Wind Energy Association, 122 C Street NW, Washington DC 20002).

Wind Energy Resource Atlas of the United States (prepared by Pacific Northwest Laboratory for the U.S. Department of Energy, DOE/CH10093-4, March 1987: available from AWEA, 122 C Street NW, Washington DC 20002).

Wind Characteristics: An Analysis for the Generation of Wind Power (J. Rohatgi and V. Nelson, 1994: available from Alternative Energy Institute, West Texas A&M University, Canyon, Texas 79016).

An Assessment of the Available Windy Land Area and Wind Energy Potential in the Contiguous United States (D. Elliott, L. Wendell and G. Gower, Report No. PNL-7789, 1991: available from National Technical Information Service, 5285 Port Royal Road, Springfield, VA 22151).

Powering the Midwest: Renewable Electricity for the Economy and the Environment (Union of Concerned Scientists (UCS), 1993: available from UCS, 26 Church Street, Cambridge, MA 02238).

Utility Wind Systems

The Meteorological Aspects of Siting Large Wind Turbines (T. Hiester and W. Pennell, Report No. PNL-2522, 1981: available from National Technical Information Service, 5285 Port Royal Road, Springfield, VA 22151).

Planning Your First Wind Power Project: A Primer for Utilities: Everything You Need to Know to Bring your First Wind Power Plant On-Line (K. Conover and E. Davis, Report No. EPRI TR-104398, 1994: available from the EPRI Distribution Center, P.O. Box 23205, Pleasant Hill, CA 94523).

Siting Guidelines for Utility Application of Wind Turbines (W. Pennell, Report No. EPRI AP-2795, 1983: available from Electric Power Research Institute, Research Reports Center, Box 50490, Palo Alto, CA 94303).

Status of Wake and Array Loss Research (Dennis Elliott, presented at AWEA's Windpower '91 Conference, Palm Springs, CA, September 1991: available from AWEA, 122 C Street, NW, Washington DC 20002).

Small Wind Systems

A Siting Handbook for Small Wind Energy Conversion Systems (H. Wegley, J. Ramsdell, M. Orgill and R. Drake, Report No. PNL-2521 Rev. 1, 1980: available from National Technical Information Service, 5285 Port Royal Road, Springfield, VA 22151).

Understanding Your Wind Resource (Newell Thomas, American Wind Energy Association (AWEA): available from AWEA, 122 C Street NW, Washington, DC 20002).

International Wind Resource Assessments

Caribbean and Central America Wind Energy Assessment (D. Elliott, presented at AWEA's Windpower '87 Conference, San Francisco, CA, October, 1987: available from AWEA, 122 C Street NW, Washington, DC 20002).

Mexico Wind Resource Assessment Project (M. Schwartz and D. Elliott, presented at AWEA's Windpower '95 Conference, Washington, DC, March, 1995: available from AWEA, 122 C Street NW, Washington, DC 20002).

European Wind Atlas (I. Troen and E.L. Petersen, 1989: available from Department of Meteorology and Wind Energy, Risø National Laboratory, Roskilde, Denmark).

World-Wide Wind Resource Assessment (N.J. Cherry, D.L. Elliott, and C.I. Aspliden, presented at AWEA's Wind Workshop V Conference, Washington, DC, October 1981: available from AWEA, 122 C Street NW, Washington, DC 20002).

18C

Availability of Renewable Resources: Biomass Energy

Marjorie A. Franklin
Franklin Associates, Ltd.

Frank Kreith
Consulting Engineer

Introduction

Biomass can be defined as organic matter available on a renewable basis for conversion to energy and heat. It includes agricultural and wood residues, crops grown especially for biomass, animal waste, and the organic portions of municipal solid waste. Biomass has a potential for many applications and is probably the oldest form of energy used by man. The Union of Concerned Scientists (UCS) claims that about 833 megawatts (MW) of biomass electricity generating capacity is currently available in the Midwest alone, most of it in Michigan, Minnesota, Ohio, and Wisconsin, using primarily residues from forest product industries and municipal solid wastes (see Table 18C.1).

The American Paper Institute (API) (U.S. Pulp and Paper Industries Energy Use, New York, 1992) has estimated that today the paper and pulp industry is the leading user of wood and wood wastes for energy production in the United States. When wood is processed at lumber, pulp, and paper mills, large quantities of residue are created (e.g., wood scraps, bark, and sawdust). The pulp and paper industry is using these residues to meet its own energy needs, and API has estimated that these industries generated, in 1991, about 1.4 quadrillion Btu, or 1.6% of the U.S. energy supply. In addition, some power plants, particularly in California, use wood chips as primary or supplementary fuel, but no specific data on the cost and efficiency of those plants are available.

Since almost all of the waste at large lumber, pulp, and paper mills is already used for energy production, little expansion for the use of this resource is likely. For expansion of biomass as a fuel source, special plantations would have to be developed or residues from logging and agricultural activities would have to be collected. At present, it has been estimated that wood chips might be expected to cost anywhere from $26 per dry ton to $38 per dry ton ($1.50 to $2.20 per million Btu). By comparison, coal was recently sold to electric utilities at prices ranging from $1.25 per million Btu in Minnesota to $1.06 per million Btu in Michigan. Moreover, if the market for fuel wood were to expand, prices could be expected to rise, as has been observed in California. The cost of collecting and delivering forest residues in California was reported to vary from $25 to 45 per dry ton, but in time of a fuel shortage, the spot market price has recently risen to as high as $54 per ton (PG&E Biomass Qualifying Facilities, Lessons Learned, Scoping Study Phase I, San Ramon, California, PG&E, 1991).

0-8493-2514-5/96/$0.00+$.50

TABLE 18C.1 Biomass Electric Capacity (megawatts)

State	Crop Residues	Landfill Methane	Municipal Solid Waste	Wood and Mill Residue	Total
Illinois	0.0	22.2	0.0	0.0	22.2
Indiana	0.0	0.0	0.0	0.0	0.0
Iowa	17.8	0.2	4.0	0.0	22.0
Kansas	0.0	0.0	0.0	0.0	0.0
Michigan	0.0	8.4	99.9	255.3	363.6
Minnesota	0.0	0.2	113.4	10.0	123.6
Missouri	0.0	0.0	0.0	0.0	0.0
Nebraska	0.0	0.0	0.0	0.0	0.0
North Dakota	9.0	0.0	0.0	0.0	9.0
Ohio	0.0	0.0	114.8	29.5	144.3
South Dakota	0.0	0.0	0.0	0.0	0.0
Wisconsin	0.0	19.8	13.5	114.6	147.9
Midwest	26.8	50.8	345.6	409.4	832.6

Source: J. Hamrin and N. Rader, *Investing in the Future: A Regulator's Guide to Renewables (Appendix),* prepared for the National Association of Regulatory Utility Commissioners, Washington, D.C. (draft, 1992).

Although opportunities exist to expand biomass-derived energy resources, there is little evidence that commercial activities will develop at a sufficiently rapid pace to generate appreciable additional biomass energy within the 10- to 15-year time frame envisioned for this handbook. As noted by the UCS, in the next decade, the price of biomass resources is not likely to be competitive with that of coal or natural gas. Moreover, some types of wood residues and energy crops are troublesome to burn in existing power plants. Despite the availability and potential of biomass feedstocks, near-term prospects for appreciable commercial penetration appear dim without less expensive growing, collecting, and hauling methods and more efficient biomass power plant technology. This is probably best illustrated in graph, prepared by the UCS, shown here in Figure 18C.1. In one graph, the potential supply is plotted as a function of cost for conventional power plant technology using biomass as a fuel. It can be seen that there is practically no potential for conversion with energy crops below a cost of 7 cents/kWh. For biomass wastes and residues, shown in the second graph, the minimal cost is on the order of 5 cents/kWh, but no appreciable amount of power can be generated at costs less than 8 cents/kWh. Obviously, these costs are in excess of what coal-fired or natural gas–fired power plants can provide.

An extensive review of the technical and economic status of wood energy feedstock production has been conducted at the Oak Ridge National Laboratory (Technical and Economic Status of Wood Energy Feedstock Production by Robert D. Perlack and Lynn L. Wright, *Energy,* vol. 20, #4, pp. 279–284, 1995). This review updates a previous evaluation conducted at Oak Ridge National Laboratory in 1987 to evaluate the economics of producing wood energy feedstocks using short rotation forestry techniques. The earlier review concluded that a delivered cost (growing, harvesting, handling, and hauling) of wood feedstocks would range between $3.00 and $4.10 per GJ (1987 dollars) using available planting materials and harvesting technology. The review also concluded that these costs could be reduced to under $2 per GJ by the year 2000 if advances from tree breeding and genetics, crop management, and harvesting technology research were adopted.

The 1995 review concludes that the 1987 findings are still valid but that the estimates of technical progress and cost reduction were overly optimistic. These optimistic projections about technical progress and cost reduction assumed significantly faster research progress and higher conventional fuel costs, which did not occur during the 1980s and 1990s. The 1995 review still projects that producing wood feedstock at delivered costs under $2 per GJ in 1994 dollars is possible but concludes that this goal will not likely be met before the year 2010. The review further estimates that current delivered costs range from about $2.30 per GJ in the Pacific Northwest to a high of $3.30 per GJ in the Midwest corn belt. Realistic estimates of current costs are between $2.60 and

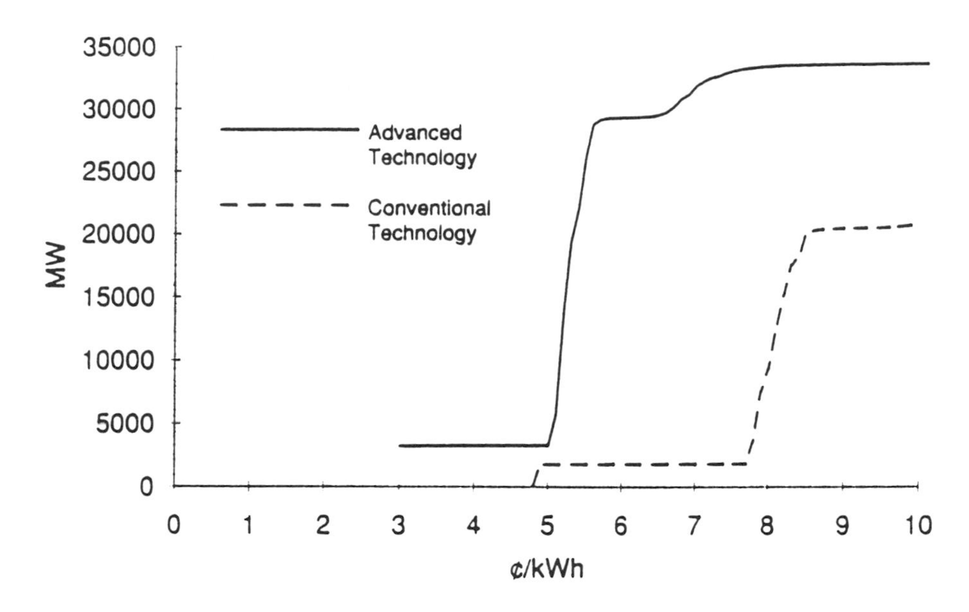

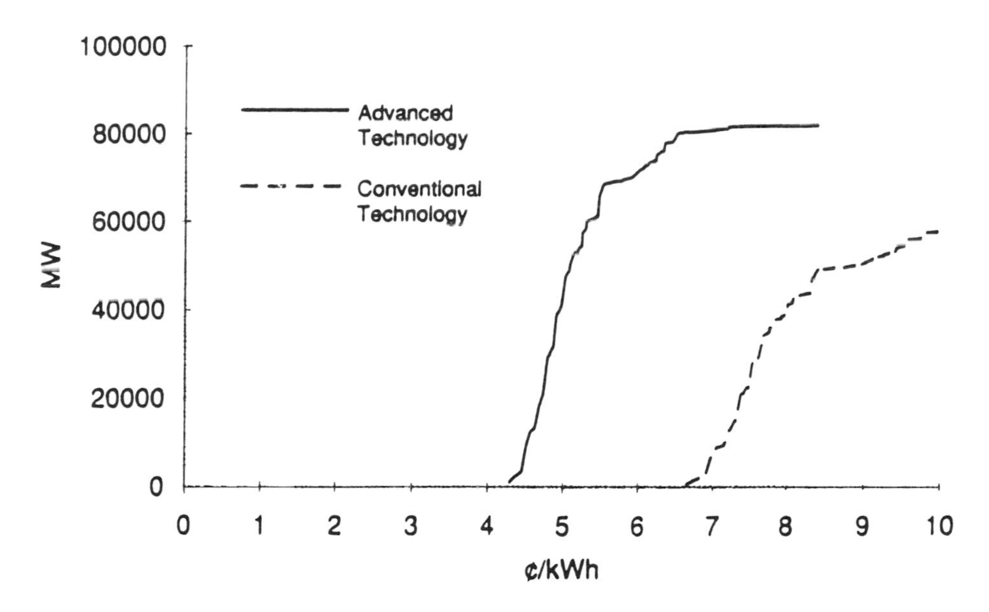

FIGURE 18C.1 Supply versus cost for energy crops and for biomass wastes and residues with advanced and conventional technology.

$2.90 per GJ in 1994 dollars. At current biomass prices with available conversion technology, it does not appear that energy generation from feedstocks using short rotation forestry techniques is competitive, nor will it be competitive within the time frame of 10 years targeted for the presentations in this handbook. Consequently, it is suggested that the reader follow the current literature, in case progress should occur more rapidly than current estimates indicate.

The only kind of biomass technology that has achieved market penetration and is likely to continue its growth in the next decade is production of electric power from municipal solid waste.

Even though the cost of utilizing municipal solid waste for power production is higher than that of power plants fired by natural gas or coal, there is a growing need to dispose of these wastes. Thus, the cost of electric energy from waste-to-energy plant is partially offset by savings in tipping fees for landfills. Despite vigorous efforts in recycling and source reduction, the best estimates from EPA and industry indicate that substantial amounts of wastes, of the order of 150 million tons per year by 2000, will have to be disposed of either in landfills or in incinerators, and thus assure a continued supply of fuel for the future.

18C.1 Characterization of Municipal Solid Waste (MSW) in the United States

This section has been extracted from the most recent in a series of reports released by the U.S. Environmental Protection Agency (EPA) to characterize MSW in the United States (U.S. Environmental Protection Agency. Characterization of Municipal Solid Waste: 1995 Update. (EPA 530-R-96-011). March 1996.). It characterizes the national waste stream based on data through 1994 and makes projections using three possible scenarios for recycling/recovery in the year 2010 (30, 35, and 40%). Recovery scenarios depend on local situations and states, and local communities can set their own goals.

Identifying the components of the municipal solid waste (MSW) stream is an important step toward addressing the issues associated with using it for energy generation. MSW characterizations, which analyze the quantity and composition of the municipal solid waste stream, involve estimating how much MSW is generated, recycled, combusted, and disposed of in landfills. This section characterizes the municipal solid waste stream of *the nation as a whole.* Local and regional variations are not addressed, but suggestions for use of the information in this report by local planners are included in Chapter 1 of the EPA report.

The method used for the characterization of MSW for this section estimates the waste stream on a nationwide basis by a "material flows methodology." EPA's Office of Solid Waste and its predecessors in the Public Health Service sponsored work in the 1960s and early 1970s to develop this methodology, which is based on production data (by weight) for the materials and products in the waste stream, with adjustments for imports, exports, and product lifetimes.

Materials and Products in MSW

In 1994, generation of municipal solid waste totaled 209 million tons. A breakdown by weight of the *materials* generated in MSW in 1994 is shown in Table 18C.2. Paper and paperboard products are the largest component of municipal solid waste by weight (39% of generation) and yard trimmings are the second largest component (15% of generation). Five of the remaining materials in MSW—glass, metals, plastic, wood, and food wastes—range between 6 and 9% each by weight of total MSW generated. Other materials in MSW, including rubber and leather, textiles, and small amounts of miscellaneous wastes, made up approximately 9% of MSW in 1994.

Most of the materials in MSW have some level of recovery for recycling or composting. This is illustrated for 1994 in Table 18C.2. Since each material category (except for food wastes and yard trimmings) is made up of many different products, some of which may not be recovered at all, the overall recovery rate for any particular material will be lower than recovery rates for some products within the materials category.

The highest recovery rate shown in Table 18C.2 is that for nonferrous metals other than aluminum (66% of generation). This is because the lead in lead–acid batteries is recovered at very high rates. Aluminum is recovered at approximately 38% of generation overall, even though aluminum cans are recovered at rates above 60%. Likewise, the overall recovery rate for paper and paperboard is 35%, even though corrugated containers are recovered at rates above 50%.

The many products in MSW are grouped into three main categories: durable goods (for example, appliances), nondurable goods (for example, newspapers), and containers and packaging

TABLE 18C.2 Generation and Recovery of Materials in MSW, 1994 (in millions of tons and percentage of generation of each material)

	Weight Generated	Weight Recovered	Percent of Generation
Paper and paperboard	81.3	28.7	35.3
Glass	13.3	3.1	23.4
Metals			
Ferrous metals	11.5	3.7	32.3
Aluminum	3.1	1.2	37.6
Other nonferrous metals	1.2	0.8	66.1
Total metals	15.8	5.7	35.9
Plastics	19.8	0.9	4.7
Rubber and Leather	6.4	0.5	7.1
Textiles	6.6	0.8	11.7
Wood	14.6	1.4	9.8
Other materials	3.6	0.8	20.9
Total Materials in Products	161.3	41.8	25.9
Other Wastes			
Food Wastes	14.1	0.5	3.4
Yard Trimmings	30.6	7.0	22.9
Miscellaneous Inorganic Wastes	3.1	Neg.	Neg.
Total Other Wastes	47.8	7.5	15.7
Total Municipal Solid Waste	209.1	49.3	23.6

Note: Includes wastes from residential, commercial, and institutional sources. Neg. = Less than 50,000 tons or 0.05 percent. Numbers in this table have been rounded to the first decimal place.

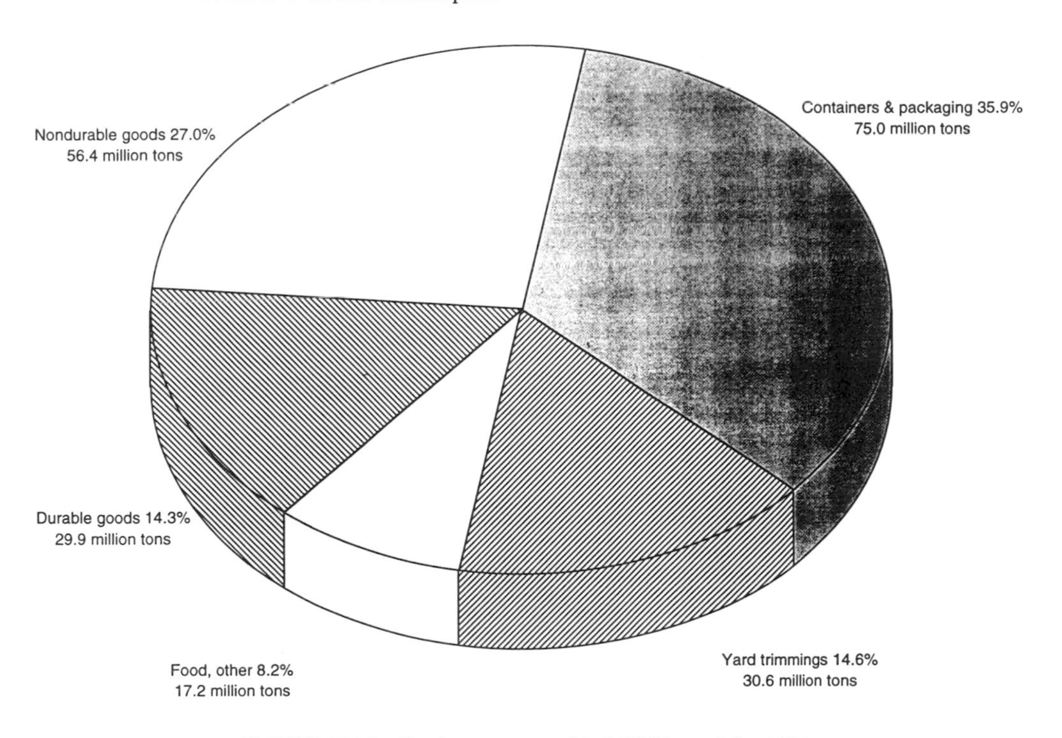

FIGURE 18C.2 Products generated in MSW by weight, 1994.

(Figure 18C.2). The materials in MSW are generally made up of products from each category. There are exceptions, however. The durable goods category contains no paper and paperboard. The nondurable goods category includes only small amounts of metals and essentially no glass or wood. The containers and packaging category includes only very small amounts of rubber, leather, and textiles.

Generation and recovery of MSW, broken down by materials, are shown in Table 18C.2. Overall, the materials in durable goods were recovered at a rate of approximately 15% in 1994. The nonferrous metals were recovered at a rate of approximately 66% because of the high rate of recovery of lead–acid batteries. The recovery of these batteries also accounts for the high rate of recovery of "other materials," which are the nonlead components of the batteries. Considerable amounts of ferrous metals are recovered from appliances in the durables category, and some rubber is recovered from tires.

Overall, recovery in the nondurable goods category was estimated to be 22% in 1994. In this category, large amounts of newspapers, office papers, and some other paper products are recovered. Recovery from the containers and packaging category is the highest of these categories—33% of generation. Aluminum packaging was recovered at 55% in 1994 (mostly aluminum beverage cans), and steel (mostly cans) was recovered at over 51%. Paper and paperboard packaging recovery was estimated at 45% overall in 1994, with corrugated containers accounting for most of that tonnage. Glass containers were estimated to have been recovered at 26% overall, while wood packaging (mostly pallets) was estimated to have been recovered at 14% of generation. Plastic containers and packaging were estimated to have been recovered at an overall rate of 8% in 1994, with most of the recovered plastics being soft drink bottles and milk and water bottles.

Management of MSW

The breakdown of how much waste went to recycling and composting, combustion, and landfills in 1994 is shown in Figure 18C.3. Recovery of materials for recycling and composting was estimated to have been 49 million tons, or 24% of generation, in 1994. Combustion of MSW (nearly all with energy recovery) was estimated to have been 33 million tons, or 16% of generation, in 1994. The remainder, 127 million tons of MSW (61% of generation), was assumed to have been landfilled (although small amounts may have been littered or self-disposed, e.g., on farms). Recovery rates have increased from 13% in 1988 to 17% in 1990 to 24% in 1994 (Figure 18C.4).

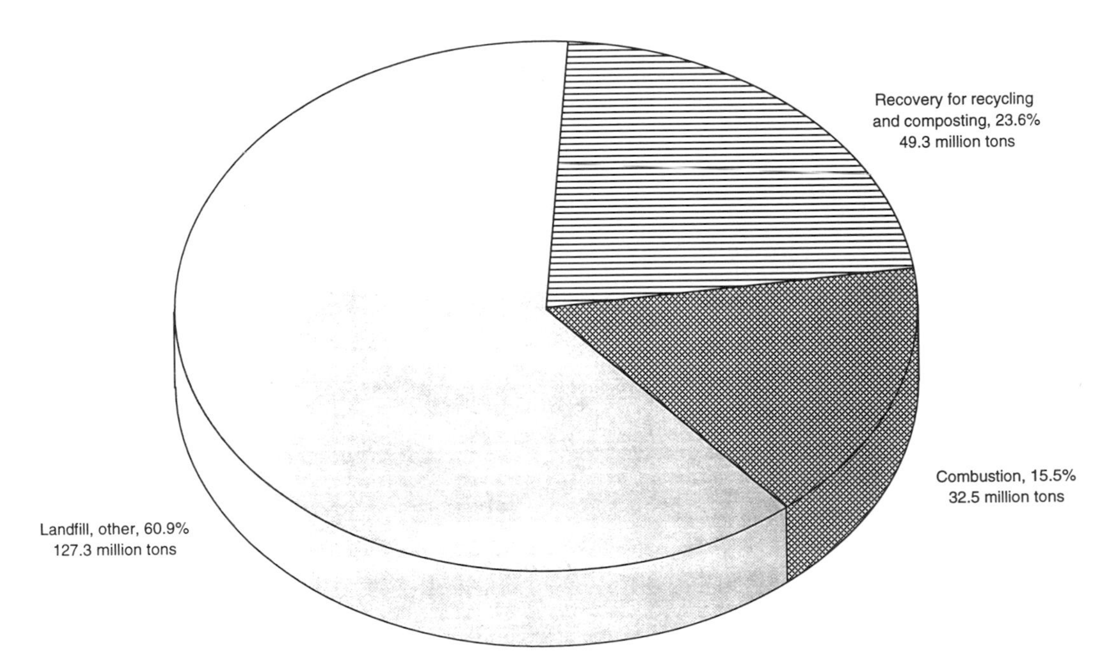

FIGURE 18C.3 Management of MSW in the United States, 1994.

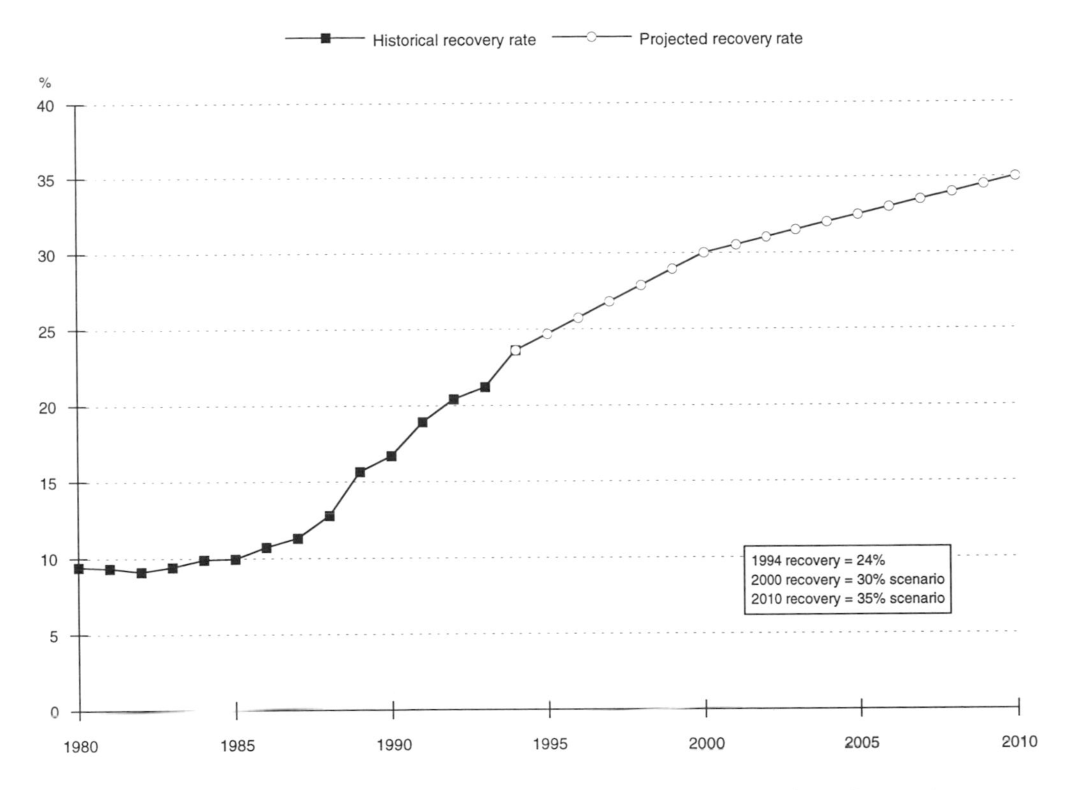

FIGURE 18C.4 Recovery for recycling and composting in percent of total MSW generation.

The EPA study looked at a range of recovery scenarios from 30% to 40% nationwide for the year 2010. A midrange projected scenario of 35% in the year 2010 was used to illustrate the effects of recovery on future municipal solid waste management. To achieve this level of recovery, it was assumed that local, state, and federal agencies will continue to emphasize recycling and composting as a priority; that industries will continue to make the necessary investments in recovery and utilization of materials; that state and local governments will continue to expand programs designed to keep yard trimmings out of landfills; and that most U.S. citizens will have access to some sort of recovery program by the year 2010.

Additional Perspectives on MSW

Per capita generation of MSW is an important parameter used by solid waste management planners. During the period 1960 to 1994, per capita generation of MSW increased steadily from 2.7 pounds per person per day to 4.4 pounds per person per day. During the period of 1994 to 2010, per capita generation of products (including packaging) is projected to continue to increase if present trends continue. The per capita generation of yard trimmings, however, is projected to decline if current source reduction activities at the state and local levels continue. Nevertheless per capita generation could increase from 4.4 pounds per person per day in 1994 to 4.8 pounds per person per day in 2010.

Residential and commercial sources of MSW are included here to characterize their composition. Commercial sources include stores, warehouses, institutions such as schools, and some industrial sites where packaging is generated. The source where the MSW is generated is highly relevant to management techniques, including its collection for waste-to-energy combustion.

Estimates of residential and commercial generation of MSW were made for EPA. Residential wastes (including wastes from multifamily dwellings) are estimated to be 55 to 65% of total

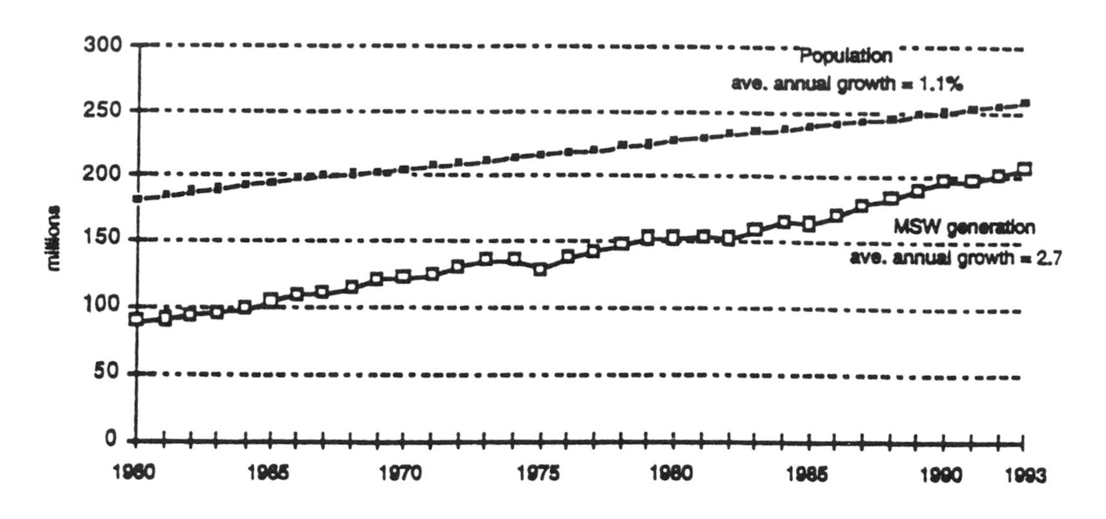

FIGURE 18C.5 U.S. population and municipal solid waste generation, 1960 to 1993.

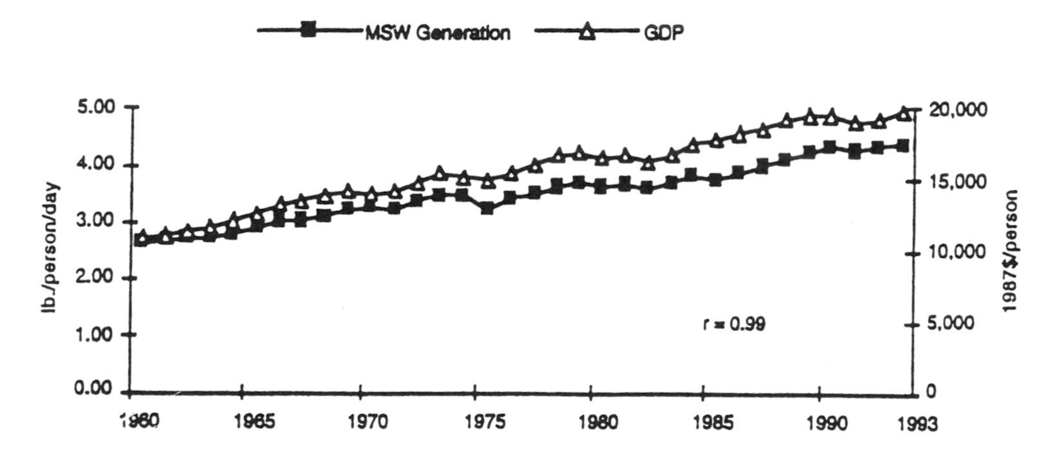

FIGURE 18C.6 Municipal solid waste generation and gross domestic product, 1960 to 1993.

generation, with commercial wastes ranging between 35 and 45% of generation. Local and regional factors such as climate and level of commercial activity contribute to the variations.

Several factors affect future MSW generation. Increasing population clearly contributes to increasing generation of MSW nationwide. In statistical language, the correlation coefficient *(r)* between MSW generation and population from 1960 to 1993 is .99, a high degree of correlation.

Population is not the only factor leading to increased MSW generation; historical trends show that MSW generation has been increasing more rapidly than population (Figure 18C.5). While average annual population growth over the last 33-year period was 1.1%, average annual growth of MSW generation was 2.7%. In other words, per capita generation of MSW increased over the historical period.

Many reasons have been suggested for the growth in per capita MSW generation, such as changes in lifestyles, more two-income wage earners in households, smaller households, and changes in the workplace (especially in offices). It seems clear that many of these reasons are related to changes in the level of economic activity, which has been generally upward except for occasional recessions. A plot of per capita MSW generation and economic activity as measured by gross domestic product (GDP) (in 1987 dollars per capita) as shown in Figure 18C.6. During the 33-year period, MSW per capita generation increased by 65% while GDP (in constant dollars) on a per capita basis

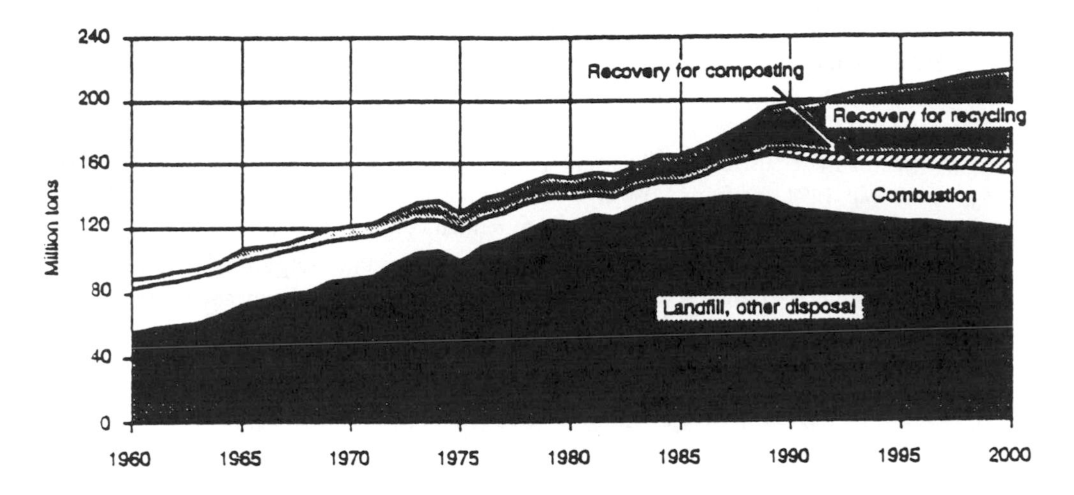

FIGURE 18C.7 Municipal solid waste management, 1960 to 2000.

increased by 82%. The correlation coefficient *(r)* between per capita MSW generation and per capita GDP is .99, a strong positive correlation.

On the basis of this preliminary analysis, it appears that population growth accounts for a portion of the increase in MSW generation, but that economic activity (and perhaps other factors such as household size) contributes to the increase over and above population growth.

Trends in MSW Generation, Recovery, and Discards

Generation of municipal solid waste grew steadily between 1960 and 1994, from 88 million to 209 million tons per year (Figure 18C.7). Per capita generation of MSW increased from 2.7 pounds per person per day in 1960 to 4.4 pound per person per day in 1994. Projected per capita MSW generation in the year 2010 is 4.8 pounds per person per day (262 million tons).

Combustors handled an estimated 30% of MSW generated in 1960, mostly through incinerators with no energy recovery and no air pollution controls. In the 1960s and 1970s, combustion dropped steadily as the old incinerators were closed, reaching a low of less than 10% of MSW generated by 1980, then increasing to approximately 16% of MSW in 1990. Between 1990 and 1994, combustion remained around 16% of MSW generation. All major new facilities have energy recovery and are designed to meet air pollution standards. The EPA report projects that the tonnage of MSW combusted will increase to 38 million tons slightly by the year 2010.

Summary

The generation of municipal solid waste continues to increase steadily, both in overall tonnage and in pounds per capita. There is some evidence that recycling, composting, and source reduction measures, particularly efforts to keep yard trimmings out of the waste management system and reduce packaging waste, are leading to a decline in the percentage of MSW being sent for disposal to waste-to-energy combustion and landfill facilities. Major findings of the EPA study include the following:

- In 1994, 209 million tons, or 4.4 pounds per person per day, of MSW were generated. After materials recovery for recycling and composting, discards were 3.4 pounds per person per day. Virtually all of these discards were either combusted or sent to landfills.
- EPA projects that the per capita generation rate will increase by the year 2010 to 4.8 pounds per person per day.

- Even with significant source reduction efforts, generation of MSW is projected to increase to 262 million tons in the year 2010. However, discards to combustion facilities or landfills are projected to decline from 160 million tons in 1994 to 156 million tons in 2000, then increase to 170 million tons.
- Recovery of materials for recycling and composting was estimated to be 24% of MSW generated in 1994, up from 17% in 1990. Combustion facilities managed 16% of total generation, and this percentage is expected to remain constant or decline.
- Between 1990 and 1994, recovery of materials for recycling and composting increased from 33 million tons to 49 million tons, an increase of 48%. Recovery of paper and paperboard accounted for over half of this increased tonnage. Yard trimmings for composting contributed the next largest increase in tonnage recovered.

19

Active Solar Heating Systems

T. Agami Reddy
Texas A & M University

Introduction

This section defines the scope of the entire chapter and presents a brief overview of the types of applications that solar thermal energy can potentially satisfy.

Motivation and Scope

Successful solar system design is an iterative process involving consideration of many technical, practical, reliability, cost, code, and environmental considerations (Mueller Associates, 1985). The success of a project involves identification of and intelligent selection among trade-offs, for which a proper understanding of goals, objectives, and constraints is essential. Given the limited experience available in the solar field, it is advisable to keep solar systems as simple as possible and not be lured by the promise of higher efficiency offered by more complex systems. Because of the location-specific variability of the solar resource, solar systems offer certain design complexities and concerns not encountered in traditional energy systems.

The objective of this chapter is to provide energy professionals with a fundamental working knowledge of the scientific and engineering principles of solar collectors and solar systems relevant to both the prefeasibility study and the feasibility study of a solar project. Conventional equipment such as heat exchangers, pumps, and piping layout is but briefly described. Because of space

0-8493-2514-5/96/$0.00+$.50

limitations, certain equations/correlations had to be omitted, and proper justice could not be given to several concepts and design approaches. Effort has been made to provide the reader with pertinent references to textbooks, manuals, and research papers.

A detailed design of solar systems requires in-depth knowledge and experience in (i) the use of specially developed computer programs for detailed simulation of solar system performance, (ii) designing conventional equipment, controls, and hydronic systems, (iii) practical aspects of equipment installation, and (iv) economic analysis. These aspects are not addressed here, given the limited scope of this chapter. Readers interested in acquiring such details can consult manuals such as SERI (1989) or Mueller Associates (1985).

Design of large solar installations is a length process. The process is much less involved when a small domestic hot water system or unitary solar equipment or single solar appliances such as solar stills, solar cookers, or solar dryers are to be installed. Not only do such appliances differ in engineering construction from region to region, there are also standardized commercially available units whose designs are already more or less optimized by the manufacturers, normally as a result of previous experimentation, both technical or otherwise. Such equipment is not described in this chapter for want for space.

This chapter is not meant to be a primer on solar thermal systems; rather, it is a reasonably self-contained chapter that classifies and briefly describes the major application areas of solar thermal systems, presents the basic theory of solar collector performance along with that of specific sub-components, describes the various design approaches by which long-term or annual collector performance can be predicted in order to size solar systems, and addresses the synergism between system thermal performance and economic considerations, which determines system design and sizing. Case studies of a few solar thermal systems are also included.

This chapter is limited to nonelectricity applications and also avoids passive solar heating and cooling of buildings, as these issues are covered in other chapters of this handbook. The design concepts described in this chapter are applicable to domestic water heating, swimming pool heating, active space heating, industrial process heat, convective drying systems, and solar cooling systems.

Brief Overview of Energy Use by Sector

Though solar radiation is freely available, its widespread acceptance is strongly dictated by the equipment cost required to convert it to forms suitable for human needs. These costs are influenced by considerations such as energy type (electricity or thermal), temperature level (low, medium, or high), relationship between demand and solar availability, and status of technological maturity. Table 19.1 provides a simple guide to energy use patterns in the United States, which are generally reflective of those in developed countries. In terms of direct thermal applications, we note that space heat and hot water in the residential and commercial sectors account for about 14% and 6.7% respectively of the total energy use. These needs can well be met by flat-plate solar collectors whose technology is mature and whose installation and operation are relatively simple. In the industrial sector, about 1.2% of the total energy use is below 100°C, an application again well matched to flat-plate solar collectors; about 4.7% between 100 and 170°C, a temperature range appropriate for stationary concentrating collectors, and 3.6% between 177 and 288°C, a range suitable for tracking collectors. All these generic collector types will be addressed in this chapter.

Energy needs in developing countries are very much different than those in the developed countries and, moreover, seem to vary significantly from country to country (see Table 19.2). For example, energy use for cooking is a major end use in India, while energy use in commercial buildings is very low both in India and Brazil. Also, energy use for process heat applications is high in both countries, with it being higher in Brazil. Another distinctive feature is that while all the energy use in the United States is from commercial fuels (see Table 19.3), a large fraction of the residential/commercial energy use in developing countries is from traditional sources of energy,

TABLE 19.1 Energy Consumption of the United States in 1977 by End Use (Units of 10^{18} J/yr. Numbers in parentheses show percentage of total)

Sector	Total Energy Use	Application	Energy Use
Transportation	21.2 (26.2%)		
Residential	18.2 (22.5%)	Space heat	8.7 (10.8%)
		Hot water	2.5 (3.1%)
		Air-conditioning	1.2 (1.5%)
		Refrigerators and freezers	2.4 (3.0%)
		Lights	1.1 (1.4%)
		Other	2.3 (2.8%)
Commercial	11.7 (14.5%)	Space heat	5.2 (6.4%)
		Air-conditioning	2.5 (3.1%)
		Hot water	0.26 (0.3%)
		Lights	2.5 (3.1%)
		Other	1.2 (1.5%)
Industrial process heat	12.0 (16.0%)	Below 100°C	1.0 (1.2%)
		100–177°C	3.8 (4.7%)
		177–288°C	2.9 (3.6%)
		288–593°C	2.6 (3.2%)
		593–1,090°C	1.0 (1.2%)
		above 1,090°C	1.6 (2.0%)
Other industrial	16.8 (20.8%)		
Total	80.8 (100%)		

Source: Rabl, 1985.

TABLE 19.2 Percentage Energy Use by Service in Two Developing Countries

		Brazil		India	
Sector	Application	% of Total	% Energy Use	% of Total	% Energy Use
Residential		14.3		47.0	
	Cooking		12.2		42.5
	Lighting		0.7		4.2
	Appliances		1.4		0.3
Commercial		3.5		2.0	
	Cooling		0.9		0.1
	Lights		1.2		0.04
	Appliances		1.4		0.06
Industrial		44.7		35.0	
	Process heat		40.3		23.1
	Motor drive		3.7		11.1
	Lighting/others		0.7		0.8
Transport		30.6		11.0	
Agriculture		4.8		5.0	
Others		2.1			
Per capita use (GJ)		43.4		11.7	

Source: Adapted from OTA, 1991.

for example firewood. Hence the successful introduction of solar thermal devices and systems will not only impact conventional energy use (as it does in developed countries), it will alleviate many of the ills associated with energy use in developing countries (deforestation, for example) and uplift the general standard of living by providing services (and jobs) not available earlier. These are some of the extraneous reasons why solar energy (and renewable energy in general) is still being ardently pursued in developing countries, while countries such as the United States seem to have developed a skeptical attitude, at least for the near future.

TABLE 19.3 Total Delivered Energy in Three Different Regions of the World in 1985 (Units: 10^{18} J/yr)

		Latin America	India/China	U.S.
Residential/commercial				
	Commercial	2.3	7.3	16.8
	Traditional	2.6	4.7	—
Industrial				
	Commercial	4.1	13.0	16.4
	Traditional	0.8	0.2	—
Total				
	Commercial	10.1	22.2	51.8
	Traditional	3.4	4.8	—

Source: OTA, 1991.

Types of Thermal Loads

One of the crucial design aspects to consider while designing solar thermal systems is the temporal match of the load and the energy delivered from the solar system. Energy requirements may not be constant over the annual seasonal cycle. Process loads are essentially constant over the year, with domestic hot water loads exhibiting a small seasonal dependence while space heating and space cooling loads exhibit strong seasonal variations. Since there is no space heating load during the summer months, a solar system applied to that load could be inoperative during these months. Adding domestic water or cooling load to the system would allow a more uniform operation of the solar system throughout the year. To generalize, thermal loads normally encountered in solar engineering fall into four categories:

1. Those that are dependent on weather conditions and do not require energy conversion (for example, space heating),
2. Those that are used for direct thermal applications and are independent of weather conditions (for example, process loads and domestic hot water),
3. Those that depend on weather conditions and do not require energy conversion (for example, space cooling),
4. Those that are independent of weather conditions but require energy conversion (for example, motive power applications that are constant over the year).

9.1 Solar Collectors

Collector Types

A solar thermal collector is a heat exchanger that converts radiant solar energy into heat. In essence this consists of a receiver that absorbs the solar radiation and then transfers the thermal energy to a working fluid. Because of the nature of the radiant energy (its spectral characteristics, its diurnal and seasonal variability, changes in diffuse to global fraction, etc.) as well as the different types of applications for which solar thermal energy can be used, the analysis and design of solar collectors present unique and unconventional problems in heat transfer, optics, and material science. The classification of solar collectors can be made according to the type of working fluid (water, air, or oils) or the type of solar receiver used (nontracking or tracking).

Most commonly used working fluids are water (glycol being added for freeze protection) and air. Table 19.4 identifies the relative advantages and potential disadvantages of air and liquid collectors and associated systems. Because of the poorer heat transfer characteristics of air with the solar absorber, the air collector may operate at a higher temperature than a liquid-filled

TABLE 19.4 Advantages and Disadvantages of Liquid and Air Systems

Characteristics	Liquid	Air
Efficiency	Collectors generally more efficient for a given temperature difference	Collectors generally operate at slightly lower efficiency.
System configuration	Can be readily combined with service hot water and cooling systems	Space heat can be supplied directly but does not adapt easily to cooling. Can preheat hot water.
Freeze protection	May require antifreeze and heat exchangers that add cost and reduce efficiency	None needed
Maintenance	Precautions must be taken against leakage, corrosion and boiling.	Low maintenance requirements. Leaks repaired readily with duct tape, but leaks may be difficult to find.
Space requirements	Insulated pipes take up nominal space and are more convenient to install in existing buildings.	Duct work and rock storage units are bulky, but ducting is a standard HVAC installation technique.
Operation	Less energy required to pump liquids	More energy required by blowers to move air; noisier operation
Cost	Collectors cost more	Storage costs more
State of the art	Has received considerable attention from solar industry	Has received less attention from solar industry

Source: SERI, 1989.

TABLE 19.5 Types of Solar Thermal Collectors

Nontracking Collectors	Tracking Collectors
Basic flat-plate	Parabolic troughs
Flat-plate enhanced with side reflectors or V-troughs	Fresnel reflectors
Tubular collectors	Paraboloids
Compound parabolic concentrators (CPCs)	Heliostats with central receivers

collector, resulting in greater thermal losses and, consequently, a lower efficiency. The choice of the working fluid is usually dictated by the application. For example, air collectors are suitable for space heating and convective drying applications, while liquid collectors are the obvious choice for domestic and industrial hot water applications. In certain high-temperature applications, special types of oils are used that provide better heat transfer characteristics.

The second criterion of collector classification is according to the presence of a mechanism to track the sun throughout the day and year in either a continuous or discreet fashion (see Table 19.5). The stationary flat-plate collectors are rigidly mounted, facing toward the equator with a tilt angle from the horizontal roughly equal to the latitude of the location for optimal year-round operation. The compound parabolic concentrators (CPCs) can be designed either as completely stationary devices or as devices that need seasonal adjustments only. On the other hand, Fresnel reflectors, paraboloids, and heliostats need two-axis tracking. Parabolic troughs have one axis tracking either along the east–west direction or the north–south direction. These collector types are described by Kreider (1979) and Rabl (1985).

A third classification criterion is to distinguish between nonconcentrating and concentrating collectors. The main reason for using concentrating collectors is not that *more energy* can be collected but that the thermal energy is obtained at higher temperatures. This is done by decreasing the area from which heat losses occur (called the receiver area) with respect to the aperture area (i.e., the area that intercepts the solar radiation). The ratio of the aperture to receiver area is called the *concentration ratio.*

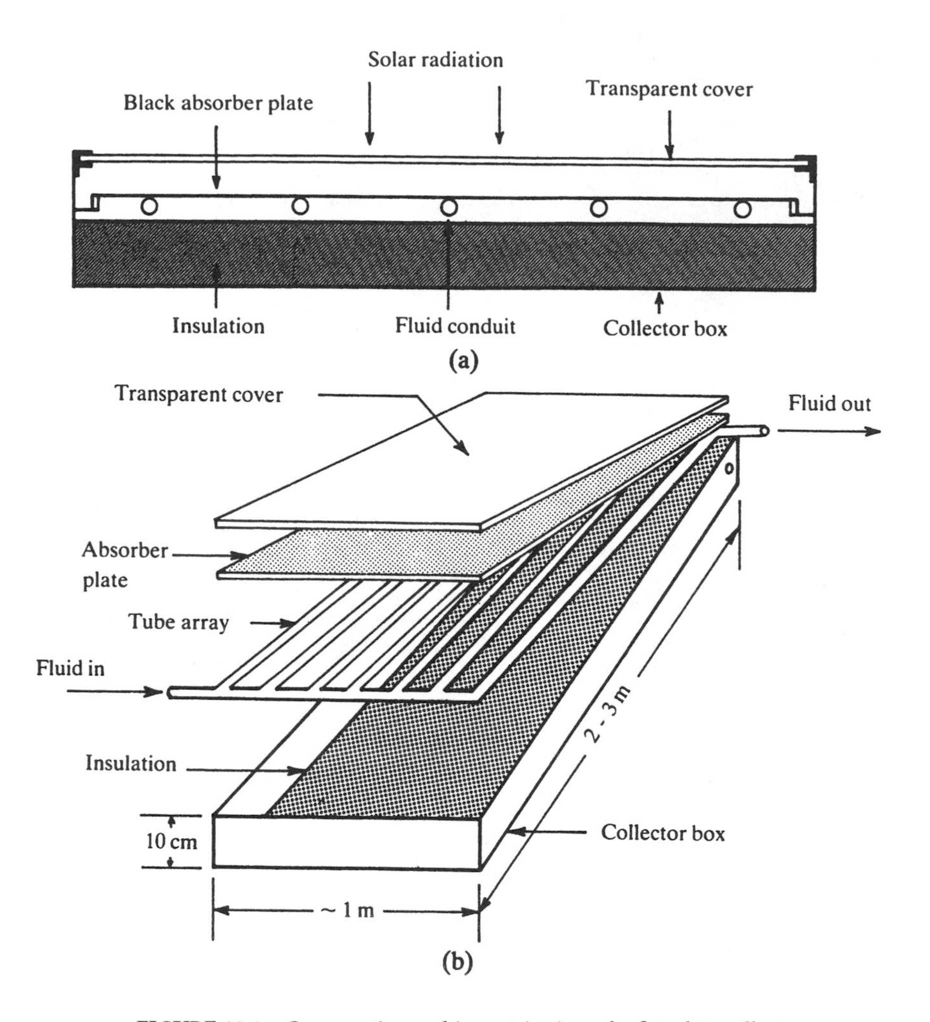

FIGURE 19.1 Cross-section and isometric view of a flat-plate collector.

Flat-Plate Collectors

Description

The flat-plate collector is the most common conversion device in operation today, since it is most economical and appropriate for delivering energy at temperatures up to about 100°C. The construction of flat-plate collectors is relatively simple, and many commercial models are available.

Figure 19.1 shows the physical arrangements of the major components of a conventional flat-plate collector with a liquid working fluid. The blackened absorber is heated by radiation admitted via the transparent cover. Thermal losses to the surroundings from the absorber are contained by the cover, which acts as a black body to the infrared radiation (this effect is called the *greenhouse* effect), and by insulation provided under the absorber plate. Passages attached to the absorber are filled with a circulating fluid, which extracts energy from the hot absorber. The simplicity of the overall device makes for long service life.

The absorber is the most complex portion of the flat-plate collector, and a great variety of configurations are currently available for liquid and air collectors. Figure 19.2 illustrates some of these concepts in absorber design for both liquid and air absorbers. Conventional materials are copper, aluminum, and steel. The absorber is either painted with a dull black paint or can be coated with a *selective surface* to improve performance (see "Improvements to Flat-Plate Collector Performance"

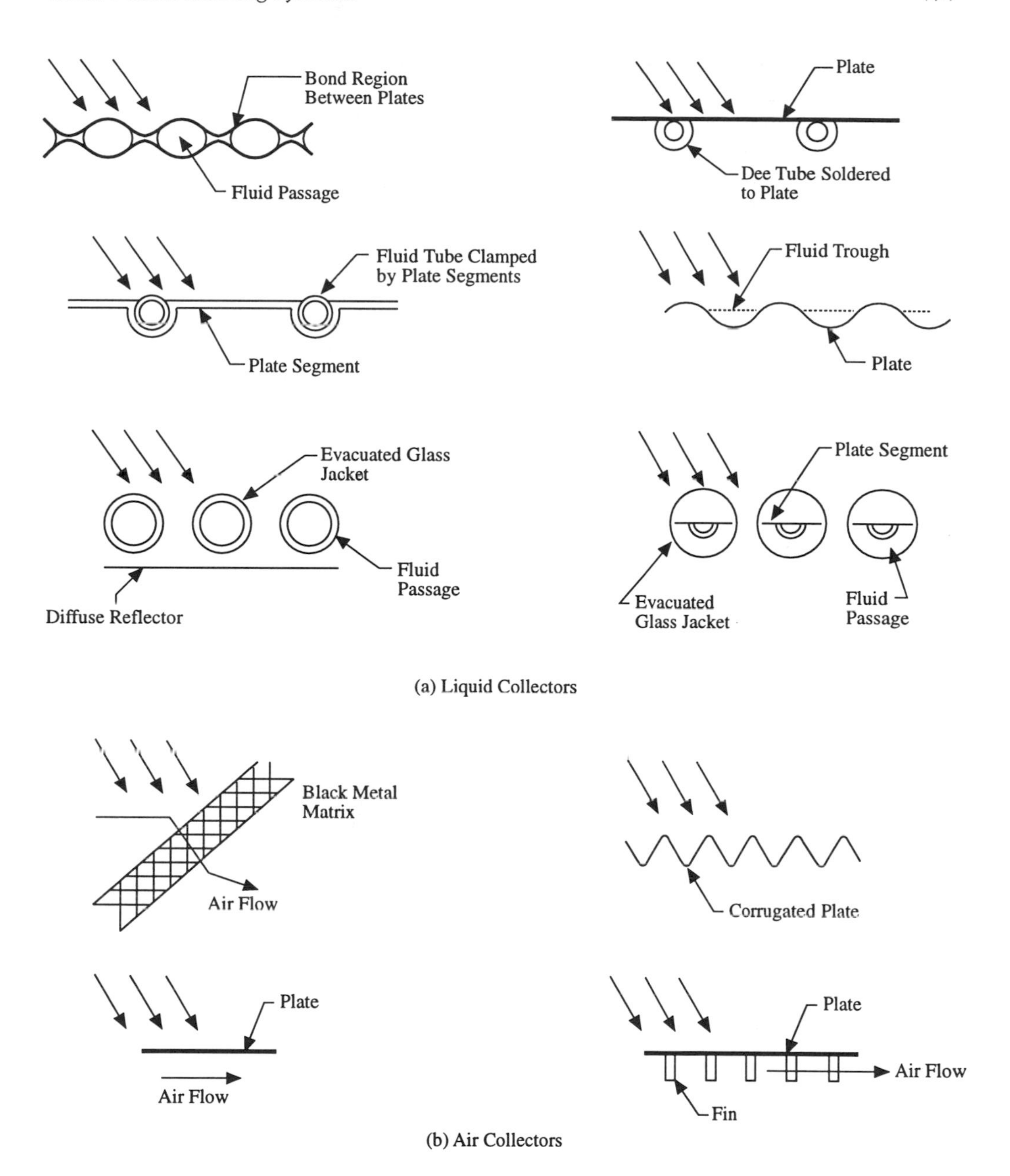

FIGURE 19.2 Typical flat-plate absorber configurations.

for more details). Bonded plates having internal passageways perform well as absorber plates because the hydraulic passageways can be designed for optimal fluid and thermal performance. Such collectors are called *roll-bond* collectors. Another common absorber consists of tubes soldered or brazed to a single metal sheet, and mechanical attachments of the tubes to the plate have also been employed. This type of collector is called a *tube-and-sheet* collector. Heat pipe collectors have also been developed, though these are not as widespread as the previous two types. The so-called *trickle type* of flat-plate collector, with the fluid flowing directly over the corrugated absorber plate, dispenses entirely with fluid passageways. Tubular collectors have also been used because of the relative ease by which air can be evacuated from such collectors, thereby reducing convective heat losses from the absorber to the ambient air.

The absorber in an air collector normally requires a larger surface than in a liquid collector because of the poorer heat transfer coefficients of the flowing air stream. Roughness elements and producing turbulence by way of devices such as expanded metal foil, wool, and overlapping plates have been used as a means for increasing the heat transfer from the absorber to the working fluid. Another approach to enhance heat transfer is to use packed beds of expanded metal foils or matrices between the glazing and the bottom plate.

Modeling

A particular modeling approach and the corresponding degree of complexity in the model are dictated by the objective as well as by experience gained from past simulation work. For example, it has been found that transient collector behavior has insignificant influence when one is interested in determining the long-term performance of a solar thermal system. For complex systems or systems meant for nonstandard applications, detailed modeling and careful simulation of system operation are a must initially, and simplifications in component models and system operation can subsequently be made. However, in the case of solar thermal systems, many of the possible applications have been studied to date and a backlog of experience is available not only concerning system configurations but also with reference to the degree of component model complexity.

Because of low collector time constants (about 5 min.), heat capacity effects are usually small. Then the instantaneous (or hourly, because radiation data are normally available in hourly time increments only) steady-state useful energy q_C in watts delivered by a solar flat-plate collector of surface area A_C is given by

$$q_C = A_C F' \left[I_T \eta_0 - U_L \left(T_{Cm} - T_a \right) \right]^+ \tag{19.1}$$

where F' is the plate efficiency factor, which is a measure of how good the heat transfer is between the fluid and the absorber plate; η_0 is the optical efficiency, or the product of the transmittance and absorptance of the cover and absorber of the collector; U_L is the overall heat loss coefficient of the collector, which is dependent on collector design only and is normally expressed in $W/(m^2\,°C)$; T_{Cm} is the *mean* fluid temperature in the collector (in °C); and I_T is the radiation intensity on the plane of the collector (in W/m^2). The + sign denotes that negative values are to be set to zero, which physically implies that the collector should not be operated when q_C is negative (i.e., when the collector loses more heat than it can collect).

However, because T_{Cm} is not a convenient quantity to use, it is more appropriate to express collector performance in terms of the fluid *inlet* temperature to the collector (T_{Ci}). This equation is known as the classical Hottel–Whillier–Bliss (HWB) equation and is most widely used to predict instantaneous collector performance:

$$q_C = A_C F_R \left[I_T \eta_0 - U_L \left(T_{Ci} - T_a \right) \right]^+ \tag{19.2}$$

where F_R is called the heat removal factor and is a measure of the solar collector performance as a heat exchanger, since it can be interpreted as the ratio of actual heat transfer to the maximum possible heat transfer. It is related to F′ by

$$\frac{F_R}{F'} = \frac{(mc_p)_C}{A_C F' U_L} \left\{ 1 - \exp\left[-\frac{A_C U_L F'}{(mc_p)_C} \right] \right\} \tag{19.3}$$

where m_C is the total fluid flow rate through the collectors and c_{pc} is the specific heat of the fluid flowing through the collector. The variation of (F_R/F') with $[(mC_p)_C/A_C U_L F']$ is shown graphically

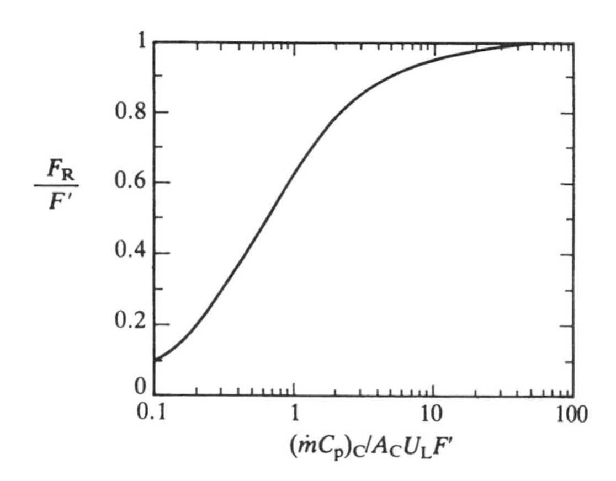

FIGURE 19.3 Variation of F_R/F' as a function of $[(mc_p)_c/(A_c U_L F')]$ (Duffie and Beckman, 1980).

in Figure 19.3. Note the asymptotic behavior of the plot, which suggests that increasing the fluid flow rate more than a certain amount results in little improvement in F_R (and hence in q_C) while causing a quadratic increase in the pressure drop.

Factors influencing solar collector performance are of three types: (i) constructional, that is, related to collector design and materials used, (ii) climatic, and (iii) operational, that is, fluid temperature, flow rate, and so on. The plate efficiency factor F' is a factor that depends on the physical constructional features and is essentially a constant for a given liquid collector. (This is not true for air collectors, which require more careful analysis.) Operational features involve changes in m_C and T_{Ci}. While changes in m_C affect F_R as per Eq. (19.3), we note from Eq. (19.2) that to enhance q_C, T_{Ci} needs to be kept as low as possible. For solar collectors that are operated under more or less constant flow rates, specifying $F_R\,\eta_0$ and $F_R\,U_L$ is adequate to predict collector performance under varying climatic conditions.

There are a number of procedures by which collectors have been tested. The most common is a *steady-state procedure*, where transient effects due to collector heat capacity are minimized by performing tests only during periods when radiation and ambient temperature are steady. The procedure involves simultaneous and accurate measurements of the mass flow rate, the inlet and outlet temperatures of the collector fluid, and the ambient conditions (incident solar radiation, air temperature, and wind speed). The most widely used test procedure is the ASHRAE Standard 93-77 (1978), whose test setup is shown in Figure 19.4. Though a solar simulator can be used to perform indoor testing, outdoor testing is always more realistic and less expensive. The procedure can be used for nonconcentrating collectors using air or liquid as the working fluid (but not two phase mixtures) that have a single inlet and a single outlet and contain no integral thermal storage.

Steady-state procedures have been in use for a relatively long period and though the basis is very simple the engineering setup is relatively expensive (see Figure 19.4). From an overall heat balance on the collector fluid and from Eq. (19.2), the expressions for the instantaneous collector efficiency under normal solar incidence are

$$\eta_C \equiv \frac{q_C}{A_C I_T} = \frac{\left(mc_p\right)_C \left(T_{Co} - T_{Ci}\right)}{A_C I_T} \tag{19.4}$$

$$= \left[F_R \eta_n - F_R U_L \left(\frac{T_{Ci} - T_a}{I_T}\right)\right] \tag{19.5}$$

where η_n is the optical efficiency at normal solar incidence.

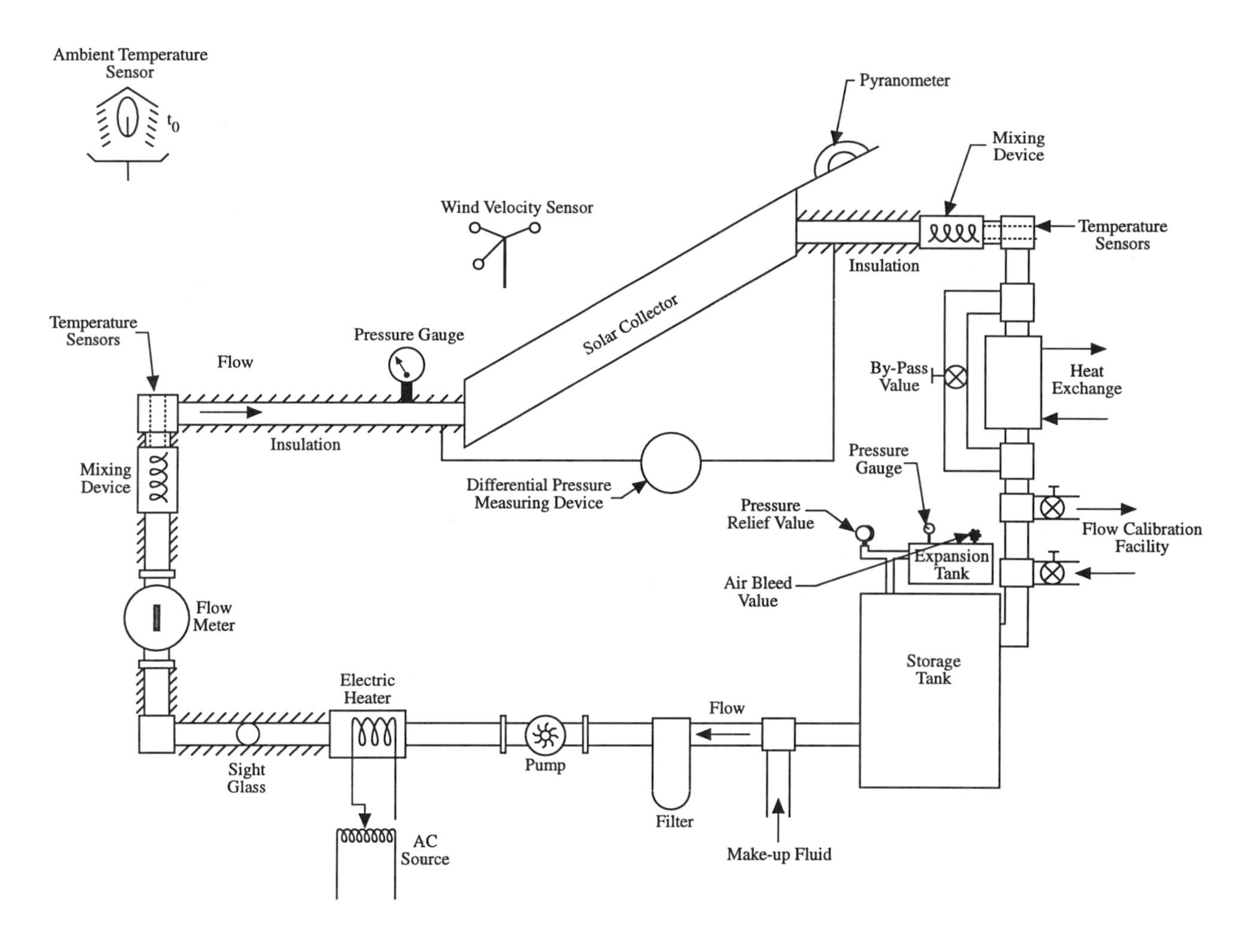

FIGURE 19.4 Set up for testing liquid collectors according to ASHRAE Standard 93-72.

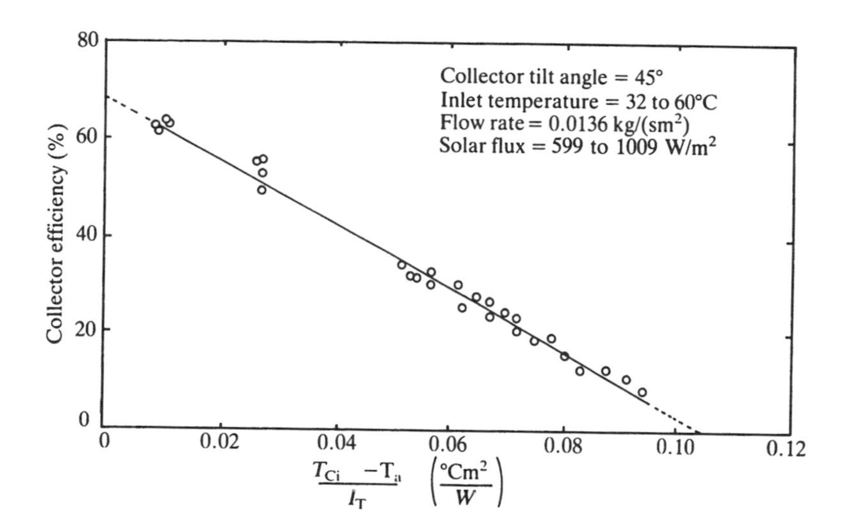

FIGURE 19.5 Thermal efficiency curve for a double glazed flat-plate liquid collector (ASHRAE 1978). Test conducted outdoors on a 1.2 m by 1.25 m panel with 10.2 cm of glass fiber back insulation and a flat copper absorber with black coating of emissivity of 0.97.

From the test data, points of η_c against reduced temperature $[(T_{Ci} - T_a) / I_T]$ are plotted as shown in Figure 19.5. Then a linear fit is made to these data points by regression, from which the values of $F_R \eta_n$ and $F_R U_L$ are easily deduced. It will be noted that if the reduced term were to be taken as $[(T_{Cm} - T_a)/I_T]$, estimates of $F'\eta_n$ and $F'U_L$ would be correspondingly obtained.

Incidence Angle Modifier

The optical efficiency η_0 depends on the collector configuration and varies with the angle of incidence as well as with the relative values of diffuse and beam radiation. The incidence angle modifier is defined as $K_\eta \equiv (\eta_0/\eta_n)$. For flat plate collectors with 1 or 2 glass covers, K_η is almost unchanged up to incidence angles of 60°, after which it abruptly drops to zero.

A simple way to model the variation of K_η with incidence angle for flat plate collectors is to specify η_n, the optical efficiency of the collector at normal beam incidence, to assume the entire radiation to be beam, and to use the following expression for the angular dependence (ASHRAE, 1978)

$$K_\eta = 1 + b_0\left(\frac{1}{\cos\theta} - 1\right) \tag{19.6}$$

where θ is the solar angle of incidence on the collector plate (in degrees) and b_0 is a constant called the incidence angle modifier coefficient. Plotting K_η against $[(1/\cos\theta)-1]$ results in linear plots (see Figure 19.6), thus justifying the use of Eq. (19.6). We note that for one-glass and two-glass covers, approximate values of b_0 are –0.10 and –0.17 respectively.

In case the diffuse solar fraction is high, one needs to distinguish between beam, diffuse, and ground-reflected components. Diffuse radiation, by its very nature, has no single incidence angle. One simple was is to assume an equivalent incidence angle of 60° for diffuse and ground-reflected components. One would then use Eq. (19.6) for the beam component along with its corresponding value of θ and account for the contribution of diffuse and ground reflected components by assuming a value of θ = 60° in Eq. (19.6). For more accurate estimation, one can use the relationship between the effective diffuse solar incidence angle versus collector tilt given in Duffie and Beckman (1980). It should be noted that the preceding equation gives misleading results with incidence angles close to 90°. An alternative functional form for the incidence angle modifier for both flat-plate and concentrating collectors has been proposed by Rabl (1981).

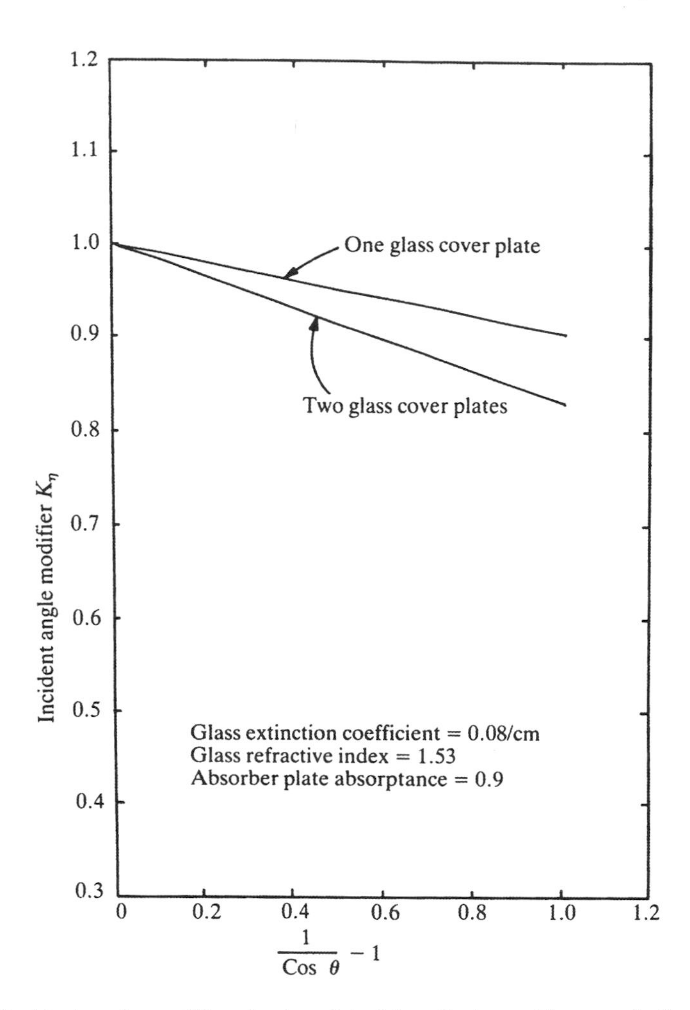

FIGURE 19.6 Incident angle modifiers for two flat-plate collectors with non-selective coating on the absorber. (Adapted from ASHRAE, 1978.)

Example 19.1. From the thermal efficiency curve given in Figure 19.5, determine the performance parameters of the corresponding solar collector.

Extrapolating the curve yields y-intercept = 0.69, x-intercept = 0.105 (m^2 °C/W). Since the reduced temperature in Figure 19.5 is in terms of the inlet fluid temperature to the collector, Eq. (19.5) yields $F_R \eta_n = 0.69$ and $F_R U_L = 0.69/0.105 = 6.57$ W/(m^2 °C). Alternatively, the collector parameters in terms of the plate efficiency factor can be deduced. From Figure 19.5, the collector area = 1.22 × 1.25 = 1.525 m^2, while the flow rate (m/A_C) = 0.0136 kg/(s m^2). From Eq. (19.3),

$$F'/F_R = -(0.0136 \times 4190/6.57)\ \ln\left[1 - 6.57/(0.0136 \times 4190)\right] = 1.0625$$

Thus $F'U_L = 6.57 \times 1.0625 = 6.98$ W/(m^2 °C) and $F'\eta_n = 0.69 \times 1.0625 = 0.733$.

Example 19.2. How would the optical efficiency be effected at a solar incidence angle of 60° for a flat plate collector with two glass covers?

Assume a value of $b_0 = -0.17$. From Eq. (19.6), $K_\eta = 0.83$. Thus

$$F_R \eta_0 = F_R \eta_n K_\eta = 0.69 \times 0.83 = 0.57$$

Other Collector Characteristics

There are three collector characteristics that a comprehensive collector testing process should also address. The collector *time constant* is a measure that determines how intermittent sunshine affects collector performance and is useful in defining an operating control strategy for the collector array that avoids instability. Collector performance is usually enhanced if collector time constants are kept low. ASHRAE 93-77 also includes a method for determining this value. Commercial collectors usually have time constants of about 5 minutes or less, and this justifies the use of the HWB model (see Eq. 19.2).

Another quantity to be determined from collector tests is the collector *stagnation temperature*. This is the equilibrium temperature reached by the absorber plate when no heat is being extracted from the collector. Determining the maximum stagnation temperature, which occurs under high I_T and T_a values, is useful in order to safeguard against reduced collector life due to thermal damage to collectors (namely irreversible thermal expansion, sagging of covers, physical deterioration, optical changes, etc.) in the field when not in use. Though the stagnation temperature could be estimated from Eq. (19.2) by setting $q_C = 0$ and solving for T_{Ci}, it is better to perform actual tests on collectors before field installation.

The third collector characteristic of interest is the *pressure drop* across the collector for different fluid flow rates. This is an important consideration for liquid collectors, and more so for air collectors, in order to keep parasitic energy consumption (namely electricity to drive pumps and blowers) to a minimum in large collector arrays.

Improvements to Flat-Plate Collector Performance

There are a number of ways by which the performance of the basic flat-plate collectors can be improved. One way is to enhance optical efficiency by treatment of the glass cover thereby reducing reflection and enhancing performance. As much as a 4% increase has been reported (Anderson, 1977). Low-iron glass can also reduce solar absorption losses by a few percent.

These improvements are modest compared to possible improvements from reducing losses from the absorber plate. Essentially, the infrared upward reradiation losses from the heated absorber plate have to be decreased. One could use a second glass cover to reduce the losses, albeit at the expense of higher cost and lower optical efficiency. Usually for water heating applications, radiation accounts for about two-thirds of the losses from the absorber to the cover with convective losses making up the rest (conduction is less than about 5%). The most widely used manner of reducing these radiation losses is to use selective surfaces whose emissivity varies with wavelength (as against matte-black painted absorbers, which are essentially gray bodies). Note that 98% of the solar spectrum is at wavelengths less than 3.0 μm, whereas less than 1% of the black body radiation from a 200°C surface is at wavelengths less than 3.0 μm. Thus selective surfaces for solar collectors should have high solar absorptance (i.e., low reflectance in the solar spectrum) and low long-wave emittance (i.e., high reflectance in the long-wave spectrum). The spectral reflectance of some commonly used selective surfaces is shown in Figure 19.7. Several commercial collectors for water heating or low-pressure steam (for absorption cooling or process heat applications) are available that use selective surfaces.

Another technique to simultaneously reduce both convective and radiative losses between the absorber and the transparent cover is to use honeycomb material (Hollands, 1965). The honeycomb material can be reflective or transparent (the latter is more common) and should be sized properly. Glass honeycombs have had some success in reducing losses in high-temperature concentrating receivers, but plastics are usually recommended for use in flat-plate collectors. Because of the poor thermal aging properties, honeycomb flat-plate collectors have had little commercial success. Currently the most promising kind seems to be the simplest (both in terms of analysis and construction), namely collectors using horizontal rectangular slats (Meyer, 1978). Convection can be entirely suppressed provided the slats with the proper aspect ratio are used.

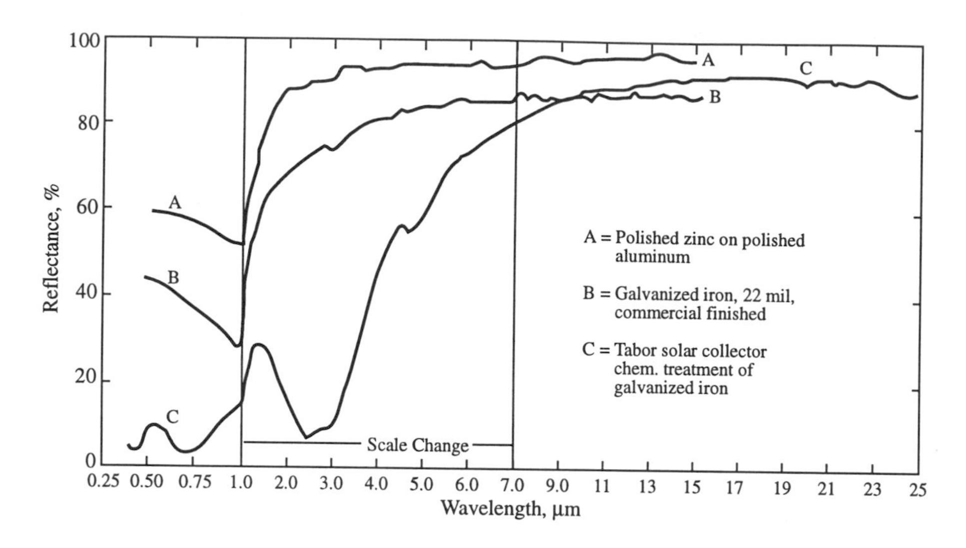

FIGURE 19.7 Spectral reflectance of several surfaces (Source: Edwards et al., 1960).

Finally, collector output can be enhanced by using side reflectors, for instance a sheet of anodized aluminum. The justification in using these is their low cost and simplicity. For instance, a reflector placed in front of a tilted collector cannot but increase collector performance because losses are unchanged and more solar radiation is intercepted by the collector. Reflectors in other geometries may cast a shadow on the collector and reduce performance. Note also that reflectors would produce rather nonuniform illumination over the day and during the year, which, though not a problem in thermal collectors, may drastically penalize the electric output of photovoltaic modules. Whether reflectors are cost-effective depends on the particular circumstances and practical questions such as aesthetics and space availability. The complexity involved in the analysis of collectors with planar reflectors can be reduced by assuming the reflector to be long compared to its width and treating the problem in two dimensions only. How optical performance of solar collectors are affected by side planar reflectors is discussed in several papers, for example Larson (1980) and Chiam (1981).

Other Collector Types

Evacuated Tubular Collectors

One method of obtaining temperatures between 100°C and 200°C is to use evacuated tubular collectors. The advantage in creating and being able to maintain a vacuum is that convection losses between glazing and absorber can be eliminated. There are different possible arrangements of configuring evacuated tubular collectors. Two designs are shown in Figure 19.8. The first is like a small flat-plate collector with the liquid to be heated making one pass through the collector tube. The second uses an all-glass construction with the glass absorber tube being coated selectively. The fluid being heated passes up the middle of the absorber tube and then back through the annulus. Evacuated tubes can collect both direct and diffuse radiation and do not require tracking. Glass breakage and leaking joints due to thermal expansion are some of the problems which have been experienced with such collector types. Various reflector shapes (like flat-plate, V-groove, circular, cylindrical, involute, etc.) placed behind the tubes are often used to usefully collect some of the solar energy, which may otherwise be lost, thus providing a small amount of concentration.

Compound Parabolic Concentrators (CPCs)

The CPC collector, discovered in 1966, consists of parabolic reflectors that funnel radiation from aperture to absorber rather than focusing it. The right and left halves belong to different parabolas (hence the name *compound*) with the edges of the receiver being the foci of the opposite parabola

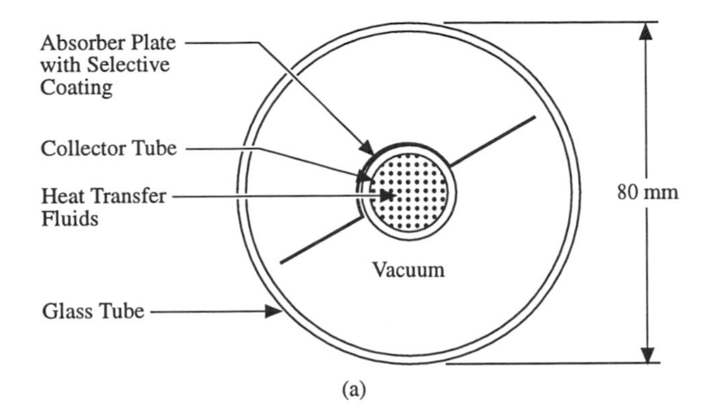

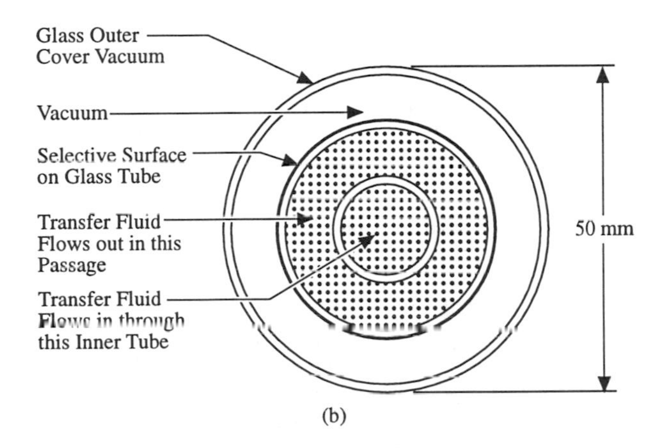

FIGURE 19.8 Evacuated tubular collectors (Source: Charters and Pryor, 1982).

(see Fig. 19.9). It has been proven that such collectors are *ideal* in that any solar ray, be it beam or diffuse, incident on the aperture within the acceptance angle will reach the absorber while all others will bounce back to and fro and re-emerge through the aperture. CPCs are also called *nonimaging* concentrators because they do not form clearly defined images of the solar disk on the absorber surface as achieved in classical concentrators. CPCs can be designed both as low-concentration devices with large acceptance angles or as high-concentration devices with small acceptance angles. CPCs with low concentration ratios (of about 2) and with east–west axes can be operated as stationary devices throughout the year or at most with seasonal adjustments only. CPCs, unlike other concentrators, are able to collect all the beam and a large portion of the diffuse radiation. Also they do not require highly specular surfaces and can thus better tolerate dust and degradation. A typical module made up of several CPCs is shown in Figure 19.10. The absorber surface is located at the bottom of the trough, and a glass cover may also be used to encase the entire module. CPCs show considerable promise for water heating close to the boiling point and for low-pressure steam applications. Further details about the different types of absorber and receiver shapes used, the effect of truncation of the receiver and the optics, can be found in Rabl (1985).

9.2 Long-Term Performance of Solar Collectors

Effect of Day-to-Day Changes in Solar Insolation

Instantaneous or hourly performance of solar collectors has been discussed in "Flat-Plate Collectors." For example, one would be tempted to use the HWB equation (Eq. 19.2) to predict long-term

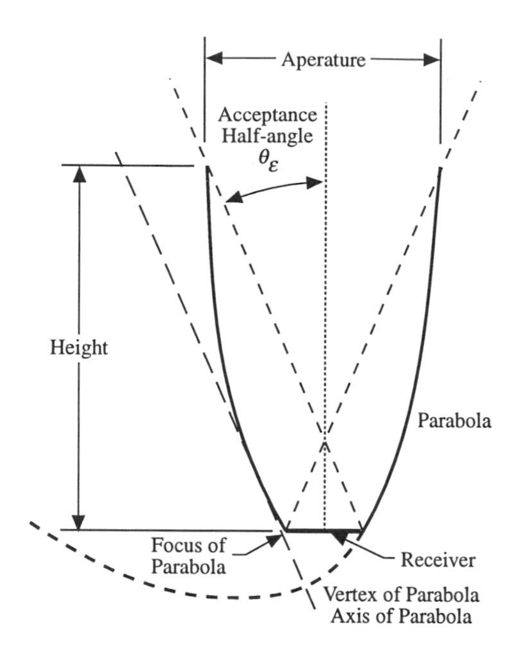

FIGURE 19.9 Cross-section of a symmetrical non-truncated CPC (Source: Duffie and Beckman, 1980).

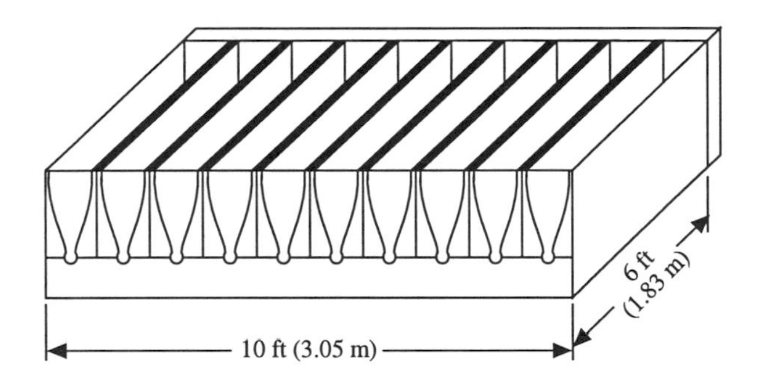

FIGURE 19.10 A CPC collector module (Source: SERI, 1989).

collector performance at a prespecified and constant fluid inlet temperature T_{Ci} merely by assuming average hourly values of I_T and T_a. Such a procedure would be erroneous and lead to underestimation of collector output because of the presence of the control function, which implies that collectors are turned on only when $q_C > 0$, that is, when radiation I_T exceeds a certain critical value I_C. This critical radiation value is found by setting q_C in Eq. (19.2) to zero:

$$I_C = U_L(T_{Ci} - T_a)/\eta_0 \tag{19.7a}$$

To be more rigorous, a small increment δ to account for pumping power and stability of controls can also be included if needed by modifying the equation to

$$I_C = U_L(T_{Ci} + \delta - T_a)/\eta_0 \tag{19.7b}$$

Then, Eq. (19.2) can be rewritten in terms of I_C as

$$q_C = A_C F_R \eta_0 [I_T - I_C]^+ \tag{19.8}$$

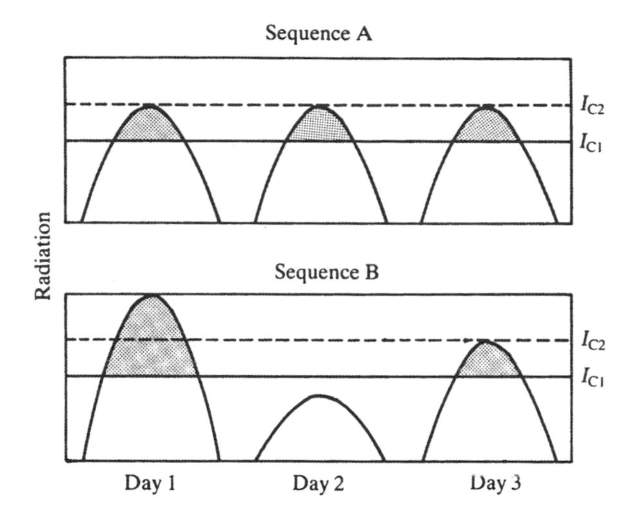

FIGURE 19.11 Effect of radiation distribution on collector long-term performance (Source: Klein, 1978).

Why one cannot simply assume a mean value of I_T in order to predict the mean value of q_C will be illustrated by the following simple concept (Klein, 1978). Consider the three identical day sequences shown in sequence A of Figure 19.11. If I_{C1} is the critical radiation intensity, and if it is constant over the whole day, the useful energy collected by the collector is represented by the sum of the shaded areas. If a higher critical radiation value shown as I_{C2} in Figure 19.11 is selected, we note that no useful energy is collected at all. Actual weather sequences would not look like that in sequence A but rather like that in sequence B, which is comprised of an excellent, a poor, and an average day. Even if both sequences have the same average radiation over 3 days, a collector subjected to sequence B will collect useful energy when the critical radiation is I_{C2}. Thus, neglecting the variation of radiation intensity from day to day over the long term and dealing with mean values would result in an underestimation of collector performance.

Loads are to a certain extent repetitive from day to day over a season or even the year. Consequently, one can also expect collectors to be subjected to a known diurnal repetitive pattern or mode of operation, that is, the collector inlet temperature T_{Ci} has a known repetitive pattern.

Individual Hourly Utilizability

In this mode, T_{Ci} is assumed to vary over the day but has the same variation for all the days over a period of N days (where $N = 30$ days for monthly and $N = 365$ for yearly periods). Then from Eq. (19.8), *total* useful energy collected over N days during individual hour i of the day is

$$q_{CN}(i) = A_C F_R \overline{\eta}_0 \overline{I}_{Ti} \sum_{i=1}^{N} \frac{\left[I_{Ti} - I_C\right]^+}{\overline{I}_{Ti}} \tag{19.9}$$

Let us define the radiation ratio

$$X_i = I_{Ti} / \overline{I}_{Ti} \tag{19.10}$$

and the critical radiation ratio

$$X_C = I_C / \overline{I}_{Ti}$$

The modified HWB equation (Eq. 19.8) can be rewritten as

$$q_{CN}(i) = A_C F_R \overline{\eta}_0 \overline{I}_{Ti} N \phi_i(x_c) \quad \textbf{(19.11)}$$

where the individual hourly utilizability factor ϕ_i is identified as

$$\phi_i(X_C) = \frac{1}{N} \sum^{N} (X_i - X_C)^+ \quad \textbf{(19.12)}$$

Thus ϕ_i can be considered to be the fraction of the incident solar radiation that can be converted to useful heat by an ideal collector (i.e., whose $F_R \eta_0 = 1$). The utilizability factor is thus a *radiation statistic* in the sense that it depends solely on the radiation values at the specific location. As such, it is in no way dependent on the solar collector itself. Only after the radiation statistics have been applied is a collector dependent significance attached to X_C.

Hourly utilizability curves on a *monthly* basis that are independent of location were generated by Liu and Jordan (1963) over 30 years ago for flat-plate collectors (see Fig. 19.12). The key climatic parameter which permits generalization is the *monthly clearness index* $\overline{K}$ of the location defined as

$$\overline{K} = \overline{H}/\overline{H}_0 \quad \textbf{(19.13)}$$

where $\overline{H}$ is the monthly mean daily global radiation on the horizontal surface and $\overline{H}_0$ is the monthly mean daily extraterrestrial radiation on a horizontal surface.

Extensive tables giving monthly values of $\overline{K}$ for several different locations worldwide can be found in several books, for example, Duffie and Beckman (1980) or Reddy (1987). The curves apply to equator-facing tilted collectors with the effect of collector tilt accounted for by the factor $\overline{R}_{b,T}$ which is the ratio of the monthly mean daily extraterrestrial radiation on the tilted collector to that on a horizontal surface. Monthly mean daily calculations can be made using the 15th of the month, though better accuracy is achieved using slightly different dates (Reddy, 1987). Clark et al. (1983), working from measured data from several U.S. cities, have proposed the following correlation for individual hourly utilizability over monthly time scales applicable to *flat-plate collectors only*:

$$\phi_i = 0 \qquad \text{for } X_C \geq X_{\max} \quad \textbf{(19.14)}$$

$$= (1 - X_C/X_{\max})^2 \qquad \text{for } X_{\max} = 2$$

$$= \left| |a| - \left[a^2 + (1+2a)(1 - X_C/X_{\max})^2 \right]^{1/2} \right| \qquad \text{otherwise}$$

where

$$a = (X_{\max} - 1)(2 - X_{\max}) \quad \textbf{(19.15)}$$

and

$$X_{\max} = 1.85 + 0.169\ (\overline{r}_T/\overline{k}^2) - 0.0696\ \cos\beta/\overline{k}^2 - 0.981\ \overline{k}/(\cos\ \delta)^2 \quad \textbf{(19.16)}$$

where $\overline{k}$ is the monthly mean *hourly* clearness index for the particular hour, δ is the solar declination, β is the tilt angle of the collector plane with respect to the horizontal, and $\overline{r}_T$ is the ratio

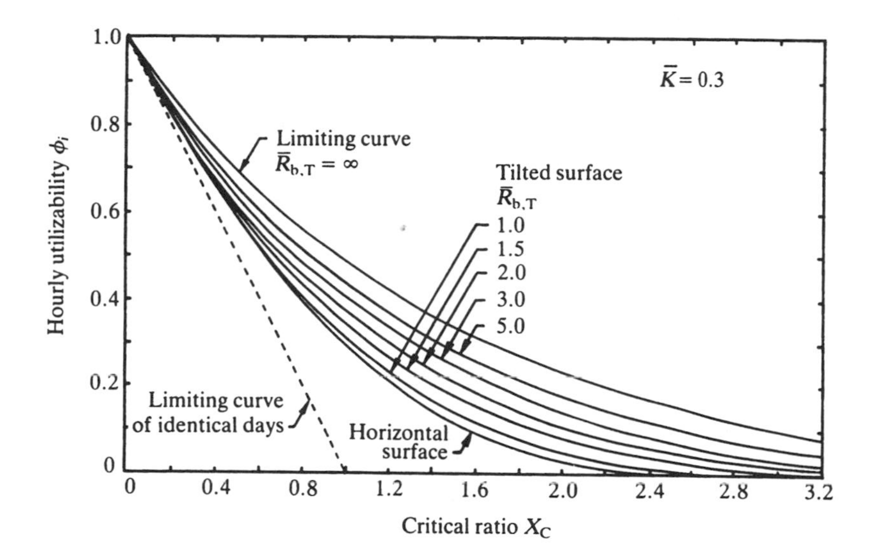

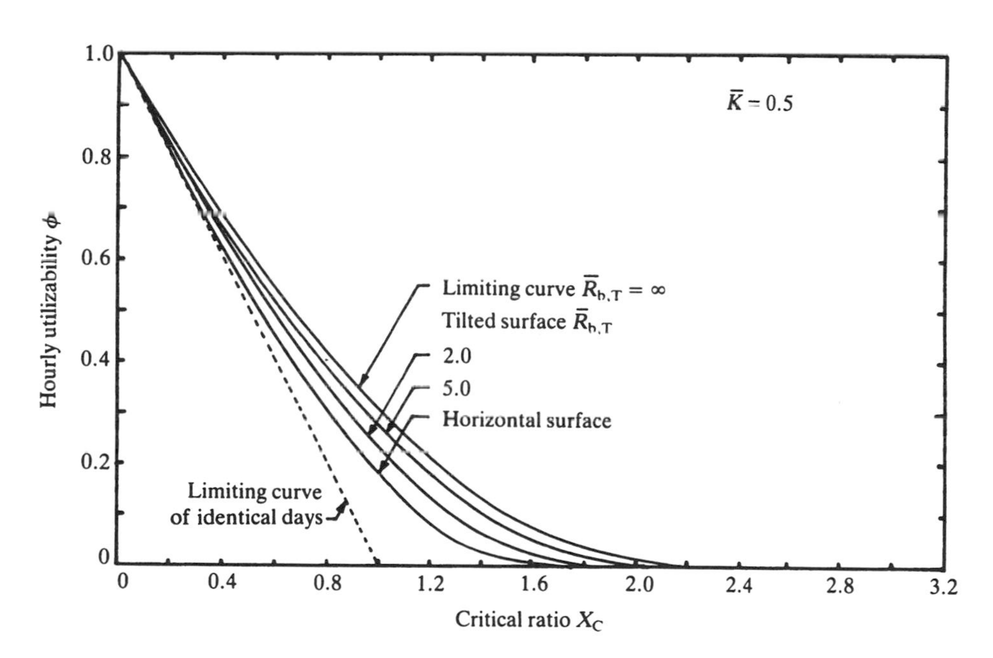

FIGURE 19.12(a) Generalized hourly utilizability curves of Liu and Jordan (1963) for three different monthly mean clearness indices $\bar{K}$.

of monthly average hourly global radiation on a tilted surface to that on a horizontal surface for that particular hour. For an isotropic sky assumption, $\bar{r}_T$ is given by

$$\bar{r}_T = \left(1 - \frac{\bar{I}_d}{\bar{I}}\right) r_{b,T} + \left(\frac{1 + \cos\beta}{2}\right)\frac{\bar{I}_d}{\bar{I}} + \left(\frac{1 - \cos\beta}{2}\right)\rho \tag{19.17}$$

where $\bar{I}_d$ and $\bar{I}$ are the hourly diffuse and global radiation on the horizontal surface, $r_{b,T}$ is the ratio of hourly beam radiation on the tilted surface to that on a horizontal surface (this is a purely astronomical quantity and can be calculated accurately from geometric considerations), and ρ is the ground albedo.

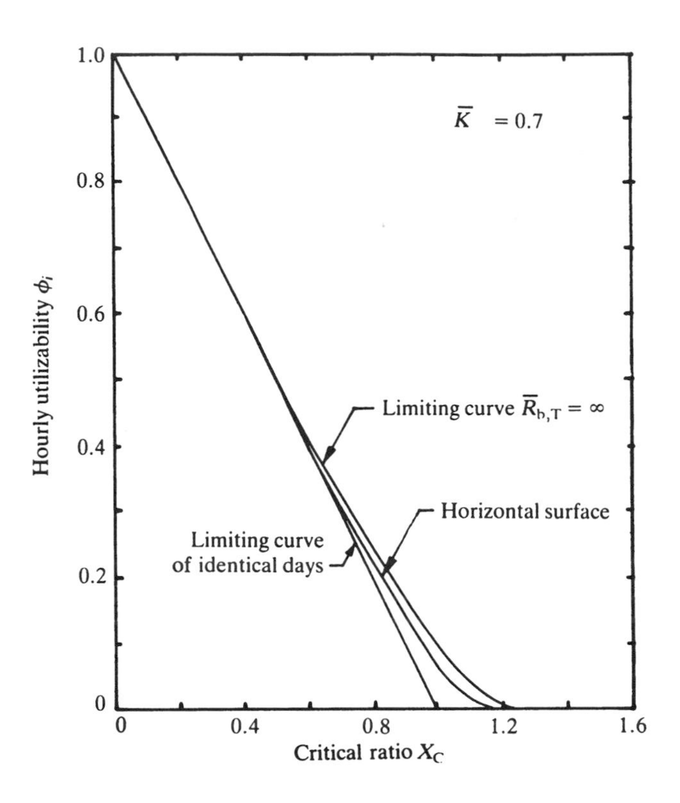

FIGURE 19.12(b)

Example 19.3. Compute the total energy collected during 11:30 to 12:30 for the month of September in New York, NY (latitude: 40.75°N, T_a = 20°C) by a flat-plate solar collector of 5 m^2 area having zero tilt. The collector performance parameters are $F_R\,\eta_0$ = 0.54 and $F_R\,U_L$ = 3.21 W/(m^2 °C) and the collector inlet temperature is 80°C. The corresponding hourly mean clearness index $\bar{k}$ is 0.44, and the monthly mean hourly radiation on a horizontal surface $\bar{I}_{Ti}$ (11:30 to 12:30) is 6.0 MJ/(m^2 h).

From Eq. (19.7a), critical radiation I_C = 3.21 × (80 – 20)/0.54 = 356.7 W/m^2 = 1.28 MJ/(m^2 h). For the average day of September, solar declination δ = 2.2°. Also, because the collector is horizontal $\bar{r}_T$ = 1 and β = 0. Thus from Eq. (19.16)

$$X_{max} = 1.85 + 0.169/0.44^2 - 0.0696/0.44^2 - 0.981 \times 0.44/(\cos\ 2.2)^2 = 1.93.$$

Also, from Eq. (19.15), a = (1.93 – 1)/(2 – 1.93) = 13.29.

The critical radiation ratio X_C = 1.28/1.93 = 0.663.

Because $X_C < X_{max}$, from Eq. (19.14) we have

$$\phi_i(X_C) = \left|13.29 - \left[13.29^2 + (1 + 2 \times 13.29)(1 - 0.663/1.93)^2\right]^{1/2}\right| = |13.29 - 13.73| = 0.44$$

Finally, the total energy collected is given by Eq. (19.11)

$$q_{CN}(11:30 \text{ to } 12:30) = 5 \times 0.54 \times 6.0 \times 30 \times 0.44 = 214 \text{ MJ/h}$$

Daily Utilizability

Basis

In this mode, T_{Ci}, and hence the critical radiation level, is assumed constant during all hours of the day. The *total* useful energy over N days that can be collected by solar collectors operated all day over n hours is given by

$$Q_{CN} = A_C F_R \overline{\eta}_0 \overline{H}_T N \overline{\phi} \tag{19.18}$$

where $\overline{H}_T$ is the average daily global radiation on the collector surface, and $\overline{\phi}$ (called Phibar) is the daily utilizability factor, defined as

$$\overline{\phi} = \sum^{N}\sum^{n}(I_T - I_C)^+ \Big/ \sum^{N}\sum^{n} I_T = \frac{1}{Nn}\sum^{N}\sum^{n}(X_i - X_C)^+ \tag{19.19}$$

Generalized correlations have been developed both at monthly time scales and for annual time scales based on the parameter $\overline{K}$. Generalized (i.e., location and month independent) correlations for $\overline{\phi}$ on a *monthly* time scale have been proposed by Theilacker and Klein (1980). These are strictly applicable for flat-plate collectors only. Collares-Pereira and Rabl (1979) have also proposed generalized correlations for $\overline{\phi}$ on a monthly time scale which, though a little more tedious to use are applicable to concentrating collectors as well. The reader may refer to Rabl (1985) or Reddy (1987) for complete expressions.

Monthly Time Scales

The Phibar method of determining the daily utilizability fraction proposed by Theilacker and Klein (1980) correlates $\overline{\phi}$ to the following factors:

1. A geometry factor $\overline{R}_T/\overline{r}_{T,noon}$, which incorporates the effects of collector orientation, location, and time of year. $\overline{R}_T$ is the ratio of monthly average global radiation on the tilted surface to that on a horizontal surface. $\overline{r}_{T,noon}$ is the ratio of radiation at noon on the tilted surface to that on a horizontal surface for the average day of the month. Geometrically, $\overline{r}_{T,noon}$ is a measure of the maximum height of the radiation curve over the day, whereas R_T is a measure of the enclosed area. Generally the value $(\overline{R}_T/\overline{r}_{T,noon})$ is between 0.9 and 1.5.
2. A dimensionless critical radiation level $\overline{X}_{C,K}$ where

$$\overline{X}_{C,K} = I_C / \overline{I}_{T,noon} \tag{19.20}$$

with $\overline{I}_{T,noon}$, the radiation intensity on the tilted surface at noon, given by

$$\overline{I}_{T,noon} = \overline{r}_{noon}\overline{r}_{T,noon}\overline{H} \tag{19.21}$$

where $\overline{r}_{noon}$ is the ratio of radiation at noon to the daily global radiation on a horizontal surface during the mean day of the month which can be calculated from the following correlation proposed by Liu and Jordan (1960):

$$r(W) = \frac{I(W)}{H} = \frac{\pi}{24}(a + b\cos W)\frac{(\cos W - \cos W_S)}{\left(\sin W_S - \frac{\pi}{180}W_S\cos W_S\right)} \tag{19.22}$$

with

$$a = 0.409 + 0.5016 \sin(W_s - 60)$$

$$b = 0.6609 - 0.4767 \sin(W_s - 60)$$

where W is the hour angle corresponding to the midpoint of the hour (in degrees) and W_S is the sunset hour angle given by

$$\cos W_S = -\tan L \tan \delta \tag{19.23}$$

where L is the latitude of the location. The fraction r is the ratio of hourly to daily global radiation on a horizontal surface. The factors $\bar{r}_{T,\text{noon}}$ and $\bar{r}_{\text{noon}}$ can be determined from Eqs. (19.17) and (19.22) respectively with $W = 0°$.

The Theilacker and Klein correlation for the daily utilizability for equator-facing flat-plate collectors is

$$\bar{\phi}(X_{C,K}) = \exp\left\{\left[a' + b'\left(\bar{r}_{T,\text{noon}}/\bar{R}_T\right)\right]\left[X_{C,K} + c'\ X_{C,K}^2\right]\right\} \tag{19.24}$$

where

$$\begin{aligned} a' &= 7.476 - 20.00\ \bar{K} + 11.188\ \bar{K}^2 \\ b' &= -8.562 + 18.679\ \bar{K} - 9.948\ \bar{K}^2 \\ c' &= -0.722 + 2.426\ \bar{K} + 0.439\ \bar{K}^2 \end{aligned} \tag{19.25}$$

How $\bar{\phi}$ varies with the critical radiation ratio $\bar{X}_{C,K}$ for three different values of $\bar{K}$ is shown in Figure 19.13.

Example 19.4. A flat-plate collector operated horizontally at Fort Worth, Texas ($L = 32.75°$N), has a surface area of 20 m^2. It is used to heat 10 kg/min. of water entering the collector at a constant temperature of 80°C each day from 6 A.M. to 6 P.M. The collector performance parameters are $F_R\eta_0 = 0.70$ and $F_R U_L = 5.0$ W/(m^2 °C). Use Klein's correlation to compute the energy collected by the solar collectors during September. Assume $\bar{H} = 18.28$ MJ/(m^2 d), $\bar{K} = 0.57$ and $\bar{T}_a = 25$°C. Assume the mean sunset hour angle for September to be 90°.

The critical radiation is calculated first:

$$I_C = (5/0.7) \times (80 - 25) = 393\ \text{W/m}^2 = 1.414\ \text{MJ}/(\text{m}^2\text{h})$$

For a horizontal surface, $\bar{R}_T = \bar{r}_{T,\text{noon}} = 1$. From Eq. (19.22), $r(W = 0) = \frac{\pi}{24}(a + b) = 0.140$. Klein's critical radiation ratio (Eq. 19.20) $\bar{X}_{C,K} = 1.414/(18.28 \times 0.140) = 0.553$. From Eq. (19.24), $\bar{\phi} = 0.318$. Finally, from Eq. (19.18), the total monthly energy collected by the solar collectors is $Q_{CM} = 20 \times 0.7 \times 30 \times 0.318 \times 18.28 = 2.44$ GJ/month.

Yearly Time Scales

Generalized expressions for the *yearly* average energy delivered by the principal collector types with constant radiation threshold (i.e., when the fluid inlet temperature is constant for all hours during the day over the entire year) have been developed by Rabl (1981) based on data from several U.S. locations. The correlations are basically quadratic of the form

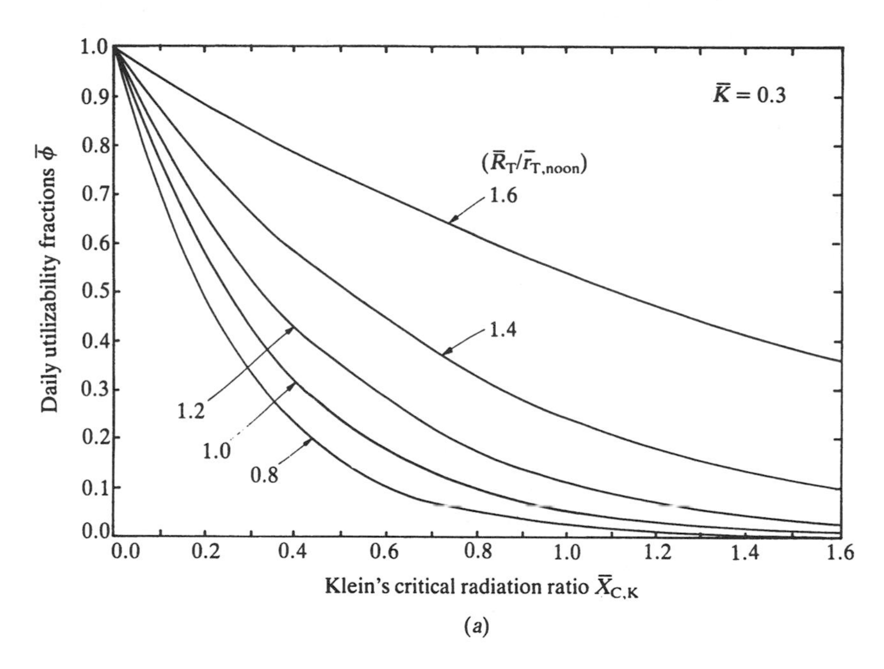

FIGURE 19.13(a) Generalized daily utilizability curves of Theilacker and Klein (1980) for three different $\bar{K}$ values.

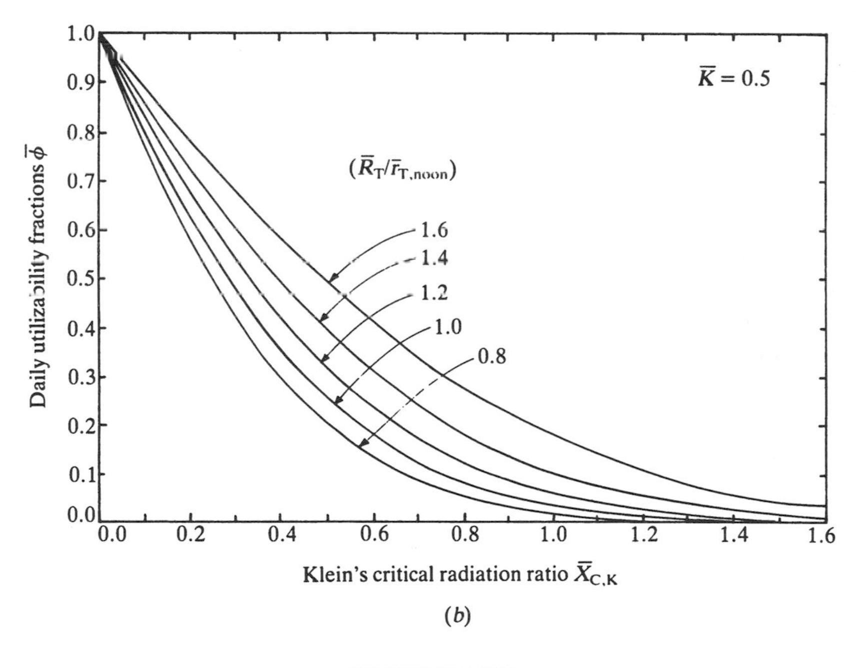

FIGURE 19.13(b)

$$\frac{Q_{CY}}{A_C F_R \eta_n} = \tilde{a} + \tilde{b} I_C + \tilde{c} I_C^2 \tag{19.26}$$

where the coefficients $\tilde{a}$, $\tilde{b}$, and $\tilde{c}$ are functions of collector type and/or tracking mode, climate, and in some cases, latitude. The complete expressions as revised by Gordon and Rabl (1982) are given in Reddy (1987). Note that the yearly *daytime* average value of T_a should be used to determine

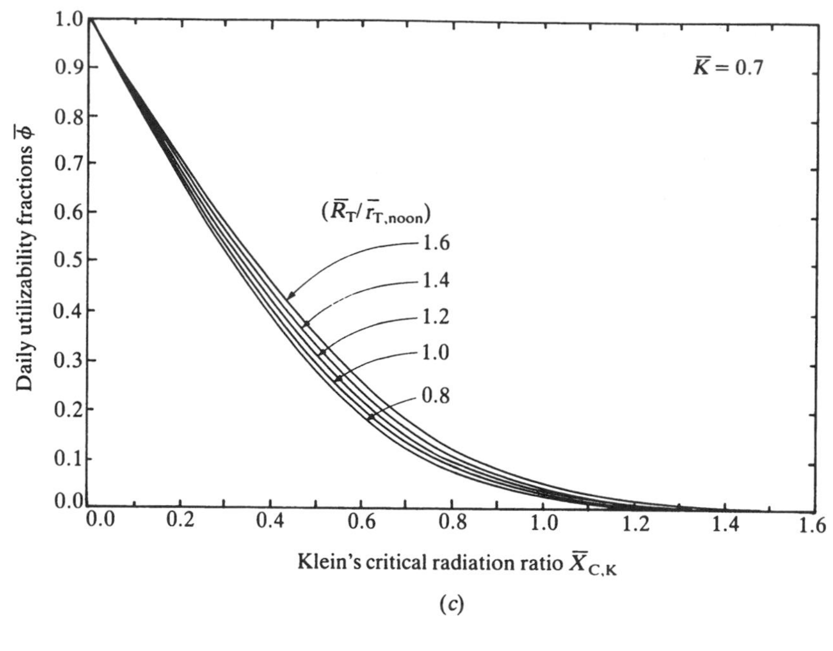

FIGURE 19.13(c)

I_C. If this is not available, the yearly mean *daily* average value can be used. Plots of Q_{CY} versus I_C for flat-plate collectors that face the equator with tilt equal to the latitude are shown in Figure 19.14. The solar radiation enters these expressions as $\tilde{I}_{bn}$, the annual average beam radiation at normal incidence. This can be estimated from the following correlation

$$\tilde{I}_{bn} = 1.37\ \tilde{K} - 0.34 \tag{19.27}$$

where $\tilde{I}_{bn}$ is in kW/m^2 and $\tilde{K}$ is the annual average clearness index of the location. Values of $\tilde{K}$ for several locations worldwide are given in Reddy (1987).

This correlation is strictly valid for latitudes ranging from 25° to 48°. If used for lower latitudes, the correlation is said to lead to overprediction. Hence, it is recommended that for such lower latitudes a value of 25° be used to compute Q_{Cy}.

A direct comparison of the yearly performance of different collector types is given in Figure 19.15 (from Rabl, 1981). A latitude of 35°N is assumed and plots of Q_{Cy} vs. $(T_{Ci} - T_a)$ have been generated in a sunny climate with I_{bn} = 0.6 kW/m^2. Relevant collector performance data are given in Figure 19.15. The cross-over point between flat-plate and concentrating collectors is approximately 25°C above ambient temperature whether the climate is sunny or cloudy.

19.3 Solar Systems

Classification

Solar thermal systems can be divided into two categories: standalone or solar supplemented. They can be further classified by means of energy collection as active or passive, by their use as residential or industrial. Further, they can be divided by collector type into liquid or air systems, and by the type of storage they use into seasonal or daily systems.

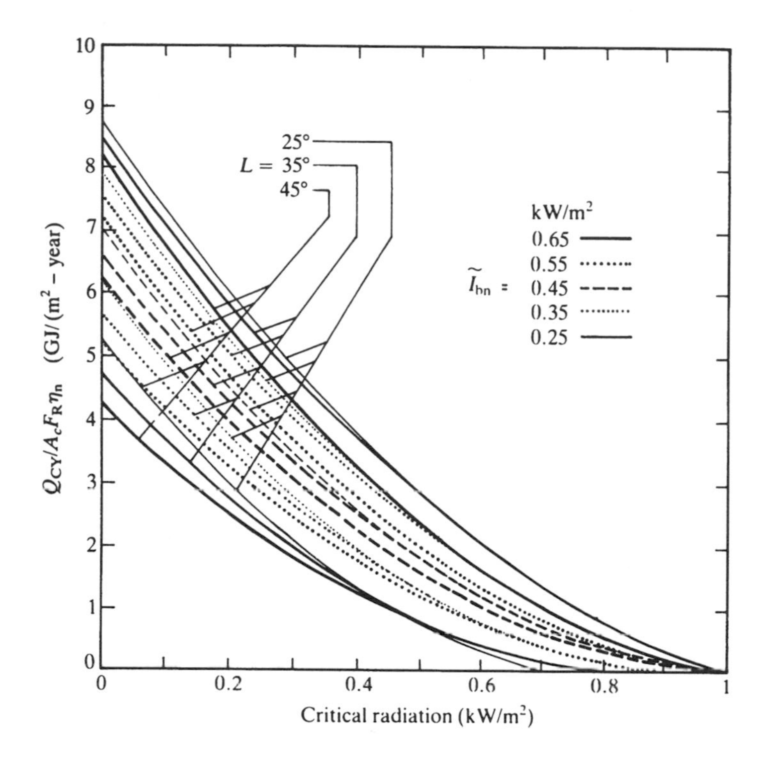

FIGURE 19.14 Yearly total energy delivered by flat-plate collectors with tilt equal to latitude (Source: Gordon and Rabl, 1982).

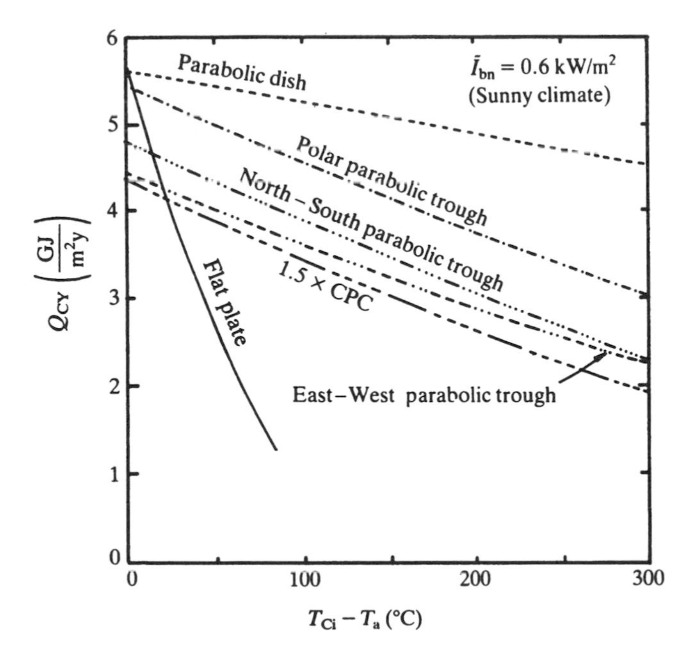

FIGURE 19.15 Figure illustrating the comparative performance (yearly collectible energy) of different collector types as a function of the difference between collector inlet temperature and ambient collector performance parameters $F'\eta_0$ and $F'U_L$ in W/(m² °C) are: flat plate (0.70 and 5.0), CPC (0.60 and 0.75), parabolic trough (0.65 and 0.67), and parabolic dish (0.61 and 0.27) (Source: Rabl, 1981).

Standalone and Solar Supplemented Systems

Standalone systems are systems in which solar energy is the only source of energy input used to meet the required load. Such systems are normally designed for applications where a certain amount of tolerance is permissible concerning the load requirement; in other words, where it is not absolutely imperative that the specified load be met each and every instant. This leniency is generally admissible in the case of certain residential and agricultural applications. The primary reasons for using such systems are their low cost and simplicity of operation.

Solar-supplemented systems, widely used for both industrial and residential purposes, are those in which solar energy supplies part of the required heat load, the rest being met by an auxiliary source of heat input. Due to the daily variations in incident solar radiation, the portion of the required heat load supplied by the solar energy system may vary from day to day. However, the auxiliary source is so designed that at any instant it is capable of meeting the remainder of the required heat load. It is normal practice to incorporate an auxiliary heat source large enough to supply the entire heat load required. Thus, the benefit in the solar subsystem is not in its capacity credit (i.e., not that a smaller capacity conventional system can be used), but rather that a part of the conventional fuel consumption is displaced. The solar subsystem thus acts as a fuel economizer.

Solar-supplemented energy systems will be the primary focus of this chapter. Designing such systems has acquired a certain firm scientific rationale, and the underlying methodologies have reached a certain maturity and diversity, which may satisfy professionals from allied fields. On the other hand, unitary solar apparatus are not discussed here, since these are designed and sized based on local requirements, material availability, construction practices, and practical experience. Simple rules of thumb based on prior experimentation are usually resorted to for designing such systems.

Active and Passive Systems

Active systems are those systems that need electric pumps or blowers to collect solar energy. It is evident that the amount of solar energy collected should be more than the electrical energy used. Active systems are invariably used for industrial applications and for most domestic and commercial applications as well. *Passive systems* are those systems that collect or use solar energy without direct recourse to any source of conventional power, such as electricity, to aid in the collection. Thus, either such systems operate by natural thermosyphon (for example domestic water heating systems) between collector, storage, and load or, in the case of space heating, the architecture of the building is such as to favor optimal use of solar energy. Use of a passive system for space heating applications, however, in no way precludes the use of a back-up auxiliary system.

Residential and Industrial Systems

Basically, the principles and the components used in these two types of systems are alike, the difference being in the load distribution, control strategies, and relative importance of the components with respect to each other. Whereas *residential* loads have sharp peaks in the early morning or in the evening and have significant seasonal variations, industrial loads tend to be fairly uniform over the year. Constant loads favor the use of solar energy because good equipment utilization can be achieved. Because of differences in load distribution, the role played by the storage differs for both applications. Residential loads often occur at times when solar radiation is no longer available. Thus the collector and the storage subsystems interact in a mode without heat withdrawal from the storage. Finally, for economic reasons, many residential systems are designed to operate by natural thermosyphon, in which case no pumps or controls are needed.

On the other hand, for *industrial and commercial* applications, there is no a priori relationship between the time dependence of the load and the period of sunshine. Moreover, a high reliability has to be assured, so the solar system will have to be combined with a conventional system. Very often, a significant portion of the load can be directly supplied by the solar system even without storage. Another option is to use buffer storage for short periods, on the order of a few hours, in case of discontinuous batch process loads. Thus, the proper design of the storage component has to be given adequate consideration. At present, due to economic constraints as well as the fact that proper awareness of the various installations and operational difficulties associated with larger solar thermal systems is still lacking, solar thermal systems are normally designed either (i) with the no-storage option, or (ii) with buffer storage where a small fraction of the total heat demand is only supplied by the solar system.

Liquid and Air Collectors

Although air has been the primary fluid for space heating and drying applications, solar air heating systems have until recently been relegated to second place, mainly as a result of the engineering difficulties associated with such systems. Also, applications involving hot air are probably less common than those needing hot water. Air systems for space heating are well described by Lof (1981).

Even with liquid solar collectors, various configurations are possible, and these can be classified basically as *nontracking* (which include flat-plate collectors and compound parabolic concentrators) or *tracking* collectors (which include various types of concentrating collectors). For low-grade thermal heat, for which solar energy is most suited, flat-plate collectors are far more appropriate than concentrating collectors, not only because of their lower cost but also because of their higher thermal efficiencies at low temperature levels. Moreover, their operation and maintenance costs are lower. Finally, for locations having a high fraction of diffuse radiation, as in the tropics, flat-plate collectors are considered to be thermally superior because they can make use of diffuse radiation as well as beam radiation. Although the system design methodologies presented in this chapter explicitly assume flat plate collector systems, these design approaches can be equally used with concentrating collectors.

Daily and Seasonal Storage

By *daily storage* is meant systems having capacities equivalent to at most a few days of demand (i.e., just enough to tide over day-to-day climatic fluctuations). In *seasonal storage*, solar energy is stored during the summer for use in winter. Industrial demand loads, which are more or less uniform over the year, are badly suited for seasonal-storage systems. This is also true of air-conditioning for domestic and commercial applications because the load is maximum when solar radiation is also maximum, and vice versa. The present-day economics of seasonal storage units do not usually make such systems an economical proposition except for community heating in cold climates.

Closed-Loop and Open-Loop Systems

The two possible configurations of solar thermal systems with daily storage are classified as closed-loop or open-loop systems. Though different authors define these differently, we shall define these as follows. A *closed-loop system* has been defined as a circuit in which the performance of the solar collector is directly dependent on the storage temperature. Figure 19.16 gives a schematic of a closed-loop system in which the fluid circulating in the collectors does not mix with the fluid supplying thermal energy to the load. Thus, these two subsystems are distinct in the sense that any combination of fluids (water or air) is theoretically feasible (a heat exchanger, as shown in the figure, is of course imperative when the fluids are different). However, in practice, only water–water, water–air, or air–air combinations are used. From the point of system performance, the storage temperature normally varies over the day and, consequently, so does collector performance. Closed-loop system configurations have been widely used to date for domestic hot water and space heating applications. The flow rate per unit collector area is generally around 50 kg/(h m^2) for liquid collectors. The storage volume makes about 5 to 10 passes through the collector during a typical sunny day, and this is why such systems are called *multipass* systems. The temperature rise for each pass is small, of the order of 2 to 5°C for systems with circulating pumps and about 10°C for thermosyphon systems. An expansion tank and a check valve to prevent reverse thermosyphoning at nights, although not shown in the figure, are essential for such system configurations.

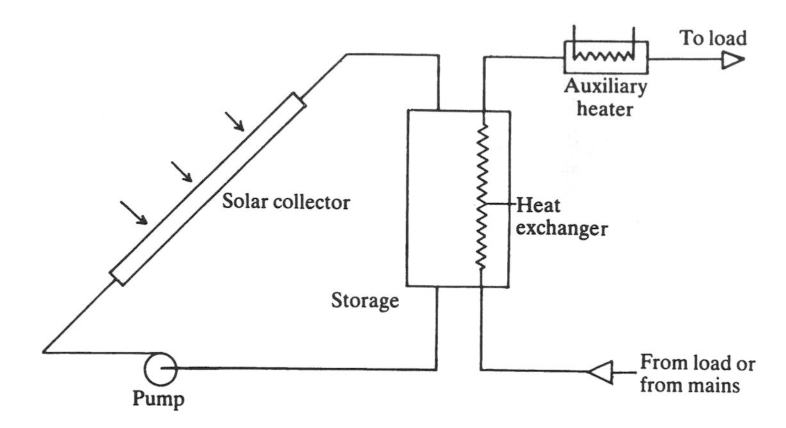

FIGURE 19.16 Schematic of a closed-loop solar system.

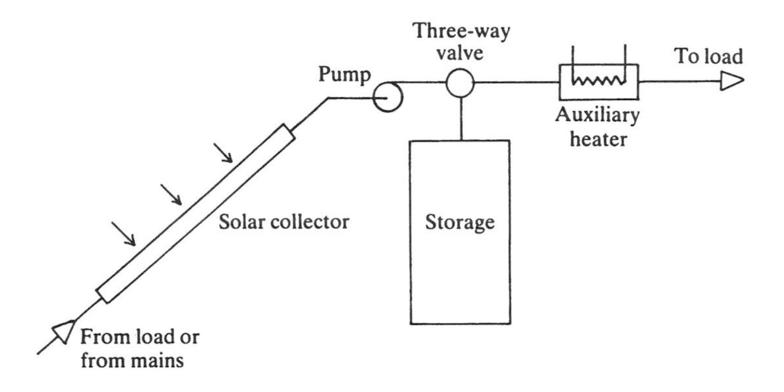

FIGURE 19.17 Schematic of an open-loop solar system.

Figure 19.17 illustrates one of the possible configurations of *open-loop systems*. Open-loop systems are defined as systems in which the collector performance is independent of the storage temperature. The working fluid may be rejected (or a heat recuperator can be used) if contaminants are picked up during its passage through the load. Alternatively, the working fluid could be directly recalculated back to the entrance of the solar collector field. In all these open-loop configurations, the collector is subject to a given or known inlet temperature specified by the load requirements.

If the working fluid is water, instead of having a continuous flow rate (in which case the outlet temperature of the water will vary with insolation), a solenoid valve can be placed just at the exit of the collector, set so as to open when the desired temperature level of the fluid in the collector is reached. The water is then discharged into storage, and fresh water is taken into the collector. The solar collector will thus operate in a discontinuous manner, but this will ensure that the temperature in the storage is always at the desired level. An alternative way of ensuring uniform collector outlet temperature is to vary the flow rate according to the incident radiation. One can collect a couple of percent more energy than with constant rate single-pass designs (Gordon and Zarmi, 1985). However, this entails changing the flow rate of the pump more or less continuously, which is injurious to the pump and results in reduced life. Of all the three variants of the open-loop configuration, the first one, namely the single-pass open-loop solar thermal system configuration with constant flow rate and without a solenoid valve, is the most common.

As stated earlier, closed-loop systems are appropriate for domestic applications. Until recently, industrial process heat systems were also designed as large solar domestic hot water systems with high collector flow rates and with the storage tank volume making several passes per day through the collectors. Consequently, the storage tank tends to be fairly well mixed. Also the tank must be strong enough to withstand the high pressure from the water mains. The open-loop single-pass configuration, wherein the required average daily fluid flow is circulated just once through the collectors with the collector inlet temperature at its lowest value, has been found to be able to deliver as much as 40% more yearly energy for industrial process heat applications than the multipass designs (Collares-Pereira et al., 1984). Finally, in a closed-loop system where an equal amount of fresh water is introduced into storage whenever a certain amount of hot water is drawn off by the load, it is not possible to extract the entire amount of thermal energy contained in storage since the storage temperature is continuously reduced due to mixing. This *partial depletion effect* in the storage tank is not experienced in open-loop systems. The penalty in yearly energy delivery ranges typically from about 10% for daytime-only loads to around 30% for nighttime-only loads compared to a closed-loop multipass system where the storage is depleted every day. Other advantages of open-loop systems are (i) the storage tank need not be pressurized (and hence is less costly), and (ii) the pump size and parasitic power can be lowered.

A final note of caution is required. The single-pass design is not recommended for *variable* loads. The tank size is based on yearly daily load volumes, and efficient use of storage requires near-total depletion of the daily collected energy each day. If the load draw is markedly lower than its average value, the storage would get full relatively early the next day and solar collection would cease. It is because industrial loads tend to be more uniform, both during the day and over the year, than domestic applications that the single-pass open-loop configuration is recommended for such applications.

Description of a Typical Closed-Loop System

Figure 19.18 illustrates a typical closed-loop solar-supplemented liquid heating system. The useful energy is often (but not always) delivered to the storage tank via a collector–heat exchanger, which separates the collector fluid stream and the storage fluid. Such an arrangement is necessary either for antifreeze protection or to avoid corrosion of the collectors by untreated water containing gases and salts. A safety relief valve is provided because the system piping is normally nonpressurized, and any steam produced in the solar collectors will be let off from this valve. When this happens, energy dumping is said to take place. Fluid from storage is withdrawn and made to flow through the load–heat exchanger when the load calls for heat. Whenever possible, one should withdraw fluid directly from the storage and pass it through the load, and avoid incorporating the load–heat exchanger, since it introduces additional thermal penalties and involves extra equipment and additional parasitic power use. Heat is withdrawn from the storage tank at the top and reinjected at the bottom in order to derive maximum benefit from the thermal stratification that occurs in the storage tank. A bypass circuit is incorporated prior to the load heat exchanger and comes into play

1. when there is no heat in the storage tank (i.e., storage temperature T_S is less than the fluid temperature entering the load heat exchanger T_{Xi})
2. when T_S is such that the temperature of the fluid leaving the load heat exchanger is greater than that required by the load (i.e., $T_{Xo} > T_{Li}$, in which case the three-way valve bypasses part of the flow so that $T_{Xo} = T_{Li}$). The bypass arrangement is thus a differential control device which is said to modulate the flow such that the above condition is met. Another operational strategy for maintaining $T_{Xo} = T_{Li}$ is to operate the pump in a "bang-bang" fashion (i.e., by short cycling the pump). Such an operation is not advisable, however, since it would lead to premature pump failure.

An auxiliary heater of the *topping-up type* supplies just enough heat to raise T_{Xo} to T_{Li}. After passing through the load, the fluid (which can be either water or air) can be recirculated or, in case of liquid contamination through the load, fresh liquid can be introduced. The auxiliary heater

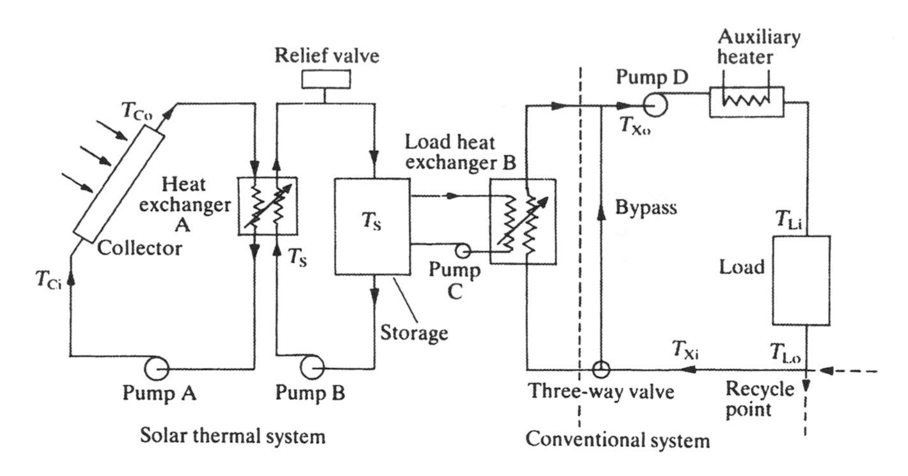

FIGURE 19.18 Schematic of a typical closed-loop system with auxiliary heater placed in series (also referred to as a topping-up type).

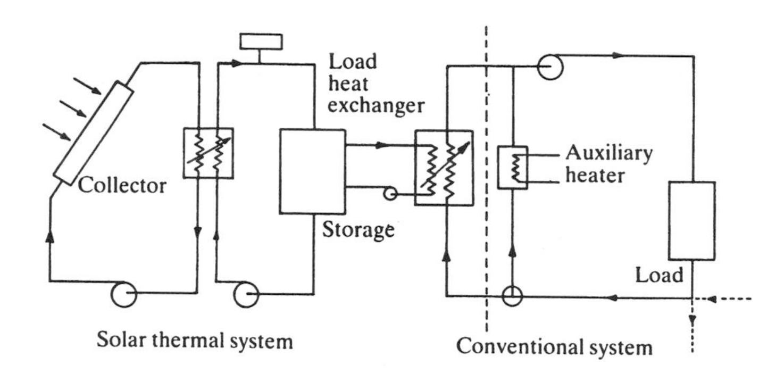

FIGURE 19.19 Schematic of a typical closed-loop system with auxiliary heater placed in parallel (also referred to as an all-or-nothing type).

can also be placed in parallel with the load (see Fig. 19.19), in which case it is called an *all-or-nothing type*. Although such an arrangement is thermally less efficient than the topping-up type, this type is widely used during the solar retrofit of heating systems because it involves little mechanical modifications or alterations to the auxiliary heater itself.

It is obvious that there could also be solar-supplemented energy systems that do not include a storage element in the system. Figure 19.20 shows such a system configuration with the auxiliary heater installed in series. The operation of such systems is not very different from that of systems with storage, the primary difference being that whenever instantaneous solar energy collection exceeds load requirements (i.e., $T_{Co} > T_{Li}$), energy dumping takes place. It is obvious that by definition there cannot be a closed-loop, no-storage solar thermal system. Solar thermal systems without storage are easier to construct and operate, and even though they may be effective for 8 to 10 hours a day, they are appropriate for applications such as process heat in industry.

Active closed-loop solar systems as described earlier are widely used for service hot water systems, that is, for domestic hot water and process heat applications as well as for space heat. There are different variants to this generic configuration. A system without the collector–heat exchanger is referred to having collectors *directly coupled* to the storage tank (as against *indirect coupling* as in Fig. 19.16). For domestic hot water systems, the system can be simplified by placing the auxiliary heater (which is simply an electric heater) directly inside the storage tank. One would like to maintain stratification in the tank so that the coolest fluid is at the bottom of the storage tank,

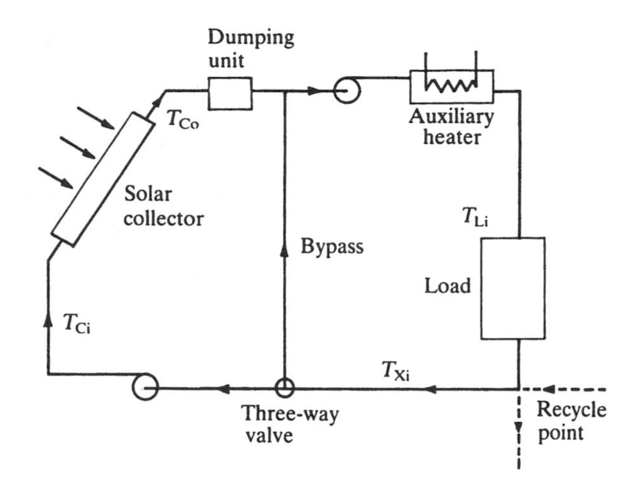

FIGURE 19.20 Simple solar thermal system without storage.

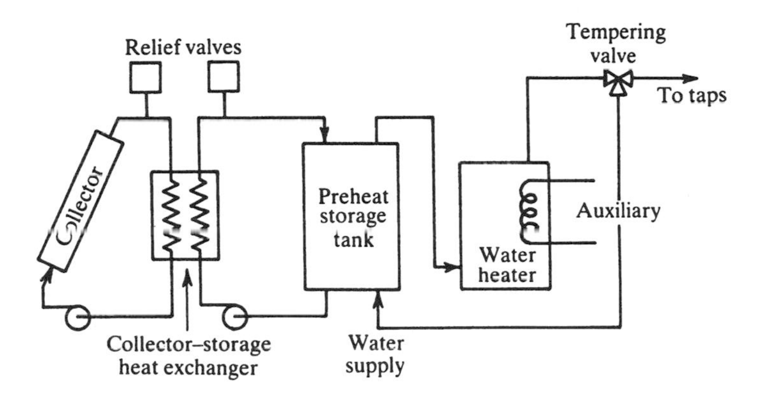

FIGURE 19.21 Schematic of a standard domestic hot-water system with double tank arrangement (Source: Duffie and Beckman, 1980).

thereby enhancing collection efficiency. Consequently, the electric heater is placed at about the upper third portion of the tank so as to assure good collection efficiency while assuring adequate hot water supply to the load. A more efficient but expensive option is widely used in the United States: the *double tank system*, shown in Figure 19.21. Here the functions of solar storage and auxiliary heating are separated, with the solar tank acting as a preheater for the conventional gas or electric unit. Note that a further system simplification can be achieved for domestic applications by placing the load heat exchanger directly inside the storage tank. In certain cases, one can even eliminate the heat exchanger completely.

Another system configuration is the *drain-back* (also called drain-out) system, where the collectors are emptied each time the solar system shuts off. Thus the system invariably loses collector fluid at least once, and often several times, each day. No collector–heat exchanger is needed, and freeze protection is inherent in such a configuration. However, careful piping design and installation, as well as a two-speed pump, are needed for the system to work properly (Newton and Gilman, 1977). The drain-back configuration may be either open (vented to atmosphere) or closed (for better corrosion protection). Long-term experience in the United States with the drain-back system has shown it to be very reliable if engineered properly. A third type of system configuration is the *drain-down* system, where the fluid from the collector array is removed only when adverse conditions, such as freezing or boiling, occur. This design is used when freezing ambient temperatures are only infrequently encountered.

Active solar systems of the type described above are mostly used in countries such as the United States and Canada. Countries such as Australia, India, and Israel (where freezing is rare) usually prefer thermosyphon systems. No circulating pump is needed, the fluid circulation being driven by density difference between the cooler water in the inlet pipe and the storage tank and the hotter water in the outlet pipe of the collector and the storage tank. The low fluid flow in thermosyphon systems enhances thermal stratification in the storage tank. The system is usually fail-proof, and a study by Liu and Fanney (1980) reported that a thermosyphon system performed better than several pumped service hot water systems. If operated properly, thermosyphon and active solar systems are comparable in their thermal performance. A major constraint in installing thermosyphon systems in already existing residences is the requirement that the bottom of the storage tank be at least 20 cm or more higher than the top of the solar collector in order to avoid reverse thermosyphoning at night. To overcome this, spring-loaded one-way valves have been used, but with mixed success.

Controls

There are basically five categories to be considered when designing automatic controls (Mueller Associates, 1985): (i) collection to storage, (ii) storage to load, (iii) auxiliary energy to load, (iv) miscellaneous (i.e., heat dumping, freeze protection, overheating, etc.), and (v) alarms. The three major control system components are sensors, controllers, and actuating devices. Sensors are used to detect conditions (such as temperatures, pressures, etc.). Controllers receive output from the sensors, select a course of action, and signal a system component to adjust the condition. Actuated devices are components such as pumps, valves, and alarms that execute controller commands and regulate the system.

The sensors for the controls must be set, operated, and located correctly if the solar system is to collect solar energy effectively, reduce operating time, wear and tear of active components, and minimize auxiliary and parasitic energy use. Moreover, sensors also need to be calibrated frequently. For diagnostic purposes, it may be advisable to add extra sensors and data acquisition equipment in order to verify system operation and keep track of long-term system operation. Potential problems can be then rectified in time. The reader may refer to manuals by Mueller Associates (1985) or by SERI (1989) for more details on controls pertaining to solar energy systems.

Though single-point temperature controllers or solar-cell-activated controls have been used for activated solar collectors, the best way to do so is by differential temperature controllers. Temperature sensors are used to measure the fluid temperature at collector outlet and at the bottom of the storage tank. When the difference is greater than a set amount, say 5°C, then the controller turns the pump on. If the pump is running and the temperature difference falls below another preset value, say 1°C, the controller stops the pump. The temperature deadband between switching-off and reactivating levels should be set with care, since too high a deadband would adversely affect collection efficiency and too low a value would result in short cycling of the collector pump. Figure 19.22 taken from CSU (1980), shows typical diurnal temperature variations of the liquids at collector exit T_1 and in the storage bottom T_3 as a result of heat withdrawal and/or heat losses from the storage. At about 8:30 A.M., $T_1 > T_3$ and, since there is no flow in the collector, T_1 increases rapidly until the difference $(T_1 - T_3)$ reaches the preset activation level (shown as point 1). The collector pump A comes on, and liquid circulation through the collector begins. Because of this cold water surge, T_1 decreases, resulting in a drop of $(T_1 - T_3)$ to the preset deactivating level (shown as point 2). The pump switches off, and liquid flow through the collectors stops. Gradually T_1 increases again, and so on. The number of on-off cycles at system start-up depends on solar intensity, fluid flow rate, volume of water in the collector loop, and the differential controller setting. A similar phenomenon of cycling also occurs in the afternoon. However, the error introduced in solar collector long-term performance predictions by neglecting this cycling effect in the modeling equations is usually small.

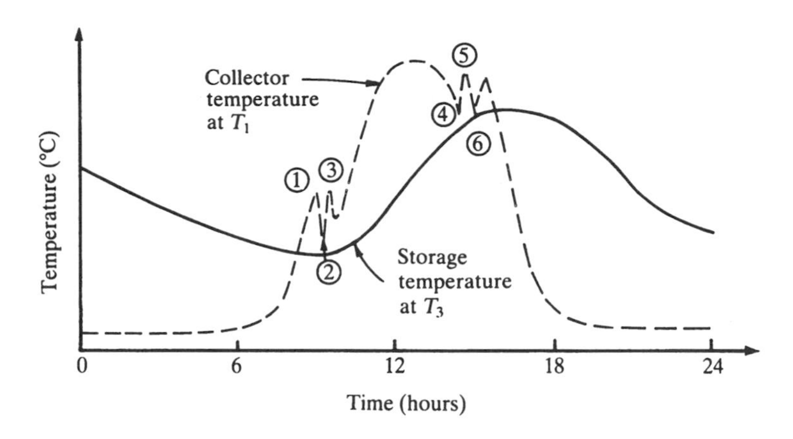

FIGURE 19.22 Typical diurnal variation of collector and storage temperatures (Source: CSU, 1980).

Corrections to Collector Performance Parameters

Combined Collector–Heat Exchanger Performance

The use of the heat exchanger A in Figure 19.18 imposes a penalty on the performance of the solar system because T_{Ci} is always higher than T_S, thereby decreasing q_C (see Fig. 19.23). The collector–heat exchanger can be implicitly accounted for by suitably modifying the collector performance parameters. Recall from basic heat transfer the concept of heat exchanger effectiveness E defined as the ratio of the actual heat transfer rate to the maximum possible heat transfer rate, that is,

$$E = (mc_p)_a (T_{ai} - T_{ao}) / (mc_p)_{\min} (T_{ai} - T_{bi}) \quad \textbf{(19.28a)}$$

$$= (mc_p)_h (T_{bo} - T_{bi}) / (mc_p)_{\min} (T_{ai} - T_{bi}) \quad \textbf{(19.28b)}$$

where $(mC_p)_x$ = capacitance rate of fluid X (with $X = a$ for the warmer fluid, or $X = b$ for the cooler fluid) and $(mC_p)_{\min}$ is the lower heat capacitance value of either stream. The advantage of this modeling approach is that, to a good approximation, E can be considered constant in spite of variations in temperature levels provided the mass flow rates of both fluids remain constant. Thus knowing the two flow rates, E, T_{ai}, and T_{bi}, both the exit fluid temperatures can be conveniently deduced. De Winter (1975) has shown that the combined performance of the solar collector and the heat exchanger can be conveniently modeled by replacing the collector heat removal factor F_R by a combined collector–exchanger heat removal factor F'_R such that

$$\frac{F'_R}{F_R} = \left[1 + \frac{F_R U_L A_C}{(mc_p)_C} \left\{ \frac{(mc_p)_C}{E_A (mc_p)\min} - 1 \right\} \right]^{-1} \quad \textbf{(19.29)}$$

where $(mc_p)_c$ is the capacitance rate of the fluid through the collector and E_A is the effectiveness of heat exchanger A. The variation of F'_R/F_R is shown in Figure 19.24. The plots exhibit the same type of asymptotic behavior with mass flow rate as in Figure 19.3.

The design of the collector–heat exchanger also requires care if the penalty imposed by it on the solar collection is to be minimized. Using a large heat exchanger increases the effectiveness and lowers this penalty; that is, the ratio (F_R'/F_R) is high, but the associated initial and operating costs may be higher. Both these considerations need to be balanced for optimum design (see Fig. 19.25).

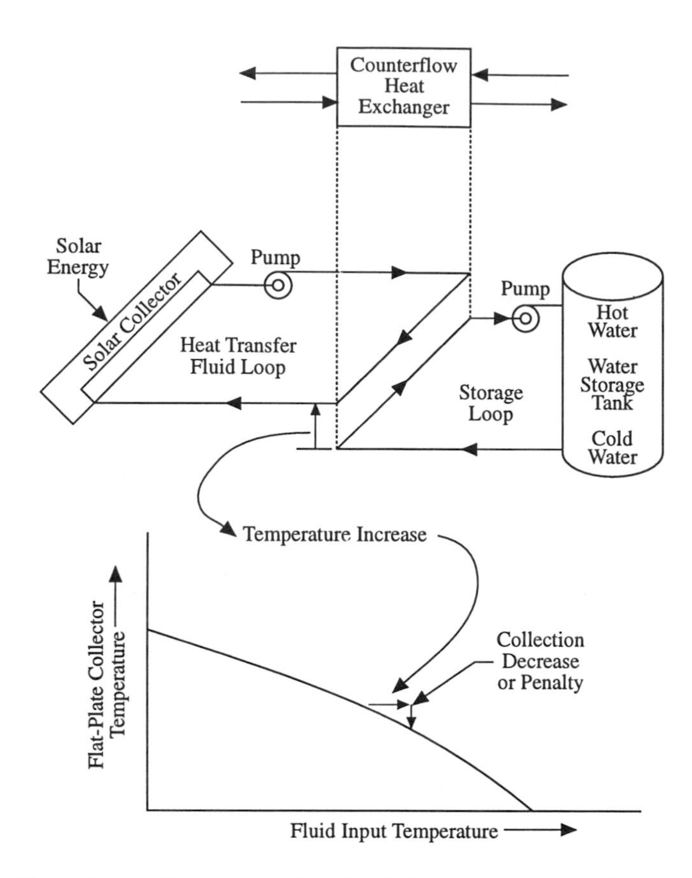

FIGURE 19.23 Heat collection decrease caused by double-loop heat exchangers (Source: Cole et al., 1979).

Optimum heat exchanger area A_X can be found from the following equation proposed by Cole et al. (1979):

$$A_X = A_C \left[\frac{F_R U_L C_C}{U_X C_X} \right]^{1/2} \quad \textbf{(19.30)}$$

where A_C is the collector area, C_C is the cost per unit collector area, C_X is the cost per unit heat exchanger area, and U_X is the heat loss per unit area of the heat exchanger.

Collector Piping and Shading Losses

Other corrections that can be applied to collector performance parameters include those for thermal losses from the piping (or from ducts) between the collection subsystem and the storage unit. Beckman (1978) has shown that these losses can be conveniently taken into consideration by suitably modifying the η_n and U_L terms of the solar collectors as follows:

$$\frac{\eta_n'}{\eta_n} = \left[1 + \frac{u_d A_0}{(mc_p)_C} \right]^{-1} \quad \text{and} \quad \frac{U_L'}{U_L} = \frac{1 - \dfrac{U_d A_i}{(mc_p)_C} + \dfrac{U_d (A_i + A_0)}{A_C F_R U_L}}{1 + \dfrac{U_d A_0}{(mc_p)_C}} \quad \textbf{(19.31)}$$

where U_d is the heat coefficient from the pipe or duct, A_0 is the heat loss area of the outlet pipe or duct, and A_i is the heat loss area of the inlet pipe or duct.

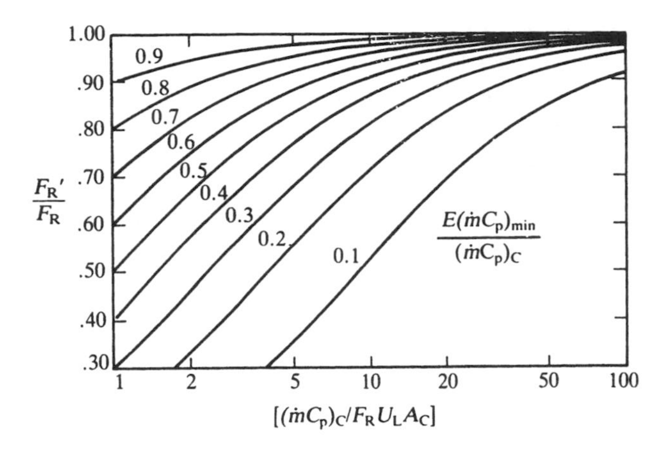

FIGURE 19.24 Variation of collector-heat exchanger correction factor (Source: Duffie and Beckman, 1980).

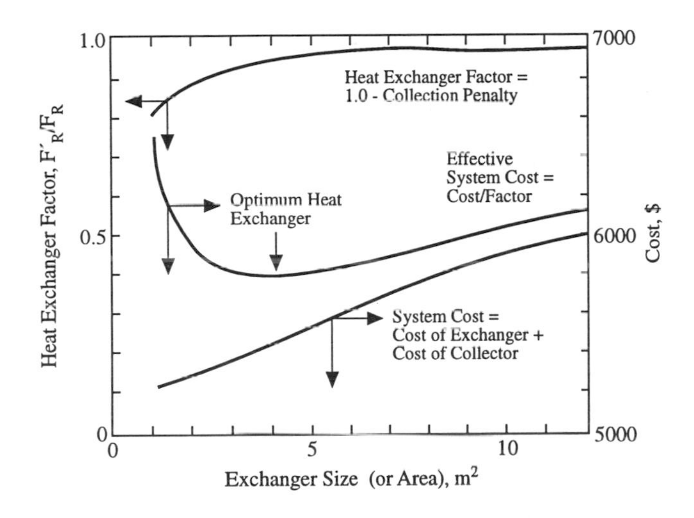

FIGURE 19.25 Typical heat-exchanger optimization plot (Source: Cole et al., 1979).

When large collector arrays are mounted on flat roofs or level ground, multiple rows of collectors are usually arranged in a sawtooth fashion. These multiple rows must be spaced so that they do not shade each other at low sun angles. Unlimited space is rarely available, and it is desirable to space the rows as close as possible to minimize piping and to keep land costs low. Some amount of shading, especially during early mornings and late evenings during the winter months is generally acceptable. Detailed analysis of shading losses is cumbersome though not difficult and equations presented in standard text books such as Duffie and Beckman (1980) can be used directly.

Thermal Storage Systems

Low-temperature solar thermal energy can be stored in liquids, solids, or phase change materials (PCMs). Water is the most frequently used liquid storage medium because of its low cost and high specific heat. The most widely used solid storage medium is rocks (usually of uniform circular size 25 to 40 mm in diameter). PCM storage is much less bulky because of the high latent heat of the PCM material, but this technology has yet to become economical and safe for widespread use.

Water storage would be the obvious choice when liquid collectors are used to supply hot water to a load. When hot air is required (for space heat or for convective drying), one has two options: an air collector with a pebble-bed storage or a system with liquid collectors, water storage, and a load heat exchanger to transfer heat from the hot water to the working air stream. Though a number of solar air systems have been designed and operated successfully (mainly for space heating), water storage is very often the medium selected. Water has twice the heat capacity of rock, so water storage tanks will be smaller than rock-bed containers. Moreover, rock storage systems require higher parasitic energy to operate, have higher installation costs, and require more sophisticated controls. Water storage permits simultaneous charging and discharging while such an operation is not possible for rock storage systems. The various types of materials used as containers for water and rock-bed storage and the types of design, installation, and operation details one needs to take care of in such storage systems are described by Mueller Associates (1985) and SERI (1989).

Sensible storage systems, whether water or rock-bed, exhibit a certain amount of thermal stratification. Standard textbooks present relevant equations to model such effects. In the case of active closed-loop multipass hot water systems, storage stratification effects can be neglected for long-term system performance with little loss of accuracy. Moreover, this leads to conservative system design (i.e., solar contribution is underpredicted if stratification is neglected). A designer who wishes to account for the effect of stratification in the water storage can resort to a formulation by Phillips and Dave (1982), who showed that this effect can be fairly well modeled by introducing a *stratification coefficient* (which is a system constant that needs to be determined only once) and treating the storage subsystem as fully mixed. However, this approach is limited to the specific case of no (or very little) heat withdrawal from storage during the collection period. Even when water storage systems are highly stratified, simulation studies seem to indicate that modeling storage as a one-dimensional plug-flow three-node heat transfer problem yields satisfactory results of long-term solar system performance.

The thermal losses q_w from the storage tank can be modeled as

$$q_w = (UA_S)(T_S - T_{env}) \quad \textbf{(19.32)}$$

where (UA_S) is the storage overall heat loss per unit temperature difference and T_{env} is the temperature of the air surrounding the storage tank. Note that (UA_S) depends (i) on the storage size, which is a parameter to be sized during system design, and (ii) on the configuration of the storage tank (i.e., on the length by diameter ratio in case of a cylindrical tank). For storage tanks, this ratio is normally in the range of 1.0 to 2.0.

Solar System Simulation

A system model is nothing but an assembly of appropriate component modeling equations that are to be solved over time subject to certain forcing functions (i.e., the meteorological data and load data). The resulting set of simultaneous equations can be solved either analytically or numerically.

The analytical method of resolution is appropriate, or possible, only for simplified system configurations and operating conditions. This approach has had some success in the analysis and design of open-loop systems (refer to Reddy, 1987, and Gordon and Rabl, 1986, for more details). On the other hand, numerical simulation can be performed for any system configuration and operating strategy, however complex. However, this is time-consuming and expensive in computer time and requires a high level of operator expertise.

We shall illustrate the approach of numerical simulation by considering the simple solar system shown in Figure 19.18. Assuming a fully mixed storage tank, the instantaneous energy balance equation is

$$(Mc_p)_S \, (dT_S/dt) = q_C - q_u - q_w \quad \textbf{(19.33)}$$

where q_C is the useful energy delivered by the solar collector (given by Eq. 19.2

q_w is the thermal loss from the storage tank (given by Eq. 19.32)

q_u is the useful heat transferred through the load heat exchanger, which can be determined as follows:

The maximum hourly rate of energy transfer through the load heat exchanger is

$$q_{max} = E_B \left(mc_p\right)_{min} \left(T_S - T_{Xi}\right) \delta_L \tag{19.34}$$

where δ_L is a control function whose value is either 1 or 0 depending upon whether there is a heat demand or not. Since q_{max} can be greater than the amount of thermal energy q_L actually required by the load, the bypass arrangement can be conveniently modeled as

$$q_u = \min\left(q_{max}, q_L\right) \tag{19.35}$$

where

$$q_L = \left(mc_p\right)_L \left(T_{Li} - T_{xi}\right) \tag{19.36}$$

for water heating and industrial process heat loads. Space heating and cooling loads can be conveniently determined by one of the several variants of the bin-type methods (ASHRAE, 1985).

The amount of energy q_{aux} supplied by a topping-up type of auxiliary heater is

$$q_{aux} = q_L - q_u \tag{19.37}$$

Assuming $T_{env} = T_a$, Eq. (19.33) can be expanded into

$$\left(Mc_p\right)_S \frac{dT_S}{dt} - A_C F_R\left[I_T \eta_0 - U_L\left(T_S - T_a\right)\right]^+ - \left(mc_p\right)_S \left(T_S - T_{Xi}\right)\delta_L \quad \left(UA\right)_S\left(T_S - T_a\right) \tag{19.38}$$

The presence of control functions and time dependence of I_T and T_a prevent a general analytical treatment, though, as mentioned earlier, specific cases can be handled. The numerical approach involves expressing this differential equation in finite difference form. After rearranging, one gets

$$-(Ua)_S\left(T_{S,b} - T_a\right)\}T_{S,f} = T_{S,b} + \frac{\Delta t}{\left(Mc_p\right)_S}\left\{A_C F_R\left[I_T \eta_0 - U_L\left(T_{S,b} - T_a\right)\right]^+ - \left(mc_p\right)_S\left(T_{S,b} - T_{X,i}\right)\delta_L \tag{19.39}$$

where $T_{S,b}$ and $T_{S,f}$ are the storage temperatures at the beginning and the end of the time step Δt. The time step is sufficiently small (say 1 hour) that I_T and T_a can be assumed constant. This equation is repeatedly used over the time period in question (day, month, or year), and the total energy supplied by the collector or to the load can be estimated.

Such methods of simulation, referred as *stepwise* steady-state simulations, implicitly assume that the solar thermal system operates in a steady-state manner during one time step, at the end of which it undergoes an abrupt change in operating conditions as a result of changes in the forcing functions, and thereby attains a new steady-state operating level. Although in reality, the system performance varies smoothly over time and is consequently different from that outlined earlier, it has been found that, in most cases, taking time steps of the order of 1 hour yields acceptable results of long-term performance.

The objective of solar-supplemented energy systems is to displace part of the conventional fuel consumption of the auxiliary heater. The index used to represent the contribution of the solar thermal system is the *solar fraction*, which is the fraction of the total energy required by the load that is supplied by the solar system. The solar fraction could be expressed over any time scale, with month and year being the most common. Two commonly used definitions of the monthly solar fraction are

1. thermal solar fraction:

$$f_Y = Q_{UM}/Q_{LM} = 1 - Q_{aux,M}/Q_{LM} \quad \textbf{(19.40)}$$

where Q_{UM} is the monthly total thermal energy supplied by the solar system
Q_{LM} is the monthly total thermal requirements of the load
$Q_{aux,M}$ is the monthly total auxiliary energy consumed

2. energy solar fraction (i.e., thermal plus parasitic energy):

$$f'_M = Q'_{UM}/Q'_{LM} \quad \textbf{(19.41)}$$

where Q_{UM}' is Q_{UM} *minus* the parasitic energy consumed by the solar system and Q_{LM}' is Q_{LM} *plus* the parasitic energy consumed by the load.

Example 19.5. Simulate the closed-loop solar thermal system shown in Figure 19.18 for each hour of a day assuming both collector and load heat exchangers to be absent (i.e., $E_A = E_B = 1$). Assume the following data as input for the simulation: $A_C = 10$ m², $F_RU_L = 5.0$ W/m² °C), $F_R\eta_0 = 0.7$, $(Mc_p) = 2.0$ MJ/°C and $(UA)_S = 3$ W/°C. Water is withdrawn to meet a load from 9 A.M. to 7 P.M. (solar time) at a constant rate of 60 kg/h and is replenished from the mains at a temperature of 25°C. The storage temperature at the start (i.e., at 6 A.M.) is 40°C, and the environment temperature is equal to the ambient temperature. The temperature of the water entering the load should not exceed 55°C. The hourly values of the solar radiation on the plane of the collector are given in column 2 of Table 19.6 and the ambient temperature is assumed constant over the day and equal to 25°C. The variation of the optical efficiency with angle of incidence can be neglected.

TABLE 19.6 Simulation Results of Example 19.5

(1) Solar time (h)	(2) I_T (MJ/m²h)	(3) $T_{s,f}$ (°C)	(4) q_C (MJ/h)	(5) q_w (MJ/h)	(6) q_{max} (MJ/h)	(7) q_L (MJ/h)	(8) q_U (MJ/h)	(9) q_{aux} (MJ/h)
Start		40.00						
6-7	0.37	39.92	0.00	0.16	0.00	0.00	0.00	0.00
7-8	0.95	41.82	3.96	0.16	0.00	0.00	0.00	0.00
8-9	1.54	45.61	7.75	0.18	0.00	0.00	0.00	0.00
9-10	2.00	48.05	10.29	0.22	5.18	7.54	5.18	2.36
10-11	2.27	50.90	11.74	0.25	5.79	7.54	5.79	1.75
11-12	2.46	53.78	12.56	0.28	6.51	7.54	6.51	1.03
12-13	2.50	56.17	12.32	0.31	7.24	7.54	7.24	0.31
13-14	2.24	57.26	10.07	0.34	7.84	7.54	7.54	0.00
14-15	2.12	57.84	9.03	0.35	8.11	7.54	7.54	0.00
15-16	1.37	55.73	3.68	0.35	8.25	7.54	7.54	0.00
16-17	0.76	51.79	0.00	0.33	7.72	7.54	7.54	0.00
17-18	0.23	48.28	0.00	0.29	6.73	7.54	6.73	0.81
18-19	0.00	45.23	0.00	0.25	5.85	7.54	5.85	1.69
Total	18.81	—	81.41	3.48	69.24	75.40	67.48	7.94

The results of the simulation are given in Table 19.6. The following equations should permit the reader to verify for himself the results obtained. Simulating the system entails solving the following equations in the sequence given here:

Column 4. Useful energy delivered by the collector (Eq. 19.2)

$$q_C = 10 \times [0.7 \times I_T - 5 \times (3{,}600/10^6) \times (T_{S,b} - 25)]^+ \qquad (MJ/h)$$

The term $(3{,}600/10^6)$ is introduced to convert W/m^2 (the units in which I_T is expressed) into MJ/(h m^2). Note that $T_{S,b}$ is taken to be equal to $T_{S,f}$ of the final hour.

Column 5. Thermal losses from the storage tank (Eq. 19.32)

$$q_w = 3 \times (3{,}600/10^6)\,(T_{S,b} - 25) \qquad (MJ/h)$$

Column 6. The maximum rate of energy that can be transferred from the load can be calculated from Eq. (19.34)

$$q_{max} = 60 \times (4{,}190/10^6)\,(T_{S,b} - 25) \qquad (MJ/h)$$

Column 7. The thermal energy required by the load (from Eq. 19.36)

$$q_L = 60 \times (4{,}190/10^6)\,(55 - 25) = 7.54\ MJ/h$$

Column 8. The actual amount of heat withdrawn from storage (Eq. 19.35)

$$q_u = \min[\text{column 6, column 7}]$$

Column 9. The amount of energy supplied by the auxiliary heater (Eq. 19.37)

$$q_{aux} = \text{column 7} - \text{column 8}$$

The final storage temperature $T_{S,f}$ is now calculated from Eq. (19.39)

$$T_{S,f} = T_{S,b} + [\text{column 4} - \text{column 8} - \text{column 5}]/2.0$$

From Table 19.6, we note that the solar collector efficiency over the entire day is [81.41/(18.81 × 10)] = 0.43. The corresponding daily solar fraction = (67.48/75.40) = 0.895.

19.4 Solar System Sizing Methodology

Sizing of solar systems primarily involves determining the collector area and storage size that are most cost effective. Stand-alone and solar-supplemented systems have to be treated separately since the basic design problem is somewhat different. The interested reader can refer to Gordon (1987) for sizing standalone systems.

Solar-Supplemented Systems

Production Functions

Because of the annual variation of incident solar radiation, it is not normally economical to size a solar subsystem such that it provides 100% of the heat demand. Most solar energy systems follow

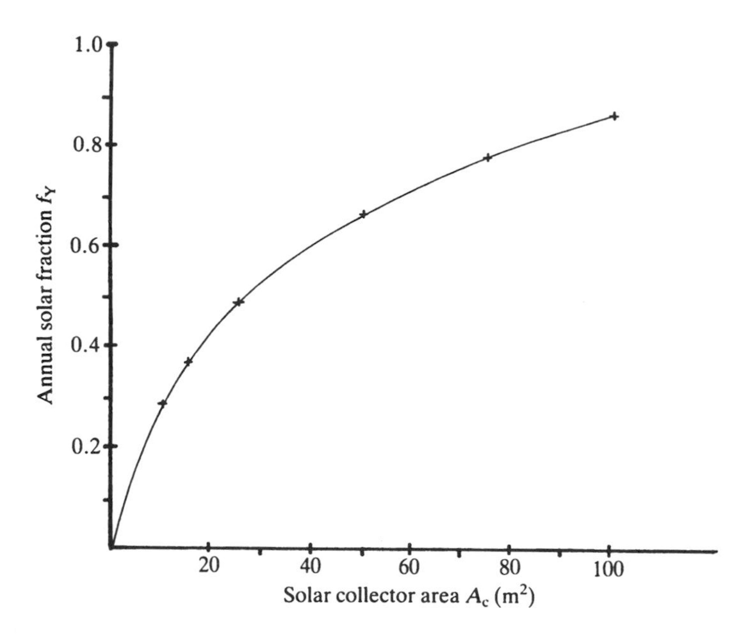

FIGURE 19.26 A typical solar system production function (see Example 19.6).

the *law of diminishing returns.* This implies that increasing the size of the solar collector subsystem results in a less than proportional increase in the annual fuel savings (or alternatively, in the annual solar fraction).

Any model has two types of variables: exogenous and endogenous. The *exogenous parameters* are also called the input variables, and these in turn may be of two kinds. *Variable* exogenous parameters are the collector area A_C, the collector performance parameters $F_R\eta_n$ and F_RU_L, the collector tilt, the thermal storage capacity $(Mc_p)_S$, the heat exchanger size, and the control strategies of the solar thermal system. On the other hand, the climatic data specified by radiation and the ambient temperature, as well as the end-use thermal demand characteristics, are called *constrained* exogenous parameters because they are imposed externally and cannot be changed. The *endogenous* parameters are the output parameters whose values are to be determined, the annual solar fraction being one of the parameters most often sought.

Figure 19.26 illustrates the law of diminishing results. The annual solar fraction f_Y is seen to increase with collector area but at a decreasing rate and at a certain point will reach saturation. Variation of any of the other exogenous parameters also exhibits a similar trend. The technical relationship between f_Y and one or several variable exogenous parameters for a given location is called the *yearly production function.*

It is only for certain simple types of solar thermal systems that an analytical expression for the production can be deduced directly from theoretical considerations. The most common approach is to carry out computer simulations of the particular system (solar plus auxiliary) over the complete year for several combinations of values of the exogenous parameters. The production function can subsequently be determined by an empirical curve fit to these discrete sets of points.

Example 19.6 Kreider (1979) gives the following expression for the production function of an industrial solar water heater for a certain location:

$$f_Y = Q_{UY}/Q_{LY} = \left(0.35 - \frac{F_RU_L}{100F_R\eta_n}\right) \ln\left(1 + \frac{20F_R\eta_nA_C}{Q_{LY}}\right) \qquad \mathbf{(19.42)}$$

where Q_{UY} is the thermal energy delivered by the solar thermal system over the year in GJ/y; Q_{LY} is the yearly thermal load demand, also in GJ/y; and $F_R U_L$ is in W/(m^2 °C). Note that only certain solar system exogenous parameters figure explicitly in this expression, thereby implying that other exogenous parameters (for example, storage volume) have not been varied during the study. As an illustration, let us assume the following nominal values: $Q_{LY} = 100$ GJ/y, $F_R U_L = 2.0$ W/(m^2 °C), and $F_R \eta_n = 0.7$. For a 1% increase in collector area A_C, the corresponding percentage increase in Q_{UY} (called elasticity) can be determined:

$$\frac{dQ_{UY}}{Q_{UY}} = \frac{dA_C}{A_C}\left[\left(\frac{Q_{LY}}{20F_R\eta_n A_C}+1\right)\ln\left(1+\frac{20F_R\eta_n A_C}{Q_{LY}}\right)\right]^{-1} \quad (19.43)$$

From this, we obtain the expression for *marginal productivity*

$$\frac{dQ_{UY}}{dA_C} = \frac{Q_{UY}}{A_C}\left[\left(\frac{Q_{LY}}{20F_R\eta_n A_C}+1\right)\ln\left(1+\frac{20F_R\eta_n A_C}{Q_{LY}}\right)\right]^{-1} \quad (19.44)$$

Numerical values can be obtained from the preceding expression. Though Q_{UY} increases with A_C, the marginal productivity of Q_{UY} goes on decreasing with increasing A_C, thus illustrating the law of diminishing returns. A qualitative explanation of this phenomenon is that as A_C increases, the mean operating temperature level of the collector increases, thus leading to decreasing solar collection rates. Figure 19.26 illustrates the variation of f_Y with A_C as given by Eq. (19.42) when the preceding numerical values are used.

The objective of the sizing study in its widest perspective is to determine, for a given specific thermal end use, the size and configuration of the solar subsystem that results in the most economical operation of the entire system. This economical optimum can be determined using the production function along with an appropriate economic analysis. Several authors—for example, Duffie and Beckman (1980) or Rabl (1985)—have presented fairly rigorous methodologies of economic analysis, but a simple approach is adequate to illustrate the concepts and for preliminary system sizing.

Simplified Economic Analysis

It is widely recognized that *discounted cash flow analysis* is most appropriate for applications such as sizing an energy system. This analysis takes into account both the initial cost incurred during the installation of the system and the annual running costs over its entire life span.

The economic objective function for optimal system selection can be expressed in terms of either the energy cost incurred or the energy savings. These two approaches are basically similar and differ in the sense that the objective function of the former has to be minimized while that of the latter has to be maximized. In our analysis, we shall consider the latter approach, which can further be subdivided into the following two methods:

1. Present worth or life cycle savings, wherein all running costs are discounted to the beginning of the first year of operation of the system.
2. Annualized life cycle savings, wherein the initial expenditure incurred at the start as well as the running costs over the life of the installation are expressed as a yearly mean value.

The *present worth* method is used here. Also, our primary consideration being to size the solar subsystem, it is much simpler to limit our consideration to all the *extra* expenses and savings incurred as a result of incorporating the subsystem. Thus, the costs associated with the conventional and auxiliary systems are not included in the analysis. Finally, in order to permit generality of usage, the simplified economic analysis methodology ignores the effect of income tax, which not only varies from country to country but also depends on the income slab of the enterprise.

1. *Initial cost:* Neglecting effects such as down-payment and salvage value, the initial cost C_I' of the *extra* solar installation can be very simply stated as

$$C_I' = C_b + A_C C_a + M_S C_s \tag{19.45}$$

where C_b is the cost of components that are independent of collector area and storage size (like pumps and controls), C_a is the cost of components that are linearly dependent on collector area (like collector costs, heat exchangers, piping, etc.), and C_S is the cost of components linearly dependent on storage mass. Note that the expression can be expanded to explicitly contain as many exogenous parameters as need to be optimized.

2. *Operating costs:* These costs are spread over n years of the life span of the system. They comprise the running costs, the maintenance and replacement costs, the insurance costs, and the savings.

The **running costs** C_F' of the solar subsystem comprise the power (normally electricity) required for its pumps, blowers, and controls. The costs associated with this parasitic energy are supposed to be paid at the end of each year. Thus the present worth (i.e., the yearly cost discounted to the beginning of the first year) is

$$\begin{aligned} C_F' &= \frac{C_{F1}(1+j_e)}{(1+i)} + \frac{C_{F1}(1+j_e)^2}{(1+i)^2} + \cdots + \frac{C_{F1}(1+j_e)^n}{(1+i)^n} \\ &= \frac{C_{F1}}{(1+i_e')} + \frac{C_{F1}}{(1+i_e')^2} + \cdots + \frac{C_{F1}}{(1+i_e')^n} = C_{F1}\left[\frac{1-(1+i_e')^{-n}}{i_e'}\right] \end{aligned} \tag{19.46a}$$

where C_{F1} is the operational cost over the first year, which could be expressed in terms of the solar thermal system exogenous parameters, or alternatively in terms of the total yearly thermal energy delivered by the system, and i_e' is the effective annual discount rate for electricity:

$$C_F' = nC_{F1} \tag{19.46b}$$

where i and j_e are the annual discount rate and the annual electricity cost inflation rate respectively. They are assumed to be constant over n years of the life of the system. For the special case when $i_e' = 0$,

$$i_e' = (i - j_e)/(1 + j_e) \tag{19.47}$$

Annual parasitic energy is difficult to estimate properly, since its magnitude will depend on the number of hours during which the solar system operates during the year. Normally for space heating applications, parasitic energy will be of the order of 5 to 7% of the total energy delivered by a solar liquid system, while for air system it could be of the order of 10 to 15%. How to determine number of hours of solar system operation is discussed by Mitchell et al. (1981) and Reddy (1987).

The **maintenance and replacement costs** are also supposed to be paid at the end of each year. The present worth over n years is

$$C'_M = C_{M1}\left[\frac{1-\left(\frac{1+M}{1+i'}\right)^n}{(i'-M)}\right] \quad \textbf{(19.48a)}$$

where i' is the effective discount rate given by

$$i' = (i-j)/(1+j) \quad \textbf{(19.48b)}$$

with j being the annual inflation rate that is assumed constant over n years. C_{M1} is the maintenance cost over the first year and M is the corresponding annual rate of *increase* of maintenance, also assumed constant over n years. For the special case when both M and i' are equal to zero

$$C'_M = n\,C_{M1} \quad \textbf{(19.49)}$$

The **insurance costs** are supposed to be paid at the beginning of each year. The present worth over n years is

$$C'_A = C_{A1}\left[\frac{1-\left(\frac{1-A}{1+i'}\right)^n}{(A+i')}\right](1+i') \quad \textbf{(19.50)}$$

where C_{A1} is the sum paid at the beginning of the first year and A is the annual rate of *decrease* in insurance (assumed constant over n years).

The present worth of the **savings** in conventional energy due to the solar subsystem is

$$E' = \frac{1}{\eta_{\text{Aux}}} f_Y Q_{LY} C_f \left[\frac{1-\left(1+i'_f\right)^{-n}}{i'_f}\right] \quad \textbf{(19.51)}$$

where Q_{LY} is the annual thermal demand of energy to be met, η_{Aux} is the thermal efficiency of the auxiliary heater (for oil burners, this is around 0.8), C_f is the unit cost of conventional energy paid at the end of the first year, and i_f' is the effective annual discount rate for the fuel (supposed constant over n years). This is given by

$$i'_f = (i-j_f)/(1+j_f) \quad \textbf{(19.52)}$$

where j_f is the annual rate of increase in the cost of conventional fuel, assumed constant over n years.

Discounted total savings are given by

$$C'_S = E' - (C'_I + C'_F + C'_M + C'_A) \quad \textbf{(19.53)}$$

If all the economic parameters are specified, this equation reduces to

$$C'_S = f\,(A_C, M_s)$$

In other words, C_S' is a function of the variable exogenous parameters whose optimum values are being sought. The objective of the economic study is to determine the set of exogenous parameters whose combination results in C_S' being maximized. In view of the uncertainty in all the economic and other parameters that influence C_S', is seems unwise to invest too much time and effort in trying to estimate f_Y too accurately in a preliminary sizing study.

Example 19.7. For the production function given in Example 19.6, determine the optimal collector area A_C for C_S' to be maximum given the following economic data: $n = 15$ years, $c_b = \$1,000$, $C_a = \$150/m^2$, $j_f = j_e = 0.10$, $j = 0.08$, $i = 0.10$, $C_{F1} = 0.05\ Q_{UY}C_f$, $C_{M1} = 0.05\ C_I'$, $C_{A1} = 0.02\ C_I'$, $M = 0.05$, $A = 0.05$, $C_f = \$0.10/kWh = \$27.8/GJ$, and $\eta_{Aux} = 1.0$.

Then $i' = 0.0185$, $i_e' = i_f' = 0$, $C_F' = 20.85\ Q_{UY}$, $C_M' = 0.919\ C_I'$, $C_A' = 0.193\ C_I'$, and $E' = 417.0\ Q_{UY}$.

Thus

$$\begin{aligned} C_S' &= 417\ Q_{UY} - C_I' - 20.85\ Q_{UY} - 0.919\ C_I' - 0.193\ C_I' \\ &= 12{,}733.4 \times \ln(1 + 0.14\ A_C) - 2{,}112 - 316.8\ A_C \end{aligned} \tag{19.54}$$

The maximum value of C_S' is found when $(dC_S'/dA_C) = 0$. From Eq. (19.54), we find that C_S' is maximum when $A_C = 33\ m^2$. The corresponding solar fraction is $f_Y = 55.5\%$. Figure 19.27 illustrates the variation of C_S' in the framework of the preceding numerical example for two different values of C_f. Curve 1 is for $C_f = 0.1$/kWh (as assumed earlier), and $C_f = 0.01$/kWh (curve 2) corresponds to the case when cost of conventional energy is low. In the latter case, C_S' is negative for all values of A_C, thereby indicating that there is no incentive at all in using a solar system.

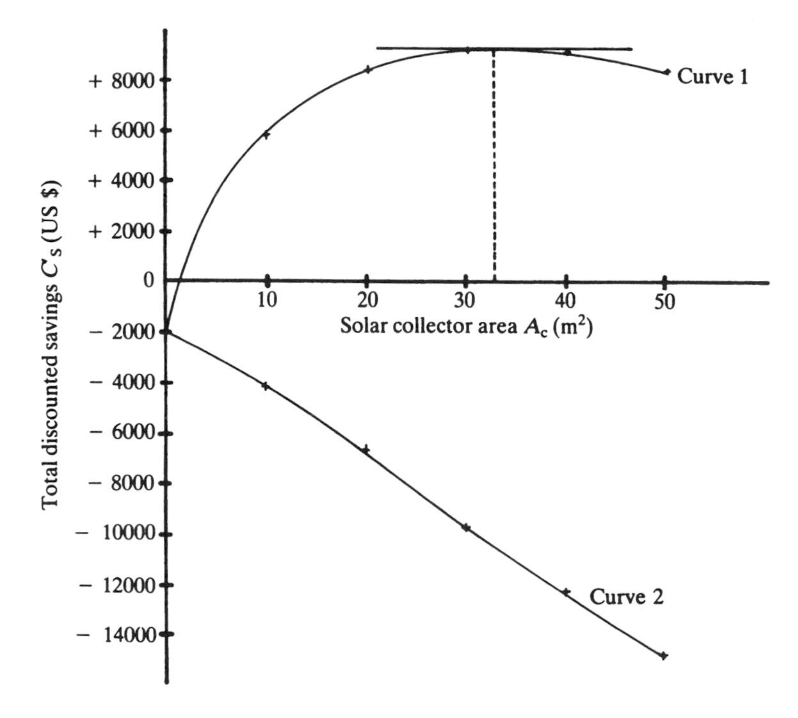

FIGURE 19.27 Variation of total discounted savings with collector area for two different values of conventional fuel unit cost C_f (see Example 19.7). Curve 1 is for $C_f = \$0.1$/kWh and Curve 2 is for $C_f = \$0.01$/kWh.

Another way of presenting the results of an economic analysis would be to determine the minimum number of years for which $C_S' > 0$. Solving Eq. (19.54) iteratively, we get a payback with cash flow discounting of 5.5 years. Such an analysis is often useful for managerial decision-making purposes.

19.5 Solar System Design Methods

Classification

Design methods may be separated into three generic classes. The *simple* category, usually associated with the prefeasibility study phase involves quick manual calculations of solar collector/system performance and rule-of-thumb engineering estimates. For example, the generalized yearly correlations proposed by Rabl (1981) and described in Section 19.2 could be conveniently used for year-round, more or less constant loads. The approach is directly valid for open-loop solar systems, while it could also be used for closed-loop systems if an *average* collector inlet temperature could be determined. A simple manner of selecting this temperature $\overline{T}_m$ for domestic closed-loop multipass systems is to assume the following empirical relation:

$$\overline{T}_m = T_{mains}/3 + (2/3)\,T_{set} \qquad \textbf{(19.55)}$$

where T_{mains} is the average annual supply temperature and T_{set} is the required hot water temperature (about 60 to 80°C in most cases).

These manual methods often use general guidelines, graphs, and/or tables for sizing and performance evaluation. The designer should have a certain amount of knowledge and experience in solar system design in order to make pertinent assumptions and simplifications regarding the operation of the particular system.

Mid-level design methods are resorted to during the feasibility phase of a project. The main focus of this chapter has been toward this level, and a few of these design methods will be presented in this section. A personal computer is best suited to these design methods because they could be conveniently programmed to suit the designer's tastes and purpose (spreadsheet programs, or better still one of the numerous equation-solver software packages, are most convenient). Alternatively, commercially available software packages such as f-chart (Beckman et al., 1977) could also be used for certain specific system configurations.

Detailed design methods involve performing hourly simulations of the solar system over the entire year from which accurate optimization of solar collector and other equipment can be performed. Several simulation programs for active solar energy systems are available, TRNSYS (Klein et al., 1975, 1979) developed at the University of Wisconsin—Madison being perhaps the best known. This public-domain software has technical support and is being constantly upgraded. TRNSYS contains simulation models of numerous subsystem components (solar radiation, solar equipment, loads, mechanical equipment, controls, etc.) that comprise a solar energy system. A user can conveniently hook up components representative of a particular solar system to be analyzed and then simulate that system's performance at a level of detail that the user selects. Thus TRNSYS provides the design with large flexibility, diversity, and convenience of usage.

As pointed out by Rabl (1985), the detailed computer simulations approach, though a valuable tool, has several problems. Judgment is needed both in the selection of the input and in the evaluation of the output. The very flexibility of big simulation programs has drawbacks. So many variables must be specified by the user that errors in interpretation or specification are common. Also, learning how to use the program is a time-consuming task. Because of the numerous system variables to be optimized, the program may have to be run for numerous sets of combinations, which adds to expense and time. The inexperienced user can be easily misled by the second-order

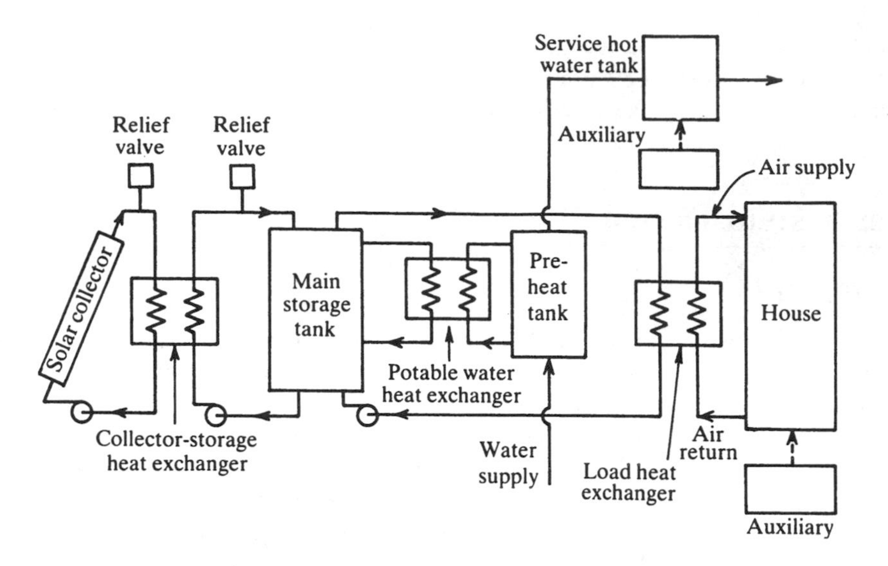

FIGURE 19.28 Schematic of the standard space heating liquid system configuration for the f-chart method (Source: Duffie and Beckman, 1980).

details while missing first-order effects. For example, uncertainties in load, solar radiation, and economic variables are usually very large, and long-term performance simulation results are only accurate to within a certain degree. Nevertheless, detailed simulation programs, if properly used by experienced designers, can provide valuable information on system design and optimization aspects at the final stages of a project design.

There are basically three types of mid-level design approaches: the empirical correlation approach, the analytical approach, and the one-day repetitive methods (described fully in Reddy, 1987). We shall illustrate their use by means of specific applications.

Active Space Heating

The solar system configuration for this particular application has become more or less standardized. For example, for a liquid system, one would use the system shown in Figure 19.28. One of the most widely used design methods is the f-chart method (Beckman et al., 1977; Duffie and Beckman, 1980), which is applicable for standardized liquid and air heating systems as well as for standardized domestic hot water systems. The f-chart method basically involves using a simple algebraic correlation that has been deduced from numerous TRNSYS simulation runs of these standard solar systems subject to a wide range of climates and solar system parameters. Correlations were developed between monthly solar fractions and two easily calculated dimensionless variables X and Y, where

$$X = \left(A_C\, F_R'\, U_L \left(T_{\text{Ref}} - \overline{T}_a\right)\Delta t\right) / Q_{LM} \quad \textbf{(19.56)}$$

$$Y = A_C F_R' \overline{\eta}_0\, \overline{H}_T N / Q_{LM} \quad \textbf{(19.57)}$$

where A_C = collector area (m^2)
F_R' = collector-heat exchanger heat removal factor (given by Eq. 19.29)
U_L = collector overall loss coefficient (W/(m^2 °C))
Δt = total number of seconds in the month = 3,600 × 24 × N = 86,400 × N
$\overline{T}_a$ = monthly average ambient temperature (°C)

T_{Ref} = an empirically derived reference temperature, taken as 100°C
Q_{LM} = monthly total heating load for space heating and/or hot water (J)
$\overline{H}_T$ = monthly average daily radiation incident on the collector surface per unit area (J/m²)
N = number of days in the month
$\overline{\eta}_o$ = monthly average collector optical efficiency

The dimensionless variable X is the ratio of reference collector losses over the entire month to the monthly total heat load; the variable Y is the ratio of the monthly total solar energy absorbed by the collectors to the monthly total heat load. It will be noted that the collector area and its performance parameters are the predominant exogenous variables that appear in these expressions. For changes in secondary exogenous parameters, the following corrective terms X_C and Y_C should be applied for liquid systems:

1. for changes in storage capacity:

$$X_C/X = \left(\text{actual storage capacity}/\text{standard storage capacity}\right)^{-0.25} \quad \textbf{(19.58)}$$

   where the standard storage volume is 75 L/m² of collector area.

2. for changes in heat exchanger size:

$$Y_C/Y = 0.39 + 0.65\ \exp\left[-\left(0.139\ (UA)_B/\left(E_L\left(mc_p\right)_{min}\right)\right)\right] \quad \textbf{(19.59)}$$

The monthly solar fraction for liquid space heating can then be determined from the following empirical correlation:

$$f_M = 1.029\,Y - 0.065\,X - 0.245\,Y^2 + 0.0018\,X^2 + 0.0215\,Y^3 \quad \textbf{(19.60)}$$

subject to the conditions that $0 \le X \le 15$ and $0 \le Y \le 3$. This empirical correlation is shown graphically in Figure 19.29.

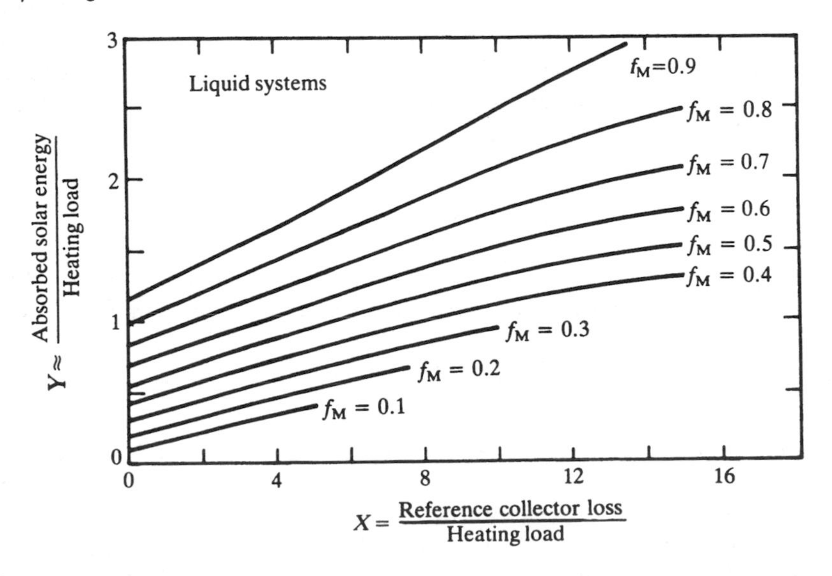

FIGURE 19.29 The f-chart correlation for liquid system configuration (Source: Duffie and Beckman, 1980).

A similar correlation has also been proposed for space heating systems using air collectors and pebble-bed storage. The procedure for exploiting the preceding empirical correlations is as follows. For a predetermined location, specified by its 12 monthly radiation and ambient temperature values, Eq. (19.60) is repeatedly used for each month of the year for a particular set of variable exogenous parameters. The monthly solar fraction f_M and thence the annual thermal energy delivered by the solar thermal system are easily deduced. Subsequently, the entire procedure is repeated for different values and combinations of variable exogenous parameters. Finally, an economic analysis is performed to determine optimal sizes of various solar system components. Care must be exercised that the exogenous parameters considered are not outside the range of validity of the f-chart empirical correlations.

Example 19.8. (adapted from Duffie and Beckman, 1980). A solar heating system is to be designed for Madison, Wisconsin (latitude 43°N) using one-cover collectors with $F_R\eta_n = 0.74$ and $F_R U_L = 4$ W/(m² °C). The collector faces south with a slope of 60° from the horizontal. The average daily radiation on the tilted surface in January is 12.9 MJ/m², and the average ambient temperature is –7°C. The heat load is 36 GJ for space heating and hot water. The collector–heat exchanger correction factor is 0.97 and the ratio of monthly average to normal incidence optical efficiency is 0.96. Calculate the energy delivered by the solar system in January if 50 m² of collector area is to be used.

From Eqs. (19.56) and (19.57), with $A_C = 50$ m²,

$$X = 4.0 \times 0.97 \times \left[100 - \left(-7\right)\right] \times 31 \times 86{,}400 \times 50 / \left(36 \times 10^9\right) = 1.54$$

$$Y = 0.74 \times 0.97 \times 0.96 \times 12.9 \times 10^6 \times 31 \times 50 / \left(36 \times 10^9\right) = 0.38$$

From Eq. (19.60), the solar fraction for January is $f_M = 0.26$. Thus the useful energy delivered by the solar system = 0.26 × 36 = 9.4 GJ.

In an effort to reduce the tediousness involved in having to perform 12 monthly calculations, two analogous approaches that enable the annual solar fraction to be determined directly have been developed by Barley and Winn (1978) and Lameiro and Bendt (1979). These involve the computation of a few site-specific empirical coefficients, thereby rendering the approach less general. For example, the *relative-area* method suggested by Barley and Winn enable the designer to directly calculate the annual solar fraction of the corresponding system using four site-specific empirical coefficients. The approach involves curve fits to simulation results of the f-chart method for specific locations in order to deduce a correlation such as:

$$f = c_1 + c_2 \ln\left(A/A_{0.5}\right) \tag{19.61}$$

where c_1 and c_2 are location-specific parameters that are tabulated for several United States locations, and $A_{0.5}$ is the collector area corresponding to an annual solar fraction of 0.5 given by

$$A_{0.5} = A_S\left(UA\right)/\left(F_R'\eta_0 - F_R'U_L Z\right) \tag{19.62}$$

where A_S and Z are two more location specific parameters, UA is the overall heat loss coefficient of the building, and $F_R'\eta_0$ and $F_R'U_L$ are the corresponding solar collector performance parameters corrected for the effect of the collector–heat exchanger.

Barley and Winn also proposed a simplified economic life-cycle analysis whereby the optimal collector area could be determined directly. Another well-known approach is the *Solar Load Radio* (SLR) method for sizing residential space heating systems (Hunn, 1980).

Domestic Water Heating

The f-chart correlation (Eq. 19.60) can also be used to predict the monthly solar fraction for domestic hot water systems represented by Figure 19.21 provided the water mains temperature T_{mains} is between 5 and 20°C and the minimum acceptable hot water temperature drawn from the storage for end use (called the set water temperature T_w) is between 50 and 70°C. Further, the dimensionless parameter X must be corrected by the following ratio

$$X_w/X = \left(11.6 + 1.8T_w + 3.86\ T_{mains} - 2.32\overline{T}_a\right)/\left(100 - \overline{T}_a\right) \quad \textbf{(19.63)}$$

In case the domestic hot water load is much smaller than the space heat load, it is recommended that Eq. (19.60) be used without the above correction.

Industrial Process Heat

As discussed in "Description of a Typical Closed-Loop System," two types of solar systems for industrial process heat are currently used: the closed-loop multipass systems (with an added distinction that the auxiliary heater may be placed either in series or in parallel (see Figs. 19.18 and 19.19) and the open-loop singlepass system. How such systems can be designed will be described next.

Closed-Loop Multipass Systems

Auxiliary Heater in Parallel. The Phibar-f chart method (Klein and Beckman, 1979, Duffie and Beckman, 1980, Reddy, 1987) is a generalization of the f-chart method in the sense that no restrictions need be imposed on the temperature limits of the heated fluid in the solar thermal system. However, three basic criteria for the thermal load have to be satisfied for the Phibar-f chart method to be applicable: (i) the thermal load must be constant and uniform over each day and for at least a month, (ii) the thermal energy supplied to the load must be above a minimum temperature that completely specifies the temperature level of operation of the load, and (iii) either there is no conversion efficiency in the load (as in the case of hot water usage) or the efficiency of conversion is constant (either because the load temperature level is constant or because the conversion efficiency is independent of the load temperature level). The approach is strictly applicable to solar systems with the auxiliary heater in parallel (Fig. 19.19).

A typical application for the Phibar-f chart method is absorption air-conditioning. The hot water inlet temperature from the collectors to the generator must be above a minimum temperature level (say, 80°C) for the system to use solar heat. If the solar fluid temperature is less (even by a small amount), the entire energy to heat up the water to 80°C is supplied by the auxiliary system.

As a result of continuous interaction between storage and collector in a closed-loop system, the variation of the storage temperature (and hence the fluid inlet temperature to the collectors) over the day and over the month is undetermined. The Phibar-f chart method implicitly takes this into account and reduces these temperature fluctuations down to a monthly mean equivalent storage temperature $\overline{T}_S$. The determination of this temperature in conjunction with the daily utilizability approach is the basis of the design approach.

The basic empirical correlation of the Phibar-f chart method, shown graphically in Figure 19.30, is as follows:

$$f_M = Y\overline{\phi} - a\left[\exp\left(bf_M\right) - 1\right]\left[1 - \exp\left(cX\right)\right] \quad \textbf{(19.64)}$$

with $0 < X < 20$ and $0 < Y < 1.6$, and $\overline{\phi}$ is the Klein daily utilizability fraction described in "Daily Utilizability" and given by Eq. (19.24). Y is given by Eq. (19.57), and X is now slightly different from Eq. (19.56) and is defined as:

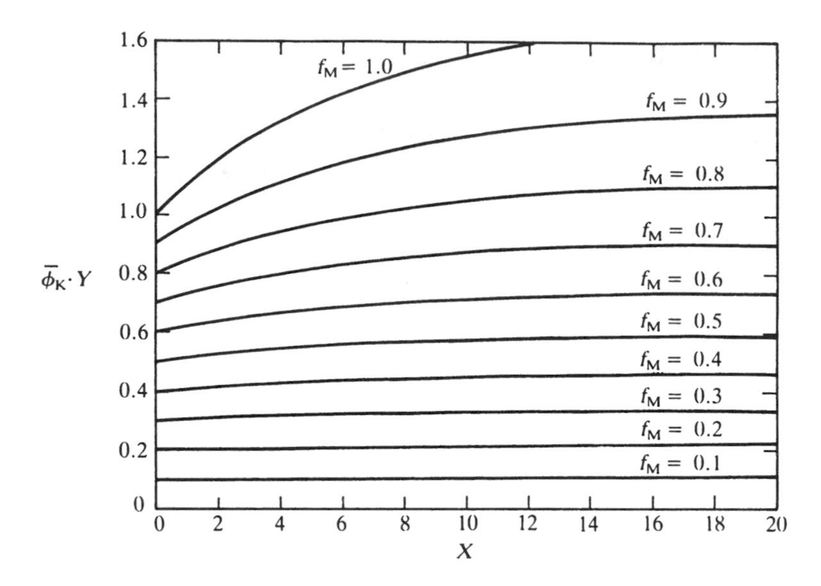

FIGURE 19.30 The Phibar-f chart correlation for a storage capacity of 350 kJ/m² and for a 12 hour per day thermal load (Source: Duffie and Beckman, 1980).

$$X = A_C F_R\, U_L\, \Delta t\, (100°C)/Q_{LM} \tag{19.65}$$

The values of the constants a, b, and c are given by the following:

1. for an end-use load operating between 6 A.M. and 6 P.M. every day of the month,

$$a = 0.015\left[\left(Mc_p\right)_S / 350\,\text{kJ}/\left(\text{m}^2\,°\text{C}\right)\right]^{-0.76} \quad \text{for} \quad 175 \le \left[\left(Mc_p\right)_S / A_C\right] \le 1{,}400\ \text{kJ}/\left(\text{m}^2\,°\text{C}\right) \tag{19.66}$$

and $b = 3.85$ and $c = -0.15$

2. for an end-use load operating 24 hours per day over the entire month,

$$a = 0.043 \text{ only for } \left[\left(Mc_p\right)_S / A_C\right] = 350\ \text{kJ}/\left(\text{m}^2\,°\text{C}\right),\ b = 2.81, \text{ and } c = -0.18 \tag{19.67}$$

It will be noted that $(Y\bar{\phi})$ denotes the maximum solar fraction that would have resulted had T_{Ci}, the inlet temperature to the collector, been equal to T_{Li} throughout the month. The term in Eq. (19.64) that is subtracted from $(Y\bar{\phi})$ represents the decrease in the solar fraction as a result of $T_{Ci} > T_{Xi}$. The solar fraction computed from Eq. (19.64) has to be corrected for the effect of thermal losses from the storage as well as the presence of the load–heat exchanger, both of which will decrease the solar fraction. For complete details, refer to Duffie and Beckman (1980) or Reddy (1987). Note that Eq. (19.64) needs to be solved for f_m in an iterative manner.

Auxiliary Heater in Series. The Phibar-f chart method has also been modified to include solar systems with the auxiliary heater in series as shown in Figure 19.18. This configuration leads to higher solar fractions but retrofit to existing systems may be more costly.

In this case, the empirical correlation given by Eq. (19.64) has been modified by Braun et al. (1983) as follows:

$$f_M = Y\ \bar{\phi} - a\left[\exp\left(bf_M\right) - 1\right]\left[1 - \exp(cX)\right]\ \exp(-1.959Z) \tag{19.68}$$

with $Z = Q_{LM}/(C_L \times 100°C)$ and (i) when there is no load-heat exchanger, C_L is the monthly total load heat capacitance, which is the product of the monthly total mass of water used and the specific heat capacity of water, and (ii) when there is a load-heat exchanger present $C_L = E_L \times C_{min}$, where E_L is the effectiveness of the load-heat exchanger and C_{min} is the monthly total heat capacitance, which is the lesser of the two fluids rates across the load heat exchanger.

The modified Phibar-f chart is similar to the original method in respect to load uniformity on a day-to-day basis over the month and in assuming no conversion efficiency. The interested may refer to Braun et al. (1983) or Reddy (1987) for complete details.

Open-Loop Single-Pass Systems

The advantages offered by open-loop single-pass systems over closed-loop multipass systems for meeting constant loads has been described in "Closed-Loop and Open-Loop Systems." Because industrial loads operate during the entire sun-up hours or even for 24 hours daily, the simplest solar thermal system is one with no heat storage (Fig. 19.20). A sizable portion (between 25 and 70%) of the day-time thermal load can be supplied by such systems and consequently, the sizing of such systems will be described below (Gordon and Rabl, 1982). We shall assume that T_{Li} and T_{Xi} are constant for all hours during system operation. Because no storage is provided, excess solar energy collection (whenever $T_{Ci} > T_{Li}$) will have to be dumped out.

The maximum collector area $\hat{A}_C$ for which energy dumping does not occur at any time of the year can be found from the following instantaneous heat balance equation:

$$P_L = \hat{A}_C \hat{F}_R \left[I_{max} \eta_n - U_L (T_{Ci} - T_a) \right] \quad \textbf{(19.69)}$$

where P_L, the instantaneous thermal heat demand of the load (say, in kW) is given by

$$P_L = m_L c_p (T_{Li} - T_{Xi}) \quad \textbf{(19.70)}$$

and F_R is the heat removal factor of the collector field when its surface area is $\hat{A}_C$. Since $\hat{A}_C$ is as yet unknown, the value of $\hat{F}_R$ is also undetermined. (Note that though the *total* fluid flow rate is known, the flow rate per unit collector area is not known.) Recall that the plate efficiency factor F' for liquid collectors can be assumed constant and independent of fluid flow rate per unit collector area. Equation (19.69) can be expressed in terms of critical radiation level I_C:

$$P_L = \hat{A}_C \hat{F}_R \eta_n (I_{max} - I_C) \quad \textbf{(19.71a)}$$

or

$$\hat{A}_C \hat{F}_R \eta_n = P_L / (I_{max} - I_C) \quad \textbf{(19.71b)}$$

Substituting Eq. (19.3) in lieu of F_R and rearranging yields

$$\hat{A}_C = -(m_L c_p / F' U_L) \ \ln\left[1 - P_L U_L / \left(\eta_n (I_{max} - I_C) \, m_L c_p\right)\right] \quad \textbf{(19.72)}$$

If the actual collector area A_C exceeds this value, dumping will occur as soon as the radiation intensity reaches a value I_D, whose value is determined from the following heat balance:

$$P_L = A_C F_R \eta_n (I_D - I_C) \quad \textbf{(19.73a)}$$

Hence

$$I_D = I_C + P_L/(A_C F_R \eta_n) \quad \textbf{(19.73b)}$$

Note that the value of I_D decreases with increasing collector area A_C, thereby indicating that increasing amounts of solar energy will have to be dumped out.

Since the solar thermal system is operational during the entire sunshine hours of the year, the yearly total energy collected can be directly determined by the Rabl correlation given by Eq. (19.26). Similarly, the yearly total solar energy collected by the solar system which has got to be dumped out is

$$Q_{DY} = A_C F_R \eta_n \left(\tilde{a} + \tilde{b} \cdot I_D + \tilde{c} \cdot I_D^2\right) \quad \textbf{(19.74)}$$

The yearly total solar energy delivered to the load is

$$\begin{aligned} Q_{UY} &= Q_{CY} - Q_{DY} \\ &= A_C F_R \eta_n \left[\tilde{b} \cdot (I_C - I_D) + \tilde{c} \cdot (I_C^2 - I_D^2)\right] \quad \textbf{(19.75)} \\ &= -(\tilde{b} + 2\tilde{c} I_C) P_L - \tilde{c} P_L^2 / (A_C F_R \eta_n) \end{aligned}$$

$$= -(\tilde{b} + 2\tilde{c} I_C) P_L - \tilde{c} P_L^2 / (A_C F_R \eta_n) \quad \textbf{(19.76)}$$

Replacing the value of F_R given by Eq. (19.3), the annual production function in terms of A_C is

$$Q_{UY} = -(\tilde{b} + 2\tilde{c} I_C) P_L - \frac{\tilde{c} P_L^2}{\left(\dfrac{F'\eta_n}{F'U_L}\right) \cdot (m_L c_p) \left[1 - \exp\left(-\dfrac{F'U_L A_C}{m_L c_p}\right)\right]} \quad \textbf{(19.77)}$$

subject to the condition that $A_C > \hat{A}_C$. If the thermal load is not needed during all days of the year due to holidays or maintenance shut-down, the production function can be reduced proportionally. This is illustrated in the following example.

Example 19.8. Obtain the annual production function of an open-loop solar thermal system without storage that is to be set up in Boston, Massachusetts according to the following load specifications: industrial hot water load for 12 hours a day (6 A.M. to 6 P.M.) and during 290 days a year, mass flow rate $m_L = 0.25$ kg/s, required inlet temperature $T_{Li} = 60$°C. Contaminants are picked up in the load, so that all used water is to be rejected and fresh water at ambient temperature is taken in. Flat-plate collectors with tilt equal to latitude with the following parameters are used $F'\eta_n = 0.75$ and $F'U_L = 5.5$ W/(m² °C). The latitude of Boston is 42.36°N = 0.739 radians. The yearly $\bar{K} = 0.45$ and $\bar{T}_a = 10.9$°C. Use the following Gordon and Rabl (1981) correlation:

$$\begin{aligned} Q_{CY}/A_c F_R \eta_n = &\left[(5.215 + 6.973\, I_{bn}) + (-5.412 + 4.293\, I_{bn})L + (1.403 - 0.899\, I_{bn})L^2\right] + \\ &\left[(-18.596 - 5.931\, I_{bn}) + (15.468 + 18.845\, I_{bn})L + (-0.164 - 35.510\, I_{bn})L^2\right] I_C + \\ &\left[(-14.601 - 3.570\, I_{bn}) + (13.675 - 15.549\, I_{bn})L + (-1.620 + 30.564\, I_{bn})L^2\right] I_C^2 \end{aligned}$$

From Eq. (19.27), $\tilde{I}_{bn} = 1.37 \times 0.45 - 0.34 = 0.276$ kW/m². The critical radiation level $I_C = 0$, since $T_{Ci} = T_a$. Consequently, Eq. (19.26), using the above expression reduces to

$$Q_{CY}/(A_C F_R \eta_n) = 5.215 + 6.973 \times 0.276 + (-5.412 + 4.293 \times 0.276) \times 0.739 + (1.403 - 0.899 \times 0.276) \times 0.739^2 = 4.646 \text{ GJ}/(\text{m}^2 \text{ y}).$$

The expression for the dumped out energy is found from Eq. (19.74) and the previous expression by replacing I_C by I_D:

$$\begin{aligned} Q_{CY,\text{dump}}/(A_C F_R \eta_n) = {} & 4.646 + [(-18.596 - 5.931 \times 0.276) + (15.468 + 18.845 \times 0.276) \times 0.739 + \\ & (-0.164 - 35.510 \times 0.276) \times 0.739^2]\ I_D + [(14.601 - 3.57 \times 0.276) + \\ & (-13.675 - 15.549 \times 0.276) \times 0.739 + (1.620 + 30.564 \times 0.276) \times 0.739^2]\ I_D^2 \\ = {} & 4.646 - 10.40\ I_D + 5.83\ I_D^2\ (\text{GJ}/\text{m}^2\ \text{y}) \end{aligned}$$

The thermal energy demand $P_L = 0.25 \times 4.19 \times (60 - 10.9) = 51.43$ kW.

The annual production function is

$$\begin{aligned} Q_{UY} \times (365/290) = {} & -(-10.40 + 2 \cdot \tilde{c} \times 0) \times 51.43 \\ & -(5.83 \times 51.43^2)/\{(0.75/5.5)(0.25 \times 4.19)[1 - \exp[-(5.5 \times A_C)/(0.25 \times 4190)]]\} \end{aligned}$$

or $Q_{UY} = 424.96 - 85.78/[1 - \exp(-A_C/190.45)]\ (\text{GJ/y})$

Complete details as well as how this approach can be extended to solar systems with storage (see Fig. 19.31) can be found in Rabl (1985) or Reddy (1987).

19.6 Design Recommendations and Costs

Design Recommendations

As mentioned earlier, design methods reduce computational effort compared to detailed computer simulations. Even with this decrease, the problem of optimal system design and sizing remains formidable because of

a. The presence of several solar thermal system configuration alternatives.
b. The determination of optimal component sizes for a given system.
c. The presence of certain technical and economic constraints.
d. The choice of proper climatic, technical, and economic input parameters.
e. The need to perform sensitivity analysis of both technical and economic parameters.

For most practical design work, a judicious mix of theoretical expertise and practical acumen is essential. Proper focus right from the start on the important input variables as well as the restriction of the normal range of variation would lead to a great decrease in design time and effort several examples of successful case studies and system design recommendations are described in the published literature (see, for example, Kutcher et al., 1982).

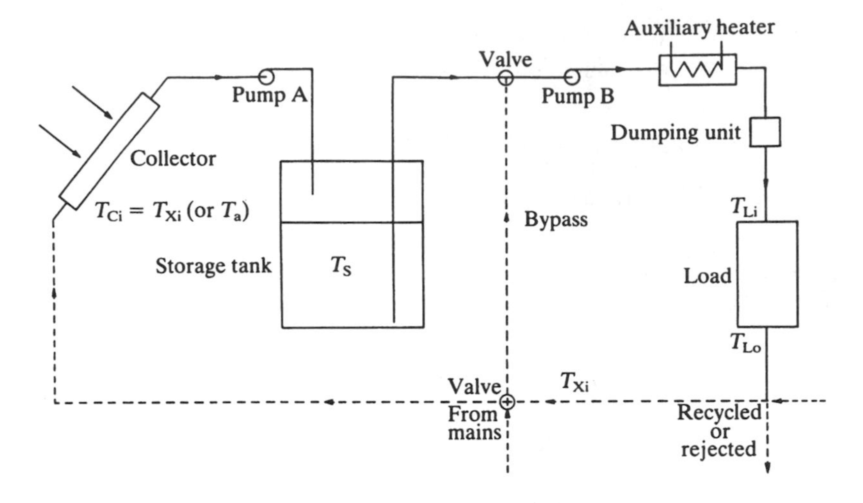

FIGURE 19.31 Open-loop solar industrial hot-water system with storage.

TABLE 19.7 Percentage of Total System Cost by Component

Cost component	Percentage range
Collectors	15–30
Collector installation	5–10
Collector support structure	5–20*
Storage tanks	5–7
Piping and specialties	10–30
Pumps	1–3
Heat exchangers	0–5**
Chiller	5–10
Miscellaneous	2–10
Instrumentation	1–3
Insulation	2–8
Control subsystem	4–9
Electrical	2–6

* For collectors mounted directly on a tilted roof
** For systems without heat exchangers.
From Mueller Associates, 1985.

Solar System Costs

How the individual components of the solar system contribute to the total cost can be gauged from Table 19.7. We note that collectors constitute the major fraction (from 15 to 30%), thus suggesting that collectors should be selected and sized with great care. Piping costs are next with other collector-related costs like installation and support structure being also important.

Figure 19.32 depicts cost ranges for three of the most promising solar technologies, swimming pool heaters, solar water heaters, and industrial heaters for the years 1980, 1990, and 2000. How passive heating compares with these technologies is also to be noted. These costs have been obtained from surveys (at least three experts were consulted for each technology) and from literature searches.

Swimming pool heaters are less expensive than domestic water heaters because they operate at relatively low temperatures, about 27°C, and use inexpensive unglazed flat-plate collectors (polymer collectors are often used). The cost of solar pool heating ranges from \$10 to \$17/GJ. The constant dollar cost has not changed significantly since 1980, although performance and reliability have

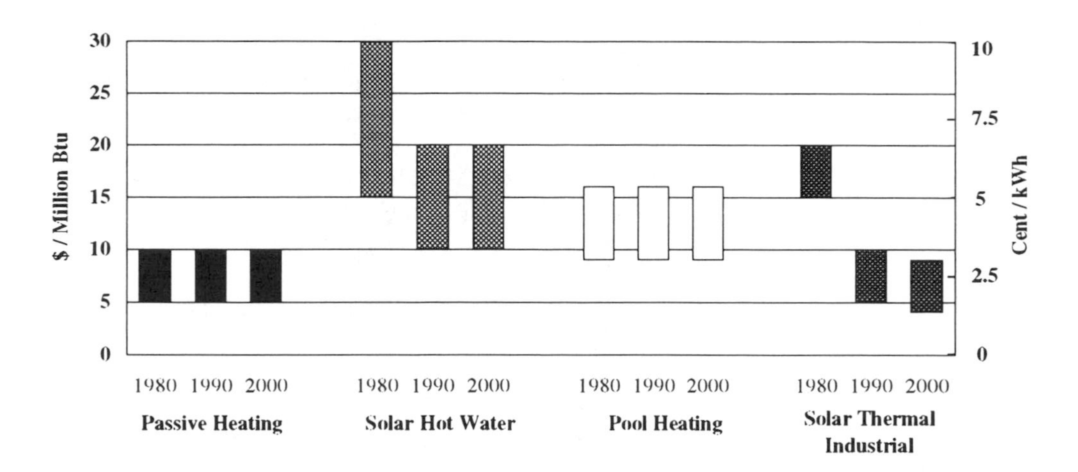

FIGURE 19.32 Cost ranges for solar thermal technologies (Source: Larson et al., 1992).

improved. Costs are not expected to change over the next decade unless strong market growth brings this down. More than 75% of the energy cost is due to initial investment, which pool owners, who are often affluent, can afford to meet. Because swimming pools are located in favorable solar climates, there is often a good match between demand and solar energy.

Solar water heating is now a very reliable and mature technology. Such systems are available in a variety of designs. About 70% of the energy cost is for equipment, 20% for installation, and 10% for operation and maintenance. The costs have dropped from $15 to $30 per GJ in 1980 to $10 to $20 per GJ in 1990. Trends indicate that demand may increase over the next decade. Recent new utility demand side management (DSM) programs are encouraging solar water heating as a means to reduce demand peaks. Consequently, solar water heating may become more economically attractive in the future.

Solar thermal costs for industrial applications are lower than the previous two applications as a result of the dramatic drop from 1980 to 1990 (see Fig. 19.32) from a range of $15 to $20 per GJ to $5 to $10 per GJ as well as due to economy of scale. They are expected to decline still further. About 80% of the energy cost is due to initial investment while 20% is due to operation and maintenance. One-axis parabolic trough is currently the most cost-effective of the several types of concentrating systems currently available.

References

Anderson, B. (1977). *Solar Energy: Fundamentals in Building Design*, McGraw-Hill, New York.

ASHRAE Standard 93-77 (1978). *Methods of Testing to Determine the Thermal Performance of Solar Collectors*, American Society of Heating, Refrigeration and Air Conditioning Engineers, New York.

ASHRAE (1985). *Fundamentals*, American Society of Heating, Refrigeration and Air Conditioning Engineers, New York.

Barley, C.D. and Winn, C.B. (1978). Optimal sizing of solar collectors by the method of relative areas, *Solar Energy*, 21, p. 279.

Beckman, W.A. (1978). Technical note: Duct and pipe losses in solar energy systems, *Solar Energy*, 21, p. 531.

Beckman, W.A., Klein, S.A. and Duffie, J.A. (1977). *Solar Heating Design by the f-Chart Method*, Wiley Interscience, New York.

Braun, J.E., Klein, S.A. and Pearson, K.A. (1983). An improved design method for solar water heating systems, *Solar Energy*, 31, p. 597.

Charters, W.W.S. and Pryor, T.L. (1982). *An Introduction to the Installation of Solar Energy Systems*, Victoria Solar Energy Council, Melbourne, Australia.

Chiam, H.F. (1981). Planar concentrators for flat-plate solar collectors, *Solar Energy*, 26, p. 503.

Clark, D.R., Klein, S.A. and Beckman, W.A. (1983). Algorithm for evaluating the hourly radiation utilizability function, *ASME Journal of Solar Energy Eng.*, 105, p. 281.

Cole, R.L., Nield, K.J., Rohde, R.R. and Wolosewicz, R.M. (eds.) (1979). *Design and Installation Manual for Thermal Energy Storage*, ANL-79-15, Argonne National Laboratory, Argonne, IL.

Collares-Pereira, M. and Rabl, A. (1979). Derivation of method for predicting the long-term average energy delivery of solar collectors, *Solar Energy*, 23, p. 223.

Collares-Pereira, M., Gordon, J.M., Rabl, A. and Zarmi, Y. (1984). Design and optimization of solar industrial hot water systems with storage, *Solar Energy*, 32, p. 121.

CSU (1980). *Solar Heating and Cooling of Residential Buildings—Design of Systems*, manual prepared by the Solar Energy Applications Laboratory, Colorado State University.

de Winter, F. (1975). Heat exchanger penalties in double loop solar water heating systems, *Solar Energy*, 17, p. 335.

Duffie, J.A. and Beckman, W.A. (1980). *Solar Engineering of Thermal Processes*, Wiley Interscience, New York.

Edwards, D.K., Nelson, K.E., Roddick, R.D. and Gier, J.T. (1960). Basic Studies on the Use of Solar Energy, Report no. 60-93, Dept. of Engineering, Univ. of California at Los Angeles, CA.

Gordon, J.M. and Rabl, A. (1982). Design, analysis and optimization of solar industrial process heat plants without storage, *Solar Energy*, 28, p.519.

Gordon, J.M. and Zarmi, Y. (1985). An analytic model for the long-term performance of solar thermal systems with well-mixed storage, *Solar Energy*, 35, p. 55.

Gordon, J.M. and Rabl, A. (1986). Design of solar industrial process heat systems, in *Reviews of Renewable Energy Sources*, M.S. Sodha, S.S. Mathur, M.A.S. Malik and T.C. Kandpal (eds.), Ch. 6, Wiley Eastern, New Delhi.

Gordon, J.M. (1987). Optimal sizing of stand-alone photovoltaic systems, *Solar Cells*, 20, p. 295.

Hollands, K.G.T. (1965). Honeycomb devices in flat-plate solar collectors, *Solar Energy*, 9, p. 159.

Hunn, B.D. (1980). A simplified method for sizing active solar space heating systems, in *Solar Energy Technology Handbook, Part B: Applications, System Design and Economics*, W.C. Dickinson and P.N Cheremisinoff (eds.), Ch. 44, p. 639, Marcel Dekker, New York.

Klein, S.A., Cooper, P.I., Freeman, T.L., Beekman, D.M., Beckman, W.A. and Duffie, J.A. (1975). A method of simulation of solar processes and its applications, *Solar Energy*, 17, p. 29.

Klein, S.A. (1978). Calculation of flat-plate collector utilizability, *Solar Energy*, 21, p. 393.

Klein, S.A. et al. (1979). TRNSYS-A Transient System Simulation User's Manual, University of Wisconsin-Madison Engineering Experiment Station Report 38-10.

Klein, S.A. and Beckman, W.A. (1979). A general design method for closed-loop solar energy systems, *Solar Energy*, 22, p. 269.

Kreider, J.F. (1979). *Medium and High Temperature Solar Processes*, Academic Press, New York.

Kutcher, C.F., Davenport, R.L., Dougherty, D.A., Gee, R.C., Masterson, P.M. and May, E.K. (1982). Design Approaches for Solar Industrial Process Heat Systems, SERI/TR-253-1356, Solar Energy Research Institute, Golden, CO.

Lameiro, G.F. and Bendt, P. (1978). The GFL method for designing solar energy space heating and domestic hot water systems, Proc. American Solar Energy Society Conf., Vol. 2, p. 113, Denver, CO.

Larson, D.C. (1980). Optimization of flat-plate collector flat mirror system, *Solar Energy*, 24, p. 203.

Larson, R.W., Vignola, F. and West, R. (1992). *Economics of Solar Energy Technologies*, American Solar Energy Society Report, Boulder, CO.

Liu, B.Y.H. and Jordan, R.C. (1960). The inter-relationship and characteristic distribution of direct, diffuse and total solar radiation, *Solar Energy*, 4, p. 1.

Liu, B.Y.H. and Jordan, R.C. (1963). A rational procedure for predicting the long-term average performance of flat-plate solar energy collectors, *Solar Energy*, 7, p. 53.

Liu, S.T. and Fanney, A.H. (1980). Comparing experimental and computer-predicted performance for solar hot water systems, *ASHRAE Journal*, 22, No. 5, p. 34.

Löf, G.O.G. (1981). Air based solar systems for space heating, in *Solar Energy Handbook*, J.F. Kreider and F. Kreith (eds.), McGraw-Hill, New York.

Meyer, B.A. (1978). Natural convection heat transfer in small and moderate aspect ratio enclosures—An application to flat-plate collectors, in *Thermal Storage and Heat Transfer in Solar Energy Systems*, F. Kreith, R. Boehm, J. Mitchell and R. Bannerot (eds.), American Society of Mechanical Engineers, New York.

Mitchell, J.C., Theilacker, J.C. and Klein, S.A. (1981). Technical note: Calculation of monthly average collector operating time and parasitic energy requirements, *Solar Energy*, 26, p. 555.

Mueller Associates (1985). *Active Solar Thermal Design Manual*, funded by U.S. DOE (no. EG-77-C-01-4042), SERI(XY-2-02046-1) and ASHRAE (project no. 40), Baltimore, MD.

Newton, A.B. and Gilman, S.H. (1981). *Solar Collector Performance Manual*, funded by U.S. DOE (no. EG-77-C-01-4042), SERI(XH-9-8265-1) and ASHRAE (project no. 32, Task 3).

OTA (1991). Office of Technology Assessment, U.S. Congress, Washington, D.C.

Phillips, W.F. and Dave, R.N. (1982). Effect of stratification on the performance of liquid-based solar heating systems, *Solar Energy*, 29, p. 111.

Rabl, A. (1981). Yearly average performance of the principal solar collector types, *Solar Energy*, 27, p. 215.

Rabl, A. (1985). *Active Solar Collectors and Their Applications*, Oxford University Press, New York.

Reddy, T.A. (1987). *The Design and Sizing of Active Solar Thermal Systems*, Oxford University Press, Oxford, U.K.

SERI (1989). *Engineering Principles and Concepts for Active Solar Systems*, Hemisphere Publishing Company, New York.

Theilacker, J.C. and Klein, S.A. (1980). Improvements in the utilizability relationships, American Section of the International Solar Energy Society Meeting Proceedings, P. 271, Phoenix, AZ.

20

Passive Solar Heating, Cooling and Daylighting

Jeffrey H. Morehouse
University of South Carolina

Introduction

Passive systems are defined, quite generally, as systems in which the thermal energy flow is by natural means, that is, by conduction, radiation, and natural convection. In the context of solar systems for heating buildings, a **passive heating system** is one in which the sun's radiant energy is converted to heat upon absorption by the system, usually after transmission through glazing; the heat is then transferred to thermal storage by natural means or used to directly heat the building; and transfer from thermal storage to the building is also by natural means. **Passive cooling systems** use natural energy flows to transfer heat to the environmental sinks: the ground, air, and sky.

If any one of these major heat transfer paths employs a forced convection path, such as a pump or fan to force flow of a heat transfer fluid, then the system is not truly passive, and it is referred to as having an **active** component or subsystem. **Hybrid systems**—either for heating or cooling—are ones in which there is a parallel path for energy flow that is active, or ones in which the distribution of heat to or from thermal storage is by active means. By convention, the use of movable insulation, even though moved by a motor, does not convert a system from the passive category to active.

The use of the sun's radiant energy for the natural illumination of a building's interior spaces is called **daylighting**. Daylighting design approaches use both solar beam radiation (referred to as **sunlight**) and the diffuse radiation scattered by the atmosphere (referred to as **skylight**) as sources for interior lighting, with historical design emphasis being on utilizing skylight.

0-8493-2514-5/96/$0.00+$.50

Distinction Between a Passive System and Energy Conservation

Energy conservation features are designed to reduce the energy required to thermally condition a building—either heating energy or cooling energy. Thus the use of insulation to reduce either heating or cooling loads is neither a passive solar measure nor a natural cooling measure—it simply reduces the amount of heating and cooling energy required of the passive solar, natural cooling, or auxiliary heat or cooling supply system. Similarly, the use of window shading or window placement to reduce solar gains and thus reduce summer cooling loads is an energy conservation measure. Conversely, the placement of windows to enhance solar gains to offset winter heating loads and/or to provide daylighting is clearly a passive solar feature. Passive features increase the use of solar energy for heating and daylight, plus the use of ambient "coolth" for cooling. The use of building fins to reduce wind scouring of the building surface is energy conservation, but the use of a thermal chimney to draw air through the building to provide cooling is a passive cooling effect, and so forth.

Key Elements of Economic Consideration

The distinction between passive and energy conservation is not a critical one, since the economic calculation is the same in both cases—a trade-off between the life cycle cost of the energy saved (performance) and the life cycle cost of the initial investment, operating, and maintenance costs, (cost).

Performance—Net Energy Savings

The net annual energy saved by the installation of the passive system is the key performance parameter to be determined. The basis for calculating the economics of solar energy is to compare it against a "normal" building; thus, the actual difference in cost of fuel is the difference in auxiliary energy that would be used with and without solar. Thus the energy saved rather than energy delivered, energy collected, useful energy, or some other energy measure, must be determined. The distinction between energy delivered to the building and energy saved may be illustrated as follows: in a building with major solar gains, and with the auxiliary heating thermostat set at 18°C, the solar heat supply may be sufficient to maintain the temperature in the 18 to 24°C range during much of the time when a nonsolar building would remain at 18°C. The extra energy delivered by solar to keep the building above 18°C is a "nonsaved" bonus over and above the energy saved by not using the auxiliary.

Cost—Over and Above "Normal" Construction

The other significant part of the economic trade-off involves the cost of construction of the passive building and also of the "normal" building against which it is to be compared. The convention, adopted from the economics used for active solar systems, is to define a "solar add-on cost." Again, this is a difficult definition in the case of most passive designs because the building is significantly altered compared to typical construction. In the case of a one-to-one replacement of one wall for another, the methodology is relatively straightforward. However, in other cases it becomes more complex (and more arbitrary) and involves assumptions and simulations concerning the typical construction building.

General System Application Status and Costs (ASES, 1992)

Since 1980 over one-quarter million buildings in the U.S. have been constructed or retrofitted with passive features. Passive heating finds application primarily in single-family dwellings and secondarily in small commercial buildings. Since large commercial buildings often require more cooling than heating, daylighting features, which reduce lighting loads and the associated cooling loads, are usually more appropriate.

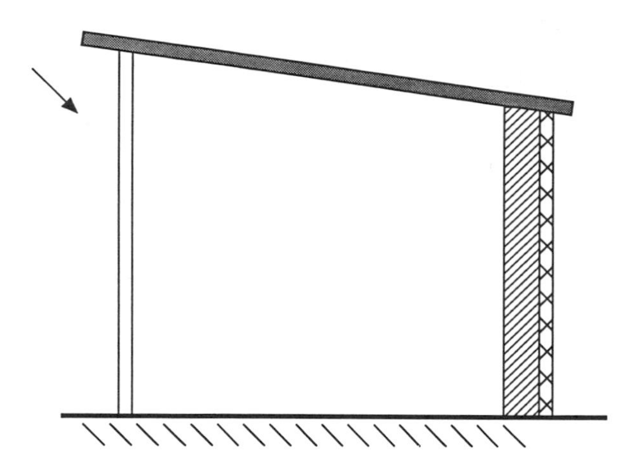

FIGURE 20.1 Direct gain.

For a typical passive design in a favorable climate, passive heating might supply up to one-third of a home's original load at a cost of $5 to 10 per million Btu net energy saved. As for daylighting, an appropriately designed daylighting system can supply lighting at a cost of 2.5 to 5¢ per kWh.

20.1 Passive Solar Heating Design Fundamentals

Passive solar heating concepts have usually been described with reference to their physical configuration. A more general method of cataloging the various approaches is to distinguish three general categories: direct gain, indirect, and isolated. Most of the physical configurations of passive heating systems are seen to fit within one of these three categories.

For direct gain (Fig. 20.1) sunlight enters the heated space, is converted to heat at absorbing surfaces, and is dispersed throughout the space and to the various enclosing surfaces and room contents.

For indirect category systems, sunlight is absorbed and stored by a mass interposed between the glazing and the conditioned space. The conditioned space is partially enclosed and bounded by the thermal storage mass, so a strong natural (and uncontrolled) thermal coupling is achieved. Examples of the indirect approach are the thermal storage wall, the thermal storage roof, and the northerly room of an attached sunspace.

In the thermal storage wall (Figure 20.2), sunlight penetrates glazing and is absorbed and converted to heat at a wall surface interposed between the glazing and the heated space. The wall is usually masonry (Trombe wall) or containers filled with water (water wall), although it might contain phase-change material. The attached sunspace (Figure 20.3) is actually a two-zone combination of direct gain and thermal storage wall. Sunlight enters and heats a direct gain southerly "sunspace" and also heats a mass wall separating the northerly buffered space, which is heated indirectly. The "sunspace" is frequently used as a greenhouse, in which case the system is called an "attached greenhouse." The thermal storage roof (Figure 20.4) is similar to the thermal storage wall except that the interposed thermal storage mass is located on the building roof.

The isolated category concept is an indirect system, except that there is a distinct thermal separation (by means of either insulation or physical separation) between the thermal storage and the heated space. The convective loop, as depicted in Fig. 20.5, is in this category and is often used to heat domestic water. It is most akin to conventional active systems in that there is a separate collector and separate thermal storage. It is a passive approach, however, because the thermal

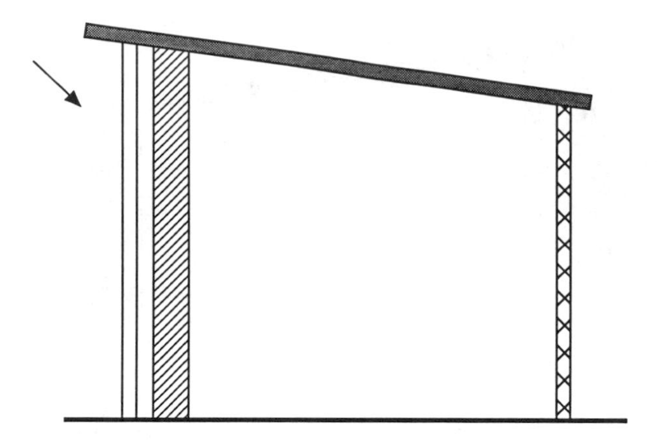

FIGURE 20.2 Thermal storage wall.

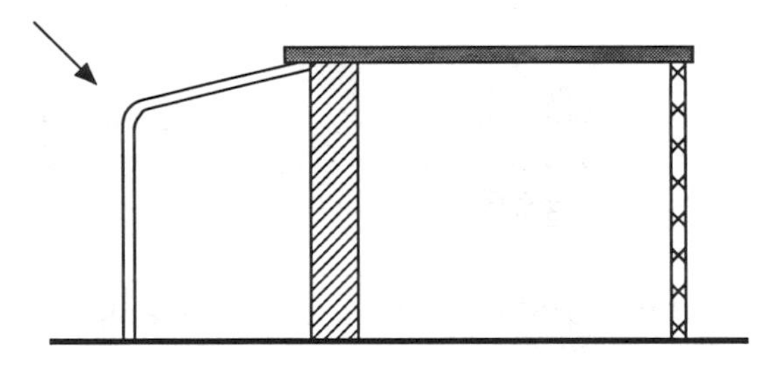

FIGURE 20.3 Attached sunspace.

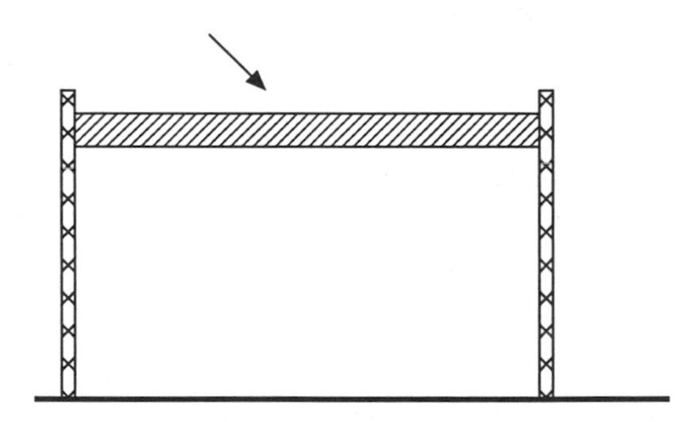

FIGURE 20.4 Thermal storage roof.

energy flow is by natural convection. The thermal storage wall, thermal storage roof, and attached sunspace approaches can also be made into isolated systems by insulating between thermal storage and the heated space.

Passive Design Approaches

Design of a passive heating system involves selection of the passive feature type(s), estimation of thermal performance, and cost estimation; ideally, a cost/performance optimization would be performed. Owner and architect ideas usually establish the passive feature type, with size and cost estimation generally available. However, the thermal performance of a passive heating system has to be calculated.

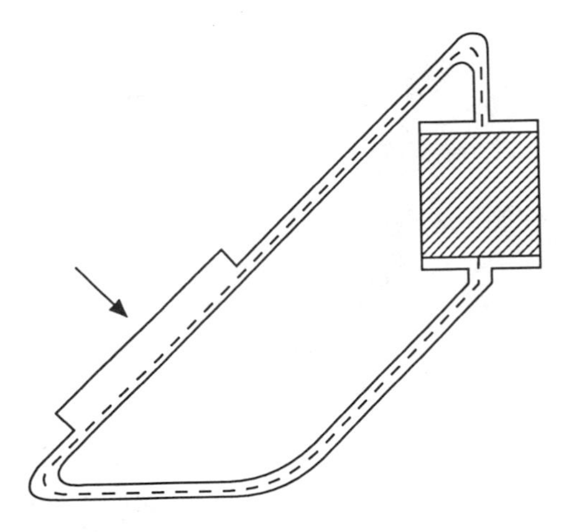

FIGURE 20.5 Convective loop.

There are several "levels" of methods that can be used to estimate the thermal performance of passive designs. The first level involves a rule of thumb and/or generalized calculation to get a starting estimate for size and/or annual performance. A second-level method involves climate, building, and passive system details, which allow annual performance determination, plus some sensitivity to passive system design changes. Third-level methods involve periodic calculations (hourly, monthly) of performance and permit more detailed variations of climatic, building, and passive solar system design parameters.

These levels of design methods have a basis of commonality in that they all are derived from correlations of a multitude of computer simulations of passive systems (PSDH, 1980, 1984). As a result, many passive design approaches reference a similar set of defined terms:

- A_p, solar projected area, m^2 (ft^2): the net south-facing passive solar glazing area projected onto a vertical plane.
- NLC, net building load coefficient, kJ/CDD (Btu/FDD): net load of the nonsolar portion of the building per day per degree of indoor–outdoor temperature difference. The CDD and FDD terms refer to Centigrade and Fahrenheit degree days, respectively.
- Q_{net}, net reference load, Wh (Btu): heat loss from nonsolar portion of building as calculated by

$$Q_{net} = \text{NLC} \times (\text{No. of degree days}) \tag{20.1}$$

- LCR, load collector ratio, kJ/m^2 CDD (Btu/ft^2 FDD): ratio of NLC to A_p,

$$\text{LCR} = \text{NLC}/A_p \tag{20.2}$$

- SSF, solar savings fraction, %: percentage reduction in required auxiliary heating relative to net reference load,

$$\text{SSF} = 1 - \frac{\text{Auxiliary heat required } (Q_{aux})}{\text{Net reference load } (Q_{net})} \tag{20.3}$$

so, using Eq. (20.1),

$$\text{Auxiliary heat required, } Q_{aux} = (1 - \text{SSF}) \times \text{NLC} \times (\text{No. of degree days}). \tag{20.4}$$

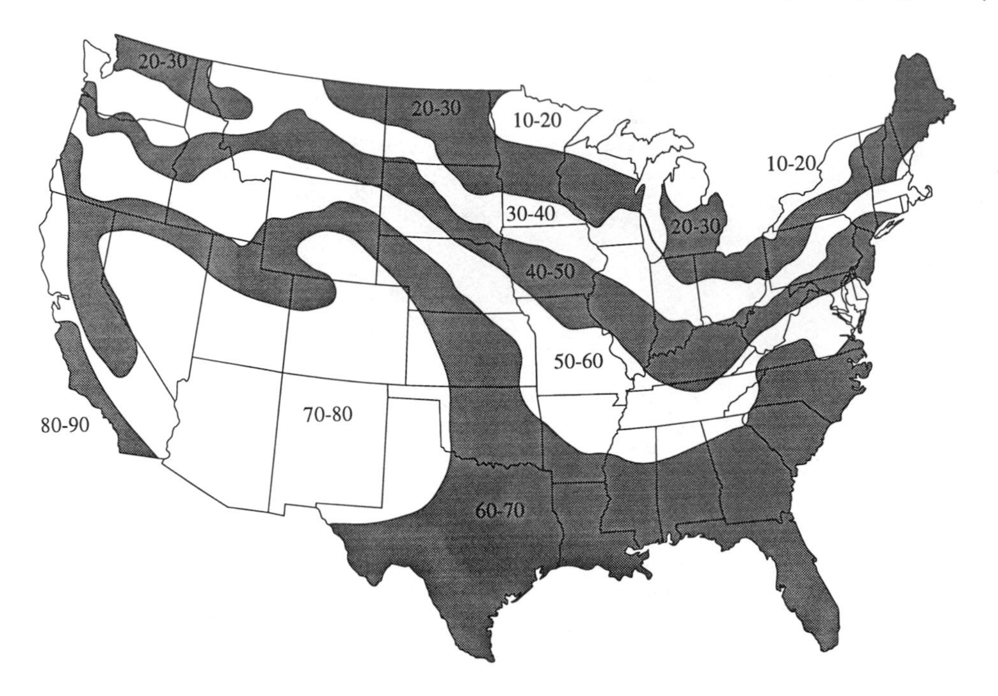

FIGURE 20.6 Starting-point values of solar savings fraction (SSF) in percent. *Source:* PSDH, 1984.

The amount of auxiliary heat required is often a basis of comparison between possible solar designs as well as being the basis for determining building energy operating costs. Thus, many of the passive design methods are based on determining SSF, NLC, and the number of degree days in order to calculate the auxiliary heat required for a particular passive system by using Eq. (20.4).

The First Level: Generalized Methods

A first estimate or starting value is needed to begin the overall passive system design process. Generalized methods and rules of thumb have been developed to generate initial values for solar aperture size, storage size, solar savings fraction, auxiliary heat required, and other size and performance characteristics.

Load

A rule of thumb used in conventional building design is that a design heating load of 120 to 160 kJ/CDD per m^2 of floor area (6 to 8 Btu/FDD ft^2) is considered an energy conservative design. Reducing these nonsolar values by 20% to solarize the proposed south-facing solar wall gives rule-of-thumb NLC values per unit of floor area:

$$\text{NLC/Floor area} = 100 \text{ to } 130 \text{ kJ/CDD m}^2 \left(4.8 \text{ to } 6.4 \text{ Btu/FDD ft}^2\right) \tag{20.5}$$

Solar Savings Fraction

A method of getting starting-point values for the solar savings fraction is presented in Figure 20.6 (PSDH, 1984). The map values represent optimum SSF in % for a particular set of conservation and passive-solar costs for different climates across the United States.

LCR

Another rule of thumb is associated with "good" values of LCR (PSDH, 1984). Different values are used depending on whether the design is for a "cold" or "warm" climate:

$$\text{"Good" LCR} = \begin{cases} \text{For cold climate: } 410\ \text{kJ/m}^2\ \text{CDD}\ \left(20\ \text{Btu/ft}^2\ \text{FDD}\right) \\ \text{For warm climate: } 610\ \text{kJ/m}^2\ \text{CDD}\ \left(30\ \text{Btu/ft}^2\ \text{FDD}\right) \end{cases} \quad \textbf{(20.6)}$$

Storage

Rules of thumb for thermal mass storage relate storage material total heat capacity to the solar projected area (PSDH, 1984). The use of the storage mass is to provide for heating on cloudy days and to regulate sunny day room air temperature swing. When the thermal mass directly absorbs the solar radiation, each square meter of the projected glazing area requires enough mass to store 613 kJ/C. If the storage material is not in direct sunlight, but heated from room air only, then four times as much mass is needed.

In a room with a directly sunlight heated storage mass, the room air temperature swing will be approximately one-half the storage mass temperature swing. For room air heated storage, the air temperature swing is twice that of the storage mass.

Example 20.1. A Denver, Colorado, building is to have a floor area of 195 m^2 (2100 ft^2). Determine rule-of-thumb size and performance characteristics.

Solution: From Equation (20.5) the NLC is estimated as

$$\begin{aligned} \text{NLC} &= \left(115\ \text{kJ/CDD m}^?\right) \times \left(195\ \text{m}^?\right) \\ &= 22{,}400\ \text{kJ/CDD}\ \left(11{,}800\ \text{Btu/FDD}\right) \end{aligned}$$

Using the "cold" LCR value and Equation (20.2), the passive solar projected area is

$$\begin{aligned} A_p &= \text{NLC/LCR} = \left(22{,}400\ \text{kJ/CDD}\right) / \left(410\ \text{kJ/m}^2\ \text{CDD}\right) \\ &= 54.7\ \text{m}^2\ \left(588\ \text{ft}^2\right) \end{aligned}$$

Locating Denver on the map of Figure 20.6 gives an SSF in the 70 to 80% range (use 75%). An annual °C-degree-day value can be found in city climate tables (NCDC, 1992; PSDH, 1984), and is 3,491 CDD (6,283 FDD) for Denver. Thus, the auxiliary heat required, Q_{aux}, is found using Eq. (20.4):

$$\begin{aligned} Q_{aux} &= (1-.75)\left(22{,}400\ \text{kJ/CDD}\right)\left(3{,}491\ \text{CDD}\right) \\ &= 19{,}600\ \text{MJ}\ \left(18.5 \times 10^6\ \text{Btu}\right)\ \text{annually} \end{aligned}$$

The thermal storage can be sized using directly solar heated and/or room air heated mass by using the projected area. Assuming brick with a specific heat capacity of 840 J/kg °C; the storage mass is found by

$$A_p \times (613\ \text{kJ/C}) = m \times (840\ \text{J/kg°C})$$

$$m_d = 40{,}000\ \text{kg}\ (88{,}000\ \text{lbm}) \quad \text{Direct sun}$$

$$\text{or} \qquad m_a = 160{,}000\ \text{kg}\ (351{,}000\ \text{lbm}) \quad \text{Air heated}$$

A more location-dependent set of rules of thumb is presented in (PSDH, 1980). The first rule of thumb relates solar projected area as a percentage of floor area to solar savings fraction, with and without night insulation of the solar glazing:

A solar projected area of (R1)% to (R2)% of the floor area can be expected to produce an SSF in (location) of (S1)% to (S2)%, or, if R9 night insulation is used, of (S3)% to (S4)%.

where the values of R1, R2, S1, S2, S3 and S4 are found using Table 20.1 for the location. The thermal storage mass rule of thumb is again related to the solar projected area:

A thermal storage wall should have 14 kg × SSF(%) of water or 71 kg × SSF (%) of masonry for each square meter of solar projected area. For a direct gain space, the mass above should be used with a surface area of at least three times the solar projected area, and masonry no thicker than 10–15 cm. If the mass is located in back rooms, then four times the above mass is needed.

Example 20.2. Determine size and performance passive solar characteristics with the location-dependent set of rules of thumb for the house of the previous example.

Solution: Using Table 20.1 with the 195 m^2 house in Denver yields

$$\text{Solar projected area} = 12\% \text{ to } 23\% \text{ of floor area}$$

$$= 23.4\ \text{m}^2 \text{ to } 44.9\ \text{m}^2$$

$$\text{SSF (no night insulation)} = 27\% \text{ to } 43\%$$

$$\text{SSF (R9 night insulation)} = 47\% \text{ to } 74\%$$

Using the rule of thumb for the thermal storage mass:

$$m = 14\ \text{kg} \times 73\% \times 44.9\ \text{m}^2$$

$$= 46{,}000\ \text{kg}\ (101{,}000\ \text{lbm}) \quad \text{Thermal wall or direct gain}$$

Comparing the results of this example to those of the previous example, the various rules of thumb can be seen to produce "roughly" similar answers. General system cost and performance information can be generated with results from rule-of-thumb calculations, but a more detailed level of information is needed to determine passive system type (direct gain, thermal wall, sunspace), size, performance, and costs.

The Second Level: LCR Method

The LCR method was developed by calculating the annual SSF for 94 reference passive solar systems for 219 U.S. and Canadian locations over a range of LCR values. (*Note:* The solar load ratio (SLR) method was used to make these calculations, and this SLR method is described in the next section as the third-level method.) Tables were constructed for each city with LCR versus SSF listed for each of the 94 reference passive systems. The LCR method is useful for making estimates of the annual performance of specific types of passive system(s) combinations. The method consists of the following steps (PSDH, 1984):

1. Determine building parameters:
    a. Building load coefficient, NLC
    b. Solar projected area, A_p
    c. Load collector ratio, LCR = NLC/A_p

TABLE 20.1 Values to Be Used in the Glazing Area and SSF Relations Rules of Thumb

City	B1	B2	S1	S2	S3	S4
Birmingham, Alabama	.09	.18	22	37	34	58
Mobile, Alabama	.06	.12	26	44	34	60
Montgomery, Alabama	.07	.15	24	41	34	59
Phoenix, Arizona	.06	.12	37	60	48	75
Prescott, Arizona	.10	.20	29	48	44	72
Tucson, Arizona	.06	.12	35	57	45	73
Winslow, Arizona	.12	.24	30	47	48	74
Yuma, Arizona	.04	.09	43	66	51	78
Fort Smith, Arkansas	.10	.20	24	39	38	64
Little Rock, Arkansas	.10	.19	23	38	37	62
Bakersfield, California	.08	.15	31	50	42	67
Baggett, California	.07	.15	35	56	46	73
Fresno, California	.09	.17	29	46	41	65
Long Beach, California	.05	.10	35	58	44	72
Los Angeles, California	.05	.09	36	58	44	72
Mount Shasta, California	.11	.21	24	38	42	67
Needles, California	.06	.12	39	61	49	76
Oakland, California	.07	.15	35	55	46	72
Red Bluff, California	.09	.18	29	46	41	65
Sacramento, California	.09	.18	29	47	41	66
San Diego, California	.04	.09	37	61	46	74
San Francisco, California	.06	.13	34	54	45	71
Santa Maria, California	.05	.11	31	53	42	69
Colorado Springs, Colorado	.12	.24	27	42	47	74
Denver, Colorado	.12	.23	27	43	47	74
Eagle, Colorado	.14	.29	25	35	53	77
Grand Junction, Colorado	.13	.27	29	43	50	76
Pueblo, Colorado	.11	.23	29	45	48	75
Hartford, Connecticut	.17	.35	14	19	40	64
Wilmington, Delaware	.15	.29	19	30	39	63
Washington, DC	.12	.23	18	28	37	61
Apalachicola, Florida	.05	.10	28	47	36	61
Daytona Beach, Florida	.04	.08	30	51	36	63
Jacksonville, Florida	.05	.09	27	47	35	62
Miami, Florida	.01	.02	27	48	31	54
Orlando, Florida	.03	.06	30	52	37	63
Tallahassee, Florida	.05	.11	26	45	35	60
Tampa, Florida	.03	.06	30	52	36	63
West Palm Beach, Florida	.01	.03	30	51	34	59
Atlanta, Georgia	.06	.17	22	36	34	58
Augusta, Georgia	.06	.16	24	40	35	60
Macon, Georgia	.07	.15	25	41	35	59
Savannah, Georgia	.06	.13	25	43	35	60
Boise, Idaho	.14	.28	27	38	48	71
Lewiston, Idaho	.15	.29	22	29	44	65
Pocatello, Idaho	.13	.26	25	35	51	74
Chicago, Illinois	.17	.35	17	23	43	67
Moline, Illinois	.20	.39	17	22	46	70
Springfield, Illinois	.15	.30	19	26	42	67
Evansville, Indiana	.14	.27	19	29	37	61
Fort Wayne, Indiana	.16	.33	13	17	37	60
Indianapolis, Indiana	.14	.28	15	21	37	60
South Bend, Indiana	.18	.35	12	15	39	61
Burlington, Iowa	.18	.36	20	27	47	71
Des Moines, Iowa	.21	.43	19	25	58	75
Mason City, Iowa	.22	.44	18	19	56	79
Sioux City, Iowa	.23	.46	20	24	53	76

TABLE 20.1 (continued) Values to Be Used in the Glazing Area and SSF Relations Rules of Thumb

City	B1	B2	S1	S2	S3	S4
Dodge City, Kansas	.12	.23	27	42	46	73
Goodland, Kansas	.13	.27	26	39	47	74
Topeka, Kansas	.14	.26	24	35	45	71
Wichita, Kansas	.14	.26	26	41	45	72
Lexington, Kentucky	.13	.27	17	26	35	58
Louisville, Kentucky	.13	.27	18	27	35	59
Baton Rouge, Louisiana	.06	.12	26	43	34	59
Lake Charles, Louisiana	.06	.11	24	41	32	57
New Orleans, Louisiana	.05	.11	27	46	35	61
Shreveport, Louisiana	.08	.15	26	43	36	61
Caribou, Maine	.25	.30	- N	R -	53	74
Portland, Maine	.17	.34	14	17	45	69
Baltimore, Maryland	.14	.27	19	30	38	62
Boston, Massachusetts	.15	.29	17	25	40	64
Alpena, Michigan	.21	.42	- N	R -	47	69
Detroit, Michigan	.17	.34	13	17	39	61
Flint, Michigan	.15	.31	11	12	40	62
Grand Rapids, Michigan	.19	.38	12	13	39	61
Sault Ste. Marie, Michigan	.25	.50	- N	R -	50	70
Traverse City, Michigan	.18	.36	- N	R -	42	62
Duluth, Minnesota	.25	.50	- N	R -	50	70
International Falls, Minnesota	.25	.50	- N	R -	47	66
Minneapolis-St. Paul, Minnesota	.25	.50	- N	R -	55	76
Rochester, Minnesota	.24	.49	- N	R -	54	76
Jackson, Mississippi	.06	.15	24	48	34	59
Meridian, Mississippi	.08	.15	23	39	34	58
Columbia, Missouri	.13	.26	20	30	41	66
Kansas City, Missouri	.14	.29	22	32	44	70
Saint Louis, Missouri	.15	.29	21	33	41	65
Springfield, Missouri	.13	.26	22	34	40	65
Billings, Montana	.16	.32	24	31	53	76
Cut Bank, Montana	.24	.49	22	23	62	81
Dillon, Montana	.16	.32	24	32	54	77
Glasgow, Montana	.25	.50	- N	R -	55	75
Great Falls, Montana	.18	.37	23	26	56	77
Helena, Montana	.20	.39	21	25	55	77
Lewistown, Montana	.19	.38	21	25	54	76
Miles City, Montana	.23	.47	21	23	60	80
Missoula, Montana	.18	.36	15	16	47	68
Grand Island, Nebraska	.18	.36	24	33	51	76
North Omaha, Nebraska	.20	.48	21	29	51	76
North Platte, Nebraska	.17	.34	25	36	50	76
Scottsbluff, Nebraska	.16	.31	24	36	49	74
Elko, Nevada	.12	.25	27	39	52	76
Ely, Nevada	.12	.23	27	41	50	77
Las Vegas, Nevada	.09	.18	35	56	48	75
Lovelock, Nevada	.13	.25	32	48	53	78
Reno, Nevada	.11	.22	31	48	49	76
Tonopah, Nevada	.11	.23	31	48	51	77
Winnemucca, Nevada	.13	.26	28	42	49	75
Concord, New Hampshire	.17	.34	13	15	45	68
Newark, New Jersey	.13	.25	19	29	39	64
Albuquerque, New Mexico	.11	.22	29	47	46	73
Clayton, New Mexico	.10	.20	28	45	45	73
Farmington, New Mexico	.12	.24	29	45	49	76
Los Alamos, New Mexico	.11	.22	25	40	44	72

TABLE 20.1 (continued) Values to Be Used in the Glazing Area and SSF Relations Rules of Thumb

City	B1	B2	S1	S2	S3	S4
Roswell, New Mexico	.10	.19	30	49	45	73
Truth or Consequences, New Mexico	.09	.17	32	51	46	73
Tucumcari, New Mexico	.10	.20	30	48	45	73
Zuni, New Mexico	.11	.21	27	43	45	73
Albany, New York	.21	.41	13	15	43	66
Binghamton, New York	.15	.30	- N	R -	35	56
Buffalo, New York	.19	.37	- N	R -	36	57
Massena, New York	.25	.50	- N	R -	50	71
New York (Central Park), NY	.15	.30	16	25	36	59
Rochester, New York	.18	.37	- N	R -	37	58
Syracuse, New York	.19	.38	- N	R -	37	59
Asheville, North Carolina	.10	.20	21	35	36	61
Cape Hatteras, North Carolina	.09	.17	24	40	36	60
Charlotte, North Carolina	.08	.17	23	38	36	60
Greensboro, North Carolina	.10	.20	23	37	37	63
Raleigh-Durham, North Carolina	.09	.19	22	37	36	61
Bismarck, North Dakota	.25	.50	- N	R -	56	77
Fargo, North Dakota	.25	.50	- N	R -	51	72
Minot, North Dakota	.25	.50	- N	R -	52	72
Akron-Canton, Ohio	.15	.31	12	16	35	57
Cincinnati, Ohio	.12	.24	15	23	35	57
Cleveland, Ohio	.15	.31	11	14	34	55
Columbus, Ohio	.14	.28	13	18	35	57
Dayton, Ohio	.14	.28	14	20	36	59
Toledo, Ohio	.17	.34	13	17	38	61
Youngstown, Ohio	.16	.32	- N	R -	34	54
Oklahoma City, Oklahoma	.11	.22	25	41	41	67
Tulsa, Oklahoma	.11	.22	24	38	40	65
Astoria, Oregon	.09	.19	21	34	37	60
Burns, Oregon	.13	.25	23	32	47	71
Medford, Oregon	.12	.24	21	32	38	60
North Bend, Oregon	.09	.17	25	42	38	64
Pendleton, Oregon	.14	.27	22	30	43	64
Portland, Oregon	.13	.26	21	31	38	60
Redmond, Oregon	.13	.27	26	38	47	71
Salem, Oregon	.12	.24	21	32	37	59
Allentown, Pennsylvania	.15	.29	16	24	39	63
Erie, Pennsylvania	.17	.34	- N	R -	35	55
Harrisburg, Pennsylvania	.13	.26	17	26	38	62
Philadelphia, Pennsylvania	.15	.29	19	29	38	62
Pittsburgh, Pennsylvania	.14	.28	12	16	33	55
Wilkes-Barre-Scranton, PA	.16	.32	13	18	37	60
Providence, Rhode Island	.15	.30	17	24	40	64
Charleston, South Carolina	.07	.14	25	41	34	59
Columbia, South Carolina	.08	.17	25	41	36	61
Greenville-Spartanburg, SC	.08	.17	23	38	36	60
Huron, South Dakota	.25	.50	- N	R -	58	79
Pierre, South Dakota	.22	.43	21	23	58	80
Rapid City, South Dakota	.15	.30	23	32	51	76
Sioux Falls, South Dakota	.22	.45	18	19	57	79
Chattanooga, Tennessee	.09	.19	19	32	33	56
Knoxville, Tennessee	.09	.18	20	33	33	56
Memphis, Tennessee	.09	.19	22	36	36	60
Nashville, Tennessee	.10	.21	19	30	33	55
Abilene, Texas	.09	.18	29	47	41	68
Amarillo, Texas	.11	.22	29	46	45	72

TABLE 20.1 (continued) Values to Be Used in the Glazing Area and SSF Relations Rules of Thumb

City	B1	B2	S1	S2	S3	S4
Austin, Texas	.06	.13	27	46	37	63
Brownsville, Texas	.03	.06	27	46	32	57
Corpus Christi, Texas	.05	.09	29	49	36	63
Dallas, Texas	.08	.17	27	44	38	64
Del Rio, Texas	.06	.12	30	50	39	66
El Paso, Texas	.09	.17	32	53	45	72
Forth Worth, Texas	.09	.17	26	44	38	64
Houston, Texas	.06	.11	25	43	34	59
Laredo, Texas	.05	.09	31	52	39	64
Lubbock, Texas	.09	.19	30	49	44	72
Lufkin, Texas	.07	.14	26	43	35	61
Midland-Odessa, Texas	.09	.18	32	52	44	72
Port Arthur, Texas	.06	.11	26	44	34	60
San Angelo, Texas	.08	.15	29	48	40	67
San Antonio, Texas	.06	.12	28	48	38	64
Sherman, Texas	.10	.20	25	41	38	64
Waco, Texas	.06	.15	27	45	38	64
Wichita Falls, Texas	.10	.20	27	45	41	67
Bryce Canyon, Utah	.13	.25	26	39	52	78
Cedar City, Utah	.12	.24	28	43	48	75
Salt Lake City, Utah	.13	.26	27	39	48	72
Burlington, Vermont	.22	.43	- N	R -	46	68
Norfolk, Virginia	.09	.19	23	38	37	62
Richmond, Virginia	.11	.22	21	34	37	61
Roanoke, Virginia	.11	.23	21	34	37	61
Olympia, Washington	.12	.23	20	29	38	59
Seattle-Tacoma, Washington	.11	.22	21	30	39	59
Spokane, Washington	.20	.39	20	24	48	68
Yakima, Washington	.18	.36	24	31	49	70
Charleston, West Virginia	.13	.25	16	24	32	54
Huntington, West Virginia	.13	.25	17	27	34	57
Eau Claire, Wisconsin	.25	.50	- N	R -	53	75
Green Bay, Wisconsin	.23	.46	- N	R -	53	75
La Crosse, Wisconsin	.21	.43	- N	R -	52	75
Madison, Wisconsin	.20	.40	15	17	51	74
Milwaukee, Wisconsin	.18	.35	15	18	48	71
Casper, Wyoming	.13	.26	27	39	53	78
Cheyenne, Wyoming	.11	.21	25	39	47	74
Rock Springs, Wyoming	.14	.28	26	38	54	79
Sheridan, Wyoming	.16	.31	22	30	52	75
	Canada					
Edmonton, Alberta	.25	.50	- N	R -	54	72
Suffield, Alberta	.25	.50	28	30	67	85
Nanaimo, British Columbia	.13	.26	26	35	45	66
Vancouver, British Columbia	.13	.26	20	28	48	60
Winnipeg, Manitoba	.25	.50	- N	R -	54	74
Dartmouth, Nova Scotia	.14	.28	17	24	45	70
Moosonee, Ontario	.25	.50	- N	R -	48	67
Ottawa, Ontario	.25	.50	- N	R -	59	80
Toronto, Ontario	.18	.36	17	23	44	68
Normandie, Quebec	.25	.50	- N	R -	54	74

Note: -NR- means Not Recommended
Source: PSDH, 1980.

2. Find the short designation of the reference system closest to the passive system design (Table 20.2).
3. Enter the LCR Tables (Table 20.3):
    a. Find the city
    b. Find the reference system listing
    c. Determine annual SSF by interpolation using the LCR value from above
    d. Note the annual heating degree days (No. of degree of days)
4. Calculate the annual auxiliary heat required:

$$\text{Auxiliary heat required} = (1 - \text{SSF}) \times \text{NLC} \times (\text{No. of degree days})$$

If more than one reference solar system is being used, then find the "aperture area weighted" SSF for the combination. Determine each individual reference system SSF using the total aperture area LCR, then take the "area weighted" average of the individual SSFs.

The LCR method allows no variation from the 94 reference passive designs. To treat off-reference designs, sensitivity curves have been produced that illustrate the effect on SSF of varying one or two design variables. These curves were produced for six cities, chosen to represent wide geographical and climatological ranges. Several of these sensitivity curves are presented in Figure 20.7.

While the complete LCR Table (PSDH, 1984) includes 219 locations, Table 20.3 only includes the six "representative" cities, purely due to space restrictions. However, Table 20.2 includes the description of all 94 reference systems for use both with the LCR method and with the SLR method described later.

Example 20.3. The previously used 2,100 ft^2 building with NLC = 11,800 Btu/FDD is preliminarily designed to be located in Medford, Oregon, with 180 ft^2 of 12" thick vented Trombe wall and 130 ft^2 of direct gain, both systems having double glazing, nighttime insulation, and 30 Btu/ft^2 thermal storage capacity. Determine the annual auxiliary energy needed by this design.

Solution: Step 1 yields:

$$\text{NLC} = 11{,}800 \text{ Btu/FDD}$$

$$A_p = 180 + 130 = 320 \text{ ft}^2$$

$$\text{LCR} = 11{,}800/320 = 36.8 \text{ Btu/FDD ft}^2$$

Step 2 yields: From Table 20.2 the short designations for the appropriate systems are

TWD4 (Trombe wall)
DGA3 (Direct gain)

Step 3 yields: From Table 20.3 for Medford, Oregon, with LCR = 36.8

TWD4: SSF (TW) = 0.42
DGA3: SSF (DG) = 0.37

Determine the "weighted area" average SSF:

$$SSF = \frac{180\,(0.42) + 130\,(0.37)}{320} = 0.39$$

TABLE 20.2 Designations and Characteristics for 94 Reference Systems

(a) Overall System Characteristics

Masonry properties	
thermal conductivity (k)	
sunspace floor	0.5 Btu/hr/ft/°F
all other masonry	1.0 Btu/hr/ft/°F
density (Q)	150 lb/ft³
specific heat (c)	0.2 Btu/lb/°F
infrared emittance of normal surface	0.9
infrared emittance of selective surface	0.1
Solar absorptances	
waterwall	1.0
masonry, Trombe wall	1.0
direct gain and sunspace	0.8
sunspace: water containers	0.9
lightweight common wall	0.7
other lightweight surfaces	0.3
Glazing properties	
transmission characteristics	diffuse
orientation	due south
index of refraction	1.526
extinction coefficient	0.5 $inch^{-1}$
thickness of each pane	one-eighth inch
gap between panes	one-half inch
ared emittance	0.9
Control range	
room temperature	65 to 75°F
sunspace temperature	45 to 95°F
internal heat generation	0
Thermocirculation vents (when used)	
vent area/projected area (sum of both upper and lower vents)	0.06
height between vents	8 ft
reverse flow	none
Nighttime insulation (when used)	
thermal resistance	R9
in place, solar time	5:30 P.M. to 7:30 A.M.
Solar radiation assumptions	
shading	none
ground diffuse reflectance	0.3

(b) Direct-Gain (DG) System Types

Designation	Thermal Storage Capacity* (in Btu/ft²/°F)	Mass Thickness* (inches)	Mass-Area-to-Glazing-Area Ratio	No. of Glazings	Nighttime Insulation
A1	30	2	6	2	no
A2	30	2	6	3	no
A3	30	2	6	2	yes
B1	45	6	3	2	no
B2	45	6	3	3	no
B3	45	6	3	2	yes
C1	60	4	6	2	no

TABLE 20.2 (continued) Designations and Characteristics for 94 Reference Systems

Designation	Thermal Storage Capacity* (in Btu/ft²/°F)	Mass Thickness* (inches)	Mass-Area-to-Glazing-Area Ratio	No. of Glazings	Nighttime Insulation
C2	60	4	6	3	no
C3	60	4	6	2	yes

(c) Vented Trombe-Wall (TW) System Types

Designation	Thermal Storage Capacity* (Btu/ft²/°F)	Wall Thickness* (inches)	ρck (Btu²/hr/ft⁴/°F²)	No. of Glazings	Wall Surface	Nighttime Insulation
A1	15	6	30	2	normal	no
A2	22.5	9	30	2	normal	no
A3	30	12	30	2	normal	no
A4	45	18	30	2	normal	no
B1	15	6	15	2	normal	no
B2	22.5	9	15	2	normal	no
B3	30	12	15	2	normal	no
B4	45	18	15	2	normal	no
C1	15	6	7.5	2	normal	no
C2	22.5	9	7.5	2	normal	no
C3	30	12	7.5	2	normal	no
C4	45	18	7.5	2	normal	no
D1	30	12	30	1	normal	no
D2	30	12	30	3	normal	no
D3	30	12	30	1	normal	yes
D4	30	12	30	2	normal	yes
D5	30	12	30	3	normal	yes
E1	30	12	30	1	selective	no
E2	30	12	30	2	selective	no
E3	30	12	30	1	selective	yes
E4	30	12	30	2	selective	yes

(d) Unvented Trombe-Wall (TW) System Types

Designation	Thermal Storage Capacity* (Btu/ft²/°F)	Wall Thickness* (inches)	ρck (Btu²/hr/ft⁴/°F²)	No. of Glazings	Wall Surface	Nighttime Insulation
F1	15	6	30	2	normal	no
F2	22.5	9	30	2	normal	no
F3	30	12	30	2	normal	no
F4	45	18	30	2	normal	no
G1	15	6	15	2	normal	no
G2	22.5	9	15	2	normal	no
G3	30	12	15	2	normal	no
G4	45	18	15	2	normal	no
H1	15	6	7.5	2	normal	no
H2	22.5	9	7.5	2	normal	no
H3	30	12	7.5	2	normal	no
H4	45	18	7.5	2	normal	no
I1	30	12	30	1	normal	no
I2	30	12	30	3	normal	no
I3	30	12	30	1	normal	yes
I4	30	12	30	2	normal	yes
I5	30	12	30	3	normal	yes
J1	30	12	30	1	selective	no
J2	30	12	30	2	selective	no
J3	30	12	30	1	selective	yes
J4	30	12	30	2	selective	yes

TABLE 20.2 (continued) Designations and Characteristics for 94 Reference Systems

(e) Waterwall (WW) System Types

Designation	Thermal Storage Capacity* (in Btu/ft²/°F)	Wall Thickness (inches)	No. of Glazings	Wall Surface	Nighttime Insulation
A1	15.6	3	2	normal	no
A2	31.2	6	2	normal	no
A3	46.8	9	2	normal	no
A4	62.4	12	2	normal	no
A5	93.6	18	2	normal	no
A6	124.8	24	2	normal	no
B1	46.8	9	1	normal	no
B2	46.8	9	3	normal	no
B3	46.8	9	1	normal	yes
B4	46.8	9	2	normal	yes
B5	46.8	9	3	normal	yes
C1	46.8	9	1	selective	no
C2	46.8	9	2	selective	no
C3	46.8	9	1	selective	yes
C4	46.8	9	2	selective	yes

(f) Sunspace (SS) System Types

Designation	Type	Tilt (degrees)	Common Wall	End Walls	Nighttime Insulation
A1	attached	50	masonry	opaque	no
A2	attached	50	masonry	opaque	yes
A3	attached	50	masonry	glazed	no
A4	attached	50	masonry	glazed	yes
A5	attached	50	insulated	opaque	no
A6	attached	50	insulated	opaque	yes
A7	attached	50	insulated	glazed	no
A8	attached	50	insulated	glazed	yes
B1	attached	90/30	masonry	opaque	no
B2	attached	90/30	masonry	opaque	yes
B3	attached	90/30	masonry	glazed	no
B4	attached	90/30	masonry	glazed	yes
B5	attached	90/30	insulated	opaque	no
B6	attached	90/30	insulated	opaque	yes
B7	attached	90/30	insulated	glazed	no
B8	attached	90/30	insulated	glazed	yes
C1	semienclosed	90	masonry	common	no
C2	semienclosed	90	masonry	common	yes
C3	semienclosed	90	insulated	common	no
C4	semienclosed	90	insulated	common	yes
D1	semienclosed	50	masonry	common	no
D2	semienclosed	50	masonry	common	yes
D3	semienclosed	50	insulated	common	no
D4	semienclosed	50	insulated	common	yes
E1	semienclosed	90/30	masonry	common	no
E2	semienclosed	90/30	masonry	common	yes
E3	semienclosed	90/30	insulated	common	no
E4	semienclosed	90/30	insulated	common	yes

* The thermal storage capacity is per unit of projected area, or, equivalently, the quantity ρck. The wall thickness is listed only as an appropriate guide by assuming $\rho c = 30$ Btu/ft³/°F.

Source: PSDH, 1984.

TABLE 20.3 LCR Tables for Six Representative Cities (Albuquerque, Boston, Madison, Medford, Nashville, Santa Maria)

SSF	.10	.20	.30	.40	.50	.60	.70	.80	.90
Santa Maria, California								3053 DD	
WW A1	1776	240	119	73	50	35	25	18	12
WW A2	617	259	154	103	74	54	39	28	19
WW A3	523	261	164	114	82	61	45	33	22
WW A4	482	260	169	119	87	65	48	35	24
WW A5	461	263	175	125	92	69	52	38	26
WW A6	447	263	177	128	95	72	54	40	27
WW B1	556	220	128	85	60	43	32	23	15
WW B2	462	256	168	119	88	66	49	36	25
WW B3	542	315	211	151	112	85	64	47	32
WW B4	455	283	197	144	109	83	63	47	32
WW B5	414	263	184	136	103	79	60	45	31
WW C1	569	330	221	159	118	89	67	49	33
WW C2	478	288	197	143	107	81	61	45	31
WW C3	483	318	228	170	130	100	77	57	40
WW C4	426	280	200	149	114	88	68	51	35
TW A1	1515	227	113	70	48	34	24	17	11
TW A2	625	234	134	89	63	46	33	24	16
TW A3	508	231	140	95	68	50	37	27	18
TW A4	431	217	137	95	69	51	38	28	19
TW B1	859	212	112	71	49	35	25	18	12
TW B2	502	209	124	83	59	43	32	23	15
TW B3	438	201	123	84	60	44	33	24	16
TW B4	400	104	112	76	55	40	30	22	14
TW C1	568	188	105	69	48	35	25	18	12
TW C2	435	178	105	70	50	36	27	19	13
TW C3	413	165	97	64	46	33	25	18	12
TW C4	426	146	82	54	38	27	20	14	10
TW D1	403	170	101	67	48	35	25	18	12
TW D2	488	242	152	105	76	57	42	31	21
TW D3	509	271	175	123	90	67	50	36	25
TW D4	464	266	177	127	94	71	53	39	27
TW D5	425	250	169	122	91	69	52	38	26
TW E1	581	309	199	140	102	76	57	42	28
TW E2	512	283	186	132	97	73	55	40	27
TW E3	537	328	225	164	123	94	71	53	36
TW E4	466	287	199	145	109	83	63	47	32
TW F1	713	198	107	68	47	34	25	18	12
TW F2	455	199	120	81	58	42	31	22	15
TW F3	378	190	120	83	60	45	33	24	16
TW F4	311	169	110	77	57	42	32	23	16
TW G1	450	170	98	65	46	33	24	17	12
TW G2	331	163	102	70	51	38	28	20	14
TW G3	278	147	94	66	48	36	27	20	13
TW G4	222	120	78	55	40	30	22	16	11
TW H1	295	137	84	57	41	30	22	16	11
TW H2	226	118	75	52	38	28	21	15	10
TW H3	187	99	64	44	33	24	18	13	9
TW H4	143	75	48	33	24	18	14	10	7
TW I1	318	144	88	59	42	31	23	16	11
TW I2	377	203	132	93	68	51	38	28	19
TW I3	404	226	149	106	78	58	44	32	22
TW I4	387	230	156	113	84	64	48	36	24
TW I5	370	226	155	113	85	65	49	36	25

TABLE 20.3 (continued) LCR Tables for Six Representative Cities (Albuquerque, Boston, Madison, Medford, Nashville, Santa Maria)

SSF	.10	.20	.30	.40	.50	.60	.70	.80	.90
TW J1	483	271	179	127	94	71	53	39	26
TW J2	422	246	165	119	88	67	50	37	25
TW J3	446	283	199	146	111	85	65	48	33
TW J4	400	254	178	132	100	77	58	43	30
DG A1	392	188	117	79	55	38	26	16	7
DG A2	389	190	121	85	61	45	32	22	14
DG A3	443	220	142	102	77	58	44	31	19
DG B1	384	191	122	86	64	48	35	24	13
DG B2	394	196	127	91	69	53	40	29	19
DG B3	445	222	145	105	80	62	49	37	25
DG C1	451	225	146	104	78	61	47	34	21
DG C2	453	226	148	106	80	63	49	37	25
DG C3	509	254	167	121	92	73	58	45	31
SS A1	1171	396	220	142	98	69	49	34	22
SS A2	1028	468	283	190	135	98	71	50	33
SS A3	1174	380	209	133	91	64	45	31	20
SS A4	1077	481	289	193	136	98	71	50	32
SS A5	1896	400	204	127	86	60	42	29	18
SS A6	1030	468	283	190	135	97	71	50	32
SS A7	2199	359	178	109	72	50	35	24	15
SS A8	1089	478	285	190	133	96	69	48	31
SS B1	802	298	170	111	77	55	40	28	18
SS B2	785	366	224	152	108	79	57	41	27
SS B3	770	287	163	106	74	52	37	26	17
SS B4	790	368	224	152	108	78	57	40	26
SS B5	1022	271	144	91	62	44	31	22	14
SS B6	750	356	219	149	106	77	56	40	26
SS B7	937	242	127	80	54	38	27	19	12
SS B8	750	352	215	146	103	75	55	39	25
SS C1	481	232	144	99	71	52	39	28	19
SS C2	482	262	170	120	88	66	49	36	24
SS C3	487	185	107	71	50	36	27	19	13
SS C4	473	235	147	102	74	55	41	30	20
SS D1	1107	477	282	188	132	95	68	48	31
SS D2	928	511	332	232	169	125	92	66	43
SS D3	1353	449	248	160	110	78	56	39	25
SS D4	946	500	319	222	160	117	86	61	40
SS E1	838	378	227	153	108	78	56	40	26
SS E2	766	419	272	190	138	102	75	54	36
SS E3	973	322	178	115	79	56	40	28	18
SS E4	780	393	247	170	122	89	65	47	31
Albuquerque, New Mexico								4292 DD	
WW A1	1052	130	62	38	25	18	13	9	6
WW A2	354	144	84	56	39	29	21	15	10
WW A3	300	146	90	62	45	33	24	18	12
WW A4	276	146	93	65	47	35	26	19	13
WW A5	264	148	97	69	50	38	28	21	14
WW A6	256	148	99	70	52	39	30	22	15
WW B1	293	111	63	41	28	20	15	11	7
WW B2	270	147	96	67	49	37	28	20	14
WW B3	314	179	119	84	62	47	35	26	18
WW B4	275	169	116	85	64	49	37	28	19
WW B5	252	159	110	81	61	47	36	27	19
WW C1	333	190	126	89	66	50	38	28	19
WW C2	287	171	115	83	62	47	36	27	18
WW C3	293	191	136	101	77	59	46	34	24

TABLE 20.3 (continued) LCR Tables for Six Representative Cities (Albuquerque, Boston, Madison, Medford, Nashville, Santa Maria)

SSF	.10	.20	.30	.40	.50	.60	.70	.80	.90
WW C4	264	172	122	91	69	54	41	31	22
TW A1	900	124	60	37	25	17	12	9	6
TW A2	361	130	73	48	33	24	18	13	8
TW A3	293	129	77	52	37	27	20	15	10
TW A4	249	123	76	52	38	28	21	15	10
TW B1	502	117	60	38	26	18	13	9	6
TW B2	291	118	68	45	32	23	17	12	8
TW B3	254	114	68	46	33	24	18	13	9
TW B4	233	104	63	42	30	22	16	12	8
TW C1	332	106	58	37	26	19	14	10	6
TW C2	255	101	58	39	27	20	15	11	7
TW C3	243	94	54	36	25	18	13	10	7
TW C4	254	84	46	30	21	15	11	8	5
TW D1	213	86	50	33	23	17	12	9	6
TW D2	287	139	86	59	43	32	24	17	12
TW D3	294	153	97	68	49	37	27	20	14
TW D4	281	158	104	74	55	41	31	23	16
TW D5	260	151	101	73	54	41	31	23	16
TW E1	339	177	113	78	57	43	32	23	16
TW E2	308	168	109	77	56	42	32	23	16
TW E3	323	195	133	96	72	55	42	31	21
TW E4	287	175	120	88	66	50	38	28	20
TW F1	409	108	57	36	24	17	13	9	6
TW F2	260	110	65	43	31	22	17	12	8
TW F3	216	106	66	45	33	24	10	13	9
TW F4	178	95	61	42	31	23	17	13	9
TW G1	256	93	53	34	24	17	13	9	6
TW G2	189	91	56	38	27	20	15	11	7
TW G3	159	82	52	36	26	20	15	11	7
TW G4	128	68	43	30	22	16	12	9	6
TW H1	168	76	45	31	22	16	12	9	6
TW H2	130	66	41	29	21	15	11	8	6
TW H3	108	56	35	25	8	13	10	7	5
TW H4	83	42	27	19	13	10	7	5	4
TW I1	166	73	43	29	20	15	11	8	5
TW I2	221	117	75	52	30	28	21	16	11
TW I3	234	128	83	59	43	32	24	10	12
TW I4	234	137	92	66	49	37	28	21	14
TW I5	226	136	93	67	50	38	29	22	15
TW J1	282	156	102	72	53	40	30	22	15
TW J2	254	146	97	69	51	39	29	22	15
TW J3	269	169	118	86	65	50	38	29	20
TW J4	247	155	106	80	60	46	35	26	18
DG A1	211	97	57	36	22	13	5	—	—
DG A2	227	107	67	46	32	23	16	10	5
DG A3	274	131	83	59	44	34	25	18	10
DG B1	210	97	60	42	30	21	13	6	—
DG B2	232	110	69	49	37	28	21	14	8
DG B3	277	134	85	61	47	37	28	21	14
DG C1	253	120	74	53	39	30	22	14	—
DG C2	271	130	82	59	45	35	26	19	12
DG C3	318	155	96	71	54	43	34	26	18
SS A1	591	187	101	64	44	31	22	16	10
SS A2	531	232	137	92	65	47	34	25	16
SS A3	566	170	90	56	38	27	19	13	8
SS A4	537	230	135	89	63	45	33	23	15
SS A5	980	187	92	56	37	26	18	13	8

TABLE 20.3 (continued) LCR Tables for Six Representative Cities (Albuquerque, Boston, Madison, Medford, Nashville, Santa Maria)

SSF	.10	.20	.30	.40	.50	.60	.70	.80	.90
SS A6	529	231	136	91	64	47	34	24	16
SS A7	1103	158	74	44	29	20	14	10	6
SS A8	540	226	131	87	61	44	32	23	15
SS B1	403	141	78	50	35	25	18	13	8
SS B2	412	186	111	75	53	39	28	20	14
SS B3	372	130	71	46	31	22	16	11	7
SS B4	403	181	106	72	51	37	27	20	13
SS B5	518	127	65	40	27	19	13	9	6
SS B6	390	179	106	73	52	38	28	20	13
SS B7	457	108	54	33	22	16	11	8	5
SS B8	379	171	102	69	49	35	26	19	12
SS C1	270	126	77	52	37	27	20	15	10
SS C2	282	150	97	68	49	37	28	20	14
SS C3	276	101	57	37	26	19	14	10	7
SS C4	277	135	83	57	41	31	23	17	11
SS D1	548	225	130	85	59	43	31	22	14
SS D2	474	253	162	113	82	61	45	33	22
SS D3	683	212	113	72	49	35	25	17	11
SS D4	484	248	156	107	77	57	42	30	20
SS E1	410	176	103	68	48	35	25	18	12
SS E2	390	208	133	92	67	50	37	27	18
SS E3	487	151	80	51	35	25	18	12	8
SS E4	400	195	120	82	59	43	32	23	15
Nashville, Tennessee								3696 DD	
WW A1	588	60	24	13	8	5	3	2	1
WW A2	192	70	38	23	15	11	7	5	3
WW A3	161	72	42	27	18	13	9	6	4
WW A4	148	72	43	29	20	14	10	7	5
WW A5	141	74	46	31	22	16	11	8	5
WW A6	137	74	47	32	22	16	12	8	5
WW B1	135	41	19	10	6	3	2	—	—
WW B2	152	78	48	33	23	17	12	9	6
WW B3	179	97	61	42	30	22	16	12	8
WW B4	164	97	65	46	34	25	19	14	9
WW B5	153	93	63	45	33	25	19	14	9
WW C1	193	105	67	46	33	24	18	13	8
WW C2	169	97	63	44	32	24	18	13	8
WW C3	181	115	79	58	43	33	25	18	12
WW C4	164	104	72	53	39	30	23	17	11
TW A1	509	59	25	13	8	5	3	2	1
TW A2	199	64	33	20	13	9	6	4	3
TW A3	160	65	36	23	15	11	8	5	3
TW A4	136	62	36	23	16	11	8	6	4
TW B1	282	57	26	15	9	6	4	3	2
TW B2	161	59	32	20	13	9	6	4	3
TW B3	141	58	32	21	14	10	7	5	3
TW B4	131	54	30	19	13	9	7	5	3
TW C1	188	53	27	16	10	7	5	3	2
TW C2	144	52	28	18	12	8	6	4	2
TW C3	139	49	27	17	11	8	5	4	2
TW C4	149	45	23	14	9	7	5	3	2
TW D1	99	33	16	9	5	3	2	1	—
TW D2	164	75	44	29	20	14	10	7	5
TW D3	167	82	49	33	23	17	12	8	5
TW D4	168	91	58	40	29	21	15	11	7
TW D5	160	89	58	40	29	22	16	12	8

TABLE 20.3 (continued) LCR Tables for Six Representative Cities (Albuquerque, Boston, Madison, Medford, Nashville, Santa Maria)

SSF	.10	.20	.30	.40	.50	.60	.70	.80	.90
TW E1	198	98	59	40	28	20	15	10	7
TW E2	182	95	59	40	29	21	15	11	7
TW E3	197	115	76	54	39	29	22	16	11
TW E4	178	105	70	50	37	27	20	15	10
TW F1	221	50	23	13	8	5	4	2	1
TW F2	139	53	29	18	12	8	6	4	2
TW F3	116	52	30	19	13	9	7	5	3
TW F4	96	47	28	19	13	9	7	5	3
TW G1	137	44	22	13	9	6	4	3	2
TW G2	101	44	25	16	11	8	5	4	2
TW G3	86	41	24	16	11	8	6	4	2
TW G4	69	34	21	14	10	7	5	3	2
TW H1	89	36	20	13	8	6	4	3	2
TW H2	69	33	19	12	9	6	4	3	2
TW H3	59	28	17	11	8	5	4	3	2
TW H4	46	22	13	9	6	4	3	2	1
TW I1	74	26	13	7	4	2	1	—	—
TW I2	125	62	38	25	18	13	9	7	4
TW I3	133	69	43	29	20	15	11	8	5
TW I4	139	78	51	35	26	19	14	10	7
TW I5	137	80	53	37	27	20	15	11	7
TW J1	164	86	54	36	26	19	14	10	6
TW J2	150	82	53	36	26	19	14	10	7
TW J3	165	101	68	49	36	27	20	15	10
TW J4	133	93	63	46	34	25	19	11	10
DG A1	98	34	—	—	—	—	—	—	—
DG A2	130	55	31	19	11	6	—	—	—
DG A3	173	78	47	32	23	16	11	7	2
DG B1	100	36	17	—	—	—	—	—	—
DG B2	134	58	33	22	15	10	6	—	—
DG B3	177	81	49	33	24	18	14	10	6
DG C1	131	52	28	17	9	—	—	—	—
DG C2	161	71	42	28	20	14	10	6	—
DG C3	205	94	57	39	29	22	17	12	8
SS A1	351	100	50	29	19	13	9	6	4
SS A2	328	135	76	49	33	24	17	12	8
SS A3	330	87	41	24	15	10	6	4	2
SS A4	331	133	74	47	32	22	16	11	7
SS A5	595	98	43	24	15	10	7	4	2
SS A6	324	132	75	48	32	23	16	11	7
SS A7	668	79	32	17	10	6	4	2	1
SS A8	330	129	71	45	30	21	15	10	6
SS B1	236	74	38	23	15	10	7	5	3
SS B2	258	110	63	41	28	20	14	10	6
SS B3	212	65	32	19	12	8	5	3	2
SS B4	251	105	60	39	27	19	13	9	6
SS B5	307	65	30	17	10	7	4	3	2
SS B6	241	104	60	39	27	19	14	10	6
SS B7	264	52	23	12	7	5	3	2	—
SS B8	233	98	56	36	25	17	12	9	5
SS C1	141	60	33	21	14	10	7	5	3
SS C2	161	81	50	33	23	17	12	9	6
SS C3	149	48	25	15	10	7	4	3	2
SS C4	160	73	43	28	19	14	10	7	5
SS D1	317	119	64	39	26	18	13	8	5
SS D2	287	147	90	61	43	31	23	16	10
SS D3	405	113	55	33	21	14	10	6	4

TABLE 20.3 (continued) LCR Tables for Six Representative Cities (Albuquerque, Boston, Madison, Medford, Nashville, Santa Maria)

SSF	.10	.20	.30	.40	.50	.60	.70	.80	.90
SS D4	295	144	87	58	40	29	21	15	10
SS E1	229	89	48	29	19	13	9	6	4
SS E2	233	118	72	48	34	24	18	12	8
SS E3	283	77	37	22	14	9	6	4	2
SS E4	242	111	65	43	29	21	15	11	7
Medford, Oregon								4930 DD	
WW A1	708	64	24	11	—	—	—	—	—
WW A2	212	73	38	22	13	7	3	—	—
WW A3	174	75	41	25	16	9	5	2	—
WW A4	158	74	43	27	17	11	6	3	1
WW A5	149	75	45	29	19	12	7	4	2
WW A6	144	75	46	30	20	13	8	4	2
WW B1	154	43	16	—	—	—	—	—	—
WW B2	162	80	48	31	21	14	9	6	3
WW B3	190	100	62	41	28	19	13	8	5
WW B4	171	99	65	45	32	23	16	11	7
WW B5	160	95	63	45	32	23	17	12	7
WW C1	205	108	67	45	31	21	15	10	6
WW C2	178	99	63	43	30	22	15	10	6
WW C3	189	117	80	57	42	31	23	16	10
WW C4	170	106	72	52	38	28	21	15	9
TW A1	607	63	25	12	5	—	—	—	—
TW A2	222	68	33	19	11	6	2	—	—
TW A3	175	67	36	21	13	8	4	2	—
TW A4	147	64	36	22	14	9	5	3	1
TW B1	327	61	27	14	7	3	—	—	—
TW B2	178	62	32	19	12	7	4	2	—
TW B3	154	60	33	20	12	8	4	2	1
TW B4	143	56	31	19	12	8	5	2	1
TW C1	212	56	27	15	9	5	2	—	—
TW C2	159	55	28	17	11	7	4	2	—
TW C3	154	52	27	16	10	6	4	2	1
TW C4	167	48	24	14	9	5	3	2	—
TW D1	112	34	14	—	—	—	—	—	—
TW D2	177	77	44	28	18	12	8	5	3
TW D3	180	85	50	32	21	14	9	6	3
TW D4	177	93	58	39	27	19	13	9	5
TW D5	168	92	58	40	28	20	14	10	6
TW E1	213	101	60	39	26	18	12	8	4
TW E2	194	98	59	39	27	19	13	9	5
TW E3	208	118	77	53	38	27	20	13	8
TW E4	186	108	71	49	36	26	19	13	8
TW F1	256	53	23	12	5	—	—	—	—
TW F2	153	56	29	17	10	5	2	—	—
TW F3	125	54	30	18	11	7	3	1	—
TW F4	102	48	28	18	11	7	4	2	1
TW G1	153	46	22	12	7	—	—	—	—
TW G2	109	46	25	15	9	5	3	1	—
TW G3	92	42	24	15	9	6	3	2	—
TW G4	74	35	20	13	8	5	3	2	—
TW H1	97	38	20	12	7	4	1	—	—
TW H2	75	34	19	12	7	5	3	1	—
TW H3	63	29	17	10	7	4	3	1	—
TW H4	49	23	13	8	5	3	2	1	—
TW I1	83	27	10	—	—	—	—	—	—
TW I2	133	64	38	24	16	11	7	4	2

TABLE 20.3 (continued) LCR Tables for Six Representative Cities (Albuquerque, Boston, Madison, Medford, Nashville, Santa Maria)

SSF	.10	.20	.30	.40	.50	.60	.70	.80	.90
TW I3	142	71	43	28	19	13	9	5	3
TW I4	146	80	51	35	25	17	12	8	5
TW I5	144	82	53	37	26	19	13	9	6
TW J1	175	89	54	36	24	17	11	7	4
TW J2	158	85	53	36	25	18	12	8	5
TW J3	173	103	69	48	35	26	18	13	8
TW J4	160	96	64	45	33	24	17	12	8
DG A1	110	35	—	—	—	—	—	—	—
DG A2	142	58	32	18	9	—	—	—	—
DG A3	187	82	48	32	22	15	9	5	—
DG B1	110	40	15	—	—	—	—	—	—
DG B2	146	61	35	21	13	7	—	—	—
DG B3	193	84	51	34	24	17	12	7	3
DG C1	144	57	29	13	—	—	—	—	—
DG C2	177	75	44	28	19	12	6	—	—
DG C3	224	98	60	41	29	21	14	10	5
SS A1	415	110	51	28	16	9	4	2	—
SS A2	372	146	79	48	31	21	14	8	5
SS A3	397	96	42	21	10	—	—	—	—
SS A4	379	144	76	46	29	19	12	7	4
SS A5	732	111	45	23	12	5	—	—	—
SS A6	368	143	77	47	30	20	13	8	4
SS A7	846	90	33	14	—	—	—	—	—
SS A8	379	140	73	44	27	17	11	6	3
SS B1	274	81	38	21	12	6	2	—	—
SS B2	288	117	65	40	26	18	12	7	4
SS B3	249	71	33	17	8	—	—	—	—
SS B4	282	113	62	38	25	16	11	7	4
SS B5	368	72	30	15	7	—	—	—	—
SS B6	269	111	62	30	25	17	11	7	4
SS B7	323	58	23	10	—	—	—	—	—
SS B8	262	106	57	35	23	15	9	6	3
SS C1	153	62	33	19	11	5	—	—	—
SS C2	172	83	50	32	22	15	10	6	3
SS C3	166	51	24	13	7	3	—	—	—
SS C4	173	76	43	27	18	12	8	5	3
SS D1	367	129	65	37	22	13	7	3	1
SS D2	318	156	92	60	40	27	18	12	7
SS D3	480	124	57	31	18	10	5	2	—
SS D4	328	153	89	57	38	26	17	11	6
SS E1	262	95	48	27	15	7	—	—	—
SS E2	257	124	73	47	31	21	14	9	5
SS E3	334	84	38	20	10	4	—	—	—
SS E4	269	118	67	42	27	18	12	7	4
Boston, Massachusetts								5621 DD	
WW A1	368	28	9	—	—	—	—	—	—
WW A2	119	41	20	12	7	5	3	2	—
WW A3	101	43	24	15	10	6	4	3	1
WW A4	93	44	26	16	11	7	5	3	2
WW A5	89	45	27	18	12	8	6	4	2
WW A6	87	46	28	19	13	9	6	4	3
WW B1	59	—	—	—	—	—	—	—	—
WW B2	103	52	31	21	15	10	7	5	3
WW B3	123	66	41	28	20	14	10	7	5
WW B4	118	70	46	33	24	18	13	9	6
WW B5	113	69	46	33	25	18	14	10	7

TABLE 20.3 (continued) LCR Tables for Six Representative Cities (Albuquerque, Boston, Madison, Medford, Nashville, Santa Maria)

SSF	.10	.20	.30	.40	.50	.60	.70	.80	.90
WW C1	135	72	46	31	22	16	12	8	5
WW C2	121	68	44	31	22	16	12	9	6
WW C3	136	86	60	44	33	25	19	14	9
WW C4	124	78	54	40	30	23	17	12	8
TW A1	324	30	11	4	—	—	—	—	—
TW A2	126	37	18	10	6	4	2	1	—
TW A3	102	39	21	13	8	5	3	2	1
TW A4	88	38	22	14	9	6	4	3	2
TW B1	180	32	13	7	4	2	—	—	—
TW B2	104	36	19	11	7	5	3	2	1
TW B3	92	36	19	12	8	5	3	2	1
TW B4	86	34	19	12	8	5	4	2	1
TW C1	122	32	15	9	5	3	2	1	—
TW C2	95	33	17	10	7	4	3	2	1
TW C3	93	31	16	10	6	4	3	2	1
TW C4	102	29	15	9	6	4	3	2	1
TW D1	45	—	—	—	—	—	—	—	—
TW D2	112	49	28	18	12	9	6	4	3
TW D3	113	54	32	21	15	10	7	5	3
TW D4	121	64	41	28	20	15	11	8	5
TW D5	118	66	42	30	21	16	12	8	6
TW E1	138	67	40	27	18	13	9	7	4
TW E2	130	66	41	28	20	14	10	7	5
TW E3	146	84	56	39	29	21	16	11	8
TW E4	133	78	52	37	27	20	15	11	7
TW F1	134	25	10	4	—	—	—	—	—
TW F2	86	30	16	9	5	3	2	1	—
TW F3	72	31	17	11	7	4	3	2	1
TW F4	61	29	17	11	7	5	3	2	1
TW G1	83	24	11	6	3	2	—	—	—
TW G2	63	26	14	9	5	4	2	1	—
TW G3	54	25	14	9	6	4	3	2	1
TW G4	45	21	12	8	5	4	3	2	1
TW H1	54	21	11	6	4	2	1	—	—
TW H2	44	20	11	7	5	3	2	1	—
TW H3	38	17	10	6	4	3	2	1	—
TW H4	30	14	8	5	3	2	2	1	—
TW I1	30	—	—	—	—	—	—	—	—
TW I2	84	41	24	16	11	8	6	4	2
TW I3	91	46	28	19	13	9	7	5	3
TW I4	100	56	36	25	18	13	10	7	5
TW I5	101	58	38	27	20	15	11	8	5
TW J1	114	59	37	25	17	12	9	6	4
TW J2	107	58	37	25	18	13	10	7	4
TW J3	123	75	51	36	27	20	15	11	7
TW J4	115	70	47	34	25	19	14	10	7
DG A1	43	—	—	—	—	—	—	—	—
DG A2	85	34	18	9	—	—	—	—	—
DG A3	125	56	33	22	16	11	7	4	—
DG B1	44	—	—	—	—	—	—	—	—
DG B2	87	36	20	12	7	—	—	—	—
DG B3	129	58	35	24	17	13	9	6	3
DG C1	71	23	—	—	—	—	—	—	—
DG C2	109	47	27	17	12	8	4	—	—
DG C3	151	68	41	28	21	16	12	8	5
SS A1	230	61	29	16	10	6	4	2	1

TABLE 20.3 (continued) LCR Tables for Six Representative Cities (Albuquerque, Boston, Madison, Medford, Nashville, Santa Maria)

SSF	.10	.20	.30	.40	.50	.60	.70	.80	.90
SS A2	231	93	52	33	22	15	11	7	5
SS A3	205	48	20	10	4	—	—	—	—
SS A4	229	90	49	31	20	14	9	6	4
SS A5	389	58	23	11	6	3	—	—	—
SS A6	226	91	50	32	21	15	10	7	4
SS A7	420	40	12	—	—	—	—	—	—
SS A8	226	86	46	28	19	12	8	6	3
SS B1	151	44	21	12	7	4	2	1	—
SS B2	183	77	43	28	19	13	9	6	4
SS B3	129	36	16	8	3	—	—	—	—
SS B4	176	73	41	26	17	12	8	6	4
SS B5	193	36	15	7	3	—	—	—	—
SS B6	169	72	41	26	18	12	9	6	4
SS B7	157	25	7	—	—	—	—	—	—
SS B8	160	66	37	23	16	11	7	5	3
SS C1	84	33	17	10	6	4	2	1	—
SS C2	110	54	33	22	15	11	8	5	3
SS C3	91	26	12	7	4	2	—	—	—
SS C4	109	48	28	18	12	9	6	4	3
SS D1	206	73	38	22	14	9	5	3	2
SS D2	203	103	63	42	29	21	15	10	6
SS D3	264	69	32	18	10	6	4	2	1
SS D4	208	100	60	39	27	19	14	9	6
SS E1	140	51	25	14	8	4	2	—	—
SS E2	161	80	48	32	22	15	11	7	5
SS E3	177	44	19	10	5	2	—	—	—
SS E4	166	75	43	28	19	13	9	6	4
Madison, Wisconsin								7730 DD	
WW A1	278	—	—	—	—	—	—	—	—
WW A2	91	27	12	—	—	—	—	—	—
WW A3	77	30	15	8	3	—	—	—	—
WW A4	72	32	17	10	5	—	—	—	—
WW A5	69	33	19	11	7	4	—	—	—
WW A6	67	34	19	12	7	4	2	—	—
WW B1	—	—	—	—	—	—	—	—	—
WW B2	84	41	24	15	10	7	5	3	2
WW B3	102	53	32	21	15	10	7	5	3
WW B4	101	59	39	27	19	14	10	7	5
WW B5	98	59	39	28	20	15	11	8	5
WW C1	113	59	37	25	17	12	8	6	3
WW C2	103	57	37	25	18	13	9	6	4
WW C3	119	75	51	37	28	21	15	11	7
WW C4	109	68	47	34	25	19	14	10	7
TW A1	249	16	—	—	—	—	—	—	—
TW A2	97	26	11	4	[illegible]	—	—	—	—
TW A3	79	28	13	7	3	—	—	—	—
TW A4	69	28	15	9	5	3	—	—	—
TW B1	139	20	5	—	—	—	—	—	—
TW B2	81	26	12	6	3	—	—	—	—
TW B3	72	27	13	7	4	2	—	—	—
TW B4	69	26	13	8	5	3	1	—	—
TW C1	96	23	10	4	—	—	—	—	—
TW C2	76	25	12	7	4	2	—	—	—
TW C3	75	24	12	7	4	2	1	—	—
TW C4	84	23	11	6	4	2	1	—	—
TW D1	—	—	—	—	—	—	—	—	—

TABLE 20.3 (continued) LCR Tables for Six Representative Cities (Albuquerque, Boston, Madison, Medford, Nashville, Santa Maria)

SSF	.10	.20	.30	.40	.50	.60	.70	.80	.90
TW D2	91	39	22	13	9	6	4	2	1
TW D3	93	43	25	16	10	7	5	3	1
TW D4	103	54	34	23	16	12	8	6	4
TW D5	102	56	36	25	18	13	10	7	4
TW E1	115	54	32	21	14	10	7	4	3
TW E2	110	55	34	22	16	11	8	5	3
TW E3	126	72	47	33	24	18	13	9	6
TW E4	116	68	45	32	23	17	13	9	6
TW F1	99	13	—	—	—	—	—	—	—
TW F2	65	20	8	—	—	—	—	—	—
TW F3	55	22	11	5	—	—	—	—	—
TW F4	47	21	11	7	4	2	—	—	—
TW G1	61	14	—	—	—	—	—	—	—
TW G2	47	18	8	4	—	—	—	—	—
TW G3	42	18	9	5	3	—	—	—	—
TW G4	35	16	9	5	3	2	—	—	—
TW H1	41	13	6	—	—	—	—	—	—
TW H2	34	14	7	4	2	—	—	—	—
TW H3	29	13	7	4	2	1	—	—	—
TW H4	24	10	6	3	2	1	—	—	—
TW I1	—	—	—	—	—	—	—	—	—
TW I2	68	32	18	12	8	5	3	2	1
TW I3	75	37	22	14	10	7	4	3	2
TW I4	85	47	30	21	15	11	8	5	3
TW I5	87	50	33	23	16	12	9	6	4
TW J1	95	48	29	19	13	9	6	4	3
TW J2	91	48	30	21	14	10	7	5	3
TW J3	106	65	43	31	23	17	12	9	6
TW J4	100	61	41	29	21	16	12	9	6
DG A1	—	—	—	—	—	—	—	—	—
DG A2	68	25	11	—	—	—	—	—	—
DG A3	109	47	28	18	12	8	5	—	—
DG B1	—	—	—	—	—	—	—	—	—
DG B2	70	27	14	6	—	—	—	—	—
DG B3	114	50	30	20	14	10	7	4	—
DG C1	47	—	—	—	—	—	—	—	—
DG C2	91	37	21	13	7	—	—	—	—
DG C3	133	59	35	24	17	13	9	6	3
SS A1	192	47	20	9	3	—	—	—	—
SS A2	200	78	42	26	17	12	8	5	3
SS A3	166	32	—	—	—	—	—	—	—
SS A4	197	74	39	23	15	10	6	4	2
SS A5	329	42	13	—	—	—	—	—	—
SS A6	195	75	40	25	16	11	7	5	3
SS A7	349	22	—	—	—	—	—	—	—
SS A8	192	69	36	21	13	8	5	3	2
SS B1	122	32	13	5	—	—	—	—	—
SS B2	158	64	36	22	15	10	7	5	3
SS B3	100	22	—	—	—	—	—	—	—
SS B4	150	60	33	29	13	9	6	4	2
SS B5	156	24	—	—	—	—	—	—	—
SS B6	145	59	33	20	13	9	6	4	2
SS B7	122	—	—	—	—	—	—	—	—
SS B8	136	54	29	18	11	7	5	3	2
SS C1	61	20	7	—	—	—	—	—	—
SS C2	90	43	25	16	11	7	5	3	2
SS C3	67	16	—	—	—	—	—	—	—
SS C4	90	38	22	13	9	6	4	2	1

TABLE 20.3 (continued) LCR Tables for Six Representative Cities (Albuquerque, Boston, Madison, Medford, Nashville, Santa Maria)

SSF	.10	.20	.30	.40	.50	.60	.70	.80	.90
SS D1	169	56	26	13	6	—	—	—	—
SS D2	175	86	51	34	23	16	11	7	5
SS D3	221	52	21	10	—	—	—	—	—
SS D4	179	84	49	32	21	15	10	7	4
SS E1	108	34	12	—	—	—	—	—	—
SS E2	135	65	38	24	16	11	7	5	3
SS E3	141	29	8	—	—	—	—	—	—
SS E4	140	61	34	21	14	9	6	4	2

Source: PSDH, 1984.

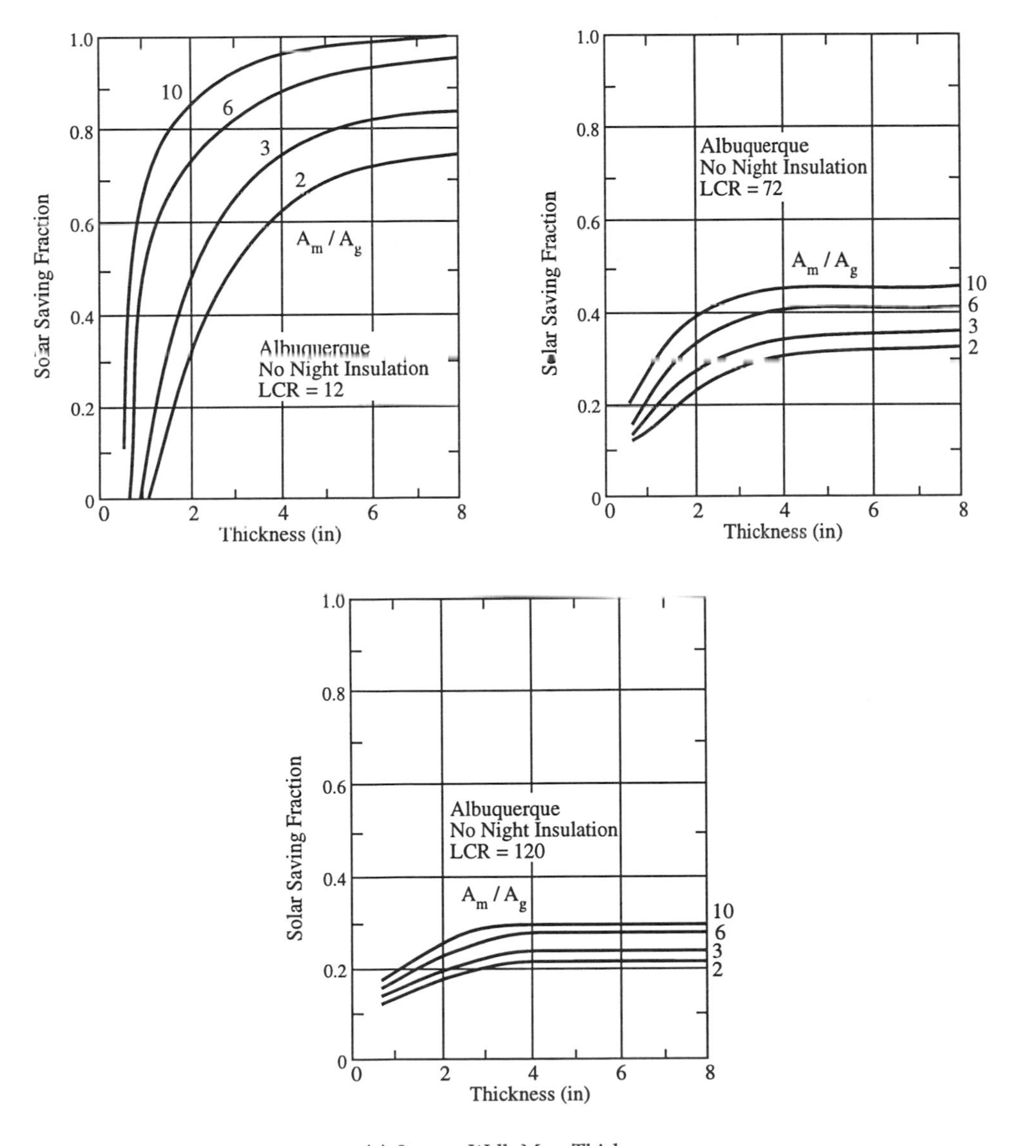

(a) Storage Wall: Mass Thickness

FIGURE 20.7 Sensitivity of SSF to off-reference conditions. *Source:* PSDH, 1984.

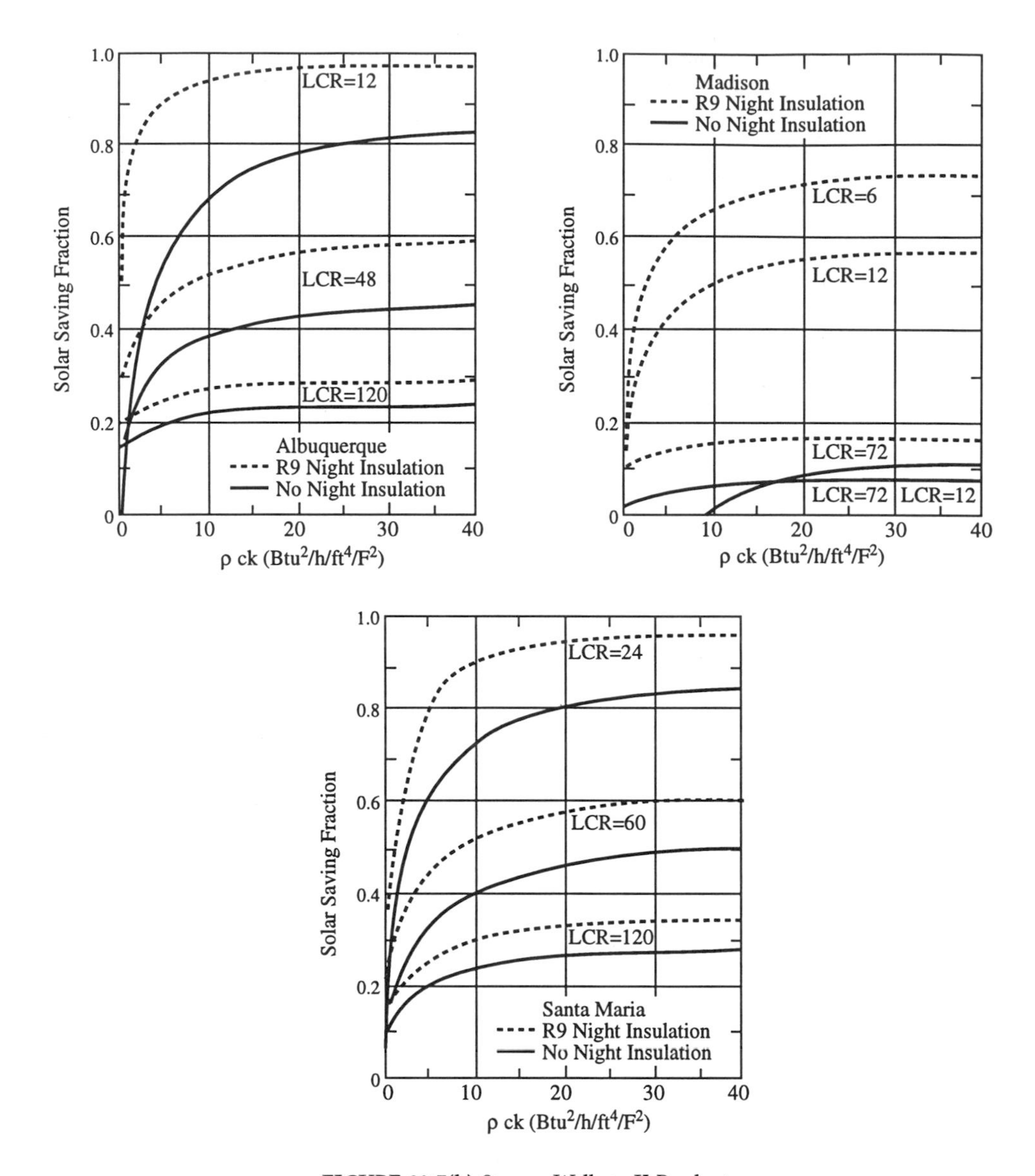

FIGURE 20.7(b) Storage Wall: ρ cK Product

Step 4 yields: Using Eq. (20.4) and reading 4,930 FDD from Table 20.3

$$Q_{aux} = (1 - 0.39) \times 11{,}800 \text{ Btu} \times 4{,}930 \text{ FDD}$$
$$= 35.5 \times 10^6 \text{ Btu annually}$$

Using the reference system characteristics yields the thermal storage size:

- Trombe wall ($\rho ck = 30$, concrete properties from Table 20.2c):

$$\text{m(TW)} = \text{density} \times \text{area} \times \text{thickness}$$
$$= 150 \text{ lbm/ft}^3 \times 180 \text{ ft}^2 \times 1 \text{ ft}$$
$$= 27{,}000 \text{ lbm}$$

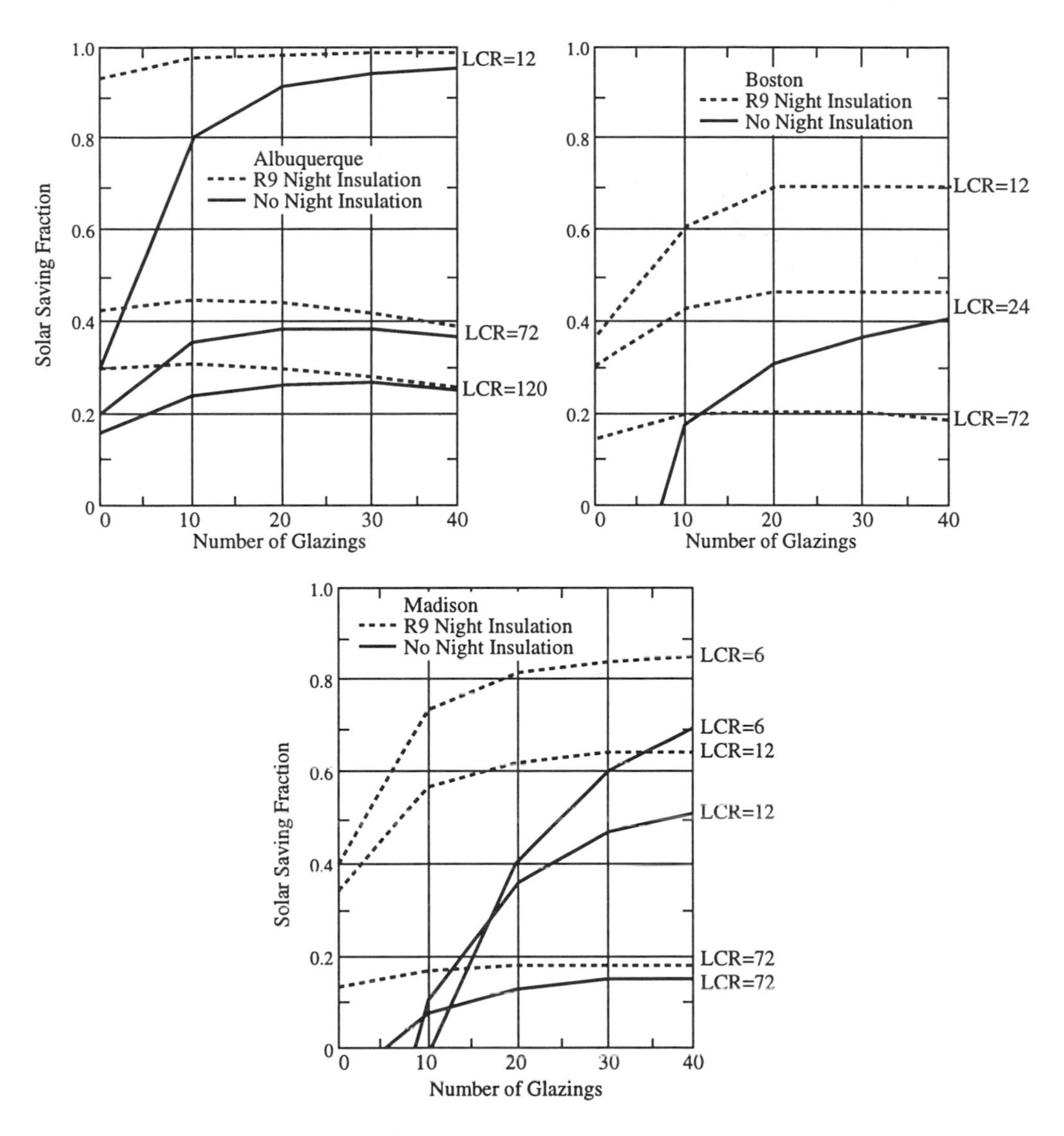

FIGURE 20.7(c) Storage Wall: Number of Glazings

- Direct gain ($\rho ck = 30$, concrete properties), using mass area–to–glazing area ratio of 6:

$$\text{mass area} = 6 \times 130 = 780 \text{ ft}^2 \text{ of 2" thick concrete}$$

$$m(DG) = 150 \text{ lbm/ft}^3 \times 780 \text{ ft}^2 \times 1/6 \text{ ft}$$

$$= 19{,}500 \text{ lbm}$$

Thus, using the LCR method allows a basic design of passive system types for the 94 reference systems, and the resulting annual performance. A bit more design variation can be obtained by using the sensitivity curves of Figure 20.7 to modify the SSF of a particular reference system. For instance, the direct gain system SSF just presented would increase by approximately 0.03 (to 0.40) if the mass-glazing-area ratio (assumed 6) were increased to 10, and would decrease by about 0.04 (to 0.33) if the mass-glazing-area ratio were decreased to 3. This information provides a designer with quantitative information for making trade-offs.

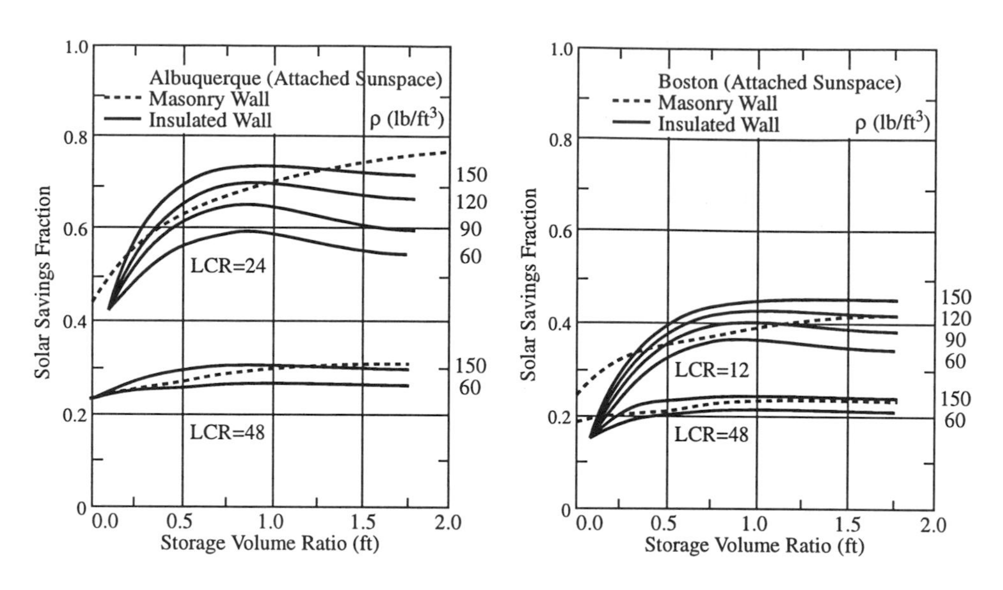

FIGURE 20.7(d) Sunspace: Storage Volume to Projected Area Ratio

The Third Level: SLR Method

The solar-load ratio (SLR) method calculates monthly performance, and the terms and values used are monthly based. The method allows the use of specific location weather data and the 94 reference design systems (Table 20.2). In addition, the sensitivity curves (Figure 20.7) can again be used to define performance outside the reference design systems. The result of the SLR method is the determination of the monthly heating auxiliary energy required, which is then summed to give the annual requirement for auxiliary heating energy. Generally, the SLR method gives annual values within ±3% of detailed simulation results, but the monthly values may vary more (PSDH, 1984; Duffie and Beckman, 1991). Thus, the monthly LCR method is more "accurate" than the rule-of-thumb methods, plus providing the designer with system performance on a month-by-month basis.

The SLR method uses equations and correlation parameters for each of the 94 reference systems combined with the insolation absorbed by the system, the monthly degree days, and the system's LCR to determine the monthly SSF. These correlation parameters are listed in Table 20.4 as A, B, C, D, R, G, H, and LCRs for each reference system (PSDH, 1984). The correlation equations are

$$\text{SSF} = 1 - K\,(1 - F) \tag{20.7}$$

where

$$K = 1 + \text{G}/\text{LCR} \tag{20.8}$$

$$F = \begin{cases} AX, & \text{when } X < R \\ B - C\exp(-DX), & \text{when } X > R \end{cases} \tag{20.9}$$

$$X = \frac{S/DD - (\text{LCRs})\;H}{(\text{LCR})K} \tag{20.10}$$

and X is called the generalized solar load ratio. The term S is the monthly insolation absorbed by the system per unit of solar projected area. Monthly average daily insolation data on a vertical south facing surface can be found and/or calculated using various sources (PSDH, 1984; McQusiton

TABLE 20.4 SLR Correlation Parameters for the 94 Reference Systems

Type	A	B	C	D	R	G	H	LCRs	STDV
WW A1	0.0000	1.0000	.9172	.4841	–9.0000	0.00	1.17	13.0	.053
WW A2	0.0000	1.0000	.9833	.7603	–9.0000	0.00	.92	13.0	.046
WW A3	0.0000	1.0000	1.0171	.8852	–9.0000	0.00	.85	13.0	.040
WW A4	0.0000	1.0000	1.0395	.9569	–9.0000	0.00	.81	13.0	.037
WW A5	0.0000	1.0000	1.0604	1.0387	–9.0000	0.00	.78	13.0	.034
WW A6	0.0000	1.0000	1.0735	1.0827	–9.0000	0.00	.76	13.0	.033
WW B1	0.0000	1.0000	.9754	.5518	–9.0000	0.00	.92	22.0	.051
WW B2	0.0000	1.0000	1.0487	1.0851	–9.0000	0.00	.78	9.2	.036
WW B3	0.0000	1.0000	1.0673	1.0087	–9.0000	0.00	.95	8.9	.038
WW B4	0.0000	1.0000	1.1028	1.1811	–9.0000	0.00	.74	5.8	.034
WW B5	0.0000	1.0000	1.1146	1.2771	–9.0000	0.00	.56	4.5	.032
WW C1	0.0000	1.0000	1.0667	1.0437	–9.0000	0.00	.62	12.0	.038
WW C2	0.0000	1.0000	1.0846	1.1482	–9.0000	0.00	.59	8.7	.035
WW C3	0.0000	1.0000	1.1419	1.1756	–9.0000	0.00	.28	5.5	.033
WW C4	0.0000	1.0000	1.1401	1.2378	–9.0000	0.00	.23	4.3	.032
TW A1	0.0000	1.0000	.9194	.4601	–9.0000	0.00	1.11	13.0	.048
TW A2	0.0000	1.0000	.9680	.6318	–9.0000	0.00	.92	13.0	.043
TW A3	0.0000	1.0000	.9964	.7123	–9.0000	0.00	.85	13.0	.038
TW A4	0.0000	1.0000	1.0190	.7332	–9.0000	0.00	.79	13.0	.032
TW B1	0.0000	1.0000	.9364	.4777	–9.0000	0.00	1.01	13.0	.045
TW B2	0.0000	1.0000	.9821	.6020	–9.0000	0.00	.85	13.0	.038
TW B3	0.0000	1.0000	.9980	.6191	–9.0000	0.00	.80	13.0	.033
TW B4	0.0000	1.0000	.9981	.5615	–9.0000	0.00	.76	13.0	.028
TW C1	0.0000	1.0000	.9558	.4709	–9.0000	0.00	.89	13.0	.039
TW C2	0.0000	1.0000	.9788	.4964	–9.0000	0.00	.79	13.0	.033
TW C3	0.0000	1.0000	.9760	.4519	–9.0000	0.00	.76	13.0	.029
TW C4	0.0000	1.0000	.9588	.3612	–9.0000	0.00	.73	13.0	.026
TW D1	0.0000	1.0000	.9842	.4418	–9.0000	0.00	.89	22.0	.040
TW D2	0.0000	1.0000	1.0150	.8994	–9.0000	0.00	.80	9.2	.036
TW D3	0.0000	1.0000	1.0346	.7810	–9.0000	0.00	1.08	8.9	.036
TW D4	0.0000	1.0000	1.0606	.9770	–9.0000	0.00	.85	5.8	.035
TW D5	0.0000	1.0000	1.0721	1.0718	–9.0000	0.00	.61	4.5	.033
TW E1	0.0000	1.0000	1.0345	.8753	–9.0000	0.00	.68	12.0	.037
TW E2	0.0000	1.0000	1.0476	1.0050	–9.0000	0.00	.66	8.7	.035
TW E3	0.0000	1.0000	1.0919	1.0739	9.0000	0.00	.61	5.5	.034
TW E4	0.0000	1.0000	1.0971	1.1429	–9.0000	0.00	.47	4.3	.033
TW F1	0.0000	1.0000	.9430	.4744	–9.0000	0.00	1.09	13.0	.047
TW F2	0.0000	1.0000	.9900	.6053	–9.0000	0.00	.93	13.0	.041
TW F3	0.0000	1.0000	1.0189	.6502	–9.0000	0.00	.86	13.0	.036
TW F4	0.0000	1.0000	1.0419	.6258	–9.0000	0.00	.80	13.0	.032
TW G1	0.0000	1.0000	.9693	.4714	–9.0000	0.00	1.01	13.0	.042
TW G2	0.0000	1.0000	1.0133	.5462	–9.0000	0.00	.88	13.0	.035
TW G3	0.0000	1.0000	1.0325	.5269	–9.0000	0.00	.82	13.0	.031
TW G4	0.0000	1.0000	1.0401	.4400	–9.0000	0.00	.77	13.0	.030
TW H1	0.0000	1.0000	1.0002	.4356	–9.0000	0.00	.93	13.0	.034
TW H2	0.0000	1.0000	1.0280	.4151	–9.0000	0.00	.83	13.0	.030
TW H3	0.0000	1.0000	1.0327	.3522	–9.0000	0.00	.78	13.0	.029
TW H4	0.0000	1.0000	1.0287	.2600	–9.0000	0.00	.74	13.0	.024
TW I1	0.0000	1.0000	.9974	.4036	–9.0000	0.00	.91	22.0	.038
TW I2	0.0000	1.0000	1.0386	.8313	–9.0000	0.00	.80	9.2	.034
TW I3	0.0000	1.0000	1.0514	.6886	–9.0000	0.00	1.01	8.9	.034
TW I4	0.0000	1.0000	1.0781	.8952	–9.0000	0.00	.82	5.8	.032
TW I5	0.0000	1.0000	1.0902	1.0284	–9.0000	0.00	.65	4.5	.032
TW J1	0.0000	1.0000	1.0537	.8227	–9.0000	0.00	.65	12.0	.037
TW J2	0.0000	1.0000	1.0677	.9312	–9.0000	0.00	.62	8.7	.035
TW J3	0.0000	1.0000	1.1153	.9831	–9.0000	0.00	.44	5.5	.034
TW J4	0.0000	1.0000	1.1154	1.0607	–9.0000	0.00	.38	4.3	.033
DG A1	.5650	1.0090	1.0440	.7175	.3931	9.36	0.00	0.0	.046

TABLE 20.4 (continued) SLR Correlation Parameters for the 94 Reference Systems

Type	A	B	C	D	R	G	H	LCRs	STDV
DG A2	.5906	1.0060	1.0650	.8099	.4681	5.28	0.00	0.0	.039
DG A3	.5442	.9715	1.1300	.9273	.7068	2.64	0.00	0.0	.036
DG B1	.5739	.9948	1.2510	1.0610	.7905	9.60	0.00	0.0	.042
DG B2	.6180	1.0000	1.2760	1.1560	.7528	5.52	0.00	0.0	.035
DG B3	.5601	.9839	1.3520	1.1510	.8879	2.38	0.00	0.0	.032
DG C1	.6344	.9887	1.5270	1.4380	.8632	9.60	0.00	0.0	.039
DG C2	.6763	.9994	1.4000	1.3940	.7604	5.28	0.00	0.0	.033
DG C3	.6182	.9859	1.5660	1.4370	.8990	2.40	0.00	0.0	.031
SS A1	0.0000	1.0000	.9587	.4770	–9.0000	0.00	.83	18.6	.027
SS A2	0.0000	1.0000	.9982	.6614	–9.0000	0.00	.77	10.4	.026
SS A3	0.0000	1.0000	.9552	.4230	–9.0000	0.00	.83	23.6	.030
SS A4	0.0000	1.0000	.9956	.6277	–9.0000	0.00	.80	12.4	.026
SS A5	0.0000	1.0000	.9300	.4041	–9.0000	0.00	.96	18.6	.031
SS A6	0.0000	1.0000	.9981	.6660	–9.0000	0.00	.86	10.4	.028
SS A7	0.0000	1.0000	.9219	.3225	–9.0000	0.00	.96	23.6	.035
SS A8	0.0000	1.0000	.9922	.6173	–9.0000	0.00	.90	12.4	.028
SS B1	0.0000	1.0000	.9683	.4954	–9.0000	0.00	.84	16.3	.028
SS B2	0.0000	1.0000	1.0029	.6802	–9.0000	0.00	.74	8.5	.026
SS B3	0.0000	1.0000	.9689	.4685	–9.0000	0.00	.82	19.3	.029
SS B4	0.0000	1.0000	1.0029	.6641	–9.0000	0.00	.76	9.7	.026
SS B5	0.0000	1.0000	.9408	.3866	–9.0000	0.00	.97	16.3	.030
SS B6	0.0000	1.0000	1.0068	.6778	–9.0000	0.00	.84	8.5	.028
SS B7	0.0000	1.0000	.9395	.3363	–9.0000	0.00	.95	19.3	.032
SS B8	0.0000	1.0000	1.0047	.6469	–9.0000	0.00	.87	9.7	.027
SS C1	0.0000	1.0000	1.0087	.7683	–9.0000	0.00	.76	16.3	.025
SS C2	0.0000	1.0000	1.0412	.9281	–9.0000	0.00	.78	10.0	.027
SS C3	0.0000	1.0000	.9699	.5106	–9.0000	0.00	.79	16.3	.024
SS C4	0.0000	1.0000	1.0152	.7523	–9.0000	0.00	.81	10.0	.025
SS D1	0.0000	1.0000	.9889	.6643	–9.0000	0.00	.84	17.8	.028
SS D2	0.0000	1.0000	1.0493	.8753	–9.0000	0.00	.70	9.9	.028
SS D3	0.0000	1.0000	.9570	.5285	–9.0000	0.00	.90	17.8	.029
SS D4	0.0000	1.0000	1.0356	.8142	–9.0000	0.00	.73	9.9	.028
SS E1	0.0000	1.0000	.9968	.7004	–9.0000	0.00	.77	19.6	.027
SS E2	0.0000	1.0000	1.0468	.9054	–9.0000	0.00	.76	10.8	.027
SS E3	0.0000	1.0000	.9565	.4827	–9.0000	0.00	.81	19.6	.028
SS E4	0.0000	1.0000	1.0214	.7694	–9.0000	0.00	.79	10.8	.027

Source: PSDH, 1984.

and Parker, 1994) and the S term can be determined by multiplying by a transmission and an absorption factor and the number of days in the month. Absorption factors for all systems are close to 0.96 (PSDH, 1984), whereas the transmission is approximately 0.9 for single glazing, 0.8 for double glazing, and 0.7 for triple glazing.

Example 20.4. For the vented, 180 ft², double glazed with night insulation, 12" thick Trombe wall system (TWD4) in the NLC = 11,800 Btu/FDD house in Medford, Oregon, determine the auxiliary energy required in January.

Solution: Weather data for Medford, Oregon (PSDH, 1984), yields for January ($N = 31$, days): daily vertical surface insolation = 565 Btu/ft² and 880 FDD, so $S = (31)(565)(0.8)(0.96) =$ 13,452 Btu/ft² month

The LCR = NLC/A_p = 11,800/180 = 65.6 Btu/FDD ft²

From Table 4 at TWD4: A = 0, B = 1, C = 1.0606, D = 0.977, R = –9, G = 0, H = 0.85, LCRs = 5.8 Btu/FDD ft².

Substituting into Eq. (20.7) gives

$$K = 1 + 0/65.6 = 1$$

Eq. (20.9) gives

$$X = \frac{13{,}452/880 - (5.8 \times 0.85)}{65.6 \times 1} = 0.16$$

Eq. (20.8) gives

$$F = 1 - 1.0606\ e^{-0.977 \times 0.16} = 0.09$$

and Eq. (20.6) gives

$$SSF = 1 - 1\ (1 - 0.09) = 0.09$$

The January auxiliary energy required can be calculated using Eq. (20.4):

$$\begin{aligned} Q_{aux}(\text{Jan}) &= (1 - \text{SSF}) \times \text{NLC} \times (\text{No. of degree days}) \\ &= (1 - 0.09) \times 11{,}800 \times 880 \\ &= 9{,}450{,}000\ \text{Btu} \end{aligned}$$

As mentioned, the use of sensitivity curves (PSDH, 1984) as in Figure 20.7 will allow SSF to be determined for many off-reference system design conditions, such as storage mass, number of glazings, and other more esoteric parameters. Also, the use of multiple passive system types within one building would be approached as was done with the LCR method: a combined area LCR would be used to calculate the SSF for each type system individually, and then a weighted-area (aperture) average SSF would be determined for the building.

20.2 Passive Cooling Design Fundamentals

Passive cooling systems are designed to use natural means to transfer heat from buildings, including convection/ventilation, evaporation, radiation, and conduction. However, the most important element in both passive and conventional cooling design is to prevent heat from entering the building in the first place. Cooling conservation techniques involve building surface colors, insulations, special window glazings, overhangs and orientation, and numerous other architectural/engineering features, most of which are treated in Chapter 9.

Solar Control

Controlling the solar energy input to reduce the cooling load is usually a passive design concern because solar input may be needed for other purposes, such as daylighting throughout the year and/or heating during the winter. The solar control is normally "designed in" via the shading of the solar windows, where direct radiation is desired for winter heating and needs to be excluded during the cooling season.

The shading control of the windows can be of various types and "controllability," ranging from drapes and blinds, use of deciduous trees, to the commonly used overhangs and vertical louvers.

TABLE 20.5 South-Facing Window Overhang Rule of Thumb

$$\text{Length of the Overhang} = \frac{\text{Window Height}}{F}$$

(a) Overhang Factors

North Latitude	F*
28	5.6–11.1
32	4.0–6.3
36	3.0–4.5
40	2.5–3.4
44	2.0–2.7
48	1.7–2.2
52	1.5–1.8
56	1.3–1.5

(b) Roof Overhang Geometry

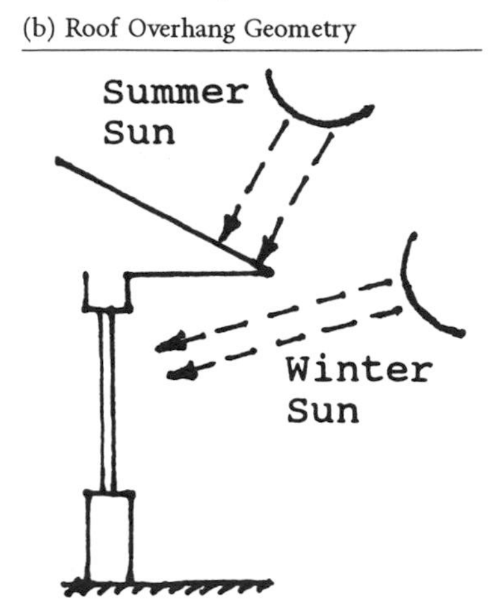

Properly sized overhangs shade out hot summer sun but allow winter sun (which is lower in the sky) to penetrate windows.

* Select a factor according to your latitude. Higher values provide complete shading at noon on June 21; lower values, until August 1.
Source: Halacy, 1984.

Table 20.5 provides a rule of thumb for determining proper south-facing window overhang for both winter heat and summer shade. Technical details on calculating shading from various devices and orientations are found in ASHRAE (1989) and Olgyay and Olgyay (1976).

Natural Convection/Ventilation

Movement of air provides cooling comfort through convection and evaporation from human skin. ASHRAE (1989) places the comfort limit at 79°F for an air velocity of up to 50 ft/min (fpm), 82°F for 160 fpm, and 85°F for 200 fpm. To determine whether or not comfort conditions can be obtained, a designer must calculate the volumetric flow rate, Q, which is passing through the occupied space. Using the cross-sectional area, A_x, of the space and the room air velocity, V_a, required, the flow is determined by

$$Q = A_x V_a \tag{20.11}$$

The proper placement of windows, "narrow" building shape, and open landscaping can enhance natural wind flow to provide ventilation. The air flow rate through open windows for wind-driven ventilation is given by ASHRAE (1989):

$$Q = C_v V_w A_w \tag{20.12}$$

where Q = air flow rate, m^3/s
A_w = free area of inlet opening, m^2
V_w = wind velocity, m/s
C_v = effectiveness of opening = 0.5 to 0.6 for wind perpendicular to opening, and 0.25 to 0.35 for wind diagonal to opening

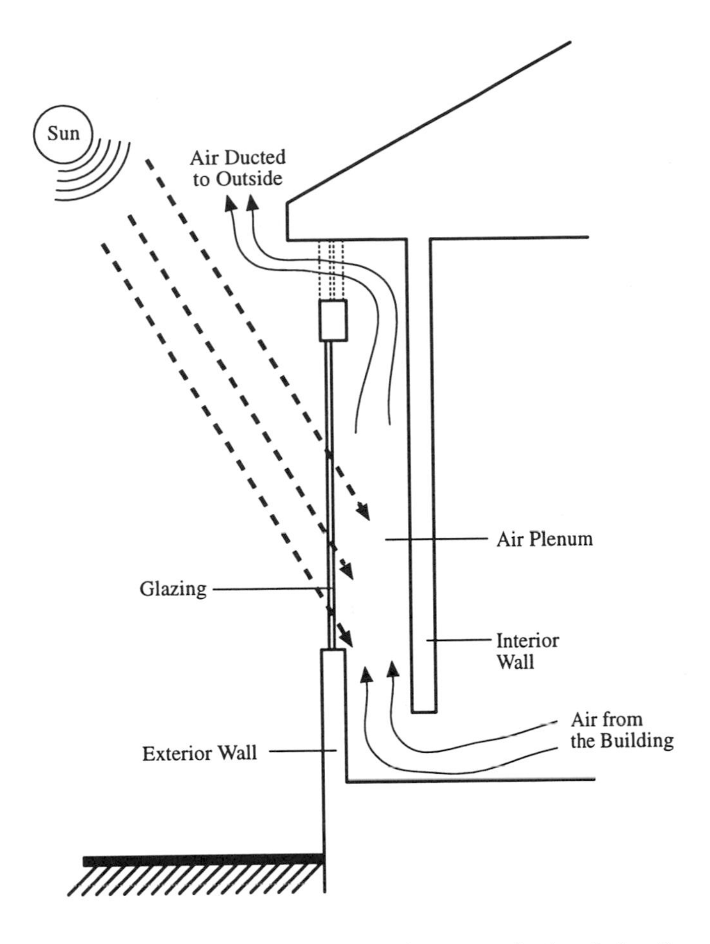

FIGURE 20.8 The stack-effect/solar chimney concept to induce convection/ventilation. *Source:* PSDH, 1980.

The stack effect can induce ventilation when warm air rises to the top of a structure and exhausts outside, while cooler outside air enters the structure to replace it. Figure 20.8 illustrates the solar chimney concept, which can easily be adapted to a thermal storage wall system. The greatest stack-effect flow rate is produced by maximizing the stack height and air temperature in the stack, as shown by

$$Q = 0.116\, A_j \sqrt{h(T_S - T_o)} \tag{20.13}$$

where Q = stack flow rate, m^3/s

A_j = area of inlets or outlets (whichever is smaller), m^2

h = inlet to outlet height, m

T_s = average temperature in stack, °C

T_o = outdoor air temperature, °C

If inlet or outlet area is twice the other, the flow rate will increase by 25%, and by 35% if the areas' ratio is 3:1 or larger.

Another approach to passive convective/ventilative cooling involves using cooler night air to reduce the temperature of the building and/or a storage mass. Thus, the building/storage mass is prepared to accept part of the heat load during the hotter daytime. This type of convective system can also be combined with evaporative and radiative modes of heat transfer, utilizing air and/or water as the convective fluid.

Evaporation

When air with less than 100% relative humidity moves over a water surface, the evaporation of water causes both the air and the water itself to cool. The lowest temperature that can be reached by this direct evaporative cooling effect is the wet-bulb temperature of the air. This wet-bulb temperature is directly related to the relative humidity, with lower wet-bulb temperature associated with lower relative humidity. Thus, dry air (low relative humidity) has a low wet-bulb temperature and will undergo a large temperature drop with evaporative cooling. Humid air (high relative humidity) can only be slightly cooled evaporatively. The wet-bulb temperature for various relative humidity and air temperature conditions can be found via the "psychrometric chart" available in most thermodynamic texts. Normally, an evaporative cooling process does not cool the air all the way down to the wet-bulb temperature, but only part of the way. To get the maximum temperature decrease it is necessary to have a large water surface area in contact with the air for a long time. Interior ponds and fountain sprays can provide this air–water contact area.

The use of water sprays and ponds on roofs provides cooling primarily via evaporation. The use of roof ponds as a thermal cooling mass has been tried by several investigators (Hay and Tellot, 1969; Marlatt et al., 1984) and is often linked with nighttime radiative cooling.

The hybrid system involving a fan and wetted mat, the "swamp cooler," is by far the most widely used evaporative cooling technology. Direct, indirect, and combined evaporative cooling system design features are described in ASHRAE (1989, 1991). Work in Australia (Close, 1968) investigated rock storage beds that were chilled using evaporatively cooled night air. Room air was then circulated through the bed during the day to provide space cooling.

Radiative Cooling

All warm objects emit thermal infrared radiation; the hotter the body, the more energy it emits. A passive cooling scheme is to use the cooler night sky as a sink for thermal radiation emitted by a warm storage mass, thus chilling the mass for cooling use the next day. The net radiative cooling rate, Q_R, for a horizontal unit surface (ASHRAE, 1989) is

$$Q_r = \varepsilon\sigma\left(T_{\text{body}}^4 - T_{sky}^4\right) \qquad \textbf{(20.14)}$$

where Q_r = net radiative cooling rate, W/m² (Btu/h ft²)
ε = surface emissivity fraction (usually 0.9 for water)
σ = 5.67×10^{-8} W/m² K⁴ (1.714×10^{-9} Btu/h ft² R⁴)
T_{body} = warm body temperature, Kelvin (Rankine)
T_{sky} = effective sky temperature, Kelvin (Rankine)

The effective sky temperature depends primarily upon atmospheric water vapor. However, the monthly average air–sky temperature difference has been determined (Martin and Berdahl, 1984) and Figure 20.9 presents these values for July (in °F) for the United States.

Example 20.5. Estimate the overnight cooling possible for a 12 m², 80°F thermal storage roof during July in Los Angeles.

Solution: Assume the roof storage unit is black with $\varepsilon = 0.9$. From Figure 20.9, $T_{\text{air}} - T_{\text{sky}}$ is approximately 10°F for Los Angeles. From weather data for LA airport, (PSDH, 1984; ASHRAE, 1989), the July average temperature is 69°F with a range of 15°F. Assuming night temperatures vary from the average (69°F) down to half the daily range (15/2), then the average nighttime temperature is chosen as 69 – (1/2)(15/2) = 65°F. So, T_{sky} = 65 – 10 = 55°F. From Eq. (20.14),

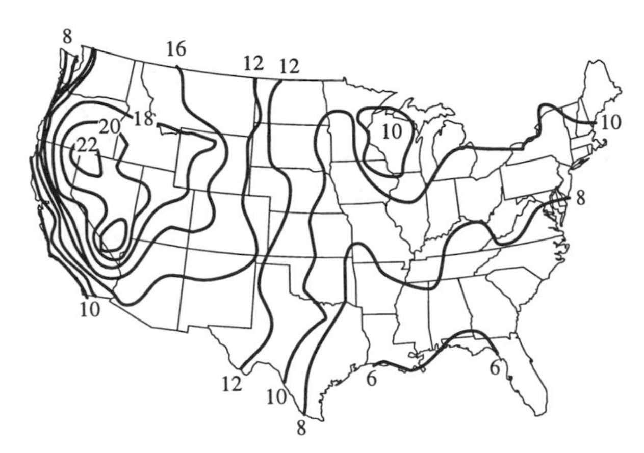

FIGURE 20.9 Average monthly sky temperature depression ($T_{air} - T_{sky}$) for July in °F. *Source:* Adapted from Martin and Berdahl, 1984.

$$Q_r = 0.9\left(1.714\times10^{-9}\right)\left[\left(460+80\right)^4-\left(460+55\right)^4\right]$$
$$= 22.7 \text{ Btu/h ft}^2$$

For a 10-hour night and 12 m^2 (129 ft^2) roof area,

$$\text{Total radiative cooling} = 22.7\ (10)(129)$$
$$= 29{,}300 \text{ Btu}$$

Note: This does not include the convective cooling possible which can be approximated, for still air, (ASHRAE, 1989) by

$$\text{Total } Q_{conv} = hA\left(T_{roof} - T_{air}\right)(\text{Time})$$
$$= 5(129)(80-55)(10)$$
$$= 161{,}000 \text{ Btu}$$

convection is usually the more dominant mode of nighttime cooling.

20.3 Daylighting Design Fundamentals

In the last century, electric lighting was considered an alternative technology to **daylighting**. Today the situation is reversed, primarily due to the economics of energy use and conservation. However, there are good physiological and psychological reasons for using daylight as an illuminant. The quality of daylight, composed of **sunlight** and **skylight**, matches the human eye's response, thus permitting lower light levels for task comfort, better color rendering, and clearer object discrimination (Robbins, 1986).

Determination of the daylighting available at a given location in a building space at a given time is important to evaluate the reduction possible in electric lighting and the associated impact on heating and cooling loads. Daylight provides about 110 lm/W, fluorescent lamps about 75 lm/W, and incandescent lamps about 20 lm/W; thus, daylighting causes only 1/2 to 1/5 the heating that electric lighting does, significantly reducing the building cooling load.

Aperture controls such as blinds and drapes are used to moderate the amount of daylight entering the space, as are the architectural features of the building itself (glazing type, area, and orientation; overhangs and wingwalls; lightshelves; etc.). Electric lighting dimming controls are used to adjust the electric light level based on the quantity of the daylighting. With these two types of controls (aperture and lighting), the electrical lighting and cooling energy use and demand, as well as cooling system sizing, can be reduced. However, the determination of the daylighting position and time **illuminance** value within the space is required before energy usage and demand reduction calculations can be made.

Daylighting is provided through a variety of glazing features which can be grouped as **sidelighting** (light enters via the side of the space) and **toplighting** (light enters from the ceiling area). Figure 20.10 presents several architectural forms producing sidelighting and toplighting, with the dashed lines representing illuminance distributions within the space. The calculation of workplane illuminance depends on whether sidelighting and/or toplighting features are used, and the combined illuminance values are additive.

Sun–Window Geometry

The solar illuminance on a vertical or horizontal window depends on the position of the sun relative to that window. In the method described here, the sun and sky illuminance values are determined using the sun's altitude angle (α) and the sun–window azimuth angle difference (ϕ). These angles need to be determined for the particular time of day, day of year, and window placement under investigation.

Solar Altitude Angle (α). The solar altitude angle is the angle swept out by a person's arm when pointing to the horizon directly below the sun and then raising the arm to point at the sun. The equation to calculate solar altitude, α, is

$$\sin\alpha = \cos L\ \cos\delta\ \cos H + \sin L\ \sin\delta \tag{20.15}$$

where

L = local latitude (degrees)
δ = earth-sun declination (degrees) given by

$$\delta = 23.45\ \sin\left[360\ (n-81)/365\right] \tag{20.16}$$

n = day number of the year
H = hour angle (degrees) given by

$$H = \frac{(12\text{ noon} - \text{time})(\text{in minutes})}{4};\quad (+\text{ morning},\ -\text{ afternoon}) \tag{20.17}$$

Sun–Window Azimuth Angle Difference (ϕ). The difference between the sun's azimuth and the window's azimuth needs to be calculated for vertical window illuminance. The window's azimuth angle, γ_w, is determined by which way it faces, as measured from south (east of south is positive, westward is negative). The solar azimuth angle, γ_s, is calculated:

$$\sin\gamma_s = \frac{\cos\delta\ \sin H}{\cos\alpha} \tag{20.18}$$

The sun–window azimuth angle difference, ϕ, is given by the absolute value of the difference between γ_s and γ_w:

$$\phi = |\gamma_s - \gamma_w| \tag{20.19}$$

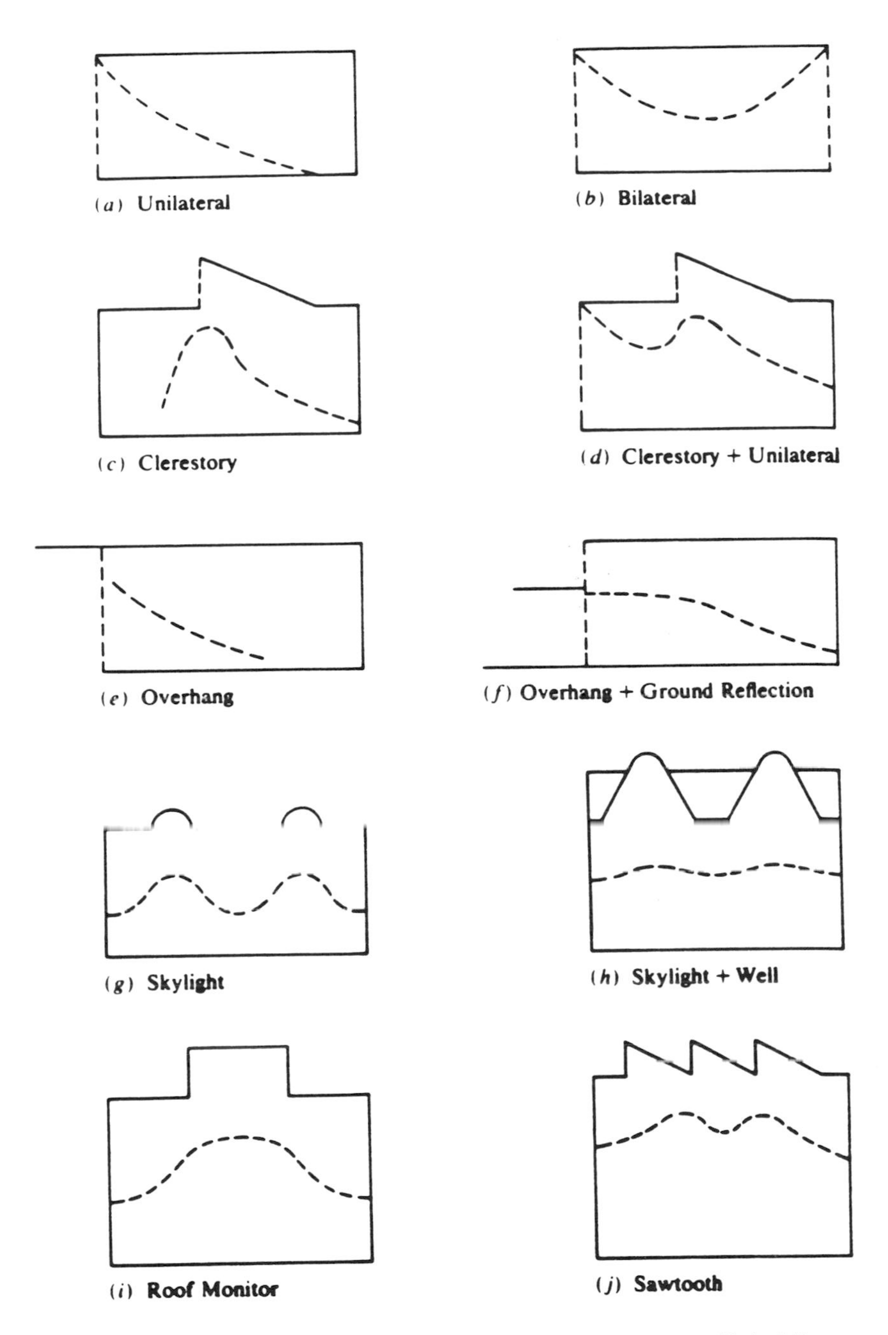

FIGURE 20.10 Example of sidelighting and toplighting architectural features (dashed lines represent illuminance distributions). *Source:* Murdoch, 1985.

Lumen Method of Sidelighting (Vertical Windows)

The lumen method of sidelighting calculates interior horizontal illuminance at three points, as shown in Figure 20.11, at the 30-inch (0.76 m) work plane on the room-and-window centerline. A vertical window is assumed to extend from 36 inches (0.91 m) above the floor to the ceiling. The method accounts for both direct and ground-reflected sunlight and skylight, so both horizontal and vertical illuminances from sun and sky are needed. The steps in the lumen method of sidelighting are presented next.

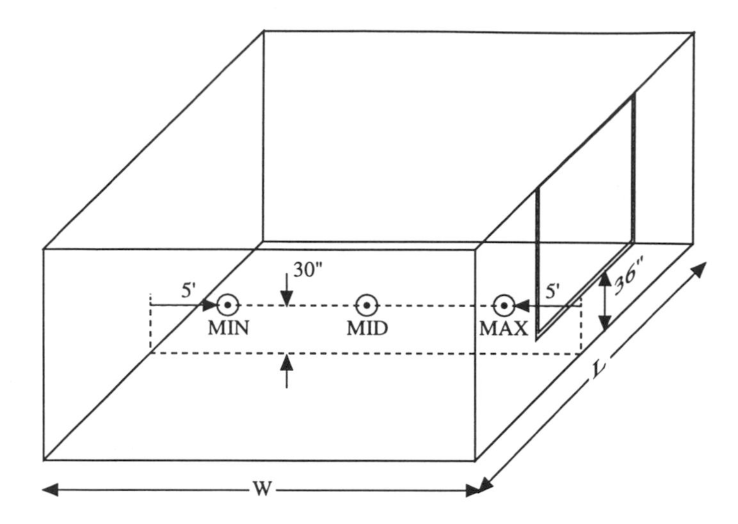

FIGURE 20.11 Location of illumination points within the room (along centerline of window) determined by lumen method of sidelighting..

The incident direct and ground-reflected window illuminance are normally found for both cloudy and clear days for representative days during the year and times during the day. Thus, the interior illumination due to sidelighting and skylighting can then be examined for effectiveness throughout the year.

Step 1: Incident Direct Sky and Sun Illuminances

The solar altitude and sun–window azimuth angle difference are calculated for the latitude, date, and time. Using these two angles, the total illuminance on the window (E_{sw}) can be determined from the direct sun illuminance (E_{uw}) plus the direct sky illuminance (E_{kw}) from the appropriate graph in Figure 20.12.

Step 2: Incident Ground-Reflected Illuminance

The sun illuminance on the ground (E_{ug}), plus the overcast or clear sky illuminance (E_{kg}) on the ground, make up the total horizontal illuminance on the ground surface (E_{sg}). A fraction of the ground surface illuminance is then considered diffusely reflected onto the vertical window surface (E_{gw}), where *gw* indicates from the ground to the window.

The horizontal ground illuminances can be determined using Figure 20.13, where the clear sky plus direct sun case and the overcast sky case are functions of solar altitude. The fraction of the ground illuminance diffusely reflected onto the window depends on the reflectivity (ρ) of the ground surface (see Table 20.6) and the window-to-ground surface geometry.

If the ground surface is considered uniformly reflective from the window outward, then

$$E_{gw} = \frac{\rho \, E_{sg}}{2} \tag{20.20}$$

A more complicated ground-reflection case is illustrated in Figure 20.14, with multiple "strips" of differently reflecting ground being handled using the angles to the window, where a strip's illuminance is calculated,

$$E_{gw(\text{strip})} = \frac{\rho_{\text{strip}} E_{sg}}{2} \left(\cos\,\theta_1 - \cos\,\theta_2\right) \tag{20.21}$$

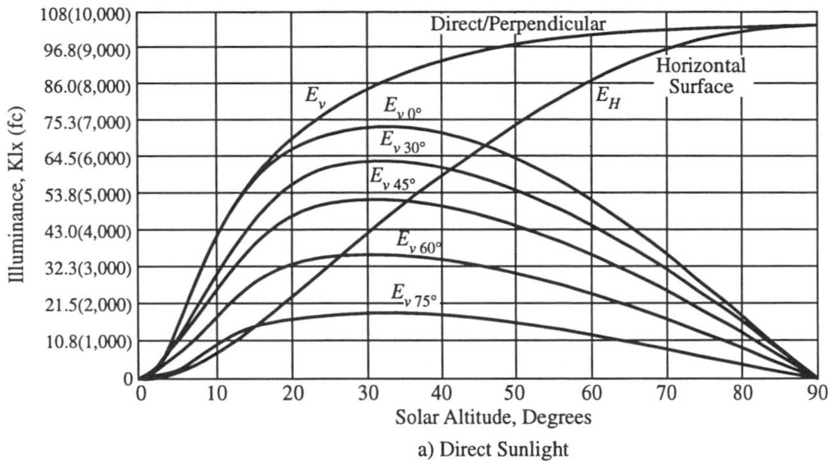

a) Direct Sunlight

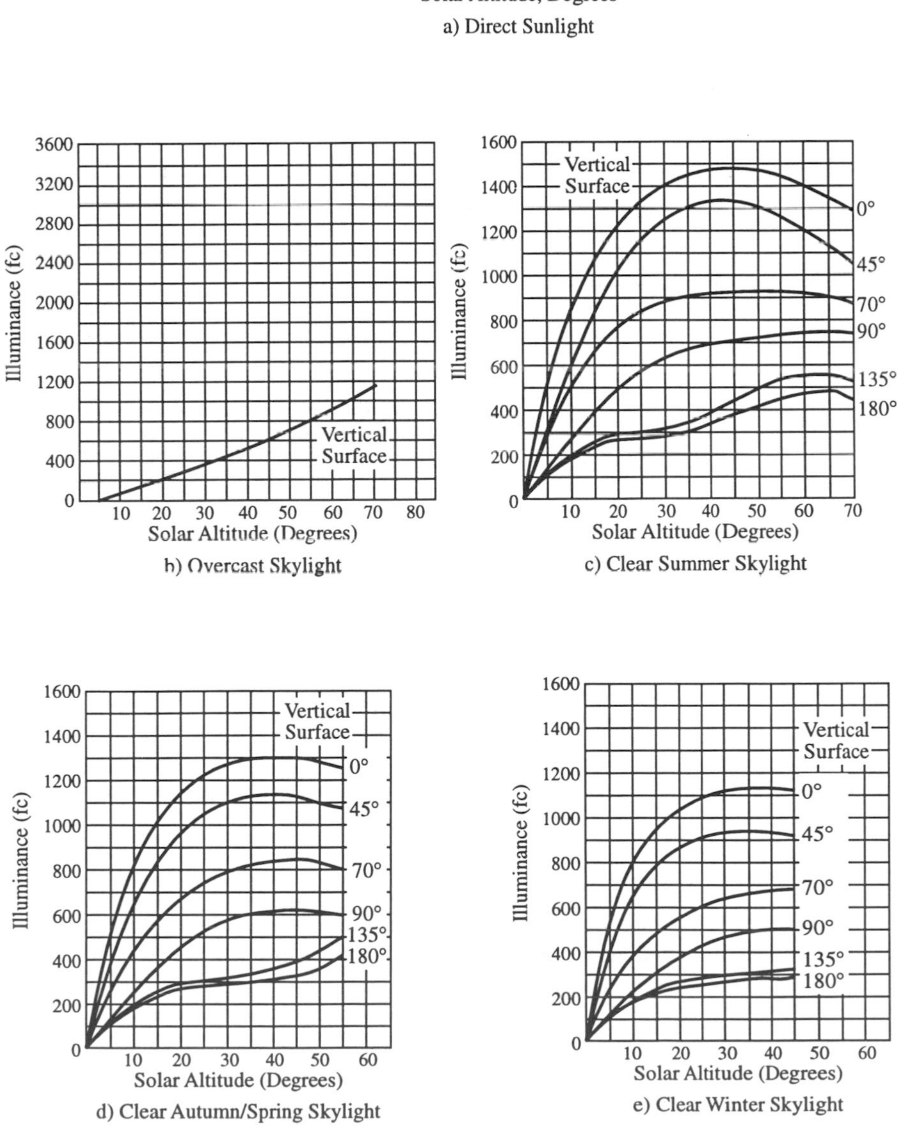

b) Overcast Skylight

c) Clear Summer Skylight

d) Clear Autumn/Spring Skylight

e) Clear Winter Skylight

FIGURE 20.12 Vertical illuminance from (a) direct sunlight and (b–e) skylight, for various sun–window azimuth angle differences. *Source:* IES, 1979.

TABLE 20.6 Ground Reflectivities

Material	ρ(%)
Cement	27
Concrete	20–40
Asphalt	7–14
Earth	10
Grass	6–20
Vegetation	25
Snow	70
Red brick	30
Gravel	15
White paint	55–75

Source: Murdoch, 1985.

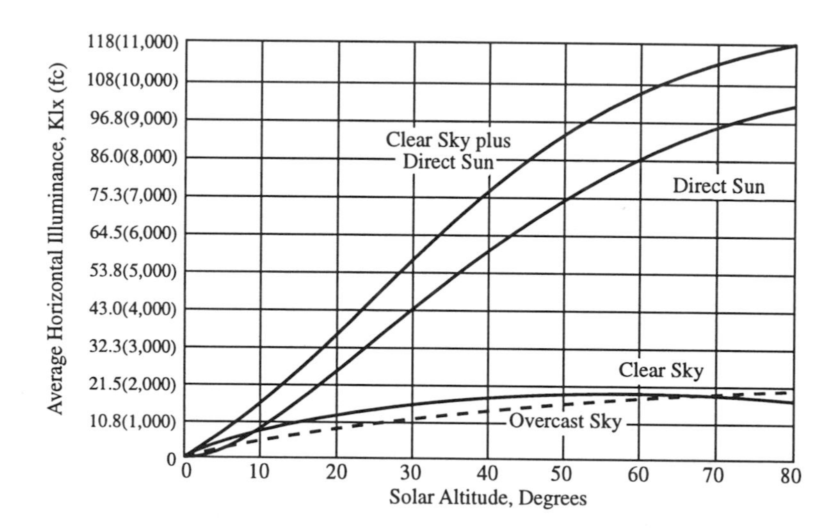

FIGURE 20.13 Horizontal illuminance for overcast sky, clear sky, direct sun, and clear sky plus direct sun. *Source:* Murdoch, 1985.

and the total reflected is the sum of the strip illuminances:

$$E_{gw} = \frac{E_{sg}}{2}\left[\rho_1\left(\cos 0 - \cos \theta_1\right) + \rho_2\left(\cos \theta_1 - \cos \theta_2\right) + \cdots + \rho_n\left(\cos \theta_{n-1} - \cos 90\right)\right] \quad \textbf{(20.22)}$$

Step 3: Luminous Flux Entering Space

The direct sky–sun and ground-reflected **luminous fluxes** entering the building are attenuated by the transmissivity of the window. Table 20.7 presents the transmittance fraction (τ) of several window glasses. The fluxes entering the space are

$$\phi_{sw} = E_{sw}\tau\ A_w \quad \textbf{(20.23)}$$

$$\phi_{gw} = E_{gw}\tau\ A_w$$

where A_w is the glass area.

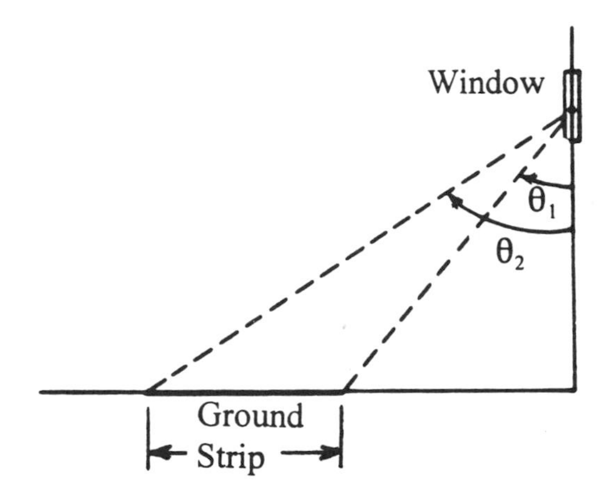

FIGURE 20.14 Geometry for ground "strips." *Source:* Murdoch, 1985.

TABLE 20.7 Glass Transmittances

Glass	Thickness (in.)	τ(%)
Clear	$\frac{1}{8}$	89
Clear	$\frac{3}{16}$	88
Clear	$\frac{1}{4}$	87
Clear	$\frac{5}{16}$	86
Grey	$\frac{1}{8}$	61
Grey	$\frac{3}{16}$	51
Grey	$\frac{1}{4}$	44
Grey	$\frac{5}{16}$	35
Bronze	$\frac{1}{8}$	68
Bronze	$\frac{3}{16}$	59
Bronze	$\frac{1}{4}$	52
Bronze	$\frac{5}{16}$	44
Thermopane	$\frac{1}{8}$	80
Thermopane	$\frac{3}{16}$	79
Thermopane	$\frac{1}{4}$	77

Source: Murdoch, 1985.

Step 4: Light Loss Factor

The light loss factor (K_m) accounts for the depreciation of luminous flux due to dirt on the window (WDD, window dirt depreciation) and room surface dirt (RSDD, room surface dirt depreciation). WDD depends on how often the window is cleaned, but a 6-month average for offices is 0.83 and for factories is 0.71 (Murdoch, 1985).

The RSDD is a more complex calculation involving time between cleanings, the direct–indirect flux distribution, and room proportions. However, for rooms cleaned regularly, RSDD is around 0.94 and for once-a-year-cleaned dirty rooms, the RSDD would be around 0.84.

The light loss factor is the product of the preceding two fractions:

$$K_m = (\text{WDD})\,(\text{RSDD}) \tag{20.24}$$

Step 5: Work-Plane Illuminances

As discussed earlier, Figure 20.11 illustrates the location of the work-plane illuminances determined with this lumen method of sidelighting. The three illuminances (max, mid, min) are determined using two coefficients of utilization, the *C* factor and *K* factor. The *C* factor depends on room length and width and wall reflectance. The K factor depends on ceiling–floor height, room width, and wall reflectance. Table 20.8 presents C and K values for the three incoming fluxes: sun plus clear sky, overcast sky, and ground reflected. Assumed ceiling and floor reflectances are given for this case with no window controls (shades, blinds, overhangs, etc.). These further window control complexities can be found in IES (1984), LOF (1976), and others. A reflectance of 70% represents light-colored walls, with 30% representing darker walls.

The work-plane luminous fluxes are calculated using

$$E_{sp} = \phi_{sw} C_s K_s K_m \qquad (20.25)$$

$$E_{gp} = \phi_{gw} C_g K_g K_m$$

where the sp and gp refer to the sky-to-work-plane and ground-to-work-plane illuminances.

Example 20.6. Determine the clear sky illuminances for a 30-ft-long, 30-ft-wide, 10-ft-high room with a 20-ft-long window with a 3-ft sill. The window faces 10°E of South, the building is at 32°N latitude, and it is January 15 at 2 P.M. The ground cover outside is grass, the glass is ¼-inch clear, and the walls light colored.

Solution: Following the steps in the "sidelighting" method:

Step 1: With $L = 32$, $n = 15$, $H = (12–14)60/4 = –30$,

$$\delta = 23.45 \ \sin\left[360(15-81)/365\right] = -21.3°$$

Then, Eq. (20.14) yields $\alpha = 41.7°$, Eq. (20.17) yields $\gamma_s = -38.7°$, and Eq. (20.18) yields $\phi = |-38.7 - (+10)| = 48.7°$.

From Figure (20.12) with $\infty = 41.7°$ and $\phi = 48.7°$

(a) For clear sky (winter, no sun): $E_{kw} = 875$ fc.

(b) For direct sun: $E_{uw} = 4{,}100$ fc.

(c) Total clear sky plus direct: $E_{sw} = 4{,}975$ fc.

(*Note*: A high E_{uw} value probably indicates a glare situation!)

Step 2: Horizontal illuminances from Figure 20.13: $E_{sg} = 7{,}400$ fc

Then Eq. (20.19) yields, with $\rho_{grass} = 0.06$, $E_{gw} = 222$ fc

Step 3: From Eq. (20.22), with $\tau = 0.87$ and $A_w = 140$ ft^2,

$$\phi_{sw} = 4{,}975\ (0.87)(140) = 605{,}955 \text{ lm}$$

$$\phi_{gw} = 222\ (0.87)(140) = 27{,}040 \text{ lm}$$

Step 4: For a clean office room:

$$K_m = (0.83)(0.94) = 0.78$$

Step 5: From Table 20.8, for 30' width, 30' length, 10' ceiling, and wall reflectivity 70%,

(a) Clear sky:

C_s, max = 0.0137; K_s, max = 0.125

C_s, mid = 0.0062; K_s, mid = 0.110

C_s, min = 0.0047; K_s, min = 0.107

TABLE 20.8a C and K Factors For No Window Controls For: (a) Overcast Sky

Illumination by Overcast Sky
C: Coefficient of Utilization

Room Length		20'		30'		40'	
Wall Reflectance		70%	30%	70%	30%	70%	30%
	Room Width						
Max	20'	.0276	.0251	.0191	.0173	.0143	.0137
	30'	.0272	.0248	.0188	.0172	.0137	.0131
	40'	.0269	.0246	.0182	.0171	.0133	.0130
Mid	20'	.0159	.0177	.0101	.0087	.0081	.0071
	30'	.0058	.0050	.0054	.0040	.0034	.0033
	40'	.0039	.0027	.0030	.0023	.0022	.0019
Min	20'	.0087	.0053	.0063	.0043	.0050	.0037
	30'	.0032	.0019	.0029	.0017	.0020	.0014
	40'	.0019	.0009	.0016	.0009	.0012	.0008

K: Coefficient of Utilization

Ceiling Height		8'		10'		12'		14'	
Wall Reflectance		70%	30%	70%	30%	70%	30%	70%	30%
	Room Width								
Max	20'	.125	.129	.121	.123	.111	.111	.0991	.0973
	30'	.122	.131	.122	.121	.111	.111	.0945	.0973
	40'	.145	.133	.131	.126	.111	.111	.0973	.0982
Mid	20'	.0908	.0982	.107	.115	.111	.111	.105	.122
	30'	.156	.102	.0939	.113	.111	.111	.121	.134
	40'	.106	.0948	.123	.107	.111	.111	.135	.127
Min	20'	.0908	.102	.0951	.114	.111	.111	.118	.134
	30'	.0924	.119	.101	.114	.111	.111	.125	.126
	40'	.111	.0926	.125	.109	.111	.111	.133	.130

Source: IES, 1979.

TABLE 20.8b C and K Factors For No Window Controls For: (b) Clear Sky

Illumination by Clear Sky
C: Coefficient of Utilization

Room Length		20'		30'		40'	
Wall Reflectance		70%	30%	70%	30%	70%	30%
	Room Width						
Max	20'	.0206	.0173	.0143	.0123	.0110	.0098
	30'	.0203	.0173	.0137	.0120	.0098	.0092
	40'	.0200	.0168	.0131	.0119	.0096	.0091
Mid	20'	.0153	.0104	.0100	.0079	.0083	.0067
	30'	.0082	.0054	.0062	.0043	.0046	.0037
	40'	.0052	.0032	.0040	.0028	.0029	.0023
Min	20'	.0106	.0060	.0079	.0049	.0067	.0043
	30'	.0054	.0028	.0047	.0023	.0032	.0021
	40'	.0031	.0014	.0027	.0013	.0021	.0012

K: Coefficient of Utilization

Ceiling Height		8'		10'		12'		14'	
Wall Reflectance		70%	30%	70%	30%	70%	30%	70%	30%
	Room Width								
Max	20'	.145	.155	.129	.132	.111	.111	.101	.0982
	30'	.141	.149	.125	.130	.111	.111	.0954	.101
	40'	.157	.157	.135	.134	.111	.111	.0964	.0991
Mid	20'	.110	.128	.116	.126	.111	.111	.103	.108
	30'	.106	.125	.110	.129	.111	.111	.112	.120
	40'	.117	.118	.122	.118	.111	.111	.123	.122
Min	20'	.105	.129	.112	.130	.111	.111	.111	.116
	30'	.0994	.144	.107	.126	.111	.111	.107	.124
	40'	.119	.116	.130	.118	.111	.111	.120	.118

Source: IES, 1979.

TABLE 20.8c C and K Factors For No Window Controls For: (c) Groun Illumination (Ceiling Reflectance, 80%; Floor Reflectance, 30%)

Ground Illumination

C: Coefficient of Utilization

Room Length		20'		30'		40'	
Wall Reflectance		70%	30%	70%	30%	70%	30%
	Room Width						
Max	20'	.0147	.0112	.0102	.0088	.0081	.0071
Max	30'	.0141	.0012	.0098	.0088	.0077	.0070
Max	40'	.0137	.0112	.0093	.0086	.0072	.0069
Mid	20'	.0128	.0090	.0094	.0071	.0073	.0060
Mid	30'	.0083	.0057	.0062	.0048	.0050	.0041
Mid	40'	.0055	.0037	.0044	.0033	.0042	.0026
Min	20'	.0106	.0071	.0082	.0054	.0067	.0044
Min	30'	.0051	.0026	.0041	.0023	.0033	.0021
Min	40'	.0029	.0018	.0026	.0012	.0022	.0011

K: Coefficient of Utilization

Ceiling Height		8'		10'		12'		14'	
Wall Reflectance		70%	30%	70%	30%	70%	30%	70%	30%
	Room Width								
Max	20'	.124	.206	.140	.135	.111	.111	.0909	.0859
Max	30'	.182	.188	.140	.143	.111	.111	.0918	.0878
Max	40'	.124	.182	.140	.142	.111	.111	.0936	.0879
Mid	20'	.123	.145	.122	.129	.111	.111	.100	.0945
Mid	30'	.0966	.104	.107	.112	.111	.111	.110	.105
Mid	40'	.0790	.0786	.0999	.106	.111	.111	.118	.118
Min	20'	.0994	.108	.110	.114	.111	.111	.107	.104
Min	30'	.0816	.0822	.0984	.105	.111	.111	.121	.116
Min	40'	.0700	.0656	.0946	.0986	.111	.111	.125	.132

Source: IES, 1979.

(b) Ground reflected:

C_g, max = 0.0098; K_g, max = 0.140

C_g, mid = 0.0062; K_g mid = 0.107

C_g, min = 0.0041; K_g min = 0.0984

Then using Eq. (20.24),

$$E_{sp}, \text{ max} = 605{,}955\ (0.0137)(0.125)\ (0.78) = 809 \text{ fc}$$

$$E_{sp}, \text{ mid} = 605{,}955\ (0.0062)(0.110)\ (0.78) = 322 \text{ fc}$$

$$E_{sp}, \text{ min} = 605{,}955\ (0.0047)(0.107)\ (0.78) = 238 \text{ fc}$$

$$E_{gp}, \text{ max} = 27{,}040\ (0.0098)(0.140)\ (0.78) = 29 \text{ fc}$$

$$E_{gp}, \text{ mid} = 27{,}040\ (0.0062)(0.107)\ (0.78) = 14 \text{ fc}$$

$$E_{gp}, \text{ min} = 27{,}040\ (0.0041)(0.984)\ (0.78) = 9 \text{ fc}$$

Thus,

$$E_{\text{max}} = 838 \text{ fc}$$

$$E_{\text{mid}} = 336 \text{ fc}$$

$$E_{\text{min}} = 247 \text{ fc}$$

Lumen Method of Skylighting

The lumen method of skylighting calculates the average illuminance at the interior work plane provided by horizontal skylights mounted on the roof. The procedure for skylighting is generally the same as that described earlier for sidelighting. As with windows, the illuminance from both overcast sky and clear sky plus sun cases are determined for days in different seasons and times of the day, and a judgment is then made as to the number and size of skylights and any controls needed.

The procedure is presented in four steps: (1) finding the horizontal illuminance on the outside of the skylight; (2) calculating the effective transmittance through the skylight and its well; (3) figuring the light loss factor and the utilization coefficient; and (4) calculating illuminance on the work plane.

Step 1: Horizontal Sky and Sun Illuminances

The illuminance value for an overcast sky or a clear sky plus sun situation can be determined from Figure 20.13 knowing only the solar altitude.

Step 2: Net Skylight Transmittance

The transmittance of the skylight is determined by the transmittance of the skylight cover(s), the reflective efficiency of the skylight well, the net-to-gross skylight area, and the transmittance of any light-control devices (lenses, louvers, etc.).

The transmittance for several flat-sheet plastic materials used in skylight domes is presented in Table 20.9. To get the effective dome transmittance (T_D) from the flat-plate transmittance (T_F) value (AAMA, 1977), use

$$T_D = 1.25\ T_F(1.18 - 0.416\ T_F) \tag{20.26}$$

TABLE 20.9 Flat-Plate Plastic Material Transmittance for Skylights

Type	Thickness (in.)	Transmittance (%)
Transparent	$\frac{1}{8}-\frac{3}{16}$	92
Dense translucent	$\frac{1}{8}$	32
Dense translucent	$\frac{3}{16}$	24
Medium translucent	$\frac{1}{8}$	56
Medium translucent	$\frac{3}{16}$	52
Light translucent	$\frac{1}{8}$	72
Light translucent	$\frac{3}{16}$	68

Source: Murdoch, 1985.

If a double-domed skylight is used, then the single-dome transmittances are combined as follows (Pierson, 1962):

$$T_D = \frac{T_1\ T_2}{T_1 + T_2 - T_1\ T_2} \tag{20.27}$$

If the diffuse and direct transmittances for solar radiation are available for the skylight glazing material, it is possible to follow this procedure and determine diffuse and direct dome transmittances separately. However, this difference is usually not a significant factor in the overall calculations.

The efficiency of the skylight well (N_w) is the fraction of the luminous flux from the dome that enters the room from the well. The well index (WI) is a geometric index given by

$$\mathrm{WI} = \frac{h(w+l)}{2wl} \tag{20.28}$$

and is used with the well-wall reflectance value in Figure 20.15 to determine N_w.

With T_D and N_w determined, the net skylight transmittance for the skylight and well is given by:

$$T_n = T_D N_W R_A T_C \tag{20.29}$$

where R_A = ratio of net to gross skylight areas
T_C = transmittance of any light-controlling devices

Step 3: Light Loss Factor and Utilization Coefficient

The light loss factor (K_m) is again defined as the product of the room surface dirt depreciation (RSDD) and the skylight direct depreciation (SDD) fractions, similar to Eq. (20.24). Following the sidelighting case, the RSDD value for clean rooms is around 0.94, and 0.84 for dirty rooms. Without specific data indicating otherwise, the SDD fraction is often taken as 0.75 for office buildings and 0.65 for industrial areas.

The fraction of the luminous flux on the skylight that reaches the work plane (K_u) is the product of the net transmittance (T_n) and the room coefficient of utilization (RCU). Dietz et al. (1981) developed RCU equations for office and warehouse interiors with ceiling, wall, and floor reflectances of 75%, 50%, and 30%, and 50%, 30%, and 20%, respectively.

$$\mathrm{RCU} = \frac{1}{1 + A(\mathrm{RCR})^B}, \quad \text{if RCR} < 8 \tag{20.30}$$

where $A = 0.0288$ and $B = 1.560$ (offices)
$A = 0.0995$ and $B = 1.087$ (warehouses)

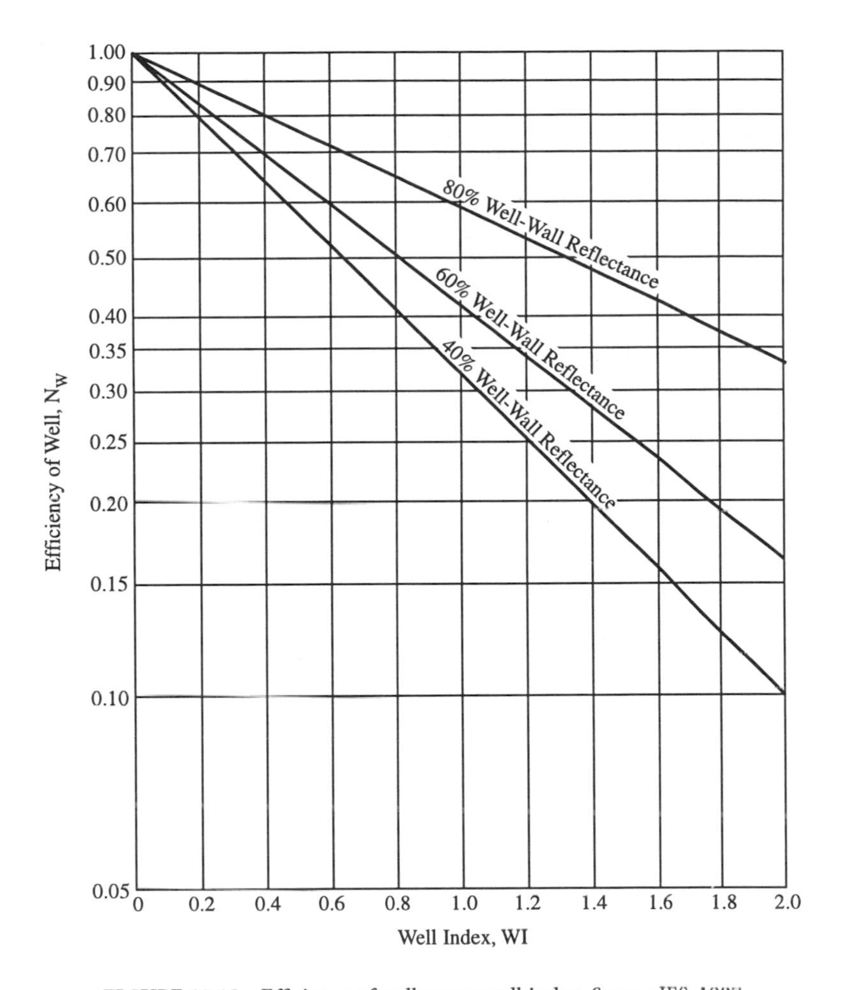

FIGURE 20.15 Efficiency of well versus well index. *Source:* IES, 1987.

and room cavity ratio (RCR) is given by

$$\text{RCR} = \frac{5h_c(l+w)}{1w} \tag{20.31}$$

with h_c the ceiling height above the work plane.

The RCU is then multiplied by the previously determined T_n to give the fraction through the skylight and room:

$$K_u = T_n \, (\text{RCU}) \tag{20.32}$$

Step 4: Work-Plane Illuminance

The illuminance at the work plane (E_{TWP}) is given by

$$E_{TWP} = E_H \left(\frac{A_T}{A_{WP}} \right) K_u K_m \tag{20.33}$$

where E_H is the horizontal overcast or clear sky plus sun illuminance from Step 1, A_T is total gross area of the skylights (number of skylights times skylight gross), and A_{WP} is the work-plane area (generally room length times width). Note that in Eq. (20.33), it is also possible to fix the E_{TWP} at some desired value and determine the skylight area required.

Rules of thumb for skylight placement for uniform illumination include 4 to 8% of roof area and spacing less than 1½ times ceiling-to-work-plane distance between skylights (Murdoch, 1984).

Example 20.7. Determine the work-plane "clear sky" illuminance for a 30' × 30' × 10' office with 75% ceiling, 50% wall, and 30% floor reflectance with four 4' × 4' double-domed skylights at 2 P.M. on January 15th. The skylight well is 1' deep at 32° latitude with 60% reflectance walls, and the outer and inner dome plastic transmittances are 0.85 and 0.45, respectively. The net skylight area is 90%.

Solution: Follow the four steps in the lumen method for skylighting.

Step 1: Use Figure 20.13 with the solar altitude of 41.7° (previous example) for the clear sky plus sun curve to get horizontal illuminance:

$$E_H = 7{,}400 \text{ fc}$$

Step 2: To get net transmittance (T_n), use Eq. (20.26) to get domed transmittances,

$$T_1 = 1.25(0.85)\left[1.18 - 0.416(0.85)\right] = 0.89$$

$$T_2\ (T_F = 0.45) = 0.56$$

and Eq. (20.26) to get total dome transmittance

$$T_D = \frac{(0.89)\ (0.56)}{(0.89) + (0.56) - (0.89)\ (0.56)} = 0.52$$

To get well efficiency, use WI = 0.25 from Eq. (20.28) with 60% wall reflectance in Figure 20.15 to give N_w = 0.80. With R_A = 0.90, use Eq. (20.29) to calculate net transmittance: T_n = (0.52)(0.80)(90)(1.0) = 0.37.

Step 3: The light loss factor is assumed to be from "typical" values in Eq. (20.23): K_m = (0.94)(0.75) = 0.70. The room utilization coefficient is determined using Eqs. (20.29) and (20.30):

$$\text{RCR} = \frac{5(7.5)\ (30 + 30)}{(30)\ (30)} = 2.5$$

$$\text{RCU} = \left[1 + 0.0288(2.5)^{1.560}\right]^{-1} = 0.89$$

and Eq. (20.32) yields K_u = (0.37)(0.89) = 0.33.

Step 4: The work-plane illuminance is given by Eq. (20.33):

$$E_{TWP} = 7{,}400\left[\frac{4(16)}{30(30)}\right]\ 0.33(0.70)$$

$$E_{TWP} = 122 \text{ fc}$$

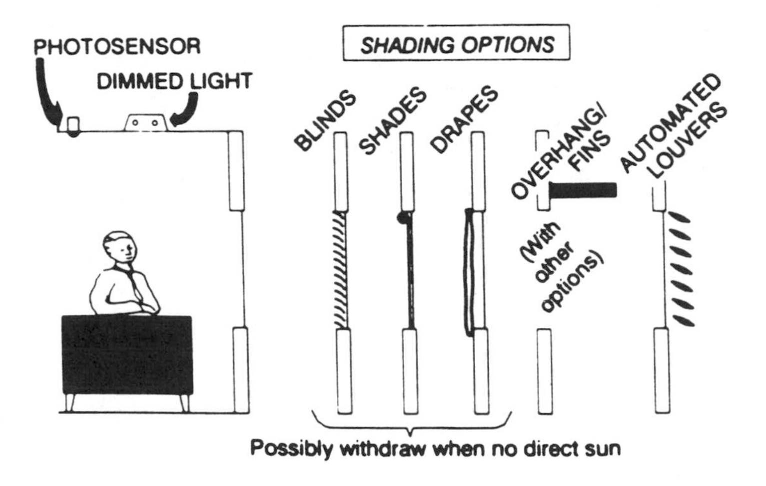

FIGURE 20.16 Daylighting system controls. *Source:* Rundquist, 1991.

Daylighting Controls and Economics

Controls, both aperture and lighting, directly affect the efficacy of the daylighting system. As shown in Figure 20.16, aperture controls can be architectural (overhangs, light shelves, etc.) and/or window shading devices (blinds, automated louvers, etc.). The aperture controls generally moderate the sunlight entering the space to maximize/minimize solar thermal gain, permit the proper amount of light for visibility, and prevent glare and beam radiation onto the workplane. Photosensor control of electric lighting allows the dimming (or shutting off) of the lights in proportion to the amount of available daylighting illuminance.

Dimming or shutting off electric lighting when daylighting illuminance levels are sufficient reduces both the lighting electrical load and the cooling load, and may even reduce the required installed cooling capacity for the building. In most cases, increasing the solar gain for daylighting purposes, with daylighting controls, saves more in electrical lighting energy and the cooling energy associated with the lighting than is incurred with the added solar gain (Rundquist, 1991). The net change of space heating or cooling load can be calculated using conventional methods once the daylighting solar gain and its associated lighting energy reduction have been determined.

In determining the annual energy savings from daylighting (E_T), the annual lighting energy saved from daylighting (E_L) is added with the reduction in cooling system energy (ΔE_C) and with the negative of the heating system energy increase (ΔE_H):

$$E_T = E_L + \Delta E_C - \Delta E_H \tag{20.34}$$

A simple approach to estimating the heating and cooling energy changes associated with the lighting energy reduction is by using the fraction of the year associated with the cooling or heating season (f_c, f_H) and the seasonal COP of the cooling or heating equipment. Thus, Eq. (20.34) can be expressed as

$$\begin{aligned} E_T &= E_L + \frac{f_C E_L}{\text{COP}_C} - \frac{f_H E_L}{\text{COP}_H} \\ E_T &= E_L\left(1 + \frac{f_C}{\text{COP}_C} - \frac{f_H}{\text{COP}_H}\right) \end{aligned} \tag{20.35}$$

It should be noted that the increased solar gain due to daylighting has not been included here but would reduce summer savings and increase winter savings. If it is assumed that the increased wintertime daylighting solar gain approximately offsets the reduced lighting heat gain, then the last term in Eq. (20.35) becomes negligible.

To determine the annual lighting energy saved (E_L), calculations using the lumen method described earlier should be performed on a monthly basis for both clear and overcast days for the space under investigation. Monthly weather data for the site would then be used to prorate clear and overcast lighting energy demands monthly. Subtracting the calculated (controlled) daylighting illuminance from the design illuminance leaves the supplementary lighting needed, which determines the lighting energy required.

This procedure has been computerized and includes many details of controls, daylighting methods, weather, and heating and cooling load calculations. ASHRAE (1989) lists many of the methods and simulation techniques currently used with daylighting and its associated energy effects.

Economic Effects

The economic benefit of daylighting is directly tied to the reduction in lighting electrical energy operating costs. Also, lower cooling system operating costs are possible due to the reduced electrical lighting load. The reduction of lighting and cooling system power during peak demand periods could also beneficially affect demand charges.

The reduction of the design cooling load through the use of daylighting can also lead to the reduction of installed or first-cost cooling system dollars. Normally, an automatic lighting control system is required to take advantage of the reduced lighting/cooling effect, and the control system cost minus any cooling system cost savings must be expressed as a "net" first cost. A payback time for the lighting control system ("net" or not) can be calculated from the ratio of first costs to yearly operating savings. In some cases, these paybacks for daylighting controls have been found to be in the range of 1 to 5 years for office building spaces (Rundquist, 1991).

Example 20.8. A 30' × 20' space has a photosensor dimmer control with installed lighting density of 2.0 W/ft². The required workplace illuminance is 60 fc and the available daylighting illuminance is calculated as 40 fc on the summer peak afternoon. Determine the effect on the cooling system (adapted from Rundquist, 1991).

Solution: The lighting power reduction is (2.0 W/ft²)(30 × 20) ft² × (40 fc/60 fc) = 800 W. The space cooling load would also be reduced by this amount (assuming CLF = 1.0):

$$\frac{800 \text{ W} \times 3.413 \text{ Btuh/W}}{12{,}000 \text{ Btuh/ton}} = 0.23 \text{ ton}$$

Assuming 1½ tons nominally installed for 600 ft² of space at $2,200/ton, the 0.23-ton reduction is "worth" 0.23 × $2,200/ton = $506. The lighting controls cost about $1/ft² of controlled area, so the net installed first cost is

$$\text{Net first cost} = \$600 \text{ controls} - \$500 \text{ A/C savings} = \$100$$

Assuming the day-to-monthly-to-annual illuminance calculations gave a 30% reduction in annual lighting, the associated operating savings can be determined. Lighting energy savings are

$$E_L = 0.30 \times 2.0 \text{ W/ft}^2 \times 600 \text{ ft}^2 \times 2{,}500 \text{ h/yr} = 900 \text{ kWh}$$

Using Eq. (20.35) to also include cooling energy saved due to lighting reduction (with COP_c = 2.5, f_c = 0.5, and neglecting heating) gives

$$E_T = 900 \left(1 + \frac{0.5}{2.5}\right) = 1{,}080 \text{ kWh}$$

At \$0.10 kWh, the operating costs savings are \$0.10/kWh × 1,080 kWh = \$108.

Thus, the simple payback is approximately 1 year (100/108) for the "net" situation, and a little over 5½ years (600/108) against the "controls" cost alone. It should also be noted that the 800 W lighting electrical reduction at peak hours, with an associated cooling energy reduction of 800 W/2.5 COP = 320 W, provides a peak demand reduction for the space of 1.1 kW, which can be used as a "first-cost savings" to offset control system costs.

20.4 Defining Terms

Active system: A system employing a forced (pump or fan) convection heat transfer fluid flow.

Daylighting: The use of the sun's radiant energy for illumination of a building's interior space.

Hybrid system: A system with parallel passive and active flow systems or one using forced convection flow to distribute from thermal storage.

Illuminance: The density of luminous flux incident on a unit surface. Illuminance is calculated by dividing the luminous flux (in lumens) by the surface area (m^2, ft^2). Units are lux (lx) (lumens/m^2) in SI and footcandles (fc) (lumens/ft^2) in English systems.

Luminous flux: The time rate of flow of luminous energy (lumens). A lumen (lm) is the rate which luminous energy from a 1 candela (cd) intensity source is incident on a 1 m^2 surface 1 m from the source.

Passive cooling system: A system using natural energy flows to transfer heat to the environmental sinks (ground, air, and sky).

Passive heating system: A system in which the sun's radiant energy is converted to heat by absorption in the system, and the heat is distributed by naturally occurring processes.

Sidelighting: Daylighting by light entering through the wall/side of a space.

Skylight: The diffuse solar radiation from a clear or overcast sky, excluding the direct radiation from the sun.

Sunlight: The direct solar radiation from the sun.

Toplighting: Daylighting by light entering through the ceiling area of a space.

References

AAMA. 1977. Voluntary Standard Procedure for Calculating Skylight Annual Energy Balance, Architectural Aluminum Manufacturers Association Publication 1602.1.1977, Chicago, IL.

Anderson, B. and Wells, M. 1981. *Passive Solar Energy*, Brick House Publishing, Andover, MA.

Anderson, E.E. 1983. *Fundamentals of Solar Energy Conversion*, Addison-Wesley, Reading MA.

ASES, 1992. *Economics of Solar Energy Technologies*. Edited by R. Larson, F. Vignola, and R. West, American Solar Energy Society, December 1992.

ASHRAE Handbook. 1989. *Fundamentals*, I-P Edition, American Society of Heating, Refrigerating and Air-Conditioning Engineers, Atlanta, GA.

ASHRAE Handbook. 1991. *Heating, Ventilating, and Air-Conditioning Applications*, I-P Edition, American Society of Heating, Refrigerating and Air-Conditioning Engineers, Atlanta, GA.

Close, D.J., Dunkle, R.V., and Robeson, K.A. 1968. Design and Performance of a Thermal Storage Air Conditioning System. *Mech. and Chem. Eng. Trans.*, Institute Eng. Australia, MC4, 45.

Design Guidelines for the Passive Solar Home. 1980. Pacific Power and Light Pamphlet 6163 9/80 10M ML.

Dietz, P., Murdoch, J., Pokoski, J. and Boyle, J. 1981. A Skylight Energy Balance Analysis Procedure, *Journal of the Illum. Eng. Soc.*, October.

Duffie, J.A. and Beckman, W.A. 1991. *Solar Engineering of Thermal Processes*, 2nd ed., Wiley, New York.

Halacy, D.S. 1984. *Home Energy*, Rodale Press, Emmaus, PA.

Hay, H. and Yellott, J. 1969. Natural Air Conditioning with Roof Ponds and Movable Insulation. *ASHRAE Transactions* 75(1):165–77.

IES, 1987. *Lighting Handbook, Application Volume.* Illumination Engineering Society, New York.

Lebens, R.M. 1980. *Passive Solar Heating Design.* Applied Science Publishers Ltd., London, England.

LOF. 1976. *How to Predict Interior Daylight Illumination*, Libbey-Owens-Ford Co., Toledo, OH.

Martin, M. and Berdahl, P. 1984. Characteristics of Infrared Sky Radiation in the United States. *Solar Energy* 33(314):321–36.

Marlatt, W., Murray, C. and Squire, S. 1984. *Roof Pond Systems Energy Technology Engineering Center.* Rockwell International Report No. ETEC 6, April.

McQuiston, P.C. and Parker, J.D. 1994. *Heating, Ventilating, and Air Conditioning*, 4th ed., Wiley, New York.

Murdoch, J.B. 1985. *Illumination Engineering—From Edison's Lamp to the Laser.* Macmillan, New York.

NCDC, 1992. National Climactic Data Center. *Climatography of the U.S. #81.* Federal Building, Asheville, NC 28801.

Olgyay, A. and Olgyay, V. 1967. *Solar Control and Shading Devices.* Princeton University Press, Princeton, New Jersey.

PSDH, 1980. *Passive Solar Design Handbook.* Volume One: *Passive Solar Design Concepts*, DOE/CS-0127/1, March 1980. Prepared by Total Environmental Action, Inc. (B. Anderson, C. Michal, P. Temple, and D. Lewis); Volume Two: *Passive Solar Design Analysis*, DOE/CS-0127/2, January 1980. Prepared by Los Alamos Scientific Laboratory (J.D. Balcomb, D. Barley, R. McFarland, J. Perry, W. Wray and S. Noll). U.S. Department of Energy, Washington, DC.

PSDH, 1984. *Passive Solar Design Handbook.* Part One: Total Environmental Action, Inc., Part Two: Los Alamos Scientific Laboratory, Part Three: Los Alamos National Laboratory. Van Nostrand Reinhold, New York.

Phillips, R.O. 1980. Making the Best Use of Daylight in Buildings. In *Solar Energy Applications in the Design of Buildings*, ed. H.J. Cowen, pp. 95–120. Applied Science Publishers London, England.

Pierson, O. 1962. *Acrylics for the Architectural Control of Solar Energy.* Rohm and Haas, Philadelphia, PA.

Robbins, C.L. 1986. *Daylighting—Design and Analysis.* Van Nostrand, New York.

Rundquist, R.A. 1991. Daylighting Controls: Orphan of HVAC Design. *ASHRAE J.* November:30–34.

For Further Information

The most complete reference for passive system heating design is still the *Passive Solar Design Handbook*, all three parts. *Solar Today* magazine, published by the American Solar Energy Society, is the most available source for current practice designs and economics, as well as a source for passive system equipment suppliers.

The *ASHRAE Handbook of Fundamentals*, Chapter 28, is a general introduction to passive cooling techniques and calculations, with an emphasis on evaporative cooling. *Passive Solar Buildings* and *Passive Cooling*, published by MIT Press, contain a large variety of techniques and details concerning passive system designs and economics.

The Illumination Engineering Society's Lighting Handbook, Applications presents the basis for and details of daylighting and artificial lighting design techniques. However, most texts on illumination present simplified format daylighting procedures.

Passive Solar Design Strategies: Guidelines for Homebuilders (Passive Solar Industries Council, Washington, DC, 1989) presents a user-friendly approach to passive solar design.

21A

Solar Thermal Power and Industrial Heat: Solar Thermal Power

Lorin L. Vant-Hull
University of Houston

Introduction

Definition

A concentrating solar thermal power system, using a reflective (or sometimes refractive) optical system, concentrates the direct beam component of the solar radiation, allowing the solar radiation to be converted into sensible heat over a wide range of temperatures. This heat can be transformed directly into mechanical energy, via a heat engine or turbine, and used as shaft power or to generate electricity. Alternatively the heat can be used directly as process heat or in a cogeneration plant.

Advantages

Solar thermal power plants have numerous unique advantages. Fueled by the sun, they emit no carbon dioxide, nitrous oxides, volatile organic compounds, or other pollutants, and make no net contribution to the earth's heat burden. Comprising simple optical and thermal–mechanical systems, they require relatively low technology (compared to nuclear or photovoltaic power) and provide more jobs per dollar invested or MW capacity than most other alternatives (Horst, 1979).

Compared to wind or photovoltaic electric plants, solar thermal offers the advantage of easy and low-cost storage of the intermediate energy. Any excess heat over current demand can easily be stored (e.g., as a tank of hot fluid) and delivered on demand to a turbine or process, even at night. This capability can enhance the value of the system by transferring collected energy to times

0-8493-2514-5/96/$0.00+$.50

of peak demand, thus making the plant dispatchable and eligible for peaking credit on a utility grid. It can also allow the annual average capacity factor to be increased from about 25% (typical of run-of-the-sun operation) to 50% or more. As an added advantage, these thermal systems can easily be hybridized by simply adding a fossil fuel burner to heat either the heat transfer/storage fluid or the final working fluid on those occasions when the solar resource does not meet the demand. More complex hybrids involving combined cycles, topping cycles, or superheaters are also possible (Bohn, Williams, and Price, 1995; Noble et al., 1995).

Thermal electric generating systems are relatively standardized and well defined, and, through grid connections, can service a very large market. In contrast, each process heat application is unique and the solar collector must be on-site. Previous studies have shown that solar thermal can meet the requirements of process heat applications over a wide range of temperatures and plant sizes (Munjal, 1985). However, the arguments presented above have channeled current efforts into the development of solar thermal electric power stations, STEPS. There is a market for 5 to 25 kWe individual modules or small groups of such modules, but they must be exceedingly reliable or the time and cost required to travel to the sites for maintenance becomes prohibitive. Thus, in this chapter we will concentrate on systems in the 20 to 200 MWe range.

Environmental Issues

In contrast to conventional fossil plants, a STEPS plant produces no on-site pollution during operation. Although intensive in materials, the total pollution generated in production of a STEPS plant, from mining to installation, is less than that produced by an oil burning plant in 6 months of operation (Vant-Hull, 1992).

International agreements to limit worldwide CO_2 production to 1990 levels resulted from the Rio conference in 1992. Although enhanced efficiency of energy use will play an important role in meeting this goal, many public utility commissions and utilities recognize that solar thermal electricity offers a near-term, utility-compatible, CO_2-free alternative to fossil-fired plants, which produce about two pounds of CO_2 per kWh of electricity produced.

21A.1 Classification of Concentrating Solar Thermal Power Systems

Purpose of Concentrating

The efficiency of conversion of heat into electricity increases with increasing heat-source temperature for constant heat-sink temperature. (The thermodynamic basis for this is discussed in Chapter 2 and later in this chapter under "Heat Engines.") Therefore, it is desirable to make available the heat from a solar energy absorber at an elevated temperature. Although there are solar thermal electric generating systems operating at lower temperatures (e.g., solar ponds), it is generally considered that the heat should be available at a temperature of at least 300°C for efficient electric power generation. The only way that temperatures this high can be achieved with solar collectors is by concentrating the solar radiation onto an absorber with an area much smaller than that of the beam receptor. Concentrating also requires tracking of the sun. Concentrating solar collectors are considered in this chapter.

General Requirements for Concentrating Systems

Figure 21A.1 is a schematic diagram of a solar thermal electric power system. This generalized system consists of five subsystems:

Concentrator
Receiver

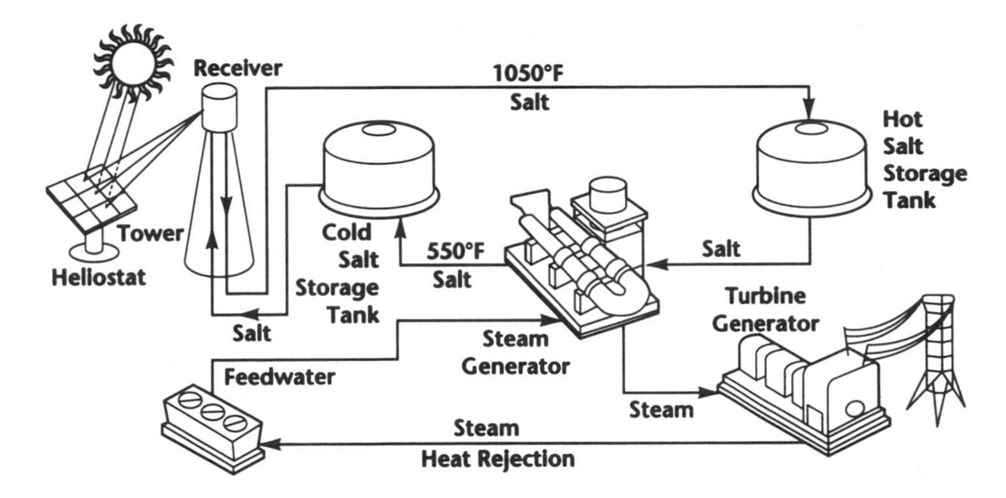

FIGURE 21A.1 Schematic diagram of a solar thermal electric power plant illustrating the basic subsystems and their configuration. The specific system illustrated is a molten salt central receiver system. *Source*: Burch, B.D., 1993. *Solar Thermal Electric, Five-Year Program Plan, FY 1993 Through 1997.* DOE, Washington, DC.

Storage
Heat engine
Electricity generation and transmission

The principles, types, and general design methods for concentrators and receivers—the uniquely solar aspects of the system—are discussed here. Heat storage is covered in Chapter 16, and not here, except as it applies to specific system design. Heat engine efficiency is included here, as well as being developed in Chapter 2. The electrical subsystem does not differ in essence from that of any electric power generation process, so it is not discussed here.

Practical, cost-effective concentrating collectors/thermal engines have a number of general requirements:

- Good direct beam sunlight with little cloud cover or haze
- Avoid refractive optics
- Good quality specular reflectors
- Accurate (1 mrad or better) solar tracking
- Total added beam dispersion comparable to the angle subtended by the sun
- Principles of **concentrator optics** rather than diffraction-limited imaging optics used in design
- Use of low-cost mass-produced components
- Tracking module size determined by economic design considerations, typically over 50 m^2
- Reliable, high-temperature, receiver/engine designed for frequent (at least daily) thermal cycles
- Solar qualification of all off-the-shelf components
- Elimination of, or extreme care with, single-point-failure elements (e.g., provide parallel computers on uninterruptable power to compute sun location and tracking angles)

Classification

There are three primary classification schemes for concentrating solar systems: one based on the location of the receiver, one on the tracking mode employed, and the third on the receiver configuration. They are commonly used together.

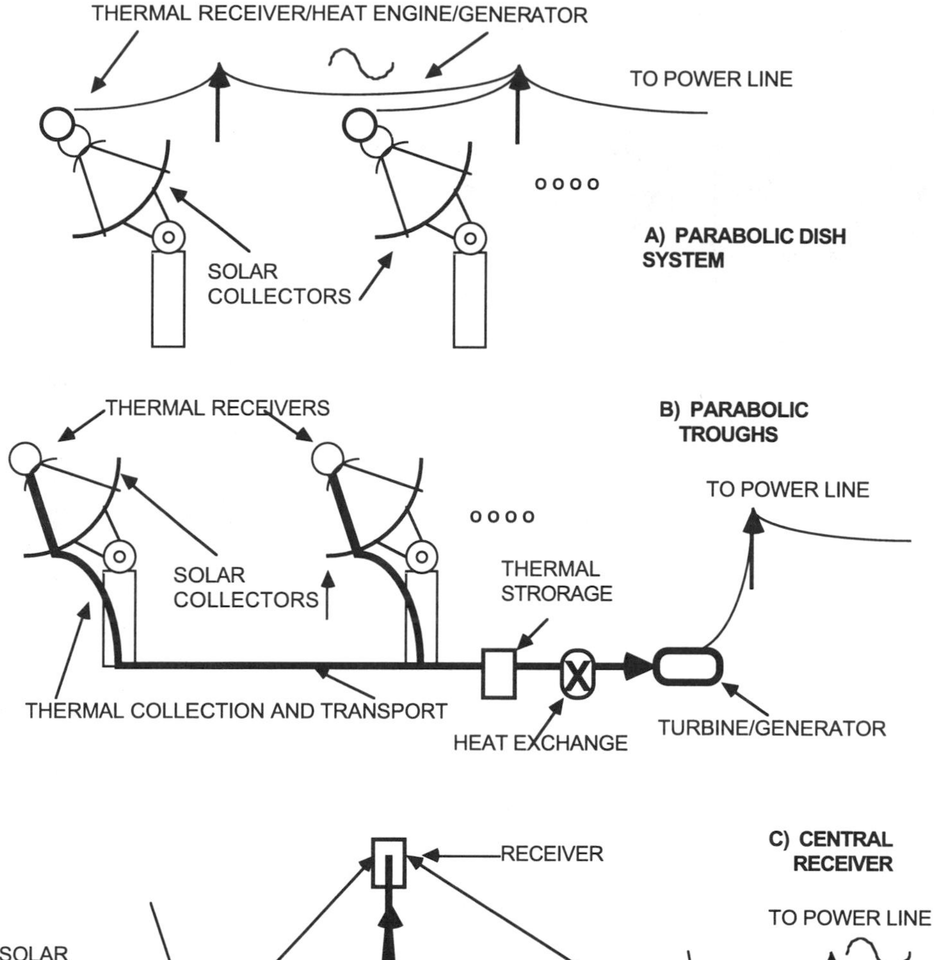

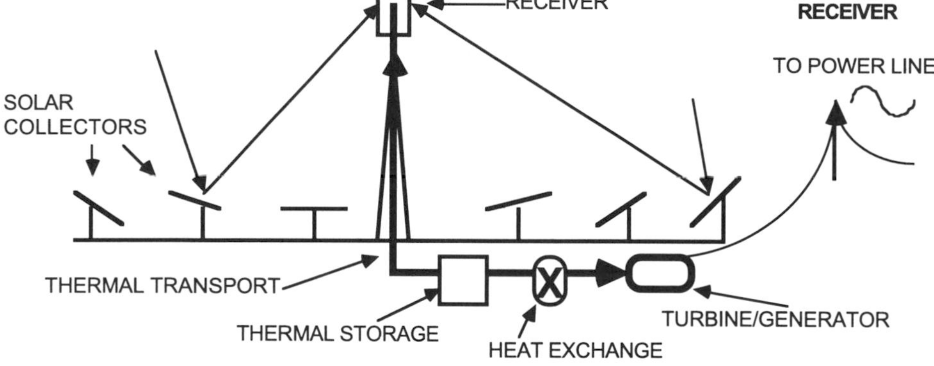

FIGURE 21A.2 Alternative placement of receivers (and heat engines) in (A) a distributed receiver system using individual heat engines, typical of a dish Stirling or dish Brayton system, (B) a distributed receiver system using a central heat engine typical of trough systems, and (C) a central receiver system.

Receiver Location

In either case the solar energy is absorbed at the receiver, converted to heat, and transported to a storage unit or directly to a heat engine (Fig. 21A.2).

Distributed Receivers. Distributed receiver systems consist of small collector modules (dishes or troughs of aperture 25–250 m^2), each of which illuminates its own receiver. Distributed receiver systems may employ a heat engine/generator unit with each module (typical of two-axis tracking dish systems). Alternatively, they may use an extensive piping network to bring the thermal energy

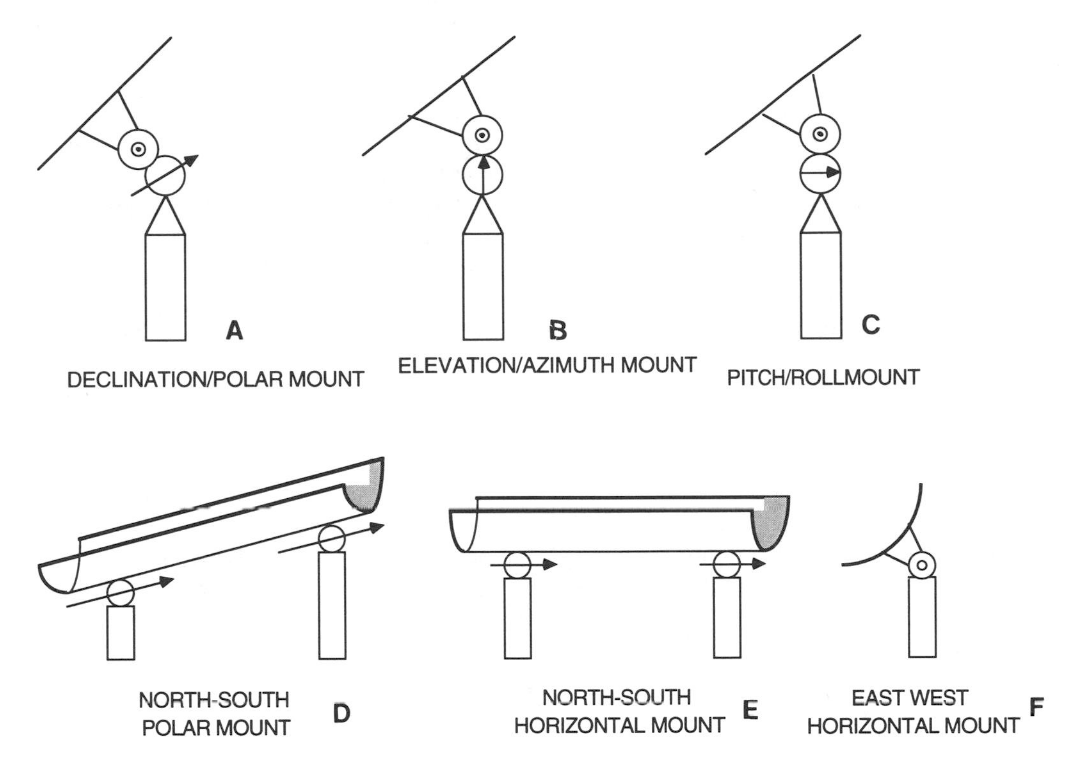

FIGURE 21A.3 The primary tracking configurations are defined by the rotation axis. Two-axis tracking: (A) declination over polar axis, (B) elevation over azimuth, (C) pitch over roll. One-axis tracking. (D) north–south polar axis; collector rotates at a steady 15°/hour (incidence angle equals solar declination, <23.5°), (E) north–south horizontal, (F) east–west horizontal mount. The arrow indicates an axis in the plane of the figure, and the ⊙ indicates an axis perpendicular to this plane.

from each single module or series string or loop of modules to a central point where a turbine generator and perhaps a thermal storage unit is located (typical of one-axis tracking trough systems).

Central Receivers. In a central receiver system many individual collector modules (heliostats) redirect the sunlight to a single elevated central receiver. The **heat transfer fluid** is pumped through a riser, a receiver, and a downcomer to the thermal storage unit and/or the heat engine, which are usually located on the ground near the receiver support tower (due to weight considerations). However, a direct-coupled Stirling or Brayton cycle engine (with no thermal storage or heat transfer loop) can be colocated with the receiver.

Solar Tracking

Two-Axis Tracking Systems. To achieve complete (perfect) tracking of the sun throughout the year requires motion around two axes (Fig. 21A.3). There are many choices of axes. The two most favored are *declination over polar* (rotate at 15°/hour around polar axis and adjust declination daily) and *elevation over azimuth* (rotate about azimuth axis and adjust elevation axis to match solar motion). Others such as *azimuth over elevation* or *pitch over roll* become unwieldy for large modules and so are seldom used.

Parabolic Dishes. For a monolithic module, two-axis tracking offers the obvious advantages of keeping the full aperture normal to the sun at all times and maintaining a constant flux distribution at the receiver. This system allows two axes of concentration, so paraxial rays can map a parabolic concentrator surface into a "point" in the focal zone (but note Eq. 21A.3). Such concentrators are commonly called dishes, or parabolic dishes, or may be referred to as point focus systems.

Central Receivers. If a parabolic dish is aimed at the sun, the reflector cut into facets, and the facets projected to the ground along straight lines emanating from the focal point, the only effect on the performance would be the requirement of a larger receiver to capture the light projected from the more distant facets (compare Eq. 21A.3 and 21A.4). When the sun moved, it would be necessary to change the orientation of the facets so that the reflected light would still be directed at the receiver. Because the facets would not move around the field, as they would if a new projection from the receiver were used at each instant, some parts of facets would be shaded at any time other than that corresponding to the original projection. In the central receiver concept, these facets are called heliostats (each of which must track on two axes), and the "dish" is increased in area to several square kilometers, corresponding to a focal length, or receiver height, of a hundred meters or more. Two-axis tracking provides many of the advantages of parabolic dishes, and additional economies of scale accrue. As the energy is collected to a central region by optical transmission, no piping network is required.

One-Axis Tracking Systems. Complete solar tracking is given up in order to achieve an economy or simplicity resulting from single-axis tracking (i.e., only one tracking motor, drive, and sensor is required). The resulting system can only focus on one axis; that is, paraxial rays can map the paraboloid collector into a line parallel to the tracking axis (but note Eq. 21A.3) resulting in geometric concentration approximately equal to the square root of the concentration of a comparable two-axis tracking system (note Eqs. 21A.5 and 21A.6). Such concentrators are commonly called troughs, or parabolic troughs, or they may be referred to as line focus systems.

The tracking axis of the troughs can be oriented east–west or north–south: the N–S axis also allows that the axis be horizontal or tilted parallel to the polar axis. Again, each approach has obvious advantages and disadvantages which will be considered in more detail in Chapter 21B. The polar axis mount is more expensive, but it keeps the aperture more nearly normal to the sun and tracks at a uniform 15°/hour. The N–S horizontal axis performs poorly in winter when the effective aperture is reduced by the cosine of the angle of incidence of the sunlight on the aperture (the cosine factor), but well in the morning and afternoon, while the E–W axis performs well near noon all year but poorly in the morning and afternoon when the cosine factor is small.

Nontracking Systems. Stationary collectors can achieve some concentration by use of CPCs (compound parabolic concentrators), but as the concentration is limited to 5 or 10, only low temperatures (<200°C) can be achieved. Such systems are of little or no interest for power applications.

Receiver Configuration

The configuration selected for the receiver defines many parameters of the design, may place limits or constraints on some parameters, and may play an important role in plant operation and maintenance procedures and costs (Fig. 21A.4).

Cavity Receivers. A cavity receiver locates its heat absorbing structure within an insulated enclosure behind an aperture plate, which is sized to admit the majority of the concentrated sunlight to the cavity but protects the structure from spilled radiation. It is thus somewhat protected from the environment. Light entering the aperture is subjected to multiple reflections, enhancing the effective absorptivity of the cavity. As radiation losses can only occur through the aperture, the absorbing area of the receiver can be enlarged without increasing the radiative loss, but convective losses are increased because of the larger heated area. Cavity receivers are thus well suited for high-temperature gas-cooled receivers where the poor heat transfer characteristics require a low flux density on the heat exchanger surface. The operative concentration is defined by the aperture area rather than the heat exchanger area. Of course by definition of a cavity, α/ε is approximately unity (see Eq. 21A.2).

Disadvantages of cavity receivers are that they tend to be large and heavy, and the aperture restricts the field of view of the receiver. Thus, the angle, ψ_{rim}, between the normal to the aperture

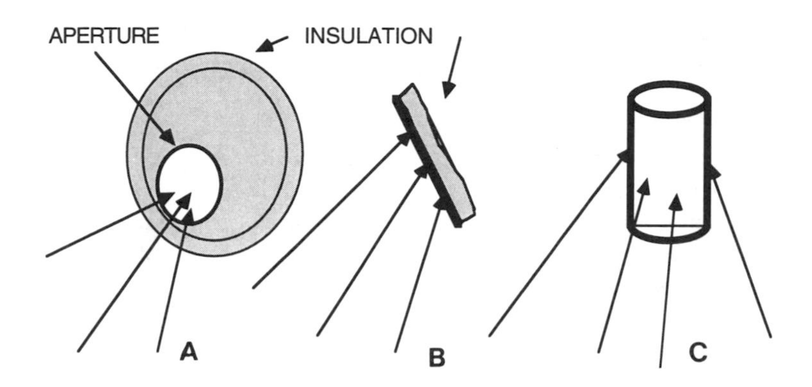

FIGURE 21A.4 Receiver configurations: (A) cavity; (B) flat plate–planar; (C) external.

(which may face downward or be tilted toward the north) and the boundary of the collector must be less than 90° and is more typically restricted to <75°.

Planar (flat) Receivers. Essentially, the aperture of a cavity receiver is replaced with a planar heat exchange surface. The flux distribution on the focal plane may need to be adjusted and the heat exchanger enlarged to handle the incident energy, but the back side can be insulated, reducing losses. The main advantages of a planar receiver are simplicity and low cost. Restrictions of the field of view are the same as for a cavity.

External Receivers. A simple tubular receiver in a trough system is an example of an external receiver. It allows light to reach the receiver from all directions, so ψ_{rim} can extend to 120° or more. External receivers are relatively simple and lightweight because they require no enclosure. By selecting a high-performance heat transfer medium, the allowable flux density on the receiver can be made as high as 850 kW/m^2 (molten salt) or 1,000 kW/m^2 (liquid sodium). This allows the receiver area to be reduced nearly to that of the aperture of a comparable cavity receiver. Removal of the restriction on ψ_{rim} set by an aperture gives the system designer more freedom in optimizing the overall system. Thus, the receiver tower of a central receiver system using an external receiver need only be ~70% as tall as if a cavity receiver is specified, and so costs about half as much.

Disadvantages of an external receiver are that the reflected fraction, $(1 - \alpha)$, of the incident light is lost, the receiver is exposed to the elements, and if the heat transfer fluid can freeze the receiver must be drained, and reheated prior to filling, as a protective measure.

21A.2 Recent Commercially Directed Developments

In 1972, under the Research Applied to National Needs (RANN) program, the National Science Foundation (NSF) initiated programs directed toward the eventual commercialization of large-scale STEPS plants. The initial feasibility studies identified three primary alternative systems: (1) arrays of parabolic trough collectors connected in series by heat transfer piping to a central turbine, (2) fields of heliostats transmitting energy optically to a central receiver and from there as heat to a turbine, and (3) point focus dishes with individually mounted organic Rankine, Stirling, or Brayton cycle engines.

Improved design and mass production of the concentrator, selection and development of the most cost-effective receiver, and reduction of thermal loss at the receiver were recognized as requirements for commercial success of any solar thermal system. In addition, the reliability of the rotating joints between trough modules and reliability of the individual dish engines were recognized as particular problems and considerable development effort was directed toward refining these items. In contrast, the central receiver was not modular and so required a large investment to provide any sort of an operational facility required for test or development.

Each of these issues was addressed in the subsequent DoE solar program and reported in a series of annual meetings (Anonymous, 1985; Diver, 1986; Tyner, 1987; Couch, 1989). A series of modular trough test systems were commissioned, built, and tested. Several engine manufacturers were commissioned to design and build specialized solar engines for use with parabolic dishes. a 5 MW thermal central receiver test facility was built by Sandia near Albuquerque, New Mexico, and operated to test receivers and heliostats. Eventually this facility expanded to a test facility for essentially all solar thermal components, including pumps and valves, innovative collector systems, dish receivers and engines, and complete trough assemblies. Sizable federally supported development programs were also in place in France, Germany, Spain, and Japan, with smaller but complimentary programs in a multitude of other countries (Grasse, Hertlein, and Winter, 1991).

In the early 1980s all of this effort had yielded three primary results, which showed the direction of development required to achieve reliable and commercially viable dish, central receiver, or trough systems.

- Experimental Stirling, Brayton, and organic Rankine heat engines had been built and operated with solar collectors with good results, achieving a record system efficiency of 30% (Droher, 1986). However, there was still a problem with reliability of the prototype receiver/engine complex. Consequently, a long-term DoE-financed cooperative program was put in place to develop a parabolic dish system comprising a low-cost and effective concentrator and an efficient receiver coupled to a reliable Stirling, Brayton, or organic Rankine heat engine.
- Eight central receiver system experiments were performed worldwide. The most successful of these, partly because of its larger scale, was Solar One, a largely DoE-financed, 10 MWe (35 MW thermal), pilot-plant-scale prototype of a central receiver system built near Barstow, California, shown in Figure 21A.5. It comprised a cylindrical, 7.1-m-diameter by 13.5-m-tall, once-through-to-superheated steam tube-wall receiver mounted 76 m above a field of 1,818 heliostats, each of area 39 m^2. There was a working oil–rock thermocline thermal storage system, but it was rarely used: The 500°C steam was usually fed directly to the dual-admission, multiple-extraction turbine, which was connected to the Southern California Edison (So Cal Ed) grid. This system met essentially all its design goals, and after a 5-year test and evaluation phase, So Cal Ed operated the plant in a pure power production mode for an additional 3 years (Radosevich, 1988).

In addition to a few small experimental trough systems (Coolidge, IEA), a series of eight solar electric generating (SEGS) trough-electric systems ranging from 13 to 80 MWe (total capacity, 355 MWe) were built commercially in the Barstow, California area by Luz International Limited (LUZ) (see Chapter 21B).

Current Issues

The collector modules of point-focusing distributed receiver systems may consist of a single-element monolithic dish, a multifaceted dish, a Fresnel mirror, articulated Fresnel mirror elements on a partially tracking platform ("Slats", etc.), or arrays of any of these. In contrast, the point-focusing central receiver system consists of fixed articulated Fresnel mirror elements (large two-axis-tracking heliostats) with a single receiver at the focal zone, which is elevated above the plane containing the heliostats.

Probably a dozen versions of each type above have been proposed. Many of these have been built and tested, and most perform reasonably well (Meinel and Meinel, 1976). However, commercial viability is the ultimate objective of all this work, and it requires much more than reasonable performance. The system must be constructed and operated at low cost and must effectively collect and efficiently convert the solar energy to high-temperature heat. Although no solar thermal system

FIGURE 21A.5 View of Solar One showing a portion of the heliostat field, the tower, and the external receiver. The receiver is 7.1 m in diameter and 15 m tall, elevated 75 m above the plane of the heliostats. The white region below the receiver is a set of four targets for beam alignment. Solar Two will reuse the tower and collector, but the receiver will be only 5.06 m in diameter and 6.21 m tall. *Source:* McDonnell Douglas Astronautics Company, Huntington Beach, CA.

has as yet been proven commercially viable, with the possible exception of the later Luz systems (Lotker, 1991), several show great promise for commercial viability, many are possible candidates for commercial viability, and several have proven to be complete failures for commercial viability as defined earlier.

Summary of Progress and Plans

In light of these issues, it is interesting to consider the approach that has been followed by DoE and its associated National Laboratories such as Aerospace, Sandia, and SERI (now NREL) over the last few decades. Following 1985, this program was developed with the assistance and general consensus of the Solar Energy Industries Association.

Work of 1972–1985—Develop and Test Lots of Systems. The basic idea seemed to be that any concept with a strong advocate deserved an opportunity to prove itself, at least through the prototype or pilot plant stage. The result was that there were literally hundreds of concepts and variations in different states of development, and there was no definitive system in place for eliminating any of them from further development or funding.

Work of 1985–1995—Select and Advance Systems with the Best Potential to Become Commercially Viable. Through competitive design studies and prototype development and testing, the most promising of the many solar thermal systems were evaluated and further work on a number of them was terminated or developmental support was put off until specific issues were resolved. Some specific projects terminated for cause were these: linear central receiver showed no cost advantages over the point-focus concept and offered much lower concentration and efficiency; the solar gridiron or large spherical bowl–tracking receiver concept, had an inefficient and not inexpensive collector (the bowl), and the tracking receiver was complicated and suffered from extreme flux variations; and in several variations of Slats systems incorporating multielement venetian blind–like collectors, the multitude of parts made construction and maintenance expensive and the performance was generally not particularly good outside a laboratory-like environment.

Those systems selected as showing high potential to become commercially viable were subjected to rigorous evaluation, and programs were carried out to mitigate the items posing the greatest technical or perceptual risk. For example, NREL instituted a program to understand the degradation of reflector surfaces and to develop long-lived high-reflectivity polymer mirrors (Short, 1988, Kennedy and Jorgensen, 1994), whereas Sandia funded prototype low-cost heliostat development and developed and tested receiver concepts and other components (Mancini, 1994). Efforts such as these had the effect of reentering the logarithmic cost/experience curve at a lower-cost position without the extreme cost of producing hundreds or thousands of components. Simultaneously, perceived risks were mitigated by demonstrating the performance of components such as molten salt pumps and valves in a simulated solar environment (Mancini, 1994) or by identifying commercial operations that used components in a similar fashion with good reliability.

Work of 1995–2000—Precommercial Demonstration of Best Systems. During the current period there will be demonstrations of prototypes of several systems deemed to have high potential to become commercially viable. These programs will proceed under cost-sharing cooperative research agreements between DoE and industry (joint ventures). Examples of such joint ventures are assistance in reducing operating and maintenance costs for the several operating "commercial" SEGS plants, and extending the "lessons learned" to other solar technologies; construction of Solar Two as a second-generation central receiver plant utilizing the tower and heliostat field of Solar One; a program to build and install five prototype 7.5 kWe dish Stirling systems at utility sites around the country in 1996; and two joint ventures to develop 25 kWe dish Stirling systems (DoE, 1994).

Federal cost sharing with industry is essential at this stage in development in order to mitigate the risks associated with building precommercial prototype systems at a large enough scale that realistic life, performance, and cost evaluations can be made prior to commercial development. Specific risks addressed in the joint venture projects are technology risk (performance), construction risk (cost or schedule overruns), operating risk (breakdown/unavailability of product), regulatory risk (disallowance of utility investments, expiration of favorable tax legislation), and financial risk (perception of these risks—real or imaginary—can increase financing costs remarkably [DoE, 1992]).

Federal participation in such joint ventures is clearly in the national interest in order to achieve the advantages mentioned in the introduction.

Work of 2000+—Install Commercial Systems by Thousands. Assuming the successful completion of the cooperative agreements and the continuation of technology development work at the National Laboratories, all technical and perceived risks associated with the commercial implementation of both dish Stirling and central receiver systems should be resolved. Also, trough systems will be better positioned for economical production of energy. At this point the market will have available several new, nonpolluting, low-risk, reasonable-cost solar systems that should be reliable and free of the risk of fuel-cost increases or shortages. The interest utilities and the solar industry have shown in participating with DoE in the 50% cost-shared joint ventures and cooperative research agreements during the 1990s suggests that the utilities will be more than willing partners in the program to add solar thermal electric power to their portfolio of plants and thereby reduce both system risk and the total cost of energy.

1995 Status

Currently there are no commercial STEPS plants in construction.

The eight existing SEGS plants are being operated commercially by operating agents responsible to the plant investors. Many of the companies that contributed expertise or components to the original SEGS plants are cooperating to continue the development of a fourth-generation SEGS plant (see Chapter 21B).

All of the experimental central receiver systems were decommissioned in the mid-1980s (Grasse, Hertlein, and Winter, 1991; Kolb et al., 1994). The accumulated information was incorporated in a cost-shared study funded by DoE, directed by several utilities, and carried out by major A&E firms, utility suppliers, and the solar industry: the "Utility Solar Central Receiver Study," or the "Utility Study" (Hillesland and Weber, 1990). After evaluating, on a system basis, all of the available options, this group unanimously agreed that the central receiver system most suitable for development by the U.S. utilities should have a configuration very similar to that of Solar One, but should use a molten sodium/potassium nitrate salt, known as "draw salt," as both the receiver coolant and as a storage medium. They further recommended that a first system test be carried out by refurbishing Solar One. Solar Two, designed for 10 MWe, would include a new receiver, heat transfer system, 30 MWe-h salt storage (about 100 MWh of heat at 1,050°F), a steam generator, a new control computer, and 10,000 m^2 of additional heliostats. This demonstration of the technology should be followed immediately by a commitment by the consortium to build a series of three to five 100 MWe demonstration plants to be followed by full-scale 200 MWe commercial plants.

This program is currently well under way as a joint venture, funded equally by a utility-based consortium headed by So Cal Ed and by the DoE. Under consortium management, the design and construction of Solar Two is complete and acceptance testing is scheduled to start in June 1996. Enhanced capabilities of the new receiver will allow it to operate with only one-third the heat-absorbing area of the Solar One receiver, and the turbine will always operate from stored heat, protecting the turbine generator from solar transients and assuring dispatchability of the plant. Solar Two is described more fully later in this chapter.

A parallel program in Europe directed by the International Energy Agency, known as Phoebus, took a somewhat different tack. The IEA assumed that the primary market for European STEPS plants would be for export to sun-rich third-world nations. Consequently they were interested in a system that would operate well in such an environment. Citing enhanced reliability and safety as the most important criteria in this environment, they selected a high-temperature air-cooled direct-absorption solar receiver, a ceramic brick thermal storage unit, and a high-temperature gas turbine. Concentrated solar radiation is incident on a thick honeycomb-like grid structure. Air is drawn in parallel through the grid, absorbing the heat and keeping the external surface relatively cool, reducing thermal loss. After the heated air passes through the thermal store, the subatmospheric-pressure depleted air is recompressed by a fan and rejected to the atmosphere, or it may be injected at the circumference of the receiver to control convection and retrieve residual heat.

The system will use two alternate stores: one store can be charged by the solar heated gas at subatmospheric pressure; meanwhile, the other fully charged store can be discharged using pressurized air so that the turbine will operate more efficiently. Successful component tests have resulted in a design for a 30 MWe unit (De Laquil et al., 1990). Plans to initiate construction in Jordan were terminated by the Iraqi invasion of Kuwait in 1991, but development has continued.

Many lessons have been learned from a brief list of the generic problems that have plagued the early attempts at the commercialization of solar thermal systems over the past few decades: these include failure of the reflective surface, poor optical performance, poor thermal performance, poor engine performance, short life of components, high component repair/replacement cost, too many parts, manufacturing process not viable for low-cost production, system not designed-to-cost, failure of critical components, excessive maintenance requirements, demise of production/service corporation, and so on. Each of these issues has been addressed, and the remaining problems will be mitigated by the programs reported in "Summary of Progress and Plans".

In addition to the problems of developing a reliable new energy resource, the solar community has been faced with severe competition from low-cost nonrenewable fossil fuels. Additional barriers are disproportionate taxes (sales tax, income tax on construction and operations salaries, property tax, etc.) on solar energy collection components compared to the tax-free treatment of fossil fuel energy costs, lax regulation of pollutants released by fossil fuel producers and power plants, high interest rates on the capital-intensive solar plants, general public utility commission rules favoring lowest-first-cost installations, and lack of recognition that the solar facilities provided a system with a zero-cost noninterdictable fuel supply compared to the uncertainties of fossil fuels 20 or 30 years into the future.

21A.3 Heat Engines

Systems of high concentration are used to achieve high efficiency at high outlet temperature from the solar concentrator. The motivation for this follows directly from the properties of the heat engine or process which will use the collected energy, so it is worthwhile to consider them briefly before proceeding. (A more detailed discussion may be found in Chapter 2.)

"Power" results from operating a practical irreversible thermodynamic cycle. The efficiency of such a cycle can be expressed as

$$\eta_p \sim \eta_{he}\eta_c \sim (0.3 \text{ to } 0.7)\left[1 - T_c(\mathrm{K})/T_h(\mathrm{K})\right] \quad \textbf{(21A.1)}$$

Here $T(\mathrm{K})$ reminds us that the exhaust temperature, T_c, and the operating temperature, T_h, must be expressed in absolute temperature, either Kelvin or Rankine, η_c is the usual Carnot cycle efficiency, and η_{he} is an estimate of the relative efficiency of a well-designed practical heat engine with the larger values accruing to larger, higher inlet temperature engines of a given type (Grasse, Hertlein, and Winter, 1991). Clearly a high operating temperature is advantageous. In fact, the inlet temperature T_h must be greater than 300°C = 573 K for reasonable (33%) practical heat engine efficiency. A more typical value for a utility-scale steam Rankine cycle engine is 550°C, while efficient Stirling, Brayton, or gas turbine cycles may require inlet temperatures of 1,000°C. Process heat applications also cover this entire temperature range, and process efficiency also generally drops off rapidly with reduced temperature.

21A.4 Basic Principles of "Concentrator Optics" Systems

Introduction

For a concentrating solar system to be cost-effective, it must deliver a high system efficiency. Generally the selected application defines a temperature range, and as indicated in the previous section, most solar thermal applications lie in the range 300 to 1,000°C. System size plays an important role in defining the technology: a single-dish electric system can provide 5 to 25 kWe, single-trough thermal loops can provide heat for one MWe, and a central receiver can be effective at 10 MWe. Larger requirements can be met by arrays of dishes, multiple trough loops, or larger central receivers—up to 200 MWe. The smaller systems require automation, reliability, and colocation to prevent operation and maintenance (O&M) costs from dominating power production costs, while larger systems also benefit from economies of scale.

In any high-temperature concentrating system the receiver is the focal point of the design effort. Because radiation and convection losses are proportional to the area of the receiver, it should be as small as possible (high concentration). The receiver must also be large enough to intercept and absorb a large fraction of the solar energy directed toward it by the collector, but larger receivers are more costly, reduce the concentration, and result in higher thermal losses. Alternatively the designer can specify a more perfect optical system to allow a reduction in the size of the focal region and the receiver, but this will increase the cost of the collector. At very high concentrations or temperatures, the design may also be limited by materials properties of the receiver or of the heat transfer medium. A guiding principle that resolves many issues and is a starting point for any design analysis is the angular subtended by the sun.

Beam Errors

The receiver in a concentrating system must be large enough to accept the direct beam energy from the sun, after the beam has been degraded by the atmosphere and the optical system. At the earth's surface, the half angle of the solar disk, $\alpha = 0.267° = 0.00465$ radian = 4.65 mr. As a result, the design of concentrating optical systems requires an approach much different from that of diffraction limited imaging optical systems. The objective is *not* to form an image of the sun, but rather to capture as much of its energy as feasible on a receiver, which is kept small to minimize thermal losses.

The energy distribution in the solar beam may be approximated by a Gaussian solar radiance, $L = L_0\left[\exp\left(\xi^2/2\sigma_s^2\right)\right]$ with $\sigma_s \sim 0.6\alpha$ (so the 9.3 mr diameter of the solar disk is about $3.3\sigma_s$) and ξ representing the angle measured from the center of the sun. The effect of mechanical, optical, or tracking errors can each be expressed in terms of an angular beam error (note that reflection doubles the effect of any error associated with a change in the surface orientation). Angular beam errors may be incorporated by adding σ_e, which represents the standard deviation of their combined effect, in quadrature to σ_s, to obtain $\sigma_d = \sqrt{\left(\sigma_s^2 + \sigma_e^2\right)}$. This σ_d represents a degraded sun which will produce an equivalent image on ideal optics. So long as σ_e is less than σ_s, σ_d is not significantly enlarged over its minimum value. As σ_s^2 is about 7.8 mr^2, σ_e^2 can be comprised of four 1.4 mr error sources, or seven 1 mr sources, or any other combination that does not exceed the goal. A reasonable design can usually maintain each error source well below 1 mr. A more sophisticated analysis can proceed by expanding each element of the error budget in terms of moments of the Hermite polynomials (Walzel, Lipps, and Vant-Hull, 1977), which are the orthogonal polynomial set for the Gaussian function, but this added complication is only called for in final design analysis.

Thermal Losses and Concentration

Thermal losses at the receiver are primarily from radiation and convection, so the thermal efficiency, η_{th} (= 1 – thermal loss/solar input), may be expressed through

$$\eta_{th} = 1 - A_{\text{receiver}}\left[\varepsilon\sigma\left(T^4 - T_a^4\right) + c\left(T - T_a\right)^{1.25}\right] \Big/ G_b\, A_{\text{collector}}\, \eta_{\text{col}}\eta_{\text{int}}\alpha$$

$$= 1 - \left[\sigma\left(T^4 - T_a^4\right) + (c/\varepsilon)\left(T - T_a\right)^{1.25}\right] \Big/ G_b\, C\, \eta_{\text{col}}\eta_{\text{int}} \qquad \textbf{(21A.2)}$$

$$C = \alpha\ A_{\text{collector}}/\varepsilon\ A_{\text{receiver}} = C_g\ \alpha/\varepsilon$$

Here T_a is the ambient temperature (Kelvin), ε is the emissivity of the receiver at the temperature T, σ is the Stephan Boltzmann constant, 5.67×10^{-8} W/m^2 K^4, and c is the temperature independent convective loss coefficient, typically 1 to 25 W/m^2 K$^{1.25}$. Also, α is the absorptivity for direct beam solar radiation (G_b) and η_{col} is the collector efficiency, which includes reflectivity (typically 0.85–0.95), cosine of angle of incidence effects (typically 0.7–0.9), and any factor to account for shading of mirror surfaces or blocking of reflected sunlight (typically 0.9–1.0). It is conventional to use the actual mirrored area of the collector for $A_{collector}$. The fraction of reflected light that is intercepted by the receiver is η_{int}, representing scattering and spillage losses. We see that the geometric concentration C_g plays an important role in suppressing the effects of thermal losses.

Sizing the Receiver

A brief table (Table 21A.1) of the one and two-dimensional Gaussian integrals will be useful in estimating the required receiver size, as it represents the fraction of the incident energy reflected from *each* collector element which will be captured by a receiver of a given angular size (ξ), as seen from that collector element (i.e., the receiver radius is p tan ξ where p is the slant range to the receiver). Accepting the degraded sun parameter, σ_d, as a measure of system quality, we define $\kappa = \xi/\sigma_d$. If we choose ξ to give $\kappa = 1.5$ for a boundary element, κ will be 3 for an internal element at half the slant range, p, to the receiver, and so on. Thus, it is important to remember that the system interception is a sum over the component interceptions given in this table.

TABLE 21A.1 Table of Gaussian Integrals (interception factors): 1D, tabulate [erf($\kappa^2/2$)]; 2D, tabulate [1 – exp ($\kappa^2/2$)]

$\kappa = \xi/\sigma_d$ →	0	1.0	1.5	2.0	2.5	3.0	4.0
One-dimensional Gaussian	0.000	0.683	0.866	0.954	0.988	0.997	0.9999
Two-dimensional Gaussian	0.000	0.393	0.675	0.865	0.957	0.989	0.9996

Clearly, closer elements are more effective at delivering energy to a given receiver. For example, we may require that the receiver intercept over 86% of the energy reflected from the boundary of the collector (more distant elements would have a still lower interception and so not be cost effective). An interception of 86% requires a value for κ of 1.5 for a one-dimensional concentrator such as a trough, and $\kappa = 2$ for a two-dimensional concentrator such as a parabolic dish. For an element closer in at half the slant range, κ will be doubled, giving interceptions for such elements of 99.7% and 99.96% respectively.

We can now write an expression for r, the required receiver half width (trough, 1D Gaussian) or radius (dish or central receiver, 2D Gaussian) based on the value of ξ, interpolated from the Table 21A.1, which gives the minimum acceptable interception factor for the extreme rays (and a higher value for all others). In terms of the rim angle, ψ_{rim}, and the focal length, f, we have

$$r = \xi p_{rim} = \kappa\sigma_d 2f/(1+\cos\Psi_{rim}) \quad \text{for a parabolic reflector} \tag{21A.3}$$

$$r = \xi p_{rim} = \kappa\sigma_d f/\cos\Psi_{rim} \quad \text{for a Fresnel reflector or central receiver} \tag{21A.4}$$

Here p is the distance from each mirror element to the receiver: p_{rim} can become $\geq 3f$ for large rim angles (>110° in Eq. 21A.3, >71° in Eq. 21A.4). A Fresnel reflector is made by projecting parabolic mirror elements onto a planar surface.

At this point it is reasonable to point out that standard refractive optics are not practical for most solar applications, both because they suffer from chromatic aberrations and because the large size of the required components leads to unreasonably thick and heavy lenses. Even though the optical quality of cast lenses would be adequate, the material cost is prohibitive. (Fresnel refractors, which remain thin even in large sizes, may be used in certain cases). Also, for high concentration systems operating at elevated temperatures, it is difficult to achieve a ratio of solar absorbtivity to infrared emissivity very much greater than unity, particularly if the requirements of long life and highest possible absorptivity are imposed. It is important to note that a geometrical concentration of only 10/1 means that a 1% loss in absorptivity requires a decrease in emissivity of over 10% to compensate. Consequently, in high-concentration systems a high α is sought and little attention is paid to the value of ε. (But for the SEGS moderate-concentration parabolic trough systems Luz used a proprietary absorber coating operating in an antireflection-coated Pyrex vacuum jacket to achieve an α/ε of about 5 at the operating temperature of 300°C.)

Geometric Concentration

The geometric concentration C_g = collector aperture/receiver area, can be immediately written for the principle systems. If we assume open cylindrical or spherical receivers, note that the collector radius R is given by $p_{rim} \sin\psi_{rim}$, and refer to Eqs. (21A.3) and (21A.4), we can immediately write for a one-axis (linear trough) system,

$$C_{g1} \sim 2RL/2\pi rL = R/\pi r = p_{rim}\sin\Psi_{rim}/p_{rim}\pi\kappa\sigma_d = \sin\Psi_{rim}/\pi\kappa\sigma_d \tag{21A.5}$$

and for a two-axis (parabolic dish) system,

$$C_{g2} \sim \pi R^2/4\pi r^2 = p_{rim}\sin^2\Psi_{rim}/(2\kappa\sigma_d p_{rim})^2 = (\sin\Psi_{rim}/2\kappa\sigma_d)^2 = (C_{g1}\pi/2)^2 \tag{21A.6}$$

Additional factors of order 2 may result from specifying flat or partially insulated receivers, and a Fresnel collector will bypass a fraction of the light, resulting in a lower concentration.

A most important fact to note about these results is that the concentration is a function of $\kappa\sigma_d$ and of ψ_{rim} only, and not of the system size (i.e., not of the focal length f). Thus, system accuracy (σ_e, incorporated into σ_d), system geometry (ψ_{rim}), and system size (f) are nearly independent variables under the control of the system designer, as is the rim element interception which defines κ. For instance, the module focal length (f) and the aperture width ($p_{rim} \sin\psi_{rim}$) can be selected to provide the most cost-effective system providing sufficient system accuracy. The rim angle is usually in the range 45 to 135°, which gives sin ψ_{rim} between 0.7 and 1.0, so it does not have a strong effect on the concentration and thus is available for other design considerations.

Although these same equations apply to Fresnel optical systems, it is necessary to include an additional factor to represent the fraction of sunlight redirected toward the receiver by the Fresnel elements. This will be less than unity due to bypassed light, or the blocking of reflected sunlight by the adjacent Fresnel ring. In central receiver systems a portion of the light may also be blocked from reaching the reflecting surface by adjacent elements (shading); consequently, the designer may increase the separation of elements (allowing the fraction of light *bypassed* to increase to 50% in the center, and 85% near the boundary) in order to achieve good performance for a wide range of sun positions. To retain the same mirror area p_{rim} must be increased, causing an increase in the receiver size and the focal length to achieve an optimum design.

Finally, it seems clear that the principles of concentration optics require a mindset much different from that involved in the design of diffraction limited imaging optics such as cameras or telescopes, where the quality of the image and freedom from optical aberrations are the primary objectives.

21A.5 Design and Performance of Concentrating STEPS

Two-Axis Tracking Central Receiver (Solar Tower) Systems

Basic Design Issues

A solar tower system is essentially a large-scale two-axis tracking parabolic concentrator approximated by a faceted Fresnel mirror concentrator. It is capable of efficiently and economically redirecting 50 to over 1,000 MW of solar energy to a central receiver and converting it to high-quality heat for a utility-style turbine, or for a process heat application. The receiver is located at the focal point of the Fresnel collector, 100 to 300 m above the ground. The Fresnel collector consists of 2,000 to 20,000 large (typically 50–150 m^2) stationary two-axis tracking nearly flat mirrors (heliostats). Each heliostat is computer actuated so that the normal to its reflective surface bisects the angle between the sun and the receiver. Thus, it redirects the sun's energy to the receiver. This "optical transmission" feature solves the problem of transporting the solar energy incident on several km^2 to an efficient commercial scale turbine generator. The multitude of nominally focused heliostats can easily provide a concentration of over 500 suns (~500 kW/m^2) for over 90% of the redirected energy.

As the collector represents about 40% of the direct capital cost, the heliostats are a critical component of the system and so must be designed for low-cost mass production. Fortunately the central receiver system places few additional restrictions on the designer other than requiring that the reflected beam error from all causes be smaller than, or comparable to, the solar divergence angle (i.e., $\sigma_e \leq \sigma_e \sim \alpha/2 \sim 2.3$ mr). Other desirable features are high reflectivity and an ability to withstand expected winds and nominal earthquakes.

Considerable effort has been devoted in the past decade to the development of lower-cost heliostats. Development, along with construction of a few prototypes, has led to confident reductions by a factor of two or so in the estimated cost of production runs of a few thousand "conventional" glass–metal heliostats. It has also resulted in the development of the "stretched membrane" heliostat. This low-cost design comprises two aluminum or steel membranes welded in tension to a metal ring, like a large tambourine. A very slight vacuum causes the front surface to develop a curvature corresponding to a focal length variable from 100 to over 1,000 m, and thin silvered glass or plastic film is bonded to the front surface to form a mirror. Several 50 m^2 prototypes have been successfully tested. Good optical quality is readily achieved. Properly silvered glass mirrors have a demonstrated life of 30 years, although care must be taken with the process. In contrast, the silvered polymer reflector, ECP 305, developed jointly by NREL and the 3M Company provides over 90% reflectivity and a life expectancy exceeding 5 years before in-field replacement is required (Kennedy and Jorgensen, 1994). Current research is directed to innovative approaches to produce equally reflective advanced films with a projected life exceeding 10 years, and a lower production cost (Kennedy and Jorgensen, 1994).

As a corollary to the comments following Eq. (21A.6), it is desirable to pack the heliostats as closely as possible to allow a small receiver and high concentration. However, the heliostats are individual tracking Fresnel segments subject to complex performance factors, which should be optimized over the sunlit hours of the year. Their effective area is reduced by the cosine of the angle of incidence ι; the annual average of cos ι varies from about 0.9 two tower heights north of the tower to about 0.7 two tower heights south of the tower. Typical preliminary designs place about two-thirds of the heliostats north of the tower to take advantage of this factor, although overall system optimization tends to reduce the unbalance in response to the resulting excess flux density on the north side of the receiver. It is cost-effective to accept the lower performance associated with moving heliostats to the south field in order to ameliorate the high flux areas on the receiver, rather than to increase receiver costs and thermal loses by making the receiver larger.

If the heliostats are spaced too closely, their corners may collide, so "mechanical limits" preclude pedestal spacing closer than the maximum dimension (i.e., the diagonal or diameter of the heliostat, D_m). This causes a significant disadvantage for heliostats with aspect ratios (height/width) very different from unity.

The next most rigid limitation on spacing of heliostats is the blocking of reflected light from reaching the receiver due to a neighbor nearer the central tower. Simple trigonometry shows that no blocking can occur if the radial heliostat spacing exceeds $H \cos \iota / \cos \psi$, where H is the heliostat height, ι is the angle of incidence on the mirror, and $\psi = \arctan R/f$, with R the radial distance from the tower and f the focal length of the Fresnel system, essentially the receiver elevation. If the annual average value $\langle \cos \iota \rangle$ is used for each heliostat relatively little blocking will occur.

The final effect to consider is shading of the incoming light by neighboring heliostats. This occurs predominantly at low sun angles and in the central regions of the field where the blocking condition would allow close spacing ($\cos \psi \sim 1$). The shadows move during the day and year, as do the heliostats' orientations, so no simple rules of thumb can be given, particularly since the heliostats are not uniformly spaced due to the issues just discussed.

An optimization procedure based on cost-effective use of heliostats (Lipps and Vant-Hull, 1978) gives a required separation for close in heliostats with $\psi < 45°$ of $\sim 2.1\ D_m$ for square heliostats and $\sim 1.7\ D_m$ for round heliostats. For larger ψ, the blocking constraint dominates, so the radial spacing increases as $\cos \langle \iota \rangle / \cos \psi$. This leads to the idea of "staggering" circles of heliostats spaced at half the required radial spacing, since the nominal azimuthal separation ΔZ, defined by shading, allows the heliostats in the circle behind to "peak through" at the receiver and experience little blocking. For such a field at 35° latitude a typical optimized field using square low-cost heliostats has heliostat spacing which may be represented in terms of the receiver elevation angle, $\theta = \pi/2 - \psi$, and with θ expressed in radians, by (Vant-Hull, 1991)

$$\Delta R/D_m = 1.009/\Theta - 0.0630 + 0.4803\Theta = 5.078 \quad \text{at } \Theta = 0.2\ rad\ (13°)$$

$$\Delta Z/D_m = 2.1700 - 0.6598\ \Theta + 1.2470\ \Theta^2 = 2.087 \quad \text{at } \Theta = 0.2\ rad\ (13°)$$

Finally, the boundary of the field is defined by a combination of decreased interception due to the large distance to the receiver, and increases in land and wiring costs due to the large distance to the receiver, and increases in land and wiring costs due to the large spacing between heliostats to prevent blocking. Also, for large systems, atmospheric absorption between the heliostat and the receiver over 1 to 2 km becomes appreciable (5% on clear days).

The receiver must efficiently absorb the incident energy, convert it to heat at 500°C to 1,000°C, and survive for 20 to 30 years in the high-flux environment. It is also subject to one daily thermal cycle from ambient to operating temperature, and many smaller cycles due to cloud transit and so on. Thus stress fatigue as well as corrosion by the high-temperature heat transfer medium are important design issues.

The "Utility Study" (1985–1988)

In the "Utility Study" (Hillesland and Weber, 1990) an alternative was sought to the pressurized water steam heat transfer medium used in Solar One. The Solar One receiver had an allowable flux limit of only 350 kW/m^2, and there were problems associated with storing the heat (heat exchange from high-pressure steam to heat transfer oil and later back to steam).For the Utility Study, liquid sodium (Na) and a molten salt (draw salt, $Na0_{.6}K0_{.4}NO_3$) were the primary candidates for the heat transfer and storage medium due to their low vapor pressure and excellent heat transfer characteristics. In addition, there is significant experience and background with these materials from the Liquid Metal Fast Breeder Reactor program, from industrial experience with draw salt, and from solar receiver tests of the past decade involving both materials (Grasse, Hertlein, and Winter, 1991). Na provided a higher allowable flux limit (1,000 kW/m^2) compared to draw salt (850 kW/m^2). However, the higher cost of sodium requires that the storage unit use an alternative medium such as draw salt, so a sodium receiver requires an intermediate heat exchanger. The cost and added complexity associated with this heat exchanger plus the fire hazard posed by any leakage of the sodium metal resulted in a clear decision to use draw salt for both the receiver and storage (Hillesland and Weber, 1990).

The choice of an external receiver vs. a cavity was based partly on the lower system cost due to fewer restrictions posed by the external receiver. The larger field of view allowed a shorter tower with less piping and lower pumping costs. In addition, the cavity itself was more costly and because of the large tubing lengths required could not reasonably be scaled up beyond about 300 MWth. An auxiliary study showed that a system with a single 1000 MWth unit was substantially more economic and simpler than coupling three 333 MWth cavity receivers. Thus "nitrate was added to the sodium," and a unanimous recommendation was made by the three utility-led teams that a utility-based consortium be formed.

Solar Two

This consortium would begin with the Utility Study design for a 100 MWe first commercial system using a south-biased heliostat field illuminating an external molten salt receiver. This system also incorporated 3 to 6 hours of thermal storage using draw salt as both the heat transfer medium and the storage medium. A further recommendation was made to refurbish Solar One as a 42 MWth molten salt system: Solar Two. Solar Two would reduce the risk inherent in scaling up from existing molten salt receiver experience (5–10 MWth) to 300 to 500 MWth typical of a first 100 MWe commercial design. It would also provide operating experience with the external molten salt receiver and heat transfer system, including the steam generator. As the draw salt crystallizes at about 220°C, heat tracing of all the molten salt pipes and valves is an important issue for such a system, and particular attention is being given to reliability and to the reduction of parasitic loads associated with the heat tracing at Solar Two.

Solar Two is a scaled prototype of a 100 MWe precommercial system that would have a solar multiple of 1.3 to 1.8, allowing a 3- to 6-hour storage unit to be filled on most clear summer days while operating the turbine at full load. This produces a capacity factor of 0.40 to 0.56 and should receive full credit as a peaking plant, being available on demand during 85% of peak load hours. The Solar Two consortium was organized with the intentions that, following successful operation of Solar Two, consortium members would order 3 to 6 such 100 MWe plants in a single procurement in order to achieve significant "economies of scale," resulting in a busbar cost of 8 to 10 cents/kWh (in 1988$) from these dispatchable plants. As the first step in this venture a consortium has been formed (led by So Cal Ed) and construction of Solar Two is complete. Dedication of the plant occurred on June 5, 1996 with 5 hours of operation of both the solar plant and the turbine-generation at over 70% of rated output.

The receiver design for Solar Two is based on the design requirements for a 100 MWe system. Thus it will allow reliable scale-up. The cylindrical external receiver consists of 2 circuits of 12 panels each. Warm salt will enter each circuit on the north side (where the flux density is ~800 kW/m^2) and pass through the panels in serpentine flow (thereby eliminating interpanel risers) and exit at 565°C (1,050°F) on the south. A single crossover between the east and west circuits is used to balance the morning and afternoon asymmetries in the heat load on the two circuits. At the 565°C outlet temperature the allowable flux is limited to only ~350 kW/m^2 to assure that the salt surface film stays below the salt corrosion limit temperature of 600°C (1,112°F). The heliostat field has been tailored to assure that the receiver flux limitations are satisfied. In addition, multiple aim points will be computed on a real-time basis and computed flux density maps will be used to identify areas of excess flux and to select appropriate heliostats to defocus. In addition, the few hundred heliostats to be used for preheating of the dry receiver panels prior to establishing salt flow will be selected and aimed using the same programming (Vant-Hull and Pitman, 1990).

The receiver will experience one complete thermal cycle per day (~10,000 in its 30-year design life), but may also see several less severe cycles each day due to cloud passages and so on. As the tubes in each panel are only heated from the front, thermal stress fatigue is an important design parameter. Extensive tests and analysis has shown that 316H stainless steel has adequate corrosion resistance and can withstand the expected thermal stress fatigue. Use of thin-wall tubing (1.2 mm) and relatively high flow rates results in a very high allowable flux and therefore a smaller, more efficient receiver. Unlike Solar One, the tubes are not welded together to form a tube wall, but are supported from the headers and held in place against transverse motion by tube clips, which can slide vertically to allow free thermal expansion. These design features have produced a receiver with conservative but very high flux capability that is fully scalable to a 100 MWe design.

Estimation of Design Parameters for a Central Receiver System

As an example of the use of the design equations, we will analyze a "first commercial central receiver design," as developed during the Utility Study and detailed earlier. The first decision is the application, which in this case is clearly to power a conventional 100 MWe utility steam turbine. A nominal allowance for parasitic power of 10% includes the cold salt pumps (6.2 MWe), hot salt pumps (0.7 MWe), feedwater pumps (2.2 MWe), heliostat field (0.8 MWe), and miscellaneous loads (3 MWe), not all of which operate simultaneously. A solar multiple (ratio of receiver output to turbine inlet power ratings) of 1.8 was chosen to enable filling thermal storage while generating rated power on the design day. With a reheat turbine efficiency of 43.2%, the required receiver design point thermal output power, DPTOP = (100 + 10)1.8/0.432 = 468 MW thermal. Many of the design parameters for this system are listed in Table 21A.2.

Receiver thermal losses depend on the receiver area and the local surface temperature. A first estimate of the receiver area can be obtained from the allowable flux density limit. Due to end effects, and the reduced allowable at the high-temperature outlet (on the south side), the average allowable is reduced from the specified 825 kW/m^2 to about 440 kW/m^2, resulting in a required receiver area of 468 MW/0.44 MW/m^2 = 1,064 m^2. For the first commercial design a cylindrical receiver with an aspect ratio of 1.3 was selected to allow the highly peaked flux distribution to be leveled by selection of a vertical distribution of aim points, resulting in a receiver radius of 16.1 m.

To provide the required 1,000°F steam to the turbine, the salt storage temperature is set at 1,050°F (565°C), which must be the receiver outlet temperature. The draw salt begins to freeze at 430°F (220°C), safely below the feedwater preheat outlet temperature of 455°F, so the cold salt storage tank temperature is set at 500°F (288°C), which defines the receiver inlet temperature. Converting to Kelvin, and allowing for a flux-dependent 100 to 200 K temperature rise through the tube wall, results in a receiver surface temperature ranging from 750 K to 850 K. Using the average of these for T, 300 K for the ambient (T_a), 0.88 for ε (typical of rough oxidized steel), and a nominal value of $c = 5$ W/m^2 K$^{1.25}$, the thermal loss from Eq. (21A.2) is

$$\begin{aligned} TL &= A_{\text{receiver}}\left[\varepsilon\sigma\left(T^4 - T_a^4\right) + c\left(T - T_a\right)^{1.25}\right] \\ &= 1,056\left[0.88\cdot 5.67\cdot 10^{-8}\cdot\left(800^4 - 300^4\right) + 5\left(800 - 300\right)^{1.25}\right] \\ &= 1,056\left[20,000 + 11,600\right] = 31.6\ \text{MW} \end{aligned}$$

The resulting thermal efficiency is

$$\begin{aligned} \eta_{\text{th}} &= 1 - \text{TL/incident power} = 1 - \text{TL}/\left(\text{DPTOP } + \text{TL}\right) \\ &= 1 - 31.6/\left(468 + 31.6\right) = 0.937 - 93.7\% \end{aligned}$$

TABLE 21A.2 Solar Two and First Commercial Plant from Utility Study

	Solar Two	First Commercial Utility Study‡
Design point power (electric)	10 MW	100 MW
Scheduled acceptance test	June 1996	~2,000
Location	15 km E of Barstow CA, 35°N, 117°W	Arizona, Nevada, or Southern California
Site elevation	593 m	400–800 m
Heliostat field	Glass/metal	Glass/metal or membrane
Configuration	Radial stagger	Radial stager
Reflective area	71,000 m² @ 39 m²	884,000 m² @ 150 m²
(supplement to Solar One field)	+ 10,000 m² @ 95 m²	
Constraint on field (system trade)	South field favored to reduce peak flux on north side of receiver.	
Ground coverage	27% avg., 40 to 18%	25% avg., 40 to 16%
Collector area (land)	0.3 km², ~75 acres	3.54 km², ~875 acres
Receiver	External, salt in tube	External, salt in tube
Focal height above heliostats	73.4 m	185 m
Diameter	5.05 m	16.1 m
Length	6.21 m	21.0 m
Heat absorbing area	98.6 m²	1,068 m²
Design point power (absorbed thermal)	42.2 MW	468 MW
Absorbed flux density, Maximum	850 kW/m²*	820 kW/m²*
Average	430 kW/m²*	440 kW/m²*
Thermal storage unit	Hot and cold tank	Hot and cold tank
Medium	Draw salt	Draw salt
Rating (thermal)	3 h @ 35 MW	6 h @ 260 MW
Weight of salt required	1.3 million kg	16 million kg
Turbogenerator rating (nominal, thermal)	35 MW	260 MW
Turbogenerator rating (nominal, electric)	12.5 MW	110 MW
Solar multiple (receiver/turbine)	1.2	1.8
Heat engine efficiency	35% rated	42.3% rated
Power generation efficiency†	5.8% annual avg.	14.6% annual avg.

* Solar Two final receiver design is more aggressive than the preliminary commercial design, which was completed several years earlier.

† estimated as electrical output/(annual insolation · reflective area)

‡ *Source*: 1988. *Arizona Public Service Utility Solar Central Receiver Study*, Report No. DOE/AL/38741-1, November 1988.

The required collector area is calculated at the design point via absorbed solar power = DPTOP + TL = $G\rho A_c(\text{CSB } T\eta A)\alpha$, so

$$A_c = (468 + 31.6)\cdot 10^6 / \left(950 \cdot 0.92 \cdot (0.83 \cdot 0.92 \cdot 0.93 \cdot 0.99)\cdot 0.92\right) = 884{,}000 \text{ m}^2$$

Here G is the design point beam insolation (= 950 W/m² reflector); ρ the average heliostat reflectivity; CSB the combined effect of cosine, shading, and blocking losses; T the transmission loss from the heliostat to the receiver; η the receiver interception factor; A the heliostat availability; and α the receiver absorptivity. Rightfully, the term in parenthesis should be computed for each heliostat and then averaged, but using field average values for each item is a reasonable approximation.

The land area for the field is computed from the ground coverage F, and the collector area A_c, with a nominal 3% addition for roads and the power block, thus

$$A_f = (1.03)(A_c/F) = (1.03)(884{,}000/0.25) = 3.64 \text{ km}^2 = 1.4 \text{ sq. miles} = 900 \text{ acres}$$

Although the field is biased toward the south to minimize north to south variations in the receiver flux, the south radius is still only about 70% of the north field radius, approximated by replacing π by 2.5

$$A_f = \pi R^2 - (0.3R \cdot \pi R/2) \cong 2.5\ R^2$$

so

$$R = \left(A_f/2.5\right)^{0.5} = (3{,}640{,}000/2.5)^{0.5} = 1{,}210\ m$$

In this case, the range of receiver elevation angles from the edge of the field ($\theta_{rim} = 90° - \psi_{rim}$) is

$$\text{in the north,} \quad \theta_{rim} = \arctan 185/1{,}210 = 8.7°$$

$$\text{due south,} \quad \theta_{rim} = \arctan 185/0.7 \cdot 1210 = 12.3°$$

The optimum rim angle ψ_{rim} will depend on the relative cost of heliostats and will be decreased if an exceedingly high flux density is required (a higher tower allows a higher heliostat density and so a smaller field with a higher concentration). Typical values range about 2.5° around the preceding figures.

As a check, the concentration can now be calculated. The beam accuracy for this heliostat is described by $\sigma_e = 2.63$ mr. To this it is reasonable to add (in quadrature) 1 mr contributions due to focusing errors and to off-axis aberrations in addition to the 2.32 mr contribution from the sun to obtain $\sigma_d = (6.92 + 1 + 1 + 5.38)^{0.5} = 14.3^{0.5} = 3.78$ mr.

If we wish the extreme heliostat to have an interception factor of 0.865 we must choose $\kappa = 2$ for a spherical receiver, or 1.5 for a cylindrical receiver. For a central receiver with a cylindrical receiver, Eq. (21A.4) gives

$$r = \xi p_{rim} = \kappa\sigma_d f/\cos\Psi_{rim} = 1.5 \cdot 0.00378 \cdot 185\ \text{m}/\cos 81.3 = 6.85\ \text{m}$$

(which is less than the receiver radius of 8.08 m selected based on the allowable flux limitation).

From Eq. (21A.6), the maximum concentration for a receiver radius of 6.85 m is

$$C_{g2} = \left(\sin\Psi_{rim}/2\kappa\sigma_d\right)^2 = (\sin 81.3/2 \cdot 1.5 \cdot 0.00378)^2 = 87^2 = 7{,}600.$$

For a central receiver system, this must be reduced by the collector efficiency (in parenthesis in the equation for A_c, ~0.703) and the ground coverage ratio of 0.25, giving $C_{g2} = 1{,}335$. Introducing the other loss factors and the insolation of 950 W/m^2, this corresponds to an absorbed flux density of 1 MW/m^2, somewhat above the allowable, which explains why a larger receiver was specified in the design study.

It is clear from the design equations that the optical characteristics of the design will scale linearly, and the focal length, *f*, is the natural scaling parameter. Thus, the field and receiver dimensions are proportional to *f*, while the field and receiver area and the absorbed power are proportional to f^2. In contrast, the beam errors and flux density on the receiver are independent of *f*. Changing the latitude, insolation characteristics, relative heliostat size or accuracy (σ_e), or relative component costs have a more complicated effect on the design and will require a new system optimization.

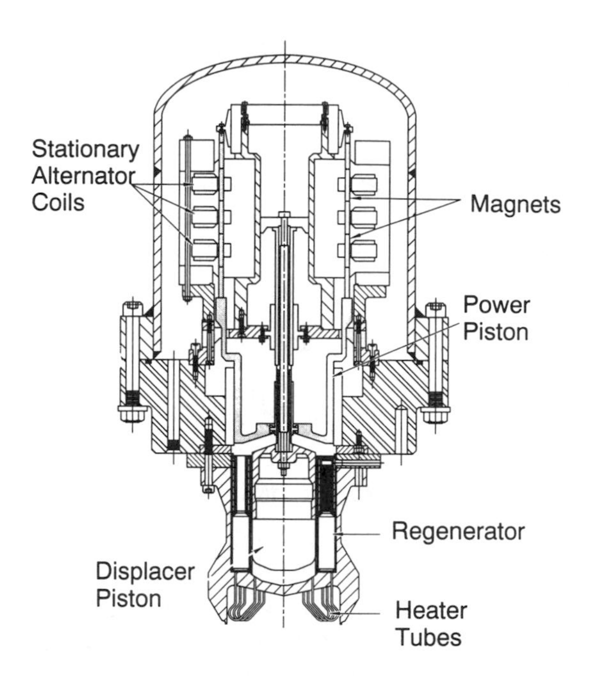

FIGURE 21A.6(A) (A) A line drawing of the Cummins/Sunpower free piston Stirling engine showing the moving parts (the power piston and the displacer), and the internal electric generator. (B) Line drawing of the United Stirling AB 4-95 MKII solarized kinematic Stirling engine showing also the crankcase and drive shaft required to couple to the external electric generator. *Source*: Stine, W.B., Diver, R.B., 1994. *A Compendium of Solar Dish/Stirling Technology*, Technical Report SAND93-7026, Sandia National Laboratories.

Two-Axis Tracking Distributed Receiver (Parabolic Dish) Systems

Modularity

Parabolic dish systems are intrinsically modular. A single dish can have a paraboloidal reflective area ranging from 10 m^2 to 300 m^2, although most are between 30 and 120 m^2. Each such reflector illuminates its own receiver to form a concentrating module, with a thermal rating of 25 to 100 kW. The unavailability of reliable and compact heat engines in this range before the 1980s mandated that early dish systems use a piping arrangement to collect 0.1 to 20 MW of thermal energy to a central location for process heat or electric generation. Two-axis tracking makes the interconnecting piping more complicated (3D flexible joints), and unlike the thermal loops used in trough systems, insulated hot and cold lines need to run from each collector to the ground and then to the pedestal of the next collector.

Efforts of the last decade have led to development of several candidate Stirling or Brayton cycle engines in the 5 to 25 kWe range that have been particularly modified for solar applications (solarized) and are poised for commercial development (Figure 21A.6). With such an engine mounted on each concentrator–receiver module, one has a truly modular solar electric system. A single unit (or a few) can be installed as an electric resource at a remote site or at a location on the demand side of an overloaded power line. Alternatively, thousands of units can be installed at one site feeding electricity to a major installation or into a power grid.

Dish Electric System Elements

A dish electric system consists of four primary subsystems, which are largely independent, but must be designed to work together effectively. The subsystems, shown in Figure 21A.2(a), are

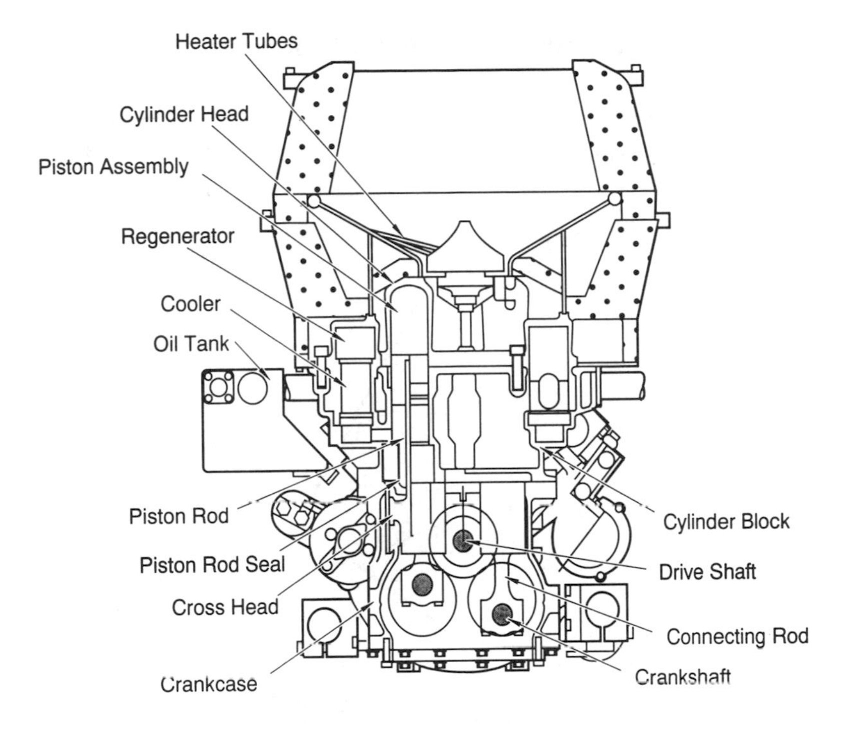

FIGURE 21A.6(B)

Concentrator, comprising a foundation/pedestal, tracking system, receiver support structure, parabolic substrate, and reflective surface

Receiver, comprising an aperture, absorbing surface, thermal interface, heat transfer medium, containment, insulation, and interface to heat engine

Heat engine, comprising a thermal interface, working fluid, piston, regenerator, waste heat rejection, casing, and generator interface

Electric interface, comprising the electrical output from the heat engine, frequency conversion to DC or 60 Hz, combination of signals from many engines, transformation to power line voltage, and conditioning of the output power

Control system, which includes such elements as collector tracking algorithms, safety interlocks, receiver and heat engine controls and interfaces, and dispatch/operation of the electric interface system

The last two items in this list are crucial to the success of any dish system but are rather outside the scope of this chapter. Good solutions seem to be available for the electric interface using solid-state electronic switching, and so on. The heat engine will be discussed with respect to its interaction with the receiver, but in no more detail than required. The working fluid in most such engines is high-pressure hydrogen or helium operating at 600°C to 800°C. Kinematic Stirling motors couple a power piston and a displacer cylinder via a crankshaft or a swash-plate. This rotating member drives a standard electric generator (Fig. 21A.6B). Alternatively the free piston Stirling motor uses no crankshaft and employs moving magnets mounted directly to the piston or displacer to couple to internal or external coils, producing the generated EMF directly (Stine and Diver, 1994) (Fig. 21A.6A). Thus, the heat engine–generator assembly can be delivered as factory-sealed hermetic units with no external drive shaft. Additional internal seals and bearings may be eliminated by use of close tolerances, linear motions, bellows, and leaf springs, resulting in the potential for very high reliability.

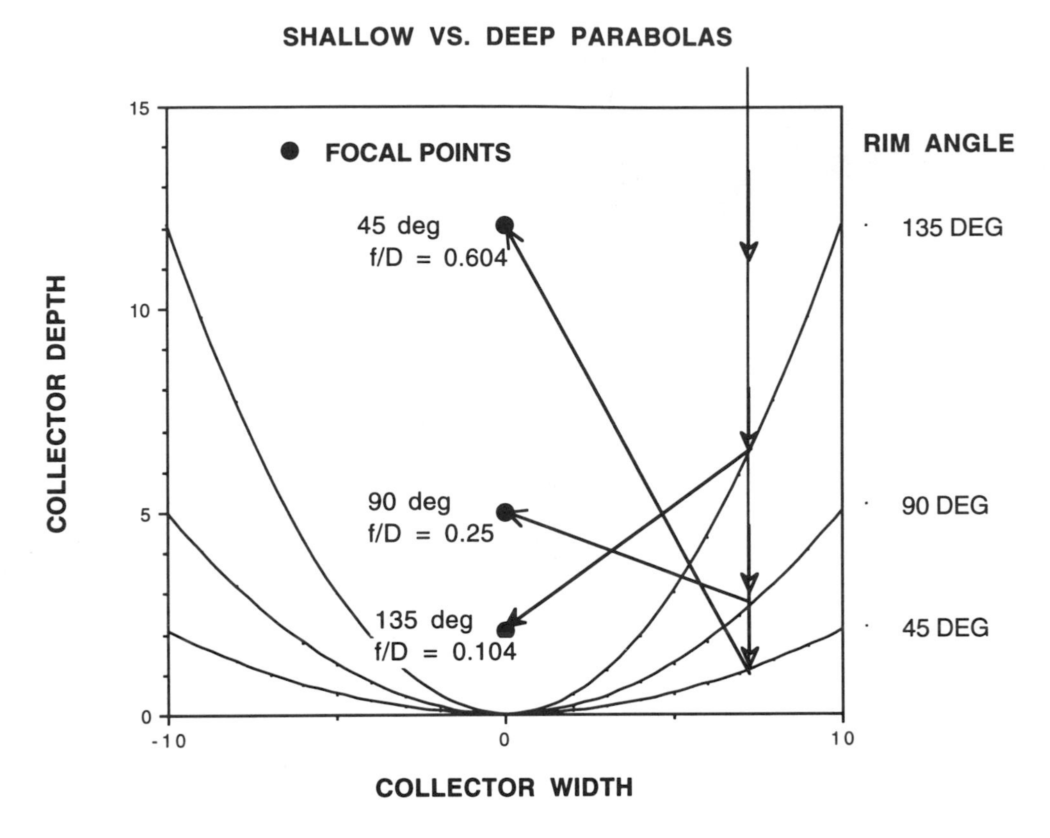

FIGURE 21A.7 Comparison of large *f/D* ratio (shallow) dishes suitable for use with a cavity or secondary concentrator, and small *f/D* ratio (deep) dishes.

Concentrator Design

Dish electric systems can produce high concentrations and do not require a thermal transport system, so they can easily produce the 600°C to 800°C working temperature required by an efficient heat engine. At these temperatures radiation losses from an external receiver are severe, so cavity receivers are usually specified. This limits the *f/D* ratio of the parabola to be substantially greater than 0.25 ($\psi_{rim} < 90°$). Deeper dishes (Fig. 21A.7) use their surface area less efficiently (cos ι effect), require more curvature of their reflectors, and are more cumbersome because of the three-dimensional support structure required. Consequently, the cavity imposed limitation to larger *f/D* ratios is not a significant restriction on the parabolic concentrators.

Stretched membrane dishes can take several forms. A rigid structure can be formed by stretching a thin aluminum or stainless steel membrane over both faces of a ring about one-tenth-diameter thick. After being bonded in place, a slight vacuum is applied, causing the surface to elasticity deform into an approximate parabola. For deeper dishes, it may be necessary to *plastically* deform the surface as well. This is accomplished either by applying air pressure to the "open" surface, or by "hydroforming," wherein the dish is filled with water to specified depths as a vacuum is applied to the underside, resulting in plastic deformation to a near ideal parabolic shape. Dishes 7 to 17 m in diameter have been built, achieving good accuracy (i.e., a good "optical figure" matching the desired parabola quite well). As an alternative, a number of smaller facets (a meter or two in diameter) can be mounted on a parabolic space frame to achieve the same effect. The smaller facets do not require plastic deformation and are much easier to handle than the larger monolithic dishes. However, they do bypass some sunlight and require individual alignment and focusing. The focusing in either case is generally accomplished by adjusting the air pressure between the two membranes. One advantage of such systems is that they can be easily and rapidly defocused in time of need by changing the internal pressure. An alternative design uses formed metal triangular or trapezoidal elements called gores, joined together to form a circular parabolic reflector.

Each of these types of mirrors is lustered by bonding to it either thin, somewhat flexible, silvered low-iron glass (91–93% reflectivity), an aluminized plastic (80–90% solar reflectivity), or a silvered acrylic film (94% reflectivity). The glass is somewhat fragile, but such reflectors have very good reflectivity and long lives. The plastic films are easier to apply, but lifetime and price have been serious problems. Extensive research and cooperative development programs are in place to increase the projected life from 5 to over 10 years at over 90% reflectivity, but such a lifetime has not yet been demonstrated (Kennedy and Jorgensen, 1994).

The alternative to the stretched membrane is the old standby, metal–glass facets similar to the heliostats at Solar One. In this application the support must have a parabolic shape, and the facets (unless very small) are also shaped to a shallow spherical curvature. Ten to a hundred facets are mounted to a space frame or steel truss assembly to form the dish, and the engine is supported above it by further trusses, or a tripod, just as for the faceted membrane concentrator. The mirrors can be self-supporting thick or laminated glass facets, or they can be formed metal sheets, lustered as described earlier.

The layout of an array of dish electric systems is very simple. The electric cabling is relatively inexpensive (compared to heat transport piping), so there is little incentive to crowd the units together. The only optical interaction between units is shading, resulting in a pedestal spacing of 3 to 4 diameters E–W, and 2 to 3 diameters N–S.

Of course, the site must be selected to provide good direct beam sunlight. As thermal storage beyond a few minutes is difficult to incorporate into these modules, it is important that the load match the insolation availability or that there be a backup energy source. The cavity receiver and the high operating temperature make this an ideal candidate for a fossil-hybrid system. One option is to heat the absorbing surface directly with a specially designed gas fired burner inside the cavity; an alternative that has been demonstrated (Christiansen, 1995) is to use a secondary heat absorbing surface for the gas burner.

Receiver Design

Earlier generations of dish receivers used pumped heat transfer fluids carried in tubes through the concentrated flux area and then to the central heat engine. More recent "direct illumination receivers" (DIRs) consist essentially of an enclosure or cavity constructed around the head or heat exchanger of the heat engine. The incident solar flux is redistributed somewhat by the cavity, but both the incident solar and the cavity infrared radiation impinge directly on the thermal absorbing surface of the heat engine. In either design, the heat absorbing surface is exposed to rapid fluctuations in flux and temperature—with time due to variations in insolation, and in space due to the distribution of the incident flux.

An alternative is to bathe the heat transfer surface of the heat engine in a saturated high-temperature alkali metal vapor (typically Na or NaK). The vapor condenses on any cooler region it encounters, typically the head of the heat engine, delivering thermal energy upon demand at rates up to 800 kW/m^2 and with a temperature variation of only a few degrees. The condensed liquid metal is returned to the evaporating side of the receiver (refluxed) via gravity flow. Such reflux receivers can use either pool boiling or the wicking action familiar from heat-pipe technology to transform the solar energy incident on their absorbing surface into latent heat in the high-temperature, high-pressure alkali metal vapor. Typically the absorber is a hemispherical shell serving as the back wall of the cavity, and is mounted directly to the Stirling or Brayton cycle heat engine.

If a Brayton cycle heat engine is employed, typically the working fluid (air) is heated in a volumetric receiver. In such a cavity type receiver a matrix of hot metal or ceramic fins or wires is directly heated by exposure to concentrated solar flux. The absorbed heat is transferred to the working fluid as it is drawn through the matrix.

Current Efforts to Commercialize Dish Electric Systems

Currently Industry and the DoE are participating in three industry-led cost-shared programs to develop specific commercially oriented dish engine systems (DoE, 1994, 1995).

Dish Stirling Joint Ventures. In 1988 Cummins Power Generation Inc. (Cummins) determined to enter the market with a 5 to 10 kWe dish/Stirling system directed at remote applications (e.g., in developing countries). To this end, Cummins initiated development of solarized Stirling motors integrated with a heat-pipe receiver developed by Thermacore, Inc. They also acquired rights to the La Jet dish concentrator, a lightweight, 460 ft^2 (43 m^2), 24-facet, stretched membrane collector supported by a space frame assembly and mounted on a declination-over-polar-axis drive. Seven hundred of these collectors comprised the 1980s Warner Springs thermal collection field. In 1991 Cummins entered into an $18.3 million, 5-year cost-sharing arrangement with DoE to commercialize their dish Stirling system design. Supported by Sandia, Thermacore has tested several versions of heat pipe receivers for thousands of on-sun hours, and Cummins is working out problems on a first-generation 7 kWe free piston Stirling alternator. 1995 plans call for placing four second-generation 7 kWe systems at various test sites and cooperating utilities around the United States. In 1996 Cummins plans to manufacture 25 of the 7 kWe systems and begin their distribution to remote sites. Cummins is expected to produce 50 systems in 1997 at a projected price estimated to be in the $12,000/kWe range.

Thus, with a relatively small injection of cash and the substantial design and testing support of Sandia, DoE will help Cummins bring small-scale standalone solar thermal electric plants to the market. It is not clear whether the political, technical, or financial support is more important in this case. It is clear that the success of such a development has as much benefit to the country as to the company, as it will bring jobs, increase exports, reduce worldwide CO_2, and aid development of remote regions of the earth. Thus, it is appropriate that the federal government assist the company in overcoming the risks (both real and perceived) associated with such a commercial development.

The Utility Scale Joint Venture Program (USJVP). In addition to a reliable 7 to 9 kWe standalone dish electric system, there are applications for a grid connected dish electric system: the demand side of overloaded transmission lines, rapid capacity addition, large arrays, and so on. In this case economy of scale wins over ease of handling and convenience, so a nominal 25 kWe system using a nominal 100 m^2 collector has been selected for development. In this arena, DoE is participating in two 50% cost shared joint ventures, contributing technical expertise and about $18 million to each.

The Cummins Power Generation Inc./Thermacore Inc. USJVP. Cummins is integrating the system design for this joint venture initiated in January 1994, and will produce the 25 kWe Stirling engine. It is currently developing a free piston engine with the assistance of the Clever Fellows Innovation Consortium and is retaining as a backup a kinematic engine developed and tested on-sun by Aisin-Seiki of Japan. Thermacore will provide heat pipe receivers rated at 90 kW thermal, following a design that has already been tested on-sun at Sandia. Based on life-cycle costs, Cummins has selected a concentrator constructed from trapezoidal metal gores joined to form a full parabolic dish, and lustered by thin silvered glass, rather than the faceted stretched membrane collector used in the DSJVP. The current activity will lead to construction of prototypical samples of each system element and their integration for a full system test in 1996.

The Science Applications International Corp. (SAIC)/Stirling Thermal Motors (STM)/Detroit Diesel Corp. (DDC) USJVP. This joint venture was initiated in November 1994, with SAIC as the system integrator. SAIC will also supply the concentrator, a 16-facet, 970 ft^2 (90 m^2), stretched

FIGURE 21A.8 Photograph of an operating dish Stirling system. The SAIC 16 facet 90 m^2 stretched membrane dish is shown driving the Stirling Thermal Motors 24 kWm heat engine, which produces 20 kWe at 1,800 rpm. *Source*: Roger Davenport, Science Applications International Corporation (SAIC).

membrane collector supported by a space frame assembly and mounted on an elevation-over-azimuth drive. It is lustered by 0.7-mm-thick silvered glass mirrors giving a reflectivity of 90%. STM will provide both a directly illuminated receiver (DIR) and a third-generation kinematic Stirling engine. The engine is based on the previously tested STM 4-120 four-cylinder engine. It uses a variable-angle swash-plate to allow control of the operating cycle and a pressurized crankcase to minimize seal problems. A gas-fired hybrid receiver that uses the DIR heater head is also being developed. DDC will manufacture the cold parts of the engine, as well as assist in design for manufacturability decisions.

Phase 1 is scheduled for completion in 1995 and includes on-sun operation of the dish Stirling system shown in Figure 21A.8, extended laboratory testing of 10 Stirling engines, and fabrication of the hybrid/DIR receiver. Phase 2 will involve fabrication of an alternate receiver design (perhaps a heat pipe receiver) and selection, redesign and fabrication of five complete dish Stirling systems for testing at utility sites.

References

Anonymous, 1988. Arizona Public Service Utility Solar Recoiler Study, Report No. DOE/AL/38741-1, November.

Anonymous, 1985. Proceedings of the Department of Energy Solar Central Receiver Annual Meeting, SAND85-8241, Sandia National Laboratories, Albuquerque, NM.

Bohn, M., Williams, T., and Price, H., 1995. *Combined Cycle Power Tower*, Proceedings of the ASME Solar Energy Division Meeting, March 19–24, Maui, Hawaii.

Burch, B.D., 1993. *Solar Thermal Electric, Five Year Program Plan, FY 1993 Through 1997*. Department of Energy, Washington, D.C.

Christiansen, J., 1995. Technical Status of Cummins' Dish/Stirling Systems, SOLTEC 95, San Antonio, TX.

Couch, W.A., ed., 1989. Proceedings of the Annual Solar Thermal Technology Research and Development Conference, SAND89-0463, Sandia National Laboratories, Albuquerque, NM.

De Laquil, P. et al., 1990. *PHOEBUS Project: 30 MWe Solar Central Receiver Plant Conceptual Design*, ASME, Solar Engineering—1990, p.25–30, Twelfth Annual ASME Intl. Solar Energy Conference, Miami, FL.

Diver, R.B., ed.,1986. Proceedings of the Solar Thermal Technology Conference, SAND86-0536, Sandia National Laboratories, Livermore, CA, N. T. I. C.

DoE, 1992. Solar 2000, A Collaborative Strategy, Dept. of Energy, Office of Solar Energy Conversion, Washington, DC.

DoE, 1994. *Solar Thermal Electric; Annual Summary, Fiscal Year 1994*, Dept. of Energy, Washington, DC.

DoE, 1995. *Solar Thermal Electric; Quarterly Progress Reports, Jan 1995, April 1995*, Dept. of Energy, Washington, DC.

Droher, J.J., 1986. *Performance of the Vanguard Solar Dish-Stirling Engine Module*, Technical Report EPRI AP-4608, Electric Power Research Institute, Palo Alto, CA.

Grasse, W., Hertlein, H.P., and Winter, C.-J., 1991. Thermal Solar Power Plants Experience. In *Solar Power Plants*, ed. C.-J. Winter, R.L. Sizmann, and L.L. Vant-Hull, pp.215–282. Springer-Verlag, Heidelberg, New York.

Hillesland, T., and Weber, E.R., 1990. Utilities' Study of Solar Central Receivers. In *Solar Thermal Technology—Research Development and Applications*, Proc. 4th Intl. Symposium, Santa Fe, NM, pp.165–176, ed. B.P. Gupta and W.H. Traugott, Hemisphere Publishing Corporation, New York, NY.

Horst, B., 1979. *Regional Labor Cost and Manhour Commitments for Selected Solar Technologies*, Science Applications, Inc., Sept. 1979.

Kennedy, C., and Jorgensen, G., 1994. State-of-the-Art Low-Cost Solar Reflector Materials. Presented at the 8th International Vacuum Web Coating Conference, Las Vegas, Nov. 6–8, 1994. Also available as NREL technical report TP-471-7022. National Renewable Energy Laboratory, Golden, CO.

Kolb, G.J., Chavez, J.M., Klimas, P.C., Meinecke, W., Becker, M., and Kiera, M., 1994. Evaluation of Second Generation Central Receiver Technologies. In Proceedings of the 1994 ASME International Solar Energy Conference, March 27–30, San Francisco, CA. ASME, New York.

Lipps, F.W., and Vant-Hull, L.L., 1978. A Cellwise Method for the Optimization of Large Central Receiver Systems, *Solar Energy*, 20:505–516.

Lotker, M., 1991. *Barriers to Commercialization of Large-Scale Solar Electricity: Lessons Learned from the LUZ Experience*, Contractor Report SAND91-7014, Sandia National Laboratories, Albuquerque, NM.

Mancini, T.R., 1994. *An Overview: Component Development for Solar Thermal Systems, SAND 94-2431*, Proceedings of the 7th International Symposium on Solar Thermal Concentrating Technologies, Sept 26–30, 1994, Moscow, Russia.

Meinel, A.B., and Meinel, M. P., 1976. *Applied Solar Energy, an Introduction*, Addison-Wesley, Reading, MA.

Munjal, P.K., 1985. *Solar Central Receiver Preliminary Design Studies*, Report ATR-86(5836)-IND (DE85011385), The Aerospace Corp., El Segundo, CA.

Noble, J., Olan, R., White, M., Kesseli, J., Bohn, M., School, K., and Becker, E., 1995. *Test Results from a 10 kWe Solar/Natural Gas Hybrid Pool Boiler Receiver*, Proceedings of the ASME Solar Energy Division Meeting, March 19–24, Maui, Hawaii.

Radosevich, L.G., 1988. *Final Report on the Power Production Phase of the 10 MWe Solar Thermal Central Receiver Pilot Plant*, SNLL Technical Report SAND87-8022, Sandia National Laboratories, Albuquerque, NM.

Short, W.D., 1988. *Optical Goals for Polymeric Film Reflectors*, SERI/SP-253-3383; National Renewable Energy Laboratory, Golden, CO.

Stine, W.B., and Diver, R. B., 1994. *A Compendium of Solar Dish/Stirling Technology*, Technical Report SAND93-7026, Sandia National Laboratories, Albuquerque, NM.

Tyner, C.E., ed., 1987. Proceedings of the Solar Thermal Technology Conference, SAND87-1258, Sandia National Laboratories, Albuquerque, NM.

Vant-Hull, L.L., 1991. Concentrator Optics. In *Solar Power Plants*, ed. C.-J. Winter, R.L. Sizmann, and L.L. Vant-Hull, pp.215–282. Springer-Verlag, Heidelberg, New York.

Vant-Hull, L.L., 1992. Solar Thermal Electricity: An Environmentally Benign and Viable Alternative. *Perspectives in Energy*, 2:157–166.

Vant-Hull, L.L., and Pitman, C.L., 1990. Static and Dynamic Response of a Heliostat Field to Flux Density Limitations on a Central Receiver. In *Solar Engineering*, ed. J.T. Beard and M.A. Ebadian, Book No. H00581-1990, ASME, New York.

Walzel, M.D., Lipps, F.W., and Vant-Hull, L.L., 1977. A Solar Flux Density Calculation for a Solar Tower Concentrator Using a Two-Dimensional Hermite Function Expansion. *Solar Energy*, 19:239–253.

21B

Solar Thermal Power and Industrial Heat: Parabolic Trough Concentrating Collectors Component and System Design

Randy C. Gee

E. Kenneth May

Industrial Solar Technology Corporation

Introduction

Parabolic trough collectors are the most developed and deployed type of solar concentrator. The advanced status of trough technology results mainly because it is simpler than other concentrator systems: troughs track the sun in one axis (not two), and receiver temperatures are generally more moderate than in collector types operating at higher concentrations. The inherent modularity of parabolic troughs is another important advantage. Troughs can be configured in any appropriate collector area to meet the desired load, and their simple arrangement in parallel rows eases field layout and minimizes interconnecting piping.

Parabolic troughs are also the most flexible of solar heat technologies. Their potential range of application covers heating hot water for domestic, commercial, or industrial uses; heating air; providing building heat or industrial drying; providing heat to absorption chillers or desiccant

0-8493-2514-5/96/$0.00+$.50

cooling systems; and generating steam for a host of commercial and industrial applications, as well as for the production of electricity. In addition, troughs can be economically deployed on a much smaller scale than a field of heliostats, while the development of parabolic dish systems is directed primarily toward electric-only applications.

The design of parabolic trough components and systems is described in this chapter. Methods and design tools are presented that emphasize practical ways to evaluate and analyze the performance of trough components and systems. Readers interested in additional details regarding the optics, heat transfer, performance testing, system performance prediction, or system economic analysis may consult the many texts and reports that have been referenced.

21B.1 Parabolic Trough Fundamentals

The goal in designing a parabolic trough is to create a mechanical structure that economically concentrates the solar energy falling on the large reflector area onto the smaller receiver area. In reducing the area of the receiver, thermal losses are reduced, so allowing efficient operation at high temperatures. Concentration ratios (i.e., reflector aperture area/receiver surface area) are typically in the range of 15 to 30. Optical concentration to this level requires the trough concentrators to rotate to follow the apparent motion of the sun.

Figure 21B.1 (Harrigan, 1981) shows the principal parts of a parabolic trough. The reflector has a parabolic contour and is made of or covered with a material that has a high specular reflectance. The receiver is a black-colored steel absorber tube surrounded by a cylindrical glass envelope to reduce thermal losses. Generally a selective surface is used on the absorber tube to further reduce thermal losses. The parabolic concentrator reflects direct solar radiation onto the receiver located at the focus of the parabola. As the absorber tube is heated by solar radiation, a fluid flows within the tube and withdraws the heat.

To keep the concentrated sunlight on the line of the receiver throughout the day, the collector must track the sun. A drive system is employed to provide the necessary one-dimensional rotation. Sometimes an electric motor and gearbox combination is used as the powering mechanism (as shown), some designs use hydraulic drives, and some use multirow drive systems in which a single motor and linear actuator are used to orient several parallel rows of collectors with an interconnecting wire rope/pulley system. Pylons between each trough module support them off the ground and allow for their rotation as they track the sun. The pylons are normally fastened to anchor bolts that are embedded in concrete foundations. Most trough systems are ground mounted, so drilled piers are often used for the foundations.

To accommodate the movement of the receiver relative to the stationary piping on each end of the collector rows, a flexible connection is needed. Typically, flexible hoses or ball joints are used, although rotating unions have also been employed. The flexible connections also allow for the linear growth of the absorber tube due to thermal expansion.

21B.2 Tracking of Parabolic Troughs

Parabolic troughs are often installed so that their axes of rotation are oriented either north–south or east–west. However, any orientation can be suitable, and the choice is often dictated by the orientation of the land or roof areas on which the troughs are installed. Orientation affects the performance of the troughs, especially on a seasonal basis. Seasonal variations in collector output for troughs oriented north–south can be quite large. Three to four times more energy can be delivered during summer months than during winter months. For troughs oriented east–west, seasonal variations in energy delivery are generally much smaller, typically less than 50%, and reflect the seasonal variation in the amount of available insolation.

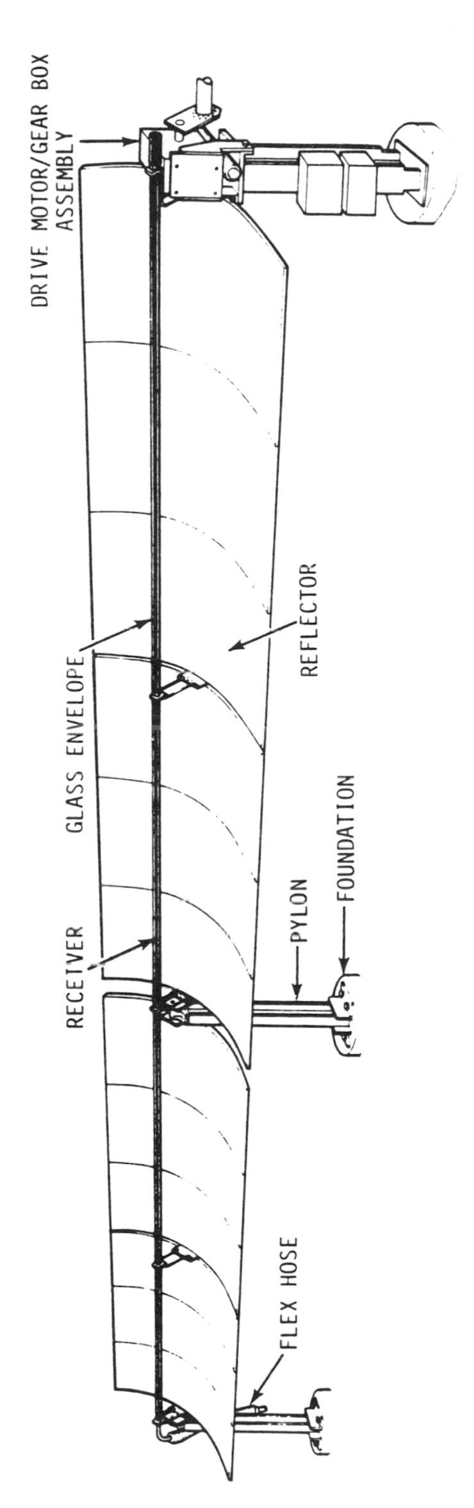

FIGURE 21B.1 Typical line-focusing collector design. *Source:* Harrigan, 1981.

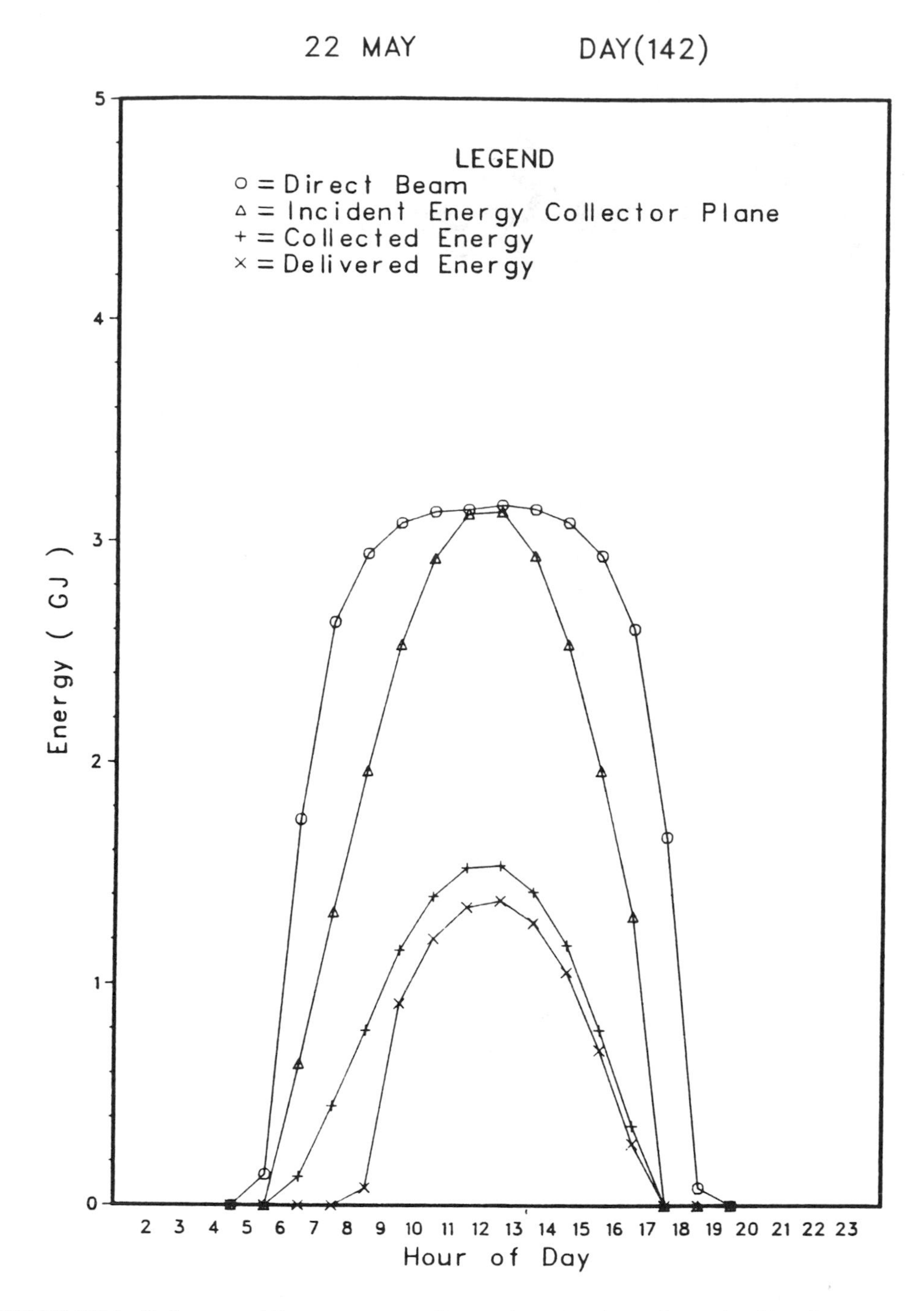

FIGURE 21B.2 Daily energy delivery: east–west collector orientation. *Source*: Lewandowski et al., 1984, p. 51.

Considering the output of a parabolic trough system on a daily basis, a trough system oriented east–west will achieve normal incidence (when it is directly facing the sun) every day at solar noon, when its thermal output will peak. Solar noon is that time of day at a particular location when the sun is directly due south. As shown in Figure 21B.2 (Lewandowski et al., 1984, p. 51), the profile of energy delivery with time is a bell-shaped curve that closely follows the pattern of the solar radiation incident on the collector plane. In contrast, a north–south trough only achieves normal incidence in the summer at times early in the morning and late in the evening. The system will

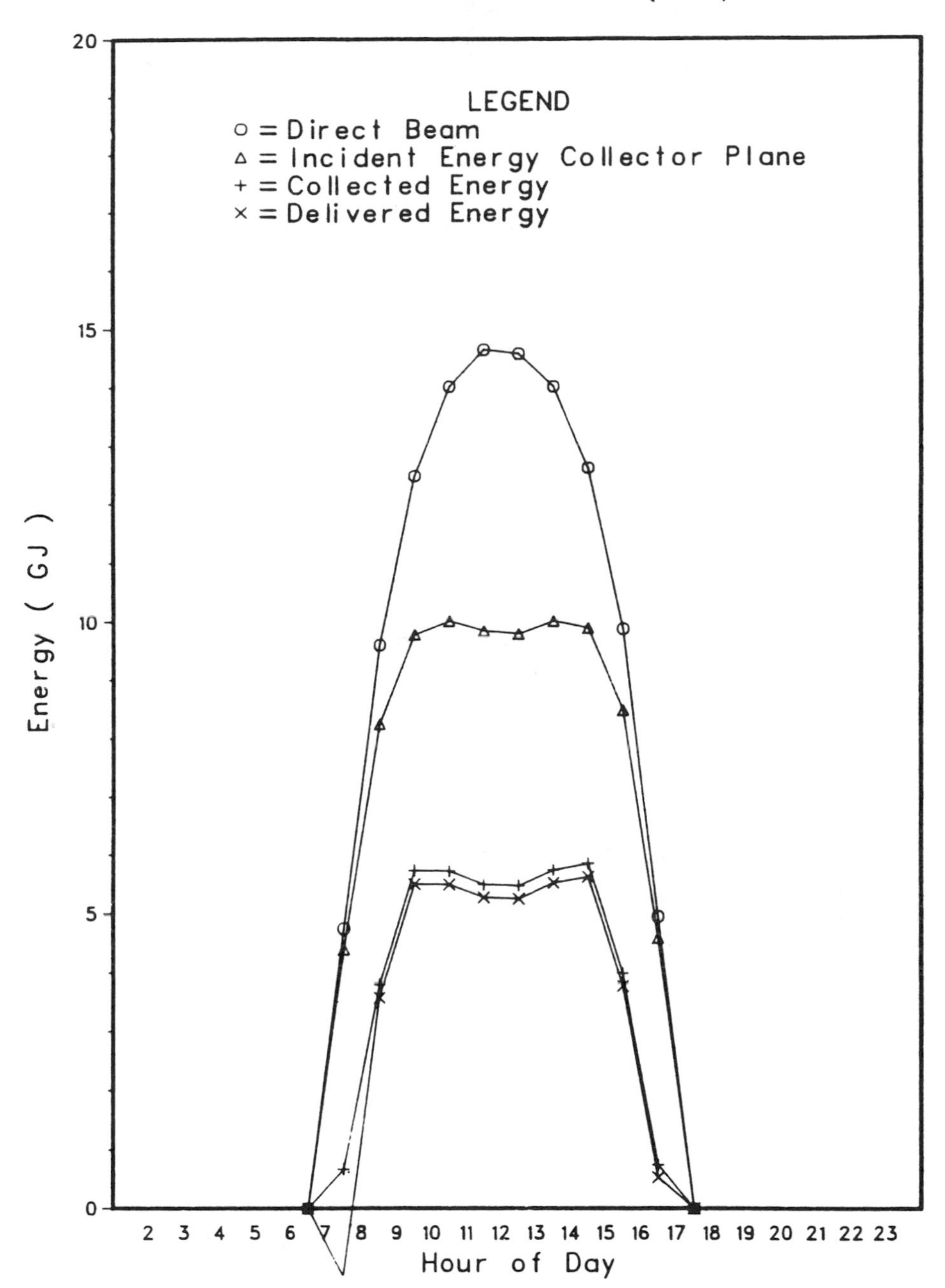

FIGURE 21B.3 Daily energy delivery: north–south collector orientation. *Source*: Lewandowski et al., 1984, p. 21.

not peak in output at these times, however, because the large air mass of the atmosphere at low sun elevations attenuates the intensity of the solar radiation. As shown in Figure 21B.3 (Lewandowski et al., 1984, p. 21), the energy delivery profile is an inverted, U-shaped pattern that more closely follows the shape of the direct-beam insolation profile than a trough system oriented east–west. Since the incident angle between the sun and the plane of the trough increases around noon, a system oriented north–south achieves peak performance twice during the day, several hours before and after solar noon. For systems not oriented along north–south or east–west lines, thermal performance is somewhere between the two extremes. Figure 21B.4 (Lewandowski et al., 1984, p. 57) illustrates the output of a system oriented 25° west of north. This configuration favors performance in the afternoon. For the day shown, thermal output peaks at about 2 P.M. standard time.

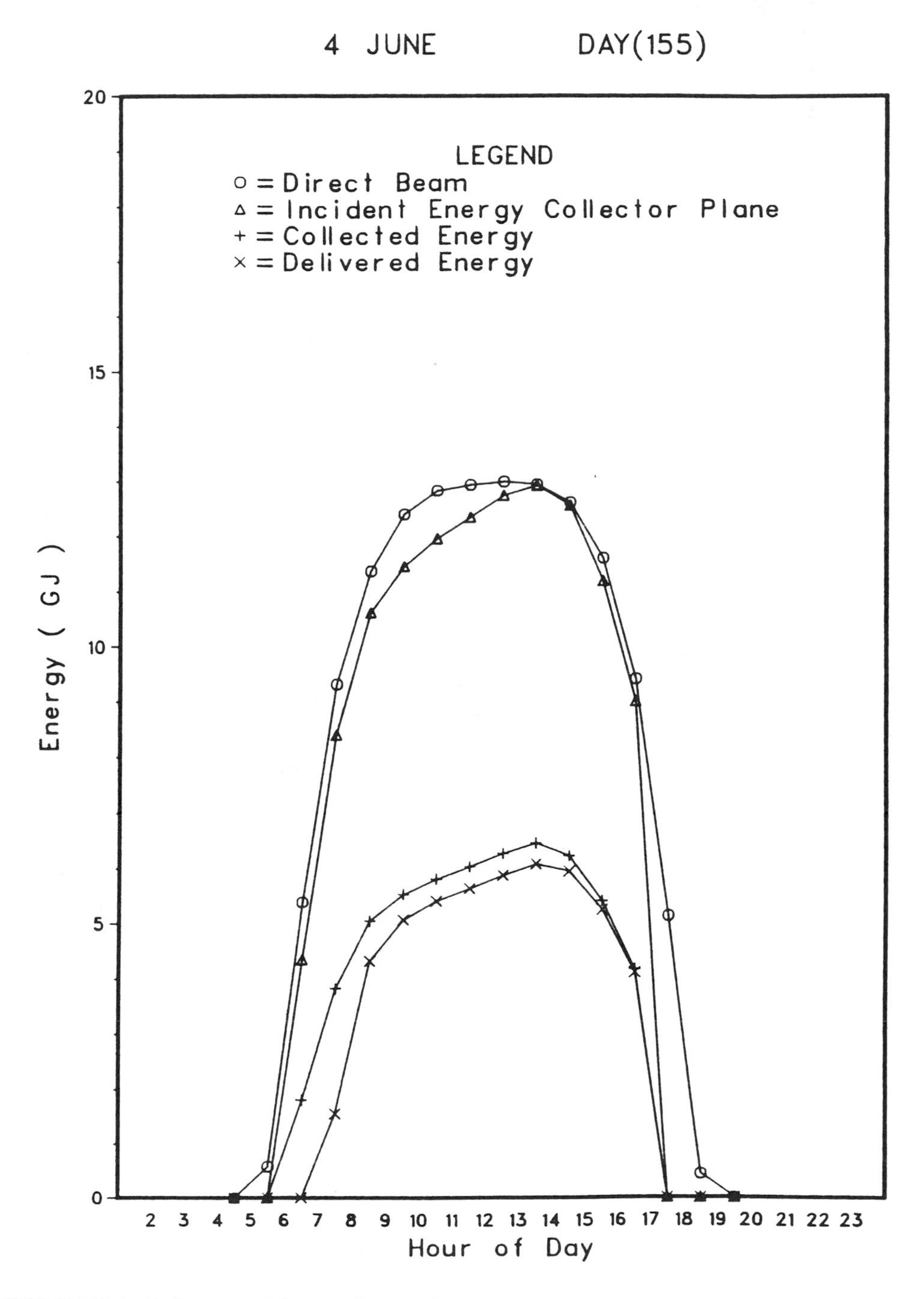

FIGURE 21B.4 Daily energy delivery: off-axis collector orientation. *Source*: Lewandowski et al., 1984, p. 57.

A single-axis parabolic trough rotates about its axis so that the sun is contained within the plane that extends normal to the collector's aperture and through the receiver. For horizontally mounted troughs,* the concentrator tracking angle (0° = faceup) can be calculated as

$$\text{Tracking angle} = \tan^{-1}\left[\frac{\cos\theta_{\text{axis}}\cos AZ\,\cos EL - \sin\theta_{\text{axis}}\sin AZ\,\cos EL}{\sin EL}\right] \quad \textbf{(21B.1)}$$

where AZ and EL define the sun's position, and θ_{axis} defines the direction of the rotational axis of the troughs, (θ_{axis} = 0° for east–west, and 90° for north–south).

This equation reduces to a simpler form for east–west and north–south rotational axis troughs mounted horizontally.

$$\text{N/S track angle} = \tan^{-1}\left(-\sin AZ/\tan EL\right) \quad \textbf{(21B.2)}$$

$$\text{E/W track angle} = \tan^{-1}\left(\cos AZ/\tan EL\right) \quad \textbf{(21B.3)}$$

Since parabolic troughs typically track in only one axis, the collector aperture is not always normal to the sun. Energy delivery at nonnormal incidence is corrected using the cosine of the incidence angle that the rays of the sun make with the normal to the collector aperture.

For horizontal parabolic troughs, the cosine factor at any given time can be conveniently calculated based on the sun's position (Lewandowski et al., 1984):

$$\text{East–west cosine of incident angle} = \left[1-\cos^2(EL)\,\sin^2(AZ)\right]^{1/2} \quad \textbf{(21B.4)}$$

$$\text{North–south cosine of incident angle} = \left[1-\cos^2(EL)\,\cos^2(AZ)\right]^{1/2} \quad \textbf{(21B.5)}$$

In general, troughs oriented north–south have significantly more available insolation (lower cosine losses) in the summer than do troughs oriented east–west. Over a full year, north–south troughs also have lower cosine losses than do east–west troughs. A comparison of energy incident in the plane of troughs oriented north–south and east–west, as a percentage of direct normal radiation, is shown in Table 21B.1 for several U.S. cities (NREL, 1994).

TABLE 21B.1 Available Insolation versus Direct Normal Radiation

City	North–South	East–West
Phoenix, Arizona	0.88	0.76
Boulder, Colorado	0.85	0.76
Sacramento, California	0.89	0.75
Miami, Florida	0.90	0.77

Note that cosine effects are generally less than 15% for north–south orientations and less than 25% for east–west orientations. This typically results in better overall energy production for a north–south collector field than for an east–west field. However, in situations where wintertime performance is most important or where seasonal variations in energy output should be minimized, an east–west orientation may be preferred.

*Tracking angle equations for horizontal and nonhorizontal troughs are provided in Gee (1982).

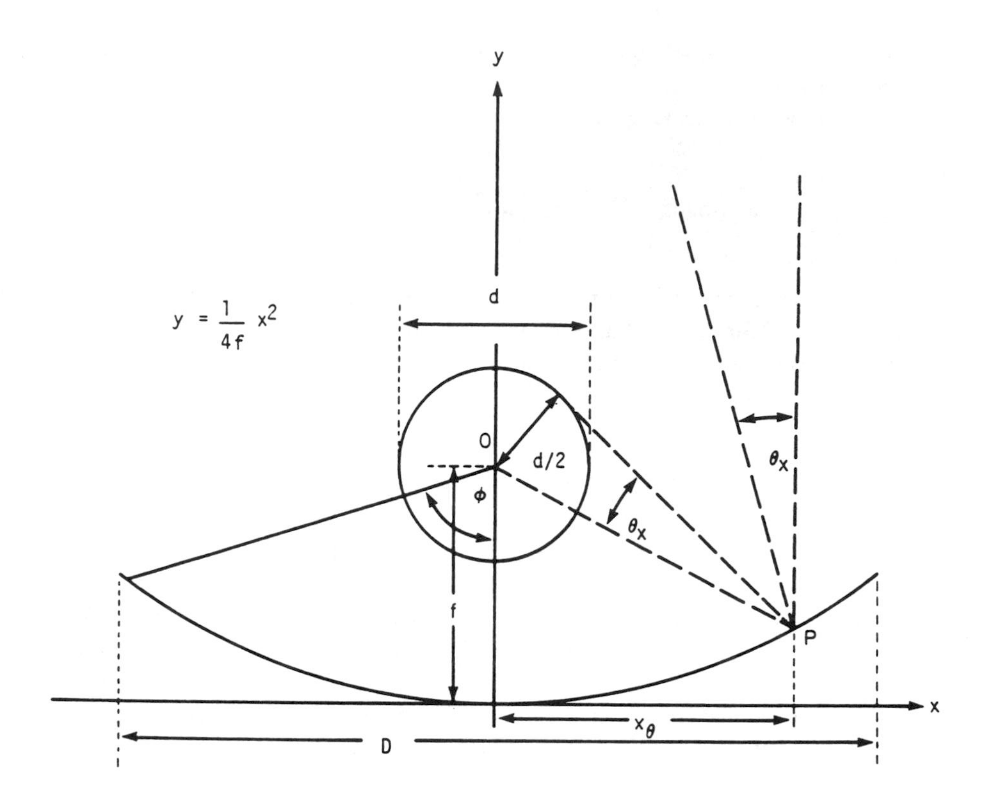

FIGURE 21B.5 Geometric relations for a parabolic trough concentrator. *Source*: Bendt et al., 1979.

21B.3 Concentrator Design

The concentrator intercepts and reflects the sun's rays to the receiver. The surface of the concentrator has a parabolic cross section so that all rays of light parallel to its axis reflect from the surface to a focal line. The basic geometrical relation that defines the shape of the concentrator is

$$y = x^2/(4f) \tag{21B.6}$$

where f = focal length of concentrator.

Figure 21B.5 (Bendt et al., 1979) defines the geometric relationships for a parabolic concentrator. D is the concentrator aperture width, d is the absorber tube diameter, and ϕ is the concentrator rim angle. The rim angle can be calculated from the following relationship:

$$D/(4f) = \tan(\phi/2) \tag{21B.7}$$

The geometric concentration ratio C is given as

$$C = D/(\pi d) \tag{21B.8}$$

Most parabolic troughs use concentrators with rim angles of between 70° and 110°, as this range provides a balance between concentrator optical and structural requirements. Rim angles near the low end of the range minimize concentrator arc length but have higher average concentrator-to-receiver distances. Rim angles near the upper end of the range minimize average concentrator-to-receiver distances but require longer concentrator arc lengths. Fortunately, the optimum choice of

FIGURE 21B.6 Small-scale trough (BSAR solar). *Source*: SERI, 1990.

rim angle is so broad that the choice of rim angle for a trough can be determined by other considerations, such as mechanical strength and ease of manufacturing.

A number of different designs have been developed for parabolic trough concentrators. All must provide rotational rigidity to resist wind loads. To achieve this, some designs use torque tubes, some use monocoque construction, and others use diagonal rods across the front aperture. Concentrator size varies considerably too; from small ½ × 2-meter modules (for example BSAR Solar, shown in Figure 21B.6 with monocoque construction), to medium-sized 2.3 × 6.1-meter modules (Industrial Solar Technology, shown in Fig. 21B.7 with diagonal rods), to the large 5.8 × 12-meter modules (Luz LS-3, shown in Fig. 21B.8, which uses a torque tube).

The reflective surface used on trough concentrators must have a high degree of specularity. Two materials are used extensively—silvered glass and metallized acrylic films. Silvered glass is durable and abrasion resistant, but is susceptible to breakage due to its brittle nature. Metallized films are flexible (not brittle) and lower in cost, but are less abrasion resistant and are still being actively developed to improve lifetimes. Both silvered glass and silvered acrylic films can achieve specular reflectance higher than 90%. Aluminized films are also available at lower cost, but specular reflectance falls in the mid-80% range. In addition to the optical error introduced by the finite specularity of the reflective surface, a number of other optical errors contribute to widening of the reflected beam at the receiver, such as concentrator slope error, tracking error and receiver mislocation errors. Generally, these errors can be characterized by Gaussian distributions, so the total amount of beam spreading can be calculated as the root mean square (rms) value of all the optical errors. Since the sun itself is not a point source of light, the sun's finite size is often approximated with an rms width (σ_{sun}) of 2.8 mrad. However, for line-focus optics, the apparent width of the sun increases with incidence angle* and a good average all-day rms width is 5.0 mrad.

Total beam spreading (σ_{total}) is the rms total of the optical errors and the sun shape expressed as

$$\sigma_{total} = \left(\sigma^2_{optical} + \sigma^2_{sun}\right)^{1/2} \qquad \textbf{(21B.9)}$$

*Bendt et al. (1979) show that the apparent width of the sun increases as $1/\cos\theta$ with incidence angle θ.

FIGURE 21B.7 Medium size trough: IST system supplying process hot water for the California Correctional Institution in Tehachapi, California. *Source*: Becker et al., 1995.

$$\sigma^2_{optical} = 4\sigma^2_{con} + \sigma^2_{track} + \lambda(\theta)\left(4\sigma^2_{con} + \sigma^2_{spec}\right) + \sigma^2_{disp} + \sigma^2_{spec} \tag{21B.10}$$

where

σ_{con} = rms angular deviation of concentrator from perfect parabola (slope error)
σ_{track} = rms angular spread due to sun tracking error
σ_{spec} = rms angular spread of reflected beam due to imperfect specularity of reflector
σ_{disp} = equivalent rms angular spread, which accounts for the imperfect placement of the receiver
$\lambda(\theta)$ = coefficient that accounts for the rim-angle-dependent contribution of longitudinal mirror errors to transverse beam spreading*

The intercept factor (the fraction of the reflected beam that is intercepted by the receiver) can be calculated using σ_{total}, the concentrator rim angle ϕ, and the concentration ratio C as shown in Figure 21B.9. Note that the intercept factor depends on the product σ_{total} C, not on C and σ_{total} separately. If σ_{total} has been estimated, various concentration ratios (i.e., receiver sizes) can be evaluated for the resulting impact on intercept factor. A typical design value for the intercept factor is 95%; that is, losses due to spillover amount to about 5%.

21B.4 Receiver Design

The receiver of a parabolic trough is typically a steel tube surrounded by a cylindrical borosilicate glass tube. Carbon steel is typically sufficient, but sometimes stainless steel is used for high-

*A value of 0.1 is recommended for $\lambda(\theta)$ for concentrators with rim angles between 80° and 100°.

FIGURE 21B.8 Large-scale trough: The Luz LS-3 collector. *Source*: Becker et al., 1995.

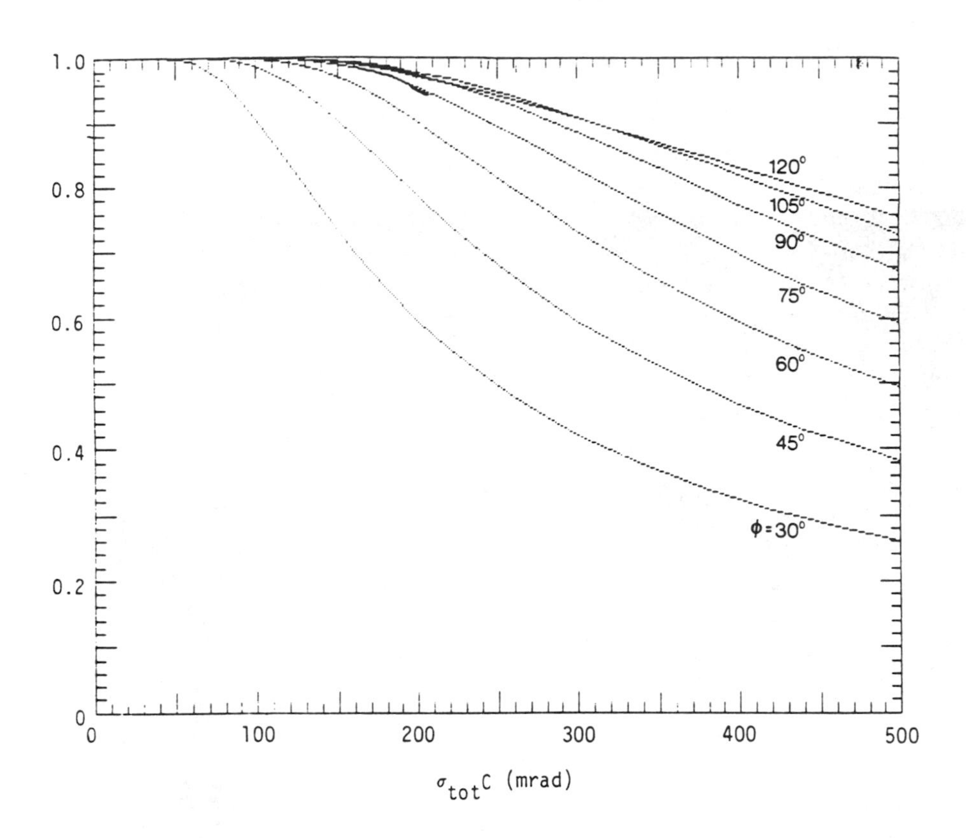

FIGURE 21B.9 Intercept factor vs. σ_{tot} for different rim angles. *Source*: Bendt et al., 1979.

temperature operation, particularly if the receiver is evacuated. The steel tube is normally coated with a selective surface to maximize absorptance of concentrated sunlight while minimizing thermal losses of the receiver. A popular selective coating is black chrome, although alternatives such as black nickel and cermet coatings have recently been developed by trough manufacturers. Selective surfaces give the absorber tube a high solar absorptance (around 0.95) and a low thermal emittance (around 0.15 to 0.30 depending on temperature). Cermet coatings have especially low thermal emittances and excellent thermal stability at temperatures as high as 400°C. Black nickel has a higher absorptance than the cermet coating and is lower in cost, but in an air environment is generally limited to operating temperatures below 300°C.

The glass cylinder that surrounds the absorber tube reduces the convective heat loss from the receiver. The receiver glazing diameter is sized to minimize the conductive/convective losses. Too small a glass diameter (small gap between the absorber tube and the surrounding glass) results in excessive conduction losses, whereas too large a glass diameter (large gap) results in excessive convective losses. The optimal gap is a function of the Rayleigh Number and so varies with temperature and the gap dimension, but is typically about 0.8 cm (Ratzel and Simpson, 1979, p. 23). Of course, the gap size must be large enough to ensure that the absorber tube cannot contact the outer glass because of geometrical errors, such as bowing of the absorber tube or glazing.

One way to further reduce heat loss from the receiver is to evacuate the space between the absorber tube and the glazing. This entirely eliminates conductive/convective losses between the absorber and glazing, thereby significantly reducing the overall heat loss from the receiver, although at the expense of a large increase in manufacturing costs. In a typical design, glass-to-metal seals are used to provide a vacuum-tight seal between the glass annulus and a convoluted metal bellows that accommodates the differential thermal expansion of the absorber tube relative to the glazing. Chemical getters in the receiver annulus absorb reactive gases that slowly evolve from the absorber during operation. Figure 21B.10 shows the design of the Luz LS-3 receiver employed on the later SEGS plants.

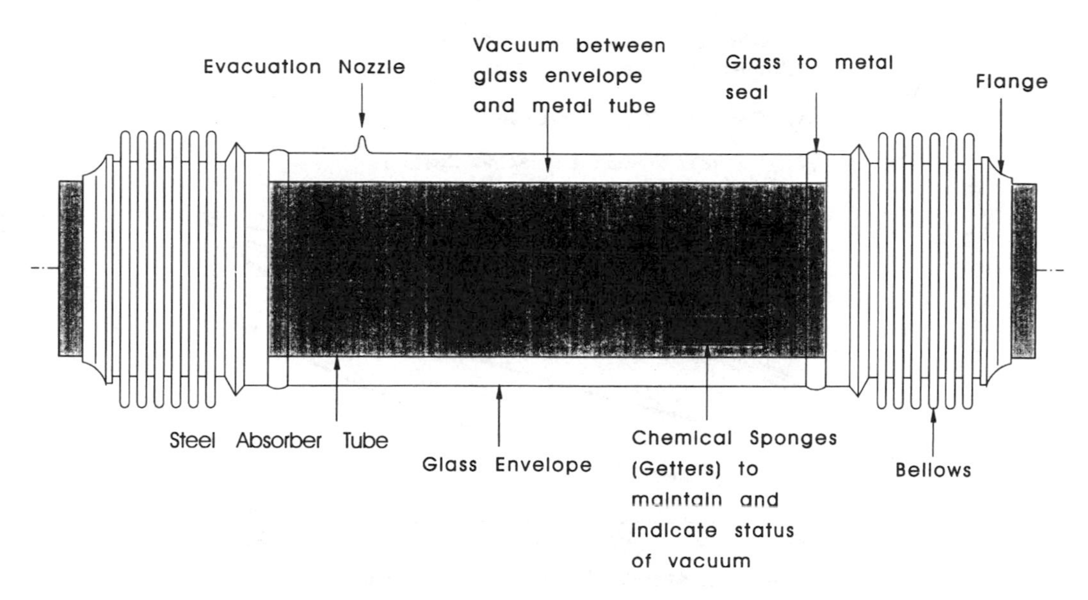

FIGURE 21B.10 Details of LS-3 receiver assembly. *Source*: Becker et al., 1995.

Prior to the development of an evacuated receiver and a cermet selective coating by Luz, the operating temperature range of parabolic trough collectors was limited to about 300°C. Above this temperature, most other selective coatings, including black chrome, degrade in an atmosphere of air. In addition, heat losses from a receiver with an air annulus operating above 300°C greatly reduce the collector efficiency. Driven by the desire to increase heat engine efficiency, Luz technologists pushed the outlet temperature of their later SEGS plants to 390°C. This was possible because the development of a reliable evacuated receiver greatly reduced thermal losses and the cermet selective coating was stable at high operating temperatures.

Heat losses from parabolic trough collectors can be calculated from the efficiency test data that is normally available from the manufacturer. The data shown in Figures 21B.11 and 21B.12, however, provide calculated thermal loss rates (Gee, 1980) for a variety of different receiver designs. These results are based on thermal modeling of trough receivers, so actual receiver losses may be somewhat higher due to thermal loss mechanisms that have not been accounted for, such as additional losses from supports or expansion bellows. The so-called "reference" receiver has an absorber tube diameter of 2.54 cm and a selective coating with an emittance of 0.15 at 100°C and 0.25 at 300°C, varying linearly between and beyond these limits. The receiver glazing has an emittance of 0.9. While the absorber tube diameter is fixed, the receiver glazing diameter is sized to minimize the conductive/convective losses. In addition to the reference receiver, four other lower-loss receiver designs are shown:

- Selective surface emittance reduced to 0.05 at 100°C and to 0.15 at 300°C (linear change with temperature)
- Receiver glazing emittance reduced to 0.15 (i.e., heat mirror coating)
- Xenon back-filled annulus, to reduce thermal conduction/convection
- Evacuated receiver

The heat-loss coefficients are shown as a function of average absorber tube temperature in Figure 21B.11. Relating heat loss to average absorber tube temperature eliminates the need to specify the fluid inlet or outlet temperatures, flow rate, and fluid properties. Also, the heat-loss coefficient is expressed in terms of absorber tube surface area,

$$U_L = \frac{\text{Heat loss rate}}{\left(T_{\text{abs}} - T_{\text{amb}}\right) A_{\text{abs}}} \tag{21B.11}$$

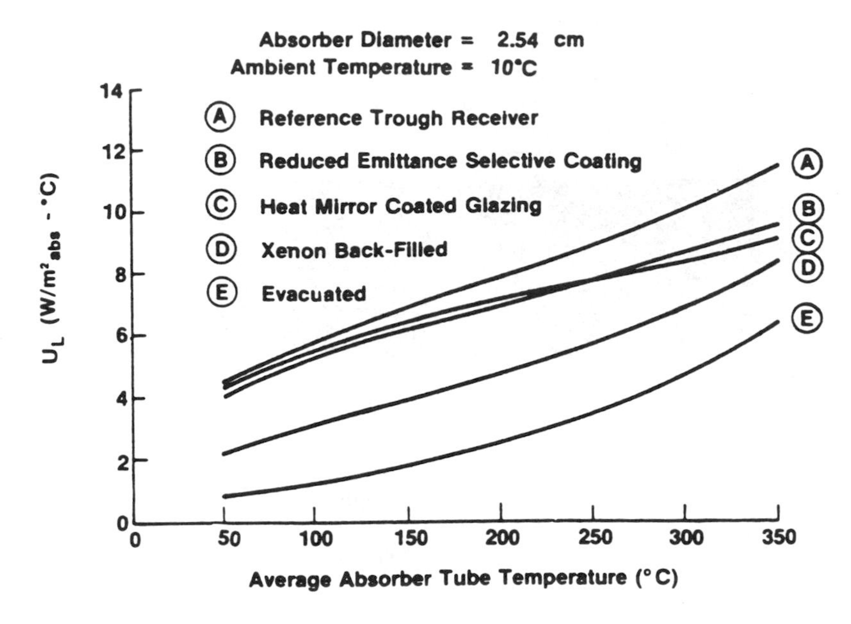

FIGURE 21B.11 Absorber heat loss coefficients vs. temperature. *Source*: Gee, 1980.

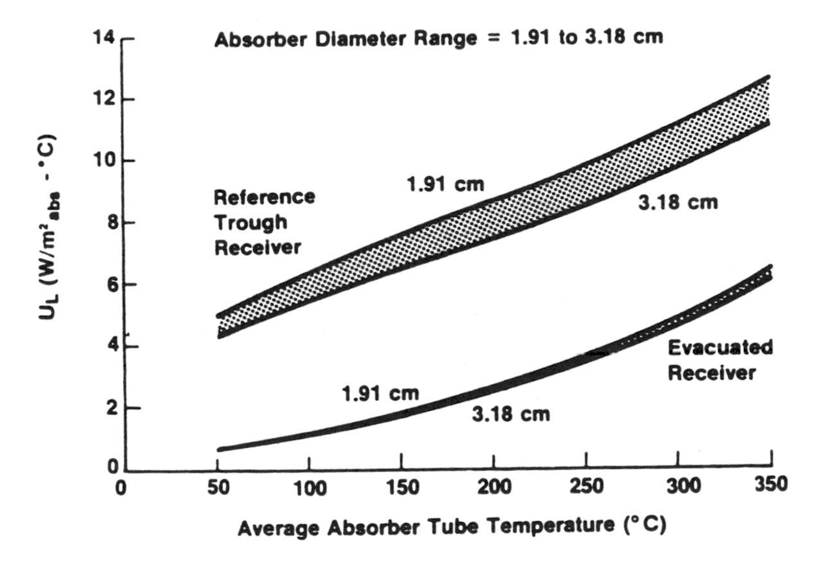

FIGURE 21B.12 Heat-loss coefficient variance with absorber diameter. *Source*: Gee, 1980 .

because on this basis the heat-loss coefficients are applicable over a wide range of absorber diameters, as shown in Figure 21B.12. The effect of diameter is larger for the reference receiver than for the evacuated receiver, because conductive/convective heat losses increase by slightly less than the increase in receiver diameter, whereas radiation losses are very closely a direct function of diameter. Larger-diameter absorbers result in smaller heat losses per unit absorber area, although the absolute heat loss rate is greater.

The graphs shown in Figures 21B.11 and 21B.12 are based on a constant outside ambient temperature of 10°C. This assumption is not nearly as restrictive as might be assumed. The heat loss coefficients do vary with ambient temperature, but the variance is especially small when the coefficients are expressed as a function of absorber temperature itself, rather than as a function of ΔT (absorber temperature above ambient) (Gee, 1981).

The transmittance of sunlight through a standard borosilicate glazing is typically about 0.91. This optical loss is very significant and largely arises from the reflectance losses of the glazing. Cost-effective techniques have been developed to reduce these reflectance losses using antireflective coatings. The two main techniques are phase separation and solgel coating. Phase separation involves producing a graded index coating at the surface of the glass through a process of heat treatment and etching (McCollister and Pettit, 1983). More recently, solgel coatings have been developed to provide antireflective properties. Solgel coatings can be applied at room temperature, but heat treatment at a temperature of about 500°C is needed to densify the coating and produce the desired properties. This is a lower temperature requirement than for the phase separation technique and offers a distinct advantage. Solgel research was pioneered at Sandia National Laboratories (Ashley, 1984) and has been commercialized by IST (1995). Overall solar-weighted transmittance can be improved to about 0.96 with this technique.

Determining the best size for the receiver involves trading off thermal losses against optical losses. As the absorber tube diameter is decreased, heat losses decrease, but so does the fraction of the incident solar radiation that is intercepted by the receiver. To properly predict the optimal concentration ratio, the impact of optical errors and thermal losses must be considered on an all-day basis, not just instantaneously. A concentration ratio optimization procedure that incorporates all-day affects has been developed specifically for parabolic troughs (Bendt et al., 1979). For most situations, the optimization is very straightforward and requires only two principal quantities,* Q_L/η_0 and σ_{tot}, which can be graphically related to the optimum concentration ratio as shown in Figure 21B.13. The first quantity is shown along the abscissa of the graph and is the heat-loss rate of the receiver per unit of absorber-tube surface area divided by the optical efficiency of the collector. Values for Q_L may be obtained from Figure 21B.11 for various types of receivers. The second quantity, σ_{tot}, is the total all-day average rms beam spread resulting from all sources of optical error and was defined earlier.

Note from Figure 21B.13 that the optimum concentration ratio is quite dependent on receiver heat-loss rate. For high temperatures (and, therefore, high Q_L values), concentration ratios of 20 to 30 are optimal. For lower temperature operation, concentration ratios should be significantly lower—in the 10 to 20 range. The value of σ_{tot} further defines the optimum concentration ratio within this range.

Examples

(A) Using Figure 21B.11, at an average operating temperature of 290°C (550°F), the heat-loss rate from a typical receiver is about 9.5 W/m^2 °C. The driving force for heat transfer is the difference between the absorber tube and ambient temperatures, 280°C, so that the overall heat loss rate is 2,660 W/m^2. For an assumed optical efficiency of 0.75, the x-axis value of Figure 21B.13 is 3,550 W/m$^2_{abs}$. A good all-day σ_{tot} value for a very optically precise parabolic trough is 8 mrad. (This is based on σ_{con} = 2.5 mrad, σ_{spec} = 2 mrad, σ_{track} = 2 mrad and σ_{disp} = 2 mrad.) For these values, the optimum concentration ratio is shown in Figure 21B.13 to be about 29.

(B) For a less optically precise parabolic trough operating at lower temperatures, Figure 21B.13 shows that a significantly lower concentration ratio should be used. For example, at 150°C the heat-loss rate from a typical receiver is about 924 W/m$^2_{abs}$ (140 × 6.6). Again, assuming an optical efficiency of 0.75, the x-axis value of Figure 21B.13 is 1,230 W/m^2. Choosing a σ_{tot} value of 14 mrad as an estimate of reduced-cost, lower-precision trough optics, the optimal concentration ratio is shown to be around 14. This is about half the concentration ratio recommended for the optically precise, high-temperature trough. Thus, a larger receiver by almost a factor of two is recommended for this lower-temperature receiver design.

*The sensitivity of concentration ratio optimization to other variables (e.g., geographic latitude, annual irradiation, and receiver shading) is very small, typically resulting in a concentration ratio change of less than one.

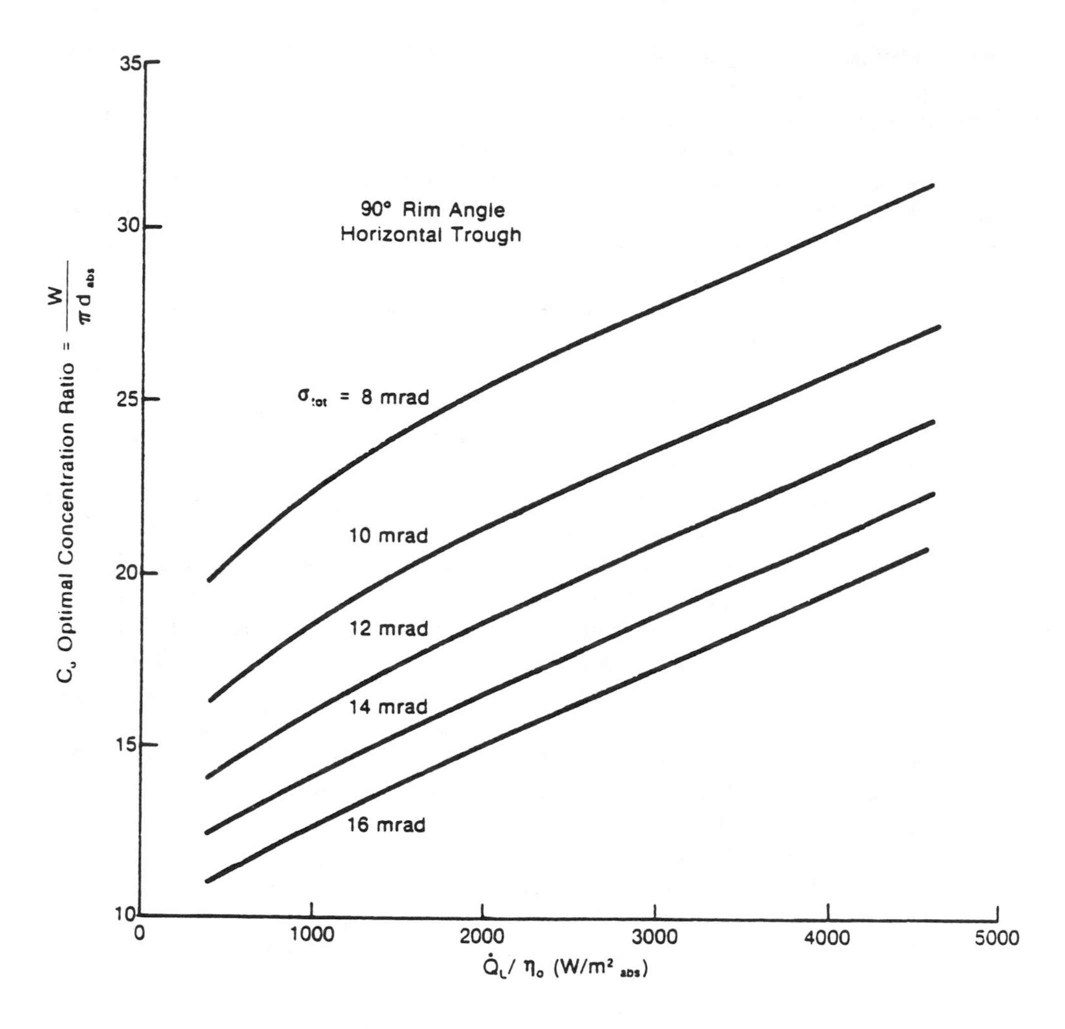

FIGURE 21B.13 Optimum concentration ratios vs. receiver heat loss. *Source*: Gee, 1983.

21B.5 Trackers, Drives, and Flexible Connections

Apart from the concentrator and receiver, several other parabolic trough components greatly affect the overall operation, performance, and cost of the system. These components include sun trackers, drives, and flexible connections. There are many different design approaches for each of these items.

Sun Trackers

Sun trackers are sensing elements used to orient the position of a solar collector toward the sun. There are three major types of sun trackers: computer trackers, aperture-based trackers, and flux-line trackers. A computer tracker uses a clock input to compute the sun's position and initiates collector rotation to this computed angle. Although quite simple in theory, this type of tracker requires an accurate method of determining the collector's angular position as it is rotated. The most commonly used device of this type is an optical shaft encoder. However, because high-resolution optical shaft encoders are expensive, computer tracking is not widely used.

The other two tracking methods, aperture-based tracking and flux-line tracking, utilize a simple feedback loop to track the sun. An error signal is generated by the sun-tracker sensor if the collector is not correctly pointed at the sun. When the error signal exceeds a threshold, the collector drive system initiates collector rotation until the tracker electronics are once again satisfied. Aperture-based trackers and flux-line trackers differ principally in where the sun tracking sensors are located. An aperture-based sun tracker can be mounted anywhere on the collector (usually near the center of the collector row) and rotates with it. The two sensors of a shadow-band tracker are separated by a partition or shadowing strip, which shades one of the sensors if the tracker is not pointed directly at the sun. The sensors produce an error signal when both sensors are not equally illuminated. The sensors of flux-line sun trackers are located at or near the receiver, and are sensitive to concentrated flux. As with aperture-based trackers, if the collector is mispointed, an error signal is generated, which in turn controls the rotation of the collector.

Each type of sun tracker has its advantages and disadvantages. Aperture-based trackers rely on a fixed relationship between the focal line and the tracking sensor. If the relationship of the sensor changes with respect to the focal line (e.g., because of deformation of the concentrator or sagging of the receiver tube), an aperture-based tracker cannot compensate with an adjusted tracking angle. A flux-line tracker can compensate for these effects because flux-line trackers actually sense the position of the concentrated flux relative to the receiver. Another disadvantage of aperture-based trackers is that they must be very carefully aligned in two planes: parallel with the rotational axis of the trough, and perpendicular to the collector's aperture plane. Flux-line trackers are much easier to install, since they are simply mounted to the receiver. A disadvantage associated with flux-line trackers is that the tracking angle is set based on only the relatively small concentrator area that the sun sensors are viewing. If this area is not representative of the overall concentrator (e.g., there is localized concentrator deformation, excessive dirt or dust soiling), the receiver may not be in optimum focus.

For a well-designed and well-maintained collector, both types of sun trackers are adequate if they are installed and adjusted correctly. The merits of individual sun trackers should therefore be judged principally on tracking accuracy, cost, durability, reliability, and control features.

Drives

Typically, troughs are arranged in parallel rows with several trough modules connected together in the row. At the center of each row of modules, a drive subsystem provides for rotation in order to track the sun. Various types of drive subsystems have been used. Many early designs used electric motor-driven gear reducers. Some, like the Luz LS-3 concentrator modules, use hydraulic drives.

The costs associated with drive and control systems can be reduced by connecting together large collector areas. For example, Luz-developed parabolic troughs progressed from 128 m^2/drive for the LS-1, to 235 m^2/drive for the LS-2, and then to 545 m^2/drive for the LS-3. Another way to reduce trough drive and control costs is by linking parallel rows of collectors together to share the same drive subsystem. This works because when individual collector rows are all parallel, they all need to be oriented at the same tracking angle throughout the day. Although a number of multirow drive systems can be conceptualized, Industrial Solar Technology has developed one that uses wire rope to interconnect collector rows via large pulleys located at the center of each row of collectors. A motorized linear actuator provides for translation of the wire rope, which in turn rotates all the pulleys and interconnected rows together. The IST design allows up to 500 m^2 of collector area to be driven with a single motor and linear actuator.

Flexible Connections

Flexible connections are needed at the ends of trough collector rows to accommodate the rotation of the concentrators as they track the sun throughout the day. Two main types of flexible connections have been used: flexible hoses (flex-hoses) and ball joints.

Flex-hoses are made of an inner hose and an outer stripwound metal hose. The inner hose carries the fluid and is generally made of a long convoluted bellows surrounded by a metal braid. The outer hose has an interlocking design so that it bends only so far, thereby defining a minimum bending radius. This minimum radius prevents overstressing of the inner hose, which would

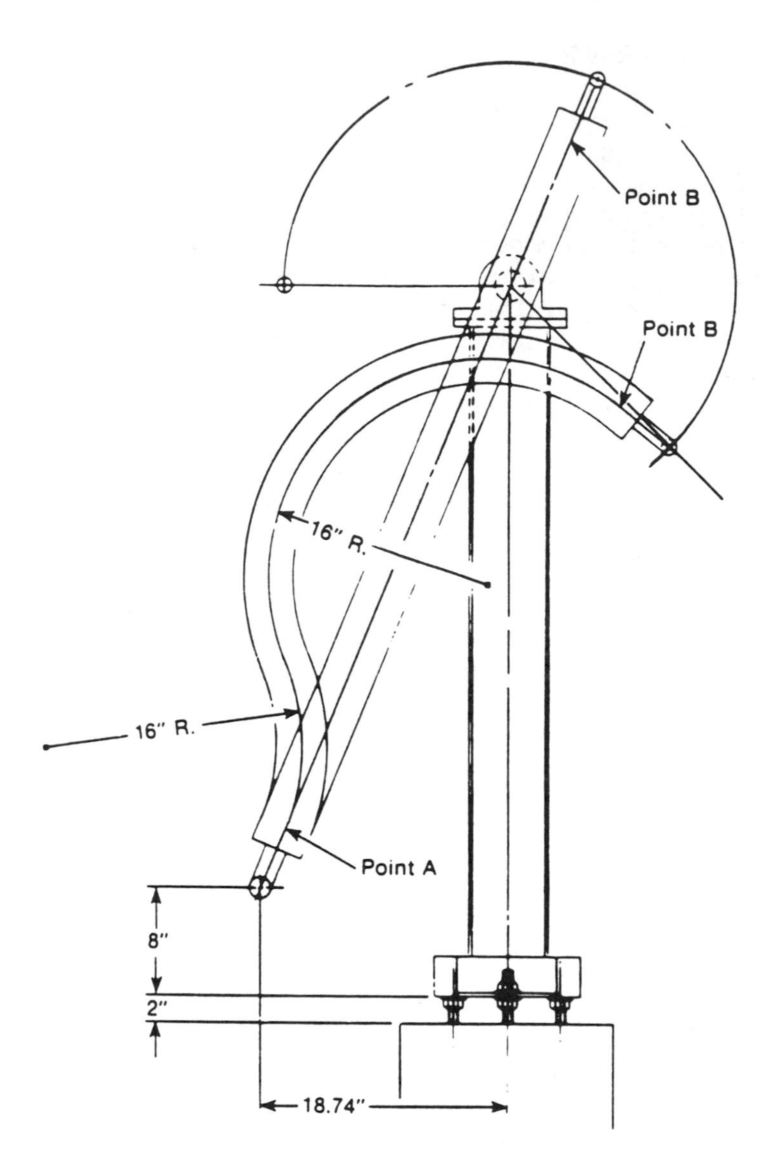

FIGURE 21B.14 Flex-hose installation. *Source:* Kutscher et al., 1982.

otherwise lead to fatigue failure, resulting in a leak. Flexible insulation between the inner and outer hoses reduces thermal loss from the assembly.

Another important flex-hose design requirement is ensuring that they are installed to flex in one dimension only. Torsional stress can cause early failure; thus, imposing such stresses during installation and operation should be avoided (Boyd, 1980). An example of correct flex-hose installation is shown in Figure 21B.14.

Ball joints offer an alternative to flex-hoses. KJC Operating Company has developed a design using commercial components that comprises two ball joints, connected by an insulated fixed-length pipe. One ball joint is located at the end of the receiver, and the other is located at the connection to the fixed field piping. The ball joints slip at these two locations to allow for the rotation of the receiver as the concentrator tracks, as well as to accommodate the thermal expansion of the receiver. KJC has determined that this design is more reliable than the flex-hose it replaces and also significantly reduces their solar field pressure drop and pumping power requirements (KJC, 1994).

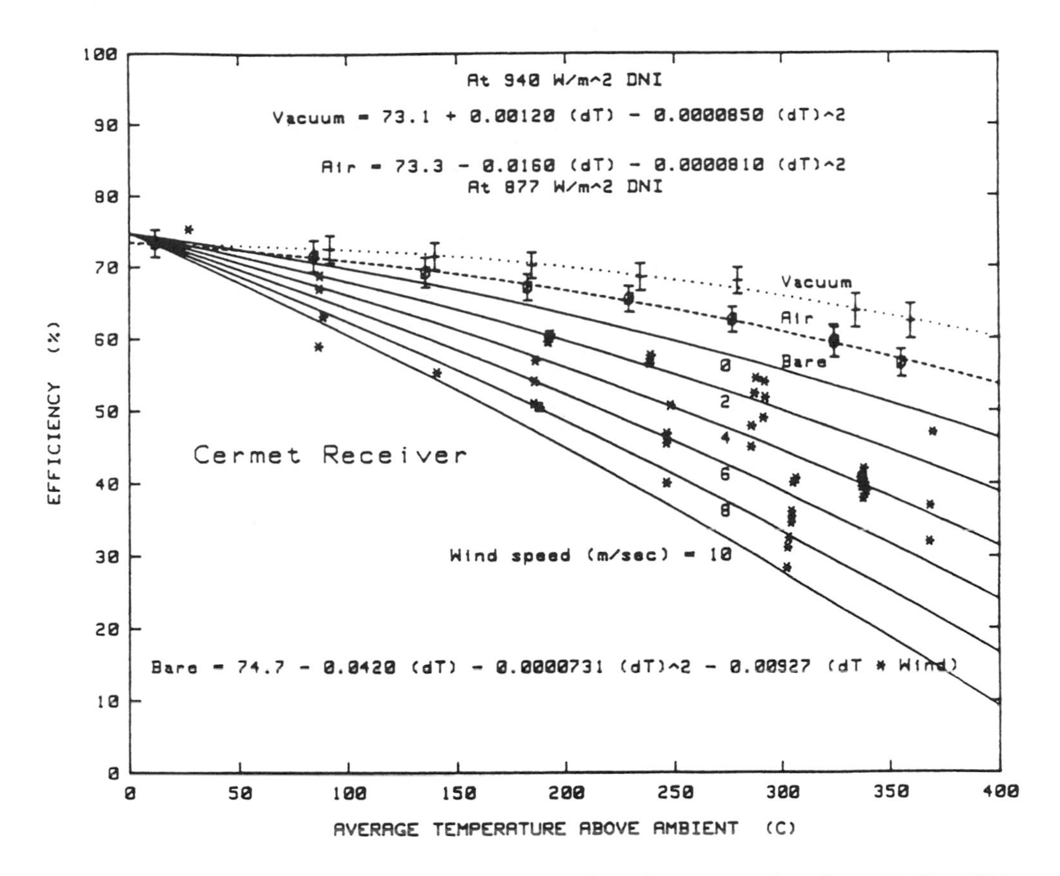

FIGURE 21B.15 SEGS LS-2 efficiency vs. temperature and wind—cermet receiver. *Source*: Dudley, 1994.

21B.6 Instantaneous Measured Efficiency

Sandia National Laboratories in Albuquerque has conducted most of the efficiency testing for parabolic troughs. Sandia's AZTRAK rotating platform allows the incidence angle of the tested collector to be controlled and incorporates instrumentation to accurately measure energy collection of the test piece. Sandia recently tested two parabolic trough collectors using its rotating turntable system: the Luz LS-2, and the IST collector.

The LS-2 modules use silvered glass mirrors and an evacuated receiver with an antireflective-coated glazing and a cermet (graded ceramic—metal) selective coating. The LS-2 has an aperture width of 5 meters, and the absorber tube is 70 mm in diameter. Sandia National Laboratories test results for the LS-2 are shown in Figure 21B.15. In this figure, efficiency is shown as a function of ΔT for the level of beam irradiance specified on the graph. Sandia measured collector performance with three different receiver configurations: the normal evacuated receiver, a nonevacuated receiver (air in the annulus), and a bare absorber tube (glazing removed). Without the glazing, wind speed significantly affects collector performance and efficiency decreases greatly.

The Sandia efficiency test results for the LS-2 with an evacuated receiver can be related to irradiance and ΔT in the most useful form of the efficiency equation for parabolic troughs*:

$$\eta = 0.731 + 0.01128\ \Delta T/I - 0.000799\ \Delta T^2/I \tag{21B.12}$$

*This form of the efficiency equation was obtained by factoring out the 940 W/m² irradiance that was present when the Sandia tests were conducted.

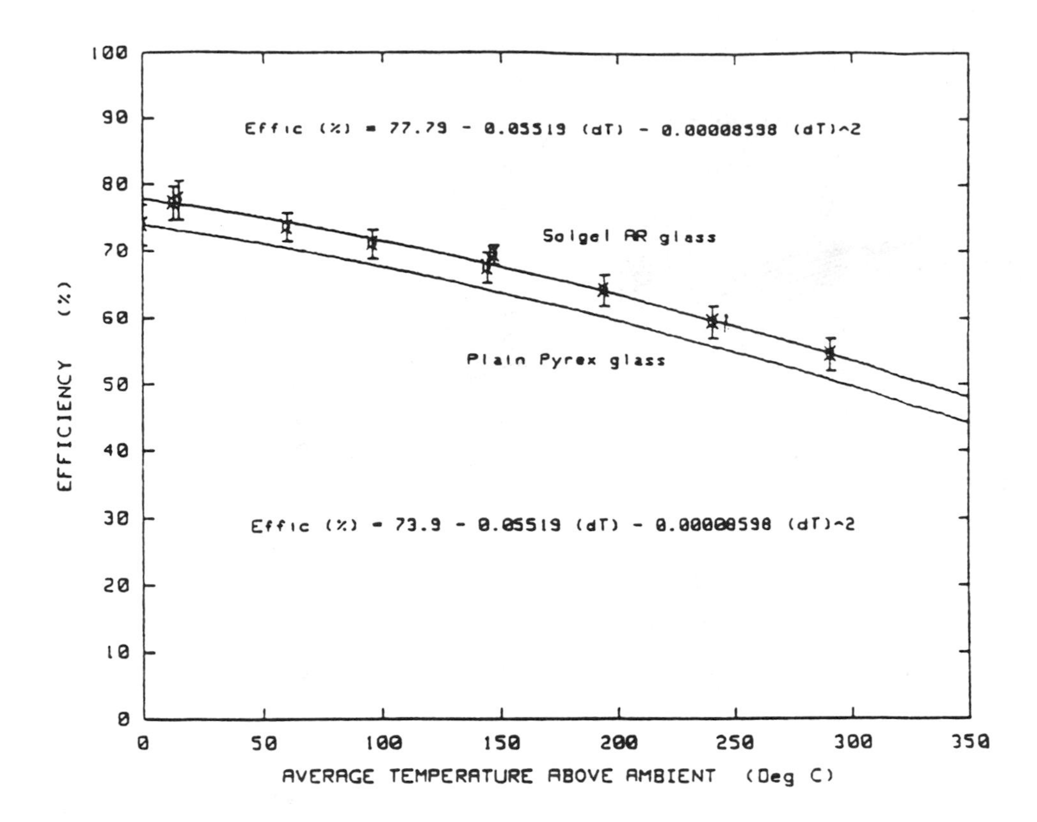

FIGURE 21B.16 IST black nickel efficiency comparison—solgel glass/pyrex glass. *Source*: Sandia National Laboratories, 1995.

Readers interested in additional test results, including incidence angle modifier results, are referred to Dudley and Evans (1994).

The IST module uses silvered acrylic reflective film (3M ECP-305), and a nonevacuated receiver with an antireflective coated glazing and a black nickel selective coating. The aperture width is 2.3 meters, and the absorber tube is 51 mm in diameter. Sandia test results for this collector are shown in Figure 21B.16, where again efficiency is shown as a function of ΔT for a high beam irradiance (above 900 W/m^2). Two efficiency curves are shown. The upper curve has solgel-coated receiver glazings, and the lower curve uses uncoated standard borosilicate glazings.

Based on the full set of data from the Sandia tests, the efficiency of the IST collector can be characterized as

$$\eta = 0.762 - 0.2125\ \Delta T/I - 0.001672\ \Delta T^2/I \tag{21B.13}$$

Again, incidence angle modifiers and further test results are provided in the Sandia test report (Sandia, 1995).

21B.7 Annual Collector Output

The average annual energy delivery of a parabolic trough collector system depends on many variables: collector efficiency characteristics, system configuration, load temperature and load profile, storage tank size (if any), collector field layout, piping network size and layout, and site characteristics such as insolation. Accounting for all these details is often handled using a computer simulation. This approach is recommended for detailed design, when accuracy requirements are

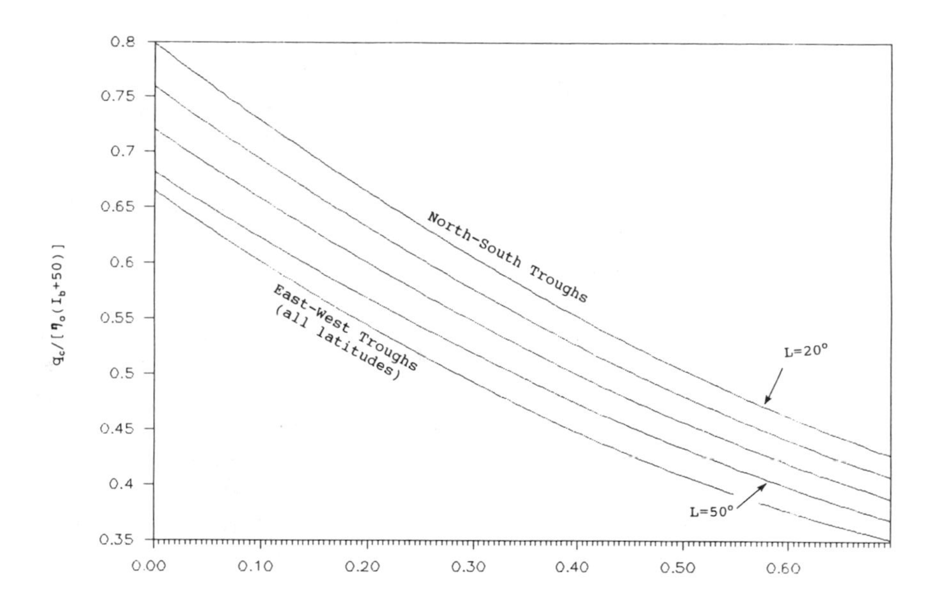

FIGURE 21B.17 Graphical determination of yearly average collection rate. *Source*: Courtesy Industrial Solar Technology.

highest and when system-level affects need to be accurately assessed. Fortunately, however, only a few variables dominate the annual output of well designed trough systems, and simple methods may be used to calculate trough collector performance in the field when estimates are sufficient—such as for feasibility analysis

The variables that dominate trough energy output are

- The optical efficiency and heat-loss coefficient of the collector
- The average collector operating temperature
- The average annual direct normal irradiance at the site

Empirical performance correlations have been derived, using these dominant variables, based on thousands of detailed computer simulations (Kutscher et al., 1982). For horizontally mounted parabolic troughs, these empirical correlations permit a graphical determination of average energy delivery, as shown in Figure 21B.17. Alternatively, performance can be calculated using the equations shown below the graph.

To use this simple estimation technique, we must first calculate the value noted on the x-axis of the graph, a term referred to as the Intensity Ratio. The Intensity Ratio is the critical intensity of the collector (the irradiance at which heat loss just balances heat gain) divided by the average intensity of the sun's irradiance during daylight hours.

$$\text{Intensity Ratio} = U_L(T_{\text{collector}} - T_{\text{ambient}})/\eta_0\, I_{\text{annual}}$$

where

η_0 = collector optical efficiency
U_L = heat loss coefficient,* slope of collector efficiency curve, (W/m^2 °C)
I_{annual} = average annual direct normal irradiation (W/m^2)
$T_{\text{collector}}$ = average collector operating temperature (°C)
T_{ambient} = average annual daytime ambient temperature (°C)

*If a second-order efficiency equation is used, calculate U_L at the expected collector operating temperature; that is, $U_L = U_{L,1} + U_{L,2}(T_{\text{collector}} - T_{\text{ambient}})$.

The collector optical efficiency and heat loss coefficient can be obtained from test data, or the collector efficiency equation. The long-term average annual direct normal irradiance, I_{annual}, can be estimated from the insolation map shown in Figure 21B.18. Of course, if better long-term irradiance data are available, they should be used instead. Note that the average irradiance value is based on a daytime average (using 4,380 hours of annual daylight hours per year or an average of 12 hours per day per year). The average daytime ambient temperature can generally be obtained from reference texts or an atlas.

The average collector temperature is not generally known, but an estimate will normally suffice for troughs because efficiency is relatively insensitive to temperature. For systems with thermal storage delivering a typical load fraction of around 50%, a reasonable estimate for the average collector operating temperature is the required load temperature. The following example illustrates the basis for this estimate. Consider a hot water system, in which the solar fraction is designed to be about 50%. That is, the solar system is expected to deliver about half the hot water needed over a year, with the rest provided from a fossil fuel backup system. If the water supply temperature averages 20°C and the desired use temperature of the hot water is 70°C, we anticipate an average supply temperature of 45°C. If the approach temperature for the heat exchanger is 10°C, then the collector inlet temperature is 55°C. If an average temperature differential across the solar field is 30°C, then the collector outlet temperature is 85°C and the average collector temperature is 70°C. Using this example as a basis, it is apparent that the average collector temperature will increase at load fractions greater than 50% and will approach the load inlet temperature as the load fraction declines to zero.

With steam systems without storage, the average collector operating temperature will be somewhat above the steam saturation temperature. A reasonable approximation of the average collector operating temperature is 30°C above the steam saturation temperature.

After calculating the Intensity Ratio, determine the ordinate (y-axis value) from the graph for the latitude of the site and the assumed orientation of the trough rotational axis. The ordinate is then multiplied by the product of the collector optical efficiency and an average irradiance level (plus 50 W/m^2)* to arrive at q_c, the average annual energy collection rate per unit of collector area (in units of W/m^2). Multiplying q_c by 4,380 (number of daylight hours in a year) and the collector area (in m^2) gives an estimate of the annual energy collection (in watthours).

A number of other correction factors can be applied to this result to improve accuracy, such as storage tank modifiers, heat exchanger modifiers, shading loss modifiers, and piping loss modifiers. Details of these additional corrections can be obtained in Kutscher et al. (1982).

21B.8 System Sizing and Performance Estimates

An analysis of the economic feasibility of a parabolic trough installation requires sizing the solar system, estimating annual energy output and its value, and comparing this revenue stream with the installed cost of the system. A computer simulation that accurately models the thermal characteristics of the solar system and uses hourly weather data is an essential tool to fully engineer and optimize a large-scale solar system. However, the information provided here combines performance estimates with some guidelines to allow a preliminary feasibility analysis to be conducted.

The previous section provided a means of estimating the annual output of a parabolic trough. Average annual weather data are used in making these calculations. However, the procedure does not give collector output on a given day, nor does it account for heat losses in delivering the collected energy from the solar collectors to the load. Heat losses, both when the system is operating and when it is cooling down overnight, are a function of the thermal mass of the solar system, which includes the collectors, the piping system from the collectors to the load and any auxiliary equipment, such as a storage tank or heat exchanger, and the degree of pipe insulation, as well as ambient and system operating temperatures. Some guidelines to calculate system heat losses are

*The extra 50 W/m^2 in this term has been shown to increase the accuracy of the empirical fit to the data.

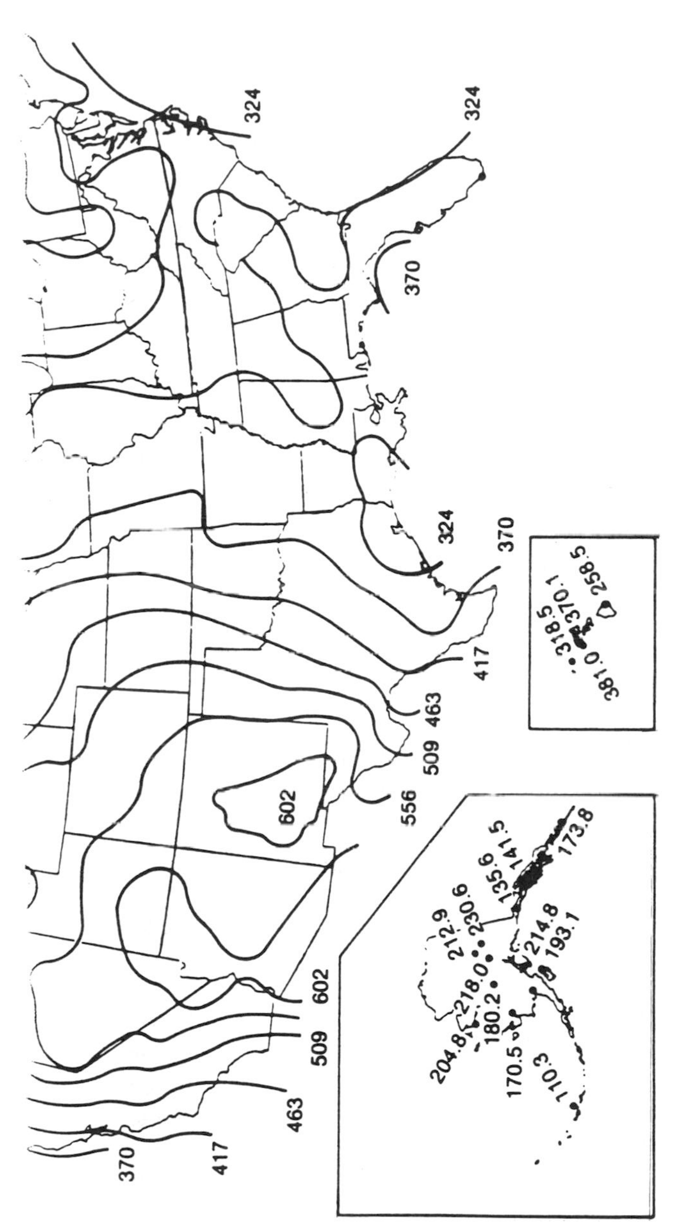

FIGURE 21B.18 Average direct normal irradiance during daylight hours. *Source:* Kutscher et al., 1982.

provided in Lewandowski et al. (1984). Based on this report, for a well-designed piping system, where the load is close (up to 100 m) to the solar collector field, the thermal mass of the piping (not including the absorber tube) using an aqueous heat transfer fluid should be no greater than 3 kJ/°C m^2 of collector aperture area, and the heat loss coefficient should be no greater than 0.4 kJ/°C m^2 h. For a system using a heat transfer oil, the upper design value for the heat capacity of the piping is 2 kJ/°C m^2, and the heat loss coefficient is 0.5 kJ/°C m^2 h.

The peak thermal output of the collectors can be determined from the collector efficiency equation. The optical efficiency should be multiplied by a factor to account for the effects of dirt. The rate at which dirt accumulates on the collector surface is very site dependent (Morris, 1982). However, a reasonable value for the "dirt factor" of the reflective surface might be 0.92—an average value for a washing cycle of a month or so. Peak collector output occurs at normal incidence, when cosine losses are zero, and at a solar radiation level around 1,000 W/m^2.

As an example, assume a solar system is proposed for an industrial application where the hot water load is 25,000 gallons per day (94,750 Lpd). The system is to be sized to meet the minimum energy demand, so that the potential of the solar system is fully utilized. It will be oriented on an axis running east–west to provide the most uniform energy delivery on an annual basis. The water is used at a temperature of 50°C, and in the summer it enters the system at 15°C. Solar collectors with the operating characteristics of the IST trough are to be used. Average ambient temperature during the summer day is 25°C. The temperature of the solar system varies with the thermal energy storage tank, but assume that the average tank temperature is 35°C, such that the collector inlet is 45°C, the collector temperature outlet is 75°C, and the average collector temperature is 60°C.

From this information, the minimum thermal load is

$$94{,}750 \times 4{,}180 \times (50 - 15) = 13.9\,\text{GJ/day}$$

Based on the IST collector efficiency equation, with a dirt factor of 0.92, average collector temperature of 60°C and ambient temperature of 25°C, the peak collector efficiency is about 69%.

For U.S. latitudes, output over a clear day when energy production is not interrupted by clouds is equivalent to about 6 hours of peak energy delivery for an east–west system, and about 9 hours for a north–south system. Hence, the required area for this solar system oriented east–west, based on 6 hours of peak energy delivery is

$$13.9 \times 10^9 / (0.69 \times 1{,}000 \times 3{,}600 \times 6) = 933 \text{ m}^2$$

Heat loss from the piping system for this collector field, based on the heat loss data just provided is about

$$933 \times 0.4 \times (60 - 25) = 13{,}060 \text{ kJ/h}$$

This instantaneous heat loss is only about 0.5% of the peak delivery rate. However, this heat loss is relatively constant while the system is operating, so as a fraction of energy delivery, the percentage loss increases when the system is operating at nonpeak conditions. If the system ran 3,000 hours per year, and the average annual daytime temperature was 20°C, then the annual heat loss from the system piping when it was operating would be

$$933 \times 0.4 \times (60 - 20) \times 3{,}000 = 45 \text{ GJ/yr } (12{,}440 \text{ kWh/yr})$$

A more significant heat-loss mechanism is the loss due to cooldown. It can be assumed, somewhat conservatively, that the system cools to ambient temperature each night. Say that this temperature is 10°C on average and the system cycles 300 times per year. Then, heat loss from the transport piping system is

$$300 \times 933 \times 3 \times (60 - 10) = 42 \text{ GJ } (11{,}660 \text{ kWh})$$

In addition, the absorber tube piping and fluid will also cool. The heat capacity of the absorber tube is about 4 kJ/°C m^2 h, so heat loss from the absorber adds about 56 GJ to the system losses. Using these approximations, heat loss can be subtracted from the collected energy to get a better estimate of energy delivery. Even for a system operating at low temperatures, thermal losses that total 143 GJ are around 5% of the collector output. Such losses can require an increase in the necessary collector area to meet the thermal load.

Determining accurate load data for an application can be a challenge. It is unusual when the load data are readily available from existing instrumentation, such as when the thermal output of the boiler that the solar system is to supplement is monitored continuously. More commonly, the only monitored data at a plant will be the monthly energy bills from the meter serving the entire site. For some applications, this limited information can be sufficient. If the meter records energy use for building heating and hot water, then the utility bills in the summer (when there is no space heating load) can provide information on the amount of energy used to produce hot water. If the solar system is sized to meet the minimum load, then this information, together with an estimate of the efficiency of the conventional heating system, is all that is needed. In other instances, to supplement utility bills, literature data are sometimes available that will provide estimates of the thermal inputs for various industrial processes. For some applications, it may be necessary to measure the thermal load for a particular process to develop the information needed to engineer a solar system design. Nonintrusive ultrasonic flow meters are particularly useful for measuring flow rates to determine energy use. These and other energy measuring devices can be connected to compact data acquisition systems to record information over extended hours of operation.

If load data is deduced from utility bills, estimates of the conventional heating source efficiency must be made to determine how much of the fuel source is converted into useable energy. Boilers rarely operate at design conditions, and for applications such as DHW, they frequently cycle on and off—an operating mode that greatly increases standby losses. System efficiency of conventional heating equipment in delivering energy to the load is typically in the range of 50 to 70%.

Once a preliminary size of the solar system has been defined, aspects such as the layout of the collector field on the system site, the routing of piping in the solar system loop, and the interface with the thermal load must be specified. Based on this level of detail, system performance and cost estimates can be refined. The next step in the engineering process precedes construction and involves the production of actual system drawings once an agreement to build the solar system has been reached. Such drawings would show structural work, such as pylon supports, the surrounding solar field fence, and any auxiliary solar system buildings, together with plumbing and electrical drawings. Following the construction period, an essential part of an installation is a training period for O&M personnel, complete with system documentation and manuals.

21B.9 System Configuration

The piping in a parabolic trough system, including the absorber, is typically made of steel. This material is more compatible with the high-temperature capability of parabolic trough systems, and in large sizes steel piping is less expensive than copper. Steel-pipe joining methods can be through threaded fittings or welding.

The use of steel piping is one reason why trough systems are almost always closed piping loops that transfer energy to the load through heat exchange. Using water (or glycol/water mixtures in freezing climates) as the collector fluid, deaeration occurs as the fluid is heated, and the addition of chemical inhibitors protects the steel piping from corrosion. Similarly, the use of an organic

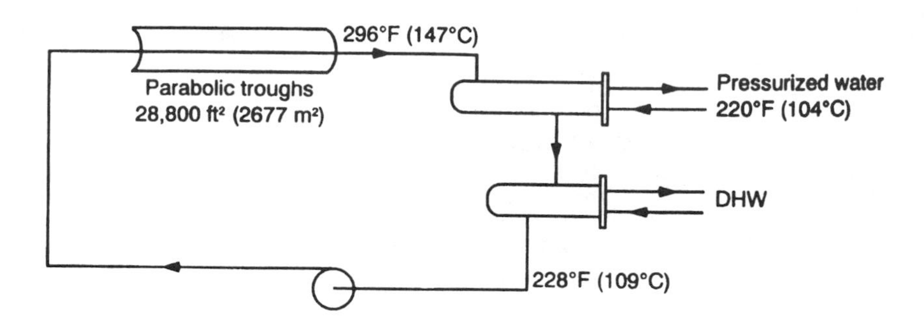

FIGURE 21B.19 Energy transfer through direct heat exchange. *Source*: SEIA, 1995.

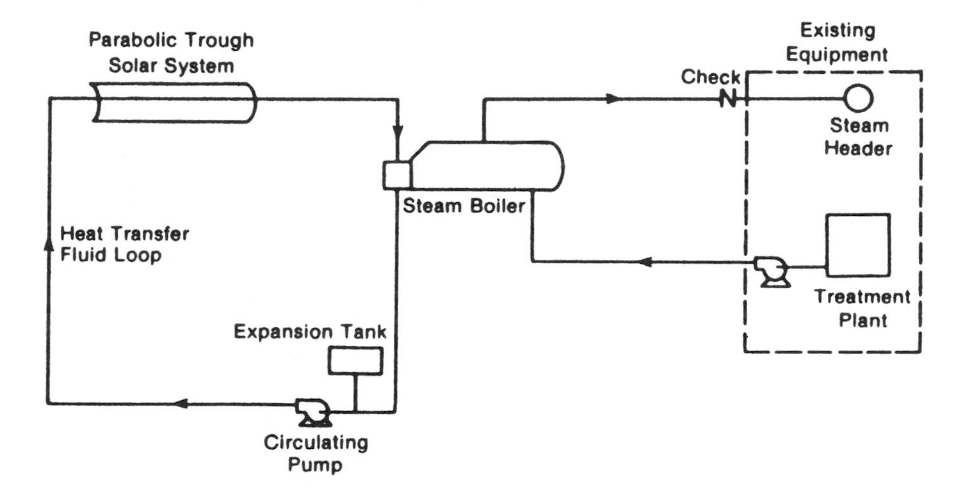

FIGURE 21B.20 Solar steam generating system. *Source*: Kutscher et al., 1982.

heat transfer fluid provides a corrosion-resistant environment for the steel piping. Using a closed system and good quality fill water also eliminates the scaling on the inside of pipes that can reduce heat transfer rates in closed systems.

Another major determinant of parabolic trough system design is whether thermal energy storage is incorporated. Without thermal storage, energy delivery to the load is determined entirely by the output of the solar field. This will vary throughout the day, responding to solar input levels and the incident angle to the sun. The inlet temperature of the solar field will be tied closely to the load temperature through the heat exchanger; the temperature differential across the solar field varies with the energy collected.

Two system configurations where thermal energy storage is not incorporated are shown in Figures 21B.19 and 21B.20. Figure 21B.19 shows the solar system exchanging heat directly to the load through a heat exchanger. The load could be a pressurized water loop, used to heat a chemical process for instance, or air, used for a drying application. Figure 21B.20 illustrates the production of steam. The collector fluid boils boiler feedwater in a unfired steam generator. Steam generated in this manner is delivered to the load through a check valve as long as the pressure in the generator is greater than the line pressure. An alternative means of generating steam that has been proposed (Murphy and May, 1982) is to boil water directly in the trough absorber tubes. This concept is also proposed as the basis for the next generation of Luz-type electric generating plants.

Without energy storage, the solar energy delivery potential of the solar collectors cannot exceed the thermal requirements of the load, unless the solar system output is modulated by defocussing

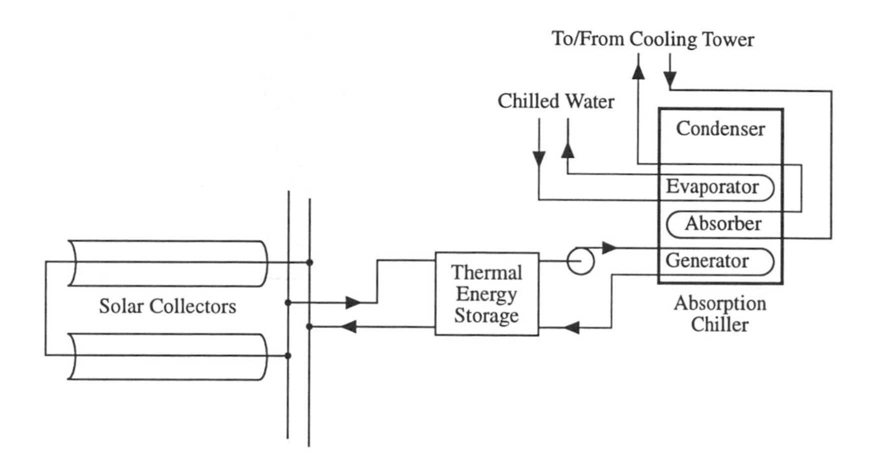

FIGURE 21B.21 Integration of parabolic troughs with absorption chiller. *Source*: Courtesy Industrial Solar Technology.

part to the field. Economically, this action is usually justified only for short time periods, since taking part of the solar system out of service is wasting the potential of a capital resource.

From a design point of view, sizing a solar system without thermal storage is very straightforward. The maximum output of the solar field, with suitable allowances for collector soiling and field thermal losses, is equated to the thermal load. The solar field area is rounded off to a practical size based on the configuration of the solar modules, and a check is made to see that the collectors will fit in the available land area.

Thermal energy storage in a solar system can be relatively short-term or "buffer" storage ranging from a fraction to several hours duration, or long-term storage where the need is to supply heat on a 24-hour basis, or even to store heat collected over a day or a weekend when a plant is shut down. Incorporating thermal energy storage involves an economic trade-off between the value of the additional energy delivered to the load and the economies of scale resulting from the installation of a larger solar system than the nonstorage case, against the cost and thermal penalties of incorporating storage. Charging storage always requires that the solar system run at a temperature higher than the nonstorage case, thus degrading the performance of the solar system. However, without thermal storage, the size of the solar field can be limited to a size that is too small to be economically justified.

Short-term thermal storage can even out fluctuations in thermal output and can continue energy delivery during short periods of cloud cover. Such storage can be particularly valuable when running equipment, such as an absorption chiller, that will run inefficiently if subject to frequent start-ups and shutdowns. The integration of the solar system with storage and a chiller is shown in Figure 21B.21.

Thermal storage for DHW applications is typically very economical. Increasing the load fraction supplied by solar energy more than compensates for the cost of storage and the small reduction in thermal performance due to increased system operating temperatures. Performance degradation of trough systems is not large at the relatively low operating temperature of a DHW system. A system configuration for DHW applications is shown in Figure 21B.22. An unpressurized tank is filled with water. Heat from the solar field is transferred to the tank through copper coils at the bottom. A trough system can efficiently increase the temperature in the tank to just below the boiling point. Heat is extracted from the tank using a second set of copper coils placed at the top of the tank through which domestic water flows. If necessary, the outlet water stream can be tempered before it is delivered to the load.

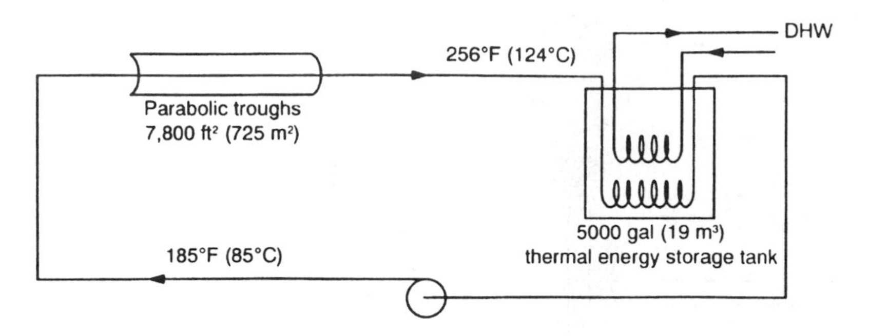

FIGURE 21B.22 DHW system configuration. *Source*: SEIA, 1995.

Energy storage systems operating at temperatures above the boiling point of water are much more costly than those operating below this temperature. For aqueous storage media, a pressure vessel must be used, and pressures increase rapidly with temperature. An organic heat transfer oil is an alternate storage medium that does not require a pressure vessel. However, even the cost for the fluid with the lowest temperature capability is several dollars per gallon, and approximately twice the volume compared to water is required for a given amount of storage because such oils have greatly reduced specific heat and lower density than water. Another cost of using higher-temperature storage results from the thermal penalties involved. The solar field will operate at a high absolute temperature, higher than the load temperature, and higher still to be able to charge storage. Consequently, for higher-temperature energy storage systems the efficiency of energy collection is reduced and thermal losses, both while the system is operating and as a result of the overnight cooldown, become a more significant fraction of the energy collected.

Using the same simplified analysis as before, but for a solar system using an organic heat transfer fluid and operating at an average temperature of 200°C, the peak thermal efficiency of the collectors is about 61%. This is over 10% less than when operating at the lower temperature.* The size of the solar field to meet the peak energy demand is

$$13.9 \times 10^9/(0.61 \times 1,000 \times 3,600 \times 6) = 1,050 \text{ m}^2$$

Heat loss from the piping when the system is operating is

$$1,050 \times 3,000 \times 0.5 \times (200 - 20) = 284 \text{ GJ/yr}$$

The thermal capacity of the receiver tube containing oil is about 3.3 kJ/°C m², so the cooldown heat losses for the absorber tube and piping are

$$1,050 \times 300 \times (2 + 3.3) \times (200 - 20) = 301 \text{ GJ/yr}$$

Hence, heat losses from this system total about 585 GJ/yr, about four times as much as from the system operating at a lower temperature. Using the stated assumptions, this is about 20% of the delivered energy. (In fact, the stated assumptions are conservative. At higher-temperature operations, more pipe insulation and smaller-diameter pipes would be used to reduce heat losses, and pipes do not cool to ambient temperature overnight.) Nevertheless, the longer-term storage of energy at temperatures above the boiling point of water is generally only justified when displacing high-cost fuels.

Analyzing and optimizing a system with thermal storage is more complicated than the nonstorage case. The calculation involves a trade-off between storage size and energy revenue. Often the

*On an annual basis, collector thermal energy delivery drops even more, well over 20%.

most economic storage size will not collect all the available energy. For short periods the solar system can be shut down because the upper temperature limit of the storage tank is reached. For high-temperature storage systems, another trade-off involves the size of the storage tank and how high to limit the maximum storage temperature.

Once the basic area of the solar collectors and the configuration with the thermal load have been established, the layout of the solar field needs to be defined. Within any possible site constraints, the goal of the layout is to maximize revenue while minimizing installed cost.

As noted previously, the design orientation of the collectors is influenced most by the timing of the thermal load. Orientation also influences the row-to-row distance, since for a given shading loss collectors oriented along an east–west axis can be packed closer together than collectors oriented north–south. Determining row-to-row spacing, which is also a function of latitude, trades off shading against the increased capital cost of longer pipes and increased operating losses due to greater heat losses and electric power consumed for pumping.

Collectors are linked together with a discrete number of modules in a row. Considerations of pressure drop and pumping power determine how many collector rows are connected in series, and how many delta-temperature (delta-T) loops comprise the entire solar field. Possible collector arrangements are in U-loops so that the inlet and outlet headers are at the same end of the field. Another configuration is with one or more collector rows in a straight line with headers at either end. In very large solar fields, such as those used for solar electric systems, piping can be arranged to service collectors on each side of the headers. The aerial view shown in Figure 21B.23 of the Solar Electric Generating Systems (SEGS-III through SEGS-VII) at Kramer Junction, California, illustrates the collectors configured in many parallel rows. The statistics listed in Table 21B.2 for the SEGS-VIII plant at Harper Lake California (Lotker, 1989) give some idea of the size of these facilities. A schematic of the solar power plant is shown in Figure 21B.24.

TABLE 21B.2 Characteristics of SEGS-VIII

Net electrical power	80 MW
Net electric output	252,750 MWh/yr
Collector area	464,340 m^2
Capacity factor (solar and natural gas)	35%
Capacity factor (solar alone)	26.5%
Thermal electric efficiency	38.1%

There are eight individual solar modules, model LS-3, in each solar collector assembly (SCA), comprising 545 m^2 of collector area, that track the sun using a single controller and drive system. Six SCAs form a flow loop, of which there are 142 in the entire field. Each individual flow loop would be a fairly large-scale process heating system!

For systems with multiple parallel flow paths, the piping should be sized to try and deliver equal flows through each delta-T string. However, some flow imbalance is acceptable, as long as the string with the least flow does not exceed the maximum system limit, whether such a limit is defined by the boiling point, the mechanical limits of the piping system, or the thermal limit of the selective surface. Reverse return piping to achieve flow balancing is inadvisable for trough systems. The increase in piping length, resulting in greater installation and running costs from increased thermal loss and pumping power, is not justifiable. If it is necessary to restrict flow in loops closest to the supply lines, then throttling valves can be incorporated at the inlet to the loop.

21B.10 Operating and Maintenance

Modern trough systems are designed to achieve extreme reliability under automatic operation, and to minimize O&M costs. Long-life corrosion-resistant materials are used for the concentrator structure. Bearings are permanently lubricated. Electronic controllers have a long history of reliable

FIGURE 21B.23 Overview of KJC plants, SEGS-III to SEGS-VII. *Source*: Becker et al., 1995.

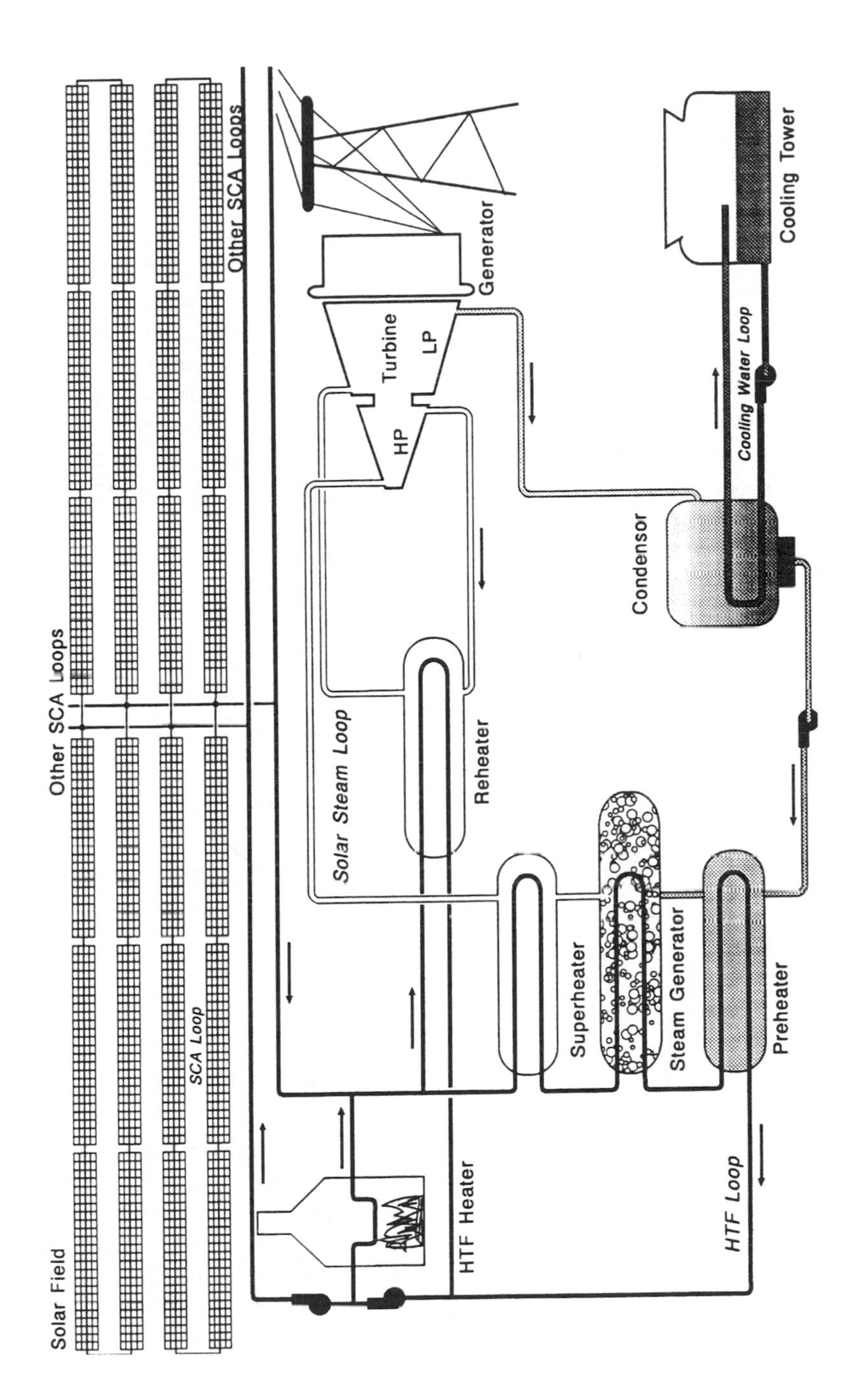

FIGURE 2 B.24 Schematic of SEGS Plant. *Source:* Becker et al., 1995.

operation. Systems are instrumented to provide indications of suboptimal performance. Most maintenance tasks are preventative in nature. Regular inspections check the focus of the collectors, mechanical components, and flexible hoses. Motors and gears are lubricated at annual or semiannual intervals.

The most significant maintenance task involves washing the reflectors to maintain system performance. Washing frequency depends on the trade-off between the cost of washing and the value of the additional revenue that results. For a SEGS plant operating during the summer peak, washing may occur at weekly intervals. For a process heat system of much smaller scale, the washing interval may be 1 to 2 months.

21B.11 Economic Analysis

In the last round of the DOE demonstration projects in the early 1980s, two 50,000 ft² solar systems were installed for prices in the range of \$55/ft² (\$600/m²) (Green, 1983). Few solar systems have been installed since that time to provide a large basis for comparison. However, the systems that have been installed in recent years indicate that considerable strides have been made to reduce costs and to improve system performance and reliability. A 2,667 m² solar system installed by IST in Tehachapi, California, in 1990 had a turn-key system price of \$23/ft² (\$248/m²). IST estimates that such a system built in 1995 as part of a large production run, could be installed for slightly less than the above price. Costs would decline somewhat for larger systems and increase for smaller or more complicated systems, such as for steam generation or incorporating thermal storage.

In 1989, Luz proposed to build and operate 80 MW solar electric plants with the cost parameters listed in Table 21B.3 (Lotker, 1989). The estimated costs are deduced from a range of data published by Luz.

TABLE 21B.3 Solar Power Systems Costs

Net electrical power	80 MW
Total system cost	\$168 million
Interconnection and land	6 million
Collector cost (est.)	70 million
O&M cost	3.75 million/yr
Natural gas cost	3.50 per million Btu
Natural gas cost (est.)	2.8 million/yr
Electricity delivered	252,750 MWh/yr

Revenues

Considering the operation of a solar thermal system, the value of the solar energy delivered depends on the cost of the displaced conventional fuel. Presently, a typical cost for natural gas is \$4 per million Btu (\$0.014/kWh), for #2 fuel oil is \$0.80/gallon (\$5.77 per million Btu, \$0.020/kWh), and for electricity used for heating purposes is \$0.06/kWh (\$17.60 per million Btu). The value of the delivered solar energy also depends on the efficiency with which the conventional fuel is burned. Boilers typically do not operate at design conditions, and for applications such as DHW, they cycle on and off frequently, increasing standby losses. System efficiency of conventional heating equipment in delivering energy to the load is typically in the range of 50 to 70%. Consequently, the value of solar energy displacing natural gas and fuel oil (burned at 60% efficiency), and electricity (used at 100% efficiency), based on these costs is \$0.023, \$0.033, and \$0.06/kWh, respectively.

System revenues will vary after the first year of operation depending on whether fuel prices increase or not. For the following example, assume that the general inflation rate is 3% per year and that energy prices increase at 1% above the general rate of inflation.

The proposed solar system is to be located in the southwest of the United States, and the collector area is 2,677 m² (28,800 ft²). The system installed cost is \$21.50/ft², so the total price is \$619,200. Subtracted from system revenues are costs for maintaining the system, insurance for both liability and property damage, and taxes. All these costs decline on a unit area basis as systems get larger. For the example system, O&M and insurance costs are about \$0.35/ft², or about \$10,000 per year. Such costs are assumed to increase in the future at the general rate of inflation.

Based on system revenues and costs, a cash flow table for the solar system over its life can be constructed. The cash flow analysis is conducted over 20 years, even though the life of the solar installation should exceed this time period. Tables 21B.4 and 21B.5 show systems displacing natural gas and electricity. Table 21B.5 is truncated after 10 years for brevity.

TABLE 21B.4 Cash Flow for a Solar System Displacing Natural Gas

Year	Revenue	Expenses	Net Income	Recovery of Investment
1	59,086	10,000	49,086	–570,114
2	61,467	10,300	51,167	–518,948
3	63,944	10,609	53,335	–465,613
4	66,521	10,927	55,594	–410,019
5	69,202	11,255	57,946	–352,073
6	71,990	11,593	60,398	–291,675
7	74,892	11,941	62,951	–228,724
8	77,910	12,299	65,611	–163,113
9	81,049	12,668	68,382	–94,731
10	84,316	13,048	71,268	–23,463
11	87,714	13,439	74,275	50,811
12	91,249	13,842	77,406	128,217
13	94,926	14,258	80,668	208,886
14	98,751	14,685	84,066	292,952
15	102,731	15,126	87,605	380,557
16	106,871	15,580	91,291	471,848
17	111,178	16,047	95,131	566,979
18	115,659	16,528	99,130	666,110
19	120,320	17,024	103,295	769,405
20	125,168	17,535	107,633	877,038

Internal Rate of Return over 20 years, 8.8%

TABLE 21B.5 Cash Flow for a Solar System Displacing Electricity

Year	Revenue	Expenses	Net Income	Recovery of Investment
1	151,934	10,000	141,934	–477,266
2	158,057	10,300	147,757	–329,508
3	164,427	10,609	153,818	–175,690
4	171,053	10,927	160,126	–15,564
5	177,947	11,255	166,692	151,128
6	185,118	11,593	173,525	324,653
7	192,578	11,941	180,638	505,291
8	200,339	12,299	188,041	693,331
9	208,413	12,668	195,745	889,077
10	216,812	13,048	203,764	1,092,841

Internal Rate of Return over 20 years, 26.6%

The revenue stream from a system displacing natural gas is much lower than that from a system replacing electricity. The last column shows how the initial investment is repaid. Using a simple measure of economic viability, the simple payback of the system displacing natural gas before taxes is 11 years. Using this criterion, the system is unlikely to be funded. However, when the value of the net revenue stream is compared against the initial investment, the internal rate of return before taxes is 8.8%. This is reasonable when compared against investments such as government bonds, although those investments carry less risk. Since solar systems can be depreciated over 6 years, and currently are eligible for a 10% investment tax credit, the after-tax rate of return will be slightly higher than the rate before taxes. This increased rate could be attractive under a third-party ownership scenario. The way this works is that a third-party investor would purchase the solar system from the installer. The investor would receive the income from the solar system, less a discount on energy cost allowed the system user, in return for providing a location for the solar system and purchasing the thermal output. If an energy purchase agreement can be negotiated with the system user and the discounted revenue stream is sufficient to attract outside investment, it is possible that the system could go ahead.

For applications where the fuel displaced is electricity, an investment in a solar system should be very attractive. Before taxes, the simple payback is 4 years and the internal rate of return is 26%.

Overall, these examples illustrate the difficulty in marketing solar energy systems in the United States when the predominant fuel is natural gas. Even under the most favorable climatic conditions, returns are only marginally attractive for investors. Market penetration will increase, however, as industry drives down costs and energy prices increase in the future.

The economics of a parabolic trough system generating electricity is illustrated in the Table 21B.6 based on the data in Table 21B.2. Such plants are well matched to the power needs of utilities, since much of the power generated coincides with the peak power demands. Compared to a process heat system the numbers are very large. Revenue is based on an aggregate electric rate of $0.10/kWh. Expenses are the total of O&M costs, and the costs for natural gas that provides about 25% of the total energy delivered by the plant. Using these assumptions, the simple payback on the system is about 7 years, and the before-tax internal rate of return on the investment over its 30-year life is 14.4%. Again, this is quite attractive. However, currently payments by utilities is less than the $0.10/kWh figure used, so the analysis is more equivalent to markets outside the United States where power rates are higher.

TABLE 21B.6 Cash Flow for a Solar Electric System ($1,000)

Year	Revenue	Expenses	Net Income	Recovery of Investment
1	25,275	6,575	18,700	−155,300
2	26,294	6,772	19,521	−135,779
3	27,353	6,975	20,378	−115,401
4	28,456	7,185	21,271	−94,130
5	29,602	7,400	22,202	−71,928
6	30,795	7,622	23,173	−48,755
7	32,036	7,851	24,185	−24,569
8	33,327	8,086	25,241	672
9	34,670	8,329	26,341	27,013
10	36,068	8,579	27,489	54,502
Internal Rate of Return over 30 years, 14.43%				

References

Ashley, C.S. and Reed, S.T. *Sol-Gel Derived AR Coatings for Solar Receivers.* Sandia National Laboratories, Albuquerque, NM, SAND84-0662, September 1984.

Becker, M., Gupta, B., Meinecke, W., and Bohn, M. *Solar Energy Concentrating Systems: Applications and Technologies.* C.F. Mueller, Heidelberg, 1995.

Bendt, P., Rabl, A., Gaul, H., and Reed, K. *Optical Analysis and Optimization of Line-Focus Solar Collectors.* Solar Energy Research Institute, Golden, CO, SERI/TR 34-092, September 1979.

Boyd, W.E. *Insulated Metal Hose for Tracking Receiver Application.* Proceedings of the Line-Focus Solar Thermal Energy Technology Development Seminar, Sandia National Laboratories, Albuquerque, NM, September 9–11, 1980.

Dudley, V.E. and Evans, L.R. *Test Results: SEGS LS-2 Solar Collector,* Sandia National Laboratories, Albuquerque, NM, SAND94-1884, December 1994.

Gee, R. *Long-Term Average Performance Benefits of Parabolic Trough Improvements.* Solar Energy Research Institute, Golden, CO, SERI/TR-632-439, March 1980.

Gee, R. *Near-Term Improvements in Parabolic Troughs: An Economic and Performance Assessment.* Solar Energy Research Institute, Golden, CO, SERI/TR-632-870, August 1981.

Gee, R. *An Experimental Performance Evaluation of Line-Focus Sun Trackers.* Solar Energy Research Institute, Golden, CO, SERI/TR-632-646, May 1982.

Gee, R. *Low-Temperature IPH Parabolic Troughs: Design Variations and Cost-Reduction Potential.* Solar Energy Research Institute, Golden, CO, SERI/TR-253-1662, August 1983.

Green, S.T. *Solar IPH System for Caterpillar Tractor Company at San Leandro, California.* Proceedings of the Distributed Solar Collector Summary Conference—Technology and Applications, Sandia National Laboratories, Albuquerque, NM, SAND83-0137C, March 1983.

Harrigan, R.W. *Handbook for the Conceptual Design of Parabolic Trough Solar Energy Systems—Process Heat Applications.* Sandia National Laboratories, Albuquerque, NM, SAND81-0763, July 1981.

Industrial Solar Technology (IST), *Final Report—Development of an Advanced Parabolic Trough Receiver,* draft version submitted to Sandia National Laboratories, Albuquerque, NM, July 1995.

KJC Operating Company. *O&M Cost Reductions in Solar Thermal Electric Plants.* Interim Report on Project Status, Sandia National Laboratories, Albuquerque, NM, September 1, 1994.

Kutscher, C., Davenport, R., Dougherty, D., Gee, R., and May, E. *Design Approaches for Solar Industrial Process Heat Systems.* Solar Energy Research Institute, Golden, CO, SERI/TR-253-1356, August 1982.

Lewandowski, A., Gee, R., and May, K. *Industrial Process Heat Data Analysis and Evaluation. Vol.* 1. Solar Energy Research Institute, SERI/TR-253-2161, July 1984.

Lotker, M. *Solar Energy for Utility Peaking/Intermediate Load Duty: A Cost-Effective Option for the 1990s.* The 1989 Electric Utility Business Environment Conference and Exhibition, March 29, 1989.

Morris, V.L. *Final Report—Solar Collector Materials Exposure to the IPH Site Environment.* Sandia National Laboratories, Albuquerque, SAND81-7028, January 1982.

McCollister, H.L. and Pettit, R.B. Anti-reflection Pyrex Envelopes for Parabolic Solar Collectors. *J. of Solar Energy Eng. Vol.* 105, pp.425–429, November 1983.

Murphy, L.M. and May, E.K. *Steam Generation in Line-Focus Solar Collectors: A Comparative Assessment of Thermal Performance, Operating Stability, and Cost Issues.* Solar Energy Research Institute, Golden, CO, SERI/TR-632-1311, 1982.

NREL (National Renewable Energy Laboratory). *Solar Radiation Data Manual for Flat-Plate and Concentrating Collectors,* April 1994, NREL, Golden, CO, NREL/TP-463-5607.

Ratzel, C.R. and Simpson, C.E. *Heat Loss Reduction Techniques for Annular Solar Receiver Designs.* Sandia National Laboratories, Albuquerque, NM, SAND78-1769, February 1979.

Ratzel, C.R. and Sisson, C.E. *Annular Solar Receiver Thermal Characteristics.* Sandia National Laboratories, Albuquerque, NM, SAND79-1010, February 1979.

Sandia National Laboratories (Sandia). *Test Results: Industrial Solar Technology Parabolic Trough Solar Collector,* draft version dated April 1994, Sandia National Laboratories, Albuquerque, NM, SAND94-1116, November 1995.

SERI (Solar Energy Research Institute). *Science & Technology in Review.* Golden, CO, Solar Energy Research Institute, Summer, 1990.

SEIA (Solar Energy Industries Association). *Catalog of Successfully Operating Solar Process Heat Systems.* Solar Energy Industries Association, Washington, DC, 1995.

22

Photovoltaic Solar Energy

Jerrold H. Krenz
University of Colorado

Introduction

Although the modern silicon photovoltaic cell was invented in 1954, the use of arrays of photovoltaic cells to complement conventional electrical generation systems (fossil, nuclear, and hydro) was not proposed until the energy crisis of the 1970s. From one perspective, the operation of an array of photovoltaic cells is deceptively simple; place the array in direct sunlight and connect an electrical load to its terminals. There are no moving parts for a nonfocused array, a reasonably long life can be expected, and the pollution of an operating system is negligible (manufacturing and disposal are not as benign). Why then are not arrays of photovoltaic cells being used more extensively to supplement conventional electric power plants? The answer can be summed up in a single word—cost.

The high cost of arrays of photovoltaic cells (relative to conventional power plants) is a consequence of the cost of constituent materials and the fabrication processes needed to manufacture photovoltaic cells. These cells are semiconductor devices that are related to semiconductor integrated circuits of electronic systems. While the cost of electronic integrated circuits has been rapidly declining, the decline in the recent cost of photovoltaic arrays has been much slower. This should not be too surprising. An important factor in the cost reduction of semiconductor devices has been their miniaturization (a 1 μm scale is now common). The number of devices that can be incorporated on a single integrated circuit has been steadily increasing—approximately doubling every two years. Hence, fewer integrated circuits tend to be required for a particular application. (Compare today's nearly "empty" personal computer with the earliest models packed full of integrated circuits.) The cost of many integrated circuit chips has changed only marginally, although each chip accomplishes considerably more than its predecessors. For example, each improved memory chip has resulted in a quadrupling of the number of bits that it can store. Over

0-8493-2514-5/96/$0.00+$.50

a period of only two decades, memory chips have increased from 4 Kb (4,096 bits) to 4 Mb (4,194,304 bits), while their price has changed very little. The thousand-fold increase in stored bits has decreased the cost of a stored bit to approximately 0.1% of its earlier value.

Miniaturization has played a key role in producing integrated circuits for electronic applications, but it is not useful for reducing the cost of photovoltaic cells, since for a given fabrication technology, it is the surface area of a photovoltaic cell that determines the amount of sunlight it intercepts (the number of photons per second it can potentially utilize). An increase in conversion efficiency does reduce the area of a cell or an array required to produce a specified electrical output power for a given illumination level. However, the greatest increase in conversion efficiency that might be anticipated for large-scale systems is a doubling of their present 10 to 14% efficiencies. Smaller cells can be used with solar concentrating systems, but they require higher power cells rather than miniaturized devices similar to those of integrated circuits.

Many of the fabrication techniques developed by the semiconductor industry for producing integrated circuits have resulted in improvement of photovoltaic cells and in a reduction of their costs. Nevertheless, a further reduction in cost by a factor of two or three from the mid-1990s cost of approximately $5 to $10/W is considered necessary before photovoltaic cells become competitive for some utility applications (a factor of 10 may be required for centralized power generation). This implies that the cost of a photovoltaic array would be comparable to the cost of a high-quality window of the same area. There is considerable debate as to the likelihood of achieving a factor of ten cost reduction. For example, can such a cost reduction be achieved through an improvement of existing manufacturing techniques or will entirely new manufacturing techniques be needed? If significant cost reductions are not achieved, electrical energy produced by photovoltaics will not compete with conventionally produced electrical energy in centralized applications in the next 10 to 15 years.

These considerations do not preclude the use of photovoltaic systems for numerous applications. Photovoltaic power can be economical when utility-distributed electric power is either not readily available or not cost-effective. Photovoltaic systems are being used to power remote radio and microwave communication systems as well as remote weather monitoring stations. In rural villages, photovoltaic systems are used to power small lighting systems, operate television sets, and power small refrigerators. Residential structures distant from an existing utility distribution grid can be powered by a photovoltaic system, thus eliminating the need for a costly utility extension. Even within a high-voltage, high-power utility system, the use of sectionalizing switches powered by photovoltaics avoids the need for costly step-down transformers. Moreover, the switches can also operate during power outages. Simple self-contained photovoltaic systems, such as a residential security light, avoid the cost of a buried power line that might otherwise be required. Thus, for many applications, a photovoltaic system is presently an economic alternative.

Utility-produced electrical energy is relatively cheap and has tended to not be used very efficiently. Since electrical energy produced by photovoltaic systems is presently much more expensive, it is imperative that it be used to power efficient energy-using systems. An economic analysis based on both the cost of the photovoltaic system and the energy consuming system is necessary. The added cost of a high-efficiency consuming system needs to be weighed against the avoided photovoltaic costs (the larger photovoltaic system that would be required for a less efficient system). Hence, a systems perspective is imperative.

22.1 Photovoltaic Principles

Historical Perspective

The photoconductivity of selenium (a change of conductance in the presence of light) was discovered by Willoughby Smith in 1873. Three years later W.G. Adams and R.E. Day observed a photovoltaic effect (an induced voltage) in selenium.[1] These discoveries were soon followed by the development of the selenium photovoltaic cell—the first device to convert solar energy directly to electrical energy. Until recently, selenium photovoltaic cells, even though less than 1% efficient, had been extensively used for photographic light meters. It was not until after the invention of the point contact and the junction transistor at Bell Telephone Laboratories that modern photovoltaic cells capable of much higher conversion efficiencies were developed.

The point-contact transistor was invented by John Bardeen, W.H. Brittain, and William Shockley, an invention for which they shared the 1956 Nobel Prize in Physics.[2] The point contact transistor of 1947 was followed by the invention of the much improved junction transistor in 1950. The study of semiconductors, in particular the behavior of free electrons and holes in semiconductors, led to the invention of the photovoltaic junction diode in 1953.[3,4] Even the earliest silicon photovoltaic cells, when placed in direct sunlight, had conversion efficiencies as high as 4%—an improvement by more than a factor of four over selenium diodes. The basic theory of the operation of a photovoltaic junction diode, published in 1955, built on the theory developed for semiconductor transistors.[5] As a result of the theoretical developments and the maturing of the technological base established for producing semiconductor transistors, photovoltaic cells with 10% efficiencies were constructed as early as 1957.

Photon Interactions

The operation of a photovoltaic cell depends on a photon interaction with valence electrons of the semiconductor from which it is fabricated.

$$E_{\text{photon}} = h\nu = hc/\lambda, \text{ J} \tag{22.1}$$

where

h = Planck's constant, 6.625×10^{-34} J s
ν = frequency, Hertz
c = velocity of light, 3×10^{8} m/s
λ = wavelength, m

In this equation, the energy of a photon, E_{photon}, is expressed in joules (J). For semiconductor applications, it is generally more convenient to express photon energies in electron volts (eV), that is, to divide the energy expressed in joules by the electronic charge, q.

$$E_{\text{photon-eV}} = hc/q\lambda = 1.242/\lambda_{\mu m}, \text{ eV} \tag{22.2}$$

where

q = electronic charge, 1.6×10^{-19} C
$\lambda_{\mu m}$ = wavelength, μm

A photon with a wavelength of 0.5 μm (a wavelength near the center of the visible spectrum), has an energy of approximately 2.5 eV (Figure 22.1). For a silicon semiconductor at 20°C, a photon energy of 1.12 eV is sufficient to liberate a valence electron from its valence energy band. The free electron and hole generated by the photon interaction can result in a device current.

The first photovoltaic application occurred in 1955, when an array of photovoltaic cells was installed for powering a repeater of a rural telephone system. Although the test was successful from a technical perspective, the high cost of the array made the system uncompetitive with alternative power sources. With the advent of artificial satellites (the first of these was Sputnik in 1957), arrays of photovoltaic cells became more widely used. The first solar-powered satellite was Vanguard I, launched in 1958—its radio transmitter, powered by silicon photovoltaic cells, operated for

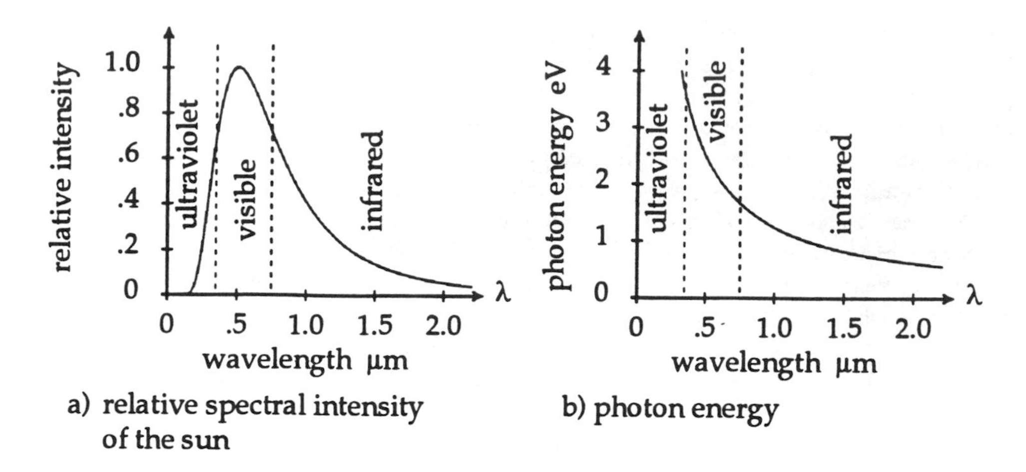

FIGURE 22.1 Relative spectral density of the sun outside the earth's atmosphere and photon energy.

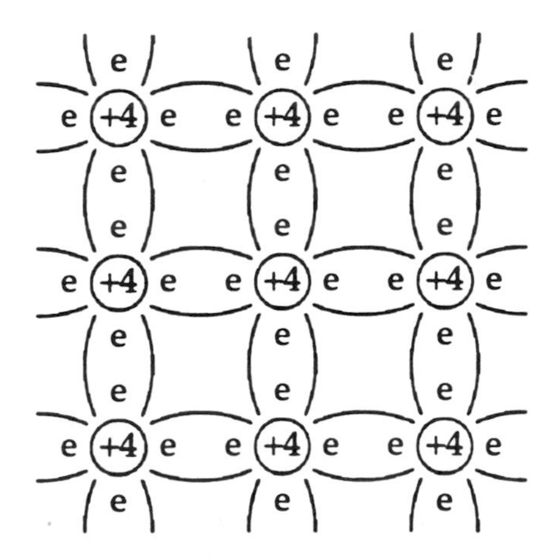

FIGURE 22.2 Two-dimensional representation of a silicon crystal.

approximately 8 years. Despite their high cost (early costs were as high as $100,000/W), photovoltaic cells for satellite applications were more economic than all alternatives. Arrays of photovoltaic cells power all present-day communication satellites. Although the production of arrays of photovoltaic cells for satellites has resulted in a steady decline in their cost, even at the time of the energy crisis of the 1970s, their costs were prohibitive for terrestrial applications.

Semiconductors

Although several types of semiconductors are available and have been used for fabricating photovoltaic cells, silicon is used for most of the presently available cells. Silicon (atomic number 14) has an outer electron shell of four valence electrons. Thus, it is neither a good conductor nor a good insulator. In its crystalline form, silicon atoms form covalent bonds by sharing valence electrons with their four neighboring atoms. As indicated in the two-dimensional representation of Figure 22.2, the nucleus and the inner electronic shells have a net positive charge consisting of four electronic charges. For the physical crystal, the atoms are arranged in a three-dimensional tetrahedral pattern, each atom being equally distant from its four neighbors.

The energy necessary for an electron to escape from the valence bond of a silicon crystal at 20°C is 1.12 eV. An electron that so escapes a valence bond is referred to as a *free electron*, since it is available for conduction. On the average, each valence electron has a thermal energy of $\frac{3}{2}kT$, or approximately 0.038 eV at room temperature, but owing to the Maxwellian velocity distribution of electrons, a few electrons will have sufficient energy to escape their valence bonds. Each free electron leaves behind a vacancy or hole in the valence bond. This allows a valence electron of an adjacent bond to readily move into this vacancy and, in the process, leave behind a new vacancy. In the presence of an electric field, both the free electrons and the valence electrons experience a force and tend to drift in the direction of the force. The net effect is a drift of the vacancy or hole in a direction opposite to that of the valence electrons. The drift of a hole is therefore in the same direction as that of a positive charge. Rather than consider the motion of numerous valence electrons, it is more convenient to deal with the motion of the hole, that is, the "missing" electron. An electric field gives rise to a drift of holes in the direction of the field and a drift of free electrons in the opposite direction. Since free electrons and holes are formed in pairs, a pure or intrinsic semiconductor material has equal number densities of holes and free electrons.

The current of a semiconductor depends upon the density of the carriers as well as the mobility of the carriers (the free electrons and holes). While electron–hole pairs are continuously generated within the crystal, the concurrent recombination of free electrons and holes results in an equilibrium value for their density. Recombination occurs when a free electron literally "falls" into a hole and in the process gives up its potential energy. Since the generation rate of carriers depends upon thermal energy, their density is temperature dependent. For silicon, the equilibrium density of free electrons and holes, n_i, is $1.5 \times 10^{16}/\text{m}^3$ at 20°C. This is an exceedingly small density compared to that of the valence electrons: one electron–hole pair for each approximately 10^{13} valence electrons. Incident light with photon energies greater than the escape energy (also known as the band-gap energy) for valence electrons (1.12 eV for silicon) can generate additional electron–hole pairs, which in turn increases the conductivity of the semiconductor. This effect, photoconductivity, is used in many light-sensing devices. An energy conversion cell, however, requires a more complex semiconductor configuration.

Extrinsic semiconductors contain selected impurities that are introduced by the process of doping. In doped semiconductors, conduction can be predominantly the result of either free electrons or holes, rather than due to equal quantities of the two carriers. For example, in the case in which an atom of a silicon crystal has been replaced with an atom with five valence electrons such as antimony, arsenic, or phosphorus (Fig. 22.3a), four of the five valence electrons complete covalent bonds with adjacent silicon atoms and the fifth or excess electron has, at room temperature, sufficient energy to become a free electron. The free electron in this case, however, does not leave behind a hole. An impurity atom of this type (5 valence electrons) is referred to as a *donor atom*. If sufficient donor atoms are present in the crystal (a density much larger than n_i), the density of free electrons will greatly exceed that of the holes, and as a result, conduction will be primarily due to free electrons. A semiconductor with donor impurities is referred to as an n-type semiconductor (n standing for negative).

A semiconductor in which the current is due primarily to holes can be formed by the addition of impurities with three valence electrons, such as boron, aluminum, or indium (Figure 22.3b). This impurity atom has insufficient valence electrons to complete four valence bonds. The incomplete bond results in a hole that can contribute to the semiconductor current. Impurity atoms with three valence electrons accept an additional valence electron and are hence referred to as *acceptor atoms*. A sufficient density of acceptor atoms (a density much larger than n_i) results in a semiconductor in which holes are the dominant carriers. Doping with acceptor atoms results in a p-type (positive) semiconductor.

The Semiconductor Junction Diode

A photovoltaic cell relies on a semiconductor diode to convert photon energy to electrical energy. A diode is formed by a transition from p-type to n-type semiconductor material. For charge carriers, free electrons and holes, to readily migrate from one region to the other, the diode must

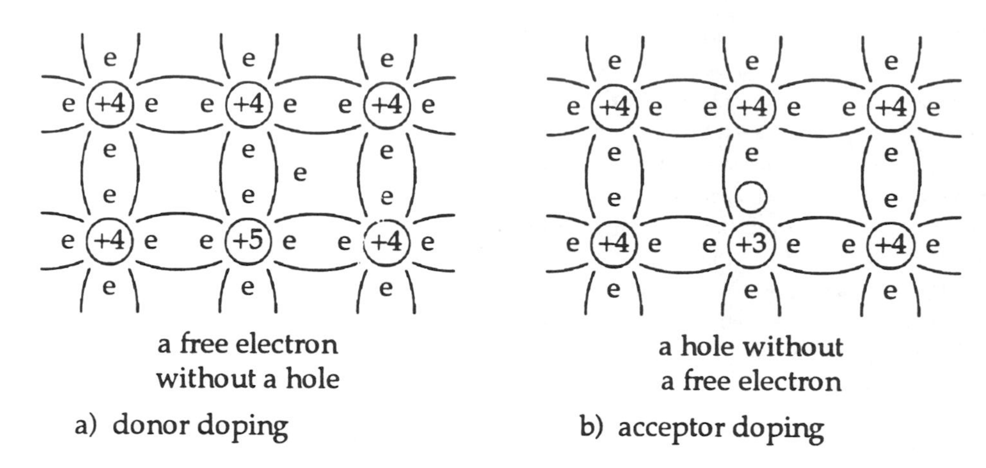

FIGURE 22.3 Doping of a semiconductor

be formed from a single crystal. Figure 22.4(a) illustrates a diode junction and Figure 22.4(b) indicates the doping concentration on each side of a *p–n* junction (the densities of donor and acceptor atoms need not be equal). While light energy would normally be incident on the left face of the diode shown, the behavior of a diode in the absence of light will first be considered. Recombination occurs in the region of the junction, as indicated in Figure 22.4(c). Free electrons from the *n*-type material (left side) tend to diffuse into the *p*-type region where they readily recombine because of the very large concentration of holes. Also, holes from the right side of the junction tend to diffuse into the *n*-type material and rapidly recombine with the numerous free electrons. These movements of free electrons and holes result in a net charge density in the region of the junction. Free electrons moving from the left side of the junction leave a net positive charge behind; the migrating holes leave a net negative charge behind on the right side of the junction.

The charge distribution gives rise to an electric field and hence a potential difference across the junction (the zero for the potential profile indicated in Fig. 22.4d is arbitrary). The potential across the junction tends to retard the diffusion of carriers across the junction—an eventual equilibrium condition prevails. In addition to the free electrons of the *n*-type semiconductor and the holes of the *p*-type (the majority carriers) the few holes of the *n*-type and the few free electrons of the *p*-type semiconductor (the minority carriers) must also be considered. Minority carriers tend to cross the junction readily since the potential profile is such that it enhances their migration. Therefore, the potential for zero current allows only the diffusion across the junction of sufficient majority carriers to cancel the current of minority carriers.

The potential difference across the junction is not directly available to an external circuit. For zero current (a dark condition), the contact potential differences of the semiconductor–metal junctions cancel the diode junction potential and result in a zero terminal voltage. An externally applied potential difference modifies the junction potential. For a basic semiconductor junction, a small positive potential decreases the potential barrier and hence allows a greater quantity of majority carriers to cross the junction. On the other hand, a negative potential increases the potential barrier and decreases the current due to majority carriers but leaves the current due to the minority carriers essentially unaffected. The diode current, I_D, is the sum of the current due to majority carriers and that due to minority carriers. For moderately large negative potentials, only minority carriers cross the junction. The minority current depends primarily upon the generation rate of minority carriers and is hence determined by the temperature and the semiconductor material. For a given temperature, however, it is roughly a constant, $-I_0$, which is independent of the junction potential. The current due to the majority carriers depends upon the potential profile, since only free electrons and holes with energies in excess of the barrier potential will cross the junction. As a result of the carriers tending to have a Maxwellian velocity distribution, a current

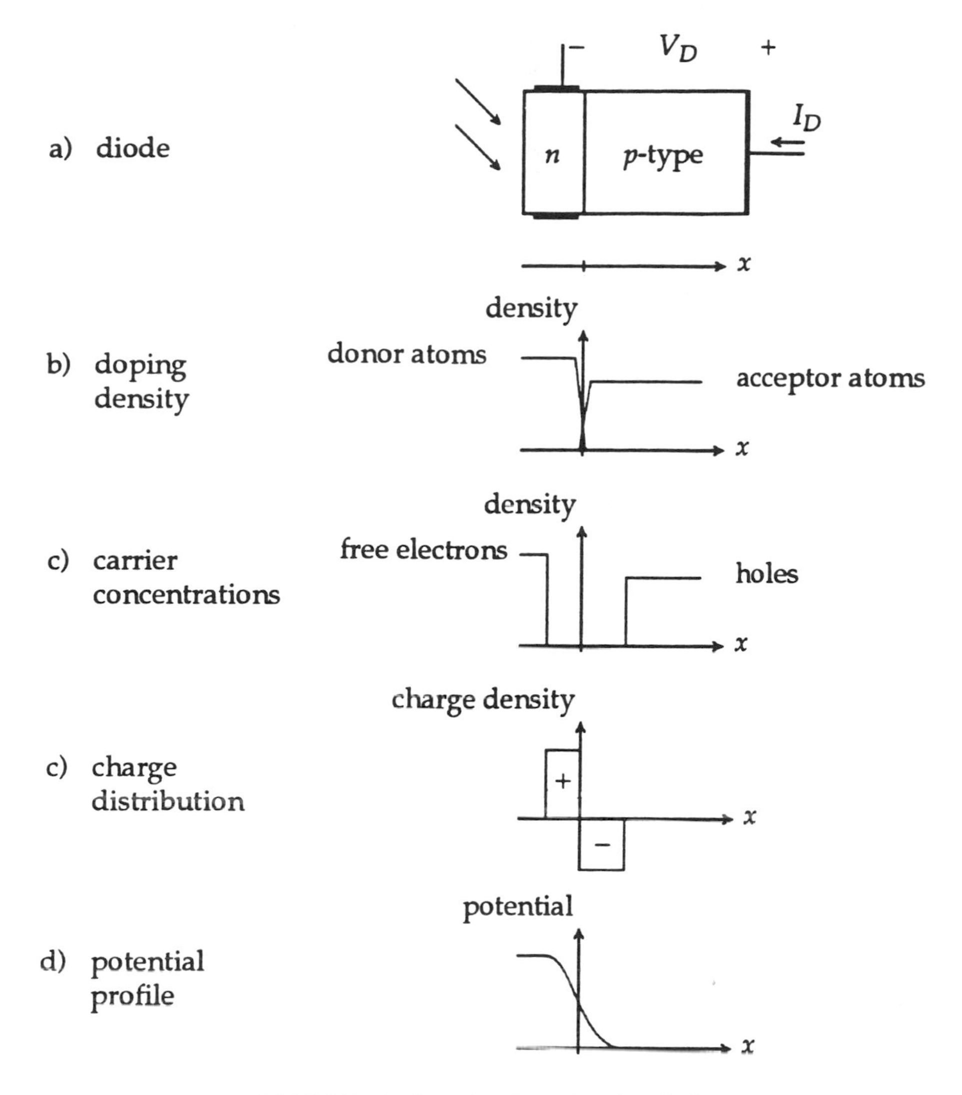

FIGURE 22.4 A semiconductor junction diode.

with an exponential dependence on the potential is obtained. The net result (a rigorous derivation is beyond the scope of this treatment) is given by the following expression.

$$I_D = \underbrace{I_0 e^{qV_D/nkT}}_{\text{majority carriers}} - \underbrace{I_0}_{\text{minority carriers}} \tag{22.3}$$

where

V_D = diode voltage, V
k = Boltzmann constant, 1.38×10^{-23} J/K
T = absolute temperature, K
n = dimensionless ideality factor

The constant I_0 multiplying the exponential term of the majority current carriers results in a zero current for a zero voltage. Figure 22.5 is a plot of the current–voltage characteristic of a junction diode in which the exponential voltage dependence results in a very rapid increase of current with

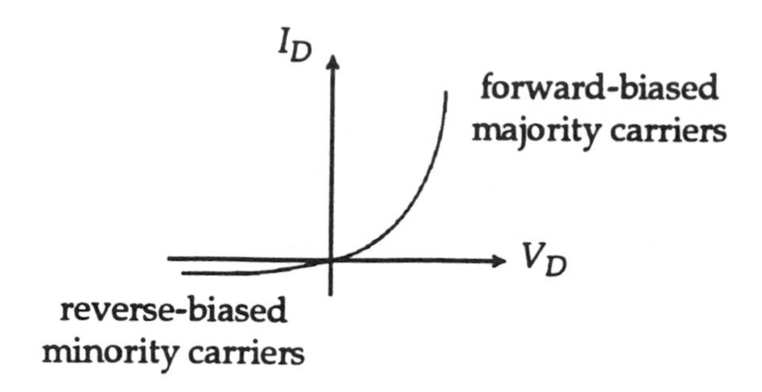

FIGURE 22.5 Current versus voltage characteristic of a junction diode. Only the region for $|V_D|$ small is shown.

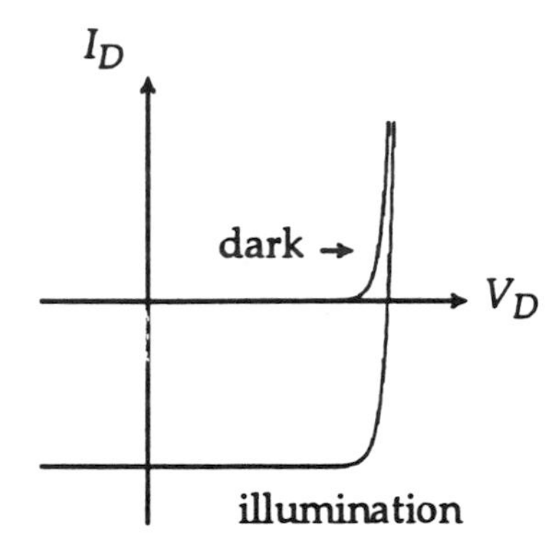

FIGURE 22.6 Current-versus-voltage characteristic of a photovoltaic cell.

voltage. The current $-I_0$ for reverse bias is frequently negligible compared to the forward current. Appreciable forward current for a silicon diode of interest requires a voltage of 0.6 to 0.7 V.

The Photovoltaic Junction Diode

Incident radiation with photons having energies equal to or in excess of the energy required to liberate valence electrons will generate additional electron–hole pairs in the semiconductor. Ideally, the *n*-type region of the diode of Figure 22.4(a) is sufficiently thin that the incident light is absorbed in the *p*-type region immediately adjacent to the junction. Equal quantities of majority and minority carriers (free electrons and holes, respectively, for the *p*-type material) will be generated. Since large quantities of majority carriers are present even in the absence of light, the excess majority carriers will have little effect. Without light, however, only a few minority carriers are present, and therefore, the current due to the photon-generated minority carriers of the incident radiation may be significant. Since minority carriers readily cross the junction, a minority carrier current proportional to the photon-generation rate of minority carriers results. This gives rise to the current-versus-voltage characteristic of Figure 22.6.

An elementary circuit model of a solar cell consists of a photon-dependent current source, I_p, which is in parallel with a junction diode (Figure 22.7). An expression for the load current, I_L, is readily obtained.

$$\begin{aligned} I_L &= I_p - I_D \\ &= I_p - I_0 e^{qV_D/nkT} + I_0 \end{aligned} \tag{22.4}$$

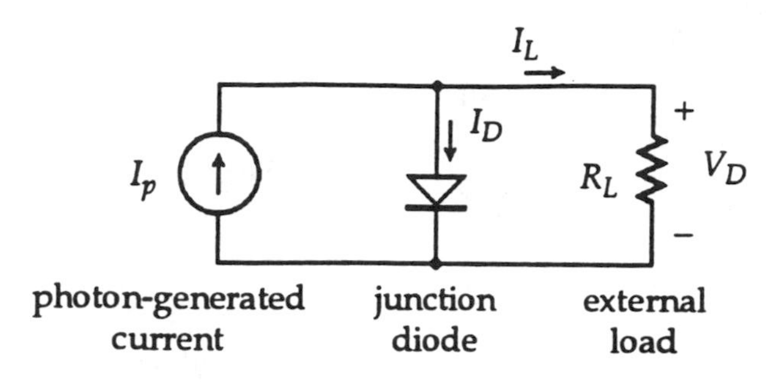

FIGURE 22.7 An idealized circuit model of a photovoltaic cell.

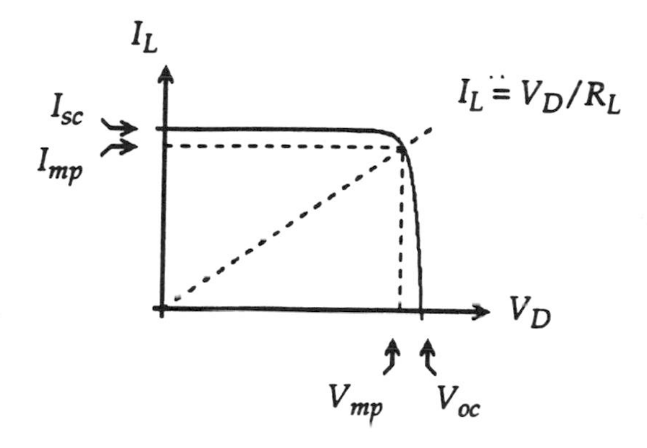

FIGURE 22.8 Load current of a photovoltaic cell.

where I_p = photon generated current.

A second current–voltage relationship is determined by the value of the load resistance.

$$I_L = V_D/R_L \tag{22.5}$$

Knowing the output voltage and current, the output power, P, is obtained.

$$P = I_L V_D \tag{22.6}$$

For a given light intensity, the output power depends upon the value of the load resistance. Often of interest is the maximum obtainable output power, P_{max}, the corresponding voltage and current being $V_{L\,mp}$ and $I_{L\,mp}$, respectively.

$$P_{max} = I_{mp} V_{mp} \tag{22.7}$$

The maximum power as indicated in Figure 22.8, represents the area of the rectangle formed by $I_{L\,mp}$ and $V_{D\,mp}$, which is clearly less than the product of the open-circuit voltage, V_{oc}, and the short-circuit current, I_{sc}.

$$P_{max} < I_{sc} V_{oc} \tag{22.8}$$

An expression for the open-circuit voltage, V_{oc}, may be obtained by setting the load current, I_L, equal to zero.

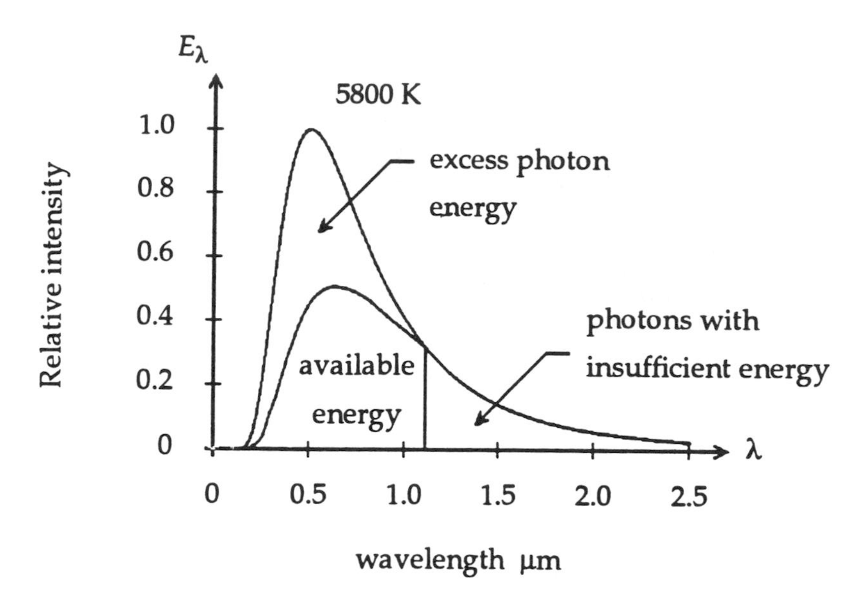

FIGURE 22.9 Energy utilization for a silicon photocell as a function of wavelength.

$$0 = I_p - I_0 e^{qV_{oc}/nkT} + I_0$$
$$V_{oc} = (nkT/q)\ln(I_p/I_0 + 1) \tag{22.9}$$

For a useful photovoltaic diode, the photon current, I_p, is much larger than I_0. Therefore, the 1 of the logarithm of Eq. (22.9) may be neglected.

$$V_{oc} = (nkT/q)\ln(I_p/I_0) \tag{22.10}$$

The open-circuit voltage of a typical silicon photovoltaic cell is only about 0.6 V. Hence, it is generally necessary to connect cells in series to achieve potentials adequate for most applications.

Optimizing the performance of solar cells is the subject of a now classic paper by Loferski.[6] Since only photons with an energy in excess of that required to generate free electron–hole pairs, the band-gap energy (E_g), contribute to the photon current, the spectral distribution of the incident radiation is important. Figure 22.9 shows the spectral energy density distribution for a 5,800 K blackbody radiator, a distribution that is approximately that of the sun. While the incident radiation is reduced for several wavelengths at the earth's surface due to atmospheric absorption, the distribution of Figure 22.9 will be used to obtain an approximate result. Since a photon energy of 1.12 eV is required to generate free electron–hole pairs in silicon (the band-gap energy), photons with wavelengths greater than 1.1 μm have insufficient energy to generate carriers. The area under the wavelength distribution curve to the right of 1.1 μm is therefore proportional to the energy that cannot be utilized by a photovoltaic cell; approximately 23% of the total energy is lost due to this effect. While all photons with wavelengths less than 1.1 μm can generate carriers, not all the energy of the individual photons is utilized. Since each photon generally produces only one free electron–hole pair, the excess energy of a photon serves only to heat the semiconductor.

A quantitative measure of the effectiveness of photons with energies in excess of E_g is readily obtained. If E_λ is the spectral energy density of the incident photons, $E_\lambda d\lambda$ is the energy in the range $d\lambda$. The corresponding number of photons in this differential range of wavelengths, dn, may be obtained by dividing $E_\lambda d\lambda$ by the energy of one photon, namely, $h\nu$.

$$dn = \frac{E_\lambda d\lambda}{h\nu} = \frac{\lambda E_\lambda d\lambda}{hc} \tag{22.11}$$

The useful energy of each photon is E_g, the energy necessary for an electron to break its valence bond.

$$E_g dn = \frac{E_g \lambda E_\lambda d\lambda}{hc} \tag{22.12}$$

The gap energy may be related to the wavelength of a photon.

$$E_g = \frac{hc}{\lambda_g}, \ E_g dn = \left(\frac{\lambda}{\lambda_g}\right) E_\lambda d\lambda \tag{22.13}$$

This energy distribution, valid for $\lambda \le \lambda_g$, is indicated in Figure 22.9. When this effect is taken into account, only approximately 44% of the incident radiation energy can be potentially utilized by a silicon photovoltaic cell. While semiconductors with larger band-gap energies tend to make fuller use of the energy of those photons which interact, a greater proportion of the incident photons have insufficient energy to interact. Conversely, lower band-gap energies allow more photons to interact but less fully utilize the energy of the reacting photons. Based upon this consideration alone, the optimum band-gap energy is on the order of 1.5 eV, somewhat greater than that of silicon.

For typical photovoltaic cells, the output voltage is considerably less than the band-gap potential, typically about half to two-thirds of the band-gap energy expressed in electron volts. Therefore, the theoretical efficiency of converting photon energy to electrical energy is only about 22%. (Loferski gives a theoretical value of 19.2% for silicon.) While efficiencies in excess of this theoretical efficiency have been achieved in the laboratory, efficiencies of commercially available cells tend to be in the range of 12 to 16%.

Fabrication of Single-Crystal Silicon Photovoltaic Cells

The single-crystal silicon cell, produced by processes similar to those used to manufacture electronic integrated circuits, is the most widely produced photovoltaic cell.[7] Quartzite, a pure form of sand (SiO_2), is used as the primary feed stock. It is reduced through a carbon reaction in a high-temperature furnace to produce metallurgical-grade silicon. The metallurgical-grade silicon is then treated with hydrogen chloride to form trichlorosilane ($SiHCl_3$). A fractional distillation process is then used to remove impurities. The next step, a hydrogen reduction reaction of the purified trichlorosilane, yields electronic-grade silicon, a polycrystalline material.

The growth of a single crystal from the polycrystalline silicon is accomplished by the Czochralski Process. A seed crystal of silicon is brought into contact with a silicon melt (silicon has a melting point of 1,415°C). The seed is then slowly withdrawn from the melt so that the silicon that adheres to the crystal forms a single crystal as it solidifies. By slowly withdrawing the crystal, the growth is continued, forming a cylindrical crystal, called a boule, with a length of 1 m or more and a diameter of 10 cm or more. In order to form *p*-type silicon, required for one region of the photovoltaic cell, an acceptor impurity, generally boron, is added to the silicon melt before the crystal is withdrawn.

The boule, after it is cooled, is sliced into thin wafers. These round wafers, having a thickness of less than 1 mm, account for the round structure that characterizes many photovoltaic cells. Alternatively, the boule can first be cut lengthwise to produce a square cell. The minimum wafer thickness is limited by the physical properties of crystalline silicon; thicknesses of 0.25 mm or more are common for photovoltaic cells. As a result of the slicing operation, approximately half of the original silicon crystal is lost as sawdust. The wafer is chemically etched, lapped to produce flat surfaces, and then cleaned.

The wafer, as a result of the acceptor atoms added to the melt from which it was grown, is *p*-type silicon. The top surface (that exposed to light) must now be doped with a sufficient density of donor atoms to neutralize the effect of the acceptor atoms and result in a net donor doping; an *n*-type semiconductor is formed. This thin *n*-type region forms a junction diode with the *p*-type semiconductor wafer. Phosphorus, a common donor, is typically introduced by forming a phosphorus oxide on the surface of the wafer (generally on both sides). At an elevated temperature (800–900°C), a diffusion of the phosphorus atoms into the crystal occurs. After the diffusion, it is common to remove the remaining oxide surface layer.

Metallic contacts to both the thin *n*-type semiconductor, the "top" surface exposed to light, and the *p*-type wafer are necessary. A typical photovoltaic cell with a diameter of 10 cm, when placed in bright sunlight, will produce an electrical power of approximately 1 W. Hence, for a cell voltage of 0.5 V, the current is 2 A. As a consequence of this large current, a low-resistance connection to the semiconductor is necessary—even small resistive losses will significantly reduce the efficiency of a photovoltaic cell. Screen printed metal pastes are generally used to form a metal contact. For the front surface, the conductors are deposited in a grid configuration, while for the back surface, either a solid layer or a dense grid is used.

The formation of a module in which the cells are connected together to produce a desired voltage and current is the final step. This involves soldering interconnecting wires to the individual cells. The cells are then encapsulated, that is, sealed within a weatherproof enclosure, and mounted in a rigid supporting frame.

These numerous intricate semiconductor processing steps, most of which are carried out at elevated temperatures in controlled atmosphere furnaces, result in the high cost of photovoltaic cells. Automated batch processes are used to reduce costs, but for most applications costs are still considerably higher than those of competing schemes for producing electrical power.

Other Photovoltaic Cells

An alternative to growing single-crystal silicon with the Czochralski process is casting silicon directly into rectangular ingots. These ingots are then sliced into wafers, producing rectangular cells, which fit into modules more efficiently than round wafers. Unfortunately, casting produces polycrystalline silicon (silicon consisting of small crystallites or grains), which typically results in less efficient photovoltaic cells than those fabricated from single-crystal silicon. An alternative process that eliminates slicing and the concurrent silicon loss is the direct growth of thin ribbons of silicon. A single-crystal silicon ribbon can be obtained by pulling silicon from a melt through a narrow graphite die. The thin ribbon, less than a millimeter thick, is then cut into wafers for producing cells. However, as a result of impurities introduced by the graphite die, the silicon is generally of a lower quality than that produced by the Czochralski process.

Elements in groups III and V of the periodic table can also be used to produce single-crystal semiconductors.[8] A common III–V semiconductor is gallium arsenide. Each atom of gallium (valence 3) has bonds with four equally distant atoms of arsenic (valence 5). Similarly, each atom of arsenic has bonds with four equally distant gallium atoms. Through a sharing of valence electrons, all valence bonds are complete—a crystal has equal numbers of gallium and arsenic atoms. Gallium arsenide can be doped with either donor or acceptor atoms to form *n*- or *p*-type semiconductors. The band-gap energy of gallium arsenide, 1.42 eV, is very close to that which results in the maximum photon utilization efficiency. Furthermore, gallium arsenide, as well as most other III–V semiconductors, has a much larger photon absorption coefficient than silicon. As a consequence, photovoltaic cells fabricated from III–V semiconductors need be only a few microns thick, whereas single-crystal silicon cell designs usually require a thickness in excess of 100 μm or more. Experimental photovoltaic cells of gallium arsenide and related materials have been tested with efficiencies in excess of 30%. Other III–V semiconductors suitable for photovoltaic cells include gallium arsenide crystals with aluminum, indium, and phosphorus atom substitutions as well as indium phosphide. High efficiencies have been achieved with III–V semiconductors, but the cost of these cells is considerably greater than that of single-crystal silicon cells.

An alternative approach that is expected to significantly reduce the cost of photovoltaic cells is the use of thin-film cells.[9,10] A polycrystalline semiconductor film may be deposited on a substrate such as glass by a variety of techniques, which include chemical vapor deposition, electrodeposition, evaporation screen painting, spraying, and spattering. Layers are deposited in succession, first a conducting contact layer followed by two different semiconductor layers on which is deposited a second conducting layer. The two polycrystalline semiconductor layers form what is known as a heterojunction device. Experimental photovoltaic cells made from cadmium telluride and copper indium diselenide semiconductors have achieved conversion efficiencies of 16% and 17%, respectively.

Also promising are thin-film photovoltaic cells made from amorphous silicon.[11,12] Hydrogenated amorphous silicon (a-Si:H), which contains hydrogen atoms bonded to the silicon atoms, has many crystal-like properties. It can be doped to produce *n*- and *p*-type semiconductors from which a junction can be formed. Experimental photovoltaic cells fabricated using a-Si:H have achieved efficiencies of 13%. Owing to their low cost, these cells are extensively used for powering watches and calculators (in terms of the number of cells, the largest current application of photovoltaic cells). Unfortunately, amorphous silicon cells degrade with time—stabilizing at 65 to 85% of their initial efficiency. Alternative stabilized structures need to be developed to minimize this effect.

An advantage of thin-film cells is that their cost is expected to be considerably less than single-crystal silicon cells. It is anticipated that very large area photovoltaic cells (1 m^2 or larger) will eventually be produced. Using an automated manufacturing process, these cells will then be segmented into smaller cells and interconnected to form a complete module.

Module Interconnections of Photovoltaic Cells

Series- and parallel-connected photovoltaic cells are generally utilized to achieve a desired load voltage and current. The interconnection of identical cells equally illuminated does not pose a problem. If n identical cells are connected in series, a voltage n times that of a single cell is obtained. Similarly, if n cells are connected in parallel, a current that is n times that of a single cell results.

As a result of manufacturing tolerances, however, the cells of a module will not be identical. Furthermore, the performance of individual cells may degrade differently during operation. Shading or soiling will also reduce a cell's output. Not only will the overall performance of a module of cells be degraded due to an energy loss of a particular cell, but, depending on the interconnections between cells, a much larger loss may occur, as the other cells may no longer function optimally.

Consider the case of the series connection of two photovoltaic cells with different output currents (Figure 22.10). The resultant current versus voltage of the two cells, the dashed characteristic of Figure 22.10, is obtained by adding the individual cell voltages for each value of current. This characteristic, it will be noted, is essentially that of a series connection of two of the low output cells, ②. This effect is important because for a series connection of several cells, the performance of the lowest output cell determines the performance of the entire array of cells.

For two parallel-connected cells, the overall current is obtained by adding the currents of the individual cells (Fig. 22.11). While a current loss results due to a poorly functioning cell, ②, the terminal voltage remains nearly unchanged. Hence the loss is essentially that due to the poorly functioning cell—the effect on the behavior of the other cell, ①, is a slightly reduced voltage.

As a result of the small voltage obtained from an individual cell, 0.5 to 0.6 V, it is necessary to connect cells in series to obtain the voltages required for most applications. For example, to charge a 12 V lead–acid storage battery, as many as 36 cells may be required to provide an adequate charging voltage (approximately 14.5 V), overcome the losses of a regulator circuit, and function for less than a maximum of solar illumination intensity. A combination of series and parallel connections is generally utilized to minimize the effect of a cell with a low output. The array of

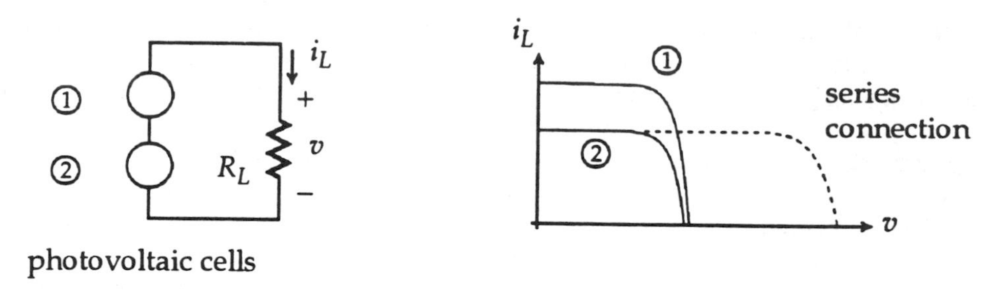

FIGURE 22.10 A series connection of two photovoltaic cells.

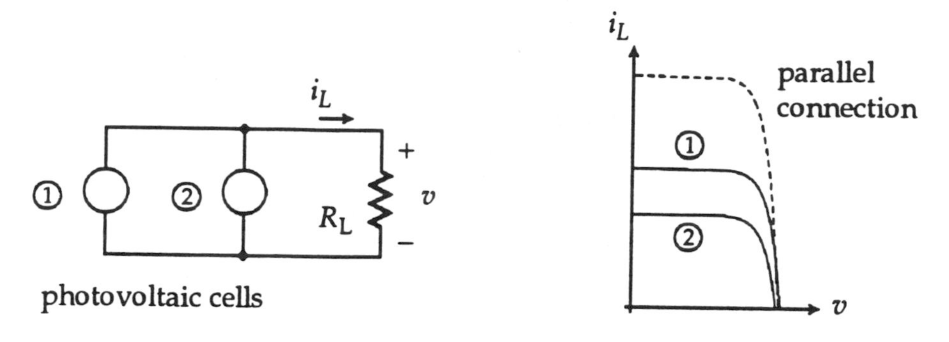

FIGURE 22.11 A parallel connection of two photovoltaic cells.

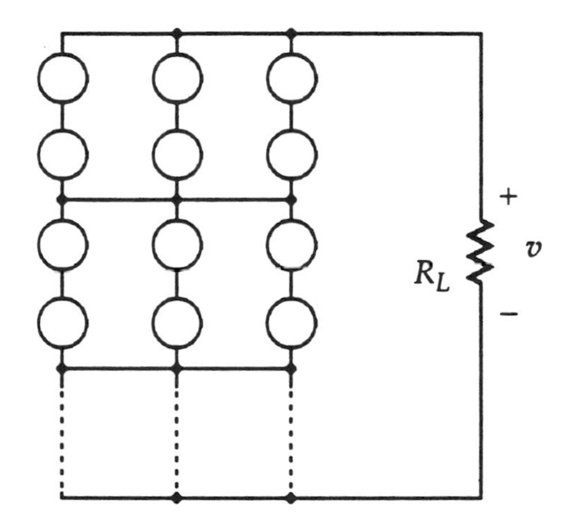

FIGURE 22.12 Series-parallel-series connections of photovoltaic cells.

Figure 22.12 has a series connection of two cells that are connected in parallel with two other sets of series-connected cells. These cells are then connected in series with other similar combinations to obtain a desired load voltage. A circuit such as that of Figure 22.12 (other combinations of series and parallel connections are also used) tends to reduce the effect of a poorly functioning cell on the overall performance of the module.

A complete failure of a single cell is also possible (a broken lead or the cell completely covered). For this situation, the cell may become an open circuit or a junction diode without a photon-produced current (Fig. 22.7 with $I_p = 0$). If the cell is part of a series circuit, it will prevent the other series-connected cells from producing a current. Depending on the circuit, a reverse bias

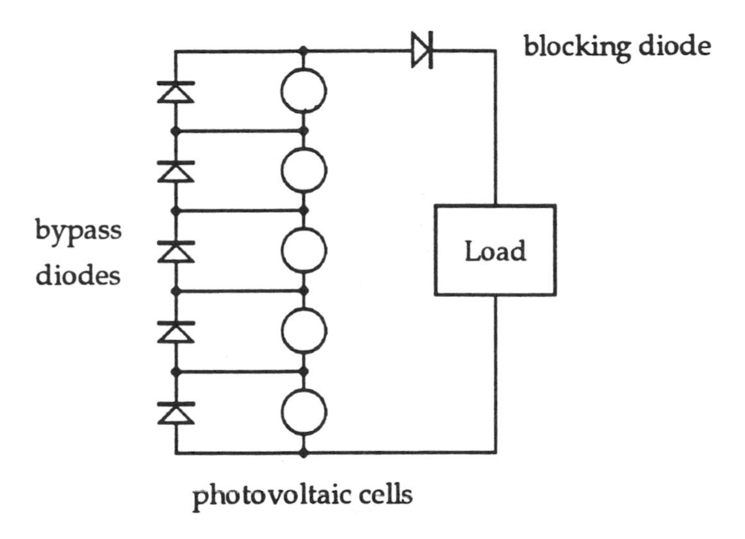

FIGURE 22.13 Series-connected photovoltaic cells with bypass diodes.

voltage may develop across the faulty cell, which could cause a destructive breakdown of the cell. Bypass diodes (conventional silicon rectifier diodes) are utilized to prevent this from occurring (Fig. 22.13). For normal operation, the bypass diodes will be reverse biased and their currents will be negligible. If a cell fails, however, the diode of that cell will provide a current path. The array continues to function, but its overall voltage will be reduced by the loss of the voltage of the faulty cell as well as by the voltage developed across the bypass diode (approximately 0.7 V). A blocking diode in series with the cells is also indicated in Figure 22.13. This diode will prevent a load such as a battery from resulting in a reverse current of the photovoltaic cells when they are not illuminated. Unfortunately, the blocking diode reduces the load voltage by approximately 0.7 V.

As a result of the series and parallel connections of the photovoltaic cells of a module, the overall performance of the module tends to be degraded by a poorly functioning cell or cells. While the degradation is greater than that associated with the loss of a particular cell, a design using a combination of interconnections and bypass and blocking diodes tends to minimize the performance degradation of the overall module.

22.2 Electrical Energy Usage

Expectations Arising from the Energy Crisis of the 1970s

Prior to the oil embargo of 1973, energy consumption in the United States, as well as the rest of the world, had been steadily increasing. Furthermore, an increasing fraction of the primary energy inputs were being transformed to electrical energy. Electrical energy currently accounts for over one-third of the overall primary energy usage in the United States. An important benefit associated with energy supplied by an electric utility is its instant availability (in essence, it is only necessary to "plug in" to the nearest electrical outlet). As a consequence, an electrical utility system must have sufficient generating capacity to meet a peak in demand—if the system is unable to meet a peak demand, a system failure (a blackout) occurs.

Before 1970, trends in electrical energy usage and generating capacity were highly predictable. Each decade since 1900, without exception, usage and capacity had doubled; that is, they increased, on average, 7% each year. Hence, if one assumed that this trend was an unalterable property of the electrical system (as experts of the 1970s mistakenly assumed), a prediction of future quantities was straightforward—each decade a doubling should occur. To continue this trend over a period

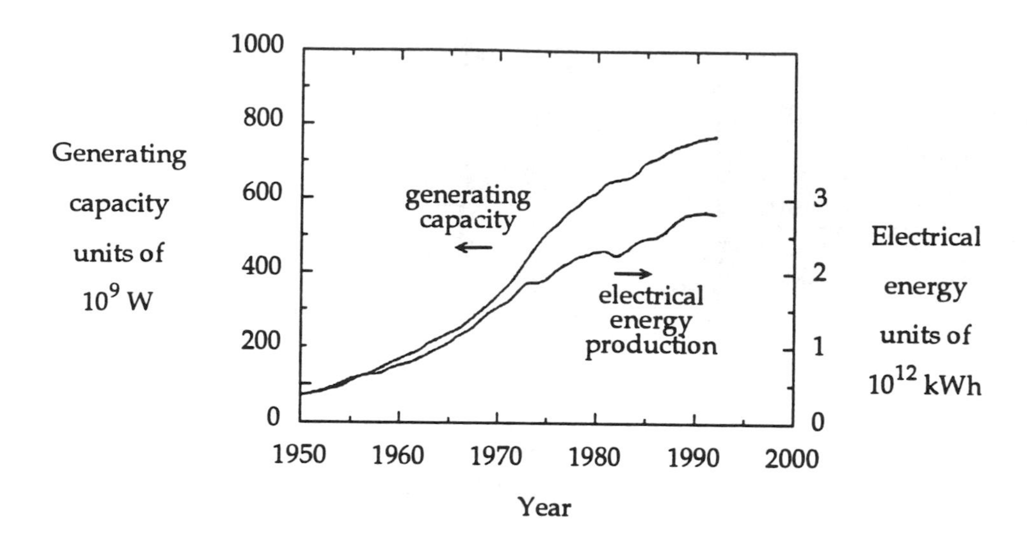

FIGURE 22.14 U.S. utility electrical generating capacity and energy production.[13-15]

of one decade, one new unit of capacity would need to be added for each existing unit (assuming no failures or retirements of old units). For a two-decade interval, a second doubling would occur, that is a capacity of four units would be needed for each unit at the beginning of the period. As can be seen from the data of Figure 22.14, however, the trend did not continue after 1973.[13-15] While electrical energy consumption has continued to increase, that for 1990 was "only" 83% greater than that of 1970, considerably less than the 300% increase corresponding to a doubling each decade. The increase over the most recent decade (1983–1993) was 25%, considerably less than the 100% corresponding to a doubling.

Forecasts done in the early 1970s for future electrical generating capacity are given in Table 22.1. The forecast errors for 1985 are inordinate, from 29 to 53% too high. Given the long lead times required for constructing electric power plants, a generating capacity no greater than 800 GW can be expected for the year 2000. Hence, the projections of Table 22.1 for the year 2000 are 135 to 181% too high.

A large part of the prediction errors was due to a clouded vision of the potential of nuclear power plants, a vision of the atom supplying over half of the nation's electricity by the end of the century. The Atomic Energy Commission in 1974 forecast a nuclear generating capacity of 260 GW in 1985 and 1,200 GW in 2000.[16] These forecasts were even more inaccurate than those for overall generating capacity. In 1994, the installed nuclear generating capacity was 99 GW.[15] Furthermore, a decline in capacity is expected over the next several years as a result of aging reactors being retired.

TABLE 22.1 Early 1970s Projections of U.S. Electrical Generating Capacity[16]

	GW*	
	1985	2000
Actual generating capacity (1972, 400 GW)	698	
Department of the Interior (1972)	915	1,880
Federal Power Commission (1974)	1,070	2,250
Atomic Energy Commission (1974)	903	2,220

* 1 GW is equal to 10^9 W, approximately the generating capacity of a large electric power plant.

Photovoltaic Expectations

Starting in the 1970s and continuing through the present, there have been numerous projections of the extent to which photovoltaic cells will be used to produce electrical energy. A joint National Science Foundation and National Aeronautics and Space Administration (NSF/NASA) study in 1972 envisioned that "a significant input" will occur from photovoltaic systems on buildings between 1985 and 1995 and that large industrial central power stations will be installed by the year 2000.[17] In 1974, following the oil embargo of 1973, hearings were held by the Subcommittee on Energy of the Committee on Science and Astronautics of the U.S. House of Representatives.[18] Presentations of expert witnesses (Loferski, Ralph, Rappaport, and Goldsmith) were extremely optimistic regarding the energy producing potential of photovoltaic cells. Loferski, for example, predicted a cost of 30¢/W in 1983 based on a yearly production rate of 200 MW. Numerous books published during the 1970s asserted expectations of low-cost photovoltaic systems supplying a significant fraction of the electrical energy used in the United States.[19-21]

In 1979 the American Physical Society (APS) Study Group published *Solar Photovoltaic Energy Conversion*.[22] As stated by Ehrenreich and Martin, "To obtain conclusions that would be substantially free of preconceived opinions, the study group was to be chosen primarily from scientists who possessed appropriate disciplinary backgrounds but were not involved in photovoltaics in a major way."[23] The APS study concluded that the photovoltaic module costs would have to be reduced to 10 to 40¢/W_p (peak, 1975 dollars) in order for electricity produced by photovoltaic systems to be economic. However, even if this cost goal were to be achieved, they concluded that photovoltaic systems were unlikely to produce more than 1% of the U.S. electrical energy in 2000, as 30 years or more would be required for a maturing of the necessary technologies.

Another totally unrealistic proposal for the use of photovoltaics to produce electrical energy is now all but forgotten. In 1968, Glaser proposed using satellites with a mammoth photovoltaic array (100 km^2) that would transmit the energy to earth with microwave beams.[24,25] This proposed system was to provide nearly continuous supply of electrical power—an interruption would occur only for a short daily interval when the satellite passed through the earth's shadow. The proposal received considerable attention during the 1970s. However, a study published in 1981 by the National Academy of Sciences was extremely pessimistic regarding the feasibility of this concept.[26] As a result of the *Challenger* accident and continued problems with the space shuttle, satellite power stations are now a past dream. Not only are lower-cost photovoltaic cells required, it is also necessary to lift satellites into a stationary orbit at a cost that is only a very small fraction of present costs. Photovoltaic power-producing systems appear to be destined to remain earth based for the foreseeable future.

Many recent projections continue to imply that photovoltaic cells will "soon" produce electrical energy that will be competitive with that produced by fossil-fueled power plants.[27,28] Even the titles of published articles, such as "Photovoltaics: Unlimited Electrical Energy from the Sun,"[29] have been misleading. The term "unlimited" carries nearly the same connotation as that once put forth by the early nuclear power industry, "too cheap to meter." Realistic appraisals of photovoltaic energy systems are still lacking.

Photovoltaic Power Demonstration Systems

Utility demonstration systems ranging from peak powers of 15 kW to several megawatts have established the technical feasibility of photovoltaic cells supplying electrical power to a utility grid.[30-33] For the systems tested, the land requirements are modest. For example, a 1 MW photovoltaic system with an overall system efficiency of 10% requires a cell area of approximately 10^4 m^2 (incident insolation of 1 kW/m^2). Accounting for a cell packing factor (associated with round cells) and for the additional area needed for a tracking system, the land area remains modest—a few times 10^4 m^2.

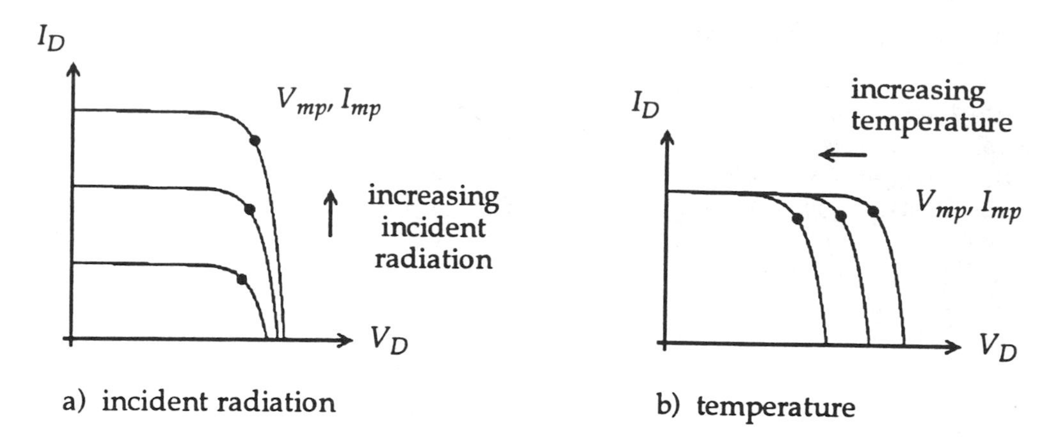

FIGURE 22.15 Variation of a photovoltaic cell's current-versus-voltage characteristic as a function of incident radiation and temperature.

In order to achieve usable voltages and currents, photovoltaic cells are connected together to form modules which are then connected in segments to comprise an array. For a photovoltaic cell to produce its maximum output power, the current and voltage of a cell must be optimized (the maximum power quantities, I_{mp} and V_{mp} of Fig. 22.8). The current-versus-voltage characteristic of a photovoltaic cell, however, depends on both the incident radiation and on the temperature of the cell (Fig. 22.15). For each level of incident radiation and temperature, an optimal voltage and current exists for which the output power of a cell is a maximum. This implies that the equivalent load resistance connected to a cell, V_{mp}/I_{mp}, must be changed as the incident radiation and temperature changes.

The optimal overall current and voltage of an array of photovoltaic cells also depends on the incident radiation and temperature, both of which may vary across the array. A malfunctioning cell or cells will also affect the current-versus-voltage characteristic of the array. Hence, an electrical power conversion unit (PCU) is required, which results in an electrical load that corresponds to a maximum output power. In addition, for utility applications, it is necessary to convert the DC power of the array to 3-phase AC power (Fig. 22.16). The operation of the power conditioning unit relies on power-electronic semiconductor switching devices to first convert the variable output voltage of the photovoltaic array to a constant DC voltage and then to convert this voltage to 3-phase (60 Hz in North America) AC power. The maximum power tracking is achieved by an electronic control system (generally microprocessor controlled), which continuously varies the load current by small increments to determine the corresponding change in the array voltage. Through this process, the optimal array current and voltage are obtained. The response of the control system is sufficiently rapid to track changing conditions such as the effect of a passing cloud. While conversion efficiencies in excess of 90% are achieved for large power conversion units, this implies an effective reduction of the photovoltaic cell efficiency. For example, the efficiency of a nominal 15% photovoltaic module is reduced to 13.5% when supplying electrical power to a utility grid through a 90% efficient power conversion unit.

Several utility-connected photovoltaic systems have been installed to ascertain the feasibility of providing utility power from arrays of photovoltaic cells. There does not appear to be any significant technical obstacle—it is only economic competitiveness that precludes a wide-scale usage of photovoltaic arrays. A brief summary of several demonstration projects follows.

Hesperia-Lugo: This installation was located in the Mojave desert near Hesperia, California. The 1 MW_p facility, installed in 1981, consisted of single-crystal silicon photovoltaic cells. Flat-plate modules are mounted on two-axis trackers to maximize the electrical output power. While the measured peak power of the system was less than its design value of 1 MW (based on a "standardized" insolation of 1 kW/m^2), it has performed well—having an

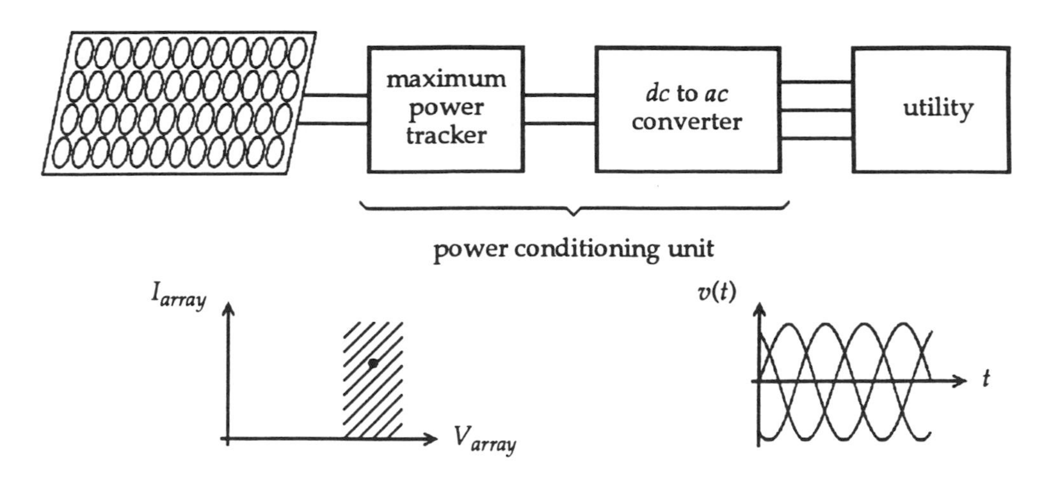

FIGURE 22.16 A utility photovoltaic system.

availability of over 96% and capacity factor of 36% in 1989. The silicon photovoltaic modules have been removed owing to their value being much greater than can be justified by the electrical energy they produced.

Carrisa Plains: This facility, installed in 1984 and 1985, was located in San Luis Obispo County, California.[34,35] It consisted of a 5.2 MW_p array of single-crystal silicon photovoltaic cells. Two-axis trackers, with mirrors mounted on the sides of the modules (2× magnification), were used to enhance the system's output power. While the initial peak output power was 5.2 MW, its output had dropped to 3.0 MW by 1990. The decline in efficiency was attributed to an oxidation of the encapsulation, a result of the elevated temperatures caused by the mirrors. This plant, the largest facility to date (1995), was also dismantled in order to sell its photovoltaic collectors.

City of Austin PV300 System: This is a smaller facility, 300 kW_p, that was installed in 1987. It uses single-crystal silicon photovoltaic cells and a single-axis (north–south) tracker. It has produced a measured peak AC output power of 250 kW.

3M PV300 System: This is a 300 kW_p single-crystal photovoltaic system installed outside of Austin, Texas, in 1990. A 22× linear Fresenal lens concentrating system is used in conjunction with two-axis trackers.

PVUSA: Photovoltaics for Utility Scale Applications is a cooperative demonstration facility designed to test commercial and precommercial photovoltaic systems.[36-38] While the main test site is located in Davis, California, the program also supports facilities at host utility sites. Two types of demonstration projects are being considered. Emerging module technologies (EMT) consist of unproved but promising photovoltaic technologies—20 kW_p units are being tested. Higher-power, utility-scale (US) systems based on more mature technologies are being tested at a 200 kW_p or higher power levels. In addition to single-crystal silicon cells, other photovoltaic technologies, including amorphous, polycrystalline, and thin-film silicon cells, as well as nonsilicon cells, are being tested. A goal of the project is to arrive at detailed cost estimates for producing electrical energy from various photovoltaic systems.

15 kW Amorphous Silicon Photovoltaic System: This system was installed in 1988 in Orlando, Florida.[39-41] The photovoltaic array is mounted at a fixed tilt of 25° to the South—tracking is not used. Over a 3-year period (1988–1991) the overall performance of the system declined by 25%, so that by 1991, the peak power efficiency was only 4%.

Kobern-Gondorf: This facility, constructed in 1987, is located in the lower Moselle region of Germany.[42] Its function is similar to that of PVUSA, that is, to test various types and sizes of solar arrays.

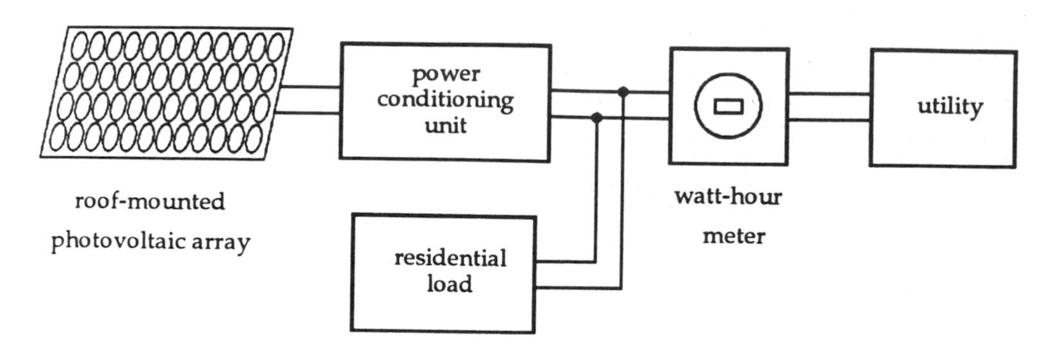

FIGURE 22.17 A residential utility-connected photovoltaic system.

Other systems have also been proposed. Detailed cost estimates for three alternative utility-scale (50 MW) installations have been prepared by Stolte et al.[43] Two of the proposed systems are based on high-concentration cells (500:1 and 325:1) having conversion efficiencies of 27.4%. The third system is based on flat-plate collectors (no tracking) using copper indium diselenide cells with an efficiency of 13%. An electrical energy cost of 11 to 13¢/kWh is projected for these systems. Costs depend on achieving efficiencies of the production units that are equal to those of experimental laboratory cells and on modules having a 30-year life expectancy. Since no production experience is presently available to compare the production cost estimates, the energy cost estimates must be treated as tentative.

An alternative to very large utility-type photovoltaic systems is small utility-connected distributed systems mounted on individual residences (or commercial buildings).[44,45] Electrical energy produced by the photovoltaic arrays is supplied directly to the residence; if the residential demand is less than the array output, energy is delivered to the utility-grid system (Fig. 22.17). No energy storage is needed and to the extent that there is a favorable correlation between the demand and the available energy from the photovoltaic array, a utility benefit occurs. Photovoltaic arrays can generally be mounted with a fixed orientation on roofs.

A research and demonstration project in Gardner, Massachusetts, involves 30 residential systems.[46,47] Single-crystal silicon cells are mounted in 4 × 6-ft aluminum modules and have a peak power of 220 W (11% efficient cells). Each residence has ten photovoltaic units, resulting in a peak capacity of 2.2 kW. A maximum-power tracking inverter (power conditioning unit) is used to supply 240 V AC electrical power to the residence. When the array output power exceeds the residential demand, power is fed back to the utility through a conventional watt-hour meter. Since the meter turns backward for this condition, the home-owner receives credit for the excess electrical energy at the same rate for which energy is purchased. Alternative metering schemes, which might, for example, give additional credit for power during peak demand periods, are possible, but they would increase the complexity of the system. Based on 2 years of data, the average amount of photovoltaic-produced electrical energy was approximately 2,200 kWh/year, approximately one-third of the yearly electrical energy used by a residence. While an evaluation of the interaction of the residential systems with the utility system was carried out, data on the cost of the photovoltaic system were not included in published articles.[46,47] Only the technical feasibility of this type of residential system was established—it was not claimed to be a reasonable economic investment.

Another example of a proposed grid-connected residential photovoltaic system is project "Megawatt" in Switzerland.[48] The intent of this program is to install 333 roof mounted systems of 3 kW. The initial installation phase in 1989 consisted of only 10 units and was followed by a second phase of 100 units installed in 1990; still further increases have followed.

While operating, arrays of photovoltaic cells produce no pollutants or greenhouse gases. As with all manufacturing processes, both the production and the disposal of photovoltaic systems has an environmental impact.[49] The reduction of quartzite (silicon dioxide) with carbon results in carbon dioxide emissions. In addition, the direct and indirect energy inputs associated with the production of photovoltaic systems results in emissions of greenhouse gases as well as other pollutants. Undesirable emissions also occur as a result of the semiconductor processing steps used to produce photovoltaic cells. Most of these processes are similar to those presently used extensively by the microelectronics industry. Emissions associated with this industry seem to be well monitored and controlled. Even with increased production rates, as would be required for the large-scale utilization of photovoltaic systems, the environmental impact is expected to be much less than that of both fossil-fueled and nuclear-powered generating systems.

22.3 Stand-Alone Photovoltaic Systems

Introduction

Although grid-connected photovoltaic systems have not been cost-effective, standalone systems for some applications in which utility power is not readily available can be cost-effective. These applications include not only remote sites for which a utility extension would be prohibitively expensive, but also low-power electricity needs in which the cost of the photovoltaic system is less than the cost of a utility connection. An example of the latter situation is aircraft warning lights on high-voltage electric transmission towers. Even though standalone applications are unlikely to produce "significant" amounts of electrical energy (relative to that produced by utilities), they can result in an economical means of obtaining electrical power.[50]

The simplest standalone photovoltaic system consists of a photovoltaic array connected directly to an electrical load. An example is pumping water from a well for livestock. Since pumping occurs only when sufficient solar insolation is available, the pumped water is stored and used when needed. Photovoltaic-driven water pumps may be considered the modern equivalent of the windmill-driven water pumps that were common on farms at the beginning of the 20th century. Another application is the solar-powered calculator, designed to operate at ambient light levels. Calculators, which generally use amorphous-silicon photovoltaic cells, represent the largest application in terms of the number of cells used.

For most standalone photovoltaic systems, battery storage of electrical energy is required. In addition, an auxiliary power system such as an internal combustion engine driving an electric generator may be required to provide electrical energy during extended intervals of inadequate solar insolation. While many electrical loads can be supplied by DC power from a photovoltaic-battery system, other loads such as induction motors require AC power. Therefore, a DC-to-AC inverter, which can also transform the voltage level, may be needed. These components are indicated in Figure 22.18.

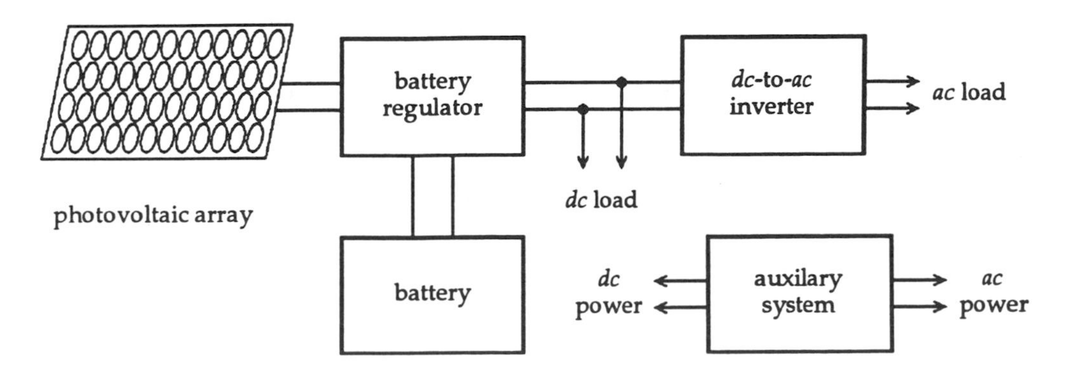

FIGURE 22.18 A stand-alone photovoltaic system.

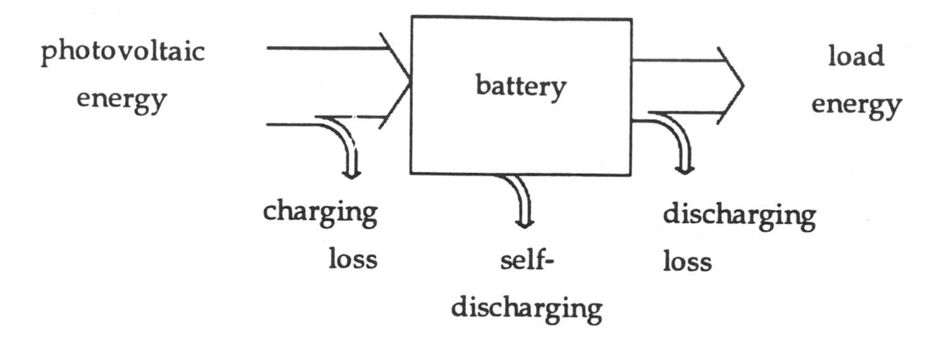

FIGURE 22.19 Battery energy storage.

For utility applications, a tracking system is generally used to orient the photovoltaic array so as to obtain the maximum energy output. For small photovoltaic systems, however, a fixed photovoltaic array orientation is generally used. While this simplifies the mounting of the photovoltaic array, the daily energy output is reduced and a significant seasonal variation occurs. A south-facing photovoltaic collector (northern hemisphere) is desired, optimized for noontime collection. Since the altitude angle of the array affects the seasonal energy variation, this angle must be optimized to produce the greatest daily energy output when the demand is greatest. If, for example, the greatest electrical demand corresponds to the period of minimal solar insolation, such as for most lighting applications, the altitude angle should be optimized for winter collection. The net result is that the daily energy collection in the summer for this orientation is less than it would be if the array were oriented for optimum summer collection. In effect, the potential energy capacity of the photovoltaic system is not fully utilized in the summer, since the capacity of the system corresponds to the period of greatest demand and least solar insolation.

Battery storage, as indicated in Figure 22.18, is needed to provide electrical energy throughout a 24-hour period as well as to provide electrical energy during extended intervals of inadequate solar insolation (cloud cover).[51-53] In addition, a battery system will provide intermittent load powers in excess of the output power of the photovoltaic array. Rechargeable batteries suitable for photovoltaic systems include lead–acid batteries similar to those used for automobiles and nickel–cadmium batteries of the type frequently used for portable electronic equipment. Since lead–acid batteries, per unit of energy capacity, cost considerably less than nickel–cadmium batteries, they are generally used for other than low-power photovoltaic systems. Batteries introduce an additional energy loss of the system. A storage battery relies on a reversible chemical reaction in which electrical energy is stored as chemical energy—the overall reaction is not completely reversible. A charge efficiency, also referred to as a coulomb efficiency, expresses the loss of electric charge. Typical lead–acid batteries have a charge efficiency of 80 to 90%; the precise value depends on the charging and discharging currents. Furthermore, batteries have a self-discharge loss, a slow leakage of charge within the cell that occurs even when the battery is not used (Fig. 22.19). An additional energy loss occurs as a result of a variation in the battery terminal voltage—a higher voltage is needed to charge the battery than that obtained when the battery is producing a load current (Fig. 22.20). The net result is that for typical photovoltaic installations the average energy storage efficiency of a lead–acid battery is only about 70%.

The voltage of a battery, when being charged or discharged, provides an approximate indication of its charge status. When being charged, the terminal voltage increases fairly abruptly at full charge. At this point, a regulator will either remove the battery from the circuit of the photovoltaic array (a series cut-off switch) or provide a shunt path for the current of the photovoltaic array. For both cases, photovoltaic power is lost—it is dissipated by the cells of the photovoltaic array for a series cut-off-type regulator or dissipated by the shunt circuit of a shunt-type regulator. The discharge voltage is used to determine the condition at which the battery approaches being completely

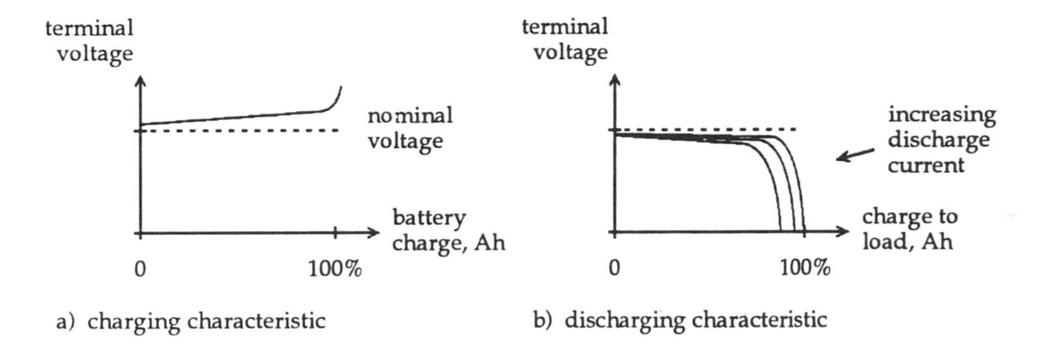

FIGURE 22.20 Charging and discharging voltages of a battery.

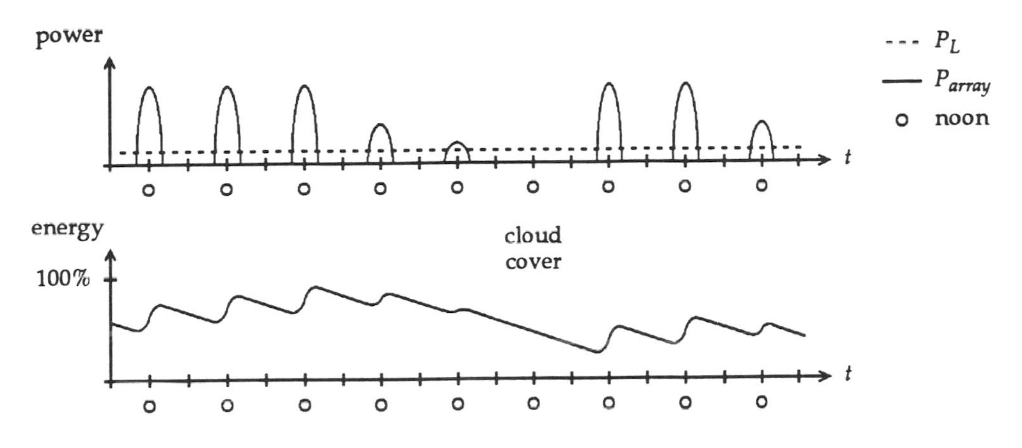

FIGURE 22.21 A photovoltaic system with energy storage.

discharged. When the terminal voltage falls below some established minimum voltage, the regulator will disconnect the load from the battery. Without a regulator circuit, either an excessive charging or discharging of the battery will occur; both conditions tend to reduce the battery life. Hence, a well-designed regulator circuit is important.[54-58]

Batteries are rated in terms of their charge capacity, which is expressed in the hybrid unit of ampere-hours, Ah. Since the dimension of amperes is coulombs per second, C/s, one ampere-hour is equal to a charge of 3,600 coulombs (1 Ah = 3,600 C). The charge capacity of a battery depends on the discharge rate, that is, the load current, and is generally specified for a relatively small load current. The terminal potential of a battery depends on the number of cells. Each cell of a lead–acid battery has a nominal terminal voltage of 2 V, so a battery with six cells, as is typical for automobiles, has a nominal terminal voltage of 12 V. Hence, for an automobile-type lead–acid battery with a charge capacity of 50 Ah, the stored energy capacity is 600 Wh (50 Ah × 12 V). Such a battery will supply a load current of 1 A at a potential of approximately 12 V for 50 h (a power of 12 W for 50 h). Several storage batteries are needed for many applications where the stored energy of a single battery is inadequate.

To illustrate the behavior of a photovoltaic system with energy storage, consider the case of Figure 22.21 for a constant load power (such as a telephone repeater). No losses will be assumed. Without a photovoltaic system, the discharge rate of the battery by an electrical load power of P_L would be constant in time.

$$E_{\text{battery}} = E_{\text{initial}} - P_L t \tag{22.14}$$

If the load power is expressed in watts and the time t in hours, the energy quantities, $E_{battery}$ and $E_{initial}$, are in watt-hours, Wh. During the interval of no photovoltaic produced power, a linear decrease in battery energy occurs, P_L watt-hours per hour, Wh/h. The daily energy input to the battery is the time integral of the output power of the photovoltaic array. For an hourly time scale, this is the area under the photovoltaic output power curve.

Battery losses may be taken into account by introducing an average charging efficiency, η_C, a discharging efficiency, η_D, and a power loss, P_{loss}, associated with the self-discharge of the battery (the charge leakage rate times the nominal potential of the battery). If the power supplied by the photovoltaic array is P_{array} and the time-dependent load power is P_L, the power supplied to the battery is the difference of these quantities. This results in the following equation for the battery energy:

$$P_{battery} = \begin{cases} \eta_C\left(P_{array} - P_L\right) & \text{for } P_{array} - P_L > 0 \\ \left(P_{array} - P_L\right)/\eta_D & \text{otherwise.} \end{cases}$$

$$E_{battery} = E_{initial} + \int_0^t P_{battery}\,dt' - \int_0^t P_L\,dt' \tag{22.15}$$

It should be emphasized that the efficiencies are average quantities that depend not only on the current levels, but also the general state of the battery. In addition, the battery is disconnected from the circuit when it is fully charged and when it approaches being completely discharged.

An important factor associated with a standalone photovoltaic system is *availability*, the fraction of the time for which the photovoltaic system can supply its electrical load. This depends on the size of the photovoltaic array (its peak capacity), the size of the battery storage system (its charge capacity and nominal terminal voltage), and the efficiency of the regulator system. For a high availability, the system needs to be designed for operation with minimum anticipated daily solar insolation levels. To increase the availability of a system, the size of the photovoltaic array and battery must be increased, thus increasing the cost of the system.

To achieve a high availability without having an excessively large photovoltaic-battery system, an auxiliary energy system is needed. The most common system is an internal combustion engine (gasoline-, diesel-, or propane-fueled) driving an electric generator. Since the auxiliary system can be used to rapidly recharge the batteries, it needs to operate over only relatively short and infrequent intervals. An auxiliary system can also be used to provide infrequently needed high-load powers that may be excessive for the photovoltaic system.

Residential Stand-Alone Photovoltaic Systems

Based on present electric utility rates and photovoltaic costs, a standalone photovoltaic system cannot be economically justified if a utility connection is readily available. However, for a remote residential site that requires a lengthy utility extension, a photovoltaic system may prove an economical alternative to extending the utility line.[59,60] In addition, a utility extension may not be possible for some extremely isolated sites.

A utility extension involves an initial capital investment even though it may be "paid for" on an amortized monthly schedule by a residential customer. Therefore, a direct comparison of the capital investment required for a photovoltaic system with that required for a utility extension that would otherwise be necessary is appropriate. It should be noted that the yearly operating and maintenance costs of the photovoltaic system are unlikely to exceed the yearly electricity charges of the utility. Hence a comparison of capital costs is sufficient for making a wise economic decision. Since the cost of a utility extension can be $5,000 to 9,000/km, the cost of a several-kilometer utility extension will generally exceed the cost of a residential photovoltaic system.[60,61]

At first glance, a cost comparison seems straightforward. However, unless the photovoltaic system has very large power capacity, it will not be equivalent to a utility connection, which provides, from the perspective of residential electrical demands, a nearly unlimited peak power capacity. Regardless of the photovoltaic system installed, using it to provide electric heating is not economic. In addition, for a photovoltaic system to be economic, electricity needs to be used considerably more efficiently than it is for a utility connection. Life-style changes associated with being more aware of excessive electricity usage will be advantageous. In addition, a photovoltaic system will require periodic maintenance; cleaning collectors, monitoring and replacing batteries, and attending to equipment failures. The homeowner will be responsible for maintenance, whereas the failure of a line extension is the responsibility of the utility.

The capacity of a utility extension is much greater than the demand of a single residence. Therefore, the eventual connection of additional residences to an extension could occur—the added cost (transformer and residential drops) would be small. While the extension cost for a single residence might justify a photovoltaic system, the cost of a utility extension divided among several residences might not justify individual standalone photovoltaic systems. Hence, the possibility of additional connections to a utility extension needs to be taken into account, otherwise one could end up with a residence having a large (and costly) photovoltaic system and being located next to a utility extension put in by neighbors.

The first step in designing a residential standalone system involves a recognition of the cost of electrical energy produced by a photovoltaic array. The daily winter solar insolation incident on an optimally oriented photovoltaic array will amount to 3 to 6 kWh/m^2. The lower value, 3 kWh/m^2, corresponds to an average value for the northern continental United States, while the higher value corresponds to the sunny southwestern regions. Suppose the photovoltaic collector has an efficiency of 14% and the average efficiency of the rest of the system, including batteries, is 70%. This results in an effective overall average array efficiency of approximately 10%. Hence, for a daily incident energy of 3 kWh/m^2, the daily available electrical energy for a one square meter photovoltaic array is 0.3 kWh. If the photovoltaic array cost is \$1,000/m^2 (\$7.14/W$_p$ based on a standard test condition of 1,000 W/m^2) and the yearly amortization rate is 10% of the initial value (for example, 8%/year interest on a 20-year loan), the daily "cost" of the system is 27.4¢ for each square meter of photovoltaic array. This translates to an electrical energy cost of 91¢/kWh. For a 6 kWh/m^2 incident radiation value, the cost is 45.5¢/kWh. The amortized cost of the rest of the system as well as operating and maintenance costs need to be added to these figures. Although higher quantities of energy may be obtained in the summer, the additional energy may not be needed and thus has no value.

Residential electrical energy consumption depends on many variables: residence size, number and ages of occupants, location (which determines climate and expected hours of sunshine), and life styles. Therefore, even if a statistical average of electricity usage were available (which might be interpreted as corresponding to a "typical" residence), this would be of little value when designing a photovoltaic standalone system for a particular residence and its occupants. Variations between residences are too great. Furthermore, statistics for electrical energy usage are based on utility-supplied electrical energy and energy costs, 7 to 15¢/kWh in the United States, are relatively low. Also, the peak power capacity of a utility connection is seldom a constraint for a residential customer. At present, there is only a minor incentive for a residential consumer to devote much attention to the electrical energy efficiency of appliances or to be concerned with energy-type trade-offs, that is, choosing between electricity or a fuel such as natural gas for a particular need.

A listing or residential electrical energy usage is indicated in Table 22.2. A moderate size residence with a floor area of 100 to 200 m^2, an occupancy of two or three individuals, and a moderate climate (neither excessively harsh winters nor extremely hot summers) has been assumed. Even for the estimates of this table, a large range in energy usage is apparent. (A slightly different perspective is given by Strong and Scheller.[61]) Excluding electrical energy that might be used for heating and cooling, monthly energy usage is from 250 to 1,000 kWh. The lower estimate excludes

TABLE 22.2 Utility-Supplied Residential Electrical energy Consumption

	Monthly kWh	Alternative
Refrigerator and freezer	80–150	Separate high-efficiency units or propane absorption units
Cooking: range and oven	50–100	Propane stove or more efficient alternative cooking appliances
Two 2,500 W burners		
Two 1,500 W burners		
2,500 W oven		
2,500 W broiler		
Domestic water heater	0–400	Propane- or heating-oil-fueled heater
Dish washer	0–60	
Clothes washer	20–50	
Clothes dryer	0–150	Propane heated with electric drive for the blower and tumbler
Heating system—mechanical water or air circulation	20–60*	Superinsulated structure that needs very little heat circulation
Lighting	30–100	High-efficiency compact fluorescent units
Computer	0–20	Energy-efficient notebook computer
Television and VCR	15–30	
Air conditioning	†	Superinsulated structure to minimize cooling requirements
Small appliances	20–200	
Space heater	0–100	Superinsulated structure will minimize need.

* Heating season only

† Cooling season only. A small room air-conditioner, 3 to 4 hours of daily operation, requires approximately 100 kWh/month.

electric domestic water heating and is premised on having moderately efficient appliances and on a minimum of wasteful habits. Even for 250 kWh/month, photovoltaic-produced electrical energy would be very expensive—$114 to 228/month (6 or 3 kWh/m^2 solar insolation, respectively) based on previously introduced estimates, which include only the cost of the photovoltaic array.

Given the very high cost of photovoltaic-produced electricity, it is imperative to use electricity very efficiently. In essence, the design goal for a residence relying on a photovoltaic system is to provide energy-produced amenities comparable to those available for a utility-supplied residence—not comparable amounts of electrical energy. Significant investments that reduce electrical energy needs can be justified. In particular, the investment necessary to reduce energy usage by 1 kWh needs to be compared with the photovoltaic investment that would otherwise be needed to supply the 1 kWh of electrical energy. As a result of the high cost of photovoltaic systems, efficiency investments considerably greater than those that can be economically justified for utility-supplied electricity can be justified.

Efficient usage of electrical energy has received considerable attention because through efficient usage (produced by investments to improve efficiencies), construction of additional utility generating facilities can be deferred.[62-65] The same types of efficiency improvements are also appropriate for a photovoltaic-supplied residence, and even the most expensive of proposed improvements will tend to be economic.

One important example is the refrigerator—nearly every U.S. kitchen is equipped with a refrigerator-freezer combined unit, and an additional freezer may be available for long-term storage. Although the energy efficiency of refrigerators has been improving (improved insulation and more efficient compressors) and considerable potential exists for future improvements, the common combined refrigerator-freezer unit is particularly inefficient, since the heat that leaks into the refrigerator (≈5°C) is extracted from the freezer at a temperature of –15 to –20°C. Not only is a larger refrigeration unit needed (its coefficient of performance is smaller than if the heat were extracted from the refrigerator compartment), moisture condenses on the evaporator in the freezer, requiring frequent defrosting. If separate units are used, or if a combined refrigerator-freezer has two refrigerator units, a significant efficiency improvement can be realized. For example, a high-efficiency 450-liter refrigerator-freezer designed for photovoltaic systems has a rated energy consumption of only 22 kWh/month.[66] A further efficiency improvement for a photovoltaic system would be to add thermal mass within the refrigerator to store "cold." Cooling could then be powered directly from the photovoltaic array thus eliminating battery storage (and hence loss) for this task. Absorption refrigeration systems fueled by propane are a nonelectric alternative.

Similar consideration should be given to the other items of Table 22.2. Instead of electricity, propane could be used for cooking. Alternatively, this may not be desirable in a well-insulated, reasonably air-tight structure. Electric appliances, including the electric kettle and insulated individual cooking pots with built-in heating elements, are more efficient than range-top cooking. A small, well-insulated countertop oven can frequently substitute for the less efficient large oven of a range. Domestic hot water may be obtained from a solar collector and/or the fuel used for heating (propane or heating oil)—powering an electric heater with photovoltaic-produced electricity is not economic. Dishwashing can be done by hand, but a high-temperature dishwasher is more hygienic. Improved motors and pumps and running a dishwasher only when full reduce its electric energy demand. Electricity used by a clothes washer can be reduced with improved agitating systems that move a minimum quantity of water. A clothes dryer can be heated with propane. With conventional residential construction, a forced distribution of heat is necessary—hot-air vents or hot-water radiators are generally located along the outer perimeter of the structure. In a superinsulated structure, natural air circulation is adequate, since "cold spots" do not tend to occur. Since fluorescent lamps are three to four times more energy efficient than incandescent lamps, they offer a significant energy saving.

Water pumping (from a well) may be needed, since it is unlikely that utility-supplied water would be available when electric power is not. A water pump can be powered directly from a photovoltaic array during times of high solar insolation, since the water can be stored in a large tank. Either a small water pump or a gravity-feed system can be used to distribute the water as needed.

The design of a photovoltaic system requires the selection of energy-efficient appliances and lighting fixtures. Based on the smaller energy quantities of Table 22.2, it is possible to reduce the monthly energy demand to 100 kWh. Using the parameters already introduced, a photovoltaic array of 11 m^2 (100 kWh/month = 3.3 kWh/day) would be needed for a 10% efficient system and a solar insolation of 3 kWh/day. This implies a photovoltaic array cost of approximately $11,000. For a 6 kWh/day solar insolation, the cost would be approximately $5,500. It should be noted that the 250 kWh/month consumption rate based on conventional energy usage patterns, would require an investment of approximately $27,800 (3 kWh/day solar insolation).

In order to supply electrical energy over a few-day interval of no solar insolation, several lead–acid batteries are needed. If, for example, 10 series-connected 12 V batteries are used, the resultant potential of 120 V may be used to directly power many conventional 120 V appliances—including special DC fluorescent lamp fixtures that are available. Batteries increase both the initial investment and the operating costs of the system. A suitable 12 V, 180 Ah industrial deep-cycle lead–acid battery will cost approximately $600; that is, an investment of $6,000 is required for a 10-battery system (Table 22.3). The energy storage capacity of the system, 21.6 kWh, is sufficient for 6 days of operation. For a battery lifetime rating of 1,800 charge/recharge cycles (80% depletion), 31,100 kWh of electrical energy can be stored before the battery needs to be replaced. This implies an average stored energy cost of 19¢/kWh—a significant added cost. If storage is needed for all energy (100 kWh/month), the battery would last over 25 years. Both the recycling of the batteries and the smelting of the lead needed to replace that lost result in significant emissions of lead compounds.[67] Thus, although photovoltaic systems don't produce pollutant emissions during their operation, they are not as environmentally clean as purported by some.

A DC-to-AC inverter will also be needed for those appliances that cannot use DC power, such as a microwave oven or an induction motor. An inverter reaches its peak efficiency (over 90% for some units) at its rated output power while having a significant standby loss for no load. Hence, an inverter needs to be turned on, either manually or automatically, only when needed. Small electronic appliances (stereo, TV, VCR) may also need AC power. A small dedicated inverter is desirable (alternatively, 12 or 24 V DC electronic appliances intended for mobile homes could be used).

TABLE 22.3 Electric Storage Batteries

Manufacturer and Model	Model Number	List Price ($)	Shallow/ Deep Cycle (S/D)	Nominal Capacity (Ah)	Nominal Voltage (V)	Depth of Discharge (%)	Life (Cycles)	Energy Delivered (kWh)	Cost ($/kWh)
GNB Absolyte	638	63	S	42	6	50	1,000	126	0.50
	1260	114	S	59	12	50	1,000	359	0.32
	6-35A09	654	S	202	12	50	3,000	3,636	0.18
	3-75A25	1,168	S	1,300	6	50	3,000	11,700	0.10
Exide Tubular Modular	6E95-5	600	S	192	12	15	4,100	1,417	0.42
						20	3,900	1,797	0.33
	6E120-9	930	S	538	12	15	4,100	3,970	0.23
						20	3,900	5,036	0.19
	3E120-21	945	S	1,346	6	15	4,100	4,967	0.19
						20	3,900	6,299	0.15
Delco-Remy Photovoltaic	2000	78	S	105	12	10	1,800	227	0.34
						15	1,250	236	0.33
						20	850	214	0.36
Globe Solar Reserve Gel Cell	3SRC-125G	181	S	125	6	10	2,000	150	1.21
	SRC-250G	100	S	250	2	10	2,000	100	1.00
	SRC-375G	150	S	375	2	10	2,000	150	1.00
Globe	GC12-800-38	100	S	80	12	20	1,500	288	0.35
			D	80	12	80	250	240	0.42
GNB Absolyte	638	63	D	40	6	80	500	96	0.65
	1260	114	D	56	12	80	500	269	0.42
	6-35A09	654	D	185	12	80	1,500	2,664	0.25
	3-75A25	1,168	D	1,190	6	80	1,500	8,568	0.14
Surrette	CH-375	231	D	375	6	80	1,400	2,520	0.09
	NS-29	522	D	490	6	80	1,400	3,293	0.16
	NS-33	813	D	564	8	80	1,400	5,053	0.16
Exide Tubular Modular	6E95-5	600	D	180	12	80	1,800	3,110	0.19
	6E120-9	930	D	360	12	80	1,800	6,221	0.15
	3E120-21	945	D	1,250	6	80	1,800	10,800	0.09

Source: From Strong, S.J. and Scheller, W.G., *The Solar Electric House: Energy for the Environmentally Responsive, Energy-Independent Home*, Sustainability Press, Still River, MA, 1993, with permission.

To complete a system, an auxiliary power source is required for extended intervals of inadequate solar insolation. Most likely, this will be either a diesel- or propane-fueled internal combustion engine driving an electric generator for charging batteries and possibly supplying AC power for heavy loads. Ideally, a photovoltaic system will require very little electrical energy supplied by the auxiliary system. Since the auxiliary system is needed to assure an uninterrupted supply of electrical energy, its capacity must be adequate to supply the daily demand—as if the photovoltaic array were not present. To supply a daily energy demand of 3.3 kWh (100 kWh/month), a 3 kW generator would recharge the batteries in less than 2 hours (70% battery energy efficiency).

The electrical energy produced by the auxiliary system results in an additional expense—approximately 20¢/kWh based on fuel costs and a comparable amount for engine maintenance and replacement (or rebuilding). When designing the photovoltaic system, a balancing of the cost of the energy produced by the photovoltaic modules making up an array with the cost of the electrical energy obtained from the auxiliary system is needed. If it is anticipated that the cost of photovoltaic modules will decrease in time, it is not necessary to include all modules in an initial installation. The auxiliary system could be relied upon to provide a large portion of the electrical demand. A small benefit occurs in that the auxiliary system can be programmed to run during intervals of peak electric demand thus reducing battery losses.

Several residences with photovoltaic systems are described in the literature.[50,61,68] The details provided can be used to gain a perspective of the technical choices available and of the costs.

Other Photovoltaic Stand-Alone Applications

Pumping water for livestock and irrigation is an application for which numerous photovoltaic systems have been installed throughout the world.[68-73] These systems are cost-effective for remote sites at which the cost of a utility extension is high. Photovoltaic-powered pumping systems installed in areas that do not have utility-supplied electric power have been found to be more economic than pumps powered by alternative means.

The simplest pumping systems are for shallow-water sources (shallow wells, springs, etc.), in which suction-type pumps may be used. These pumps can readily be fitted with a DC motor, thus eliminating the need for a DC-to-AC inverter. For larger pumping depths (up to a maximum of 30 m), jet pumps fitted with DC motors can be used. Although jet pumps require a second pipe that goes down the well (the pumping jet is at the bottom of the well), the pump and motor are at the top of the well and can hence be readily serviced.

Submersible pumps tend to be more efficient than jet pumps, and they can be used for very deep wells. The entire unit, an electric motor and a multistage centrifugal pump, is housed in a small-diameter assembly that is attached to the end of the drop pipe at the bottom of the well. A waterproof cable is used to supply electric power to the motor. If a standard, readily available, AC pump is used, a DC-to-AC inverter along with battery storage to provide the starting surge current is needed. Less readily available DC submersible pumps eliminate the need for an inverter and batteries. Several Wyoming installations, all using DC pumps, are described by Chowdhury et al.[60]

Another type of pump suitable for deep wells with a low recharge rate are pump jacks (this type of pump is common for stripper oil wells). An above-ground DC motor can be used to drive the mechanical system attached to the "sucker rod" that connects to a piston pump at the bottom of the well. Either batteries or an electronic control system is needed for the starting surge current of the pump.

A basic photovoltaic module consisting of a single series-connected string of cells can be used to directly charge a lead-acid storage battery (Fig. 22.22). A module of 36 cells will produce a load voltage of approximately 18 V, an open-circuit voltage in excess of 20 V, and a current of 2 to 2.5 A (40–50 W_p rating). The photovoltaic module indicated in Figure 22.22 is connected directly to the 12 V battery through a blocking diode. As a result of the current versus voltage characteristic of a photovoltaic cell, the charging current will be only slightly less than the short-circuit current of a

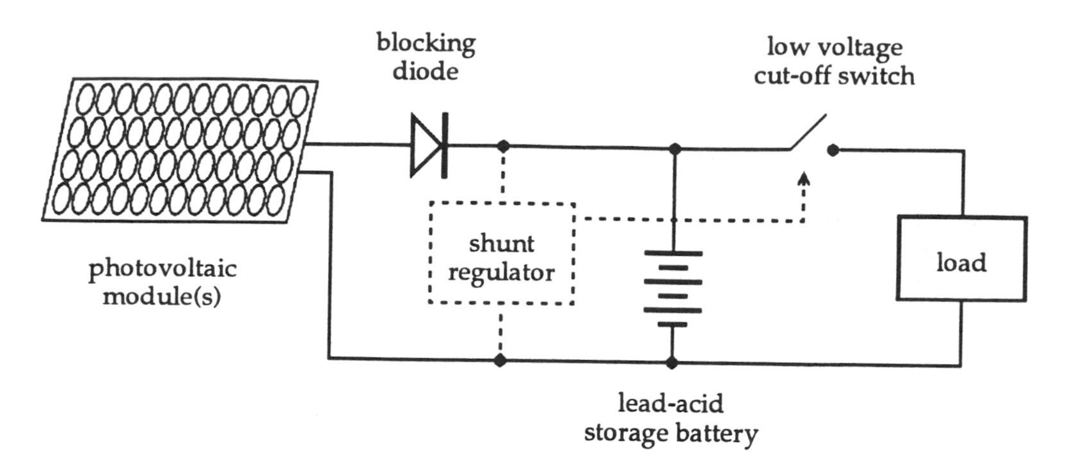

FIGURE 22.22 An elementary stand-alone photovoltaic system.

cell for a battery voltage of 13 to 14 V. A power loss occurs because for a 36-cell module, the voltage is less than that for maximum power. For a small system, the corresponding energy loss is considered acceptable, since it mitigates the need for a more complex electronic controller circuit.

An optional shunt-type regulator is indicated in Figure 22.22. For a very low-cost unit, the regulator might be omitted. Since the battery potential increases when it becomes fully charged the current of the photovoltaic module will tend to decrease, thus tending to reduce the charging current (self-regulation). The potentials of both the battery and the photovoltaic cells, however, are dependent on temperature. Hence, an electronic regulator, which adds to the cost of the unit, may be preferable—it will not only prevent the battery from being overcharged but it can also disconnect the load from the circuit when the battery voltage falls below a set minimum value.

A basic system consisting of a single photovoltaic module and battery (with or without a regulator) can be used to operate various low-power DC loads.[74] For example, consider an outdoor lighting application. A self-contained pole-mounted photovoltaic module and lamp that has an enclosed battery at its base eliminates the need for a buried utility connection. A single 40 W_p photovoltaic module will provide a daily battery charge of 6 to 12 Ah (depending on the solar insolation) to a 12 V battery. This is an energy of 72 to 144 Wh. Hence, this unit could power an 18 W low-pressure sodium lamp for 4 to 8 hours each night. A higher power lamp could be used if its operating time is restricted (2 to 4 hours for a 36 W lamp). By using additional photovoltaic modules and batteries, higher power lamps and/or longer operating times can be achieved.

The lighting system can be modified to function as a security alarm by adding a motion detector and audio alarm. Alternatively, a motion detector could simply be used to turn on the light when needed rather than having the light on for several hours each night. This would significantly reduce the required size of the photovoltaic system and hence its cost.

A similar photovoltaic system could be used to power roadside warning lights and signals. Not needing a utility connection, portable photovoltaic systems are ideally suited for fulfilling temporary needs. Photovoltaic units can be used for remote monitoring of road conditions and traffic. Emergency telephones using a radio link can be located along highways without regard to the availability of electric and telephone utility connections.

Remote monitoring of weather conditions and stream or river flows is another application well suited for photovoltaic systems. Besides providing a continuous data record, such monitoring systems can be used, for example, to warn of impending flood conditions. An example of remote monitoring is the photovoltaic-powered water-flow meters installed by the City of Austin Water Department.[75]

Photovoltaic-powered remote telecommunication repeater amplifiers (for telephone and other services) are extensively used in Australia—arrays having a total capacity of over 2 MW_p have been installed.[76] The high availability required by a telecommunication system is achieved with a large storage battery capacity.

Cathodic protection is required for oil and gas pipelines, wells, storage tanks, and other metallic structures in contact with the earth. A photovoltaic system can be used to provide an electrical current that counteracts the natural electrochemical earth current in order to reduce corrosion.

Approximately one-third of the world's population does not have utility-supplied electric power. Since a substantial portion of this rural population is located in regions of relatively high levels of solar insolation, photovoltaic systems are particularly well suited to provide electricity in these regions.[77-80] Without electric power, kerosene lamps and candles are the main source of the limited low-quality nighttime illumination available. A basic photovoltaic system of Figure 22.22 with a single photovoltaic module of 40 to 50 W_p can be used for a residential lighting system powering one or two 18 W fluorescent lamps for a few hours each evening. This system could also be used to power a small 12 V radio or television set.

A photovoltaic system might be used to power a small refrigerator—a typical need would be for the storage of vaccines and other medical supplies as well as providing ice for cold packs. Consider a 450-liter efficient refrigerator (suggested for the preceding residential application) that has an energy requirement of 22 kWh/month.[66] An average daily energy of 0.73 kWh is therefore required. For a tropical region with an average daily insolation of 6 kWh/m^2, an energy of 0.84 kWh/m^2 is obtained from a 14% efficient photovoltaic module. If the regulator and battery are 70% efficient, the effective energy output of the collector is 0.59 kWh/m^2. Hence, a module area of 1.24 m^2 is required to provide the 0.73 kWh required by the refrigerator. For a photovoltaic module cost of $1,000/m^2, this implies an investment of $1,240.

A 12 V battery with a charge storage capacity of 60.8 Ah (an energy of 0.73 kWh) will provide the energy storage for one day of operation. Therefore, a battery with a charge storage capacity of 180 Ah (Table 22.3) would provide, based on a discharge of 80%, slightly more than 2 days (2.37 days) of operation. The battery system must be sized taking into account the expected variability of the daily solar insolation. A 360 Ah battery, for example, will provide nearly 5 days (4.74 days) of operation. For an application where reliability is essential, such as a medical refrigerator, a storage capacity of at least this size would be needed.

Lighting for streets and paths along with community buildings can also be provided by basic photovoltaic systems. An important need in medical clinics is small refrigerators to store vaccines and to produce ice packs.[81,82] Since a very reliable system is required, a conservatively designed photovoltaic system with a large battery storage capacity is required. The photovoltaic array of this system, however, can be used to charge the batteries of an auxiliary system when the batteries of the refrigerator are fully charged.

Although at first glance it would seem surprising, electric utilities find photovoltaic systems of value for supplying low-power electrical needs (a few watts) adjacent to high-voltage transmission systems (many megawatts).[83-86] Low-power needs have generally been supplied by the extension of a single-phase distribution line (potential up to approximately 15 kV) or through an expensive transformer connection to a high-voltage line. The typical cost, for example, for a transformer connection to a 115 kV line is over $13,000. Hence, for an application requiring, on the average, only several watts of electrical power, a photovoltaic system is less expensive than a connection to the utility power system.

One particularly important utility photovoltaic application is the powering of sectionalizing switches of high-voltage transmission and distribution lines. These switches are used for removing electric power from sections of high-voltage lines. It is necessary that these motor-driven switches be battery powered so that they can be operated during a system power outage or when the line is disconnected. Since battery storage is needed, a photovoltaic array and controller can substitute for a battery charger powered by the utility connection. The average power required for the occasional operation of the switch, the remote terminal, and the radio control system is small. Hence, a few standard photovoltaic modules that directly charge a 12 V lead–acid battery are adequate.

Aircraft warning beacons for high-voltage transmission towers can also be powered by a photovoltaic system—again avoiding an otherwise expensive utility extension or transformer. While the power requirement of the photovoltaic array (a few hundred peak watts) is greater than for a sectionalizing switch, the utility tower can readily accommodate the larger photovoltaic array.

Conclusion

An operating single-crystal silicon photovoltaic cell was demonstrated in 1953, and its theory of operation was relatively well understood before the end of that decade. Manufacturing techniques required for single-crystal silicon devices have greatly improved over the past 40 years. The fabrication of single-crystal silicon photovoltaic cells can be classified as a relatively mature technology. Even so, the cost of photovoltaic modules is still too high for the electricity they produce to be competitive with conventional utility-generated electrical energy. However, these modules have proven economic for many standalone applications.

In order to reduce the cost of photovoltaic modules, a considerable effort has been expended on developing thin-film cells. A photovoltaic cell formed from a thin film of amorphous silicon deposited on a supporting substrate was introduced in 1974. After over 20 years, only a limited theoretical understanding of amorphous silicon is available. A similar situation exists for other thin-film cells, such as those of cadmium telluride and copper indium–diselenide. The elaborate experimental techniques used to improve the performance of thin-film cells have been based mainly on empirical observations. Perhaps, the structure of these cells is too complex for a better theoretical understanding.

The potential for thin-film photovoltaic cells is extremely difficult to assess, since what may be important developments have been obscured by corporate proprietary policies. Since there is not a free exchange of information, it is impossible for an impartial observer to evaluate, based on relevant data, what could be overly optimistic predictions of possible future developments.

References

1. Wolf, M., Photovoltaic solar energy conversion systems, in *Solar Energy Handbook*, Kreider, J.F. and Kreith, F., Eds., McGraw-Hill, New York, 24-1 to 24-35, 1981.
2. Shockley, W., The path to the conception of the junction transistor, *IEEE Transactions on Electron Devices*, ED-23(7), 597–620, 1976.
3. Chapin, D.M., Fuller, C.S. and Pearson, G.L., A new silicon p-n junction photocell for converting solar radiation into electrical power, *Journal of Applied Physics*, 25(5), 676–677, 1954.
4. Smits, F.M., History of silicon solar cells, *IEEE Transactions on Electron Devices*, ED-23(7), 640–643, 1976.
5. Prince, M.B., Silicon solar energy converters, *Journal of Applied Physics*, 26(5), 534–540, 1955.
6. Loferski, J.J., Theoretical considerations governing the choice of the optimum semiconductor for photovoltaic solar energy conversion, Journal of Applied Physics, 27(7), 777–784, 1956.
7. Schwartz, R.J., Photovoltaic power generation, *Proceedings of the IEEE*, 81(3), 355–364, 1993.
8. Sze, S.M., *Physics of Semiconductor Devices*, 2nd Ed. John Wiley and Sons, New York.
9. Zweibel, K., Thin-film photovoltaic cells, *American Scientist*, 81(4), 362–369, 1993.
10. Carlson, D.E., Low-cost power from thin-film photovoltaics, in *Electricity: Efficient End-Use and New Generation Technologies, and Their Planning Implications*, Johansson, T.B., Bodulund, B. and Williams, R.H., Eds., Lund University Press, Lund, Sweden, 1989.
11. Carlson, D.E., Overview of amorphous silicon photovoltaic module development, *Solar Cells*, 30(1–4), 277–283, 1991.
12. Carlson, D.E., Amorphous-silicon solar cells, *IEEE Transactions of Electron Devices*, 36(12), 2775–2780, 1989.

13. Dupree, W., Enzer, H., Miller, S. and Hillier, D., *Energy Perspectives 2,* Department of the Interior, Washington, DC, 1976.
14. Energy Information Administration, Monthly Energy Review, Department of Energy, Washington, DC, 1995.
15. Energy Information Administration, *Electric Power Monthly,* DOE/EIA-0226(95/04), Department of Energy, Washington, D.C. 1995.
16. Friedlander, G.D., Energy's hazy future, IEEE Spectrum, 12(5), 32–43, 1975.
17. NSF/NASA Solar Energy Panel, *An Assessment of Solar Energy as a National Energy Resource,* National Science Foundation, Washington, DC, 1972.
18. U.S. Congress (93rd), Hearings before the Subcommittee on Energy of the Committee on Science and Astronautics, U.S. House of Representatives, U.S. Government Printing Office, Washington, DC, 1974.
19. Williams, J.R., *Solar Energy: Technology and Applications,* Ann Arbor Science Publishers, Ann Arbor, MI, 1974.
20. Deudney, D. and Flavin, C., *Renewable Energy: The Power to Choose,* W.W. Norton, New York, 1983.
21. Hayes, D., *Rays of Hope: The Transition to a Post-Petroleum World,* W.W. Norton, New York, 1977.
22. Ehrenreich, H., *Principal Conclusions of the American Physical Society Study Group on Solar Photovoltaic Energy Conversion,* American Physical Society, New York, 1979.
23. Ehrenreich, H. and Martin, J.H., Solar photovoltaic energy, *Physics Today,* 32(9), 25–32, 1979.
24. Glaser, P.E., Satellite solar power station, *Solar Energy,* 12(3), 353–361, 1969.
25. Glaser, P.E., Power from the sun: Its future, *Science,* 162(3856), 857–861, 1968.
26. National Research Council, Committee on Satellite Power Systems, Environmental Studies Board, Commission on Natural Resources, Electric Power from Orbit: A Critique of a Satellite Power System, National Academy Press, Washington, DC, 1981.
27. Hubbard, H. M., Photovoltaics today and tomorrow, *Science,* 244(4902), 297–304, 1989.
28. Stone, J.L., SOLAR 2000: The next critical step towards large-scale commercialization of photovoltaics in the United States, *Solar Energy Materials and Solar Cells,* 34(1–4), 41–49, 1994.
29. Stone, J.L., Photovoltaics: Unlimited electrical energy from the sun, *Physics Today,* 46(9), 22–29, 1993.
30. Ramakumar, R., Photovoltaic systems, *Proceedings of the IEEE,* 81(3), 365–377, 1993.
31. Conover, K., Photovoltaic operation and maintenance evaluation, *IEEE Transactions on Energy Conversion,* 5(2), 279–283, 1990.
32. Schaefer, J.C., Review of photovoltaic power plant performance and economics, *IEEE Transactions on Energy Conversion,* 5(2), 232–238, 1990.
33. Durand, S.J., Bowling, D.R. and Risser, V.V., Lessons learned from testing utility connected PV systems, *Twenty-First IEEE Photovoltaic Specialists Conference—1990, IEEE,* New York, 909–913, 1990.
34. Wenger, H.J., Schaefer, J., Rosenthal, A., Hammond, B. and Schlueter, L., Decline of the Carrisa Plains PV power plant: The impact of concentrating sunlight on flat plates, *Twenty-Second IEEE Photovoltaic Specialists Conference—1991,* IEEE, New York, 586–592, 1991.
35. Wenger, H.J., Jennings, C. and Innaucci, J.J., Carrisa Plains PV power plant performance, *Twenty-First IEEE Photovoltaic Specialists Conference—1990,* IEEE, New York, 844–849, 1990.
36. Candelario, T.R., Hester, S.L., Townsend, T.U. and Shipman, D.J., PVUSA—Performance, experience, and cost, *Twenty-Second IEEE Photovoltaic Specialists Conference—1990,* IEEE, New York, 493–500, 1990.
37. Hester, S.L., Townsend, T.U., Clements, W.T. and Stolte, W.J., PVUSA: Lessons learned from startup and early operation, *Twenty-First IEEE Photovoltaic Specialists Conference—1990,* IEEE, New York, 937–943, 1990.

38. Matlin, R. and Lenskold, R., PVUSA programme pushes photovoltaics into the 1990s, *Modern Power Systems,* 9(11), 75–78, 1989.
39. Atmaram, G.H., Marion, B. and Herig, C., Three years performance and reliability of a 15 kWp amorphous silicon photovoltaic system, *Twenty-Second IEEE Photovoltaic Specialists Conference—1991,* IEEE, New York, 600–607, 1991.
40. Atmaram, G.H., Marion, B. and Herig, C., Performance and reliability of a 15 kWp amorphous silicon photovoltaic system, *Twenty-First IEEE Photovoltaic Specialists Conference—1990,* IEEE, New York, 821–830, 1990.
41. Atmaram, G.H., Marion, Z.B. and Herig, C., First year performance of a 15 kWp amorphous silicon photovoltaic system, *IEEE Transactions on Energy Conversion,* 5(2), 290–298, 1990.
42. Beyer, L., Dietrich, B., Pottbrock, R. and Lotfi, A., Solar modules tested at Kobern-Gondorf, *Modern Power Systems,* 10(11), 81–85, 1990.
43. Stolte, W.J., Whisnant, R.A. and McGowin, C.R., Cost of energy from utility-scale PV systems, Advances in *Solar Energy: An Annual Review of Research and Development,* 9, 245–203.
44. Goodman, F.R. Jr., DeMeo, E.A. and Zavadil, R.M., Residential photovoltaics in the United States: Status and outlook, *Solar Energy Materials and Solar Cells,* 35, 375–386, 1994.
45. Russell, M.C. and Kern, E.C., Lessons learned with residential photovoltaic systems, *Twenty-First IEEE Photovoltaic Specialists Conference—1990,* IEEE, New York, 898–901, 1990.
46. Bzura, J.J., Performance of grid-connected photovoltaic systems on residences and commercial buildings in New England, *IEEE Transactions on Energy Conversion,* 7(1), 79–82, 1992.
47. Bzura, J.J., The New England Electric Photovoltaic Systems Research and Demonstration Project, *IEEE Transactions on Energy Conversion,* 5(2), 284–289, 1990.
48. Real, M. and Ludi, H., Project megawatt: Experience with photovoltaics in Switzerland, *Twenty-Second IEEE Photovoltaic Specialists Conference—1991,* IEEE, New York, 574–575, 1991.
49. Baumann, A.E., Environmental costs of photovoltaics, *IEEE Proceedings—A,* 140(1), 76–80, 1993.
50. Wenham, S.R., Green, M.A. and Watt, M.E., *Applied Photovoltaics,* Centre for Photovoltaic Devices and Systems, University of NSW, New South Wales, Australia, undated.
51. Alminauskas, V., Performance evaluation of lead acid batteries for use with solar panels, *Twenty-Third IEEE Photovoltaic Specialists Conference—1993,* IEEE, New York, 1258–1263, 1993.
52. Stevens, J., Kratochvil, J. and Harrington, S., Field investigation of the relationship between battery size and PV system performance, *Twenty-Third IEEE Photovoltaic Specialists Conference—1993,* IEEE, New York, 1163–1169, 1993.
53. Khousam, K.Y. and Khousam, L., Optimum matching of a photovoltaic array to a storage battery, *Twenty-Second IEEE Photovoltaic Specialists Conference—1992,* IEEE, New York, 706–711, 1992.
54. Snyman, D.B. and Enslin, J.H.R., Simplified maximum power point controller for PV installations, *Twenty-Third IEEE Photovoltaic Specialists Conference—1993,* IEEE, New York, 1240–1245, 1993.
55. Khouzam, K.Y., Khouzam, L. and Groumpos, P., Optimum matching of ohmic loads to the photovoltaic array, *Solar Energy,* 46(2), 101–108, 1991.
56. Dunlop, J., Bower, W. and Harrington, S., Performance of battery charge controllers: First year test report, *Twenty-Second IEEE Photovoltaic Specialists Conference—1992,* IEEE, New York, 640–645, 1992.
57. Wolf, S.M.M. and Enslin, J.H.R., Economical, PV maximum power point tracking regulator with simplistic controller, *Twenty-Fourth IEEE Annual Power Electronics Specialists Conference—1993,* IEEE, New York, 581–587, 1993.
58. Won, C.-Y., Kim, D.-H., Kim, S.-C., Kim, W.-S. and Kim, H.-S., A new maximum power point tracker of photovoltaic arrays using a fuzzy controller, *Twenty-Fifth IEEE Annual Power Electronics Specialists Conference—1994,* IEEE, New York, 396–403, 1994.

59. DeLaune, J., The economics of remote PV for a Wisconsin utility, *Solar Today*, 8(5), 23–25, 1994.
60. Chowdhury, B.H., Ula, S. and Stokes, K., Photovoltaic-powered water pumping—design, and implementation: Case studies in Wyoming, *IEEE Transactions on Energy Conversion*, 8(4), 646–651, 1993.
61. Strong, S.J. and Scheller, W.G., *The Solar Electric House: Energy for the Environmentally Responsive, Energy-Independent Home*, Sustainability Press, Still River, MA, 1993.
62. Fickett, A.P., Gellings, C.W. and Lovins, A.B., Efficient use of electricity, *Scientific American*, 263(3), 65–74, 1990.
63. Johansson, T.B., Bodlund, B. and Williams, R.H. (Eds.), *Electricity: Efficient End-Use and New Generation Technologies, and Their Planning Implications*, Lund University Press, Lund, Sweden, 1989.
64. Nørgård, J.S., Low electricity appliances—Options for the future, in *Electricity: Efficient End-Use and New Generation Technologies, and Their Planning Implications*, Johansson, T.B., Bodlund, B. and Williams, R.H., Eds., Lund University Press, Lund, Sweden, 125–172, 1989.
65. McGowan, T., Energy-efficient lighting, in *Electricity: Efficient End-Use and New Generation Technologies, and Their Planning Implications*, Johansson, T.B., Bodlund, B. and Williams, R.H., Eds., Lund University Press, Lund Sweden, 59–88, 1989.
66. Perry, T.S., "Green" refrigerators, *IEEE Spectrum*, 31(8), 25–30, 1994.
67. Lave, L.B., Hendrickson, C.T. and McMichael, F.C., Environmental implications of electric cars, *Science*, 268(5213), 993–995, 1995.
68. Crane, R., Twelve years of living with photovoltaic power, *Twenty-Second IEEE Photovoltaic Specialists Conference—1991*, IEEE, New York, 717–721, 1991.
69. Dankoff, W., Pumping water, *Solar Age*, 9(2), 28–34, 1984.
70. Malukatin, S., Water from the African sun, *IEEE Spectrum*, 34(10), 40–43, 1994.
71. Zaki, A.M., Eskander, M.N. and Elewa, M.M., Control of maximum-efficiency PV-generator irrigation system using PLC, *Renewable Energy*, 4(4), 447–453, 1994.
72. Chaurey, A., Sadaphal, P.M. and Tyaqi, D., Experiences with SPV water pumping systems for rural applications in India, *Renewable Energy*, 3(8), 961–964, 1993.
73. Bazarov, B.A., Strebkov, D.S., Shaimerdangulyev, G., Abdylkhekimov, E.S. and Charyev, K., Solar photovoltaic water-lifting systems (Creation and experience of operation in conditions of the Turkmen SSR), *Applied Solar Energy*, 25(3), 69–72, 1989.
74. Interstate Renewable Energy Council, *Procurement Guide for Renewable Energy Systems*, Interstate Renewable Energy Council, Albany, NY, Undated.
75. Hoffner, J.E., Ragsdale, K. and Libby, L., Photovoltaics vs. utility line extension in Austin, Texas, *Twenty-Third IEEE Photovoltaic Specialists Conference—1993*, IEEE, New York, 1125–1128, 1993.
76. Anon., Telecom taps into solar power, *Electrical Engineer*, 68(7), 22–23, 1991.
77. Lysen, E.H., Photovolts for villages, *IEEE Spectrum*, 31(10), 34–39, 1994.
78. Bhattacharya, T.K., Homemade watts for rural India, *IEEE Spectrum*, 31(10), 44–46, 1994.
79. Singh, N.P. and Bhargava, B., Photovoltaic powered lights for Indian villages, *Twenty-First IEEE Photovoltaic Specialists Conference—1990*, IEEE, New York, 991–993, 1990.
80. Panggabean, L.M., Four years Sukatani, "The solar village Indonesia," *Solar Energy Materials and Solar Cells*, 34(1–4), 387–394, 1994.
81. Kilfooyle, D., Marion, B. and Ventre, G., Lessons learned from photovoltaic vaccine refrigerators, *Twenty-First IEEE Photovoltaic Specialists Conference—1990*, IEEE, New York, 985–990, 1990.
82. Salameh, Z.M. and Lynch, W.A., Single-stage dual priority regulator for photovoltaic systems, *Solar Energy*, 48(6), 349–351, 1992.
83. Bigger, J.E., Kern, E.C. Jr. and Russell, M.C., Powering T&D sectionalizing switches with photovoltaics: A low cost option, *IEEE Transactions on Energy Conversion*, 8(2), 165–171, 1993.

84. Bigger, J.E., Kern, E.C. Jr. and Russell, M.C., Cost-effective photovoltaic applications for electric utilities, *Twenty-Second IEEE Photovoltaic Specialists Conference—1991,* IEEE, New York, 486–492, 1991.
85. Alzieu, J., Cunff, Y.L., Marie, J. and Moller, Y., Two examples of photovoltaic generators on a high-voltage overhead line, *Twenty-Second IEEE Photovoltaic Specialists Conference—1991,* IEEE, New York, 994–998, 1991.
86. Jennings, C., PG&E's cost-effective photovoltaic installations, *Twenty-Second IEEE Photovoltaic Specialists Conference—1991,* IEEE, New York, 914–918, 1991.

23

Wind Power

David M. Eggleston
DME Engineering

Introduction

Wind power is a renewable energy resource for which the technology has matured enough to be very close to competing with conventional sources of electricity in areas with good wind resources.[1] Wind electric generation technology passed from the research and development realm to commercial industry in the 1980s during wind development in California[2] and Denmark. Wind plants are now routinely providing power, integrated with coal, nuclear, and other conventional generating plants, into the utility grids in many countries around the world.

23.1 The Industry

In regions with favorable conditions for wind plant development, such as California in the early 1980s and Germany and India in the early 1990s, the growth rate in installed capacity has been rapid, as is shown later. From an established base, wind plant development in North America,

0-8493-2514-5/96/$0.00+$.50

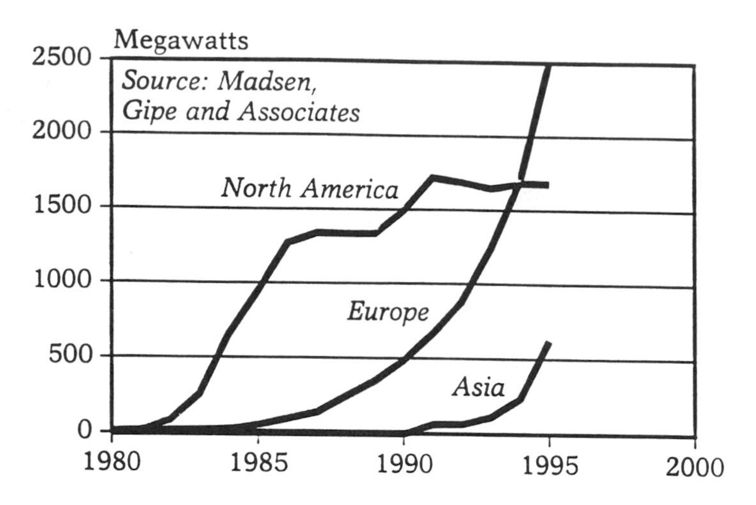

FIGURE 23.1 Wind energy generating capacity by region, 1980–1995. *Source*: World Watch Institute, *Vital Signs 1996*, With permission.

Europe, and Asia exhibited a worldwide growth rate of 25% in 1994, and a total of 3,730 MW of nameplate capacity had been installed by the end of that year.[3] Rapid growth has continued through 1995. As noted in Chapter 18B, there are no technological barriers to wind energy providing 20% of the world's electricity. Yearly energy production from wind plants worldwide reached 6 TWh (terawatt-hours, 10^{12} or million-million watt-hours) for 1994, and was predicted to reach 7 TWh by 1995.[4]

Although market uncertainty has slowed wind plant capacity growth in California and development in the United States, wind plants are being installed outside of California in Minnesota, Washington, Wyoming, Texas, Wisconsin, Maine, and other states. New plants have been installed in Germany, the United Kingdom, the Netherlands, Spain, India, Ireland, and China. India installed at least 114 MW in 1994 and had reached 300 MW by June 1995.[5] These installations have taken place during a time of low fossil fuel prices. This growth has been made possible by improved understanding of the wind environment, by manufacturing and technology advances, by series production runs of wind turbines of high reliability, and by the gradual increase in machine size and corresponding reduction in maintenance and operating cost.

The record of total installed wind plant capacity is shown in Figure 23.1. Europe has now exceeded North America in installed capacity, and strong growth is continuing in Europe and has spread to India. Joint venture wind plant development has been initiated in China and in Eastern European countries, and plans for wind plant development are being finalized in South America, especially in Brazil and Argentina.

Electric power production from wind plants in Europe and North America is compared in Figure 23.2. The yearly production of Europe has exceeded that of all of North America and is increasing strongly. Germany added 139 MW of capacity in the first half of 1995, reaching a total of more than 754 MW installed.[6]

Interest in wind plants has resulted in wind-powered refrigeration plants for the fishing industry in Egypt, in large turbines producing power on the Golan Heights in Israel, and in power for islands in the Atlantic and Pacific.

Small wind turbines are supplying power off-grid in remote villages, to a reported 100,000 nomadic herders in Inner and Outer Mongolia, and to villages in Morocco, Mexico, and other developing countries. In the developed world small turbines are providing power for remote dwellings and for battery charging on boats. Small turbines are also providing power to unmanned communications facilities on mountain tops and in other remote sites, often having extreme winds and severe environmental conditions, such as in Antarctica. Small grid-connected wind plants are also contributing to the power requirements of homes and farms.

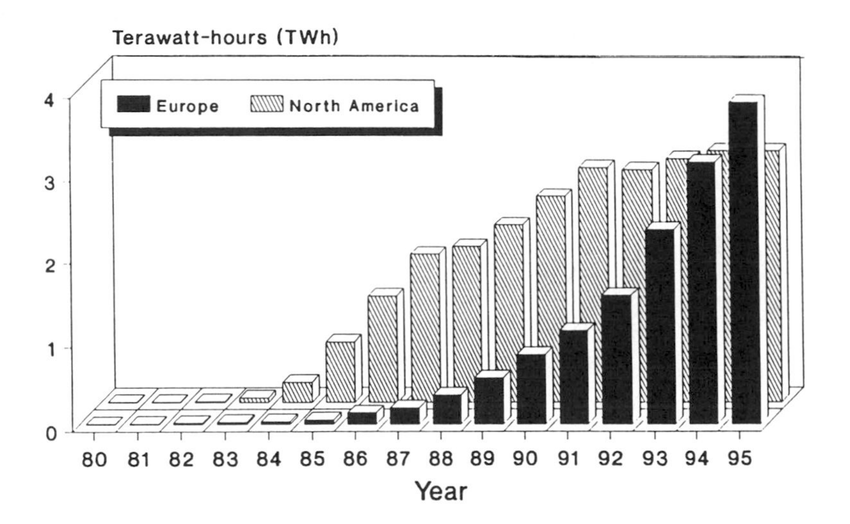

FIGURE 23.2 North American and European wind generation. *Source:* Gipe, Paul. *Wind Power Comes of Age*, p. 13. Wiley, 1995.

23.2 Wind Inhomogeneity

While calculations are often based upon a homogeneous steady wind, the actual winds at a site are more complex. Figure 23.3 shows the idealized homogeneous wind in (a), followed by a wind having uniform vertical wind shear (b), and finally by a turbulent wind (c). Since the rotor slices through these turbulent, unsteady winds, the blades experience changing velocities, which depend upon the wind speed and the rotor RPM. These changing velocities result in turbulent loads and stochastic motions of the blades, which can lead to material fatigue. Rotor blades designed for a 30-year life must anticipate these conditions and be capable of withstanding these loads and motions. Some wind turbine manufacturers specify the turbulence level their machines can withstand, downrating the life in severe environments. The best wind plant sites have high energy but low turbulence, and wind resource work as in Chapter 18B must include measurement of turbulence.

23.3 Energy Content of Wind

A wind velocity V_0 carries a power density, *p*, of

$$P = 1/2\ \rho V_0^3,\ \mathrm{W/m^2} \tag{23.1}$$

Thus the power density is directly proportional to air density ρ, and to the cube of wind velocity V_0. The dependence upon the cube of wind velocity has many important practical implications. The sensitivity of the power content to errors in velocity is large. For example, if a site has an expected wind speed of 7 m/s but this turns out in practice to be only 6 m/s, the power will be only 63% of that expected. Thus the accuracy of measurement in the wind resource work of Chapter 18B must be stressed.

Matching a generator system to a rotor is also made difficult by the cubic dependence on wind velocity. Obtaining high efficiency in low winds may require multiple generators and/or generators with windings for several different pole arrangements.

The density of air varies according to the ideal gas law as $\rho = p/RT$ as described in Chapter 18B, Equations (2) and (3), decreasing with altitude and with temperature. Therefore the most energy is obtained from a wind plant at low altitude and in cold conditions. Suitable reductions in power

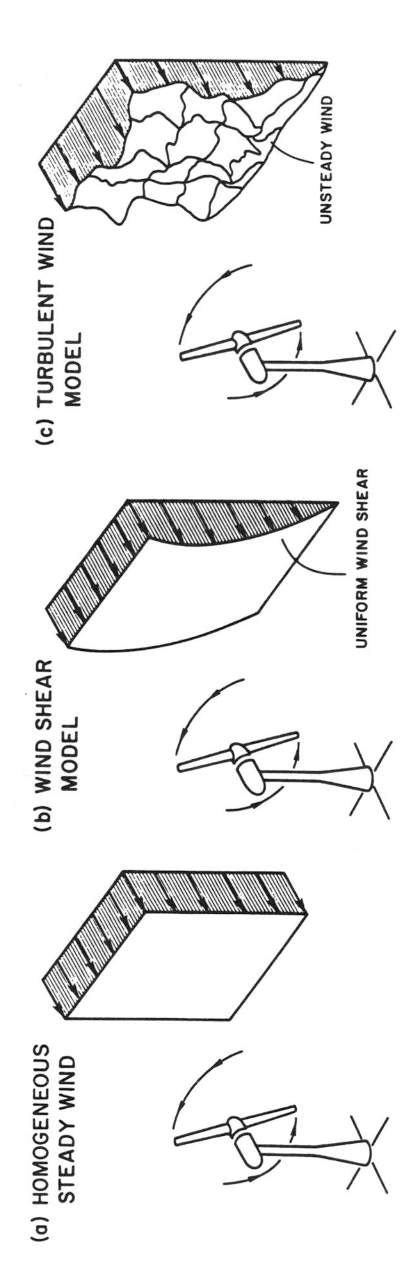

FIGURE 23.3 Wind models. *Source:* Eggleston, D.M., and Stoddard, F.S. *Wind Turbine Engineering Design*. Van Nostrand Reinhold, 1987.

estimates must be made for altitudes above sea level and for warm climates. Significant changes in power output of a wind plant can occur due to changes in air density.

23.4 Wind Plant Characteristics

Wind power from one machine is not a constant energy source, and power from it cannot be obtained on demand. At most locations on the earth wind power is of low energy density, so that a large area of wind must be intercepted in order to harness a large amount of power. As noted in Chapter 18B, the dependence of wind energy on the cube of wind speed makes wind prospecting a necessary activity in the development of any wind plant. It is necessary to search for the most continually windy sites. However, once a very windy site has been identified, the economic viability of establishing a wind plant on the site can be evaluated. The ultimate practical concern is whether a wind plant development provides the energy needed and gives promise of a good rate of return on the investment, other aspects of the development being also acceptable.

Technologies for inexpensively storing power are maturing, and their development promises to greatly improve the load-matching and dispatchability of wind power. Lead–acid storage batteries, although expensive, have been used to carry over power from high to low wind periods in California. Compressed-air energy storage, although inefficient, has also been used in utility applications to store base-load power for later use, and could be applied to wind plants. Combination with a gas turbine is another promising method of guaranteeing power availability of a wind plant.

23.5 Mechanics of Wind Power

Utilizing the Rankine–Froude actuator-disk theory, the amount of power extractable by a rotor can be determined and optimized. Consider a uniform, homogeneous wind of velocity V_0. If the wind is not retarded, no power can be obtained. If the wind is slowed to zero speed, again no power can be extracted, since there is no flow through the rotor. Somewhere in between these conditions lies a maximum. The maximum turns out to occur when the wind is slowed to $^2/_3\ V_0$ by the time it reaches the rotor plane, and is further slowed to $^1/_3\ V_0$ far downstream. It can be shown that the maximum power that can be thus extracted is $16/27 \approx 0.59259$ or about 59% of the kinetic energy of the wind. The changes in wind velocity and pressure are shown in Figure 23.4(a) and (b).[7]

When it is operating in this condition the rotor has a thrust coefficient (the ratio of the thrust force to the product of the dynamic pressure times the rotor area), $C_T = T/qA = 8/9$, where T is the thrust force, $q = {}^1\!/_2 \rho V_0^2$ is the dynamic pressure, and A is the rotor area. Since a flat plate has a drag coefficient D/qA of approximately 1.17 (this varies with the Reynolds Number), a working wind rotor experiences a thrust force about 76% as large as that on a circular flat plate of diameter equal to that of the rotor. The yaw bearing must withstand this load, and the tower structure and foundation must be able to react to the large overturning moment thus generated. Even larger loads are possible in other operating conditions, as will be shown.

Since air under these conditions acts as an incompressible fluid, the cylinder of air intercepted by the rotor must expand as the wind slows down. The theory implies that the diameter of the stream increases from 18.4% smaller than the rotor far upstream to 41.4% larger than the rotor far downstream. When a large array of wind turbines is laid out, the velocity deficit from the upstream machines detracts from the power that can be developed by machines downstream, an effect called *array losses*. For this reason, machines must be located an adequate distance apart. This separation allows turbulent mixing to re-establish the wind velocity downstream.

Wind plants have been built with a wide variety of cross-wind and downwind spacings. Generally, on sites of limited area, cross-wind spacings of $2D$ to $3D$ and downwind spacings of $8D$ to $10D$ (where D represents the rotor diameter) are about the minimum suitable for a wind plant.

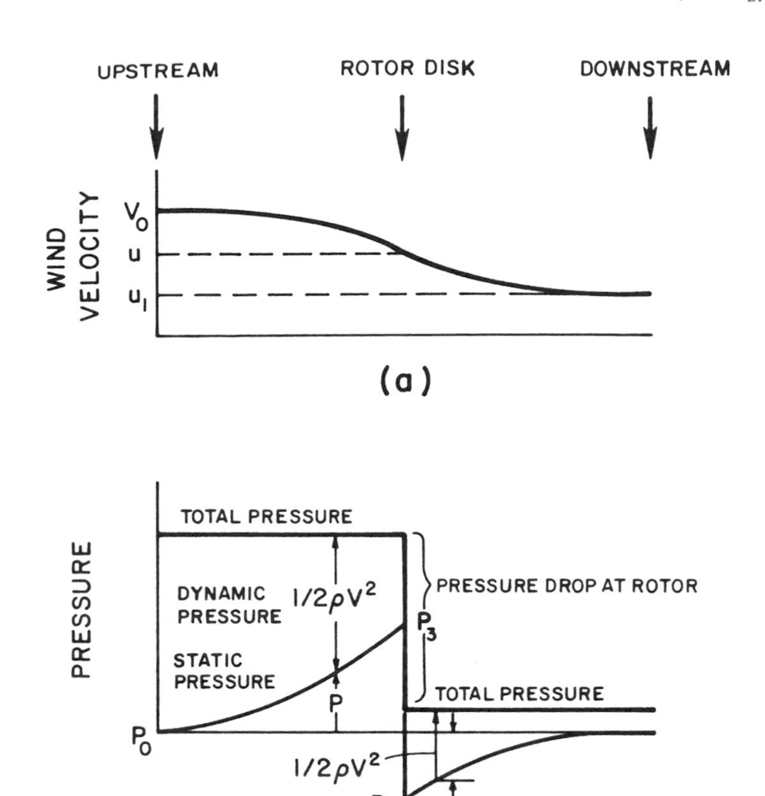

FIGURE 23.4 Actuator disk model. (a) Velocity variation; (b) pressure relationships. *Source:* Eggleston and Stoddard. *Wind Turbine Engineering Design.* Van Nostrand Reinhold, 1987.

On large unrestricted areas, larger spacings are desirable. With greater spacing, the further reduction of array losses must be balanced against the cost of extra wiring, grading, and land costs in placing the turbines farther apart.

For a wind turbine, the principle of conservation of rotational momentum requires that, if a rotor acquires a rotational momentum in one direction, the air that caused it to turn must have an equal and opposite rotational momentum in the other direction. The principle gives rise to a rotation of the wake behind the rotor. The kinetic energy of rotation is wasted energy and lowers the efficiency of the rotor. The amount of energy lost depends on the *solidity* of the rotor, which may be roughly defined as the percentage of the rotor area that is occupied by blades.[8] Thus the more the rotor disk is covered with blades, the greater the rotational kinetic energy that is wasted. Modern wind turbine rotors are generally of low solidity, to keep the amount of energy lost to rotational motion of the wake small.

23.6 Types of Machines

In engineering design there are many ways of satisfying the requirements of long life and efficient extraction of energy from the wind, and many different styles of wind machine have been tried, as shown in Figure 23.5. Of these, the propeller or horizontal axis wind turbine (HAWT) with two or three blades and the Darrieus or vertical axis wind turbine (VAWT) with two or three blades have evolved beyond the prototype stage and are routinely used for generating electrical power. Some HAWT machines with one blade are also made commercially.

WIND MACHINES

Taxonomy

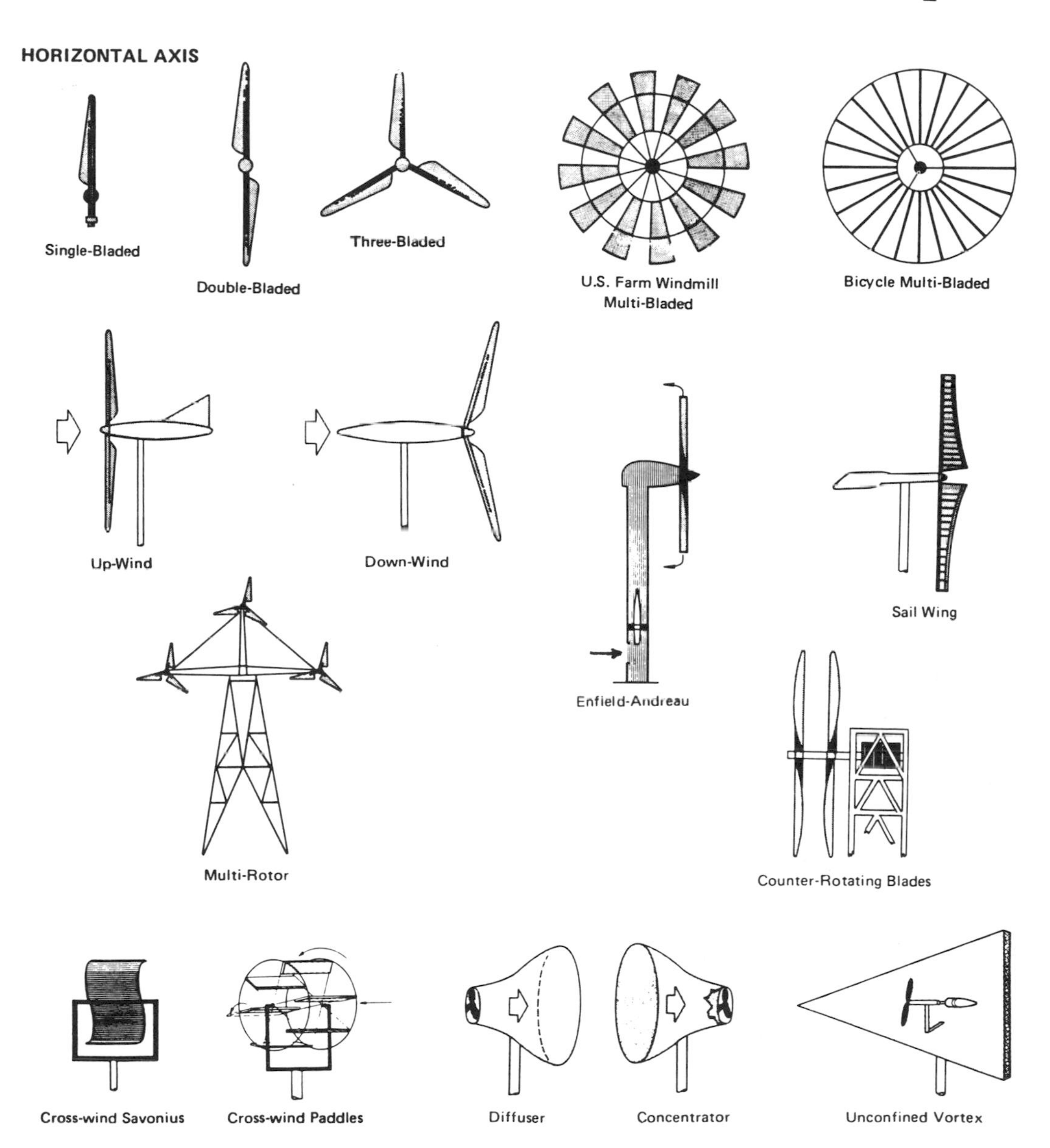

FIGURE 23.5(a) Wind turbine configurations.

One of the most fundamental distinctions in wind machines is whether they operate on a *drag* principle or on a *lift* principle. Wind machines operating on a lift principle can generate a great deal more power than those operating on a drag principle. This fact was first discovered in connection with sailing ships when it was found that more power was generated and speeds were greater while sailing with a cross wind (reaching) than with a tail wind (running).

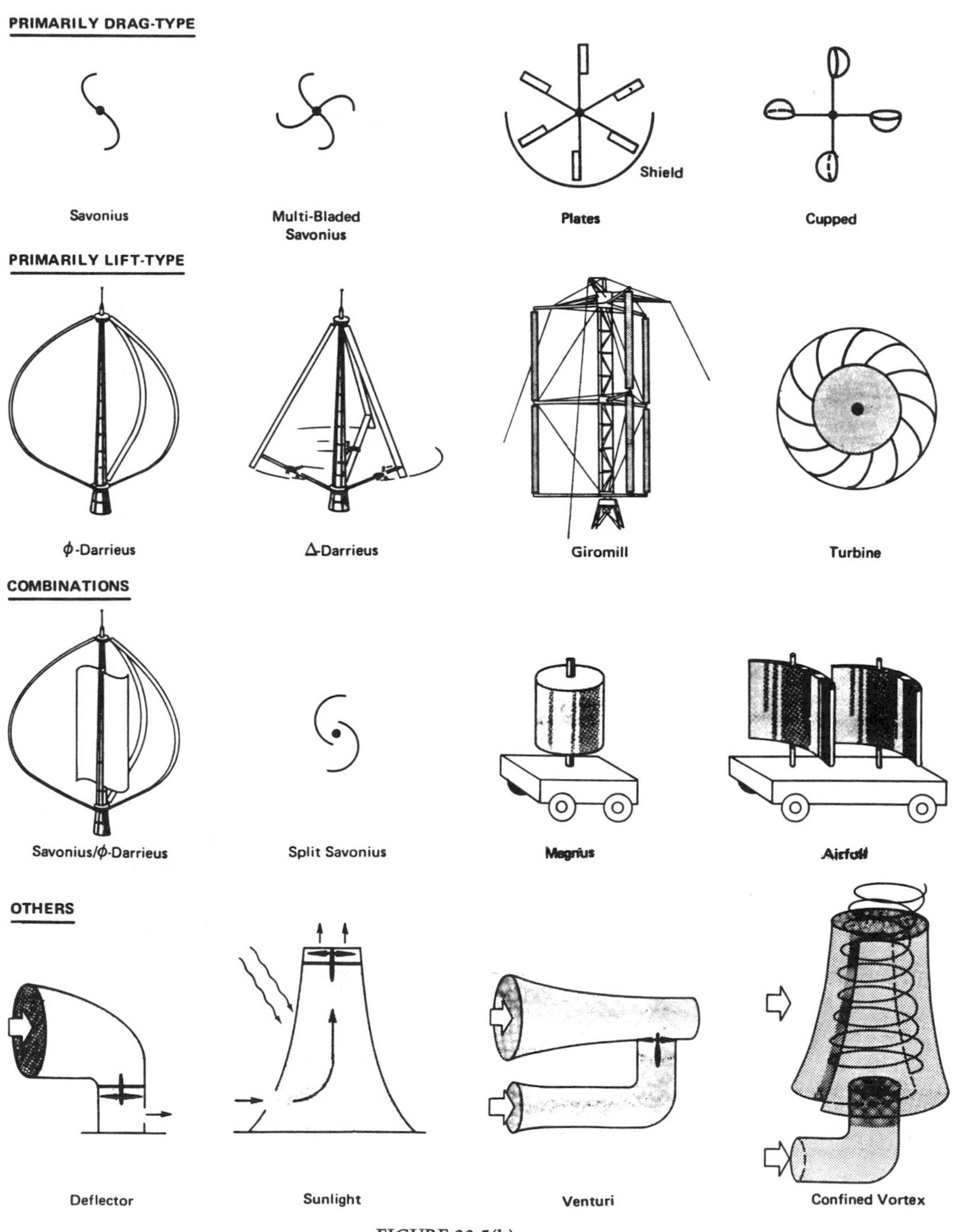

FIGURE 23.5(b)

Consider a drag device of area A and drag coefficient C_D moving with the wind at velocity V. The drag force on the plate multiplied by the velocity of the plate represents power achieved, P, and is given by

$$P = 0.5\,\rho\left(V_0 - V\right)^2 C_D\,A\,V = 0.5\,\rho\,V_0^3\,A \times \left(1 - x\right)^2 C_D \tag{23.2}$$

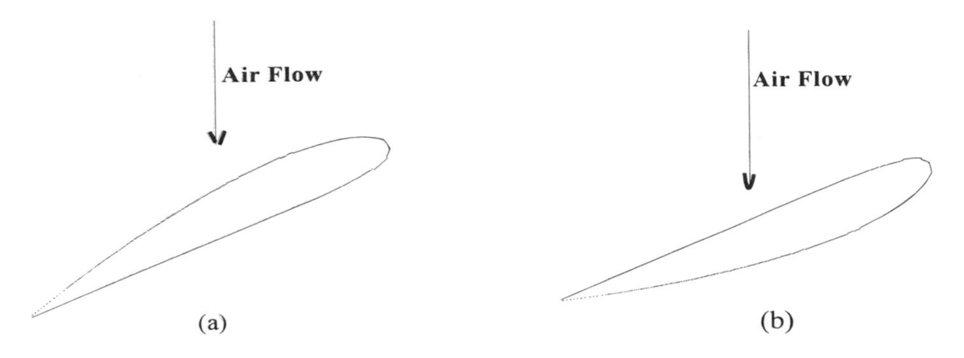

FIGURE 23.6 Rotor shapes. (a) Propeller; (b) wind turbine.

where $x = V/V_0$. If this power is optimized with respect to x,[9] it is found that the maximum power is generated when $x = 1/3$. In fact, the ratio $x = V/V_0$, called the *speed ratio*, is a fundamental parameter in wind turbine performance. Thus a drag-based machine operates most effectively at a speed ratio $x = 1/3$.

A lifting surface of equal area A with lift/drag ratio $\varepsilon = C_L/C_D$ can attain values of ε between 10 and 100. Let the maximum lift and drag coefficients $C_{L_{MAX}}$ and $C_{D_{MAX}}$, which are close to unity, be taken to be exactly one. If the power generated by a lifting surface is optimized, and it is found that the maximum power for a lift device is larger than that for a drag device by a factor $\varepsilon^2 = (C_L/C_D)^2$. Since even moderate airfoils attain $(C_L/C_D) = 10$, a lift device can produce potentially 100 times as much power as a drag device. For this reason, machines based on a lift principle can always generate vastly more power than can drag devices, so the most efficient commercial machines are invariably based upon a lift principle. The optimal value of x for a lift device is $x = \frac{2}{3}\varepsilon$. Thus a lift device with $\varepsilon = 10$ would produce maximum power when operated at a speed ratio of 6.67. This speed ratio is 20 times what would be optimal for the drag device.

23.7 Theory of Operation

VAWTs rely on the average lift component in the direction of blade motion during an unsteady cyclic process to generate power. While steady flow processes generally enable higher machine efficiencies than unsteady flows, horizontal-axis wind machines also experience unsteady flows as a result of wind shear, turbulence, yaw orientation, and other conditions. Thus all wind machines must to a certain extent operate in unsteady flow.

The HAWT rotor bears a superficial resemblance to a propeller, as shown in Figure 23.6. A propeller must have power applied, while a wind turbine rotor must absorb power from the wind. A propeller would make an extremely inefficient wind turbine rotor, because if the propeller of Figure 23.6(a) were turned upside down, its trailing edge would correspond to the wind turbine rotor leading edge and the airfoil shape would be backward.

HAWT blades use airfoil shapes that vary with radius r arranged in a planform and with a twist distribution determined by design. The design specifications for a wind turbine rotor and for the airfoils used are quite different from those for a propeller or an airplane wing. The relative wind at any radius r on the rotor depends upon the wind speed, the RPM, and the amount the wind has been slowed prior to reaching the rotor. The wind velocity at the rotor plane will have been slowed by an amount v and is taken as $V_0(1-a)$, where the fractional slowing $a = v/V_0$ is called the *axial interference factor*. Over a section of rotor blade of length Δr the *lift* force L and the *drag* force D are felt by the blade. The component of L in the direction of blade rotation generates power, as shown in Figure 23.7(a), while the components perpendicular to the rotor plane generate *thrust* T. The length from the *trailing edge* to the *leading edge* along the line joining the two extreme points is called the airfoil *chord*. Here θ is the blade *twist* angle, ϕ is the *relative wind angle*, and α is the *angle of attack*. In Europe, a different reference for angle of attack, the *zero lift line*, is commonly

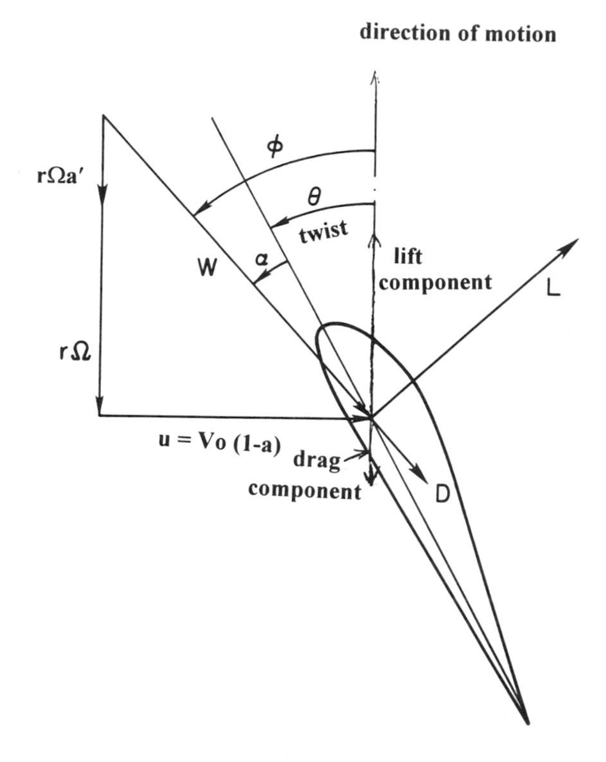

FIGURE 23.7(a) Lift, drag, and relative wind. *Source:* Eggleston, D.M., and Stoddard, F.S. *Wind Turbine Engineering Design*. Van Nostrand Reinhold, 1987.

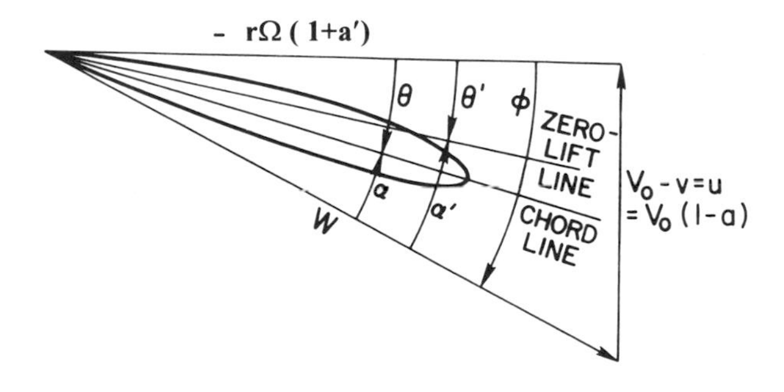

FIGURE 23.7(b) The two angle of attack references. *Source:* Eggleston, D.M., and Stoddard, F.S. *Wind Turbine Engineering Design*. Van Nostrand Reinhold, 1987.

used, and the correspondingly different angle of attack is denoted θ' as shown in Figure 23.7(b). When airfoil lift is plotted using the zero lift line as reference, the lift coefficient versus angle of attack curve always goes through the origin.

Since during operation the relative wind angle generally goes from near zero at the tip to 90 degrees at the root, the twist angle θ must be near zero at the tip and large at the root in order to generate the maximum amount of power.

A rotor has been called an *airscrew*,[10] a very appropriate description. A screw has a *pitch*, which is the amount the screw advances with each turn. A rotor acts as an airscrew in that if it is rotating freely, it will have a rotational speed that bears a fixed ratio to the wind speed. This *zero-slip speed* is almost exactly the speed for an anemometer propeller as in Chapter 18B. A wind turbine rotor, however, is loaded by driving a generator and must generate a torque by means of the difference between its RPM and its zero slip speed. The effective pitch under load is represented by the *tip speed ratio* X, and a fixed-pitch rotor will have an optimal tip speed ratio. If the generator load goes to zero, such as when the utility line connection is lost, the rotor will rapidly attempt to reach its zero-slip speed, which may easily be in excess of its safe survival RPM. Therefore, all wind turbines must have a control to prevent overspeed in case of a loss-of-line fault.

There is a whole set of conditions in which a horizontal-axis rotor may operate, and viewing these as a continuum yields considerable insight. Figure 23.8 shows the different operating states, together with their *thrust coefficient* C_T. As the pitch angle θ' with respect to the zero-lift line is varied, shaft power must be applied in (a), would be absorbed in (c), and must again be applied in (d). A wind turbine with a fixed-pitch rotor driving an induction generator may pass through the states (a) through (c) as it comes on line. Note that in the vortex ring state prior to entering the propeller brake state, the thrust coefficient can reach values as high as 2.0.

When a wind turbine rotor is designed or tested, the performance is represented as a dimensionless *power coefficient* $C_P = P/0.5\rho V_0^3 A$. The power coefficient is plotted versus tip speed ratio X, in Figure 23.9 for several different types of wind machines. Conditions where X is small represent either low RPM of the rotor or high wind speeds, with corresponding high angles of attack for the airfoils. Where the curve reaches a maximum is the best performance or design point of the rotor. It again goes to zero at the zero-slip condition.

The Glauert Optimum Rotor Theory, which neglects the effects of drag, allows defining the planform and twist to yield conditions for an optimum rotor (i.e., maximum power generation for a particular tip speed ratio). If the maximum of the peak of the C_P vs. X curve is very sharp, the rotor will fail to maximize the power except at the maximum point. By distributing the optimum condition along the blade, the peak can be made broader.[11] By operating a rotor at variable speed utilizing blade pitch control or equivalent, it is possible to control the RPM and keep the rotor operating near the best tip speed ratio. Several modern turbines, such as the Enercon E-40 series and the Kenetech VS-33 and 45, operate in a variable-speed mode. It is not clear yet that variable speed operation results in enough extra energy capture to pay for the complex mechanisms or power electronics required.[12]

Darrieus-style machines make use of the fact that a symmetric airfoil oriented and moving tangentially generates a lift component in its direction of motion under a wide variety of relative wind angles, especially when going cross-wind on the upwind side, as shown in Figure 23.10. The blade torque has a strong cyclic behavior. The wake of the central column of the rotor affects the blades on the downwind side. The flow is rather complex and unsteady, but good performance has been demonstrated. A variety of VAWT prototypes have been built and tested. These include several variable-geometry machines. The 4 kW ASI/Pinson and the McDonnell 40 kW giromills used cyclic pitch to improve performance, and the 125 kW 25-m.-dia. Musgrove Arrow VAWT went from an H to a double-arrow configuration in high winds to control input power. A series of Darrieus prototype machines was developed in Canada and at the Sandia Labs in Albuquerque. The 4 MW Eolé machine at Cap-Chat, Quebec in Canada is the largest VAWT ever constructed. Only the ϕ Darrieus has reached series production in any significant numbers.

A Darrieus-style machine relies on a symmetrical airfoil. This is somewhat of a disadvantage, since a well-designed cambered (convex shaped) airfoil can always generate the same lift with less drag than that of a corresponding symmetric airfoil. However, research at Sandia Labs in the late 1980s resulted in development of airfoils specifically to optimize VAWT performance. On the other hand, under high-wind conditions the Darrieus may retain performance it would ordinarily lose from the rapidly changing angle of attack and lift due to a phenomenon called *dynamic stall*,[13] which also affects horizontal-axis rotors. The Darrieus machine has the advantage of not requiring devices to orient it to face the wind and of allowing the generator to be near ground level for easy access. Disadvantages include having much of the rotor at a low height, where the wind velocity is low due to wind shear and where wind turbulence is great. On a ridge with negative wind shear in proximity to the ground, blades being near the ground may not be a disadvantage. Although

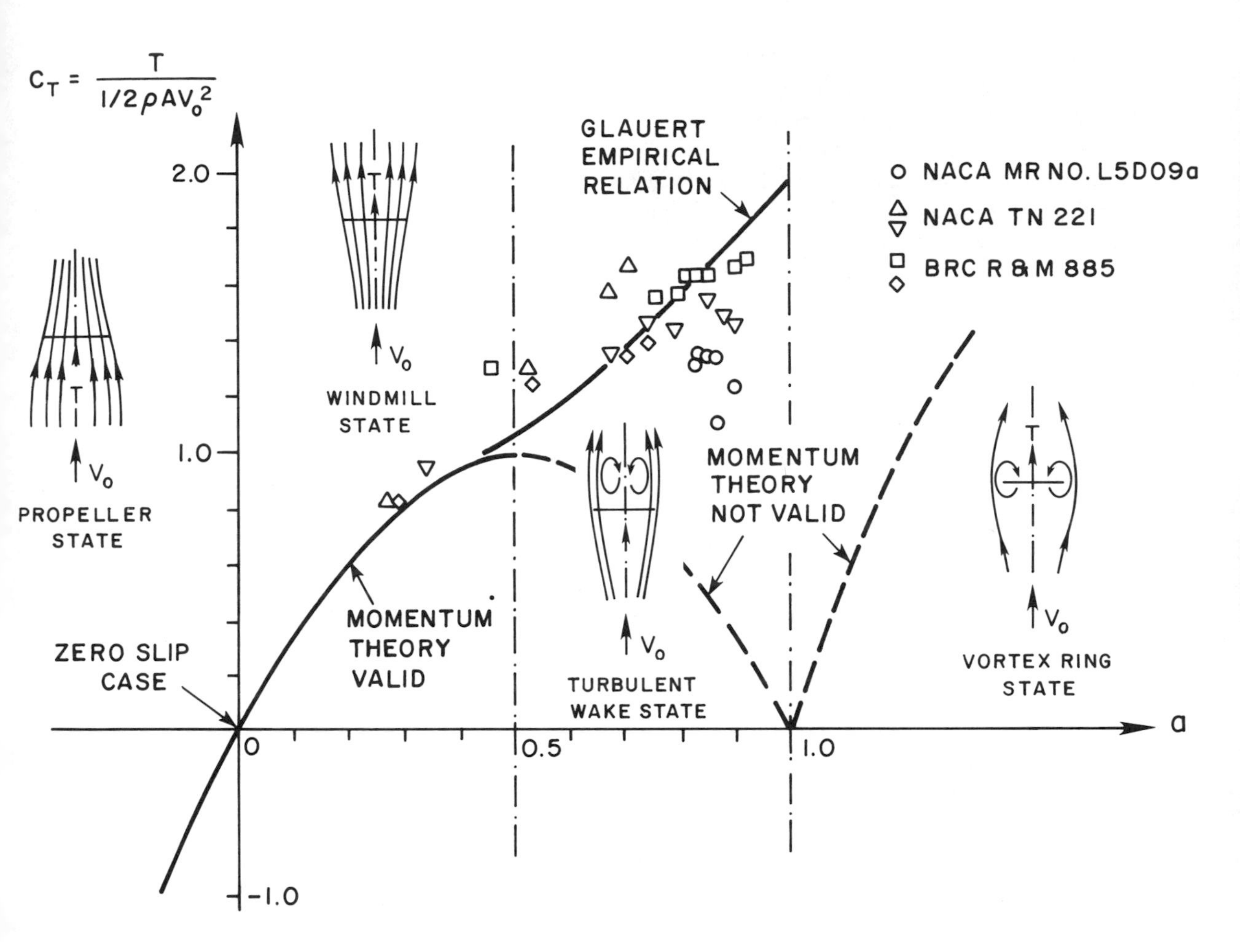

FIGURE 23.8 Thrust coefficient and rotor states. *Source:* Eggleston, D.M., and Stoddard, F.S. *Wind Turbine Engineering Design.* Van Nostrand Reinhold, 1987.

both vertical-axis and horizontal-axis machines have achieved commercial success, the horizontal-axis or "propeller"-type machines are far more numerous. It remains to be seen whether vertical-axis machines can gain a large market share.

23.8 Rotors

Rotors for horizontal-axis turbines may have a rigid hub or may include a degree of flexibility, most commonly achieved by the use of a hinge pin. The latter are commonly referred to as teetered rotors. Single-bladed rotors have been demonstrated, balanced with a counterweight, but generally have to operate at high tip speeds in order to retain efficiency and can generate high noise levels. Two and three-bladed rotors are common. Four-bladed rotors, like their ancestors in windmills in the Netherlands and the United Kingdom, were utilized in wind turbines built by Terrence Mehrkam in the United States, but have not survived past the 1980s. The low-speed Gaviotas MV2E,[14] designed for tropical water pumping, uses 5 blades, and 8 to 24 blades are required in fantails and in American-style farm windmills to achieve high starting torque. Medium to large utility wind turbines have either 2 or 3 blades, since little is gained in maximum efficiency for low-solidity rotors in going to larger numbers of blades[15] and cost is reduced with fewer blades.

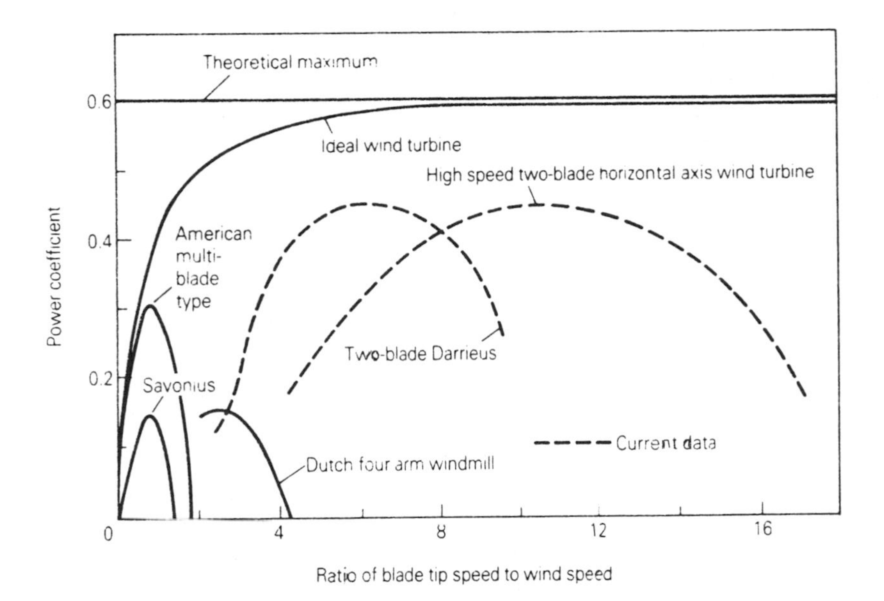

FIGURE 23.9 Wind turbine power coefficients.

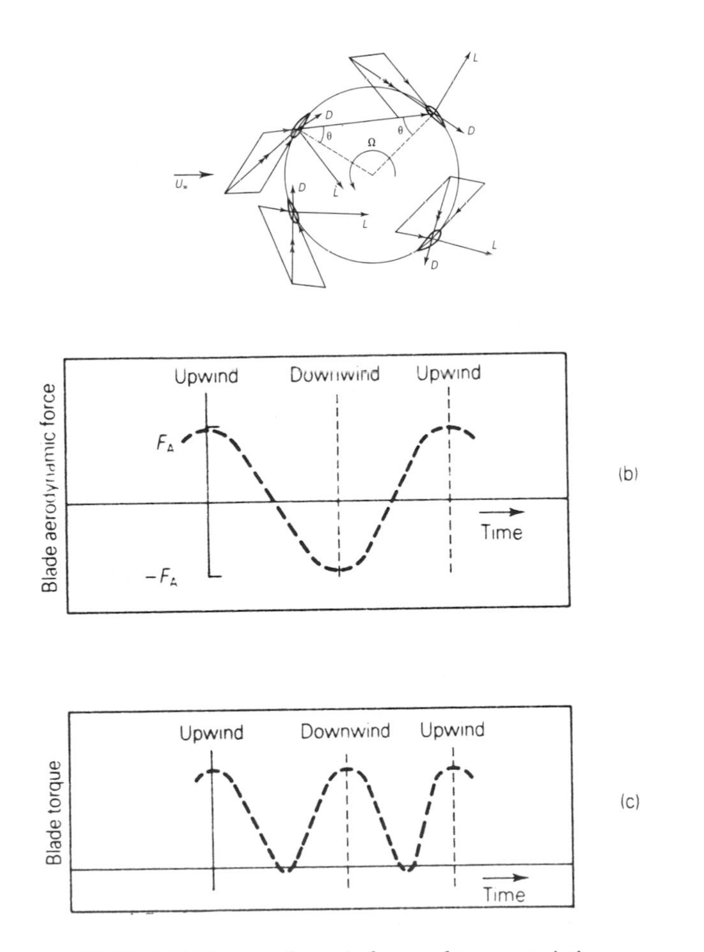

FIGURE 23.10 Aerodynamic force and torque variation.

The most successful blades have been made from fiber-reinforced polyester (FRP) and laminated wood. Glass fiber-reinforced polyester (GFRP) blades are very common among large utility wind turbines. Metal blades have been used, but aluminum has not fared well due to its relatively poor fatigue properties, and metal blades can cause serious disturbances to radio and TV reception.[16]

Early wind turbines utilized airfoils developed for airplanes and sailplanes. The design requirements for wind turbines are quite different from those for airplanes, so in some conditions aircraft airfoils gave inferior results. Airplanes must avoid the stall condition, whereas wind turbine blades operate partially or large stalled much of the time. Wind turbines also benefit from high lift and greater thickness (for increased strength) near the root. Development of a series of special airfoils for horizontal-axis wind turbines was undertaken by aerodynamicist James Tangler from the U.S. SERI (now NREL) and designed by Dan Somers of Airfoils, Inc.[17] Tests have confirmed improved performance of rotors using the NREL airfoils,[18] and airfoils especially designed for wind turbines are now in use worldwide.

The rotors of modern ϕ-type vertical-axis machines have been constructed of GRFP pultrusions or aluminum extrusions, anchored at top and bottom. Any lateral stiffeners for the blades need to be aerodynamically faired, since they otherwise can generate very large drag torques. Vertical axis turbines have a wide variety of vibrational modes and frequencies, and good structural dynamic modeling is important in understanding and avoiding operating resonances. Wide-range variable-speed operation would be difficult. Airfoils need to be tailored for VAWT operation.

23.9 Performance

The performance of a wind turbine is given by its *power curve*, the power output in kW as a function of wind speed. Typical power curves for stall-controlled and pitch-controlled rotors are shown in Figure 23.11. The wind speed at which the machine will begin to produce power is called the *cut-in* speed. The power rises with wind speed until the *rated power* speed is reached. A stall-controlled rotor will then begin to limit power passively by progressive stalling of sections of the blades. The shape of the power curve after achieving rated power is a result of this blade twist design process.

Variable-pitch machines control the maximum power very near the rated value, all the way up to the *cut-out* speed. Operation above cut-out would result in long-term structural damage to the rotor. Variable pitch allows more control of power and less thrust force on the tower in very high winds, with the complete feathering of the blades as an option.

The power curve is calculated during design and measured experimentally during prototype testing. The wind speed at which rated power is reached varies among machines, and must be matched to the wind spectrum of the site. The Carter 300 machine is unusual in that it reaches rated power at a very high wind speed of 18.8 m/s, while most machines reach rated power by about 12 to 14 m/s. European wind power incentives are sometimes indexed to the power produced at 12 m/s instead of the rated capacity of the machine. This is a more basic way of expressing machine performance and serves as a means of useful comparison between machines.

The *capacity factor* is the ratio of the actual output to the rated capacity over some period, and is an important measure of both machine performance and the wind spectrum of the site. Well-sited and managed wind plants routinely achieve capacity factors of 0.25 or above, and in exceptional sites reach as high as 0.50.

Another method of expressing machine performance is the power produced per unit area of the rotor, kWh/m^2 yr. The higher the value the more productive the machine and rotor. This parameter is dependent upon the winds at the site as well as machine performance. Values as high as 1869 kWh/m^2 yr have been reported for a Vestas V 225 kW machine in 1994,[19] although values near 1,000 are more common in the mid-1990s. Machine performance figures can be found in the publications of the California Energy Commission in the U.S. and in WindStats for Europe.[20]

The ideal energy produced by a wind power plant may be estimated by multiplying the site wind-speed frequency spectrum by the machine power curve. The actual power received at a

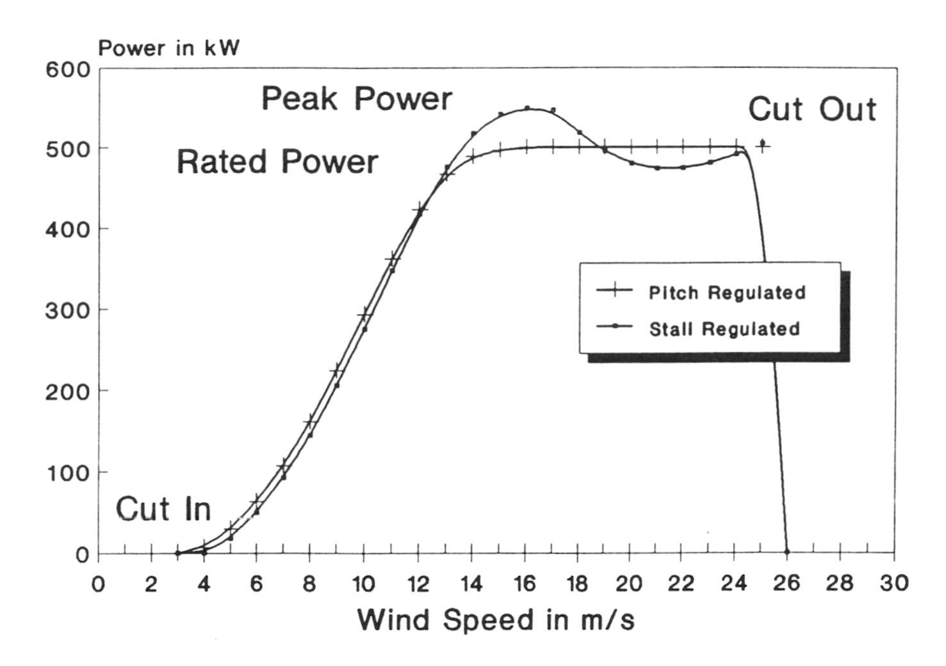

FIGURE 23.11 Sample power curves for 500 kW wind turbines. These turbines are quite different, although both have the same rating. One is 37 m (120 ft) in diameter and regulates power by aerodynamic stall. The other is 39 m (130 ft) in diameter and regulates power by varying blade pitch. *Source:* Gipe, Paul. *Wind Power Comes of Age*. Wiley, 1995.

substation is reduced from this estimate by many factors, including machine misalignment with the wind, line losses, array losses, forced outages, blade soiling, and so on. Procedures for correcting machine output predictions are given in Figure 23.12.

23.10 Yaw Control

Horizontal-axis machines must be "winded," that is, turned to face the wind for starting and low-wind conditions. Historically, this was first done manually, then automatically using fantails, which were small rotors oriented perpendicular to the main rotor. If the wind direction changed, the fantail would sense the crosswind and drive a gear system to turn the mill until the crosswind was reduced. The fantail was a very effective mechanical feedback control system having the advantage of yielding low yaw rates and guaranteeing elimination of any steady yaw error. Use of the fantail was carried over into several first-generation European turbines installed in California, including the Danish Riisager-Vindmoller and the German Aeroman machines, but has since lost favor. On large wind turbines an active yaw drive, using a signal from a wind direction sensor above the nacelle, has generally been the most convenient method of generating the large torques required. Usually the turbine is restrained in yaw by friction plates until the wind direction changes enough to require a new rotor orientation.

Some turbines have locked the nacelle in yaw until the wind direction changed, and others have been allowed to yaw freely once faced into the wind. Locking the nacelle to the tower can result in very large loads being transmitted that can cause tower fatigue failures. Depending upon rotor parameters, some rotors have an inherent aerodynamic stability causing them to track the wind direction,[21] even if they are facing upwind. Designers have utilized this stability to allow the rotor to yaw freely over a limited range to minimize tower loads, even though upwind.

Small machines, requiring simplicity, often use *tail vanes* for yaw orientation. Tail vanes tend to give very high yaw rates, aggravating rotor gyroscopic stresses, especially in rigid, two-bladed rotors.

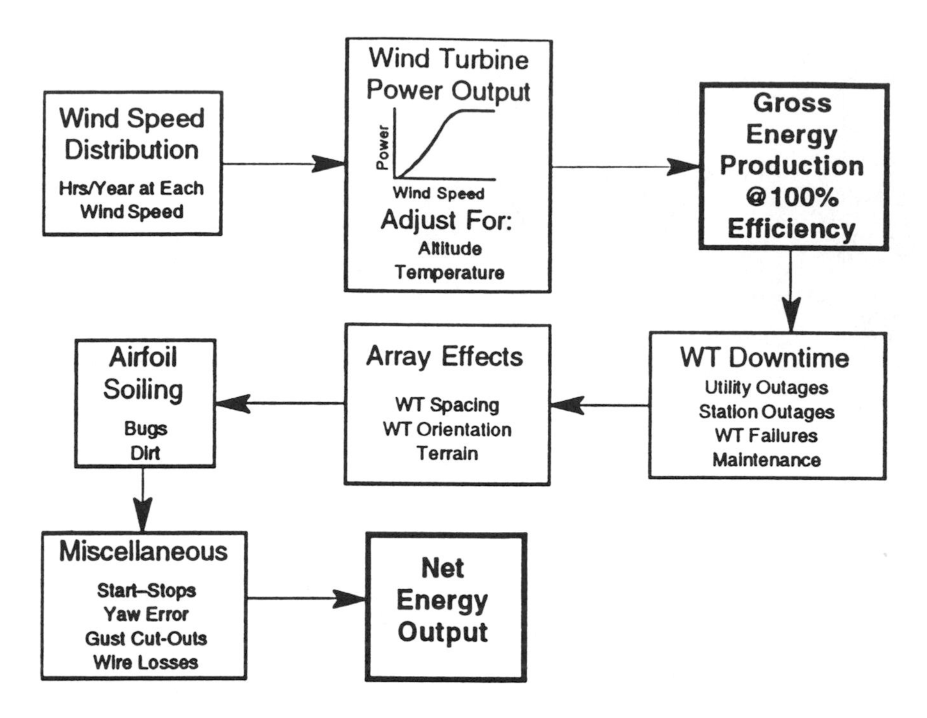

FIGURE 23.12 Flow chart for predicting net annual energy production from a wind power station.

For larger machines, tail vanes become too large to be practical. In the modern era, downwind machines using free yaw often experienced sudden yaw motions, which can lead to very high gyroscopic stresses on the rotor.

In very high winds, horizontal-axis machines may be yawed out of the wind for protection. One very large prototype machine, the 1.5 MW Gamma 60 in Italy, uses yaw control for both power limiting and overspeed control. It has experienced an overspeed condition resulting from inability to yaw out of the wind quickly enough, but was not damaged.

23.11 Generators

Induction generators (sometimes described as asynchronous) are by far the most common types used in utility-class wind turbines. They have the advantages of simplicity, reliability, and low cost, and can be used as motors to accelerate the rotor up to speed for starting. Disadvantages are that they must draw power from the utility to provide their magnetizing current and therefore do not operate at unity power factor. Capacitor banks are necessary to compensate their reactive power requirements, but can also be used to provide dynamic braking.[22] Induction generators having windings for both 6 or 4 poles are sometimes used to provide two-speed operation. The 6-pole winding is used at low RPM for starting and low winds, and the 4-pole winding for full power at the higher wind speeds.

Standard induction generators have only a small speed variation, or *slip*, between actual and synchronous speed. Special *high-slip* generators are used on Vestas V-600 machines to allow a limited range of variable-speed operation.

The efficiency of induction generators varies considerably with load, and a goal of design is to operate at the highest efficiencies possible. Designers sometimes use multiple generators, either of the same or different ratings, to better match the power from a rotor. Early Danish mid-sized wind turbines often used a small generator in light winds and a larger generator for stronger winds. Several 1,000 kW prototypes currently under test use four 250 kW generators to match rotor power with wind speed to achieve their total rating.

Generator winding insulation deteriorates with overtemperature, overspeed, mechanical abrasion and with sudden transients, as well as deterioration due to age. Each time power is applied the windings will flex. Generators for wind turbine applications should be designed to resist these effects. Corrosion protection must be provided, especially in adverse environments such as deserts or seacoasts. Resistive thermal detectors (RTDs) are usually installed in windings during manufacture to provide inputs for the turbine control system to protect against overheating.

Synchronous generators are used in a few large machines. They are very stiff electrically and require careful speed matching to be brought on line, but are capable of high efficiency.

Small machines often utilize permanent magnet alternators and convert the output to DC for charging batteries. Some home- and farm-sized machines carry the process further by adding a *synchronous inverter* so the machine can be tied into a utility grid. Permanent-magnet alternators have the disadvantage that their fields cannot be turned off in case of faults.[23]

Enercon in Germany has pioneered a line of variable-speed, transmissionless (no gearbox) wind turbines using a large multipole ring generator and power electronics for AC-DC-AC conversion. Whether this type of design will prove superior to multispeed induction machines with transmissions remains to be determined.

23.12 Power Control

A fundamental problem in wind turbines is restraining the maximum power when the generator(s) have reached full load. Power control may be achieved using variable pitch, stall-control, ailerons, and flaps, and by yawing the machine out of the wind. On commercial-sized machines, flaps were first suggested in the development of the early (1941–1945) Smith–Putnam 1.2 MW wind turbine.[24] Spoilers have been used on a medium-sized Danish design, and ailerons have recently been used on one model of the Zond Z-40. However, in northern regions where moderate to severe icing can occur, machines must be designed and operated such that icing cannot disable the power control system.

All airfoil surfaces experience the *stall* condition in which lift decreases and drag increases when the angle of attack exceeds an angle of about 14 degrees. For stall-controlled machines the rotor, which normally runs at constant speed, is designed to progressively stall from the root out to the tip, somewhat as airplane wings with *washout* (i.e., the tip twisted downward so the wing root stalls prior to the tip). This arrangement allows the rotor to gradually lose effectiveness as the wind speed increases, thereby preventing generator overload. Another advantage of stall-controlled rotors is low *gust sensitivity*, in that, if a gust occurs, the blade automatically stalls to limit the torque spike due to the gust. However, stall does not limit the thrust load.

In a rotor having *full span pitch* control, the whole blade is rotated to a different pitch angle in order to control power output. Thus pitch can be considered as the addition of a constant to the twist distribution $\theta(r)$ at every point along the blade. A blade can be rotated either in the direction of increasing θ (toward what is called *feathering* in an airplane propeller, when θ becomes 90 degrees) or in the direction of decreasing θ, which increases the angle of attack α toward stall.

Pitch control allows the power input to the generator to be cut off at exactly the desired level, resulting in a flat power curve all the way from the rated wind speed out to the cut-out wind speed, when the machine is shut down for protection.

In case of a sudden gust, the angle of attack quickly increases. If the blade is not near stall, such a gust can result in torque spikes as high as 100% over the average torque. Because pitch rates are often limited, it is usually much faster to pitch to stall than to pitch to feather. The control system for a variable-pitch machine can be designed to operate in a stall-controlled mode, thus obtaining the limited gust sensitivity of the stall-controlled rotor. As average wind speeds increase above the cut-out wind speed, power may be brought to zero by pitching toward feather.

By allowing some variation in rotor speed, a gust can be allowed to speed up the rotor and avoid overloading the generator. This can be accomplished by a variety of options, but typically for

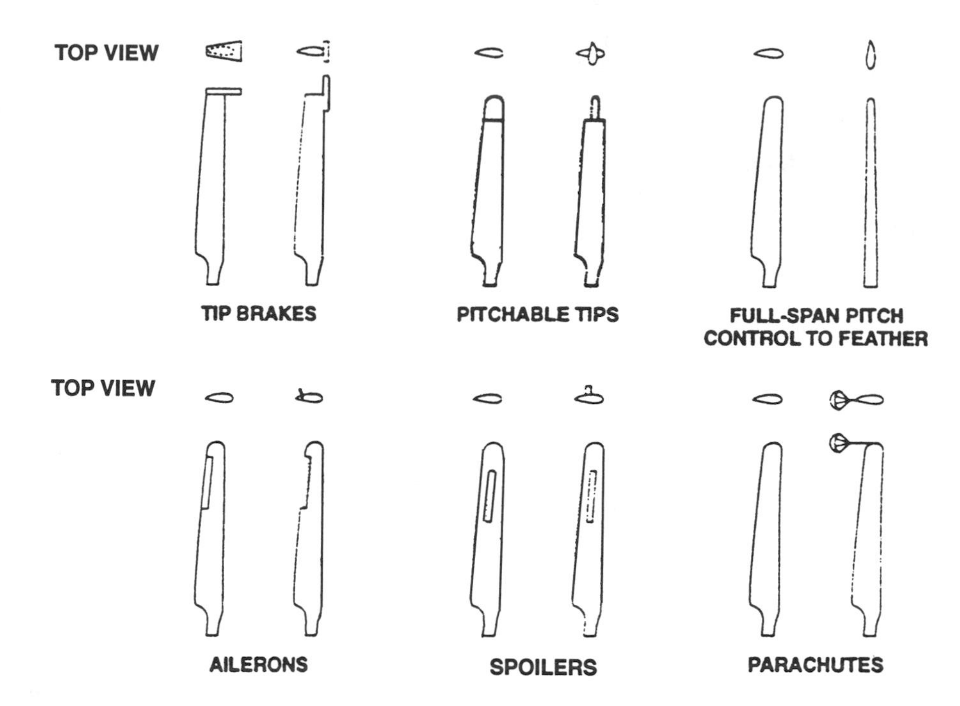

FIGURE 23.13 Overspeed control. Means to limit power and control speed of rotor under emergency conditions. *Source:* Gipe, Paul. *Wind Power Comes of Age.* Wiley, 1995.

induction generators an AC-DC-AC power electronics system is used. Recently Vestas has incorporated a high-slip induction generator to provide a modest speed variation in their V-600 series. Alternatively, as in the case of the Enercon E-40, a large multipole generator is driven directly by the rotor, thereby eliminating the gearbox.

23.13 Overspeed Control

Another significant challenge for designers is to control of rotor speed when the generator load is lost. This must generally be done very quickly, as a rotor at full power generates a torque that is high in relation to its inertia. For medium-sized machines, it may take only a few seconds to reach destructive speeds, so the net rotor torque must be reduced fast. As noted earlier for pitch-controlled rotors with rate-limited pitching mechanisms, pitching toward stall is usually much faster than pitching toward feather.

Several styles of aerodynamic overspeed controls are shown in Figure 23.13. For stall-controlled machines a typical solution has been pitchable tips. A short section of the blade including the tip pitches to 90 degrees, destroying lift and creating extra drag to slow the rotor. These are usually centrifugally activated. Another solution is tip brakes, which release to swing out and cause increased drag. Ailerons and spoilers have also been used. In each case, it is important that the mechanism return the blade to normal operation when the wind subsides without the intervention of an operator. On a two- or three-bladed rotor, it is necessary that the emergency control mechanisms on each blade work simultaneously to avoid unbalanced action. In icing environments, these tip mechanisms are likely to fail, and other overspeed control methods are recommended.

Yet another solution for overspeed is to slow the rotor using mechanical brakes. Brakes are often installed on the high-speed shaft, since smaller and less expensive brakes can be used there. Brakes on the high-speed shaft cannot slow the rotor if the transmission or the couplings fail, and brakes on the low-speed shaft are desirable. Brakes can wear, can overheat, and can fail. Worn or overheated brakes may not have the capacity to stop a rotor in high winds. The Tacke TW 300 and 600 do not have pitching tips, but utilize disk brakes on both the low- and high-speed shafts to provide redundancy.

Application of brakes must be done with precise controls, since a sudden high torque conducted through the transmission can overload the gears and couplings.

23.14 Towers

For horizontal-axis turbines, towers hold the rotor up where wind speeds are higher and turbulence lower. Tower heights have increased with the development of larger machines, so tower height is usually about equal to the rotor diameter. The tower and its foundation must withstand all loads from the torque and thrust of the rotor, even during storm winds.

Towers may be either *hard*, where the tower resonant frequencies are above those of the rotor and machinery, or *soft*, where the tower resonant frequencies are below that of the rotor and machinery. Soft towers are almost universal among contemporary turbines due to weight and cost considerations.

Lattice towers are inexpensive and generally require simpler and less costly foundations. However, they have the disadvantage of providing perching places for birds, requiring periodic retorquing of bolts, a laborious process, and of having very many parts that generally have to be assembled in the field near the site.

Tubular towers have the advantage of providing a sheltered environment so that the machines can be serviced in any weather; having no places for birds to perch; providing shelter for the ladder, cables, instruments and controls; and being nestable for convenient shipping and assembly on site. Kenetech Windpower has developed a hybrid tower that uses a tubular tower on a spread lattice tower foundation.

23.15 Cranes and Rigging

Cranes are most often used for erection, although wind turbines may be erected by many other means, including tilt-up towers, gin poles, helicopters, and even elevator mechanisms built into the tower. The cost of major repair can depend heavily on crane rental costs, but these are usually omitted from cost of energy calculations.[25] Provision must be made for getting large cranes to the site, including suitable access roads. Wind turbines must be designed for low-cost maintenance and repair in order that the cost of energy over a 20- to 30-year life can be minimized.

23.16 Noise

The components of a wind turbine generate some level of acoustic noise. Noise, generally defined as unwanted sound, is produced by air flows past various parts of the rotor, the meshing of gears, generators, hydraulic pumps, transformers, power electronics, forced-air cooling systems, and other sources. Noise can be understood in terms of noise sources, noise propagation, and noise perception. Noise generally decays with distance from the source. For turbines installed in the vicinity of dwellings the effects of noise must be considered, while in remote locations having no human habitation low levels of noise would ordinarily be of no significance.

The scientific study of acoustic noise is highly developed, as are techniques for noise measurement. Much noise research has been done on propellers and helicopter rotors, especially in connection with military applications, and methods of predicting rotor noise are also quite well established. Wind turbines are routinely evaluated for noise emissions, and their measured noise level is usually available from the manufacturer or from certification agencies. Many studies of wind plant noise have also been conducted.[26]

At any location, some level of *ambient noise* exists. The wind itself causes noise in trees and shrubs that increases rapidly with wind velocity. If noise from wind turbines is not tonal in nature and its average level is below the ambient noise level, it generally cannot be perceived by humans.

Of special interest is the fact that if the noise has a strong low-frequency component, it may penetrate buildings. It may even cause a lightly constructed building to resonate. This type of noise can sometimes be perceived within a building, while outside the dwelling it is masked by the ambient noise. Likewise, noise that is masked by the ambient noise level on a windy ridge can be transmitted and be annoying in a quiet valley. Noise can be studied in terms of *frequency spectrum*, and both propagation and perception depend upon the frequencies at which significant energy is transmitted.

Noise can be divided into *narrow band* noise such as *harmonics* from the *blade passage frequency* or gear noise, and *broad band noise* such as from aerodynamic or mechanical impulses, *boundary layer interaction*, or *vortex shedding*, especially at the trailing edge.[27]

Aerodynamic noise from wind turbine rotors comes from many sources, but that associated with the *tip shape* and *speed*, and with the *thickness of the trailing edge* is usually the most significant. Machines designed for low tip speeds are generally much less noisy than those operating at higher tip speeds. Most modern machines limit the tip speed to about 60 m/s (134 mph) to obtain a low noise signature, although some performance may thereby be sacrificed. *Upwind* rotors are generally less noisy than *downwind* rotors, since passage of the blades through the tower wake can generate impulsive loads and resulting low-frequency harmonics of blade passage noise. In locations where noise is critical, one manufacturer can program the turbine controller to minimize rotor noise, at the expense of losing of the order of 2% of the rotor power.

Machinery noise inside the nacelle can be reduced by grinding and polishing of gear surfaces and by use of vibration-absorbing mounts. Use of noise-absorbing materials can dampen nacelle noise, including sympathetic vibration of flexible panels. Air vents in the nacelle should preferably be oriented upward so that nacelle noise is not inadvertently focused toward the ground.

Damage to the tip area or to the leading or trailing edge of blades from impacts with hail or other objects can emit very annoying noise requiring prompt repair.[28] Wind turbine siting and environmental impact regulations, where they exist, often specify allowable noise levels or distance offsets from dwellings for wind plants.

23.17 Avian Issues

Wind plants require structures, including towers, utility poles, transmission lines, transformers, substations, and maintenance buildings. A novel element is the moving structure, the rotor. All structures can affect avian life. Many birds are injured or killed by running into structures, and many more are killed by motor vehicles and hunters.[29] The rotors of wind turbines constitute yet another hazard for bird life. Critical issues for wind plants include legal prohibitions, such as those regarding the Migratory Bird Treaty Act and the Endangered Species Act in the United States.

Birds of prey, or *raptors*, seem to be the most vulnerable to wind turbine rotors. Their behavior in focusing on their prey can lead them into the presence of rotor blades. Various methods of reducing raptor/rotor conflicts have been suggested, including reducing or eliminating raptor prey near wind plants, painting blades to make them more visible to birds, and installing noise sources disliked by birds. Studies have focused on critical issues such as the mortality rate of birds near wind turbines, the nature of avian–wind turbine interaction, and wind plant siting in relation to bird flyways. Since many birds migrate at night, radar has been used to track them. Research in these areas is continuing.

Avian authorities, such as the National Audobon Society and Birdlife International, have taken a keen interest in the environmental effects of wind plants on birds. Since wind turbines help reduce utility plant emissions, including greenhouse gases, which can have long-term effects on bird habitat, wind plants have both positive and negative effects on bird life. After detailed study, several bird organizations have publicly approved wind plant developments.[30] A cautious approach, based on studying bird interaction problems of an initial phase of a wind plant before approving the next phase of expansion, has been recommended. Wind plant developers must be sensitive to avian issues and attempt to minimize adverse effects on bird life.

23.18 Commercial Machines

A variety of commercial utility grade wind turbines is available, and newer, larger models are undergoing prototype testing. The data presented in Table 23.1 was obtained from manufacturers publications and technical papers and reports. Most major manufacturers have prototype commercial wind turbines in the 700 to 1,500 kW range intended for marketing in the 1997–2000 time frame.

Figure 23.14 shows a diagram of the Vestas V-39 500 kW turbine nacelle. Figure 23.15 shows the aileron-controlled version of the Zond Z-40, 550 kW turbine now installed near Ft. Davis, Texas. A similar model having full-span pitch control and a tubular tower is to be installed in New England. A number of current designs, including the Carter 300, the Zond Z-40, and the Tacke TW 300 and TW 600 machines utilize an integrated drive train, combining the rotor shaft and bearings, the transmission, the generator, the brakes, and the housing that attaches to the yaw bearing on the tower, all in one unit. This design is intended to eliminate the bed plate, provide accurate alignment of subsystems, and minimize cost.

23.19 Wind Turbine Procurement

To purchase turbines for a wind plant, specifications must be developed and vendors surveyed. Considerations in the choice of a turbine include

- Frequency spectrum of expected winds at the site.
- Turbulence level of winds at site.
- Machine power curve.
- Machine cost.
- Delivery schedule.
- Warrantees and guarantees.
- Machine test and operating experience.
- Machine maintenance and operating costs.
- Machine suitability for site environmental conditions.
- Machine noise signature/proximity of human habitation.
- Soil conditions; tower foundation requirements.
- Machine power quality.
- Machine commissioning procedures.

These specifications must cover all the particulars of machine performance, liability in case of failure or need for repair, and operation and maintenance costs over the entire life of the machine. Denmark and Germany, among other countries, have wind turbine certification programs. These programs provide a basic certification, giving buyers some assurance of machine durability and performance beyond manufacturer warrantees. Most countries do not yet have certification programs for wind turbines, although emerging markets, such as India, have underscored the importance of such certificates as a uniform method of assurance in machine procurement.

23.20 Wind Energy Cost

Costs of wind energy have declined from levels near $0.20/kWh in the early 1980s to current contracted costs of less than $0.05/kWh in 1995. Methodologies for determining cost of wind energy take several different approaches.[31] Whatever method to calculate costs is used, all the assumptions on which it is based must be clearly stated. Comparing costs, especially those for different energy technologies, must be done very carefully.

TABLE 23.1

Make	Model	Rating, kW	Rated Pwr @ m/s	Hub Ht Std, m	Power, kW 12 m/s, hub	Rotor Diameters	RPM	GenSpeed	No. of Blades	Type	Generator Type
Vestas	V29-225	225	14	31	215.5	29	30.5/40.5	Two	3	U	Ind
Vestas	V39-500	500	16	39		39	30		3	U	Ind
Vestas	V39-600	600	17	40.5	476	40	32.5	–5%>+5%	3	U	Ind
Vestas	V42-600	600	17	40.5	510	42	30.0	–5%>+5%	3	U	Ind
Vestas	V44-600	600	17	40.5	514	44	28	–5%>+5%	3	U	Ind
Kenetech	KVS-33	300				33		V			
Kenetech	KVS-46										
Enercon	E-40	550				40	18.3/6.41	V	3	U	Multipole
Enercon											
Enercon											
Zond	Z-40	550	12.6	30, 40, 50	495	40	29		3	U	Ind
Zond	Z-40	550	12.6	30, 40, 50	495	40	29		3	U	Ind
Zond	Z-46					46					
AWT		26				26					
AOC	15-50	50				15					
Bonus											
Nortank											
Nordex											
Nedwind											
Tacke	TW 600	200/600	14.5	50.52	470	43	18/27				Ind 6/4p
Tacke	TW 200	100/300	14	40	270	33	23.1/34.6				Ind 6/4
Micon	M1500	150/600	14	40/46	520	43	18/27	Two	3	U	Ind 6/4
Micon	M700	40/225	16	30/36	187	29.8	25/37.5	Two	3	U	Ind 6/4
Micon	M750	100/400	15	36	308	31	23.8/35.7	Two	3	U	Ind 6/4

Vestas	V29-225	Low	VP	VP	4	25	56		5,000	9,000	12,000		46.3/61.5
Vestas	V39-500	Low	VP	VP	4.5	25	54	75	6,000	18,000	26,000		61.3
Vestas	V39-600	High	VP	VP	4.5	30	54	75	8,400	17,300	23,200		68.1
Vestas	V42-600	High	VP	VP	4.5	25	50	70	8,400	17,300	23,200		66
Vestas	V44-600	High	VP	VP	4.5	20	43	60	8,400	17,300	23,200		64.5
Kenetech	KVS-33												
Kenetech	KVS-46												
Enercon	E-40	N/A	VP	VP									
Enercon													
Enercon													
Zond	Z-40	Low	VP	VP	4	29							60.7
Zond	Z-40	Low	Aileron	Aileron	4	29							60.7
Zond	Z-46												
AWT													
AOC	15-50												
Bonus													
Nortank													
Nordex													
Nedwind													
Tacke	TW 600	Low	S	BL & BH	3	25	65		5,500	29,000	60,000	99	40.5/60.8
Tacke	TW 200	Low	S	BL & BH	3	25	65		6,000	20,000	26,000	95	40/59.8
Micon	M1500	Low	S	TB	3.5	25	69		13,000	19,000	40,000		40.5/60.8
Micon	M700	Low	S	TB	4	25	70		5,000	8,500	12,000		39/58.5
Micon	M750	Low	S	TB	4	25	69		6,000	12,000	21,000		38.6/57.9

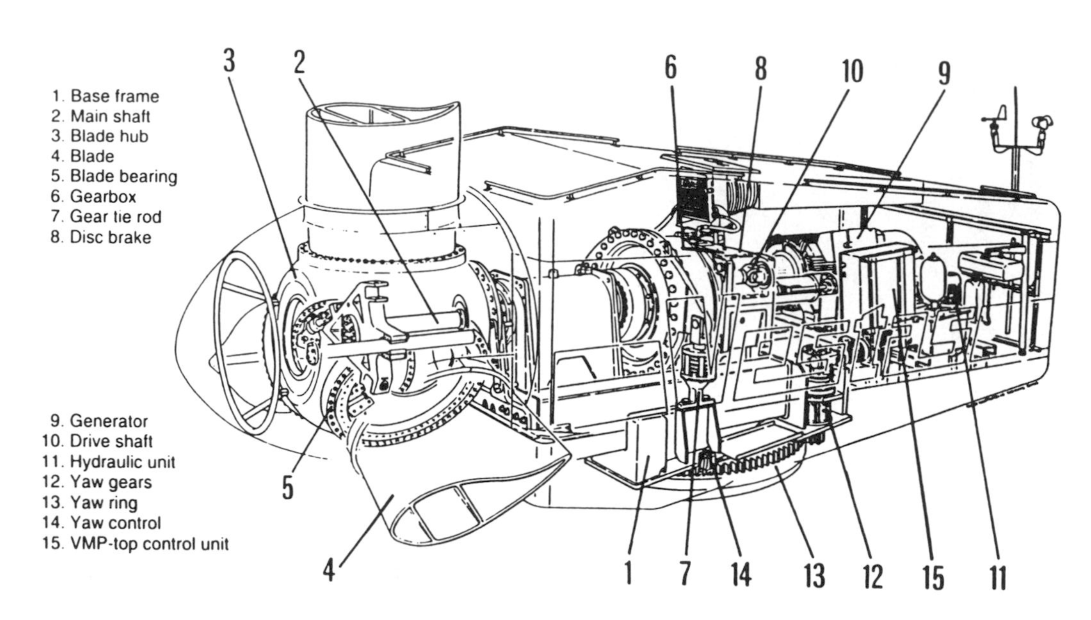

FIGURE 23.14 Vestas V-39 Nacelle.

FIGURE 23.15 Zond Z-40.

The best way to determine costs of wind energy is through cash flow analyses using actual costs of a wind plant. However, actual costs are usually considered proprietary. Thus estimates of all the costs involved must be made. There are several ways of dealing with inflation.[32] One simple method is to ignore it, giving costs in *constant* or *real* dollars. Estimates calculated with constant dollars will be lower than those that consider inflation. If estimates for the inflation rate are made, costs in a long-term project can be expressed in *nominal* dollars, which include the effects of both real cost escalation *and* cost inflation due to reduced dollar purchasing power. If the nominal costs that increase over time are converted into an equivalent constant value, as in an annuity, a *levelized* cost may be determined.[33]

23.21 Wind Plant Development

Four ingredients are necessary for the development of a wind plant to both generate and sell wind power: rights to build and operate the plant on a windy site, a power sales agreement (which usually includes transmission facilities for delivering the power), project financing, and means for project construction.

Land

The first step for a wind plant project is to locate a prospective site with good wind resources and a suitable transmission line nearby. The second step is to obtain rights to develop a wind plant on the land.

The necessary structures and roads for a wind plant typically occupy only about 5% of the land area. Effective wind plant development usually has little long-term effect on farming, ranching, or other compatible land uses.

If the developer succeeds in getting a land agreement, he or she will usually want to begin a wind resource evaluation on the land as described in Chapter 18B and should then plan activities to enlist the support of local authorities and communities, and plan to satisfy any requirements for permits, environmental assessments, and protection of archaeological sites. Any new taxes to be expected as a result of the development must be considered. The developer should attempt to cooperate with local game, wildlife, and bird authorities. Known bird migration routes should be avoided.

In land negotiations the developer and the landowner have a number of objectives. The agreement must attempt to satisfy these objectives and still provide both parties with an attractive business arrangement.[34] The landowner should be compensated to share in the success of the project and be protected in case of project failure. This agreement must continue to protect and effectively serve the interests of both parties for the entire term of the project, usually 30 years or longer. It may be renewable for additional periods on some carefully chosen basis. It is important that the agreement between developer and landowner be crafted to avoid or efficiently resolve confrontational situations that could result in losses to both parties.

A common method of compensating the landowner is payment of a production royalty. For this method there must be a power production reporting system that guarantees the landowner knowledge of how much power was produced during each payment period on the landowner's property. The royalty rate will depend on the sales price of the power produced.

Issues to be covered in this agreement include

- Length of time the agreement will be in effect.
- Payment of royalties and fees.
- Reporting arrangements to ensure that the amount of power produced on the landowner's property is accurately measured.
- Protection of the resource owner from legal and financial liabilities.
- Arrangements covering on-site interconnection and substations.

- Allowance for assignment by the developer and for financing by the landowner.
- Arrangements covering default and termination, including derelict projects and site restoration.
- Arrangements regarding payment of taxes resulting from the development.
- Arrangements to protect the land from unnecessary road construction, erosion, adverse effects on wildlife, to restore land damaged during construction, and to provide for other concurrent uses of the land.

The best relationship between developer and landowner is one of mutual respect and cooperation toward their both profiting from a successful project.

Power Sales Agreement

There must be a buyer for wind-generating energy to be sold, and there must be some means of conveying the power produced to said buyer. The usual commitment to buy wind-generated electrical power is by means of a power sales agreement (PSA) with a utility. The utility usually owns and operates the transmission lines and the power distribution system.

The PSA sets forth terms and conditions under which the utility will purchase power from a wind plant. It needs to be a carefully thought-out document, balancing the needs and obligations of the utility, the wind plant developer, and the financiers of the project. For the developer, it defines the sales price for the power sold and the minimum acceptable power quality, and imposes many constraints and obligations on the seller. For the utility, it defines the amount and quality of the power to be purchased, the variation of price with time of day to meet the utility's load requirements, and the obligations the utility wants to satisfy legally and financially, and imposes a host of financial and legal protections in case of default or disagreements. Utility lawyers work continually to develop language to protect the utility from almost any conceivable problem. Thus the PSA is weighted to satisfy the utility's requirements and protect its interests in case of difficulties.

Of particular import to the seller are the conditions under which the utility can declare an outage. In this case the utility may cause the wind plant to completely or partially shut down. During such outage, the seller's income is temporarily halted, even if the wind is blowing hard and the plant is capable of full output. With good cooperation between the seller and the utility, planned outages can be scheduled at times of expected low winds, and unplanned outages can be corrected quickly.

Financing

The financing of a wind plant will depend upon how the project is envisioned. The plant can be conceived and built by a developer with the intent of retaining ownership and selling power to the utility. A utility may indicate an interest in eventual ownership of the plant or in taking an equity in the plant, or the utility can finance and own the plant with design and execution done by a consulting firm under their direction. In the case considered here the plant is developed and financed with the independent power producer intending to retain and sell power to the utility through a PSA. A typical arrangement for financing a project will be shown in the following example.

Construction

The developer is responsible for planning and executing the project in the most efficient manner. This means getting together all the elements of the project, choosing reliable partners, and managing the work skillfully. As an example of planning, Ken Karas, president and CEO of Zond Systems, Inc., has graciously provided a complete example of a typical project plan. Figure 23.16 shows the development and financing schedule for a 50 MW project.[35] The developer must have a site in mind and begin the project 4 years prior to the start of operation. Since wind plants are modular, construction can take place within a single year. The equity financing begins 1 year prior

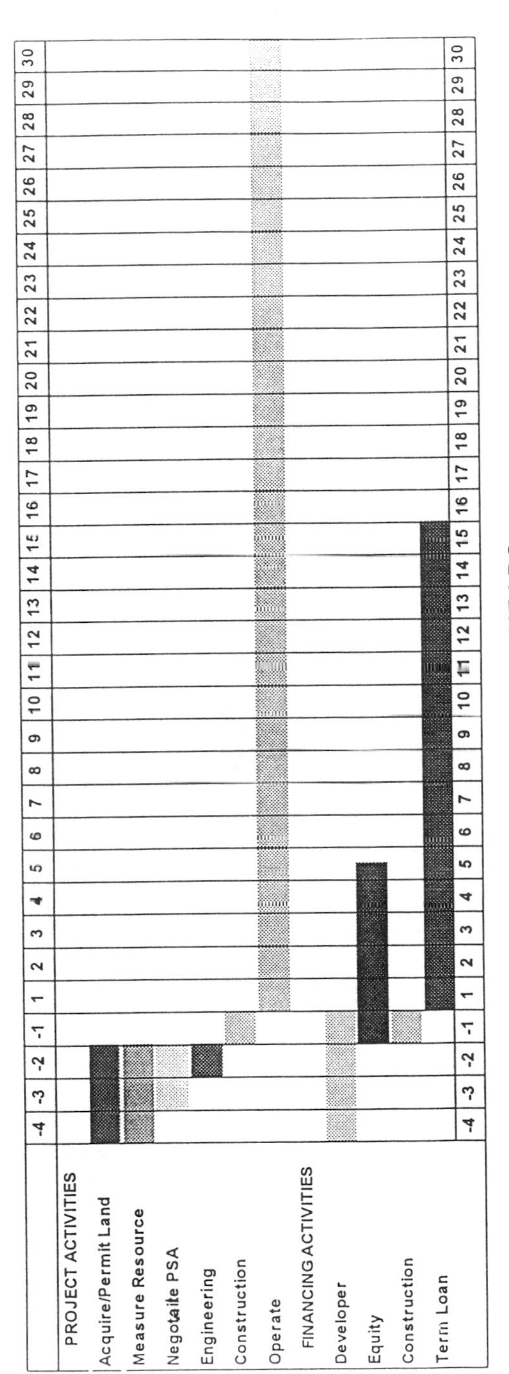

FIGURE 23.16 Fifty-megawatt project: development and financing schedule.

Assumptions:	
Capacity:	50MW
Cost:	$48mm ($950/kW)
Inflation:	4% p.a.
Energy Rate Escalator:	3% p.a.
Capacity Factor:	30%
Energy Rates:	1994 1999 2004 2009 2014 2019 2024
Nominal ¢/kWh	4.5 5.2 6.0 7.0 8.1 9.4 10.9
Levelized ¢/kWh	5.5 @ 11% discount rate
Equity:	30% of project cost
Equity Return:	18% (after tax)
Debt:	70% of project cost
Debt Terms:	
Term:	15 years
Rate:	9.5%
Coverage Ratio	140%
Reserves	6 months
Production Tax Credit:	1.56¢ per kWh

FIGURE 23.17 Project economic information.

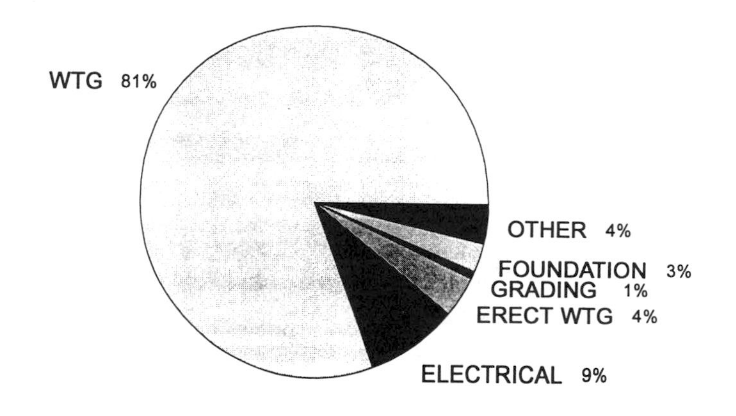

FIGURE 23.18 Project construction costs.

to operation, and lasts for 4.5 years thereafter, while the long-term debt begins on completion of the plant and extends over a period of 15 years. The plant is designed with a 30-year life, and income from the plant then continues for the remaining 15 years.

The assumptions made in laying out the project are shown in Figure 23.17. The nominal energy rates are assumed to increase at 3% per year, starting at 4.5 cents/kWh and reaching 10.9 cents/kWh in 2024, resulting in 5.5 cents/kWh levelized over the 30-year period using an 11% discount rate. The equity is taken to be 30%.

Project construction costs are given in Figure 23.18. The wind turbine generators are assumed to cost 81% of the total. Project operating costs are shown in Figure 23.19, taking the year 2000 as a representative year in the life of the project. The cost breakdown by year is shown in graphical form in Figure 23.20 and in tabular form in Figure 23.21. The largest income rate occurs in the last few years of the project, when all debts have been serviced.

No particular wind plant development is likely to follow this example in all particulars, but the cash flows give a good general idea of how a wind plant project can be accomplished. The interests of the utility, the equity holders, the long-term debt holders, and the project manager are all apparent in the interpretation of the results.

($-000's)	1994	1999	2004	2009	2014	2019	2024
Power Revenues	5,938	6,883	7,980	9,251	10,724	12,432	14,412
Interest on Reserves	0	165	165	165	0	0	0
Total Revenues	5,938	7,049	8,145	9,416	10,724	12,432	14,412
O&M Expense	546	664	808	983	1,196	1,456	1,771
Land Expense	178	207	239	278	322	373	432
Insurance	118	144	175	213	259	315	383
Property Taxes	550	204	165	165	0	0	0
Administration	104	119	136	157	142	173	211
Management Fee	119	138	160	185	214	249	288
Total Expenses	1,616	1,474	1,683	1,980	2,134	2,565	3,085
Pre-debt Cash Flow	4,322	5,574	6,462	7,436	8,590	9,867	11,327
Debt Service	3,087	3,982	4,616	4,579	0	0	0
Reserve Funding	617	0	0	0	0	0	0
Cash Available for Distribution	617	1,458	1,745	2,790	8,725	10,032	11,492
Equity Distribution	563	1,452	1,684	2,606	7,835	8,999	10,300
Developer Distribution	54	140	162	251	756	868	997

FIGURE 23.19 Project economic results.

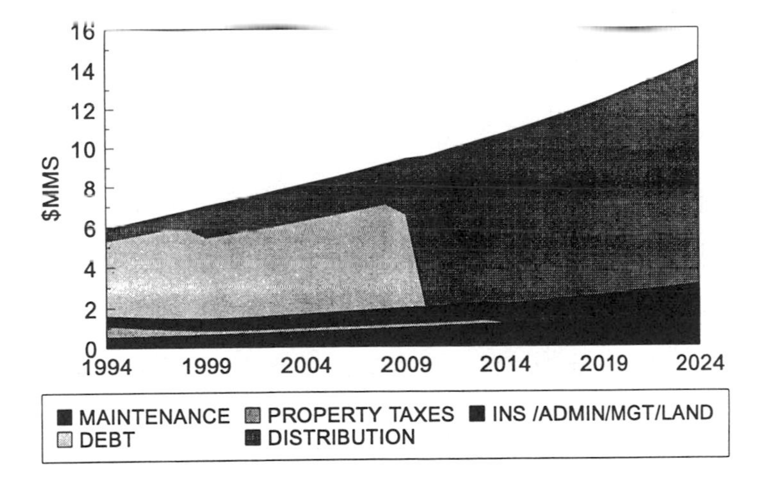

FIGURE 23.20 Project operating costs.

($-000's)	Year 2000	% of Total Revenues	¢/kWh
Power Revenues	7,090	98%	5.4
Interest on Reserves	165	2%	0.1
Total Revenues	7,255	100%	5.5
O&M Expense	691	10%	0.5
Land Expense	213	3%	0.2
Insurance	149	2%	0.1
Property Taxes	176	2%	0.1
Administration	122	2%	0.1
Management Fee	142	2%	0.1
Total Expenses	1,493	21%	1.1
Pre-debt Cash Flow	5,762	79%	4.4
Debt Service	4,116	57%	3.1
Reserve Funding	0	0%	0.0
Cash Available for Distribution	1,646	23%	1.2
Equity Distribution	1,498	21%	1.1
Developer Distribution	148	2%	0.1

FIGURE 23.21 Project economic results overtime.

References and Endnotes

1. NSP Awards Next 100 MW Project to Zond Systems, *Wind Energy Weekly*, v. 14, n. 652, June 26, 1995. p. 1, "the cost of energy for the project is estimated to be 3 cents/kWh levelized over 30 years."
2. About half the machines installed in the 1980s in California were of Danish and other European design.
3. Madsen, Birger T., 1995, Market Update, *WindStats*, v. 8, n. 2, pp. 1–3.
4. Gipe, Paul, 1995, *Wind Energy Comes of Age*, Wiley, p. 2.
5. Madsen, Birger T., reporting a statement by S. Krishna Kumar, Minister for Non Conventional Energy Sources, India, in *WindStats*, v. 8, n. 2, p. 3.
6. Ibid.
7. For details, see Eggleston, D.M., and Stoddard, F.S., 1987, *Wind Turbine Engineering Design*, Van Nostrand Reinhold.
8. Solidity is the ratio of the blade area to the total swept area. Blade twist is ignored in calculating solidity. A number of different definitions of solidity are in use. The local radius is called r and the tip radius R, while B is the number of blades. The *local* solidity, $\sigma_\ell = Bc\Delta r/2\pi r\Delta r = Bc/2\pi r$ is the fraction of the annulus occupied by blades. The *rectangular* solidity α assumes blades of constant chord c all the way to the axis, giving $B(Rc)/\pi R^2$. The *total* solidity is the ratio of the total blade area to the swept area, or $B\int c(r)dr/\pi R^2$.
9. Analysis done originally by Wilson, R.E., and Lissaman, P.B.S., 1974, *Applied Aerodynamics of Wind Power Machines*, Oregon State University NSF report. See also Rohatgi, J.S., and Nelson, V., 1994, *Wind Characteristics, an Analysis for the Generation of Wind Power*, Alternative Energy Institute, West Texas A&M University.
10. Early theories of the screw propeller were developed in connection with ship propulsion by Rankine and W. Froude, and refined by R.E. Froude. See Durand, W.F., Ed., *Aerodynamic Theory, Vol. IV, Division L (by H. Glauert), Airplane Propellers*, pp. 170–360. See also Glauert, H., *The Analysis of Experimental Results in the Windmill Brake and Vortex Ring States of an Airscrew*, Brit. R & M 1026, 1926.

11. For details, see Eggleston, D.M., and Stoddard, F.S., 1987, *Wind Turbine Engineering Design*, Van Nostrand Reinhold, p. 54, Dual Optimum.
12. Gipe, P., Searching for variable speed performance improvements, *WindStats*, v. 8, n. 4, Autumn, 1995, pp. 1–2.
13. Carr, L.W., McAlister, K.W., and McCroskey, W.J., 1977, Analysis of the development of dynamic stall based on oscillating airfoil experiments. NASA Ames Research Center, Moffett Field, CA, Jan. 1977. In a phone contact Dale Berg of Sandia Lab's reports effects in high winds on the DOE 34m turbine at Bushland, TX most likely due to dynamic stall.
14. Las Gaviotas Tropical Windmill, Double Effect. "Las Gaviotas", Bogota 5, D.E. Colombia S.A., P.O. Box A.A. 18261. This little wind pump can operate in winds as low as 3 mph to pump small quantities of water from wells up to 22.8 m (75 ft) in depth.
15. Miller, Rene, and Dugundji, J. et al., 1978, *Wind Energy Conversion*, MIT Aeroelastic and Structures Research Lab TR-184-8, Aerodynamics of horizontal axis wind turbines, Cat. UC-60, Sept. See also Eggleston & Stoddard, 1987, *Wind Turbine Engineering Design*, Van Nostrand Reinhold, p. 79.
16. Spera, D.A., ed. 1994, *Wind Turbine Technology*, Chapter 9, by Sengupta, D.L., & T.B.A. Senior, ASME Press, p. 447–486.
17. Tangler, J.L., and Somers, D.M., 1985, "Advanced Airfoils for HAWT's," *Proceedings, Wind Power '85 Conference*, American Wind Energy Association, pp. 45–51.
18. Tangler, J., Smith, B., et al., 1990, SERI thin-airfoil blade atmospheric performance test: final results, *Proceedings, Windpower '90*, AWE, Sept. 24–28, Washington, DC, pp. 118–125.
19. H.P. Klein and B. Canter, Eurowin reports improved turbine performance, *WindStats*, v. 8, n. 3, 1995.
20. Ibid.
21. For details, see Eggleston, D.M., and Stoddard, F.S., 1987, *Wind Turbine Engineering Design*, Van Nostrand Reinhold, pp. 205–211.
22. Scheda, F.A., and Helmick, C.G., 1985, Harnessing self-excitation for wind turbine induction generators. *Proceedings, Fourth ASME Wind Energy Symposium*, February 17–21, Dallas, TX.
23. For details, see Eggleston, D.M., and Stoddard, F.S., 1987, *Wind Turbine Engineering Design*, Van Nostrand Reinhold, p. 317.
24. Putnam, Palmer C., 1948, *Power from the Wind*, Cost reductions by means of major modifications of the 1945 design, flaps, Van Nostrand Reinhold, p. 202.
25. See Gipe, Paul, 1995, *Wind Energy Comes of Age*, Chapter 4, Death knell of the giants, Wiley, pp. 96–115.
26. Hubbard, H.H., and Shepherd, K.P., 1994, *Wind Turbine Acoustics*, Chapter 7 of *Wind Turbine Technology*, edited by D.A. Spera, ASME Press. See also, Moroz, E., *Experimental and Theoretical Characterization of Acoustic Noise from a Yaw Controlled Teetered Rotor Wind Turbine*, M.S. Thesis, Univ. of Texas, El Paso, July, 1995.
27. Ibid, p. 338.
28. Comment by T.U. Electric executive J.B. Headrick, 1995, regarding operational experience of a demonstration wind plant, at EPRI/DOE Turbine Verification Program Review, Ft. Davis, TX, Sept. 28, 1995.
29. Joost van Kastersen, 1995, Bird Society Satisfied, *Wind Power Monthly*, v. 11, n. 9, Sept. 1995, pp. 23–24.
30. Davidson, Ros, 1995, Gaining a better understanding, *Wind Power Monthly*, v. 11, n. 6, June, 1995.
31. See Karas, K.C., 1992, Wind energy: what does it really cost?, *Proceedings, AWEA Windpower '92*, Seattle, WA, pp. 157–166.
32. Ibid, p. 160.

33. For a more detailed discussion, see Gipe, Paul, 1995, *Wind Energy Comes of Age*, Chapter 7, Wind energy's declining costs, Wiley, pp. 226–248.
34. Tinkerman, Randall M., 1994, "Wind Rights Acquisition: Lessons From Altamont Pass," *AWEA Proceedings, Windpower '94*, May 10–13, Minneapolis, MN, pp. 314–345.
35. Karas, Ken, 1994, "Wind Power Economics," AWEA Wall Street Workshop, presentation slides, Nov. 10.

24

Waste-to-Energy Combustion

Charles O. Velzy
Private Consultant

Roger E. Hecklinger
Roy F. Weston, Inc.

Introduction

One of the most serious issues facing urbanized areas today is development of cost-effective environmentally acceptable disposal of the community's solid waste. The solid waste generated in a community may be collected by private companies or governmental entities, or portions by both, but the assurance that the waste is ultimately disposed of in an environmentally safe manner is a governmental responsibility.

Solid waste management has become a major concern in the United States, largely because of the declining number of landfills available and the increasing concern with environmental problems. These developments have emphasized the possibility of utilizing municipal solid waste, which, for all practical purposes is a renewable commodity, for the generation of electricity. An analysis by Penner and Richards[49a] showed that incineration of municipal waste, even after 30% of the waste was recycled, could provide as much electric power as eight large nuclear or coal generating stations by the end of the century. Their analysis further concluded that this could provide 1 to 2% of the total electric energy needs in the United States at prices competitive with coal fired base load power plants.

The basic technology for modern waste-to-energy combustion was developed in Europe during the 1960s and 1970s. This technology, which has been modified and improved since its development, has been widely implemented in the United States. However, despite the fact that incineration of solid waste can decrease its volume ninefold and ameliorate the final waste disposal into landfills, the full potential of utilizing solid waste for energy production is not being realized because of widespread fears regarding environmental pollution. In preparing this chapter, the realities of the situation have been taken into account and the discussion emphasizes the prevention of pollution

0-8493-2514-5/96/$0.00+$.50

as much or more than the production of power from waste. Waste-to-energy combustion in modern facilities with adequate environmental safeguards and careful monitoring has been shown to be a safe and cost-effective technology that is likely to increase in importance during the next decade.

Two conditions usually point to the use of combustion processes in treating municipal solid waste prior to ultimate disposal: the waste is collected in an urbanized area with little or no conveniently located land for siting of sanitary landfills (need for volume reduction); and markets exist for energy recovered from the combustion process, and possibly for reclaimed materials, with the energy attractively priced.

Modern waste-to-energy (WTE) plants reflect significant advances that have been made in addressing the technical and practical difficulties of material handling, combustion control, and flue gas clean-up. In the early days of waste incineration, when air pollution regulations were undemanding or nonexistent, relatively simple, fixed-grate plants operating on a single- or two-shift basis were common. However, with increasingly stringent air pollution control regulations, more complex plants requiring continuous operation are now being built.

24.1 Waste Quantities and Characteristics

Municipal solid waste (MSW) as used herein refers to solid waste collected from residences and commercial and institutional waste. It does not include industrial waste, which is another problem and varies widely in quantity and characteristics, depending on the industry and specific industrial plant. Changes in packaging practices and improvements in the general standard of living have resulted in significant increases in the quantities of solid waste generated over the past 50 years. Additionally, increasing emphasis on and participation in the recycling of wastes by local communities has resulted in significant variations in quantity and characteristics of MSW at the local community level. All of these factors must be considered when planning a WTE facility. Chapter 18C gives more information on waste availability.

Waste Quantities

In the United States approximately 160 million tons of MSW were generated in 1983, increasing to 207 million tons in 1993.[23] MSW generation is projected to increase to 218 million tons by the year 2000. At the local level the quantity of solid waste generated varies geographically, daily, and seasonally, according to the effectiveness of the local recycling initiatives, and differences in socio-economic conditions.[56]

Over the past 30 years, numerous studies by EPA,[21] APWA,[1] and others[60] have indicated that urbanized areas in this country generate approximately 2.0 pounds per capita per day of MSW from residences and another 2.0 pounds per capita per day from commercial and institutional facilities, on a national average basis. Thus, a typical community of 100,000 inhabitants would generate about 200 tons per day of gross MSW discards averaged over a 1-year time period.

These projections are subject to adjustments related to specific community characteristics. Thus, communities in the south, with longer active growing seasons than those in the north, tend to produce and collect more yard waste. Recent requirements for on-site disposal and/or composting of yard waste is changing this variable. Rural communities tend to produce less waste per capita than highly urbanized areas because of their greater potential for on-site waste disposal. In the past, the communities in the north tended to produce more waste in the winter due to the prevalence of heating of homes with solid fuels, which produced large quantities of ashes for disposal. Variations in yard wastes and ashes produced from home heating with solid fuels are also examples of variations in MSW quantities related to seasonal effects. Seasonal variations in MSW generation have been noted to range ±15% from the average, while daily variations in waste collections may range up to ±50% from the average, depending largely on number of collections per week. Daily variations in waste quantities are more important in designing certain plant components, while geographic and seasonal effects are more important in establishing plant size.

Waste quantities are also affected by the effectiveness of local recycling initiatives and by socio-economic conditions. EPA studies have indicated an increase, nationally, in waste recycling from 6.6% in 1960–1970, to 10.7% in 1986, and 21.7% in 1995.[22,23] The national recycling rate is projected to increase to about 30% by the year 2000, at which time it is expected to level off. A community in New England[6] projected a 14.0% drop in MSW generation between 1989 and 1991 due to recycling activities in the community, and a 6.0% drop due to the recession during this period in this region of the country. A 15% drop in MSW collections was noted in a Long Island community during a recessionary period in 1972. The impact of recycling on MSW generation should be considered in plant sizing, while the impact of recessionary periods on plant economics may require specific consideration during project planning.

Waste Characteristics

It is important in approaching the design of a WTE facility that one consider the potential variations in both physical and chemical composition of MSW. Historically, one of the most troublesome areas in WTE plants has been materials handling systems. To successfully select materials handling system components and design an integrated process, one must have adequate information on the variability and extremes of the physical size and shape the solid waste facility must have, the bulk density and angle of repose of the material, and the variation in noncombustible content. This information generally is not available from published surveys and reports, and can only be secured through inspection of the MSW in the field. If materials handling facilities for refuse feeding and residue handling are not large enough and oriented properly to pass the largest bulky items in the MSW, or large enough and rugged enough to handle the quantities of materials required to meet plant design capacity, the plant will experience expensive periods of down time and might have to be derated.

In the design of the furnace/boiler portion of WTE facilities the refuse characteristics are the calorific value, moisture content, proportion of noncombustibles, and other components (such as heavy metals, chlorine, and sulfur) whose presence during combustion will result in the need for flue gas clean-up. The capacity of a WTE furnace boiler is roughly inversely proportional to the calorific or heating value of the waste. Table 24.1 illustrates the variation in waste characteristics that has been observed in studies defining the average solid waste composition in the United States over the past 30 years.

TABLE 24.1 Waste Characteristics (in percent)

		U.S. Average				
	Oceanside, NY[35]	U.S. EPA[62]	U.S. EPA, 1986[22]		U.S. EPA, 1993[23]	
	1966–1967	1977	Generated	After Recycle	Generated	After Recycle
Paper Materials	32.72–53.33	35.0	41.0	35.6	37.6	31.7
Plastics	2.45–8.82	3.8	6.5	7.3	9.3	11.5
Rubber & Leather	—	—	2.5	2.8	3.0	3.6
Textiles	2.24–3.97	4.3	1.8	2.0	2.9	3.3
Garbage (or Organics)	7.23–16.70	14.9	7.9	8.9	6.7	8.5
Wood & Lumber	1.22–6.58	3.8	3.7	4.1	6.6	7.7
Yard Wastes*	0.26–33.33	16.3	17.9	20.1	15.9	16.2
Non-combustible	22.47–14.36	21.9	18.7	19.2	18.0	17.5
Total		100.0	100.0	100.0	100.0	100.0
Recycled	—	6.6	10.7		21.7	

* Including grass, dirt, and leaves.
Note: See indicated references for futher detail.

TABLE 24.2 Variation in Heat Content of MSW

	15		25	
Noncombustible, %	Comb. %	Heat Cont. Btu/lb	Comb. %	Heat Cont. Btu/lb
Moisture %				
20	65	6.125	55	5.225
30	55	5.225	45	4.275
40	45	4.275	35	3.325
50	35	3.225	25	2.375

As indicated in Table 24.1, approximately 40 to 45% of the combustible fraction of MSW is composed of cellulosic material such as paper and wood. This percentage has remained relatively constant over the past 30 years, even after taking into account the threefold increase in recycling percentage achieved over this period. The remainder of the combustible content is composed of various types of plastics, rubber, and leather. The heat released by burning cellulose is approximately 8,000 Btu/lb (on a dry basis), while that released by plastics, rubber, and leather is significantly higher on a per pound, dry basis. Heat released by burning garbage (on a dry basis) is only slightly less than cellulose. However, the moisture content of garbage has been observed to range from 50 to 75%, by weight, while that of the cellulosic fraction of MSW usually ranges from 15 to 30%.

In recent years, it has been observed that the higher heating value (HHV) of the combustible portion of MSW (moisture and ash free) averages about 9,400 Btu/lb. Considering the recent changes in MSW composition following recycling (increase in plastics while cellulosic material has remained relatively constant), this moisture and ash free HHV has probably increased to 9,500 Btu/lb. Taking 9,500 Btu/lb as the moisture and ash free heat content of MSW, the following Table 24.2 illustrates the variation in as-received heat content that one could expect in MSW with moisture content ranging from 20 to 50% by weight and noncombustible content ranging from 15 to 25% by weight.

Moisture content is a highly important and also a highly variable characteristic of waste materials. The moisture content of MSW has been observed to vary from 15 to 70%. This variation may be due, for example, to seasonal variations in precipitation, the nature of the waste (e.g., grass clippings vs. paper) and the method of storage and collection (e.g., open versus closed containers/trucks). Thus, after a heavy rain, the moisture content of the solid waste may be so high that it may be difficult to sustain combustion. The combustion of solid waste usually can proceed without supplementary fuel when the heat value is greater than 3,500 to 4,000 Btu/lb (8,140 to 9,300 kJ/kg). This type of variation in MSW composition must be considered in the design of WTE facilities.

The ultimate or elemental analysis of the combustible portion of the MSW refers to the chemical analysis of the waste for carbon, hydrogen, oxygen, sulfur, chlorine, and nitrogen. This information is used to estimate the heat content of waste, moisture, and ash free; to predict the composition of the flue gases; and, from the last three elements (sulfur, chlorine, and nitrogen), to assess the possible impact of waste combustion on air pollution. A typical analysis of solid waste is presented in Table 24.3.

24.2 Design of WTE Facilities

The primary function of a WTE facility is to reduce solid waste to an inert residue with minimum adverse impact on the environment. Thermal efficiency, in terms of maximizing the capture of energy liberated in the combustion process, is of secondary importance. WTE facilities are usually classified as mass burn systems, or refuse-derived fuel (RDF) systems.

TABLE 24.3 Analysis of Solid Waste

	Percent by Weight	
	West Europe*[,58]	U.S.[19,57]
Proximate Analysis		
Combustible	42.1	50.3
Water	31.0	25.2
Ash and inert material	26.9	24.5
Total	100	100
Ultimate (Elemental) Analysis of Combustibles		
Carbon	51.1	50.9
Hydrogen	7.1	6.8
Oxygen	40.1	40.3
Nitrogen	1.2	1.0
Sulfur	0.5	0.4
Chlorine	—	0.6
Total	100	100

* Gross heat value, as fired = 9,000 kJ/kg.

General Features

Types of Facilities

Mass-burn systems are large facilities (usually over 200 tons per day) that burn, as received, unprocessed MSW, which is extremely heterogeneous. Most mass burn systems burn the waste in a single combustion chamber under conditions of excess air (i.e., more than is needed to complete combustion) (see Figure 24.1). The waste is burned on a sloping, moving grate, which helps agitate the MSW and mixes it with combustion air. Many different proprietary grate systems exist.

In refuse-derived fuel (RDF) systems, usually large facilities, the MSW is first processed (see Figure 24.2) by mechanical means to produce a more homogeneous material prior to introduction into a furnace/boiler. Several types of RDF can be made—coarse, fluff, powder, and densified. These differ in complexity and horsepower requirements of the waste processing facilities, size of particle produced, and whether or not the material is compacted under pressure into pellets, briquettes, or similar forms. The coarse type of RDF is the most common form produced at this time.

RDF can be burned in one of two types of boilers. It can be used as the sole or primary fuel in dedicated boilers (see Figure 24.3) or it can be cofired with conventional fossil fuels in existing industrial or utility boilers. One advantage of these systems is that RDF can be produced at one location for use at a nearby off-site boiler, allowing for flexibility in locating processing facilities. Also, some materials, such as steel and glass, can be recovered for recycling during the initial processing step.

Mass-burn and RDF systems together account for about 90% of the current and planned incineration capacity. Modular units, described briefly later, make up the other 10% of incineration capacity.

Operation and Capacity

The capacity to be provided in a facility is a function of the number of shifts (one, two, or three) the plant is to operate; the area and population to be served; and the rate of refuse production for the population served. A small plant (100 tons per day) without energy recovery might be operated one shift per day. For capacities above 400 tons per day, or any plant with energy recovery, economic and/or equipment operating considerations usually dictate three-shift operation.

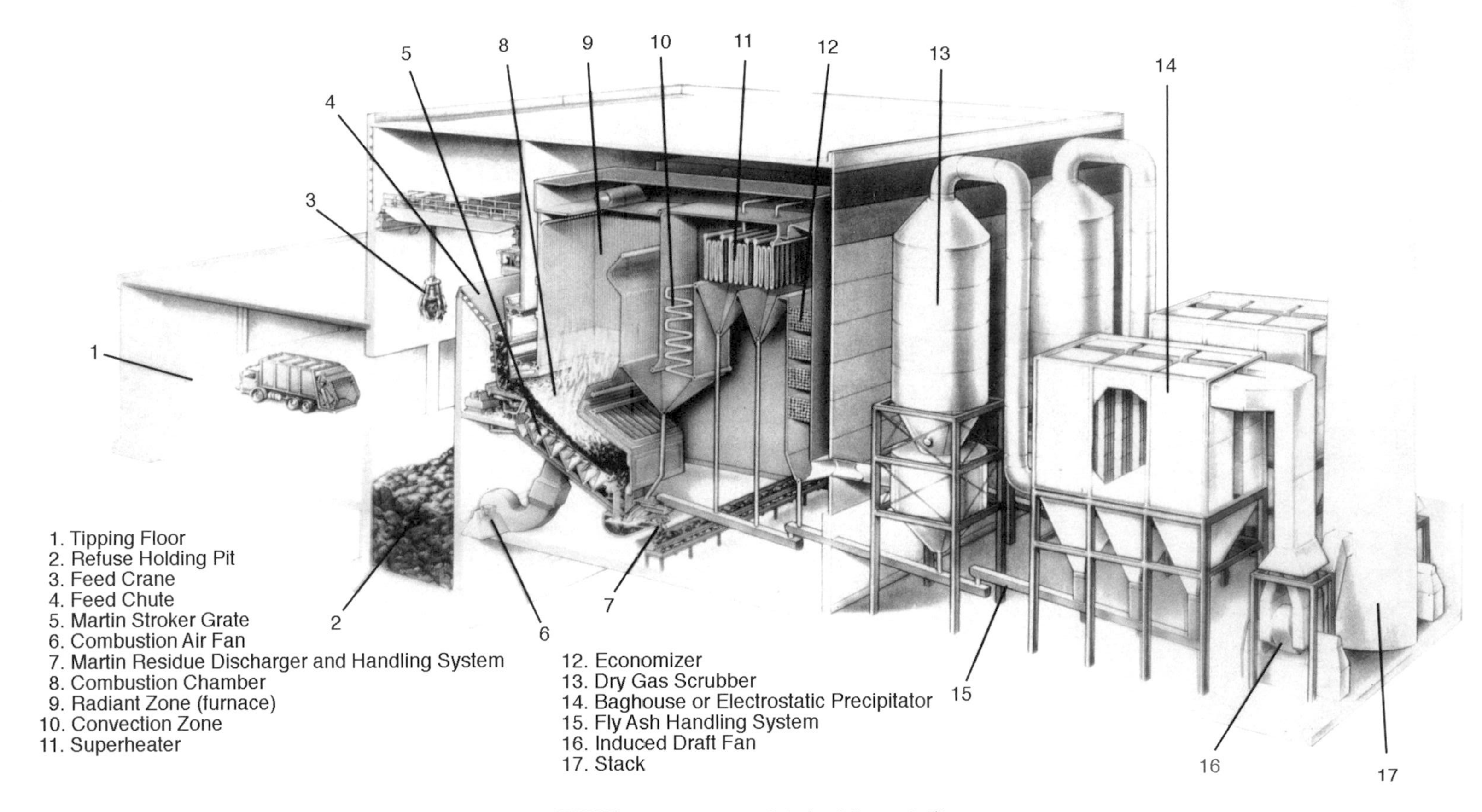

FIGURE 24.1 A typical Ogden Martin facility.

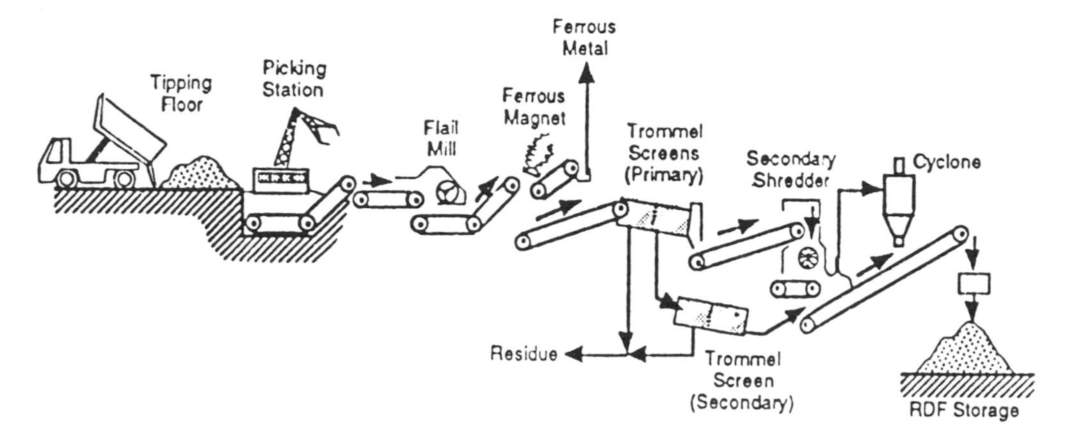

FIGURE 24.2 RDF processing system.

If collection records, preferably by weight, are available for the community, forecasts for determining required plant capacity can be made with reasonable accuracy. If records are not available, refuse quantities for establishing plant size may be approximated by assuming a refuse generation rate of 4.0 pounds per capita per day when there is little or no waste from industry. Of course, in planning for plant capacity, the impact of local recycling activities on both quantity and characteristics of MSW must be considered as discussed earlier.

Other factors must be taken into account in establishing the size or capacity of a facility. Should the facility serve only one community, or should it be regional and serve several communities? What are the possible benefits of economies of scale? What is the impact of the cost of hauling refuse to a central point on overall project economics? There is substantial evidence available at present to show that implementation becomes much more difficult as the number of separate political jurisdictions is increased. Imposition of regional plans on local jurisdictions to achieve economies of scale, where it cannot be conclusively demonstrated that such regional plans make sound economic sense based on the total cost of the solid waste management plan, including the cost of transporting the solid waste to the regional facility, is, at best, unwise. Economies of scale in these projects have tended to be illusory, while haul costs to gather sufficient waste together to achieve the economies of scale have tended to be ignored in developing total project economics.

Siting

One of the key issues to face in implementing a WTE project is locating a site for the facility. Since MSW is usually delivered to these plants by truck, inevitably there will be substantial truck traffic in the vicinity of the plants. The equipment and processes used in these plants are industrial in nature. They are generally noisy at the source and tend to produce dust and odors. These facts indicate that it is desirable to site such plants in industrial or commercial areas.[41] It has been contended, as cited in a 1989 OTA report,[61] "that sites are sometimes selected to avoid middle- and higher-income neighborhoods that have sufficient resources to fight such development."

Plants should be located near major highways to minimize the impact from increased truck traffic. As shown by operating WTE plants in Europe and the United States, it is possible to control all nuisance conditions by proper attention to the details of plant design. The local impact of truck deliveries to the plant can be minimized by providing sufficient length of access road so that refuse truck queuing does not take place on public highways. Odors and noise can be confined to the plant building. Odors and fugitive dust can be destroyed by collecting plant air and using it for combustion air supply. Noise should be attenuated at the source to maintain healthful working conditions. In all cases, there is no need to adversely impact the surrounding neighborhood. Proper attention to architectural treatment can result in a structure that is pleasing in appearance.

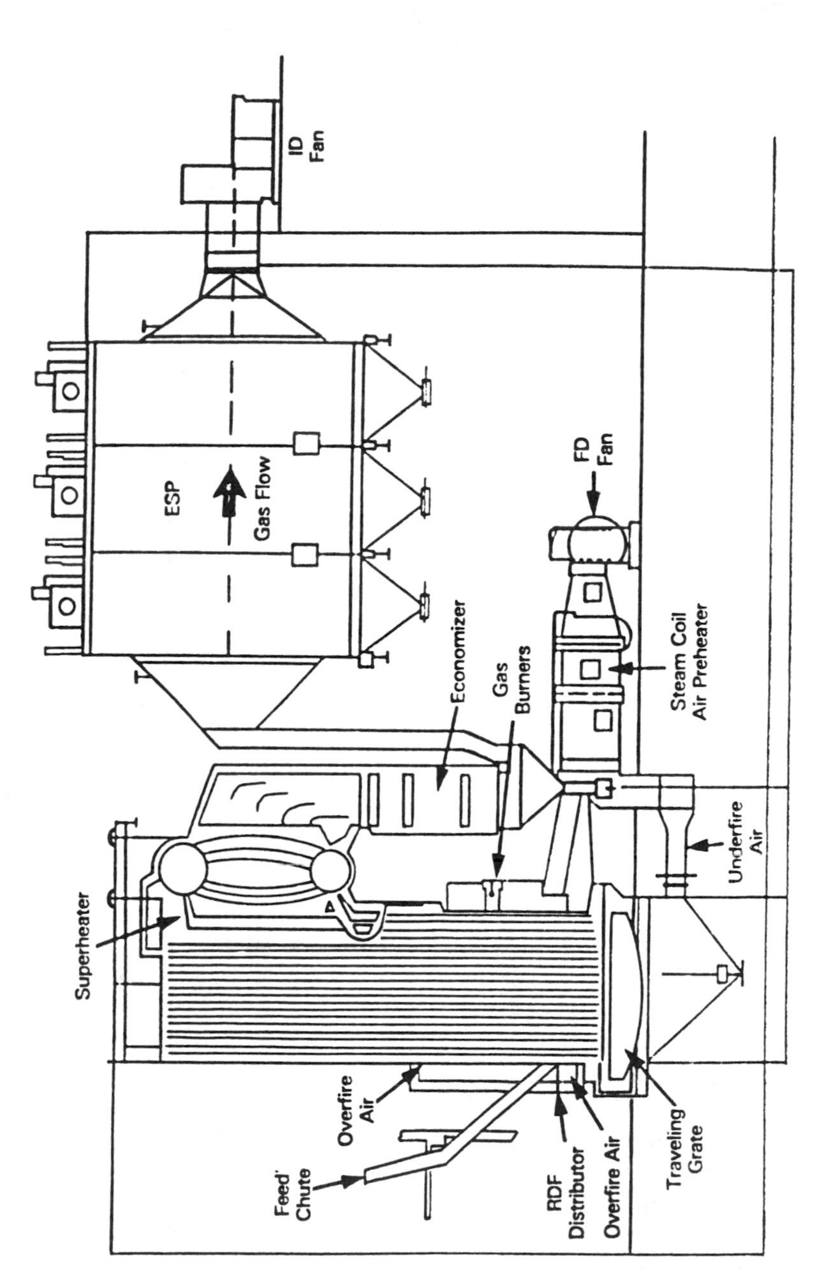

FIGURE 24.3 Albany RDF fired boiler.

Since considerable vertical distance is frequently required in the passage of MSW through a mass burn WTE plant, there is an advantage in a sloping or hillside site. Collection trucks can deliver MSW at the higher elevation, while residue trucks operate at the lower elevation, requiring minimum site regrading. This consideration does not generally apply to RDF plants.

Feed Handling

Refuse Receipt, Processing, and Storage

Scales, preferably integrated into an automated record keeping system, should be provided to record the weight of MSW delivered to the WTE plant. Either the entire tipping area or individual tipping positions should be enclosed so as to control potential nuisance conditions in the vicinity of the plant, such as blowing papers, dust, and/or odors. The number of tipping positions provided should take into consideration the peak number of trucks expected per hour at the facility and should be located so the trucks have adequate time and room to maneuver to and from the dumping positions while minimizing queue time.

Collections usually are limited to one 8-hour daily shift 5 days per week (sometimes with partial weekend collection) while burning will usually be continuous, so ample storage must be provided. This usually requires 2 to 3 days of refuse storage at most WTE plants. Seasonal and cyclic variations should also be a consideration in establishing plant storage requirements.

Refuse storage in large mass burn plants is normally in long, relatively narrow and deep pits either extending along the front of the furnaces, or split in two halves extending from either side of the front end of the furnaces. If the storage pit is over 25 feet in width, it will generally be necessary to rehandle the refuse dumped from the trucks. In smaller mass burn plants and in RDF plants, floor dumping and storage of the MSW either on the dump floor or in shallow, relatively wide pits is common practice.

When computing the dimensions required for storage of as received MSW, the required volume may be determined based on an MSW bulk density of from 400 to 600 lb/yd^3 (240 to 360 kg/m^3).[1,56a] Other factors to consider in sizing and laying out refuse storage facilities in WTE plants is that refuse flows very poorly and can maintain an angle of repose greater than 90°. Thus, MSW is commonly stacked in storage facilities to maximize storage capacity.

Sizing a refuse storage pit requires use of empirical data, judgment, and knowledge of plant layout and operations. The process is best explained through an example. The following assumptions are made:

Plant capacity = 600 tons/day (T/D), 24 hr/day, 7 days/week.
Refuse collection = 5 days/week in trucks holding 5 tons of waste.
Fifty percent of the waste is delivered in 2 peak periods of 90 min. each day.
Average turnaround time per truck of 10 min = 6 trucks per hour at each tipping position.
Bulk density of waste stored in pit = 500 lb/yd^3 (4 yd^3/ton of waste).
3 days storage to be provided in pit.
Pit depth is 35 ft from tipping floor (usually varies from 30 to 45 ft).
Pit width is 35 ft (usually varies from 35 to 50 ft).
Distance from pit bottom to charging floor is 100 ft (this is a function of plant layout and equipment arrangement).

The pit should be at least long enough to provide sufficient truck tipping positions on one side so that the trucks are not unduly delayed in discharging their waste (tipping) into the pit. It has been found in practice that it takes an average of 10 mins for the trucks to perform the tipping operation. Also, each tipping position must provide at least 13.5 ft of unobstructed width for this operation. The furnaces are located near the high wall on the opposite side of the pit. A 600 T/D plant would include two furnace units. Experience indicates that the pit length should be at least 70 ft to accommodate these two units. The length required to provide the proper number of tipping positions is determined as follows:

Daily waste delivery: 600 tons × 7 days/5 days = 840 tons/day
Number of trucks/day: 840 tons × 1 day/5 tons/truck = 168 trucks/day
Trucks/peak period: 162/2 × 1/2 peaks = 42 trucks/peak period
Trucks/hr during peak: 42 × 60/90 = 28 trucks/hr
Tipping positions needed: 28/6 trucks @ position = 4.7; say 5 tipping positions

Provide 20 ft for each tipping position to allow for convenient truck access and space for armored building support columns. Thus, the pit length would be 100 ft. This is greater than the 70 ft minimum required to accommodate the furnaces.

Pit storage capacity is computed by calculating the entire volume of the pit below the tipping floor and half the volume above the tipping floor. Half the pit width above the tipping floor allows the grapple access to move the waste away from the tipping positions and reflects the fact that the waste will maintain a vertical face when stacked. The adequacy of the pit to hold 3 days of plant operating waste is checked as follows:

Volume required: 600 T/D × 4 yd^3/T × 3 days = 7200 yd^3
Storage below tipping floor: 35 × 35 × 100 × 1/27 = 2537 yd^3
Storage above tipping floor: (along furnace wall) 7200 – 4537 = 2663 yd^3
Height of storage along furnace wall: 35/2 × 100 × H + 2663 × 27; H = 71,901/1750 = 41.1 ft

Thus, 3 days' waste storage can be accommodated in this pit by filling the pit below the tipping floor and stacking waste for half the width of the pit to a height of just over 41 ft along the furnace wall of the pit, which rises 65 ft above the tipping floor. This is a satisfactory design.

At most newer and more successful RDF plants, after receiving the MSW in a floor dump type of operation, the MSW is loaded onto conveyors that carry the material to flail mills or trommels with bag-breaking blades. These facilities break apart the bags containing the waste, allowing glass and some metals to be separated from the remaining MSW. The separated MSW, primarily the light combustible fraction, is then reduced in size. Removal of the glass prior to the size reduction process alleviates the problem, experienced in earlier plants, of contamination of combustible material with glass shards.

Processes to produce powdered fuel or RDF fuel pellets, although interesting, have not been developed to a state of commercial availability. Some limited work is still being undertaken to improve the economics and operability of such systems. However it appears that commercial availability, if this ever occurs, is many years in the future. A process to produce RDF by "hydro pumping," after being attempted in two full-scale plants, was not commercially successful.

Refuse Feeding

Batch feeding of MSW, practiced in the past in mass-burn plants, is undesirable and is not practiced in modern plants. In the larger mass burn plants, the solid waste is usually moved from the storage pit to a charging hopper by a traveling bridge crane and a grapple or orange-peel type of bucket. The grapple or bucket size is established by a duty cycle analysis, taking into account the quantity of material that must be moved from the pit to the furnaces, the distances over which the material must be moved, allowable crane speeds, and the need to rehandle (mixing and/or stacking) material in the pit. Buckets or grapples can range in size from 1.5 to 8 yd^3 (1 to 6 m^3) capacity and larger.

The crane used in this service should be capable of meeting the severest of duty requirements.[49] The load lifting capability is established by adding to the bucket or grapple weight, 1.5 times the volumetric capacity of the bucket times a density of MSW of 600 to 800 lb/yd^3 (360 to 480 kg/m^3).[35] In the past, the crane has been operated from an air conditioned cab mounted on the bridge. However, in many European plants, and with increasing frequency in the United States, crane operation is being centralized in a fixed control room, usually located at the charging floor elevation and either over the tipping positions opposite the charging hoppers or in the vicinity of the charging hoppers.

In modern mass-burn plants the MSW is deposited from the crane grapple into a charging hopper. The charging hopper, which is built large enough to prevent spillage on the charging floor and with slopes steep enough to prevent bridging, is placed on top of a vertical feed chute that discharges the MSW into the furnace. The feed chute may be constructed of water cooled steel plates or steel plates lined with smooth refractory material. The chute is normally at least 4 ft (1.2 m) wide, to pass large objects with a minimum of bridging, and 8 to 14 ft (2.4 to 4.2 m) long. It is kept full of refuse to prevent uncontrolled admission of air into the furnace. The refuse is fed from the bottom of the feed chute into the furnace by a portion of the mechanical grate, or by a ram. The ram generally provides better control of the rate of feed into the furnace than the older technique of using a portion of the mechanical grate for refuse feed.

In RDF plants, conveyors, live bottom bins, and pneumatic handling of the size-reduced MSW combustible material has been utilized. The fuel material is usually blown into these furnaces, where it is either burned on spreader stokers or partially burned while in suspension, with combustion being completed on grates at the bottom of the furnace. These fuel feeding systems are generally more complex than the mass burn systems and have experienced problems related to properly sizing the equipment, plugging of the feeding system, and higher than expected maintenance.

Combustion Furnaces

Combustion Principles

Combustion is the rapid oxidation of combustible substances with release of heat. Oxygen is the sole supporter of combustion. Carbon and hydrogen are by far the most important of the combustible substances. These two elements occur either in a free or combined state in all fuels—solid, liquid, and gaseous, Sulfur is the only other element considered to be combustible. In combustion of MSW, sulfur is a minor constituent with regard to heating value. However, it is a concern in design of the air pollution control equipment. The only source of oxygen considered here will be the oxygen in the air around us.

Table 24.4 displays the elements and compounds that play a part in the combustion process. The elemental and molecular weights displayed are approximate values which are sufficient for combustion calculations. Nitrogen is listed as chemical nitrogen N_2, with a molecular weight of 28.0 and as atmospheric nitrogen, N_{2atm}, which is a calculated figure to account for trace constituents of dry air. Water occurs as a vapor in air and in the products of combustion and as a liquid or vapor constituent of MSW fuel.

TABLE 24.4 Elements and Compounds Encountered in Combustion[33a]

Substance	Molecular Symbol	Molecular Weight	Form	Density (lb per ft³)
Carbon	C	12.0	Solid	—
Hydrogen	H_2	2.0	Gas	0.0053
Sulfur	S	32.1	Solid	—
Carbon Monoxide	CO	28.0	Gas	0.0780
Methane	CH_4	16.0	Gas	0.0424
Acetylene	C_2H_2	26.0	Gas	0.0697
Ethylene	C_2H_4	28.0	Gas	0.0746
Ethane	C_2H_6	30.1	Gas	0.0803
Oxygen	O_2	32.0	Gas	0.0846
Nitrogen	N_2	28.0	Gas	0.0744
Nitrogen-Atmos.	N_{2atm}	28.2	Gas	0.0748
Dry Air		29.0	Gas	0.0766
Carbon Dioxide	CO_2	44.0	Gas	0.1170
Water	H_2O	18.0	Gas/liquid	0.0476
Sulfur Dioxide	SO_2	64.1	Gas	0.1733

A U.S. standard atmosphere of dry air has been defined as a mechanical mixture of 20.947% O_2, 78.086% N_2, 0.934% Ar (argon), and 0.033% CO_2 by volume.[15] The percentages of argon and carbon dioxide in air can be combined with chemical nitrogen to develop the following compositions of dry air by volume and by weight that can be used for combustion calculations:

Constituent	% by Volume	% by Weight
Oxygen, O_2	20.95	23.14
Atmospheric nitrogen, N_{2atm}	79.05	76.86

Atmospheric air also contains some water vapor. The level of water vapor in air, or its humidity, is a function of atmospheric conditions. It is measured by wet and dry bulb thermometer readings and a psychrometric chart. If specific data are not known, the American Boiler Manufacturers Association recommends a standard of 0.13 pounds of water per pound of dry air, which corresponds to 60% relative humidity and a dry bulb temperature of 80°F.

Table 24.5 displays the chemical reactions of combustion. These reactions result in complete combustion; that is, the elements and compounds unite with all the oxygen with which they are capable of entering into combination. In actuality, combustion is a more complex process in which heat in the combustion chamber causes intermediate reactions leading up to complete combustion. An example of intermediate steps to complete combustion would be when carbon reacts with oxygen to form carbon monoxide and, later in the combustion process, the carbon monoxide reacts with more oxygen to form carbon dioxide. The combined reaction produces precisely the same result as if an atom of carbon combined with a molecule of oxygen to form a molecule of carbon dioxide in the initial reaction. An effectively controlled combustion process results in well less than 0.1% of the carbon in the fuel leaving the combustion chamber as carbon monoxide; and the remaining 99.9+% of the carbon in the fuel leaves the combustion process as carbon dioxide. It should also be noted with regard to Table 24.5 that some of the sulfur in a fuel may combust to SO_3 rather than SO_2 with a markedly higher release of heat. However, it is known that only a small portion of the sulfur will combust to SO_3 and some of the sulfur in fuel may be in the form of pyrites (FeS_2), which do not combust at all. Therefore, only the SO_2 reaction is given.

TABLE 24.5 Chemical Reactions of Combustion[33a]

Combustible	Reaction
Carbon	$C + O_2 = CO_2$
Hydrogen	$2 H_2 + O_2 = 2 H_2O$
Sulfur	$S + O_2 = SO_2$
Carbon Monoxide	$2 CO + O_2 = 2 CO_2$
Methane	$CH_4 + 2 O_2 = CO_2 + 2 H_2O$
Acetylene	$2 C_2H_2 + 5 O_2 = 4 CO_2 + 2 H_2O$
Ethylene	$C_2H_4 + 3 O_2 = 2 CO_2 + 2 H_2O$
Ethane	$2 C_2H_6 + 7 O_2 = 4 CO_2 + 6 H_2O$

Factors directly affecting furnace design are the moisture and the combustible content of the solid waste to be burnt and the volatility of the material to be burnt. The means for temperature control and sizing of flues and other plant elements should be based on design parameters that result in large sizes. Combustion controls should provide satisfactory operation for loads below the maximum rated capacity of the units.

The combustible portion of MSW is composed largely of cellulose and similar materials originating from wood, mixed with appreciable amounts of plastics and rubber, as well as some fats, oils, and waxes. The heat released by burning dry cellulose is approximately 8,000 Btu/lb, while that released by certain plastics, rubber, fats, oils, and so on, may be as high as 17,000 Btu/lb. If

TABLE 24.6 Heat of Combustion[33a]

Combustible	Molecular Symbol	Heating Value (Btu per Pound)	
		Gross	Net
Carbon	C	14,100	14,100
Hydrogen	H_2	61,100	51,600
Sulfur	S	3,980	3,980
Carbon Monoxide	CO	4,350	4,350
Methane	CH_4	23,900	21,500
Acetylene	C_2H_2	21,500	20,800
Ethylene	C_2H_4	21,600	20,300
Ethane	C_2H_6	22,300	20,400

cellulose and plastics, rubber, oil, and fat exist in the ratio of 5:1 in MSW, the heat content of the dry combustible matter only is approximately 9,500 Btu/lb.

The heat released in combustion of basic combustible substances is displayed in Table 24.6. The heating value of a substance can be expressed either as higher (or gross) heating value or as lower (or net) heating value. The higher heating value takes into account the fact that water vapor formed or evaporated in the process of combustion includes the latent heat of vaporization, which could be recovered if the products of combustion are reduced in temperature sufficiently to condense the water vapor to liquid water. The lower heating value is predicated on the assumption that the latent heat of vaporization will not be recovered from the products of combustion.

The heat released during combustion may be determined in a bomb calorimeter, a device with a metal container (bomb) immersed in a water jacket. The MSW sample is burned with a known quantity of oxygen, and the heat released is determined by measuring the increase in temperature of the water in the water jacket. Since the bomb calorimeter is cooled to near ambient conditions, the heat recovery measured includes the latent heat of vaporization as the products of combustion are cooled and condensed in the bomb. That is, the bomb calorimeter inherently measures higher heating value (HHV). It has been customary in the United States to express heating value as HHV. In Europe and elsewhere, heating value is frequently expressed as the lower heating value (LHV).

Heating value can be converted from HHV to LHV if weight decimal percentages of moisture and hydrogen (other than the hydrogen in moisture) in the fuel are known, using the following formula:

$$LHV_{Btu/lb} = HIV_{Btu/lb} - \left[\%H_2O + \left(9 \times \%H_2\right)\right] \times \left(1{,}050\ Btu/lb\right) \tag{24.1}$$

$$LHV_{J/kg} = HHV_{J/kg} - 2440 \times \left[\left(9 \times \%H_2\right) + \%H_2O\right]\ KJ/kg \tag{24.2}$$

For example (using data from Table 24.9),

$$LHV_{Btu/lb} = HHV_{Btu/lb} - \left[\%H_2O + \left(9 \times \%H_2\right)\right] \times 1{,}050\ Btu/lb$$

$$LHV_{Btu/lb} = 4{,}940 - \left[0.30 + \left(9 \times 0.047\right)\right] \times 1{,}050$$

$$LHV_{Btu/lb} = 4{,}940 - \left[0.30 + 0.42\right] \times 1{,}050$$

$$LHV_{Btu/lb} = 4{,}940 - 756 = 4{,}184\ Btu/lb$$

Another method for determining the approximate higher heating value for MSW is to perform an ultimate analysis and then apply Dulong's formula:

$$\text{HHV}\left(\text{Btu/lb}\right)=14{,}544\ C+62{,}028\left(H_2-\frac{O_2}{8}\right)+4050\,S \quad \textbf{(24.3)}$$

where C, H_2, O_2, and S represent the decimal proportionate parts by weight of carbon, hydrogen, oxygen, and sulfur in the fuel. The term $O_2/8$ is a correction used to account for hydrogen which is already combined with oxygen in the form of water. For example (using data from Table 24.9)

$$\text{HHV}_{(\text{Btu/lb})}=14{,}544\ \text{C}+62{,}028\left(\text{H}_2-\frac{\text{O}_2}{8}\right)+4{,}050\,\text{S}$$

$$\text{HHV}=14{,}544\times 0.257+62{,}028\left(0.047-\frac{0.21}{8}\right)+4{,}050\times 0.001$$

$$\text{HHV}=3{,}738+62{,}028\left(0.047-0.026\right)+4.0$$

$$\text{HHV}=3{,}738+62{,}028\times 0.021+4.0$$

$$\text{HHV}=3{,}738+1{,}303+4=5{,}045\ \text{Btu/lb}$$

$$\text{versus: }\ 9{,}500\times 0.52=4{,}940\ \text{Btu/lb}$$

The American Society for Testing and Materials (ASTM) publishes methods for determining the ultimate analysis of solid fuels such as MSW.[2] The ultimate analysis of a fuel is developed through measures of carbon, hydrogen, sulfur, nitrogen, ash, and moisture content. Oxygen is normally determined "by difference"; that is, once the percentages of the other components are measured, the remaining material is assumed to be oxygen. For solid fuels, such as MSW, it is frequently desirable to determine the proximate analysis of the fuel. The procedure for determining the proximate analysis is also prescribed by ASTM.[2] The qualities of the fuel measured in percentage by weight are moisture, volatile matter, fixed carbon, and ash. This provides an indication of combustion characteristics of a solid fuel. As a solid fuel is heated to combustion, first the moisture in the fuel evaporates, then some of the combustible constituents volatilize (gasify) and combust as a gas with oxygen, and the remaining combustible constituents remain as fixed carbon in a solid state and combust with oxygen to form carbon dioxide. The material remaining after combustion is complete is the ash. MSW, with a high percentage of volatiles and a low percentage of fixed carbon, burns with much flame.

Table 24.7 displays ignition temperatures for combustible substances. The ignition temperature is the temperature to which the combustible substance must be raised before it will unite in chemical combination with oxygen. Thus, the temperature must be reached and oxygen must be present for combustion to take place. Ignition temperatures are not fixed temperatures for a given substance. The actual ignition temperature is influenced by combustion chamber configuration, oxygen fuel ratio, and synergistic effect of multiple combustible substances. The ignition temperature of MSW is the ignition temperature of its fixed carbon component. The volatile components of MSW are gasified but not ignited before the ignition temperature is attained.

The oxygen, nitrogen, and air data displayed in Table 24.8 represent the weight of air theoretically required to completely combust one pound of a combustible substance. The weight of oxygen required is the ratio of molecular weight of oxygen to molecular weight of the combustion constituent as displayed in Table 24.5. The weights of nitrogen and air required are calculated from the percentage by weight constituents of dry air. In actuality, to achieve complete combustion, air in excess of the theoretical requirement is required for complete combustion to increase the likelihood that all of the combustible substances are joined with sufficient oxygen to complete combustion. The level of excess air required in the combustion of MSW depends on the configuration of

TABLE 24.7 Ignition Temperatures[57a]

Combustible	Molecular Symbol	Ignition Temp. °F
Carbon (Fixed)	C	650
Hydrogen	H_2	1,080
Sulfur	S	470
Carbon Monoxide	CO	1,170
Methane	CH_4	1,270
Acetylene	C_2H_2	700
Ethylene	C_2H_4	960
Ethane	C_2H_6	1,020

TABLE 24.8 Theoretical Combustion Air[33a]

		Pounds Per Pound of Combustible					
		Required for Combustion			Products of Combustion		
Combustible	Molecular Symbol	O_2	N_{atm}	Air	CO_2	H_2O	N_{atm}
Carbon		2.67	8.87	11.54	3.67		8.87
Hydrogen		8.00	26.57	34.57		9.00	26.57
Sulfur		1.00	3.32	4.32	2.00	SO_2	3.32
Carbon Monoxide		0.57	1.89	2.46	1.57		1.89
Methane		4.00	13.29	17.29	2.75	2.25	13.29
Acetylene		3.08	10.23	13.31	3.38	0.70	10.23
Ethylene		3.43	11.39	14.82	3.14	1.29	11.39
Ethane		3.73	12.39	16.12	2.93	1.80	12.39

the combustion chamber, the nature of the fuel firing equipment, and the effectiveness of mixing combustion air with the MSW. An excess air level of 80% is commonly associated with combustion of MSW in modern WTE facilities.

Excess air serves to dilute and thereby reduce the temperature of the products of combustion. The reduction of temperature tends to reduce the heat energy available for useful work. Therefore, the actual excess air used in the combustion process is a balance between the desire to achieve complete combustion and the objective of maximizing the heat energy available for useful work.

It is frequently useful to know the temperature attained by combustion. The heat released during combustion heats the products of combustion to a calculable temperature. It must be understood that the calculation procedure presented here assumes complete combustion and that no heat is lost to the surrounding environment. Thus, it is a temperature that is useful to compare one combustion process with another. The heat available for heating the products of combustion is the lower heating value of the fuel. The increase in temperature is the lower heating value divided by the mean specific heat of the products of combustion. The mean specific heat is a function of the constituent products of combustion ($W_{P.C.}$) and the temperature. To approximate the theoretical temperature attainable, one can use a specific heat of 0.55 Btu/lb per °F for water vapor (W_{H_2O}) and 0.28 Btu/lb per °F for the other gaseous products of combustion ($W_{P.C.} - W_{H_2O}$). Thus, the formula approximating the temperature attained during combustion is

$$T_{comb} = T + \frac{LHV_{Btu/lb}}{0.55W_{H_2O} + 0.28\left(W_{P.C.} - W_{H_2O}\right)} \qquad \textbf{(24.4)}$$

For example (using data from Table 24.9)

TABLE 24.9 Sample Calculation for MSW

Ultimate Analysis		Theoretical Air		Products of Combustion							
Substance	Fraction % By Weight	lb/lb		CO_2 Incl. SO_2		H_2O		N_2		O_2	
(1)	(2)	(3)	(2) × (3)	(4)	(2) × (4)	(5)	(2) × (5)	(6)	(2) × (6)	(7)	(7) × (3)
Moisture	0.300					× 1.00	0.30				
Carbon	0.257	× 11.54	2.97	× 3.67	0.92			× 8.87	2.28		
Hydrogen	0.047	× 34.57	1.62			× 9.00	0.42	× 26.57	1.25		
Sulfur	0.001	× 4.32	0.04	× 2.00	0.02			× 3.32	0.03		
Nitrogen	0.005							× 1.00	0.01		
Oxygen	0.210	÷ 0.2314	–[(2) ÷ (3)] = –0.94					× 0.7686	(2 ÷ 3) × (6) = –0.72		
Ash	0.180										
Total	1.000		3.69		0.94		0.72		2.85		
HHV 4,940 Btu/lb											
80% Excess Air			2.95					× 0.7686	2.27	× 0.2314	0.68
Total Dry Air			6.64								
Moisture in Air		× 0.013	0.09				0.09				
Total Products of Combustion					0.94		0.81		5.12		0.68

Total Products of Combustion		
CO_2	0.94	0.125
H_2O	0.81	0.107
N_2	5.12	0.678
O_2	0.68	0.090
Total	7.55 lb/lb fuel	1.000

LHV = 4,940 – [0.30 + (9 × 0.047)] × (1,050)
= 4,184 Btu/lb

Temperature Developed in Combustion = 60 + 4,184/[(0.81 × 0.55) + (7.55 – 0.81)0.28]
= 1,850°F

$$T_{comb} = T_{ambient} + \frac{LHV_{Btu/lb}}{0.55\ W_{H_2O} + 0.28\left(W_{P.C.} - W_{H_2O}\right)}$$

If the ambient temperature is assumed to be 60°F, then

$$T_{comb} = 60 + \frac{4{,}184}{0.55 \times 0.81 + 0.28(7.55 - 0.81)}$$

$$T_{comb} = 60 + \frac{4{,}184}{0.45 + 1.89} = 60 + \frac{4{,}184}{2.34}$$

$$T_{comb} = 1{,}848 \cong 1{,}850°F$$

Typical combustion calculations are provided in Table 24.9 for MSW to determine the products of the combustion process. Each of the combustible substances combines and completely combusts with oxygen as displayed in Table 24.5. The weight ratio of oxygen to the combustible substance is the ratio of molecular weights. Table 24.8 displays the weight or volume of oxygen theoretically required for complete combustion of one pound of the combustible substance. Sulfur dioxide from combustion of sulfur in fuel is combined with CO_2 in the sample calculation as a matter of convenience. If desired, a separate column can be prepared for sulfur dioxide in the products of combustion. Oxygen in the fuel combines with the combustible substances in the fuel, thereby reducing the quantity of air required to achieve complete combustion. The sample calculation uses the weight percentages of oxygen to reduce the theoretical air requirements and the nitrogen in the products of combustion. The decimal percentage of excess air is multiplied by the total theoretical air requirement to establish the weight of excess air and the total air requirement including excess air.

Furnaces

While the general principles of a modern incinerator burning as-received MSW are common to all types, the specific solid waste combustion process is rather complex. The waste is heated by contact with hot combustion gases or preheated air, and by radiation from the furnace walls. Drying occurs in a temperature range of 50 to 150°C. At higher temperatures, volatile matter is formed by complicated thermal decomposition reactions. This volatile matter is generally combustible and, after ignition, produces flames. The remaining material is further degased and burns much more slowly. In an RDF furnace (see Figure 24.3), most of the volatile matter and some of the fixed carbon is burned in suspension while the remaining fixed carbon is combusted on a grate at the bottom of the furnace.

The complexity of the combustion of solid waste streams results from the nature of the decomposition and burning reactions and their association with heat transfer, air flow, and diffusion. In most incinerators, combustion takes place while the solids are supported on and conveyed by a grate. Since the early 1960s, most MSW incinerators have incorporated one of a number of available proprietary grate systems that allow continuous feed of unscreened waste into and movement through furnaces with integral boiler facilities. The grate performs several functions: provides support for the refuse, admits underfire air through openings in the grate surface, transports the solid waste from feed mechanism to ash quench, agitates the bed, and serves to redistribute the burning mass.

The basic design factors which determine furnace capacity are grate area and furnace volume. Also, the available capacity and method of introducing both underfire and overfire air will influence, to a lesser extent, furnace capacity. Required grate area, in a conservative design, is normally determined by limiting the burning rate to between 60 and 70 lb/ft^2 hr (290 and 340 kg/m^2 hr) of grate area.[1] This is based on limiting the heat release rate loading on the grate to 250,000 and 300,000 Btu/ft^2 of grate hr (2.8 to 3.4 GJ/m^2 hr).

Furnace volume required is established by the rate of heat release from the fuel. Thus, furnace volume is generally established by using heat release rates ranging from 12,500 to 20,000 Btu/ft^3/hr (450 to 750 MJ/m^3/hr), with the lower heat release rate being more desirable from the standpoint of developing a conservative design. A conservative approach to design in this area is desirable because of probable periodic operation above design capacity to meet short-term higher than normal refuse collections and possible receipt of high-heat-content waste.

Water wall units burning as-received MSW have been built as small as 75 to 100 tons per day (68 to 91 tonnes/day) capacity. However, the cost per ton of rated capacity of such units is relatively high. A more common unit size for both mass burn and RDF furnaces is 250 to 300 tons/day (225 to 270 tonnes/day), while water wall mass fired units have been built as large as 750 to 1,200 tons/day (675 to 1,090 tonnes/day) capacity.[4]

The primary objective of a mechanical grate in a mass burn furnace is to convey the refuse from the point of feed through the burning zone to the point of residue discharge with a proper depth of fuel and sufficient retention time to achieve complete combustion. The refuse bed should be greatly agitated so as to enhance combustion. However, the agitation should not be so pronounced that particulate emissions are unreasonably increased. The rate of movement of the grate or its parts should be adjustable to meet varying conditions or needs in the furnace.

In the United States over the past 20 years, several types of mechanical grates have been used in continuous feed furnaces burning as-received MSW. These include traveling grates (see Fig. 24.4), reciprocating grates (see Fig. 24.5), rocking grates (see Fig. 24.6), rolling grates (see Fig. 24.7), and rotary kilns (both water cooled and refractory lined). The traveling grate conveys the refuse through the furnace on the grate surface. Stirring is accomplished by building the grate in two or more sections, with drop between sections to agitate the material. The reciprocating grates, rocking grates, and rolling grates agitate and move the refuse material through the furnace by the movement of the grate elements and the incline of the grate bed. Additional agitation is obtained, particularly in the reciprocating grate, by substantial drops in elevation between grate sections. The rotary kiln slowly rotates to tumble the refuse material, which is conveyed through the inside of the cylinder. The combustor is inclined from the horizontal so that gravity assists in moving the material through the unit.

Other grate systems have been developed in Europe for burning as-received MSW, some of which are currently being utilized in plants being constructed or in operation in the United States. The Volund incinerator (Danish) (see Fig. 24.8) uses a slowly rotating, refractory-lined cylinder or kiln, which is fed by a two-section (drying and ignition) reciprocating grate. Refuse passes through the kiln and residue is discharged to a water quench when combustion is completed. The rolling grate, or so-called Dusseldorf System (see Fig. 24.7), uses a series of 5 or more rotating cylindrical grates, or drums, placed at a slope of about 30 degrees.[53] The refuse is conveyed by the surface of the drums, which rotate in the direction of refuse flow, and is agitated as it tumbles from drum to drum. Underfire air is introduced through the surface of the drums. Both the Von Roll and the Martin grates use a reciprocating motion to push the refuse material through the furnace. However, in the Martin grate (see Fig. 24.9), the grate surface slopes steeply down from the feed end of the furnace to the ash discharge end and the grate sections push the refuse uphill against the flow of waste, causing a gentle tumbling and agitation of the fuel bed.

Another variable feature in the various grate designs is the percentage of open area to allow for passage of underfire air.[66] These air openings vary from approximately 2 to 30% of the grate surface area. The smaller air openings tend to limit the quantity of siftings dropping through the grates and create a pressure drop that assists in controlling the point of introduction of underfire air. RDF grates generally have a smaller percentage of air openings. Larger air openings make control of underfire air more difficult but allow for continuous removal of fine material, which could interfere with the combustion process, from the fuel bed.

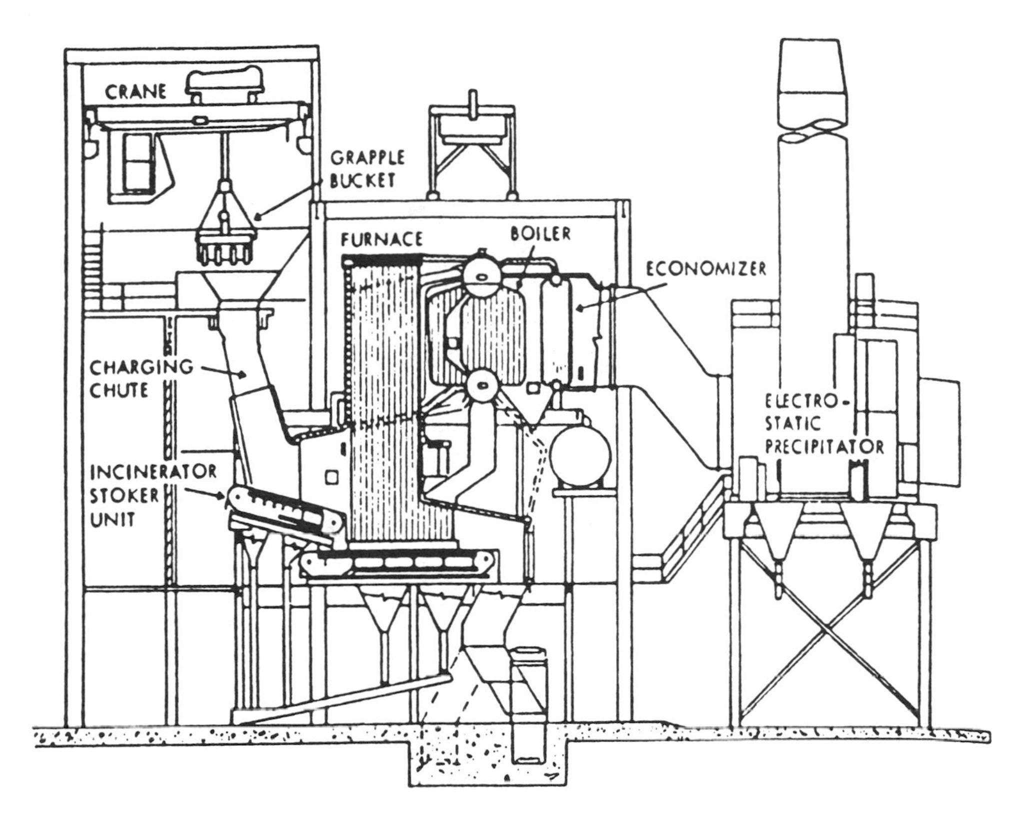

FIGURE 24.4 Traveling grate system. *Source:* Braun, Metzger, and Vogg.

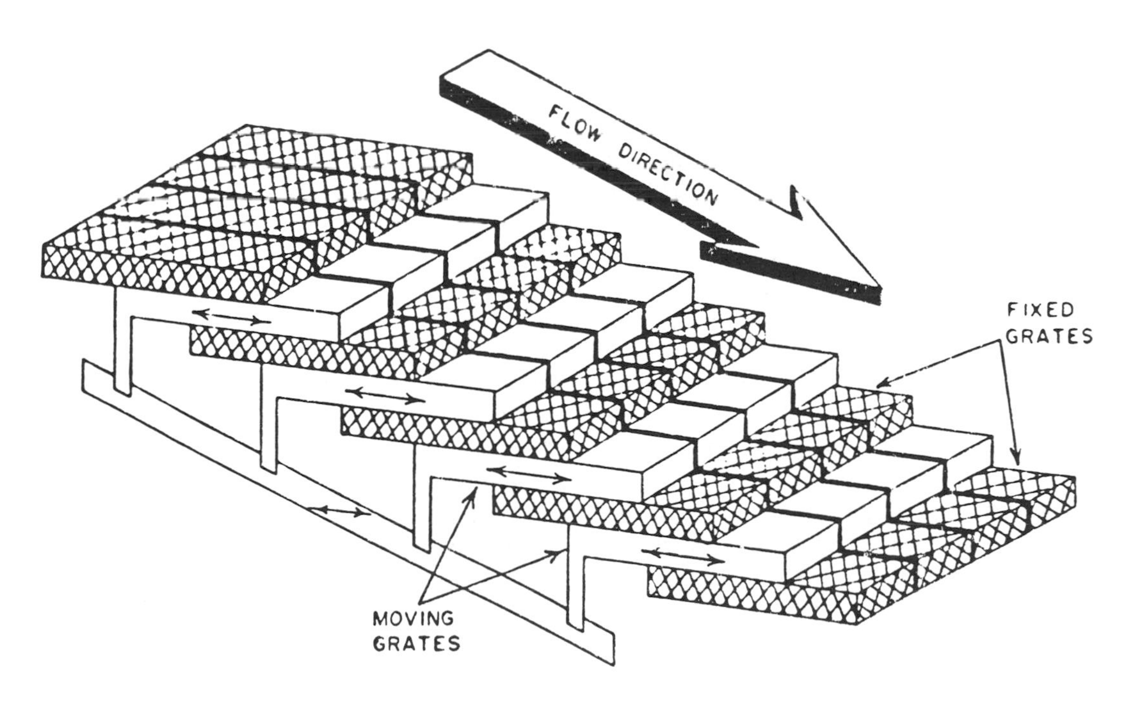

FIGURE 24.5 Reciprocating grates.

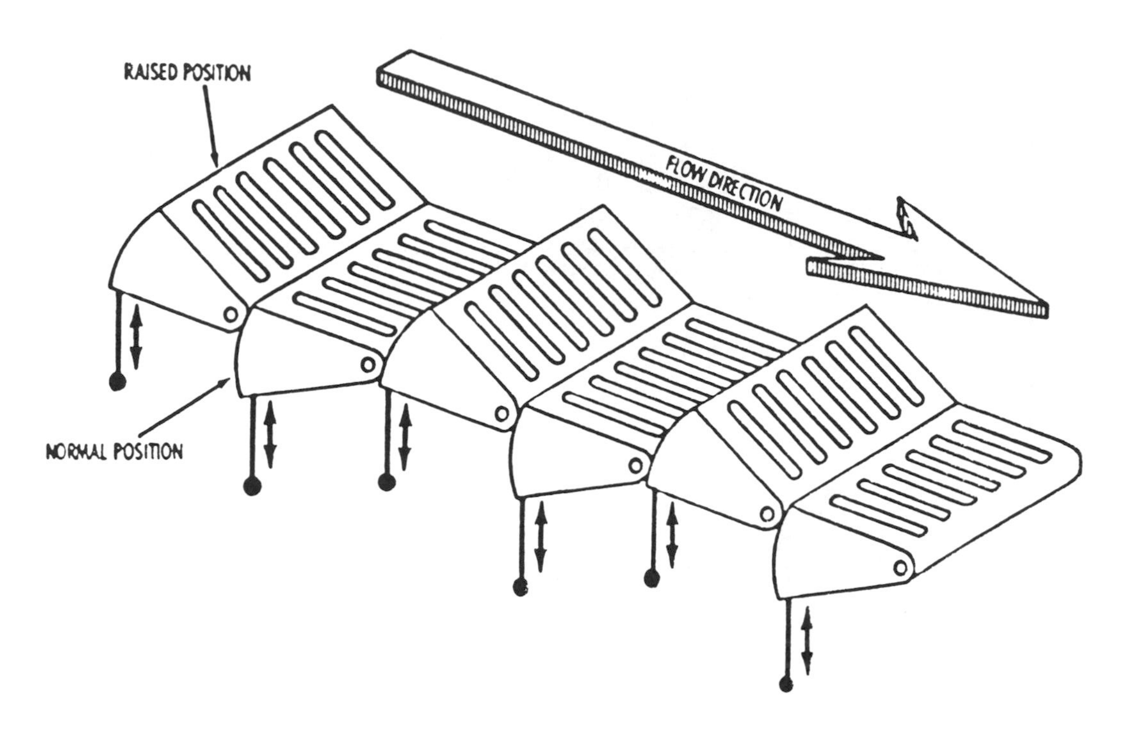

FIGURE 24.6 Rocking grates.

Furnace configuration is largely dictated by the type of grate used. In the continuous feed mechanical grate system, the furnace is rectangular in plan and the height is dependent upon the volume required by the limiting rate of heat release cited earlier. An optimum furnace configuration would provide sufficient volume for retention of gases in the high-temperature zone of maximum fuel volatilization long enough to ensure complete combustion, and would be arranged so that the entire volume is effectively utilized. Temperatures are usually high enough with present-day refuse for proper combustion. Turbulence should be provided by a properly designed overfire air system.

With present-day mass fired water wall furnaces, the use of refractories in furnace construction has been minimized but not eliminated. Refractory materials may be used to line charging chutes, provide a transition enclosure between the top of the grates and the bottom of the water walls, a protective coating on the water wall tubes, and an insulating layer between the hot gases and the metal walls of flues downstream of the primary combustion chamber. Refractory brick used in a charging chute must be able to withstand high temperatures, flame impingement, thermal shock, slagging, spalling, and abrasion. The protective coating on the water wall tubes must be relatively dense castable material with a relatively high heat conductivity.[67] Insulating refractories used in flues downstream from the boilers, on the other hand, should have a low heat conductivity.

Refractories are generally classified according to their physical and chemical properties, such as resistance to chemical attack, hardness, strength, heat conductivity, porosity, and thermal expansion.[46] The material may be cast in brick in a variety of shapes and laid up with air-setting or thermal-setting mortar, or may be used in a moldable or plastic form. Material used in incinerator construction includes "high duty" and "superduty" fireclay brick, phosphate-bonded alumina material, and silicon carbide, among others. In selecting the proper materials for application in this type of service, because the variety of materials is so great and the conditions of service so varied and severe, [16] advice of a recognized manufacturer should be sought.

As indicated in the section on combustion calculations, the combustion process requires oxygen to complete the reactions involved in the burning process. The air that must be delivered in the furnace to supply the exact amount of oxygen required for completion of combustion is called the

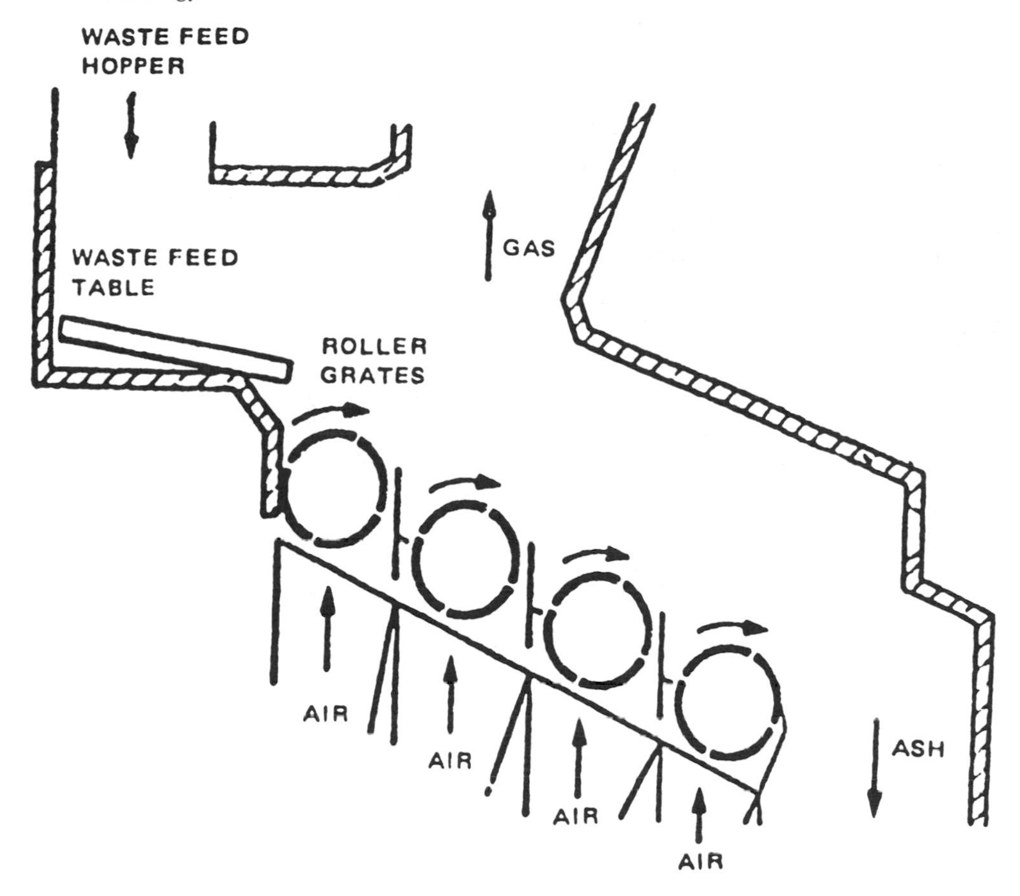

FIGURE 24.7 VKW system. *Source:* Braun, Metzger, and Vogg.

stoichiometric air requirement. Additional air supplied to the furnace is called excess air and is usually expressed as a percentage of the stoichiometric requirements.

The total air supply capacity in an incinerator must be greater than the stoichiometric requirement for combustion because of imperfect mixing and to assist in controlling temperatures, particularly with dry, high-heat-content refuse. The total combustion air requirements can range from 6 to 8 lb of air/lb of refuse for mass fired water wall furnaces.

In the modern mass-burn mechanical grate furnace chamber, at least two blower systems should be provided to supply combustion air to the furnace—one for underfire or undergrate air and the other for overfire air. Underfire air, admitted to the furnace from under the grates and through the fuel bed, is used to supply primary air to the combustion process and to cool the grates.

Overfire air may be introduced in two levels. Air introduced at the first level, called secondary air, immediately above the fuel bed, is used to promote turbulence and mixing, and to complete the combustion of volatile gases driven off the bed of burning solid waste. The second row of nozzles, which are generally located higher in the furnace wall, allow the introduction of air, called tertiary air, into the furnace to promote additional mixing of gases and for temperature control.

Blower capacities should be divided so that the underfire air blower is capable of furnishing half to two-thirds of the total calculated combustion air requirements, while the overfire blower should have a capacity of somewhat less than half of the total calculated air requirements. Setting these capacities requires some judgment related to assessing how great a variation is anticipated in refuse heat contents during plant operation. Dampers should be provided on fan inlets and on air distribution ducts for control purposes.

Pressures on underfire air systems in mass burn units for most U.S. types of grates will normally range from 2 to 5 inches of water. European grates frequently require a higher pressure. The pressure

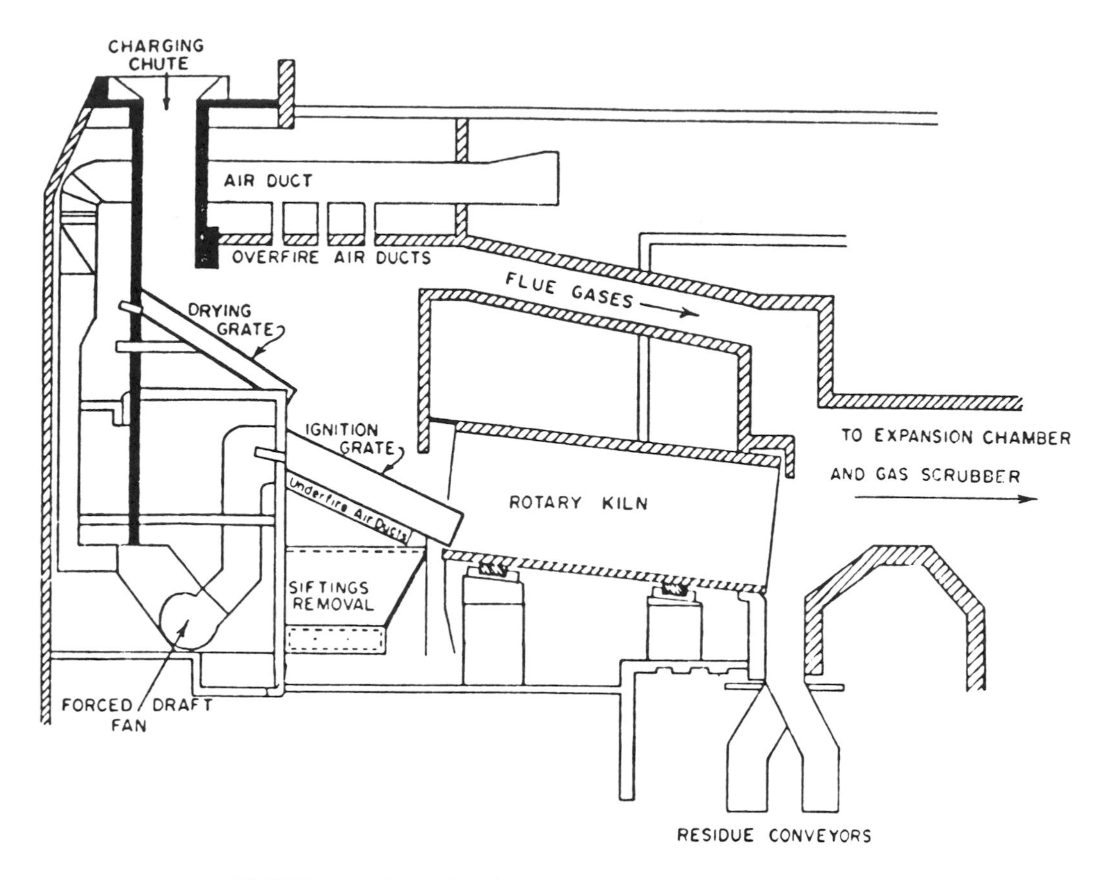

FIGURE 24.8 Rotary kiln furnace. *Source:* Braun, Metzger and Vogg.

on the overfire air should be high enough that the air, when introduced into the furnace, produces adequate turbulence without impinging on the opposite wall. This is normally accomplished by the use of numerous relatively small ($1^1/_2$ inches to 3 inches in diameter) nozzles at pressures of 20 inches of water or higher.

In an RDF fired, spreader–stoker type of unit, the combustible material is generally introduced through several air-swept spouts in the front water wall, is partially burned in suspension, and then falls onto a grate on which combustion is completed as the partially burned material is conveyed to the residue discharge under the front water wall face of the furnace. Densified RDF can also be burned in such units. These units can generally handle a coarser RDF than the so-called full-suspension burning units. The RDF can furnish all the combustible input to the system, or it may be cofired with a fossil fuel, generally coal.

The full-suspension combustion concept was originally proposed so that finely shredded combustible material from MSW could be burned in existing utility boilers. In this way, the expense of constructing a boiler would be mitigated and 10 to 15% of the fossil fuel normally consumed by the utility would be displaced (saved) by burning the RDF. This has been the least successful of the system types due to problems related to the additional handling and greater power required to achieve a finer shred. Also, some utility boilers seemed to experience a greater tendency for slag formation in the boiler when RDF was added. While the concept initially anticipated that the RDF would completely burn in suspension, experience to date indicates that this does not occur. Accordingly, dump grates became a necessity in such furnace boiler units to allow for completion of combustion prior to water quenching of the residue.

Some combustion air in RDF fired units is introduced with the fuel through the air-swept feed spouts to distribute the fuel on the grate. Additional air is introduced into the furnace higher in

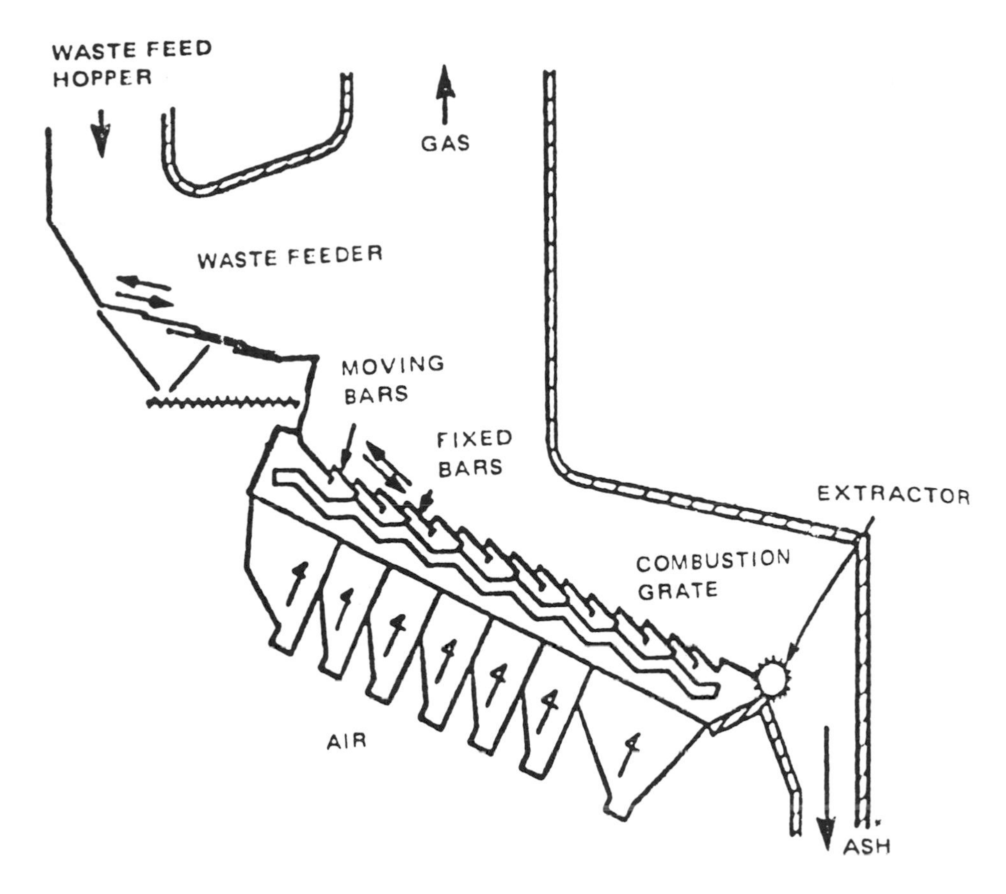

FIGURE 24.9 Martin system. *Source:* Braun, Metzger, and Vogg.

the water wall area to enhance turbulence and mixing in the unit and/or to control temperatures. This additional combustion air supply is similar to the tertiary air utilized in the mass burn units.

Boilers

Substantial quantities of heat energy may be recovered during the thermal destruction of the combustible portions of MSW. Systems that have been successfully used to recover this energy include mass fired refractory combustion chambers followed by a convection boiler section; a mass fired water wall unit where the water wall furnace enclosure forms an integral part of the boiler system; an RDF semisuspension fired spreader stoker/boiler unit; and an RDF suspension burning utility type of boiler. Each system has apparent advantages and disadvantages.

Refractory Furnace with Waste Heat Boiler

In a refractory furnace waste heat boiler unit, energy extraction efficiencies are generally lower, assuming the same boiler outlet temperatures, than with the other systems. Approximately 50 to 60% of the heat generated in the combustion process may be recovered with such systems. These units can produce approximately 2 lb steam/lb of normal MSW (heat content = 4,500 Btu/lb), versus 3 or more lb/lb MSW in mass fired water wall units. This lower efficiency of steam generation is caused by larger heat losses due to higher excess air quantities needed with such units to control furnace temperatures so that furnace refractories are not damaged. However, the boilers in such units, if properly designed and operated, generally are less susceptible to boiler tube metal wastage problems than the other systems listed earlier.

Mass Fired Water Wall Units

Mass fired water wall units are the most widely utilized type of heat recovery unit in this field today. In this type of unit, the primary combustion chamber is fabricated from closely spaced steel tubes through which water circulates. This water wall lined, primary combustion chamber incorporated into the overall boiler system is followed by a convection type of boiler surface. It has been found desirable in these plants to coat a substantial height of the primary combustion chamber, subject to higher temperatures and flame impingement, with a thin coating of a silicon carbide type of refractory material, and to limit average gas velocities to under 15 ft/s (4.5 m/s) in this portion of the furnace. Gas velocities entering the boiler convection bank should be less than 30 ft/s (9.0 m/s).[67] Efficiency of heat recovery in such units has been found to range generally from 65 to 70%, with steam production usually about 3 lb of steam/lb of normal as-received MSW. Water table studies have been found to be very useful in the larger units to check on combinations of furnace configuration and location of overfire combustion of air nozzles.[24a]

RDF Fired Water Wall Units

As pointed out earlier, RDF may be burned in a semisuspension fired spreader stoker/boiler unit where the RDF is introduced through several air-swept spouts in the front water wall, partially burns in suspension, and then falls on a grate on which combustion is completed. Alternatively, a full-suspension type of furnace/boiler can be used wherein the RDF is blown into the furnace through nozzles located one-half to two-thirds up the height of the water wall furnace enclosure. In this type of unit most of the RDF, usually composed of smaller sized particles than in the semisuspension fired unit, burns in suspension. However, it has been found necessary in this type of unit to install dump grates in the bottom of the water wall enclosed primary combustion chamber to retain the substantial amount of unburned material. In both these types of units the water wall lined primary combustion enclosure (furnace) may be followed by a superheater (usually), a convection boiler heat transfer surface, and (sometimes) an economizer surface.

Efficiencies of RDF fired boilers generally range from 60 to 80% of the heat input from the RDF. Steam production from RDF would normally be expected to be somewhat greater than 3 lb of steam/lb of RDF. However, when one takes into account the combustible material lost in the processing of as-received MSW to produce the RDF, steam production normally will fall below 3 lb of steam/lb of as-received MSW.[33]

If the energy recovered from the combustion of as-received MSW or RDF is to be used to produce electricity, some superheating is desirable, if not necessary. Since boiler tube metal wastage in these plants is, at least partially, a function of tube metal temperature[8] and steam is a less efficient cooling medium than water, superheater surface is more prone to metal wastage problems than other areas of boiler tubing. Tube metal temperatures, above which metal wastage can be a significant operational problem, are generally thought to range from 650° to 850°F (345° to 455°C). These temperatures are lower than those for maximum efficiency of electrical generation by steam driven turbines. It is desirable to consider this in facility design to reduce plant downtime and minimize maintenance costs.

Residue Handling and Disposal

The residue from a well-designed, well-operated mass fired incinerator burning as-received refuse will include the noncombustible material in the MSW and usually somewhat less than 3% of the combustibles. The nature of this material will vary from relatively fine, light ash, burned tin cans, and partly melted glass to large, bulky items such as lawn mowers and bicycles.

In most modern WTE plants bottom ash residue is discharged from the end of the furnace grate through a chute into a trough filled with water. Removal from the trough may be either by a ram discharger onto a conveyor or by a flight conveyor to an elevated storage hopper from which it is discharged to a truck. If a water-filled trough with a flight conveyor is used, normally two troughs are provided, arranged so that the residue can be discharged through either trough. The second trough serves as a standby. Fly ash, residue collected in the air pollution control equipment downstream of the furnace/boiler, is usually returned back to and mixed with the bottom ash.

A key feature in the design of ash discharge facilities is provision for sealing the discharge end of the furnace to prevent uncontrollable admission of air. This seal is usually provided by carrying the ash discharge chute at least 6 inches (15 cm) below the water surface in the receiving trough. In the design of the conveyor mechanism, the proportions should be large because the material frequently contains bulky metal items and wire, potential causes of jamming. Also, the residue material tends to be extremely abrasive.

Residue is usually taken to a landfill for final disposal. Many modern facilities dispose of their residue at monofills (landfills that accept WTE plant residue only). The volume of material remaining for ultimate disposal will range from 5 to 15% of that received at the plant. Many plants currently operating in the United States that weigh MSW received at the plant and residue discharged from the furnaces, indicate that the weight of MSW is only reduced by from 50 to 60%. However, as much as one-third of the residue weight in these plants may be attributed to incomplete drainage of the material prior to its discharge into the final transportation container. The ram-type ash discharger used in European and some of the new, large U.S. plants generally achieves much better dewatering of residues than older water-filled trough, ash drag residue handling systems.

The main components of ash are inert materials of low solubility such as silicates, clay, and sand. Aluminum, calcium, chlorine, iron, selenium, sodium, and zinc are major elements in all particles and, along with carbon, can comprise over 10% by weight of the ash.[20]

A broad range of trace metals and organic compounds may be found in fly and bottom ash. Data on ash composition are difficult to compare, however, because they reflect different types and sizes of facilities, unknown sample collection methodology and sample size at each facility, interlab variation in testing procedures (even using the same test), and the heterogeneous nature of MSW itself. In addition, the presence of a substance in ash does not mean that it will enter the environment. Its fate depends on its solubility, how the ash is managed, and whether the ash is subject to conditions that cause leaching.[61]

Metals tend to be distributed differently in fly and bottom ash. Most volatile and semivolatile metals, such as arsenic, mercury, lead, cadmium, and zinc, tend to be more concentrated or "enriched" in fly ash.[55,64] Less volatile metals, such as aluminum, chromium, iron, nickel, and tin, typically are concentrated in bottom ash.[54,55]

Organic chemicals also exhibit differing distributions. Dioxin/furan and polychlorinated biphenyls (PCBs) tend to be enriched in fly ash, while other chemicals such as polycyclic aromatic hydrocarbons (PAHs) and phthalates tend to be concentrated in bottom ash.[64] Concentrations of dioxins/furans in fly ash exhibit a wide range, but they are significantly lower in ash from modern facilities than in ash from older incinerators.[30,65,73]

From a regulatory standpoint a number of different testing procedures have been developed and utilized by regulatory agencies over the past several years in an attempt to predict the behavior of MSW residues deposited in landfills. Most of these methods were developed to predict leaching characteristics of residues deposited in landfills with raw or as-received MSW. Test results using these methods have been quite variable. However, as pointed out earlier, most modern WTE facilities dispose of their residue in monofills.[61] Tests of leachate from such monofills indicate metals concentrations below Extraction Procedure (EP) toxicity limits, and in most cases below U.S. drinking water standards.[61] Most test data show little or no leaching of organic chemicals.[31,61] The regulatory requirements for ash residues is an area still being hotly debated between WTE plant operators and organizations such as the Environmental Defense Fund (EDF).

24.3 Air Pollution Control Facilities

Potential emissions from the burning of MSW may be broadly classified into particulates, gaseous emissions, products of incomplete combustion (PICs) (e.g., carbon monoxide and hydrocarbons),

TABLE 24.10 Gaseous Emission Factors for Municipal Refuse Incinerators (lb/ton)[10,13,27,61,69]

	New York Incinerators 1968–1969	Test Results U.S. Plants 1971–1978	Martin Plants 1984–1986	EPA Data Base Tests Through 1988	
				Mass Burn	RDF
Carbon Monoxide	—	3.7–9.3	0.2	0.06–16.2	1.0–5.2
Nitrogen Oxides	—	0.5–2.2	5.0–6.0	0.5–4.5	2.3–3.2
Hydrocarbons	0.1–22.1	1.1	0.015–0.006	0.01–0.1	0.005–0.01
Hydrochloric Acid	1.4–8.6	4.6–14.5	5.0–0.2	0.05–5.7	0.02–9.3
Sulfur Oxides	1.3–8.0	0.8–2.2	1.0–2.0	0.05–4.8	0.05–2.3

and trace emissions. Particulates have been a matter of concern and regulatory agency attention for some time. The initial concern was from the standpoint of reducing gross emissions that were both an aesthetic and a potential public health problem. Current interest and concern, since the initial problem has largely been solved, is directed toward better control of submicron-size particles[68] and other potential pollutants.

Major Constituents

Gaseous emissions, such as CO, SO_2, and NO_x, are not generally felt to be a major problem in an incineration plant. However, control measures may be required at a specific site if it is located in an area designated by EPA as a nonattainment area for that particular gaseous emission.

Common gaseous emission factors, based on tests at a number of energy-from-waste-plants, are shown in Table 24.10. High carbon monoxide and hydrocarbon emissions are caused by incomplete combustion and/or upsets in combustion conditions. High nitrogen oxide emissions are generally caused by high combustion temperatures and/or high nitrogen content in the waste. Hydrogen chloride (and hydrogen fluoride) and sulfur oxides, on the other hand, are directly a function of the chlorine (fluorine) and sulfur content in the fuel. The highest emissions, cited in Table 24.10, are from older, poorly controlled plants without significant pollution control equipment.

Carbon monoxide and hydrocarbon emissions are best controlled by maintaining proper combustion conditions. Nitrogen oxide emissions are controlled by ammonia injection or by use of combustion control techniques such as limitation of combustion temperatures or recirculation of flue gases. Hydrogen chloride (and hydrogen fluoride) and sulfur oxides are best removed by acid gas washing or adsorbing devices using chemical treatment. Note in the last column of Table 24.10 that recent successful attempts to limit hydrocarbon emissions by improving combustion conditions and raising furnace operating temperatures seem to have resulted in increasing the level of NO_x emissions.

Thermodynamic equilibrium considerations indicate that under excess air conditions and with temperatures of 800°C (1,472°F) and higher, maintained in a completely mixed reactor for a suitable period of time, emissions of organic or hydrocarbon compounds should be at nondetectable levels. However, measurements at operating plants, particularly those constructed prior to the early to mid-1980s, indicated significant emissions of trace organic or hydrocarbon compounds, some of which are toxic. These tests indicated that the basic objective of combustion control, thorough mixing of combustion products with oxygen at a temperature that is sufficiently high to provide for the rapid destruction of all organic or hydrocarbon compounds, had not been achieved in these early WTE plants.

If the fuel, or the gas driven off of the fuel bed, is not adequately mixed with air, fuel-rich pockets will exist containing relatively high levels of hydrocarbons, which then can be carried out of the combustion system. Kinetic considerations indicate that such hydrocarbons can be destroyed rapidly in the presence of oxygen at elevated temperatures. Also, if too much combustion air is introduced into the combustion chamber, either in total or in a particular area of the chamber, temperatures will be reduced, combustion reactions can be quenched, and hydrocarbons carried out of the combustion system. Achieving the goal of proper combustion control, destruction of all hydrocarbon compounds to form carbon dioxide and water, will minimize emission of potentially toxic substances as well as other compounds that may be precursors and capable of forming toxic compounds downstream in cooler regions of the boiler.

Table 24.10 shows that the hydrocarbons can vary significantly, frequently over relatively short periods of time, based on measurements at older incineration plants. The highest levels shown in this table occurred in one of the older plants and no doubt indicate very poor combustion conditions. Tests at modern WTE plants indicate consistently low levels of hydrocarbons, which are indicative of good combustion control. In modern, well-designed and operated plants, photochemical oxidants and PAH are in concentrations too low to cause any known adverse health effects. Tests[50,59] for other substances that might be of concern, such as polychlorinated biphenyls (PCBs), generally have found levels discharged to the atmosphere to be so low as to have a negligible impact on the environment.

Upset conditions in energy-from-waste plants can lead to local air-deficient conditions resulting in the emission of organic compounds. PAH are formed during fuel-rich combustion as a consequence of free radical reactions in the high temperature flame zone. In addition, it is found that in the presence of water cooled surfaces, such as found in oil fired home-heating furnaces, a high fraction of the polycyclic compounds are oxygenated. Similar free radical reactions probably take place in fuel-rich zones of incinerator flames yielding PAH, oxygenated compounds such as phenols, and, in the presence of chlorine, some dioxin/furan. The argument for the synthesis of dioxin/furan at temperatures of 400 to 800°F is supported by the increase in the concentration of these pollutants across a heat recovery boiler downstream of the refractory lined combustion chamber of an incinerator.

This free radical mechanism appears to be the dominant source of dioxin/furan in incinerators. These compounds may also be present as contaminants in a number of chemicals, most notably chlorinated phenols and polychlorinated biphenyls (PCBs). Their presence in MSW results from the use of these chemicals, discontinued in some cases, as fungicides and bactericides for the phenol derivatives, or the use of PCBs as heat exchanger and capacitor fluids. These compounds are expected to survive in a furnace combustion chamber only if large excesses in the local air flow cool the gases to below the decomposition or reaction temperature. Dioxin/furan can also be produced by condensation reactions involving the chlorinated phenols and biphenyls. The observed formation of dioxin/furan when fly ash from MSW incinerators is heated to temperatures of 480°F (250°C) suggest such catalyzed condensation reactions of chlorinated phenols. PCBs are precursors to furan, and pyrolysis of PCBs in laboratory reactors at elevated temperatures for a few seconds has yielded furan.

Trace Metals

Trace metals are not destroyed during incineration, and the composition of wastes therefore provides, on a statistical basis, the measure of the total inorganic residue. The distribution of trace metals between bottom ash and ash carried over to the air pollution control device is dependent upon the design and operation of the incinerator and the composition of the feed. The amount of ash carried up and with the flue gases discharged from a burning refuse bed increases with increasing underfire air rate and with bed agitation. Modular incinerators (described later in "Status of Other Technologies") with low underfire air flow rates tend to have lower particulate emissions than conventional mass burn units and RDF units for this reason. In addition, the amount of ash carried from the combustion chamber will be influenced by the particle size of the inorganic content of the MSW.

The distribution of trace metals between the different components of refuse has a strong influence on the fate of the trace metals. For example, TiO_2 used as a pigment in paper products, has a particle size of about 0.2 μm and will tend to be carried off by the flue gases passing through the refuse bed, whereas TiO_2 present in glass will remain with the glass in the bottom ash. Up to 20%

TABLE 24.11 Trace Metal Emissions Test Results[14,32,47,48,69]

			Particulate Removal Efficiency 99+%				
						Dry Scrubber, Fab. Filter	
	Japanese Plant Uncontrolled	Braintree Mass. Part. Rem. Eff. 74+%	German Plants	Japanese Plants	Tulsa, OK	Marion Co., OR	Pilot Plant Canada
Arsenic (AS) (lb/Ton × 10^{-3})	<0.4	0.125	0.09	<0.0016	—	—	0.00033–0.00064
Beryllium (Be) (lb/Ton × 10^{-3})	<0.3	0.00027	0.002	<0.0016	0.000025	0.000021	—
Cadmium (Cd) (lb/Ton × 10^{-3})	0.7	1.30	0.25	0.11	—	—	ND–0.006
Chromium (Cr) (lb/Ton × 10^{-3})	16.0	0.34	0.185	0.026	—	—	ND–0.016
Lead (Pb) (lb/Ton × 10^{-3})	17.0	42.2	10.0	0.1	3.5	0.29	ND–0.08
Mercury (Hg) (lb/Ton × 10^{-3})							
on particulates	0.5	0.11	0.067	0.03	—	—	—
vapor phase	0.8	4.38*	—	0.90	3.5	2.9	0.16–9.83
Selenium (Se) (lb/Ton × 10^{-3})	<0.3	—	—	<0.0016	—	—	—
Particulates (lb/Ton)	25.7	1.3	0.5	0.19	0.13	0.16	<0.01

* Total on particulate and vapor phase.

of the inorganic content of the waste will be entrained from burning refuse beds to form fly ash particles. The remainder will end up in the residue.

Volatile metals and their compounds, usually present in trace amounts in the feed, will vaporize from the refuse and condense in the cooler portions of a furnace either as an ultra fine aerosol (size less than 1 μm) or on the surface of the fly ash, preferentially on the surface of the finer ash particles. A large fraction of certain elements in the feed, such as mercury, will be volatilized.

Elements such as sodium (Na), lead (Pb), Zinc (Zn), and Cadmium (Cd), will be distributed between the volatiles and the residue in amounts that depend on the chemical and physical form in which the elements are present. For example, sodium in glass will be retained in the residue but that in common salt will partially disassociate and be discharged with the emission gases.

Some of the data on metal emissions that are available from tests on resource recovery plants is shown in Table 24.11. Note that the emission of trace metals can be dramatically limited at WTE plants by the use of high-efficiency particulate control devices that are installed on modern facilities.

While sampling for metal emissions is fairly well established, in order to obtain enough sample to analyze at highly controlled sources, samples times are extremely long, sometimes over eight hours using the U.S. EPA Method 5 sample train (relatively low sample rate). Several studies[25,26,39] of incinerator emissions in the United States in the 1970s concluded that "municipal incinerators can be major sources of Cd, Zn, Sb and possibly Sn… ." This conclusion is based on the relative concentration of these materials in the total suspended particulate catch. However, two of the three plants tested in these studies utilized inefficient air pollution control facilities. Thus, particulate emissions in these plants were relatively high when compared to the German and Japanese plant data in Table 24.11, which is similar to emission data from most modern mass burn and RDF waste-to-energy plants in the United States.[61] Note also in Table 24.11 that, as the efficiency of a particulate control improves, trace metal emissions generally decrease, and in most cases decrease significantly. Even though there is ample evidence from test data[17] to indicate that heavy metals tend to concentrate on the finer particulates, there is also evidence from test results to show that at high particulate removal efficiency (99%±), high trace metal removal (99%±) is achieved. Thus, the conclusion in these studies of incinerators quoted earlier is not valid for WTE plants utilizing efficient air pollution control devices.

TABLE 24.12 Lime Addition with Baghouse,[32] Mercury Concentrations (μg/km³ @ 8% O_2)

Operation Dry System	Inlet	Outlet	%Removal
230°F(110°C)	440	40	90.9
260°F(125°C)	480	23	97.9
285°F(140°C)	320	20	93.8
390°F(>200°C)	450	610	0
Wet-Dry System			
140°C	290	10	94.7
140°C←Recycle	350	19	94.7

Volatiles

Since mercury is a very volatile metal, it exists in vapor phase at temperatures as low as 68 to 122°F (20–50°C). Several studies have indicated that sufficient cooling of the flue gas (typically below 140°C, based on test results conducted to date) and a highly efficient particulate removal system to remove the particles on which the mercury has been adsorbed[5,7,74] are both required to achieve high mercury removal. High mercury removal has been obtained for the scrubber/fabric filter system, provided that the flue gas is adequately cooled (see Table 24.12).

Particulate Control

Electrostatic precipitators were the most commonly used device of gas cleaning for particulate emission control in municipal incinerators, as they can be designed to achieve very high collection efficiencies (99% or higher) and their operating costs are moderate. However, the investment cost and volumetric space required are high. Some precipitators that have been installed on solid waste incinerators incorporated paper char traps to prevent this material from being emitted. Charred paper has large area, low density, low resistivity, and low adhesion on electrodes.

Fabric filters can operate at high efficiency, even in the submicron particle size range. They were not widely used in solid waste incineration plants until the late 1980s because of their relatively high investment and operating costs, and limited life at high temperatures. However, with the increasing emphasis of regulatory agencies on acid gas control and an apparent trend to lower particulate emission levels, baghouses, usually in conjunction with addition of lime, are used more widely in the U.S. on municipal waste incinerators.[6] This is considered by the EPA as an element in Best Available Control Technology (BACT).

Test results (see Table 24.13) for municipal solid waste incinerators equipped with the conventional two-field electrostatic precipitator (EP) have shown a wide range of particulate emissions, varying from 50 to 300 mg/Nm³ (0.02 to 0.13 gr/dscf).* The three- and four-field electrostatic precipitators are achieving emissions of 20 to 75 mg/Nm³ (0.009 to 0.03 gr/dscf).[61] An emission level below 20 mg/Nm³ (0.009 gr/dscf) is technically possible and has been demonstrated at Zurich, where lime addition was retrofitted on a plant with an existing two-field electrostatic precipitator. However, a high capital cost may usually be anticipated to construct a precipitator with a sufficient number of fields and adequate treatment area to consistently achieve such performance.

The scrubber/fabric filter control systems (DS, BH) have been shown to be capable of operating at a particulate emission level of 20 mg/Nm³ (0.009 gr/dscf) and lower (see Table 24.13). The material selected for the filter bags can have an important effect on filtering efficiency and the emission level thus achieved. In general, test results to date for the scrubber/fabric indicate lower particulate emissions than those for electrostatic precipitators on WTE plants. However, in general, electrostatic precipitators have not been designed to meet emission levels as low as those specified

*The Symbol Nm³ means m³ at "normal" conditions: dry, 1 atm, 0°C.

TABLE 24.13 Particulate Emissions From Refuse Fired Plants[70]

	Particulates (gr/dscf)* @ 12% CO_2
Plant G (1983); EP	0.0321
Plant T (1984); DS, BH	0.012
Plant M (1984); DS, EP	0.0104
Plant W (1985); DS, BH	0.004
Plant P (1985); EP	0.0163
Plant T (1986); EP	0.007
Plant M (1986); DS, BH	0.007

* Grains per dry standard cubic foot.

for fabric filter installations. Electrostatic precipitators following spray drying absorbers in Europe have been tested at particulate emission levels of 1 to 8 mg/Nm^3 (0.00045–0.0036 gr/dscf). The reliability and overall economics of the various control processes must be considered when making a selection of equipment to meet these very low emission control requirements. Data are available[61] on emission levels for approximately 30 different specific elements, many of them heavy metals. Elements found to occur in stack emission from municipal waste incinerators are lead, chromium, cadmium, arsenic, zinc, antimony, mercury, molybdenum, calcium, vanadium, aluminum, magnesium, barium, potassium, strontium, sodium, manganese, cobalt, copper, silver, iron, titanium, boron, phosphorus, tin, and others.

Since the condensation point for metals such as lead, cadmium, chromium, and zinc is above 570°F (300°C), the removal efficiency for such metals is highly dependent on the particulate removal efficiency. Some metal compounds, particularly chlorides such as $AsCl_3$ at 252°F (122°C) and $SnCl_4$ at 212°F (100°C), have condensation points below 300°C. For such compounds, particulate collection temperatures will be a factor in collection efficiency. High removal (over 99%) has been observed for most metals for highly efficient (over 99%) particulate removal systems operating at appropriate temperatures. Many existing facilities have relatively inefficient particulate control equipment and emissions of metals are relatively high in such facilities (see Table 24.11).

Acid Gas Control

Initial efforts at acid gas control used wet collectors. However, this type of flue gas cleaning equipment is subject to problems such as corrosion, erosion, generation of acidic waste water, wet plumes, and, not least, high operating cost. Because of these problems, various semiwet and dry methods of cleaning flue gases have been developed and installed. These methods of gas treatment are based on the injection of slurried or powdered lime, limestone, or dolomite; adsorption; and absorption; followed by chemical conversion.[61] Since the reactivity of these lime materials is rather low, a multiple of the stoichiometric quantity is normally required to obtain a satisfactory cleaning effect. High removal efficiencies can be achieved for HCl, but reduction of SO_2 and SO_3 is more difficult to achieve and maintain. Slaked lime is highly reactive and stoichiometric ratios of 1.2 to 1.7 have been used for 97 to 99% HCl removals and 60 to 90% SO_2 reductions, depending on operating conditions and particulate collector (fabric filters having demonstrated higher removal efficiencies than electrostatic precipitators). Lime addition for acid gas control is the second element of EPA's BACT.

The important operating parameters for such equipment are flue gas temperature and composition, contact time, relative velocity of particles and gas stream, and possible activation of particles. See Tables 24.12, 24.14, and 24.15 for operating results achieved using dry and semidry lime injection followed by a baghouse for removal of heavy metal and trace organic pollutants from incinerator emissions.

TABLE 24.14 Lime Addition with Baghouse, Percent Removal of Organics[32]

	Dry System				Wet/Dry System	
	110°C	125°C	140°C	200°C	140°C	140°C Recycle
CB	95	98	98	62	>99	99
PCB	72	>99	>99	54	>99	>99
PAH	84	82	84	98	>99	79
CP	97	99	99	56	99	96

TABLE 24.15 Lime Addition with Baghouse, Metal Concentrations[9,32] (μg/km³ @ 8% O_2)

Metal	Inlet	Outlet	Removal
Zinc	77,000–108,000	5–10	96→99.99
Cadmium	1000–3500	1.0–0.6	96→99.96
Lead	34,000–44,000	1–6	95→99.98
Chromium	1400–3100	0.2–1	>99.92
Nickel	700–2500	0.4–2	>99.81
Arsenic	80–150	0.02–0.07	>99.95
Antimony	800–2200	0.2–0.6	>99.92
Mercury	190–480	10–610	0→90

Lime injection into a scrubber/fabric filter system has resulted in removal efficiencies of 90 to 99% for HCl and 70 to 90% SO_2, provided that the flue gas temperature and the stoichiometric ratio for lime addition are suitable. This combination of processes has reduced HCl levels below 20 ppm and SO_2 to levels below 40 ppm for MSW waste-to-energy plants. This technology has also been extensively used in other applications for acid gas removal. The scrubber/electrostatic precipitator combination has been shown to provide about 90% HCl removal, but typically less SO_2 removal (about 50%). Some tests in Europe have demonstrated 60 to 85% removals of SO_2. Since electrostatic precipitators and baghouses alone are not effective in removing HCl and SO_2, lime injection into the furnace has been tested with some success (about 50 to 70% efficiency).

Some sampling to determine HF removal has been reported. In general, HF removal normally follows HCl removal (i.e., is usually over 90 to 95%).

A NO_x removal system was installed on a MSW WTE plant in California. The State of California has demanded selective noncatalytic reduction of NO_x for other planned WTE plants.

Organic Compound Control

Organic compounds for which emission data are available include polychlorinated dibenzodioxins (PCDDs), polychlorinated dibenzofurans (PCDFs), chlorobenzenes (CBs), chlorophenols (CPs), polycyclic-aromatic hydrocarbons (PAHs), and PCBs. A number of other organic compounds, including aldehydes, chlorinated alkanes, and phthalic acid esters, have also been identified in specific testing programs. Since dioxin/furan emissions have generated the most interest over the past several years, there are more data for these compounds, in particular for the tetrahomologues, and especially the 2,3,7,8-substituted isomers. The other compounds have been less frequently reported in the literature. The reason for this emphasis is the toxicity of dioxin/furan to laboratory animals and the perceived risk to humans.

The available test data clearly show that dioxin/furan exit the boilers and, depending on the emission control devices employed, some fraction enters the atmosphere either as gases or adsorbed onto particulates. In addition, the solids remaining behind in the form of fly ash and bottom ash

TABLE 24.16 Summary of Average Total PCDD/PCDF Concentrations from MSW Combustion in Modern Plants (ng/Nm^3, dry, at 12% CO_2)[27,45,47,48]

	Total PCDD + PCDF
Peekskill, NY	
E.P. Only (1985)	100.25
Wurzburg, FRG	
Dry Scrub.–Baghouse (1985)	49.95
Tulsa, OK	
E.P. Only (1986)	34.45
Marion Co., OR	
Dry Scrub.–Baghouse (1986)	1.55

contain most of the same compounds, and these become another potential source of discharge to the environment.

Emission data for total dioxin/furan generally fall into three main categories:

- Low emissions, in the range of 20 to 130 ng/Nm^3
- Medium emissions, from 130 to 1,000 ng/Nm^3
- High emissions, over 1,000 ng/Nm^3

Average dioxin/furan emission from older plants to ranges about 500 to 1,000 ng/Nm^3. The lower emission levels tend to be associated with newer, well-operated mass fired facilities such as water well plants, and with modular, starved or controlled air types of incinerators (see Table 24.16). In most test programs, adequate operating data were not collected to correlate emissions with operations. Researchers in the field theorize that combustion conditions can play a role in minimizing emissions, and several studies[57,61,65] were conducted in Canada and the United States to define that role more exactly.

Emissions from MSW combustion contain small amounts of many different dioxin/furan isomers. While individual dioxin/furan isomers have widely differing toxicities, the 2,3,7,7-TCDD isomer, present as a small proportion of the total dioxin/furan, is of greatest known toxicological concern. Based on animal studies it has been generally concluded that other 2,3,7,8-substituted dioxin/furan isomers, in addition to the 2,3,7,8-TCDD, are also likely to be of toxicological concern. A method for expressing the relative overall toxicological impact of all dioxin/furan isomers, as so-called "2,3,7,8-TCDD toxic equivalents," was developed in the mid-1980s[2a] and has been used by the EPA intermittently in its regulatory efforts since this time.

In this method, emissions are sampled, extracted, and analyzed for all constituent isomers of dioxin/furan. A system of toxicity weighting factors from the existing toxicological data (based almost entirely on animal studies) is applied to each constituent dioxin/furan isomer and the results are summed to arrive at the 2,3,7,8-TCDD toxic equivalent. An example of dioxin/furan test results expressed as 2,3,7,8-TCDD toxic equivalents, using three different systems of weighting factors, is shown in Table 24.17.

Recently, emission control systems consisting of a scrubber/fabric filter have been evaluated for dioxin emissions.[32] Dioxin removal efficiencies exceeding 99% were obtained, which resulted in dioxin concentrations at the stack that approach the detection limit of the sampling and analytical equipment currently available. Emissions of furan followed a similar range of values as dioxin with the scrubber/high efficiency particulate removal combination reducing furan to very low or non-detectable levels.

Some limited data on emissions of CB, CP, PCB, and PAH are available. Most sampling programs for dioxin/furan have unfortunately neglected to analyze for these compounds. Maximum levels from Canadian studies[32] are as included in Table 24.18 along with some data from tests on U.S. plants.[31] The scrubber/fabric filter technology generally demonstrated removal rates of 80 to 99% for these compounds in these studies.

TABLE 24.17 Toxic Equivalent Emissions by U.S. EPA, Swedish, and California Methods[29,45,47,48,76]

Facility	Toxic Equivalents ng/Nm^3 at 12% CO_2		
	U.S. EPA	Swedish	California
Peekskill, NY	1.62	3.83	9.73
Tulsa, OK	0.7	1.74	4.75
Wurzburg, FRG	0.37	0.81	2.11
Marion Co., OR	0.11	0.16	0.29
From WHO Workshop; Naples, Italy			
Max. from Avg. Oper.	25.0	52.78	134.5
Achievable with no acid gas cleaning	0.9	2.2	5.94

TABLE 24.18 Organic Emissions (ng/Nm^3)[3,32]

Chemical Emitted	U.S. Plants	Canadian Pilot Plant	
	Before Particulate Removal	Before Particulate Removal	After Scrubber/Fabric Filter
CB	10,000 to 500,000	17,000	3,000
CP	22,000 to 80,000	30,000	8,000
PCB	—	700	Non-detectable (ND)
PAH	ND to 5,600,000	30,000	130

Very few studies report on other organic products in the flue gas. Some data from tests on older plants have been reported for aldehydes and certain other volatile hydrocarbons.[3] Such data are not available for newer plants.

The conventional combustion gas measurements include CO, total hydrocarbons (THCs), CO_2, and H_2O. Both CO and THC have been of interest as potential surrogates for dioxin/furan emissions; however, no strong correlations have been found in previous studies. In fact, very few studies have attempted to determine CO and dioxin/furan emission data for several operating conditions on the same incinerator to develop a correlation. On the other hand, several authors have attempted to correlate CO and dioxin/furan data obtained from several different facilities. From such comparisons, it appears that low CO levels (below 100 ppm) are associated with low dioxin/furan emissions.[36] High CO levels of several 100 ppm and even over 1,000 ppm have been associated with high dioxin/furan emissions. During poor or upset combustion conditions, CO levels of several thousand ppm have been observed and THC levels have risen from a typical 1 to 5 ppm to 100 ppm and more. Since one of the measures available to incinerator operators to optimize combustion control is to minimize CO production, one would assume from these general correlations that this would also tend to minimize dioxin/furan emissions, along with emissions of other trace chlorinated hydrocarbons. THC is not as useful as CO as an indicator of proper combustion because of problems in sampling to consistently obtain a representative sample for analysis at the analytical instrument.

24.4 Performance

Mass-burning of MSW is the most highly developed and commercially proven combustion process presently available for reducing the volume of MSW prior to ultimate disposal of residuals on the land, and for extracting energy from the waste.[67] Hundreds of such plants, incorporating various grate systems and boiler concepts, which differ in details of design, construction, and quality of operation, have been built throughout the world since the mid-1960s. Mass burn systems are generally furnished with a guaranteed availability of 85%, while in practice availabilities of 90 to 95% have been achieved.[61] Availability cannot approach 100% because standard maintenance practice requires periodic shutdowns. The newest mass burn facilities seem capable of achieving high reliability, based on their performance so far in Europe, Japan, and the United States.

RDF systems generally have not been as reliable as mass burn systems because of the greater complexity of their processing systems.[61] Since the early 1970s approximately two dozen RDF plants were built and placed in operation in this country, producing RDF with some related materials. Many of these plants had frequent and substantial technical problems and needed significant modifications. During this period about 10% of the mass burn plants built were subsequently closed (after operating for 10 to 20 years), while approximately 50% of the RDF plants were closed. Of the RDF plants remaining in operation, at least half were in start-up for periods lasting several years. Many plants were substantially retrofitted, and some remained in operation at only a fraction of their original design capacity.[71]

Numerous "trial burns" have been conducted between the mid-1970s and the mid 1980s in existing large utility-type boiler units. Results were generally satisfactory with respect to combustion conditions, but mixed results were experienced with respect to plant operation and maintenance conditions. Consequently, these operations were not pursued beyond the "trial burn" stage.

Nonetheless, some RDF systems have operated reliably once start-up problems were overcome. New RDF facilities have performed better, but it remains to be seen how reliable and economical they will be over time.[61]

24.5 Costs

It is extremely difficult to obtain accurate, consistent, and comparable WTE plant construction cost data from which to develop information which might be useful in predicting a planned new plant's construction cost during the study state of a project. However, a 1988 study[51] has developed such data (appropriate for the time frame of mid-1987), which is confirmed in general by this author's personal experience. This study indicates that for the upper 90% confidence limit for the smallest facility, and the largest facility, the construction costs would range as indicated below:

1. A small modular combustion unit with a waste heat boiler and a capacity of less than 250 TPD—$68,000 and $40,000 per ton of daily MSW processing capability. (In most instances such plants don't incorporate the same degree of equipment redundancy, and/or the same quality of equipment as the larger plants.)
2. A small refractory wall furnace with waste heat boiler and dry scrubbers of between 200 and 500 TPD capacity—$90,000 and $70,000 per ton of daily MSW processing capability.
3. A small, field erected, water wall congeneration or electric generation facility with dry scrubbers of between 500 and 1,500 TPD capacity—$112,000 and $85,000 per ton of daily MSW processing capability.
4. A large, field erected, water wall congeneration or electric generation facility with a dry scrubber between 2,000 and 3,000 TPD capacity—$129,000 and $112,000 per ton of daily MSW processing capacity.

In this study, [51] the construction costs were said to include the vendor quote for construction plus contingency, utility interconnection expenses, and any identified allowances clearly associated with the construction price. All other costs, such as land acquisition, interest during construction, development costs, and management fees were not considered or included, where known, due to their highly project specific nature.

The following specific observations were made by the authors at the conclusion of this study:[51]

1. Capital construction price decreases with increasing size within size ranges and increases with a higher-value energy product.
2. Price is also affected by the construction, procurement, and air pollution control methods employed.
3. Refuse derived fuel and mass burning water wall facilities are competitively priced with each other.

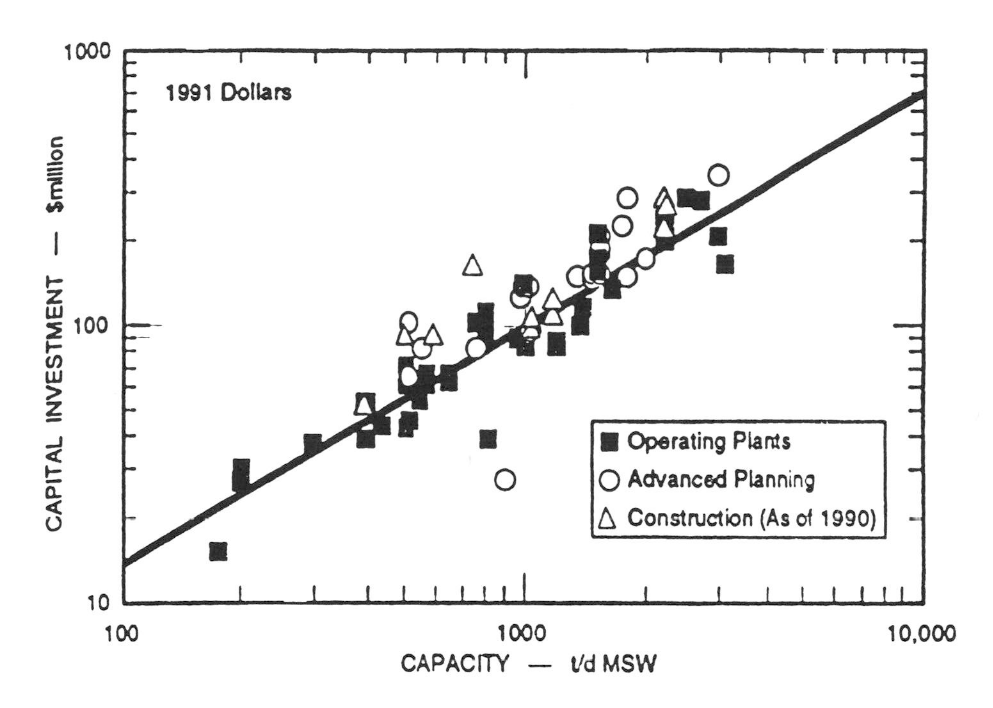

FIGURE 24.10 Effect of plant capacity on capital costs for mass burn facilities (electricity only), excluding costs associated with collection (e.g., trucks).

The effect of plant capacity on capital costs for mass burn plants is shown on Fig. 24.10. Capital costs for other types of plants are similar. However, reliable information from which to estimate capital costs for RDF plants is not as readily available as for mass burn plants for several reasons.[71] First, many RDF plants, as noted earlier, have undergone significant retrofit programs after initial construction. This raises a question as to what the capital costs would have been if the plant, as originally built, had incorporated all successful retrofit changes. Second, some RDF plant construction costs only include the cost of the MSW processing plant, while other costs include both MSW processing plant and boiler. The similarity of RDF and mass burn water wall facilities noted in item 3 may indicate a comparison between RDF/boiler systems and mass fired/water wall boiler systems.

With respect to operating costs and/or tipping fees, information is even more difficult to obtain from which to develop costs for planning purposes. Costs cited in the literature from 1989 through 1994 range from $40 to $80 per ton of daily rated MSW processing capability.[18,40,44,70,71] Tipping fees on Long Island, which has generally high labor rates, high power costs, and very long hauls for residue disposal, have been noted to range up to $110 per ton of daily rated MSW processing capability.[24] Plants in other parts of the country where plant operating cost elements are significantly lower have been found to have tipping fees closer to $40 per ton.[6] Thus, tipping fees for a specific facility would have to be developed based on cost factors for that specific plant.

The effect of plant capacity on O&M costs for mass burn plants is shown on Fig. 24.11. Information is so limited on other types of plants, and the costs are so dependent on local conditions that we do not feel that curves developed for other types of plants would be useful.

24.6 Status of Other Technologies

Several other technologies have been used to a small extent to burn MSW and beneficially use the energy produced in the combustion process. Their use in the future depends on numerous factors,

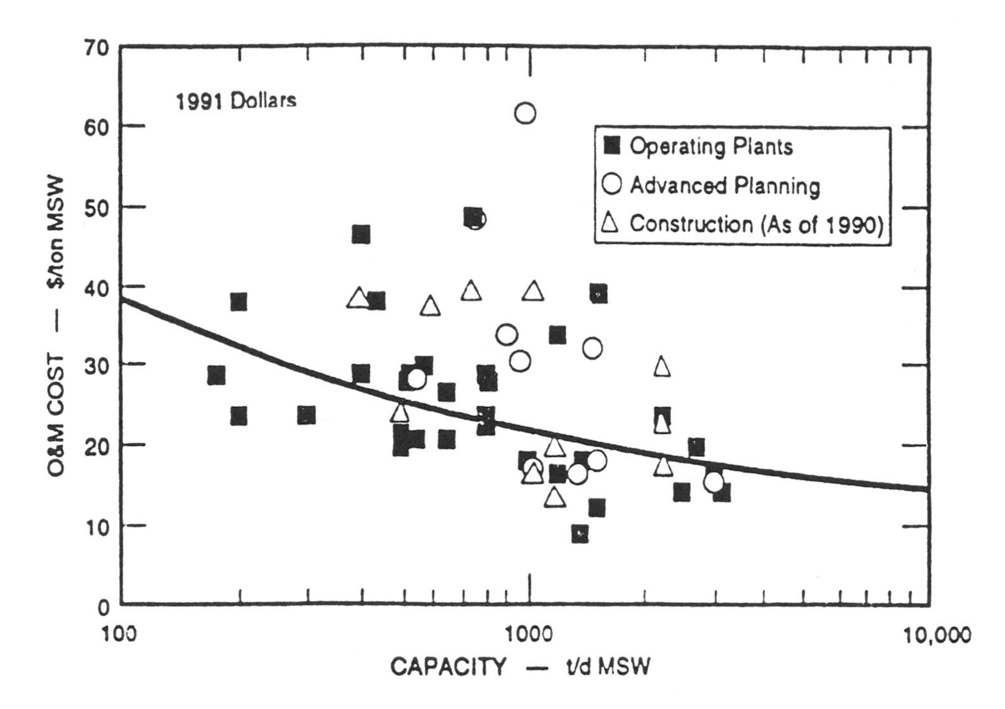

FIGURE 24.11 Effect of plant capacity on O&M costs for mass-burn facilities (electricity-producing only), excluding costs associated with collection (e.g., trucks).

perhaps the most important of which is full-scale demonstration of successful, reliable operation, after which total operational costs are shown to be competitive with mass burn and/or RDF combustion costs.

Modular Systems

Modular systems, generally utilized in smaller plants, are assemblies of factory-prefabricated major components joined together in the field to form a total operational system. They have been built in individual unit sizes up to just over 100 tons per day capacity, combined into plants of just over 300 tons per day capacity. Modular systems are similar to mass burn systems in that they combust unprocessed MSW, but they feature two combustion chambers, and the MSW is charged into the system with a hydraulic ram.

The primary chamber of a modular system is usually operated in a slightly oxygen deficient ("starved air") environment. The volatile portion of the MSW is vaporized in this chamber and the resulting gases flow into the secondary chamber. The secondary chamber operates in an excess air condition, and combustion of the gases driven off the MSW is essentially completed in this chamber.

One advantage of these units, as indicated in the section on costs, is low cost. Another advantage is that factory prefabrication of major system components can result in shortening of the field construction time. One disadvantage of the two-chamber modular system is that waste burn-out in the residue is not always complete, which increases ash quantities and reduces the efficiency of energy recovery.[63] Energy recovery efficiency is also reduced due to generally higher "excess air" levels carried in these units. Also, combustion control is generally less effective in this type of unit, increasing the possibility of discharge of trace organic emissions. As pointed out earlier, these types of units generally utilize a lower quality of equipment and include less redundancy than larger mass fired water wall and RDF WTE plants. Modular plants are responsible for about 10% of the total MSW burned at this time in the United States.

Fluidized Beds

Fluidized bed combustion (FBC) differs from mass-burn and RDF combustion in that the fuel is burned in "fluid suspension"—entrained along with particles of sand in an upward flow of turbulent air at a temperature controlled to 1,500 to 1,600°F (816–971°C). To date, it has been used primarily to burn sewage sludge, industrial waste, and coal. Currently, several facilities are in advanced planning stages or construction for combustion of RDF.

"Bubbling" FBC designs retain the material near the bottom of the furnace, while "circulating" designs allow bed material to move upward and then be returned near the bottom of the bed for further combustion. The reason for the interest in this combustion technique to burn RDF is the potential for these designs to provide more consistent combustion because of the extreme turbulence and the proximity of the RDF waste particles to the hot sand particles.[43] Such systems also require lower combustion temperatures than mass-burn and current RDF systems.

Pyrolysis

Pyrolysis is the chemical decomposition of a substance by heat in the absence of oxygen. It generally occurs at relatively low temperatures (900 to 1,100°F, compared with 1,800°F for mass burn). The heterogeneous nature of MSW makes pyrolysis reactions complex and difficult to control. Besides producing larger quantities of solid residues that must be managed for ultimate disposal, pyrolysis produces liquid tar and/or gases that are potentially marketable energy forms. The quality of the fuel products depends on the material fed into the reactor (e.g., moisture, ash, cellulose, trace constituent content) and operating conditions (e.g., temperature and particle size).[61]

Several pyrolysis facilities were built in the 1970s with grants from EPA. These facilities were unable to produce quality fuels in substantial quantities. No one in the United States has yet successfully developed and applied the pyrolysis technology to MSW combustion. However, the use of pyrolysis for MSW management still attracts attention in other countries.

Hydropulping

Hydropulping is a process of producing RDF from MSW while the waste is immersed in water. The waste is shredded in a machine similar to a large blender. Much of the noncombustible material is separated from the shredded material, and the remaining material was dewatered to about 50% water content in large presses. The dewatered combustible material is then burned in large water boilers.

Two full-size plants were built in this country and operated for a period of several years. The plants were finally shut down and either dismantled or converted to other processes due to higher than anticipated maintenance costs and lower than anticipated energy production and income.

24.7 Future Issues and Trends

It has been demonstrated by actual experience that modern mass burn and RDF fired MSW WTE plants can be designed and operated with reasonable assurance of continuous service and without adversely affecting nearby neighborhood property values. Operating experience with these plants has been largely satisfactory, as has been demonstrated by a recent survey of citizens living in neighborhoods with WTE plants located nearby.[37] Allegations that WTE plant sites are situated adjacent to neighborhoods of low-income, disadvantaged, or minority populations ignore the specific technical siting criteria outlined earlier (i.e., adjacent to major highways, low land cost, industrial type area, etc.) which are generally followed in siting such facilities. Such areas frequently are closer to low-income neighborhoods than to middle- and higher-income neighborhoods.[61]

Two other institutional issues impacting WTE, flow control and classification of ash residues, have worked their way through our court system, with the Supreme Court rendering important decisions in 1994. With respect to flow control, the Court found that a local community could not force the MSW from that community to be taken to a specific facility such as a WTE plant.[24] The Court's decision with respect to ash residues found that ash residue from combustion of MSW is not exempt from the rules and regulations for hazardous waste. Thus, the MSW residue must be tested and, if found to be hazardous, must be disposed of in secure landfills[24] at about ten times the cost of disposal in monofills. The actual impact of this ruling has not as yet been determined, although the testing requirements alone will increase the cost of operation of WTE plants.

Another issue facing WTE plants is the uncertain future of regulatory requirements, both from the standpoint of legislation and from that of the regulatory agencies. Over the past 10 years legislation has been passed by Congress calling for Best Available Control Technology (BACT), then Lowest Achievable Emission Requirements (LAER), and then, most recently, Maximum Achievable Control Technology (MACT). The impact of this legislation, each calling for significant reductions in allowable emissions (absent any indication of the existence of a significant public health problem or benefit), has been to require extensive retrofits of existing plants and addition of equipment to proposed new plants, all at substantial expense without proven benefit and, in many cases, without prior proof of operational viability.

Several positive actions are occurring in the field. Thus, most project developers have recognized the desirability of implementing a proactive program early in the project planning process to involve the public, particularly those in the vicinity of the proposed facility.[56] Also, the potential for materials recycling, which had been overenthusiastically embraced a number of years ago (state recycling goals as high as 70%, with some local communities projecting 100% of MSW management through recycling), is gradually being recognized.[41] Franklin Associates[23] projects an increase in the recycling rate of from 22% in 1993 to 30% by the year 2000. Much of this increase in recycling is to come from increases in recovery of paper materials and diversion of yard wastes to composting. The impact of these changes in waste composition on energy available at WTE plants will be minimal, with the reduction in moisture content due to diversion of yard wastes being a positive factor.

Kiser,[37] a representative from the WTE industry trade association, noted in early 1994, "As we speak, there are seven projects under construction with a design capacity of 8,830 tons a day, so the industry is doing well in that regard." Another article in January 1995[42] indicated, "In spite of public resistance and regulatory challenges, $2 billion was spent to construct or purchase new incinerators in 1994. In addition, $500 million was spent to upgrade incineration equipment." Franklin Associates[23] projects an increase in MSW disposed of in WTE plants of 1.1 million tons per year from 1993 to the year 2000 (32,900,000 tons per year in 1993, to 34,000,000 tons per year in 2000). This would indicate the addition of new plant capacity of about 3,550 tons per day of rated capacity.

Thus, the need to generate electric energy and safely manage the MSW generated by modern civilization, particularly in the vicinity of major metropolitan areas, together with the proven performance of modern WTE plants, indicate that this technology will be utilized to dispose of a portion of this country's solid waste and provide electricity.

References

1. American Public Works Association, *Municipal Refuse Disposal*, 3rd edition, Public Administration Service, Chicago, IL, 1970.
2. American Society for Testing and Materials (ASTM), *Annual Book of ASTM Standards*, Section 5, Vol. 05.05, *Fuels: Coal and Coke*, Philadelphia, PA.

2a. Barnes, D., "Rump Session." *Chemosphere*, Vol. 14, No. 6-7, June–July 1985, pp. 987–989.

3. Battelle Columbus Labs, "Characterization of Stack Emissions from Municipal Refuse-to-Energy Systems." National Technical Information Service, PB87-110482, October, 1982.
4. Beltz, P.R., Engdahl, R.B., and Dartoy, J., "Evaluation of European Refuse-Fired Energy Systems Design Practices, Summary, Conclusions and Inventory," prepared for U.S. EPA by Battelle Laboratories, Columbus, OH, 1979.

5. Bergstrom, J.G.T., "Mercury Behavior in Flue Gases," *Waste Management and Research*, Vol. 4, No. 1, March 1986.
6. Bilmes, J.S., "Impact of the Recession and Recycling on Solid Waste Processing Facilities in New England," *Proceedings 1992 National Waste Processing Conference*, Detroit, MI, May 17–20, ASME, New York, pp. 351–360.
7. Braun, H., Metzger, M., and Vogg, H., *Zur Problematik der Quecksilber-Abscheidung aus Rauchgasen von Mullverbrennungsanlagen*, Vol. 1, Teil 2.
8. Bryers, R.W. (Ed.), *Ash Deposits and Corrosion due to Impurities in Combustion Gases*, Hemisphere Publishing Corp., New York, 1978.
9. Carlsson, K., "Heavy Metals from Energy-from-Waste Plants—Comparison of Gas Cleaning Systems," *Waste Management & Research*, Vol. 4, No. 1, March 1986.
10. Carrotti, A.A., and Smith, R.A., *Gaseous Emissions from Municipal Incinerators*. USEPA Publication No. SW-18C, 1974.
11. Charles, M.A., and Kiser, J.V.L., "Waste-to-Energy: Benefits Beyond Waste Disposal." *Solid Waste Technologies*, Vol. 8, No. 7, p. 12, 1995.
12. Churney, K.L., Ledford, Jr., A.E., Bruce, S.S., and Domalski, E.S., "The Chlorine Content of Municipal Solid Waste from Baltimore County, MD and Brooklyn, NY," National Bureau of Standards, NBSIR 85-3218; April 1985.
13. Cooper Engineers, Inc., "Air Emissions Tests of a Deutsche Babcock Anlagen, Dry Scrubber System at the Munich North Refuse-Fired Power Plant," February 1985.
14. Cooper Engineers, Inc., "Air Emissions and Performance Testing of a Dry Scrubber (Quench Reactor). Dry Venturi and Fabric Filter System Operating on Flue Gas From Combustion of Municipal Solid Waste in Japan," West County Agency of Contra Costa County Waste Co-Disposal/Energy Recovery Project, May 1985.
15. Lide, D.R., *CRC Handbook of Chemistry and Physics*, 71st ed., CRC Press, Boca Raton, FL, 1990.
16. Criss, G.H., and Olsen, R.A., "The Chemistry of Incinerator Slags and Their Compatibility with Fireclay and High Alumina Refractories," *Proceedings 1968 National Incinerator Conference, ASME*, New York, pp. 53–68.
17. Dannecker, W., "Schadstoffmessungen bei Mullverbrennungsanlagen," *VGB Krazftwekstechnick* Vol. 63, March 3, 1983.
18. Davis, C.F., "Annual Snapshot of Six Large Scale RDF Projects," *Proceedings 1992 National Waste Processing Conference*, Detroit, MI, May 17–20, ASME, New York, pp. 75–88.
19. Domalski, E.S., Ledford, A.E., Jr., Bruce, S.S., and Churney, K.L. "The Chlorine Content of Municipal Solid Waste from Baltimore County, Maryland, and Brooklyn, NY," *Proceedings 1986 National Waste Processing Conference*, Denver, CO, June 1–4, ASME, New York, pp. 435–448.
20. Eighmy, T.T., Collins, M.R., DiPietro, J.V., and Guay, M.A., "Factors Affecting Inorganic Leaching Phenomena from Incineration Residues." Paper presented at Conference on Municipal Solid Waste Technology, San Diego, CA, January 30–February 1, 1989.
21. Franklin Associates, Ltd., "Characterization of Municipal Solid Waste in the United States, 1960–2000," report prepared for U.S. EPA, NTIS No. PB87-17323, Prairie Village, KS, July 25, 1986.
22. Franklin Associates, Ltd., "Characterization of Municipal Solid Waste in the United States, 1960 to 2000 (Update 1988)," report prepared for U.S. EPA, Office of Solid Waste and Emergency Response, Prairie Village, KS, March 30, 1988.
23. Franklin Associates, Ltd., "Characterization of Municipal Solid Waste in the United States. (1994 Update)," report prepared for U.S. EPA Office of Solid Waste and Emergency Response, Prairie Village, KS, November 1994.
24. Freeman, D., "Waste-to-Energy Fights the Fires of Opposition," *World Wastes*, Vol. 37, No. 1, January 1994, p. 24.

24a. Fryling, G.R. (Ed.), *Combustion Engineering*, Combustion Engineering, Inc., New York, 1966.

25. Greenberg, R.R., Zoller, W.H., and Gordon, G.E., "Composition and Size Distributions of Particles Released in Refuse Incineration," *Environmental Science and Technology*, Vol. 12, No. 5, May 1978. ACS, pp. 566–573.
26. Greenberg, R.R., Gordon, G.E., Zoller, W.H., Jacko, R.B., Neuendorf, D.W., and Yost, K.J., "Composition of Particles from the Nicosia Municipal Incinerator," *Environmental Science and Technology*, Vol. 12, No. 11, November 1978. ACS, pp. 1329–1332.
27. Hahn, J.L., VonDemfange, H.P., Zurlinden, R.A., Stianche, K.F., and Seelinger, R.W. (Ogden Martin Systems, Inc.), and Weiand, H., Schetter, G., Spichal, P., and Martin, J.E. (Martin GmbH)" 1986 National Waste Processing Conference, Denver, CO, June 1–4, 1986. ASME, New York.
28. Hahn, J.L., "Dioxin Emissions from Modern, Mass-Fired, Stoker/Boiler for Use in Waste-to-Energy Risk Assessments." Paper presented at APCA Meeting, Minneapolis, MN, June 1986.
29. Hahn, J.L., and Sussman, D.B., "Dioxin Emissions from Modern, Mass-Fired, Stoker/Boilers with Advanced Air Pollution Control Equipment." Presented at Dioxin '86, Fukuoka, Japan, September 1986.
30. Hahn, J.L., and Sofaer, D.S., "Variability of NO_x Emissions from Modern Mass-Fired Resource Recovery Facilities." Paper No. 88-21.7 presented at 81st Annual Meeting of Air Pollution Control Association, Dallas, TX, June 1988.
31. Hasselriis, F., "The Environmental and Health Impact of Waste Combustion—The Rush to Judgment versus Getting the Facts First," *Proceedings 1992 National Waste Processing Conference*, Detroit, MI, May 17–20, ASME, New York, pp. 361–369.
32. Hay, D.J., Finkelstein, A., Klicius, R. (Environment Canada), and Marenlette, L. (Flakt Canada, Ltd.), "The National Incinerator Testing and Evaluation Program: An Assessment of (A) Two-Stage Incineration (B) Pilot Scale Emission Control," Report EPS 3/UP/2, September 1986.
33. Hecklinger, R.S., "The Relative Value of Energy Derived From Municipal Refuse," *Journal of Energy Resources Technology*, Vol. 101, pp. 251–259, 1979.
33a. Hecklinger, R.S., Chapter on "Combustion" in *The Engineering Handbook*, CRC Press, Inc., Boca Raton, FL, 1996.
34. Junk, G.A., and Ford, C.S., "A Review of Organic Emissions from Selected Combustion Processes," *Chemosphere*, Vol. 9, pp. 187–230, 1980.
35. Kaiser, E.R., Zeit, C.D., and McCaffery, J.B., "Municipal Incinerator Refuse and Residue," *Proceedings 1968 National Incinerator Conference*, ASME, New York, pp. 142–153.
36. Kilgroe, J.D., Lanier, W.S., and Von Alten, T.R., "Development of Good Combustion Practice for Municipal Waste Combustors," *Proceedings 1992 National Waste Processing Conference*, Detroit, MI, May 17–20, ASME, New York, pp. 145–156.
37. Kiser, J.V.L., Charles, M.A., Bridges, J.P., Carr, G.L., and Velzy, C.O., "Waste-to-Energy: Citizens Respond to Plants in Their Neighborhoods," *Solid Waste Technologies*, Vol. 8, No. 3, May/June 1994, p. 18.
38. Konhelm, C.S., and Koehler, S.N., "Plant Sitings and Property Values," *Waste Age*, April 1986, p. 95.
39. Law, S.L., and Gordon, G.E., "Sources of Metals in Municipal Incinerator Emissions," *Environmental Science and Technology*, Vol. 13, No. 4, April 1979, ACS, pp. 432–438.
40. Leavitt, C., "Calculating the Costs of Waste Management," *World Wastes*, Vol. 37, No. 4, April 1994, p. 42.
41. Michaels, A., "Solid Waste Forum," *Public Works*, March 1995, p. 72.
42. Miller, R.K., "Industry Execs See Bright Future for Incineration," *World Wastes*, Vol. 38, No. 1, January 1995, p. 12.
43. Minott, D.H., "Operating Principles and Environmental Performance of Fluid-Bed Energy Recovery Facilities," Paper No. 88-21.9 presented at 81st Annual Meeting of Air Pollution Control Association, Dallas, TX, June 19–24, 1989.

44. Murdoch, J.D. and Gay, J.L., "Material Recovery with Incineration, Monmouth County, N.J.," Proceedings, 27th Annual International Solid Waste Exposition, Tulsa, OK, August 14–17, 1989, SWANA, Silver Springs, MD (Pub. #GR-0028) p. 329.
45. NYS Department of Environmental Conservation Bureau of Toxic Air Sampling, Division of Air Resources, "Preliminary Report on Westchester RESCO RRF," January 8, 1986.
46. Norton, F.H., "Refractories," *Marks Standard Handbook for Mechanical Engineers*, 8th ed., McGraw-Hill, New York, 1978, pp. 6-172 and 6.173.
47. Ogden Projects, Inc., "Environmental Test Report, Walter B. Hall Resource Recovery Facility," October 20, 1986.
48. Ogden Projects, Inc., "Environmental Test Report, Marion County Solid Waste-to-Energy Facility," December 5, 1986.
49. O'Malley, W.R., "Special Factors Involved in Specifying Incinerator Cranes," Proceedings 1968 National Incinerator Conference, ASME, New York, pp. 211–215.

49a. Penner, S.S. and Richards, M.B., "Estimates of Growth Rates for Municipal-Waste Incineration and Environmental Controls Costs for Coal Utilization in the U.S." *Energy: The International Journal*, Vol. 14, p. 961, 1989.

50. Richard, J.J., and Junk, G.A., "Polychlorinated Biphenyls in Effluents from Combustion of Coal/Refuse," *Environmental Science and Technology*, Vol. 15, No. 9, September 1981, ACS, pp. 1095–1100.
51. Rigo, H.G., and Conley, A.D., "Waste-to-Energy Facility Capital Costs," *Proceedings 1988 National Waste Processing Conference*, Philadelphia, PA, May 1–4, ASME, New York, pp. 23–28.
52. Rigo, H.G. and Chandler, A.J., "Managing Metals in MSW: Based on Sound Science or Myth?" *Solid Waste Technologies*, Vol. 8, No. 3, May/June 1994, p. 25.
53. Rogus, C.A., "An Appraisal of Refuse Incineration in Western Europe," *Proceedings 1966 National Incinerator Conference*, ASME, New York, pp. 114–123.
54. Sawell, S.E., Bridle, T.R., and Constable, T.W., "Leachability of Organic and Inorganic Contaminants in Ashes from Lime-Based Air Pollution Control Devices on a Municipal Waste Incinerator," Paper No. 87-94A presented at 80th Annual Meeting of Air Pollution Control Association, New York, June 21–26, 1987.
55. Sawell, S.E., and Constable, T.W., "NITEP Phase IIB: Assessment of Contaminant Leachability from the Residues of a Mass Burning Incinerator," Vol. VI of National Incinerator Testing and Evaluation Program, The Combustion Characterization of Mass Burning Incinerator Technology, Quebec City, Toronto, Canada, Environment Canada, August 1988.
56. Scarlett, T., "WTE Report Focuses on Socioeconomics," *World Wastes*, Vol. 37, No. 5, May 1994, p. 14.

56a. Scherrer, R., and Oberlaender, B., "Refuse Pit Storage Requirements," *Proceedings 1990 National Waste Processing Conference*, Long Beach, CA, June 3–6, ASME, New York, pp. 135–142.

57. Seeker, W.R., Lanier, W.S., and Heap, M.P., "Municipal Waste Combustion Study: Combustion Control of Organic Emissions," EPA, Research Triangle Park, NC, January 1987.

57a. Stultz, S.C. and Kitto, J.B. (Eds.), *Steam: Its Generation and Use*, 40th ed., The Babcock and Wilcox Co., Barberton, OH, 1992.

58. Suess, M.J., et al., "Solid Waste Management: Selected Topics," WHO Regional Office for Europe, Copenhagen, Denmark, 1985.
59. Timm, C.M., "Sampling Survey Related to Possible Emission of Polychlorinated Biphenyls from the Incineration of Domestic Refuse," NTIS Pub. No. PB-251-285, 1975.
60. U.S. Congress, Office of Technology Assessment, *Facing America's Trash, What Next for Municipal Solid Waste*, Chapter 3, Generation and Composition of MSW, OTA-A-424, Washington, DC, U.S. Government Printing Office, October, 1989.

61. U.S. Congress, Office of Technology Assessment, *Facing America's Trash, What Next for Municipal Solid Waste,* Chapter 6, Incineration, OTA-A-424, Washington, DC, U.S. Government Printing Office, October 1989.
62. U.S. Environmental Protection Agency, *Fourth Report to Congress, Resources Recovery and Waste Reduction,* Report No. SW-600, Washington, DC, 1977.
63. U.S. Environmental Protection Agency, *Municipal Waste Combustion Study, Report to Congress,* EPA/530-SW-87-021a, Washington, DC, 1987.
64. U.S. Environmental Protection Agency, "Characterization of Municipal Waste Combustor Ashes and Leachates from Municipal Solid Waste Landfills, Monofills and Codisposal Sites," prepared by NUS Corp. for Office of Solid Waste and Emergency Response, EPA/530-SW-87-028A, Washington, DC, October 1987.
65. U.S. Environmental Protection Agency, "Municipal Waste Combustion Multipollutant Study, Characterization Emission Test Report," Office of Air Quality Planning and Standards, EMB Report No. 87-MIN-04, Research Triangle Park, NC, September 1988.
66. Velzy, C.O., "The Enigma of Incinerator Design," *Incinerator and Solid Waste Technology,* ASME, New York, pp. 65–74.
67. Velzy, C.O., "30 Years of Refuse Fired Boiler Experience," presented at Engineering Foundation Conference, Franklin Pierce College, July 17, 1978.
68. Velzy, C.O., "Trace Emissions in Resource Recovery—Problems, Issues and Possible Control Techniques," *Proceedings of the DOE-ANL Workshop; Energy from Municipal Waste: State-of-the-Art and Emerging Technologies,* Argonne National Laboratory Report ANL/CNSV-TM-137, February 1984.
69. Velzy, C.O., "Standards and Control of Trace Emissions from Refuse-Fired Facilities." *Municipal Solid Waste as a Utility Fuel, EPRI Conference Proceedings,* Madison, WI, November 20–22, 1985.
70. Velzy, C.O., "U.S. Experience in Combustion of Municipal Solid Waste," paper presented at APCA Specialty Conference on Regulatory Approaches for Control of Air Pollutants, Atlanta, GA, February 20, 1987.
71. Velzy, C.O., "Considerations in Planning for an Energy-from-Waste Incinerator," paper presented at New Techniques and Practical Advice for Problems in Municipal Waste Management, Toronto, Ontario, Canada, April 1–2, 1987.
72. Visalli, J.R., "Environmental Impact Considerations in Recycling Solid Waste," *Journal of Resource Management and Technology,* Vol. 14, No. 4, pp. 241–245, December 1985.
73. Visalli, J.R., "A Comparison of Dioxin, Furan and Combustion Gas Data from Test Programs at Three MSW Incinerators," *Journal of the Air Pollution Control Association,* Vol. 37, No. 12, pp. 1451–1463, December 1987.
74. Vogg, H., Braun, H., Metzger, M., and Schneider, J., "The Specific Role of Cadmium and Mercury in Municipal Solid Waste Incineration," *Waste Management and Research,* Vol. 4, No. 1, March 1986.
75. Vogg, H., Metzger, M., Stieglitz, L., "Recent Findings on the Formation and Decomposition of PCDD/PCDF in Solid Municipal Waste Incineration," ISWA/WHO, Specialized Seminar, Copenhagen, Denmark, January 1987.
76. World Health Organization, "Report on PCDD and PCDF Emissions from Incinerators for Municipal Sewage, Sludge and Solid Waste—Evaluation of Human Exposure," from WHO Workshop, Naples, Italy, March 1986.

Index